A New Pattern Comprehensive Textbook of

Quantitative Aptitude

Maths in Moments

Vetted by a Panel of Experts

A Universal Resource for

NRA (CET), SSC, RAILWAYS, BANKS,
CAT, MAT, SNAP, XAT, CLAT, CDS, PSUs, UPSC & PCS

Satish Kumar
Under Secretary to Govt. of India

ARIHANT PUBLICATIONS (INDIA) LIMITED

arihant

ARIHANT PUBLICATIONS (INDIA) LIMITED

Note : *The views expressed by the author in the book are his own and not that of Government*

卐 **Administrative & Production Offices**

Regd. Office

'Ramchhaya' 4577/15, Agarwal Road, Darya Ganj, New Delhi -110002
Tele: 011- 47630600, 43518550

卐 **Head Office**

Kalindi, TP Nagar, Meerut (UP) - 250002
Tel: 0121-7156203, 7156204

卐 **Sales & Support Offices**

Agra, Ahmedabad, Bengaluru, Bareilly, Chennai, Delhi, Guwahati, Hyderabad, Jaipur, Jhansi, Kolkata, Lucknow, Nagpur & Pune.

卐 **ISBN** 978-93-26196-97-0

卐 **PRICE** ₹895.00

PO No. : TXT-59-T070845-5-26

Published by Arihant Publications (India) Ltd.

For further information about the books published by Arihant, log on to www.arihantbooks.com or e-mail at info@arihantbooks.com

Follow us on   

Dedicated To

My Parents

Shri Jagannath Singh
Smt. Tara Devi

My Sons

Shivansh & Saransh

About the Author

Shri Satish Kumar, born in village- Lodipur, P.O.- Pataria, in the District of Jehanabad (Bihar) in 1977, had cracked many competitive examinations like Bank PO, Railway, etc. before he joined the Ministry of Housing and Urban Affairs as Assistant Section Officer in the year 2003 in pursuance of results of Staff Selection Commission (Combined Graduate Level) Examination -2000, securing 8th rank across the country in the first attempt. Thereafter, he secured 3rd rank in Limited Departmental Competitive Examination-2006 (meant for the post of Section Officer), conducted by the Union Public Service Commission. Further, he cracked Civil Services (Main) examinations three times in 2007, 2008 and 2009.

From 2011 to 2014, he served in Ministry of Health and Family Welfare, Government of India as Section Officer and dealt with policy matters of Central Government Health Scheme. He also completed the World Health Organisation (WHO) In-Country Fellowship Training Program on Public Private Partnership in Health. Moreover, he contributed a lot in preparation of National Action Plan for Children (NAPC)- 2016, a key document, released by the Ministry of Women and Child Development in 2017 when he was dealing with the matters pertaining to Child Welfare in the capacity of Under Secretary to the Government of India. He participated in several conferences on Child Welfare issues persistent in India.

Presently, he is working as Under Secretary to the Government of India in the Department of Personnel and Training, Ministry of Personnel, Public Grievances and Pensions, North Block, New Delhi.

Shri Kumar has recently completed a two year regular Post Graduate Diploma in Management (Financial Management) Programme, equivalent to MBA, from the National Institute of Financial Management(NIFM), now renamed as Arun Jaitley NIFM, an autonomous body of the Ministry of Finance, Department of Expenditure, Government of India. He also participated in an Executive Program in Financial Management and Governance at University of California Riverside Extension, USA. His dissertation titled as 'Reforms, their impacts vis-à-vis contemporary issues and challenges in Public Sector Banks of India', may be identified to be germane to the present banking scenario in the country.

He started his career with an NGO and subsequently switched over teaching line, when he was only 21. He founded 'FAO Educational Campus' in his home district 'Jehanabad' in 1999 and taught thousands of students aspiring for government sector jobs and admission into precious national level institutions. A lot of students, now working in different Government organization and top corporate houses, have been benefitted hitherto by the benevolent initiative of Shri Kumar. He has sui generis pedagogy for diverse competitive examinations like SSC, Banks, Railways, CDS, CAT, MAT, etc. His several articles on Mathematics have been published in a magazine 'Geo Child', India and he also earned laurels from the intellectual community in the country.

Preface

At the very outset, I would like to explain as to what necessitated bringing out a new pattern textbook on Quantitative Aptitude when a plethora of books are available in market claiming to be full of short tricks (hot tricks/modern tricks!) to solve the problems in jiffy. However, as they claim, these so called tricks are not the tricks in true sense, these are only the results of the algebraic solution of the arithmetic questions. Moreover, these books are devoid of scientific approaches compelling the aspirants to remember the tricks without understanding the logic and rationale behind them. The real game starts when the aspirants waste their precious time in the examination hall in just selecting the suitable/applicable tricks on the basis of what they have memorized from the so called short-trick books, for solving the question/s. As the concept for solving the problems is not explained in these books even the master practitioner from these books fails to solve the questions even for a slight change in pattern or language.

I was very surprised when some students apprised me of the books in which question-wise tricks have been given. I also noticed that most of the students were capable to solve a particular question by so called tricks but not the same question in subjective mode. The dependency on the tricks and devoid of basic understandings were the main reasons.

It will not be out of context to mention that most of the books do not cover the entire syllabus of the competitive examinations and students are bound to buy/consult more than one book of mathematics to cater to their need. Also, solutions given in these books lack conceptual clarity and the questions based on the same concept are not placed together. They claim to include the tricks to solve the question quickly, but ironically, they don't include chapters even on fundamental calculations. Obviously, without conceptual clarity and without knowing the real tricks/methods for quick calculation, we cannot solve the questions swiftly.

In the above scenario when aspirants are groping in the dark for a comprehensive book on mathematics for competitive examinations which can cover all the nitty gritty of solving the problems with conceptual clarity, I decided to write an exhaustive book with scientific methods for solving questions promptly. Before penning this book an academic panel comprising the heads of certain institutions for competitive examinations, experienced Maths-teachers, aspirants and other eminent faculties, was constituted. The Panel discussed the problems being faced by aspirants in Quantitative Aptitude section at mental as well as psychological levels in detail. On recommendations of the panel, this book has been finalized. This book has also been vetted by them. It is based on **'one unit/chapter - one theme'.**

It took years of painstaking research to write this encyclopedic book. This book covers all the topics which are being asked contemporarily in diverse competitive examinations, follows an innovative, systematic and inspirational alternative conceptual structure which also provides opportunity to the aspirants to choose the best method to solve the problem, among various methods of solutions given, as per their choice and academic background. Needless to mention that the suitability of the methods depends upon the academic background and interest of the aspirants.

Questions included under the 'Illustrations' have been solved using 2-3 alternative approaches which shall not only widen the arena of thinking level of the aspirants but also provide opportunity to them to select the best approach. The problems under 'Work Out' which are similar to the illustrated problems have been solved with the best effective method only due to space constraints. However, alternative methods of solutions to the problems under 'Work Out' have been provided wherein it has been felt necessary.

As regards structures of this text book, it has been divided into 11 units and idea behind this division is 'different concept – different unit'. First unit augments your Calculation Ability, Second unit is related to certain Operations of Numbers and their Application, Third unit enumerates the wide application of Percentage, Fourth unit is based on the application of the concept of Ratio, Fifth unit is relating to application of both concepts i.e. Percentage and Ratio. Sixth unit is all about Mensuration, Seventh unit is Data-Interpretation, Eighth is Trigonometry, Ninth is Algebra and Tenth and Eleventh are Geometry and Co-ordinate Geometry respectively.

Each chapter/section commences with learning objectives and structures(Inside). Further, each chapter includes analysis of nature of questions as well as their solutions with alternative approaches. I can say with aplomb that most of the questions in upcoming competitive exams will not be out of the ambit of this book directly or indirectly.

Furthermore, **for the following reasons, the aspirant should consult this book:**

- This text book is suitable for all objective as well as subjective competitive examinations. In DSSSB, UP Lower Sub-ordinate exams, etc. subjective solutions are required. It is also useful for the secondary school students who are keen to build the foundation for future competitive examinations by widening their arena of thinking.
- It includes almost all chapters required as per the latest trends and patterns of the competitive examinations. 3000+ illustrations and 2500+ problems (fully solved) have been provided so as to cover all aspects associated with the chapters. Questions asked in recent years (based on memory in case of online examinations) have also been included. Thus, there is no need for the aspirants to consult any other books even any practice book, once they get hold of this book.
- For the ease of understanding and to avoid overlapping of concepts, this book has been divided into 11 units, each has different concept than others. Concepts to solve basic level to high difficulty level problems have been envisaged.
- Each unit of this text book has unique central theme/concept which is applicable for all the questions associated with the chapter/unit. Hence, there would be no stress on the aspirants in selecting the applicable approach.
- Each chapter/section commences with learning objectives and structures so that aspirants can test themselves as to what he gained after completion of the chapter.
- In arithmetic section almost all of the questions have been solved without algebraic equations to accelerate speed in solving the problems. Needless to mention that algebraic solutions require comparatively more time.

- The concepts associated with arithmetic section have been crafted/interpreted in such a way that problems of all the eight chapters under Unit- III (viz. Percentage, Profit & Loss, Discount, SI, CI, Time & Work, Work & Wages and Pipes & Cistern), Unit-IV (viz. Ratio, Proportion & Variation, Problems based on Ages, Partnership and Alligation/Mixture, Unit- V (viz. TSD, Average Speed, Relative Speed, To & Fro motion, Races & Games, Circular Motion, Problems based on trains, Boat & stream & Escalator) can be solved in moments without pen and paper if the aspirants get command over the 'concepts of 'Percentage' and 'Ratio'.
- A lot of questions under 'Illustrations', have been solved with 2 – 3 approaches so that the aspirants may choose the best method which suits them according to their educational background and interest. Certain questions have been solved with 5-6 different methods.
- It teaches us how to solve problems in moments with diverse inspirational approaches stressing on pragmatic aspects in lieu of academic aspects in coherent and lucid way. With these innovative and inspirational techniques, the aspirants shall be able to solve most of the questions in moments without pen and paper.
- It recognizes the sorts of questions which seek specific approaches and can easily/promptly be solved without pen and paper or/and through the given options. The time-factor in the examination hall has been the main causes for this endeavor.
- This book is also suitable for the students of non-mathematical background as diagrams have been used for better understanding.
- Critical facts, generic version of the questions and observations have been provided in the boxes/shaded portion to revise them quickly just before the examinations.
- This consummate book obviates the requirement of the coaching institutions as well as necessity of different previous years' question banks for respective exams to the greater extent. Obviously, it is like a boon for those working students who have no time for the institutes.

I hope this book would help the aspirants in cracking the various competitive examinations culminating in employment in Government sector and also in admissions in the precious institutes. Efforts have been made to achieve 6-sigma criteria. However, it cannot be denied possibility of any error. Therefore, dear students and faculty friends are requested to bring to my notice, any error they encounter while going through this book, so that the next edition can be error free.

Needless to say that improvement is a dynamic process and obviously valuable suggestions/comments of the readers are most welcome. I will be obliged if the aspirants share their suggestion/comments, observations, idea, query or doubt regarding this book at: mathsinmoments@gmail.com.

I wish the aspirants all the best for their bright career.

Author

Acknowledgement

(i) The successful completion of this consummate book gives me immense satisfaction, which would, in deed, be imperfect without referring to the personality who motivated me, encouraged me and guided me consistently.

(ii) I express my sincere gratitude and thanks to Smt. Rashmi Saxena Sahni (IRS-84), Principal Director General of Income Tax and my Guruji Shri G.N. Singh, Joint Secretary, Ministry of Agriculture and Farmers' Welfare, Government of India, Shri Prakash Kr. Tamrakar (Director, M/o SJ & E, GOI) and Dr. A. K. Sharan, Professor, Arun Jaitley National Institute of Financial Management for their motivation and guidance at macro and micro levels.

(iii) I convey my gratefulness and special thanks to Hon'be Judge Shri Virender Kumar Bansal, Director, Delhi Judicial Academy and Shri B.K.Goel, Member, CAG State Audit Advisory Board, GOI & Ex-Member, Haryana-SCPCR for their invariable support, guidance and critical suggestions without which I would not be able to complete this book in totality.

(iv) I am very grateful to all the members of the expert panel who spared their priceless time for sharing not only their experiences but also the innovative ideas as to how to secure maximum possible marks in Quantitative Aptitude in minimum span. The recommendations of the Panel shaped this book. The Panel also vetted the book.

(v) A special mention here is of Shri Sudama Prasad (Asstt. Audit Officer, O/o CAG) and Shri Arvind Kumar Suraj, Under Secretary, M/o Finance who spared their valuable time whenever I required in connection of writing this book.

(vi) I deeply acknowledge the support of my Brothers- Mr. Nitish Kumar(ITO) and Mr. Ritish Kumar(Custom Deptt.) and Sisters- Mrs. Kalawati Kumari (SI-Bihar Police) and Mrs. Jully Kumari (Block Panchayat Raj Officer-BPRO), who had been inspiring me constantly to write a very good book of Mathematics for diverse competitive examinations. They shared their difficulties in Maths in details, experiences during preparation for the exams. I also recognize the contribution of my son Mr. Shivansh to the greater extent who always used to say that '*Papa, you teach me far far better than my teachers, then why not you teach my all friends even my teachers*'. It gave me implicit strength to write this book. Mr. Kranti (pursuing MBBS, AIIMS), my nephew, had been inspiring me to do something for the unprivileged/weaker section in education arena. Shri Arun Kumar, Sr.SE, Railway and Shri Krishna Narayan Prasad, BPRO(both brothers in laws) contributed a lot in writing this book.

(vii) I acknowledge immensely to my all students specially Mr. Prasun (BOB-PO), Mrs. Jyoti(Corporate Sector), Mr. Sunil Kr. Meena (Lecturer), Mrs. Sajal (CSS), Mr. Ajay Kumar (Rly), Mr. Kamlesh Kumar(Rly) and many more........., who appreciated my teaching style and given me the motivation to write this monumental book of Mathematics.

(viii) I convey my thanks to my all senior officers and colleagues who supported and motivated me for this august work.

(ix) I am very grateful to Director, Shri Deepesh Jain, Shri Keshav Agarwal, Shri Ram Kishor CA and Priya Rai (Project Coordinator) all the staff of Arihant Publication for publishing as well as technical inputs. Without their supports and hard works, publication was not possible in such a lucid and limpid ways.

(x) Last but not least, I communicate my sincere and heartfelt thanks to my parents, close friends especially Mr. Santosh Kumar (IAS), Mr. Suraj Patel (IAS), Mr. Rajesh Kumar (IRS), Mr. Amit Kumar (IRS), Mr. Sanjay Kumar Meena (Suptt. GST), Dr. Sushil Kr. Vimal, Dr. Sunil Kr. Singh, Dr. Anand T. Gudivada, Shri Manoj Manu (Under Secretary, SSC), Shri Rajbir Singh (Under Secretary, DoPT), Mr. B.K.Singh (CSS), Mr. Unni Krishnan T (CSS), Mr. Rahul Kumar(BDO-Bihar), Mr. Vishal Vaibhav (Dy. Commandant, CRPF), Mrs. Suman Parmar (CSS), Mr. Ratnakar Yadav (CSS), Mrs Swati Jatav(CSS), Mr. Ramji Lal Jangir(CSS, DoPT), Mr. P.K.Barnwal(CSSS), Mr. Mukesh Kumar & Mr. Ashutosh Kumar (both- IIT-ISM), Mr. Sumit Kishore (Advocate, Supreme Court), Mrs. Priyanka Singh (Deoghar), Mrs Nisha Singh (Sr. Manager, IOB) and all DoPT Officers (DoPTians) who supported me in each and every step of this precious life.

For Information to **DEAR ASPIRANTS**

- Certain questions have been indicated as were asked in more than one competitive examinations. It is quite possibility that there may be slightly changes in the figures and languages of the question. However, there is no change in the conception/pattern associated with.
- In certain examinations like CAT, few questions don't have options. However, in order to bring uniformity in the fabric/pattern of the book, options have been created against such questions.
- Certain questions are based on the memory, hence, it may be possibility of slightly changes from the original questions.
- Problems under Illustrations have been solved in many ways, however, due to space constraints the problems under Work Out have been solved with the best effective approach.
- The author is of the view that the difficulty level of the questions depends upon the clarity of the conceptions associated with them. Since all the concepts have been elaborated immensely in limpid way in this book, no need to divide/present the questions under headings of different difficulty levels.

NRA : Changing the recruitment process...

In order to mitigate the hardship to candidates seeking Government jobs by providing the same platform to all candidates, and with a view to set a new standard of equity and inclusiveness in recruitment, Government has set up the National Recruitment Agency (NRA) vide order dated 28.08.2020. The mandate of NRA is to conduct first level screening of candidates through computer based online Common Eligibility Test (CET) at Graduate, Higher Secondary and Matriculate levels for jobs in Government sector for which recruitment is presently carried out through the Staff Selection Commission (SSC), Railway Recruitment Boards (RRBs) and the Institute of Banking Personnel Selection (IBPS).

NRA will conduct only preliminary screening of candidates and the final selection shall be done through separate specialized tests/examinations to be conducted by respective recruitment agencies viz. SSC, RRBs and IBPS. Based on the score obtained in CET, candidates may appear in the aforesaid specialized exams/tests. The score obtained by candidates in CET may also be used by the State Governments, the public sector and the private sector in their recruitments, based on a Memorandum of Understanding entered into with NRA.

Initially, a CET for each of the three levels (the graduate, the 12th pass and the 10th pass) shall be held bi-annually and NRA shall enhance the frequency of the CET at each level in a planned manner. There shall be no restriction on the number of attempts to be taken by a candidate to appear in the CET. Relaxation in the upper age limit shall be given to candidates of SC/ST/OBC and other categories as per the extant policy of the Government. As per the reply to a Lok Sabha Question, given in February, 2021, every year, approximately 2.5 crore candidates appear in multiple recruitment examinations for approximately 1.25 lakh vacancies.

Contents

Unit 1
Fundamentals

Addition and Subtraction • Multiplication
• Unit digit, last two digits and No. of Trailing Zeros of a Product/(Base)^{Exponent}
• Division • How to Compute Remainder
• Concept of Factor and Factor-Theorem • Comparison of Fractions
• Square and Cube • Square- Root, Cube-Root and *n*th-Root
• Indices and Surds • Simplification

01

Addition and Subtraction

Learning OBJECTIVES

After completing this chapter, aspirants will be able to

- Understand the significance of this chapter as the mother of all arithmetic chapters;
- Learn how to compute addition and subtraction of Whole numbers, Decimal numbers, Fractions, Recurring decimals, a series of natural numbers and numbers/terms given in a special/unique series;
- Understand the meaning of Recurring decimal;
- Realise difference between conventional approaches and alternative approaches;
- Understand the philosophy of the division of this chapter in various parts;
- Analyse the trend of the questions being asked in various competitive examinations;
- Solve the questions in moments;
- Grip on concepts helping to solve the questions without pen and paper;
- Learn new concepts/fundamentals;
- Apply the pragmatic approaches in lieu of academic approaches;
- Ponder on the alternative methods; and
- Know many more interesting and useful facts.

Inside

- *Introduction*
- *Addition and Subtraction of Whole Numbers*
- *9's Rule*
- *Addition and Subtraction of Decimal Numbers*
- *Addition and Subtraction of Fractions*
- *Recurring Decimal and their Addition and Subtraction*
- *Sum of Series of Natural Numbers*
- *Unique Series based on Addition and Subtraction*
- ....
- .... *& Many More*
- *Work Out with Solutions*

Introduction

You would appreciate to know that all the arithmetic problems are associated with only two operations - Addition and Subtraction. You may contend that multiplication, division, etc., are also instruments to solve the arithmetic questions. In this regard, it will not be out of context to mention that these operations are nothing but diverse versions of the addition and subtraction *viz.* multiplication is the shortest form of addition, division is the shortage form of subtraction, square is the another form of multiplication, etc.

Now, I think, no need to mention the importance of this chapter.

The techniques envisaged in this chapter will obviously reduce the time taken in solving the questions related to almost all arithmetical chapters. This chapter is very critical especially for banking services examinations wherein questions are based on direct applications of the addition and subtraction. The direct applications of addition and subtraction are also seen in the chapters like Simplification, Data Interpretations, etc., to the greater extent.

Addition and Subtraction may have diverse dimensions like

A. Addition and Subtraction of Whole numbers
B. Addition and Subtraction of Decimal numbers
C. Addition and Subtraction of Fractions
D. Addition and Subtraction of Recurring decimals
E. Addition of series of Natural numbers
F. Unique series based on Addition and Subtraction:
 (i) Series seeking no special concept(s)
 (ii) Based on sum of natural numbers/their squares/cubes
G. Addition of reciprocals of consecutive natural numbers

A. Addition and Subtraction of Whole Numbers

Type I. When the given whole numbers are added

$$69375 + 8621 + 3425 + 132 = ?$$

***Step* I** Start adding from extreme right digits (i.e. unit place digits)

⇒ 2 + 5 + 1 + 5 = ①❸

Here, only extreme right digit of the sum i.e. ❸ has to be written and the remaining portion i.e.① will be carried over the sum of ten's place digits.

***Step* II** Add ten's place digits

⇒ 3 + 2 + 2 + 7 + ① = ①❺

Here, ❺ will be written and the remaining portion i.e. ① will be carried over the sum of hundred's place digits.

***Step* III** Add hundred's place digits

⇒ 1 + 4 + 6 + 3 + ① = ①❺

Here, ❺ will be written and the remaining portion i.e. ① will be carried over the sum of thousand's place digits.

***Step* IV** Add thousand's place digits

⇒ 3 + 8 + 9 + ① = ②❶

Here, ❶ will be written and the remaining portion i.e.② will be carried over the sum of ten thousand's place digits.

***Step* V** Add ten thousand's place digits:

⇒ 6 + ② = ❽

Now, arrange the dark circles from right to left starting from Step I, i.e.❽❶❺❺❸. Thus, the sum of given numbers = 81553.

Note *These steps should be done mentally.*

Type II. When addition and subtraction both operations are required

$$76385 - 43216 + 2164 - 437 + 8 = ?$$

Conventional Approach

***Step* I** Add the numbers with positive sign and append + sign before the resultant.

⇒ 76385 + 2164 + 8 = (+) 78557

***Step* II** Add the numbers with negative sign and append – sign before the resultant.

⇒ 43216 + 437 = (–) 43653

***Step* III** Subtract smaller from the bigger and append the sign of bigger before the outcome.

⇒ 78557 – 43653 = (+) 34904

Limitation : This approach requires much space as well as much time.

Alternative Approach

We add the unit's place digits in Step I, then ten's place digits in Step II, then hundred's place digits in Step III and so on. We process the same with following conditions.

A.	If sum of the digits in the same place is	(+) single digit	Write the same digit.
B.	......do......	(+) double digit	Write extreme right digit and the remaining portion will be carried over the next sum of digits of the same place.
C.	......do......	(–) single digit	Subtract the digit(ignoring sign) from 10 and write the resultant. Further, (–1) will be carried over the next sum of digits of the same place.
D.	......do......	(–) double digit	Subtract the digit(ignoring sign) from just next multiple of 10 and write the resultant. If subtraction is made from *nth* multiple of 10, (–*n*) be carried over the next sum of digits of the same place. *For example,* if subtraction is made from 20, (–2) will be carried over.

Note *The similar process is adopted in case sum of the digits in the same place is either in (+) three digits or (–) three digits.*

Take one example, 76385 – 43216 + 2164 – 437 + 8 = ?

***Step* I** Add unit place digits with signs.

⇒ + 8 – 7 + 4 – 6 + 5 = ❹

***Step* II** Add ten's place digits with signs.

⇒ – 3 + 6 – 1 + 8 = ①⓿

Here, ⓿ will be written and the remaining portion i.e. ① will be carried over the sum of hundred's place digits.

Step **III** Add hundred's place digits with signs.

$\Rightarrow -4+1-2+3+$ ① $=-1$ but ❾

We will write ❾ as we have to subtract 1 from multiple of 10, which is just above 1.

Since 1 is subtracted from 10, (– ①) will be carried over the sum of thousand's place digits.

Step **IV** Add thousand's place digits with signs.

$\Rightarrow +2-3+6+$ (– ①) = ❹

Step **V** Add ten thousand's place digits with signs.

$\Rightarrow \quad -4+7=$ ❸

All the above steps can be done mentally.

Now, arrange the dark circles from right to left starting from Step I i.e. ❸❹❾⓿❹.

Thus, the required answer = 34904.

➘ Problem 1 $74156 - ? - 321 - 20 + 520 = 69894$

[BOM(PO)-08]

Sol. We should alienate the terms from the question mark (?). In other words, all the terms one side and question mark another side.

Now, $\quad 74156 - 321 - 20 + 520 - 69894 = ?$

Note *You are advised to assume all the terms one side and question mark(?) another side without shifting the term(s) either on Left Hand Side(LHS) or Right Hand Side(RHS) in actual so as to save time.*

Step **I** $\quad -4+0-0-1+6=\underline{1}$

Step **II** $\quad -9+2-2-2+5=-6$, i.e. $(10-6=\underline{4})$

Step **III** $\quad -8+5-3+1+(-1)=-6$, i.e. $(10-6=\underline{4})$

Step **IV** $\quad -9+4+(-1)=-6$, i.e. $(10-6=\underline{4})$

Step **V** $\quad -6+7+(-1)=\underline{0}$

Hence, required answer = 04441

It is also important to note that you may reduce the number of terms by solving some terms orally. *For example,* you may write the question as 74156 – 341 + 520 – 69894 = ? to save few moments.

➘ Problem 2 $-4798 + 231 - ? - 378 = -8273$

Sol. ***Step*** **I** $\quad +3-8+1-8=-12$, i.e. $(20-12=\underline{8})$

Step **II** $\quad 7-7+3-9+(-2)=-8$, i.e. $(10-8=\underline{2})$

Step **III** $\quad +2-3+2-7+(-1)=-7$, i.e. $(10-7=\underline{3})$

Step **IV** $\quad 8-4+(-1)=3$

Hence, required answer = 3328

➘ Problem 3 $3765 - 346 + 1870 - ? = 6524$

Sol. ***Step*** **I** $\quad -4+0-6+5=-5$, i.e. $(10-5=\underline{5})$

Step **II** $\quad -2+7-4+6+(-1)=6$

Step **III** $\quad -5+8-3+7=7$

Step **IV** $\quad -6+1+3=-2$, i.e. $(10-2=\underline{8})$

It may be observed from all the above questions except last one that the last step yields positive number whereas the last question yields negative number in last step. In these circumstances, you should subtract the final result i.e., 8765 from the number formed by appending zeros as many as number of steps/number of digits existing in a number to the greater extent, on right side of 1, 2, ... as the case may be. 1 if last digit is subtracted from 10, 2 if subtraction from 20 and so on. The outcome on subtraction with the negative sign will be the final answer.

Hence, final answer = – (10000 – 8765) = – 1235.

➘ Problem 4 $-6369 + 1236 - 5696 + 327 = ?$

Sol. ***Step*** **I** $\quad +7-6+6-9=-2$, i.e. $(10-2=\underline{8})$

Step **II** $\quad +2-9+3-6+(-1)=-11$, i.e. $(20-11=\underline{9})$

Step **III** $\quad +3-6+2-3+(-2)=-6$, i.e. $(10-6=\underline{4})$

Step **IV** $\quad -5+1-6+(-1)=-11$, i.e. $(20-11=9)$

Hence, required answer = – (20000 – 9498) = – 10502.

All the steps can be done mentally, so try yourself.

9's Rule

The sum of digits seen in all numbers to be added, leaves the same remainder as the sum of the digits of resultant leaves in case both are divided by 9.

For example, 13 + 28 + 137 = 178

If we divide (1 + 3 + 2 + 8 + 1 + 3 + 7) by 9, we get 7 as remainder.

Further, if we divide (1 + 7 + 8) by 9, we get 7 as remainder.

This property is generally known as 9's Rule. We may also find the remainder in the following manners.

(a) $13+28+137 \Rightarrow 4+10+11 \Rightarrow 4+1+0+1+1=7$

(7 will leave 7 as remainder if it is divided by 9)

(b) $13+28+137 \Rightarrow 4+(10-9)+(11-9)=7$

(we have considered only remainders)

(c) ~~1~~ + 3 + 2 + ~~8~~ + 1 + 3 + 7 ⇒ ~~5~~ + ~~1~~ + ~~3~~ + 7 = 7

You are free to prefer any mode whichever suits you best.

Observation This rule has significance only when we desire to check the correctness of the answer, otherwise it is not so useful in the competitive examinations as it will consume your precious moments in the examination hall.

Take one example:

➘ Problem 5 $74156 - ? - 321 - 20 + 520 = 69894$

Sol. $74156 - 321 - 20 + 520 - 69894 = ?$

$\Rightarrow \quad 23-6-2+7-36$ (on addition of digits)

$\Rightarrow \quad 5-6-2+7-0$

[considering only remainders if they are divided by 9]

$\Rightarrow 4$

Let the options be 3451, 4441, 5401 and 4531.

Now, proceed through options.

(a) 3451 $\Rightarrow (3+4+5+1) \div 9 \Rightarrow$ remainder = 4

(b) 4441 $\Rightarrow (4+4+4+1) \div 9 \Rightarrow$ remainder = 4

(c) 5401 $\Rightarrow (5+4+0+1) \div 9 \Rightarrow$ remainder = 1

(d) 4531 $\Rightarrow (4+5+3+1) \div 9 \Rightarrow$ remainder = 4

It appears that either (a) or (b) or (d) is the correct answer as the sum of digits seen in all numbers leaves the same remainder as the sum of the digits of resultant leaves in case both are divided by 9. It connotes that 9's Rule should be avoided despite being an interesting method.

B. Addition and Subtraction of Decimal Numbers

***Case* I** When number of digits after decimal point is same in all the given numbers, we simply solve the problems as has been done in case of sum of whole numbers assuming there are no decimal points in any number. Thereafter, we append a decimal point on the resultant from the extreme right hand side after the same number of digits as each number has after decimal point. For example:

Problem 6 792.02 + 101.32 − 306.76 = ?

Sol. Assume there is no decimal points in any number.

Step I $-6+2+2=-2$, i.e. $(10-2=\underline{8})$

Step II $-7+3+0+(-1)=-5$, i.e. $(10-5=\underline{5})$

Step III $-6+1+2+(-1)=-4$, i.e. $(10-4=\underline{6})$

Step IV $-0+0+9+(-1)=8$

Step V $-3+1+7=5$

Hence, result = 58658

Now, append decimal point to the resultant after two digits from the extreme right as each number has two digits after decimal point. Thus, the required answer = 586.58

Case II When number of digits after decimal point is not same in all the given numbers, we make them equal by appending required number of zeros on their extreme right. Then, we simply solve the problems as has been done in case of sum of whole numbers assuming there are no decimal points in any number.

Thereafter, we append a decimal point on the resultant from the extreme right after the same number of digits as each number has after decimal point after equalisation.

Problem 7 4 + 4.44 + 44.4 + 4.04 + 444 = ? ***[BOB(PO)-07]***

Sol. Given, expression = 4.00 + 4.44 + 44.40 + 4.04 + 444.00

(We have equalised the number of digits after the decimal point in each number)

Now, we solve on the above line assuming there is no decimal point in any number.

Then, we get the resultant as 50088.

Then, required answer = 500.88

C. Addition and Subtraction of Fractions

When an object is divided into a number of equal parts, then each part is called a fraction.

For example Two fifth of an object may be written as 2/5. 2 is called numerator*(N)* and 5 is called denominator *(D)*. There are three types of fractions.

(i) **Proper fraction** When numerator < denominator, *viz.* 2/5, 7/9, etc.

(ii) **Improper fraction**
When numerator > denominator, *viz.* 6/5, 10/9, etc.

(iii) **Mixed fraction** It has a whole number plus a fraction, *viz.* $12\frac{1}{2}$, $8\frac{1}{3}$, etc.

Addition and Subtraction of fractions with/without whole numbers

Types	Addition	Subtraction
1. *Whole number and Proper fraction*	⊙ $6+\frac{3}{5}=?$ ***Conventional approach*** $\frac{6}{1}+\frac{3}{5}=\frac{\left(6\times\frac{5}{1}\right)+\left(3\times\frac{5}{5}\right)}{\text{LCM of 1 and 5 i.e. 5}}$ $=\frac{33}{5}=6\frac{3}{5}$ ***Alternatively*** Delete only '+' sign. Hence, $6+\frac{3}{5}=6\frac{3}{5}$ Similarly, $17+\frac{19}{29}=17\frac{19}{29}$	⊙ $6-\frac{3}{5}=?$ ***Conventional approach*** $\frac{6}{1}-\frac{3}{5}=\frac{\left(6\times\frac{5}{1}\right)-\left(3\times\frac{5}{5}\right)}{\text{LCM of 1 and 5 i.e. 5}}=\frac{27}{5}=5\frac{2}{5}$ ***Alternatively*** $6-\frac{3}{5}=\mathbf{5}+\left(1-\frac{3}{5}\right)=5+\frac{5-3}{5}$ $\left(\text{i.e. }5+\frac{D-N}{D}\right)$ Hence, $(6-1)\frac{D-N}{D}$ Similarly, $17-\frac{19}{29}=(17-1)\frac{29-19}{29}=16\frac{10}{29}$

Types	Addition	Subtraction
2. *Proper Fraction and Proper fraction(s) When D(s) are same*	$\frac{3}{7}+\frac{2}{7}=\frac{3+2}{7}=\frac{5}{7}$ $\frac{8}{9}+\frac{2}{9}+\frac{7}{9}=\frac{8+2+7}{9}=\frac{17}{9}=1\frac{8}{9}$	$\frac{3}{7}-\frac{2}{7}=\frac{3-2}{7}=\frac{1}{7}$ $\frac{8}{9}-\frac{2}{9}-\frac{7}{9}=\frac{8-2-7}{9}=\frac{-1}{9}=-\frac{1}{9}$
3. *Proper fraction and Proper fraction(s) When one D(s) is/are multiple(s) of other D(s)*	$\frac{1}{2}+\frac{3}{8}=?$ ***Conventional Approach*** $\frac{\left(1\times\frac{8}{2}\right)+\left(3\times\frac{8}{8}\right)}{\text{LCM of 2 and 8 i.e. 8}}=\frac{7}{8}$ ***Alternatively,*** Make both Ds equal by multiplying *N* and *D* by the same number and proceed as type-2. $\frac{1\times4}{2\times4}+\frac{3}{8}=\frac{4}{8}+\frac{3}{8}=\frac{7}{8}$ $\frac{5}{8}+\frac{1}{2}+\frac{3}{4}=\frac{5}{8}+\frac{1\cdot4}{2\cdot4}+\frac{3\cdot2}{4\cdot2}=\frac{15}{8}=1\frac{7}{8}$	$\frac{1}{2}-\frac{3}{8}=?$ ***Conventional Approach*** $\frac{\left(1\times\frac{8}{2}\right)-\left(3\times\frac{8}{8}\right)}{\text{LCM of 2 and 8 i.e. 8}}=\frac{1}{8}$ ***Alternatively,*** Make both Ds equal by multiplying *N* and *D* by the same number and proceed as type-2. $\frac{1\times4}{2\times4}-\frac{3}{8}=\frac{4}{8}-\frac{3}{8}=\frac{1}{8}$ $\frac{5}{8}-\frac{1}{2}-\frac{3}{4}=\frac{5}{8}-\frac{1\cdot4}{2\cdot4}-\frac{3\cdot2}{4\cdot2}=\frac{-5}{8}$
4. *Proper Fraction and Proper fraction(s) When one D is not multiple of another*	$\frac{2}{3}+\frac{1}{5}=?$ ***Conventional Approach*** $\frac{\left(2\times\frac{15}{3}\right)+\left(1\times\frac{15}{5}\right)}{\text{LCM of 3 and 5 i.e. 15}}=\frac{13}{15}$ ***Alternatively,*** $\frac{\text{Sum of cross multiplication}}{\text{Multiplication of both Ds}}=\frac{2\cdot5+1\cdot3}{3\cdot5}=\frac{13}{15}$ [it is applicable for type-3 also]. $\frac{7}{11}+\frac{3}{4}=\frac{7\cdot4+3\cdot11}{11\cdot4}=\frac{61}{44}=1\frac{17}{44}$	$\frac{2}{3}-\frac{1}{5}=?$ ***Conventional Approach*** $\frac{\left(2\times\frac{15}{3}\right)-\left(1\times\frac{15}{5}\right)}{\text{LCM of 3 and 5 i.e. 15}}=\frac{7}{15}$ ***Alternatively,*** $\frac{\text{Left dia. multiplication}-\text{Right dia. multiplication}}{\text{Multiplication of both Ds}}$ (dia. ⇒ Diagonally) $=\frac{2\cdot5-1\cdot3}{3\cdot5}=\frac{7}{15}$ [it is applicable for type-3 also]. $\frac{7}{11}-\frac{3}{4}=\frac{7\cdot4-3\cdot11}{11\cdot4}=\frac{-5}{44}$
5. *Whole number and mixed fraction*	$18+13\frac{1}{17}=?$ ***Conventional Approach*** $\frac{18}{1}+\frac{222}{17}=\frac{\left(18\times\frac{17}{1}\right)+\left(222\times\frac{17}{17}\right)}{\text{LCM of 1 and 17 i.e. 17}}=\frac{528}{17}=31\frac{1}{17}$ ***Alternatively,*** $18+13\frac{1}{17}=(18+13)\,\frac{1}{17}=31\frac{1}{17}$ $8+12\frac{10}{13}=(8+12)\,\frac{10}{13}=20\frac{10}{13}$	$18-13\frac{1}{17}=?$ ***Conventional Approach*** $\frac{18}{1}-\frac{222}{17}=\frac{\left(18\times\frac{17}{1}\right)-\left(222\times\frac{17}{17}\right)}{\text{LCM of 1 and 17 i.e. 17}}$ $=\frac{84}{17}=4\frac{16}{17}$ ***Alternatively,*** $18-13\frac{1}{17}=18-\left(13+\frac{1}{17}\right)=(18-13)-\frac{1}{17}$ $=5-\frac{1}{17}=4\frac{16}{17}$ [see type-1] $8-12\frac{10}{13}=8-\left(12+\frac{10}{13}\right)(8-12)-\frac{10}{13}$ $=-4-\frac{10}{13}=-\left(4+\frac{10}{13}\right)=-4\frac{10}{13}$

Types	Addition	Subtraction

6. *Proper fraction and mixed fraction*

Addition

$\frac{1}{2} + 18\frac{3}{7} = ?$

Conventional Approach

$$\frac{1}{2} + \frac{129}{7} = \frac{\left(1 \times \frac{14}{2}\right) + \left(129 \times \frac{14}{7}\right)}{\text{LCM of 2 and 7 i.e. 14}} = \frac{265}{14} = 18\frac{13}{14}$$

Alternatively,

Add whole numbers and fractions separately.

$$0\frac{1}{2} + 18\frac{3}{7} = (0 + 18) + \left(\frac{1}{2} + \frac{3}{7}\right)$$

$$= 18 + \frac{7 \cdot 1 + 2 \cdot 3}{2 \cdot 7} = 18\frac{13}{14} \quad \text{[see type-4]}$$

$$13\frac{3}{4} + \frac{11}{12} = 13\frac{3}{4} + 0\frac{11}{12} = (13 + 0) + \left(\frac{3}{4} + \frac{11}{12}\right)$$

$$= 13 + \left(\frac{3 \cdot 3}{4 \cdot 3} + \frac{11}{12}\right) \quad \text{[see type-3]}$$

$$= 13 + \frac{20}{12} = 13 + 1\frac{2}{3} = 14\frac{2}{3}$$

Subtraction

$\frac{1}{2} - 18\frac{3}{7} = ?$

Conventional Approach

$$\frac{1}{2} - \frac{129}{7} = \frac{\left(1 \times \frac{14}{2}\right) - \left(129 \times \frac{14}{7}\right)}{\text{LCM of 2 and 7 i.e. 14}} = -\frac{251}{14} = -17\frac{13}{14}$$

Alternatively,

Process whole numbers and fractions separately.

$$0\frac{1}{2} - 18\frac{3}{7} = (0 - 18) + \left(\frac{1}{2} - \frac{3}{7}\right) = -18 + \frac{7 \cdot 1 - 2.3}{2 \cdot 7}$$

[like type-4]

$$= -18 + \frac{1}{14} = -\left(18 - \frac{1}{14}\right) = -17\frac{13}{14} \quad \text{[see type-1]}$$

$$13\frac{3}{4} - \frac{11}{12} = 13\frac{3}{4} - 0\frac{11}{12} = (13 - 0) + \left(\frac{3}{4} - \frac{11}{12}\right)$$

$$= 13 + \left(\frac{3 \cdot 3}{4 \cdot 3} - \frac{11}{12}\right) \quad \text{[see type-3]}$$

$$= 13 + \frac{-2}{12} = 13 - \frac{1}{6} = 12\frac{5}{6}$$

Addition

$18\frac{3}{8} + 29\frac{13}{15} = ?$

Conventional Approach

$$\frac{147}{8} + \frac{448}{15} = \frac{\left(147 \times \frac{120}{8}\right) + \left(448 \times \frac{120}{15}\right)}{\text{LCM of 8 and 15 i.e. 120}}$$

$$= \frac{5789}{120} = 48\frac{29}{120}$$

Alternatively,

Add whole numbers and fractions separately.

$$18\frac{3}{8} + 29\frac{13}{15} = (18 + 29) + \left(\frac{3}{8} + \frac{13}{15}\right)$$

$$= 47 + \frac{15 \cdot 3 + 8 \cdot 13}{8 \cdot 15} \quad \text{[see type-4]}$$

$$= 47 + \frac{149}{120} = 47 + 1\frac{29}{120} = 48\frac{29}{120}$$

Subtraction

$18\frac{3}{8} - 29\frac{13}{15} = ?$

Conventional Approach,

$$\frac{147}{8} - \frac{448}{15} = \frac{\left(147 \times \frac{120}{8}\right) - \left(448 \times \frac{120}{15}\right)}{\text{LCM of 8 and 15 i.e. 120}}$$

$$= \frac{-1379}{120} = -11\frac{59}{120}$$

Alternatively,

Process whole numbers and fractions separately.

$$18\frac{3}{8} - 29\frac{13}{15} = (18 - 29) + \left(\frac{3}{8} - \frac{13}{15}\right)$$

$$= -11 + \frac{15 \cdot 3 - 8 \cdot 13}{8 \cdot 15} \quad \text{[see type-4]}$$

$$= -11 - \frac{59}{120} = -\left(11 + \frac{59}{120}\right) = -11\frac{59}{120}$$

7. *Mixed fraction and mixed fraction(s)*

Addition

$$113\frac{7}{8} + 56\frac{1}{3} + 59\frac{5}{16} = ?$$

$$= (113 + 56 + 59) + \left(\frac{7}{8} + \frac{1}{3} + \frac{5}{16}\right)$$

$$= 228 + \left(\frac{7 \cdot 2}{8 \cdot 2} + \frac{5}{16} + \frac{1}{3}\right) \quad \text{[see type-3]}$$

$$= 228 + \left(\frac{19}{16} + \frac{1}{3}\right) = 228 + \left(1\frac{3}{16} + \frac{1}{3}\right)$$

$$= 228 + 1 + \left(\frac{3}{16} + \frac{1}{3}\right) = 229 + \frac{3 \cdot 3 + 1 \cdot 16}{16 \cdot 3}$$

[see type-4]

$$= 229 + \frac{25}{48} = 229\frac{25}{48}$$

Alternatively,

$$113\frac{7 \cdot 6}{8 \cdot 6} + 56\frac{1 \cdot 16}{3 \cdot 16} + 59\frac{5 \cdot 3}{16 \cdot 3} \quad \text{[see type-3]}$$

$$= 228\frac{73}{48} = 228 + 1\frac{25}{48} = 229\frac{25}{48}$$

Subtraction

$$113\frac{7}{8} - 56\frac{1}{3} - 59\frac{5}{16} = ?$$

$$= (113 - 56 - 59) + \left(\frac{7}{8} - \frac{1}{3} - \frac{5}{16}\right)$$

$$= -2 + \left(\frac{7 \cdot 2}{8 \cdot 2} - \frac{5}{16} - \frac{1}{3}\right) \quad \text{[see type-3]}$$

$$= -2 + \left(\frac{9}{16} - \frac{1}{3}\right) = -2 + \left(\frac{9 \cdot 3 - 1 \cdot 16}{16 \cdot 3}\right)$$

$$= -2 + \left(\frac{11}{48}\right) = -\left(2 - \frac{11}{48}\right) = -1\frac{37}{48} \quad \text{[see type-1]}$$

Alternatively,

$$113\frac{7 \cdot 6}{8 \cdot 6} - 56\frac{1 \cdot 16}{3 \cdot 16} - 59\frac{5 \cdot 3}{16 \cdot 3} \quad \text{[see type-3]}$$

$$= (113 - 56 - 59) + \frac{11}{48} = -2 + \left(\frac{11}{48}\right) = -\left(2 - \frac{11}{48}\right)$$

$$= -1\frac{37}{48} \quad \text{[see type-1]}$$

Now solve few questions without pen and papers.

Problem 8 $7\frac{8}{11}+19\frac{1}{11}-8\frac{7}{55}-?=10\frac{6}{11}$

Sol. Alienate the terms from the question mark (?). In other words, all the terms one side and question mark another side as has been done in case of addition and subtraction of whole numbers. But, it should be mentally without writing.

$$\Rightarrow 7\frac{8}{11}+19\frac{1}{11}-8\frac{7}{55}-10\frac{6}{11}=?$$

$$\Rightarrow (7+19-8-10)+\left(\frac{8}{11}+\frac{1}{11}-\frac{7}{55}-\frac{6}{11}\right)$$

$$=8+\left(\frac{8}{11}+\frac{1}{11}-\frac{6}{11}-\frac{7}{55}\right)$$

$$=8+\left(\frac{3}{11}-\frac{7}{55}\right)$$

$$=8+\left(\frac{3\cdot 5}{11\cdot 5}-\frac{7}{55}\right)=8\frac{8}{55}$$

Problem 9 $3\frac{6}{7}-6\frac{1}{4}+5\frac{1}{3}=?$ ***[Andhra Bank(PO)-11]***

Sol. $(3-6+5)+\left(\frac{6}{7}-\frac{1}{4}+\frac{1}{3}\right)=2+\left(\frac{6\cdot 4-1\cdot 7}{7\cdot 4}+\frac{1}{3}\right)$

$$=2+\left(\frac{17}{28}+\frac{1}{3}\right)$$

$$=2+\left(\frac{17\cdot 3+1\cdot 28}{28\cdot 3}\right)=2\frac{79}{84}$$

Problem 10 $4\frac{1}{2}+6\frac{2}{3}+5\frac{1}{3}=?$ ***[NABARD (Asstt. Mngr)-02]***

Sol. $(4+6+5)\left(\frac{1}{2}+\frac{2}{3}+\frac{1}{3}\right)=15+\left(\frac{1}{2}+\frac{3}{3}\right)$

$$=15+1+\frac{1}{2}=16\frac{1}{2}$$

D. Recurring Decimal and their Addition and Subtraction

A Recurring Decimal (also known as Repeating Decimal or Circulating decimal) is a decimal which contains a digit or set of digits after the decimal point that repeat without end. The portion which repeats is generally shown by dots over the first and last digits of the repeating pattern/set. It is also shown by a line (bar) over the pattern. Sometimes, repeating pattern is shown within a bracket too.

For example, $\frac{1}{3}=0.3333\ldots\ldots=0.\dot{3}$ or $0.\overline{3}$ or $0.(3)$

$\frac{9}{11}=0.818181\ldots\ldots\ldots=0.\dot{8}\dot{1}$ or $\overline{81}$ or $0.(81)$

$\frac{22}{7}=3.142857142857\ldots\ldots=3.\dot{1}4285\dot{7}$

or $3.\overline{142857}$ or $3.(142857)$

Before elaborating the process of addition and subtraction, we will discuss 'Conversion of Decimal into fractions'.

Observe some examples

1. Convert $3.\dot{3}$ into fraction.

$\Rightarrow \quad 3.\dot{3}=3.3333\ldots$

Let the value of 3.3333 .... be x, then

$x=3.3333\ldots$ ...(i)

$10x=33.3333\ldots$ ...(ii)

[multiplying both sides by 10]

On subtracting Eq. (i) from Eq. (ii), we get $9x=30$

$\Rightarrow \quad x=\frac{30}{9}=\frac{10}{3}$

2. Convert $0.\overline{81}$ into fraction.

Let $0.818181\ldots=x$ ...(i)

Then, $10x=8.18181\ldots\ldots$ ...(ii)

Now, it is not possible to obtain integer value if we subtract the Eq. (i) from Eq. (ii).

Thus, we may multiply Eq. (i) by 100.

$\Rightarrow \quad 100x=81.8181\ldots$ ...(iii)

Now, on subtraction of Eq. (i) from Eq. (iii), we get

$100x-x=(81.8181\ldots)-(0.818181\ldots)$

$\Rightarrow \quad 99x=81 \Rightarrow x=\frac{9}{11}$

3. Convert $0.5\dot{3}$ into fraction.

Let $0.5333333\ldots.=x$ ...(i)

Then, $10x=5.333333\ldots\ldots$ ...(ii)

Now, it is not possible to obtain integer value if we subtract the Eq. (i) from Eq. (ii).

Thus, we may multiply Eq. (i) by 100.

$\Rightarrow \quad 100x=53.333333\ldots$ ...(iii)

Now, on subtraction of Eq. (ii) from Eq. (iii), we get

$100x-10x=(53.333333\ldots.)-(5.333333\ldots\ldots)$

$\Rightarrow \quad 90x=48 \Rightarrow x=\frac{8}{15}$

Observation All the above proceedings conclude that we may write the fraction directly by adopting a fixed pattern as below.

Required fraction = {Given number ignoring decimal as well as bar − number(ignoring decimal) consisting of all digits except bar-digit(s)} ÷ {Number formed by appending zero(s) on 1 as many as number of digits after the decimal point − number formed by appending zero(s) on 1 as many as number of digits having no bar after the decimal point}

Note *If there is no any digit having no bar after decimal point, we append nothing on 1. It means, we subtract only 1.*

Now, convert $0.5\dot{3}$ into fraction.

Required fraction $=\frac{053-05}{100-10}=\frac{48}{90}=\frac{8}{15}$

Problem 11 $0.\overline{001}$ is equal to *[SSC (CGL-T-I)-10]*

Sol. Required fraction $= \dfrac{0001-0}{1000-1} = \dfrac{1}{999}$

Problem 12 $2.8\overline{768}$ is equal to *[SSC (CPO-SI)-06]*

Sol. Required fraction $= \dfrac{28768-28}{10000-10} = \dfrac{28740}{9990} = \dfrac{958}{333} = 2\dfrac{292}{333}$

Alternatively,

We may convert the given number in fraction without considering the whole number.

Now, convert $0.8\overline{768}$ into fraction. Thereafter 2 will be added in that fraction.

$$0.8\overline{768} = \frac{08768-8}{10000-10} = \frac{8760}{9990} = \frac{292}{333}$$

Hence, the required answer $= 2 + \dfrac{292}{333} = 2\dfrac{292}{333}$. Hence, place the whole number (left earlier) just before the fraction obtained.

Note *No doubt, we should adhere to the alternative approach in such types of cases so as to avoid mistake while converting the improper fraction into mixed fraction.*

Addition and Subtraction of Recurring Decimals

Take few examples to understand the concepts.

$0.34\overline{67} + 0.13\overline{33}$ is equal to:

[SSC (CGL-PT)-02, LIC (AAO)-09, CDS-19]

$$\Rightarrow \frac{03467-034}{10000-100} + \frac{01333-013}{10000-100} = \frac{4753}{10000-100}$$

$$= \frac{4801-48}{10000-100} = 0.48\overline{01}$$

[also, you may divide 4753 by 9900 to get $0.48\overline{01}$]

Alternatively,

```
      0.3467676767...
  (+) 0.1333333333...
  -------------------
      0.4801010100...
```

Clearly, repetition starts from a set of 01 onwards. However, you observe that the last digit of sum is 0 (in lieu of 1 as should have been), which does not seem to be in order. In this regard, it is to note that last digit 0 will automatically become 1 when given numbers are expanded further.

Thus, required answer $= 0.48\overline{01}$

Remark

(i) You may take hints from the given options as to how much expansion of recurring decimals should be appropriate to ascertain the pattern.

(ii) Students are advised to adhere to the alternative approach to get the answer quickly.

Problem 13 $8.3\dot{1} + 0.\dot{6} + 0.00\dot{2}$ is equal to

Sol.

```
      8.311111....
      0.666666....
  (+) 0.002222....
  ----------------
      8.979999....
```

Hence, required answer $= 8{\cdot}97\dot{9}$

Problem 14 $3.\overline{87} - 2.\overline{59} = ?$

Sol.

```
      3.878787....
  (-) 2.595959....
  ----------------
      1.282828....
```

Hence, required answer $= 1.\overline{28}$

Problem 15 $3.\overline{36} - 2.\overline{05} + 1.\overline{33}$ equals

Sol.

```
      3.363636...
  (-) 2.050505....
  ----------------
      1.313131....
  (+) 1.333333....
  ----------------
      2.646464....
```

Hence, required answer $= 2.\overline{64}$

Problem 16 The difference between $5.\overline{76}$ and $2.\dot{3}$ is *[SSC (CISF-ASI)-10]*

Sol.

```
      5.767676....
  (-) 2.333333....
  ----------------
      3.434343....
```

Hence, required answer $= 3.\overline{43}$

Problem 17 $4.2\overline{69} + 10.\overline{321}$ is

Sol.

```
       4.269696969696969....
  (+) 10.321321321321321....
  --------------------------
      14.591018291018290....
```

Hence, required answer $= 14.5\,\overline{910182}$

E. Sum of Series of Natural Numbers

Problems based on Series of Natural Numbers

If the series are in Arithmetic Progression (AP), the sum of given series $= \dfrac{n}{2}[a + l]$. AP series refers to a series with equal difference between any two consecutive terms/numbers, detailed in chapter ' Progression'.

Where, n = total number of terms, a = initial term and l = last term. Term simply refers to the number.

Then, Sum of first n natural numbers i.e.

$$(1 + 2 + 3 + \ldots + n) = \frac{n}{2}[a + l]$$

$$= \frac{n}{2}[1 + n] = \frac{n(n+1)}{2}$$

Problem 18 Find out the sum of first 100 natural numbers . **[CLAT-16]**

Sol. Required sum $= \frac{n}{2}[a + l] = \frac{100}{2}[1 + 100] = 5050$

[$\because$ 1 is an initial term(a) and 100 is the last term (l)]

Alternatively,

Required sum $= \frac{n(n+1)}{2} = \frac{100(100+1)}{2}$

$= 50 \times 101 = 5050$

Problem 19 Find out the sum of all natural numbers upto 36 starting from 10.

Sol. $(10 + 11 + 12 + \ldots + 36) = ?$

Then, required sum $= \frac{n}{2}[a + l] = \frac{27}{2}[10 + 36] = 621$

[$\because a = 10, l = 36, n = (36 - 10) + 1 = 27$]

Alternatively,

Required sum $= (1 + 2 + 3 + \ldots + 36) - (1 + 2 + 3 + \ldots + 9)$

$= \frac{36(36+1)}{2} - \frac{9(9+1)}{2}$

$= 666 - 45 = 621$

Problem 20 Find out the sum of all natural numbers between 22 and 38.

Sol. $(23 + 24 + 25 + \ldots + 37) = ?$

Then, required sum $= \frac{n}{2}[a + l] = \frac{15}{2}[23 + 37] = 450$

[$\because a = 23, l = 37$ and $n = (37 - 23) + 1 = 15$]

Alternatively,

Required sum $= (1 + 2 + 3 + \ldots + 37) - (1 + 2 + 3 + \ldots + 22)$

$= \frac{37(37+1)}{2} - \frac{22(22+1)}{2}$

$= 703 - 253 = 450$

Problem 21 Find out the sum of 10 consecutive natural numbers starting from 15.

$(15 + 16 + 17 + \ldots + 24) = ?$

Sol. Then, required sum $= \frac{n}{2}[a + l] = \frac{10}{2}[15 + 24] = 195$

[$\because a = 15, l = 15 + (10 - 1) = 24$ and $n = 10$]

Alternatively I,

Required sum $= \frac{n}{2}[a + l] = \frac{n}{2}[a + a + (n - 1)]$

$= \frac{n}{2}[2a + (n - 1)] = \frac{10}{2}[2 \cdot 15 + (10 - 1)] = 195$

Alternatively II,

Required sum $= (1 + 2 + 3 + \ldots + 24) - (1 + 2 + 3 + \ldots + 14)$

$= \frac{24(24+1)}{2} - \frac{14(14+1)}{2} = 300 - 105 = 195$

Problem 22 Find out the sum of 10 consecutive natural numbers ending with 28.

$(19 + 20 + 21 + \ldots + 28) = ?$

Sol. Then, required sum $= \frac{n}{2}[a + l] = \frac{10}{2}[19 + 28] = 235$

[$\because a = 28 - (10 - 1) = 19, l = 28$ and $n = 10$]

Alternatively I,

Required sum $= \frac{n}{2}[a + l] = \frac{n}{2}[l - (n - 1) + l] = \frac{n}{2}[2l - (n - 1)]$

$= \frac{10}{2}[2 \cdot 28 - (10 - 1)] = 235$

Alternatively II,

Required sum $= (1 + 2 + 3 + \ldots + 28) - (1 + 2 + 3 + \ldots + 18)$

$= \frac{28(28+1)}{2} - \frac{18(18+1)}{2}$

$= 406 - 171 = 235$

In brief,

Sum of n consecutive natural numbers starting from a

$= \frac{n}{2}[2a + (n - 1)]$

and Sum of n consecutive natural numbers ending with term l

$= \frac{n}{2}[2l - (n - 1)]$

Problem 23 Find the sum of first 12 multiples of 7. **[CLAT-13]**

Sol. Clearly, $(7 \times 1) + (7 \times 2) + (7 \times 3) + \ldots + (7 \times 12) = ?$

Then, required sum $= \frac{n}{2}[a + l] = \frac{12}{2}[7 + 84] = 546$

Alternatively,

$(7 \times 1) + (7 \times 2) + (7 \times 3) + \ldots + (7 \times 12)$

$= 7(1 + 2 + 3 + \ldots + 12) = 7 \times \frac{12(12+1)}{2}$

$= 546$

Problem 24 Find the sum of all natural numbers between 100 and 200 which are divisible by 6.

Sol. Clearly, 1st term = 102 and last term is 198.

Then, $n = \{(198 \div 6) - (102 \div 6)\} + 1 = 17$

[$l = a + (n - 1)d \Rightarrow 198 = 102 + (n - 1)6$

$\Rightarrow n = 17$, detailed in chapter 'Progression'.

Here, only common sense has been applied.]

Now, required sum $= \frac{n}{2}[a + l] = \frac{17}{2}[102 + 198] = 2550$

Alternatively,

$(102 + 108 + 114 + \ldots + 198) = ?$

$= 6(17 + 18 + \ldots + 33)$

$= 6\{(1 + 2 + 3 + \ldots + 33) - (1 + 2 + 3 + \ldots + 16)\}$

$= 6\left\{\frac{33(33+1)}{2} - \frac{16(16+1)}{2}\right\}$

$= 6 \times 425 = 2550$

Problem 25 Given, $1 + 2 + 3 + 4 + \ldots + 10 = 55$, then $6 + 12 + 18 + 24 + \ldots + 60 = ?$ **[SSC(CGL-PT)-04]**

Sol. Given expression $= 6 + 12 + 18 + 24 + \ldots + 60$

$= 6(1 + 2 + 3 + 4 + \ldots + 10)$

$= 6 \times 55 = 330$

Problem 26 Expression $1+2+3....+49+50+49+....+3+2+1=?$

Sol. Given expression $= (1+2+3....+49)$
$+(49+.....+3+2+1)+50$
$= 2\times\dfrac{49(49+1)}{2}+50 = 50(49+1)$
$= (50)^2 = 2500$

Observation

$1+2+3+.....+n+(n-1)+(n-2)+...+3+2+1=n^2$

Problems based on Series of Odd Numbers

If the series are in Arithmetic Progression (AP), the sum of given series $=\dfrac{n}{2}[a+l]$ or $\dfrac{n}{2}[2a+(n-1)d]$

Where, n = total number of terms, a = initial term, l = last term and d = common difference. Term simply refers to the number. The details are in the chapter 'Progression'.

Then, sum of first n odd numbers i.e., $(1+3+5+...+n$ terms)

$$=\frac{n}{2}[2a+(n-1)d]=\frac{n}{2}[2\cdot1+(n-1)2]=n^2.$$

Problem 27 Find the sum of first 12 odd numbers.

Sol. Required sum $= n^2 = (12)^2 = 144$

Alternatively,

Obviously, the expression is $(1+3+5+......+23)$

Here, $a=1$, $l=2n-1=(2\times12)-1=23$

Required sum $=\dfrac{(l+a)(l-a+2)}{4}=\dfrac{(23+1)(23-1+2)}{4}=144$

Problem 28 Expression: $1+3+5+....+99=?$

Sol. Before solving this question, build a concept to find total number of odd numbers in a series starting from 1 and the last odd number is known.

For example, What is the total number of odd numbers in the series 1, 3, ..., 9 ?

Here, total number of odd numbers $=\dfrac{\text{Last odd number}+1}{2}$

$=\dfrac{9+1}{2}=5$

Similarly, total number of odd numbers in the given expression $=\dfrac{99+1}{2}=50.$

Now, required sum $= n^2 = (50)^2 = 2500$

Note *We may find the number of odd numbers given in an AP series with the help of a formula, $l=a+(n-1)d$, detailed in chapter 'Progression'.*

Alternatively I,

$$\text{Sum}=\frac{n}{2}[a+l]=\frac{50}{2}[1+99]=2500$$

Or

$$\text{Sum}=\frac{n}{2}[2a+(n-1)d],$$

where, a = initial term, l = last term, n = number of total terms, d = difference between two consecutive terms i.e. common difference.

$$=\frac{50}{2}[2\cdot1+(50-1)2]=2500$$

[This approach is applicable for all AP series.]

Alternatively II,

$$\text{Sum}=\frac{(l+a)(l-a+2)}{4}=\frac{(99+1)(99-1+2)}{4}=2500$$

Alternatively III,

$$\text{Sum}=\frac{(\text{number just after last odd number})^2-(\text{number just before initial odd number})^2}{4}$$

$$=\frac{(100)^2-(0)^2}{4}=2500$$

Problem 29 Find the sum of all odd numbers upto 37 starting from 17.

Sol. Given expression, $17+19+21+....+37$

$\Rightarrow (1+3+5+....+37)-(1+3+5+....+15)$

$=\left\{n=\dfrac{(37+1)}{2}\right\}^2-\left\{n=\dfrac{(15+1)}{2}\right\}^2$

$= 361-64 = 297$

Alternatively I, Here, $a=17$, $l=37$ and

$$n=\frac{\text{last odd number}-\text{initial odd number}}{2}+1$$

$$=\frac{37-17}{2}+1=11$$

Now, sum $=\dfrac{n}{2}[a+l]=\dfrac{11}{2}[17+37]=297$

Alternatively II,

$$\text{Sum}=\frac{(l+a)(l-a+2)}{4}$$

$$=\frac{(37+17)(37-17+2)}{4}=297$$

Alternatively III,

$$\text{Sum}=\frac{(\text{number just after last odd number})^2-(\text{number just before initial odd number})^2}{4}$$

$$=\frac{(38)^2-(16)^2}{4}=\frac{(38+16)(38-16)}{4}=297$$

Problem 30 Find the sum of all odd numbers between 19 and 100.

Sol. Given expression, 21 + 23 + 25 + ... + 99

$\Rightarrow (1+3+5+\ldots+99)-(1+3+5+\ldots+19)$

$$=\left\{n=\frac{(99+1)}{2}\right\}^2-\left\{n=\frac{(19+1)}{2}\right\}^2$$

$$=2500-100=2400$$

Alternatively I, Here, $a=21$, $l=99$

and $n=\dfrac{\text{last odd number} - \text{initial odd number}}{2}+1$

$$=\frac{99-21}{2}+1=40$$

Now, sum $=\dfrac{n}{2}[a+l]=\dfrac{40}{2}[21+99]=2400$

Alternatively II,

$$\text{Sum}=\frac{(l+a)(l-a+2)}{4}=\frac{(99+21)(99-21+2)}{4}=2400$$

Alternatively III,

$$\text{Sum}=\frac{(\text{number just after last odd number})^2-(\text{number just before initial odd number})^2}{4}$$

$$=\frac{(100)^2-(20)^2}{4}=\frac{(100+20)(100-20)}{4}=2400$$

Problem 31 Find the sum of 24 consecutive odd numbers starting from 13.

Sol. Given expression, 13 + 15 + 17 + ... upto 24 terms

Clearly, there are $=\dfrac{11+1}{2}$ i.e. 6 odd numbers upto 11.

It implies that there are 6 odd numbers before 13 starting from 1.

Hence, required sum = {sum of first (24 + 6) odd numbers} − (sum of first 6 odd numbers)

$$=(30)^2-(6)^2=900-36=864$$

Alternatively I,

Here, $a=13$, $n=24$ and $l=a+(n-1)d=13+(24-1)2=59$

Now, sum $=\dfrac{n}{2}[a+l]=\dfrac{24}{2}[13+59]=864$

Alternatively II,

$$\text{Sum}=\frac{(l+a)(l-a+2)}{4}=\frac{(59+13)(59-13+2)}{4}=864$$

Alternatively III,

$$\text{Sum}=\frac{(\text{number just after last odd number})^2-(\text{number just before initial odd number})^2}{4}$$

$$=\frac{(60)^2-(12)^2}{4}=\frac{(60+12)(60-12)}{4}$$

$$=864$$

Problems Based on Series of Even Numbers

If the series are in Arithmetic Progression (AP), the sum of given series $=\dfrac{n}{2}[a+l]$ or $\dfrac{n}{2}[2a+(n-1)d]$

Where, n = total number of terms, a = initial term, l = last term and d = common difference. Term simply refers to the number. The details are in the chapter 'Progression'.

Then, sum of first n even numbers i.e.

$$(2+4+6+\ldots+n \text{ terms})=\frac{n}{2}[2a+(n-1)d]$$

$$=\frac{n}{2}[2\cdot 2+(n-1)2]=n(n+1)$$

Problem 32 Find the sum of first 25 even numbers.

Sol. Required sum $=n(n+1)=25(25+1)=650$

Alternatively I,

Expression is (2 + 4 + 6 + ... + 50)

$$\Rightarrow\quad 2(1+2+3+\ldots+25)=2\times\frac{25\times 26}{2}=650$$

Alternatively II,

Obviously, the expression is (2 + 4 + 6 + ... 50)

Here, $a=2$, $l=2n=2\times 25=50$

Then, required sum $=\dfrac{(l+a)(l-a+2)}{4}=\dfrac{(50+2)(50-2+2)}{4}$

$$=650$$

Alternatively III,

Required sum $=\dfrac{n}{2}[a+l]=\dfrac{25}{2}[2+50]=650$

Alternatively IV,

$$\text{Sum}=\frac{(\text{number just after last even number})^2-(\text{number just before initial even number})^2}{4}$$

$$=\frac{(51)^2-(1)^2}{4}=\frac{(51+1)(51-1)}{4}=650$$

Problem 33 $2+4+6+\ldots+104=?$

Sol. The total number of even numbers in a series starting from 2 and the last even number is known $=\dfrac{\text{Last even number}}{2}$

$$=\frac{104}{2}=52$$

Now, required sum $=n(n+1)=52(52+1)=52\times 53=2756$

Note We may find the number of even numbers given in a AP series with the help of a formula, $l=a+(n-1)d$, detailed in chapter 'Progression'.

Alternatively I,

$$2+4+6+\ldots+104=2(1+2+3+\ldots+52)$$

$$=2\times\frac{52\times 53}{2}=2756$$

Alternatively II,

Sum $=\dfrac{n}{2}[a+l]=\dfrac{52}{2}[2+104]=2756$

Or Sum $=\dfrac{n}{2}[2a+(n-1)d]$,

where a = initial term, l = last term, n = number of total terms, d = difference between two consecutive terms i.e, common difference.

$$=\frac{52}{2}[2\cdot 2+(52-1)2]=2756$$

[This approach is applicable for all AP series.]

Alternatively III, Sum $= \dfrac{(l+a)(l-a+2)}{4}$

$= \dfrac{(104+2)(104-2+2)}{4} = 2756$

Alternatively IV,

$$\text{Sum} = \frac{(\text{number just after last even number})^2 - (\text{number just before initial even number})^2}{4}$$

$$= \frac{(105)^2 - (1)^2}{4} = \frac{(105+1)(105-1)}{4} = 2756$$

Problem 34 Find the sum of all even numbers upto 50 starting from 20.

Sol. Given expression, $20 + 22 + 24 + ... + 50$

$\Rightarrow (2 + 4 + 6 .... + 50) - (2 + 4 + 6 + ... + 18)$

$= (n = 25)(25 + 1) - (n = 9)(9 + 1) = 650 - 90 = 560$

Alternatively I,

Given expression, $20 + 22 + 24 + ... + 50$

$\Rightarrow (2 + 4 + 6 .... + 50) - (2 + 4 + 6 + ... + 18)$

$= 2\{(1 + 2 + 3 + ... + 25) - (1 + 2 + 3 + ... + 9)\}$

$= 2\left\{\dfrac{25 \times 26}{2} - \dfrac{9 \times 10}{2}\right\} = 560$

Alternatively II, Here, $a = 20$, $l = 50$ and

$$n = \frac{\text{last even number} - \text{initial even number}}{2} + 1$$

$$= \frac{50-20}{2} + 1 = 16$$

Now, sum $= \dfrac{n}{2}[a + l] = \dfrac{16}{2}[20 + 50] = 560$

Alternatively III,

$$\text{Sum} = \frac{(l+a)(l-a+2)}{4} = \frac{(50+20)(50-20+2)}{4} = 560$$

Alternatively IV,

$$\text{Sum} = \frac{(\text{number just after last even number})^2 - (\text{number just before initial even number})^2}{4}$$

$$= \frac{(51)^2 - (19)^2}{4} = \frac{(51+19)(51-19)}{4} = 560$$

Problem 35 Find the sum of all even numbers between 101 and 240.

Sol. Expression $(102 + 104 + 106 + ... + 238)$

$$\text{Sum} = \frac{(l+a)(l-a+2)}{4} = \frac{(238+102)(238-102+2)}{4} = 11730$$

Please solve with other approaches too.

Problem 36 Find the sum of 15 consecutive even numbers starting from 22.

Sol. Here, $a = 22$, $l = a + (n-1)d = 22 + (15-1)2 = 50$

$$\text{Sum} = \frac{(l+a)(l-a+2)}{4} = \frac{(50+22)(50-22+2)}{4} = 540$$

Please solve with other approaches too.

Observations

(i) Aspirants are advised to stress on the model, sum $= \dfrac{(l+a)(l-a+2)}{4}$ as it is very convenient and applicable also for both odd and even series.

(ii) This model is applicable for only ascending order. In case of a series of negative odd/even numbers, we will arrange first in ascending order, if required. Then, the above model will be applied.

For example What is the sum of the series $-64, -66, -68, ..., -100$? *[XAT-15]*

Ascending order : $-100, ..., -68, -66, -64$

Now, sum $= \dfrac{(l+a)(l-a+2)}{4}$

$= \dfrac{(-64-100)(-64+100+2)}{4} = -1558$

Or Given expression, $-\{64 + 66 + 68 + ... + 100\}$

$= -\left\{\dfrac{(100+64)(100-64+2)}{4}\right\} = -1558$

Sum of the Squares of Numbers Given in a Certain Series

(i) Sum of squares of first n natural numbers $= \dfrac{n(n+1)(2n+1)}{6}$

(ii) Sum of squares of first n even numbers $= \dfrac{2n(n+1)(2n+1)}{3}$

(iii) Sum of squares of first n odd numbers $= \dfrac{n(2n+1)(2n-1)}{3}$

Problem 37 $5^2 + 6^2 + ... + (10)^2 + (20)^2 = ?$ *[SSC (CPO-SI)-03]*

Sol. Required sum

$= \{1^2 + 2^2 + ... + (10)^2\} - \{1^2 + 2^2 + ... + 4^2\} + (20)^2$

$= \dfrac{10(10+1)(2\cdot 10+1)}{6} - \dfrac{4(4+1)(2\cdot 4+1)}{6} + 400$

$= 385 - 30 + 400 = 755$

Problem 38 Given that $1^2 + 2^2 + 3^2 + ... + (20)^2 = 2870$, the value of $\{2^2 + 4^2 + 6^2 + ... + (40)^2\} = ?$

Sol. $[2^2 + 4^2 + 6^2 .... + (40)^2] = 2^2\{1^2 + 2^2 + 3^2 .... + (20)^2\}$

$= 4 \times 2870$

$= 11480$

Problem 39 Find the sum of squares of all natural numbers between 13 and 20.

Or

Find the sum of integers which are perfect square and lying between 190 and 365.

Sol. Clearly, $(14)^2 + (15)^2 + (16)^2 \ldots + (19)^2 = ?$

Required sum $= \{1^2 + 2^2 + \ldots + (19)^2\} - \{1^2 + 2^2 + \ldots + (13)^2\}$

$= \dfrac{19(19+1)(2\cdot 9+1)}{6} - \dfrac{13(13+1)(2\cdot 13+1)}{6}$

$= 2470 - 819 = 1651$

Problem 40 Find the sum of squares of first 15 multiples of 8.

Sol. Clearly, $(8\times 1)^2 + (8\times 2)^2 + (8\times 3)^2 + \ldots + (8\times 15)^2 = ?$

$= 8^2\{1^2 + 2^2 + \ldots + (15)^2\}$

$= 8^2 \times \dfrac{15(15+1)(2\cdot 15+1)}{6} = 79360$

Problem 41 Find the sum of squares of first 10 odd numbers.

Sol. Required sum $= \dfrac{n(2n+1)(2n-1)}{3} = \dfrac{10(2\cdot 10+1)(2\cdot 10-1)}{3}$

$= 1330$

Problem 42 Find the sum of squares of all odd numbers between 13 and 25.

Sol. Clearly, $(15)^2 + (17)^2 + (19)^2 + \ldots + (23)^2 = ?$

$= \{(1)^2 + (3)^2 + (5)^2 + \ldots + (23)^2\} - \{(1)^2 + (3)^2 + (5)^2 + \ldots + (13)^2\}$

$= \dfrac{12(2\cdot 12+1)(2\cdot 12-1)}{3} - \dfrac{7(2\cdot 7+1)(2\cdot 7-1)}{3}$

$\left[\because \text{number of terms in 1st part} = \dfrac{23+1}{2} = 12 \text{ and that in 2nd part} = \dfrac{13+1}{2} = 7\right]$

$= 2300 - 455 = 1845$

Problem 43 Find the value of $4 + 36 + 100 + \ldots + 324$.

Sol. $4\{1 + 9 + 25 + \ldots + 81\}$

$= 4\{(1)^2 + (3)^2 + (5)^2 + \ldots + (9)^2\}$

$= 4 \times \dfrac{5(2\cdot 5+1)(2\cdot 5-1)}{3}$ $\left[\because n = \dfrac{9+1}{2} = 5\right]$

$= 660$

Problem 44 Find the value of $0.5 + 4.5 + 12.5 + \ldots + 144.5$.

Sol. $\dfrac{1}{2}[1 + 9 + 25 + \ldots + 289]$

$\Rightarrow \dfrac{1}{2}[(1)^2 + (3)^2 + (5)^2 + \ldots + (17)^2]$

$= \dfrac{1}{2} \times \dfrac{9(2\cdot 9+1)(2\cdot 9-1)}{3}$ $\left[\because n = \dfrac{17+1}{2} = 9\right]$

$= 484.5$

Problem 45 Find the sum of the squares of first 20 even numbers.

Sol. Required sum $= \dfrac{2n(n+1)(2n+1)}{3} = \dfrac{2\times 20(20+1)(2\cdot 20+1)}{3}$

$= 11480$

Problem 46 Find the sum of the squares of all even numbers between 13 and 30.

Sol. Clearly, sum $= (14)^2 + (16)^2 + (18)^2 + \ldots + (28)^2$

$= \{(2)^2 + (4)^2 + (6)^2 + \ldots + (28)^2\} - \{(2)^2 + (4)^2 + (6)^2 + \ldots + (12)^2\}$

$= \dfrac{2\times 14(14+1)(2\cdot 14+1)}{3} - \dfrac{2\times 6(6+1)(2\cdot 6+1)}{3}$

$\left[\because \text{number of terms in 1st part} = \dfrac{28}{2} = 14 \text{ and that in 2nd part} = \dfrac{12}{2} = 6\right]$

$= 4060 - 364 = 3696$

Sum of the Cubes of Numbers Given in a Certain Series

(i) Sum of cubes of first n natural numbers $= \left\{\dfrac{n(n+1)}{2}\right\}^2$

(ii) Sum of cubes of first n even numbers $= 2n^2(n+1)^2$

(iii) Sum of cubes of first n odd numbers $= n^2(2n^2 - 1)$

Problem 47 $5^3 + 6^3 + \ldots + (10)^3 = ?$

Sol. Required sum $= \{1^3 + 2^3 + \ldots + (10)^3\} - \{1^3 + 2^3 + \ldots + 4^3\}$

$= \left\{\dfrac{10(10+1)}{2}\right\}^2 - \left\{\dfrac{4(4+1)}{2}\right\}^2$

$= 3025 - 100 = 2925$

Problem 48 If $1^3 + 2^3 + \ldots + 9^3 = 2025$, then $(0.11)^3 + (0.22)^3 + \ldots + (0.99)^3 = ?$

Sol. $(0.11)^3 + (0.22)^3 + \ldots + (0.99)^3$

$= (0.11)^3\ \{1^3 + 2^3 + \ldots + 9^3\}$

$= (0.11)^3 \times 2025$

$= 2.695275$

Problem 49 If $1^3 + 2^3 + \ldots + (10)^3 = 3025$, then $4 + 32 + 108 + \ldots + 4000 = ?$ *[SSC(CGL-T-I)-11, SSC (CGL-PT)-02]*

Sol. $4 + 32 + 108 + \ldots + 4000$

$\Rightarrow 4\{1^3 + 2^3 + \ldots + (10)^3\} = 4 \times 3025 = 12100$

Problem 50 Find the sum of all perfect cubes lying between 10 and 10000.

Sol. Clearly, $3^3 + 4^3 + \ldots + (21)^3$

$= \{1^3 + 2^3 + \ldots + (21)^3\} - \{1^3 + 2^3\}$

$= \left\{\dfrac{21(21+1)}{2}\right\}^2 - 9$

$= 53361 - 9 = 53352$

Problem 51 Find the sum of cubes of first 12 odd numbers.

Sol. Required sum $= n^2(2n^2 - 1)$

$= (12)^2\ \{2(12)^2 - 1\} = 41328$

Problem 52 Find the value of

$$(11)^3 + (13)^3 + (15)^3 + \ldots + (23)^3.$$

Sol. Clearly, sum $= \{1^3 + 3^3 + \ldots + (23)^3\} - \{1^3 + 3^3 + \ldots + (9)^3\}$

$$= (12)^2 \{2 (12)^2 - 1\} - (5)^2 \{2 (5)^2 - 1\}$$

$$\left[\because \text{number of terms in 1st part} = \frac{23+1}{2} = 12 \right.$$

$$\left.\text{and that in 2nd part} = \frac{9+1}{2} = 5\right]$$

$$= 41328 - 1225 = 40103$$

Problem 53 Find the value of

$$2 + 54 + 250 + \ldots + 2662.$$

Sol. $2 + 54 + 250 + \ldots + 2662 = 2\{1^3 + 3^3 + \ldots + (11)^3\}$

$$= 2 \times 6^2\{2 \cdot 6^2 - 1\} \qquad [\because n = 6]$$

$$= 5112$$

Problem 54 Find the value of

$$171.5 + 364.5 + 665.5 + \ldots + 2456.5.$$

Sol. $\frac{1}{2}\{343 + 729 + 1331 + \ldots + 4913\}$

$$\Rightarrow \frac{1}{2}\{7^3 + 9^3 + (11)^3 \ldots + (17)^3\}$$

$$\Rightarrow \frac{1}{2}[\{1^3 + 3^3 + \ldots + (17)^3\} - \{1^3 + 3^3 + (5)^3\}]$$

$$= \frac{1}{2}[9^2\{2 \cdot 9^2 - 1\} - 3^2\{2 \cdot 3^2 - 1\}]$$

$$\left[\because \text{number of terms in 1st part} = \frac{17+1}{2} = 9 \right.$$

$$\left.\text{and that in 2nd part} = \frac{5+1}{2} = 3\right]$$

$$= \frac{1}{2}\{13041 - 153\} = 6444$$

Problem 55 Find the sum of cubes of first eight even numbers.

Sol. Sum $= 2n^2(n + 1)^2 = 2 \times 8^2(8 + 1)^2 = 10368$

Problem 56 Find the value of

$$(10)^3 + (12)^3 + (14)^3 + \ldots + (22)^3.$$

SSC (CGL-T-II)-17

Sol. Clearly, sum $= \{2^3 + 4^3 + 6^3 + \ldots + (22)^3\} - \{2^3 + 4^3 + \ldots + 8^3\}$

$$= 2(11)^2 (11 + 1)^2 - 2 \cdot 4^2(4+1)^2$$

$$\left[\because \text{number of terms in 1st part} = \frac{22}{2} = 11 \right.$$

$$\left.\text{and that in 2nd part} = \frac{8}{2} = 4\right]$$

$$= 34848 - 800 = 34048$$

Problem 57 Find the value of $2\frac{2}{3} + 21\frac{1}{3} + 72 + \ldots + 333\frac{1}{3}$.

Sol. $\frac{1}{3}\{8 + 64 + 216 + \ldots + 1000\}$

$$\Rightarrow \frac{1}{3}\{2^3 + 4^3 + 6^3 + \ldots + (10)^3\} = \frac{1}{3}\{2 \cdot 5^2(5+1)^2\} = 600$$

F. Unique Series based on Addition and Subtraction

Here, Unique series refers to the series which follow a definite pattern, however, it appears that they don't follow so at first instance. Hence, such types of questions have been divided into two categories, (I) : Unique series based on logical thinking only which seek no any specific concepts. (II) : Unique series based on sum of natural numbers, their squares/cubes.

I. Unique Series Seeking No Specific Concepts

Problem 58 Find the sum of first 20 terms of a series

$$\frac{1}{5 \cdot 6} + \frac{1}{6 \cdot 7} + \frac{1}{7 \cdot 8} + \ldots\ldots$$

[SSC (CGL-PT)-08]

Sol. Given, 1st term $= \frac{1}{5 \cdot 6} = \frac{1}{5} - \frac{1}{6} = \frac{1}{(1+4)} - \frac{1}{(1+5)}$

2nd term $= \frac{1}{6 \cdot 7} = \frac{1}{6} - \frac{1}{7} = \frac{1}{(2+4)} - \frac{1}{(2+5)}$

and 3rd term $= \frac{1}{7 \cdot 8} = \frac{1}{7} - \frac{1}{8} = \frac{1}{(3+4)} - \frac{1}{(3+5)}$ ... and so on.

Then, 20th term $= \frac{1}{(20+4)} - \frac{1}{(20+5)}$

Now, the expression is $\frac{1}{5} - \frac{1}{6} + \frac{1}{6} - \frac{1}{7} + \frac{1}{7} - \frac{1}{8} + \ldots + \frac{1}{24} - \frac{1}{25}$

$$= \frac{1}{5} - \frac{1}{25} = \frac{4}{25} = 0.16$$

Problem 59 $\frac{1}{20} + \frac{1}{30} + \frac{1}{42} + \frac{1}{56} + \frac{1}{72} + \frac{1}{90} + \frac{1}{110} + \frac{1}{132}$ is equal to

Sol. $\left(\frac{1}{4} - \frac{1}{5}\right) + \left(\frac{1}{5} - \frac{1}{6}\right) + \left(\frac{1}{6} - \frac{1}{7}\right) \ldots + \left(\frac{1}{11} - \frac{1}{12}\right) = \frac{1}{4} - \frac{1}{12} = \frac{1}{6}$

Problem 60 $\frac{1}{1 \cdot 4} + \frac{1}{4 \cdot 7} + \frac{1}{7 \cdot 10} + \frac{1}{10 \cdot 13} + \frac{1}{13.16} = ?$

[SSC (CGL-PT)-07]

Sol. $\frac{1}{3}\left\{\left(\frac{1}{1} - \frac{1}{4}\right) + \left(\frac{1}{4} - \frac{1}{7}\right) + \left(\frac{1}{7} - \frac{1}{10}\right) + \left(\frac{1}{10} - \frac{1}{13}\right) + \left(\frac{1}{13} - \frac{1}{16}\right)\right\}$

$$\Rightarrow \frac{1}{3}\left\{\frac{1}{1} - \frac{1}{16}\right\} = \frac{1}{3} \times \frac{15}{16} = \frac{5}{16}$$

Problem 61 $\frac{1}{1 \cdot 2 \cdot 3} + \frac{1}{2 \cdot 3 \cdot 4} + \frac{1}{3 \cdot 4 \cdot 5} + \frac{1}{4 \cdot 5 \cdot 6} = ?$

Sol. $\frac{1}{6}\left\{1 + \frac{1}{4} + \frac{1}{10} + \frac{1}{20}\right\} = \frac{1}{6} \times \frac{28}{20} = \frac{7}{30}$

Problem 62 $\frac{3}{1^2 \cdot 2^2} + \frac{5}{2^2 \cdot 3^2} + \frac{7}{3^2 \cdot 4^2} + \ldots + \frac{19}{9^2 \cdot (10)^2} = ?$

Sol. $\left(\frac{1}{1^2} - \frac{1}{2^2}\right) + \left(\frac{1}{2^2} - \frac{1}{3^2}\right) + \left(\frac{1}{3^2} - \frac{1}{4^2}\right) + \ldots + \left(\frac{1}{9^2} - \frac{1}{(10)^2}\right)$

$$= \left(\frac{1}{1^2} - \frac{1}{(10)^2}\right) = 1 - \frac{1}{100} = \frac{99}{100}$$

Problem 63 $\frac{1}{10}+\frac{1}{15}+\frac{1}{21}+\ldots+\frac{1}{120}$ is equal to

Sol. $2\left\{\frac{1}{20}+\frac{1}{30}+\frac{1}{42}+\ldots+\frac{1}{240}\right\}$

$\Rightarrow 2\left\{\left(\frac{1}{4}-\frac{1}{5}\right)+\left(\frac{1}{5}-\frac{1}{6}\right)+\left(\frac{1}{6}-\frac{1}{7}\right)\ldots+\left(\frac{1}{15}-\frac{1}{16}\right)\right\}$

$\Rightarrow 2\left\{\frac{1}{4}-\frac{1}{16}\right\}=2\times\frac{3}{16}=\frac{3}{8}$

Problem 64 $\frac{10}{36}+\frac{14}{144}+\frac{18}{400}+\ldots+\frac{38}{8100}=?$

Sol. $\Rightarrow 2\left\{\frac{5}{36}+\frac{7}{144}+\frac{9}{400}+\ldots+\frac{19}{8100}\right\}$

$\Rightarrow 2\left\{\frac{5}{2^2\cdot 3^2}+\frac{7}{3^2\cdot 4^2}+\frac{9}{4^2\cdot 5^2}+\ldots+\frac{19}{9^2\cdot(10)^2}\right\}$

$=2\left\{\left(\frac{1}{2^2}-\frac{1}{3^2}\right)+\left(\frac{1}{3^2}-\frac{1}{4^2}\right)+\left(\frac{1}{4^2}-\frac{1}{5^2}\right)+\ldots+\left(\frac{1}{9^2}-\frac{1}{(10)^2}\right)\right\}$

$=2\left(\frac{1}{2^2}-\frac{1}{(10)^2}\right)=2\times\frac{24}{100}=\frac{12}{25}$

Problem 65 $\frac{1}{14}+\frac{3}{56}+\frac{1}{24}+\frac{1}{30}+\ldots+\frac{1}{44}$ is equal to

Sol. $\Rightarrow \frac{3}{42}+\frac{3}{56}+\frac{3}{72}+\frac{3}{90}+\ldots+\frac{3}{132}$

$\Rightarrow 3\left\{\frac{1}{42}+\frac{1}{56}+\frac{1}{72}+\frac{1}{90}+\ldots+\frac{1}{132}\right\}$

$=3\left\{\left(\frac{1}{6}-\frac{1}{7}\right)+\left(\frac{1}{7}-\frac{1}{8}\right)+\left(\frac{1}{8}-\frac{1}{9}\right)+\ldots+\left(\frac{1}{11}-\frac{1}{12}\right)\right\}$

$=3\left\{\frac{1}{6}-\frac{1}{12}\right\}=3\times\frac{1}{12}=\frac{1}{4}$

Problem 66 $\frac{1}{75}+\frac{1}{105}+\frac{1}{140}+\ldots+\frac{1}{275}$ is equal to

Sol. $\Rightarrow \frac{2}{5}\left\{\frac{1}{30}+\frac{1}{42}+\frac{1}{56}+\ldots+\frac{1}{110}\right\}$

$\Rightarrow \frac{2}{5}\left\{\left(\frac{1}{5}-\frac{1}{6}\right)+\left(\frac{1}{6}-\frac{1}{7}\right)+\left(\frac{1}{7}-\frac{1}{8}\right)\ldots+\left(\frac{1}{10}-\frac{1}{11}\right)\right\}$

$=\frac{2}{5}\left\{\frac{1}{5}-\frac{1}{11}\right\}=\frac{2}{5}\times\frac{6}{55}=\frac{12}{275}$

Problem 67 $\frac{1}{4\cdot 5}+\frac{1}{4\cdot 11}+\frac{1}{7\cdot 11}+\ldots+\frac{1}{16\cdot 35}=?$

Sol. $\frac{2}{2\cdot 4\cdot 5}+\frac{2}{2\cdot 4\cdot 11}+\frac{2}{2\cdot 7\cdot 11}+\ldots+\frac{2}{2\cdot 16\cdot 35}$

$=\frac{2}{5\cdot 8}+\frac{2}{8\cdot 11}+\frac{2}{11\cdot 14}+\ldots+\frac{2}{32\cdot 35}$

$=\frac{2}{3}\left\{\left(\frac{1}{5}-\frac{1}{8}\right)+\left(\frac{1}{8}-\frac{1}{11}\right)+\left(\frac{1}{11}-\frac{1}{14}\right)+\ldots+\left(\frac{1}{32}-\frac{1}{35}\right)\right\}$

$=\frac{2}{3}\left\{\frac{1}{5}-\frac{1}{35}\right\}=\frac{2}{3}\times\frac{6}{35}=\frac{4}{35}$

Problem 68 $\frac{7}{8\cdot 9}+\frac{9}{8\cdot 25}+\frac{11}{18\cdot 25}+\frac{13}{18\cdot 49}+\ldots+\frac{21}{50\cdot 121}=?$

Sol. $2\left\{\frac{7}{9\cdot 16}+\frac{9}{16\cdot 25}+\frac{11}{25\cdot 36}+\frac{13}{36\cdot 49}+\ldots+\frac{21}{100\cdot 121}\right\}$

$=2\left\{\left(\frac{1}{9}-\frac{1}{16}\right)+\left(\frac{1}{16}-\frac{1}{25}\right)+\left(\frac{1}{25}-\frac{1}{36}\right)\ldots+\left(\frac{1}{100}-\frac{1}{121}\right)\right\}$

$=2\left\{\frac{1}{9}-\frac{1}{121}\right\}=2\times\frac{112}{121\cdot 9}=\frac{224}{1089}$

Problem 69 $\frac{1}{3}+\frac{1}{15}+\frac{1}{35}+\frac{1}{63}+\frac{1}{99}+\frac{1}{143}$ is equal to

Sol. $\frac{1}{1\cdot 3}+\frac{1}{3\cdot 5}+\frac{1}{5\cdot 7}+\frac{1}{7\cdot 9}+\frac{1}{9\cdot 11}+\frac{1}{11\cdot 13}$

$\Rightarrow \frac{1}{2}\left\{\left(\frac{1}{1}-\frac{1}{3}\right)+\left(\frac{1}{3}-\frac{1}{5}\right)+\left(\frac{1}{5}-\frac{1}{7}\right)+\ldots+\left(\frac{1}{11}-\frac{1}{13}\right)\right\}$

$=\frac{1}{2}\left\{\frac{1}{1}-\frac{1}{13}\right\}=\frac{1}{2}\times\frac{12}{13}=\frac{6}{13}$

Problem 70 Find the value of the series

$\frac{1}{2}+\frac{1}{3}-\frac{1}{4}-\frac{1}{2}-\frac{1}{3}+\frac{1}{4}+\frac{1}{2}+\frac{1}{3}-\frac{1}{4}-\frac{1}{2}$ ... upto 52 terms.

Sol. Clearly, the value of first 6 terms = 0

Value of further 6 terms = 0

Then, value upto 48th term = 8 × 0 = 0

Now, the series starting from 49th term to 52nd term will be

$\Rightarrow \frac{1}{2}+\frac{1}{3}-\frac{1}{4}-\frac{1}{2}$

Then, its value $=\frac{1}{12}$

Problem 71 Find the value of the series

$\frac{1}{2}-\frac{1}{3}+\frac{1}{6}+\frac{1}{2}-\frac{1}{3}+\frac{1}{6}+\frac{1}{2}-\frac{1}{3}$ .... upto 22 terms.

Sol. Clearly, the value of first 3 terms $=\frac{1}{2}-\frac{1}{3}+\frac{1}{6}=\frac{1}{3}$

Then, value of first 21 terms $=7\times\frac{1}{3}=\frac{7}{3}$ and 22nd term $=\frac{1}{2}$

Hence, value upto 22 terms $=\frac{7}{3}+\frac{1}{2}=2\frac{1}{3}+\frac{1}{2}=2\frac{5}{6}$

Problem 72 Find the value of the series $-5+7-9+11-13+15$.... upto 50 tems.

Sol. Since each even-place number is '+', 50th term will also be positive. Now, arrange the expression in pairs like,

$(-5+7)+(-9+11)+(-13+15)$........ upto 25 pairs.

Hence, required answer = 2 × 25 = 50

Problem 73 The sum $1(1!)+2(2!)+3(3!)+\ldots+n(n!)$ equals

(a) $3(n!)+n-3$ (b) $(n+1)!-(n-1)!$
(c) $(n+1)!-1$ (d) $2(n!)-2n-1$

Sol. 1st term $=1(1!)=(2-1)\,1!=2\cdot 1!-1!=2!-1!$

2nd term $=2(2!)=(3-1)\,2!=3\cdot 2!-2!=3!-2!$

3rd term $=3(3!)=(4-1)\,3!=4\cdot 3!-3!=4!-3!$

....

nth term $=n(n!)=(n+1-1)n!=(n+1)\cdot n!-n!=(n+1)!-n!$

The required sum $=2!-1!+3!-2!+4!-3!+\ldots+(n+1)!-n!$
$=(n+1)!-1!$

Problem 74 Find the value of $+3-5+7-9+11-13\ldots$ upto 15 terms.

Sol. Without considering the sign, 15th term $= a + (n-1)d$

$= 3 + (15-1)2 = 31$

Since each odd place number is '+', 15th term will also be positive.

Now, arrange the expression in pairs like,

$(3-5)+(7-9)+(11-13)+\ldots$ upto 7 pairs + 15th term

$= (-2\times 7) + 31 = 17$

[***Note*** If number of terms is small, you may solve the problem directly by writing all the terms. Here, you may also do.]

Problem 75 $1+\frac{1}{2}+\frac{1}{2\cdot 3}+\frac{1}{2\cdot 3\cdot 4}+\frac{1}{2\cdot 3\cdot 4\cdot 5}+\ldots = ?$

Sol. $\Rightarrow 1+\frac{1}{1\cdot 2}+\frac{1}{1\cdot 2\cdot 3}+\frac{1}{1\cdot 2\cdot 3\cdot 4}+\frac{1}{1\cdot 2\cdot 3\cdot 4\cdot 5}+\ldots = ?$

$\Rightarrow 1+\frac{1}{2!}+\frac{1}{3!}+\frac{1}{4!}+\frac{1}{5!}+\ldots$

$\Rightarrow e-1$ [where, $e = 1+\frac{1}{1!}+\frac{1}{2!}+\frac{1}{3!}+\ldots$]

$= 2.71828 - 1 = 1.71828$

Problem 76 The sum of first n terms of the given series $1^2+2\cdot 2^2+3^2+2\cdot 4^2+5^2+2\cdot 6^2\ldots$ is $\frac{n(n+1)^2}{2}$, when n is even. When n is odd, the sum will be

(a) $\frac{n(n+1)^2}{2}$ (b) $\frac{n^2(n+1)}{2}$

(c) $n(n+1)^2$ (d) None of these

Sol. When n is an odd number, $(n-1)$ will obviously be even.

Now, sum of $(n-1)$ terms of the series

$$= \frac{(n-1)(n-1+1)^2}{2} = \frac{n^2(n-1)}{2}$$

Also, nth term $= n^2$ as n is odd.

Now, sum of n terms = sum of $(n-1)$ terms + nth term

$$= \frac{n^2(n-1)}{2} + n^2 = \frac{n^2(n+1)}{2}$$

Alternatively,

Sum of one term = 1

Sum of 3 terms = 18

Sum of 5 terms = 75 ...... and so on.

Now, check as to which option yields 1, 18, ... on putting $n = 1, 3$ respectively.

II. Unique Series based on Sum of Natural Numbers/their Squares/Cubes

Problem 77 $1\cdot 2+2\cdot 3+3\cdot 4+\ldots+24\cdot 25 = ?$

Sol. Clearly, 1st term = 1 × (1 more than 1)

2nd term = 2 × (1 more than 2)

3rd term = 3 × (1 more than 3)....... and so on

Then, nth term $= n \times$ (1 more than n) $= n(n+1) = n^2 + n$

Now, 1st term $= 1^2 + 1$ $[\because n = 1]$

2nd term $= 2^2 + 2$ $[\because n = 2]$

3rd term $= 3^2 + 3$ $[\because n = 3]$

....

24th term $= (24)^2 + 24$

$\therefore$ Required sum $= \{1^2+2^2+3^2+\ldots+(24)^2\} + \{1+2+3+\ldots+24\}$

$$= \frac{24(24+1)(2\cdot 24+1)}{6} + \frac{24(24+1)}{2} = 4900 + 300 = 5200$$

Remarks *In these sorts of questions, you have to split the terms in the form of (+) or (–) or both before processing further.*

Problem 78 $1\cdot 3+2\cdot 4+3\cdot 5+\ldots+18\cdot 20 = ?$

Sol. Clearly, nth term $= n \times$ (2 more than n) $= n(n+2) = n^2 + 2n$

Now, 1st term $= 1^2 + 2\cdot 1$

2nd term $= 2^2 + 2\cdot 2$

3rd term $= 3^2 + 2\cdot 3$

....

18th term $= (18)^2 + 2\cdot 18$

$\therefore$ Required sum $= \{1^2+2^2+3^2+\ldots+(18)^2\} + 2\{1+2+3+\ldots+18\}$

$$= \frac{18(18+1)(2\cdot 18+1)}{6} + \left\{2\times\frac{18(18+1)}{2}\right\} = 2109 + 342 = 2451$$

Problem 79 $7\cdot 9+8\cdot 10+9\cdot 11+\ldots+21\cdot 23 = ?$

Sol. Clearly, 1st term $= (1+6)\times(1+8)$

2nd term $= (2+6)\times(2+8)$

3rd term $= (3+6)\times(3+8)$

.... and so on

Then, nth term $= (n+6)\times(n+8) = n^2 + 14n + 48$

Now, 1st term $= 1^2 + 14\cdot 1 + 48$ $\{\because n = 1\}$

2nd term $= 2^2 + 14\cdot 2 + 48$

3rd term $= 3^2 + 14\cdot 3 + 48$

....

15th term $= (15)^2 + 14\cdot 15 + 48$

$[\because (n+6) = 21 \Rightarrow n = 6$ or $(n+8) = 23 \Rightarrow n = 15]$

$\therefore$ Required sum $= \{1^2+2^2+3^2+\ldots+(15)^2\} + 14\{1+2+3+\ldots+15\} + (48\times 15)$

$$= \frac{15(15+1)(2\cdot 15+1)}{6} + \left\{14\times\frac{15(15+1)}{2}\right\} + (48\times 15)$$

$= 1240 + 1680 + 720 = 3640$

Problem 80 $1\cdot 2\cdot 3+2\cdot 3\cdot 4+3\cdot 4\cdot 5+\ldots+12\cdot 13\cdot 14 = ?$

Sol. Clearly, nth term $= n\times(n+1)(n+2) = n^3 + 3n^2 + 2n$

Now, 1st term $= 1^3 + 3\cdot 1^2 + 2\cdot 1$

2nd term $= 2^3 + 3\cdot 2^2 + 2\cdot 2$

3rd term $= 3^3 + 3\cdot 3^2 + 2\cdot 3$

....

12th term $= (12)^3 + 3\cdot(12)^2 + 2\cdot 12$

$\therefore$ Required sum $= \{1^3+2^3+3^3+\ldots+(12)^3\} + 3\{1^2+2^2+3^2+\ldots+(12)^2\} + 2\{1+2+3+\ldots+12\}$

$$= \left\{\frac{12(12+1)}{2}\right\}^2 + 3\left\{\frac{12(12+1)(2\cdot 12+1)}{6}\right\} + \left\{2\times\frac{12(12+1)}{2}\right\}$$

$= 6084 + 1950 + 156 = 8190$

➘ Problem 81 $1\cdot 0+2\cdot 1+3\cdot 4+4\cdot 9+\ldots+10\cdot 81=?$

Sol. Clearly, nth term $= n\times(n-1)^2 = n^3-2n^2+n$

Now, you may proceed directly as below

Required sum $=\{1^3+2^3+3^3+\ldots+(10)^3\}$
$-2\{1^2+2^2+3^2+\ldots+(10)^2\}+\{1+2+3+\ldots+10\}$

$$=\left\{\frac{10(10+1)}{2}\right\}^2-2\left\{\frac{10(10+1)(2\cdot 10+1)}{6}\right\}+\left\{\frac{10(10+1)}{2}\right\}$$

$= 3025 - 770 + 55 = 2310$

➘ Problem 82

$$1\cdot 4+2\cdot 7+3\cdot 12+4\cdot 19+\ldots.+20\cdot 403=?$$

Sol. Clearly, nth term $= n\times(n^2+3) = n^3+3n$

Now, you may proceed directly as below:

Required sum $=\{1^3+2^3+3^3+\ldots+(20)^3\}$
$+3\{1+2+3+\ldots+20\}$

$$=\left\{\frac{20(20+1)}{2}\right\}^2+3\left\{\frac{20(20+1)}{2}\right\}$$

$= 44100 + 630 = 44730$

➘ Problem 83 Let $t_1, t_2, \ldots$ be real numbers such that $t_1+t_2+\ldots+t_n=2n^2+9n+13$, for every positive integer $n\geq 2$. If $t_k=103$, then k equals ***[CAT-18]***

Sol. Given, $t_1+t_2+\ldots+t_n=2n^2+9n+13$

Then, $t_1+t_2+\ldots+t_{n-1}=2(n-1)^2+9(n-1)+13$

Subtracting the latter equation from the former we get,

$t_n = 2n^2-2n^2+4n-2+9n-9n+9+13-13$
$= 4n+7$

Then, $t_k = 4k+7$

$\Rightarrow\quad 103 = 4k+7 \Rightarrow k-24$

➘ Problem 84 The sum of $(n+1)$ terms of $\frac{1}{1}+\frac{1}{1+2}+\frac{1}{1+2+3}+\ldots.$ is

(a) $\frac{n}{n+1}$ (b) $\frac{2n}{n+1}$

(c) $\frac{2}{n(n+1)}$ (d) $\frac{2(n+1)}{n+2}$

Sol. Clearly, nth term $=\frac{1}{1+2+3+\ldots+n}=\frac{1}{\frac{n(n+1)}{2}}$

$$=\frac{2}{n(n+1)}=2\left(\frac{1}{n}-\frac{1}{n+1}\right)$$

Hence, 1st term $=2\left(\frac{1}{1}-\frac{1}{1+1}\right)=2\left(\frac{1}{1}-\frac{1}{2}\right)$

2nd term $=2\left(\frac{1}{2}-\frac{1}{3}\right)$

3rd term $=2\left(\frac{1}{3}-\frac{1}{4}\right)$

....

$(n+1)$th term $=2\left(\frac{1}{n+1}-\frac{1}{n+2}\right)$

$\therefore$ Required sum $=2\left[\left(1+\frac{1}{2}+\frac{1}{3}+\ldots+\frac{1}{n+1}\right)-\left(\frac{1}{2}+\frac{1}{3}+\ldots+\frac{1}{n+1}+\frac{1}{n+2}\right)\right]$

$$=2\left[1-\frac{1}{n+2}\right]=\frac{2(n+1)}{n+2}$$

➘ Problem 85 Find the value of the following series upto 20 terms $\{4^2-1^2\}+\{5^2-2^2\}+\{6^2-3^2\}+\{7^2-4^2\}+\ldots.$

Sol. Clearly, nth term $=(n+3)^2-n^2=6n+9$

$\therefore$ Required sum $=6\{1+2+3+\ldots+20\}+9\cdot 20$

$$=6\left\{\frac{20(20+1)}{2}\right\}+180=1260+180=1440$$

Alternatively,

Given expression $15+21+27+33+\ldots$ upto 20 terms.

$\therefore$ Required sum $=\frac{n}{2}\{2a+(n-1)d\}=\frac{20}{2}\{2\cdot 15+(20-1)6\}$

$= 1440$

➘ Problem 86 Find the value of the following series $4^2-1^2+5^2-2^2+6^2-3^2+\ldots$ upto 21 terms.

Sol. Clearly, 10 pairs can be formed taking 1st two terms together, further two terms together and so on. But, 21st term will be single.

Now, value of 10 pairs $= 15+21+27+\ldots.$ upto 10 terms

$$=\frac{10}{2}\{2\cdot 15+(10-1)6\}=420$$

Further, 1st term $=4^2$, 3rd term $=5^2$, ... then, 21st term $=(14^2)$

Since each odd- place term is '+', 21st term will also be positive.

Hence, 21st term $=+(14^2)=+196$

$\therefore$ Required answer $=420+190-616$

➘ Problem 87 $\{4^3+1^3\}+\{6^3+3^3\}+\{8^3+5^3\}\ldots$upto 12 terms = ?

Sol. Clearly, 1st term $=\{4^3+1^3\}=\{(2\cdot 1+2)^3+(2\cdot 1-1)^3\}$

2nd term $=\{(2\cdot \mathbf{2}+2)^3+(2\cdot \mathbf{2}-1)^3\}$

3rd term $=\{(2\cdot \mathbf{3}+2)^3+(2\cdot \mathbf{3}-1)^3\}$

....and so on.

Then, nth term $=\{(2\cdot \boldsymbol{n}+2)^3+(2\cdot \boldsymbol{n}-1)^3\}$

$= 8n^3+8+24n^2+24n+8n^3-1-12n^2+6n$

$= 16n^3+12n^2+30n+7$

$\therefore$ Required sum $=16\{1^3+2^3+3^3+\ldots+(12)^3\}$
$+12\{1^2+2^2+3^2+..+(12)^2\}$
$+30\{1+2+3+\ldots+12\}+\{7\times 12\}$

$$=16\left\{\frac{12(12+1)}{2}\right\}^2+12\left\{\frac{12(12+1)(2\cdot 12+1)}{6}\right\}+30\left\{\frac{12(12+1)}{2}\right\}+84$$

$= 97344+7800+2340+84=107568$

➘ Problem 88 $7+20+39+64+\ldots.$ upto 13 terms = ?

Sol. Clearly, 1st term = 7

2nd term = 7 + 13

3rd term = 7 + 13 + 19 .... and so on

Then, nth term = 7 + 13 + 19 + ... upto n terms.

$$= \frac{n}{2}\{2a + (n-1)d\}$$

$$= \frac{n}{2}\{2 \cdot 7 + (n-1)6\} = 7n + 3n^2 - 3n = 3n^2 + 4n$$

Now, required sum $= 3\{1^2 + 2^2 + 3^2 + \dots + (13)^2\} + 4\{1 + 2 + 3 + \dots + 13\}$

$$= 3\left\{\frac{13(13+1)(2 \cdot 13 + 1)}{6}\right\} + 4\left\{\frac{13(13+1)}{2}\right\}$$

$$= 2457 + 364 = 2821.$$

Problem 89 The sum of $(n-1)$ terms of $1 + (1+3) + (1+3+5) + \dots$ is

(a) $\frac{n(n+1)(2n+1)}{6}$ (b) $\frac{n^2(n+1)}{4}$

(c) $\frac{n(n-1)(2n-1)}{6}$ (d) n^2

Sol. Clearly, nth term = (1 + 3 + 5 ... + n terms) = n^2

Hence, 1st term = 1^2

2nd term = 2^2

3rd term = 3^2

....

$(n-1)$th term = $(n-1)^2$

Required sum $= 1^2 + 2^2 + \dots + (n-1)^2$

$$= \frac{(n-1)(n)(2n-2+1)}{6} = \frac{n(n-1)(2n-1)}{6}$$

G. Addition of Reciprocals of Consecutive Natural Numbers

Estimate the value of $\frac{1}{n} + \frac{1}{n+1} + \frac{1}{n+2} + \dots + \frac{1}{n+10}$, where n is a natural number.

If all the terms were $\frac{1}{n+10}$,

then sum $= \left(\frac{1}{n+10} \times \text{number of terms}\right)$ i.e. $\left(\frac{11}{n+10}\right)$.

Since the value of $\frac{1}{n+10}$ is least among all the given terms, the actual sum will be more than $\left(\frac{11}{n+10}\right)$.

Similarly, if all the terms were $\frac{1}{n}$,

then sum $= \left(\frac{1}{n} \times \text{number of terms}\right)$ i.e. $\left(\frac{11}{n}\right)$.

Since, the value of $\frac{1}{n}$ is greatest among all the given terms, the actual sum will be less than $\left(\frac{11}{n}\right)$.

It also holds good that

(i) Value of the expression is more than (Value of middle term × number of terms) i.e. more than $\frac{11}{(n+5)}$. Here, the value of middle term of the expression having n terms refers to $\left(\frac{n+1}{2}\right)$th term if n is odd and does $\left\{\text{average of } \left(\frac{n}{2}\right)\text{th term and } \left(\frac{n}{2}+1\right)\text{th term}\right\}$ if n is even.

(ii) The sum of the reciprocals of consecutive natural numbers is never an integer whether started from 1 or not.

Now, take some examples

Problem 90 If A is the sum of the reciprocals of the consecutive integers from 101 to 200 (both inclusive), which of the following is true?

(a) $\frac{1}{2} < A < 1$ (b) $1 < A < 2$

(c) $\frac{5}{4} < A < \frac{3}{2}$ (d) None of these

Sol. (a) As given, $A = \frac{1}{101} + \frac{1}{102} + \frac{1}{103} + .. + \frac{1}{200}$

If all the terms were $\frac{1}{100}$. Then, sum $= \left(\frac{1}{100} \times 100\right) = 1$.

Since $\frac{1}{100}$ is greater than each of the given terms, the actual sum will be less than 1. Further, if all the terms were $\frac{1}{200}$.

Then, sum $= \left(\frac{1}{200} \times 100\right) = \frac{1}{2}$. Since $\frac{1}{200}$ is least among all the given terms, the actual sum will be greater than $\frac{1}{2}$.

Hence, option(a) is correct.

Problem 91 What is the closest value of sum of the reciprocals of the consecutive integers from 20 to 30 (both inclusive)?

(a) $\frac{2}{5}$ (b) $\frac{1}{4}$ (c) $\frac{9}{20}$ (d) $\frac{3}{4}$

Sol. (c) Sum $= \frac{1}{20} + \frac{1}{21} + \frac{1}{22} + \dots + \frac{1}{30}$

Clearly, on the above pattern, the required sum will be greater than $\left(\frac{1}{30} \times 11\right)$ but lesser than $\left(\frac{1}{20} \times 11\right)$.

Hence, $\frac{11}{30} < \text{sum} < \frac{11}{20}$

However, values of option(a) and (c) are lying between $\frac{11}{30}$ and $\frac{11}{20}$.

Further, the required sum shall be greater than the (Value of middle term × number of terms)

i.e. greater than $\frac{11}{25}$.

Since the value of option(c) is greater than $\frac{11}{25}$, the correct option is (c).

WORKOUT

1. The value of $\{1^3 + 2^3 + 3^3 + \ldots + (15)^3\} - \{1 + 2 + 3 + \ldots + 15\}$ is **[MAT-2000]**
(a) 14280 (b) 14400
(c) 12280 (d) 13280

2. Find $1^3 + 2^3 + 3^3 + \ldots + (15)^3$. **[MAT-98, SSC (DEO)-09]**
(a) 11025 (b) 13400
(c) 900 (d) 14400

3. $\frac{1}{2}+\frac{1}{6}+\frac{1}{12}+\frac{1}{20}+\frac{1}{30}+\frac{1}{42}+\frac{1}{56}+\frac{1}{72}+\frac{1}{90}+\frac{1}{110}+\frac{1}{132}$ is equal to
(a) $\frac{10}{99}$ (b) $\frac{11}{131}$
(c) $\frac{11}{12}$ (d) $\frac{12}{55}$

4. Express $0.\overline{39} + 0.\overline{53} + 0.\overline{8}$ as a rational number of the form $\frac{p}{q}$.
(a) $\frac{20}{99}$ (b) $\frac{20}{11}$
(c) $\frac{10}{11}$ (d) $\frac{12}{53}$

5. $\frac{3}{4}+\frac{5}{36}+\frac{7}{144}+\ldots+\frac{17}{5184}+\frac{19}{8100}$ is equal to
(a) $\frac{13}{99}$ (b) $\frac{99}{100}$
(c) $\frac{10}{111}$ (d) $\frac{19}{21}$

6. Express $0.\overline{13} + 0.\overline{45} + 1.\overline{12}$ as a rational number of the form $\frac{p}{q}$.
(a) $\frac{13}{99}$ (b) $\frac{99}{100}$
(c) $\frac{169}{99}$ (d) $\frac{19}{21}$

7. $4^2 + 5^2 + 6^2 + \ldots + (12)^2 = ?$
(a) 842 (b) 1440
(c) 636 (d) 1328

8. $\frac{1}{1\cdot 2}+\frac{1}{2\cdot 3}+\frac{1}{3\cdot 4}+\ldots+\frac{1}{99\cdot 100}$ is equal to **[SSC (CGL-T-II)-10]**
(a) $\frac{100}{99}$ (b) $\frac{99}{100}$
(c) $\frac{1}{9900}$ (d) $\frac{1000}{99}$

9. The sum $11^2 + 12^2 + \ldots. + 21^2 = ?$ **[SSC (CGL-T-II)-11]**
(a) 2926 (b) 3017
(c) 3215 (d) 3311

10. Find the value of
$$\left(1-\frac{1}{n+1}\right)+\left(1-\frac{2}{n+1}\right)+\left(1-\frac{3}{n+1}\right)+\ldots+\left(1-\frac{n}{n+1}\right).$$
[SSC (CGL-T-II)-13]
(a) n (b) $\frac{n}{2}$
(c) $(n+1)$ (d) $\frac{(n+1)}{2}$

11. If $1^3 + 2^3 + \ldots + (10)^3 = 3025$, then find the value of $2^3 + 4^3 + 6^3 + \ldots + (20)^3 = ?$
(a) 6050 (b) 9075
(c) 12100 (d) 24200

12. Find the sum of the first five terms of the following series
$\frac{1}{1\cdot 4}+\frac{1}{4\cdot 7}+\frac{1}{7\cdot 10}+\ldots.$
(a) $\frac{5}{16}$ (b) $\frac{9}{32}$
(c) $\frac{7}{16}$ (d) $\frac{1}{210}$

13. If $1^2 + 2^2 + 3^2 + \ldots. + x^2 = \frac{x(x+1)(2x+1)}{6}$, then $1^2 + 3^2 + 5^2 + \ldots + (19)^2$ is equal to
(a) 1330 (b) 2100
(c) 2485 (d) 2500

14. The number of terms in the series $1 + 3 + 5 + 7 + \ldots + 73 + 75$ is
(a) 28 (b) 30
(c) 36 (d) 38

15. How many terms of the series "1 + 2 + 3 + ..." add to 5050?
(a) 50 (b) 51 (c) 100 (d) 101

16. When simplified, the sum
$\frac{1}{2}+\frac{1}{6}+\frac{1}{12}+\frac{1}{20}+\frac{1}{30}+\ldots+\frac{1}{n(n+1)}$ is equal to
[SSC (SO-Audit)-06, NICL (AO)-13]
(a) $\frac{n}{(n+1)}$ (b) $\frac{1}{(n+1)}$
(c) $\frac{1}{n}$ (d) $\frac{2(n-1)}{n}$

17. $1^2 - 2^2 + 3^2 - 4^2 + \ldots - (10)^2$ is equal to **[SSC (SO-Audit) 06, SSC (ASI) 10]**
(a) 45 (b) −45
(c) −54 (d) −55

18. $(1 + 3 + 5 + 7 + 9 + \ldots + 99)$ is **[SSC (CGL-PT)-07, 08]**
(a) 2050 (b) 2500
(c) 2005 (d) 2002

19. Given that, $1^2 + 2^2 + 3^2 + \ldots + n^2 = \frac{n(n+1)(2n+1)}{6}$, then $10^2 + 11^2 + 12^2 + \ldots + (20)^2$ is equal to: **[SSC (CGL-PT)-08]**
(a) 2626 (b) 2585
(c) 3747 (d) 2555

20. The sum of the series (1 +0.6 +0.06 + 0.006 + 0.0006 + ...) is **[SSC (CGL-T-I)-10]**
(a) $1\frac{2}{3}$ (b) $1\frac{1}{3}$
(c) $2\frac{1}{3}$ (d) $2\frac{2}{3}$

21. Expression $1 + 2 + 3 + \ldots + 79 + 80 + 79 + \ldots + 3 + 2 + 1 = ?$
(a) 3200 (b) 2800
(c) 3600 (d) 6400

22. Find out the sum of all natural numbers between 35 and 55.
(a) 855 (b) 280
(c) 960 (d) 756

23. Find the sum of all odd numbers between 20 and 101.
(a) 2400 (b) 2800
(c) 3600 (d) 6400

24. $13\cdot 15 + 14\cdot 16 + 15\cdot 17 + \ldots + 31\cdot 33 = ?$
(a) 10602 (b) 10620
(c) 12206 (d) 12364

25. 5 + 13 + 24 + 38 + ... upto 19 terms = ?
(a) 2460 (b) 4360
(c) 4370 (d) 4400

26. Find the value of the series −5 + 7 − 9 + 11 − 13 + 15 ... upto 60 terms.
(a) 60 (b) 70 (c) 120 (d) 80

27. $1\cdot 2\cdot 3 + 2\cdot 3\cdot 4 + 3\cdot 4\cdot 5 + \ldots. + 14\cdot 15\cdot 16 = ?$
(a) 14260 (b) 16800
(c) 36200 (d) 14280

28. Sum of the series $\frac{2}{3}+\frac{8}{9}+\frac{26}{27}+\frac{80}{81}+\ldots$ to n terms is

(a) $n-\frac{1}{2}(3^n-1)$ (b) $n+\frac{1}{2}(3^n-1)$

(c) $n+\frac{1}{2}(1-3^{-n})$ (d) $n+\frac{1}{2}(3^{-n}-1)$

29. The sum $1+3-5+7+9-11+13+15-17+\ldots+61+63-65$ is equal to *[LIC (AAO)-11]*

(a) 319 (b) 330 (c) 341 (d) 451

30. If $x=\frac{1}{2}+\frac{1}{6}+\frac{1}{12}+\frac{1}{20}+\frac{1}{30}+\frac{1}{42}+\frac{1}{56}+\frac{1}{72}$, then value of $\frac{1}{x}$ is closest to *[LIC (AAO)-11]*

(a) 1.1 (b) 1 (c) 0.9 (d) 0.8

31. $2\frac{1}{8}-1\frac{1}{16}=?+1\frac{1}{32}-1\frac{9}{64}$ *[IOB (PO)-11]*

(a) $2\frac{9}{32}$ (b) $1\frac{9}{64}$ (c) $2\frac{5}{32}$ (d) $1\frac{11}{64}$

(e) None of these

32. $3\frac{6}{7}-6\frac{1}{4}+5\frac{1}{3}=?$ *[Andhra Bank (PO)-11 P & S Bank (PO)-11]*

(a) $1\frac{65}{84}$ (b) $8\frac{1}{84}$ (c) $2\frac{79}{84}$ (d) $5\frac{47}{84}$

(e) None of these

33. $1\frac{1}{4}+1\frac{1}{6}-1\frac{1}{8}=?+1\frac{1}{12}$ *[Corporation Bank (PO)-11]*

(a) $\frac{5}{24}$ (b) $\frac{7}{24}$ (c) $\frac{5}{12}$ (d) $\frac{7}{12}$

(e) None of these

34. $441.01-232.99+1649.99=?+1225.92$ *[Corporation Bank (PO)-11]*

(a) 600 (b) 630 (c) 660 (d) 690

(e) 720

35. $454.58-376.89+121.45-95.42=?$ *[Syndicate Bank (PO)-2010]*

(a) 102.22 (b) 103.72

(c) 91.72 (d) 92.32

(e) None of these

36. $3\frac{3}{4}+4\frac{2}{5}-3\frac{1}{8}=?$ *[Indian Bank (PO)-10, SBI (PO-Pre)-18]*

(a) $4\frac{1}{40}$ (b) $5\frac{1}{40}$ (c) $6\frac{1}{40}$ (d) $5\frac{3}{40}$

(e) None of these

37. $726.34+888.12-?=1001.88$ *[IDBI Bank (PO)-09]*

(a) 621.58 (b) 602.64

(c) 654.54 (d) 618.78

(e) None of these

38. The sum of the series $1\cdot3^2+2\cdot5^2+3\cdot7^2+\ldots$ upto 20 terms is

(a) 188090 (b) 189080

(c) 199080 (d) None of these

39. Find the sum of cubes of first nine even numbers.

(a) 12400 (b) 12800

(c) 16200 (d) 17200

40. If $R=1+\frac{1}{100}+\frac{1}{107}+\frac{1}{114}+\ldots+\frac{1}{149}$, then closest value of R is

(a) $\frac{53}{50}$ (b) $\frac{100}{149}$ (c) $1\frac{1}{4}$ (d) 2.37

41. What is the value of the following expression? *[CAT-2000]*

$$\frac{1}{2^2-1}+\frac{1}{4^2-1}+\frac{1}{6^2-1}+\ldots+\frac{1}{20^2-1}$$

(a) $\frac{9}{19}$ (b) $\frac{10}{19}$ (c) $\frac{10}{21}$ (d) $\frac{11}{21}$

42. Find the sum

$$\sqrt{1+\frac{1}{1^2}+\frac{1}{2^2}}+\sqrt{1+\frac{1}{2^2}+\frac{1}{3^2}}+\ldots+\sqrt{1+\frac{1}{2007^2}+\frac{1}{2008^2}}$$

[CAT-08]

(a) $2008-\frac{1}{2008}$ (b) $2007-\frac{1}{2007}$

(c) $2007-\frac{1}{2008}$ (d) $2008-\frac{1}{2007}$

(e) $2008-\frac{1}{2009}$

43. What is the sum of first 40 terms of $1+3+4+5+7+7+10+9+\ldots$? *[SSC (CGL-T-II)-17]*

(a) 1010 (b) 1115 (c) 1030 (d) 1031

44. What is the value of

$$\frac{1}{3\times7}+\frac{1}{7\times11}+\frac{1}{11\times15}+\ldots+\frac{1}{899\times903}?$$

[SSC (CGL-T-II)-17]

(a) $\frac{21}{509}$ (b) $\frac{18}{403}$ (c) $\frac{25}{301}$ (d) $\frac{29}{31}$

45. $5466.97-3245.01+1122.99=?+2309.99$ *[SBI (PO)-17]*

(a) 1130 (b) 1000 (c) 1100 (d) 1030

(e) 1060

46. $302.46+395.72-123.47=?$ *[IBPS (Specialist Officer) 13]*

(a) 576.77 (b) 547.17

(c) 547.77 (d) 574.71

(e) 577.71

47. $\frac{1}{2}+\frac{1}{5}+\frac{1}{8}+\frac{1}{11}+\frac{1}{20}+\frac{1}{41}+\frac{1}{110}+\frac{1}{1640}$ is equal to *[Oriental Ince. Company (AAO)-12]*

(a) 0.97 (b) 0.99 (c) 1 (d) 1.23

48. A child was asked to add first few natural numbers (that is $1+2+3+\ldots$) so long his patience permitted. As he stopped, he gave the sum as 575. When the teacher declared the result wrong, the child discovered he had missed one number in the sequence during addition. The number he missed was

(a) less than 10 (b) 10

(c) 15 (d) more than 15

49. Let S_n denote the sum of the squares of the first n odd natural numbers. If $S_n=533n$, find the value of n.

(a) 18 (b) 20 (c) 24 (d) 30

50. N, the set of natural numbers, is divided into subsets $A_1=(1)$, $A_2=(2,3)$, $A_3=(4,5,6)$, $A_4=(7,8,9,10)$ and so on. What is the sum of the elements of the subset A_{50}? *[CAT-09]*

(a) 42455 (b) 61250

(c) 62525 (d) 65525

51. The value of the sum $7\cdot11+11\cdot15+15\cdot19+\ldots+95\cdot99$ is *[CAT-18]*

(a) 80707 (b) 80773

(c) 80730 (d) 80751

52. The natural numbers are divided into groups as (1), (2, 3, 4), (5, 6, 7, 8, 9)... and so on. Then, the sum of the numbers in the 15th group is equal to *(CAT-21)*

(a) 7471 (b) 6090

(c) 4941 (d) 6119

Answer Key

1. (a)	**2.** (d)	**3.** (c)	**4.** (b)	**5.** (b)	**6.** (c)	**7.** (c)	**8.** (b)	**9.** (a)	**10.** (b)
11. (d)	**12.** (a)	**13.** (a)	**14.** (d)	**15.** (c)	**16.** (a)	**17.** (d)	**18.** (b)	**19.** (b)	**20.** (a)
21. (d)	**22.** (a)	**23.** (a)	**24.** (a)	**25.** (c)	**26.** (a)	**27.** (d)	**28.** (d)	**29.** (a)	**30.** (a)
31. (d)	**32.** (c)	**33.** (a)	**34.** (b)	**35.** (b)	**36.** (b)	**37.** (e)	**38.** (a)	**39.** (c)	**40.** (a)
41. (c)	**42.** (a)	**43.** (c)	**44.** (c)	**45.** (d)	**46.** (d)	**47.** (c)	**48.** (d)	**49.** (b)	**50.** (c)
51. (a)	**52** (d)								

Hints and SOLUTIONS

1. $\left\{\frac{15(15+1)}{2}\right\}^2 - \left\{\frac{15(15+1)}{2}\right\} = 14400 - 120 = 14280$

2. $\left\{\frac{15(15+1)}{2}\right\}^2 = 14400$

3. $\left(\frac{1}{1}-\frac{1}{2}\right)+\left(\frac{1}{2}-\frac{1}{3}\right)+\left(\frac{1}{3}-\frac{1}{4}\right)+\left(\frac{1}{4}-\frac{1}{5}\right)+\left(\frac{1}{5}-\frac{1}{6}\right)+\left(\frac{1}{6}-\frac{1}{7}\right)$

$+ \ldots + \left(\frac{1}{11}-\frac{1}{12}\right)$

$= \frac{1}{1} - \frac{1}{12} = \frac{11}{12}$

4. $\frac{039-0}{100-1} + \frac{053-0}{100-1} + \frac{08-0}{10-1} = \frac{180}{99} = \frac{20}{11}$

Alternatively,

0.393939.....
0.535353.....
(+) 0.888888......
1.818180.....

Hence, $1.\overline{81} = 1\frac{81-0}{100-1} = 1\frac{9}{11}$

$= \frac{20}{11}$

5. $\left\{\frac{3}{1^2\cdot 2^2} + \frac{5}{2^2\cdot 3^2} + \frac{7}{3^2\cdot 4^2} + \frac{9}{4^2\cdot 5^2} + \ldots + \frac{19}{9^2\cdot(10)^2}\right\}$

$= \left\{\left(\frac{1}{1^2}-\frac{1}{2^2}\right)+\left(\frac{1}{2^2}-\frac{1}{3^2}\right)+\left(\frac{1}{3^2}-\frac{1}{4^2}\right)+\left(\frac{1}{4^2}-\frac{1}{5^2}\right)+ \ldots + \left(\frac{1}{9^2}-\frac{1}{10^2}\right)\right\}$

$= \frac{1}{(1)^2} - \frac{1}{(10)^2} = \frac{100-1}{100} = \frac{99}{100}$

6. $\frac{013-0}{100-1} + \frac{045-0}{100-1} + \frac{112-1}{100-1} = \frac{169}{99}$

7. Required sum $= \{1^2 + 2^2 + \ldots + (12)^2\} - \{1^2 + 2^2 + 3^2\}$

$= \frac{12(12+1)(2\cdot 12+1)}{6} - 14$

$= 650 - 14 = 636$

8. $\left(\frac{1}{1}-\frac{1}{2}\right)+\left(\frac{1}{2}-\frac{1}{3}\right)+\left(\frac{1}{3}-\frac{1}{4}\right)+\ldots+\left(\frac{1}{99}-\frac{1}{100}\right)$

$= \frac{1}{1} - \frac{1}{100} = \frac{99}{100}$

9. Required sum $= \{1^2 + 2^2 + \ldots + (21)^2\} - \{1^2 + 2^2 + \ldots + 10^2\}$

$= \frac{21(21+1)(2\cdot 21+1)}{6} - \frac{10(10+1)(2\cdot 10+1)}{6}$

$= 3311 - 385 = 2926$

10. Obviously, there are n terms.

Then, required value $= (1 + 1 + 1 + \ldots$ upto n terms$)$

$- \left(\frac{1}{n+1} + \frac{2}{n+1} + \frac{3}{n+1} + \ldots + \frac{n}{n+1}\right)$

$= n - \left(\frac{1+2+3+\ldots+n}{n+1}\right) = n - \left(\frac{\frac{n(n+1)}{2}}{n+1}\right) = n - \frac{n}{2} = \frac{n}{2}$

11. $2^3 + 4^3 + 6^3 \ldots + (20)^3$

$= (1\cdot 2)^3 + (2\cdot 2)^3 + (2\cdot 3)^3 + \ldots + (2\cdot 10)^3$

$= 2^3\{1^3 + 2^3 + 3^3 + \ldots + (10)^3\}$

$= 8 \times 3025$

$= 24200$

12. $\frac{1}{1\cdot 4} + \frac{1}{4\cdot 7} + \frac{1}{7\cdot 10} + \frac{1}{10\cdot 13} + \frac{1}{13\cdot 16} = ?$

$= \frac{1}{3}\left\{\left(\frac{1}{1}-\frac{1}{4}\right)+\left(\frac{1}{4}-\frac{1}{7}\right)+\left(\frac{1}{7}-\frac{1}{10}\right)+\left(\frac{1}{10}-\frac{1}{13}\right)+\left(\frac{1}{13}-\frac{1}{16}\right)\right\}$

$= \frac{1}{3}\left\{\frac{1}{1} - \frac{1}{16}\right\} = \frac{1}{3} \times \frac{15}{16} = \frac{5}{16}$

13. Required sum

$= \{1^2 + 2^2 + 3^2 + \ldots + 20^2\} - \{2^2 + 4^2 + 6^2 + \ldots + 20^2\}$

$= \frac{20(20+1)(2\cdot 20+1)}{6} - 4\{1^2 + 2^2 + 3^2 + \ldots + 10^2\}$

$= 2870 - 4\left\{\frac{10(10+1)(2\cdot 10+1)}{6}\right\}$

$= 2870 - 1540$

$= 1330$

Alternatively,

Number of terms $= \frac{19+1}{2} = 10$

We know,

and sum of squares of first n odd numbers $= \frac{n(2n+1)(2n-1)}{3}$

Then, sum of squares of first 10 odd numbers

$$= \frac{10(2\cdot 10+1)(2\cdot 10-1)}{3} = 1330$$

14. Required number of terms $= \frac{75+1}{2} = 38$

15. Clearly, $\frac{n(n+1)}{2} = 5050 \Rightarrow n(n+1) = 10100 \Rightarrow n = 100.$

16. $\left(\frac{1}{1}-\frac{1}{2}\right)+\left(\frac{1}{2}-\frac{1}{3}\right)+\left(\frac{1}{3}-\frac{1}{4}\right)+\left(\frac{1}{4}-\frac{1}{5}\right)+\left(\frac{1}{5}-\frac{1}{6}\right)+\ldots+$

$$\left(\frac{1}{n}-\frac{1}{(n+1)}\right) = \frac{1}{1}-\frac{1}{(n+1)} = \frac{n}{(n+1)}$$

17. Required value

$= \{1^2+3^2+5^2+\ldots+9^2\} - \{2^2+4^2+6^2+..+10^2\}$

We know, sum of squares of first n odd numbers

$$= \frac{n(2n+1)(2n-1)}{3}$$

and sum of squares of first n even numbers

$$= \frac{2n(n+1)(2n+1)}{3}$$

Then, required answer $= \frac{5(2\cdot 5+1)(2\cdot 5-1)}{3}$

$$- \frac{2\cdot 5(5+1)(2\cdot 5+1)}{3}$$

$$= 165 - 220 = -55$$

18. Here, $n = \frac{99+1}{2} = 50$

Required sum $= n^2 = (50)^2 = 2500$

19. Required value $= \{1^2+2^2+3^2+\ldots+20^2\} - \{1^2+2^2+3^2+\ldots+9^2\}$

$$= \frac{20(20+1)(2\cdot 20+1)}{6} - \frac{9(9+1)(2\cdot 9+1)}{6}$$

$$= 2870 - 285 = 2585$$

20. $1 + 0.6 + 0.06 + 0.006 + 0.0006 + \ldots = 1.666666\ldots = 1.\overline{6}$

$$= 1\frac{6-0}{10-1} = 1\frac{2}{3}$$

21. Given expression $= (80)^2 = 6400$

22. $(36+37+38+\ldots+54) = ?$

Then, required sum $= \frac{n}{2}[a+l] = \frac{19}{2}[36+54] = 855$

$[\because a = 36,\ l = 54$ and $n = (54-36)+1 = 19]$

23. Clearly, $a = 21$ and $l = 99$, then,

$$\text{Sum} = \frac{(l+a)(l-a+2)}{4} = \frac{(99+21)(99-21+2)}{4} = 2400$$

24. Clearly, 1st term $= (1+12)\times(1+14)$

2nd term $= (2+12)\times(2+14)$

3rd term $= (3+12)\times(3+14)$

.... and so on

Then, nth term $= (n+12)\times(n+14) = n^2 + 26n + 168$

Now, 1st term $= 1^2 + 26\cdot 1 + 168$ $[\because n = 1]$

2nd term $= 2^2 + 26\cdot 2 + 168$

3rd term $= 3^2 + 26\cdot 3 + 168$

....

19th term $= (19)^2 + 26\cdot 19 + 168$

$\{\because (n+12) = 31 \Rightarrow n = 19$
or $(n+14) = 33 \Rightarrow n = 19\}$

$\therefore$ Required sum $= \{1^2+2^2+3^2+\ldots+(19)^2\}$
$+ 26\{1+2+3+\ldots+19\} + (168\times 19)$

$$= \frac{19(19+1)(2\cdot 19+1)}{6} + \left\{26\times\frac{19(19+1)}{2}\right\} + (168\times 19)$$

$$= 2470 + 4940 + 3192 = 10602$$

25. Clearly, 1st term = 5

2nd term = 5 + 8

3rd term = 5 + 8 + 11

.... and so on

Then, nth term = 5 + 8 + 11 + ... upto n terms.

$$= \frac{n}{2}\{2a+(n-1)d\} = \frac{n}{2}\{2\cdot 5+(n-1)3\}$$

$$= 5n + \frac{3}{2}n^2 - \frac{3}{2}n = \frac{3}{2}n^2 + \frac{7}{2}n$$

Now, required sum

$$= \frac{3}{2}\{1^2+2^2+3^2+\ldots+(19)^2\} + \frac{7}{2}\{1+2+3+\ldots+19\}$$

$$= \frac{3}{2}\left\{\frac{19(19+1)(2\cdot 19+1)}{6}\right\} + \frac{7}{2}\left\{\frac{19(19+1)}{2}\right\}$$

$$= 3705 + 665 = 4370$$

26. Since each even-place number is '+', 60th term will also be positive. Now, arrange the expression in pairs like

$(-5+7)+(-9+11)+(-13+15)$ ...upto 30 pairs.

Hence, required answer $= 2\times 30 = 60$

27. Clearly, nth term $= n\times(n+1)(n+2) = n^3 + 3n^2 + 2n$

Now, 1st term $= 1^3 + 3\cdot 1^2 + 2\cdot 1$

2nd term $= 2^3 + 3\cdot 2^2 + 2\cdot 2$

3rd term $= 3^3 + 3\cdot 3^2 + 2\cdot 3$

....

14th term $= (14)^3 + 3\cdot(14)^2 + 2\cdot 14$

$\therefore$ Required sum $= \{1^3+2^3+3^3+\ldots+(14)^3\}$
$+ 3\{1^2+2^2+3^2+\ldots+(14)^2\} + 2\{1+2+3+\ldots+14\}$

$$= \left\{\frac{14(14+1)}{2}\right\}^2 + 3\left\{\frac{14(14+1)(2\cdot 14+1)}{6}\right\} + \left\{2\times\frac{14(14+1)}{2}\right\}$$

$$= 11025 + 3045 + 210 = 14280$$

28. Clearly, nth term $= \frac{1}{3^n}(3^n - 1) = 1 - \frac{1}{3^n}$

Hence, 1st term $= 1 - \left(\frac{1}{3}\right)^1$

2nd term $= 1 - \left(\frac{1}{3}\right)^2$

3rd term $= 1 - \left(\frac{1}{3}\right)^3$

....

nth term $= 1 - \left(\frac{1}{3}\right)^n$

Required sum = (1+ 1+ ...upto n terms)

$$-\left\{\frac{1}{3}+\left(\frac{1}{3}\right)^2+\left(\frac{1}{3}\right)^3+\ldots+\left(\frac{1}{3}\right)^n\right\}$$

$$= n - \left\{\frac{\frac{1}{3}\left\{1-\left(\frac{1}{3}\right)^n\right\}}{1-\frac{1}{3}}\right\} \text{\{In GP, } S_n = \frac{a(1-r^n)}{1-r},$$

detailed in chapter 'Progression'}

$$= n - \frac{1}{2}\left\{1-\left(\frac{1}{3}\right)^n\right\} = n - \frac{1}{2}\{1-3^{-n}\}$$

$$= n + \frac{1}{2}(3^{-n}-1)$$

29. If all are positive, then sum $= \frac{(l+a)(l-a+2)}{4}$

$$= \frac{(65+1)(65-1+2)}{4} = 1089$$

Now, (– 5 – 11 – 17 – ... – 65)

$$= \frac{n}{2}(a+l) = \frac{11}{2}(-5-65) = -385$$

$[\because t_n = a + (n-1)d \Rightarrow -65 = -5 + (n-1)(-6)$
$\Rightarrow n - 1 = 10 \Rightarrow n = 11]$

Since negative numbers have been taken earlier also as positive, the required answer = 1089 – (385 + 385) = 319

30. $x = \left(\frac{1}{1}-\frac{1}{2}\right)+\left(\frac{1}{2}-\frac{1}{3}\right)+\left(\frac{1}{3}-\frac{1}{4}\right)+\ldots+\left(\frac{1}{8}-\frac{1}{9}\right)$

$$\Rightarrow x = \frac{1}{1}-\frac{1}{9} \Rightarrow x = \frac{8}{9} \Rightarrow \frac{1}{x} = \frac{9}{8} = 1.125$$

Hence, option (a) is correct.

31. $2\frac{1}{8}-1\frac{1}{16}-1\frac{1}{32}+1\frac{9}{64}=?$

$$\Rightarrow (2-1-1+1)+\left(\frac{1}{8}-\frac{1}{16}-\frac{1}{32}+\frac{9}{64}\right)$$

$$= 1+\left(\frac{1\cdot 8}{8\cdot 8}-\frac{1\cdot 4}{16\cdot 4}-\frac{1\cdot 2}{32\cdot 2}+\frac{9}{64}\right) = 1\frac{11}{64}$$

32. $(3-6+5)+\left(\frac{6}{7}-\frac{1}{4}+\frac{1}{3}\right)$

$$= 2+\left(\frac{6\cdot 12}{7\cdot 12}-\frac{1\cdot 21}{4\cdot 21}+\frac{1\cdot 28}{3\cdot 28}\right) = 2\frac{79}{84}$$

33. $1\frac{1}{4}+1\frac{1}{6}-1\frac{1}{8}-1\frac{1}{12}=?$

$$\Rightarrow (1+1-1-1)+\left(\frac{1}{4}-\frac{1}{8}+\frac{1}{6}-\frac{1}{12}\right)$$

$$\Rightarrow 0+\left(\frac{1}{4}-\frac{1}{8}\right)+\left(\frac{1}{6}-\frac{1}{12}\right) = \left(\frac{1\cdot 2}{4\cdot 2}-\frac{1}{8}\right)+\left(\frac{1\cdot 2}{6\cdot 2}-\frac{1}{12}\right)$$

$$= \frac{1}{8}+\frac{1}{12} = \frac{1\cdot 3}{8\cdot 3}+\frac{1\cdot 2}{12\cdot 2} = \frac{5}{24}$$

34. 441.01 – 232.99 + 1649.99 – 1225.92 = ?

= 632.09

i.e. 630 (approx.)

35. 454.58 – 376.89 + 121.45 – 95.42 = 103.72

36. $(3+4-3)+\left(\frac{3}{4}+\frac{2}{5}-\frac{1}{8}\right)$

$$= 4+\left(\frac{3}{4}-\frac{1}{8}+\frac{2}{5}\right) = 4+\left(\frac{3\cdot 2}{4\cdot 2}-\frac{1}{8}+\frac{2}{5}\right) = 4+\left(\frac{5}{8}+\frac{2}{5}\right)$$

$$= 4+\left(\frac{5\cdot 5+2\cdot 8}{8\cdot 5}\right) = 4+\left(\frac{41}{40}\right) = 4+1\frac{1}{40} = 5\frac{1}{40}$$

37. 726.34 + 888.12 – 1001.88 = 612.58

38. Clearly, nth term $= n(2n+1)^2 = 4n^3 + 4\cdot n^2 + n$

Hence, 1st term $= 4\cdot 1^3 + 4\cdot 1^2 + 1$

2nd term $= 4\cdot 2^3 + 4\cdot 2^2 + 2$

3rd term $= 4\cdot 3^3 + 4\cdot 3^2 + 3$

....

20th term $= 4\cdot 20^3 + 4\cdot 20^2 + 20$

Required sum $= 4(1^3 + 2^3 + 3^3 + \ldots + 20^3)$
$+ 4(1^2 + 2^2 + 3^2 + \ldots + 20^2) + (1 + 2 + 3 + \ldots + 20)$

$$= 4\left\{\frac{20(20+1)}{2}\right\}^2 + 4\left\{\frac{20(20+1)(2\cdot 20+1)}{6}\right\} + \left\{\frac{20(20+1)}{2}\right\}$$

$$= 176400 + 11480 + 210 = 188090$$

39. Sum $= 2n^2(n+1)^2 = 2\times 9^2(9+1)^2 = 16200$

40. Clearly, $R - 1 = \frac{1}{100}+\frac{1}{107}+\frac{1}{114}+\ldots+\frac{1}{149}$

Here, number of terms = 8

If all the terms were $\frac{1}{100}$, then $R - 1 = \left(\frac{1}{100}\times 8\right) = \frac{2}{25}$.

Since $\frac{1}{100}$ is greatest among all the given terms, the actual value of $(R-1)$ will be less than $\frac{2}{25}$.

Further, if all the terms were $\frac{1}{149}$,

then $R - 1 = \left(\frac{1}{149}\times 8\right) = \frac{8}{149}$.

Since $\frac{1}{149}$ is least among all the given terms, the actual value of $(R-1)$ will be greater than $\frac{8}{149}$.

Hence, $\frac{8}{149} < (R-1) < \frac{2}{25}$

$$\Rightarrow \frac{8}{149}+1 < R < \frac{2}{25}+1$$

$$\Rightarrow \frac{157}{149} < R < \frac{27}{25}$$

Only option(a) is satisfying this condition.

41. $\frac{1}{2^2-1}+\frac{1}{4^2-1}+\frac{1}{6^2-1}+\ldots+\frac{1}{20^2-1}$

$$= \frac{1}{1\cdot 3}+\frac{1}{3\cdot 5}+\frac{1}{5\cdot 7}+\ldots+\frac{1}{19\cdot 21}$$

$$= \frac{1}{2}\left\{\left(\frac{1}{1}-\frac{1}{3}\right)+\left(\frac{1}{3}-\frac{1}{5}\right)+\left(\frac{1}{5}-\frac{1}{7}\right)+\ldots+\left(\frac{1}{19}-\frac{1}{21}\right)\right\}$$

$$= \frac{1}{2}\left\{\frac{1}{1}-\frac{1}{21}\right\} = \frac{1}{2}\times\frac{20}{21} = \frac{10}{21}$$

42. 1st term $= \sqrt{1+\frac{1}{1^2}+\frac{1}{2^2}} = \sqrt{\frac{9}{4}} = \frac{3}{2} = 2-\frac{1}{2}$

Sum of first two terms $= \frac{3}{2} + \sqrt{1+\frac{1}{2^2}+\frac{1}{3^2}} = \frac{3}{2}+\frac{7}{6}$

$= \frac{8}{3} = 3-\frac{1}{3}$

Similarly, sum of first three terms $= 4-\frac{1}{4}$

Sum of first 2007 terms $= 2008 - \frac{1}{2008}$

43. Required sum $= 1+3+4+5+7+7+10+9+\ldots$ upto 40 terms

$= (1+4+7+10+13+\ldots$ upto 20 terms$) + (3+5+7+9+\ldots$ upto 20 terms$)$

$= \frac{20}{2}\{2\cdot 1+(20-1)3\}+[21^2-1]$

$= 590+440 = 1030$

44. $\frac{1}{4}\left\{\left(\frac{1}{3}-\frac{1}{7}\right)+\left(\frac{1}{7}-\frac{1}{11}\right)+\left(\frac{1}{11}-\frac{1}{15}\right)+\ldots+\left(\frac{1}{899}-\frac{1}{903}\right)\right\}$

$= \frac{1}{4}\left\{\frac{1}{3}-\frac{1}{903}\right\} = \frac{75}{903} = \frac{25}{301}$

45. ? = 5467 − 3245 + 1123 − 2310 ≈ 1030

46. ?= 302.46 + 395.72 − 123.47 = 574.71

47. $\frac{1}{2}+\frac{1}{5}+\frac{1}{8}+\frac{1}{11}+\frac{1}{20}+\frac{1}{41}+\left(\frac{1}{10}-\frac{1}{11}\right)+\left(\frac{1}{40}-\frac{1}{41}\right)$

$= \frac{1}{2}+\frac{1}{5}+\frac{1}{8}+\frac{1}{20}+\frac{1}{10}+\frac{1}{40} = \frac{40}{40} = 1$

48. Let he added first n numbers.

Then, sum of first n natural numbers $= \frac{n(n+1)}{2}$

$\Rightarrow \quad 575 = \frac{n(n+1)}{2}$

If we assume, $n = 34$, then $\frac{n(n+1)}{2} = 595$

and if we assume, $n = 33$, then $\frac{n(n+1)}{2} = 561$

It may be construed from the above that he added first 34 natural numbers but missed a number which is equal to (595 − 575).

Hence, option (d) is correct.

49. Sum of squares of first n odd numbers $= \frac{n(2n+1)(2n-1)}{3}$

$\Rightarrow \quad 533n = \frac{n(2n+1)(2n-1)}{3}$

$\Rightarrow \quad 1599n = n(2n+1)(2n-1)$

$\Rightarrow \quad 1599 = 4n^2-1$

$\Rightarrow \quad 4n^2 = 1600$

$\Rightarrow \quad n = 20$

50. Natural numbers covered upto A_1 = first 1 natural number

Natural numbers covered upto A_2 = first $\left\{\frac{2(2+1)}{2}\right\}$ natural numbers

Natural numbers covered upto A_3 = first $\left\{\frac{3(3+1)}{2}\right\}$ natural numbers

Natural numbers covered upto A_4 = first $\left\{\frac{4(4+1)}{2}\right\}$ natural numbers

∴ Natural numbers covered upto A_{49} = first $\left\{\frac{49(49+1)}{2}\right\}$ natural numbers i.e. first1225 natural numbers

Then, A_{50} = (1226, 1227, 1228, ...upto 50th term)

∴ Required sum $= \frac{50}{2}\times(2\cdot 1226+49\cdot 1)$

$= 62525$

51. 1st term $= (1\cdot 4+3)\times(1\cdot 4+7)$

2nd term $= (2\cdot 4+3)\times(2\cdot 4+7)$

....

nth term $= (n\cdot 4+3)\times(n\cdot 4+7)$

$= 16n^2+40n+21$

Now, 1st term $= 16\cdot 1^2+40\cdot 1+21$

2nd term $= 16\cdot 2^2+40\cdot 2+21$

....

23rd term $= 16\cdot 23^2+40\cdot 23+21$

[∵ 95 is the 23rd term in AP series, 7, 11, 15, 19, ...]

Then, required sum $= 16\{1^2+2^2+3^2+\ldots+(23)^2\} + 40\{1+2+3+\ldots+23\}+21\cdot 23$

$= 16\left\{\frac{23(23+1)(2\cdot 23+1)}{6}\right\}+40\left\{\frac{23(23+1)}{2}\right\}+483$

$= 80707$

52. Last terms of 1st, 2nd, 3rd ...15th group are $1^2, 2^2, 3^2, \ldots 15^2$ respectively and they have 1, 3, 5, ..., 29 terms respectively.

Hence, 15th group is (197, 198, ..., 225)

Then, the required sum $= \frac{29}{2}[197+225] = 6119$

02

Multiplication

Learning OBJECTIVES

After completing this chapter, aspirants will be able to

- Understand the significance of this chapter;
- Learn how to compute multiplication of Whole numbers, Decimal numbers, Fractions, Recurring decimals, and numbers/terms given in a special/unique series with unique approaches;
- Realise difference between conventional approaches and alternative approaches;
- Understand the philosophy of the division of this chapter in various parts;
- Analyse the trend of the questions being asked in various competitive examinations;
- Solve the questions in moments;
- Grip on concepts helping to solve the questions without pen and paper;
- Learn new concepts/fundamentals;
- Apply the pragmatic approaches in lieu of academic approaches; and
- Ponder on the alternative methods.

Introduction

Needless to say that there is no chapter of arithmetic where applications of the multiplication don't exist. The techniques envisaged in this chapter will reduce the time taken in solving the questions related to almost all arithmetical chapters.

This chapter is very critical especially for Banking services examinations wherein questions are based on direct applications of the multiplications. In addition, the core process to solve the questions asked there, are based on the multiplication. Directly or indirectly almost all the chapters are associated with the process of multiplication. The applications of multiplication are seen in the chapters like Simplification, Data Interpretation, etc., to the greater extent.

In simple words, multiplication is the shortest process of addition of repeated given numbers. The number which is to be repeated is called the multiplicand and the times by which multiplicand will be repeated is called multiplier.

Inside

- *Multiplication of Whole Numbers : General Rules*
- *Multiplication of Fractions/ Whole Numbers – Fractions*
- *Multiplication of Decimal Numbers*
- *Multiplication of Recurring Decimals*
- *Unique Series Based on Multiplication*
- ....
- .... *& many more*
- *Work Out with Solutions*

For example $6+6+6=18=6\times 3$

Here, 6 is called multiplicand and 3 is called multiplier. 18 is called product.

Multiplication may have diverse dimensions

A. Multiplication of **Whole numbers**
B. Multiplication of **Fractions**
C. Multiplication of **Decimal numbers**
D. Multiplication of **Recurring decimals**
E. **Unique series** based on Multiplication

A. Multiplication of Whole Numbers : General Rules

(i) Multiplication of Numbers Each Having Two Digits

$$\begin{array}{rr} a & b \\ \times c & d \\ \hline ad & bd \\ ac \quad bc & \\ \hline ac \mid ad+bc \mid bd \end{array}$$

We may arrive directly at the last line to save the time in the examination hall as well as paper/space provided on the question paper.

For example

$$\begin{array}{r} 34 \\ \times\ 56 \\ \hline \end{array}$$

$5\cdot 3$	$6\cdot 3$ + $5\cdot 4$	$6\cdot 4$	
15	38	24	
15	40 (i.e. 38 + 2)	4	Taken only unit digit (4) of 1st right term (24) and remaining portion(2) has been carried over left.
19 (i.e. 15 + 4)	0	4	Taken only unit digit(0) of 2nd right term(40) and remaining portion(4) has been carried over left.

Here, product is **1904**.

Note *Each term from the extreme right will have only unit digit (except the extreme left term) and remaining portion of the product is carried over left.*

Now, try yourself to find the product of following pairs of numbers mentally or in one line/ step.

Problem 1 $65\times 49=?$
2 $78\times 45=?$
3 $39\times 86=?$
4 $56\times 73=?$

(ii) Multiplication of Numbers Each Having Three Digits

$$\begin{array}{rrrrr} & & a & b & c \\ & & d & e & f \\ \hline & & af & bf & cf \\ & ae & be & ce & \\ ad & bd & cd & & \\ \hline \end{array}$$

$$ad \mid ae+bd \mid af+be+cd \mid bf+ce \mid cf$$

For example

$$\begin{array}{r} 743 \\ \times 256 \\ \hline \end{array}$$

$7\cdot 2$	$2\cdot 4+7\cdot 5$	$7\cdot 6+2\cdot 3+4\cdot 5$	$4\cdot 6+5\cdot 3$	$3\cdot 6$	
14	43	68	39	18	
19	0	2	0	8	On carrying over left

Here, product is **190208.**

Problem 5 $785\times 39=?$

Sol. Solve by equalizing number of digits in both numbers like below

$$\begin{array}{r} 785 \\ \times 039 \\ \hline \end{array}$$

Answer is 30615.

(iii) Multiplication of Numbers Each Having Four Digits

$$\begin{array}{rrrrrrr} & & & a & b & c & d \\ & & & e & f & g & h \\ \hline & & & ah & bh & ch & dh \\ & & ag & bg & cg & dg & \\ & af & bf & cf & df & & \\ ae & be & ce & de & & & \\ \hline \end{array}$$

$$ae \mid af+be \mid ag+bf+ce \mid ah+bg+cf+de \mid bh+cg+df \mid ch+dg \mid dh$$

For example

$$\begin{array}{r} 2135 \\ \times 6894 \\ \hline \end{array}$$

$2\cdot 6$	$2\cdot 8+1\cdot 6$	$2\cdot 9+6\cdot 3+1\cdot 8$	$2\cdot 4+6\cdot 5+1\cdot 9+3\cdot 8$	$1\cdot 4+8\cdot 5+3\cdot 9$	$3\cdot 4+9\cdot 5$	$5\cdot 4$	
12	22	44	71	71	57	20	
14	7	1	8	6	9	0	On carrying over left

Here, product is **14718690.**

Try yourself to solve the following in one line/ step.

Problem 6 $9683 \times 125 = ?$

[*Hint* : 9683×0125]

Ans. 1210375

Problem 7 $6837 \times 2358 = ?$

Ans. 16121646

Problem 8 $9863 \times 7985 = ?$

Ans. 78756055

Problem 9 $8663 \times 73 = ?$

[*Hint* : 8663×0073]

Ans. 632399

(iv) Multiplication of Numbers Each Having Five Digits

$$\begin{array}{ccccc} a & b & c & d & e \\ & f & g & h & i & k \end{array}$$

$$af \mid ag + bf \mid cf + ah + bg \mid ai + fd + gc + bh \mid ak + ef + gd + bi + ch \mid bk + ge + hd + ci \mid ck + eh + di \mid dk + ie \mid ek$$

Try yourself to solve the followings in one line/step

Problem 10 $23682 \times 18795 = ?$

Ans. 445103190

Problem 11 $68321 \times 8397 = ?$

Ans. 573091437

Problem 12 $37852 \times 957 = ?$

Ans. 36224364

Observation The above patterns are applicable in all types of multiplications as it is the shortest form of the conventional method. However, there may be many cases wherein finding product is so easy that it can be done mentally without using pen and paper. **We will discuss case by case.**

Case I When both numbers (i.e. multiplicand and multiplier) are closer to 10

There may be following possibilities

(a) both are less than 10

(b) both are above 10

(c) one is less than 10 and another is above 10

Here, 10 may be termed as the basc number.

Case I (a) When both are less than 10

For example $7 \times 6 = ?$

***Step* I** Hide anyone figure with your finger (say, 7), it means only 6 is appearing.

***Step* II** The number by which appearing number is less than the base number(10), should be subtracted from the hidden number.

$\Rightarrow \quad 7 - (10 - 6) = 3$

***Step* III** Find the product of the numbers by which both numbers are less than the base number 10 and write the extreme right digit (i.e. unit digit) of the product from the right hand side. The remaining portion, if any, of the product will be carried over the left figure. Here, left figure means the outcome of the step(II).

$\Rightarrow \quad \text{Product} = (10 - 7) \times (10 - 6) = 12$

Here, we will take digit 2 and write from the extreme right. Then, remaining portion (i.e.1) will be carried over left, hence, $(3 + 1)$.

Now, required answer $= (3 + 1) \mid (2) = 42$.

The above process may be depicted ut-infra.

Step I	Step II	Step III
Hide any one figure	Hidden figure – (difference between base number and appearing figure)	Product of differences of numbers from the base number
7	$7 - (10 - 6)$	$(10 - 7) \times (10 - 6)$
	3	12
		Only unit digit (i.e. 2) of the product is to be depicted just below the line. The remaining (i.e.1) is to be carried over left.
	$(3 + 1)$	2
	Required answer = **42**	

Take another example $7 \times 8 = ?$

Step I	Step II	Step III
Hide any one figure	Hidden figure – (difference between base number and appearing figure)	Product of differences of numbers from the base number
8	$8 - (10 - 7)$	$(10 - 8) \times (10 - 7)$
	5	6
		Only unit digit (i.e. 6) of the product is to be depicted just below the line. The remaining (i.e.nothing) is to be carried over left.
	5	6
	Required answer = **56**	

Try yourself to solve the following

Problem 13 $6 \times 9 = ?$

14 $8 \times 6 = ?$

15 $7 \times 9 = ?$

***Case* I (b)** When both are above 10

All the processes cited above will remain unchanged, only negative sign will turn into positive.

***Step* I** Hide anyone figure.

***Step* II** The number by which appearing number is above the base number 10, should be added to the hidden number.

***Step* III** Find the product of the numbers by which both numbers are above the base number 10 and write the extreme right digit (i.e. unit digit) of the product from the right hand side. The remaining portion, if any, of the product will be carried over the left figure. Here, left figure means the outcome of the step (II).

The operation may be carried out as below in case of 13×14.

Step (I)	Step (II)	Step (III)
Hide any one figure	Hidden figure + (difference between base number and appearing figure)	Product of differences of numbers from the base number
13	13 + (14 – 10)	(13 – 10) × (14 – 10)
	17	12
		Only unit digit (i.e. 2) of the product is to be depicted just below the line. The remaining portion (i.e. 1) is to be carried over left.
	(17 + 1)	2
	18	2
	Required answer = **182**	

➘ Problem 16 $15 \times 16 = ?$

Step (I)	Step (II)	Step (III)
Hide any one figure	Hidden figure + (difference between base number and appearing figure)	Product of differences of numbers from the base number
15	15 + (16 – 10)	(15 – 10) × (16 – 10)
	21	30
	(21 + 3)	0
	24	0
	Required answer = **240**	

Try yourself to solve the following without using pen and paper.

➘ Problem 17 $16 \times 19 = ?$

18 $18 \times 16 = ?$

19 $17 \times 12 = ?$

***Case* I (c)** When one number is less than 10 and another is above 10

For example $16 \times 7 = ?$

Here, the process for finding the product is slightly different from the above.

***Step* I** Hide anyone figure either 16 or 7(say, 16).

***Step* II** The number by which appearing number is **less** than the base number(10), should be **subtracted** from the hidden number.

$\Rightarrow \quad 16 - (10 - 7) = 13$

Or

***Step* I** Hide the figure 7 instead of 16.

***Step* II** The number by which appearing number is **above** the base number(10), should be **added** to the hidden number.

$\Rightarrow \quad 7 + (16 - 10) = 13$

Catch words : Less ⇒ Subtract and Above ⇒ Add

***Step* III** Subtract the product of the numbers by which both numbers are less/more than the base number 10, from either 10 or its multiple which is just above the product. If it is subtracted from 1st multiple of 10 (i.e. 10), subtract 1 from the outcome of the step (ii). In case, it is subtracted from 2nd multiple (i.e. 20), subtract 2 from the outcome of the step (ii) and so on.

Here, product is $(16 - 10) \times (10 - 7)$ i.e. 18. Thus, it should be subtracted from the 2nd multiple of 10 as 2nd multiple is just above the 18. Thereafter, subtract 2 from the outcome of the step(II).

Now, required answer is $(13 - 2) \mid (20 - 18) = 112$

The above process may be depicted, infra.

Step I	Step II		Step III
Hide any one figure	Hidden figure – (difference between base number and appearing figure) *Or* Hidden figure + (difference between base number and appearing figure) **(Remarks** If appearing figure is less than the base number, subtract the difference from hidden figure. In case of above the base figure, add the difference to hidden figure.)		(10 or its multiple) – (Product of differences of numbers from the base number). Then, subtract 1, 2, ... as case may be, from the outcome of step (II).
If 16 is hidden	16 – (10 – 7) = 13	In both cases of hiding, outcome will be same (i.e. 13). You have to hide only one.	(2 × 10) – [(16 – 10) × (10 – 7)]
If 7 is hidden	7 + (16 – 10) = 13		
	13		2
	(13 – 2)		2
	11		2
	Required answer = **112**		

Problem 20 $8 \times 12 = ?$

Step I	Step II	Step III
Hide any one figure	Hidden figure – (difference between base number and appearing figure) *Or* Hidden figure + (difference between base number and appearing figure)	(10 or its multiple) – (Product of differences of numbers from the base number). Then, subtract 1, 2, .. as case may be, from the outcome of step (II).
8	8 + (12 – 10) [∵ appearing figure is above the base number.]	(*1* × 10) – {(10 – 8) × (12 – 10)}
	10	6
	(10 – *1*)	6
	9	6
	Required answer = **96**	

Observations The outcome of step (II) in each case may be obtained by [Sum of both figures – Base number]. Hence, we may find the product of figures given in all the above three cases as following.

For 7 × 6

(7 + 6 – 10)	(10 – 7) (10 – 6)
3	12
(3 + 1)	2
42	

For 13 × 14

(13 + 14 – 10)	(13 – 10) (14 – 10)
17	12
(17 + 1)	2
182	

For 16 × 7

(16 + 7 – 10)	(*2* × 10) – (16 – 10) (10 – 7)
13	2
(13 – *2*)	2
11	2
112	

Try yourself to solve the following

Problem 21 $16 \times 9 = ?$

22 $8 \times 14 = ?$

23 $19 \times 7 = ?$

Case II When both numbers (i.e. multiplicand and multiplier) are closer to 20

There may be following possibilities like Case I

(a) both are less than 20

(b) both are above 20

(c) one is less than 20 and another is above 20

The process elaborated for Case I is applicable here also. Only we double the outcome of the step (II) as the base number given here (i.e. 20) is two times the previous base number (i.e.10). In case the base number is 30, we triples the outcome of step (II) and so on.

***Case* II (a)** When both are less than 20

For example $15 \times 17 = ?$

Step I	Step II	Step III
Hide any one figure	2 × {Hidden figure – (difference between base number and appearing figure)}	Product of differences of numbers from the base number
15	2 × {15 – (20 – 17)}	(20 – 15) × (20 – 17)
	24	15
		Only unit digit (i.e. 5) of the product is to be depicted just below the line. The remaining (i.e. nothing) is to be carried over left.
	(24+1)	5
	Required answer = **255**	

Note *It may also be solved considering the base number as 10.*

***Case* II (b)** When both are above 20

For example $23 \times 28 = ?$

Step I	Step II	Step III
Hide any one figure	2 × {Hidden figure + (difference between base number and appearing figure)}	Product of differences of numbers from the base number
23	2 × {23 + (28 – 20)}	(23 – 20) × (28 – 20)
	62	24
		Only unit digit (i.e. 4) of the product is to be depicted just below the line. The remaining (i.e. nothing) is to be carried over left.
	(62 + 2)	4
	Required answer = **644**	

Case **II (c)** When one number is less than 20 and another is above 20

For example 14 × 27 = ?

Step I	Step II	Step III
Hide any one figure	2 × {Hidden figure – (difference between base number and appearing figure)} *Or* 2 × {Hidden figure + (difference between base number and appearing figure)} **(Remarks :** If appearing figure is less than the base number, subtract the difference from hidden figure. In case of above the base figure, add the difference to hidden figure.)	(10 or its multiple) – (Product of differences of numbers from the base number). Then, subtract 1, 2, ... as case may be, from the outcome of step (II).
14	2 × {14 + (27 – 20)}	(*5* × 10) – {(20 – 14) × (27 – 20)}
	42	8
	(42 – *5*)	8
	37	8
	Required answer = **378**	

Observations The outcome of step (II) in each case may be obtained by (Sum of both figures – base number). Illustrations will be the same as given for cases I(a), I(b) and I(c).

Try yourself to solve the followings without using pen and paper.

Problem 24 18 × 19 = ?

25 24 × 29 = ?

26 27 × 18 = ?

Case III When both numbers (i.e. multiplicand and multiplier) are closer to 30

Needless to say that here the base number is 30. This case will follow the same pattern as given above.

Case **III (a)** 27 × 21 = ?

Let the first figure be hidden here, also for Cases III(b) and III(c).

3 × {27 (30 – 21)} *Or* 3 × (27 + 21 – 30)	(30 – 27) (30 – 21)
54	27
(54 + 2)	7
Required answer = **567**	

Case **III (b)** For 34 × 38

3 × {34 + (38 – 30)} *Or* 3 × (34 + 38 – 30)	(34 – 30) (38 – 30)
126	32
(126 + 3)	2
Required answer = **1292**	

Case **III (c)** For 26 × 36

3 × {26 + (36 – 30)} *Or* 3 × (26 + 36 – 30)	(*3* × 10) – {(30 – 26) (36 – 30)}
96	6
(96 – *3*)	6
93	6
Required answer = **936**	

Now, Try yourself to solve the following without using pen and paper.

Problem 27 26 × 29 = ?

28 33 × 38 = ?

29 29 × 37 = ?

Note *You ought to take some examples considering base numbers as 40 , 50, 60 and so on and solve mentally without using pen and paper. Some examples may be 44 × 48 = ?, 56 × 53 = ?, 75 × 62 = ?, 89 × 76 = ?, 54 × 67 = ?.*

Case IV When both numbers (i.e. multiplicand and multiplier) are closer to 100.

The pattern would be same as discussed above, however, the last box/space meant for the product will house two digits and surplus portion will be carried over the left. In case the base number is 1000, the last box will house three digits and surplus portion will be carried over.

Coding for memory (number of zeros ⇒ number of digits in the last box meant for product.) *viz.*

10 or its multiple ⇒ one digit in product-box

100 or its multiple ⇒ two digits in product-box

1000 or its multiple ⇒ three digits in product-box

Case **IV (a)** 92 × 98 = ?

Let first figure be hidden here, also for case IV(b) and IV(c).

{92 – (100 – 98)}	(100 – 92) (100 – 98)
90	16
Required answer = **9016**	

***Case* IV (b)** For 109×112

{109 + (112 – 100)}	(109 – 100) (112 – 100)
121	108
(121 + 1)	08
Required answer = **12208**	

***Case* IV (c)** For 84×106

Here, the product of differences of both figures from the base number will be subtracted from either 100 or its multiple.

In case of subtraction from 1st multiple, we will subtract 1 from the left box. In case, from 2nd multiple, 2 will be subtracted from the left box and so on.

{84 + (106 – 100)}	(*1* × 100) – {(100 – 84) (106 – 100)}
90	(*1* × 100) – 96
(90 – *1*)	4
89	04, as this box must contain two digits
Required answer = **8904**	

Now, try yourself to solve the following without using pen and paper.

Problem 30 $91 \times 99 = ?$

31 $107 \times 117 = ?$

32 $94 \times 116 = ?$

Case V When both numbers (i.e. multiplicand and multiplier) are closer to multiple of 100 *viz.* 200/300/400/......

For examples :

***Case* V (a)** $312 \times 315 = ?$

Clearly, the base number is 300.

3 × {312 + (315 – 300)}	(312 – 300) (315 – 300)
981	180
(981 + 1)	80
Required answer = **98280**	

***Case* V (b)** $418 \times 382 = ?$

Clearly base number is 400.

4 × {418 – (400 – 382)}	(*4* × 100) – {(418 – 400) (400 – 384)}
1600	(*4* × 100) – 324
(1600 – *4*)	76
Required answer = **159676**	

Now, try yourself to find the product of the following without using pen and paper.

Problem 33 $229 \times 195 = ?$

34 $338 \times 313 = ?$

35 $429 \times 371 = ?$

Case VI When both numbers (i.e. multiplicand and multiplier) are closer to 1000 or its multiple

For examples :

***Case* VI (a)** $987 \times 992 = ?$

{987 – (1000 – 992)}	(1000 – 987) (1000 – 992)
979	104 (this box will contain three digits as base number i.e. 1000 has three zeros)
979	104
Required answer = **979104**	

***Case* VI (b)** $1018 \times 992 = ?$

{1018 – (1000 – 992)}	(*1* × 1000) – {(1018 – 1000)(1000 – 992)}
1010	(*1* × 1000) – 144
(1010 – *1*)	856
Required answer = **1009856**	

Try yourself to find the product of the following without using pen and paper.

Problem 36 $983 \times 1012 = ?$

37 $1017 \times 1014 = ?$

38 $984 \times 986 = ?$

Observations All the approaches envisaged above, have been worked out from the following expression:

(a) $(x - a)(x - b) = x(x - a - b) + ab$

(b) $(x + a)(x + b) = x(x + a + b) + ab$

(c) $(x - a)(x + b) = x(x + b - a) - ab$

It is germane to mention that we are now capable to compute the product of those numbers which are closer than each other, in other words, which have common base number, in few seconds. Now consider the multiplication of 28 and 78. Obviously, both the numbers are not closer to each other. Then, 50 may be assumed as base number.

5 × {28 + (78 – 50)}	(*62* × 10) – {(50 – 28) (78 – 50)}
280	4
(280 – *62*)	4
218	4
Required answer = **2184**	

It seems that this method is very boring and will take much time. Hence, it is appropriate to adhere to the general rules meant for the multiplication, given in beginning, for the multiplication of these types of numbers.

Case VII When sum of unit digits is 10 and remaining portions of both the numbers are same

$$\begin{array}{r} a\,b \\ \times\, a\,c \\ \hline a^2 \,|\, a\,(b+c) \,|\, bc \end{array}$$

But, given that $b + c = 10$.

Then, required product = $a^2 | 10a | bc$

[∵ 0, suffixed with middle term, will accommodate the left digit of bc and the remaining portion of middle term will be carried over the left.]

Hence, required product = $a(a+1) | bc$

For example

***Case* VII (a)** $76 \times 74 = ?$

7(7 + 1)	6 × 4
56	24
Required answer = **5624**	

***Case* VII (b)** $112 \times 118 = ?$

11(11 + 1)	2 × 8
132	16
Required answer = **13216**	

***Case* VII (c)** $81 \times 89 = ?$

8(8 + 1)	1 × 9
72	09 (this box will contain always two digits as the middle term is suffixed by 0.)
Required answer = **7209**	

Try yourself to solve the following without using pen and paper.

Problem 39 $95 \times 95 = ?$

40 $128 \times 122 = ?$

41 $1002 \times 1008 = ?$

Case VIII When sum of tens place digits is 10 and unit digits are same

$$\begin{array}{r} a\,b \\ \times\, c\,d \\ \hline ac | b(a+c) | b^2 \end{array}$$

But, given that $a + c = 10$.

Then, the required product = $ac | 10b | b^2$

[obviously, 0, suffixed with the middle term, will accommodate the left digit of b^2 and the remaining portion of middle term will be carried over the left.]

Hence, the required product = $ac + b | b^2$

For example

***Case* VIII (a)** $43 \times 63 = ?$

(4 × 6) + 3	3 × 3
27	09, this box will contain always two digits for the above reason.
Required answer = **2709**	

***Case* VIII (b)** $78 \times 38 = ?$

(7 × 3) + 8	8 × 8
29	64
Required answer = **2964**	

Try yourself to solve the following without using pen and paper.

Problem 42 $75 \times 35 = ?$

43 $18 \times 98 = ?$

44 $46 \times 66 = ?$

Case IX When sum of numbers each formed with last two digits of multiplicand and multiplier, is a multiple of 10 and the remaining portions (i.e. numbers after deleting last two digits) are same.

For example $1512 \times 1508 = ?$

1st Part	2nd part
(remaining portion)^2 + {(remaining portion) × sum mentioned above (i.e. a multiple of 10)/100}	Product of numbers formed with last two digits.
$(15)^2 + \left\{15 \times \frac{(12+8)}{100}\right\}$	12 × 8
228	0096
Required answer = **2280096**	

Points to be Noted

(i) 2nd part must be in four digits. If actual product is not in four digits, append the required number of zeros before the product as done in this case (<u>00</u>96)

(ii) If 1st part is whole number, just write down (228, as done here). In case, it is mixed number(i.e. whole number + fraction), convert the fraction into decimal number, thereafter, delete the decimal (*viz.* take 6 instead of 0.6). Now, add this number (say, 6) to extreme left digit of the four digit product and if the sum exceeds 9, follow the carry-over left rule.

Take Some More Examples

(a) $3517 \times 3513 = ?$

1st Part	2nd part
(remaining portion)^2 + {(remaining portion) × sum mentioned above (i.e. a multiple of 10)/100}	Product of numbers formed with last two digits.
$(35)^2 + \left\{35 \times \frac{(17+13)}{100}\right\}$	17 × 13
1235.5	0221
1235 (5)	**0**221
1235 (5 + **0**)	221
Required answer = **12355221**	

(b) $1372 \times 1348 = ?$

1st Part	2nd part
(remaining portion)^2 + {(remaining portion) × sum mentioned above (i.e. a multiple of 10)/100}	Product of numbers formed with last two digits.
$(13)^2 + \left\{13 \times \frac{(72+48)}{100}\right\}$	72×48
184.6	3456
184 (6)	3456
184 (6 + 3)	456
Required answer = **1849456**	

(c) $1246 \times 1244 = ?$

1st Part	2nd part
(remaining portion)^2 + {(remaining portion) × sum mentioned above (i.e. a multiple of 10)/100}	Product of numbers formed with last two digits.
$(12)^2 + \left\{12 \times \frac{(46+44)}{100}\right\}$	46×44
154.8	2024
154 (8)	2024
154 (8 + 2)	024
154 (10)	024
1550	024
Required answer = **1550024**	

Try yourself to solve the followings

Problem 45 1742×1718

46 2717×2733

47 1512×1523 *

Note *This conception also works when the sum of numbers each formed with last two digits of multiplicand and multiplier, is not a multiple of 10 and the remaining portions (i.e. numbers after deleting last two digits) are same. It may be illustrated as below.*

For example 817×818

1st part	2nd part
(remaining portion)^2 + {(remaining portion) × sum mentioned above (i.e. a multiple of 10)/100}	Product of numbers formed with last two digits.
$(8)^2 + \left\{8 \times \frac{(17+18)}{100}\right\}$	17×18
66.8	0306
66 (8)	0306
66(8 + 0)	306
Required answer = **668306**	

Case X When a number is to be multiplied by 5^n, where n = natural numbers.

For example $198 \times 5^2 = ?$

$$\Rightarrow \quad 198 \times \frac{100}{2^2} = \frac{19800}{2^2} = 4950$$

It may be construed from the above that you should divide the given number (i.e. number other than 5^n) by 2^n after appending 'n' zeros at end of that number. Needless to say that dividing a number by 2^n may be easier than multiplying that number by 5^n in some cases.

Take another example : $362 \times 125 = ?$

$$\Rightarrow \quad 362 \times 5^3 = \frac{362000}{2^3} = 45250$$

Try yourself to find the product of the followings

Problem 48 $479 \times 625 = ?$

49 $3125 \times 9876 = ?$

50 $125 \times 6666 = ?$

Case XI When a number is to be multiplied by 75

For example $639 \times 75 = ?$

$$\Rightarrow \quad 639 \times \frac{300}{4} = \frac{191700}{4} = 47925$$

Note *We write 75 as $\frac{300}{4}$ for the convenience.*

Now, solve the followings

Problem 51 $283 \times 75 = ?$

52 $3125 \times 75 = ?$

Case XII When a number is to be multiplied by a number closer to 100, 1000, 10000 and so on

For example

(a) $378 \times 93 = ?$

$= 378 \times (100 - 7) = 37800 - 2646 = 35154$

(b) $367496 \times 1012 = ?$

$= 367496 \times (1000 + 12) = 367496000 + 4409952$

$= 371905952$

(c) $9999 \times 9999 = ?$

$= 9999 \times (10000 - 1) = 99990000 - 9999 = 99980001$

Observation We fetch the multiplier into form of $\{(10^n) \pm x\}$ for the convenience.

Try yourself to solve the followings

Problem 53 $479 \times 989 = ?$

54 $3125 \times 9992 = ?$

55 $1958 \times 9999987 = ?$

Case XIII When a number is to be multiplied by 11, 111, 1111, ...

For example

(a) $1732 \times 11 = ?$

Step I Append one zero at both sides of the multiplicand. In case the multiplier is 111, append two zeros on both sides and in case of 1111, append three zeros on both sides of the multiplicand and so on.

***Step* II** Add two consecutive digits from the extreme right and again add the two consecutive digits from the right leaving only last digit of the previous group of consecutive digits. Further, add two consecutive digits from the right leaving only last digit of the just previous group of consecutive digits. Repeat the same until the extreme left digit (say, 0) is included once in a group of two consecutive digits.

Note

(a) If multiplier is of 2 digits, we make a group of two consecutive digits and in case of 3 digits, a group of three consecutive digits is made and so on.

(b) 'Rule of carry over left' will be taken care after every sum of consecutive digits,

Now, the operation can be done as below for 1732 × 11.

***Step* I** 1732 ⇒ 017320

***Step* II** 0173<u>20</u> ⇒ 0 + 2 = 2

017<u>32</u>0 ⇒ 2 + 3 = 5

01<u>73</u>20 ⇒ 3 + 7 = 10

0<u>17</u>320 ⇒ 7 + 1 = 8

<u>01</u>7320 ⇒ 1 + 0 = 1

Now, write from the extreme right hand side like:

1	8	10	5	2	
1	(8 + 1)	0	5	2	Adhering to 'carry over left rule'

Required product = 19052

(b) 539 × 1111 = ?

***Step* I** 539 ⇒ 000539000

***Step* II** 00053<u>9000</u> ⇒ 0 + 0 + 0 + 9 = 9

0005<u>3900</u>0 ⇒ 0 + 0 + 9 + 3 = 12

000<u>5390</u>00 ⇒ 0 + 9 + 3 + 5 = 17

00<u>0539</u>000 ⇒ 9 + 3 + 5 + 0 = 17

0<u>0053</u>9000 ⇒ 3 + 5 + 0 + 0 = 8

<u>0005</u>39000 ⇒ 5 + 0 + 0 + 0 = 5

Adhering to 'carry over left rule' and writing from the extreme right hand side, we get 598829

Hence, product is **598829**.

(c) 67 × 444 = ?

⇒ 67 × 4 × 111 = either (67 × 111) × 4

or (67 × 4) × 111

On the above pattern, 67 × 111 = 7437

Then, required product = 7437 × 4 = **29748**

In my opinion the aspirants ought to first multiply by 1111.... .Then multiply by the digit associated with 1111....(i.e. 4). The same has been done above. It will rather forestall the mistake to be committed during the application of 'carry over left rule' number of times as it will entail requirement of application of 'carry over left rule' lesser times.

Try yourself to find the product of the followings

Problem 56 5679 × 111 = ?

57 2879 × 777777 = ?

Case XIV When a number is to be multiplied by a number upto 30.

The best way is to proceed through tables as it saves time to the greater extent. Thus, aspirants are supposed to get by heart the tables upto 30. More tables, more convenient, it is a catch line for the operation of multiplication.

For example 2385 × 19 = ?

19 × 2	19 × 3	19 × 8	19 × 5	
38	57	152	95	
45	3	1	5	Adhering to 'carry over left rule'

Now take one example of multiplication 94 × 96 = ?

(i) **Through General rule**

9 · 9	9 · 4 + 9 · 6	6 · 4
81	90	24
90	2	4

Required answer is **9024**.

(ii) Under case 'When sum of unit digits is 10 and remaining portions of both the numbers are same'.

9(9 + 1)	6 × 4
90	24
9024	

(iii) Under case 'When both numbers (i.e. multiplicand and multiplier) are closer to 100'.

{94 – (100 – 96)}	{100 – 94} {100 – 96}
90	24
9024	

(iv) Under case 'When a number is to be multiplied by a number closer to 100, 1000, 10000 and so on'.

94 × 96 = 94 × (100 – 4) = 9400 – 376

= 9024

Perusal of the section 'Multiplication of whole numbers' reveals that there are so many approaches to compute the product of two given numbers. Thus, in beginning, aspirants should ponder upon diverse techniques to find out product of two given numbers. The above is shown as only illustration.

It is also germane to clarify that diverse techniques may be derived for the multiplication, however, the same has not been done in this chapter deliberately keeping in view the practicality of their applications.

B. Multiplication of Fractions/ Whole Numbers-Fractions

General Rules

Type I	
Generic version	$a \times \frac{b}{c} = \frac{ab}{c}$
Example	$6 \times \frac{2}{5} = \frac{12}{5} = 2\frac{2}{5}$

Type II	
Generic version	$\frac{a}{b} \times \frac{c}{d} = \frac{ac}{bd}$
Example	$\frac{1}{6} \times \frac{7}{8} = \frac{1 \times 7}{6 \times 8} = \frac{7}{48}$

Type III	
Generic version	$a\frac{b}{c} \times d = \left(a + \frac{b}{c}\right) \times d = ad + \left(\frac{b}{c} \times d\right)$
Example	$19\frac{2}{3} \times 6 = (19 \times 6) + \left(\frac{2}{3} \times 6\right)$ $= 114 + 4 = 118$

Type IV	
Generic version	$a\frac{b}{c} \times \frac{d}{e} = \left(a \times \frac{d}{e}\right) + \left(\frac{b}{c} \times \frac{d}{e}\right)$
Example	$15\frac{2}{3} \times \frac{4}{5} = \left(15 \times \frac{4}{5}\right) + \left(\frac{2}{3} \times \frac{4}{5}\right)$ $= 12 + \frac{8}{15} = 12\frac{8}{15}$

Type V	
Generic version	$\frac{b}{c} \times d\frac{e}{f} = (a \times d) + \left(\frac{b}{c} \times \frac{e}{f}\right) + \left(\frac{b}{c} \times d\right) + \left(a \times \frac{e}{f}\right)$
Example	$16\frac{2}{3} \times 14\frac{7}{8}$ $= (16 \times 14) + \left(\frac{2}{3} \times \frac{7}{8}\right) + \left(\frac{2}{3} \times 14\right) + \left(16 \times \frac{7}{8}\right)$ $= 224 + \frac{7}{12} + 9\frac{1}{3} + 14 = 247\frac{11}{12}$

Special Rule for Special Cases

Case I	When integral part of both mixed fractions are same and the sum of fractional parts is 1.
Generic Version	$a\frac{b}{c} \times a\frac{d}{e} = a(a+1) + \left(\frac{b}{c} \times \frac{d}{e}\right)$
Example	$6\frac{2}{7} \times 6\frac{5}{7} = 6(6+1) + \left(\frac{2}{7} \times \frac{5}{7}\right) = 42\frac{10}{49}$
Case II	When integral parts of both mixed numbers are same and the sum of fractional part is $\frac{1}{2}$.
Generic Version	$a\frac{b}{c} \times a\frac{d}{e} = a^2 + \frac{a}{2} + \left(\frac{b}{c} \times \frac{d}{e}\right)$
Example	$16\frac{2}{5} \times 16\frac{1}{10} = (16)^2 + \frac{16}{2} + \left(\frac{2}{5} \times \frac{1}{10}\right) = 264\frac{1}{25}$

Problem 58 $5\frac{7}{9} \times 8\frac{4}{5} \times 9\frac{2}{3} = ?$ ***[Corporation Bank(PO)-09]***

(a) 490 (b) 590
(c) 540 (d) 460
(e) 520

Sol. Take any two terms, thereafter multiply the product thus obtained with the 3rd term.

$$\Rightarrow 5\frac{7}{9} \times 8\frac{4}{5} = (5 \times 8) + \left(\frac{7}{9} \times \frac{4}{5}\right) + \left(8 \times \frac{7}{9}\right) + \left(5 \times \frac{4}{5}\right)$$

$$= 40 + \frac{28}{45} + \frac{56}{9} + 4$$

$$= 44 + \frac{28}{45} + 6\frac{2}{9}$$

$$= 50\frac{38}{45}$$

Now, $50\frac{38}{45} \times 9\frac{2}{3} = (50 \times 9) + \left(\frac{38}{45} \times \frac{2}{3}\right) + \left(\frac{38}{45} \times 9\right) + \left(50 \times \frac{2}{3}\right)$

$$= 450 + \frac{76}{135} + 7\frac{3}{5} + 33\frac{1}{3}$$

$$= 490 + \frac{76}{135} + \frac{3}{5} + \frac{1}{3} = 490 \text{ (approx.)}$$

Hence, correct option is (a).

Alternatively,

$5.8 \times 8.8 \times 9.7 = 5.8 \times \{85.4\} = 495.3 = 490$ (approx.)

Hence, correct option is (a).

Problem 59 If * represents a number, then the value of * in $5 \times \frac{3}{*} \times 3\frac{1}{2} = 19$ is ***[SSC(CGL-PT)-05]***

(a) 7 (b) 4
(c) 6 (d) 2

Sol. $5\frac{3}{*} \times \frac{7}{2} = 19 \Rightarrow 5\frac{3}{*} = \frac{38}{7} = 5\frac{3}{7}$

Obviously, * = 7

Alternatively,

Let * = x

$$\Rightarrow 5\frac{3}{x} \times \frac{7}{2} = 19 \Rightarrow \frac{5x+3}{x} = \frac{38}{7}$$

$$\Rightarrow x = 7$$

C. Multiplication of Decimal Numbers

The approaches elaborated for multiplication of whole numbers will also be employed in case of multiplication of decimal numbers. In other words, we multiply decimal numbers ignoring the decimal points by the same approaches. Thereafter, decimal point is placed in the product after the certain number of digits from the right end, equal to total number of digits existing on the right side of decimal point in all given numbers which are to be multiplied.

For example (a) $6.39 \times 0.611 = ?$

***Step* I** Find the product ignoring the decimal points of both numbers.

Now, $639 \times 611 = ?$

$(6)^2 + \left\{6 \times \frac{(39+11)}{100}\right\}$	39×11
39	0429
390429	

[you are free to find the product by other approaches as per your choice]

***Step* II** Compute number of digits existing after decimal point in each number.

In case of 6.39 , number of digits existing after decimal point = 2 and in case of 0.611, number of digits existing after decimal point = 3

Thus, total number of digits after decimal points = 2 + 3 = 5

Now, put the decimal point after 5 digits from the right end of the product.

Hence, required product is **3.90429**.

Note *In case total number of existing digits after decimal points is greater than the number of digits of the product, we put zero(s) on left end of the product to equalise their number.*

Aspirant should mind the place of points (below and centre) as indicated below

2.5 (i.e. two and half) and $2 \cdot 5$ (i.e. two multiplied by 5)

(b) $0.0639 \times 0.0611 = ?$

Product of 639 and 611 = 390429

Clearly, total number of digits after decimal points = (4 + 4) = 8, which is greater than the number of digits in the product(i.e. 6) by 2. Hence, we will append two zeros in left end of the product and place the decimal point at extreme left end of the product. Hence, required product = **0.00390429**

(c) $3.6785 \times 1000 = ?$

When the decimal number is to be multiplied by (10^n), we shift the decimal point n digits ahead towards the right end of the number.

Hence, required product = 3678.5

Problem 60 $25.05 \times 123.95 + 388.999 \times 15.001 = ?$ ***[SBI(Associates-PO)-11]***

(a) 9000 (b) 8950
(c) 8935 (d) 8975
(e) 8995

Sol. $25 \times 124 + 399 \times 15 = 3100 + 5835 = 8935$

[If one option had been 'None of these', the exact computation would have been required]

Problem 61 $23.999 \times 9.004 \times 16.997 = ?$

(a) 3200 (b) 4100
(c) 2700 (d) 3700
(e) 4500

Sol. $24 \times 9 \times 17 = 216 \times 17 = 3672$

Hence, option(d) is correct.

Problem 62 $8.88 \times 88.8 \times 88 = ?$

(a) 68301.142 (b) 79391.642
(c) 65365.824 (d) 76218.414
(e) None of these

Sol. $8 \times 8 \times 8 \times [(111 \times 111) \times 11]$

[without considering the decimal points]

$= 512 \times 12321 \times 11$

$= 5632 \times 12321 = 69391872$

Now, total number of digits after decimal points = (2 + 1 + 0) = 3

Hence, required product = 69391.872

Thus, option (e) is correct.

Problem 63 $14.4 \times 16.5 \times 8 = ?$ ***[IDBI Bank(PO)-09]***

(a) 1908 (b) 1900.8
(c) 1912.4 (d) 1924.8
(e) None of these

Sol. $144 \times 165 \times 8$ {Ignoring decimal points)

$= 144 \times 1320 = 190080$

Then, required product = 1900.80

D. Multiplication of Recurring Decimals

Initially, we convert recurring decimals into fractions and then multiply as usual. Take some examples

Problem 64 $63.2\overline{31} \times 26 = ?$

Sol. $63\frac{(231-2)}{(1000-10)} \times 26 = 63\frac{229}{990} \times 26 = (63 \times 26) + \left(\frac{229}{990} \times 26\right)$

$= 1638 + \frac{5954}{990} = 1638 + 6.0\overline{14} = 1644.0\overline{14}$

Alternatively,

$63.2\overline{31} \times 26 = 63.2313131... \times 26 = 1644.01414$

(clearly, repetition starts from 14 onwards)

$\therefore$ Required answer $= 1644.0\overline{14}$

Problem 65 $37.2\overline{56} \times 100 = ?$

Sol. $37\frac{(256-2)}{(1000-10)} \times 100 = 37\frac{254}{(1000-10)} \times 100$

$= 37\frac{254}{990} \times 100 = 3700 + \frac{25400}{990} = 3700 + 25\frac{65}{99}$

$= 3700 + 25.\overline{65} = 3725.\overline{65}$

Alternatively,

$37.25656565.... \times 100 = 3725.6565...$

$= 3725.\overline{65}$

Problem 66 $2.3\overline{6} \times 4.\overline{5} = ?$

Sol. $2\frac{(36-3)}{(100-10)} \times 4\frac{(5-0)}{(10-1)}$

$$= \frac{213}{90} \times \frac{41}{9} = 10.78148$$

E. Unique Series Based on Multiplication

Problem 67 $\left(1-\frac{1}{2}\right)\left(1-\frac{1}{3}\right)\left(1-\frac{1}{4}\right)....\left(1-\frac{1}{n-1}\right)\left(1-\frac{1}{n}\right) = ?$

[SSC(CGL-PT) 08]

Sol. $\frac{1}{2} \times \frac{2}{3} \times \frac{3}{4} \times .... \times \frac{n-2}{n-1} \times \frac{n-1}{n} = \frac{1}{n}$

Problem 68 The value of $\left(1+\frac{1}{x}\right)\left(1+\frac{1}{x+1}\right)\left(1+\frac{1}{x+2}\right)\left(1+\frac{1}{x+3}\right)$ is

Sol. $\left(\frac{x+1}{x}\right)\left(\frac{x+2}{x+1}\right)\left(\frac{x+3}{x+2}\right)\left(\frac{x+4}{x+3}\right) = \frac{x+4}{x}$

Problem 69 $\left(1+\frac{1}{7}\right)\left(1+\frac{1}{8}\right)\left(1+\frac{1}{9}\right)...\left(1+\frac{1}{19}\right) = ?$

Sol. $\frac{8}{7} \times \frac{9}{8} \times \frac{10}{9} \times .... \times \frac{20}{19} = \frac{20}{7} = 2\frac{6}{7}$

Problem 70 $\left(2+\frac{2}{7}\right)\left(3+\frac{3}{8}\right)\left(4+\frac{4}{9}\right)\left(5+\frac{5}{10}\right)....\left(9+\frac{9}{14}\right) = ?$

Sol. $2\left(1+\frac{1}{7}\right) \times 3\left(1+\frac{1}{8}\right) \times 4\left(1+\frac{1}{9}\right) \times 5\left(1+\frac{1}{10}\right) \times .... \times 9\left(1+\frac{1}{14}\right)$

$$= 2\cdot3\cdot4\cdot5\cdot6\cdot7\cdot8\cdot9 \times \left(1+\frac{1}{7}\right)\left(1+\frac{1}{8}\right)\left(1+\frac{1}{9}\right) \times ... \times \left(1+\frac{1}{14}\right)$$

$$= 362880 \times \frac{8}{7} \times \frac{9}{8} \times \frac{10}{9} \times .... \times \frac{15}{14}$$

$$= 362880 \times \frac{15}{7} = 777600$$

Problem 71 $\left(1+\frac{1}{2}\right)\left(1+\frac{1}{3}\right)\left(1+\frac{1}{4}\right)....\left(1+\frac{1}{120}\right) = ?$

[SSC(CGL-PT)-03]

Sol. $\frac{3}{2} \times \frac{4}{3} \times \frac{5}{4} \times ... \times \frac{121}{120} = \frac{121}{2} = 60.5$

Problem 72 $(1\cdot2\cdot3\cdot....\cdot9)(11\cdot12\cdot13\cdot....\cdot19)(21\cdot22\cdot23\cdot....\cdot29) .... (71\cdot72\cdot73\cdot....\cdot79) = ?$

Sol. $(1\cdot2\cdot3\cdot...\cdot9) \times \frac{10}{10} \times (11\cdot12\cdot13\cdot...\cdot19) \times \frac{20}{20} \times (21\cdot22\cdot23\cdot...\cdot29) .... \times \frac{70}{70} \times (71\cdot72\cdot73\cdot...\cdot79)$

$$= \frac{79!}{10\cdot20\cdot30...70} = \frac{79!}{5040\cdot(10)^7} = \frac{79!}{504\cdot(10)^8}$$

Problem 73 $\left(1-\frac{1}{3^2}\right)\left(1-\frac{1}{4^2}\right)\left(1-\frac{1}{5^2}\right)...\left(1-\frac{1}{(11)^2}\right)\left(1-\frac{1}{(12)^2}\right) = ?$

[SSC(CGL-Mains)-04]

Sol. $\left(\frac{3^2-1}{3^2}\right)\left(\frac{4^2-1}{4^2}\right)\left(\frac{5^2-1}{5^2}\right)...\left(\frac{(11)^2-1}{(11)^2}\right)\left(\frac{(12)^2-1}{(12)^2}\right)$

$$= \left\{\frac{(3-1)(3+1)}{3^2}\right\} \times \left\{\frac{(4-1)(4+1)}{4^2}\right\} \times \left\{\frac{(5-1)(5+1)}{5^2}\right\} .... \left\{\frac{(11-1)(11+1)}{(11)^2}\right\}\left\{\frac{(12-1)(12+1)}{(12)^2}\right\}$$

$$= \left\{\frac{2}{3} \times \frac{4}{3}\right\} \times \left\{\frac{3}{4} \times \frac{5}{4}\right\} \times \left\{\frac{4}{5} \times \frac{6}{5}\right\} \times ... \times \left\{\frac{10}{11} \times \frac{12}{11}\right\} \times \left\{\frac{11}{12} \times \frac{13}{12}\right\}$$

$$= \frac{2}{3} \times \frac{13}{12} = \frac{13}{18}$$

Problem 74 Consider a sequence S whose nth term T_n is defined as $1+\frac{3}{n}$, where $n = 1, 2, ...$ Find the product of all the consecutive terms of S starting from the 4th term to the 60th term.

Sol. 4th term $= 1+\frac{3}{4} = \frac{7}{4}$

5th term $= 1+\frac{3}{5} = \frac{8}{5}$

6th term $= 1+\frac{3}{6} = \frac{9}{6}$

....

59th term $= 1+\frac{3}{59} = \frac{62}{59}$

60th term $= 1+\frac{3}{60} = \frac{63}{60}$

Now, the required product

$$= \frac{7}{4} \times \frac{8}{5} \times \frac{9}{6} \times \frac{10}{7} \times \frac{11}{8} \times \frac{12}{9} \times ... \frac{60}{57} \times \frac{61}{58} \times \frac{62}{59} \times \frac{63}{60}$$

$$= \frac{61 \times 62 \times 63}{4 \times 5 \times 6} = 1985.55$$

WORKOUT

1. $96.84 \times 172.3 = ?$

(a) 13115.532 (b) 16685.532
(c) 16785.532 (d) 16185.532

2. $8.7 \times 5.8 \times 3.51 = ?$

(a) 177.1136 (b) 177.2146
(c) 157.1146 (d) 177.1146

3. $1.016 \times 10.17 = ?$

(a) 103.3272 (b) 107.1172
(c) 107.1142 (d) 107.1192

4. $1.712 \times 17.08 = ?$

(a) 29.24096 (b) 29.24196
(c) 27.24096 (d) 27.24096

5. $1958 \times 999997 = ?$

(a) 1957894126 (b) 1757994126
(c) 1954994126 (d) 1957994126

6. $2987 \times 1111 = ?$

(a) 3317557 (b) 3318567
(c) 3318547 (d) 3318557

7. $2\frac{3}{10} \times 4\frac{6}{7} \times 7\frac{1}{2} = ?$ ***[Indian Bank(PO)-10]***

(a) 68 (b) 72
(c) 93 (d) 84
(e) 101

8. For what value of *, statement $= \left[\frac{*}{21} \times \frac{*}{189}\right] = 1$ is correct?

(a) 3969 (b) 147
(c) 63 (d) 21

9. $999\frac{98}{99} \times 99$ is equal to ***[SSC(DEO/LDC)-10]***

(a) 98999 (b) 97891
(c) 97999 (d) 89999

10. $282 \times 82 \times 0.2 = ?$ ***[BOM(AO)-09]***

(a) 4624.8
(b) 4734.6
(c) 4604.4
(d) 4324.2
(e) None of these

11. $117.95 \times 8.017 \times 4.98 = ?$ ***[RBI(Grade-B Officer)-09]***

(a) 4670 (b) 4780
(c) 4840 (d) 4720
(e) 4800

12. $14.4 \times 16.5 \times 8 = ?$ ***[IDBI(PO)-09]***

(a) 1908.0
(b) 1900.8
(c) 1912.4
(d) 1924.8
(e) None of these

13. $12.564 \times 22.009 \times 17.932 = ?$ ***[Indian Bank(PO)-10]***

(a) 4901 (b) 4895
(c) 4800 (d) 4950
(e) 4350

14. $54.327 \times 357.2 \times 0.0057$ is the same as ***[Hotel Mgmt.-97]***

(a) $5.4327 \times 3.572 \times 5.7$
(b) $5.4327 \times 3.572 \times .57$
(c) $54327 \times 3572 \times 0.0000057$
(d) None of these

15. $0.\overline{09} \times 7.\overline{3}$ is equal to ***[SSC(CGL-PT)-03]***

(a) $0.\overline{6}$ (b) $0.\overline{657}$
(c) $0.6\overline{7}$ (d) $0.6\overline{57}$

16. $1.\overline{2} \times 0.\overline{03} = ?$ ***[SSC(CPO-SI)-09]***

(a) $0.\overline{04}$ (b) $0.\overline{036}$
(c) $1.\overline{13}$ (d) $0.\overline{037}$

17. $\left(1-\frac{1}{3}\right)\left(1-\frac{1}{4}\right)\left(1-\frac{1}{5}\right)....\left(1-\frac{1}{99}\right)\left(1-\frac{1}{100}\right) = ?$ ***[SSC(CGL-PT)-03,08]***

(a) $\frac{2}{99}$ (b) $\frac{1}{25}$ (c) $\frac{1}{50}$ (d) $\frac{1}{100}$

18. If $\left(1+\frac{1}{2}\right)\left(1+\frac{1}{4}\right)\left(1+\frac{1}{6}\right)\left(1+\frac{1}{8}\right)\left(1-\frac{1}{3}\right)\left(1-\frac{1}{5}\right)\left(1-\frac{1}{7}\right) = 1+\frac{1}{x}$, then what is the value of x? ***[SSC(CGL-T-II)-17]***

(a) 6 (b) 8
(c) 5 (d) 7

19. $\left(1+\frac{1}{5}\right)\left(1+\frac{1}{6}\right)\left(1+\frac{1}{7}\right)....\left(1+\frac{1}{21}\right) = ?$

(a) 9.6 (b) 4.4
(c) 2.21 (d) 21.5

20. $\left(2-\frac{1}{3}\right)\left(2-\frac{3}{5}\right)\left(2-\frac{5}{7}\right)...\left(2-\frac{997}{999}\right)$ is equal to ***[SSC(SAS)-10]***

(a) $\frac{1001}{999}$ (b) $\frac{999}{1001}$
(c) $\frac{1001}{3}$ (d) $\frac{5}{1001}$

21. Consider the multiplication $999 \times abc = def132$ in decimal notation, where a, b, c, d, e and f are digits. What are the values of a, b, c, d, e and f respectively? ***[CDS-19]***

(a) 6,6,8,6,8,7 (b) 8,6,8,6,7,8
(c) 6,8,8,7,8,6 (d) 8,6,8,8,6,7

Answer Key

1. (b)	2. (d)	3. (a)	4. (a)	5. (d)	6. (d)	7. (d)	8. (c)	9. (a)	10. (a)
11. (d)	12. (b)	13. (d)	14. (a)	15. (a)	16. (d)	17. (c)	18. (b)	19. (b)	20. (c)
21. (d)									

Hints and SOLUTIONS

1. We have,

$$\begin{array}{r} a\ b\ c\ d \\ \times\ e\ f\ g\ h \\ \hline ae \mid af + be \mid ag + bf + ce \mid ah + bg + cf + de \mid bh \\ + cg + df \mid ch + dg \mid dh \end{array}$$

$$\begin{array}{r} 9684 \\ \times 1723 \\ \hline \end{array}$$

9·1	9·7 + 1·6	9·2 + 1·8 + 6·7	9·3 + 1·4 + 6·2 + 7·8	6·3 + 7·4 + 8·2	8·3 + 2·4	4·3	
9	69	68	99	62	32	12	
16	6	8	5	5	3	2	On carrying over left

Hence, required product = 16685.532

2. Multiply first two terms, thereafter multiply the resultant with the third one. We may use the pattern:

$8.7 \times 5.8 = 50.46$

Then,
$$\begin{array}{r} 5046 \\ \times 0351 \\ \hline \end{array}$$

Try with the above approach. Answer is 177.1146

3. Compute the value of 1016×1017 as below:

{1016 + (1017 − 1000)}	{1016 − 1000} {1017 − 1000}
1033	272 (this box will contain three digits as the base number i.e. 1000 has three zeros)
1033	272
1033272	

Then, the required answer = 103.3272

4.

1st part	2nd part
(remaining portion)^2 + {(remaining portion) × sum mentioned above (i.e. a multiple of 10)/100}	Product of numbers formed with last two digits.
$(17)^2 + \left\{17 \times \dfrac{(12 + 08)}{100}\right\}$	12×8
292.4	0096
292 (4)	0096
292 (4 + 0)	096
2924096	

Alternatively, you may compute on the pattern as below

$$\begin{array}{r} a\ b\ c\ d \\ \times\ e\ f\ g\ h \\ \hline ae \mid af + be \mid ag + bf + ce \mid ah + bg + cf + de \mid bh \\ + cg + df \mid ch + dg \mid dh \end{array}$$

Hence, the required answer = 29.24096

5. $1958 \times 999997 = 1958 \times (1000000 - 3)$

$= 1958000000 - 5874 = 1957994126$

6. ***Step* I** 2987 ⇒ 0002987000

***Step* II** 000298**7000** ⇒ 0 + 0 + 0 + 7 = 7

00029**8700**0 ⇒ 0 + 0 + 7 + 8 = 15

0002**9870**00 ⇒ 0 + 7 + 8 + 9 = 24

000**2987**000 ⇒ 7 + 8 + 9 + 2 = 26

00**0298**7000 ⇒ 8 + 9 + 2 + 0 = 19

0**0029**87000 ⇒ 9 + 2 + 0 + 0 = 11

0002987000 ⇒ 2 + 0 + 0 + 0 = 2

Adhering to 'carry over left rule' and writing from the extreme right hand side, we get 3318557. Hence, the product is 3318557.

7. $2\frac{3}{10} \times 4\frac{6}{7} = (2 \times 4) + \left(\frac{3}{10} \times \frac{6}{7}\right) + \left(2 \times \frac{6}{7}\right) + \left(4 \times \frac{3}{10}\right)$

$= 8 + \frac{9}{35} + 1\frac{5}{7} + 1\frac{1}{5}$

$= 10 + \frac{41}{35} = 11\frac{6}{35}$

Now, $11\frac{6}{35} \times 7\frac{1}{2} = (11 \times 7) + \left(\frac{6}{35} \times \frac{1}{2}\right) + \left(7 \times \frac{6}{35}\right) + \left(11 \times \frac{1}{2}\right)$

$= 77 + \frac{3}{35} + 1\frac{1}{5} + 5\frac{1}{2}$

$= 83 + \frac{2}{7} + \frac{1}{2} = 83\frac{11}{14} \approx 84$

Hence, option(d) is correct.

Alternatively $2.3 \times 4.8 \times 7.5 \approx 84$

8. Clearly, the product of * and * must be 21×189 to get 1 after simplification.

$\Rightarrow \quad * \times * = 21 \times 189$

$\Rightarrow \quad * \times * = 3 \times 7 \times 27 \times 7 = 9 \times 9 \times 7 \times 7$

$\Rightarrow \quad * = 9 \times 7 = 63$

9. $999\frac{98}{99} \times 99 = (999 \times 99) + \left\{\frac{98}{99} \times 99\right\}$

$= (99900 - 999) + 98 = 98999$

10. $82 \times 2 = 164$

Then,
$$\begin{array}{r} 282 \\ \times\ 164 \\ \hline \end{array}$$

2·1	2·6 + 1·8	2·4 + 1·2 + 8·6	8·4 + 6·2	2·4	
2	20	58	44	8	
4	6	2	4	8	On carrying over left

Hence, the required answer = 4624.8

11. Approx value of $117.95 \times 8.017 \times 4.98$

$= 118 \times 8 \times 5 = 118 \times 40 = 4720$

12. $16.5 \times 8 = 132$

Now,
$$\begin{array}{r} 132 \\ \times 144 \\ \hline \end{array}$$

1·1	1·4+ 1·3	1·4+1·2 +3·4	3·4+ 4·2	2·4	
1	7	18	20	8	
1	9	0	0	8	On carrying over left

Hence, the required answer = 1900.8

13. $12.564 \times 22.009 \times 17.932 \approx 12.5 \times 22 \times 18$ i.e. 4950

Hence, correct option is (d).

14. Total number of digits existing on the right side of the decimal points w.r.t expression given in question $= 3 + 1 + 4 = 8$

Now, ascertain as to which option has total of 8 digits after decimal points. Hence, option (a) is correct.

15. $0.\overline{09} \times 7.\overline{3} = \dfrac{9-0}{100-1} \times 7\dfrac{3}{(10-1)}$

$= \dfrac{1}{11} \times \dfrac{22}{3}$

$= 0.66666 = 0.\overline{6}$

Alternatively,

$0.\overline{09} \times 7.\overline{3} = 0.090909 \times 7.3333 = 0.6666\ldots = 0.\overline{6}$

16. $1\dfrac{(2-0)}{(10-1)} \times \dfrac{(3-0)}{(100-1)}$

$= \dfrac{11}{9} \times \dfrac{3}{99} = \dfrac{1}{27} = 0.\overline{037}$

Alternatively,

$1.\overline{2} \times 0.\overline{03} = 1.222222 \times 0.030303 = 0.\overline{037}$

17. $\dfrac{2}{3} \times \dfrac{3}{4} \times \dfrac{4}{5} \ldots \times \dfrac{98}{99} \times \dfrac{99}{100} = \dfrac{2}{100} = \dfrac{1}{50}$

18. Given, $\left[\left(1+\frac{1}{2}\right)\left(1-\frac{1}{3}\right)\right] \times \left[\left(1+\frac{1}{4}\right)\left(1-\frac{1}{5}\right)\right]$

$\times \left[\left(1+\frac{1}{6}\right)\left(1-\frac{1}{7}\right)\right]\left(1+\frac{1}{8}\right) = 1 + \frac{1}{x}$

$\Rightarrow \quad \left(1+\frac{1}{8}\right) = 1 + \frac{1}{x} \Rightarrow x = 8$

19. $\dfrac{6}{5} \times \dfrac{7}{6} \times \dfrac{8}{7} \times \ldots \times \dfrac{22}{21} = \dfrac{22}{5} = 4.4$

20. Given expression:

$\left(2-\frac{1}{3}\right)\left(2-\frac{3}{5}\right)\left(2-\frac{5}{7}\right)\ldots\left(2-\frac{997}{999}\right)$

$= \dfrac{5}{3} \times \dfrac{7}{5} \times \dfrac{9}{7} \times \ldots \times \dfrac{1001}{999} = \dfrac{1001}{3}$

21. Given, $999 \times abc = def132$

$\Rightarrow (1000 - 1) \times abc = def132$

$$\Rightarrow \begin{array}{r} a\,b\,c\,0\,0\,0 \\ -\,a\,b\,c \\ \hline d\,e\,f\,1\,3\,2 \end{array}$$

Clearly, $c = 8$, $b = 6$ and $a = 8$ (for the lower row) and $c = 7$, $b = 6$ and $a = 8$ (for the upper row). Values of a, b and c (for upper row) are actually values of d, e and f respectively. Hence, option (d) is correct.

Alternatively, $\quad 999 \times abc = def132$

$\Rightarrow \quad (1000 - 1) \times abc = def000 + 132$

$\Rightarrow \quad (abc)1000 - abc = (def + 1)1000 - 868$

On comparing LHS and RHS, we get $a = 8$, $b = 6$, $c = 8$

and $\quad d = a = 8$, $e = b = 6$

and $\quad f = c - 1 = 8 - 1 = 7$.

03

Unit Digit, Last Two Digits and No. of Trailing Zeros of a Product/(Base)^{Exponent}

A. How to Compute 'Unit Digit' of a Product/(Base)^{Exponent}

Aspirants are supposed to be capable to find the digit in units place of the following patterns to handle any type of questions wherein unit digit is asked to be found.

I. $x \times y \times z$, where x, y and z are integers.

II. $(x)^n$, where x is an integer and n be a natural number.

Pattern (I) $x \times y \times z$, where x, y and z are integers.

1. Find the Unit Digit of ($14 \times 16 \times 297$).

Multiply any two unit digits, *viz.* $4 \times 6 = 2\boxed{4}$

Now, multiply the unit digit (i.e. 4) of the product thus obtained, with the unit digit of third number.

$\Rightarrow \qquad 4 \times 7 = 2\boxed{8}$

Obviously, 8 will be the unit digit of ($14 \times 16 \times 297$).

2. Find the Unit Place Digit of ($367 \times 6259 \times 43518 \times 82$).

***Step* I** We multiply unit digits of first two terms.

i.e. $7 \times 9 = 6\boxed{3}$

***Step* II** Unit digit of previous product is multiplied with the unit digit of third term.

i.e. $3 \times 8 = 2\boxed{4}$

***Step* III** Unit digit of just previous product is multiplied with unit digit of the fourth term

i.e. $4 \times 2 = \boxed{8}$

Hence, unit digit of $367 \times 6259 \times 43518 \times 82$ is 8.

Observation The whole process can be done mentally in few seconds.

Inside

- *How to compute 'Unit Digit' of a Product/(Base)^{Exponent}*
- *How to Compute 'Last Two Digits' of a Product/(Base)^{Exponent}*
- *How to Compute 'Number of Trailing Zeros of a Product/(Base)^{Exponent}*
- ....
- .... *& Many More*
- *Work Out with Solutions*

3. Find the Unit Digit of the Product of all the Odd Prime Numbers.

Obviously, we have to find unit digit of

$$(3\times 5\times 7\times 11\times 13\times \ldots)$$

Now, $3\times 5 = 1\boxed{5}$

$5\times 7 = 3\boxed{5}$

$5\times 11 = 5\boxed{5}$

....

....

It may be construed from the above that when 5 is multiplied with any odd number, it always yields 5 as unit digit. Thus, required unit digit = 5.

Pattern (II) $(x)^n$, where *x* is an integer and *n* be a natural number

4. Find is the unit digit of $(372)^{378}$.

Obviously, the unit digit of the product/resultant will depend on the unit digit of the base.

Now, look at the following trends

Expressions	Unit digit
$(372)^1$	$2^1 \Rightarrow \mathbf{2}$
$(372)^2$	$2^2 \Rightarrow 4^*$
$(372)^3$	$2^3 \Rightarrow 8$
$(372)^4$	$2^4 \Rightarrow 6$
$(372)^5$	$2^5 \Rightarrow \mathbf{2}$
$(372)^6$	$2^6 \Rightarrow 4^*$
$(372)^7$	$2^7 \Rightarrow 8$
$(372)^8$	$2^8 \Rightarrow 6$
$(372)^9$	$2^9 \Rightarrow \mathbf{2}$
$(372)^{10}$	$2^{10} \Rightarrow 4^*$

The above trends reveal that the unit digit repeats itself after every 4 steps or say, after every 4th power. In other words, Unit digit of $(372)^1$ = Unit digit of $(372)^{1+4}$ = Unit digit of $(372)^{1+4+4}$ and so on. Similarly, Unit digit of $(372)^2$ = Unit digit of $(372)^{2+4}$ = Unit digit of $(372)^{2+4+4}$ and so on.

Further, if we have to find the unit digit of $(372)^{10}$, we may say that the unit digit of $(372)^{10}$ is the same as that of $(372)^2$. It means exponent 10 can be replaced by 2. Since unit digit repeats itself after every four steps, we may simply divide 10 by 4, the remainder so obtained i.e. 2 will replace the original exponent. It is also clarified that if the exponent is divisible exactly be 4, the original exponent will be replaced by 4 itself. It may be corroborated by the fact that unit digit of $(372)^4$ = unit digit of $(372)^8$, here, 8 is exactly divisible by 4.

Now, come on the example: Unit digit of $(372)^{378}$ = ?

Find the remainder dividing 378 by 4.

Here, remainder will be 2.

Then, the required unit digit $= (2)^2 = 4$

Note *We get the same remainder when a number formed with last two digits of the exponent is divided by 4 as we divide the exponent by 4. Therefore, to find remainder or check divisibility, consider the same. In the above context, Remainder if $378 \div 4$ = Remainder if $78 \div 4$.*

5. Find the Unit Digit of $(62)^{140}$.

Clearly, 140 is exactly divisible by 4.

$\therefore$ Required unit digit = unit digit of $(2)^4 = 6$

Now, have a look at the trends if unit digit of the base is 3.

Expressions	Unit digit
$(\ldots 3)^1$	3
$(\ldots 3)^2$	9
$(\ldots 3)^3$	7
$(\ldots 3)^4$	1
$(\ldots 3)^5$	3
$(\ldots 3)^6$	9
$(\ldots 3)^7$	7
$(\ldots 3)^8$	1
$(\ldots 3)^9$	3
$(\ldots 3)^{10}$	9

Here, the same unit digit appears after every four steps. Hence, the approach employed in case of 2 as unit digit of base number, will also be applicable here.

6. Find the Unit Digit of $(123)^{43}$.

Remainder if 43 is divided by 4 = 3

Then, required unit digit = unit digit of $(3)^3 = 7$

7. Find the Unit Digit of $(948723)^{56324}$.

Clearly, 56324 is exactly divisible by 4.

$\therefore$ Required unit digit = unit digit of $(3)^4 = 1$

Further, if we look at trends in respect of base numbers having 7 or 8 as unit digit, the above pattern will be observed. Obviously, we will adhere to the same approach.

The above whole concept may be summarised as below

1. The base numbers having 2/3/7/8 as unit digit follow the same patterns i.e. after every 4 steps, the same unit digit appears.
2. Exponent is always divided by 4 for the above reason.
3. If remainder is 1/2/3, the required unit digit will be unit digit of $(2 \text{ or } 3 \text{ or } 7 \text{ or } 8)^1 / (2 \text{ or } 3 \text{ or } 7 \text{ or } 8)^2 / (2 \text{ or } 3 \text{ or } 7 \text{ or } 8)^3$ respectively.
4. If the exponent is exactly divisible by 4, the required unit digit will be unit digit of $(2 \text{ or } 3 \text{ or } 7 \text{ or } 8)^4$.

8. Find the Unit Digit of $(3867)^{340}$.

Clearly, 340 is exactly divisible by 4,

$\therefore$ Required unit digit = Unit digit of $(7)^4 = 1$

9. Find the unit digit of $(368)^{123}$.

Remainder if 123 is divided by 4 = 3

$\therefore$ Required unit digit = Unit digit of $(8)^3 = 2$

Now, have a look at trends with regard to the base numbers having 1/5/6/0 as unit digit.

Expressions	Unit digit	Expressions	Unit digit
$(...1)^1$	1	$(...5)^1$	5
$(...1)^2$	1	$(...5)^2$	5
$(...1)^3$	1	$(...5)^3$	5
$(...1)^4$	1	$(...5)^4$	5
$(...6)^1$	6	$(...0)^1$	0
$(...6)^2$	6	$(...0)^2$	0
$(...6)^3$	6	$(...0)^3$	0
$(...6)^4$	6	$(...0)^4$	0

Observation The base numbers having 1/5/6/0 as unit digit follow the same patterns i.e. the required unit digit will always be 1/5/6/0 irrespective of value of exponent. In other words, unit digit of the base number is the required unit digit of the expression, if the base number ends with 1/5/6/0.

10. Find the Unit Digit of $(361)^{132}$.

The required unit digit = 1 as base number ends with 1.

Now, have a look at trends with regard to the base numbers having 4/9 as unit digit.

Expressions	Unit digit	Expressions	Unit digit
$(...4)^1$	4	$(...9)^1$	9
$(...4)^2$	6	$(...9)^2$	1
$(...4)^3$	4	$(...9)^3$	9
$(...4)^4$	6	$(...9)^4$	1
$(...4)^5$	4	$(...9)^5$	9

Observations

1. The base numbers having 4/9 as unit digit follow the same patterns i.e. after every 2 steps, the same unit digit appears.
2. If exponent is an odd number, the required unit digit will be unit digit of $(4)^1/(9)^1$ as the case may be. In case of even number that will be unit digit of $(4)^2/(9)^2$. In other words, the required unit digit will be 4/9, if exponent to (....4)/(....9) is an odd number. In case of even number, the required unit digit will be 6/1.

11. Find the Unit Digit of $(1369)^{35}$.

Here, the exponent is an odd number.

$\therefore$ Required unit digit = Unit digit of $(9)^1 = 9$

12. Find the Unit Digit of $(64)^{68}$.

Here, the exponent is an even number.

$\therefore$ Required unit digit = Unit digit of $(4)^2 = 6$

Note *For any natural number n, n^5 has the same unit digit as n has.*

Illustrations

Problem 1 **The digit in unit's place of the product $(2153)^{167}$ is**

Sol. Required answer = Unit digit of $3^3 = 7$

[$\because$ 167 ÷ 4 leaves the remainder 3]

Problem 2 **Find the unit digit in the product $(4387)^{245} \times (621)^{72}$** ***[SSC(CGL-T-I)-11]***

Sol. Unit digit of $(4387)^{245}$ = Unit digit of $(7)^1$

[$\because$ 245 ÷ 4 leaves the remainder 1]

= 7

Further, unit digit of $(621)^{72} = 1$

[$\because$ Base number having 1 as unit digit always yields 1 as unit digit of the product, irrespective of value of exponent]

$\therefore$ Required unit digit = 7 × 1 = 7

Problem 3 **The last digit of $(1001)^{2008} + 1002$ is** ***[SSC(CGL-T-I)-11]***

Sol. Unit digit of $(1001)^{2008}$ = 1 and unit digit of 1002 = 2

On addition, we get 3 as last digit of the sum.

$\therefore$ Required answer = 3.

Problem 4 **The unit digit in the sum of $(124)^{372} + (124)^{373}$ is** ***[SSC(CGL-T-I)-11]***

Sol. Unit digit of $(124)^{372}$ = Unit digit of $(4)^2 = 6$

[$\because$ 372 is an even number]

and unit digit of $(124)^{373}$ = Unit digit of $(4)^1 = 4$

[$\because$ 373 is an odd number]

$\therefore$ Required unit digit = Unit digit of sum of 4 and 6 = 0

Problem 5 **X and Y are the digits at the unit's place of the numbers $(480X)$ and $(789Y)$ where $X \neq Y$. However, the digits at the unit's place of the numbers $(408X)^{63}$ and $(789Y)^{85}$ are the same. What will be the possible value(s) of $(X + Y)$?** ***[XAT-18]***

Sol.

$(408X)^{63}$		$(789Y)^{85}$	
If X = 0	Unit digit = 0	If Y = 0	Unit digit = 0
If X = 1	Unit digit = 1	If Y = 1	Unit digit = 1
If X = 2*	Unit digit = 8	If Y = 2#	Unit digit = 2
If X = 3$	Unit digit = 7	IF Y = 3@	Unit digit = 3
If X = 4	Unit digit = 4	If Y = 4	Unit digit = 4
If X = 5	Unit digit = 5	If Y = 5	Unit digit = 5
If X = 6	Unit digit = 6	If Y = 6	Unit digit = 6
If X = 7@	Unit digit = 3	If Y = 7$	Unit digit = 7
If X = 8#	Unit digit = 2	If Y = 8*	Unit digit = 8
If X = 9	Unit digit = 9	If Y = 9	Unit digit = 9

Since $X \neq Y$, possible pairs of (X, Y) are (2, 8), (3, 7), (7, 3) and (8, 2).

Clearly, $X + Y = 10$

Problem 6 Find the unit digit of $(2153)^{9!} + (388)^{12!}$

Sol. It is to note that 9! is read as factorial of 9 i.e. $9\times8\times7\times6\times5\times4\times3\times2\times1$

Now, unit digit of $(2153)^{9!}$ = unit digit of $3^4 = 1$

[∵ 9! is exactly divisible by 4.]

and unit digit of $(388)^{12!}$ = unit digit of $8^4 = 6$

[∵ 12! is exactly divisible by 4.]

Hence, required unit digit = 1 + 6 = 7

Problem 7 Find the unit digit of $1^6 + 2^6 + 3^6 + \ldots + (100)^6$.

Sol. Unit digit of $1^6 = 1$

Unit digit of 2^6 = Unit digit of $2^2 = 4$

Unit digit of 3^6 = Unit digit of $3^2 = 9$

Unit digit of $4^6 = 6$

Unit digit of $5^6 = 5$

Unit digit of $6^6 = 6$

Unit digit of 7^6 = Unit digit of $7^2 = 9$

Unit digit of 8^6 = Unit digit of $8^2 = 4$

Unit digit of $9^6 = 1$

Unit digit of $(10)^6 = 0$

Hence, unit digit of $\{1^6 + 2^6 + 3^6 \ldots + (10)^6\}$ = Unit digit of $\{1+4+9+6+5+6+9+4+1+0\} = 5$

Obviously, the unit digit of sum of further ten terms (i.e. $\{(11)^6 + (12)^6 + (13)^6 + \ldots + (20)^6\}$ will also be 5 as each of them has the same unit digit as was in first 10 terms.

Now, required unit digit = unit digit of (10 × 5) = 0.

Problem 8 Find the unit digit of $1^2 + 2^2 + 3^2 + \ldots + (49)^2$.

Sol. Try on the above pattern.

Alternatively,

Sum of square of n natural number

$$= \frac{n(n+1)(2n+1)}{6} = \frac{49\times50\times99}{6} = 40425$$

Hence, required unit digit = 5

Problem 9 Unit's digit of $[7^{(7)^7} + 777^{777}]$ is

[Oriental Insc. Co. (AAO)-12]

Sol. Unit digit of $7^{(7)^7}$ = Unit digit of $7^3 = 3$

[∵ $7^7 \div 4$ leaves 3 as remainder. You may learn 'how to compute remainder' in a separate chapter]

and unit digit of 777^{777} = unit digit of $7^1 = 7$

[∵ 777 ÷ 4 leaves 1 as remainder]

Then, required unit's digit = unit digit of (3 + 7) = 0

Problem 10 If $5^x - 3^y = 13438$ and $5^{x-1} + 3^{y+1} = 9686$, then $x + y$ equals ***[CAT-19]***

Sol. Obviously, unit digit of 5^{x-1} will be 5 for all positive integral value of x provided $x \ge 2$.

Now, $5^{x-1} + 3^{y+1} = 9686$

$\Rightarrow$ $(\ldots5) + 3^{y+1} = 9686$

∴ Unit digit of $3^{y+1} = 1$

It implies that the value of $y + 1$ will be in form of $4k$ as 3^{4k} ends with 1.

Further, we have $3^4 = 81$, $3^8 = 6561$, 3^{12} = more than 9686.

Clearly, the value of 3^{y+1} cannot be 3^{12} or more as $5^{x-1} + 3^{y+1} < 3^{12}$.

Hence, the value of 3^{y+1} may either be 81 or 6581.

Then, value of 5^{x-1} = 9686 – 81 (i.e. 9605) or 9686 – 6561 (i.e. 3125)

Furthermore, the value of 5^{x-1}cannot be 9605 for any integral value x.

Thus, $5^{x-1} = 3125$

then $3^{y+1} = 6561$

$\Rightarrow$ $5^{x-1} = 5^5$

and $3^{y+1} = 3^8$

$\Rightarrow$ $x = 6$ and $y = 7$

Then, $x + y = 13$

Note *The rationale behind not proceeding through 1st equation is that we cannot estimate, at the first instance, the value of x and y as each may take larger value.*

B. How to Compute 'Last Two Digits' of a Product/(Base)^{Exponent}

Pattern (I) When the expression is in the form of $x \times y \times z \times \ldots$

Problem 11 Find the last two digits of $356 \times 42 \times 3687$.

Sol. Multiply any two terms(say, 356 and 42) to the extent of two digits only from the right.

```
   356
 × 042
......52
```

[Recall the general rule for multiplication ...... (5 × 2 + 4 × 6| (6 × 2)]

Now, multiply (...52) with 3rd term to the extent of two digits from the right with the help of general rule for multiplication.

```
   ..52
 × 3687
 .... 24
```

Hence, required answer = 24

Problem 12 Find the last two digits of $7526 \times 3548 \times 351 \times 695$.

Sol.

```
   7526
 × 3548
 .....48
  × 351
 .....48
  × 695
 ......60
```

Required answer = 60

Pattern (II) When the expression is in the form of (base)^{Exponent}

Case I **When unit digit of the base is 1**

Problem 13 Find the last two digits of $(681)^{63}$.

Sol. Here, unit digit of base is 1, then unit digit of the final product be 1 and tens digit of the final product = Last digit of the product of tens digit of base and unit digit of the exponent

$= 8 \times 3 = 2\boxed{4}$

$\therefore$ Last two digits of $(671)^{63} = 41$

Observation We may find the last two digits of a^b with the help of above concept provided the last digit of the base is 1.

Problem 14 Find the last two digits of $(861)^{36965}$.

Sol. Unit digit = 1

Tens digit = Last digit of $6 \times 5 = 3\boxed{0}$

$\therefore$ Required answer = 01

Note *It may be noted that the above concept will not be applicable in case unit digit of base is other than 1. Thus, we may discuss the different concepts case-wise.*

Case II **When unit digit of the base is 3/7/9**

We ought to try to convert the base-number ending with 3/7/9 into the numbers ending with 1 by splitting of exponent. Thereafter, we will adhere to the pattern as given in case-I. It will not be out of context to mention that $(3)^4$, $(7)^4$ and $(9)^2$ end with 1.

Now, take examples

Problem 15 What are the last two digits of $(353)^{689}$.

Sol. $(353)^{689} = [(353)^4]^{172} \times (353)^1 = [(353)^2 \times (353)^2]^{172} \times 353$

$= [(..09 \times ..09)]^{172} \times 353 = (...81)^{172} \times 353$

Now, apply the concept meant for the base number ending in 1 in case of $(...81)^{172}$.

$= (...61) \times 353 = (.......33)$

Hence, required answer = 33

Problem 16 Find the last two digits of $(4687)^{235}$.

Sol. $(4687)^{235} = [(4687)^4]^{58} \times (4687)^3$

$= [(4687)^2 \times (4687)^2]^{58} \times (4687)^3$

$= [(..69 \times ..69)]^{58} \times (4687)^2 \times (4687)^1$

$= (...61)^{58} \times [(.......69) \times (4687)]$

$= (........81) \times (...03) = (........43)$

Hence, required answer = 43.

Problem 17 Find the last two digits of $(59)^{127}$.

Sol. $(59)^{127} = [(59)^2]^{63} \times 59$

$= [59 \times 59]^{63} \times 59 = (......81)^{63} \times 59$

$= (...41) \times (59) = (...19)$

Hence, required answer = 19

Case III **When unit digit of the base is 2/4/6/8**

We find the last two digit of $(...2/4/6/8)^{\text{any integer}}$ with the help of the following facts

(i) 2^{10} ends with 24.

(ii) (...24) raised to an even power always ends with 76.

(iii) (...24) raised to an odd power always ends with 24.

(iv) 76 is only one even two-digit number which raised to any power gives the last two digits as 76. Also,
when 76 is multiplied with 2^n, the last two digits of the product remain the same as the last two digits of 2^n provided $n > 1$.

Now, try to find the last two digits of $(42)^{379}$ on the basis of the above facts.

$\Rightarrow \quad (42)^{379} = (2 \times 21)^{379} = (2)^{379} \times (21)^{379}$

$= [(2)^{10}]^{37} \times \{(2)^9 \times (21)^{379}\}$

$= (..24)^{37} \times \{(-12) \times (...81) = (...24) \times (..72) = (..28)$

Hence, required answer = 28

Problem 18 Find the last two digits of $(2)^{788}$.

Sol. $(2)^{788} = [(2)^{10}]^{78} \times (2)^8$

$= (...24)^{78} \times (2)^8 = (...76) \times (...56) = (...56)$

Hence, required answer = 56

Problem 19 Find the last two digits of $(6434)^{325}$.

Sol. $(6434)^{325} = (2 \times 3217)^{325} = (2)^{325} \times (3217)^{325}$

$= [(2)^{10}]^{32} \times (2)^5 \times (3217)^{325}$

$= (...24)^{32} \times 32 \times [(3217)^4]^{81} \times (3217)$

$= (...76) \times 32 \times [(3217)^2 \times (3217)^2]^{81} \times (3217)$

$= (...76) \times 32 \times [(..89) \times (..89)]^{81} \times (3217)$

$= (..32) \times (.21)^{81} \times (3217)$

$= (..32) \times (.21) \times (3217)$

$= (...72) \times (3217) = (..24)$

Hence, required answer = 24

Problem 20 Find the last two digits of $(736)^{215}$.

Sol. $(736)^{215} = (2^5 \times 23)^{215} = 2^{1075} \times (23)^{215}$

$= (2^{10})^{107} \times 2^5 \times (23)^{215}$

$= (.24)^{107} \times 2^5 \times [(23)^4]^{53} \times (23)^3$

$= (...24) \times 32 \times (...29 \times ...29)^{53} \times (23)^3$

$= (...68) \times (...41)^{53} \times 23^3$

$= (...68) \times (...21) \times 23^3$

$= (...28) \times (...67) = ...76.$

Hence, required answer = 76

Problem 21 Find the last two digit of $(78)^{21}$.

Sol. $(78)^{21} = (2 \times 39)^{21}$

$= (2)^{21} \times (39)^{21}$

$= (2^{10})^2 \times 2^1 \times [(39)^2]^{10} \times (39)^1$
$= (.24)^2 \times 2 \times (.21)^{10} \times 39$
$= (..76) \times 2 \times (..01) \times 39$
$= (..52) \times (..39) = (.28)$

Hence, required answer = 28

Observation We ought to split the given base of $(\text{base})^{\text{exponent}}$ in format of $\{(2)^{\text{maximum possible integer}} \times (\text{number ends with odd number})\}$.

Case IV When unit digit of the base is 5

The last two digits of the product are either 25 or 75. Mind the chart given below.

Tens digit of Base	Power/ Exponent	Second last digit	Last digit
Odd	Odd	7	5
Odd	Even	2	5
Even	Odd	2	5
Even	Even	2	5

Remark *The above chart reveals that if tens digit of base and power both are odd, the last two digits will be 75, otherwise, in all combinations, it will be 25.*

Take an example

Problem 22 Find the last two digit of $(785)^{241}$.

Sol. Here, tens digit of the base is 8 i.e. an even number and exponent is an odd number. Hence, the required last two digits = 25.

Other Critical Facts

The last two digits of x^2, $(50-x)^2$, $(50+x)^2$, $(100-x)^2$, $(100+x)^2$, ... will always be same. For example, last two digits of $(14)^2$, $(36)^2$, $(64)^2$, $(86)^2$ and $(114)^2$ will be the same i.e. 96. It may be used whenever it is required in the problem.

C. How to Compute Number of Trailing Zeros of a Product/ $(\text{Base})^{\text{Exponent}}$

Take one example

Problem 23 The numbers 1, 2, 3, 4, ..., 100 are multiplied together. The number of zeros at the end (on the right) of the product must be. ***[SSC(DEO)-09]***

Sol. One zero can be obtained by multiplying the number having 5 as unit digit with an even number. Since there are such 10 numbers(*viz.* 5,15, ... ,95) in the given series, 10 zeros can be obtained. Here, number of even numbers is not less than the number of numbers end with 5. It is critical to note that we may derive number of number 2 from prime factorisation of the given number(s) and obviously, every number 2 will accommodate a number ending with 5 to fetch one zero. **For illustration:** 48 is equivalent to four even numbers as $48 = 2 \times 2 \times 2 \times 2 \times 3$, hence, 48 will accommodate 4 numbers ending with 5 and will fetch four zeros. However, 25 may be written as 5×5, it means one additional zero is obtained when multiplied by an even number. Also, 50 may be written as $5 \times 5 \times 2$, it means one additional zero is obtained. Further, $75 = 5 \times 5 \times 3$, i.e. one more zero is added.

Hence, total number of zeros as of now

$= 10 + 1 + 1 + 1 = 13$

Further, the number ending with zero(s) will add additional number of zero(s). Here, such types of numbers are 10, 20, 30, 40, 50, 60, 70, 80, 90, 100. Obviously, they will fetch 11 zeros.

Thus, required number of zeros = 13 + 11 = 24.

Alternatively,

The above concept may be abridged as below

Number of multiples of 5 up to 100 $= \frac{100}{5} = 20$

Number of multiples of 5^2 up to 100 $= \frac{100}{25} = 4$

$\therefore$ Total number of zeros = 20 + 4 = 24.

Note *Finding number of multiples of 5^3 is not possible up to 100, but we shall consider in other cases, if possible.*

Generic Version

Number of zeros in the end of a product of numbers in these types of series, where adequate number of even numbers or number 2(derived from the prime factorization of the given numbers) exists = number of multiples of 5^1 + number of multiples of 5^2 + number of multiples of 5^3 and so on.

Problem 24 The number 1, 2, 3, 4, ..., 1000 are multiplied together. The number of zeros at right end of the product must be ***[SSC(CGL-PT)-05]***

Sol. Required number of zeros = number of multiples of 5^1 + number of multiples of 5^2 + number of multiples of 5^3 + number of multiples of 5^4

$$= \frac{1000}{5} + \frac{1000}{5^2} + \frac{1000}{5^3} + \frac{1000}{5^4}$$

$= 200 + 40 + 8 + 1 = 249$

Note *In case of decimal number, we take only integer,* e.g. $\frac{1000}{5^4} = 1.6 \Rightarrow 1$

Problem 25 Find the number of zeros in the product $5 \times 15 \times 25 \times 35 \times \ldots \times 95 \times 8 \times 12$.

Sol. We have already discussed that we may derive number of number 2 from prime factorisation of the given number(s) and obviously, every number 2, thus obtained, will accommodate a number ending with 5 to fetch one zero.

Now, $8 \times 12 = 2 \times 2 \times 2 \times 2 \times 2 \times 3$

The above 5 number 2 will accommodate only 5 given numbers end with 5 to fetch 5 zeros.

Hence, required number of zeros = 5

Problem 26 The number 2, 4, 6, 8, 10 .... 196, 198, 200 are multiplied together. The number of zeros at right end of the product is

Sol. Number of multiples of 5 in the series
= 20 i.e. [10, 20, 30, ..., 200]
Number of multiples of 5^2 in the series
= 4 i.e. [50, 100, 150 and 200]
Number of multiples of 5^3 in the series = 0
∴ Total number of zeros
= 20 + 4 = 24

Alternatively,
Number of trailing zeros in such type of product, where adequate number of even numbers or number 2(derived from the prime factorisation of the given numbers) exists = (sum of powers of bases having 5 as a factor) + (sum of powers of bases having 5^2 as a factor) + (sum of powers of bases having 5^3 as a factor) + ... so on.
Here, bases 5, 10, 15, 20, ..., 45, 50 have 5 as a factor.
Bases 25 and 50 have 5^2 as a factor.
Then, required number of trailing zeros
$= (5 + 10 + 15 + \dots + 45 + 50) + \{1(25 + 50)\}$
= 275 + 75 = 350

Problem 27 Find the number of zeros at the right end of $1^1 \times 2^2 \times 3^3 \times 4^4 \times \dots \times 50^{50}$.

Sol. Number of fives can be obtained from the following terms
$5^5, 10^{10}, 15^{15}, 20^{20}, 25^{25}, 30^{30}, 35^{35}, 40^{40}, 45^{45}, 50^{50}$
$\Rightarrow 5^5, (2\cdot5)^{10}, (3\cdot5)^{15}, (4\cdot5)^{20}, (5\cdot5)^{25}, (5\cdot6)^{30}, (5\cdot7)^{35}, (5\cdot8)^{40}, (5\cdot9)^{45}, (5\cdot5\cdot2)^{50}$
Clearly, number of fives in the product
= 5 + 10 + 15 + 20 + (25 + 25) + 30 + 35 + 40 + 45 + (50 + 50) = 350
Since number of 2's is more than number of 5's in the given sequence, the required number of zeros = 350.

Problem 28 Find the number of zeros at the right end of $5^5 \times 10^{10} \times 15^{15} \times \dots \times 50^{50}$.

Sol. Number of 5's from the above = 350
In order to fetch 350 zeros, number of 2's must be atleast 350.
Here, number of 2's can be derived from the following terms
$10^{10}, 20^{20}, 30^{30}, 40^{40}, 50^{50}$
$\Rightarrow (2\cdot5)^{10}, (2^2\cdot5)^{20}, (5\cdot2\cdot3)^{30}, (2^3\cdot5)^{40}, (5\cdot5\cdot2)^{50}$
Clearly, number of 2's = 10 + 40 + 30 + 120 + 50 = 250
It implies that only 250 sets of (2, 5) can be formed.
Hence, required number of zeros = 250.

WORKOUT

1. What is the last digit in the expansion of $(3)^{4798}$. *[CDS-10]*
(a) 1 (b) 3 (c) 7 (d) 9

2. What is the last digit in the expansion of $(2457)^{754}$ *[CDS-09, SSC (SO-Comm)-07, SSC (CPO)-03]*
(a) 8 (b) 3 (c) 7 (d) 9

3. The unit's digit in the product of $(7)^{35} \times (3)^{71} \times (11)^{55}$ is
(a) 1 (b) 3 (c) 9 (d) 7

4. The unit digit in $3 \times 38 \times 537 \times 1256$ is *[SSC(CGL-T-II)-13]*
(a) 8 (b) 3 (c) 4 (d) 6

5. The digit in unit's place of the product $81 \times 82 \times 83 \times \dots \times 89$ is
(a) 0 (b) 2
(c) 8 (d) 6

6. The digit in unit's place of the product $(2153)^{167}$ is
(a) 1 (b) 3
(c) 7 (d) 9

7. The unit digit in the product $(2464)^{1793} \times (615)^{317} \times (131)^{491}$ is *[SSC(CPO-SI)-04, SSC(CGL-T-II)-17, IIFT-12]*
(a) 0 (b) 3 (c) 2 (d) 5

8. The digit in the unit's place of $[(251)^{98} + (21)^{29} - (106)^{100} + (705)^{35} - (16)^4 + 259]$ is *[SSC(CGL-T-I)-11, SSC(CGL-PT)-2000]*
(a) 1 (b) 4 (c) 5 (d) 6

9. One's digit of the number $(22)^{23}$ is *[SSC(CGL-T-I)-11, SSC(CPO-SI)-08]*
(a) 8 (b) 4 (c) 2 (d) 6

10. What is the unit digit of the sum of first 111 whole numbers ? *[SSC(CGL-T-II)-17]*
(a) 4 (b) 6
(c) 5 (d) 0

11. What is the unit digit of $1^5 + 2^5 + 3^5 + \dots + 20^5$? *[SSC(CGL-T-II)-17]*
(a) 4 (b) 2
(c) 5 (d) 0

12. The unit digit of $(13)^{2003}$ is *[New India Assur.(AO)-09]*
(a) 1 (b) 3 (c) 7 (d) 9

13. The unit's digit of $34563^{20359} + 2358^{784}$ is
(a) 1 (b) 3 (c) 7 (d) 9

14. Find the last two digits of $256 \times 34 \times 241$.
(a) 21 (b) 64 (c) 01 (d) 41

15. What are the last two digits of $(4891)^{3415632}$?
(a) 81 (b) 61 (c) 01 (d) 21

16. What are the last two digits of $(483)^{302}$?
(a) 39 (b) 49 (c) 89 (d) 41

17. Find the last two digits of $(364)^{457}$.
(a) 04 (b) 44 (c) 54 (d) 94

18. Find the last two digit of $(78)^{941}$.
(a) 68 (b) 08 (c) 18 (d) 28

19. Find the last two digit of $(835)^{8791}$.
(a) 55 (b) 45 (c) 75 (d) 05

20. Find the last two digit of $(7)^{2008}$. *[CAT-08, RMLU-15]*
(a) 21 (b) 61
(c) 01 (d) 41
(e) 81

21. Find the number of zeros at the right end of 21!.
(a) 4 (b) 5
(c) 6 (d) 7

22. Find the number of zeros at the right end of 150!.
(a) 44 (b) 35
(c) 37 (d) 47

23. Find the number of zeros at the right end of $1^1 \times 2^2 \times 3^3 \times 4^4 \times \ldots \times 150^{150}$.
(a) 2175 (b) 2825
(c) 2975 (d) 2505

24. What is the unit digit of $58^{58^{58^{58}}}$?
(a) 2 (b) 8 (c) 4 (d) 6

25. The rightmost non-zero digit of the number 30^{2720} is *[CAT-05]*
(a) 1 (b) 3 (c) 7 (d) 9

26. If 10^n divides $6^{23} \times 75^9 \times 105^2$, then what is the largest value of n? *[CDS-19]*
(a) 20 (b) 22 (c) 23 (d) 28

Answer Key

1. (d)	2. (d)	3. (a)	4. (a)	5. (a)	6. (c)	7. (a)	8. (b)	9. (a)	10. (c)
11. (d)	12. (c)	13. (b)	14. (b)	15. (a)	16. (c)	17. (a)	18. (d)	19. (c)	20. (c)
21. (a)	22. (c)	23 (c)	24 (d)	25 (a)	26 (a)				

Hints and SOLUTIONS

1. Required answer = Unit digit of 3^2
[$\because$ 4798 ÷ 4 leaves the remainder 2]
= 9

2. Required answer = Unit digit of 7^2
[$\because$ 754 ÷ 4 leaves the remainder 2]
= 9

3. Unit digit of $(7)^{35}$ = Unit digit of $(7)^3$
[$\because$ 35 ÷ 4 leaves the remainder 3]
= 3
Further, unit digit of $(3)^{71}$ = Unit digit of $(3)^3$
[$\because$ 71 ÷ 4 leaves the remainder 3]
= 7
and unit digit of $(11)^{55}$ = 1
Then, required unit digit = Unit digit of $(3 \times 7 \times 1) = 1$

4. Required unit digit = Unit digit of $(3 \times 8 \times 7 \times 6)$
= Unit digit of $(4 \times 2) = 8$

5. Required unit digit = Unit digit of $(1 \times 2 \times 3 \ldots \times 9)$
= Unit digit of $(1 \times 2 \times 3) \times (4 \times 5 \times 6) \times (7 \times 8 \times 9)$
= Unit digit of $(6 \times 0 \times 4) = 0$

6. Required answer = Unit digit of 3^3
[$\because$ 167 ÷ 4 leaves the remainder 3]
= 7

7. Unit digit of $(2464)^{1793}$ = Unit digit of $(4)^1$ [$\because$ 1793 is an odd number]
= 4
Further, unit digit of $(615)^{317} = 5$
and unit digit of $(131)^{491} = 1$
Then, required unit digit = Unit digit of $(4 \times 5 \times 1) = 0$

8. Required unit digit
$= 1 + 1 - 6 + 5 - 6 + 9 = 4$

9. Required answer = Unit digit of 2^3
[$\because$ 23 ÷ 4 leaves the remainder 3]
= 8

10. Sum of 111 whole numbers
$= 0 + 1 + 2 + 3 + \ldots + 110$
$= \frac{110 \times 111}{2}$
$= 55 \times 111$
$\therefore$ Required unit digit $= 5 \times 1 = 5$

11. We know, for any natural number n, n^5 has the same unit digit as n has.
Hence, required unit digit = Unit digit of $(1 + 2 + 3 + \ldots + 9 + 0 + 1 + 2 + 3 + \ldots + 9 + 0)$
= Unit digit of 90 = 0

12. Required answer = Unit digit of 3^3
[$\because$ 2003 ÷ 4 leaves the remainder 3]
= 7

13. Unit digit of 34563^{20359}
= Unit digit of 3^3
[$\because$ 20359 ÷ 4 leaves the remainder 3]
= 7
and Unit digit of 2358^{784}
= Unit digit of 8^4
[$\because$ 784 ÷ 4 leaves the remainder 0]
= 6
Then, required unit digit
= Unit digit of (7 + 6) = 3

14.
$$\begin{array}{r} 256 \\ \times\, 034 \\ \hline \ldots.04 \end{array}$$
Now, multiply (.......04) with 3rd term to the extent of two digits from the right with the help of general rule for multiplication.
$$\begin{array}{r} \ldots\ldots04 \\ \times\, 241 \\ \hline \ldots.64 \end{array}$$
Hence, required answer = 64

15. Unit digit = 1
Tens digit = Last digit of $9 \times 2 = 8$
$\therefore$ Required answer = 81

16. $(483)^{302} = [(483)^4]^{75} \times (483)^2$
$= [(483)^2 \times (483)^2]^{75} \times \{483 \times 483\}$
$= [(..89 \times ..89)]^{75} \times (..89)$
$= (..21)^{75} \times (..89)$
$= (\ldots 01) \times (\ldots 89)$
$= (\ldots 89)$
Hence, required answer = 89

17. $(364)^{457} = (2^2 \times 91)^{457} = (2)^{914} \times (91)^{457}$

$= [(2)^{10}]^{91} \times \{(2)^4 \times (91)^{457}\}$

$= (...24)^{91} \times \{(16) \times (...31)$

$= (...24) \times (...96)$

$= (...04)$

Hence, required answer = 04

18. $(78)^{941} = (2 \times 39)^{941} = (2)^{941} \times (39)^{941}$

$= (2^{10})^{94} \times 2^1 \times [(39)^2]^{470} \times (39)^1$

$= (..24)^{94} \times 2 \times (...21)^{470} \times 39$

$= (..76) \times 2 \times (..01) \times 39$

$= (...52) \times (..39) = (..28)$

Hence, required answer = 28

19. Here, tens digit of the base is 3 i.e. an odd number and exponent is also an odd number. Hence, the required last two digits = 75.

20. $(7)^{2008} = [(7)^4]^{502} = (49 \times 49)^{502}$

$= (..01)^{502} = 01$

Hence, required answer = 01

21. Required product $= 21 \times 20 \times 19 \times 18 \times 17 \times ... \times 2 \times 1$

Number of multiples of 5 in the series = 4

Hence, required number of zeros at the right end of 21! = 4

22. Required product $= 150 \times 149 \times 148 \times ... \times 3 \times 2 \times 1$

Number of multiples of 5 in the series $= 150 \div 5 = 30$

Number of multiples of 5^2 in the series $= 150 \div 5^2$

$= 6$

Number of multiples of 5^3 in the series $= 150 \div 5^3 = 1$

$\therefore$ Total number of zeros

$= 30 + 6 + 1$

$= 37$

23. Number of trailing zeros = (Sum of powers of bases having 5 as a factor) + (Sum of powers of bases having 5^2 as a factor) + (Sum of powers of bases having 5^3 as a factor) + ... so on.

Here, bases 5, 10, 15, 20, ..., 145, 150 have 5 as a factor. Bases 25, 50, 75, 100, 125 and 150 have 5^2 as a factor and the base 125 has 5^3 as a factor.

Then, required number of trailing zeros $= (5 + 10 + 15 + ... + 145 + 150)$

$= (5 + 10 + 15 + ... + 145 + 150)$
$+ (25 + 50 + 75 + 100 + 125 + 150)$
$+ 125 = 2325 + 525 + 125 = 2975$

24. $58^{58^{58^{58}}} = 58^{58^{58 \cdot 58 \cdot 58 \cdot 58 \cdot 58 ... \text{ 58 times}}}$

$= 58^{58 \cdot 58 \cdot 58 \cdot ...(58 \cdot 58 \cdot 58 \cdot 58 \cdot 58 ... \text{58 times}) \text{ times}}$

$= (58)^{\text{a multiple of 4}}$

[$\because 58 \cdot 58$ is exactly divisible by 4]

Now, unit digit of $(58)^{\text{a multiple of 4}}$

= Unit digit of 8^4

[$\because$ power $\div$ 4 leaves the remainder 0]

$= 6$

25. $30^{2720} = (30^4)^{680}$

$= (810000)^{680}$

Clearly, the right most digit is 0 and the right most non-zero digit is 1 as the unit digit of $(...1)^n = 1$

26. $6^{23} \times 75^9 \times 105^2 = 2^{23} \times 3^{23} \times 3^9 \times 5^{18} \times 3^2 \times 5^2 \times 7^2 = 2^{23} \times 3^{34} \times 5^{20} \times 7^2$

$\therefore$ Number of trailing zeros = 20

[$\because$ maximum number of base 5 which can be combined with 2's (bases) so as to fetch zeros = 20]

Hence, the largest value of $n = 20$.

04

Division

Learning OBJECTIVES

After completing this chapter, aspirants will be able to

- Understand the significance of this chapter in almost all the arithmetic chapters;
- Learn how to perform divisions based on Whole numbers, Decimal numbers, Fractions, Recurring decimals;
- Test the divisibility by certain prime numbers and composite numbers;
- Understand the concept of division by multiple divisors;
- Realise themselves very comfortable in division by a number closer to 100, 1000, 10000 and so on;
- Realise themselves very comfortable in division by some specific numbers;
- Mind the problems where some known approaches can be applied after some alternation in form/pattern of problems;
- Understand the philosophy of the division of this chapter in various parts;
- Analyse the trend of the questions being asked in various competitive examinations;
- Solve the questions in moments;
- Grip on concepts helping to solve the questions without pen and paper;
- Learn new concepts/fundamentals;
- Apply the pragmatic approaches in lieu of academic approaches; and
- Ponder on the alternative methods.

Inside

Introduction

It is one of the basic operations in Mathematics, which may be termed as shortest form of subtraction. This chapter plays critical roles in all the competitive exams like CAT, SSC, IBPS, Railways, etc.

The techniques envisaged in this chapter will obviously reduce the time taken in solving the questions related to almost all arithmetical chapters. This chapter is very critical not only for Banking services examinations wherein questions are based on direct applications of the Division, but also for other major competitive exams.

The direct applications of division are also seen in the chapters like Simplification, Data Interpretations, etc. to the greater extent.

Division may have diverse dimensions like

A. **Test of Divisibility.**
B. Division based on **Whole numbers**
C. Division based on **Decimal numbers**
D. Division based on **Recurring decimals**
E. Division based on **Fractions.**

A. Test of Divisibility

'Test of Divisibility' means a procedure to test a number as to whether it is divisible by a certain number or not without dividing that number in actual.

- **Divisibility by 2** Any number of which last digit is divisible by 2, is also divisible by 2.

  For example 352<u>4</u> is divisible by 2 as last digit 4 is divisible by 2.

  153<u>0</u> is divisible by 2 as last digit 0 is divisible by 2. (∵ quotient = 0)

  1523<u>5</u> is not divisible by 2 as last digit 5 is not divisible by 2.

- **Divisibility by 4** A number is divisible by 4 if the number formed by its last two digits is divisible by 4.

  For example 3<u>468</u> ✓ [∵ 68 is divisible by 4]
  6846<u>34</u> ✗ [∵ 34 is not divisible by 4]

- **Divisibility by 8** A number is divisible by 8 if the number formed by its last three digits is divisible by 8.

  For example 36<u>840</u> ✓ [∵ 840 is divisible by 8]
  46<u>340</u> ✗ [∵ 340 is not divisible by 8]

- **Divisibility by 16** A number is divisible by 16 if the number formed by its last four digits is divisible by 16.

  For example 634<u>2638</u> ✗ [∵ 2638 is not divisible by 16]
  5638<u>2080</u> ✓ [∵ 2080 is divisible by 16]

- **Divisibility by 3** A number is divisible by 3 if the sum of its all digits is divisible by 3.

  For example 8625 ✓ [∵ (8 + 6 + 2 + 5) i.e. 21 is divisible by 3]

  3269 ✗ [∵ (3 + 2 + 6 + 9) i.e. 20 is not divisible by 3]

- **Divisibility by 6** Since, 6 = 2 × 3, the number divisible by 2 and 3 each will obviously be divisible by 6. Thus, we should test the number applying the rules meant for divisibility by 2 and 3 individually.

  For example 4653 ✗ [∵ last digit is not divisible by 2. Further, no need to check its divisibility by 3]

  5648 ✗ [last digit is divisible by 2, but (5 + 6 + 4 + 8) is not divisible by 3]

  2466 ✓ [last digit is divisible by 2 and also (2 + 4 + 6 + 6) is divisible by 3]

- **Divisibility by 9** A number is divisible by 9 if the sum of its all digits is divisible by 9.

  For example 6354 ✓
  [∵ (6 + 3 + 5 + 4) i.e. 18 is divisible by 9]

  36254 ✗ [∵ (3 + 6 + 2 + 5 + 4) i.e. 20 is not divisible by 9]

- **Divisibility by 12** Since 12 = 3 × 4, the number divisible by 3 and 4 each will obviously be divisible by 12. Thus, we should test the number applying the rules meant for divisibility by 3 and 4 individually.

  For example 138636 ✓ [∵ (1 + 3 + 8 + 6 + 3 + 6) i.e. 27 is divisible by 3 and also the number formed with last two digits is divisible by 4]

  632154 ✗ [sum of its all digits is divisible by 3 but the number formed with last two digits is not divisible by 4]

- **Divisibility by 18** Since 18 = 2 × 9, the number divisible by 2 and 9 each will obviously be divisible by 18. Thus, we should test the number applying the rules meant for divisibility by 2 and 9 individually.

  For example 38916 ✓ [∵ last digit is divisible by 2 and (3 + 8 + 9 + 1 + 6) is divisible by 9]

  135622 ✗ [∵ last digit is divisible by 2 but (1 + 3 + 5 + 6 + 2 + 2) is not divisible by 9]

- **Divisibility by 5** Any number of which last digit is either 0 or 5, is divisible by 5.

  For example 30580 ✓, 46325 ✓, 4350 ✓, 103 ✗

- **Divisibility by 10** Any number having last digit as 0, is divisible by 10.

  For example 430 ✓, 205 ✗

- **Divisibility by 15** Since 15 = 3 × 5, the number divisible by 3 and 5 each will obviously be divisible by 15. Thus, we should test the number applying the rules meant for divisibility by 3 and 5 individually.

  For example 681435 ✓ [∵ (6 + 8 + 1 + 4 + 3 + 5) i.e. 27 is divisible by 3 and the last digit of the number is 0.]

  334795 ✗ [∵ (3 + 3 + 4 + 7 + 9 + 5) is not divisible by 3. Further, no need to test by 5.]

- **Divisibility by 25** A number is divisible by 25 if the number formed by its last two digits is divisible by 25.

  For example 15850 ✓ [∵ 50 is divisible by 25]
  32400 ✓ [∵ 00 is divisible by 25]
  24560 ✗ [∵ 60 is not divisible by 25]

- **Divisibility by 125** A number is divisible by 125 if the number formed by its last three digits is divisible by 125.

  For example 15750 ✓ [∵ 750 is divisible by 125]
  324000 ✓ [∵ 000 is divisible by 125]
  24560 ✗ [∵ 560 is not divisible by 125]

- **Divisibility by 11** If the difference between sum of digits at odd places and that at even places is divisible by 11, the number is also divisible by 11.

  For example 563497

Sum of digits at odd places = 7 + 4 + 6 = 17

and sum of digits at even places= 9 + 3 + 5 = 17

Then, difference between both sum = 17 − 17 = 0

Since 0 is divisible by 11, the number 563497 is divisible by 11.

- **Divisibility by 7, 11 and 13** If the difference between the number formed with the last three digits of the given number and the number existing with remaining digits is divisible by 7, the given number will be divisible by 7. In case the aforesaid difference is divisible by 11, the given number will be divisible by 11 and also in case the difference is divisible by 13, the given number will be divisible by 13.

For example Test for 646321.

646**<u>321</u>** ⇒ 646 − 321 = 325

Since the 325 is divisible neither by 7 nor by 11, the number 646321 will not be divisible by either 7 or 11. However, 646321 will be divisible by 13 as 325 is divisible by 13.

52**<u>738</u>** ⇒ 738 − 52 = 686

Since the 686 is divisible by 7, 52738 will be divisible by 7. But 52738 is not divisible by either 11 or 13 as 686 is divisible by neither 11 nor 13.

- **Divisibility by 14** Since 14 = 2 × 7, the number divisible by both 2 and 7 will be divisible by 14. Thus, we should test the number applying the rules meant for divisibility by 2 and 7 individually.

For example 15862 ✓

Since the last digit is divisible by 2, 15862 is divisible by 2.

Further, 15**<u>862</u>** ⇒ 862 − 15 = 847

Since, 847 is divisible by 7, 15862 will be divisible by 7.

632154 ✗

Since the last digit is divisible by 2, 632154 is divisible by 2.

Further, 632**<u>154</u>** ⇒ 632 − 154 = 478

Since 478 is not divisible by 7, 632154 will not be divisible by 7.

- **Divisibility by 17, 19, 23, 29, 31, 37, 41, …….. prime numbers**

The technique which is to be elaborated is based on osculation approach. We may write

17 ⇒ 17 × 3 = 51 [just 1 more than <u>5</u>0]

19 ⇒ 19 × 1 = 19 [just 1 less than <u>2</u>0]

23 ⇒ 23 × 3 = 69 [just 1 less than <u>7</u>0]

29 ⇒ 29 × 1 = 29 [just 1 less than <u>3</u>0]

31 ⇒ 31 × 1 = 31 [just 1 more than <u>3</u>0]

37 ⇒ 37 × 3 = 111 [just 1 more than <u>11</u>0]

41 ⇒ 41 × 1 = 41 [just 1 more than <u>4</u>0] and so on.

In case of 17, 19, 23, 29, 31,…… last digit will be multiplied by 5, 2, 7, 3, 3, ……..respectively as the base numbers i.e. 50, 20, 70, 30, 30 ….. are 5 times, two times, seven times, three times, three times,…… of 10 respectively. Further, the product thus obtained will be subtracted from/added to the number existing with the digits except the last one. Since the base number i.e. 50 (in case of 17) is less than 51, subtraction is made otherwise addition, *viz.* in case of 19, 23, 29, the product will be added to. This process is repeated until you feel comfortable to divide the resultant by the desired number.

Now, understand the above approach through an example: Check as to whether 552279 is divisible by 17, 19, 23, ………, Concept of positive and negative osculators have been left deliberately so as to avoid complexity in understanding.

- **By 17** 552279 ⇒ 55227 − (5 × 9) = 5518**<u>2</u>** [Five times the last digit of the given number has been subtracted from the number existing with the digits except the last one. Since 50 is less than 51, subtraction has been made.]

⇒ 5518 − (5 × 2) = 550**<u>8</u>**

⇒ 550 − (5 × 8) = 51**<u>0</u>**

⇒ 51 − (5 × 0) = 51

Since 51 is divisible by 17, the given number 552279 is divisible by 17.

- **By 19** 552279 ⇒ 55227 + (2 × 9) = 5524**<u>5</u>** [Two times the last digit of the given number has been added to the number existing with the digits except the last one as the base number is 20. Since 20 is more than 19, addition has been made.]

⇒ 5524 + (2 × 5) = 553**<u>4</u>**

⇒ 553 + (2 × 4) = 56**<u>1</u>**

⇒ 56 + (2 × 1) = 58

Since 58 is not divisible by 19, the given number 552279 is not divisible by 19.

- **By 23** 552279 ⇒ 55227 + (7 × 9) = 5529**<u>0</u>** [7 times the last digit of the given number has been added to the number existing with the digits except the last one as the base number is 70. Since 70 is more than 69, addition has been made.]

⇒ 5529 + (7 × 0) = 552**<u>9</u>**

⇒ 552 + (7 × 9) = 61**<u>5</u>**

⇒ 61 + (7 × 5) = 9**<u>6</u>**

⇒ 9 + (7 × 6) = 51

Since 51 is not divisible by 23, the given number 552279 is not divisible by 23.

Now, test the divisibility of the given number by 29, 31, 37, ………… on the above line. The above approach is also applicable for 7, 11 and 13.

[∵ 7 ⇒ 7 × 3 = 21 (just 1 more than 20), 11 ⇒ 11 × 1 = 11 (just 1 more than 10) and 13 ⇒ 13 × 3 = 39 (just 1 less than 40)].

Illustrations

Problem 1 If * is a digit such that 5824* is divisible by 11, then * equals

[SSC(CGL-PT)-2000, SSC(CPO)-04, 08, CDS-08]

Sol. Sum of digits at odd places = * + 2 + 5 = 7 + *

and sum of digits at even places = 4 + 8 = 12

Clearly, * = 5 makes the difference between both sums divisible by 11.

Problem 2 If $34849xy$ is divisible by 80, $x + y = ?$

[SSC(CGL-T-I)-17]

Sol. Since $80 = 5 \times 16$, the number divisible by both 5 and 16 will be divisible by 80. Thus, we should test the number applying the rules meant for divisibility by 5 and 16 individually.

$34849xy$ is divisible by 5 only when $y = 0$ or 5. But, in case the number is divisible by 16, the number should not have 5 as last digit. Hence, $y = 0$.

Further, in case $34849xy$ is divisible by 16, the number formed with last four digits must be divisible by 16.

At $x = 6$, $49\,x\,0$ becomes divisible by 16

[$\because$ 4960 is divisible by 16]

Now, $x + y = 6 + 0 = 6$

Problem 3 Find the minimum value of $x + y$ if $4638x7y$ is divisible by 12.

Sol. Since $4638x7y$ is divisible by 12, it is also divisible by both 3 and 4.

For divisibility by 4, $7y$ must be divisible by 4. Obviously, at $y = 2$ or 6, $7y$ is divisible by 4.

But we have to take minimum value of y, hence, $y = 2$.

Further, for divisibility by 3, $(4 + 6 + 3 + 8 + x + 7 + 2)$ i.e. $(30 + x)$ must be divisible by 3.

At $x = 0, 3, 6$ and 9, $(30 + x)$ is divisible by 3. But we have to take minimum value of x, hence, $x = 0$.

Now, $x + y = 0 + 2 = 2$

Problem 4 Which of the following is divisible by 165?

(a) 41352705 (b) 362175
(c) 263585 (d) 7263710

Sol. Since $165 = 3 \times 5 \times 11$, the number divisible by 3, 5 and 11 each, will be divisible by 165. Thus, we should test the numbers applying the rules meant for divisibility by 3, 5 and 11 individually.

Option(a) $41352705 \Rightarrow (4 + 1 + 3 + 5 + 2 + 7 + 0 + 5)$ is divisible by 3. Hence, the number is divisible by 3.

Further, the number has 5 as its last digit. Hence, divisible by 5.

Now, $41352\underline{\mathbf{705}} \Rightarrow 41352 - 705 = 40647$

$40\underline{\mathbf{647}} \Rightarrow 647 - 40 = 607$

Since 607 is not divisible by 11, 41352705 is not divisible by 11

Option(b) $362175 \Rightarrow (3 + 6 + 2 + 1 + 7 + 5)$ is divisible by 3, hence, the number is divisible by 3.

Further, the number has 5 as its last digit. Hence, divisible by 5.

Now, $362\underline{\mathbf{175}} \Rightarrow 362 - 175 = 187$

Since 187 is divisible by 11, 362175 is divisible by 11.

Alternatively,

The difference between $(5 + 1 + 6)$ and $(7 + 2 + 3)$ i.e. 0 is divisible by 11, the number is divisible by 11.

Hence, option(b) is correct.

Problem 5 What is the maximum value of $(x - y)$ if $36789x7y$ is exactly divisibly by 55. (x and y are not the same).

Sol. Since $55 = 5 \times 11$, the number divisible by both 5 and 11 will be divisible by 55. Thus, we should test the number applying the rules meant for divisibility by 5 and 11 individually.

$36789x7y$ is divisible by 5 only when $y = 0$ or 5.

Now, difference between $(y + x + 8 + 6)$ and $(7 + 9 + 7 + 3)$ must be divisible by 11. Obviously, $(x + y) = 1, 12, 23, 34$........

In case $(x + y) = 1$, maximum value of $x - y = 1 - 0 = 1$

[$\because y = 0$ or 5 and we can't take $y = 5$ as $(x + y) = 1$]

Further, in case $(x + y) = 12$, the maximum value of $x - y = 7 - 5 = 2$ [$\because y = 0$ or 5 and we can't take $y = 0$ as in doing so the value of x will be in a two-digit number.]

Also, in case $(x + y) = 23$ or more, the value of x will be in a two-digit number at $y = 0$ or 5.

Hence, required answer = 2.

Critical Facts Regarding Divisibility

(i) A sum of n consecutive whole numbers is always divisible by n.

(ii) The product of 3 consecutive natural numbers the first of which is even, is divisible by 24.

(iii) The difference between any number and the number formed with same digits but in reverse order, is always divisible by 9.

For examples (a) $63 - 36 = 27$, divisible by 9

(b) $532 - 235 = 297$, divisible by 9.

(iv) Any number written in form of $\{(10)^n - 1\}$ is divisible by 9.

(v) For any natural number, $(n^3 - n)$ is divisible by 6.

(vi) The product of any n consecutive numbers is divisible by $n!$.

(vii) If x and y are divisible by d individually, then their sum and their difference are also divisible by d.

(viii) If a number A is divisible by B and B is divisible by C, then A is divisible by C also.

(ix) If a number is divisible by many prime numbers individually, then it must be necessarily divisible by the product of those prime numbers.

(x) $a^4 - b^4$ is always divisible by 8, if a and b are two odd integers.

(xi) $n^2(n^2 - 1)$ is always divisible by 12, if n is a whole number greater than 1.

(xii) $x^n - a^n$ is exactly divisible by $(x - a)$, whether n is odd or even integer.

(xiii) $x^n - a^n$ is exactly divisible by $(x + a)$, when n is an even integer.

(xiv) $x^n + a^n$ is exactly divisible by $(x + a)$, when n is an odd integer.

(xv) $x^n + a^n$ is never divisible by $(x - a)$.

Illustrations

Problem 6 If a number is divisible by 11 and 13, then it must be necessarily

(a) divisible by $(11 + 13)$ (b) divisible by $(13 - 11)$

(c) divisible by (11×13) (d) 429

Sol. (c) We know that if a number is divisible by certain prime numbers individually, then it must be necessarily divisible by the product of those prime numbers. Hence, option (c) is correct.

Problem 7 If a and b are two odd positive integers, by which of the following integers is $(a^4 - b^4)$ always divisible? ***[SSC(CGL-T-I) -11]***

(a) 3 (b) 6 (c) 8 (d) 12

Sol. We know that, $(a^4 - b^4)$ is always divisible by 8 if a and b are two odd integers. Hence, option (c) is correct.

B. Division Based on Whole Numbers

I General Rules

Type I Division of a whole number by a two-digit whole number

Understand the concept through an example *viz.* $46785 \div 62 = ?$

Step **I** For sake of convenience it is assumed that the divisor is 60 at the first instance i.e. a multiple of 10 just below the divisor. Draw a line after one digit from the extreme right as 60 has only one zero.

6	4678	5
②		

Step **II** We divide 46 by 6. Then, quotient = 7, remainder = 4 and obviously the next dividend will be 47.

Now, write 7 as quotient as depicted below. But, the actual dividend would be [47 − ② × 7)] i.e. 33 as ② was not considered earlier while dividing.

6	46 ^{4}7 8	5
○	**7**	

Step **III** We divide 33 by 6. Then, quotient = 5, remainder = 3 and obviously the next dividend will be 38. Now, write 5 as quotient just after the earlier quotient as depicted below. But, the actual dividend would be [38 − (② × 5)] i.e. 28 as ② was not considered earlier while dividing.

6	467 ^{3}8	5
②	**75**	

Step **IV** We divide 28 by 6. Then, quotient = 4, remainder = 4 and obviously the next dividend will be 45.
Now, write 4 as quotient just after earlier quotients as depicted below. But the actual dividend = [45 − ② × 4)] i.e. 37. Since the dividend 37 is lesser than the divisor 62, 37 will be as remainder.
Mind the placement of 37.

6	4678	^{4}5
②	7 54	37

Hence, required quotient = 754 and remainder 37.

Note *In case we get actual dividend negative at any step, we should reduce the quotient by one and repeat, if required, till we get the positive actual dividend.*

Take another Example $379521 \div 57 = ?$

Step **I** For sake of convenience it is assumed that the divisor is 50 at the first instance i.e. a multiple of 10 just below the divisor. Draw a line after one digit from the extreme right as 50 has only one zero.

5	3 7 9 5 2	1
⑦		

Step **II** We divide 37 by 5. Then, quotient = 7, remainder = 2 and obviously the next dividend will be 29. But the actual dividend would be [29 − (⑦ × 7)] i.e. − 20 as ② was not considered earlier while dividing. To avoid the negative resultant, we will take quotient one less than the earlier, hence, 6. Then, the remainder will be 7 and the next dividend will be 79. Accordingly, the actual dividend = [79 − (⑦ × 6)] = 37

5	37 ^{7}9 5 2	1
⑦	**6**	

Step **III** We divide 37 by 5. Then, quotient = 7, remainder = 2 and obviously the next dividend will be 25. But the actual dividend would be [25 − (⑦ × 7)] i.e. − 24.
To avoid the negative resultant, we will take quotient one less than the earlier, hence, 6. Then, the remainder will be 7 and the next dividend will be 75. Accordingly, the actual dividend = [75 − (⑦ × 6)] = 33

5	379 ^{7}5 2	1
⑦	**66**	

Step **IV** We divide 33 by 5. Then, quotient = 6, remainder = 3 and obviously the next dividend will be 32. But the actual dividend would be [32 − (⑦ × 6)] i.e. − 10.

To avoid the negative resultant, we will take quotient one less than the earlier, hence, 5. Then, the remainder will be 8 and the next dividend will be 82. Accordingly, the actual dividend = [82 − (⑦ × 5)] = 47

5	3795 ^{8}2	1
⑦	**665**	

***Step* V** We divide 47 by 5. Then, quotient = 9, remainder = 2 and obviously the next dividend will be 21. But the actual dividend would be [21 – (⑦ × 9)] i.e. – 42. To avoid the negative resultant, we will take quotient one less than the earlier, hence, 8.

Then, the remainder will be 7 and the next dividend will be 71. Accordingly, the actual dividend = [71 – (⑦ × 8)]. Since, the dividend 15 is lesser than the divisor 57, 15 will be as remainder. Mind the placement of 15.

5	37952	$^{2}1$
⑦	**6658**	15

Hence, required quotient = 6658 and remainder = 15.

Type II Division of a whole number by a three-digit whole number

***Case* I** When divisor is comparatively smaller like 123, 128, 157, etc.

In this case we should adopt the same approach as has been applied in case of a two-digit divisor. We will assume 123 as <u>12</u> 3, 128 as <u>12</u> 8 and so on.

For Example 64325 ÷ 123 = ?

***Step* I** For sake of convenience it is assumed that the divisor is 120 at the first instance i.e. a multiple of 10 just below the divisor. Draw a line after one digit from the extreme right as 120 has only one zero.

12	6 4 3 2	5
③		

***Step* II** We divide 64 by 12. Then, quotient = 5, remainder = 4 and obviously the next dividend will be 43. Now, write 5 as quotient as depicted below.
But, the actual dividend would be [43 – (③ × 5)] i.e. 28 as ③ was not considered earlier while dividing.

12	64 $^{4}3$ 2	5
③	**5**	

***Step* III** We divide 28 by 12. Then, quotient = 2, remainder = 4 and obviously the next dividend will be 42. Now, write 2 as quotient just after earlier quotient as depicted below. But, the actual dividend would be [42 – (③ × 2)] i.e. 36.

12	643 $^{4}2$	5
③	**52**	

***Step* IV** We divide 36 by 12. Then, quotient = 3, remainder = 0 and obviously the next dividend will be 05.

But the actual dividend would be [05 – (③ × 3)] i.e. – 4. To avoid the negative resultant, we will take quotient one less than the earlier, hence, 2.

Then, the remainder will be 12 and the next dividend will be 125. Accordingly, the actual dividend = [125 – (③ × 2)] = 119.

Since, the dividend 119 is lesser than the divisor 123, 119 will be as remainder. Mind the placement of 119.

12	6432	$^{0}5$
③	**522**	119

Finally, quotient = 522 and remainder = 119.

***Case* II** When divisor is comparatively bigger like 523, 736, etc.

For Example 803215 ÷ 523 = ?

In these types of cases, given divisor is assumed as a multiple of 100 just below the divisor.

***Step* I** For sake of convenience it is assumed that the divisor is 500 at the first instance i.e. a multiple of 100 just below the divisor. Draw a line after two digits from the extreme right as 100 or 500 has two zeros.

5	8 0 3 2	1 5
23		

***Step* II** We divide 8 by 5. Then, quotient = 1, remainder = 3 and obviously the next dividend will be 30. Now, write 1 as quotient as depicted below. But, the actual dividend would be [30 – (Quotient × 1st digit of 23)] i.e. [30 – (Quotient × 1st digit of 23)] i.e. 30 – (1 × 2) = 28.

5	8 $^{3}0$ 3 2	1 5
23	**1**	

***Step* III** We divide 28 by 5. Then, quotient = 5, remainder = 3 and obviously the next dividend will be 33. But, the actual dividend would be $\left\{33 - \begin{pmatrix}23\\ \times\\ 15\end{pmatrix}\right\}$

i.e. [33 – (2 · 5 + 3 · 1)], hence, 20.

5	80 $^{3}3$ 2	1 5
23	**15**	

***Step* IV** We divide 20 by 5. Then, Quotient = 4, remainder = 0 and obviously the next dividend will be 02. But the actual dividend would be $\left\{02 - \begin{pmatrix}23\\ \times\\ 54\end{pmatrix}\right\}$

i.e. [33 – (2 · 4 + 3 · 5)], hence, negative value. To avoid the negative resultant, we will take quotient one less than the earlier, hence, 3. Then, the remainder will be 5 and the next dividend will be 52.

Accordingly, the actual dividend = $\left\{52 - \begin{pmatrix}23\\ \times\\ 53\end{pmatrix}\right\}$ i.e.

[52 – (2 · 3 + 3 · 5)], hence, 31.

5	803 $^{0}2$	1 5
23	**153**	

***Step* V** We divide 31 by 5. Then, quotient = 6, remainder = 1 and obviously the next dividend will be 11. But, the actual dividend would be negative. To avoid the negative resultant, we will take quotient one less than the earlier, hence, 5.

Then, the remainder will be 6 and the next dividend will be 61. Accordingly, the actual dividend $= \left\{61 - \begin{pmatrix} 23 \\ \times \\ 35 \end{pmatrix}\right\}$

i.e. $[61 - (2 \cdot 5 + 3 \cdot 3)]$, hence, 42.

5	8032	$^{1}15$
23	**1535**	

***Step* VI** We divide 42 by 5. Then, quotient = 8, remainder = 2 and obviously the next dividend will be 25. But the actual dividend would be negative. To avoid the negative resultant, we will take quotient one less than the earlier, hence, 7. Then, the remainder will be 7 and the next dividend will be 75. Accordingly, the actual dividend $= \left\{75 - \begin{pmatrix} 23 \\ \times \\ 57 \end{pmatrix}\right\}$

i.e. $[75 - (2 \cdot 7 + 3 \cdot 5)]$, hence, 46.

5	8032	$1\ {}^{2}5$
23	**1535**	**7**

***Step* VII** We divide 46 by 5. Then, quotient = 9, remainder = 1. However, to avoid the negative resultant, we will take quotient one less than the earlier, hence, 8.

Hence, $803215 \div 523 = 1535.78$.

Note *If you intend to find out the remainder, the required remainder = The remainder as envisaged in step-V suffixing the last two digits of given original dividend* $- \left\{10 \times \begin{pmatrix} 23 \\ \times \\ 35 \end{pmatrix}\right\}$

– [last digit of 23 × last quotient as concluded in step-V]

$\Rightarrow 6\underline{15} - 10\,(2 \cdot 5 + 3 \cdot 3) - (3 \times 5) = 410$

Hence, quotient = 1535 and remainder = 410.

Type III Division of a whole number by a four-digit whole number

For Example $893625 \div 1243 = ?$

Obviously, the format will be as below as we have to assume the divisor in form of a multiple of 1000.

1	8 9 3	6 2 5
243		

***Step* I** We divide 8 by 1. Then, quotient = 8, remainder = 0 and obviously the next dividend will be 09. But, the actual dividend would be [09 – (Quotient × 1st digit of 243)] i.e. $9 - (8 \times 2)$, hence, a negative value. To avoid the negative resultant, we will take quotient one less than the earlier, hence, 7. Then, the remainder will be 1 and the next dividend will be 19. Accordingly, the actual dividend $= 19 - (7 \times 2) = 5$.

1	8 $^{0}9$ 3	6 2 5
243	**7**	

***Step* II** We divide 5 by 1. However, to avoid the negative resultant/dividend, we will take quotient 4 less than what should be, hence, 1. Then, the remainder will be 4 and the next dividend will be 43. Accordingly, the actual dividend $= \left\{43 - \begin{pmatrix} 24 \\ \times \\ 71 \end{pmatrix}\right\}$

i.e. $[43 - (2 \cdot 1 + 4 \cdot 7)]$, hence, 13. Here, 24 is the first two digits of 243 and 71 is the quotient obtained so far.

1	89 $^{0}3$	6 2 5
243		

***Step* III** We divide 13 by 1. However, to avoid the negative resultant/dividend, we will take quotient 5 less than what should be, hence, 8. Then, the remainder will be 5 and the next dividend will be 5625.

1	893	$^{0}6$ 2 5
243	**718**	

Now, required remainder = 5625 – 100 [sum of cross multiplication of 243 and last three digits of quotients] – 10 [sum of cross multiplication of two right digits of 243 and last two digits of quotients] – [product of last digit of 243 and last digit of quotient]

$$= 5625 - \left\{100 \times \begin{pmatrix} 243 \\ \times \\ 718 \end{pmatrix}\right\} - \left\{10 \times \begin{pmatrix} 43 \\ \times \\ 18 \end{pmatrix}\right\} - \{3 \times 8\}$$

$$= 5625 - 100\{2 \cdot 8 + 3 \cdot 7 + 4 \cdot 1\} - 10\{4 \cdot 8 + 3 \cdot 1\} - \{3 \times 8\}$$

$$= 5625 - 4100 - 350 - 24 = 1151$$

Hence, required quotient = 718 and remainder = 1151.

Note *Quotient may be obtained in decimal if we continue to process as above. The required quotient = 718.92*

Observations The above approaches may bother you if you don't practice much. However, if you are not feeling convenient despite much practice, leave this technique and stress on conventional /alternative approaches.

Try yourself

➘ Problem 8 $25346 \div 49$

Sol. Q → 517 and R → 13

➘ Problem 9 $25346 \div 59$

Sol. Q → 429 and R → 35

Problem 10 35342 ÷ 49

Sol. Q → 721 and R → 13

Problem 11 853461 ÷ 67

Sol. Q → 12738 and R → 15

Problem 12 756849 ÷ 78

Sol. Q → 9703 and R → 15

Problem 13 25346 ÷ 137

Sol. Q → 185 and R → 1

Problem 14 2534 ÷ 151

Sol. Q → 16 and R → 118

Problem 15 35342 ÷ 539

Sol. Q → 65 and R → 307

Problem 16 85346 ÷ 603

Sol. Q → 141 and R → 323

Problem 17 75681 ÷ 1324

Sol. Q → 57 and R → 213

II Division by Multiple Divisors

The divisor may be of two terms and more than two terms in the form of multiplication.

For example $\frac{4687}{3\times 7}, \frac{89321}{2\times 3\times 11}$, etc.

In such cases, divide the dividend by the first divisor and write the quotient below the dividend and remainder on the right of the quotient. Further, divide the quotient so obtained by the next divisor and write the quotient and remainder in the above manner. We continue this process till division has been performed by all of the divisors.

Divisors (D)	Dividend (4687)	Remainders (R)
$D_1 \Rightarrow 3$	1562	$1 \Rightarrow R_1$
$D_2 \Rightarrow 7$	223	$1 \Rightarrow R_2$

Clearly, quotient = 223 and remainder = $1 + 1 \cdot 3 = 4$

Generic Version

Required remainder in case of two divisors

$$= R_1 + R_2 \cdot D_1$$

In case of 3 divisors, required Remainder

$$= R_1 + R_2 \cdot D_1 + R_3 \cdot D_1 \cdot D_2$$

and in case of 4 divisors, remainder

$$= R_1 + R_2 \cdot D_1 + R_3 \cdot D_1 \cdot D_2 + R_4 \cdot D_1 \cdot D_2 \cdot D_3$$

and so on.

Needless to say that the last dividend will be the final/required dividend.

Now, take another example $\frac{89321}{2\times 3\times 11}$

Divisors (D)	Dividend (89321)	Remainders (R)
$D_1 \Rightarrow 2$	44660	$1 \Rightarrow R_1$
$D_2 \Rightarrow 3$	14886	$2 \Rightarrow R_2$
$D_3 \Rightarrow 11$	1353	$3 \Rightarrow R_3$

Now, final quotient = 1353

and final remainder

$$= R_1 + R_2 \cdot D_1 + R_3 \cdot D_1 \cdot D_2$$
$$= 1 + 2 \cdot 2 + 3 \cdot 2 \cdot 3 = 23$$

Hence, $\frac{89321}{2\times 3\times 11} = 1353 \frac{23}{2\times 3\times 11} = 1353.35$

Now, take one more example $\frac{836255}{216} = ?$

We should split the divisor/denominator in the product form like (12×18), (24×9), $(4\times 6\times 9)$, (8×27) etc as per our choice which suits better. Now, you have to proceed on the above pattern.

Observations You may divide the given dividend easily by 12, 18, 24, 9, 4, 6, 2, etc. If you are well versed with their tables but division by 108 seems to be somewhat difficult. However, this difficulty may be dispelled when we learn the division by a number closer to 100, 1000, ...and so on.

III Division by a Number Closer to 100, 1000, 10000 and so on

***Case* I When divisor is less than 100, 1000, 10000 ..... and so on.**

Consider an example 78523 ÷ 98

Imagine we have to divide the given number by 100 instead of 98. Then, quotient would be 785 and remainder be 23. In order to alienate the remainder from the quotient, a vertical line after two digits from the extreme right may be drawn. In case the divisor is closer to 1000, this line may be drawn after three digits from the right and so on.

785 | 23

But, it is true that the actual divisor is 98, it implies that every time (here, 785 times) the remainder equal to 2 will be accumulated in addition to remainder 23. Hence, 2×785 i.e. 1570 in addition to 23.

Further, it is assumed that 1570 is divided by 100 instead of 98, the quotient = 15 and remainder = 70. Also, in this case, remainder would be (15×2) i.e. 30 over 70.

15 | 70

Now, we should stop the process as 30 when divided by 100 yields no quotient and at this stage no question arises as to how much remainder is to be accumulated over 30.

From above, total quotient = 785 + 15 + 0 = 800

and total remainder = 23 + 70 + 30 = 123

Then, required answer $= 800\frac{123}{98} = 800 + 1\frac{25}{98} = 801\frac{25}{98}$

The approach elaborated above may be depicted as below

785	23
15	70 [Quotient × number by which the divisor is less than 100 i.e. 785 × 2, hence, 1570]
0	30 [Quotient × number by which the divisor is less than 100 i.e. 15 × 2, hence, 30]
800	123

Note *(i) We continue the process till no digit remains on left side of the line.*

(ii) The numbers to be written on right side of the line must be in two digits.

Then, required answer $= 800\frac{123}{98} = 800 + 1\frac{25}{98} = 801\frac{25}{98}$

Take another example: 362587 ÷ 93

3625	87	
253	75	[∵ 3625 × 7 = 25375]
17	71	[∵ 253 × 7 = 1771]
1	19	[17 × 7 = 119]
0	07	[1 × 7 = 07]
3896	259	

Then, required answer

$= 3896\frac{259}{93} = 3896 + 2\frac{73}{93} = 3898\frac{73}{93}$

Remarks *If the sum of remainders i.e. sum of numbers on right side is a large number, you may adopt the same process for* $\frac{\text{sum of remainders}}{\text{divisor}}$ *as is for* $\frac{\text{original dividend}}{\text{divisor}}$.

Take other example 6359632 ÷ 982

Here, the divisor is 18 less than 1000, thus we should draw a line after three digits from the right end of the dividend. Also, keep in mind that the right side numbers should be in three digits.

6359	632	
114	462	[∵ 6359 × 18 = 114462]
2	052	[∵ 114 × 18 = 2052]
0	036	[∵ 2 × 18 = 36]
6475	1182	

Hence, required answer

$= 6475\frac{1182}{982} = 6475 + 1\frac{200}{982}$

$= 6476\frac{100}{491}$

Case **II When divisor is more than 100, 1000, 10000 .....and so on.**

The process meant for the cases wherein the divisors are less than 100, 1000, ... is applicable here also but the difference is that we add both sides of line in those cases whereas here we have to subtract and add alternately from 2nd row. Rest operations are same.

For example 78523 ÷ 102

Case I	**78523 ÷ 98**	**Case II**	**78523 ÷ 102**
785	23	785	23
15	70 [∵ 785 × 2 = 1570]	(−) 15	(−) 70 [∵ 785 × 2 = 1570]
0	30 [∵ 15 × 2 = 30]	(+) 0	(+) 30 [∵ 15 × 2 = 30]
800	123	770	(−)17
Then, required answer $= 800\frac{123}{98} = 800 + 1\frac{25}{98} = 801\frac{25}{98}$		Then, required answer $= 770\frac{-17}{102} = 770 - \frac{17}{102} = 769\frac{5}{6}$	

Another example 365281 ÷ 1017 = ?

365	281
(−)6	(−) 205
(+) 0	(+) 102
359	(+)178

Then, required answer $= 359\frac{178}{1017}$

IV. Division by Some Specific Numbers

By 5ⁿ

(i) $2367 \div 5^2 = 2367 \times \frac{4}{100} = 94.68$

(ii) $36213 \div 5^3 = 36213 \times \frac{8}{1000} = 289.704$

(iii) $1267 \div 5^4 = 1267 \times \frac{16}{10000} = 2.0272$

By 35 $567 \div 35 = 567 \times \frac{2}{70} = \frac{1134}{7 \times 10} = \frac{162}{10} = 16.2$

By 45 $8657 \div 45 = 8657 \times \frac{2}{90} = \frac{17314}{9 \times 10} = \frac{1923.7}{10} = 192.37$

V. Problems Wherein the Above Approaches can be Applied After Some Alternation in Form/Pattern of Problems

(i) $\frac{638752}{253} = ? \Rightarrow \frac{638752 \times 4}{253 \times 4} = \frac{2555008}{1012}$

Now, apply the concept 'When the divisor is more than 100, 1000, 10000 .....and so on.

The answer is $2524\frac{720}{1012}$ i.e. $2524\frac{180}{253}$.

(ii) $\dfrac{123876529}{1648} = ?$

$\Rightarrow \dfrac{123876529}{16 \times 103}$

Now, apply the concept "Division by multiple Divisors". The concept "When the divisor is more than 100, 1000, 10000 .....and so on" may be applied.

The answer is $75167 \dfrac{1313}{1648}$ i.e. 75167.8

(iii) $\dfrac{36325}{168} = ?$

$\Rightarrow \dfrac{36325 \times 6}{168 \times 6} = \dfrac{217950}{1008}$

Now, apply the concept "When the divisor is more than 100, 1000, 10000 .....and so on".

The answer is $216 \dfrac{222}{1008}$ i.e. $216 \dfrac{37}{168}$.

Aspirants are Supposed to

(i) memorise the tables upto 25 so as to divide the given number directly in few seconds.

(ii) split the divisor into product form, where it is possible, in such a way that you are capable to divide the dividends by each divisor/term and then process accordingly.

(iii) multiply the divisor by a suitable number so as to bring it near to 100, 1000, ... if it is possible. The dividend should also be multiplied by the same number. Thereafter, process accordingly.

C. Division of/by Decimal Number(s)

For example $\dfrac{6.936}{1.7}$

We divide numerator(i.e. dividend) by denominator (i.e. divisor) ignoring the decimal point(s).

$\Rightarrow \dfrac{6936}{17} = 408$

Here, number of digits after decimal point in numerator $= 3$ (say, x)

and number of digits after decimal point in denominator $= 1$ (say, y)

Then, difference between x and $y = 3 - 1 = 2$

Now, put decimal point after two digits from the right end of the resultant as $x > y$. In case $x < y$, put two zeros on the resultant on the right end.

Here, $x > y$, then required answer = 4.08.

Take another example $\dfrac{3161.1}{0.0123} = ?$

$\Rightarrow \dfrac{31611}{00123} = 257$ [Ignoring decimal points]

Here, number of digits after decimal point in numerator $= 1$ (say, x)

and number of digits after decimal point in denominator $= 4$ (say, y)

Then, difference between x and $y = 4 - 1 = 3$

Now, put three zeros on the resultant at the right end as $x < y$.

Hence, required answer = 257000

Other examples

(i) $\dfrac{188.945}{23} = ?$

$\Rightarrow \dfrac{188945}{23}$ [Ignoring decimal point]

$= 8215$

Clearly, $x = 3$ and $y = 0$ then difference between x and $y = 3 - 0 = 3$.

Now, put decimal point after three digits from the right end of the resultant as $x > y$.

Hence, required answer = 8.215.

(ii) $\dfrac{575575}{0.91} = ?$

$\Rightarrow \dfrac{575575}{091}$ [Ignoring decimal point]

$= 6325$

Clearly, $x = 0$ and $y = 2$ then difference between x and $y = 2 - 0 = 2$.

Now, put two zeros on the resultant at the right end as $x < y$.

Hence, required answer = 632500

(iii) $\dfrac{3.7426}{1000} = ?$

When a decimal number is divided by $(10)^n$, the decimal point shifts n digits towards left end.

Hence, required answer = 0.0037426

D. Division of Recurring Decimals

Build the concepts through examples:

(i) $0.01\overline{2} \div 100 = ?$

$\Rightarrow 0.0122222........ \div 100 = 0.000122222........$

$= 0.0001\overline{2}$

(ii) $4.8\overline{62} \div 4 = ?$

$\Rightarrow 4.8626262............ \div 4 = 1.2156565........$

$= 1.21\overline{56}$

(iii) $24.36\overline{12} \div 21 = ?$

$\Rightarrow 24.36121212121212......... \div 21$

$= 1.160057720057720.....$

$= 1.16\overline{005772}$

E. Division of/by Fraction(s)

Types	Patterns	Examples
I	$\frac{a}{b} \div c = ?$ $\Rightarrow \frac{a}{b} \div \frac{c}{1}$ $= \frac{\text{Left diagonally multiplication}}{\text{Right diagonally multiplication}}$ $= \frac{a \cdot 1}{b \cdot c} = \frac{a}{bc}$	(a) $\frac{2}{3} \div 5 = ?$ $\Rightarrow \frac{2}{3} \div \frac{5}{1} = \frac{2 \cdot 1}{3 \cdot 5} = \frac{2}{15}$ (b) $\frac{4}{7} \div 8 = ?$ $\Rightarrow \frac{4 \cdot 1}{7 \cdot 8} = \frac{1}{14}$
II	$\frac{a}{b} \div \frac{c}{d} = ?$ $= \frac{\text{Left diagonally multiplication}}{\text{Right diagonally multiplication}}$ $= \frac{a \cdot d}{b \cdot c} = \frac{ad}{bc}$	(a) $\frac{4}{9} \div \frac{17}{23} = ?$ $\Rightarrow \frac{4 \cdot 23}{9 \cdot 17} = \frac{92}{153}$ (b) $13 \div \frac{4}{5} = ?$ $\Rightarrow \frac{13}{1} \div \frac{4}{5} = \frac{13 \cdot 5}{4 \cdot 1} = 16\frac{1}{4}$
III	$a\frac{b}{c} \div d = ?$ $\Rightarrow \left(a + \frac{b}{c}\right) \div d$ $= (a \div d) + \left(\frac{b}{c} \div d\right)$	$18\frac{1}{2} \div 5 = ?$ $\Rightarrow (18 \div 5) + \left(\frac{1}{2} \div 5\right)$ $\Rightarrow 3\frac{3}{5} + \frac{1}{10} = 3\frac{3 \cdot 2}{5 \cdot 2} + \frac{1}{10}$ [Recall the addition method] $= 3\frac{7}{10}$
IV	$a\frac{b}{c} \div \frac{d}{e} = ?$ $\Rightarrow \left(a + \frac{b}{c}\right) \div \frac{d}{e}$ $= \left(a \div \frac{d}{e}\right) + \left(\frac{b}{c} \div \frac{d}{e}\right)$	$112\frac{2}{7} \div \frac{3}{5} = ?$ $\Rightarrow \left(112 \div \frac{3}{5}\right) + \left(\frac{2}{7} \div \frac{3}{5}\right)$ $= \frac{112 \cdot 5}{1 \cdot 3} + \frac{2 \cdot 5}{3 \cdot 7}$ $= 186\frac{2}{3} + \frac{2 \cdot 5}{3 \cdot 7}$ $= 186\frac{2 \cdot 7}{3 \cdot 7} + \frac{2 \cdot 5}{3 \cdot 7} = 186\frac{24}{21}$ $= 186 + 1\frac{1}{7} = 187\frac{1}{7}$

Illustrations

➘ Problem 18 $4\frac{2}{3} \div \frac{1}{3} \div \frac{2}{5} = ?$

Sol. $\Rightarrow \left(4\frac{2}{3} \div \frac{1}{3}\right) \div \frac{2}{5} = \left\{\left(4 \div \frac{1}{3}\right) + \left(\frac{2}{3} \div \frac{1}{3}\right)\right\} \div \frac{2}{5}$

$$= \{12 + 2\} \div \frac{2}{5} = 14 \div \frac{2}{5} = 35$$

Alternatively

$$4\frac{2}{3} \div \frac{1}{3} \div \frac{2}{5} = \frac{4\frac{2}{3}}{\frac{1}{3} \times \frac{2}{5}} \qquad \left[\because A \div B \div C = \frac{A}{BC}\right]$$

$$= 4\frac{2}{3} \div \frac{2}{15} = \left(4 \div \frac{2}{15}\right) + \left(\frac{2}{3} \div \frac{2}{15}\right) = 30 + 5 = 35$$

➘ Problem 19 $\frac{272}{68} \div \frac{17}{?} = \frac{12}{17}$

Sol. $\frac{272 \times ?}{17 \times 68} = \frac{12}{17} \Rightarrow 272 \times ? = 12 \times 68 \Rightarrow ? = \frac{12 \times 68}{272} = 3$

➘ Problem 20 $\frac{50}{0.2} \div \frac{0.5}{5} = ?$

Sol. $\frac{50 \times 5}{0.2 \times 0.5} = 2500$

➘ Problem 21 $4\frac{16}{17} \times 1\frac{11}{16} \div \frac{7}{38} = ?$ **[BOI(PO)-08]**

Sol. $4\frac{16}{17} \times \left\{1\frac{11}{16} \div \frac{7}{38}\right\} = 4\frac{16}{17} \times \left\{\left(1 \div \frac{7}{38}\right) + \left(\frac{11}{16} \div \frac{7}{38}\right)\right\}$

$$= 4\frac{16}{17} \times \left\{\left(\frac{38}{7}\right) + \left(\frac{11 \cdot 38}{16 \cdot 7}\right)\right\} = 4\frac{16}{17} \times \left\{\left(\frac{38 \cdot 16}{7 \cdot 16}\right) + \left(\frac{11 \cdot 38}{16 \cdot 7}\right)\right\}$$

$$= 4\frac{16}{17} \times \left(\frac{38 \cdot 27}{7 \cdot 16}\right) = \left(4 \times \frac{38 \cdot 27}{7 \cdot 16}\right) + \left(\frac{16}{17} \times \frac{38 \cdot 27}{7 \cdot 16}\right)$$

$$= \left(\frac{19 \cdot 27}{7 \cdot 2}\right) + \left(\frac{38 \cdot 27}{17 \cdot 7}\right) = \left(\frac{19 \cdot 27 \cdot 17}{7 \cdot 2 \cdot 17}\right) + \left(\frac{38 \cdot 27 \cdot 2}{17 \cdot 7 \cdot 2}\right)$$

$$= \frac{19 \cdot 21 \cdot 27}{17 \cdot 7 \cdot 2} = \frac{1539}{34} = 45\frac{9}{34}$$

Alternatively,

$$\frac{84}{17} \times \frac{27}{16} \div \frac{7}{38} = ?$$

$$= \frac{84}{17} \times \frac{27}{16} \times \frac{38}{7} = \frac{81 \times 19}{34}$$

$$= \frac{1539}{34} = 45\frac{9}{34}$$

➘ Problem 22 Find the value of * in the following.

$$1\frac{2}{3} \div \frac{2}{7} \times \frac{*}{7} = 1\frac{1}{4} \times \frac{2}{3} \div \frac{1}{6}$$

Sol. $\frac{5}{3} \div \frac{2}{7} \times \frac{*}{7} = \frac{5}{4} \times \frac{2}{3} \div \frac{1}{6}$

$$\Rightarrow \frac{5}{3} \times \frac{7}{2} \times \frac{*}{7} = \frac{5}{4} \times \frac{2}{3} \times 6$$

$$\Rightarrow * = 6$$

F. Successive Division

It has been covered under Number System.

WORKOUT

1. $456.23 \div 18 = ?$
(a) 25.34 (b) 35.1
(c) 25.92 (d) None of these

2. $9999 \div 99 \div 9 = ?$ (approx. value) ***[Andhra Bank(PO)-11]***
(a) 18 (b) 15
(c) 6 (d) 11
(e) 20

3. $5554.999 \div 50.007 = ?$ (approx. value)
(a) 110 (b) 150
(c) 200 (d) 50
(e) 125

4. Find the remainder in case of $\frac{86347}{2 \times 3 \times 13}$.
(a) 1 (b) 15
(c) 6 (d) 71

5. $475283 \div 1013 = ?$
(a) $469\frac{178}{1013}$ (b) $469\frac{186}{1013}$
(c) $469\frac{176}{1013}$ (d) $469\frac{196}{1013}$

6. $3942 \div 64 \div 3 = ?$ (approx. value) ***[Indian Bank(PO)-10]***
(a) 29 (b) 32 (c) 21 (d) 17
(e) 11

7. $74844 \div ? = 54 \times 63$
(a) 34 (b) 42 (c) 22 (d) 54
(e) None of these

8. $4.56 \div 625 = ?$
(a) 0.02366 (b) 0.007296
(c) 0.01256 (d) None of these

9. $3.10\overline{12} \div 12 = ?$
(a) $0.258\overline{334}$ (b) $0.257\overline{434}$
(c) $0.258\overline{434}$ (d) None of these

10. Which one of the following number is divisible by 11? ***[CDS-13]***
(a) 45678940 (b) 93455120
(c) 87524398 (d) 54857266

11. Which one of the following is correct ?
The number 222222 is
(a) divisible by 3 but not divisible by 7
(b) divisible by 3 and 7 but not divisible by 11
(c) divisible by 2 and 7 but not divisible by 11
(d) divisible by 3, 7 and 11

12. If the sum of the digits of an even number is divisible by 9, then that number is always divisible by
(a) 24 (b) 12 (c) 18 (d) 27

13. If $738A6A$ is divisible by 11, then the value of A is ***[SSC(CGL-T-II)-12]***
(a) 6 (b) 3 (c) 9 (d) 1

14. On dividing 4996 by a certain number, the quotient is 62 and the remainder is 36. What is the divisor? ***[CDS-08]***
(a) 80 (b) 85 (c) 90 (d) 95

15. A number G236G0 can be divided by 36, if G is ***[SNAP-11]***
(a) 8 (b) 6
(c) 1
(d) More than one values are possible

16. The number 58129745812974 is divisible by ***[CDS-12]***
(a) 4 (b) 8
(c) 11 (d) None of these

17. How many even integers n, where $100 \le n \le 200$, are divisible neither by 7 nor by 9? ***[CAT-03]***
(a) 40 (b) 37
(c) 39 (d) 38

18. By adding x to 1254934, the resulting number becomes divisible by 11, while adding y to 1254934 makes the resulting number divisible by 3. Which one of the following is the set of values for x and y?
(a) $x = 1, y = 1$ (b) $x = 1, y = -1$
(c) $x = -1, y = 1$ (d) $x = -1, y = -1$

19. If the nine-digit number $708x6y8z9$ is divisible by 99, then what is the value of $x + y + z$? ***[SSC(CGL-T-I)-19]***
(a) 27 (b) 5 (c) 16 (d) 9

20. When 335 is added to $5A7$, the result is $8B2$. $8B2$ is divisible by 3. What is the largest possible value of A? ***[SSC(CGL-T-II)-13]***
(a) 8 (b) 4 (c) 2 (d) 1

21. Which among the following is the largest four digit number that is divisible by 88? ***[CDS-11]***
(a) 9988 (b) 9944
(c) 9966 (d) 8888

22. The largest integer that divides product of any four consecutive integers is ***[CDS-11]***
(a) 4 (b) 6 (c) 12 (d) 24

23. If a number $774958A96B$ is to be divided by 8 and 9, the respective values of A and B will be ***[CAT-96]***
(a) 7, 8 (b) 8, 0
(c) 5, 8 (d) None of these

24. What is the value of M and N respectively if $M39048458N$ is divisible by 8 and 11, where M and N are single digit integers? ***[MAT-04]***
(a) 7, 8 (b) 8, 6 (c) 6, 4 (d) 5, 4

25. What least value must be given to ⊙ so that the number 84705⊙2 is divisible by 9? ***[CDS-09]***
(a) 0 (b) 1 (c) 2 (d) 3

26. In a division operation, the divisor is 5 times the quotient and twice the remainder. If the remainder is 15, then what is the dividend? ***[CDS-07, SSC (Matric-Mains)-2000]***
(a) 175 (b) 185 (c) 195 (d) 250

27. In a long division sum the quotient consists of two digits. If the dividend is 40051 and the two remainders are 173 and 294, find the divisor.
(a) 479 (b) 430 (c) 58 (d) 532

28. If the 6-digit numbers $x35624$ and $1257y4$ are divisible by 11 and 72 respectively, then what is the value of $(5x - 2y)$? ***[SSC(CGL-T-I)-19]***
(a) 14 (b) 12
(c) 10 (d) 13

29. The number of integers x such that $0.25 < 2^x < 200$ and $2^x + 2$ is perfectly divisible by either 3 or 4 is ***[CAT-18]***
(a) 14 (b) 8
(c) 5 (d) 13

Answer Key

1. (a)	2. (d)	3. (a)	4. (a)	5. (b)	6. (c)	7. (c)	8. (b)	9. (c)	10. (b)
11. (d)	12. (c)	13. (c)	14. (a)	15. (a)	16. (c)	17. (c)	18. (b)	19. (c)	20. (b)
21. (b)	22. (d)	23. (b)	24. (c)	25. (b)	26. (c)	27. (a)	28. (a)	29. (c)	

Hints and SOLUTIONS

1. Divide directly with help of the table of 18. Option (a) is correct.

2. $9999 \div 99 \div 9 = 101 \div 9$
$= 11.22$
$= 11 \text{(approx.)}$

Alternatively, $10000 \div 100 \div 10 = 10 \approx 11$

3. $5555 \div 50 = 111.10 \approx 110$

4.

Divisors (D)	Dividend (86347)	Remainders (R)
$D_1 \Rightarrow 2$	43173	$1 \Rightarrow R_1$
$D_2 \Rightarrow 3$	14391	$0 \Rightarrow R_2$
$D_3 \Rightarrow 13$	1107	$0 \Rightarrow R_3$

Remainder
$= R_1 + R_2 \cdot D_1 + R_3 \cdot D_1 \cdot D_2$
$= 1 + 0 \cdot 2 + 0 \cdot 2 \cdot 3 = 1$

5.

475	283
(–) 6	(–) 175
(+) 0	(+) 078
469	(+)186

Then, required answer $= 469 \frac{186}{1013}$

6. $3942 \div 64 \div 3 = \frac{3942}{8 \times 8 \times 3}$

Divisors (D)	Dividend (3942)	Remainders (R)
$D_1 \Rightarrow 8$	492	......
$D_2 \Rightarrow 8$	61	......
$D_3 \Rightarrow 3$	20	....

Clearly, quotient is approx. 20. Hence, option(c) is correct.

Note No need to write remainders as we intend to get approx. value.

7. $? = \frac{74844}{6 \times 9 \times 9 \times 7}$

Divisors (D)	Dividend (74844)	Remainders (R)
$D_1 \Rightarrow 6$	12474	0
$D_2 \Rightarrow 9$	1386	0
$D_3 \Rightarrow 9$	154	0
$D_4 \Rightarrow 7$	22	0

Hence, option(c) is correct.

8. $4.56 \div 625 = 4.56 \div 5^4$
$= 4.56 \times \frac{16}{10000} = \frac{72.96}{10000} = 0.007296$

9. $3.10121212 \div 12 = 0.258\overline{434}$

10. From option (a), 45678940 ⇒
Sum of digits at odd places
$= 0 + 9 + 7 + 5 = 21$
and sum of digits at even places
$= 4 + 8 + 6 + 4 = 22$
Then, their difference $= 22 - 21 = 1$, which is not divisible by 11.
From option (b) 93455120 ⇒
Sum of digits at odd places
$= 0 + 1 + 5 + 3 = 9$
and sum of digits at even places
$= 2 + 5 + 4 + 9$
$= 20$
Then, their difference $= 20 - 9$
$= 11$
which is divisible by 11.
Hence, option (b) is the correct answer.

Alternatively,
(a)45678<u>940</u> $\Rightarrow 45678 - 940 = 44738$
44<u>738</u> $\Rightarrow 738 - 44 = 694$, which is not divisible by 11.
(b) 93455<u>120</u> $\Rightarrow 93455 - 120 = 93335$
93<u>335</u> $\Rightarrow 335 - 93 = 242$, which is divisible by 11.
Hence, 93455120 is divisible by 11.

11. Sum of all digits of 222222 = 12, which is divisible by 3. Hence, 222222 is divisible by 3.
Further, 222<u>222</u> $\Rightarrow 222 - 222 = 0$, which is divisible by both of 7 and 11. Hence, 222222 is divisible by both 7 and 11.
Hence, option(d) is correct.

12. It is obvious that an even number is always divisible by 2. Further, the sum of digits of an even number is divisible by 9, it means, that number is divisible by both of 9 and 2. Also, $18 = 2 \times 9$. Hence, that number is always divisible by 18.

13. Sum of digits at odd places
$= A + A + 3$
$= 2A + 3$
and sum of digits at even places
$= 6 + 8 + 7 = 21$
Then, their difference
$= 21 - (2A + 3)$
$= 18 - 2A$
Clearly, at $A = 9$, the difference i.e. $18 - 2A$ will be divisible by 11. Hence, option (c) is correct.

14. Dividend = (Divisor × Quotient) + Remainder
$\Rightarrow 4996 = (\text{Divisor} \times 62) + 36$
$\Rightarrow \text{Divisor} = [(4996 - 36) \div 62]$
$= 80$

15. $36 = 4 \times 9$
A number is divisible by 4, if the number formed with the last two digits of that number is divisible by 4. Hence, value of G may be 0, 2, 4, 6, 8.

Further, a number is divisible by 9, if sum of all digits is divisible by 9.
Now, sum of digits

$= G + 2 + 3 + 6 + G + 0$

$= 11 + 2G$

In order to make $11 + 2G$ divisible by 9 we should put value of G as 8.

Thus, required value of G = 8

16. (a) From option (a) 4 ⇒ Since the number formed with last two digits i.e. 74 is not divisible by 4, the given number is not divisible by 4.

From option (b) 8 ⇒ Since the number formed with last three digits i.e. 974 is not divisible by 8, the given number is not divisible by 8.

From option (c) 11 ⇒ Sum of digits at odd places

$= 4 + 9 + 1 + 5 + 7 + 2 + 8 = 36$

and sum of digits at even places

$= 7 + 2 + 8 + 4 + 9 + 1 + 5$

$= 36$

Then, their difference $= 36 - 36 = 0$, which is divisible by 11.

Hence, the given number is divisible by 11.

17. Number of even integers divisible by 7 = 7 [i.e. 112, 126, 140, 154, 168, 182, 196]

and number of even integers divisible by 9 = 6

[i.e. 108, 126, 144, 162, 180, 198]

Further, number of even integers divisible by both 9 and 7 = 1 [i.e. 126].

Hence, number of even integers divisible by either 7 or 9

$= (7 + 6) - 1 = 12$

Then, number of even integers divisible neither by 7 nor by 9

= (Number of even integers from 100 to 200) − 12 = 51 − 12 = 39

18. Sum of digits at odd places

$= 4 + 9 + 5 + 1 = 19$

and sum of digits at even places

$= 3 + 4 + 2 = 9$

Then, their difference

$= 19 - 9 = 10$,

which lacks 1 to become divisible by 11.

Clearly, if we add 1 to the given number, the sum of digits at odd places will become 20 resulting in the above difference becomes divisible by 11.

Hence, $x = 1$

Further, sum of all digits of 1254934 = 28 , which may become divisible by 3 if we add (−1) to it.

Hence, $y = -1$

Thus, option (b) is correct.

19. Clearly, $708x6y8z9$ must be divisible by both 11 and 9.

For 11

Sum of digits at odd places

$= 9 + 8 + 6 + 8 + 7 = 38$

and sum of digits at even places

$= z + y + x + 0 = x + y + z$

Then, their difference must be divisible by 11.

In case $x + y + z = 5$, their difference will be divisible by 11

In case $x + y + z = 16$, their difference will be divisible by 11

In case $x + y + z = 27$, their difference will be divisible by 11

In case $x + y + z = 38$, their difference will be divisible by 11. In this situation the values of at least one of x, y and z will not be in single digit. Hence, the same is not possible.

For 9

$(7 + 0 + 8 + x + 6 + y + 8 + z + 9)$ i.e. $38 + x + y + z$ must be divisible by 9. It is possible only if $(x + y + z) = 7$ or 16 or 23 or 32 or .....

Since 16 is common for both circumstances, $x + y + z = 16$

20. $335 + 5A7 = 8B2$

Since $8B2$ is divisible by 3, sum of its digits must be divisible by 3.

$\Rightarrow \quad 8 + B + 2 = 10 + B$

At $B = 2$, 5 and 8, $(10 + B)$ is divisible by 3.

Hence, largest possible value of $B = 8$

Then, largest possible value of

$A = 8 - 3 - 1 = 4$

[∵ 1 was to be carried over left, $5 + 7 = 12$]

Alternatively

You may proceed through options.

(a) $8 \Rightarrow 335 + 5\underline{8}7 \neq 8B2$

(b) $4 \Rightarrow 335 + 5\underline{4}7 = 8B2$

and here $B = 8$

Hence, option (b) is correct.

21. Since $88 = 8 \times 11$, the number divisible by 8 and 11 each, will be divisible by 88. Thus, we should test the numbers applying the rules meant for divisibility by 8 and 11 individually.

Option (a), $9\underline{988} \Rightarrow 988 - 9 = 979$, which is divisible by 11.

Hence, 9988 is divisible by 11.

Further, the number formed with last three digits of the number i.e. 988 is not divisible by 8. Hence, 9988 is not divisible by 8.

(b) $9\underline{944} \Rightarrow 944 - 9 = 935$, which is divisible by 11.

Hence, 9944 is divisible by 11.

Further, the number formed with last three digits of the number i.e. 944 is divisible by 8. Hence, 9944 is divisible by 8.

Thus, option (b) is correct.

22. We know, the product of any n consecutive numbers is divisible by $n!$. Here, $n = 4$, then the desired divisor $= 4! = 4 \times 3 \times 2 \times 1 = 24$.

Since 24 is largest among the options, option (d) is correct.

23. A number is divisible by 8, if the number formed with the last three digits of that number is divisible by 8. Hence, value of B may be 0 or 8.

Further, a number is divisible by 9, if sum of all digits is divisible by 9.
Now, sum of digits

$= B + 6 + 9 + A + 8 + 5 + 9 + 4 + 7 + 7$

$= 55 + A + B$.

In order to make $55 + A + B$ divisible by 9, the value of $(A + B)$ must be 8 or 17 or 23, ...

If $B = 0$, $A = 8$, 17, ...

If B = 8, $A = 0$, 9, 15, ...

Since the value of each of A and B is in one digit, the respective values of A and B = 8, 0 or 0, 8

Thus, option(b) is correct.

24. Sum of digits at odd places

$= N + 5 + 8 + 0 + 3$

$= 16 + N$

and sum of digits at even places

$= 8 + 4 + 4 + 9 + M$

$= 25 + M$

Then, their difference

$= (25 + M) - (16 + N)$

$= 9 + (M - N)$

Clearly, at $(M - N) = 2$, the difference will be divisible by 11. Hence, options (b) and (c) seem to be correct at first instance.

But, the given number should also be divisible by 8.

Option (b) $\Rightarrow N = 6$. Since the number formed with last three digits i.e. 586 is not divisible by 8, the given number cannot be divisible by 8.

Option (c) $\Rightarrow N = 4$. Since the number formed with last three digits i.e. 584 is divisible by 8, the given number is divisible by 8.

Hence, option(c) is correct.

25. A number is divisible by 9, if sum of all digits is divisible by 9.

Now, sum of digits

$= 8 + 4 + 7 + 0 + 5 + \odot + 2 = 26 + \odot$

In order to make $26 + \odot$ divisibly by 9, we should add 1, 10, 19, ... Since we require least value of single digit, option (b) is correct.

26. Divisor $= 2 \times 15 = 30$

Then, quotient $= 30 \div 5 = 6$

Now, Dividend

= (Divisor × Quotient) + Remainder

$= (30 \times 6) + 15 = 195$

27. D) 40051 (

P

1731

Q

294

Clearly, $P = 4005 - 173$

$= 3832$

and $Q = 1731 - 294$

$= 1437$

It is obvious that the divisor 'D' will be HCF of 3832 and 1437.

Hence, value of the divisor = 479

28. For $x35624$,

Sum of digits at odd places

$= 4 + 6 + 3 = 13$

and sum of digits at even places

$= 2 + 5 + x = 7 + x$

Then, their difference must be divisible by 11 as the given number is divisible by 11.

At $x = 6$, difference is 0, which is divisible by 11.

For $1257y4$,

Since it is divisible by 72, it must be divisible by both 8 and 9.

If $1257y4$ is divisible by 9, $(1 + 2 + 5 + 7 + y + 4)$ must be divisible by 9. Obviously, y must be 8. Further, the number $7y4$ must be divisible by 8 if the given number is divisible by 8. At $y = 8$, $7y4$ is divisible by 8. Hence, at $y = 8$, the given number is divisible by 72.

Now, $(5x - 2y) = (5 \cdot 6 - 2 \cdot 8) = 14$

29. Given, $\frac{1}{4} < 2^x < 200$

Clearly, value of x

$= -2, -1, 0, 1, 2, 3, 4, 5, 6, 7$

Further, $2^x + 2$ will be divisible by 3 only if x is 0 or even positive integers.

Hence, $x = 0, 2, 4, 6$

$2^x + 2$ will be divisible by 4 only if x is 1.

Then, number of possible values of $x = 5$ (i.e. 0, 1, 2, 4, 6)

05

How to Compute Remainder

This chapter is basically the extension of chapter 'Division'. We will stress on diverse cases of divisions. If $a, b, c, \ldots\ldots$ are divided by d and $a_r, b_r, c_r, \ldots\ldots\ldots$ are remainders respectively , then:

(A) $\frac{a \times b \times c \times \ldots}{d}$ leaves the remainder what is left by $\frac{a_r \times b_r \times c_r \times \ldots}{d}$

For example What is remainder if $(10 \times 15 \times 18)$ is divided by 7.

Clearly, $\frac{10}{7}$ leaves the remainder 3, $\frac{15}{7}$ leaves 1 and $\frac{18}{7}$ leaves 4.

Then, $\frac{3 \times 1 \times 4}{7}$ will leave the remainder 5.

Hence, $(10 \times 15 \times 18)$ will leave remainder 5 if it is divided by 7.

(B) $\frac{a + b + c + \ldots}{d}$ leaves the remainder what is left by $\frac{a_r + b_r + c_r + \ldots}{d}$

For example Find the remainder if $(10 + 15 + 18)$ is divided by 7.

Clearly, $\frac{10}{7}$ leaves the remainder 3, $\frac{15}{7}$ leaves 1 and $\frac{18}{7}$ leaves 4.

Then, $\frac{3 + 1 + 4}{7}$ will leave the remainder 1.

Hence, $(10 + 15 + 18)$ will leave remainder 1 if it is divided by 7.

(C) $\frac{a - b}{d}$ leaves the remainder what is outcome on subtracting the remainder obtained by dividing b by d from the remainder obtained by dividing a by d. In case the outcome is negative, the required remainder is (d + negative figure).

For example The remainder left by

$\frac{128 - 76}{12}$ = (Remainder left by 128/12) – (Remainder left by 76/12)

$= 8 - 4 = 4$

Inside

- *Brief Introduction*
- *Computation of Remainder in Diverse Cases with Illustrations*
- ....
- .... *& Many More*
- *Work Out with Solutions*

Another example Remainder in case of $\frac{147-86}{9}$

= (Remainder left by 147/9) – (Remainder left by 86/9)

= 3 – 5 = – 2

Since the outcome is negative, the required remainder = 9 + (– 2) = 7

(D) If $\frac{a}{d}$ leaves the remainder r, then $\frac{a}{\text{any factor of } d}$ will leave the same remainder as the remainder left by $\frac{r}{\text{that factor}}$.

For example A number, when divided by 114 leaves remainder 21. If the same number is divided by 19, then the remainder will be: ***[SSC(CGL-T-I)-10]***

Since 19 is a factor of 114, the required remainder = remainder in case of $\frac{21}{19} = 2$.

Illustrations Based on Fact A

Problem 1 Find the remainder for $\frac{86 \times 113 \times 219 \times 415 \times 320}{17}$

Sol. The required remainder = Remainder in case of $\frac{1 \times 11 \times 15 \times 7 \times 14}{17}$

[∵ 1, 11, 15, 7 and 14 are respective remainders]

= Remainder in case of $\frac{165 \times 98}{17}$ = Remainder in case of $\frac{12 \times 13}{17}$

[∵ 12 and 13 are remainders for 165/17 and 98/17 respectively]

= Remainder in case of $\frac{156}{17} = 3$.

Problem 2 Find the remainder when $(9)^{648}$ is divided by 5.

Sol. $\frac{9 \times 9 \times 9 \times 9 \times \ldots 648 \text{ times}}{5}$

$\Rightarrow \frac{4 \times 4 \times 4 \times 4 \times \ldots 648 \text{ times}}{5}$ [∵ 9/5 leaves 4 as remainder]

[here, finding multiplication of 4s (648 times) is very difficult, thus we should combine two or more terms for sake of convenience . If we combine two terms we will get 1 as remainder and multiplication of 1s (324 times) is very easy.]

$\Rightarrow \frac{(4 \times 4) \times (4 \times 4) \times \ldots 324 \text{ pairs}}{5}$

$\Rightarrow \frac{1 \times 1 \times 1 \times 1 \times \ldots 324 \text{ times}}{5}$

[∵ (4 × 4)/5 leaves 1 as remainder]

$\Rightarrow \frac{1}{5}$

Then, remainder = 1 if 1 is divided by 5.

Hence, required remainder = 1.

Alternatively,

If dividend is in form of $(\text{base})^{\text{exponent}}$, we should rewrite it in such a way that new base leaves the remainder 1 when divided by the given divisor.

$\Rightarrow \frac{(9)^{648}}{5} = \frac{(9^2)^{324}}{5} \Rightarrow \frac{(1)^{324}}{5}$ [∵ $\frac{(9)^2}{5}$ gives remainder 1]

$\Rightarrow 1$

Hence, required remainder = 1.

[**Remarks** If the base (here, 9) is greater than the divisor (here, 5), we should write remainder as the base for sake of convenience and proceed on the above line, *viz.* $\frac{(9)^{648}}{5} \Rightarrow \frac{(4)^{648}}{5}$

$\Rightarrow \frac{(4^2)^{324}}{5} \Rightarrow \frac{(1)^{324}}{5}$ [∵ $\frac{(4)^2}{5}$ gives remainder 1]

$\Rightarrow 1$

Problem 3 Find the remainder for $\frac{(5)^{344}}{31}$.

Sol. Clearly, $\frac{(5)^3}{31}$ gives remainder 1. Now, we should rewrite the dividend as below:

$\Rightarrow \frac{(5^3)^{114} \times 5^2}{31} \Rightarrow \frac{(1)^{114} \times 5^2}{31}$

$\Rightarrow \frac{1 \times 25}{31} \Rightarrow 25$

Hence, required remainder = 25.

Problem 4 Find the remainder for $\frac{(3)^{149}}{121}$.

Sol. Clearly, $\frac{(3)^5}{121}$ gives remainder 1. Now, we should rewrite the dividend as below:

$\Rightarrow \frac{(3^5)^{29} \times 3^4}{121} \Rightarrow \frac{(1)^{29} \times 3^4}{121}$

$\Rightarrow \frac{1 \times 81}{121} \Rightarrow 81$

Hence, required remainder = 81.

Problem 5 Find the remainder for $\frac{(2292)^{5555}}{7}$.

Sol. $\frac{(2292)^{5555}}{7} \Rightarrow \frac{(3)^{5555}}{7}$ [∵ 2292/7 leaves 3 as remainder]

$\Rightarrow \frac{(3^6)^{925} \times 3^5}{7}$ [∵ $\frac{(3)^6}{7}$ leaves remainder 1]

$\Rightarrow \frac{(1)^{925} \times 5}{7}$ [∵ $\frac{(3)^5}{7}$ leaves remainder 5]

$\Rightarrow 5$

Hence, required remainder = 5.

Problem 6 When a positive integer is divided by d, the remainder is 15. When ten times of the same number is divided by d, the remainder is 6. The least possible value of d is: ***[SSC (CGL-T-I)-19]***

Sol. Clearly, $\frac{15 \times 10}{d}$ will leave the remainder 6.

It implies that (150 – 6) i.e. 144 is divisible by d.

Since the remainder, in the first case, is 15, the value of d must be greater than 15. Further, we have to find least value of d, a factor of 144 just above 15 will be the required answer. Obviously, 16 is the least possible value of d.

Problem 7 Find the remainder for $\frac{(79)^{439}}{78}$.

Sol. $\frac{(79)^{439}}{78} \Rightarrow \frac{(1)^{439}}{78}$ [$\because$ 79/78 leaves remainder 1]

$\Rightarrow 1$

Problem 8 Find the remainder for $\frac{(83)^{112}}{84}$.

Sol. $\frac{(83)^{112}}{84} \Rightarrow \frac{(83^2)^{56}}{84}$ [$\because \frac{(83)^2}{84}$ leaves remainder 1]

$\Rightarrow \frac{(1)^{56}}{84} \Rightarrow 1.$

Problem 9 Find the remainder for $\frac{(83)^{143}}{84}$.

Sol. Clearly, $\frac{(83)^2}{84}$ leaves remainder 1. Then, we should rewrite the dividend as below:

$\Rightarrow \frac{\{(83)^2\}^{71} \times 83}{84} \Rightarrow \frac{(1)^{71} \times 83}{84} \Rightarrow \frac{1 \times 83}{84} \Rightarrow 83.$

Observations from the above

(i) Remainder in case of $\frac{(n+1)^{\text{odd or even}}}{n}$ is always 1.

(ii) Remainder in case of $\frac{(n)^{\text{even}}}{(n+1)}$ is always 1.

(iii) Remainder in case of $\frac{(n)^{\text{odd}}}{(n+1)}$ is always n.

Problem 10 If $(17)^{200}$ is divided by 18, the remainder is:

Sol. $\frac{(17)^{200}}{18} \Rightarrow \frac{(17^2)^{100}}{18}$ [$\because \frac{(17)^2}{18}$ leaves remainder 1]

$\Rightarrow \frac{(1)^{100}}{18} \Rightarrow 1.$

Alternatively,

Remainder in case of $\frac{(n)^{\text{even}}}{(n+1)}$ is always 1.

Hence, the required answer is 1.

Problem 11 When $(2)^{33}$ is divided by 10, the remainder will be **[SSC (CGL-T-I)-11]**

Sol. $\frac{(2)^{33}}{10} \Rightarrow \frac{(2^5)^6 \times 2^3}{10}$

$\Rightarrow \frac{(2)^6 \times 2^3}{10} \Rightarrow \frac{512}{10} \Rightarrow 2.$

Alternatively,

Obviously, the unit digit of $(2)^{33}$ will be the required remainder as the divisor ends with 0.

$\Rightarrow$ 33/4 gives 1 as remainder

Then, unit digit = $2^1 = 2$ [Recall the concept of unit digit]

Hence, required answer = 2.

Problem 12 The remainder when $(3)^{21}$ is divided by 5 is **[SSC(CGL-T-I)-11]**

Sol. $\frac{(3)^{21}}{5} \Rightarrow \frac{(3^4)^5 \times 3}{5}$ [$\because \frac{(3)^4}{5}$ leaves remainder 1]

$\Rightarrow \frac{(1)^5 \times 3}{5} \Rightarrow 3.$

Problem 13 When n is divisible by 5, the remainder is 2. What is the remainder when n^2 is divided by 5? **[SSC(CGL-T-I)-11]**

Sol. $\frac{(n)^2}{5} \Rightarrow \frac{n \times n}{5}$

$\Rightarrow \frac{\text{Remainder in case of } \frac{n}{5} \times \text{Remainder in case of } \frac{n}{5}}{5}$

$\Rightarrow \frac{2 \times 2}{5} \Rightarrow 4$

Problem 14 When $32^{33^{34}}$ is divided by 11, the remainder will be?

Sol. $\frac{32^{33^{34}}}{11}$

$\Rightarrow \frac{32^{\text{an odd number}}}{11}$ [$(\text{odd})^{\text{even}}$ is always an odd number]

$\Rightarrow \frac{(32)^{\text{even}} \times (32)^1}{11}$

$\Rightarrow \frac{\{(32)^2\}^{\text{even or odd}} \times (32)^1}{11}$ [$\because \frac{(32)^2}{11}$ leaves remainder 1]

$\Rightarrow \frac{(1)^{\text{even or odd}} \times (32)^1}{11} \Rightarrow 10$

Illustrations Based on Fact B

Problem 15 $(7^{19} + 2)$ is divided by 6, the remainder is

Sol. Remainder in case of $(7^{19} \div 6) = 1$

and remainder in case of $(2 \div 6) = 2$

Then, required remainder = reminder in case of $\{(1 + 2) \div 6\} = 3.$

Problem 16 What is the remainder left after dividing 1! + 2! + 3! + ...+ 100! by 7? **[MAT-03]**

Sol. Clearly, 7! + 8! + 9! + ...+ 100! will exactly be divisibly by 7 as each term has 7 as a factor. Hence, no remainder.

Further, remainder in case of $\{(1! + 2! + 3! + \ldots\ldots + 6!) \div 7\}$

= Remainder in case of $\{(R \text{ on } \frac{1!}{7} + R \text{ on } \frac{2!}{7} + \ldots + R \text{ on } \frac{6!}{7}) \div 7\}$

= Remainder in case of $\{(1 + 2 + 6 + 3 + 1 + 6) \div 7\} = 5.$

Illustrations Based on Fact C

Problem 17 $(9^6 - 11)$ when divided by 8 would leave a remainder of

Sol. Remainder in case of $\frac{(9^6 - 11)}{8}$

= Remainder in case of $(9^6 \div 8)$ – remainder in case of $(11 \div 8)$

$= 1 - 3 = -2$

Since the outcome is negative, the required remainder

$= 8 + (-2) = 6.$

Illustrations Based on Fact D

Problem 18 A number, when divided by 899, leaves remainder 63. What will be the remainder if the same number is divided by 29? *SSC(CGL-PT)-08*

Sol. Required remainder = Remainder in case of $(63 \div 29)$ as 29 is a factor of 899 = 5.

Problem 19 A number divided by 56 gives 29 as remainder. If the same number is divided by 8, the remainder is:

Sol. Required remainder = Remainder in case of $(29 \div 8)$ as 8 is a factor of 56 = 5.

Problem 20 A number divided by 64 gives 7 as remainder. If a number 9 more than the former is divided by 8, the remainder is:

Sol. The remainder if the same number as earlier is divided by 8 = Remainder in case of $(7 \div 8)$ as 8 is a factor of 64 = 7

Then, remainder in case of {(earlier number + 9) ÷ 8} =

$$\left\{\frac{\text{Remainder in case of (earlier number} \div 8) + \text{Remainder in case of } (9 \div 8)}{8}\right\}$$

= Remainder in case of $\left(\frac{7+1}{8}\right) = 0.$

Other Critical Facts

(i) The remainder in case $\{a^{E(n)} \div n\}$ is always 1, where $E(n)$ is Euler number of n and a and n are co-prime to each other. Euler number of n refers to the number of numbers which are less than n and co-prime to n. For example, Euler number of 10 is 4 as each of 1, 3, 7, 9 is lesser than 10 and co-prime to 10.

Euler number of n

$$= E(n) = n \times \left(1 - \frac{1}{\text{1st prime factor}}\right) \times \left(1 - \frac{1}{\text{2nd prime factor}}\right) \times \left(1 - \frac{1}{\text{3rd prime factor}}\right) \times \ldots\ldots$$

Since 10 has two prime factors i.e. 2 and 5,

$$E(10) = 10 \times \left(1 - \frac{1}{2}\right) \times \left(1 - \frac{1}{5}\right) = 4$$

For example The remainder in case of $\{17^8 \div 20\}$ is 1 as 8 is Euler number of 20 and also, 17 and 20 are co-prime to each other.

(ii) Since Euler number of a prime number is always 1 less than the number, the remainder in case $\{a^n \div (n+1)\}$ shall be 1 if $(n + 1)$ is a prime number and a and $(n + 1)$ are co-prime to each other.

For example Find the remainder if 2^{1040} is divided by 131 ?

$$\frac{2^{1040}}{131} \Rightarrow \frac{(2^{130})^8}{131} \Rightarrow \frac{(1)^8}{131}$$

$\left[\because \frac{2^{130}}{131}\right.$ leaves the remainder 1 adhering to the above fact.]

$\Rightarrow 1$

Hence, required remainder = 1

(iii) The remainder in case of $\{n! \div (n+1)\}$ is always n if $(n + 1)$ is a prime number. Also, the remainder in case $\{(n-1)! \div (n+1)\}$ is always 1, if $(n + 1)$ is a prime number.

For example The remainder in case of $\frac{18!}{19}$ will be 18 and the remainder in case of $\frac{17!}{19}$ will be 1.

(iv) The remainder obtained when a prime number greater than 6 is divided by 6, is either 1 or 5.

(v) Rules of divisibility can be employed for computation of the remainder. For example: (a) remainder in case of (given number ÷ 4) = remainder in case of {(number formed with last two digits of that number) ÷ 4} (b) remainder in case of (given number ÷ 8) = remainder in case of {(number formed with last three digits of that number) ÷ 8} (c) remainder in case of (given number ÷ 9) = remainder in case of {(sum of digits of that number) ÷ 9}, etc.

WORKOUT

1. What is the remainder when $(1235 \times 4523 \times 2451)$ is divided by 12 ? ***[CDS-14]***

(a) 1 (b) 3 (c) 5 (d) 7

2. Let $n! = 1 \times 2 \times 3 \times \ldots \times n$ for integer $n \geq 1$. If $p = 1! + (2 \times 2!) + (3 \times 3!) + \ldots + (10 \times 10!)$, then $p + 2$ when divided by 11! leaves a remainder of ***[CAT-05]***

(a) 0 (b) 10 (c) 7 (d) 1

3. $(9^6 + 7)$, when divided by 8, would have a remainder of

(a) 0 (b) 5
(c) 6 (d) None of these

4. What is the remainder when 4^{1012} is divided by 7 ?

(a) 4 (b) 1
(c) 2 (d) 3

5. If a positive integer leaves remainder 28 when divided by 143, then what is the remainder obtained on dividing the same number by 13 ? ***[CDS-12, MAT-07, SSC (SAS)-10 SSC(CGL-T-II)-10, CAT-98]***

(a) 0 (b) 9
(c) 2 (d) 10

6. The remainder when $(7)^{84}$ is divided by 342 is ***[MAT-03, CAT-99]***

(a) 0 (b) 1 (c) 49 (d) 342

7. What is the remainder when the number $(4444)^{4444}$ is divided by 9? ***[CDS-17]***

(a) 4 (b) 6 (c) 7 (d) 8

8. What is the remainder when $7^{74} - 5^{74}$ is divided by 4 ? ***[CAT-09]***

(a) 0 (b) 1
(c) 2 (d) None of these

9. $(67^{67} + 67)$ is divided by 68, the remainder is ***[SSC(CGL-T-II)-12]***

(a) 1 (b) 63 (c) 66 (d) 67

10. When a positive integer n is divided by 5, the remainder is 2. What is the remainder when the number $3n$ is divided by 5 ? ***[CDS-11, 09, SSC(CGL-T-II)-11]***

(a) 1 (b) 2 (c) 3 (d) 4

11. When 2^{256} is divided by 17, the remainder would be ***[CAT-02]***

(a) 1
(b) 16
(c) 14
(d) None of the above

12. The remainder on dividing given integers a and b by 7 are, respectively 5 and 4. What is the remainder when ab is divided by 7?

(a) 3 (b) 4 (c) 5 (d) 6

13. What is the remainder when 4^{96} is divided by 6 ? ***[CAT-03]***

(a) 0 (b) 2 (c) 3 (d) 4

14. The numbers 1 to 29 are written side by side as follows:

1234567891011.......2829

If the number is divided by 9, then what is the remainder?

(a) 0 (b) 1
(c) 3 (d) None of these

15. A hundred digit number is formed by writing first 54 natural numbers in front of each other as 12345678910111213......... .Find the remainder when this number is divided by 8. ***[CAT-98]***

(a) 1 (b) 7 (c) 2 (d) 0

16. What is the remainder if $(19^{20} \div 20^{19})$ is divided by 7? ***[XAT-20]***

(a) 1 (b) 3 (c) 0 (d) 5

Answer Key

1. (b)	2. (d)	3. (a)	4. (a)	5. (c)	6. (b)	7. (c)	8. (a)	9. (c)	10. (a)
11. (a)	12. (d)	13. (d)	14. (c)	15. (a)	16. (d)				

Hints and SOLUTIONS

1. The required remainder = Remainder in case of $\frac{11 \times 11 \times 3}{12}$

[$\because$ 11, 11 and 3 are respective remainders]

$= \text{Remainder in case of } \frac{11 \times 33}{12}$

$= \text{Remainder in case of } \frac{11 \times 9}{12}$

[$\because$ 11 and 9 are remainders for 11/12 and 33/12 respectively]

$= \text{Remainder in case of } \frac{99}{12} = 3$

2. $n \times n! = (n + 1 - 1) \times n! = (n + 1)! - n!$

Then, $1 \times 1! = (1 + 1)! - 1! = 2! - 1!$ and $2 \times 2! = (2 + 1)! - 2! = 3! - 2!$ and so on.

Now, $p = 1! + (2 \times 2!) + (3 \times 3!) + \ldots + (10 \times 10!)$

$= (2! - 1!) + (3! - 2!) + (4! - 3!) + \ldots + (11! - 10!)$

$= (11! - 1!)$

$\therefore p + 2 = (11! - 1!) + 2$

$= (11! + 1)$

Obviously, $(11! + 1)$ will leave remainder 1 when divided by 11!.

3. Remainder in case of $(9^6 \div 8) = 1$ and remainder in case of $(7 \div 8) = 7$

Then, required remainder = remainder in case of $\{(1 + 7) \div 8\} = 0$

4. Clearly, $\frac{(4)^3}{7}$ leaves remainder 1.

Now, we should rewrite the dividend as below

$\Rightarrow \frac{(4^3)^{337} \times 4^1}{7}$

$\Rightarrow \frac{(1)^{337} \times 4^1}{7}$

$\Rightarrow \frac{1 \times 4}{7} \Rightarrow 4$

Hence, required remainder = 4

5. Required remainder = Remainder in case of $(28 \div 13)$ as 13 is a factor of 143 = 2.

6. $\frac{(7)^{84}}{342} \Rightarrow \frac{(7^3)^{28}}{342}$

$[\because \frac{(7)^3}{342}$ leaves remainder 1]

$\Rightarrow \frac{(1)^{28}}{5} \Rightarrow 1$

7. $\frac{(4444)^{4444}}{9}$

$\Rightarrow \frac{(7)^{4444}}{9}$

$[\because \frac{4444}{9}$ leaves the remainder 7]

$\Rightarrow \frac{(7^3)^{1481} \times 7}{9}$

$\Rightarrow \frac{(1)^{1481} \times 7}{9}$

$[\because \frac{7^3}{9}$ leaves the remainder 1]

$\Rightarrow \frac{1 \times 7}{9} \Rightarrow 7$

8. Remainder in case of $(7^{74} \div 4) = 1$

$[\because \frac{(7)^{74}}{4} \Rightarrow \frac{(7^2)^{37}}{4} \Rightarrow \frac{(1)^{37}}{4}$ $(\because \frac{(7)^2}{4}$ leaves the remainder 1) $\Rightarrow 1]$

Remainder in case of $(5^{74} \div 4) = 1$

$[\because \frac{(5)^{74}}{4} \Rightarrow \frac{(1)^{74}}{4}$ $(\because \frac{5}{4}$ leaves the remainder 1) $\Rightarrow 1]$

Then, required remainder = 1 − 1 = 0

9. Remainder in case of $(67^{67} \div 68) = 67$

$[\because$ Remainder in case of $\frac{(n)^{odd}}{(n+1)}$ is always $n]$

and remainder in case of $(67 \div 68) = 67$

Then, required remainder

= remainder in case of $\{(67 + 67) \div 68\}$

= 66

10. Remainder in case of $\frac{3 \times n}{5}$

$= \frac{\text{Remainder in case of } (3 \div 5) \times \text{Remainder in case of } (n \div 5)}{5}$

$\Rightarrow$ Remainder in case of $\frac{3 \times 2}{5} = 1$

Alternatively,

You may take a suitable number (say, 7) and find the remainder dividing (3×7) by 5.

11. $\frac{(2)^{256}}{17} \Rightarrow \frac{(2^{16})^{16}}{17} \Rightarrow \frac{(1)^{16}}{17}$

$[\because \frac{2^{16}}{17}$ leaves remainder 1 as remainder in case $\{a^n \div (n+1)\}$ is 1 if $(n+1)$ is a prime number and a and $(n+1)$ are co-prime to each other]

$\Rightarrow 1$

12. Remainder in case of $\frac{a \times b}{7}$

= Remainder in case of $\frac{\text{Remainder in case of } (a \div 7) \times \text{Remainder in case of } (b \div 7)}{7}$

= Remainder in case of $\frac{5 \times 4}{7} = 6$

Alternatively,

You may take a suitable number (say, $a = 12$, $b = 11$) and find the remainder dividing (12×11) by 7.

13. 4^1 leaves remainder 4 if it is divided by 6.

4^2 leaves remainder 4 if it is divided by 6.

4^3 leaves remainder 4 if it is divided by 6.

4^4 leaves remainder 4 if it is divided by 6.

4^5 leaves remainder 4 if it is divided by 6.

4^6 leaves remainder 4 if it is divided by 6.

....

....

Then, 4^{96} will also leave remainder 4 if it is divided by 6.

14. Sum of digits from numbers 1 to 10 = 46

Sum of digits from numbers 11 to 20 = 56

Sum of digits from numbers 21 to 29 = 63

$\therefore$ Sum of digits in the given number = 46 + 56 + 63 = 165

Then, remainder in case 165 is divided by 9 = 3

Hence, the required answer = 3

15. Clearly, the last three digits will be 545. Recall the concept of divisibility of a number by 8. The same rule is applicable here too.

Remainder in case of $(545 \div 8) = 1$

Hence, the required remainder = 1

16. For remainder in case of $(19^{20} \div 7)$

$\frac{(19^2)^{10}}{7} \Rightarrow \frac{(4)^{10}}{7}$

$[\because \frac{(19)^2}{7}$ leaves the remainder 4]

$\Rightarrow \frac{(4^3)^3 \times 4}{7} \Rightarrow \frac{(1)^3 \times 4}{7}$

$[\because \frac{(4)^3}{7}$ leaves the remainder 1]

$\Rightarrow \frac{1 \times 4}{7} \Rightarrow 4$

Hence, remainder in case of $(19^{20} \div 7) = 4$

For remainder in case of $(20^{19} \div 7)$:

$\frac{(20^2)^9 \times 20}{7} \Rightarrow \frac{(1)^9 \times 20}{7}$

$[\because \frac{(20)^2}{7}$ leaves the remainder 1]

$\Rightarrow \frac{1 \times 20}{7} \Rightarrow 6$

Hence, remainder in case of $(20^{19} \div 7) = 6$

Then, required remainder

= 4 − 6

= − 2

Since the outcome is negative, the required remainder

= 7 + (− 2) = 5.

06

Concept of Factor and Factor-Theorem

Learning OBJECTIVES

After completing this chapter, aspirants will be able to

- Understand the significance of this chapter in almost all the arithmetic chapters;
- Learn how to compute number of odd/even factors of a number and their sum;
- Know the highest power of a number which can divide a given number exactly;
- Be familiar with Factor-Theorem with its applications;
- Solve diverse types of question on divisibility;
- Observe the problems where some known approaches can be applied after some alternation in form/pattern of problems;
- Understand the philosophy of the division of this chapter in various parts;
- Analyse the trend of the questions being asked in various competitive examinations;
- Solve the questions in moments;
- Grip on concepts helping to solve the questions without pen and paper;
- Learn new concepts/fundamentals;
- Apply the pragmatic approaches in lieu of academic approaches; and
- Ponder on the alternative methods.

Inside

- *Factors of a Number : Introduction*
- *How to find Number of Factors/Divisors of a Composite Number*
- *Sum of Factors of a Number*
- *Computation of Number of Odd Factors of a Number*
- *Computation of Number of Even Factors of a Number*
- *Computation of Highest Power of a Number Dividing a Factorial Number*
- *Factor-Theorem*
- ....
- .... *& Many More*
- *Work Out with Solutions*

A. Factors of a Number : Introduction

A number which divides a given number is called factor of that given number and the given number is called a multiple of that factor.

For example 18 is divisible by 1, 2, 3, 6, 9 and 18. Then, it is said that 1, 2, 3, 6, 9 and 18 are factors of 18 and 18 is a multiple of 1, 2, 3, 6, 9 and 18.

Note *1 and the given number itself (here, 1 and 18) are called improper factors and 2, 3, 6 and 9 are called proper factors.*

Prime Factorisation When a natural number is expressed as the product of prime numbers, the factorisation of the number is called its prime factorisation.

For example

(a) $48 \Rightarrow 2 \times 2 \times 2 \times 2 \times 3$

(b) $36 \Rightarrow 2 \times 2 \times 3 \times 3$

How to find Number of Factors/Divisors of a Composite Number

At outset, the given composite number must be expressed in form of $a^p \times b^q \times c^r \times ....$, where $a, b, c, ...$ are prime numbers. Then, the number of factors will be

$$(p+1)\cdot(q+1)\cdot(r+1)..........$$

For example What is number of factors of 100 ?

$\Rightarrow$ $100 = 2^2 \times 5^2$

Now, total number of factors of $100 = (2+1)\cdot(2+1) = 9$

Note *Perfect square number has odd number of factors. Also, the sum of all factors of a perfect square number is always odd.*

Illustrations

Problem 1 Find the number of factors of 320.

Sol. $320 = 2^6 \times 5^1$

Now, total number of factors of $320 = (6+1)\cdot(1+1) = 14$

Problem 2 Find the number of factors of 1800.

Sol. $1800 = 2^3 \times 3^2 \times 5^2$

Now, total number of factors of 1800

$= (3+1)\cdot(2+1)\cdot(2+1) = 36$

Problem 3 When 200 is divided by a positive integer x, the remainder is 8. How many values of x are there ? ***[SSC(CGL-T-I)-19]***

Sol. Clearly, (200 – 8) i.e., 192 is divisible by x. Since, the remainder is 8, x must be greater than 8. Further, $192 = 2^6 \times 3^1$

Then, number of factors of $192 = (6+1)\cdot(1+1) = 14$

Since factors 1, 2, 3, 4, 6, 8 are not greater than 8, x may take (14 – 6) i.e. 8 values.

Sum of Factors of a Number

If the number is in form of $a^p \times b^q \times c^r \times ....$, (where $a, b, c, ...$ are prime numbers), the sum of all factors will be $\frac{\{a^{(p+1)}-1\}\times\{b^{(q+1)}-1\}\times\{c^{(r+1)}-1\}\times...}{(a-1)(b-1)(c-1)...}$

Problem 4 Find the sum of all factors of 72.

Sol. $72 = 2^3 \times 3^2$

Now, required sum $= \frac{[2^{(3+1)}-1]\times[3^{(2+1)}-1]}{(2-1)(3-1)} = \frac{15 \times 26}{1 \times 2} = 195$

Problem 5 Find the sum of all factors of 140.

Sol. $140 = 2^2 \times 5^1 \times 7^1$

Now, required sum $= \frac{[2^{(2+1)}-1]\times[5^{(1+1)}-1]\times[7^{(1+1)}-1]}{(2-1)(5-1)(7-1)}$

$= \frac{7 \times 24 \times 48}{1 \times 4 \times 6} = 336$

Computation of Number of Odd Factors of a Number

If the number is in form of $a^p \times b^q \times c^r \times ...$, (where a is an even prime number and $b, c, ...$ are odd prime numbers), the number of odd factors $= (q+1)(r+1)...$

Problem 6 What is the number of odd factors of 120 ?

Sol. $120 = 2^3 \times 3^1 \times 5^1$

Here, 3 and 5 are odd prime numbers. Then, number of odd factors of $120 = (1+1)\cdot(1+1) = 4$

Problem 7 What is the number of odd factors of 1700 ?

Sol. $1700 = 2^2 \times 5^2 \times (17)^1$

Here, 5 and 17 are odd prime numbers. Then, number of odd factors of $1700 = (2+1)\cdot(1+1) = 6$

Computation of Number of Even Factors of a Number

If the number is in form of $a^p \times b^q \times c^r \times ...$, (where a is an even prime number and $b, c, ...$ are odd prime numbers), the number of even factors

$$= p(q+1)(r+1)....$$

Alternatively,

Number of even factors = Total number of factors – Total number of odd factors

Problem 8 What is number of even factors of 100?

Sol. $100 = 2^2 \times 5^2$

Now, total number of even factors of $100 = 2(2+1) = 6$

Alternatively,

$\Rightarrow$ $100 = 2^2 \times 5^2$

Now, total number of factors of $100 = (2+1)\cdot(2+1) = 9$

and total number of odd factors $= (2+1) = 3$

Hence, total number of even factors $= 9 - 3 = 6$

Problem 9 What is number of even factors of 540 ?

Sol. $540 = 2^2 \times 3^3 \times 5^1$

Now, total number of even factors of $540 = 2(3+1)(1+1) = 16$

Alternatively,

$\Rightarrow$ $540 = 2^2 \times 3^3 \times 5^1$

Now, total number of factors of

$540 = (2+1)\cdot(3+1)\cdot(1+1) = 24$

and total number of odd factors $= (3+1)(1+1) = 8$

Hence, total number of even factors $= 24 - 8 = 16$

B. Computation of Highest Power of a Number Dividing a Factorial Number

Recall the topic "Finding number of zeros in the end of a product" wherein we have arrived at conclusion that number of 5's in 100! may be found in the following way

$$\Rightarrow \quad \frac{100}{5}+\frac{100}{5^2}+\frac{100}{5^3} \Rightarrow 20+4+0=24$$

Hence, there are twenty four 5's in 100!. It implies that 5^{24} may divide 100! exactly. In other words, 24 is the maximum power to 5 which may divide 100! exactly.
The above process is also applicable/true for the other prime numbers.

For example Find the largest power of 7 that can divide 1500! exactly.

$$\text{Total number of 7 in } 1500! = \frac{1500}{7}+\frac{1500}{7^2}+\frac{1500}{7^3}+\frac{1500}{7^4}$$

$$= 214+30+4+0=248$$

It implies that 7^{248} may divide 1500! exactly. Hence, required answer = 248

Alternatively,

7	1500
7	214
7	30
7	4
	0

Remarks *Divide successive quotients till we get 0 as last quotient.*
Here, total number of 7's = 214 + 30 + 4 + 0 = 248
Hence, required answer = 248

Illustrations

Problem 10 What is the highest power of 15 which can divide 90! exactly.

Or

If $(15)^n$ divides 90! exactly, then what is the maximum value of n?

Sol. 15 = 3 × 5

Now, for number of 5's in 90!

5	90
5	18
5	3
	0

Hence, number of 5's in 90! = 18 + 3 = 21

Furthor, for number of 3's in 90! →

3	90
3	30
3	10
3	3
3	1
	0

Hence, number of 3's in 90! = 30 + 10 + 3 + 1 = 44
Here, there are twenty one 5's and forty four 3's.
In order to obtain 15, one 5 and one 3 are required; hence, only twenty one 15's can be obtained. It implies that $(15)^{21}$ will divide 90! exactly.
Hence, required answer = 21.

Observation If divisor is in form of b^n and b is in form of a product of all different prime numbers, then maximum value of n depends on the number of greatest prime number. The logic behind it is that number of greatest prime number will be lesser than that of any smaller prime number.

For example In this question no need to compute number of 3's, computation of number of 5's will suffice.

Problem 11 135! is exactly divisible by $(30)^n$.
What is the maximum value of n?

Sol. 30 = 2 × 3 × 5

Now, for number of 5's in 135! ⇒

5	135
5	27
5	5
5	1
	0

Hence, number of 5's = 27 + 5 + 1 + 0 = 33
Here, no need to compute number of 2's and 3's as 30 is a product of all different prime numbers. Thus, the maximum value of n = 33.

Problem 12 What maximum power of 175 will exactly divide 100!?

Sol. 175 = 5 × 5 × 7. Here, 175 is not a product of all different prime numbers.
Now, for number of 7's in 100! ⇒

7	100
7	14
7	2
	0

Now, for number of 7's in 100! = 14 + 2 + 0 = 16
Further, for number of 5's in 100! =

5	100
5	20
5	4
	0

Hence, number of 5's in 100! = 20 + 4 + 0 = 24
Then, number of (5 × 5) = (24 ÷ 2) = 12
Clearly, in order to obtain 175, one (5 × 5) and one 7 are required and we have only twelve (5 × 5). Thus, we can obtain only twelve 175's.

Problem 13 What is the maximum value of n if 8^n divides 139! exactly ?

Sol. $8 = 2 \times 2 \times 2$

Number of 2's in 139! ⇒

2	139
2	69
2	34
2	17
2	8
2	4
2	2
2	1
	0

Hence, number of 2's in 139!

$= 69 + 34 + 17 + 8 + 4 + 2 + 1 + 0 = 135.$

But, we have to find power of 8. It means, we should combine three 2's to obtain one 8.

Thus, maximum power of 8 = (135 ÷ 3) = 45

∴ Maximum value of n = 45

C. Factor-Theorem

(i) $x^n - a^n$ is exactly divisible by $(x - a)$, whether n is odd or even integer. Hence, $(x - a)$ is a factor of $x^n - a^n$.

(ii) $x^n - a^n$ is exactly divisible by $(x + a)$, when n is an even integer. Hence, $(x + a)$ is a factor of $x^n - a^n$.

(iii) $x^n + a^n$ is exactly divisible by $(x + a)$, when n is an odd integer. Hence, $(x + a)$ is a factor of $x^n + a^n$.

(iv) $x^n + a^n$ is never divisible by $(x - a)$. Hence, $(x - a)$ may never be a factor of $x^n + a^n$.

Explanations of the above facts have been detailed in section 'Algebra'.

Illustrations based on the above facts

Problem 14 $(49)^{15} - 1$ is exactly divisible by

(a) 50 (b) 51

(c) 29 (d) 8

Sol. (d) $(49)^{15} - 1 = (49)^{15} - (1)^{15}$

We know, $x^n - a^n$ is exactly divisible by $(x - a)$, whether n is odd or even integer. Hence, $(49)^{15} - (1)^{15}$ will exactly be divisible by (49 – 1). Further, a number divisible by d, is also divisible by all of factors of d. It means, $(49)^{15} - 1$ is not only divisible by 48 but also by all factors of 48. Thus, option (d) is correct.

On the above line, the number $[(37)^{371} - (26)^{371}$ is exactly divisible by (37 – 26) i.e., 11 .... asked in **[IIFT- 19]**

Problem 15 A common factor of $[(13)^7 + (11)^7]$ and $[(13)^5 + (11)^5]$ is

(a) 24 (b) $(13)^5 + (11)^5$

(c) $(13)^2 + (11)^2$ (d) None of these

Sol. $(13)^7 + (11)^7$ will exactly be divisible by (13 + 11) as 7 is an odd integer. Further, $(13)^5 + (11)^5$ will also be divisible by (13 + 11) exactly as 5 is also an odd integer. Hence, common factor = (13 + 11) = 24

Problem 16 $2^{16} - 1$ is exactly divisible by **[SSC(CGL-T-I)-11]**

(a) 11 (b) 13

(c) 17 (d) 19

Sol. Since, $x^n - a^n$ is exactly divisible by $(x - a)$ and $(x + a)$) in case n is an even integer, $2^{16} - 1^{16}$ will be divisible by both (2 – 1) and (2 + 1). But, 3 is not in options.

Now, $2^{16} - 1$ may be written as $(2^4)^4 - 1$ i.e., $16^4 - 1^4$.

Obviously, $16^4 - 1^4$ will exactly be divisible by both (16 – 1) and (16 + 1) as 4 is an even integer. Hence, option (c) is correct.

Alternatively,

You may find the value of $2^{16} - 1$ and test its divisibility by the given options.

Problem 17 $(19)^5 + (21)^5$ is divisible by **[CDS-13]**

(a) Only 10 (b) Only 20

(c) Both 10 and 20 (d) Neither 10 nor 20

Sol. We know, $x^n + a^n$ is exactly divisible by $(x + a)$, when n is an odd integer. It means, $(19)^5 + (21)^5$ is divisible by (19 + 21) i.e. by 40. Further, a number divisible by d, is also divisible by all of factors of d. It implies that $(19)^5 + (21)^5$ is also divisible by 20 and 10.

Hence, option (c) is correct.

Miscellaneous Problems Based on Divisibility

Problem 18 $4^{61} + 4^{62} + 4^{63} + 4^{64}$ is divisible by

(a) 3 (b) 10

(c) 11 (d) 13

Sol. $4^{61} + 4^{62} + 4^{63} + 4^{64} = 4^{61}(1 + 4^1 + 4^2 + 4^3)$

$= 4^{61} \times 85 = 4 \times 4^{60} \times 85 = 340 \times 4^{60}$,

which will obviously be divisible by 10 as it ends with 0.

Problem 19 If n is a whole number greater than 1, then $n^2(n^2 - 1)$ is always divisible by **[SSC(CPO-SI)-05, CDS-14]**

(a) 16 (b) 12

(c) 10 (d) 8

Sol. In such types of problems we should test the divisibility of given term by putting appropriate values of n.

When, $n = 2$, the value of $n^2(n^2 - 1) = 12$

When, $n = 3$, the value of $n^2(n^2 - 1) = 72$

When, $n = 4$, the value of $n^2(n^2 - 1) = 240$

When, $n = 5$, the value of $n^2(n^2 - 1) = 600$

Conclusion may be drawn from the above that all the values of term are divisible by 12. Hence, $n^2(n^2 - 1)$ is always divisibly by 12.

Problem 20 If b is the largest square divisor of c and a^2 divides c, then which one of the following is correct (where, a, b and c are integers)? *[CDS-13]*

(a) b divides a (b) a does not divide b
(c) a divides b (d) a and b are co-prime

Sol. Obviously, a^2 is either equal to or less than b.
Since, b is the largest square divisor, $c = bx$ {where, x is not a whole square number}.
Further, a^2 divides c. Hence, a^2 divides bx. It implies that a divides b because x, being not a whole square number, is not divisible by a^2.

Problem 21 x is a positive integer such that $x^2 + 12$ is exactly divisible by x. Find all the possible values of x.

Sol. Since $x^2 + 12$ is exactly divisible by x, 12 must be divisible by x. Hence, the possible values of $x = 1, 2, 3, 4, 6$ and 12.

WORKOUT

1. What is the number of divisors of 360? *[CDS-14]*
(a) 12 (b) 18
(c) 24 (d) None of these

2. Which one of the following has least number of divisors?
(a) 88 (b) 91
(c) 96 (d) 99

3. For a positive integer n, define $d(n)$ = The number of positive divisors of n. What is the value of $d[d\{d(12)\}]$?
(a) 1 (b) 2
(c) 4 (d) None of these

4. The highest power of 9 dividing 99! completely is
(a) 20 (b) 24
(c) 22 (d) 11

5. What is the greatest power of 5 which can divide 80! exactly?
(a) 15 (b) 16
(c) 19 (d) 13

6. Which one of the following will completely divide $5^{71} + 5^{72} + 5^{73}$? *[SSC(CGL-T-I)-11, SSC (CGL-T-II)-20]*
(a) 150 (b) 160
(c) 155 (d) 30

7. $3^{25} + 3^{26} + 3^{27} + 3^{28}$ is divisible by
(a) 11 (b) 16
(c) 25 (d) 30

8. Which one of the following is correct?
$a^n + b^n$ is divisible by $a - b$
(a) for all integral value of n
(b) when n is an even integer
(c) when n is an odd integer
(d) for no integral value of n

9. The number $6n^2 + 6n$ for natural number n is always divisible by *[MAT-05]*
(a) 6 (b) 18
(c) 12 (d) 6 and 12

10. For two positive integers a and b, if $(a + b)^{a+b}$ is divisible by 500, then the least possible value of $a \times b$, is *[XAT-16]*
(a) 8
(b) 9
(c) 12
(d) 10
(e) None of the above

11. The number of positive integers n in the range $12 \le n \le 40$ such that the product $(n - 1)(n - 2) \dots 3 \cdot 2 \cdot 1$ is not divisible by n is
(a) 5 (b) 7
(c) 13 (d) 14

12. The remainder, when $(15^{23} + 23^{23})$ is divided by 19, is *[CAT-04]*
(a) 0 (b) 4
(c) 15 (d) 18

13. If $x = (16^3 + 17^3 + 18^3 + 19^3)$, then x divided by 70 leaves a remainder of *[CAT-05]*
(a) 0 (b) 1 (c) 69 (d) 35

14. The largest number amongst the following that will perfectly divide $101^{100} - 1$ is
(a) 100 (b) 10000
(c) 100^{100} (d) 100000

15. What is the remainder when $(17^{23} + 23^{23} + 29^{23})$ is divided by 23?
(a) 0 (b) 1 (c) 21 (d) 22

16. $7^{6n} - 6^{6n}$, where n is an integer > 0, is divisible by *[CAT-02]*
(a) 13 (b) 127
(c) 559 (d) None of these

17. For the product $n(n + 1)(2n + 1)$, $n \in N$, which one of the following is not necessarily true? *[CAT-95]*
(a) It is even
(b) Divisible by 3
(c) Divisible by the sum of the squares of first n natural numbers
(d) Never divisible by 237

18. If $22^3 + 23^3 + 24^3 + \dots + 87^3 + 88^3$ is divided by 110, then the remainder will be
(a) 0 (b) 1
(c) 55 (d) 44

Answer Key

1. (c)	2. (b)	3. (d)	4. (b)	5. (c)	6. (c)	7. (d)	8. (d)	9. (d)	10. (b)
11. (b)	12. (a)	13. (a)	14. (b)	15. (a)	16. (b)	17. (d)	18. (c)		

Hints and SOLUTIONS

1. $360 = 2^3 \times 3^2 \times 5^1$

Then, total number of divisors of 360

$= (3+1)\cdot(2+1)\cdot(1+1) = 24$

2. (a) $88 \Rightarrow 2^3 \times 11^1$, then number of factors $= (3+1)(1+1) = 12$

(b) $91 \Rightarrow 7^1 \times 13^1$, then number of factors $= (1+1)(1+1) = 4$

(c) $96 \Rightarrow 2^5 \times 3^1$, then number of factors $= (5+1)(1+1) = 12$

(d) $99 \Rightarrow 3^2 \times 11^1$, then number of factors $= (2+1)(1+1) = 6$

Hence, 91 has the least number of divisors.

3. $d[d\{d(12)\}] = d[d\{6\}]$

[$\because$ positive divisors of 12 are 1, 2, 3, 4, 6 and 12]

$= d\,[4]$ [$\because$ positive divisors of 6 are 1, 2, 3 and 6]

$= 3$ [$\because$ positive divisors of 4 are 1, 2 and 4]

4. $9 = 3 \times 3$

For number of 3's in 99!

3	99
3	33
3	11
3	3
3	1
	0

Hence, number of 3's in 99! = 33 + 11 + 3 + 1 + 0 = 48

But, we have to find power of 9. It means, we should combine two 3's to obtain one 9.

Thus, maximum power of $9 = (48 \div 2) = 24$

5.

5	80
5	16
5	3
	0

Hence, greatest power of 5 = 16 + 3 + 0 = 19

6. $5^{71} + 5^{72} + 5^{73} = 5^{71}(1 + 5^1 + 5^2)$

$= 5^{71} \times 31 = 5^{70} \times 5^1 \times 31$

$= 5^{70} \times 155,$

which is obviously exactly divisible by 155.

7. $3^{25} + 3^{26} + 3^{27} + 3^{28} = 3^{25}(1 + 3^1 + 3^2 + 3^3)$

$= 3^{25} \times 40 = 3 \times 3^{24} \times 40$

$= 3^{24} \times 120,$

which is exactly divisible by 30 as 120 is divisible by 30.

8. We know, $x^n + a^n$ is never divisible by $(x - a)$.

Hence, option (d) is correct.

9. $6n^2 + 6n = 6(n^2 + n)$

At $n = 1$, $6(n^2 + n) = 6(1^2 + 1) = 12$, which is divisible by 12.

Further, a number divisible by d, is also divisible by all of factors of d. It implies that $6(n^2 + n)$ is also divisible by 6, a factor of 12.

Hence, option (d) is correct.

10. $(a+b)^{a+b}$ is divisible by 500, it implies that base $(a+b)$ must be a multiple of 10. If we take minimum value of $(a+b)$ i.e. 10, then 10^{10} would be divisible by 500.

Thus, the least value of $ab = 1 \times 9 = 9$.

11. Product $(n-1)(n-2)$ .... $3 \cdot 2 \cdot 1$ will not be divisible by n only when n is a prime number.

Now, number of prime numbers from 12 to 40 = 7 [i.e. 13, 17, 19, 23, 29, 31, 37]

Hence, required answer = 7

12. We know, $x^n + a^n$ is exactly divisible by $(x + a)$, when n is an odd integer. Hence, $(15^{23} + 23^{23})$ is divisible by $(15 + 23)$ i.e. by 38. Further, a number divisible by d, is also divisible by all of factors of d. It implies that $(15^{23} + 23^{23})$ is also divisible by 19. Hence, remainder = 0

13. We know, $x^n + a^n$ is exactly divisible by $(x + a)$, when n is an odd integer. Hence, $(17^3 + 18^3)$ is divisible by $(17 + 18)$ i.e. by 35. Also, $(16^3 + 19^3)$ is divisible by $(16 + 19)$ i.e. by 35. It means $(16^3 + 17^3 + 18^3 + 19^3)$ is divisible by 35.

Further, $(16^3 + 17^3 + 18^3 + 19^3)$ will have 0 as unit digit as $(\ldots 6 + \ldots 3 + \ldots 2 + \ldots 9) = \ldots .0$.

From above, $(16^3 + 17^3 + 18^3 + 19^3)$ is divisible by both 35 and 10. It implies that $(16^3 + 17^3 + 18^3 + 19^3)$ is a multiple of LCM of 35 and 10 and the fact that their LCM is 70.

Hence, $(16^3 + 17^3 + 18^3 + 19^3)$ is divisible by 70. Then, remainder = 0.

Alternatively,

You may find the cubes of 16, 17, 18 and 19 and divide their sum by 70.

14. $101^{100} = (100 + 1)^{100}$

$$= {}^{100}C_0 \cdot 100^{100-0} \cdot 1^0 + {}^{100}C_1 \cdot 100^{100-1} \cdot 1^1 + {}^{100}C_2 \cdot 100^{100-2} \cdot 1^2 + \ldots + {}^{100}C_{99} \cdot 100^{100-99} \cdot 1^{99} + {}^{100}C_{100} \cdot 100^{100-100} \cdot 1^{100}$$

[Recall binomial theorem, $(a+b)^n = \Sigma_{k=0}^{n} \left\langle {n \atop k} \right\rangle a^{n-k} b^k$,

where $\left\langle {n \atop k} \right\rangle$ refers to n_{c_k}]

$$= {}^{100}C_0 \cdot 100^{100} + {}^{100}C_1 \cdot 100^{99} + {}^{100}C_2 \cdot 100^{98} + \ldots + {}^{100}C_{99} \cdot 100^1 + 1$$

Now, $101^{100} - 1 = {}^{100}C_0 \cdot 100^{100} + {}^{100}C_1 \cdot 100^{99} + {}^{100}C_2 \cdot 100^{98} + \ldots + {}^{100}C_{99} \cdot 100^1$

Clearly, each of the above terms is divisible by 10, 100, 1000, 10000. Since 10000 is the greatest, option(b) is correct.

15. Remainder in case of $\frac{(17^{23} + 23^{23} + 29^{23})}{23}$ = Remainder in case of [{Remainder in case of $\frac{(17^{23} + 29^{23})}{23}$ +remainder in case of $\frac{(23^{23})}{23}$} ÷ 23]

Now, remainder in case of $\frac{(17^{23} + 29^{23})}{23} = 0$

[$\because x^n + a^n$ is exactly divisible by $(x + a)$, when n is an odd integer. Hence, $(17^{23} + 29^{23})$ is divisible by (17 + 29). i.e. by 46. It implies that $(17^{23} + 29^{23})$ is also divisible by 23, a factor of 46]

Further, remainder in case of $\frac{(23^{23})}{23} = 0$

Then, required remainder = remainder in case of [{0 + 0} ÷ 23] = 0

16. $7^{6n} - 6^{6n} = (7^{3n})^2 - (6^{3n})^2$

For $n = 1$, $(7^{3n})^2 - (6^{3n})^2 = (7^3)^2 - (6^3)^2$

$= (343 + 216)(343 - 216) = 559 \times 127$

Hence, $7^{6n} - 6^{6n}$ is divisible by 127.

17. Obviously, at $n = 237$, the product is divisible by 237. Hence, option (d) is correct.

18. Given dividend $= (22^3 + 88^3) + (23^3 + 87^3) + (24^3 + 86^3) + \dots + (54^3 + 56^3) + 55^3$

We know, $x^n + a^n$ is exactly divisible by $(x + a)$, when n is an odd integer. It implies that all the above terms except 55^3 are divisible by 110.

Now, the remainder if 55^3 is divided by 110 = 55

Hence, the required remainder = 55

Alternatively,

Given expression $= \left\{\frac{88(88+1)}{2}\right\}^2 - \left\{\frac{21(21+1)}{2}\right\}^2$

$= 15335056 - 53361 = 15281695$

Now, remainder in case 15281695 ÷ 110 = 55

07 Comparison of Fractions

Critical Relevant Properties of Proper Fractions

(a) In case of same denominator, the other having highest numerator will have the greatest value. In other words, higher numerator, higher value.

For example $\frac{17}{23} > \frac{15}{23} > \frac{11}{23} > \frac{10}{23}$

(b) In case of same numerator, the fraction having lowest denominator will have the greatest value i.e. lower denominator, higher value.

For example $\frac{5}{11} > \frac{5}{17} > \frac{5}{24} > \frac{5}{37}$

(c) If the difference between the numerator and the denominator of given fractions each is same, the fraction having highest numerator or denominator will have highest value i.e. higher numerator or denominator, higher value.

For example $\frac{6}{43} > \frac{5}{42} > \frac{3}{40} > \frac{2}{39}$

(d) $\frac{x}{y}$ will be greater than $\frac{m}{n}$, if $xn > my$. Conversely, if $xn < my$, $\frac{x}{y}$ will be smaller than $\frac{m}{n}$.

For example $\frac{6}{13} > \frac{5}{12}$ $[\because (6 \times 12) > (5 \times 13)]$

Now, take some examples for better understanding of the applications of properties of the fraction given above. The other methods are also given.

Problem 1. If the fractions $\frac{1}{2}, \frac{2}{3}, \frac{5}{9}, \frac{6}{13}$ and $\frac{7}{9}$ are arranged in ascending order of their values, which one will be fourth? *[UBI (Clerk)-08]*

(a) $\frac{2}{3}$ (b) $\frac{6}{13}$

(c) $\frac{5}{9}$ (d) $\frac{7}{9}$

Inside

- *Critical Relevant Properties of Proper Fractions*
- *Methods for Comparison of Fractions with Illustration*
- ....
- .... *& Many More*
- *Work Out with Solutions*

Method I Through Conversion of Fractions into Decimals

$\frac{1}{2} = 0.50, \frac{2}{3} = 0.66, \frac{5}{9} = 0.56, \frac{6}{13} = 0.46, \frac{7}{9} = 0.78$

Hence, ascending order $\Rightarrow \frac{6}{13} < \frac{1}{2} < \frac{5}{9} < \frac{2}{3} < \frac{7}{9}$

Clearly, the fourth fraction $= \frac{2}{3}$

Method II Through Cross Multiplication

Take any two fractions for comparison (say, $\frac{1}{2}$ and $\frac{2}{3}$)

$\Rightarrow \quad \frac{1}{2} >< \frac{2}{3}$

$\Rightarrow \quad (1 \times 3) < (2 \times 2) \Rightarrow 3 < 4$

$\therefore \quad \frac{1}{2} < \frac{2}{3}$ ....(i)

Further, take $\frac{2}{3}$ and $\frac{5}{9}$

$\Rightarrow \quad 18 > 15$

$\therefore \quad \frac{2}{3} > \frac{5}{9}$ ...(ii)

Now, take $\frac{1}{2}$ and $\frac{5}{9}$

$\Rightarrow \quad 9 < 10$

$\therefore \quad \frac{1}{2} < \frac{5}{9}$ .... (iii)

From the above 3 results,

$\frac{2}{3} > \frac{5}{9} > \frac{1}{2}$ ...(iv)

Now, take $\frac{1}{2}$ and $\frac{6}{13}$

$\Rightarrow \quad 13 > 12$

$\therefore \quad \frac{1}{2} > \frac{6}{13}$ ...(v)

Further, take $\frac{7}{9}$ and $\frac{1}{2}$

$\Rightarrow \quad 14 > 9$

$\therefore \quad \frac{7}{9} > \frac{1}{2}$ ...(vi)

Now, take $\frac{2}{3}$ and $\frac{7}{9}$

$\Rightarrow \quad 18 < 21$

$\therefore \quad \frac{2}{3} < \frac{7}{9}$ ...(vii)

From the results of Eqs. (iv), (v) and (vii)

$\frac{6}{13} < \frac{1}{2} < \frac{5}{9} < \frac{2}{3} < \frac{7}{9}$

Hence, fourth fraction $= \frac{2}{3}$

Method III Through Equalising the Denominators

Find the LCM of given denominators i.e. 234

Now, $\frac{1}{2} = \frac{1 \times 117}{2 \times 117} = \frac{117}{234}$

$\frac{2}{3} = \frac{2 \times 78}{3 \times 78} = \frac{156}{234}$

$\frac{5}{9} = \frac{5 \times 26}{9 \times 26} = \frac{130}{234}$

$\frac{6}{13} = \frac{6 \times 18}{13 \times 18} = \frac{108}{234}$

$\frac{7}{9} = \frac{7 \times 26}{9 \times 26} = \frac{182}{234}$

Since $108 < 117 < 130 < 156 < 182$,

$\frac{6}{13} < \frac{1}{2} < \frac{5}{9} < \frac{2}{3} < \frac{7}{9}$

Hence, fourth fraction $= \frac{2}{3}$

Method IV Through Equalising the Numerators

Take LCM of the numerators i.e. 210

Now, $\frac{1}{2} = \frac{1 \times 210}{2 \times 210} = \frac{210}{420}$

$\frac{2}{3} = \frac{2 \times 105}{3 \times 105} = \frac{210}{315}$

$\frac{5}{9} = \frac{5 \times 42}{9 \times 42} = \frac{210}{378}$

$\frac{6}{13} = \frac{6 \times 35}{13 \times 35} = \frac{210}{455}$

$\frac{7}{9} = \frac{7 \times 30}{9 \times 30} = \frac{210}{270}$

Since $270 < 315 < 378 < 420 < 455$,

$\frac{6}{13} < \frac{1}{2} < \frac{5}{9} < \frac{2}{3} < \frac{7}{9}$

Hence, fourth fraction $= \frac{2}{3}$

Method V Through Multiplication by LCM of Denominators

Here, LCM of denominators = 234

Now, $\frac{1}{2} \Rightarrow \frac{1}{2} \times 234 = 117$

$\frac{2}{3} \Rightarrow \frac{2}{3} \times 234 = 156$

$\frac{5}{9} \Rightarrow \frac{5}{9} \times 234 = 130$

$\frac{6}{13} \Rightarrow \frac{6}{13} \times 234 = 108$

$$\frac{7}{9} \Rightarrow \frac{7}{9} \times 234 = 182$$

Since $108 < 117 < 130 < 156 < 182$,

$$\frac{6}{13} < \frac{1}{2} < \frac{5}{9} < \frac{2}{3} < \frac{7}{9}$$

Hence, fourth fraction $= \frac{2}{3}$

Method VI Through Downward Cancellation

$$\frac{1}{2} \Rightarrow \frac{1}{2}$$

$$\frac{2}{3} \Rightarrow \frac{1}{1.5}$$

$$\frac{5}{9} \Rightarrow \frac{1}{1.8}$$

$$\frac{6}{13} \Rightarrow \frac{1}{2.1}$$

$$\frac{7}{9} \Rightarrow \frac{1}{1.2}$$

Since $2.1 > 2 > 1.8 > 1.5 > 1.2$,

$$\frac{6}{13} < \frac{1}{2} < \frac{5}{9} < \frac{2}{3} < \frac{7}{9}$$

Hence, fourth fraction $= \frac{2}{3}$

Method VII Through Equalising the Differences between Numerator and Denominator of every given Fraction

Here, differences between numerator and denominator of given fractions are 1, 1, 4, 7 and 2.

Now, their LCM = 28

Now, obtain the quotient by dividing LCM (i.e. 28) by respective differences and then multiply the numerator and denominator each with this quotient so as to get same differences as under

Obviously, quotients are:

$28 \div 1 = 28$, $28 \div 1 = 28$,

$28 \div 4 = 7$, $28 \div 7 = 4$,

$28 \div 2 = 14$

Now,
$$\frac{1}{2} = \frac{1 \times 28}{2 \times 28} = \frac{28}{56}$$

$$\frac{2}{3} = \frac{2 \times 28}{3 \times 28} = \frac{56}{84}$$

$$\frac{5}{9} = \frac{5 \times 7}{9 \times 7} = \frac{35}{63}$$

$$\frac{6}{13} = \frac{6 \times 4}{13 \times 4} = \frac{24}{52}$$

$$\frac{7}{9} = \frac{7 \times 14}{9 \times 14} = \frac{98}{126}$$

Clearly, the differences are same, then the fraction having higher numerator or denominator, will have higher value.

i.e. $\frac{7}{9} > \frac{2}{3} > \frac{5}{9} > \frac{1}{2} > \frac{6}{13}$

or $\frac{6}{13} < \frac{1}{2} < \frac{5}{9} < \frac{2}{3} < \frac{7}{9}$

Hence, fourth fraction $= \frac{2}{3}$

Alternatively,

The above process may be curtailed as under

LCM of differences = 28

Now, multiply the numerators with the quotients obtained by dividing LCM by respective differences.

Clearly, quotients are : 28, 28, 7 , 4 , 14

Now,
$$\frac{1}{2} = 1 \times 28 = 28$$

$$\frac{2}{3} = 2 \times 28 = 56$$

$$\frac{5}{9} = 5 \times 7 = 35$$

$$\frac{6}{13} = 6 \times 4 = 24$$

$$\frac{7}{9} = 7 \times 14 = 98$$

Since, $24 < 28 < 35 < 56 < 98$,

$$\frac{6}{13} < \frac{1}{2} < \frac{5}{9} < \frac{2}{3} < \frac{7}{9}$$

Hence, fourth fraction $= \frac{2}{3}$

Method VIII Through Dividing the Numerators by Respective Differences between Numerator and Denominator

i.e. $\frac{N}{D-N}$, not $\frac{N}{N-D}$.

Now,
$$\frac{1}{2} \Rightarrow 1 \div 1 = 1$$

$$\frac{2}{3} \Rightarrow 2 \div 1 = 2$$

$$\frac{5}{9} \Rightarrow 5 \div 4 = 1.25$$

$$\frac{6}{13} \Rightarrow 6 \div 7 = 0.85$$

$$\frac{7}{9} \Rightarrow 7 \div 2 = 3.5$$

Since $0.85 < 1 < 1.25 < 2 < 3.5$,

$$\frac{6}{13} < \frac{1}{2} < \frac{5}{9} < \frac{2}{3} < \frac{7}{9}$$

Hence, fourth fraction $= \frac{2}{3}$

Now, take an another example for this method.

Problem 2. Which of the following is the smallest fraction? *[SSC (CGL-PT)-02]*

(a) $\frac{8}{25}$ (b) $\frac{7}{23}$
(c) $\frac{11}{23}$ (d) $\frac{14}{53}$

Sol. Given fractions $= \frac{8}{25}, \frac{7}{23}, \frac{11}{23}, \frac{14}{53}$

On dividing numerator by respective differences between numerator and denominator

$\Rightarrow \frac{8}{(25-8)}, \frac{7}{(23-7)}, \frac{11}{(23-11)}, \frac{14}{(53-14)}$

$\Rightarrow \frac{8}{17}, \frac{7}{16}, \frac{11}{12}, \frac{14}{39}$

However, it is difficult to find their values in decimals. Thus, you are advised to repeat the process of division.

Then,

$\Rightarrow \frac{8}{9}, \frac{7}{9}, \frac{11}{1}, \frac{14}{25}$

If you are still facing the difficulties to find their values, you may repeat the process further. But, the third fraction will converge into negative figure $\left(\frac{11}{1-11}\right)$ on further processing. Thus, we should stop the process.

$\Rightarrow$ 0 .9, 0.7, 11, 0.56

Since 0.56 is the smallest, $\frac{14}{53}$ is the smallest fraction.

It may be construed from the above that whenever any fraction converts into improper fraction, we should stop the further processing. In improper fraction, numerator is greater than denominator.

Aspirants must adopt this innovative technique as far as possible as it is also very useful in data interpretation.

WORKOUT

1. Which of the following fractions is the smallest?
(a) $\frac{7}{6}$ (b) $\frac{7}{9}$ (c) $\frac{4}{5}$ (d) $\frac{5}{7}$

2. The least among the following fractions *[SSC (CGL-T-I)- 2010]*
(a) $\frac{15}{16}$ (b) $\frac{19}{20}$
(c) $\frac{24}{25}$ (d) $\frac{34}{35}$

3. Which of the following are in descending order of their value? *[RRB-02, MAT-01]*
(a) $\frac{5}{9}, \frac{7}{11}, \frac{8}{15}, \frac{11}{17}$ (b) $\frac{5}{9}, \frac{8}{15}, \frac{11}{17}, \frac{7}{11}$
(c) $\frac{11}{17}, \frac{7}{11}, \frac{8}{15}, \frac{5}{9}$ (d) $\frac{11}{17}, \frac{7}{11}, \frac{5}{9}, \frac{8}{15}$

4. If the fractions $\frac{2}{5}, \frac{3}{8}, \frac{4}{9}, \frac{5}{13}$ and $\frac{6}{11}$ are arranged in ascending order of their values, which one will be fourth? *[SBI (PO)-08]*
(a) $\frac{4}{9}$ (b) $\frac{5}{13}$ (c) $\frac{3}{8}$ (d) $\frac{2}{5}$
(e) None of these

5. If all the fractions $\frac{3}{5}, \frac{1}{8}, \frac{8}{11}, \frac{4}{9}, \frac{2}{7}, \frac{5}{7}$ and $\frac{5}{12}$ are arranged in descending order of their values, which one will be third?
(a) $\frac{1}{8}$ (b) $\frac{4}{9}$
(c) $\frac{5}{12}$ (d) $\frac{8}{11}$
(e) None of these

6. Which of the following are in descending order of their value?
(a) $\frac{3}{5}, \frac{7}{9}, \frac{6}{7}$ (b) $\frac{7}{9}, \frac{3}{5}, \frac{6}{7}$
(c) $\frac{7}{9}, \frac{6}{7}, \frac{3}{5}$ (d) $\frac{3}{5}, \frac{6}{7}, \frac{7}{9}$
(e) None of these

7. Which of the following are in descending order of their value?
(a) $\frac{19}{21}, \frac{11}{14}, \frac{16}{19}$ (b) $\frac{19}{21}, \frac{16}{19}, \frac{11}{14}$
(c) $\frac{11}{14}, \frac{16}{19}, \frac{19}{21}$ (d) $\frac{11}{14}, \frac{19}{21}, \frac{16}{19}$
(e) None of these

8. The greatest among the following fractions
(a) $\frac{1}{3}$ (b) $\frac{2}{5}$
(c) $\frac{3}{7}$ (d) $\frac{3}{5}$
(e) None of these

9. The least among the following fractions
(a) $\frac{13}{16}$ (b) $\frac{15}{19}$
(c) $\frac{17}{21}$ (d) $\frac{7}{8}$
(e) None of these

10. Which of the following fractions are in descending order ? *[Indian Bank(RMO)-10]*
(a) $\frac{5}{12}, \frac{2}{7}, \frac{3}{11}, \frac{2}{9}, \frac{1}{5}$ (b) $\frac{2}{7}, \frac{2}{9}, \frac{3}{11}, \frac{1}{5}, \frac{5}{12}$
(c) $\frac{1}{5}, \frac{2}{9}, \frac{3}{11}, \frac{2}{7}, \frac{5}{12}$
(d) $\frac{5}{12}, \frac{3}{11}, \frac{2}{7}, \frac{2}{9}, \frac{1}{5}$
(e) None of these

Answer Key

1. (d)	2. (a)	3. (d)	4. (a)	5. (e)	6. (e)	7. (b)	8. (d)	9. (b)	10. (a)

Hints and SOLUTIONS

1. Since improper fraction i.e. $\frac{7}{6}$ can not be smaller than any proper fraction, we have to compare among the remaining three.

$\frac{7}{9}, \frac{4}{5}, \frac{5}{7}$

$\Rightarrow \frac{7}{(9-7)}, \frac{4}{(5-4)}, \frac{5}{(7-5)}$

$\Rightarrow$ 3.5, 4, 2.5

Since 2.5 is smallest, $\frac{5}{7}$ is the smallest fraction.

2. $\frac{15}{(16-15)}, \frac{19}{(20-19)}, \frac{24}{(25-24)}, \frac{34}{(35-34)}$

$\Rightarrow$ 15, 19, 24, 34

Hence, $\frac{15}{16}$ is the least fraction.

Alternatively,

Since the difference between numerator and denominator of each fraction is same, the fraction having least numerator will have least value.

Hence, least fraction is $\frac{15}{16}$.

3. Given fractions

$= \frac{5}{9}, \frac{7}{11}, \frac{8}{15}, \frac{11}{17}$

$\Rightarrow \frac{5}{(9-5)}, \frac{7}{(11-7)}, \frac{8}{(15-8)}, \frac{11}{(17-11)}$

$\Rightarrow \frac{5}{4}, \frac{7}{4}, \frac{8}{7}, \frac{11}{6}$

$\Rightarrow$ 1.2, 1.7, 1.1, 1.8

Then, descending order is

$\frac{11}{17} > \frac{7}{11} > \frac{5}{9} > \frac{8}{15}$

Hence, option (d) is correct.

4. $\frac{2}{5}, \frac{3}{8}, \frac{4}{9}, \frac{5}{13}, \frac{6}{11}$

$\Rightarrow \frac{2}{3}, \frac{3}{5}, \frac{4}{5}, \frac{5}{8}, \frac{6}{5}$

$\Rightarrow$ 0.66, 0.60, 0.80, 0.62, 1.2

Then, ascending order

$\Rightarrow$ 0.60, 0.62, 0.66, 0.80, 1.2

Hence, fourth fraction = $\frac{4}{9}$

5. $\frac{3}{5}, \frac{1}{8}, \frac{8}{11}, \frac{4}{9}, \frac{2}{7}, \frac{5}{7}, \frac{5}{12}$

$\Rightarrow \frac{3}{2}, \frac{1}{7}, \frac{8}{3}, \frac{4}{5}, \frac{2}{5}, \frac{5}{2}, \frac{5}{7}$

$\Rightarrow$ 1.5, 0.14, 2.66, 0.80, 0.40, 2.50, 0.71

Descending order

= 2.66, 2.50, 1.5, 0.80, 0.71, 0.40, 0.14

Hence, third fraction = $\frac{3}{5}$

6. Given fractions $\frac{3}{5}, \frac{7}{9}, \frac{6}{7}$

$\Rightarrow \frac{3}{2}, \frac{7}{2}, \frac{6}{1} \Rightarrow$ 1.5, 3.5, 6

Descending order $\Rightarrow$ 6, 3.5, 1.5

Hence, required answer = $\frac{6}{7}, \frac{7}{9}, \frac{3}{5}$

7. Given fractions

$= \frac{19}{21}, \frac{11}{14}, \frac{16}{19} \Rightarrow \frac{19}{2}, \frac{11}{3}, \frac{16}{3}$

$\Rightarrow$ 9.5, 3.66, 5.33

Descending order $\Rightarrow$ 9.5, 5.33, 3.66

Hence, required answer = $\frac{19}{21}, \frac{16}{19}, \frac{11}{14}$

8. Given fractions

$= \frac{1}{3}, \frac{2}{5}, \frac{3}{7}, \frac{3}{5} \Rightarrow \frac{1}{2}, \frac{2}{3}, \frac{3}{4}, \frac{3}{2}$

$\Rightarrow$ 0.50, 0.66, 0.75, 1.5

Hence, the greatest fraction = $\frac{3}{5}$

9. Given fractions

$= \frac{13}{16}, \frac{15}{19}, \frac{17}{21}, \frac{7}{8}$

$\Rightarrow \frac{13}{3}, \frac{15}{4}, \frac{17}{4}, \frac{7}{1}$

$\Rightarrow$ 4.33, 3.75, 4.25, 7

Hence, the least fraction = $\frac{15}{19}$

10. Given fraction

$= \frac{5}{12}, \frac{2}{7}, \frac{3}{11}, \frac{2}{9}, \frac{1}{5}$

$\Rightarrow \frac{5}{7}, \frac{2}{5}, \frac{3}{8}, \frac{2}{7}, \frac{1}{4}$

$\Rightarrow \frac{5}{2}, \frac{2}{3}, \frac{3}{5}, \frac{2}{5}, \frac{1}{3}$

$\Rightarrow$ 2.5, 0.66, 0.6, 0.4, 0.33

It is in descending order. Hence, option (a) is correct.

08

Square and Cube

Learning OBJECTIVES

After completing this chapter, aspirants will be able to

- Understand the significance of this chapter in numerical ability;
- Learn how to compute square and cube of Whole numbers, Decimal numbers and Fractions;
- Understand the concept of square and cube through very unique approaches;
- Realise themselves very comfortable in squaring some specific numbers as well as numbers nearer to multiple of 100;
- Observe that only two techniques have been emphasised for squaring any number;
- Understand the philosophy of the division of this chapter in various parts;
- Compute square and cube in moments;
- Grip on concepts helping to compute square and cube without pen and paper;
- Learn new concepts/fundamentals;
- Apply the pragmatic approaches in lieu of academic approaches; and
- Ponder on the alternative methods.

Inside

Introduction

The techniques envisaged in this chapter will obviously reduce the time taken in solving the questions related to arithmetical chapters. This chapter is very critical especially for Banking services examinations wherein questions are based on direct applications of the square and cubes of numbers. The direct applications of square and cube are seen in the chapters like Simplification, Data Interpretations, etc. to the greater extent.

Square

If a number is multiplied with itself, the resultant is called the square of that number. There are diverse ways to find the square of a number with the limitations that one method is applicable for a particular case, another is applicable for another case and so on.

Here, we will stress on only two techniques with help of which you would be capable to find the square of any number. This technique has only one prerequisite that you have the squares of 1 to 25 and fact that most of aspirants know the same.

Square may have diverse dimensions like:

I. Square of Whole numbers

II. Square of Decimal numbers

III. Square of Fractions.

I. Square of Whole Numbers

Rules for squaring (1 to 25)

Aspirants are requested to get squares of 1 to 25 by heart instead of applying other technique(s) so as to save time in examination hall.

Rules for squaring (26 to 75)

(Number – 25)	Square of difference from 50
	This box contains always 2 digits. In case of more than 2 digits, the remaining will be carried over left. Further, in case of lesser, we append 0 before that digit.

Examples

(a) $(43)^2 = ?$

(43 – 25)	$(50 - 43)^2$
1849	

(b) $(58)^2 = ?$

(58 – 25)	$(58 - 50)^2$
3364	

(c) $(74)^2 = ?$

(74 – 25)	$(74 - 50)^2$
49	576
(49 + 5)	76
5476	

(d) $(52)^2 = ?$

(52 – 25)	$(52 - 50)^2$
27	4
27	04
2704	

Rules for Squaring (76 and above)

Recall the techniques 'When both numbers (i.e. multiplicand and multiplier) are closer to 100 and its multiple' or closer to 1000 and its multiple and so on as envisaged in chapter 'Multiplication'.

***Case* I** When the base number is 100.

(a) $(87)^2 = 87 \times 87$

{87 – (100 – 87)}	{100 – 87} {100 – 87}
74	169
7569	

(b) $(117)^2 = 117 \times 117$

{117 + (117 – 100)}	{117 – 100} {117 – 100}
134	289
13689	

(c) $(157)^2 = 157 \times 157$

{157 + (157 – 100)}	$(57)^2$
214	$(57 - 25) \mid (57 - 50)^2$ [applying the square rule meant for (26 – 75)]
214	3249
24649	

***Case* II** When the base number is 200/300/400, ...

(a) $(186)^2 = 186 \times 186$

2{186 – (200 – 186)}	{200 – 186} {200 – 186}
344	196
34596	

(b) $(247)^2 = 247 \times 247$

2{247 + (247 – 200)}	$(47)^2$
588	$(47 - 25) \mid (50 - 47)^2$ [applying the square rule meant for (26 -75)]
588	2209
61009	

(c) $(768)^2 = 768 \times 768$

8{768 - (800 – 768)}	$(32)^2$
5888	1024
589824	

***Case* III** When the base number is 1000 or its multiple.

$(972)^2 = 972 \times 972$

{ 972 - (1000 – 972)}	{1000 – 972} {1000 – 972}
944	784
944784	

Observation

(i) If the number formed with last two digits of the number whose square is required, is more than 50, we take next multiple of 100 as base number, otherwise existing multiple for sake of convenience, *viz.* 768 ⇒ 800, 732 ⇒ 700, 972 ⇒ 1000, 246 ⇒ 200, etc.

(ii) If you feel inconvenience with the above method, confine yourself on one-liner method meant for multiplication, explained in chapter 'Multiplication'.

For example: (a) $(823)^2 = ?$

$\Rightarrow$ 823
× 823
677329

(b) $(1347)^2 = ?$

$\Rightarrow$ 1347
× 1347
1814409

Some Specific Cases

(i) **When a number ending with 5.**

***Step* I** Multiply the number leaving unit digit (i.e. 5) with the number 1 higher than it.

***Step* II** Append 25 on the right side of the above product. It will be the required square.

For examples

(a) $(75)^2 \Rightarrow 7(7+1)|\,25 \Rightarrow 5625$

(b) $(135)^2 \Rightarrow 13(13+1)|\,25 \Rightarrow 18225$

(ii) **When a number ending with 25.**

***Step* I** Multiply the number leaving last two digits (i.e. 25) with a number formed by suffixing 5 to it.

***Step* II** Append 625 on the right side of the above product.

For examples

(a) $(625)^2 \Rightarrow 6 \times (65)|\,625 \Rightarrow 390625$

(b) $(1325)^2 \Rightarrow 13 \times (135)|\,625 \Rightarrow 1755625$

(iii) **When the number is formed with repetition of a digit .**

(a) $(1111)^2 = ?$

***Step* I** Write 1234 from left to right end as there are four 1's.

***Step* II** Thereafter write in descending order i.e. 321.

Hence, required square = <u>1234321</u>.

(b) $(11111)^2$ = <u>123454321</u>.

(c) If $n^2 = 12345678987654321$, what is n? ***[CAT-99]***

Clearly, $n = 111111111$

The above technique is also applicable in square of a number with repetition of any single digit after some alternations in format. **For example:**

(d) $(3333)^2 = 3^2 \times (1111)^2 = 9 \times 1234321 = 11108889$

(e) $(777)^2 = 7^2 \times (111)^2 = 49 \times 12321 = 603729$

(f) $(999)^2 = 9^2 \times (111)^2 = 81 \times 12321 = 998001$

Alternatively I

Here, there are three (say n) times 9. Now, write $(n-1)$ times 9 followed by 8 and thereafter $(n-1)$ times 0 and finally 1.

Hence, <u>99</u> 8 <u>00</u> 1

Alternatively II

Multiply 999 with 999 as elaborated in the chapter 'Multiplication' under head 'When a number is to be multiplied by a number closer to 100, 1000, 10000 and so on.

$\Rightarrow 999 \times (1000 - 1) = 999000 - 999$
$= 998001$

II. Square of Decimal Numbers

We square the given number assuming that there is no decimal point. Thereafter, we append the decimal point on the resultant after as many digits from right extreme as twice the number of digits existing after decimal point in the given number.

For example

(a) $(4\cdot 6)^2 = ? \Rightarrow (46)^2 \Rightarrow$

(46 – 25)	$(50-46)^2$
2116	

Number of digits after decimal point in given number = 1

Then, two times of 1 = 2

Further, put the decimal point on the resultant after 2 digits from the extreme right to get the answer.

Hence, required answer = 21.16

(b) $(4\cdot 23)^2 = ? \Rightarrow (423)^2 \Rightarrow$

4(423 + 23)	$(23)^2$
1784	529
178929	

Hence, required answer = 17.8929

III. Square of Fractions

***Type* I** $\left(\frac{A}{B}\right)^2 = \frac{A^2}{B^2}$

Example $\left(\frac{2}{3}\right)^2 = \frac{2^2}{3^2} = \frac{4}{9}$

***Type* II** $\left(A\frac{B}{C}\right)^2 = (A)^2 + \left(\frac{B}{C}\right)^2 + 2\cdot A\cdot\frac{B}{C}$

Example $\left(4\frac{2}{5}\right)^2 = (4)^2 + \left(\frac{2}{5}\right)^2 + 2\cdot 4\cdot\frac{2}{5}$

$$= 16 + \frac{4}{25} + \frac{16}{5}$$

$$= 16 + \frac{4}{25} + 3\frac{1}{5}$$

$$= 19 + \frac{4}{25} + \frac{1}{5}$$

$$= 19 + \frac{9}{25} = 19\frac{9}{25}$$

Try yourself to find the value of the followings

Problem 1 $(83)^2$ Answer: 6889

2 $(189)^2$ Answer: 35721

3 $(691)^2$ Answer: 477481

4 $(17.25)^2$ Answer: 297.5625

5 $(6666.66)^2$ Answer: 44444355.5556

6 $\left(19\frac{2}{7}\right)^2$ Answer: $371\frac{46}{49}$

Cube

If a number is multiplied two times with itself, the resultant is called the cube of that number.

Generally, cube of very large numbers are not required in competitive exams, that is why we will stress on the method suitable for a two-digit number only. Further, aspirants are requested to get cubes of 1 to 12 by heart instead of applying any technique(s) so as to save time in examination hall.

Cube may have diverse dimensions like

I. Cube of Whole numbers.

II. Cube of Decimal numbers.

III. Cube of Fractions.

I. Cube of Whole Numbers

$(ab)^3 \Rightarrow$

a^3	$3a^2b$	$3b^2a$	b^3
a^3	$a^3 \cdot \frac{b}{a} + 2 \cdot a^3 \cdot \frac{b}{a}$	$a^3 \cdot \frac{b}{a} \cdot \frac{b}{a} + 2 \cdot a^3 \cdot \frac{b}{a} \cdot \frac{b}{a}$	b^3
a^3	**Initial term (i.e. a^3) $\times \frac{b}{a}$ + its twice**	**Previous term (i.e. $a^3 \cdot \frac{b}{a}$) $\times \frac{b}{a}$ + its twice**	b^3

We will follow this pattern so as to obtain the cube.

Examples

(i) $(32)^3 = ?$

Clearly, $a = 3$ and $b = 2 \Rightarrow$

a^3	Initial term (i.e. a^3) $\times \frac{b}{a}$ + its twice	Previous term (i.e. $a^3 \cdot \frac{b}{a}$) $\times \frac{b}{a}$ + its twice	b^3
$3^3 = 27$	$27 \times \frac{2}{3}$ (i.e. 18) + 36	$18 \times \frac{2}{3}$ (i.e. 12) + 24	$2^3 = 8$
27	54	36	8
Now, we have to keep only one digit in all boxes/columns except the last. The remaining will be carried over the left.			
32	7	6	8

Hence, required cube = 32768

Observation The above pattern is convenient only if you have command over cubes of 1 to 9.

(ii) $(74)^3 = ?$

Clearly, $a = 7$ and $b = 4$

$\Rightarrow$

a^3	Initial term (i.e. a^3) $\times \frac{b}{a}$ + its twice	Previous term (i.e. $a^3 \cdot \frac{b}{a}$) $\times \frac{b}{a}$ + its twice	b^3
$7^3 = 343$	$343 \times \frac{4}{7}$ (i.e. 196)+293	$196 \times \frac{4}{7}$ (i.e. 112)+224	$4^3 = 64$
343	588	336	64
405	2	2	4

Hence, required cube = 405224

(iii) $(123)^3 = ?$

Clearly, $a = 12$ and $b = 3$

$\Rightarrow$

a^3	Initial term (i.e. a^3) $\times \frac{b}{a}$ + its twice	Previous term (i.e. $a^3 \cdot \frac{b}{a}$) $\times \frac{b}{a}$ + its twice	b^3
$12^3 = 1728$	$1728 \times \frac{3}{12}$ (i.e. 432) + 864	$432 \times \frac{3}{12}$ (i.e. 108) + 216	$3^3 = 27$
1728	1296	324	27
1860	8	6	7

Hence, required cube = 1860867

II. Cube of Decimal Numbers

We cube the given number assuming that there is no decimal point. Thereafter, we append the decimal point on the resultant after as many digits from right extreme as thrice the number of digits existing after decimal point in the given number.

For example $(4.6)^3 = ?$

$\Rightarrow$ $(46)^3 = ?$

Clearly, $a = 4$ and $b = 6$

a^3	Initial term (i.e. a^3) $\times \frac{b}{a}$ + its twice	Previous term (i.e. $a^3 \cdot \frac{b}{a}$) $\times \frac{b}{a}$ + its twice	b^3
$4^3 = 64$	$64 \times \frac{6}{4}$ (i.e. 96)+192	$96 \times \frac{6}{4}$ (i.e. 144)+288	$6^3 = 216$
64	288	432	216
97	3	3	6

Hence, cube of 46 = 97336

Here, number of digits after decimal point in given number = 1

Then, its thrice = 3

Further, put the decimal point on the resultant after 3 digits from the extreme right to get the answer.

Hence, required answer = 97.336

III. Cube of Fractions

***Type* I** $\left(\frac{A}{B}\right)^3 = \frac{A^3}{B^3}$

Example $\left(\frac{2}{3}\right)^3 = \frac{2^3}{3^3} = \frac{8}{27}$

***Type* II** $\left(A\frac{B}{C}\right)^3 = \left(A + \frac{B}{C}\right)^3 = (A)^3 + \left(\frac{B}{C}\right)^3 + 3 \cdot A \cdot \frac{B}{C}\left(A + \frac{B}{C}\right)$

Example $\left(4\frac{2}{3}\right)^3 = (4)^3 + \left(\frac{2}{3}\right)^3 + 3 \cdot 4 \cdot \frac{2}{3}\left(4 + \frac{2}{3}\right)$

$= 64 + \frac{8}{27} + 32 + \frac{16}{3} = 96 + \frac{8}{27} + 5\frac{1}{3}$

$= 101 + \frac{17}{27} = 101\frac{17}{27}$

Try yourself to find the value of the followings:

➘ Problem 7 $(83)^3$ Answer: 571787

8 $(6.7)^3$ Answer: 300.763

9 $(11.3)^3$ Answer:1442.897

10 $\left(9\frac{2}{7}\right)^3$ Answer: $800\frac{225}{343}$

WORKOUT

1. $(12.51)^4 = ?$ (approx.) ***[SBI(PO)-08]***

(a) 22300 (b) 23400
(c) 24500 (d) 25600
(e) 26700

2. $(56)^2 \times 2.5385 = ?$ (approx.)

(a) 4740 (b) 5570
(c) 6600 (d) 7960
(e) 8880

3. $(52.36)^3 = ?$ (approx.) ***[UBI(PO)-08]***

(a) 129600 (b) 138850
(c) 143550 (d) 165790
(e) 177370

4. $(34.34)^2 + (5.96)^2 - (23.09)^2 = ?$ (approx.) ***[Corporation Bank(PO)-11]***

(a) 510
(b) 540
(c) 620
(d) 680
(e) 650

5. $(11.7)^3 = ?$

(a) 1502.123
(b) 1601.613
(c) 1420.123
(d) 1331.243

6. $\left(24\frac{2}{3}\right)^2 = ?$

(a) $608\frac{2}{9}$ (b) $608\frac{4}{9}$
(c) $606\frac{2}{9}$ (d) $609\frac{2}{9}$

7. $\left(8\frac{2}{5}\right)^3 = ?$

(a) $572\frac{91}{125}$ (b) $582\frac{83}{125}$
(c) $592\frac{88}{125}$ (d) $592\frac{88}{105}$

Answer Key

1. (c)	2. (d)	3. (c)	4. (d)	5. (b)	6. (b)	7. (c)			

Hints and SOLUTIONS

1. $(12.51)^4 = (12.51)^2 \times (12.51)^2 \approx (12.5)^2 \times (12.5)^2$

i.e. $(125)^2 \times (125)^2$ if we ignore decimals for the time being.

Now, $(125)^2 \Rightarrow 1 \times (15) \mid 625 \Rightarrow 15625$

Then, $(125)^2 \times (125)^2 = (15625)^2$

Further, $(15625)^2 \Rightarrow 156 \times (1565) \mid 625 \Rightarrow 244140625$

Answer is 24414.0625 if we take decimals into consideration.

Hence, the correct option is (c).

2. $(56)^2 \Rightarrow$

(56 – 25)	$(56 - 50)^2$
31	36
3136	

Then, $(56)^2 \times 2.5385 \approx 3136 \times 2.5$ i.e. 7840

[we divided 3136 by 4 and then appended 0 on the resultant as $2.5 = \frac{10}{4}$.

Alternatively, we can multiply 3136 by 25 with the help of table of 25 and place the decimal after one digit from the right side of the product]

Hence, option (d) is correct.

3. $(52.36)^3 \approx (52)^3$

Now, $(52)^3 \Rightarrow$

a^3	Initial term (i.e. a^3) $\times \frac{b}{a}$ + its twice	Previous term (i.e. $a^3 \cdot \frac{b}{a}) \times \frac{b}{a}$ + its twice	b^3
$5^3 = 125$	$125 \times \frac{2}{5}$ (i.e. 50) +100	$50 \times \frac{2}{5}$ (i.e. 20) + 40	$2^3 = 8$
125	150	60	8
140	6	0	8

i.e. $(52)^3 = 140608$

Hence, option(c) is correct.

4. $(34.34)^2 + (5.96)^2 - (23.09)^2 \approx (34)^2 + (6)^2 - (23)^2$

i.e. 1156 + 36 – 529, hence, 663

[**Note:** $(34)^2 \Rightarrow (34 - 25) \mid (50 - 34)^2 \Rightarrow 9 \mid 256 \Rightarrow 1156$]

Hence, option (d) is correct.

5. $(11.7)^3 \Rightarrow (117)^3$ if we ignore decimals for the time being.

Now, $(117)^3 \Rightarrow$

$11^3 = 1331$	$1331 \times \frac{7}{11}$ (i.e. 847) + 1694	$847 \times \frac{7}{11}$ (i.e. 539) + 1078	$7^3 = 343$
1331	2541	1617	343
1601	6	1	3

i.e. $(117)^3 = 1601613$

Then, $(11.7)^3 = 1601.613$

Hence, option (b) is correct.

6. $\left(24\frac{2}{3}\right)^2 = (24)^2 + \left(\frac{2}{3}\right)^2 + 2 \cdot 24 \cdot \frac{2}{3}$

$= 576 + \frac{4}{9} + 32$

$= 608 + \frac{4}{9} = 608\frac{4}{9}$

Hence, option (b) is correct.

7. $\left(8\frac{2}{5}\right)^3 = (8)^3 + \left(\frac{2}{5}\right)^3 + 3 \cdot 8 \cdot \frac{2}{5}\left(8 + \frac{2}{5}\right)$

$= 512 + \frac{8}{125} + 76\frac{4}{5} + 3\frac{21}{25}$

$= (512 + 76 + 3) + \frac{8 + 100 + 105}{125}$

$= 591 + 1\frac{88}{125}$

$= 592\frac{88}{125}$

Hence, option (c) is correct.

09

Square-Root, Cube-Root and *n*th Root

Learning OBJECTIVES

After completing this chapter, aspirants will be able to

- Understand the significance of this chapter in numerical ability;
- Learn how to compute square-root, cube-root and *n*th root of Whole numbers and Decimal numbers;
- Understand the concept of square-root, cube-root and *n*th root through diverse approaches;
- Realise themselves very comfortable in computation of square-root, cube-root and *n*th root through estimation method;
- Understand the philosophy of the division of this chapter in various parts;
- Compute square-root, cube-root and *n*th root of a number in moments;
- Grip on concepts helping to compute square-root, cube-root and *n*th root of a number without pen and paper;
- Learn new concepts/fundamentals;
- Apply the pragmatic approaches in lieu of academic approaches; and
- Ponder on the alternative methods.

Introduction

The techniques envisaged in this chapter will obviously reduce the time taken in solving the questions related to arithmetical chapters. This chapter is very critical especially for Banking services examinations wherein questions are based on direct applications of the square-root, cube-root and *n*th root of numbers. Their direct applications are also seen in the chapters like Simplification, etc. to the greater extent.

Inside

A. Square Root

When the square of a number is equal to another number, the former is called square-root of latter. Every positive integer has two square-roots. To illustrate, square-roots of 64 are + 8 and − 8. Square-root may have diverse dimensions like

I. Square Root of Whole numbers

II. Square Root of Decimal numbers

I. Square-Root of Whole Numbers

Traditionally, there are two methods for obtaining the square-root namely

(i) Prime Factorisation Method

(ii) Division Method

Apart from the above, we will also discuss other approaches to compute the square root of a number with a view to saving time in examination hall.

Prime Factorisation Method

Firstly, we express the given number in form of product of prime numbers. Then, we take one prime number out of two same prime numbers because the square root of a number (say $n \times n$) is n.

Take one example $\sqrt{9216} = ?$

2	9216
2	4608
2	2304
2	1152
2	576
2	288
2	144
2	72
2	36
2	18
3	9
	3

Clearly, $9216 = \underline{2 \times 2} \times \underline{2 \times 2} \times \underline{2 \times 2} \times \underline{2 \times 2} \times \underline{2 \times 2} \times \underline{3 \times 3}$

Then, square root of $9216 = 2 \times 2 \times 2 \times 2 \times 2 \times 3 = 96$

[Technically, ± 96]

Take another example $\sqrt{128} = ?$

2	128
2	64
2	32
2	16
2	8
2	4
	2

Clearly, $128 = \underline{2 \times 2} \times \underline{2 \times 2} \times \underline{2 \times 2} \times 2$

$= \underline{2 \times 2} \times \underline{2 \times 2} \times \underline{2 \times 2} \times \underline{\sqrt{2} \times \sqrt{2}}$

Then, square root of $128 = 2 \times 2 \times 2 \times \sqrt{2} = 8\sqrt{2}$

[Technically, $\pm 8\sqrt{2}$]

Division Method

Take one example: $\sqrt{9216} = ?$

***Step* I** Make pairs from right to left end. It is also possible that the extreme left digit may either be part of a pair or not, *viz.* $\overline{92}\,\overline{16}$, $1\overline{63}\,\overline{46}$, *etc.*

***Step* II** Start the process by dividing extreme left pair/digit by a suitable divisor in order to obtain quotient same as divisor.

9	$\overline{92}\,\overline{16}$	9
	81	

***Step* III** Now, add the quotient to the divisor. Then, the next dividend will be the next pair prefixing the previous remainder i.e. 1116.

9	$\overline{92}\,\overline{16}$	9
9	81	
18	1116	

***Step* IV** Now, append a suitable digit with the sum 18 so that the product of that digit and the number so obtained is equal or less than the dividend.

9	$\overline{92}\,\overline{16}$	96
9	81	
186	1116	
6	1116	
	xxxx	

***Step* V** If the number whose square root is to be obtained is large number, we proceed on the above line until all the pairs get exhausted. In case a remainder is left, annex two zeros to it and append decimal point to the quotient. At every step we will append two zeros to the remainder. The quotient thus obtained would be the required square root of the given number.

Take another example: $\sqrt{500} = ?$

2	$5\,\overline{00}$	22.36...
2	4	
42	100	
2	84	
443	1600	
3	1329	
4466	27100	
6	26796	

Continue, if required in the question.

Other Approach

Before discussing the approach we should mind the following facts.

Perfect square number ending in	Last digit of its square-root
1	1 or 9, *viz.* $\sqrt{14641} = 121$, $\sqrt{6241} = 79$
4	2 or 8, *viz.* $\sqrt{484} = 22$, $\sqrt{64} = 8$
6	4 or 6, *viz.* $\sqrt{576} = 24$, $\sqrt{1296} = 36$
9	3 or 7, *viz.* $\sqrt{1849} = 43$, $\sqrt{3249} = 57$
5	5, *viz.* $\sqrt{625} = 25$
00	0, *viz.* $\sqrt{57600} = 240$

Observations

(i) A perfect square number doesn't have 2,3,7 and 8 as its last digit.

[Which of (2222, 11664, 343343, 220347) are not perfect squares: - **CDS(2017)**]

(ii) It is seen from the above that the sum of possible unit digits of square-roots in case of perfect square numbers ending in 1/4/6/9 is 10.

(iii) If the square number has n digits, its square-root has either $\frac{n}{2}$ or $\frac{n+1}{2}$ digits.

Square-Root of a Perfect Square Number

[Nature of options of the question reveals as to whether the given number is perfect square or not]

Problem 1 What is the square root of 4624?

Sol. Make pairs from right to left end as $\overline{46}\,\overline{24}$. It is obvious that it will have a 2-digit square root ($\because n/2 = 4/2 = 2$). Further, its square root ends either in 2 or 8 as the given number ends in 4.

46 is lying between 6^2 and 7^2. We have to take lower base i.e. 6 as extreme left of the square root.

Hence, the square root of 4624 is either 62 or 68.

Now, find a number between 62 and 68 ending in 5 i.e. 65. The reason behind selection of a mid number ending in 5 is that we can easily compute its square, technique for squaring such numbers has already been elaborated in the topic 'Square'.

Clearly, $(65)^2 = 6\,(6+1)\,|\,25 = 4225$

$\because 4624 > (65)^2$

$\therefore$ Square root of $4624 > 65$

Hence, required square root = 68, not 62.

Alternatively

6	$\overline{46}\,\overline{24}$	6
6	36	
12 ⊙	1024	

Put either 2 or 8 in ⊙ to get 1024. Clearly, ⊙ will take 8 as $128 \times 8 = 1024$.

Hence, required square root of 4624 = 68

Note *This Modified Division Method saves time as we have to put maximum two digits instead of many digits one by one as is done in conventional division method.*

Problem 2 What is the square-root of 5041?

Sol. $5041 \Rightarrow \overline{50}\,\overline{41}$

It is obvious that it will have a 2-digit square root ($\because n/2 = 4/2 = 2$). Further, its square root ends either in 1 or 9 as the given number ends in 1.

But, $7^2 < 50 < 8^2$

Hence, the square root of 5041 is either 71 or 79.

Now, $5041 < (75)^2$

$\therefore$ Square root of $5041 < 75$

Hence, required square root = 71

Alternatively

With help of Modified Division Method explained in previous question.

7	$\overline{50}\,\overline{41}$	7
7	49	
14 ⊙	141	

Put either 1 or 9 in ⊙ to get 141. Clearly, ⊙ will take 1 as $141 \times 1 = 141$.

Hence, required square root of 5041 = 71

Problem 3 $\sqrt{18769} = ?$

Sol. Obviously, $1\,\overline{87}\,\overline{69}$ will have 3-digit square root ending in 3 or 7.

$1\,\overline{87}\,\overline{69}$ may be written as $\overline{187}\,\overline{69}$ with a view to comparing 187 to avoid complexity involved in comparing 1 and 87 separately.

But, $13^2 < 187 < 14^2$

Hence, the square root of 18769 is either 133 or 137.

Now, $18769 > (135)^2$

$\therefore$ Square root of $18769 > 135$

Hence, required square root = 137

Alternatively

With help of Modified Division Method explained in previous questions.

Problem 4 $\sqrt{107584} = ?$

Sol. Obviously, $\overline{1075}\,\overline{84}$ will have 3-digit square root ending in 2 or 8.

But, $32^2 < 1075 < 33^2$

Hence, the square root of 107584 is either 322 or 328.

Since $107584 > (325)^2$, square root of $107584 > 325$.

Hence, required square root = 328

Alternatively

With the help of Modified Division Method explained in previous questions.

Problem 5 $\sqrt{751689} = ?$

Sol. Obviously, $\overline{7516}\,\overline{89}$ will have 3-digit square root ending in 3 or 7.

We may estimate quickly that 7516 is close to the square to 80 as $8^2 < 75 < 9^2$. But, not known exactly.

However, the closest number $= \dfrac{80 + \dfrac{7516}{80}}{2} = 87$ (approx.)

[explained in next sub-topic]

But, $86^2 < 7516 < 87^2$.

$\therefore$ Square root = 863 or 867

Now, $751689 > (865)^2$

$\therefore$ Square-root of 751689 > 865

Hence, required square root = 867

Alternatively

With the help of Modified Division Method explained in previous questions.

Square Root of any Number Even Perfect Square Number

Problem 6 Take one example: What is the square root of 600 ?

Sol. $\Rightarrow 6\overline{00}$

Clearly, its square root has two digits and the extreme left digit is 2.

Let the square root is 20. (you are free to take any figure, but dividing 600 by 20 is very convenient.)

Then, average of 20 and $\dfrac{600}{20} = \dfrac{20 + 30}{2} = 25$

Here, 25 is the approximate square root. For more accuracy, we may repeat the process.

Average of 25 and $\dfrac{600}{25} = \dfrac{25 + 24}{2} = 24.50$

For more accuracy, average of 24.5 and $\dfrac{600}{24.5}$

$= \dfrac{24.5 + 24.49}{2} = 24.49$

Here, 24.49 is the accurate value of square root as 24.5 and $\dfrac{600}{24.5}$ (i.e. 24.49) are almost equal. Hence, we should stop the process of repetition. It is important to note that process of repetition depends upon the demand of the question. If question demands approx. value, we should process accordingly.

This approach may be called **'Estimation Method'**. It is very useful in computing the approx. value of square root as asked in banking examinations frequently.

Alternatively I

> Let the number whose square is just below the given number (N) be B. Then, the square root of $N = B + \dfrac{N - B^2}{2B + 1}$

Clearly, $(24)^2 < 600 < (25)^2$

Then, $\sqrt{600} = 24 + \dfrac{600 - (24)^2}{2 \cdot 24 + 1} = 24 + \dfrac{24}{49} = 24.49$

Alternatively II

$\sqrt{600} = \sqrt{6 \times 100} = \sqrt{6} \times 10$

$= 2.449 \times 10 = 24.49$

Alternatively III

By Division Method, already detailed.

Problem 7 $\sqrt{7921} = ?$

Sol. $\Rightarrow \overline{79}\,\overline{21}$

Clearly, its square root has two digits and the extreme left is 8.

Let the square root is 80 (you are free to take any figure, but dividing 7921 by 80 is very convenient)

Then, average of 80 and $\dfrac{7921}{80} = \dfrac{80 + 99.01}{2} = 89.5$

Hence, 89.5 is the approximate square root.

For more accuracy, we may repeat the process.

Then, average of 89.5 and $\dfrac{7921}{89.5} = \dfrac{89.5 + 88.5}{2} = 89$

Hence, 89 is the accurate value of square root as 89.5 and $\dfrac{7921}{89}$ (i.e. 89) are almost equal. Hence, we should stop the process of repetition.

Further,

If we assume square root as 90, then average of 90 and $\dfrac{7921}{90} = \dfrac{90 + 88.01}{2} = 89.005$

Hence, 89 is the square root of 7921.

Note *Here, 79 is nearer to 9^2 in comparison to 8^2. It implies that 80 is far away from the actual square root of 7921 but 90 is nearer to it. Hence, we should estimate the square root as 90 instead of 80 so as to save the steps/time.*

Alternatively,

You may compute square root by other approaches as applied in previous question.

Problem 8 $\sqrt{11986} = ?$

Sol. Obviously, $\overline{1}\,\overline{19}\,\overline{86}$ will have 3-digit square root nearer to 110 as square of 11 is near to 119.

Now, average of 110 and $\dfrac{11986}{110} = \dfrac{110 + 108.96}{2} = 109.48$

Hence, 109.48 is the square root of 11986.

Since 110 and 108.96 are closer to each other, 109.48 is the accurate value of square root of 11986.

Problem 9 $\sqrt{896889} = ?$

Sol. Obviously, $\overline{89}\,\overline{68}\,\overline{89}$ will have 3-digit square root nearer to 900 as square of 90 is near to 8968.

Now, average of 900 and $\dfrac{896889}{900} = \dfrac{900 + 996.54}{2}$

$= 948.27$(approx.)

Since 900 and 996.54 are not so closer, 948.27 is not the actual square root. It is the approx. value of square root. For more accuracy, you may repeat the process. The actual square root is 947.04.

Try yourself to find the approximate value of the followings.

Problem 10 $\sqrt{123456}$ Answer: 351.36 (accurate value)

11 $\sqrt{23569}$ Answer: 153.52 (accurate value)

12 $\sqrt{78912}$ Answer: 280.91 (accurate value)

13 $\sqrt{88209}$ Answer: 297 (accurate value)

14 $\sqrt{1261129}$ Answer: 1123 (accurate value)

II. Square Root of Decimal Numbers

***Step* I** Ignore the decimal point and compute the square-root of the given number.

***Step* II** If the left side to decimal point has n digits then we should append decimal point on the resultant after $n/2$ digits from left end. In case $n/2$ does not yield integer, append decimal point after integer next to $n/2$.

Alternatively,

Take only integer part into consideration and compute its square root.

Examples

Problem 15 $\sqrt{104.04} = ?$

Sol. The square root of 10404 = 102

Here, digits before decimal point = 3

Then, $3/2 = 1.5 \Rightarrow 2$(next integer)

Thus, we should append decimal point after 2 digits from the left end.

$\therefore$ Required square-root = 10.2

Alternatively

Clearly, $1\overline{04}$ has square root in two digits and the left digit is 1.

Let the square root of 104 be 10.

Then, average of 10 and $\frac{104}{10} = \frac{10 + 10.4}{2} = 10.2$

Since 10 and 10.4 are closer to each other, 10.2 is the accurate value of square root of 104 or 104.04.

Problem 16 $\sqrt{90.25} = ?$

Sol. The square root of 9025 = 95

Here, digits before decimal point = 2

Then, 2/2 = 1 (integer itself)

Thus, we should append decimal point after 1 digit from the left end.

$\therefore$ Required square-root = 9.5

Alternatively

Clearly, $\overline{90}$ has square root in one digit which is nearer to 9.

Let the square root of 90 be 9.

Then, average of 9 and $\frac{90}{9} = \frac{9 + 10}{2} = 9.5$

Since 9 and 10 are closer to each other, 9.5 is the accurate value of square root of 90 or 90.25.

Problem 17 $\sqrt{180.23} = ?$

Sol. Clearly, $1\overline{80}$ has square root in two digits and the left digit is 1.

Let the square root of 180 be 15 as dividing 180 by 15 is so easy.

Then, average of 15 and $\frac{180}{15} = \frac{15 + 12}{2} = 13.5$

Since 15 and 12 are not so closer to each other, 13.5 is the approx value of square root of 180 or 180.23.

Try yourself to find the approximate value of the followings.

Problem 18 $\sqrt{25623.56}$ **Answer : 160.07(accurate value)**

19 $\sqrt{1231.152}$ **Answer : 35.08 (accurate value)**

B. Cube Root

The cube-root of any number (say $n \times n \times n$) is n.

It is denoted by $\sqrt[3]{x}$ or $x^{\frac{1}{3}}$, where x is a number whose cube-root is required. Cube-root may have diverse dimensions like:

I. Cube root of Whole numbers

II. Cube root of Decimal numbers

I. Cube Root of Whole Numbers

Generally, the cube root of a number is obtained by the Prime Factorisation Method. We may take few examples.

Problem 20 $\sqrt[3]{64} = ?$

Sol.

2	64
2	32
2	16
2	8
2	4
	2

Clearly, $64 = 2 \times 2 \times 2 \times 2 \times 2 \times 2$

Now, make groups of three same primes and then find product of primes taking one out of each group.

$$64 = \underline{2 \times 2 \times 2} \times \underline{2 \times 2 \times 2}$$

Hence, $\sqrt[3]{64} = 2$ (from one group) $\times$ 2 (from another group) = 4

Problem 21 $\sqrt[3]{216} = ?$

Sol.

2	216
2	108
2	54
3	27
3	9
	3

Clearly, $216 = 2 \times 2 \times 2 \times 3 \times 3 \times 3$

Now, make groups of three same primes and then find product of primes taking one out of each group.

$$216 = \underline{2 \times 2 \times 2} \times \underline{3 \times 3 \times 3}$$

Hence, $\sqrt[3]{64} = 2 \times 3 = 6$

Problem 22 $\sqrt[3]{108} = ?$

Sol.

2	108
2	54
3	27
3	9
	3

Clearly, $108 = 2 \times 2 \times 3 \times 3 \times 3$

Now, make groups of three same primes and then find product of primes taking one out of each group.

$$108 = 2 \times 2 \times \underline{3 \times 3 \times 3}$$

Hence, $\sqrt[3]{108} = \sqrt[3]{2 \times 2} \times 3 = 3\sqrt[3]{4}$

Apart from the above Prime Factorisation Method, we will discuss another approach to compute cube root of an exact cube in moments. Before discussing the same, we should mind the following facts.

Exact cubes ending in	Last digit of its cube root
1	1
2	8
3	7
4	4
5	5
6	6
7	3
8	2
9	9
0	0

Observations If unit digit of an exact cube is 1/4/5/6/9/0, its cube root has the same unit digit whereas in case of 2/3/7/8 as unit digit of exact cube, its cube root has complement of 10 as unit digit.

Cube Root of Exact Cubes having Maximum of 6 Digits

Nature of options of the question reveals as to whether the given number is an exact cube or not.

Take few examples

Problem 23 What is the cube root of 79507.

Sol. Step I Make a group of three consecutive digits from extreme right (i.e. $79\,\overline{507}$). Obviously, the unit digit of the cube root would be 3 as the given number has 7 as unit digit.

Step II Try to find two such numbers, the number formed with remaining digits(i.e. 79) lies between their cubes.

Here, $4^3 < 79 < 5^3$.

Step III The lower base i.e. 4 will be tens digit of the cube root. In case the number formed with digits other than last three digits from extreme right, is a cube of a number (say n), we write n as tens digit of the required cube root.

Hence, required cube root = 43

Problem 24 $\sqrt[3]{1728} = ?$

Sol. $1\,\overline{728}$

Since the number ends with 8, the unit digit of its cube root would be 2.

Further, 1 is cube of 1. Then, 1 will be tens digit of cube root.

Thus, required cube root = 12.

Problem 25 $\sqrt[3]{830584} = ?$

Sol. $830\,\overline{584}$

Since the number ends with 4, the unit digit of its cube root would be 4.

Further, $9^3 < 830 < 10^3$

Thus, required cube root = 94

Cube Root of Exact Cubes having 7/8/9 Digits

Obviously, the cube root of a number having 7/8/9 digits will have 3 digits. Let its left digit be L, middle be M and right digit be R. The process for finding L and R will be same as indicated in case of exact cubes having maximum of 6 digits.

For example $\sqrt[3]{2146689} = ? \Rightarrow 2\,\overline{146}\,\overline{689}$

Here, $R = 9$ [$\because$ unit digit of given number is 9]

$\Rightarrow$ and $L = 1$ [$\because 1^3 < 2 < 2^3$]

Now, proceed for finding value of M.

Subtract R^3 from the given number and delete the last digit of the resultant.

$\Rightarrow 2146689 - 9^3 = 2145960 \Rightarrow 214596$ (After deletion of last digit)

Further, it is also fact that the unit digit of $(3R^2M)$ would be same as unit digit of 214596.

$\Rightarrow$ Unit digit of $(3 \cdot 9^2 \cdot M)$ = Unit digit of 214596

$\Rightarrow M = 2$

Hence, required cube root = LMR = 129.

Problem 26 $\sqrt[3]{432081216} = ?$

Sol. $\Rightarrow 432\,\overline{081}\,\overline{216}$

Here, $R = 6$ [$\because$ unit digit of given number is 6]

and $L = 7$ [$\because 7^3 < 432 < 8^3$]

Now, proceed for finding value of M:

Subtract R^3 from the given number and delete the last digit of the resultant.

$\Rightarrow 432081216 - 6^3 = 432081000$

$\Rightarrow 43208100$ (After deletion of last digit)

Further, it is also fact that the unit digit of $(3R^2M)$ would be same as unit digit of 43208100.

$\Rightarrow$ Unit digit of $(3 \cdot 6^2 \cdot M)$ = Unit digit of 43208100

$\Rightarrow M = 5$ or 0.

Hence, required cube root = LMR = 756 or 706.

Now, match with the options given. Correct answer is 756.

Problem 27 $\sqrt[3]{381078125} = ?$

Sol. $\Rightarrow 381\,\overline{078}\,\overline{125}$

Here, $R = 5$ [$\because$ unit digit of given number is 5]

and $L = 7$ [$\because 7^3 < 381 < 8^3$]

Now, subtract R^3 from the given number and delete the last digit of the resultant.

$\Rightarrow 381078125 - 5^3 = 381078000$

$\Rightarrow 38107800$ (after deletion of last digit)

Further, it is also fact that the unit digit of $(3R^2M)$ would be same as unit digit of 38107800.

$\Rightarrow$ Unit digit of $(3 \cdot 5^2 \cdot M)$ = unit digit of 38107800

$\Rightarrow M = 2$ or 4 or 6 or 8

Hence, required cube root = LMR = 725 or 745 or 765 or 785

Now, match with the options given.

Correct answer is 725.

Cube Root of any Number Even Exact Cube

We will discuss an another approach other than the above.

Problem 28 $\sqrt[3]{1061228} = ?$

Sol. ⇒ $1\ \overline{061}\ \overline{228}$

Clearly, its cube root has 3 digits and the left (extreme) digit is 1.

Let the cube root is 100 (you are free to take any figure starting with 1, but dividing a number by 100 is very convenient.)

Then, $1061228 \div 100 = 10612.28$

Further, $10612.28 \div 100 = 106.1228$

Now, average of 100, 100 and 106.1228

$$= \frac{100 + 100 + 106.1228}{3} = 102.04$$

[Alternatively, actual average = average (assumed) + $\frac{\text{sum of values obtained by subtracting the assumed average from the given each value}}{\text{number of given vaules}}$, detailed in chapter 'Average'.

Hence, required average

$$= 100 + \frac{0 + 0 + 6.1228}{3}$$

$$= 100 + 2.04 = 102.04]$$

Since 100, 100 and 106.1228 are not so closer to each other, 102.04 can only be approx. value of cube root, not the accurate value.

For more accuracy, we should start the process with 102 (not necessarily with 102.04) to make divisions easier.

$$1061228 \div 102 = 10404.2$$

Further, $10404.2 \div 102 = 102.0019$

Now, avcrago of 102, 102 and 102.0019

$$= 102 + \frac{0.0019}{3}$$

$$= 102.0006$$

Since 102, 102 and 102.0019 are closer to each other, the accurate value of cube root = 102.0006

Problem 29 $\sqrt[3]{141584} = ?$

Sol. $\Rightarrow 141\ \overline{584}$

Clearly, its cube-root has 2-digits and the left (extreme) digit is 5.

Let the cube root is 50 (you are free to take any figure starting with 5, but dividing a number by 50 is very convenient)

Then, $141584 \div 50 = 2831.68$

Further, $2831.68 \div 50 = 56.6336$

Now, average of 50, 50 and 56.6336

$$= 50 + \frac{6.6336}{3} = 52.21$$

Since 50, 50 and 56.6336 are not so closer to each other, 52.21 is not the accurate value of the required cube root.

It is only approximate value.

For more accuracy, We should start the process with 52 (not necessarily with 52.21).

The accurate value of cube root = 52.12.

Try yourself to find the approximate value of the followings

Problem 30 $\sqrt[3]{140608}$ Answer 52 (accurate value)

31 $\sqrt[3]{5735339}$ Answer 179 (accurate value)

32 $\sqrt[3]{12356789}$ Answer 231.19(accurate value)

33 $\sqrt[3]{32961}$ Answer 32.06 (accurate value)

II. Cube Root of Decimal Numbers

Problem 34 $\sqrt[3]{103.823} = ?$

Sol. If the options given indicate that 103.823 is an exact cube, then ignore decimal point. Hence, 103 823 ⇒ $103\ \overline{823}$

Since the number ends with 3, the unit digit of its cube root would be 7.

Further, $4^3 < 103 < 5^3$

Thus, cube root of 103823 = 47.

Now, we should append decimal point after one digit from the left extreme as integer value of cube root of the number before the decimal point (here, 103) is in single digit.

Then, cube root of 103.823 = 4.7

If we are not able to identify whether the given number is exact cube or not, we start the process considering only the number before the decimal point (here, 103).

Let the cube root of 103 be 4 as $4^3 < 103 < 5^3$.

Then, $103 \div 4 = 25.75$

Further, $25.75 \div 4 = 6.44$

Now, average of 4, 4 and 6.44

$$= 4 + \frac{2.44}{3} = 4.8$$

Since 4, 4 and 6.44 are not so closer, 4.8 cannot be accurate value of the required cube root, however, it is approximate value. If we proceed further, we may get the exact value of cube root.

Alternatively

We may process ignoring the decimal point ⇒ $103\ \overline{823}$

Clearly, its cube root has 2-digits and the left digit is 4 as $4^3 < 103 < 5^3$.

Let the cube-root of 103823 be 40.

Then, $103823 \div 40 = 2595.56$

Further, $2595.56 \div 40 = 64.88$

Now, average of 40, 40 and 64.88

$$= 40 + \frac{24.88}{3}$$

$$= 48.29$$

Since 40, 40 and 64.88 are not so closer, 48.29 is not the accurate value of cube root of 103823, however the approximate value.

Hence, approximate value of cube root of 103.823 = 4.829

Problem 35 $\sqrt[3]{154.85} = ?$

Sol. Let the cube root of 154 be 5 as $5^3 < 154 < 6^3$.

Then, $154 \div 5 = 30.8$

Further, $30.8 \div 5 = 6.16$

Now, average of 5, 5 and 6.16 $= 5 + \frac{1.16}{3}$

$= 5.39$

5.39 is only approximate value, not accurate as 5, 5 and 6.16 are not so closer to each other.

If we proceed further, we may get the exact value of cube root.

Problem 36 Find the cube root of 0.037.

[SSC(FCI-Mains)-12]

Sol. $\Rightarrow 0.\overline{037} = 0.037037........$

Ignore decimal point $\Rightarrow$ 0037

Let the cube root of 0037 be 3 as $3^3 < 37 < 4^3$

Then, $37 \div 3 = 12.33$

Further, $12.33 \div 3 = 4.11$

Now, average of 3, 3 and 4.11

$= 3 + \frac{1.11}{3} = 3.37$

Hence, approx. value of cube-root of 0037

$= 3.37$

Then, approx. value of cube-root of 0.037

$= 0.337 = 0.3$

Try yourself to find the approximate value of the followings.

Problem 37 $\sqrt[3]{25013.36}$ Answer 29.24(accurate value)

38 $\sqrt[3]{1465.31}$ Answer 11.35(accurate value)

C. *n*th Root of a Number

The similar procedure is adhered to as has been done in case of square root or cube root.

For examples

Problem 39 Find the fourth root of 256.

Sol. Prime Factorisation Method:

2	256
2	128
2	64
2	32
2	16
2	8
2	4
	2

Clearly, $256 = \underline{2 \times 2 \times 2 \times 2} \times \underline{2 \times 2 \times 2 \times 2}$

Then, $\sqrt[4]{256} = 2 \times 2 = 4$

Estimation-Method

Let the 4th root of 256 be 4.

Then, $256 \div 4 = 64$

Again, $64 \div 4 = 16$

Further, $16 \div 4 = 4$

Now, average of 4, 4, 4 and 4 = 4

Since 4, 4, 4 and 4 are equal, the exact value of 4th root of 256 = 4

Problem 40 What is 5th root of 5153632 ?

Sol. Prime Factorisation Method

2	5153632
2	2576816
2	1288408
2	644204
2	322102
11	161051
11	14641
11	1331
11	121
	11

Clearly, $5153632 = \underline{2 \times 2 \times 2 \times 2 \times 2} \times \underline{11 \times 11 \times 11 \times 11 \times 11}$

Then, $\sqrt[5]{5153632} = 2 \times 11 = 22$

Estimation-Method $\Rightarrow 5\overline{153632}$

Clearly, its 5th root has 2-digits and the left one is 2 as $2^5 < 51 < 3^5$.

Let the 5th root of 5153632 be 20.

Then, $5153632 \div 20 = 257681.6$

$\Rightarrow 257681.6 \div 20 = 12884.08$

$\Rightarrow 12884.08 \div 20 = 644.204$

$\Rightarrow 644.204 \div 20 = 32.2102$

Now, average of 20, 20, 20, 20 and 32.21 $= 20 + \frac{12.21}{5} = 22.42$

Since 20, 20, 20, 20 and 32.21 are not closer to each other, the approx. value of 5th root = 22.42

For more accuracy, we may repeat the process.

Problem 41 What is the 4th root of 25204.7376?

Sol. We may process ignoring the decimal point $\Rightarrow = 2\overline{5204}$

Clearly, its 4th root has 2-digits and the left digit is 1 as $1^4 < 2 < 2^4$

Let the 4th root of 25204 be 10.

Then, $25204 \div 10 = 2520.4$

Further, $2520.4 \div 10 = 252.04$

Again, $252.04 \div 10 = 25.204$

Now, average of 10, 10, 10 and 25.204 $= 10 + \frac{15.204}{4} = 13.8$

Since 10, 10, 10 and 25.204 are not closer to each other, the approx. value of 4th root of 25204.7376 = 13.8

For more accuracy, repeat the process with 13.8 or 13.

$\Rightarrow 25204 \div 13 = 1938.77$

$\Rightarrow 1938.77 \div 13 = 149.13$

$\Rightarrow 149.13 \div 13 = 11.47$

Now, average of 13, 13, 13 and 11.47 $= 13 + \frac{-1.53}{4} = 12.6$

Since 13, 13, 13 and 11.47 are closer to each other, the accurate value of 4th root of 25204.7376 = 12.6

Try yourself to find the approximate value of the followings.

Problem 42 $\sqrt[4]{4562}$ Answer 8.21 (accurate value)

43 $\sqrt[5]{1465.31}$ Answer 4.29 (accurate value)

WORKOUT

1. $\sqrt[3]{41132} = ?$ (approximate value)
(a) 35 (b) 42 (c) 22 (d) 29
(e) 45

2. $\sqrt{729} \times \sqrt{338} = ?$ (approximate value) ***[NABARD(PO)-08]***
(a) 470 (b) 482 (c) 521 (d) 530
(e) 496

3. $\sqrt{97344} = ?$
[Punjab and Sind Bank(PO)-08]
(a) 302 (b) 322 (c) 292 (d) 342
(e) 312

4. $\sqrt[2]{2230} = ?$ (approximate value)
(a) 54 (b) 59 (c) 41 (d) 37
(e) 47

5. $\sqrt{8000} = ?$ (approximate value)
[Bank of Maharashtra(AO)-09]
(a) 76 (b) 89
(c) 65 (d) 97
(e) 58

6. $\sqrt[3]{84900} = ?$ (approximate value)
(a) 56 (b) 44
(c) 67 (d) 33
(e) 21

7. What is the approximate value of $\sqrt{237.93}$?
(a) 17 (b) 21
(c) 37 (d) 15.4
(e) 21

8. $\sqrt[3]{1560896} = ?$
(a) 116 (b) 126
(c) 136 (d) 156
(e) 106

9. $\sqrt[3]{367.13} = ?$ (approximate value)
(a) 7.2 (b) 7.7 (c) 8.1 (d) 8.2
(e) 7.9

10. What is the approx. value of 6th root of 85766600.19 ?
(a) 19
(b) 25
(c) 21
(d) 28
(e) 27

Answer Key

1. (a)	2. (e)	3. (e)	4. (e)	5. (b)	6. (b)	7. (d)	8. (a)	9. (a)	10. (c)

Hints and SOLUTIONS

1. Since the given options are approximate values of cube root of 41132, we cannot apply the rule(s) meant for computing cube root of a exact cube number. Hence, we should apply the approach which is meant for finding cube root of any number even exact cube.

Given number: $41\overline{132}$

Clearly, its cube root has 2-digits and the left (extreme) digit is 3 as $3^3 < 41 < 4^3$.

Let the cube root is 30 (you are free to take any figure starting with 3, but dividing a number by 30 is very convenient.)

Then, $41132 \div 30 = 1371.06$

Further, $1371.06 \div 30 = 45.70$

Now, average of 30, 30 and 45.70

$= 30 + \dfrac{15.70}{3}$

$= 35.23$

Hence, option (a) is correct. No need to proceed further for accurate value of cube root.

Alternatively, Clearly, its cube-root has 2-digits and the left (extreme) digit is 3 as $3^3 < 41 < 4^3$.

Since only one option starts with 3, option(a) is correct.

2. $729 \Rightarrow 7\,\overline{29}$

Clearly, its square-root has 2-digits and the left (extreme) digit is 2 as $2^2 < 7 < 3^2$.

Let the square-root is 30. Then, average of 30 and $(729 \div 30) = 27.15$

Hence, $\sqrt{729} = 27$ [you can directly write $\sqrt{729} = 27$ if you are aware that square of 27 be 729]

Now, $338 \Rightarrow 3\,\overline{38}$

Clearly, its square-root has 2-digits and the left (extreme) digit is 1 as $1^2 < 3 < 2^2$.

Let the square root is 20. Then, average of 20 and $(338 \div 20) = 18.45$

Hence, $\sqrt{338} \approx 18.4$

Now, $27 \times 18.4 = 496.8$

Hence, option(e) is correct.

3. It appears from the options that 97344 is a perfect square number. Hence, you may apply the technique meant for the same.

Obviously, $9\,\overline{73}\,\overline{44}$ will have 3-digit square root ending in 2 or 8.

Further, $9\,\overline{73}\,\overline{44}$ may be written as $\overline{973}\,\overline{44}$.

But, $31^2 < 973 < 32^2$

Hence, the square root of 97344 is either 312 or 318.

$\therefore$ Option(e) is correct.

4. $2230 = 22\,\overline{30}$

Clearly, its square-root has 2-digits and the left (extreme) digit is 4 as $4^2 < 22 < 5^2$.

Let the square-root is 50. Then, average of 50 and $(2230 \div 50) = 47.3$

Hence, option(e) is correct.

5. $8000 = 80\,\overline{00}$

Clearly, its square root has 2-digits and the left (extreme) digit is 8 as $8^2 < 80 < 9^2$.

Since only one option starts with 8, option(b) is correct.

6. Given number, $84\,\overline{900}$

Clearly, its cube root has 2 digits and the left (extreme) digit is 4 as $4^3 < 84 < 5^3$.

Since only one option starts with 4, option(b) is correct.

7. Clearly, $2\overline{37}$ has square root in 2-digits and the left digit is 1 as $1^2 < 2 < 2^2$.

Let the square root of 237 be 15.

Then, average of 15 and $\dfrac{237}{15} =$ average of 15 and 15.8 = 15.4

Since 15 and 15.8 are closer to each other, 15.4 is the accurate value of square-root of 237.

Hence, option(d) is correct.

8. It is clear from the options that 1560896 is a perfect cube number.

$\Rightarrow \overline{1560}\,\overline{896}$

Here, $R = 6$

[$\because$ unit digit of given number is 6]

and $L = 1$ [$\because 1^3 = 1$]

Now, subtract R^3 from the given number and delete the last digit of the resultant.

$\Rightarrow 1560896 - 6^3 = 1560680$

$\Rightarrow$ 156068 (after deletion of last digit)

Further, it is also fact that the unit digit of $(3R^2M)$ would be same as unit digit of 156068.

$\Rightarrow$ Unit digit of $(3 \cdot 6^2 \cdot M)$

= Unit digit of 156068

$\Rightarrow \quad M = 1$ or 6

Hence, required cube-root

$= LMR = 116$ or 166.

Hence, option(a) is correct.

9. Let the cube-root of 367 be 7 as $7^3 < 367 < 8^3$.

Then, $367 \div 7 = 52.4$

Further, $52.4 \div 7 = 7.48$

Now, average of 7, 7 and 7.48

$= 7 + \dfrac{0.48}{3}$

$= 7.16$

Since 7, 7 and 7.48 are closer to each other, the accurate value of cube-root of 367 is 7.16.

Hence, option(a) is correct.

10. We may process ignoring the decimal point $\Rightarrow 85\,\overline{766600}$

Clearly, its 6th root has 2. digits and the left digit is 2 as $2^6 < 85 < 3^6$.

Let the 6th root of 85766600 be 20.

Then, $85766600 \div 20 = 4288330$

$4288330 \div 20 = 214416.5$

$214416.5 \div 20 = 10720.8$

$10720.8 \div 20 = 536.04$

$536.04 \div 20 = 26.8$

Now, average of 20, 20, 20, 20, 20 and $26.8 = 20 + \dfrac{6.8}{6}$

$= 21.13$

Since 20, 20, 20, 20, 20 and 26.8 are relatively closer to each other, the accurate value of 6th root of 85766600 will be almost equal to 21. Hence, option(c) is correct.

10

Indices and Surds

Learning OBJECTIVES

After completing this chapter, aspirants will be able to

- Understand the significance of this chapter in numerical ability;
- Comprehend the meaning of Index and base;
- Understand Laws of Indices;
- Learn meaning of Surd and its sorts with examples;
- Learn how to compare the magnitudes of surds;
- Understand the philosophy of the categorisation of the questions;
- Analyse the trend of the questions being asked in various competitive examinations;
- Recognise the types of question which can be easily solved through the given options;
- Grip on concepts helping to solve the questions based on Indices/Surds in moments;
- Learn new concepts/fundamentals;
- Apply the pragmatic approaches in lieu of academic approaches; and
- Ponder on the alternative methods.

Inside

Introduction

This chapter is, in principle, extension of the concept of square/cube and nth root of a number. No doubt, it is very critical for the competitive examinations as almost all the competitive examinations like CAT, MAT, SSC, Banks, Railway, LIC etc. cover this topic. We will discuss both the concepts 'Indices and Surds' in detail.

A. Indices

$a^n = a \times a \times a \times \ldots . n$ times. Here, n is called 'index' or 'exponent' and a is called 'base'. We read a^n as 'a to the power n' or 'a raised to the power n'.

$base \Leftarrow a^{n \Rightarrow exponent/\, index}$

Laws of Indices

Keeping in view the pragmatic approach and building your concept, a need is felt to give an example before each of laws of Exponent/Indices.

(a) $2^5 \times 2^3 = \underline{2 \times 2 \times 2 \times 2 \times 2} \times \underline{2 \times 2 \times 2} = 2^8$

It implies that $\boldsymbol{a^m \times a^n = a^{m+n}}$

Similarly, $\boldsymbol{a^m \times a^n \times a^p = a^{m+n+p}}$

(b) $2^5 \div 2^3 = \frac{2\times2\times2\times2\times2}{2\times2\times2} = 2^2$

It implies that $\boldsymbol{a^m \div a^n = a^{m-n}}$

Similarly, $\boldsymbol{a^m \div a^n \div a^p = a^{m-n-p}}$

(c) $2^3 \div 2^3 = \frac{2\times2\times2}{2\times2\times2} = 1$

It implies that $a^m \div a^m = 1$

$\Rightarrow \qquad a^{m-m} = 1 \Rightarrow \boldsymbol{a^0 = 1}$

(d) $(2^3)^4 = 2^3 \times 2^3 \times 2^3 \times 2^3 = 2^{12}$

It implies that $\boldsymbol{(a^m)^n = a^{mn}}$

(e) $2^{3^4} = 2^{3\times3\times3\times3} = 2^{81}$

It implies that $\boldsymbol{a^{m^n} = (a)^{m^n}}$

From (d) and (e), $\boldsymbol{(a^m)^n \neq a^{m^n}}$

(f) $(2\times3)^4 = (2\times3)\times(2\times3)\times(2\times3)\times(2\times3)$

$= \underline{2\times2\times2\times2\times3\times3\times3\times3} = 2^4\times3^4$

It implies that $\boldsymbol{(ab)^m = a^m \cdot b^m}$

Similarly, $\boldsymbol{(a^m \cdot b^m)^n = a^{mn} \cdot b^{mn}}$

In addition to above, some facts, given below, are also critical.

(i) $\boldsymbol{a^n = \frac{1}{a^{-n}}}$ or $\boldsymbol{a^{-n} = \frac{1}{a^n}}$

e.g. $2^3 = \frac{1}{2^{-3}}$ or $2^{-3} = \frac{1}{2^3}$

(ii) **If $\boldsymbol{a^x = a^y}$, then $\boldsymbol{x = y}$**

e.g. if $4^x = 4^7$, then $x = 7$

Also, **if $\boldsymbol{a^x = b^x}$, then $\boldsymbol{a = b}$**

e.g. if $a^7 = 4^7$, then $a = 4$

(iii) **If $\boldsymbol{a^x = m}$, then $\boldsymbol{a = m^{\frac{1}{x}}}$**

e.g. if $a^3 = 8$, then $a = 8^{\frac{1}{3}} = \sqrt[3]{8} = \sqrt[3]{2\times2\times2} = 2$

(iv) **If $\boldsymbol{a^{\frac{1}{x}} = m}$, then $\boldsymbol{a = m^x}$**

e.g. $a^{\frac{1}{3}} = 3$, then $a = 3^3 = 27$

(v) **If $\boldsymbol{a^x = m^y}$, then $\boldsymbol{a = m^{\frac{y}{x}}}$ and $\boldsymbol{m = a^{\frac{x}{y}}}$**

e.g. if $a^2 = 4^6$, then $a = 4^{\frac{6}{2}} = 4^3 = 64$

Further, if $(64)^2 = m^6$, then $m = (64)^{\frac{2}{6}} = (64)^{\frac{1}{3}}$

$= \sqrt[3]{64} = \sqrt[3]{4\times4\times4} = 4$

B. Surds

Before discussing surds, it is imperative to know the rational and irrational numbers. A number of the form p/q is called a rational number, where p and q are integers, prime to each other and $q \neq 0$.

e.g. : $2\left(i.e. \frac{2}{1}\right), \frac{2}{3}, \frac{-3}{4}$, etc.

On the other hand, a number which is not rational is called irrational number. It means an irrational number cannot be expressed in the form p/q, where p and q are integers, prime to each other and $q \neq 0$. e.g.: $\sqrt{2}, \sqrt{3}$ etc.

When a root of a rational number 'a' (i.e. value of $\sqrt[n]{a}$) cannot be exactly obtained, then this root is called a surd. e.g.: $\sqrt{2}, \sqrt[3]{4}, \sqrt[3]{10}$, etc. But, $\sqrt{4}$ and $\sqrt[3]{125}$ are not surds as their values are 2 and 5 respectively. It may be construed from the above that all surds are irrational numbers but its converse is not true.

$\sqrt[n]{a}$ is called 'surd' $\Rightarrow \sqrt[3]{7}$

a is called 'radicand' $\Rightarrow 7$

and n is called 'order of the surd' $\Rightarrow 3$

Also, the surds of second order are known as 'quadratic surds', *viz.* $\sqrt{5}, \sqrt{7}$, etc. The surds of third order are known as 'cubic surds', *viz.* $\sqrt[3]{3}, \sqrt[3]{11}$, etc. and the surds of fourth order are known as 'biquadratic surds', *viz.* $\sqrt[4]{3}, \sqrt[4]{15}, 5^{\frac{7}{4}}$, etc.

Pure Surd A surd having no rational factor is called Pure Surd. e.g. $\sqrt{10}$, $\sqrt[3]{7}$, etc.

Mixed Surd The product of a rational number and a surd is known as Mixed Surd. e.g. $5\sqrt{2}, 7\sqrt[3]{16}, \sqrt{40}$ (i.e. $2\sqrt{10}$), etc.

Similar Surd Surds having the same irrational factor are called similar surds or like surds. e.g. $\sqrt{2}$, $5\sqrt{2}$ and $\sqrt{8}$ (i.e. $5\sqrt{2}$) are similar surds.

Conjugate or Complementary Surds The conjugate surd of a surd $(\sqrt{a}+\sqrt{b})$ is $(\sqrt{a}-\sqrt{b})$ and *vice-versa*. Also, the conjugate surd of a surd $(a+\sqrt{b})$ is $(a-\sqrt{b})$ and *vice-versa*. It is important to note that their product is a rational number.e.g. $(\sqrt{5}+\sqrt{7})$ and $(\sqrt{5}-\sqrt{7})$ are conjugate to each other.

Rationalisation of Surds The process of multiplying a given surd by an another surd in order to obtain the product as a rational number, is called rationalisation of the given surd.

Examples: $3\sqrt{5}$ (i.e. given surd) $\times \sqrt{5} = 15$

and $(3+2\sqrt{2})$ (i.e. given surd) $\times (3-2\sqrt{2}) = 9-8 = 1$

Here, $\sqrt{5}$ is a rationalizing factor of $3\sqrt{5}$ and $(3-2\sqrt{2})$ is a rationalizing factor of $(3+2\sqrt{2})$.

Note

(i) The rationalising factor of $a^{\frac{1}{n}}$ *is* $a^{\left(1-\frac{1}{n}\right)}$.
(ii) The rationalising factor of $(\sqrt{a}+\sqrt{b})$ *is* $(\sqrt{a}-\sqrt{b})$ *and vice-versa.*

Illustrations

1. The rationalising factor of $3\sqrt{3}$ is *[SSC(CPO)-03]*
$\Rightarrow \sqrt{3}$

2. Find the value of $\frac{\sqrt{5}-\sqrt{3}}{\sqrt{5}+\sqrt{3}}$. *[RRB(ASM)/Ajmer-08]*

$$\Rightarrow \frac{\sqrt{5}-\sqrt{3}}{\sqrt{5}+\sqrt{3}} \times \frac{\sqrt{5}-\sqrt{3}}{\sqrt{5}-\sqrt{3}}$$

{Rationalising the denominator}

$$= \frac{(\sqrt{5})^2+(\sqrt{3})^2-2\cdot\sqrt{5}\,\sqrt{3}}{(\sqrt{5})^2-(\sqrt{3})^2} = \frac{8-2\sqrt{15}}{5-3} = 4-\sqrt{15}$$

Some Critical Facts

(i) Only similar surds can be added or subtracted.

For example:

- $7\sqrt{2}+9\sqrt{2}=16\sqrt{2}$
- $8\sqrt{3}-5\sqrt{3}=3\sqrt{3}$
- $18\sqrt{2}-\sqrt{50}=18\sqrt{2}-\sqrt{2\times 25}$
  $=18\sqrt{2}-5\sqrt{2}=13\sqrt{2}$
- $\sqrt{8}-\sqrt{4}-\sqrt{2}=?$ *[SSC (CGL-PT)-2000]*
  $\Rightarrow \sqrt{2\times 4}-\sqrt{4}-\sqrt{2}=2\sqrt{2}-2-\sqrt{2}=(2\sqrt{2}-\sqrt{2})-2$
  $=\sqrt{2}-2=\sqrt{2}-\sqrt{2}\cdot\sqrt{2}=\sqrt{2}(1-\sqrt{2})$
- $7\sqrt{3}+10\sqrt{5}=7\sqrt{3}+10\sqrt{5}$ ($\because$ surds are not similar)

(ii) Surds of same order are multiplied or divided as per following laws:

- $\sqrt[n]{a}\times\sqrt[n]{b}=\sqrt[n]{ab}$, e.g. $\sqrt[5]{3}\times\sqrt[5]{7}=\sqrt[5]{21}$
- $\sqrt[n]{a}\div\sqrt[n]{b}=\sqrt[n]{\frac{a}{b}}$, e.g. $\sqrt[3]{6}\div\sqrt[3]{2}=\sqrt[3]{\frac{6}{2}}=\sqrt[3]{3}$

(iii) In case of same order, the greater radicand, the greater surd.

- $\boldsymbol{\sqrt[n]{a}>\sqrt[n]{b}}$ **if** $\boldsymbol{a>b}$
- $\boldsymbol{\sqrt[n]{a}<\sqrt[n]{b}}$ **if** $\boldsymbol{a<b}$ e.g. $\sqrt[3]{7}>\sqrt[3]{5}$ as $7>5$

(iv) **If** $\boldsymbol{a+\sqrt{b}=c+\sqrt{d}}$ **or** $\boldsymbol{a-\sqrt{b}=c-\sqrt{d}}$**, then** $\boldsymbol{a=c}$ **and** $\boldsymbol{b=d}$**.**

e.g. Find the values of rational numbers a and b if $\frac{4+3\sqrt{5}}{4-3\sqrt{5}}=a+b\sqrt{5}$

$$\Rightarrow \frac{4+3\sqrt{5}}{4-3\sqrt{5}}=\frac{4+3\sqrt{5}}{4-3\sqrt{5}}\times\frac{4+3\sqrt{5}}{4+3\sqrt{5}}$$

{Rationalising the denominator}

$$=\frac{(4)^2+(3\sqrt{5})^2+2\cdot 4\cdot 3\sqrt{5}}{(4)^2-(3\sqrt{5})^2}=\frac{61+24\sqrt{5}}{-29}$$

$$=\frac{61}{-29}+\frac{24\sqrt{5}}{-29}=a+b\sqrt{5} \quad \text{(as given)}$$

$$\Rightarrow a=\frac{-61}{29} \text{ and } b=\frac{-24}{29}$$

(v) **If** $\boldsymbol{a+\sqrt{b}=c+\sqrt{d}}$ **then** $\boldsymbol{a-\sqrt{b}=c-\sqrt{d}}$ **and vice-versa.**

(vi) $\boldsymbol{(\sqrt[n]{a})^n=a}$, e.g. $(\sqrt[3]{7})^3=7$

(vii) $\boldsymbol{\sqrt[m]{\sqrt[n]{a}}=\sqrt[mn]{a}=\sqrt[n]{\sqrt[m]{a}}}$, e.g. $\sqrt[4]{\sqrt[3]{5}}=\sqrt[12]{5}=\sqrt[3]{\sqrt[4]{5}}$,

(viii) $\boldsymbol{(\sqrt[n]{a})^m=\sqrt[n]{a^m}}$, e.g. $(\sqrt[3]{11})^5=\sqrt[3]{(11)^5}$

(ix) $\boldsymbol{\sqrt[n]{a}=\sqrt[nm]{a^m}}$, e.g. $\sqrt[4]{3}=\sqrt[4\cdot5]{3^5}=\sqrt[20]{3^5}$

Comparison of Magnitudes of Surds

As we are aware that 'the greater radicand, the greater surd' in case of same order. Thus, we ought to try to equalize the orders of different given surds so as to compare them. Further, the best way for equalizing the orders is to make each order equal to LCM of all given orders by multiplying it with an appropriate number.

Illustrations

- Which of the following is the greatest ?
$\sqrt{2}$, $\sqrt[3]{3}$ and $\sqrt[4]{4}$ *[RRB(ASM)-Ajmer-01, CAT-06, SSC(SAS)-10, SSC(CGL-T-II)-10]*

Given surds are: $\sqrt[2]{2}$, $\sqrt[3]{3}$ and $\sqrt[4]{4}$

LCM of orders 2, 3 and 4 = 12

Now, we have to make each order equal to 12 by multiplying it with an appropriate number.

$$\sqrt[2]{2}\Rightarrow\sqrt[2\cdot6]{2^6}=\sqrt[12]{64}$$
$$\sqrt[3]{3}\Rightarrow\sqrt[3\cdot4]{3^4}=\sqrt[12]{81}$$
$$\sqrt[4]{4}\Rightarrow\sqrt[4\cdot3]{4^3}=\sqrt[12]{64}$$

In case of same order, 'the greater radicand, the greater surd', hence, $\sqrt[12]{81}$ is greatest. The corresponding surd is $\sqrt[3]{3}$.

- Which is greater : $(\sqrt{15}-\sqrt{13})$ or $(\sqrt{19}-\sqrt{17})$

$$(\sqrt{15}-\sqrt{13})=\frac{(\sqrt{15}-\sqrt{13})}{(\sqrt{15}+\sqrt{13})}\times(\sqrt{15}+\sqrt{13})$$
$$=\frac{2}{(\sqrt{15}+\sqrt{13})}$$

and $(\sqrt{19}-\sqrt{17})=\frac{(\sqrt{19}-\sqrt{17})}{(\sqrt{19}+\sqrt{17})}\times(\sqrt{19}+\sqrt{17})=\frac{2}{(\sqrt{19}+\sqrt{17})}$

Since $(\sqrt{19}+\sqrt{17})>(\sqrt{15}+\sqrt{13})$, $\frac{2}{(\sqrt{15}+\sqrt{13})}>\frac{2}{(\sqrt{19}+\sqrt{17})}$

Hence, $(\sqrt{15}-\sqrt{13})>(\sqrt{19}-\sqrt{17})$

It may be construed from the above that if the differences between numbers appeared in form of square roots of each term are positive and equal, the term having lowest number will have the greatest value and the term having highest number will have lowest value. Here, difference refers 1st number minus 2nd number.

In case of negative and equal differences, the term having lowest number will have the lowest value and the term having highest number will have the greatest value.

This concept can also be employed for the numbers appearing in form of nth root.

Here, $15 - 13 = 19 - 17 = +2$

Another example

$$(\sqrt[3]{10} - \sqrt[3]{13}) < (\sqrt[3]{25} - \sqrt[3]{28}) < (\sqrt[3]{26} - \sqrt[3]{29})$$

It is also important to note that if the sum of the two numbers appeared in form of nth-root under each term, are equal, then the term having lowest difference between its numbers will have greatest value. On the contrary, the term having highest difference will have lowest value.

Compare $\sqrt[n]{50} + \sqrt[n]{40}$, $\sqrt[n]{60} + \sqrt[n]{30}$ and $\sqrt[n]{20} + \sqrt[n]{70}$. Here, $50 + 40 = 60 + 30 = 70 + 20$, but, the difference between 50 and 40 is 10, that between 60 and 30 is 30 and that between 20 and 70 is 50.

Since 1st term has lowest difference, it will have greatest value. We may arrange them in ascending or descending order considering the quantum of differences.

Hence, descending order:

$$\sqrt[n]{50} + \sqrt[n]{40},\ \sqrt[n]{60} + \sqrt[n]{30},\ \sqrt[n]{20} + \sqrt[n]{70}$$

Which of the following statement(s) is/are true?

I. $(\sqrt{11} + \sqrt{7}) < (\sqrt{10} + \sqrt{8})$

II. $(\sqrt{17} + \sqrt{11}) > (\sqrt{15} + \sqrt{13})$ ***[SSC(CGL-II)-17]***

Statement(I): Sum are equal i.e. 18. Since $(10 - 8)$ is lesser than $(11 - 7)$, $(\sqrt{10} + \sqrt{8}) > (\sqrt{11} + \sqrt{7})$. Hence, Statement(I) is correct.

Statement(II): Sum are equal i.e. 28. Since $(15 - 13)$ is lesser than $(17 - 11)$, $(\sqrt{15} + \sqrt{13}) > (\sqrt{17} + \sqrt{11})$.

Hence, Statement(II) is not true.

∴ Only Statement(I) is correct.

Problems Based on Indices and Surds

Type I

$$\sqrt[n]{a\sqrt[n]{a\sqrt[n]{a\sqrt[n]{a\ldots\ldots}}}} = ?$$

Let $\sqrt[n]{a\sqrt[n]{a\sqrt[n]{a\sqrt[n]{a\sqrt[n]{a\ldots\ldots}}}}} = x$

⇒ $a\sqrt[n]{a\sqrt[n]{a\sqrt[n]{a\ldots\ldots}}} = x^n$ (∵ both sides raised to the power n)

⇒ $a \cdot x = x^n$

⇒ $a = x^{n-1}$

⇒ $x = \sqrt[n-1]{a} = a^{\frac{1}{(n-1)}}$

Hence, $\sqrt[n]{a\sqrt[n]{a\sqrt[n]{a\sqrt[n]{a\ldots\ldots}}}} = \sqrt[n-1]{a} = a^{\frac{1}{(n-1)}}$

Problem 1 $\sqrt{5\sqrt{5\sqrt{5\ldots\ldots\ldots}}} = ?$

Sol. Let $\sqrt{5\sqrt{5\sqrt{5\ldots\ldots\ldots}}} = x$

⇒ $5\sqrt{5\sqrt{5\ldots\ldots\ldots}} = x^2$ {Squaring both sides}

⇒ $5 \times x = x^2$

⇒ $x = 5$

Alternatively

$\sqrt[2]{5\sqrt[2]{5\sqrt[2]{5\ldots\ldots\ldots}}} = 5^{\frac{1}{(2-1)}} = 5$

Problem 2 $\sqrt[3]{5\sqrt[3]{5\sqrt[3]{5\sqrt[3]{5\ldots\ldots\ldots}}}} = ?$

Sol. Let $\sqrt[3]{5\sqrt[3]{5\sqrt[3]{5\sqrt[3]{5\ldots\ldots\ldots}}}} = x$

⇒ $5\sqrt[3]{5\sqrt[3]{5\sqrt[3]{5\ldots\ldots\ldots}}} = x^3$ {cubing both sides}

⇒ $5 \times x = x^3$

⇒ $5 = x^2$

⇒ $x = \sqrt{5}$

Alternatively

$\sqrt[3]{5\sqrt[3]{5\sqrt[3]{5\sqrt[3]{5\ldots\ldots\ldots}}}} = \sqrt[3-1]{5} = \sqrt{5}$

Type II

$$\sqrt[n]{a\sqrt[n]{a\sqrt[n]{a}}} = ?$$

⇒ $\sqrt[n]{a\sqrt[n]{a\sqrt[n]{a}}} = \sqrt[n]{a\sqrt[n]{a \cdot a^{\frac{1}{n}}}} = \sqrt[n]{a\sqrt[n]{a^{\frac{n+1}{n}}}}$

$= \sqrt[n]{a \cdot a^{\frac{n+1}{n^2}}} = \sqrt[n]{a^{\frac{n^2+n+1}{n^2}}} = a^{\frac{n^2+n+1}{n^3}} = a^{\frac{(n^3-1)}{(n-1)n^3}}$

Hence, in case of 3 times of a, the resultant $= a^{\frac{(n^3-1)}{(n-1)n^3}}$

Similarly, in case of 4 times of a, the resultant $= a^{\frac{(n^4-1)}{(n-1)n^4}}$

and in case of 5 times of a, the resultant $= a^{\frac{(n^5-1)}{(n-1)n^5}}$ and so on.

Problem 3 $\sqrt{2\sqrt{2\sqrt{2\sqrt{2\sqrt{2}}}}} = ?$ ***[SSC(SAS)-10]***

Sol. ⇒ $\sqrt{2\sqrt{2\sqrt{2\sqrt{2\sqrt{2}}}}} = \sqrt{2\sqrt{2\sqrt{2\sqrt{2 \cdot 2^{\frac{1}{2}}}}}} = \sqrt{2\sqrt{2\sqrt{2\sqrt{2^{1+\frac{1}{2}}}}}}$

$$= \sqrt{2\sqrt{2\sqrt{2 \cdot 2^{\frac{3}{2 \cdot 2}}}}} = \sqrt{2\sqrt{2\sqrt{2^{1+\frac{3}{4}}}}} = \sqrt{2\sqrt{2 \cdot 2^{\frac{7}{4 \cdot 2}}}}$$

$$= \sqrt{2\sqrt{2^{1+\frac{7}{8}}}} = \sqrt{2 \cdot 2^{\frac{15}{8 \cdot 2}}} = \sqrt{2^{1+\frac{15}{16}}} = \sqrt{2^{\frac{31}{16 \cdot 2}}} = 2^{\frac{31}{32}}$$

Alternatively

Since a (say, 2) has appeared 5 times, the required answer

$$= a^{\frac{(n^5 - 1)}{(n-1)n^5}}, \text{ where } n \text{ be } 2. = 2^{\frac{(2^5 - 1)}{(2-1)2^5}} = 2^{\frac{31}{32}}$$

Problem 4 $\sqrt[5]{7\sqrt[5]{7\sqrt[5]{7\sqrt[5]{7}}}} = ?$

Sol. Since a (say, 7) has appeared 4 times, the required answer

$$= a^{\frac{(n^4 - 1)}{(n-1)n^4}}, \text{ where } n \text{ be } 5 = 7^{\frac{(5^4 - 1)}{(5-1)5^4}} = 7^{\frac{156}{625}}$$

Type III

Problem 5 $\sqrt{6 + \sqrt{6 + \sqrt{6 + \sqrt{6 + \ldots\ldots\ldots}}}} = ?$

[CDS-13, SSC(SAS)-10, SSC(CGL-T-II)-12, 13]

Sol. Let $\sqrt{6 + \sqrt{6 + \sqrt{6} + \sqrt{6 + \ldots\ldots\ldots}}} = x$

Then, $\sqrt{6 + x} = x$

$\Rightarrow \quad 6 + x = x^2$ {Squaring both sides}

$\Rightarrow \quad x^2 - x - 6 = 0$

$\Rightarrow \quad x^2 - 3x + 2x - 6 = 0$

$\Rightarrow \quad x(x - 3) + 2(x - 3) = 0 \Rightarrow x = -2$, and 3

The value of $\sqrt{6 + \sqrt{6 + \sqrt{6 + \sqrt{6 + \ldots\ldots\ldots}}}}$ cannot be negative as value of $\sqrt{6 + \text{positive number}}$ cannot be negative.

Hence, required answer = 3

Alternatively-I,

Split the appearing number in two parts in such a way that their difference is 1 and their product is the number itself.

If '+' sign appears in the given term, then the required answer = greater part and in case of '–' sign, the required answer = smaller part

Here, appearing number i.e. $6 = 2 \times 3$

Since there is positive sign, the required answer = greater part = 3

Alternatively II,

If '+' sign appears, then the required answer

$$= \frac{\sqrt{(4 \cdot \text{appeared number}) + 1} + 1}{2}$$

In case of negative sign, the required answer

$$= \frac{\sqrt{(4 \cdot \text{appeared number}) + 1} - 1}{2}$$

Here, the required answer $= \dfrac{\sqrt{(4 \cdot \text{appeared number}) + 1} + 1}{2}$

$$= \frac{\sqrt{(4 \cdot 6) + 1} + 1}{2} = 3$$

Problem 6 $\sqrt{6 - \sqrt{6 - \sqrt{6 - \sqrt{6 - \ldots\ldots\ldots}}}} = ?$

Sol. $\Rightarrow \quad 6 = 2 \times 3$

Since there is negative sign, the required answer = smaller part = 2

You may try with another method too as given above.

Problem 7 $\sqrt{2 + \sqrt{2 + \sqrt{2 + \sqrt{2} + \ldots\ldots\ldots}}} = ?$

[SSC(CGL-PT)-08, CDS-19]

Sol. $\Rightarrow \quad 2 = 1 \times 2$

Since there is positive sign, the required answer = greater part = 2

You may try with another method too as given in box.

Problem 8 $\sqrt{1 + \sqrt{1 + \sqrt{1 + \sqrt{1 + \ldots\ldots\ldots}}}} = ?$ **[CDS-17]**

Sol. Let $\sqrt{1 + \sqrt{1 + \sqrt{1 + \sqrt{1 + \ldots\ldots\ldots}}}} = x$

Then, $\sqrt{1 + x} = x$

$\Rightarrow \quad 1 + x = x^2$ {Squaring both sides}

$\Rightarrow \quad x^2 - x - 1 = 0$

$\Rightarrow \quad x = \dfrac{-(-1) \pm \sqrt{(-1)^2 - 4 \cdot 1 \cdot (-1)}}{2 \cdot 1}$

$\Rightarrow \quad x = \dfrac{1 \pm \sqrt{5}}{2 \cdot 1}$

$\Rightarrow \quad x = \dfrac{1 + \sqrt{5}}{2}$ and $x = \dfrac{1 - \sqrt{5}}{2}$

The value of $\sqrt{1 + \sqrt{1 + \sqrt{1 + \sqrt{1 + \ldots\ldots}}}}$ cannot be negative as value of $\sqrt{1 + \text{positive number}}$ cannot be negative.

Hence, $x = \dfrac{1 + \sqrt{5}}{2} = \dfrac{1 + 2.236}{2} = 1.61$

Alternatively I

$\Rightarrow \quad 1 = 1.6 \times 0.6$ (approx.)

Since there is positive sign, the required answer = greater part = 1.6

Alternatively II

Here, the required answer $= \dfrac{\sqrt{(4 \cdot \text{appeared number}) + 1} + 1}{2}$

$$= \frac{\sqrt{(4 \cdot 1) + 1} + 1}{2} = \frac{1 + 2.236}{2} = 1.61$$

Problem 9 $\sqrt{8 - \sqrt{8 - \sqrt{8 - \sqrt{8 - \ldots\ldots\ldots}}}} = ?$

Sol. Here, the required answer $= \dfrac{\sqrt{(4 \cdot \text{appeared number}) + 1} - 1}{2}$

$$= \frac{\sqrt{(4 \cdot 8) + 1} - 1}{2} = \frac{\sqrt{33} - 1}{2} = 2.37$$

{You may compute value of $\sqrt{33}$ quickly with the help of Estimation Method, elaborated in chapter 'Square- Root, Cube-Root and nth-Root'.}

Type IV

➘ Problem 10 $\sqrt{-\sqrt{3}+\sqrt{3+8\sqrt{7+4\sqrt{3}}}}=?$

Sol. $\Rightarrow\sqrt{-\sqrt{3}+\sqrt{3+8\sqrt{7+4\sqrt{3}}}}$

$$=\sqrt{-\sqrt{3}+\sqrt{3+8\sqrt{2^2+(\sqrt{3})^2+2\cdot 2\cdot\sqrt{3}}}}$$

$$=\sqrt{-\sqrt{3}+\sqrt{3+8\sqrt{(2+\sqrt{3})^2}}}$$

$[\because a^2+b^2+2ab=(a+b)^2]$

$$=\sqrt{-\sqrt{3}+\sqrt{3+16+8\sqrt{3}}}$$

$$=\sqrt{-\sqrt{3}+\sqrt{4^2+(\sqrt{3})^2+2\cdot 4\cdot\sqrt{3}}}$$

$$=\sqrt{-\sqrt{3}+\sqrt{(4+\sqrt{3})^2}}=\sqrt{-\sqrt{3}+(4+\sqrt{3})}=2$$

➘ Problem 11 If $p=999$, then $\sqrt[3]{p(p^2+3p+3)+1}=?$

[SSS(CGL-PT)-02, 08, SSC(CGL-T-I)-11]

Sol. $\Rightarrow\sqrt[3]{p(p^2+3p+3)+1}=\sqrt[3]{p^3+3p^2+3p+1}$

$$=\sqrt[3]{(p+1)^3}=p+1=999+1=1000$$

➘ Problem 12 $\sqrt{(0.798)^2+0.404\times 0.798+(0.202)^2}+1=?$

Sol. $\sqrt{(0.798)^2+0.404\times 0.798+(0.202)^2}+1$

$$=\sqrt{(0.798)^2+2\times 0.202\times 0.798+(0.202)^2}+1$$

$$=\sqrt{(0.798+0.202)^2}+1=0.798+0.202+1=2$$

➘ Problem 13 $\sqrt{8-2\sqrt{15}}=?$

Sol. $\Rightarrow\sqrt{8-2\sqrt{15}}=\sqrt{(\sqrt{5})^2+(\sqrt{3})^2-2\cdot\sqrt{5}\cdot\sqrt{3}}$

$$=\sqrt{(\sqrt{5}-\sqrt{3})^2}=\sqrt{5}-\sqrt{3}$$

➘ Problem 14 $\dfrac{\sqrt{4-\sqrt{7}}}{\sqrt{8+3\sqrt{7}-2\sqrt{2}}}=?$

Sol. $\Rightarrow\dfrac{\sqrt{4-\sqrt{7}}}{\sqrt{8+3\sqrt{7}}-2\sqrt{2}}=\dfrac{\sqrt{\dfrac{(\sqrt{7})^2+(1)^2-2\sqrt{7}\cdot 1}{2}}}{\sqrt{\dfrac{(\sqrt{7})^2+(3)^2+2\cdot\sqrt{7}\cdot 3}{2}}-2\sqrt{2}}$

$$=\frac{\sqrt{\dfrac{(\sqrt{7}-1)^2}{2}}}{\sqrt{\dfrac{(\sqrt{7}+3)^2}{2}}-2\sqrt{2}}=\frac{\dfrac{(\sqrt{7}-1)}{\sqrt{2}}}{\dfrac{(\sqrt{7}+3)}{\sqrt{2}}-2\sqrt{2}}=\frac{(\sqrt{7}-1)}{(\sqrt{7}+3)-4}=1$$

Type V

➘ Problem 15 The greatest of $\sqrt{2}$, $\sqrt[6]{3}$, $\sqrt[3]{4}$ and $\sqrt[4]{5}$ is :

[SSC(CGL-PT)-05, SSC(CGL-T-II)-17]

Sol. Given surds are: $\sqrt[2]{2}$, $\sqrt[6]{3}$, $\sqrt[3]{4}$ and $\sqrt[4]{5}$

LCM of orders 2, 6, 3 and 4 = 12

Now, we have to make each order equal to 12 by multiplying it with an appropriate number.

$$\sqrt[2]{2}\Rightarrow\sqrt[2\cdot 6]{2^6}=\sqrt[12]{64}$$

$$\sqrt[6]{3}\Rightarrow\sqrt[6\cdot 2]{3^2}=\sqrt[12]{9}$$

$$\sqrt[3]{4}\Rightarrow\sqrt[3\cdot 4]{4^4}=\sqrt[12]{256}$$

$$\sqrt[4]{5}\Rightarrow\sqrt[4\cdot 3]{5^3}=\sqrt[12]{125}$$

In case of same order, 'the greater radicand, the greater surd', hence, $\sqrt[12]{256}$ is greatest. The corresponding surd is $\sqrt[3]{4}$.

➘ Problem 16 Which is the greatest among: $(\sqrt{19}-\sqrt{17})$, $(\sqrt{13}-\sqrt{11})$, $(\sqrt{7}-\sqrt{5})$ and $(\sqrt{5}-\sqrt{3})$

[SSC(CGL-PT)-05, SSC(CGL-T-II)-10]

Sol. Here, differences between numbers appeared under square roots of each term are positive and equal, i.e.

$19-17=13-11=7-5=5-3=+2$

Then, the term having lowest number will have the greatest value. Hence, $(\sqrt{5}-\sqrt{3})$ is the greatest term as it has lowest number 3.

➘ Problem 17 Which is the greatest among $(\sqrt[5]{41}-\sqrt[5]{46})$, $(\sqrt[5]{33}-\sqrt[5]{38})$, $(\sqrt[5]{39}-\sqrt[5]{44})$ and $(\sqrt[5]{28}-\sqrt[5]{33})$

Sol. We know, in case of negative and equal differences, the term having the lowest number will have the lowest value and the term having the highest number will have the greatest value. Since $(\sqrt[5]{41}-\sqrt[5]{46})$ has the highest number (say, 46), it will have the greatest value.

➘ Problem 18 Arrange the following terms in an ascending order

$$(\sqrt[3]{21}-\sqrt[3]{24}),\ (\sqrt[3]{8}-\sqrt[3]{11}),\ (\sqrt[3]{19}-\sqrt[3]{22})\text{ and }(\sqrt[3]{27}-\sqrt[3]{30})$$

Sol. We know, in case of negative and equal differences, the term having the lowest number will have the lowest value and the term having the highest number will have the greatest value. Now, we may arrange in ascending order considering 'the lower number, the lower value'.

Hence, the ascending order

$$(\sqrt[3]{8}-\sqrt[3]{11}),\ (\sqrt[3]{19}-\sqrt[3]{22}),\ (\sqrt[3]{21}-\sqrt[3]{24}),\ (\sqrt[3]{27}-\sqrt[3]{30})$$

Type VI

➘ Problem 19 $\{(-2)^{-2}\}^{-2}=?$

Sol. $\Rightarrow\quad\{(-2)^{-2}\}^{-2}=(-2)^{-2(-2)}=(-2)^4=16$

➘ Problem 20 $\left[\left\{\left(-\dfrac{1}{2}\right)^2\right\}^{-2}\right]^{-1}=?$

[SSC (10+2-LDC/DEO)-10]

Sol. $\Rightarrow\left[\left\{\left(-\dfrac{1}{2}\right)^2\right\}^{-2}\right]^{-1}=\left\{\left(-\dfrac{1}{2}\right)^2\right\}^{-2\cdot(-1)}$

$$=\left(-\frac{1}{2}\right)^{2\cdot 2}=\left(-\frac{1}{2}\right)^4=\frac{1}{16}$$

Problem 21 $\left\{\left(\sqrt[5]{x^{-\frac{3}{5}}}\right)^{-\frac{5}{3}}\right\}^5 = ?$ *[SSC(CGL-T-I)-10]*

Sol. $\Rightarrow \left\{\left(\sqrt[5]{x^{-\frac{3}{5}}}\right)^{-\frac{5}{3}}\right\}^5 = \left(\sqrt[5]{x^{-\frac{3}{5}}}\right)^{-\frac{5\cdot 5}{3}} = x^{-\frac{3}{5\cdot 5}\times\left(-\frac{5\cdot 5}{3}\right)} = x$

Problem 22 $\left(\frac{1}{2}\right)^{-\frac{1}{2}} = ?$

Sol. $\Rightarrow \left(\frac{1}{2}\right)^{-\frac{1}{2}} = \left(\frac{1}{2^1}\right)^{-\frac{1}{2}} = (2^{-1})^{-\frac{1}{2}} = (2)^{\frac{1}{2}} = \sqrt{2}$

Alternatively

$\Rightarrow \left(\frac{1}{2}\right)^{-\frac{1}{2}} = \frac{1}{\left(\frac{1}{2}\right)^{\frac{1}{2}}} = \left(\frac{1}{\frac{1}{2}}\right)^{\frac{1}{2}} = (2)^{\frac{1}{2}} = \sqrt{2}$

Problem 23 $(0.04)^{-1.5} = ?$

Sol. $\Rightarrow (0.04)^{-1.5} = \left(\frac{1}{25}\right)^{-1.5}$

$= (25^{-1})^{-1.5} = (25)^{1.5} = (5^2)^{1.5} = 5^3 = 125$

Type VII

Problem 24 $(64)^{\frac{-2}{3}} \times \left(\frac{1}{4}\right)^{-2} = ?$

Sol. $\Rightarrow (64)^{\frac{-2}{3}} \times \left(\frac{1}{4}\right)^{-2} = (4^3)^{\frac{-2}{3}} \times (4^{-1})^{-2} = 4^{-2} \times 4^2 = 4^0 = 1$

Problem 25 $\left(\frac{1}{2}\right)^{-2} \times \left(\frac{1}{3}\right)^{-2} \times \left(\frac{1}{4}\right)^{-2} = ?$

Sol. $\left(\frac{1}{2}\right)^{-2} \times \left(\frac{1}{3}\right)^{-2} \times \left(\frac{1}{4}\right)^{-2} = \left(\frac{1}{2}\times\frac{1}{3}\times\frac{1}{4}\right)^{-2}$

$[\because a^m \cdot b^m \cdot c^m = (abc)^m]$

$= \left(\frac{1}{24}\right)^{-2} = (24^{-1})^{-2} = 24^2 = 576$

Problem 26 $(\sqrt{72} - \sqrt{18}) \div \sqrt{12} = ?$

Sol. $\Rightarrow (\sqrt{72} - \sqrt{18}) \div \sqrt{12} = (\sqrt{2\cdot 36} - \sqrt{2\cdot 9}) \div \sqrt{3\cdot 4}$

$= (6\sqrt{2} - 3\sqrt{2}) \div 2\sqrt{3} = 3\sqrt{2} \div 2\sqrt{3}$

$= (\sqrt{3}\times\sqrt{3}\times\sqrt{2}) \div (\sqrt{2}\times\sqrt{2}\times\sqrt{3}) = \sqrt{\frac{3}{2}}$

Problem 27 $(16)^{\frac{3}{2}} + (16)^{\frac{-3}{2}} = ?$

Sol. $\Rightarrow (16)^{\frac{3}{2}} + (16)^{\frac{-3}{2}} = (4^2)^{\frac{3}{2}} + (4^2)^{\frac{-3}{2}}$

$= 4^3 + 4^{-3} = 64 + \frac{1}{64} = 64\frac{1}{64}$

Problem 28 $(\sqrt[3]{\sqrt[6]{5^9}})^4 \times (\sqrt[6]{\sqrt[3]{5^9}})^4 = ?$

Sol. $\Rightarrow (\sqrt[3]{\sqrt[6]{5^9}})^4 \times (\sqrt[6]{\sqrt[3]{5^9}})^4 = (\sqrt[3\cdot 6]{5^9})^4 \times (\sqrt[6\cdot 3]{5^9})^4 \quad (\because \sqrt[m]{\sqrt[n]{a}} = \sqrt[mn]{a})$

$\Rightarrow \left(5^{\frac{9}{18}}\right)^4 \times \left(5^{\frac{9}{18}}\right)^4 = 5^2 \times 5^2 = 5^4 = 625$

Problem 29 $\sqrt[3]{3.5} + \sqrt[3]{2.5}\,[(\sqrt[3]{3.5})^2 - \sqrt[3]{8.75} + (\sqrt[3]{2.5})^2] = ?$

Sol. $\Rightarrow (\sqrt[3]{3.5} + \sqrt[3]{2.5})\,[(\sqrt[3]{3.5})^2 - \sqrt[3]{8.75} + (\sqrt[3]{2.5})^2]$

$= (\sqrt[3]{3.5} + \sqrt[3]{2.5})\,[(\sqrt[3]{3.5})^2 - (\sqrt[3]{3.5}\times\sqrt[3]{2.5}) + (\sqrt[3]{2.5})^2]$

$[\because (a+b)(a^2 - ab + b^2) = a^3 + b^3]$

$= (\sqrt[3]{3.5})^3 + (\sqrt[3]{2.5})^3 = 3.5 + 2.5 = 6$

Problem 30 $(3 + 2\sqrt{2})^{-3} + (3 - 2\sqrt{2})^{-3} = ?$

Sol. $\Rightarrow (3 + 2\sqrt{2})^{-3} + (3 - 2\sqrt{2})^{-3}$

$= \frac{1}{(3+2\sqrt{2})^3} + \frac{1}{(3-2\sqrt{2})^3} = \frac{(3-2\sqrt{2})^3 + (3+2\sqrt{2})^3}{(3+2\sqrt{2})^3 \times (3-2\sqrt{2})^3}$

$= \frac{[(3-2\sqrt{2}) + (3+2\sqrt{2})]\,[(3-2\sqrt{2})^2 - (3-2\sqrt{2})(3+2\sqrt{2}) + (3+2\sqrt{2})^2]}{[(3+2\sqrt{2})\cdot(3-2\sqrt{2})]^3}$

$[\because a^3 + b^3 = (a+b)(a^2 - ab + b^2)]$

$= \frac{6[17 - 12\sqrt{2} - 1 + 17 + 12\sqrt{2}]}{[1]^3} = 6\times 33 = 198$

Problem 31 $(2\sqrt{392} - 21) + (\sqrt{8} - 7)^2 = (?)^2$ *[Corporation Bank(PO)-11]*

Sol. $\Rightarrow (2\sqrt{392} - 21) + (\sqrt{8} - 7)^2 = (?)^2$

$\Rightarrow (2\sqrt{49\cdot 8} - 21) + 8 + 49 - 2\cdot 7\cdot\sqrt{8} = (?)^2$

$\Rightarrow 14\sqrt{8} - 21 + 8 + 49 - 2\cdot 7\cdot\sqrt{8} = (?)^2$

$\Rightarrow 36 = (?)^2 \Rightarrow (6)^2 = (?)^2 \Rightarrow ? = 6$

Problem 32 The square root of $\left(\frac{\sqrt{3}+\sqrt{2}}{\sqrt{3}-\sqrt{2}}\right)$ is *[SSC(CGL-T-I)- 11]*

Sol. $\left(\frac{\sqrt{3}+\sqrt{2}}{\sqrt{3}-\sqrt{2}}\right) = \left(\frac{\sqrt{3}+\sqrt{2}}{\sqrt{3}-\sqrt{2}} \times \frac{\sqrt{3}+\sqrt{2}}{\sqrt{3}+\sqrt{2}}\right)$

{Rationalising the denominator}

$= \frac{(\sqrt{3}+\sqrt{2})^2}{9-8} = (\sqrt{3}+\sqrt{2})^2$

Then, square root of $(\sqrt{3}+\sqrt{2})^2 = \sqrt{3}+\sqrt{2}$

Problem 33 Find the value of $\left(2^{\frac{1}{4}} - 1\right)\left(2^{\frac{3}{4}} + 2^{\frac{1}{2}} + 2^{\frac{1}{4}} + 1\right)$ *[NIFT-03]*

Sol. $(2^{\frac{1}{4}} - 1)(2^{\frac{3}{4}} + 2^{\frac{1}{2}} + 2^{\frac{1}{4}} + 1)$

$= 2^{\frac{1}{4}}\cdot 2^{\frac{3}{4}} + 2^{\frac{1}{4}}\cdot 2^{\frac{1}{2}} + 2^{\frac{1}{4}}\cdot 2^{\frac{1}{4}} + 2^{\frac{1}{4}}\cdot 1 - 2^{\frac{3}{4}} - 2^{\frac{1}{2}} - 2^{\frac{1}{4}} - 1$

$= 2 + 2^{\frac{3}{4}} + 2^{\frac{1}{2}} + 2^{\frac{1}{4}} - 2^{\frac{3}{4}} - 2^{\frac{1}{2}} - 2^{\frac{1}{4}} - 1 = 2 - 1 = 1$

Problem 34 $(28+10\sqrt{3})^{\frac{1}{2}}-(7-4\sqrt{3})^{\frac{-1}{2}}=?$

Sol. $\Rightarrow (28+10\sqrt{3})^{\frac{1}{2}}-(7-4\sqrt{3})^{\frac{-1}{2}}$

$$=[(5)^2+2\cdot5\cdot\sqrt{3}+(\sqrt{3})^2]^{\frac{1}{2}}-[(2)^2-2\cdot2\cdot\sqrt{3}+(\sqrt{3})^2]^{\frac{-1}{2}}$$

$$=[(5+\sqrt{3})^2]^{\frac{1}{2}}-[(2-\sqrt{3})^2]^{\frac{-1}{2}}=5+\sqrt{3}-\frac{1}{(2-\sqrt{3})}$$

$$=5+\sqrt{3}-\frac{1}{(2-\sqrt{3})}\times\frac{(2+\sqrt{3})}{(2+\sqrt{3})}$$

$$=5+\sqrt{3}-(2+\sqrt{3})=3$$

Problem 35 $\sqrt{\frac{19+8\sqrt{3}}{7-4\sqrt{3}}}=?$

Sol. $\Rightarrow \sqrt{\frac{(4)^2+2\cdot4\cdot\sqrt{3}+(\sqrt{3})^2}{(2)^2-2\cdot2\cdot\sqrt{3}+(\sqrt{3})^2}}=\sqrt{\frac{(4+\sqrt{3})^2}{(2-\sqrt{3})^2}}=\frac{4+\sqrt{3}}{2-\sqrt{3}}$

$$=\frac{4+\sqrt{3}}{2-\sqrt{3}}\times\frac{2+\sqrt{3}}{2+\sqrt{3}}$$

$$=\frac{8+4\sqrt{3}+2\sqrt{3}+3}{4-3}$$

$$=11+6\sqrt{3}$$

Type VIII

Problem 36 If $3^{x+8}=27^{2x+1}$, then $x=?$

Sol. Given, $3^{x+8}=(3^3)^{2x+1}$

$\Rightarrow\ 3^{x+8}=3^{6x+3}$

$\Rightarrow\ x+8=6x+3$ $\{\because$ If $a^x=a^y$, then $x=y\}$

$\Rightarrow\ 5x=5\ \therefore\ x=1$

Problem 37 If $\sqrt{1-\frac{x^3}{100}}=\frac{3}{5}$, then $x=?$ ***[SSC(CGL-T-I)- 11]***

Sol. Given, $\sqrt{1-\frac{x^3}{100}}=\frac{3}{5}\Rightarrow1-\frac{x^3}{100}=\left(\frac{3}{5}\right)^2$

$\Rightarrow\ 1-\frac{x^3}{100}=\frac{9}{25}\Rightarrow\frac{x^3}{100}=\frac{16}{25}\Rightarrow x^3=64$

$\Rightarrow\ x^3=4^3\ \therefore\ x=4$

Problem 38 $(0.64)^4\div(0.512)^3\times(0.8)^4=(0.8)^{?+3}$ ***[IOB (PO)-11]***

Sol. Given, $[(0.8^2)^4\div(0.8^3)^3]\times(0.8)^4=(0.8)^{?+3}$

$\Rightarrow\ (0.8)^{2\cdot4-3\cdot3+4}=(0.8)^{?+3}\Rightarrow(0.8)^3=(0.8)^{?+3}$

$\Rightarrow\ 3=?+3\Rightarrow?=0$

Problem 39 $8^{1.1}\times4^{2.7}\times2^{3.3}=2^?$

Sol. $\Rightarrow\ (2^3)^{1.1}\times(2^2)^{2.7}\times2^{3.3}=2^?$

$\Rightarrow\ 2^{3.3+5.4+3.3}=2^?$

$\Rightarrow\ 2^{12}=2^?\Rightarrow?=12$

Problem 40 $\frac{\sqrt{3+x}+\sqrt{3-x}}{\sqrt{3+x}-\sqrt{3-x}}=2$, then $x=?$ ***[SSC (CGL-T-I)- 10]***

Sol. Given, $\frac{\sqrt{3+x}+\sqrt{3-x}}{\sqrt{3+x}-\sqrt{3-x}}=2$

$\Rightarrow\ \frac{\sqrt{3+x}+\sqrt{3-x}}{\sqrt{3+x}-\sqrt{3-x}}\times\frac{\sqrt{3+x}+\sqrt{3-x}}{\sqrt{3+x}+\sqrt{3-x}}=2$

$\Rightarrow\ \frac{(\sqrt{3+x})^2+(\sqrt{3-x})^2+2\cdot\sqrt{3+x}\times\sqrt{3-x}}{(\sqrt{3+x})^2-(\sqrt{3-x})^2}=2$

$\Rightarrow\ \frac{3+x+3-x+2\cdot\sqrt{9-x^2}}{3+x-3+x}=2$

$\Rightarrow\ 6+2\cdot\sqrt{9-x^2}=4x$

$\Rightarrow\ 2\cdot\sqrt{9-x^2}=4x-6\Rightarrow\sqrt{9-x^2}=2x-3$

$\Rightarrow\ 9-x^2=(2x-3)^2$ {Squaring both sides}

$\Rightarrow\ 9-x^2=4x^2+9-12x\Rightarrow5x^2-12x=0$

$\Rightarrow\ 5x^2=12x\Rightarrow x=\frac{12}{5}$

Problem 41 If $x^{x\sqrt{x}}=(x\sqrt{x})^x$, then $x=?$

Sol. Given, $x^{x\cdot x^{\frac{1}{2}}}=\left(x\cdot x^{\frac{1}{2}}\right)^x$

$\Rightarrow\ x^{x^{\frac{3}{2}}}=\left(x^{\frac{3}{2}}\right)^x\Rightarrow x^{x^{\frac{3}{2}}}=x^{\frac{3x}{2}}\Rightarrow x^{\frac{3}{2}}=\frac{3x}{2}$

$\Rightarrow\ \frac{x^{\frac{3}{2}}}{x}=\frac{3}{2}\Rightarrow x^{\frac{1}{2}}=\frac{3}{2}\Rightarrow x=\left(\frac{3}{2}\right)^2=\frac{9}{4}$

Problem 42 If $9\sqrt{x}=\sqrt{12}+\sqrt{147}$, then $x=?$

Sol. Given, $9\sqrt{x}=\sqrt{12}+\sqrt{147}$

$\Rightarrow\ 9\sqrt{x}=2\sqrt{3}+7\sqrt{3}\Rightarrow9\sqrt{x}=9\sqrt{3}\Rightarrow x=3$

Problem 43 If $5^{\sqrt{x}}+12^{\sqrt{x}}=13^{\sqrt{x}}$, then $x=?$

Sol. Clearly, $5^2+12^2=13^2$

Then, $\sqrt{x}=2$

$\therefore\ x=4$

Alternatively,

Proceed through options if given.

Problem 44 If $\sqrt{2^x}=256$, then $x=?$

Sol. Given, $\sqrt{2^x}=256$

$\Rightarrow\ (2^x)^{\frac{1}{2}}=2^8\Rightarrow\frac{x}{2}=8\Rightarrow x=16$

Problem 45 If $2^{z-3}-2^{z-4}+2^{z+5}=1026$, then $z=?$

Sol. Given, $2^{z-3}-2^{z-4}+2^{z+5}=1026$

$\Rightarrow\ 2^z\left(\frac{1}{2^3}-\frac{1}{2^4}+2^5\right)=1026$

$\Rightarrow\ 2^z\left(\frac{1}{8}-\frac{1}{16}+32\right)=1026$

$\Rightarrow \quad 2^z\left(\frac{513}{16}\right) = 1026 \Rightarrow 2^z = 1026 \times \frac{16}{513} = 32$

$\Rightarrow \quad 2^z = 2^5 \Rightarrow z = 5$

Problem 46 If $2^{2x-y} = 16$ and $2^{x+y} = 32$, the value of xy is

Sol. Given, $2^{2x-y} = 2^4 \Rightarrow 2x - y = 4$ ...(i)

Further, $2^{x+y} = 2^5 \Rightarrow x + y = 5$ ...(ii)

From Eqs. (i) and (ii), we get $x = 3$ and $y = 2$

Then, $xy = 3 \cdot 2 = 6$

Problem 47 If $2^{x-1} + 2^{x+3} = 272$, then $x = ?$

Sol. Given, $2^{x-1} + 2^{x+3} = 272$,

$\Rightarrow \quad 2^{x-1}(1 + 2^4) = 272$

$\Rightarrow \quad 2^{x-1} = (272 \div 17) = 16$

$\Rightarrow \quad 2^{x-1} = 2^4 \Rightarrow x - 1 = 4 \Rightarrow x = 5$

Problem 48 If $\left(\frac{a}{b}\right)^{x-1} = \left(\frac{b}{a}\right)^{x-3}$, then $x = ?$ ***[MBA-03]***

Sol. Given, $\left(\frac{a}{b}\right)^{x-1} = \left(\frac{b}{a}\right)^{x-3} \Rightarrow \left(\frac{a}{b}\right)^{x-1} = \left(\frac{1}{\frac{a}{b}}\right)^{x-3}$

$\Rightarrow \quad \left(\frac{a}{b}\right)^{x-1} = \left\{\left(\frac{a}{b}\right)^{-1}\right\}^{x-3} \Rightarrow \left(\frac{a}{b}\right)^{x-1} = \left(\frac{a}{b}\right)^{3-x}$

$\Rightarrow \quad x - 1 = 3 - x \Rightarrow 2x = 4$

$\therefore \quad x = 2$

Problem 49 If $\sqrt{4x-9} + \sqrt{4x+9} = 5 + \sqrt{7}$, then $x = ?$ ***[SSC(CGL T I) 11]***

Sol. Given, $\sqrt{4x-9} + \sqrt{4x+9} = 5 + \sqrt{7}$

$\Rightarrow (\sqrt{4x-9} + \sqrt{4x+9})^2 = (5 + \sqrt{7})^2$ {squaring both sides}

$\Rightarrow (4x - 9 + 4x + 9 + 2 \cdot \sqrt{16x^2 - 81})$

$= 25 + 7 + 2 \cdot 5 \cdot \sqrt{7}$

$\Rightarrow 8x + 2\sqrt{16x^2 - 81} = 32 + 10\sqrt{7}$

$\Rightarrow \quad \sqrt{16x^2 - 81} = 16 + 5\sqrt{7} - 4x$

$\Rightarrow \quad 16x^2 - 81 = 256 + 175 + 160\sqrt{7}$

$+ 16x^2 - 128x - 40x\sqrt{7}$

{squaring both sides}

$\Rightarrow \quad x(128 + 40\sqrt{7}) = 512 + 160\sqrt{7}$

$\Rightarrow \quad x(128 + 40\sqrt{7}) = 4(128 + 40\sqrt{7})$

$\Rightarrow \quad x = 4$

Alternatively I Proceed through options by putting value of x in the given equation.

Alternatively II We have to ensure as to whether value of x for $\sqrt{4x-9} = 5$ is equal to that of x for $\sqrt{4x+9} = \sqrt{7}$. If not, check as to whether value of x for $\sqrt{4x-9} = \sqrt{7}$ is equal to that of x for $\sqrt{4x+9} = 5$.

In second case, both the equations yield $x = 4$. Hence, required answer = 4.

Problem 50 If $\sqrt{\frac{x-a}{x-b}} + \frac{a}{x} = \sqrt{\frac{x-b}{x-a}} + \frac{b}{x}$, $b \neq a$, then $x = ?$

[SSC(CGL-T-I)- 12]

Sol. Given, $\sqrt{\frac{x-a}{x-b}} + \frac{a}{x} = \sqrt{\frac{x-b}{x-a}} + \frac{b}{x}$

$\Rightarrow \quad \sqrt{\frac{x-a}{x-b}} - \sqrt{\frac{x-b}{x-a}} = \frac{b}{x} - \frac{a}{x}$

$\Rightarrow \quad \frac{x-a-x+b}{\sqrt{(x-b)(x-a)}} = \frac{b-a}{x}$

$\Rightarrow \quad x = \sqrt{(x-b)(x-a)}$

$\Rightarrow \quad x^2 = x^2 - x(a+b) + ab$ {squaring both sides}

$\Rightarrow \quad x(a+b) = ab \Rightarrow x = \frac{ab}{(a+b)}$

Type IX

Problem 51 If $\sqrt{15} = 3.88$, then $\sqrt{\frac{5}{3}} = ?$

Sol. $\Rightarrow \sqrt{\frac{5}{3}} = \sqrt{\frac{5 \times 3}{3 \times 3}} = \frac{\sqrt{15}}{3} = \frac{3.88}{3} = 1.29\bar{3}$

Problem 52 Assume $\sqrt{13} = 3.605$ (approx.) and $\sqrt{130} = 11.40$ (approx.), then find the value of $\sqrt{1.3} + \sqrt{1300} + \sqrt{0.013}$

Sol. $\Rightarrow \sqrt{1.3} + \sqrt{1300} + \sqrt{0.013} = \sqrt{\frac{130}{100}} + \sqrt{13 \times 100} + \sqrt{\frac{130}{10000}}$

$= \frac{\sqrt{130}}{10} + (\sqrt{13} \times 10) + \frac{\sqrt{130}}{100}$

$= \frac{11.40}{10} + (3.605 \times 10) + \frac{11.40}{100}$

$= 1.140 + 36.050 + 0.114 = 37.304$

Problem 53 The value of $\sqrt{32} - \sqrt{128} + \sqrt{50}$ correct to 3 places of decimals is:

Sol. $\sqrt{32} - \sqrt{128} + \sqrt{50} = \sqrt{2 \times 16} - \sqrt{2 \times 64} + \sqrt{2 \times 25}$

$= 4\sqrt{2} - 8\sqrt{2} + 5\sqrt{2} = \sqrt{2} = 1.414$

Problem 54 Given $\sqrt{5} = 2.236$, $\sqrt{3} = 1.732$, then $\frac{1}{\sqrt{5} + \sqrt{3}} = ?$

Sol. $\Rightarrow \frac{1}{\sqrt{5}+\sqrt{3}} = \frac{1}{\sqrt{5}+\sqrt{3}} \times \frac{\sqrt{5}-\sqrt{3}}{\sqrt{5}-\sqrt{3}} = \frac{\sqrt{5}-\sqrt{3}}{5-3}$

$= \frac{2.236 - 1.732}{2} = \frac{0.504}{2} = 0.252$

Problem 55 Given, $\sqrt{3} = 1.732$, then $\sqrt{147} - \frac{1}{4}\sqrt{48} - \sqrt{75} = ?$ ***[SSC(CGL-PT)- 06, SSC(CPO)-08, SSC(10+2, LDC/DEO)-10]***

Sol. $\Rightarrow \sqrt{147} - \frac{1}{4}\sqrt{48} - \sqrt{75} = \sqrt{3 \times 49} - \frac{1}{4}\sqrt{3 \times 16} - \sqrt{3 \times 25}$

$= 7\sqrt{3} - \left(\frac{1}{4} \times 4\sqrt{3}\right) - 5\sqrt{3} = \sqrt{3} = 1.732$

➘ Problem 56 If $\sqrt{2} = 1.4142$, find the value of

$$2\sqrt{2} + \sqrt{2} + \frac{1}{2+\sqrt{2}} + \frac{1}{\sqrt{2}-2}$$ **[SSC(CGL-T-I)- 11]**

Sol. $2\sqrt{2} + \sqrt{2} + \frac{1}{2+\sqrt{2}} + \frac{1}{\sqrt{2}-2}$

$$\Rightarrow \quad 3\sqrt{2} + \frac{\sqrt{2}-2+\sqrt{2}+2}{(\sqrt{2}+2)(\sqrt{2}-2)}$$

$$= 3\sqrt{2} + \frac{2\sqrt{2}}{2-4} = 3\sqrt{2} - \sqrt{2}$$

$$\Rightarrow \quad 2\sqrt{2} = 2 \times 1.4142 = 2.8284$$

Type X

➘ Problem 57 The value of $\frac{3\sqrt{2}}{\sqrt{3}+\sqrt{6}} - \frac{4\sqrt{3}}{\sqrt{6}+\sqrt{2}} + \frac{\sqrt{6}}{\sqrt{3}+\sqrt{2}}$ is **[SSC(CGL-PT)- 02, SSC(CPO)-04, 05, 07, SSC(CGL-T-I)-11]**

Sol. 1st term $= \frac{3\sqrt{2}}{\sqrt{3}+\sqrt{6}} = \frac{3\sqrt{2}}{\sqrt{6}+\sqrt{3}} \times \frac{\sqrt{6}-\sqrt{3}}{\sqrt{6}-\sqrt{3}}$

$$= \frac{3\sqrt{12}-3\sqrt{6}}{6-3} = \sqrt{12}-\sqrt{6}$$

2nd term $= \frac{4\sqrt{3}}{\sqrt{6}+\sqrt{2}} = \frac{4\sqrt{3}}{\sqrt{6}+\sqrt{2}} \times \frac{\sqrt{6}-\sqrt{2}}{\sqrt{6}-\sqrt{2}}$

$$= \frac{4\sqrt{18}-4\sqrt{6}}{6-2} = \sqrt{18}-\sqrt{6}$$

3rd term $= \frac{\sqrt{6}}{\sqrt{3}+\sqrt{2}} = \frac{\sqrt{6}}{\sqrt{3}+\sqrt{2}} \times \frac{\sqrt{3}-\sqrt{2}}{\sqrt{3}-\sqrt{2}}$

$$= \frac{\sqrt{18}-\sqrt{12}}{3-2} = \sqrt{18}-\sqrt{12}$$

Now, $\frac{3\sqrt{3}}{\sqrt{3}+\sqrt{6}} - \frac{4\sqrt{3}}{\sqrt{6}+\sqrt{2}} + \frac{\sqrt{6}}{\sqrt{3}+\sqrt{2}}$

$$= \sqrt{12}-\sqrt{6}-\sqrt{18}+\sqrt{6}-\sqrt{18}-\sqrt{12} = 0$$

Alternatively

$$\Rightarrow \quad \frac{3\sqrt{2}}{\sqrt{3}+\sqrt{6}} - \frac{4\sqrt{3}}{\sqrt{6}+\sqrt{2}} + \frac{\sqrt{6}}{\sqrt{3}+\sqrt{2}}$$

$$= \frac{3\sqrt{2}}{\sqrt{3}(1+\sqrt{2})} - \frac{4\sqrt{3}}{\sqrt{2}(\sqrt{3}+1)} + \frac{\sqrt{6}}{\sqrt{3}+\sqrt{2}}$$

$$= \frac{\sqrt{6}}{(1+\sqrt{2})} - \frac{2\sqrt{6}}{(\sqrt{3}+1)} + \frac{\sqrt{6}}{\sqrt{3}+\sqrt{2}}$$

Now rationalise the denominators of each term and solve.

➘ Problem 58 $3 + \frac{1}{\sqrt{4}+\sqrt{5}} + \frac{1}{\sqrt{5}+\sqrt{6}} + \frac{1}{\sqrt{6}+\sqrt{7}} + \frac{1}{\sqrt{7}+\sqrt{8}} + \frac{1}{\sqrt{8}+\sqrt{9}} = ?$

Sol. 2nd term $= \frac{1}{\sqrt{4}+\sqrt{5}} = \frac{1}{\sqrt{5}+\sqrt{4}} \times \frac{\sqrt{5}-\sqrt{4}}{\sqrt{5}-\sqrt{4}} = \sqrt{5}-\sqrt{4}$

3rd term $= \frac{1}{\sqrt{5}+\sqrt{6}} = \frac{1}{\sqrt{6}+\sqrt{5}} \times \frac{\sqrt{6}-\sqrt{5}}{\sqrt{6}-\sqrt{5}} = \sqrt{6}-\sqrt{5}$ and so on.

Now,

$$3 + \frac{1}{\sqrt{4}+\sqrt{5}} + \frac{1}{\sqrt{5}+\sqrt{6}} + \frac{1}{\sqrt{6}+\sqrt{7}} + \frac{1}{\sqrt{7}+\sqrt{8}} + \frac{1}{\sqrt{8}+\sqrt{9}}$$

$$= 3 + \sqrt{5}-\sqrt{4}+\sqrt{6}-\sqrt{5}+\sqrt{7}-\sqrt{6}+\sqrt{8}-\sqrt{7} + \sqrt{9}-\sqrt{8} = 3-\sqrt{4}+\sqrt{9} = 3-2+3 = 4$$

➘ Problem 59

$$\frac{1}{\sqrt{2}+1} + \frac{1}{\sqrt{3}+\sqrt{2}} + \frac{1}{\sqrt{4}+\sqrt{3}} + \ldots + \frac{1}{\sqrt{100}+\sqrt{99}} = ?$$

[SSC(MTS)-11, SSC(SAS)-10, CDS -08, MAT-05]

Sol. 1st term $= \frac{1}{\sqrt{2}+1} = \frac{1}{\sqrt{2}+\sqrt{1}} \times \frac{\sqrt{2}-\sqrt{1}}{\sqrt{2}-\sqrt{1}} = \sqrt{2}\ \sqrt{1}$

2nd term $= \frac{1}{\sqrt{3}+\sqrt{2}} = \frac{1}{\sqrt{3}+\sqrt{2}} \times \frac{\sqrt{3}-\sqrt{2}}{\sqrt{3}-\sqrt{2}} = \sqrt{3}-\sqrt{2}$

and so on.

Now, $\frac{1}{\sqrt{2}+1} + \frac{1}{\sqrt{3}+\sqrt{2}} + \frac{1}{\sqrt{4}+\sqrt{3}} + \ldots + \frac{1}{\sqrt{100}+\sqrt{99}}$

$$= \sqrt{2}-\sqrt{1}+\sqrt{3}-\sqrt{2}+\sqrt{4}-\sqrt{3}+\ldots+\sqrt{100}-\sqrt{99}$$

$$= -\sqrt{1}+\sqrt{100} = -1+10 = 9$$

➘ Problem 60 $\frac{1}{\sqrt{12-\sqrt{140}}} - \frac{1}{\sqrt{8-\sqrt{60}}} - \frac{2}{\sqrt{10+\sqrt{84}}} = ?$

Sol.

$$\frac{1}{\sqrt{(\sqrt{7})^2+(\sqrt{5})^2-2\cdot\sqrt{7}\cdot\sqrt{5}}} - \frac{1}{\sqrt{(\sqrt{5})^2+(\sqrt{3})^2-2\cdot\sqrt{5}\cdot\sqrt{3}}} - \frac{2}{\sqrt{(\sqrt{7})^2+(\sqrt{3})^2+4\cdot\sqrt{7}\cdot\sqrt{3}}}$$

$$= \frac{1}{\sqrt{(\sqrt{7}-\sqrt{5})^2}} - \frac{1}{\sqrt{(\sqrt{5}-\sqrt{3})^2}} - \frac{2}{\sqrt{(\sqrt{7}+\sqrt{3})^2}}$$

$$= \frac{1}{\sqrt{7}-\sqrt{5}} - \frac{1}{\sqrt{5}-\sqrt{3}} - \frac{2}{\sqrt{7}+\sqrt{3}}$$

$$= \left\{\frac{1}{\sqrt{7}-\sqrt{5}} \times \frac{\sqrt{7}+\sqrt{5}}{\sqrt{7}+\sqrt{5}}\right\} - \left\{\frac{1}{\sqrt{5}-\sqrt{3}} \times \frac{\sqrt{5}+\sqrt{3}}{\sqrt{5}+\sqrt{3}}\right\} - \left\{\frac{2}{\sqrt{7}+\sqrt{3}} \times \frac{\sqrt{7}-\sqrt{3}}{\sqrt{7}-\sqrt{3}}\right\}$$

$$= \frac{\sqrt{7}+\sqrt{5}}{7-5} - \frac{\sqrt{5}+\sqrt{3}}{5-3} - \frac{2(\sqrt{7}-\sqrt{3})}{7-3}$$

$$= \frac{\sqrt{7}+\sqrt{5}-\sqrt{5}-\sqrt{3}-\sqrt{7}+\sqrt{3}}{2} = 0$$

➘ Problem 61 The value of

$$\frac{2+\sqrt{3}}{2-\sqrt{3}} + \frac{2-\sqrt{3}}{2+\sqrt{3}} + \frac{\sqrt{3}+1}{\sqrt{3}-1}$$ is **[SSC(CGL-T-I)-11]**

Sol. $\Rightarrow \left(\frac{2+\sqrt{3}}{2-\sqrt{3}} + \frac{2-\sqrt{3}}{2+\sqrt{3}}\right) + \frac{\sqrt{3}+1}{\sqrt{3}-1}$

$$= \left(\frac{4+3+2\cdot2\sqrt{3}+4+3-2\cdot2\sqrt{3}}{4-3}\right) + \left(\frac{\sqrt{3}+1}{\sqrt{3}-1} \times \frac{\sqrt{3}+1}{\sqrt{3}+1}\right)$$

$$= 14 + \left(\frac{3+1+2\cdot\sqrt{3}\cdot1}{3-1}\right)$$

$$= 14+2+\sqrt{3} = 16+\sqrt{3}$$

Problem 62 $\frac{2}{\sqrt{5}+\sqrt{3}}+\frac{3}{\sqrt{6}-\sqrt{3}}+\frac{1}{\sqrt{6}+\sqrt{5}}$ is equal to

Sol. $\Rightarrow \frac{2}{\sqrt{5}+\sqrt{3}}+\frac{3}{\sqrt{6}-\sqrt{3}}+\frac{1}{\sqrt{6}+\sqrt{5}}$

$$=\left(\frac{2}{\sqrt{5}+\sqrt{3}}\times\frac{\sqrt{5}-\sqrt{3}}{\sqrt{5}-\sqrt{3}}\right)+\left(\frac{3}{\sqrt{6}-\sqrt{3}}\times\frac{\sqrt{6}+\sqrt{3}}{\sqrt{6}+\sqrt{3}}\right)+\left(\frac{1}{\sqrt{6}+\sqrt{5}}\times\frac{\sqrt{6}-\sqrt{5}}{\sqrt{6}-\sqrt{5}}\right)$$

$$=\sqrt{5}-\sqrt{3}+\sqrt{6}+\sqrt{3}+\sqrt{6}-\sqrt{5}=2\sqrt{6}$$

Problem 63 $\frac{1}{1+\sqrt{2}+\sqrt{3}}+\frac{1}{1-\sqrt{2}+\sqrt{3}}=?$

Sol. $\Rightarrow \frac{1}{(1+\sqrt{3})+\sqrt{2}}+\frac{1}{(1+\sqrt{3})-\sqrt{2}}$

$$=\frac{(1+\sqrt{3})-\sqrt{2}+(1+\sqrt{3})+\sqrt{2}}{(1+\sqrt{3})^2-(\sqrt{2})^2}$$

$$\Rightarrow \frac{2(1+\sqrt{3})}{4+2\sqrt{3}-2}=\frac{2(1+\sqrt{3})}{2(1+\sqrt{3})}=1$$

Type XI

Problem 64 If $x=y^a$, $y=z^b$ and $z=x^c$, then $abc=?$

Sol. Given, $z=x^c$

$\Rightarrow \quad z=(y^a)^c \Rightarrow z=y^{ac} \Rightarrow z=(z^b)^{ac}=z^{abc}$

$\Rightarrow \quad z^1=z^{abc} \Rightarrow abc=1$

Problem 65 Given that $10^{0.48}=x$, $10^{0.70}=y$ and $x^z=y^2$, then the value of z is close to

Sol. Given, $x^z=y^2$

$\Rightarrow \quad (10^{0.48})^z=(10^{0.70})^2 \Rightarrow 10^{0.48z}=10^{0.70\times 2}$

$\Rightarrow \quad 0.48z=0.70\times 2 \Rightarrow z=2.9$ (approx.)

Problem 66 If $a^x=b$, $b^y=c$ and $c^z=a$, then $\frac{1}{xyz}=?$

Sol. Given, $b=a^x$

$\Rightarrow \quad b=(c^z)^x \Rightarrow b=c^{zx} \Rightarrow b=(b^y)^{zx}=b^{xyz}$

$\Rightarrow \quad b^1=b^{xyz} \Rightarrow xyz=1$

Then, $\quad \frac{1}{xyz}=\frac{1}{1}=1$

Problem 67 If $3^x=9^y=81^z$ and $\frac{1}{2x}+\frac{1}{2y}+\frac{1}{3z}=\frac{17}{36}$ then $y=?$

Sol. Given, $\quad 3^x=9^y=81^z$

$\Rightarrow \quad 3^x-(3^2)^y=(3^4)^z \Rightarrow x-2y=4z$

Now, $\frac{1}{2x}+\frac{1}{2y}+\frac{1}{3z}=\frac{17}{36}\Rightarrow\frac{1}{2\cdot 2y}+\frac{1}{2y}+\frac{1}{3\cdot\frac{y}{2}}=\frac{17}{36}$

$\Rightarrow \quad \frac{1}{4y}+\frac{1}{2y}+\frac{2}{3y}=\frac{17}{36}$

$\Rightarrow \quad \frac{3+6+8}{12y}=\frac{17}{36}\Rightarrow y=3$

Problem 68 If $3^{2x}=5^{-y}=15^z$ then $\frac{1}{2x}-\frac{1}{y}-\frac{1}{z}=?$

Sol. Let $k=3^{2x}=5^{-y}=15^z$

Then, $\quad 3^{2x}=k\Rightarrow 3=k^{\frac{1}{2x}}$ {If $a^x=m$, then $a=m^{\frac{1}{x}}$}

and $\quad 5^{-y}=k\Rightarrow 5=k^{\frac{-1}{y}}$ and $15^z=k\Rightarrow 15=k^{\frac{1}{z}}$

Now, $\quad 3\times 5=15 \Rightarrow k^{\frac{1}{2x}}\times k^{\frac{-1}{y}}=k^{\frac{1}{z}}$

$\Rightarrow \quad k^{\frac{1}{2x}+\left(\frac{-1}{y}\right)}=k^{\frac{1}{z}}\Rightarrow \frac{1}{2x}+\left(\frac{-1}{y}\right)=\frac{1}{z}$

$\Rightarrow \quad \frac{1}{2x}-\frac{1}{y}-\frac{1}{z}=0$

Problem 69 If $a^x=b^y=c^z$ and $b^2=ac$, then $y=?$ ***[CDS-19]***

Sol. Let $k=a^x=b^y=c^z$

Then, $\quad a^x=k\Rightarrow a=k^{\frac{1}{x}}$ {If $a^x=m$, then $a=m^{\frac{1}{x}}$}

and $\quad b^y=k\Rightarrow b=k^{\frac{1}{y}}$ and $c^z=k\Rightarrow c=k^{\frac{1}{z}}$

As given, $b^2=ac$

$\Rightarrow \quad \left(k^{\frac{1}{y}}\right)^2=k^{\frac{1}{x}}\cdot k^{\frac{1}{z}}\Rightarrow k^{\frac{2}{y}}=k^{\frac{1}{x}}\cdot k^{\frac{1}{z}}$

$\Rightarrow \quad \frac{2}{y}=\frac{1}{x}+\frac{1}{z}\Rightarrow\frac{2}{y}=\frac{z+x}{xz}\Rightarrow y=\frac{2xz}{x+z}$

Type XII

Problem 70 $\frac{1}{1+a^{(n-m)}}+\frac{1}{1+a^{(m-n)}}=?$

Sol. $\Rightarrow \frac{1}{1+a^{(n-m)}}+\frac{1}{1+a^{(m-n)}}=\frac{1}{1+\frac{a^n}{a^m}}+\frac{1}{1+\frac{a^m}{a^n}}$

$$=\frac{a^m}{a^m+a^n}+\frac{a^n}{a^n+a^m}=\frac{a^m+a^n}{a^n+a^m}=1$$

Problem 71 $\frac{1}{1+x^{(b-a)}+x^{(c-a)}}+\frac{1}{1+x^{(a-b)}+x^{(c-b)}}+\frac{1}{1+x^{(b-c)}+x^{(a-c)}}=?$

Sol. $\frac{1}{1+\frac{x^b}{x^a}+\frac{x^c}{x^a}}+\frac{1}{1+\frac{x^a}{x^b}+\frac{x^c}{x^b}}+\frac{1}{1+\frac{x^b}{x^c}+\frac{x^a}{x^c}}$

$$=\frac{1}{\frac{x^a+x^b+x^c}{x^a}}+\frac{1}{\frac{x^b+x^a+x^c}{x^b}}+\frac{1}{\frac{x^c+x^b+x^a}{x^c}}$$

$$=\frac{x^a}{x^a+x^b+x^c}+\frac{x^b}{x^a+x^b+x^c}+\frac{x^c}{x^a+x^b+x^c}=\frac{x^a+x^b+x^c}{x^a+x^b+x^c}=1$$

➘ Problem 72 If $abc = 1$, then $\dfrac{1}{1+a+b^{-1}} + \dfrac{1}{1+b+c^{-1}} + \dfrac{1}{1+c+a^{-1}} = ?$

Sol. $\Rightarrow \dfrac{1}{1+a+b^{-1}} + \dfrac{1}{1+b+c^{-1}} + \dfrac{1}{1+c+a^{-1}}$

$= \dfrac{1}{1+a+\frac{1}{b}} + \dfrac{1}{1+b+\frac{1}{c}} + \dfrac{1}{1+c+\frac{1}{a}}$

$= \dfrac{b}{b+ab+1} + \dfrac{1}{1+b+ab} + \dfrac{1}{1+\frac{1}{ab}+\frac{1}{a}}$

$\left\{\because abc = 1 \Rightarrow \dfrac{1}{c} = ab \text{ and } c = \dfrac{1}{ab}\right\}$

$= \dfrac{b}{b+ab+1} + \dfrac{1}{1+b+ab} + \dfrac{ab}{ab+1+b}$

$= \dfrac{b+1+ab}{b+ab+1} = 1$

Type XIII

➘ Problem 73 $\left(\dfrac{x^c}{x^a}\right)^{c+a} \cdot \left(\dfrac{x^a}{x^b}\right)^{a+b} \cdot \left(\dfrac{x^b}{x^c}\right)^{b+c} - 2 = ?$

Sol. $(x^{c-a})^{c+a} \cdot (x^{a-b})^{a+b} \cdot (x^{b-c})^{b+c} - 2$

$= x^{c^2-a^2} \cdot x^{a^2-b^2} \cdot x^{b^2-c^2} - 2$

$= x^{c^2-a^2+a^2-b^2+b^2-c^2} - 2$

$= x^0 - 2 = 1 - 2 = -1$

➘ Problem 74 $\left(\dfrac{x^b}{x^c}\right)^{a} \cdot \left(\dfrac{x^c}{x^a}\right)^{b} \cdot \left(\dfrac{x^a}{x^b}\right)^{c} - \left(\dfrac{x^a}{x^b}\right)^{a^2+ab+b^2}$

$\cdot \left(\dfrac{x^b}{x^c}\right)^{b^2+bc+c^2} \left(\dfrac{x^c}{x^a}\right)^{c^2+ac+a^2} + \left(\dfrac{x^a}{x^b}\right)^{0} = ?$

Sol. $\Rightarrow (x^{b-c})^a \cdot (x^{c-a})^b \cdot (x^{a-b})^c - (x^{a-b})^{a^2+ab+b^2}$

$(x^{b-c})^{b^2+bc+c^2} (x^{c-a})^{c^2+ac+a^2} + 1$

$= x^{ab-ac} \cdot x^{bc-ab} \cdot x^{ac-bc} - x^{a^3-b^3+b^3-c^3+c^3-a^3} + 1$

$= x^{ab-ac+bc-ab+ac-bc} - x^0 + 1$

$= 1 - 1 + 1 = 1$

➘ Problem 75 $\left(\dfrac{x^b}{x^c}\right)^{b+c-a} \cdot \left(\dfrac{x^c}{x^a}\right)^{c+a-b} \cdot \left(\dfrac{x^a}{x^b}\right)^{a+b-c} = ?$

[LIC (AAO)-03]

Sol. $\Rightarrow (x^{b-c})^{b+c-a} \cdot (x^{c-a})^{c+a-b} \cdot (x^{a-b})^{a+b-c}$

$= x^{b^2+bc-ab-cb-c^2+ac} \cdot x^{c^2+ac-bc-ac-a^2+ab}$

$\cdot x^{a^2+ab-ac-ab-b^2+bc}$

$= x^{b^2+bc-ab-cb-c^2+ac+c^2+ac-bc-ac-a^2}$

$^{+ab+a^2+ab-ac-ab-b^2+bc}$

$= x^0 = 1$

Type XIV

➘ Problem 76 If $\dfrac{4\sqrt{3}+5\sqrt{2}}{\sqrt{48}+\sqrt{18}} = a + b\sqrt{6}$, values of a and b respectively are ***[SSC(CGL-T-I)-11, SSC (CGL-T-II)-19]***

Sol. $\Rightarrow \dfrac{4\sqrt{3}+5\sqrt{2}}{\sqrt{48}+\sqrt{18}} = \dfrac{4\sqrt{3}+5\sqrt{2}}{4\sqrt{3}+3\sqrt{2}}$

$= \dfrac{4\sqrt{3}+5\sqrt{2}}{4\sqrt{3}+3\sqrt{2}} \times \dfrac{4\sqrt{3}-3\sqrt{2}}{4\sqrt{3}-3\sqrt{2}}$ {Rationalising the denominator}

$= \dfrac{48 - 12\sqrt{6} + 20\sqrt{6} - 30}{(4\sqrt{3})^2 - (3\sqrt{2})^2}$

$= \dfrac{18 + 8\sqrt{6}}{48 - 18} = \dfrac{18}{30} + \dfrac{8\sqrt{6}}{30}$

$= \dfrac{3}{5} + \dfrac{4}{15}\sqrt{6} = a + b\sqrt{6}$ (given)

$\Rightarrow \quad a = \dfrac{3}{5}$ and $b = \dfrac{4}{15}$

WORKOUT

1. If $9^{2x-1} - 81^{x-1} = 1944$, then x is **[CAT-17]**

(a) 3 (b) $\frac{9}{4}$
(c) $\frac{4}{9}$ (d) $\frac{1}{3}$

2. $\sqrt{8 + \sqrt{57 + \sqrt{38 + \sqrt{108 + \sqrt{169}}}}} = ?$ **[SSC(CGL-T-I)-11]**

(a) 4 (b) 6
(c) 8 (d) 10

3. Consider the following in respect of the numbers $\sqrt{2}$, $\sqrt[3]{3}$ and $\sqrt[6]{6}$

I. $\sqrt[6]{6}$ is the greatest number.
II. $\sqrt{2}$ is the smallest number.

Which of the above statements is/are correct ? **[CDS-14, 07]**

(a) Only I
(b) Only II
(c) Both I and II
(d) Neither I nor II

4. $\sqrt{24^4} + 224 = ? \times 20^2$

(a) 20 (b) 4
(c) 2 (d) 16
(e) None of these

5. If $(x)^{\frac{1}{m}} = (y)^{\frac{1}{n}} = (z)^{\frac{1}{p}}$ and $xyz = 1$, then what is the value of $m + n + p$?

(a) 0 (b) 1
(c) 2 (d) -1

6. Which of the following is not the reciprocal of $\left(\frac{2}{3}\right)^4$?

(a) $\left(\frac{3}{2}\right)^4$ (b) $\left(\frac{2}{3}\right)^{-4}$
(c) $\left(\frac{3}{2}\right)^{-4}$ (d) $\frac{3^4}{2^4}$

7. Which is the smallest number among the following ? **[CDS-13]**

(a) $[(5^{-2})^{-2}]^{-2}$ (b) $[(5^{-2})^{2}]^{-2}$
(c) $[(2^{-5})^{-2}]^{-2}$ (d) $[(2^{-5})^{2}]^{-2}$

8. $(2 \times 3)^3 \div (4 \times 9)^2 \times (27 \times 8)^2 = (6)^?$ **[Syndicate Bank(PO)-10]**

(a) 5 (b) 6
(c) 3 (d) 8
(e) None of these

9. If $\sqrt{7\sqrt{7\sqrt{7\sqrt{7 \ldots \infty}}}} = (343)^{y-1}$, then y is equal to **[SSC(CGL-T-II)-10]**

(a) $\frac{2}{3}$ (b) 1
(c) $\frac{4}{3}$ (d) $\frac{3}{4}$

10. If $\sqrt{1 + \frac{93}{196}} = 1 + \frac{x}{14}$, then what does x equal to ?

(a) 1 (b) 2
(c) 3 (d) 4

11. $\sqrt{7 + \sqrt{7 - \sqrt{7 + \sqrt{7 - \ldots\ldots\ldots \infty}}}} = ?$ **[IIFT-12]**

(a) 1 (b) 2
(c) 3 (d) 4

12. $[(3\sqrt{8} + \sqrt{8}) \times (8\sqrt{8} + 7\sqrt{8})] - 98 = ?$ **[IBPS(CWE)-11]**

(a) $2\sqrt{8}$ (b) $8\sqrt{8}$
(c) 382 (d) 386
(e) None of these

13. $16 \times 8^{n+2} = 2^m$, then m is equal to **[CDS-13]**

(a) $n + 8$
(b) $2n + 10$
(c) $3n + 2$
(d) $3n + 10$

14. $(8)^7 \times (2)^6 \div (8)^{2.4} = (8)^?$ **[Andhra Bank(PO)-09]**

(a) 10.6 (b) 9.6
(c) 8.6 (d) 6.6
(e) None of these

15. The simplified value of $6\sqrt{\frac{1}{(27)^{\frac{2}{3}}} + \frac{1}{(8)^{\frac{2}{3}}}}$ is

(a) $6\sqrt{35}$ (b) $6\sqrt{6}$
(c) $\sqrt{13}$ (d) $\frac{6}{\sqrt{13}}$

16. If $x = \{a + \sqrt{a^2 + b^3}\}^{\frac{1}{3}} + \{a - \sqrt{a^2 + b^3}\}^{\frac{1}{3}}$, then what is the value of $x^3 + 3bx - 2a$?

(a) $2a^3$ (b) $-2a^3$
(c) 1 (d) 0

17. The simplest value of the expression $\left\{\frac{4^{p+\frac{1}{4}} \times \sqrt{2 \times 2^p}}{2 \times \sqrt{2^{-p}}}\right\}^{\frac{1}{p}}$ is: **[IIFT-15]**

(a) 4 (b) 8
(c) 4^p (d) 8^p
(e) None of these

18. $\frac{1}{\sqrt[3]{4} + \sqrt[3]{2} + 1}$ is equal to **[SSC(SAS)-10]**

(a) $\sqrt[3]{2} + 1$ (b) $\sqrt[3]{2} - 1$
(c) $-\sqrt[3]{2} - 1$ (d) $1 - \sqrt[3]{2}$

19. $\sqrt{4^2 \times 8 + 9^2 - 40} = ?$ **[Corporation Bank(PO)-11]**

(a) 169 (b) 13
(c) 14 (d) 196
(e) None of these

20. If $\sqrt{10 + \sqrt[3]{x}} = 4$, then what is the value of x ? **[CDS-12]**

(a) 150 (b) 216
(c) 316 (d) 450

21. If $a = 2 + \sqrt{3}$, then what is the value of $(a^2 + a^{-2})$?

(a) 12 (b) 14
(c) 16 (d) 18

22. If $196x^4 = x^6$, then x^3 is equal to which one of the following ?

(a) $\frac{x^6}{14}$ (b) $14x^4$
(c) $\frac{x^2}{14}$ (d) $14x^2$

23. If $a^x = b$, $b^y = c$ and $xyz = 1$, then $c^z = ?$

(a) a (b) b
(c) ab (d) $\frac{a}{b}$

24. If $3^{x+y} = 81$ and $81^{x-y} = 3$, then the value of $x = ?$

(a) $\frac{17}{16}$ (b) $\frac{17}{8}$
(c) $\frac{17}{4}$ (d) $\frac{15}{4}$

25. $2\sqrt[3]{40} - 4\sqrt[3]{320} + 3\sqrt[3]{625} - 3\sqrt[3]{5}$ is equal to **[SSC(CGL-T-II)-11]**

(a) $-2\sqrt[3]{340}$ (b) 0
(c) $\sqrt[3]{340}$ (d) $\sqrt[3]{660}$

26. $\sqrt{4+\sqrt{4-\sqrt{4+\sqrt{4-\ldots\ldots\infty}}}} = ?$ *[CAT-05]*

(a) 3
(b) $\frac{\sqrt{13}-1}{2}$
(c) $\frac{\sqrt{13}+1}{2}$
(d) $\sqrt{13}$

27. p and q are positive numbers such that $p^q = q^p$ and $q = 9p$. The value of p is: *[XAT-13]*

(a) $\sqrt{9}$ (b) $\sqrt[6]{9}$
(c) $\sqrt[9]{9}$ (d) $\sqrt[8]{9}$
(e) $\sqrt[3]{9}$

28. If $a^x = b^y = c^z$ and $abc = 1$, then $xy + yz + zx = ?$

(a) xyz (b) $x + y + z$
(c) 0 (d) 1

29. If $(3.7)^x = (0.037)^y = 10000$, then what is the value of $\frac{1}{x} - \frac{1}{y}$?

(a) 1 (b) 2
(c) $\frac{1}{2}$ (d) $\frac{1}{4}$

30. What are the values of x and y that satisfy both the equations ?

$2^{0.7x} \cdot 3^{-1.25y} = \frac{8\sqrt{6}}{27}$

$4^{0.3x} \cdot 9^{0.2y} = 8 \cdot (81)^{\frac{1}{5}}$ *[CAT-06]*

(a) $x = 2$, $y = 5$
(b) $x = 2.5$, $y = 6$
(c) $x = 3$, $y = 5$
(d) $x = 3$, $y = 4$
(e) $x = 5$, $y = 2$

31. $\sqrt[3]{2^4\sqrt{2^{-5}\sqrt{2^6}}} = ?$

(a) 2 (b) 1
(c) 3 (d) 4

32. The value of $\sqrt{2+\sqrt{3}} + \sqrt{2-\sqrt{3}}$ is *[LIC(AAO)-11]*

(a) $\sqrt{6}$ (b) $2\sqrt{2}$
(c) $2\sqrt{3}$ (d) 6

33. What is $\frac{\sqrt{5}+\sqrt{3}}{\sqrt{5}-\sqrt{3}} + \frac{\sqrt{5}-\sqrt{3}}{\sqrt{5}+\sqrt{3}}$ equal to?

(a) 16 (b) 8
(c) 4 (d) $\sqrt{15}$

34. $2^m + 2^{1+m} = 24$, then what is the value of m ? *[CDS-11]*

(a) 8 (b) 10
(c) 3 (d) 7

35. The value of $\sqrt[6]{24-16\sqrt{2}} \times \sqrt[3]{4+2\sqrt{2}}$ is *[LIC(AAO)-12]*

(a) $\sqrt[3]{2}$ (b) $\sqrt[6]{2}$
(c) $\sqrt{2}$ (d) 2

36. The value of $\sqrt[3]{\frac{4}{9}} - \sqrt[3]{\frac{2}{9}} + \sqrt[3]{\frac{1}{9}}$ is

(a) $\sqrt[3]{3}$ (b) $\frac{1}{\sqrt[3]{3}}$
(c) $\frac{\sqrt[3]{3}}{\sqrt[3]{2}+1}$ (d) $\frac{3}{\sqrt[3]{2}+1}$

37. The value of $\sqrt{2\sqrt[3]{4\sqrt{2\sqrt[3]{4\sqrt{2\sqrt[3]{4\ldots\ldots\ldots}}}}}}$ is *[SSC(CGL-T-II)-13]*

(a) 2 (b) 2^2
(c) 2^3 (d) 2^5

38. What is $\frac{1}{\sqrt{9}-\sqrt{8}} - \frac{1}{\sqrt{8}-\sqrt{7}} + \frac{1}{\sqrt{7}-\sqrt{6}} - \frac{1}{\sqrt{6}-\sqrt{5}} + \frac{1}{\sqrt{5}-\sqrt{4}} = ?$

(a) 0 (b) 1
(c) 5 (d) $\frac{1}{3}$

39. $(36)^{\frac{1}{6}} = ?$

(a) $\sqrt{6}$ (b) $\sqrt[3]{6}$
(c) $2\sqrt{6}$ (d) 6

40. If $p^x = r^y = m$ and $r^w = p^z = n$, then which one of the following is correct ? *[CDS-10,09]*

(a) $xw = yz$ (b) $xz = yw$
(c) $x + y = w + z$ (d) $x - y = w - z$

41. If $x = t^{\frac{1}{t-1}}$ and $y = t^{\frac{t}{t-1}}$, $t > 0$, $t \neq 1$, then what is the relation between x and y? *[CDS-17]*

(a) $y^x = x^{\frac{1}{y}}$ (b) $x^{\frac{1}{y}} = y^{\frac{1}{x}}$
(c) $x^y = y^x$ (d) $x^y = y^{\frac{1}{x}}$

42. Which of the following is true ? *[CAT-97]*

(a) $7^{3^2} = (7^3)^2$ (b) $7^{3^2} > (7^3)^2$
(c) $7^{3^2} < (7^3)^2$ (d) None of these

43. If $9^x3^y = 2187$ and $2^{3x}2^{2y} - 4^{xy} = 0$, then what can be the value of $(x + y)$?

(a) 1 (b) 3
(c) 5 (d) 7

44. Which of the following statement(s) is/are true ? *[SSC(CGL-T-II)-17]*

I. $\frac{2}{3\sqrt{5}} < \frac{3}{2\sqrt{5}} < \frac{5}{4\sqrt{3}}$

II. $\frac{3}{2\sqrt{5}} < \frac{2}{3\sqrt{3}} < \frac{7}{4\sqrt{5}}$

(a) Only I (b) Only II
(c) Both I and II (d) Neither I nor II

45. If m and n are integers such that $(\sqrt{2})^{19}3^44^29^m8^n = 3^n16^m(\sqrt[4]{64})$, then m is *[CAT-19]*

(a) – 20 (b) – 12
(c) – 24 (d) – 16

Answer Key

1. (b)	2. (a)	3. (d)	4. (c)	5. (a)	6. (b)	7. (c)	8. (a)	9. (c)	10. (c)
11. (c)	12. (c)	13. (d)	14. (d)	15. (c)	16. (d)	17. (b)	18. (b)	19. (b)	20. (b)
21. (b)	22. (d)	23. (a)	24. (b)	25. (b)	26. (c)	27. (d)	28. (c)	29. (c)	30. (e)
31. (a)	32. (a)	33. (b)	34. (c)	35. (d)	36. (c)	37. (a)	38. (c)	39. (b)	40. (a)
41. (c)	42. (b)	43. (c)	44. (a)	45. (b)					

Hints and SOLUTIONS

1. Given, $9^{2x-1} - 81^{x-1} = 1944$

⇒ $9^{2x-1} - (9^2)^{x-1} = 1944 \Rightarrow 9^{2x-1} - 9^{2x-2} = 1944$

⇒ $9^{2x-2}(9-1) = 1944 \Rightarrow (3^2)^{2x-2} = 243$

⇒ $3^{4x-4} = 3^5 \Rightarrow 4x - 4 = 5 \Rightarrow x = \frac{9}{4}$

2. $\sqrt{8+\sqrt{57+\sqrt{38+\sqrt{108+13}}}} = \sqrt{8+\sqrt{57+\sqrt{38+11}}}$

$= \sqrt{8+\sqrt{57+7}} = \sqrt{8+8} = 4$

3. Given surds are: $\sqrt[2]{2}, \sqrt[3]{3}$ and $\sqrt[6]{6}$

LCM of orders 2, 3, and 6 = 6

Now, we have to make each order equal to 6 by multiplying it with an appropriate number.

$\sqrt[2]{2} \Rightarrow \sqrt[2\cdot3]{2^3} = \sqrt[6]{8}$

$\sqrt[3]{3} \Rightarrow \sqrt[3\cdot2]{3^2} = \sqrt[6]{9}$

$\sqrt[6]{6} \Rightarrow \sqrt[6]{6} = \sqrt[6]{6}$

In case of same order, 'the greater radicand, the greater surd', hence, $\sqrt[6]{9}$ is greatest. The corresponding surd is $\sqrt[3]{3}$. It is also seen that $\sqrt[6]{6}$ is the smallest surd.

Hence, option (d) is correct.

4. $\sqrt{24^4} + 224 = ? \times 20^2 \Rightarrow (24^4)^{\frac{1}{2}} + 224 = ? \times 20^2$

⇒ $576 + 224 = ? \times 400 \Rightarrow ? = 2$

5. Let $k = (x)^{\frac{1}{m}} = (y)^{\frac{1}{n}} = (z)^{\frac{1}{p}}$

Then, $(x)^{\frac{1}{m}} = k \Rightarrow x = k^m$ {If $a^x = m$, then $a = m^{\frac{1}{x}}$}

and $(y)^{\frac{1}{n}} = k \Rightarrow y = k^n$ and $(z)^{\frac{1}{p}} = k \Rightarrow z = k^p$

As given, $xyz = 1$

⇒ $k^m \cdot k^n \cdot k^p = k^0 \Rightarrow m + n + p = 0$

6. Reciprocal of $\left(\frac{2}{3}\right)^4 = \frac{1}{\left(\frac{2}{3}\right)^4} = \frac{3^4}{2^4}$

Reciprocal of $\left(\frac{2}{3}\right)^4 = \frac{1}{\left(\frac{2}{3}\right)^4} = \frac{1^4}{\left(\frac{2}{3}\right)^4} = \left(\frac{1}{\frac{2}{3}}\right)^4 = \left(\frac{3}{2}\right)^4$

Also, reciprocal of $\left(\frac{2}{3}\right)^4 = \frac{1}{\left(\frac{2}{3}\right)^4} = \left(\frac{2}{3}\right)^{-4}$

Hence, option (b) is correct answer.

7. (a) ⇒ $[(5^{-2})^{-2}]^{-2} = 5^{(-2)\cdot(-2)\cdot(-2)} = (5^{-2})^4$

(b) ⇒ $[(5^{-2})^{-2}]^{2} = 5^{(-2)\cdot(-2)\cdot(2)} = (5^{2})^4$

(c) ⇒ $[(2^{-5})^{-2}]^{-2} = 2^{(-5)\cdot(-2)\cdot(-2)} = (2^{-5})^4$

(d) ⇒ $[(2^{-5})^{2}]^{-2} = 2^{(-5)\cdot(2)\cdot(-2)} = (2^{5})^4$

Since 2^{-5} is the smallest, $[(2^{-5})^{-2}]^{-2}$ is the smallest surd.

8. $(2\times3)^3 \div (4\times9)^2 \times (27\times8)^2 = (6)^?$

⇒ $(2\times3)^3 \div (2^2\times3^2)^2 \times (3^3\times2^3)^2 = (6)^?$

⇒ $(2\times3)^3 \div (2\times3)^{2\cdot2} \times (2\times3)^{2\cdot3} = (6)^?$

⇒ $(2\times3)^{3-4+6} = (6)^? \Rightarrow (6)^5 = (6)^?$

⇒ ? = 5

9. $\sqrt{7\sqrt{7\sqrt{7\sqrt{7\ldots\infty}}}} = (343)^{y-1}$

⇒ $7^{\frac{1}{(2-1)}} = (343)^{y-1}$

⇒ $7 = (7^3)^{y-1}$

⇒ $3y - 3 = 1 \Rightarrow y = \frac{4}{3}$

10. Given, $\sqrt{1+\frac{93}{196}} = 1 + \frac{x}{14} \Rightarrow \sqrt{\frac{289}{196}} = 1 + \frac{x}{14}$

⇒ $\frac{17}{14} = 1 + \frac{x}{14} \Rightarrow \frac{3}{14} = \frac{x}{14} \Rightarrow x = 3$

11. Let $\sqrt{7+\sqrt{7-\sqrt{7+\sqrt{7-\ldots\ldots\infty}}}} = x$

⇒ $\sqrt{7+\sqrt{7-x}} = x$

⇒ $7 + \sqrt{7-x} = x^2$

Proceeding through options we get $x = 3$.

12. $[(3\sqrt{8}+\sqrt{8})\times(8\sqrt{8}+7\sqrt{8})] - 98 = [\sqrt{8}(3+1)\times\sqrt{8}(8+7)] - 98$

$= [4\sqrt{8}\times15\sqrt{8}] - 98 = [60\times8] - 98 = 382$

13. $16\times8^{n+2} = 2^m \Rightarrow 2^4\times(2^3)^{n+2} - 2^m$

⇒ $2^{4+3n+6} = 2^m \Rightarrow m = 3n + 10$

14. $(8)^7 \times \left(8^{\frac{1}{3}}\right)^6 \div (8)^{2.4} = (8)^? \Rightarrow (8)^{7+2-2.4} = (8)^?$

⇒ $(8)^{6.6} = (8)^?$

⇒ ?= 6.6

15. $6\times\sqrt{\frac{1}{(27)^{\frac{2}{3}}} + \frac{1}{(8)^{\frac{2}{3}}}} = 6\times\sqrt{\frac{1}{(3^3)^{\frac{2}{3}}} + \frac{1}{(2^3)^{\frac{2}{3}}}}$

$= 6\times\sqrt{\frac{1}{9}+\frac{1}{4}} = 6\times\sqrt{\frac{13}{36}} = \sqrt{13}$

16. Given, $x = \{a+\sqrt{a^2+b^3}\}^{\frac{1}{3}} + \{a-\sqrt{a^2+b^3}\}^{\frac{1}{3}}$

⇒ $x^3 = a + \sqrt{a^2+b^3} + a - \sqrt{a^2+b^3} + 3\{a+\sqrt{a^2+b^3}\}^{\frac{1}{3}} \cdot \{a-\sqrt{a^2+b^3}\}^{\frac{1}{3}} [\{a+\sqrt{a^2+b^3}\}^{\frac{1}{3}} + \{a-\sqrt{a^2+b^3}\}^{\frac{1}{3}}]$

{cubing both sides}

⇒ $x^3 = 2a + 3\{a^2 - a^2 - b^3\}^{\frac{1}{3}} [\{a+\sqrt{a^2+b^3}\}^{\frac{1}{3}} + \{a-\sqrt{a^2+b^3}\}^{\frac{1}{3}}]$

⇒ $x^3 = 2a - 3bx \Rightarrow x^3 + 3bx - 2a = 0$

17. $\left\{\dfrac{4^{p+\frac{1}{4}} \times \sqrt{2\times 2^p}}{2\times\sqrt{2^{-p}}}\right\}^{\frac{1}{p}} = \left\{\dfrac{(2^2)^{p+\frac{1}{4}} \times (2^{p+1})^{\frac{1}{2}}}{2\times(2^{-p})^{\frac{1}{2}}}\right\}^{\frac{1}{p}}$

$= \left\{\dfrac{2^{2p+\frac{1}{2}+\frac{p}{2}+\frac{1}{2}}}{2^{1-\frac{p}{2}}}\right\}^{\frac{1}{p}} = \left\{2^{2p+\frac{1}{2}+\frac{p}{2}+\frac{1}{2}-1+\frac{p}{2}}\right\}^{\frac{1}{p}} = \{2^{3p}\}^{\frac{1}{p}} = 2^3 = 8$

18. $\dfrac{1}{\sqrt[3]{4}+\sqrt[3]{2}+1} = \dfrac{1}{(2)^{\frac{2}{3}}+(2)^{\frac{1}{3}}+1}$

$= \dfrac{\{(2)^{\frac{1}{3}}-1\}}{\{(2)^{\frac{1}{3}}-1\}\{(2)^{\frac{2}{3}}+(2)^{\frac{1}{3}}\cdot 1+(1)^2\}}$

$= \dfrac{\{(2)^{\frac{1}{3}}-1\}}{\left\{2^{\frac{1}{3}}\right\}^3-\{1\}^3}$ $\{\because (a-b)(a^2+ab+b^2)=a^3-b^3\}$

$= \dfrac{\{(2)^{\frac{1}{3}}-1\}}{2-1} = \sqrt[3]{2}-1$

19. $\sqrt{4^2\times 8+9^2-40} = \sqrt{128+81-40} = \sqrt{169} = 13$

20. Given, $\sqrt{10+\sqrt[3]{x}} = 4$

$\Rightarrow \quad 10+\sqrt[3]{x} = 16$ {squaring both sides}

$\Rightarrow \quad \sqrt[3]{x} = 6$

$\Rightarrow \quad x = 6^3$ {cubing both sides}

$\Rightarrow \quad x = 216$

21. $(a^2+a^{-2}) = \left(a^2+\dfrac{1}{a^2}\right) = \left(a+\dfrac{1}{a}\right)^2-2$

$= \left(2+\sqrt{3}+\dfrac{1}{2+\sqrt{3}}\right)^2-2 = \left(\dfrac{4+3+4\sqrt{3}+1}{2+\sqrt{3}}\right)^2-2$

$= \left\{\dfrac{4(2+\sqrt{3})}{2+\sqrt{3}}\right\}^2-2 = 16-2 = 14$

22. Given, $196x^4 = x^6 \Rightarrow \sqrt{196x^4} = \sqrt{x^6} \Rightarrow 14x^2 = x^3$

23. Given, $b^y = c \Rightarrow (a^x)^y = c \Rightarrow a^{xy} = c$

$\Rightarrow \quad (a^{xy})^z = (c)^z \Rightarrow a^{xyz} = c^z$

$\Rightarrow \quad a^1 = c^z$ $\{\because xyz = 1\}$

Hence, $\quad c^z = a$

24. Given, $3^{x+y} = 81 \Rightarrow 3^{x+y} = 3^4$

$\Rightarrow \quad x+y = 4$ ...(i)

Also, $81^{x-y} = 3 \Rightarrow (3^4)^{x-y} = 3^1$

$\Rightarrow \quad x-y = \dfrac{1}{4}$ ...(ii)

On adding Eqs. (i) and (ii), we get $2x = \dfrac{17}{4} \Rightarrow x = \dfrac{17}{8}$

25. $2\sqrt[3]{40}-4\sqrt[3]{320}+3\sqrt[3]{625}-3\sqrt[3]{5}$

$= 2\sqrt[3]{5\times 2^3}-4\sqrt[3]{5\times 4^3}+3\sqrt[3]{5\times 5^3}-3\sqrt[3]{5}$

$= 2\cdot 2\sqrt[3]{5}-4\cdot 4\sqrt[3]{5}+3\cdot 5\sqrt[3]{5}-3\sqrt[3]{5} = 0$

26. Let $\sqrt{4+\sqrt{4-\sqrt{4+\sqrt{4-\ldots\ldots\infty}}}} = x$

$\Rightarrow \quad \sqrt{4+\sqrt{4-x}} = x \Rightarrow 4+\sqrt{4-x} = x^2$

$\Rightarrow \quad 4-x = (x^2-4)^2$

Here, options are 3, 1.30, 2.30 and 3.60

Proceeding through options, we get $x = 2.30$. Hence, option (c) is correct.

27. Given, $p^q = q^p \Rightarrow p^{9p} = (9p)^p$ $\{\because q = 9p\}$

$\Rightarrow \quad (p^9)^p = (9p)^p \Rightarrow p^9 = 9p \Rightarrow p^8 = 9 \Rightarrow p = \sqrt[8]{9}$

28. Let $k = a^x = b^y = c^z$

Then, $\quad a^x = k \Rightarrow a = k^{\frac{1}{x}}$ {If $a^x = m$, then $a = m^{\frac{1}{x}}$}

and $\quad b^y = k \Rightarrow b = k^{\frac{1}{y}}$ and $c^z = k \Rightarrow c = k^{\frac{1}{z}}$

As given, $abc = 1$

$\Rightarrow \quad k^{\frac{1}{x}}\cdot k^{\frac{1}{z}}\cdot k^{\frac{1}{y}} = k^0 \Rightarrow \dfrac{1}{x}+\dfrac{1}{y}+\dfrac{1}{z} = 0 \Rightarrow \dfrac{yz+xz+xy}{xyz} = 0$

$\Rightarrow \quad xy+yz+zx = 0$

29. Given, $(3.7)^x = 10000 \Rightarrow 3.7 = (10000)^{\frac{1}{x}} = (100)^{\frac{2}{x}}$

and $(0.037)^y = 10000 \Rightarrow 0.037 = (10000)^{\frac{1}{y}}$

$\Rightarrow \quad 3.7 = (100)^{\frac{2}{y}}\times 100 = (100)^{\frac{2}{y}+1}$

From the above, $(100)^{\frac{2}{x}} = (100)^{\frac{2}{y}+1}$

$\Rightarrow \quad \dfrac{2}{x} = \dfrac{2}{y}+1 \Rightarrow \dfrac{2}{x}-\dfrac{2}{y} = 1 \Rightarrow \dfrac{1}{x}-\dfrac{1}{y} = \dfrac{1}{2}$

30. Given, $2^{0.7x}\cdot 3^{-1.25y} = \dfrac{8\sqrt{6}}{27}$

$\Rightarrow \quad 2^{0.7x}\cdot 3^{-1.25y} = (2)^3\times(2\cdot 3)^{\frac{1}{2}}\times(3)^{-3}$

$\Rightarrow \quad 2^{0.7x}\cdot 3^{-1.25y} = (2)^{\frac{7}{2}}\times(3)^{\frac{-5}{2}}$

Hence, $0.7x = \dfrac{7}{2}$ and $-1.25y = \dfrac{-5}{2} \Rightarrow x = 5$ and $y = 2$

Further, $\quad 4^{0.3x}\cdot 9^{0.2y} = 8\cdot(81)^{\frac{1}{5}}$

$\Rightarrow \quad (2^2)^{0.3x}\cdot(3^2)^{0.2y} = 2^3\cdot(3^4)^{\frac{1}{5}}$

$\Rightarrow \quad (2)^{0.6x}\cdot(3)^{0.4y} = 2^3\cdot(3)^{\frac{4}{5}}$

Hence, $0.6x = 3$ and $0.4y = \dfrac{4}{5} \Rightarrow x = 5$ and $y = 2$

Thus, option (e) is correct.

31. $\sqrt[3]{2^4\sqrt{2^{-5}\sqrt{2^6}}} = \sqrt[3]{2^4\sqrt{2^{-5}\cdot 2^{\frac{6}{2}}}} = \sqrt[3]{2^4\sqrt{2^{-2}}}$

$= \sqrt[3]{2^4\cdot 2^{-\frac{2}{2}}} = \sqrt[3]{2^3} = 2^{\frac{3}{3}} = 2$

Alternatively,

$(6\div 2)+(-5) = -2$

$(-2\div 2)+4 = 3 \Rightarrow (3\div 3) = 1$

Then, required answer $= 2^1 = 2$

32. $\sqrt{2+\sqrt{3}}+\sqrt{2-\sqrt{3}}$

$$=\sqrt{\frac{(\sqrt{3})^2+(1)^2+2\cdot\sqrt{3}\cdot 1}{2}}+\sqrt{\frac{(\sqrt{3})^2+(1)^2-2\cdot\sqrt{3}\cdot 1}{2}}$$

$$=\sqrt{\frac{(\sqrt{3}+1)^2}{2}}+\sqrt{\frac{(\sqrt{3}-1)^2}{2}}=\frac{(\sqrt{3}+1)+(\sqrt{3}-1)}{\sqrt{2}}$$

$$=\frac{2\sqrt{3}}{\sqrt{2}}=\sqrt{2}\cdot\sqrt{3}=\sqrt{6}$$

33. $\dfrac{\sqrt{5}+\sqrt{3}}{\sqrt{5}-\sqrt{3}}+\dfrac{\sqrt{5}-\sqrt{3}}{\sqrt{5}+\sqrt{3}}$

$$=\left(\frac{5+3+2\cdot\sqrt{5}\cdot\sqrt{3}+5+3-2\cdot\sqrt{5}\cdot\sqrt{3}}{5-3}\right)=8$$

34. Given, $2^m+2^{1+m}=24 \Rightarrow 2^m(1+2)=24$

$\Rightarrow \quad 2^m=8 \Rightarrow 2^m=2^3 \Rightarrow m=3$

35. $\sqrt[6]{24-16\sqrt{2}}\times\sqrt[3]{4+2\sqrt{2}}$

$$=\sqrt[6]{4^2+(2\sqrt{2})^2-2\cdot 4\cdot 2\sqrt{2}}\times\sqrt[3]{4+2\sqrt{2}}$$

$$=\sqrt[6]{(4-2\sqrt{2})^2}\times\sqrt[3]{4+2\sqrt{2}}=\sqrt[3]{4-2\sqrt{2}}\times\sqrt[3]{4+2\sqrt{2}}$$

$$=\sqrt[3]{4^2-(2\sqrt{2})^2}=\sqrt[3]{8}=2$$

36. $\sqrt[3]{\dfrac{4}{9}}-\sqrt[3]{\dfrac{2}{9}}+\sqrt[3]{\dfrac{1}{9}}=\dfrac{\sqrt[3]{4}-\sqrt[3]{2}+\sqrt[3]{1}}{\sqrt[3]{9}}=\dfrac{\left(2^{\frac{1}{3}}\right)^2-2^{\frac{1}{3\cdot 1}}+(1)^2}{\sqrt[3]{9}}$

$$=\frac{\left(2^{\frac{1}{3}}+1\right)\left(\left(2^{\frac{1}{3}}\right)^2-2^{\frac{1}{3\cdot 1}}+(1)^2\right)}{\left(2^{\frac{1}{3}}+1\right)\sqrt[3]{9}}=\frac{\left(2^{\frac{1}{3}}\right)^3+(1)^3}{\left(2^{\frac{1}{3}}+1\right)\sqrt[3]{9}}$$

$\{\because (a+b)(a^2-ab+b^2)=a^3+b^3\}$

$$=\frac{3}{(\sqrt[3]{2}+1)\sqrt[3]{3^2}}=\frac{3^{1-\frac{2}{3}}}{(\sqrt[3]{2}+1)}=\frac{\sqrt[3]{3}}{\sqrt[3]{2}+1}$$

37. Let $\sqrt{2\sqrt[3]{4\sqrt{2\sqrt[3]{4\sqrt{2\sqrt[3]{4\ldots\ldots}}}}}}=x$

$\Rightarrow \quad \sqrt{2\sqrt[3]{4x}}=x$

$\Rightarrow \quad 2\cdot\sqrt[3]{4x}=x^2$ {squaring both sides}

$\Rightarrow \quad 8\times 4x=x^6$ {cubing both sides}

$\Rightarrow \quad 32=x^5 \Rightarrow 2^5=x^5 \Rightarrow x=2$

38. 1st term $=\dfrac{1}{\sqrt{9}-\sqrt{8}}=\dfrac{1}{\sqrt{9}-\sqrt{8}}\times\dfrac{\sqrt{9}+\sqrt{8}}{\sqrt{9}+\sqrt{8}}=\sqrt{9}+\sqrt{8}$

2nd term $=\dfrac{1}{\sqrt{8}-\sqrt{7}}=\dfrac{1}{\sqrt{8}-\sqrt{7}}\times\dfrac{\sqrt{8}+\sqrt{7}}{\sqrt{8}+\sqrt{7}}=\sqrt{8}+\sqrt{7}$

and so on.

Now, $\dfrac{1}{\sqrt{9}-\sqrt{8}}-\dfrac{1}{\sqrt{8}-\sqrt{7}}+\dfrac{1}{\sqrt{7}-\sqrt{6}}-\dfrac{1}{\sqrt{6}-\sqrt{5}}+\dfrac{1}{\sqrt{5}-\sqrt{4}}$

$=\sqrt{9}+\sqrt{8}-(\sqrt{8}+\sqrt{7})+\sqrt{7}+\sqrt{6}-(\sqrt{6}+\sqrt{5})+\sqrt{5}+\sqrt{4}$

$=\sqrt{9}+\sqrt{8}-\sqrt{8}-\sqrt{7}+\sqrt{7}+\sqrt{6}-\sqrt{6}-\sqrt{5})+\sqrt{5}+\sqrt{4}$

$=\sqrt{9}+\sqrt{4}=3+2=5$

39. $(36)^{\frac{1}{6}}=(6^2)^{\frac{1}{6}}=6^{2\cdot\frac{1}{6}}=\sqrt[3]{6}$

40. Given, $p^x=r^y$

$\Rightarrow \quad (p^x)^w=(r^y)^w \Rightarrow p^{xw}=r^{yw}$ ...(i)

Further given, $r^w=p^z$

$\Rightarrow \quad (r^w)^y=(p^z)^y \Rightarrow r^{wy}=p^{zy}$ ...(ii)

From Eqs. (i) and (ii), we get $p^{xw}=p^{zy} \Rightarrow xw=yz$

Hence, option (a) is correct.

41. Given, $y=\dfrac{t}{t^{t-1}} \Rightarrow y=t^{\frac{1}{t-1}\times t} \Rightarrow y=x^t$

$\Rightarrow \quad y^{t^{\frac{1}{t-1}}}=(x^t)^{t^{\frac{1}{t-1}}} \Rightarrow y^x=(x)^{t\times t^{\frac{1}{t-1}}}$

$\Rightarrow \quad y^x=(x)^{t^{1+\frac{1}{t-1}}} \Rightarrow y^x=x^y$

Hence, option (c) is correct.

42. $7^{3^2}=7^9$ and $(7^3)^2=7^6$. Hence, option (b) is correct.

43. Given, $9^x3^y=2187 \Rightarrow 3^{2x}\cdot 3^y=3^7 \Rightarrow 2x+y=7 \Rightarrow y=7-2x$

Also given, $2^{3x}2^{2y}-4^{xy}=0 \Rightarrow 2^{3x}2^{2y}=2^{2xy} \Rightarrow 3x+2y=2xy$

$3x+2(7-2x)=2x(7-2x)$ $\{\because y=7-2x\}$

$\Rightarrow \quad 14-x=14x-4x^2 \Rightarrow 4x^2-15x+14=0$

$\Rightarrow \quad 4x^2-7x-8x+14=0 \Rightarrow x(4x-7)-2(4x-7)=0$

$\Rightarrow \quad x=2$ and $x=\dfrac{7}{4}$

From above, $2x+y=7 \Rightarrow x+y=7-x$

$\Rightarrow \quad x+y=7-2$ and $7-\dfrac{7}{4}$ i.e. 5 and 5.25

44. Statement (I): $\dfrac{2}{3\sqrt{5}}<\dfrac{3}{2\sqrt{5}}<\dfrac{5}{4\sqrt{3}}$

Here, $\dfrac{2}{3\sqrt{5}}<\dfrac{3}{2\sqrt{5}}$ as $\dfrac{2}{3}<\dfrac{3}{2}$

(both have been divided by $\sqrt{5}$)

Also, $\dfrac{3}{2\sqrt{5}}<\dfrac{5}{4\sqrt{3}}$ as $12\sqrt{3}<10\sqrt{5}$

Hence, Statement (I) is true.

Statement (II): $\dfrac{3}{2\sqrt{5}}<\dfrac{2}{3\sqrt{3}}<\dfrac{7}{4\sqrt{5}}$

Here, $\dfrac{3}{2\sqrt{5}}\nless\dfrac{2}{3\sqrt{3}}$ as $9\sqrt{3}\nless 4\sqrt{5}$

$\therefore$ Option (a) is correct.

45. Given, $(\sqrt{2})^{19}\,3^4 4^2 9^m 8^n=3^n 16^m(\sqrt[4]{64})$

$\Rightarrow \quad 2^{\frac{19}{2}}\times 3^4\times 2^4\times 3^{2m}\times 2^{3n}=3^n\times 2^{4m}\times 2^{\frac{3}{2}}$

$\Rightarrow \quad 2^{\left(\frac{19}{2}+4+3n\right)}\times 3^{(2m+4)}=2^{\left(\frac{3}{2}+4m\right)}\times 3^n$

$\Rightarrow \quad \dfrac{19}{2}+4+3n=\dfrac{3}{2}+4m$ and $2m+4=n$

(comparing the powers of the same bases)

$\Rightarrow \quad \dfrac{19}{2}+4+3(2m+4)=\dfrac{3}{2}+4m$

(putting the value of n)

$\Rightarrow \quad m=-12$

11

Simplification

Learning OBJECTIVES

After completing this chapter, aspirants will be able to

- Learn meaning of Simplification;
- Understand the significance of this chapter in numerical ability;
- Comprehend the applications of mathematical operations together;
- Understand fundamentals of operations to be done in a sequence;
- Understand the philosophy of the categorisation of the questions;
- Analyse the trend of the questions being asked in various competitive examinations;
- Recognise the types of question which can be easily solved through the given options;
- Grip on concepts helping to solve the questions based on simplification in moments;
- Learn some new concepts/fundamentals;
- Apply the pragmatic approaches in lieu of academic approaches; and
- Ponder on the alternative methods.

Introduction

This topic is very critical for the competitive examinations as almost all the competitive examinations like CAT, MAT, SSC, Banks, Railway, LIC etc. cover this topic. Further, it will not be out of context to say that Simplification-Problems based on surds and indices, simplification of unique series have been elaborated under the chapter 'Indices and Surds'.

Simplification means expression in a simpler form. In order to simplify the given expression, first of all vinculum/bar must be removed and thereafter all the brackets must be removed strictly in the order ( ), { }, [ ] .

After the execution of the above operations, we must execute the operations 'of ', '÷', '×', '+', '–' strictly in order.

Note

(i) 'of' means multiplication but operated before division.

(ii) '+' and '–' may be executed simultaneously or their order may be interchanged.

Rule of 'V-BODMAS'

The above order may be remembered as:

V	Vinculum/Bar	⇓
B	Brackets	
O	Of	
D	Division	
M	Multiplication	
A	Addition	
S	Subtraction	

The aspirants are supposed to apply the techniques, meant for fast calculation as discussed earlier, in this chapter.

Illustrations

Type I

Problems based on mathematical operations and expressed in the linear mode.

Problem 1 $(800 \div 64) \times (1296 \div 36) = ?$

Sol. $\Rightarrow (800 \div 64) \times (1296 \div 36) = \frac{800}{64} \times \frac{1296}{36} = \frac{100}{8} \times \frac{1296}{36}$

$= 12.5 \times 36 = 450$

Problem 2 $\frac{1}{8}$ of $\frac{2}{3}$ of $\frac{3}{5}$ of $1715 = ?$ *[SBI –Associates(PO)-11)]*

Sol. $\frac{1}{8} \times \frac{2}{3} \times \frac{3}{5} \times 1715 = \frac{343}{4} = 85.75$

Problem 3 $25.05 \times 123.95 + 388.999 \times 15.001 = ?$

Sol. Since the options were in whole numbers, thus we should also make the expression in whole numbers for calculation.

$\Rightarrow 25 \times 124 + 389 \times 15$

[Now, apply approach meant for multiplication]

$= 3100 + 5835 = 8935$

Problem 4 $(23.1)^2 + (48.6)^2 - (39.8)^2 = ? + 1147.69$

[IBPS (CWE)-11, SBI(PO-Pre.)-17, IBPS(PO-Pre.)-19]

Sol. $533.61 + 2361.96 - 1584.04 = ? + 1147.69$

[Applying tricks for computation of square]

$\Rightarrow 533.61 + 2361.96 - 1584.04 - 1147.69 = ?$

[Now, observe the alternative approach given under 'When addition and subtraction both operations are required' in chapter 'Addition and subtraction'.]

$\Rightarrow ? = 163.84$

Problem 5 $\sqrt{11449} \times \sqrt{6241} - (54)^2 = \sqrt{?} + (74)^2$

Sol. $\Rightarrow (107 \times 79) - 2916 - 5476 = \sqrt{?}$

[Applying tricks for computation of square-root and square]

$\Rightarrow 8453 - 2916 - 5476 = \sqrt{?}$

$\Rightarrow 61 = \sqrt{?}$

$\Rightarrow ? = (61)^2$

$= 3721$

Problem 6 $\sqrt{400} + \sqrt{0.0400} + \sqrt{0.000004} = ?$

[SSC (CGL-PT)-04, SSC(CPO) – 09, SSC(CGL-T-II)-12]

Sol. $\sqrt{400} + \sqrt{0.0400} + \sqrt{0.000004}$

$= 20 + 0.2 + 0.002 = 20 + 0.200 + 0.002 = 20.202$

Problem 7 $\frac{458}{32} \div \frac{19}{789} \times \frac{232}{78} = ?$ *[Corporation Bank(PO)-11]*

Sol. $\frac{458}{32} \div \frac{19}{789} \times \frac{232}{78} = \frac{458}{32} \times \frac{789}{19} \times \frac{232}{78} = ?$

$\Rightarrow 14.31 \times 41.53 \times 2.97 = 1765$ (approx.)

Note *You are supposed to find the answer in approximate value, not exactly, in the exams in case of such types of questions.*

Problem 8 $\sqrt{7378} \times \sqrt{1330} \div \sqrt{660} = ?$ (not to calculate exactly)

Sol. $\sqrt{7378} \times \sqrt{1330} \div \sqrt{660} = 86 \times 36 \div 26 = 119$ (approx.)

[Applying Estimation method for computation of square-root]

Problem 9 $\sqrt{5400} \times \sqrt{2120} \div \sqrt{460} = ?$ (not to calculate exactly)

Sol. $\sqrt{5400} \times \sqrt{2120} \div \sqrt{460} = 74 \times 46 \div 22 = 160$ (approx.)

[Applying Estimation method for computation of square-root]

Problem 10 $(21)^2 - 3717 \div 59 = ? \times 8$

Sol. $(21)^2 - 3717 \div 59 = ? \times 8$

$\Rightarrow 441 - 3717 \div 59 = ? \times 8$

$\Rightarrow 441 - 63 = ? \times 8$

$\Rightarrow ? = \frac{378}{8} = 47.25$

Problem 11 $4\frac{1}{2} \times 4\frac{1}{3} - 8\frac{1}{3} \div 5\frac{2}{3} = ?$ *[SBI(PO-Pre.)-18]*

Sol. $4\frac{1}{2} \times 4\frac{1}{3} - 8\frac{1}{3} \div 5\frac{2}{3} = 4\frac{1}{2} \times 4\frac{1}{3} - \frac{25}{3} \times \frac{3}{17}$

$= 19\frac{1}{2} - 1\frac{8}{17} = 18\frac{1}{34}$

Problem 12 Find the value of

$18 - [10 - \{6 - (7 - \overline{9 - 4})\}]$.

Sol. $18 - [10 - \{6 - (7 - \overline{9 - 4})\}]$

$= 18 - [10 - \{6 - (7 - 5)\}]$

$= 18 - [10 - \{6 - 2\}]$

$= 18 - [10 - 4]$

$= 18 - 6$

$= 12$

Problem 13 $1 \div [1 + 1 \div \{1 + 1 \div (1 + 1 \div 2)\}] = ?$

[SSC(CGL-PT)-99, 01, 02, SBI(PO-Pre.)-18]

Sol. $\Rightarrow 1 \div [1 + 1 \div \{1 + 1 \div (1 + 1 \div 2)\}]$

$= 1 \div [1 + 1 \div \{1 + 1 \div \left(1 + \frac{1}{2}\right)\}]$

$$= 1 \div [1 + 1 \div \{1 + 1 \div \frac{3}{2}\}]$$

$$= 1 \div [1 + 1 \div \left\{1 + \frac{2}{3}\right\}] = 1 \div [1 + 1 \div \frac{5}{3}]$$

$$= 1 \div \left[1 + \frac{3}{5}\right] = 1 \div \frac{8}{5} = \frac{5}{8}$$

Problem 14 $b - [b - (a + b) - \{b - (b - \overline{a - b})\} + 2a] = ?$

Sol. $b - [b - (a + b) - \{b - (b - \overline{a - b})\} + 2a]$

$$= b - [b - (a + b) - \{b - (b - a + b)\} + 2a]$$
$$= b - [b - a - b - \{b - b + a - b\} + 2a]$$
$$= b - [-a - \{a - b\} + 2a]$$
$$= b - [-a - a + b + 2a]$$
$$= b - [b]$$
$$= b - b = 0$$

Type II

Problems based on mathematical operations and expressed in a ladder shape.

Problem 15 $3 + \cfrac{5}{3 + \cfrac{2}{3 + \cfrac{2}{2 + \cfrac{1}{2}}}} = ?$

Sol. $3 + \cfrac{5}{3 + \cfrac{2}{3 + \cfrac{2}{2 + \cfrac{1}{2}}}} = 3 + \cfrac{5}{3 + \cfrac{2}{3 + \left(2 \times \frac{2}{5}\right)}}$

$$= 3 + \cfrac{5}{3 + \cfrac{2}{\frac{19}{5}}} = 3 + \cfrac{5}{3 + \frac{10}{19}}$$

$$= 3 + \cfrac{5}{\frac{67}{19}} = 3 + \frac{95}{67} = 3 + 1\frac{28}{67} = 4\frac{28}{67}$$

Problem 16 If a, b and c are positive integers such that $\cfrac{1}{a + \cfrac{1}{b + \cfrac{1}{c + \cfrac{1}{2}}}} = \frac{16}{23}$, then what is the mean of a, b and c? *[CDS-19]*

$$\frac{16}{23} = \frac{1}{\frac{23}{16}} = \cfrac{1}{1 + \frac{7}{16}} = \cfrac{1}{1 + \left(\cfrac{1}{\frac{16}{7}}\right)}$$

$$= \cfrac{1}{1 + \cfrac{1}{2 + \frac{2}{7}}} = \cfrac{1}{1 + \cfrac{1}{2 + \left(\cfrac{1}{\frac{7}{2}}\right)}}$$

$$= \cfrac{1}{1 + \cfrac{1}{2 + \left(\cfrac{1}{3 + \frac{1}{2}}\right)}} = \cfrac{1}{1 + \cfrac{1}{2 + \cfrac{1}{3 + \cfrac{1}{2}}}} = \cfrac{1}{a + \cfrac{1}{b + \cfrac{1}{c + \cfrac{1}{2}}}}$$

Clearly, $a = 1$, $b = 2$ and $c = 3$

Then, their mean $= \frac{1 + 2 + 3}{3} = 2$

Problem 17 $\dfrac{\frac{1}{3} \div \frac{1}{3} \times \frac{1}{3}}{\frac{1}{3} \div \frac{1}{3} \text{ of } \frac{1}{3}} - \frac{1}{9} = ?$

Sol. $\dfrac{\frac{1}{3} \div \frac{1}{3} \times \frac{1}{3}}{\frac{1}{3} \div \frac{1}{3} \text{ of } \frac{1}{3}} - \frac{1}{9} = \dfrac{\frac{1}{3} \times \frac{3}{1} \times \frac{1}{3}}{\frac{1}{3} \div \frac{1}{9}} - \frac{1}{9} = \dfrac{\frac{1}{3}}{\frac{1}{3} \times \frac{9}{1}} - \frac{1}{9}$

$$= \dfrac{\frac{1}{3}}{\frac{1}{3} \times \frac{9}{1}} - \frac{1}{9} = \frac{1}{3} \div \frac{3}{1} - \frac{1}{9} = \frac{1}{3} \times \frac{1}{3} - \frac{1}{9} = \frac{1}{9} - \frac{1}{9} = 0$$

Problem 18

$$\frac{7}{5 - 2\frac{2}{3}} \div \dfrac{3 - \cfrac{2}{3 - 1\frac{1}{2}}}{4 - 1\frac{1}{2}} - \frac{5}{7} \times \left[\frac{7}{10} + 1\frac{1}{5} \times \frac{3\frac{1}{3} - 2\frac{1}{2}}{2\frac{5}{21} - 2}\right] = ?$$

Sol. 1st term $= \dfrac{7}{5 - 2\frac{2}{3}} = \dfrac{7}{2\frac{1}{3}} = 7 \times \frac{3}{7} = 3$

2nd term $= \dfrac{3 - \cfrac{2}{3 - 1\frac{1}{2}}}{4 - 1\frac{1}{2}} = \dfrac{3 - \cfrac{2}{1\frac{1}{2}}}{2\frac{1}{2}} = \dfrac{3 - \frac{4}{3}}{\frac{5}{2}} = \frac{5}{3} \times \frac{2}{5} = \frac{2}{3}$

3rd term $= \frac{5}{7}$

4th term $= \left[\frac{7}{10} + 1\frac{1}{5} \times \dfrac{3\frac{1}{3} - 2\frac{1}{2}}{2\frac{5}{21} - 2}\right] = \left[\frac{7}{10} + 1\frac{1}{5} \times \dfrac{\frac{5}{6}}{\frac{5}{21}}\right]$

$$= \left[\frac{7}{10} + \frac{6}{5} \times \frac{5}{6} \times \frac{21}{5}\right] = \left[\frac{7}{10} + \frac{21}{5}\right] = \frac{49}{10}$$

Now, expression $= 3 \div \frac{2}{3} - \frac{5}{7} \times \frac{49}{10} = 3 \times \frac{3}{2} - \frac{5}{7} \times \frac{49}{10}$

$$= \frac{9}{2} - \frac{7}{2} = 1$$

Problem 19 If $\cfrac{2x}{1 + \cfrac{1}{1 + \cfrac{x}{1 - x}}} = 1$, then $x = ?$

Sol. Given, $\cfrac{2x}{1 + \cfrac{1}{1 + \cfrac{x}{1 - x}}} = 1$

$$\Rightarrow \quad \frac{2x}{1+\dfrac{1}{\dfrac{1-x+x}{1-x}}} = 1 \Rightarrow \frac{2x}{1+\dfrac{1}{\dfrac{1}{1-x}}} = 1 \Rightarrow \frac{2x}{1+1-x} = 1$$

$$\Rightarrow \quad 2x = 2 - x \Rightarrow 3x = 2$$

$$\therefore \quad x = \frac{2}{3}$$

Type III

Problems based on mathematical operations, where the symbol has specific meaning.

Problem 20 If $a * b = 2(a+b)$, then $5 * 2 = ?$

Sol. Given, $a * b = 2(a+b) \Rightarrow 5 * 2 = 2(5+2) = 14$

Problem 21 If $x * y = x^2 + y^2 - xy$, then, $11 * 13 = ?$ *[SSC(CGL-T-I)-11]*

Sol. Given, $x * y = x^2 + y^2 - xy$

$$\Rightarrow \quad 11 * 13 = 11^2 + 13^2 - 11 \cdot 13 = 147$$

Type IV

Simplification based on Algebraic Identities.

[Students are advised to peruse the Algebraic Identities given under 'Algebra' before solving the question given below]

Problem 22 $\dfrac{(0.03)^2 - (0.01)^2}{0.03 - 0.01} = ?$

Sol. $\dfrac{(0.03)^2 - (0.01)^2}{0.03 - 0.01} = \dfrac{(0.03+0.01)(0.03-0.01)}{(0.03-0.01)}$ $[\because a^2 - b^2 = (a+b)(a-b)]$

$$= (0.03 + 0.01) = 0.04$$

Problem 23 $\dfrac{(2.697 - 0.498)^2 + (2.697 + 0.498)^2}{2.697 \times 2.697 + 0.498 \times 0.498} = ?$ *[SSC(CGL-PT)-08, CDS-17]*

Sol. Let $2.697 = a$ and $0.498 = b$

Then, given expression $= \dfrac{(a-b)^2 + (a+b)^2}{a^2 + b^2} = \dfrac{2(a^2 + b^2)}{a^2 + b^2} = 2$

Problem 24 $\dfrac{137 \times 137 + 137 \times 133 + 133 \times 133}{137 \times 137 \times 137 - 133 \times 133 \times 133} = ?$

Sol. Given expression $= \dfrac{a^2 + ab + b^2}{a^3 - b^3}$ [Assuming $a = 137$ and $b = 133$]

$$= \frac{a^2 + ab + b^2}{(a-b)(a^2 + ab + b^2)} = \frac{1}{(a-b)} = \frac{1}{(137-133)} = \frac{1}{4}$$

Problem 25

$$\frac{\frac{1}{3}\cdot\frac{1}{3}\cdot\frac{1}{3} + \frac{1}{4}\cdot\frac{1}{4}\cdot\frac{1}{4} - 3\cdot\frac{1}{3}\cdot\frac{1}{4}\cdot\frac{1}{5} + \frac{1}{5}\cdot\frac{1}{5}\cdot\frac{1}{5}}{\frac{1}{3}\cdot\frac{1}{3} + \frac{1}{4}\cdot\frac{1}{4} + \frac{1}{5}\cdot\frac{1}{5} - \left(\frac{1}{3}\cdot\frac{1}{4} + \frac{1}{4}\cdot\frac{1}{5} + \frac{1}{3}\cdot\frac{1}{5}\right)} = ?$$

[SSC(CGL-PT)-04, CDS-12]

Sol. Given expression $= \dfrac{a^3 + b^3 - 3abc + c^3}{a^2 + b^2 + c^2 - (ab + bc + ca)}$

[Assuming $a = \frac{1}{3}, b = \frac{1}{4}$ and $c = \frac{1}{5}$]

$$= \frac{(a+b+c)(a^2 + b^2 + c^2 - ab - bc - ca)}{(a^2 + b^2 + c^2 - ab - bc - ca)}$$

$$= (a+b+c) = \left(\frac{1}{3} + \frac{1}{4} + \frac{1}{5}\right) = \frac{47}{60}$$

Problem 26 $(0.98)^3 + (0.02)^3 + 3 \times 0.98 \times 0.02 - 1 = ?$

Sol. Given expression $= a^3 + b^3 + 3ab - 1$

[Assuming $a = 0.98, b = 0.02$]

$$= a^3 + b^3 + 3ab(a+b) - 1 \quad [\because a + b = 1]$$

$$= (a+b)^3 - 1$$

$$= (1)^3 - 1 = 0$$

Problem 27 $\dfrac{\left(\frac{6}{5}\right)^4 + \left(\frac{5}{6}\right)^4 + 1}{\left(\frac{6}{5}\right)^2 + \left(\frac{5}{6}\right)^2 + 1} = ?$

Sol. Given expression $= \dfrac{(a)^4 + (b)^4 + 1}{(a)^2 + (b)^2 + 1}$ [Assuming $a = \frac{6}{5}, b = \frac{5}{6}$]

$$= \frac{(a^2)^2 + (b^2)^2 + 1}{(a)^2 + (b)^2 + 1}$$

$$= \frac{(a^2 + b^2)^2 - 2 \cdot a^2 \cdot b^2 + 1}{(a)^2 + (b)^2 + 1}$$

$[\because x^2 + y^2 = (x+y)^2 - 2xy]$

$$= \frac{(a^2 + b^2)^2 - 2 \cdot (ab)^2 + 1}{(a)^2 + (b)^2 + 1} = \frac{(a^2 + b^2)^2 - 2 \cdot 1^2 + 1}{(a)^2 + (b)^2 + 1}$$

$[\because ab = \frac{6}{5} \times \frac{5}{6} = 1]$

$$= \frac{(a^2 + b^2)^2 - 1}{a^2 + b^2 + 1} = \frac{(a^2 + b^2)^2 - (1)^2}{a^2 + b^2 + 1}$$

$$= \frac{(a^2 + b^2 + 1)(a^2 + b^2 - 1)}{(a^2 + b^2 + 1)}$$

$$= (a^2 + b^2 - 1) = \left(\frac{6}{5}\right)^2 + \left(\frac{5}{6}\right)^2 - 1 = \frac{36}{25} + \frac{25}{36} - 1$$

$$= \frac{1021}{900} = 1\frac{121}{900}$$

Problem 28

$$\frac{(1.7)^3 + (2.2)^3 + (3.1)^3 - 3 \times 1.7 \times 2.2 \times 3.1}{(0.5)^2 + (0.9)^2 + (1.4)^2} = ?$$

Sol. Given expression $= \dfrac{a^3 + b^3 + c^3 - 3abc}{(b-a)^2 + (c-b)^2 + (c-a)^2}$

[Assuming $a = 1.7, b = 2.2$ and $c = 3.1$]

$$= \frac{\frac{1}{2}(a+b+c)\{(a-b)^2 + (b-c)^2 + (c-a)^2\}}{\{(a-b)^2 + (b-c)^2 + (c-a)^2\}}$$

$$= \frac{1}{2}(a+b+c) = \frac{1}{2}(1.7 + 2.2 + 3.1) = 3.5$$

Problem 29 $\dfrac{(0.25)^3 + (0.43)^3 - (0.68)^3}{3 \cdot (0.25)\,(0.43)\,(0.68)} = ?$

Sol. Given, expression $= \dfrac{a^3 + b^3 + c^3}{3ab\,(-c)}$

[Assuming $a = 0.25$, $b = 0.43$ and $c = -0.68$]

$= \dfrac{3abc}{3ab(-c)}$ $\quad [\because a^3 + b^3 + c^3 = 3abc$, if $a + b + c = 0]$

$= -1$

Problem 30 $\dfrac{(a-b)^3 + (b-c)^3 + (c-a)^3}{5 \cdot (a-b)\,(b-c)\,(c-a)} = ?$

Sol. Given expression $= \dfrac{(a-b)^3 + (b-c)^3 + (c-a)^3}{5 \cdot (a-b)\,(b-c)\,(c-a)}$

$= \dfrac{3(a-b)\,(b-c)\,(c-a)}{5(a-b)(b-c)\,(c-a)} = \dfrac{3}{5}$

$[\because x^3 + y^3 + z^3 = 3xyz$, if $x + y + z = 0$. Here, $(a-b) + (b-c) + (c-a) = 0]$

Problem 31 $\dfrac{1.\bar{3} \times 1.\bar{3} \times 1.\bar{3} - 1}{1.\bar{3} + 1.\bar{3} \times 1.\bar{3} + 1} = ?$

Sol. Given expression $= \dfrac{1.\bar{3} \times 1.\bar{3} \times 1.\bar{3} - 1}{1.\bar{3} + 1.\bar{3} \times 1.\bar{3} + 1}$

$= \dfrac{(1.\bar{3})^3 - (1)^3}{1.\bar{3} \times 1 + 1.\bar{3} \times 1.\bar{3} + 1 \times 1}$

$= \dfrac{(a)^3 - (b)^3}{ab + a^2 + b^2}$ $\quad$ [Assuming $a = 1.\bar{3}$, $b = 1$]

$= \dfrac{(a-b)\,(a^2 + ab + b^2)}{ab + a^2 + b^2}$

$= a - b = 1.\bar{3} - 1 = .\bar{3} = \dfrac{3}{9} = \dfrac{1}{3}$

Type V

Where commonality is found between the numerator and the denominator

Problem 32 $\dfrac{(3.07)^2 + (0.0193)^2}{(0.307)^2 + (0.00193)^2} = ?$

Sol. $\dfrac{(3.07)^2 + (0.0193)^2}{(0.307)^2 + (0.00193)^2} = \dfrac{(3.07)^2 + (0.0193)^2}{\left(\dfrac{3.07}{10}\right)^2 + \left(\dfrac{0.0193}{10}\right)^2}$

$= \dfrac{(3.07)^2 + (0.0193)^2}{\dfrac{1}{100}\{(3.07)^2 + (0.0193)^2\}} = 100$

Problem 33 $\dfrac{(0.22)^3 + (0.11)^3 + (0.32)^3}{(0.66)^3 + (0.96)^3 + (0.33)^3} = ?$

Sol. $\Rightarrow \dfrac{(0.22)^3 + (0.11)^3 + (0.32)^3}{(0.66)^3 + (0.96)^3 + (0.33)^3}$

$= \dfrac{(0.22)^3 + (0.11)^3 + (0.32)^3}{3^3\{(0.22)^3 + (0.11)^3 + (0.32)^3\}} = \dfrac{1}{27}$

Problem 34 $\left\{\dfrac{1 \cdot 2 \cdot 4 + 2 \cdot 4 \cdot 8 + 3 \cdot 6 \cdot 12 + \ldots}{1 \cdot 3 \cdot 9 + 2 \cdot 6 \cdot 18 + 3 \cdot 9 \cdot 27 + \ldots}\right\}^{\frac{1}{3}} = ?$

Sol. $\Rightarrow \left\{\dfrac{1 \cdot 2 \cdot 4 + 2 \cdot 4 \cdot 8 + 3 \cdot 6 \cdot 12 + \ldots}{1 \cdot 3 \cdot 9 + 2 \cdot 6 \cdot 18 + 3 \cdot 9 \cdot 27 + \ldots}\right\}^{\frac{1}{3}}$

$= \left\{\dfrac{8\,(1 + 8 + 27 \ldots}{27(1 + 8 + 27 \ldots}\right\}^{\frac{1}{3}} = \left\{\dfrac{8}{27}\right\}^{\frac{1}{3}} = \left\{\left(\dfrac{2}{3}\right)^3\right\}^{\frac{1}{3}} = \dfrac{2}{3}$

Problem 35

$\left\{\dfrac{(1.331)^{-1} + (1.331)^{-2} + \ldots + (1.331)^{-6}}{(1.331)^{-2} + (1.331)^{-3} + \ldots + (1.331)^{-7}}\right\}^{\frac{1}{3}} \div 1.1 = ?$

Sol. Given expression,

$\left\{\dfrac{(1.331)^{-1} + (1.331)^{-2} + \ldots\ldots\ldots\ldots + (1.331)^{-6}}{(1.331)^{-2} + (1.331)^{-3} + \ldots\ldots\ldots\ldots + (1.331)^{-7}}\right\}^{\frac{1}{3}} \div 1.1$

$= \left\{\dfrac{1}{(1.331)^{-1}} \times \dfrac{(1.331)^{-1} + (1.331)^{-2} + \ldots + (1.331)^{-6}}{(1.331)^{-1} + (1.331)^{-2} + \ldots + (1.331)^{-6}}\right\}^{\frac{1}{3}} \div 1.1$

$= \left\{\dfrac{1}{(1.331)^{-1}}\right\}^{\frac{1}{3}} \div 1.1 = \{1.331\}^{\frac{1}{3}} \div 1.1$

$= \{(1.1)^3\}^{\frac{1}{3}} \div 1.1$

$= 1.1 \div 1.1 = 1$

Type VI

Problems based on mathematical operations having modulus.

Absolute value/Modulus of a number indicate the distance of the given number on the number line from 0 irrespective of its direction. It implies that the absolute value/modulus of a number cannot be negative as direction of the number from the zero has been ignored.

In other words, 'absolute value' or 'modulus' of a real number x is the non-negative value of x regardless of its sign.

For example, absolute value of -5 or modules of $-5 = |-5| = 5$. $|-5|$ may be written as 'abs(5)'.

Other examples: Modulus (in short, Mod) of

(a) $3 - 5 = |3 - 5| = |-2| = 2$

(b) $|4| = 4$

Problem 36 $\dfrac{8 - [5 - (-3 + 2)] \div 2}{|5 - 3| - |5 - 3| \div 3} = ?$

Sol. $\dfrac{8 - [5 - (-3 + 2)] \div 2}{|5 - 3| - |5 - 8| \div 3}$

$= \dfrac{8 - [5 - (-1)] \div 2}{|2| - |-3| \div 3} = \dfrac{8 - [5 + 1] \div 2}{2 - 3 \div 3}$

$= \dfrac{8 - 6 \div 2}{2 - 3 \div 3} = \dfrac{8 - 3}{2 - 1} = 5$

Problem 37 $\dfrac{9|3-5|-5|4| \div 10}{-3(5)-2\times 4 \div 2} = ?$

Sol. $\dfrac{9|3-5|-5|4|\div 10}{-3(5)-2\times 4\div 2} = \dfrac{9|-2|-5|4|\div 10}{(-15)-8\div 2}$

$= \dfrac{9\cdot 2-5\cdot 4\div 10}{(-15)-4}$

$= \dfrac{18-2}{-19} = \dfrac{-16}{19}$

Problem 38 $\dfrac{|3-5|}{\text{Maximum value of } 18-|4x-5|} = ?$ For the maximum value of $18-|4x-5|$

Sol. The value of $18-|4x-5|$ will take maximum only when the value of $|4x-5|$ is minimum. Further, the minimum value of any $|k|$ is zero. Hence, maximum value of

$18-|4x-5| = 18-0 = 18$

Now, $\dfrac{|3-5|}{\text{Maximum value of } 18-|4x-5|} = \dfrac{2}{18} = \dfrac{1}{9}$

Problem 39 Greatest number of two digits × minimum value of $|5x-7|-11 = ?$

Sol. For the minimum value of $|5x-7|-11$:

The value of $|5x-7|-11$ will take minimum only when the value of $|5x-7|$ is minimum. Further, the minimum possible value of $|5x-7| = 0$. Hence, the minimum value of

$|5x-7|-11 = 0-11 = -11$

Now, Greatest number of two digits × minimum value of

$|5x-7|-11 = 99\times(-11)$

$= -1089$

Type VII

Simplification based on mathematical operations having recurring decimals.

Problem 40 $8.3\overline{1} + 0.\overline{6} + 0.00\overline{2} - (0.\overline{09}\times 7.\overline{3}) = ?$

[SSC (CGL-T-I)-12, SSC (CGL-T-II)-20]

Sol.

8.311111...
0.666666...
(+) 0.002222..
8.979999...

Now, $0.\overline{09}\times 7.\overline{3} = 0.090909...\times 7.3333 = 0.6666 = 0.\overline{6}$

Then, 8.979999...
(−) 0.666666...
8.313333

Hence, required answer $= 8.31\overline{3}$

Problem 41. $3 + 1.\overline{2}\times 0.\overline{03} - (0.\overline{4}\div 2) = ?$

Sol. $1.\overline{2}\times 0.\overline{03} = 1\dfrac{(2-0)}{(10-1)}\times\dfrac{(3-0)}{(100-1)} = \dfrac{11}{9}\times\dfrac{3}{99} = \dfrac{1}{27} = 0.\overline{037}$

and $(0.\overline{4}\div 2) = 0.444444...\div 2 = 0.222222 = 0.\overline{2}$

Then, $3 + 0.\overline{037} - 0.\overline{2} = 3.037037037... - 0.222222222....$

$= 2.814814815... = 2.\overline{814}$

Type VIII

Simplification with certain stipulations.

Problem 42 If $\dfrac{a}{b} = \dfrac{3}{4}$ and $8a + 5b = 22$, then $a = ?$

[RRB-02]

Given, $\dfrac{a}{b} = \dfrac{3}{4} \Rightarrow 3b = 4a$

$\Rightarrow$ $5b = \dfrac{4}{3}\times 5a = \dfrac{20a}{3}$

Now, $8a + 5b = 22$

$\Rightarrow$ $8a + \dfrac{20a}{3} = 22$

$\Rightarrow$ $\dfrac{44a}{3} = 22 \Rightarrow a = 1.5$

Alternatively I

You may proceed through given options. Let options be 6, 1.5, ...

(a) $6 \Rightarrow$ If $a = 6$, then $b = 8$ as $\dfrac{a}{b} = \dfrac{3}{4}$. Further, $8\cdot 6 + 5\cdot 8 \neq 22$

(b) $1.5 \Rightarrow$ If $a = 1.5$, then $b = 2$ as $\dfrac{a}{b} = \dfrac{3}{4}$.

Further, $8\cdot 1.5 + 5\cdot 2 = 22$

Hence, option(b) is correct.

Alternatively-II

Given, $\dfrac{a}{b} = \dfrac{3}{4} \Rightarrow \dfrac{a}{3} = \dfrac{b}{4} = k$(let)

$\Rightarrow$ $a = 3k$ and $b = 4k$

Now, $8a + 5b = 22$

$\Rightarrow$ $8\cdot 3k + 5\cdot 4k = 22$

$\Rightarrow$ $44k = 22 \Rightarrow k = \dfrac{1}{2}$

Then, $a = 3k = 3\cdot\dfrac{1}{2} = 1.5$

Problem 43 If $\dfrac{a}{3} = \dfrac{b}{4} = \dfrac{c}{7}$, then $\dfrac{a+b+c}{c} = ?$

Sol. Given, $\dfrac{a}{3} = \dfrac{b}{4} = \dfrac{c}{7}$

$\Rightarrow$ $\dfrac{a}{3} = \dfrac{b}{4} = \dfrac{c}{7} = k$ (let)

$\Rightarrow$ $a = 3k$, $b = 4k$ and $c = 7k$

Now, $\dfrac{a+b+c}{c} = \dfrac{3k+4k+7k}{7k} = \dfrac{14k}{7k} = 2$

Alternatively

Given, $\dfrac{a}{3} = \dfrac{b}{4} = \dfrac{c}{7} \Rightarrow \dfrac{a}{c} = \dfrac{3}{7}$ and $\dfrac{b}{c} = \dfrac{4}{7}$

Now, $\dfrac{a+b+c}{c} = \dfrac{a}{c} + \dfrac{b}{c} + 1 = \dfrac{3}{7} + \dfrac{4}{7} + 1 = 2$

Problem 44 If $\dfrac{x}{y} = \dfrac{3}{4}$, then $\dfrac{6}{7} + \dfrac{y-x}{y+x} = ?$

Sol. $\dfrac{6}{7} + \dfrac{y-x}{y+x} = \dfrac{6}{7} + \dfrac{\frac{y-x}{y}}{\frac{y+x}{y}} = \dfrac{6}{7} + \dfrac{1-\frac{x}{y}}{1+\frac{x}{y}}$

$$= \frac{6}{7} + \frac{1-\frac{3}{4}}{1+\frac{3}{4}} = \frac{6}{7} + \frac{\frac{1}{4}}{\frac{7}{4}} = \frac{6}{7} + \frac{1}{7} = 1$$

Alternatively

Given, $\frac{x}{y} = \frac{3}{4} \Rightarrow \frac{x}{3} = \frac{y}{4} = k$ (let)

$\Rightarrow$ $x = 3k$ and $y = 4k$

Now, $\frac{6}{7} + \frac{y-x}{y+x} = \frac{6}{7} + \frac{4k-3k}{4k+3k} = \frac{6}{7} + \frac{1}{7} = 1$

[**Note**: you may simply put $x = 3$ and $y = 4$ in the expression to find its value. These values may be wrong, but this approach is workable for finding value of the given expression quickly. Hence, we should have no hesitation to adopt the same]

Problem 45 If $\frac{a}{b} = \frac{3}{2}$, then $\frac{2a+5b}{2a-5b} = ?$ ***[SSC(CGL-PT)-06]***

Sol. We may put $a = 3$ and $b = 2$ in the given expression for finding its value quickly despite the fact that values of a and b may be wrong.

$\Rightarrow$ $\frac{2a+5b}{2a-5b} = \frac{2\cdot3+5\cdot2}{2\cdot3-5\cdot2} = \frac{16}{-4} = -4$

Alternatively

Given, $\frac{a}{b} = \frac{3}{2} \Rightarrow 2a = 3b$

Now, $\frac{2a+5b}{2a-5b} = \frac{3b+5b}{3b-5b}$

$= \frac{8b}{-2b} = -4$

Problem 46 If $\frac{x}{y} = \frac{a+2}{a-2}$, then $\frac{x^2-y^2}{x^2+y^2} = ?$

Sol. $\frac{x^2-y^2}{x^2+y^2} = \frac{(a+2)^2-(a-2)^2}{(a+2)^2+(a-2)^2}$ [Putting $x = a+2$ and $y = a-2$]

$= \frac{8a}{2a^2+8} = \frac{4a}{a^2+4}$

Problem 47 If $\frac{a}{b} = \frac{2}{3}$ and $\frac{b}{c} = \frac{9}{10}$ then $\frac{c^2-a^2}{c^2+a^2} = ?$

Sol. $\Rightarrow \frac{a}{b} \times \frac{b}{c} = \frac{2}{3} \times \frac{9}{10} \Rightarrow \frac{a}{c} = \frac{3}{5}$

Given, expression $= \frac{c^2-a^2}{c^2+a^2} = \frac{5^2-3^2}{5^2+3^2}$

[Putting $a = 3$ and $c = 5$]

$= \frac{16}{34} = \frac{8}{17}$

Alternatively, you may compute $a:c$ if $a:b$ and $b:c$ are known, by the process as detailed in chapter 'RATIO, PROPORTATION AND VARIATION'.

WORKOUT

1. If $\frac{50}{*} = \frac{*}{12\frac{1}{2}}$, then $* = ?$

(a) $\frac{25}{2}$ (b) $\frac{4}{25}$ (c) 4 (d) 25

2. $\frac{725\times725\times725+371\times371\times371}{725\times725-725\times371+371\times371} = ?$ ***[CDS-13, 12]***

(a) 9610 (b) 1960
(c) 1096 (d) 1016

3. The value of $\frac{\frac{1}{2}\div\frac{1}{2}\text{ of }\frac{1}{2}}{\frac{1}{2}+\frac{1}{2}\text{ of }\frac{1}{2}}$ is

(a) $\frac{11}{3}$ (b) $\frac{8}{3}$ (c) 3 (d) 1

4. Let $y = \cfrac{1}{2+\cfrac{1}{3+\cfrac{1}{2+\cfrac{1}{3+\ldots}}}}$, then what is the value of y? ***[CAT-04]***

(a) $\frac{\sqrt{13}+3}{2}$ (b) $\frac{\sqrt{13}-3}{2}$
(c) $\frac{\sqrt{15}+3}{2}$ (d) $\frac{\sqrt{15}-3}{2}$

5. If $x = 111\ldots1$ (20 digits), $y = 333\ldots3$ (10 digits) and $z = 222\ldots2$ (10 digits), then what is $\frac{x-y^2}{z}$ equal to? ***[CDS-17]***

(a) 1 (b) 2
(c) 3 (d) $\frac{1}{2}$

6. $\frac{(0.337+0.126)^2-(0.337-0.126)^2}{0.337\times0.126} = ?$

(a) 4 (b) 0.211
(c) 0.469 (d) 0.4246

7. $\left|\frac{3}{5}\times\frac{-5}{10}\right| = \left|\frac{3}{5}\right| \times \ldots.$

(a) $\frac{-5}{8}$ (b) $\frac{4}{8}$
(c) $\frac{8}{5}$ (d) $\frac{3}{8}$

8. If $\frac{x+y}{z} = 2$, then what is the value of $\frac{y}{y-z} + \frac{x}{x-z}$? ***[SSC(CGL-T-II)-17]***

(a) 1
(b) 2
(c) 0
(d) – 1

9. What is the simplified form of $\frac{x^2-3x+2}{x^3-8} \div \frac{x^2-9}{x^2+7x+12} \times \frac{x^3+2x^2+4x}{x^2+3x-4}$?

(a) $\frac{x}{(x-3)}$ (b) $\frac{(x-2)}{(x-3)}$
(c) $\frac{x}{(x+3)}$ (d) $\frac{(x+3)}{(x+4)}$

10. $98 \div 14 \times 49 - 294 = (?)^2$ ***[Syndicate Bank (PO)-10]***

(a) – 14 (b) $\sqrt{7}$
(c) – 7 (d) 196
(e) None of these

11. $\frac{(0.75)^3}{1-0.75}+[0.75+(0.75)^2+1]$ is

(a) 1 (b) 2
(c) 3 (d) 4

12. What is $\frac{x^2-3x+2}{x^2-5x+6} \div \frac{x^2-5x+4}{x^2-7x+12}$ equal to **[CDS-11]**

(a) $\frac{x+3}{x-3}$ (b) 1
(c) $\frac{x+1}{x-1}$ (d) 2

Directions (Q.Nos. 13-15) *What approximate value should come at the place of question mark (?) in the following questions?* **[SBI(PO)-17]**

13. $13.03^2 + ? + 21.998 \times 4.012 = 298.998$ **[IBPS (PO-Pre)-19]**

(a) 32 (b) 42
(c) 52 (d) 62
(e) 72

14. $\sqrt{454+985} - ?^2 \div 18.752 = 18.9001$

(a) 19 (b) 18
(c) 21 (d) 25
(e) 15

15. $\sqrt{33125} \times \sqrt{2600} - (83.01)^2 = ?^2 + (36.99)^2$

(a) 19 (b) 28
(c) 21 (d) 25
(e) 32

16. $\frac{256 \times 256 - 144 \times 144}{112} = ?$ **[SSC(CGL-T-I)-10]**

(a) 420 (b) 400
(c) 360 (d) 320

17. $\frac{(0.67)^3 - 0.001}{(0.67)^2 + 0.067 + 0.01} = ?$

(a) 0.57 (b) 0.68
(c) 0.0067 (d) 1.34

18. What is $\frac{(x^2+y^2)(x-y)-(x-y)^3}{x^2y-xy^2}$ equal to **[CDS-13]**

(a) 1 (b) 2
(c) 4 (d) – 2

19. $(999)^2 - (998)^2$ is equal to **[MAT-08]**

(a) 1 (b) 999
(c) 1997 (d) 998

20. $\frac{(0.137+0.098)^2+(0.137-0.098)^2}{0.137 \times 0.137 + 0.098 \times 0.098}$ is simplified to

(a) 2.0 (b) 0.2
(c) 0.235 (d) 0.039

21. $(21)^2 - 3717 + 59 = ? \times 8$ **[IOB (PO)-11]**

(a) 43.75 (b) 42.25
(c) 45.75 (d) 47.25
(e) None of these

22. What is $\frac{x^8+4}{x^4+2x^2+2}$ equal to?

(a) $x^4 + 2x^2 - 2$
(b) $x^4 - 2x^2 + 2$
(c) $x^4 - 2x^2 - 2$
(d) Cannot be determined

23. $\frac{1}{8}$ of $\frac{2}{3}$ of $\frac{3}{5}$ of 1715 = ? **[SBI-Associates (PO)-11]**

(a) 80 (b) 85
(c) 90 (d) 95
(e) 75

24. What is the value of $3 \times 0.3 \times 0.03 \times 0.003 \times 0.0003 \times 30$? **[CDS-12]**

(a) $(0.09)^3$ (b) $(0.009)^3$
(c) $(0.0009)^3$ (d) None of these

25. $9999 + 99 + 9 = ?$ **[Andhra Bank (PO)-11]**

(a) 18 (b) 15
(c) 6 (d) 11
(e) 20

26. What is the value of x in

$$1+\frac{1}{1+\left\{\frac{1}{1+\frac{1}{x}}\right\}}=\frac{11}{7}$$

(a) 1 (b) 3
(c) $\frac{1}{2}$ (d) $\frac{7}{11}$

27. Find the value of

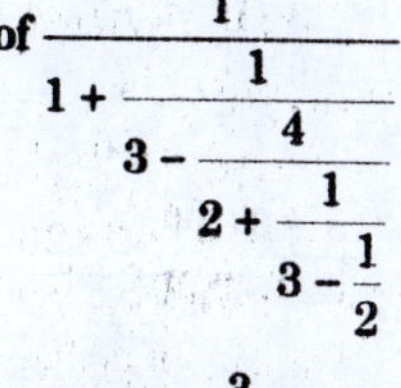

$$\frac{1}{1+\frac{1}{3-\frac{4}{2+\frac{1}{3-\frac{1}{2}}}}}+\frac{3}{3-\frac{4}{3+\frac{1}{2-0.\bar{3}}}}$$

(a) $2\frac{11}{119}$ (b) $3\frac{11}{119}$
(c) $2\frac{19}{119}$ (d) $2\frac{21}{119}$

28. $\frac{(2.3)^3 - 0.027}{(2.3)^2 + 0.69 + 0.09} = ?$ **[CDS-11, 08]**

(a) 0.3 (b) 2.3
(c) 1 (d) 2

29. If $\frac{a}{x}+\frac{y}{b}=1$ and $\frac{b}{y}+\frac{z}{c}=1$, then $\frac{x}{a}+\frac{c}{z}=?$

(a) 1 (b) 2 (c) 3 (d) 4

30. If $x * y = (x+3)^2(y-1)$ then $5 * 4 = ?$ **[SSC (CGL-T-1)-11]**

(a) 192 (b) 182
(c) 180 (d) 172

31. Suppose $p * q = 2p + 2q - pq$, where p, q are natural numbers. If $8 * x = 4$, then what is the value of x?

(a) 1 (b) 2
(c) 3 (d) 4

32. $\sqrt{\frac{(0.35)^2+0.70+1}{2.25}+0.19} = ?$ **[CDS-17]**

(a) 1 (b) 2
(c) 3 (d) 4

33. If $\frac{a}{b}=\frac{1}{3}, \frac{b}{c}=2, \frac{c}{d}=\frac{1}{2}, \frac{d}{e}=3$ and $\frac{e}{f}=\frac{1}{4}$, then what is the value of $\frac{abc}{def}=?$ **[CAT-06]**

(a) $\frac{3}{4}$ (b) $\frac{27}{8}$ (c) $\frac{3}{8}$ (d) $\frac{27}{4}$
(e) $\frac{1}{4}$

34. Value of A in the expression : $5 + 12 \times 10 \div \frac{120}{240} = A \times 10$, is **[CLAT-14]**

(a) 11 (b) 24.5
(c) 34 (d) 6.5

35. $\frac{2}{2+\frac{2}{3+\frac{2}{3+\frac{2}{3}}}\times 0.39} = ?$

(a) $\frac{100}{111}$ (b) $\frac{127}{128}$ (c) $\frac{3}{8}$ (d) $\frac{27}{41}$

Answer Key

1. (d)	2. (c)	3. (b)	4. (d)	5. (a)	6. (a)	7. (b)	8. (b)	9. (a)	10. (c)
11. (d)	12. (b)	13. (b)	14. (a)	15. (e)	16. (b)	17. (a)	18. (b)	19. (c)	20. (a)
21. (d)	22. (b)	23. (b)	24. (b)	25. (d)	26. (b)	27. (c)	28. (d)	29. (a)	30. (a)
31. (b)	32. (a)	33. (c)	34. (b)	35. (a)					

Hints and SOLUTIONS

1. Given, $\frac{50}{*} = \frac{*}{12\frac{1}{2}} \Rightarrow (*)^2 = 2 \times 2 \times 12\frac{1}{2} \times 12\frac{1}{2}$

$\Rightarrow \quad * = 2 \times 12\frac{1}{2} = 25$

2. Given expression $= \frac{725 \times 725 \times 725 + 371 \times 371 \times 371}{725 \times 725 - 725 \times 371 + 371 \times 371}$

$= \frac{(a)^3 + (b)^3}{a^2 - ab + b^2}$ [Assuming $a = 725$, $b = 371$]

$= \frac{(a+b)(a^2 - ab + b^2)}{a^2 - ab + b^2} = a + b = 725 + 371 = 1096$

3. $\frac{\frac{1}{2} \div \frac{1}{2} \text{ of } \frac{1}{2}}{\frac{1}{2} + \frac{1}{2} \text{ of } \frac{1}{2}} = \frac{\frac{1}{2} \div \frac{1}{4}}{\frac{1}{2} + \frac{1}{4}} = \frac{\frac{1}{2} \times \frac{4}{1}}{\frac{3}{4}} = 2 \times \frac{4}{3} = \frac{8}{3}$

4. Given, $y = \cfrac{1}{2 + \cfrac{1}{3 + \cfrac{1}{2 + \cfrac{1}{3 + \ldots}}}}$

$\Rightarrow \quad y = \cfrac{1}{2 + \cfrac{1}{3 + y}} \Rightarrow y = \cfrac{1}{\cfrac{7 + 2y}{3 + y}} = \frac{3 + y}{7 + 2y}$

$\Rightarrow \quad 7y + 2y^2 = 3 + y \Rightarrow 2y^2 + 6y - 3 = 0$

$\Rightarrow \quad y = \frac{-6 \pm \sqrt{6^2 - 4 \cdot 2(-3)}}{2 \cdot 2}$

$= \frac{-6 \pm \sqrt{60}}{2.2} = \frac{-3 \pm \sqrt{15}}{2}$

Since y is a positive value, $y = \frac{-3 + \sqrt{15}}{2} = \frac{\sqrt{15} - 3}{2}$

5. $x = 111..1$(20 digits)

$\Rightarrow \quad x = \frac{999...9 \text{ (20 digits)}}{9} = \frac{10^{20} - 1}{9}$

$y = 333...3$ (10 digits)

$\Rightarrow \quad y = 3\{111...1 \text{ (10 digits)}\} = 3 \times \frac{10^{10} - 1}{9} = \frac{10^{10} - 1}{3}$

$z = 222...2$ (10 digits)

$\Rightarrow \quad y = 2\{111...1 \text{ (10 digits)}\} = \frac{2(10^{10} - 1)}{9}$

Now, $\frac{x - y^2}{z} = \frac{\frac{10^{20} - 1}{9} - \left(\frac{10^{10} - 1}{3}\right)^2}{\frac{2(10^{10} - 1)}{9}}$

$= \frac{(10^{20} - 1) - (10^{10} - 1)^2}{2(10^{10} - 1)}$

$= \frac{(10^{10} + 1)(10^{10} - 1) - (10^{10} - 1)^2}{2(10^{10} - 1)}$

$= \frac{(10^{10} + 1) - (10^{10} - 1)}{2} = 1$

6. Let $0.337 = a$ and $0.126 = b$

Then, given expression $= \frac{(a + b)^2 - (a - b)^2}{ab}$

$= \frac{4ab}{ab} = 4$

7. $\left|\frac{3}{5} \times \frac{-5}{10}\right| = \left|\frac{3}{5}\right| \times \left|\frac{-5}{10}\right| = \left|\frac{3}{5}\right| \times \frac{1}{2} = \left|\frac{3}{5}\right| \times \frac{4}{8}$

Hence, required answer is (b).

8. Given, $\frac{x + y}{z} = 2 \Rightarrow z = \frac{x + y}{2}$

Now, $\frac{y}{y - z} + \frac{x}{x - z} = \cfrac{y}{y - \cfrac{x + y}{2}} + \cfrac{x}{x - \cfrac{x + y}{2}}$

$= \frac{2y}{y - x} + \frac{2x}{x - y}$

$= \frac{2y}{y - x} - \frac{2x}{y - x} = \frac{2(y - x)}{y - x} = 2$

9. 1st term $= \frac{x^2 - 3x + 2}{x^3 - 8} = \frac{x^2 - 2x - x + 2}{(x - 2)(x^2 + 2x + 4)}$

$= \frac{x(x - 2) - 1(x - 2)}{(x - 2)(x^2 + 2x + 4)}$

$= \frac{(x - 2)(x - 1)}{(x - 2)(x^2 + 2x + 4)} = \frac{(x - 1)}{(x^2 + 2x + 4)}$

2nd term $= \frac{x^2 - 9}{x^2 + 7x + 12} = \frac{(x - 3)(x + 3)}{x^2 + 3x + 4x + 12}$

$$= \frac{(x-3)(x+3)}{x(x+3)+4(x+3)} = \frac{(x-3)(x+3)}{(x+3)(x+4)} = \frac{(x-3)}{(x+4)}$$

3rd term $= \frac{x^3+2x^2+4x}{x^2+3x-4} = \frac{x(x^2+2x+4)}{x^2+4x-x-4}$

$$= \frac{x(x^2+2x+4)}{x(x+4)-1(x+4)} = \frac{x(x^2+2x+4)}{(x+4)(x-1)}$$

Now, expression $= \frac{(x-1)}{(x^2+2x+4)} \div \frac{(x-3)}{(x+4)} \times \frac{x(x^2+2x+4)}{(x+4)(x-1)}$

$$= \frac{(x-1)}{(x^2+2x+4)} \times \frac{(x+4)}{(x-3)} \times \frac{x(x^2+2x+4)}{(x+4)(x-1)} = \frac{x}{(x-3)}$$

10. $98 \div 14 \times 49 - 294 = (?)^2$

$\Rightarrow \quad 98 \times \frac{1}{14} \times 49 - 294 = (?)^2$

$\Rightarrow \quad 343 - 294 = (?)^2$

$\Rightarrow \quad 49 = (?)^2$

$\Rightarrow \quad ? = \pm 7$

Hence, option (c) is correct.

11. $\frac{(0.75)^3}{1-0.75} + [0.75 + (0.75)^2 + 1]$

$$= \frac{(0.75)^3 + (1-0.75)[0.75 \cdot 1 + (0.75)^2 + 1]}{1-0.75}$$

$$= \frac{(0.75)^3 + (1^3 - 0.75^3)}{1-0.75} \quad [\because (a-b)(a^2+ab+b^2) = a^3 - b^3]$$

$$= \frac{1^3}{1-0.75} = 4$$

12. $\frac{x^2-3x+2}{x^2-5x+6} \div \frac{x^2-5x+4}{x^2-7x+12}$

$$= \frac{x^2-2x-x+2}{x^2-3x-2x+6} \div \frac{x^2-4x-x+4}{x^2-4x-3x+12}$$

$$= \frac{x(x-2) \quad 1(x-2)}{x(x-3)-2(x-3)} \div \frac{x(x-4)-1(x-4)}{x(x-4)-3(x-4)}$$

$$= \frac{(x-2)(x-1)}{(x-3)(x-2)} \div \frac{(x-4)(x-1)}{(x-4)(x-3)} = \frac{(x-1)}{(x-3)} \div \frac{(x-1)}{(x-3)} = 1$$

13. $13.03^2 + ? + 21.998 \times 4.012 = 298.998$

$\Rightarrow \quad 13^2 + ? + 22 \times 4 = 299$ (making approximation)

$\Rightarrow \quad ? = 299 - 169 - 88 = 42$

14. $\sqrt{454 + 985} - ?^2 \div 18.752 = 18.9001$

$\Rightarrow \quad \sqrt{1439} - ?^2 \div 18.752 = 18.9001$

$\Rightarrow \quad 38 - ?^2 \div 19 = 19$ (making approximation)

[Applying Estimation method for computation of square-root]

$\Rightarrow \quad 38 - 19 = (?^2 \div 19)$

$\Rightarrow \quad ?^2 = 19 \times 19$

$\Rightarrow \quad ? - 19$

15. $\sqrt{33125} \times \sqrt{2600} - (83.01)^2 = ?^2 + (36.99)^2$

$\Rightarrow \quad 182 \times 51 - (83)^2 + ?^2 + (37)^2$ (making approximation)

$\Rightarrow \quad 9282 - 6889 = ?^2 + 1369$

$\Rightarrow \quad ?^2 = 1024$

$\Rightarrow \quad ? = 32$

16. $\frac{256 \times 256 - 144 \times 144}{112} = \frac{(a)^2 - (b)^2}{a-b}$

[Assuming $a = 256$ and $b = 144$]

$$= \frac{(a+b)(a-b)}{a-b} = a + b = 256 + 144 = 400$$

17. $\frac{(0.67)^3 - (0.1)^3}{(0.67)^2 + 0.67 \times 0.1 + (0.1)^2} = \frac{a^3 - b^3}{(a^2+ab+b^2)}$

[Assuming $a = 0.67$ and $b = 0.1$]

$$= \frac{(a-b)(a^2+ab+b^2)}{(a^2+ab+b^2)} = a - b = 0.67 - 0.1 = 0.57$$

18. $\frac{(x^2+y^2)(x-y) - (x-y)^3}{x^2y - xy^2}$

$$= \frac{x^3 + xy^2 - x^2y - y^3 - \{x^3 - y^3 - 3x^2y + 3y^2x\}}{x^2y - xy^2}$$

$$= \frac{x^3 + xy^2 - x^2y - y^3 - x^3 + y^3 + 3x^2y - 3y^2x\}}{x^2y - xy^2}$$

$$= \frac{2(x^2y - xy^2)}{x^2y - xy^2} = 2$$

19. $(999)^2 - (998)^2 = (999 + 998)(999 - 998) = 1997$

20. Let $0.137 = a$ and $0.098 = b$

Then, given expression $= \frac{(a+b)^2 + (a-b)^2}{(a)^2 + (b)^2} = \frac{2(a^2+b^2)}{(a^2+b^2)} = 2$

21. Given, $(21)^2 - 3717 \div 59 = ? \times 8$

$\Rightarrow \quad 441 - 3717 \div 59 = ? \times 8 \Rightarrow 441 - 3717 \times \frac{1}{59} = ? \times 8$

$\Rightarrow \quad 441 - 63 = ? \times 8 \Rightarrow ? = \frac{378}{8} = 47.25$

22. $\frac{x^8+4}{x^4+2x^2+2} = \frac{x^8+4x^4+4-4x^4}{x^4+2x^2+2} = \frac{(x^4+2)^2 - (2x^2)^2}{x^4+2x^2+2}$

$$= \frac{(x^4+2x^2+2)(x^4-2x^2+2)}{x^4+2x^2+2}$$

$$= (x^4 - 2x^2 + 2)$$

23. $\frac{1}{8}$ of $\frac{2}{3}$ of $\frac{3}{5}$ of $1715 = \frac{1}{8} \times \frac{2}{3} \times \frac{3}{5} \times 1715 = 85.75 = 85$ (approx.)

24. Given expression $= 3 \times 3 \times (10)^{-1} \times 3 \times (10)^{-2} \times 3 \times (10)^{-3} \times 3 \times (10)^{-4} \times 3 \times (10)^1$

$$= (3)^6 \times (10)^{-9}$$

$$= (3^2)^3 \times (10^{-3})^3 = (3^2 \times 10^{-3})^3$$

$$= (0.009)^3$$

25. $9999 \div 99 \div 9 = 9999 \times \frac{1}{99} \times \frac{1}{9} = 11.22 = 11$ (approx.)

26. Given, $1 + \cfrac{1}{1 + \left\{\cfrac{1}{1 + \cfrac{1}{x}}\right\}} = \frac{11}{7} \Rightarrow 1 + \cfrac{1}{1 + \left\{\cfrac{1}{\cfrac{1+x}{x}}\right\}} = \frac{11}{7}$

$\Rightarrow \quad 1 + \cfrac{1.}{1 + \left\{\cfrac{x}{1+x}\right\}} = \frac{11}{7} \Rightarrow 1 + \cfrac{1}{\cfrac{1+x+x}{1+x}} = \frac{11}{7}$

$$\Rightarrow \quad 1+\frac{1+x}{1+2x}=\frac{11}{7}$$

$$\Rightarrow \quad \frac{1+2x+1+x}{1+2x}=\frac{11}{7}$$

$$\Rightarrow \quad \frac{2+3x}{1+2x}=\frac{11}{7}$$

$$\Rightarrow \quad 14+21x=11+22x \Rightarrow x=3$$

27. $$\frac{1}{1+\cfrac{1}{3-\cfrac{4}{2+\cfrac{1}{3-\cfrac{1}{2}}}}}+\frac{3}{3-\cfrac{4}{3+\cfrac{1}{2-0.\overline{3}}}}$$

$$=\frac{1}{1+\cfrac{1}{3-\cfrac{4}{2+\cfrac{2}{5}}}}+\frac{3}{3-\cfrac{4}{3+\cfrac{1}{2-\cfrac{1}{3}}}}$$

$$=\frac{1}{1+\cfrac{1}{3-\cfrac{5}{3}}}+\frac{3}{3-\cfrac{4}{3+\cfrac{3}{5}}}=\frac{1}{1+\cfrac{3}{4}}+\frac{3}{3-\cfrac{10}{9}}$$

$$=\frac{4}{7}+\frac{27}{17}=\frac{257}{119}=2\frac{19}{119}$$

28. $$\frac{(2.3)^3-0.027}{(2.3)^2+0.69+0.09}=\frac{(2.3)^3-(0.3)^3}{(2.3)^2+2.3\times 0.3+(0.3)^2}$$

$$=\frac{(2.3-0.3)\{(2.3)^2+2.3\times 0.3+(0.3)^2\}}{(2.3)^2+2.3\times 0.3+(0.3)^2}=(2.3-0.3)=2$$

29. Given, $\frac{a}{x}+\frac{y}{b}=1 \Rightarrow \frac{a}{x}=1-\frac{y}{b}=\frac{b-y}{b} \Rightarrow \frac{x}{a}=\frac{b}{b-y}$

Also given, $\frac{b}{y}+\frac{z}{c}=1$

$$\Rightarrow \frac{z}{c}=1-\frac{b}{y}=\frac{y-b}{y} \Rightarrow \frac{c}{z}=\frac{y}{y-b}$$

Now, $\frac{x}{a}+\frac{c}{z}=\frac{b}{b-y}+\frac{y}{y-b}=\frac{b}{b-y}-\frac{y}{(b-y)}=\frac{(b-y)}{(b-y)}=1$

30. Given, $x * y=(x+3)^2(y-1)$

$$\Rightarrow \quad 5 * 4=(5+3)^2(4-1)=64\cdot 3=192$$

31. Given, $p * q=2p+2q-pq$

Then, $8 * x=2\cdot 8+2\cdot x-8\cdot x$

$$\Rightarrow \quad 4=16-6x \Rightarrow 6x=12 \Rightarrow x=2$$

32. $$\sqrt{\frac{(0.35)^2+0.70+1}{2.25}+0.19}$$

$$=\sqrt{\frac{(0.35)^2+2\cdot 0.35\cdot 1+1^2}{2.25}+0.19}$$

$$=\sqrt{\frac{(0.35+1)^2}{2.25}+0.19}=\sqrt{\frac{1.35\times 1.35}{2.25}+0.19}$$

$$=\sqrt{(0.9\times 0.9)+0.19}=\sqrt{0.81+0.19}=1$$

33. $\frac{a}{b}\times\frac{b}{c}\times\frac{c}{d}=\frac{1}{3}\times 2\times\frac{1}{2}\Rightarrow\frac{a}{d}=\frac{1}{3}$

$$\frac{b}{c}\times\frac{c}{d}\times\frac{d}{e}=2\times\frac{1}{2}\times 3\Rightarrow\frac{b}{e}=3$$

$$\frac{c}{d}\times\frac{d}{e}\times\frac{e}{f}=\frac{1}{2}\times 3\times\frac{1}{4}\Rightarrow\frac{c}{f}=\frac{3}{8}$$

Now, $\frac{abc}{def}=\frac{a}{d}\times\frac{b}{e}\times\frac{c}{f}=\frac{1}{3}\times 3\times\frac{3}{8}=\frac{3}{8}$

34. Given expression $5+12\times 10\div\frac{120}{240}=A\times 10$

$$\Rightarrow 5+12\times 10\times 2=A\times 10 \Rightarrow 245=A\times 10 \Rightarrow A=24.5$$

35. $$\frac{2}{2+\cfrac{2}{3+\cfrac{2}{3+\cfrac{2}{3}}}\times 0.39}=\frac{2}{2+\cfrac{2}{3+\cfrac{2}{\cfrac{11}{3}}}\times 0.39}$$

$$=\frac{2}{2+\cfrac{2}{3+\cfrac{6}{11}}\times 0.39}=\frac{2}{2+\cfrac{2}{\cfrac{39}{11}}\times 0.39}$$

$$=\frac{2}{2+\left(\cfrac{22}{39}\times 0.39\right)}=\frac{2}{2+\cfrac{22}{100}}$$

$$=\frac{2}{2+\cfrac{11}{50}}=2\div\frac{111}{50}$$

$$=2\times\frac{50}{111}=\frac{100}{111}$$

Unit 2
Numbers, Series and Way of Counting

- Number System • LCM & HCF/GCD
- Average • Progressions
- Permutation and Combination
- Probability

12

Number System

Learning OBJECTIVES

After completing this chapter, aspirants will be able to

- Learn diverse types of number system;
- Understand the significance of binary system in the computer field;
- Convert one base to another base meant for number systems;
- Understand the fundamentals of mathematical operations for binary system;
- Learn the difference between place-value and face-value of a digit in numeral;
- Learn various types of numbers;
- Recognise interesting properties of numbers;
- Understand the philosophy of the categorization of the questions;
- Analyse the trend of the questions being asked in various competitive examinations;
- Recognise the types of question which can be easily solved through the given options;
- Grip on concepts helping to solve the questions based on number-system in moments;
- Learn new concepts/fundamentals;
- Apply the pragmatic approaches in lieu of academic approaches; and
- Ponder on the alternative methods.

Inside

- *Brief Introduction*
- *Major Types of Number System in Brief*
- *Conversion of a Base Into Decimal*
- *Conversion of Decimal Into Another Base*
- *Conversion of a Base (Other Than Decimal) to Another Base (Other Than Decimal)*
- *Mathematical Operations in Binary System*
- *Place-Value (Local Value) v/s Face-Value of a Digit in a Numeral*
- *Place Value and Face-Value of Digits in a Decimal Number*
- *General Classification of Numbers*
- *Some Special Numbers*
- *Interesting Properties of Numbers*
- *Diverse Types of Problems Including Puzzles*
- ....
- .... *& Many More*
- *Work Out with Solutions*

Introduction

This topic is very critical for the competitive examinations as almost all the competitive examinations like CAT, MAT, SSC, Banks, Railway, LIC etc. cover this topic. The number system is basically refers to the number of symbols used to represent numbers. With their symbols, each number system is able to represent any quantity. 'Base' is defined as the total number of digits available in the number system.

A. Major Types of Number Systems in Brief

Number-system	Base	Digits	Remarks
Decimal	10	0, 1, 2, 3, 4, 5, 6, 7, 8, 9	This is the most common number system, we will discuss in detail under this chapter. A group of symbols representing a number is called **numeral**, e.g. 542. A decimal number is written as ... $_{10}$, e.g. 523_{10}, but as per trend, 10 is not indicated.
Binary	2	0, 1	Each binary digit is called as **bit**. Actually bit is a short form of binary digit. A group of 4 bit is called **nibble** and a group of 8 bit is called as **byte**. Value of digit depends upon the position of the digit in the number. A binary number is written as ....$_2$, e.g. 1001_2
Octal	8	0, 1, 2, 3, 4, 5, 6, 7	Three binary digits are sufficient to represent any octal number in binary because $2^3 = 8$ and the fact that there are only 8 digits in octal number. An octal number is written as ....$_8$, e.g. 65_8.
Hexa-decimal	16	0, 1, 2, 3, 4, 5, 6, 7, 8, 9, A, B, C, D, E, F	Due to combination of numeric digits and alphabets, this number system is also called **alphanumeric number system**, widely used in computer system. A hexa-decimal number is written as ....$_{16}$, e.g. $C3_{16}$. Clearly, $A = 10$, $B = 11$, $C = 12$, $D = 13$, $E = 14$ and $F = 15$. On the above analogy, four binary digits are sufficient to represent any hexa-decimal number in binary.

Other number system like ternary(or trinary), Quaternary, Quinary, Senary are far less common.

Since the first of the digits is 0, the maximum value of a single digit is equal to one less than the value of the base. For illustration, maximum value of a digit in decimal number system = 10 – 1 = 9 , similarly, in octal system, that is (8 – 1) hence, 7.

B. Conversion of a Base into Decimal

Generic Version :

(a) $(\ldots\ldots w\,x\,y\,z)_{\text{base}} = \ldots\ldots + w\cdot(\text{base})^3 + x\cdot(\text{base})^2 + y\cdot(\text{base})^1 + z\cdot(\text{base})^0$

(b) $(\ldots\ldots v\,w.\,x\,y\,z)_{\text{base}} = \ldots\ldots + v\cdot(\text{base})^1 + w\cdot(\text{base})^0 + x\cdot(\text{base})^{-1} + y\cdot(\text{base})^{-2} + z\cdot(\text{base})^{-3}$

For examples,

(i) $11101_2 = (?)_{10}$

$11101_2 \Rightarrow 1\cdot 2^4 + 1\cdot 2^3 + 1\cdot 2^2 + 0\cdot 2^1 + 1\cdot 2^0$

$= 16 + 8 + 4 + 0 + 1 = 29$

Hence, $11101_2 = 29_{10}$

(ii) $111.01_2 = (?)_{10}$

$111.01_2 \Rightarrow 1\cdot 2^2 + 1\cdot 2^1 + 1\cdot 2^0 + 0\cdot 2^{-1} + 1\cdot 2^{-2}$

$= 4 + 2 + 1 + 0 + 1/4 = 7.25$

Hence, $111.01_2 = 7.25_{10}$

(iii) $475_8 = (?)_{10}$

$475_8 \Rightarrow 4\cdot 8^2 + 7\cdot 8^1 + 5\cdot 8^0 = 256 + 56 + 5 = 317$

Hence, $475_8 = 317_{10}$

(iv) $1CA_{16} = (?)_{10}$

$1CA_{16} \Rightarrow 1\cdot 16^2 + C\cdot 16^1 + A\cdot 16^0$

$= 1\cdot 256 + 12\cdot 16 + 10\cdot 1$

$= 256 + 192 + 10 = 458$

Hence, $1CA_{16} = 458_{10}$

(v) $1243_6 = (?)_{10}$

$1243_6 \Rightarrow 1\cdot 6^3 + 2\cdot 6^2 + 4\cdot 6^1 + 3\cdot 6^0$

$= 1\cdot 216 + 2\cdot 36 + 4\cdot 6 + 3\cdot 1$

$= 216 + 72 + 24 + 3 = 315$

Hence, $1243_6 = 315_{10}$

(vi) $2B.F4_{16} = (?)_{10}$

$2B.F4_{16} \Rightarrow 2\cdot 16^1 + B\cdot 16^0 + F\cdot 16^{-1} + 4\cdot 16^{-2}$

$= 2\cdot 16 + 11\cdot 1 + 15\cdot \frac{1}{16} + 4\cdot \frac{1}{16^2}$

$= 32 + 11 + 0.9375 + 0.015625 = 43.953125$

Hence, $2B.F4_{16} = 43.953125_{10}$

C. Conversion of Decimal into Another Base

We apply **Division - Remainder Technique** to convert decimal to another base. For example

(a) $29_{10} = ?_2$

2	29	Remainders
	14	1
	7	0
	3	1
	1	1
	0	1

Now, write the remainder-digits from bottom to top i.e. 11101.

Hence, $29_{10} = 11101_2$

(b) $458_{10} = ?_{16}$

16	458	Remainders
	28	10 (i.e. A)
	1	12 (i.e. C)
	0	1

Now, write the remainder-digits from bottom to top i.e. 1CA.

Hence, $458_{10} = 1CA_{16}$

D. Conversion of a Base (other than decimal) to Another Base (other than decimal)

***Step* I** Convert the given base in decimal.

***Step* II** Convert the decimal so obtained in the desired base.

For Examples

(a) $11101_2 = (?)_8$

$11101_2 \Rightarrow 1\cdot 2^4 + 1\cdot 2^3 + 1\cdot 2^2 + 0\cdot 2^1 + 1\cdot 2^0$

$= 16 + 8 + 4 + 0 + 1 = 29$

Hence, $11101_2 = 29_{10}$

Now,

8	29	Remainders
	3	5
	0	3

It means $29_{10} = 35_8$

Hence, $11101_2 = 35_8$

Alternatively

(For binary to octal conversion)

Divide the given binary into groups of 3, starting from the extreme right. We may prefix 0, if required, for our convenience. Here, groups are 011 and 101.

Now, $\underline{011} \Rightarrow 0\cdot 2^2 + 1\cdot 2^1 + 1\cdot 2^0 = 3$

and $\underline{101} \Rightarrow 1\cdot 2^2 + 0\cdot 2^1 + 1\cdot 2^0 = 5$

Hence, $11101_2 = 35_8$

(b) $111001010_2 = ?_{16}$

We may convert firstly from binary to decimal and then decimal to hexa decimal.

Alternatively

(For binary to hexa decimal conversion)

Divide the given binary into groups of 4, starting from the extreme right. We may prefix 0, if required, for our convenience. Here, groups are 0001, 1100 and 1010

Now, $\underline{0001} \Rightarrow 0\cdot 2^3 + 0\cdot 2^2 + 0\cdot 2^1 + 1\cdot 2^0 = 1$

and $\underline{1100} \Rightarrow 1\cdot 2^3 + 1\cdot 2^2 + 0\cdot 2^1 + 0\cdot 2^0 = 12 = C$

and $\underline{1010} \Rightarrow 1\cdot 2^3 + 0\cdot 2^2 + 1\cdot 2^1 + 0\cdot 2^0 = 10 = A$

Hence, $111001010_2 = 1CA_{16}$

(c) $35_8 = ?_2$

We may convert firstly from octal to decimal and then decimal to binary.

Alternatively,

(For octal to binary conversion)

$3 \Rightarrow 011_2$ (we have to write always in a set of 3 digits)

$5 \Rightarrow 101_2$

Now, combine the groups like 011 101

Hence, $35_8 = 11101_2$

(d) $1CA_{16} = ?_2$

We may convert firstly from hexa decimal to decimal and then decimal to binary.

Alternatively

(For hexa decimal to binary conversion)

$1 \Rightarrow 0001_2$ (we have to write always in a set of 4 digits)

C (i.e. 12) $\Rightarrow 1100_2$

A (i.e. 10) $\Rightarrow 1010_2$

Now, combine the groups like 0001 1100 1010

Hence, $1CA_{16} = 111001010_2$

E. Mathematical Operations in Binary System

Rules for Addition

0 + 0 = 0, 0 + 1 = 1, 1 + 0 = 1 and 1 + 1 = 1 0, 1 will be carried over immediate left column sum in the same way as in the decimal number system.

For examples (a):

$$\begin{array}{r} ^{1}\quad\; \\ 1\,0\,0\,1 \\ +1\,1\,0\,1 \\ \hline 1\,0\,1\,1\,0 \end{array}$$

Explanation

1st column 1 + 1 = 10, 0 has been written and 1 is carried over the immediate left column sum.

2nd column 0 + 0 + 1 = 1

3rd column 0 + 1 = 1

4th column 1 + 1 = 1 0 (no scope for carrying over)

Alternatively

Convert into decimal system and then add, thereafter convert the decimal number so obtained in binary system again.

$1001_2 = 9_{10}$

and $1101_2 = 13_{10}$

Now, $9_{10} + 13_{10} = 22_{10}$

Further, $22_{10} = 10110_2$

(b)

$$\begin{array}{r} 1\,1\,1 \\ +1\,1\,0 \\ \hline 1\,1\,0\,1 \end{array}$$

Explanation

1st column 1 + 0 = 1

2nd column 1 + 1 = 1 0, 0 has been written and 1 is carried over the immediate left column sum.

3rd column 1 + 1 + 1 = 1 0 + 1 = 1 1

Alternatively

Convert into decimal system and then add, thereafter convert the decimal number so obtained in binary system again.

$111_2 = 7_{10}$ and $110_2 = 6_{10}$

Now, $7_{10} + 6_{10} = 13_{10}$

Further, $13_{10} = 1101_2$

Rules for Subtraction

$0 - 0 = 0$, $1 - 0 = 1$, $1 - 1 = 0$ and $0 - 1 = 1$ (borrowing 1 from immediate left upper digit) or $10 - 1 = 1$. Also, that upper digit reduces by 1 and the remaining process is similar as is done in decimal system.

For examples (a):

$$\begin{array}{r} 1\,1\,0\,1 \\ -\,1\,0\,0\,1 \\ \hline 0\,1\,0\,0 \end{array}$$

Explanation

1st column $1 - 1 = 0$

2nd column $0 - 0 = 0$

3rd column $1 - 0 = 1$

4th column $1 - 1 = 0$

Alternatively

Convert into decimal system and then subtract, thereafter convert the decimal number so obtained in binary system again.

$1101_2 = 13_{10}$ and $1001_2 = 9_{10}$

Now, $13_{10} - 9_{10} = 4_{10}$

Further, $4_{10} = 100_2$

(b)

$$\begin{array}{r} 1\,1\,0 \\ -\,1\,0\,1 \\ \hline 0\,0\,1 = 1 \end{array}$$

Explanation

1st column $0 - 1 = 1$

2nd column $(1 - 1) - 0 = 0$

3rd column $1 - 1 = 0$

Alternatively

Convert into decimal system and then subtract, thereafter convert the decimal number so obtained in binary system again.

$110_2 = 6_{10}$ and $101_2 = 5_{10}$

Now, $6_{10} - 5_{10} = 1_{10}$

Further, $1_{10} = 1_2$

For multiplication and division of binary numbers, we should convert the given numbers into decimal numbers and then again convert the number so obtained in binary system to save time and to avoid any confusion. Rules for multiplication and division have not been discussed deliberately.

F. Place Value (Local Value) v/s Face-Value of a Digit in a Numeral

In fact, the decimal system is a positional-value system in which the value of a digit in a number depends on its position.

For example : in 542, 5 represents 500 (i.e. 5×10^2), 4 represents 40 (i.e. 4×10^1) and 2 represents 2 (i.e. 4×10^0) units. Here, 500, 40 and 2 are place- values of 5, 4 and 2 respectively.

On the other hand, the Face-Value of a digit in a number is the value of the digit itself irrespective of its position/location. It means, in 542, face value 2 is 2, that of 4 is 4 and that of 5 is 5.

The above discussion may be depicted as below.

Digits in Number (542)	Face-value	Place-value
2	2	$2 \times 1 = 2$
4	4	$4 \times 10 = 40$
5	5	$5 \times 100 = 500$

We may also write 542 as $(5 \times 100) + (4 \times 10) + (2 \times 1)$.

On the above line, a two digit numbers may be expressed as $10x + y$ (where x and y are tens and unit digits respectively). Similarly, a three digit number may be expressed as $100x + 10y + z$ (where x, y and z are hundreds, tens and unit digits respectively).

Note

(i) Difference between any number and the number obtained by writing the digits in reverse order is exactly divisible by 9. In case of a two-digit number, the aforesaid difference is the product of the difference between both digits and 9.

(ii) Further, in case of a three-digit number, the difference(d) between the original number and the number obtained by writing the digits in reverse order is the product of the difference between unit and hundreds digits and 99, hence the difference(d) is also divisible by 99.

(iii) Sum of any two digit number and the number obtained by writing the digits in reverse order is exactly divisible by 11. **[SSC (CGL-PT)-03]**

(iv) If a number of two digits is x times the sum of its digits, then the number formed by interchanging the digits is equal to the sum of the digits multiplied by $(11 - x)$.

G. Place Value and Face-value of Digits in a Decimal Number

Take a decimal number 23.124.

Digits	2	3	.	1	2	4
Place-Value	$2 \times 10 = 20$	$3 \times 1 = 3$		$1 \times \frac{1}{10} = 0.1$	$2 \times \frac{1}{100} = 0.02$	$4 \times \frac{1}{1000} = 0.004$

Note *The Face-values of 2, 3, 1, 4 are 2, 3, 1, 4 respectively.*

H. General Classification of Numbers

Natural Numbers

Counting numbers i.e.1, 2, 3, 4 .... are called natural numbers.

Whole Numbers

When we include 'zero' in the natural numbers, it is known as whole numbers, e.g. 0, 1, 2, 3, 4 .... It implies that each of natural numbers is a whole number but its converse is not true. Also, 0 is the only whole number which is not a natural number.

It is also true that every square number is of the form $5n$ or $(5n - 1)$ or $(5n + 1)$ where, n is a whole number.

Integers

All natural numbers, 0 and negatives of counting numbers together form the set of integers, e.g (..... ,– 4, – 3, – 2, – 1,.... 0, 1, 2, 3, 4........)

Positive Integers $\Rightarrow$ 1, 2, 3, 4 ...

Negative Integers $\Rightarrow$ – 1, – 2, – 3, – 4 ...

Neither positive nor negative integer $\Rightarrow$ 0

Non-negative integers $\Rightarrow$ 0, 1, 2, 3, 4 ...

Non-positive integers $\Rightarrow$ 0, – 1, – 2, – 3, – 4 ...

Note

(i) Any integer can be expressed in one of the three forms $3k, 3k+1, 3k+2$, where k is an integer.

(ii) Every square integer (except 1) is either a multiple of 4 or greater than a multiple of 4 by unity.

(iii) The product of two square integers is a square integer.

(iv) If the product of n positive integers is n^n, then their sum is never less than n^2.

Even Numbers

All the natural numbers which are divisible by 2 are known as even numbers. For example: 2, 4, 6, 8,....... . Every even number greater that 4 can be expressed as the sum of two odd prime numbers. e.g. $6 = 3 + 3$, $8 = 3 + 5$, $10 = 3 + 7$, $12 = 5 + 7$ and so on.

Odd Numbers

All the natural numbers which are not divisible by 2 are known as odd numbers. e.g. 1, 3, 5, 7...

Prime Numbers

A number other than 1 is called a prime number if it is divisible only by 1 and itself, e.g. 2, 3, 5, 7, 11 .... Also, 2 is the only even prime number and 3 is the lowest odd prime number.

How to Test Whether a Number is Prime or not

Take an integer just higher than the approximate square root of the number which is to be tested. If the number to be tested is not divisible by any of the prime numbers below that integer, then it is a prime number, otherwise, a composite number.

Check as to whether 251 is prime or not

Approximate value of square-root of 251 = 15. ...

Now, integer just higher than 15. ... = 16

or

You may write as $251 < 16^2$ in lieu of above process. It implies that you have to express the number to be tested always as just less than the square of a suitable number.

Further, prime numbers below 16 are: 2, 3, 5, 7, 11 and 13. Clearly, 251 is not divisible by any of 2, 3, 5, 7, 11 and 13. Hence, 251 is a prime number.

Composite Numbers

A number other than 1, which is not a prime number is called composite number, e.g. 4, 6, 8, 9, 10, 12, etc. It may be construed that every composite number is a natural number.

Note *Every natural number except 1 is either prime or composite number.*

Twin Primes

Two prime numbers are said to be twin primes if there is only one composite number between them, e.g. 3 and 5, 5 and 7, 11 and 13, etc.

Prime Triplet

A set of three consecutive prime numbers differing by 2, is called prime triplet, e.g. 3, 5, 7. It is the only prime triplet.

Co-Prime Numbers

Two natural numbers are said to be co-prime if they have no common factor other than 1. It implies that the Highest Common Factor(HCF) of co-prime numbers is always 1, e.g. 3 and 4, 12 and 25, 11 and 23, etc.

Facts Relating to Prime Numbers

(i) A prime number greater than or equal to 5 leaves the remainder either 1 or 5 when it is divided by 6.

(ii) $(p^2 - 1)$ is always divisible by 24 if p is a prime number not lesser than 5.

(iii) $(p^4 - 1)$ is divisible by 2400 if p is a prime number greater than 5.

(iv) Total number of prime numbers between 1 and 100 is 25 out of which number of two-digit prime numbers is 21. Between 100 and 200, there are 21 prime numbers. *[CDS-17]*

(v) Every prime number of the form $3k + 1$ can be represented in the form $6m + 1$ (where, k and m are integers), when k is an even. *[CDS-13]*

Rational and Irrational Numbers

A number of the form p/q is called a rational number, where p and q are integers, prime to each other and $q \neq 0$.

e.g. $2\left(\text{i.e. } \frac{2}{1}\right), \frac{2}{3}, \frac{-3}{4}$, etc.

On the other hand, a number which is not rational is called irrational number. It means an irrational number cannot be expressed in the form p/q, where p and q are integers, prime to each other and $q \neq 0$.

e.g. $\sqrt{2}$, $\sqrt{3}$ etc.

Recurring decimal like $2.\dot{3}$ can be expressed in form p/q, thus they are rational. Also, every integer is a rational number.

There are an infinite number of rational and irrational numbers between any two rational numbers.

Sum or Product of two irrational numbers may be rational or irrational.

Examples $2\sqrt{3} + \sqrt{5} = 2\sqrt{3} + \sqrt{5} \rightarrow$ irrational

$(7 + 2\sqrt{3}) + (-2\sqrt{3}) = 7 \rightarrow$ rational

Further, $\sqrt{5} \times \sqrt{2} = \sqrt{10} \rightarrow$ irrational

and $\sqrt{8} \times \sqrt{2} = \sqrt{16} = 4 \rightarrow$ rational

Sum or product of two rational numbers is a rational number.

Examples $2 + \frac{1}{3} = \frac{7}{3}, 2 \times \frac{1}{3} = \frac{2}{3}$

Sum or product of a rational number and an irrational number is an irrational number.

Examples $2 + \sqrt{5} = 2 + \sqrt{5} \rightarrow$ irrational

$2 \times \sqrt{5} = 2\sqrt{5} \rightarrow$ irrational

Real Numbers

Real numbers include both rational and irrational numbers. Hence, every rational number is a real number.

Complex Numbers

Complex numbers include real numbers and imaginary number such as square-root of negative number.

I. Classification of Numbers on Basis of the Sum of their Factors

Perfect Numbers

A number is called a perfect number if the sum of its all factors is equal to twice the number, e.g. 6, 28, 496, 8128, etc. Here, sum of all factors of 6 = 1 + 2 + 3 + 6 = 12, which is twice the number 6. It is germane to mention that only 17 perfect numbers have been discovered as on date.

Deficient Numbers

A number is known as deficient number if the sum of its all factors is less than twice the number, e.g. 8, 10, etc.

Abundant Numbers

A number is abundant if the sum of its all factors is greater than twice the number, e.g. 12, 18, etc.

J. Some Special Numbers

Ramanujan Numbers

The numbers which may be expressed as the sum of two cubes in two different ways are known as Ramanujan numbers, e.g. $1729 = (1)^3 + (12)^3 = (9)^3 + (10)^3$

$$4104 = (2)^3 + (16)^3 = (9)^3 + (15)^3$$

$$13832 = (2)^3 + (24)^3 = (18)^3 + (20)^3$$

$$20683 = (10)^3 + (27)^3 = (19)^3 + (24)^3$$

Other property of $1729 \Rightarrow (1 + 7 + 2 + 9 = 19) \times 91 = 1729$

Smith Numbers

If sum of the digits of a composite integer is equal to the sum of the digits of its prime factors, that composite integer is known as Smith number.

e.g. 22, 85, etc.

$22 \Rightarrow \{2 + 2 = 4 = 2 + 1 + 1$ (2 and 11 are prime factors of 22)$\}$

$85 \Rightarrow \{8 + 5 = 13 = 1 + 7 + 5$ (17 and 5 are prime factors of 85)$\}$

Narcissistic Numbers

If the number is equal to the sum of cubes of its digits, it is known as Narcissistic number.

e.g. 153, 371, etc.

$153 \Rightarrow 153 = 1^3 + 5^3 + 3^3$ and $371 = 3^3 + 7^3 + 1^3$

It is also fact that $153 = 1! + 2! + 3! + 4! + 5!$

Happy and Unhappy Numbers

A positive integer is replaced by the sum of the squares of its digits and this process is repeated until the number equals one. The numbers for which this process ends in 1 are known as happy numbers, otherwise unhappy numbers (or sad numbers).

e.g. 7, 10, 13, etc. are happy numbers. 12 is sad number.

$7 \Rightarrow 7^2$ (i.e. 49) $\Rightarrow 4^2 + 9^2$ (i.e. 97) $\Rightarrow 9^2 + 7^2$ (i.e. 130)

$\Rightarrow 1^2 + 3^2 + 0^2$ (i.e., 10) $\Rightarrow 1^2 + 0^2$ (i.e. 1)

Happy Couple Numbers

If two consecutive positive integers are happy, they are known as happy couple numbers. e.g. 31 and 32, 301 and 302, etc.

Pan-digital Numbers

The number where all the digits i.e. from 0 to 9 exist once. e.g. 7165283904. It is to note that it is square of 84648 also.

Armstrong Number A number which can be expressed as sum of some power of its digits e.g. 370, 407, 153, etc.

$370 \Rightarrow 370 = 3^3 + 7^3 + 0^3$

$407 \Rightarrow 407 = 4^3 + 0^3 + 7^3$

Extraordinary Numbers

A number is said to be extraordinary, if the number $wxyz$ is equal to $w^x y^z$.
e.g. 2592 as $2592 = 2^5 \cdot 9^2$

Curious Numbers

When the number is equal to the sum of the factorials of its digits, the number is said to be curious number.
e.g. 2, 145, etc.

$$2 \Rightarrow 2 = 2!$$
$$145 \Rightarrow 145 = 1! + 4! + 5!$$

Cheap Numbers

The positive integers formed with either 1 or 2 or 3 only.
e.g. 1, 2, 3, 11, 22, 333, etc.

Cheaper Numbers

The positive integers formed with any two of 1, 2 and 3.
e.g. 12, 23, 32, 112, 232, 313, etc.

Cheapest Numbers

The positive integers formed with all of 1, 2 and 3. e.g. 123, 1123, etc.

Interesting Properties of Numbers

(i) There are only 7 numbers which are divisible by all the natural numbers lesser than their square roots. e.g. 2, 3, 4, 6, 8, 12, 24.

(ii) Digits of product of two same/different numbers are in reverse order of the digits of their sum. e.g.

$9 + 9 = 18$ and $9 \times 9 = 81$	$24 + 3 = 27$ and $24 \times 3 = 72$
$47 + 2 = 49$	$497 + 2 = 499$
and $47 \times 2 = 94$	and $497 \times 2 = 994$

Try for more ...

(iii)
$$1 = 1^2$$
$$1 + 3 = 2^2$$
$$1 + 3 + 5 = 3^2$$
$$1 + 3 + 5 + 7 = 4^2 \text{ and so on.}$$

(iv) $1^3 + 2^3 + 3^3 + 4^3 + 5^3 + 6^3 + \ldots\ldots$
$$1^3 = 1^2$$
$$1^3 + 2^3 = (1 + 2)^2$$
$$1^3 + 2^3 + 3^3 = (1 + 2 + 3)^2$$
$$1^3 + 2^3 + 3^3 + 4^3 = (1 + 2 + 3 + 4)^2$$
and so on.

(v) $1 + 3 + 5 + 7 + 9 + 11 + 13 + 15 + 17 + 19 + \ldots$
$$\Rightarrow 1 + (3 + 5) + (7 + 9 + 11) + (13 + 15 + 17 + 19) + \ldots$$
$$= 1^3 \quad + 2^3 \quad + 3^3 \quad + 4^3 + \text{so on.}$$

(vi)
$$37 \times 3 = 111$$
$$37 \times 6 = 222$$
$$37 \times 9 = 333$$
$$37 \times 12 = 444$$
$$37 \times 15 = 555$$
and so on.

(vii) $(14)^2 + (87)^2 = (41)^2 + (78)^2$
Here, $14 \Rightarrow 41$ and $87 \Rightarrow 78$ (reverse order)
$$(15)^2 + (75)^2 = (51)^2 + (57)^2$$
$$(17)^2 + (84)^2 = (71)^2 + (48)^2$$
$$(26)^2 + (97)^2 = (62)^2 + (79)^2$$

(viii) $153 = 1^3 + 5^3 + 3^3$
$$370 = 3^3 + 7^3 + 0^3$$
$$371 = 3^3 + 7^3 + 1^3$$
$$407 = 4^3 + 0^3 + 7^3$$

(ix) $(12)^2 = 144$
$$(21)^2 = 441$$
$$(13)^2 = 169$$
$$(31)^2 = 961$$

K. Diverse Types of Problems

Problems Based on Different Number-System

Problem 1 Two numbers in the base system B are 121_B and 11_B. The sum of these two numbers in decimal system is 20. Find the value of 1010_B in decimal system?

Sol. We know $(\ldots w\, x\, y\, z)_{\text{base}} = \{\ldots\ldots + w \cdot (\text{base})^3 + x \cdot (\text{base})^2 + y \cdot (\text{base})^1 + z \cdot (\text{base})^0\}_{10}$

Here, $121_B = \{1 \cdot (B)^2 + 2 \cdot (B)^1 + 1 \cdot (B)^0\}_{10}$
$= B^2 + 2B + 1$

and $11_B = \{1 \cdot (B)^1 + 1 \cdot (B)^0\}_{10} = B + 1$

As given, $B^2 + 2B + 1 + B + 1 = 20$
$$\Rightarrow B^2 + 3B - 18 = 0$$
$$\Rightarrow B^2 + 6B - 3B - 18 = 0$$
$$\Rightarrow B(B + 6) - 3(B + 6) = 0$$
$$\Rightarrow B = 3 \quad [\because -6 \text{ cannot be a base}]$$

Now, $1010_B = \{1 \cdot (3)^3 + 0 \cdot (3)^2 + 1 \cdot (3)^1 + 0 \cdot (3)^0\}_{10}$
$= 30_{10}$

Problem 2 Consider the expression: $(xxx)_b = x^3$, where b is the base, and x is any digit of base b. Find the value of b. **[XAT-13]**

Sol. Here, $(xxx)_b = \{x \cdot (b)^2 + x \cdot (b)^1 + x \cdot (b)^0\}_{10} = x^3$ (as given)
$$\Rightarrow x\{(b)^2 + (b)^1 + (b)^0\} = x^3$$
$$\Rightarrow b^2 + b + 1 = x^2$$
$$\Rightarrow x > b$$

But it is the fact that x cannot be greater than the base b.
Hence, there is no possible value of b.

Problem 3 A certain number written in a certain base is 144. Which of the following is always true?

I. Square root of the number written in the same base is 12.

II. If the base is increased by 2, the number becomes 100.

(a) Only I (b) Only II

(c) Both I and II (d) Neither I nor II

Sol. Since, there is 4 in the number, the base is not lesser than 5.

Let the base be b, then $(144)_b$

$$= \{1\cdot(b)^2 + 4\cdot(b)^1 + 4\cdot(b)^0\}_{10}$$

$$= \{(b+2)^2\}_{10}$$

Now, square root of $\{(b+2)^2\}_{10} = (b+2)_{10}$

Then, $(b+2)_{10} = (?)_b$

b	$b+2$	Remainders
	1	2
	0	1

Hence, $(b+2)_{10} = (12)_b$

$\therefore$ Statement I is true.

Further, $\{(b+2)^2\}_{10} = (?)_{b+2}$

$b+2$	$(b+2)^2$	Remainders
	$b+2$	0
	1	0
	0	1

Hence, $\{(b+2)^2\}_{10} = (100)_{b+2}$

$\therefore$ Statement II is true.

Thus, option(c) is correct.

Problem 4 A manager is not used to work in the decimal system. She says that there are 100 employees in the office of which 24 are males and 32 are females. Which number system does the manager use ?

Sol. Let she uses B - base number system.

Then, $(100)_B = (24)_B + (32)_B$

$\Rightarrow \quad 1\cdot B^2 + 0\cdot B^1 + 0\cdot B^1 = 2\cdot B^1 + 4\cdot B^0 + 3\cdot B^1 + 2\cdot B^0$

$\Rightarrow \quad B^2 = 2B + 4 + 3B + 2$

$\Rightarrow \quad B^2 - 5B - 6 = 0$

$\Rightarrow \quad B^2 - 6B + B - 6 = 0$

$\Rightarrow \quad B(B-6) + 1(B-6) \Rightarrow B = 6 \text{ or } -1$

Since −1 cannot be the base of any number system, the manager is using 6-base number system.

Problems Based on Successive Division

Problem 5 A number being successively divided by 2, 3 and 6 leaves remainders 1, 2 and 4 respectively. Find the respective remainders if the order of divisors be reversed.

Sol.

2	A	Remainders(R)
3	B	$1 \Rightarrow R_1$(1st remainder)
6	C	$2 \Rightarrow R_2$
	D	$4 \Rightarrow R_3$

Clearly, D may be any number. If D is minimum, A would be minimum and in case of maximum value of D, A be maximum.

Let D be 0.

Then, $C = 0 \times 6 + 4 = 4$

and $B = 4 \times 3 + 2 = 14$

and $A = 14 \times 2 + 1 = 29$, it is the least dividend.

Alternatively

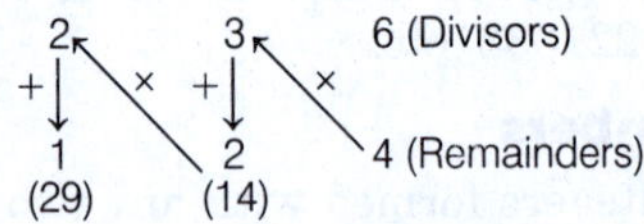

Step I $\Rightarrow 4 \times 3 + 2 = 14$

Step II $\Rightarrow 14 \times 2 + 1 = 29$

Now, 29 is to be divided by 6, 3 and 2 respectively.

6	29	Remainders(R)
3	4	$5 \Rightarrow R_1$
2	1	$1 \Rightarrow R_2$
	0	$1 \Rightarrow R_3$

Hence, required remainders are 5, 1 and 1 respectively.

Problem 6 A number was divided successively in order by 4, 5 and 6. The remainders were respectively 2, 3 and 4. The number is ***[SSC(CPO)-1997, SSC(CGL-T-II)-10]***

Sol. The options given were 214, 476, 954, 908.

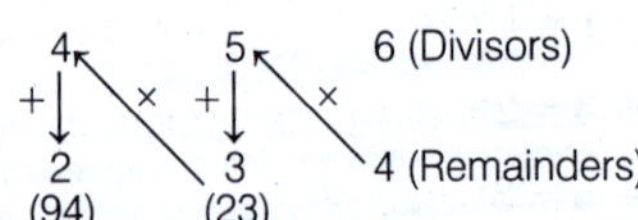

Then, the least dividend $= (5 \times 4 + 3) \times 4 + 2 = 94$

The next higher dividend $= (4 \times 5 \times 6) \times 1 + 94 = 214$

Hence, option (a) is correct.

If required in other questions, we will proceed as below:

$$\begin{cases} \text{the next higher dividend} = (4 \times 5 \times 6) \times 2 + 94 = 334 \\ \text{the next higher dividend} = (4 \times 5 \times 6) \times 3 + 94 = 454 \end{cases}$$

Alternatively

Proceed from options and check which option yields 2, 3, 4 as remainders on dividing successively in order by 4, 5 and 6.

Problem 7 When a number is divided by another number, the remainder is 41. If divisor factors are used then remainders are 2, 1 and 3 respectively. While dividing from factors, 2nd and 3rd divisors are 4 and 7 respectively. Find the divisor.

Sol. Let the 1st divisor (i.e. a factor of actual divisor) be x.

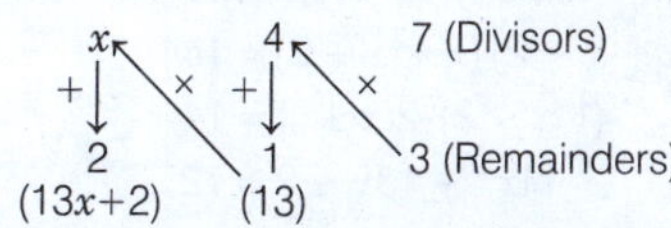

$\therefore$ Least dividend $= (4 \times 3 + 1)\cdot x + 2 = 13x + 2$

At $x = 1$,

one possible Divisor $= (1 \times 4 \times 7) = 28$

Dividend $= (13 \times 1 + 2) = 15$

Then, remainder = 15, not 41

At $x = 2$,

one possible Divisor $= (2 \times 4 \times 7) = 56$

Dividend $= (13 \times 2 + 2) = 28$

Then, remainder = 28, not 41

At $x = 3$,

one possible Divisor $= (3 \times 4 \times 7) = 84$

Dividend $= (13 \times 3 + 2) = 41$

Then, remainder = 41

Hence, actual divisor = 84.

Problem 8 Two different numbers when divided by a certain divisor leave remainders 547 and 343 respectively. When the sum of the two numbers is divided by the same divisor, the remainder is 211. What is the divisor ?

Sol. We know that if a, b, c, ... are divided by d and a_r, b_r, c_r, ... are remainders respectively , then $\frac{a + b + c + \dots}{d}$ leaves the same remainder what is left by $\frac{a_r + b_r + c_r + \dots}{d}$. Recall the concept of 'computation of remainder' described earlier.

Here, the sum of both given numbers will yield the samo remainder what would be left by (547 + 349) when divided by the same divisor. But, given that remainder is 211. It implies that the divisor is either {(547 + 349) − 211} or one of its factors. Hence, either 685 or one of its factors.

It is also mentioned that since the remainders are 547, 349 and 211, the divisor will obviously be greater that these remainders. Hence, the actual divisor is 685 itself, not one of its factors.

Problem 9 How many numbers lie between 100 and 1000 which when successively divided by 4, 5 and 6 leaves the respective remainders of 1, 2 and 3.

Sol.

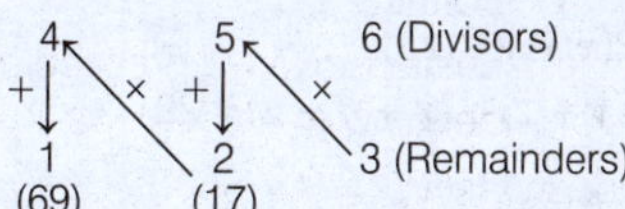

Then, the least dividend $= (5 \times 3 + 2) \times 4 + 1 = 69$

The next higher dividend $= (4 \times 5 \times 6) \times 1 + 69 = 189$

Next higher dividend $= (4 \times 5 \times 6) \times 2 + 69 = 309$

Next higher dividend $= (4 \times 5 \times 6) \times 3 + 69 = 429$

Next higher dividend $= (4 \times 5 \times 6) \times 7 + 69 = 909$

Hence, there are 7 such numbers.

Alternatively

$1000 - 69 = 931$

Since $931 \div (4 \times 5 \times 6)$ yields 7 as remainder, the required answer = 7

Problems Based on Place Value and Face Value of Digits in a Number

Problem 10 The difference between the place values of 7 and 3 in the number 527435 is ***[RRB-01]***

Digits in Number(527435)	Place-value
5	5×1
3	$3 \times 10 = 30$
4	4×100
7	$7 \times 1000 = 7000$
2	2×10000
5	5×100000

Now, the required difference = 7000 − 30 = 6970.

Problems Based on Digits of a Number/Changes in their Positions

Problem 11 In a three digit number the digit in the unit's place is twice the digit in the ten's place and 1.5 times the digit in the hundred's place. If the sum of all three digits of the number is 13, what is the number ? ***[SBI(PO)-08]***

Sol. Let ten's place digit be x.

Then, unit's place digit $= 2x$

and hundred's place digit $= 2x \times \frac{2}{3} = \frac{4x}{3}$

As given, $x + 2x + \frac{4x}{3} = 13$

$\Rightarrow \quad 13x = 13 \times 3$

$\therefore \quad x = 3$

Then, unit digit $= 2 \times 3 = 6$ and hundreds digit $= \frac{4\cdot 3}{3} = 4$

Hence, required number = 436

Alternatively

Proceed through given options to check as to which option is satisfying the conditions.

Problem 12 If the digits of a two digit number are interchanged, the newly formed number is more than the original number by 18, and sum of the digits is 8, then what was the original number?

Sol. Let the unit digit be y and tens digit be x, then number

$= 10x + y$

In case of mutual interchanges of digits, the new number

$= 10y + x$

As given, $(10y + x) - (10x + y) = 18$

$\Rightarrow \quad 9y - 9x = 18 \Rightarrow y - x = 2$ ...(i)

Further, $x + y = 8$, as given ...(ii)

On solving Eqs. (i) and (ii), we get $y = 5$ and $x = 3$

Hence, original number $= 10x + y = 10\cdot 3 + 5 = 35$

Alternatively I

We know, the difference between a two digit number and the number obtained by writing the digits in reverse order is the product of the difference between both digits and 9.

Hence, (difference between both digits) × 9 = 18
Difference between both digits = 2
Further, sum of both digits = 8, as given
Obviously, digits are 3 and 5.
Since, the original number is lesser than newly formed number, the tens digit would be lesser. Hence, the original number = 35

Alternatively II

Proceed through options. Let options be 26, 35, ...

(a) 26 ⇒ 2 + 6 = 8, but 62 − 26 ≠ 18
(b) 35 ⇒ 3 + 5 = 8 and 53 − 35 = 18

Hence, option(b) is correct.

Problem 13 **A two digit number is five times the sum of digits. If 9 is added to the number, the digits are interchanged. Find the sum of digits.**

Sol. Let unit digit be y and tens digit be x, then original number $= 10x + y$

Given, $10x + y = 5(x + y)$
$\Rightarrow \quad 5x - 4y = 0 \Rightarrow x = 0.8y$
Further, $10x + y + 9 = 10y + x$ [∵ digits are interchanged]
$\Rightarrow \quad 10x - x + y - 10y = -9 \Rightarrow 9x - 9y = -9$
$\Rightarrow \quad x - y = -1 \Rightarrow 0.8y - y = -1$
$\Rightarrow \quad y = 5$
Then, $x = 0.8 \times 5 = 4$
∴ Sum of digits = 4 + 5 = 9

Problem 14 **The sum of a two digit number and the number obtained by writing the digits in reverse order is a perfect square number. The number of such numbers is** ***[SSC(MTS)-11]***

Sol. Let the unit digit be y and tens digit be x, then number $= 10x + y$

Given, $10x + y + 10y + x =$ a perfect square number
$\Rightarrow \quad 11x + 11y =$ a perfect square number
$\Rightarrow \quad 11(x + y) =$ a perfect square number

Clearly, $(x + y)$ should be 11 so as to make $11(x + y)$ a perfect square number.

x + y	y + x
1 + 10	10 + 1
2 + 9	9 + 2
3 + 8	8 + 3
4 + 7	7 + 4
5 + 6	6 + 5

Here, x and y are digits, thus, $x = 10$ or $y = 10$ are not possible as 10 is a number, not a digit. Now, number of such numbers = 8.

Problem 15 **The unit digit of a two digit number is greater than tens digit by 2 and the product of that number and sum of its digits is 144. The number is**

Sol. Let the unit digit be y and tens digit be x, then number $= 10x + y$

As given, $y - x = 2 \Rightarrow y = x + 2$

Further given, $(10x + y)(x + y) = 144$
$\Rightarrow \quad (10x + x + 2)(x + x + 2) = 144$
$\Rightarrow \quad (11x + 2) \cdot 2(x + 1) = 144$
$\Rightarrow \quad 11x^2 + 13x + 2 = 72$
$\Rightarrow \quad 11x^2 + 13x - 70 = 0$
$\Rightarrow \quad 11x^2 - 22x + 35x - 70 = 0$
$\Rightarrow \quad 11x(x - 2) + 35(x - 2) = 0$
$\Rightarrow \quad x = 2$ or $-35/11$

Since, unit digit cannot be negative, the value of $x = 2$.
Then, $y = 2 + 2 = 4$
Hence, the number = 24

Alternatively

Proceed through options. Let options be 46, 24, 42, 26.

(a) 46 ⇒ 6 − 4 = 2 but 46 × (4 + 6) ≠ 144
(b) 24 ⇒ 4 − 2 = 2 and 24 × (2 + 4) = 144

Hence, option(b) is correct.

Problem 16 **The tens digit of a two digit number is less than unit digit by 2 and the sum of three times of the number and 6/7 of the number obtained by writing digits in reverse order is 108. Find the sum of digits.**

Sol. Let unit digit be y and tens be x. Then, original number $= 10x + y$

Given, $3(10x + y) + \frac{6}{7}(10y + x) = 108$
$\Rightarrow \quad 10x + y + \frac{20y}{7} + \frac{2x}{7} = 36$
$\Rightarrow \quad 72x + 27y = 36 \times 7$
$\Rightarrow \quad 8x + 3y = 28$ ... (i)
Also given, $y - x = 2$ ... (ii)
From Eqs. (i) and (ii), we get $x = 2$ and $y = 4$
∴ Sum of digits = 2 + 4 = 6

Problem 17 **The sum of all possible two-digit numbers formed from three different one-digit natural numbers when divided by the sum of the original numbers is equal to**

Sol. Let the one-digit natural numbers be x, y and z.

Sum of all possible two-digit numbers
$= (10x + y) + (10y + z) + (10x + z) + (10y + x) + (10z + x) + (10z + y)$
$= 22(x + y + z)$

Now, $22(x + y + z) \div (x + y + z) = 22$

Alternatively

Assume, three one-digit numbers are 2, 3 and 4.
Sum of possible two digit numbers
= 23 + 24 + 34 + 32 + 42 + 43 = 198
Now, 198 ÷ (2 + 3 + 4) = 22

Problem 18 **A number consists of four digits having 8 in the unit's place. If the digit to the extreme left is shifted to immediate right to the unit's place, keeping all other digits as they are, the new number formed exceeds the original number by 1305. Find the original number.**

Sol. Let the original number be $xyz8$.

New number after shifting = $yz8x$

As given, $yz8x - xyz8 = 1305$

$$\Rightarrow \begin{array}{r} y\,z\,8\,x \\ (-)\,x\,y\,z\,8 \\ \hline 1\,3\,0\,5 \end{array}$$

Clearly, $x = 3$ $[13 - 8 = 5]$

$z = 7$ $[7 - 7 = 0]$

$y = 4$ $[7 - 4 = 3]$

Hence, the original number = $xyz8$ = 3478

Alternatively

The best way is to proceed through options.

Problem 19 **If we reverse the digits of a two digit number, then the difference between the original number and new number is 36, the difference between the digits is ?**

Sol. Let unit digit be y and tens be x. Then, original number $= 10x + y$

Number obtained by reversing the digits = $10y + x$

As given, $(10x + y) - (10y + x) = 36$

$\Rightarrow$ $9x - 9y = 36$

$\therefore$ $x - y = 4$

Alternatively

We know, the difference between a two digit number and the number obtained by writing the digits in reverse order is the product of the difference between both digits and 9.

Hence, (difference between both digits) × 9 = 36

$\therefore$ Difference between both digits = 4

Problem 20 **The sum of units digit and hundreds digit of a three digit number is one-digit number. If this number is written in reverse order, then find out the possible sum of tens digits of both the original and new numbers.**

Sol. Let the options are 18, 13, 17 and 28.

It is obvious that tens digit will remain unchanged even after writing digits in reverse order.

Also, sum of tens digits of both original and changed numbers will always be even because odd + odd = even and even + even = even.

Thus, options (b) and (c) are incorrect.

As regards option (d), sum of tens digits of both numbers cannot be greater than 18 (i.e. 9 + 9) as tens digit may have maximum value of 9. Hence, option (d) is also incorrect.

In view of the above, option (a) is correct.

Problem 21 **The difference between a 3-digit number and an another 3-digit number formed with the same digits in reverse order is 495. Find the minimum possible sum of all the digits.**

Sol. Original number = $100x + 10y + z$

(Let z, y and x be unit, tens and hundreds digits respectively)

Then, new number = $100z + 10y + x$

Difference = $100x + 10y + z - 100z - 10y - x$

$= 99x - 99z = 495$ (as given)

$\Rightarrow$ $x - z = 5$

Possible values of (x, z) = (5, 0), (6, 1), (7, 2), (8, 3), (9, 4)

Obviously, (5, 0) is not admissible since if unit digit of the original number becomes 0, then on reversal, the new number would be of two digits only.

Further, tens digit (y), may be 0, 1, 2 ... 9.

Now, minimum possible value of $(x + y + z) = 6 + 0 + 1 = 7$.

Problem 22 **The number obtained by interchanging the two digits of a two-digit number is lesser than the original number by 54. If the sum of the digits is 12, then what is the original number?** ***[IDBI(PO)- 09]***

Sol. Difference between a two digit number and the number obtained by writing the digits in reverse order = (difference between both digits) × 9 = 54

$\Rightarrow$ Difference between both digits = 6

Further, sum of both digits = 12

Then, the digits are $\frac{12 + 6}{2}$ and $\frac{12 - 6}{2}$ i.e. 9 and 3.

Then, the original number = 93 or 39.

Since the original number is greater than newly formed number, the tens digit would be greater. Hence, the original number = 93.

Alternatively

The best way is to proceed through options.

Problem 23 **A three-digit number had digits in strictly descending order and divisible by 10. By changing the places of the digits, a new three-digit number is constructed in such a way that the new number is also divisiblo by 10. The difference between the original number and the new number is divisible by 40. How many numbers will satisfy all these conditions ?** ***[XAT-15]***

Sol. Let tens and hundreds digits of original number be y and x respectively. Since the original number is divisible by 10, its unit digit must be 0.

Then, original number = $100x + 10y + 0$

Further, the number formed by changing the places of digits is also divisible by 10, it implies that only tens and hundreds digits have been interchanged.

Hence, new number = $100y + 10x + 0$

It is obvious that the original number is greater than the new number because the digits of the former are in descending order.

Then, their difference = $100x + 10y + 0 - (100y + 10x + 0)$

$= 90x - 90y$

$= 90(x - y)$

Since $90(x - y)$ is divisible by 40, $(x - y)$ must be divisibly by 4. Further, the maximum possible value of $x - y$ is 8 as digit x may take maximum value of 9 and y may take minimum value of 1.

The possible combinations of x and y

= (9, 1), (9, 5), (8, 4), (7, 3), (6, 2), (5, 1).

Hence, there are 6 numbers i.e. 910, 950, 840, 730, 620 and 510 which are satisfying the given conditions.

Problems Based on Relation Between Two/Three Numbers

Problems related to numbers based on ratio have been detailed in chapter 'RATIO, PROPORTION AND VARIATION'. However, one question is provided here for observation.

Problem 24 Two natural numbers are in the ratio 3 : 5 and their product is 2160. The smaller of the numbers is *[SSC (CGL-T-I)-10]*

Sol. Let the numbers be $3x$ and $5x$.

Given, $3x \times 5x = 2160$

$\Rightarrow x^2 = 144 \Rightarrow x = 12$

Then, smaller number $= 3x = 3 \times 12 = 36$

Alternatively I

As given, $(3 \times 5)\text{ unit}^2 = 2400 \Rightarrow \text{unit}^2 = 144$

$\therefore$ 1 unit = 12

Then, smaller number = 3 units $= 3 \times 12 = 36$

Alternatively II

Proceed through options. Let options be 24, 36, ...

(a) $24 \Rightarrow 3$ units = 24, then 5 units = 40
and their product $= 24 \times 40 \neq 2160$

(b) $36 \Rightarrow 3$ units = 36, then 5 units = 60
and their product $= 36 \times 60 = 2160$

Hence, option(b) is correct.

Problem 25 The sum of two numbers is 22 and sum of their squares is 404. Find their product.

Sol. Let numbers be x and y. Then, $x + y = 22$

Also given, $x^2 + y^2 = 404 \Rightarrow (x+y)^2 - 2xy = 404$

$\Rightarrow (22)^2 - 2xy = 404 \Rightarrow 80 = 2xy$

$\therefore xy = 40$

Hence, product of numbers = 40

Problem 26 I multiplied a natural number by 18 and another by 21 and added the products, which one of the following could be the sum? *[SSC(CGL-T-1)-11]*

Sol. Given options are 2007, 2008, 2006 and 2002.
Since both the products are multiples of 3, their sum must be a multiple of 3. Only option (a) is divisible by 3.
Hence, option(a) is correct.

Problem 27 The sum of two numbers is 24 and their product is 143. The sum of their squares is

Sol. Let numbers be x and y.

Given, $x + y = 24$ and $xy = 143$

Further, $x^2 + y^2 = (x+y)^2 - 2xy$

$= (24)^2 - 2 \cdot 143 = 576 - 286 = 290$

Problem 28 Sum of two numbers is 40 and their product is 375. What will be the sum of their reciprocals?

Sol. Let numbers be x and y.

Given, $x + y = 40$ and $xy = 375$

Now, $\frac{1}{x} + \frac{1}{y} = \frac{y+x}{xy} = \frac{40}{375} = \frac{8}{75}$

Problem 29 The product of two numbers is 11520 and their quotient is 9/5. Find the difference of two numbers.

Sol. Let numbers be x and y.

Given, $\frac{x}{y} = \frac{9}{5} \Rightarrow y = \frac{5x}{9}$

Further given, $xy = 11520 \Rightarrow x \times \frac{5x}{9} = 11520$

$\Rightarrow x^2 = 20736 \Rightarrow x = 144$

Then, $y = \frac{5 \times 144}{9} = 80$

Now, difference between x and $y = 144 - 80 = 64$

Problem 30 Two numbers x and y $(x > y)$ are such that their sum is equal to three times their difference. Then, the value of $\frac{3xy}{2(x^2 - y^2)}$ will be

Sol. Let numbers be x and y.

Given, $x + y = 3(x - y) \Rightarrow x - 3x = -3y - y$

$\Rightarrow -2x = -4y \Rightarrow x = 2y$

Now, $\frac{3xy}{2(x^2 - y^2)} = \frac{3 \cdot 2y \cdot y}{2\{(2y)^2 - y^2\}} = \frac{6y^2}{6y^2} = 1$

Alternatively

Take the numbers satisfying the given conditions. Obviously, they may be 1 and 2. Since $x > y$, $x = 2$ and $y = 1$

Now, $\frac{3xy}{2(x^2 - y^2)} = \frac{3 \cdot 2 \cdot 1}{2(2^2 - 1)} = 1$

Problem 31 If the sum of two numbers be multiplied by each number separately, the product so obtained are 247 and 114. The sum of the numbers is *[SSC(CGL-T-I)-11]*

Sol. Obviously, sum of both numbers is common in both products. It is also fact that 19 is common in both products (19×13) and (19×6). It implies that 19 is the sum of both numbers.

Problem 32 Find the higher of the two successive natural numbers if the square of the sum of those numbers exceeds the sum of their squares by 40.

Sol. Let the numbers be x and $(x + 1)$.

As given, $(x + x + 1)^2 = x^2 + (x+1)^2 + 40$

$\Rightarrow 4x^2 + 4x + 1 = x^2 + x^2 + 2x + 1 + 40$

$\Rightarrow 2x^2 + 2x - 40 = 0 \Rightarrow x^2 + x - 20 = 0$

$\Rightarrow x^2 + 5x - 4x - 20 = 0 \Rightarrow x(x+5) - 4(x+5) = 0$

$\Rightarrow (x-4)(x+5) = 0 \Rightarrow x = 4$ and -5

$\Rightarrow x = 4$ because a natural number cannot be negative.

Then, higher natural number $= x + 1 = 4 + 1 = 5$

Alternatively

The best way is to proceed through options. Let options be 3, 5, ...

(a) 3 $\Rightarrow$ then smaller number = 2 and $(2 + 3)^2$

$\neq 2^2 + 3^2 + 40$

(b) 5 $\Rightarrow$ then smaller number = 4 and $(4 + 5)^2$

$= 4^2 + 5^2 + 40$

Hence, option(b) is correct.

Problem 33 **The product of a two digit number and a number formed with the same digits written in reverse order, is equal to 1462. Find the lower number.**

Sol. The best way is to proceed through options. Let the options be 23, 34, ...

(a) 23 $\Rightarrow 23 \times 32 \neq 1462$ (b) 34 $\Rightarrow 34 \times 43 = 1462$

Hence, 34 is the lower number.

Problem 34 **If we divide a two-digit number by a number formed with the same digits written in reverse order, we get 2 as quotient and 18 as remainder. If we add 9 to that number, we get square of the sum of both digits. Find the number.**

Sol. The best way is to proceed through options. Let the options be 52, 72, ...

(a) 52 $\Rightarrow 52 \div 25$ does not yield 2 as quotient and 18 as remainder.

(b) 72 $\Rightarrow 72 \div 27$ yields 2 as quotient and 18 as remainder.

Further, $(72 + 9) = (7 + 2)^2$

Hence, option(b) is correct.

Problems Based on Calculation Made by Mistake

Problem 35 **A student was asked to divide a number by 6 and add 12 to the quotient. He, however, first added 12 to the number and then divided it by 6, getting 112 as the answer. The correct answer should have been**

Sol. Let the number be x.

Then, $\frac{x + 12}{6} = 112$

$\Rightarrow$ $x + 12 = 672 \Rightarrow x = 660$

Now, the correct answer $= \frac{660}{6} + 12 = 122$

Alternatively

The student by mistake added 12 to the number first and then divided by 6. It implies that the actual quotient should have been less than 112 by $(12 \div 6)$ i.e. 110.

Now, the correct answer = 110 + 12 = 122

Problem 36 **A boy was asked to multiply a number by 12. By mistake he multiplied by 21 and got his answer 63 more than the correct answer. The number to be multiplied was.**

Sol. Let the required number be x.

Given, $21x - 12x = 63 \Rightarrow x = 7$

Alternatively I

Required number $= \frac{63}{21 - 12} = 7$

Alternatively II

You may proceed through options. Let options be 8, 7, ...

(a) 8 $\Rightarrow 8 \times 21 - 8 \times 12 \neq 63$

(b) 7 $\Rightarrow 7 \times 21 - 7 \times 12 = 63$

Hence, option(b) is correct.

Problem 37 **A candidate in an examination was asked to find 5/14 of a certain number. By mistake he found 5/4 of it. Thus, his answer was 25 more than the correct answer. The number was**

Sol. Let the number be x.

As given, $\frac{5x}{4} - \frac{5x}{14} = 25 \Rightarrow x = 28$

Alternatively I

The required number $= \frac{25}{\frac{5}{4} - \frac{5}{14}} = 28$

Alternatively II

You may proceed through options. Let options be 56, 28, ...

(a) 56 $\Rightarrow 56 \times \frac{5}{4} - 56 \times \frac{5}{14} \neq 25$

(a) 28 $\Rightarrow 28 \times \frac{5}{4} - 28 \times \frac{5}{14} = 25$

Hence, option(b) is correct.

Problem 38 **A boy multiplies 987 by a certain number and obtains 559981 as his answer. If in the answer, both 9's are wrong but the other digits are correct, then the correct answer will be**

Sol. Let options be 553681, 555181, 555681 and 556581

$987 = 3 \times 7 \times 47.$

Thus, the required number (i.e. correct product) must be divisible by each one of 3, 7 and 47.

Now, proceed through option.

(a) 553681 $\rightarrow$ not divisible by 3.

(b) 555181 $\rightarrow$ not divisible by 3.

(c) 555681 $\rightarrow$ divisible by 3,7 and 47. [see the rules of divisibility]

Hence, option(c) is correct.

Problem 39 **In doing a division of a question with zero remainder, a candidate took 12 as divisor instead of 21. The quotient by him was 35. The correct quotient is**

Sol. Dividend $= 12 \times 35 + 0 = 420$

Then, correct quotient $= \frac{420}{21} = 20$

Alternatively I

The required number $= 35 \times \frac{12}{21} = 20$

Finding Desired Number Like Perfect Cube/Square/Exactly Divisible by Certain Number

Problem 40 **By what least number should 4320 be multiplied so as to obtain a number which is a perfect cube?** *[SSC(CPO)-10, SSC(CGL-T-1)-10]*

Sol.

2	4320
2	2160
2	1080
2	540
2	270
3	135
3	45
3	15
	5

Clearly, $4320 = 2 \times 2 \times 2 \times 2 \times 2 \times 3 \times 3 \times 3 \times 5$

Obviously, 4320 should be multiplied by $2 \times 5 \times 5$ so as to obtain a perfect cube number.

$\Rightarrow$ $\underline{2 \times 2 \times 2} \times \underline{2 \times 2 \times ❷} \times \underline{3 \times 3 \times 3} \times \underline{5 \times ❺ \times ❺}$

Problem 41 **The least number to be subtracted from 36798 to get a number which is exactly divisible by 78 is**

Sol.

78) 36798 (471
$\underline{36738}$
××60

Clearly, we should subtract 60 from the dividend in order to make it exactly divisible by 78.

Problem 42 **The largest number of five digits, which is a perfect square, is** ***[SSC(CGL-PT)-08]***

Sol.

3	9 $\overline{99}\,\overline{99}$	316
3	9	
61	× 99	
1	61	
626	3899	
6	3756	
	143	

Clearly, we should subtract 143 from the 99999 to make the resultant a perfect square.

Hence, the required answer = 99999 − 143 = 99856

Alternatively

Recall the 'Estimation Method' meant for computing square root under chapter 'Square-root, cube-root and *n*th-root'.

Largest number of five digits = 99999

Obviously, 9 $\overline{99}\,\overline{99}$ will have a three digit square-root with the extreme left digit 3. Let the square-root of 99999 be 300.

Then, average of 300 and (99999 ÷ 300)

$$= \frac{300 + 333.33}{2}$$

$$= 316.66$$

Hence, the approx. value of square-root of 99999 = 316

Now, the number whose square-root is 316

$$= 316 \times 316$$

$$= 99856$$

It reveals that 99856 is the largest number of five digits which is a perfect square number.

Problem 43 **The smallest natural number by which 3000 must be divided to make the quotient a perfect cube.**

Sol. $3000 = 2 \times 2 \times 2 \times 5 \times 5 \times 5 \times 3$

It is apparent that if we divide the above expression by 3, the quotient will become a perfect cube.

Hence, required answer = 3

Problem 44 **The smallest number which should be added to the number 8958 so that the result is a perfect square.**

Sol.

9	$\overline{89}\,\overline{58}$	95
9	81	
185	× 858	
5	925	
190	− 67	

The above process reveals that 67 is required to add with the given number so as to make the resultant a perfect square.

Alternatively

Let the square root be 90.

$$\text{Then, estimated square root} = \frac{90 + \dfrac{8958}{90}}{2}$$

$$= \frac{90 + 99.5}{2}$$

$$= 94.8 \text{ (approx.)}$$

The next integer square root = 95

The number whose square root is 95 = 95 × 95

= 9025

∴ Required number to be added = 9025 − 8958 = 67

Problem 45 **The missing digit (*) in 1276*, so that the number is a perfect square, is**

Sol.

1	1 $\overline{27}\,\overline{6*}$	113
1	1	
21	× 27	
1	21	
223	× 66*	
3	669	
	×××	

Clearly, * is 9.

Alternatively

Obviously, 1 $\overline{27}\,\overline{6*}$ will have a three digit square-root with the extreme left digit 1.

Let the square root of 1276* be 100.

Then, average of 100 and (1276* ÷ 100) $= \dfrac{100 + 127.6*}{2}$

$= 113.8$

Hence, the approx. value of square root of 1276* = 113

Now, we may check as to which digit may be in place of *.

113 × 113 = 12769

We may check through 114 × 114, if required.

Clearly, * is 9.

Problem 46 For what value(s) of a is $x+\frac{1}{4}\sqrt{x}+a^2$ a perfect square?

Sol. Given expression $= x+\frac{1}{4}\sqrt{x}+a^2=(\sqrt{x})^2+2\cdot\frac{1}{8}\cdot\sqrt{x}+\left(\frac{1}{8}\right)^2$

$$=\left(\sqrt{x}+\frac{1}{8}\right)^2 \rightarrow \text{a perfect square}$$

Clearly, $a^2=\left(\frac{1}{8}\right)^2 \Rightarrow a=\pm\frac{1}{8}$

Problems Based on Components of Division

Problem 47 In a question on division, the divisor is 7 times the quotient and 3 times the remainder. If the remainder is 28, then the dividend is

Sol. Divisor = 3 × remainder = 3 × 28 = 84

Further, Divisor = 7 × quotient

$\Rightarrow$ 84 = 7 × quotient $\Rightarrow$ quotient = 12

Now, Dividend = Divisor × Quotient + Remainder

= 84 × 12 + 28 = 1008 + 28 = 1036

Problem 48 7 is added to a certain number, the sum is multiplied by 5, the product is divided by 9 and 3 is subtracted from the quotient. Thus, if the remainder left is 12, what was the original number ?

Sol. Quotient = 12 + 3 = 15

The dividend = 15 × 9 = 135

The sum of 7 and the unknown number = 135 ÷ 5 = 27

Then, the unknown number = 27 − 7 = 20

Alternatively

You may proceed through options. Let options be 30, 20, ...

(a) $30 \Rightarrow [\{(30+7)\times 5\}\div 9]-3\neq 12$

(b) $20 \Rightarrow [\{(20+7)\times 5\}\div 9]-3=12]$

Hence, option(b) is correct.

Problem 49 A number when divided by the sum of 555 and 445 gives two times their difference as quotient and 30 as the remainder. The number is

Sol. Dividend = Divisor × Quotient + Remainder

= (555 + 445) × 2(555 − 445) + 30 = 220030

Problem 50 A number when divided by a divisor leaves a remainder of 16. When twice the original number is divided by the same divisor, the remainder is 9. Then, the divisor is

Sol. Let the original dividend be n and divisor be d. Then,

Remainder in case of $\frac{2\times n}{d}$ = Remainder in case of $\frac{\text{remainder in case of } \frac{2}{d} \times \text{remainder in case of } \frac{n}{d}}{d}$

[Recall the concept of 'computation of remainder' under chapter 'Division'.]

= Remainder in case of $\frac{2\times 16}{d}$

[∵ Remainder is 16, the divisor is greater than 16. Hence, remainder in case of $\frac{2}{d}=2$]

= Remainder in case of $\frac{32}{d}$ = 9 (given)

Clearly, the divisor is either (32 − 9) or one of its factors. It is also fact that the divisor is greater than 16.

Thus, the required divisor = 23.

Alternatively

The best way is to proceed through options. Let options be 25, 23,...

(a) 25 $\Rightarrow$ Suppose the dividend be 41. Here, remainder is 16. Further, in case of (2 × 41) ÷ 25, the remainder is not equal to 9.

(b) 23 $\Rightarrow$ Suppose the dividend be 39.Here, remainder is 16. Further, in case of (2 × 39) ÷ 23, the remainder is equal to 9.

Hence, option(b) is correct.

Problem 51 235 when divided by a certain divisor leaves the remainder 4. When 472 is divided by the same divisor, the remainder obtained is 10. Also, if the sum of 235 and 472 is divided by the same divisor, the remainder is 3. What is the divisor ?

Sol. Let the original dividend be n and divisor be d. Then,

Remainder in case of $\frac{235+472}{d}$ = Remainder in case of $\frac{\text{remainder in case of } \frac{235}{d} + \text{remainder in case of } \frac{472}{d}}{d}$

[Recall the concept of 'computation of remainder' under chapter 'Division'.]

= Remainder in case of $\frac{4+10}{d}$

[∵ Remainder is 10 in one case, the divisor is greater than 10]

= Remainder in case of $\frac{14}{d}$ = 3 (given)

Clearly, the divisor is either (14 − 3) or one of its factors. It is also fact that the divisor is greater than 10.

Thus, the required divisor = 11.

Alternatively

You may proceed through options. Let options be 15, 11,......

(a) 15 $\Rightarrow$ Remainder in case of (235 ÷ 15) is not equal to 4.

(b) 11 $\Rightarrow$ Remainders in case of (235 ÷ 11), (472 ÷ 11), {235 + 472) ÷ 11} are 4, 10 and 3 respectively as have been given in questions.

Hence, option(b) is correct.

Nature of Expression/Value of the Number Hidden in Expression

Problem 52 If a and b are odd numbers, then which of the following is even ? ***[SSC(CGL-T-1)-11]***

Sol. The given options are

(a) $a+b+ab$ (b) $a+b-1$

(c) $a+b+1$ (d) $a+b+2ab$

It is important to note that:
(i) Odd + Odd = Even,
Odd + Even = Odd,
Even + Even = Even
(ii) Odd × Odd = Odd, Odd × Even = Even,
Even × Even = Even
(iii) Odd – 1 = Even or 0, Even – 1 = Odd
(iv) Odd + 1 = Even, Even + 1 = Odd
[Here, odd refers to odd number and even refers to even number]

Options

(a) $\Rightarrow a + b + ab \Rightarrow$ odd + odd + odd · odd = even + odd = odd
(b) $\Rightarrow a + b - 1 \Rightarrow$ odd + odd – 1 = even – 1 = odd
(c) $\Rightarrow a + b + 1 \Rightarrow$ odd + odd + 1 = even + 1 = odd
(d) $\Rightarrow a + b + 2ab \Rightarrow$ odd + odd + 2 · odd · odd = even + 2 · odd = even + even · odd = even + even = even

Hence, option(d) is correct.

Alternatively

In these types of cases, aspirants are advised to assume the values of a and b to arrive at conclusion.

Problem 53 Find the largest number which exactly divides every number of the form $(n^3 - n)(n - 2)$, where n is a natural number greater than 2.

Sol. At $n = 3$, $(n^3 - n)(n - 2) = (3^3 - 3)(3 - 2) = 24$
At $n = 4$, $(n^3 - n)(n - 2) = (4^3 - 4)(4 - 2) = 120$
At $n = 5$, $(n^3 - n)(n - 2) = (5^3 - 5)(5 - 2) = 360$
The above trend reveals that the largest number which can exactly divide every number of the form $(n^3 - n)(n - 2)$, is 24.

Problem 54 If m and $n(n > 1)$ are whole numbers such that $m^n = 121$, the value of $(m - 1)^{n+1}$ is **[SSC(CPO)- 08]**

Sol. Given, $m^n = 121 \Rightarrow m^n = 11^2 \Rightarrow m = 11$ and $n = 2$
Now, $(m - 1)^{n+1} = (11 - 1)^{2+1} = 1000$

Problem 55 If a and b are two distinct natural numbers, which one of the following is true ?

Sol. Given options are
(a) $\sqrt{a+b} > \sqrt{a} + \sqrt{b}$ (b) $\sqrt{a+b} = \sqrt{a} + \sqrt{b}$
(c) $ab = 1$ (d) $\sqrt{a+b} < \sqrt{a} + \sqrt{b}$
Let $a = 4$ and $b = 9$, then options
(a) $\sqrt{a+b} > \sqrt{a} + \sqrt{b} \Rightarrow \sqrt{4+9} > \sqrt{4} + \sqrt{9} \Rightarrow$ false
(b) $\sqrt{a+b} = \sqrt{a} + \sqrt{b} \Rightarrow \sqrt{4+9} = \sqrt{4} + \sqrt{9} \Rightarrow$ false
(c) $ab = 1 \Rightarrow 4 \times 9 = 1 \Rightarrow$ false
(d) $\sqrt{a+b} < \sqrt{a} + \sqrt{b} \Rightarrow \sqrt{4+9} < \sqrt{4} + \sqrt{9} \Rightarrow$ true
Hence, option(d) is correct.

Problem 56 The greatest whole number by which the expression $n^4 + 6n^3 + 11n^2 + 6n + 24$ is divisible for every natural number n, is

Sol. Let $n = 1$, then $n^4 + 6n^3 + 11n^2 + 6n + 24$
$= 1^4 + 6 \cdot 1^3 + 11 \cdot 1^2 + 6 \cdot 1 + 24 = 48$, which is divisible by 48.
Since we have taken value of n the lowest, the required greatest whole number = 48

Problem 57 x is a positive integer such that $x^2 + 12$ is exactly divisible by x. Find all possible values of x.

Sol. $x^2 + 12$ is exactly divisible by x. It is also fact that x^2 is divisibly by x. Then, it is obvious that 12 would also be exactly divisible by x. Then, value of $x = 1, 2, 3, 4, 6$ and 12.

Problem 58 If the sum of the numbers $(21x)^2$ and x^3 is divisible by 7, then the value of x is

Sol. Given, $(21x)^2 + x^3$ is divisible by 7. It is also true that $(21x)^2$ is divisible by 7. Then, it is obvious that x^3 would be divisible by 7 and it is possible only if $x = 7$.

Computation of Value of Number if Relation Between its Parts/ Number-its Part is Given

Problem 59 Find a number, one-seventh of which exceeds its eleventh part by 100. **[SSC(CGL-T-1)-11]**

Sol. $\left(\frac{1}{7} - \frac{1}{11}\right)$ of the number = 100
$\Rightarrow \frac{4}{77}$ of the number = 100
$\Rightarrow$ Number $= \frac{100 \times 77}{4} = 1925$

Problem 60 If three-fifth of sixty per cent of a number is 36, the number is?

Sol. Given, $\frac{3}{5}$ of 60% of the number = 36
$\Rightarrow$ 36% of the number = 36
$\therefore$ 100% of the number = 100

Alternatively

You may proceed through options. Let options be 200, 100, ...
(a) $200 \Rightarrow 3/5$ of 60% of $200 \neq 36$
(b) $100 \Rightarrow 3/5$ of 60% of 100 = 36
Hence, option(b) is correct.

Problem 61 Three-fifth of the square of a certain number is 126.15. What is the number?

Sol. Let number be x.
As given, $\frac{3}{5} \times x^2 = 126.15 \Rightarrow x^2 = \frac{126.15 \times 5}{3} = 210.25$
$\therefore$ $x = \sqrt{210.25} = 14.5$

Alternatively

You may proceed through options. Let options be 15, 14.5, ...
(a) $15 \Rightarrow 3/5$ of $(15)^2 \neq 126.15$
(b) $14.5 \Rightarrow 3/5$ of $(14.5)^2 = 126.15$
Hence, option(b) is correct.

Problem 62 When 60 is subtracted from 60% of a number, the result is 60. The number is ?

Sol. 60% of the number – 60 = 60
$\Rightarrow$ 60% of the number = 120
$\Rightarrow$ 100% of the number = 200

Alternatively

You may proceed through options. Let options be 120, 200, ...

(a) 120 ⇒ 60% of 120 − 60 ≠ 60

(b) 200 ⇒ 60% of 200 − 60 = 60

Hence, option(b) is correct.

Problem 63 The difference between a number and $\frac{2}{5}$th of the number is 30. The number is? *[RRB(ECRC)-08]*

Sol. $\left(1-\frac{2}{5}\right)$ of the number = 30

⇒ $\frac{3}{5}$ of the number = 30

⇒ The number = 50

Alternatively

You may proceed through options. Let options be 75, 50, ...

(a) 75 ⇒ 75 − 2 / 5 of 75 ≠ 30

(b) 50 ⇒ 50 − 2/5 of 50 = 30

Hence, option(b) is correct.

Problem 64 $\frac{1}{2}$ of $\frac{3}{4}$ of a number is $2\frac{1}{2}$ of 10. What is the number?

Sol. $\frac{1}{2}\times\frac{3}{4}\times$ number $= 10\times\frac{5}{2}$

⇒ Number $=\frac{25\times 2\times 4}{3}=66\frac{2}{3}$

Alternatively

You may proceed through options. Let options be 50, $66\frac{2}{3}$, ...

(a) 50 ⇒ $\frac{1}{2}$ of $\frac{3}{4}$ of $50\neq 2\frac{1}{2}$ of 10

(b) $66\frac{2}{3}\Rightarrow\frac{1}{2}$ of $\frac{3}{4}$ of $66\frac{2}{3}=2\frac{1}{2}$ of 10

Hence, option(b) is correct.

Problem 65 A number which when multiplied by 11 becomes as much above 180 as it was originally below 180. Find the number.

Sol. As per question, number × 11 − 180 = 180 − Number

⇒ 12 × Number = 360 ⇒ Number = 30

Alternatively

You may proceed through options.

Let options be 20, 30, ...

(a) 20 ⇒ 20 × 11 − 180 ≠ 180 − 20

(b) 30 ⇒ 30 × 11 − 180 = 180 − 30

Hence, option(b) is correct.

Problem 66 A number of two digits has 3 for its unit's digit and the sum of digits is 1/7 of the number itself. The number is? *[LIC-03]*

Sol. Let tens digit be x.

Then, the number = $10x + 3$

Further, $(x + 3) = 1/7$ of $(10x + 3)$

⇒ $7x + 21 = 10x + 3$

⇒ $3x = 18$

⇒ $x = 6$

Hence, the number = $10x + 3 = 10\cdot 6 + 3 = 63$

Alternatively

You may proceed through options. Let options be 43, 63, ...

(a) 43 ⇒ (4 + 3) ≠ 1/7 of 43

(b) 63 ⇒ (6 + 3) = 1/7 of 63

Hence, option(b) is correct.

Problems Based on Prime Numbers

Problem 67 Which one of the following is a prime number? *[CDS-11]*

Sol. Options given are 161, 173, 171, 221.

(a) 161 ⇒ $161 < (13)^2$

The prime numbers below 13 = 2, 3, 5, 7, 11.

Since 161 is divisible by 7, it is not a prime number.

(b) 173 ⇒ $173 < (14)^2$

The prime numbers below 14 = 2, 3, 5, 7, 11, 13.

Since 173 is not divisible by any of the numbers 2, 3, 5, 7, 11, 13, it is a prime number.

Problem 68 Let p denote the product $2\cdot 3\cdot 5 \ldots\ldots 59\cdot 61$ of all primes from 2 to 61. Consider the sequence $p + n(2\leq n\leq 59)$. What is the number of primes in this sequence (where n is a natural number)?

Sol. Clearly, product p ends with 0 as $2\cdot 5 = 10$

In case n be an even number, $p + n = (\ldots 0)$ + even number, which is also divisible by other than 1 and itself.

In case n be an odd number(say 3),

$p + n = (2\cdot 3\cdot 5\ldots 59\cdot 61) + 3 = 3(2\cdot 5\ldots 59\cdot 61 + 1)$, which is also divisible by 3.

Similarly, for all values of n $(2\leq n\leq 59)$ as odd, $p + n$ will also be divisible by other than 1 and itself.

It implies that there is no prime number in the aforementioned sequence.

Problem 69 The sum of three prime numbers is 54 and the difference between two numbers is 6 then one of the number is

Sol. Let options given be 23, 31, 37, 41.

Let the 1st prime number be x.

Then, 2nd prime number = $x + 6$

and 3rd prime number = y

As given, $x + x + 6 + y = 54$

⇒ $2x + y = 48$

In case y be an odd prime number, x would be in fraction. Hence, y must be an even prime number and it is the fact that 2 is only even prime number.

Then, $2x + 2 = 48$

∴ $x = 23$

Thus, 1st prime number = 23

2nd prime number = 23 + 6 = 29

3rd prime number = 2

∴ Correct option is (a).

Problem 70 The difference between two numbers is a prime number and also, their sum is a prime number. If higher number is 20, then how many possible values of the lower number may be?

Sol. Let the lower number be x.

Given that $(20 - x)$ and $(20 + x)$ both are prime numbers.

Clearly, at $x = 3, 9, 17$ the above condition is satisfied. Hence, there may be 3 possible values of the lower number.

Problem 71 There are four prime numbers in ascending order. Product of the first three numbers is 385 and that of last three numbers is 1001. Find out the greatest prime number.

Sol. $385 = 5 \times 7 \times 11$ and $1001 = 7 \times 11 \times 13$

Clearly, numbers are : 5, 7, 11 and 13.

Hence, the greatest number = 13

Problem 72 Let $P = [2, 3, 4, \ldots\ldots, 100]$ and $Q = \{101, 102, 103, \ldots\ldots., 200\}$. How many elements of Q are there such that they don't have any element of P as a factor ?

Sol. Clearly, we have to find the number of prime numbers from the elements of Q as they will have no factor from the elements of P. No need to mention that the factors of a prime number is 1 and number itself. Since there are 21 prime numbers from 101 to 200, the required answer is 21.

Puzzles on Numbers

Problem 73 The number of trees in each row of a garden is equal to the total number of rows in the garden. After 111 trees have been uprooted in a storm, there remain 10914 trees in the garden. The number of rows of trees in the garden is **[SSC(CPO)-07]**

Sol. Let number of rows be x.

Then, each row will have x trees.

Now, total number of trees $= x \cdot x = x^2$

As given, $x^2 - 111 = 10914 \Rightarrow x^2 = 11025$

$\therefore \quad x = 105$

Alternatively

You may write directly as number of rows $= \sqrt{10914 + 111} = 105$

Problem 74 In a classroom there are certain number of benches. If 6 students are made to sit on a bench, then to accommodate all of them, one more bench is needed. However, if 7 students are made to sit on a bench, then after accommodating all of them, space for 5 students is left. What is the total number of students in the class?

Sol. Let number of benches be B.

Case I Number of students $= 6B + 6$

Case II Number of students $= 7B - 5$

Obviously, $6B + 6 = 7B - 5 \Rightarrow -B = -11$

$\therefore \quad B = 11$

Hence, number of students

$= 6B + 6 = (6 \times 11) + 6 = 72$

Alternatively

Proceed through options. Let options be 30, 72, ...

(a) 30

$\Rightarrow$ ***Case* I** Number of benches $= (30 - 6) \div 6 = 4$

Further, ***Case* II** number of students $= 7 \cdot 4 - 5 \neq 30$

(b) 72

$\Rightarrow$ ***Case* I** Number of benches $= (72 - 6) \div 6 = 11$

Further, ***Case* II** number of students $= 7 \cdot 11 - 5 = 72$

Hence, option (b) is correct.

Problem 75 In a class, $\frac{3}{5}$ of the students are girls and rest are boys. If $\frac{2}{9}$ of the girls and $\frac{1}{4}$ of the boys are absent, what part of the total number of students are present?

Sol. Absent Girls $= \frac{3}{5} \times \frac{2}{9} = \frac{2}{15}$ of total number of students

and absent boys $= \left(1 - \frac{3}{5}\right) \times \frac{1}{4} = \frac{2}{5} \times \frac{1}{4} = \frac{1}{10}$ of total number of students

Then, total absent students $= \frac{2}{15} + \frac{1}{10} = \frac{7}{30}$ of total number of students

$\therefore$ Present students $= 1 - \frac{7}{30} = \frac{23}{30}$ of the total number of students

Problem 76 A classroom has equal number of boys and girls. Eight girls left to play Kho-Kho, leaving twice as many boys as girls in the classroom. What was the total number of girls and boys present initially?

Sol. Let number of boys and girls each initially be equal to x.

As given, $2(x - 8) = x \Rightarrow 2x - 16 = x \Rightarrow x = 16$

$\therefore$ Total number of students initially present $= 16 + 16 = 32$

Alternatively

Proceed through options. Let options be 24, 32, ...

(a) $24 \Rightarrow 2\{(24 \div 2) - 8\} \neq (24 \div 2)$

(b) $32 \Rightarrow 2\{(32 \div 2) - 8\} = (32 \div 2)$

Hence, option(b) is correct.

Problem 77 A lady sells the first customer, the half the apples in her stock and half an apple, half the remaining stock and half an apple to the second customer and in same manner, she sells apples to the third and fourth customer. At last she has 15 apples with her. How many apples did she have initially? **[SSC (Asstt.-Mains)-95, CAT-08]**

Sol. Let number of apples initially be x.

Number of apples sold to 1st customer $= \frac{x}{2} + \frac{1}{2}$

Remaining apples $= x - \left(\frac{x}{2} + \frac{1}{2}\right) = \frac{x}{2} - \frac{1}{2} = \frac{x - 1}{2}$

Number of apples sold to 2nd customer $= \left(\frac{x - 1}{2}\right) \times \frac{1}{2} + \frac{1}{2}$

Remaining apples $= \frac{x-1}{2} - \left(\frac{x-1}{4} + \frac{1}{2}\right) = \frac{x-1}{4} - \frac{1}{2}$

Number of apples sold to 3rd customer $= \left(\frac{x-1}{4} - \frac{1}{2}\right) \times \frac{1}{2} + \frac{1}{2}$

$= \frac{x-1}{8} - \frac{1}{4} + \frac{1}{2} = \frac{x-1}{8} + \frac{1}{4}$

Remaining apples $= \left(\frac{x-1}{4} - \frac{1}{2}\right) - \left(\frac{x-1}{8} + \frac{1}{4}\right) = \frac{x-1}{8} - \frac{3}{4}$

Number of apples sold to 4th customer $= \left(\frac{x-1}{8} - \frac{3}{4}\right) \times \frac{1}{2} + \frac{1}{2}$

$= \frac{x-1}{16} - \frac{3}{8} + \frac{1}{2} = \frac{x-1}{16} + \frac{1}{8}$

Remaining apples $= \left(\frac{x-1}{8} - \frac{3}{4}\right) - \left(\frac{x-1}{16} + \frac{1}{8}\right)$

$= \frac{x-1}{16} - \frac{7}{8}$

As given, $\frac{x-1}{16} - \frac{7}{8} = 15$

$\Rightarrow \frac{x-1}{16} = \frac{127}{8}$

$\Rightarrow x - 1 = 254$

$\therefore x = 255$

Alternatively I

Number of apples left with lady in end = 15

Number of apples left with lady before selling to 4th customer

$= \left(15 + \frac{1}{2}\right)2 = 31$

Number of apples left with lady before selling to 3rd customer

$= \left(31 + \frac{1}{2}\right)2 = 63$

Number of apples left with lady before selling to 2nd customer

$= \left(63 + \frac{1}{2}\right)2 = 127$

Number of apples left with lady before selling to 1st customer

$= \left(127 + \frac{1}{2}\right)2 = 255$

Alternatively II

Proceed through options. Let options be 225, 255, ...

(a) $225 \Rightarrow (225 \div 2) + 1/2 = 113 \rightarrow$ 1st customer

$(225 - 113) \div 2 + 1/2 = 56.5 \rightarrow$ 2nd customer

$(112 - 56.5) \div 2 + \frac{1}{2} = 28.25 \rightarrow$ 3rd customer

$(55.5 - 28.25) \div 2 + 1/2 = 14.125 \rightarrow$ 4th customer

Then, remaining apples $= 27.25 - 14.125 = 13.125 \neq 15$

(b) $255 \Rightarrow (255 \div 2) + 1/2 = 128 \rightarrow$ 1st customer

$(255 - 128) \div 2 + 1/2 = 64 \rightarrow$ 2nd customer

$(127 - 64) \div 2 + 1/2 = 32 \rightarrow$ 3rd customer

$(63 - 32) \div 2 + 1/2 = 16 \rightarrow$ 4th customer

Then, remaining apples $= 31 - 16 = 15$

Hence, option(b) is correct.

Problem 78 In a certain office 1/3rd of the workers are women, 1/2 of the women are married and 1/3rd of the married women have children. If ¾ of the men are married and 2/3rd of the married men have children, what part of the workers are without children?

Sol.

Descriptions	Women	Men
Part of total number of workers	$\frac{1}{3}$	$1 - \frac{1}{3} = \frac{2}{3}$
Married	$\frac{1}{3} \times \frac{1}{2} = \frac{1}{6}$	$\frac{2}{3} \times \frac{3}{4} = \frac{1}{2}$
Having children	$\frac{1}{6} \times \frac{1}{3} = \frac{1}{18}$	$\frac{1}{2} \times \frac{2}{3} = \frac{1}{3}$
Without children	$\frac{1}{3} - \frac{1}{18} = \frac{5}{18}$	$\frac{2}{3} - \frac{1}{3} = \frac{1}{3}$

Hence, part of workers without children $= \frac{5}{18} + \frac{1}{3} = \frac{11}{18}$

Directions (Problem Nos. 79-81) A young girl Roopa leaves home with x flowers, goes to the bank of a nearby river. On the bank of the river, there are four places of worship, standing in a row. She dips all the x flowers into the river, the number of flowers doubles. Then, she enters the first place of worship, offers y flowers to the deity. She dips the remaining flowers into the river and again the number of flowers doubles. She goes to the second place of worship, offers y flowers to the deity. She dips the remaining flowers into the river and again the number of flowers doubles. She goes to the third place of worship, offers y flowers to the deity. She dips the remaining flowers into the river and again the number of flowers doubles. She goes to the fourth place of worship, offers y flowers to the deity. Now, she is left with no flowers in hand. *[CAT-99]*

Problem 79 If Roopa leaves home with 30 flowers, the number of flowers she offers to each deity is

Sol. Number of flowers she has before offering y flowers to the deity at 1st place $= 30 \times 2 = 60$

[$\because$ number of flowers double when she dips them in river]

Number of flowers she has before offering y flowers to the deity at 2nd place $= 2(60 - y)$

Number of flowers she has before offering y flowers to the deity at 3rd place $= 2(120 - 2y - y) = 240 - 6y$

Number of flowers she has before offering y flowers to the deity at 4th place $= 2(240 - 6y - y) = 480 - 14y$

Number of flowers she has after offering y flowers to the deity at 4th place $= 480 - 14y - y = 480 - 15y$

According to the question, $480 - 15y = 0 \Rightarrow y = 32$

Hence, required answer = 32

Alternatively

Number of flowers she has before offering y flowers to the deity at 4th place $= y$

Number of flowers she has before offering y flowers to the deity at 3rd place $= y + \frac{y}{2} = \frac{3y}{2}$

Number of flowers she has before offering y flowers to the deity at 2nd place $= y + \frac{3y}{4} = \frac{7y}{4}$

Number of flowers she has before offering y flowers to the deity at 1st place $= y + \frac{7y}{8} = \frac{15y}{8}$

Then, number of flowers she has at the time of leaving home

$= \frac{15y}{8 \times 2} \Rightarrow 30 = \frac{15y}{16} \Rightarrow y = 32$

Problem 80 The minimum number of flowers that could be offered to each deity is

Sol. From the above, number of flowers she has at the time of leaving home $= \frac{15y}{16} \Rightarrow x = \frac{15y}{16} \Rightarrow y = \frac{16x}{15}$

The value of y will be integer as well as minimum only when $x = 15$.

At $x = 15$, $y = 16$

Hence, minimum number of flowers that could be offered to each deity = 16

Problem 81 The minimum number of flowers with which Roopa leaves home is

Sol. From the above, number of flowers she has at the time of leaving home $= \frac{15y}{16}$

$$\Rightarrow \quad x = \frac{15y}{16}$$

Clearly, at $y = 16$, the value of x is integer as well as minimum.

Then, at $y = 16$, $x = \frac{15 \times 16}{16} = 15$

Problem 82 An intelligence agency forms a code of two distinct digits selected from 0, 1, 2, ..., 9 such that the first digit of the code is non-zero. The code handwritten on a slip, can however potentially create confusion when read upside down. For example, the code 91 may appear as 16. How many codes are there for which no such confusion can arise ? ***[CAT-93, 03]***

Sol. Number of codes formed with two distinct codes not started with zero $= 9 \times 9 = 81$

The codes which can create confusions are 16-91, 18-81, 19-61, 68-89, 86-98.

Hence, required number of codes $= 81 - 10 = 71$

Problem 83 From a group of boys and girls 15 girls leave. Then two boys are left for each girl. After this 45 boys leave. There are then 5 girls for each boy. Find the number of girls in the beginning.

Sol. Let number of boys be x and that of girls be y in the beginning

Case I When 15 girls leave the group

$$x = 2(y - 15)$$

$$\Rightarrow \quad x - 2y = -30 \quad \ldots\text{(i)}$$

Case II When 45 boys leave the group

$$5(x - 45) = y - 15$$

$$\Rightarrow \quad 5x - y = 210 \quad \ldots\text{(ii)}$$

From Eqs. (i) and (ii) we get, $y = 40$

Hence, number of girls in beginning = 40

Alternatively

Proceed through options. Let options be 60, 40, ...

(a) 60 $\Rightarrow$ number of boys before 45 boys leave $= 2(60 - 15) = 90$

Further, after 45 boys leave, $5(90 - 45) \neq (60 - 15)$

(b) 40 $\Rightarrow$ number of boys before 45 boys leave $= 2(40 - 15) = 50$

Further, after 45 boys leave, $5(50 - 45) = (40 - 15)$

Hence, option(b) is correct.

Problem 84 On a 2 km road, a total of 201 trees are planted on each side of the road at equal distances. How many such trees in all will be planted on both sides of a 50 km road such that the distance between two consecutive trees is the same as that of the consecutive trees on the 2 km road?

Sol. Number of trees on each side of the 2 km road $= 201 = 200 + 1$

Now, 2000 m ÷ 200 = 10 m

Hence, trees have been planted at equal distance of 10 m.

> For better understanding, suppose you have to plant a tree on a path 50 m long at equal distance of 10 m, then you will plant {(50 ÷ 10) + 1} trees.

Further, number of trees on each side of the 50 km road

$= (50000 \div 10) + 1 = 5001$

Then, number of trees on both sides of the 50 km road

$= 2 \times 5001$

$= 10002$

Problem 85 The soldiers in a regiment can be drawn up in a hollow square 4 deep. If these soldiers can be drawn up in the shape of solid square, the number of soldiers along a side reduced by 16. Find the number of soldiers.

Sol. Let the number of soldiers along a side be x (1st case).

Clearly, total number of soldiers $= x^2 - (x - 8)^2$

Now, number of soldiers along a side be $x - 16$ (if arranged in a solid square). Then, total number of soldiers $= (x - 16)^2$

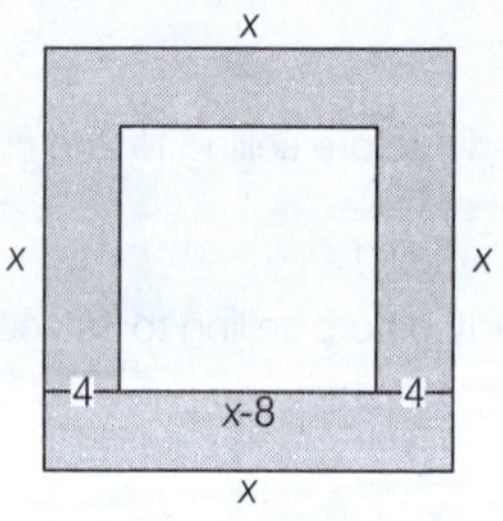

But, number of soldiers in both cases are equal

$$\therefore \quad x^2 - (x - 8)^2 = (x - 16)^2$$

$$\Rightarrow \quad x^2 - x^2 + 16x - 64 = x^2 - 32x + 256$$

$$\Rightarrow \quad x^2 - 48x + 320 = 0$$

$$\Rightarrow \quad x^2 - 40x - 8x + 320 = 0$$

$$\Rightarrow \quad x(x - 40) - 8(x - 40) = 0$$

$$\Rightarrow \quad (x - 40)(x - 8) = 0$$

$$\Rightarrow \quad x = 40 \text{ and } x = 8 \text{ (it is not possible)}$$

$\therefore$ Total number of soldiers $= (x - 16)^2 = (40 - 16)^2 = 576$

Alternatively

Proceed through options. Let options be 529, 576, 676 and 625.

(a) 529 $\Rightarrow (23)^2 \Rightarrow$ It means number of soldiers along a side of solid square = 23.

Thus, number of soldiers along a side of hollow square $= 23 + 16 = 39$

Now, $(39)^2 - \{39 - (2 \times 4)\}^2 \neq 529$

(b) 576 $\Rightarrow (24)^2 \Rightarrow$ It means, number of soldiers along a side of solid square = 24.

Thus, number of soldiers along a side of hollow square

$= 24 + 16 = 40$

Now, $(40)^2 - \{40 - (2 \times 4)\}^2 = 576$

Hence, option(b) is correct.

Problem 86 In a school, students were called for the Flag Hoisting ceremony on August 15. After the ceremony, small boxes of sweets were distributed among the students. In each class, the student with roll no. 1 got one box of sweets, student with roll no. 2 got 2 boxes of sweets and so on. In class III, a total of 1200 boxes of sweets were distributed. By mistake, one of the students of Class III got double the sweets he was entitled to get. Identify the roll number of the student who got twice as many boxes of sweets as compared to his entitlement. **[IIFT-14]**

Sol. Let there are n students in class III and x be roll no. of the student who got twice as many as compared to his/her entitlement.

Then, total number of boxes of sweets distributed

$$= \frac{n(n+1)}{2} + x = 1200$$

$\Rightarrow \quad n(n+1) = 2400 - 2x$

Obviously, n will be lesser than 50 and $(2400 - 2x)$ will be lesser than 2400 and will be even as $2x$ is even.

Now, by **hit and trial method**

$49 \times 50 = 2450$ which is greater than 2400.

$48 \times 49 = 2352 = 2400 - 48 = 2400 - 2 \cdot 24$

Hence, $x = 24$

Alternatively,

Proceed through options. Let options be 22, 24, 30 and 28.

(a) $22 \Rightarrow 2400 - 2x = 2400 - 2 \cdot 22 = 2356$, which is not a product of two consecutive natural numbers.

(b) $24 \Rightarrow 2400 - 2x = 2400 - 2 \cdot 24 = 2352$ which is a product of two consecutive natural numbers.

Hence, option(b) is correct.

Problem 87 N persons stand on the circumference of a circle at a distinct points. Each possible pair of persons, not standing next to each other, sings a 2 min song one pair after the other. If the total time taken for singing is 28 min, what is N? **[CAT-04]**

Sol. Recall the concept of number of handshakes when each shakes hand with each other. In this case, number of handshakes $= \frac{N(N-1)}{2}$

On the above line, number of occurrence of singing when each sings with another one $= \frac{N(N-1)}{2}$

But, a person cannot sing with his/her two adjacent persons, then number of occurrence of singing $= \frac{N(N-3)}{2}$

Or

Number of possible pairs $= {}^NC_2 - N = \frac{N!}{2!(N-2)!} - N$

$$= \frac{N(N-1)}{2} - N = \frac{N(N-3)}{2}$$

[Concepts of combination have been detailed in chapter 'Permutation and Combination']

[**Alternatively** you may recall the concept of number of diagonals in a polygon, given in chapter 'Geometry'.]

Now, total time taken in singing $= \frac{N(N-3)}{2} \times 2$ min

$\Rightarrow \quad 28 = \frac{N(N-3)}{2} \times 2$

$\Rightarrow \quad 28 = N(N-3) = 7(7-3)$

Hence, $\quad N = 7$

Problem 88 The integers 1, 2, ...40 are written on a blackboard. The following operation is then repeated 39 times. In each repetition, any two numbers, say a and b, currently on the blackboard are erased and a new number $a + b - 1$ is written. What will be the number left on the board at the end? **[CAT-08]**

Sol. Initially, sum of 1, 2, ... $40 = \frac{40 \times 41}{2} = 820$

After erasing any two numbers and writing a number 1 less than their sum, the new sum of the sequence

$= 820 - 1 = 819$

It is obvious that after every operation, sum of the numbers appeared on board reduces by 1.

Then, after 39 operation, the number left on the board

$= 820 - 39 = 781$

Problem 89 Amitabh picks a random integer between 1 and 999, double it and gives the result to Sashi. Each time Sashi gets a number from Amitabh, he adds 50 to the number and gives the result back to Amitabh, who doubles the number again. The first person, whose result is more than 1000 loses the game. Let x be the smallest initial number that results in a win for Amitabh. The sum of the digits of x is **[XAT-14]**

Sol.

Numbers with Amitabh	Numbers with Shashi	Round
$2x$	$2x + 50$	I
$2(2x + 50)$	$4x + 150$	II
$8x + 300$	$8x + 350$	III
$16x + 700$	$16x + 750$	IV
$32x + 1500$	$32x + 1550$	V
......	......	VI

It is seen from the above that in Vth round Amitabh has the number greater than 1000. It implies that he is losing the game. But, he wins, hence, the game is over in IVth round.

For winning, the number with Amitabh must be lesser than 1000.

Thus, $\quad 16x + 700 < 1000$

$\Rightarrow \quad 16x < 300 \Rightarrow x < 18.75$

Since, Shashi loses the game, he has the number greater than 1000.

Hence, $\quad 16x + 750 > 1000$

$\Rightarrow \quad 16x > 250 \Rightarrow x > 15.625$

From the above, x must be 16, 17 and 18.

Hence, sum of digits of smallest number $= 1 + 6 = 7$

Problem 90 A girl counted in the following way on the fingers of her left hand; she started by calling the thumb 1, the index finger 2, the middle finger 3, the ring finger 4, the little finger 5 and then reversed direction calling the ring finger 6, the middle finger 7 and so on. She counted upto 1994. She ended counting on which finger? *[MAT-08, CAT-93]*

Sol.

Thumb	Index	Middle	Ring	Little
1 →	2 →	3 →	4 →	5
9	← 8	← 7	← 6	
	10 →	11 →	12 →	13
17	← 16	← 15	← 14	
	18 →	19 →	20 →	21
	← 24	← 23	← 22	

The above counting pattern reveals that every multiple of 8 falls on index finger and moves towards thumb. Then, last multiple of 8 upto 1994 = 1992. Hence, 1992 will come on index finger, 1993 on thumb and 1994 on index finger.

Thus, the required answer is index finger.

Problem 91 A shop sells bags in three sizes: small, medium and large. A large bag costs ₹ 1000, a medium bag costs ₹ 200 and a small bag costs ₹ 50. Three buyers-Ashish, Banti and Chintu independently buy some numbers of these types of bags. The respective amounts spent by Ashish, Banti and Chintu are equal. Put together, the shop sells 1 large bag, 15 small bags and some medium bags of these three buyers. What is the minimum number of medium bags that the shop sells to them? *[XAT-20]*

Sol. Since any one of three persons buys one large bag, the minimum amount spent by each of them is ₹ 1000.

Then, total amount spent(minimum) = $3 \times 1000 = 3000$, out of which one large bag costing ₹ 1000 and 15 small bags costing ₹ 750 are bought. The remaining amount i.e. ₹ 1250 or more is to be spent on the purchase of medium bags.

Out of ₹ 1250, 6 medium bags can be purchased and obviously, ₹ 50 will be left. In order to purchase 7th medium bag, ₹ 150 more is required which can evenly be shared by each of them at the rate of ₹ 50. Hence, minimum number of medium bags the shop sells = 7. Here, each of them has spent ₹ 1050.

Problems Based on Fractions

Problem 92 The denominator of a fraction is 3 more than the numerator. If numerator as well as denominator is increased by 4, the fraction becomes 4/5. What was the original fraction ? *[SBI(PO)-99]*

Sol. Let the numerator be x.

Then, denominator $= x + 3$

Further, $\dfrac{x+4}{x+3+4} = \dfrac{4}{5}$

$\Rightarrow 5x + 20 = 4x + 28 \Rightarrow x = 8$

Hence, original fraction $= \dfrac{8}{8+3} = \dfrac{8}{11}$

Alternatively

The best way is to proceed through options. Let options be $\dfrac{5}{8}, \dfrac{8}{11}, \ldots$

(a) $\dfrac{5}{8} \Rightarrow 8$ is 3 more than 5. Further, $\dfrac{5+4}{8+4} \neq \dfrac{4}{5}$

(b) $\dfrac{8}{11} \Rightarrow 11$ is 3 more than 8. Further, $\dfrac{8+4}{11+4} = \dfrac{4}{5}$

Hence, option(b) is correct.

Problem 93 The numerator and denominator of a fraction are in ratio 2 : 3. If 6 is subtracted from the numerator, the result is a fraction that has a value 2/3 of the original fraction. The numerator of the original fraction is

Sol. Let the fraction be $\dfrac{2x}{3x}$.

As given, $\dfrac{2x-6}{3x} = \dfrac{2x}{3x} \times \dfrac{2}{3}$

$\Rightarrow 18x - 54 = 12x$

$\Rightarrow 6x = 54$

$\Rightarrow x = 9$

Hence, numerator $= 2x = 2 \cdot 9 = 18$

Alternatively

The best way is to proceed through options. Let options be 16, 18, ..

(a) $16 \Rightarrow$ fraction $= \dfrac{16}{24}$.

Further, $\dfrac{16-6}{24} \neq \dfrac{16}{24} \times \dfrac{2}{3}$

(b) $18 \Rightarrow$ fraction $= \dfrac{18}{27}$.

Further, $\dfrac{18-6}{27} = \dfrac{18}{27} \times \dfrac{2}{3}$

Hence, option(b) is correct.

Problem 94 If the numerator of a fraction is increased by 2 and denominator is increased by 3, the fraction becomes 7/9 and if both the numerator as well as denominator are decreased by 1, the fraction becomes 4/5. What is the original fraction ?

Sol. Let the fraction be $\dfrac{x}{y}$.

Then, $\dfrac{x+2}{y+3} = \dfrac{7}{9}$

$\Rightarrow 9x + 18 = 7y + 21$

$\Rightarrow 9x - 7y = 3 \quad \ldots(i)$

Further, $\dfrac{x-1}{y-1} = \dfrac{4}{5}$

$\Rightarrow 5x - 5 = 4y - 4$

$\Rightarrow 5x - 4y = 1 \quad \ldots(ii)$

From Eqs. (i) and (ii), we get $x = 5$ and $y = 6$

Hence, original fraction $= \dfrac{5}{6}$

Alternatively

The best way is to proceed through options. Let options be $\frac{9}{11}, \frac{5}{6}, \ldots$

(a) $\frac{9}{11} \Rightarrow \frac{9+2}{11+3} \neq \frac{7}{9}$ (b) $\frac{5}{6} \Rightarrow \frac{5+2}{6+3} = \frac{7}{9}$.

Further, $\frac{5-1}{6-1} = \frac{4}{5}$

Hence, option (b) is correct.

Problem 95 By how much does $\frac{6}{\frac{7}{8}}$ exceed $\frac{\frac{6}{7}}{8}$?

Sol. Given numbers are $6 \times \frac{8}{7}$ and $\frac{6}{7 \times 8}$ i.e., $6\frac{6}{7}$ and $\frac{3}{28}$

Then, required answer $= 6\frac{6}{7} - \frac{3}{28} = 6\frac{24}{28} - \frac{3}{28} = 6\frac{21}{28} = 6\frac{3}{4}$

Problem 96 If the difference between the reciprocal of a positive proper fraction and the fraction itself be 9/20, then the fraction is

Sol. Let the fraction be y/x [where $y < x$ as it is a proper fraction]

Then, $\frac{x}{y} - \frac{y}{x} = \frac{9}{20}$

$\Rightarrow \frac{x^2 - y^2}{xy} = \frac{9}{20}$

$\Rightarrow \frac{(x+y)(x-y)}{xy} = \frac{9}{20}$

$\Rightarrow \frac{(5+4)(5-4)}{5 \cdot 4} = \frac{9}{20}$

$\Rightarrow x = 5$ and $y = 4$

Hence, fraction $= \frac{4}{5}$

Alternatively

The best way is to proceed through options. Let options be $\frac{3}{5}, \frac{4}{5}, \ldots$

(a) $\frac{3}{5} \Rightarrow \frac{5}{3} - \frac{3}{5} \neq \frac{9}{20}$ (b) $\frac{4}{5} \Rightarrow \frac{5}{4} - \frac{4}{5} = \frac{9}{20}$

Hence, option(b) is correct.

Miscellaneous

Problem 97 If p and q represent digits, what is the possible maximum value of d in the statement $5p9 + 327 + 2d8 = 1114$? ***[SSC(CGL-T-1)-10]***

Sol.

```
 5 p 9
 3 2 7
 2 d 8
1 1 1 4
```

Clearly, $p + d$ must be 7 to obtain tens digit of the resultant as 1.

If $p = 0$, then d would be maximum.

Hence, possible maximum value of $d = 7 - 0 = 7$

Problem 98 If '∇' stands for the operation adding first number to twice the second number. Find the value of $(1 \nabla 2) \nabla 3$.

Sol. $(1 \nabla 2) = 1 + 2 \cdot 2 = 5$

Then, $(5 \nabla 3) = 5 + 2 \cdot 3 = 11$

Problem 99 A three-digit number $4a3$ is added to another three-digit number 984 to give a four-digit number $13b7$, which is divisible by 11. Find the value of $2a + b$.

Sol.

```
  4 a 3
  9 8 4
1 3 b 7
```

Clearly, if $a = 1$, then $b = 9$ and if $a = 2$, then $b = 0$ and 1 will be carried over the left, which has not been allowed here as given that $4 + 9 = 13$.

$\therefore$ $a = 1$ and $b = 9$ [$\because$ 1397 is also divisible by 11]

Now, $2a + b = 2 \cdot 1 + 9 = 11$

Problem 100 There are 6 friends and all have to have hand-shake with all without repetition. How many hand shakes possible?

Sol. When one person shakes hand with another, it implies that both shake hands at that point of time.

Hence, number of handshakes $= \frac{6 \times 5}{2} = 15$

Note *In such types of questions, you may directly put the value of number of persons in place of n in expression* $\frac{n(n-1)}{2}$.

Alternatively

Number of handshakes $= {}^nC_2 = {}^6C_2$

$= \frac{6 \times 5}{2} = 15$

[Concepts of combination have been detailed in chapter 'Permutation and Combination']

Problem 101 In a family there are some boys and girls. All girls told that they are having equal number of brothers and sisters and boys told that they are having twice the number of sisters than brothers. How many boys and girls present in that family?

Sol. Let number of girls be x and that of boys be y.

Each of girls is saying that she has equal number of brothers and sisters. It means, she is excluding herself while counting.

Then, $(x - 1) = y \Rightarrow x - y = 1$ ...(i)

Further, each of boys (excluding himself) is saying that he has twice the number of sisters than brothers.

Then, $2(y - 1) = x$

$\Rightarrow x - 2y = -2$ ...(ii)

From Eqs. (i) and (ii), we get $y = 3$ and $x = 4$

Alternatively

The best way is to proceed through options. Let options be:

(a) 3 girls and 4 boys

(b) 4 girls and 3 boys

(c) 2 boys and 5 girls

(d) 5 boys and 2 girls

(a) 3 girls and 4 boys $\Rightarrow (3-1) \neq 4$

(b) 4 girls and 3 boys $\Rightarrow (4-1) = 3$ and $2(3-1) = 4$

Hence, option(b) is correct.

Problem 102 In a group of 26 persons, a total of 14 take tea while 10 take tea but not coffee. How many persons in this group take coffee but not tea. *[SSC(CGL-T-I)-12, CAT-18]*

Sol. Persons who take tea = 14

Persons who take tea not coffee = 10 i.e. persons who take only tea

It means, (14 – 10) persons i.e. 4 take both tea and coffee.

Now, persons who take coffee but not tea i.e. who take only coffee = 26 – 10 – 4 = 12

Problem 103 400 students were admitted to the 2018-19 MBA batch. 200 of them did not choose 'Business Statistics'. 100 of them did not choose 'International Management'. There were 80 students Who did not choose any of the two subjects. Find the number of students who choose both Business Statistics and International Management. *[IIFT-19]*

Sol. Clearly, $B + C + M + n = 400$

$\Rightarrow \quad B + C = 400 - 200 = 200 \quad [\because M + n = 200 \text{ (given)}]$

Also, $B + n = 100$ (given)

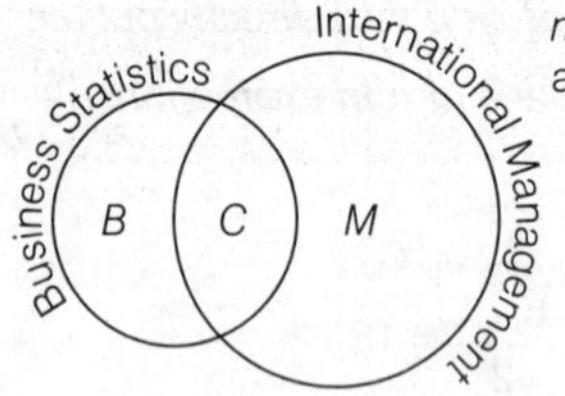

$n = 80$, who did not choose any of the two subjects

Now, $B = 100 - n = 100 - 80 = 20$

Then, $C = 200 - B = 200 - 20a = 180$

Hence, number of students who chose both subjects = 180.

Note *The concepts for such types of questions are detailed in chapter 'Percentage' under head 'Problems based on Examination'.*

Problem 104 In an examination a student scores 4 marks for every correct answer and losses 1 mark for every wrong answer. If he attempts all 75 questions and scores 125 marks, the number of questions he attempts correctly is? *[SSC (CGL-T-I)-11]*

Sol. Let the number of correct answer be x.

As given, $(x \times 4) - (75 - x) \times 1 = 125$

$\Rightarrow \quad 4x - 75 + x = 125$

$\Rightarrow \quad 5x = 200$

$\therefore \quad x = 40$

Alternatively

Proceed through options. Let options be 35, 40, ...

(a) $35 \Rightarrow 35 \cdot 4 - (75 - 35) \cdot 1 \neq 125$

(b) $40 \Rightarrow 40 \cdot 4 - (75 - 40) \cdot 1 = 125$

Hence, option(b) is correct.

Problem 105 A positive integer, which when added to 1000, gives a sum which is greater than when it is multiplied by 1000. This positive integer is?

Sol. Clearly, $(1000 + 1) > 1 \times 1000$

Hence, required positive integer = 1

Problem 106 $\frac{5}{9}$th of a number is equal to 25% of the second number. The second number is equal to $\frac{1}{4}$th of the third number. The value of third number is 2960. What is the 30% of the first number? *[IOB(PO)-11]*

Sol. Given, 3rd number = 2960

$$\text{2nd number} = 2960 \times \frac{1}{4} = 740$$

Also given, $\frac{5}{9}$ of 1st number = 25% of the 2nd number

$\Rightarrow \frac{5}{9}$ of 1st number = 25% of 740

$\Rightarrow \frac{5}{9}$ of 1st number = 185

$\Rightarrow$ 1st number $= 185 \times \frac{9}{5}$

$\Rightarrow$ 30% of 1st number $= 185 \times \frac{9}{5} \times \frac{30}{100}$

$= 99.9$

Problem 107 Thrice the square of a natural number decreased by 4 times the number is equal to 50 more than the number. The number is

Sol. Let the number be x.

Then, $3x^2 - 4x = x + 50$

$\Rightarrow \quad 3x^2 - 5x - 50 = 0$

$\Rightarrow \quad 3x^2 - 15x + 10x - 50 = 0$

$\Rightarrow \quad 3x(x - 5) + 10(x - 5) = 0$

$\Rightarrow \quad (x - 5)(3x + 10) = 0$

$\Rightarrow \quad x = 5$ and $\frac{-10}{3}$

Since, $\frac{-10}{3}$ is not a natural number, the required number = 5

Alternatively

The best way is to proceed through options. Let options be 4, 5,...

(a) $4 \Rightarrow 3 \cdot 4^2 - 4 \cdot 4 \neq 4 + 50$

(b) $5 \Rightarrow 3 \cdot 5^2 - 4 \cdot 5 = 5 + 50$

Hence, option (b) is correct.

WORKOUT

1. Two numbers 297_B and 792_B, belong to base B number system. If the first number is a factor of the second number, then the value of B is ***[XAT-14]***

(a) 11 (b) 12 (c) 15 (d) 17
(e) 19

2. A person gives half the lemons what he has, to another person but takes one from what he gives. Further he repeats the same with 30 persons more and at last he has 2 lemons. How many lemons did he have initially.

(a) 2 (b) 90
(c) 240 (d) 61

3. A number when divided successively by 4 and 5 leaves remainders 1 and 4 respectively, when it is successively divided by 5 and 4, then the respective remainders will be

(a) 1, 2 (b) 2, 3
(c) 3, 2 (d) 4, 1

4. On dividing a number by 9, the remainder is 8. The quotient so obtained when divided by 11, leaves the remainder 9. Now, the quotient so obtained when divided by 13, leaves the remainder 8. Find the remainder when the given number is divided by 1287.

(a) 752 (b) 229 (c) 881 (d) 781

5. The product of two natural numbers is 17. Then, the sum of the reciprocals of their squares is

(a) $\frac{1}{289}$ (b) $\frac{289}{290}$
(c) $\frac{290}{289}$ (d) 289

6. The smallest value of n, for which $2n + 1$ is not a prime number, is

(a) 3 (b) 4
(c) 5 (d) None of these

[Hotel Management - 03]

7. The multiplication of a three-digit number $XY5$ with digit Z yields $X215$. What is $X + Y + Z$ equal to ***[CDS-14]***

(a) 13 (b) 15 (c) 17 (d) 18

8. A man has a certain number of mangoes. He sells half of what he had and one more to A, half the remainder and one more to B, half the remainder and one more to C, half the remainder and one more to D, by which time he has only one left. Find how many he had at first?

(a) 76 (b) 46
(c) 85 (d) 72

9. The sum of two numbers, one of which is one-third of the other is 36. The smaller number is

(a) 6 (b) 7
(c) 8 (d) 9

10. How many whole numbers between 100 and 800 contain the digit 2 ? ***[XAT-13]***

(a) 200 (b) 214
(c) 220 (d) 240
(e) 248

11. If such numbers which are divisible by 5 and also those which have 5 as one of the digits are eliminated from the numbers 1 to 60, how many numbers would remain ?

(a) 40 (b) 47
(c) 53 (d) 45

12. A certain number of two digits is three times the sum of its digits. If 45 is added to the number, then the digits will be reversed. What is the sum of the squares of the two digits of the number ? ***[CDS-14]***

(a) 41 (b) 53
(c) 45 (d) 64

13. Three prime numbers are such that the product of first two numbers is 551 and that of last two numbers is 1073. Find their sum.

(a) 75 (b) 81
(c) 85 (d) 89

14. The students present in an auditorium are asked to stand in rows. If there were 4 students extra in a row there would be 4 less rows. However, if 2 students were less in a row, there would be 4 more rows. The number of students present in the auditorium is: ***[MAT-10]***

(a) 80 (b) 96
(c) 100 (d) 128

15. $(BE)^2 = MPB$, where B, E, M and P are distinct integers, then M = ? ***[CAT-98]***

(a) 2 (b) 3
(c) 9 (d) None of these

16. If the numbers q, $q + 2$ and $q + 6$ are all prime, then what can be the value of $3q + 9$?

(a) Only 18 (b) Only 42
(c) Only 60 (d) Both (b) and (c)

17. The sum of the place values of 3 in the number 503535 is

(a) 3300 (b) 0.6
(c) 60 (d) 3030

18. Find the smallest number which, when multiplied by 9, gives the product as 1 followed by a certain number of 7's only.

(a) 16428 (b) 19753
(c) 17777 (d) None of these

19. For a positive integer x, define $f(x)$ such that $f(x + a) = f(x \times a)$, where a is an integer and $f(1) = 4$. If the value of $f(1003) = k$, then the value of k will be ***[XAT-15]***

(a) 1003 (b) 1004
(c) 1005 (d) 1006
(e) None of these

20. What is the value of x for which x, $(x + 1)$ and $(x + 3)$ are all prime numbers ? ***[CDS-10]***

(a) 1 (b) 2
(c) 3 (d) None of these

21. ₹ 6500 were divided equally among a certain number of persons. Had there been 15 more persons each would have got ₹ 30 less. Find the original number of persons ?

(a) 45 (b) 50
(c) 55 (d) 48

22. In the board meeting of a FMCG company, everybody present in the meeting shakes hand with everybody else. If the total number of handshakes is 78, the number of members who attended the board meeting is? *[IIFT-15]*

(a) 7 (b) 9
(c) 11 (d) 13
(e) None of these

23. A number is interesting, if on adding the sum of the digits of the number and the product of the digits of the number, the result is equal to the number. What fraction of numbers between 10 and 100 (both 10 and 100 included) is interesting? *[XAT-13]*

(a) 0.1 (b) 0.11
(c) 0.16 (d) 0.22
(e) None of these

24. In a number system, the product of 44 and 11 is 3414. The number 3111 of this system, when converted to the decimal number system becomes *[CAT-01]*

(a) 406 (b) 1086
(c) 213 (d) 691

25. Consider the following statements: *[CDS-13]*

I. To obtain prime numbers less than 121, we are to reject all the multiplies of 2, 3, 5 and 7.

II. Every composite number less than 121 is divisible by a prime number less than 11.

Which of the statements given above is/are correct ?

(a) Only I
(b) Only II
(c) Both I and II
(d) Neither I nor II

26. Two times a two-digit number is 9 times the number obtained by reversing the digits and sum of the digits is 9. The number is *[MAT-10]*

(a) 72 (b) 81
(c) 63 (d) 54

27. The sum of all the digits of the numbers from 1 to 100 is *[SSC(CGL-T-II)-10]*

(a) 5050 (b) 903
(c) 901 (d) 900

28. Suppose, the seed of any positive integer n is defined as follows
Seed(n) = n, if $n < 10$
= seed ($s(n)$), otherwise
where, $s(n)$ indicates the sum of digits of n, for example,
seed(7) = 7, seed(248) = seed (2 + 4 + 8) = seed(14) = seed (1 + 4) = seed(5) = 5 etc.

How many positive integers n, such that $n < 500$, will have seed (n) = 9? *[CAT-08]*

(a) 39 (b) 72
(c) 81 (d) 108
(e) 55

29. If r and s are any real numbers such that $0 \le s \le 1$ and $r + s = 1$, then what is the maximum value of the product ?

(a) 1 (b) $\frac{3}{4}$
(c) $\frac{1}{2}$ (d) $\frac{1}{4}$

30. If the difference of the squares of two natural numbers is 37, then the sum of the squares of the numbers will be *[SSC(SAS)-10]*

(a) 690 (b) 687
(c) 685 (d) 672

31. Z is the product of first 31 natural numbers . If $X = Z + 1$, then the numbers of primes among $X + 1$, $X + 2$, ... $X + 29$, $X + 30$ is *[IIFT-12]*

(a) 30
(b) 2
(c) Cannot be determined
(d) None of the above

32. Given that $-1 \le v \le 1$, $-2 \le u \le -0.5$ and $-2 \le z \le -0.5$, and $w = \frac{vz}{u}$, then which of the following is necessarily true ? *[CAT-03]*

(a) $-0.5 \le w \le 2$ (b) $-4 \le w \le 4$
(c) $-4 \le w \le 2$ (d) $-2 \le w \le -0.5$

33. Each boy contributed rupees equal to the number of girls and each girl contributed rupees equal to the number of boys in a class of 60 students. If the total amount thus collected is ₹ 1600, how many boys are there in the class? *[LIC(AAO)-10]*

(a) 30 (b) 25 (c) 50
(d) Data inadequate

34. The smallest positive prime(say p) such that $2^p - 1$ is not a prime, is *[CDS-13]*

(a) 5 (b) 11
(c) 17 (d) 29

35. At the first stop on his route, a driver unloaded 2/5 of the packages in his van. After he unloaded another three packages at his next stop, ½ of the original number of packages remained. How many packages were in the van before the first delivery ?

(a) 25
(b) 30
(c) 10
(d) 36

36. If $f(x) = ax + b$, a and b are positive real numbers and $f(f(x)) = 9x + 8$, then value of $a + b$ is *[XAT-17]*

(a) 3 (b) 4
(c) 5 (d) 6
(e) None of these

37. A group of 630 children is arranged in rows for a group photograph session. Each row contains three fewer children than the row in front of it. What number of rows is not possible ? *[CAT-06]*

(a) 3 (b) 4
(c) 5 (d) 6
(e) 7

38. A General, while arranging his men, who were 6000 in number, in the form of a square, found that there were 71 men left over. How many were arranged in each row ? *[SSC(CGL-T-II)-10]*

(a) 73 (b) 77
(c) 87 (d) 93

39. The product of two alternate odd integers exceeds three times the smaller by 12. What is the larger number ?

(a) 3 (b) 5
(c) 7 (d) 9

40. If x is a positive number, then which of the following fractions has the greatest value ?

(a) $\frac{x}{x}$ (b) $\frac{x+1}{x}$
(c) $\frac{x}{x+1}$ (d) $\frac{x+2}{x+3}$

41. The prime number 1999 can be written as $a^2 - b^2$, where a and b are natural numbers. Then, the value of $a^2 + b^2$ is **[LIC(AAO)-11]**
(a) 1998000 (b) 1998001
(c) 1999000 (d) 1999001

42. The sum of seven consecutive even numbers of set A is 378. What is the sum of a different set of four consecutive numbers whose lowest number is 23 less than the mean of set A? **Corporation Bank(PO)-11**
(a) 132 (b) 146
(c) 156 (d) 124
(e) None of these

43. The digits of a three-digit number A are written in the reverse order to form another three digit number B. If $B > A$ and $B - A$ is perfectly divisible by 7, then which of the following is necessarily true? **[CAT-05]**
(a) $100 < A < 299$ (b) $106 < A < 305$
(c) $112 < A < 311$ (d) $118 < A < 317$

44. Consider the following statements: If p is a prime such that $p + 2$ is also a prime, then **CDS-11**
I. $p(p+2)+1$ is a perfect square.
II. 12 is a divisor of $p + (p+2)$, if $p > 3$.
Which of the above statements given is/are correct?
(a) Only I (b) Only II
(c) Both I and II (d) Neither I nor II

45. A man has 1044 candles. After burning, he can make a new candle from 9 stubs left behind. Find the maximum number of candles that can be made.
(a) 116 (b) 120
(c) 130 (d) 140

46. If k is an integer and 0.0010101×10^k is greater than 1000, what is the least possible value of k? **[IIFT-12]**
(a) 4 (b) 5
(c) 6 (d) 7

47. $11001_2 - 1011_2 = ?$
(a) 1110_2 (b) 1101_2
(c) 1100_2 (d) 110_2

48. The sum of a number and its reciprocal is thrice the difference of the number and its reciprocal. The number is **[SNAP-11]**
(a) $\pm\sqrt{2}$ (b) $\pm\frac{1}{\sqrt{2}}$
(c) $\pm\frac{1}{\sqrt{3}}$ (d) $\pm\sqrt{3}$

49. A test has 50 questions. A student scores 1 mark for a correct answer, –1/3 for a wrong answer and –1/6 for not attempting a question. If the net score of a student is 32, the number of questions answered wrongly by that student cannot be less than **[CAT-03]**
(a) 6 (b) 12
(c) 3 (d) 9

50. If m and n are natural numbers, then $\sqrt[m]{n}$ is **[CDS-13]**
(a) always irrational
(b) irrational unless n is the mth power of an integer
(c) irrational unless m is the nth power of an integer
(d) irrational unless m and n are co-prime

51. How many odd numbered pages are there in a book of 1089 pages?
(a) 545 (b) 544
(c) 546 (d) 542

52. a, b, c are integers $|a| \neq |b| \neq |c|$ and $-10 \le a, b, c \le 10$. What will be the maximum possible value of $[abc - (a+b+c)]$? **[XAT-16]**
(a) 524 (b) 693
(c) 731 (d) 970
(e) None of these

53. Let x and y be positive integers such that x is prime and y is composite. Then, **[CAT-03]**
(a) $y - x$ cannot be an even number
(b) xy cannot be an even integer
(c) $\frac{x+y}{x}$ cannot be an even integer
(d) None of the above statements is true

54. The number zero (0) is surrounded by the same 2-digit number on both (left and right) the sides; for example, 25025, 67067, etc. The largest number that always divides such a number is
(a) 7 (b) 11
(c) 13 (d) 1001

55. In a group of buffaloes and ducks, the number of legs are 24 more than twice the number of heads. What is the number of buffaloes in the group?
(a) 6 (b) 12
(c) 8 (d) None of these

56. If k is a positive integer, then every square integer is of the form **[CDS-13]**
(a) Only $4k$ (b) $4k$ or $4k+3$
(c) $4k+1$ or $4k+3$ (d) $4k$ or $4k+1$

57. The sum of the possible values of X in the equation $|X+7| + |X-8| = 16$ is **[XAT-14]**
(a) 0 (b) 1
(c) 2 (d) 3
(e) None of these

58. If N, $(N+2)$ and $(N+4)$ are prime numbers, then the number of possible solutions for N are
(a) 1 (b) 2
(c) 3 (d) None of these

59. An Army Commander wishing to draw up his 5180 men in the form of a solid square found that he had 4 men less. If he could get four more men and form the solid square, the number of men in the front row is?
(a) 72 (b) 68
(c) 78 (d) 82

60. For a positive integer n, let P_n denote the product of the digits of n and S_n denote the sum of digits of n. The number of integers between 10 and 1000 for which $P_n + S_n = n$ is **[CAT-05]**
(a) 81 (b) 16
(c) 18 (d) 9

61. $475_8 = ?_6$
(a) 10135_6 (b) 1245_6
(c) 5425_6 (d) 3215_6

62. The pair of rational numbers that lies between 1/4 and 3/4 is
(a) $\frac{262}{1000}, \frac{752}{1000}$ (b) $\frac{24}{100}, \frac{74}{100}$
(c) $\frac{9}{40}, \frac{31}{40}$ (d) $\frac{252}{1000}, \frac{748}{1000}$

63. Three friends divided some bullets equally. After all of them shot 4 bullets, the total number of bullets remaining is equal to the bullets each had after division. Find the original number of bullets?
(a) 15 (b) 18
(c) 20 (d) 17

64. Let T be the set of integers {3, 11, 19, 27, ...451, 459, 467} and S be a subset of T such that the sum of no two elements of S is 470. The maximum possible number of elements in S is

(a) 32 (b) 28
(c) 29 (d) 30 **[CAT-03]**

65. If a number of two digits is k times the sum of its digits, then the number formed by interchanging the digits is the sum of the digits multiplied by **[LIC(AAO)-10]**

(a) $9 + k$ (b) $10 + k$
(c) $11 - k$ (d) $k - 1$

66. $11101010_2 = (......)_{16}$

(a) EA_{16} (b) CA_{16}
(c) $1A_{16}$ (d) $E3_{16}$

67. A sequence of 4 digits, when considered as a number in base 10 is four times the number it represents in base 6. What is the sum of the digits of the sequence ?

(a) 9 (b) 8
(c) 6 (d) 7

68. The two digit number, which when divided by sum of the digits and product of the digits respectively, the remainder is same and the difference of quotients is one, is

(a) 14 (b) 32
(c) 23 (d) 41

69. One of a group of swans, 7/2 times the square root of the number are playing on the shore of the pond. The two remaining are inside the pond. What is the total number of swans ?

(a) 10 (b) 16 (c) 12 (d) 14

70. The number of employees in Obelix Menhir company is a prime number and is less than 300. The ratio of the number of employees who are graduates and above, to that of employees who are not, can possibly be **[CAT-06]**

(a) 101 : 88 (b) 87 : 100
(c) 110 : 111 (d) 85 : 98
(e) 97 : 84

71. If the product of the integers a, b, c and d is 3094 and if $1 < a < b < c < d$, then what is the product of b and c? **[IIFT-13]**

(a) 26 (b) 91
(c) 133 (d) 221

72. If the product of n positive real numbers is unity, then their sum is necessarily **[CAT-03]**

(a) a multiple of n
(b) equal to $n + 1/n$
(c) never less than n
(d) a positive integer

73. A three-digit number is divisible by 11 and has its digit in the unit's place equal to 1. The number is 297 more than the number obtained by reversing the digits. What is the number? **[CDS-10]**

(a) 121 (b) 231
(c) 451 (d) 561

74. If a, b and c are distinct natural numbers less than 25, what is the maximum possible value of $|a-b| + |b-c| - |c-a|$?

(a) 44 (b) 46
(c) 23 (d) 21

75. One-fourth of a herd of camels was seen in the forest. Twice the square root of the herd had gone to mountains and the remaining 15 camels were seen on the bank of the river. Find the total number of camels?

(a) 32 (b) 36
(c) 35 (d) 38

76. p, q and r are three non-negative integers such that $p + q + r = 10$. The maximum value of $pq + qr + pr + pqr$ is **[XAT-13]**

(a) ≥ 40 and < 50
(b) ≥ 50 and < 60
(c) ≥ 60 and < 70
(d) ≥ 70 and < 80
(e) ≥ 80 and < 90

77. If $|b| \ge 1$ and $x = -|a|b$, then which one of the following is necessarily true? **[CAT-03]**

(a) $a - xb < 0$ (b) $a - xb \ge 0$
(c) $a - xb > 0$ (d) $a - xb \le 0$

78. If sum of an integer and its reciprocal is 10/3, then the integer is equal to **[CDS-12]**

(a) 3 (b) 6
(c) 12 (d) 13

79. Which of the following integers is the square of an integer of every integer n?

(a) $n^2 + 1$ (b) $n^2 + n$
(c) $n^2 + 2n$ (d) $n^2 + 2n + 1$

80. A real number x satisfying $1 - \frac{1}{n} < x \le 3 + \frac{1}{n}$, for every positive integer , is best described by **[CAT-03]**

(a) $1 < x < 4$ (b) $1 < x \le 3$
(c) $0 < x \le 4$ (d) $1 \le x \le 3$

81. Which one of the following numbers is not a square of any natural number?

(a) 5041 (b) 9852
(c) 1936 (d) 6241

82. If a, $a + 2$ and $a + 4$ are prime numbers, then the number of possible solutions for a is **[CAT-03]**

(a) one
(b) two
(c) three
(d) more than three

83. What is the least fraction which, when added to or subtracted from $\frac{29}{12} + \frac{15}{16}$, will make the result a whole number?

(a) $\frac{21}{38}$ (b) $\frac{31}{38}$
(c) $\frac{31}{48}$ (d) $\frac{17}{48}$

84. If a, b, c and d are four different positive integers selected from 1 to 25, then the highest possible value of $[(a + b) + (c + d)] / [(a + b) + (c - d)]$ would be **[XAT-15]**

(a) 47 (b) 49
(c) 51 (d) 96
(e) None of these

85. When you reverse the digits of the number 13, the number increases by 18. How many other two-digit numbers increase by 18 when their digits are reversed ? **[CAT-06]**

(a) 5 (b) 6
(c) 7 (d) 8
(e) 10

86. What is $26^2 + 97^2$ equal to

(a) $27^2 + 93^2$ (b) $34^2 + 93^2$
(c) $82^2 + 41^2$ (d) $79^2 + 62^2$

87. If x is positive even integer and y is negative odd integer, then x^y is **[CDS-13]**

(a) odd integer
(b) even integer
(c) rational number
(d) None of the above

88. The sum of four consecutive two-digit odd numbers, when divided by 10, becomes a perfect square. Which of the following can possibly be one of these four numbers? **[CAT-06]**

(a) 21 (b) 25 (c) 41 (d) 67
(e) 73

89. If the numerator and the denominator of a proper fraction are increased by the same quantity, then the resulting fraction is

(a) always greater than the original fraction
(b) always less than the original fraction
(c) always equal to the original fraction
(d) None of the above

90. Let a, b, c be distinct digits. Consider a two-digit number 'ab' and a three-digit number 'ccb', both defined under the usual decimal number system, if $(ab)^2 = ccb > 300$, then the value of b is **[CAT-99]**

(a) 1 (b) 0 (c) 5 (d) 6

91. If the product of three consecutive positive integers is 15600 then the sum of squares of these integers is **[CAT-17]**

(a) 1777 (b) 1785 (c) 1875 (d) 1877

92. How many numbers are there between 500 and 600 in which 9 occurs only once ? **[MAT-09]**

(a) 19 (b) 18 (c) 20 (d) 21

93. Which one of the following is neither prime not composite number ? **[CDS-12]**

(a) 1 (b) 2
(c) 3 (d) None of these

94. For every positive real number $\left[\frac{x}{2}\right] + \left[\frac{x+1}{2}\right] = \ldots\ldots$

Where [ ] is the greatest integer function.

(a) 0 (b) 1
(c) $[x + 1]$ (d) $[x]$

95. Consider four digit numbers for which first two digits are equal and the last two digits are also equal. How many such numbers are perfect squares ? **[CAT-07]**

(a) 1 (b) 3
(c) 2 (d) 4
(e) 0

96. The values of the numbers 2^{2004} and 5^{2004} are written one after another. How many digits are there in all? **[CAT-11]**

(a) 4008 (b) 2003
(c) 2004 (d) None of these

97. A printer numbers the pages of a book starting with 1 and uses 3189 digits in all. How many pages does the book have ?

(a) 1074 (b) 1075
(c) 1000 (d) None of these

98. A positive whole number M less than 100 is represented in base 2 notation, base 3 notation and base 5 notation. It is found that in all three cases the last digit is 1, while in exactly two out of the three cases the leading digit is 1. Then M equals **[CAT-03]**

(a) 31 (b) 63 (c) 75 (d) 91

99. In a company, each employee gives a gift to every other employee. If the number of gifts is 132, then the number of employees in the company is

(a) 11 (b) 13 (c) 12 (d) 8

100. A rich merchant had collected many gold coins. He did not want anybody to know about him. One day, his wife asked, "How many gold coins do we have?" After pausing a moment, he replied, "Well ! if we divide the coins into two unequal numbers, then 48 times the difference between the two numbers equals the difference between the squares of the two numbers". The wife looked puzzled. Can you help the merchant's wife by finding out how many gold coins the merchant has ? **[CAT-02]**

(a) 96 (b) 53
(c) 43 (d) None of these

101. Suppose n is an integer such that the sum of the digits of n is 2 and $10^{10} < n < 10^{11}$. The number of different values of n is **[CAT-14]**

(a) 11 (b) 10
(c) 9 (d) 8

102. P, Q and R are three consecutive odd numbers in ascending order. If the value of three times P is 3 less than two times R, find the value of R. **[CAT-97]**

(a) 5 (b) 7
(c) 9 (d) 11

103. How many 100-digit positive numbers are there? **[SSC(CGL-T-II)-17]**

(a) 9×10^{99} (b) 9×10^{100}
(c) 10100 (d) 11×10^{98}

104. If $a_1 = 1$ and $a_{n+1} - 3a_n + 2 = 4n$ for every positive integer, then a_{100} equals **[CAT-05]**

(a) $3^{99} - 200$ (b) $3^{99} + 200$
(c) $3^{100} - 200$ (d) $3^{100} + 200$

105. In a 4-digit number, the sum of the first two digits is equal to that of the last digits. The sum of the first and last digits is equal to the third digit. Finally, the sum of the second and fourth digits is twice the sum of the other two digits. What is the third digit of the number? **[CAT-01]**

(a) 5 (b) 8
(c) 1 (d) 4

106. How many pairs of positive integers m, n satisfy $\frac{1}{m} + \frac{4}{n} = \frac{1}{12}$, where n is an odd integer less than 60 ? **[CAT-07]**

(a) 3 (b) 6
(c) 4 (d) 7
(e) 5

107. A three-digit number is eleven times the two-digit number formed by using the hundred's and the unit's digit of the three-digit number respectively, in the ten's and unit's place of the two-digit number. If the difference between the digit in ten's place and the digit in hundred's place is 1, then what is the digit in the unit's place ?

(a) 2 (b) 3
(c) 1 (d) 4

108. The owner of a local jewellery store hired 3 watchmen to guard his diamond, but a thief still got in and stole some diamonds. On the way out, the thief met each watchman, one at a time. To each he gave 1/2 of the diamonds he had then and 2 more besides. He escaped with one diamond. How many did he steal originally? **[CAT-02]**

(a) 40
(b) 36
(c) 25
(d) None of the above

109. In a tournament, there are n teams $T_1, T_2, \ldots\ldots T_n$ with $n > 5$. Each team consists of k players, $k > 3$. The following pairs of teams have one player in common:

T_1 and T_2, T_2 and T_3, ... , T_{n-1} and T_n & T_n and T_1.

No other pairs of teams has any player in common. How many players are participating in the tournament considering all the n teams together ? **[CAT-07]**

(a) $(n - 1)(k - 1)$
(b) $n(k - 1)$
(c) $k(n - 1)$
(d) $n(k - 2)$
(e) $k(n - 2)$

110. If $N = (11^{p+7})(7^{q-2})(5^{r+1})(3^s)$ is a perfect cube, where p, q, r and s are positive integers, then the smallest value of $p + q + r + s$ is **[XAT-17]**

(a) 5 (b) 6
(c) 7 (d) 8
(e) 9

111. A calculator has two memory buttons- P and Q. Value 1 is initially stored in both memory locations. The following sequence of steps is carried out five times:

I. Add 1 to Q
II. Multiply P and Q
III. Store the result in P

What is the value stored in memory location P after this process ?

(a) 120 (b) 450
(c) 600 (d) 720

112. Arun can climb a coconut tree by 1.5 feet by each lift; however he slips 0.5 feet every time he makes the next lift. How many individual lifts he will have to reach the top of the coconut tree of 18.5 feet ? **[CLAT-11]**

(a) 20 (b) 19 (c) 18 (d) 17

113. In a six-digit number, the sixth, that is the rightmost digit, is the sum of the first three digits, the fifth digit is the sum of first two digit, the third digit is equal to the first digit, the second digit is twice the first digit and the fourth digit is the sum of fifth and sixth digits. Then, the largest possible value of the fourth digit is **[CAT-19]**

(a) 8 (b) 7 (c) 6 (d) 9

114. A prime number contains the digit X at unit's place. How many such digits of X are possible? **[CDS-19]**

(a) 3 (b) 4 (c) 5 (d) 6

115. The sum of three prime numbers is 100. If one of them exceeds another by 36, then one of the numbers is

(a) 17 (b) 29
(c) 43 (d) None of these

116. The product of two positive number is 616. If the ratio of the difference of their cubes to the cube of their difference is 157 : 3, then the sum of the two numbers is: **[CAT-19]**

(a) 58 (b) 50
(c) 95 (d) 85

117. Garima had only ₹ 200, ₹ 500 and ₹ 2000 notes in her wallet. She goes to shoppers stop and purchases some dresses and gives half of her ₹ 2000 notes and in turn receives same number of ₹ 200 notes. She then goes to a restaurant and gives all her ₹ 500 notes and receives thirty ₹ 2000 notes, which increases the number of ₹ 2000 notes, she had, by 75%. If now she has fifty ₹ 200 notes, what were the original number of ₹ 2000 and ₹ 200 notes she had at the start? **[IIFT-18]**

(a) 60, 10
(b) 60,15
(c) 80,10
(d) 80,15

118. A box contains 6 cricket balls, 5 tennis balls and 4 rubber balls. Of these, some balls are defective. The proportion of defective cricket balls is more than the proportion of defective tennis balls but less than the proportion of defective rubber balls. Moreover, the overall proportion of defective balls is twice the proportion of defective tennis balls. What best can be said about the number of defective rubber balls in the box ? **[XAT-20]**

(a) It is either 2 or 3
(b) It is exactly 2
(c) It is either 3 or 4
(d) It is exactly 3
(e) It is either 0 or 1.

119. How many three digit numbers are greater than 100 and increase by 198 when the three digits are arranged in the reverse order? **[CAT-21]**

(a) 70 (b) 80
(c) 90 (d) 95

Answer Key

1. (e)	2. (a)	3. (b)	4. (c)	5. (c)	6. (b)	7. (a)	8. (b)	9. (d)	10. (b)
11. (a)	12. (b)	13. (c)	14. (b)	15. (b)	16. (d)	17. (d)	18. (b)	19. (e)	20. (b)
21. (b)	22. (d)	23. (e)	24. (a)	25. (c)	26. (b)	27. (c)	28. (e)	29. (d)	30. (c)
31. (d)	32. (b)	33. (d)	34. (b)	35. (b)	36. (c)	37. (d)	38. (b)	39. (c)	40. (b)
41. (b)	42. (e)	43. (b)	44. (c)	45. (c)	46. (c)	47. (a)	48. (a)	49. (c)	50. (b)
51. (a)	52. (c)	53. (d)	54. (d)	55. (b)	56. (d)	57. (b)	58. (a)	59. (a)	60. (d)
61. (b)	62. (d)	63. (b)	64. (d)	65. (c)	66. (a)	67. (b)	68. (b)	69. (b)	70. (e)
71. (b)	72. (c)	73. (c)	74. (a)	75. (b)	76. (c)	77. (b)	78. (a)	79. (d)	80. (c)
81. (b)	82. (a)	83. (d)	84. (c)	85. (b)	86. (d)	87. (c)	88. (c)	89. (a)	90. (a)
91. (d)	92. (b)	93. (a)	94. (d)	95. (a)	96. (d)	97. (a)	98. (d)	99. (c)	100. (d)
101. (a)	102. (c)	103. (a)	104. (c)	105. (a)	106. (a)	107. (c)	108. (b)	109. (b)	110. (e)
111. (d)	112. (c)	113. (b)	114. (d)	115. (d)	116. (b)	117. (c)	118. (d)	119. (a)	

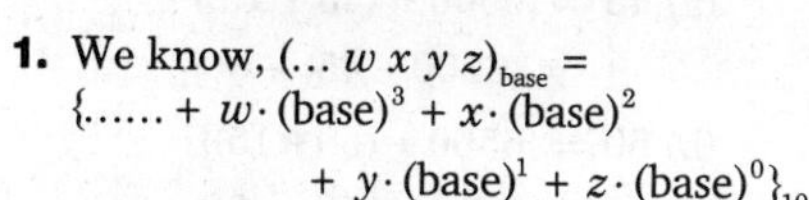

Hints and SOLUTIONS

1. We know, $(\ldots w\,x\,y\,z)_{base} = \{\ldots\ldots + w\cdot(\text{base})^3 + x\cdot(\text{base})^2 + y\cdot(\text{base})^1 + z\cdot(\text{base})^0\}_{10}$

Here,

$297_B = \{2\cdot(B)^2 + 9\cdot(B)^1 + 7\cdot(B)^0\}_{10}$
$= 2B^2 + 9B + 2$
$= (B+1)(2B+7)$

and

$792_B = \{7\cdot(B)^2 + 9\cdot(B)^1 + 2\cdot(B)^0\}_{10}$
$= 7B^2 + 9B + 2$
$= (B+1)(7B+2)$

Since the first number is a factor of the second number, $\{(7B+2) \div (2B+7)\}$ must be an integer and it is possible at $B = 19$ considering the given options.

Alternatively, you may proceed through options. Convert the given numbers in decimal system putting the value of base from the options.

2. Given, number of lemons left with him in end = 2

Then, number of lemons left before giving to 31st person
$= (2-1)\cdot 2 = 2$

Then, number of lemons left before giving to 30th person
$= (2-1)\cdot 2 = 2$

Then, number of lemons left before giving to 29th person
$= (2-1)\cdot 2 = 2$ and so on.

Then, number of lemons left before giving to 1st person $(2-1)\cdot 2 = 2$

Hence, number of lemons he had initially = 2

Alternatively, you may proceed through options.

3.
4 ↖ × 5
+↓ ↘ ↓
1 4

Then, required dividend
$= 4\times 4 + 1 = 17$

Now, divide 17 by 5 and 4 successively.

5	17	Remainders (R)
4	3	$2 \Rightarrow R_1$
	0	$3 \Rightarrow R_2$

Hence, required remainders are 2 and 3 respectively.

4.
9 ↖ × 11 ↖ × 13 (Divisors)
+↓ ↘ +↓ ↘
8 9 8 (Remainders)
(881) (97)

The least dividend
$= (11\times 8 + 9)\times 9 + 8$
$= 881$

Obviously, we will get 881 as remainder if 881 is divided by 1287.

5. $17 = 1\times 17$

It implies that numbers are 1 and 17.

Now, $\frac{1}{1^2} + \frac{1}{17^2} = 1 + \frac{1}{289} = \frac{290}{289}$

6. Proceed through options,

At $n = 3$, $2n + 1 = 2\times 3 + 1 = 7 \Rightarrow$ Prime number

At $n = 4$, $2n + 1 = 2\times 4 + 1 = 9 \Rightarrow$ not a prime number

Hence, option(b) is correct.

7.
$$\begin{array}{r} X\,Y\,5 \\ \times Z \\ \hline X\,2\,1\,5 \end{array}$$

Here, the value of Z may be 1, 3, 5, 7 or 9. But only 9 can satisfy it. You may see as below:

$$\begin{array}{r} 1\,3\,5 \\ \times 9 \\ \hline 1\,2\,1\,5 \end{array}$$

Clearly, $X = 1$, $Y = 3$ and $Z = 9$

Then, $X + Y + Z = 1 + 3 + 9$
$= 13$

8. The number of mangoes left with him in end = 1

Then, number of mangoes left before selling to
$D = (1+1)\times 2 = 4$

Then, number of mangoes left before selling to
$C = (4+1)\times 2 = 10$

Then, number of mangoes left before selling to
$B = (10+1)\times 2 = 22$

Then, number of mangoes left before selling to
$A = (22+1)\times 2 = 46$

Alternatively, you may proceed through options.

9. Let smaller number be x.

Then, greater number = $3x$

As given, $x + 3x = 36 \Rightarrow x = 9$

10. It is obvious that digit '2' can be in any of the three places- unit, tens or hundreds.

Number of numbers having digit 2 as unit digit like

102, 112, 122, ...782, 792
$= 10\times 7 = 70$

Number of numbers having digit 2 as tens digit like 120, 121, 122(already considered),

123 to 129, 220, 221, ...729
$= 9\times 7 = 63$

Number of numbers having digit 2 as hundreds digit like

200, 202, ...299.

Out of these, 220 to 229 will be excluded (i.e.10 numbers) as has already been considered.

Further, 202,212,222,232,....292 will be excluded (i.e. 10 − 1 = 9 numbers)

Hence, total such numbers
$= 100 - 10 - 9 = 81$

Finally, number of whole numbers containing the digit 2
$= 70 + 63 + 81 = 214$

11. Numbers to be eliminated are: 5, 10, 15, 20, 25, 30, 35, 40, 45, 50, 51 to 59, 60.

Here, number of numbers to be eliminated = 20

Then, required answer
$= 60 - 20 = 40$

12. Let unit digit be y and tens digit be x, then original number $= 10x + y$.

Given, $10x + y = 3(x + y)$

$\Rightarrow \quad 7x = 2y \Rightarrow x = \frac{2}{7}y$

Further, $10x + y + 45 = 10y + x$

($\because$ digits are interchanged)

$\Rightarrow \quad 10x - x + y - 10y = -45$

$\Rightarrow \quad 9x - 9y = -45$

$\Rightarrow \quad x - y = -5$

$\Rightarrow \quad \frac{2}{7}y - y = -5$

$\Rightarrow \quad -5y = -35$

$\Rightarrow \quad y = 7$

Then, $x = \frac{2}{7} \times 7 = 2$

Now, sum of squares of digits

$= 2^2 + 7^2 = 53$

13. $551 = 19 \times 29$

and $1073 = 29 \times 37$

Clearly, numbers are 19, 29 and 37.

Now, their sum = 19 + 29 + 37 = 85

14. Let there be r rows and each row contains x students.

Then, $(r - 4)(x + 4) = xr$

$\Rightarrow xr + 4r - 4x - 16 = xr$

$\Rightarrow r - x = 4$ ...(i)

Further, $(r + 4)(x - 2) = xr$

$\Rightarrow xr - 2r + 4x - 8 = xr$

$\Rightarrow -r + 2x = 4$ ...(ii)

From Eqs. (i) and (ii), we get $x = 8$ and $r = 12$

Then, number of students

$= rx = 12 \cdot 8 = 96$

Alternatively

You may proceed through options.

(a) 80 ⇒

Number of rows (r)	Number of students in each row(x)
2	40
4	20
5	16
16	5
20	4
40	2

No row of the above table is satisfying $(r - 4)(x + 4) = 80$

(b) 96 ⇒

Number of rows (r)	Number of students in each row (x)
2	48
3	32
4	24
6	16
8	12
12	8
16	6
24	4
32	3
48	2

The row 12 8 of the above table is satisfying

(i) $(r - 4)(x + 4) = 96$

(ii) $(r + 4)(x - 2) = 96$

Hence, option(b) is correct.

15. It is the fact that the value of B cannot be 2, 3, 7 or 8 as a perfect square number(say, MPB) has no unit digit like 2, 3, 7 or 8. Further, the value of B cannot be 0 as $(0E)^2$ will not yield a three-digit number. Also, the value of BE must be less than 32 as $(32)^2$ yields a four-digit number. It implies that the value of B will not be greater than 3. Hence, possible value of $B = 1$

Further, the value of E should be such that its square ends with 1 and such number is 9.

$\therefore BE = 19 \Rightarrow (19)^2 = 361$

⇒ Value of $M = 3$

16. Proceed through options:

(a) $18 \Rightarrow 3q + 9 = 18 \Rightarrow q = 3$, then prime numbers are 3, 5 and 9 but 9 is not a prime number.

(b) $42 \Rightarrow 3q + 9 = 42 \Rightarrow q = 11$, then prime numbers are 11, 13 and 17, all are primes.

(c) $60 \Rightarrow 3q + 9 = 60 \Rightarrow q = 17$, then prime numbers are 17, 19 and 23, all are primes.

Hence, option(d) is correct.

17. Place value of 3 (second from right extreme)

$= 3 \times 10 = 30$

and place value of 3 (fourth from right extreme)

$= 3 \times 1000 = 3000$

Then, their sum

$= 30 + 3000 = 3030$

18. The best way is to proceed through options.

(a) $16428 \Rightarrow 16428 \times 9 \neq 1777\ldots$

(b) $19753 \Rightarrow 19753 \times 9 = 177777$

Hence, option(b) is correct.

19. Given, $f(x + a) = f(x \times a)$

$\Rightarrow f(1002 + 1) = f(1002 \times 1)$

$\Rightarrow f(1003) = f(1002)$

Further, $f(1002) = f(1001 + 1)$

$= f(1001 \times 1)$

$= f(1001)$

Hence, $f(1003) = f(1002) = f(1001)$

$= \ldots\ldots = f(1) = 4$

20. At $x = 2$, x, $(x + 1)$ and $(x + 3)$ all are primes. Hence, option(b) is correct.

21. The best way is to proceed through options.

(a) $45 \Rightarrow \{6500 \div (45 + 15)\}$

$\neq \{6500 \div 45\} - 30$

(b) $50 \Rightarrow \{6500 \div (50 + 15)\}$

$= \{6500 \div 50\} - 30$

Hence, option(b) is correct.

22. Let number of members be n. Then, total number of handshakes

$= \frac{n(n-1)}{2}$

$\Rightarrow 78 = \frac{n(n-1)}{2}$

$\Rightarrow n(n - 1) = 156$

$\Rightarrow 13(13 - 1) = 156$

$\Rightarrow n = 13$

23. Let unit digit and tens digit be y and x respectively.

As given,

$x + y + xy = 10x + y$

$\Rightarrow xy - 9x = 0$

$\Rightarrow x(y - 9) = 0$

$\Rightarrow y - 9 = 0$

$\Rightarrow y = 9$

It is clear that the possible numbers are 19, 29, 39, 49, 59, 69, 79, 89, 99 i.e. total 9 numbers out of (10 to 100 i.e. 91 numbers)

Thus, required fraction

$= 9/91 = 0.0989$

Hence, option(e) is correct.

24. Let the base of the number system be B.

Then, $(3414)_B = 484$

$[\because 44 \times 11 = 484]$

$\Rightarrow 3B^3 + 4 \cdot B^2 + 1B + 4 = 484$

$\Rightarrow B = 5$

Now, $(3111)_5 = (?)_{10}$

Clearly,

$(3111)_5 = 3 \cdot 5^3 + 1 \cdot 5^2 + 1 \cdot 5 + 1$

$= 406$

25. Recall the concept "How to test whether a number is prime or not'. Both the statements are correct.

26. The best way is to proceed through options.

(a) $72 \Rightarrow 7 + 2 = 9$ and $2 \times 72 \neq 9 \times 27$

(b) $81 \Rightarrow 8 + 1 = 9$ and $2 \times 81 = 9 \times 18$

Hence, option (b) is correct.

27. Sum of digits of numbers from 1 to 9 $= \frac{9 \times 10}{2} = 45$

Sum of unit digits of numbers from 10 to 99

$= 9(0 + 1 + 2 + 3 + \dots + 9)$

$= 9 \times \frac{9 \times 10}{2}$

$= 405$

Sum of ten's digits of numbers from 10 to 99

$= 10(1 + 2 + 3 + \dots + 9)$

$= 10 \times \frac{9 \times 10}{2} = 450$

Sum of digits of number

$100 = 1 + 0 + 0 = 1$

Now, sum of all digits of the numbers from 1 to 100

$= 45 + 405 + 450 + 1 = 901$

28. It is true that the number having sum of its digits equal to a multiple of 9, is divisible by 9. Further, there are 55 (i.e. $500 \div 9 = 55.55$) numbers less than 500 which are divisible by 9. Obviously, their seed $(n) = 9$

29. Given, $r + s = 1$

For the maximum product, the difference between r and s should be minimum. Hence, for the maximum product, $r = s = 1/2$

Then, $rs = 1/2 \times 1/2 = \frac{1}{4}$

30. Let numbers be x and y, where $x > y$.

As given, $x^2 - y^2 = 37$

$\Rightarrow (x + y)(x - y) = 37 \times 1$

Hence, $x + y = 37$ and $x - y = 1$

On solving we get $x = 19$ and $y = 18$

Then, $x^2 + y^2 = 18^2 + 19^2 = 685$

31. Given, $X = Z + 1$

Then, $X + 1 = Z + 2$

$= (1 \cdot 2 \cdot 3 \cdot 4 \cdot 5 \cdot 6 \cdot \dots \cdot 31) + 2$

$X + 2 = (1 \cdot 2 \cdot 3 \cdot 4 \cdot 5 \cdot 6 \cdot \dots \cdot 31) + 3$

$X + 3 = (1 \cdot 2 \cdot 3 \cdot 4 \cdot 5 \cdot 6 \cdot \dots \cdot 31) + 4$

$X + 4 = (1 \cdot 2 \cdot 3 \cdot 4 \cdot 5 \cdot 6 \cdot \dots \cdot 31) + 5$

....

.... so on.

Clearly, $X + 1$ is divisible by 2 as $X + 1 = 2\{(1 \cdot 2 \cdot 3 \cdot 4 \cdot 5 \cdot 6 \cdot \dots \cdot 31) + 1\}$ and $X + 2$ is divisible by 3 as $X + 2 = 3\{(1 \cdot 2 \cdot 3 \cdot 4 \cdot 5 \cdot 6 \cdot \dots \cdot 31) + 1\}$ and so on.

Hence, $X + 1, X + 2, \dots\dots X + 29, X + 30$ all are composite numbers.

Thus, option(d) is correct.

32. For obtaining positive value of w, value of vz must be negative as u is always negative. Further, to maximise the value of w, the numerical value of numerator (i.e. vz) must be maximum and value of denominator (i.e. u) u should be least.

Hence, $u = -0.5$, $v = 1$, $z = -2$

Then, maximum value of

$w = \frac{vz}{u} = \frac{1 \times (-2)}{-0.5} = 4$

Further, for obtaining negative value of w, value of vz must be positive as u is always negative. Also, in order to get minimum value of w, the numerical value of vz must be maximum and that of u should be minimum. To get numerical value of vz maximum, we should take values of v and z minimum(i.e. both in negative).

Hence, $u = -0.5$, $v = -1$, $z = -2$

Then, minimum value of

$w = \frac{vz}{u} = \frac{(-1) \times (-2)}{-0.5} = -4$

Thus, $-4 \le w \le 4$

33. Let number of boys be B. Then, number of girls $= 60 - B$

As given,

$B(60 - B) + (60 - B)B = 1600$

$\Rightarrow B(60 - B) = 800$

$\Rightarrow B^2 - 60B + 800 = 0$

$\Rightarrow B^2 - 40B - 20B + 800 = 0$

$\Rightarrow B(B - 40) - 20(B - 40) = 0$

$\Rightarrow B = 20$ or 40

Then, number of girls = 40 or 20

Hence, option(d) is correct.

34. Proceed through given options.

(a) $5 \Rightarrow 2^5 - 1 = 31$, which is a prime number.

(b) $11 \Rightarrow 2^{11} - 1 = 2047$, which is not a prime number as it is divisible by 23.

Hence, option(b) is correct.

35. The best way is to proceed through options.

(a) $25 \Rightarrow$ number of packages remained in van

$= 25 - \left\{25 \times \frac{2}{5} + 3\right\} \ne 1/2$ of 25.

(b) $30 \Rightarrow$ number of packages remained in van

$= 30 - \left\{30 \times \frac{2}{5} + 3\right\} = 1/2$ of 30

Thus, option (b) is correct.

36. Given, $f(x) = ax + b$

Then, $f(f(x)) = f(ax + b)$

$= a(ax + b) + b$

$= 9x + 8$ (as given)

$\Rightarrow a^2x + ab + b = 9x + 8$

$\Rightarrow a^2x + b(a + 1) = 9x + 8$

$\Rightarrow a^2 = 9$ and $b(a + 1) = 8$

$\Rightarrow a = 3$

and $b(3 + 1) = 8$

i.e., $b = 2$

Then, $a + b = 3 + 2 = 5$

37. Proceed through options. Let there be n children in the first/front row.

Option (a)

$n + (n - 3) + (n - 6) = 630$

$\Rightarrow n = 213$, possible value

Option (b)

$n + (n - 3) + (n - 6) + (n - 9) = 630$

$\Rightarrow n = 162$, possible value

Option (c)

$n + (n - 3) + (n - 6) + (n - 9) + (n - 12) = 630$

$\Rightarrow n = 132$, possible value

Option (d) $n + (n - 3) + (n - 6) + \dots + (n - 15) = 630$

$\Rightarrow n = 112.5$, not possible as number of children cannot be in fraction.

Option (e) $n + (n - 3) + (n - 6) + \dots + (n - 18) = 630$

$\Rightarrow n = 99$, possible value

Hence, option(d) is correct.

Alternatively

Clearly, number of children in each row makes an arithmetic progression with common difference 3. Let there be n rows and last row contains a children.

Then, total number of children

$= \frac{n}{2}\{2a + (n - 1)d\}$

$\Rightarrow 630 = \frac{n}{2}\{2a + (n - 1)3\}$

Now, find the value of a putting the values of options in the above equation.

(a) $3 \Rightarrow 630 = \frac{3}{2}\{2a + (3 - 1)3\}$

$\Rightarrow a = 207$, possible value

Similarly, process with remaining options. Only option(d) yields the value of a in fraction, which is not possible.

Hence, option (d) is correct.

38. Number of men arranged in each row

$= \sqrt{\text{Number of men arranged in form of a square}}$

$= \sqrt{6000 - 71} = 77$

39. Let first odd integer be x. Then, another be $x + 4$.

As given, $x(x + 4) = 3x + 12$

$\Rightarrow \quad x^2 + 4x - 3x - 12 = 0$

$\Rightarrow \quad x(x + 4) - 3(x + 4) = 0$

$\Rightarrow \quad (x - 3)(x + 4) = 0$

$\Rightarrow \quad x = 3 \qquad [\because x \neq -4]$

Hence, larger number = 3 + 4 = 7

Alternatively, the best way is to proceed through options.

40. (a) $\frac{x}{x} \Rightarrow 1$

(b) $\frac{x+1}{x} \Rightarrow 1 + \frac{1}{x}$ = greater than 1

(c) $\frac{x}{x+1} \Rightarrow \frac{x+1-1}{x+1} = \frac{x+1}{x+1} - \frac{1}{x+1}$

$= 1 - \frac{1}{x+1}$

= less than 1

(d) $\frac{x+2}{x+3} \Rightarrow \frac{x+3-1}{x+3} = \frac{x+3}{x+3} - \frac{1}{x+3}$

$= 1 - \frac{1}{x+3}$

= less than 1

Hence, option(b) is correct.

Alternatively, In such types of questions, we should put value of x (assume, 1) in options. Then, values of options be 1, 2, 1/2 ,3/4. Hence, option(b) has greatest value.

41. Given, $a^2 - b^2 = 1999$

$\Rightarrow (a - b)(a + b) = 1 \times 1999$

$\Rightarrow \quad a - b = 1$

and $\quad a + b = 1999$

On solving, we get

$a = 1000$

and $\quad b = 999$

Then, $a^2 + b^2 = 1000^2 + 999^2$

$= 1998001$

42. Sum of numbers of another set
= [{(378 ÷ 7) − 23} i.e. 31] +

32 + 33 + 34 = 130

43. Let unit, tens and hundreds digits be z, y and x respectively, where x and z are not zero. If z or x be zero, the new number will be of two digits only, and it contradicts the condition envisaged in the question.

Then, the original number

$A = 100x + 10y + z$

and new number $B = 100z + 10y + x$

Now, $B - A = 100z + 10y + x$

$\qquad - 100x - 10y - z$

$= 99z - 99x$

$= 99(z - x)$

$(B - A)$ will be divisible by 7 only when $(z - x) = 7$

Hence, value of $(z, x) = (8, 1)$ and $(9, 2)$, y can have any value from 0 to 9.

Thus, possible values of A are 108, 118, 128, 138, ... 198, 209, 219, ,..........289, 299.

∴ Hence, option (b) is correct.

44. Let value of p be 5. Then, $p(p + 2) + 1 = 5(5 + 2) + 1 = 36$, hence a perfect square.

Further,
$p + (p + 2) = 5 + (5 + 2) = 12$, divisible by 12.

Hence, option(c) is correct.

Note *You are free to take any value of p but it must be greater than 3 for the second statement.*

45. In case 1044 candles burn, number of candles made from stubs

= 1044/9 = 116

Further, in case 116 candles burn, number of candles made from stubs = 116/9 i.e. 12 candles {8 stubs will remain}

Further, in case 12 candles burn, number of candles made from stubs = 12/9 i.e. 1 candle {3 stubs will remain}

Obviously, from (8 stubs + 3 stubs), one more candle can be made.

Hence, maximum number of candles those can be made

= 116 + 12 + 1 + 1 = 130

46. $0.0010101 \times 10^6 = 0001010.1$, which is greater than 1000.

Hence, option(c) is correct.

47.
```
  1 1 0 0 1
 -1 0 1 1
 ---------
  0 1 1 1 0
```

Explanation

1st column: 1 − 1 = 0

2nd column: 0 − 1 = 1

3rd column: 1 − 0 = 1

4th column: 0 − 1 = 1

5th column: (1 − 1) = 0

Alternatively you may convert given binary numbers into decimal system and then subtract, thereafter convert the decimal number so obtained in binary system again.

48. Let the number be x.

As given, $x + \frac{1}{x} = 3\left(x - \frac{1}{x}\right)$

$\Rightarrow \quad x - 3x = \frac{-3}{x} - \frac{1}{x}$

$\Rightarrow \quad -2x^2 = -4$

$\Rightarrow \quad x^2 = 2, x = \pm\sqrt{2}$

Hence, option(a) is correct.

49. Let the number of questions answered correctly be x, wrongly be y and not attempted be z.

Then, $x + y + z = 50 \qquad$ ...(i)

Further, net score $= x - \frac{y}{3} - \frac{z}{6} = 32$

$\Rightarrow \quad 6x - 2y - z = 192 \qquad$ ...(ii)

Adding Eqs. (i) and (ii), we get

$7x - y = 242$

$\Rightarrow \quad x = \frac{242 + y}{7}$

Obviously, x, y and z are positive integers.

Then, at $y = 3$, we get $x = 35$ and at $y = 10$, we get $x = 36$

Hence, the number of questions answered wrongly (i.e. y) by that student cannot be less than 3.

Note *Values of x, y and z respectively are either 35, 3, 12 or 36, 10, 4.*

50. Option(b) is correct as $\sqrt[m]{a^m} = a$

51. Number of odd numbered pages

$= \frac{1089 + 1}{2} = 545$

52. We have to compute maximum possible value so we will stress upon the value of abc. Greater the value of abc, greater the value of the given expression.

The maximum value of

$|a| \times |b| \times |c| = 10 \times 9 \times 8 = 720$

Further, in order to get maximum value of expression, we should minimize the value of $(a + b + c)$ to the greater extent and it may be feasible when $a = -10$, $b = -9$ and $c = 8$ {the values of a, b, c may be interchanged}. In case values of a, b, c are kept negative, their product will be negative.

Hence, maximum possible value of $[abc - (a + b + c)]$

$= 720 - \{-10 - 9 + 8\} = 731$

53. Let $x = 2$ and $y = 10$

Then,

(a) $y - x = 10 - 2$
$= 8$, an even integer

(b) $xy = 2 \times 10$
$= 20$, an even integer

(c) $\frac{2+10}{2} = 6$, an even integer

Hence, option(d) is correct.

54. Obviously, 1001 will be the largest number.

55. Let number of buffaloes be x and that of ducks be y.

Then, number of heads $= x + y$

and number of legs $= 4x + 2y$

As given,

$4x + 2y = 2(x + y) + 24$

$\Rightarrow \quad 2x = 24$

$\Rightarrow \quad x = 12$

56. We know, every square integer (except 1) is either a multiple of 4 or greater than a multiple of 4 by unity.

Hence, option(d) is correct.

57. Given, $|x + 7| + |x - 8| = 16$

Taking positive values,

$x + 7 + x - 8 = 16$

$\Rightarrow \quad 2x - 1 = 16$

$\Rightarrow \quad x = 8.5$

Taking negative values,

$-x - 7 - x + 8 = 16$

$[\because |-(x+7)| + |-(x-8)| = 16]$

$\Rightarrow \quad -2x + 1 = 16$

$\Rightarrow \quad x = -7.5$

Hence, sum of possible values of $x = 8.5 - 7.5 = 1$

Note *In case one is taken positive and another as negative, LHS ≠ RHS.*

58. When $N = 3$, N, $(N + 2)$ and $(N + 4)$ all are primes. Hence, option(a) is correct.

59. Number of soldiers required to form a solid square $= 5180 + 4 = 5184$

Then, number of row in front row or any row $= \sqrt{5184} = 72$

60. Clearly, n can be a two digit or three digit.

In case of two digits numbers,

Let x and y be tens and unit digit respectively.

Then, $n = 10x + y$, $P_n = xy$ and $S_n = x + y$

As given, $P_n + S_n = n$

$\Rightarrow \quad xy + x + y = 10x + y$

$\Rightarrow \quad x(y + 1) = 10x \Rightarrow y = 9$

Hence, possible value of numbers are 19, 29, 39,............99, i.e. total number of such integers = 9

In case of three digits numbers,

Let unit, tens and hundreds digits be z, y and x respectively,

Then, $n = 100x + 10y + z$, $P_n = xyz$ and $S_n = x + y + z$

As given, $P_n + S_n = n$

$\Rightarrow xyz + x + y + z = 100x + 10y + z$

$\Rightarrow \quad xyz = 99x + 9y$

$\Rightarrow \quad z = \frac{99}{y} + \frac{9}{x}$

Since z, y and x are digits, the value of z cannot be in single digit if $z = \frac{99}{y} + \frac{9}{x}$ is considered.

Hence, there is no number of 3 digits.

Thus, option(d) is correct.

61. $475_8 \Rightarrow 4 \cdot 8^2 + 7 \cdot 8^1 + 5 \cdot 8^0$
$= 256 + 56 + 5 = 317$

Hence, $\quad 475_8 = 317_{10}$

Now,

6	317	Remainders
	52	5
	8	4
	1	2

Hence, $475_8 = 1245_6$

62. Clearly, $\frac{252}{1000}, \frac{748}{1000}$ i.e. 0.252, 0.748 lies between 1/4 (i.e. 0.250) and 3/4 (i.e. 0.750).

Hence, option(d) is correct.

63. Let they have x bullets each initially.

Then, total number of bullets remaining after shot

$= 3(x - 4) = x$ (given)

$\Rightarrow \quad 2x = 12 \Rightarrow x = 6$

Hence, original number of bullets

$= 6 \times 3 = 18$

Alternatively

The best way is to proceed through options.

(a) 15 ⇒ total number of bullets remaining after shot
$= 3\{(15 \div 3) - 4\} \neq (15 \div 3)$

(b) 18 ⇒ Total number of bullets remaining after shot
$= 3\{(18 \div 3) - 4\} = (18 \div 3)$

Thus, option(b) is correct.

64. Obviously, T is a set of 59 integers because $n = \frac{467 - 3}{8} + 1$

Further, 1st term (i.e. 3) + 59th term (i.e. 467) = 470

and 2nd term (i.e. 11) + 58th term (i.e. 459) = 470

....

29th term + 31st term = 470

Clearly, 30th term will be single having value $3 + (30 - 1)8$ i.e. 235.

In case we can take one number each from the 29 pairs along with 30th term, the sum of any two elements will never be 470.

Hence, maximum possible number of elements in

$S = 29 + 1 = 30$

65. Already given in relevant box/ shaded portion. Option(c) is correct.

66. Divide the given binary into groups of 4, starting from the extreme right. We may prefix 0, if required, for our convenience. Here, groups are 1110 and 1010

Now, 1110_3

$\Rightarrow 1 \cdot 2^3 + 1 \cdot 2^2 + 1 \cdot 2^1 + 0 \cdot 2^0 = 14 = E$

and $1010 \Rightarrow 1 \cdot 2^3 + 0 \cdot 2^2 + 1 \cdot 2^1 + 0 \cdot 2^0$
$= 10 = A$

Hence, $\quad 11101010_2 = EA_{16}$

67. Let the sequence be $wxyz$.

Then,

$(wxyz)_{10} = 1000w + 100x + 10y + z$

and $(wxyz)_6 = 216w + 36x + 6y + z$

As given, $1000w + 100x + 10y + z$
$= 4(216w + 36x + 6y + z)$

$\Rightarrow \quad 136w = 44x + 14y + 3z$

It can be observed from this equation that:

(i) Since the digits in the number system with base 6 be 0, 1, 2, 3, 4, 5. The value of w may be (1 to 5), but not 0. Values of other variables may be (0 to 5).

(ii) At any value of w from 1 to 5, the value of LHS will always be even. Also, values of each of $44x$ and $14y$ will be even.

(iii) If LHS is even, $44x$ is even and $14y$ is even , then value of z must be either 0 or 2 or 4.

(iv) If we take maximum value of x, y and z, the value of RHS will be $44 \cdot 5 + 14 \cdot 5 + 3 \cdot 4 = 302$. In this situation, value of w cannot be more than 2. Hence, its value may be either 1 or 2.

Keeping in view the above observations, we may solve the equation with trial and error method: At $w = 1$, $x = 2$, $y = 3$ and $z = 2$, the above equation is satisfied.

Then, required sum

$= 1 + 2 + 3 + 2 = 8$

68. The best way is to proceed through options.

(a) $14 \Rightarrow$ Remainders in case of $\{14 \div (1 + 4)\}$ and $\{14 \div (1 \times 4)\}$ are not same.

(b) $32 \Rightarrow$ Remainders in case of $\{32 \div (3 + 2)\}$ and $\{32 \div (3 \times 2)\}$ are same and difference between quotients is one.

Hence, option(b) is correct.

69. The best way is to proceed through options.

(a) $10 \Rightarrow \frac{7}{2} \times \sqrt{10} + 2 \neq 10$

(b) $16 \Rightarrow \frac{7}{2} \times \sqrt{16} + 2 = 16$

Hence, option(b) is correct.

70. Proceed through options.

(a) $101 : 88 \Rightarrow 101 + 88 = 189$, not a prime number

(b) $87 : 100 \Rightarrow 87 + 100 = 187$, not a prime number

(c) $110 : 111 \Rightarrow 110 + 111 = 221$, not a prime number

(d) $85 : 98 \Rightarrow 85 + 98 = 183$, not a prime number

(e) $97 : 84 \Rightarrow 97 + 84 = 181$, a prime number

Hence, option(e) is correct.

71. $3094 = 2 \times 7 \times 13 \times 17$

Hence, $a = 2$, $b = 7$, $c = 13$ and $d = 17$

Now, $bc = 7 \times 13 = 91$

72. Let two numbers be 3 and 1/3.

Hence, $n = 2$

Here, $3 \times \frac{1}{3} = 1$

Then, $3 + \frac{1}{3} > 2$ (i.e. n)

[Hence, options(a), (b) and (d) - eliminated]

Further, let three numbers be 2, 3/5 and 5/6. Hence, $n = 3$

Here, $2 \times \frac{3}{5} \times \frac{5}{6} = 1$

Then, $2 + \frac{3}{5} + \frac{5}{6} > 3$ (i.e., n) {Hence, options(a), (b) and (d) - eliminated}

Thus, option(c) is correct.

73. Proceed through options.

(a) $121 \Rightarrow$ divisible by 11, but $121 - 121 \neq 297$

(b) $231 \Rightarrow$ divisible by 11, but $231 - 132 \neq 297$

(c) $451 \Rightarrow$ divisible by 11 and $451 - 154 = 297$

Hence, option(c) is correct.

74. Clearly, the values of 1st term and 2nd term must be maximum and the value of third term should be minimum. It is possible only when a and c are closer to each other and are far away from b.

Hence, one possibility is $a = 1$, $c = 2$ and $b = 24$

Then, $|a - b| + |b - c| - |c - a|$

$= |1 - 24| + |24 - 2| - |2 - 1| = 44$

Second possibility is $a = 24$, $c = 23$ and $b = 1$

Then, $|a - b| + |b - c| - |c - a|$

$= |24 - 1| + |1 - 23| - |23 - 24| = 44$

Hence, option(a) is correct.

75. The best way is to proceed through options.

(a) $32 \Rightarrow 32 - \{(32 \div 4) + 2\sqrt{32}\} \neq 15$

(b) $36 \Rightarrow 36 - \{(36 \div 4) + 2\sqrt{36}\} = 15$

Hence, option(b) is correct.

76. (c) It is obvious that $pq + qr + pr + pqr$ will have maximum value if p, q and r are as close as possible. Since, they are integers, their values may be 3,3 and 4 respectively. The order of integers does not affect the value of the expression.

Now, $pq + qr + pr + pqr$

$= 3 \cdot 3 + 3 \cdot 4 + 4 \cdot 3 + 3 \cdot 3 \cdot 4$

$= 69$

Hence, option(c) is correct.

77. $a - xb = a - \{-|a| b\} b$

$= a + |a| b^2$

It is obvious that b^2 will always have positive value and the minimum value of b^2 be 1 as $|b| \geq 1$.

In case we take value of a positive, value of $a + |a| b^2$ will be positive at any value of b^2. But, in case of negative value of a and minimum value of $|a| b^2$, $a + |a| b^2$ will equal to 0. Further, the value of $a + |a| b^2$, will increase from 0 if the value of b^2 increases from 1 in case of negative value of a.

Hence, $a - xb \geq 0$

78. $\frac{10}{3} = 3\frac{1}{3} = 3 + \frac{1}{3}$

Hence, the integer is 3.

79. Obviously, $(n + 1)^2 = n^2 + 2n + 1$

Hence, option(d) is correct.

80. If we take minimum value of n i.e. 1, we get

$$1 - \frac{1}{1} < x \leq 3 + \frac{1}{1}$$

$\Rightarrow \quad 0 < x \leq 4$

Further, in case we take maximum value of n i.e. ∞, we get

$$1 - \frac{1}{\infty} < x \leq 3 + \frac{1}{\infty}$$

$\Rightarrow \quad 1 - 0 < x \leq 3 + 0$

$\Rightarrow \quad 1 < x \leq 3$

From the above, it is possibility of $0 < x \leq 4$.

81. We know that the number ending with 2, 3, 7, 8 cannot be square of a natural number. Hence, option(b) is correct.

82. Only one set of such prime numbers exists viz., 3, 5, 7.

Hence, option(a) is correct.

83. $\frac{29}{12} + \frac{15}{16} = 2\frac{5}{12} + \frac{15}{16}$

$= 2 + \frac{65}{48} = 3 + \frac{17}{48}$

In order to make the result a whole number we may either subtract $\frac{17}{48}$ or add $\frac{31}{48}$. It is also fact that $\frac{17}{48}$ is lesser than $\frac{31}{48}$.

Hence, option(d) is correct.

84. Let $a + b + c = x$

Then, the value of expression

$= \frac{x + d}{x - d} = \frac{(x - d) + 2d}{x - d}$

$= 1 + \frac{2d}{x - d}$

The value of $\frac{2d}{x - d}$ will be maximum only when value of d is maximum and value of denominator i.e. $(x - d)$ is minimum. Further, value of d cannot be greater than 25.

Hence, maximum value of

$$\frac{2d}{x - d} = \frac{2 \times 25}{26 - 25} = 50$$

Then, maximum value of

$$1 + \frac{2d}{x - d} = 1 + 50 = 51$$

85. Let tens and unit digit be x and y respectively, where $x, y \neq 0$.

Then, original number $= 10x + y$

and new number $= 10y + x$

As given, $10y + x - (10x + y) = 18$

$\Rightarrow 9y - 9x = 18 \Rightarrow y - x = 2$

The possible values of $y, x = (3,1), (4, 2), (5, 3), (6, 4), (7, 5), (8, 6), (9, 7)$.

Then, original numbers are 13, 24, 35, 46, 57, 68 and 79.

$\therefore$ Other than 13, there are 6 such numbers.

Hence, option(b) is correct.

86. Observe the 'Interesting properties of numbers' detailed in this chapter. Option(d) is correct.

87. Obviously, x^y will be a rational number as $a^{-n} = \frac{1}{a^n}$

88. Let four consecutive numbers are m ,$m + 2$, $m + 4$ and $m + 6$.

As given,

$m + m + 2 + m + 4 + m + 6 = 10 \times n^2$

Where, n is a natural number.

$\Rightarrow \quad 4m + 12 = 10 \times n^2$

$\Rightarrow \quad 2m + 6 = 5n^2$

$\Rightarrow \quad m = \frac{5n^2 - 6}{2}$

Let us consider value of n as 1, 2, 3, ...

At $n = 1$, m is not integer.

At $n = 2$, m is 7. Then, numbers are 7, 9, 11, 13, but these numbers are not in options.

At $n = 3$, m is not integer.

At $n = 4$, $m = 37$. Then, numbers are 37, 39, 41, 43 and the fact that 41 is available in options. Obviously, no need to process further.

Hence, option(c) is correct.

89. If the difference between the numerator and the denominator of given fractions each is same, the fraction having highest numerator or denominator will have highest value. The concept has been detailed in chapter 'Comparison of Fractions'.

Hence, option (a) is correct.

90. Clearly, the value of ab must be less than 32 as $(32)^2$ yields a four-digit number.

Since $(ab)^2 > 300$, the value of ab must be greater than 17.

Hence, value of ab is between 17 and 32.

Further, the value of b should be such that its square ends with the same value and such values are 0,1, 5, 6.

It implies that the possible values of $ab = 20, 21, 25, 26, 30, 31$

Now, arrive at the answer with trial and error method:

$(20)^2 = 400, (21)^2 = 441,$

Clearly, $ab = 21$ i.e, $b = 1$

91. Recall the estimation method to compute the cube root of a number.

$\Rightarrow \overline{015}\ \overline{600}$

Clearly, its cube-root has 2 digits and the left (extreme) digit is 2.

Let the cube-root is 20 (you are free to take any figure starting with 2, but dividing a number by 20 is relatively convenient)

Then, $15600 \div 20 = 780$

Further, $780 \div 20 = 39$

Now, average of 20, 20 and

$39 = 20 + \frac{19}{3} = 26.33$

Hence, approx. value of cube-root $= 26$

It implies that all the three consecutive integers are closer to 26. Further, 15600 is exactly divisible by 25 as a number formed with last two digits is divisible by 25.

We may write, $15600 = 25 \times 624$

$= 25 \times 24 \times 26$

$= 24 \times 25 \times 26$

Clearly, the numbers are 24, 25 and 26.

Now, sum of their squares

$= 576 + 625 + 676$

$= 1877$

92. Required numbers are 509, 519, 529, 539, 549, 559, 569, 579, 589, 590 to 598.

Hence, number of such numbers = 18

93. It is true that every natural number except 1 is either prime or composite number. Hence, option(a) is correct.

94. $\left[\frac{x}{2}\right] + \left[\frac{x+1}{2}\right] = \frac{x}{2} + \frac{x+1}{2}$

$= \frac{2x+1}{2} = x + \frac{1}{2} = [x]$

as [ ] is the greatest integer function.

95. It is true that any number of the form $aabb$ is divisible by 11. Since $aabb$ is a perfect square, it is a multiple of 121 also.

Hence, the possible numbers are : $121 \times 1^2, 121 \times 2^2, 121 \times 3^2, ...$

Only 121×8^2 is in form of $aabb$. Hence, it is the desired perfect square number.

$\therefore$ Hence, option (a) is correct.

96. Number of digits when 2^1 and 5^1 are written one after another i.e. $25 = (1 + 1) = 2$

Number of digits when 2^2 and 5^2 are written one after another i.e. 425 $= (2 + 1) = 3$

Number of digits when 2^3 and 5^3 are written one after another i.e. $8125 = (3 + 1) = 4$

On the above pattern, number of digits when 2^{2004} and 5^{2004} are written one after another

$= (2004 + 1) = 2005$

97. Total number of 1-digit numbers $= 9$, then total digit

$= 9 \cdot 1 = 9$

Total number of 2-digit numbers $= (99 - 9)$, then total digit

$= 90 \cdot 2 = 180$

Total number of 3-digit numbers $= (999 - 99)$, then total digit

$= 900 \cdot 3 = 2700$

Now, total digits used

$= 9 + 180 + 2700$

$= 2889$

Hence, only (3189 – 2889) i.e. 300 digits have been used for 4-digit numbers.

Then, number of 4-digit numbers for which 300 digits have been used

$= (300 \div 4) = 75$

Thus, total number of pages

$= 9 + 90 + 900 + 75$

$= 1074$

98. Obviously, the number M will leave 1 as remainder when divided by 2, 3 and 5 individually as in all three cases last digit is 1. Hence, option(b) and option(c) are eliminated. Now, check the remaining options.

(a) $31_{10} = 11111_2 = 1011_3 = 111_5$

(d) $91_{10} = 1011011_2 = 10101_3 = 331_5$

Obviously, option(d) is correct.

Alternatively

Since the last digit is 1, the number M should leave remainder 1 when it is divided by each of 2, 3 and 5.

Hence, the value of M = LCM of 2, 3

and $5+1=30+1=31$

Next higher value of M

$= 2\cdot 30+1=61$

Next higher value of M

$= 3\cdot 30+1=91$

But, only 31 and 91 are among the given options.

(a) $(31)_{10} = (11111)_2 = (1011)_3 = (111)_5$. Here, all the cases have leading digit as 1, which contradicts the stipulation given in the question. Hence, option(d) is correct.

99. Let number of employees be x.

Since each employee gives a gift to every other employee, it means no one gives a gift to himself/herself.

Hence, total number of gifts $= x(x-1) = 132$

$\Rightarrow \quad x^2 - x - 132 = 0$

$\Rightarrow \quad x^2 - 12x + 11x - 132 = 0$

$\Rightarrow \quad x(x-12) + 11(x-12) = 0$

$\Rightarrow \quad (x-12)(x+11) = 0$

$\Rightarrow x = 12$ and -11 [number of employees cannot be in negative figure]

Hence, number of employees = 12

Alternatively The aspirants should proceed through options.

100. Let the unequal numbers be x and y.

Then, $48(x-y) = x^2 - y^2$

$\Rightarrow \quad x + y = 48$

Hence, option(d) is correct.

101. $10^{10} = 10{,}000{,}000{,}000$

Since sum of digits is 2, any one of 10 zeroes can be replaced by 1. In doing so, the value of the number will be between 10^{10} and 10^{11}.

Clearly, we may form 10 different number by replacing the zero by 1.

Further, if we replace 1 of 10, 000,000,000 by 2, the sum of digits will also be 2 and the number will be between 10^{10} and 10^{11}.

Hence, number of different values of

$n = 10 + 1 = 11$

102. Since P, Q and R are three consecutive odd numbers,

$R = P + 4$

$\Rightarrow \quad 3R = 3P + 12$

$\Rightarrow \quad 3R = (2R - 3) + 12$ [$\because 3P = 2R - 3$ given]

$\Rightarrow \quad R = 9$

103. Number of 2-digit positive numbers

$= (99 - 10) + 1 = 90 = 9 \times 10^1$

Number of 3-digit positive numbers

$= (999 - 100) + 1 = 900 = 9 \times 10^2$

$\therefore$ Number of 100-digit positive numbers $= 9 \times 10^{99}$

104. Given, $a_{n+1} - 3a_n + 2 = 4n$

$\Rightarrow \quad a_{n+1} = 3a_n + 4n - 2$

Now, at $n = 1$, $a_2 = 3a_1 + 4\cdot 1 - 2$

$= 3\cdot 1 + 2 = 3^2 - 2\cdot 2 = 5$

At $n = 2$, $a_3 = 3a_2 + 4\cdot 2 - 2$

$= 3\cdot 5 + 6 = 3^3 - 2\cdot 3 = 21$

At $n = 3$, $a_4 = 3a_3 + 4\cdot 3 - 2$

$= 3\cdot 21 + 10$

$= 3^4 - 2\cdot 4 = 73$

It is clear from the above that

$a_n = 3^n - 2n$

Hence, $a_{100} = 3^{100} - 200$

105. Let the 1st, 2nd, 3rd and 4th digits be a, b, c and d respectively.

As given, $a + b = c + d$, $a + d = c$ and $b + d = 2(a + c)$

Clearly, $(a + d) + (b + d) = 2a + 3c$

$\Rightarrow \quad c + d + 2d = 2(c - d) + 3c$

$\Rightarrow \quad 5d = 4c$

$\Rightarrow \quad c = \frac{5}{4}d$

Since c and d both are integers, the possible value of d is either 4 or 8.

At $d = 4$, $c = 5$ and at $d = 8$, $c = 10$. But value of c cannot be in two digits. Hence, $c = 5$

106. Given, $\frac{1}{m} + \frac{4}{n} = \frac{1}{12}$

$\Rightarrow \quad \frac{1}{m} = \frac{1}{12} - \frac{4}{n} = \frac{n-48}{12n}$

$\Rightarrow \quad m = \frac{12n}{n-48}$

Since m and n are positive integers, n must be greater than 48. The possible values of n (less than 60 and odd) = 49, 51, 53, 55, 57 and 59.

At $n = 49, 51, 57$, m has integral values.

Thus, option(a) is correct.

107. Let the three-digit number be $100x + 10y + z$ where x, y and z are hundred's, ten's and unit digit respectively.

As given,

$100x + 10y + z = 11(10x + z)$

$\Rightarrow \quad 10y = 10x + 10z$

$\Rightarrow \quad y = x + z$

$\Rightarrow \quad y - x = z$

$\Rightarrow \quad 1 = z$

Note *if we take $(x - y) = 1$, then value of z will be -1, not possible*

Hence, option (c) is correct.

108. The number of diamonds the thief has before he gave diamonds to the 3rd watchman $= (1 + 2) \times 2$

$= 6$

The number of diamonds the thief has before he gave diamonds to the 2nd watchman $= (6 + 2) \times 2$

$= 16$

The number of diamonds the thief has before he gave diamonds to the 1st watchman $= (16 + 2) \times 2 = 36$

Hence, the thief originally steals 36 diamonds.

109. It is clear that each team is a constituent of two pairs of teams. It implies that each team shares 2 players with other two teams (viz. in case of T_1 & T_2 and T_2 & T_3 one player of T_2 is common for T_1 and T_3. Then, each team has $(k - 2)$ players, which has not been shared. Accordingly, for n teams, number of players $= n(k - 2)$.

Further, for n teams, number of pairs of teams $= n$ and each of n pairs of teams has 1 player in common.

Then, total number of players

$= n(k - 2) + n = n(k - 2 + 1)$

$= n(k - 1)$

110. N can be a perfect cube only when each of four given terms is a perfect cube. Further, each base number i.e. each of the given prime numbers should have power as 0, 3, 6, 9, ... if each term be a perfect cube. It is also given that $(p, q, r, s) > 0$

Now, (11^{p+7}) will be a perfect cube if $p = 2, 5, 8, \ldots\ldots$

(7^{q-2}) will be a perfect cube if $q = 2, 5, 8, \ldots\ldots$

(5^{r+1}) will be a perfect cube if $r = 2, 5, 8, \ldots\ldots$

(3^s) will be a perfect cube if $s = 3, 6, 9 \ldots\ldots$

Hence, smallest value of

$p + q + r + s = 2 + 2 + 2 + 3 = 9$

111.

Step	P	Q	New Q	$P \times$ New Q	New P
I	1	1	$1+1=2$	1×2	2
II	2	2	$2+1=3$	2×3	6
III	6	3	$3+1=4$	6×4	24
IV	24	4	$4+1=5$	24×5	120
V	120	5	$5+1=6$	120×6	720

Hence, option(d) is correct.

112. The net height he climbs in one lift = 15 − 0.5 = 10 ft

Then, he can climb (18.5 − 1.5) ft in 17 lifts.

It is obvious that after 17 lifts, he can climb 1.5 feet in 18th lift and at this juncture, no question arises of his slipping as he reaches the top of the tree.

Hence, option(c) is correct.

113. Let the number be $abcdef$.

As given, $f = a + b + c$, $e = a + b$, $c = a$, $b = 2a$, $d = e + f$

Now, $d = e + f = (a + b) + (a + b + c)$

$= (a + 2a) + (a + 2a + a) = 7a$

The maximum possible value of a will be 1 as d and a both are digits. Hence, maximum value of $d = 7 \times 1 = 7$.

114. The number having 1, 3, 5, 7, 9 as unit digit, may be prime numbers. It is also fact that 2 is also prime number. Hence, the correct option is 5 + 1 = 6.

115. Let the prime numbers be x, $x + 36$ and y.

As given, $x + x + 36 + y = 100 \Rightarrow 2x + y = 64$

Obviously, $2x$ shall be even number as {2 × (either even or odd)} is always even.

If $2x$ is even, then y will necessarily be even as even + even = even. Y, a prime number, is even number, it implies that $y = 2$ as 2 is only even prime number.

Putting the value of y in equation $2x + y = 64$, we get $x = 31$

Hence, the prime numbers are 2, 31 and 67.

∴ Hence, option (d) is correct.

116. Let the numbers be a and b.

Given, $\dfrac{a^3 - b^3}{(a-b)^3} = \dfrac{157}{3}$

$\Rightarrow \dfrac{(a-b)(a^2 + b^2 + ab)}{(a-b)^3} = \dfrac{157}{3}$

$\Rightarrow \dfrac{(a^2 + b^2 + ab)}{(a-b)^2} = \dfrac{157}{3}$

$\Rightarrow \dfrac{(a-b)^2 + 3ab}{(a-b)^2} = \dfrac{157}{3}$

$\Rightarrow 1 + \dfrac{3ab}{(a-b)^2} = \dfrac{157}{3}$

$\Rightarrow \dfrac{3 \times 616}{(a-b)^2} = \dfrac{154}{3}$

$\Rightarrow (a-b)^2 = 36$

$\Rightarrow (a+b)^2 - 4ab = 36$

$\Rightarrow (a+b)^2 = 36 + 4 \cdot 616$

$\Rightarrow a + b = 50$

117. Let Garima had n ₹ 2000 notes and x ₹ 200 in beginning.

After shopping, she will have $n/2$ ₹ 2000 notes, $(x + n/2)$ ₹200 notes and the same number of ₹ 500 notes what she had in beginning.

After transaction in restaurant, she will have $(n/2 + 30)$ ₹2000 notes and fifty ₹ 200 notes.

As given, $(n/2 + 30)$ is 175% of $n/2 \Rightarrow n = 80$

Further, $(x + n/2) = 50$

$\Rightarrow x + 80/2 = 50$

$\Rightarrow x = 10$

Hence, option(c) is correct.

118. Let c cricket balls, t tennis balls and r rubber balls are defective.

As given, $\dfrac{r}{4} > \dfrac{c}{6} > \dfrac{t}{5}$

and $\dfrac{r + c + t}{6 + 5 + 4} = 2 \times \dfrac{t}{5}$

$\Rightarrow c = 5t - r$

If $r = 1$, $t = 1$, then $c = 4$, but not satisfying $\dfrac{r}{4} > \dfrac{c}{6} > \dfrac{t}{5}$

If $r = 1$, $t = 2$, then $c = 9$, but the value of c cannot exceed 6. If we increase value of t beyond 2, the value of c continues to exceed 6.

Hence, value of t cannot be more than 1.

If $r = 2$, $t = 1$, then $c = 3$, but not satisfying $\dfrac{r}{4} > \dfrac{c}{6} > \dfrac{t}{5}$

If $r = 2$, $t = 2$, then $c = 8$, the value of c cannot exceed 6.

Hence, value of t cannot be more than 1.

If $r = 3$, $t = 1$, then $c = 2$, also satisfying $\dfrac{r}{4} > \dfrac{c}{6} > \dfrac{t}{5}$

If $r = 3$, $t = 2$, then $c = 7$, the value of c cannot exceed 6.

Hence, value of t cannot be more than 1.

If $r = 4$, $t = 1$, then $c = 1$, but not satisfying $\dfrac{r}{4} > \dfrac{c}{6} > \dfrac{t}{5}$

If $r = 4$, $t = 2$, then $c = 6$, but not satisfying $\dfrac{r}{4} > \dfrac{c}{6} > \dfrac{t}{5}$

If $r = 4$, $t = 3$, then $c = 11$, the value of c cannot exceed 6.

Hence, value of t cannot be more than 2.

Thus, option (d) is correct.

119. Let z, y and x be unit, tens and hundreds digits respectively. Then, original number $= 100x + 10y + z$

As given,

$100x + 10y + z + 198 = 100z + 10y + x$

$\Rightarrow 99x - 99z = -198$

$\Rightarrow z - x = 2$

Hence, the possible combinations are :

1 − 3, 2 − 4, 3 − 5, 4 − 6, 5 − 7, 6 − 8, 7 − 9

Obviously, each combination will have 10 numbers. Thus, required answer = 70

13

LCM, HCF/GCD

Learning OBJECTIVES

After completing this chapter, aspirants will be able to

- Understand the meaning of multiple and factor;
- Differentiate between LCM and HCF and learn the alternative approaches to find LCM and HCF in addition to conservative approaches;
- Observe the critical properties related to LCM as well as HCF;
- Apply these properties while solving the questions;
- Observe the concepts of remainders associated with LCM and HCF with the diverse cases along with illustrations;
- Learn how to find LCM and HCF of decimal numbers, fractions and algebraic expression;
- Learn how to compute LCM and HCF of the numbers expressed as $(\text{Base})^{\text{exponent}}$/ Product of $(\text{Base})^{\text{exponent}}$/ product of prime/composite and $(\text{Base})^{\text{exponent}}$;
- Solve the questions seeking concepts of both LCM and HCF;
- Observe the Pragmatic applications of the concepts of HCF and LCM;
- Realise how the wordings of the questions are analised;
- Observe the relation among LCM, HCF and product of numbers;
- Solve most of the questions without pen and paper as critical properties have been given;
- Feel comfortable to choose the appropriate approaches out of so many approaches as per his/her choice;
- Observe the diverse sorts of categories of the questions;
- Understand the philosophy of the division of this chapter in various parts;
- Analyse the trend of the questions being asked in various competitive examinations;
- Recognise the types of questions which can be easily solved through the given options;

Inside

- *Brief introduction*
- *Concept of LCM and HCF, their Critical Properties and Various Approaches for their Computation*
- *Problems Based on LCM/HCF with or without Association with Concepts of Remainders*
- *LCM/HCF of Fractions, Decimal Numbers, Expression Given in Form of $(\text{Base})^{\text{exponent}}$, Product of Prime/Composite and $(\text{Base})^{\text{exponent}}$ and Algebraic Expression*
- *Application of Concepts of Both LCM and HCF Simultaneously*
- *Pragmatic Application of Concepts of LCM and HCF*
- *Application of Concepts of LCM in Case of Circular Motion as well as in Comparison of Fractions*
- ....
- .... *& Many More*
- *Work Out with Solutions*

- Solve the questions in moments;
- Grip on concepts helping to solve the questions without pen and paper;
- Learn new concepts/fundamentals to the greater extent;
- Apply the pragmatic approaches in lieu of academic approaches; and
- Ponder on the alternative methods.

Introduction

This chapter is not so complicated if you are well versed with the real meanings of LCM and HCF and relation between them. The practical problems before the aspirants in this chapter is that they don't understand the questions in true spirit. That is why this chapter has been discussed category-wise. Alternative approaches have been developed so that most of the questions can be solved without pen and paper. Further, properties related to LCM and HCF has been discussed so that the questions are solved without adopting a long processes.

No doubt, this chapter plays a vital role in securing marks/ranks in the competitive examinations like CAT, MAT, SSC, CDS, NDA, Bank, LIC, PSU, Railway, etc. I hope, aspirants will realise themselves totally different after perusal of this chapter and obviously this will accelerate their efforts to achieve the goal.

I. Concept of Least Common Multiple (LCM)

Least Common Multiple (LCM) of two or more than two numbers/expressions is the least number/expression which is exactly divisible by all these numbers/expressions. To understand the concept of LCM in better way, consider an example:

'What is the LCM of 4, 6 and 8 ?'

Multiples of 4 = {4, 8 , 12, 16 , 20, <u>24</u>, 28, 32, 36, 40, 44, <u>48</u>, 52,......}

Multiples of 6 = {6, 12, 18, <u>24</u>, 30, 36, 42, <u>48</u>, 54, 60, ...}

Multiples of 8 = {8, 16, <u>24</u>, 32, 40, 48, 56, 64, 72, ... }

Here, common multiples = {24, 48, ...}

Least common multiples among {24, 48, ...} = 24

Hence, LCM = 24

Before discussing the diverse approaches, it will be appropriate to describe the conventional methods for computing the LCM. Generally, there are two methods for the same.

A. Factor-Method

we will comprehend this method through an example. Suppose you have to find LCM of 24, 36 and 40. First, we split the numbers in form of product of prime numbers.

$$24 = \boxed{2} \times \boxed{2} \times \boxed{2} \times \boxed{3}$$
$$36 = \boxed{2} \times \boxed{2} \times \boxed{3} \times \boxed{3}$$
$$40 = \boxed{2} \times \boxed{2} \times \boxed{2} \times 5$$

Now, take common factor from the above and multiply them.

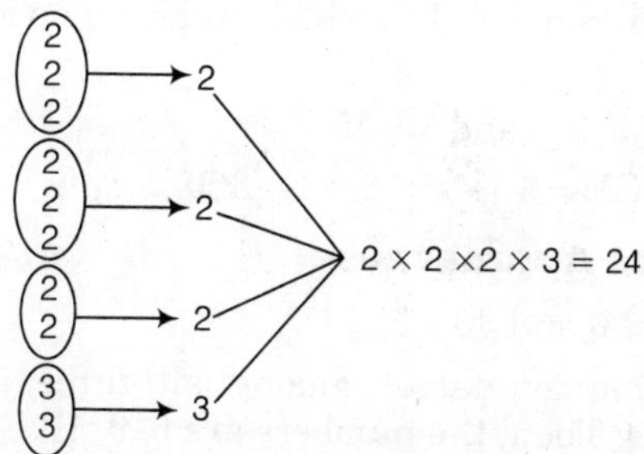

Further, you have to multiply 24 with the remaining factors which are common neither in two expressions nor in all three.

Hence, LCM = 24 × 3 × 5 = 360

Now, find the LCM of 28, 70 and 154 on the above line.

$$28 = \boxed{2} \times 2 \times \boxed{7}$$
$$70 = \boxed{2} \times 5 \times \boxed{7}$$
$$154 = \boxed{2} \times \boxed{7} \times 11$$

∴ Required answer = 2 × 7 × 2 × 5 × 11 = 1540

B. Division-Method

Find the LCM of 24, 36 and 40.

2	24, 36, 40
2	12, 18, 20
2	6, 9, 10
3	3, 9, 5
	1, 3, 5

∴ Required LCM = 2 × 2 × 2 × 3 × 1 × 3 × 5 = 360

Now, find the LCM of 28, 70 and 154 on the above line.

2	28, 70, 154
7	14, 35, 77
	2, 5, 11

∴ Required LCM = 2 × 7 × 2 × 5 × 11 = 1540

Alternative Approaches-I

You will agree with the view that both the methods may take some more time than what is required in diverse competitive exams. Hence, a pragmatic approach has been devised to compute LCM. Build your concept through examples: What is LCM of 24, 36 and 40 ?

Greatest common factor among all three at the first instance = 4. It is to note that if you take greatest common factor as 2 by mistake, there will be no effect on the final value of LCM and indeed there is a lot of possibility of these sorts of mistakes on selection at the first instance.

Then, the numbers are 6, 9, 10.

Now, greatest common factor = 3 {only 6 and 9 have been considered as 10 is not a multiple of 3}

Then, the numbers are 2, 3, 10.

Further, greatest common factor = 2 {only 2 and 10 have been considered as 3 is not a multiple of 2}

Then, numbers are 1, 3 , 5 which will have no any common factor.

Now, LCM of 1, 3 and 5 = 15

The final LCM = $15 \times 2 \times 3 \times 4 = 360$

Alternative Approaches-II

LCM of 24, 36 and 40 = ?

Greatest common factor among all three at the first instance = 4. Then, the numbers are 6, 9, 10.

Now, ascertain as to which least multiple of the highest number(i.e.10) is exactly divisible by the remaining numbers (i.e. 6 and 9). Obviously, it will be 90.

Then, required LCM = $90 \times 4 = 360$

Note *In case one or more of the remaining numbers is/are factor(s) of the highest number (say, 10), it is not required to ensure as to whether that least multiple of highest number is divisible by them or not as a factor of any number divides that number exactly.*

Now, find the LCM of 28, 70, and 154.

Greatest common factor at first instance = 14 (if you take 7, no problem).

Then, the numbers are 2, 5, 11.

Since all of 2, 5 and 11 are prime numbers, the required LCM will be the same as their product i.e. $2 \times 5 \times 11 = 110$

∴ Final LCM = $110 \times 14 = 1540$

Find the LCM of 12, 18 and 24.

If numbers are such that you know their tables, you may find the LCM directly. Here, highest number is 24 and also the fact that 12 is a factor of 24. Thus, you have check as to which least multiple of 24 is exactly divisible by 18 also. Obviously, it is 72.

∴ Required LCM = 72

We may try this question with **Alternative Approaches I** Let greatest common factor be 2 [Actually it is 6, however, 2 is taken so as to ascertain there is no effect on value of LCM in case of such types of mistakes]

Then, the numbers are 6, 9, 12.

Now, greatest common factor = 3

Then, the numbers are 2, 3, 4.

Now, greatest common factor = 2

Then, the numbers are 1, 3, 2 which will have no common factor. Now, LCM of 1, 3, 2 = 6

∴ Required LCM = $6 \times 2 \times 3 \times 2 = 72$

Observations Aspirants should adhere to the Alternative Approaches-I

Critical Properties of LCM

(i) The LCM of the given numbers and its multiples are exactly divisible by all of the given numbers including their factors.

e.g. The LCM of 4, 6 and 8 = 24. Then, 24 and its multiples like 48, 72, 96, ... will exactly be divisible by 4, 6 and 8 and also by all factors of 4, 6 and 8.

(ii) The LCM of two or more numbers is greater than or equal to the largest of those numbers.

(iii) LCM of prime numbers is equal to their product.

Illustrations

Problem Based on LCM but not Associated with the Concept of Remainders

Type I

When LCM may be or may not be altered by either division or multiplication or addition or subtraction.

Problem 1 Find the least number which is exactly divisible by 12, 15, 20 and 54.

Sol. As we know, LCM is the least number which is exactly divisible by the set of given numbers.

∴ Required number = LCM of 12, 15, 20 and 54 = 540

Problem 2 What will be the least number which when doubled will be exactly divisible by 12, 18, 21 and 30?

Sol. The least number exactly divisible by 12, 18, 21 and 30 = LCM of 12, 18, 21 and 30 = 1260

Obviously, the required number should be half the LCM so that it becomes equal to value of LCM when it is doubled.

∴ Required number = $\frac{1260}{2} = 630$

Problem 3 The smallest number which when diminished by 7, is divisible by 12, 16, 18, 21 and 28 ***[LIC(AAO)-03]***

Sol. The smallest number exactly divisible by 12, 16, 18, 21 and 28 = LCM of 12, 16, 18, 21 and 28 = 1008. The desired number must be greater than the smallest number by 7 so that its value becomes equal to value of smallest number when it is diminished by 7.

Thus, desired number = 1008 + 7 = 1015

Problem 4 The least numbers which should be added to 57 so as to make the sum exactly divisible by 5, 6, 4 and 3, is

Sol. The smallest number exactly divisible by 5, 6, 4 and 3 = LCM of 5, 6, 4 and 3 = 60. The desired number must be 57 less than the smallest number so that its value becomes equal to value of smallest number when it is added to 57.

Thus, desired number = 60 − 57 = 3

Problem 5 What will be the least number which when divided by 7, is divisible by 12, 15 and 20.

Sol. The least number exactly divisible by 12, 15 and 20 = LCM of 12, 15 and 20 = 60. Obviously, the required number should be 7 times the LCM, so that it becomes equal to value of LCM when it is divided by 7. Hence, required number = $60 \times 7 = 420$

Observations We act in opposite direction of the operations given in the questions. If in question, multiplication is written, we do division and we do addition when subtraction is written in the questions and *vice-versa.*

Type II

Where a certain multiple of LCM is required.

Problem 6 The greatest number of four digits which is divisible by 15, 25, 40 and 75 is ? *[SSC(CGL-PT)-03, SSC(CGL-T-II)-13]*

Sol. The least number divisible by 15, 25, 40 and 75 = LCM of 15, 25, 40 and 75 = 600. But, 600 is not a four-digit number, however, its multiples may be four-digit numbers which will obviously be divisible by the given numbers. Now, we may proceed through the given options. The option having the greatest value among the options divisible by 600, will be correct option. Let the options be 9000, 9400, 9600 and 9800. Clearly, 9000 and 9600 are divisible by 600. Hence, required number = 9600 as it is greater than 9000.

Alternatively

The greatest number of four digits = 9999, then it is obvious that the required number will be near to it. We divide 9999 by 600.

```
600)9999(16
    600
    3999
    3600
    ×399
```

If we subtract the remainder 399 from 9999, the resultant would exactly be divisible by 600 i.e. individually by each of 15, 25, 40 and 75. Here, the resultant(i.e. 9600) is a four-digit number.

Further, if we add (600 − 399) i.e. 201 to 9999, the resultant would also be exactly divisible by 600 i.e. individually by each of 15, 25, 40 and 75. Here, the resultant(10200) is a five-digit number. As per question, the required number should be of four digits. Hence, the required number = 9999 − 399 = 9600

Problem 7 Find the least number of 5 digits, which is exactly divisible by 15, 25, 40 and 75.

Sol. LCM of 15, 25, 40 and 75 = 600. But, 600 is not a five-digit number, however, its multiples may be five-digit numbers which will obviously be divisible by the given numbers. Now,

```
600)10000(16
     600
     4000
     3600
     ×400
```

If we add (600 − 400) i.e. 200 to 10000, the resultant would be exactly divisible by 600 i.e. individually by each of given numbers.

Here, the resultant (i.e. 10200) is a five-digit number.

Further, if we subtract 400 from 10000, the resultant would also be divisible by 600 i.e. individually by each of 15, 25, 40 and 75. But, the resultant (i.e. 9600) is a four-digit number, which is not desirable in the question.

Hence, required number = 10000 + (600 − 400) = 10200

Problem 8 The number nearest to 43582 divisible by each of 25, 50 and 75 is: *[SSC(CPO)-07]*

Sol. LCM of 25, 50 and 75 = 150

```
150)43582(290
    300
    1358
    1350
    ×××82
```

Clearly, (43582 − 82) i.e. 43500 is exactly divisible by 150 i.e. individually by each of 25, 50 and 75.

Further, 43582 + (150 − 82) = 43650, is also divisible by 150 i.e. individually by each of 25, 50 and 75.

It is clear that 43650 is closer to 43582 in comparison to 43500. Hence, required number = 43650.

Problem 9 The least number which is a perfect square and is divisible by each of the numbers 21, 36 and 66.

Sol. $21 = 3 \times 7$, $36 = 2 \times 2 \times 3 \times 3$ and

$66 = 2 \times 3 \times 11$

Now, LCM $= 3 \times 3 \times 2 \times 2 \times 7 \times 11$

In order to make $\underline{3 \times 3} \times \underline{2 \times 2} \times \underline{7 \times 11}$ a perfect square,

we must multiple it by 7×11.

∴ Perfect square number $= \underline{3 \times 3} \times \underline{2 \times 2} \times \underline{7 \times 7} \times \underline{11 \times 11}$

$= 213444$

Problem 10 Find the number of numbers lying between 500 and 1000, which are divisible by 12, 15, 20 and 24.

Sol. Least number divisible by each of 12, 15, 20 and 24 = LCM 12, 15, 20 and 24 = 120. It is fact that the multiples of 120 will also be divisible by each of 12, 15, 20 and 24. Such multiples are 120, 240, 360, 480, 500, 720, 840, 960, 1080, ...

Now, numbers lying between 500 and 1000 ⇒ 600, 720, 840, 960

∴ Number of such numbers = 4

Problem 11 Find the number of numbers between 1000 and 10000, which are divisible by 15, 21 and 35 and are odd also.

Sol. The least number divisible by 15, 21 and 35 = 105

The number just above 1000, divisible by given numbers = 1050 and the number just below 10000, divisible by given numbers = 9975 {we can find 9975 through the process mentioned earlier}

Now, numbers divisible by 105 ⇒ 1050, 1155, 1260 -------- 9975

Total number of such numbers $= \frac{9975 - 1050}{105} + 1 = 85 + 1 = 86$

(43 even and 43 odd as the series starts with even number and ends with odd)

∴ Required number = 43

Note *You may apply AP series concept to compute number of numbers.* $9975 = 1050 + (n-1)105 \Rightarrow n - 1 = 85 \Rightarrow n = 86$

Type III

Finding number(s) or relation between them if their LCM is known.

Problem 12 The LCM of two numbers is 48. The numbers are in the ratio 2 : 3. The sum of the numbers is *[SSC(CGL-PT)- 03, SSC(MTS)-11]*

Sol. Let the numbers are $2x$ and $3x$.

$\text{LCM} = 2 \times 3 \times x = 6x$

As given, $6x = 48$

$\therefore \quad x = 8$

Then, sum of numbers $= 2x + 3x = 5x = 5 \times 8 = 40$

Alternatively

LCM of 2 and 3 = 6

As given, 6 units = 48

$\therefore \quad (2 + 3)$ units = 40

Problem 13 Product of two co-prime numbers is 117. Their LCM should be

(a) 1 (b) 117
(c) equal to their HCF (d) Cannot be calculated

Sol. 117 may be written as

1×117
3×39
9×13

Clearly, 1 and 117 & 9 and 13 are co-primes. Hence, their LCM = 117

Alternatively

We know, LCM × HCF = 1st number × 2nd number

and HCF of co-prime numbers is always 1. {These will be discussed later in this Chapter}

Then, LCM × 1 = 117

$\therefore \quad$ LCM = 117

Problem 14 LCM of two prime numbers x and $y (x > y)$ is 161. The value of $3y - x$ is *[SSC (CGL-T-II)-13, CDS-21]*

Sol. We know, LCM of prime numbers is equal to their product.

Hence, $161 = x \times y$

$\Rightarrow \quad 1 \times 161$ or $7 \times 23 = x \times y$

In case of (1×161), 1 is not prime number.

But in case of (7×23), 7 and 23 both are prime numbers.

$\therefore \quad x = 23$ and $y = 7$ $\quad [\because x > y$ as given$]$

Now, $3y - x = (3 \times 7) - 23 = -2$

Alternatively

We know, LCM × HCF = 1st number × 2nd number

and HCF of prime numbers is always 1.

Then, $\text{LCM} \times \text{HCF} = x \times y$

$\Rightarrow \quad 161 \times 1 = xy$

$\Rightarrow \quad 1 \times 161$ or $7 \times 23 = x \times y$

In case of (1×161), 1 is not prime number.

But in case of (7×23), 7 and 23 both are prime numbers.

$\therefore \quad x = 23$ and $y = 7$ $\quad [\because x > y$ as given$]$

Now, $3y - x = (3 \times 7) - 23 = -2$

Problem 15 The LCM of two multiples of 12 is 1056. If one of them is 132, find another.

Sol. Given, one number $= 132 = 12 \times 11$

Clearly, second number $= 12 \times x$

$\{\because$ both numbers are multiples of 12$\}$

Now, their LCM $= 12 \times x \times 11 = 1056$ (given)

$\therefore \quad x = 8$

Then, second number $= 12 \times 8$

$= 96$

Alternatively

Proceed through options. Let options be 12, 96, 72, 132

(a) 12 ⇒ It is multiple of 12, but LCM of 12 and 132 ≠ 1056

(b) 96 ⇒ It is multiple of 12 and LCM of 96 and 132 = 1056

Hence, option(b) is correct.

Concept of Remainders Associated with LCM

Case I **Remainders are same when a number is divided by a set of numbers.**

Problem 16 What will be the least number which when divided by 12,15,20 and 54 leaves 4 as remainder in each case?

Sol. Least number exactly divisible by 12, 15, 20 and 54 = LCM of 12, 15, 20 and 54 = 540. If we increase 540 by 4, obviously, the new number i.e. 544 will leave 4 as remainder in each case.

∴ Required number = 544

Alternatively

Proceed through options. Let the options be 450, 544, 454, 540. Now, ascertain as to which option is satisfying the conditions.

(a) 450 ⇒ 450 ÷ 12, remainder ≠ 4

(b) 544 ⇒ 544 ÷ 12, remainder = 4; 544 ÷ 15, remainder = 4;
544 ÷ 20, remainder = 4;
and 544 ÷ 54, remainder = 4

Hence, option (b) is correct.

Problem 17 Find the greatest number of five digits which when divided by 3, 5, 8, 12 have 2 as remainder. *[SSC(CGL-PT)-02, CLAT-18]*

Sol. LCM of 3,5,8 and 12 = 120. Now, we divide five-digit greatest number i.e. 99999 by 120 and get 39 as remainder. It implies that (99999 – 39) i.e. 99960 is the greatest number of five digits which is exactly divisible by 3, 5, 8 and 12.

In order to get 2 as remainder in each case, we should add 2 to 99960. Then, the required greatest number

$= 99960 + 2$

$= 99962$

Alternatively

Proceed through options. Let the options be 99999, 99942, 99960, 99962

(a) 99999 ⇒ 99999 ÷ 3, remainder ≠ 2

(b) 999842 ⇒ It leaves 2 as remainder in each case when divided by each of 3, 5, 8 and 12.

But we observe that there is some greater number in the options which may also satisfy the given condition. Thus, continue.

(c) 99960 ⇒ It leaves 0 as remainder in each case when divided by each of 3, 5, 8 and 12.

(d) 99962 ⇒ It also leaves 2 as remainder in each case when divided by each of 3, 5, 8 and 12.

Here, option (b) and (d) both are satisfying the conditions. Since the value of option(d) is greater, this is the correct option.

Problem 18 Assume, N be a six-digit least number, which when divided by 4, 6, 10 and 15 leaves 2 as remainder in each case. Find the sum of digits of number N. ***[SSC(CGL-PT)-03, 19]***

Sol. LCM of 4, 6, 10 and 15 = 60. Now, we divide six-digit least number i.e. 100000 by 60 and get 40 as remainder. It implies that {100000 + (60 − 40)} i.e. 100020 is the least number of six digits which is exactly divisible by 4, 6, 10 and 15.

In order to get 2 as remainder in each case, we should add 2 to 100020. Then, the required least number (N)

= 100020 + 2 = 100022

Now , the sum of digits = 1 + 0 + 0 + 0 + 2 + 2 = 5

Problem 19 The least number which should be added to 4312 so that the sum leaves the remainder 5 in each case when divided by 36, 48, 60 and 96, is ?

Sol. The least number divisible by 36, 48, 60 and 96 = LCM of 36, 48, 60 and 96 = 1440. Now, we get 1432 as remainder when we divide 4312 by 1440. Obviously, if we add 8 (i.e.1440 − 1432) to 4312, the resultant will be divisible by 1440. However to get 5 as remainder, we should add 5 more i.e. 8 + 5.

∴ Required number = 13.

Problem 20 The least number which when divided by 5, 6, 7 and 8 leaves a remainder 3, but when divided by 9 leaves no remainder, is ***[LIC(AAO)-03]***

Sol. The least number divisible by 5, 6, 7 and 8 = LCM of 5, 6, 7 and 8 = 840.

Since the multiples of LCM are also divisible by the given set of numbers, 840K will be divisible by them. Then, the required number will be in form of (840K + 3), where K has an integer value. It is also given that when (840K + 3) is divided by 9 leaves no remainder.

If we divide (840K + 3) by 9, quotient = 93K and remainder = 3K + 3. Since there is no remainder, (3K + 3) must be divisible by 9 and it is possible at K = 2. Here, 2 is minimum value, other values may be 5, 8, etc.

Hence, required number = (840K + 3) = (840 × 2) + 3 = 1683

Alternatively

Proceed through options and ascertain as to which option is satisfying the conditions.

Problem 21 The least multiple of 13, which on dividing by 4, 5, 6, 7 and 8 leaves remainder 2 in each case is

Sol. On the above line, the required form of the number = {K(LCM of 4, 5, 6, 7 and 8) + 2} = 840K + 2

At K = 3, (840K + 2) becomes a multiple of 13.

Then, the required number = (840 × 3) + 2 = 2500

Alternatively

Proceed through options and ascertain as to which option is satisfying the conditions.

Case II **Remainders are different when a number is divided by a set of numbers, but the differences between the divisors and the respective remainders are equal.**

Problem 22 The least number, which when divided by 48, 60, 72,108 and 140 leaves 38, 50, 62, 98 and 130 as remainders respectively, is : ***[SSC (CGL-T-II)-20]***

Sol. The least number divisible by 48, 60, 72, 108 and 140 = LCM of 48, 60, 72, 108 and 140 = 15120

Since, 15120 is divisible by 48, we must subtract 10 from 15120 to get remainder 10 less than 48 i.e, 38. Needless to say that the earlier quotient will be diminished by 1. It also holds good for all other cases where the remainders are to be obtained 10 less than the divisors.

Hence, required number = 15120 − 10 = 15110

Observations The required number = (LCM of given set of numbers)−(difference between a divisor and the respective remainder)

Problem 23 What is the least number of four digits which when divided by 12, 15 and 18 leave the remainders 7, 10 and 13 respectively?

Sol. LCM of 12, 15 and 18 = 180. Now, we get 100 as remainder when we divide the least number of four digits (i.e. 1000) by 180. Obviously, {1000 + (180 − 100)} i.e., 1080 is the least number of four digits divisible by 12, 15 and 18.

Since the differences between the divisors and the respective remainders are equal [∵ 12 − 7 = 15 − 10 = 18 − 13 = 5], we will subtract 5 from 1080.

Hence, required number

= 1080 − 5 = 1075

Alternatively

Proceed through options and ascertain as to which option is satisfying the conditions.

Problem 24 What is the least number which when divided by 9, 12, and 15 leaves the remainders 1, 4 and 7 respectively, but exactly divisible by 11?

Sol. LCM of 9, 12 and 15 = 180

Since the differences between the divisors and the respective remainders are equal {i.e. 9 − 1 = 12 − 4 = 15 − 7 = 8}, the required number will be in form of (180K − 8).

At K = 2, (180K − 8) is divisible by 11.

Hence, required number

= (180 × 2) − 8 = 352

Alternatively

Proceed through options and ascertain as to which option is satisfying the conditions.

Case III **Remainders are different when a number is divided by a set of numbers, but the difference between the divisors and the respective remainders are not equal.**

Problem 25 What is the least number which when divided by 11 and 7, leaves the remainders 2 and 1 respectively?

Sol. Obviously, the number = $11k + 2 = 7m + 1$, where k and m both are integers.

$\Rightarrow \quad 11k = 7m - 1 \Rightarrow k = \dfrac{7m-1}{11}$

Now, we put the integer value of m in above equation starting from 1, to get the integer value of k.

At $m = 8$, k becomes 5

$\therefore$ Required number = $\{(11 \times 5) + 2\}$ or $\overline{\{(7 \times 8) + 1\}}$ i.e., 57.

Remark

In context of above question, the higher number satisfying the same conditions will be in form of [z(LCM of 11 and 7) + 57], where z is also an integer. e.g., the next higher number = 1(11 × 7) + 57 = 134. Further, the next higher number = 2(11 × 7) + 57 = 211 and so on.]

II. Concept of HCF Or GCD

HCF of two or more than two numbers/expressions is the greatest number which can divide all these numbers/expressions exactly. Now, Build your concept through an example. HCF of 12, 18, 24 = ?

Factors of 12 = [1, 2, 3, 4, 6, 12]

Factors of 18 = [1, 2, 3, 6, 9, 18]

Factors of 24 = [1, 2, 3, 4, 6, 8, 12, 24]

Here, Common Factors = [1, 2, 3, 6]

and Highest among Common Factors [1, 2, 3, 6] = 6

Hence, 6 is Highest Common Factor (HCF) or Greatest Common Divisor (GCD).

Before discussing an ideal approach, Conventional Methods i.e. Factor Method and Division-Method are being described for computation of HCF.

Factor-Method

Split the given numbers into prime factors.

$12 = \boxed{2} \times 2 \times \boxed{3}$
$18 = \boxed{2} \times 3 \times \boxed{3}$
$24 = \boxed{2} \times 2 \times 2 \times \boxed{3}$

Now, find the product of factors, which are common in all three given numbers.

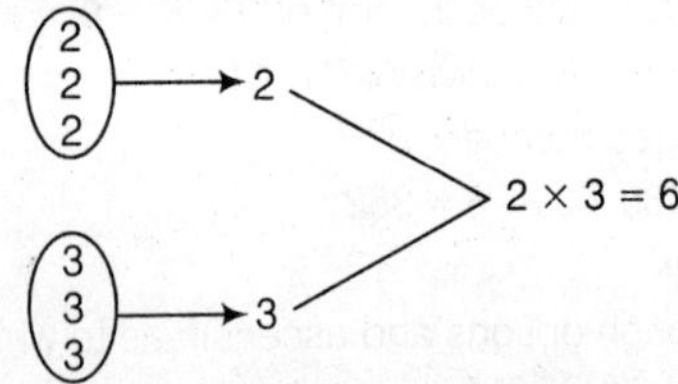

$\therefore$ Required HCF = 2 × 3 = 6

Division-Method

Take the two smallest numbers from the given numbers and divide the larger of them by the smaller and then divide the divisor by the remainder thus obtained. Further divide the just previous divisor by the remainder obtained. Repeat the process until the remainder is zero.

The last divisor would be the required HCF of two smaller numbers. If given numbers are more than two, divide the third number by the HCF of previous two smaller number and process the same until the remainder is zero. Here, the last divisor would be the HCF of all three numbers. For the four numbers, the same process is adopted.

```
12)18(1
   12
   ----
 × 6)12(2
     12
     ----
     × ×
```

HCF of 12 and 18 = 6

```
6)24(4
  24
  ---
  × ×
```

HCF of 12, 18 and 24 = 6

Alternative Approach

As observed from the above that the HCF is a factor of all possible differences between two numbers out of the given numbers. Thus, to find the HCF quickly, we divide the given numbers by the lowest possible difference of two numbers so as to test the divisibility. If the given numbers are divisible by this lowest difference, the difference itself would be the required HCF.

Further, in case all of the numbers are not divisible by the lowest possible difference, one of its factors will be the required HCF. In this situation, we should test the divisibility of given numbers by the factors of lowest possible difference one by one (greatest factor to lowest factor). The factor which divides all the given numbers will be the required HCF. Thereafter, no need to test the divisibility by the factors lower than the factor dividing all the given numbers exactly. Take some examples to build the skill.

(a) **HCF of 84, 105, 231 = ?**

Difference between 84 and 105 = 21

and Difference between 105 and 231 = 126

and Difference between 84 and 231 = 147

Here, there are three differences between two numbers are possible. Now, the lowest difference = 21. You may avoid to find the 3rd difference as it will be larger due to being a sum of previous two differences.

Now, check as to whether 21 divides all of 84, 105 and 231 exactly. Yes, it divides.

$\therefore$ Required HCF = 21

(b) **HCF of 175, 420 and 470 = ?**

Difference between 175 and 420 = 245

Difference between 420 and 470 = 50

Hence, the lowest difference = 50 [3rd difference has not been computed deliberately]

Now, check as to whether 50 divides all of 175, 420 and 470. Clearly, it is not dividing exactly.

Now, check as to which factor of 50 is dividing 175, 420 and 470 exactly and you have to try with greatest factor to lowest factor.

25 ⇒ it is also not dividing all the given numbers exactly.

5 ⇒ it is dividing all the given numbers exactly.

∴ Required HCF = 5

Critical Properties Related to HCF

(i) HCF is the highest factor common in all the given numbers. It may be construed that all the given numbers are multiples of their HCF.

(ii) The HCF of two or more numbers is either smaller or equal to the smallest of those numbers.

(iii) The HCF of prime numbers is always 1 as 1 is the highest factor common in all.

(iv) HCF is always a factor of LCM.

(v) HCF is a factor of all possible differences between two numbers out of the given numbers.

For Examples

(a) HCF of 10 and 16 is 2, which is a factor of (16 – 10).

(b) HCF of 12, 16 and 24 is 4, which is a factor of (16 – 12), (24 – 16) and (24 – 12).

(c) HCF of 15, 25 and 35 is 5 which is a factor of (25 – 15), (35 – 25) and (35 – 15)

It is also observed from the above three examples that the HCF of all possible differences between two numbers out of given numbers is not always equal to actual HCF of the given numbers. In example (b), Actual HCF = 4 = HCF of differences 4, 8 and 12, whereas in example (c) Actual HCF = 5, but the HCF of differences 10, 10 and 20 is 10.

(vi) The HCF of the sum of two numbers and their LCM is equal to the HCF of these two numbers.

Let numbers be 12 and 16. Then, their sum = 28 and their LCM = 48

Now, HCF of 28 and 48 will be equal to HCF of 12 and 16 i.e. 4

(vii) The HCF of the quotients obtained by dividing two or more numbers by their HCF is always 1.

HCF of 12, 18 and 27 = 3, then HCF of 12/3, 18/3 and 27/3 will always be 1.

(viii) If sum or product of certain pairs of two numbers is equal and also their HCF is same, then HCF will be the least number not only for the pair concerned, but also among all the given pairs of numbers. This concept is applicable in case of three numbers too.

(ix) If a and b are co-primes, the HCF of their sum and their difference will always be either 1 or 2.

(x) If (a and b) and (a and c) are two pairs of co-primes, then HCF of a and bc is equal to 1. In other words, a and bc will be co-primes.

(xi) The product of two numbers is equal to the product of their LCM and their HCF.

Let the numbers be 8 and 12. Then, their LCM = 24 and their HCF = 4

Clearly, $8 \times 12 = 24 \times 4$

Hence, 1st Number × 2nd Number = LCM × HCF

[This is applicable only in case of two numbers]

(xii) Product of n numbers = LCM × $(\text{HCF})^{n-1}$. Here, HCF refers to HCF of each possible pair out of given numbers and it is applicable only when HCFs of all possible pairs are same.

e.g. The HCF of any two numbers out of three numbers is 4 and LCM of all three numbers is 240. What is the product of all three numbers ?

Required product

$$= \text{LCM} \times (\text{HCF})^{n-1} = 240 \times (4)^{3-1} = 3840$$

(xiii) If numbers $n_1, n_2, n_3 \ldots\ldots n_x$ leave the remainders $r_1, r_2, r_3, \ldots\ldots\ldots r_x$ respectively when each of them is divided by the same number D, the HCF of $(n_1 - r_1), (n_2 - r_2), (n_3 - r_3) \ldots\ldots\ldots (n_x - r_x)$ will be the greatest value of D.

(xiv) If numbers $n_1, n_2, n_3, \ldots\ldots n_x$ leave the same remainder (but, not known) when each of them is divided by the same number D, the HCF of all possible differences between two numbers out of given numbers will be the greatest value of D.

In case of two numbers, the difference between them will be the greatest value of D. However, if this difference is greater than the smaller number, the factor of the difference just below the smaller number will be the greatest value of D.

(xv) The HCF of $(a^m - 1)$ and $(a^n - 1)$ is $(a^{\text{HCF of } m \text{ and } n} - 1)$

e.g. HCF of $(2^6 - 1)$ and $(2^4 - 1) = (2^{\text{HCF of 6 and 4}} - 1)$

$$= (2^2 - 1) = 3$$

Illustrations

Problem Based on HCF but not Associated with the Concept of Remainders

Type I

When HCF is to be computed only.

Problem 26 The greatest number that exactly divides 156, 221 and 364.

Sol. HCF is the greatest number which divides the given set of numbers exactly.

Hence, required number = HCF of 156, 221 and 364 = 13

Type II

Finding number(s) or relation between them if their HCF is known.

Problem 27 Three numbers are in the ratio 1 : 2 : 3 and their HCF is 12. The numbers are.

Sol. Let the numbers be x, $2x$ and $3x$.

Then, their HCF = Product of factors common in all three = x

As given, $x = 12$

$\therefore$ Numbers are (1×12), (2×12) and (3×12) i.e 12, 24 and 36.

Alternatively I

HCF of 1, 2 and 3 = 1

Given, 1 unit $\Rightarrow$ 12

$\therefore$ 2 units = 24 and 3 units = 36

Alternatively II

Proceed through options (check as to which option has numbers in given ratio and also satisfying the condition that their HCF is 12).

Problem 28 The product of two numbers is 2028 and their HCF is 13. The number of such pairs is:

[SSC(CGL-T-I)-11, SSC(CPO)-03]

Sol. Let the numbers be $13x$ and $13y$ as given numbers are multiples of their HCF. Further, it is fact that HCF of the quotients obtained by dividing two or more numbers by their HCF is always 1. It implies that x and y are co-prime numbers.

As given, $13x \times 13y = 2028$

$\Rightarrow \quad xy = 12$

Now, positive values of x and y may be as below:

x	1	2	3	4	6	12
y	12	6	4	3	2	1

Here, dark portion is only repetition of earlier pairs. Hence, the same has been left as order of numbers is not required at all in this question. It is clear from the above that there are three pairs satisfying that their product is 12. However, only two pairs i.e. (1, 12) and (3, 4) are co-primes. Hence, required number of pairs = 2

Note *Numbers are* $(13 \times 1, 13 \times 12)$ *and* $(13 \times 3, 13 \times 4)$

Alternatively

Product of quotients (say, x and y) obtained by dividing the given two numbers by their HCF $= \dfrac{\text{Given actual product}}{\text{HCF} \times \text{HCF}}$

$$\Rightarrow \quad xy = \frac{2028}{13 \times 13} = 12$$

Now, co-primes with product 12 are (1, 12) and (3, 4). It means, there are two such pairs.

The above concept is also applicable for more than two numbers.

Hence, product of quotients (say x, y, z ...... nth term) obtained by dividing the n given numbers by their HCF $= \dfrac{\text{Given actual product}}{(\text{HCF})^n}$

Problem 29 The product of two numbers is 4107 and their HCF is 37. The greatest number is:

Sol. Let the numbers be $37x$ and $37y$, where x and y are co-primes.

As given, $37x \times 37y = 4107$

$\Rightarrow xy = \dfrac{4107}{37 \times 37} = 3$. The possible value of x and y are (1, 3) or (3, 1). Since one is repetition of another, we take only one pair as order of numbers is not required at all. Hence, the required numbers are $(37 \times 1, 37 \times 3)$ i.e. 37 and 111.

Now, the greatest number = 111. You may directly write the greatest number = $3 \times 37 = 111$ without finding the value of smaller number.

Alternatively I

Product of quotients (say, x and y) obtained by dividing the given two numbers by their HCF $= \dfrac{\text{Given actual product}}{\text{HCF} \times \text{HCF}}$

$$\Rightarrow \quad xy = \frac{4107}{37 \times 37} = 3$$

Now, pair of co-primes with product 3 is (1,3).

Hence, greatest number = $3 \times 37 = 111$

Alternatively II

We know, if products of each pair of two numbers are the same and also HCF of each pair is same, the HCF is the least number not only for the pair concerned, but also among all the possible pairs of numbers.

Hence, least number = HCF = 37

Then, greatest number = $4107 \div 37 = 111$

Problem 30 The sum of two numbers is 528 and their HCF is 33. The number of such pairs is ?

[SSC(CGL-PT)-08, MAT-04, SSC (CGL-T-II-19]

Sol. Let the numbers be $33x$ and $33y$, where x and y are co-primes.

As given, $33x + 33y = 528$

$\Rightarrow \quad x + y = 16$

Now, pairs of co-primes with sum 16 are (1, 15), (3, 13), (5, 11) and (7, 9).

$\therefore$ Required number of such pairs = 4

Alternatively

$$x + y = \frac{\text{Actual sum of numbers}}{\text{HCF}} = \frac{528}{33} = 16$$

Now, pairs of co-primes with sum 16 are (1, 15), (3, 13), (5, 11) and (7, 9).

$\therefore$ Required number of such pairs = 4

Observations Sum of quotients (say x, y, z, ... nth term) obtained by dividing the n given numbers by their HCF $= \dfrac{\text{Actual sum of numbers}}{\text{HCF}}$

Problem 31 The HCF of two numbers are 14 and their difference is 28. How many such pairs are possible?

Sol. Let the numbers be $14x$ and $14y$ where x and y are co-primes.

As given, $\quad 14x - 14y = 28$

$\Rightarrow \quad x - y = 2$

Now, pairs of co-primes with difference 2 are (3,1), (5,3),------

Thus, There are infinite number of such pairs.

Problem 32 The sum of two numbers is 384 and their HCF is 48. Find their difference.

Sol. Let the numbers be $48x$ and $48y$, where x and y are co-primes.

As given, $48x + 48y = 384$

$\Rightarrow \quad x + y = 8$

Now, pairs of co-primes with sum 8 are (1, 7), (3, 5).

$\therefore$ Required numbers are $(48 \times 1, 48 \times 7)$ and $(48 \times 3, 48 \times 5)$.

Hence, their differences are 288 and 96.

Alternatively

Let the numbers be $48x$ and $48y$, where x and y are co-primes.

As given $48x + 48y = 384$

$\Rightarrow \quad x + y = 8$

Now, pairs of co-primes with sum 8 are (1, 7), (3, 5).

Clearly, difference in 1st pair = 6 and that in second pair = 2

$\therefore$ Actual differences $= 48 \times 6$ and 48×2 i.e. 288 and 96.

Concept of Remainders Associated with HCF

Case I Remainders are same(known) when a number divides the given set of numbers.

Problem 33 **What will be the greatest number that divides 1356, 1868 and 2764 leaving 12 as remainder in each case?** ***[Delhi Police-07]***

Sol. Given that each of 1356, 1868 and 2764 leaves the same remainder i.e. 12 when divided by a number. It implies that (1356 − 12), (1868 − 12), (2764 − 12) will be exactly divisible by that number. Now, the greatest number which can divide 1344, 1856 and 2752 = their HCF = 64.

Alternatively

You may proceed through options so as to check as to which option is satisfying the condition(s) given in the question.

Case II Remainders are not same (but known) when a number divides the given set of numbers.

Problem 34 **Find the greatest number that divides 3026 and 5053 leaving the remainders 11 and 13 respectively.** ***[SSC(CPO)-06, CLAT-13]***

Sol. Given that 3026 and 5053 leaves the remainders 11 and 13 respectively when they are divided by a number. It implies that (3026 − 11) and (5053 − 13) will be exactly divisible by that number. Now, the greatest number which can divide 3015 and 5040 = their HCF = 45.

Alternatively

You may proceed through options so as to check as to which option is satisfying the condition(s) given in the question.

Case III Remainders are same(but not known) when a number divides the given set of numbers.

Problem 35 **Find the greatest number that will divide 43, 91 and 183 so as to leave the same remainder in each case.** ***[LIC(AAO)-03, CLAT-18]***

Sol. Recall the properties related to HCF "If numbers $n_1, n_2, n_3, \dots n_x$ leave the same remainder (but, not known) when they all are divided by the same number D, the HCF of all possible differences between two numbers out of given numbers will be the greatest value of D."

Here, the greatest number = HCF of (91 − 43), (183 − 91) and (183 − 43) = HCF of 48, 92 and 140 = 4

Alternatively

You may proceed through options so as to check as to which option is satisfying the condition(s) given in the question.

Problem 36 **When 2270, 3739 and 6677 are divided by a certain number of three digits, the remainders are same. Find the remainder.**

Sol. The greatest number which divides all the three numbers leaving the same remainder

= HCF of (3739 − 2270), (6677 − 3739) and (6677 − 2270)

= HCF of 1469, 2938 and 4407 = 1469

As given, the divisor should be a 3-digit number. It is obvious that a 3-digit factor of 1469 can also satisfy the conditions as the number which is a divisible by a certain number, is also divisible by its all factors.

Now, $1469 = 13 \times 113$

Thus, the required divisor = 113 (a 3-digit number)

Further, required remainder = Remainder obtained in case anyone of given numbers is divided by 113 = 10

III. LCM and HCF of Fractions

$$\text{LCM of fractions} = \frac{\text{LCM of numerators}}{\text{HCF of denominators}}$$

$$\text{HCF of fractions} = \frac{\text{HCF of numerators}}{\text{LCM of denominators}}$$

[It is important to note that LCM of fractions may be either a fraction or an integer but the HCF of fractions is always a fraction]

Problem 37 **Find the LCM of $\frac{2}{5}, \frac{8}{15}$ and $\frac{12}{25}$.**

Sol. Required LCM $= \frac{\text{LCM of 2, 8 and 12}}{\text{HCF of 5,15 and 25}}$

$= \frac{24}{5} = 4\frac{4}{5}$

Problem 38 **Find the HCF of $\frac{3}{16}, \frac{5}{12}$ and $\frac{7}{18}$.** ***[RRB-Ajmer(ASM)-01]***

Sol. Required HCF $= \frac{\text{HCF of 3, 5 and 7}}{\text{LCM of 16, 12 and 18}}$

$= \frac{1}{144}$

Problem 39 **The smallest fraction, which each of $\frac{6}{7}, \frac{5}{14}, \frac{10}{21}$ will divide exactly is :**

Sol. The LCM of the given fractions is the smallest fraction, which is exactly divisible by all of them.

$\therefore$ Required fraction = LCM of $\frac{6}{7}, \frac{5}{14}$ and $\frac{10}{21}$

$= \frac{\text{LCM of 6, 5 and 10}}{\text{HCF of 7, 14 and 21}}$

$= \frac{30}{7} = 4\frac{2}{7}$

IV. LCM and HCF of Decimal Numbers

Steps for Computing LCM and HCF

(i) First of all equalise the number of digits after the decimal point by suffixing zero(s), wherever is required.

(ii) Thereafter treat these numbers as integers (i.e. without decimal points)

(iii) Find LCM or HCF of these integers and put the decimal point leaving as many digits from its extreme right as there are in each of the numbers (which have been equalised in terms of number of digits after decimal point) after the decimal point.

Problem 40 Find the LCM of 1.7, 0.68 and 1.02. *[CLAT-14]*

Sol. Given numbers = 1.70, 0.68 and 1.02

[we equalised number of digits after decimal point by suffixing one zero in the first number]

Now, LCM of 170, 068 and 102 = 1020

Since there are two digits after decimal point in every number after equalisation in terms of number of digits after decimal point, we put decimal points after two digits from the right extreme.

∴ Required LCM = 10.20

Problem 41 The HCF of 1.75, 5.6 and 7 is: *[Hotel Mgmt - 02]*

Sol. Given numbers: 1.75, 5.60 and 7.00

Now HCF of 175, 560 and 700 = 35

∴ Required HCF = 0.35

LCM and HCF of the numbers expressed as (Base)^{exponent} / Product of (Base)^{exponent} / product of prime/composite and (Base)^{exponent}

S.No.	Cases	Numbers/terms	LCM	HCF
1.	**(Base)^{exponent}** When bases are primes and same	$3^7, 3^5, 3^6$	The **base with maximum power** i.e. 3^7	The **base with minimum power** i.e. 3^5
		$2^{-2}, 2^5, 2^3$ -HCF = ? *[SSC(CPO)-97]*	2^5	2^{-2}
2.	**(Base)^{exponent}** When all/some bases are composite	$4^3, 2^4, 2^5$	Split the composite bases into primes. $(2^2)^3, 2^4, 2^5 \Rightarrow 2^6, 2^4, 2^5$ LCM = The **base with maximum power** i.e. 2^6 or 4^3	Split the composite bases into primes. $(2^2)^3, 2^4, 2^5 \Rightarrow 2^6, 2^4, 2^5$ HCF = The **base with minimum power** i.e. 2^4
		$3^7, 9^6, 3^2$	Given terms: $3^7, (3\times 3)^6, 3^2$ $\Rightarrow 3^7, 3^{12}, 3^2$ LCM = 3^{12} or 9^6	Given terms: $3^7, 3^{12}, 3^2$ HCF = 3^2
3.	**Product of (Base)^{exponent}** When bases are primes/composites	$2^2\times 3^7, 3^2\times 5^3, 2^5\times 3^4\times 7^2$	All of the bases are already in primes, thus, no need to split. **LCM = Product of all the different bases with maximum powers.** $= 2^5\times 3^7\times 5^3\times 7^2$	All of the bases are already in primes, thus, no need to split. **HCF = Product of bases (common in all terms) with minimum powers** $= 3^2$
		$4^3\times 2^2\times 3^4, 2^7\times 5^2\times 3^2, 6^9\times 5^3, 2^7\times 5^4\times 3^{15}$ *[RRB-03]*	Split the composite bases into primes $(2^2)^3\times 2^2\times 3^4, 2^7\times 5^2\times 3^2$ $(2\times 3)^9\times 5^3, 2^7\times 5^4\times 3^{15}$ $\Rightarrow 2^8\times 3^4, 2^7\times 5^2\times 3^2, 2^9\times 3^9\times 5^3, 2^7\times 5^4\times 3^{15}$ **LCM = Product of all the different bases with maximum powers** $= 2^9\times 3^{15}\times 5^4$	Split the composite bases into primes $(2^2)^3\times 2^2\times 3^4, 2^7\times 5^2\times 3^2$ $(2\times 3)^9\times 5^3, 2^7\times 5^4\times 3^{15}$ $\Rightarrow 2^8\times 3^4, 2^7\times 5^2\times 3^2, 2^9\times 3^9\times 5^3, 2^7\times 5^4\times 3^{15}$ **HCF = Product of bases (common in all terms) with minimum powers** $= 2^7\times 3^2$
4.	**Product of primes/composites and (Base)^{exponent}**	$3\times 7^3\times 5, 5^4\times 7^3\times 9, 7^2\times 5$	Split the composite bases/numbers into primes. Terms: $3\times 7^3\times 5, 5^4\times 7^3\times 3^2, 7^2\times 5$ **LCM = Product of all the different bases with maximum powers** $= 3^2\times 5^4\times 7^3$	Split the composite bases/numbers into primes. Terms: $3\times 7^3\times 5, 5^4\times 7^3\times 3^2, 7^2\times 5$ **HCF = Product of bases (common in all terms) with minimum powers** $= 7^2\times 5$
		$5\times 7^2\times 11, 2^3\times 22\times 3, 18\times 11\times 9$	Split the composite bases/numbers into primes. Terms $5\times 7^2\times 11, 2^4\times 11\times 3, 2\times 11\times 3^4$ LCM = $2^4\times 3^4\times 5\times 7^2\times 11$	Split the composite bases/numbers into primes. Terms $5\times 7^2\times 11, 2^4\times 11\times 3, 2\times 11\times 3^4$ HCF = 11

Note *Approaches for computing LCM and HCF of the numbers expressed in products of prime numbers have already been elaborated under respective 'FACTOR-METHOD'.*

Take one more example

If $p = 2^3 \cdot 3^{10} \cdot 5$ and $Q = 2^5 \cdot 3 \cdot 7$, then HCF of P and Q is: ***[SSC(10+2-DEO/LDC)-11]***

HCF = Product of bases (common in all terms) with minimum powers $= 2^3 \cdot 3$

V. LCM and HCF of Algebraic Expressions

We split the expressions into form of the products of factors so as to find the common factor(s), if any. In other words, we factorise the expressions. So far as the concepts are concerned, those are the same as meant for LCM and HCF of the numbers expressed as $(\text{Base})^{\text{exponent}}$/ Product of $(\text{Base})^{\text{exponent}}$/ Product of prime/composite and $(\text{Base})^{\text{exponent}}$. Aspirants are advised to command over the algebraic identities so as to split the expressions.

Problem 42 Find the LCM and HCF of $(x-1)^{17}(x+2)^{13}(x+3)^4$ and $(x+2)^2(x+3)^7(x+1)^8$.

Sol. We know, LCM = Product of all the different bases with maximum powers $= (x-1)^{17}(x+2)^{13}(x+3)^7(x+1)^8$

Further, HCF = Product of bases (common in all terms) with minimum powers $= (x+2)^2(x+3)^4$.

Problem 43 Find the LCM of $(x+5)^2(x+1)^3(x-3)$ and $(x+1)^2(x-3)^2(x+5)$. ***[RRB-Bhubneshwar(ASM)-02]***

Sol. We know, LCM = Product of all the different bases with maximum powers $= (x+5)^2(x+1)^3(x-3)^2$.

Problem 44 What is the HCF of $a^2b^4 + 2a^2b^2$ and $(ab)^7 - 4a^2b^9$? ***[CDS-13]***

Sol. $a^2b^4 + 2a^2b^2 = a^2b^2(b^2+2)$

and $(ab)^7 - 4a^2b^9 = a^2b^2(a^5b^5 - 4b^7)$

Now, HCF = Product of bases (common in all terms) with minimum powers $= a^2b^2$.

Problem 45 The HCF of $(x^4 - y^4)$ and $(x^6 - y^6)$ is:

Sol. 1st term $= (x^4 - y^4) = \{(x^2)^2 - (y^2)^2\} = (x^2 - y^2)(x^2 + y^2)$

$= (x+y)(x-y)(x^2+y^2)$

2nd term $= (x^6 - y^6) = \{(x^3)^2 - (y^3)^2\}$

$= (x^3 - y^3)(x^3 + y^3)$

$= (x-y)(x^2+y^2+xy)(x+y)(x^2+y^2-xy)$

Now, HCF = Product of bases (common in all terms) with minimum powers $= (x+y)(x-y) = x^2 - y^2$

Problem 46 If $(x+2)$ is the HCF of $x^2 + ax + b$ and $x^2 + cx + d$ {where, $a \neq c$ and $b \neq d$}, then which one of the following is correct?

(a) $a + b = c + d$ (b) $2a + b = 2c + d$

(c) $b + 2c = 2a + d$ (d) $b - 2c = 2a - d$

Sol. We know that the HCF of the given numbers divides them exactly. Hence, both the term leave 0 as remainders when divided by $(x+2)$. Now, recall the Remainder Theorem, obviously, $x = -2$.

$x^2 + ax + b \Rightarrow (-2)^2 + a(-2) + b = 0 \Rightarrow 4 - 2a + b = 0$ ... (i)

$x^2 + cx + d \Rightarrow (-2)^2 + c(-2) + d = 0 \Rightarrow 4 - 2c + d = 0$ ... (ii)

From Eqs. (i) and (ii), we have $4 - 2a + b = 4 - 2c + d$

$\Rightarrow \quad b + 2c = 2a + d$

VI. Problems Seeking Concepts of Both LCM and HCF

Problem 47 The LCM of three different numbers is 120. Which of the following can't be their HCF?

(a) 8 (b) 12

(c) 24 (d) 35

Sol. HCF is a factor of LCM and 35 is not a factor of 120. Hence, 35 can't be HCF of those three numbers.

Problem 48 The ratio of two numbers is 3 : 4 and their HCF is 4. Their LCM is? ***[SSC(CGL-PT)-02, SSC(CGL-T-II)-10]***

Sol. Let the numbers be $3x$ and $4x$. Then, their HCF $= x = 4$ (as given)

Now, LCM of $3x$ and $4x = 3 \times 4 \times x = 12x = 12 \times 4 = 48$

Alternatively

HCF of 3 and 4 = 1 and LCM of 3 and 4 = 12

Given that, 1unit = 4, then 12 units = 48

Problem 49 Three numbers are in the ratio 3 : 4 : 5 and their LCM is 2400. Their HCF is? ***[MAT-03, SSC(SAS)-10]***

Sol. Let the numbers be $3x$, $4x$ and $5x$.

Then, LCM $= 60x = 2400$ (as given)

$\Rightarrow \quad x = 40$

Now, HCF of $3x$, $4x$ and $5x = x = 40$

$\therefore$ Required answer = 40

Alternatively

LCM of 3, 4 and 5 = 60 and their HCF = 1

Given, 60 units = 2400

$\therefore$ 1 unit = 40

Problem 50 The HCF of two numbers is 11 and their LCM is 7700. If one of the numbers is 275, then the other is

Sol. We know, 1st number × 2nd number = LCM × HCF

$\Rightarrow 275 \times$ 2nd number $= 7700 \times 11$

$\Rightarrow$ 2nd number $= \dfrac{7700 \times 11}{275} = 308$

Problem 51 The HCF and LCM of two numbers are 11 and 385 respectively. If one number lies between 75 and 125, then that number is?

Sol. Let the number be $11x$ and $11y$, where x and y are co-primes.

Now, their LCM $= 11xy = 385 \Rightarrow xy = 35$

Now, pairs of co-primes with product 35 are: (1, 35) and (5, 7)

Thus, numbers are : $(11\times 1, 11\times 35)$ and $(11\times 5, 11\times 7)$
$= (11, 385)$ and $(55, 77)$
Thus, the number lying between 75 and 125 = 77

Alternatively

LCM × HCF = 1st number × 2nd number
$\Rightarrow$ $385\times 11 =$ 1st number × 2nd number
$\Rightarrow$ $55\times 7\times 11 =$ 1st number × 2nd number
$\Rightarrow$ $11\times 5\times \overline{11\times 7} =$ 1st number × 2nd number

Clearly, that number is 77 [$\because$ 77 lies between 75 and 125 and HCF and LCM of (55 and 77) are 11 and 385 respectively]

Problem 52 **Two numbers, both greater than 29, have HCF 29 and LCM 4147. The sum of the numbers is?**

Sol. Let the number be $29x$ and $29y$, where x and y are co-primes.

Then, their LCM $= 29xy = 4147$ (as given) $\Rightarrow xy = 143$

Now, pairs of co-primes with product 143 are: (1, 143) and (11, 13).

Thus, numbers are $(29\times 1, 29\times 143)$ and $(29\times 11, 29\times 13)$

Since both numbers are greater than 29, the suitable pair of numbers $= 29\times 11$ and 29×13 i.e. 319 and 377.

Now, their sum = 319 + 377 = 696

Alternatively

We have, LCM × HCF = 1st number × 2nd number
$\Rightarrow$ $4147\times 29 =$ 1st number × 2nd number
$\Rightarrow$ $29\times 11\times 13\times 29 =$ 1st number × 2nd number
$\Rightarrow$ $29\times 11\times 29\times 13 =$ 1st number × 2nd number

Clearly, numbers are $(29\times 11, 29\times 13)$

$\therefore$ Their sum = 319 + 377 = 696

Problem 53 **The HCF and LCM of two numbers are 21 and 84 respectively. If their ratio is 1 : 4, then the larger number is?** ***[MAT-97]***

Sol. Let the numbers be x and $4x$.

We have, LCM × HCF = 1st number × 2nd number
$\Rightarrow$ $84\times 21 = x\times 4x$
$\Rightarrow$ $21\times 21 = x^2 \Rightarrow 21 = x$

Now, largest number $= 4x = 4\times 21 = 84$

Alternatively I

We have, LCM × HCF = 1st number × 2nd number
$\Rightarrow$ $84\times 21 =$ 1st number × 2nd number
$\Rightarrow$ $21\times 84 =$ 1st number × 2nd number

Clearly, 1st number : 2nd number = 21 : 84 = 1 : 4

Hence, the largest number = 84

Alternatively II

Proceed through options as to which option satisfies the given condition(s).

Problem 54 **The LCM of two numbers is 495 and their HCF is 5. If the sum of the numbers is 100, then their difference is?**

Sol. Let the numbers be $5x$ and $5y$, where x and y are co-primes.

Now, LCM $= 5xy = 495$
$\Rightarrow$ $xy = 99$

Further, sum of numbers $= 5x + 5y = 100 \Rightarrow (x + y) = 20$

We have, $(x-y)^2 = (x+y)^2 - 4xy = (20)^2 - 4\cdot 99 = 4$
$\Rightarrow$ $x - y = 2 \Rightarrow 5x - 5y = 2\times 5 = 10$

Hence, difference in numbers = 10

Alternatively I

Let the numbers be $5x$ and $5y$, where x and y are co-primes.

Then, LCM $= 5xy = 495 \Rightarrow xy = 99$

Now, pairs of co-primes with product 99 are (1, 99) and (9, 11).

Thus, numbers are $(1\times 5, 99\times 5)$ and $(9\times 5, 11\times 5)$.

Since sum of numbers is 100, the numbers are (45, 55).

Hence, their difference = 55 − 45 = 10

Alternatively II

We have, LCM × HCF = 1st number × 2nd number
$\Rightarrow$ $495\times 5 =$ 1st number × 2nd number
$\Rightarrow$ $5\times 9\times 5\times 11 =$ 1st number × 2nd number

Since sum of (5×9) and (5×11) is 100 and their HCF is 5, their difference = 55 − 45 = 10

Alternatively III

Proceed through options. Let options be 20, 10, ...

(a) 20 $\Rightarrow$ If their difference is 20, numbers are 40 and 60 as their sum is 100. But neither their HCF is 5 nor their LCM is 495.

(b) 10 $\Rightarrow$ If their difference is 10, numbers are 45 and 55 as their sum is 100. Also, their HCF is 5 and their LCM is 495.

Hence, option(b) is correct.

Problem 55 **The HCF of the sum of two numbers and their LCM is 13. Find the HCF of the numbers.**

Sol. Let the numbers be Ha and Hb, where H is their HCF and a and b are co-primes.

Then, their sum $= Ha + Hb = H(a+b)$ and their LCM $= Hab$

We know that $(a+b)$ and ab are co-primes if a and b are co-primes.

$\therefore$ HCF of $H(a+b)$ and $Hab = H = 13$ (as given)

Hence, HCF of the numbers = 13

Alternatively

Recall the properties of HCF "The HCF of the sum of two numbers and their LCM is equal to the HCF of these two numbers.

Hence, required answer = 13

Problem 56 **The LCM of two numbers is 12 times their HCF. If one of the numbers is 16 and the sum of HCF and LCM is 52, the other number is?** ***[SSC(CGL-T-I)-12, MAT-18]***

Sol. Let the HCF of numbers be H. Then, their LCM $= 12H$.

As given, $H + 12H = 52 \Rightarrow H = 4$

Hence, HCF = 4 and LCM $= 4\times 12 = 48$

Now, LCM × HCF = 1st number × 2nd number
$\Rightarrow$ $48\times 4 = 16\times$ 2nd number
$\Rightarrow$ 2nd number = 12

Alternatively

Proceed through options. Let options be 28, 12, 18 and 13.

(a) 28 $\Rightarrow$ LCM of 16 and 28 = 112, which is not 12 times the HCF of 16 and 28.

(b) 12 $\Rightarrow$ LCM of 16 and 12 = 48, which is 12 times the HCF of 16 and 28.

Hence, option (b) is correct.

Problem 57 The HCF of two numbers is 8. Which one of the following can never be their LCM ?

(a) 24 (b) 48 (c) 56 (d) 60

Sol. We know that HCF is a factor of LCM. Clearly, 8 is not a factor of 60.

Hence, 60 can never be LCM.

Problem 58 The HCF of two numbers is 23 and the other two factors of their LCM are 13 and 14. The larger of the two numbers is:

Sol. Let the numbers be $23x$ and $23y$, where x and y are co-primes.

Then, LCM = $23xy$

Given that 13 and 14 are other two factors of LCM. It means, one possible LCM may be $23 \times 13 \times 14$. Hence, one pair of numbers are (23×13) and (23×14).

Then, the larger number = $23 \times 14 = 322$

[If it is in option, it is the correct answer, otherwise the larger number may be greater than 322. e.g. Numbers may be $(23 \times 13, 23 \times 28)$, $(23 \times 13, 23 \times 42)$ and so on.]

Problem 59 The product of LCM and HCF of two numbers is 24. Find the numbers if their difference is 2. ***[SSC(CGL-PT)-99, 06, 08]***

Sol. We have, LCM × HCF = 1st number × 2nd number

$\Rightarrow$ 24 = 1st Number × 2nd number

$\Rightarrow 1 \times 24$ or 2×12 or 3×8 or 4×6 = 1st number × 2nd number

The required pair is (4, 6) as difference between 4 and 6 is 2.

Alternatively

Proceed through options. Let options be 8 and 6, 6 and 4, 8 and 10 and 2 and 4.

(a) 8 and 6 $\Rightarrow$ Difference is 2, but product of their LCM and HCF $\neq 24$

(b) 6 and 4 $\Rightarrow$ Difference is 2 and also product of their LCM and HCF = 24

Hence, option(b) is correct.

Problem 60 The HCF and LCM of two numbers are 8 and 1728 respectively. How many such pairs are possible? ***[SSC(MTS)-11]***

Sol. Let the numbers be $8x$ and $8y$, where x and y are co-primes.

As given, LCM = 1728 $\Rightarrow$ $8xy = 1728$

$\therefore$ $xy = 216$

Now, pairs of co-primes with product 216 are : (1, 216) and (8, 27)

Thus, required number of such pairs = 2

Problem 61 The HCF and LCM of 77, 99 and x are 11 and 3465 respectively, then minimum value of x = ?

(a) 11 (b) 55 (c) 45 (d) 35

Sol. $77 = 11 \times 7$

$99 = 11 \times 3 \times 3$

$x = 11 \times a$ [$\because$ HCF is a factor common in all numbers]

Now, put the value of a:

At $a = 1$, LCM = $11 \times 7 \times 3 \times 3 \neq 3465$

At $a = 5$, LCM = $11 \times 7 \times 3 \times 3 \times 5 = 3465$ [we took $a = 1$ and 5 as options (a) and (b) indicate so].

$\therefore$ Minimum value of $x = 11 \times 5 = 55$

Alternatively

Proceed through options.

(a) 11 $\Rightarrow$ HCF of 77, 99 and 11 is 11, but their LCM $\neq 3465$

(b) 55 $\Rightarrow$ HCF of 77, 99 and 55 is 11 and also their LCM = 3465

Hence, option(b) is correct.

Problem 62 LCM of two numbers is 120 and their HCF is 10. Then, which one of the following may be sum of the numbers?

(a) 140 (b) 80 (c) 60 (d) 70

Sol. Let the numbers be $10x$ and $10y$, where x and y are co-primes.

Then, their LCM = $10xy = 120$ (given)

$\Rightarrow$ $xy = 12$

Now, pairs of co-primes with product 12 are : (1,12) and (3,4).

Thus, numbers are $(10 \times 1, 10 \times 12)$ and $(10 \times 3, 10 \times 4)$.

Clearly, sum of numbers is either 130 or 70.

Hence, option (d) is correct.

Alternatively

We know that "The HCF of the sum of two numbers and their LCM is equal to the HCF of these two numbers.

Now, ensure that the HCF of 120 and value of which option is 10. Clearly, HCF of 120 and 70 is 10.

Hence, option (d) is correct.

Problem 63 The sum of two numbers is 36 and their HCF and LCM are 3 and 105 respectively. Find the sum of their reciprocals. ***[SSC(CGL-T-I)-10, SSC (10+2-DEO/LDC)-10, SSC(SAS)-10]***

Sol. Let the number be $3x$ and $3y$, where x and y are co-primes.

Their LCM = $3xy = 105$ (given)

Also, $3x + 3y = 36$ (given)

$\Rightarrow$ $x + y = 12$

Now, sum of their reciprocals = $\frac{1}{3x} + \frac{1}{3y} = \frac{y + x}{3xy} = \frac{12}{105} = \frac{4}{35}$

Problem 64 HCF of 3240, 3600 and a third number is 36 and their LCM is $2^4 \times 3^5 \times 5^2 \times 7^2$. The third numbers is: ***[SSC(CGL-PT)-99, MAT-01, MAT-05, LIC(AAO)-12]***

Sol. $3240 = 2 \times 2 \times 2 \times 3 \times 3 \times 3 \times 3 \times 5 = 2^3 \times 3^4 \times 5$

$3600 = 2 \times 2 \times 2 \times 2 \times 3 \times 3 \times 5 \times 5 = 2^4 \times 3^2 \times 5^2$

Third number = ?

The third number must have 36 (i.e. $2^2 \times 3^2$) as its factor because HCF is a factor common in all given numbers. Further, since LCM is the product of all the different bases with maximum powers, the third number must have 3^5 and 7^2 as its factors. Clearly, 3^2 can be accommodated under 3^5.

Hence, third number = $2^2 \times 3^5 \times 7^2$

Alternatively

You may proceed through options. You have to ascertain as to which option with two given numbers is yielding LCM and HCF as $2^4 \times 3^5 \times 5^2 \times 7^2$ and 36 respectively.

Problem 65 **Sum of two numbers is 64 and the sum and the difference of LCM and HCF of these numbers are 128 and 112 respectively. Find the possible smaller number?**

Sol. Given, LCM + HCF = 128 and LCM − HCF = 112

Then, LCM = 120 and HCF = 8

Let the number be $8x$ and $8y$, where x and y are co-primes.

Then, their LCM = $8xy = 120 \Rightarrow xy = 15$

∴ Pairs of co-primes with product 15 = (1, 15), (3, 5)

Hence, numbers are $(1 \times 8, 15 \times 8)$, $(3 \times 8, 5 \times 8)$

i.e. (8, 120) and (24, 40)

∴ Required number = 8

Alternatively

You may proceed through options. Let options be 26, 8, 16 and 20.

(a) $26 \Rightarrow$ then greater number = 64 − 26 = 38 but LCM + HCF ≠ 128

(b) $8 \Rightarrow$ then greater number = 64 − 8 = 56 and LCM + HCF = 128 and LCM − HCF = 112

Hence, option(b) is correct.

Problem 66 **The product of four numbers is 69120 and HCF of each possible pair is 4. Find the greatest number of 4 digits which when divided by each of the four numbers leaves the remainder 3 in each case.**

Sol. Let the numbers be $4a$, $4b$, $4c$ and $4d$ where a, b, c and d are all co-primes to each other.

As given, $4a \times 4b \times 4c \times 4d = 69120$

$\Rightarrow \quad 4abcd = 1080$

It means, their LCM = $4abcd = 1080$

Now, we divide the greatest number of 4 digits i.e. 9999 by 1080 and get 279 as remainder.

Hence, required answer = (9999 − 279) + 3 = 9723

Problem 67 **The LCM of four numbers is 420 and HCF of each possible pair is 2. Find the product of all four numbers.**

Sol. Let the numbers be $2a$, $2b$, $2c$ and $2d$ where a, b, c and d are all co-primes to each other.

Then, their LCM = $2abcd = 420$ (given)

Further, product of all numbers = $2a \times 2b \times 2c \times 2d$

$= 8 \times 2abcd = 8 \times 420 = 3360$

Alternatively

We know, product of n numbers = LCM × $(\text{HCF})^{n-1}$.

Where, HCF refers to HCF of each possible pair out of given numbers.

$= 420 \times (2)^{4-1} = 3360$

VII. Pragmatic Applications of the Concepts of HCF

Problem 68 **An officer was appointed on maximum daily wages on a contract money of ₹ 4956. But on being absent for some days, he was paid ₹ 3894. For how many days was he absent?**

Sol. Since an officer has been appointed on maximum daily wages, we should find the maximum number which may divide both the numbers 4956 and 3894. Obviously, that number = their HCF = 354. Hence, his daily wages = ₹ 354

∴ Number of days for which he was appointed

= 4956/354 = 14 days

and Number of days for which he was paid

= 3894/354 = 11 days.

Hence, number of days for which he was absent

= 14 − 11 = 3 days.

Problem 69 **Find the side of the largest possible square slabs which can be paved on the floor of a room 2 m 50 cm long and 1m 50 cm broad. Also, find the number of such slabs to pave the floor.** *[LIC(AAO)-07]*

Sol. Clearly, HCF of 2.50 m and 1.50 m is the side of the largest possible square slabs i.e. 50 cm.

Then, number of slabs = $\frac{\text{Area of Floor}}{\text{Area of one slab}}$

$= \frac{2.50 \times 1.50}{0.50 \times 0.50} = 15$

Problem 70 **The maximum number of students among the 1001 pens and 910 pencils can be distributed in such a way that each student gets the same number of pens and same number of pencils is**

Sol. Required number of students

= HCF of 1001 and 910 = 91

Problem 71 **A salesman purchased some pens for ₹ 84 and sold some pens for ₹ 35 at no profit and no loss. Find the minimum number of pens left with him.**

Sol. Clearly, the salesman will have minimum number of pens when the cost of each pen is maximum.

It means, maximum cost of each pen = HCF of 84 and 35 = ₹ 7

Then, minimum number of pens he may purchase = 84/7 = 12

And maximum number of pens he may sell = 35/7 = 5

Thus, minimum number of pens left with him = 12 − 5 = 7

Problem 72 **The greatest possible length which can be used to measure exactly the lengths 7 m, 3 m 85 cm and 12 m 95 cm is:** *[RRB-03]*

Sol. Clearly, the greatest possible length

= HCF of 700 cm, 385 cm and 1295 cm

= 35 cm.

Problem 73 **132 mango trees, 204 banana trees and 228 apple trees are to be planted in rows such that each row contains the same number of trees of one variety only. Find the minimum number of rows in which the above trees may be planted.**

Sol. Number of trees in each row = HCF of 132, 204 and 228 = 12

Hence, required number of rows = $\frac{132}{12} + \frac{204}{12} + \frac{228}{12}$

= 11 + 17 + 19 = 47

Problem 74 There are 408 boys and 312 girls in a school which are to be divided into equal sections of either boys or girls alone. Find the total number of sections thus formed: *[LIC(AAO)-07]*

Sol. Number of boys or girls that can be accommodated in one section = HCF of 312 and 408 = 24

$\therefore$ Total number of sections $= \frac{408}{24} + \frac{312}{24} = 17 + 13 = 30$

Problem 75 In a school 105 boys and 56 girls are to be divided into the largest possible equal classes, so that there are equal number of boys and girls in each class. Find the number of classes.

Sol. Required number of classes = HCF of 105 and 56 = 7

Problem 76 There are four fields measuring 345, 120, 225 and 240 square meters in which vegetable-bunds having equal sizes are to be made. If length of the each bund is 6 m. Find the maximum breadth of the bunds.

Sol. Clearly, area (size) of each bund

= HCF of 345, 120, 225 and 240 = 15 m^2

Then, breadth $= \frac{\text{area}}{\text{length}} = \frac{15}{6} = 2.5\text{m}$

VIII. Pragmatic Applications of the Concepts of LCM

Problem 77 If the students of a class can be grouped exactly into 6 or 8 or 10, then the minimum number of students in the class must be

Sol. Clearly, minimum number of students = LCM of 6, 8 and 10 = 120

Problem 78 Three bells ring simultaneously at 11am. They ring at regular intervals of 20 min, 30 min and 40 min, respectively. The time when all the three ring together next is *[SSC(CGL-T-I)-11, SSC(CGL-T-II)-11, CLAT-16]*

Sol. LCM of 20, 30 and 40 = 120. Hence, 120 is the least number which is exactly divisible by 20, 30 and 40. It means, after 120 min all the three bells will ring simultaneously i.e. at 1PM.

Problem 79 A General can draw up his soldiers in the rows of 10, 15 or 18 soldiers and he can also draw them up in the form of a solid square. Find the least number of soldiers with the General. *[SSC(CGL-PT)-07]*

Sol. $10 = 2 \times 5$

$15 = 3 \times 5$

$18 = 2 \times 3 \times 3$

Now, LCM of 10, 15 and 18 $= 2 \times 3 \times 5 \times 3$

With this number of soldiers (i.e. $2 \times 3 \times 5 \times 3$) the General can draw up his soldiers in the rows of 10, 15 or 18, but he cannot draw them up in form of solid square. Hence, we require a perfect square number which should also be a multiple of the LCM and obviously, we can get the same by multiplying LCM by (2×5). Thus, the required number of soldiers

$= 2 \times 3 \times 5 \times 3 \times (2 \times 5) = 900$

Problem 80 The students of a school can stand in a hollow square 12, 15 or 18 deep and also into a solid square. Find the number of students.

Sol. LCM of 12, 15 and 18 $= 2 \times 2 \times 3 \times 3 \times 5$

Clearly, we must multiply LCM by 5 to make the resultant a square.

Hence, required number of students

$= \underline{2 \times 2} \times \underline{3 \times 3} \times \underline{5 \times 5} = 900$

Problem 81 Two wheels having 18 and 30 teeth respectively revolve together. If the smaller wheel makes 40 revolutions per min, then find the number of revolutions made by the larger wheel per minute.

Sol. LCM of 18 and 30 = 90

It means, two particular teeth will again be together after 90 teeth and till this event, in a particular time, number of revolutions made by the smaller wheel $= \frac{90}{18} = 5$, while number of revolutions made by the larger wheel $= \frac{90}{30} = 3$.

As given, $5 \Rightarrow 40 \therefore 3 \Rightarrow 24$

Thus, number of revolutions made by larger wheel per minute = 24

Problem 82 Two wheels having 24 and 30 teeth respectively revolve together. How many times will the two particular teeth be together in a period of 4 hours provided the smaller wheel makes 75 revolutions per minute?

Sol. LCM of 24 and 30 = 120

It means, two particular teeth will again be together after 120 teeth and till this event number of revolutions made by the smaller wheel $= \frac{120}{24} = 5$, while number of revolutions made by the larger wheel $= \frac{120}{30} = 4$

Clearly, both the particular teeth will be together after 5 revolutions made by the smaller wheel or after 4 revolutions made by the larger wheel.

Hence, both particular teeth will be together $(75 \div 5)$ times in 1 min.

Then, in 4 h, both will be together $(75 \div 5) \times 240$ i.e. 3600 times

* **Applications of the concepts of LCM in case of circular motion.**

  Detailed in the chapter 'Time and Distance'.

* **Applications of the concepts of LCM in comparing fractions.**

  Detailed in the chapter 'Comparison of Fractions'.

WORKOUT

1. What is the number of integral solutions of the equations HCF $(a, b) = 5$ and $a + b = 65$ **[CDS-14, 12]**
(a) Exactly one (b) Infinitely many
(c) Less than 65 (d) None

2. If a and b be positive integers, then what is HCF $\left\{\frac{a}{\text{HCF}(a, b)}, \frac{b}{\text{HCF}(a, b)}\right\}$ equal to ?
(a) a (b) b
(c) 1 (d) $\frac{a}{\text{HCF}(a, b)}$

3. The HCF of $2^{-7}, 2^{-8}$ and 2^{-5} is
(a) 2^{-7} (b) 2^{-8}
(c) 2^{-5} (d) None of these

4. For any integers 'a' and 'b' with HCF$(a, b) = 1$, what is HCF $(a + b, a - b)$ equal to ?
(a) It is always 1 (b) It is always 2
(c) Either 1 or 2 (d) None of these

5. For any integer n, what is HCF $(22n + 7, 3n + 10)$ equal to ?
(a) n (b) 11 **[CDS-14]**
(c) 1 (d) None of these

6. In a fire range, 4 shooters are firing at their respective targets. The first, the second, the third and the fourth shooter hit the target once is every 5 *s*, 6 *s*, 7 *s* and 8 *s* respectively. If all of them hit their target at 9 : 00 am , when will they hit their target together again ?
(a) 9 : 04 am (b) 9 : 08 am
(c) 9 : 14 am (d) None of these

7. What is the HCF of $8(x^5 - x^3 + x)$ and $28(x^6 + 1)$? **[CDS-14]**
(a) $4(x^4 - x^2 + 1)$ (b) $2(x^4 - x^2 + 1)$
(c) $(x^4 - x^2 + 1)$ (d) None of these

8. An Army commander wishing to draw up his 5180 men in the form of a solid square found that he had 4 men less. If he could get four more and form the solid square, the number of men in the front row is: **[MAT- 08]**
(a) 72 (b) 68 (c) 78 (d) 82

9. Find the least number which on being divided by 5, 6, 8, 9 and 12 leaves in each case a remainder 1 but when divided by 13 leaves no remainder.
(a) 3504 (b) 3601
(c) 361 (d) None of these

10. An ascending series of numbers satisfies the following conditions: **[XAT-15]**
I. When divided by 3, 4, 5 and 6 the numbers leave *a* remainder of 2.
II. When divided by 11, the numbers have no remainder.
The 6th number in this series will be:
(a) 242 (b) 2882 (c) 3542 (d) 4202
(e) None of these

11. 5 bells tolling together and toll at intervals of 2, 4, 6, 8 and 10 s respectively. How many times do the five bells toll together in 20 min?
(a) 10 (b) 11 (c) 12 (d) 15

12. If for integers a, b and c, if HCF $(a, b) = 1$ and HCF$(a, c) = 1$, then which one of the following is correct ? **[CDS-13]**
(a) HCF$(a, bc) = 1$ (b) HCF$(a, bc) = a$
(c) HCF$(a, bc) = b$ (d) None of these

13. A rod is cut into 3 equal parts. The resulting portions are then cut into 12, 18 and 32 equal parts respectively. If each of the resulting portions have integer length, the minimum length of the rod is: **[IIFT-13]**
(a) 6912 units (b) 864 units
(c) 288 units (d) 240 units

14. The HCF of two numbers of the same number of digits is 45 and their LCM is 540. Find the greatest number.
(a) 180 (b) 135 (c) 240 (d) 360

15. Consider those numbers between 300 and 400 such that when each number is divided by 6, 9 and 12, it leaves 4 as remainder in each case. What is the sum of the numbers ? **[CDS- 10]**
(a) 692 (b) 764 (c) 1080 (d) 1092

16. The HCF and LCM of two numbers are 21 and 4641 respectively. If one of the numbers lies between 200 and 300, then the two numbers are
(a) 273, 357 (b) 273, 361
(c) 273, 359 (d) 273, 363

17. Three Vice Presidents (VP) regularly visit the plant on different days. Due to labour unrest, VP(HR) regularly visits the plant after a gap of 2 days. VP(operations) regularly visits the plant after a gap of 3 days. VP(Sales) regularly visits the plant after a gap of 6 days. The VPs do not deviate from their regular schedules. CEO of the company meets the VPs when all the three VPs come to plant together. CEO is on leave from January 5th to January 28th, 2012. Last time CEO met the VPs on January 3, 2012. When is the next time CEO will meet all the VPs? **[XAT-12]**
(a) February 6, 2012
(b) February 7, 2012
(c) February 8, 2012
(d) February 9, 2012
(e) None of the above

18. The LCM of $\frac{1}{3}, \frac{5}{6}, \frac{2}{9}, \frac{4}{27}$ is **[LIC(AAO)-09]**
(a) $\frac{1}{54}$ (b) $\frac{10}{27}$ (c) $\frac{20}{3}$ (d) $\frac{27}{4}$

19. The numbers 11284 and 7655 when divided by a certain 3-digit number, give equal remainders. The 3-digit number is. **[SSC(SAS)-10]**
(a) 190 (b) 191
(c) 192 (d) 193

20. The LCM of three different numbers is 150. Which of the following cannot be their HCF ?
(a) 15 (b) 25 (c) 55 (d) 50

21. A person has to completely put each of three liquids 403 L of petrol, 465 L of diesel and 496 L of Mobil oil in bottles of equal size without mixing any of the above three types of liquids. What is the least possible number of bottles required ? **[IAS(Pre.)-07]**
(a) 34 (b) 44
(c) 46 (d) None of these

22. Consider the following statements
I. The HCF of $x + y$ and $x^{10} - y^{10}$ is $x + y$
II. The HCF of $x + y$ and $x^{10} + y^{10}$ is $x + y$.

III. The HCF of $x - y$ and $x^{10} + y^{10}$ is $x - y$.

IV. The HCF of $x - y$ and $x^{10} - y^{10}$ is $x - y$.

Which of the statements given above are correct ?

(a) I and II (b) II and III

(c) I and IV (d) II and IV

23. A general while arranging his men, who were 6000 in number, in the form of a square, found that there were 71 men left over. How many were arranged in each row ? ***[SSC(CGL-T-II)-10]***

(a) 73 (b) 77 (c) 87 (d) 93

24. What is the LCM of $a^3b - ab^3$, $a^3b^2 + a^2b^3$ and $ab(a + b)$?

(a) $a^2b^2(a^2 - b^2)$ (b) $ab(a^2 - b^2)$

(c) $a^2b^2 + ab^3$ (d) $a^3b^3(a^2 - b^2)$

25. When a number is divided by 2, 3, 4, 5 or 6 remainder in each case is 1. But the number is exactly divisible by 7. If the number lies between 250 and 350, the sum of digits of the number will be ***[LIC(AAO)-11]***

(a) 4 (b) 5 (c) 7 (d) 10

26. The LCM of two numbers is 2376 while their HCF is 33. If one of the numbers is 297, then the other number is ***[CDS-13]***

(a) 216 (b) 264 (c) 642 (d) 792

27. The LCM of $(x^3 - x^2 - 2x)$ and $(x^3 + x^2)$ is ***[CDS-13]***

(a) $(x^3 - x^2 - 2x)$ (b) $x^2 + 2x$

(c) $(x^4 - x^3 - 2x^2)$ (d) $x - 2$

28. When asked for his taxi number, the driver replied, "If you divide the number of my taxi by 2, 3, 4, 5, 6 each time you will find a remainder of one. But, if you divide it by 11, the remainder is zero. You will also not find any other driver with a taxi having a lower number who can say the same". What is the taxi number ?

(a) 121 (b) 1001 (c) 1881 (d) 781

29. What is the greatest number that divides 13850 and 17030 and leaves a remainder 17 ?

(a) 477 (b) 159 (c) 107 (d) 87

30. If $(x - 6)$ is the HCF of $(x^2 - 2x - 24)$ and $(x^2 - kx - 6)$, then what is the value of k? ***[CDS-12]***

(a) 3 (b) 5 (c) 6 (d) 8

31. 21 mango trees, 42 apple trees and 56 orange trees have to be planted in rows such that each row contains the same number of trees of one variety only. Find the minimum number of rows in which the above trees may be planted ?

(a) 3 (b) 15 (c) 17 (d) 20

32. The product of two numbers is 6912 and their GCD is 24. What is their LCM ?

(a) 280 (b) 286 (c) 288 (d) 296

33. What is the LCM of 2/3, 7/9 and 14/15 ?

(a) 7/3 (b) 14/3

(c) 2/3 (d) 1/3

34. The least number which when divided by 5, 6, 7 and 8 leaves a remainder 3 is

(a) 423 (b) 843

(c) 1683 (d) 2523

35. If HCF of three numbers 144, x and 192 is 12, then the number x cannot be:

(a) 180 (b) 84

(c) 60 (d) 48

36. What is the HCF of 3.0, 1.2 and 0.06 ?

(a) 0.6 (b) 0.06 (c) 6.0 (d) 6.06

37. What is the least number of square tiles required to pave the floor of a room 9 m 99 cm long and 4 m 7 cm broad ? ***[CDS-11]***

(a) 247 (b) 277 (c) 297 (d) 307

38. The HCF of two polynomials $p(x)$ and $q(x)$ is $2x(x + 2)$ and LCM is $24x(x + 2)^2(x - 2)$. If $p(x) = 8x^3 + 32x^2 + 32x$, then what is $q(x)$ equal to? ***[CDS-11]***

(a) $4x^3 - 16x$ (b) $6x^3 - 24x$

(c) $12x^3 + 24x$ (d) $12x^3 - 24x$

39. If $(x + k)$ is the HCF of $ax^2 + ax + b$ and $x^2 + cx + d$, then what is the value of k ?

(a) $\frac{b + d}{a + c}$ (b) $\frac{a + b}{c + d}$

(c) $\frac{a - b}{c - d}$ (d) None of these

40. What is the sum of the digits of the least number which when divided by 52, leaves 33 as remainder, when divided by 78, leaves 59 as remainder and when divided by 117, leaves 98 as remainder ? ***[CDS-10, 09, MAT-04, FMS-11, LIC(AAO)-12]***

(a) 17 (b) 18

(c) 19 (d) 21

41. Let p, q and r be natural numbers. If m is their LCM and n is their HCF, consider the following:

I. $mn = pqr$, if each of p, q and r is prime.

II. $mn = pqr$, if p, q and r are relatively prime in pairs.

Which of the above statements is/are correct ?

(a) Only I (b) Only II

(c) Both I and II (d) Neither I nor II

42. What is the greatest number which divides 392, 486 and 627 so as to leave the same remainder in each case ?

(a) 47 (b) 43

(c) 37 (d) 34

43. LCM of two numbers is 16 times their HCF. The sum of LCM and HCF is 850. If one number is 50, then what is the other number ?

(a) 800 (b) 1200

(c) 1600 (d) 2400

44. A person has four iron bars whose lengths are 24 m, 36 m, 48 m, and 72 m respectively. This person wants to cut pieces of same length from each of four bars. What is the least number of total pieces if he is to cut without any wastage ?

(a) 10 (b) 15

(c) 20 (d) 25

45. Consider the following statements:

I. If $a = bc$ with HCF$(b, c) = 1$, then HCF(c, bd) = HCF (c, d)

II. If $a = bc$ with HCF$(b, c) = 1$, then LCM(a, d) = LCM (c, bd)

Which of the above statement(s) is/are correct ? ***[CDS-17]***

(a) I only (b) II only

(c) Both I and II (d) Neither I nor II

46. N is the largest two-digit number, which when divided by 3, 4 and 6 leaves the remainder 1, 2 and 4 respectively. What is the remainder when N is divided by 5? ***[SSC(CGL-T-II)-17]***

(a) 4 (b) 2 (c) 0 (d) 1

Answer Key

1. (c)	2. (c)	3. (b)	4. (c)	5. (c)	6. (c)	7. (a)	8. (a)	9. (b)	10. (c)
11. (a)	12. (a)	13. (b)	14. (a)	15. (a)	16. (a)	17. (c)	18. (c)	19. (b)	20. (c)
21. (b)	22. (c)	23. (b)	24. (a)	25. (a)	26. (b)	27. (c)	28. (a)	29. (b)	30. (b)
31. (c)	32. (c)	33. (b)	34. (b)	35. (d)	36. (b)	37. (c)	38. (b)	39. (d)	40. (a)
41. (c)	42. (a)	43. (a)	44. (b)	45. (a)	46. (a)				

Hints and SOLUTIONS

1. Given that sum of two numbers is 65 and their HCF is 5. Then, how many such pairs are possible?

Let the numbers be $5x$ and $5y$, where x and y are co-prime.

As given, $5x + 5y = 65$

$\Rightarrow \quad x + y = 13$

Now, pairs of co-primes with sum 13 are (1, 12), (2, 11),(3, 10), (4, 9), (5, 8) and (6, 7).

$\therefore$ Required number of such pairs = 6

Hence, number of solutions is less than 65.

2. Recall the properties related to HCF 'The HCF of the quotients obtained by dividing two or more numbers by their HCF is always 1'.

Hence, option(c) is correct.

3. In case of same prime-base, the base with minimum power is the required HCF.

Here, HCF = 2^{-8}.

4. Recall the properties related to HCF 'if a and b are co-primes, the HCF of their sum and their difference will always be either 1 or 2'. Here, a and b are with HCF 1, it means, they are co-primes.

5. Since there is no factor common in both terms, their HCF will be 1.

6. LCM of 5, 6, 7 and 8 = 840

Hence, they all will hit the target again together after 840 seconds i.e. at 9 : 14 am.

7. $8(x^5 - x^3 + x)$

$= 2 \times 2 \times 2 \times x(x^4 - x^2 + 1)$

and $28(x^6 + 1)$

$= 2 \times 2 \times 7 \times \{(x^2)^3 + (1^2)^3\}$

$= 2 \times 2 \times 7 \times (x^2 + 1)(x^4 - x^2 + 1)$

Now, HCF = Product of bases (common in all terms) with minimum powers

$= 2 \times 2 \times (x^4 - x^2 + 1) = 4(x^4 - x^2 + 1)$

8. Clearly, with (5180 + 4) men a solid square can be drawn up. Then, square root of 5184 = 72.

Hence, the number of men in the front(i.e. first) row = 72

9. The least number divisible by 5, 6, 8, 9 and 12 = LCM of 5, 6, 8, 9 and 12 = 360.

Since the multiples of LCM are also divisible by the given set of numbers, $360K$ will be divisible by them. Then, the required number will be in form of $(360K + 1)$, where K has an integer value. It is also given that when $(360K + 1)$ is divided by 13 leaves no remainder.

If we divide $(360K + 1)$ by 13, quotient = $27K$ and remainder = $9K + 1$. Since there is no remainder, $(9K + 1)$ must be divisible by 13 and it is possible at $K = 10$. Here, 10 is minimum value.

Hence, required number

$= (360K + 1) = (360 \times 10) + 1 = 3601$

Alternatively: You may check the options as to which one is satisfying the conditions.

10. The least number divisible by 3, 4, 5 and 6 = LCM of 3, 4, 5 and 6 = 60.

Since the multiples of LCM are also divisible by the given set of numbers, $60K$ will be divisible by them. Then, the required number will be in form of $(60K + 2)$, where K has an integer value. It is also given that when $(60K + 2)$ is divided by 11 leaves no remainder.

If we divide $(60K + 2)$ by 11, quotient = $5K$ and remainder = $5K + 2$. Since there is no remainder, $(5K + 2)$ must be divisible by 11 and it is possible at $K = 4$. Here, 4 is minimum value of K. Since K is an integer, the next values of $K = (4 + 11), (15 + 11), (26 + 11), (37 + 11), (48 + 11)$ and so on.

Hence, required number

$= \{(60 \times 59) + 2)\}$

$= 3542$

11. LCM of 2, 4, 6, 8 and 10 = 120.

Hence, 120 is the least number which is exactly divisible by 2, 4, 6, 8 and 10. It means, after 120 s all the five bells will ring simultaneously.

Clearly, in 2 min $\Rightarrow$ 1 time

$\therefore$ In 20 min = 10 times.

12. Recall the properties related to HCF 'if (a and b) and (a and c) are two pairs of co-primes, then HCF of a and bc is equal to 1.

13. Each portion of the rod must be divisible by 12, 18 and 32. Hence, minimum length of the one portion = LCM of 12, 18 and 32 = 288

Then, minimum length of the rod

$= 3 \times 288 = 864$ units

14. LCM $\times$ HCF = 1st number $\times$ 2nd number

$\Rightarrow 45 \times 540 =$ 1st number $\times$ 2nd number

$\Rightarrow 3 \times 3 \times 5 \times 2 \times 2 \times 3 \times 3 \times 3 \times 5$

= 1st number $\times$ 2nd number

$\Rightarrow \underline{3 \times 3 \times 5 \times 2 \times 2 \times 3 \times 3 \times 3 \times 5}$

= 1st number $\times$ 2nd number

$\Rightarrow (180) \times (135) = $ 1st number × 2nd number [∵ we have brought both numbers in 3 digits]

Hence, option(a) is correct.

15. The least number exactly divisible by 6, 9 and 12 = 36

Now, we have to find the numbers between 300 and 400 which are divisible by 36 i.e. individually by each of 6, 9 and 12.

Clearly, 324 and 360 are multiples of 36.

Since remainders are 4 in each case, the required numbers will be (324 + 4) and (360 + 4).

Hence, sum of those numbers

= 328 + 364 = 692

16. LCM × HCF = 1st number × 2nd number

$\Rightarrow 21 \times 4641$ = 1st number × 2nd number

$\Rightarrow 3 \times 7 \times 3 \times 13 \times 119$ = 1st number × 2nd number

$\Rightarrow (3 \times 7 \times 13) \times (3 \times 119)$ = 1st number × 2nd number

$\Rightarrow 273 \times 357$ = 1st number × 2nd number

Hence, option(a) is correct.

17. As given, VP(HR) visits the plant at the interval of 3 days, VP(operations) visits the plant the interval of 4 days and VP(Sales) visits the plant the interval of 6 days. Hence, they all meet at interval of {LCM of 3, 4 and 6} days. It means they all meet at interval of 12 days.

The next meeting will be 12 days beyond the last meeting day i.e. on (3 + 12) January 2012, then on (3 + 12 + 12) January 2012 and next on 8 February 2012.

Since CEO was on leave from January 5th to January 28th, he meets all the VPs on February 8, 2012.

18. Required LCM

$$= \frac{\text{LCM of numerators}}{\text{HCF of denominators}} = \frac{20}{3}$$

19. Recall the properties related to HCF. In case two numbers are divided by a number D and the remainders are equal, the difference between them will be the greatest value of D. However, if this difference is greater than the smaller number, the factor of the difference just below the smaller number will be the greatest value of D.

Here, 11284 and 7655 can give equal remainders when divided by their difference (11284 − 7655) i.e. by 3629. But the divisor should be a 3-digit number as sought in the question. It is true that both the numbers will also be divisible by the factors of 3629. 3629 = 19 × 191

Hence, the required answer = 191

20. We know that, HCF is always a factor of LCM.

Here, 55 is not a factor of 150.

Hence, option(c) is correct.

21. The maximum capacity of a bottle = HCF of 403, 465 and 496 = 31L.

Then, number of bottles required

= (403 ÷ 31) + (465 ÷ 31) + (496 ÷ 31)

= 13 + 15 + 16 = 44

22. Recall the concept of Remainder/Factor theorem. Clearly, $x + y$ is a factor of $x^{10} - y^{10}$ and $x - y$ is also a factor of $x^{10} - y^{10}$

Then, option(c) is correct.

23. Clearly, with (6000 − 71) men a solid square can be drawn up.

Then, square root of 5929 = 77

Hence, the number of men in the each row = 77.

24. $a^3b - ab^3 = ab(a^2 - b^2)$

$= ab(a + b)(a - b)$

and $a^3b^2 + a^2b^3 = a^2b^2(a + b)$

and $ab(a + b)$

Now, LCM = Product of all the different bases with maximum powers $= a^2b^2(a + b)(a - b)$

$= a^2b^2(a^2 - b^2)$

25. The least number divisible by 2, 3, 4, 5 and 6 = LCM of 2, 3, 4, 5 and 6 = 60.

Since the multiples of LCM are also divisible by the given set of numbers, $60K$ will be divisible by them. Then, the required number will be in form of $(60K + 1)$, where K has an integer value. It is also given that when $(60K + 1)$ is divided by 7 leaves no remainder.

If we divide $(60K + 1)$ by 7, quotient $= 8K$ and remainder $= 4K + 1$. Since there is no remainder, $(4K + 1)$ must be divisible by 7 and it is possible at $K = 5$. Here, 5 is minimum value of K.

Since K is an integer, the next values of $K = (5 + 7), (12 + 7)$ and so on.

Hence, such numbers $= \{(60 \times 5) + 1\}, \{(60 \times 12) + 1\}$ and so on.

Since we require the number between 250 and 350, the required number $= \{(60 \times 5) + 1\} = 301$

Now, sum of digits of the number

= 3 + 0 + 1 = 4

26. LCM × HCF = 1st number × 2nd number

$\Rightarrow 2376 \times 33 = 297 \times$ 2nd number

$\Rightarrow$ 2nd number = 264

27. $(x^3 - x^2 - 2x) = x(x^2 - x - 2)$

$= x(x^2 - 2x + x - 2)$

$= x\{x(x - 2) + 1(x - 2)\}$

$= x(x - 2)(x + 1)$

Further, $(x^3 + x^2) = x^2(x + 1)$

LCM = Product of all the different bases with maximum powers

$= x^2(x - 2)(x + 1) = (x^4 - x^3 - 2x^2)$

28. The least number divisible by 2, 3, 4, 5 and 6 = LCM of 2, 3, 4, 5 and 6 = 60

Since the multiples of LCM are also divisible by the given set of numbers, $60K$ will be divisible by them. Then, the required number will be in form of $(60K + 1)$, where K has an integer value. It is also given that when $(60K + 1)$ is divided by 11, the remainder is zero. Clearly, if we divide $(60K + 1)$ by 11, quotient $= 5K$ and remainder $= 5K + 1$. Since the remainder is 0, $(5K + 1)$ must be divisible by 11 and it is possible at $K = 2$. Here, 2 is minimum value, other values may be 13, 24, etc.

Hence, required number

$= (60K + 1) = (60 \times 2) + 1 = 121$

Alternatively

The best way is to proceed through option.

29. Required number = HCF of (13850 − 17) and (17030) = 159

30. Since $(x - 6)$ is the HCF of both expressions it is also a factor of both expressions. Recall the remainder theorem, at $x = 6$, the value of

$(x^2 - kx - 6) = 0$

$\Rightarrow \quad 6^2 - 6k - 6 = 0 \Rightarrow k = 5$

31. Number of trees in each row = HCF of 21, 42 and 56 = 7

Hence, required number of rows

$$= \frac{21}{7} + \frac{42}{7} + \frac{56}{7} = 3 + 6 + 8 = 17$$

32. LCM × HCF = 1st number × 2nd number

$\Rightarrow$ LCM × 24 = 6912

$\Rightarrow$ LCM = 6912 ÷ 24 = 288

33. $\text{LCM} = \dfrac{\text{LCM of numerators}}{\text{HCF of denominators}}$

$= \dfrac{\text{LCM of 2, 7 and 14}}{\text{HCF of 3, 9 and 15}}$

$= \dfrac{14}{3}$

34. Required number = (LCM of 5, 6, 7 and 8) + 3 = 840 + 3 = 843

35. Factors of 144 and 192 is

$144 = 12 \times 2 \times 2 \times 3$

$192 = 12 \times 2 \times 2 \times 2 \times 2$

$x = 12 \times ?$

Given, 12 is a highest factor common in all three. Hence, the number which is a multiple of 12, can be the value of x with the stipulation that other factor(s) of x should not be common in all three.

In case of 48, $x = 12 \times 2 \times 2$. Clearly, (2×2) is common in all three in addition to 12.

Hence, 48 cannot be the value of x. Needless to say that, in case $x = 48$, the HCF will be 48, not 12.

36. Given numbers are 3.00, 1.20 and 0.06

Now HCF of 300, 120 and 006 = 6

Then, required HCF = 0.06

37. Since we require least number of tiles, the size of tiles must be maximum.

∴ Side of the tiles = HCF of 999 cm and 407 cm = 37 cm

Now, number of tiles

$= \dfrac{999 \times 407}{37 \times 37} = 297$

38. LCM × HCF

= 1st number(polynomial) × 2nd number (polynomial)

$\Rightarrow 24x(x+2)^2(x-2) \times 2x(x+2)$

$= (8x^3 + 32x^2 + 32x) \times$ 2nd polynomial

$\Rightarrow$ 2nd polynomial

$= \dfrac{24x(x+2)^2(x-2) \times 2x(x+2)}{(8x^3 + 32x^2 + 32x)}$

$= \dfrac{48x^2(x+2)^3(x-2)}{8x(x^2+4x+4)}$

$= \dfrac{6x(x+2)^3(x-2)}{(x+2)^2}$

$= 6x(x+2)(x-2)$

$= 6x(x^2-4)$

$= 6x^3 - 24x$

39. Since, $(x + k)$ is a factor common in $ax^2 + ax + b$ and $x^2 + cx + d$, at $x = -k$, value of both expressions each will be 0.

Recall the remainder/factor theorem.

Hence, $a(-k)^2 + a(-k) + b = (-k)^2 + c(-k) + d$

$\Rightarrow k^2(a-1) - k(a-c) + b - d = 0$

$\Rightarrow k = \dfrac{-(c-a) \pm \sqrt{(c-a)^2 - 4(a-1)(b-d)}}{2(a-1)}$

Hence, option(d) is correct.

40. Recall the concept associated with LCM "'Remainders are different when a number is divided by a set of numbers, but the differences between the divisors and the respective remainders are equal".

The required number = (LCM of given set of numbers) − (difference between a divisor and respective remainder)

= (LCM of 52, 78 and 117) − (52 − 33)

= 468 − 19 = 449

Then sum of digits of 449

= 4 + 4 + 9 = 17

41. Statement I $\Rightarrow$ HCF of prime numbers is always 1 and LCM of prime numbers is equal to their product.

Hence, LCM × HCF = their products i.e. $mn = pqr$. It is a true statement.

Statement II $\Rightarrow$ Recall the properties related to HCF, Product of n numbers = LCM × $(\text{HCF})^{n-1}$, where HCF refers to HCF of each possible pair. Here, it is given that all are relatively prime in pairs, it means, HCF of each pair is 1.

Then, $pqr = mn$. Hence, statement II is also true.

Obviously, option(c) is correct.

42. Recall the properties related to HCF "If numbers $n_1, n_2, n_3, \ldots n_x$ leave the same remainder (but, not known) when each of them is divided by the same number D, the HCF of all possible differences between two numbers out of given numbers will be the greatest value of D."

Here, the greatest number = HCF of (486− 392), (627 − 486) and (627 − 392) = HCF of 94, 141 and 235 = 47

43. Let the HCF of numbers be H. Then, their LCM = $16H$.

As given, $H + 16H = 850$

$\Rightarrow H = 50$

Hence, HCF = 50 and LCM = 50 × 16 = 800

Now, LCM × HCF = 1st number × 2nd number

$\Rightarrow$ 800 × 50 = 50 × 2nd number

$\Rightarrow$ 2nd number = 800

44. Since the person requires least number of pieces, he should make the pieces in maximum possible size.

Now, HCF of 24, 36, 48 and 72 = 12m

Then, number of total pieces

= (24 ÷ 12) + (36 ÷ 12) + (48 ÷ 12) + (72 ÷ 12)

= 2 + 3 + 4 + 6 = 15

45. Statement (I) Since HCF$(b, c) = 1$, b and c are co-primes. Then, HCF (c, bd) will depend upon the availability of common factor(s) between c and d. Hence, statement(I) is correct.

Statement (II) Since HCF$(b, c) = 1$, b and c are co-primes. Now, LCM (a, d) = LCM(bc, d). But, it is not necessarily true that LCM(bc, d) = LCM(c, bd) {*viz.* LCM $(3 \cdot 20, 3) \neq$ LCM$(20, 3 \cdot 3)$}

Hence, option(a) is correct.

46. LCM of 3, 4 and 6 = 12

Since the differences between the divisors and the respective remainders are equal {i.e. 3 − 1 = 4 − 2 = 6 − 4 = 2), the required number will be in form of $(12K - 2)$.

At $K = 8$, $(12K - 2)$ will be the greatest two-digit number.

Hence, the greatest two-digit number = $12 \cdot 8 - 2 = 94$

Now, remainder in case of (94 ÷ 5) = 4

14

Average

Learning OBJECTIVES

After completing this chapter, aspirants will be able to

- Understand the meaning of average;
- Observe the properties of average with illustrations;
- Realise how the average of a given data set alters when one or more values are included therein;
- Realise how the average of a given data set alters when one or more values are excluded;
- Analyse the alteration in average of the data set when some values are taken incorrectly;
- Comprehend the concept of Weighted Average;
- Identify the types of problems based on cricket;
- Identify the questions which can also be solved by the Alligation-Rule;
- Solve most of the questions without pen and paper as unique understanding/fundamentals/concepts have been developed;
- Compute the average of series of the numbers, their squares and their cubes. Recognise the fundamental approaches which is applicable for the questions;
- Feel comfortable to choose the appropriate approaches out of so many approaches as per his/her choice;
- Comprehend the solutions of complicated questions easily as usages of appropriate words have been made while solving the problems;
- Observe the diverse sorts of categories of the questions;
- Learn how to solve the maximum questions with the concepts associated with 'Ratio';
- Understand the philosophy of the division of this chapter in various parts;
- Analyse the trend of the questions being asked in various competitive examinations. Recognise the types of question which can be easily solved through the given options;
- Solve the questions in moments;
- Grip on concepts helping to solve the questions without pen and paper;
- Learn new concepts/ fundamentals to the greater extent;
- Apply the pragmatic approaches in lieu of academic approaches; and
- Ponder on the alternative methods.

Inside

- *Brief Introduction*
- *Properties of Average and Problems Based Theorem*
- *Concept of Inclusion and Exclusion of Number(s) in a given set of Numbers.*
- *Alteration in Average Due to Computational Errors*
- *Concept of Weighted Average*
- *Problems Based on Cricket*
- *Concept of Average Speed*
- *Average of the Diverse Series of Numbers*
- *Average of the Squares and Cubes of Numbers*
- *Miscellaneous*
- *..............................*
- *.............................. & Many More*
- *Work Out with Solutions*

Introduction

This chapter is very easy, I agree with. However, it can take some more time than what is required in the various competitive examinations. Keeping in mind this fact, a unique approach has been developed so that most of the questions can be solved without pen and paper.

The concept of Alligation has been taken into account to the some extent in this chapter to solve the questions quickly. Needless to say that it has already been mentioned in the chapter 'Alligation/Mixture' that Alligation-Rule is a substitute of Weighted Average method.

No doubt, this chapter plays a vital role in securing marks/ranks in the **competitive** examinations like CAT, MAT, SSC, CDS, NDA, Bank, LIC, PSU, Railway, etc. I hope, aspirants will realize themselves totally different after perusal of this chapter and obviously this will accelerate their efforts to achieve the goal.

There are several different meanings of 'average' but the most general is a measure of central tendency, which means measure of 'central' value of the data set. In general, data set means a list/series of values/numbers. The central value may be assumed in place of given each value. In other words, the average is the number which distributes the sum evenly and smoothes out the variation of the numbers.

Means, Medians and Modes are kinds of averages, however, the term average generally refers to an arithmetic mean. Average is calculated by dividing the sum of values in the set by their number. Hence,

$$\text{Average} = \frac{\text{sum of values of the data set}}{\text{number of values}}$$

Or, Sum of all values = number of values × average

For Example Find the average of 3, 9, 7 and 5.

$$\text{Average} = \frac{\text{sum of 3, 9, 7 and 5}}{\text{number of given values}}$$

$$= \frac{3+9+7+5}{4} = 6$$

Properties of Average

(A) Average of any data set always lies between highest and lowest values of that set.

For example in the above question,

3 < 6 < 9

↓ ↓

Lowest value, Highest value

(B) If each of the given values of a data set is replaced by the average, there will be no change in their sum.

Considering a data set (3, 9, 7 and 5)

S.No.	Values/ numbers	average	Difference between numbers and average i.e. (average – number)
1.	3	6	(+ 3)
2.	9	6	(– 3)
3.	7	6	(– 1)
4.	5	6	(+ 1)
Total	**24**	**24**	

The above chart also reveals that the sum of differences between average and numbers above the average (i.e. 3 + 1) is equal to sum of differences between average and numbers below the average (i.e. 3 + 1). In other words, the net surplus i.e., 4 owing to numbers being above the average is equal to the net deficit i.e. 4 due to numbers being below the average.

(C) If each number/ value of a given data set is increased by m then average will increase by m. However, if the average is increased by a certain value 'm' then it is not necessary that each number/value has been increased by m, it may be or it may not be. However, it is true that the sum of those numbers/values will increase by the product of 'm' and number of numbers/values. Similarly, it happens accordingly in case of reduction.

For example Let the numbers be 3, 9 , 7 and 5 as above and every number is increased by 5, then numbers be $(3+5), (9+5), (7+5)$ and $(5+5)$. Now, their average $= \frac{8+14+12+10}{4} = 11$, which is 5 more than earlier average(6).

On the other hand, if average increases by 5, (say, $6+5$), the set of numbers may be following or different-

3, 19, 5, 17 # 1, 7, 27, 9 # 2, 18, 13, 11

A lot of sets may be crafted ascertaining their sum is (4×11). No doubt, the earlier sum i.e. 24 has increased by the product of increment(i.e. 5) and number of numbers(i.e. 4). The same pattern will follow in case of decrease in each number/value of a data set.

(D) If each number/value of a data set is multiplied by a certain value 'm', then their average will become m times the previous average. Further, keep in mind that if average becomes m times previous average, then it is not necessary that each number/value has been multiplied by m, it may be or may not be. However, their sum will become m times the earlier sum. Similarly, it happens accordingly in case of division.

From the above discussions at points (c) and (d), we may construe that-

(i) If each number/value of a data set is increased/ decreased/ multiplied/divided by a certain value 'm', then average will change accordingly.

(ii) If average of a data set is increased/decreased by a certain value 'm' then sum of all numbers/values will increase or decrease by the product of 'm' and number of numbers/values irrespective of changes in individual number/value.

(iii) If average of a data set is multiplied/divided by a certain value 'm', then sum of all numbers/values becomes m times/ $(1/m)$ times of earlier sum irrespective of changes in individual number/value.

Illustrations

Problems Based on Properties (A) of the Average

Problem 1 Find out the average of 2.397, 0.385, 10.50 and 55.13.

(a) 57.49 (b) 67.35
(c) 0.385 (d) 17.103

Sol. Since, the average of the given data set lies between the highest and lowest values, option(d) is correct.

Alternatively

$$\text{Average} = \frac{\text{sum of values of the data set}}{\text{number of values}}$$

$$= \frac{2.397 + 0.385 + 10.50 + 55.13}{4} = 17.103$$

Problems Based on Properties (B) of the Average

Problem 2 A student was asked to find the arithmetic mean of the numbers 3, 11, 7, 9, 15, 13, 8, 19, 17, 21, 14 and x. He found the mean to be 12. What should be number in place of x? ***[SSC(SO)-03]***

Sol. $3 - 12 = -9$

$11 - 12 = -1$

$7 - 12 = -5$

$9 - 12 = -3$

$15 - 12 = +3$

$13 - 12 = +1$

$8 - 12 = -4$

$19 - 12 = +7$

$17 - 12 = +5$

$21 - 12 - +9$

$14 - 12 = +2$

$(x - 12) = ?$

Here, sum of surpluses = 27 and sum of deficits = 22

We know that, sum of surpluses is equal to sum of deficits. Here, we require 5 more as deficit and it is possible only when x would be 5 less than the average. Hence, $x = 7$.

Note *The whole process can be done mentally without pen and paper*

Alternatively

$$\text{Average} = \frac{\text{sum of values of the data set}}{\text{number of values}}$$

$$\Rightarrow \quad 12 = \frac{3 + 11 + 7 + 9 + 15 + 13 + 8 + 19 + 17 + 21 + 14 + x}{12}$$

$$\Rightarrow \quad 144 = 137 + x \Rightarrow x = 7$$

Problem 3 A grocer has a sale of ₹ 6435, ₹ 6927, ₹ 6855, ₹ 7230 and ₹ 6562 for 5 consecutive months. How much sale must he have in the sixth month so that he gets an average sale of ₹ 6500.

Sol. Let he has a sale of ₹ x in sixth month.

$6435 - 6500 = -65$

$6927 - 6500 = +427$

$6855 - 6500 = +355$

$7230 - 6500 = +730$

$6562 - 6500 = +62$

$(x - 6500) = ?$

Sum of surpluses = 1574 and sum of deficits = 65

Since, sum of surpluses is equal to sum of deficits, we require 1509 more as deficit and it is possible only when x would be 1509 less than the average. Hence, $x = 6500 - 1509 = 4991$.

Alternatively

$$\text{Average} = \frac{\text{sum of values of the data set}}{\text{number of values}}$$

$$\Rightarrow \quad 6500 = \frac{6435 + 6927 + 6855 + 7230 + 6562 + x}{6}$$

$$\Rightarrow \quad 39000 = 34009 + x$$

$$\Rightarrow \quad x = 4991.$$

Problem 4 The average of 8 numbers is 20. The average of first two numbers is $15\frac{1}{2}$ and that of the next three is $21\frac{1}{3}$. If the sixth number be less than the seventh and eighth numbers by 4 and 7 respectively. Then the eighth number is?

Sol. Assume, each of 8 numbers be 20.

Each of first two numbers is $4\frac{1}{2}$ less than 20. Then, deficits = 9

Further, each of next three numbers is $1\frac{1}{3}$ more than 20, hence, surpluses = 4

Thus, up to first five numbers total deficits

$= 9 - 4 = 5$

But we know, total surpluses = total deficits, therefore, next three numbers (i.e. 6th, 7th, 8th) should fetch together a surplus of 5.

Then, 6th number + 7th number + 8th number

$= (20 \times 3) + 5 = 65$

$\Rightarrow$ 6th number + (6th number + 4) + 8th number = 65

$\Rightarrow$ (8th number − 7) + (8th number − 7 + 4) + 8th number = 65

$\Rightarrow \quad 3 \times$ 8th number = 65 + 10

$\therefore \quad$ 8th number = 25

Alternatively

We know, sum of all values = number of values × average

$\therefore$ Sum of first two numbers $= 2 \times 15\frac{1}{2} = 31$

and sum of further next three numbers $= 21\frac{1}{3} \times 3 = 64$

and sum of all 8 numbers $= 20 \times 8 = 160$

Then, sum of 6th , 7th and 8th numbers

$= 160 - (31 + 64) = 65$

Now, 6th number + 7th number + 8th number = 65

$\Rightarrow$ 6th number + (6th number + 4) + 8th number = 65

$\Rightarrow$ (8th number − 7) + (8th number − 7 + 4)+ 8th number = 65

$\Rightarrow$ 3 × 8th number = 65 + 10

$\therefore$ 8th number = 25

Problem 5 **The average of 20 numbers is 12. The average of the first 12 numbers is 11 and that of the next 7 numbers is 10. The last number is what?**

Sol. Assume each number be 12 as their average is 12.

Total deficit in case of first 12 numbers $= 12 \times (12 - 11) = 12$

and total deficit in case of next 7 numbers $= 7 \times (12 - 10) = 14$

Now, total deficits = total surpluses = 12 + 14 = 26 and this surplus will obviously shift to the remaining number(s) i.e., last number. Hence, last number = 12 + 26 = 38.

Alternatively I

Assume each number be 12 as their average is 12. However, each of first 12 numbers gives 1 to last number and also each of next 7 numbers gives 2 to last number.

Hence, the last number $= 12 + (12 \times 1) + (7 \times 2) = 38$.

Alternatively II

Sum of all 20 numbers $= 20 \times 12 = 240$

and sum of first 12 numbers $= 12 \times 11 = 132$

and sum of next 7 numbers beyond 12 numbers $= 7 \times 10 = 70$

Then, 20th number being the last number

$= 240 - (132 + 70) = 38$

Problem 6 **The average of 11 numbers is 10.8. If the average of the first six numbers is 10.4 and that of last six number is 11.5, then the middle number is**

[SSC(CGL-T-II)-13, SSC(CGL-PT)-2000, 01, 02, MAT-03, 01]

Sol. Since the average of all 11 numbers is 10.8, we may assume each of them is 10.8.

Total deficit in case of first 6 numbers $= 6(10.8 - 10.4) = 2.4$

and total surplus in case of last 6 numbers $= 6(11.5 - 10.8) = 4.2$

Clearly, net effect(surplus) $= 4.2 - 2.4 = 1.8$ which is on account of counting of middle number twice.

Thus, the middle number $= 10.8 + 1.8 = 12.6$

Alternatively

Sum of all 11 numbers $= 11 \times 10.8 = 118.8$

and sum of first 6 numbers $= 6 \times 10.4 = 62.4$

and sum of last 6 numbers $= 6 \times 11.5 = 69$

It is clear that the middle number has been taken into consideration twice instead of taking once.

Hence, the middle number being 6th number

$= (62.4 + 69) - 118.8 = 12.6$

Problem 7 **The average of 25 results is 18. That of first 12 is 14 and of last 12 is 17. What is thirteenth result?**

[RRB-Ajmer(ASM)- 01]

Sol. Since the average of all 25 results is 18, we may assume each of them is 18.

Deficit in case of first 12 numbers $= 12 \times (18 - 14) = 48$

and deficit in case of last 12 numbers $= 12 \times (18 - 17) = 12$

Now, total surpluses = total deficit = 48 + 12 = 60 and this surplus will move to the remaining number(s) i.e., 13th result.

$\therefore$ 13th number $= 18 + 60 = 78$

Alternatively

Sum of all 25 results $= 25 \times 18 = 450$

and sum of first 12 results $= 12 \times 14 = 168$

and sum of last 12 results $= 12 \times 17 = 204$

Then, 13th result $= 450 - (168 + 204) = 78$

Problem 8 **Six persons went to a hotel for taking their meals. Five of them spent ₹ 32 each on their meals while the sixth person spent ₹ 80 more than the average expenditure of all the six. Total money spent by all the persons is:**

Sol. Since expenditures incurred by the sixth person is 80 more than the average expenditure of the all, a surplus will be of 80.

Total deficits = total surpluses = 80.

Clearly, the deficits of 80 will move to the remaining 5 persons.

$\therefore$ Deficit per person $= 80 \div 5 = 16$. But, the average expenditures of the 5 persons after deficit = 32

$\therefore$ Actual average expenditures of all $= 32 + 16 = 48$

Hence, total expenditures $= 6 \times 48 =$ ₹ 288.

Alternatively

Let the average expenditures of all the six be ₹ x.

Then, total expenditures $= 6 \times x = 6x$

As given, $6x = (32 \times 5) + (x + 80)$

$\Rightarrow$ $5x = 240 \Rightarrow x = 48$

Hence, total expenditures $= 6x = 6 \times 48 =$ ₹ 288.

Problem 9 **The average of 20 numbers is zero. Of them, at the most, how many may be greater than zero.**

[Hotel Management - 02]

Sol. It is possible that surplus in respect of 19 numbers may be equal to deficit in respect of one number. Thus, 19 numbers may be greater than zero.

Problems Based on Properties (C) of the Average

Problem 10 **David obtained 76, 65, 82, 67 and 85 marks (out of 100) in English, Mathematics, Physics, Chemistry and Biology. What are his average marks?**

Sol. Assume average is near to 70. Then, we may write:

76 as 70 + (6)

65 as 70 + (− 5)

82 as 70 + (12)

67 as 70 + (− 3)

85 as 70 + (15)

Now, the average of 6, – 5, 12, – 3 and

$$15 = \frac{6 - 5 + 12 - 3 + 15}{5} = 5$$

It is obvious that each number (i.e., 6, – 5, 12, – 3 and 15) should be increased by 70 so as to get the original given numbers i.e. 76, 65, ... Further, we know that when each number/ value of a given data set is increased by m then average will increase by m.

$\therefore$ Required average = 5 + 70 = 75

It is construed from the above that the required average will be equal to the sum of assumed average and average of the numbers obtained on subtraction of assumed average from each of given numbers. The whole process can be made mentally without pen and paper. Indeed, this technique is very useful in data-analysis/interpretation.

You are free to take any convenient number as assumed average. Suppose you have taken 80 instead of 70.

Then, average of the numbers obtained on subtraction of assumed average from each given numbers

$$= \frac{-4 - 15 + 2 - 13 + 5}{5} = -5$$

Now, required average = assumed average

+ (– 5) = 80 – 5 = 75

Problem 11 **Of the 3 numbers, the second is twice the first and thrice the third. If the average of the three numbers is 44, then the largest number is?**

[SSC(CGL-T-II) -12, SSC(CGL-PT)-08, MAT- 2000, 99]

Sol. Let the first number be 100. Then, 2nd number = 200 and 3rd number = 66.66

Now, you have to find their average. Assume their average be 100.

Now, subtract the assumed average from each of the given numbers and find the average of those differences.

$\therefore$ Average = assumed average

$$+ \frac{(100 - 100) + (200 - 100) + (66.66 - 100)}{3}$$

$$= 100 + \frac{66.66}{3} = 122.22$$

As given, $122.22 \Rightarrow 44$

$\therefore$ $200 \Rightarrow 72$

Alternatively

Let the first number be x.

Then, second number = $2x$ and third number = $\frac{2x}{3}$

Now, $\text{average} = \frac{x + 2x + \frac{2x}{3}}{3} = 44$

$\Rightarrow$ $\frac{11x}{9} = 44 \Rightarrow x = 36$

$\therefore$ Largest number = $2x = 2 \times 36 = 72$

Problem 12 **The average marks of a group of 20 students is reduced by 3 when a student, whose score is 75, is replaced by a new student. What is the score of new student?**

Sol. Here, average is reduced by 3, thus the sum of marks obtained by 20 students will be reduced by (3×20). Obviously, it is due to replacement of one old student by a new one. It implies that a student scoring 75 marks has been replaced by a new student whose score is 60 less than 75. Hence, the score of new student = 75 – 60 = 15.

Alternatively

Let, initial average marks of 20 students are x.

Then, sum of their marks = $20 \times x = 20x$

As per question, $20x - 75$ + score of new student = $20(x - 3)$

$\Rightarrow$ score of new student = – 60 + 75 = 15

Problem 13 **The average age of 8 men is increased by 2 years when two of them whose ages are 21 years and 23 years are replaced by two new men. The average age of the two new men is?**

Sol. Here, average age of 8 men is increased by 2 years. Thus, the sum of their ages will increase by (2×8) years and this increment is on account of replacement. It means, the sum of ages of two new men is 16 years more than sum of ages of two old men replaced.

Hence, sum of ages of the two new men

= 21 + 23 + 16 = 60 years

$\therefore$ Average age of the two new men

= 60 ÷ 2 = 30 years

Alternatively

Let initial average age of all 8 men is x.

Then, sum of their ages = $8 \times x = 8x$

As per question,

$8x - 21 - 23$ + sum of ages of two new men= $8(x + 2)$

$\Rightarrow$ Sum of ages of two new men = 21 + 23 + 16 = 60

$\therefore$ Average age of the two new men = 60 ÷ 2 = 30 years

Problem 14 **The present average age of family of four members is 36 years. If the present age of the youngest member of the family be 12 years, the average age of the family at birth of the youngest member was?**

Sol. At birth of youngest member, each of the members was 12 years younger. It means, average age of all four at birth of youngest one was 12 years less than the present average age. Hence, average age at birth of the youngest member

= 36 – 12 = 24 years

$\therefore$ Sum of ages of all four persons at birth of youngest

= 24 × 4 = 96

But in reality, there were only 3 members at birth of youngest. Hence, average age of the family at birth of the youngest member = 96 ÷ 3 = 32 years

Alternatively

Let the average age of the family at birth of the youngest member was x. Then, sum of their ages = $3x + 0$

[$\because$ one is new born]

After 12 years, when the youngest one attains the age of 12, sum of their ages = $3x + 0 + (4 \times 12)$

As given, sum of their present ages = 36×4

Hence, $3x + 0 + (4 \times 12) = 36 \times 4$

$\Rightarrow$ x = 32 years

Problem 15 **The average of some numbers is 54.6. If 75% of the numbers are increased by 5.6 each and the rest are decreased by 8.4 each, then what is the average of the numbers so obtained ?**

Sol. Ratio of numbers increased to numbers decreased = 3 : 1

Then, overall effect on the average

$$= \frac{3(+5.6) + 1(-8.4)}{3+1} = +2.1$$

Hence, the increased average = 54.6 + 2.1

= 56.7

Problems Based on Properties (D) of the Average

Problem 16 **The average of 17 numbers is 3.25. If each number is multiplied by 16, then the average of the new set of numbers is**

Sol. We know that when each number/value of a data set is multiplied by a certain value 'm', then their average will become m times the previous average.

∴ Required average = 3.25 × 16 = 52

Problem 17 **The average of 80 numbers is 75. If each number becomes 40% of itself, then the new average will be ?**

Sol. Required average = 40% of 75 = 30

Problem 18 **The average of first nine prime numbers is $11\frac{1}{9}$. What will be average if each number is divided by 10?**

Sol. Required average $= 11\frac{1}{9} \times \frac{1}{10} = 1\frac{1}{9}$

Problem 19 **The arithmetic mean of 100 observations is 24. 6 is added to each of the observations and then each of them is multiplied by 2.5. Find the new arithmetic mean.** *[SSC(CGL-T-II)-13]*

Sol. We know, if each number/ value of a given data set is increased by m then average will increase by m. Further, If each number/value of a data set is multiplied by a certain value 'n', then their average will become n times.

In view of the above, new mean = (24 + 6) × 2.5 = 75

Concept of Inclusion of Number(s) in a Given Set of Numbers

We have seen that the average of 3, 9, 7 and 5 is 6. Now, a 5th number is included in this set resulting in the average (overall) increases by 1.5, then you have to find the value of 5th number.

Each of previous four numbers may be assumed as 6 because their average is 6. Later, due to inclusion of 5th number, the average of all the five numbers becomes 7.5, it means we may again assume that each number is 7.5. Now, peruse the chart.

Before inclusion of 5th number		After inclusion of 5th number	
1st number	6	7.5 ⇒	6 + 1.5
2nd number	6	7.5 ⇒	6 + 1.5
3rd number	6	7.5 ⇒	6 + 1.5
4th number	6	7.5 ⇒	6 + 1.5
5th number	No exists	7.5 ⇒	7.5

Clearly, total increment over the previous sum

= (4 × 1.5) + 7.5 = 13.5

and it happens due to inclusion of fifth number.

∴ Value of fifth number = 13.5

In other words, we may conclude that the fifth number gives/ sacrifices 1.5 to each of previous four numbers to maintain the average of 7.5 keeping 7.5 for itself. Hence, the fifth number = 7.5 + (4 × 1.5) = 13.5

Generic Version

Value of the number included later = Value of final average + (Number of previous numbers × increment in previous average).

Now, take another example for building the concepts

"There is a set of four numbers 4,10, 8 and 6. The overall average decreases by 1 when one number is included in the set. Find the fifth number."

Each of previous four numbers may be assumed as 7 because their average is 7. Later, due to inclusion of 5th number the average of all the five numbers becomes 6, it means we may again assume that each number is 6. Now peruse the chart.

Before inclusion of 5th number		After inclusion of 5th number	
1st number	7	6 ⇒	7 − 1
2nd number	7	6 ⇒	7 − 1
3rd number	7	6 ⇒	7 − 1
4th number	7	6 ⇒	7 − 1
5th number	No exists	6 ⇒	6

Clearly, the fifth number having its certain value takes 1 from each of previous four numbers (i.e. 4 × 1) for being 6. Hence, the fifth number = 6 − (4 × 1) = 2

Generic Version

Value of the number included later = Value of final average − (Number of previous numbers × decrease in previous average).

Alternatively

***Case* I** Sum of all four previous numbers = 6 × 4 = 24

and sum of all five numbers = (6 + 1.5) × 5 = 37.5

∴ Value of 5th number = 37.5 − 24 = 13.5

***Case* II** Sum of all four previous numbers = 7 × 4 = 28

and sum of all five numbers = (7 − 1) × 5 = 30

∴ Value of 5th number = 30 − 28 = 2

Problem 20 The average age of 36 students in a group is 14 years. When teacher's age is included to it, the average age increases by one. What is the teacher's age in years? *[RBI-03, CLAT-10]*

Sol. Clearly, teacher's age made the average of students increase by one maintaining his/her own average age of 15.

Thus, teacher's age = 15 + (36 × 1) = 51 years

Alternatively

Age of the teacher = (37 × 15) − (36 × 14) = 555 − 504

= 51 years

Remarks *Aspirants are advised to avoid the alternative method as it may lead mistakes during multiplications and subtraction specially in case of large figures.*

Problem 21 The average age of 6 sons of a family is 8 years. If the ages of their parents are included the average age becomes 22 years. Find out the age of mother, if father is 8 years older than mother.

Sol. Clearly, ages of parents made the age of sons increase by 14 years maintaining their average age of 22.

Thus, parent's age = (22 × 2) + (6 × 14) = 44 + 84 = 128

But it is also given that father is 8 years older than mother,

$\therefore$ Age of mother = $\frac{128-8}{2}$ = 60 years.

Alternatively

Parents' age = (8 × 22) − (6 × 8) = 128

But it is also given that father is 8 years older than mother,

$\therefore$ Age of mother = $\frac{128-8}{2}$ = 60 years

Problem 22 3 years ago, the average age of a family of 5 members was 17 years. A baby having been born, the average age of the family is the same today. The present age of the baby is? *[SSC(CGL-PT)-04, SBI(PO)-17]*

Sol. At present each member of family has been grown up by 3 years, thus average should have become 3 years more than 17 years i.e. 20 years. However, due to inclusion of newly born baby, the present average remains 17 i.e. 3 less than 20. Now, this question reads: 'the average age of a family of 5 members is 20 years, but due to inclusion of a baby, the overall average becomes 3 years less than 20 years, what is the age of the baby?'

Clearly, the age of the baby made the average age of members of the family decrease by 3 years for maintaining his/her average age of 17. Hence, the present age of the baby

= 17 − (5 × 3) = 2 years

Alternatively

Sum of ages of family members 3 years ago = 17 × 5 = 85

Since at present, each member of family has been grown up by 3 years, sum of their ages at present = 85 + (5 × 3) = 100 years.

Further, given that average of all family-members including baby is 17. Then, sum of their ages (including baby)

= 17 × 6 = 102

$\therefore$ Present age of the baby = 102 − 100 = 2 years

Problems Based on Inclusion of Number(s) in Specific Cases

Problem 23 The average age of 20 students in a school is 16.25 years. One boy, aged 15 years is replaced by 3 newcomers resulting the new averages of all students becomes 17 years. Find out the average of all three newcomers.

Sol. Assume, only one newcomer comes in place of the student left, maintaining the average of 17.

Then, age of this new comer = 15 + (20 × 0.75)

= 30 years

But in fact there are two more newcomers, who are also maintaining the average of 17. It means sum of the ages of these two newcomers = 17 × 2 = 34 years

Average of all three newcomers = $\frac{30+34}{3}$ = 10 + 11.33

= 21.33 years

Note *It cannot be said definitely that the age of 1st newcomer is 30 years but it is certain that the sum of all three will be (30 + 34) years.*

Alternatively,

Sum of ages of all 20 old students = 20 × 16.25 = 325

Then, sum of ages of 19 old students when one leaves

= 325 − 15 = 310

Since the average age of 19 old students and 3 newcomers is 17 years, sum of ages of all 22 students = 17 × 22 = 374

$\therefore$ Sum of ages of 3 newcomers = 374 − 310 = 64

Hence, average of all three newcomers = 64 ÷ 3

= 21.33 years

Problem 24 The average age of 30 boys in a class is 15 years. One boy, aged 20 years, left the class but two new boys came in his place whose ages differ by 5 years. If the average age of all the boys now in the class becomes 15 years, the age of the younger newcomer is

Sol. Assume, only one newcomer comes in place of the student left, maintaining the average of 15 same as earlier. Then, age of this new comer = 20 years. But in fact, there is one more newcomer, who are also maintaining the average of 15. It means, age of this newcomer is 15 years.

Keeping in view the above and the fact that their ages differ by 5 years, it may be construed that the age of younger newcomer is 15 years.

Alternatively,

Sum of ages of all 30 old boys = 30 × 15 = 450

Then, sum of ages of 29 old boys when one leaves

= 450 − 20 = 430

Since the average age of 29 old boys and 2 newcomers is 15 years, sum of ages of all 31 boys = 15 × 31 = 465

$\therefore$ Sum of ages of both newcomers = 465 − 430 = 35

Further, the difference between ages of both newcomers is 5 years, the age of the younger newcomer = $\frac{35-5}{2}$ = 15 years

Concept of Exclusion of Number(s) from a Given set of Numbers

Let there be a set of 2, 5, 6 and 7. Now a number is excluded from this set resulting the average of remaining numbers increases by 1. Find out the value of the number excluded?

Since average of the set is 5, we may assume each number be 5.

$$2 \Rightarrow 5$$
$$5 \Rightarrow 5$$
$$6 \Rightarrow 5$$
$$7 \Rightarrow 5$$

When, one number is excluded from the set, the average of all three remaining numbers increases by 1. It implies that the sum of all remaining numbers increases by (1×3) and in fact this increment is on account of curtailment of 3 from the number to be excluded. But, we have assumed the number to be excluded as 5. Thus, the excluded number $= 5 - (1 \times 3) = 2$

Generic Version

> Value of the number excluded = Value of previous average − (Number of remaining numbers × increase in previous average).

Now, take **another example :** From a set of 3, 5, 10 and 12, one number is excluded resulting in new average decreases by 1.5. Find the excluded number.

Since average of the set is 7.5, we may assume each number be 7.5.

$$3 \Rightarrow 7.5$$
$$5 \Rightarrow 7.5$$
$$10 \Rightarrow 7.5$$
$$12 \Rightarrow 7.5$$

When one number is excluded from the set, the average of all three remaining numbers decreases by 1.5. It implies that the sum of all remaining numbers decreases by (1.5×3). It may be construed that the number to be excluded takes 1.5 each from the remaining numbers to make them decrease by 1.5 and thereafter it leaves the set. It means, the number to be excluded is (1.5×3) more than the average value of the all numbers including the number to be excluded later. But, we have assumed the number to be excluded as 7.5 since the overall average was 7.5 before the exclusion. Thus, the excluded number $= 7.5 + (1.5 \times 3) = 12$.

Generic Version

> Value of the number excluded
> = Value of previous average + (Number of remaining numbers × decrease in previous average).
>
> In case of exclusion of two or more numbers, we will act accordingly on the same line. An illustration is also given below.

Problem 25 **The average of five numbers is 27, if one number is excluded, the average becomes 25. The excluded number is?**

Sol. The average of remaining four numbers is decreased by 2. It implies that decrease in their sum i.e., (2×4) has been accumulated towards the excluded number and the fact that assumed value of excluded number is 27.

$\therefore$ Excluded number $= 27 + 8 = 35$

Note *You may directly write on pattern of the generic version Excluded number = Value of previous average + (Number of remaining numbers × decrease in previous average)*

$$= 27 + (4 \times 2) = 35$$

Alternatively

Sum of all five numbers $= 5 \times 27 = 135$

And sum of all four numbers after one's exclusion

$$= 4 \times 25 = 100$$

$\therefore$ Value of excluded number $= 135 - 100 = 35$

Problem 26 **The average of marks of 28 students in mathematics was 50. 8 students left the school, then this average increased by 5. What is the average of marks obtained by the students who left the school ?**

[SSC(CGL-PT)-05, SSC(SAS)-10]

Sol. The average of remaining 20 numbers is increased by 5, resulting in their sum will increase by 100. This increment is actually against the curtailment in sum of marks obtained by 8 students who left the school. Here, sum of 8 students marks is 50×8 i.e. 400. Then, actual marks of 8 students

$$= 400 - 100 = 300$$

$$\therefore \quad \text{Average} = \frac{300}{8} = 37.5$$

Alternatively

Sum of marks of all 28 students $= 50 \times 28$

$$= 1400$$

and sum of all 20 remaining students

$$= 20 \times (50 + 5) = 1100$$

$\therefore$ Sum of marks of 8 students who left the school

$$= 1400 - 1100 = 300$$

Then, their average marks $= 300 \div 8 = 37.5$

Problem 27 **The batting average of 40 innings of a cricket player is 50 runs. His highest score exceeds his lowest score by 172 runs. If these two innings are excluded, the average of the remaining 38 innings is 48. His highest score was:**

[SSC(CGL-T-II) -12, MAT-09, SSC(Asstt.-Mains)-03, SSC(SAS)-10]

Sol. We know, value of the number excluded = value of previous average + (number of remaining numbers × decrease in previous average).

$\Rightarrow$ Sum of scores of two innings to be excluded

$$= \{(2 \times 50) + (38 \times 2)\}$$
$$= 176$$

Then, highest score $= \dfrac{(176 - 172)}{2} + 172 = 174$

Alternatively

Sum of the scores in two innings excluded later

$$= (50 \times 40) - (38 \times 48) = 176$$

Then, highest score $= \dfrac{(176 - 172)}{2} + 172 = 174$

Alteration in Average due to Computational Errors

Problem 28 The mean of 100 items was 16. Later on it was discovered that an item 16 was misread as 61 and another item 43 was misread as 34. It was also found that the number of items was 90 and not 100. Then, what is the correct mean?

[SSC(CGL-T-II)-11, IBPS(CWE)-11, SSC(CGL-T-II)-17, CLAT-16]

Sol. $(61+34)-(16+43)=95-59=36$ has been taken more than the original sum while computing the average on account of misreading. It implies that the given average is 36/100 i.e. 0.36 more than correct mean. Thus, the correct mean of 100 items = 46 − 0.36 = 45.64.

Further, 100 items has been taken instead of 90 items. Thus, the correct mean $= 45.64 \times \frac{100}{90} = 50.7$

Problem 29 In an examination, the average of marks was found to be 50. For deducting marks for computational errors, the marks of 100 candidates had to be changed from 90 to 60 each and so average of marks came down to 45. The total number of candidates, who appeared at the examination was?

Sol. Let, the total number of students be x.

Total sum before correction $= 50x$.

Then, correct sum $= 50x - 100(90-60)$

$\therefore$ Correct average $= \frac{50x - 3000}{x} = 45$

$\Rightarrow 50x - 3000 = 45x$

$\Rightarrow 5x = 3000 \Rightarrow x = 600$

Alternatively

Total marks deduction $= 100(90-60) = 3000$

Since the overall average of marks of candidates has decreased by 5 and the fact that total deduction is 3000, number of total candidates $= 3000 \div 5 = 600$.

Concept of Weighted Average

When average of groups having different number of same type elements is computed, it is called weighted average. Suppose, there are x groups having $n_1, n_2, n_3 \dots n_x$ elements with averages $a_1, a_2, a_3, a_4 \dots a_x$ respectively. Then, weighted average

$$= \frac{n_1a_1 + n_2a_2 + n_3a_3 + n_4a_4 + \dots\dots + n_xa_x}{n_1 + n_2 + n_3 + n_4 + \dots\dots + n_x}$$

Note *If we take ratio of $n_1, n_2, n_3 \dots n_x$ instead of their real value, there will be no effect on the value of weighted average.*

Illustrations

Problem 30 The average age of 20 boys in a class is 12 years. 5 new boys are admitted to the class whose average age is 7 years. The average age of the boys in the class becomes?

Sol. New average age $= \frac{(20 \times 12) + (5 \times 7)}{20 + 5} = 11$ years

Alternatively I

20 boys ⇒ 12 years i.e. (7 + 5) years

5 boys ⇒ 7 years

Assume that the average age of 20 boys is 7 years and given that average age of 5 boys is 7 years. Then, average age of (20 + 5) boys will be 7 years. But in fact, average of 20 boys is 5 greater than the assumed average age(i.e. 7).

Hence, (20×5) years will be distributed among all 25 boys.

$\therefore$ Final average age $= 7 + (100 \div 25) = 11$ years

Alternatively II

Let the overall average be x.

Using Alligation-Rule,

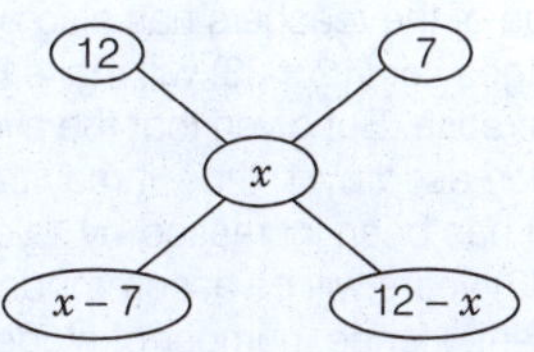

$\therefore$ Ratio of number of old boys to that of new boys $= (x-7):(12-x)$

$\Rightarrow (x-7):(12-x) = 20:5$

$\Rightarrow 48 - 4x = x - 7$

$\Rightarrow x = 11$

Hence, overall average age of the class = 11 years

Problem 31 In a school, the average age of students is 6 years and the average age of 12 teachers is 40 years. If he average age of the combined group of all the teacher and students is 7 years, then the number of students is?

Sol. Let the number of students be x.

Weighted average $= \frac{(x \times 6) + (12 \times 40)}{x + 12}$

$\Rightarrow 7 = \frac{(x \times 6) + (12 \times 40)}{x + 12}$

$\Rightarrow 7(x + 12) = (x \times 6) + (12 \times 40)$

$\Rightarrow 7x - 6x = (12 \times 40) - (12 \times 7)$

$\Rightarrow x = 396$

Alternatively I

By applying the Alligation-Rule

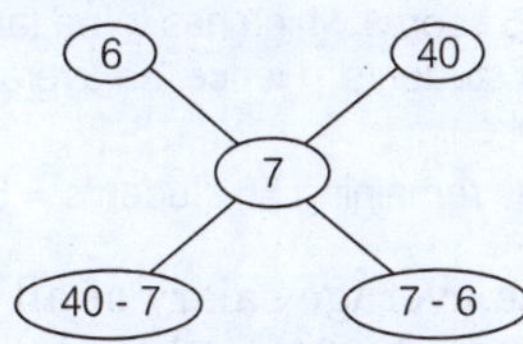

$\therefore$ Ratio of number of students to that of teachers = 33 : 1

As given, 1 unit = 12

$\therefore$ 33 units $= 33 \times 12 = 396$

Hence, number of students = 396

Alternatively II

Students	7 − 1	6
	7 − 1	
	7 − 1	
	and so on	
Teachers	7 + 33	40
	7 + 33	
	7 + 33	
	Upto 12 terms	

The average age of the teachers has also become 7 years, then the surplus of ages i.e. (12×33) will make the average age of the students increase. But given that the average age of students has increased by 1, thus, it may be assumed that age of each student has been increased by 1 year.
Further, (12×33) years will be accommodated for increasing 1 year per student. Hence, number of students $= 12 \times 33 = 396$.

Problem 32 **The arithmetic mean of the scores of a group of students in a test was 52. The brightest 20% of them secured a mean score of 80 and the dullest 25% a mean score of 31. The mean score of remaining 55% is?**

Sol. Let mean score of the remaining 55% be x.

Then, the weighted average $= \dfrac{(20 \times 80) + (25 \times 31) + (55 \times x)}{20 + 25 + 55}$

$$\Rightarrow \quad 52 = \frac{(20 \times 80) + (25 \times 31) + (55 \times x)}{20 + 25 + 55}$$

$$\Rightarrow \quad 5200 = 1600 + 775 + 55x$$

$$\Rightarrow \quad 55x = 2825$$

$$\therefore \quad x = 51.36$$

Alternatively

Let the total number of students be 100.

Clearly, 20 students are scoring above the average i.e. by $(80 - 52)$

$\therefore$ Sum of scores by 20 students $= 28 \times 20 = 560$ (surplus)

Further, 25 students are scoring below the average i.e. by $(52 - 31)$

$\therefore$ Sum of scores by 25 students $= 25 \times 21 = 525$ (deficit)

Now, we may conclude that $(20 + 25)$ students have a surplus of $(560 - 525)$ i.e. 35 scores which has to be taken from scores of the remaining 55 students. Hence, an average of $(35 \div 55)$ i.e. 0.64 has to be taken.

Thus, mean of the remaining 55 students $= 52 - 0.64 = 51.36$

Problem 33 **The average salary of all 40 employees in an organization is ₹ 6000/month and number of officers is 1/3 of number of clerks. Find out the average salary of all the officers.**

Sol. Ratio of number of officers to number of clerks = 1 : 3

$\therefore$ Number of officers $= 40 \times \dfrac{1}{4} = 10$

Then, number of clerks $= 40 - 10 = 30$

Here, we cannot find the surplus or deficit. Thus, we may say that data is insufficient.

Alternatively

Combined(weighted) average salary and individual number of officers and clerks are known. Then, in order to know the average salary of all officers, we must have the information as to what is the average salary of clerks. Hence, given data is insufficient.

Problems Based on Cricket

Problem 34 **A cricketer has a certain average of runs for his 8 innings. In the 9th innings, he scores 100 runs, thereby increases his average by 9 runs. His new average of run is?** ***[SSC(CGL-T-II)-13, SSC(CPO)-08, CLAT-11]***

Sol. Let the average of runs for 8 innings be x.

Total sum $= 8 \times x = 8x$

Total runs in 9 innings $= 8x + 100$

Average for 9 innings $= \dfrac{8x + 100}{9}$

As given, $\dfrac{8x + 100}{9} = x + 9$

$$\Rightarrow \quad 8x + 100 = 9x + 81$$

$$\therefore \quad x = 19$$

$$\therefore \quad \text{New average} = 19 + 9 = 28.$$

Alternatively I

1st inning	? + 9 (final average)
2nd inning	
3rd inning	
... upto 8th inning	
9th inning	100 runs

The average for 8 innings has been increased by 9. It means 72 (8×9) runs has been taken from 9th inning's run(100). Now, runs left for 9th inning is $(100 - 72)$ i.e. 28. Obviously, the average for 1st 8 innings and run for 9th inning will be equal. Hence, new average = 28.

(In this question, the old average for 8 innings $= 28 - 9 = 19$)

Alternatively II

Recall the 'Concept of Inclusion of number(s) in a given set of numbers'. Clearly, 9th inning has made the average for 8 innings increase by 9 keeping final average run of x (let) for itself.

Thus, run made in 9th inning $= x + (9 \times 8)$

$$\Rightarrow \quad 100 = x + 72 \Rightarrow x = 28$$

Problem 35 **A cricketer scores 8 runs in 12th inning resulting in, overall average decreases by 1. Find out the average of all innings before 12th inning.** ***[SSC(CGL-T-II)-17]***

Sol. Let, the average of 1st 11 innings be x.

Then, sum of runs in total of 12 innings $= 11x + 8$

$\therefore$ Average for 12 innings $= \dfrac{11x + 8}{12} = (x - 1)$

$$\Rightarrow \quad 11x + 8 = 12x - 12$$

$$\Rightarrow \quad x = 20$$

Alternatively I

The overall average for 11 innings decreases by 1. It means each of first 11 innings has given 1 run to 12th inning.

$\therefore$ Total given runs $= 11 \times 1 = 11$ runs.

Now run for 12th inning $= 11 + 8 = 19$

Obviously, the average run for 1st 11 innings and run for 12th inning will be equal. Hence, new average for all 12 innings = 19

$\therefore$ Average for all innings before 12th inning $= 19 + 1 = 20$

Alternatively II

Recall the 'Concept of Inclusion of number(s) in a given set of numbers'. Clearly, 12th inning has made the average for 11 innings decrease by 1 maintaining its own final average run of x (let).

Thus, run made in 12th inning $= x - (11 \times 1)$

$\Rightarrow \quad 8 = x - 11$

$\Rightarrow \quad x = 19$

Hence, final average for 12 innings = 19

$\therefore$ Average for all innings before 12th inning $= 19 + 1 = 20$

Problem 36 The bowling average of a cricketer was 12.4. He improves his bowling average by 0.2 points when he takes 5 wickets for 26 runs in his last match. The number of wickets taken by him before the last match was?

Sol. Let the number of wickets taken by the cricketer before the last match be x.

It is given that the cricketer takes 1 wicket by giving 12.4 runs through bowling.

$\therefore$ Total runs given $= 12.4x$

Then, including last match, total runs given $= 12.4x + 26$

Now, $\quad \text{average} = \dfrac{\text{total run given}}{\text{total wicket taken}}$

$\Rightarrow \quad (12.4 - 0.2) = \dfrac{12.4x + 26}{x + 5}$

$\Rightarrow \quad 12.4x + 26 = 12.2 + 61$

$\Rightarrow \quad 0.2x = 35$

$\therefore \quad x = 175$

Alternatively

The bowling average reduces to (12.4 – 0.2) and it is also true for the last 5 wickets.

$\therefore$ Sum of runs $= (12.4 - 0.2) \times 5 = 61$

Real savings $= 61 - 26 = 35$ which is due to decrease in each of previous bowlings by 0.2 points.

$\therefore$ Number of wickets before the last match $= \dfrac{35}{0.2} = 175$

[$\because$ 0.2 runs are reduced/saved, when 1 wicket is taken

$\therefore$ 35 runs will be saved when $\dfrac{35}{0.2}$ wickets are taken]

Concept of Average Speed

Detailed in the chapter 'Time and Distance'.

Average of the series of numbers

If the series are in Arithmetic Progression (AP), the sum of given series $= \dfrac{n}{2}[a + l]$

Where, n = total number of terms, a = initial term and l = last term. Term simply refers to the number.

$\therefore$ Average of the given AP series

$$= \frac{\text{sum of terms}}{\text{total number of terms}} = \frac{\frac{n}{2}[a + l]}{n}$$

$$= \frac{a + l}{2} = \frac{\text{1st term} + \text{last term}}{2}$$

This conception is applicable in all types of problems based on AP series. Now, we will discuss the questions category wise.

Problems Based on Series of Natural Numbers

Problem 37 Find out the average of first 24 natural numbers.

Sol. Required average $= \dfrac{a + l}{2} = \dfrac{1 + 24}{2} = 12.5$

[$\because$ 1 is an initial term(a) and 24 is the last term(l)]

Problem 38 Find out the average of all natural numbers upto 36 starting from 10.

Sol. Required average $= \dfrac{a + l}{2} = \dfrac{10 + 36}{2} = 23$ $\quad$ [$\because a = 10, l = 36$]

Problem 39 Find out the average of all natural numbers between 22 and 38.

Sol. Required average $= \dfrac{a + l}{2} = \dfrac{23 + 37}{2} = 30$

[$\because a = 23$, $l = 37$ as word, 'between' is in the question]

Remarks *only one model : average* $= \dfrac{a + l}{2}$ *has been applied in above all three cases*

Problem 40 Find out the average of 10 consecutive natural numbers starting from 15.

Sol. Required average $= \dfrac{15 + 24}{2} = 19.5$

[$\because a = 15$, $l = 24$, series is 15,16,17, ... 24]

Or

$$\text{Average} = \frac{15 + 24}{2} = \frac{15 + (15 + 9)}{2}$$

$$= \frac{(15 + 15) + 9}{2} = 15 + \frac{9}{2} = a + \frac{(n - 1)}{2}$$

[where, a = initial term and n = number of consecutive natural numbers]

Or

$$\text{Average} = \frac{15 + 24}{2} = \frac{(24 - 9) + 24}{2}$$

$$= \frac{(24 + 24) - 9}{2} = 24 - \frac{9}{2} = 1 - \frac{(n - 1)}{2}$$

{Where, l = last term and n = number of consecutive natural numbers}

In brief,

Average of n consecutive natural numbers starting from a
$= a + \frac{(n-1)}{2}$ or, $a = \text{average} - \frac{(n-1)}{2}$, and Average of
n consecutive natural numbers ending with term l
$= l - \frac{(n-1)}{2}$ or, $l = \text{average} + \frac{(n-1)}{2}$

Problem 41 Find out the 3rd term of a series of 9 consecutive natural numbers of which average is 27.

Sol. We know, $a = \text{average} - \frac{(n-1)}{2} = 27 - \frac{(9-1)}{2} = 23$

Then, 3rd term $= 23 + 2 = 25$

Problem 42 The average of 7 consecutive numbers is 20. The largest of these numbers is

Sol. Obviously, the last number of the series will be the largest number.

We have, $l = \text{average} + \frac{(n-1)}{2} = 20 + \frac{(7-1)}{2} = 23$

Problem 43 The average of 7 consecutive numbers is P. If the next three numbers are also added, the average shall *[IIFT-13]*

(a) remain unchanged (b) increase by 1
(c) increase by 1.5 (d) increase by 2

Sol. Average of 7 consecutive numbers $= a + \frac{(7-1)}{2} = a + 3$, where a is an initial number.

Further, Average of 10 consecutive numbers
$= a + \frac{(10-1)}{2} = a + 4.5$

It is clear that the latter average will increase by 1.5.

Problems Based on Series of Even Numbers

Problem 44 Find out the average of first 24 even numbers.

Problem 45 Find the average of all even numbers upto 46 starting from 16.

Problem 46 Find the average of all even numbers between 36 and 49.

Solutions

44. Average $= \frac{a+l}{2} = \frac{2+48}{2} = 25$

[Series is 2, 4, ... 48, $a = 2$, $l = 48$]

Remarks *You may observe from the above that the required average is always $(n+1)$, where n = number of terms in the series*

45. Required average $= \frac{a+l}{2} = \frac{16+46}{2} = 31$ [$\because a = 16, l = 46$]

46. Required average $= \frac{a+l}{2} = \frac{38+48}{2} = 43$ [$\because a = 38, l = 48$]

Remarks *Only one model average $= \frac{a+l}{2}$ has been applied in above all the three cases*

Problem 47 Find out the average of 7 consecutive even numbers starting from 18.

Sol. Here, 1st term $= a = 18$, Last term $= l = 18 + (2 \times 6)$ or $2 \cdot 6 = 30$

Then, average $= \frac{18+30}{2} = 24$

Or

Average $= \frac{18+30}{2} = \frac{18+(18+12)}{2}$
$= \frac{(18+18)+12}{2} = 18 + 6 = a + (n-1)$

[where, a = initial term and n = number of consecutive even numbers]

Or

Average $= \frac{18+30}{2} = \frac{(30-12)+30}{2}$
$= \frac{(30+30)-12}{2} = 30 - 6 = l - (n-1)$

[where, l = last term and n = number of consecutive even numbers]

In brief,

Average of n consecutive even numbers starting from a
$= a + (n-1)$ or, $a = \text{average} - (n-1)$
and Average of n consecutive even numbers ending with term l
$= l - (n-1)$ or, $l = \text{average} + (n-1)$.

Note *The above both outcomes suit also for the n consecutive odd numbers.*

Problems Based on Series of Odd Numbers

Problem 48 Find out the average of first 29 odd numbers.

Problem 49 Find out the average of all odd numbers upto 37 starting from 17.

Problem 50 Find out the average of all odd numbers between 36 and 57.

Solutions

48. Required average $= \frac{a+l}{2} = \frac{1+57}{2} = 29$

[$\because$ series is 1,3,5,......... 57, $a = 1$ and $l = 57$]

Remarks *You may say, the required average will always be n where n = number of terms in the series*

49. Required average $= \frac{a+l}{2} = \frac{17+37}{2} = 27$ [$\because a = 17, l = 37$]

50. Required average $= \frac{a+l}{2} = \frac{37+55}{2} = 46$ [$\because a = 37$, $l = 55$ as word 'between' is in the question]

Remarks *Only one model average $= \frac{a+l}{2}$ has been applied in above all three cases.*

Problem 51 Find out the average of 7 consecutive odd numbers starting from 97.

Sol. Required average

$= a + (n - 1) = 97 + (7 - 1) = 103$

Problem 52 The average of four consecutive odd numbers is 12. Which will be lowest odd number ?

[NABARD (Asstt. Mngr)- 02]

Sol. Obviously, the initial number will be the lowest one.

Then, a = average $- (n - 1)$

$= 12 - (4 - 1) = 9$

Problem 53 The sum of three consecutive odd numbers is 38 more than the average of these numbers. What is the first of these numbers ?

Sol. 1st term = average $- (n - 1)$ = average $- (3 - 1)$

$\Rightarrow$ 1st term $= m - 2$, where, m = their average.

Then, 2nd term $= (m - 2) + 2 = m$

[$\because$ 2nd will be 2 more than 1st in this odd number series]

and 3rd i.e. last term $= m + 2$

Now, sum of these three terms $= 3m$

As given, $3m = m + 38$

$\Rightarrow$ $2m = 38$

$\therefore$ $m = 19$

$\therefore$ 1st term $= 19 - 2 = 17$

Alternatively

Let the average of all three numbers be m.

Then, their sum $= 3m$

As given, $3m = m + 38$

$\Rightarrow$ $2m = 38$

$\therefore$ $m = 19$

Now, 1st term $= a$ = average $- (n - 1)$

$= 19 - (3 - 1) = 17$

Problem 54 The average of five consecutive odd numbers is 61. What is the difference between the highest and lowest numbers ?

[Bank (PO)-03, SSC(CGL-T-II)-17]

Sol. Lowest number = 1st term $= a$ = average $- (n - 1)$

$= 61 - (5 - 1) = 57$

and Highest number = last term $= l$ = average $+ (n - 1)$

$= 61 + (5 - 1) = 65$

Now, required difference $= 65 - 57 = 8$

Problem 55 The average of five consecutive odd numbers is m. If the next three odd numbers are also included, then what is the increase in the average ?

[SSC(CGL-T-I)-19]

Sol. Average of 5 consecutive odd numbers = initial number $+ (5 - 1)$

and average of 8 consecutive odd numbers = initial number $+ (8 - 1)$

Since initial numbers in this case are same, the increase in the average is $(7 - 4)$ i.e. 3.

Average of the Squares of Numbers

(i) Average of squares of first n natural numbers $= \dfrac{(n+1)(2n+1)}{6}$

(ii) Average of squares of first n even numbers $= \dfrac{2(n+1)(2n+1)}{3}$

(iii) Average of square of first n odd numbers $= \dfrac{4n^2 - 1}{3}$

Problem 56 Find out the average of squares of first 22 even numbers ?

Sol. Required average

$$= \frac{2(n+1)(2n+1)}{3} = \frac{2(22+1)(2 \cdot 22 + 1)}{3} = \frac{46 \times 45}{3} = 690$$

Problem 57 What is the average of squares of first 10 odd numbers ?

Sol. Required average $= \dfrac{4n^2 - 1}{3} = \dfrac{4(10)^2 - 1}{3} = 133$

Problem 58 What is the average of squares of even numbers between 13 and 39 ?

Sol. In such types of questions where the above formulae are not applicable, we try to find their sum as detailed in the chapter 'Addition and Subtraction'. Thereafter, we may divide the sum by number of numbers so as to obtain the average.

Average of the Cubes of Numbers

(i) Average of cubes of first n natural numbers $= \dfrac{n(n+1)^2}{4}$

(ii) Average of cubes of first n even numbers $= 2n(n+1)^2$

(iii) Average of cubes of first n odd numbers $= n(2n^2 - 1)$

Problem 59 What is the average of cubes of first 8 odd numbers ?

Sol. Required average $= n(2n^2 - 1) = 8(2 \cdot 8^2 - 1) = 1016$

Miscellaneous

Problem 60 Three numbers A, B and C are in ratio 1 : 2 : 3 and their average is 600. A is increased by 10% and B is decreased by 20%, then by how much percent must C be increased so as to increase 5% in their average ?

Sol. Average of 1, 2 and 3 units = 2

As given, 2 units $= 600 \Rightarrow B$

$\therefore$ 1 unit $= 300 \Rightarrow A$

and 3 units $= 900 \Rightarrow C$

In order to increase overall average by 5%, their sum should be increased by 5%. It means, the sum should be increased by (5% of 1800) i.e. by 90.

Now, increment in A = 10% of 300 = 30
and decrease in B = 20% of 600 = 120
Hence, net effect on the sum = + 30 − 120 = − 90
Obviously, number C must be increased by 180 so as to make the sum increase by 90.
Now, required percent age = $\left(\frac{180}{900} \times 100\right)\% = 20\%$

Alternatively

Let number C is increased by x%.
In order to increase overall average by 5%, their sum should be increased by 5%.

Hence, $(1 \times 110\%) + (2 \times 80\%) + \{3 \times (100 + x)\%\}$
$= (1 + 2 + 3) \times 105\%$
$\Rightarrow 110 + 160 + 300 + 3x = 6 \times 105$
$\Rightarrow 3x = 630 - 570 = 60$
$\therefore x = 20$

Thus, number C must be increased by 20% so as to make the average increase by 5%.

Problem 61 The average weight of 3 men A, B and C is 84 kg. Another man D joins the group and the average now becomes 80 kg. If another man E, whose weight is 3 kg more than that of D, replaces A, then the average weight of B, C, D and E becomes 79 kg. The weight of A is

[Bank(PO)- 03, CLAT-17]

Sol. We know, value of the number included later = Value of final average − (Number of previous numbers × decrease in previous average).

Hence, weight of $D = 80 - (3 \times 4) = 68$ kg
Then, weight of E = 3 kg more than D = 71 kg
The average of B, C, D and E is 79
$\therefore$ Average of B and $C = 79 + \frac{(79 - 68) + (79 - 71)}{2} = 88.5$
But, average of A, B and C = 84
$\therefore$ Weight of $A = 84 - (88.5 - 84) - (88.5 - 84) = 75$ kg

Problem 62 There are four natural numbers. The average of any three numbers is added in fourth number and in this way, numbers 29, 23, 21 and 17 are obtained. Find the smallest natural number.

Sol. Let numbers be a, b, c and d.

As given, $\frac{a + b + c}{3} + d = 29$

$\Rightarrow a + b + c + 3d = 87$ ...(i)
Similarly, $a + b + d + 3c = 69$ ...(ii)
and $a + c + d + 3b = 63$ ... (iii)
and $b + c + d + 3a = 51$ ... (iv)

Adding all four Eqs. we get, $6(a + b + c + d) = 270$
$\Rightarrow (a + b + c + d) = 45$
Putting this value in Eq. (i) we get, $45 + 2d = 87$
$\Rightarrow d = 21$
Similarly, putting this value in Eqs. (ii), (iii) and (iv), we get $c = 12$, $b = 9$ and $a = 3$
Hence, numbers are 3, 9, 12 and 21. Then, smallest number = 3

Problem 63 Two statements have been given below the questions as (A) and (B).

Mark answer(a) if question can be answered using A alone but not using B alone

Mark answer(b) if question can be answered using B alone but not using A alone

Mark answer(c) if question can be answered using A and B together but not using either A or B alone

Mark answer(d) if question cannot be answered even using A and B together.

The average weight of a class of 100 students is 45 kg. The class consists of two Sections I and II, each with 50 students. The average weight, WI of Section I is smaller than the average weight WII of Section II. If the heaviest student, say Deepak, of Section II is moved to Section I and the lightest student, say Poonam, of Section I is moved to Section II, then the average weights of the two sections are switched, i.e. the average weight of Section I becomes WII and that to Section II becomes WI. What is the weight of Poonam?

(A) WII − WI = 1.0

(B) moving Deepak from Section II to I (without any move from I to II) makes the average weights of the two sections equal.

[CAT-07]

Sol. (c) Weighted average = $\frac{n_1 w_{\text{I}} + n_2 w_{\text{II}}}{n_1 + n_2}$

$\Rightarrow 45 = \frac{50 w_{\text{I}} + 50 w_{\text{II}}}{50 + 50}$

$\Rightarrow w_{\text{I}} + w_{\text{II}} = 90$ where, $w_{\text{II}} > w_{\text{I}}$

Further, if Deepak and Poonam exchange their sections, then w_{I} becomes w_{II} and vice-versa. In both cases, the difference in weights of Deepak and Poonam = $50(w_{\text{II}} - w_{\text{I}})$

Suppose we have information that $w_{\text{II}} - w_{\text{I}} = 1.0$ then, we can find the value of difference between the weights of Deepak and Poonam. But we cannot find the weight of Poonam.

Further, if information as envisaged in Statement (B) is used, then after shift of Deepak from Section II to Section I,

average weight of Section I = $\frac{50 w_{\text{I}} + \text{weight of Deepak}}{51}$

and average weight of Section II = $\frac{50 w_{\text{II}} - \text{weight of Deepak}}{49}$

As given, $\frac{50\, w_{\text{I}} + \text{weight of Deepak}}{51} = \frac{50 w_{\text{II}} - \text{weight of Deepak}}{49}$

If the values of W_{I} and W_{II} are not known, we cannot find the weight of Deepak. But we can find the value of w_{I} and w_{II} from the equations (i) $w_{\text{I}} + w_{\text{II}} = 90$ and (ii) $w_{\text{II}} - w_{\text{I}} = 1.0$. No doubt, we can find the weight of Poonam if we find the weight of Deepak as difference between their weigths is $50(w_{\text{II}} - w_{\text{I}})$ i.e. (50×1).

Conclusion The question can be answered using A and B together but not using either A or B alone.

WORKOUT

1. There are two Classes A and B. The average weight of the students in Class A is 40 kg. The average weight of the students in Class B is 60 kg. A student, whose weight is x kg left A and joined B. As a result, the average weight of A, as well as that of B decreased. Which of the following must be true?

(a) $35 < x \le 40$ (b) $40 < x < 60$
(c) $60 < x < 65$ (d) $30 < x \le 35$

2. If the average of A and B is 30, the average of C and D is 20, then which of the following is/are correct?

I. The average of B and C must be greater than 25.

II. The average of A and D must be less than 25.

Select the correct answer using the codes given below: ***[CDS-14]***

(a) Only I (b) Only II
(c) Either I or II (d) Neither I nor II

3. While calculating the average of a batsman as 36 in 100 matches that he played, one of the score 90 was incorrectly noted as 40. The percentage error is

(a) 0.6% (b) 1.36%
(c) 1.34% (d) 1.21%

4. The mean of 50 numbers is 30. Later it was discovered that two entries were wrongly entered as 82 and 13 instead of 28 and 31. Find the correct mean. ***[SSC(CGL-T-I)-11, CSAT-19]***

(a) 29.28 (b) 30.82
(c) 30.72 (d) 48.53

5. The average score of Rahul, Manish and Suresh is 63. Rahul's score is 15 less than Ajay and 10 more than Manish. If Ajay scored 30 marks more than the average score of Rahul, Manish and Suresh, what is the sum of Manish's and Suresh's scores ? ***[Allahabad Bank(PO)-10]***

(a) 120 (b) 111 (c) 117
(d) Cannot be determined
(e) None of the above

6. A batsman scores 80 runs in his sixth inning and thus increases his average by 5. What is the average after six innings?

(a) 50 (b) 55
(c) 60 (d) 65

7. The average marks of a student in 8 subjects is 87. Of these, the highest marks are 2 more than the one next in value. If these two subjects are eliminated, the average marks of the remaining subjets is 85. What is the highest score? ***[MAT-12, 09]***

(a) 91 (b) 94
(c) 89 (d) 96

8. The average of 8 men, 7 women and 1 boy is 45 years. If the average age of 8 men is 48 years and that of 7 women is 46 years, then the age of the boy is

(a) 14 years (b) 15 years
(c) 16 years (d) 18 years

9. The average of 5 consecutive natural numbers is m. If the next three natural numbers are also included, how much more than m will the average of these 8 numbers be? ***[SSC (CPO)-06, CDS-19]***

(a) 1.5 (b) 2
(c) 3 (d) 2.5

10. The average weight of a class of 15 boys and 10 girls is 38.4 kg. If the average weight of the boys is 40 kg, then what is the average weight of the girls ? ***[CDS-13, IBPS(PO-Pre.)-18]***

(a) 36.5 kg (b) 35 kg
(c) 36 kg (d) 35.6 kg

11. The average monthly salary of employees, consisting of officers and workers of an organization is ₹ 3000. The average salary of an officer is ₹ 10000 while that of a worker is ₹ 2000 per month. If there are total 400 employees in the organisation, find the number of officers. ***[MAT-12, 07, SSC(SAS)-10]***

(a) 50 (b) 60
(c) 80 (d) 40

12. The average age of a couple is 25 years. The average age of the family just after the birth of the first child was 18 years. The average age of the family just after second child was born was 15 years. The average age of the family after the third and the fourth children (who are twins) were born was 12 years. If the present average age of the family of six persons is 16 years, how old is the eldest child? ***[CAT-09]***

(a) 6 years (b) 7 years
(c) 8 years (d) 9 years

13. Nine numbers are written in ascending order. The middle number is the average of the nine numbers. The average of the first five larger numbers is 68 and that of the five smaller numbers is 44. What is the sum of all nine numbers?

(a) 450 (b) 501
(c) 504 (d) 540

14. The average age of 3 children in a family is 20% of the average age of the father and the eldest child. The total age of the mother and the youngest child is 39 years. If the father's age is 26 years, what is the age of the second child?

(a) 20 years
(b) 18 years
(c) 15 years
(d) Cannot be determined

15. Prof. Suman takes a number of quizzes for a course. All the quizzes are out of 100. A student can get an A grade in the course, if the average of her scores is more than or equal to 90. Grade B is awarded to a student, if the average of her scores is between 87 and 89 (both included). If the average is below 87, the student gets a C grade. Ramesh is preparing for the last quiz and he realizes that he will score a minimum of 97 to get an A grade. After the quiz, he realize that he will score 70, and he will just manage a grade B. How many quizzes did Prof. Suman take ? *[XAT-14]*

(a) 6 (b) 7
(c) 8 (d) 9

16. The mean weight of 150 students in a class is 60 kg. The mean weight of boys is 70 kg and that of girls is 55 kg. What is the number of boys in the class ? *[CDS-12, 10, 19]*

(a) 50 (b) 60
(c) 75 (d) 100

17. The average of 5 consecutive odd numbers A, B, C, D and E is 41. What is the product of A and E ?

(a) 1677 (b) 1517
(c) 1665 (d) 1591

18. The average salary of all the staff in an office of a corporate house is ₹ 5000. The average salary of the officers is ₹ 14000 and that of the rest is ₹ 4000. If the total number of staff is 500, the number of officers is

[SSC(CGL-T-II)-12, IBPS(PO-Pre.)-18]

(a) 10 (b) 15
(c) 25 (d) 50

19. The average salary of 100 employees in an office is ₹ 16000 per month. The management decided to raise salary of every employee by 5% but stopped a transport allowance of ₹ 800 per month which was paid earlier to every employee. What will be the new average monthly salary ? *[IAS(Pre.)- 07]*

(a) ₹ 16000
(b) ₹ 16500
(c) ₹ 16800
(d) Cannot be known since data are insufficient.

20. The population of a state increased from 100 million to 169 million in two decades. What is the average increase in population per decade ?

(a) 35% (b) 32%
(c) 34.5% (d) 69%

21. Out of 250 observations, the first 100 observations have mean 5 and the average of the remaining 150 observations is 25/3. What is the average of the whole group of observations ?

[CDS-11, 10, MAT-08, CLAT-14]

(a) 6 (b) 7
(c) 8 (d) 9

22. Average rainfall on Monday, Tuesday, Wednesday and Thursday is 420.5 cm and average on Tuesday, Wednesday , Thursday and Friday is 440.5 cm. If the ratio of rainfall for Monday and Friday is 20 : 21, find the rainfall in cm on Monday and Friday. *[SSC(CGL-T-II)-11]*

(a) 1800, 1890
(b) 1600, 1680
(c) 1700, 1470
(d) 1682, 1762

23. The average marks of the students in four Sections A, B, C and D of a school is 60%. The average marks of the students of A, B, C and D individually are 45%, 50%, 72% and 80% respectively. If the average marks of the students of Sections A and B together is 48% and that of the students of B and C together is 60%, what is the ratio of the number of students in Section A and D? *[MAT-12]*

(a) 2 : 3 (b) 4 : 3
(c) 5 : 3 (d) 3 : 5

24. The age of the captain of a cricket team of 11 players is 25 yearss and the wicketkeeper is 3 years older than the captain. If the ages of these two are excluded, the average age of the remaining players of the team becomes 1 years less than the average age of the whole team. What is the average age of the whole team ?

(a) 20 (b) 21
(c) 23 (d) 22

25. Four numbers are written in a row. The average of first two numbers is 7, that of middle two numbers is 2.3 and that of last two numbers is 8.4. The average of first and last number is *[LIC(AAO)-11]*

(a) 5.9
(b) 10.7
(c) 13.1
(d) Cannot be determined

26. The average (arithmetic mean) of x and y is 40. If $z = 10$, then what is the average of x, y and z ?

(a) $16\frac{2}{3}$ (b) 30
(c) 25 (d) $17\frac{1}{2}$

27. There are 30 students in a class. The average age of the first 10 students is 12.5 years. The average age of the next 20 students is 13.1 yearss. The average age of the whole class is? *[MAT-09, NABARD(PO)-09]*

(a) 12.5 years (b) 12.7 years
(c) 12.8 years (d) 12.9 years

28. The average weight of 29 students in a class is 48 kg. If the weight of the teacher is included, the average weight rises by 500 gm. Find the weight of the teacher.

(a) 57 kg (b) 60 kg
(c) 65 kg (d) 63 kg

29. The weight of the body, calculated as the average of seven different experiments, is 53.735 g. The average of the first three experiments is 54.005 g. The fourth was greater than the fifth by 0.004 g and the average of sixth and seventh was 0.010 g less than the average of the first three. Find the weight of the body in fourth experiment.

(a) 53.072 g (b) 63.072 g
(c) 53. 726 g (d) 56. 72 g

30. There are four friends. The average score in unit test of the first three is 15 and that of the last three is 16. If the score of the last friend is 19, then first friend's score is what per cent of average of the last three ?

(a) 66.66%
(b) 300%
(c) 33.33%
(d) None of the above

31. The average of four consecutive even numbers is 27. The largest of these numbers is *[SSC(SAS)-10]*

(a) 28 (b) 26
(c) 32 (d) 30

32. The average weight of 45 students in a class is 52 kg. 5 of them whose average weight is 48 kg leave the class and other 5 students whose average weight is 54 kg join the class. What is the new average weight (in kg) of the class ?

(a) 52.6 (b) $52\frac{2}{3}$
(c) $52\frac{1}{3}$ (d) None of these

33. The average monthly expenditures of a family was ₹ 2200 during the first 3 months, ₹ 2250 during the next 4 months and ₹ 3120 during the last 5 months of a year. If the total savings during the year were ₹ 1260, then the average monthly income was

(a) ₹ 2605 (b) ₹ 2805
(c) ₹ 2705 (d) ₹ 2905

34. The average of 5 consecutive integers starting with 'm' is n. What is the average of 6 consecutive integers starting with $(m + 2)$?

(a) $\frac{2n+5}{2}$ (b) $(n + 2)$
(c) $(n + 3)$ (d) $\frac{2n+9}{2}$

35. A team of 8 persons joins in a shooting competition. The best marksman scored 85 points. If he had scored 92 points, the average score for the team would have been 84. The number of points, the team scored was *[LIC(AAO)-09]*

(a) 645 (b) 665 (c) 588 (d) 672

36. In a class, there are 20 boys whose average age is decreased by 2 months, when one boy age 18 years is replaced by a new boy. The age of the new boy is

(a) 14 years 8 months
(b) 16 years 4 months
(c) 15 years
(d) 17 years 10 months

37. The average of marks obtained by 120 students was 35. If the average of passed candidates was 39 and that of failed candidates was 15. The number of candidates passed in the examination was

(a) 100 (b) 110
(c) 120 (d) 80

38. The average age of a woman and her daughter is 42 years. The ratio of their ages is 2 : 1 respectively. What is the daughter's age ?

(a) 28 years (b) 18 years
(c) 52 years (d) 32 years

39. Ten years ago, the ages of the members of a joint family of eight people added upto 231 years. Three years later, one member died at the age of 60 years and a child was born during the same year. After another three years, one more member died, again at 60, and a child was born during the same year. The current average age of this eight member joint family is nearest to *[CAT-07]*

(a) 24 years (b) 23 years
(c) 22 years (d) 21 years
(e) 25 years

40. If the average of 6 consecutive even numbers is 25. What is the difference between the highest and lowest numbers ? *[SSC(CGL-T-II)-13]*

(a) 8 (b) 10
(c) 12 (d) 14

41. The average monthly income of a person in a certain family of 5 is ₹ 1000. What will be the monthly average income of a person in the same family if the income of one person will be increased by ₹ 12000 per year? *[IAS(Pre.)- 97]*

(a) ₹ 1200 (b) ₹ 1600
(c) ₹ 2000 (d) ₹ 3400

42. The average temperature from Monday to Thursday is 48° and from Tuesday to Friday is 52°. If the temperature on Monday is 42°, what was it on Friday ? *[MAT-09, 05]*

(a) 55° (b) 52°
(c) 58° (d) 51°

43. The average of first three numbers is double of the fourth number. If the average of all the four numbers is 12, find the 4th number. *[SSC(CGL-T-II)-13*

(a) $\frac{48}{7}$ (b) 16
(c) 20 (d) $\frac{18}{7}$

44. A class consists of 20 boys and 30 girls. In the mid-semester examination, the average score of the girls was 5 higher than that of the boys. In the final exam, however, the average score of the girls dropped by 3 while the average score of the entire class increased by 2. The increase in the average score of the boys is *[CAT-17]*

(a) 9.5 (b) 10
(c) 4.5 (d) 6

45. The total marks obtained by a student in Physics, Chemistry and Mathematics together is 120 more than the marks obtained by him in Chemistry. What is the average marks obtained by him in Physics and Mathematics together ?

(a) 60
(b) 120
(c) 40
(d) Cannot be determined
(e) None of the above

46. The mean monthly salary paid to 75 workers in a factory is ₹ 5680. The mean salary of 25 of them is ₹ 5400 and that of 30 others is ₹ 5700. The mean salary of the remaining workers is

(a) ₹ 5000 (b) ₹ 7000
(c) ₹ 6000 (d) ₹ 8000

47. The average of 5 numbers is 306.4. The average of the first two numbers is 431 and the average of the last two numbers is 214.5. What is the third number ?

(a) 108
(b) 52
(c) 321
(d) Cannot be determined
(e) None of the above

48. The monthly income of a family of four earning members was ₹ 15130. One of the daughters in the family got married and left home, so the average monthly income of the family came down to ₹ 14660. What is the monthly income of the married daughter ?

(a) ₹ 15350
(b) ₹ 12000
(c) ₹ 16540
(d) Cannot be determined
(e) None of the above

49. A batsman has a certain average of runs for 12 innings. In the 13th inning, he scores 96 runs thereby increasing his average by 5 runs. What is his average after the 13th inning ? *[MAT-08, SSC(CGL-T-II)-10]*

(a) 64 (b) 48
(c) 36 (d) 72

50. The mean of the marks obtained by 100 students is 60. If the marks obtained by one of the students was incorrectly

calculated as 75, whereas the actual marks obtained by him was 65, what is the correct mean of the marks obtained by the students ?
[NABARD(PO)-08, IOB(PO)-09, Syndicate Bank(PO)-10]

(a) 59 (b) 58.50 (c) 50
(d) Cannot be determined
(e) None of the above

51. In a mathematics examination the average score of the students who passed is x and those who failed is y. If the average score of all students who appeared in the examination is z, then percentage of students who failed was **[LIC(AAO)-12]**

(a) $\frac{100xy}{z^2}$ (b) $\frac{100(y-z)}{(x-y)}$
(c) $\frac{100(x-y)}{(x-z)}$ (d) $\frac{100(x-z)}{(x-y)}$

52. The mean of 25 observations was found to be 78.4. But later on it was found that 96 was misread as 69. The correct mean is
[MAT-07, SSC(CGL-T-II)-17]

(a) 79.48 (b) 76.54
(c) 81.32 (d) 78.4

53. A set of consecutive positive integers beginning with 1 is written on the blackboard. A student came along and erased one number. The average of remaining numbers is $35\frac{7}{17}$. What was the number erased ? **[CAT-01]**

(a) 8
(b) 7
(c) 9
(d) None of the above

54. In a class of five students, the average weight of the 4 lightest students is 40 kg, average weight of the 4 heaviest students is 45 kg. What is the difference between the maximum and minimum possible average weight overall ?

(a) 2.8 kg (b) 3.2 kg
(c) 3 kg (d) 4 kg

55. In an apartment complex, the number of people aged 51 years and above is 30 and there are at most 39 people whose ages are below 51 years. The average age of all the people in the apartment complex is 38 years. What is the largest possible average age, in years, of the people whose ages are below 51 years? **[CAT-18]**

(a) 27 (b) 28 (c) 26 (d) 25

56. The average of 30 integers is 5. Among these 30 integers, there are exactly 20 which do not exceed 5. What is the highest possible value of the average of these 20 integers ? **[CAT-19]**

(a) 4 (b) 3.5 (c) 4.5 (d) 5

57. The average of the first and the last number of a set of five two digit integers is its middle number. Second number is 3/2 of the first number. Sum of first three numbers is 85. Middle number is M and the average of all five numbers is A. What can be the values of A and M provided the fourth number is 40.
[IBPS(PO-Mains)-18]

(a) 25, 34 (b) 35, 35
(c) 28, 40 (d) 28, 32
(e) 34, 37

58. There is a sequence of 11 consecutive odd numbers. If the average of first 7 numbers is x, then find the average of all the 11 integers. **[MAT-18]**

(a) $x + 3$
(b) $x + 4$
(c) $x + 7$
(d) x

59. The average age of a number of persons in a group was calculated as 35 years, which was 2.5 years more than the correct average as there was an error in recording the age of two persons as 38.5 years and 40 years instead of 29 years and 22 years respectively. The number of persons in the group was **[SSC(CGL-T-I)-19]**

(a) 11 (b) 12
(c) 15 (d) 13

60. Ramesh and Gautam are among 22 students who write an examination. Ramesh scores 82.5. The average score of the 21 students other than Gautam is 62. The average score of all the 22 students is one more than the average score of the 21 students other than Ramesh. The score of Gautam is **[CAT-19]**

(a) 49 (b) 48
(c) 51 (d) 53

61. The mean of all 4-digit even natural numbers of the form '$aabb$' where $a > 0$ is **[CAT-20]**

(a) 5050
(b) 4466
(c) 5544
(d) 4864

Answer Key

1. (b)	2. (d)	3. (b)	4. (a)	5. (b)	6. (b)	7. (b)	8. (a)	9. (a)	10. (c)
11. (a)	12. (d)	13. (c)	14. (d)	15. (d)	16. (a)	17. (c)	18. (d)	19. (a)	20. (c)
21. (b)	22. (b)	23. (b)	24. (d)	25. (c)	26. (b)	27. (d)	28. (d)	29. (a)	30. (d)
31. (d)	32. (b)	33. (c)	34. (a)	35. (b)	36. (a)	37. (a)	38. (a)	39. (a)	40. (b)
41. (a)	42. (c)	43. (a)	44. (a)	45. (a)	46. (c)	47. (e)	48. (c)	49. (c)	50. (e)
51. (d)	52. (a)	53. (b)	54. (c)	55. (b)	56. (c)	57. (b)	58. (b)	59. (a)	60. (c)
61. (c)									

Hints and SOLUTIONS

1. A student left Class A resulting in the average weight of the Class A decreased. It implies that his weight must be more than the average weight of the Class A i.e. $x > 40$. Further, after he joined the Class B, the average weight of Class B decreased. It implies that his weight must be less than the average weight of the Class B i.e. $x < 60$. Hence, $40 < x < 60$.

2. Clearly, $(A + B) = 2 \times 30 = 60$ and $(C + D) = 2 \times 20 = 40$

Since individual values of A, B, C and D are not specified, we cannot arrive at conclusions given as I and II. Hence, option(d) is correct.

3. The correct sum of scores should have been 50 more than what was taken incorrectly.

Then, correct average

$$= 36 + \frac{50}{100} = 36.5$$

$\therefore$ Error percent

$$= \left(\frac{0.5}{36.5} \times 100\right)\% = 1.36\%$$

4. $(82 + 13) - (28 + 31) = 95 - 59 = 36$ has been taken more than the original sum while computing the average. It implies that the given average is 36/50 i.e. 0.72 more than correct mean/ average.

Thus, correct mean

$$= 30 - 0.72 = 29.28.$$

5. Sum of scores of (Rahul + Manish + Suresh) $= 63 \times 3 = 189$

Ajay's score $= 30 + 63 = 93$

and Rahul's score $= 93 - 15 = 78$

Now, sum of Manish's and Suresh's scores $= 189 - 78 = 111$.

6. Recall the 'Concept of Inclusion of number(s) in a given set of numbers'. Clearly, 6th inning has made the average for 5 innings increase by 5 maintaining its own final average run of x (let).

Thus, run made in 6th inning

$$= x + (5 \times 5)$$

$$\Rightarrow \quad 80 = x + 25$$

$$\Rightarrow \quad x = 55$$

7. Sum of the marks in two subjects eliminated later

$$= (87 \times 8) - (85 \times 6)$$

$$= 696 - 510 = 186$$

Then, highest score

$$= \frac{(186 - 2)}{2} + 2 = 94$$

8. Total surpluses in case of 8 men

$$= 8 \times (48 - 45) = 24$$

and total surpluses in case of 7 women $= 7(46 - 45) = 7$

Now, total deficits = total surpluses

$$= 24 + 7 = 31$$

This deficit will be accommodated against age of the boy.

Hence, age of the boy

$$= 45 - 31 = 14 \text{ years}$$

9. Average of 5 natural number

$$= \frac{a + l}{2} = \frac{1 + 5}{2} = \frac{6}{2} = 3$$

[assuming $a = 1$]

Now, average of 8 consecutive natural number $= \dfrac{1 + 8}{2} = 4.5$

Required answer $= 4.5 - 3 = 1.5$.

10. Let average weight of girls be x kg. Now, applying the Alligation-Rule:

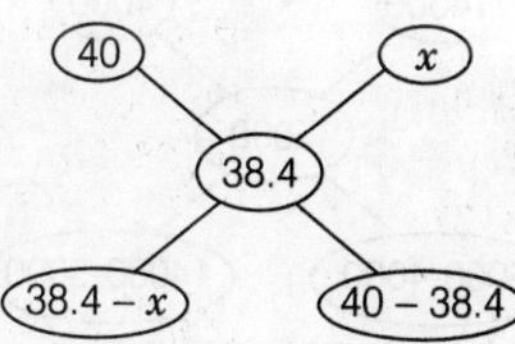

$\therefore$ Ratio of number of boys to that of girls $= (38.4 - x) : 1.6$

As given, $(38.4 - x) : 1.6 = 15 : 10$ (given)

$$\Rightarrow \quad 4.8 = 76.8 - 2x$$

$$\Rightarrow \quad x = 36 \text{ kg.}$$

Alternatively

Total surpluses in respect of boys

$$= 15(40 - 38.4) = 24$$

obviously, total deficits = 24, which will be accommodated among 10 girls. Thus, per girl deficit

$$= 24 \div 10 = 2.4$$

Then, average weight of girls

$$= 38.4 - 2.4 = 36 \text{ kg.}$$

11. By applying Alligation-Rule,

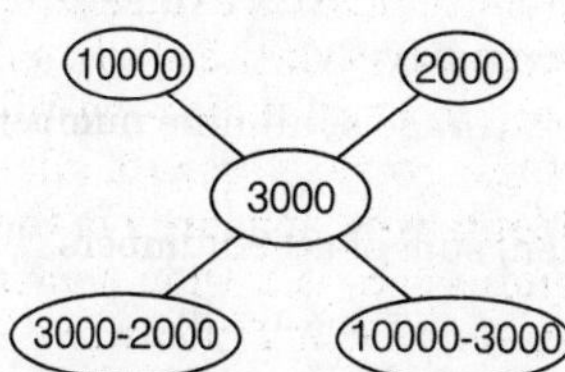

$\therefore$ Ratio of number of officers to that of workers $= 1000 : 7000 = 1 : 7$

As given, $(1 + 7)$ units $= 400$

$\therefore$ 1 unit $= 50$

12. Sum of ages of the couple in the beginning $= 2 \times 25 = 50$

Then, sum of ages of the couple just after the birth of 1st child $= 18 \times 3 = 54$. It implies that 1st child was born when sum of ages of couple was 54 as the age of child was 0 that time. It may be interpreted that after 2 years 1st child was born as each of couple has grown up by 2 years.

Sum of ages of the (couple + 1st child) just after the birth of 2nd child $= 15 \times 4 = 60$. It means the 2nd child was born when they all three have grown up by further 2 years $[\because (60 - 54) \div 3]$

Further, sum of ages of the (couple + 2 children) just after the birth of 3rd and 4th children $= 12 \times 6 = 72$. It implies that each of (couple + 2 children) has grown up by further 3 years. $[\because (72 - 60) \div 4]$

Now, the sum of present ages of the 6 family members $= 16 \times 6 = 96$. It means each of family members has grown up by 4 years. $[\because (96 - 72) \div 6]$

Hence, present age of the eldest child $= 2 + 3 + 4 = 9$ years

13. Since the average of all 9 numbers is x (let), we may assume each of them is x.

Total deficit in case of first 5 numbers $= 5(x - 44)$

and total surplus in case of last 5 numbers $= 5(68 - x)$

Clearly, net effect(surplus)

$= 5(68 - x - x + 44)$
$= (560 - 10x)$

which is on account of counting of middle number twice. Thus, the middle number $= x + (560 - 10x)$

$= 560 - 9x = x$ (given)

$\therefore \quad x = 56$

Then, sum of all nine numbers

$= 56 \times 9 = 504$

Alternatively

Let average of all nine numbers be x.

Then, sum of all 9 numbers

$= 9 \times x = 9x$

and sum of 5 larger numbers

$= 5 \times 68 = 340$

and sum of 5 smaller numbers

$= 5 \times 44 = 220$

It is clear that the middle number has been taken into consideration twice instead of taking once.

Hence, the middle number being 5th number

$= (340 + 220) - 9x = x$ (given)

$\Rightarrow \quad x = 56$

Then, sum of all nine numbers

$= 56 \times 9 = 504$

14. Let average age of the father and the eldest child be x years.

Then, sum of ages of the father and the eldest child $= 2x$

Hence, age of the eldest child

$= 2x - 26$

Further, average age of 3 children = 20% of $x = x/5$

Then, sum of ages of all 3 children

$= 3x/5$

$\Rightarrow$ Age of eldest child + age of second child + age of youngest child $= 3x/5$

$\Rightarrow (2x - 26)$ + age of second child + age of youngest child $= 3x/5$

Clearly, we cannot find the age of second child with the help of given data. Hence, option(d) is correct.

15. It is clear that when Ramesh scores 97 in last quiz, his overall average is 90 whereas when he scores 70 in last quiz, his overall average is 87.It means when his scores in last quiz increase by (97 – 70), his overall average increases by (90 – 87). Then, it may be construed that 3 marks are added per quiz to accommodate scores of 27. Hence, number of quizzes $= 27 \div 3 = 9$.

16. Let, number of boys be x.

Total surpluses in case of boys

$= x(70 - 60) = 10x$

obviously, total deficits $= 10x$, which will be accommodated among 10 girls. Thus, per girl deficit

$= 10x \div (150 - x)$

Then, average weight of girls

$= 60 - \{10x \div (150 - x)\} = 55$ (given)

$\Rightarrow \quad 750 - 5x = 10x$

$\Rightarrow \quad x = 50$

Alternatively

Let, number of boys be x. Now, applying the Alligation-Rule

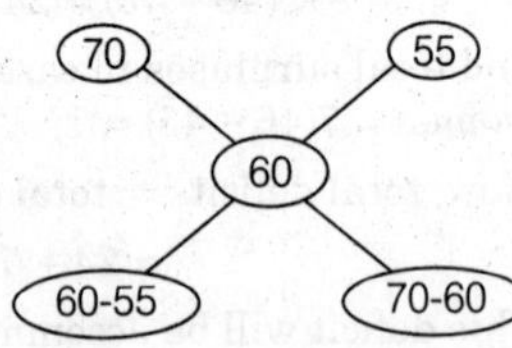

$\therefore$ Ratio of number of boys to that of girls $= 5 : 10 = 1 : 2$

As given, $(1 + 2)$ units $= 150$

$\therefore$ 1 unit = 50

17. We know, first number(a) = average $- (n - 1)$

$\Rightarrow \quad A = 41 - (5 - 1) = 37$

and last number(l)

$=$ average $+ (n - 1)$

$\Rightarrow \quad E = 41 + (5 - 1) = 45$

Now, $A \times E = 37 \times 45 = 1665$

18. By applying Alligation-Rule,

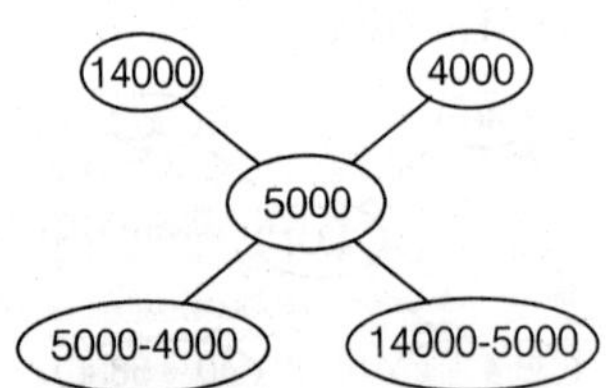

$\therefore$ Ratio of number of officers to that of rest staff $= 1000 : 9000 = 1 : 9$

As given, $(1 + 9)$ units $= 500$

$\therefore$ 1 unit = 50

19. We know, if each number/ value of a given data set is increased by m then average will increase by m. Here, salary of every employee has been increased by 5%, the average will become 105% of 16000 i.e., 16800.

Simultaneously, transport allowance of each employee amounting to ₹ 800 per month has been stopped. Further, we know, if each number/ value of a given data set is decreased by m then average will also decrease by m. Then, new average monthly salary $= 16800 - 800 = 16000$.

20. Increase percent in population in 2 decades = 69%

$\therefore$ Average increase in the population $= 69\% \div 2 = 34.5\%$

21. Weighted average

$$= \frac{(100 \times 5) + (150 \times 25/3)}{100 + 150}$$

$$= \frac{1750}{250} = 7$$

22. Sum of rainfall on Tuesday, Wednesday and Thursday in the 1st case $= (420.5 \times 4)$ – rainfall on Monday

Further, in 2nd case, sum of rainfall on Tuesday, Wednesday and Thursday $= (440.5 \times 4)$ – rainfall on Friday

Hence,

(420.5×4) – rainfall on Monday

$= (440.5 \times 4)$ – rainfall on Friday

$\therefore$ Rainfall on Friday – Rainfall on Monday = 80

As given, $(21 - 20)$ unit = 80

$\therefore \quad$ 21 units = 1680

and 20 units = 1600

Hence, option (b) is correct.

23. By applying Alligation-Rule,

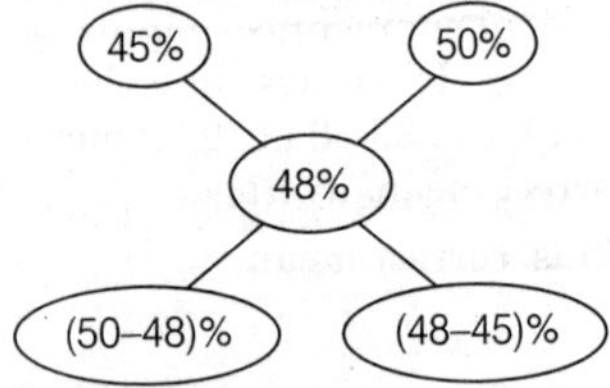

$\therefore$ Ratio of number of student in Section A to that in Section $B = 2 : 3$

Similarly, applying the Alligation – Rule, we obtain:

Ratio of number of student in Section B to that in Section C

$= (72 - 60) : (60 - 50) = 6 : 5$

On combination, ratio of students in Sections A, B and $C = 4 : 6 : 5$

Let, ratio of students in Sections A, B, C and $D = 4 : 6 : 5 : x$.

Then, as per question,

$(4 \times 45\%) + (6 \times 50\%) + (5 \times 72\%)$
$+ (x \times 80\%)$
$= \{(4 + 6 + 5 + x) \times 60\%\}$

$\Rightarrow 180 + 300 + 360 + 80x$
$= 900 + 60x$
$\Rightarrow \quad 20x = 60$
$\Rightarrow \quad x = 3$

Hence, ratio of the number of students in Sections A and $D = 4 : 3$

24. We know, value of the number excluded = Value of previous average + (Number of remaining numbers × decrease in previous average).

Let the initial average of whole team be x years.

∴ Sum of ages of both players excluded $= \{(2 \times x) + (9 \times 1)\}$

$\Rightarrow 25 + (25 + 3) = 2x + 9 \Rightarrow x = 22$

25. Sum of first two numbers (say A and B) $= 2 \times 7 = 14$

Sum of middle two numbers (say B and C) $= 2 \times 2.3 = 4.6$

and sum of last two numbers (say C and D) $= 2 \times 8.4 = 16.8$

On addition,

$(A + D + 2B + 2C) = 35.4$

$\Rightarrow A + D = 35.4 - 2(4.6) = 26.2$

Then, required average

$= 26.2 \div 2 = 13.1$

26. Sum of x and $y = 2 \times 40 = 80$

Then, sum of x, y and $z = 80 + 10 = 90$

∴ Their average $= 90 \div 3 = 30$

27. 20 students $\Rightarrow 12.5 + 0.6$

10 students $\Rightarrow 12.5$

Obviously, (20×0.6) is to be distributed among 30 students.

Then, final average

$= 12.5 + \dfrac{(20 \times 0.6)}{30}$

$= 12.5 + 0.4 = 12.9$ years

Alternatively

Weighted average

$= \dfrac{(10 \times 12.5) + (20 \times 13.1)}{(10 + 20)} = 12.9$

Hence, average age of the whole class = 12.9 years

28. Clearly, teacher's weight made the average of students increase by 1/2 kg maintaining his own average weight of $48\frac{1}{2}$ kg.

Thus, teacher's weight

$= 48\frac{1}{2} + (29 \times 1/2) = 63$ kg

29. Sum of first five experiments = sum of all seven − sum of 6th and 7th $= (7 \times 53.735) - 2(54.005 - 0.010)$

$= 268.155$

Further, sum of 4th and 5th experiments = sum of first five − sum of first three

$= 268.155 - (3 \times 54.005) = 106.14$

Since 4th is 0.004 gm greater than 5th, value of fourth experiment

$= \dfrac{106.14 - 0.004}{2} + 0.004 = 53.072$ g

30. Sum of scores of first three friends $= 15 \times 3 = 45$

∴ Sum of scores of all four friends

$= 45 + 19 = 64$

But, sum of scores of last three friends $= 16 \times 3 = 48$

Then, score of 1st friend

$= 64 - 48 = 16$

Clearly, score of 1st friend i.e. 16 is 100% of the average of the last three friends. Hence, option (d) is correct.

31. Largest number $= l =$ average $+ (n - 1) = 27 + (4 - 1) = 30$

32. Due to exchange, increase in sum of weights $= 5(54 - 48) = 30$, which is to be accommodated among 45 students.

Hence, new average weight of the class $= 52 + \dfrac{30}{45} = 52\frac{2}{3}$ kg.

33. Weighted average

$= \dfrac{(2200 \times 3) + (2250 \times 4) + (3120 \times 5)}{(3 + 4 + 5)}$

$= 2600$

Hence, average monthly expenditures = ₹ 2600

Further, average monthly savings

$= 1260 \div 12 =$ ₹ 105

Then, average monthly income

$= 2600 + 105 =$ ₹ 2705

34. We know, average of n consecutive natural numbers starting from $a = a + \dfrac{(n-1)}{2}$

Here, $n = m + \dfrac{(5-1)}{2} \Rightarrow n = m + 2$

Now, average of 6 consecutive integers $= (m + 2) + \dfrac{(6-1)}{2}$

$= n + \dfrac{5}{2} \qquad [\because n = m + 2]$

$= \dfrac{2n + 5}{2}$

35. In case, the best marksman scores 7 points more, the average will become $\frac{7}{8}$ more than the original average. Hence, original average

$= 84 - \dfrac{7}{8} = \dfrac{665}{8}$

Now, total points scored by the team

$= \dfrac{665}{8} \times 8 = 665$

36. Here, average is reduced by 2 months, thus the sum of ages of 20 boys will be reduced by (2×20) months. Obviously, it is due to replacement of one old boy by a new one. It implies that a boy aging 18 years has been replaced by a new boy whose age is (2×20) months less than 18 years. Hence, the age of new boy = 18 years − 40 months

= 14 years 8 months

37. By applying Alligation-Rule,

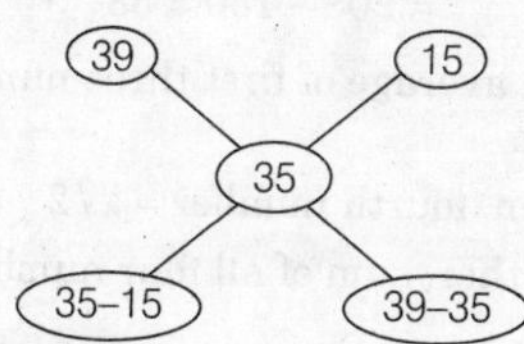

∴ Ratio of number of passed candidates to that of failed candidates $= 20 : 4 = 5 : 1$

As given, $(5 + 1)$ units = 120

∴ 5 units = 100.

38. Sum of their ages $= 42 \times 2 = 84$

As given, $(2 + 1)$ units = 84

∴ 1 unit = 28 years

39. The sum of present ages of all 8 members, if two adults at the age of 60 were not died

$= 231 + (10 \times 8) = 311$ years

But, two adult members have died at the age of 60 each, hence, a loss of 120 years is realised. But, there is no change in number of family members as two children were born when they had died.

Hence, actual sum of ages of all 8 members at present

$= 311 - (2 \times 60) = 191$

Then, their average

$= 191 \div 8 = 23.875$

= 24 years (approx.)

40. Lowest number = 1st term $= a$

= average $- (n - 1)$

$= 25 - (6 - 1) = 20$

and Highest number = last term = l

$= \text{average} + (n-1)$

$= 25 + (6-1) = 30$

Now, required difference

$= 30 - 20 = 10$

41. Monthly increment of only one person $= 12000 \div 12 = 1000$

Which is to be accommodated amongst all the family members.

Hence, new monthly average income

$= 1000 + \dfrac{1000}{5} = ₹\, 1200$

42. Sum of temperatures on M, T, W and Th $= 4 \times 48 = 192°$

$\therefore$ Sum of temperatures on T, W and Th $= 192 - 42 = 150°$

Further, sum of temperatures on T, W, Th and F $= 52 \times 4 = 208°$

$\therefore$ Temperature on F

$= 208 - 150 = 58°$

43. Let, average of first three numbers be x.

Then, fourth number $= x/2$

Further, sum of all four numbers

$= 4 \times 12$

Thus, $4 \times 12 = 3x + x/2$

$\Rightarrow \quad \dfrac{x}{2} = \dfrac{48}{7}$

44. In the final examination: overall increase in score $= (20 + 30) \times 2 = 100$, decrease in score of the girls $= 30 \times 3 = 90$ and increase in score of boys = ?

Clearly, increase in score of boys − decrease in score of girls = overall increase

$\Rightarrow \quad ? - 90 = +\, 100$

$\Rightarrow \quad ? = 190$

Then, increase in the average score of the boys $= 190 \div 20 = 9.5$

Hence, option(a) is correct.

45. Given, Marks in $(P + C + M)$ = Marks in $C + 120$

$\Rightarrow$ Marks in $(P + M) = 120$

$\Rightarrow$ Average marks in P and M

$= 120 \div 2 = 60$

46. Total deficits in case of 25 workers $= 25 \times 280 = 7000$

and total surpluses in case of 30 workers $= 30 \times 20 = 600$

Hence, net effect (deficits) = 6400. Then, total surpluses will also be of ₹ 6400 as total deficits = total surpluses. Surpluses of ₹ 6400 are to be accommodated among the remaining 20 workers. It means, surplus per worker

$= 6400 \div 20$

$= 320$

Then, mean salary of the remaining 20 workers $= 5680 + 320 = ₹\, 6000$

Note *You may directly write that mean salary of the remaining 20 workers*

$$= \frac{(75 \times 5680) - (25 \times 5400) - (30 \times 5700)}{20}$$

$= ₹\, 6000$

47. The third number

$= (306.4 \times 5) - (2 \times 431) - (2 \times 214.5)$

$= 241$

48. We know, value of the number excluded = Value of previous average + (Number of remaining numbers × decrease in previous average).

Here, required answer

$= 15130 + (3 \times 470)$

$= ₹\, 16540$

49. Recall the 'Concept of Inclusion of number(s) in a given set of numbers'. Clearly, 13th inning has made the average for 12 innings increase by 5 maintaining its own final average run of x (let).

Thus, run made in 13th inning

$= x + (12 \times 5)$

$\Rightarrow \quad 96 = x + 60$

$\Rightarrow \quad x = 36$

50. Clearly, 10 has been taken more than the original sum by mistake while computing the average. It implies that the given average is 10/100 i.e., 0.1 more than correct mean/ average. Thus, correct mean of 100 students $= 60 - 0.1 = 59.9$

51. By applying Alligation-Rule,

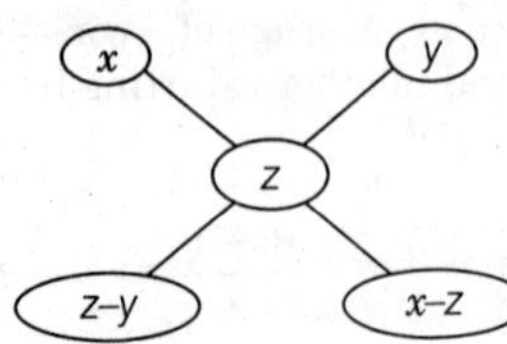

$\therefore$ Ratio of number of passed students to that of failed students

$= (z - y) : (x - z)$

Now, total students

$= z - y + x - z$

$= x - y$

$\therefore$ Percentage of failed students

$= \left\{ \dfrac{(x - z)}{(x - y)} \times 100 \right\} \%$

Note *We have assumed $x > y$. If we assume $x < y$, the percentage will be same.*

52. The correct sum of observations should have been (96 - 69) i.e. 27 more than what was taken incorrectly.

Then, correct average

$= 78.4 + \dfrac{27}{25} = 79.48$

53. Let there are n consecutive numbers starting with 1 and the number erased by the student be x.

Then, average of $(n - 1)$ numbers

$$= \frac{\frac{n(n+1)}{2} - x}{n - 1} = 35\frac{7}{17}$$

$$\Rightarrow \frac{\frac{n(n+1)}{2} - x}{n - 1} = \frac{602}{7}$$

$$\Rightarrow \frac{n(n+1) - 2x}{2(n-1)} = \frac{602}{17}$$

Now, proceed through options.

(a) at $x = 8$, n is not an integer.

(b) at $x = 7$, n is an integer (i.e. 69)

Hence, option(b) is correct.

54. Let the names of the students be a, b, c, d and e who are in ascending order of weights.

Since the average weight of a, b, c, d is 40 kg, $a + b + c + d = 40 \times 4$

$= 160$ kg

Then, total weight of all students

$= 160 + e$

Further, $b + c + d + e = 45 \times 4 = 180$

Then, total weight of all students

$= 180 + a$

Hence, $160 + e = 180 + a$

$\Rightarrow \quad e = 20 + a$

Now, the highest possible value of $a = 40$ as the average weight of a, b, c and d is 40 $(a < b < c < d)$.

$\therefore \quad e = 40 + 20 = 60$

Then, maximum possible average weight $= \dfrac{160 + 60}{5} = 44$ kg

Similarly, the least value of e will be 45 as the average weight of b, c, d and e is 45 $(b < c < d < e)$.

Then, least value of $a = 45 - 20 = 25$

∴ Minimum possible average weight

$= \frac{180+25}{5} = 41\text{kg}$

Thus, the required difference

$= 44 - 41 = 3\text{kg}$

55. In order to get largest possible average age of the people whose ages are below 51 years, the average age of the people aging 51 years and above must be minimum and the same may be 51 years.

Let number of people aging below 51 years be x.

Then, average age of x people

$= \frac{38(30+x) - 51.30}{x}$

$= \frac{38x - 390}{x} = 38 - \frac{390}{x}$

Average age of x people will be maximum if x takes maximum value and the maximum possible value of x is 39.

∴ Required average age

$= 38 - \frac{390}{39} = 28$

56. Given that 20 integers do not exceed 5. It implies that each of the remaining 10 integers is greater than 5. In order to get maximum possible value of average of 20 integers not exceeding 5, the value of each of remaining 10 integers must be minimum and it is possibility that each be 6.

Then, sum of 20 integers not exceeding 5 = $(30 \times 5) - (10 \times 6) = 90$

∴ Required average

$= 90 \div 20 = 4.5$

57. Let the 1st number be x. Then, 2nd number $= \frac{3x}{2}$

Now, 3rd number i.e. middle number

$= 85 - x - \frac{3x}{2} = 85 - \frac{5x}{2}$

Since the average of the first and the last number is the middle number, it implies that the sum of both numbers

$= 2 \times \left(85 - \frac{5x}{2}\right) = 170 - 5x$

Now, average of all five numbers

$= \frac{\text{sum of 1st and last number} + \text{2nd number} + \text{middle number} + \text{fourth number}}{5}$

$= \frac{170 - 5x + \frac{3x}{2} + 85 - \frac{5x}{2} + 40}{5}$

$= 59 - \frac{6x}{5}$

Now, proceed through options

(a) If average is 25, then

$59 - \frac{6x}{5} = 25$

$\Rightarrow x = 28.33$, not possible as all are integers.

(b) If average is 35, then

$59 - \frac{6x}{5} = 35$

$\Rightarrow \quad x = 20$

Then, middle number

$= 85 - \frac{5 \times 20}{2} = 35$, matching with option.

Hence, option(b) is correct.

58. Average of 7 consecutive odd numbers starting from a

$= a + (7 - 1)$

$\Rightarrow \quad x = a + 6$

$\Rightarrow \quad a = x - 6$

Then, average of 11 consecutive odd numbers starting from $(x - 6)$

$= (x - 6) + (11 - 1)$

$= x + 4$

59. Clearly, $(38.5 + 40 - 29 - 22)$ i.e. 27.5 years more have been taken into consideration while computing the average resulting in excess of 2.5 years over the correct average. It may be construed that the age of each person has been taken 2.5 years more than their actual age.

Hence, total number of persons

$= 27.5 \div 2.5 = 11$

60. Let the score of Gautam be x.

Then, sum of scores of 21 students other than Gautam

$= 21 \times 62$

Then, sum of all 22 students

$= (21 \times 62) + \text{score of Gautam}$

$= (21 \times 62) + x$

∴ Average score of 22 students

$= \frac{(21 \times 62) + x}{22}$

As per question, $\frac{(21 \times 62) + x}{22} - 1$

= Average score of the 21 students other than Ramesh

$= \frac{(21 \times 62) + x}{22} - 1$

$= \frac{(21 \times 62) + x - 82.5}{21}$

On solving, $x = 51$

61. The possible even numbers are:

1100, 1122, 1144, 1166, 1188, 2200, 2222, ... , 9988

Clearly, mean of underlined numbers = mid term = 1144

Similarly, mean of next five numbers = 2244

....

Mean of last five numbers = 9944

Finally, overall mean = mid term of series : 1144, 2244, 3344,...... , 9944 = 5544

15

Progressions

Learning OBJECTIVES

After completing this chapter, aspirants will be able to

- Understand the significance of this chapter in diverse competitive exams;
- Be familiar with different types of progressions with examples along with their properties;
- Be familiar with concepts of Arithmetic Mean, Geometric Mean and Harmonic Mean and relation among them;
- Know the concept of Arithmetico-Geometric sequence and its applications;
- Realise themselves competent to solve all types of problems based on progressions;
- Understand the philosophy of division of 'Illustrations' in many parts;
- Identify the problems which can easily be solved through options;
- Realise themselves competent to solve the questions in unique way;
- Analyse the trend of the questions being asked in various competitive examinations;
- Solve the questions in moments;
- Learn some new concepts/fundamentals;
- Apply the pragmatic approaches in lieu of academic approaches; and
- Ponder on the alternative methods.

Inside

- *Introduction*
- *Arithmetic Progression(AP) and it's Properties, sum of terms, arithmetic Mean(AM)*
- *Geometric Progression(GP) and it's Properties, Sum of Terms, Geometric Mean(GM)*
- *Harmonic Progression(HP) and Harmonic Mean(HM)*
- *Arithmetico-Geometric Sequence*
- *Relation Between AM, GM and HM*
- *Illustrations*
- ....
- .... *& Many More*
- *Work Out with Solutions*

Introduction

This chapter is very critical for the aspirants appearing in CAT, MAT, IIFT, Bank-PO, SSC, etc. Basically, the concepts of the progression seek the logical interpretation of the situations involved in the questions. This chapter may be called as extension of the chapter' Number-System' as both the chapters have similar logics to the greater extent. Sequences that follow a definite pattern are known as progressions. There are three kinds of progression:

(A) Arithmetic Progression(AP)

(B) Geometric Progression(GP)

(C) Harmonic Progression(HP)

In addition to above, we will also discuss about 'Arithmetico-geometric sequence'.

(A) Arithmetic Progression(AP)

A sequence of quantities is said to be Arithmetic Progression if each term differs from its preceding term by a constant. This constant is known as 'Common Difference' of the Arithmetic Progression.

For examples Each of the following sequences is AP.

(i) 5, 8, 11, 14, ... (ii) 5, 1, –3, –7, ...

(iii) –10, –8, –6, –4, ...

Now, observe the 1st progression 5, 8, 11, 14, ...

(a) 5, 8, 11, 14, etc., are the terms of the progression.

(b) 5 is the initial term(notation: a).

(c) The common difference(notation: d) is (8 – 5) or (11 – 8) or (14 – 11) i.e. 3.

(d) 2nd term may be written as $(a + d)$, 3rd term as $(a + 2d)$ and 4th term as $(a + 3d)$. If we observe the coefficient of d for a particular term, we find that it is one less than the rank of that term from the left in the progression. Hence, ***n*th term of this progression be $\{a + (n - 1)d\}$**. This nth term of the progression is known as its general term. If the general term is l, then $\boldsymbol{l = \{a + (n - 1)\, d\}}$.

(e) **The *n*th term from the end of an AP = $\{l - (n - 1)d\}$**, where l is the last term, and d is the common difference.

If we observe the 2nd and 3rd progression, common differences are – 4 and 2, respectively.

Sum of terms of AP

Consider the 1st AP.

$$\text{1st term} = 5 = a + 0 \cdot d$$
$$\text{2nd term} = 8 = a + 1d$$
$$\text{3rd term} = 11 = a + 2d$$
$$\text{4th term} = 14 = a + 3d$$

....

....

$$n\text{th term} = a + (n - 1)d$$

$$\text{Sum of } n \text{ terms} = (a + a + a + \ldots + n\text{th term}) + d[0 + 1 + 2 + 3 + \ldots + (n - 1)]$$

$$= na + d \cdot \frac{(n-1)(n-1+1)}{2}$$

$$= \frac{n}{2}[2a + (n - 1)d]$$

Or

$\frac{n}{2}(a + l)$, where l is the last term of AP.

[illegible] even numbers, their squares, cubes, etc., have been discussed under the chapter 'Addition and Subtraction'.

It is important to note that:

(i) Sum of 1st term and the last term, that of 2nd term and 2nd last term, that of 3rd term and 3rd last term and so on, are equal. It implies that; sum of nth term and mth term = sum of pth term and $(m + n - p)$th term.

(ii) Sum of all terms of an AP = number of terms × Average

$$\text{and average} = \frac{\text{1st term + last term}}{2}$$
$$= \frac{\text{2nd term + 2nd last term}}{2}$$
$$= \frac{\text{3rd term + 3rd last term}}{2} \text{ or so on.}$$

It also holds true that if number of terms of an AP is odd, the middle term is its average, hence middle term

$$= \frac{\text{1st term + last term}}{2}$$
$$= \frac{\text{2nd term + 2nd last term}}{2}$$
$$= \frac{\text{3rd term + 3rd last term}}{2} \text{ or so on.}$$

And in case of even number, average of AP = average of $\frac{n}{2}$ th term and $\left(\frac{n}{2} + 1\right)$th term.

(iii) In certain cases wherein some numbers are repeated with changed sign, the sum of n_1 terms is equal to sum of n_2 terms.

For examples

(a) –15, –10, –5, 0, 5, 10, 15, 20, 25, 30, ...

(b) 15, 10, 5, 0, –5, –10, –15, –20, ...

In both examples (a) and (b), sum of first 3 terms is equal to that of first 4 terms. We may write as $S_3 = S_4$ for sake of convenience.

Also, $S_1 = S_6$ and $S_2 = S_5$

It also holds good that sum of (1 + 6 or 2 + 5 or 3 + 4) terms, i.e. $S_7 = 0$.

(c) –49, –35, –21, –7, 7, 21, 35, 49, 63, 77, 91, ...

(d) 49, 35, 21, 7, –7, –21, –35, –49, –63, –77, –91, ...

Here [i.e. for (c) & (d)], $S_1 = S_7, S_2 = S_6$ and $S_3 = S_5$

Also, sum of (1 + 7 or 2 + 6 or 3 + 5) terms, i.e. $S_8 = 0$.

It may be construed from the above that:

- If $S_n = S_m$, then $S_{m+n} = 0$
- In case of AP wherein 0 is a term, sum of terms opted in odd number = sum of terms opted in even number whereas in the AP wherein 0 is not a term, sum of terms opted in odd number = sum of terms opted in odd number or sum of terms opted in even number

(iv) If S_1, S_2 and S_3 be the sum of first n, $2n$ and $3n$ terms respectively of an AP, then $S_3 = 3(S_2 - S_1)$.

Properties of Arithmetic Progression

(i) If a constant is added to or subtracted from each term of an AP, the resulting sequence is also an AP with the same common difference as before.

(ii) If each term of an AP is multiplied or divided by a non-zero constant, the resulting sequence is also an AP with new common difference, equal to either product of old common difference and that constant or the quotient obtained by dividing the old common difference by that constant as the case may be.

(iii) If two arithmetic progressions are added or subtracted termwise, then the sequence containing the resulting term shall be an AP.

(iv) If three quantities are in AP, then the middle term is equal to sum of other two terms only if the middle term is 0.

(v) If $\frac{1}{a+b}, \frac{1}{c+a}, \frac{1}{b+c}$ are in AP, then a^2, b^2 and c^2 are also in AP.

Arithmetic Mean (AM)

Arithmetic mean of an AP containing n terms

$$= \frac{\text{Sum of terms}}{\text{Number of terms}}$$

$$= \frac{\frac{n}{2}[2a+(n-1)d]}{n} = \frac{[2a+(n-1)d]}{2},$$

where a is initial term and d is common difference.

Or

$$\text{AM} = \frac{\frac{n}{2}(a+l)}{n} = \frac{(a+l)}{2},$$

where a is initial term and l is the last term.

Alternatively

$$\text{AM} = \frac{(\text{1st term} + \text{last term})}{2}$$

$$= \frac{(\text{2nd term} + \text{2nd last term})}{2}$$

$$= \frac{(\text{3rd term} + \text{3rd last term})}{2}$$

$$= \frac{m\text{th term} + m\text{th last term}}{2}$$

For sake of convenience, 1st term and last term, 2nd term and 2nd last term, ... may be termed as corresponding pairs.

Remarks

(i) If number of terms in an AP is odd, the middle term is its AM.

(ii) If number of terms in an AP is even, then average of $\frac{n}{2}$ th term and $\left(\frac{n}{2}+1\right)$th term is its AM.

(iii) Number of terms × AM = Sum of terms

(iv) Since sum of 1st term and the last term, that of 2nd term and 2nd last term, that of 3rd term and 3rd last term and so on, are equal, Arithmetic Mean of these corresponding pairs are equal.

How to insert n numbers between two given numbers so that the resulting sequence comes out to be an AP

Let the given numbers be a and b and we like to insert n numbers $k_1, k_2, k_3, \ldots\ldots, k_n$ between them in such a way that $a, k_1, k_2, k_3, \ldots\ldots, k_n, b$ be an AP.

The resulting sequence : $a, k_1, k_2, k_3, \ldots\ldots, k_n, b$

$$\text{1st term} = a$$
$$k_1 = a + d$$
$$k_2 = a + 2d$$
$$k_3 = a + 3d$$
$$\ldots\ldots\ldots\ldots$$
$$\ldots\ldots\ldots\ldots$$
$$k_n = a + nd$$
$$\text{Last term} = b$$

Obviously, $k_n + d = b$

$\Rightarrow$ $a + nd + d = b$

$\Rightarrow$ $d(n+1) = b - a$

$\Rightarrow$ $d = \frac{b-a}{n+1}$

Hence, $k_1 = a + \left(\frac{b-a}{n+1}\right), k_2 = a + 2\left(\frac{b-a}{n+1}\right),$

$$k_3 = a + 3\left(\frac{b-a}{n+1}\right), \ldots\ldots\ldots, k_n = a + n\left(\frac{b-a}{n+1}\right).$$

Generic version: If n AMs are inserted, then rth AM

$$= k_r = a + r\left(\frac{b-a}{n+1}\right), r \le n$$

In case $n = 1$, the number inserted

$$= a + \left(\frac{b-a}{1+1}\right) = \frac{a+b}{2}.$$

It implies that, the number inserted is the AM of the given numbers a and b. Similarly, k_1 is AM of a and k_2, k_2 is AM of k_1 and k_3 and so on.

Miscellaneous

If we have to assume 3 terms in AP, we should assume them as $(a-d)$, a and $(a+d)$ for sake of convenience. We are free to assume them as $a, (a+d)$ and $(a+2d)$.

In case of 4 terms: $(a-3d), (a-d), (a+d)$ and $(a+3d)$

In case of 5 terms: $(a-2d), (a-d), a, (a+d)$ and $(a+2d)$

In case of 6 terms:

$(a-5d), (a-3d), (a-d), (a+d)(a+3d)$ and $(a+5d)$.

(B) Geometric Progression (GP)

A sequence of non-zero quantities is said to be Geometric Progression if ratio of any term to its preceding term is a constant. This constant factor is known as 'Common Ratio' of the Geometric Progression.

For examples Each of the following sequences is GP.

(i) 3, 6, 12, 24, ...

(ii) 0.2, 0.02, 0.002, 0.0002, ...

(iii) $\frac{-1}{2}, \frac{1}{4}, \frac{-1}{8}, \frac{1}{16}, ...$

Now, observe the 1st progression 3, 6, 12, 24, ...

(a) 3, 6, 12, 24, etc., are the terms of the progression.

(b) 3 is the initial term(notation: a).

(c) The common ratio (notation: r) is $(6 \div 3)$ or $(12 \div 6)$ or $(24 \div 12)$, i.e. 2.

(d) 2nd term may be written as ar, 3rd term as ar^2 and 4th term as ar^3. If we observe the power of r for a particular term, we find that it is one less than the rank of that term from the left in the progression. Hence, **nth term of this progression be ar^{n-1}**. This nth term of the progression is known as its general term. If the general term is l, then $l = ar^{n-1}$

If we observe the 2nd and 3rd progression, common ratio are 0.1 and $\frac{-1}{2}$, respectively.

Sum of Finite and Infinite Terms of GP

Let there be n terms and their sum is S_n.

Then, $S_n = a + ar + ar^2 + ar^3 + ... + ar^{n-1}$ ...(i)

$\Rightarrow \quad r \cdot S_n = ar + ar^2 + ar^3 + ar^4 + ... + ar^{n-1} + ar^n$ ...(ii)

Subtracting Eq. (ii) from Eq. (i), we get

$$S_n - r \cdot S_n = a - ar^n \quad \Rightarrow \quad S_n = \frac{a(1-r^n)}{1-r}$$

Or $$S_n = \frac{a(r^n - 1)}{r - 1}$$

where $r \neq 1$, otherwise, S_n is undefined.

Both results yield the sum of n terms, however, it is convenient to apply $S_n = \frac{a(1-r^n)}{1-r}$ if $-1 < r < 1$ and

$$S_n = \frac{a(r^n - 1)}{r - 1} \text{ if } r > 1 \text{ or } r < -1.$$

Now, consider $S_n = \frac{a(r^n - 1)}{r - 1}$

$$\Rightarrow \quad S_n = \frac{ar^n - a}{r - 1} = \frac{ar^{n-1} \cdot r - a}{r - 1}$$

$$\Rightarrow \quad S_n = \frac{lr - a}{r - 1}, \text{ where } l \text{ is the last term.}$$

Further, from the above, $S_n = \frac{a(1-r^n)}{1-r}$

$$\Rightarrow \quad S_n = \frac{a}{1-r} - \frac{ar^n}{1-r}$$

In case n is ∞, r^n becomes 0, if $-1 < r < 1$.

Then, $$S_\infty = \frac{a}{1-r} - \frac{a \cdot 0}{1-r} = \frac{a}{1-r}$$

Hence, **the sum to infinity of a geometric progression with initial term a and common ratio $r = \frac{a}{1-r}$, where** $-1 < r < 1$.

Properties of Geometric Progression

(i) If each term of a GP is multiplied or divided by a non-zero constant, the resulting sequence is also a GP with the same common ratio as before.

(ii) If two geometric progressions are multiplied or divided term-wise, then the sequence containing the resulting term shall be a GP with the new common ratio. This new common ratio is equal to either product of both common ratios or the quotient obtained by dividing one by another common ratio, as the case may be.

(iii) Product of 1st term and the last term, that of 2nd term and 2nd last term, that of 3rd term and 3rd last term and so on, are equal.

(iv) If number of terms in a geometric sequence is odd, then square of the middle term is equal to either product of first and the last term or product of second term and second last term or product of third term and third last term or so on.

In case of even number of terms, the product of $\frac{n}{2}$ th term and $\left(\frac{n}{2} + 1\right)$th term is equal to either product of first and the last term or product of second term and second last term or product of third term and third last term or so on.

(v) The product of mth and nth terms of a geometric sequence is equal to square of $\left(\frac{m+n}{2}\right)$th term, if $\frac{m+n}{2}$ is integer. In case $\frac{m+n}{2}$ is not an integer, the product of mth and nth terms of a geometric sequence is equal to product of $\left\{\left(\frac{m+n}{2}\right) - 0.5\right\}$th term and $\left\{\left(\frac{m+n}{2}\right) + 0.5\right\}$th term.

(vi) If a, b, c are in GP, then a^2, b^2 and c^2 are also in GP.

(vii) If a, b, c are in GP, then their logarithms will be in AP.

(viii) If pth, qth, rth and sth terms of an AP be in GP, then $(p-q), (q-r), (r-s)$ will be in GP.

(ix) If $a = b = c$, then a, b, c will be in both AP and GP.

Geometric Mean(GM)

If three quantities are in GP, the middle one is said to be 'Geometric Mean' of the other two quantities.

Let a, G, b be in GP, then common ratio $= \frac{G}{a} = \frac{b}{G}$

$$\Rightarrow \quad G^2 = ab \Rightarrow G = \pm\sqrt{ab}$$

Hence, GM of a and $b = \sqrt{ab}$

In general, the GM of n non-zero positive numbers is nth root of their product.

For example: GM of a, b, c, d is $\sqrt[4]{abcd}$.

Note *Generally, GM of negative numbers is not calculated. However, if all the given numbers are negative, then GM on n numbers* $= -(|\text{product of } n \text{ numbers}|)^{\frac{1}{n}}$. *For example, GM of* $-1, -3, -9$ *is* -3.

How to insert n numbers between two given numbers so that the resulting sequence comes out to be a GP

Let the given numbers be a and b and we like to insert n numbers $k_1, k_2, k_3, \ldots\ldots\ldots k_n$, between them in such a way that $a, k_1, k_2, k_3, \ldots\ldots\ldots; k_n, b$ be a GP.

The resulting sequence : $a, k_1, k_2, k_3, \ldots\ldots, k_n, b$

$$\text{1st term} = a$$
$$k_1 = ar$$
$$k_2 = ar^2$$
$$k_3 = ar^3$$
$$\ldots\ldots\ldots\ldots$$
$$\ldots\ldots\ldots\ldots$$
$$k_n = ar^n$$
$$\text{Last term} = b$$

Obviously, $\quad ar^n \cdot r = b$

$$\Rightarrow \quad ar^{n+1} = b$$

$$\Rightarrow \quad r^{n+1} = \frac{b}{a} \Rightarrow r = \left(\frac{b}{a}\right)^{\frac{1}{n+1}}$$

Hence, $k_1 = a\left(\frac{b}{a}\right)^{\frac{1}{n+1}}, k_2 = a\left(\frac{b}{a}\right)^{\frac{2}{n+1}}, k_3 = a\left(\frac{b}{a}\right)^{\frac{3}{n+1}}, \ldots\ldots\ldots,$

$$k_n = a\left(\frac{b}{a}\right)^{\frac{n}{n+1}}$$

Generic version: If n GMs are inserted, then rth GM

$$= k_r = a\left(\frac{b}{a}\right)^{\frac{r}{n+1}}, r \le n$$

Clearly, k_1 is GM of a and k_2, k_2 is GM of k_1 and k_3 and so on.

Miscellaneous

If we have to assume 3 terms in GP, we should assume them as $\frac{a}{r}$, a and ar for sake of convenience. We are free to assume them as a, ar and ar^2.

In case of 4 terms: $\frac{a}{r^3}, \frac{a}{r}, ar$ and ar^3.

In case of 5 terms: $\frac{a}{r^2}, \frac{a}{r}, a, ar$ and ar^2.

(C) Harmonic Progression (HP)

If the sequence $a, b, c, d, \ldots\ldots$ is an AP, then the sequence with their reciprocals is said to be Harmonic Progression(HP). Hence, $\frac{1}{a}, \frac{1}{b}, \frac{1}{c}, \frac{1}{d}, \ldots\ldots\ldots\ldots\ldots$ is a HP.

Example $\frac{1}{3}, \frac{1}{7}, \frac{1}{11}, \frac{1}{15}, \ldots\ldots\ldots$ is a HP as 3, 7, 11, 15, ... is an AP.

The converse is also true. It means, if $a, b, c, d, \ldots\ldots\ldots$ is a HP, then $\frac{1}{a}, \frac{1}{b}, \frac{1}{c}, \frac{1}{d}, \ldots\ldots\ldots\ldots\ldots$ is an AP.

Sum of terms of a HP There is no general formula to find the sum of any number of terms in Harmonic Progression.

Harmonic Mean (HM)

If three quantities are in HP, the middle one is said to be 'Harmonic Mean' of the other two quantities.

Let a, H, b be in HP, then H is HM of a and b.

Obviously, $\frac{1}{a}, \frac{1}{H}, \frac{1}{b}$ will be in AP.

Then, $$\frac{1}{H} - \frac{1}{a} = \frac{1}{b} - \frac{1}{H}$$

$$\Rightarrow \quad \frac{2}{H} = \frac{a+b}{ab}$$

$$\Rightarrow \quad H = \frac{2ab}{a+b}$$

Hence, HM of two quantities a and b

$$= \frac{2ab}{a+b}$$

In general, harmonic mean of $x_1, x_2, x_3, \ldots\ldots\ldots x_n$

$$= \frac{n}{\frac{1}{x_1} + \frac{1}{x_2} + \frac{1}{x_3} + \ldots\ldots + \frac{1}{x_n}}$$

Note

(i) Most of the questions related to HP are solved by applying the properties/rules of the corresponding AP.

(ii) If a, b, c be in HP, then $a^2 + c^2 > 2b^2$.

(D) Arithmetico-Geometric Sequence

This is the sequence which contains terms as product of arithmetic progression and geometric progression term by term. If $a_1, a_2, a_3, \ldots a_n$ is a arithmetic progression and $g_1, g_2, g_3, \ldots\ldots g_n$ is a geometric progression, then $a_1g_1, a_2g_2, a_3g_3, \ldots\ldots\ldots a_ng_n$ is an 'Arithmetico-geometric sequence'.

Let a and g be initial terms of AP and GP respectively and d and r are common difference and common ratio respectively. Then, the form of the Arithmetico-geometric sequence will be as under:

1st term: $ag = ag$

2nd term: $(a + d)gr = (ag + dg)r$

3rd term: $(a + 2d)gr^2 = (ag + 2dg)r^2$

4th term: $(a + 3d)gr^3 = (ag + 3dg)r^3$

....

....

nth term: $\{a + (n - 1)d\}\, gr^{n-1} = \{ag + (n - 1)dg\}r^{n-1}$

$$S_n = ag + (ag + dg)r + (ag + 2dg)r^2 + \ldots + \{ag + (n - 1)dg\}r^{n-1} \quad \ldots \text{(i)}$$

$$r \cdot S_n = agr + (ag + dg)r^2 + (ag + 2dg)r^3 + \ldots + \{ag + (n - 1)dg\}r^n \quad \ldots \text{(ii)}$$

Subtracting Eq. (ii)from (i), we get

$$S_n - r \cdot S_n = ag + dgr + dgr^2 + dgr^3 + \ldots\ldots + dgr^{n-1} - \{a + (n - 1)d\}gr^n$$

$$\Rightarrow S_n(1 - r) = ag + dgr(1 + r + r^2 + \ldots\ldots + r^{n-2}) - \{a + (n - 1)d\}\, gr^n$$

$$\Rightarrow S_n(1 - r) = ag + dgr\left\{\frac{1(1 - r^{n-1})}{1 - r}\right\} - \{a + (n - 1\}d\}\, gr^n$$

$$\Rightarrow \mathbf{S_n = \frac{ag}{1 - r} + \frac{dgr(1 - r^{n-1})}{(1 - r)^2} - \frac{\{a + (n - 1)d\}gr^n}{1 - r}}$$

In case $n = \infty$, $S_n = \dfrac{ag}{1 - r} + \dfrac{dgr}{(1 - r)^2}$ where, $-1 < r < 1$

Relation between AM, GM and HM

1. Let A, G and H be Arithmetic mean, Geometric mean and Harmonic mean of two positive quantities a and b respectively.

   Then, $A \times H = \dfrac{a + b}{2} \times \dfrac{2ab}{a + b} = ab = (\sqrt{ab})^2 = G^2$

   Hence, $\mathbf{AH = G^2}$

2. For real positive numbers $x_1, x_2, x_3, \ldots\ldots\ldots x_n$, **AM ≥ GM ≥ HM**

   [i.e. $\dfrac{x_1 + x_2 + x_3 + \ldots + x_n}{n} \geq \sqrt[n]{x_1\, x_2\, x_3 \ldots\ldots x_n}$ $\geq \dfrac{n}{\dfrac{1}{x_1} + \dfrac{1}{x_2} + \dfrac{1}{x_3} + \ldots + \dfrac{1}{x_n}}$]. It is also important to note that **AM = GM = HM, when $x_1 = x_2 = x_3 = \ldots\ldots = x_n$.**

Illustrations

Problems Based on AP

Problem 1 Which term of the AP 5, 9, 13, 17, ... is 81 ? ***[SSC(CGL-PT)-08]***

Sol. Let nth term be 81.

$\Rightarrow \quad 81 = a + (n - 1)d$

$\Rightarrow \quad 81 = 5 + (n - 1)4 \qquad [\because a = 5 \text{ and } d = 9 - 5 = 4]$

$\Rightarrow \quad 76 = (n - 1)4$

$\Rightarrow \quad n - 1 = 19 \Rightarrow n = 20$

Hence, 20th term of the AP is 81.

Alternatively

Let nth term be 81.

Obviously, nth term exceeds the 1st term by $(n - 1)$ times the common difference. Hence, $81 - 5 = (n - 1)4 \Rightarrow n = 20$

Problem 2 For what value of n are the nth terms of the following two AP's the same ?

(a) 13, 19, 25, ...

(b) 69, 68, 67, ...

Sol. nth term of 1st AP $= 13 + (n - 1)6 \qquad \{\because a = 13 \text{ and } d = 6\}$

and nth term of 2nd AP $= 69 + (n - 1)(-1)$

$[\because a = 69 \text{ and } d = -1]$

As given, $13 + (n - 1)6 = 69 + (n - 1)(-1)$

$\Rightarrow \quad 6n + 7 = -n + 70$

$\Rightarrow \quad 7n = 63$

$\therefore \quad n = 9$

Hence, 9th term of both the progressions are same.

Problem 3 If 11th and 15th terms of an AP are -30 and -46 respectively, then what is the 5th term ?

Sol. 11th term $= a + 10d$

$\Rightarrow \quad -30 = a + 10d$

Also, $\quad -46 = a + 14d$

$\Rightarrow \quad -46 = a + 10d + 4d$

$\Rightarrow \quad -46 = -30 + 4d$

$\Rightarrow \quad 4d = -16$

$\Rightarrow \quad d = -4$

On substituting $d = -4$ in equation $-30 = a + 10d$, we get

$-30 = a + 10(-4)$

$\Rightarrow \quad a = -30 + 40 = 10$

Now, 5th term $= 10 + 4(-4) = -6$

Alternatively

Given, 11th term = – 30 and 15th term = – 46

Clearly, the 15th term exceeds the 11th term by – 16 and this is due to addition of (15 – 11) times the common difference.

Hence, $4d = -16$

$\Rightarrow \quad d = -4$

Further, the 11th term will exceed the 5th term by (11 – 5) times the common difference. Hence, 11th term = 5th term + 6 (– 4).

Then, 5th term = – 30 + 24 = – 6

Note *This process can be done mentally without pen and paper in moments.*

Problem 4 The 5th term of an AP is zero, then what is the ratio of 98th term to 36th term?

Sol. It is obvious that the 98th term will exceed the 5th terms by (98 – 5) times the common difference. Since 5th term is 0, 98th term = $93d$. Also, the 36th term exceeds the 5th term by (36 – 5) times the common difference. Since 5th term is 0, 36th term = $31d$

Now, required ratio = $93d : 31d = 3 : 1$

Problem 5 If m times the mth term of an AP is equal to n times the nth term and $m \neq n$, then its $(m+n)$th term is

Sol. Obviously, mth term = $a + (m-1)d$ and nth term = $a + (n-1)d$

According to question, $m\{a + (m-1)d\} = n\{a + (n-1)d\}$

$\Rightarrow \quad a(m-n) + (m^2 - m - n^2 + n)d = 0$

$\Rightarrow \quad a(m-n) + \{(m^2 - n^2) - (m-n)\}d = 0$

$\Rightarrow \quad (m-n)\{a + (m+n-1)d\} = 0$

$\Rightarrow \quad \{a + (m+n-1)d\} = 0 \qquad \{\because m - n \neq 0\}$

$\Rightarrow \quad (m+n)$th term = 0

Problem 6 Which term of the AP 98, 93, 88, ... is the first negative term ?

Sol. Let nth term be the first negative term.

Obviously, nth term will exceed the 1st term by $(n-1)$ times the common difference.

Hence, nth term = $98 + (n-1)(-5)$

$\Rightarrow \quad 98 + (n-1)(-5) < 0$

$\Rightarrow \quad 103 - 5n < 0$

$\Rightarrow \quad 103 < 5n$

$\Rightarrow \quad 20.6 < n$

Hence, 21st term onwards will be negative. In other words, 21st term is the first negative term.

Problem 7 Find the 10th term from the end of the AP 4, 9, 14, ..., 254.

Sol. 10th term from the end = $\{l - (n-1)d\} = 254 - (10-1)5 = 209$

Alternatively

Let the progression be 254, 249, ..., 14, 9, 4.

Then, 10th term will exceed the 1st term by 9 times the common difference. Here, common difference

= 249 – 254 = – 5

Hence, 10th term = 254 + (10 – 1) (– 5) = 209

Problem 8 Sum of three numbers in AP is 33 and their product is 1155. Find the 3rd number.

Sol. Let the numbers be $(a-d)$, a and $(a+d)$.

Then, $(a-d) + a + (a+d) = 33$

$\Rightarrow \quad 3a = 33 \Rightarrow a = 11$

Further, $(a-d) \times a \times (a+d) = 1155$

$\Rightarrow \quad (a^2 - d^2)a = 1155$

$\Rightarrow \quad (11^2 - d^2)11 = 1155$

$\Rightarrow \quad (11^2 - d^2) = 105$

$\Rightarrow \quad d^2 = 16$

$\Rightarrow \quad d = \pm 4$

Hence, 3rd number be {11 + (4)} or {11 + (– 4)} i.e. 15 or 7

Note *Numbers are (7, 11, 15) or (15, 11, 7)*

Alternatively

Let the numbers be $(a-d)$, a and $(a+d)$..

Then, $(a-d) + a + (a+d) = 33$

$\Rightarrow \quad 3a = 33$

$\Rightarrow \quad a = 11$

Now, proceed through options. Let options be 17, 15, 13, 14.

(a) 17 ⇒ numbers be 5, 11, 17 and their product ≠ 1155.

(b) 15 ⇒ numbers be 7, 11, 15 and their product =1155

Hence, option(b) is correct.

Problem 9 The number of common terms in the two sequences 17, 21, 25, ..., 417 and 16, 21, 26, ..., 466 is ***[CAT-08, 19]***

Sol. The common difference of the 1st sequence = 4

and the common difference for the 2nd sequence = 5

Also, the first common term for the both sequences = 21

Obviously, the next common terms will as 41, 61, 81 , ... with the common difference of 20 as LCM of 4 and 5 is 20.

Hence, the sequence of common terms be 21, 41, 61, 81, ..., 401.

Then, $\quad 401 = 21 + (n-1)20$

$\Rightarrow \quad n - 1 = 19 \Rightarrow n = 20$

Thus, number of common terms in both the given sequences = 20

Problem 10 If $a_1, a_2, a_3, \dots \dots \dots a_{n+1}$ are in AP, then the value of $\frac{1}{a_1a_2} + \frac{1}{a_2a_3} + \dots \dots + \frac{1}{a_na_{n+1}} = ?$

(a) $\frac{n-1}{a_1a_{n+1}}$ (b) $\frac{1}{a_1a_{n+1}}$

(c) $\frac{n+1}{a_1a_{n+1}}$ (d) $\frac{n}{a_1a_{n+1}}$

Sol. Given expression: $\frac{1}{a_1a_2} + \frac{1}{a_2a_3} + \dots + \frac{1}{a_na_{n+1}}$

$$= \frac{1}{d}\left\{\frac{a_2 - a_1}{a_1a_2} + \frac{a_3 - a_2}{a_2a_3} + \dots \frac{a_{n+1} - a_n}{a_na_{n+1}}\right\}$$

[let d be the common difference]

$$= \frac{1}{d}\left\{\frac{1}{a_1} - \frac{1}{a_2} + \frac{1}{a_2} - \frac{1}{a_3} \ldots + \frac{1}{a_n} - \frac{1}{a_{n+1}}\right\}$$

$$= \frac{1}{d}\left\{\frac{1}{a_1} - \frac{1}{a_{n+1}}\right\} = \frac{1}{d}\left\{\frac{a_{n+1} - a_1}{a_1 a_{n+1}}\right\}$$

$$= \frac{1}{d}\left\{\frac{nd}{a_1 a_{n+1}}\right\}$$

[Obviously, term a_{n+1} will exceed the term a_1 by nd]

$$= \frac{n}{a_1 a_{n+1}}$$

Problems Based on Concept of Arithmetic Mean (AM)

Problem 11 The four arithmetic means between 3 and 23 are

(a) 5, 9, 11, 13 (b) 7, 11, 15, 19

(c) 5, 11, 15, 22 (d) 7, 15, 19, 21

Sol. Let four AMs k_1, k_2, k_3 and k_4 be inserted between 3 and 23.

The resulting sequence : 3, k_1, k_2, k_3, k_4, 23

Now, $k_1 = 3 + \left(\frac{23-3}{4+1}\right)$, $k_2 = 3 + 2\left(\frac{23-3}{4+1}\right)$, $k_3 = 3 + 3\left(\frac{23-3}{4+1}\right)$

and $k_4 = 3 + 4\left(\frac{23-3}{4+1}\right)$ i.e. 7, 11, 15, 19

Alternatively

You may proceed through options. You may check as to which option when inserted between 3 and 23 yields an AP.

Problem 12 After inserting n Arithmetic Means between 2 and 38, the sum of the resulting progression is 200. The value of n is

Sol. The resulting sequence be: 2, 1st AM, 2nd AM, 3rd AM, ..., 38

Now, sum $= \frac{n}{2}\{a + l\}$

$\Rightarrow \quad 200 = \frac{n}{2}(2 + 38)$

$\Rightarrow \quad n = 10$

Hence, total number of terms in the arithmetic sequence 2, ..., 38 is 10.

$\therefore$ Number of arithmetic means inserted = 10 − 2 = 8

Problems Based on Sum of Terms of AP

Problem 13 If the 5th and 12th terms of an AP are − 4 and − 18 respectively, find the sum of first 20 terms of the AP.

Sol. Given, 5th term = − 4 and 12th term = − 18

Clearly, the 12th term exceeds the 5th term by − 14 and this is due to addition of (12 − 5) times the common difference.

Hence, $7d = -14$

$\Rightarrow \quad d = -2$

Further, the 5th term will exceed the 1st term by (5 − 1) times the common difference. Hence, 5th term $= a + 4(-2)$

$\Rightarrow \quad -4 = a - 8 \Rightarrow a = 4$

Now, sum of first 20 terms $= \frac{n}{2}\{2a + (n-1)d\}$

$$= \frac{20}{2}\{2\cdot 4 + (20-1)(-2)\} = -300$$

Alternatively

Given, 5th term = − 4 and 12th term = − 18

Clearly, the 12th term exceeds the 5th term by − 14 and this is due to addition of (12 − 5) times the common difference.

Hence, $7d = -14$

$\Rightarrow \quad d = -2$

Further, the 20th term will exceed the 12th term by (20 − 12) times the common difference. Hence, 20th term

$= -18 + 8(-2) = -34$

Similarly, the 5th term will exceed the 1st term by (5 − 1) times the common difference. Hence, $-4 = a + 4(-2)$

$\Rightarrow \quad a = 4$

Now, required sum $= \frac{n}{2}\{a + l\} = \frac{20}{2}\{4 + (-34)\} = -300$

Problem 14 If the nth term of an AP is $(2n + 1)$, find the sum of first n terms of the AP.

Sol. 1st term $= 2\cdot 1 + 1 = 3$

2nd term $= 2\cdot 2 + 1 = 5$

Hence, $a = 3, d = 5 - 3 = 2$

Now, sum of first n terms $= \frac{n}{2}\{2a + (n-1)d\}$

$$= \frac{n}{2}\{2\cdot 3 + (n-1)2\} = \frac{n}{2}\{2n + 4\} = n^2 + 2n$$

Alternatively

1st term $= 2\cdot 1 + 1$

2nd term $= 2\cdot 2 + 1$

3rd term $= 2\cdot 3 + 1$

....

nth term $= 2\cdot n + 1$

Sum $= 2(1 + 2 + 3 + \ldots + n) + (1 + 1 + 1 + \ldots n \text{ times})$

$= 2 \times \frac{n}{2}(n+1) + n = n^2 + 2n$

Problem 15 Find the number of terms of the AP 63, 60, 57, ... so that their sum is 693.

Sol. Here, $a = 63, d = -3$

Let number of terms be n.

Then, sum of first n terms $= \frac{n}{2}\{2a + (n-1)d\}$

$\Rightarrow \quad 693 = \frac{n}{2}\{2a + (n-1)d\}$

$\Rightarrow \quad 693 = \frac{n}{2}\{2\cdot 63 + (n-1)(-3)\}$ $\quad [\because a = 63, d = -3]$

$\Rightarrow \quad 1386 = 126n - 3n^2 + 3n$

$\Rightarrow \quad 3n^2 - 129n + 1386 = 0$

$\Rightarrow \quad 3n^2 - 66n - 63n + 1386 = 0$

$\Rightarrow \quad 3n(n-22) - 63(n-22) = 0$

$\Rightarrow \quad (3n - 63)(n - 22) = 0$

$\Rightarrow \quad n = 21$ or 22 {22nd term is 0}

Problem 16 In an AP, the sum of first n terms is $\left\{\frac{3n^2}{2}+\frac{5n}{2}\right\}$. Find its 25th term.

Sol. Clearly, sum of first 24 terms $=\left\{\frac{3(24)^2}{2}+\frac{5\cdot 24}{2}\right\}=924$

and sum of first 25 terms $=\left\{\frac{3(25)^2}{2}+\frac{5\cdot 25}{2}\right\}=1000$

Hence, 25th term $=1000-924=76$

Problem 17 If $a_1, a_2, a_3, \ldots\ldots\ldots\ldots\ a_n$ be an AP and S_1, S_2 and S_3 be the sum of first n, $2n$ and $3n$ terms respectively, then $S_3-S_2-S_1$ is equal to (where a is the first term and d is common difference).

(a) $3a-2n-d$ (b) $a(n+2d)$

(c) $3a+2nd$ (d) $2n^2d$

Sol. $S_3-S_2-S_1=\frac{3n}{2}\{2a+(3n-1)d\}-\frac{2n}{2}\{2a+(2n-1)d\}-\frac{n}{2}\{2a+(n-1)d\}$

$$=3an+3nd\left(\frac{3n-1}{2}\right)-2an-nd(2n-1)-an-nd\left(\frac{n-1}{2}\right)$$

$$=nd\left[\left(\frac{9n-3}{2}\right)-(2n-1)-\left(\frac{n-1}{2}\right)\right]$$

$$=nd\left(\frac{4n}{2}\right)$$

$$=2n^2d$$

Problem 18 The sum of 3rd and 15th elements of an arithmetic progression is equal to the sum of 6th, 11th and 13th elements of the same progression. Then, which element of the series should necessarily be equal to zero? **[CAT-03]**

Sol. Let the first term and common difference be a and d respectively.

As given, 3rd term + 15th term = 6th term + 11th term + 13 term

$\Rightarrow$ $(a+2d)+(a+14d)=(a+5d)+(a+10d)+(a+12d)$

$\Rightarrow$ $16d=a+27d$

$\Rightarrow$ $a+11d=0$

$\Rightarrow$ 12th term = 0

Alternatively,

We know, sum of nth term and mth term = Sum of pth term and $(m+n-p)$th term.

Here, 3rd term + 15th term = 6th term + 12 (i.e. 3 + 15 – 6)th term.

As given, 3rd term + 15th term = 6th term + 11th term + 13th term

Hence, 12th term = 11th term + 13th term

It is fact that 11th, 12th and 13th terms are in AP and the middle term is equal to sum of other two terms. It is possible only when the middle term i.e. 12th term = 0.

Problem 19 If the sum of the first 11 terms of an arithmetic progression equals that of the first 19 terms, then what is the sum of the first 30 terms? **[CAT-04]**

Sol. Let the first term and common difference be a and d respectively.

As given, $S_{11}=S_{19}$

$\Rightarrow$ $\frac{11}{2}\{2a+(11-1)d\}=\frac{19}{2}\{2a+(19-1)d\}$

$\Rightarrow$ $22a+110d=38a+342d$

$\Rightarrow$ $16a+232d=0$

$\Rightarrow$ $2a+29d=0$

Now, $S_{30}=\frac{30}{2}\{2a+(30-1)d\}=15(2a+29d)=15\times 0=0$

Alternatively

We know, if $S_n=S_m$, then $S_{m+n}=0$

As given, $S_{11}=S_{19}$. Then, $S_{19+11}=0$

Hence, $S_{30}=0$

Problem 20 Consider the sequence of numbers $a_1, a_2, a_3, \ldots\ldots\ldots$ to infinity where $a_1=81.33$ and $a_2=-19$ and $a_j=a_{j-1}-a_{j-2}$ for $j\geq 3$. What is the sum of the first 6002 terms of this sequence? **[CAT-13]**

Sol. The terms are as below:

1st term $=a_1=81.33$

2nd term $=a_2=-19$

3rd term $=a_3=a_{3-1}-a_{3-2}=a_2-a_1$

4th term $=a_4=a_{4-1}-a_{4-2}=a_3-a_2=a_2-a_1-a_2=-a_1$

5th term $=a_5=a_{5-1}-a_{5-2}=a_4-a_3=-a_1-a_2+a_1=-a_2$

6th term $=a_6=a_{6-1}-a_{6-2}=a_5-a_4=-a_2+a_1$

On the above pattern,

7th term $=a_1$, 8th term $=a_2$, 9th term $=a_2-a_1$

10th term $=-a_1$, 11th term $=-a_2$,

12th term $=-a_2+a_1$ and so on.

Now, sum of first six terms

$$=a_1+a_2+a_2-a_1-a_1-a_2-a_2+a_1=0$$

and sum of next six terms = 0

Also, sum of next six terms = 0 and so on.

Then, sum of first 6000 terms = 0

Now, sum of 6002 terms = Sum of first 6000 terms + 6001st term + 6002nd term = 0 + 81.33 + (– 19) = 62.33

[$\because$ 6001st term = 81.33 and 6002nd term = – 19 as seen from the pattern]

Problems Based on GP

Problem 21 By adding the same constant to each of 31, 7, -1 a geometric progression results. The common ratio is **[SSC(CGL-PT)-2000]**

Sol. Let the constant be x.

Then, $(31+x)$, $(7+x)$ and $(-1+x)$ be in GP.

$\therefore$ $\frac{(7+x)}{(31+x)}=\frac{(-1+x)}{(7+x)}$

$\Rightarrow$ $49+x^2+14x=x^2+30x-31$

$\Rightarrow$ $16x=80$

$\Rightarrow$ $x=5$

Then, common ratio $=\frac{(7+x)}{(31+x)}=\frac{(7+5)}{(31+5)}=\frac{12}{36}=\frac{1}{3}$

Problem 22 If 5th term of a GP is $\frac{1}{3}$ and 9th term is $\frac{16}{243}$, then 4th term will be

Sol. It is obvious that 9th term will be $r^{(9-5)}$ times the 5th term.

Hence, $\frac{16}{243} = r^4 \cdot \frac{1}{3} \Rightarrow r = \frac{2}{3}$

Further, 5th term will be r times the 4th term.

Then, 4th term $= \frac{1}{3} \div \frac{2}{3} = \frac{1}{2}$

Problem 23 If the 4th, 7th and 10th terms of a GP be a, b, c respectively, then the relation between a, b, c is

Sol. Let the first term and the common ratio be k and r respectively.

As given, $k \cdot r^3 = a$, $k \cdot r^6 = b$ and $k \cdot r^9 = c$

Clearly, $b^2 = ac$

Alternatively I

We know, if number of terms in a geometric sequence is odd, then square of the middle term is equal to product of first and the last term or product of second term and second last term or product of third term and third last term and so on.

Here, 7th term may be treated as the middle term if we assume, 4th term as the first term and 10th be the last term.

Then, square of middle term = 4th term × 10th term

$\Rightarrow \quad b^2 = ac$

Alternatively II

We know, the product of mth and nth terms of a geometric sequence is equal to square of $\left(\frac{m+n}{2}\right)$th term, if $\frac{m+n}{2}$ is integer.

Hence, 4th term × 10th term = square of $\left(\frac{4+10}{2}\right)$th term

$\Rightarrow \quad ac = b^2$

Problem 24 If $(p+q)$th term of a GP be m and $(p-q)$th term be n, then the pth term will be (in terms of m and n)?

Sol. We know that the product of mth and nth terms of a geometric sequence is equal to square of $\left(\frac{m+n}{2}\right)$th term, if $\frac{m+n}{2}$ is integer.

Here, $(p+q)$th term × $(p-q)$th term = square of $\left(\frac{p+q+p-q}{2}\right)$th term

$\Rightarrow \quad mn = (p\text{th term})^2$

$\Rightarrow \quad p\text{th term} = \sqrt{mn}$

Alternatively

Since $(p+q)$th term and $(p-q)$th term are equidistant from pth term, pth term will be their GM. Hence, pth term $= \sqrt{mn}$

Problem 25 If every term of a GP with positive term is the sum of its two previous terms, then the common ratio of the series is?

Sol. Let the 1st term and the common ratio be a and r respectively.

Then, nth term $= ar^{n-1}$, $(n-1)$th term $= ar^{n-2}$

and $(n-2)$th term $= ar^{n-3}$

As given, $ar^{n-1} = ar^{n-2} + ar^{n-3}$

$\Rightarrow \quad \frac{1}{r} = \frac{1}{r^2} + \frac{1}{r^3}$

$\Rightarrow \quad 1 = \frac{1}{r} + \frac{1}{r^2}$

$\Rightarrow \quad r^2 - r - 1 = 0$

$\Rightarrow \quad r = \frac{1 \pm \sqrt{(-1)^2 - 4 \cdot 1(-1)}}{2 \cdot 1} = \frac{1 \pm \sqrt{5}}{2} = \frac{1+\sqrt{5}}{2}$

[∵ every term is positive]

Problem 26 If pth, qth and rth terms of a GP are a, b, c respectively, then $a^{q-r} \cdot b^{r-p} \cdot c^{p-q}$ is equal to

(a) 0 (b) 1
(c) abc (d) pqr

Sol. Let 1st term and the common ratio be A and r respectively.

Then, $a = Ar^{p-1}$, $b = Ar^{q-1}$ and $c = Ar^{r-1}$

Now, $a^{q-r} \cdot b^{r-p} \cdot c^{p-q} = (Ar^{p-1})^{q-r} \cdot (Ar^{q-1})^{r-p} \cdot (Ar^{r-1})^{p-q}$

$= A^0 \cdot r^0 = 1$

Problems Based on Concept of GM

Problem 27 The geometric mean of variables 32, 4, 8, x and 2 is 8. What is the value of variable x? **[CDS-12]**

Sol. $\text{GM} = \sqrt[5]{32 \cdot 4 \cdot 8 \cdot x \cdot 2}$

$\Rightarrow \quad 8 = \sqrt[5]{2^5 \cdot 2^5 \cdot x \cdot 2}$

$\Rightarrow \quad 8 = 4 \cdot \sqrt[5]{2x} \Rightarrow 2 = \sqrt[5]{2x}$

$\Rightarrow \quad 2^5 = 2x \Rightarrow x = 16$

Problem 28 If five GMs are inserted between 486 and 2/3, then 4th GM will be:

Sol. We know, if n GMs are inserted, then rth GM

$= k_r = a\left(\frac{b}{a}\right)^{\frac{r}{n+1}}, r \le n$

$\therefore$ 4th GM $= a\left(\frac{b}{a}\right)^{\frac{4}{5+1}}$

$= 486 \times \left(\frac{2/3}{486}\right)^{\frac{4}{5+1}} = 6$

Alternatively

The resulting sequence : 486, 1st GM, 2nd GM, 3rd GM, 4th Gm, 5th GM, 2/3.

Total number of terms = 7

Now, 7th term $= 486 \times r^6$

$\Rightarrow \quad \frac{2}{3} = 486 \times r^6$

$\Rightarrow \quad r^6 = 343 \times 3$

$\Rightarrow \quad r = \frac{1}{3}$

Now, 4th GM i.e. 5th term $= 486 \times r^4 = 486 \times \left(\frac{1}{3}\right)^4 = 6$

Problems Based on Sum of Terms of GP

Problem 29 The sum of first 8 terms of a geometric progression is 6560 and the common ratio is 3. The first term is

Sol. We have, $S_n = \dfrac{a(r^n - 1)}{r-1}$ if $r > 1$ or $r < -1$

$\Rightarrow \quad S_8 = \dfrac{a(3^8 - 1)}{3-1} \Rightarrow 6560 = \dfrac{6560a}{3-1} \Rightarrow a = 2$

Problem 30 Consider the set of numbers $\{1, 3, 3^2, 3^3, \ldots\ldots\ldots\ldots 3^{100}\}$. The ratio of the last number and the sum of the remaining numbers is closest to *[XAT-16]*

Sol. Sum of terms $1, 3, 3^2, 3^3, \ldots 3^{99} = \dfrac{1(3^{100} - 1)}{3-1} = \dfrac{(3^{100} - 1)}{2}$

$[\because a = 1, r = 3 \text{ and } r > 1]$

Now, required ratio $= \dfrac{3^{100}}{\left(\dfrac{3^{100} - 1}{2}\right)} = \dfrac{2 \times 3^{100}}{3^{100} - 1} = 2$ (approx.)

Problem 31 A man starts going for morning walk every day. The distance walked by him on the first day was 2 kms. Everyday he walks half of the distance walked on the previous day. What can be the maximum total distance walked by him in his life time? *[SBI(PO)-01]*

Sol. Total distance walked $= 2 + 1 + 1/2 + 1/4 + \ldots \infty$

Here, $a = 2$ and $r = 1/2$

Then, total distance walked $= \dfrac{a}{1-r} = \dfrac{2}{1 - \dfrac{1}{2}} = 4$ km

Problem 32 If the sum of n terms of a GP is 255 and nth term is 128 and common ratio is 2, then the first term will be

Sol. Let the initial term and the common ratio be a and r respectively.

As given, nth term = 128

$\Rightarrow \quad ar^{n-1} = 128 \Rightarrow ar^{n-1} \times r = 128r \Rightarrow ar^n = 128r$

Now, sum of n terms of a GP $= \dfrac{a(r^n - 1)}{r-1} \Rightarrow 255 = \dfrac{ar^n - a}{r-1}$

$\Rightarrow \quad 255 = \dfrac{128\,r - a}{r-1} \Rightarrow 255 = \dfrac{128 \cdot 2 - a}{2-1} \Rightarrow a = 1$

Problem 33 The first term of a GP is 7, the last term is 448 and sum of all terms is 889, then the common ratio is

Sol. Let the initial term and the common ratio be a and r respectively,

We have, $S_n = \dfrac{lr - a}{r-1}$, where l is the last term

$\Rightarrow \quad 889 = \dfrac{448 \cdot r - 7}{r-1} \Rightarrow 889r - 889 = 448r - 7$

$\Rightarrow \quad 441r = 882 \Rightarrow r = 2$

Problems Based on Combination of AP and GP

Problem 34 Terms a, 1, b are in arithmetic progression and terms 1, a, b are in geometric progression. Find a and b given $a \neq b$. *[SSC(FCI-Asstt. Grd-III)-13]*

Sol. Since a, 1, b are in AP, $1 - a = b - 1 \Rightarrow a + b = 2$

Also, 1, a, b are in GP. Then, $a = \sqrt{1b} \Rightarrow a^2 = b$

Clearly, $a + a^2 = 2$

$\Rightarrow \quad a^2 + a - 2 = 0$

$\Rightarrow \quad a^2 + 2a - a - 2 = 0$

$\Rightarrow \quad a(a + 2) - 1(a + 2) = 0$

$\Rightarrow \quad (a - 1)(a + 2) = 0 \Rightarrow a = 1 \text{ or } -2$

Then, $b = (2 - 1)$ or $\{2 - (-2)\}$ i.e. 1 or 4.

$\therefore$ Values of a, b are either (1, 1) or (–2, 4).

Hence, the required answer is (–2, 4) as $a \neq b$.

Problem 35 If p, q and r are three unequal numbers such that p, q and r are in AP and $p, r - q, q - p$ are in GP, then $p : q : r$ is equal to *[IIFT-15]*

Sol. Let the common difference be d.

Then, $p = q - d$ and $r = q + d$

Since $p, r - q, q - p$ are in GP, $\dfrac{r-q}{p} = \dfrac{q-p}{r-q}$

$\Rightarrow \quad \dfrac{d}{p} = \dfrac{d}{d} \Rightarrow p = d$

Hence, $p = d, q = 2d$ and $r = 3d$

$\therefore p : q : r = d : 2d : 3d = 1 : 2 : 3$

Alternatively

The best way is to proceed through options. Let options be 2 : 3 : 4, 1 : 2 : 3, 3 : 2 : 1 and 1 : 3 : 4.

(a) $2 : 3 : 4 \Rightarrow 3 - 2 = 4 - 3$ i.e. p, q, r are in AP.

But, $\dfrac{r-q}{p} \neq \dfrac{q-p}{r-q}$

(b) $1 : 2 : 3 \Rightarrow 2 - 1 = 3 - 2$ i.e. p, q, r are in AP.

Also, $\dfrac{r-q}{p} = \dfrac{q-p}{r-q}$

Hence, option(b) is correct.

Problems Based on HP, HM

Problem 36 The fifth term of the HP, $2, 2\dfrac{1}{2}, 3\dfrac{1}{3} \ldots$ will be

Sol. Since $2, \dfrac{5}{2}, \dfrac{10}{3}, \ldots$ are in HP, $\dfrac{1}{2}, \dfrac{2}{5}, \dfrac{3}{10} \ldots\ldots$ will be in AP.

Clearly, in arithmetic sequence, the initial term, $a = 1/2$ and common difference, $d = \dfrac{-1}{10}$

Then, 5th term of the arithmetic sequence $= \dfrac{1}{2} + 4\left(\dfrac{-1}{10}\right) = \dfrac{1}{10}$

Now, fifth term of HP sequence = Reciprocal of $\dfrac{1}{10} = 10$

Problem 37 If 5th term of a HP is $\frac{1}{45}$ and 11th term is $\frac{1}{69}$, then its 16th term will be

Sol. Obviously, 5th and 11th term of the corresponding AP be 45 and 69 respectively.

Clearly, the 11th term exceeds the 5th term by 24 and this is due to addition of (11 – 5) times the common difference.

Hence, $6d = 24 \Rightarrow d = 4$

Further, the 16th term will exceed the 11th term by (16 – 11) times the common difference.

Hence, 16th term = 11th term + 5(4) = 69 + 20 = 89

$\therefore$ 16th term of the Harmonic sequence $= \frac{1}{89}$

Problem 38 In a HP, pth term is q and the q th term is p. Then, pq th term is

Sol. Obviously, pth and qth term of the corresponding AP be $\frac{1}{q}$ and $\frac{1}{p}$ respectively.

Let the 1st term and the common difference of arithmetic sequence be a and d respectively.

Then, $\frac{1}{q} = a + (p-1)d$ and $\frac{1}{p} = a + (q-1)d$

$\Rightarrow \frac{1}{q} - \frac{1}{p} = d(p-q) \Rightarrow d = \frac{1}{pq}$

Putting the value of d in Eq. $\frac{1}{q} = a + (p-1)d$, we get

$\frac{1}{q} = a + (p-1)d \Rightarrow \frac{1}{q} = a + (p-1)\frac{1}{pq}$

$\Rightarrow \frac{1}{q} = a + \frac{1}{q} - \frac{1}{pq} \Rightarrow a = \frac{1}{pq}$

Now, pqth term of AP $= a + (pq-1)d = \frac{1}{pq} + (pq-1)\frac{1}{pq}$

$= \frac{1}{pq} + 1 - \frac{1}{pq} = 1$

Then, pqth term of HP = reciprocal of 1 = 1

Problem 39 If H is the harmonic mean between p and q, then the value of $\frac{H}{p} + \frac{H}{q}$ is

Sol. Harmonic mean of p and $q = \frac{2pq}{p+q}$

Now, $\frac{H}{p} + \frac{H}{q} = H \times \frac{p+q}{pq}$

$= \frac{2pq}{p+q} \times \frac{p+q}{pq} = 2$

Problem 40 The sixth HM between 3 and $\frac{6}{13}$ is

Sol. Clearly, terms of corresponding AP be $\frac{1}{3}$ and $\frac{13}{6}$.

We know, if n AM's are inserted between a and b, then nth AM

$= a + n\left(\frac{b-a}{n+1}\right)$

Here, 6th AM $= \frac{1}{3} + 6\left(\frac{\frac{13}{6} - \frac{1}{3}}{6+1}\right) = \frac{120}{63}$

$\therefore$ 6th HM $= \frac{63}{120}$

Problems Based on Relation Between AM, GM and HM

Problem 41 If $a^{\frac{1}{x}} = b^{\frac{1}{y}} = c^{\frac{1}{z}}$ and a, b, c are in GP, then x, y, z will be in:

Sol. Let $a^{\frac{1}{x}} = b^{\frac{1}{y}} = c^{\frac{1}{z}} = k$

Then, $a = k^x$, $b = k^y$ and $c = k^z$

Since a, b, c are in GP, $k^y / k^x = k^z / k^y$

$\Rightarrow k^{2y} = k^{x+z} \Rightarrow 2y = x + z \Rightarrow y = \frac{x+z}{2}$

Hence, x, y, z are in AP.

Problem 42 If the ratio of HM and GM between two number a and b is 4 : 5, then the ratio of the two numbers will be:

Sol. Given, $\frac{2ab}{a+b} : \sqrt{ab} = 4 : 5 \Rightarrow \frac{a+b}{2\sqrt{ab}} = \frac{5}{4}$

$\Rightarrow \frac{a+b+2\sqrt{ab}}{a+b-2\sqrt{ab}} = \frac{5+4}{5-4}$

$\Rightarrow \frac{(\sqrt{a}+\sqrt{b})^2}{(\sqrt{a}-\sqrt{b})^2} = \frac{9}{1} \Rightarrow \frac{\sqrt{a}+\sqrt{b}}{\sqrt{a}-\sqrt{b}} = \frac{3}{1}$

$\Rightarrow \frac{\sqrt{a}+\sqrt{b}+\sqrt{a}-\sqrt{b}}{\sqrt{a}+\sqrt{b}-\sqrt{a}+\sqrt{b}} = \frac{3+1}{3-1} \Rightarrow \frac{\sqrt{a}}{\sqrt{b}} = 2$

$\Rightarrow a : b = 4 : 1$

Alternatively

The best way is to proceed through options. Let options be 1 : 2, 4 : 1, ...

(a) $1 : 2 \Rightarrow \text{HM} : \text{GM} = \frac{2\cdot1\cdot2}{1+2} : \sqrt{1\cdot2} \neq 4 : 5$

(b) $4 : 1 \Rightarrow \text{HM} : \text{GM} = \frac{2\cdot4\cdot1}{4+1} : \sqrt{4\cdot1} = \frac{8}{5} : 2 = 4 : 5$

Hence, option(b) is correct.

Problems based on Arithmetico-geometric Sequence

Problem 43 $1 + \frac{3}{2} + \frac{5}{2^2} + \frac{7}{2^3} + \ldots \infty$ is equal to :

Sol. Given expression: $1 + \frac{3}{2} + \frac{5}{2^2} + \frac{7}{2^3} + \ldots \infty$

1st term $= 1\cdot1$

2nd term $= 3\cdot\frac{1}{2}$

3rd term $= 5\cdot\frac{1}{2^2}$

....

Hence, the given series is an arithmetico-geometric series.

Here, $a = 1, g = 1, d = (3-1) = 2, r = \frac{1}{2} \div 1 = \frac{1}{2}$

Now, $S_\infty = \frac{ag}{1-r} + \frac{dgr}{(1-r)^2} = \frac{1 \cdot 1}{1-\frac{1}{2}} + \frac{2 \cdot 1 \cdot \frac{1}{2}}{\left(1-\frac{1}{2}\right)^2} = 6$

Alternatively

Let $S = 1 + \frac{3}{2} + \frac{5}{2^2} + \frac{7}{2^3} + \dots \infty$ ... (i)

Then, $\frac{S}{2} = \frac{1}{2} + \frac{3}{2^2} + \frac{5}{2^3} + \dots \infty$ ...(ii)

Subtracting Eqs. (ii) from (i) we get,

$S - \frac{S}{2} = 1 + \left\{1 + \frac{1}{2} + \frac{1}{2^2} + \dots\dots \infty\right\}$

$\Rightarrow \frac{S}{2} = 1 + \frac{a}{1-r}$ where, $a = 1$ and $r = \frac{1}{2}$

$\Rightarrow \frac{S}{2} = 1 + \frac{1}{1-\frac{1}{2}} = 3 \Rightarrow S = 6$

Problem 44 The value of $2^{\frac{1}{4}} \cdot 4^{\frac{1}{8}} \cdot 8^{\frac{1}{16}} \dots \infty$ is

Sol. The given expression: $2^{\frac{1}{4}} \cdot 4^{\frac{1}{8}} \cdot 8^{\frac{1}{16}} \dots \infty$

$= 2^{\frac{1}{4} + \frac{2}{8} + \frac{3}{16} + \dots \infty}$

Clearly, $\frac{1}{4} + \frac{2}{8} + \frac{3}{16} + \dots \infty$ is an arithmetico-geometric series.

1st term $= 1 \cdot \frac{1}{4}$ 2nd term $= 2 \cdot \frac{1}{8}$

3rd term $= 3 \cdot \frac{1}{16}$

....

Here, $a = 1, g = \frac{1}{4}, d = 1, r = \frac{1}{2}$

Then, $\frac{1}{4} + \frac{2}{8} + \frac{3}{16} + \dots \infty = \frac{ag}{1-r} + \frac{dgr}{(1-r)^2} = \frac{1 \cdot \frac{1}{4}}{1-\frac{1}{2}} + \frac{1 \cdot \frac{1}{4} \cdot \frac{1}{2}}{\left(1-\frac{1}{2}\right)^2} = 1$

Now, $2^{\frac{1}{4} + \frac{2}{8} + \frac{3}{16} + \dots\dots \infty} = 2^1 = 2$

Alternatively

The given expression: $2^{\frac{1}{4}} \cdot 4^{\frac{1}{8}} \cdot 8^{\frac{1}{16}} \dots \infty = 2^{\frac{1}{4} + \frac{2}{8} + \frac{3}{16} + \frac{4}{32}} \dots \infty$

Let $S = \frac{1}{4} + \frac{2}{8} + \frac{3}{16} + \frac{4}{32} \dots \infty$ ...(i)

Then, $\frac{S}{2} = \frac{1}{8} + \frac{2}{16} + \frac{3}{32} + \dots \infty$ ...(ii)

Subtracting Eqs. (ii) from (i) we get,

$S - \frac{S}{2} = \frac{1}{4} + \frac{1}{8} + \frac{1}{16} + \dots \infty \Rightarrow \frac{S}{2} = \frac{a}{1-r}$ where, $a = \frac{1}{4}$ and $r = \frac{1}{2}$

$\Rightarrow \frac{S}{2} = \frac{\frac{1}{4}}{1-\frac{1}{2}} = 1/2$

$\Rightarrow S = 1$

Then, $2^S = 2^1 = 2$

WORKOUT

1. If the 8th term of an AP is 31 and its 15th term is 16 more than the 11th term, find the 13th term.

(a) 51 (b) 52
(c) 54 (d) 61

2. How many terms are there in the AP 3, 7, 11, ...,51 ?

(a) 23 (b) 16
(c) 13 (d) None of these

3. If a, b, c are in AP, then $\frac{(a-c)^2}{(b^2 - ac)} = ?$

(a) 1 (b) 2
(c) 3 (d) 4

4. Which term of the AP 3, 8, 13, 18, ... will be 55 more than its 20th term? ***[SSC(CGL-PT)-06]***

(a) 32nd (b) 31st
(c) 29th (d) 28th

5. Let S_n denotes the sum of n terms of an AP. If $S_{2n} = 3S_n$, then ratio $\frac{S_{3n}}{S_n} = ?$

(a) 10 (b) 6 (c) 8 (d) 4

6. If the 4th term of an arithmetic progression is 14 and 12th term is 70, then the first term is

(a) – 10 (b) – 7 (c) 7 (d) 10

7. If twice the 11th term of an AP is equal to 7 times of its 21st term, then its 25th term is equal to

(a) 24 (b) 120
(c) 0 (d) None of these

8. If the 10th term of the sequence $a, a-b, a-2b, a-3b, \dots\dots$ is 20 and the 20th term is 10, then the xth term of the series is ***[SSC(CPO)-06]***

(a) $10 - x$ (b) $20 - x$
(c) $29 - x$ (d) $30 - x$

9. If the sum of two extreme numbers of AP with four terms is 8 and product of remaining two middle terms is 15, then the greatest number of the series will be

(a) 5 (b) 7
(c) 9 (d) 11

10. Find the sum of all three-digit natural numbers which are multiples of 7.

(a) 45612 (b) 52682
(c) 72256 (d) 70336

11. If p, q, r are in geometric progression, then which is true among the followings? ***[SSC(CGL-T-I)-12]***

(a) $q = \frac{p+r}{2}$ (b) $p^2 = qr$
(c) $q = \sqrt{pr}$ (d) $\frac{p}{r} = \frac{r}{q}$

12. If $a^x = b^y = c^z$ and a, b, c are in GP, then x, y, z will be in

(a) AP (b) GP
(c) HP (d) None of these

13. What is the sum of the following series ? **[XAT-15]**

$-64, -66, -68, \ldots\ldots, -100$

(a) -1458 (b) -1558
(c) -1568 (d) -1664
(e) None of these

14. If mth term of series 63, 65, 67, ... and 3, 10, 17, ... be equal, then m = ?

(a) 15 (b) 13
(c) 12 (d) 11

15. x, 17, $3x - y^2 - 2$ and $3x + y^2 - 30$ are four consecutive terms of an increasing arithmetic sequence. Then, sum of the four numbers is divisible by **[XAT-14]**

(a) 2 (b) 3
(c) 5 (d) 7
(e) 11

16. If the arithmetic, geometric and harmonic means between two distinct real numbers be A, G and H respectively, then the relation between them is

(a) $A^2 = GH$ (b) $H^2 = AG$
(c) $G = AH$ (d) $G^2 = AH$

17. If $1 + 10 + 10^2 + \ldots\ldots$ upto n terms $= \frac{10^n - 1}{9}$, then the sum of the series $4 + 44 + 444 + \ldots$ upto n term is **[SSC(CPO)-07]**

(a) $\frac{4(10^n - 1)}{9} - \frac{4n}{9}$
(b) $\frac{4(10^n - 1)}{81} - \frac{4n}{9}$
(c) $\frac{40(10^n - 1)}{81} - \frac{4n}{9}$
(d) $\frac{40(10^n - 1)}{9} - \frac{4n}{9}$

18. If AM and HM of two numbers are 36 and 25 respectively, then what is GM of the same numbers ?

(a) 30 (b) 32 (c) 90 (d) 45

19. The sum of n arithmetic means between a and b, is

(a) $\frac{n(a+b)}{2}$ (b) $n(a + b)$
(c) $\frac{(n+1)(a+b)}{2}$ (d) $(n + 1)(a + b)$

20. The sum of three consecutive numbers in GP is 21 and the sum of their square is 189. The product of the number is

(a) 72 (b) 216
(c) 108 (d) 144

21. If $x, 1, z$ are in AP and $x, 2, z$ are in GP, then $x, 4, z$ will be in

(a) AP (b) GP
(c) HP (d) None of these

22. In the four numbers first three are in GP and last three are in AP, whose common difference is 6. If the 1st and last numbers are same, then first will be ?

(a) 2 (b) 4
(c) 6 (d) 8

23. If $a_1, a_2, a_3, \ldots\ldots\ldots\ldots a_n$ are in AP, where $a_i > 0$ for all i, then the value of $\frac{1}{\sqrt{a_1} + \sqrt{a_2}} + \frac{1}{\sqrt{a_2} + \sqrt{a_3}} + \ldots + \frac{1}{\sqrt{a_{n-1}} + \sqrt{a_n}} = ?$

(a) $\frac{n-1}{\sqrt{a_1} + \sqrt{a_n}}$ (b) $\frac{n+1}{\sqrt{a_1} + \sqrt{a_n}}$
(c) $\frac{n-1}{\sqrt{a_1} - \sqrt{a_n}}$ (d) $\frac{n+1}{\sqrt{a_1} - \sqrt{a_n}}$

24. If the third term of a GP is 4, then the product of its first 5 terms is

(a) 64 (b) 256
(c) 1024 (d) None of these

25. Let T_r be the rth term of an AP for $r = 1, 2, 3, \ldots\ldots$ If for some positive integers m, n we have $T_m = \frac{1}{n}$ and $T_n = \frac{1}{m}$, then T_{mn} equals

(a) $\frac{1}{mn}$ (b) $\frac{1}{m} + \frac{1}{n}$
(c) 0 (d) 1

26. If the sum of the first $2n$ terms of 2, 5, 8, ... is equal to the sum of first n terms of 57, 59, 61, ... then n is equal to

(a) 10 (b) 11
(c) 12 (d) 13

27. The middle term of an arithmetic series 3, 7, 11, ...,147 is **[RRB (P.way, Kolkata)-2000]**

(a) 71 (b) 75
(c) 79 (d) 83

28. The first two terms of a geometric progression add upto 12. The sum of third and fourth terms is 48. If the terms of the geometric progression are alternately positive and negative , then the first term is

(a) 4 (b) -4 (c) -12 (d) 12

29. If a, b, c are in AP, then $\frac{1}{bc}, \frac{1}{ca}, \frac{1}{ab}$ will be in

(a) AP (b) GP
(c) HP (d) None of these

30. If A_1 and A_2 are the two AMs between two numbers a and b and G_1 and G_2 be two GMs between same two numbers, then $\frac{A_1 + A_2}{G_1 G_2}$ is

(a) $\frac{a+b}{ab}$ (b) $\frac{a+b}{2ab}$
(c) $\frac{2ab}{a+b}$ (d) $\frac{ab}{a+b}$

31. $1 + \frac{4}{5} + \frac{7}{5^2} + \frac{10}{5^3} + \ldots \infty$ is equal to

(a) $\frac{3}{16}$ (b) $\frac{35}{8}$ (c) $\frac{35}{4}$ (d) $\frac{35}{16}$

32. What is the value of $\frac{1}{0.2} + \frac{1}{0.02} + \frac{1}{0.002} + \ldots\ldots\ldots$upto 9 terms ? **[SSC(CGL-T-II)-17]**

(a) 222222222 (b) 111111111
(c) 555555555 (d) 525252525

33. If the arithmetic, geometric and harmonic means between two distinct real numbers be A, G and H respectively, then the relation between them is:

(a) $A > G > H$ (b) $A > G < H$
(c) $H > G > A$ (d) $G > A > H$

34. The sum of an infinite geometric series is 3. A series, which is formed by squares of its terms, have the sum also 3. What is the common ratio ?

(a) 1/2 (b) 3
(c) 4 (d) 2

35. If the sum of n terms of an AP is $nA + n^2B$, where A, B are constants, then its common difference will be ?

(a) $A - B$ (b) $A + B$
(c) $2A$ (d) $2B$

36. The sum of first ten terms of an AP is 155 and the sum of first two terms of a GP is 9. The first term of the AP is equal to common ratio of the GP and the first term of the

GP is equal to the common difference of the AP. Which can be the AP as per the given conditions ?

(a) 2, 4, 6, 8, ...

(b) 2, 5, 8 , 11, ...

(c) 25/2, 79/6, 83/6, ...

(d) both (b) and (c)

37. In a geometric progression consisting of positive terms, each term equals the sum of the next two terms. Then, the common ratio of this progression is equal to

(a) $\sqrt{5}$ (b) $\frac{(\sqrt{5}-1)}{2}$

(c) $\frac{(1-\sqrt{5})}{2}$ (d) $\frac{\sqrt{5}}{2}$

38. If the pth, qth and rth terms of an arithmetic sequence are a, b and c respectively, then the value of $a(q-r)+b(r-p)+c(p-q)=?$

(a) 1 (b) – 1

(c) $\frac{1}{2}$ (d) 0

39. The harmonic mean of $\frac{a}{1-ab}$ and $\frac{a}{1+ab}$ is

(a) a (b) $\frac{1}{1-a^2b^2}$

(c) $\frac{a}{1-a^2b^2}$ (d) None of these

40. If sum of n terms of an AP is $3n^2+5n$ and $T_m=164$, then $m=?$

(a) 26

(b) 27

(c) 28

(d) None of the above

41. Fifth term of a GP is 2, then the product of its 9 terms is

(a) 256 (b) 512

(c) 1024 (d) None of these

42. There are 15 terms in an arithmetic progression. Its first term is 5 and their sum is 390. The middle term is

(a) 32 (b) 23

(c) 26 (d) 29

43. If the square of the 7th term of an arithmetic progression with positive common difference equals the product of the 3rd and 17th terms, then the ratio of the first term to the common difference is **[CAT- 17]**

(a) 2 : 3 (b) 3 : 2

(c) 3 : 4 (d) 4 : 3

44. The infinite sum $1+\frac{4}{7}+\frac{9}{7^2}+\frac{16}{7^3}+\frac{25}{7^4}+\ldots\ldots$ equals

(a) $\frac{27}{14}$ (b) $\frac{21}{13}$

(c) $\frac{49}{27}$ (d) $\frac{256}{147}$

45. Aman is given ₹ 158 in one rupee denominations. He has been asked to allocate them into a number of bags such that any amount required between ₹ 1 and ₹ 158 can be given by handing out a certain number of bags without opening them. What is the minimum number of bags required ? **[MAT-17]**

(a) 11

(b) 12

(c) 13

(d) None of the above

46. If $(2n+1)+(2n+3)+(2n+5)+\ldots+(2n+47)=5280$, then what is the value of $1+2+3+\ldots+n$? **[CAT-19]**

(a) 4851

(b) 3052

(c) 5412

(d) None of the above

Answer Key

1. (a)	2. (c)	3. (d)	4. (b)	5. (b)	6. (b)	7. (c)	8. (d)	9. (b)	10. (d)
11. (c)	12. (c)	13. (b)	14. (b)	15. (a)	16. (d)	17. (c)	18. (a)	19. (a)	20. (b)
21. (c)	22. (d)	23. (a)	24. (d)	25. (d)	26. (b)	27. (b)	28. (c)	29. (a)	30. (a)
31. (d)	32. (c)	33. (a)	34. (a)	35. (d)	36. (d)	37. (b)	38. (d)	39. (a)	40. (b)
41. (b)	42. (c)	43. (a)	44. (c)	45. (b)	46. (a)				

Hints and SOLUTIONS

1. As given, 15th term exceeds the 11th term by 16. It implies that the increase in 11th term is due to addition of (15 – 11) times the common difference.

Hence, $d = 4$.

Obviously, 13th term will exceed the 8th term by (13 – 8) times the common difference.

$\therefore$ 13th term $= 31 + 5 \cdot 4 = 51$

2. Let nth term be 51.

Obviously, nth term exceeds the 1st term by $(n - 1)$ times the common difference.

Hence, $51 - 3 = (n - 1)4$

$\Rightarrow \quad n = 13$

3. You may assume any value of a, b and c. Let a, b and c be 2, 4 and 6 respectively.

Then, $\dfrac{(a-c)^2}{(b^2 - ac)} = \dfrac{(2-6)^2}{(4^2 - 2 \cdot 6)} = \dfrac{16}{4} = 4$

4. Since the required term is 55 more than the 20th term, the required term will be (55 ÷ 5) terms ahead of 20th term as common difference is 5. Hence, the required term is (20 + 11) i.e. 31st term.

5. Wo know, if S_n, S_{2n} and S_{3n} be the sum of first n, $2n$ and $3n$ terms respectively of an AP, then

$S_{3n} = 3(S_{2n} - S_n)$

$\Rightarrow S_{3n} = 3(3S_n - S_n)$

$\{\because S_{2n} = 3S_n$ as given$\}$

$\Rightarrow \quad S_{3n} = 6S_n \Rightarrow \dfrac{S_{3n}}{S_n} = 6$

6. As given, 12th term exceeds the 4th term by 56. It implies that the increase in 4th term is due to addition of (12 – 4) times the common difference. Hence, $8d = 56$, $\therefore d = 7$

Obviously, 4th term will exceed the 1st term by (4 – 1)times the common difference.

$\therefore$ 1st term $= 14 - 3 \cdot 7 = -7$

7. Let the first term and the common difference be a and d respectively.

As given, $2\{a + 10d)\} = 7\{a + 20d\}$

$\Rightarrow 5a + 120d = 0$

$\Rightarrow a + 24d = 0$

$\Rightarrow$ 25th term $= 0$

8. Clearly, the sequence is in AP as the first term is a and the common difference is $-b$.

It is true that 20th term exceeds 10th term by (20 – 10) times the common difference.

$\Rightarrow 10 = 20 + 10d \Rightarrow d = -1$

Further, 10th term exceed the first term by (10 – 1) times the common difference.

$\Rightarrow 20 = a + 9(-1) \Rightarrow a = 29$

Now, x th term

$= a + (x - 1)(-1) = 30 - x$

9. Let the numbers be $a - 3d$, $a - d$, $a + d$, $a + 3d$.

As given, $a - 3d + a + 3d = 8$

$\Rightarrow a = 4$

Further, $(a - d)(a + d) = 15$

$\Rightarrow a^2 - d^2 = 15 \Rightarrow 4^2 - d^2 = 15$

$\Rightarrow d = \pm 1$

Now, at $d = +1$, the greatest term

$= a + 3d = 4 + 3 \cdot 1 = 7$

and at $d = -1$, the greatest term

$= a - 3d = 4 - 3(-1) = 7$

Hence, option(b) is correct.

10. The smallest number of three digits divisible by $7 = 105$ and the greatest number of three digits divisible by 7 = 994

Hence, AP be 105, 112, 119, 126, …., 994.

Then, number of terms

$= \left\{\dfrac{994}{7} - \dfrac{105}{7}\right\} + 1$

$= 142 - 15 + 1 = 128$

Now, the required sum

$= \dfrac{n}{2}\{a + l\} = \dfrac{128}{2}\{105 + 994\}$

$= 70336$

11. Since p, q, r be in GP, then common ratio $= \dfrac{q}{p} = \dfrac{r}{q} \Rightarrow q = \sqrt{pr}$

Alternatively

we know, if three quantities are in GP, the middle one is said to be 'Geometric Mean' of the other two quantities.

The GM of p and $r = \sqrt{pr}$.

Hence, $q = \sqrt{pr}$

12. Let $a^x = b^y = c^z = k$

Then, $a = k^{\frac{1}{x}}$, $b = k^{\frac{1}{y}}$ and $c = k^{\frac{1}{z}}$

Since a, b, c are in GP,

$$k^{\frac{1}{y}} / k^{\frac{1}{x}} = k^{\frac{1}{z}} / k^{\frac{1}{y}}$$

$\Rightarrow \quad k^{\frac{2}{y}} = k^{\frac{1}{x} + \frac{1}{z}}$

$\Rightarrow \quad \dfrac{2}{y} = \dfrac{1}{x} + \dfrac{1}{z}$

$\Rightarrow \quad y = \dfrac{2xz}{x + z}$

Hence, x, y, z are in HP.

13. Let – 100 be nth term of the series.

Obviously, – 100 will exceed – 64 by $(n - 1)$ times the common difference.

Here, common difference = – 2

$\therefore \quad -100 - (-64) = (n - 1)(-2)$

$\Rightarrow n = 19$

Now, the required sum $= \dfrac{n}{2}\{a + l\}$

$= \dfrac{19}{2}\{-64 + (-100)\} = -1558$

14. As given,

$63 + (m - 1)2 = 3 + (m - 1)7$

$\Rightarrow 63 - 2 - 3 + 7 = 7m - 2m$

$\Rightarrow 65 = 5m$

$\Rightarrow m = 13$

15. Since terms are in AP,

$17 - x = (3x - y^2 - 2) - 17$

$\Rightarrow y^2 - 4x = -36$ … (i)

Further,

$(3x + y^2 - 30) - (3x - y^2 - 2)$

$= (3x - y^2 - 2) - 17$

$\Rightarrow 3y^2 - 3x = 9$

$\Rightarrow \quad y^2 - x = 3$ … (ii)

Solving Eqs (i) and (ii), we get

$x = 13$, $y^2 = 16$

Now, sum of terms

$= 13 + 17 + (3 \cdot 13 - 16 - 2)$
$+ (3 \cdot 13 + 16 - 30)$

$= 76$, which is divisible by 2.

Hence, option(a) is correct.

16. Clearly, option(d) is correct.

17. Given expression 4 + 44 + 444 + … ….. upto n term.

$= \dfrac{4}{9}(9 + 99 + 999 + \ldots$ to n terms$)$

$= \frac{4}{9}\{(10-1)+(100-1)+(1000-1)+$ .....to n terms)

$= \frac{4}{9}\{(10+10^2+\ldots$ upto n terms$)-n\}$

$= \frac{4}{9}\times\frac{10(10^n-1)}{10-1}-\frac{4}{9}n$

$= \frac{40(10^n-1)}{81}-\frac{4n}{9}$

18. We know, $AH = G^2$

$\Rightarrow \quad 36\times 25 = G^2$

$\Rightarrow \quad G = 30$

19. The arithmetic sequence be a, 1st AM, 2nd AM, ..., b

Then, sum $= \frac{n}{2}\{a+l\} = \frac{n}{2}(a+b)$

20. Let the numbers be $\frac{a}{r}$, a and ar.

As given, $\frac{a}{r}+a+ar = 21$

$\Rightarrow \frac{a}{r}(1+r+r^2) = 21$

$\Rightarrow \left(\frac{a}{r}\right)^2(1+r+r^2)^2 = 441 \quad \ldots(i)$

[squaring both sides]

Further, $\left(\frac{a}{r}\right)^2 + a^2 + a^2r^2 = 189$

$\Rightarrow \left(\frac{a}{r}\right)^2(1+r^2+r^4) = 189 \quad \ldots(ii)$

Dividing Eq. (i) by Eq. (ii), we get

$$\frac{(1+r+r^2)^2}{(1+r^2+r^4)} = \frac{441}{189}$$

$$\Rightarrow \frac{(1+r+r^2)^2}{(1+2r^2+r^4)-r^2} = \frac{7}{3}$$

$$\Rightarrow \frac{(1+r+r^2)^2}{(1+r^2)^2-r^2} = \frac{7}{3}$$

$$\Rightarrow \frac{(1+r+r^2)^2}{(1+r^2+r)(1+r^2-r)} = \frac{7}{3}$$

$\Rightarrow 3(1+r^2+r) = 7r^2-7r+7 = 0$

$\Rightarrow 2r^2-5r+2 = 0 \Rightarrow r = 2$ or $1/2$

Then, at $r=2$, $\frac{a}{2}(1+2+2^2) = 21$

$\Rightarrow \quad a = 6$

and at $r = \frac{1}{2}$, $2a\left\{1+\frac{1}{2}+\left(\frac{1}{2}\right)^2\right\} = 21$

$\Rightarrow a = 6$

Now, product of numbers

$= \frac{a}{r}\times a\times ar = a^3$

$= 6^3 = 216$

Alternatively

$\frac{a}{r}(1+r+r^2) = 3(1+2+4)$

and $\left(\frac{a}{r}\right)^2(1+r^2+r^4) = 9(1+4+16)$

Hence, $r = 2$ and $\frac{a}{r} = 3$

$\Rightarrow a = 6$

Now, product of numbers

$= \frac{a}{r}\times a\times ar$

$= a^3 = 6^3 = 216$

21. Since x, 1, z are in AP,

$2\cdot 1 = x+z \quad \ldots(i)$

Further, x, 2, z are in GP,

$4 = xz \quad \ldots(ii)$

Dividing Eqs. (ii) by (i) we get,

$\frac{xz}{x+z} = 2$

$\Rightarrow \quad \frac{2xz}{x+z} = 4$

It implies that 4 is harmonic mean of x and z. Hence, x, 4, z are in HP.

22. Let the numbers be $\frac{a}{r}$, a, ar and $\frac{a}{r}$ as 1st and last terms are same.

Since a, ar and $\frac{a}{r}$ are in AP,

$2ar = a+\frac{a}{r}$

$\Rightarrow \quad 2r = 1+\frac{1}{r}$

$\Rightarrow \quad r = \frac{-1}{2}$ or 1

Further, $ar - a = 6$

$\Rightarrow a(r-1) = 6$

At $r = 1$, $0 = 6$ (not possible)

and at $r = \frac{-1}{2}$, $a = -4$

Hence, $a = -4$ and $r = \frac{-1}{2}$

Then, the first term $= \frac{a}{r} = \frac{-4}{\frac{-1}{2}} = 8$

23. $\frac{1}{\sqrt{a_1}+\sqrt{a_2}}+\frac{1}{\sqrt{a_2}+\sqrt{a_3}}+\ldots+\frac{1}{\sqrt{a_{n-1}}+\sqrt{a_n}}$

$= \frac{1}{\sqrt{a_2}+\sqrt{a_1}}+\frac{1}{\sqrt{a_3}+\sqrt{a_2}}+\ldots+\frac{1}{\sqrt{a_n}+\sqrt{a_{n-1}}}$

$= \frac{\sqrt{a_2}-\sqrt{a_1}}{a_2-a_1}+\frac{\sqrt{a_3}-\sqrt{a_2}}{a_3-a_2}+\ldots+\frac{\sqrt{a_n}-\sqrt{a_{n-1}}}{a_n-a_{n-1}}$

$= \frac{\sqrt{a_2}-\sqrt{a_1}}{d}+\frac{\sqrt{a_3}-\sqrt{a_2}}{d}+\ldots+\frac{\sqrt{a_n}-\sqrt{a_{n-1}}}{d}$

[where d is common difference]

$= \frac{1}{d}\{\sqrt{a_2}-\sqrt{a_1}+\sqrt{a_3}-\sqrt{a_2}+\ldots\ldots+\sqrt{a_n}-\sqrt{a_{n-1}}\}$

$= \frac{1}{d}\{\sqrt{a_n}-\sqrt{a_1}\}$

$= \frac{1}{d}\times\frac{\sqrt{a_n}-\sqrt{a_1}}{\sqrt{a_n}+\sqrt{a_1}}\times\sqrt{a_n}+\sqrt{a_1}$

$= \frac{1}{d}\times\frac{a_n-a_1}{\sqrt{a_n}+\sqrt{a_1}}$

$= \frac{1}{d}\times\frac{(n-1)d}{\sqrt{a_n}+\sqrt{a_1}} = \frac{(n-1)}{\sqrt{a_n}+\sqrt{a_1}}$

[$\because$ nth term exceeds 1st term by $(n-1)d$]

24. Let the 1st term and the common ratio be a and r respectively.

Then, 3rd term $= ar^2 = 4$

Now, $a\cdot ar\cdot ar^2\cdot ar^3\cdot ar^4$

$= a^5r^{10} = (ar^2)^5 = 4^5$

Hence, option(d) is correct.

25. Let the first term and the common difference be a and d respectively.

Then, $\quad T_m = \frac{1}{n} = a+(m-1)d$

$\Rightarrow \frac{1}{mn} = \frac{a+(m-1)d}{m} \quad \ldots(i)$

and $\quad T_n = \frac{1}{m} = a+(n-1)d$

$\Rightarrow \frac{1}{mn} = \frac{a+(n-1)d}{n} \quad \ldots(ii)$

From Eqs. (i) and (ii),

$\frac{a+(m-1)d}{m} = \frac{a+(n-1)d}{n}$

$\Rightarrow an+mnd-nd = am+mnd-dm$

$\Rightarrow \quad dm-nd = am-an$

$\Rightarrow \quad d(m-n) = a(m-n)$

$\Rightarrow \quad d = a$

Putting the value of d in Eq. (i), we get

$\frac{1}{mn} = \frac{a+(m-1)a}{m}$

$\Rightarrow \quad \frac{1}{mn} = a \Rightarrow mn = \frac{1}{a}$

Now, $T_{mn} = a+(mn-1)d$

$= a + \left(\frac{1}{a} - 1\right) a$

$= a + 1 - a = 1$

26. Clearly, $\frac{2n}{2}[2 \cdot 2 + (2n-1)3]$

$= \frac{n}{2}[2 \cdot 57 + (n-1)2]$

$\Rightarrow 8 + 12n - 6 = 114 + 2n - 2$

$\Rightarrow \quad 10n = 110$

$\Rightarrow \quad n = 11$

27. The middle term

$= \frac{\text{1st term} + \text{last term}}{2}$

$= \frac{3 + 147}{2} = 75$

28. Let the GP be: $a, ar, ar^2, ar^3, ar^4, \ldots$

As given, $a + ar = 12$

$\Rightarrow \quad a(1 + r) = 12$

$\Rightarrow \quad (1 + r) = 12/a$

Also, $ar^2 + ar^3 = 48$

$\Rightarrow \quad ar^2(1 + r) = 48$

$\Rightarrow \quad ar^2 \times \frac{12}{a} = 48$

$\Rightarrow r = \pm 2$

At $r = 2$, $(1 + 2) = \frac{12}{a}$

$\Rightarrow a = 4$

At $r = -2$, $(1 - 2) = \frac{12}{a}$

$\Rightarrow a = -12$

Clearly, at $r = +2$, GP be 4, 8, 16, 32 and at $r = -2$, GP be -12, 24, -48, 96

Since terms are alternately positive and negative, the first term is (-12).

29. We know, if each term of an AP is divided by a non-zero constant, the resulting sequence is also an AP with new common difference.

Hence, $\frac{a}{abc}, \frac{b}{abc}, \frac{c}{abc}$ will also be in AP i.e. $\frac{1}{bc}, \frac{1}{ca}, \frac{1}{ab}$ are in AP.

30. Clearly, $A_1 = a + \left(\frac{b-a}{2+1}\right)$

and $A_2 = a + 2\left(\frac{b-a}{2+1}\right)$

Further, $G_1 = a\left(\frac{b}{a}\right)^{\frac{1}{3}}$

and $G_2 = a\left(\frac{b}{a}\right)^{\frac{2}{3}}$

Now,

$$\frac{A_1 + A_2}{G_1 G_2} = \frac{a + \left(\frac{b-a}{2+1}\right) + a + 2\left(\frac{b-a}{2+1}\right)}{a\left(\frac{b}{a}\right)^{\frac{1}{3}} \times a\left(\frac{b}{a}\right)^{\frac{2}{3}}}$$

$$= \frac{a + b}{ab}$$

31. Given expression:

$$1 + \frac{4}{5} + \frac{7}{5^2} + \frac{10}{5^3} + \ldots\ldots \infty$$

1st term $= 1 \cdot 1$

2nd term $= 4 \cdot \frac{1}{5}$

3rd term $= 7 \cdot \frac{1}{5^2}$

....

Hence, the given series is an arithmetico geometric series.

Here, $a = 1$, $g = 1$, $d = (4 - 1) = 3$,

$r = \frac{1}{5} \div 1 = \frac{1}{5}$

Now, $S_\infty = \frac{ag}{1 - r} + \frac{dgr}{(1 - r)^2}$

$$= \frac{1 \cdot 1}{1 - \frac{1}{5}} + \frac{3 \cdot 1 \cdot \frac{1}{5}}{\left(1 - \frac{1}{5}\right)^2} = \frac{35}{16}$$

32. Here, $a = \frac{1}{0.2}$

and $\quad r = \frac{1}{0.02} \div \frac{1}{0.2} = 10$

We have, $S_n = \frac{a(r^n - 1)}{r - 1}$,

if $r > 1$ or $r < -1$

$\Rightarrow \quad S_9 = \frac{\frac{1}{0.2}(10^9 - 1)}{10 - 1}$

$= \frac{999999999}{0.2 \times 9}$

$= 555555555$

33. Clearly, option(a) is correct.

34. Given, $3 = a + ar + ar^2 + ar^3 + \ldots$ upto ∞ terms

Also given, $3 = a^2 + a^2r^2 + a^2r^4 + \ldots$upto ∞ terms

We know, the sum to infinity of a geometric progression with initial term a and common ratio $r = \frac{a}{1 - r}$, where $-1 < r < 1$.

Hence, $3 = \frac{a}{1 - r}$

and $\quad 3 = \frac{a^2}{1 - r^2}$

$\Rightarrow \quad a = 3(1 - r)$

and $\quad a^2 = 3(1 - r^2)$

$\Rightarrow \quad a^2 = 9(r^2 + 1 - 2r)$

and $\quad a^2 = 3(1 - r^2)$

$\Rightarrow 9(r^2 + 1 - 2r) = 3(1 - r^2)$

$\Rightarrow 3r^2 + 3 - 6r = 1 - r^2$

$\Rightarrow \quad 2r^2 - 3r + 1 = 0$

$\Rightarrow \quad r = 1$ or $1/2$

At $r = 1$, $a = 0$, which is not possible.

Alternatively, $3 \neq \frac{a}{1 - 0}$

Hence, $r = \frac{1}{2}$

35. Sum of first one term

$= 1 \cdot A + 1^2 \cdot B$

$= A + B$

Sum of first two terms

$= 2 \cdot A + 2^2 \cdot B$

$= 2A + 4B$

Sum of first three terms

$= 3 \cdot A + 3^2 \cdot B = 3A + 9B$

Sum of first four terms

$= 4 \cdot A + 4^2 \cdot B = 4A + 16B$

....

Hence, the sequence be: $(A + B)$, $\{(2A + 4B) - (A + B)\}$, $\{(3A + 9B) - (2A + 4B)\}$, ...

i.e. $(A + B), (A + 3B), (A + 5B), \ldots$

Then, common difference

$= (A + 3B) - (A + B) = 2B$

36. Let the first term and the common difference of the AP be a and d respectively.

Then, $155 = \frac{10}{2}\{2a + (10 - 1)d\}$

$\Rightarrow \quad 2a + 9d = 31$

Further, first term of GP $= d$ and common ratio $= a$

Then, $d + da = 9$

$\Rightarrow \quad d = \frac{9}{1 + a}$

Putting the value of d in equation $2a + 9d = 31$, we get $a = 2, \frac{25}{2}$,

Then, at $a = 2$, $d = 3$ and

at $a = \frac{25}{2}$, $d = \frac{2}{3}$

Hence, AP be: (i) 2, $(2 + 3)$, $(2 + 2 \cdot 3)$, ...i.e. 2, 5, 8, 11, ...

(ii) $\frac{25}{2}, \left(\frac{25}{2}+\frac{2}{3}\right), \left(\frac{25}{2}+2\cdot\frac{2}{3}\right), \ldots$

i.e. $\frac{25}{2}, \frac{79}{6}, \frac{83}{6}, \ldots$

Alternatively, you may proceed through options.

37. Let the GP be: $a, ar, ar^2, ar^3, ar^4, \ldots$

As given, $a = ar + ar^2$,

$ar = ar^2 + ar^3$ and so on.

$\Rightarrow \quad 1 = r + r^2$

$\Rightarrow r^2 + r - 1 = 0$

$\Rightarrow \quad r = \frac{(\sqrt{5}-1)}{2}$ or $\frac{(-\sqrt{5}-1)}{2}$

Since all the terms are positive, value of r cannot be negative.

Hence, option(b) is correct.

38. Let initial term and the common difference be x and d respectively.

Then, pth term $= a = x + (p-1)d$,

qth term $= b = x + (q-1)d$

and rth term $= c = x + (r-1)d$

Now, $a(q-r) + b(r-p) + c(p-q)$

$= a(q-r) + b(r-p) + c(p-q)$

$= \{x + (p-1)d\}$

$(q-r) + \{x + (q-1)d\}$

$(r-p) + \{x + (r-1)d\}(p-q) = 0$

39. $\text{HM} = \dfrac{2 \times \dfrac{a}{1-ab} \times \dfrac{a}{1+ab}}{\dfrac{a}{1-ab} + \dfrac{a}{1+ab}} = a$

40. Sum of $(m-1)$ terms

$= 3(m-1)^2 + 5(m-1)$

and sum of m terms $= 3m^2 + 5m$

Then, mth term i.e.

$T_m = \{3m^2 + 5m\}$

$- \{3(m-1)^2 + 5(m-1)\}$

$\Rightarrow 164 = 3m^2 + 5m - 3m^2 + 6m$

$- 3 - 5m + 5$

$\Rightarrow 164 = 6m + 2$

$\Rightarrow \quad m = 27$

41. Let the initial term and the common ratio be a and r respectively.

Then, GP be $a, ar, ar^2, ar^3, ar^4, \ldots, ar^8$

As given, $ar^4 = 2$

Then, product of 9 terms

$= a^9 r^{(1+2+3 \cdots\cdots + 8)}$

$= a^9 r^{36} = (ar^4)^9$

$= 2^9 = 512$

42. Clearly,

$390 = \frac{15}{2}\{5 + l\} \left[\because \text{Sum} = \frac{n}{2}\{a + l\}\right]$

$\Rightarrow \quad l = 47$

We know, if number of terms of an AP is odd, the middle term is its average, hence, middle term

$= \frac{\text{1st term} + \text{last term}}{2}$

$\therefore$ Middle term $= \frac{5+47}{2} = 26$

43. Let the first term and the common difference be a and d respectively.

As given,

$(a+6d)^2 = (a+2d)(a+16d)$

$\Rightarrow a^2 + 12ad + 36d^2$

$= a^2 + 18ad + 32d^2$

$\Rightarrow \quad 4d^2 = 6ad$

$\Rightarrow \quad 4d = 6a$

$\Rightarrow \quad a : d = 4 : 6 = 2 : 3$

Alternatively, you may proceed through options.

44. Let $S_n = 1 + \frac{4}{7} + \frac{9}{7^2} + \frac{16}{7^3} + \frac{25}{7^4} + \ldots$

Then, $\frac{S_n}{7} = \frac{1}{7} + \frac{4}{7^2} + \frac{9}{7^3} + \frac{16}{7^4} + \ldots$

Now,

$S_n - \frac{S_n}{7} = 1 + \frac{3}{7} + \frac{5}{7^2} + \frac{7}{7^3} + \frac{9}{7^4} + \ldots$

$\Rightarrow \quad \frac{6S_n}{7} = 1 + \frac{3}{7} + \frac{5}{7^2} + \frac{7}{7^3} + \frac{9}{7^4} + \ldots$

$\Rightarrow \quad \frac{6S_n}{7\cdot 7} = \frac{1}{7} + \frac{3}{7^2} + \frac{5}{7^3} + \frac{7}{7^4} + \ldots$

Then,

$\frac{6S_n}{7} - \frac{6S_n}{7\cdot 7} = 1 + \frac{2}{7} + \frac{2}{7^2}$

$+ \frac{2}{7^3} + \frac{2}{7^4} + \ldots$

$\Rightarrow \frac{36S_n}{49} - 1 = \frac{2}{7} + \frac{2}{7^2} + \frac{2}{7^3} + \frac{2}{7^4} + \ldots$

$\Rightarrow \frac{36S_n}{49} - 1 = \left(\dfrac{\frac{2}{7}}{1 - \frac{1}{7}}\right)$

$\left\{\because S_\infty = \frac{a}{1-r}, \text{ where } -1 < r < 1\right\}$

$\Rightarrow \frac{36S_n}{49} - 1 = \frac{1}{3}$

$\Rightarrow \quad \frac{36S_n}{49} = \frac{4}{3}$

$\Rightarrow \quad S_n = \frac{49}{27}$

45. Let the 1st bag contains one coin/note and the second bag contains two coins, then upto ₹ 3 (i.e. 1 + 2) can be covered. To illustrate, if any person seeks ₹ 1, 1st bag may be handed over. In case he seeks ₹ 2, 2nd bag may be given to him. In case he demands ₹ 3, both bags may be handed over to him.

Further, if the 3rd bag contains 4 coins, all the three bags can manage any amount required from 1 to (1 + 2 + 4). On the above pattern, 4th bag must contain 8 coins to manage the amount from ₹ 1 to ₹ 15, 5th bag must contain 16 coins to manage the amount from ₹ 1 to ₹ 31 and so on.

Hence, the sequence is : 1, 2, 4, 8, 16, 32, 64, 128,....... i.e. GP series.

We know, $S_n = \frac{a(r^n - 1)}{r - 1}$ if $r > 1$

$\Rightarrow \quad S_n = \frac{1(2^n - 1)}{2 - 1} = 2^n - 1$

If $n = 7$, any amount from ₹ 1 to ₹ $(2^7 - 1)$ i.e. ₹ 127 can be managed. Obviously, the 7th bag will contain 64 coins. But we have to manage the amount from ₹ 1 to ₹ 158. Then, 8th bag should contain 128 coins , but 127 coins have already been utilized. Thus, the remaining coins i.e. 31 must be allocated in such a way that any amount from ₹ 128 to ₹ 158 can be managed.

We have observed that first 5 bags are able to manage an amount from ₹ 1 to ₹ 31. Therefore, 31 coins may further be allocated in another 5 bags which will contain 1, 2 , 4, 8 and 16 coins respectively.

Hence, total number of bags

$= 7 + 5 = 12$

46. Number of terms $= \frac{1 + 47}{2} = 24$

As given,

$(2n+1) + (2n+3) + (2n+5)$

$+ \ldots + (2n+47) = 5280$

$\Rightarrow 24 \times 2n + (1 + 3 + 5 + \ldots + 47)$

$= 5280$

$\Rightarrow \quad 48n + 24^2 = 5280$

$\because$ Sum of first n odd numbers $= n^2$

$\Rightarrow \quad n = 98$

Then, $1 + 2 + 3 + \ldots + 98$

$= \frac{98 \times 99}{2} = 4851$

16

Permutation and Combination

Learning OBJECTIVES

After completing this chapter, aspirants will be able to

- Understand meaning of Permutation and Combination;
- Differentiate between Concepts of Permutation and Combination;
- Find number of ways without counting in actual;
- Understand the concept of Circular Permutation with reference to clock-wise and anti-clockwise directions;
- Visualise the critical cases associated with the concepts of Permutation and Combination through a chart with illustrations;
- Comprehend the logic applicable for restricted arrangements/Selections;
- Understand the philosophy of the division of this chapter in various cases;
- Analyse the trend of the questions being asked in various competitive examinations;
- Solve the questions in moments;
- Grip on concepts helping to solve the questions without pen and paper;
- Learn some new concepts/fundamentals;
- Apply the pragmatic approaches in lieu of academic approaches; and
- Ponder on the alternative methods.

Inside

- *Introduction Including Difference Between the Concepts of Permutation and Combination*
- *Applicability of Concepts of Permutation and Combination in Diverse Cases Through a Chart.*
- *Illustrations based on Permutation Interalia Including Restricted and Circular Permutation*
- *Illustrations Based on Combination*
- ....
- .... *& Many More*
- *Work Out with Solutions*

Introduction

This chapter is very critical for the aspirants appearing in CAT, MAT, IIFT, Bank-PO, SSC, etc. Basically, the concepts of Permutation and Combination seek the logical interpretation of the situations involved in the questions, thus the aspirant must stress upon systematic thinking/approach. Also, command over this chapter will lead you in the chapter 'Probability' as the concepts associated with 'Permutation and Combination' are widely applied there. In fact, these concepts enable us in determining the number of different ways of arranging and selecting objects without listing them in actual.

Before discussing the concepts in detail, we must know the meaning of the words 'Permutation' and 'Combination'.

A permutation is an arrangement in a definite order of a number of objects taken all or some at a time. In other words, if n number of given objects are arranged in different orders taking all or some at a time, each of these arrangements is known as PERMUTATION. The number of permutation of n different objects taken r of them at a time is expressed by ${}^nP_r$ where P stands for Permutation.

Each of different selections/groups/collections formed by taking all or some of a number of given objects at time is called COMBINATION. The number of combinations of n different things taking r of them at a time is denoted by ${}^nC_r$ where C stands for Combination.

Please keep in mind that arrangement has importance of order of objects but selection/collection/group has no importance of their order. It may be said that number of selections in which order of taken objects holds no importance is determined by 'Combination' and where it holds importance is determined by Permutation. Thus, we should have no objection if permutation is termed as ordered selection/group/collection.

Now we will discuss the concepts of Permutation and Combination through illustrations elaborated in the following chart.

Permutation	Combination

***Case* I When all the objects are distinct and all are taken at a time**

{Where repetition is not allowed}	{Where repetition is not allowed}
Find the number of 3 letter words, with or without meaning, which can be formed out of the letters of the word 'HOW', where repetition of the letters is not allowed.	Find the number of groups/selections consisting of all three letters of the word 'HOW', where repetition of the letters is not allowed.

$\left.\begin{matrix}\text{HOW}\\ \text{HWO}\\ \text{OWH}\\ \text{OHW}\\ \text{WHO}\\ \text{WOH}\end{matrix}\right\} \Rightarrow$ when we write all the possible words in actual

Hence, required number of words = 6

Alternatively-I,

There will be as many words as there are ways of filling in 3 vacant places O O O by the given three letters without allowing repetition.

The first place can be filled in 3 different ways by anyone of the given 3 letters(H, O,W). Following which the second place can be filled in 2 different ways by anyone of the remaining 2 letters. Following which the third place can be filled in one way. Thus, the number of ways in which three places can be filled is : $3 \times 2 \times 1$ i.e, 6

The above can be represented by a diagram, given below:

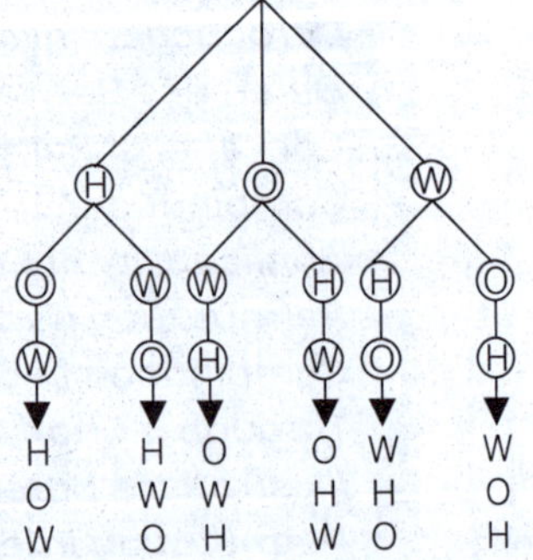

Alternatively II

From the above, total number of ways/3-letter words $= 3 \times 2 \times 1 = 3!$ $\{\because n! = n \times (n-1) \times (n-2) \times \ldots \times 1\}$

Observation The number of permutations of n different objects taken all at a time $= n! =$ product of n consecutive numbers in descending order starting from n.

Since all the possible words, to be made, will have same letters, the number of such groups = 1. It implies that number of combinations of n different objects taking all at a time is always 1, where repetition is not allowed.

Hence, ${}^nC_n = 1$, where, n in upper part of C is the total number of different objects and n in lower part of C is number of objects taken at a time. The details may be seen in case II.

Note *(i) Read $n!$ as 'factorial of n' or 'n factorial'. (ii) $0! = 1$*

{Where repetition is allowed}

Find the number of 3 letter words, with or without meaning, which can be formed out of the letters of the word 'HOW', where repetition of the letters is allowed.

Here, three vacant places OOO are to be filled in. The first place can be filled in 3 different ways by anyone of the given three letters. Following which 2nd place can be filled in 3 different ways by anyone of the given three letters. Following which 3rd place can be filled in 3 different ways by anyone of the given three letters.

Thus, total number of ways $= 3 \times 3 \times 3 = 27$

Alternatively,

The number of permutation of n different objects taken r at a time, where repetition is allowed $= n^r$

Here, total number of ways $= 3^3 = 27$

{Where repetition is allowed}

Find the number of selections of 4 letters which can be formed out of four letters A, B, C and D where repetition of the letters is allowed

$$\text{Required number of selection} = \frac{\{r + (n-1)\}!}{r!\,(n-1)!}$$

where, $\begin{cases} n = \text{total number of different objects} \\ \text{and } r = \text{number of objects taken at a time} \end{cases}$

$$= \frac{\{4 + (4-1)\}!}{4!\,(4-1)!} = \frac{7!}{4!\,3!} = 35$$

Observation The number of combinations of r objects which can be formed out of n different objects, where repetition is allowed

$$= \frac{\text{Product of } r \text{ consecutive numbers in descending order starting from } (n+r-1)}{\text{Product of } r \text{ consecutive number in descending order starting from } r}$$

Here, $n + r - 1 = 4 + 4 - 1 = 7$

$\therefore$ Required number of combinations $= \dfrac{7 \times 6 \times 5 \times 4}{4 \times 3 \times 2 \times 1} = 35$

Case II When all the objects are distinct and some of them are taken at a time

{Where repetition is not allowed}

How many 3-digit numbers can be formed from the digits 1, 2, 3, 4, 5, 6 and 7 assuming that repetition of digits is not allowed.

The number of permutations of n different things taken r at a time, where repetition is not allowed is denoted by ${}^nP_r$. The value of

$${}^nP_r = \frac{n!}{(n-r)!}$$

Here, $n = 7$ and $r = 3$

Then, required number of permutations $= {}^7P_3 = \dfrac{7!}{(7-3)!}$

$$= \frac{7 \times 6 \times 5 \times 4 \times 3 \times 2 \times 1}{4 \times 3 \times 2 \times 1} = 210$$

Observation Required number of permutations = Product of r consecutive numbers in descending order starting from n.

$= 7 \times 6 \times 5 = 210$

Remarks

If we take n objects out of n different objects allowing no repetition (as has been done in 1st question of case-I), number of permutations $= {}^nP_n = \dfrac{n!}{(n-n)!} = \dfrac{n!}{0!} = n!$

{Where repetition is not allowed}

How many combinations can be formed by taking 3 objects out of 7 different objects A, B, C, D, E, F and G at a time, where repetition of objects is not allowed.

The number of combinations of n different things taken r at a time, where repetition is not allowed is denoted by ${}^nC_r$. The value of

$${}^nC_r = \frac{n!}{r!\,(n-r)!}$$

Here, $n = 7$ and $r = 3$

Then, required number of combinations $= {}^7C_3 = \dfrac{7!}{3!\,(7-3)!}$

$$= \frac{7 \times 6 \times 5 \times 4 \times 3 \times 2 \times 1}{3 \times 2 \times 1 \times 4 \times 3 \times 2 \times 1} = 35$$

Observation Required number of combinations

$$= \frac{\text{Product of } r \text{ consecutive numbers in descending order starting from } n}{\text{Product of } r \text{ consecutive numbers in descending order starting from } r}$$

Mind the followings :

(i) ${}^nC_0 = \dfrac{n!}{0!\,(n-0)!} = \dfrac{n!}{1 \cdot n!} = 1$, hence, $\boxed{{}^nC_0 = 1}$

(ii) ${}^nC_n = \dfrac{n!}{n!\,(n-n)!} = \dfrac{n!}{n!\,1} = 1$, hence, $\boxed{{}^nC_n = 1}$

(iii) ${}^nC_r = \dfrac{n!}{r!\,(n-r)!} = \dfrac{n!}{(n-r)!\,\{n-(n-r)\}!}$

$= {}^nC_{n-r}$, hence, $\boxed{{}^nC_r = {}^nC_{n-r}}$

(iv) ${}^nC_r = \dfrac{n!}{r!\,(n-r)!} = \dfrac{{}^nP_r}{r!}$

{Where repetition is allowed}

How many 3-digit numbers can be formed from the digits 1, 2, 3, 4, 5, 6 and 7 assuming that repetition of digits is allowed.

The number of permutations of n different things taken r at a time, where repetition is allowed = n^r

Here, $n = 7$ and $r = 3$

Then, required number of permutations = $7^3 = 343$

Observation Required number of permutations $= n \times n \times \ldots r$ times.

{Where repetition is allowed}

How many combinations can be formed by taking 3 objects out of 7 different objects A, B, C, D, E, F and G at a time, where repetition of objects is allowed.

$$\text{Required number of combinations} = \frac{\{r + (n-1)\}!}{r!\,(n-1)!}$$

where, $\begin{Bmatrix} n = \text{total number of different objects} \\ \text{and } r = \text{number of objects taken at a time} \end{Bmatrix}$

Here, $n = 7$ and $r = 3$

$$\text{Then, required number} = \frac{\{3 + (7-1)\}!}{3!\,(7-1)!} = \frac{9!}{3!\,6!} = 84$$

Observation The number of combinations of r objects which can be formed out of n different objects, where repetition is allowed

$$= \frac{\text{Product of } r \text{ consecutive numbers in descending order starting from } (n + r - 1)}{\text{Product of } r \text{ consecutive numbers in descending order starting from } r}$$

Here, $n + r - 1 = 7 + 3 - 1 = 9$

$$\therefore \text{Required number of combinations} = \frac{9 \times 8 \times 7}{3 \times 2 \times 1} = 84$$

***Case* III When all the objects are not distinct and all are taken at a time**

In how many different ways can the letters of the word 'ARRANGE' be arranged ?

The number of permutations of n objects taken all at a time, where k_1 objects are of first kind, k_2 objects are of second kind, ...k_m objects are of mth kind and rest, if any, are all different

$$= \frac{n!}{k_1!\,k_2!\ldots\ldots k_m!}$$

Here, there are 7 letters out of which there are 2 As and 2 Rs and rest are all different.

$$\therefore \text{Required number of ways} = \frac{7!}{2!\,2!} = 1260$$

***Case* III When all the objects are not distinct and all or some of them are taken at a time**

A basket contains 3 apples and 2 mangoes. Find the number of ways a person can select some or all of fruits.

1 A
2 A
3 A
1 M
2 M
1 A + 1 M
2 A + 1 M
3 A + 1 M
1 A + 2 M
2 A + 2 M
3 A + 2 M

clearly, total number of ways

$$= \begin{Bmatrix} 3 \text{ (i.e. total number of apples} + 2 \text{(i.e. total number of mangoes} \\ + \\ 3 \text{ (i.e. total number of apples} \times 2 \text{ (i.e. total number of mangoes} \end{Bmatrix}$$

$= 11$

Observation

Total number of ways $= 3 + 2 + 3 \times 2 = 3 \times 2 + 3 + 2 + 1 - 1$

$= 3(2 + 1) + 1(2 + 1) - 1 = (3 + 1)(2 + 1) - 1$

$= \{(\text{number of apples} + 1) \times (\text{number of mangoes} + 1)\} - 1$

On the same pattern, total number of selections of some or all things i.e. atleast one out of $(p + q + r + \ldots)$ where p are alike of one kind, q are alike of 2nd kind, r are alike are 3rd kind and so on.

$= \{(p + 1)(q + 1)(r + 1)\ldots\ldots\} - 1$

Case IV When all the objects are not distinct and some of them are taken at a time

Find the number of permutations of the letters of the word 'PAPERS' taking three of them at a time.	There are 3 apples, 2 oranges and 2 mangoes in a basket and a person has to select 4 fruits at a time. In how many ways can he do so ?
The permutations formed from given 6 letters (i.e., 2 Ps and 1 each of A, E, R, S) can be divided into two sets as below: (i) 2 alike + 1 different (ii) all three are different **Now, proceed for 1st set(2 alike + 1 different)** Each of 4 letters A, E, R, S can be taken with 2 Ps. Then, number of such groups = 4 Further, number of permutations in each group = $\frac{3!}{2!} = 3$ {∵ each group has 2 alike letters. Recall case-III} Hence, total number of permutations in those 4 groups $= 4 \times 3 = 12$ **Further, proceed for 2nd set(all three are different):** There are five different letters i.e., P, A, E, R, S. Then, number of permutations of these 5 letters taking 3 at a time = ${}^5P_3$ $= \frac{5!}{(5-3)!} = 60$ Thus, required number of permutations = 12 + 60 = 72	The following sets of selections may be made. (i) 3 alike + 1 different (ii) 2 alike of any kind + 2 others different (iii) 2 alike of one kind + 2 alike of another kind **Now, proceed for 1st set (3 alike + 1 different)** It means 3 apples + 1 fruit from any of the remaining two kinds of fruits. Hence, number of ways = 2 **Proceed for 2nd set (2 alike + 2 others different)** It means, 2 alike may be taken from any of three kinds of fruits i.e., in 3 ways. As regards selection of 2 others different, those may be taken from remaining two kinds of fruits (i.e., other than the kind from which 2 alike has been selected), one from each kind. Hence, only one way. In other words, if we select 2 apples, we will select one orange and one mango and so on. Hence, total number of ways for 2nd set = 3 × 1 = 3 **Proceed for 3rd set (2 alike of one kind + 2 alike of another kind)** We have to select 2 alike from one kind of fruits and 2 alike from any one of remaining two kinds of fruits. Hence, a total of 3 ways. Thus, required number of ways = 2 + 3 + 3 = 8

***Case* V Restricted arrangements (Permutation)**	***Case* V Restricted selections (Combination)**
How many words of 3 letters with or without meaning can be formed out of 5 letters A, B, C, D and E if (i) letters A and B are always included (ii) letters B and C are never included (iii) letters A and B both are together (iv) letters A and B both are never together	A delegation of four doctors is to be selected out of 10 doctors. In how many ways can the delegation be selected if (i) all doctors are willing (ii) two particular doctors have to be included (iii) two particular doctors are not willing to be in delegation (iv) two particular doctors are not willing to be together in delegation.
• The number of arrangements of n distinct objects taking r at a time when e particular objects are (a) always included = ${}^{n-e}C_{r-e} \times r!$ (b) never included = ${}^{n-e}C_r \times r!$ (c) all together = $(r-e+1) \cdot e! \cdot {}^{n-e}P_{r-e}$ • The number of arrangements of n distinct objects taking all at a time when e particular objects are all together $= (n-e+1)! \cdot e!$ • The number of arrangements of n distinct objects taking all at a time when e particular objects are all together, k another particular objects are all together and so on $= (n-e-k-\dots+1+1+\dots)! \cdot e! \cdot k! \dots$ • The number of arrangements of n objects taking all at a time when e particular objects are all together $= \frac{(n-e+1)! \cdot e!}{m! \cdot t! \dots}$, if m objects are alike of one kind, t objects are alike of another kind and so on.	• The number of combinations of n different objects taking r objects at a time when e particular objects are: (a) always included = ${}^{n-e}C_{r-e}$ (b) never included = ${}^{n-e}C_r$ Now, the solutions are (i) Required answer = ${}^nC_r = {}^{10}C_4 = \frac{10 \times 9 \times 8 \times 7}{4 \times 3 \times 2 \times 1} = 210$ (ii) In case two particular doctors are included, we need to select only 2 out of remaining 8. Then, number of ways = ${}^8C_2 = 28$

Now, the solutions are

(i) Required answer $= {}^{n-e}C_{r-e} \times r! = {}^{5-2}C_{3-2} \times 3!$

$\{\because n = 5, e = 2, r = 3\}$

$$= {}^3C_1 \times 3! = \frac{3!}{1!\,(3-1)!} \times 3! = 18$$

(ii) When letters B and C are not included, then we have to select all three from the remaining three letters A, D and E. Thus, total number of ways $= {}^3C_3 = 1$. Further, these three letters can be arranged in 3! ways.

Thus, total number of ways $= 1 \times 3! = 6$

Alternatively,

Required number of ways $= {}^{n-e}C_r \times r! = {}^{5-2}C_3 \times 3!$

$\{\because n = 5, e = 2, r = 3\}$

$$= {}^3C_3 \times 3! = \frac{3!}{3!\,(3-3)!} \times 3! = 6$$

(iii) Required number of ways $= (r - e + 1) \cdot e! \cdot {}^{n-e}P_{r-e}$

$\{\because n = 5, e = 2, r = 3\}$

$$= (3 - 2 + 1) \cdot 2! \cdot {}^{5-2}P_{3-2} = 2 \cdot 2 \cdot 3 = 12$$

(iv) Required number of ways = number of ways in which letters A and B are included–number of ways in which letters A and B both are together = 18 − 12 = 6

Now, in how many different ways can the letters of the word 'SOFTWARE' be arranged in such a way that the vowels come together? ***[Canara Bank(PO)-03]***

Required number of ways $= (n - e + 1)!\, e!$

$= (8 - 3 + 1)!\, 3! = 4320$

$\{\because$ total number of letters $= n = 8$, total number of vowels $= e = 3\}$

Further consider the following question

In how many different ways can the letters of the word 'SOFTWARE' be arranged in such a way that the vowels come together and consonants S and T also come together ?

Sol: There are 3 vowels. If we assume all vowels as 1 letter and consonants S and T as 1 letter, then total number of letters = 1 + 1 + 1 (i.e. F) + 1 (i.e. W) + 1 (i.e., R) = 5, which can be arranged in 5! ways. Further, 3 vowels can be arranged in 3! ways and 2 consonants(S and T) can also be arranged in 2! ways.

Hence, required number of ways $= 5! \times 3! \times 2! = 1440$

Alternatively,

Required number of ways $= (n - e - k - \ldots + 1 + 1 + \ldots)! \cdot e!\, k! \ldots = (8 - 3 - 2 + 1 + 1)! \cdot 3!\, 2! = 5! \times 3! \times 2! = 1440$

Alternatively,

Required number of ways $= {}^{n-e}C_{r-e} = {}^{10-2}C_{4-2} = 28$

$\{\because n = 10, e = 2, r = 4\}$

(iii) Since two particular doctors are not to be included, we need to select all four out of remaining 8.

Then, number of ways $= {}^8C_4 = 70$

Alternatively,

Required number of ways $= {}^{n-e}C_r = {}^{10-2}C_4 = 70$

$\{\because n = 10, e = 2, r = 4\}$

(iv) Total number of ways for selecting 4 out of 10 = 210 and total number of ways in which two particulars are selected = 28

Then, number of ways in which both are not included together = 210 − 28 = 182

Case **VI Other Restricted arrangements (Permutation)**	***Case*** **VI Other Restricted combinations**
3 women and 4 men are to be seated in a row in such a way that the women don't sit together. In how many different ways can it be done?	A person has 4 friends. In how many ways can he invite one or more of them on a tea-party ?
If there are m men and w women (where, $m + 1 \geq w$) and they have to sit in a row in such a way that no two women sit together, then total number of such arrangements $= {}^{m+1}P_w \cdot m!$	• Number of combinations when 1 or more things out of n different things are taken i.e., some or all of them are taken $= {}^nC_1 + {}^nC_2 + {}^nC_3 + \ldots\ldots + {}^nC_n = 2^n - 1$ • Number of combinations when zero or more things out of n different things are taken $= {}^nC_0 + {}^nC_1 + {}^nC_2 + {}^nC_3 + \ldots\ldots + {}^nC_n = 2^n$ • Number of ways of selection of r things out of n identical things = 1 • Number of ways of selection of zero or more things from a group of n identical things $= (n + 1)$ • Number of ways of selections of k consecutive objects out of n things in a row $= (n - k + 1)$

Here, required number of ways = ${}^{4+1}P_3 \cdot 4! = 1440$

Alternatively

4 men can be seated in a row in 4! ways. Obviously, five places(□) would be created viz. {□ M □ M □ M □ M □} in which three women can be arranged in ${}^5P_3$ways.

Hence, total number of ways in which no two women sit together

$$= 4! \times {}^5P_3 = 4 \times 3 \times 2 \times 1 \times 5 \times 4 \times 3 = 1440$$

***Case* VII Circular Permutation**

(i) In how many ways can 6 boys stand in a circle ?

(ii) Find the number of ways in which 8 different flowers can be strung together to make a garland in such a way that three particular flowers are always together ?

In circular permutation, clockwise and anti-clockwise order of arrangements are considered as distinct permutations. There may be two types of circular permutation.

Type I *When distinction is made between clockwise and anti-clockwise arrangements, then total number of circular permutations of n different objects = (n − 1)!*

Needless to say that in circular permutation, objects are arranged in the form of circle. Now, take 3 different objects(i.e. A, B, C) to understand in the better way.

Total number of linear arrangements = 3! = 6 {i.e. ABC, ACB, BAC, BCA, CAB, CBA}. But there is no difference among ABC, CAB and BCA when they are arranged along a circle, only positional changes have been made maintaining the same order with respect to clockwise direction as well as anti-clockwise direction, shown as below:

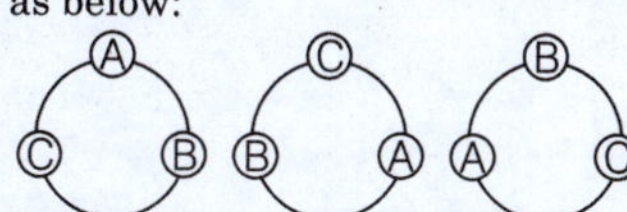

Further, there is no difference among ACB, BAC, CBA on the above line.

It is clear that only two different arrangements (ABC and ACB) may be made with respect to both directions as shown below:

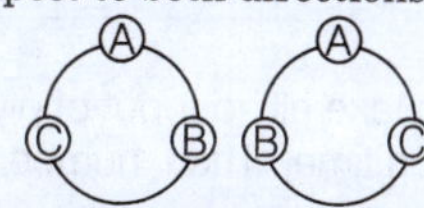

It is also seen that here distinction has been considered between clockwise and anti-clockwise arrangements.

Hence, these 2 different arrangements can be obtained on dividing 6 by 3. In other words, number of circular permutations $= \dfrac{n!}{n}$

$$= \frac{n \cdot (n-1)!}{n} = (n-1)!$$

Sol.(i) Required number of arrangements = (6 − 1)! = 120

Type II *When distinction between clockwise and anti-clockwise arrangements is not considered then total number of circular permutations of n different objects = (n − 1)!/ 2*

When distinction between clockwise and anti-clockwise arrangements is not considered, observations can be made from both directions and obviously they will be same. Now, take the above example. If we consider the distinction between clockwise and anti-clockwise arrangements, arrangements are as below:

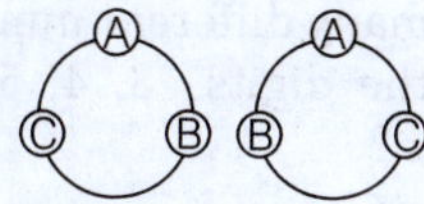

Solution: Required number of ways = $2^n - 1 = 2^4 - 1 = 15$

***Case* VII Division into Groups**

(i) Find the number of ways in which 8 different things can be divided into two groups of 5 things and 3 things.

(ii) Find the number of ways in which 8 different things can be divided into two equal groups.

(iii) Find the number of ways in which 8 different prizes can be distributed equally between 2 boys.

- The number of ways in which $(m + n)$ different things can be divided into two groups, one contains m things and another contains n things

$$= \{{}^{m+n}C_m\} \times \{{}^nC_n\} = \frac{(m+n)!}{m! \times n!}$$

$$\text{Or, } \{{}^{m+n}C_n\} \times \{{}^mC_m\} = \frac{(m+n)!}{m! \times n!}$$

On the above line,

- The number of ways in which $(m + n + p)$ different things can be divided into three groups of m, n and p things

$$= \{{}^{m+n+p}C_m\} \times \{{}^{n+p}C_n\} \times \{{}^pC_p\} = \frac{(m+n+p)!}{m! \times n! \times p!}$$

- Number of ways of dividing mn different things into m equal groups, each group containing n things

$$= \frac{(mn)!}{(n! \times n! \times \ldots m \text{ times}) \times m!}$$

- Number of ways in which mn distinct things can be distributed equally to m groups, each group containing n things

$$= \frac{(mn)!}{(n! \times n! \times \ldots m \text{ times})}$$

{∵ in event of distribution, order is considered}

Note

*(i) Number of ways in which **n distinct things** can be distributed to r different persons, each person can get any number of things =* r^n

*(ii) Number of ways in which **n identical things** can be distributed to r different persons, each person can get any number of things including zero =* ${}^{n+r-1}C_{r-1}$

*(iii) Number of ways in which **n identical things** can be distributed to r different persons, each person gets atleast one thing =* ${}^{n-1}C_{r-1}$

Sol. (i) Required number of ways $= \dfrac{(5+3)!}{5! \times 3!} = 56$

Sol. (ii) Required number of ways $= \dfrac{(2 \cdot 4)!}{4! \times 4! \times 2!} = 35$

Sol. (iii) Required number of ways $= \dfrac{(2 \cdot 4)!}{4! \times 4!} = 70$

Clearly , 1st arrangement with reference to clockwise = ABC

and 2nd arrangement with reference to clockwise = ACB

If we consider no distinction between clockwise and anti-clockwise arrangements, both arrangements appear to one arrangement. *Viz.*

1st arrangement with reference to clockwise = ABC

and 2nd arrangement with reference to anticlockwise = ABC

It is construed from the above that two circular permutations may be counted as one. Thus, total number of circular permutations of n different objects = $(n-1)!/2$

Examples of this type are a necklace made from different colored metal/stones, a garland made from different flowers, etc. If the necklace on the left is turned over we get the arrangement on the right, hence, distinction between clockwise and anti-clockwise arrangements is not considered.

Sol. (ii) Here, difference between clock-wise and anti-clockwise arrangements is not to be considered. If we assume three particular flowers as one flower , then total number of flowers = $8-3+1=6$, which can be arranged in $(6-1)!/2$ ways i.e. in 60 ways.

Further, those 3 particular flowers can be arranged in 3! ways i.e. 6 ways.

Hence, required number of ways = $60 \times 6 = 360$.

Alternative approach

Since there is no any end point in circular permutations, we may assume one thing fixed and the remaining can be arranged accordingly, *viz.* number of circular permutations of n different objects = $(n-1)!$, if distinction is made between clockwise and anti-clockwise arrangements. When distinction between clockwise and anti-clockwise arrangements is not considered then total number of circular permutations of n different objects = $(n-1)!/2$

On the above analogy

- Number of circular permutations of n different objects taken r at time in case clock-wise and anticlockwise orders are different $= \frac{^nP_r}{r}$ and in case clockwise and anti-clockwise orders are not different = $\frac{^nP_r}{2r}$

Application of the concept of combination in Geometry will be discussed in the Unit - IX (Geometry).

Problems Based on Permutation

Case A When all objects are distinct

Problem 1 In how many different ways can the letters of the word 'BLOATING' be arranged ? ***[Maharashtra Bank(PO)- 08]***

Sol. Clearly, we have to take all letters out of given 8 distinct letters at a time . Then, number of ways

$$= {}^8P_8 = \frac{8!}{(8-8)!}$$

$$= 8! = 40320$$

Alternatively,

Required number of ways = Product of 8 consecutive numbers in descending order starting from 8

$$= 8 \times 7 \times 6 \times 5 \times 4 \times 3 \times 2 \times 1 = 40320$$

Problem 2 Four persons are to be seated in a row to watch a dance programme and 4 chairs are for them. In how many ways can they take their seats ?

Sol. Clearly, we have to take all four out of given 4 distinct objects/persons at a time . Then, number of ways = ${}^4P_4$

$$= \frac{4!}{(4-4)!} = 4! = 24$$

Alternatively,

Required number of ways = product of 4 consecutive numbers in descending order starting from 4 = $4 \times 3 \times 2 \times 1 = 24$

Problem 3 In an almirah there is a space for 4 clothes. In how many ways can it be filled with 7 different clothes ?

Sol. Required number of ways = ${}^7P_4 = \frac{7!}{(7-4)!} = 840$

Alternatively,

Required number of ways = Product of 4 consecutive numbers in descending order starting from 7 = $7 \times 6 \times 5 \times 4 = 840$

Problem 4 How many different numbers of 3 digits can be formed with the digits 3, 4, 5, 6 and 7 without repetition of digits ?

Sol. Required number of ways = $^5P_3 = \frac{5!}{(5-3)!} = 60$

Alternatively,

Required number of ways = Product of 3 consecutive numbers in descending order starting from 5 = $5 \times 4 \times 3 = 60$

Problem 5 How many different numbers of 3 digits can be formed with the digits 3, 4, 5, 6 and 7 if repetition of digits is allowed?

Sol. Required number of ways = $n^r = 5^3 = 125$

Alternatively,

Required number of ways = Product of three 5 s

$= 5 \times 5 \times 5 = 125$

OR

On the above line, observe one question A lock is equipped with digit-system, which can be opened with a four digit code. There are four rings indicating digits 0 to 9. How many maximum attempts would have to be made before getting the right code? ***[MAT-19]***

Sol. Obviously, the correct code will have four digits out of 0, 1, 2, ..., 9. Repetition of the digits is also possible.

Then, total number of 4-digit codes = 10^4

$\therefore$ Maximum number of attempts before the getting the right code

$= 10^4 - 1$

Problem 6 How many words containing only the vowels taken from the letters of a word 'DAUGHTER' can be formed?

Sol. Number of vowels in 'DAUGHTER' = 3

Then, required number of desired words

$= {^3P_3} = \frac{3!}{(3-3)!} = 3! = 6$

Alternatively,

Required number of desired words = Product of 3 consecutive numbers in descending order starting from 3 = $3 \times 2 \times 1 = 6$

Problem 7 How many words ending with the letter M can be formed with the letters of the word 'GAMES' taken all together, repetition not being allowed?

Sol. Since letter M is fixed as last letter of the desired words, we have to take 4 letters out of remaining 4 letters at a time.

$\therefore$ Required number of ways = $^4P_4 = \frac{4!}{(4-4)!} = 4! = 24$

On the above line, 120 words can be formed with the word CUSTOM with a condition that the word should begin with M. ***[SNAP-18]***

Problem 8 How many 5-digit numbers can be formed with digits 2, 3, 4, 5 and 6 such that 5 is always at hundred's place?

Sol. Since digit 5 is fixed at hundreds digit of the desired numbers, we have to take 4 digits out of remaining 4 digits at a time.

$\therefore$ Required number of ways = $^4P_4 = \frac{4!}{(4-4)!} = 4! = 24$

Problem 9 How many 3-digit numbers can be formed from the digits 2, 3, 5, 6, 7 and 9, which are divisible by 5 and none of digits is repeated? ***[SSC(CGL-PT)-2000]***

Sol. The number divisible by 5 must have either 0 or 5 at as unit digit. Here, 5 is to be fixed as unit digit. Hence, we have to take remaining 2 digits out of remaining 5 digits at a time.

$\therefore$ Required number of numbers = $^5P_2 = \frac{5!}{(5-2)!} = 20$

Problem 10 How many five-digit numbers can be formed with digits 3, 4, 5, 6 and 7 which are divisible by 2?

Sol. Obviously, the desired numbers must have either 4 or 6 as unit digit.

In case of 4 at unit place, the remaining 4 digits can be arranged in $^4P_4$ i.e., in 24 ways.

Further, in case of 6 at unit place, the remaining 4 digits can also be arranged in $^4P_4$ i.e., in 24 ways.

Hence, total number of desired numbers = 24 + 24 = 48

Problem 11 How many numbers can be formed with digits 3, 4, 5 and 6 without repetition of digits, which are greater than 400.

Sol. Obviously, the desired numbers be either of three digits starting with 4, 5, 6 or of the four digits starting with any digit.

Number of 3-digit numbers starting with 4 = $^3P_2 = 6$

{$\because$ remaining two digits can be taken from remaining 3 digits}

Similarly, number of 3-digit numbers starting with 5 = 6

and number of 3-digit numbers starting with 6 = 5

Further, number of 4-digit numbers = $^4P_4 = 24$

Thus, total number of required numbers

$= 6 + 6 + 6 + 24$

$= 42$

Alternatively,

Total number of 3-digit numbers formed with 3, 4, 5 and 6

$= {^4P_3} = 24$

and number of 3-digit numbers formed with 3, 4, 5 and 6 which starts with 3 = $^3P_2 = 6$

Hence, total number of 3-digit numbers not starting with 3 = 24 − 6 = 18

Further, number of 4-digit numbers = $^4P_4 = 24$

Thus, total number of required numbers = 18 + 24

= 42

Problem 12 How many 3-digit numbers can be formed with digits 0, 2, 3 and 4, if repetition of digits is not allowed?

Sol. Number of 3-digit numbers when we take 3 digits out of 4 given digits at a time = $^4P_3 = 24$

Further, number of 3-digits numbers started with 0 = $^3P_2 = 6$

{$\because$ in case of 0 at hundred place, tens and unit place can be filled with the remaining 3 digits}

Hence, required number of desired numbers = 24 − 6 = 18

Case B When all objects are not distinct

Problem 13 **In how many different ways can the letters of the word 'PADDLED' be arranged ?** ***[Corporation Bank(PO)- 06]***

Sol. The word 'PADDLED' consists of seven letters out of which the letter *D* is repeated thrice.

∵ Required number of ways

$$= \frac{7!}{3!} = 7 \times 6 \times 5 \times 4$$
$$= 840$$

Problem 14 **In how many different ways can the letters of the word 'RUMOUR' be arranged ?**

Sol. Total number of letters = 6, number of letter *U* = 2 and number of letter *R* = 2

Then, required number of arrangements $= \frac{6!}{2!2!}$
$= 180$

Problem 15 **All letters of the word 'AGAIN' are permuted in all possible ways and the words so formed (with or without meaning) are written as in a dictionary. Then, the 50th word is** ***[MAT-11]***

Sol. Obviously, in dictionary A will come first, then G, I and N respectively.

Number of words started with letter *A* = 4! = 24

{∵ If *A* is the first letter, remaining 4 will be arranged in 4! ways}

Number of words started with letter $G = \frac{4!}{2!} = 12$ {∵ there is 2 As}

Number of words started with letter $I = \frac{4!}{2!} = 12$ {∵ there is 2 As}

Now, the word after (24 + 12 + 12) words i.e., 49th word will start with *N* and that will be NAAGI (mind the order of letters).

Then, 50th word = NAAIG

Case C Restricted/Conditional arrangements

Problem 16 **Arrange the letters of word 'AUCTION' in such a way that the vowels always occur together. Find the number of ways.**

Sol. If we assume the vowels of the given word (i.e., *A*, *U*, *I*, *O*) as one letter, then total number of letters= 4 { i.e. CTN(AUIO)}

Clearly, these 4 letters can be arranged in 4! ways.

Further, all four vowels can also be arranged in 4! ways

Hence, required number of ways = 4! × 4! = 576

Alternatively,

We know, number of arrangements of *n* distinct objects taking all at a time when e particular objects are all together

$= (n - e + 1)! \cdot e!$

Here, $\begin{cases} n = \text{number of objects} = 7 \\ e = \text{number of objects to be taken togehter} = 4 \\ i.e., \text{ number of vowels} \end{cases}$

$= (7 - 4 + 1)! \cdot 4!$

$= 840$

Problem 17 **In how many different ways can the letters of the word 'CORPORATION' be arranged so that the vowels come together ?** ***[SBI(PO)- 03]***

Sol. If we assume, the vowels of the given word (i.e., *O*, *O*, *A*, *I*, *O*) as one letter, then total number of letters = 7 { i.e., CRPRTN(OOAIO)}

Clearly, out of these 7 letters, *R* comes twice.

∴ Number of arrangements taking all 7 letters at a time

$$= \frac{7!}{2!} = 2520$$

Further, Out of 5 vowels, O comes thrice.

∴ Number of arrangements taking all 5 vowels at a time

$$= \frac{5!}{3!} = 20$$

Hence, required number of ways = 2520 × 20 = 50400

Alternatively,

We know, number of arrangements of *n* objects taking all at a time when *e* particular objects are all together $= \frac{(n-e+1)! \cdot e!}{m! \cdot t! \cdot \ldots\ldots}$,

if *m* objects are alike of one kind, *t* objects are alike of another kind and so on.

Here, *n* = number of letters of the given word = 11

e = number of vowels = 5

m = 2 (∵ *R* comes twice)

t = 3 (∵ *O* comes thrice)

Then, required number of ways $= \frac{(n-e+1)! \cdot e!}{m! \cdot t!}$

$$= \frac{(11-5+1)! \cdot 5!}{2! \cdot 3!} = \frac{7! \cdot 5!}{2! \cdot 3!} = 50400$$

Problem 18 **In how many different ways can the letters of the word 'ADJUST' be arranged so that the vowels never come together ?**

Sol. Number of ways taking all 6 letters at a time = 6! = 720

Number of arrangements of 6 letters in which both vowels *A* and *U* come together $= (n - e + 1)! \cdot e! = (6 - 2 + 1)! \cdot 2! = 5! \cdot 2!$ $= 240$

Hence, required number of ways = 720 − 240 = 480

Problem 19 **In how many different ways can the letters of the word 'DETAIL' be arranged in such a way that the vowels occupy only the odd position.**

Sol. There are three vowels (*E*, *A* and *I*), which are to be placed at odd places i.e., at 1st , 3rd and 5th position and obviously, these three vowel can be arranged in 3! ways.

Further, the remaining three constants at even places can be arranged in 3! ways.

Hence, required number of ways = 3! × 3! = 36

Problem 20 **2 girls and 4 boys are to be seated in a row in such a way that the girls don't sit together . In how many different ways can it be done ?**

Sol. Obviously, 6 different persons can sit in 6! ways i.e., in 720 ways.

Number of arrangements in which two girls sit together

$= (n - e + 1)! e! = (6 - 2 + 1)! \cdot 2! = 5! \cdot 2! = 240$

Hence, number of required ways = 720 − 240 = 480

Alternatively-I,

4 boys can be seated in a row in 4! ways. Obviously, five places(□) would be created *viz.* {□ B □ B □ B □ B □ } in which two girls may be arranged in $^5P_2$ways.

Hence, total number of ways in which no two girls sit together

$= 4! \times {}^5P_2 = 4 \times 3 \times 2 \times 1 \times 5 \times 4 = 480$

Alternatively-II

We know, if there are m men and w women (where, $m + 1 \geq w$) and they have to sit in a row in such a way that no two women sit together, then total number of such arrangements $= {}^{m+1}P_w \cdot m!$

On the same line, required number of ways $= {}^{4+1}P_2 \cdot 4! = 480$

Problem 21 There are 10 persons $P1$, $P2$, ... $P9$, $P10$. Out of these 10 persons, 5 persons are to be arranged in a line such that in each arrangement $P1$ must occur whereas $P4$ and $P5$ do not occur. The number of such possible arrangements are: *[MAT-14, 19]*

Sol. Since two persons should not be there in the line, we should take 5 persons out of remaining 8 persons. But, one person must be there in the line, hence, we have to take another 4 persons out of 7 persons. Then, number of ways to select 4 persons out of 7 persons $= {}^7C_4$. Further, 5 persons in each line can be arranged in 5! Ways.

Hence, required number of possible arrangements $= {}^7C_4 \times 5!$

Alternatively,

We know, number of arrangements of n distinct objects taking r at a time when e particular objects are always included

$= {}^{n-e}C_{r-e} \times r!$

Here, $n = 8$ {∵ two persons are not to be selected}, $e = 1$, $r = 5$

∴ Required number of possible arrangements $= {}^{8-1}C_{5-1} \times 5!$

$= {}^7C_4 \times 5!$

Problem 22 In how many ways can 10 boys $a_1, a_2\ a_3 \ldots a_{10}$ be seated in a row such that a_2 always sits before a_5 and a_5 always sits before a_7? *[MAT-13]*

Sol. (1)--(2)---(3)--(4)--(5)---(6)····(7)······(8)···(9)······(10)·· >

Given that a_2 always sits before a_5 and a_5 always sits before a_7, it implies that:

a_7 can't sit on seat number (1) and (2).

a_5 can't sit on seat number (1) and (10).

and a_2 can't sit on seat number (9) and (10).

Hence, the order of sitting of a_2, a_5 and $a_7 \Rightarrow a_2 \ldots a_5 \ldots a_7$

Now, the following cases need to be considered:

Case (i) When a_2 at seat number1, a_5 can sit either on seat number 2 or 3 or 4 or 5 or 6 or 7 or 8 or 9

If a_5 is at seat number 2, a_7 can sit in 8 ways.

If a_5 is at seat number 3, a_7 can sit in 7 ways.

If a_5 is at seat number 4, a_7 can sit in 6 ways.

If a_5 is at seat number 5, a_7 can sit in 5 ways.

If a_5 is at seat number 6, a_7 can sit in 4 ways.

If a_5 is at seat number 7, a_7 can sit in 3 ways.

If a_5 is at seat number 8, a_7 can sit in 2 ways.

If a_5 is at seat number 9, a_7 can sit in 1 way.

Hence, total number of ways in this case

$= (8 + 7 + 6 + \ldots + 2 + 1) = 36$

Case (ii) When a_2 at seat number 2, a_5 can sit either on seat number 3 or 4 or 5 or 6 or 7 or 8 or 9

On the above pattern, total number of ways in this case

$= (7 + 6 + \ldots + 2 + 1) = 28$

Obviously, there will be 8 cases as a_2 can sit either on seat number 1 or 2 or 3 or 4 or 5 or 6 or 7 or 8.

On the above pattern, total number of ways of sitting of a_2, a_5 and a_7 in given order i.e., $a_2 \ldots a_5 \ldots a_7$

$= 36 + 28 + 21 + 15 + 10 + 6 + 3 + 1 = 120$

Further, the remaining 10 seats will be occupied by 7 boys other than a_2, a_5 and a_7 in 7! Ways.

Thus, required number of ways $= 120 \times 7! = 604800$

Alternatively,

Obviously, the order of sitting of a_2, a_5, and$a_7 \Rightarrow a_2 \ldots a_5 \ldots a_7$

Number of ways of arrangements of a_2, a_5 and a_7 assuming other seven are seated at fixed seats for time being $= 3! = 6$

These 6 arrangements will be

(i) $\underline{a_2\ a_5\ a_7}\ a_1\ a_3\ a_4\ a_6\ a_8\ a_9\ a_{10}$

(ii) $\underline{a_2\ a_7\ a_5}\ a_1\ a_3\ a_4\ a_6\ a_8\ a_9\ a_{10}$

(iii) $\underline{a_5\ a_2\ a_7}\ a_1\ a_3\ a_4\ a_6\ a_8\ a_9\ a_{10}$

(iv) $\underline{a_5\ a_7\ a_2}\ a_1\ a_3\ a_4\ a_6\ a_8\ a_9\ a_{10}$

(v) $\underline{a_7\ a_5\ a_2}\ a_1\ a_3\ a_4\ a_6\ a_8\ a_9\ a_{10}$

(vi) $\underline{a_7\ a_2\ a_5}\ a_1\ a_3\ a_4\ a_6\ a_8\ a_9\ a_{10}$

Only arrangement (i) is as per given condition. It may be construed that 1/6 of the total arrangements are as per conditions/order given in the question.

Clearly, total number of arrangements of 10 persons = 10!

∴ Required number of ways = 1 / 6 of 10! = 604800

Problem 23 There are six teachers. Out of them, two are primary teachers and two are secondary teachers. They are to stand in a row so as the primary teachers, middle teachers and secondary teachers are always in a set. The number of ways in which they can do so, is:

Sol. We have to arrange in such a way that both primary teachers be together, both middle teachers be together and also both secondary teachers be together.

Obviously, these 3 pairs can be arranged in 3! ways and each pair can be arranged in 2!.

Hence, required number of ways $= 3! \times 2! \times 2! \times 2! = 48$

Alternatively,

Number of arrangements of n distinct objects taking all at a time when e particular objects are all together, k another particular objects are all together and so on

$= (n - e - k - \ldots + 1 + 1 + \ldots)! \cdot e! \cdot k! \ldots\ldots$

∴ Required number of ways

$= (6 - 2 - 2 - 2 + 1 + 1 + 1)! \cdot 2! \cdot 2! \cdot 2!$

$= 3! \times 2! \times 2! \times 2! = 48$

Problem 24 An examination paper contains 8 questions of which 4 have 3 possible answers each, 3 have 2 possible answers each and the remaining one has 5 possible answers. The total number of possible answers to all the questions is *[MAT-11, 08]*

(a) 1278 (b) 1728 (c) 1306 (d) 3240

Sol. For better understanding, take one question 'what is total number of possible answers of 2 questions if each has 3 possible answers ?'

Let questions be (*A*) and (*B*). (*A*) has three answers *m*, *n* and *q* and (*B*) has *p*, *r* and *t*.

The possible answers of 2 questions will be as below:

(A)	(B)	
m	p	1st possible answer
m	r	2nd
m	t	3rd
n	p	4th
n	r	5th
n	t	6th
q	p	7th
q	r	8th
q	t	9th

Hence, total number of possible answers of 2 questions = 9.

Generic Version

Total number of possible answers of n questions having m possible answers each = m^n

Now, come at the question: there are three parts of examination paper, first part contains 4 questions each having 3 possible answers, 2nd part contains 3 questions each having 2 possible answers and 3rd has 1 question with 5 possible answers. Clearly, 1st part can be answered in 3^4 ways, 2nd in 2^3 ways and 3rd in 5^1 ways

Hence, whole examination paper can be answered in $3^4 \times 2^3 \times 5^1$ ways i.e., 3240 ways.

Case D Circular Permutation

Problem 25 A company CEO invited nine persons for a business meeting, where the host will be seated at a circular table. How many different arrangements are possible if two invitees *X* and *Y* be seated on either side of the host CEO ?

Sol. Total persons for the meeting = 10

Assume *X*, *Y* and CEO as one person for the time being. Then, total number of persons = 10 – 3 + 1 = 8

These 8 persons can be arranged in (8 – 1)! ways. Further, *X*, *Y* and CEO may be arranged in two ways i.e., *X*- CEO - *Y* and *Y*-CEO -*X* .

Hence, required number of ways

$= (8-1)! \times 2 = 5040 \times 2 = 10080$

Problem 26 Find the number of ways in which ten different flowers can be strung together to make a garland in such a way that three particular flowers are always together.

Sol. Here, difference between clock-wise and anti clock-wise arrangements is not to be considered.

If we assume three particular flowers as one flower , then total number of flowers = 10 – 3 + 1 = 8 , which can be arranged in (8 – 1)!/2 ways i.e., in 2520 ways.

Further, those 3 particular flowers can be arranged in 3! ways i.e., 6 ways.

Hence, required number of ways = 2520 × 6 = 15120.

Problems Based on Combinations

Problem 27 In how many ways can a cricket team consisting of 11 players be made out of a batch of 17 players ?

Sol. Number of ways of selecting 11 players out of 17 players at a time = $^{17}C_{11} = \dfrac{17 \times 16 \times 15 \times 14 \times 13 \times 12}{6 \times 5 \times 4 \times 3 \times 2 \times 1} = 12376$

Problem 28 How many triangles can be formed with 6 non-collinear points lying in a plane ?

Sol. In order to draw/form a triangle, three points are required. It implies that we have to select 3 points out of given 6 points at a time. Then, required number of ways/triangles

$= {}^6C_3 = \dfrac{6 \times 5 \times 4}{3 \times 2 \times 1} = 20$

Problem 29 How many combinations of 3 objects out of 5 different objects can be formed if the repetition of objects is allowed.

Sol. Required number of ways

$= \dfrac{\text{Product of } r \text{ consecutive numbers in descending order starting from } (n + r - 1)}{\text{Product of } r \text{ consecutive numbers in descending order starting from } r}$

$= \dfrac{7 \times 6 \times 5}{3 \times 2 \times 1} \{\because n = 5 \text{ and } r = 3\} = 35$

Problem 30 In how many different ways a group of 4 men and 4 women be formed out of 7 men and 8 women ? ***[Canara Bank(PO)-03]***

Sol. Number of ways of selecting 4 men out of 7 men = $^7C_4 = 35$

and Number of ways of selecting 4 women out of 8 women

$= {}^8C_4 = 70$

Then, required number of ways = 35 × 70 = 2450

Problem 31 In how many ways can 5 questions be selected from a question paper consisting of 10 questions such that it always contains question number 2 and 3.

Sol. Since two questions are fixed to be taken, only 3 questions are to be selected out of remaining 8 questions.

Then, required number of ways = $^8C_3 = 56$

Alternatively,

Here, $n = 10$, $r = 5$ and $p = 2$

Then, required number of ways

$= {}^{n-p}C_{r-p} = {}^{10-2}C_{5-2} = {}^8C_3 = 56$

Problem 32 A student is to answer 10 out of 13 questions in an examination such that he must choose atleast 4 from first five questions. The number of choices available to him is ***[MAT-13]***

Sol. **Case-I** When he chooses 4 from first 5 questions

Then, number of ways of selecting 4 out of first 5 = $^5C_4 = 5$

and number of ways of selecting 6 out of remaining 8

$= {^8C_6} = 28$

Hence, total number of ways = $5 \times 28 = 140$

Case-II When he chooses 5 from first 5 questions

Then, number of ways of selecting 5 out of first 5 = $^5C_5 = 1$

and number of ways of selecting 5 out of remaining 8

$= {^8C_5} = 56$

Hence, total number of ways = $1 \times 56 = 56$

$\therefore$ Required number of ways = $140 + 56 = 196$

Problem 33 **In a tennis tournament, eight players must be divided into four teams of two players each. The number of different ways in which this can be done is**

Sol. We know, number of ways of dividing mn different things in m groups containing n things each

$$= \frac{(mn)!}{(n! \times n! \times \ldots m \text{ times}) \times m!}$$

$$\therefore \text{Required number of ways} = \frac{8!}{2! \times 2! \times 2! \times 2! \times 4!}$$

$\{\because m = 4, n = 2\}$

$= 105$

Problem 34 **A committee of 3 experts is to be selected out of a pannel of 7 persons. Three of them are engineers, three are managers and one is both engineer and manager. In how many ways can the committee be selected, if it must have atleast one engineer and one manager ?**

Sol. There may be 3 cases.

Case-I 1 out of 3 engineers and 2 out of 3 managers

Case-II 2 out of 3 engineers and 1 out of 3 managers

Case-III 1 person who is both engineer and manager and 2 out of remaining 6 persons.

Now, total number of ways for Case-I = $^3C_1 \times {^3C_2} = 9$

total number of ways for Case-II = $^3C_2 \times {^3C_1} = 9$

and total number of ways for Case-III = $1 \times {^6C_2} = 15$

Hence, required number of ways = $9 + 9 + 15 = 33$

Problem 35 **Five balls of different colours are to be placed in three different boxes such that any box contains atleast one ball. What is the maximum number of different ways in which this can be done ?**

[IAS(Pre.) - 07, MAT-10]

Sol. There may be two cases.

Case-I (2, 2, 1) – (2, 1, 2) – (1, 2, 2)

Case-II (3, 1, 1) – (1, 3, 1) – (1, 1, 3)

If we take case I (2, 2, 1), then number of ways = $\frac{5!}{2! \cdot 2! \cdot 1!} = 30$

$\{\because$ Number of ways in which $(m + n + p)$ different things can be divided into three groups containing m, n and p things respectively = $\frac{(m+n+p)!}{m! \times n! \times p!}\}$

Obviously, total number of ways for case-I = $30 \times 3 = 90$

Similarly, for the case-II, total number of ways

$$= 3 \times \frac{5!}{3! \cdot 1! \cdot 1!} = 60$$

Hence, required number of ways = $90 + 60 = 150$

Alternatively,

Required number of ways = $r^n - {^rC_1}(r-1)^n + {^rC_2}(r-2)^n$

$= 3^5 - {^3C_1}(3-1)^5 + {^3C_2}(3-2)^5 \quad \{\because n = 5 \text{ and } r = 3\}$

$= 243 - 3 \cdot 32 + 3 \cdot 1 = 150$

Problem 36 **There are 5 different green dyes, 4 different blue dyes and 3 different red dyes. How many combinations of dyes can be chosen taking atleast one green and one blue dye ?**

Sol. Number of ways of selecting at least one green dye out of 5 green dyes = $2^n - 1 = 2^5 - 1 = 31$

$\{\because$ Number of combinations when 1 or more things out of n different things are taken $= 2^n - 1\}$

Similarly, number of ways of selecting at least one blue dye out of 4 blue dyes = $2^n - 1 = 2^4 - 1 = 15$

and Number of ways of selecting at least one red dye or no red dye out of 3 red dyes = $2^n = 2^3 = 8$

$\{\because$ Number of combinations when zero or more things out of n different things are taken $= 2^n\}$

Hence, required number of ways = $31 \times 15 \times 8 = 3720$

Problem 37 **The number of ways in which a committee of 3 ladies and 4 gentlemen can be appointed out of 8 ladies and 7 gentlemen, if Mrs *X* refuses to serve in a committee of which Mr. *Y* is a member, is?**

[MAT-09, IIFT-14, IAS(Pre.) - 03]

Sol. Number of ways of selecting 3 ladies and 4 gentlemen out of 8 ladies and 7 gentlemen = $^8C_3 \times {^7C_4} = 56 \times 35 = 1960$

and number of ways of selecting 3 ladies and 4 gentlemen in which Mrs X and Mr. Y both are included

$= {^7C_2} \times {^6C_3} = 21 \times 20 = 420$

$\therefore$ Number of ways in which Mrs X and Mr. Y don't serve together = $1960 - 420 = 1540$

Alternatively,

Case-I When Mr. *Y* is a member, Mrs *X* is not a member

$\therefore$ Number of ways = $^6C_3 \times {^7C_3} = 20 \times 35 = 700$

Case-II When Mr. *Y* is not a member, Mrs *X* may be a member

$\therefore$ Number of ways = $^6C_4 \times {^8C_3} = 15 \times 56 = 840$

Hence, required number of ways = $700 + 840 = 1540$

Problem 38 **In a sports meet for senior citizens organised by the Rotary Club in Kolkata, 9 married couples participated in table tennis mixed double event. The number of ways in which the mixed double team can be made, so that no husband and wife play in the same set, is:** ***[IIFT-13]***

Sol. Number of ways of selecting 2 men out of 9 men = $^9C_2$

According to the question, since two men have already been selected, their wives cannot be selected. It implies that only 7 women left from which 2 can be selected in $^7C_2$ ways.

Hence, number of ways of selection of 2 men and 2 women
$= {}^9C_2 \times {}^7C_2$

Obviously, 2 men and 2 women shall form 2 teams of mixed double in 2 ways {*viz.* $(m_1 + w_1)$ against $(m_2 + w_2)$ and $(m_1 + w_2)$ against $(m_2 + w_1)$}

Hence, required number of ways $= {}^9C_2 \times {}^7C_2 \times 2 = 1512$

Problem 39 A student is required to answer 6 out of 10 questions in an examination. The questions are divided into two groups, each containing 5 questions. She is not allowed to attempt more than 4 questions from each group. The number of different ways in which the student can choose the 6 questions is ***[IIFT-12]***

Sol. The selection can be made as follows :

Case-I 2 questions from 1st group and 4 questions from 2nd group

Case-II 3 questions from 1st group and 3 questions from 2nd group

Case-III 4 questions from 1st group and 2 questions from 2nd group

The number of ways for Case-I $= {}^5C_2 \times {}^5C_4 = 10 \times 5 = 50$

Number of ways for Case-II $= {}^5C_3 \times {}^5C_3 = 10 \times 10 = 100$

and number of ways for Case-III $= {}^5C_4 \times {}^5C_2 = 5 \times 10 = 50$

Hence, required number of ways $= 50 + 100 + 50 = 200$

Problem 40 Eight points lie on the circumference of a circle. The difference between the number of triangles and the number of quadrilaterals that can be formed by connecting these points is ***[IIFT-12]***

Sol. It is obvious that 3 points are required to make one triangle and 4 points for one quadrilateral.

Hence, number of triangles $= {}^8C_3 = 56$

and number of quadrilaterals $= {}^8C_4 = 70$

$\therefore$ Required answer $= 70 - 56 = 14$

Problem 41 There are 12 towns grouped into four zones with 3 towns/zone. It is intended to connect the towns with telephone lines such that, every two towns are connected with three direct lines if they belong to the same zone and with only one direct line otherwise. How many direct telephone lines are required ? ***[CAT-03]***

Sol. Number of single direct lines among 12 towns $= {}^{12}C_2 = 66$

Further, number of direct lines in each zone in addition to those which have already been included in 66 $= (3 \times 3 - 3) = 6$

$\therefore$ number of such direct lines in 4 zones $= 6 \times 4 = 24$

Hence, required number of direct lines $= 66 + 24 = 90$

Problem 42 In a chess competition involving some boys and girls of a school, every student had to play exactly one game with every other student. It was found that in 45 games both the players were girls, and in 190 games both were boys. The number of games in which one player was a boy and the other was a girl is: ***[CAT-05]***

Sol. Let number of boys be x and that of girls be y.

As given, ${}^xC_2 = 190 \Rightarrow \dfrac{x!}{2! \times (x-2)!} = 190$

$\Rightarrow \dfrac{x \times (x-1) \times (x-2)!}{2! \times (x-2)!} = 190$

$\Rightarrow x(x-1) = 380 \Rightarrow x = 20$

Also, ${}^yC_2 = 45 \Rightarrow y = 10$

Hence, total number of players $= x + y = 20 + 10 = 30$

Now, required number of games $= {}^{20}C_1 \times {}^{10}C_1 = 20 \times 10 = 200$

Problem 43 Out of 5 women and 4 men, a committee of three members is to be formed in such a way that at least one member is a woman. In how many different ways can this be done ?

Sol. There may be three cases as under

Case-I $1W + 2M$

Case-II $2W + 1M$

Case-III $3W + 0M$

Number of ways for Case-I $= {}^5C_1 \times {}^4C_2 = 5 \times 6 = 30$

Number of ways for Case-II $= {}^5C_2 \times {}^4C_1 = 10 \times 4 = 40$

Number of ways for Case-III $= {}^5C_3 = 10$

Hence, required number of ways $= 30 + 40 + 10 = 80$

Problems 44 - 45 A committee of five members is to be formed out of 5 Professors, 6 Teachers and 3 Readers. In how many different ways can it be done if?

44. The committee should consist of 2 Professors, 2 teachers and 1 Reader.

45. The committee should include all the 3 Readers. ***[IDBI(PO)-09]***

Sol. **(Problems 44 - 45)**

44. Required number of ways $= {}^5C_2 \times {}^6C_2 \times {}^3C_1$
$= 10 \times 15 \times 3 = 450$

45. Required number of ways $= {}^3C_3 \times {}^{(6+5)}C_2 = 1 \times 55 = 55$

Problem 46 At an election, a voter may vote for any number of candidates, not greater than the number to be elected. There are 10 candidates and 4 are be elected. If a voter votes for at least one candidate, then the number of ways in which he can vote is:

Sol. Required number of ways $= {}^{10}C_1 + {}^{10}C_2 + {}^{10}C_3 + {}^{10}C_4$
$= 10 + 45 + 120 + 210 = 385$

Problem 47 From 6 different novels and 3 different dictionaries, 4 novels and 1 dictionary are to be selected and arranged in a row on the shelf so that the dictionary is always in the middle. Then, the number of such arrangements is

Sol. Number of ways of selecting 4 novels out of 6 $= {}^6C_4 = 15$

and number of ways of selecting 1 dictionary out of 3 $= {}^3C_1 = 3$

Since dictionary is to be kept in middle, the 4 novels can be arranged in 4! ways.

Hence, required number of ways $= 15 \times 3 \times 4! = 1080$

Problem 48 There are two urns. Urn A has 3 distinct red balls and Urn B has 9 distinct blue balls. From each urn two balls are taken out at random and then transferred to the other. The number of ways in which this can be done is

Sol. Required number of ways = ${}^3C_2 \times {}^9C_2 = 3 \times 36 = 108$

Problem 49 Assuming the balls to be identical except for difference in colours, the number of ways in which one or more balls can be selected from 10 white, 9 green and 7 black balls is

Sol. We know, total number of selections of some or all things i.e., atleast one out of $(p + q + r + \dots\dots)$ where p are alike of one kind, q are alike of 2nd kind, r are alike are 3rd kind and so on.
$= \{(p+1)(q+1)(r+1)\dots\} - 1$

$\therefore$ Required number of ways $= \{(10+1)(9+1)(7+1)\} - 1 = 879$

Problem 50 A square is divided into 9 identical smaller squares. Six identical balls are to be placed in these smaller squares such that each of the three rows gets at least one ball (one ball in one square only). In how many different ways can this be done ?

Sol. Number of ways of placing six balls in 9 squares = ${}^9C_6 = 84$

Number of ways of placing six balls in 6 squares (leaving top row empty) = ${}^6C_6 = 1$

Number of ways of placing six balls in 6 squares (leaving middle row empty) = ${}^6C_6 = 1$

Number of ways of placing six balls in 6 squares (leaving bottom row empty) = ${}^6C_6 = 1$

$\therefore$ Number of ways of placing 6 balls in such a way that each row gets at least one ball $= 84 - 1 - 1 - 1 = 81$

Problem 51 Each of 8 identical balls is to placed in the squares shown in the figure given below in a horizontal direction such that one horizontal row contains 6 balls and the other horizontal row contains 2 balls. In how many maximum different ways can this be done ?

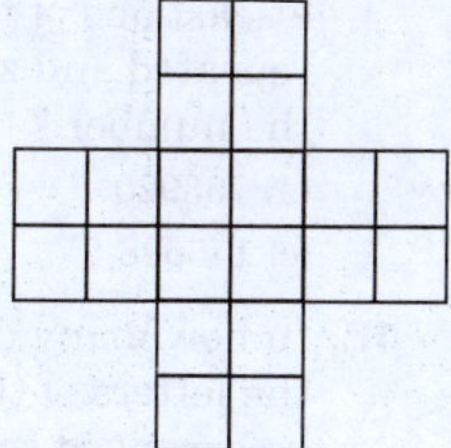

Sol. There may be four cases

(i) 6 balls in row-III and 2 balls in row -IV

(ii) 6 balls in row-IV and 2 balls in row -III

(iii) 6 balls in row-III and 2 balls in row -I or II or V or VI

(iv) 6 balls in row-IV and 2 balls in row -I or II or V or VI

Now, number of ways for Case(i) = ${}^6C_6 \times {}^6C_2 = 15$

Number of ways for Case(ii) = ${}^6C_6 \times {}^6C_2 = 15$

Number of ways for Case(iii) = $({}^6C_6 \times {}^2C_2) + ({}^6C_6 \times {}^2C_2)$
$+ ({}^6C_6 \times {}^2C_2) + ({}^6C_6 \times {}^2C_2) = 4$

Number of ways for case(iv)
$= ({}^6C_6 \times {}^2C_2) + ({}^6C_6 \times {}^2C_2) + ({}^6C_6 \times {}^2C_2) + ({}^6C_6 \times {}^2C_2) = 4$

Then, required number of ways $= 15 + 15 + 4 + 4$
$= 38$

Problem 52 Amit has five friends: 3 girls and 2 boys. Amit's wife also has 5 friends: 3 boys and 2 girls. In how many maximum number of different ways can they invite 2 boys and 2 girls such that two of them are Amit's friends and two are his wife's ? ***[IAS(Pre.) - 07]***

Sol. There may be three cases.

Case (i) 2 boys (Amit's) + 2 girls(Amit's wife's)

Case (ii) 2 boys (Amit's wife's) + 2 girls(Amit's)

Case (iii) 1 boy (Amit's) + 1 boy (Amit's wife's) + 1 girl (Amit's) + 1 girl(Amit's wife's)

Now, number of ways for case(i) = ${}^2C_2 \times {}^2C_2$
$= 1 \times 1 = 1$

Number of ways for case(ii) = ${}^3C_2 \times {}^3C_2$
$= 3 \times 3 = 9$

and Number of ways for case(iii) = ${}^2C_1 \times {}^3C_1 \times {}^3C_1 \times {}^2C_1$
$= 36$

Then, required number of ways $= 1 + 9 + 36$
$= 46$

Problem 53 Find the number of natural number solutions of $a + b + c = 12$

Sol. We may read this question as "Find the number of ways in which n identical things can be distributed to r different persons, each person gets atleast one thing."

Here, $n = 12$ and $r = 3$

Then, required number of solutions
$= {}^{n-1}C_{r-1} = {}^{12-1}C_{3-1} = {}^{11}C_2$

Problem 54 What is the number of positive integral solutions of $a + b + c = 12$ if $a \geq 2$, $b \geq 3$ and $c \geq 4$?

Sol. We may read this question as "Find the number of ways in which 12 identical things can be distributed among 3 persons a, b and c such that a gets atleast 2 things, b atleast 3 things and c atleast 4 things.

If a gets 2, b gets 3 and c gets 4 things initially, then 3 things will be left out and these 3 things can be distributed further among them in such a way that each of them can get any number of things (including 0). Obviously, this can be done in ${}^{n+r-1}C_{r-1}$ ways.

Since $n = 3$ and $r = 3$, number of ways = ${}^5C_2$

WORKOUT

1. 5 students are to be seated on 5 seats. In how many different ways can they be seated ?

(a) 120 (b) 20
(c) 60 (d) 72
(e) None of these

2. The total number of 8-digit landline telephone numbers that can be formed having atleast one of their digits repeated, is: ***[IIFT-14]***

(a) 98185600 (b) 97428800
(c) 100000000 (d) None of these

3. In a certain city, all telephone numbers have 6-digits, the first 2-digits always being 41 or 42 or 46 or 62 or 64. The number of telephone numbers having all six 6-digits distinct is ***[MAT-11]***

(a) 9200 (b) 7200
(c) 8400 (d) 1200

4. If the letters of word 'SACHIN' are arranged in all possible ways and these words are written out as in dictionary, then the word SACHIN appears at serial number ***[IAS(Pre.-07)]***

(a) 601 (b) 600
(c) 603 (d) 602

5. How many integers, greater than 999 but not greater than 4000, can be formed with the digits 0, 1, 2, 3 and 4 if repetition of digits is allowed ?

(a) 499 (b) 500
(c) 375 (d) 376
(e) 501 ***[CAT-08]***

6. 10 different letters of an alphabet are given. Words with 5 letters are formed from these given letters. Then, the number of words which have atleast one letter repeated is

(a) 69760 (b) 30240
(c) 99748 (d) None of these

7. On a shelf there are 4 books on Economics, 3 books on Management and 4 books on Statistics. In how many different ways can the books are arranged so that the books on Economics are kept together ?

(a) 967680 (b) 120960
(c) 5040 (d) 40320
(e) None of these

8. In how many different ways can the letters of the word 'ARMOUR' be arranged ?
[Syndicate Bank(PO)-10, Allahabad Bank(PO)-10, IOB(PO)-09, SBI(PO)-08]

(a) 720 (b) 300
(c) 640 (d) 350
(e) None of these

9. Three boys and three girls are to be seated around a table in a circle. Among the boys, X does not want any girl adjacent to him and the girl Y does not want any boy adjacent to her. How many such arrangements are possible ? ***[MAT-10, 09]***

(a) 8 (b) 4 (c) 6 (d) 2

10. How many ways are there to arrange the letters in the word 'GARDEN' with the vowels in alphabetical order ?

(a) 120 (b) 480
(c) 360 (d) 240

11. A new flag has to be designed with six vertical stripes using some or all of the colours yellow, green, blue and red. Then the number of ways in which this can be done such that no two adjacent stripes have the same colour is ***[CAT-04]***

(a) 12×81 (b) 16×192
(c) 20×125 (d) 24×216

12. 2 men and 1 woman board a bus in which 5 seats are vacant. One of these five seats is reserved for ladies. A woman may or may not sit on the seat reserved for ladies but a man cannot sit on the seat reserved for ladies. In how many different ways can the five seats be occupied by these three passengers? ***[IAS(Pre.)-05]***

(a) 12 (b) 24
(c) 36 (d) 48

13. A family consists of a grandfather, 5 sons and daughters and 8 grandchildren. They are to be seated in a row for dinner. The grandchildren wish to occupy the 4 seats at each end and grandfather refuses to have a grandchild on either side of him. The number of ways in which the family can be made to sit is:

(a) 11360 (b) 11520
(c) 21530 (d) None of these

14. Out of 8 consonants and 5 vowels, how many words can be made, each containing 4 consonants and 3 vowels ? ***[IIFT-13]***

(a) 700 (b) 504000
(c) 3528000 (d) 7056000

15. In how many different ways can 6 players be arranged in a line such that two of them, Ajit and Mukherjee, are never together ?

(a) 120 (b) 240
(c) 360 (d) 480

16. How many different words can be formed by jumbling the letters in the word MISSISSIPPI in which no two S are adjacent ?

(a) 7350 (b) 1050
(c) 3750 (d) 6250

17. What is the number of 6-digit telephone numbers in a city if atleast one of their digits is repeated and zero cannot initiate, the number ? ***[MAT-10]***

(a) 763920 (b) 453621
(c) 145698 (d) 781243

18. In how many different ways can the letters of the word 'DESIGN' be arranged so that the vowels are at the two ends ? ***[UBI(PO)-09]***

(a) 48 (b) 72 (c) 36 (d) 24
(e) None of these

19. Total number of 4-digit odd numbers that can be formed using 0, 1, 2, 3, 5, 7(using repetition allowed) are

(a) 216 (b) 375
(c) 400 (d) 720

20. A number lock on a suitcase has 3 wheels each labelled with 10 digits from 0 to 9. If opening of the lock is a particular sequence of 3 digits with no repeats, how many such sequences will be possible ? ***[MAT- 09]***

(a) 720 (b) 760
(c) 680 (d) 780

21. Number greater than 1000 but less than 4000 is formed using the digits 0, 1, 2, 3, 4 (repetition allowed) is:

(a) 125 (b) 105
(c) 375 (d) 625

22. How many three digit positive integers with digits x, y and z in hundred's, ten's and unit's place respectively, exist such that $x < y, z < y$ and $x \neq 0$? ***[CAT-03]***

(a) 245 (b) 285
(c) 240 (d) 320

23. The number of ways in which 6 men and 5 women can dine at a round table, if no two women are to sit together is given by :

(a) $6! \times 5!$ (b) $5! \times 4!$
(c) 30 (d) $7! \times 5!$

24. There are 6 boxes numbered 1, 2, ...6. Each box is to be filled up either with a red or green ball in such a way that at least 1 box contains a green ball and the boxes containing green balls are consecutively numbered. The total number of ways in which this can be done is ***[CAT-03, MAT-10]***

(a) 5 (b) 21
(c) 33 (d) 60

25. A string of three English letters is formed as per the following rules:

(A) The 1st letter is any vowel.
(B) The 2nd letter is m, n or p.
(C) If the 2nd letter is m then the third letter is any vowel which is different from the 1st letter.
(D) If the 2nd letter is n then the 3rd letter is e or u.
(E) if the 2nd letter is p then the 3rd letter is the same as the 1st letter.

Now answer the followings

(i) How many strings of letters can possible be formed using the above rules ?

(a) 40 (b) 45 (c) 30 (d) 35

(ii) How many strings of letters can possibly be formed using the above rules such that the third letter of the string is e? ***[CAT-03]***

(a) 8 (b) 9
(c) 10 (d) 11

26. How many numbers are there in all from 6000 to 6999 (Both 6000 and 6999 included) having at least one of their digits repeated ? ***[IAS(Pre.)-06]***

(a) 216 (b) 356
(c) 496 (d) 504

27. There are 6 tasks and 6 persons. Task 1 cannot be assigned either to person 1 or to person 2; task 2 must be assigned to either person 3 or person 4. Every person is to be assigned one task. In how many ways can the assignment be done ***[CAT-06]***

(a) 144 (b) 180
(c) 192 (d) 360
(e) 716

28. What is the number of ways in which 5 identical balls can be distributed among 10 identical boxes, if not more than one ball can go into a box? ***[MAT-10]***

(a) 204 (b) 230
(c) 250 (d) None of these

29. There are 6 persons-A, B, C, D, E and F. They are to be seated in a row such that B never sits anywhere ahead of A and C never sits anywhere ahead of B. In how many different ways can this be done? ***[IAS(Pre.)-05]***

(a) 120 (b) 240
(c) 360 (d) 480

30. In how many ways can six different rings be worn on 4 fingers of one hand ?

(a) 10 (b) 12
(c) 4^6 (d) 16

31. In a question paper, there are four multiple choice questions. Each question has five choices with only one choice for its correct answer. What is the total number of ways in which a candidate will not get all the four answers correct ? ***[IAS(Pre.)-06]***

(a) 19 (b) 120
(c) 624 (d) 1024

32. From a group of 6 men and 4 women a committee of 4 persons is to be formed. In how many different ways can it be done so that the committee has at least 2 men?

(a) 210 (b) 225
(c) 195 (d) 185
(e) None of these

33. In how many different ways can 10 identical green balls and 7 identical blue balls be arranged in a line in such a way that no two blue balls be together ?

(a) 720 (b) 2880
(c) 330 (d) 3125

34. Each of 2 women and 3 men is to occupy one chair out of 8 chairs, each of which is numbered from 1 to 8. First, women are to be occupy any two chairs from those numbered 1 to 4; and then the 3 men would occupy any three chairs out of the remaining 6 chairs. What is the maximum number of different ways in which this can be done ?

(a) 40 (b) 132
(c) 1440 (d) 3660

35. Find the number of triangles formed in an octagon?

(a) 20 (b) 80 (c) 30 (d) 56

36. If ${}^nP_4 = 56 \times {}^{n-2}C_2$, then $n = ?$

(a) 7 (b) 8
(c) 9 (d) -9

[RRB-Kolkata(P.Way)-2000]

37. Three dice (each having six faces with each face having one number from 1 to 6) are rolled. What is the number of possible outcomes such that at least one dice shows the number 2 ? ***[IAS(Pre.)-07]***

(a) 36
(b) 81
(c) 91
(d) 116

38. There are 4 candidates for the post of a lecturer in Mathematics and one is to be selected by votes of 5 men. The number of ways in which the votes can be given is:

(a) 1048
(b) 1072
(c) 1024
(d) None of the above

39. There are three prizes to be distributed among five students. If each student can get any number of prizes, then this can be done in:

(a)125 ways (b) 81 ways
(c) 75 ways (d) 27 ways

40. 8 identical gifts to be distributed among 4 boys so that each boy gets at least one medal. In how many ways can this be done?

(a) 25 (b) 84
(c) 35 (d) 21

41. There are three prizes to be distributed among five students. If no student gets more than one prize, then this can be done in

(a) 10 ways (b) 30 ways
(c) 60 ways (d) 80 ways

42. In how many ways can 7 identical erasers be distributed among 4 kids in such a way that each kid gets atleast one eraser but no body gets more than three erasers? *[CAT- 17]*

(a) 20 (b) 16
(c) 14 (d) 15

43. In a meeting of 7 members of a family the family-head does not intend to sit next to the one particular family member. In how many ways can all be seated around a circular table ?

(a) 480 (b) 720
(c) 120 (d) None of these

44. Out of 11 members of a family 4 are males and rest females. The family hired three taxis on rent to see a museum. The members have to sit in the cars in such a way that not more than 4 members are in any one car and each car has at least one male member. In how many different ways the members can travel ?

(a) 658 (b) 126
(c) 140 (d) 532
(e) None of these

45. Find the number of ways in which 4 balls can be selected out of 50 identical balls.

(a) $^{50}C_4$ (b) 4^{50}
(c) 1 (d) None of these

46. In how many ways can 5 males and 5 females be seated around a circular table so that no two males are together ?

(a) 720 (b) 2880
(c) 380 (d) 3125

47. From 5 consonants and 4 vowels, ... words can be formed using 3 consonants and 2 vowels.

(a) 6500 (b) 7200
(c) 6200 (d) 5200

48. On red flag, three white flags and two blue flags are arranged in a line such that (I) no two adjacent flags are of the same colour. (II) the flags at the two ends of the line are of different colour. In how many ways can the flags be arranged ?

(a) 7 (b) 3
(c) 2 (d) 6

49. A man has 9 friends: 4 boys and 5 girls. In how many ways can he invite them, if there have to be 3 exactly girls in the invitees ? *[CLAT- 15]*

(a) 320 (b) 160 (c) 80 (d) 200

50. A man has 5 friends and his wife has 4 friends. They want to invite either of their friends, one or more to a party. In how many ways can they do so ? *[MAT-18]*

(a) 9 (b) 18
(c) 31 (d) 46

51. How many numbers with two or more digits can be formed with the digits 1, 2, 3, 4, 5, 6, 7, 8, 9 so that in every such number, each digit is used at most once and the digits appear in the ascending order ? *[CAT-18]*

(a) 320 (b) 160
(c) 502 (d) 520

Answer Key

1. (a)	2. (a)	3. (c)	4. (a)	5. (d)	6. (a)	7. (a)	8. (e)	9. (b)	10. (c)
11. (a)	12. (c)	13. (d)	14. (c)	15. (d)	16. (a)	17. (a)	18. (a)	19. (d)	20. (a)
21. (c)	22. (c)	23. (a)	24. (b)	25. (i) d (ii) c	26. (c)	27. (a)	28. (d)	29. (a)	30. (c)
31. (c)	32. (d)	33. (c)	34. (c)	35. (d)	36. (b)	37. (c)	38. (c)	39. (a)	40. (c)
41. (a)	42. (b)	43. (a)	44. (a)	45. (c)	46. (b)	47. (b)	48. (d)	49. (b)	50. (d)
51. (c)									

Hints and SOLUTIONS

1. Required number of ways
$= 5 \times 4 \times 3 \times 2 \times 1 = 120$

2. Number of 8-digit telephone numbers when repetition of digits is allowed $= 10^8$ $\{\because n = 10$ (i.e. 0 to 9) and $r = 8\}$
and number of 8-digit telephone numbers when repetition of digits is not allowed $= {}^{10}P_8 = 1814400$
Then, required number
$= 10^8 - 1814400 = 98185600$

3. If first 2 digits of 6-digit telephone numbers are fixed, the remaining 4 digits can be taken out of remaining 8 digits i.e., in ${}^8P_4$ ways.
Since there are 5 types of telephone numbers starting with 41, 42, 46, 62 and 64, total number of telephone numbers having all 6 digits distinct
$= 5 \times {}^8P_4 = 5 \times 8 \times 7 \times 6 \times 5 = 8400$

4. Alphabetical order is A, C, H, I, N and S.
Number of words starting with A
$= 5! = 120$
$\{\because$ in case A is fixed, remaining five will be arranged in different ways$\}$
On the above line,
Number of words starting with C
$= 5! = 120$
Number of words starting with H
$= 5! = 120$
Number of words starting with I
$= 5! = 120$
Number of words starting with N
$= 5! = 120$
Hence, number of words started with either A or C or H or I or N $= 600$
Clearly, 601st word will start with S and other letters will be arranged as per dictionary. Then, 601st word will be 'SACHIN'.

5. The numbers will range from 1000 to 4000. But, for the sake of convenience, we first consider the numbers from 1000 to 3999. The desired numbers will be of 4 digits
⇒ □□□□
Clearly, 1st box(extreme left) can be filled by either 1 or 2 or 3, i.e., in 3 ways.
2nd box (i.e., Hundreds place) can be filled by either 0 or 1 or 2 or 3 or 4 i.e. in 5 ways.
3rd box(i.e., Tens place) can be filled by either 0 or 1 or 2 or 3 or 4 i.e. in 5 ways.
and 4th box(i.e., unit place) can be filled by either 0 or 1 or 2 or 3 or 4 i.e., in 5 ways.
Hence, total number of numbers
$= 3 \times 5 \times 5 \times 5 = 375$
But we have to include 4000 also, then required number of numbers
$= 375 + 1 = 376$

6. Number of permutations of 10 different things taken 5 at a time, where repetition is allowed $= 10^5$
Number of permutations of 10 different things taken 5 at a time, where repetition is not allowed
$= {}^{10}P_5$
$\therefore$ required number of ways
$= 10^5 - {}^{10}P_5$
$= 100000 - (10 \times 9 \times 8 \times 7 \times 6)$
$= 100000 - 30240 = 69760$

7. Here, total number of books
$= 1 + 3 + 4 = 8$
$\{\because$ we assumed 4 books on Economics as one book for the time being$\}$
Now, number of ways in which 8 books can be arranged $= 8! = 40320$
Obviously, 4 books on Economics can be arranged in 4! Ways.
Thus, required number of ways
$= 40320 \times 4! = 967680$

8. Here, number of letters = 6 and letter 'R' comes twice.
$\therefore$ Required number of ways
$= \dfrac{6!}{2!} = 360$

9. Obviously, $G1$ and $G2$ can interchange their positions. Also, $B1$ and $B2$ can interchange their positions.

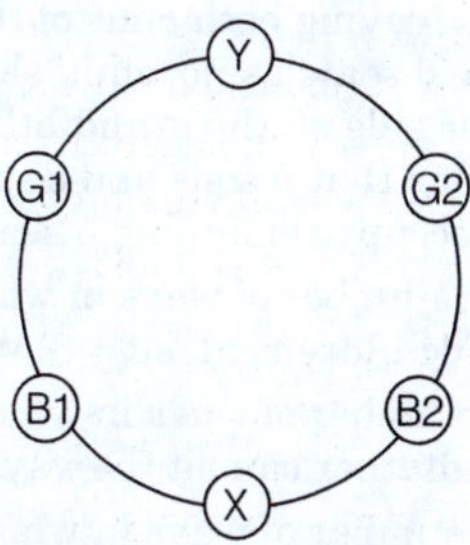

Hence, required number of arrangements $= 2 \times 2 = 4$

10. Clearly, in half of the arrangements, A will be before E and in half , E will be before A.
Hence, required number of ways
$= 6! \div 2 = 360$

11. We can choose any one of the given 4 colours for the 1st strip. Then, any one of the remaining three colours, can be chosen for the second strip as no two adjacent stripes have the same colour. Further, 3rd strip can be anyone of 4 colours except that has been used in 2nd strip. 4th strip can be anyone of 4 colours except that has been used in 3rd strip and so on.
Thus, required number of ways
$= 4 \times 3 \times 3 \times 3 \times 3 \times 3 = 12 \times 81$

12. There may be two cases.
Case-I When the lady sits on the reserved seat.
Then, 2 men can sit in ${}^4P_2$ ways.
$\therefore$ Total number of ways
$= 1 \times {}^4P_2 = 12$
Case-II When the lady does not sit on the reserved seat.
Then, the lady can sit either on the seat out of remaining 4 seats i.e., in 4 ways.
and 2 men can sit in ${}^3P_2$ ways as number of seats left for men
$= 4 - 1 = 3$
$\therefore$ Total number of ways
$= 4 \times {}^3P_2 = 24$
Thus, required number of ways
$= 12 + 24 = 36$

13. Clearly, there are (1 + 5 + 8) members in the family.

Given that 4 grand children occupied at one end and 4 occupied at another end. Then, grandfather can sit on anyone of the middle 6 seats leaving both ends of the row of these 6 seats as no child should be either side of the grandfather. It is obvious that 5 sons and daughter will occupy remaining 5 seats.

Now, number of ways in which grandchildren can sit = 8! Ways

and number of ways in which grandfather can sit = 4 ways

and number of ways in which 5 sons and daughter can sit = 5! ways

Hence, required number of ways $= 8! \times 4 \times 5! = 19353600$

14. Number of ways of selecting 4 consonants out of 8 consonants

$$= {}^8C_4 = 70$$

and Number of ways of selecting 3 vowels out of 5 vowels $= {}^5C_3 = 10$

and Number of ways of arranging the 7 letters = 7!

Then, required number of words

$$= 70 \times 10 \times 7! = 3528000$$

15. Number of ways of arranging 6 players = 6!

If we assume Ajit and Mukherjee as one player, then number of players = 5 and these 5 players can be arranged in 5! ways.

Further, number of ways in which Ajit and Mukherjee can be arranged = 2!

Hence, number of arrangements in which Ajit and Mukherjee are together = 5! × 2!

∴ Required number of ways

$$= 6! - (5! \times 2!) = 480$$

16. Total number of letters in the given word = 11

Number of $S = 4$

Then, other than S, seven letters M, I, I, I, P, P, I can be arranged in $\frac{7!}{2! \cdot 4!}$ ways.

$\{\because$ number of $P = 2$ and number of $I = 4\}$

Further, four S can be placed in 8 spaces in ${}^8C_4$ ways.

Thus, required number of ways

$$= \frac{7!}{2! \cdot 4!} \times {}^8C_4$$

$$= 7 \times 5 \times 3 \times 70 = 7350$$

17. Total number of 6-digit (telephone)numbers

= 999999 − 99999

= 900000, not initiated by zero

Number of 6-digit telephone numbers in which zero can not initiate the number and no digit is repeated = numbers formed by taking 6 digits from 0 to 9 − numbers formed fixing 0 in beginning and taking 5 from 1 to 9

$$= {}^{10}P_6 - {}^9P_5 = 10 \times 9 \times 8 \times 7 \times 6 \times 5 - 9 \times 8 \times 7 \times 6 \times 5$$

= 136080

Then, required number

$$= 900000 - 136080 = 763920$$

18. There are two vowels (i.e. E and I) which are to be placed at two ends. Then, remaining 4 consonants can be arranged in 4! ways. Further, two vowels can be arranged in 2! ways.

Hence, required number of ways

$$= 4! \times 2! = 48$$

19. The 4-digit odd numbers ⇒ □□□□

Clearly, 1st box(extreme left) can be filled by either 1 or 2 or 3 or 5 or 7 i.e., in 5 ways.

2nd box (i.e., Hundreds place) can be filled by either 0 or 1 or 2 or 3 or 5 or 7 i.e. in 6 ways.

3rd box(i.e., Tens place) can be filled by either 0 or 1 or 2 or 3 or 5 or 7 i.e. in 6 ways.

and 4th box(i.e., unit place) can be filled by either 1 or 3 or 5 or 7 as desired numbers be odd i.e. in 4 ways.

Hence, required number of odd numbers = 5 × 6 × 6 × 4 = 720

20. Required number of sequences

$$= {}^{10}P_3 = 10 \times 9 \times 8 = 720$$

21. The desired numbers will be of 4 digits ⇒ □□□□

Clearly, 1st box(extreme left) can be filled by either 1 or 2 or 3 i.e., in 3 ways.

2nd box (i.e. Hundreds place) can be filled by either 0 or 1 or 2 or 3 or 4 i.e. in 5 ways.

3rd box(i.e. Tens place) can be filled by either 0 or 1 or 2 or 3 or 4 i.e. in 5 ways.

and 4th box(i.e. unit place) can be filled by either 0 or 1 or 2 or 3 or 4 i.e. in 5 ways.

Hence, required number of numbers

$$= 3 \times 5 \times 5 \times 5 = 375$$

22. Clearly, possible value of $x = 1, 2, 3, \ldots\ldots 8$ $\{\because x \neq 0\}$

Possible value of $y = 2, 3, 4, \ldots 9$ $\{\because x < y\}$

Possible value of $z = 0, 1, 2, \ldots 8$ $\{\because z < y\}$

Now, observe the following:

x	y	z	Number of numbers
1	2	0, 1	1 × 1 × 2 = 2
1, 2	3	0, 1, 2	2 × 1 × 3 = 6
1, 2, 3	4	0, 1, 2, 3	3 × 1 × 4 = 12
1, 2, 3, 4	5	0, 1, 2, 3, 4	4 × 1 × 5 = 20
1, 2, 3, 4, 5	6	0, 1, 2, 3, 4, 5	5 × 1 × 6 = 30
1, 2, 3, 4, 5, 6	7	0, 1, 2, 3, 4, 5, 6	6 × 1 × 7 = 42
1, 2, 3, 4, 5, 6, 7	8	0, 1, 2, 3, 4, 5, 6, 7	7 × 1 × 8 = 56
1, 2, 3, 4, 5, 6, 7, 8	9	0, 1, 2, 3, 4, 5, 6, 7, 8	8 × 1 × 9 = 72
		Total	240

23. Number of circular permutations $= (n-1)!$

Hence, number of ways of arranging 6 men around a table = (6 − 1)! = 5!

Considering the condition that no 2 women are to sit together, 5 women can be adjusted against 6 blank places, shown in the figure in 6! ways.

Hence, required number of ways = 5! × 6!

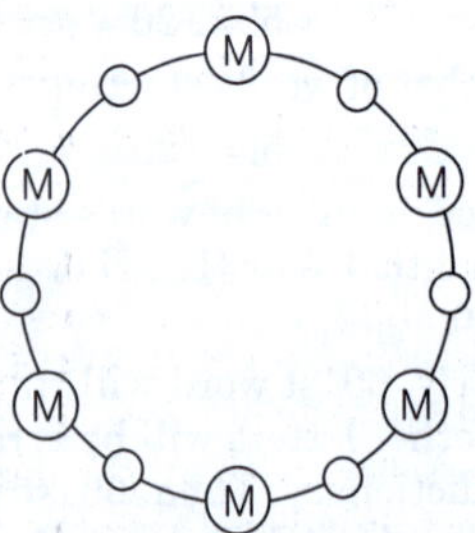

24. Given, at least one box contains a green ball.

Now, number of ways of filling 1 green ball = 6 {1, 2, 3, 4, 5, 6}

In case of 2 green balls, number of ways of filling = 5

{(1, 2), (2, 3), (3, 4), (4, 5), (5, 6)}

In case of 3 green balls, number of ways of filling = 4

{(1, 2, 3), (2, 3, 4), (3, 4, 5), (4, 5, 6)}
In case of 4 green balls, number of ways of filling = 3
{(1, 2, 3, 4), (2, 3, 4, 5), (3, 4, 5, 6)}
In case of 5 green balls, number of ways of filling = 2
{(1, 2, 3, 4, 5), (2, 3, 4, 5,6)}
In case of 6 green balls, number of ways of filling = 1 {(1, 2, 3, 4, 5, 6)}

Thus, required number of ways
= 6 + 5 + 4 + 3 + 2 + 1 = 21

25 (i).

1st letter	2nd letter	3rd letter	Number of ways
a, e, i, o, u	*m*	Any vowel except that has been taken as 1st letter	5 × 1 × 4 = 20
a, e, i, o, u	*n*	*e* or *u*	5 × 1 × 2 = 10
a, e, i, o, u	*p*	Same as 1st letter, so no further need to select the 3rd once 1st is selected.	5 × 1 × 1 = 5

Thus, required number of possible strings of letters = 20 + 10 + 5 = 35

25. (ii).

1st letter	2nd letter	3rd letter	Number of ways
a, i, o, u	*m*	*e*	4 × 1 × 1 = 4
a, e, i, o, u	*n*	*e*	5 × 1 × 1 = 5
e	*p*	*e*	1 × 1 × 1 = 1

Thus, required number of ways
= 4 + 5 + 1 = 10

26. Since the first digit is 6, the remaining three digits can be selected out of 0, 1, 2, 3, 4, 5, 7, 8 and 9 in ${}^9C_3$ ways in order to form the numbers with all the digits distinct. Further, these last three digits can be arranged in 3! ways.

Hence, total number of numbers in which all digits are distinct

$= {}^9C_3 \times 3! = 84 \times 6 = 504$

OR

1st digit is fixed as 6. 2nd digit can be either of remaining 9 digits, 3rd digit can be either of remaining 8 digits and 4th i.e., last digit can be either of remaining 7 digits.

Hence, total number of numbers without repetition of digits

$= 1 \times 9 \times 8 \times 7 = 504$

∴ Required number of ways
= {total number of numbers from 6000 to 6999 } – {total number of numbers in which all digits are distinct}
= 1000 – 504 = 496

27. Given, task 2 is to be assigned to either person 3 or person 4. Thus, it can be done in 2 ways.

According to the question, Task 1 cannot be assigned either to person 1 or to person 2. It implies that Task 1 can be assigned anyone except person1, person 2 and the person who has been assigned task 2 already. i.e., Task 1 can be assigned in 3 ways.

Further, remaining 4 tasks can be assigned to remaining 4 persons in 4! ways.

Hence, required number of ways
$= 2 \times 3 \times 4! = 144$

28. Required number of ways
$= {}^{10}C_5 = 252$

29. Obviously, the order of sitting of A, B and $C \Rightarrow C \dots B \dots A$

Number of ways of arrangements of A, B and C assuming D, E and F are seated at fixed seats for time being = 3! = 6

These 6 arrangements will be

(i) *ABCDEF* (ii) *ACBDEF*
(iii) *BACDEF* (iv) *BCADEF*
(v) *CABDEF* (vi) *CBADEF*

Only arrangement (vi) is as per given condition. It may be construed that 1/6 of the total arrangements are as per conditions/order given in the question.

Clearly, total number of arrangements of 6 persons = 6! = 720

∴ required number of ways = 1/6 of 720 = 120

30. Obviously, each ring can go to one of the 4 fingers and this can be done in 4 ways.

Hence, required number of ways
$= 4 \times 4 \times 4 \times 4 \times 4 \times 4 = 4^6$

31. Since each question can be answered in 5 ways, the question paper can be answered in $5 \times 5 \times 5 \times 5$ i.e. in 625 ways.

Number of ways in which a candidate will get all the four answers correct = 1

∴ Total number of ways in which a candidate will not get all the four answers correct = 625 – 1 = 624

32. There may be three cases as under:

Case-I $2M + 2W$
Case-II $3M + 1W$
Case-III $4M + 0W$

Number of ways for Case-I
$= {}^6C_2 \times {}^4C_2 = 15 \times 6 = 90$

Number of ways for Case-II
$= {}^6C_3 \times {}^4C_1 = 20 \times 4 = 80$

Number of ways for Case-III
$= {}^6C_4 = 15$

Hence, required number of ways
= 90 + 80 + 15 = 185

33. Clearly, 11 places/slots will be created if we place 10 green balls in a line { i.e. ☐ G ☐ G ☐ ....G ☐}.
We have to put 7 blue balls against these 11 places/slots created. Hence, required number of ways

$${}^{11}C_7 = 330$$

(Here, order is not important as balls are identical).

34. Number of ways for women = ${}^4P_2$
Number of ways for men = ${}^6P_3$
Then, required number of ways
$= {}^4P_2 \times {}^6P_3 = 1440$

35. Since a triangle can be formed by selecting any three of the eight vertices of the octagon, required number of ways/triangles = ${}^8C_3 = 56$

36. Given, ${}^nP_4 = 56 \times {}^{n-2}P_2$

$\Rightarrow \quad \dfrac{n!}{(n-4)!} = 56 \times \dfrac{(n-2)!}{(n-2-2)!}$

$\Rightarrow n \times (n-1) \times (n-2)! = 56 \times (n-2)!$

$\Rightarrow \quad n \times (n-1) = 56$

$\Rightarrow \quad n \times (n-1) = 8 \times (8-1)$

Hence, $n = 8$

37. When three dice are rolled, any figure out of 1 to 6 may be appeared on any dice. Hence, total possible ways = 6 × 6 × 6 = 216

Number of ways in which '2' does not appear on any dice = 5 × 5 × 5
= 125

Then, number of possible outcomes in which atleast one dice shows '2'
= 216 – 125
= 91

38. Obviously, each man can vote for one of the 4 candidates and this can be done in 4 ways.

Hence, required number of ways
= 4 × 4 × 4 × 4 × 4 = 1024

39. Required number of ways = $5^3 = 125$

40. We know, number of ways in which n identical things can be distributed to r different persons, each person gets at least one thing $= {}^{n-1}C_{r-1}$

Hence, required number of ways $= {}^{8-1}C_{4-1} = {}^7C_3 = 35$

41. The required number of ways $= {}^5C_3 = 10$

42. We know, number of ways in which n identical things can be distributed to r different persons, each person gets at least one thing $= {}^{n-1}C_{r-1}$

Then, number of ways in which each person can get atleast one eraser $= {}^{7-1}C_{4-1} = {}^6C_3 = 20$

Further, number of ways in which any one can get more than 3 erasers = 4 {i.e. (4, 1, 1, 1), (1, 4, 1, 1), (1, 1, 4, 1) and (1, 1, 1, 4)}.

Hence, required number of ways = 20 – 4 = 16

43. Number of arrangements in case of circular permutation $= (7-1)! = 6!$

If we assume the family-head and that particular family member as one member for time being, then number of arrangements $= (6-1)! = 5!$

Further, these two members can be arranged in 2! Ways.

Hence, number of arrangements in which the family-head and that particular family member be together $= 5! \times 2!$

$\therefore$ required number of ways $= 6! - (5! \times 2!) = 5! \times 4 = 480$

44. (*a*)

	Number of passengers	Possible Combinations	Number of ways
Taxi-I	4	$1m + 3f$ or $2m + 2f$	$({}^4C_1 \times {}^7C_3) + ({}^4C_2 \times {}^7C_2) = (4 \cdot 35) + (6 \cdot 21) = 266$
Tax-II	4	$2m + 2f$ or $1m + 3f$	$({}^4C_2 \times {}^7C_2) + ({}^4C_1 \times {}^7C_3) = (6 \cdot 21) + (4 \cdot 35) = 266$
Tax-III	3	$1m + 2f$ or $2m + 1f$	$({}^4C_1 \times {}^7C_2 + ({}^4C_2 \times {}^7C_1) = (4 \cdot 21) + (6 \cdot 7) = 126$
		Required number of ways	266 + 266 + 126 = 658

45. Since all the balls are identical, number of ways to select 4 balls = 1

46. Number of ways of arranging 5 males around a table $= (5-1)! = 4!$

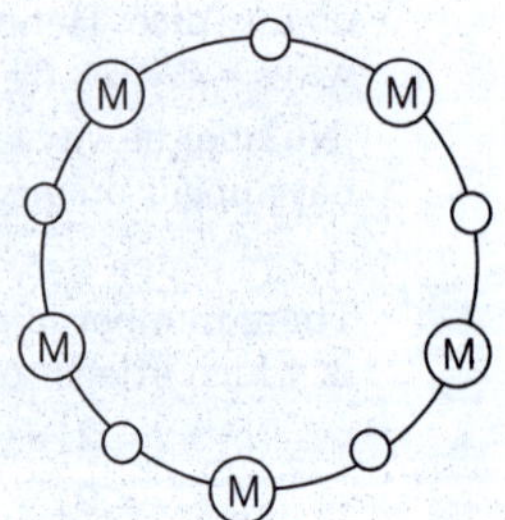

Considering the condition that no two males are to sit together, 5 females can be adjusted against 5 blank places, shown in the figure, in 5! ways.

Hence, required number of ways $= 4! \times 5!$
$= 2880$

47. Number of selection of 3 consonants out of 5 $= {}^5C_3$ and number of selection of 2 vowels out of 4 $= {}^4C_2$

$\therefore$ Number of selections each having 5 letters $= {}^5C_3 \times {}^4C_2$

Further, each selection can be arranged in 5! Ways.

$\therefore$ Required number of words $= {}^5C_3 \times {}^4C_2 \times 5!$
$= 7200$

48. There may be two cases:(a) □ W □ W □ W (b) W □ W □ W □, where W connotes white flag.

In each case, □ can be occupied by two blue and one red flags.

Clearly, two blue and one red flags can occupy three □ in $\frac{3!}{2!}$ i.e. 3 ways.

Hence, total number of ways (i.e. for both cases) = 3 + 3 = 6 ways.

49. Number of ways for inviting 3 girls out of 5 girls $= {}^5C_3 = 10$

He is free to invite either 0 or 1 or 2 or 3 or 4 boys out of 4 boys. Then, total number of ways for inviting boy(s) $= 2^4 = 16$

{$\because$ number of combinations when zero or more things out of n different things are taken $= 2^n$}

Hence, required number of ways $= 10 \times 16 = 160$

50. Total number of ways the man can invite his friends
$= {}^5C_1 + {}^5C_2 + {}^5C_3 + {}^5C_4 + {}^5C_5$

We know, ${}^nC_1 + {}^nC_2 + {}^nC_3 + \ldots + {}^nC_n = 2^n - 1$

Hence, total number of ways the man can invite
$= 2^5 - 1 = 31$

Similarly, total number of ways his wife can invite her friends $= 2^4 - 1 = 15$

Then, required number of ways = 31 + 15
= 46

51. Number of n-digit numbers using digits out of given 9 digits $= {}^9P_n$. It is true that only one number out of ${}^9P_n$ numbers will be in the ascending order and we require the same. Hence, we have to take only one number from ${}^9P_n$, $n = 2, 3, \ldots, 9$. It implies that we are talking about the concept of Combination.

Now, required number of numbers
$= {}^9C_2 + {}^9C_3 + {}^9C_4 + \ldots + {}^9C_9$

We know, ${}^nC_1 + {}^nC_2 + {}^nC_3 + \ldots + {}^nC_n = 2^n - 1$

$\therefore$ Required answer $= (2^9 - 1) - {}^9C_1$
$= 512 - 1 - 9$
$= 502$

17

Probability

Learning OBJECTIVES

After completing this chapter, aspirants will be able to

- Understand the meaning of Probability and importance of this chapter;
- Know relevant terms related to concept of Probability;
- Understand the relation between Event and Sample Space;
- Differentiate between methods meant for conjunction 'OR' and 'AND';
- Comprehend the concept of mutually exclusive events and non-mutually exclusive events;
- Understand the philosophy of the division of this chapter in various cases;
- Analyse the trend of the questions being asked in various competitive examinations;
- Solve the questions in moments;
- Learn new concepts/fundamentals;
- Apply the pragmatic approaches in lieu of academic approaches; and
- Ponder on the alternative methods.

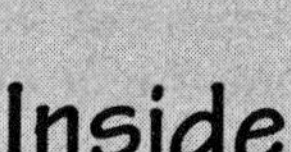

Inside

Introduction

This chapter plays a vital role so as to secure marks in competitive examinations like CAT, MAT, IIFT, Bank-PO, SSC, etc. It is as logical in nature as the topic 'Permutation and Combination' and in order to command over this chapter, the aspirants should have been well versed with the concepts of 'Permutation and Combination'.

Probability is the measure of the likeliness that an event will occur. The value of probability lies between 0 and 1 where 0 indicates impossibility and 1 connotes certainty. If the value of probability is less, it is least chance that the event will occur and *vice-versa.* The concept of the probability is applied in various domain like Physics, Chemistry, IT, economics, Statistics, Management, weather forecasting, etc.

Some Relevant Terms

Experiment : An operation which can produce some well defined outcomes.

Random Experiment : An experiment in which all possible outcomes are known and the exact output cannot be predicted in advance, is called random experiment.

Example Tossing a fair coin, rolling of an unbiased dice etc. Most of the questions are based on the random experiment.

Sample Space The set of all possible outcomes in a random experiment is called a Sample Space(S).

Event Any subset of a sample space is called an Event(E).

Probability of occurrence of an Event [$P(E)$]

$$= \frac{\text{Number of ways in which desired event occurs}}{\text{Total number of possible outcomes}}$$

Hence, $P(E) = \frac{n(E)}{n(S)}$

Understanding of Basic Concepts Through Examples

(i) **In a single throw of a die, what is the probability of getting 4?**

Here, getting 4 is an event and it will occur only when 4 will appear on the top face of the die. Clearly, number of ways in which desired event occurs i.e. $n(E) = 1$.

It is also noted that in a throw of a die, we may get either 1 or 2 or 3 or 4 or 5 or 6 i.e., total number of possible outcomes, $n(S) = 6$

Now, probability of getting $4 = \frac{n(E)}{n(S)} = \frac{1}{6}$

(ii) **In a single throw of a die, what is the probability of getting a number greater than 3?**

Getting either 4 or 5 or 6 is the event and it will occur when either 4 or 5 or 6 appears on the top face of the die. It implies that the desired event occurs when any one of these three figures comes on the top face. Hence, number of ways in which desired event occurs i.e. $n(E) = 3$. We may also write $n(E) = {}^3C_1$

{Recall, ${}^nC_r = \frac{n!}{r!(n-r)!}$}

In other words, $n(E) =$

Total availability of desired figure/item $(s)_{C_{\text{number of figure/item(s) desired in actual}}}$

$= (3 \text{ i. e. } 4, 5, 6)_{C_{1 (\text{i.e., either 4 or 5 or 6})}}$

Further, number of ways in which any one out of six figures will appear on the top face of the die $= {}^6C_1$ i.e. total number of possible outcomes, $n(S)$.

Then, $P(E) = \frac{n(E)}{n(S)} = \frac{{}^3C_1}{{}^6C_1} = \frac{3}{6} = \frac{1}{2}$

Remarks

We may write $n(E) = {}^1C_1$ and $n(S) = {}^6C_1$ for the 1st question

Alternatively,

Probability of getting 4 = 1/6

Probability of getting 5 = 1/6

Probability of getting 6 = 1/6

Probability of getting either 4 or 5 or 6 $= 1/6 + 1/6 + 1/6 = 1/2$

Remarks

in case of 'OR' we simply add the values of probabilities of the relevant events

(iii) **In a throw of a coin, find the probability of getting a tail.**

Number of ways in which desired event (i.e., getting a tail) occurs i.e. $n(E) = {}^1C_1$

In other words, $n(E)$

= total availability of desired items

$(S)_{C_{\text{number of items desired in actual}}} = {}^1C_1$

and Number of ways of getting anyone out of head and tail i.e. $n(S) = {}^2C_1$

Then, $P(E) = \frac{n(E)}{n(S)} = \frac{{}^1C_1}{{}^2C_1} = \frac{1}{2}$

{You can just write 1/2 as value of probability ignoring the Combination-format. But my intention is to develop your understanding which will help you in solving the complicated problems}

(iv) **A bag contains 3 white and 2 black balls. Find the probability of drawing a white ball at random.**

Getting a white ball is the event and it will occur when either 1st white ball or 2nd white ball or 3rd white ball is drawn. Hence, $n(E) = {}^3C_1$

In other words, $n(E)$

= total availability of desired items

$(S)_{C_{\text{number of items desired in actual}}} = {}^3C_1$

Further, number of ways to select one out of all five balls $= {}^5C_1$ i.e. total number of possible outcomes, $n(S)$.

Then, $P(E) = \frac{n(E)}{n(S)} = \frac{{}^3C_1}{{}^5C_1} = \frac{3}{5}$

Concept of Sample Space in Diverse Cases

Case A unbiased coin(s)

	Sample Space [$n(S)$] i.e. total number of possible outcomes	Possible outcomes
When a coin is tossed	${}^2C_1 = 2$	$[H, T]$
When two coins are tossed simultaneously	$({}^2C_1)^2 = 2^2 = 4$	$[(H,H), (H,T)$ $(T,H), (T,T)]$
When n coins are tossed simultaneously	$({}^2C_1)^n = 2^n$	Depends upon the value of n.

***Case* B** unbiased die/dice		
When a die is thrown	${}^6C_1 = 6$	{1, 2, 3, 4, 5, 6}
When two dice are thrown simultaneously	$({}^6C_1)^2 = 6^2 = 36$	[(1, 1), (1, 2), (1, 3), (1, 4), (1, 5), (1, 6), (2, 1), (2, 2), ...]
When n dice are thrown simultaneously	$({}^6C_1)^n = 6^n$	Depends upon the value of n.
***Case* C** 52-card deck		
When one card is drawn	${}^{52}C_1 = 52$	All 52 cards
When two cards are drawn	${}^{52}C_2 = 26 \times 51$	Different combinations such as (ace, queen), (ace, king), ...
When n cards are drawn	${}^{52}C_n$	Depends upon value of n.

Details of 52-card deck

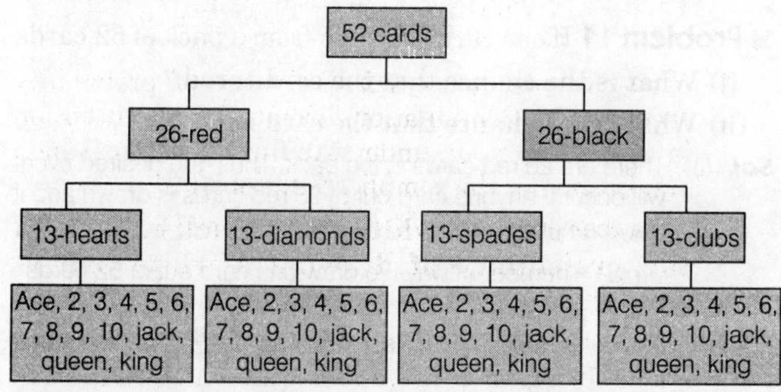

Hearts, Diamonds, Spades and Clubs are known as suits. Hence, a deck has 4 suits and each suit has 13 cards (i.e. ace, 2, 3, 4, 5, 6, 7, 8, 9, 10, jack, queen and king). Aces, kings, queens and jacks are called face cards, hence there are 16 face cards in the deck of 52 playing cards.

Problems Based Directly on the Basic Concept of Probability $\left\{P(E) = \dfrac{n(E)}{n(S)}\right\}$

Problem 1 Three unbiased coins are tossed. What is the probability of getting two tails ?

Sol. There are three heads and three tails. We can take two tails out of three available tails in ${}^3C_2$ ways. Hence, $n(E) = {}^3C_2 = 3$

Further, total number of possible outcomes $= n(S) = 2^3 = 8$

Now, $P(E) = \dfrac{n(E)}{n(S)} = \dfrac{3}{8}$

Alternatively,

Events = [(T, T, H), (T, H, T), (H, T, T)]

Hence, $n(E) = 3$

Now, $P(E) = \dfrac{n(E)}{n(S)} = \dfrac{3}{8}$

Problem 2 Three unbiased coins are tossed. What is the probability of getting at least two heads ?

Sol. Probability of getting 2 heads $= \dfrac{{}^3C_2}{2^3} = \dfrac{3}{8}$

and Probability of getting 3 heads $= \dfrac{{}^3C_3}{2^3} = \dfrac{1}{8}$

$\therefore$ Probability of getting at least two heads $= \dfrac{3}{8} + \dfrac{1}{8} = \dfrac{1}{2}$

Alternatively,

Here, $n(E) = 4$ [(T, H, H), (H, T, H), (H, H, T) (H, H, H)]

Then, $P(E) = \dfrac{n(E)}{n(S)} = \dfrac{4}{8} = \dfrac{1}{2}$

Problem 3 Three unbiased coins are tossed. What is the probability of getting at most two tails?

Sol. Probability of getting all three tails $= \dfrac{{}^3C_3}{2^3} = \dfrac{1}{8}$

Then, probability of getting at most two tails $= 1 - \dfrac{1}{8} = \dfrac{7}{8}$

Alternatively I,

$n(E) = 7$ [(T, H, H), (H, T, H), (H, H, T), (H, H, H), (H, T, T), (T, T, H), (T, H, T)]

and $n(S) = 2^3 = 8$

Then, probability of getting at most two tails $= \dfrac{n(E)}{n(S)} = \dfrac{7}{8}$

Alternatively II,

$(E) = {}^3C_0 + {}^3C_1 + {}^3C_2 = 1 + 3 + 3 = 7$

and $n(S) = 2^3 = 8$

Then, probability of getting at most two tails $= \dfrac{n(E)}{n(S)} = \dfrac{7}{8}$

Problem 4 What is the probability of getting a sum of 9 from two throws of a die ? ***[MBA-02]***

Sol. Obviously, the desired event will occur if any one set out of [(3, 6), (4, 5), (5, 4), (6, 3)] appears and it may occur in ${}^4C_1$ ways.

Hence, $n(E) = {}^4C_1$

Then, probability of getting a sum of 9 $= \dfrac{n(E)}{n(S)} = \dfrac{{}^4C_1}{6^2} = \dfrac{4}{36} = \dfrac{1}{9}$

Problem 5 Two dice are thrown simultaneously. What is the probability of getting two numbers whose product is even ?

Sol. The desired event will occur if anyone set out of [(1, 2), (1, 4), (1, 6), (2, 1), (2, 2), (2, 3), (2, 4), (2, 5), (2, 6), (3, 2), (3, 4), (3, 6), (4, 1), (4, 2), (4, 3), (4, 4), (4, 5), (4, 6), (5, 2), (5, 4), (5, 6), (6, 1), (6, 2), (6, 3), (6, 4), (6, 5), (6, 6)] appears after both dice are thrown simultaneously and obviously it may occur in ${}^{27}C_1$ ways.

Hence, $n(E) = {}^{27}C_1 = 27$

and $n(S) = 6^2 = 36$

$\therefore$ $P(E) = \dfrac{27}{36} = \dfrac{3}{4}$

Alternatively

You may count the number of products which are not even. Number of such products is 9.

Then, $n(E) = 36 - 9 = 27$

Hence, $P(E) = \frac{27}{36} = \frac{3}{4}$

Problem 6 **Four dice are thrown simultaneously. What is the probability of getting the same faces ?**

Sol. The desired event will occur if anyone set out of [(1, 1, 1, 1), (2, 2, 2, 2), (3, 3, 3, 3), (4, 4, 4, 4), (5, 5, 5, 5), (6, 6, 6, 6)] appears and obviously, it may occur in $^6C_1$ ways.

Hence, $n(E) = {^6C_1} = 6$

and $n(S) = 6^4 = 1296$

$\therefore$ $P(E) = \frac{6}{1296} = \frac{1}{216}$

Problem 7 **Two dice are thrown simultaneously. What is the probability of getting the numbers whose difference is 2 ?**

Sol. The desired event will occur if anyone set out of [(1, 3), (2, 4), (3, 1), (3, 5), (4, 2), (4, 6), (5, 3), (6, 4)] appears and obviously, it may occur in $^8C_1$ ways.

Hence, $n(E) = {^8C_1} = 8$ and $n(S) = 6^2 = 36$

$\therefore$ $P(E) = \frac{8}{36} = \frac{2}{9}$

Problem 8 **In a class there are 50 boys and 40 girls. What is the probability that the topper be a boy?**

Sol. The desired event will occur if anyone out of 50 boys becomes topper and obviously, it may occur in $^{50}C_1$ways.

Hence, $n(E) = {^{50}C_1} = 50$

and $n(S) = 50 + 40 = 90$

$\therefore$ $P(E) = \frac{50}{90} = \frac{5}{9}$

Problem 9 **The letters of the word 'NUMBER' are arranged at random. What is the probability for each case?**

(i) there are exactly 4 letters between U and E

(ii) all vowels are together

(iii) all vowels are not together

(iv) all consonants are together

(v) all consonants are not together

Sol. (i) $U - - - - E$ or $E - - - - U$

Obviously, 4 letters between them will be arranged in 4! ways. Thus, $n(E) = 2 \times 4!$

and $n(S) = 6!$ {$\because$ 6 letters can be arranged in 6! ways}

$\therefore$ $P(E) = \frac{2 \times 4!}{6!} = \frac{1}{15}$

(ii) There are two vowels which may be termed as a single letter, then, number of letters = 5 and it can be arranged in 5! ways. It is also fact that the two vowels kept together interchange their positions.

Thus, $n(E) = 5! \times 2$

and $n(S) = 6!$

$\therefore$ $P(E) = \frac{5! \times 2}{6!} = \frac{1}{3}$

(iii) Probability that all vowels are not together
$= 1 -$ Probability of being all vowels together
$= 1 - 1/3 = 2/3$

(iv) On the same line as done for (ii), $n(E) = 3! \times 4!$ and $n(S) = 6!$

$\therefore$ $P(E) = \frac{3! \times 4!}{6!} = \frac{1}{5}$

(v) Probability of being all consonant not together
$= 1 - 1/5 = 4/5$

Problem 10 **What is the chance that a leap year, selected at random, will contain fifty-three Sundays ?**

Sol. Leap year $\Rightarrow$ 366 days = 52 weeks + 2 days
= 52 Sundays + (Sat, Sun) or (Sun, Mon) or (Mon, Tues) or (Tues, Wed) or (Wed, Thu) or (Thu, Fri) or (Fri, Sat)

Clearly, probability of being Sunday in 2 days $= \frac{n(E)}{n(S)} = 2/7$

Problem 11 **If one card is drawn from a pack of 52 cards**

(i) What is the chance that the card is red ?

(ii) What is the chance that the card is a king?

Sol. (i) There are 26 red cards in the pack. Thus, the desired event will occur if anyone card out of 26 red cards is drawn and it may be happened in $^{26}C_1$ ways. Hence, $n(E) = {^{26}C_1}$

$n(S)$ = number of ways to draw one card out of 52 cards $= {^{52}C_1}$

$\therefore$ $P(E) = \frac{^{26}C_1}{^{52}C_1} = \frac{1}{2}$

(ii) There are 4 kings.

Thus, $n(E) = {^4C_1}$

$\therefore$ $P(E) = \frac{^4C_1}{^{52}C_1} = \frac{1}{13}$

Problem 12 **If three cards are drawn from a pack of 52 cards, what is the chance that all are queens ?**

Sol. There are 4 queens and the desire event will occur when 3 out of 4 queens will be drawn and it may be happened in $^4C_3$ ways. Hence, $n(E) = {^4C_3}$

and $n(S)$ = number of ways to draw 3 cards out of 52 cards $= {^{52}C_3}$

$\therefore$ $P(E) = \frac{^4C_3}{^{52}C_3} = \frac{4 \times 3 \times 2}{52 \times 51 \times 50} = \frac{1}{5525}$

Problem 13 **M and N stand in a ring with 8 other persons. If the arrangement of the 10 persons is at random, find the chance that there are exactly 3 persons between M and N.** ***[MAT-10]***

Sol. Obviously, there will be two situations if M is at any point:

(a) N will be 3 points ahead(say, clockwise)

(b) N will be 3 points behind(say, anti-clockwise)

Obviously, the desire event will occur in 2 ways as there are two situations. We may say, $n(E) = {}^2C_1$ {taking one out of two situations}.

and $n(S) = 9$

i.e., {$M\,N$, $M \cdot N$, $M \cdot\cdot N$, $M \cdots N$, $M \cdots\cdot N$, $M \cdots\cdots N$, $M \cdots\cdots\cdot N$, $M \cdots\cdots\cdot\cdot N$, $M \cdots\cdots\cdots N$

⇒ One point indicates one person, two indicate two persons and so on}

∴ $P(E) = \dfrac{2}{9}$

Alternatively,

Given there are 3 persons between M and N. Further, M and N can be arranged in 2! ways as they can interchange their positions.

Also, remaining 8 persons (including 3 persons between M and N) can be arranged in 8! ways.

Hence, number of ways of arranging 3 person between M and $N = 2! \times 8!$

{Alternatively, number of ways of arranging 3 person between M and $N = {}^8C_3 \times 3! \times 5! \times 2! = 8! \times 2!$}

and total number of arrangements = $(10 - 1)! = 9!$ {Recall the concept of circular permutation}

∴ Required probability $= \dfrac{2! \times 8!}{9!} = \dfrac{2}{9}$

Problem 14 9 identical balls are put in three different boxes. What is probability of putting balls in the boxes in equal numbers ?

Sol. We know, number of ways in which n identical things can be distributed to r different persons, each person can get any number of things including zero $= {}^{n+r-1}C_{r-1}$

Here, $n(S) = {}^{9+3-1}C_{3-1} = {}^{11}C_2$

Further, $n(E) = 1$ {∵ each box is to have 3 balls, which can be done in 1 way}

∴ $P(E) = \dfrac{1}{{}^{11}C_2} = \dfrac{1}{55}$

Problem 15 There are six letters and each of them corresponds an envelope. They are placed in the envelopes at random. What is probability of getting all the letters in the right envelopes?

Sol. Event ⇒ All the letters in the right envelopes.

Then, $n(E) = 1$ {∵ there is only one way to place all letters in their corresponding envelopes}.

and $n(s) = {}^6P_6 = 6!$

∴ $P(E) = \dfrac{1}{6!} = \dfrac{1}{720}$

Problem 16 For the above question, what is the probability of not having all letters in right envelopes ?

Sol. $P(E) = 1 - \dfrac{1}{6!} = 1 - \dfrac{1}{720} = \dfrac{719}{720}$

Problem 17 A natural number is selected at random and then divided by 8. What is the probability that the remainder is an even number ?

Sol. Any natural number when divided by 8 may yield remainders as 0, 1, 2, 3, 4, 5, 6 and 7. Out of these 8 figures, only three are even.

Hence, $n(E) = {}^3C_1$ and $n(S) = {}^8C_1$

Then, $P(E) = \dfrac{{}^3C_1}{{}^8C_1} = \dfrac{3}{8}$

Problem 18 The digits 1, 2, 3, 4, 5 and 6 are written at random in order to form a six digit number. What is the probability that it is divisible by 4?

Sol. The number having last two digits as 12, 16, 24, 32, 36, 52, 56, 64 will only be divisible by 4. It means, there are 8 ways of choosing such last two digits. Further, corresponding to each of these ways, the remaining 4 digits can be arranged in ${}^4P_4$ ways.

Hence, number of six-digit numbers divisible by 4

$= 8 \times {}^4P_4 = 8 \times 4!$

Then, $n(E) = {}^nC_r = {}^{8 \times 4!}C_1 = 8 \times 4!$

and $n(S)$ = number of numbers formed with all six digits = 6!

∴ Required probability $= \dfrac{8 \times 4!}{6!} = \dfrac{4}{15}$

Problem 19 In a box, there are 3 red and 7 black caps. One cap is picked up randomly. What is the probability that it is not a red cap ?

[RBI-Note Printing PVT. Ltd- 03]

Sol. Probability of being not red cap = Probability of being black cap

$= \dfrac{n(E)}{n(S)} = \dfrac{{}^7C_1}{{}^{10}C_1} = \dfrac{7}{10}$

Alternatively,

Probability of being red cap $= \dfrac{n(E)}{n(S)} = \dfrac{{}^3C_1}{{}^{10}C_1} = \dfrac{3}{10}$

Then, probability of being not red cap $= 1 - \dfrac{3}{10} = \dfrac{7}{10}$

Problem 20 One-third of 12 mangoes got rotten. If 4 mangoes are taken out randomly, what is the probability that no mango is rotten?

Sol. As per question, 8 mangoes are not rotten.

Then, probability that no mango is rotten

$= \dfrac{n(E)}{n(S)} = \dfrac{{}^8C_4}{{}^{12}C_4} = \dfrac{70}{495} = \dfrac{14}{99}$

Problem 21 In a box there are 8 red, 7 blue and 6 green balls, one ball is picked up randomly. What is the probability that is neither red nor green?

Sol. Probability of being blue ball $= \dfrac{n(E)}{n(S)} = \dfrac{{}^7C_1}{{}^{21}C_1} = \dfrac{7}{21} = \dfrac{1}{3}$

Problem 22 An urn contains 5 red, 3 green, 2 blue and 4 yellow marbles. If four marbles are picked at random, what is the probability that none is red ? ***[BOM(Agri. Officer)-09]***

Sol. The desired event will occur if 4 marbles are taken out from all marbles other than red and it will happen in ${}^9C_4$ ways.

Clearly, $n(S) = {}^{14}C_4$

Then, probability that none is red $= \dfrac{{}^9C_4}{{}^{14}C_4} = \dfrac{18}{143}$

Problem 23 If 3 balls are picked up at random from a basket containing 3 blue, 2 green and 5 red balls, what is the probability that at least one is red ? ***[BOM(PO)-07]***

Sol. Probability of getting 3 non-red balls $= \dfrac{{}^{(3+2)}C_3}{{}^{10}C_3} = \dfrac{1}{12}$

Then, probability of getting at least one red $= 1 - 1/12 = 11/12$

Problem 24 A bag contains 10 green and 5 red balls. Two balls are chosen at random and first one is found to be red. What is the probability that the second one is also red ?

Sol. $n(E) =$ selection of one out of remaining 4 red balls $= {}^4C_1$

{∵ one already found to be red}

and $n(S) =$ selection of one out of remaining 14 balls $= {}^{14}C_1$

Then, probability that the second one is red $= \dfrac{{}^4C_1}{{}^{14}C_1} = \dfrac{2}{7}$

Directions (Q. Nos. 25 and 26) The details of students in a school having only three classes are as under:

Classes	Boys	Girls
A	50	----
B	----	30
C	----	40

Note

(a) *Probability of selecting a girl from Class A out of all girls in all classes is 3/10.*

(b) *Probability of selecting a boy from Class B is equal to that of selecting a boy from Class C.*

(c) *Probability of selecting a boy from all students in all classes is 31/51.* ***[IBPS(PO-Mains)-19]***

Problem 25 If 10% of boys from Class *B* and 20% of girls from Class *A* scored greater than 90% in an examination, then what is the difference between number of boys and girls who secured 90% or less than 90% marks.

Problem 26 What is the difference between average number of girls in Classes *B* and *C* and median of boys in the school?

Sol. Consider note (*a*)

Let the number of girls in Class *A* be *x*.

Then, probability of selecting a girl from Class *A* out of all girls in all classes $= \dfrac{{}^xC_1}{{}^{(x+30+40)}C_1} = \dfrac{x}{70+x} = \dfrac{3}{10}$

$\Rightarrow \quad x = 30$

Consider note (*b*)

Let number of boys in Clases *B* and *C* be *y* and *z* respectively.

As given, $\dfrac{y}{y+30} = \dfrac{z}{z+40}$

$\Rightarrow \quad yz + 40y = yz + 30z$

$\Rightarrow \quad y : z = 3 : 4$

Consider note (*c*)

Probability of selecting a boy from all students in all classes

$= \dfrac{50+y+z}{50+y+z+100} = \dfrac{31}{51}$

$\Rightarrow \quad \dfrac{50+(y+z)}{150+(y+z)} = \dfrac{31}{51}$

$\Rightarrow \quad 20(y+z) = 31 \cdot 150 - 50 \cdot 51$

$\Rightarrow \quad y + z = 105$

Since, $y : z = 3 : 4$, $y = 45$ and $z = 60$.

Now, the given information is as under

Classes	Boys	Girls
A	50	30
B	45	30
C	60	40

25 Required difference = difference between (80% of 45) and (66.66% of 30) = 16

26 Average of girls in Classes *B* and *C*

$= (30 + 40) \div 2 = 35$

and median of observations (45, 50, 60 -----arranged in ascending order) = middle observation = 50

Then, required difference = 50 − 35 = 15

Concept of Conjunction *'OR'*

(in case of mutually exclusive events)

Two events are said to be mutually exclusive if $E_1 \cap E_2 = \phi$. In other words a set of events is mutually exclusive when there is no common outcomes. If $E(A) =$ 2, 3, 4 and $E(B) =$ 1, 5, 6, then they are said to be mutually exclusive.

We simply replace the conjunction *'OR'* by the sign of addition whenever *'OR'* connects two or more events directly or indirectly.

Problem 27 The probability of a horse *A* winning a race is 1/5 and the probability of a horse *B* winning the same race is 1/4. What is the probability that either *A* or *B* wins a race ?

Sol. Here, both the events are mutually exclusive as both the horses cannot win the race simultaneously.

Now, probability of winning the race by either *A* **or** *B*

= Probability of winning the race by Horse *A* + Probability of winning the race by Horse *B*

$= 1/5 + 1/4 = 9/20$

Problem 28 An urn contains 5 red, 3 green, 2 blue and 4 yellow marbles. If two marbles are picked at random, what is the probability that either both are red or both are green? *[BOM(Agri. Officer)-09]*

Sol. Probability of getting both marbles red $= \frac{n(E_1)}{n(S)} = \frac{^5C_2}{^{14}C_2} = \frac{10}{91}$

and probability of getting both marbles green

$= \frac{n(E_2)}{n(S)} = \frac{^3C_2}{^{14}C_2} = \frac{3}{91}$

Remarks *No need to explain the above as we have, discussed earlier at greater length.*
*Then, probability of getting either both balls red **or** both green* $= \frac{10}{91} + \frac{3}{91} = \frac{13}{91} = \frac{1}{7}$

Problem 29 Three unbiased coins are tossed. What is the probability of getting at least two heads ?

Sol. Probability of getting 2 heads $= \frac{^3C_2}{2^3} = \frac{3}{8}$

and probability of getting 3 heads $= \frac{^3C_3}{2^3} = \frac{1}{8}$

∴ Probability of getting at least two heads

= Probability of getting either two heads <u>or</u> three heads

$= \frac{3}{8} + \frac{1}{8} = \frac{1}{2}$

Remarks *The same question has been discussed earlier also, but in different ways*

Problem 30 An urn contains 5 white and 3 black balls. Two balls are drawn at random. Find the probability that they are of the same color.

Sol. Probability of getting both balls white $= \frac{n(E_1)}{n(S)} = \frac{^5C_2}{^8C_2} = \frac{5}{14}$

and probability of getting both balls black $= \frac{n(E_2)}{n(S)}$

$= \frac{^3C_2}{^8C_2} = \frac{3}{28}$

Then, probability that both the balls are of the same color = Probability of getting both balls white <u>or</u> probability of getting both balls black $= \frac{5}{14} + \frac{3}{28} = \frac{13}{28}$

Problem 31 From an ordinary pack a card is drawn at random. What is the probability that it is either a king or a queen ?

Sol. Probability of getting a king $= \frac{n(E_1)}{n(S)} = \frac{^4C_1}{^{52}C_1} = \frac{1}{13}$

and probability of getting a queen $= \frac{n(E_2)}{n(S)} = \frac{^4C_1}{^{52}C_1} = \frac{1}{13}$

Then, probability of getting either a king <u>or</u> a queen

$= \frac{1}{13} + \frac{1}{13} = \frac{2}{13}$

Concept of Conjunction 'OR'
(in case of non- mutually exclusive events)

Problem 32 20 tickets are numbered from 1 to 20. One ticket is drawn at random. Find the probability of the ticket to have a

(i) number less than 10 or an odd integer.

(ii) number divisible by 3 or 4.

Sol. (i) Numbers less than 10 = 1, 2, 3, 4, 5, 6, 7, 8, 9

Odd integers = 1, 3, 5, 7, 9, 11, 13, 15, 17, 19

Common integers = 1, 3, 5, 7, 9

Since there are some common integers between both the events, they cannot be said to be mutually exclusive, hence, non- mutually exclusive events.

Now, probability of getting the number less than 10

$= \frac{^9C_1}{^{20}C_1} = \frac{9}{20}$

and probability of getting odd number $= \frac{^{10}C_1}{^{20}C_1} = \frac{1}{2}$

and probability of getting common number$= \frac{^5C_1}{^{20}C_1} = \frac{1}{4}$

Then, the probability of the ticket to have a number less than 10 or an odd integer $= \frac{9}{20} + \frac{1}{2} - \frac{1}{4} = \frac{7}{10}$

It is important to note that

In case of mutually exclusive events we only add, but in case of non-mutually exclusive events we minus also. We may recall the concept of statistics,
$n(A \cup B) = n(A) + n(B) - n(A \cap B)$.
Here, $P(A \cup B) = P(A) + P(B) - P(A \cap B)$.

(ii) Numbers divisible by 3 = 3, 6, 9, 12, 15, 18

and numbers divisible by 4 = 4, 8, 12, 16, 20

Common between them = 12

Now, probability of getting the number divisible by 3

$= \frac{^6C_1}{^{20}C_1} = \frac{6}{20}$

and probability of getting the number divisible by 4

$= \frac{^5C_1}{^{20}C_1} = \frac{5}{20}$

and probability of getting common number(i.e., divisible by both 3 and 4) $= \frac{^1C_1}{^{20}C_1} = \frac{1}{20}$

Then, probability of getting the number either divisible by 3 or 4 $= \frac{6}{20} + \frac{5}{20} - \frac{1}{20} = \frac{1}{2}$

Problem 33 Two cards are drawn at random from a pack of 52 cards. What is the probability that either both are red or both are kings ?

Sol. We know, there are 26 red cards and 4 kings(i.e. 2 red + 2 black) in a pack of cards.

Now, probability of getting both cards red $= \frac{^{26}C_2}{^{52}C_2} = \frac{25}{102}$

and probability of getting both cards kings $= \frac{^4C_2}{^{52}C_2} = \frac{1}{221}$

and probability of getting common cards (i.e., both are red kings)

$$= \frac{^2C_2}{^{52}C_2} = \frac{1}{1326}$$

Then, probability that either both cards are red or both are kings

$$= \frac{25}{102} + \frac{1}{221} - \frac{1}{1326}$$

$$= \frac{325 + 6 - 1}{1326} = \frac{330}{1326} = \frac{55}{221}$$

Problem 34 A chartered accountant applied for a job in two Firms X and Y. The ability of his being selected in Firm X is 0.7 and being rejected at Y is 0.5 and the probability of atleast one of his applications being rejected is 0.6. What is the probability that he will be selected atleast one of the firms ? ***[MAT-08]***

Sol. Probability of being rejected in Firm X = 1 – 0.7 = 0.3

Probability of being rejected in Firm Y = 0.5

and Probability of being rejected in atleast one of the firms = Probability of being rejected either in Firm X or Y or both = 0.6

Now, probability of being rejected either in Firm X or Y or both

= 0.3 + 0.5 – probability of being rejected in both firms

⇒ 0.6 = 0.3 + 0.5 – probability of being rejected in both firms

⇒ probability of being rejected in both firms = 0.2

∴ probability of being selected in atleast one of the firms

= 1 – 02 = 0.8

Alternatively

Probability of being selected in Firm X = 0.7

Probability of being selected in Firm Y = 1 – 0.5 = 0.5

and Probability of being rejected in atleast one of the Firms = 0.6

Then, probability of being selected in both Firms X and Y

= 1 – 0.6 = 0.4

∴ Probability of being selected in atleast one of the Firms = Probability of being selected either in Firm X or Y or both = Probability of being selected in firm X + Probability of being selected in Firm Y – Probability of being selected in both firms X and Y = 0.7 + 0.5 – 0.4 = 0.8

Concept of conjunction *'AND'*

Type I

When occurrence of event depends upon single source

We multiply the number of ways of selection for 1st, 2nd, 3rd, ... parts of the event to compute the value of $n(E)$ if parts of the event is connected with *'AND'* directly or indirectly.

Problem 35 From a pack of 52 cards two are drawn at random. Find the chance that one is a queen and another is jack.

Sol. Event ⇒ getting 1 queen *AND* one jack.

∴ $n(E)$ = number of ways of choosing 1 queen out of 4 *AND* 1 jack out of 4 = $^4C_1 \times {^4C_1}$

and $n(S)$ = number of ways of choosing 2 out of 52 = $^{52}C_2$

Then, probability of getting 1 queen and 1 jack

$$= \frac{^4C_1 \times {^4C_1}}{^{52}C_2} = \frac{8}{663}$$

Problem 36 Four cards are drawn from a deck of 52 cards at random. What is the chance that they all are of different suits ?

Sol. We know, there are 4 suits namely hearts, diamond, spades and clubs and each has 13 cards.

Event ⇒ getting one card from hearts *AND* one card from diamonds *AND* one card from spades *AND* one card from clubs.

Hence, $n(E) = {^{13}C_1} \times {^{13}C_1} \times {^{13}C_1} \times {^{13}C_1}$

and $n(S) = {^{52}C_4}$

Then, required probability $= \frac{^{13}C_1 \times {^{13}C_1} \times {^{13}C_1} \times {^{13}C_1}}{^{52}C_4} = \frac{2197}{20825}$

Problem 37 In game of a bridge, what is the probability that one hand will have four kings.

Sol. We know, in a game of bridge, one hand holds 13 cards.

$n(E)$ = number of ways of choosing 4 kings out of 4 kings *AND* 9 out of remaining 48 cards = $^4C_4 \times {^{48}C_9}$

and $n(S) = {^{52}C_{13}}$

Then, required probability $= \frac{^4C_4 \times {^{48}C_9}}{^{52}C_{13}} = \frac{11}{4165}$

Problem 38 Four cards are drawn from a deck of 52 cards at random. What is the chance that these four will have just two jacks ?

Sol. $n(E)$ = number of ways of choosing 2 out of 4 jacks AND 2 out of remaining 48 cards = $^4C_2 \times {^{48}C_2}$

and $n(S) = {^{52}C_4}$

Then, required probability $= \frac{^4C_2 \times {^{48}C_2}}{^{52}C_4}$

Directions (Q. Nos. 39 – 41) Consider the given information and answer the following questions.

There are some identical green balls and blue balls (number of blue balls > number of green balls) in a Bag -M. Two balls are drawn, then

I. Probability of getting both balls green is A.

II. Probability of getting one green and one blue is 32/69.

III. In case certain number of blue balls and green balls of another Bag -N are taken out and then added to Bag-M, the probability of getting both balls green from Bag-M is 3/29 provided number of blue balls taken out from Bag-N is 2 more than that of green balls.

[SBI(PO-Mains)-19]

Problem 39 What were the total number of balls in Bag-M in beginning ?

Problem 40 What is the value of $2A$?

Problem 41 What is the number of green balls taken out from the Bag-N ?

Sol. (Q. Nos. 39-41) Let there are x blue balls and y green balls in the bag-M. Then, probability of getting both balls green

$$= \frac{^yC_2}{^{(x+y)}C_2} = A \text{ (given)}$$

Further, probability of getting one green and one blue ball

$$= \frac{^xC_1 \times {^yC_1}}{^{(x+y)}C_2} = \frac{32}{69}$$

$$\Rightarrow \frac{2 \times x \times y}{(x+y)(x+y-1)} = \frac{32}{3 \times 23}$$

$$\Rightarrow \frac{x \times y}{(x+y)(x+y-1)} = \frac{16 \times 8}{3 \times 8 \times 23} = \frac{16 \times 8}{24 \times 23}$$

Clearly, $(x + y) = 24$ and $xy = 16 \times 8$

Hence, $x = 16$ and $y = 8$ $\{\because x > y\}$

Further, let g green balls and $(g + 2)$ blue balls are drawn from Bag-N and the same are added to Bag-M.

Then, number of blue balls in Bag-$M = 16 + g + 2$

and number of green balls in Bag-$N = 8 + g$

Hence, total number of balls in Bag-$M = 26 + 2g$

As per question,

Probability of getting both balls green $= \frac{3}{29}$

$$\Rightarrow \frac{^{(8+g)}C_2}{^{(26+2g)}C_2} = \frac{3}{29}$$

$$\Rightarrow \frac{(8+g)(8+g-1)}{(26+2g)(26+2g-1)} = \frac{30 \times 3}{30 \times 29} = \frac{10 \times 9}{30 \times 29}$$

Clearly, $26 + 2g = 30$ or $(8 + g) = 10 \Rightarrow g = 2$

39. Total number of balls $= 16 + 8 = 24$

40. $A = \frac{^yC_2}{^{(x+y)}C_2}$ (from above)

$$\Rightarrow A = \frac{^8C_2}{^{(16+8)}C_2} = \frac{8(8-1)}{24(24-1)} = \frac{7}{69}$$

$$\Rightarrow 2A = \frac{14}{69}$$

41. Since $g = 2$, two green balls were taken out from Bag-N.

Type II

When occurrence of event depends upon more than one source

We multiply the value of individual probability if parts of the event is connected with *'AND'*.

Problem 42 Two dice are thrown. What is the probability of getting both numbers prime ?

Sol. Event ⇒ Getting a prime number from 1st die *AND* a prime number from 2nd die.

Probability of getting a prime number from 1st die $= \frac{^3C_1}{^6C_1} = \frac{1}{2}$

$\{\because$ prime numbers are : 2, 3, 5$\}$

and Probability of getting a prime number from 2nd die

$$= \frac{^3C_1}{^6C_1} = \frac{1}{2}$$

Then, probability of getting both the numbers prime

$$= 1/2 \times 1/2 = 1/4$$

Alternatively,

$n(E) = 9$ [(2, 2), (2, 3), (2, 5), (3,2), (3, 3), (3, 5), (5,2), (5, 3), (5, 5)]

and $n(S) = 6^2 = 36$

Then, required probability $= \frac{9}{36} = \frac{1}{4}$

Remarks *The above approach has already been discussed earlier*

Problem 43 One card is drawn randomly from each of two decks of cards. What is the chance that both of them are red ?

Sol. On the above pattern, probability of getting both the cards red

= (Probability of getting a red card from 1st deck)

× (Probability of getting a red card from 2nd deck)

$$= \frac{^{26}C_1}{^{52}C_1} \times \frac{^{26}C_1}{^{52}C_1} = \frac{1}{2} \times \frac{1}{2} = \frac{1}{4}$$

Problem 44 There are 3 red and 4 white balls in one bag. The second bag contains 4 black and 5 green balls. A ball is taken out from each bag. What is the probability that one ball is red and another green ?

Sol. Event ⇒ getting one ball red *AND* another green.

Probability of getting one red ball from the 1st bag

$$= \frac{^3C_1}{^{(3+4)}C_1} = \frac{3}{7}$$

and probability of getting one green ball from the 2nd bag

$$= \frac{^5C_1}{^{(4+5)}C_1} = \frac{5}{9}$$

Then, probability of getting one ball red and another green

$$= \frac{3}{7} \times \frac{5}{9} = \frac{5}{21}$$

Problem 45 The probabilities of solving a question by two students are 1/3 and 1/5. What is the probability that the question will be solved?

Sol. Probability that the question will be solved by anyone

= Probability that 1st student will solve + probability that 2nd student will solve − probability that both will solve the question

$$= \frac{1}{3} + \frac{1}{5} - \frac{1}{3} \cdot \frac{1}{5} = \frac{7}{15}$$

Alternatively,

The probability that 1st student will not solve the problem

$$= 1 - \frac{1}{3} = \frac{2}{3}$$

and Probability that 2nd student will not solve the question
$= 1 - \frac{1}{5} = \frac{4}{5}$

Then, probability that both will not solve the question
$= \frac{2}{3} \times \frac{4}{5} = \frac{8}{15}$

$\therefore$ Probability that the question will be solved by anyone
$= 1 - \frac{8}{15} = \frac{7}{15}$

Remarks *In order to make the events mutually exclusive, their contrary have been considered.*

Problem 46 What is the probability of getting at least one figure of 1 while throwing two dice?

Sol. Probability of getting figures(except 1) from 1st die = 5/6
Probability of getting figures(except 1) from 2nd die = 5/6
Then, probability of getting a figure(except 1) from 1st die *AND* a figure(except 1) from 2nd die = $5/6 \times 5/6 = \frac{25}{36}$
Then, probability of getting at least one figure of 1 = $1 - \frac{25}{36} = \frac{11}{36}$

Alternatively
$n(S) = 6^2 = 36$
$n(E) = 11$ {(1, 1), (1, 2), (1, 3), (1, 4), (1, 5), (1, 6), (2, 1) (3, 1), (4, 1), (5, 1), (6, 1)}
Then, probability of getting at least one figure of 1 = $\frac{11}{36}$

Problems Based on Combination of Conjunction *'OR'* and *'AND'*

{Recall 'OR' means '+' and 'AND' means '×'}

Problem 47 What is the probability of getting one odd and another composite number while throwing two dice?

Sol. Event $\Rightarrow$ Getting one number odd and another composite. This can be got as below

(i) an odd number on 1st die *AND* a composite number on 2nd die

OR

(ii) a composite number on 1st die *AND* an odd number on 2nd die

Probability in case(i)= $\frac{^3C_1}{^6C_1} \times \frac{^2C_1}{^6C_1} = 1/6$

and probability in case(ii)= $\frac{^2C_1}{^6C_1} \times \frac{^3C_1}{^6C_1} = 1/6$

Then, required probability = 1/6 + 1/6
{$\because$ both events/cases are mutually exclusive}
= 1/3

Alternatively, you may double the value of probability of any one case to get the answer.

Problem 48 One card is drawn randomly from each of two packs of cards. What is the probability of getting both cards of different colors?

Sol. Event $\Rightarrow$

(i) Getting a black card from 1st pack *AND* a red card from 2nd pack

OR

(ii) Getting a red card from 1st pack *AND* a black card from 2nd pack

Probability in case(i) = $\frac{^{26}C_1}{^{52}C_1} \times \frac{^{26}C_1}{^{52}C_1} = 1/4$

and probability in case(ii) = $\frac{^{26}C_1}{^{52}C_1} \times \frac{^{26}C_1}{^{52}C_1} = 1/4$

Then, required probability = 1/4 + 1/4
{$\because$ both events/cases are mutually exclusive}
= 1/2

Alternatively, required probability = 2 × Probability of getting a black card from 1st pack AND a red card from 2nd pack = 2 × 1/4 = 1/2

Problem 49 Ram speaks the truth in 70% cases and Jyoti in 80% of the cases. What is the probability that they are contradicting each other in speaking the same fact?

Sol. There will be situations of contradiction, if one speaks truth whereas another a lie. Obviously, there will be two situations as below:

(i) Ram speaks truth *AND* Jyoti a lie

'OR'

(ii) Jyoti speaks truth *AND* Ram a lie

Now, probability for 1st situation = $\frac{70}{100} \times \frac{(100-80)}{100} = \frac{7}{50}$

and probability for 2nd situation = $\frac{80}{100} \times \frac{(100-70)}{100} = \frac{6}{25}$

Since both situations/events are mutually exclusive,
required probability = $\frac{7}{50} + \frac{6}{25} = \frac{19}{50} = 38\%$

Problem 50 Two persons are to be selected from a group of 8 men and 6 women. What is the probability that at least one of the selected persons is a woman?

Sol. Event $\Rightarrow$

(i) getting one woman *AND* one man

OR

(ii) getting one woman *AND* one woman

Now probability for case(i) = $\frac{^6C_1 \times {^8C_1}}{^{14}C_2}$

and probability for case(ii) = $\frac{^6C_2}{^{14}C_2}$

$\therefore$ Required probability = $\frac{^6C_1 \times {^8C_1}}{^{14}C_2} + \frac{^6C_2}{^{14}C_2}$

{$\because$ both events/cases are mutually exclusive}

$= \frac{9}{13}$

Alternatively,

Probability that no woman is selected = $\frac{^8C_2}{^{14}C_2} = \frac{4}{13}$

$\therefore$ Probability of selecting at least one woman= $1 - \frac{4}{13} = \frac{9}{13}$

Problem 51 There are two bags. One contains 4 white and 6 green balls. Another bag contains 5 white and 4 yellow balls. 2 balls are drawn from one of the bags at random. What is the probability of getting the balls of different colors?

Sol. Probability of getting 1st bag = 1/2

and probability of getting 2nd bag = 1/2

Probability of getting two balls of different colors (i.e., white and green) from the 1st bag $= \frac{{}^4C_1 \times {}^6C_1}{{}^{10}C_2} = \frac{8}{15}$

and probability of getting two balls of different colors (i.e., white and yellow) from the 2nd bag $= \frac{{}^5C_1 \times {}^4C_1}{{}^{10}C_2} = \frac{5}{9}$

Then, required probability $= \left(1/2 \times \frac{8}{15}\right) + \left(1/2 \times \frac{5}{9}\right) = \frac{49}{90}$

Problem 52 The probabilities of solving a question by two students are 1/3 and 1/5. What is the probability that the question will be solved?

Sol. This question has been discussed earlier, however, we will discuss with another approach.

Event ⇒

1st student solves the question *AND* 2nd student does not solve

OR

2nd student solves the question *AND* 1st student does not solve

OR

1st student solves the question *AND* 2nd student does also.

Then, required probability

$$= \left\{\frac{1}{3} \times \left(1 - \frac{1}{5}\right)\right\} + \left\{\frac{1}{5} \times \left(1 - \frac{1}{3}\right)\right\} + \frac{1}{3} \cdot \frac{1}{5} = \frac{7}{15}$$

Problem 53 A boy and a girl are to be appeared in an interview for two vacancies. The probability of selection of the boy is 1/3 and that of the girl is 1/4. What is the probability that only one of them is selected? **[MAT-05, SNAP-18]**

Sol. Event ⇒ Selection of boy <u>AND</u> not selection of girl

OR

Selection of girl <u>AND</u> not selection of boy

Then, required probability $= \left\{\frac{1}{3} \times \left(1 - \frac{1}{4}\right)\right\} + \left\{\frac{1}{4} \times \left(1 - \frac{1}{3}\right)\right\} = \frac{5}{12}$

Problem 54 There are three boxes. First box contains 4 red and 3 green balls, second contains 2 red and 1 green balls and third contains 3 red and 2 green balls. One ball is drawn from each of three boxes at random. What is the probability of getting 2 red and 1 green ball?

Sol. Event ⇒

1 red from 1st box <u>AND</u> 1 red from 2nd box <u>AND</u> 1 green from 3rd box

OR

1 red from 1st box <u>AND</u> 1 green from 2nd box <u>AND</u> 1 red from 3rd box

OR

1 green from 1st box *AND* 1 red from 2nd box *AND* 1 red from 3rd box

Probability for 1st case $= \frac{{}^4C_1}{{}^7C_1} \times \frac{{}^2C_1}{{}^3C_1} \times \frac{{}^2C_1}{{}^5C_1} = \frac{16}{105}$

and Probability for 2nd case $= \frac{{}^4C_1}{{}^7C_1} \times \frac{{}^1C_1}{{}^3C_1} \times \frac{{}^3C_1}{{}^5C_1} = \frac{4}{35}$

and Probability for 3rd case $= \frac{{}^3C_1}{{}^7C_1} \times \frac{{}^2C_1}{{}^3C_1} \times \frac{{}^3C_1}{{}^5C_1} = \frac{6}{35}$

Then, required probability $= \frac{16}{105} + \frac{4}{35} + \frac{6}{35} = \frac{46}{105}$

Problem 55 A letter is taken out from a word 'GUESSES' and another letter is taken out from the word 'SUCCESS'. What is the probability that they are the same letters?

Sol.

GUESSES ⇒	G	U	EE	SSS	Letters G and C are not common
SUCCESS ⇒	CC	U	E	SSS	

Probability of getting U $= \frac{{}^1C_1}{{}^7C_1} \times \frac{{}^1C_1}{{}^7C_1} = \frac{1}{49}$

Probability of getting E $= \frac{{}^2C_1}{{}^7C_1} \times \frac{{}^1C_1}{{}^7C_1} = \frac{2}{49}$

Probability of getting S $= \frac{{}^3C_1}{{}^7C_1} \times \frac{{}^3C_1}{{}^7C_1} = \frac{9}{49}$

Then, required probability $= \frac{1}{49} + \frac{2}{49} + \frac{9}{49} = \frac{12}{49}$

Problems Based on Conditional Probability

Problem 56 What is the probability of the occurrence of a number greater than 1 if only odd numbers appear on the top face of a die while it is thrown?

Sol. $n(E) = 2$ {i.e. 3, 5}

and $n(S) = 3$ {i.e. 1, 3, 5}

Hence, $P(E) = 2/3$

Problem 57 A family has three children. What is the chance that all three are girls if it is known that one of them is a girl?

Sol. $n(E) =$ {i.e. *GGG*}

and $n(S) = 3$ {i.e. *GGG*, *GBG*, *GBB*}

Hence, $P(E) = 1/3$

Problem 58 3 items are to be selected out of 5 items. What is probability that a particular item will be selected if an another item has already been selected?

Sol. Since one item has already been selected, desired 2 items can be selected out of remaining 4 items. Thus, $n(S) = {}^4C_2 = 6$

It is obvious that the desired particular item may be combined with either of remaining 3 items i.e. in 3 ways.

Hence, $P(E) = 3/6 = 1/2$

Concept of Probability in Case of Repeated Trials

If p is the probability of success in one trial, then probability of m successes in n trials
$$= {}^nC_m \times p^m \times (1-p)^{n-m}$$

Problem 59 A coin is tossed four times. What is the probability of getting exactly two tails?

'OR'

A coin is tossed four times. What is the probability of getting two heads and two tails ?'

Sol. Here, $n = 4, m = 2$

and p = probability of getting head in one trial = 1/2

Then, required probability $= {}^nC_m \times p^m \times (1-p)^{n-m}$

$$= {}^4C_2 \times \left(\frac{1}{2}\right)^2 \times \left(1-\frac{1}{2}\right)^{4-2} = \frac{3}{8}$$

Alternatively,

For the sake of convenience, the above question may be interpreted that 4 coins are tossed. No doubt, both are conveying the same sense.

Since, there are 4 heads and 4 tails, $n(E) = {}^4C_2 = 6$

and $n(S) = (2)^4 = 16$

Then, $P(E) = \frac{6}{16} = \frac{3}{8}$

Problem 60 Two dice are thrown simultaneously. What is probability of getting exactly 3 aces on both in 5 trials ?

Sol. Here, $n = 5, m = 3$ {∵ 3 times • – • required}

and p = Probability of getting • – • in one trial $= \frac{1}{6} \times \frac{1}{6} = \frac{1}{36}$

Then, required probability $= {}^nC_m \times p^m \times (1-p)^{n-m}$

$$= {}^5C_3 \times \left(\frac{1}{36}\right)^3 \times \left(1-\frac{1}{36}\right)^{5-3}$$

$$= 12250 \times \left(\frac{1}{36}\right)^5$$

Problem 61 A boy tosses a coin 4 times. What is probability of getting at least one tail ?

Sol. Here, $n = 4$, m = either 1 or 2 or 3 or 4

and p = probability of getting tail in one trial = 1/2

Then, required probability

$$= \left\{{}^4C_1 \times \left(\frac{1}{2}\right)^1 \times \left(1-\frac{1}{2}\right)^{4-1}\right\} + \left\{{}^4C_2 \times \left(\frac{1}{2}\right)^2 \times \left(1-\frac{1}{2}\right)^{4-2}\right\}$$
$$+ \left\{{}^4C_3 \times \left(\frac{1}{2}\right)^3 \times \left(1-\frac{1}{2}\right)^{4-3}\right\} + \left\{{}^4C_4 \times \left(\frac{1}{2}\right)^4 \times \left(1-\frac{1}{2}\right)^{4-4}\right\}$$

$$= \frac{1}{4} + \frac{3}{8} + \frac{1}{4} + \frac{1}{16} = \frac{15}{16}$$

Alternatively I,

Probability of getting one tail $= \frac{{}^4C_1}{2^4} = \frac{1}{4}$

Probability of getting two tails $= \frac{{}^4C_2}{2^4} = \frac{3}{8}$

Probability of getting three tails $= \frac{{}^4C_3}{2^4} = \frac{1}{4}$

Probability of getting four tails $= \frac{{}^4C_4}{2^4} = \frac{1}{16}$

Then, required probability $= \frac{1}{4} + \frac{3}{8} + \frac{1}{4} + \frac{1}{16} = \frac{15}{16}$

Alternatively II,

Probability of getting four heads $= \frac{{}^4C_4}{2^4} = \frac{1}{16}$

Then, probability of getting at least one tail $= 1 - \frac{1}{16} = \frac{15}{16}$

Problem 62 A shooter can hit a target once in two shots. What is the probability the target is hit if he fires two shots?

Sol. Here, $n = 2$

m = either 1 or 2

and p = probability of hitting the target in one trial = 1/2

Then, required probability

$$= \left\{{}^2C_1 \times \left(\frac{1}{2}\right)^1 \times \left(1-\frac{1}{2}\right)^{2-1}\right\} + \left\{{}^2C_2 \times \left(\frac{1}{2}\right)^2 \times \left(1-\frac{1}{2}\right)^{2-2}\right\} = \frac{3}{4}$$

Problem 63 An urn contains 8 yellow and 4 green balls. A ball is drawn and replaced. What is the probability that there be 3 green and 1 yellow in 4 drawings ?

Sol. **We may restructure the question as** What is the probability of getting only 1 yellow ball in 4 drawings ? The reason is that if we get only 1 yellow ball, 3 balls will obviously be green and *vice-versa*.

Now, $n = 4$

$m = 1$

and p = probability of getting 1 yellow ball in one trial

$$= \frac{8}{(8+4)} = \frac{2}{3}$$

Then, required probability $= {}^nC_m \times p^m \times (1-p)^{n-m}$

$$= {}^4C_1 \times \left(\frac{2}{3}\right)^1 \times \left(1-\frac{2}{3}\right)^{4-1} = \frac{8}{81}$$

Note *We may also solve it taking 3 green balls as reference. Then, $n = 4$, $m = 3$ and $p = 4/12 = 1/3$*

WORKOUT

1. A bag contains 3 black, 4 white and 5 red balls. Two balls are taken out at random. Find the probability that both of them are black or white.

(a) 3/22 (b) 7/12 (c) 1/12 (d) 5/22

2. Varun throws two unbiased dice together and gets a sum of 7. If his friend Tarun, then throws the same two dice, what is the probability that the sum is less than 7 ? ***[MAT-12]***

(a) 1/6 (b) 7/12 (c) 1/2 (d) 5/12

3. The probability that in household LPG will last 60 days or more is 0.8 and that it will last at most 90 days is 0.6. The probability that the LPG will last at most 60 to 90 days is: ***[IIFT-12]***

(a) 0.40 (b) 0.50
(c) 0.75 (d) None of these

4. An urn contains 5 white, 4 black and 3 yellow balls. Four balls are drawn at random. Find the probability that all of them are of the same colour.

(a) 2/23 (b) 2/165 (c) 2/63 (d) 5/122

5. Ashish is studying late into the night and is hungry. He opens his mother's snack cupboard without switching on the lights, knowing that his mother has kept 10 packets of chips and biscuits in the cupboard. He pulls out 3 packets from the cupboard and all of them turn out to be chips. What is the probability that the snack cupboard contains 1 packet of biscuits and 9 packets of chips ? ***[IIFT-12]***

(a) 6/55 (b) 12/73 (c) 14/55 (d) 7/50

6. Three identical dice are rolled. What is the probability that same number will appear on each of them?

(a) 1/6 (b) 7/36 (c) 1/12 (d) 1/36

7. Out of five girls and three boys, four children are to be randomly selected for a quiz contest. What is the probability that all selected children are girls ?

(a) 1/14 (b) 1/7
(c) 5/17 (d) 2/17
(e) None of these

8. The probability that a randomly chosen positive divisor of 10^{29} is an integer multiple of 10^{23} is $\frac{a^2}{b^2}$, then $b - a$ would be ***[XAT-14]***

(a) 8 (b) 15 (c) 21 (d) 23

9. There are four boxes. Each box contains two balls: one red and one blue. You draw one ball from each of the four boxes. What is the probability of drawing at least one red ball ? ***[CAT-10]***

(a) $\frac{1}{2}$ (b) $\frac{1}{4}$ (c) $\frac{1}{16}$ (d) $\frac{15}{16}$

10. Two cubes have their faces painted either red or blue. The first cube has 5 red faces and 1 blue face. When two cubes are rolled simultaneously the probability that the two top faces show the same colour is ½. How many red faces are there on the second cube ? ***[MAT-13]***

(a) 3 (b) 2 (c) 4 (d) 5

11. A die is thrown 3 times. If getting a six is considered a success, find the probability of at least two successes.

(a) 1/16 (b) 7/12 (c) 2/27 (d) 5/12

12. Out of 20 consecutive positive integers, two are chosen at random. The probability that their sum is odd, is

(a) 10/19 (b) 1/20 (c) 19/20 (d) 9/19

13. The answer sheets of 5 engineering students can be checked by anyone of 9 professors. What is the probability that all the 5 answer sheets are checked by exactly 2 professors? ***[IIFT-12]***

(a) 20/2187 (b) 40/2187
(c) 40/729 (d) None of these

14. If the probability that A will live 15 yr is 7/8 and that B will live 15 yr is 9/10, then what is the probability that both will live after 15 yr ?

(a) 1/20 (b) 63/80
(c) 1/5 (d) None of these

15. The probability that a student is not a swimmer is 1/5. Then, the probability that out of the five students, four are swimmer is:

(a) $^5C_4\left(\frac{4}{5}\right)^2\left(\frac{1}{5}\right)$ (b) $\left(\frac{4}{5}\right)^4\left(\frac{1}{5}\right)$

(c) $^5C_4\left(\frac{4}{5}\right)^4\left(\frac{1}{5}\right)$ (d) None of these

16. The probability of different faces of a biased die to appear are as follows

Face number	1	2	3	4	5	6
Probability	0.1	0.32	0.21	0.15	0.05	0.17

The die is thrown and it is known that either face number 1 or 2 will appear. The probability of the face number 1 to appear is

(a) 2/23 (b) 5/21
(c) 1/2 (d) 5/18

17. The probability that a leap year selected at random contains either 53 Sundays or 53 Mondays, is ***[SNAP-11]***

(a) 17/53 (b) 1/53
(c) 3/7 (d) None of these

18. Two fair dice are thrown. Given that, the sum of the dice is less than or equal to 4, find the probability that only one dice shows two. ***[MAT-14,19]***

(a) 1/4 (b) 1/2
(c) 2/3 (d) 1/3

19. Six playing cards are lying face down on a table, two of them are kings. Two cards are drawn at random. Let a denote the probability that atleast one of the cards drawn is a king and b denote the probability of not drawing a king. The ratio a/b is ***[XAT-13]***

(a) ≥ 0.25 and < 0.5
(b) ≥ 0.5 and < 0.75
(c) ≥ 0.75 and < 0.1
(d) > 1.0 and < 1.25
(e) ≥ 1.25

20. A class consists of 100 students; 25 of them are girls and 75 boys; 20 of them are rich and the remaining poor; 40 of them are fair-complexioned. The probability of selecting a fair complexioned rich girl is:
(a) 0.05 (b) 0.04
(c) 0.02 (d) 0.08

21. Three groups *A*, *B* and *C* are contesting for a position on the Board of Directors of a company. The probability of their winning are 0.5, 0.3 and 0.2 respectively. If the group *A* wins, then the probability of introducing a new product is 0.7 and the corresponding probabilities for group *B* and *C* are 0.6 and 0.5 respectively. The probability that the new product will be introduced is
(a) 0.52 (b) 0.74
(c) 0.63 (d) None of these

22. Five persons entered the lift cabin on the ground floor of an 8-floor house. Suppose that each of them, independently and with equal probability, can leave the cabin at any floor beginning with the first. Find the probability of all the five persons leaving at different floors.
(a) 121/512 (b) 15/64
(c) 11/512 (d) 105/512

23. Two packs of cards are thoroughly mixed and shuffled and two cards are drawn at random, one after the other. What is the probability that both of them are jacks ? ***[IAS(Pre.)- 96]***
(a) 1/13 (b) 2/13
(c) 7/1339 (d) 1/169

24. Ten cars are parked in a row, what is the probability that there are exactly five cars between the particular two cars ?
(a) $\frac{8 \times 5!}{10!}$ (b) $\frac{4 \times 5! \times 2!}{10!}$
(c) $\frac{8 \times 8!}{10!}$ (d) $\frac{16 \times 8!}{10!}$

25. 12 defective pens are accidentally mixed with 132 good ones. It is not possible to just look at a pen and tell whether or not it is defective. One pen is taken out at random from this lot. Determine the probability that the pen taken out is a good one. ***[CLAT-16]***
(a) 11/12 (b) 10/12
(c) 7/12 (d) 9/12

26. A fair die is rolled once. The probability of an odd number less than 5 turns up and probability of a number greater than or equal to 3 turns up are respectively.
(a) 1/6 (b) 5/12
(c) 1/2 (d) 2/3

27. There are 100 students in a college class of which 36 are boys studying Statistics and 13 girls are not studying Statistics. If there are 55 girls in all, the probability that a boy picked up at random is not studying Statistics is
(a) 1/5 (b) 2/5
(c) 3/5 (d) 4/5

28. A box contains 6 red balls, 7 green balls and 5 blue balls. Each ball is of a different size. One ball is selected at random. The probability that the ball selected is the smallest red ball is ***[CAT-93]***
(a) 1/18 (b) 1/3
(c) 1/6 (d) 2/3

29. An urn contains 2 white and 2 black balls. A ball is drawn at random. If it is white it is not replaced into the urn, otherwise it is replaced along with another ball of the same colour. The process is repeated. Find the probability that the third ball drawn is black.
(a) 23/30 (b) 7/15
(c) 1/12 (d) 7/30

30. It has been found that if *A* and *B* play a game 12 times, *A* wins 6 times, *B* wins 4 times and they draw twice. *A* and *B* take part in a series of 3 games. The probability that they will win alternatively is ***[MAT-13]***
(a) 5/12 (b) 5/36
(c) 1/6 (d) 3/4

31. 3 digits are chosen at randon from 1, 2, 3, 4, 5, 6, 7, 8 and 9 without repeating any digit. What is the probability that their product is odd ? ***[IAS(Pre.)- 06]***
(a) 2/3 (b) 5/108
(c) 5/42 (d) 7/48

32. The probability that an event *A* happens in one trial of an experiment is 0.4. Three independent trials of the experiment are performed. Find the probability that the event *A* happens at least once.
(a) 4/75 (b) 98/125
(c) 1/201 (d) 7/125

33. *A* and *B* play a game where each is asked to select a number from 1 to 16. If the two numbers match, both of them win a prize. Find the probability that they will not win a prize in a single trail.
(a) 15/16 (b) 12/17
(c) 14/15 (d) 12/13

34. An anti-aircraft gun can take a maximum of four shots at an enemy plane moving away from it. The probability of hitting the plane at the first , second, third and fourth shot are 0.4, 0.3, 0.2 and 0.1 respectively. What is the probability that the gun hits the plane ?
(a) 0.025 (b) 0.1236
(c) 0.6958 (d) 0.6976

35. A shopkeeper received a pack of 15 pens, out of which 4 were defective. The shopkeeper decided to examine every pen one by one selecting a pen at random. The pens examined are not put back. What is the probability that ninth one examined is the last defective pen ? ***[CAT-13]***
(a) $\frac{11}{195}$ (b) $\frac{16}{195}$ (c) $\frac{8}{195}$ (d) $\frac{17}{195}$

36. A classroom has 3 electric lamps. From a collection of 10 electric bulbs of which 6 are good, 3 are selected at random and put in the lamps. Find the probability that all lamps are burning.
(a) 1/8 (b) 1/4 (c) 1/2 (d) 1/6

37. A Bag *A* contains 2 white and 3 red balls and the Bag *B* contains 4 white and 5 red balls. One ball is drawn at random from a randomly chosen bag and is found to be red. The probability that it was drawn from the Bag *B* was ?
(a) $\frac{5}{14}$ (b) $\frac{5}{16}$
(c) $\frac{5}{18}$ (d) $\frac{25}{52}$

38. A person draws a card from a pack of 52, replaces it and shuffles it. He continues doing it until he draws a heart. What is the probability that he has to make 3 trials? ***[MAT-14, 19]***

(a) 9/64 (b) 3/64
(c) 5/64 (d) 1/64

39. A man takes a step forward with probability 0.4 and backward with probability as 0.6. Find the probability that at the end of eleven steps he is one step away from the starting point.

(a) $462 \times \left(\frac{3}{25}\right)^5$ (b) $462 \times \left(\frac{6}{25}\right)^5$
(c) $462 \times \left(\frac{2}{25}\right)^5$ (d) $462 \times \left(\frac{1}{25}\right)^5$

40. Eight chits are numbered 1 to 8. Three are drawn one by one with replacement. The probability that the least number on any selected chit is 6, is

(a) $1-\left(\frac{3}{4}\right)^3$ (b) $\left(\frac{3}{4}\right)^3$
(c) $\left(\frac{3}{8}\right)^3$ (d) $1-\left(\frac{3}{8}\right)^3$

41. A can hit the target 4 times in 5 shots; B, 3 times in 4 shots and C twice in 3 shots. They fire one each. What is the probability that atleast two shots hit the target ?

(a) 13/30
(b) 5/6
(c) 11/40
(d) None of the above

42. A purse contains 4 copper coins, 3 silver coins and a second purse contains 6 copper coins and 2 silver coins. A coin is taken out of any purse at random. The probability that it is copper coin is:

(a) $\frac{4}{7}$ (b) $\frac{3}{4}$
(c) $\frac{3}{7}$ (d) $\frac{37}{56}$

43. A five digit number is formed by the digits 1, 2, 3, 4, 5 without repetition. Find the probability that the number formed is divisible by 4.

(a) $\frac{4}{7}$ (b) $\frac{3}{4}$
(c) $\frac{1}{7}$ (d) $\frac{1}{5}$

44. An unbiased die with faces marked 1, 2, 3, 4, 5 and 6 is rolled four times. Out of four face values obtained, the probability that minimum face value is not less than 2 and maximum face value is not greater than 5 is:

(a) $\frac{16}{81}$ (b) $\frac{1}{81}$
(c) $\frac{80}{81}$ (d) $\frac{65}{81}$

45. Three faces of a fair die are yellow, two faces red and one blue. The die is tossed three times. The probability that the colors, yellow, red and blue appear in the first, second and third tosses respectively is

(a) $\frac{5}{36}$ (b) $\frac{1}{36}$
(c) $\frac{3}{7}$ (d) $\frac{17}{36}$

46. Two events A and B have probabilities 0.25 and 0.50 respectively. The probability that both A and B occur simultaneously is 0.12. Then, the probability that neither A nor B occurs is ***[MAT-11]***

(a) 0.38 (b) 0.13 (c) 0.63 (d) 0.37

47. A has 3 shares in a lottery containing 3 prizes and 9 blanks. B has two shares in a lottery containing 2 prizes and 6 blanks. Compare their chances of success.

(a) 145 : 362 (b) 952 : 715
(c) 123 : 213 (d) 145 : 716

48. A committee consists of 9 experts taken from three institutions A, B and C of which 2 are from A, 3 from B and 4 from C. If three experts resign, then the probability that they belong to different institutions is

(a) 1/729 (b) 1/24
(c) 1/21 (d) 2/7

49. In a factory where toys are manufactured, Machines A, B and C produce 25%, 35% and 40% of the total toys respectively. Of their output, 5%, 4% and 2% respectively are defective toys. If a toy drawn at random is found to be defective, what is the probability that it is manufactured on Machine B? ***[MAT-12]***

(a) 17/69 (b) 28/69
(c) 35/69 (d) None of these

50. If four whole numbers taken at random are multiplied together, what is the probability that the last digit in the product is 1, 3, 7 or 9 ?

(a) 8/125 (b) 3/625
(c) 16/625 (d) 37/125

51. There are two Boxes A and B. A has 5 oranges and 6 apples in it and Box B contains 3 apples and 4 oranges in it. A fruit is taken from A and placed in B, after which a fruit is then transferred from B to A. What is the probability that the configuration of boxes does not change due to the transfers ?

(a) 49/88
(b) 45/88
(c) 25/44
(d) None of the above

52. An instrument manufactured by a company consists of two Parts A and B. In manufacturing Part A, 9 out of 100 are likely to be defective and in manufacturing Part B, 5 out of 100 are likely to be defective. Calculate the probability that both the instruments will not be defective.

(a) 0.91 (b) 0.86
(c) 0.95 (d) 0.83

53. A committee of three members is to be selected out of 3 men and 2 women. What is probability that the committee has at least one woman? ***[SBI(PO)-08]***

(a) 1/14 (b) 1/7
(c) 5/17 (d) 2/17
(e) None of these

54. The probability that a computer company will get a computer hardware contract is 2/3 and the probability that it will not get a software contract is 5/9. If the probability of getting atleast one contract is 4/5, what is the probability that it will get both the contract ? ***[MAT-10, 08]***

(a) 14/45 (b) 17/45
(c) 16/45 (d) 11/45

55. In a competition A, B and C are participating. The probability that A wins is twice that of B, the probability that B wins is twice that of C. The probability that A loses is ***[MAT-13]***

(a) 3/4 (b) 1/2
(c) 1/4 (d) None of these

Directions (Q. Nos. 56 - 60): *Study the given information carefully to answer the questions that follow.* ***[IBPS-CWE(PO/MT)-11, SBI(PO)-10, OBC(PO)-09, BOM(AO)-09]***

A urn contains 4 green, 5 blue, 2 red and 3 yellow marbles.

56. If two marbles are drawn at random, what is the probability that both are red or at least one is red ?
(a) 26/51 (b) 1/7
(c) 199/364 (d) 133/191
(e) None of these

57. If three marbles are drawn at random, what is the probability that at least one is yellow ?
(a) 1/3 (b) 165/364
(c) 199/364 (d) 3/11
(e) None of these

58. If eight marbles are drawn at random, what is the probability that there are equal numbers of marbles of each color ?
(a) 4/7 (b) 361/728
(c) 60/1001 (d) 1/1
(e) None of these

59. If three marbles are drawn at random, what is the probability that none is green ?
(a) 2/7 (b) 253/728
(c) 10/21 (d) 14/91
(e) 30/91

60. If four marbles are drawn at random, what is the probability that two are blue and two are red?
(a) 10/1001 (b) 9/14
(c) 17/364 (d) 2/7
(e) None of these

61. A point is selected at random from the interior of a circle. The probability that the point is closer to the centre than boundary of the circle is ***[MAT-13]***
(a) 3/4 (b) 1/2
(c) 1/4 (d) None of these

62. There are two urns. One contains two white balls and four red balls, the other contains three white and nine red balls. All balls are of the same shape and size. From each urn, one ball is drawn. What is the probability of getting both the balls of the same colour ? ***[CLAT-17]***
(a) 1/24 (b) 7/12 (c) 1/12 (d) 1/2

63. Namrata wants to visit four cities *A*, *B*, *C* and *D* on an official trip. The probability that she visits *A* just before *B* is
(a) 1/2 (b) 1/12 (c) 1/6 (d) 1/4

64. A die is rolled twice. What is the probability that the number in the second roll will be higher than that in the first ? ***[XAT-17]***
(a) 5/36 (b) 8/36
(c) 15/36 (d) 21/36
(e) None of these

65. Two bags *A* and *B* contain different colours of balls. Bag *A* contains 2 red balls, 4 green balls and 5 white balls whereas there are 4 red balls, x green balls and 4 white balls are in Bag *B*. Find the value of x provided the probability of drawing one red ball from Bag *A* is 2/11 less than that of drawing one white ball from Bag *B*. ***[IBPS(PO-mains)-18]***
(a) 4 (b) 3 (c) 6 (d) 5
(e) None of these

66. A physical therapist of Russian football team knows that the team will play 40% of its matches on artificial turf, this season. Because of his vast experience he knows that a football player's chance of incurring a knee injury is 50% higher if he is playing on artificial turf instead of grass. If the player's chances of a knee injury on artificial turf is 0.42, what is the probability that a football player with knee injury, incurred the injury while playing on grass ? ***[IIFT-18]***
(a) 0.336 (b) 0.28
(c) 0.5 (d) None of these

67. A number is drawn at random between 1 and 100. What is the probability that it is an odd as well as a prime number ? ***[MAT-18]***
(a) 1/4 (b) 3/25
(c) 6/25 (d) 23/100

68. A bag contains 5 red balls and some blue balls. If the probability of drawing a blue ball is three times the probability of drawing a red ball, how many total balls are there ? ***[MAT-18]***
(a) 15 (b) 25
(c) 20 (d) None of these

69. Ashok has a bag containing 40 cards, numbered with the integers from 1 to 40. No two cards are numbered with the same integer. Likewise, his sister Shilpa has another bag containing only five cards that are numbered with the integers from 1 to 5, with no integer repeating. Their mother, Latha, randomely draws one card each from Ashok's and Shilpa's bags and notes down their respective numbers. If Latha divides the number obtained from Ashok's bag by the number obtained from Shilpa's, what is the probability that the remainder will not be greater than 2 ? ***[XAT-20]***
(a) 0.8 (b) 0.73
(c) 0.94 (d) 0.87
(e) 0.91

Answer Key

1. (a)	2. (d)	3. (a)	4. (b)	5. (c)	6. (d)	7. (a)	8. (d)	9. (d)	10. (a)
11. (c)	12. (a)	13. (b)	14. (d)	15. (c)	16. (b)	17. (c)	18. (d)	19. (e)	20. (c)
21. (c)	22. (d)	23. (c)	24. (c)	25. (a)	26. (d)	27. (a)	28. (a)	29. (a)	30. (b)
31. (c)	32. (b)	33. (a)	34. (d)	35. (c)	36. (d)	37. (d)	38. (a)	39. (b)	40. (c)
41. (b)	42. (d)	43. (d)	44. (a)	45. (b)	46. (d)	47. (b)	48. (d)	49. (b)	50. (c)
51. (a)	52. (b)	53. (e)	54. (a)	55. (d)	56. (e)	57. (c)	58. (c)	59. (e)	60. (a)
61. (c)	62. (b)	63. (d)	64. (c)	65. (b)	66. (c)	67. (c)	68. (c)	69. (d)	

Hints and SOLUTIONS

1. Probability of getting both balls black $= \dfrac{n(E_1)}{n(S)} = \dfrac{^3C_2}{^{12}C_2} = \dfrac{1}{22}$

and Probability of getting both balls white $= \dfrac{n(E_2)}{n(S)} = \dfrac{^4C_2}{^{12}C_2} = \dfrac{1}{11}$

Then, probability that both of them are black or white

$$= \frac{1}{22} + \frac{1}{11} = \frac{3}{22}$$

2. Clearly, both events are independents.

$$n(E) = 15$$

{i.e. 1 − 5, 1 − 4, 1 − 3, 1 − 2, 1 − 1, 2 − 4, 2 − 3, 2 − 2, 2 − 1, 3 − 3, 3 − 2, 3 − 1, 4 − 2, 4 − 1, 5 − 1}

and $\quad n(S) = 6^2 = 36$

Hence, required probability

$$= 15/36 = 5/12$$

3. The probability that LPG will last less than 60 days

$$= 1 - 0.8 = 0.2$$

and Probability that LPG will last at most 90 days = 0.6

Hence, the probability that the LPG will last at most 60 to 90 days

$$= 0.6 \quad 0.2 = 0.4$$

4. Probability of getting all 4 balls white $= \dfrac{n(E_1)}{n(S)} = \dfrac{^5C_4}{^{12}C_4} = \dfrac{5}{495} = \dfrac{1}{99}$

and probability of getting all 4 balls black $= \dfrac{n(E_2)}{n(S)} = \dfrac{^4C_4}{^{12}C_4} = \dfrac{1}{495}$

and probability of getting all 4 balls yellow = 0

{$\because$ there are only 3 yellow balls}

Then, probability that all of them are of the same colour = Probability of getting all 4 balls either white or black or yellow

$$= \frac{1}{99} + \frac{1}{495} + 0 = \frac{2}{165}$$

5. The possible number of chips = 3 or 4 or 5 or 6 or 7 or 8 or 9 or 10

$\therefore$ Total number of cases

$$= {^3C_3} + {^4C_3} + {^5C_3} + {^6C_3} + {^7C_3} + {^8C_3} + {^9C_3} + {^{10}C_3} = 330$$

Number of desired cases $= {^9C_3} = 84$

Then, required probability

$$= \frac{84}{330} = \frac{14}{55}$$

6. $n(E) = 6$ {i.e. (1,1,1), (2,2,2),... (6,6,6)}

and $\quad n(S) = 6^3$

Then, $P(E) = \dfrac{6}{6^3} = \dfrac{1}{36}$

7. Required probability $= \dfrac{^5C_4}{^8C_4} = \dfrac{1}{14}$

8. $10^{29} = 2^{29} \times 5^{29}$

$\therefore$ Total number of positive divisors

$$= (29 + 1) \times (29 + 1)$$

$$= 900$$

Further, $10^{29} = 10^{23} \times 10^6$

$$= 10^{23} \times 2^6 \times 5^6$$

$\therefore$ Number of positive divisors, multiple of $10^{23} = (6 + 1)\cdot(6 + 1) = 49$

Now, required probability $= \dfrac{49}{900}$

$$\Rightarrow \quad \frac{a^2}{b^2} = \frac{49}{900}$$

$$\Rightarrow \quad \frac{a}{b} \quad \frac{7}{30}$$

Hence, $a = 7$ and $b = 30$

Then, $b - a = 30 - 7 = 23$

9. Probability of drawing all four blue balls = 1/2 × 1/2 × 1/2 × 1/2

$$= \frac{1}{16}$$

Then, probability of drawing at least one red ball $= 1 - \dfrac{1}{16} = \dfrac{15}{16}$

10. **For the first cube**

Probability of getting red face $= \dfrac{5}{6}$

and Probability of getting blue face $= \dfrac{1}{6}$

For the second cube

Probability of getting red face $= \dfrac{x}{6}$

{let there be x red faces on 2nd cube}

and probability of getting blue face

$$= \frac{6 - x}{6}$$

As per question,

Probability of getting red face on 1st cube *AND* probability of getting red face on 2nd cube *OR* Probability of getting blue face on 1st cube *AND* probability of getting blue face on 2nd cube = 1/2

$$\Rightarrow \left\{\frac{5}{6} \times \frac{x}{6}\right\} + \left\{\frac{1}{6} \times \frac{6 - x}{6}\right\} = \frac{1}{2}$$

$$\Rightarrow \quad 5x + 6 - x = 18$$

$$\Rightarrow \quad 4x = 12$$

$$\Rightarrow \quad x = 3$$

Hence, number of red faces on the second cube = 3

11. Here, $n = 3$

m = either 2 or 3

p = probability of getting six in one trial = 1/6

Then, required probability

$$= \left\{{^3C_2} \times \left(\frac{1}{6}\right)^2 \times \left(1 - \frac{1}{6}\right)^{3-2}\right\} + \left\{{^3C_3} \times \left(\frac{1}{6}\right)^3 \times \left(1 - \frac{1}{6}\right)^{3-3}\right\}$$

$$= \frac{2}{27}$$

12. $n(S) = {^{20}C_2} = 190$; $n(E) = 10 \times 10$

{$\because$ there are 10 odd and 10 even numbers in 20 consecutive integers and also, for the sum to be odd one number should be odd and another even}

Then, required probability

$$= \frac{10 \times 10}{190} = \frac{10}{19}$$

13. Number of ways to check 5 answer sheets by 9 professors $= 9^5$

Number of ways to select 2 professors out of 9 professors

$$= {^9C_2} = 36$$

Number of ways to check 5 answer sheet by two particular professors

$$- 2^5 = 32$$

Required probability $= \dfrac{36 \times (32 - 2)}{9^5}$

{$\because$ only one professor can check all the 5 answer sheets}

$$= \frac{40}{2187}$$

14. Probability that both will live 15 yr $=\frac{7}{8}\times\frac{9}{10}=\frac{63}{80}$

Then, probability that both will live after 15 yr $=1-\frac{63}{80}=\frac{17}{80}$

15. Here, $n=5, m=4$ and $p=$ probability of being swimmer $=1-\frac{1}{5}=\frac{4}{5}$

Then, required probability

$={}^nC_m \times p^m \times (1-p)^{n-m}$

$={}^5C_4 \times \left(\frac{4}{5}\right)^4 \times \left(1-\frac{4}{5}\right)^{5-4}$

$={}^5C_4\left(\frac{4}{5}\right)^4\left(\frac{1}{5}\right)$

16. Required probability $=\frac{0.1}{0.1+0.32}=\frac{5}{21}$

17. Leap year $\Rightarrow$ 366 days = 52 weeks + 2 days = 52 Sundays + (Sat, **Sun**) or (**Sun**, **Mon**) or (**Mon**, Tues) or (Tues, Wed) or (Wed, Thu) or (Thu, Fri) or (Fri, Sat)

Clearly, probability of being either 53 Sundays or 53 Mondays

$=\frac{n(E)}{n(S)}=\frac{3}{7}$

18. $n(S)=6$ {i.e. (1,1), (1,2), (1, 3), (2,1),(3, 1)(2,2)}

and $n(E)=2$ {i.e. (1,2), (2,1)}

$\therefore \quad P(E)=2/6=1/3$

19. Probability of getting none of 2 cards a king $=b=\frac{{}^4C_2}{{}^6C_2}=\frac{2}{5}$

$\therefore$ Probability of getting at least one king $=a=1-\frac{2}{5}=\frac{3}{5}$

Now, $\frac{a}{b}=\frac{\frac{3}{5}}{\frac{2}{5}}=\frac{3}{2}=1.5>1.25$

Hence, option(e) is correct.

20. Probability that a girl is selected

$=\frac{25}{100}=0.25$

Probability that a rich person is selected $=\frac{20}{100}=0.2$

Probability that a fair complexioned person is selected $=\frac{40}{100}=0.4$

Then, probability of selecting a rich *AND* fair complexioned *AND* girl

$=0.25\times 0.2\times 0.4=0.02$

Hence, required probability = 0.02

21. Required probability

$=0.5\times 0.7+0.3\times 0.6+0.2\times 0.5$

$=0.63$

22. Number of ways for leaving at different floors $={}^8P_5$

and total number of ways for leaving $=8^5$ {recall the concept of permutation}

Then, required probability

$=\frac{{}^8P_5}{8^5}=\frac{105}{512}$

23. Total number of cards

$=2\times 52=104$

and total number of jacks $=2\times 4=8$

Probability of getting a jack in first draw $=\frac{{}^8C_1}{{}^{104}C_1}=\frac{1}{13}$

and probability of getting a jack in second draw $=\frac{{}^7C_1}{{}^{103}C_1}=\frac{7}{103}$

Then, required probability

$=\frac{1}{13}\times\frac{7}{103}=\frac{7}{1339}$

24. Since there are 5 cars between two particular cars, the positions of two particular cars must be at 1 and 7, 2 and 8, 3 and 9 and 4 and 10 imagining that 1, 2, 3, ...,10 are positions of the cars where they are to be parked.

Further, these two particular cars can be arranged in 2! ways as they can interchange their positions.

Also, remaining 8 cars (including 5 cars between two particulars cars) can be arranged in 8! ways.

Hence, number of ways of arranging 5 cars between two particular cars

$=4\times 2!\times 8!$

{Alternatively, number of ways of arranging 5 cars between two particular cars

$=4\times {}^8C_5\times 5!\times 3!\times 2!$

$=4\times 8!\times 2!$}

and number of total arrangement

$=10!$

$\therefore$ Required probability $=\frac{4\times 2!\times 8!}{10!}$

$=\frac{8\times 8!}{10!}$

25. Required probability $=\frac{{}^{132}C_1}{{}^{144}C_1}=\frac{11}{12}$

{$\because$ total number of pens $=132+12=144$}

26. **Case (i)** $n(E)=2$ {i.e. 1, 3} and $n(S)=6$. Then, $P(E)=2/6=1/3$

Case (ii) $n(E)=4$ {i.e. 3, 4, 5, 6}

and $\quad n(S)=6$

Then, $\quad P(E)=4/6=2/3$

27. Total number of boys $=100-55=45$

Number of boys studying statistics $=36$

$\therefore$ Number of boys not studying statistics $=45-36=9$

Then, required probability

$=\frac{9}{45}=\frac{1}{5}$

28. Probability of getting red ball irrespective of sizes $=\frac{6}{6+7+5}=\frac{1}{3}$

Then, probability of getting smallest red ball $=\frac{1}{3\times 6}=\frac{1}{18}$

29. **Case(i)** $W+W+B$

Case (ii) $W+B+B$

Case(iii) $B+B+B$

Case(iv) $B+W+B$

Probability in case(i)

$=\frac{{}^2C_1}{{}^4C_1}\times\frac{{}^1C_1}{{}^3C_1}\times\frac{{}^2C_1}{{}^2C_1}=\frac{1}{6}$

Probability in case(ii)

$=\frac{{}^2C_1}{{}^4C_1}\times\frac{{}^2C_1}{{}^3C_1}\times\frac{{}^3C_1}{{}^4C_1}=\frac{1}{4}$

Probability in case(iii)

$=\frac{{}^2C_1}{{}^4C_1}\times\frac{{}^3C_1}{{}^5C_1}\times\frac{{}^4C_1}{{}^6C_1}=\frac{1}{5}$

Probability in case(iv)

$=\frac{{}^2C_1}{{}^4C_1}\times\frac{{}^2C_1}{{}^5C_1}\times\frac{{}^3C_1}{{}^4C_1}=\frac{3}{20}$

$\therefore$ Required probability

$=\frac{1}{6}+\frac{1}{4}+\frac{1}{5}+\frac{3}{20}=\frac{23}{30}$

30. Probability that A wins

$=6/12=1/2$

and probability that B wins

$=4/12=1/3$

There will be two situations

Case (i) A wins then B wins and then A wins

Case (ii) B wins then B wins and then B wins

Probability for case (i)

$=\frac{1}{2}\times\frac{1}{3}\times\frac{1}{2}=\frac{1}{12}$

and probability for case (ii)

$= \frac{1}{3} \times \frac{1}{2} \times \frac{1}{3} = \frac{1}{18}$

Then, required probability

$= \frac{1}{12} + \frac{1}{18}$

$= \frac{5}{36}$

31. It is obvious that all three digits must be odd to yield the odd product. Here, there are 5 odd digits.

Required probability $= \frac{^5C_3}{^9C_3} = \frac{5}{42}$

32. Probability of not happening in 3 trials

$= {}^3C_3 \times (1-0.4)^3 \times \{1-(1-0.4)\}^{3-3}$

$= \frac{27}{125}$

Then, probability of happening

$= 1 - \frac{27}{125} = \frac{98}{125}$

33. Event ⇒ winning the prize

$n(E) = 16$ {i.e. 1-1, 2-2, 3-3, ...,16-16}

and $n(S) = 16^2 = 256$

∴ Probability that they win the prize

$= \frac{16}{256} = \frac{1}{16}$

Then, probability that they will not win the prize $= 1 - \frac{1}{16} = \frac{15}{16}$

34. The probability of not hitting in 1st shot $= 1 - 0.4 = 0.6$

The probability of not hitting in 2nd shot $= 1 - 0.3 = 0.7$

The probability of not hitting in 3rd shot $= 1 - 0.2 = 0.8$

The probability of not hitting in 4th shot $= 1 - 0.1 = 0.9$

Hence, probability of not hitting the plane $= 0.6 \times 0.7 \times 0.8 \times 0.9$

$= 0.3024$

Then, probability of hitting the plane $= 1 - 0.3024 = 0.6976$

35. Probability of getting exactly 3 defectives in the examination of first 8 pens $= \frac{^4C_3 \times {}^{11}C_5}{^{15}C_8}$

and probability of getting ninth one defective $= \frac{1}{7}$

{∵ 8 have already been examined}

Hence, required probability

$= \frac{^4C_3 \times {}^{11}C_5}{^{15}C_8} \times \frac{1}{7} = \frac{8}{195}$

Alternatively,

Required probability

$= \frac{^8C_3 \times {}^4C_3}{^{15}C_8} \times \frac{1}{(15-3)} = \frac{8}{195}$

36. $n(E) = {}^6C_3$ and $n(S) = {}^{10}C_3$

∴ Required probability

$= \frac{^6C_3}{^{10}C_3} = \frac{6 \times 5 \times 4}{10 \times 9 \times 8} = \frac{1}{6}$

37. Probability of getting a red ball from Bag $A = 3/5$

and probability of getting a red ball from Bag $B = 5/9$

Then, required probability

$= \frac{5/9}{\frac{3}{5} + \frac{5}{9}} = \frac{25}{52}$

38. It is obvious that he does not get a heart in 1st and 2nd trials. But in 3rd trial, he gets a heart.

Probability in 1st trial $= \frac{52-13}{52}$

{∵ there are 13 hearts in a pack of 52}

$= \frac{39}{52} = \frac{3}{4}$

Probability in 2nd trial $= \frac{39}{52} = \frac{3}{4}$

Probability in 3rd trial $= \frac{13}{52} = \frac{1}{4}$

{∵ he gets the desired card}

∴ Required probability

$= \frac{3}{4} \times \frac{3}{4} \times \frac{1}{4} = \frac{9}{64}$

39. Probability of one step forward

$= {}^{11}C_6 \times (0.4)^6 \times (1-0.4)^{11-6}$

and probability of one step backward

$= {}^{11}C_6 \times (0.6)^6 \times (1-0.6)^{11-6}$

Then, required probability

$= \{{}^{11}C_6 \times (0.4)^6 \times (1-0.4)^{11-6}\}$
$\quad + \{{}^{11}C_6 \times (0.6)^6 \times (1-0.6)^{11-6}\}$

$= {}^{11}C_6 \times (0.4)^5 \times (0.6)^5 \{0.4 + 0.6\}$

$= 462 \times \left(\frac{6}{25}\right)^5$

40. Here, $n = 3$, $m = 3$

and p = probability of getting either 6 or 7 or 8 $= \frac{3}{8}$

Then, required probability

$= {}^nC_m \times p^m \times (1-p)^{n-m}$

$= {}^3C_3 \times \left(\frac{3}{8}\right)^3 \times \left(1 - \frac{3}{8}\right)^{3-3} = \left(\frac{3}{8}\right)^3$

41. Probability that A hits the target in one shot $= 4/5$

Probability that B hits the target in one shot $= 3/4$

Probability that C hits the target in one shot $= 2/3$

Now,

(i) probability that all three hit the target in one shot

$= \frac{4}{5} \times \frac{3}{4} \times \frac{2}{3} = \frac{2}{5}$

(ii) probability that A and B hit the target but C does not

$= \frac{4}{5} \times \frac{3}{4} \times \left(1 - \frac{2}{3}\right) = \frac{1}{5}$

(iii) probability that B and C hit the target but A does not

$= \frac{3}{4} \times \frac{2}{3} \times \left(1 - \frac{4}{5}\right) = \frac{1}{10}$

(iv) probability that C and A hit the target but B does not

$= \frac{2}{3} \times \frac{4}{5} \times \left(1 - \frac{3}{4}\right) = \frac{2}{15}$

Atleast two shots hit the target means either any two or all three shots hit the target.

∴ Required probability

$= \frac{2}{5} + \frac{1}{5} + \frac{1}{10} + \frac{2}{15} = \frac{5}{6}$

42. Required probability

$= \left\{1/2 \times \frac{4}{(4+3)}\right\} + \left\{1/2 \times \frac{6}{(6+2)}\right\}$

$= \frac{37}{56}$

43. Total number of numbers formed without repetition $= {}^5P_5 = 120$

We know, if the number formed with last two digits of a given number is divisible by 4, the given number is divisible by 4.

Here, the numbers having last two digits as 12, 24, 32, 52 will obviously be divisible by 4. Then, number of such numbers $= 3! \times 4 = 24$

Hence, required probability

$= \frac{24}{120} = \frac{1}{5}$

44. Here, $n = 4$, $m = 4$

and p = probability of getting either 2 or 3 or 4 or 5 in one trial

$= \frac{4}{6} = \frac{2}{3}$

Then, required probability

$= {}^nC_m \times p^m \times (1-p)^{n-m}$

$$= {}^4C_4 \times \left(\frac{2}{3}\right)^4 \times \left(1-\frac{2}{3}\right)^{4-4} = \frac{16}{81}$$

Alternatively,

Required probability

$$= \frac{{}^4C_1 \times {}^4C_1 \times {}^4C_1 \times {}^4C_1}{6^4} = \frac{16}{81}$$

45. Required probability = probability of getting yellow face in first toss × probability of getting red face in second toss × probability of getting blue face in third toss

$$= \frac{3}{6} \times \frac{2}{6} \times \frac{1}{6} = \frac{1}{36}$$

46. Probability that either A or B or both occurs

$$= 0.25 + 0.50 - 0.12 = 0.63$$

Then, the probability that neither A nor B occurs $= 1 - 0.63 = 0.37$

47. A will not succeed if his all three shares (tickets) are of blanks.

$\therefore$ Probability that A will not succeed

$$= \frac{{}^9C_3}{{}^{12}C_3} = \frac{9 \times 8 \times 7}{12 \times 11 \times 10} = \frac{21}{55}$$

Hence, probability of success of A

$$= 1 - \frac{21}{55} = \frac{34}{55}$$

Further, B will not succeed if his both shares (tickets) are of blanks.

$\therefore$ Probability that B will not succeed

$$= \frac{{}^6C_2}{{}^8C_2} = \frac{6 \times 5}{8 \times 7} = \frac{15}{28}$$

Hence, probability of success of B

$$= 1 - \frac{15}{28} = \frac{13}{28}$$

$\therefore$ Required ratio

$$= \frac{34}{55} : \frac{13}{28} = 952 : 715$$

Alternatively,

Winning cases for A

: WLL, WWL, WWW

{W- Winning ticket, L- Loosing ticket}

Then, probability of success of A

$$= \frac{{}^3C_1 \times {}^9C_2}{{}^{12}C_3} + \frac{{}^3C_2 \times {}^9C_1}{{}^{12}C_3} + \frac{{}^3C_3 \times {}^9C_0}{{}^{12}C_3} = \frac{34}{55}$$

On the above line, probability of success of $B = \frac{13}{28}$

$\therefore$ Required ratio $= \frac{34}{55} : \frac{13}{28} = 952 : 715$

48. Probability that one resigns from A *AND* second one from B *AND* third one from C

$$= \frac{{}^2C_1 \times {}^3C_1 \times {}^4C_1}{{}^9C_3} = 2/7$$

49. Number of toys manufactured by Machine A = 25% and number of defective toys = 5% of 25% = 1.25%

Number of toys manufactured by Machine B = 35% and number of defective toys = 4% of 35% = 1.4%

Number of toys manufactured by Machine C = 40% and number of defective toys = 2% of 40% = 0.8%

Then, required probability

$$= \frac{1.4}{1.25 + 1.4 + 0.8} = \frac{28}{69}$$

50. It is obvious that the numbers must have 1, 3, 7 or 9 in units place to yield a product having 1, 3, 7 or 9 as units digit.

Then, probability that unit digit of any one of the four numbers be 1, 3, 7 or 9 $= \frac{4}{10}$

Hence, required probability

$$= \left(\frac{4}{10}\right)^4 = \frac{16}{625}$$

51. The configuration of boxes cannot be changed in following two cases

Case (i) When one orange is transferred from Box A to B and B to A

Case (ii) When one apple is transferred from Box A to B and B to A

Probability for case(i) $= \frac{5}{11} \times \frac{5}{8}$ {not 5/7 as there are 3 apples and 5 oranges in Box B at the time of selection from it}

and probability for case(ii) $= \frac{6}{11} \times \frac{4}{8}$

{not 3/7 as there are 4 apples and 4 oranges in Box B at the time of selection from it}

$\therefore$ Required probability

$$= \frac{5}{11} \times \frac{5}{8} + \frac{6}{11} \times \frac{4}{8} = \frac{49}{88}$$

52. Probability that Part A is non-defective $= 1 - 0.09 = 0.91$

and probability that part B is non-defective $= 1 - 0.05 = 0.95$

Then, probability that both parts are non-defective

$$= 0.91 \times 0.95 = 0.86$$

53. Probability that all three are men

$$= \frac{{}^3C_3}{{}^5C_3} = \frac{1}{10}$$

$\therefore$ Probability that atleast one be woman $= 1 - \frac{1}{10} = \frac{9}{10}$

54. Given, probability of getting hardware contract = 2/3

and probability of getting software contract $= 1 - 5/9 = 4/9$

and probability of getting atleast one contract = 4/5

Now, probability of getting atleast one contract = Probability of getting either hardware or software or both contract = Probability of getting one hardware contract + Probability of getting software contract – Probability of getting both contracts

$\Rightarrow \frac{4}{5} = \frac{2}{3} + \frac{4}{9} -$ Probability of getting both contracts

$\Rightarrow$ Probability of getting both contracts $= \frac{10}{9} - \frac{4}{5} = \frac{14}{45}$

55. Let the probability that C wins $= x$

Then, probability of winning of B $= 2x$

and probability of winning of $A = 4x$

Since, sum of all probability is 1, $x + 2x + 4x = 1 \Rightarrow x = 1/7$

Then, probability of winning of A

$$= \frac{4}{7}$$

$\therefore$ Probability that A loses the competition $= 1 - 4/7 = 3/7$

56. Probability of getting one red *AND* one non-red

$$= \frac{{}^2C_1 \times {}^{(14-2)}C_1}{{}^{14}C_2} = \frac{24}{91}$$

and probability of getting two red

$$= \frac{{}^2C_2}{{}^{14}C_2} = \frac{1}{91}$$

$\therefore$ Probability of getting both marbles red or at least one red

$$= \frac{24}{91} + \frac{1}{91} = \frac{25}{91}$$

Alternatively,

Probability of getting both the marbles not being red

$$= \frac{{}^{(14-2)}C_2}{{}^{14}C_2} = \frac{66}{91}$$

$\therefore$ Probability of getting at least one red $= 1 - \frac{66}{91} = \frac{25}{91}$

57. Probability of getting all three marbles not being yellow

$$= \frac{^{(14-3)}C_3}{^{14}C_3} = \frac{165}{364}$$

$\therefore$ Probability of getting at least one yellow $= 1 - \frac{165}{364} = \frac{199}{364}$

58. Obviously, we have to find probability of getting 2 green *AND* 2 blue, *AND* 2 red *AND* 2 yellow marbles.

Then, required probability

$$= \frac{^4C_2 \times {^5C_2} \times {^2C_2} \times {^3C_2}}{^{14}C_8}$$

$$= \frac{180}{3003} = \frac{60}{1001}$$

59. Probability that none is green

$$= \frac{^{(14-4)}C_3}{^{14}C_3} = \frac{30}{91}$$

60. Probability of getting two blue *AND* two red marbles

$$= \frac{^5C_2 \times {^2C_2}}{^{14}C_4} = \frac{10}{1001}$$

61. A point can be said to be closer to the centre of a circle of radius r than its boundary when it lies within area of the concentric circle of radius $r/2$.

Then, required probability

$$= \frac{\text{area of circle of radius } r/2}{\text{area of the circle}}$$

$$= \frac{\pi\left(\frac{r}{2}\right)^2}{\pi r^2} = \frac{1}{4}$$

62. Event

$\Rightarrow$ 1 white from 1st urn *AND* 1 white from 2nd urn

OR

1 red from 1st urn *AND* 1 red from 2nd urn

Probability for 1st case

$$= \frac{^2C_1}{^6C_1} \times \frac{^3C_1}{^{12}C_1} = \frac{1}{12}$$

and probability for 2nd case

$$= \frac{^4C_1}{^6C_1} \times \frac{^9C_1}{^{12}C_1} = \frac{1}{2}$$

Then, required probability

$$= \frac{1}{12} + \frac{1}{2} = \frac{7}{12}$$

63. Assuming (AB) as a single term/city, number of cities = 3, which can be arranged in 3! ways. Hence, $n(E) = 3!$

Clearly, $n(S) = 4!$

$\therefore$ Required probability $= 3!/4! = 1/4$

64. Obviously, the desired event will occur if any one set out of [(1, 2), (1, 3), (1, 4), (1, 5), (1, 6), (2, 3), (2, 4), ... i.e., total number = 5 + 4 + 3 + 2 + 1 = 15] appears and it may occur in $^{15}C_1$ ways.

Then, required probability

$$= \frac{^{15}C_1}{6^2} = \frac{15}{36}$$

65. Probability of getting one red ball from Bag $A = \frac{^2C_1}{^{(2+4+5)}C_1}$

$$= \frac{2}{11}$$

and probability of getting one white ball from Bag B

$$= \frac{^4C_1}{^{(4+x+4)}C_1} = \frac{4}{8+x}$$

As given, $\frac{2}{11} = \frac{4}{8+x} - \frac{2}{11} \Rightarrow x = 3$

66. Number of matches to be played on artificial turf = 40%

and probability of being injured on artificial turf = 0.42

Further, number of matches to be played on grass= (100 − 40)% = 60%

and probability of being injured on grass $= 0.42 \times \frac{2}{3} = 0.28$

{$\because$ probability of being injured on artificial turf is 150% of that on grass}

Then, required probability

$$= \frac{60\% \times 0.28}{(40\% \times 0.42) + (60\% \times 0.28)} = 0.5$$

Alternatively,

Ratio of matches to be played on artificial turf and grass = **2** : 3 and ratio of probability of being injured on turf and grass = **3** : 2.

Clearly, $(2 \times 3) = (3 \times 2)$

Then, the required probability

$$= \frac{1}{1+1} = 0.5$$

67. We know that, there are 25 prime numbers between 1 and 100 out of which only 2 is even number.

Hence, the required probability

$$= \frac{24}{100} = \frac{6}{25}$$

68. Let there are n blue balls in that bag.

As given, $\frac{n}{5+n} = 3 \times \frac{5}{5+n}$

$\Rightarrow \quad n = 15$

Hence, total number of balls

$= 5 + 15 = 20$

69. Latha draws one card from Ashok's bag and one card from Shilpa's. Hence, total number of ways of selection $= {^{40}C_1} \times {^5C_1} = 200$

If Latha obtains a card, from Shilpa's bag, numbered with 1, the remainder will be 0 in case the integer indicated on the card, obtained from Ashok's bag is divided by 1 and of course it can be done in 40 ways (like 1/1, 2/1, 3/1…., 40/1).

Similarly, if the card drawn from Shilpa's bag shows 2, the remainder will either be 1 or 0. Hence, number of ways = 40 (like ½, 2/2, 3/2,……. 40/2). Further, if Shilpa's card indicates 3, the remainder will either be 1 or 2 or 0 and number of ways will be 40.

In case Shilpa's card shows 4, the remainder will either be 1 or 2 or 3 or 0. Since the remainder should not be greater than 2, the number of ways will be (40 − 10) i.e., 30 because every 4th number started from 3 (i.e., 3, 7, 11, …. 39) will yield 3 as remainder and such 10 remainders are to be eliminated.

Likewise, if Shilpa's card shows 5, the remainder will either be 1 or 2 or 3 or 4 or 0. Here, we have to eliminate remainders 3 and 4. Obviously, a block of 1st five consecutive integers will yield two remainders as 3 and 4. Further, next block of five consecutive integers will yield two remainders as 3 and 4 and so on. Thus, we can eliminate such 16 ways. Then, number of ways

$= 40 - 16 = 24$

Then, the required probability

$$= \frac{40 + 40 + 40 + 30 + 24}{200}$$

$= 0.87$

Alternatively,

We may focus on only two cards numbered with 4 and 5, obtained from Shilpa's bag. In case Shilpa's card shows 4, 10 ways will be rejected. Likewise, if Shilpa's card shows 5, 16 ways will be rejected. Then, required probability

$$= 1 - \frac{10+16}{200} = 0.87$$

Unit 3

Percentage and Its Extensive Applications

- Percentage
- Profit and Loss
- Discount
- Simple Interest
- Compound Interest
- Time and Work
- Work and Wages
- Pipes and Cisterns

18

Percentage

Learning OBJECTIVES

After completing this chapter, aspirants will be able to :

- Understand the meaning of percentage and concept of base-value for percentage determination;
- Understand the pattern of conversion of fraction into percentage and *vice-versa*; the pattern of conversion of ratio into percentage and *vice-versa*; and the pattern of conversion of decimals into percentage and *vice-versa*;
- Recognise the significance of fraction to Percentage Conversion Table in solution of questions based on percentage;
- Realise that the fraction to Percentage Conversion Table is also a critical instrument for other chapters like Number System, Profit and Loss, Simple Interest, Compound Interest, Time and Work, Time and Distance, Data Interpretation, etc.;
- Express one number as percentage of another one;
- Solve the question based on pattern of Missing Figure;
- Understand the conception of 'Product Constancy' in diverse cases;
- Compute the effect on number in case of Successive Increase/Decrease/ Combination of both;
- Analyse the effect on product in case of changes in factor(s);
- Understand the application of the percentage in diverse sections like Examinations, Elections, Numbers, Income and Expenditures etc.;
- Understand the philosophy of the division of this chapter in various parts;
- Analyse the trend of the questions being asked in various competitive examinations;
- Recognise the types of questions which can be easily solved through the given options and solve the questions in moments;
- Grip the concepts helping to solve the questions without pen and paper;
- Learn new concepts/fundamentals;
- Apply the pragmatic approaches in lieu of academic approaches; and
- Ponder on the alternative methods.

Inside

Introduction

Percentage means parts per hundred. The word comes from the latin phrase per centum which means per hundred. In other words, we may say that a percent is a ratio whose second term is 100. It is denoted by a symbol '%'.

This chapter is very critical as a lot of questions envisaged in the most of the chapters are based on the concept of the percentage. It is imperative to mention that the central concepts associated with the chapters, *viz.* Profit and Loss, Discount, Simple interest, Compound Interest, Time and Work, Pipes and Cisterns, Work and Wages are based on the conception of percentage. Hence, these chapters may broadly be termed as extension of the conception of percentage and its applications. Further, the acquaintance with conception of Percentage also leads in field of Financial Management as well as day to day professional life. In addition, chapters Time and Distance, Data Interpretation, etc. also require the conception of percentage.

Conversion of Fraction Into Term of Percentage and *Vice-Versa*

Suppose you have to convert $\frac{3}{4}$ into term of percentage.

Literally, $\frac{3}{4}$ means, out of 4 $\Rightarrow$ 3

Then, out of 1 $\Rightarrow \frac{3}{4}$ and out of 100 $\Rightarrow \frac{3}{4} \times 100 = 75$

Here, we evaluated the numerator term with reference to 100 as denominator. It means out of 100, we are obtaining 75 and it is denoted as 75%. Thus, $\frac{3}{4} = 75\%$.

Considering the above facts, we may say that in order to convert a fraction into term of percentage, we should multiply the fraction by 100. Conversely, for converting the percentage term into fraction, we should divide the percentage term by 100.

$$\text{Fraction} \times 100 = \text{Percentage}$$

or, $$\text{Percentage} \div 100 = \text{Fraction}$$

Problem 1. Convert $\frac{5}{6}$ into percentage.

Sol. $\frac{5}{6} \times 100 = 83.33\%$

Problem 2. Convert 35% into fraction.

Sol. $\frac{35}{100} = \frac{7}{20}$

Conversion of Ratio Into Term of Percentage and *Vice-Versa*

We write the given ratio in term of fraction and then convert the same in percentage term. Conversely, in order to convert the percentage into ratio, we obtain fraction and then write in term of ratio. For Example

Problem 3. Convert 2 : 3 into percentage.

Sol. Clearly, 2 : 3 means $\frac{2}{3}$

Then, $\frac{2}{3} = \frac{2}{3} \times 100\% = 66.66\%$

Problem 4. Convert 46% into term of ratio.

Sol. $\frac{46}{100} = \frac{23}{50} = 23 : 50$

Conversion of Decimals Into Term of Percentage and *Vice-Versa*

Since Decimals have technically the same meaning as fraction, in order to convert a decimal into term of percentage, we multiply the decimal by 100. Conversely, for converting the percentage term into decimal, we divide the percentage term by 100. For Example

Problem 5. Convert 39% into decimals.

Sol. $\frac{39}{100} = 0.39$

Problem 6. Convert 0.45 into term of percentage.

Sol. $0.45 \times 100 = 45\%$

Percentage of a Quantity

Suppose, you have to find the value of 5% of 326. It means you have to find the numerator term with reference to 326 as denominator. Given, out of 100 $\Rightarrow$ 5

Then, out of 326 $\Rightarrow \frac{5}{100} \times 326 = 16.3$

In another way, 5% of 326 = 326 × 5%

$$= 326 \times \frac{5}{100} = 16.3$$

Obviously, we should proceed through second approach to find the required value.

Problem 7. Find the value of 14.28% of 854.

Sol. $14.28\% \text{ of } 854 = 854 \times \frac{14.28}{100}$

It seems to be tough to calculate or say, it will take much time to arrive at final answer. If you had memorized/known that 14.28% is equal to $\frac{1}{7}$, you would have solved in moments as under

$\Rightarrow \quad 14.28\% \text{ of } 854 = 854 \times \frac{1}{7}$

$= 122$

Problem 8. Find the value of 6.66% of 345.

Sol. $6.66\% \text{ of } 345 = 345 \times \frac{6.66}{100}$

Oh, again tough to calculate.

Alternatively,

$6.66\% \text{ of } 345 = 345 \times \frac{1}{15} = 23 \quad \left[\because 6.66\% = \frac{1}{15}\right]$

Observations The problems like Q. No. 7 and 8 are compelling us to memorise a basic chart depicting the value of a fraction in term of percentage.

Fraction to Percentage Conversion and *Vice-Versa* Table

	NUMERATORS								
		1	2	3	4	5	6	7	8
	1	100							
	2	50							
	3	33.33	66.66						
	4	25		75					
	5	20	40	60	80				
D	6	16.66	33.33			83.33			
E	7	14.28	28.57	42.86	57.14	71.43	85.71		
N	8	12.5		37.5		62.5		87.5	
O	9	11.11	22.22		44.44	55.55		77.77	88.88
M	10	10							
I	11	9.09	18.18	27.27	36.36	45.45	54.54	63.63	
N	12	8.33				41.66		58.33	
A	15	6.66	13.33		26.66				
T	16	6.25		18.75		31.25			
O	18	5.55							
R	20	5							
S	25	4	8	12					
	30	3.33							
	35	2.85	5.71	8.57					
	40	2.5		7.5					
	45	2.22							

Figures in white Background are in Percentage term

Aspirants are requested to get it by heart as it is very important tool for solving the problems envisaged in chapters like Number System, Profit and Loss, Simple Interest, Compound Interest, Time and Work, Time and Distance, Data Interpretation, etc.

Problem 9. $16\frac{2}{3}\%$ of 600 gm $- 33\frac{1}{3}\%$ of 180 gm = ?

Sol. Observe the Fraction to Percentage Conversion Table [column 2 and row 7] where in figure 16.66 $\left(\text{i.e. } 16\frac{2}{3}\right)$ in percentage has been written. For this figure, numerator is 1 and denominator is 6.

It means $16\frac{2}{3}\% = \frac{1}{6}$.

In the same way, $33\frac{1}{3}\% = \frac{1}{3}$.

Now, $600 \times \frac{1}{6} - 180 \times \frac{1}{3} = 100 - 60$

$= 40$ gm.

Problem 10. 83.33% of 684 + 10.23 = ?

Sol. 83.33% of $684 + 10.23 = 684 \times \frac{5}{6} + 10.23$

$= 570 + 10.23 = 580.23$

Problem 11. $93\frac{1}{3}\%$ of 465 = ?

Sol. It is not given in chart and it is also fact that if Percentage Conversion Table is filled up with so many figures, it is not possible to retain in mind for a long period. Now, we will try to solve through another approach.

$93\frac{1}{3}\%$ of $465 = 465 \times (100 - 6.66)\%$

$= 465 \times 100\% - 465 \times 6.66\%$

$= 465 - 465 \times \frac{1}{15} = 434$

In other words, it may be said that the required value is $465 \times \frac{14}{15}$

Now try to solve quickly without pen and paper, 87.5% of 800 = ?

Problem 12. 131.25% of 3376 = ?

Sol. 131.25% of $3376 = (100\% + 31.25\%)$ of 3376

$= 3376 \times 100\% + 3376 \times 31.25\%$

$= 3376 + 3376 \times \frac{5}{16}$

$= 3376 + 1055 = 4431$

Problem 13. 12% of 493 − 5.02 = ?

Sol. $493 \times \frac{12}{100} - 5.02 = 59.16 - 5.02 = 54.14$

{Here we did not write 12% of 493 as $493 \times \frac{1}{8.33}$ because it is difficult to divide 493 by 8.33 in comparison to multiply 493 by 12 and then divided by 100. Thus, don't be rigid to convert percentage into fraction all times.}

Problem 14. 64.4% of 856 = ?

Sol. 64.4% of $856 = 856 \times \frac{64.4}{100} = 551.264$

[If you don't have multiplication trick in mind, it will take some more time. Needless to say that tricks have been given under Multiplication chapter]. Figure 64.4 is also not appeared in the Percentage Conversion Table. We may try to solve it by another approach.

It is very easy to find out 100%, 10%, 1%, 0.1%, 0.01%, 0.001% of any number. Our approach will depend upon these values.

64.4% of 856 = [50% + 10% + (4 × 1%) + (4 × 0.1%)] of 856

$\Rightarrow$ 50% of 856 = 428

10% of 856 = 85.6

1% of 856 = 8.56

$\therefore$ 4% of 856 = 4 × 8.56 = 34.24

0.1% of 856 = 0.856

$\therefore$ 0.4% of 856 = 4 × 0.856 = 3.424

On addition, **551.264**

{The whole process can be done mentally without pen and paper, try please}

Problem 15. 43.2% of 1096 = ?

Sol. $\{4 \times 10\% + 3 \times 1\% + 2 \times 0.1\%\}$ of 1096

$= 4 \times 109.6 + 3 \times 10.96 + 2 \times 1.096$

$= 438.4 + 32.88 + 2.192 = 473.472$

[you may find the answer from options much before the final step of the process. Hence, no need to go ahead]

Expressing one Number as Percentage of Another One

e.g., $\frac{1}{2}$ is what percent of $\frac{1}{3}$?

Given, out of $\frac{1}{3} \Rightarrow \frac{1}{2}$

Then, out of $100 \Rightarrow \frac{\frac{1}{2}}{\frac{1}{3}} \times 100 = 150$

Hence, $\frac{1}{2}$ is 150% of $\frac{1}{3}$

Observation

Required percentage

$$= \left(\frac{\text{Number to be expressed in percentage}}{\text{another number}} \times 100 \right) \%$$

Problem 16. What percent of ₹ 13.50 is 27 paise?

Sol. We should convert both the quantities in the same unit. Here, ₹ 13.50 = 1350 paise. Then, required percentage =

$$\left(\frac{\text{Number to be expressed in percentage}}{\text{another number}} \times 100 \right) \%$$

$$= \left(\frac{27}{1350} \times 100 \right) \% = 2\%$$

Hence, 2% of ₹ 13.50 is 27 paise.

Pattern of Missing Figure

E.g. ? % of 16 = 0.48

We may say that the question is "what percent of 16 is 0.48". The spirit of the question is as same as that of the above question.

Then, required percentage $= \left(\frac{0.48}{16} \times 100 \right) \% = 3\%$

Hence, 3% of 16 is 0.48.

Alternatively

We should bring the question in this following form, if required, **Number ×? = single number or product of number but** not in form of addition of subtraction.

Then, required answer/Missing figure

$$= \left\{ \frac{\text{Number (single or in product form only) of opposite side of ?}}{\text{number attached with ?}} \times 100 \right\}$$

$$= \frac{0.48}{16} \times 100 = 3$$

Problem 17. 0.35% of ? = 0.252

Sol. Missing figure $= \frac{0.252}{0.35} \times 100 = 72$

Problem 18. 14% of 250 × ?% of 150 = 840

Sol. Missing figure $= \frac{840}{250 \times \frac{14}{100} \times 150} \times 100 = 16$

Problem 19. 5083 + 25% of ? + 289 = 6385.5

[PNB-PO-10]

Sol. 5083 + 25% of ? + 289 = 6385.5

$\Rightarrow$ 25% of ? = 6385.5 − 5083 − 289

= 1013.5

Now, missing figure $= \frac{1013.5}{25} \times 100 = 4054$

Problem 20. 36% of 365 + ? % of 56.2 = 156.69

[IDBI-PO-10]

Sol. $\Rightarrow$?% of 56.2 $= 156.69 - 365 \times \frac{36}{100}$

$\Rightarrow$?% of 56.2 $= 156.69 - 730 \times \frac{18}{100}$

[$\because$ it is easier to multiply by 18]

$\Rightarrow$? % of 56.2 = 156.69 − 131.40

= 25.29

Now, missing figure $= \frac{25.29}{56.2} \times 100 = 45$

Base Value for Percentage Determination

We cannot find out the percentage without knowing the base on which it is to be calculated. Thus, whenever you are asked to find out the percentage, you have to try first to find out the answer to the question 'Percentage of what base?. Take some examples

Problem 21. *B* got 20% marks less than *A*. What percent marks did *A* get more than *B*?

Sol. Let marks obtained by *A* be 100.

$20\% \text{ of } 100 = 100 \times \frac{20}{100} = 20$ [Here, base value is 100]

Then, marks obtained by

B = 100 − 20 = 80

Now, the question is that by how much percent is *A*'s marks more than *B* ? Obviously, we have to find the percentage value on base 80.

$\because$ When *B*'s marks is 80, *A*'s marks is 20 more than *B*'s

$\therefore$ .... 1, ... $\frac{20}{80}$ ....

$\therefore$ .... 100, ... $\frac{20}{80} \times 100 = 25$

Hence, *A*'s marks is 25% more than *B*'s.

Problem 22. The income of *X* is 20% more than that of *Y*. By how much percent is income of *Y* less than *X*'s income?

[SSC(CGL-T-I)-08-02-01-2000, CPO-10/06/05/03, CDS-10]

Sol. Let the income of Y be 100.

Then, income of $X = 120$

$\because$ When X's income is 120, Y's income is 20 less than X's.

$\therefore$ .... 1, ... $\frac{20}{120}$ ....

$\therefore$ .... 100, ... $\frac{20}{120} \times 100 = 16.66$

Hence, income of Y is 16.66% less than that of X.

Observations

1. More or increase %

$$= \left\{\frac{\text{Less or Decrease\%}}{100 - \text{Less or Decrease\%}} \times 100\right\} \%$$

$$\left\{\text{i.e.} \frac{20}{(100-20)} \times 100\right\} \%$$

2. Less or decrease %

$$= \left\{\frac{\text{More or increase \%}}{(100 + \text{More or Increase\%})} \times 100\right\} \%$$

$$\left\{\text{i.e.} \frac{20}{(100+20)} \times 100\right\} \%$$

Alternative approach (Fraction method)

1. Given, less % = 20% = $\frac{1}{5}\left(\text{say}, \frac{n}{d}\right)$ then,

More or increase % = $\left[\frac{1}{(5-1)} \times 100\right]$ %

$$\left[\text{say}, \left\{\frac{n}{(d-n)} \times 100\right\} \%\right]$$

2. Given, more % = 20% = $\frac{1}{5}\left(\text{say}, \frac{n}{d}\right)$ then,

Less or decrease % = $\left[\frac{1}{(5+1)} \times 100\right]$ %

$$\left[\text{say}, \left\{\frac{n}{(d+n)} \times 100\right\} \%\right]$$

[The details of this alternative approach have also been given under topic 'Concept of Product constancy'.]

Problem 23. A number is increased by 40%, then by how much percent should the resultant be reduced to get back to the original value?

Sol. Required answer = $\left[\frac{40}{(100+40)} \times 100\right]\% = 28.57\%$

Alternatively

$$40\% = \frac{2}{5}$$

$\therefore$ Required answer = $\left[\frac{2}{(5+2)} \times 100\right]\% = 28.57\%$

Problem 24. At the end of year 1998, Shepard bought nine dozen goats. Henceforth, every year he added $p\%$ of the goats at the beginning of the year and sold $q\%$ of the goats at the end of the year where $p > 0$ and $q > 0$. If Shepard had nine dozen goats at the end of the year, after making the sales for that year, which of the following is true? ***[CAT-03]***

(a) $p = q$ (b) $p < q$

(c) $p > q$ (d) $p = \frac{q}{2}$

Sol. If a number is increased by x%, then the resultant must be reduced by $\left[\frac{x}{(100+x)} \times 100\right]$% to get back to the original value. Here, x will always be greater than $\frac{x}{(100+x)} \times 100$.

Since the number of goats remains the same, $p > q$.

Problem 25. First and second numbers are less than a third number by 30% and 37% respectively. The second number is less than the first by : ***[SSC(CGL-T-1)-11, BOM (PO)-08, SSC(CGL-T-II)-17, MAT-18]***

Sol. Let 3rd number be 100.

Then, Ist number = 70 and 2nd number = 63

Clearly, the second number is 7 less than first number (base value).

$\because$ $70 \Rightarrow 7$

$\therefore$ $100 \Rightarrow 10$

Hence, the second number is 10% less than the first number.

Alternatively

$$\text{less \%} = \left[\frac{\text{less}}{\text{original quantity}} \times 100\right]\%$$

$$= \left[\frac{7}{70} \times 100\right]\% = 10\%$$

[The base for calculating the percentage change is always the original quantity unless otherwise instructed]

Problem 26. Number A is 30% less than B and 25% more than C. By how much percent is C less than B?

Sol. Let the number B be 100.

Then, $A = 100 - 30 = 70$, which is 25% more than number C.

It means C is $\left[\frac{25}{125} \times 100\right]$% i.e. 20% less than A.

Hence, $C = (100 - 20)\%$ of $70 = 56$

Clearly, C is 44% less than number B.

Alternatively

Let the number B be 100.

Then, $A = 100 - 30 = 70$, which is 25% more than number C.

It means, when A is 125, C is 100.

In case A is 70, C will be $\frac{100}{125} \times 70$ i.e. 56

Clearly, C is 44% less than number B.

Problem 27. C's income is 20% more than B's and B's income is 25% more than A's. By how much percentage is C's income more than A's? *[SSC(CGL-PT)-99, CDS-11]*

Sol. Let A's income be 100.

Then B's income = 125

and C's income = 120% of 125 = $125 \times \frac{120}{100} = 150$

Clearly, C's income is 50% more than A's.

Concept of Product Constancy

Suppose, we buy 100 pens at the rate of ₹ 5 per pen. Then, total expenditure = number of pens × rate = 100×5 = ₹ 500. Now, rate is increased by 25%, however, we are inclined not to increase our expenditure.

It means we will buy less number of pens with the same expenditure. Number of pens after rate increase

$$= \frac{500}{5 \times 125\%} = 80$$

i.e. 20% reduction in number of pens.

Here, number of pens × price = expenditure

$$100 \times 5 = 500 \text{ (Before price hike)}$$

$$80 \times 6.25 = 500 \text{ (After price hike)}$$

We may say that here, expenditure is a product of number of pens and price. In both cases, expenditures (product) are kept same by reduction in number of pens after increase in price.

In case the price of pen is reduced by 20%, we will buy more pens with the same expenditure.

Then, number of pens = $\frac{500}{5 \times 80\%} = 125$ i.e. 25% increase in number of pens.

Observations

(a) If product (i.e. expenditure) is kept constant and one factor (i.e. price) is increased by 25% (say, x%), the other factor (i.e. number of pens) will reduce by $\left[\frac{25}{(100+25)} \times 100\right]\% \left[\text{say}, \left\{\frac{x}{(100+x)} \times 100\right\}\%\right]$

[Note : x will always be greater than $\left\{\frac{x}{(100+x)} \times 100\right\}\%$]

(b) Keeping the product constant if one factor is increased by $\frac{n}{d}$, the other factor will reduce by $\left[\frac{n}{(d+n)} \times 100\right]\%$.

(c) It also holds good that if product is kept constant and one factor (i.e. price) is decreased by x% the other factor will reduce by $\left[\frac{x}{(100-x)} \times 100\right]\%$.

[Note x will always be lesser than $\left\{\frac{x}{(100-x)} \times 100\right\}\%$]

(d) Also, keeping the product constant if one factor is decreased by $\frac{n}{d}$, the other factor will increase by $\left[\frac{n}{(d-n)} \times 100\right]\%$.

(e) This concept of product constancy is also applicable in case of following products :

(i) one number × another number = product of numbers

(ii) rate × number = price

(iii) price × quantity = expenditure

(iv) speed × time = distance

(v) rate × time = cost

(vi) time × rate = wages

(vii) average × number of quantities = total value

(viii) efficiency × time = work

(ix) one dimension × another dimension (with/without any constant term) = area of rectangle/triangle/ quadrilateral/cone etc.

(x) consumption or sales × tax = revenue and other such product.

Illustrations

Problem 28. The base of a triangle is increased by 10%. By what percent should its height be reduced so as to keep its area constant.

Sol. ⇒ Since area of the triangle is a product of base and height with a constant term $\left(\frac{1}{2}\right)$. If one factor i.e. base is increased by 10%, another factor i.e. height must be reduced by $\left\{\frac{x}{(100+x)} \times 100\right\}\%$ so as to keep its product (i.e. area) constant.

Hence, required answer = $\left\{\frac{10}{(100+10)} \times 100\right\}\% = 9.09\%$

Alternatively I

Here, base is increased by 10% i.e. by $\frac{1}{10}$, then height must be reduced by $\left\{\frac{1}{(10+1)} \times 100\right\}\%$ so as to keep its product (i.e. area) constant.

Hence, required answer = 9.09%

Alternatively II

Let base be 10 and height be 10 units

Then, area of triangle = $\frac{1}{2} \times 10 \times 10 = 50$ sq. unit.

Now, base becomes 11 from 10.

Then, $50 = \frac{1}{2} \times 11 \times \text{height}$

⇒ height = 9.09

It means, reduction in height

= 10 − 9.09 = 0.91

Then, reduction % $= \left(\frac{0.91}{10} \times 100\right)\% = 9.1\%$

[difference in answer is on account of taking only two digits after decimal]

Alternatively III

You may assume the dimensions of triangle in x and y terms also to solve the question.

Problem 29. The length of a rectangle is increased by 60%. By what percent would the width have to be decreased so as to maintain the same area? ***[MAT-03]***

Sol. Required answer $= \left\{\frac{60}{(100+60)} \times 100\right\}\% = 37.5\%$

Try yourself through another method as mentioned in earlier question.

Problem 30. The price of rice is reduced by 2%. How many kilogram of rice can now be bought for the money which was sufficient to buy 49 kg of rice earlier?

Sol. Here, product i.e. expenditure is also constant.

Then, increase % in quantity of rice $= \left\{\frac{2}{(100-2)} \times 100\right\}\%$

$= \frac{100}{49}\%$

It means, quantity of rice after reduction in price

$= 49 + \frac{100}{49}\%$ of $49 = 49 + 1 = 50$ kg.

Problem 31. A reduction of 20% in the price of sugar enables me to purchase 5 kg more for ₹ 600. Find the price of sugar per kg before the reduction of price. ***[SSC (CGL-T-I)-11, MAT-14, 19]***

Sol. Increase % in quantity of sugar $= \left\{\frac{20}{(100-20)} \times 100\right\}\% = 25\%$

But given, 25% is equal to 5 kg.

Then, 100% will be equal to 20 kg. When we say that the consumption is increased by 100%, it means consumption has been increased by 20 kg. Hence, the earlier consumption in question was 20 kg and total expenditure thereon was ₹ 600.

Then, earlier rate of sugar $= \frac{600}{20} =$ ₹ 30 /kg.

Alternatively I

Increase % in quantity of sugar $= \left\{\frac{20}{(100-20)} \times 100\right\}\% = 25\%$

I will get 25% extra quantity of sugar of which monetary value will be 25% of 600 = ₹ 150.

Obviously, the price of 5 kg will be ₹ 150. Hence, earlier rate of sugar $= \frac{150}{5} =$ ₹ 30/kg.

Alternatively II

Let the earlier price was ₹ p/kg. Then, consumption $= \frac{600}{p}$

After reduction in price, present consumption $= \frac{600}{p \times 80\%} = \frac{750}{p}$

But, given that, $\frac{750}{p} - \frac{600}{p} = 5$

$\Rightarrow \quad 150 = 5p$

$\therefore \quad p =$ ₹ 30

Alternatively III

Let the earlier price was ₹ p/kg. Then, consumption $= \frac{600}{p}$

After reduction in price, present consumption

$= \frac{600}{p \times (100-20)\%}$

$= \frac{600 \times 100}{p(100-20)}$

But, given that $\frac{600 \times 100}{p(100-20)} - \frac{600}{p} = 5$

$\Rightarrow \quad \frac{600 \times 100 - 600(100-20)}{p(100-20)} = 5$

$\Rightarrow \quad p = \frac{600 \times 20}{5(100-20)} =$ ₹ 30

Initial price $= \frac{600 \times 20}{5\,(100-20)}$

Then, final price $= \frac{600 \times 20}{5\,(100-20)} \times \frac{(100-20)}{100} = \frac{600 \times 20}{5 \times 100}$

The generic formula may be derived from the above explanation.

Initial price $= \frac{\text{expenditure} \times \text{decrease \% in price}}{\text{change in quantity} \times (100 - \text{decrease\% in price})}$

Or, $\frac{\text{expenditure} \times \text{increase \% in price}}{\text{change in quantity} \times (100 + \text{increase \% in price})}$

Final price (increased or reduced)

$= \frac{\text{expenditure} \times \text{increase or decrease \% in price}}{\text{change in quantity} \times 100}$

Remarks

If formula for final price is kept in mind, you will be able to find the initial price with help of the relation between final price and initial price. The relation is as under

Final price = initial price × (100 + increase % in price)%

Or, Final price = initial price × (100 − decrease % in price)%

For the above question, final price $= \frac{600 \times 20}{5 \times 100} = 24$

Then, 24 = initial price × (100 − 20)%

$\Rightarrow$ Initial price $= 24 \times \frac{5}{4} =$ ₹ 30

Successive Increase/Decrease/ Combination of Both

Understand the concept through an example. A number is increased by 20%, further, it is increased by 30%. Find out the percentage increase in that number.

Initially, original number has been increased by 20%. Further, this increased number is increased by 30%, it means original number and the increment both will be increased by 30%.

Thus, total increment = 20% + 30% + 30% of 20%

$$= 20\% + 30\% + 20\% \times \frac{30}{100} = \left\{20 + 30 + \frac{20 \times 30}{100}\right\}\% = 56\%$$

Now, we may institutionalise this concept as under keeping in mind '+' for increase and '−' for decrease.

(a) If $x\%$ and $y\%$ are two successive increments, then

$$\text{Effect on original number} = \left\{+x + y + \frac{xy}{100}\right\}\%$$

(b) If $x\%$ is increment and $y\%$ is decline, then

$$\text{Effect on original number} = \left\{+x - y - \frac{xy}{100}\right\}\%$$

(c) If $x\%$ is decline and $y\%$ is increment, then

$$\text{Effect on original number} = \left\{-x + y - \frac{xy}{100}\right\}\%$$

(d) If $x\%$ and $y\%$ are two successive declines, then

$$\text{Effect on original number} = \left\{-x - y + \frac{xy}{100}\right\}\%$$

Note

(i) If outcome is positive, the original number increases and in case of negative outcome, the original number decreases.

(ii) When a number is decreased by x% and further decreased by x%, the number by which the decreased number further decreases is x% less than the number by which the number decreases initially.

Let a number (say, 80) is decreased by 20% and further decreased by 20%. Clearly, the number initially decreases by 16 and further decreases by 12.8. Then, 12.8 will be 20% less than 16.

This also holds good in case a number is increased by x% and further increased by x%. It means the number by which the increased number further increases will be x% greater than the number by which the number increases initially.

(iii) If sum of values of each set of successive increases is equal, the set, having maximum product of given values, augments the original number more in comparison to other sets. Hence, more product, more augmentation.

Per contra, if sum of values of each set of successive decreases is equal, the set, having least product of given values, diminishes the original number more in comparison to other sets. Hence, less product more decrease.

This conception is applicable only for the set of two successive increases/decreases. However, you may write a given value (say, x) as a set of x and 0 for application of this approach. The same approach may be adopted for the successive rates of interest, successive discounts, etc.

Problem 32. Salary of a person is increased by 20%, then it is decreased by 20%. Change in his salary is :

Sol. Effect on salary $= \left\{+20 - 20 - \frac{20 \times 20}{100}\right\}\% = -4\%$

Negative outcome means his salary is decreased by 4%.

Problem 33. If the price of a book is first decreased by 25% and then increased by 20%, then the net change in the price will be?

Sol. Effect on price $= \left\{-25 + 20 - \frac{25 \times 20}{100}\right\}\% = -10\%$

Negative outcome indicates that the price has been decreased by 10%.

Problem 34. A number is decreased by 10% and then increased by 10%. The number so obtained is 10 less than original number. What was the original number?

Sol. Effect on original number $= \left\{-10 + 10 - \frac{10 \times 10}{100}\right\}\% = -1\%$

Negative outcomes shows that the original number has been decreased by 1%.

Given, 1% = 10, then 100% = 1000

Hence, original number = 1000

Problem 35. A number is increased by 20% and then again by 20%. By what per cent should the increased number be reduced so as to get back the original number? ***[SSC (CGL-PT)-04]***

Sol. Effect on original number

$$= \left\{+20 + 20 + \frac{20 \times 20}{100}\right\}\% = +44\%$$

Hence, the original number has been increased by 44%.

Then, required reduction %

$$= \left\{\frac{44}{(100 + 44)} \times 100\right\}\%$$

$$= 30\frac{5}{9}\%$$

Problem 36. A number is increased by 10% and then again by 20%. Further, decreased by 25%. Find out the effect on the original number.

Sol. Firstly, we should take any two figures as per our convenience to work out the effect.

Effect on taking two figures 10% and 20%

$$= \left\{+10 + 20 + \frac{10 \times 20}{100}\right\}\%$$

$$= +32\%$$

Now, effect on taking the outcome (i.e. + 32%) and the remaining figure (−25%)

$$= \left\{+32 - 25 - \frac{32 \times 25}{100}\right\}\% = -1\%$$

Hence, the original number will decrease by 1%.

Problem 37. Samyak's expenditure is first decreased by 20% and again increased by 25%. The result is same as Manu's expenditure increased by 20% and then reduced by 35%. Find the ratio of Samyak's expenditure to that of Manu.

Sol. Effect on Samyak's expenditure $= \left\{-20 + 25 - \frac{20 \times 25}{100}\right\}\% = 0\%$

It means, Samyak's altered expenditure is 100% of the original.

Now, effect on Manu's expenditure

$$= \left\{+20 - 35 - \frac{20 \times 35}{100}\right\}\% = -22\%$$

It means Manu's expenditure becomes 78% of his original expenditure.

Given, 100% of Samyak's expenditure

= 78% of Manu's expenditure.

⇒ Samyak's expenditure : Manu's expenditure

= 78 : 100 = 39 : 50

Problem 38. After receiving two successive raises, Ajitha's salary became equal to $\frac{15}{8}$ times of her initial salary. By how much percent was the salary raised the first time if the second raise was twice as high (in per cent) as the first? ***[MAT-14]***

Sol. Clearly, effect on salary after two successive raises = 87.5%

Let the first time raise be x%. Then, 2nd time raise = $2x$%

Then, $\left\{+x + 2x + \frac{x \times 2x}{100}\right\}\% = 87.5\%$

$$\Rightarrow \quad 3x + \frac{x^2}{50} = 87.5$$

$$\Rightarrow \quad 150x + x^2 = 4375$$

$$\Rightarrow \quad x^2 - 25x + 175x - 4375 = 0$$

$$\Rightarrow \quad x(x - 25) + 175(x - 25) = 0$$

$$\Rightarrow \quad (x - 25)(x + 175) = 0$$

Hence, $x = 25\%$

Alternatively

The best way is to proceed through the given options. Suppose options are 30%, 25%, etc.

(a) 30% $\Rightarrow \left\{+30 + 60 + \frac{30 \times 60}{100}\right\}\% \neq +87.5\%$

(b) 25% $\Rightarrow \left\{+25 + 50 + \frac{25 \times 50}{100}\right\}\% = +87.5\%$

Hence, option (b) is correct.

Problem 39. The owner of an art shop conducts his business in the following manner. Every once in a while he raises his prices by x% then a while later he reduces all the new prices by x%. After one such up-down cycle, the price of a paining decreased by ₹ 441. After a second up-down cycle, the painting was sold for ₹ 1944.81. What was the original price of the painting (in ₹)? ***[CAT-01]***

Sol. Effect on the original price of the painting after one up-down cycle

$$= \left\{+x - x - \frac{x \times x}{100}\right\}\% = -\frac{x^2}{100}\%$$

Clearly, $\frac{x^2}{100}$% of the original price of painting = 441

In second up-down cycle, the original price reduced by 441, will further be reduced by $\frac{x^2}{100}$%.

i.e. $(p - 441)\left(100 - \frac{x^2}{100}\right)\% = 1944.81$, where p is the original price of the painting.

$$\Rightarrow \quad 100\% \text{ of } p - 100\% \text{ of } 441 - \frac{x^2}{100}\% \text{ of } p + \frac{x^2}{100}\% \text{ of } 441 = 1944.81$$

$$\Rightarrow \quad p - 441 - 441 + \frac{(441)^2}{p} = 1944.81$$

$$\Rightarrow \quad p^2 - 2826.81p + (441)^2 = 0$$

$$\Rightarrow \quad p = 2756.25 \text{ and } 70.56$$

But $p \neq 70.56$ as it cannot be reduced by 441 in the first up-down cycle.

Hence, $p = 2756.25$

Alternatively

Clearly, the original price is initially decreases by $\frac{x^2}{100}$% and further decreases by $\frac{x^2}{100}$% i.e. by same percentage.

Now, proceed through options. Let option be 2500, 2756.25, etc.

(a) 2500 ⇒ 2500 initially decreases by 441 i.e. by 17.64%. The decreased number i.e. 2059 further decreases to 1944.81, hence decreases by 5.5%. Since 17.64% ≠ 5.5%, option (a) is incorrect.

(b) 2756.25 ⇒ 2756.25 initially decreases by 441 i.e. by 16%. The decreased number i.e. 2315.25 further decreases to 1944.81. hence, decreases by 16%. Since decrease percentage in 1st cycle is equal to that in 2nd cycle, option (b) is correct.

Problem 40. A, B, C, D and E are five employees working in a company. In two successive years, each of them got hikes in his salary as follows

A : p% and $(p + 1)$%

B : $(p + 2)$% and $(p - 1)$%

C : $(p + 3)$% and $(p - 2)$%

D : $(p + 4)$% and $(p - 3)$%

E : $(p + 5)$% and $(p - 4)$%

If all of them have the same salary at the end of two years, who got the least hike in his salary? ***[XAT-19]***

Sol. Here, sum of values of each set is equal i.e. $2p + 1$. We know, if sum of values of each set of successive increases is equal, the set, having maximum product of given values, augments the original number more in comparison to other sets. Hence, more product, more augmentation.

Further, for any value of p, the product of given two increments for employee A will be maximum and that for employees E, per contra, will be minimum. Hence, increment percentage for A will maximum and that for E will be minimum. Eventually, salaries of all employees are same at the end of two years. It implies that the salary of E was, in beginning, maximum among all. Thus, the actual hike in salary was least for E.

Effect on Product in Case of Changes in Factor(s)

There will be the same concept as envisaged under topic 'Successive increase/decrease/combination of both'. However, for better understanding, the concept is recapitulated here in other words. Needless to say that Product may be of two factors or more.

Examples of two factors of a product have been given under 'Concept of Product Constancy' in detail, *viz.* price × quantity = expenditure, speed × time = distance, rate × time = cost, etc.

Example of three factors of a product may also be seen in 3-*D* figures like cylinder, cone, prism, pyramid, etc. *viz* volume of a cylinder = π × radius × radius × height, product of any three numbers, etc.

For the product of two factors

(a) When one factor is increased by $x\%$ and another by $y\%$, the effect on their product

$$= \left\{ + x + y + \frac{xy}{100} \right\} \%$$

(b) When one factor is increased by $x\%$ and another is decreased by $y\%$, the effect on their product

$$= \left\{ + x - y - \frac{xy}{100} \right\} \%$$

(c) When one factor is decreased by $x\%$ and another is increased by $y\%$, the effect on their product

$$= \left\{ - x + y - \frac{xy}{100} \right\} \%$$

(d) When one factor is decreased by $x\%$ and another by $y\%$, the effect on their product

$$= \left\{ - x - y + \frac{xy}{100} \right\} \%$$

For the product of three factors

If one factor is increased by $x\%$, second factor is decreased by $y\%$ and third factor is increased by $z\%$, the effect on product will be as under :

Effect on taking any two figures with their sign (say,

$$+ x, - y) = \left\{ + x - y - \frac{xy}{100} \right\} \% = r\% \text{ (let)}$$

Now, we have to take the resultant with sign (i.e. $+ r$) and the remaining figure with sign (i.e. $+ z$) to compute the final effect.

$$\text{Then, final effect} = \left\{ + r + z + \frac{rz}{100} \right\} \%$$

The same pattern is applicable for more than three factors.

Note

(a) If outcome is positive, the product increases and in case of negative outcome, the product decreases.

(b) If one factor is increased or decreased, the product will be changed by the same magnitude.

Take Some Examples

Problem 41. The price of a T.V. set is decreased by 25% as a result of which the sale increased by 20%. What will be the effect on the total revenue of the shop? *[Bank (PO)-03, MAT-09, CLAT-19]*

Sol. Here, revenue (i.e. product) = price × sale

$$\text{Effect on total revenue} = \left\{ -25 + 20 - \frac{25 \times 20}{100} \right\} \% = -10\%$$

Alternatively I

Let the initial price be ₹ 100/set and the sale was 100 sets.

Then, initial revenue = 100 × 100 = ₹ 10000

Changed price = ₹ 75 and changed sale = 120

Then, changed revenue = 75 × 120 = ₹ 9000

∴ Reduction in revenue = 10000 − 9000 = 1000

$$\text{Then, reduction } \% = \left\{ \frac{1000}{10000} \times 100 \right\} \% = 10\%$$

Alternatively II

Changed price = 75% of the original price = 0.75

and changed sale = 120% of the original sale = 1.20.

Thus, changed revenue = 0.75 × 1.20

= 0.9 i.e. a loss of 0.1 and 0.1 means 10%.

Alternatively III

$$\text{Effect on revenue} = \left[\left\{ \left(\frac{100-25}{100} \times \frac{(100+20)}{100} \right) - 1 \right\} \times 100 \right] \%$$

$$= [\{0.9 - 1\} \times 100]\% = -10\%.$$

[you have to append '−' in case of decrease and '+' in case of increase for the above pattern]

Problem 42. What happens to the area of a square when its side is halved. Its area will : *[RRB-03]*

Sol. Here, side of the square is decreased by 50% and area = side × side. Hence, both figures are same i.e. − 50%

$$\text{Now, effect on area} = \left\{ -50 - 50 + \frac{50 \times 50}{100} \right\} \% = -75\%$$

Thus, area will reduce by 75%. In other words, its area will become one-fourth.

Problem 43. An error of 2% in excess is made while measuring the side of a square. The percentage of error in the calculating area of the square is :

$$\textbf{Sol. } \text{Effect on area} = \left\{ + 2 + 2 + \frac{2 \times 2}{100} \right\} \% = + 4.04\%$$

Hence, percentage of error = 4.04% (excess).

Problem 44. The length and breadth of a room are decreased by 10% and 20% respectively. But, height is increased by 25%. Find out the effect on its volume.

Sol. Three figures are : −10%, −20% and + 25%.

Effect on taking any two figures (say, −10% and −20%)

$$= \left\{ -10 - 20 + \frac{10 \times 20}{100} \right\} \% = -28\%$$

Now, effect on taking the resultant (i.e. −28%) and the remaining figure (i.e. +25%) $= \left\{ -28 + 25 - \frac{28 \times 25}{100} \right\} \% = -10\%$

Hence, the volume will decrease by 10%.

Problem 45. The price of sugar is increased by 25%. By how much percent should the family reduce the consumption so as to allow increment of only 12.5% in the earlier expenditure? *[SSC (CGL-T-I)-11, SNAP-18]*

Sol. Here, expenditure is a product of price and consumption.

Now, effect on expenditure $= \left\{+25 - x - \frac{25 \times x}{100}\right\}\%$

[Let reduction in consumption be x%]

$\Rightarrow \quad +12.5\% = \left\{+25 - \frac{5x}{4}\right\}\% \Rightarrow \frac{5x}{4} = 12.5 \Rightarrow x = 10$

Hence, the family should reduce their consumption by 10%.

Alternatively

Reduction in consumption keeping the expenditure constant

$= \left\{\frac{25}{(100+25)} \times 100\right\}\% = 20\%$

In case consumption is reduced by 20%, there will be no effect on expenditure. But, given that expenditure is allowed to be increased by 12.5%.

Then, effect on consumption $= \left\{-20 + 12.5 - \frac{20 \times 12.5}{100}\right\}\%$

$= -10\%$

Hence, the family should reduce their consumption by 10%.

Problem 46. Two ships A and B cover the same distance by using different types of fuel and A covers 25% more distance per litre of fuel than B does. The fuel meant for A is 5% costlier than that for B. By how much percent is expenditure incurred on fuel for Ship A less than expenditure on fuel for Ship B?

Sol. Suppose, Ship B covers 100 km/L of fuel, then Ship A will cover 125 km/L of fuel. Further, let the distance be 500 km (assumed for sake of convenience)

Then, fuel consumed by Ship B = 5 L and that by Ship A = 4 L.

Suppose, fuel for Ship B is at ₹ 100/L then, total expenditure on fuel for Ship B = 100 × 5 = ₹ 500

and total expenditure on fuel for Ship A = 105 × 4 = ₹ 420

Now, saving (i.e. less in expenditure) = ₹ 500 − ₹ 420 = ₹ 80

percent less in expenditure $= \left(\frac{80}{500} \times 100\right)\% = 16\%$

Alternatively I

Ship A covers 25% more distance per L than B does, it means Ship A will consume fuel 20% $\left(\text{i.e. } \frac{25}{125} \times 100\right)$ less than B does.

But fuel used by Ship A is 5% costlier than Ship B does. Then, effect on expenditure $= \left\{-20 + 5 - \frac{20 \times 5}{100}\right\}\% = -16\%$

It means, expenditure incurred on fuel for Ship A is 16% less than expenditure on fuel for Ship B.

Alternatively II

Given, $\quad \left\{+25 - x - \frac{25 \times x}{100}\right\}\% = 5\%$

$\Rightarrow \quad \left\{+25 - \frac{5x}{4}\right\}\% = 5\%$

$\Rightarrow \quad \frac{5x}{4} = 20 \Rightarrow x = 16\%$

Problems Based on Population

Before attempting the questions under this section, revise the concept of 'Successive Increase/Decrease/ combination of both' and also observe the concept of Compound Interest given in the chapter 'Compound Interest'.

Problem 47. The population of a town increases every year by 4%. If its present population is 50000, then after 2 yr, it will be? *[SSC (CGL-PT)-08]*

Sol. Given that, the present population increases by 4% in first year and again by 4% in second year.

Then, effect on original population

$= \left\{+4 + 4 + \frac{4 \times 4}{100}\right\}\% = 8.16\%$

It means the present population will increase by 8.16% after 2 yr. Then, the population after 2 yr = 108.16% of 50000 = 54080

Alternatively I

The population after one year

$= 104\% \text{ of } 50000 = 50000 \times \frac{104}{100} = 52000$

Then population after two years

$= 104\% \text{ of } 52000 = 52000 \times \frac{104}{100} = 54080$

{**Note** In case of decline, you have to subtract the decline % from 100%}

Alternatively II

The final population $= 50000\left(1 + \frac{4}{100}\right)\left(1 + \frac{4}{100}\right) = 54080$

Note *In case of decline you have to append '–' instead of '+'. This is the based on the concept of Compound Interest. Here, principal may be assumed as 50000 and rate of interest may be assumed as 4%. The details of logic behind this conception have been given at the chapter 'Compound Interest'.*

Alternatively III

Compound interest = $\boxed{\text{Cumulative simple interest}}$.

$\boxed{\text{given rate of interest}^2} = \boxed{4+4} \cdot \boxed{4}^2 = 8.16\%$

Amount = 108.16% of principal

{please refer to the chapter 'Compound Interest'}

Similarly, the final population = 108.16% of the original population = 108.16% of 50000 = 54080

Remarks *Aspirants should adopt the first or last approach so as to solve the Problems in easy way as the case may be.*

Problem 48. The population of a village has increased annually at the rate of 25%. If at end of 3 yrs it is 10000, the population in the beginning of the first year was?

Sol. Effect on original population after two years

$= \left\{+25 + 25 + \frac{25 \times 25}{100}\right\}\% = +56.25\%$

Then, final effect after three years

$= \left\{+56.25 + 25 + \frac{56.25 \times 25}{100}\right\}\% = +95.3125\%$

Hence, the original population has increased by 95.3125%.

Thus, present population = 195.3125% of the original population

$\Rightarrow$ 10000 = 195.3125% of the original population

$\therefore$ Original population $= 10000 \times \frac{100}{195.3125} = 5120$

Alternatively

Compound interest =

Cumulative simple interest $\cdot 3r^2 \mid r^3 = +\ 95.3125\%$

{please refer to the chapter 'Compound Interest'}

Hence, the original population has increased by 95.3125%.

Thus, present population = 195.3125% of the original population

$\Rightarrow$ 10000 = 195.3125% of the original population

$\therefore$ Original population $= 10000 \times \frac{100}{195.3125} = 5120$

[please refer to the chapter 'Compound Interest']

Problem 49. **The population of a variety of tiny bush in an experimental field increased by 10% in the first year, increased by 8% in the second year but decreased by 10% in the third year. If the present number of bushes in the experimental field is 26730, then the number of bushes in the beginning was ?**

Sol. Effect on population on taking any two figures (say, +10%, –10%) $= \left\{+10 - 10 - \frac{10 \times 10}{100}\right\}\% = -1\%$

Then, final effect $= \left\{+8 - 1 - \frac{8 \times 1}{100}\right\}\% = +\ 6.92\%$

'+' indicates that the final population of bushes has increased by 6.92%. Thus, present population = 106.92% of the original population.

$\Rightarrow$ 26730 = 106.92% of the original population

$\therefore$ Original population of bushes $= 26730 \times \frac{100}{106.92} = 25000$

Alternatively

Final population

$$= \text{original population}\left(1 + \frac{10}{100}\right)\left(1 + \frac{8}{100}\right)\left(1 - \frac{10}{100}\right)$$

$$\Rightarrow 26730 = \text{original population}\left(1 + \frac{10}{100}\right)\left(1 + \frac{8}{100}\right)\left(1 - \frac{10}{100}\right)$$

$\Rightarrow$ 26730 = Original population × 1.1 × 1.08 × 0.9

$\therefore$ Original population = 25000

Problem 50. **The number of inhabitants in a town increases at a certain rate percentage. The number at present is 375000 and the number 3 yrs ago was 352947. Find the rate percent?**

Sol. Final population = original population $\times \left(1 + \frac{r}{100}\right)^3$

$$\Rightarrow \quad 375000 = 352947\left(1 + \frac{r}{100}\right)^3$$

$$\Rightarrow \quad \frac{375000}{352947} = \left(1 + \frac{r}{100}\right)^3$$

$$\Rightarrow \quad \frac{125000}{117649} = \left(1 + \frac{r}{100}\right)^3$$

$$\Rightarrow \quad \left(\frac{50}{49}\right)^3 = \left(1 + \frac{r}{100}\right)^3$$

$$\Rightarrow \quad \frac{50}{49} = 1 + \frac{r}{100}$$

$$\Rightarrow \quad r = 2\frac{2}{49}\%$$

Alternatively

The approaches for solving these types of questions in moments have been detailed in chapter 'Compound Interest'. Due to space constraints, here, the whole approach/process is not possible to be given.

Problem 51. **In a town, the population was 8000. In one year, male population increased by 10% and female population increased by 8% but the total population increased by 9%. The number of males in the town was?**

Sol. Let the male population was p, then female population $= 8000 - p$

Given, 110% of p + 108% of $(8000 - p)$ = 109% of 8000

$\Rightarrow \quad 1.1p + 1.08\,(8000 - p) = 1.09 \times 8000$

$\Rightarrow \quad 0.02p = 80$

$\therefore \quad p = 4000$

Alternatively

Let the male population was x and the female population was y.

According to the question,

10% of x + 8% of y = 9% of $(x + y)$

$\Rightarrow \quad 10x + 8y = 9x + 9y$

$\Rightarrow \quad 10x - 9x = 9y - 8y$

$\Rightarrow \quad x\,(10 - 9) = y(9 - 8)$

$\Rightarrow \quad \frac{x}{y} = \frac{9 - 8}{10 - 9}$

This above process may be depicted as under

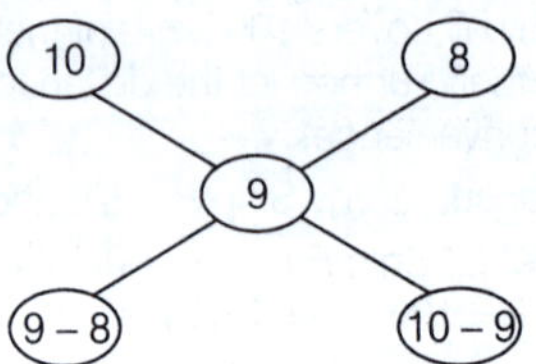

$x : y = (9 - 8) : (10 - 9) = 1 : 1$

[This pattern is very useful in chapter 'Mixture/Alligation' and also, details have been given thereunder]

Hence, the male population $= 8000 \times \frac{1}{(1 + 1)}$

$= 4000$

Problem 52. **The population of a town was 5600 three years back but it is 7000 right now. What would be the population after three years if the growth-rate has been constant and has been compounding annually.**

[SSC (CGL-T-I)-12, MAT-18]

Sol. Increase in population over three years

$= 7000 - 5600 = 1400$

Thus, increment % $= \frac{1400}{5600} \times 100 = 25\%$

Then, population after 3 yrs from now

= 125% of 7000

= 8750

Problem 53. Out of adult population of city, 60% men and 20% women are married. Find out the percentage of married adult population assuming that no man marries with more than one woman and *vice-versa*.

Sol. Let the number of adult men be 100.

Then, number of married adult men = 60% of 100 = 60

Obviously, number of married adult women = 60 as one man marries with one woman and *vice-versa*. But given that number of married woman i.e. 60 is 20% of the total adult woman population.

Hence, total adult woman population = 300

Then, total adult population of the city = 100 + 300 = 400

and total married adult = 60 + 60 = 120

Thus, married adult percent = $\left(\frac{120}{400} \times 100\right)\% = 30\%$

Alternatively,

Given, number of married adult men = number of married adult women

$\Rightarrow$ 60% of adult men = 20% of adult women

Clearly, number of adult men : number of adult women = 1 : 3

Then, married adult % = $\frac{(1 \times 60\%) + (3 \times 20\%)}{(1+3)} \times 100 = 30\%$

Problems Based on Examination

Problem 54. In an examination, a student who gets 20% of the maximum marks fails by 5 marks. Another student who scores 30% of the maximum marks gets 20 marks more than the passing marks. The necessary percentage required for passing is : **[SSC (CGL-PT)-2000, SSC (CGL-T-I)-19]**

Sol. Clearly, the difference between both students' marks = 5 + 20 = 25 and difference of marks in terms of percentage = 30% − 20% = 10%.

Hence, 10% of the maximum marks = 25

$\therefore$ Maximum marks = 250

Now, marks obtained by 1st student = 20% of 250 = 50

But he fails by 5 marks, it means the passing marks = 50 + 5 = 55

Then, passing marks percent = $\left(\frac{55}{250} \times 100\right)\% = 22\%$

Alternatively

From the above, the maximum marks = 250

Given that 1st student gets 20% marks but fails by 5 marks and 5 marks is equal to 2% of the maximum marks. Hence, passing marks % = 20% + 2% = 22%

Or

The second student gets 30% marks, which is 20 marks more than the passing marks and 20 marks is equal to 8% of the maximum marks. Hence, passing marks % = 30% − 8% = 22%.

Problem 55. A candidate who gets 25% marks fails by 20 marks but another candidate who gets 45% marks gets $28\frac{4}{7}\%$ more than the passing marks. Find the maximum marks.

Sol. Let the maximum marks be x.

Passing marks for 1st candidate = (25% of x) + 20

and the passing marks for the second candidate = 45% of $x - 28\frac{4}{7}$% of passing marks = 45% of $x - 28\frac{4}{7}$% of (25% of x + 20)

Obviously, (25% of x) + 20 = 45% of $x - 28\frac{4}{7}$% of (25% of x + 20)

$$\Rightarrow \quad 0 = 20\% \text{ of } x - 20 - \frac{2x}{28} - \frac{40}{7}$$

$$\Rightarrow \quad \frac{9x}{70} = \frac{180}{7} \Rightarrow x = 200$$

Alternatively

You may proceed through options also.

Problem 56. In an examination, one student scores 9 marks more than another, which is equal to 56% of total sum of marks obtained by both. Marks obtained by both are : **[SSC (CGL-PT)-04]**

Sol. Since, one gets 56% of total sum of marks obtained by both students, another will get 44% of the sum. Then, difference in their marks = 12%, but, the actual difference is 9.

It means, 12% $\Rightarrow$ 9

Then, 100% $\Rightarrow$ 75

Hence, total sum of their marks is 75 and difference between them is 9.

Now, you may proceed through options. However, details are being given as under

Marks obtained by 1st student = $\frac{(75-9)}{2} + 9 = 42$ and by another = $\frac{(75-9)}{2} = 33$

Alternatively

You may proceed through options also.

Problem 57. At an examination in which maximum marks are 500, A got 10% less than B, B got 25% more than C, C got 20% less than D. If A got 360 marks, what percent of marks was obtained by D?

Sol. Let the marks obtained by $D = x$

Marks obtained by C = 80% of x = $0.8x$

Marks obtained by B = 125% of $0.8x = x$

Marks obtained by A = 90% of $x = 0.9x$

Given, $0.9x = 360$

$\Rightarrow$ $x = 400$

% of marks obtained by $D = \left(\frac{400}{500} \times 100\right)\% = 80\%$

Alternatively I

Marks of $D \times \frac{(100-20)}{100} \times \frac{(100+25)}{100} \times \frac{(100-10)}{100}$ = Marks of A = 360

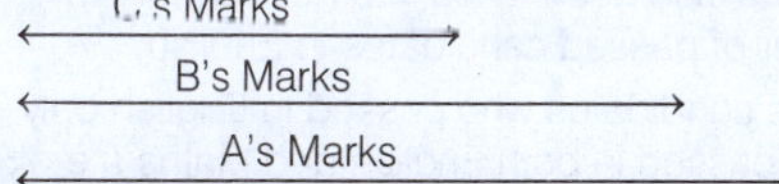

Then, marks of $D = \frac{360 \times 100 \times 100 \times 100}{80 \times 125 \times 90} = 400$

$\therefore$ Percentage of marks obtained by $D = \left(\frac{400}{500} \times 100\right)\%$

$= 80\%$

Alternatively II

Effect on marks of D while taking two figures (say, –20, +25)

$= \left\{-20 + 25 - \frac{20 \times 25}{100}\right\}\%$

$= 0\%$

Final effect $= \left\{-10 + 0 - \frac{10 \times 0}{100}\right\}\%$

$= -10\%$

Hence, marks of D has been reduced by 10%. Obviously, his/her reduced marks will be equal to A's marks

$\Rightarrow$ 90% of marks of $D = 360 \Rightarrow D$'s marks $= 400$

$\therefore$ Percentage of marks obtained by $D = \left(\frac{400}{500} \times 100\right)\%$

$= 80\%$

Problem 58. In an examination, passing marks is 50%. Divansh gets marks 20% less than pass marks and Shivansh gets 10% more than the Divansh's. What per cent less than the sum of their marks should Saransh get to pass that examination?

Sol. Let the total marks be 100.

Then, passing marks = 50

Marks obtained by Divansh = 80% of 50 = 40

and Marks obtained by Shivansh = 110% of 40 = 44

Sum of marks obtained by them = 40 + 44 = 84

Since passing marks is 50, Saransh should get 34 marks less than sum of marks obtained by Divansh and Shivansh to pass the examination.

Then, required percentage $= \left(\frac{34}{84} \times 100\right)\%$

$= 40.48\%$

Application of Set-theory

(Applicable Also in Cases Other Than Examination)

Problem 59. In an examination, 80% of the candidates passed in English and 85% in Mathematics; while 75% passed in both English and Mathematics. If 45 candidates failed in both these subjects, then the total number of candidates was ?

[SSC (CPO)-08, IBPS-13, CLAT-19 (Partly)]

Sol. Given, passed in English = 80%

Passed in Maths = 85%

Passed in both English and Maths = 75%

It implies that, number of candidates passed in **English only** is only 5% as 75% has also been taken into account while computing number of passed candidates in English.

In other words, the candidates who passed in English only (i.e. 5%) and who passed in both English and Maths (i.e. 75%), both categories of candidates have been treated as passed in English (i.e. 80%).

Similarly, number of candidate passed in

Maths only = 85% – 75% = 10%

Now, we may conclude that :

Total passed candidates either in English or in Maths or in both = Total number of students passed at least in one subject = **Passed in English only + Passed in Maths only + passed in both subjects**

$= (80\% - 75\%) + (85\% - 75\%) + 75\%$

$= 90\%$

Or

Total passed candidates either in English or in Maths or in both

$= (80\% - 75\%) + (85\% - 75\%) + 75\%$

$= 80\% + 85\% - 75\%$

= Passed in English + Passed in Maths – Passed in both subjects

The above understanding may be portrayed in a Venn-diagram as under

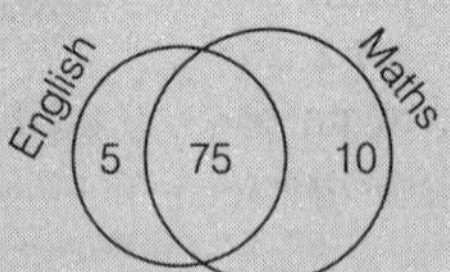

Now, come to the question :

From above, total passed either in English or in Maths or in both = 90%. Then, failed candidates = 10%

Given, 10% $\Rightarrow$ 45, then 100% $\Rightarrow$ 450

Hence, total number of students = 450

Alternatively

We may also find out the percentage of failed candidates to solve the Problem.

Failed in English = 100% – 80% = 20%

Failed in Maths = 100% – 85% = 15%

Failed in both = 100% – 75% = 25%

[It is not possible that failed in both subjects is 25% and in one subject either 20% or 15%, however, ignoring this reality another approach has been adopted to arrive at correct answer.]

Now, total failed candidates

= failed in Eng. + failed in Maths – failed in both

= 20% + 15% – 25% = 10%

Given that 45 candidates failed, it means 10% of the total candidates = 45, then, total number of students = 450.

Problem 60. In an examination 60% of the candidates passed in English and 70% passed in Maths, but 20% failed in both subject. If 2500 candidates passed in both the subjects, the number of candidates that appeared at the examination was? ***[SSC (CGL-PT) 07]***

Sol. **We must convert all the data either in passed form or in failed form without considering whether these are true in real sense or not.**

Passed in English = 60%

Passed in Maths = 70%

Passed in both English and Maths = 100% – 20% = 80%

Remarks *The above figures are not possible in real sense.*

Now, total passed candidate either in English or in Maths or in both = 60% + 70% – 80% = 50%

Given that 2500 candidates passed, it means, 50% of the total candidate appeared in the examination = 2500. Then, total candidate appeared in the examination = 5000.

Alternatively

Failed in English = (100 – 60)% = 40%

Failed in Maths = (100 – 70)% = 30%

Failed in both English and Maths = 20%

Now, total failed candidates either in English or in Maths or in both = 40% + 30% – 20% = 50%

Then, percentage of passed candidates = (100 – 50)% = 50%

Given that 2500 candidates passed, it means, 50% of the total candidate appeared in the examination = 2500. Then, total candidate appeared in the examination = 5000.

Problem 61. In an examination for two subjects namely History and Hindi, 60 students passed in History only and 80 passed in both subjects. Find out the percentage of students passed in Hindi only if total of 250 students passed in that examination.

Sol. Here, number of students passed in History only = 60 and number of students passed in both History and Hindi = 80

It means, passed in History = 60 + 80 = 140

Now, total passed = Passed in History + Passed in Hindi – Passed in both

$\Rightarrow$ 250 = 140 + Passed in Hindi – 80

$\Rightarrow$ Passed in Hindi = 190

Then, passed in Hindi only

= Passed in Hindi – Passed in both = 190 – 80 = 110

Now, required percentage $= \left(\frac{110}{250} \times 100\right)\% = 44\%$

Alternatively

Total passed candidates either in History or in Hindi or in both = Passed in History only + Passed in Hindi only + Passed in both

$\Rightarrow$ 250 = 60 + Passed in Hindi only + 80

$\Rightarrow$ Passed in Hindi only = 110

Now, required percentage $= \left(\frac{110}{250} \times 100\right)\% = 44\%$

Problem 62. In a function, the president awarded 40 medals in Music, 20 medals in Maths and 30 medals in Science. There are total of 60 awardees in the function, out of which 5 persons got medals in all the three categories. Then, what percent of total awardees got medals in exactly two of these categories?

Sol. As per Set Theory, $\boldsymbol{n(A \cup B \cup C) = n(A) + n(B) + n(C) - n(A \cap B) - n(B \cap C) - n(A \cap C) + n(A \cap B \cap C)}$

$\Rightarrow 60 = 40 + 20 + 30 - n(A \cap B) - n(B \cap C) - n(A \cap C) + 5$

$\Rightarrow n(A \cap B) + n(B \cap C) + n(A \cap C) = 35$

Further, the number of awardees getting medals in exactly two of three categories $= \{n(A \cap B) - n(A \cap B \cap C))\}$

$+ \{n(B \cap C) - n(A \cap B \cap C)\} + \{n(A \cap C) - n(A \cap B \cap C)\}$

$= \boldsymbol{n(A \cap B) + n(B \cap C) + n(A \cap C) - 3n(A \cap B \cap C)}$

$= 35 - 3 \times 5 = 20$

$\therefore$ Required % $= \left\{\frac{20}{60} \times 100\right\}\% = 33.33\%$

Remarks *Aspirants must memorise the above both formulae so as to solve the question quickly*

Directions (Problem Nos. 63 and 64) *These questions are based on the following information*

There are three different cable channel namely Ahead, Luck and Bang. In a survey, it was found that 85% of viewers respond to Bang, 20% to Luck and 30% to Ahead. 20% of viewers respond to exactly two channels and 5% to none.

Problem 63. What percentage of the viewers responded to all three?

Problem 64. Assuming 20% respond to ahead and Bang and 16% respond to Bang and Luck, what is the percentage of viewers who watch only Luck? ***[CAT-95]***

63. Let number of viewers responding to Ahead = $n(A)$

Number of viewers responding to Bang = $n(B)$

Number of viewers responding to Luck = $n(C)$

As per Set Theory, $n(A \cup B \cup C) = n(A) + n(B) + n(C) - n(A \cap B) - n(B \cap C) - n(A \cap C) + n(A \cap B \cap C)$

$\Rightarrow 100 - 5 = 30 + 85 + 20 - \{n(A \cap B) + n(B \cap C) + n(A \cap C) - n(A \cap B \cap C)\}$

$\Rightarrow \quad 95 = 135 - [\{n(A \cap B) - n(A \cap B \cap C)\} + \{n(B \cap C) - n(A \cap B \cap C)\} + \{n(A \cap C) - n(A \cap B \cap C)\} + 3n(A \cap B \cap C)] + n(A \cap B \cap C)$

$\Rightarrow \quad -40 = -[20 + 3n(A \cap B \cap C)] + n(A \cap B \cap C)$

$\Rightarrow \quad -20 = -2n(A \cap B \cap C)$

$\rightarrow \quad n(A \cap B \cap C) = 10$

64. We have, $n(A \cup B \cup C) = n(A) + n(B) + n(C) - n(A \cap B) - n(B \cap C) - n(A \cap C) + n(A \cap B \cap C)$

$\Rightarrow \quad 95 = 30 + 85 + 20 - 20 - 16 - n(A \cap C) + 10$

$\Rightarrow n(A \cap C) = 14$

Now, number of viewers watching only Luck channel = Viewers watching Luck – viewers watching exactly Bang and Luck – viewers watching exactly Ahead and Luck – viewers watching all three

$= 20 - \{n(B \cap C) - n(A \cap B \cap C)\} - \{n(A \cap C) - n(A \cap B \cap C)\} - n(A \cap B \cap C)$

$= 20 - \{16 - 10\} - \{14 - 10\} - 10$

$= 20 - 6 - 4 - 10 = 0$

Problem 65. Out of 105 students taking an examination English and Mathematics, 80 students pass in English, 75 students pass in Mathematics and 10 students fail in both the subjects. What is the percentage of students passed in only one subject?

Sol. Number of students failed in English = 105 – 80 = 25

and number of students failed in Maths = 105 – 75 = 30

Given, number of students failed in both subjects = 10

Hence, number of students failed either in English or in Maths or in both = number of students failed at least in one subject

$= 25 + 30 - 10 = 45$

It means, number of students passed in only one subject

$= 45 - 10 = 35$

$\therefore$ Required % $= \left(\frac{35}{105} \times 100\right)\% = 33.33\%$

Alternatively

You may try with reference to number of passing students in lieu of number of failed student as taken above.

The above pattern is also applicable in types of given Problems (infra).

Problem 66. In a school, 40% of the students play football and 50% play cricket. If 18% of the students neither play football nor cricket, the percentage of students playing both is :

Sol. Number of students playing either football or cricket or both = Students playing football + Students playing cricket – Students playing both

$\Rightarrow$ (100 – 18)% = 40% + 50% – Students playing both

$\Rightarrow$ Students playing both = 8%

Problem 67. 70% of the students who joined XLRI last year play football, 75% play cricket, 80% play basketball and 85% play carrom. The minimum percentage of students who play all four games is : ***[XAT, 13]***

Sol. Given that 70% play football. It means, percentage of students not playing football = 30%

Also, 75% play cricket. It means, percentage of students not playing cricket = 25%

On the above line, percentage of students not playing basketball = 20% and percentage of students not playing carrom = 15%

Hence, maximum percentage of students not playing at least one game = (30 + 25 + 20 + 15)% = 90%. Then, obviously the remaining 10% of students play all the four games.

Alternatively

Percentage of players playing all given games = Sum of percentage of players playing different games – {(number of games – 1) × 100}%

Here, the required percentage = (70 + 75 + 80 + 85)% – {(4 – 1) × 100}% = 310% – 300% = 10%

Problems Based on Election

Before understanding the gravity of the problems based on Election, we must know that :

(a) There is a voter list in which eligible voters are enrolled. Out of enrolled voters, some voters cast their votes and it is possible that out of votes casted/polled, some votes may be invalid due to the technical error or other reasons. Counting of valid votes is made on which winning is decided. However, some times, analysis of winner or loser is done in terms of number of enrolled voters, casted votes etc.

(b) The winner wins the election by a majority of votes with reference to his/her closest contester and it is convention. However, in Maths, the reference may be changed and the same is specified in the question.

Problem 68. In an election between two candidates, the candidate getting 60% of the votes polled, is elected by a majority of 14000 votes. The number of votes obtained by the winning candidates is :

Sol. Votes in favour of the winner = 60%

and votes in favour of loser = 40%

Then, difference between their votes = 20%

But, given that the winner wins the election by a majority of 14000 votes, it means the difference between their votes is 14000.

Obviously, 20% of the votes polled = 14000

$\therefore$ 60% of the votes polled = 42000

Hence, number of votes obtained by the winner = 42000

Problem 69. 10% of the voters did not cast their vote in an election between two candidates. 10% of the votes polled were found invalid. The successful candidate got 54% of the valid votes and won by a majority of 1620 votes. The number of voters enrolled on the voter list was? ***[SSC (CGL-PT) 03]***

Sol. Let number of voters enrolled be 100.

Number of votes polled = 100 – 10 = 90

Number of valid votes = 90% of 90 = 81

Since winner gets 54% of the valid votes, the loser will get 46% of the valid votes. Hence, difference between their votes = 8% of the valid votes = 8% of 81 = 6.48

Clearly, when difference is 6.48, number of voters enrolled is 100

$\therefore$ When difference is 1620, number of voters enrolled

$= \frac{100}{6.48} \times 1620 = 25000$

Remarks We may compute valid votes, votes polled on the same pattern for the above question. e.g.

For Calculation of Valid votes

When difference is 6.48, number of valid votes is 81

$\therefore$ When difference is 1620, number of valid votes

$= \frac{81}{6.48} \times 1620 = 20250$

Alternatively

Effect on number of voters enrolled

$= \left\{-10 - 10 + \frac{10 \times 10}{100}\right\}\% = -19\%$

Hence, percentage of valid votes = (100 – 19)% = 81% of voter enrolled = 81% of 25000 = 20250

For calculation of votes polled

When difference is 6.48, number of votes polled is 90

$\therefore$ When difference is 1620, number of votes polled

$= \frac{90}{6.48} \times 1620 = 22500$

Problem 70. 8% of the people eligible to vote are between 18 and 21 years of age. In an election, 85% of those eligible to vote, who were between 18 and 21, actually voted. In that election, the number of persons between 18 and 21, who actually voted, was what percent of those eligible to vote? ***[RRB-98]***

Sol. Let the eligible voters be 100.

Voters between 18 and 21 yr of age = 8% of 100 = 8

Then, number of voters between 18 and 21 yr of age, who actually votes = 85% of 8 = 6.80

Hence, required percentage = 6.8%

Alternatively

Required percentage = 85% of 8% = $8\% \times \frac{85}{100}$ = 6.8%

Problem 71. In an election between 3 candidates x, y and z, x gets 20% more votes than y does and y gets $47\frac{1}{17}\%$ more than z does. If x wins the election by a majority of 5000 with reference to y. Then, find out the number of voters enrolled on voter list if 20% of voters did not cast their votes and 10% votes polled were declared as invalid.

Sol. Votes got by y = 100 (let)

Then, votes got by x = 120% of 100 = 120

and votes got by $z = 100 \times \frac{100}{\left(100 + 47\frac{1}{17}\right)} = 68$

If a number becomes n after increment by a%, the original number will be $n \times \frac{100}{(100+a)}$. In case of decrease, the original number = $n \times \frac{100}{(100-a)}$.

Evidence :

Let the original number be m, then the new number (n) = $(100 + a)$% of m in case of increment.

$\Rightarrow$ new number $(n) = m \times \frac{(100+a)}{100}$

$\Rightarrow \quad m = n \times \frac{100}{(100+a)}$

Similarly, in case of decline, $m = n \times \frac{100}{(100-a)}$.

The same concept has been applied to get number of votes by Z. You will also observe the same concept in almost all the arithmetic topics.

Now, difference between number of votes obtained by x and y = 20

$\because$ If difference is 20, total number of valid votes is 288 (i.e. 100 + 120 + 68)

$\therefore$ When difference is of 5000, total number of valid votes.

$= \frac{288}{20} \times 5000 = 72000$

Then, total casted votes = $72000 \times \frac{100}{(100-10)} = 80000$

Further, total enrolled voters = $80000 \times \frac{100}{(100-20)} = 100000$

Problem 72. In an election between two contestants, 20% of voters enrolled on voter list didn't cast their votes and 10% votes were declared invalid. A candidate got 50% of number of the total voters and won the election by a majority of 14000. Find the total valid votes as well as total casted votes.

Sol. Let the voters enrolled be 100.

Casted votes = 80% of 100 = 80

Valid votes = 90% of 80 = 72

Votes in favour of winner = 50% of 100 = 50

Then, votes in favour of loser = 72 − 50 = 22

Now, difference between votes obtained by them

= 50 − 22 = 28

$\because$ If difference is 28, total number of valid votes is 72

$\therefore$ When difference is of 14000, total number of valid votes

$= \frac{72}{28} \times 14000 = 36000$

Further, if difference is 28, total number of casted votes is 80

$\therefore$ When difference is of 14000, total number of casted votes

$= \frac{80}{28} \times 14000 = 40000$

Alternatively

Total casted votes = $36000 \times \frac{100}{(100-10)}$

= 40000

Problem 73. $\frac{2}{5}$ of the voters promise to vote for P and the rest promised to vote for Q. Of these, on the last day 15% of the voters went back of their promise to vote for P and 25% of voters went back of their promise to vote for Q and P lost by 2 votes. Then, the total number of voters is : ***[CAT-95]***

Sol. Given, voters promise to vote for P = 40% of the total voters and voters promise to vote for Q = 60% of the total voters.

Then, voters actually voted for P = {(100 − 15)% of 40%} + {25% of 60%} = 49%

Then, voters actually voted for Q = (100 − 49)% = 51%

Difference between number of votes got by them = 2%

Clearly, 2% of the total votes = 2 (given)

$\therefore$ Total number of votes = 100

Problems Based on Income, Expenditure and Savings

Problem 74. Rohit spends 40% of his salary on food, 20% on house rent, 10% on entertainment and 10% on conveyance. If his savings at the end of a month are ₹ 1500, then monthly salary is : ***[SSC (CGL-PT) 03, Andhra Bank (MA) 09]***

Sol. Total expenditure = 40% + 20% + 10% + 10% = 80%

Then, savings = 100% − 80% = 20%

Obviously, 20% of monthly salary (i.e. savings) is equal to ₹ 1500.

Then, monthly salary = 1500 × 5 = ₹ 7500

Problem 75. From the salary of an officer, 10% deducted as house rent, 20% of rest he spends on conveyance, 20% of the rest he pays as income tax and 10% of the balance he spends on clothes. Then, he is left with ₹ 15552. Find his salary. ***[LIC (AAO) 03, MAT-11, 09, 07]***

Sol. Let the salary be ₹ 100.

Expenditure on house rent = 10% of 100 = 10

Then, balance amount = 100 – 10 = 90

Now, expenditure on conveyance = 20% of 90 = 18

Then, balance amount = 90 – 18 = 72

Further, expenditure on income tax = 20% of 72 = 14.4

Then, balance amount = 72 – 14.4 = 57.6

Finally, expenditure on clothes = 10% of 57.6 = 5.76

Then, balance amount = 57.6 – 5.76 = 51.84

It reveals that is case balance is ₹ 51.84, total salary is ₹ 100

∴ When balance is ₹ 15552, total salary

$$= \frac{100}{51.84} \times 15552 = 30000$$

Alternatively

Let the salary be x.

Then, $x \times \frac{90}{100} \times \frac{80}{100} \times \frac{80}{100} \times \frac{90}{100} = ₹\ 15552$

∴ $x = 30000$

Alternatively II

Effect on salary while taking two figures (say, –10, and –20)

$$= \left\{-10 - 20 + \frac{10 \times 20}{100}\right\}\%$$

$$= -28\%$$

Further, effect while taking third figure (say, –20) with the above resultant

$$= \left\{-28 - 20 + \frac{28 \times 20}{100}\right\}\% = -42.4\%$$

Again, effect while taking fourth figure (say, –10) with just the above resultant $= \left\{-42.4 - 10 + \frac{42.4 \times 10}{100}\right\}\% = -48.16\%$

Hence, the salary will reduce by 48.16% due to expenditure on various components.

Then savings = 100% – 48.16% = 51.84%

Obviously, 51.84% of total salary = ₹ 15552 (as given)

∴ Total salary $= \frac{100}{51.84} \times 15552 = 30000$

Problem 76. A man spends 80% of his income. Later, his income is increased by 20% but he increases his expenditure by 10% only. Find the percentage increase in his savings.

Sol. Let the initial income be ₹ 100.

As per question, expenditure = 80% of 100 = 80

Then, savings = 100 – 80 = 20

After increment, income becomes ₹ 120 and expenditure becomes 110% of 80 i.e. 88

Then, savings = 120 – 88 = 32

Clearly, increase in savings = 32 – 20 = 12

∴ Percentage increase $= \left(\frac{12}{20} \times 100\right)\% = 60\%$

Problem 77. In a family, the ratio of expenses to savings is 5 : 3. If income and expenses are increased by 50% and 40% respectively, there would be a surplus of ₹ 8400 in savings. The increased savings is :

Sol. Income = Expenditure + Savings = 5 + 3 = 8 units

Now,

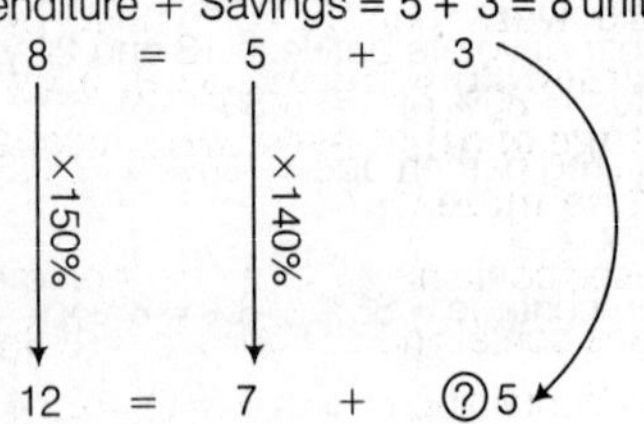

Clearly, increase in saving = 5 – 3 = 2 units = ₹ 8400 (as given)

Then, 5 units = 21000

Hence, the increased savings = ₹ 21000

Problems Based on Number(s)

Problem 78. The sum of two numbers is $\frac{28}{25}$ of the 1st number. The second number is what percent of the first? ***[Hotel Mgmt-97]***

Sol. Let 1st number be 100.

Then, sum of both numbers $= 100 \times \frac{28}{25} = 112$

Thus, second number = 112 – 100 = 12

Clearly, 2nd number is 12% of the first number.

Alternatively

Since sum of two numbers is 112% i.e. $\left(\frac{28}{25} \times 100\right)\%$ of the 1st number, the second number is 12% of the first number.

Problem 79. Three-fifth of a number is 30 more than 50% of that number. What is 80% of that number? ***[NABARD (Asst. Mngt 02)]***

Sol. Let the number be x.

Given, $\frac{3x}{5} - (50\% \text{ of } x) = 30$

$\Rightarrow \frac{3x}{5} - \frac{x}{2} = 30 \Rightarrow \frac{x}{10} = 30$

∴ $x = 300$

Now, 80% of 300 $= \left(300 \times \frac{80}{100}\right) = 240$

Alternatively

Needless to say that three-fifth means 60%. Since (60-50)% of that number is equal to 30, 80% of that number = 240.

Problem 80. The difference of two numbers is 20% of the larger number. If the smaller number is 20, the larger number is :

Sol. Given that the difference of two numbers is 20% of the larger number, it means difference between the larger number and its one-fifth is equal to the smaller number.

Hence, (100 – 20) % of larger number = 20,

then larger number = 25

Alternatively I

Let the larger number be x.

Given, $x - 20 = \frac{x}{5} \Rightarrow \frac{4x}{5} = 20 \Rightarrow x = 25$

Alternatively II

The best way is to proceed through given options.

Problem 81. In a recent survey, 40% houses contained two or more were people. Of those houses containing only one person 25% were having only a male. What is the percentage of all houses, which contains exactly one female and no males?

Sol. Given, houses containing 2 or more people = 40%

Then, houses containing one person = (100 − 40)% = 60%

and houses containing only one male = 25% of 60% = 15%

Thus, houses containing exactly one female and no males

= 60% − 15% = 45%

Problem 82. A student multiplied a number by $\frac{3}{5}$ instead of $\frac{5}{3}$. What is the percentage error in the calculation?

Sol. Let the number be 1.

Given, error $= 1 \times \frac{5}{3} - 1 \times \frac{3}{5} = \frac{16}{15}$,

$\therefore$ Error % $= \left(\frac{\frac{16}{15} \times 100}{\frac{5}{3}}\right)\% = 64\%$

Alternatively,

$\frac{3}{5} = 60\%$ and $\frac{5}{3} = 166.66\%$

Then, error = 166.66% − 60% = 106.66%

$\therefore$ Error % $= \left(\frac{106.66}{166.66} \times 100\right)\% = 64\%$

Problem 83. If $x\%$ of y is 100 and $y\%$ of z is 200, then the relation between x and z is :

Sol. Given, $x\%$ of $y = 100$

$\Rightarrow \frac{xy}{100} = 100 \Rightarrow y = \frac{10000}{x}$

Also given, $y\%$ of $z = 200$

$\Rightarrow yz = 100 \times 200 \Rightarrow y = \frac{20000}{z}$

Clearly, $\frac{10000}{x} = \frac{20000}{z}$

$\Rightarrow z = 2x$

Alternatively

Given, $x\%$ of $y = 100 \Rightarrow y\%$ of $x = 100$

Further, $y\%$ of $z = 200$

Obviously, $z = 2x$

Problem 84. In a school, there are 20 more boys than girls. If the number of boys is increased by 10% and number of girls is also increased by 15%, there would be 19 more boys than girls. Find out the number of students in that school.

Sol. Let the number of girls be x.

Then, number of boys = $x + 20$

As per question, 110% of $(x + 20)$ − 115% of $x = 19$

$\Rightarrow \quad -5\%$ of $x = 19 - 22$

$\Rightarrow \quad x = 60$

Hence, number of girls = 60 and number of boys

= 60 + 20 = 80

Then, total number of students = 60 + 80 = 140

Alternatively,

Given that number of boys is increased by 10%. It means, there will be difference of (20 + 10% of 20) i.e. 22 between number of boys and girls in case number of girls is also increased by the same figure (i.e. 10%) as in case of boys. But actual difference is 19 and it is due to increment in number of girls by 15% instead of 10%. Thus, we may construe that (15 − 10)% of number of girls = (22 − 19), then number of girls = 60.

Now, total students = 60 + (60 + 20) = 140

Problem 85. The numerator of a fraction has been increased by 25% and its denominator has been decreased by 20%. What is the effect on the original fraction?

Sol. Let the fraction be $\frac{x}{y}$.

After alternation in numerator and denominator the fraction becomes $\frac{125\% \text{ of } x}{80\% \text{ of } y} = \frac{25x}{16y}$

Increase in value of original fraction $= \frac{25x}{16y} - \frac{x}{y} = \frac{9x}{16y}$

Then, percentage increase $= \left(\frac{\frac{9x}{16y}}{\frac{x}{y}} \times 100\right)\% = 56.25\%$

Alternatively,

Effect on the fraction in case of alternation (in %) in numerator or denominator or in both =

$$\left[\begin{Bmatrix}\text{(Alternation in numerator with sign)} \\ -\text{ (Alternation in denominator with sign)}\end{Bmatrix} \times \left\{\frac{100}{100 + \text{(Alternation in denominator with sign)}}\right\}\right]\%$$

Explanation :

(a) If numerator is increased by $x\%$ and denominator is increased by $y\%$, the effect on value of the fraction

$$= \left[\{(+x) - (+y)\} \times \frac{100}{(100 + (+y))}\right]\%$$

(b) If numerator is increased by $x\%$ and denominator is decreased by $y\%$, the effect on value of the fraction

$$= \left[\{(+x) - (-y)\} \times \frac{100}{(100 + (-y))}\right]\%$$

(c) If numerator is decreased by $x\%$ and denominator is increased by $y\%$, the effect on value of the fraction

$$= \left[\{(-x) - (+y)\} \times \frac{100}{\{100 + (+y)\}}\right]\%$$

(d) If numerator is decreased by $x\%$ and denominator is decreased by $y\%$, the effect on value of the fraction

$$= \left[\{(-x) - (-y)\} \times \frac{100}{\{100 + (-y)\}}\right]\%$$

Note

(i) If outcome is positive, the original fraction increases and in case of negative outcome, the original fraction decreases.

(ii) If only one of the components (either numerator or denominator) is altered, we can use 0% as alternation for the unchanged component. ***For example****. If the denominator of a fraction is increased by 20%, the effect on the value of the fraction will be* $\left[\{(\pm 0)-(+20)\}\times\frac{100}{\{100+(+20)\}}\right]=-16.66\%$.

Hence, the value of original fraction has been reduced by 16.66%.

For the above question, the effect on the fraction

$$=\left[\{(+25)-(-20)\}\times\frac{100}{\{100+(-20)\}}\right]\%$$

$$=\left[45\times\frac{100}{80}\right]\%=+56.25\%$$

Since the outcome is positive, the value of fraction has been increased by 56.25%

Problem 86. A fraction is squared and then its numerator is increased by 20% and denominator is decreased by 25% resulting in the new fraction becomes $\frac{2}{5}$ of original fraction. The difference between numerator and denominator is what?

Sol. Let the fraction be $\frac{x}{y}$.

Given, $\frac{120\% \text{ of } x^2}{75\% \text{ of } y^2}=\frac{2}{5}\times\frac{x}{y}$

$\Rightarrow \frac{8x^2}{5y^2}=\frac{2}{5}\times\frac{x}{y} \Rightarrow \frac{x}{y}=\frac{1}{4}$

Hence, difference between numerator and denominator $=4-1=3$

Problem 87. When 40% of number A is subtracted from number B, B becomes 55% of its previous value, then which of the following is definitely true?

(a) $A<B$ (b) $B=A$

(c) $A>B$ (d) $A\geq B$

Sol. Clearly, 40% of $A=(100-55)\%$ of B. It means, A is definitely greater than B.

Miscellaneous Problems

Problem 88. In a contract, amount of work has been increased by 20%, then by how much percent must the number of workers be increased to complete the work in time if new workers have 60% more efficiency than the existing workers?

Sol. It is obvious that in order to complete 20% more work, 20% more workers having same efficiency as existing workers's are required. But given that new worker have 25% more efficiency than the existing workers. Then, required percentage

$$=\left(20\times\frac{100}{160}\right)\%=12.5\%$$

Problem 89. A salesman's commission is 5% on all sales upto ₹ 10000 and 4% on all sales exceeding this. He remits ₹ 31100 to his parent company after deducting his commission. Find the total sales.

[RRB-01, SSC (CGL-Mains) 03, IIFT-15]

Sol. Let the total sales be ₹ x.

Then, commission upto sales of ₹ 10000 = 5% of 10000 = 500 and commission beyond sales of ₹ 10000 = 4% of $(x-10000)$

According to question,

$$x-\{500+4\% \text{ of } (x-10000)\}=31100$$

$$\Rightarrow x-\frac{x}{25}=31100+100$$

$$\Rightarrow \frac{24x}{25}=31200 \Rightarrow x=32500$$

Alternatively

The best way is to proceed through given options.

Problem 90. Fresh fruit contains 68% water and dry fruit contains 20% water. How much dry fruit can be obtained from 100 kg of fresh fruits?

Sol. From 100 kg of fresh fruit, 32 kg of pulp may be obtained as it contains 68% water. Now, assume that x kg of dry fruit may be obtained from 100 kg fresh fruits.

Then, quantity of pulp in x kg of dry fruit = 80% of $x=\frac{4x}{5}$

Obviously, $32=\frac{4x}{5}$ [both are quantity of pulp in 100 kg of fresh fruits] $\Rightarrow x=40$ kg

Alternatively I

From 100 kg of fresh fruit, 32 kg of pulp may be obtained as it contains 68% water.

∵ From 80 kg pulp, 100 kg of dry fruits may be obtained.

∴ From 32 kg pulp, $\left(\frac{100}{80}\times 32\right)$ kg of dry fruits may be obtained i.e. 40 kg.

Alternatively II

Quantity of dry fruits $=100\times\frac{(100-68)}{100}\times\frac{100}{(100-20)}=40$ kg

Problem 91. A man bought goods worth ₹ 40, out of which 40 paise went on sales tax on taxable purchases. If the tax rate is 8%, then what was the cost of tax free items?

Sol. Given, 8% of cost of taxable items = ₹ 0.4

∴ 100% of cost of taxable items = ₹ 5

Hence, cost of tax free items = ₹ 40 − ₹ 5 = ₹ 35

Problem 92. 40% of the women are more than 30 years old and 80% of the women are less than or equal to 55 years old. 25% of all of the women play cricket. If 25% of the women above the age of 55 yrs play cricket, find out the percentage of cricket players who are less than or equal to 55 yrs.

Sol. Let the number of women be 100.

Given, number of women less than or equal to 55 yrs = 80

∴ Number of women above the age of 55 years = 20

Now, number of cricket players being less than or equal to age of 55

= Total number of cricket players – number of cricket players above age of 55

= 25% of 100 – 25% of 20

= 25 – 5 = 20

∴ Required percentage $= \left(\frac{20}{25} \times 100\right)\% = 80\%$

Problem 93. A spider climbed 62.5% of the height of the pole in one hour and in the next hour it covered 12.5% of the remaining height. If the height of the pole is 192 m, then distance climbed in second hour is :

Sol. As the end of 1st hour the remaining height of the pole

= (100 – 62.5)% = 37.5%

Then, distance covered in 2nd hour

= 12.5% of 37.5% of 192 = 9 m.

Problem 94. The pressure of gas is inversely proportional to its volume and directly proportional to temperature. If volume is decreased by 20% and temperature is increased by 25%, then what will be the effect on the original pressure?

Sol. Given, $p \propto \frac{1}{v}$ and $p \propto t$

[where, p is pressure, v be the volume and t be temperature]

On combining, $p \propto \frac{t}{v}$

$\Rightarrow \quad p = k \cdot \frac{t}{v}$ [where, k is constant]

Now, new $p = k \cdot \frac{125\% \text{ of } t}{80\% \text{ of } v} = k \cdot \frac{25t}{16v}$

Then, difference between new and original pressure

$= k \cdot \frac{25t}{16v} - k \cdot \frac{t}{v} = k \cdot \frac{9t}{16v}$

∴ Percentage difference $= \left(\frac{k \cdot \frac{9t}{16v}}{k \cdot \frac{t}{v}} \times 100\right)\% = 56.25\%$

Hence, the original pressure will increase by 56.25%.

Alternatively

We derived, $p = k \cdot \frac{t}{v}$

Here, numerator (t) has been increased by 25% and denominator (v) has been decreased by 20%, then, effect on value of the fraction (i.e. p)

$= \left[\{(+25) - (-20)\} \times \frac{100}{\{100 + (-20)\}}\right]\%$

$= \left[45 \times \frac{100}{80}\right]\% = +56.25\%$

Since the outcome is positive, the value of fraction (i.e., p) has been increased by 56.25%.

WORKOUT

1. The value of machine is ₹ 200000. At the end of every year its value reduces at the rate of 2% of that at the beginning of the year. Find its value at the end of third year.

(a) 178238.4 (b) 148238.4
(c) 188238.4 (d) None of these

2. If m% of (m) + n% of (n), = 2% of (mn), then what percent of m is n? ***[CDS-14, CDS-17]***

(a) 50%
(b) 75%
(c) 100%
(d) Cannot be determined

3. When 30% of one number is subtracted from another number, the second number reduces to its four-fifth. What is the ratio of first to the second number?

(a) 4 : 7 (b) 3 : 2 (c) 2 : 5
(d) Cannot be determined
(e) None of these

4. 400 students took the mock test. 60% of the boys and 80% of the girls cleared the cut off in the test. If the total percentage of students qualifying is 65%, how many girls appeared in the test? ***[MAT 2014, 19]***

(a) 100 (b) 120
(c) 150 (d) 300

5. Akash scored 73 marks in subject *A*. He scored 56% in subject *B* and *x* marks in subject *C*. Maximum marks in each subject were 150. The overall percentage marks obtained by Akash in all the three subjects together were 54%. How many marks did he score in subject *C*? ***[IBPS (PO)-12, IAS (PT)-98]***

(a) 84
(b) 86
(c) 79
(d) 73
(e) None of the above

6. Tax on commodity is decreased by 10% and thereby its consumption increases by 8%. Find the increase or decrease in the revenue obtained from the commodity.

(a) 2.8% decrease (b) 2.1% increase
(c) 1.8% decrease (d) We cannot say

7. If 50% of ($x - y$) = 40% of ($x + y$), then what percent of x is y?

(a) $10\frac{1}{9}\%$ (b) $11\frac{1}{9}\%$ (c) $13\frac{1}{9}\%$ (d) $21\frac{1}{9}\%$

8. In an election between two candidates, one got 72% of the total valid votes. 25% of the total votes are invalid. If the total votes are 8200, what is the number of valid votes the other person got? ***[OBC (PO)-08, PNB (PO)-09, UPPSC (Revenue Inspector)-15]***

(a) 1835 (b) 1722
(c) 2050 (d) 4428
(e) None of these

9. The number of votes not cast for the Praja Party increased by 25% in the National General Election over those not cast for it in the previous Assemble Polls and the Praja Party lost by a majority twice as large as by which it had won the Assembly Polls. If a total of 260000 people voted each time, how many voted for the Praja Party in the previous Assembly polls? ***[CAT-94]***

(a) 110000 (b) 150000
(c) 140000 (d) 120000

10. One-fourth of 60% of a number is equal to two-fifths of 20% of another number. What is the ratio of the first number to second?

(a) 4 : 7
(b) 8 : 13
(c) 5 : 9
(d) cannot be determined
(e) None of these

11. Out of five questions in a paper, $\frac{1}{20}$th of students answered all questions and $\frac{1}{20}$ th none. Of the rest, $\frac{1}{4}$th answered only four and $\frac{1}{5}$th answered only one question. If 24.5% of the total number of student answered only three questions and 200 answered only two, what was the total number of students?

(a) 840 (b) 860
(c) 800 (d) None of these

12. To an examination, a candidate needs 40% marks. All questions carry equal marks. A candidate just passed by getting 10 answers correct by attempting 15 of the total questions. How many questions are there in the examination?

(a) 2 (b) 30 (c) 40 (d) 45

13. The present population of a village is 5500. If the number of males increases by 11% and the number of females increases by 20%, then the population will become 6330. What is the present population of females in the village?

(a) 3000 (b) 2500
(c) 3500 (d) 2000

14. A person who has a certain amount with him goes to market. He can buy 50 oranges or 40 mangoes. He retains 10% of the amount for taxi fares and buys 20 mangoes and of the balance he purchases oranges. Number of oranges he can purchase is : ***[CAT-95]***

(a) 36 (b) 40
(c) 15 (d) 20

15. Ram ordered for 6 black toys and some additional brown toys. The prices of a black toy is 2.5 times that of brown toy. While preparing the bill, the clerk interchanged the number of black toys which increased the bill by 45%. Find the number of brown toys.

(a) 14 (b) 16
(c) 15 (d) 24

16. What is the number whose 20% is 30% of 40?

(a) 90 (b) 60
(c) 80 (d) 50

17. On a test consisting of 75 questions carrying one mark each, Samir answered 75% of the first 40 questions correctly. What approximate percent of the other 35 questions does he need to answer correctly to score 80% on the entire test? ***[NABARD (PO)-08, MAT-12]***

(a) 90 (b) 75
(c) 86 (d) 60
(e) 58

18. In a town 25% families own a phone and 15% own a car, 65% families own neither a phone nor a car. 2000 families own both a car and a phone.
Consider the following statements in this regard:
I. 10% families own both a car and a phone
II. 35% families own either a car or a phone
III. 40000 families live in the town.
Which of the above statements are correct? ***[IAS (PT)-99, 01]***

(a) I and II (b) I and III
(c) II and III (d) I, II and III

19. There are two types of employees in Sun Metals, general graduates and engineers. 40% of the employees in Sun Metals are general graduates, and 75% of the engineers earn more than ₹ 5 lakh/yr. If 50% of the organisation's employees earn more than ₹ 5 lakh/yr, what proportion of the general graduates employed by the organisation earn ₹ 5 lakh or less? ***[XAT-10]***

(a) $\frac{3}{5}$ (b) $\frac{3}{4}$
(c) $\frac{1}{2}$ (d) $\frac{2}{5}$
(e) None of these

20. A motorist reduces his distance covered annually (in km) by x% when the price of petrol increased by y%. Find the increase percent in his annual petrol bill.

(a) $\left\{-x + y - \frac{xy}{100}\right\}$ %
(b) $\left\{x - y - \frac{xy}{100}\right\}$ %
(c) $\left\{\frac{xy}{100}\right\}$ %
(d) $\left\{\frac{2(x + y)}{100}\right\}$ %

21. X, Y and Z had taken a dinner together. The cost of the meal of Z was 20% more than that of Y and the cost of the meal of X was $\frac{5}{6}$ as much as the cost of the meal of Z. If Y paid ₹ 100, then what was the total amount that all of them had paid? ***[CDS-13]***

(a) ₹ 285 (b) ₹ 355
(c) ₹ 300 (d) None of these

22. What is 5% of 50% of 500? ***[CDS-12]***

(a) 12.5 (b) 25
(c) 1.25 (d) 6.25

23. The pre-paid recharge of Airtel gives 21% less talktime than the same price pre-paid of Vodafone. The post-paid talktime of Airtel is 12% more than its pre-paid recharge, having the same price. Further, the post-paid talktime of same price of Vodafone is 15% less than its pre-paid recharge. How many percent less/more talktime can one get from the Airtel post-paid service compared to the post-paid service of Vodafone? ***[IIFT-15]***

(a) 3.9% more (b) 4.7% less
(c) 4.7% more (d) 2.8% less

24. A building worth ₹ 1331000 is constructed on a land worth ₹ 729000. If the land appreciates at 10% per annum and the building depreciates at 10% per annum, find the time after which the values of land and the building will be the same.

(a) 3 yr (b) 4 yr
(c) 2.5 yr (d) 4 yr

25. p% of a number P is q% more than r% of the number R. If the difference between P and R is r% of R and if the sum of P and R is 210, then which of the following statements is always true?

(a) $P = 110, R = 100$
(b) $P = 220, R = 200$
(c) $P = 3300, R = 3000$
(d) All of the above

26. In a class of 110 students, x students take both Mathematics and Statics, $2x + 20$ students take Mathematics and $2x + 30$ students take Statistics. There are no students who take neither Mathematics nor Statistics. X is equal to what percent of total students?

(a) 18.18% (b) 20%
(c) 25% (d) 30.3%

27. A salesman offers a commission of x% on the first sale of worth ₹ 3000 and y% on the part of sale exceeding it. He gives ₹ 1100 as commission on sale of ₹ 7000 and ₹ 1660 as commission on sale of ₹ 11000. Find the value of x and y respectively: ***[SIDBI Bank Officer-14]***

(a) 18% and 14% (b) 18% and 12%
(c) 14% and 16% (d) 16% and 14%
(e) None of these

28. Suppose that an equal number of persons are born on each day. What will be the percentage of persons whose birthday will fall on 29th February?

(a) 0.741 (b) 0.273
(c) 0.068 (d) None of these

29. The price of sugar has been increased by 32% but a family reduced its consumption so that the expenditure on it was only 10% more than before. If the early monthly consumption was 10 kg, what is the monthly consumption now?

(a) 18.33 kg (b) 8.33 kg
(c) 16.66 kg (d) 12.5 kg

30. A person saves 6% of his income. 2 yrs later, his income shoots up by 15% but his savings remain the same. Find the hike in his expenditure?
[MAT-14, 19, SSC(CGL-T-I-19), SSC (CGL-T-II)-19]

(a) 13.65% (b) 12.45%
(c) 14.85% (d) 15.95%

31. The radius of the base of a right circular cone is increased by 15% keeping the height fixed. The volume of the cone will be increased by:

(a) 30% (b) 31%
(c) 32.25% (d) 34.75%

32. 45% of 224 × ?% of 120 = 8104.32

(a) 67 (b) 62 (c) 59 (d) 71
(e) None of these

33. 18.5% of 425 + 16.2% of 388 = ?

(a) 141 (b) 224
(c) 116 (d) 183
(e) 215

34. An interview panel found that a candidate has given a wrong data about his height. While filling up his form he filled up 20% more than his actual height. His actual height is 5 feet 2 inches. By what approximate percent should he reduce his height to get actual height? ***[LIC (AAO)-13, MAT-09]***

(a) 15 (b) 14 (c) 18 (d) 17
(e) None of these

35. The number of males per hundred females in a country was 120. At the next census, the total population increased by 5% and the female population increased by 10%. Find the increased percent in the male population.

(a) 0.83% (b) 62.5%
(c) 16.66% (d) 33.33%

36. A salesman's terms were changed from a flat commission of 5% on all his sales to a fixed salary of ₹ 1000 plus 2.5% commission on all sales exceeding ₹ 4000. If his remuneration as per new scheme was ₹ 600 more than the first scheme, what were his sales worth?

(a) ₹ 11000 (b) ₹ 12000
(c) ₹ 16000 (d) ₹ 17000

37. Rubina decided to donate 16% of her monthly salary to an NGO. On the day of donation she changed her mind and donated ₹ 6567, which was 75% of what she had decided earlier. How much is Rubina's monthly salary (in ₹)?

(a) 8756 (b) 54725
(c) 656700 (d) 45696
(e) None of these

38. 53% of a number is 358 less than the square of 26. What is the value of three-fourth of 23% of that number?

(a) 101 (b) 109.5
(c) 113 (d) 103.5
(e) None of these

39. In a survey of political preferences, 78% of those asked were in favour of at least one of the proposals I, II and III, 50% of those asked favoured proposal I, 30% favoured proposal II and 20% favoured proposal III. If 5% of those asked favoured all three of the proposals, what percentage of those asked favoured more than one of the three proposals?
[CAT-99, IIFT-18]

(a) 10 (b) 12
(c) 17 (d) 22

40. If 80% of A = 20% of B and $B = 5x$% of A, then the value of x is ***[CLAT-14]***

(a) 75 (b) 80
(c) 90 (d) 85

41. Village A has a population of 6800, which is decreasing at the rate of 120/yr. Village B has a population of 4200, which is increasing at the rate of 80/yr. In how many years will the population of the two villages be equal?

(a) 9 yrs (b) 11 yrs
(c) 13 yrs (d) 16 yrs

42. A man spends 80% of his income. With the increase in the cost of living his expenditure increase by 37.5% and his income increases by $16\frac{2}{3}$%, find his present savings.

(a) 5.71%
(b) 1.59%
(c) 4.935
(d) None of the above

43. If 7% of the total quantity of wheat is lost in grinding when a country has to import 6 million tonne but when 5.2% is lost it can export 3 million tonne, then what is the quantity of wheat grown in the country? *[MAT-13, 12, 08]*

(a) 600 m tonne (b) 500 m tonne
(c) 400 m tonne (d) 700 m tonne

44. If the numerator of a fraction is increased by 500% and the denominator is increased by 300%, the resultant fraction is $1\frac{1}{17}$. What was the original fraction?
[UBI (PO) 08, SBI (PO) 08, PNB (PO)09, IOB (PO) 09, NABARD (PO) 09, Corporation Bank (PO) 10]

(a) $\frac{12}{17}$ (b) $\frac{13}{17}$ (c) $\frac{3}{7}$ (d) $\frac{4}{11}$
(e) None of these

45. In a class of 40 students and 8 teachers, each student got sweets that are 20% of the total number of students and each teacher got sweets that are 25% of the total number of students. How many sweets were there?

(a) 420 (b) 400 (c) 320 (d) 360
(e) None of these

46. If a% of x is equal to b% of y, then c% of y is what per cent of x?

(a) c% (b) abc% (c) $\frac{ac}{b}$% (d) $\frac{bc}{a}$%

47. Amar, Akbar and Antony are friends, being looked after by a matron Farah. Amar weighs 50% more than Akbar and Antony weighs 25% less than Amar. Farah weighs a third of the combined weight of the three boys. All four together weighs 232 kg. The correct arrangement of the person in the ascending order of their weights is: *[IAS (PT)-99]*

(a) Antony, Akbar, Farah, Amar
(b) Antony, Akbar, Amar, Farah
(c) Akbar, Antony, Amar, Farah
(d) Akbar, Antony, Faray, Amar

48. Gauri went to the stationers and bought items worth ₹ 25, out of which 30 paise went on sales tax on taxable purchases. If the tax rate was 6%, then what was the cost of tax free items (in ₹) *[MAT-12]*

(a) 15 (b) 15.70 (c) 19.70 (d) 20

49. A city has a population of 300000 out of which 180000 are males. 50% of the population is literate. If 70% of the males are literate, the number of literate females is : *[IAS (Pre)-01, MAT-06, 18]*

(a) 24000 (b) 30000
(c) 54000 (d) 60000

50. In an election, 10% of the people in the voter's list did not participate. 60 votes were declared invalid. There are only two candidates A and B. A defeated B by 308 votes. It was found that 46% of the people listed in the voter's list voted for A. Find out the total number of votes polled. *[SSC (CGL/Mains)-2000, MAT-06]*

(a) 10126 (b) 11160
(c) 11300 (d) 103500

51. At a college entrance examination, each candidate is admitted or rejected according to whether he has passed or failed the test. Of the candidates who are really capable, 80% pass the tests and of the incapable, 25% pass the tests. Given that 40% of the candidates are really capable, then the proportion of capable college students is about ?

(a) 73% (b) 70%
(c) 68% (d) 75%

52. Out of a total of 120 musicians in a club, 5% can play all the three instruments, guitar, violin and flute. It so happens that the number of musicians who can play any two and only two of the above instruments is 30. The number of muscians who can play the guitar alone is 40. What is the total number of those who can play violin alone or flute alone? *[IAS (PT)-95, MAT-06, CSAT-14]*

(a) 45 (b) 44 (c) 38 (d) 30

53. Peter got 30% of the maximum marks in an examination and failed by 10 marks. However, Paul who took the same examination got 40% of the total marks and got 15 more than the passing marks in the examination. What were the passing marks in the examination?

(a) 35 (b) 250
(c) 75 (d) 85

54. $\frac{5}{9}$th part of the population in a village are males. If 30% of the males are married, the percentage of unmarried females in the total population is *[CDS-17]*

(a) $20\frac{2}{9}$ (b) $27\frac{2}{9}$
(c) $27\frac{7}{9}$ (d) $29\frac{2}{9}$

55. In an examination, 40% marks are required to pass. A obtains 10% less than the number of marks required to pass. B obtains $11\frac{1}{9}$% less than A and C obtained $41\frac{3}{17}$% less than the number of marks obtained by A and B together. What marks did C get?

(a) 50 (b) 40
(c) 35 (d) 45

56. A man's income is increased by ₹ 1200 and at the same time, the rate of tax to be paid is reduced from 12% to 10%. He now pays the same amount of tax as before. What is his increased income, if 20% of his income is exempted from tax in both cases? *[MAT-10, SSC (CGL-T-II)-17]*

(a) ₹ 6300 (b) ₹ 7200
(c) ₹ 4500 (d) ₹ 4500

57. Ravi invests 50% of his monthly savings in fixed deposits. 30% of the rest of his savings is invested in stocks and the rest goes into Ravi's saving bank account. If the total amount deposited by him in the bank (for savings account and fixed deposits) is ₹ 59500, then Ravi's total monthly savings (in ₹) is : *[CAT-17]*

(a) 70000
(b) 87000
(c) 84000
(d) None of these

58. The number of girls appearing for an admission test is twice the number of boys. If 30% of the girls and 45% of the boys get admission, the percentage of candidates who do not get admission is : *[CAT-17]*

(a) 35
(b) 50
(c) 60
(d) 65

59. The income of Amala is 20% more than that of Bimala and 20% less than that of Kamala. If Kamala's income goes down by 4% and Bimala's goes up by 10%, then the percentage by which Kamala's income would exceed Bimala's is nearest to : ***[CAT-19]***

(a) 31 (b) 28 (c) 32 (d) 29

60. In a class, 60% of the students are girls and the rest are boys. There are 30 more girls than boys. If 68% of the students, including 30 boys, pass an examination, the percentage of the girls who do not pass is : ***[CAT-19]***

(a) 20% (b) 50%
(c) 30% (d) None of these

61. A business group has 3 companies *X*, *Y*, *Z* and a Trust *P* which is engaged in charitable activities. Each group company has to donate 5% of its own funds to the Trust, excluding the loan which the company has taken from other companies of the group. *X* has given a loan to *Y* which is equivalent to 10% of the funds of *Y*. After receiving the loan, *Y* has funds which are two times the funds of *Z*. If *Z* gave ₹ 10000 as donation to the Trust *P*, how much is the approximate contribution of Company *Y* to the Trust *P*? ***[IIFT-18]***

(a) 17000 (b) 18000
(c) 19000 (d) 20000

62. Meena scores 40% in an examination and after review, even though her score is increased by 50%, she fails by 35 marks. If her post-review scored is increased by 20%, she will have 7 marks more than the passing score. The percentage score needed for passing the examination is : ***[CAT-19]***

(a) 75 (b) 60 (c) 70 (d) 80

63. Nailini has received a total of 600 messages from four friends Anita, Bina, Chaitra and Divya, Bina and Divya have respectively sent 30% and 20% of these messages, while Anita has sent an equal number of messages as Chaitra. Moreover, Nalini finds that of Anita's, Bina's, Chaitra's and Divya's message, 60%, 40%, 80% and 50% respectively are jokes. What percentage of the jokes, received by Nalini, have been sent neither by Divya nor by Bina? ***[XAT-20]***

(a) 65.12 (b) 61.4 (c) 38.6 (d) 57
(e) 34.88

Answer key

1. (c)	2. (c)	3. (e)	4. (a)	5. (b)	6. (a)	7. (b)	8. (b)	9. (c)	10. (e)
11. (c)	12. (a)	13. (b)	14. (d)	15. (c)	16. (b)	17. (c)	18. (c)	19. (e)	20. (a)
21. (d)	22. (a)	23. (a)	24. (a)	25. (a)	26. (a)	27. (a)	28. (c)	29. (b)	30. (d)
31. (c)	32. (a)	33. (a)	34. (d)	35. (a)	36. (b)	37. (b)	38. (d)	39. (c)	40. (b)
41. (c)	42. (a)	43. (b)	44. (a)	45. (b)	46. (c)	47. (d)	48. (d)	49. (a)	50. (b)
51. (c)	52. (b)	53. (d)	54. (c)	55. (b)	56. (b)	57. (a)	58. (d)	59. (a)	60. (a)
61. (b)	62. (c)	63. (b)							

Hints and SOLUTIONS

1. Value at the end of third year

$= 200000 \times \left(1 - \frac{2}{100}\right)^3$

$= 200000 \times \frac{49}{50} \times \frac{49}{50} \times \frac{49}{50}$

$= ₹\ 188238.4$

Alternatively

Try yourself through another method as already given in illustrations.

2. Given, $\frac{m^2}{100} + \frac{n^2}{100} = \frac{2mn}{100}$

$\Rightarrow \quad m^2 + n^2 = 2mn$

$\Rightarrow \quad m^2 + n^2 - 2mn = 0$

$\Rightarrow \quad (m - n)^2 = 0$

$\Rightarrow \quad m = n$

Hence, 100% of *m* is *n*.

3. Clearly, 30% of 1st number

$= (100 - 80)\%$ of the 2nd number

$\Rightarrow$ 1st number : 2nd number = 2 : 3

4.

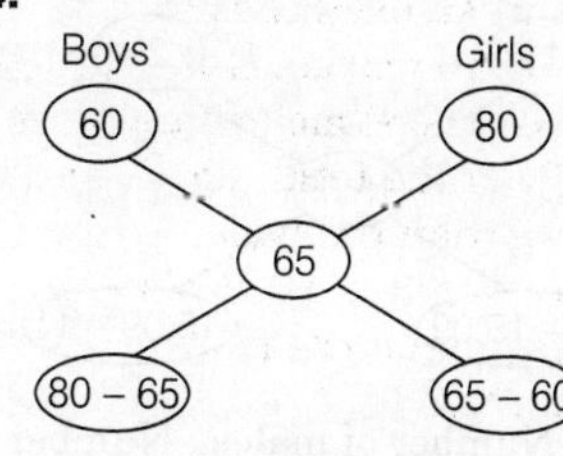

Obviously, number of boys : Number of girls = $(80 - 65)\% : (65 - 60)\% = 3 : 1$

Hence, number of girls appeared in exam $= 400 \times \frac{1}{(1+3)} = 100$

[We will substitute the method, shown in last line of solution, in chapter 'Ratio and Proportion' to compute quickly].

5. Score in subject *A* = 73, in subject *B* = 56% of 150 = 84 and in subject *C* = ?

Now, $73 + 84 + ? = 3 \times 54\%$ of 150

$\Rightarrow \quad ? = 86$

6. Here, revenue = tax × consumption

Now, observe the concept 'Effect on product in case of changes in factor(s)'. Then, effect on revenue

$= \left\{-10 + 8 - \frac{10 \times 8}{100}\right\}\% = -2.8\%$

Since outcome is negative, the revenue from the commodity will decrease by 2.8%.

7. Given, 50% of $(x - y)$ = 40% of $(x + y)$

$\Rightarrow \quad 0.5x - 0.5y = 0.4x + 0.4y$

$\Rightarrow \quad 0.1x = 0.9y \Rightarrow x = 9y$

Now, ? % of $x = y$

Then, following the pattern of missing figure, required percent

$= \frac{100y}{x} = \frac{100y}{9y} \quad [\because x = 9y]$

$= 11\frac{1}{9}\%$

8. Given, invalid votes = 25% of the total votes

∴ Valid votes = 75% of the total votes = 75% of 8200

Then, votes got by the other person = valid votes – votes got by first candidate = 75% of 8200 – 72% of 75% of 8200

= 75% of 8200 – 54% of 8200

= 21% of 8200

= 1722

Alternatively,

Valid votes = 75% of total votes

Votes got by A = 72% of 75% = 54% of total votes

Hence, votes got by B

= (75 – 54)% of total votes

= 21% of 8200 = 1722

9. Let number of votes not cast for Praja Party be x in the assembly polls. Then, number of votes got by Praja Party in assembly polls = $260000 - x$

Further, number of votes not cast for Praja Party in the General National Election (GNE) = $1.25x$

∴ Number of votes got by Praja Party in GNE = $260000 - 1.25x$

According to the question,

$2[(260000 - x) - x]$

$= 1.25x - (260000 - 1.25x)$

$\Rightarrow \quad 520000 - 4x = 2.5x - 260000$

$\Rightarrow \quad 6.5x = 780000$

$\Rightarrow \quad x = 120000$

Thus, number of votes got by Praja Party in assembly polls

= 260000 – 120000 = 140000

10. Given, $\frac{1}{4} \times 60\%$ of first number

$= \frac{2}{5} \times 20\%$ of second number

$\Rightarrow$ first number : second number = 8 : 15

11. Given, number of students who answered all 5 questions = 5%

Number of students who answered 0 question = 5%

Number of students who answered 4 questions

$= \frac{1}{4}$ of (100 – 5 – 5)% = 22.5%

Number of students who answered 1 question

$= \frac{1}{5}$ of (100 – 5 – 5)% = 18%

Number of students who answered 3 question = 24.5%

Obviously, number of students who answered only 2 questions,

= (100 – 5 – 5 – 22.5 – 18 – 24.5)% = 25%

As given, 25% of the students = 200

∴ Total number of students = 800

12. Given, 10 questions carry 40 % marks. It means 25 questions will carry 100% marks. Hence, total number of questions in exam. = 25.

13. Let the female population is p, then male population = $5500 - p$

Given,

111% of $(5500 - p)$ + 120% of $p = 6330$

$\Rightarrow$ 111% of 5500 + 9% of $p = 6330$

$\Rightarrow \quad 0.9p = 225$

$\therefore \quad p = 2500$

Alternatively I

Increase in population

$= \left(\frac{830}{5500} \times 100\right)\% = 15.09\%$

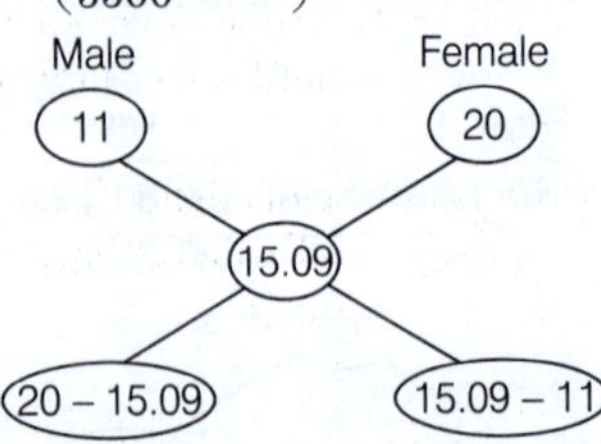

Hence, Number of males : Number of females = (20 – 15.09) : (15.09 – 11)

= 4.91 : 4.09

Then, the female population

$= 5500 \times \frac{4.09}{(4.91 + 4.09)} = 2500$

Alternatively II

You may proceed through options also.

(a) 3000 ⇒ 120% of 3000 + 111% of 2500 ≠ 6330

(b) 2500 ⇒ 120% of 2500 + 111% of 3000 = 6330

Hence, option (b) is correct.

14. Let the person goes to market keeping ₹ 100.

Expenditure on purchasing

= 90% of 100 = ₹ 90

Given, price of 50 Oranges

= price of 40 Mangoes = ₹ 100

Hence, price of each Orange

$= \frac{100}{50}$ = ₹ 2 and that of each mango $= \frac{100}{40} = 2.50$

Now, expenditure on purchase of 20 mangoes = 20 × 2.50 = ₹ 50

Then, balance = 90 – 50 = ₹ 40

∴ Number of Oranges he bought

$= \frac{40}{2} = 20$

Alternatively

Given, price of 50 oranges = price of 40 mangoes. If the person intends to buy both items, he can buy 25 oranges and 20 mangoes with the same amount. Firstly, he has bought 20 mangoes, then with the balance amount he can buy 25 oranges provided no provision of taxi fares is allowed. In case 10% is kept for fares, number of oranges will be reduced by (10 × 2)%. Hence, number of oranges he bought = 80% of 25 = 20.

15. Let number of brown toys be x and price of each brown toy is ₹ 1

Then, initial bill before mistake

$= (6 \times 2.50) + (x \times 1) = 15 + x$

Due to interchanges in prices, bill amount = $(6 \times 1) + (x \times 2.5)$

$= 6 + 2.5x$

Given, $(6 + 2.5x)$ = 145% of $(15 + x)$

$\Rightarrow \quad 6 - \frac{87}{4} = \frac{29x}{20} - 2.5x$

$\Rightarrow \quad \frac{63}{4} = \frac{21x}{20}$

$\Rightarrow \quad x = 15$

Alternatively

The best way is to proceed through options.

16. 20% of ? = 30% of 40

⇒ We know from the 'pattern of Missing figure' that missing figure

$$= \left(\frac{\text{Number (single or in product form only) of opposite side of ?}}{\text{number attached with ?}} \times 100 \right)$$

$$= \frac{0.3 \times 40}{20} \times 100 = 60$$

17. (c) Since each question carries one mark, Samir must need to answer correctly of 80% of 75 questions i.e. 60 questions in order to score 80% marks. He answered only 75% of 40 i.e. 30 questions correctly. Now, he need to answer 30 questions correctly out of remaining 35 questions.

Hence, required percentage

$$= \left(\frac{30}{35} \times 100 \right)\% = 85.71\%$$

= 86% (approx.)

Alternatively

75% of 40 + x% of 35 = 80% of 75

⇒ 30 + x% of 35 = 60

$$\Rightarrow \frac{35x}{100} = 30$$

⇒ $x = 85.71\% = 86\%$ (approx.)

18. Given, 65% families own neither a phone nor a car. It may be construed from this statement that 35% families own either a phone or a car or both.

Then, 35% = Number of families with a phone + Number of families with a car − Number of families with both a car and a phone.

⇒ 35% = 25% + 15% − Number of families with both a car and a phone

⇒ Number of families with both a car and a phone = 5%

But given, 5% = 2000

∴ 100% = 40000

Hence, 40000 families live in the town.

Thus, option (c) is correct.

Alternatively

65% = (100 − 25)% + (100 − 15)% − number of families not having both = 95%

It means, number of families with both a car and phone = (100 − 95)% = 5%. But given, 5% = 2000

∴ 100% = 40000

{**Remarks** Assume, passed = having and failed = not having. Details are in illustrations}

19. Let the total employees be 100.

Number of general graduates = 40

Then, number of engineers = 60

Engineers getting more than ₹ 5 lakh/yr = 75% of 60 = 45

Also given, 50% of the total employees i.e. total of 50 are getting more than ₹ 5 lakh/yr. It means only 5 employees under category of general graduates are getting more than ₹ 5 lakh/yr.

Hence, proportion of the general graduates getting ₹ 5 lakh or less

$$= \frac{(40-5)}{40} = \frac{7}{8}.$$

20. We know, Distance covered × Price of petrol = expenses(bill)

Now, observe the concept 'Effect on producet in case of changes in factor(s)'. Then, effect on bill

$$= \left\{ -x + y - \frac{xy}{100} \right\}\%$$ Note that the outcome may be either positive depending upon values of x and y.

21. Given, cost of meal of Y be ₹ 100.

Then, cost of meal of $Z = 120$

and cost of meal of $X = 5/6$ of $120 = 100$

∴ Total cost of meals

= 100 + 120 + 100

= ₹ 320

22. 5% of 50% of 500

$$= 500 \times \frac{5}{100} \times \frac{50}{100} = 12.5$$

23. Let the person intends to spend ₹ x. Also assume that in case of pre-paid, talktime of Vodafone = 100 min.

Then, with the same price (i.e. x), talktime of Airtel = 21% less then 100 = 79 min.

Further, talktime of vodafone in case of postpaid = 15% less than 100 = 85 min

and talktime of Airtel in case of Postpaid = 12% more than 79

= 88.48 min

∴ Required %

$$= \left(\frac{88.48 - 85}{85} \times 100 \right)\%$$

= 4.09%

Hence, the person gets 4.09% more talktime from the Airtel postpaid service compared to the post-paid service of Vodafone. Closest option is (a).

24. Let after t years, their values are same.

$$\Rightarrow 1331000 \left(1 - \frac{10}{100}\right)^t = 729000 \left(1 + \frac{10}{100}\right)^t$$

$$\Rightarrow 1331000\,(0.9)^t = 729000\,(1.1)^t$$

$$\Rightarrow \left(\frac{0.9}{1.1}\right)^t = \frac{7290000}{1331000}$$

$$\Rightarrow \left(\frac{0.9}{1.1}\right)^t = \frac{729}{1331}$$

$$\Rightarrow \left(\frac{0.9}{1.1}\right)^t = \left(\frac{9}{11}\right)^3$$

$$\Rightarrow \left(\frac{0.9}{1.1}\right)^t = \left(\frac{0.9}{1.1}\right)^3$$

⇒ $t = 3$ yrs

Alternatively

The best way is to proceed through options.

25. Option (a) is correct as the sum of P and R is 210 and the same fact is given in the question.

26. Given, $110 = (2x + 20) + (2x + 30) - x$

⇒ $110 = 3x + 50$

⇒ $x = 20$

$$\therefore \text{Required } \% = \left(\frac{20}{110} \times 100 \right)\%$$

= 18.18%

27. Given, x% of 3000 + y% of (7000 − 3000) = ₹ 1100

⇒ $30x + 40y = 1100$ ...(i)

Also given, x% of 3000 + y% of (11000 − 3000) = ₹ 1660

⇒ $30x + 80y = 1660$ ...(ii)

On subtracting Eq. (i) from Eq. (ii), we get, $40y = 560$

⇒ $y = 14\%$

Putting the value of y in Eq. (i), we get $x = 18\%$

Alternatively

Given that due to increase in sales from ₹ 7000 to ₹ 11000 (i.e. ₹ 4000), commission is increased from 1100 to 1660 (i.e. 560). It is clear that this increment in commission is due to application of y% on the increased sales (i.e. 4000) and the fact that ₹ 560 is 14% of 4000. Hence, $y = 14\%$

Now, in case of sales of ₹ 7000, commission at the rate of 14% for sales of ₹ 4000 = 560. Then, commission at the rate of x% for sales of first ₹ 3000 = 1100 − 560 = 540. Obviously, 540 is 18% of 3000. It means, x = 18%.

28. We know that 29th February falls in a leap year and a leap year comes once in four years. It is also fact that a non-leap year has 365 days while a leap year has 366 days.

Let each day one person is born. Then, in 4 years, number of person born = (3 × 365) + 366 = 1461

Then, required percentage

$$= \left(\frac{1}{1461} \times 100\right)\% = 0.068\%$$

29. Let the family has reduced the consumption by x%.

Here, expenditure is a product of price and consumption. Observing the concept of 'Effect on product in case of changes in factor(s)', effect on expenditure $= \left\{+32 - x - \frac{32 \times x}{100}\right\}\%$

$$\Rightarrow \quad +10\% = \left\{+32 - \frac{33x}{25}\right\}\%$$

$$\Rightarrow \quad \frac{33x}{25} = 22$$

$$\Rightarrow \quad x = \frac{22 \times 25}{33} = 16.66\%$$

Hence, the consumption has been reduced by 16.66%. Then, the present consumption = 83.33% of 10 = 8.33 kg.

Alternatively

Reduction in consumption keeping the expenditure constant

$$= \left\{\frac{32}{(100+32)} \times 100\right\}\%$$

$$= 24.24\%$$

In case consumption is reduced by 24.24%, there will be no effect on expenditure. But, given that expenditure is allowed to be increased by 10%.

Then, effect on consumption

$$= \left\{-24.24 + 10 - \frac{24.24 \times 10}{100}\right\}\%$$

$$= -16.66\%$$

Hence, the consumption has been reduced by 16.66%. Then, the present consumption = 83.33% of 10 = 8.33 kg.

30. Given, income (100) = expenditure (94) + savings (6)

After alternation ⇒ income (115) = expenditure (?) + savings (6)

Then, expenditure = 115 − 6 = 109

Hence, hike in expenditure

$$= 109 - 94 = 15$$

Then, hike % $= \left(\frac{15}{94} \times 100\right)\%$

$$= 15.95\%$$

31. We know that volume of the cone $= \frac{1}{3}\pi r^2 h$ i.e. its volume is a product of three factors with a constant term. Here, only radius has been increased, it means two factors (say, r, r) have been changed. Now, observe the concept 'Effect on product in case of changes in factor(s)'.

Then, effect on volume

$$= \left\{+15 + 15 + \frac{15 \times 15}{100}\right\}\%$$

$$= +32.25\%$$

Since outcome is positive, volume of the cone will be increased by 32.25%

32. Following the pattern of missing figure, ? $= \frac{8104.32}{0.45 \times 224 \times 120} \times 100$

$$= 67$$

33. 18.5% of 425 + 16.2% of 388

= 78.62 + 62.85 = 141.47 = 141 (approx.)

34. Observe the concept of 'Base value for percentage determination', required answer

$$= \left\{\frac{20}{(100+20)} \times 100\right\}\%$$

= 16.66% = 17% (approx.)

35. Let the initial population be 220 [i.e. 100 females + 120 males]

Then, increased population

= 105% of 220

= 231

and increased female population

= 110% of 100

= 110

Thus, increased male population

= 231 − 110 = 121

∴ % increase

$$= \left\{\frac{(121-120)}{120} \times 100\right\}\% = 0.83\%$$

Alternatively

Ratio of male to female

= 120 : 100 = 6 : 5

Let Male population increases by x%. Then,

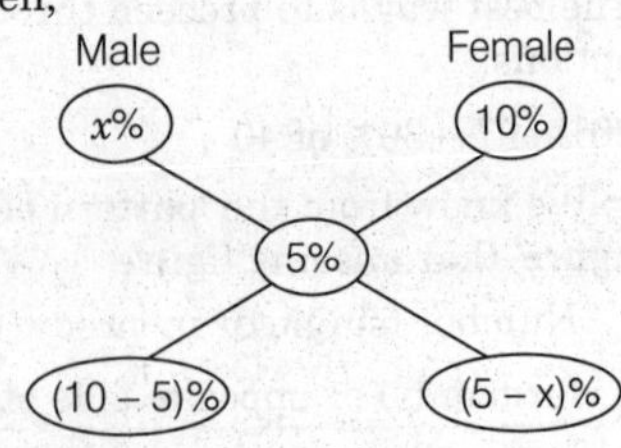

Obviously, Male : Female

= (10 − 5) : (5 − x)

= 6 : 5

$$\Rightarrow \quad 25 = 30 - 6x$$

$$\Rightarrow \quad x = \frac{5}{6}$$

Hence, % increase in Male population

$$= \frac{5}{6}\% = 0.83\%$$

[**Remarks** Another approach has also been given in chapter' Ratio and Proportion in addition to the above approach in detail]

36. Let sales worth were ₹ x.

Given, Commission in new scheme = Commission in old scheme + ₹ 600

⇒ 1000 + 2.5% of (x − 4000) = 5% of x + 600

⇒ 2.5% of x = 1000 − 600 − 100

⇒ x = 12000

Alternatively

You may proceed through options. Ensure that which option satisfies the condition that Commission in new scheme = Commission in old scheme + ₹ 600

(a) 11000 ⇒ 1000 + 2.5% of (11000 − 4000) ≠ 5% of 11000 + 600

(b) 12000 ⇒ 1000 + 2.5% of (12000 − 4000) = 5% of 12000 + 600

Hence, option (b) is correct.

37. Given, 75% of 16% of monthly salary = ₹ 6567

⇒ 12% of monthly salary = ₹ 6567

∴ Monthly salary = ₹ 54725

38. Given, 53% of the number

= (26 × 26) − 358

$$\Rightarrow \frac{23 \times 3}{4}\% \text{ of that number}$$

$$= \frac{318}{53} \times \frac{69}{4}$$

= 103.5

39. $n(A \cup B \cup C) = n(A) + n(B) + n(C) - n(A \cap B) - n(B \cap C) - n(A \cap C) + n(A \cap B \cap C)$

$\Rightarrow 78 = 50 + 30 + 20 - [n(A \cap B) + n(B \cap C) + n(A \cap C) + n(A \cap B \cap C)]$

$\Rightarrow 78 = 100 - [\{n(A \cap B) - n(A \cap B \cap C)\}] + \{n(B \cap C) - n(A \cap B \cap C)\} + \{n(A \cap C)\} - n(A \cap B \cap C)\} + 3n(A \cap B \cap C)] + n(A \cap B \cap C)$

$\Rightarrow 78 = 100 - [\{n(A \cap B) - n(A \cap B \cap C)\}] + \{n(B \cap C) - n(A \cap B \cap C)\} + \{n(A \cap C)\} - n(A \cap B \cap C)\} - 2n(A \cap B \cap C)\}$

$\Rightarrow [\{n(A \cap B) - n(A \cap B \cap C)\} + \{n(B \cap C) - n(A \cap B \cap C)\} + \{n(A \cap C)\} - n(A \cap B \cap C)\}]$

$= 100 - 78 - 2 \times 5 = 12$

$\Rightarrow$ Number of persons favouring exactly two of three proposals = 12

Then, percentage of those favouring more than one of the three proposals

= Number of persons favouring exactly two of three proposals + Number of persons favouring all three of the proposals

$= 12 + 5 = 17$

Hence, option (c) is correct.

40. Given, 80% of A = 20% of $B \Rightarrow 4A = B$

Further, $B = 5x\%$ of A

$\Rightarrow 4A = 5x\%$ of A

$\Rightarrow 5x = 400$

$\Rightarrow x = 80$

41. Let the time period is x years.

Then, population of village A after x years = $6800 - 120\,x$

and population of village B after x years = $4200 + 80x$

But given, $6800 - 120x = 4200 + 80x$

$\Rightarrow 200x = 2600$

$\therefore x = 13$ yrs

Alternatively

You may proceed through options.

42. Let the income be ₹ 100.

Then, expenditure = 80

Now, increased income

= 116.66% of 1100 = ₹ 116.66

and increased expenditure

= 137.5% of 80

= ₹ 110

Thus, savings = 116.66 − 110

= ₹ 6.66

$\therefore$ Savings $= \left(\dfrac{6.66}{116.66} \times 100\right)\% = 5.71\%$

43. Clearly, difference in lost % = 1.8% and difference in production = (6 + 3) m tonne

It means, 1.8% of production = 9 m tonne

Then, the production = 500 m tonne

Hence, quantity of wheat grown in country = 500 m tonne.

44. Let the fraction be $\dfrac{x}{y}$

Then, $\dfrac{x + 5x}{y + 3y} = \dfrac{18}{17}$

$\Rightarrow \dfrac{3x}{2y} = \dfrac{18}{17}$

$\Rightarrow \dfrac{x}{y} = \dfrac{12}{17}$

Alternatively

Proceed through options.

(a) $\dfrac{12}{17} \Rightarrow \dfrac{12 + 60}{17 + 51} = \dfrac{18}{17} = 1\dfrac{1}{17}$

Hence, option (a) is correct.

45. Total number sweets = 40 × (20% of 40) + 8 × (25% of 40) = 400

46. Given, $b\%$ of $y = a\%$ of x

$\Rightarrow y = \dfrac{a}{b}$ of x

$\Rightarrow c\%$ of $y = c\%$ of $\dfrac{a}{b}$ of $x = \dfrac{ac}{b}\%$ of x

Hence, option (c) is correct.

47. Let the weight of Akbar be $100x$ kg.

Then, weight of Amar = $150x$

and weight of Antony

= 75% of $150x = 112.5x$

Further, weight of Farah

$= \dfrac{100x + 150x + 112.5x}{3}$

$= 120.83x$

Hence, option (d) is correct.

48. Given, 6% of cost of taxable items = 0.3

$\Rightarrow$ Cost of taxable items

$= 0.3 \times 16.66$

= ₹ 5

Hence, cost of tax free items

= ₹ 25 − ₹ 5

= ₹ 20.

49. Number of literate population

= 50% of 300000 = 150000

and number of literate males

= 70% of 180000 = 126000

Hence, number of literate females

= 150000 − 126000 = 24000

50. Let people in the voter's list be x.

Then, number of votes polled = 90% of $x = \dfrac{9x}{10}$

and number of valid votes $= \dfrac{9x}{10} - 60$

Number of votes got by candidate A = 46% of $x = \dfrac{23x}{50}$

Obviously, $\dfrac{23x}{50}$ = (number of votes obtained by B + 308)

$\Rightarrow \dfrac{23x}{50} = \left[\left(\dfrac{9x}{10} - 60\right) - \dfrac{23x}{50}\right] + 308$

$\Rightarrow \dfrac{x}{50} = 248$

$\Rightarrow x = 12400$

Then, number of votes polled

$= \dfrac{9x}{10} = \dfrac{9 \times 12400}{10} = 11160$

51. Capable students = 40%

and capable who passed the test

= 80% of 40% = 32%

Further, incapable students

= 100% − 40% = 60%

and incapable but passed the test

= 25% of 60% = 15%

Total passed candidates

= 32% + 15% = 47%

Then, proportion of capable students

$= \left(\dfrac{32}{47} \times 100\right)\%$

= 68% (approx.)

52. Number of those who play all three instruments = 5% of 120 = 6

Number of those who play any two and only two instruments = 30

Number of those who play guitar alone = 40

Thus, number of those who can play violin alone or flute alone

= 120 − (6 + 30 + 40)

= 44

53. Clearly, the difference between both students' marks = 10 + 15 = 25 and difference of marks in terms of percentage = 40% − 30% = 10%.

Hence, 10% of the maximum marks = 25

$\therefore$ Maximum marks = 250

Now, marks obtained by Peter

= 30% of 250 = 75,

But he fails by 10 marks, it means the passing marks = 75 + 10 = 85

54. Clearly, $\frac{4}{9}$th part of the total population are females.

Number of married females = Number of married males = 30% of $\frac{5}{9}$ of the total population = $\frac{1}{6}$ of the total population.

Then, number of unmarried females

$= \frac{4}{9} - \frac{1}{6}$ i.e. $\frac{5}{18}$th part of the total population.

Then, required percentage

$= \left(\frac{5}{18} \times 100\right)\%$

$= 27\frac{7}{9}\%$

55. Marks obtained by A = 90% of 40% of total marks = 36%

B got $= \left(100 - 11\frac{1}{9}\right)\%$ of 36% = 32% of total marks

C got $= \left(100 - 41\frac{3}{17}\right)\%$ of (36% + 32%)

= 40% of total marks.

56. Let initial income be ₹ x.

Then, taxable income = 80% of x = $0.8x$

[∵ 20% of income is exempted]

Then, earlier tax paid = 12% of 80% of x

As per question, 10% of 80% of

$(x + 1200)$ = 12% of 80% of x

⇒ 2% of x = 10% of 1200

∴ $x = 6000$

Thus, increased income

= 6000 + 1200 = ₹ 7200

Alternatively I

The rate of tax has been reduced from 12% to 10%, however, the tax amount being paid earlier is unchanged. The reason behind it is that the income has been increased by ₹ 1200. It means reduction of 2% has been accommodated in tax applicable on increased income. In other words, 2% of initial income = 10% of ₹ 1200.

Hence, initial income = 6000

Then, increased income

= 6000 + 1200

= ₹ 7200

Alternatively II

You may proceed through options.

(a) 6300 ⇒ 12% of 80% of 5100 ≠ 10% of 80% of 6300

(b) 7200 ⇒ 12% of 80% of 6000 = 10% of 80% of 7200

Hence, option (b) is correct.

57. Investment in fixed deposits = 50% of monthly savings

Investment in stock = 30% of 50% of monthly savings i.e. 15%

Investment in savings bank account

= (100 − 50 − 15)% i.e. 35% of monthly savings.

As given, (50 + 35)% of monthly savings = 59500

∴ 100% of the monthly savings

= ₹ 70000

58. Let number of girls and boys be 200 and 100 respectively.

Then, number of candidates getting no admission

= 70% of 200 + 55% of 100

= 195, which is 65% of 300.

59. Let Bimala's income be 100.

Then, Amala's income = 120

As given, 120 = 80% of Kamala's income

∴ 100% of Kamala's income = 150

Reduced income of Kamala

= 96% of 150 = 144

and increased income of Bimala = 110

Hence, required percentage

$= \left(\frac{144 - 110}{110} \times 100\right)\% \approx 31\%$

60. Clearly, 40% of the students are boys.

As given, (60 − 40)% of the student

= 30

∴ 60% of the students

= 90 = number of girls

∴ 100% of the students = 150

Number of students passed

= 68% of 150 = 102

Then, number of girls passed

= 102 − 30 = 72

Now, required percentage

$= \left(\frac{90 - 72}{90} \times 100\right)\%$

= 20%

61. 5% of the own funds of Z = ₹ 10000

⇒ 100% of the own funds of Z = 200000

Then, Y's funds (own + loan)

= 2 × 200000 = 400000

∴ Y's own funds

= 90% of 400000

= 360000

Now, contribution of Y to the trust P

= 5% of 360000 = 18000

62. Meena's score after review

= 150% of 40%

= 60%

Then, proposed score

= 120% of 60%

= 72%

Clearly, the difference between her post review score and proposed score = 35 + 7 = 42 and difference of marks in terms of percentage

= 72% − 60%

= 12%.

Hence, 12% of the maximum marks

= 42

∴ 100% of maximum marks = 350

Given that she gets 60% marks but fails by 35 marks and 35 marks is equal to 10% of the maximum marks.

Hence, passing marks %

= 60% + 10% = 70%

63.

	% of total messages	Jokes as % of messages sent by individual	Jokes as % of total sent messages
Bina	30%	40%	40% of 30% = 12%
Divya	20%	50%	50% of 20% = 10%
Anita	25%	60%	60% of 25% = 15%
Chaitra	25%	80%	80% of 25% = 20%

Required percentage = $\left\{\frac{(15 + 20)}{(12 + 10 + 15 + 20)} \times 100\right\}\% = 61.4\%$

19

Profit and Loss

Learning OBJECTIVES

After completing this chapter, aspirants will be able to :

- Understand the significance of this chapter in competitive exams, in field of finance/commerce etc., and our daily life as well;
- Understand the diverse terminology related to Profit and Loss;
- Understand how to apply the concept of percentage like product constancy, effect on number in case of Successive Increase/Decrease combination of both, etc., in this chapter as this chapter is extension of application of percentage;
- Differentiate between Profit/Loss on Cost Price (CP) and that on Selling Price (SP);
- Analyse the different approaches for solution of a question and will also be able to choose the best one for the exams;
- Realise themselves competent to solve the questions in many ways including conventional way;
- Understand how Profit/Loss alters in a defined way when CP/SP changes;
- Appreciate the diverse cases under one type of questions;
- Recognise the concept of Profit and Loss in different context like selling or purchasing rate, in terms of CP/SP etc;
- Comprehend the concept of Profit/ Loss in various types of transactions;
- Recognise how one concept works in different sorts of problems;
- Use the concept of ratio in this chapter;
- Realise the significance of generic version of the questions;
- Recognise the types of question which can be easily solved through the given options;
- Analyse the trend of the questions being asked in various competitive examinations;
- Solve the questions in moments;
- Grip the concepts helping to solve the questions without pen and paper;
- Learn new concepts/fundamentals;
- Apply the pragmatic approaches in lieu of academic approaches; and
- Ponder on the alternative methods.

Inside

Introduction

Almost all the competitive examinations like CAT, MAT, SSC, Banks, Railway, LIC, etc, cover this topic and in my opinion, it is an easy topic and requires no much complexity in logical thinking. Hence, the highest marks can be secured without much efforts.

In deed, the concepts envisaged in this chapter is based on the conceptions of the application of percentage, percentage rule, percentage conversion table, etc. Thus, we must have command over the 'Percentage' chapter, if we intend to solve the questions based on 'Profit and Loss' in moments. The application of this chapter is also very wide in our daily life. In addition, in field of finance, commerce, business it plays vital roles. It is worth mentioning that Profit and Loss are the terms related to transactions in trade as well as business.

Terminology

(a) **Cost Price (CP)** The price at which a product is purchased.

(b) **Selling Price (SP)** The price at which a product is sold.

It may be construed that the cost price of a customer is the same as selling price of the seller.

(c) **Overhead charges/Overheads** The expenses incurred on transportation, rent, salary of employees, packaging, maintenance, advertisements etc., are included under the head of 'Overhead charges'. It is to mention that overhead charges are not taken into account in case not given in the question. It is also worth mentioning that these overhead charges are added in the cost price. It means these overheads and the desired profit when added to the cost price, it becomes selling price for a trader/shopkeeper.

(d) **Profit :** If SP > CP, profit/gain is earned.

Profit = SP – CP ⇒ SP = CP + Profit ⇒ CP = SP – Profit

and profit % = profit on CP assuming CP be ₹ 100

$$= \left(\frac{\text{profit}}{\text{CP}} \times 100\right)\%$$

(e) **Loss :** If SP < CP, loss is incurred.

Loss = CP – SP

⇒ SP = CP – Loss

⇒ CP = SP + Loss

and loss % = loss on CP assuming CP be ₹ 100

$$= \left(\frac{\text{loss}}{\text{CP}} \times 100\right)\%$$

> ***Elucidation*** Profit or Loss is always computed on CP unless otherwise stated. In case we have to compute profit% or loss% on SP, the profit% will be $\left(\frac{\text{Profit}}{\text{SP}} \times 100\right)\%$ and Loss % = $\left(\frac{\text{Loss}}{\text{SP}} \times 100\right)\%$

Illustrations

Type I **When value of any two of three terms (i.e. CP, SP and Profit/Loss on CP)are given in respect of the single product.**

Problem 1. A person buys 100 cups at ₹ 10 each. On the way 10 cups are broken. He sells the remaining cups at ₹11 each. Find his loss percent.

Sol. CP = 100 × 10 = ₹ 1000

SP = (100 – 10) × 11 = ₹ 990

Now, loss = CP – SP = 1000 – 990 = 10

Hence, loss percent = $\left(\frac{\text{loss}}{\text{CP}} \times 100\right)\% = \left(\frac{10}{1000} \times 100\right)\% = 1\%$

Alternatively

When CP is ₹ 1000, loss is ₹ 10,

then in case CP be ₹ 100, loss will be ₹ 1.

Hence, loss % = 1%

Problem 2. A retailer buys a radio for ₹ 225. His overhead expenses are ₹ 15. He sells the radio for ₹ 300. Find the profit percent of the retailer.

Sol. Actual CP = CP + Overheads = ₹ 225 + ₹ 15 = ₹ 240

Given, SP = ₹ 300

Then, profit = SP – CP = 300 – 240 = 60

Hence, profit percent = $\left(\frac{\text{profit}}{\text{CP}} \times 100\right)\%$

$= \left(\frac{60}{240} \times 100\right)\% = 25\%$

Alternatively

When CP is ₹ 240, profit is ₹ 60

Then, in case CP be ₹ 100, profit will be ₹ 25.

Hence, Profit per cent = 25%

Problem 3. An item when sold for ₹ 1690 earned 30% profit on the cost price. Then, find the cost price.

[SSC (FCI-Asst. Gr.-III) 12, RRB-Ajmer (ASM) 08]

Sol. Let the CP be ₹ 100.

Then, SP on a profit of 30% = ₹ 130

Now, when SP is ₹ 130, CP is ₹ 100,

Then, in case SP be 1690, CP will be $\frac{100}{130} \times 1690$, i.e. ₹ 1300.

Alternatively I

Let CP be x.

Then, SP on 30% profit = 130% of x

Given, 130% of x = 1690

⇒ $x \times \frac{130}{100} = 1690$

⇒ x = ₹ 1300

Alternatively II

The SP is 30% more than CP. It means CP is $\left(\frac{30}{130} \times 100\right)\%$, i.e. 23.08% less than SP.

Now, 23.08% of 1690 = ₹ 390

[The above calculation can be done mentally without paper and pen like : 10% = 169, then 20% = 338

Further, 1% = 16.90, then 3% = 50.70
and 0.01% = 0.1690, then 0.8% = 1.35
Finally, 23.08% of 1690 = 338 + 50.70 + 1.35 = ₹ 390 (approx.)]
Hence, CP = 1690 − 390 = ₹ 1300
or
From above, CP is 23.08% less than SP.
Hence, CP = (100 − 23.08) % of SP = 76.92% of 1690
= ₹ 1300

Alternatively III

You have to express SP always in terms of % of CP.
Here, SP = 130% of CP
As given, 130% of CP = 1690
Then, 100% of CP = 1300

Alternatively IV

It is imperative to note that,

$$CP = SP \times \frac{100}{(100 + \text{Profit\%})}$$

or, $$CP = SP \times \frac{100}{(100 - \text{loss\%})}$$

In other way, $$SP = CP \times \frac{(100 + \text{profit\%})}{100}$$

or, $$SP = CP \times \frac{(100 - \text{loss \%})}{100}$$

Here, $$CP = SP \times \frac{100}{(100 + \text{profit\%})}$$
$$= 1690 \times \frac{100}{(100 + 30)} = ₹\ 1300$$

Alternatively V

Proceed through the options. Check as to 130% of value of which option is equal to 1690.

Observations Aspirants are advised to adopt the approaches detailed under Alternatively II and III to save the time in examination hall.

Problem 4. On selling an article for ₹ 270, there is a gain of 12.5%, what is the cost price?
[RRB (ASM-Ajmer) 08]

Sol. As given, 112.5% of CP = 270

Then, 100% of CP = $270 \times \frac{100}{112.5} = 270 \times \frac{8}{9} = 240$

[Recall the Percentage – Conversion table, 12.5% $= \frac{1}{8}$, i.e. 12.5% × 8 = 100%]

[such types of calculations have been illustrated in the chapter 'Percentage']

Alternatively

You may try with other approaches detailed earlier.

Problem 5. Find the SP at $16\frac{2}{3}$% loss if CP of an article is ₹ 618.

Sol. Here, SP is 83.33% of CP.
Given, 100% of CP = 618
Then, 83.33% of CP = $618 \times \frac{5}{6} = 515$ [Recall the Percentage – Conversion table, 16.66% × 6 = 100%]
Hence, SP = ₹ 515

Alternatively

Loss = 16.66% of 618 = $618 \times \frac{1}{6} = 103$

Then, SP = CP − Loss = 618 − 103 = ₹ 515

Problem 6. The difference between the CP and SP of an article is ₹ 240. If the profit is 20%, then find the selling price. ***[SSC (CGL-Tier-I) 11]***

Sol. Given, SP is 20% more than CP.
It means, 20% of CP = ₹ 240 (as given)
Then, 120% of CP = ₹ 1440
Hence, SP = ₹ 1440

Problem 7. Find the profit % earned by selling an article at certain price if a man incurs a loss of 20% by selling that article at half the earlier SP.

Sol. Let the initial SP be ₹ 100. Then, new SP = ₹ 50
New SP, i.e. ₹ 50 = 80% of the CP, then 100% of CP = 62.5
Now, it is clear that SP is 100 and CP is 62.5.

Then, profit % = $\left\{\frac{(100 - 62.5)}{62.5} \times 100\right\}$ % = 60%

Alternatively

Given, half of the initial SP = 80% of CP
Then, initial SP = 160% of CP
Clearly, profit% = 160% − 100% = 60%

Problem 8. If selling price of an article is 8/5 times its cost price, then how much profit percent on it?

Sol. Given, SP = 160% of CP {∵ 8/5 = 160%}
Then, profit% = 160% − 100% = 60%

Problem 9. On selling a watch for ₹ 280, a man earns 1/3 of the cost price. Determine the CP.

Sol. Given, SP = 133.33% of CP
⇒ 133.33% of CP = 280
∴ 100% of CP = ₹ 210

Type II **When value of any two of three terms (i.e. CP, SP and Profit/Loss on SP) are given in respect of the single product.**

Problem 10. A merchant buys an article for ₹ 27 and sells it at a profit of 10% of the selling price. The selling price is

Sol. Given, SP − CP = 10% of SP
⇒ CP = 90% of SP
⇒ 27 = 90% of SP
⇒ 100% of SP = ₹ 30

Alternatively I

Let SP be ₹ 100
Then, profit = 10% of 100 = 10
∴ CP = SP − Profit = 100 − 10 = ₹ 90
∵ If CP is ₹ 90, SP be ₹ 100
∴ When CP be 27, SP = ₹ 30

Alternatively II

Proceed through options (let options be 32, 30, 28, 25)

(a) $32 \Rightarrow 32 - (10\% \text{ of } 32) \neq 27$ (b) $30 \Rightarrow 30 - (10\% \text{ of } 30) = 27$

Hence, option (b) is correct.

Type III **Concept of Profit and Loss in context of purchasing and selling rate for the products**

Type IIIA

Problem 11. If I would have purchased 11 articles for ₹ 10 and sold all the articles at the rate of 10 for ₹ 11, the profit percent would have been?

[SSC (CGL-PT) 02, UPPSC (Revenue Inspector) 15, SSC (CGL-T-II) 17]

Sol. Given, CP of 11 articles = ₹ 10, then CP of 1 article $= \frac{10}{11}$

Further, SP of 10 articles = ₹ 11, then SP of 1 article $= \frac{11}{10}$

Now, $\text{profit} = SP - CP = \frac{11}{10} - \frac{10}{11} = \frac{21}{110}$

Then, $\text{profit } \% = \left(\frac{\frac{21}{110}}{\frac{10}{11}} \times 100\right)\% = 21\%$

Alternatively I

Let the number of articles be LCM of number of articles, i.e. 110.

Now, CP of 110 articles at the rate of 11 for ₹ 10 = ₹ 100

and SP of 110 articles at the rate of 10 for ₹ 11 = ₹ 121

Hence, profit % = 21%

Alternatively II

Write the given figures in following format.

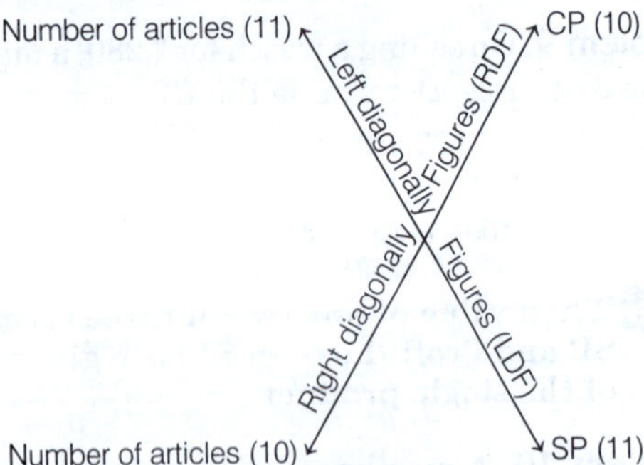

Profit or Loss%

$$= \left(\frac{\text{Product of left diagonally figures (LDF)} - \text{Product of right diagonally figures (RDF)}}{\text{Product of right diagonally figures (RDF)}} \times 100\right)\%$$

$$= \left[\frac{(11 \times 11 - 10 \times 10)}{10 \times 10} \times 100\right]\% = 21\%$$

Note

(a) *If the outcome is positive, profit is earned and in case of negative outcome, loss is incurred.*

(b) *This approach may be termed as 'Format Method' or 'Format Model' for future reference.*

Remarks *We should adhere to the 'Format Method' so as to arrive at answer quickly.*

Problem 12. A man purchases some eggs at 3 for ₹ 5 and sold them 5 of ₹ 12. Thus, he gained ₹ 143 in all. The number of eggs he bought is

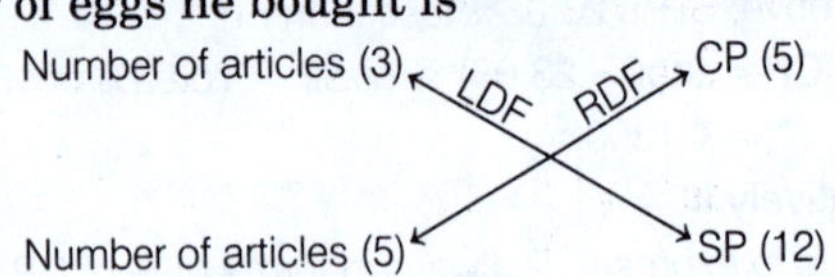

Sol. Then Profit or Loss percent

$$= \left[\frac{(\text{Product of LDF} - \text{Product of RDF})}{\text{Product of RDF}} \times 100\right]\%$$

$$\left[\frac{(3 \times 12 - 5 \times 5)}{5 \times 5} \times 100\right]\% = 44\%$$

As given, 44% of CP = ₹ 143

Then, 100% of CP = ₹ 325

Now, number of eggs bought for ₹ 325 at 3 for ₹ 5 = 195

Alternatively

Let number of eggs bought be the LCM of number of eggs i.e. 15.

Then, CP of 15 eggs = ₹ 25 and SP of 25 eggs = ₹ 36

∴ Profit = 36 − 25 = ₹ 11

Now, when profit is ₹ 11, number of eggs bought is 15.

Then, in case profit is ₹ 143, number of eggs = 195

Type IIIB

Problem 13. If the cost price of 10 chairs is equal to the selling price of 16 articles, find loss%.

[SSC (Investigator) 10, SSC (CGL-PT) 2000, SSC (CGL-T-I)-11]

Sol. Let CP and SP each be ₹ 1

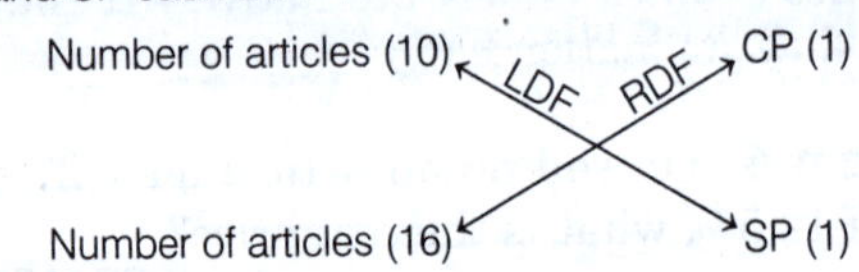

Then, Profit or Loss percent

$$= \left[\frac{(10 \times 1 - 16 \times 1)}{16 \times 1} \times 100\right]\% = -37.5\%$$

Since the outcome is negative, a loss of 37.5% is incurred.

Alternatively

$$\text{Profit or Loss percent} = \left[\frac{\text{Number of items purchased} - \text{Number of items sold}}{\text{Number of items sold}} \times 100\right]\%$$

When CP = SP.

Note *If the outcome is positive, profit is earned and in case of negative out come, loss is incurred*

$$\text{Here, Profit or Loss } \% = \left[\frac{(10 - 16)}{16} \times 100\right]\%$$

$$= -37.5\%$$

Since the outcome is negative, a loss of 37.5% is incurred.

Alternatively II

Let the CP of SP be ₹ 80. [∵ LCM of number of articles = 80]

(You are free to take any figure instead of 80. The reason behind taking 80, i.e. LCM of 10 and 16 is that it is least and exactly divisible by both figures.)

Then, CP per chair $= \frac{80}{10} =$ ₹ 8 and SP per chair $= \frac{80}{16} =$ ₹ 5

Now, loss % $= \left(\frac{3}{8} \times 100\right)\% = 37.5\%$

Problem 14. Oranges are bought at the rate of 10 for ₹ 25 and sold at the rate of 9 for ₹ 25. The profit is

Sol. Profit or Loss % $= \left[\frac{\text{(Number of items purchased} - \text{number of items sold)}}{\text{number of items sold}} \times 100\right]\%,$

when CP = SP.

Now, Profit or Loss % $= \left[\frac{(10-9)}{9} \times 100\right]\% = 11\frac{1}{9}\%$

Since the outcome is positive, a profit of $11\frac{1}{9}\%$ is earned.

Alternatively

Number of articles (10) —LDF— SP (25)

Number of articles (9) —RDF— CP (25)

Profit or Loss % $= \left[\frac{(25 \times 10 - 25 \times 9)}{25 \times 9} \times 100\right]\% = 11\frac{1}{9}\%$

Since the outcome is positive, profit is earned.

Problem 15. The cost price of 20 articles is the same as the selling price of x articles. If the profit is 25%, then the value of x is

Sol. Profit or Loss % $= \left[\frac{\text{Number of items purchased} - \text{number of items sold}}{\text{number of items sold}} \times 100\right]\%,$

when CP = SP.

$\Rightarrow \quad +25\% = \left[\frac{(20-x)}{x} \times 100\right]\%$

$\Rightarrow \quad 25x = 2000 - 100x \Rightarrow x = 16$

Alternatively

You may also solve with the help of 'Formal Method'.

Type III C

Problem 16. A man sold 20 apples for certain money and gained 20%. How many apples did he buy for the same amount?

Sol. Given, SP is 20% more than CP.

Then, CP is $\left(\frac{20}{120} \times 100\right)\%$ i.e. 16.66% less than SP.

Let SP be ₹ 100. Then, CP = 100 − 16.66 = 83.33.

Clearly, for ₹ 83.33, he buys 20 apples.

∴ For ₹ 100 he will buy 24 apples.

Alternatively I

Given, 120% of CP = SP (let ₹ 100)

Then, 100% of CP = ₹ 83.33

Clearly, for ₹ 83.33, he buys 20 apples

∴ For ₹ 100 he will buy 24 apples.

Alternatively II

Number of items bought for the given (same) amount

$= \frac{(100 + \text{profit}\%)}{100} \times \text{number of items sold}$

or $\frac{(100 - \text{loss}\%)}{100} \times \text{number of items sold}$

and Number of items sold for the given (same) amount

$= \frac{100}{(100 + \text{Profit}\%)} \times \text{number of items bought}$

or, $\frac{100}{(100 - \text{loss}\%)} \times \text{number of items bought}$

Here, number of items bought for the given amount

$= \frac{(100 + \text{profit}\%)}{100} \times \text{number of items sold}$

$= \frac{(100+20)}{100} \times 20 = 24$

Alternatively III

Let he buys x apples for a certain money (let, ₹ 1)

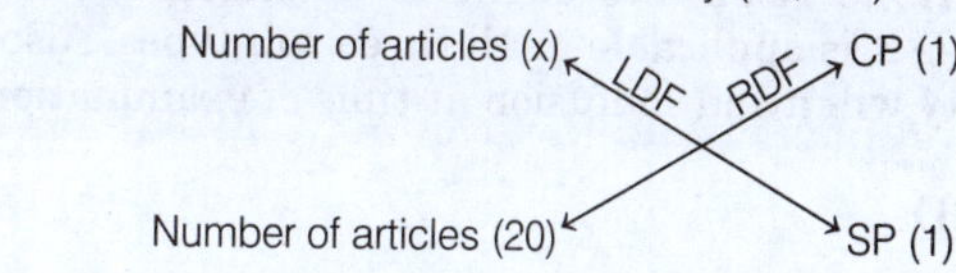

Profit or Loss % $= \left[\frac{(x \times 1 - 20 \times 1)}{20 \times 1} \times 100\right]\%$

$\Rightarrow \quad +20\% = (5x - 100)\%$

$\Rightarrow \quad x = 24.$

Alternatively IV

Proceed through options (Suppose options are 20, 22, 24, 28)

(a) 20 ⇒ no profit, no loss

(b) 22 $\Rightarrow \left[\frac{(22-20)}{20} \times 100\right]\% \neq 20\%$

(c) 24 $\Rightarrow \left[\frac{(24-20)}{20} \times 100\right]\% = 20\%$

[Recall, Profit or Loss % $= \left\{\frac{\text{Number of items purchased} - \text{number of items sold}}{\text{number of items sold}} \times 100\right\}\%$, when CP = SP]

Hence, correct option is (c).

Problem 17. By selling 15 pens for a rupee, a man loses 20%. How many pens did he buy for a rupee?

Sol. You are free to adhere to any one of the five approaches given in just the above question. Adhering to the one approach,

Number of items bought the given (same) amount

$= \frac{(100 - \text{loss}\%)}{100} \times \text{number of items sold}$

$= \frac{(100-20)}{100} \times 15 = 12$

Problem 18. A vendor bought toffees at 6 for a rupee. How many for a rupee must he sell to gain 20%?

Sol. Here, you are also free to adhere to any one of the five approaches given before. Adhering to the one approach,

Number of items sold for the given (same) amount

$$= \frac{100}{(100 + \text{profit}\%)} \times \text{number of items bought}$$

$$= \frac{100}{(100 + 20)} \times 6 = 5$$

Secondly,

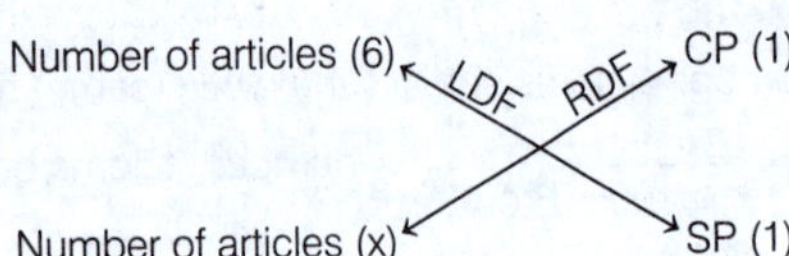

$$\text{Profit or Loss \%} = \left[\frac{(6 \times 1 - x \times 1)}{x \times 1} \times 100\right]\%$$

$$\Rightarrow \quad +20\% = \left[\frac{(6 - x)}{x} \times 100\right]\%$$

$$\Rightarrow \quad +20x = (600 - 100x)$$

$$\Rightarrow \quad x = 5$$

> ***Observations*** Aspirants ought to emphasis on 'Format Method' as it is applicable in all three sub-types. Also, this uniformity will avoid confusion at time of examination.

Type III D

Problem 19. A loss of 20% is incurred when 6 articles are sold for a rupee. To gain 20%, how many articles should be sold for a rupee?

[SSC (SO-Audit)-05, RRB-Bhubneshwar (ASM)-02]

Sol. Number of articles bought for the given (i.e. ₹ 1) in case of a loss of 20%

$$= \frac{(100 - \text{loss}\%)}{100} \times 6$$

Further, to gain 20%, number of items sold for the given (₹ 1)

$$= \frac{100}{(100 + \text{profit }\%)} \times \text{number of items bought}$$

$$= \frac{100}{(100 + \text{profit }\%)} \times \frac{(100 - \text{loss }\%)}{100} \times 6$$

$$= \frac{(100 - \text{loss }\%)}{(100 + \text{profit }\%)} \times 6$$

$$= \frac{(100 - \text{loss }\%)}{(100 + \text{profit }\%)} \times \text{number of articles sold}$$

Hence, number of articles to be sold

$$= \frac{(100 - \text{loss }\%)}{(100 + \text{profit }\%)} \times \text{number of articles sold}$$

Generic version

> Number of articles to be sold
>
> $$= \text{Number of articles sold} \times \frac{(100 + \text{given } P\%) \text{ or } (100 - \text{given } L\%)}{(100 + \text{desired } P\%) \text{ or } (100 - \text{desired } L\%)}$$

For this question, number of articles to be sold

$$= 6 \times \frac{(100 - \text{given } L\%)}{(100 + \text{desired } P\%)}$$

$$= 6 \times \frac{(100 - 20)}{(100 + 20)} = 4$$

Alternatively I

(100% + desired P%), i.e. 120% ⇒ 6

Then, (100% – given L%), i.e. 80% ⇒ 4

Generic version

> (100% + desired P%) or (100% – desired L%) ⇒ Number of articles sold
>
> Then, (100% + given P%) or (100% – given L%) ⇒ ?

Alternatively II

Initially, we will find number of articles purchased for a rupee applying 'Format Model.'

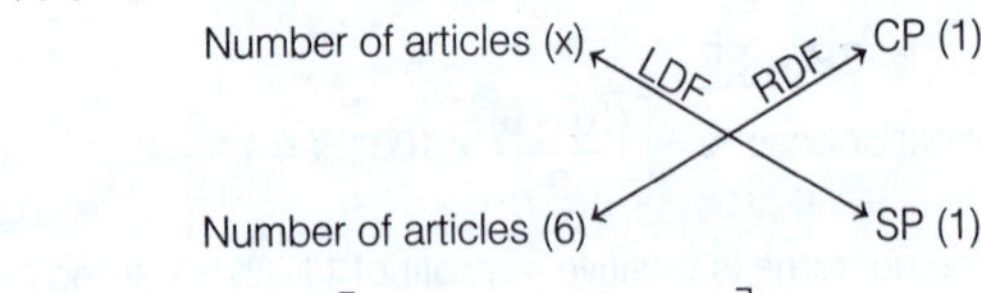

$$\text{Profit or Loss \%} = \left[\frac{(x \times 1 - 6 \times 1)}{6 \times 1} \times 100\right]\%$$

$$\Rightarrow \quad -20\% = \left[\frac{(x - 6)}{6} \times 100\right]\%$$

$$\Rightarrow \quad -120 = (100x - 600)$$

$$\Rightarrow \quad x = 4.8$$

Further, we will find number of articles to be sold to gain 20%.

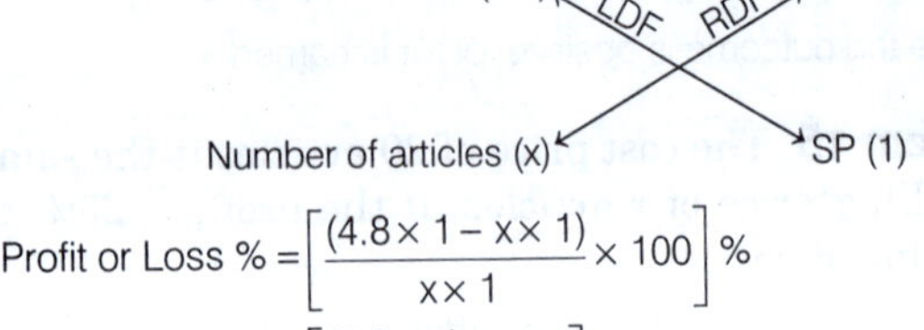

$$\text{Profit or Loss \%} = \left[\frac{(4.8 \times 1 - x \times 1)}{x \times 1} \times 100\right]\%$$

$$\Rightarrow \quad +20\% = \left[\frac{(4.8 - x)}{x} \times 100\right]\%$$

$$\Rightarrow \quad +20x = (480 - 100x)$$

$$\Rightarrow \quad x = 4$$

Problem 20. By selling 24 articles for a rupee a man gain 20%. To lose 10%, how many articles should be sold for a rupee?

Sol. On the line explained in previous question,

(100% – desired L%), i.e., 90% ⇒ 24

Then, (100% + given P%), i.e. 120% ⇒ 32

(Other methods as explained before, are also applicable)

Problem 21. By selling 18 articles for a rupee, a vendor earns 25%. How many articles should be sold so as to gain 50%?

Sol. Here, (100% + desired P%), i.e. 150% ⇒ 18

Then, (100% + given P%), i.e. 125% = 15.

(Other methods as explained before, are also applicable)

Problem 22. A man makes a profit of 20% when he sells 5 articles for ₹ 20. Find his profit or loss percent if he sells 8 articles for ₹ 25.

Sol. Number of articles he buys for ₹ 20

$$= \frac{(100 + \text{profit }\%)}{100} \times \text{number of items sold}$$

$$= \frac{120}{100} \times 5 = 6$$

Now, apply 'Format Model'

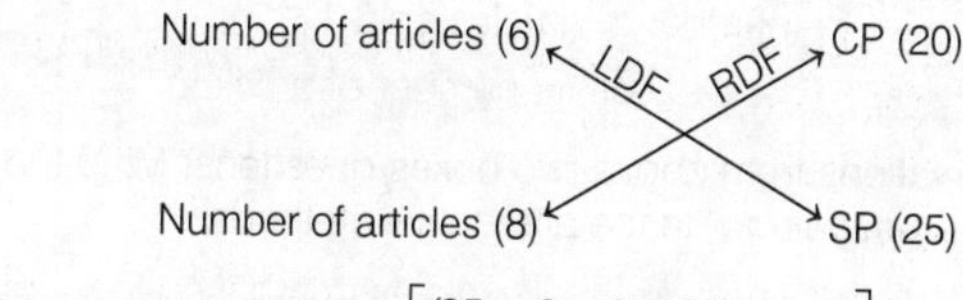

$$\text{Profit or Loss percent} = \left[\frac{(25 \times 6 - 8 \times 20)}{8 \times 20} \times 100\right]\%$$

$$= -6\frac{1}{4}\%$$

Since the outcome is negative, a loss of $6\frac{1}{4}$% is incurred.

Alternatively

Let the desired profit be P%.

Then, $(100 + P)\% = 5$

$\therefore \quad (100 + 20)\% = \dfrac{5}{(100 + P)} \times 120 = \dfrac{32}{5}$

(Given, for ₹ 25, 8 articles are sold. Then, for ₹ 20, $\frac{32}{5}$ articles will be sold. Mind that the amount must be same for this approach as you observed in previous questions.)

$\Rightarrow \quad 3000 = 3200 + 32P$

$\therefore \quad P = -6\frac{1}{4}\%$

Hence, a loss of $6\frac{1}{4}$% is incurred.

Problem 23. By selling 30 articles for a rupee, a man earns P% profit and by selling 28 articles for a rupee, he earns $(P + 10)$% profit. Find his purchasing rate per rupee.

Sol. Here, $(100 + P + 10)\% \Rightarrow 30$

Then, $(100 + P)\%$

$\Rightarrow \quad \dfrac{30}{(110 + P)} \times (100 + P) = 28$ (given)

$\Rightarrow \quad 1500 + 15P = 1540 + 14P$

$\Rightarrow \quad P = 40\%$

Now, we have to compute the purchasing rate if he earns 40% profit by selling 30 articles for a rupee.

Obviously, number of article purchased for ₹ 1

$$= \frac{(100 + 40)}{100} \times 30 = 42$$

Alternatively,

Proceed through the options (let options are 40, 42, 45, 36).

(a) $40 \Rightarrow \dfrac{(40 - 30)}{30} \times 100 = 33.33\%$

and $\dfrac{(40 - 28)}{28} \times 100 = 42.86\%$

Clearly, $42.86\% \neq (33.33 + 10)\%$

(b) $42 \Rightarrow \dfrac{(42 - 30)}{30} \times 100 = 40\%$

and $\dfrac{(42 - 28)}{28} \times 100 = 50\%$

Clearly, $50\% = (40 + 10)\%$

Hence, option (b) is correct.

Type IV Relation between Profit/Loss on CP and that on SP

Problem 24. If a man estimates his loss as 20% of the selling price, then find his loss percent.

Sol. Clearly, you have to find the loss % on CP.

Let the SP be ₹ 100.

Then, loss = 20% of 100 = 20

$\therefore$ CP = 100 + 20 = 120

Now, at 120 as CP, loss is 20.

$\therefore$ At 100 as CP, Loss $= \dfrac{20}{120} \times 100 = \dfrac{50}{3} = 16\dfrac{2}{3}$

Hence, loss percent on CP $= 16\frac{2}{3}\%$

Alternatively

Recall the problem envisaged in 'Percentage' chapter : "Price of sugar is increased by 20% then by how much percent the consumption must be reduced to maintain the earlier budget".

Answer is $\left(\dfrac{20}{100 + 20} \times 100\right)\% = \dfrac{50}{3}\%$

$$= 16\frac{2}{3}\%$$

Here, you have to compare the question with the percentage question, but read loss of 20% as increase of 20% and percent of profit as reduction percentage, i.e. read with opposite meaning.

It means,

$$\text{loss \% on CP} = \left(\frac{\text{loss \% on SP}}{100 + \text{loss \% on SP}} \times 100\right)\%$$

Or

$$\text{Profit \% on CP} = \left(\frac{\text{profit \% on SP}}{100 - \text{profit \% on SP}} \times 100\right)\%$$

Here, loss % on CP $= \left[\dfrac{20}{120} \times 100\right]\% = \dfrac{50}{3}\% = 16\dfrac{2}{3}\%$

Problem 25. If the ,percentage of profit calculated on selling price of an article is 20. Percentage of profit calculated of cost price will be

[SSC (Investigator)-10, SSC (SAS)-10, SSC (CGL-T-I)-19]

Sol. Profit % on CP $= \left[\dfrac{20}{(100 - 20)} \times 100\right]\% = 25\%$

Problem 26. The profit percentage on the cost price of an article is 25%. Find the profit percent on the selling price.

Sol. Let the CP be ₹ 100.

Then, SP on 25% profit = 125

$\because$ At ₹ 125 as SP, profit is ₹ 25.

$\therefore$ At ₹ 100 s SP, Profit = $\left(\frac{25}{125} \times 100\right)\% = 20\%$

> **It means** profit % on SP = $\left(\frac{\text{Profit \% on CP}}{100 + \text{Profit \% on CP}} \times 100\right)\%$
>
> Or Loss % on SP = $\left(\frac{\text{loss \% on CP}}{100 - \text{loss \% on CP}} \times 100\right)\%$
>
> **Reference** Base value for percentage determination (given in chapter 'Percentage').

Problem 27. The profit percentage of traders X and Y is same (say, 25%) on selling the articles for ₹ 2400 each. Find the difference between their profits if X calculates his profit on selling price and Y calculates on CP.

Sol. Profit of X = $2400 \times \frac{25}{100} = 600$

$\because$ SP of Y at 25% profit is ₹ 2400.

$\therefore$ CP of Y = $2400 \times \frac{100}{125} = 1920$

Hence, profit of Y = 2400 − 1920 = 480

Then, difference between their profits = 600 − 480 = 120.

Alternatively

Profit of X = 25% on SP

and profit of Y = 25% on CP = $\left[\frac{25}{(100+25)} \times 100\right]$ % on SP

= 20% on SP

$\therefore$ Difference between their profits = (25 − 20)% of 2400 = 120

Type V **When Profit/Loss is given in terms of SP/CP of some articles**

A. In terms of SP

Problem 28. By selling 144 hens Mahesh suffered a loss equal to the selling price of 6 hens. Find his loss percent.
[SSC (CGL-PT)-07, UPPSC (Revenue Inspector)-15]

Sol. Let SP of 144 hens be ₹ 144.

Then, SP of 6 hens = ₹ 6 (given as loss)

$\therefore$ CP = 144 + 6 = 150

$\therefore$ Loss % = $\frac{6 \times 100}{150} = 4\%$

Alternatively

$$\text{Loss \%} = \left[\frac{6}{(144+6)} \times 100\right]\% = 4\%$$

Generic Version

> $$\text{Loss \%} = \left(\frac{\text{Loss in number of articles}}{(\text{number of articles sold} + \text{Loss in number of articles})} \times 100\right)\%$$

Problem 29. By selling 60 articles a vendor gains the selling price of 15 articles. Find his gain percentage.
[SSC (CPO)-11, SSC (CGL-T-II)-12, 13]

Sol. Let SP of 60 articles be ₹ 60.

Then, SP of 15 articles = ₹ 15 (given as profit)

$\therefore$ CP = SP − Profit = 60 − 15 = 45

Now, profit % = $\left(\frac{15}{45} \times 100\right)\% = 33\frac{1}{3}\%$

Alternatively

$$\text{Profit \%} = \left[\frac{15}{(60-15)} \times 100\right]\% = 33\frac{1}{3}\%$$

Generic Version

> $$\text{Profit \%} = \left(\frac{\text{Profit in number of articles}}{(\text{number of articles sold} - \text{Profit in number of articles})} \times 100\right)\%$$

Observations from above two boxes/questions: Mind the sign while proceeding with the alternative methods.

Problem 30. Mohan bought 25 books for ₹ 2000 and sold them at a profit equal to the selling price of 5 books. The selling price of one book is

Sol. Let the SP of one book be x.

Then, SP of 25 books = $25x$ and SP of 5 books = $5x$ (given as profit)

Given that, CP of 25 books = 2000

We know, profit = SP − CP

$\Rightarrow$ $5x = 25x - 2000$

$\Rightarrow$ x = ₹ 100

Alternatively I

Clearly, Mohan gained the selling price of 5 books by selling 25 books.

Thus, his profit% = $\left(\frac{5}{25-5} \times 100\right)\% = 25\%$

Now, CP of 1 book = $\frac{2000}{25} = 80$

$\therefore$ SP of 1 book at a profit of 25% = $80 \times \frac{125}{100}$ = ₹ 100

Alternatively II

Proceed through options (let options be 120,100,80,200).

(a) 120 $\Rightarrow$ Total SP = 120 × 25 (i.e., 3000), CP = 2000 (given)

Hence, profit (i.e. 3000 − 2000) ≠ SP of 5 books (i.e. 5 × 120)

(b) 100 $\Rightarrow$ Total SP = 100 × 25 (i.e. 2500). CP = 2000 (given)

Hence, profit (i.e. 2500 − 2000) = SP of 5 books (i.e. 5 × 100)

Thus, option (b) is correct.

B. In terms of CP

Problem 31. By selling 33 m of cloth a person gains the cost of 11 m. Find the gain percentage.

Sol. Let CP of 33 m be ₹ 33.

Then, CP of 11 m = ₹ 11 (given as profit)

Thus, profit% = $\frac{11 \times 100}{33} = 33\frac{1}{3}\%$

Alternatively

$$\text{Gain \%} = \left(\frac{11}{33} \times 100\right)\% = 33\frac{1}{3}\%$$

Generic Version

> $$\text{Profit\%} = \left(\frac{\text{Profit in number of articles}}{\text{number of articles sold}} \times 100\right)\%$$

Problem 32. By selling 35 mangoes a grocer suffered a loss equal to the cost price of 7 mangoes. Find his loss percent.

Sol. Clearly, loss% = $\left(\frac{7}{35}\times 100\right)\% = 20\%$

Generic version

$$\text{Loss\%} = \left(\frac{\text{Loss in number of articles}}{\text{number of articles sold}}\times 100\right)\%$$

Problem 33. A book-seller bought 200 textbooks for ₹ 12000. He wanted to sell them at a profit so that he got 20 books free. At what profit percent should he sell them.

Sol. Given, CP of 200 books = 12000

Then, profit = CP of 20 books = 1200

Now, profit% = $\left(\frac{1200}{12000}\times 100\right)\% = 10\%$

Alternatively I

The book-seller intends to make a profit equal to CP of 20 books by selling 200 books. Since 20 is 10% of 200, the required profit is 10%.

Remarks *No need to use the figure ₹ 12000.*

Alternatively II

You may also proceed through options (say, 10, 20, 30, 40).

(a) 10 ⇒ Profit = 10% of 12000 = 1200, which is equal to cost price of 20 books.

Hence, option (a) is correct.

Problem 34. By selling 14 watches of equal cost price at the rate of ₹ 450 each, there is a profit equal to the cost price of 4 watches. What is the cost price of watch?

[SSC (10 + 2)/LDC/DEO-08, SSC (CGL-T-II)-13]

Sol. Let CP of each watch be ₹ x.

Then, total CP = $14x$ and total SP = 450 × 14 = 6300

As given, $6300 - 14x = 4x$

$\therefore \quad x = \frac{6300}{18} = 350$

Alternatively

Profit% = $\left(\frac{4}{14}\times 10\right)\% = \frac{200}{7}\%$

Given, SP of each watch on a profit of $\frac{200}{7}\%$ = ₹ 450

$\therefore$ CP of each watch = $450 \times \frac{100}{\left(100+\frac{200}{7}\right)} = 350$

Type VI **When Profit/Loss is numerically equal to SP/CP**

Problem 35. If the profit percent got on selling an article is numerically equal to its cost price (in ₹) and the selling price is ₹ 39, then the cost price will be

Sol. Let the cost price be ₹ x.

Then, profit = $x\%$

Now, $\quad \text{SP} = x \times \frac{(100+x)}{100} = 39$ (given)

$\Rightarrow \quad 100x + x^2 = 3900$

$\Rightarrow \quad x^2 + 100x - 3900 = 0$

$\Rightarrow \quad x^2 + 130x - 30x - 3900 = 0$

$\Rightarrow \quad x(x+130) - 30(x+130) = 0$

$\Rightarrow \quad x = -130$ (not possible) or $x = 30$

Hence, cost price is ₹ 30.

Alternatively I

$$\text{CP} = (10\sqrt{25+\text{SP}}) - 50 \text{ (in case of profit)}$$

Here, $\quad \text{CP} = (10\sqrt{25+39}) - 50 = 80 - 50 = 30$

Alternatively II

The best way is to proceed through options. (Say, options be 20, 30, 22, 40)

(a) 20 ⇒ 120% of 20 ≠ 39 (b) 30 ⇒ 130% of 30 = 39

Hence, option (b) is correct.

Problem 36. If the loss percent incurred on selling an article is numerically equal to its cost price in rupees. Then, find its CP if SP is ₹ 24.

Sol. Conventional approach as same as applied in just above question.

Alternatively I

$$\text{CP} = 50 \pm (10\sqrt{25-\text{SP}}) \text{ (in case of loss)}$$

Here, $\quad \text{CP} = 50 \pm (10\sqrt{25-24})$

$\Rightarrow \quad \text{CP} = (50+10)$ or $(50-10)$ ⇒ CP = 60 or 40

Alternatively II

Proceed through options. (Say, options be 30, 40, 50, 70)

(a) 30 ⇒ (100 – 30)% of 30 ≠ 24

(b) 40 ⇒ (100 – 40)% of 40 = 24

Hence, option (b) is correct.

Type VII **Calculation of Profit or Loss, when ratio of CPs/ratio of CP and SP is given and *vice-versa*.**

Problem 37. The ratio of cost price and selling price is 5 : 4, the loss percent is

Sol. Assume, CP = ₹ 5 and SP = ₹ 4

Then, loss = ₹ 1

Now, loss% = $\left[\frac{1}{5}\times 100\right]\% = 20\%$

Problem 38. If the ratio of the CP and SP of an article be 5 : 6, the percentage of gain is

Sol. On the above pattern, gain% = $\left(\frac{1}{5}\times 100\right)\% = 20\%$

Problem 39. The cost price : selling prices of an article is $a : b$, if b is 200% of a, then percentage of profit on cost price is

Sol. Clearly, $a : b = 1 : 2$

Then, profit % = $\left(\frac{1}{1}\times 100\right)\% = 100\%$

Problem 40. The cash difference between selling prices of an article at a profit of 4% and 6% is ₹ 3. The ratio of the two selling prices is *[SSC (CPO)-03, SSC (CGL-T-I)-10]*

Sol. Clearly, first SP = 104% of CP and second SP = 106% of CP

It means, (106 – 104)% of CP = 3

∴ 100% of CP = 150

Hence, CP = 150

Now, ratio of both selling prices

= 104% of 150 : 106% of 150 = 52 : 53

Alternatively

No need to use the difference of ₹ 3 as both SPs have been calculated on the same CP.

Then, their ratio = 104 : 106 = 52 : 53

Problem 41. Partha earns 15% on an investment but loses 10% on another investment. If the ratio of two investments is 3 : 5, then find the combined loss percent. *[SSC (SO-commercial)-06]*

Sol. Let the first investment be ₹ 3 and 2nd be ₹ 5.

Thus, her total investment = ₹ 8.

Now, profit on first investment = $\frac{3 \times 15}{100} = 0.45$

and loss on second investment = $\frac{5 \times 10}{100} = 0.50$

Clearly, combined loss = (0.50 – 0.45) = 0.05

Then, combined loss % = $\left[\frac{0.05}{8} \times 100\right]\% = 0.625\%$

Alternatively

Profit ⇒ 3 × 15% = 45% and loss ⇒ 5 × 10% = 50%

Clearly, loss = 5%

∴ Combined loss% = $\left[\frac{5}{(3+5)}\right]\% = 0.625\%$

Problem 42. If an article is sold at 200% profit, then the ratio of its cost price to its selling price will be

Sol. Let CP be 100. Then, SP = 100 + 200 = 300

Hence, required ratio = 100 : 300 = 1 : 3

Problem 43. The ratio of the cost prices of the three articles is 1 : 2 : 3 and those are sold in number of 2 : 3 : 5 at profit of 10%, 15% and 20%, respectively. Find the overall profit percentage.

Sol. Let cost prices be x, $2x$ and $3x$ and their numbers are $2y$, $3y$ and $5y$, respectively.

Then, total CP = $(x \times 2y) + (2x \times 3y) + (3x \times 5y) = 23xy$

Now, overall profit = 10% of $(x \times 2y)$ + 15% of $(2x \times 3y)$ + 20% of $(3x \times 5y) = 0.2xy + 0.9xy + 3xy = 4.1xy = 4.1xy$

Then, overall profit% = $\left(\frac{4.1xy}{23xy} \times 100\right)\% = 17.82\%$

Alternatively

Given, ratio of cost prices = 1 : 2 : 3

and ratio of their numbers = 2 : 3 : 5

Hence, ratio of their CPs = (1 × 2) : (2 × 3) : (3 × 5) = 2 : 6 : 15

Now, profit = (2 × 10%) + (6 × 15%) + (15 × 20%) = 410%

∴ Overall profit % = $\frac{410\%}{(2+6+15)} = 17.82\%$

Problem 44. The ratio of profit percent on three articles is 2 : 4 : 3 and that of their selling prices is 11 : 18 : 23. If CP of article A is ₹ 200 and the profit percent on article B is 20%, then what is the overall profit percentage?

Sol. Given, ratio of profits (on A, B and C) = 2 : 4 : 3

Also, given that profit on article B = 20% (i.e. 4 units = 20%)

Then, profit are 10%, 20% and 15% respectively.

It is also known that CP of A is ₹ 200.

Then, SP of A at a profit of 10% = 110% of 200 = 220

Now, ratio of selling prices = 11 : 18 : 23

It means, 11 units = 220

∴ One unit = 20

Clearly, there selling prices are 220, 360 (i.e. 18 × 20) and 460 (i.e. 23 × 20).

Further, CP of A = SP × $\frac{100}{(100+10)}$

$= 220 \times \frac{100}{110} = 200$

and CP of B = $360 \times \frac{100}{120} = 300$

and CP of C = $460 \times \frac{100}{115} = 400$

Now, overall profit = 10% of 200 + 20% of 300 + 15% of 400

= 20 + 60 + 60 = 140

and total CP = 200 + 300 + 400 = 900

∴ Profit % = $\left(\frac{140}{900} \times 100\right)\% = 15.56\%$

Alternatively

Clearly, profits are 10%, 20% and 15%, respectively.

Now, profit % on their SPs

$= \left(\frac{10}{110} \times 100\right)\%, \left(\frac{20}{120} \times 100\right)\%$ and $\left(\frac{15}{115} \times 100\right)\%$

$= \frac{100}{11}\%, \frac{50}{3}\%$ and $\frac{300}{23}\%$

Also given, ratio of their SPs = 11 : 18 : 23

Then, profits there on

$= \frac{100}{11}\%$ of 11, $\frac{50}{3}\%$ of 18 and $\frac{300}{23}\%$ of 23 = 1, 3 and 3

Then, their CPs = (11 – 1), (18 – 3) and (23 – 3) = 10, 15 and 20.

Now, overall profit% = $\left[\frac{(1+3+3)}{(10+15+20)} \times 100\right]\%$

= 15.56%

Type VIII **Uses of false weights/metre rod, error in weights/length etc.**

Problem 45. A shopkeeper professes to sell his goods at cost price but uses a weight of 800 gm instead of kilogram. Thus,he makes a profit of

Sol. It is very clear that the shopkeeper earns a profit equal to cost price of 200 gm by investing the money equal to cost price of 800 gm only.

Let the cost price of 1000 g of that goods be ₹ 1000.

Then, CP = invested money = cost price of 800 gm = ₹ 800

and profit = cost price of 200 gm = ₹ 200.

Hence, profit % = $\left(\frac{200}{800} \times 100\right) = 25\%$

Generic version

$$\text{Gain\%} = \left[\frac{\text{Error}}{(\text{True value} - \text{Error})} \times 100\right]\%$$

Or $$\text{Gain\%} = \left[\frac{\text{Error}}{\text{Weight / Length for which money is invested}} \times 100\right]\%$$

Note If error is given in the percentage term, then gain%

$$= \left[\frac{\text{error in percentage}}{(100\% - \text{error\%})} \times 100\right]\%$$

Problem 46. A dishonest dealer professes to sell his goods at cost price. But he uses a false weight and thus gains $6\frac{18}{47}\%$. For a kilogram, he uses a weight of **[LIC (AAO)-03]**

Sol. Let error be x gm.

Then, $\text{gain \%} = \left[\frac{\text{error}}{(\text{true value} - \text{error})} \times 100\right]\%$

$\Rightarrow \quad 6\frac{18}{47}\% = \left[\frac{x}{(100 - x)} \times 100\right]\%$

$\Rightarrow \quad \frac{300}{47} = \frac{100x}{(1000 - x)}$

$\Rightarrow \quad 47x = 3000 - 3x \quad \Rightarrow \quad x = 60$

Hence, the dealer uses (1000 – 60) gm instead of 1 kg.

Alternatively

Proceed through the options (Say, options be 950, 940, 930, 900)

(a) $950 \Rightarrow \left(\frac{50}{950} \times 100\right)\% \neq 6\frac{18}{47}\%$

(b) $940 \Rightarrow \left(\frac{60}{940} \times 100\right)\% = 6\frac{18}{47}\%$

Hence, option (b) is correct.

Problem 47. A shopkeeper cheats to the extent of 10% while buying as well as selling, by using false weights. Find his total gain% if he professes to sell his article at the cost price. **[Bank PO-03]**

Sol. The shopkeeper cheats to the extent of 10% while buying, it means when he buys 1 kg, he actually takes 1100 gm. Needless to say that he pays only for 1000 gm (assume price of goods per kg be ₹ 1000). While selling he also cheats by 10%, it means when he sells 1 kg, he actually gives 900 gm. Finally, he has to sell what he has (i.e. 1100 gm).

$\because$ 900 gm appears to be 1000 gm.

$\therefore$ 1100 gm will appear to be $\frac{1000}{900} \times 1100$ i.e. $\frac{11000}{9}$ gm.

Hence, the amount he will get by selling 1100 gm

$= ₹\,\frac{11000}{9}$ [$\because$ Price of 1 gm = ₹ 1]

Now, profit = SP – CP = $\frac{11000}{9} - 1000 = \frac{2000}{9}$

Then, profit % = $\left(\frac{\frac{2000}{9}}{1000} \times 100\right)\% = \frac{200}{9}\% = 22.22\%$

Alternatively

Generic Version

Profit or Gain%

$$= \left\{\frac{\text{Actually weight or length acquired} - \text{Actually weight or length sold}}{\text{Actually weight or length sold}} \times 100\right\}\%$$

Note *In case the outcome is negative, loss in incurred and in case of positive, profit is earned.*

Here, the shopkeeper acquires 1100 gm instead of 1000 gm and sells 900 gm in lieu of 1000 gm.

Hence, profit% = $\left(\frac{1100 - 900}{900} \times 100\right)\% = \frac{200}{9}\% = 22.22\%$

Problem 48. A business man professes to sell his articles at the cost price but he cheats to the extent of 10% while buying as well as selling. Find his actual gain per cent.

Sol. Here, it is not specified that the cheating is made by using false weight. It means he buys an article of ₹ 110 at only ₹ 100 by cheating. Further, he sells this article having value of ₹ 110 at ₹ 121 (i.e. 110% of 110) as he again cheats by 10%.

It is clear that he invests only ₹ 100 and acquires ₹ 121. Hence, his gain% = 21%.

Alternatively

$$\text{Gain\%} = \left[+10 + 10 + \frac{10 \times 10}{100}\right]\% = 21\%$$

[Recall the concept of 'Successive Increasing/Decreasing combination of both' envisaged in the chapter 'Percentage']

Observation Aspirants are advised to stress on the spirit of the above both questions and mind their differences.

Problem49. A shopkeeper sells wheat at 10% profit and also uses a weight of 800 g instead of kilogram weight. Find his actual gain percent.

Sol. Profit due to faulty weight = $\left[\frac{200}{800} \times 100\right]\% = 25\%$

Then, his actual gain per cent = $\left[+25 + 10 + \frac{25 \times 10}{100}\right]\%$

$= 37.5\%$

Problem 50. A trader sells rice at a profit of 20% and uses weights which are 10% less than the correct weight. The total gain earned by him is **[CLAT-17]**

Sol. Profit due to faulty weight

$$= \left[\frac{10\% \text{ of the weight}}{(100 - 10)\% \text{ of the weight}} \times 100\right]\% = 11.11\%$$

Then, his actual gain %

$$= \left[+20 + 11.11 + \frac{20 \times 11.11}{100}\right]\% = 33.33\%$$

Problem 51. **A busines man sells the goods at 15% less, but he uses a weight of 20 gm instead of 25 gm. Find his Profit or Loss%.**

Sol. Profit due to faulty weight $= \left[\frac{5}{20} \times 100\right]\% = 25\%$

Then, his actual gain% or loss per cent

$$= \left[+25 - 15 - \frac{25 \times 15}{100}\right]\% = +6.25\%$$

Since outcome is positive, a profit of 6.25% is earned.

Problem 52 **A shopkeeper's 1 kg wheat weighs 1100 gm. He purchases wheat from a supplier using faulty weight at a rate which is same as his selling price in the market. Find his gain%.**

Sol. Let the market price of wheat is ₹ 1000 per kg.

Then, his CP = ₹ 1000, if he buys 1 kg he gets 1100 gm for ₹ 1000. It means his profit is equal to price of 100 gm i.e. ₹ 100.

Hence, profit% $= \left\{\frac{100}{1000} \times 100\right\}\% = 10\%$

Alternatively

$$\text{Gain\%} = \left[\frac{\text{Error}}{\text{Weight / Length for which money is invested}} \times 100\right]\%$$

$$= \left[\frac{100}{1000} \times 100\right]\% = 10\%$$

{The shopkeeper invests money for 1000 gm not for 1100 gm.}

Problem 53. **A trader uses a weighing balance that shows 1250 gm for a kilogram. Find his profit% if he professes to sells goods at cost price.**

Sol. Gain percent $= \left[\frac{\text{Error}}{\text{Weight / Length for which money is invested}} \times 100\right]\%$

$$= \left[\frac{250}{1000} \times 100\right]\% = 25\%$$

Type IX Goods/Article passing through various hands at certain profit or loss

Problem 54. **A purchased an article and sold it to B at a profit of 25% and B sold it at a loss of 10% and C paid ₹ 675 for it. For how much A purchased it (in ₹)?**

[SSC (FCI-Asstt. Gr.III)-12, SSC (CGL-T-II)-10, SSC (CGL-T-II)-11, 13, RRB (ASM)-Kolkata-10, CLAT-12, CLAT-18]

Sol. Let the CP of A be ₹ 100.

Then, SP of A = 125% of 100 = 125 = CP of B

Now, SP of B = 90% of 125 = 112.5 = CP of C

If CP of C is ₹ 112.5, CP of A is ₹ 100

then in case CP of C is 675, CP of $A = \frac{100}{112.5} \times 675 = 600$

Alternatively I

CP of $A \times 125\% \times 90\%$ = CP of C

$\Rightarrow$ CP of $A \times 125\% \times 90\% = 675$

$\Rightarrow$ CP of $A = \frac{675}{1.25 \times 0.9} = 600$

(Recall the concept associated with the problems based on population, given in the chapter 'Percentage')

Alternatively II

(Recall the concept of 'Successive Increase/Decrease/ combination of both' envisaged in the chapter 'Percentage')

Here, the effect on the initial value (say, effect on CP of A)

$$= \left[+25 - 10 - \frac{25 \times 10}{100}\right]\% = +12.5\%$$

Hence, the final value i.e. CP of C is 12.5% greater than CP of A.

It means, 112.5% of (CP of A) = CP of C

$\Rightarrow$ 112.5% of (CP of A) = 675

$\Rightarrow$ 100% of CP of A = ₹ 600

Thus, CP of A = ₹ 600

Alternatively III

You may proceed through the options. (Say, options be 500, 600, 800, 850)

(a) 500 $\Rightarrow 500 \times 125\% \times 90\% \neq 675$

(b) 600 $\Rightarrow 600 \times 125\% \times 90\% = 675$

Hence, option (b) is correct.

Or

Ensure through options that 112.5% of value of which option is equal to 675.

Problem 55. **A purchased a clock for ₹ 350 and sold it to B at 30% profit and B sold to C at a loss of 20% and C sold to D at a profit of 25%. Find the cost price of D.**

Sol. You are free to adhere to any one of four approaches elaborated above.

Effect on CP of A when two figures are taken

(say, + 30%, −20%) $= \left[+30 - 20 - \frac{30 \times 20}{100}\right]\% = +4\%$

Now, final effect on CP of $A = \left[+4 + 25 + \frac{4 \times 25}{100}\right]\% = +30\%$

Hence, 130% of CP of A = CP of D.

$\therefore$ CP of D = 130% of 350 = ₹ 455

Problem 56. **An article passing through two hands is finally sold at a loss of 10% at the original cost price. If the first dealer makes a profit of 20%, then the loss percent incurred by the second is**

[SSC (FCI-Asstt. (Gr.-III)-12]

Sol. Let the second dealer suffered a loss of x%.

Then, $\left[+20 - x - \frac{(20 \times x)}{100}\right]\% = -10\%$

$\Rightarrow \frac{-6x}{5} = -30 \Rightarrow x = 25\%$

Hence, loss incurred by the second dealer = 25%

Problem 57. **A sold a tap-recorder to B for ₹ 4860 at a loss of 19%. Again. B sold it to C at a price that would give A a profit of 17%. The gain of B is**

Sol. Let the CP of A be ₹ 100.

Then, CP of B = 81% of 100 = 81

and SP of $B = 117\%$ of $100 = 117$

Hence, gain of $B = 117 - 81 = 36$

Now, gain% $= \left[\frac{36}{81} \times 100\right]\% = 44.44\%$

Remarks *No need to use the given amount i.e. ₹4860.*

Problem 58. A sells an article to B at a profit of 10%. B sells the article back to A at a loss of 10%. In this transaction,

(a) A neither loses nor gains

(b) A makes a profit of 11%

(c) A makes a profit of 20%

(d) B loses 20% **[SSC (CGL-PT)-08, 2000]**

Sol. Let CP of A be ₹ 100.

Then, SP of $A = 110\%$ of $100 = 110 =$ CP of B

Further, SP of $B = 90\%$ of $110 = 99 =$ CP of A

Clearly, total profit made by $A = (110 - 100) + (100 - 99) = 11$

Hence, his total profit % = 11%

Alternatively

Effect on CP of $A = \left[+10 - 10 - \frac{10 \times 10}{100}\right]\% = -1\%$

It means A repurchases the goods at 1% less than his original CP. i.e. he makes a profit of 1%. Also, he has made already a profit of 10% during the first transaction.

Hence, his total profit% = 10% + 1% = 11%

Problem 59. A bought an article X for ₹ 8000 and an article Y for ₹ 2000. He sold them to B and B sold them to C, each making a profit of 25% on X and each incurring a loss of 10% on Y. If A were to sell them to C directly at the same selling price as that at which B had sold to C, what percent would A have gained on the whole?

Sol. Profit of A on article X in case the article is sold directly to C by A

$= \left[+25 + 25 + \frac{25 \times 25}{100}\right]\%$

$= +56.25\%$

and loss of A on article Y in case the article is sold directly to C by A

$= \left[-10 - 10 + \frac{10 \times 10}{100}\right]\% = -19\%$

Now, profit on $X = 56.25\%$ of $8000 = 4500$

and loss on $Y = 19\%$ of $2000 = 380$

Clearly, overall profit $= 4500 - 380 = 4120$

$\therefore$ Profit percent $= \left[\frac{4120}{(8000 + 2000)} \times 100\right]\%$

$= 41.2\%$

Alternatively

Given, ratio of CPs of X and Y = 4 : 1

Then, profit on $X \Rightarrow 4 \times 56.25\% = 225\%$

and loss on $Y \Rightarrow 1 \times 19\% = 19\%$.

Hence, profit = 225% − 19% = 206%

$\therefore$ Overall profit % $= \left[\frac{206}{(4+1)}\right]\% = 41.2\%$

Type X Changes in SPs on account of changes in Profit/Loss and *vice-versa*

Problem 60. If a man were to sell his handcraft for ₹ 720, he would lose 25%. To gain 25%, the selling price is **[SSC (CGL-PT)-07, 2000 SSC (10+2)/LDC-DEO-12, UPPSC (Revenue Inspector)-15, SSC (CGL-T-II)-17, CLAT-11, SSC (CGL-T-I)-19]**

Sol. CP $= 720 \times \frac{100}{(100 - 25)} = 960$

Then, SP = 125% of 960 = 1200

Hence, the required SP = ₹ 1200

Alternatively

Given, (100 − 25)% of CP = ₹ 720

Then, (100 + 25)% of CP = ₹ 1200

Hence, required SP = ₹ 1200

Problem 61. By selling a bicycle for ₹ 2850, a shopkeeper gains 14%. If the profit is reduced to 8%, then the selling price will be

Sol. Given, 114% of CP = 2850

Then, 108% of CP = 2700

Hence, required SP = ₹ 2700

Problem 62. In selling an article for ₹ 76, there is a profit of 52%. If it is sold for ₹75, the profit percent will be **[SSC (CGL-T-II)-20]**

Sol. Given, 76 = 152% of CP

Then, 75 = 150% of CP

Hence, profit % = 150% − 100% = 50%

Problem 63. If a man were to sell an article for ₹ 800, he would gain $\frac{1}{4}$th of its SP. In order to gain $\frac{1}{5}$th of its CP, the SP be

Sol. Gain of $\frac{1}{4}$ of SP = Gain of 25% of SP

$= \left[\frac{25}{(100 - 25)} \times 100\right]\%$ i.e. 33.33% gain on CP

(Recall the concept of 'Relation between Profit/Loss on CP and that on SP envisaged under model Type-IV of this chapter)

Now, 133.33% of CP = ₹ 800

Then, 120% of CP = ₹ 720

Hence, required SP = ₹ 720

Alternatively Profit $= \frac{1}{4}$th of 800 = 200

Then, CP = 800 − 200 = 600

Now, SP = 120% of 600 = ₹ 720

Problem 64. When selling price of an article is reduced from ₹ 120.72 to ₹ 98.22, the loss incurred is increased by 5 percent point. What is the selling price of the article at a profit of 5%?

Sol. Obviously, 5% of the CP = (120.72 − 98.22)

Then, 100% of CP = 22.5 × 20 = 450

Now, SP to gain 5% = 105% of 450 = 472.50

Type XI Profit/Loss is changed when CP or SP or both are changed and *vice-versa*

Problem 65. A man sells his typewriter at 5% loss. If he sells it for ₹ 80 more, he will gain 5%. The cost price of the typewriter is

[SSC (CGL-PT)-07, SSC (SAS)-10, SSC (CGL-T-I)-19, CDS-19]

Sol. Let the CP be ₹ 100.

Then, SP at 5% loss = 95 and SP at 5% gain = 105.

Clearly, when the difference between both SPs is 10, CP is 100.

∴ When the difference between both SPs will be 80, CP = 800.

Hence, the CP = ₹ 800

Alternatively

←.........Loss..............|....................Profit ....→

(–6)...(–5)...(–4)...(–3)...(–2)...(–1)...(0)...(+1)...(+2)...(+3)...(+4)...(+5)...(+6)...

←..→

Clearly, difference between (–5) and (+5) is 10.

Given, 10% of CP = ₹ 80

Then, 100% of CP = ₹ 800

Problem 66. A cooker is sold at a gain of 16%. If it has been sold for ₹ 20 more, 20% would have been gained. The cost price of the cooker is

[SSC (CPO)-09, SSC (CGL-T-II)-13]

Sol. Difference between (+16) and (+20) is 4.

Given, 4% of CP = 20

Then, 100% of CP = 500

Hence, CP of the cooker = ₹ 500

Problem 67. If an article is sold at a gain of 5% instead of being sold at a loss of 5%, one gets ₹ 5 more. What is the cost price of the article?

Sol. Difference between (+5) and (–5) is 10.

Given, 10% of CP = 5

Then, 100% of CP = 50

Hence, CP of the article = ₹ 50

Problem 68. A pen is sold at a loss of 5%. If it has been sold for ₹ 60 more, 10% would have been gained. At what price must the pen be sold so as to gain 16% profit?

Sol. Difference between (–5) and (+10) is 15.

Given, 15% of CP = 60

Then, 100% of CP = 400

Hence, CP of the article = ₹ 400

Now, desired SP = 116% of 400 = ₹ 464

Problem 69. If the selling price of an article is doubled, then its loss percent is converted into equal profit percent. The loss percent on the article is

Sol. Let the loss be x% and CP be ₹ 100. Then, SP = 100 – x.

Now, difference between $(-x)$ and $(+x)$ is $2x$.

Given, $2x$% of CP = (2 SP – SP)

Then, 100% of CP = $\frac{50 \cdot SP}{x}$

Hence, CP of the article = $\frac{50 \cdot SP}{x}$

$\Rightarrow \quad 100 = \frac{50 \cdot SP}{x}$

$\Rightarrow \quad 100 = \frac{50(100-x)}{x}$

$\Rightarrow \quad 100x = 5000 - 50x$

$\Rightarrow \quad 150x = 5000$

$\Rightarrow \quad x = 33.33\%$

Alternatively I

Let the loss be x% and CP be ₹ 100. Then, SP = $(100 - x)$.

Later SP becomes double i.e. $(200 - 2x)$.

Then, profit % = $\left\{\frac{(200-2x)-100}{100} \times 100\right\}$ % = x% (given)

$\Rightarrow \quad 100 - 2x = x$

$\Rightarrow \quad x = 33.33$

Hence, loss% = 33.33%

Alternatively II

You may proceed through options (say, options be 50%, 33.33% 40%, 20%).

Let CP be ₹ 100.

(a) 50% ⇒ SP = (100 – 50) = 50

In case SP is doubled, profit% = {2 × 50 – 100}%

= 0% ≠ 50%

(b) 33.33% ⇒ SP = (100 – 33.33) = 66.66

In case SP is doubled, profit%

= {(2 × 66.66) – 100}% = 33.33%

Hence, option (b) is correct.

Problem 70. Kranti sells an article at a profit of 10%. If he had bought it at 20% less and sold at the earlier SP, how much would he have gained?

Sol. Let the CP be ₹ x.

Then, SP = 110% of $x = \frac{11x}{10}$

Now, new CP = 80% of $x = \frac{4x}{5}$

∴ Profit % = $\left\{\frac{\frac{11x}{10} - \frac{4x}{5}}{\frac{4x}{5}} \times 100\right\}$ % = 37.5%

Alternatively I

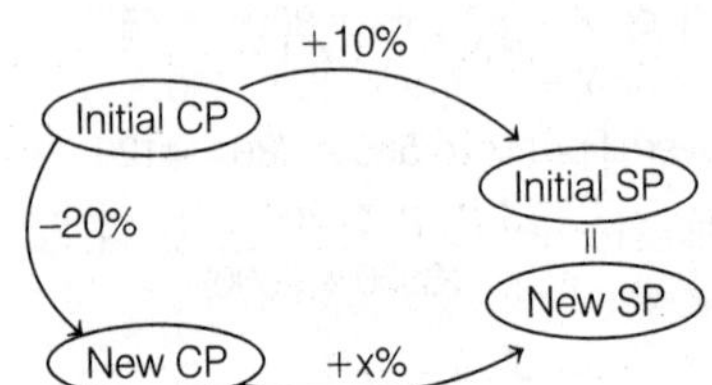

Obviously, New SP is $\left[-20 + x - \frac{20 \times x}{100}\right]$ % greater than the initial CP. Also, initial SP is 10% greater than the initial CP.

But given, Initial SP = New SP

It means, $\left[-20 + x - \frac{20 \times x}{100}\right]$ % = 10%

$$\Rightarrow \quad -20 + \frac{4x}{5} = 10$$

$$\Rightarrow \quad x = 37.5$$

Hence, gain% = 37.5%

Note *You may write directly as* $\left[-20 + x - \frac{20 \times x}{100}\right] = 10$ *and find the value of x*

Alternatively II

If Kranti had bought the article at 20% less and sold at his CP, he would have gained $\left\{\frac{20}{(100-20)} \times 100\right\}$ % i.e., 25%.

[Recall a percentage question number *A* (say, new CP) is 20% less than number *B* (say, initial CP). By how much per cent is number *B* greater than number *A*? Here, new CP is the base number on which profit is to be calculated]

But, in the question, the article is sold at 10% profit, not at its CP. It means, total profit % = $\left[+25 + 10 + \frac{25 \times 10}{100}\right]$ % = 37.25%

Note *The whole process can be done mentally without using paper and pen.*

Alternatively III

Let the initial CP be ₹ 100.

Then, initial SP = 110

Now, new CP = 80

Then, profit % = $\left\{\frac{(110-80)}{80} \times 100\right\}$ % = 37.5%

Problem 71. Richa purchased an article at $\frac{4}{5}$ of its list price and sold it at 20% more than the list price. Richa's profit percent was?

Sol. Given, Richa purchased the article at 20% less than its list price. It means, her profit % in case the article is sold at list price

$$= \left\{\frac{20}{(100-20)} \times 100\right\} \% = 25\%.$$

But, the article is sold at 20% more than the list price.

Hence, total profit %

$$= \left[+25 + 20 + \frac{25 \times 20}{100}\right] \% = 50\%$$

Problem 72. Mahesh purchased a radio at $\frac{9}{10}$ of its selling price and sold it at 8% more than its original selling price. His gain percent is?

Sol. On the above pattern, initial profit%, if the article is sold at original selling price = $\left\{\frac{10}{(100-10)} \times 100\right\}$ %

$= 11.11\%$

But, the article is sold at 8% more than the original selling price.

Hence, total profit%

$$= \left[+11.11 + 8 + \frac{11.11 \times 8}{110}\right] \% = 20\%$$

Problem 73. A book seller sells a book at a profit of 10%. If he had bought it at 4% less and sold it for ₹ 6 more, he would have gained $18\frac{3}{4}$%. The cost price of the book is **[SSC (CGL-PT) 07]**

Sol. Let the CP be ₹ x.

Then, initial SP = 110% of $x = \frac{11x}{10}$

Now, new CP = 96% of $x = \frac{24x}{25}$

and new SP = $\frac{11x}{10} + 6$

Given that new SP is $18\frac{3}{4}$% greater than new CP.

i.e. $118\frac{3}{4}\%$ of $\frac{24x}{25} = \frac{11x}{10} + 6$

$$\Rightarrow \quad \frac{57x}{50} - \frac{11x}{10} = 6$$

$$\Rightarrow \quad \frac{x}{25} = 6 \Rightarrow x = 150$$

Alternatively I

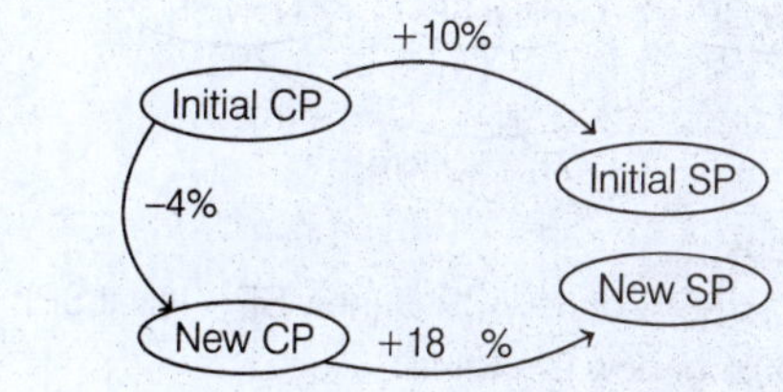

Obviously, New SP is $\left[-4 + 18\frac{3}{4} - \frac{4 \times 18\frac{3}{4}}{100}\right]$ % i.e. 14% greater than the initial CP. Also, initial SP is 10% greater than the initial CP. Hence, difference between new SP and initial SP is (14 − 10)% of CP.

But given 4% of the initial CP = ₹ 6

Then, 100% of initial CP = ₹ 150

Hence, the cost price of the book = ₹ 150

Alternatively II

Proceed through the options (say, options be 140, 150, 160, 120).

(a) 140 ⇒ Initial SP = 154, new CP = 96% of 140 = 134.4

Then, $118\frac{3}{4}\%$ of $134.4 \neq 154 + 6$

(b) 150 ⇒ initial SP = 165, new CP = 96% of 150 = 144

Then, $118\frac{3}{4}\%$ of $144 = 165 + 6$

Hence, option (b) is correct.

Alternatively III

Profit % = $\left\{\frac{4}{(100-4)} \times 100\right\}$ % = 4.16%, if seller had bought the book at 4% less than initial CP and would have sold at his initial CP.

But he sells the books at 10% profit. Then, his total profit %

$$= \left[+4.16 + 10 + \frac{4.16 \times 10}{100}\right] \% = 14.58\%$$

It means the initial SP is 14.58% more than the new CP. But it is given that total profit is 18.75%.

Obviously, (18.75 – 14.58)% of new CP = 6

{as new SP is 6 more than initial SP}

Then, 100% of new CP = 144

It is obvious that the initial CP will be 4.16% more than the new CP. Then, initial CP = 104.16% of 144 = ₹ 150.

Understand the whole process through a diagram depicted below :

I:-
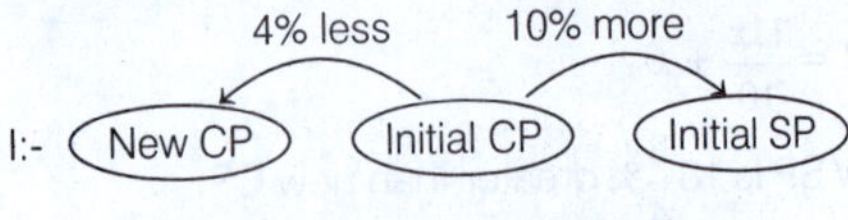

II:-
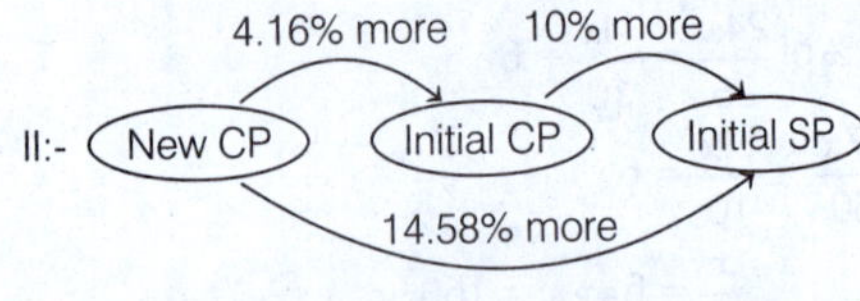

III:-
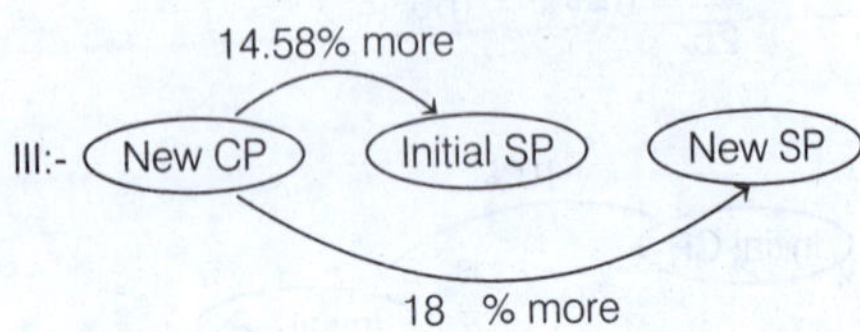

IV. It means,

(18.75 – 14.58)% of new CP = (new SP – Initial SP) = 6

Then, 100% of new CP = 144

It is obvious that the initial CP will be 4.16% more than the new CP. Then, initial CP = 104.16% of 144 = ₹ 150

Problem 74. **A person sells an article at 10% less. If he had bought it at ₹ 100 less and sold it for 25% more, he would have gained 50%. Find the initial CP.**

Sol. Let the initial CP be ₹ x.

Then, initial SP = 90% of $x = \frac{9x}{10}$

Now, new CP = $x - 100$

and new SP = 125% of $\frac{9x}{10} = \frac{9x}{8}$

Given that new SP is 50% greater than new CP.

i.e. 150% of $(x - 100) = \frac{9x}{8}$

$\Rightarrow \quad 4x - 400 = 3x$

$\Rightarrow \quad x = 400$

Alternatively I

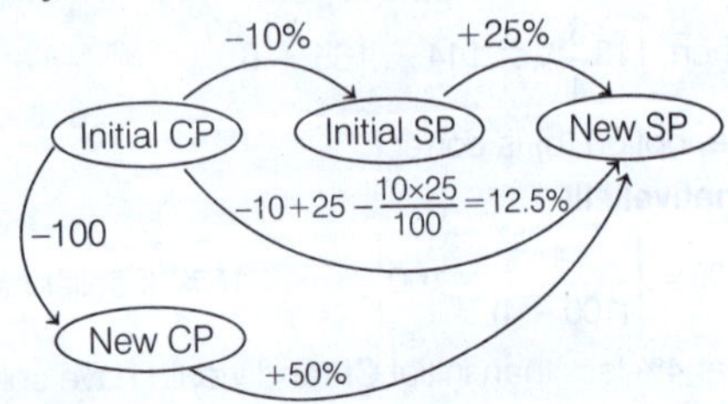

Clearly, new SP = 112.5% of initial CP = 150% of new CP

$\Rightarrow$ 112.5% of initial CP = 150% of new CP

$\Rightarrow$ 112.5% of initial CP = 150% of (initial CP – 100)

$\Rightarrow$ 37.5% of initial CP = 150

$\Rightarrow$ Initial CP = 400

Alternatively II

You may proceed through options (say, options be 300, 400, 500, 360)

(a) 300 $\Rightarrow$ initial SP = 90% of 300 = 270

New CP = 300 – 100 = 200

and new SP = 125% of 270 = 337.50

But, 150% of 200 ≠ 337.50

(b) 400 $\Rightarrow$ Initial SP = 90% of 400 = 360

New CP = 400 – 100 = 300 and new SP = 125% of 360 = 450

Clearly, 150% of 300 = 450

Hence, option (b) is correct.

Problem 75. **A person sold an article at 20% profit on the selling price. Afterwards, when the cost price reduced by 10%, then he also reduced the selling price by 10%. His percentage of profit on cost price will be**

Sol. Let the SP be ₹ 100.

Then, profit = 20 and CP = 100 – 20 = 80

Now, new CP after 10% reduction = 90% of 80 = 72

and new SP after 10% reduction = 90% of 100 = 90

Thus, profit percent = $\left\{\frac{90 - 72}{72} \times 100\right\}\% = 25\%$

Alternatively I

20% profit on SP = $\left\{\frac{20}{(100 - 20)} \times 100\right\}\%$ i.e. 25% profit on CP

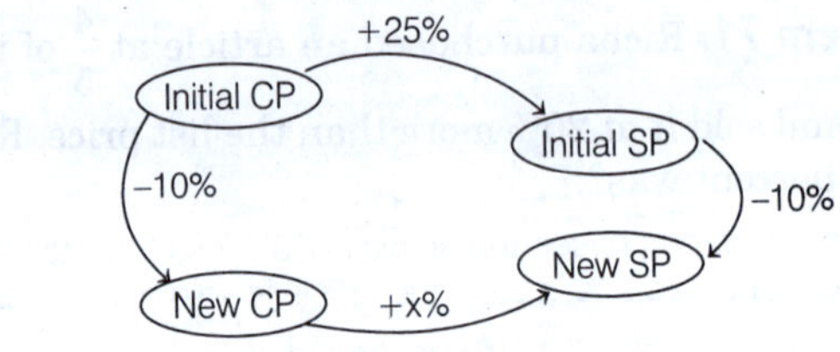

Clearly, $\left[-10 + x - \frac{10 \times x}{100}\right] = \left[+25 - 10 - \frac{25 \times 10}{100}\right]$

$\Rightarrow \quad \frac{9x}{10} = 12.5 + 10$

$\Rightarrow \quad x = 25$

Hence, profit% = 25%

Alternatively II

20% profit on SP = $\left\{\frac{20}{(100 - 20)} \times 100\right\}\%$ i.e. 25% profit on CP. If initial CP is reduced by 10%, it will be $\left\{\frac{10}{(100 - 10)} \times 100\right\}\%$ i.e. 11.11% greater than the reduced CP.

Further, selling price is also reduced by 10%. It means new SP would be $\left[+25 - 10 - \frac{25 \times 10}{100}\right]\%$ i.e. 12.5% higher than the initial CP.

Finally, profit % = $\left[+11.11 + 12.5 + \frac{11.11 \times 12.5}{100}\right]\%$

= 25%

Problem 76. A man sells his chair at a profit of 20%. If he had bought it at 25% less and sold it for ₹ 40 less, he would have gained $33\frac{1}{3}$%. The cost price of the chair is **[SSC (CGL-T-II)-10]**

Sol. Given that the chair is to be bought at 25% less. Then his profit % in case the chair is sold at its CP

$$= \left\{\frac{25}{(100-25)} \times 100\right\}\% = 33.33\%.$$

But the chair is sold at 20% more than the CP, not at its CP.
Hence, total profit%

$$= \left[+33.33 + 20 + \frac{33.33 \times 20}{100}\right]\% = 60\%$$

It is construed from the above that if the chair is bought at 25% less and sold at initial SP, the man will gain 60%. But it is also given that initial SP has been reduced by ₹ 40 resulting in his gain is only 33.33%.

It means, (60 – 33.33)% of the reduced CP = ₹ 40

∴ 100% of the reduced CP = 150 i.e. reduced CP = ₹ 150

Now, initial CP = 133.33% of 150 = ₹ 200

Alternatively I

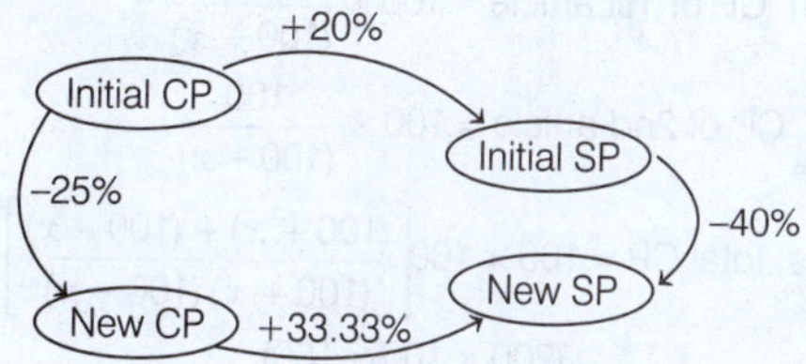

Clearly difference between 20% of initial CP and [–25 + 33.33 – (25 × 33.33)/100]% of initial CP = 40

⇒ (20 – 0)% of initial CP = 40

⇒ 100% of initial CP = 200

Hence, initial CP = ₹ 200

Alternatively II

Let initial CP be ₹ 100

Then, initial SP = ₹ 120

Now, new CP = 100 – 25 = 75

and new SP = 133.33% of 75 = 100

When difference between initial SP and new SP is ₹ (120 – 100), the initial CP is ₹ 100.

∴ In case difference is ₹ 40, the initial CP be ₹ 200.

Problem 77. An article is sold at a profit of 20%. If both the cost price and selling price were to be ₹ 20 less, the profit would be 10% more. Find the cost price of the article.

Sol. Let the initial CP be ₹ x.

Then, initial SP = 120% of $x = \frac{6x}{5}$

Now, new CP = $x - 20$

and new SP = $\frac{6x}{5} - 20$

Given that new SP is (20 + 10)% greater than new CP.

i.e. 130% of $(x - 20) = \frac{6x}{5} - 20$

$$\Rightarrow \quad \frac{x}{10} = -20 + 26 \Rightarrow x = 60$$

Alternatively I

Proceed through the options (let options be 50, 60, 70, 80)

(a) 50 ⇒ initial SP = 120% of 50 = 60

Now, 130% of (50 – 20) ≠ (60 – 20)

(b) 60 ⇒ initial SP = 120% of 60 = 72

Now, 130% of (60 – 20) = (72 – 20)

Hence option (b) is correct.

Alternatively II

Cost price = { (Initial profit % + increment in initial profit%) / increament in initial profit% × the amount by which both CP and SP have been reduced }

Here, $CP = \frac{(20+10)}{10} \times 20 = ₹ 60$

Note *This approach for solution seems to be very short. However, aspirants should avoid the same as it will be difficult to remember as well as formulate another pattern in case the amounts, by which both CP and SP are reduced, are not the same.*

Problem 78. A man sells a TV a profit of 25% of the cost. Had he sold it at a profit of 25% of the selling price, his profit would have increased by 5% of the cost price plus ₹ 100. Find the cost price of TV.

Sol. Let the initial CP be ₹ x.

Then, profit = 25% of $x = \frac{x}{4}$

and SP = $x + \frac{x}{4} = \frac{5x}{4}$

Further, profit = 25% of the selling price = $\frac{5x}{4} \times \frac{25}{100} = \frac{5x}{16}$

But given, $\frac{5x}{16} - \frac{x}{4} = \left(x \times \frac{5}{100}\right) + 100$

$$\Rightarrow \quad \frac{x}{80} = 100 \Rightarrow x = ₹ 8000$$

Alternatively

Let the CP be ₹ 100.

Then, profit = ₹ 25

and selling price = 100 + 25 = 125

Now, profit on selling price = 25% of 125 = 31.25

Difference between profit on CP and profit on SP
= 6.25 = 6.25% of CP

Further, difference between 6.25% and 5%
= 1.25%

As given, 1.25% of the CP = ₹ 100

∴ 100% of CP = 8000

Hence, CP = ₹ 8000

Type XII **Profit or Loss in Whole transaction**

Case-I ⇒ When CPs are same

Case-II ⇒ When SPs same

Case-III ⇒ When neither CPs nor SPs are same

Case-IV ⇒ When goods are sold part by part

Type XII Case I When CPs are same

Problem 79. A man bought three article for ₹ 2371.50 each. He sold one at a loss of 10%, second at a profit of 20% and third at a loss of 25%. What is his gain or loss percentage in this transaction?

Sol. Total SP = 90% of 2371.50 + 120% of 2371.50 + 75% of 2371.50 = 285% of 2371.50

and total CP = 300% of 2371.50

Thus, loss = (300 – 285)% of 2371.50

= 15% of 2371.50

Then loss % $= \left\{\dfrac{15\% \text{ of } 2371.50}{300\% \text{ of } 2371.50} \times 100\right\} = 5\%$

Alternatively

Required gain or loss per cent $= \left\{\dfrac{-10+20-25}{3}\right\}\%$

$= -5\%$

{∵ '–' connotes loss. Conversely, '+' does profit}.

Note

1. *No need to consider the given value of CP.*
2. *We simply divide the sum of given percentage of profit/loss with their symbols by number of articles. In other words, find the average.*

Problem 80. Three tables are purchased for ₹ 2500 each. First is sold at a profit of 8%, the second is at a loss of 3%. If their average selling price is ₹ 2575, find the profit percent on the third.

Sol. Average profit percent $= \left[\dfrac{(2575-2500)}{2500} \times 100\right]\%$

$= 3\%$

Now, overall gain or loss percent

$= \left[\dfrac{+8-3+x}{3}\right]\%$

$\Rightarrow \quad +3\% = \left[\dfrac{+8-3+x}{3}\right]\%$

$\Rightarrow \quad x = 4\%$

Problem 81. A person bought two articles at the same price. First is sold at 15% profit. Find the cost price of each article if the selling price of the second article is ₹ 40 more than the first's and his overall profit percent is 16% in this transaction. ***[SBI (PO-Pre)-19]***

Sol. Profit percent on the second article = (16 × 2 – 15)% = 17%

As given,

(117 – 115)% of CP of second (or each) article = ₹ 40

Then, 100% of CP of each article = ₹ 2000

Type XII Case II When SPs are same

Problem 82. A man sold two watches for ₹ 240 each. On one he gains 20% and incurs a loss of 20% on another. What is his gain or loss percent in this transaction?
[SSC (SO-Commercial), SSC (CPO)-03, SSC (CGL-T-II)-11, IBPS (CWE-PO)-13, RRB (TC)-Patna-06, CAT-96 MAT-18, CDS-19]

Sol. CP of 1st watch = ₹ 200 [∵ 120% = 240, ∴ 100% = 200]

and CP of 2nd watch = ₹ 300 [∵ 80% = 240, ∴ 100% = 300]

Thus, total CP = 200 + 300 = ₹ 500

Then, loss = 500 – (240 + 240) = 20

∴ Loss percent $= \left\{\dfrac{20}{500} \times 100\right\}\% = 4\%$

Alternatively I

Generic version of this question may be drawn as below.

The SPs of two articles are same and one has been sold at x% profit and another at x% loss. Then, find the gain or loss% in this transaction.

Let, the each SP be ₹ 100.

Then, CP of 1st article $= 100 \times \dfrac{100}{(100+x)}$

and CP of 2nd article $= 100 \times \dfrac{100}{(100-x)}$

Thus, total CP $= 100 \times 100\left[\dfrac{(100+x)+(100-x)}{(100+x)(100-x)}\right]$

$= \dfrac{(200 \times 100 \times 100)}{(100+x)(100-x)}$

Given, total SP = 100 + 100 = 200

Now, loss $= \dfrac{(200 \times 100 \times 100)}{(100+x)(100-x)} - 200$

$= \dfrac{(200 \times 100 \times 100) - (200 \times 100 \times 100) + 200x^2}{(100+x)(100-x)}$

$= \dfrac{200x^2}{(100+x)(100-x)}$

∴ Loss percent $= \left\{\dfrac{200x^2}{(100+x)(100-x)} \times \dfrac{(100+x)(100-x)}{(200 \times 100 \times 100)}\right\} \times 100\% = \dfrac{x^2}{100}\%$

Conclusion **Always a loss of $\dfrac{x^2}{100}$% is incurred.**

In context of the above question, x = 20%

∴ Loss percent $= \dfrac{(20)^2}{100}\% = 4\%$

Alternatively II

Their SPs are equal. It means,

120% (say, a) of CP of 1st article = 80% (say, b) of CP of 2nd article

Now, his returns $= \left\{\dfrac{2ab}{a+b}\right\}\%$ of total CP

$= \left[\dfrac{2 \times 120 \times 80}{(120+80)}\right]\%$ of total CP = 96% of total CP

Clearly, (100 – 96)% loss has been incurred.

Problem 83. The SPs of two articles are same. One has been sold at 10% profit and another at 20% loss. Find profit or loss percent in this transaction.

Sol. Adhering to the above approach,

$$\text{Total returns} = \left[\frac{2ab}{a+b}\right]\% \text{ of total CP}$$

$$= \left[\frac{2 \times 110 \times 80}{(110+80)}\right]\% \text{ of total CP}$$

$$= 92.63\% \text{ of total CP.}$$

Clearly, (100 – 92.63)% i.e. 7.37% loss has ben incurred.

Problem 84. A book-seller sold three books for ₹ 350 each. On 1st book he gains 20%, on 2nd book he gains 25% and incurs a loss of 10% on the third book. Find his gain or loss percent in this transaction.

Sol. Here, 120% (say, a) of CP of 1st book = 125% (say, b) of CP of 2nd book = 90% (say, c) of CP of 3rd book

$$\text{Now, his returns} = \left[\frac{3abc}{ab+bc+ca}\right]\% \text{ of total CP}$$

$$= \left[\frac{3 \times 120 \times 125 \times 90}{(120 \times 125) + (125 \times 90) + (90 \times 120)}\right]\% \text{ of the total CP}$$

$$= 109.31\% \text{ of the total CP.}$$

Hence, a profit of 9.31% has been earned in this transaction.

Problem 85. A trader sold two articles for ₹ 300 each. On one he gains 20% and gain 30% on another. Find his profit or loss in ₹.

Sol. $\text{Total returns} = \left[\frac{2ab}{a+b}\right]\% \text{ of total CP}$

$$= \left[\frac{2 \times 120 \times 130}{(120+130)}\right]\% \text{ of total CP} = 124.8\% \text{ of total CP.}$$

Clearly, total returns = 300 + 300 = 600

It means, 124.8% of total CP = ₹ 600

$\therefore$ 100% of total CP = ₹ 480.76

Hence, his profit = 600 – 480.76 = 119.24

Special case Sometimes the sum of CPs of articles are also given in addition to the fact that their SPs are equal. For Example

Problem 86. A man bought two goats for ₹ 1008. He sold one at a loss of 20% and another at a profit of 44%. If each goat was sold for the same price, the cost price of the goat which was sold at loss was

[SSC (CGL-PT)-04, SSC (CGL-T-II)-17, SSC (CGL-T-I)-19]

Sol. Clearly, 80% of CP of 1st goat = 144% of CP of 2nd goat

$\Rightarrow$ CP of 1st goat : CP of 2nd goat = 144 : 80 = 9 : 5

Given that, CP of 1st goat + CP of 2nd goat = 1008

Then, CP of 1st goat $= 1008 \times \frac{9}{(9+5)} = ₹\ 648$

Or {given, (9 + 5) units $\Rightarrow$ 1008, then, 9 units $\Rightarrow$ 648}

Note *You may directly write as CP of 1st goat: CP of 2nd goat* = (100 + 44) : (100 – 20) i.e. with reverse order of the figures.

Problem 87. A trader bought two horses for ₹ 19500. He sold one at a loss of 20% and the other at a profit of 15%. If the selling price of each horse is the same, then their cost price are respectively.

[SSC (CGL-T-I)-11, IAS (Pre)-98]

Sol. Clearly, CP of 1st horse : CP of 2nd horse

= (100 + 15) : (100 – 20)

[Please mind the reverse order of the given figures]

= 115 : 80 = 23 : 16

Then, CP of 1st horse $= 19500 \times \frac{23}{(23+16)} = 11500$

$\therefore$ CP of 2nd horse = 19500 – 11500 = 8000

Alternatively

You may proceed through options also. Ensure (i) the sum of both figures in given option is 19500 (ii) 80% of 1st figure is equal to 115% of the latter figure. The option satisfying both conditions will be the correct option.

Type XII Case III When neither CPs nor SPs are same

Problem 88. A trader purchases a watch and a wall clock for ₹ 390. He sells them making a profit of 10% on the watch and 15% on the wall clock. He earns a profit of ₹ 51.50. The difference between the original prices of the wall clock and the watch is equal to:

Sol. Let, CP of the watch be ₹ x, then CP of the wall clock = $(390 - x)$

Given, (10% of x) + 15% of $(390 - x)$ = 51.50

$\Rightarrow$ –5% of x = 51.50 – 58.50 = –7

$\Rightarrow$ x = 140

Hence, CP of watch = 140,

then CP of wall clock = 390 – 140 = 250

Then, difference between their CPs = 250 – 140 = 110

Alternatively

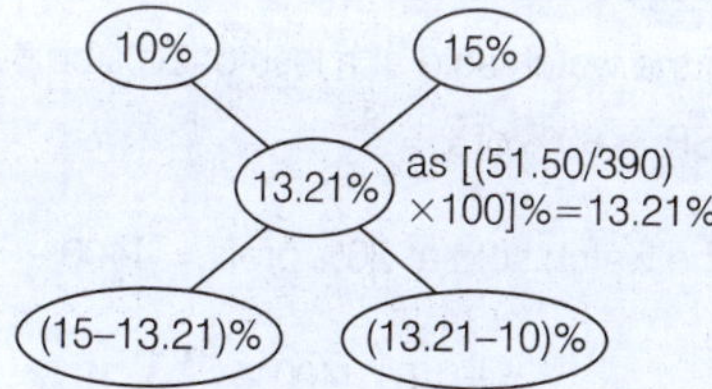

It means,

CP of the watch : CP of the wall clock

= 1.79 : 3.21

= 14 : 25

Given, (14 + 25) units $\Rightarrow$ 390

Then, (25 – 14) units $\Rightarrow$ 110

(This concept has been given in detail in the Chapter 'Alligation')

Problem 89. A businessman purchased 3 oxes and 5 horses for ₹ 6200. He sells them making a profit of 12% on oxes and 16% on horses and earns total of ₹ 930. Find the CP of one ox.

Sol. $\text{Overall profit percent} = \left[\frac{930}{6200} \times 100\right]\% = 15\%$

Now,

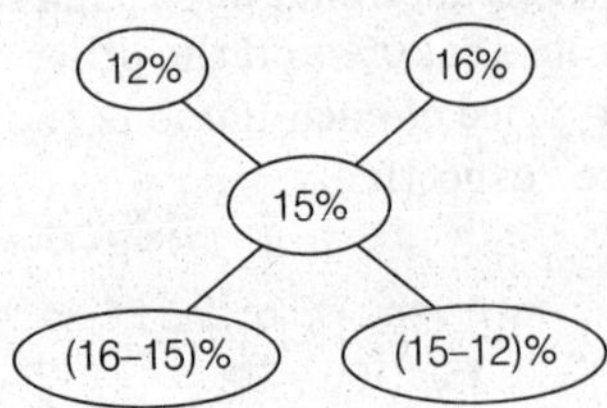

It means, CP of 3 oxes : CP of 5 horses = 1 : 3

Given, (1 + 3) units ⇒ 6200

Then, 1 unit ⇒ 1550

Hence, CP of one ox = 1550 ÷ 3 = 516.66

Problem 90. A person bought two articles A and B for ₹ 5000. He sold A at 20% profit and B at a 10% loss. He thus gained 2% on his outlays. The cost price of A was : ***[SSC (10 + 2/DEO-LDC)-08, SSC (SAS)-10, SSC (CGL-T-II)-10, 20, MAT-18, CAT-20]***

Sol.

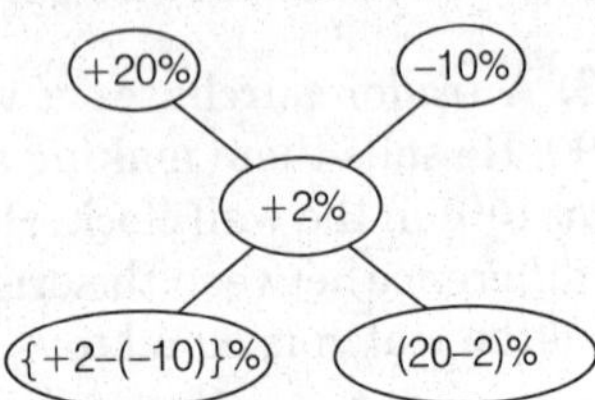

It means, CP of A : CP of B = 12 : 18 = 2 : 3

Given, (2 + 3) units ⇒ 5000

Then, 2 units ⇒ 2000

Hence, CP of A = ₹ 2000

Problem 91. A trader sold two watches together for ₹ 4400. One has been sold at 20% loss and another at 20% profit. Find the total profit or loss if the CP of the watch sold at loss is equal to 1.5 times the SP of the 2nd watch.

Sol. Let, CP of the watch sold at a loss of 20% be ₹ x.

Then, its SP = 80% of $x = \frac{4x}{5}$

∴ SP of the watch sold at 20% profit $= 4400 - \frac{4x}{5}$

Given, $x = 1.5\left(4400 - \frac{4x}{5}\right)$

⇒ $x = 6600 - 1.2x$

⇒ $x = 3000$

i.e. CP of the watch sold at a loss = 3000

Also, SP of the watch sold at profit = 3000 ÷ 1.5 = 2000

Then, its CP $= 2000 \times \frac{100}{120} = 1666.66$

Thus, loss = (3000 + 1666.66) – 4400 = ₹ 266.66

Alternatively

Let CP of the watch sold at loss be ₹ 100

Then, its SP = 80

As given, SP of the watch sold at profit $= \frac{1}{1.5} \times$ CP of the watch sold at loss $= \frac{1}{1.5} \times 100 = 66.66$

Then, CP of the watch sold at profit

$= 66.66 \times \frac{100}{120} = 55.55$

Hence, total CP = 100 + 55.55 = 155.55

and total SP = 80 + 66.66 = 146.66

Clearly, when total SP is ₹ 146.66, loss is ₹ (155.55 – 146.66).

∴ In case SP be ₹ 4400, the loss

$= \frac{8.89}{146.66} \times 4400$

= ₹ 267 (approx.).

Problem 92. A trader sold three dogs for ₹ 8000, ₹ 12000 and ₹ 6000 respectively and earned a profit of 25% on 1st dog, profit of 20% on second one. Find the CP of 3rd dog if he gained 30% in the whole transaction.

Sol. Given, 125% of CP of 1st dog = ₹ 8000

Then, 100% of CP of 1st dog = 6400

Hence, CP of 1st dog = 6400

Similarly, CP of 2nd dog = 10000

Also given, total SP = 8000 + 12000 + 6000

= 26000 and total profit = 30%

Clearly, 130% of total CP = 26000

Then, 100% of total CP = 20000

Hence, CP of 3rd dog = 20000 – (6400 + 10000) = 3600

Problem 93. The sum of CPs of two articles is ₹ 1500. A man sells one at 10% profit and another at 20% profit. If he had sold 1st article at 20% profit and second at 10% profit, he would have gained ₹ 10 more. Find the difference between their cost prices.

Sol. Let, the CP of 1st article is ₹ 1000, then CP of 2nd article = 500

In case-I, total profit = (10% of 1000) + (20% of 500) = 200

In case-II, total profit = (20% of 1000) + (10% of 500) = 250

Clearly, if difference between profits is ₹ 50, difference in CPs is ₹ 500

∴ In case difference between profit is ₹ 10, difference in CPs = 100.

Note *You are free to assume any figures instead of ₹ 1000.*

Problem 94. A cycle and a radio have been sold at 20% profit and 25% loss respectively for ₹ 705. If they were sold at 25% loss and 20% profit respectively, the sum of their SPs would have been ₹ 660. Find the CP of the radio.

Sol. Let CP of the cycle be ₹ x and that of the radio be y.

Then, 120% of x + 75% of y = ₹ 705

⇒ $x \times 8\% + y \times 5\% = 47$ ...(i)

And 75% of x + 120% of y = ₹ 660

⇒ $x \times 5\% + y \times 8\% = 44$ ...(ii)

From Eqs. (i) and (ii), we get $y \times 39\% = 117$

⇒ $y = 300$

Alternatively

Given that, 120% of x + 75% of y = 705 ...(i)

and 75% of x + 125% of y = 660 ...(ii)

On addition (i and ii)

$$195\% \text{ of } (x + y) = 1365$$

$\Rightarrow \quad x + y = 700 \quad \ldots(\text{iii})$

and on subtraction (i – ii)

$$45\% \text{ of } (x - y) = 45$$

$\Rightarrow \quad x - y = 100 \quad \ldots(\text{iv})$

Now, from Eqs. (iii) and (iv), $x = 400$ and $y = 300$

Type XII Case IV When Goods are Sold Part by Part

Problem 95. I purchased 120 exercise books at the rate of ₹ 3 each and sold $\frac{1}{3}$ of them at the rate of ₹ 4 each, $\frac{1}{2}$ of them at the rate of 5 each and the rest at the cost price. My profit percent was

Sol. Total CP = $3 \times 120 = 360$

Now, SP of $\left(\frac{1}{3} \text{ or } 120\right) = 4 \times 40 = 160$

and SP of $\left(\frac{1}{2} \text{ of } 120\right) = 5 \times 60 = 300$

and SP of remaining books = $3 \times (120 - 40 - 60) = 60$

Hence, total SP = $160 + 300 + 60 = 520$

Then, profit = $520 - 360 = 160$

$\therefore$ Profit percent = $\left[\frac{160}{360} \times 100\right]\% = 44\frac{4}{9}\%$.

Alternatively

Profit on selling $\frac{1}{3}$ of the 120 = $(4 - 3) \times 40 = 40$

and profit on selling $\frac{1}{2}$ of the 120 = $(5 - 3) \times 60 = 120$

It is obvious that neither profit nor loss has been occurred on selling the remaining books.

Hence, total profit = $40 + 120 = 160$

$\therefore$ Profit percent = $\left[\frac{160}{(3 \times 120)} \times 100\right]\% = 44\frac{4}{9}\%$.

Problem 96. A cloth merchant sold half of his cloth at 20% profit, half of the remaining cloth at 20% loss and the rest was sold at his cost price. In the total transaction, his gain or loss will be:

Sol. Let CP be ₹ 100.

Then, SP of half of the cloth = 120% of (half of 100) = ₹ 60

and SP of half of the remaining = 80% of [half of (100 – 50)]

= ₹ 20

and SP of remaining cloth = ₹ 25

Thus, total SP = $60 + 20 + 25 = 105$

$\therefore$ Gain% = 5%

Alternatively

Total profit or loss percent = $\frac{1}{2}(+20\%) + \frac{1}{4}(-20\%)$

+ remaining part × (0%) = + 5%

Since the outcome is '+', profit is earned.

Note

I. '+' connotes profit and '–' does loss.

II. Remaining part has been sold at the cost price, thus, no question arises as to profit or loss on that part.

Problem 97. A shopkeeper bought 200 articles, each costing the same. He sold 30% of the articles at 20% profit and remaining at 10% profit. If the total profit made by him is ₹ 2600, find the cost price of one article.

[SSC (10+2/DEO-LDC)-12, CAT-19]

Sol. On the above pattern,

his overall profit per cent = $\frac{3}{10}(+20\%) + \frac{7}{10}(+10\%) = 13\%$

Given, 13% of CP = 2600

Then, 100% of CP = 20000

Hence, the CP of 200 articles = 20000,

then CP of each article = ₹ 100

Problem 98. Two-third of a consignment was sold at a profit of 5% and the remaining at loss of 2%. If the total profit was ₹ 400, then the value of consignment was

[SSC (SO-Commercial)-07, MAT-09]

Sol. Overall profit or loss percent = $\frac{2}{3}(+5\%) + \frac{1}{3}(-2\%) = \frac{8}{3}\%$

Given, $\frac{8}{3}\%$ of the value of consignment = 400

Then, 100% of the value of consignment = 15000

Problem 99. A person bought 50 pens for ₹ 50 each. He sold 40 of them at a loss of 5%. He wants to gain 10% on the whole. Then, his gain % on the remaining pens should be

[SSC (CPO)-11, SSC (SAS)-10]

Sol. His overall profit or loss percent = $\frac{4}{5}(-5\%) + \frac{1}{5}(+x\%)$

$\Rightarrow \quad +10\% = -4\% + \frac{x}{5}\%$

$\Rightarrow \quad x = +70$

It means, the remaining pens should be sold at 70% profit.

Problem 100. A man purchased 150 pens at the rate of ₹ 12 per pen. He sold 50 pens at a gain of 10%. The percentage gain at which he must sell the remaining pens so as to gain 15% on the whole outlay is

[SSC (CGL-T-II)-12]

Sol. His overall profit or loss percent = $\frac{1}{3}(+10\%) + \frac{2}{3}(+x\%)$

$\Rightarrow \quad +15\% = \frac{10}{3}\% + \frac{2x}{5}\%$

$\Rightarrow \quad x = +17.5$

It means, the remaining pens should be sold at 17.5% profit.

Problem 101. A shopkeeper sold 20 articles at a profit of 20% and 30 articles at a profit of 30%. If he had sold all the 50 articles at a profit of 25%, his profit would have been reduced by ₹ 24. What is the cost price of each article?

Sol. Total number of articles = $20 + 30 = 50$

Overall profit or loss percent = $\frac{2}{5}(+20\%) + \frac{3}{5}(+30\%)$

= + 26%

As given, if the all articles were sold at a profit of 25% instead of 26%, the profit would have been reduced by ₹ 24.

It means, (26 − 25)% of the CP = ₹ 24

Then, 100% of CP = 2400

Thus, CP of all 50 articles = ₹ 2400, then each will cost ₹ 48.

Problem 102. A shopkeeper bought 30 kg of wheat at the rate of ₹ 45 per kg. He sold 40 percent of the total quantity at the rate of ₹ 50 per kg. Approximately, at what price per kg should he sell the remaining quantity to make 25% overall profit? *[Andhra Bank(PO)-10]*

Sol. Let the remaining quantity is sold @ ₹ x per kg.

Then, (40%) of 50 + (60%) of x = 125% of 45

$\Rightarrow \quad 20 + 0.6x = 56.25$

$\Rightarrow \quad x = 60.41 = 60$ (approx)

Type XIII Concept of neither Profit Nor Loss in whole transaction

Case I ⇒ When SPs are equal

Case II When sum of CPs is given, but SPs are not equal

Case III ⇒ When CP of one article is given and P/L on both articles are also given.

Type XIII Case I When SPs are equal

Problem 103. X sells two articles for ₹ 4,000 each with no loss and no gain in the transaction. If one was sold at a profit of 25%, the other is sold at a loss of? *[SSC (CGL-T-I)-11]*

Sol. Given, total SP = 4000 + 4000 = 8000

and total CP = 8000 [∵ no loss and no gain in transaction]

Now, CP of 1st article sold at 25% profit = 3200

[∵ 125% = 4000 ∴ 100% = 3200]

Obviously, CP of 2nd article = 8000 − 3200 = 4800

Then, loss incurred = 4800 − 4000 = 800

$\therefore$ Loss percent $= \left[\frac{800}{4800} \times 100\right]\% = 16\frac{2}{3}\%$

Alternatively I

Given that SP of 1st article is 25% greater than its CP. It means its CP is $\left[\frac{25}{(100+25)} \times 100\right]$% less than its SP.

Hence, profit on 1st article = 20% of its SP.

Since neither loss nor gain occurs, profit on 1st article will be equal to loss on 2nd article.

Hence, a loss of 20% of SP is incurred on 2nd article.

Then, loss percent on its CP $= \left[\frac{20}{(100+20)} \times 100\right]\%$

$= 16\frac{2}{3}\%$

Recall the concept **'Relation between Profit/Loss on CP and that on SP'**, Profit percent on SP

$= \left\{\frac{\text{Profit \% on CP}}{100 + \text{Profit \% on CP}} \times 100\right\}\%$

and Loss percent on CP $= \left(\frac{\text{Loss \% on SP}}{100 + \text{Loss \% on SP}} \times 100\right)\%$

Alternatively II

CP of 1st article $= 4000 \times \frac{100}{125} = 3200$

Then, profit on the 1st article = 4000 − 3200 = 800

Clearly, profit on 1st article = loss on 2nd article = 800

[∵ no loss and no gain in transaction]

Now, CP of 2nd article = SP on 2nd article + loss on 2nd article

= 4000 + 800 = 4800

Then, loss percent $= \left[\frac{800}{4800} \times 100\right]\%$

$= 16\frac{2}{3}\%$

Problem 104. Two articles are sold at ₹ 3000 each with no loss no gain in the transaction. One has been sold at a loss of 20%, then another has been sold at a profit of

Sol. **On the above line** (adhering to the concept 'Relation between Profit/Loss on CP and that of SP'),

Loss percent on SP of 1st article

$= \left(\frac{\text{Loss \% on CP}}{100 - \text{Loss \% on CP}} \times 100\right)\%$

$= \left\{\frac{20}{(100-20)} \times 100\right\}\%$

= 25% = Profit on SP of 2nd article

Then, profit % on CP of 2nd article

$= \left(\frac{\text{Profit\% on SP}}{100 - \text{Profit \% on SP}} \times 100\right)\%$

$= \left\{\frac{25}{(100-25)} \times 100\right\}\%$

$= 33\frac{1}{3}\%$

Type XIII Case II When sum of CPs is given, but SPs are not equal

Problem 105. The total CP of two watches is ₹ 840. One is sold at a profit of 16% and other at a loss of 12%. There is no loss or gain in the whole transaction. The cost price of the watch on which the shopkeeper gains, is :

Sol. Let the CP of 1st watch (sold at profit) be ₹ x.

Then, CP of 2nd watch = $(840 - x)$

As given, 116% of x + 88% of $(840 - x)$ = 840

[∵ total CP = total SP]

$\Rightarrow \quad 28\%$ of x = 12% of 840

$\Rightarrow \quad x = \frac{12 \times 840}{28} = 360$

Alternatively I

Let the CP of 1st watch be C_1 and that of 2nd be C_2.

Given, total CP = total SP

It means, Profit on 1st watch = Loss on 2nd watch

⇒ 16% of C_1 = 12% of C_2

$\Rightarrow \quad C_1 : C_2 = 12 : 16 = 3 : 4$

As given, $\quad C_1 + C_2 = 840$

$\therefore \quad C_1 = 360$

[∵ (3 + 4) units ⇒ 840, then 3 units = 360]

Alternatively II

You may directly write

$$C_1 = \text{sum of CPs} \times \frac{\text{Profit or loss \% on 2nd article}}{(\text{profit or loss \% on 1st article} + \text{profit or loss \% on 2nd article})}$$

Or

$$C_2 = \text{sum of CPs} \times \frac{\text{Profit or loss \% on 1st article}}{(\text{profit or loss \% on 1st article} + \text{profit or loss \% on 2nd article})}$$

Here, $C_1 = 840 \times \frac{12}{(16+12)} = 360$

↘ Problem 106. A man purchases two fans for ₹ 2160. By selling one fan at a profit of 15% and the other at a loss of 9%, he neither gains nor loses in the whole transaction. Find the cost price of the each fan in ₹

Sol. $C_1 = \text{sum of CPs} \times \frac{\text{Profit or loss \% on 2nd article}}{(\text{profit or loss \% on 1st article} + \text{profit or loss \% on 2nd article})}$

$= 2160 \times \frac{9}{(15+9)} = 810$

Then, $C_2 = 2160 - 810 = 1350.$

Type XIII <u>Case III</u> When SP of one article is given and Profit/Loss on both article are also given

↘ Problem 107. A trader earns 20% profit by selling a horse for ₹ 600 and incurs a loss of 10% on selling an ox. Find the CP of the ox, if there is neither loss nor profit in the transaction.

Sol. **In case of horse,**

Given, 120% of CP = 600

then, 20% of CP = ₹ 100 (i.e. profit on horse)

In case of ox,

Loss = ₹ 100

[In case of neither loss nor profit, profit on one article = loss on another article]

Obviously, 10% of CP of ox = 100

Then, CP = 1000

Type XIV When relation between profit and loss is given in respect of one article

↘ Problem 108. The percentage of loss when an article is sold at ₹ 50 is the same as that of the profit when it is sold at ₹ 70. The above mentioned percentage of profit or loss on the article is

[SSC (CGL-PT)-05, CLAT-19, CSAT-18]

Sol. Let the CP be ₹ x.

As given, $\left[\frac{x-50}{x} \times 100\right]\% = \left[\frac{70-x}{x} \times 100\right]\%$

$\Rightarrow \quad 2x = 120 \Rightarrow x = 60$

Then, loss percent $= \left\{\frac{60-50}{60} \times 100\right\}\%$

$= 16\frac{2}{3}\%$

Hence, percentage of profit or loss on the article $= 16\frac{2}{3}\%$.

Alternatively I

Let the profit or loss per cent be x%.

Given, $50 = (100 - x)\%$ of CP and

$70 = (100 + x)\%$ of CP

i.e. $\frac{70}{50} = \frac{(100+x)\% \text{ of CP}}{(100-x)\% \text{ of CP}}$

$\Rightarrow \quad 70(100 - x) = 50(100 + x)$

$\Rightarrow \quad x = 16\frac{2}{3}\%$

Alternatively II

Let CP be ₹ x.

Then, $x - 50 = 70 - x \Rightarrow x = 60$

Then, profit or loss % $= \left[\frac{70-60}{60} \times 100\right]\%$

$= 16\frac{2}{3}\%$

Observation You may write directly as $70(100 - x) = 50(100 + x)$ and then solve.

↘ Problem 109. Profit after selling a commodity for ₹ 524 is the same as loss after selling it for ₹ 452. The cost price of the commodity is

Sol. Let the profit or loss be ₹ x.

Then, CP in 1st case = $524 - x$

and CP in 2nd case = $452 + x$

Thus, $524 - x = 452 + x$

$\Rightarrow \quad 2x = 72 \Rightarrow x = 36$

Then, $CP = 524 - 36 = 488$

Alternatively I

We know that the average of two numbers is at the same interval from the both numbers. The contextual meaning is that CP will be the average of 524 and 452.

Hence, $CP = \frac{524 + 452}{2} = 488$

Alternatively II

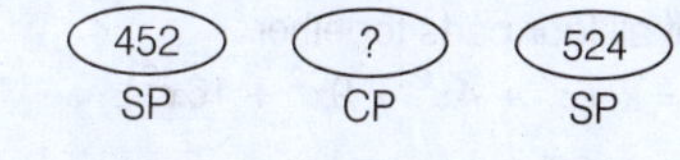

Clearly, CP – 452 (i.e. loss) = 524 – CP (i.e. profit)

$\Rightarrow \quad 2CP = 524 + 452$

$\Rightarrow \quad CP = 488$

↘ Problem 110. A person sold a TV for ₹ 9400 and he lost a particular amount. When he sold another TV of the same type at ₹ 10600, his gain was double the former loss. What was the cost price of each TV?

Sol. (9400 SP) (? CP) (10600 SP)

Clearly, $2(CP - 9400) = (10600 - CP)$

$\Rightarrow \quad 3CP = 29400$

$\therefore \quad CP = 9800$

Note *You may also proceed through options.*

Problem 111. When an article is sold for ₹ 600, a loss is incurred which is $33\frac{1}{3}\%$ less than the profit earned on selling it at ₹ 1100. Find its SP at a loss of 10%.

Sol. (600) SP (?) CP (1100) SP

As given, $(CP - 600) = 66\frac{2}{3}\%$ of $(1100 - CP)$

$\Rightarrow \quad (CP - 600) = \frac{2}{3} \times (1100 - CP)$

$\Rightarrow \quad 5CP = 2200 + 1800$

$\Rightarrow \quad CP = 800$

Now, SP at a loss of 10% = 90% of 800 = ₹ 720

Type XV Concept of loss in case of breakage of a piece of diamond/ any solid/ goods

Problem 112. The cost of a diamond varies directly as the square of its weight. A diamond is broken into four pieces and the weight of these pieces are in the ratio $1:2:3:4$. Find the price of the original piece of diamond if the total loss due to breakage is ₹ 35000.

[SSC (Asstt.-Mains)-96, 99, 2000, CAT-96]

Sol. Let the weight of four parts be x, $2x$, $3x$ and $4x$.

Then, weight of the original piece of the diamond

$= x + 2x + 3x + 4x = 10x$

As given, price of diamond $\propto$ (weight)^2

$\therefore$ Price of original piece of diamond $\propto (10x)^2$

$\Rightarrow$ Price of original piece of diamond $= k \cdot (10x)^2$

[where k is constant]

$\Rightarrow$ Price of original piece of diamond $= k \cdot 100x^2$

Similarly, price of 1st broken part $= k \cdot x^2$

Price of 2nd broken part $= k \cdot (2x)^2 = k \cdot 4x^2$

Price of 3rd broken part $= k \cdot (3x)^2 = k \cdot 9x^2$

Price of 4th broken part $= k \cdot (4x)^2 = k \cdot 16x^2$

$\therefore$ Price of all four parts together

$= k \cdot (x^2 + 4x^2 + 9x^2 + 16x^2)$

$= k \cdot 30x^2$

Now, reduction in price of original piece of diamond

$= k \cdot 100x^2 - k \cdot 30x^2 = k \cdot 70x^2$

But given, $k \cdot 70x^2$ = ₹ 35000

$\therefore \quad k \cdot 100x^2$ = ₹ 50000

Hence, the price of the original piece of diamond = ₹ 50000

Alternatively

Reduction in price of original piece of diamond

$= [(1 + 2 + 3 + 4)^2 - (1^2 + 2^2 + 3^2 + 4^2)]$ units

$= 100 - 30 = 70$ units.

Given, 70 units = 35000

$\therefore$ 100 units = 50000

Hence, the price of the original piece of diamond = ₹ 50000

Problem 113. Cost of a stone varies directly as the cube of its weight. A stone is broken into three pieces with their weights in the ratio $1:2:3$. Find the loss percent, if the loss in the total value of the stone amounts to ₹ 2700.

[SSC (CGL-T-I-16)]

Sol. On the above pattern,

Reduction in price of original piece of stone

$= [(1 + 2 + 3)^3 - (1^3 + 2^3 + 3^3)]$ units

$= 216 - 36 = 180$ units

Given, 180 units = 2700

$\therefore$ 216 units = 3240

Hence, the price of the original piece of stone = ₹ 3240

Now, loss percent $= \left[\frac{2700}{3240} \times 100\right]\% = 83\frac{1}{3}\%$

Alternatively

Reduction in price of original piece of diamond

$= [(1 + 2 + 3)^3 - (1^3 + 2^3 + 3^3)]$ units

$= 216 - 36 = 180$ units

Loss percent $= \left\{\frac{180}{216} \times 100\right\}\% = 83\frac{1}{3}\%$

Type XVI Miscellaneous problems

Problem 114. A person sells a table at a profit of 25% and a chair at a loss of $16\frac{2}{3}\%$ and the whole he gains ₹ 50. On the other hand, if he sells the table at a loss of $16\frac{2}{3}\%$ and the chair at a point of 25%, he neither gains nor loses. Find the cost price of the chair.

Sol. In the 2nd case, he neither gains nor loses. It means, Loss on table = profit on chair.

$\Rightarrow \quad 16\frac{2}{3}\%$ of CP of table (say, CT)

= 25% of CP of chair (say, CC)

$\Rightarrow \quad CT : CC = 25 : 16\frac{2}{3} = 3 : 2$

On considering 1st case,

Profit on table = 3 × 25 % = 75%

and loss on chair $= 2 \times 16\frac{2}{3}\% = 33.33\%$

Hence, combined profit in 1st case = 41.67%

$\therefore$ Combined profit per cent $= \left\{\frac{41.67}{(3+2)}\right\}\% = 8.33\%$

As given, 8.33% of (CT + CC) = ₹ 50

$\therefore$ 100% = ₹ 600

It means, CT + CC = 600

$\therefore$ CC = 240

[$\because (3 + 2)$ units $\Rightarrow 600 \therefore$ 2 units = 240]

Alternatively I

In the 2nd case he neither grains not loses. It means, Loss on table = profit on chair.

$\Rightarrow 16\frac{2}{3}\%$ on CP of table (say, CT)

= 25% of CP of chair (say, CC)

$\Rightarrow$ CT : CC $= 25 : 16\frac{2}{3} = 3:2$

Let the CT be $3x$ and CC be $2x$.

As given, 25% of $3x + \left(-16\frac{2}{3}\%\text{ of } 2x\right) = +50$

$\Rightarrow \frac{3x}{4} - \frac{x}{3} = 50$

$\Rightarrow 5x = 600$

$\Rightarrow 2x = 240$

Hence, cost of the chair = 240

Alternatively II

Let the cost price of the table be t and that of chair be c.

Given that,

+ (25% of t) – (16.66% of c) = 50 ...(i)

and – (16.66% of t) + (25% of c) = 0 ...(ii)

Now, Eq. (i) × 4

$\Rightarrow$ (100% of t) – (66.66% of c) = 200 ...(i)

and Eqn. (i) × 6

$\Rightarrow$ – (100% of t) + (150% of c) = 0 ...(ii)

On adding both Eqs. (i) and (ii), we get 83.33% of c = 200

$\therefore$ 100% of c = 240

Problem 115. **A sells his goods 20% cheaper than B and 20% dearer than C. A customer of B purchases the goods worth ₹ 600 from C. Find his profit or loss.**

Sol. Let B sells the goods for ₹ 100.

Then, SP of A = ₹ 80

and SP of $C = 80 \times \frac{100}{(100+20)}$

$= ₹\ 66\frac{2}{3}$

Now, difference between SPs of B and C

$= ₹\ 33\frac{1}{3}$

Clearly, if he buys the goods for ₹ 100, he will gain ₹ $33\frac{1}{3}$

Then, in case of buying for ₹ 600, he will gain ₹ 200.

Alternatively

SP of A is 20% dearer than that of C.

It means, SP of C is $\left[\frac{20}{120} \times 100\right]\%$ i.e. $16\frac{2}{3}\%$ cheaper than that of A.

Also given, SP of A is 20% cheaper than that of B.

The above facts may be depicted as

B .... (–20%) ...... > A ...... $\left(-16\frac{2}{3}\%\right)$...... > C

It means SP of C will be $\left[-20 - 16\frac{2}{3} + \frac{\left(20 \times 16\frac{2}{3}\right)}{100}\right]\%$ i.e. $33\frac{1}{3}\%$ cheaper than B.

Hence, the profit of the customer $= 33\frac{1}{3}\%$ of 600

= ₹ 200

Problem 116. **Sudama earns a total of 10% profit by selling a typewriter for ₹ 5000 and a bat for ₹ 4900. If he had sold the typewriter for ₹ 6000 and the bat at its CP, he would have gained total $13\frac{1}{3}\%$. Find the CP of the bat.**

Sol. Sum of their SPs in 1st case = 5000 + 4900 = 9900

It means, 110% of sum of their CPs = 9900

Then, 100% of sum of their CPs = 9000

i.e. sum of their CPs = 9000

Let the CP of the bat be ₹ x.

Then, total SPs in second case = 6000 + x

As given, $(6000 + x) = 113\frac{1}{3}\%$ of 9000

$\Rightarrow 6000 + x = 10200 \Rightarrow x = 4200$

Alternatively

Given, 110% of (sum of CPs) = 9900

Then, $113\frac{1}{3}\%$ of (sum of CPs) = 10200

Clearly, 10200 = 6000 + CP of the bat

$\therefore$ CP of the bat = 10200 – 6000 = 4200

Problem 117. **An article is sold at a profit of 120%. When a seller reduces its SP by 50%, the number of articles sold increases by 500%. Find the new profit percent.**

Sol. Let, the CP of one article be ₹ 100.

Then, its initial SP = 220% of 100 = 220

Now, number of articles sold

= 600% of 1 = 6 at the rate of ₹ (220/2)

Then, total SP = 6 × 110 = 660

And total CP = 6 × 100 = 600

Hence, new profit = 660 – 600 = 60

$\therefore$ Profit percent $= \left\{\frac{60}{600} \times 100\right\}\% = 10\%$

Problem 118. **The SP of one article is ₹ 2400 less than that of another article. If the 1st article is sold at a profit of 12% and the overall profit is 15%, then find the cost price of each provided their cost prices are same.**

Sol. Let cost price of each be ₹ x.

SP of 1st article = 112% of x

and SP of 2nd article = 112% of x + 2400

Given, [112% of x] + [112% of x + 2400] = 115% of $(x + x)$

$\Rightarrow$ 230% of x – 224% of x = 2400

$\Rightarrow$ 6% of x = 2400

$\Rightarrow$ 100% of x = 40000

Hence, cost of each article = ₹ 40000

Alternatively

1st article has been sold at a profit of 12% and second at a profit of 12% plus ₹ 2400. But, the fact is that overall profit is only 15%. It means, [(2 × 15) – (2 × 12)]% of each CP = ₹ 2400

$\Rightarrow$ 100% of each CP = ₹ 40000

Hence, cost of each article = ₹ 40000

Problem 119. A driver of auto-rickshaw makes a profit of 20% on every trip when he carries three passengers and the price of petrol is ₹ 30 per litre. Find the profit percent for the same journey if he goes for four passengers per trip and the price of petrol reduces to ₹ 24 per litre? (assuming that revenue per passenger is the same in both the cases). *[MAT-14, 19]*

Sol. Let the initial cost on consumption of petrol is ₹ 100.

Then, his revenue from 3 passengers = 120% of 100 = 120

It means, per passenger fare = ₹ 40

Now, his revenue from 4 passengers = 4 × 40 = 160 and his reduced cost on consumption of petrol is 80% of 100 (i.e. 80) as rate of petrol is reduced by 20%.

Hence, a profit of 100% is earned.

Alternatively

Clearly, increase in number of passengers = 33.33%.

Also given, earlier profit is 20%.

Then, his total profit per cent assuming no change in price of petrol

$$= \left[+ 20 + 33.33 + \frac{20 \times 33.33}{100} \right] \% = 60\%$$

But, the price of petrol is reduced by 20%. It means the driver will get $\left[\frac{20}{80} \times 100\right]$ % i.e. 25% more petrol. If he intends to keep his expenditure constant and obviously, this enhanced quantity will fetch 25% profit.

Thus, his total profit percent $= \left[+ 60 + 25 + \frac{60 \times 25}{100} \right] = 100\%$.

Problem 120. A stockist wants to make some profit by selling sugar. He contemplates about various methods. Which of the following would maximize his profit?

I. sell sugar at 10% profit.

II. use 900 g weight instead of 1 kg.

III. mix 10% impurities in sugar and selling at cost price.

IV. increase the price by 5% and reduce weights by 5%.

(a) I or III

(b) II

(c) II, III and IV

(d) profits are same *[CAT-95]*

Sol. Profit in case I ⇒ 10%

Profit in case II

$$\Rightarrow \left[\frac{100}{900} \times 100\right] \% = 11.11\%$$

Profit in case III ⇒ Impurities will appear to be sugar and its cost will fetch the gain percent. Hence, 10% profit.

Profit in case IV ⇒ profit if weights is reduced by 5%

$$= \left[\frac{5}{95} \times 100\right] \%$$

$$= 5.26\%$$

And increase in price = 5% (as given)

Then, overall profit percent

$$= \left[+ 5.26 + 5 + \frac{5.25 \times 5}{100} \right] \%$$

$$= 10.52\%$$

Hence, option (b) is correct.

WORKOUT

1. A person selling an article for ₹ 96 finds that his loss percent is one-fourth of the amount of rupees that he paid for the article. What can be the cost price? *[CDS-14]*

(a) Only ₹ 160

(b) Only ₹ 240

(c) Either ₹ 160 or ₹ 240

(d) Neither ₹ 160 nor ₹ 240

2. A carpenter wants to sell 40 chairs. If he sells them at ₹ 156, he would be able to sell all the chairs. But for every ₹ 6 increase in price, he will be left one additional unsold chair. At what selling price would he be able to maximize his profits (assuming unsold chairs remain with him)? *[MAT-14]*

(a) ₹ 198

(b) ₹ 192

(c) ₹ 204

(d) Cannot be determined

3. A shopkeeper sells sugar in such a way that the selling price of 950 g of sugar is the same as the cost price of 1 kg of sugar. What is his gain per cent? *[SSC (CGL-T-II)-10]*

(a) $5\frac{5}{19}$ (b) $5\frac{1}{5}$ (c) 5 (d) $4\frac{1}{19}$

4. Left pan of a faulty balance weighs 100 g more than its right pan. A shopkeeper keeps the weighs measure in the left pan while buying goods but keeps it in the right pan while selling his goods. He uses only 1 kg weight measure. If he sells his goods at the listed cost price, what is his gain? *[IAS (Pre.)-05]*

(a) $\frac{200}{11}\%$ (b) $\frac{200}{9}\%$

(c) $\frac{100}{9}\%$ (d) $\frac{100}{11}\%$

5. Kamya purchased an item for ₹ 46000 and sold it at a loss of 12%. With that amount she purchased another item and sold it at a gain of 12%. What was her overall gain/loss? *[Allahabad Bank (PO)-11]*

(a) Loss of ₹ 662.40

(b) Profit of ₹ 662.40

(c) Loss of ₹ 642.80

(d) Profit of ₹ 642.80

(e) None of the above

6. A man buys 4 tables and 5 chairs for ₹ 1000. If he sells the tables at 10% profit and chairs at 20% profit, he earns a profit of ₹ 120. What is the cost of one table? *[CDS-07, CDS-17]*

(a) ₹ 200 (b) ₹ 220
(c) ₹ 240 (d) ₹ 260

7. The sale price of an article including the sales tax is ₹ 616. The rate of sales tax is 10%. If the shopkeeper has made a profit of 12%, then the cost price of the article is

(a) ₹ 500 (b) ₹ 515
(c) ₹ 550 (d) ₹ 600

8. If selling price is doubled, the profit triples. Find the profit percent.

(a) $66\frac{2}{3}\%$ (b) 100%
(c) 120% (d) $105\frac{1}{3}\%$

9. The profit percentage earned by selling an article for ₹ 1920 is equal to the percentage loss incurred by selling the same article for ₹ 1280. At what price should the article be sold to make 25% profit?

(a) ₹ 2000 (b) ₹ 2400
(c) ₹2200 (d) Data inadequate
(e) None of these

10. An article is sold at a certain fixed price. By selling it at $\frac{2}{3}$ of that price, one loses 10%. The gain percent on selling it at the original price is *[SSC (CGL-T-II)-10]*

(a) 20 (b) $33\frac{1}{3}$
(c) 35 (d) 40

11. Rani bought more apples than oranges. She sells apples at ₹ 23 a piece and makes 15% profit. She sells oranges at ₹ 10 a piece and makes 25% profit. If she gets ₹ 653 after selling all the apples and oranges, then find her profit percentage. *[XAT-16]*

(a) 16.8% (b) 17.4%
(c) 17.9% (d) 18.5%
(e) 19.1%

12. A man sold two watches, each for ₹ 495. If he gained 10% on one watch and suffered a loss of 10% on the other, then what is the loss or gain percentage in the transaction?

(a) 1% gain (b) 1% loss
(c) $\frac{100}{99}\%$ loss (d) No gain no loss

13. A horse and a carriage together cost ₹ 8000. If by selling the horse at a profit of 10% and the carriage at a loss of 10% a total profit of 2.5% is made, then what is the cost price of the horse?

(a) ₹ 3000 (b) ₹ 3500
(c) ₹ 4000 (d) ₹ 5000

14. The price of jewel, passing through three hands, rises on the whole by 65%. If the first and the second sellers earned 20% and 25% profit respectively, the profit earned by the third seller is

(a) 20% (b) 15% (c) 10% (d) 5%

15. A shopkeeper sold some articles at ₹ 35 per article and gained 40% on it. What would be selling price of the each article to get 60% profit?

(a) ₹ 45 (b) ₹ 42 (c) ₹ 39 (d) ₹ 40
(e) None of these

Directions (Q.Nos. 16 and 17) *Answer the questions based on the following information.* *[CAT-96]*

A watch dealer incurs an expense of ₹ 150 for producing every watch. He also incurs an additional expenditure of ₹ 30000, which is independent of the number of watches produced. If he is able to sell a watch during the season, he sells it for ₹ 250. If he fails to do so, he has to sell each watch for ₹ 100.

16. If he is able to sell only 1200 out of 1500 watches he has made in the season, then he has made a profit of

(a) ₹ 90000 (b) ₹ 75000
(c) ₹ 45000 (d) ₹ 60000

17. If he produces 1500 watches, what is the number of watches that he must sell during the season in order to breakeven, given that he is able to sell the watches produced?

(a) 500 (b) 700
(c) 800 (d) 1000

18. Sita buys a fridge at $\frac{15}{16}$ of its original value and sells it for 10% more than its value. Then, gain per cent is

(a) 15.55% (b) 11.67%
(c) 16.67% (d) None of these

19. The cost of two articles are in the ratio 3 : 5. If there is 30% loss on the first article and 20% gain on the second article, what is overall percentage of loss or gain? *[CDS-12]*

(a) 2.25% gain (b) 5.25% loss
(c) 2% loss (d) None of these

20. What percent of selling price would be 34% of the cost price if the gross profit is 26% of the selling price? *[MAT-13, 09]*

(a) 74% (b) 25.16%
(c) 17.16% (d) 88.40%

21. By selling an article for ₹ 21, a man lost such that the percentage loss was equal to the cost price. The cost price of the article was

(a) ₹ 30 or 70
(b) ₹ 35 or 60
(c) ₹ 45
(d) ₹ 50

22. By selling an article for ₹ 98, double profit is earned than the profit that would have been obtained by selling it for ₹ 86. What is the cost price of the article?

(a) ₹ 74 (b) ₹ 75
(c) ₹ 70 (d) ₹ 68
(e) None of these

23. A loss of 19% gets converted into a profit of 17% when the selling price is increased by ₹ 162. The cost price of the article is *[SSC (CGL-T-II)-11]*

(a) ₹ 450 (b) ₹ 600
(c) ₹ 360 (d) ₹ 540

24. A man buys 200 oranges for ₹ 1000. How many oranges for ₹ 100 can be sold so that his profit percent is 25%. *[CDS-14]*

(a) 10 (b) 14
(c) 16 (d) 20

25. A shopkeeper sold an article at a profit of 17.5%. If he had bought it at 8% less and sold it at 30% profit, he would have earned ₹11.55 more as profit. Cost price of the article is *[MAT-10]*

(a) ₹ 550 (b) ₹ 675
(c) ₹ 750 (d) ₹ 1475

26. The profit earned after selling an article for ₹ 878 is the same as loss incurred after selling the article for ₹ 636. What is the cost price of the article? *[SBI(PO)-08, OBC(PO)-08 Canara Bank(PO)-09]*

(a) ₹797 (b) ₹ 787 (c) ₹ 767 (d) ₹ 757
(e) None of these

27. A man purchased 150 pens at the rate of ₹ 12 per pen. He sold 50 pens at a gain of 10%. The percentage gain at which he must sell the remaining pens so as to gain 15% on the whole outlays is. *(SSC(CGL-T-II)-11-13)*

(a) 21.5% (b) 20%
(c) 17% (d) 17.5%

28. Find the profit percent earned by selling an article for a certain money, if a business man incurs 20% loss by selling that article at 2/3 of the earlier SP.

(a) 21% (b) 25%
(c) 20% (d) 19%

29. A person bought 8 quintal of rice for certain rupees. After a week, he sold 3 quintal of rice at 10% profit, 3 quintal of rice with neither profit nor loss and 2 quintal at 5% loss. In this transaction, what is the profit? *[CDS-12]*

(a)10% (b) 20%
(c) 25% (d) None of these

30. In a certain store, the profit is 320% of the cost. If the cost increase by 25% but the selling price remains constant, approximately what percentage of the selling price is the profit?

(a) 30% (b) 70%
(c) 100% (d) 250%

31. A shopkeeper sells 100 kg of sugar partly at 10% profit and the remaining at 20% profit. If he gains 12% on the whole transaction, how much sugar he sells at 20% profit?

(a) 20 kg (b) 80 kg
(c) 60 kg (d) 75 kg

32. A shopkeeper sells his articles at their cost price but uses a faulty balance which reads 1000 g for 800 g. What is the actual profit percentage? *[CDS-14]*

(a) 25% (b) 20% (c) 40% (d) 30%

33. After selling a watch, Sultan found that he had made a loss of 10%. He also found that had he sold it for ₹ 27 more, he would have made a profit of 5%. The actual initial loss was what percentage of the profit earned, had he sold the watch for a 5% profit. *[MAT-14, 19]*

(a) 23% (b) 150%
(c) 180% (d) 200%

34. A refrigerator and a camera were sold for ₹ 12000 each. The refrigerator was sold at a loss of 20% of the cost and the camera at a gain of 20% of the cost. The entire transaction results in which one of the following?

(a) no loss no gain
(b) loss of ₹ 1000
(c) loss of ₹ 2000
(d) gain of ₹ 1000

35. Instead of a meter scale, a cloth merchant uses a 120 cm scale while buying, but uses an 80 cm scale while selling the same cloth. What is his overall profit percentage?

(a) 50% (b) 25%
(c) 40% (d) 15%

36. A person A sells a table costing ₹ 2000 to a person B and earns a profit of 6%.The person B sells it to another person C at a loss of 5%. At what price did B sell the table?

(a) ₹ 2054 (b) ₹ 2050
(c) ₹ 2024 (d) ₹ 2014

37. Aditi bought an article and spent ₹ 110 on its repairs. She then sold it to Samir at a profit of 20%. Samir sold it to Vikas at a loss of 10%. Vikas finally sold it for ₹ 1188 at a profit of 10%. How much did Aditi pay for the article? *[MAT-14, 19]*

(a) ₹ 890 (b) ₹ 1000
(c) ₹ 780 (d) ₹ 840

38. A person bought 864 articles and sold 800 of them for the price he paid for 864 articles. He sold the remaining articles at the same price per article as the other 800. The percentage gain on the entire transaction is *[LIC (AAO)-11]*

(a) 7.5% (b) 8%
(c) 8.5% (d) 9%

39. For an article the profit is 170% of the cost price. If the cost price increases by 20% but the selling price remains same, then what is the new profit percentage? *[SSC (CGL-T-I)-17]*

(a) 41 (b) 50
(c) 75 (d) 125

40. By selling 8 dozen pencils, a shopkeeper gains the selling price of 1 dozen pencils. What is the gain?

(a) $12\frac{1}{2}\%$ (b) $87\frac{1}{2}\%$
(c) $13\frac{1}{7}\%$ (d) $14\frac{2}{7}\%$

41. A man sells a book a profit of 20%. If he had bought it at 20% less and sold it for ₹ 18 less, he would have gained 25%. The cost price of the book is *[MAT-09, 03]*

(a) ₹ 80 (b) ₹ 60
(c) ₹ 70 (d) ₹ 90

42. Two items A and B are sold at a profit of 10% and 15% respectively. If the amount of profit received is the same, then the cost price of A and B may be *[RRB (TA/CA)-09, SSC (CGL-T-II)-13]*

(a) ₹ 1000 and 1500
(b) ₹ 5000 and 2000
(c) ₹ 3000 and 2000
(d) ₹ 3000 and 5000

43. A man bought a number of oranges at 3 for a rupee and an equal number at 2 for a rupee. At what price per dozen should he sell them to make a profit of 20%?

(a) ₹ 4 (b) ₹ 5
(c) ₹ 6 (d) ₹ 7

44. A house costs C rupees. Later it was sold for a profit of 25%. What is the capital gains tax if it is 50% of the profit?

(a) $\frac{C}{24}$ (b) $\frac{C}{8}$
(c) $\frac{C}{4}$ (d) $\frac{C}{2}$

45. 60% of the cost price of an article is equal to 50% of its selling price. Then, the percentage of profit or loss on the cost price is *[SSC (CGL-T-II)]*

(a) 20% loss (b) $16\frac{2}{3}\%$ profit
(c) 20% profit (d) 10% loss

46. A person sold an article for ₹ 136 and got 15% loss. Had he sold for ₹ x, he would have got a profit of 15%. Which one of the following is correct? ***[CDS-12]***

(a) $190 < x < 200$ (b) $180 < x < 190$
(c) $170 < x < 180$ (d) $160 < x < 170$

47. A student purchased a computer system and a color printer. If he sold the computer system at 10% loss and color printer at 20% gain, he would not lose anything. But if he sells the computer system at 5% gain and the color printer at 15% loss, he would lose ₹ 800 in the bargain. How much did he pay for the colour printer? ***[MAT-13, 09, 08]***

(a) ₹ 16000 (b) ₹ 8000
(c) ₹ 9000 (d) ₹ 5334

48. Harshad bought 15 pieces of DVD players at the rate of 4500 each and sold all of them at the total price of ₹ 81000. What is percent profit earned in the deal? ***[Andhra Bank (MA)-09]***

(a) $16\frac{2}{3}$ (b) 20
(c) 25 (d) 20.5
(e) None of these

49. A fruit seller buys x guavas for ₹ y and sells y guavas for ₹ x, if $x > y$, then he made ***[SSC (CGL-T-II)-12]***

(a) $\left[\frac{x^2 - y^2}{xy} \times 100\right]$ % loss
(b) $\left[\frac{x^2 - y^2}{xy} \times 100\right]$ % gain
(c) $\left[\frac{x^2 - y^2}{y^2} \times 100\right]$ % loss
(d) $\left[\frac{x^2 - y^2}{y^2} \times 100\right]$ % gain

Directions (Q. Nos. 50 and 51) *Answer the questions based on the following information.* ***[CAT-98]***

A company purchases components A and B from Germany and USA respectively. A and B form 30% and 50% of the total production cost. Current gain is 20%. Due to change in the international scenario, cost of the German mark increased by 30% and that of USA dollar increased by 22%. Due to market conditions, the selling price cannot be increased beyond 10%.

50. What is maximum current gain possible?

(a) 10% (b) 12.5%
(c) 0% (d) 7.5%

51. If the USA dollar becomes cheap by 12% over its original cost and the cost of German mark increased by 20%. What will be the gain? (The selling price is not altered).

(a) 10% (b) 15% (c) 20% (d) 7.5%

52. A merchant earns a profit of 20% by selling a basket containing 80 apples which cost of ₹ 240 but he gives one-fourth of it to his friend at cost price and sells the remaining apples. In order to earn the same profit, at what price must he sell each apple? ***[CDS-12]***

(a) ₹ 3.00 (b) ₹ 3.60
(c) ₹ 3.80 (d) ₹ 4.80

53. If 7% of the sale price of an article is equivalent to 8% of its cost price and 9% of its sale price exceeds 10% of its cost price by ₹ 1, then what is the cost price of the article

(a) ₹ 400 (b) ₹ 350
(c) ₹ 300 (d) ₹ 280

54. Manider bought two horses at ₹ 40000 each. He sold one horse at 15% gain, but had to sell the second horse at a loss. If he had suffered a loss of ₹ 3600 on the whole transaction, then the selling price of the second horse is ***[SSC (CGL-T-II)-12]***

(a) ₹ 30000 (b) ₹ 30200
(c) ₹ 30300 (d) ₹ 30400

55. A fruit-seller buys lemons at 2 for a rupee and sells them at 5 for three rupees. What is his gain percent?

(a) 10% (b) 20% (c) 25% (d) 15%

56. A manufacturer of a certain item can sell all he can produce at the selling price of ₹ 60 each. It costs him ₹ 40 in materials and labour to produce each item and he has overhead expenses of ₹ 3000 per week in order to operate that plant. The number of units he should produce and sell in order to make a profit of at least ₹ 1000 per week is ***[MAT-13, 12, 07]***

(a) 300 (b) 400
(c) 250 (d) 200

57. A retailer bought a certain number of CDs for ₹ 1800. Keeping one to himself, he sold the rest at a profit of ₹ 6 each. In total, he earned a profit of ₹ 114. The number of CDs he bought is ***[MAT-10]***

(a) 20 (b) 28
(c) 32 (d) 30

58. If selling price is tripled and cost price doubled, the profit would be 65%. What is the present profit (in percent)? ***[SSC (CGL-T-II)-17]***

(a) 10 (b) 15
(c) 20 (d) 25

59. A person bought two tables for ₹ 2200. He sells one at 5% loss and the other at 6% profit and thus on the whole he neither gains nor loses. Find the cost price of each table (in ₹).

(a) 1500, 700
(b) 1200, 1000
(c) 2000, 200
(d) 1100, 1100

60. In a market, the price of medium quality mangoes is half that of good mangoes. A shopkeeper buys 80 kg good mangoes and 40 kg medium quality mangoes from the market and then sells all these at a common price which is 10% less than the price at which he bought the good ones. His overall profit is ***[CAT-17]***

(a) 6% (b) 8%
(c) 10% (d) 12%

61. Sudha sold an article to Renu for ₹ 576 at a loss of 20%. Renu spent a sum of ₹ 224 on its transportation and sold it to Raghu at a price which would have given Sudha a profit of 24%. The percentage of gain for Renu is: ***[SSC (CGL-T-I)-19]***

(a) 13.2% (b) 11.6%
(c) 10.5% (d) 12.9%

62. On selling a pen at 5% loss and a book at 15% gain, Karim gains ₹ 7. If he sells the pen at 5% gain and the book at 10% gain, then he gains ₹ 13. The actual price of the book is ***[MAT-03, CAT-19]***

(a) ₹ 100
(b) ₹ 80
(c) ₹ 10
(d) ₹ 400

Answer key

1. (c)	2. (a)	3. (a)	4. (b)	5. (a)	6. (a)	7. (a)	8. (b)	9. (a)	10. (c)
11. (b)	12. (b)	13. (d)	14. (c)	15. (d)	16. (b)	17. (b)	18. (d)	19. (d)	20. (b)
21. (a)	22. (a)	23. (a)	24. (c)	25. (a)	26. (d)	27. (d)	28. (c)	29. (d)	30. (b)
31. (b)	32. (a)	33. (d)	34. (b)	35. (a)	36. (d)	37. (a)	38. (b)	39. (d)	40. (d)
41. (d)	42. (c)	43. (c)	44. (b)	45. (c)	46. (b)	47. (a)	48. (b)	49. (d)	50. (a)
51. (c)	52. (c)	53. (b)	54. (d)	55. (b)	56. (d)	57. (a)	58. (a)	59. (b)	60. (b)
61. (b)	62. (b)								

Hints and SOLUTIONS

1. Let the CP be ₹ x.

Loss per cent $= \left[\dfrac{(x-96)}{x} \times 100\right]\%$

$= \dfrac{x}{4}\%$ (as given)

$\Rightarrow \quad x^2 - 400x + 38400 = 0$

$\Rightarrow x^2 - 160x - 240x + 38400 = 0$

$\Rightarrow x(x-160) - 240(x-160) = 0$

$\Rightarrow \quad (x-160)(x-240) = 0$

$\Rightarrow \quad x = 160$ or 240

Hence, option (c) is correct.

Alternatively

Proceed through options

(a) $160 \Rightarrow 160 - \dfrac{160}{4}\%$ of $160 = 96$

(b) $240 \Rightarrow 240 - \dfrac{240}{4}\%$ of $240 = 96$

Hence, option (c) is correct.

2. Let the carpenter enhances the price of chair n times.

Then new SP of the chair $= 156 + 6n$

and number of chairs sold $= (40 - n)$

[∵ for every ₹ 6 increase in price, one additional chair is unsold]

Now, new revenue

$= (156 + 6n)(40 - n)$

Obviously, this revenue must be greater than the earlier revenue in order to maximise the profits.

Hence, $(156 + 6n)(40 - n) > (156 \times 40)$

$\Rightarrow (156 \times 40) + 84n - 6n^2 > (156 \times 40)$

$\Rightarrow + 84n - 6n^2 > 0$

On differentiating,

$84 - 12n = 0$

$\Rightarrow \quad n = 7$

If we put value of n exceeding 7 i.e., 8, 9, …, in the expression $(84 - 12n)$, we get negative outcome. Hence, at $n = 7$, the profit is maximum.

∴ The required *SP*

$= 156 + 6n$

$= 156 + 42$

$= 198$

3. Recall the concept 'Use of false weights/meter rod, Error in weights/length etc.'

Gain per cent

$= \left[\dfrac{\text{Error}}{\text{Weight or Length for which money is invested}} \times 100\right]\%$

$= \left[\dfrac{50}{950} \times 100\right]\% = 5\dfrac{5}{19}\%$.

4. Recall the concept 'Uses of false weights/meter rod, Error in weights/length etc.'

Profit or Gain per cent

$= \left[\dfrac{\text{Actually weight or length acquired} - \text{actually weight or length sold}}{\text{actually weight or length sold}} \times 100\right]\%$

Here, the shopkeeper acquires 1100 g instead of 1000 g and sells 900 gm in lieu of 1000 gm.

Hence, profit per cent

$= \left[\dfrac{1100 - 900}{900} \times 100\right]\%$

$= \dfrac{200}{9}\%$

$= 22.22\%$.

5. [Recall the concept of 'Successive Increasing/Decreasing/combination of both' envisaged in the chapter 'Percentage']

Overall gain or loss per cent

$= \left[-12 + 12 - \dfrac{12 \times 12}{100}\right]\%$

$= -1.44\%$

∵ '–' indicates that a loss of 1.44% is incurred.

Then, amount of loss

$= 1.44\%$ of $46000 = ₹\ 662.40$

6. Overall profit per cent

$= \left[\dfrac{120}{1000} \times 100\right]\%$

$= 12\%$

Now,

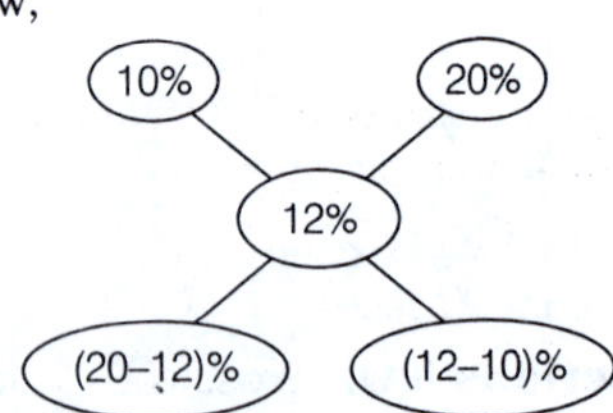

It means, CP of 4 tables : CP of 5 chairs = 8 : 2 = 4 : 1

Given, (4 + 1) units ⇒ 1000

Then, 4 unit ⇒ 800

Hence, CP of one table

= 800 ÷ 4 = 200

7. Clearly, CP × 112% × 110% = 616

⇒ CP = 500.

8. Let the profit be x% and CP be ₹ 100. Then, SP = $(100 + x)$. Later SP becomes double i.e. $(200 + 2x)$

Then, profit per cent

$$= \left[\frac{(200 + 2x) - 100}{100} \times 100\right]\%$$

$= 3x\%$ (given)

$\Rightarrow \quad 100 + 2x = 3x$

$\Rightarrow \quad x = 100$

Hence, profit per cent = 100%.

9. Let profit or loss per cent be $x\%$.

As given, $(100 + x)\%$ of CP = 1920 ...(i)

and $(100 - x)\%$ of CP = 1280 ...(ii)

On adding Eqs. (i) and (ii), we get 200% of CP = 3200

Then, 125% of CP = 2000

[∵ a profit of 25% was to earn]

Hence, SP at a profit of 25% = ₹ 2000

10. Obviously, 66.67% of the SP = 90% of the CP

Then, 100% of SP = 135% of the CP

Hence, a profit of 35% is earned.

11. Let, she buys x apples and y oranges, where $x > y$.

As given, $23x + 10y = 653$

$$\Rightarrow \quad y = \frac{653 - 23x}{10}$$

Clearly, both x and y must be integers.

At $x = 1, y = 63$,

at $x = 11, y = 40$

at $x = 21, y = 17$

Since $x > y, x = 21$ and $y = 17$

Now, cost price of each apple = 20

[∵ 115% ⇒ 23 ∴ 100% = 20]

and cost price of each orange = 8

[∵ 125% ⇒ 10 ∴ 100% = 8]

It implies that the profit on each apple = ₹ 3 and that on each orange = ₹ 2

∴ Total profit = $21 \cdot 3 + 17 \cdot 2$ = ₹ 97

and total cost = $21 \cdot 20 + 17 \cdot 8 = 556$

Now, profit percentage

$$= \left[\frac{97}{556} \times 100\right]\% = 17.4\%$$

12. Recall the concept 'Profit or Loss in whole transaction, Case II When SPs are same'.

The required loss per cent

$$= \frac{(10)^2}{100} = 1\%$$

13. Recall the concept : 'Profit or Loss in whole transaction, Case III-When neither CPs nor SPs are same'.

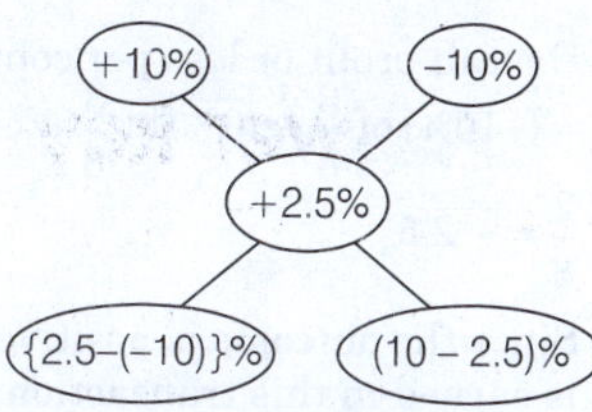

It means, CP of the horse : CP of the carriage = 12.5 : 7.5 = 5 : 3

Given, (5 + 3) units ⇒ 8000

Then, (5) units ⇒ 5000

Hence, cost price of the horse

= ₹ 5000

(This concept has been given in detail in the topic 'Alligation')

14. Rising in price till second seller

$$= \left[+ 20 + 25 + \frac{20 \times 25}{100}\right]\% = 50\%$$

Now, overall rising in price

$$= \left[+ 50 + x + \frac{50 \times x}{100}\right]\%$$

[let the profit earned by third seller be x%]

$$\Rightarrow \quad 65\% = \left[+ 50 + x + \frac{50 \times x}{1000}\right]\%$$

$$\Rightarrow \quad 65 = 50 + \frac{3x}{2} \Rightarrow x = 10$$

Hence, profit earned by third seller is 10%.

[**Note** You may proceed through options after computation of 50%. The above all steps can be made orally]

15. Recall the concept 'Changes in SPs on account of changes in profit/loss and *vice-versa*'.

Given, 140% of CP = 35

Then, 160% of CP = 40

Hence, required SP = ₹ 40

16. Selling price of 1200 watches during the season

$= 1200 \times 250 = 300000$

and selling price of remaining 300 watches in off season

$= 300 \times 100 = 30000$

∴ Total SP = 300000 + 30000

= 330000

Total cost of 1500 watches

$= (1500 \times 150) + 30000 = 255000$

Hence, profit = 330000 − 255000

= ₹ 75000

17. Let he sells x watches during the season in order to breakeven.

Now, total cost of 1500 watches

$= (1500 \times 150) + 30{,}000 = 255000$

and total SP of 1500 watches

$= 250x + 100(1500 - x)$

In case of breakeven, total cost price

= total selling price

$\Rightarrow 255000 = 250x + 150000 - 100x$

$\Rightarrow \quad 150x = 105000 \Rightarrow x = 700$

18. Let the original value of fridge is ₹ 100.

Then, her CP = $100 \times \frac{15}{16} = 93.75$

and her SP = 110

Hence, gain per cent

$$= \left[\frac{110 - 93.75}{93.75} \times 100\right]\% = 17.33\%$$

19. Recall the concept 'Calculation of Profit or Loss, when ratio of CPs/ratio of CP and SP is given and *vice-versa*.'

Loss ⇒ 3 × 30% = 90%

and profit ⇒ 5 × 20% = 100%

Clearly, profit = 10%.

∴ Combined profit per cent

$$= \left\{\frac{10\%}{(3 + 5)}\right\} = 1.25\%$$

20. Let the SP be ₹ 100.

Then, CP = 74% of 100 = 74

and 34% of the cost price

= 34% of 74 = 25.16

Clearly, 25.16 is 25.16% of SP.

21. Recall the concept 'When Profit/Loss is numerically equal to SP/CP.

CP = $50 \pm (10\sqrt{25 - \text{SP}})$ in case of loss

CP = $50 \pm (10\sqrt{25 - 21})$

$\Rightarrow \quad$ CP = (50 + 20) or (50 − 20)

$\Rightarrow \quad$ CP = 70 or 30.

Alternatively

Proceed through options

(a) 30 ⇒ 70% of 30 = 21

and 70 ⇒ 30% of 70 = 21

Hence, option (a) is correct.

[Aspirants ought to adhere to this alternative approach]

22. Let the CP be ₹ x.

As per question, $(98 - x) = 2(86 - x)$

$\Rightarrow \quad x = 172 - 98 \Rightarrow x =$ ₹ 71

23. Difference between (−19) and (+17) = 36

Clearly, 36% of CP = ₹ 162

Then 100% of CP = ₹ 450

24. Purchasing rate for ₹ 100 = 20

Number of oranges sold for the given (same) amount

$= \frac{100}{(100 + \text{Profit}\%)} \times$ number of items bought $= \frac{100}{(100 + 25)} \times 20 = 16$

Try through other methods too.

25. Recall the concept 'Profit/loss is changed when CP or SP or both are changed and *vice-versa*.

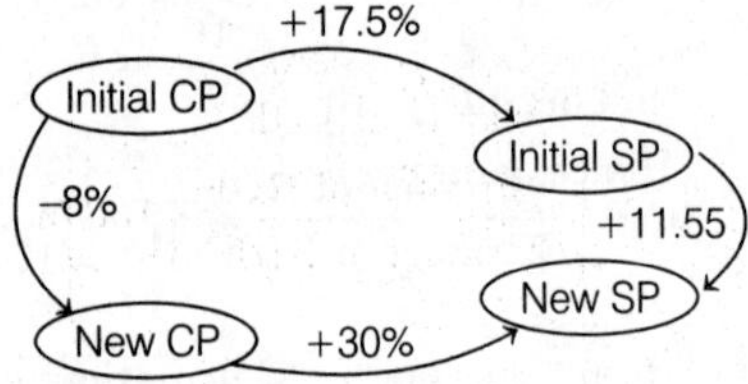

Clearly, new SP is

$\left[-8 + 30 - \frac{8 \times 30}{100}\right]\%$ i.e. 19.6%

greater than initial CP. Also, given that initial SP is 17.5% greater than initial CP. According to question, the new SP is ₹ 11.55 greater than the initial SP. It means, (19.6 – 17.5)% of initial CP = ₹ 11.55

$\Rightarrow$ 100% of initial CP = ₹ 550

Hence, cost price of the article is ₹ 550.

[**Note** you may also proceed through options]

26. We know that the average of two numbers is at the same interval from the both numbers. The contextual meaning is that CP will be the average of 524 and 452.

Hence, CP $= \frac{878 + 636}{2}$ = ₹ 757

27. Recall the concept 'When goods are sold part by part'. His overall profit or loss per cent

$= \frac{1}{3}(+10\%) + \frac{2}{3}(+x\%)$

$\Rightarrow \quad +15\% = \frac{10}{3}\% + \frac{2x}{3}\%$

$\Rightarrow \quad x = +17.5$

It means, the remaining pens should be sold at 17.5% profit.

28. Given, 2/3 of earlier SP = 80% of CP

Then, earlier SP = 120% of CP

Hence, profit per cent = 120% – 100%

= 20%

29. Recall the concept 'When goods are sold part by part'

Overall profit or loss per cent

$= (+10\%) \text{ of } \frac{3}{8} + (0\%) \text{ of } \frac{3}{8} + (-5\%) \text{ of } \frac{2}{8} = +2.5$

Since the outcome is positive, a profit is earned in this transaction.

30. Let, the CP be ₹ 100. Then, SP

= 100 + 320

= 420

Now, new CP = 125, then profit

= 420 – 125

= 295

Thus, the required percentage

$= \left[\frac{295}{420} \times 100\right]\%$

= 70% (approx.)

31. Recall the concept 'When goods are sold part by part'. Let the shopkeeper sells x kg of sugar at a profit of 10%.

Then, overall profit per cent

= [10% of x] + [20% of (100 – x)]

$\Rightarrow$ 12% of 100 = [10% of x] + [20% of (100 – x)]

$\Rightarrow 1200 = 10x + 2000 - 20x$

$\Rightarrow \quad 10x = 800$

$\Rightarrow \quad x = 80$

Hence, he sells (100 – 20) kg at a profit of 20%.

Alternatively

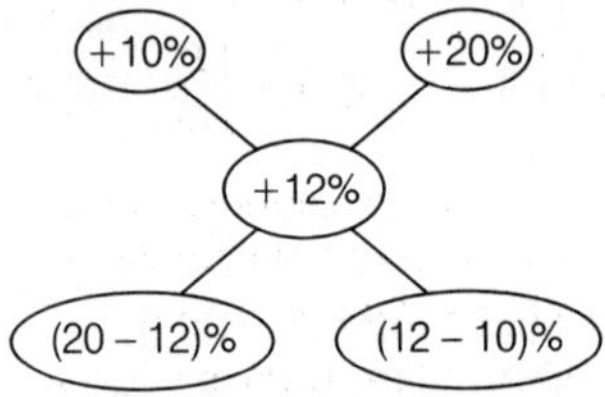

It means,

1st part sold at a profit of 10% : 2nd part sold at a profit of 20%

= 8 : 2 = 4 : 1

Given, (4 + 1) units $\Rightarrow$ 100

Then, 1 unit $\Rightarrow$ 20

Hence, the part sold at a profit of 20% = 20 kg.

32. Gain per cent

$= \left[\frac{\text{Error}}{(\text{True value} - \text{Error})} \times 100\right]\%$

$= \left[\frac{200}{800} \times 100\right]\%$

= 25%

33. Recall the concept 'Profit/loss is changed when CP or SP or both are changed and *vice-versa*'.

Clearly, difference between (–10) and (+5) is 15.

Given, 15% of CP = ₹ 27

Then, 100% of CP = ₹ 180

Now, actual loss = 10% of 180 = 18 and desired profit = 5% of 180 = 9

Hence, the actual loss is 200% of the desired profit.

Alternatively

Since the computation of profit and loss is based on the same cost price, we may find the relation between actual loss and desired profit on given figures in percentage.

Thus, actual loss (i.e. 10%) is 200% of the desired profit (i.e. 5%). No need to utilise other figure given in the question.

34. Recall the concept 'Profit or Loss in whole transation, Case II When SPs are same'.

The overall loss percent $= \frac{(20)^2}{100} = 4\%$

Now, 4% loss on CP

$= \left[\frac{4}{(100 - 4)} \times 100\right]\%$ loss on SP

$= \frac{25}{6}\%$ loss on SP

Hence, loss $= \frac{25}{6}\%$ of (12000 + 12000)

= ₹ 1000

35. Recal the concept 'Uses of false weights/meter rod, Error in weight/length etc.'.

Profit per cent

$= \left[\frac{\text{actually weight or length acquired} - \text{actually weight or length sold}}{\text{actually weight or length sold}} \times 100\right]\%$

$= \left[\frac{(120 - 80)}{80} \times 100\right]\% = 50\%$

36. Recall the concept 'Goods/Article passing through various hands at certain profit of loss'.

Now, SP for B = 2000 × 106% × 95%

= ₹ 2014

37. Recall the concept 'Goods/Article passing through various hands at certain profit or loss'.

Obviously, (CP of Aditi + 100) × 120% × 90% × 110% = SP of Vikash

$\Rightarrow$ (CP of Aditi + 110) $\times$ 120% $\times$ 90% $\times$ 110% = 1188

$\Rightarrow$ (CP of Aditi + 110) $\times 1.2 \times 0.9 \times 1.1 = 1188$

$\Rightarrow$ (CP of Aditi + 110) = 1000

$\Rightarrow$ CP of Aditi = 890.

38. The question may be rewritten as 'By selling 864 articles a person gains the selling price of 64 articles'.

Now, apply the concept 'When Profit/Loss is given in terms of SP/CP of some articles'.

Profit per cent

$$= \left[\frac{\text{Profit in number of articles}}{(\text{number of articles sold} - \text{profit in number of articles})} \times 100\right]\%$$

$$= \left[\frac{64}{(864-64)} \times 100\right]\% = 8\%$$

39. Let cost price be ₹ 100, then profit = ₹ 170.

Now, new CP = 120, then new profit = (100 + 170) – 120 = 150

$\therefore$ New profit per cent

$$= \left[\frac{150}{120} \times 100\right]\% = 125\%$$

40. Recall the concept 'When Profit/Loss is given in terms of SP/CP of some articles'.

Profit per cent

$$= \left[\frac{\text{Profit in number of articles}}{(\text{number of articles sold} - \text{profit in number of articles})} \times 100\right]\%$$

$$= \left[\frac{1}{(8-1)} \times 100\right]\%$$

$$= 14\frac{2}{7}\%$$

41. Recall the concept 'Profit/Loss is changed when CP or SP or both are changed and *vice-versa.*

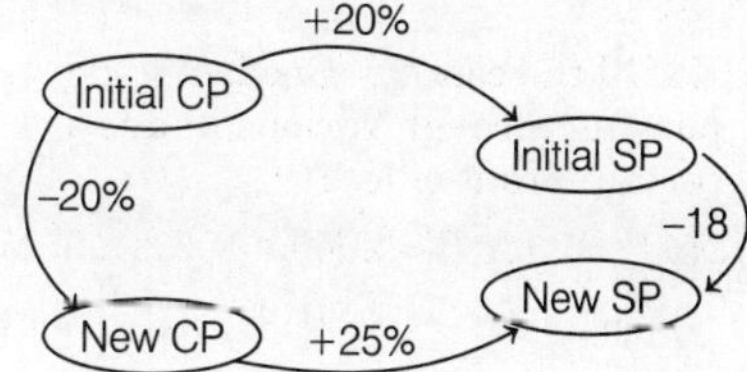

Clearly, difference between 20% of initial CP and $\left[-20 + 25 - \frac{20 \times 25}{100}\right]\%$ of initial CP = 18

$\Rightarrow$ (20 – 0)% of initial CP = 18

$\Rightarrow$ 100% of initial CP = 90

Hence, initial CP = ₹ 90.

[**Note** you may also proceed through options]

42. As given, 10% of CP of A = 15% of CP of B

$\Rightarrow$ CP of A : CP of B = 15 : 10 = 3 : 2

Hence, option (c) is correct, Please note that there will be a lot of set of CPs which are in ratio 3 : 2 like 1500 and 1000, 450 and 300, etc.

43. CP of 6 oranges at first rate = ₹ 2 and CP of 6 oranges at second rate = ₹ 3

Thus, the CP of 12 oranges = ₹ 5

Now, desired SP of 12 oranges = 120% of CP = 120% of 5 = ₹ 6.

44. Clearly, profit = 25% of $C = \frac{C}{4}$

Then, capital gains tax

$$= 50\% \text{ of } \frac{C}{4} = \frac{C}{8}$$

45. Recall the concept : 'Calculation of profit or loss, when ratio of CPs/ratio of CP and SP is given and *vice-versa*'

Given, 60% of the cost price = 50% of selling price

Then, CP : SP = 5 : 6

Assume, CP = ₹ 5

and SP = ₹ 6

Then, profit = ₹ 1

$$\therefore \text{Profit per cent} = \left\{\frac{1}{5} \times 100\right\}\% = 20\%$$

46. Recall the conception 'Changes in SPs on account of changes in profit/loss and *vice-versa.*

Given, (100 – 15)% of CP = 136

Then, (100 + 15)% of CP = 184

Hence, option (b) is correct.

47. In the 1st case, he neither gains nor loses. It means, loss on computer system = profit on color printer.

$\Rightarrow$ 10% of CP of computer system (say, CC) = 20% of CP of printer (say, CT)

$\Rightarrow$ 10% of CC = 20% of CT

$\Rightarrow$ CC : CT = 20 : 10 = 2 : 1

Let the CC be $2x$ and CT be x.

As given, (15)% of $2x$ + (–15% of x) = – 800

$\Rightarrow$ –5% of x = – 800

$\Rightarrow$ x = ₹ 16000

Thus, CP of the color printer = ₹ 16000

48. Profit per cent

$$= \left\{\frac{81000 - (15 \times 4500)}{(15 \times 4500)} \times 100\right\}\%$$

= 20%

49. Write the given figures in following format

Number of articles (x) ↖ LDF ... RDF ↗ CP (y)

Number of articles (y) ↙ ... ↘ SP (x)

Profit or Loss per cent

$$= \left[\frac{\text{Product of LDF} - \text{Product of RDF}}{\text{Product of RDF}} \times 100\right]\%$$

$$= \left[\frac{x^2 - y^2}{y^2} \times 100\right]\%$$

Its value is positive as $x > y$. It means, a profit is earned.

50. Let, the production cost be ₹ 100.

Then, share of components A and B in the production cost = 30 + 50 = 80.

$\therefore$ Share of other expenses in the production cost = 100 – 80 = 20

Clearly, selling price at a profit of 20% = 120% of 100 = ₹ 120

Now, the production cost due to international scenario

= 130% of 30 + 122% of 50 + 20 = 120

The maximum selling price = 110% of 120 = 132

$\therefore$ Maximum profit per cent

$$= \left[\frac{132 - 120}{120} \times 100\right]\%$$

= 10%

51. The production cost in the given scenario = 88% of 50 + 120% of 30 + 20 = ₹ 100

Selling price = ₹ 120

Then, profit per cent = 20%

52. Recall the concept 'When goods are sold part by part'

Now, as per question,

[0% of 1/4] + [(+x)% of 3/4] = + 20%

$\Rightarrow$ $x = 26\frac{2}{3}\%$

It means, the merchant should sell the remaining at a profit of $26\frac{2}{3}\%$.

Hence, SP of each apple

$$= 126\frac{2}{3}\% \text{ of } \left[\frac{240}{80}\right] = ₹\ 3.80$$

53. Given, 8% of CP = 7% of SP

$\Rightarrow$ SP $= \frac{8}{7}$ of CP

Also given, (9% of SP) – (10% of CP) = 1

$\Rightarrow \left(9\% \text{ of } \frac{8}{7} \text{ of CP}\right) - (10\% \text{ of CP}) = 1$

$\Rightarrow \frac{72}{700}$ of CP $- \frac{10}{100}$ of CP = 1

$\Rightarrow$ CP $= \frac{700}{2} = 350$

54. SP of second horse = total SP – SP of first horse

= (40000 + 40000 – 3600) – (115% of 40000) = 30400

55. Recall the concept 'Concept of Profit and loss in context of purchasing and selling rate for the products'.

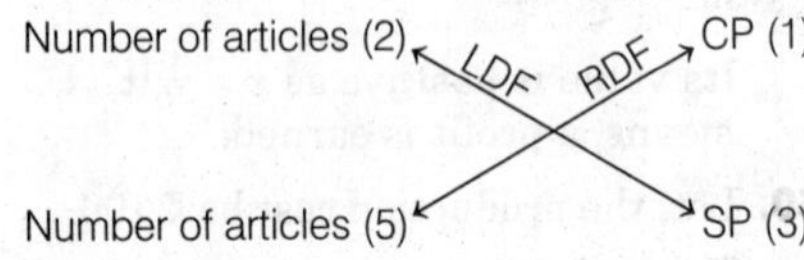

Profit or Loss per cent

$= \left[\frac{\text{Product of LDF} - \text{Product of RDF}}{\text{Product of RDF}} \times 100\right]\%$

$= \left[\frac{(2 \times 3 - 5 \times 1)}{5 \times 1} \times 100\right]\% = +\ 20\%$

Since the outcome is positive, a profit is earned.

56. Let the number of items produced a week be n.

Then, cost in production of n items

$= 40n + 3000$

and SP of n items $= 60n$

As per question, $60n$ should be 1000 more than the cost.

It means, $60n = 40n + 3000 + 1000$

$\Rightarrow n = 200$

57. He earns ₹ 6 on each CD.

Then, number of CDs sold

$= 114 \div 6 = 19$

Hence, total number of CDs he bought = 19 + 1 (kept to himself)

= 20

58. Let present CP and present SP be ₹ 100 and ₹ x respectively.

Clearly, new CP and SP be 200 and $3x$.

$\therefore$ new profit per cent

$= \left[\frac{3x - 200}{200} \times 100\right]\%$

$\Rightarrow 65 = \frac{3x - 200}{2}$

$\Rightarrow 3x = 330$

$\Rightarrow x = 110$

Then, present profit per cent

$= (110 - 100)\% = 10\%$

59. Recall the 'concept of neither profit nor loss in whole transaction Case II. When Sum of CPs is given, but SPs are not equal

C_1 = sum of CPs

$\times \frac{\text{profit or loss \% on 2nd articles}}{(\text{profit or loss \% on 1st article} + \text{profit or loss \% on 2nd article})}$

$\Rightarrow$ Cost price of 1st table

$= 2200 \times \frac{6}{(5 + 6)} = 1200$

Then, cost price of 2nd table

$= 2200 - 1200 = 1000$

Alternatively

You may proceed through options also.

(a) $\Rightarrow$ 5% of 1500 $\neq$ 6% of 700, but it is true that sum of CPs is 2200

(b) $\Rightarrow$ 5% of 1200 = 6% of 1000, but it is true that sum of CPs is 2200

Hence, option (b) is correct.

60. Let price of medium quality mangoes be ₹ x per kg and that of good ones be ₹ $2x$ per kg.

Total cost price

$= 80 \cdot 2x + 40 \cdot x = 200x$

Total selling price

$= 120\ (90\% \text{ of } 2x) = 216x$

$\therefore$ Profit per cent $= \left[\frac{16x}{200x} \times 100\right]\%$

$= 8\%$

61. CP for Renu = 576 + 224 = 800

SP for Renu = 124% of CP for Sudha

= 892.8

[$\because$ 80% of CP for Sudha = 576, then 124% of CP for Sudha = 892.8]

Then, profit per cent for Renu

$= \left[\frac{92.8}{800} \times 100\right]\% = 11.6\%$

62. Let the cost prices of pen and a book be x and y respectively.

Given, – (5% of x) + (15% of y) = 7 ...(i)

and + (5% of x) + (10% of y) = 13 ...(ii)

On adding Eqs. (i) and (ii), we get

25% of y = ₹ 20

$\therefore$ 100% of y = ₹ 80

20 Discount

Learning OBJECTIVES

After completing this chapter, aspirants will be able to :

- Understand the significance of this chapter in competitive exams, in field of finance/commerce etc. and our daily life as well;
- Understand the diverse terminology related to Discount like mark up, successive discount, etc.;
- Understand how to apply the concept of percentage like product constancy, effect on number in case of Successive Increase/Decrease/ combination of both, etc. in this chapter as this chapter is the extension of application of percentage;
- Differentiate between Mark Up and Discount;
- Analyse the different approaches for solution of a question and will also able to choose the best one for the exams;
- Realise themselves competent to solve the questions in many ways including conventional way;
- Understand how profit/loss alters in a defined way when discount changes;
- Appreciate the diverse cases under one type of questions;
- Recognise the concept of discount in different contexts like reduction in price, rebate, etc;
- Comprehend the concept of discount in various types of transactions;
- Recognise how one concept works in different sorts of problems;
- Understand the relation among profit/loss, discount and mark up;
- Use the concept of ratio in this chapter;
- Realise the significance of generic version of the questions;
- Recognise the types of question which can be easily solved through the given options;
- Analyse the trend of the questions being asked in various competitive examinations;
- Solve the questions in moments;
- Grip on concepts helping to solve the questions without pen and paper;
- Learn new concepts/fundamentals;
- Apply the pragmatic approaches in lieu of academic approaches; and
- Ponder on the alternative methods.

Inside

Introduction

Almost all the competitive examinations like CAT, MAT, SSC, Banks, Railway, LIC etc. covers this topic and in my opinion, it is an easy topic and requires no much complexity in logical thinking. Hence, the highest marks can be secured without much efforts.

In deed, the concepts envisaged in this chapter is based on the conceptions of the application of percentage, percentage-rule, percentage conversion table, etc. Thus, we must have command over the 'Percentage' chapter, if we intend to solve the questions based on 'Discounts' in moments.

The application of this chapter is also very wide in our daily life. In addition, in field of finance, commerce, business it plays vital roles.

Concept of Mark Up, Marked Price, Discount and Selling Price

Traders/shopkeepers, while selling goods add a certain percentage on the cost price so as to earn profit. This addition is called mark up and the price thus obtained is known as Marked Price/Printed Price/List Price/ labeled price.

Hence,

Marked Price = Cost Price + Mark Up

Or, Marked Price = Cost Price + Mark Up on Cost Price

Generally, goods are sold at marked price, however sometimes the traders give a discount on the marked price and sell the goods at discounted price.

Hence,

Selling Price = Marked Price – Discount on Marked Price

{Note: It is clear from the above that Mark Up is always computed on the CP whereas Discount is computed on Marked Price.}

Needless to say that there are so many tricks and techniques to solve the problems based on 'Discount'. However, we would emphasis on a core approach to solve all types of questions in this chapter. Obviously, in doing so the aspirants would save the time what is wasted in selecting the one among many tricks in the examination halls.

Now, build a concept in this regard recalling the concept of 'Successive Increase/ Decrease/ combination of both' as envisaged in the chapter 'Percentage'. Take an example: If a number (say, CP) is increased by $x\%$ (say, % Mark-Up) and further reduced by $y\%$ (say, % Discount), what will be effect on the original number.

Obviously, effect will be $\left\{+x - y - \frac{x \times y}{100}\right\}\%$. This effect may either be positive or negative. If it is positive, the original number will increase and in case of negative, the same will decrease. Here, the contextual meaning of effect is either profit or loss. Thus,

Profit or Loss%

$$= \left\{+ \textbf{Mark Up} - \textbf{Discount} - \frac{\textbf{(Mark Up} \times \textbf{Discount)}}{\textbf{100}}\right\}\%$$

It is important to note that if the outcome is positive, a profit is earned and in case of negative, a loss is incurred.

Now, discuss few examples:

I. A tradesman marks his goods at 25% above the cost price and allows purchasers a discount of $12\frac{1}{2}\%$. His profit percentage is **[SSC(CGL-PT)-08, CDS-11]**

Profit or Loss%

$$= \left\{+ \text{Mark Up} - \text{Discount} - \frac{(\text{Mark Up} \times \text{Discount})}{100}\right\}\%$$

$$= \left\{+25 - 12\frac{1}{2} - \frac{25 \times 12\frac{1}{2}}{100}\right\}\%$$

$= \{12.5 - 3.125\}\% = +9.375\%$

Since the outcome is positive, a profit of 9.375% is earned.

II. An article is marked $x\%$ above the cost price. A discount of $\frac{2x}{3}\%$ is given on the marked price. If the profit is 4% of the cost price and the value of x lies between 25 and 50, then the value of 50% of x is **[XAT-19]**

Profit or Loss%

$$= \left\{+ \text{Mark Up} - \text{Discount} - \frac{(\text{Mark Up} \times \text{Discount})}{100}\right\}\%$$

$$\Rightarrow \quad +4\% = \left\{+x - \frac{2x}{3} - \frac{x \times \frac{2x}{3}}{100}\right\}\%$$

$$\Rightarrow \quad 4 = \frac{x}{3} - \frac{x^2}{150} \Rightarrow x^2 - 50x + 600 = 0$$

$$\Rightarrow x(x - 20) - 30(x - 20) = 0 \Rightarrow x = 30 \text{ and } 20$$

Since the value of x lies between 25 and 50, $x = 30$

Then, 50% of $x = 15$

Concept of Successive Discounts

Sometimes the shopkeepers/businessmen allow successive discounts instead of a flat discount so as to attract the customers. Thus, the customers are compelled to ponder as to which type of discounts either successive or flat is beneficial to them and the same is not possible unless comparison of the flat discount rate is made with the single discount rate equivalent to the given successive discounts.

Suppose you have **to compute the single discount rate equivalent to successive discounts of 20% and 5%.**

Let the Marked Price be ₹ 100.

Then, 1st discount = 20% of 100 = ₹ 20

∴ Selling Price = 100 − 20 = ₹ 80

Now, 2nd discount = 5% of 80 = ₹ 4

∴ Selling Price = 80 − 4 = ₹ 76

Clearly, an article having marked price as ₹ 100 is sold for ₹ 76.

Hence, the single discount equivalent to two successive discounts of 20% and 5%, is (100 − 76)% i.e. 24%.

Alternatively

Now, recall the concept of 'Successive Increase/ Decrease/ combination of both' as envisaged in the chapter 'Percentage' and have an example:

"A number is decreased by 20% and further decreased by 5%, find the total effect on that number".

Effect on original number

$$= \left\{-20 - 5 + \frac{20 \times 5}{100}\right\}\%$$

$$= -24\%$$

['−' indicates that the original number has been reduced by 24%. It may be contextualized that a discount of 24% is allowed]

Hence, a single discount equivalent to successive discounts of 20% and 5% is 24%.

Problem 1. A single discount equivalent to the successive discounts of 10%, 20% and 25% is:

Sol. Taking any two discount rates on the above line,

$$\left\{-10 - 20 + \frac{10 \times 20}{100}\right\}\% = -28\%$$

Further, $\left\{-28 - 25 + \frac{28 \times 25}{100}\right\}\% = -46\%$

[here, '−' connotes the discount]

Hence, single discount equivalent to the successive discounts of 10%, 20% and 25% is 46%.

Note *Order of given successive discounts doesn't affect the value of single discount equivalent to these discounts. It means, single discount equivalent to successive discounts of 10%, 20% and 25% will be the same as the single discount equivalent to successive discounts of 10%, 25% and 20% / 20%, 10% and 25% or all possible orders.*

Problem 2. Successive discounts of 10%, 20% and 30% is equivalent to a single discount of

[SSC(CGL-T-I)-10, SSC(CGL-T-II)-11, CDS-17, 21]

Sol. You are free to take any two values (say, 20 and 30).

$$\left\{-20 - 30 + \frac{20 \times 30}{100}\right\}\% = -44\%$$

Further, $\left\{-44 - 10 + \frac{44 \times 10}{100}\right\}\% = -49.6\%$

Hence, equivalent single discount = 49.6%

Problem 3. A company offers three types of successive discounts:

(i) 25% and 15% (ii) 30% and 10% (iii) 35% and 5% which offer is the best for a customer?

(a) 1st offer

(b) 2nd offer

(c) 3rd offer

(d) Any one: all are equally good

[SSC(CGL-PT)-07, RRB Ajmer (ASM)-07]

Sol. 1st offer $\Rightarrow \left\{-25 - 15 + \frac{25 \times 15}{100}\right\}\% = -36.25\%$

2nd offer $\Rightarrow \left\{-30 - 10 + \frac{30 \times 10}{100}\right\}\% = -37\%$

3rd offer $\Rightarrow \left\{-35 - 5 + \frac{35 \times 5}{100}\right\}\% = -38.25\%$

Clearly, 3rd offer is the best as the equivalent single discount therein has the greatest value.

Alternatively I

If sum of values of discounts of different sets/offers is equal, the offer having least product of the values of given discounts is the best for the customers.

{**Note** This approach is applicable only for the sets of two successive discounts. An offer of single discount can also be written in form of a set. **For example**-an offer of 40% will be written as 40% and 0% and obviously its product will be (40 × 0).}

Here, 1st offer ⇒ 25 × 15 = 375

2nd offer ⇒ 30 × 10 = 300

3rd offer ⇒ 35 × 5 = 175

Since 3rd offer has least product value, it is the best for a customer.

Alternatively II

If sum of values of discounts of different sets/offers is equal, the offer having the highest range will be the best for the customer.

Note This approach is applicable only for the sets of two successive discounts. An offer of single discount can be written in form of a set. **For example**: an offer of 40% will be written as 40% and 0% and obviously its range will be (40 − 0).

Here, range of 1st offer ⇒ 25 − 15 = 10

Range of 2nd offer ⇒ 30 − 10 = 20

and range of 3rd offer ⇒ 35 − 5 = 30

Since 3rd offer has greatest range, it is the best for the customer.

It is important to note that (for a customer), an offer of single discount is always better than the offer consisting successive discounts provided sum of all given successive discounts is equal to that single discount. **For example** A customer will prefer to avail an offer of a discount of 40% instead of an offer consisting three successive discounts like 25%, 10% and 5%.

Problem 4. Successive discounts of $a\%$ and $b\%$ are equivalent to single discount of

(a) $(a + b)\%$ (b) $\left(\frac{a+b}{2}\right)\%$

(c) $\left(a + b - \frac{ab}{100}\right)\%$ (d) $\left(\frac{a+b}{100}\right)\%$

Sol. Here, $\left(-a - b + \frac{ab}{100}\right)\%$

Hence, equivalent single discount

$$= -\left(-a - b + \frac{ab}{100}\right)\% = \left(a + b - \frac{ab}{100}\right)\%$$

Illustrations

Case I Finding Profit or Loss when Mark Up/ Marked price and discount are given.

Problem 5. A trader's marked price is 30% above the cost price. He allows his customers 10% discount on the marked price. What profit does he make?

[SSC(CGL-PT)-03, MAT-09, CLAT-14, 18, SSC (CGL-T-II)-20]

Sol. Let the cost price be ₹ 100.

Then, marked price = 130% of 100 = 130

and discount given = 10% of 130 = 13

∴ Selling price = 130 – 13 = 117

Now, profit = SP – CP = 117 – 100 = 17

$$\therefore \quad \text{Profit } \% = \frac{17 \times 100}{100} = 17\%$$

Alternatively

Profit or Loss%

$$= \left\{+ \text{Mark Up} - \text{Discount} - \frac{(\text{Mark Up} \times \text{Discount})}{100}\right\}\%$$

$$= \left(+30 - 10 - \frac{30 \times 10}{100}\right)\% = +17\%$$

{'+' indicates that a profit of 17% is earned.}

Problem 6. A retailer buys 40 pens at the marked price of 36 pens from a wholesaler. If he sells these pens giving a discount of 1%, what is the profit percentage?

Sol. Let the Marked Price of each pen be ₹ x.

Then, his total cost price = $36x$ (for 40 pens)

He sells each pen at a discount 1%

$$\therefore \quad \text{Total SP} = 40 \times \left(x \times \frac{99}{100}\right) = 39.6x$$

Hence, Profit = $39.6x - 36x = 3.6x$

$$\therefore \quad \text{Profit } \% = \frac{3.6x \times 100}{36x} = 10\%$$

Note *You may assume marked price of each pen as ₹ 100 for prompt calculation.*

Alternatively

Given that he bought 40 pens for the marked price of 36 pens. It mean, his initial profit % = $\left\{\frac{40 - 36}{100} \times 100\right\}\% = 11\frac{1}{9}\%$. In other words, this profit % may be treated as Mark Up percentage, if no discount is allowed i.e. sold on marked price. But, a discount of 1% is allowed.

Then, effective profit %

$$= \left\{+ 11\frac{1}{9} - 1 - \frac{11\frac{1}{9} \times 1}{100}\right\}\% = +10\%$$

Since the outcome is positive, a profit of 10% is earned.

Problem 7. A man buys an article for ₹ 80 and marks it at ₹ 120. He then allows a discount of 40%. What is the loss or gain percentage?

Sol. Clearly, % mark up = 50% {∵ 80 is marked as 120}

Now, profit or loss%

$$= \left\{+ 50 - 40 - \frac{50 \times 40}{10}\right\}\% = -10\%$$

Since the outcome is negative, a loss of 10% is incurred.

Problem 8. The price of an article is raised by 30% and then two successive discounts of 10% each are allowed. Ultimately the price of the article is

[SSC(CGL-PT)-03]

Sol. Let the original price of the article be ₹ 100

Then, Marked Price = ₹ 130

Price after allowing 1st discount = 90% of 130 = 117

Further, price after allowing 2nd discount = 90% of 117 = 105.3

Clearly, the price of the article has been increased by 5.3%

Alternatively

$$\left\{- 10 - 10 + \frac{10 \times 10}{100}\right\}\% = -19\%$$

Hence, equivalent single discount = 19%.

Now, Profit or Loss%

$$= \left\{+ \text{Mark Up} - \text{Discount} - \frac{(\text{Mark Up} \times \text{Discount})}{100}\right\}\%$$

$$= \left\{+ 30 - 19 - \frac{30 \times 19}{100}\right\}\% = +5.3\%$$

Since the outcome is positive, a profit of 5.3% earned. It implies that the original price has been increased by 5.3%.

Problem 9. A trader marked his goods at 20% above the cost price. He sold half the stock at the marked price, one quarter at a discount of 20% on the marked price and the rest at a discount of 40% on the marked price. His total gain is

Sol. Let the CP of total goods be ₹ 100

Then, marked price = ₹ 120

SP of half the stock = $\frac{1}{2} \times 120$ = ₹ 60

SP of $\frac{1}{4}$ stock = 80% of $\left(\frac{1}{4} \times 120\right)$ = ₹ 24

SP of remaining $\frac{1}{4}$ stock = 60% of $\left(\frac{1}{4} \times 120\right)$ = ₹ 18

Hence, total SP = 60 + 24 + 18 = ₹ 102

∴ Gain % = (102 – 100)% = 2%

Alternatively

$\frac{1}{2}$ of stock is sold at marked price ⇒ i.e. at a profit of 20%

$\frac{1}{4}$ of stock is sold at a discount of 20%. Thus, profit or loss%

$$= \left\{+ 20 - 20 - \frac{20 \times 20}{100}\right\}\%$$

$$= -4\%$$

and $\frac{1}{4}$ of stock is sold at a discount of 40%. Thus, profit or loss%

$= \left\{+20 - 40 - \frac{20 \times 40}{100}\right\}\% = -28\%$

Now, overall profit or loss%

$= \left\{\frac{1}{2} \times (+20\%) + \frac{1}{4}(-4\%) + \frac{1}{4}(-28\%)\right\}$

{Recall the concept **'when goods are sold part by part'** in the chapter 'Profit and Loss'}

$= (10 - 1 - 7)\% = +2\%$

Thus, a profit of 2% is earned.

Problem 10. A trader marked the price of his commodity so as to include a profit of 25%. He allowed discount of 16% on the marked price. His actual profit was **[SSC(CGL-PT)-04]**

Sol. It is obvious that he should mark the price at 25% above the cost price so as to include a profit of 25%. Hence, his considered mark up percent was 25%. Now,

Profit or Loss% $= \left\{+25 - 16 - \frac{25 \times 16}{100}\right\}\% = +5\%$

Since the outcome is positive, a profit of 5% is earned.

Problem 11. A dealer buys an article at ₹ 25000 with 20% and 5% off. He spends ₹ 1000 for its repairs and sells it for ₹ 25000. What is his gain or loss percent? **[RRB-Ajmer(ASM)-07, SSC(SAS)-10, SSC(CGL-PT)-07]**

Sol. His cost price = 95% of 80% of 25000 = 19000

His total cost price = 19000 + 1000

= 20000

Then, profit% $= \left(\frac{5000}{20000} \times 100\right)\% = 25\%$

Alternatively

$\left\{-20 - 5 + \frac{20 \times 5}{10}\right\}\% = -24\%$

Hence, equivalent single discount = 24%

As given, the dealer spends 4% of the list price on repairs. It means his effective discount = 20%.

Now, his CP is 20% less than the SP (i.e. list price). Then, his SP will be $\left(\frac{20}{(100-20)} \times 100\right)\%$ i.e. 25% more than the CP.

Thus, gain % = 25%

{**Note** The all steps written above can be done mentally without pen and paper.}

Problem 12. A dealer buys a car listed at ₹ 2,00,000 at successive discounts of 5% and 10%. If he sells the car for ₹ 1,79,550, then his profit is?

Sol. $\left\{-5 - 10 + \frac{5 \times 10}{100}\right\}\% = -14.5\%$

Hence, equivalent single discount = 14.5%

∴ CP = 85.5% of 2,00,000 = 1,71,000

SP = 1, 79, 550 (as given)

∴ Profit % $= \left\{\frac{(179550 - 171000)}{171000} \times 100\right\}\%$

= 5%

Problem 13. The cost price of an article is 64% of the marked price. The gain percentage after allowing a discount of 12% on the marked price is

Sol. Let the marked price be ₹ 100.

Then, SP = 100 − 12 = 88

As given, CP = 64% of 100 = ₹ 64

∴ Profit % $= \left\{\frac{(88 - 64)}{64} \times 100\right\}\% = 37.5\%$

Alternatively

CP is 36% less than the marked price.

Hence, marked price is $\left\{\frac{36}{(100-36)} \times 100\right\}\%$ i.e. 56.25% more than the CP. It means, % Marks Up is 56.25%.

Now, Profit or Loss %

$= \left\{+5625 - 12 - \frac{56.25 \times 12}{100}\right\}\% = +37.5\%$

Since the outcome is positive, a profit of 37.5% is earned.

Problem 14. A company gives discount to his customer at 15% on the list price and this makes a profit of 19%. If the cost of production goes up by 12%, company issues a new price list in which cost of all goods have been increased by 10%. If company continues to give discount of 15% on the list price, find the profit percentage?

Sol. Let the initial list price be ₹ 100.

Then, SP = 100 − 15 = 85

and CP $= 85 \times \frac{100}{119} = \frac{500}{7}$ $\left[\text{or, } 119\% = 85 \therefore 100\% = \frac{500}{7}\right]$

As given, CP is increased by 12%, hence

New CP $= \frac{500}{7} \times \frac{112}{100}$

= ₹ 80 $\left(\text{or, } \because 100\% = \frac{500}{7} \therefore 112\% = ₹\ 80\right)$

New list price = 100 + 10 = ₹ 110

But, the discount remains same at the rate of 15%.

Then, SP $= 110 \times \frac{85}{100} = 93.5$

Thus, profit = 93.5 − 80 = 13.5

∴ Profit % $= \left\{\frac{13.5 \times 100}{80}\right\}\% = 16.875\%$

Alternatively

$\left\{+\text{Mark Up}(x) - 15 - \frac{x \times 15}{100}\right\}\% = +19\%$

$\Rightarrow \frac{17x}{20} = 34 \Rightarrow x = 40\%$

Further, list price is increased by 10%.

It means, new list price is $\left\{+40 + 10 + \frac{40 \times 10}{100}\right\}\%$ i.e. 54% above the original CP.

In this case, Profit or Loss % with the same discount

$= \left\{+54 - 15 - \frac{54 \times 15}{100}\right\}\% = +30.9\%$

Hence, a profit of 30.9% is earned.

Now, Let the original CP be ₹ 100. Then, SP = 130.9 New CP = 100 + 12 = 112 Then, Profit $= 130.9 - 112 = 18.9$ $\therefore$ Profit % $= \frac{18.9 \times 100}{112}$ $= 16.875\%$	OR	$\left\{+12 + (x) + \frac{12 \times x}{100}\right\}\%$ $= +30.9\%$ $\Rightarrow \frac{112x}{100} = 18.9$ $\Rightarrow x = 16.875\%$

Problem 15. Find the profit or loss percent, if values of discount and the mark-up over the cost price expressed in term of percentage are same.

Sol. Let discount and mark up each be x%.

Then, $\left\{+x - x - \frac{x \times x}{100}\right\}\%$ = Profit or Loss%

$\Rightarrow \frac{-x^2}{100}\%$ = Profit or Loss %

Sign '–' indicates that loss is incurred, however, we can't determine the numerical value of loss unless the value of x is known.

Case II Computation of Mark Up

Problem 16. How much percent above the cost price should a shopkeeper mark his goods so as to earn a profit of 32% after allowing a discount of 12% on the marked price? *[SSC(CGL-T-I)-12, SSC(SAS)-10, CAT-97]*

Sol. Let the CP be ₹ 100 and marked price be ₹ x.

Then, SP = 88% of x

As given, 88% of $x = 100 + 32$

$\therefore \quad x = \frac{132 \times 100}{88} = 150$

Clearly, 150 is 50% more than the CP (₹ 100).

Hence, the shopkeeper should mark 50% above the cost price.

Alternatively I

Let the Mark Up be x%. Then,

Profit or Loss%

$= \left\{+\text{Mark Up} - \text{Discount} - \frac{(\text{Mark Up} \times \text{Discount})}{100}\right\}\%$

$\Rightarrow +32 = +x - 12 - \frac{x \times 12}{100}$

$\Rightarrow \frac{88x}{100} = 44$

$\therefore \quad x = 50$

Hence, the Mark Up = 50%

Alternatively II

If discount rate and profit or loss are known then ? (given below) will be % Mark Up.
(100 – Discount)% $\Rightarrow$ (Discount + Profit)%
or (Discount – Loss)%
then 100% $\Rightarrow$?

Here, (100 – 12)% $\Rightarrow$ (12 + 32)%

Then, 100% $\Rightarrow$ 50%

Hence, the Mark Up = 50%

Alternatively III

You may proceed through options also. (Let options be 40%, 50%, ...)

Now check as to which option is yielding a profit of 32%.

(a) 40% $\Rightarrow \left\{+40 - 12 - \frac{40 \times 12}{100}\right\}\% \neq +32\%$

(b) 50% $\Rightarrow \left\{+50 - 12 - \frac{50 \times 12}{100}\right\}\% = +32\%$

Hence, option (b) is correct.

Problem 17. How much percent above the cost price should a shopkeeper mark his goods so as to incur a loss of 12% after allowing a discount of 20% on the marked price?

Sol. (100 – 20)% $\Rightarrow$ (20 – 12)%

$\therefore$ 100% $\Rightarrow$ 10%

Hence, the shopkeeper should mark 10% above the cost price.

Problem 18. The marked price of a radio is ₹ 480. The shopkeeper allows a discount of 10% and gain 8%. If no discount is allowed, his gain percent would be *[SSC(CGL-T-I)-11, SSC(CGL-T-II)-12]*

Sol. Profit or Loss%

$= \left\{+\text{Mark Up} - \text{Discount} - \frac{(\text{Mark Up} \times \text{Discount})}{100}\right\}\%$

$\Rightarrow +8\% = \left\{+\text{Mark Up}(x) - 10 - \frac{x \times 10}{100}\right\}\% \Rightarrow \frac{9x}{10} = 18 \Rightarrow x = 20$

Hence, % Mark Up = 20%

Obviously, Mark Up on CP would be his profit % when no discount is allowed.

Alternatively II

(100 – 10)% $\Rightarrow$ (10 + 8)% $\therefore$ 100% $\Rightarrow$ 20%

Problem 19. A trader allows a trade discount of 20% and a cash discount of $6\frac{1}{4}$% on the marked price of the goods and gets a net gain of 20% of the cost. By how much above the cost should the goods be marked?

Sol. $\Rightarrow \left\{-20 - 6\frac{1}{4} + \frac{20 \times 6\frac{1}{4}}{100}\right\}\% = -25\%$

Hence, equivalent single discount = 25%

Now, Profit or Loss%

$= \left\{+\text{Mark Up}(x) - 25 - \frac{x \times 25}{100}\right\}\%$

$= +20\% = \left\{+(x) - 25 - \frac{x \times 25}{100}\right\}\%$

$\Rightarrow \frac{3x}{4} = 45 \Rightarrow x = 60$

Thus, required answer = 60%

Alternatively II

From the above, equivalent single discount = 25%

Now, (100 – 25)% $\Rightarrow$ (25 + 20)%

$\therefore$ 100% $\Rightarrow$ 60%

Problem 20. A shopkeeper cheats to the extent of 10% while buying as well as selling, by using false weights. He also marks up his goods so as to get overall profit of 30%. Find the mark-up on cost price.

Sol. Here, the shopkeeper acquires 1100 gm instead of 1000 gm and sells 900 gm in lieu of 1000 gm.

Hence, Profit % $= \left\{\frac{1100-900}{900} \times 100\right\}\% = \frac{200}{9}\%$

Now, the question is that a number has been increased by $\frac{200}{9}\%$, then by how much percent should it be increased again so as to make it 30% more than the original number.

$$\left\{+\frac{200}{9} + x + \frac{x \times \frac{200}{9}}{100}\right\}\% = +30\% \Rightarrow \frac{11x}{9} = 30 - \frac{200}{9} = \frac{70}{9}$$

$\therefore x = 6.36$

Hence, % Mark Up = 6.36%

Problem 21. A trader bought some goods at a discount of 20% of the list price. He wants to mark them at such price that he can give a discount of 20% of the marked price and still makes a profit of 25%. Find the percent of the list price at which he should mark the goods.

Sol. Let the list price be ₹ 100.

Then, his cost price = 100 − 20 = 80 and profit = 25% of 80 = 20

Now, SP = 80 + 20 = 100

As given, SP is 20% less than the marked price. It means, marked price is 25% more than SP.

Hence, Marked price = 125% of 100 = 125, which is 125% of the list price.

Thus, required answer = 125%

Alternatively

For the computation of mark up :

$(100-20)\% \Rightarrow (20+25)\%$

Then, $100\% \Rightarrow 56.25\%$

Hence, the trader should mark 56.25% above the cost price. It is also given that the cost price is 20% less than the initial list price. In other words, the initial list price has been decreased by 20% and further to be increased by 56.25% so as to get desired profit.

Clearly, effect on the initial list price

$$= \left\{-20 + 56.25 - \frac{20 \times 56.25}{100}\right\}\% = +25\%$$

It means, marked price should be 25% more than the initial list price. In other words, marked price is 125% of the list price.

Case III Where Marked Price/SP/CP is to be Computed

Problem 22. A discount of $2\frac{1}{2}\%$ is given to the customer on the marked price of an article. A man bought the article for ₹ 39. The marked price of the article is

Sol. Clearly, 97.5% of marked price = ₹ 39

$\therefore$ 100% of Marked Price = ₹ 40

Hence, Marked Price = ₹ 40

Problem 23. Ravi buys an article with a discount of 25% on its marked price. He makes a profit of 10% by selling it at ₹ 660. The marked price of the article was **[SSC(CPO-SI)-07, SSC(SAS)-10]**

Sol. As given, 110% of CP = 660

$\therefore$ 100% of CP = 600

Ravi buys the article for ₹ 600 after getting 25% discount

It mean, 75% of marked price = 600

$\therefore$ 100% of marked price = 800

Hence, marked price = ₹ 800

Problem 24. An article listed at ₹ 800 is sold at successive discounts of 25% and 15%. The buyer desires to sell it off at a profit of 20% after allowing a 10% discount. What would be his list price?

Sol. $\Rightarrow \left\{-25 - 15 + \frac{25 \times 15}{100}\right\}\% = -36.25\%$

Hence, equivalent single discount = 36.25%

CP of the buyer = 63.75% of 800 = ₹ 510

Now, Profit or Loss %

$$= \left\{+\text{Mark Up }(x) - 10 - \frac{x \times 10}{100}\right\}\%$$

$$= +20\% = \left\{+(x) - 10 - \frac{x \times 10}{100}\right\}\%$$

$\Rightarrow \frac{9x}{10} = 30 \Rightarrow x = 33\frac{1}{3}$

Clearly, he should mark $33\frac{1}{3}\%$ above the CP to get a profit of 20% allowing a 10% discount.

$\therefore$ Marked price = $33\frac{1}{3}\%$ more than 510

$= 133\frac{1}{3}\%$ of 510 = ₹ 680

Alternatively

CP of the buyer $= 800 \times \left(1 - \frac{25}{100}\right)\left(1 - \frac{15}{100}\right) = 510$

{Recall the concept related to 'problems based on population' in chapter 'Percentage'}

For computation of Mark Up: $(100-10)\% \Rightarrow (10+20)\%$

Then, $100\% \Rightarrow 33.33\%$

Hence, % Mark Up = 33.33%

Then, marked price = 133.33% of 510 = ₹ 680

Try with another approaches explained in beginning

Problem 25. A dealer offers a discount of 10% on the marked price of an article and still makes a profit of 20%. If its marked price is ₹ 800, then the cost price of the article is **[SSC(CGL-PT)-02, SSC(CGL-T-I)-19]**

Sol. Profit or Loss %

$$= \left\{+\text{Mark Up }(x) - 10 - \frac{x \times 10}{100}\right\}\%$$

$$\Rightarrow +20 = +(x) - 10 - \frac{x \times 10}{100}$$

$$\Rightarrow \frac{9x}{10} = 30 \Rightarrow x = 33\frac{1}{3}$$

Clearly, the Marked Price is $33\frac{1}{3}\%$ more than the CP. It means,

$133\frac{1}{3}\%$ of CP = 800

$\therefore$ 100% of CP = 600

Alternatively

$(100 - 10)\% \Rightarrow (10 + 20)\%$

$\therefore$ $100\% \Rightarrow 33\frac{1}{3}\%$

Hence, % Mark Up = $33\frac{1}{3}\%$

Now, $133\frac{1}{3}\%$ of CP = 800

$\therefore$ 100% of CP = 600

Problem 26. An item is marked for ₹ 240 for sale. If two successive discounts of 10% and 5% are allowed on the sale price, the selling price of the article will be

Sol. $\Rightarrow \left\{-10 - 5 + \frac{10 \times 5}{100}\right\}\% = -14.5\%$

Hence, equivalent single discount = 14.5%

$\therefore$ Selling price = 85.5% of 240 = 205.20

Alternatively

SP of the article $= 240 \times \frac{90}{100} \times \frac{95}{100} = 205.20$

Problem 27. A shopkeeper earns a profit of 12% on selling a book at 10% discount on the printed price. The ratio of the cost price and the printed price of the book is ***[SSC(CGL-T-I)-10]***

Sol. $(100 - 10)\% \Rightarrow (10 + 12)\%$

$\therefore 100\% = \frac{220}{9}\%$

Hence, % Mark Up = $\frac{220}{9}\%$

Obviously, the Marked/ Printed price is $\frac{220}{9}\%$ more than CP.

Now, CP : Printed Price $= 100 : \left(100 + \frac{220}{9}\right)$

$= 900 : 1120 = 90 : 112$

$= 45 : 56$

Problem 28. In a shop, the price of all goods are marked 15% above the cost price and 5% discount is allowed on all the sales. In a certain month, the business expenses of the shop amounted to ₹ 6500 and the owner realised a profit of 6%. Find the cost price of goods sold by the shop during the month.

Sol. Profit or Loss % $= \left\{+15 - 5 - \frac{15 \times 5}{100}\right\}\%$

$= +9.25\%$

Hence, a profit of 9.25% is earned. But given that he realised only 6% profit, due to expenses of ₹ 6500.

It means (9.25 – 6)% of CP of goods = 6500

$\therefore$ 100% of CP goods = 2,00,000

Thus, Cost price of goods = ₹ 2,00,000

Problem 29. A shopkeeper sold sarees at ₹ 266 each after giving 5% discount on labelled price. Had he not given the discount, he would have earned a profit of 12% on the cost price. What was the cost price of each saree? ***[SBI(PO)-99, MAT-18]***

Sol. Each Saree is sold at ₹ 266 at a discount of 5%

$\therefore$ Marked/labeled price of each $= 266 \times \frac{100}{95} = 280$

Given that if saree had been sold without discount (i.e. at marked price), the shopkeeper would have earned 12% profit.

$\therefore$ CP $= 280 \times \frac{100}{112} =$ ₹ 250

Alternatively

CP $= 266 \times \frac{100}{(100-5)} \times \frac{100}{(100+12)} =$ ₹ 250

Problem 30. A shopkeeper gives two successive discounts of 10% and 20% on a marked price of ₹ 5000 of a bicycle. He had to give a further discount equal to 20% of his cost price on his new selling price, as a result of which he made neither a profit nor a loss. Find his cost price for the bicycle.

Sol. Single discount equivalent to successive discounts of 10% and 20% = 28%

$\left\{\because -10 - 20 + \frac{10 \times 20}{100} = -28\right\}$

Let his cost price be ₹ x.

Then, his final selling price = (100 – 28)% of 5000 –20% of x

$\because$ CP = SP

$\therefore$ (100 – 28)% of 5000 – 20% of $x = x$

$\Rightarrow 3600 = 120\%$ of $x \Rightarrow x = 3000$

Case IV How to find Discount-rate

Problem 31. A trader marked the selling price of an article at 10% above the cost price. At the time of selling he allows certain discount and suffers a loss of 1%. He allowed the discount of ***[SSC (CGL-T-II)-19]***

Sol. Profit or Loss %

$= \left\{+\text{Mark Up} - \text{Discount} - \frac{(\text{Mark Up} \times \text{Discount})}{100}\right\}\%$

$\Rightarrow -1 = \left\{+10 - d - \frac{10 \times d}{100}\right\}$

$\Rightarrow -\frac{11d}{10} = -11 \Rightarrow d = 10$

Hence, discount rate = 10%

Alternatively I

If Mark Up and profit or loss are known then? (given below) will be the discount rate.
(100 + Mark Up)% ⇒ (Mark Up – Profit)% or
(Mark Up + Loss)%
then, 100% ⇒ ?

Here, $(100 + 10)\% \Rightarrow (10 + 1)\%$

Then, $100\% \Rightarrow 10\%$

Hence, discount rate = 10%

Alternatively II

You may proceed through options also.

(Let options be 11%, 10%, ...)

Now check as to which option is yielding a loss of 1%.

(a) 11% $\Rightarrow \left\{+10-11-\frac{10\times 11}{100}\right\}\% \neq -1\%$

(b) 10% $\Rightarrow \left\{+10-10-\frac{10\times 10}{100}\right\}\% = -1\%$

Hence, option (b) is correct.

Problem 32. The marked price of a watch is ₹ 800. A shopkeeper gives two successive discounts and sells the watch at ₹ 612. If the first discount is 10%, the second discount is *[SSC(CPO-SI)-05, SSC(CGL-T-I)-17]*

Sol. A watch having marked price of ₹ 800 is sold for ₹ 612.

Hence, overall discount allowed

$$=\frac{188}{800}\times 100 = 23.5\%$$

Let the second discount be x%.

Then, $\left\{-10-x+\frac{10\times x}{100}\right\}\% = -23.5\%$

$\Rightarrow \qquad -\frac{9x}{10} = -13.5\% \Rightarrow x = 15$

Hence, the second discount is 15%.

Alternatively I

800 ...... (–10%) ......→ 720 ...... (–x%) ......→ 612

Obviously, $(100-x)\%$ of $720 = 612$

$\Rightarrow \qquad 720\times\frac{(100-x)}{100} = 612$

$\Rightarrow \qquad 7200 - 72x = 6120 \therefore x = 15$

Hence, the second discount is 15%.

Alternatively II

You may proceed through options also.

Problem 33. The marked price of a shirt and trousers are in the ratio 1:2. The shopkeeper gives 40% discount on the shirt. If the total discount on the set of the shirt and trousers is 30%, the discount offered on the trousers is

Sol. Let the marked price of shirt be ₹ x and that of trousers be ₹ $2x$.

Also, let the discount on trousers be d%.

Then,

$\Rightarrow \qquad \left(x\times\frac{40}{100}\right)+\left(2x\times\frac{d}{100}\right) = (x+2x)\times\frac{30}{100}$

$\Rightarrow \qquad \frac{2x}{5}+\frac{dx}{50} = \frac{9x}{10} \Rightarrow \frac{dx}{50} = \frac{x}{2}$

$\therefore \qquad d = 25$

Hence, the discount offered on the trousers = 25%

Alternatively I

Their Marked Price ⇒ 1 : 2

Discounts thereon ⇒ 40% : x%

Overall discount = 30%

Obviously, $(1\times 40)+(2\times x) = (1+2)30$

$\Rightarrow \qquad 2x = 50 \therefore x = 25$

Hence, the discount offered on the trousers = 25%

Alternatively II

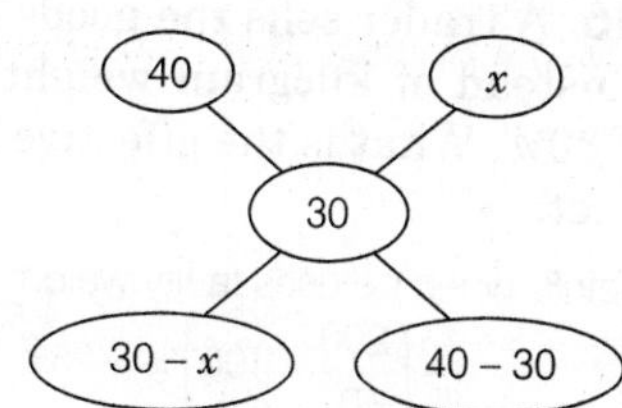

Hence, $(30-x):(40-30) = 1:2$

$\Rightarrow \qquad 60-2x = 10 \therefore x = 25$

(This concept is detailed in the chapter 'Alligation')

Problem 34. An article of cost price ₹ 8000 is marked at ₹ 11200. After allowing a discount of x% a profit of 12% is made. The value of x is

Sol. As given, $(100-x)\%$ of 11200 = 112% of 8000

$\Rightarrow 100 - x = 80 \Rightarrow x = 20$

Alternatively I

Marked price is $\left(\frac{3200}{8000}\times 100\right)\%$ i.e. 40% above the cost price.

Now, Profit or Loss% = $\left\{+40-x-\frac{40\times x}{100}\right\}\%$

$\Rightarrow \qquad +12\% = \left\{+40-\frac{7x}{5}\right\}\%$

$\Rightarrow \qquad -\frac{7x}{5} = -28$

$\therefore \qquad x = 20$

Alternatively II

Marked price is $\left(\frac{3200}{8000}\times 100\right)\%$ i.e. 40% above the cost price.

Now, (100 + 40)% ⇒ (40 – 12)%

Then, 100% ⇒ 20%

Hence, value of x = 20

Problem 35. The marked price of a chair is 40% more than its cost price and the ratio of percentage discount and percentage profit is 3 : 4. Find the discount percentage.

Sol. Now, Profit or Loss % = $\left\{+40-3x-\frac{40\times 3x}{100}\right\}\% = +4x\%$

[∵ discount : profit = 3 : 4]

$\Rightarrow \qquad +4x\% = (40-4.2x)\%$

$\Rightarrow \qquad 8.2x = 40$

$\therefore \qquad x = 4.88$

Thus, discount = $3x$% = 3 × 4.88% = 14.64%

Alternatively

Proceed through options.

{Let the options be 21%, 14.64%, ...}

(a) 21% ⇒ then profit = 28% as Discount : Profit = 3 : 4.

Now, $\left\{+40-21-\frac{40\times 21}{100}\right\}\% \neq 28\%$

(b) 14.64% ⇒ then profit = 19.52%

Now, $\left\{+40-14.64-\frac{40\times 14.64}{100}\right\}\% = 19.52\%$

Hence, option (b) is correct.

Problem 36. A trader sells the goods using a weight of 800 gram instead of kilogram weight, but he allows a discount of 30%. What is the effective discount that the customers get.

Sol. Trader's gain% when he uses faulty weight

$$= \left(\frac{200}{800} \times 100\right)\% = 25\%$$

$$\text{Effective Discount Rate} = +25 - 30 - \frac{25 \times 30}{100}$$

$$= -12.5\%$$

Hence, required answer = 12.5%

Problem 37. A shopkeeper marks the price of an article 80% above his cost price and allows a discount of $d\%$ while selling and gains .... %. If he allows a discount of $2d\%$, his profit will be ....%. Which of the following values can be filled in the same order?

I. 30, 20 II. 44, 8
III. 36, 16 IV. 60, 40

(a) Only I (b) Only II
(c) Only III (d) I and II only
(e) II and IV only *[IBPS (PO-Mains)-18]*

Sol. For 1st case:

$(100 + 80)\% \Rightarrow (80 - \text{Profit})\%$

Then, $100\% \Rightarrow \left\{\frac{80 - x}{180} \times 100\right\}\%$

(assuming profit is x%)

Hence, discount rate = $\left\{\frac{80 - x}{180} \times 100\right\}\%$

For 2nd case, discount rate

$= \left\{\frac{80 - y}{180} \times 100\right\}\%$ (assuming profit is y%)

Now, we have to check as to whether $2 \times \left\{\frac{80 - x}{180} \times 100\right\}\%$ is equal to $\left\{\frac{80 - y}{180} \times 100\right\}\%$ or not, under the given values.

In brief, whether $2(80 - x)$ is equal to $(80 - y)$ or not. Put the value of x and y, as given, to check the same.

I. $2(80 - 30) \neq (80 - 20)$
II. $2(80 - 44) = (80 - 8)$
III. $2(80 - 36) \neq (80 - 16)$
IV. $2(80 - 60) = (80 - 40)$

Hence, option (e) is correct.

Case V Concept of Difference Between Discounts

Problem 38. The difference between a discount 40% on ₹ 500 and two successive discounts of 30% and 10% on the same amount is *[SSC(CPO)-08]*

Sol. In 1st case, discount = 40%

In 2nd case, equivalent single discount = 37%

i.e. $\left(-30 - 10 + \frac{30 \times 10}{100}\right)$

Hence, difference = 3%

∴ Required answer = 3% of 500 = ₹ 15

Alternatively

If the sum of discounts in given two sets is equal, the difference in their equivalent single discounts will be $\left(\frac{\text{difference in their products}}{100}\right)\%$.

Note *It is applicable only for the sets of two successive discounts. Even a single discount may be written in form of a set. For an example: A discount of 40% may be written as 40% and 0%.*

Here, 1st set = 40% and 0%

and 2nd set = 30% and 10%

Then, required difference

$$= \left\{\frac{(30 \times 10) - (40 \times 0)}{100}\right\}\% = 3\%$$

∴ Difference (in ₹) = 3% of 500 = ₹ 15

Problem 39. The difference between equivalent discounts for two successive discounts of 30% and 20% and that for successive discounts of 40% and 10% on a certain amount in ₹ 50. Then, find that amount.

Sol. Here, the sum of discounts given in both sets are equal. Then, difference in equivalent discounts

$$= \left\{\frac{(30 \times 20) - (40 \times 10)}{100}\right\}\% = 2\%$$

As given, 2% of the amount = ₹ 50

∴ 100% of that amount = ₹ 2500

Hence, required answer = ₹ 2500

Problem 40. The difference between equivalent discount for three successive discounts of 30%, 20% and 10% and that for successive discounts of 40%, 10% and 10% on the marked price is ₹ 180. Then, find that marked price.

Sol. Case I : $\left\{-30 - 20 + \frac{30 \times 20}{100}\right\}\% = -44\%$

Further, $\left\{-44 - 10 + \frac{44 \times 10}{100}\right\}\% = -49.6\%$

Hence, equivalent single discount for 30%, 20% and 10% = 49.6%

Case II : $\left\{-40 - 10 + \frac{40 \times 10}{100}\right\}\% = -46\%$

Further, $\left\{-46 - 10 + \frac{46 \times 10}{10}\right\}\% = -51.4\%$

Hence, equivalent single discount for 40%, 10% and 10% = 51.4%

As given, (51.4 – 49.6)% of the marked price = ₹ 180

Then, 100% of the marked price = 10000

Thus, marked price = ₹ 10000

Case VI **Effect on P/L when rate of discount is changed and *vice versa*.**

Problem 41. When a shopkeeper gives 10% discount on the list price of a toy, his gain is 20%. If he had given a discount of 20%, his percentage of gain would have been *[SSC(10+2/DEO-LDC)-10, SSC(CGL-Mains)-01, SSC(CGL-T-I)-17]*

Sol. Let the list price be ₹ 100

Then, SP = 100 − 10 = 90

$\therefore$ $CP = 90 \times \frac{100}{120} = 75$

{or, 120% of CP = 90 ∴ 100% of CP = 75}

In case of a discount of 20%,

SP = 100 − 20 = 80

But CP = 75 (from the above)

Then, profit % $= \frac{(80-75)}{75} \times 100 = 6\frac{2}{3}\%$

Alternatively I

$$\text{Profit or Loss \%} = \left\{+\text{Mark Up }(x) - 10 - \frac{x \times 10}{100}\right\}\%$$

$$\Rightarrow +20 = \left\{+(x) - 10 - \frac{x \times 10}{100}\right\}$$

$$\Rightarrow \frac{9x}{10} = 30 \Rightarrow x = \frac{100}{3}$$

Hence, % mark up $= \frac{100}{3}\%$

Further, Profit or Loss%

$$= \left\{+\frac{100}{3} - 20 - \frac{\frac{100}{3} \times 20}{100}\right\}\% = \left\{13\frac{1}{3} - 6\frac{2}{3}\right\}\% = +6\frac{2}{3}\%$$

'+' indicates that a profit of $6\frac{2}{3}\%$ is earned.

Alternatively II

For the computation of Mark Up:

(100 − 10)% ⇒ (10 + 20)%

Then, $100\% \Rightarrow \frac{100}{3}\%$

Hence, % mark up $= \frac{100}{3}\%$

$$\text{Now, Profit or Loss\%} = \left\{+\frac{100}{3} - 20 - \frac{\frac{100}{3} \times 20}{100}\right\}\%$$

$$= \left\{13\frac{1}{3} - 6\frac{2}{3}\right\}\% = +6\frac{2}{3}\%$$

'+' indicates that a profit of $6\frac{2}{3}\%$ is earned.

Problem 42. A trader earns a profit of 20% by allowing 10% discount on the marked price of an article. Find the discount% allowed, if he incurs a loss of 20%.

Sol. Profit or Loss%

$$= \left\{+\text{Mark Up }(x) - 10 - \frac{x \times 10}{100}\right\}\%$$

$$\Rightarrow +20 = \left\{+\text{Mark Up }(x) - 10 - \frac{x \times 10}{100}\right\}\%$$

$$\Rightarrow \frac{9x}{10} = 30 \Rightarrow x = \frac{100}{3}\%$$

Hence, % Mark Up $= \frac{100}{3}\%$

Further, Profit or Loss%

$$= \left\{+\frac{100}{3} - \text{discount}(d) - \frac{\frac{100}{3} \times d}{100}\right\}\%$$

$$\Rightarrow -20\% = \left\{+\frac{100}{3} - \text{discount }(d) - \frac{\frac{100}{3} \times d}{100}\right\}\% \Rightarrow -\frac{4d}{3} = -\frac{160}{3}$$

$\therefore$ d = 40

Hence, discount allowed = 40%

Alternatively

(100 − 10)% ⇒ (10 + 20)%

Then, $100\% \Rightarrow \frac{100}{3}\%$

Hence, % Mark Up = 33.33%

Now, for the computation of discount rate:

(100 + 33.33)% ⇒ (33.33 + 20)%

Then, 100% ⇒ 40%

Thus, the discount rate = 40%

Case VII **Concept of Cash Discount**

Cash discount is an allowance/reduction on the amount due by the customers when they pay their bills within a stipulated period of time. The credit terms which comprise the cash discount rate and the credit period are shown in the invoice as below. $\frac{2}{10}, \frac{1}{30}, \frac{n}{60}$

Here, $\frac{2}{10}$ means that the customers will receive 2% discount if they pay this bills within 10 days of the invoice date.

$\frac{1}{30}$ means that the customers will receive 1% discount if they pay this bills within 11th to 30th day of the invoice date.

$\frac{n}{60}$ means that the full amount to be paid by the customer by 60th day of the invoice date, and no discount is allowed from 31st day onwards.

Example:

Problem 43. A trader purchases an item worth ₹ 10,000 at the credit terms: cash 5, $\frac{2}{10}, \frac{1}{30}, \frac{n}{60}$ and pays the bill on 17th day. How much should he fix its marked prices so as to gain 17% profit after allowing 10% discount.

Sol. Clearly, discount is allowed at 1% while purchasing.

Then, his CP = 99% of 10,000 = 9,900

Now, Profit or Loss%

$$= \left\{+ \text{Mark Up } (x) - 10 - \frac{x \times 10}{10}\right\}\%$$

$$\Rightarrow \quad +17 = \left\{+ \text{Mark Up } (x) - 10 - \frac{x \times 10}{100}\right\}\%$$

$$\Rightarrow \quad \frac{9x}{10} = 27 \Rightarrow x = 30\%$$

Thus, the marked price should be 30% more than its CP.

Now, the marked price = 130% of 9900 = ₹ 12870

Alternatively

From above, CP = ₹ 9900

Now, (100 − 10)% ⇒(10 + 17)%

Then, 100% ⇒30%

Hence, marked price should be 30% more than its CP.

∴ Marked price = 130% of 9900 = ₹ 12870

Miscellaneous

Problem 44. Ganu bought an article at a discount of 25% from Manu. He increases its price by ₹ 300 and sells it allowing 20% discount at a profit of ₹ 150. Find the marked price for Manu.

Sol. Let the marked price for Manu be ₹ x.

Then, CP for Ganu = 75% of $x = \frac{3x}{4}$

As given, 80% of $\left\{\frac{3x}{4} + 300\right\} = \frac{3x}{4} + 150$

$$\Rightarrow \quad \frac{3x}{5} + 240 = \frac{3x}{4} + 150$$

$$\Rightarrow \quad \frac{-3x}{20} = -90 \Rightarrow x = 600$$

Alternatively I

Ganu increases the price by ₹ 300 and also allows a discount of 20%. Thus, effective increment = 80% of 300 = 240 which includes a profit of ₹ 150 also. Obviously, ₹ (240 − 150) will be equal to 20% of the CP for Ganu. Then, 100% of the CP for Ganu = ₹ 450.

Hence, marked price for Manu = $450 \times \frac{100}{75} = 600$

Alternatively II

Ganu increases the price by ₹ 300, but he earns only a profit of ₹ 150 as he allows a discount of 20%.

Obviously, 20% of his marked price = ₹ (300 − 150)

Then, his marked price = ₹ 750

Now, his CP = 750 − 300 = 450

Then, marked price for Manu = $450 \times \frac{100}{75} = 600$

Problem 45. ***A* sells a TV to *B* at a discount of 25% on the marked price. Further, *B* sells it at a discount of 20% after increasing the cost by 40% and in doing so he earns ₹ 180 as profit. Find the initial marked price.**

Sol. Let the initial marked price of TV be ₹ x.

Then, CP for B = 75% of $x = \frac{3x}{4}$

Marked price for B = 140% of $\frac{3x}{4} = \frac{21x}{20}$

Now, SP for B = 80% of $\frac{21x}{20} = \frac{21x}{25}$

As given, $\frac{21x}{25} = \frac{3x}{4} + 180$

$$\Rightarrow \quad \frac{9x}{100} = 180 \Rightarrow x = 2000$$

Alternatively

Profit or loss% of $B = \left\{+40 - 20 - \frac{40 \times 20}{100}\right\}\% = +12\%$

Hence, B earns a profit of 12%.

As given, 12% of CP i.e. profit = ₹ 180

∴ 100% of CP = ₹ 1500

Hence, CP of B = ₹ 1500

Obviously, ₹ 1500 is a price arrived after a discount of 25%.

Hence, the initial marked price

$$= 1500 \times \frac{100}{75} = 2000$$

[or, ∵ 75% = 1500 ∴ 100% = 2000]

Problem 46. The price of a product is P. A shopkeeper raises its price by $X\%$ and then offers a discount of $Y\%$ on the raised price. The discounted price again becomes P. If Y is the difference between X and Y, then find X. ***[XAT-18]***

Sol. Since the discounted price is equal to the CP of the product, neither loss is incurred nor gain is earned.

Now, Profit or Loss% = $\left\{X - Y - \frac{X \times Y}{100}\right\}\%$

$$\Rightarrow \quad 0\% = \left\{+X - Y - \frac{X \times Y}{100}\right\}\%$$

$$\Rightarrow \quad Y - X = -\frac{X \times Y}{100}$$

$$\Rightarrow \quad \frac{X}{2} - X = -\frac{X \times \frac{X}{2}}{100} \quad \{\because Y = X - Y \Rightarrow Y = X/2\}$$

$$\Rightarrow \quad X = 100$$

Alternatively

Proceed through options. Let options be 25, 100, 20, 50.

Recall the concept associated with question: 'a number is increased by x%, by what percentage should the increased number be reduced so as to get the original number.

(a) 25% ⇒ Required reduction = $\frac{25 \times 100}{125}\% = 20\%$

But 25 − 20 ≠ 20

(b) 100% ⇒ Required reduction = $\frac{100 \times 100}{200}\% = 50\%$

and 100 − 50 = 50

Hence, option (b) is correct.

WORKOUT

1. A manufacturer marks his goods at 40% above the cost price. He allows a discount of 10% for the cash customers and 5% to the credit customers. $\frac{3}{5}$th of the goods are sold for cash and rest on credit. What is the percentage of profit when all the goods are sold and the amount realized.
   (a) 28.8% (b) 10%
   (c) 12.5% (d) 15.6%

2. A shopkeeper allows a discount of 10% on the marked price of an item but charges a sales tax of 8% on the discounted price. If the customer pays ₹ 680.40 as the price including the sales tax, then what is the marked price of the item? *[MAT-14, 19]*
   (a) ₹ 630
   (b) ₹ 700
   (c) ₹ 780
   (d) None of the above

3. A dishonest trader sells the goods using a weight of 800 gm instead of kilogram weight, but he allows a discount of 30%. What is his percentage profit if he marks up his goods by 40%.
   (a) 22.5% (b) 10%
   (c) 12.5% (d) 15%

4. As a strategy towards retention of customers, the service centre of a split AC machine manufacturer offers discount as per following rule: for the second service in a year, the customer can avail of a 10% discount, for the third and fourth servicing within a year, the customer can avail of 11% and 12% discounts respectively of the previous amount paid finally, if a customer gets more than four services within a year, he has to pay just 55% of the original servicing charges. If Rohan has availed 5 services from the same service centre in a given year, the total percentage discount availed by him is approximately *[IIFT-15]*
   (a) 16.52 (b) 20.88
   (c) 22.33 (d) 24.08

5. In respect of a bill of ₹ 10000, what is the difference between a discount of 40% and two successive discounts of 36% and 4%?
   (a) ₹ 0 (b) ₹ 144 (c) ₹ 256 (d) ₹ 400

6. Which is the profitable bargain for a consumer:
   (i) The successive discounts of 20% and 15%.
   (ii) The successive discounts of 10% and 25%.
   (a) 1st offer (b) 2nd offer
   (c) both offers (d) we cannot say

7. Which is more favourable for buyer. A discount series of 30%, 20% and 10% or a discount series of 40%, 10% and 10%?
   (a) 1st offer (b) 2nd offer
   (c) both offers (d) we cannot say

8. A merchant bought some goods at a discount of 20% of the list price. He wants to mark them at such price that he can give a discount of 20% of the marked price and still makes a profit of 20% of the selling price. The percent of the list price at which he should mark them is *[FMS-11]*
   (a) 20 (b) 100 (c) 125 (d) 80

9. One saree was purchased for ₹ 564 after getting a discount of 6% and another saree was purchased for ₹ 396 after getting a discount of 1%. Taking both the items as a single transaction, what is the percentage of discount? *[CDS-10]*
   (a) 3.5 (b) 4 (c) 7 (d) 7.5

10. A book is listed at ₹ 150, with a discount of 20%. What additional discount must be offered to bring the net price to ₹ 108. *[LIC(AAO)-09]*
    (a) 8% (b) 10% (c) 12.5% (d) 15%

11. A shopkeeper cheats to the extent of 10% while buying as well as selling, by using false weights. Also, he marks up the price of his goods by 40%. Find his total gain% if he allows a discount of 20%.
    (a) 22.5% (b) 36.9%
    (c) 32.5% (d) 15%

12. If commission of 10% is given to a customer on the marked price of an article, the gain of the trade is 20%. What will be the gain percentage of the trader if the commission is increased to 15%?
    (a) 22.5% (b) 10.5%
    (c) 13.33% (d) 16%

13. A man buys a television set which lists for ₹ 5000 at 10% discount. He gets an additional 2% discount (after the first discount) for paying cash. What does he actually pay for the set? *[CDS-13, CLAT-16]*
    (a) ₹ 4410 (b) ₹ 4400
    (c) ₹ 4000 (d) ₹ 4500

14. A garment company declared 15% discount for wholesale buyers. Mr. Sachdev bought garments from the company for ₹ 25000 after getting discount. He fixed up the selling price of garments in such a way that he earned a profit of 8% on the original company price. What is the approximate total selling price? *[MAT-10]*
    (a) ₹ 28000 (b) ₹ 29000
    (c) ₹ 31000 (d) ₹ 29500

15. In a sale, a perfume is available at a discount of 15% on the selling price. If the perfume's discounted selling price is ₹ 3675.40, what was the original selling price of the perfume? *[PNB(MT)-09]*
    (a) ₹ 4354 (b) ₹ 4824
    (c) ₹ 3324 (d) ₹ 4324
    (e) None of these

16. A pharmaceutical company manufactures 6000 strips of prescribed diabetic drugs for ₹ 800000 every month. In July, 2014, the company supplied 600 strips of medicines free to the doctors at various hospitals. Of the remaining medicines, it was able to sell 4/5th of the strips at 25% discount and the balance at the printed price of ₹ 250.

Assuming vendor's discount at the rate of a uniform 30% of the total revenue, the approximate percentage profit/ loss of the company in July, 2014 is: *[IIFT-14]*

(a) 5.5% profit (b) 4% loss
(c) 5.5% loss (d) None of these

17. The marked price of a machine is ₹ 18000. By selling it at a discount of 20%, the loss is 4%. What is the cost price of the machine?

(a) ₹ 10000 (b) ₹ 12000
(c) ₹ 14000 (d) ₹ 15000

18. An article is listed at ₹ 65. A customer bought this article for ₹ 56.16 with two successive discounts of which one is 10%. The other discount of this discount scheme that was allowed by the shopkeeper is *[MAT-10, 05]*

(a) 4% (b) 3% (c) 6% (d) 2.5%

19. Naresh purchased a TV set for ₹ 11,250 after getting discount of 10% on the labelled price. He spent ₹ 150 on transport and ₹ 800 on isntallation. At what price should it be sold so that the profit earned would be 15% if no discount was offered?

(a) ₹ 12937.5 (b) ₹ 14030
(c) ₹ 13450 (d) ₹ 15467.5
(e) None of these

20. The marked price of a piano was ₹ 15000. At time of sale, there were successive discounts of 20%, 10% and 10% respectively on it. The sale price was*[SSC(CGL-T-II)-13]*

(a) ₹ 9720 (b) ₹ 9750
(c) ₹ 9760 (d) ₹ 9780

21. Instead of a meter scale, a cloth merchant uses a 120 cm scale while buying, but uses an 80 cm scale while selling the same cloth. If he offers a discount of 20% on cash payment, what is his overall profit percentage? *[CAT-96]*

(a) 20% (b) 25% (c) 40% (d) 15%

22. By giving 25% discount a trader earns 25% profit. If he sells the item at 10% discount, what is his profit?

(a) 10% (b) 40% (c) 45% (d) 50%

23. By selling an article at 80% of its marked price, a merchant makes a loss of 12%. What will be the percent profit or loss made by the merchant if he sells the article at 95% of its marked price? *[MAT-10]*

(a) 5.5% profit (b) 1% loss
(c) 5% profit (d) 4.5% profit

24. An article is listed at ₹ 900 and two successive discounts of 8% and 8% are given on it. How much would the seller gain or loss, if he gives a single discount of 16% instead of two discounts?

(a) Gain of ₹ 4.76 (b) Loss of ₹ 5.76
(c) Gain of ₹ 5.76 (d) Loss of 4.76

25. Manoj sold an article for ₹ 15000. Had he offered a discount of 10% on the selling price, he would have earned a profit of 8%. What is the cost price? *[Andhra Bank (PO)-09]*

(a) ₹ 12500 (b) ₹ 13500
(c) ₹ 12250 (d) ₹ 13250
(e) None of these

26. Even after reducing the marked price of a transistor by ₹ 32, a shopkeeper makes a profit of 15%. If the cost price be ₹ 320, what percentage of profit would he have made if he had sold the transistor at the marked price?

(a) 25% (b) 20%
(c) 10% (d) None of these

27. The maximum retail price (MRP) of a product is 55% above its manufacturing cost. The product is sold through a retailer, who earns 23% profit on his purchase price. What is the profit percentage (expressed in nearest integer) for the manufacturer who sells his product to the retailer? The retailer gives 10% discount on MRP. *[XAT-15]*

(a) 31% (b) 22% (c) 15% (d) 13%
(e) 11%

28. Tarun bought a TV with 20% discount on the labelled price. Had he bought it with 25% discount, he would have saved ₹ 500. At what price did he buy the TV? *[SSC(CGL-T-II)-11]*

(a) ₹ 7500 (b) ₹ 8500
(c) ₹ 8000 (d) ₹ 7400

29. A dealer offers a cash discount of 20% and still makes a profit of 20%, when he further allows 16 articles to a dozen to a particularly sticky bargainer. How much percent above the cost price were his wares listed? *[CAT-94]*

(a) 100% (b) 80% (c) 75% (d) $66\frac{2}{3}\%$

30. Two shopkeepers announce the same price of ₹ 700 for a shirt. The first offers successive discounts of 30% and 6% while the second offers successive discounts of 20% and 16%. The shopkeeper that offers better discount is more of

(a) ₹ 22.40 (b) ₹ 16.80
(c) ₹ 9.80 (d) ₹ 36.40

31. A trader fixed the price of an article in such a way that by giving a rebate of 10% on the price fixed he made a profit of 15%. If the cost of the article is ₹ 72, the price fixed on it is *[IAS(Pre.)-02, SSC(SAS)-10, SSC(CGL-T-II)-10, CLAT-12]*

(a) ₹ 82.80 (b) ₹ 90.00
(c) ₹ 92.00 (d) ₹ 97.80

32. If on a marked price, the difference of selling prices with a discount of 30% and two successive discounts of 20% and 10% is ₹ 72, then the marked price is

(a) ₹ 7200 (b) ₹ 5400
(c) ₹ 3600 (d) ₹ 1800

33. Kanchan bought a clock with 25% discount on marked price. She sold it with 75% gain on the price she bought. What was her profit percent on the marked price? *[SSC(CGL-T-I)-17]*

(a) 31.25 (b) 50 (c) 56.25 (d) 60

34. On a television of brand *A* the discount is 25% and on television of brand *B* the discount is 40%. The price of *B* after discount is ₹ 2250 greater than the price of *A* after discount. What is the marked price of *A* (in ₹) if the marked price of *B* is ₹ 35000? *[SSC(CGL-T-II)-17]*

(a) 18750 (b) 21000
(c) 25000 (d) 17850

35. A tradesman gives 4% discount on the marked price and 1 article free with every 15 articles bought and still gains 35%. The marked price is more than the cost price by

(a) 16.67% (b) 20%
(c) 50% (d) 33.33%

36. The selling price of an article is ₹ 816 if the discount on it is 15%. What would be the selling price of the article if the discount on it is 25%? *[SSC(CGL-T-II)-17]*

(a) ₹ 750 (b) ₹ 720
(c) ₹ 800 (d) ₹ 700

37. By selling an article at 3/4th of the marked price, there is a gain of 25%. The ratio of the marked price and the cost price is

(a) 5 : 3 (b) 3 : 5
(c) 3 : 4 (d) 4 : 3

38. A ₹ 750 tin of cheese is offered at 8% discount and a ₹ 1250 tin of butter at 20% discount. If we buy 5 tins of cheese and 3 tins of butter, what is the effective discount we get (in %)? *[SSC(CGL-T-II)-17]*

(a) 12% (b) 15% (c) 14% (d) 16%

39. What is the maximum percentage discount that a merchant can offer on her marked price so that she ends up selling at no profit no loss, if she had initially marked her goods up by 50%?

(a) 16.67% (b) 20%
(c) 50% (d) 33.33%

40. If 60% discount is offered on the marked price, the selling price of an article becomes equal to its cost price. What is % mark up? *[SSC(CGL-T-II)-17]*

(a) 100 (b) 250 (c) 150 (d) 40

41. A shop, which sold same marked price shirts, announced an offer-if one buys three shirts then the fourth shirt is sold at a discounted price of ₹ 100 only. Patel took the offer. He left the shop with 20 shirts after paying ₹ 20000. What is the marked price of a shirt? *[XAT-17]*

(a) ₹ 1260 (b) ₹ 1300
(c) ₹ 1350 (d) ₹ 1400
(e) ₹ 1500

42. An article is marked 40% above the cost price. If discount of x% is given on the marked price of the article, then final profit of 12% is obtained. Now, if CP of a new article is ₹ 120 and x% profit is desired, then what should be the selling price of that new article? *[SBI(PO)-17]*

(a) ₹ 140 (b) ₹ 142
(c) ₹ 144 (d) ₹ 146
(e) ₹ 148

43. The marked price of an article is ₹ 740. After two successive discount of 15% and x%, it is sold for ₹ 566.10. What is the value of x? *[SSC(CGL-T-I)-19]*

(a) 12 (b) 5
(c) 10 (d) 20

44. Manu marks the price of an article ....% above his cost price and allows a discount of d% while selling and gains 20%. If he marks ....% above the cost price and allows a discount of $2d$%, his profit percent remains the same. Which of the following values can be filled in the same order?

I. 60, 100 II. 60, 140
III. 36, 72 IV. 60, 120

(a) Only I (b) Only II
(c) Only IV (d) I and II
(e) II and III

Answer key

1. (a)	2. (b)	3. (a)	4. (b)	5. (b)	6. (b)	7. (b)	8. (c)	9. (b)	10. (b)
11. (b)	12. (c)	13. (a)	14. (c)	15. (d)	16. (c)	17. (d)	18. (a)	19. (d)	20. (a)
21. (a)	22. (d)	23. (d)	24. (b)	25. (a)	26. (a)	27. (d)	28. (c)	29. (a)	30. (c)
31. (c)	32. (c)	33. (a)	34. (c)	35. (c)	36. (b)	37. (a)	38. (c)	39. (d)	40. (c)
41. (b)	42. (c)	43. (c)	44. (b)						

Hints and SOLUTIONS

1. Profit or loss % in case of cash customers

$$= \left\{+40 - 10 - \frac{40 \times 10}{10}\right\}\% = +26\%$$

and Profit or loss% in case of credit customers

$$\left\{+40 - 5 - \frac{40 \times 5}{10}\right\}\% = +33\%$$

Now, overall profit%

$$= \frac{3}{5}(+26\%) + \frac{2}{5}(+33\%) = 28.8\%$$

2. Effect on marked price of the item

$$= \left\{-10 + 8 - \frac{10 \times 8}{100}\right\}\% = -2.8\%$$

Hence, the customer bought the item at 2.8% less than the marked price (MP).

Now (100 − 2.8)% of MP = ₹ 680.40

∴ 100% of MP = 700

Thus, option (b) is correct.

3. Profit or Loss%

$$= \left\{+40 - 30 - \frac{40 \times 30}{100}\right\}\% = -2\%$$

'–' indicates that a loss of 2% is incurred.

Further, Trader's gain% when he uses faulty weight

$$= \left(\frac{200}{800} \times 100\right)\% = 25\%$$

Then, his profit or loss%

$$= \left\{+25 - 2 - \frac{25 \times 2}{100}\right\}\% = +22.5\%$$

Hence, his overall profit % is 22.5%.

4. Let the original service charges be ₹ 100. Amount paid by Rohan

For 1st service = 100

For 2nd service = 90% of 100 = 90

For 3rd service = 89% of 90 = 80.10

For 4th service = 88% of 80.10 = 70.50 (approx.)

For 5th service = 55% of 100 = 55

∴ Total amount paid

= 100 + 90 + 80.10 + 70.50 + 55

= 395.60

Then, discount availed

= 500 − 395.6 = 104.4

∴ Discount% = $\left(\frac{104.4}{500} \times 100\right)$%

= 20.88% (approx.)

5. In case of two sets of two successive discounts, if the sum of discounts in given sets is equal, the difference in their equivalent single discounts will be $\left(\frac{\text{difference in their products}}{100}\right)$%.

Here, 1st set = 40% and 0%

and 2nd set = 36% and 4%.

Then, required difference

$= \left\{\frac{(36 \times 4) - (40 \times 0)}{100}\right\}\% = 1.44\%$

∴ Difference in ₹ = 1.44% of 10000

= ₹ 144

6. We are aware of the fact that in case of the sets of two successive discounts, if sum of values of discounts of different sets is equal, the set having the highest range will be the best for the customer.

Here, range of 1st bargain

= 20 − 15 = 5

and Range of 2nd bargain

= 25 − 10 = 15

Since 2nd bargain has greater range, it is the best/profitable for the customers.

7. Case I

$\left\{-30 - 20 + \frac{30 \times 20}{100}\right\}\% = -44\%$

Further,

$\left\{-44 - 10 + \frac{44 \times 10}{100}\right\}\% = 49.6\%$

Hence, equivalent single discount for 30%, 20% and 10% = 49.6%

Case II

$\left\{-40 - 10 + \frac{40 \times 10}{100}\right\}\% = -46\%$

Further,

$\left\{-46 - 10 + \frac{46 \times 10}{100}\right\}\% = -51.4\%$

Hence, equivalent single discount for 40%, 10% and 10% = 51.4%

Obviously, 2nd discount series is more favourable for the buyer.

8. Profit is equal to 20% of the selling price, as given. It means profit is $\left\{\frac{20}{(100-20)} \times 100\right\}$% i.e. 25% of the cost price.

Now, for the computation of Mark Up:

(100 − 20)% ⇒ (20 + 25)%

Then, 100% ⇒ 56.25%

Hence, the trader should mark 56.25% above the cost price. It is also given that the cost price is 20% less than the initial list price. In other words, the initial list price has been decreased by 20% and further to be increased by 56.25% so as to get desired profit.

Clearly, effect on the initial list price

$= \left\{-20 - 56.25 - \frac{20 \times 56.25}{100}\right\}\%$

= + 25%

It means, the marked price should be 25% more than the initial list price. In other words, the marked price is 125% of the list price.

9. List Price of 1st saree

$= \left\{564 \times \frac{100}{100-6}\right\} = 600$

(or, ∵ 94% of MP = 564
∴ 100% of MP = 600)

Hence, discount on 1st saree

= 600 = 564 = 36

and List price of 2nd saree

$= 396 \times \frac{100}{(100-1)} = 400$

(or, ∵ 99% of MP = 396
∴ 100% of MP = 400)

Hence, discount on 2nd saree

= 400 − 396 = 4

Total list price = 600 + 400 = 1000

and total discount = 36 + 4 = 40

Then, overall discount%

$= \left(\frac{40}{1000} \times 100\right)\% = 4\%$

10. 150 …… (−20%) …… → 120 …… (−x%) …… → 108

Obviously, (100 − x)% of 120 = 108

$\Rightarrow \quad 120 \times \frac{(100-x)}{100} = 108$

⇒ 100 − x = 90 ⇒ x = 10

Hence, the additional discount of 10% should be allowed.

Note you may also proceed through options.

11. Profit or Gain% while using false weights

$= \left\{\frac{\text{actually weight or length acquired} - \text{actually weight or length sold}}{\text{actually weight or length sold}} \times 100\right\}\%$

Here, the shopkeeper acquires 1100 gm instead of 1000 gm and sells 900 gm in lieu of 1000 gm.

Hence, Profit%

$= \left\{\frac{1100 - 900}{900} \times 100\right\}\%$

$= \frac{200}{9}\% = 22.22\%$

Further, his profit or loss% w.r.t mark up and discount

$= \left\{+40 - 20 - \frac{40 \times 20}{100}\right\}\% = +12\%$

Now, his overall profit %

$= \left\{+22.22 + 12 + \frac{22.22 \times 12}{100}\right\}\%$

= 36.9%

12. For the computation of Mark Up:

(100 − 10)% ⇒ (10 + 20)%

Then, 100% ⇒ $\frac{100}{3}$%

Hence, % mark up = $\frac{100}{3}$%

Now, Profit or Loss%

$= \left\{+\frac{100}{3} - 15 - \frac{\frac{100}{3} \times 15}{100}\right\}\%$

$= \left\{33\frac{1}{3} - 20\right\}\%$

$= +13\frac{1}{3}\%$ or 13.33%

'+' indicates that a profit of 13.33% is earned.

13. The amount he actually paid = 98% of 90% of 5000 = ₹ 4410

14. Given, 85% of the company price

= ₹ 25000

Then, 108% of the company price

= ₹ 31764.71

Thus, option (c) is correct.

15. Given, 85% of the selling price

= ₹ 3675.40

Then, 100% of the selling price

= ₹ 4324

16. Revenue by selling 4/5 of (6000 − 600) strips at ₹ (75% of 250) per strip

= 4320 × 187.5 = 810000

and revenue by selling 1/5 of (6000 – 600) at ₹ 250 per strip

$= 1080 \times 250 = 270000$

Total revenue

$= 810000 + 270000 = 1080000$

Net revenue generation

$= 70\%$ of $1080000 = 756000$

Hence, loss

$= 800000 - 756000 = 44000$

$\therefore$ Loss % $= \left(\frac{44000}{800000} \times 100\right)\% = 5.5\%$

17. For computation of % Mark Up:

$(100 - 20)\% \Rightarrow (20 - 4)\%$

$\therefore \quad 100\% \Rightarrow 20\%$

Hence, % Mark Up = 20%

Further, 120% of CP = 18000

$\therefore \quad$ 100% of CP = 15000

Thus, option (d) is correct.

18. $65 \cdots\cdots (-10\%) \cdots\cdots \rightarrow 58.5 \cdots\cdots (-x\%) \cdots\cdots \rightarrow 56.16$

Obviously, $(100 - x)\%$ of $58.5 = 56.16$

$\Rightarrow \quad 58.5 \times \frac{(100 - x)}{100} = 56.16$

$\Rightarrow \quad 5850 - 58.5x = 5616$

$\therefore \quad x = 4$

Hence, the second discount is 4%.

19. Given, 90% of the labelled price

= ₹ 11250

Then, 100% of the labelled price

= ₹ 12500

Hence, the cost price of TV when no discount is offered = 12500

Now, total cost

$= 12500 + 150 + 800 = 13450$

$\therefore$ His desired selling price

$= 115\%$ of $13450 = 15467.50$

20. Selling price

$= 15000 \times \frac{80}{100} \times \frac{90}{100} \times \frac{90}{100} =$ ₹ 9720

21. Profit or Gain% while using false meter scale

$= \left\{\frac{\text{actually length acquired} - \text{actually length sold}}{\text{actually length sold}} \times 100\right\}\%$

Here, the merchant acquires 120 cm instead of 100 cm and sells 80 cm in lieu of 100 cm.

Hence, profit%

$= \left\{\frac{120 - 80}{80} \times 100\right\}\% = 50\%$

Then, his profit or loss% after allowing discount

$= \left\{+50 - 20 - \frac{50 \times 20}{100}\right\}\%$

$= +20\%$

22. For the computation of Mark Up:

$(100 - 25)\% \Rightarrow (25 + 25)\%$

Then, $100\% \Rightarrow \frac{200}{3}\%$

Hence, % Mark Up $= \frac{200}{3}\%$

Now, Profit or Loss%

$= \left\{+\frac{200}{3} - 10 - \frac{\frac{200}{3} \times 10}{100}\right\}\%$

$= \left\{56\frac{2}{3} - 6\frac{2}{3}\right\}\% = +50\%$

'+' indicates that a profit of 50% is earned.

23. For the computation of Mark Up:

$(100 - 20)\% \Rightarrow (20 - 12)\%$

Then, $100\% \Rightarrow 10\%$

Hence, % Mark Up = 10%

Now, Profit or Loss %

$= \left\{+10 - 5 - \frac{10 \times 5}{100}\right\}\%$

$= +4.5\%$

'+' indicates that a profit of 4.5% is earned.

24. In case of two sets of two successive discounts, if the sum of discounts in given sets is equal, the difference in their equivalent single discounts will be $\left(\frac{\text{difference in their products}}{100}\right)\%$.

Here, 1st set = 16% and 0%

and 2nd set = 8% and 8%.

Then, required difference

$= \left\{\frac{(8 \times 8) - (16 \times 0)}{100}\right\}\% = 0.64\%$

$\therefore$ Difference (in ₹) = 0.64% of 900

= ₹ 5.76

Further, since 1st set has greater range, it will be profitable for the customers. Conversely, the seller will be in relatively loss if he allows 1st set of discounts.

Hence, he is incurring a loss of ₹ 5.76.

25. For computation of % Mark Up:

$(100 - 10)\% \Rightarrow (10 + 8)\%$

Then, $100\% \Rightarrow 20\%$

Hence, Mark Up is 20%.

Obviously, 120% of the cost price

= 15000

$\therefore$ 100% of the cost price = 12500

26. If the reduction/discount had not been allowed, his additional profit would have been $\left(\frac{32}{320} \times 100\right)\% = 10\%$

Hence, his overall profit%

$= (15 + 10)\% = 25\%$.

27. Let the manufacturing cost be ₹ 100.

Then, MRP = 155

Further, SP for the retailer = 90% of 155 = 139.5, which includes 23% profit too.

Now, 123% of CP for the retailer

= 139.5

$\therefore$ 100% of CP for the retailer

= 113.41, which is SP of the manufacturer.

Hence, profit% of the manufacturer

$= (113.41 - 100)\% = 13.41\%$

$\therefore$ Option (d) is correct.

28. Given (25 – 20)% of the marked price

= ₹ 500

$\therefore$ 100% Marked price = 10000

Then, his cost price

= 80% of 10000 = ₹ 8000

29. The dealer allows 16 articles to a dozen, it implies that he is giving a discount of 25%.

Now, single discount equivalent to successive discounts of 20% and 25% = 40%

$\left\{\because -20 - 25 + \frac{20 \times 25}{100} = -40\right\}$

Now, for the computation of Mark Up:

$(100 - 40)\% \Rightarrow (40 + 20)\%$

Then, $100\% \Rightarrow 100\%$

Hence, % Mark Up = 100%

$\therefore$ Option (a) is correct.

30. In case of two sets of two successive discounts, if the sum of discounts in given sets is equal, the difference in their equivalent single discounts will be $\left(\frac{\text{difference in their products}}{100}\right)\%$.

Here, 1st set = 30% and 6%

and 2nd set = 20% and 16%.

Then, required difference

$= \left\{\frac{(20 \times 16) - (30 \times 6)}{100}\right\}\%$

$= 1.4\%$

$\therefore$ Difference in ₹ = 1.4% of 700

= ₹ 9.80

31. For computation of Mark Up:

$(100-10)\% \Rightarrow (10+15)\%$.

Then, $100\% \Rightarrow \frac{250}{9}\%$

Hence, he should fix the price $\frac{250}{9}\%$ above the cost price.

$\therefore$ Price fixed $= \left(100+\frac{250}{9}\right)\%$ of 72

= ₹ 92

32. In case of two sets of two successive discounts, if the sum of discounts in given sets is equal, the difference in their equivalent single discounts will be $\left(\frac{\text{difference in their products}}{100}\right)\%$.

Here, 1st set = 30% and 0%

and 2nd set = 20% and 10%.

Then, required difference

$= \left\{\frac{(20\times 10)-(30\times 0)}{100}\right\}\%$

= 2%

Given that $(30-28)\%$ of Marked Price = 72

$\therefore$ 100% of marked price = 3600

33. Read this question like 'A number is decreased by 25% then increased by 75%, what will be effect on the original number'.

The profit percent

$= -25+75-\frac{25\times 75}{100} = 31.25\%$

34. As given, $(100-40)\%$ of the marked price of B is 2250 greater than $(100-25)\%$ of the marked price of A.

$\Rightarrow$ 60% of 35000

= 75% of marked price of A + 2250

$\Rightarrow$ 75% of marked price of A = 18750

$\Rightarrow$ Marked price of A = ₹ 25000

35. Discount rate in context of giving 1 article free with every 15 articles bought $= \left(\frac{1}{15+1}\times 100\right)\% = 6.25\%$

Now, we will find the equivalent single discount of discounts 4% and 6.25%.

$\left\{-6.25-4+\frac{6.25\times 4}{100}\right\}\% = -10\%$

Hence, equivalent single discount

= 10%

Now, for computation of % Mark Up:

$(100-10)\% \Rightarrow (10+35)\%$,

Then, $100\% \Rightarrow 50\%$

Thus, the marked price is more than the cost price by 50%.

36. As given, $(100-15)\%$ of the marked price = 816

Then, $(100-25)\%$ of the marked price = ₹ 720

37. Let the marked price be ₹ 100.

Then, SP $= \frac{3}{4}$ of 100 = ₹ 75

Now, 125% of CP = 75

$\therefore$ 100% of CP = 60

Hence, required ratio = 100 : 60 = 5 : 3

Alternatively,

Given that a discount of 25% is given.

For computation of Mark Up:

$(100-25)\% \Rightarrow (25+25)\%$, then 100% $\Rightarrow$ 66.67%

Hence, marked price is 66.67% above the cost price.

Now, required ratio

= (100 + 66.67) : 100 = 5 : 3

38. Since $750\times 5 = 1250\times 3$, we may say that the base numbers are same.

Then, effective discount rate

$= \frac{8+20}{2} = 14\%$

39. Compare this question with a question envisaged in the chapter 'Percentage'. "A number is increased by 50%. By what percent should the increased number be reduced so as to get back the original number?"

Here, the question may be interpreted as:

"The Cost Price is increased by 50%. By what percent should the increased CP (i.e. Marked Price) be reduced/discounted so as to get back the original CP?"

Now, the required answer

$= \left(\frac{50}{(100+50)}\times 100\right)\% = 33.33\%$

40. The SP becomes CP, it means there is 0% profit.

Now, for the computation of Mark Up:

$(100-60)\%$

$\Rightarrow (60-0)\%$

Then, $100\% \Rightarrow 150\%$

Hence, % Mark Up = 150%

41. Patel buys 20 shirts out of which 5 shirts will obviously be ₹ 100 each. Hence, marked price of 15 shirts

= 20000 − 500

Then, marked price of one shirt

= 19500 ÷ 15 = ₹ 1300

42. We have, for computation of discount rate, $(100+40)\% \Rightarrow (40-12)\%$

$\therefore$ $100\% \Rightarrow 20\%$

Hence, $x = 20\%$

Then, SP at a profit of 20% = 120% of 120 = ₹ 144

43. Marked price is 740 and SP is 566.10. Hence, a discount of 23.5% is allowed.

Now, $\left\{-15-x+\frac{15\times x}{100}\right\}\% = -23.5\%$

$\Rightarrow x = 10$

44. For 1st case

$(100+\text{Mark Up})\% \Rightarrow (\text{Mark Up}-20)\%$

Then, $100\% \Rightarrow \left\{\frac{x-20}{100+x}\times 100\right\}\%$

(assuming Mark Up is x%)

Hence, discount rate

$= \left(\frac{x-20}{100+x}\times 100\right)\%$

For 2nd case, discount rate

$= \left\{\frac{y-20}{100+y}\times 100\right\}\%$

(assuming Mark Up is y%)

Now, we have to check as to whether $2\times\left\{\frac{x-20}{100+x}\times 100\right\}\%$ is equal to $\left\{\frac{y-20}{100+y}\times 100\right\}$ or not, under the given values.

In brief, whether $2\left\{\frac{x-20}{100+x}\right\}$ is equal to $\frac{y-20}{100+y}$ or not. Put the values of x and y, as given, and check the same.

I. $2\left\{\frac{60-20}{100+60}\right\} \neq \left\{\frac{100-20}{100+100}\right\}$

II. $2\left\{\frac{60-20}{100+60}\right\} = \left\{\frac{140-20}{100+140}\right\}$

III. $2\left\{\frac{36-20}{100+36}\right\} \neq \left\{\frac{72-20}{100+72}\right\}$

IV. $2\left\{\frac{60-20}{100+60}\right\} \neq \left\{\frac{120-20}{100+120}\right\}$

Hence, option (b) is correct.

21 Simple Interest

Learning OBJECTIVES

After completing this chapter, aspirants will be able to :

- Understand the significance of this chapter in competitive exams, in field of finance/commerce etc. and our daily life as well;
- Understand the diverse terminology related to interest like Principal (P), Amount (A), Simple Interest (SI), Compound Interest (CI), etc.;
- Understand how to apply the concept of percentage like 'Product Constancy', 'Effect on product in case of changes in factor(s)', etc. in this chapter as this chapter is the extension of application of percentage;
- Differentiate between Simple Interest and Compound Interest;
- Analyse the different approaches for solution of a question and will also be able to choose the best one for the exams;
- Realise themselves competent to solve the questions other than the conventional way;
- Understand how Simple Interest/ Amount alters in a defined way when rate of interest or time or both change(s);
- Appreciate the diverse cases under one approach;
- Recognise the concept of Simple Interest in different contexts like Increment in Principal , Increment in Amount, etc.;
- Comprehend the concept of Simple Interest in various types of transactions;
- Recognise how one concept works in different sorts of problems;
- Apply the rule of Alligations in this chapter;
- Use the concept of ratio in this chapter;
- Realise the significance of generic version of the questions;
- Recognise the types of questions which can be easily solved through the given options. Analyse the trend of the questions being asked in various competitive examinations;
- Solve the questions in moment;
- Grip on concepts of helping to solve the questions without pen and paper;
- Learn new concepts/fundamentals;
- Apply the pragmatic approaches in lieu of academic approaches; and
- Ponder on the alternative methods.

Inside

Introduction

No doubt, it is also an extension of application of the percentage and this chapter has utility in our financial life as well as in competitive exams like CAT, MAT, SSC, Railway, Banks, Armed forces, etc. Here, like other previous topics, you will also realise that the concepts envisaged in this topic would be different in comparison to those given elsewhere, the reason is an immense understanding of this topic adhering to one central concept.

Needless to reiterate that the understanding of the concept of percentage will lead you to secure good marks in this topic and it will not be out of context to say that 95% of the questions based on SI can be solved through little mental exercise only. It is desirable to define some critical terminology associated with this topic before proceeding on the diverse approaches.

Terminology Pertaining to Interest

(A) **Principal Money (P)** The amount of money that is borrowed/lent out.

(B) **Time (T)** The duration for which money is borrowed/lent out.

(C) **Interest (I)** The extra money,that will be paid or received, as the case may be, for the use of the principal money after a certain period is called 'Interest'.

(D) **Rate of Interest (R)** The rate at which interest is charged on principal money.

(E) **Amount (A)** The sum of the principal money and the interest accrued thereon. Hence, amount = principal money + interest

(F) **Simple Interest (SI)** The interest when it is computed uniformly only on principal money, is called 'Simple Interest'.

(G) **Compound Interest (CI)** The interest of the previous year(s)/ duration is added to the principal money while computing the 'Compound Interest'.

Now, we will discuss the simple interest only. In next chapter, the concept of compound interest will be discussed in detail.

Take one example for better understanding of the above terms.

For example Find out the SI on ₹ 500 at 6% per annum, for a period of 3 yrs.

Here, principal money (P) = ₹ 500

Time (T) = 3 yrs

Rate of Interest (R) = 6% per annum. Its literally meaning is that interest of ₹ 6 is charged on principal money of ₹ 100 for a period of 1 yr.

Now, come at the question,

$\because$ On principal money of ₹ 100 for 1 yr, the interest is ₹ 6.

$\therefore$ On principal money of ₹ 500 for 1 yr, the interest is ₹ 30.

$\therefore$ On principal money of ₹ 500 for 3 yrs, the interest is ₹ 90.

Hence, SI = ₹ 90

In other words, we may say that the principal money increases by 6% each year and obviously, in three years it will increase by 18%. This increment is nothing but the simple interest.

Hence, SI = 18% of ₹ 500 = ₹ 90.

Generic version

(Rate of interest percentage per annum) × (Time in years)
= Increment in the principal money in term of percentage
= Interest in percentage term of the principal money

Here, interest = $(R \times T)\% = (6 \times 3)\% = 18\%$

Hence, SI = 18% of the principal money
= 18% of 500 = ₹ 90.

Note *Simple interest must always be expressed in term of percentage of the principal money as interest is always calculated on the principal money with regard to simple interest. This is the central theme of this chapter.*

Take More Examples for Better Understanding

Problem 1. Find out the SI on ₹ 800 at 12% per annum for a period of 4 yrs.

Sol. SI = $(12 \times 4)\% = 48\%$

It means SI in terms of percentage of principal = 48%

Hence, SI = 48% of 800 = ₹ 384.

Problem 2. Find out the SI on ₹ 800 at 6% per annum for a period of 73 days.

Sol. Here, rate of interest per annum

$= 6\%$ and time $= \frac{73}{365} = 0.2$ yrs

Then, SI = $(6 \times 0.2)\% = 1.2\%$

Hence, SI = 1.2% of 800 = ₹ 9.60.

Problem 3. In how many years will the SI on ₹ 800 be ₹ 384 at 12% per annum?

Sol. SI = $\left(\frac{384}{800} \times 100\right)\% = 48\%$

(It is understood that SI is 48% of the principal, no need to mention the same again and again)

Now, SI = $(R \times T)\%$

$\Rightarrow$ $48\% = (12 \times T)\% \Rightarrow T = 4$ yrs

Problem 4. Find out the principal money on which SI for 4 yrs at 12% per annum is ₹ 384.

Sol. SI = $(12 \times 4)\% = 48\%$

Given, 48% of the principal money = 384

$\therefore$ 100% of the principal money = 800

Hence, principal money = ₹ 800

Problem 5. At what rate of interest per annum, the SI on ₹ 800 will be ₹ 384 for a period of 4 yrs?

Sol. $SI = \left(\frac{384}{800} \times 100\right)\% = 48\%$

Now, $SI = (R \times T)\%$

$\Rightarrow 48\% = (R \times 4)\%$

$\Rightarrow R = 12$

Hence, rate of interest per annum = 12%.

Problem 6. What amount would a man receive on a principal of ₹ 800 after 4 yrs on simple interest at the rate of 12% per annum?

Sol. SI = (12 × 4)% = 48% of the principal

Then, Amount = (100 + 48)% of the principal

Hence, amount = 148% of 800 = ₹ 1184

Observations We should adopt the above approach as it will help us solve the questions without pen and paper in moments. Further, command over the computation of percentage will save the time in the examination to the greater extent. Now, we will discuss the diverse sorts of questions being asked in various competitive exams.

Computation of R/T when SI is Given in Terms of Principal

Problem 7. The SI on a certain sum of money is 36% of the principal, find out the time, if numerical value of the rate of interest is equal to time in years.

Sol. Given, SI = 36%

$\Rightarrow 36\% = (R \times T)\%$

$\Rightarrow 36\% = (T \times T)\%$

$\Rightarrow T = \pm 6$

Since, T cannot be in negative term, the time is 6 yr.

Problem 8. The simple interest on a sum of money is 4/9 of the principal and the number of years is equal to the rate per cent per annum. The rate per annum is what? **[SSC (CGL-PT)-2000, MAT-13, LIC (AAO)-13]**

Sol. Given, $SI = \left(\frac{4}{9} \times 100\right)\%$

$\Rightarrow (R \times T)\% = \frac{400}{9}\%$

$\Rightarrow (R \times R)\% = \frac{400}{9}\%$

$\Rightarrow R = \pm 6\frac{2}{3}$

Since R cannot be negative, the rate per annum is $6\frac{2}{3}\%$.

Problem 9. If the simple interest for 6 yrs be equal to 30% of the principal, it will be equal to principal after how many years? **[SSC (CPO)-06]**

Sol. As given, 30% of the principal ⇒ 6 yrs

Then, 100% of the principal ⇒ 20 yrs.

Computation of *R*/*T* when the Principal Alters

Problem 10. A certain sum of money amounts to triple itself for a period of 20 yr. Find out the rate of interest per annum.

Sol. SI = Increment % in principal money = 200%

$\Rightarrow 200\% = (R \times T)\%$

$\Rightarrow 200\% = (R \times 20)\%$

$\Rightarrow R = 10$

Hence, rate of interest per annum = 10%.

Problem 11. A certain sum of money amounts to double itself at the rate of 5% per annum simple interest. What is the time period?

Sol. SI = Increment per cent in principal money = 100%

$\Rightarrow 100\% = (R \times T)\%$

$\Rightarrow 100\% = (5 \times T)\% \Rightarrow T = 20$

Hence, time = 20 yr.

Comparison of Increments in Principal with Reference to *R*/*T*

Problem 12. A certain sum of money becomes triple itself in 4 yrs at certain rate of interest. In how many years that sum will become seven times itself at the same rate of interest?

Sol. In first case,

SI = Increment % in principal money = 200%

$\Rightarrow 200\% = (R \times 4)\% \Rightarrow R = 50$

Hence, rate of interest per annum = 50%.

Further, in second case, SI = 600%

$\Rightarrow 600\% = (50 \times T)\% \Rightarrow T = 12$

Thus, required time = 12 yrs.

Alternatively

Increment in principal = 200% ⇒ 4 yrs

Increment in principal = 600% ⇒ 12 yrs

Problem 13. A sum of money amounts to double itself at 5% rate of interest for a certain period of time. At what rate will that sum become 5 times for the same period ?

Sol. Increment in principal = 100% ⇒ 5%

Increment in principal = 400% ⇒ 20%

Hence, required rate of interest = 20%.

Effect on SI when *R* or *T* or both is/are Altered

Problem 14. If rate of interest is increased by 30% and time is decreased by 20%, the simple interest on a certain sum of money increases by ₹ 44. Find the simple interest.

Sol. Recall the concept of "Effect on product in case of changes in factor(s)" detailed in chapter 'percentage'. We have already discussed that SI = (R × T)% of the principal.

Obviously, effect on

$$SI = \left(+30 - 20 - \frac{30 \times 20}{100}\right)\% = +4\%$$

Hence, SI will increase by 4%.

But given, 4% of the SI = ₹ 44

∴ 100% of SI = ₹ 1100.

Problem 15. Arun invested a sum of money at a certain rate of simple interest for a period of four years. Had he invested the same sum for a period of six years, the total interest earned by him would have been 50% more than the earlier interest amount. What was the rate of interest percent per annum? *[Allahabad Bank(PO)-10]*

Sol. It is obvious that time has been increased by 50%, resulting in the SI would be increased by 50% as SI is a product of rate of interest and time with a certain term. However, we cannot find the definite value of rate of interest as it will have indefinite values.

Problem 16. A sum was put at simple interest at a certain rate of interest for 3 yrs. Had it been put at 1% higher rate, it would have fetched ₹ 5100 more. The sum is *[MAT-10, CDS-21]*

Sol. Obviously, the increment in SI = (3 × 1)% of the principal money.

But given, 3% of the principal = ₹ 5100

Then, 100% of the principal ⇒ ₹ 170000.

Problem 17. A certain amount earns simple interest of ₹ 1750 after 7 yrs. Had the interest been 2% more, how much more interest would it have earned?

Sol. Obviously, the increment in SI = (2 × 7)% of the principal money.

As given, 100% of the SI = ₹ 1750

Then, 14% of the principal = ?.

The value of ? cannot be determined unless the value of the principal money is given. Hence, data is inadequate.

Problem 18. A money lender finds that due to fall in the rate of interest from 8% to 7.75%, his yearly income diminishes by ₹ 61.50. His capital is *[LIC(AAO)-09]*

Sol. Obviously, the decrease in SI = ((8 – 7.75) × 1) % of the principal money

But given, 0.25% of the principal = ₹ 61.50

Then, 100% of the principal ⇒ ₹ 24600

Hence, his capital = ₹ 24600.

Applicability of Concept of Product Constancy

Problem 19. Mr. Mehra is planning for higher education expenses of his two sons aged 15 and 12. He plans to divide ₹ 15 lakh in two equal parts and invest in two different plans such that his sons may have access to ₹ 21 lakh each when they reach the age of 21. He is looking for plans that will give him a simple interest per annum. The rates of interest of the plans for his younger son and his elder son should be : *(XAT-13)*

Sol. Principal money invested in name of each son = 15/2 = 7.5 lakh. In case of younger son, 7.5 lakh grow to ₹ 21 lakh when he attains to the age of 21 yrs i.e. in a period of (21 – 12) yrs.

Clearly, interest = $\left[\frac{(21-7.5)}{7.5} \times 100\right]\% = 180\%$

⇒ $(R \times T)\% = 180\%$

⇒ $(R \times 9)\% = 180\%$

⇒ $R = 20$

Hence, rate of interest = 20%.

In case of elder son, 7.5 lakh also grow to ₹ 21 lakhs when he attains to the age of 21yrs i.e. in a period of (21 – 15) yrs.

Clearly, interest = 180%

⇒ $(R \times T)\% = 180\%$

⇒ $(R \times 6)\% = 180\% \Rightarrow R = 30$

Hence, rate of interest = 30%

Alternatively

Since SI is a product of *R* and *T* with a certain term, the concept of Product Constancy will definitely work. Time applicable in case of elder son is 33.33% less than that applicable in case of younger son. It implies that rate applicable in case of elder son will $\left[\frac{33.33}{(100-33.33)} \times 100\right]\%$, i.e. 50% more than the rate applicable for younger son.

It is clear from the above, that *R* in case of younger son is 20%, then *R* in case of elder son will be 50% more than 20, i.e. 30%.

Effect on Amount when *R* or *T* or Both is/are Altered

Problem 20. A sum of ₹ 1800 amounts to ₹ 2000 for a period of 3 yrs. Find out the amount at 2% higher rate of interest.

Sol. Every year the principal money will increase by 2% more. Then, for a period of 3 yrs, it will increase by 6%.

Thus, increment in simple interest = 6% of 1800 = 108

Then, final amount = ₹ 2000 + ₹ 108 = ₹ 2108

Applicability of Different Rates of Interest for Different Time Periods

Problem 21. The rate of simple interest on an amount of money is 6% per annum for the first 2 yrs, 9% for next 5 yrs and 13% for period beyond 7 yrs. If the total interest on a sum at end of ten years is ₹ 7680, what is the sum? *(SIDBI Officers (Exam.)-2000, Punjab & Sind Bank(PO)-11)*

Sol. SI = (6·2 + 9·5 + 13·3)% = 96%

Given, 96% of the principal = ₹ 7680

Then, 100% of the principal = ₹ 8000

Difference in SIs on Single Principal when R is Applicable for Two Different Time Periods or T is Constant for Two Different Values of R.

Problem 22. Two equal sums of money is lent out for 8 yrs and 5 yrs at 6% per annum simple interest. Find each sum if the difference between simple interests is ₹ 270.

Sol. Obviously, difference in SI

$= [(8 \times 6) - (5 \times 6)]\%$

$= 18\%$

Given, 18% of the principal = ₹ 270

Then, 100% of the principal = ₹ 1500

Hence, each sum = ₹ 1500.

Problem 23. Two equal sums of money were invested, one at 4% and another at 4.5%. At the end of 7 yrs, the simple interest received from the latter exceeded that received from the former by ₹ 31.50. Each sum was:-

Sol. Obviously, difference in SI

$= [(7 \times 4.5) - (7 \times 4)]\%$

$= 3.5\%$

Given, 3.5% of the principal = ₹ 31.50

Then, 100% of the principal = ₹ 900

Hence, each sum = ₹ 900

Finding Values or Ratio of Parts Lent/Borrowed at Different/Same Rates for the Same Period and Overall SI is Known

Problem 24. A sum of ₹ 1550 was lent partly at 5% and partly at 8% simple interest. The total interest received after 3 yrs is ₹ 300. The ratio of money lent at 5% to that at 8% is:

Sol. Overall SI $= \left(\frac{300}{1550} \times 100\right)\% = \frac{600}{31}\%$

As given, $(5 \times 3)\%$ of first principal + $(8 \times 3)\%$ of second principal $= \frac{600}{31}\%$ of (first principal + second principal)

$\Rightarrow \quad 15\% \text{ of } P_1 + 24\% \text{ of } P_2 = \frac{600}{31}\% \text{ of } (P_1 + P_2)$

$\Rightarrow \quad \left(24 - \frac{600}{31}\right)\% \text{ of } P_2 = \left(\frac{600}{31} - 15\right)\% \text{ of } P_1$

$\Rightarrow \quad \frac{144}{31}\% \text{ of } P_2 = \frac{135}{31}\% \text{ of } P_1$

$\Rightarrow \quad 144\% \text{ of } P_2 = 135\% \text{ of } P_1$

$\Rightarrow \quad P_1 : P_2 = 144 : 135$

$= 16 : 15$

Alternatively

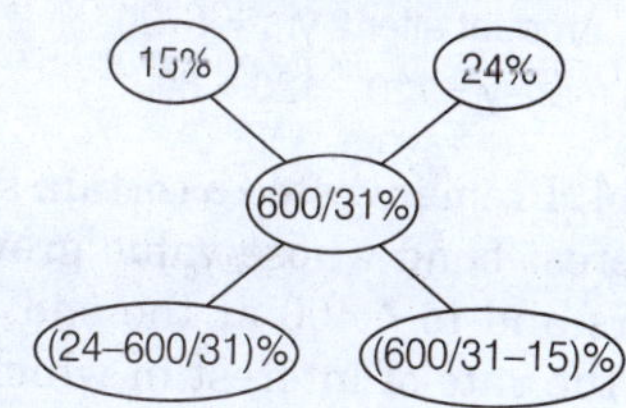

It means, $P_1 : P_2 = \frac{144}{31}\% : \frac{135}{31}\%$

$= 144 : 135 = 16 : 15$

Note *SI in terms of percentage of principal money can be taken into consideration on annum basis.*

Problem 25. Aniket deposited two parts of a sum of ₹ 25000 in different banks at rates of 15% and 18% per annum respectively. If one year he got ₹ 4050 as the total interests, what was the amount deposited at the rate of 18% per annum?

[BSRB-Patna(PO)-01, SSC(CGL-T-II)-12. MAT-10, 13, 14, 19]

Sol. SI $= \left(\frac{4050}{25000} \times 100\right)\% = 16.2\%$

As given, 15% of first principal + 18% of second principal = 16.2% of (first principal + second principal)

$\Rightarrow \quad 15\% \text{ of } P_1 + 18\% \text{ of } P_2 = 16.2\% \text{ of } (P_1 + P_2)$

$\Rightarrow \quad (18 - 16.2)\% \text{ of } P_2 = (16.2 - 15)\% P_1$

$\Rightarrow \quad 1.8\% \text{ of } P_2 = 1.2\% \text{ of } P_1$

$\Rightarrow \quad P_1 : P_2 = 18 : 12 = 3 : 2$

Then, $\quad P_2 = 25000 \times \frac{2}{(3+2)} = 10000.$

Alternatively I

Overall rate of interest $= \left(\frac{4050}{25000} \times 100\right)\% = 16.2\%$

Now,

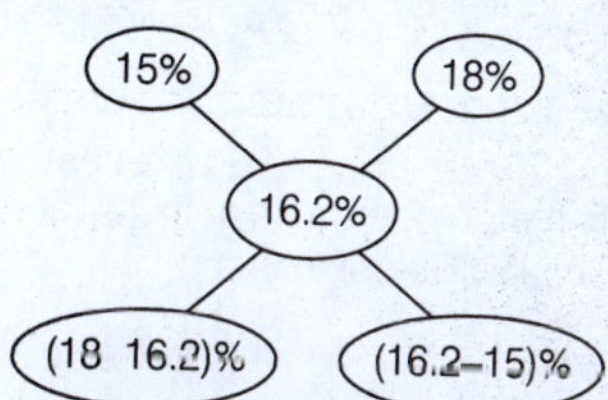

It means, $P_1 : P_2 = 1.8\% : 1.2\% = 3 : 2$

Then, $P_2 = 25000 \times \frac{2}{(3+2)} = 10000.$ (or, (3 + 2) units ⇒ 25000; Then, 2 units ⇒ 10000)

Alternatively II

You may proceed through options also. (let, options be 15000, 10000, ……).

(a) 15000 ⇒ 15% of (25000 − 15000) + 18% of 15000 ≠ 4050

(b) 10000 ⇒ 15% of (25000 − 10000) + 18% of 10000 = 4050

Hence, option (b) is correct.

Problem 26. A sum of ₹ 1000 was lent partly at 16%, partly at 14% and the remaining at 10% simple interest. The total interest received after 5 yrs is ₹ 600. If the sum lent out at 16% be P_1 and that at 10% be P_3, then which of the following is true?

Sol. Overall SI $= \left(\frac{600}{1000} \times 100\right)\% = 60\%$

As given, $(16 \times 5)\%$ of first principal + $(14 \times 5)\%$ of second principal + $(10 \times 5)\%$ of third principal

= 60% of (first principal + second principal + third principal)

$\Rightarrow 80\% \text{ of } P_1 + 70\% \text{ of } P_2 + 50\% \text{ of } P_3 = 60\% \text{ of } (P_1 + P_2 + P_3)$

$\Rightarrow \quad 8P_1 + 7P_2 + 5P_3 = 6(P_1 + P_2 + P_3) \Rightarrow 2P_1 + P_2 = P_3$

Hence, option (a) is correct.

Note *There may be infinite number of set of* $(P_1 : P_2 : P_3)$ *viz.* $(1:1:3), (1:2:4), \ldots\ldots$

Problem 27. A lends ₹ 2500 to B and a certain sum to C at the same time at 7% per annum simple interest. If after 4 yrs, A altogether receives ₹ 1120 as interest from B and C, then the sum lent to C is

Sol. Obviously, the simple interest will be 28% of the total principal money.

As given, 28% of the total principal = ₹ 1120

Then, 100% of the principal = 4000

Since first part is ₹ 2500, then second part is (4000 − 2500).

Hence the sum lent to C is ₹ 1500.

Computation of Values or Ratio of Parts Lent/Borrowed at Different/Same Rates for the Same/Different Period(s) and SI/Amount in Each is Equal

Problem 28. A person invested in all ₹ 2600 at 4%, 6% and 8% per annum simple interest. At the end of the year, he got the same interest in all three cases. The money invested at 4% is : **[SSC(CGL-PT)-03, SSC(CGL-T-I)-17,CDS-19]**

Sol. Given that, 4% of P_1 = 6% of P_2 = 8% of P_3

Then, $P_1 : P_2 : P_3 = \frac{1}{4} : \frac{1}{6} : \frac{1}{8}$

(the logic is that, higher principal × lower rate of interest = lower principal × higher rate of interest)

$\Rightarrow \quad P_1 : P_2 : P_3 = \frac{1}{4} \times 24 : \frac{1}{6} \times 24 : \frac{1}{8} \times 24 = 6 : 4 : 3$

Now, $P_1 = 2600 \times \frac{6}{(6 + 4 + 3)} = 1200$

(or, (6+ 4+ 3) units ⇒ 2600
Then, 6 units ⇒ 1200)

Alternatively

Given that, 4% of P_1 = 6% of P_2 = 8% of P_3

$\Rightarrow \quad P_1 : P_2 = 6 : 4$

and $\quad P_2 : P_3 = 4 : 3$

Then, $\quad P_1 : P_2 : P_3 = 6 : 4 : 3$

Now, (6 + 4 + 3) units ⇒ 2600

∴ 6 units ⇒ 1200.

Problem 29. The rates of simple interest in two banks HDFC and Magadh Gramin Bank are in ratio of 5:6. Find out the ratio of money invested in both banks provided the simple interest after 2 yrs fetched from those are same.

Sol. Given, (5 × 2)% of P_1 = (6 × 2)% of P_2

$\Rightarrow \quad P_1 : P_2 = 12 : 10 = 6 : 5$

Alternatively

Required ratio $= \frac{1}{5 \times 2} : \frac{1}{6 \times 2} = 6 : 5.$

Problem 30. Divide ₹ 2379 into 3 parts so that their amounts after 2yrs, 3yrs and 4 yrs respectively may be equal, the rate of interest being 5% per annum at simple interest. The first part is :

Sol. Simple interest will be (2 × 5)%, (3 × 5)% and (4 × 5)% of the respective principal money.

As given, (100 + 10)% of P_1 = (100 + 15)% of P_2 = (100 + 20)% of P_3

Then, $\quad P_1 : P_2 : P_3 = \frac{1}{110} : \frac{1}{115} : \frac{1}{120}$

$\Rightarrow \quad P_1 : P_2 : P_3 = \frac{1}{22} : \frac{1}{23} : \frac{1}{24}$

$\Rightarrow \quad P_1 : P_2 : P_3 = 276 : 264 : 253$

(On multiplying each term by LCM of denominators)

Now, $\quad P_1 = 2379 \times \frac{276}{(276 + 264 + 253)} = 828$

(or, (276+ 264+ 253) units ⇒ 2379
Then, 276 units ⇒ 828)

Alteration in Amounts when *R* or *T* or Both are Changed

Problem 31. A sum invested at 5% simple interest per annum grows to ₹ 504 in 4 yrs. The amount at 10% simple interest per annum in 2.5 yrs will grow to : **[CDS-03]**

Sol. Obviously, the amount will be (5 × 4)% above the principal money in the first case.

Now, [100 + (5 × 4)]% of the principal money = ₹ 504

Then, [100 + (10 × 2.5)]% of the principal money = ₹ 525.

Problem 32. A certain sum of money amounts to ₹ 6200 in 4 yrs at 6% per annum simple interest, in how many yrs will it amount to ₹ 7500 at the same rate of interest?

Sol. Given, ₹ 6200 = [100 + (6 × 4)]% of the principal money

Then, 7500 = 150% of the principal money

It implies that 50% of the principal is the simple interest.

Now, $\quad SI = (R \times T)\% \Rightarrow 50\% = (6 \times T)\%$

$\Rightarrow \quad T = 8.33$ yrs

Problem 33. A sum of money lent out at simple interest amounts to ₹ 720 after 2 yrs and to ₹ 1020 after a further period of 5 yrs. The sum is **[SSC (CGL-T-II)-12, 20]**

Sol. It is clear from the question that the Amount increases from ₹ 720 to ₹ 1020 on account of increment in interest in 5 yrs.

Hence, interest in 5 yrs = 1020 − 720 = 300

Then, interest in 2 yrs = ₹ 120

Also given, Amount after 2 yrs = ₹ 720

∴ Principal money = 720 − 120 = 600.

Problem 34. Prem invested a certain sum of money in a simple interest bond whose value grew to ₹ 300 at the end of 3 yrs and to ₹ 400 at the end of another 5 yrs. What was the rate of interest in which he invested his sum? **[MAT-10, SSC(Astt. Mains)-96]**

Sol. It is clear from the question that the amount increases from ₹ 300 to ₹ 400 on account of increment in interest in 5 yrs.

Hence, interest in 5 yrs = 400 − 300 = 100

Then, interest in 3 yrs = ₹ 60

Also given, amount after 3 yrs = ₹ 300

∴ Principal money = 300 − 60 = 240

Clearly, SI for 3 yrs = 25% of the principal

$\Rightarrow \quad 25\% = (R \times T)\%$

$\Rightarrow \quad 25\% = (R \times 3)\% \Rightarrow R = 8.33$

Thus, rate of interest = 8.33%.

Alternatively I

Let the rate of interest be R%.

Given, $(100 + 3R)\%$ of the principal = ₹ 300

and $(100 + 8R)\%$ of the principal = ₹ 400

$$\Rightarrow \frac{300}{400} = \frac{(100 + 3R)\% \text{ of the principal}}{(100 + 8R)\% \text{ of the principal}}$$

$$\Rightarrow \frac{300}{400} = \frac{(100 + 3R)}{(100 + 8R)}$$

$$\Rightarrow 300 + 24R = 400 + 12R \Rightarrow 12R = 100$$

$$\therefore \quad R = 8.33$$

Thus, rate of interest = 8.33%.

Note *In other way, we may express as ratio of amounts (in ₹) = ratio of amount in term of percentage of principal money.*

Alternatively II

If a sum of money at simple interest amounts to A_1 in t_1 yrs at a certain rate of interest and amounts to A_2 in t_2 yrs at the same rate of interest, then rate of interest will be $\left[\frac{100 \times (A_2 - A_1)}{A_1 t_2 - A_2 t_1}\right]\%$.

This formula has been derived from the approach given in Alternatively I.

Here, rate of interest $= \left[\frac{100 \times (400 - 300)}{300 \times 8 - 400 \times 3}\right]\% = 8.33\%$.

Problem 35. A sum of money at simple interest amounts to ₹ 9440 in 3 yrs. If the rate of interest is increased by 25%, the same sum of money amounts to ₹ 9800 in the same time. Find the rate of interest.

Sol. Let the initial rate of interest be R%.

Then, $9440 : 9800 = (100 + 3R)\% : (100 + 125\% \text{ of } 3R)\%$

$$\Rightarrow \frac{9440}{9800} = \frac{(100 + 3R)\%}{(100 + 125\% \text{ of } 3R)\%}$$

$$\Rightarrow \frac{236}{245} = \frac{(100 + 3R)}{(100 + 3.75R)}$$

$$\Rightarrow 23600 + 885R = 24500 + 735R$$

$$\Rightarrow 150R = 900 \Rightarrow R = 6$$

Hence, rate of interest = 6%.

Concept of Equal Installment for Discharging a Debt

Problem 36. What annual installment will discharge a debt of ₹ 6450 due in 4 yrs at 5% per annum simple interest? ***[SSC(CGL-T-I)-10, MAT-18; SSC (CGL-T-II)-19]***

Sol. Let, each installment be ₹ x.

Then, $\left(x + \frac{x \times 3 \times 5}{100}\right) + \left(x + \frac{x \times 2 \times 5}{100}\right) + \left(x + \frac{x \times 1 \times 5}{100}\right) + x = 6450$

$$\Rightarrow \frac{23x}{20} + \frac{22x}{20} + \frac{21x}{20} + x = 6450$$

$$\Rightarrow \frac{86x}{20} = 6450$$

$$\Rightarrow x = 1500$$

Alternatively I

$$\text{Each installment} = \frac{\text{Debt} \times 100}{(100 \times t) + [(t-1) + (t-2) + (t-3) + \dots + (t-t)] \times r}$$

where, t is time in years and r is rate of interest per annum.

∴ Each installment

$$= \frac{6450 \times 100}{(100 \times 4) + [(4-1) + (4-2) + (4-3) + (4-4)] \times 5}$$

$$= \frac{6450 \times 100}{430} = 1500$$

Alternatively II

$(100 + 105 + 110 + 115)\%$ of each installment = ₹ 6450

Then, 100% of each installment = ₹ 1500

Hence, annual installment = ₹ 1500

Problem 37. At what rate of simple interest per annum, an annual amount of ₹ 5000 discharges a debt of ₹ 15900 due in 3 yrs.

Sol. On the above pattern, obviously,

$[100 + (100 + R) + (100 + 2R)]\%$ of each installment = ₹ 15900

$\Rightarrow (300 + 3R)\%$ of each installment = ₹ 15900

$\Rightarrow$ 100% of each installment

$$= \frac{15900 \times 100}{3(100 + R)} = 5000 \text{ (as given)}$$

$$\Rightarrow 5300 = 50(100 + R)$$

$$\Rightarrow 50R = 300$$

$$\therefore R = 6$$

Hence, rate of interest per annum = 6%

Problem 38. A computer is sold either for ₹ 19200 cash or for ₹ 4800 cash down payment together with five equal monthly installments. If the rate of interest charged is 12% per annum, then the amount of each installment (nearest to a rupee) is ***[XAT-19]***

Sol. Payment to be made through installments

= 19200 − 4800 = 14400

Now, interest accrued on ₹ 14400 at 12% per annum (i.e. 1% per month) for a period of 5 months = 5% of 14400 = 720

Total amount = 14400 + 720 = 15120

Further, on the above pattern,

$(100 + 101 + 102 + 103 + 104)\% \Rightarrow$ ₹ 15120

Then, 100% $\Rightarrow$ ₹ 2964.76

Hence, each installment be ₹ 2965 (approx.).

Problem 39. A xerox machine is available for ₹ 78000 cash or for 33.33% cash down payment and 11 equal monthly installments of ₹ 4900 each. What is the rate of interest charged? *[MAT-13, 06]*

Sol. Payment to be made through installment

$= 66.66\%$ of $78000 = 52000$

Now, interest accrued on ₹ 52000 at the rate of r% per month for a period of 11 months = $11r$ % of 52000

$\therefore$ Total amount = 52000 + ($11r$ % of 52000)

Further, on the above pattern,

$$[100 + (100 + r) + (100 + 2r) + (100 + 3r) + \ldots\ldots + (100 + 10r)]\% \Rightarrow 52000 + (11r\% \text{ of } 52000)$$

$$\therefore \quad (1100 + 55r)\% \Rightarrow 52000 + (11r\% \text{ of } 52000)$$

$$\therefore \quad 100\% \Rightarrow \frac{52000 + (11r\% \text{ of } 52000)}{1100 + 55r} \times 100 = 4900$$

$$\Rightarrow \quad 52000 + 5720r = 53900 + 2695r$$

$$\Rightarrow \quad r = \frac{76}{121}$$

Then, rate of interest per annum $= \frac{76}{121} \times 12 = 7.54\%$.

Conventional method

Cost of the machine = ₹ 78000

Cost of the machine if bought by installment

$= (33.33\% \text{ of } 78000) + (4900 \times 11) =$ ₹ 79900

$\therefore$ Interest = 79900 − 78000 = ₹ 1900

The balance principal money for first month

$= 78000 - 26000 = 52000$

Balance principal money for second month

$= 52000 - 4900 = 47100$

Balance principal money for third month

$= 47100 - 4900 = 42200$

....

....

Balance principal money for eleventh month

$= 7900 - 4900 = 3000$

Sum = 52000 + 47100 + 42200 + ... + 3000 = 302500

Hence, principal money for one month = ₹ 302500

Time $= \frac{1}{12}$ yr and SI = 1900

Then, rate of interest $= \frac{1900 \times 100 \times 12}{302500} = 7.54\%$.

Concept of Repayment of Debt through Installments, each Comprising Equal Part of Principal Money

Problem 40. Ms. Debjani after her MBA graduation wants to have a start-up of her own. For this, she uses ₹ 800000 of her own saving and borrows ₹ 1200000 from a public sector bank under Mudra Scheme. As per agreement with the bank, she is supposed to repay the principal of this loan equally over the period of the loan which is 25 yrs. Two yrs after taking the first loan, she borrowed an additional loan of ₹ 800000 to finance expansion plan of her start-up. If Ms. Debjani clears all her loans in 25 yrs from the date of taking the first loan, how much total interest she has to pay on her initial borrowing? Assume simple interest at 8% per annum. *[IIFT-19]*

Sol. Money borrowed = ₹ 1200000

Principal money involved in each annual installment

$$= \frac{1200000}{25}$$

Interest involved in first installment = 8% of 1200000

(Debjani retains ₹ 1200000 for the first year)

Interest involved in second installment

$$= 8\% \text{ of } \left(1200000 - \frac{1200000}{25}\right)$$

$$= 8\% \text{ of } \frac{24}{25} \times 1200000$$

[$\because$ She retains only $\left(1200000 - \frac{1200000}{25}\right)$ as ₹ $\frac{1200000}{25}$ is paid in the first installment as principal money]

Interest involved in third installment

$$= 8\% \text{ of } \left(1200000 - \frac{2 \times 1200000}{25}\right)$$

$$= 8\% \text{ of } \frac{23}{25} \times 1200000$$

....

....

Interest involved in twenty-fifth installment

$$= 8\% \text{ of } \left(1200000 - \frac{24 \times 1200000}{25}\right)$$

$$= 8\% \text{ of } \frac{1}{25} \times 1200000$$

Hence, total interest = 8% of 1200000 + 8% of $\frac{24}{25} \times 1200000$ + 8% of $\frac{23}{25} \times 1200000 + \ldots +$ 8% of $\frac{1}{25} \times 1200000$

$$= 8\% \text{ of } \frac{25}{25} \times 1200000 + 8\% \text{ of } \frac{24}{25} \times 1200000 + 8\% \text{ of } \frac{23}{25} \times 1200000 + \ldots + 8\% \text{ of } \frac{1}{25} \times 1200000$$

$$= 8\% \text{ of } \frac{1200000}{25} (25 + 24 + 23 + \ldots + 1)$$

$$= 8\% \text{ of } \frac{1200000}{25} \times \frac{25 \times 26}{2}$$

$= 1248000$.

Generic version

(i) The total interest at the rate of r% per annum, for a debt of ₹ x, involved in n annual installments when principal amount is repaid equally with each installment over the period

$$= \left(\frac{n+1}{2}\right) \times (r\% \text{ of } x)$$

(ii) The total interest at the rate of r% per annum, for a debt of ₹ x, involved in n monthly installments when principal amount is repaid equally with each installment over the period

$$= \left(\frac{n+1}{2}\right) \times \left(\frac{r}{12}\% \text{ of } x\right)$$

Alteration in Overall SI in Case of Interchange between given Rates of Interest, Applicable for different Parts of a Sum

Problem 41. A money-lender lends a part of his money at 10% per annum and the rest at 15% per annum. His annual income is ₹ 1900. However, if he had interchanged the rate of interest on the two sums, he would have earned ₹ 200 more. The amount lent will fetch what amount at 15%? *[MAT-12]*

Sol. Let first part be ₹ x and second part be ₹ y.

As given, 10% of $x + 15\%$ of $y = ₹\,1900$ ...(i)

and 15% of $x + 10\%$ of $y = ₹\,1900 + 200$ ...(ii)

On addition Eqs. (i) and (ii),

25% of $(x + y) = 4000$

Then, $(x + y) = 16000$ ...(iii)

Further, on subtracting Eq. (i) from Eq. (ii), we get

5% of $(x - y) = 200$

Then, $(x - y) = 4000$ ...(iv)

From Eqs. (iii) and (iv), we get

$y = 6000$

Hence, the amount lent at 15% = ₹ 6000

Alternatively

Here, sum of rates of interest = 25%

Clearly, 25% of the sum of both parts

= ₹ 1900 + ₹ 2100 = ₹ 4000

Then, sum of both parts = 16000

Further, difference in rates of interest = 5%.

It means, 5% of the difference between both parts = ₹ 200

Then, difference between both parts = ₹ 4000

Since, the sum of both parts is 16000 and their difference is ₹ 4000, they will be ₹ 10000 and ₹ 6000.

Note *A part having lower rate of interest, will be bigger than the another part in case interest increases after interchange. On the other hand, the part having greater rate of interest will be bigger than the another if interest decreases after interchange.*

In the above case, first part will be ₹ 10000 and second part be ₹ 6000. Hence, the amount lent at 15% is ₹ 6000.

WORKOUT

1. A sum of money at a certain rate per annum of simple interest doubles in 5 yrs and at a different rate becomes three times in 12 yrs. The lower rate of interest per annum is

(a) 15% (b) 20%
(c) 15.75% (d) 16.67%

2. A sum of money becomes 3 times in 5 yrs at simple interest. In how many years, will the same sum become 6 times at the same rate of simple interest? *[CDS-14]*

(a) 10 (b) 12
(c) 12.5 (d) 10.5

3. A part of ₹ 38800 is lent out at 6% per six months. The rest of the amount is lent out at 5% per annum after one year. The ratio of interest after 3 yrs from the time when first amount was lent out is 5 : 4. Find the second part that was lent out at 5%. *[MAT- 14, 19]*

(a) ₹ 26600 (b) ₹ 28800
(c) ₹ 7500 (d) ₹ 28000

4. A person has left an amount of ₹ 120000 to be divided between his two sons aged 14 yrs and 12 yrs such that they get equal amount, when each attains 18 yrs of age. If the amount gets a simple interest of 5% per annum, the younger son's share at present is *[SSC (CGL-T-II)- 12]*

(a) ₹ 48,400 (b) ₹ 57,600
(c) ₹ 62,400 (d) ₹ 84,400

5. Ram had ₹ 2 lakh, part of which he lent at 15% per annum and rest at 12% per annum. Yearly interest accrued was ₹ 27600. How much did he lend at 15%?

(a) ₹ 120000 (b) ₹ 100000
(c) ₹ 80000 (d) ₹ 60000

6. A sum of money at simple interest amounts to ₹ 14160 in 3 yrs. If the rate of interest is increased to 25%, the same sum amounts to ₹ 14700 in the same time. The rate of interest is *[LIC (AAO)-11]*

(a) 22.86% (b) 25.5%
(c) 61.6% (d) 27%

7. A sum of ₹ 400 becomes ₹ 448 at simple interest in 2 yrs. In how many years will the sum of ₹ 550 amounts to ₹ 682 at the same rate?

(a) 2 (b) 3 1(c) 3.5 (d) 4

8. A person invested some amount at the rate of 12% simple interest and the remaining at 10%. He received yearly an interest of ₹ 130. Had he interchanged the amount invested, he would have received an interest of ₹ 134. How much money did he invest at the different rates ?

(a) ₹ 500 at 10% and ₹ 800 at 12%
(b) ₹ 700 at 10% and ₹ 500 at 12%
(c) ₹ 800 at 10% and ₹ 400 at 12%
(d) ₹ 700 at 10% and ₹ 600 at 12%

9. Two equal sums of money are lent at the same time at 8% and 7% per annum simple interest. The former is recovered 6 months earlier than the later and the amount in each case is ₹ 2560. The sum and the time for which the sums of money are lent out are *[MAT-05, 03, IIFT-17]*

(a) ₹ 1500, 3.5 yrs and 4 yrs
(b) ₹ 2000, 3.5 yrs and 4 yrs
(c) ₹ 2000, 4 yrs and 5.5 yrs
(d) ₹ 3000, 4 yrs and 4.5 yrs

10. A man borrows ₹ 6000 at 5% interest on reducing balance, at the start of the year. If he repays ₹ 1200 at the end of each year, find the amount of loan outstanding (in ₹) at the beginning of the third year? *[XAT-12]*

(a) ₹ 3162.75 (b) ₹ 4125
(c) ₹ 4155 (d) ₹ 5100
(e) ₹ 5355

11. At what rate percent per annum simple interest, will a sum of money triple itself in 25 yrs? *[CDS-08, 07]*

(a) 8% (b) 9%
(c) 10% (d) 12%

12. Manish borrowed a sum of ₹ 1150 from Anil at the simple rate of 6% per annum for 3 yrs. He then added some more money to the borrowed sum and lent it to Sunil for the same time at 9% per annum at simple interest. If Manish gains ₹ 274.95 by way of interest on the borrowed sum as well as his own amount from the whole transaction, then what is the sum lent by him to Sunil?

(a) ₹ 1290 (b) ₹ 1785
(c) ₹ 1285 (d) ₹ 1200
(e) None of these

13. What interest will be had on ₹ 450 in 2 yrs if an interest of ₹ 0.40 is charged on ₹ 1 for 4 yrs? *[RRB- Ajmer (ECRC)- 08]*

(a) ₹ 90 (b) ₹ 180
(c) ₹ 33 (d) None of these

14. If the simple interest on ₹ x at a rate of a% for m years is same as that on y at a rate of a^2% for m years , then $x : y$ is equal to : *[SSC (CGL-T-II)-12]*

(a) $m : a$ (b) $am : 1$
(c) $\frac{1}{m} : \frac{1}{a}$ (d) $\frac{1}{am} : 1$

15. The principal on which a simple interest of ₹ 55 will be obtained after 9 months at the rate of $3\frac{2}{3}$% per annum is

(a) ₹ 1000 (b) ₹ 1500
(c) ₹ 2000 (d) ₹ 2500

16. If ₹ 1000 be invested at interest rate of 5% and the interest be added to the principal after 10 yrs, then the number of yrs in which it will amount to ₹ 2000 is:

(a) 16.67 yrs (b) 16.25 yrs
(c) 16 yrs (d) 11 yrs

17. What amount would a man receive on a principal of ₹ 4000 after 2 yrs on simple interest at the rate of 5% per annum? *[Corporation Bank (PO)-09]*

(a) ₹ 4161 (b) ₹ 5200
(c) ₹ 4410 (d) ₹ 4100
(e) None of these

18. The ratio of the principal and its yearly amount is 8 : 9. What is the rate of simple interest per annum?

(a) 12% (b) 12.5%
(c) 10% (d) 10.5%

19. Out of a sum of ₹ 640, a part was lent at 6% simple interest and the other at 9% simple interest. If the interest on the first part after 3 yrs is equal to the interest on the second part after 6 yrs, then what is the second part? *[CDS-2008, MAT-2014, 2019]*

(a) ₹ 120 (b) ₹ 140
(c) ₹ 160 (d) ₹ 180

20. A man deposits some money in a savings bank and at the end of one year earns ₹ 15 as interest. He put in another ₹ 85 and deposits the money for another one year. After the expiry of this period he gets ₹ 420 as the sum total of the principal and the interest. What amount was originally deposited?

(a) 8% (b) 12.5%
(c) 10% (d) 5%

21. Shishupal borrowed a sum of ₹ 2000 from Abhishek at the simple rate of 7% per annum for 4 yrs. He then added some more money to the borrowed sum and lent it to Vicky for the same time at 9% per annum at simple interest. If Shishupal gains ₹ 520 by way of interest on the borrowed sum as well as his own amount from the whole transaction, then what is the sum lent by him to Vicky?

(a) ₹ 2500 (b) ₹ 3000
(c) ₹ 4500 (d) None of these

22. A person lent certain sum of money at 5% per annum simple interest and in 15 yrs the interest amounted to ₹ 250 less than the sum lent. What was the sum lent? *[SSC(CGL-T-I)-17]*

(a) ₹ 1000 (b) ₹ 1500
(c) ₹ 2400 (d) ₹ 3000

23. The sum which amounts to ₹ 364.80 in 8 yrs at 3.5% simple interest per annum is : *[CDS-11]*

(a) ₹ 285 (b) ₹ 280
(c) ₹ 275 (d) ₹ 270

24. A sum of money was borrowed at 6% per annum simple interest. At the end of first year ₹ 6800 was paid off and the rate of interest on the balance was reduced to 5% per annum. If the interest for the second year was 11/20 of the interest for the first year, what was the original sum borrowed?

(a) ₹ 17000 (b) ₹ 5000
(c) ₹ 8000 (d) ₹ 24000

25. A person invested some amount at the rate of 12% simple interest and the remaining at 10%. He received yearly an interest of ₹ 130. Had he interchanged the amounts invested, he would have received ₹ 4 more as interest. How much did he invest at 12% simple interest? *[MAT-10]*

(a) ₹ 700 (b) ₹ 500
(c) ₹ 800 (d) ₹ 400

26. A boy aged 10 yrs is left with ₹ 50000 which is under trust. The trustees invest the money at 4% per annum and pay the minor boy a sum of ₹ 1200 for his pocket money at the end of each year. The expenses of trust come out to be ₹ 300 per annum. Find the amount that will be handed over to the minor boy after he attains the age of 18 yrs.

(a) ₹ 37000 (b) ₹ 55000
(c) ₹ 54000 (d) ₹ 24000

27. ₹ 6000 becomes ₹ 7200 in 4 yrs at a certain rate of simple interest. If the rate becomes 1.5 times of itself, the amount of the same principal in 5 yrs will be :

(a) ₹ 8000 (b) ₹ 8250
(c) ₹ 9250 (d) ₹ 9000

28. A certain sum at simple interest amounts to ₹ 1350 in 5 yrs and to ₹ 1620 in 8 yrs. What is the sum? *[CDS-11]*

(a) ₹ 1000 (b) ₹ 900
(c) ₹ 700 (d) ₹ 800

29. A person lent ₹ 10000 to B for 3 yrs and ₹ 6000 to C for 4 yrs on simple interest at same rate of interest and received ₹ 5400 in all from both of them as interest. What is the rate of interest (in %)? *[SSC (CGL-T-I)-17]*

(a) 15 (b) 10 (c) 12.5 (d) 20

30. Consider the following statements.

If a sum of money is lent at simple interest, then the

I. money gets doubled in 5 yrs, if the rate of interest is $16\frac{2}{3}$%.

II. money gets doubled in 5 yrs, if the rate of interest is 20%.

III. money becomes 4 times in 10 yrs, if it gets doubled in 5 yrs.

Of these Statements :

(a) Both I and III are correct
(b) Only II is correct
(c) Only III is correct
(d) Both II and III are correct

31. A sum of money lent on simple interest triples itself in 15 yrs and 6 months. In how many years still it be doubled? *[CDS -10, SSC (CGL-T-I)-17]*

(a) 6 yrs and 3 months
(b) 7 yrs and 9 months
(c) 8 yrs and 3 months
(d) 9 yrs and 6 months

32. A man invests 3000 at the rate of 5% per annum. How much more should he invest at the rate of 8%, so that he can earn a total of 6% per annum? *[CAT-95]*

(a) ₹ 1200 (b) ₹ 1300
(c) ₹ 1500 (d) ₹ 2000

33. A person lent certain sum of money at 10% per annum simple interest. In 20 yrs the interest amounted to ₹ 500 more than the sum lent. What was the sum lent (in ₹)? *[SSC (CGL-T-I)-17]*

(a) 200 (b) 250
(c) 500 (d) 1000

34. A man borrowed ₹ 40000 at 8% simple interest per year. At the end of second year, he paid back certain amount and at the end of fifth year, he paid back ₹ 35960 and cleared the debt. What is the amount did he pay back after the second year?

(a) ₹ 16200 (b) ₹ 17400
(c) ₹ 18600 (d) None of these

35. A certain sum of money amounts to ₹ 918 in 2 yrs and ₹ 969 in 3.5 yrs at simple interest. What is the rate of interest (in %)? *[SSC (CGL-T-I)-17]*

(a) 5 (b) 4
(c) 6 (d) 8

36. Amala, Bina and Gouri invest money in the ratio 3 : 4 : 5 in fixed deposits having respective annual rates in the ratio 6 : 5 : 4. What is their total interest income (in ₹) after a year, if Bina's interest income exceeds Amala's by ₹ 25? *[CAT-19]*

(a) 6350 (b) 7250
(c) 7000 (d) 6000

37. A machine is sold either for ₹ 20000 cash or for ₹ 5000 cash down payment together with 4 equal monthly installments. If the rate of interest is charged 12% per annum, find each installment.

(a) 3200 (b) 3333.33
(c) 4415 (d) 3842.36

38. A trader owes a merchant ₹ 10028 due 1 yrs hence. The trader wants to settle the account after 3 months. If the rate of interest is 12% per annum, how much cash (in ₹) should he pay? *[MAT-18]*

(a) 9025 (b) 9200 (c) 9560 (d) 9600

39. A person invested a total amount of ₹ 15 lakh. A part of it was invested in a fixed deposit earning 6% annual interest, and the remaining amount was invested in two other deposits in the ratio 2 : 1, earning annual interest at the rates of 4% and 3% respectively. If the total annual interest income is ₹ 76000 then the amount (in ₹ lakh) invested in the fixed deposit was *[CAT-19]*

(a) 5 (b) 9
(c) 6 (d) 8

Answer key

1. (d)	2. (c)	3. (b)	4. (b)	5. (a)	6. (a)	7. (d)	8. (b)	9. (b)	10. (c)
11. (a)	12. (b)	13. (a)	14. (b)	15. (c)	16. (a)	17. (e)	18. (b)	19. (c)	20. (d)
21. (b)	22. (a)	23. (a)	24. (a)	25. (b)	26. (c)	27. (b)	28. (b)	29. (b)	30. (b)
31. (b)	32. (c)	33. (c)	34. (b)	35. (b)	36. (b)	37. (d)	38. (b)	39. (b)	

Hints and SOLUTIONS

1. In First case, SI = 100%

$\Rightarrow \quad (R \times T)\% = 100\%$

$\Rightarrow \quad (R \times 5)\% = 100\%$

$\Rightarrow \quad R = 20$

Hence, rate of interest = 20%

In second case, SI = 200%

$\Rightarrow \quad (R \times T)\% = 200\%$

$\Rightarrow \quad (R \times 12)\% = 200\%$

$\Rightarrow \quad R = 16.67$

Hence, rate of interest = 16.67%

Clearly, lower rate of interest = 16.67%.

Note The above all the process can be done without pen and paper in moments.

2. Increment in principal = 200%

$\Rightarrow$ 5 yrs

Increment in principal = 500 %

$\Rightarrow$ 12.5 yrs

3. (*b*) SI on first part for 3 yrs

$= (6 \times 2 \times 3)\%$ of P_1

SI on second part for last 2 yrs

$= (5 \times 2)\%$ of P_2

Given, $(6 \times 2 \times 3)\%$ of $P_1 : (5 \times 2)\%$ of $P_2 = 5 : 4$

$\Rightarrow \quad 50\%$ of $P_2 = 144\%$ of P_1

$\Rightarrow \quad P_1 : P_2 = 25 : 72$

$\therefore$ Second part

$$= 38800 \times \frac{72}{(25 + 72)} = 28800$$

$\left[\begin{array}{l}\text{or, } (25 + 72) \text{ units} \Rightarrow 38800 \\ \text{Then, } 72 \text{ units} \Rightarrow 28800\end{array}\right]$

4. Simple Interest will be $(4 \times 5)\%$ and $(6 \times 5)\%$ of the respective parts (principal money).

As given, $(100 + 20)\%$ of P_1

$= (100 + 30)\%$ of P_2

Then, $P_1 : P_2 = 130 : 120 = 13 : 12$

$$\therefore \quad P_2 = 120000 \times \frac{12}{(13 + 12)} = 57600$$

$\left[\begin{array}{l}\text{or, } (13 + 12) \text{ units} \Rightarrow 120000 \\ \text{Then, } 12 \text{ units} \Rightarrow 57600\end{array}\right]$

5. Overall rate of interest

$$= \left(\frac{27600}{200000} \times 100\right)\% = 13.8\%$$

Now,

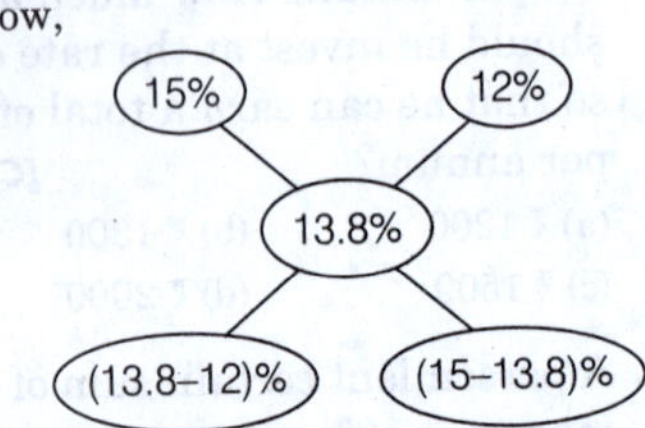

Hence, $P_1 : P_2 = 1.8\% : 1.2\% = 3 : 2$

$$\text{Then, } P_1 = 200000 \times \frac{3}{(3 + 2)} = 120000$$

$\left(\begin{array}{l}\text{or, } (3 + 2) \text{ units} \Rightarrow 200000 \\ \text{Then, } 3 \text{ units} \Rightarrow 120000\end{array}\right)$

6. Clearly, ₹ 14700 = $[100 + (25 \times 3)]\%$ of the principal

Then, ₹ 14160 = 168.57% of the principal. It implies that SI is equal to 68.57% of the principal.

Now, $\quad SI = (R \times T)\%$

$\Rightarrow \quad 68.57\% = (R \times 3)\%$

$\Rightarrow \quad R = 22.86\%$

Hence, rate of simple interest per annum = 22.86%

7. SI in first case = 12% of the principal.

Then, rate of interest per annum

$= 12\% \div 2 = 6\%$

SI in second case = 24% of the principal

Then, time = $(24\% \div 6\%)$ = 4 yrs

8. Here, difference in rates of interest = 2%.

Clearly, 2% of the difference between both parts = ₹ (134 – 130)

Then, 100% of the difference between both parts = ₹ 200

It means, the difference between money invested should be ₹ 200 and only option (b) has the same.

Alternatively

Proceed through options.

(a) (10% of 500) + (12% of 800) = 146 ≠ 130, hence, no further action to match with figure ₹ 134.

(b) (10% of 700) + (12% of 500) = 130 = 130. Further, (12% of 700) + (10% of 500) = 134

Hence, option (b) is correct.

9. Since the principals are same and amounts as well in the both cases, SIs must be same. Now, observe the options

(a) SI = $(8 \times 3.5)\% = (7 \times 4)\%$ and 128% of 1500 ≠ 2560

(b) SI = $(8 \times 3.5)\%$

$= (7 \times 4)\%$ and 128% of 2000 = 2560

Hence, option (b) is correct.

10. Loan outstanding at end of the first year = (105% of 6000) – ₹ 1200

= ₹ 5100

Further, loan outstanding at end of the second year

= (105% of 5100) – ₹ 1200

= ₹ 4155, which is also the loan outstanding at the beginning of the third year.

11. SI = Increment % in principal money = 200%

$\Rightarrow \quad 200\% = (R \times T)\%$

$\Rightarrow \quad 200\% = (R \times 25)\%$

$\Rightarrow \quad R = 8$

Hence, rate of interest per annum = 8%

12. SI paid by Manish to Anil

$= (6 \times 3)\%$ of 1150 = ₹ 207

Now, SI received by Manish

$= (9 \times 3)\%$ of the total principal

$\Rightarrow (207 + 274.95)$ = 27% of the total principal

$\Rightarrow$ 100% of total principal = ₹ 1785

Hence, sum lent by Manish to Sunil is ₹ 1785.

13. Interest on ₹ 100 for one year

$= (0.40 \times 100) \div 4$

$= 10$

Then, interest on ₹ 450 for two years

$= (10 \times 4.5) \times 2$ = ₹ 90.

Alternatively

Interest on ₹ 100 for one year

$= (0.40 \times 100) \div 4 = 10$

Hence, rate of interest per annum = 10%.

Now, SI= $(10 \times 2)\%$ of 450 = ₹ 90

14. Given, $(a \times m)\%$ of $x = (a^2 \times m^2)$ % of y

$\Rightarrow x : y = (a^2 \times m^2) : (a \times m) = am : 1$

15. (c) SI $= \left(3\frac{2}{3} \times \frac{9}{12}\right)\% = \frac{11}{4}\%$

Given, $\frac{11}{4}\%$ of the principal money $= ₹\,55$

$\therefore$ 100% of the principal money $= ₹\,2000.$

16. Amount after 10 yrs = 150% of 1000 = ₹ 1500, which will again become ₹ 2000 i.e. increment of 33.33%.

Now, SI = $(R \times T)\%$

$\Rightarrow$ 33.33% = $(5 \times T)\%$

$\Rightarrow$ $T = 6.67$ yrs

Thus, total time = 10 + 6.67 = 16.67 yrs

17. Amount = [100 + (5 × 2)]% of the principal = 110% of 4000 = ₹ 4400

Hence, option (e) is correct.

18. Clearly, yearly SI = 9 − 8 = 1 unit, which is 12.5% of the principal.

Now, SI = $(R \times T)\%$

$\Rightarrow$ 12.5% = $(R \times 1)\%$

$\Rightarrow$ $R = 12.5$

Hence, rate of simple interest per annum is 12.5%.

19. Given that, $(6 \times 3)\%$ of $P_1 = (9 \times 6)\%$ of P_2

Then, $P_1 : P_2 = 3 : 1$

Now, $P_2 = 640 \times \frac{1}{(3+1)} = 160.$

$\left(\begin{array}{l}\text{or, } (3+1) \text{ units} \Rightarrow 640\\ \text{Then, 1 unit} \Rightarrow 160\end{array}\right)$

20. Let he originally deposited ₹ P at the rate of interest of $R\%$ per annum.

Then, SI = $(R \times 1)\%$ of the principal

Given, $R\%$ of the P = ₹ 15

Then, 100% of $P = \frac{1500}{R}$

Now, new principal $= \frac{1500}{R} + 15 + 85$

According to the question, (100 + R)% of $\left(\frac{1500}{R} + 15 + 85\right) = 420$

$\Rightarrow \frac{150000}{R} + 10000 + 1500 + 100R = 42000$

$\Rightarrow \frac{150000}{R} + 100R = 30500$

$\Rightarrow \frac{1500}{R} + R = 305$

$\Rightarrow R^2 - 305R + 1500 = 0$

$\Rightarrow R^2 - 300R - 5R + 1500 = 0$

$\Rightarrow R(R - 300) - 5(R - 300) = 0$

$\Rightarrow R = 5$ and 300

Hence, rate of interest = 5% or 300%.

21. SI paid by Shishupal to Abhishek = (7 × 4%) of 2000 = ₹ 560

Now, SI received by Shishupal = (9 × 4)% of the total principal

$\Rightarrow$ (560 + 520) = 36% of the total principal

$\Rightarrow$ 100% of total principal = ₹ 3000

Hence, sum lent by Shishupal to Vicky = ₹ 3000.

22. Clearly, simple interest is (15 × 5)% of the principal money i.e. 25% less than the principal.

As given, 25% of the principal = ₹ 250

Then, 100% of the principal = ₹ 1000

23. Given, [100 + (8 × 3.5)]% of the principal = ₹ 364.80

Then, 100% of the principal = ₹ 285

24. Let the original sum be ₹ x.

Then, amount at the end of first year = 106% of x

The SI for the second year = 5% of [(106% of x) − 6800]

But given, 5% of [(106% of x) − 6800] $= \frac{11}{20}$ of 6% of x

$\Rightarrow 0.053x - 340 = 0.033x$

$\Rightarrow x = 17000$

25. Here, sum of rates of interest = 22%

Clearly, 22% of the sum of both parts = 130 + 134 = 264

Then, sum of both parts = 1200

Further, difference in rates of interest = 2%.

It means, 2% of the difference between both parts = ₹ 4

Then, difference between both parts = ₹ 200

Since the sum of both parts is ₹ 1200 and their difference is ₹ 200, they will be ₹ 700 and ₹ 500.

We know that the part having lower rate of interest, will be bigger than the another part in case interest increases after interchange.

Hence, ₹ 700 has been invested at 10% and ₹ 500 at 12% simple interest.

Hence, option (b) is correct.

26. SI per annum = 4% of 50000 = 2000

Every year savings = 2000 − 1200 − 300 = 500

Then, till his age of 18 yrs total savings = 500 × 8 = 4000

Hence, total amount to be given to the boy at end of 18 yrs = 50000 + 4000 = 54000

27. SI = 7200 − 6000 = 1200

Here, rate of interest has been increased by 50% and time by 25%.

Then, effect on SI $= \left(+\,50 + 25 + \frac{50 \times 25}{100}\right)\%$

$= +\,87.5\%$

Now, new amount = 6000 + (187.5% of 1200) = 8250

28. It is clear from the question that the amount increases from ₹ 1350 to ₹ 1620 on account of increment in interest in (8 − 5) yrs.

Hence, interest in 3 yrs = 1620 − 1350 = 270

Then, interest in 5 yrs = ₹ 450

Also given, amount after 5 yrs = ₹ 1350

$\therefore$ Principal money = 1350 − 450 = 900

29. Let, rate of interest per annum be $r\%$.

Then, SI in first case = $3r\%$ of 10000 and that in second case = $4r\%$ of 6000.

As given, $3r\%$ of 10000 + $4r\%$ of 6000 = 5400

$\Rightarrow 300r + 240r = 5400$

$\Rightarrow r = 10\%$

30. I $\Rightarrow$ SI $= \left(16\frac{2}{3} \times 5\right)\%$ of principal.

Money will get doubled when SI will be 100% of the principal. Thus, statement I is incorrect.

II $\Rightarrow$ SI = (20 × 5)% of principal. Obviously, money gets doubled as SI is 100% of the principal.

III $\Rightarrow$ SI = 300%

$\therefore r = 300\% \div 10 = 30\%$. Further, in 5 yrs. SI will be 150% of the principal. It means in 5 yrs money becomes 2.5 times not 2 times. Thus, this statement is incorrect.

Hence, option (b) is correct.

31. Increment in principal

= 200% ⇒ 15 yrs and 6 months

Increment in principal

= 100% ⇒ 7 yrs and 9 months

32.

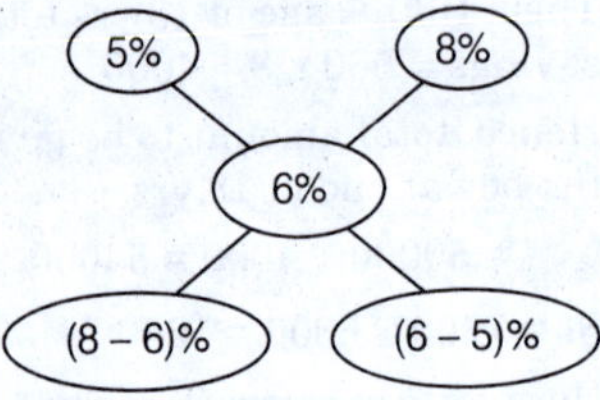

Hence, the amount invested at 5% per annum : that invested at 8% per annum = 2 : 1

As given, 2 units = ₹ 3000

∴ 1 unit = ₹ 1500

33. Clearly, SI = 200% of the principal, i.e. 100% more than the principal.

As given, 100% of the principal

= ₹ 500

Hence, principal = ₹ 500

34. At the end of second year,

Amount = [100 + (8 × 2)]% of 40000

= 46400

Let he paid ₹ x at the end of second year. Then, interest will be calculated only on $(46400 - x)$.

Hence, Amount for further three years = [100 + (8 × 3)]% of $(46400 - x)$, which is cleared by him by paying ₹ 35960.

Obviously, 124% of $(46400 - x)$

= 35960

⇒ 124% of x = 57536 – 35960 = 21576

⇒ $x = 17400$

Alternatively

Amount at the end of second year

= 116% of 40000

Now, Amount at the end of next three years i.e. at end of fifth year

= 124% of 116% of 40000

= 57536

Since some amount has been paid at the end of second year and finally paid at end of fifth year amounting to ₹ 35960.

Obviously, 124% of the paid amount at the end of second 2nd year + 35960 = 57536.

⇒ 124% of the paid amount at the end of second year = 21576

⇒ 100% of the paid amount at the end of second year = 17400

35. Clearly, interest in 1.5 yrs

= 969 – 918 = 51

Then, interest in 2 yrs = 68

∵ Amount after 2 yrs = ₹ 918

∴ Sum = 918 – 68 = 850

Further, 68 is 8% of 850.

Then, $(R \times T)\% = 8\%$

⇒ $(R \times 2)\% = 8\%$

⇒ $R = 4$

Hence, option (b) is correct.

Alternatively

You may proceed through options.

(a) 5 ⇒ If (100 + 2· 5)% of principal = 918, then (100 + 3.5 × 5)% of principal must be equal to ₹ 969, but here not so.

(b) 4 ⇒ If (100 + 2· 4)% of principal = 918, then (100 + 3.5 × 4)% of principal must be equal to ₹ 969. Here, so is this.

Hence, option (b) is correct.

36. Ratio of interest income of Amala, Bina and Gouri

= (3 × 6 × 1) : (4 × 5 × 1) : (5 × 4 × 1)

= 9 : 10 : 10

As given, (10 – 9) unit ⇒ ₹ 250

∴ (9 + 10 + 10) units ⇒ ₹ 7250

37. Payment to be made through installment = 20000 – 5000 = 15000

Now, interest accrued on ₹ 15000 at 12% per annum (i.e. 1% per month) for a period of 4 months

= 4% of 15000 = 600

Total amount

= 15000 + 600 = 15600

Further, on the earlier pattern,

(100 + 101 + 102 + 103)% ⇒ ₹ 15600

Then, 100% ⇒ ₹ 3842.36

Hence, each installment be ₹ 3842.36.

38. Clearly, the amount payable after 3 months becomes ₹ 10028 after next 9 months. The rate of interest for next 9 months will be 9% as the given rate is 12% per annum.

Hence, 109% of the amount payable after 3 months = ₹ 10028

∴ 100% of the amount payable after 3 months = ₹ 9200

39. Let an amount invested in other two deposits be ₹ x lakh.

Then, the amount invested in the fixed deposit = ₹ $(15 - x)$ lakh.

As given, 6% of $(15 - x)$ + 4% of $\frac{2}{3}x$ + 3% of $\frac{1}{3}x$ = 0.76 lakh

$\Rightarrow 90 - 6x + \frac{8}{3}x + x = 76$

$\Rightarrow x = 6$

∴ Required answer = 15 – 6

= 9 lakh

22 Compound Interest

Learning OBJECTIVES

After completing this chapter, aspirants will be able to :

- Understand the significance of this chapter in competitive exams, in field of finance/commerce etc. and in our daily life as well;
- Understand the meaning of compound interest (CI) in real sense with diverse aspects;
- Comprehend the practical meaning of interest compounded at a specific period;
- Understand how to apply the concept of percentage like 'Effect on product in case of changes in factor(s), etc. in this chapter as this chapter is the extension of application of percentage also;
- Differentiate between simple interest and compound interest with their trends;
- Ascertain the increment in principal when rate of interest or time or both are altered;
- Compare between amounts in case of CI and SI;
- Analyse the different approaches for solution of a question and will also be able to choose the best one for the exams;
- Understand the concept of tax-deduction on CI;
- Understand the fundamentals of repayment of a loan either in equal or unequal installments;
- Learn how CIs for different periods are interlinked with reference to rate of interest;
- Realise themselves competent to solve the questions other than the conventional way;
- Understand how compound interest alters in a defined way when rate of interest or time or both are change;
- Appreciate the diverse cases under one approach;
- Recognise the concept of simple interest in different contexts like increment in principals, increment in amounts, etc;
- Comprehend the concept of compound interest in various types of transactions;

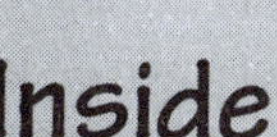

Inside

- *Introduction*
- *Computation of CI/Amount When P, R, T are Known, even When Interest is Compounded Continuously*
- *Computation of Any Term When Other Relevant Figures are Known Under Diverse Situations*
- *Comparison of Amounts with Reference to Time/Rate of Interest, even in Case of SI and CI*
- *Concept of Tax-Deduction*
- *Problems Based on Population,*
- *Concept of Equal/Unequal Installment to Discharge a Debt*
- *Concept of Periodic Investment or Recurring Deposit (Compounded Value of a Series of Payments)*
- ....
- .... *& Many More*
- *Work Out with Solutions*

- Recognise how one concept works in different sorts of problems;
- Use the concept of ratio in this chapter;
- Realise the significance of generic version of the questions;
- Recognise the types of questions which can be easily solved through the given options;
- Analyse the trend of the questions being asked in various competitive examination;
- Solve the questions in moments;
- Grip on concepts helping to solve the questions without pen and paper;
- Learn some new concepts/ fundamentals;
- Apply the pragmatic approaches in lieu of academic approaches; and
- Ponder on the alternative methods.

Introduction

Sometimes, the borrower and the lender decide to fix up a unit of time, say, yearly, half- yearly, quarterly, bimonthly or monthly to settle an account provided the interest accrued on the principal at the end of the first unit of time be added to the principal and the Amount so obtained becomes the principal for the second unit of time and so on.

The difference between Amount and money borrowed is called compound interest (CI). Thus, it is different from simple interest (SI) as in case of SI, interest is calculated uniformly at principal.

It will not be out of context to mention that this chapter is also the extension of application of percentage chapter like Profit and Loss, Discount, Simple Interest.

Suppose, the principal money is ₹ 100 and rate of interest per annum is 5%. Then,we may compare the Simple Interest and Compound Interest as following

For the period of	SI	CI
1st year	5	**5**
2nd year	5	5 + 5% of previous years' interest = 5 + (5% of 5) = **5.25**
3rd year	5	5 + 5% of last two years' interest = 5 + 5% of (5 + 5.25) = **5.5125**
4th year	5	5 + 5% of last three years' interest = 5 + 5% of (5 + 5.25 + 5.5125) = **5.788125**

From the above chart, it is observed that the SIs for the 1st, 2nd, 3rd and 4th year are same whereas the CIs are gradually increasing per year like 5 → 5.25 → 5.5125 → 5.788125 and the reason is that interest has been accrued also on the interest(s).

It may be construed from the above chart that

(i) CI for any particular year is the sum of simple interest for that (any) year and interest on the sum of previous accrued compound interests at any given rate of interest.

(ii) Point (i) may be elaborated as below:

For 1st year, difference between SI and CI = 0

For 2nd year,difference between SI and CI = r% of 1st year's CI or SI

For 3rd year, difference between SI and CI = r% of sum of previous two years' CIs

For 4th year, difference between SI and CI = r% of the sum of previous three years' CIs

(iii) $(100 + r)$% of CI for 1st yr = CI for the 2nd yr,

$(100 + r)$% of CI for 2nd yr = CI for the 3rd yr,

$(100 + r)$% of CI for 3rd year = CI for 4th year and so on.

In nutshell, CI for any particular year will be r% more than CI for its just previous year.

(iv) Case (iii) also holds good in case of cumulative CI. For example: $(100 + x)$% of cumulative CI for first two years = Cumulative CI for the next two year and $(100 + y)$ % of cumulative CI for first three years = Cumulative CI for the next three year and so on. (Here, x and y are effective rates of CI for 2 yrs and 3 yrs respectively i.e. 10.25% and 15.7625% respectively for the above chart).

In other words, if the cumulative CI for the first two years be x% of the principal, the cumulative CI for next two years will be x% more than the earlier cumulative CI. Similarly, if the cumulative CI for the first three years be y% of the principal, the cumulative CI for the next three years will be y% more than the earlier cumulative CI.

(v) It also holds good than if the effective rate of CI for the first t yrs = r%, then, effective rate of CI for next t yrs = r% more than the earlier rate of interest = $(100 + r)$% of r%.

Take a question to build the fundamentals: What is the compound interest on a sum of ₹ 1200 at the rate of 6% per annum for the period of (a) 2 yrs (b) 3 yrs (c) 4 yrs.

For the Period of Two Years

Please recall the concept 'Successive Increase/ Decrease/ combination of both' envisaged in chapter 'Percentage' as well as one question based on the concept *viz.* A number (say principal) is increased by 6% and further increased by 6%. By how much percentage will the number be increased?

Effect on the original number

$$= \left\{+6 + 6 + \frac{6 \times 6}{100}\right\}\% = +12.36\%$$

It implies that the principal money will increase by 12.36% in a period of 2 yrs . Thus, the amount (A) will be 112.36%

of the principal money. Obviously, the increment in principal i.e. 12.36% is the compound interest. In other words CI is 12.36% of the principal. Like SI, we must express the CI in term of percentage of the principal.

Hence, CI = 12.36% of 1200 = 148.32

Alternatively

From the above, it is clear that CI = $\left\{+6+6+\frac{6\times 6}{100}\right\}$ % of the principal money

$= \left\{(+6+6)+\frac{6\times 6}{100}\right\}$ % of the principal money

$= \left\{(2\times 6)+\frac{6\times 6}{100}\right\}$ % of the principal money

$= \left\{(\text{Number of years} \times \text{rate of interest})+\frac{6\times 6}{100}\right\}$ % of the principal money

$= \left\{(\text{SI for given years})+\frac{6\times 6}{100}\right\}$ % of the principal money

$= \left\{(\text{Cumulative SI})+\frac{6\times 6}{100}\right\}$ % of the principal money

= {(Cumulative SI) + 0.36}% of the principal money

CI = (Cumulative SI).$\boxed{r^2}$% of the principal money

CI = $(2\times 6)\cdot\boxed{6^2}$% of the principal money

CI = 12.36% of the principal money

Obviously, the box will contain only two digits as r^2 has been divided by 100 and in case the value of r^2 be more than two digits, the figure other than last two digits will be carried over left hand side. It is also important to note that the box will always have two digits.

For example, we have to write 04 in the box for the value of 2^2.

The above pattern may be depicted as below:

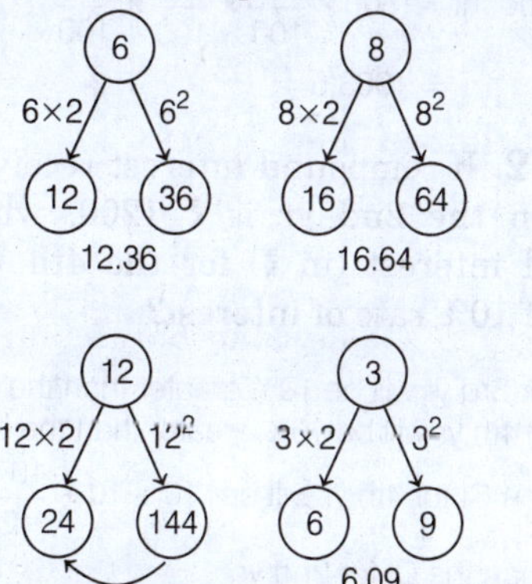

Observation: Aspirants may feel difficulty to compute the value of CI as it is a decimal number percentage of the principal. Thus, we will solve as same as given in the chapter 'Percentage' under heading 'Percentage of a quantity'. Now, 12.36% of 1200 = ?

$\Rightarrow$ {10% + (2 × 1%) + (3 × 0.1%) + (6 × 0.01%)} of 1200

= 120 + (2 × 12) + (3 × 1.2) + (6 × 0.12)

= 120 + 24 + 3.6 + 0.72 = 148.32

{you may find the answer from the options well before the final step of process. Hence, no need to go ahead till final step. Further, whole the processes/steps can be done mentally without pen and paper, try please}

For the Period of three Years

Consider a problem based on the concept of 'Successive Increase/ Decrease/ combination of both' as envisaged in chapter 'Percentage' A number is increased by 6%, again by 6% and further by 6%. By how much percent that number be increased?

Effect on number taking into consideration of only two figures

$= \left\{6+6+\frac{6\times 6}{100}\right\}\% = +12.36\%$

Then, final effect

$= \left\{+12.36+6+\frac{12.36\times 6}{100}\right\}\% = +19.1016\%$

It implies that the principal money will increase by 19.1016% in a period of 3 yrs. Thus, the amount (A) will be 119.1016% of the principal money. Obviously,the increment in principal i.e. 19.1016% is the compound interest.

Hence, CI = 19.1016% of the principal

CI = 19.1016% of 1200 = ₹ 229.22

Alternatively

CI = (Cumulative SI) . $\boxed{3r^2}\boxed{r^3}$% of the principal money

Here, CI = (Cumulative SI) . $\boxed{3\times(6)^2}\boxed{6^3}$% of the principal money

$\Rightarrow$ CI = (3 × 6) . $\boxed{108}\boxed{216}$% of the principal money

Note *Each of both boxes will have only two digits and the surplus figure(s) will be carried over left.*

$\Rightarrow$ CI = (18) . $\boxed{108+2}\boxed{16}$% of the principal money

$\Rightarrow$ CI = (18) . $\boxed{110}\boxed{16}$% of the principal money

$\Rightarrow$ CI = (18 + 1) . $\boxed{10}\boxed{16}$% of the principal money

$\Rightarrow$ CI = 19.1016% of the principal money

Hence, CI = 19.1016% of ₹ 1200 = ₹ 229.22

For the Period of Four Years

CI = 19.1016% of the principal money for the period of 3 yrs as computed above.

Then, CI for the four years on the above pattern

$= \left\{+19.1016+6+\frac{19.1016\times 6}{100}\right\}\%$

= + 26.247696%

Hence, CI = 26.247696% of the principal money

CI = 26.247696% of 1200 = ₹ 314.97

Alternatively

CI = (Cumulative SI) . $\boxed{6r^2}\boxed{4r^3}\boxed{r^4}$% of the principal money

Here, each of all three boxes will have only two digits and the surplus figure(s) will be carried over left. Details are given above.

Now, CI = (Cumulative SI) . $\boxed{6r^2}\boxed{4r^3}\boxed{r^4}$% of the principal

$\Rightarrow$ CI = (6 × 4) . $\boxed{6\cdot 6^2}\boxed{4\cdot 6^3}\boxed{6^4}$% of the principal

$\Rightarrow$ CI = (6 × 4) . 216$\boxed{864}\boxed{1296}$% of the principal

$\Rightarrow$ CI = {(6 × 4) + 2} . $\boxed{24}\boxed{76}\boxed{96}$% of the principal

$\Rightarrow$ CI = 26.247696% of 1200 = ₹ 314.97

Note *Aspirants shouldn't concentrate on finding the exact value of decimal number percent of the principal. They can say:*

12.36% → 12.4%

19.1016% → 19.1%

26.247696% → 26.2%

for the fastest calculation. Further, they can opt answer from the given alternatives. The correct alternative will either be equal or nearest to your calculation.

For the period of beyond four years

For 5 yrs, CI

$$= \left\{+ \text{CI for 4 yrs} + 6 + \frac{\text{CI for 4 yrs} \times 6}{100}\right\}\%$$

For 6 yrs, CI

$$= \left\{+ \text{CI for 5 yrs} + 6 + \frac{\text{CI for 5 yrs} \times 6}{100}\right\}\% \text{ and so on}$$

For the years given in mixed number

For example, for a period of $2\frac{1}{2}$ yrs

CI for 2 yrs = 12.36% of the principal

Given, rate of interest = 6% of per annum. It implies rate of interest for six months is only 3%.

Now, CI for $2\frac{1}{2}$ yrs = increment in principal by 3% after increment of 12.36%

Hence, CI = $\left\{+ 12.36 + 3 + \frac{12.36 \times 3}{100}\right\}$% of the principal

= 15.73% of the principal

= 15.73% of ₹ 1200 = ₹ 188.76

> ***Observation***
>
> 1. You may compute CI or amount for any period even given in mixed/decimal number like 4.2 yrs, 2.8 yrs, 3.1 yrs, etc on the above pattern.
> 2. You may solve each of the questions given above in few seconds only and more important, you may save the time if you observe the options while solving.

Illustrations

Computation of CI/Amount when *P*, *R* and *T* are known

Problem 1. Find out the amount on a sum of ₹ 1600 at 8% per annum for a period of $2\frac{1}{4}$ yrs.

Sol. CI for 2 yrs = (2 × 8) .$\boxed{8^2}$% of the principal money

= 16.64% of the principal money

Further, rate of interest for $\frac{1}{4}$ yrs = 8 ÷ 4 = 2%

Then, CI for $2\frac{1}{4}$ yrs = $\left\{16.64 + 2 + \frac{16.64 \times 2}{100}\right\}$ % of 1600

= 18.97% of 1600

= 19% (approx.) of 1600

Now, amount = 119% of 1600 = ₹ 1904 (approx.)

Obviously, it will be closer to one option which can be ticked out by you. If there is an option as 'none of these', you have to calculate exactly as below

100% of 1600 = 1600

10% of 1600 = 160

1% of 1600 = 16

0.1% of 1600 = 1.6

0.01% of 1600 = 0.16

Now, amount = 118.97% of 1600

= {100% + (2 × 10)% − 1% − (3 × 0.01)%} of 1600

= 1600 + 320 − 16 − 0.48

= 1903.52

Note *The above process seems to be long, but can be done in few moments. However, if you feel uncomfortable in this approach, it will be better to multiply 118.97% with 1600 on the same line as detailed in chapter 'Multiplication'.*

Alternatively,

Rate of interest for $\frac{1}{4}$ yr = 8 ÷ 4 = 2%

It is understood that the principal amount is increased by 8%, then by 8% and further by 2%.

Then, amount = $1600 \times \frac{108}{100} \times \frac{108}{100} \times \frac{102}{100}$

= 1903.5

Problem 2. If compound interest received on a certain amount in the 2nd yr is ₹ 1200, what will be the compound interest (in ₹) for the 4th yr on the same amount of 10% rate of interest? ***[SSC(CGL-T-1)-17]***

Sol. The CI for 3rd yr will be 10% greater than the CI for 2nd yr and the CI for 4th yr will be 10% greater than the CI for 3rd yr.

Hence, the CI for 4th yr will be $\left(10 + 10 + \frac{10 \times 10}{250}\right)$% i.e. 21% greater than the CI for 2nd yr.

Hence, the required answer = 121% of 1200

= ₹ 1452

When Interest is Compounded Periodically

Problem 3. Find the compound interest on ₹ 10000 in 2 yrs at 4% per annum, the interest being compound half yearly.

Sol. Interest is being compounded half yearly.

Thus, rate of interest for $\frac{1}{2}$ yrs = 2% (∵ 4% for 1 yr)

Also, number of $\frac{1}{2}$ yrs block in 2 yrs

i.e. frequency = 4 (say, 4 yrs for sake of convenience)

Now, CI for 2 yrs = $\left\{+2+2+\frac{2\times2}{100}\right\}$ % of the principal money

= 4.04 % of principal

∴ CI for 3 yrs = $\left\{4.04+2+\frac{2\times4.04}{100}\right\}$ % of principal

= 6.1208% of principal

Then, CI of 4 yrs = $\left\{6.1208+2+\frac{6.120\times2}{100}\right\}$ % of principal

= 8.243216% of 10000 = 824.32

Alternatively I

CI for 4 yrs = (Cumulative SI) . $\boxed{6r^2}\boxed{4r^3}\boxed{r^4}$ % of principal

= (2 × 4) . $\boxed{6\cdot2^2}\boxed{4\cdot2^3}\boxed{2^4}$ % of the principal

= (2 × 4) . $\boxed{24}\boxed{32}\boxed{16}$ % of the principal

= 8.243216% of the principal

∴ Required CI = 8.243216% of 10000 = 824.32

Alternatively II

Interest is being compounded half yearly.

Thus, rate of interest for $\frac{1}{2}$ yrs = 2% (∵ 4% for 1yr)

Also, number of $\frac{1}{2}$ yrs block in 2 yrs

= 4 (say, 4 yrs for sake of convenience)

Now, amount

$= 10000\times\frac{102}{100}\times\frac{102}{100}\times\frac{102}{100}\times\frac{102}{100} = 10824.32$

Then, CI = 10824.32 – 10000 = ₹ 824.32

Computation of CI/Amount when *P*, *R* and *T* are known and Interest is Compounded Continuously

The interest is compounded continuously means the interest is compounded infinite time in the given period.

Problem 4. ₹ 10000 in deposited in a bank at 5% per annum. Find the amount after 2 yrs, the interest being compound continuously.

Sol. Amount = Principal money × $(e)^{\frac{r}{100}\times t}$,

where e = 2.718281828,

r = rate of interest per annum

and t = time in years.

Hence, required amount

$= 10000\times(2.718281828)^{0.05\times2}$ = ₹ 11051.71

Note *Amount or CI increases when frequency of compounding increases. Hence, amount or CI in case of interest being compounded continuously > amount in case of interest being compounded daily or hourly or per second > amount in case of interest being compounded monthly > amount in case of interest being compounded bimonthly > amount in case of interest being compounded quarterly and so on.*

Finding sum or CI when Different Rates of Interest are Applicable for the Different Periods

Problem 5. A certain sum of money amounts to ₹ 28119 for a period of 3 yrs. The rate of compound interest is 3% for 1st yr, 4% for 2nd yr and 5% for 3rd yr. Find the sum.

Sol. CI for 2 yrs = $\left\{+3+4+\frac{3\times4}{100}\right\}$ % = + 7.12% of the principal.

Then, CI for 3 yrs = $\left\{+7.12+5+\frac{7.12\times5}{100}\right\}$ %

= 12.476% of principal

Hence, amount = 112.476% of the principal

As given, 112.476% of the principal = ₹ 28119

∴ 100% of the principal = ₹ 25000

Alternatively

Amount = Principal $\times\frac{103}{100}\times\frac{104}{100}\times\frac{105}{100}$

⇒ 28119 = Principal $\times\frac{103}{100}\times\frac{104}{100}\times\frac{105}{100}$

⇒ Principal = ₹ 25000

Computation of *R* or *T* when *P*, *A*, *T* or *R* are Known

Problem 6. In what time will ₹ 1000 become ₹ 1331 at 10% per annum compounded annually?

[SSC(CGL-PT)-04,SSC(SAS)-10]

Sol. Percentage increase in principal = 33.1%

Hence, CI = 33.1% of the principal

Let the options be 3 yrs, $2\frac{1}{2}$ yrs, 2 yrs and $3\frac{1}{2}$ yrs.

Now, assume that the time is 3 yrs as 3 times 10 is near about to 33.1%. But, you have to check the correctness.

For 2 yrs, CI = $\left\{+10+10+\frac{10\times10}{100}\right\}$ % of the principal money

= 21% of the principal money

For 3 yrs, CI = $\left\{+21+10+\frac{21\times10}{100}\right\}$ % of the principal

= 33.1% of the principal

Thus, answer is 3 yrs.

Alternatively I

Percentage increase in principal = 33.1%

Hence, CI = 33.1% of the principal

Now, assume that the time is 3 yrs as 3 times 10 is near about to 33.1%. But, you have to check the correctness.

CI for 3 yrs = (Cumulative SI). $\boxed{3r^2}\boxed{r^3}$% of principal money

$= (3 \times 10) . \boxed{3 \times (10)^2}\boxed{(10)^3}$% the principal

= 33.1% of the principal.

Alternatively II

We know, $A = P\left(1 + \frac{r}{100}\right)^t$

$\Rightarrow 1331 = 1000\left(1 + \frac{10}{100}\right)^t \Rightarrow \frac{1331}{1000} = \left(\frac{11}{10}\right)^t \Rightarrow \left(\frac{11}{10}\right)^3 = \left(\frac{11}{10}\right)^t$

$\Rightarrow \quad t = 3$ yrs.

Problem 7. At what rate percentage per annum will ₹ 2304 amount to ₹ 2500 in 2 yrs at compound interest?

Sol. Percentage increase in principal = 8.507%

CI for 2 yrs = 8.50% of principal.

Now proceed from options. Let options be $4\frac{1}{2}\%$, $4\frac{1}{6}\%$, $4\frac{1}{5}\%$ and $4\frac{1}{3}\%$

(a) $4.5\% \Rightarrow \left\{+ 4.5 + 4.5 + \frac{4.5 \times 4.5}{100}\right\}\% \neq +8.507\%$

(b) $4.16 \Rightarrow \left\{+ 4.16 + 4.16 + \frac{4.16 \times 4.16}{100}\right\}\% = + 8.507\%$

Hence, option (b) is correct.

Alternatively I

Let the rate of interest per annum on the principal be r%.

Now, CI for 1st yr = r% of the principal

Then, CI for 2 yrs = $\{r + (r + r\% \text{ of } r)\}$% of the principal

$= \left\{2r + \frac{r^2}{100}\right\}$% of the principal

$\therefore$ CI for 2 yrs on sum of 2304 = $2304 \times \left\{2r + \frac{r^2}{100}\right\}\%$

Then, $2304 \times \left\{2r + \frac{r^2}{100}\right\}\% = 2500 - 2304$

$\Rightarrow \quad 2304 \times \left\{2r + \frac{r^2}{100}\right\} = 19600 \Rightarrow r = 4.16\%$

Alternatively II

Amount = Principal $\times \frac{(100 + r)}{100} \times \frac{(100 + r)}{100}$

$\Rightarrow \quad 2500 = 2304 \times \left(\frac{100 + r}{100}\right)^2 \Rightarrow \frac{2500}{2304} = \left(\frac{100 + r}{100}\right)^2$

$\Rightarrow \quad \left(\frac{50}{48}\right)^2 = \left(\frac{100 + r}{100}\right)^2 \Rightarrow \frac{50}{48} = \frac{100 + r}{100} \Rightarrow r = 4.16\%$

Computation of A/P when *T*, *R* and CI are Known

Problem 8. The compound interest accrued on an amount at the end of 3 yrs at 15% per annum is ₹ 6500.52. What is the amount? ***[SBI(PO)-08, SSC(CGL-T-1)-19, CAT-20]***

Sol. CI for 3 yrs = (Cumulative SI) . $\boxed{3r^2}\boxed{r^3}$% of principal money

$= (3 \times 15) . \boxed{3 \times (15)^2}\boxed{(15)^3}$% of principal money

= 52.0875% of principal money = ₹ 6500.52 (as given)

$\Rightarrow$ 100% of the principal money = ₹ 12480

Hence, principal money = ₹ 12480

Alternatively

Amount = Principal $\times \frac{(100 + 15)}{100} \times \frac{(100 + 15)}{100} \times \frac{(100 + 15)}{100}$

$\Rightarrow \quad A = P \times 1.15 \times 1.15 \times 1.15 = 1.520875P$

Now, CI = Amount − Principal

$\Rightarrow \quad 6500.52 = 1.520875P - P \Rightarrow P = 12480$

Comparison of Amounts with Reference to Time/Rate of Interest

Problem 9. A sum of money doubles itself in 4 yrs at compound interest. It will amount to 8 times itself at the same rate of interest in ***[SSC(CGL-PT)-02, CDS-07, 21 SSC(CGL-T-II)-10, SSC(CGL-T-II)-11]***

Sol. In first 4 yrs, principal becomes $\Rightarrow$ 2 times

Then, in next 4 yrs, it will become $\Rightarrow$ 4 times

Further, in next 4 yrs, it will become $\Rightarrow$ 8 times

Hence, required time = 4 + 4 + 4 = 12 yrs

Alternatively,

If principal becomes x times itself in t yrs, it will become x^n times in $(n \times t)$ yrs at the same rate of interest. It also holds good that if principal becomes x times itself in t yrs, it will become $x^{1/n}$ times in (t/n) yrs at the same rate of interest.

Given, $x = 2$, $t = 4$ and $x^n = 8$

Then, $2^n = 8 \Rightarrow 2^n = 2^3 \Rightarrow n = 3$

Hence, required time = $(n \times t) = (3 \times 4) = 12$ yrs

Problem 10. A sum of money becomes three times itself in 5 yrs at certain rate of compound interest. In 25 yrs that sum will become how many times itself?

Sol. If 5 yrs, the principal becomes $\Rightarrow$ 3 times

In next 5 yrs, it will become $\Rightarrow$ 9 times

In next 5 yrs, it will become $\Rightarrow$ 27 times

In next 5 yrs, it will become $\Rightarrow$ 81 times

And in next 5 yrs, it will become 243 times

Hence, in total of 25 yrs, the principal will become 243 times itself.

Alternatively,

Given, $x = 3$, $t = 5$ and $(n \times t) = 25 \Rightarrow (n \times 5) = 25 \Rightarrow n = 5$

Then, $x^n = 3^5 = 243$

Hence, in total of 25 yrs, the principal will become 243 times itself.

Problem 11. A sum of money becomes 8 times in 3 yrs, if the rate is compounded annually. In how much time the same amount at the same compound interest rate will become sixteen times?

Sol. Let the principal be P.

Then, amount = $P \times (100 + r)\% \times (100 + r)\% \times (100 + r)\%$

$\Rightarrow \quad 8P = P \times (100 + r)\% \times (100 + r)\% \times (100 + r)\%$

$\Rightarrow \quad 8P \div P = (100 + r)\% \times (100 + r)\% \times (100 + r)\%$

$\Rightarrow \quad 2 \times 2 \times 2 = (100 + r)\% \times (100 + r)\% \times (100 + r)\%$

$\Rightarrow \quad 2 = (100 + r)\% \Rightarrow \frac{100 + r}{100} = 2 \Rightarrow r = 100$

Hence, rate of interest = 100%

Since rate of interest is 100%, it means the principal will be doubled every year. As given, in 3 yrs, the principal becomes 8 times. Then, in next year, the principal will become 16 times.

Thus, the required time = 4 yrs

Alternatively,

Given that the principal becomes 8 times itself in 3 yrs.

Then, it will become $8^{1/3}$ times itself in (3/3) yr.

Since every year the principal becomes 2 times, time taken to become 2^4 times itself = (1×4) yrs.

Problem 12. **A certain sum of money amounts to ₹ 9000 in 5 yrs and ₹ 13500 in 10 yrs at certain rate of compound interest. Find the sum.**

Sol. Given, in first 5 yrs the principal becomes ⇒ ₹ 9000

and in next 5 yrs the principal becomes ⇒ ₹ 13500

Here, the 2nd amount is 1.5 times the 1st amount. Then, obviously, the 1st amount will be 1.5 times the principal money.

Hence, sum = 9000 ÷ 1.5 = ₹ 6000

Problem 13. **At what rate percent per annum of compound interest, compounded annually, will a sum of money become 16 times of itself in 4 yrs?**

Sol. Given that the principal becomes 16 times itself in 4 yrs.

Then, it will become $16^{1/4}$ times itself in (4/4) yr. Hence, in 1 yr the principal becomes 2 times and it is possible only when rate of interest is 100% per annum.

Computation of Rate of Compound Interest or Principal or Both, When Amounts are Given for the Different Periods

Problem 14. **A sum of money amounts to ₹ 4840 in 2 yrs and ₹ 5324 in 3 yrs at compound interest compounded annually. The rate of interest per annum is**

[SSC(CPO)-07, SSC (SAS)-10, SSC(CGL-T-II)-11]

Sol. It is fact that compound interest is computed always on the previous Amount. Here, the CI for the 3rd yr will be computed on the Amount upto last 2 yrs i.e., on ₹ 4840.

Hence, CI for the 3rd yr = 5324 – 4840 = ₹ 484, which is 10% of ₹ 4840. It implies that the amount of ₹ 4840 has increased by 10% in only 1 yr so as to become ₹ 5324.

Thus, rate of interest = 10% per annum

Problem 15. **A sum of money amount to ₹ 5000 in 2 yrs at certain rate of interest and amounts to ₹ 5788.125 for a period of 5 yrs. The rate of interest per annum is**

Sol. Given, Amount of ₹ 5000 increases to ₹ 5788.125 in 3 yrs. Hence, increment in Amount in 3 yrs = 15.7625%

Obviously, CI for 3 yrs will be 15.7625% of the previous amount.

Now, check from given options as to which one yields 15.7625% in 3 yrs. Let given options be 4%, 5%, 6% and 8%. Then,

(a) 4% ⇒ (3 × 4) . | 3 × (4)² | (4)³ | % = 12.4864%

(b) 5% ⇒ (3 × 5) . | 3 × (5)² | (5)³ | % = 15.7625%

Hence, option (b) is correct.

Note *In case of 3 yrs, the last digit of increment percentage in principal (say, CI in term of percentage of principal) is the same as last digit of a number obtained by cubing the rate of interest. To illustrate: If rate of interest is 6%, then CI for 3 yrs = 19.1016% of the principal and $6^3 = 216$. Clearly, last digit of cube of rate of interest = last digit of CI expressed in term of percentage of principal = 6. Needless to mention that unit/last digit of cube of the numbers having different unit digits, are different from each other.*

Alternatively I

CI for 3 yrs = 15.7625% of the previous Amount. Let given options be 4%, 5%, 6% and 8%.

(a) 4% ⇒ last digit of cube of 4 ≠ last digit of 15.7625

(b) 5% ⇒ last digit of cube of 5 = last digit of 15.7625

Hence, option (b) is correct.

Alternatively II

CI for 3 yrs = 15.7625% of the previous Amount. Let given options be 4%, 5%, 6% and 8%.

It is very clear that rate of interest at 5% per annum will yield 15% at cumulative SI in 3 yrs. However, in case of compound interest, it will be little more than 15% as CI is computed also on earlier interest.

Hence, option (b) is correct.

If we have to find the sum in the above question, then

CI for 2 yrs at 5% per annum = 10.25% of the principal

Then, amount = 110.25% of the principal

Given, 110.25% of the principal = ₹ 5000

∴ 100% of the principal = ₹ 4535.15

Alternatively III

Amount of ₹ 5000 has become 1.157625 times itself in 3 yrs. Then, it will become $(1.157625)^{1/3}$ times in (3/3) yr.

Hence, in one year it becomes 1.05 times itself.

∴ Rate of interest per annum = 5%

Generic version : Rate of interest

$$= \left\{ \sqrt[\text{Intervening period in years}]{\frac{\text{bigger amount}}{\text{smaller amount}}} - 1 \right\} \times 100\%$$

Here, $r = \left\{ \sqrt[5-2]{\frac{5788.125}{5000}} - 1 \right\} \times 100 = 5\%$

Alternatively IV

Given, $5000 = P \times \left(1 + \frac{r}{100}\right)^2$

and $5788.125 = P \times \left(1 + \frac{r}{100}\right)^5$

$\Rightarrow \frac{5788.125}{5000} = \left(1 + \frac{r}{100}\right)^{5-2}$ {on dividing}

$\Rightarrow \frac{1157625}{1000000} = \left(1 + \frac{r}{100}\right)^3$

$\Rightarrow \left(\frac{105}{100}\right)^3 = \left(1 + \frac{r}{100}\right)^3$

$\Rightarrow \left(1 + \frac{5}{100}\right)^3 = \left(1 + \frac{r}{100}\right)^3$

$\Rightarrow r = 5\%$

Computation of Rate of Compound Interest or Principal or Both, When CI are Given for the Different Periods

Problem 16. In the beginning of the year 2004, a person invests some amount in a bank. In the beginning of 2007, the accumulated interest is ₹ 10000 and in the beginning of 2010, the accumulated interest becomes ₹ 25000. The interest rate is compounded annually and the annual interest rate is fixed. The principal amount is *[XAT-15]*

Sol. Let the principal be ₹ P and the effective rate of CI for the first 3 yrs = r%

Then, effective rate of CI for next 3 yrs = r% more than the earlier rate of interest = $(100 + r)$% of r%

As per question, r% of P = ₹ 10000

Further, {$(100 + r)$% of r%} of P = (25000 – 10000)

⇒ $(100 + r)$% of r% of P = 15000

⇒ $(100 + r)$% of 10000 = 15000

⇒ 100% of 10000 + r% of 10000 = 15000

⇒ $100r = 5000 \Rightarrow r = 50$

As given, r% of P = ₹ 10000

⇒ 50% of P = ₹ 10000 ⇒ P = 20000

Alternatively,

Let the cumulative CI for the first 3 yrs be r% of the principal.

Also, it is given that cumulative CI for the first 3 yrs be ₹ 10000.

Further, the cumulative CI for next 3 yrs will be r% more than the earlier cumulative CI.

⇒ 15000 = $(100 + r)$% of 10000

⇒ 15000 = 10000 + (r% of 10000)

⇒ 5000 = r% of 10000 ⇒ r = 50

Hence, cumulative CI for first 3 yrs = 50% of the principal

⇒ ₹ 10000 = 50% of the principal

Then, 100% of the principal = ₹ 20000

Concept of difference between SI and CI In percent of principal

	SI	Cumulative SI	Cumulative CI	Difference between Cumulative SI and Cumulative CI
1st yr	r	r	r	0
2nd yr	r	$2r$	(Cumulative SI). $\boxed{r^2}$	$0.\boxed{r^2}$ $\left(\text{or } \frac{r^2}{100}\right)$
3rd yr	r	$3r$	(Cumulative SI). $\boxed{3r^2}\boxed{r^3}$	$0.\boxed{3r^2}\boxed{r^3}$ $\left(\text{or } \frac{3r^2}{100}+\frac{r^3}{10000}\right)$
4th yr	r	$4r$	(Cumulative SI). $\boxed{6r^2}\boxed{4r^3}\boxed{r^4}$	$0.\boxed{6r^2}\boxed{4r^3}\boxed{r^4}$ $\left(\text{or } \frac{6r^2}{100}+\frac{4r^3}{10000}+\frac{r^4}{1000000}\right)$

(r is the rate of interest per annum)

Note *Each box will have only two digits and the surplus figure(s) will be carried over left. Also, each box will contain two digits and in case of deficiency, 0 will be appended in left hand side – this has already been discussed.*

Observation Ratio of differences between CI and SI for 2 yrs and 3 yrs at r% rate of interest

$$= \left(\frac{r^2}{100}\right)\% \text{ of the principal} : \left(\frac{3r^2}{100}+\frac{r^3}{10000}\right)\% \text{ of the principal}$$

$$= 1 : \left(3 + \frac{r}{100}\right)$$

We may also conclude that if the differences between CI and SI for 2 yrs and 3 yrs are in ratio $x : y$, then rate of interest

$$= \left(\frac{y - 3x}{x} \times 100\right)\%.$$

Now discuss few questions

Problem 17. Find out the difference between SI and CI on a sum of ₹ 1000 at 6% per annum for a period of 3 yrs and 4 yrs.

Sol. (i) From the above chart, difference between SI and CI for 3 yrs

$= 0.\boxed{3r^2}\boxed{r^3}$ % of the principal

$= 0.\boxed{3 \times 6^2}\boxed{6^3}$ % of the principal

= 1.1016% of the 1000 = ₹ 11.02

(ii) From the above chart, difference between SI and CI for 4 yrs

$= 0.\boxed{6r^2}\boxed{4r^3}\boxed{r^4}$ % of the principal

$= 0.\boxed{6 \times 6^2}\boxed{4 \times 6^3}\boxed{6^4}$ % of the principal

= 2.247696% of 1000 = ₹ 22.48

Problem 18. The difference between CI and SI on ₹ 2500 for 2 yr at 4% per annum is

[SSC(CPO)-03, SSC(CGL-T-II)-10, CLAT-14,CLAT-19]

Sol. SI for 2 yrs = (2 × 4)% of the principal

CI for 2 yrs = $\left\{+\ 4 + 4 + \frac{4 \times 4}{100}\right\}$ % of the principal money

= 8.16% of the principal money

Hence, difference between SI and CI

= 0.16% of the principal money = 0.16% of 2500 = ₹ 4

Alternatively,

From the table, difference between SI and CI for 2 yrs.

$= 0.\boxed{r^2}$ % of principal money

$= 0.\boxed{4^2}$ % of principal money

= 0.16% of the principal money

= 0.16% of 2500 = 4

Problem 19. The difference between SI and CI on a sum of ₹ 2000 for a period of $2\frac{1}{2}$ yrs at the rate of 8% per annum will be

Sol. CI for 2 yrs = (2 × 8). $\boxed{8^2}$ % of the principal

= 16.64% of the principal

Further, rate of interest for $\frac{1}{4}$ yrs = 8 ÷ 4 = 2%

Then, CI for $2\frac{1}{2}$ yrs $= \left\{16.64 + 2 + \frac{16.64 \times 2}{100}\right\}$ % of the principal

$= 18.97\%$ of the principal

and SI for $2\frac{1}{4}$ yrs $= (16 + 2)\%$ of the principal

Hence, required difference $= (18.97 - 18)\%$ of $2000 = ₹\,19.40$

Alternatively,

For the period of 2 yrs, difference $= 0.\boxed{r^2}$ % of the principal

$= 0.\boxed{(8)^2}$ % of the principal $= 0.64\%$ of the principal.

For further $\frac{1}{4}$ yrs, SI will be 2% of the principal but CI will be 2% of 16.64% of the principal over and above 2%.

Obviously, the difference for $\frac{1}{4}$ yrs = 2% of 16.64% of principal

$= 0.33\%$ of the principal

Hence, total difference $= (0.64 + 0.33)\%$ of the principal

$= 0.97\%$ of $2000 = ₹\,19.40$

Problem 20. The difference between simple interest and compound interest on ₹ 1200 for 1 yr at 10% per annum reckoned half- yearly is ***[RRB-Bhubnsr(ASM)-02]***

Sol. Interest is being compounded half yearly.

Thus, rate of interest for $\frac{1}{2}$ yrs = 5% (∵ 10% for 1 yr)

Also, number of $\frac{1}{2}$ yr blocks in 1 yr = 2

(say, 2 yrs for sake of convenience)

Now, difference between SI and CI for 2 yrs at rate of 5%

$= 0.\boxed{5^2}$ % of principal money

$= 0.25\%$ of the principal money $= 0.25\%$ of $1200 = ₹\,3$

Problem 21. The difference between SI and CI accrued on an amount of ₹ 2500 in 3 yrs is ₹ 19.0625. Find the rate of interest per annum?

Sol. Given, difference = 19.0625

Then, difference in term of percent of principal

$= \left(\frac{19.0625}{2500} \times 100\right)\% = 0.7625\%$

It holds term of good that the last digit of difference between SI and CI, given in term of percentage of principal is the same as last digit of a number obtained by cubing the rate of interest.

Now, proceed through options. Let options be 4%, 5%, 6% and 8%.

(a) 4% ⇒ last digit of cube of 4 ≠ last digit of 0.7625

(b) 5% ⇒ last digit of cube of 5 = last digit of 0.7625

Hence, option (b) is correct.

Note: In case of 3 yrs, the last digit of difference between SI and CI given in term of percentage of principal is same as the last digit of a number obtained by cubing rate of interest.

Alternatively I

Obviously, the difference between SI and CI for 3 yrs

$= 0.\boxed{3r^2}\boxed{r^3}$ % of the principal

$= 0.7625\%$ of principal (from above)

Putting the value of r from given options we get $r = 5$

Hence, rate of interest per annum = 5%

Problem 22. The difference between CI and SI accrued on a amount of ₹ 1200 in a certain time at the rate of 10% per annum is ₹ 37.20. Find the period.

Sol. CI $= \left(\frac{37.20}{1200} \times 100\right)$ % of the principal = 3.1% of the principal.

Now, check from the options (let be 2yrs, 3 yrs, 4 yrs, 5 yrs)

(a) 2 yrs ⇒ difference $= 0.\boxed{r^2}$ % of the principal

$= 0.\boxed{(10)^2}$ % of the principal

$= 1.00\%$ of the principal ≠ 3.1% of the principal

(b) 3 yrs ⇒ $0.\boxed{3 \times 10^2}\boxed{(10)^3}$ % of the principal

$= 3.1\%$ of the principal $= 3.1\%$ of the principal

Hence, option (b) is correct.

Problem 23. The difference between SI on a certain sum at the rate of 10% per annum for 2 yrs and compound interest which is compounded every 6 months is ₹ 124.05. What is the principal? ***[SBI(PO)-2000,SSC(CGL-T-II)-13]***

Sol. Interest is being compounded half yearly.

Thus, rate of interest for $\frac{1}{2}$ yrs = 5% (∵ 10% for 1 yr as given)

Also, number. of $\frac{1}{2}$ yr blocks in 2 yrs = 4

(say, 4 yrs for sake of convenience)

Now, difference between SI and CI for 4 yrs

$= 0.\boxed{6r^2}\boxed{4r^3}\boxed{r^4}$ % of the principal

$= 0.\boxed{6 \times 5^2}\boxed{4 \times 5^3}\boxed{5^4}$ % of the principal

$= 0.\boxed{150}\boxed{500}\boxed{625}$ % of the principal

$= 1.\boxed{55}\boxed{06}\boxed{25}$ % of the principal

⇒ 1.550625% of the principal = ₹ 124.05

⇒ 100% of the principal = ₹ 8000

Hence, principal money is ₹ 8000.

Alternatively

Interest is being compounded half yearly.

Thus, rate of interest for $\frac{1}{2}$ yrs = 5% (∵ 10% for 1 yr as given)

Also, number of $\frac{1}{2}$ yr blocks in 2 yrs = 4 (say, 4 yrs for sake of convenience)

As given, $P\left(1 + \frac{5}{100}\right)^4 - P - \frac{P \times 10 \times 2}{100} = 124.05$

⇒ $P\left(\frac{194481}{160000} - \frac{6}{5}\right) = 124.05 \Rightarrow P = ₹\,8000$

Problem 24. The difference of compound interest and simple interest for 3 yrs and 2 yrs are in the ratio 23 : 7 respectively. What is rate of interest per annum (in %)? ***[SSC(CGL-T-1)-17]***

Sol. Let the rate of interest be r% per annum.

Then, difference between CI and SI for 2 yrs = r% of r% of the principal $= \frac{r^2}{100}$ % of the principal.

Further, difference between CI and SI for 3 yrs = difference for 2 yrs + r% of $\left\{2r + \frac{r^2}{100}\right\}$ % of the principal

$= \frac{r^2}{100}\%$ of the principal + $\frac{2r^2}{100}\%$ of the principal

$+ \frac{r^3}{100}\%$ of the principal.

$\Rightarrow 23x = 7x + (2 \times 7x) + (r \times 7x)$ ($\because$ given ratio = 23 : 7)

$\Rightarrow 2x = 7rx$

$\Rightarrow r = \frac{2}{7} = \frac{200}{7}\%$

Alternatively,

We know, if the differences between CI and SI for 2 yrs and 3 yrs are in ration $x : y$, then rate of interest

$= \left(\frac{y - 3x}{x} \times 100\right)\%.$

Hence, required rate of interest $= \left(\frac{23 - 3 \times 7}{7} \times 100\right)\% = \frac{200}{7}\%$

Computation of Rate of Interest or Principal When SI and CI are Given for a Particular Year/Period(s)

Problem 25. The SI and CI accrued on a certain sum of money for a period of 2 yrs are ₹ 400 and ₹ 420 respectively. Find out the rate of interest per annum.

[SSC(CGL-T-II)-10]

Sol. Please refer to the chart given in the beginning of the chapter. The SI and CI for the 1st year is same, but in 2nd yr they differ as compound interest is compute also on the 1st yr interest. Here, the question may be elaborated as below for building the concept.

	SI	CI
1st yr	200	200
2nd yr	200	220
Total	400	420

Obviously, the difference of ₹ 20 between 1st yr and 2nd yr CI is due to calculation of CI at $r\%$ per annum also over the 1st yr interest. It means $r\%$ of the 1st yr interest = ₹ 20

$\Rightarrow$ $r\%$ of 200 = 20 $\Rightarrow r = 10$

Hence, rate of interest per annum = 10%

Alternatively,

We know (given below the chart in beginning) that: $(100 + r)\%$ of CI for 1st yr = CI for the 2nd yr

$\Rightarrow$ $(100 + r)\%$ of 200 = 220

$\Rightarrow$ 100% of 200 + $r\%$ of 200 = 220

$\Rightarrow$ $r\%$ of 200 = 20 $\Rightarrow$ $r = 10$

Hence, rate of interest per annum = 10%

If it is asked to find the principal in this question, then :

The SI for 1st yr is ₹ 200.

It means $(1 \times 10)\%$ of the principal = ₹ 200

$\therefore$ 100% of the principal = ₹ 2000

Problem 26. The SI and CI accrued on a certain sum of money for a period of 3 yrs are ₹ 1200 and ₹ 1261 respectively. Find out the sum and rate of interest per annum.

Sol. Let rate of interest be $r\%$.

	SI	CI
1st yr	400	400
2nd yr	400	$400 + r\%$ of $400 = 400 + 4r$
3rd yr	400	$400 + r\%$ of $\{400 + (400 + 4r)\}$
		$= 400 + 8r + \frac{r^2}{25}$

Clearly, difference between SI and CI $= 4r + 8r + \frac{r^2}{25}$

$\Rightarrow (1261 - 1200) = 12r + \frac{r^2}{25}$

$\Rightarrow 1525 = 300r + r^2$

$\Rightarrow r^2 + 300r - 1525 = 0$

$\Rightarrow r^2 + 305r - 5r - 1525 = 0$

$\Rightarrow r(r + 305) - 5(r + 305) = 0$

$\Rightarrow (r - 5)(r + 305) = 0$

$\Rightarrow r = 5$ or -305

Hence, rate of interest per annum = 5%

Further,

Obviously, 5% of the principal = ₹ 400

Then, 100% of the principal = ₹ 8000

Alternatively I

You may proceed through the options. Let options be 6%, 5%, 4% and 8%.

You have to put value of option in the above chart and ascertain as to which option makes a difference of ₹ 61.

Alternatively II

The difference between SI and CI for 3 yrs

= 0. $\boxed{3r^2}\boxed{r^3}$ % of the principal

Now, you may proceed through the option as below :

(a) 6% $\Rightarrow (3 \times 6)\%$ of the principal = ₹ 1200

$\therefore$ 100% of the principal = 6666.67

Then, difference = 0.$\boxed{3 \times 6^2}\boxed{6^3}$ % of the principal

= 1.1016% of 6666.67 $\neq$ 61

(b) 5% $\Rightarrow (3 \times 5)\%$ of the principal = ₹ 1200

$\therefore$ 100% of the principal = ₹ 8000

Then, difference = 0.$\boxed{3 \times 5^2}\boxed{5^3}$ % of the principal

= 0.7625% of 8000 = 61

Hence, option (b) is correct.

Problem 27. The simple interest on a sum of money at some rate for 3 yrs is ₹ 225 and the compound interest on the same sum of money and at the same rate for 2 yrs is ₹ 153. Find the sum and the rate per cent per annum.

	SI	CI
1st yr	75	75
2nd yr	75	78
Total	150	153

Obviously, the difference of ₹ 3 between 1st yr and 2nd yr CI is due to calculation of CI at $r\%$ per annum also over the 1st yr interest. It means $r\%$ of the first yr interest = ₹ 3.

$\Rightarrow$ $r\%$ of 75 = 3 $\Rightarrow r = 4$

Hence, rate of interest per annum = 4%

Alternatively,

We know, $(100 + r)\%$ of CI for 1st yr = CI for the 2nd yr

$\Rightarrow$ $(100 + r)\%$ of 75 = 78

$\Rightarrow$ 100% of 75 + $r\%$ of 75 = 78

$\Rightarrow$ $r\%$ of 75 = 3 $\Rightarrow r = 4$

Hence, rate of interest per annum = 4%

Further, SI for 1st yr is ₹ 75.

It means $(1 \times 4)\%$ of the principal = ₹ 75

$\therefore$ 100% of the principal = ₹ 1875

Problem 28. The SI and CI on the principal money for 2nd yr are ₹ 500 and 600 respectively. What is the rate of interest per annum?

	SI	CI
1st yr	(500)	(500)
2nd yr	500	600

Sol. Obviously, the difference of ₹ 100 between 1st yr and 2nd yr's CI is due to calculation of CI at $r\%$ per annum also over the 1st yr interest. It means $r\%$ of the 1st yr's interest = ₹ 100

$\Rightarrow$ $r\%$ of 500 = 100

$\Rightarrow$ $r = 20$

Hence, rate of interest per annum = 20%.

Alternatively,

We know,

$(100 + r)\%$ of CI for 1st year = CI for the 2nd yr

$\Rightarrow$ $(100 + r)\%$ of 500 = 600

$\Rightarrow$ $r\%$ of 500 = 100

$\Rightarrow$ $r = 20$

Hence, rate of interest per annum = 20%.

Problem 29. A sum of money is accumulating at compound interest at a certain rate of interest. If simple interest instead of compound were reckoned, the interest for first 2 yrs would be diminished by ₹ 20 and that for the first 3 yrs by ₹ 61. What is the sum? *[MAT-13]*

Sol. Given that difference between CI and SI for 2 yrs = ₹ 20 and for 3 yrs = ₹ 61. It implies that difference between them for 3rd yr = ₹ 41.

	SI	CI
1st yr	x	x
2nd yr	x	$x + 20$
3rd yr	x	$x + 41$

Obviously, tho difference of ₹ 20 between SI and CI in 2nd yr is due to calculation of CI at $r\%$ per annum also over the 1st yr's interest. It means $r\%$ of the 1st yr interest = ₹ 20

$\Rightarrow$ $r\%$ of $x = 20$

Further, the difference of ₹ 41 between SI and CI for 3rd yr is due to calculation of interest also on previous interests. It means, $r\%$ of $(x + x + 20) = 41$

$\Rightarrow$ $(r\%$ of $x) + (r\%$ of $x) + (r\%$ of $20) = 41$

$\Rightarrow$ $20 + 20 + (r\%$ of $20) = 41$

$\{\because r\%$ of $x = 20$ as obtained above$\}$

$\Rightarrow$ $(r\%$ of $20) = 1 \Rightarrow r = 5$

Hence, rate of interest = 5% per annum.

Further, $r\%$ of $x = 20$

$\Rightarrow$ 5% of $x = 20 \Rightarrow x =$ ₹ 400

Also, SI for 1st yr = 5% of the sum

$\Rightarrow$ 400 = 5% of the sum $\Rightarrow$ sum = ₹ 8000

Alternatively I,

We know that, $(100 + r)\%$ of CI for 1st yr = CI for the 2nd yr

$\Rightarrow$ $(100 + r)\%$ of $x = x + 20$

$\Rightarrow$ $r\%$ of $x = 20$

Further, $(100 + r)\%$ of CI for 2nd yr = CI for 3rd year

$\Rightarrow$ $(100 + r)\%$ of $(x + 20) = x + 41$

$\Rightarrow x + (r\%$ of $x) + 20 + (r\%$ of $20) = x + 41$

$\Rightarrow$ $x + 20 + 20 + (r\%$ of $20) = x + 41$

$\Rightarrow r = 5$

Hence, rate of interest = 5% per annum.

Further, $r\%$ of $x = 20$

$\Rightarrow$ 5% of $x = 20 \Rightarrow x =$ ₹ 400

Also, SI for 1st yr = 5% of the sum

$\Rightarrow$ 400 = 5% of the sum $\Rightarrow$ sum = ₹ 8000

Alternatively II

We know, ratio of differences between CI and SI for 2 yrs and 3 yrs at $r\%$ rate of interest = $1 : \left(3 + \frac{r}{100}\right)$

$$\Rightarrow \quad 20 : 61 = 1 : \left(3 + \frac{r}{100}\right) \Rightarrow 61 = 60 + \frac{r}{5} \Rightarrow r = 5\%$$

Obviously, the difference of ₹ 20 between SI and CI in 2nd yr is due to calculation of CI at 5% per annum also over the 1 yr's interest. It means, 5% of the 1st yr interest = ₹ 20

$\Rightarrow$ 1st yr interest = ₹ 400

Also, SI for 1st yr = 5% of the sum

$\Rightarrow$ 400 = 5% of the sum $\Rightarrow$ sum = ₹ 8000

Computation of P/A when CI for the Particular Year is Given with the Rate of Interest

Problem 30. Find the sum on which CI at 6% per annum for the 2nd yr is ₹ 318.

	SI	CI
1st yr	x	x
2nd yr	x	$x + (6\%$ of $x) = 318$

From above, $x + (6\%$ of $x) = 318 \Rightarrow 106\%$ of $x = 318$

$\Rightarrow$ 100% of $x = 300$

Obviously, 6% of the principal = ₹ 300

Then, 100% of the principal = ₹ 5000

Hence, the sum is ₹ 5000.

Alternatively I,

We know that CI for the 2nd yr is equal to the sum of SI for 1st yr and interest accrued there on at the given rate of interest. It implies that CI for the 2nd yr is 6% more than the SI for 1st yr or any yr.

Hence, 106% of SI = ₹ 318
Then, 100% of SI = ₹ 300
Further, SI for any yr = ₹ 300
⇒ 6% of the principal = ₹ 300
Then, 100% of the principal = ₹ 5000
Hence, the sum is ₹ 5000.

Alternatively II

$$\text{Principal} = \text{CI for 2nd yr} \times \frac{100}{(100+r)} \times \frac{100}{r}$$

$$\text{Now, principal} = 318 \times \frac{100}{(100+6)} \times \frac{100}{6} = ₹\,5000$$

Computation of SI, if CI for the same Period at Same Rate of Interest is Known

Problem 31. The compound interest accrued on an amount at 4% per annum for a period of 3 yrs is ₹ 624.32. Then, find the SI for the same period at the same rate of interest.

Sol. Here, CI for 3 yrs = (Cumulative SI). $\boxed{3 \times (4)^2}\boxed{4^3}$ % of the principal

⇒ CI = $(3 \times 4) \cdot \boxed{48}\boxed{64}$ % of the principal

⇒ CI = 12.4864% of the principal

As given, 12.4864% of the principal = ₹ 624.32

Then, SI i.e. (3 × 4)% of the principal = ₹ 600

{**Note** You may assume 624.32 as 625 and 12.4864 as 12.5 for sake of convenience in calculation}

Alternatively

As given, $P\left(1+\frac{4}{100}\right)^3 - P = 624.32$

$\Rightarrow P\left(\frac{17576}{15625}-1\right) = 624.32 \Rightarrow P = ₹\,5000$

Now, SI i.e. (3 × 4)% of the principal = 12% of 5000 = 600

Comparison of Amounts in case of SI and CI

Problem 32. A sum of money amounts to ₹ 3000 at 10% per annum simple interest in 2 yrs. At same rate of CI for the same period that sum would be amounted to what? ***[SSC (CGL-T-II)-19]***

Sol. SI = (2 × 10)% of the principal

and CI = $(2 \times 10) \cdot \boxed{(10)^2}$ % of the principal

CI = 21% of the principal

Now, as given, 120% of the principal = ₹ 3000

Then, 121% of the principal = ₹ 3025

Problem 33. The Amount on a certain principal at certain rate of compound interest for 2 yrs is 30% more than the Amount on the same sum at the same rate of interest for one yr at simple interest. What is the rate of interest per annum?

Sol. It is obvious that the Amount at CI for 2 yrs will exceed the Amount at SI for 1 yr by r%, where r is the rate of interest per annum.

Hence, r = 30%.

Concept of Complete yrs Required to Make the Principal *n* Times

Problem 34. The least number of complete yrs in which a sum of money put out at 20% compound interest will be more than doubled is ***[NIFT-03, CDS-11, MAT-11]***

Sol. If the concept of simple interest is considered, in 5 yrs, the sum will be doubled. Then, it is obvious that the sum will be doubled in less than 5 yrs if compound interest is considered. Now we may try with 'trail and error method'.

For 3 yrs,

CI = $(3 \times 20) . \boxed{3 \times (20)^2}\boxed{(20)^3}$ % of the principal

= 72.8% of the principal

It means, in 3 yrs, the principal will increase by only 72.8%.

For 4 yrs,

CI = $(4 \times 20) . \boxed{6 \cdot (20)^2}\boxed{4 \cdot (20)^3}\boxed{(20)^4}$ % of the principal

= 107.36% of the principal

It means, in 4 yrs, the principal will increase by 107.36%.
Hence, in 4 yrs the principal will be more than doubled.

Alternatively

From above, in 3 yrs, the principal will increase by only 72.8%.
Then, in 4 yrs, principal will increase by

$$= \left\{+72.8 + 20 + \frac{72.8 \times 20}{100}\right\}\% = \text{more than } 100\%$$

Hence, in 4 yrs the principal will be more than doubled.

Problem 35. What is the least number of complete years in which a sum of money put out at 40% annual compound interest will be more than trebled? ***[CDS-19]***

Sol. Try yourself on the above line. The answer is 4.

Concept of tax-Deduction on the Compound Interest

Problem 36. Kranti deposited ₹ 50000 in a NBFC at the rate of 10% per annum compound interest. What will he have after 4 yrs if he pays 10% tax on the interest per annum.

Sol. Effective rate of compound interest = 90% of 10% = 9%

∴ CI for period for 4 yrs

= $(4 \times 9) . \boxed{6 \cdot (9)^2}\boxed{4 \cdot (9)^3}\boxed{(9)^4}$ % of the principal

⇒ CI = 41.158161% of the principal

Hence, Amount = 141.158161% of 50000 = ₹ 70579.08

Ratio of Amounts, with reference to SI and CI, is Given and Rate of Interest to be Computed.

Problem 37. The ratio of the Amount for 3 yrs at certain rate of simple interest and that for 2 yrs at the same rate of compound interest is 10 : 9. What is the rate of interest?

Sol. Let the rate of interest be r%.

Then, Amount under SI for 3 yrs = (100 + 3r)% of the principal.

Further, CI for 2 yrs = {r + (r + r% of r)}% of the principal

$$= \left\{2r + \frac{r^2}{100}\right\}\% \text{ of the principal.}$$

Then, amount under CI $= \left\{100 + 2r + \frac{r^2}{100}\right\}$% of the principal

As given, $\frac{\left\{100 + 2r + \frac{r^2}{100}\right\}}{(100 + 3r)} = \frac{9}{10}$

$\Rightarrow \quad \frac{r^2}{10} + 20r + 1000 = 900 + 27r$

$\Rightarrow \quad r^2 - 70r + 1000 = 0 \Rightarrow r^2 - 50r - 20r + 1000 = 0$

$\Rightarrow \quad r(r - 50) - 20(r - 50) = 0 \Rightarrow r = 50 \text{ or } 20$

Hence, rate of interest = 50% or 20% per annum.

Alternatively,

You may proceed through options. Let options be 30%, 20%, 25% and 10%.

(a) 30% ⇒ Amount after 3 yrs at SI = 190% of the principal and Amount at CI for 2 yrs = 169% of the principal then, the required ratio = 190 : 169 ≠ 10 : 9

(b) 20% ⇒ Amount after 3 yrs at SI = 160% of the principal and Amount of CI for 2 yrs = 144% of the principal

Hence, the required ratio = 160 : 144 = 10 : 9

Thus, option (b) is correct.

Problems Based on Population

The problems based on the population have already been covered under chapter 'Percentage' with concepts associated with the compound interest too.

Problems Based on Devaluation/Depreciation

It is directly associated with the concept of Discount which has been given in detail in the chapter 'Discount'. However, few examples are being given for the clarity.

Problem 38. The present value of car Ciaz is 10% higher than that of car Baleno. The value of car Ciaz is depreciating by 10% per annum. Find out the depreciation percent per annum in respect of car Baleno provided values of both after 2 yrs are equal.

Sol. Let the present value of Baleno be ₹ 100.

Then, present value of Ciaz be ₹ 110.

Now, depreciation in value of Ciaz

$$= \left\{-10 - 10 + \frac{10 \times 10}{100}\right\}\% = -19\%$$

Thus, remaining value of Ciaz = (100 − 19)% of 110 = 89.1

Further, the remaining value of Baleno is also ₹ 89.1 as both have same value after 2 yrs.

Hence, decrease in value of Baleno = 100 − 89.1 = 10.9

Then, percentage decreases = 10.9%

Now, check from the options as to what percentage of depreciation will convert into 10.9% in a period of 2 yrs.

Let option be 8%, 5.6%, 7.5%, 9%

(a) 8% $\Rightarrow \left\{-8 - 8 + \frac{8 \times 8}{100}\right\}\% = -15.36\% \neq -10.9\%$

(b) 5.6% $\Rightarrow \left\{-5.6 - 5.6 + \frac{5.6 \times 5.6}{100}\right\}\% = -10.9\%$

Hence, option (b) is correct.

Problem 39. What is the present worth of a house which would be worth ₹ 50000 after 3 yrs, if it depreciates at the rate of 10%? ***[CAT-09]***

Sol. Depreciation in 2 yrs = 19% $\left[\because \left(-10 - 10 + \frac{10 \times 10}{100}\right)\% = -19\%\right]$

Now, for depreciation in 3 yrs, $\left(-19 - 10 + \frac{19 \times 10}{100}\right)\% = -27.1\%$

Hence, depreciation in 3 yrs = 27.1% of the present worth of the house

As given, (100 − 27.1)% of the present worth = ₹ 50000

⇒ 100% of the present worth = ₹ 68587.10

Finding Values or Ratio of Parts Lent/ Borrowed at Same Rates for the Different Periods and Amount in each Case is Equal

Problem 40. Divide ₹ 1165.50 between Ganu and Manu in such a way that amount of Ganu after 11.5 yrs is equal to the amount of Manu after 14.5 yrs, the interest being compounded at 10% per annum.

Sol. It is obvious that the part of Ganu is greater than that of Manu as part of Ganu takes only 11.5 yrs and that of Manu takes 14.5 yrs to become the same amount. It implies that share of Manu will be equal to that of Ganu after 3 yrs. Further, CI for 3 yrs at the rate of 10% = 33.1% of the principal.

Let options be 765.50 and 400, 665.50 and 500, 465 and 700.50, ...

Now, check through options as to one part of which option is 33.1% greater than the other part of the same option and sum of both parts is 1165.50.

(a) 756.50 and 400 ⇒ 765.50 is not 33.1% greater than 400

(b) 665.50 and 500 ⇒ 665.50 is 33.1% greater than 500

Hence, option (b) is correct.

Alternatively

Ganu's share = 33.1% more than Manu's share

⇒ Ganu's share : Manu's share = 133.1 : 100

Now divide 1165.50 in ratio 133.1 : 100

Here, (133.1 + 100) units ⇒ 1165.50

Then, 100 units ⇒ 500

Hence, Manu's share = 500

Then, Ganu's share = 1165.50 − 500 = 665.50

Problem 41. A sum of money lent out at compound interest increases in value by 50% in 5 yrs. A person wants to lend three different sums of money *X*, *Y* and *Z* for 10, 15 and 20 yrs respectively at the above rate in such a way that he gets back equal sums of money at the end of their respective periods. Find the value of *X* : *Y* : *Z*.

Sol. It may be construed from the question that each sum of money will become 1.5 times itself in a period of 5 yrs.

In 10 yrs, *X* will become (1.5 × 1.5) times.

In 15 yrs, *Y* will become (1.5 × 1.5 × 1.5) times

In 20 yrs, *Z* will become (1.5 × 1.5 × 1.5 × 1.5) times

As given, (1.5 × 1.5) times of *X* = (1.5 × 1.5 × 1.5) times of *Y*

= (1.5 × 1.5 × 1.5 × 1.5) times of *Z*

$\Rightarrow \quad X = 1.5Y = 2.25Z \Rightarrow X : Y = 1.5 : 1 = 3 : 2 = 9 : 6$

and $\quad Y : Z = 2.25 : 1.5 = 3 : 2 = 6 : 4$

Clearly, $X : Y : Z = 9 : 6 : 4$

Note *We will learn to find the ratio in the above circumstances in very easy way under chapter 'Ratio, Proportion and Variation'*

Finding Values or Ratio of Parts Lent/Borrowed at Different Rates for the Same Period and the Aggregate CI is Given

Problem 42. **Mr. Mishra invested ₹ 25000 in two fixed deposits X and Y offering compound interest at 6% per annum and 8% per annum respectively. If the total amount of interest accrued in 2 yrs through both fixed deposits is ₹ 3518, then the amount invested in scheme X is** ***[IIFT-12]***

Sol. Let the amount invested in scheme X be ₹ P.

Then, 12.36% of P + 16.64% of (25000 – P) = 3518

{∵ CI for 2 yrs 6% = 12.36% and CI for 2 yrs at the rate 8% = 16.64% of principal}

$\Rightarrow \quad -(4.28\% \text{ of } P) + 4160 = 3518$

$\Rightarrow \quad 4.28\% \text{ of } P = 642 \Rightarrow 100\% \text{ of } P = 15000$

Alternatively,

Effective rate of CI for 2 yrs $= \left(\frac{3518}{25000} \times 100\right)\% = 14.072\%$

Now, $\quad r + r + \frac{r^2}{100} = 14.07 \Rightarrow r = 6.8\%$

Then,

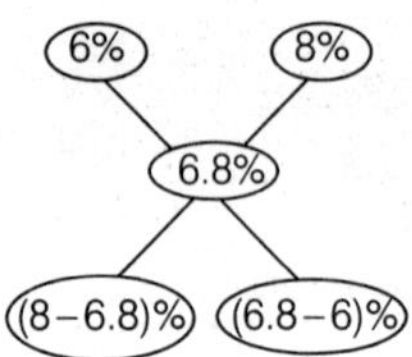

Hence, the amount invested in scheme X : Amount invested in scheme Y = 1.2 : 0.8 = 3 : 2

As given, (3 + 2) units = ₹ 25000

∴ $\quad$ 3 units = ₹ 15000

Concept of Equal Installments for Discharging a Debt

When a borrower repays the borrowed money with the interest in parts, it is known as repayment in installment. Generally, installments are equal and each installment consists of partly principal money, utilized for a certain period and interest thereon. After repayment of the 1st installment, interest is charged on the rest of the principal and so on. Discuss few problems.

Problem 43. **What equal annual payment will discharge a debt of ₹ 6620 due in 3 yrs at 10% compound interest per annum.**

Sol. Suppose a sum of ₹ p_1 has been utilized for 1 yr, p_2 has been utilized for 2 yrs and p_3 for 3 yrs. Also, let each installment be of ₹ x.

At 10% per annum, CI for 1 yr, 2 yrs and 3 yrs will be 10%, 21% and 33.1% of the principal respectively.

Obviously, x is 10% more than p_1, hence p_1 is $\left(\frac{10}{100+10} \times 100\right)\%$

i.e. 9.09% less than x. Then, $p_1 = 90.91\%$ of x

Further, x is 21% more than p_2, hence, p_2 is $\left(\frac{21}{100+21} \times 100\right)\%$

i.e. 17.36% less than x. Then, $p_2 = 82.64\%$ of x

Also, x is 33.1% more than p_3, hence p_3 is $\left(\frac{33.1}{100+33.1} \times 100\right)\%$

i.e. 24.87% less than x. Then, $p_3 = 75.13\%$ of x

Now, (90.91 + 82.64 + 75.13)% of each installment = ₹ 6620

Then, 100% of each installment = ₹ 2662

Hence, required installment = ₹ 2662

Note *You have to take only approx. value instead of exact values in decimal for sake of convenience in calculation.*

Alternatively I

Obviously, 1st installment of ₹ x will be equal to the sum of p_1 and the interest accrued thereon for 1 yr.

Hence, $\quad x = 110\%$ of p_1

$\Rightarrow p_1 = \frac{x}{110\%}$

Further, 2nd installment of ₹ x is equal to the sum of p_2 and interest accrued thereon for 2 yrs.

Hence, $\quad x = 121\%$ of p_2

$\Rightarrow \quad p_2 = \frac{x}{121\%}$

Similarly, $p_3 = \frac{x}{133.1\%}$

Now, $\quad p_1 + p_2 + p_3 = \frac{x}{110\%} + \frac{x}{121\%} + \frac{x}{133.1\%}$

$\Rightarrow \quad 6620 = \frac{100x}{110} + \frac{100x}{121} + \frac{100x}{133.1}$

$\Rightarrow \quad 6620 = \frac{100x}{110} + \frac{100x}{121} + \frac{1000x}{1331}$

$\Rightarrow \quad 6620 \times 1331 = 1210x + 1100x + 1000x$

$\Rightarrow \quad x = ₹\, 2662$

Alternatively II

$$\text{Each installment} = \frac{\text{Principal or Debt}}{\frac{1}{\left\{1+\frac{r}{100}\right\}^1} + \frac{1}{\left\{1+\frac{r}{100}\right\}^2} + \frac{1}{\left\{1+\frac{r}{100}\right\}^3} + \dots + \frac{1}{\left\{1+\frac{r}{100}\right\}^n}}$$

Where, r = Rate of interest per annum and n = number of years installments

Now, each installment

$$= \frac{6620}{\frac{1}{\left\{1+\frac{10}{100}\right\}^1} + \frac{1}{\left\{1+\frac{10}{100}\right\}^2} + \frac{1}{\left\{1+\frac{10}{100}\right\}^3}}$$

= ₹ 2662

Alternatively III

$$\text{Each installment} = \text{Borrowed money} \times \frac{\left(1+\frac{r}{100}\right)^n \times \frac{r}{100}}{\left(1+\frac{r}{100}\right)^n - 1}$$

$$OR \text{ Borrowed money} = \text{Each installment} \times \frac{\left(1+\frac{r}{100}\right)^n - 1}{\left(1+\frac{r}{100}\right)^n \times \frac{r}{100}}$$

$$\text{Hence, each installment} = 6620 \times \frac{\left(1+\frac{10}{100}\right)^3 \times \frac{10}{100}}{\left(1+\frac{10}{100}\right)^3 - 1}$$

$$= 6620 \times \frac{(1+0.1)^3 \times 0.1}{(1+0.1)^3 - 1} = ₹\, 2662$$

Problem 44. John borrowed ₹ 210000 from a bank at an interest rate of 10% per annum compounded annually. The loan was repaid in two equal installments, the first after first yr and the second after another yr. The first installment was interest of one yr plus part of the principal amount, while the second was the rest of the principal amount plus due interest thereon. Then each installment, in ₹, is : ***[CAT-18]***

Sol. $$\text{Each installment} = 210000 \times \frac{\left(1+\frac{10}{100}\right)^2 \times \frac{10}{100}}{\left(1+\frac{10}{100}\right)^2 - 1} = ₹\, 121000$$

Problem 45. The price of a sofa-set is ₹ 15000 and a customer bought it with cash down payment of ₹ 7000 and he undertakes to pay rest amount in three equal installments at 5% compound interest per annum. What is the each installment ?

Sol. After cash payment, the remaining amount

$= ₹\, 15000 - ₹\, 7000 = ₹\, 8000$

On the above line,

$$p_1 + p_2 + p_3 = \frac{x}{105\%} + \frac{x}{110.25\%} + \frac{x}{115.7625\%}$$

$$\Rightarrow \quad 8000 = \frac{100x}{105} + \frac{100x}{110.25} + \frac{100x}{115.7625}$$

$$\Rightarrow \quad x = ₹\, 2937.67 \text{ (approx.)}$$

Alternatively I

$$\text{Each installment} = \frac{8000}{\frac{1}{\left\{1+\frac{5}{100}\right\}^1} + \frac{1}{\left\{1+\frac{5}{100}\right\}^2} + \frac{1}{\left\{1+\frac{5}{100}\right\}^3}}$$

$= ₹\, 2937.67$ (approx.)

Alternatively II

$$\text{Each installment} = 8000 \times \frac{\left(1+\frac{5}{100}\right)^3 \times \frac{5}{100}}{\left(1+\frac{5}{100}\right)^3 - 1}$$

$$= 8000 \times \frac{(1+0.05)^3 \times 0.05}{(1+0.05)^3 - 1} = ₹\, 2937.67$$

Problem 46. Monika buys a Samsung's 360 litre refrigerator from M/s Coldrush Agecies for ₹ 42000. She makes a down payment of ₹ 12000 and the remaining amount in 4 equal half yearly installments. If M/s Coldrush agencies charge an interest of 10% per annum, approximately what amount Monika has to pay every six month? ***[IIFT-19]***

Sol. After cash payment, the remaining amount

$= ₹\, 42000 - ₹\, 12000 = ₹\, 30000$

Here, applicable rate of interest = 10% ÷ 2 = 5%

$$\text{Now, each installment} = 30000 \times \frac{\left(1+\frac{5}{100}\right)^4 \times \frac{5}{100}}{\left(1+\frac{5}{100}\right)^4 - 1} \approx ₹\, 8460$$

Problem 47. A sum of money is borrowed and paid back in two annual installments of ₹ 8820 each allowing 5% compound interest. Find the sum borrowed. ***[SSC (CGL-T-II)-20]***

Sol. Suppose a sum of ₹ p_1 has been utilized for 1 yr and p_2 has been utilized for 2 yrs.

Obviously, 1st installment of ₹ 8820 will be equal to the sum of p_1 and the interest accrued thereon for 1 yr.

Hence, $8820 = 105\%$ of $p_1 \Rightarrow p_1 = ₹\, 8400$

Further, 2nd installment of ₹ 8820 is equal to the sum of p_2 and interest accrued thereon for 2 yrs.

Hence, $8820 = 110.25\%$ of $p_2 \Rightarrow p_2 = ₹\, 8000$

Then, total borrowed money $= p_1 + p_2 = 8400 + 8000 = ₹\, 16400$

Alternatively I

$$\text{Each installment} = \frac{\text{Borrowed money}}{\frac{1}{\left\{1+\frac{5}{100}\right\}^1} + \frac{1}{\left\{1+\frac{5}{100}\right\}^2}}$$

$$\Rightarrow \text{Borrowed money} = \text{Each installment} \times \left[\frac{1}{\left(1+\frac{5}{100}\right)^1} + \frac{1}{\left(1+\frac{5}{100}\right)^2}\right]$$

$$= 8820 \times \left[\frac{1}{\{1+0.05\}^1} + \frac{1}{\{1+0.05\}^2}\right] = ₹\, 16400$$

Alternatively II

$$\text{Borrowed money} = \text{Each installment} \times \frac{\left(1+\frac{r}{100}\right)^n - 1}{\left(1+\frac{r}{100}\right)^n \times \frac{r}{100}}$$

$$= 8820 \times \frac{(1+0.05)^2 - 1}{(1+0.05)^2 \times 0.05} = ₹\, 16400$$

Concept of Repayment in Installments (not equal) for Discharging a Debt

Problem 48. A person borrowed ₹ 4000 at 5% per annum compounded annually. After 2 yrs, he repaid ₹2210 and then after 2 more years, he repaid the balance with interest. Find the total interest paid by him.

Sol. CI for first 2 yrs = 10.25% of 4000 = ₹ 410

Hence, the 1st installment including principal money of ₹ 1800 and interest of ₹ 410.

Then, the balance principal money = 4000 – 1800 = 2200

Now, CI for next 2 yrs = 10.25% of 2200 = 225.50

Thus, total interest paid = 410 + 225.50

= 635.50

Problem 49. A person closes his account in a bank by withdrawing ₹ 110000. 1 yr earlier, he had withdrawn ₹ 65000. 2yrs earlier, he had withdrawn ₹ 125000. How much money had he deposited in the bank at the time of opening the account 3 yrs ago if the annual interest rate was 10% compounded annually?

Sol. For the better understanding of the question, you may say that a bank borrows some money from a person and repays in three different installments.

At the end of 3rd yr, withdrawal amounting to ₹ 110000 includes the balance in bank in beginning of 3rd yr and interest there on at 10%.

Hence, 110% of the balance in bank in beginning of 3rd yr = ₹ 110000

Then, balance in bank in beginning of 3rd yr = ₹ 100000

Now, balance at end of 2nd yr before withdrawal of ₹ 65000 = 100000 + 65000 = ₹ 165000

Clearly, 110% of balance in beginning of 2nd yr = ₹ 165000

Then, balance in beginning of 2nd yr = ₹ 150000

Further, balance at end of 1st yr before withdrawal of ₹ 125000 = 150000 + 125000 = 275000

Clearly, 110% of the balance in the beginning of 1st yr = ₹ 275000

Then, balance in the beginning of 1st yr = ₹ 250000

Thus, he had opened the account with ₹ 250000.

Alternatively I

Let the amount at the time of opening of account be ₹ x.

Then, balance after 1st withdrawal = (110% of x – 125000) = $1.1x$ – 125000

Further, balance after 2nd withdrawal = {110% of ($1.1x$ – 125000)} – 65000 = $1.21x$ – 202500

And finally balance after the 3rd withdrawal

= 110% of [$1.21x$ – 202500] – 110000 = 0 {as given}

$\Rightarrow$ $1.331x - 222750 = 110000$

$\Rightarrow x =$ ₹ 250000

Alternatively II

It is obvious that each of three installments has certain sum of principal money and corresponding interest thereon. Let the principal money for 1st, 2nd and 3rd yr be p_1, p_2 and p_3 respectively.

1st installment = 110% of p_1

$\Rightarrow$ 125000 = 110% of p_1

$\Rightarrow$ $p_1 = 113636.36$

{It implies that the bank paid the customer ₹ 113636.36 as principal money (out of deposited money) and 10% interest thereon, hence total of ₹ 125000 at the end of 1st yr}

Similarly, 2nd installment = 121% of p_2

$\Rightarrow$ 65000 = 121% of p_2

$\Rightarrow$ $p_2 = 53719$

and 3rd installment = 133.1% of p_3

$\Rightarrow$ 110000 = 133.1% of p_3

$\Rightarrow$ $p_3 = 82644.62$

Hence, total amount deposited at time of opening the account

$= p_1 + p_2 + p_3$

= 113636.36 + 53719 + 82644.62

= ₹ 250000

Concept of periodic Investment of Recurring Deposit (compounded value of a series of payments)

Suppose a person invests in a Systematic Investment Plan (SIP) at 1000 per yr. It may be said that he deposits ₹ 1000 per yr in a bank. What amount will he get after 3 yrs if the rate of interest is 10% per annum and compounded annually.

There may be two cases :

***Case* I If he deposits at the end of every year**

(Popularly known as 'Ordinary annuity' or 'Regular annuity' or 'Annuity in arrear' scheme)

Clearly, the money deposited at the end of 3rd yr will attract no interest. The money deposited at end of 2nd yr will attract interest for a period of 1 yr and the money deposited at end of 1st yr will attract interest for 2 yrs.

Hence, total amount he will get = 1000 + 110% of 1000 + 121% of 1000 = ₹ 3310

Alternatively,

$$\text{Matured amount} = \text{Pereodic deposit} \times \frac{\left(1+\frac{r}{100}\right)^n - 1}{\frac{r}{100}},$$

where r = rate of interest according to the period and n = Number of investments / deposits / annuity

$$\text{Hence, matured amount} = 1000 \times \frac{\left(1+\frac{10}{100}\right)^3 - 1}{\frac{10}{100}}$$

$$= 1000 \times \frac{(1.1)^3 - 1}{0.1} = ₹\ 3310$$

***Case* II If he deposits at the beginning of every year (Popularly known as 'Due annuity' or 'Annuity in advance' scheme)**

Obviously, the money deposited at the beginning of 3rd yr will attract interest for 1 yr. The money deposited at beginning of 2nd yr will attract interest for a period of 2 yrs and the money deposited at beginning of 1st yr will attract interest for 3 yrs. Hence, total amount he will get = 110% of 1000 +121% of 1000 +133.1% of 1000 = ₹ 3641

Alternatively

Matured amount =

$$\text{Periodic deposit} \times \left[\left\{\frac{\left(1+\frac{r}{100}\right)^{n+1}-1}{\frac{r}{100}}\right\}-1\right]$$

where r = rate of interest according to the period and n = number of investments/deposits/annuity

$$\text{Hence, matured amount} = 1000 \times \left[\left\{\frac{\left(1+\frac{10}{100}\right)^{3+1}-1}{\frac{10}{100}}\right\}-1\right]$$

$$= 1000 \times \left[\left\{\frac{(1.1)^4-1}{0.1}\right\}-1\right] = ₹\ 3641$$

➘ Problem 50 How much amount should Shivansh deposit at the end of every 6 months in a bank paying 10% interest per annum in order to get ₹ 1576250 after one year six months, if interest is compounded semi annually?

Sol. Clearly, $r = 5\%$ as annual rate of interest is 10%

n = number of deposits = 3

$$\text{Now, } 1576250 = \text{Periodic deposit} \times \frac{\left(1+\frac{5}{100}\right)^3-1}{\frac{5}{100}}$$

$$\Rightarrow \text{Periodic deposit} = \frac{1576250 \times \frac{5}{100}}{\left(1+\frac{5}{100}\right)^3-1} = ₹\ 500000$$

Hence amount of deposit at end of every 6 months = ₹ 500000

Relation Between Nominal Rate of Interest and Effective Rate of Interest

Suppose the nominal rate of interest is $r\%$ per annum. Then, the effective rate of interest will be more than $r\%$ per annum, if frequency of compounding is more than 1 in a year. The effective rate of interest increases when the frequency of compounding increases.

Take an illustration

Let the nominal rate of interest is 12% per annum, then what is the effective rate of interest per annum if the interest is compounded

(i) semiannually (ii) at every four months.

(i) Rate of interest semi annually = 6%

Number of $\frac{1}{2}$ yr blocks in 1 yr i.e. frequency = 2.

The, effective rate of interest = (Cumulative SI) $\cdot \boxed{r^2}\%$

$= (2 \times 6) . \boxed{6^2}\% = 12.36\%$

(ii) Rate of interest = 4%

Number of 4 month blocks in 1 yr i.e. frequency = 3

Then, effective rate of inerest

$= (\text{Cumultative SI}) . \boxed{3r^2}\boxed{r^3}\%$

$= (\text{Cumulative SI}) . \boxed{3 \times (4)^2}\boxed{4^3}\%$

$= (3 \times 4) . \boxed{48}\boxed{64}\% = 12.4864\%$

Alternatively,

Effective rate of interest

$$= \left\{\left(1+\frac{r}{100}\right)^{\text{no. of frequency}}-1\right\} \times 100\%$$

(i) Effective rate of interest $= \left\{\left(1+\frac{6}{100}\right)^2-1\right\} \times 100\%$

$= 12.36\%$

(ii) Effective rate of interest $= \left\{\left(1+\frac{4}{100}\right)^3-1\right\} \times 100\%$

$= 12.4864\%$

Miscellaneous

➘ Problem 51 What is the total compound interest earned at the end of the three years.

I. SI earned on that amount at the same rate and period is ₹ 4500.

II. The rate of interest is 10% per annum.

III. CI for 3 years is more than the SI for that period by ₹ 465.

Which of the above statement(s) is/are necessary to answer the question?

(a) only I and II (b) only II and III
(c) only I and III (d) any two of three
(e) either II or III only

Sol. On combination of statement (I) and (II), we get, r, t and SI. Then, we may compute CI.

Further, if r is given, then we may compute the difference between SI and CI and we can obtain CI. It means statements (II) and (III) are necessary.

Also, on combination of statements (I) and (III), we get SI and difference between SI and CI. Then, we may compute CI.

Thus, option (d) is correct.

➘ Problem 52. A man borrows money at 3% per annum interest payable yearly and lends it immediately at 5% per annum interest payable half-yearly and gains thereby ₹ 165 at the end of a year. What sum of money does he borrow?

Sol. The CI to be paid by him at end of a year = 3% of the borrowed money. Further, in case of lending, $r = 2.5\%$ and $t = 2$ as interest is compounded half-yearly.

Hence, CI to be received by him at the end of the year

$= \left\{2.5 + 2.5 + \frac{2.5 \times 2.5}{100}\right\}\%$ of the lent money

$= 5.0625\%$ of the lent money

As given, $(5.0625 - 3)\%$ of the borrowed money = ₹ 165

Then, 100% of the borrowed money = ₹ 8000

Hence, he borrows ₹ 8000.

WORKOUT

1. The simple interest accrued on an amount of ₹ 84000 at the end of 3 yrs is ₹ 30240. What would be the compound interest accrued on the same amount at the same rate in the same period.
[SBI (PO)-08, BOB (AO)-08, OBC(PO)-08, Syndicate Bank (PO)-10, SBI (PO-Pre)-18, SSC (CGL-T-I)-19]
(a) ₹ 30013.95 (b) ₹ 31013.95
(c) ₹ 32013.95 (d) ₹ 33013.95
(e) ₹ 34013.95

2. The difference between CI and SI on ₹ 5000 for 2 yrs at the same rate is ₹ 72. The rate of interest per annum is :
(a) 15% (b) 12%
(c) 10% (d) 8%

3. A sum of money on compound interest amounts to ₹ 9680 in 2 yrs and to ₹ 10648 in 3 yrs. What is the rate of interest per annum?
[CDS-11]
(a) 5% (b) 10%
(c) 15% (d) 20%

4. Mukesh borrows a certain sum of money from the *ABC* Bank at 10% per annum at compound interest. The entire debt is discharged in full by Mukesh on payment of two equal amounts of ₹ 1000 each, one at the end of the 1st yr and the other at the end of the 2nd yr. What is the approximate value of the amount borrowed by him?
[MAT-14, 19]
(a) ₹ 1852 (b) ₹ 1694
(c) ₹ 1736 (d) ₹ 1792

5. A sum of ₹ 3200 invested at 10% per annum compounded quarterly amounts to ₹ 3362. Compute the time period. **[SSC (CGL-T-II)-13]**
(a) $\frac{1}{2}$ yr (b) 1 yr
(c) 2 yrs (d) $\frac{3}{4}$ yr

6. Mr. Duggal invested ₹ 20000 with the rate of interest @ 20 pcpa. The interest was compounded half-yearly for the first one year and in the next year it was compounded yearly. What will be the total interest earned at the end of 2 yrs ?
(a) ₹ 8800 (b) ₹ 9040
(c) ₹ 8040 (d) ₹ 9800
(e) None of these

7. If the rate of interest is 10% per annum and is compounded half-yearly, then the principal of ₹ 400 in $1\frac{1}{2}$ yrs will amount to :
(a) ₹ 463 (b) ₹ 463.05
(c) ₹ 463.15 (d) ₹ 463.20

8. The ratio of amount for 2 yrs under CI annually and for 1 yr under SI is 6 : 5. When the ratio of interest is same, then value of rate of interest is?
(a) 18% (b) 20% (c) 12.5% (d) 16.66%

9. Eight years after completion of your MBA degree, you start a business of your own. You invest INR 3000000 in the business that is expected to give you a return of 6% compounded annually. If the expected number of years by which your investment shall double is $72/r$, where r is the percent interest rate, the approximate expected total value of investment (in INR) from your business after 48 yrs later is : **[IIFT-14]**
(a) 24000000 (b) 36000000
(c) 48000000 (d) None of these

10. The compound interest earned by Suresh on a certain amount at the end of 2 yrs at the rate of 8 p.c.p.a was ₹ 1414.40. What was the total amount that Suresh got back at the end of 2 yrs in the form of principal plus interest earned?
(a) ₹ 9414.40 (b) ₹ 9914.40
(c) ₹ 9014.40 (d) ₹ 8914.40
(e) None of these

11. The value of a property depreciates every year by 10%. The present value of the property is ₹ 81 lakhs 2 yrs ago, its value was : **[SSC (SAS)-10]**
(a) ₹ 90 lakhs (b) ₹ 96 lakhs
(c) ₹ 100 lakhs (d) ₹ 108 lakhs

12. A sum of money put at compound interest amounts to ₹ 8820 in 2 yrs and ₹ 9261 in 3 yrs. Find the sum.
(a) ₹ 8000 (b) ₹ 7200
(c) ₹ 6500 (d) ₹ 8250

13. The simple interest on a certain sum of money for 3 yrs at 8% per annum is half the compound interest on ₹ 4000 for 2 yrs at 10% per annum. What is the sum placed on simple interest?
(a) ₹ 1550 (b) ₹ 1650
(c) ₹ 1750 (d) ₹ 2000

14. Under the Rural Housing Scheme, the DDA allotted a house to Kamal Raj for ₹ 126100. This payment is to be made in three equal annual installments. If the money is reckoned at 5% per annum compound interest, then how much is to paid by Kamal Raj in each installment?
(a) ₹ 45205 (b) ₹ 47405
(c) ₹ 46305 (d) ₹ 48505

15. If the compound interest accrued on an amount of ₹ 15000 in 2 yrs is ₹ 2496, what is the rate of interest per cent per annum?
(a) 8
(b) 10
(c) 6
(d) can't be determined
(e) None of these

16. The difference between CI and SI for 2 yrs at the rate of 10% over principal amount of ₹ x is ₹ 10. What is the value of X?
[CDS-14, MAT-10, CAT-20]
(a) ₹ 100 (b) ₹ 1000
(c) ₹ 500 (d) ₹ 5000

17. Sanju puts equal amount of money one at 10% per annum compound interest payable half-yearly and the second at a certain percent per annum compound interest payable yearly. If he gets equal amounts after 3 yrs, what is the value of the 2nd percent?
(a) 10.25% (b) 10%
(c) 9.5% (d) 8.25%

18. In how many months will ₹ 8000 yield ₹ 2648 as a compound interest at 20% per annum compounded semi-annually? *[SSC (CGL-T-II)-17]*

(a) 18 (b) 24
(c) 12 (d) 30

19. An amount of ₹ X at compound interest at 20% per annum for 3 yrs becomes Y. What is $Y:X$?

(a) 3 : 1 (b) 36 : 25
(c) 216 : 125 (d) 125 : 216

20. A man gets a simple interest of ₹ 1000 on a certain principal at the rate of 5 percent per annum in 4 yrs. What compound interest will the man get on twice the principal in 2 yrs at the same rate? *[P and S Bank (PO)-10, Corporation Bank (PO)-2011]*

(a) ₹ 1000 (b) ₹ 1005
(c) ₹ 11025 (d) ₹ 10125
(e) None of these

21. Two partners A and B together lends ₹ 84100 at 5% interest compounded annually. The amount which A gets at the end of 3 yrs is the same as what B gets at the end of 5 yrs. Determine the ratio of shares of A and B. *[SSC (SO-Audit)-97, SSC (CGL-Mains)-97, SSC (CGL-T-I)-19]*

(a) 19 : 10 (b) 169 : 144
(c) 441 : 400 (d) 13 : 10

22. Find the principal if compound interest is charged on the principal at the rate of $16\frac{2}{3}$% per annum for 2 yrs and the sum becomes ₹ 196. *[MAT-14, 19]*

(a) ₹ 140 (b) ₹ 154
(c) ₹ 150 (d) None of these

23. The compound interest on a sum for 2 yrs is ₹ 832 and the simple interest on the same sum at the same rate for the same period is ₹ 800. What is the rate of interest?

(a) 6% (b) 10%
(c) 12% (d) 8%

24. What would be the compound interest accrued on an amount of ₹ 7400 @ 13.5 per cent per annum at the end of 2 yrs ? *[IB (PO)-10, SSC (CGL-T-II)-13]*

(a) ₹ 2136.87 (b) ₹ 2306.81
(c) ₹ 2032.18 (d) ₹ 2132.87
(e) None of these

25. A money-lender borrows a certain sum of money at 3% per annum simple interest and lends it at 6% per annum compound interest, compounded half-yearly. If he gains ₹ 618 in a year, find the sum (in ₹) of money borrowed by him. *[SSC (CGL-Mains)-01, MAT-10]*

(a) 15000 (b) 16000
(c) 18000 (d) 20000

26. Three years ago your close friend had won a lottery of ₹ 1 Crore. He purchased a flat for ₹ 40 lakhs, a car for ₹ 20 lakhs and shares worth ₹ 10 lakhs. He put the remaining amount in a bank deposit that pays compound interest at the rate of 12% per annum. If today, he sells off the flat, the car and the shares at certain percentage of their original value and withdraws his entire money from the bank, the total gain in his assets is 5%. The closest approximate percentage of the original value at which he sold off the three items is : *[IIFT-13]*

(a) 60% (b) 75%
(c) 90% (d) 105%

27. A person borrowed ₹ 7500 at 16% compound interest. How much does he have to pay at the end of 2 yrs to clear the loan?

(a) ₹ 9900 (b) ₹ 10092
(c) ₹ 11000 (d) ₹ 11052

28. The difference between simple and compound interest on a sum for 2 yrs at $6\frac{2}{3}$% per annum is ₹ 14. Find the amount. *[SSC (Asstt.-Mains)-95, SSC (SO-audit)-01, CDS-17, 21]*

(a) ₹ 1425 (b) ₹ 3250
(c) ₹ 4150 (d) ₹ 3150

29. Sharma invested one-half of his savings in a bond that paid simple interest for 2 yrs and received ₹ 550 as interest. He invested the remaining in a bond that paid compound interest, interest being compounded annually, for the same 2 yrs at the same rate of interest and received ₹ 605 as interest. What was the value of his total savings before investing in these bonds? *[MAT-11]*

(a) ₹ 22000 (b) ₹ 11000
(c) ₹ 5500 (d) ₹ 2750

30. If the sum of money compounded annually becomes 1.44 times of itself in 2 yrs, then rate of interest per annum is *[SSC (CGL-T-II)-13]*

(a) 25% (b) 22%
(c) 21% (d) 20%

31. A principal sum of money is lent out at compound interest compounded annually at the rate of 20% per annum for 2 yrs. It would give ₹ 2410 more if the interest is compounded half-yearly. Find the principal sum(in ₹). *[SSC (CGL-Mains)-01, MAT-12]*

(a) 50000 (b) 100000
(c) 150000 (d) 200000

32. A, B and C invested amounts in the ratio 3 : 4 : 5 respectively. If the schemes offered compound interest at the rate of 20% per annum, 15% per annum and 10% per annum respectively, then what will be the ratio of their amounts after 1 yr? *[SSC (CGL-T-I)-17]*

(a) 3 : 15 : 25 (b) 6 : 6 : 5
(c) 36 : 46 : 55 (d) 12 : 23 : 11

33. If the difference between the SI and CI on the same principal amount at 20% for 3 yrs is ₹ 48, then the principal amount must be

(a) ₹ 650 (b) ₹ 600
(c) ₹ 375 (d) ₹ 400

34. Rohit took a loan of ₹ 20000 to purchase one LCD TV set from a finance company. He promised to make the payment after 3 yrs. The company charges compound interest at the rate of 10% per annum for the same. But, suddenly the company announces the rate of interest as 15% per annum for the last 1 yr of the loan period. What extra amount does Rohit have to pay due to this announcement of the new rate of interest?

(a) ₹ 7830 (b) ₹ 4410
(c) ₹ 6620 (d) None of these

35. Sita deposited ₹ 5000 at 10% simple interest fot 2 yrs. How much more money will Sita have in her account at the end of 2 yrs, if it is compounded semi-annually. *[SSC (CGL-T-II)-11]*

(a) ₹ 50 (b) ₹ 40 (c) ₹ 77.50 (d) ₹ 85.50

36. If the amount on a certain principal in 3 yrs at 12% rate of interest compounded annually is ₹ 12000, what will the amount (in ₹) after the 4th yrs? *[SSC (CGL-T-II)-17]*

(a) 14330 (b) 15440
(c) 13440 (d) 14550

37. What is the rate of interest (in %) if simple interest earned on a certain sum for the 3rd yrs is ₹ 2000 and compound interest earned in 2 yrs is ₹ 4160? *[SSC (CGL-T-II)-17]*

(a) 6 (b) 8
(c) 10 (d) 12

38. The difference between simple interest and compound interest at the same rate for ₹ 5000 for 2 yrs is ₹ 98. The rate of interest is:

(a) 10% (b) 14% *[CLAT-17]*
(c) 12% (d) 10.5%

39. A person wants to divide a sum of ₹ 390300 between his two sons who are 13 and 15 yrs of age respectively in such a way that their shares, if invested at 4% per annum compound interest, should produce the same amount when they becomes 18 yrs of age. Find the share of elder and younger son (in ₹) *[SSC (CGL-Mains)-2000, MAT-17]*

(a) 190300, 200000
(b) 202800, 187500
(c) 245000, 145300
(d) None of the above

40. Gopal borrows ₹ X from Ankit at 8% annual interest. He then adds Y of his own money and lends ₹ $X + Y$ to Ishan at 10% annual interest. At the end of the year, after returning Ankits dues, the net interest retained by Gopal is the same as that accrued to Ankit. On the other hand, had Gopal lent ₹ $X + 2Y$ to Ishan at 10%, then the net interest retained by him would have increased by ₹ 150. If all interests are compounded annually, then find the value of $X + Y$. *[CAT-18]*

(a) 2060 (b) 1440
(c) 6000 (d) 4000

41. What will be the compound interest on a sum ₹ 1875 after 2 yrs if the rate of interest for the 1st yr is 4% and that for the 2nd yr is 8%? *[MAT-18]*

(a) ₹ 231
(b) ₹ 341
(c) ₹ 241
(d) None of the above

42. A sum of ₹ 8000 invested at 10% per annum amounts to ₹ 9261 in a certain time, interest compounded half-yearly. What will be the compound interest (in ₹) on the same sum for the same time at double the earlier rate of interest, when interest is compounded annually? *[SSC (CGL-T-I)-19]*

(a) 2520 (b) 2480
(c) 2560 (d) 2500

43. Amal investes ₹ 12000 at 8% interest, compounded annually, and ₹ 10000 at 6% interest, compounded semi-annually, both investments being for 1 yr. Bimal invests his money at 7.5% simple interest for 1 yr. If Amal and Bimal get the same amount of interest, then the amount, in ₹ invested by Bimal is : *[CAT-19]*

(a) 20920 (b) 15440
(c) 18440 (d) 11150

Answer key

1. (e)	2. (a)	3. (b)	4. (c)	5. (a)	6. (b)	7. (b)	8. (b)	9. (c)	10. (b)
11. (c)	12. (a)	13. (c)	14. (c)	15. (a)	16. (b)	17. (a)	18. (a)	19. (c)	20. (e)
21. (c)	22. (d)	23. (d)	24. (d)	25. (d)	26. (c)	27. (b)	28. (d)	29. (d)	30. (d)
31. (b)	32. (c)	33. (c)	34. (d)	35. (c)	36. (c)	37. (b)	38. (b)	39. (b)	40. (d)
41. (a)	42. (c)	43. (a)							

Hints and SOLUTIONS

1. SI for 1 yr = 30240 ÷ 3 = 10080

Then, rate of simple interest per annum

$= \left(\frac{10080}{84000} \times 100\right)\% = 12\%$

Now, CI for 3 yrs

$= (3 \times 12) . \boxed{3 \times (12)^2} \boxed{(12)^3} \%$ of the principal

$\Rightarrow$ 40.4928% of 84000 = ₹ 34013.95

2. Difference between CI and SI in term of percentage of principal

$= \left(\frac{72}{5000} \times 100\right)\% = 1.44\%$

We know, difference between CI and SI for 2 yrs $= 0 \cdot \boxed{r^2}\%$ of the principal

Now, proceed through options

(a) $0 \cdot \boxed{r^2}\% = 0 \cdot \boxed{(12)^2}\% = 1.44\%$

Hence, option (a) is correct.

3. It is fact that compound interest is computed always on the previous amount. Here, the CI for the 3rd yr will be computed on the Amount upto last 2 yrs i.e. on ₹ 9680

Hence, CI for the 3rd yr = 10648 − 9680 = ₹ 968, which is 10% of ₹ 9680. It implies that the Amount of ₹ 9680 has increased by 10% in only one year so as to become ₹ 10648. Thus, rate of interest = 10% per annum.

4. Suppose a sum of ₹ p_1 has been utilized for 1 yr and p_2 has been utilized for 2 yrs .

Obviously, 1st installment of ₹ 1000 will be equal to the sum of p_1 and the interest accrued thereon for 1 yr.

Hence, 1000 = 110% of p_1

$\Rightarrow \quad p_1 = 909.09$

Further, 2nd installment of ₹ 1000 is equal to the sum of p_2 and interest accrued thereon for 2 yrs.

Hence, 1000 = 121% of p_2

$\Rightarrow \quad p_2 = 826.45$

Then, total borrowed money

$= p_1 + p_2$

$= 909.09 + 826.45$

= ₹ 1736 (approx)

5. Interest is compounded quarterly. Then, $r = 2.5\%$

Now, increment in principal

$= \left(\frac{162}{3200} \times 100\right)\% = 5.0625\%$

Now, proceed from options.

(a) CI for $\frac{1}{2}$ yrs

$= \left\{2.5 + 2.5 + \frac{2.5 \times 2.5}{100}\right\}\%$ of the principal = 5.0625% of the principal i.e 5.0625% increment in principal

Hence, option (a) is correct.

6. In case interest is compounded half-yearly, $r = 10\%$ and $t = 2$ yrs.

Then, CI $= (2 \times 10) . \boxed{(10)^2}\%$ of the principal = 21% of the principal

CI for the 2 yrs

$= \left\{21 + 20 + \frac{21 \times 20}{100}\right\}\%$ of the principal

= 45.2% of 20000 = ₹ 9040

7. Since interest is compounded half-yearly, $r = 5\%$ and $t = 3$ yrs

Now, CI for 3 yrs

$= (3 \times 5) . \boxed{3 \times (5)^2} \boxed{5^3}\%$ of the principal

$\Rightarrow$ CI = 15.7625 % of the principal

Hence, Amount = 115.7625% of the principal

= 115.7625% of 400 = ₹ 463.05

8. The Amount at CI for 2 yrs will exceed the Amount at SI for 1 yr by $r\%$. In other words, the amount at CI for 2 yrs will be equal to the sum of the Amount at SI for 1 yr and the interest there on at the rate of $r\%$.

Obviously, $6 = 5 + (r\% \text{ of } 5) \Rightarrow r = 20$

Hence, rate of interest = 20%

9. In (72/6 i.e. 12) yrs, the principal becomes 2 times.

Then, in (4 × 12) yrs, the principal will become 2^4 times.

Hence required amount

$= 2^4 \times 3000000 = 48000000$

10. CI for 2 yrs =16.64% of the principal

Given, 16.64% of the principal

= ₹ 1414.40

Then, Amount i.e. 116.64% of the principal = ₹ 9914.40

11. $\left\{-10 - 10 + \frac{10 \times 10}{100}\right\}\% = -19\%$

Hence, depreciation = 19% of the initial value.

As given, (100 − 19)% of initial value

= Present value

= ₹ 81 lakhs

$\therefore$ 100% of the initial value = ₹ 100 lakhs

12. It is fact that compound interest is computed always on the previous Amount. Here, the CI for the 3rd yr will be computed on the Amount upto last 2 yrs i.e. on ₹ 8820.

Hence, CI for the 3rd yr

= 9261 − 8820

= ₹ 441, which is 5% of ₹ 8820.

It implies that the amount of ₹ 8820 has increased by 5% in only 1 yr so as to become ₹ 9261.

Thus, rate of interest

= 5% per annum.

Further, CI for 2 yrs at 5% per annum = 10.25% of the principal

Then, Amount in 2 yrs =110.25% of the principal = ₹ 8820

$\therefore$ 100% of the principal = ₹ 8000

13. CI for 2 yrs = 21% of the 4000 = ₹ 840

Then, SI = 840 ÷ 2 = 420

Clearly, (3 × 8)% of the principal

= ₹ 420

$\therefore$ 100% of the principal = ₹ 1750

14. Each installment

$= 126100 \times \frac{\left(1 + \frac{5}{100}\right)^3 \times \frac{5}{100}}{\left(1 + \frac{5}{100}\right)^3 - 1}$

$= 126100 \times \frac{(1 + 0.05)^3 \times 0.05}{(1 + 0.05)^3 - 1}$

= ₹ 46305

15. CI in term of percentage of principal

$= \left(\frac{2496}{15000} \times 100\right)\% = 16.64\%$

Now, proceed through options.

(a) $8 \Rightarrow (2 \times 8) . \boxed{(8)^2}\% = 16.64\%$ of the principal

Hence, option (a) is correct.

16. For the period of 2 yrs, difference = $0.\boxed{r^2}$% of the principal = $0.\boxed{(10)^2}$% of the principal = 1% of the principal

Given, 1% of the principal = ₹ 10

Then, 100% of the principal = ₹ 1000

17. In 1st case, $r = 5\%$ and $t = 6$ yrs

Then, Amount

$$= P \times \frac{105}{100} \times \frac{105}{100} \times \frac{105}{100} \times \frac{105}{100} \times \frac{105}{100} \times \frac{105}{100}$$

In second case, $t = 3$ yrs, rate of interest = r%

Then, Amount

$$= P \times \frac{100+r}{100} \times \frac{100+r}{100} \times \frac{100+r}{100}$$

But given,

$$= P \times \frac{105}{100} \times \frac{105}{100} \times \frac{105}{100} \times \frac{105}{100} \times \frac{105}{100} \times \frac{105}{100}$$

$$= P \times \frac{100+r}{100} \times \frac{100+r}{100} \times \frac{100+r}{100}$$

$$\Rightarrow \frac{11025}{10000} \times \frac{11025}{10000} \times \frac{11025}{10000} = \frac{100+r}{100} \times \frac{100+r}{100} \times \frac{100+r}{100}$$

$$\Rightarrow \frac{11025}{10000} = \frac{100+r}{100}$$

$\Rightarrow 11025 = 10000 + 100r$

$\Rightarrow \quad r = 10.25$

Hence, second rate of interest = 10.25%

Alternatively

In first case, within 1 yr interest is to be compounded twice at the rate of 5% whereas in second case, interest will be compounded once in a year. It implies that the CI fetched in 1 yr under first case will be equal to CI fetched in 1 yr under the second case so as to maintain equality of Amounts after 3 yrs. Hence, applicable rate of interest in second case $= \left\{+5 + 5 + \frac{5 \times 5}{100}\right\}\% = 10.25\%$

18. Interest is being compounded half yearly.

Thus, rate of interest for $\frac{1}{2}$ yrs = 10%

($\because$ 20% for 1 yr)

Given, CI is 33.1% of the principal.

It is clear that the time is 3 yrs as 3 times 10 is near about to 33.1%. But, you have to check the correctness.

CI for 3 yrs = (Cumulative SI).

$\boxed{3r^2}\boxed{r^3}$% of principal money

$= (3 \times 10) . \boxed{3 \times (10)^2}\boxed{(10)^3}$% the principal

= 33.1% of the principal

Since the interest is compounded semi-annually, the required time

$= 3 \div 2 = 1.5$ yrs = 18 months

19. CI for 3 yrs =

$(3 \times 20) . \boxed{3 \times (20)^2}\boxed{(20)^3}$% of the principal

= 72.8% of the principal

Then, Amount = 172.8% of the principal

$\Rightarrow \quad Y = 172.8\%$ of X

$\Rightarrow Y : X = 172.8 : 100$

$= 1728 : 1000 = 216 : 125$

20. As given, (5×4)% of the principal i.e. SI = ₹ 1000

$\therefore$ Principal = ₹ 5000

Then, $\quad$ CI = 10.25% of (2×5000)

= ₹ 1025

21. It is obvious that the share of A is greater than that of B as share of A takes only 3 yrs and that of B takes 5 yrs to become the same Amount. It implies that share of B will be equal to that of A after 2 yrs. Further, CI for 2 yrs at the rate of 5% = 10.25% of the principal.

Hence, A's share = 10.25% more than B's share. In other words, A's share is 110.25% of B's share.

Now, A's share : B's share

= 110.25 : 100

= 11025 : 10000 = 441 : 400

22. CI for 2 yrs

$= \left\{16.67 + 16.67 + \frac{16.67 \times 16.67}{100}\right\}$% of the principal

$\Rightarrow \quad$ CI = 36.12% of the principal

Hence, Amount = 136.12% of the principal

$\Rightarrow$ 196 = 136.12% of the principal

Then, 100% of the principal = ₹ 144

23.

	SI	CI
1st yr	400	400
2nd yr	400	432
Total	800	832

Obviously, the difference of ₹ 32 between 1st yr's and 2nd yr's CI is due to calculation of CI at r% per annum also over the 1st yr's interest. It means r% of the 1st yr interest

= ₹ 32

$\Rightarrow \quad r\%$ of $400 = 32 \Rightarrow r = 8$

Hence, rate of interest per annum

= 8%

24. CI for 2 yrs

$= \left\{13.5 + 13.5 + \frac{13.5 \times 13.5}{100}\right\}$% of the principal

$\Rightarrow$ CI = 28.8225% of 7400 = ₹ 2132.87

25. The interest to be paid by him at end of a year = 3% of the borrowed money.

Further, in case of lending, $r = 3\%$ and $t = 2$ as interest is compounded half-yearly.

Hence, CI to be received by him at the end of the year = $(2 \times 3) . \boxed{3^2}$% of the lent money

= 6.09% of the lent money.

As given, $(6.09 - 3)$% of the borrowed/lent money = ₹ 618

Then, 100% of the borrowed money

= ₹ 20000

26. The money deposited in bank = ₹ 30 lakhs

CI for 3 yrs at the rate of 12% per annum = $(3 \times 12) . \boxed{3 \times (12)^2}\boxed{(12)^3}$% of the principal = 40.4928% of the principal

Then, Amount = 140.5% (approx.) of ₹ 30 lakhs = 42.15 lakhs (approx.)

Selling price of all three items

= (105% of 1 crore) − 42.15 lakhs

= 62.85 lakhs

and purchasing price of all three items = (40 + 20 + 10) = 70 lakhs

Clearly, 62.85 is approx. 90% of the 70 lakhs.

27. CI for 2 yrs = $(2 \times 16) . \boxed{(16)^2}$% of the principal money

= 34.56% of the principal money

Hence, Amount = 134.56% of 7500

= ₹ 10092

28. For the period of 2 yrs, difference

$= 0 . \boxed{r^2}$% of the principal

$= 0 . \boxed{\left\{6\frac{2}{3}\right\}^2}$% of the principal

= 0.4444% of the principal

Given, 0.4444% of the principal = ₹ 14

Then, 100% of the principal

= ₹ 3150

29.

	SI	CI
1st yr	275	275
2nd yr	275	330
Total	550	605

Obviously, the difference of ₹ 55 between 1st yr's and 2nd yr's CI is due to calculation of CI at r% per annum also over the first year's interest. It means r% of the first year's interest

$= ₹\ 55$

$\Rightarrow\ r\%$ of $275 = 55 \Rightarrow r = 20$

Hence, rate of interest per annum

$= 20\%$.

Further, SI for first year

$= 20\%$ of the sum

$\Rightarrow\ 275 = 20\%$ of the sum

$\Rightarrow$ sum $= 1375$

Thus, total savings invested in bonds

$= 1375 + 1375 = ₹\ 2750$

30. Given that the principal becomes 1.44 times itself in 2 yrs.

Then, it will become $(1.44)^{1/2}$ times itself in (2/2) yr. Hence, in 1 yr the principal becomes 1.2 times and it is possibe only when rate of interest is 20% per annum.

31. CI for 2 yrs if interest is compounded yearly $= (2 \times 20) . \boxed{(20)^2}\%$ of the principal money

CI = 44% of the principal money

In case interest is compounded half-yearly, $r = 10\%$ and $t = 4$ yrs.

Then, CI for 4 yrs = (Cumulative SI). $\boxed{6r^2}\boxed{4r^3}\boxed{r^4}$ % of the principal

$\Rightarrow$ CI $= (10 \times 4) . \boxed{6\cdot((10)^2}\boxed{4\cdot(10)^3}\boxed{(10)^4}$% of the principal

$\Rightarrow$ CI = 46.41% $\cdot$ of the principal

Difference between CI under both systems = (46.41-44)% of the principal

$\Rightarrow$ 2.41% of the principal = ₹ 2410

Then, 100% of the principal
= ₹ 100000

32. Required ratio $= 3 \times 120 : 4 \times 115 : 5 \times 110 = 36 : 46 : 55$

33. Difference between SI and CI for 3 yrs $= 0 . \boxed{3 \times (20)^2}\boxed{(20)^3}$ % of the principal

$\Rightarrow$ 12.8 % of the principal= ₹ 48

$\Rightarrow$ 100% of the principal = ₹ 375

34. CI for 3 yrs at rate of 10%
$= (3 \times 10) . \boxed{3 \times (10)^2}\boxed{(10)^3}$ % of the principal money

$\Rightarrow$ 33.1% of the principal money

After announcement of new rate for 3rd yr, CI will be as below:

CI for 2 yrs at rate of 10% = 21% of the principal

Then, CI for 3 yrs

$$= \left\{+21 + 15 + \frac{21 \times 15}{100}\right\}\%\text{ of the principal}$$

= 39.15% of the principal.

Clearly, he has to pay (39.15 – 33.1)% of 20000 more, i.e. ₹ 1210 more.

35. Interest is compounded semi-annually, then $r = 5\%$ and $t = 4$ yrs

Now, CI $= (4 \times 5) . \boxed{6\cdot 5^2}\boxed{4\cdot 5^3}\boxed{5^4}$% of the principal

$\Rightarrow$ 21.550625% of the principal

More money what she gets = (21.550625 – 20)% of 5000

= ₹ 77.50 (approx.)

36. It is fact that compound interest is computed always on the previous Amount. Here, the CI for the 4th yr will be computed on the Amount fetched in last 3 yrs i.e. on ₹ 12000.

Hence, CI for 4th yr

= 12% of 12000 = ₹ 1440

Then, Amount after 4th yr

= 12000 + 1440 = ₹ 13440

37.

	SI	CI
1st yr	2000	2000
2nd yr	2000	2160
3rd yr	2000	

Obviously, the difference of ₹ 160 between 1st yr's and 2nd yr's CI is due to calculation of CI at r% per annum also over the 1st yr's interest. It means r % of the 1st yr's interest $= ₹\ 160 \Rightarrow r\ \%$ of $2000 = 160 \Rightarrow r = 8\%$

Hence, rate of interest per annum

= 8%

38. For the period of 2 yrs, difference

$$= \frac{r^2}{100}\%\text{ of the principal}$$

$$\Rightarrow \frac{r^2}{100}\%\text{ of } 5000 = 98$$

$\Rightarrow\ r^2 = 196 \Rightarrow r = 14\%$

39. It is obvious that the share of elder son is greater than that of younger son as share of elder son takes only 3 yrs and that of younger son takes 5 yrs to become the same Amount. It implies that share of younger son will be equal to that of elder son after 2 yrs. Further, CI for 2 yrs at the rate of 4% = 8.16% of the principal.

Hence, elder son's share = 8.16% more than younger son's share. In other words, elder son's share is 108.16% of younger son's share.

Now, elder son's share : younger son's share = 108.16 : 100

= 10816 : 10000 = 676 : 625

Further, (676 + 625) units $\Rightarrow$ ₹ 390300

$\therefore$ 676 units $\Rightarrow$ 202800

Thus, share of elder son = ₹ 202800

and share of younger son

= ₹ (390300 – 202800) = ₹ 187500

40. Clearly, 2 (8% of X) = 10% of ($X + Y$) ...(i)

as the interest accrued on ($X + Y$) is twice the interest accrued on X.

Further, given that 150 + 2 (8% of X) = 10% of (X + 2Y) ...(ii)

On subtracting Eq. (ii) from Eq. (i), we get $-150 = -10\%$ of Y

$\therefore$ 100% of Y = 1500

Substituting the value of Y in Eq. (i), we get, $X = 2500$

$\therefore\ X + Y = 2500 + 1500 = ₹\ 4000$

41. $$\text{CI} = \left\{4 + 8 + \frac{4 \times 8}{100}\right\}\%\text{ of ₹ } 1875$$

= 12.32% of 1875 = ₹ 231

42. Since the interest is compounded half-yearly, rate of interest for 1/2 yr = 5%

Further, since (9261 – 8000) is 15.7625% of 8000, CI is 15.7625% of the principal.

We are familiar with the fact that CI for 3 yrs (i.e. frequency = 3) at the rate of 5% per annum is 15.7625% of the principal as CI

$= (3 \times 5) . \boxed{3 \times (5)^2}\boxed{5^3}$% of the principal.

Hence, the time period = 3/2 yrs.

If the interest is compounded yearly at double the earlier rate, $r = 20\%$.

Time remains same i.e. $t = 3/2$ yrs

Then, CI for 1.5 yrs

$$= \left\{20 + 10 + \frac{20 \times 10}{100}\right\}\%\text{ i.e. } 32\%$$

of the principal.

Thus, CI = 32% of 8000 = 2560

43. Effective rate of interest for the amount invested at 6% interest, compounded semi- annually

$= (2 \times 3) . \boxed{3^2}\% = 6.09\%$

{$\because$ Rate of interest semi- annually = 3% and frequency = 2}

As given, 8% of 12000 + 6.09% of 10000 = 7.5% of Bimal's investment

$\Rightarrow$ Bimal's investment = 20920

23

Time and Work

Learning OBJECTIVES

After completing this chapter, aspirants will be able to :

- Understand the relations among time, amount of work, work-force and efficiency;
- Apply the above relations in diverse cases;
- Realise the difference between traditional approach and new approach;
- Differentiate between concepts associated with scheduled completion of work and actual completion of work;
- Recognise how percentage method works to solve the problems quickly;
- Visualise the critical concepts through a chart;
- Comprehend the concept of negative work and their utility;
- Understand the concept of efficiency both at individual level as well as group level;
- Realise that they can solve one question in many ways;
- Realise that percentage method is easier than the fraction method;
- Understand the philosophy of the division of this chapter in various parts;
- Analyse the trend of the questions being asked in various competitive examinations;
- Recognise the types of question which can be easily solved through the given options;
- Solve the questions in moments;
- Grip on concepts helping to solve the questions without pen and paper;
- Learn new concepts/fundamentals;
- Apply the pragmatic approaches in lieu of academic approaches; and
- Ponder on the alternative methods.

Inside

Introduction

This chapter is very critical as a lot of questions are asked in diverse competitive exams like CAT, MAT, SSC, CDS, Bank-PO, Railway, etc. It has also importance in our daily life. Since concepts of percentage have been applied in this topic to the greater extent in addition to the application of the concept of ratio, aspirants are supposed to command over those concepts for solving the questions in moments.

Build Your Concepts

This chapter is the easiest chapter provided the aspirants grasp the concepts detailed here. Suppose A does a piece of work in x days and B does the same in y days, then:

work done by A in one day $= \frac{1}{x}$ units of work

and work done by B in one day $= \frac{1}{y}$ units of work.

Moreover, we may express the amount of work done as the following:

Work done by A in one day $= \frac{100}{x}$ % of work

and work done by B in one day $= \frac{100}{y}$ % of work

It is important to note that all the problems envisaged under this chapter are associated with the above concepts subject to adhering to the following relations.

(i) Time ∝ amount of work

Or

Amount of work ∝ time

(when work-force is constant)

(ii) Time $\propto \frac{1}{\text{work - force}}$

Or

Work-force $\propto \frac{1}{\text{time}}$

(when amount of work is constant)

(iii) Work-force ∝ amount of work

Or

Amount of work ∝ work-force

(when time is constant)

(iv) Efficiency $\propto \frac{1}{\text{time}}$

Or

Time $\propto \frac{1}{\text{efficiency}}$

(when amount of work is constant)

(v) Efficiency ∝ amount of work

Or

Amount of work ∝ Efficiency

(when time is constant)

(vi) Efficiency $\propto \frac{1}{\text{work - force}}$

Or

Work-force $\propto \frac{1}{\text{efficiency}}$

(when amount of work is constant)

Now, take an example for better understanding:

A can do a piece of work in 20 days while B can do the same work in 15 days. In how many days can they together finish that work?

Let the total work be of 1 unit.

Then, work done by A in one day $= \frac{1}{20}$ units

and work done by B in one day $= \frac{1}{15}$ units.

Hence, work done by $(A + B)$ in one day

$$= \frac{1}{20} + \frac{1}{15} = \frac{7}{60}$$

$\because \frac{7}{60}$ units of work are done by $(A + B)$ in 1 day.

$\therefore$ 1 unit of work will be done by $(A + B)$ in $\frac{60}{7}$ days, i.e. $8\frac{4}{7}$ days.

It implies that reciprocal of fraction of work done by them together in one day will be the required answer.

Generic Version

A can do a piece of work in x days while B can do the same work in y days. In how many days can they together finish that work?

Clearly, $(A + B)$'s one day's work $= \frac{1}{x} + \frac{1}{y} = \frac{x+y}{xy}$

Then, time required = reciprocal of $\frac{x+y}{xy} = \frac{xy}{x+y}$ days

This method may be termed as 'Fraction Method'.

Alternative approach

Needless to say that work is 100% itself.

Then, work done by A in one day $= (100 \div 20)\% = 5\%$

and work done by B in one day $= (100 \div 15)\% = 6.66\%$

Hence, work done by $(A + B)$ in one day

$$= (5 + 6.66)\% = 11.66\%.$$

Then, total time required to finish 100% of the work

$$= \frac{100}{11.66} \text{ days}$$

Now, we will simplify this fraction in the following manner.

It appears that divisor is closer to 12, hence the quotient will be approx. 8 days.

$\therefore$ Work done by them in 8 days

$$= (11.66 \times 8)\%$$
$$= \{(11 \times 8) + (0.66 \times 8)\}\%$$
$$= \{88 + (0.66 \times 6) + (0.66 \times 2)\}\%$$
$$= \{88 + 4 + 1.33\}\%$$
[$\because$ 3 times 0.66 is equal to 2]
$$= 93.33\%$$

Then, remaining work after 8 days

$$= 100\% - 93.33\% = 6.66\%$$

Further, work done by them in $\frac{1}{2}$ day = 5.83%

It may be concluded from the above that the total time required will be little more than $\left(8 + \frac{1}{2}\right)$ days. At this stage you are capable to tick out the option which is closer to $8\frac{1}{2}$.

However, if you intend to arrive at exact answer (but no need to find exactly as in most of the competitive exams options are not so much closer to each other), you have to divide 6.66 by 11.66. Obviously, the resultant would be 4/7 as dividend and divisor both are divisible by 1.66. Then, the required answer $= 8 + \frac{4}{7} = 8\frac{4}{7}$ days.

Note *The whole process can be done mentally without using pen and paper.*

This method may be termed as 'Percentage-Method', which is of course, very effective as well as time saver. We also realised from the above that we should have command also over Fraction-Percentage Conversion Table as envisaged in chapter 'Percentage'. In my opinion, aspirants should prefer percentage method to fraction method so as to arrive at option quickly. However, both methods have been elaborated for better understanding.

Case I Working together

Problem 1. *A* can do a work in 6 days and *B* in 9 days. How many days will both take together to complete the work? **[SSC(CGL-PT)-2000, CLAT-14,CDS-19]**

Sol. $\left(\begin{matrix} A \Rightarrow 16.66\% \text{ of work per day} \\ B \Rightarrow 11.11\% \text{ of work per day} \end{matrix}\right)$ hence, 27.77% per day by them.

Then, required time = (100% ÷ 27.77%) days = 3.5 days (approx.)

No need to find exact resultant, instead, observe the options. In deed, the options given were 7.5, 5.4, 3.6 and 3.

I hope you will appreciate this method.

Alternatively

$$\text{Time required} = \frac{xy}{x+y} = \frac{6 \times 9}{6+9}$$
$$= 3.6 \text{ days}$$

Problem 2. *A* can do 50% of the job in 16 days, *B* can do 1/4th of the job in 24 days. In how many days can they do 3/4th of the job working together? **[SSC(CGL-T-II)-17]**

Sol. $\left(\begin{matrix} A \Rightarrow 3.125\% \text{ of work per day} \\ B \Rightarrow 1.041\% \text{ of work per day} \end{matrix}\right)$ hence, 4.166% by $(A + B)$ per day.

Then, required time = 75% ÷ 4.166% = 18 days

Alternatively

$A \Rightarrow 32$ days and $B \Rightarrow 96$ days

Then, time taken by $(A + B)$ to finish $\frac{3}{4}$th of the job

$$= \frac{32 \times 96}{32 + 96} \times \frac{3}{4} = 18 \text{ days}$$

Problem 3. While working 7 h a day, *A* alone can complete a piece of work in 6 days and *B* alone in 8 days. In what time would they complete it together, working 8 h a day?

Sol. $\left(\begin{matrix} A \Rightarrow 16.66\% \text{ of work per day} \\ B \Rightarrow 12.5\% \text{ of work per day} \end{matrix}\right)$ hence, $(A + B) \Rightarrow 29.16\%$ per day if they work 7 h a day..

Now, time taken by them to finish the work, working 7 h per day

$$= (100\% \div 29.16\%) \text{ days} = 3.43 \text{ days.}$$

$\therefore$ By working 8 h a day, they will take $3.43 \times \frac{7}{8}$, i.e. 3 days.

Alternatively

Days taken by them to finish the work, working 7 h per day

$$= \frac{6 \times 8}{6 + 8} = 3.43 \text{ days}$$

$\therefore$ By working 8 h a day, they will take $3.43 \times \frac{7}{8}$, i.e. 3 days.

Problem 4. *A* and *B* together can do a work in 4 days. *A* alone can do the same work in 12 days. In how many days *B* alone can do the same work? **[Corporation Bank (PO)-03,CDS-17]**

Sol. $\left(\begin{matrix} A + B \Rightarrow 25\% \text{ of work per day} \\ A \Rightarrow 8.33\% \text{ of work per day} \end{matrix}\right)$ hence, 16.66% per day by *B*.

Then, time required by *B* to complete 100% of work

$$= (100\% \div 16.66\%) \text{ days} = 6 \text{ days.}$$

Alternatively Time required $= \frac{4 \times 12}{12 - 4} = 6$ days

Generic Version

$(A + B)$ can do a piece of work in x days while one of them alone can do the same work in y days. Then, time taken by another alone to finish that work = reciprocal of $\left(\frac{1}{x} - \frac{1}{y}\right) = \frac{xy}{y - x}$.

Problem 5. *A*, *B* and *C* can complete a piece of work in 24 days, 6 days and 12 days, respectively. Working together, they will complete the same work in?

Sol. $\left(\begin{matrix} A \Rightarrow 4.16\% \text{ of work per day} \\ B \Rightarrow 16.66\% \text{ of work per day} \\ C \Rightarrow 8.33\% \text{ of work per day} \end{matrix}\right)$ hence, 29.16% per day by them

Then, time required by them to complete 100% of work

= (100% ÷ 29.16%) days = 3.5 days (approx.)

Now, observe the options and tick out.

Alternatively

$$\text{Time required} = \frac{24 \times 6 \times 12}{24\cdot 6 + 6\cdot 12 + 12\cdot 24} = 3\frac{3}{7} \text{ days}$$

Generic Version

A, *B* and *C* can do a piece of work in x days, y days and z days respectively. Then, time taken by them together to finish that work = reciprocal of $\left(\frac{1}{x} + \frac{1}{y} + \frac{1}{z}\right) = \frac{xyz}{xy + yz + zx}$

Problem 6. ***A*, *B* and *C* together can completes a piece of work in 4 days. *B* and *C* can do the same work in 20 days and 25 days respectively, working alone. In how many days can *A* alone complete that work?** ***[SSC(CGL-T-II)-17]***

Sol. $\left(\begin{array}{l} A + B + C \Rightarrow 25\% \text{ of work per day} \\ B \Rightarrow 5\% \text{ of work per day} \\ C \Rightarrow 4\% \text{ of work per day} \end{array}\right)$ hence, 16% per day by *A* alone.

Then, time taken by *A* to complete the whole work

= (100% ÷ 16%) = 6.25 days.

Alternatively I

Time required

$$= \frac{4 \times 20 \times 25}{4(-20) + (-20)(-25) + (-25)4} = 6.25 \text{ days}$$

Generic Version

$(A + B + C)$ can do a piece of work in x days while *B* and *C* can finish that work in y days and z days respectively, working alone. Then, time taken by *A* alone to finish that work = Reciprocal of $\left(\frac{1}{x} - \frac{1}{y} - \frac{1}{z}\right) = \frac{xyz}{x(-y) + (-y)(-z) + (-z)x}$

Observation y and z have been replaced by $-y$ and $-z$ respectively, in $\frac{xyz}{xy + yz + zx}$.

Alternatively II

We know that, time taken by $(A + B + C)$ to finish a work $= \frac{xyz}{xy + yz + zx}$ if *A*, *B* and *C* can do a piece of work in x days, y days and z days respectively, working alone.

$$\Rightarrow \quad 4 = \frac{x \times 20 \times 25}{x\cdot 20 + 20\cdot 25 + 25\cdot x}$$

$$\Rightarrow \quad 180x + 2000 = 500x$$

$$\Rightarrow \quad x = 6.25 \text{ days}$$

Problem 7. ***A*, *B* and *C* together can finish a piece of work in $8\frac{1}{3}$ days and *B* alone can finish the same work in 25 days. In how many days can *A* and *C* together finish that work?**

Sol. $\left(\begin{array}{l} A + B + C \Rightarrow 12\% \text{ of work per day} \\ B \Rightarrow 4\% \text{ of work per day} \end{array}\right)$ hence, 8% per day by $(A + C)$.

Then, time taken by $(A + C)$ to finish the work

= (100% ÷ 8%)

= 12.5 days

Alternatively

Consider *A*, *B* and *C* as *B* and $(A + C)$. Then, time required by $(A + C)$

$$= \frac{8\frac{1}{3} \times 25}{25 - 8\frac{1}{3}} = 12.5 \text{ days.}$$

Problem 8. ***A* and *B* together can complete a work in 12 days. *A* alone can complete it in 20 days. If *B* does the work only for half a day daily, then in how many days *A* and *B* together will complete the work?** ***[RRB-03]***

Sol. $\left(\begin{array}{l} A + B \Rightarrow 8.33\% \text{ of work per day} \\ A \Rightarrow 5\% \text{ of work per day} \end{array}\right)$ hence, 3.33% per day by *B* if he works for full day. But, *B* does only for half a day. It implies that *B* does only (3.33 ÷ 2)%, i.e. 1.66% per day.

Then, per day work done by them together

= (5 + 1.66)%, i.e. 6.66%.

∴ Time required = (100% ÷ 6.66%) days

= 15 days

Alternatively

$$B\text{'s 1 day's work} = \frac{1}{12} - \frac{1}{20} = \frac{1}{30}$$

Since *B* works only for half a day daily, his daily contribution

$$= \text{half of } \frac{1}{30} = \frac{1}{60}.$$

Now, work done by them together daily

$$= \frac{1}{20} + \frac{1}{60} = \frac{1}{15}$$

Then, time required by them to complete the work = reciprocal of $\frac{1}{15}$ = 15 days

Problem 9. ***A* does 4/5 of a work in 20 days. He then calls in *B* and they together finish the remaining work in 3 days. How long *B* alone would take to do the whole work?**

Sol. $\left(\begin{array}{l} A \Rightarrow 4\% \text{ per day as he does } 80\% \text{ of work in 20 days} \\ A + B \Rightarrow 6.66\% \text{ per day as they do } 20\% \text{ of work in 3 days} \end{array}\right)$ hence, 2.66% per day by *B* alone.

Then, time taken by *B* to complete the whole work

= (100% ÷ 2.66%) = 37.5 days.

Alternatively I

A does 80% of work in 20 days, hence 4% per day. It is clear that *A* works for 23 days, i.e. he has done (23 × 4)% of work.

Hence, work done by *B* in 3 days = (100 − 92)% = 8%

Then, to complete 100% of work, *B* will take 37.5 days.

Alternatively II

A does 4/5 of a work in 20 days. It means he can complete the whole work in $20 \times \frac{5}{4}$, i.e. in 25 days.

Further, $(A + B)$ can do 1/5 of a work in 3 days. Then, they will do whole work in $3 \times \frac{5}{1}$, i.e. 15 days.

Now, time taken by B alone to finish whole work

$$= \frac{15 \times 25}{25 - 15} = 37.5 \text{ days.}$$

Problem 10. A and B can do a work in 12 days, B and C in 15 days, C and A in 20 days. If A, B and C work together, they will complete the work in?

[SSC (CGL-PT)-99, Andhra Bank (PO)- 09, SSC (SAS)-10]

Sol. Given, $\left(\begin{array}{l} A + B \Rightarrow 8.33\% \text{ of work per day} \\ B + C \Rightarrow 6.66\% \text{ of work per day} \\ C + A \Rightarrow 5\% \text{ of work per day} \end{array}\right)$

Hence, $2(A + B + C)$ do 20% of the work per day.

Then, $(A + B + C)$ will do 10% of the work per day.

$\therefore$ Time taken by $(A + B + C)$ to complete the whole work $= (100\% \div 10\%) = 10$ days.

Alternatively I

$(A + B)$'s 1 day's work $= \frac{1}{12}$

$(B + C)$'s 1 day's work $= \frac{1}{15}$

$(C + A)$'s 1 day's work $= \frac{1}{20}$

Then, $2(A + B + C)$'s 1 day's work $= \frac{1}{12} + \frac{1}{15} + \frac{1}{20} = \frac{1}{5}$

$\therefore (A + B + C)$'s 1 day's work $= \frac{1}{2\cdot 5} = \frac{1}{10}$

Hence, $(A + B + C)$ will complete the work in 10 days.

Alternatively II

Assume $(A + B)$, $(B + C)$ and $(C + A)$ as x, y and z, respectively.

Then, time taken by $(x + y + z)$ to finish the work

$$= \frac{12 \times 15 \times 20}{12\cdot 15 + 15\cdot 20 + 20\cdot 12} = 5 \text{ days}$$

Clearly, $(x + y + z) = 2(A + B + C)$

Then, time taken by $(A + B + C)$ to complete the whole work $= 5 \times 2 = 10$ days

Problem 11. A can do a piece of work in 4 h. B and C together can do it in 3 h, while A and C together can do it in 2 h. How long will B alone take to do it?

Sol. $\left.\begin{array}{l} A \Rightarrow 25\% \text{ of work per hour} \\ B + C \Rightarrow 33.33\% \text{ of work per hour} \\ A + C \Rightarrow 50\% \text{ of work per hour} \end{array}\right\}$ hence, $C \Rightarrow (50 - 25)\%$ per hour.

Then, $B \Rightarrow (33.33 - 25)\% = 8.33\%$ per hour

Or

$A + B + C \Rightarrow (25 + 33.33)\% = 58.33\%$ per hour

Then, $B \Rightarrow (58.33 - 50) = 8.33\%$ per hour

$\therefore$ Time taken by B alone to finish whole work $= (100\% \div 8.33\%)$ h $= 12$ h

Alternatively

A's one hour's work $= \frac{1}{4}$

$(B + C)$'s one hour's work $= \frac{1}{3}$

$(C + A)$'s one hour's work $= \frac{1}{2}$

Then, $(A + B + C)$'s one hour's work

$$= \frac{1}{4} + \frac{1}{3} = \frac{7}{12}$$

$\therefore$ B's one hour's work $= \frac{7}{12} - \frac{1}{2} = \frac{1}{12}$

Now, time taken by B alone to finish whole work = 12 h.

Problem 12. A can do a work in 5 days less than the time taken by B to do it. If both of them together take $11\frac{1}{9}$ days, then the time taken by B alone to do the same work (in days) is: ***[SSC (10+2 LDC /DEO)- 11, SSC (CGL-T-II)-17]***

Sol. Let B takes x days to complete the work.

Then, A will take $(x - 5)$ days.

Further, A's 1 day's work $= \frac{100}{(x - 5)}\%$

and B's 1 day's work $= \frac{100}{x}\%$

Then, work done by $(A + B)$ in 1 day $= \frac{100}{(x - 5)}\% + \frac{100}{x}\%$

$$\Rightarrow \left(100 \div 11\frac{1}{9}\right)\% = \frac{200x - 500}{x(x - 5)}\%$$

$$\Rightarrow 9 = \frac{200x - 500}{x(x - 5)}$$

$$\Rightarrow 9x^2 - 45x = 200x - 500$$

$$\Rightarrow 9x^2 - 245x + 500 = 0$$

$$\Rightarrow 9x^2 - 225x - 20x + 500 = 0$$

$$\Rightarrow 9x(x - 25) - 20(x - 25) = 0$$

$$\Rightarrow x = 25 \text{ and } \frac{20}{9}$$

(If B takes $\frac{20}{9}$ days, A will take $\left(\frac{20}{9} - 5\right)$ days i.e. negative value, hence $x = \frac{20}{9}$, not possible)

Thus, required answer = 25 days

Alternatively I

Work done by $(A + B)$ in 1 day $= \left(100 \div 11\frac{1}{9}\right)\% = 9\%$

Now, proceed through options. Let options be 15, 20, 25 and 30.

(a) $15 \Rightarrow 15(B)$, then $10(A)$ but $6.66\% + 10\% \neq 9\%$

(b) $20 \Rightarrow 20(B)$, then $15(A)$ but $5\% + 6.66\% \neq 9\%$

(c) $25 \Rightarrow 25(B)$, then $20(A)$ and $4\% + 5\% = 9\%$

Hence, option (c) is correct.

Alternatively II

Let B takes x days to complete the work.

Then, A will take $(x - 5)$ days.

Clearly, A's 1 day's work $= \frac{1}{x - 5}$

and B's 1 day's work $= \frac{1}{x}$

Further, $(A + B)$'s 1 day's work $= 1/11\frac{1}{9} = \frac{9}{100}$

Hence, $\frac{1}{x-5} + \frac{1}{x} = \frac{9}{100}$

$\Rightarrow \quad 9x^2 - 45x = 200x - 500$

$\Rightarrow \quad 9x^2 - 245x + 500 = 0$

$\Rightarrow \quad 9x^2 - 225x - 20x + 500 = 0$

$\Rightarrow \quad 9x(x-25) - 20(x-25) = 0$

$\Rightarrow \quad x = 25 \text{ and } \frac{20}{9}$

(If B takes $\frac{20}{9}$ days, A will take $\left(\frac{20}{9} - 5\right)$ days, i.e. negative value, hence $x = \frac{20}{9}$, not possible)

Thus, required answer = 25 days.

Problem 13. A and B together can finish a piece of work in 6 days. If A works for 2 days and B for 7.5 days, only 70% of the work is completed. In how many days can A alone finish the whole work?

Sol. Work done by $(A + B)$ in 1 day = 16.66%

$\therefore$ Work done by $(A + B)$ in 2 days = 33.33% (Imagining A and B worked together for 2 days)

It implies that remaining work, i.e. $(70 - 33.33)\%$ has been done by B alone in $(7.5 - 2)$ days.

Then, B's work in one day $= \frac{(70 - 33.33)\%}{(7.5 - 2)} = 6.66\%$

Thus, A's work in one day $= (16.66 - 6.66)\% = 10\%$

$\therefore$ Time taken by A to complete the whole work $= (100\% \div 10\%) = 10$ days.

Alternatively

Work done by $(A + B)$ in 2 days $= \frac{2}{6} = \frac{1}{3}$

Then, remaining work $= \frac{70}{100} - \frac{1}{3} = \frac{11}{30}$, which has been done by B in $\left(7\frac{1}{2} - 2\right)$, i.e. $5\frac{1}{2}$ days.

Hence, time taken by B to do the whole work $= 5\frac{1}{2} \times \frac{30}{11} = 15$ days.

Now, the question is : $(A + B)$ do the work in 6 days and B alone in 15 days, then A alone in?

Hence, time required by A to do the work $= \frac{6 \times 15}{15 - 6} = 10$ days.

Problem 14. In order to finish a piece of work B takes 5 days more than A does. B started to work and after 3 days C joins him, then the work is finished in next 8 days. How much time will A take to finish the work alone if all three together finish the work in 5 days?

Sol. Work done by all three together in one day = 20%

It is clear that work is completed in 11 days when B works for 11 days and C works for only 8 days.

Let A does a% of work daily, B does b% daily and C does c% daily.

Then, $\quad 11b + 8c = 100$

$\Rightarrow \quad c = \frac{100 - 11b}{8} \quad \ldots(i)$

Further, time taken by B to finish the work – 5 = time taken by A

$\Rightarrow \quad \frac{100}{b} - 5 = \frac{100}{a}$

$\Rightarrow \quad \frac{100}{b} = \frac{100}{a} + 5$

$\Rightarrow \quad b = \frac{100a}{100 + 5a} \quad \ldots(ii)$

Also, $a + b + c = 20$

$\Rightarrow \quad a + b + \frac{100 - 11b}{8} = 20$ (Putting value of c from Eq. (i))

$\Rightarrow \quad a + b - \frac{11b}{8} = \frac{15}{2} \Rightarrow a - \frac{3b}{8} = \frac{15}{2}$

$\Rightarrow \quad a - \frac{3 \times \frac{100a}{100 + 5a}}{8} = \frac{15}{2}$ (Putting value of c from Eq. (ii))

$\Rightarrow \quad a - \frac{300a}{800 + 40a} = \frac{15}{2}$

$\Rightarrow \quad 800a + 40a^2 - 300a = \frac{15}{2}(800 + 40a)$

$\Rightarrow \quad 800a + 40a^2 - 300a = 6000 + 300a$

$\Rightarrow \quad 40a^2 + 200a - 6000 = 0$

$\Rightarrow \quad a^2 + 5a - 150 = 0$

$\Rightarrow \quad a^2 + 15a - 10a - 150 = 0$

$\Rightarrow \quad a(a + 15) - 10(a + 15) = 0$

$\Rightarrow \quad a = -15 \text{ and } 10$

$\therefore \quad a = 10 \quad (\because a = -15\% \text{ not possible})$

Hence, A does 10% of work daily. Then, time taken by A alone to do the whole work $= (100\% \div 10\%) = 10$ days.

Alternatively

In such types of questions we should proceed through options. Let options be 20, 10, 8 and 15.

It is obvious that work done by all three together in one day = 20%

(a) 20 $\Rightarrow$ If A takes 20 days, B will take 25 days.

Then, C's 1 day's work $= (20 - 5 - 4)\% = 11\%$.

Further, work gets finished when B works for 11 days and C for 8 days. But $(11 \times 4)\% + (8 \times 11)\% \neq 100\%$.

(b) 10 $\Rightarrow$ If A takes 10 days, B will take 15 days. Then, C's 1 day's work $= (20 - 10 - 6.66)\% = 3.33\%$

Further, work gets finished when B works for 11 days and C for 8 days and $(11 \times 6.66)\% + (8 \times 3.33)\% = 100\%$.

Hence, option (b) is correct.

Problem 15. Saransh takes 4 days more to do a job than if Saransh and Shivansh worked together. If Shivansh worked alone, he would have required 9 days more than the time required by both together. What time will they take to do the job together?

Sol. Let they together takes x days to do the job.

Then, time taken by Saransh alone $= (x + 4)$ days and time taken by Shivansh alone $= (x + 9)$ days.

Obviously, $\quad \frac{1}{x+4} + \frac{1}{x+9} = \frac{1}{x}$

$\Rightarrow \quad x^2 + 13x + 36 = 2x^2 + 13x$

$\Rightarrow \quad x^2 = 36 \Rightarrow x = 6$ days

Alternatively I

Required time $= \sqrt{4 \times 9} = 6$ days

Generic Version :

If A and B take x days and y days respectively more than what is taken by $(A + B)$ to do a job, the time taken by $(A + B)$

$= \sqrt{\text{Product of time exceeding time taken by both together}}$

$= \sqrt{xy}$

In other words, if A and B together can do a work in d days and A alone can do the same work in $(d + x)$ days and B alone in $(d + y)$ days, then value of d will be $\sqrt{xy}$.

Alternatively II

Proceed through options. Let options be 4, 6, 8 and 10.

(a) 4 $\Rightarrow$ Time taken by Saransh and Shivansh

= 8 days and 13 days, respectively.

But, $(100 \div 8)\% + (100 \div 13)\% \neq (100 \div 4)\%$.

(b) 6 $\Rightarrow$ Time taken by Saransh and Shivansh

= 10 days and 15 days respectively

and $(100 \div 10)\% + (100 \div 15)\% = (100 \div 6)\%$.

Hence, option (b) is correct.

Problem 16. **Working together, A and B can complete a work in x days. Working alone, A takes $(8 + x)$ days and B takes $(16 + x)$ days to complete the same work. In how many days can A alone complete twice the work?**

Sol. $x = \sqrt{8 \times 16} = 8\sqrt{2}$

Then, time taken by A, working alone, to complete twice the work $= 2(8 + 8\sqrt{2}) = 16(1 + \sqrt{2})$ days.

Problem 17. **A, B and C, working alone, can do a work in $(x + 6)$ days, $(x + 8)$ days and $(x + 11)$ days, respectively. They together can do that work in x days. Working together, in how many days can they do the same work?**

Sol. Time taken by A, B and C, working together.

$$\frac{(x + 6)(x + 8)(x + 11)}{(x + 6)(x + 8) + (x + 8)(x + 11) + (x + 11)(x + 6)} = x$$

$$\Rightarrow \quad \frac{x^3 + 25x^2 + 202x + 528}{3x^2 + 50x + 202} = x$$

$$\Rightarrow \quad x^3 + 25x^2 + 202x + 528 = 3x^3 + 50x^2 + 202x$$

$$\Rightarrow \quad 2x^3 + 25x^2 - 528 = 0$$

$$\Rightarrow \quad 2x^3 - 8x^2 + 33x^2 - 132x + 132x - 528 = 0$$

$$\Rightarrow \quad 2x^2(x - 4) + 33x(x - 4) + 132(x - 4) = 0$$

$$\Rightarrow \quad (x - 4)(2x^2 + 33x + 132) = 0$$

$$\Rightarrow \quad (x - 4) = 0 \text{ and } 2x^2 + 33x + 132 = 0$$

$$\Rightarrow \quad x = 4, x = \frac{-33 \pm \sqrt{33}}{4}$$

$\Rightarrow \quad x = 4$ ($\because$ the value of x cannot be negative)

Hence, they together can do the same work in 4 days.

Observation $2x^3 + 25x^2 - 528 = 0$

$$\Rightarrow \quad 2x^3 + (6 + 8 + 11)x^2 - (6 \times 8 \times 11) = 0$$

Generic Version If A, B and C, working alone, can do a work in $(d + a)$ days, $(d + b)$ days and $(d + c)$ days respectively and they together can do that work in d days, then the positive solution of the equation $2d^3 + (a + b + c)d^2 - abc = 0$ will be the value of d.

In such types of problems, the aspirants should proceed through options instead of solving the equation. They may simply put the value of d, given in options, in the equation. The value which will satisfy the equation, will be the required answer.

Problem 18. **A and B together can do a piece of work in 30 days. A having worked for 16 days, B finishes the remaining work alone in 44 days. In how many days shall B finish the whole work alone?** ***[CBI-97, MAT 11]***

Sol. $(A + B)$'s, 1 day's work = 3.33%

Assume that A and B worked together for 16 days.

Then, work done = 16 × 3.33% = 53.33%

The remaining work i.e. (100 – 53.33)% is done by B in (44 – 16) days.

$\therefore$ 100% of work will be done by B in $\left(\frac{28}{46.66} \times 100\right)$ i.e. 60 days.

Alternatively I

Let A alone can finish the whole work in x days.

Given, time taken by $(A + B)$ to finish the whole work = 30 days.

Then, time required by B to finish the whole work alone

$$= \frac{30x}{x - 30}$$

The work gets finished when A works for 16 days and B works for 44 days.

A's 16 day's work $= \frac{16}{x}$ and B's 44 day's work $= \frac{44(x - 30)}{30x}$

Clearly, $\quad \frac{16}{x} + \frac{44(x - 30)}{30x} = 1 \Rightarrow 480 + 44x - 1320 = 30x$

$\Rightarrow \quad 14x = 840 \Rightarrow x = 60$ days

Then, time required by B to finish the whole work alone

$$= \frac{30x}{x - 30} = \frac{30 \times 60}{60 - 30} = 60 \text{ days}$$

Alternatively II

Assume that A and B worked for 16 days together. Then, amount of work done $= \frac{16}{30} = \frac{8}{15}$.

Clearly, the remaining work i.e. $1 - \frac{8}{15}\left(= \frac{7}{15}\right)$ will be done by B in (44 – 16) i.e. 28 days.

Then, time taken by B to do the whole work $= 28 \times \frac{15}{7} = 60$ days.

Problem 19. **It takes 8 days, 12 days, and 16 days for X, Y and Z respectively to complete a work. How many days will it take to complete the work if X works for 2 days, then Y works on it for until 25% of the work is left for Z to do and then Z completes the work?**

Sol. Work done by X, Y and Z per day respectively = 12.5%, 8.33% and 6.25%.

Then, required time = (2 days taken by X) + $\frac{(75\% - 2 \times 12.5\%)}{8.33\%}$ days + $\frac{25\%}{6.25\%}$ = 2 + 6 + 4 = 12 days.

Alternatively

Work done by X in 2 days $= \frac{2}{8} = \frac{1}{4}$

Work done by $Y = \left(1 - \frac{1}{4}\right) - 25\%$ of work $= \frac{1}{2}$

and time taken by Y to do $\frac{1}{2}$ of the work $= \frac{1}{2} \times 12 = 6$ days

Further, time taken by Z to do $\frac{1}{4}$ (i.e. 25%) of the work

$$= \frac{1}{4} \times 16 = 4 \text{ days.}$$

Hence, required time = 2 + 6 + 4 = 12 days.

Problem 20. **A and B can do a work in 12 days while B and C can do it in $6\frac{2}{3}$ days. After A had worked on it for 3 days and B for 4 days, C finished the work in 7 days. In how many days could B do the work separately?**

Sol. Given, $\begin{pmatrix} A + B \Rightarrow 8.33\% \text{ of work per day} \\ B + C \Rightarrow 15\% \text{ of work per day} \end{pmatrix}$

Let A and B together worked for 3 days.

Then, work done = 3 × 8.33% = 25%

But in actual, B worked for 4 days. It may be construed that B worked for one day with C.

Then, work done by $(B + C)$ in 1 day = 15%

The remaining work i.e. (100 – 25 – 15)% is done by C alone in (7 – 1) days i.e. C does 10% of work per day.

Then, B's 1 day's work = (15 – 10)% = 5%

∴ Time taken by B to do whole work = 100% – 5% = 20 days

Alternatively I

Let A and B together worked for 3 days.

Then, work done by them $= \frac{3}{12} = \frac{1}{4}$

But in actual, B worked for 4 days. It may be construed that B worked for one day with C.

Then, work done by $(B + C)$ in one day $= \frac{1}{6\frac{2}{3}} = \frac{3}{20}$.

From the above, it may be understood that the remaining work i.e. $\left(1 - \frac{1}{4} - \frac{3}{20} = \frac{3}{5}\right)$ has been completed by C alone in only (7 – 1) days. Then, time taken by C alone to do whole work $= 6 \times \frac{5}{3} = 10$ days and the fact that $(B + C)$ do the work in $6\frac{2}{3}$ days.

$$\therefore \text{Time taken by } B \text{ alone to do whole work} = \frac{10 \times 6\frac{2}{3}}{10 - 6\frac{2}{3}} = 20 \text{ days.}$$

Alternatively II

You may proceed through options also. Let options be 15, 20, 25 and 30 days.

Given, $\begin{pmatrix} A + B \Rightarrow 8.33\% \text{ of work per day} \\ B + C \Rightarrow 15\% \text{ of work per day} \end{pmatrix}$

(a) 15 : $B \rightarrow$ 6.66% per day, then C's and A's work per day respectively are (15 – 6.66)% and (8.33 – 6.66)%

then, (3 × 1.66)% + (4 × 6.66)% + (7 × 8.33)% ≠ 100%

(b) 20 : $B \rightarrow$ 5% per day, then C's and A's work per day respectively are (15 – 5)% and (8.33 – 5)%

then, (3 × 3.33)% + (4 × 5)% + (7 × 10)% = 100%

Hence, option (b) is correct.

Problem 21. **In order to finish a piece of work individually B takes 5 days more than A does and A takes 4 days more than C does. A and B can together do the work in the same time as taken by C alone. In how many days can A alone finish a piece of work?**

Sol. Let A alone takes x days to finish the work.

Then, time taken by B to do the same work = $(x + 5)$ days and time taken by C to do the same work = $(x - 4)$ days.

Now, time taken by $(A + B)$ to finish the work $= \frac{x \times (x + 5)}{x + (x + 5)}$.

Time taken by C alone $= \frac{x \times (x + 5)}{x + (x + 5)}$

$$\Rightarrow \quad x - 4 = \frac{x \times (x + 5)}{x + (x + 5)}$$

$$\Rightarrow \quad x^2 + 5x = 2x^2 - 3x - 20$$

$$\Rightarrow \quad -x^2 + 8x + 20 = 0$$

$$\Rightarrow \quad -x^2 + 10x - 2x + 20 = 0$$

$$\Rightarrow \quad -x(x - 10) - 2(x - 10) = 0$$

$$\Rightarrow \quad (-x - 2)(x - 10) = 0 \Rightarrow x = -2 \text{ and } 10$$

$$\Rightarrow \quad x = 10 \text{ days} \quad (\because \text{the time cannot be negative})$$

Alternatively

The best way is to proceed through options. Let options be 15, 10,...

(a) 15 ⇒ A takes 15 days, B takes 20 days and C takes 11 days.

But, $\frac{15 \times 20}{15 + 20} \neq 11$.

(b) 10 ⇒ A takes 10 days, B takes 15 days and C takes 6 days and $\frac{10 \times 15}{10 + 15} = 6$.

***Or* (through percentage method)**

(a) 15 ⇒ A takes 15 days (i.e. 6.66% per day), B takes 20 days (i.e. 5% per day) and C takes 11 days (i.e. 9.09% per day). But, (6.66 + 5) % ≠ 9.09%.

(b) 10 ⇒ A takes 10 days (i.e. 10% per day), B takes 15 days (i.e. 6.66% per day) and C takes 6 days (i.e. 16.66% per day). Also, (10 + 6.66)% = 16.66%.

Hence, option (b) is correct.

Problem 22. **A and B together can finish a piece of work in $8\frac{4}{7}$ days where B and C together can finish the same**

work in 6 days. In how many days can A finish that work alone provided C takes half the time in comparison to A so as to finish the work?

Sol. Work done by $(A + B)$ per day = 11.66%

Then, B will do $(11.66 - a)$% of work per day supposing A does a% of work per day.

Further, work done by $(B + C)$ per day = 16.66%.

Then, B will do $(16.66 - 2a)$% per day

($\because C$ will do twice the work in comparison to A in the same time)

Obviously, $(11.66 - a)\% = (16.66 - 2a)\% \Rightarrow a = 5\%$

Hence, A will take 20 days to do the work.

Alternatively I

Let C takes x days to finish a work. Then, time taken by $A = 2x$ days. As given, $(A + B)$ take $8\frac{4}{7}$ days to do a work. Then, time taken by B to finish same work $= \dfrac{8\frac{4}{7} \times 2x}{2x - 8\frac{4}{7}}$

Further, $(B + C)$ take 6 days to do a work. Then, time taken by B to finish same work $= \dfrac{6 \times x}{x - 6}$.

Obviously, $$\frac{8\frac{4}{7} \times 2x}{2x - 8\frac{4}{7}} = \frac{6 \times x}{x - 6}$$

$$\Rightarrow \quad \frac{20}{7}(x - 6) = 2x - \frac{60}{7}$$

$$\Rightarrow \quad \frac{6x}{7} = \frac{60}{7} \Rightarrow x = 10$$

$$\Rightarrow \quad 2x = 20 \text{ days}$$

Hence, A alone will take 20 days to do the work.

Alternatively II

You may proceed through options. Let options be 30, 20,....

(a) 30 $\Rightarrow A$ takes 30 days and C takes 15 days. But, time taken by

$$B = \frac{30 \times 8\frac{4}{7}}{30 - 8\frac{4}{7}} \neq \frac{15 \times 6}{15 - 6}$$

(b) 20 $\Rightarrow A$ takes 20 days and C takes 10 days and time taken by B

$$= \frac{20 \times 8\frac{4}{7}}{20 - 8\frac{4}{7}} = \frac{10 \times 6}{10 - 6}$$

Or (through percentage method)

Work done by $(A + B)$ per day = 11.66% and work done by $(B + C)$ per day = 16.66%.

(a) 30 $\Rightarrow A$ takes 30 days (i.e. 3.33% per day) and C takes 15 days (i.e. 6.66% per day). But, work done by B per day $= (11.66 - 3.33)\% \neq (16.66 - 6.66)\%$.

(b) 20 $\Rightarrow A$ takes 20 days (i.e. 5% per day) and C takes 10 days (i.e. 10% per day) and work done by B per day $= (11.66 - 5)\% = (16.66 - 10)\%$.

Hence, option (b) is correct.

Case II Working together but as per their schedules

Problem 23. **A alone can complete a work in 16 days and B alone in 12 days. Starting with A, they work on alternate days. The total work will be completed in?**

[SSC (CGL-PT) 04, LIC (AAO) 13]

Sol. $\left(\begin{array}{l} A \Rightarrow 6.25\% \text{ of work per day} \\ B \Rightarrow 8.33\% \text{ of work per day} \end{array}\right)$ hence, 14.58% in 2 days by them.

Then, in 4 (i.e. 2×2) days $\Rightarrow 14.58 \times 2 = 29.16\%$

In 6 (i.e. 2×3) days $\Rightarrow 14.58 \times 3 = 43.75$

....

In 12 (i.e. 2×6) days $\Rightarrow 14.58 \times 6 = 87.50\%$

In 14 (i.e. 2×7) days $\Rightarrow 14.58 \times 7 = 102.08\%$, (i.e. more than 100%).

Remarks

(We should multiply 14.53% by a suitable number so as to get the product closer or equal to 100% but not more than 100%. Thereafter, we multiply that suitable number by 2 as we have considered here an amount of work done in 2 days as one unit. The latter product is the time taken to do the amount of work equal to former product)

Hence, 14.58% × 6, i.e. 87.50% of the work is done by them in (2 × 6) i.e. 12 days.

Clearly, 12.5% of the work remains left after 12 days.

Since A starts the work, A will work on 13th day.

Then, remaining work = $(12.5 - 6.25)\% = 6.25\%$, which is to be done by B on 14th day. Obviously, B will take $\dfrac{6.25\%}{8.33\%}$ i.e. $\dfrac{3}{4}$ days

Thus, total time taken $= 12 + 1 + \dfrac{3}{4} = 13\dfrac{3}{4}$ days

Note *No need to calculate decimal value exactly. Observe the options.*

Alternatively

Work done by them in 2 days $= \dfrac{1}{16} + \dfrac{1}{12} = \dfrac{7}{48}$

Now, we should multiply $\dfrac{7}{48}$ by a suitable number so as to get the product closer or equal to 1 but not more than 1.

Here, $\dfrac{7}{48} \times 6$, i.e. $\dfrac{7}{8}$ of the work is done by them in (6×2), i.e. 12 days.

Now, remaining work after 12 days $= 1 - \dfrac{7}{8} = \dfrac{1}{8}$

Clearly, A will work on 13th day as A starts the work and the work remaining after 13 days $= \dfrac{1}{8} - \dfrac{1}{16} = \dfrac{1}{16}$, which will be done by B on 14th day.

Then, time required by B to finish $\dfrac{1}{16}$th of the work

$$= 12 \times \frac{1}{16} = \frac{3}{4} \text{ days}$$

Thus, total time taken $= 12 + 1 + \dfrac{3}{4} = 13\dfrac{3}{4}$ days

Problem 24. *A, B* and *C* can do a piece of work in 20 days, 30 days and 60 days, respectively. In how many days can *A* do the work if he is assisted by *B* and *C* on every third day?

Sol. Clearly, $\begin{pmatrix} A \Rightarrow 5\% \text{ of work per day} \\ B \Rightarrow 3.33\% \text{ of work per day} \\ C \Rightarrow 1.66\% \text{ of work per day} \end{pmatrix}$

Then, work done by *A* for first two days

$= 2 \times 5\% = 10\%$

and work done by $(A + B + C)$ on third day

$= (5 + 3.33 + 1.66)\% = 10\%$

$\therefore$ Work done by them for the first 3 days

$= (10 + 10)\% = 20\%$

Hence, $(20\% \times 5)$ i.e. 100% of work will be done by them in (5×3) i.e. 15 days.

Alternatively

Work done by *A* in first two days $= \frac{2}{20} = \frac{1}{10}$

Further, work done by $(A + B + C)$ on 3rd day

$= \frac{1}{20} + \frac{1}{30} + \frac{1}{60} = \frac{1}{10}$.

$\therefore$ Work done by them for the first 3 days $= \frac{1}{10} + \frac{1}{10} = \frac{1}{5}$

Hence, $\left(\frac{1}{5} \times 5\right)$ i.e. whole work will be done by them in (5×3) i.e. 15 days.

Problem 25. Ramesh can do a piece of work in 15 days, Sohan in 20 days and Sarita in 30 days. They start working in turns and Ramesh started the work on first day followed by Sarita on 2nd day and by Sohan on 3rd day and again by Ramesh on 4th day and so on. In how many days will the work be finished?

Sol. Clearly, $\begin{pmatrix} \text{Ramesh} \Rightarrow 6.66\% \text{ of work per day} \\ \text{Sohan} \Rightarrow 5\% \text{ of work per day} \\ \text{Sarita} \Rightarrow 3.33\% \text{ of work per day} \end{pmatrix}$

Then, work done by them together in first 3 days = 15%.

Hence, 15% × 6, i.e. 90% of the work is done by them in (3 × 6), i.e. 18 days.

Clearly, 10% of the work remains left after 18 days.

On 19th day it will be the turn of Ramesh and he will do 6.66% of the work on that day.

Then, remaining work $= (10 - 6.66)\% = 3.33\%$, which is to done by Sarita on 20th day. Obviously, Sarita will take $\frac{3.33}{3.33}$, i.e. 1 day.

Thus, total time taken = 18 + 1 + 1 = 20 days.

Alternatively

Work done by (Ramesh + Sarita + Sohan) in first 3 days

$= \frac{1}{15} + \frac{1}{30} + \frac{1}{20} = \frac{3}{20}$.

Hence, $\left(\frac{3}{20} \times 6\right)$ i.e. $\frac{9}{10}$ work will be done by them in (6×3) i.e. 18 days.

Clearly, $\frac{1}{10}$ of the work remains left after 18 days.

On 19th day it will be the turn of Ramesh and he will do $\frac{1}{15}$ of the work on that day.

Then, remaining work $= \frac{1}{10} - \frac{1}{15}$ i.e. $\frac{1}{30}$, which is to be done by Sarita on 20th day. Obviously, Sarita will take 1 day.

Thus, total time taken = 18 + 1 + 1

= 20 days.

With reference to the above problem, what proportion of the complete work is done by Ramesh? *[IIFT-18]*

Sol. Clearly, Ramesh worked for 7 days.

Then, work done by him $= 7 \times \frac{1}{15} = \frac{7}{15}$

Problem 26. *X, Y* and *Z* are three software experts, who work on upgrading the software in a number of identical systems. *X* takes a day off after every 3 days of work, *Y* takes a day off after every 4 days of work and *Z* takes a day off after every 5 days of work. Starting afresh after a common day off,

(i) *X* and *Y* working together can complete one new upgrade job in 6 days.

(ii) *Z* and *X* working together can complete two new upgrade jobs in 8 days.

(iii) *Y* and *Z* working together can complete three new upgrade jobs in 12 days.

If *X, Y* and *Z* together start afresh on a new upgrade job (after a common day off), exactly how many days will be required to complete this job? *[XAT-20]*

Sol. *X* and *Y* take together 6 days to complete one upgrade job. But, *X* takes a day off after 3 days of work and *Y* takes a day off after every 4 days of work. It implies that *X* works for 1st day, 2nd day, 3rd day, 5th day and 6th day and *Y* work for 1st day, 2nd day, 3rd day, 4th day and 6th day to complete one upgrade job.

Hence, *X* and *Y* each worked for 5 days to complete 100% of a job. Similarly, *Z* and *X* worked for 7 days and 6 days respectively to complete 200% of the same job.

And, *Y* and *Z* each worked for 10 days to complete 300% of the same job.

Since *X* and *Y* each worked for 5 days to complete 100% of a job. It implies that each can complete 10% of the job everyday. Further, *Z* and *X* worked for 7 days and 6 days respectively to complete 200% of the same job. It means, *Z* can complete $[\{200 - (6 \times 10)\}\% \div 7]\%$ i.e. 20% of the job everyday.

Now, time taken by all, working together, to complete 100% of the job = 100% ÷ (10% + 10% + 20%) = 2.5 days.

Case III Certain person(s) leave after all start the work together

Problem 27. *A* and *B* can complete a work in 15 days and 10 days, respectively. They started doing the work together but after 2 days, *B* had to leave and *A* alone completed the remaining work. The whole work was completed in?

[SSC(CGL-PT)-04, SSC(CGL-T-I)-17, MAT 18, SSC(CGL-T-I)-19]

Sol. $\begin{pmatrix} A \Rightarrow 6.66\% \text{ of work per day} \\ B \Rightarrow 10\% \text{ of work per day} \end{pmatrix}$ hence, 20% in 2 days by B.

It is clear that the remaining work i.e. 80% has been completed by A from starting to ending of the work (working along with B for 2 days and there after alone).

Now, time taken by A to finish 80% of the work

$= (80\% \div 6.66\%)$ days $= 12$ days.

Hence, the whole work was completed in 12 days.

Alternatively I

$\begin{pmatrix} A \Rightarrow 6.66\% \text{ of work per day} \\ B \Rightarrow 10\% \text{ of work per day} \end{pmatrix}$ hence, 33.33% in 2 days by them.

It is clear that the remaining work, i.e. 66.66% has been completed by A alone after first 2 days.

Now, time taken by A alone to finish 66.66% of the work

$= (66.66\% \div 6.66\%)$ days $= 10$ days.

Hence, total time taken $= 2 + 10 = 12$ days.

Alternatively II

Clearly, $\begin{pmatrix} A \Rightarrow 6.66\% \text{ of work per day} \\ B \Rightarrow 10\% \text{ of work per day} \end{pmatrix}$

Let the whole work was completed in x days. It implies that B worked for only 2 days and A worked for x days.

Then, work done by B in 2 days $= 2 \times 10\% = 20\%$ and work done by A in x days $= x \times 6.66\% = 6.66x\%$

Obviously, $20\% + 6.66x\% = 100\%$

$\Rightarrow \quad 6.66x = 80$

$\Rightarrow \quad x = 12$ days

Alternatively III

Work done by B in 2 days $= \frac{2}{10} = \frac{1}{5}$

Then, remaining work $= 1 - \frac{1}{5} = \frac{4}{5}$, which has been completed by A from starting to ending of the work (working along with B for 2 days and thereafter alone).

$\therefore$ Time taken by A to do the remaining work i.e. $\frac{4}{5}$

$= 15 \times \frac{4}{5} = 12$

Hence, the whole work was completed in 12 days.

Generic version

$\begin{pmatrix} A \Rightarrow x \text{ days} \\ B \Rightarrow y \text{ days} \end{pmatrix}$ They started together and one leaves after t days.

Then, time required in completion of whole work = time taken by the person to complete the whole work, who does not leave after certain period since work starts

$$\times \left(1 - \frac{t}{\text{time taken by the person, who leave, to complete the whole work}}\right)$$

Alternatively IV

You may proceed through options. Let options be 10, 12, 13 and 14.

Clearly, $\begin{pmatrix} A \Rightarrow 6.66\% \text{ of work per day} \\ B \Rightarrow 10\% \text{ of work per day} \end{pmatrix}$

(a) 10 $\Rightarrow$ It implies that A worked for 10 days and B worked for 2 days so as to complete whole work. However, $(10 \times 6.66\%) + (2 \times 10\%) \neq 100\%$.

(b) 12 $\Rightarrow$ It implies that A worked for 12 days and B worked for 2 days so as to complete whole work and $(12 \times 6.66\%) + (2 \times 10\%) = 100\%$.

Hence, option (b) is correct.

Alternatively V

You may proceed through options. Let options be 10, 12, 13 and 14.

(a) 10 $\Rightarrow$ It implies that A worked for 10 days and B worked for 2 days so as to complete whole work. However, $\frac{10}{15} + \frac{2}{10} \neq 1$.

(b) 12 $\Rightarrow$ It implies that A worked for 12 days and B worked for 2 days so as to complete whole work and $\frac{12}{15} + \frac{2}{10} = 1$.

Hence, option (b) is correct.

Problem 28. A and B can separately complete a piece of work in 20 days and 30 days, respectively. They work together for some time, then B left the work. If A completed the rest of work in 10 days, then B worked for? ***[SSC (CGL-PT)-08, SSC (CPO)-04, SBI (PO-Pre.)-18]***

Sol. $\begin{pmatrix} A \Rightarrow 5\% \text{ of work per day} \\ B \Rightarrow 3.33\% \text{ of work per day} \end{pmatrix}$ hence, 8.33% of work by them daily.

Now, work done by A in 10 days $= 10 \times 5\% = 50\%$.

It is obvious that the remaining work i.e. 50% has been done by them together in beginning. Then, time taken by them

$= (50\% \div 8.33\%) = 6$ days.

Hence, B worked for 6 days only.

Alternatively I

Let B leaves the work after t days. It implies that B worked for only t days and A worked for $(t + 10)$ days.

Work done by B in t days $= 3.33\% \times t = 3.33t\%$ and work done by A in $(t + 10)$ days $= 5\% \times (t + 10) = (5t + 50)\%$.

Obviously, $3.33t\% + (5t + 50)\% = 100\%$

$\Rightarrow \quad 8.33t = 50$

$\Rightarrow \quad t = 6$ days

Alternatively II

Let B leaves the work after t days.

Then, required time to complete whole work = time taken by the person to complete the whole work, who does not leave after certain period since work starts × (1 − t/time taken by the person, who leaves, to complete the whole work)

$\Rightarrow \quad 10 + t = 20\left(1 - \frac{t}{30}\right)$

$\Rightarrow \quad t + \frac{2t}{3} = 20 - 10 \Rightarrow \frac{5t}{3} = 10$

$\Rightarrow \quad t = 6$ days

Alternatively III

You may proceed through options. Let options be 4, 6, 8 and 10.

(a) 4 $\Rightarrow$ It implies that B worked for 4 days and A worked for (10 + 4) days so as to complete whole work. However, $\frac{14}{20} + \frac{4}{30} \neq 1$

Or (through percentage method)

$(14 \times 5\%) + (4 \times 3.33\%) \neq 100\%$

(b) 6 $\Rightarrow$ It implies that B worked for 6 days and A worked for (10 + 6) days so as to complete whole work and $\frac{16}{20} + \frac{6}{30} = 1.$

Or (through percentage method)

$(16 \times 5\%) + (6 \times 3.33\%) = 100\%$

Hence, option (b) is correct.

Problem 29. ***A*, *B* and *C* can separately complete a piece of work in 20 days, 25 days and 30 days, respectively. They started together doing the work but after 3 days *C* left. The whole work was completed in?**

Sol. Clearly, $\begin{pmatrix} A \Rightarrow 5\% \text{ of work per day} \\ B \Rightarrow 4\% \text{ of work per day} \\ C \Rightarrow 3.33\% \text{ of work per day} \end{pmatrix}$ hence, 10% by C in 3 days.

It is clear that the remaining work, i.e. 90% has been completed by A and B together from starting to ending of the work.

Now, time taken by $(A + B)$ to finish 90% of the work

$= [90\% \div (5 + 4)\%]$ days $= 10$ days.

Hence, the whole work was completed in 10 days.

Alternatively I

Clearly, $\begin{pmatrix} A \Rightarrow 5\% \text{ of work per day} \\ B \Rightarrow 4\% \text{ of work per day} \\ C \Rightarrow 3.33\% \text{ of work per day} \end{pmatrix}$ hence, 37% by C in 3 days.

It is clear that the remaining work i.e. 63% has been completed by $(A + B)$ after first 3 days.

Now time taken by $(A + B)$ to finish 63% of the work

$= [63 \div (5 + 4)]$ days $= 7$ days.

Hence, total time taken $= 3 + 7 = 10$ days.

Alternatively II

Clearly, $\begin{pmatrix} A \Rightarrow 5\% \text{ of work per day} \\ B \Rightarrow 4\% \text{ of work per day} \\ C \Rightarrow 3.33\% \text{ of work per day} \end{pmatrix}$

Let the whole work was completed in x days. It implies that C worked for only 3 days and A and B both worked for x days.

Then, work done by C in 3 days $= 3 \times 3.33\% = 10\%$

and work done by $(A + B)$ in x days $= x \times (5 + 4)\% = 9x\%$.

Obviously, $10\% + 9x\% = 100\%$

$\Rightarrow \quad 9x = 90 \Rightarrow x = 10$ days

Alternatively III

Work done by C in 2 days $= \frac{3}{30} = \frac{1}{10}$

Then, remaining work $= 1 - \frac{1}{10} = \frac{9}{10}$, which has been completed by $(A + B)$ from starting to ending of the work.

$\therefore$ Time taken by $(A + B)$ to do the remaining work i.e.

$$\frac{9}{10} = \frac{20 \cdot 25}{(20 + 25)} \times \frac{9}{10} = 10$$

(Recall, $A \Rightarrow x$, $B \Rightarrow y$, then time taken by them together $= \frac{xy}{x + y}$)

Hence, the whole work was completed in 10 days.

Generic version

$\begin{pmatrix} A \Rightarrow x \text{ days} \\ B \Rightarrow y \text{ days} \\ C \Rightarrow z \text{ days} \end{pmatrix}$

They started together and one leaves after t days.

Then, time required in completion of whole work = time taken by the persons (together) to complete the whole work, who do not leave after certain period since work starts

$$\times \left(1 - \frac{t}{\text{time taken by the person, who leaves, to complete the whole work}}\right)$$

Alternatively IV

You may proceed through options. Let options be 8, 10, 12 and 16.

(a) 8 $\Rightarrow$ It implies that C worked for 3 days and $(A + B)$ worked for 8 days so as to complete whole work.

However, $\quad \frac{8}{20} + \frac{8}{25} + \frac{3}{30} \neq 1.$

Or (through percentage method)

$8(5\% + 4\%) + (3 \times 3.33\%) \neq 100\%$

(b) 10 $\Rightarrow$ It implies that C worked for 3 days and $(A + B)$ worked for 10 days so as to complete whole work and

$\frac{10}{20} + \frac{10}{25} + \frac{3}{30} = 1.$

Or (through percentage method)

$10(5\% + 4\%) + (3 \times 3.33\%) = 100\%$

Hence, option (b) is correct.

Problem 30. ***A*, *B* and *C* can do a piece of work in 10 days, 15 days and 20 days, respectively. They together started the work and *A* left the work after 1 day and *B* after 3 days. In how many does will the work be finished?**

Sol. Clearly, $\begin{pmatrix} A \Rightarrow 10\% \text{ of work per day} \\ B \Rightarrow 6.66\% \text{ of work per day} \\ C \Rightarrow 5\% \text{ of work per day} \end{pmatrix}$ hence, 20% by B in 3 days.

Work done by A in 1 day + Work done by B in 3 days $= 10\% + 20\% = 30\%$

It is clear that the remaining work i.e. 70% has been completed by C from starting to ending of the work.

Now, time taken by C to finish 70% of the work $= (70\% \div 5\%)$ days $= 14$ days.

Hence, the whole work was completed in 14 days.

Alternatively I

Clearly, $\begin{pmatrix} A \Rightarrow 10\% \text{ of work per day} \\ B \Rightarrow 6.66\% \text{ of work per day} \\ C \Rightarrow 5\% \text{ of work per day} \end{pmatrix}$

Hence, $\begin{pmatrix} 21.66\% \text{ of work by all three per day} \\ \text{and } 11.66\% \text{ of work by } (B + C) \text{ per day} \end{pmatrix}$

Work done on 1st day $= 21.66\,\%$ as all three were working.

Work done from 2nd day to 3rd day $= 2 \times 11.66\%$, i.e. 23.33% as only B and C were working.

It is clear that the remaining work i.e. (100 – 21.66 – 23.33 = 55%) has been completed by C after first 3 days.

Now, time taken by C to finish 55% of the work = (55% ÷ 5%) days = 11 days.

Hence, total time taken = 3 + 11 = 14 days.

Alternatively II

$$\text{Clearly, }\begin{pmatrix} A \Rightarrow 10\% \text{ of work per day} \\ B \Rightarrow 6.66\% \text{ of work per day} \\ C \Rightarrow 5\% \text{ of work per day} \end{pmatrix}$$

Let the whole work was completed in x days. It implies that C worked for x days, A for 1 day and B for 3 days.

Obviously, $(x \times 5\%) + (1 \times 10\%) + (3 \times 6.66\%) = 100\%$

$$\Rightarrow \quad 5x = 70$$

$$\Rightarrow \quad x = 14 \text{ days}$$

Alternatively III

Work done by A in 1 day + Work done by B in 3 days

$$= \frac{1}{10} + \frac{3}{15} = \frac{3}{10}$$

Then, remaining work $= 1 - \frac{3}{10} = \frac{7}{10}$, which has been completed by C from starting to ending of the work.

$\therefore$ Time taken by C to do the remaining work i.e. $\frac{7}{10}$

$$= 20 \times \frac{7}{10} = 14$$

Hence, the whole work was completed in 14 days.

Generic Version

$\begin{pmatrix} A \Rightarrow x \text{ days} \\ B \Rightarrow y \text{ days} \\ C \Rightarrow z \text{ days} \end{pmatrix}$ They started together and any two leave after t_1 and t_2 days respectively.

Then, time required in completion of whole work = time taken by the person to complete the whole work, who does not leave after certain period since work starts

$$\times \left(1 - \frac{t_1}{\text{time taken by the person, who leaves after } t_1 \text{ days, to complete the whole work}} - \frac{t_2}{\text{time taken by the person, who leaves after } t_2 \text{ days, to complete the whole work}}\right)$$

Here, required time $= 20 \times \left(1 - \frac{1}{10} - \frac{3}{15}\right) = 14$ days

Alternatively IV

You may proceed through options. Let options be 10, 14, 16 and 18.

(a) 10 $\Rightarrow$ It implies that C worked for 10 days. A for 1 day and B for 3 days so as to complete whole work.

However, $\frac{10}{20} + \frac{1}{10} + \frac{3}{15} \neq 1$

Or (through percentage method)

$(10 \times 5\%) + (1 \times 10\%) + (3 \times 6.66\%) \neq 100\%$

(b) 14 $\Rightarrow$ It implies that C worked for 14 days, A for 1 day and B for 3 days so as to complete whole work and

$$\frac{14}{20} + \frac{1}{10} + \frac{3}{15} = 1$$

Or (through percentage method)

$$(14 \times 5\%) + (1 \times 10\%) + (3 \times 6.66\%) = 100\%$$

Hence, option (b) is correct.

Problem 31. *A, B and C complete a piece of work in 25 days, 20 days and 24 days, respectively. All work together for 2 days and then A and B left the work. C works for next* $8\frac{3}{5}$ *days and then A along with D joins C and all finish the work in next 3 days. In how many days D alone can complete the whole work?*

Sol. $\text{Clearly, }\begin{pmatrix} A \Rightarrow 4\% \text{ of work per day} \\ B \Rightarrow 5\% \text{ of work per day} \\ C \Rightarrow 4.16\% \text{ of work per day} \end{pmatrix}$

It is also clear that A works for (2 + 3) days, B for 2 days, C for $\left(2 + 8\frac{3}{5} + 3\right)$ days and D for 3 days.

Now, work done by $(A + B + C) = (5 \times 4\%)$

$$+ (2 \times 5\%) + \left(13\frac{3}{5} \times 4.16\%\right) = 86.66\%.$$

It implies that the remaining work, i.e. 13.33% has been completed by D in 3 days. Then, D can complete 100% of work in 22.5 days.

Alternatively I

Clearly, A works for (2 + 3) days, B for 2 days, C for $\left(2 + 8\frac{3}{5} + 3\right)$ days and D for 3 days.

Now, work done by $(A + B + C) = \frac{5}{25} + \frac{2}{20} + \frac{8\frac{3}{5}}{24} = \frac{13}{15}$.

It implies that the remaining work, i.e. $\frac{2}{15}$ has been completed by D in 3 days. Then, he/she can complete the whole work in $\frac{3 \times 15}{2}$ i.e. 22.5 days.

Alternatively II

You may proceed through options. Let options be 25, 22.5, 30, 40. Clearly, A works for (2 + 3) days, B for 2 days, C for $\left(2 + 8\frac{3}{5} + 3\right)$ days and D for 3 days. Then, work done by all must be equal to 1.

(a) $25 \Rightarrow \frac{5}{25} + \frac{2}{20} + \frac{8\frac{3}{5}}{24} + \frac{3}{25} \neq 1$

Or (through percentage method)

$$(5 \times 4\%) + (2 \times 5\%) + \left(13\frac{3}{5} \times 4.16\%\right) + (3 \times 4\%) \neq 100\%$$

(b) $22.5 \Rightarrow \frac{5}{25} + \frac{2}{20} + \frac{8\frac{3}{5}}{24} + \frac{3}{22.5} = 1$

Or (through percentage method)

$$(5 \times 4\%) + (2 \times 5\%) + \left(13\frac{3}{5} \times 4.16\%\right) + (3 \times 4.44\%) = 100\%$$

Hence, option (b) is correct.

Case IV Starting together some persons leave before certain time of actual completion of the work

Problem 32. A and B alone can complete a work in 9 days and 18 days, respectively. They worked together, however 3 days before the completion of the work A left. In how many days was the work completed?
[SSC (CGL-PT)-08, SSC (CPO)-08]

Sol. $\begin{pmatrix} A \Rightarrow 11.11\% \text{ of work per day} \\ B \Rightarrow 5.55\% \text{ of work per day} \end{pmatrix}$ hence, 16.66% by $(A + B)$ per day.

A has left the work 3 days before actual completion of work. It implies that B has finished the work by working last 3 days alone.

Now, work done by B in last 3 days = 3× 5.55% = 16.66%.

Thus, it is clear that the remaining work i.e. [(100 − 16.66)% = 83.33%] has been done by both together.

Then, time taken by both together to completion 83.33% of work = (83.33% ÷ 16.66%) days = 5 days.

Hence, total time required for completion of work = 5 + 3 = 8 days.

Observation

From the above, total time required = (83.33 ÷ 16.66) + 3

$$= \frac{83.33}{16.66} + 3 = \frac{133.33}{16.66} = \frac{100 + 33.33}{16.66}$$

$$= \frac{100 + 3 \times 11.11}{16.66}$$

Hence, total time required

$$= \frac{100\% + A\text{'s 3 day's work in percentage}}{\text{work done by}(A + B) \text{ per day in percentage}}$$

The above outcome may be interpreted in simple words that A's 3 day's work is a liability for both to do, in addition to 100% of the work.

Generic Version :

Case I $\begin{pmatrix} A \Rightarrow x \text{ days} \\ B \Rightarrow y \text{ days} \end{pmatrix}$ and

(They started together, but one leaves before t days of the actual completion of work.)

Then, time required in completion of whole work

$$= \left(\frac{100\% + \text{work done (in \%) in } t \text{ days by the person who left}}{\text{work done (in \%) by them together per day}} \right)$$

Case-II $\begin{pmatrix} A \Rightarrow x \text{ days} \\ B \Rightarrow y \text{ days} \\ C \Rightarrow z \text{ days} \end{pmatrix}$

and (They started together and one leaves before t days of actual completion of work.)

On the above line, time required in completion of whole work

$$= \left(\frac{100\% + \text{work done (in \%) in } t \text{ days by the person who left}}{\text{work done (in \%) by them together per day}} \right)$$

Case-III

$\begin{pmatrix} A \Rightarrow x \text{ days} \\ B \Rightarrow y \text{ days} \\ C \Rightarrow z \text{ days} \end{pmatrix}$

and (They started together and one leaves before t_1 days and another before t_2 days of actual completion of work.)

On the above line, time required in completion of whole work

$$= \left(\frac{100\% + \text{work done (in \%) in } t_1 \text{ days by the person concerned who left} + \text{work done (in \%) in } t_2 \text{ days by the person concerned who left}}{\text{work done (in \%) by them together per day}} \right)$$

Note *The above pattern works also in case of more than 3 persons.*

Alternatively

Let the time required for completion of whole work be x days. It implies that A works for $(x - 3)$ days and B for x days.

Accordingly, A's $(x - 3)$ day's work $= \frac{x - 3}{9}$

and B's x day's work $= \frac{x}{18}$

Obviously, $\frac{x - 3}{9} + \frac{x}{18} = 1$

$$\Rightarrow \frac{18(x - 3) + 9x}{9 \times 18} = 1$$

$$\Rightarrow \frac{x(18 + 9) - 18 \times 3}{9 \times 18} = 1$$

$$\Rightarrow x(18 + 9) = 9 \times 18 + 18 \times 3$$

$$\Rightarrow x = \frac{9 \times 18}{9 + 18} + \frac{18 \times 3}{9 + 18}$$

$$\Rightarrow x = \frac{9 \times 18}{9 + 18}\left(1 + \frac{3}{9}\right)$$

Generic Version :

Case-I $\begin{pmatrix} A \Rightarrow x \text{ days} \\ B \Rightarrow y \text{ days} \end{pmatrix}$ and

(they started together, but one leaves before t days of the actual completion of work.)

Then, time required in completion of whole work

$$= \frac{xy}{x + y}\left(1 + \frac{t}{\text{time taken by the person to complete whole work who leaves}}\right)$$

i.e. Time taken by both together to complete whole work

$$\times \left(1 + \frac{t}{\textbf{Time taken by the person to complete whole work who leaves}}\right)$$

Case-II $\begin{pmatrix} A \Rightarrow x \text{ days} \\ B \Rightarrow y \text{ days} \\ C \Rightarrow z \text{ days} \end{pmatrix}$ and

{they started together, but one leaves before t days of the actual completion of work.}

On the above line, time required in completion of the whole

$$\text{work} = \frac{xyz}{xy+yz+zx} \times \left(1 + \frac{t}{\text{time taken by the person to complete whole work who leaves}}\right)$$

i.e. Time taken by them together to complete the whole work

$$\times \left(1 + \frac{t}{\text{time taken by the person to complete whole work who leaves}}\right)$$

Case-III $\begin{pmatrix} A \Rightarrow x \text{ days} \\ B \Rightarrow y \text{ days} \\ C \Rightarrow z \text{ days} \end{pmatrix}$

and (They started together and one leaves before t_1 days and another before t_2 days of actual completion of work.)

On the above line, time required in completion of whole work

$$= \frac{xyz}{xy+yz+zx}\left(1 + \frac{t_1}{\text{time taken by the person to complete whole work who leaves before } t_1 \text{ days}} + \frac{t_2}{\text{time taken by the person to complete whole work who leaves before } t_2 \text{ days}}\right)$$

i.e. = Time taken by them together to complete the whole work

$$\times \left(1 + \frac{t_1}{\text{Time taken by the person to complete whole work who leaves before } t_1 \text{ days}} + \frac{t_2}{\text{Time taken by the person to complete whole work who leaves before } t_2 \text{ days}}\right)$$

Note *The above pattern works also in case of more than 3 persons.*

Here, time required in completion of whole work.

$$= \frac{9 \times 18}{9+18}\left(1 + \frac{3}{9}\right) = 8 \text{ days.}$$

Problem 33. ***A*, *B* and *C* can complete a work in 10 days, 12 days and 15 days, respectively. They started the work together. But *A* left the work before 5 days of its completion. *B* also left the work 2 days after *A* left. In how many days was the work completed?**

Sol. Clearly, B leaves the work before (5 – 2) days of completion of work. Further,

$\begin{pmatrix} A \Rightarrow 10\% \text{ of work per day} \\ B \Rightarrow 8.33\% \text{ of work per day} \\ C \Rightarrow 6.66\% \text{ of work per day} \end{pmatrix}$ hence

$\begin{pmatrix} \text{Work done by } A \text{ in 5 days} = 50\% \\ \text{Work done by } B \text{ in 3 days} = 25\% \end{pmatrix}$

Now, time required in completion of whole work

$$= \left(\frac{100\% + \text{work done (in \%) in } t_1 \text{ days by the person concerned who left} + \text{work done (in \%) in } t_2 \text{ days by the person concerned who left}}{\text{work done (in \%) by them together per day}}\right)$$

$$= \left(\frac{100\% + 50\% + 25\%}{(10 + 8.33 + 6.66)\%}\right) = 7 \text{ days}$$

Alternatively

$$\text{Required time} = \frac{xyz}{xy+yz+zx}\left(1 + \frac{t_1}{\text{time taken by the person who leaves before } t_1 \text{ days}} + \frac{t_2}{\text{time taken by the person who leaves before } t_2 \text{ days}}\right)$$

$$= \frac{10 \times 12 \times 15}{10 \cdot 12 + 12 \cdot 15 + 15 \cdot 10}\left(1 + \frac{5}{10} + \frac{3}{12}\right)$$

$$= \frac{10 \times 12 \times 15}{10 \cdot 12 + 12 \cdot 15 + 15 \cdot 10} \times \frac{7}{4} = 7 \text{ days}$$

Problem 34. ***A* and *B* alone can finish a piece of work in 20 days and 25 days, respectively. They together started the work but *B* left before some days of its completion. The whole work finished in $13\frac{1}{3}$ days. Before how many days of completion of work did *B* leave the work?**

Sol. $\begin{pmatrix} A \Rightarrow 5\% \text{ of work per day} \\ B \Rightarrow 4\% \text{ of work per day} \end{pmatrix}$ hence, 9% by $(A + B)$ per day.

Then, time required in completion of whole work

$$= \left(\frac{100\% + \text{work done (in \%) in } t \text{ days by the person who left}}{\text{work done (in \%) by them together per day}}\right)$$

$$\Rightarrow \quad \frac{40}{3} = \frac{100\% + t \cdot 4\%}{9\%}$$

$$\Rightarrow \quad 12t = 360 - 300$$

$$\Rightarrow \quad t = 5 \text{ days}$$

Alternatively I

Work done by $(A + B)$ in $13\frac{1}{3}$ days $= \frac{40}{3} \times 9\%$

$= 120\%$

Hence, extra liability = 120% – 100% = 20%, which is due to leaving the work by B.

Then, time taken by B to create liability of 20% of work at the rate of 4% per day = 5 days.

Hence, B leaves before 5 days of completion of work.

Alternatively II

$\begin{pmatrix} A \Rightarrow x \text{ days} \Rightarrow 20 \text{ days} \\ B \Rightarrow y \text{ days} \Rightarrow 25 \text{ days} \end{pmatrix}$

Time required in completion of whole work

$$= \frac{xy}{x+y}\left(1+\frac{t}{\text{Time taken by the person to complete whole work who leaves}}\right)$$

$$\Rightarrow \frac{40}{3} = \frac{20\times 25}{20+25}\left(1+\frac{t}{25}\right)$$

$$\Rightarrow 120 = 100 + 4t$$

$$\Rightarrow t = 5 \text{ days}$$

Problem 35. A man and a boy can complete a work together in 24 days. If for the last six days man alone does the work then it is completed in 26 days. How long the boy will take to complete the work alone?

Sol. Clearly, (Man + Boy)'s 1 day's work = 4.16%

Then, (Man + Boy)'s 26 day's work = 108.33%

Hence, extra liability = 108.33% – 100% = 8.33%, which is due to leaving the work by the boy before 6 days of its completion.

Then, work done by the boy per day = (8.33 ÷ 6)% = 1.38%

∴ Time taken by B to complete the work alone

= (100% ÷ 1.38%) days = 72 days.

Alternatively

Clearly, the boy leaves the work before 6 days of its actual completion. Let the boy takes x days to complete the work alone.

Then, time required in completion of whole work = time taken by both together to complete whole work

$$\times\left(1+\frac{t}{\text{time taken by the person to complete whole work who leaves}}\right)$$

$$\Rightarrow 26 = 24\times\left(1+\frac{6}{x}\right)$$

$$\Rightarrow 26-24 = \frac{24\times 6}{x}$$

$$\Rightarrow x = 72 \text{ days}$$

Case V Starting together some persons leave before certain time of Scheduled completion of the work

Problem 36. *A* and *B* alone can finish a work in 10 days and 15 days, respectively. They together started the work, then in how many days will the work be finished if *B* leaves the work before 3 days of scheduled completion of the work?

Sol. $\left(\begin{array}{l} A \Rightarrow 10\% \text{ of work per day} \\ B \Rightarrow 6.66\% \text{ of work per day}\end{array}\right)$ hence, 16.66% by $(A+B)$ per day.

Given that, B leaves the work before 3 days of scheduled completion of the work. It implies that the quantum of work to be done by $(A+B)$ in last 3 days, is done by A alone.

Clearly, work to be done by $(A+B)$ in last 3 days

= 3 × 16.66 = 50% and time taken by A to do 50% of work = (50% ÷ 10%) = 5 days

Then, total time required in completion of whole work

= [(100 ÷ 16.66) – 3] + 5 = 8 days.

Or

Total time required in completion of whole work

= [(100 – 50) ÷ 16.66] + 5 = 8 days

Alternatively I

Time to be taken by both together to finish the whole work assuming no one leaves the work $= \frac{10\times 15}{10+15} = 6$ days.

But $(A+B)$ worked for only (6 – 3) days and the remaining work is done by A.

Now, work done by $(A+B)$ in 3 days $= \frac{3}{10}+\frac{3}{15} = \frac{1}{2}$

Then, time taken by A to finish the remaining work i.e.

$\left(1-\frac{1}{2}\right) = \frac{1}{2}\times 10 = 5$ days

Thus, total time required in completion of whole work

= (6 – 3) + 5 = 8 days

Alternatively II

Scheduled time by which both together have to finish the whole work $= \frac{10\times 15}{10+15} = 6$ days.

Let the whole work is finished in x days, if B leaves before 3 days of scheduled completion of the work.

It is clear that B works for only (6 – 3) days and A works for x days.

Then, $\frac{x}{10}+\frac{3}{15} = 1$

$$\Rightarrow \frac{x}{10} = \frac{4}{5} \Rightarrow x = 8 \text{ days}$$

Generic version

Case I $\left(\begin{array}{l} A \Rightarrow x \text{ days} \\ B \Rightarrow y \text{ days}\end{array}\right)$ and

(they started together, but one leaves before t days of the scheduled completion of work.)

Then, total time required in completion of whole work

$$= \frac{xy}{x+y} + \frac{t\times\text{time taken by the person who works till end of the work}}{\text{time taken by the person to complete whole work who leaves}}$$

Or

(Time taken by both together to complete whole work $-t$) + (Time taken by the person, who works till end of the work, to finish the work what both together can do in t days)

Case-II $\left(\begin{array}{l} A \Rightarrow x \text{ days} \\ B \Rightarrow y \text{ days} \\ C \Rightarrow z \text{ days}\end{array}\right)$ and

(they started together, but one leaves before t days of the scheduled completion of work.) Then, total time required in completion of whole work =

(time taken by them together to complete whole work $-t$) + (time taken by the persons, who work till end of the work, to finish the work what all three together can do in t days)

Note *The above pattern works also in case of more than 3 persons.*

Problem 37. *A*, *B* and *C* **can finish a piece of work in 10 days, 12 days and 15 days, respectively. They together started the work but *B* leaves the work before 1 day of scheduled completion of work. In how many days will the work be finished?**

Sol. Clearly, $\begin{pmatrix} A \Rightarrow 10\% \text{ of work per day} \\ B \Rightarrow 8.33\% \text{ of work per day} \\ C \Rightarrow 6.66\% \text{ of work per day} \end{pmatrix}$

Then, total time required in completion of whole work = (time taken by them together to complete whole work $- t$) + {time taken by the persons (together), who work till end of the work, to finish the work what all three together can do in t days}

$$= \left(\frac{100\%}{(10+8.33+6.66)\%} - 1\right) + \frac{1\,(10+8.33+6.66)\%}{(10+6.66)\%}$$

$$= (4-1) + 1.5 = 4.5 \text{ days}$$

Alternatively I

Total time required in completion of whole work

= (Time taken by them together to complete whole work $- t$) + (Time taken by the persons (together), who work till end of the work, to finish the work what all three together can do in t days)

$$= \left(\frac{10\times 12\times 15}{10\cdot 12+12\cdot 15+15\cdot 10} - 1\right) + 1\left(\frac{1}{10}+\frac{1}{12}+\frac{1}{15}\right)\times\frac{10\times 15}{(10+15)}$$

$$= (4-1) + \left(\frac{1}{4}\times 6\right) = 4.5 \text{ days}$$

Alternatively II

Scheduled time by which they together have to finish the whole work $= \dfrac{10\times 12\times 15}{10\cdot 12+12\cdot 15+15\cdot 10} = 4$ days

Let the whole work is finished in x days if B leaves before 1 days of scheduled completion of the work.

It is clear that B work for only $(4-1)$ days and A and C both together work for x days.

Then, $$\frac{x}{10}+\frac{3}{12}+\frac{x}{15} = 1$$

$$\Rightarrow \quad \frac{x}{6} = \frac{3}{4}$$

$$\Rightarrow \quad x = 4.5 \text{ days}$$

Problem 38. *A*, *B* and *C* **can finish a piece of work in 10 days, 12 days and 15 days, respectively. They together started the work but *B* and *C* leave the work before 1 day and 2 days respectively of scheduled completion of work. In how many days will the work be finished?**

Sol. Clearly, $\begin{pmatrix} A \Rightarrow 10\% \text{ of work per day} \\ B \Rightarrow 8.33\% \text{ of work per day} \\ C \Rightarrow 6.66\% \text{ of work per day} \end{pmatrix}$

Scheduled time by which they together have to finish the whole work $= \dfrac{100\%}{(10+8.33+6.66)\%} = 4$ days.

It is clear that B works for only $(4-1)$ days and C for $(4-2)$ days. Work done by B and $C = 3\times 8.33 + 2\times 6.66 = 38.33\%$.

Obviously, the remaining work i.e. 61.66% work has been done by A from starting to ending of the work.

Further, time taken by A to finish 61.66% work

$= (61.66\% \div 10\%) = 6.16$ days.

Hence, total time required in completion of whole work = 6.16 days.

Alternatively

Scheduled time by which they together have to finish the whole work $= \dfrac{10\times 12\times 15}{10\cdot 12+12\cdot 15+15\cdot 12} = 4$ days.

Let the whole work is finished in x days if B and C leave the work 1 day and 2 days respectively of scheduled completion of work.

It is clear that B works for only $(4-1)$ days and C for $(4-2)$ days and A for x days.

Then, $$\frac{x}{10}+\frac{3}{12}+\frac{2}{15} = 1$$

$$\Rightarrow \quad \frac{x}{10} = \frac{37}{60}$$

$$\Rightarrow \quad x = 6\frac{1}{6} \text{ days}$$

Case VI **Starting together one leaves after sometime and another before actual completion of the work**

Problem 39. *A*, *B* and *C* **individually can finish a piece of work in 10 days, 20 days and 25 days, respectively. They together started the work but *C* left the work after 2 days of start and *B* left before 1 day of actual completion of work. In how many days will the work be finished?** ***[SSC (CGL-T-I)-17, IIFT-17]***

Sol. $\left\{\begin{matrix} A \Rightarrow 10\% \text{ of work per day} \\ B \Rightarrow 5\% \text{ of work per day} \\ C \Rightarrow 4\% \text{ of work per day} \end{matrix}\right\}$ hence, 8% by C in 2 days.

Given that, C leaves the work after 2 days of start. It implies that 8% of work is not required to be done by $(A+B)$. Further, B leaves the work before 1 day of actual completion of work. It may be construed that B's 1 day's work, i.e. 5% is an extra liability for $(A+B)$ to do, in addition to $(100-8)\%$ of the work.

Hence, total time required to complete the whole work

$$= \frac{(100-8+5)\%}{(10+5)\%} = 6.47 \text{ days.}$$

Alternatively I

Given that C leaves the work after 2 days of start. It implies that $\frac{2}{25}$ of work is not required to be done by $(A+B)$. Further, B leaves the work before 1 day of actual completion of work. It may be construed that B's 1 day's work, i.e. $\frac{1}{20}$ is an extra liability for $(A+B)$ to do, in addition to $\left(1-\frac{2}{25}\right)$ of the work.

Hence, total time required to complete the whole work

$$= \frac{10\times 20}{10+20}\left(1-\frac{2}{25}+\frac{1}{20}\right)$$

$$= \frac{20}{3}\times\frac{97}{100} = 6.47 \text{ days.}$$

(**Note** Peruse cases III and case IV for better understanding of the above pattern.)

Generic Version

$\begin{pmatrix} A \Rightarrow x \text{ days} \\ B \Rightarrow y \text{ days} \\ C \Rightarrow z \text{ days} \end{pmatrix}$ and (They started together and one leaves after t_1 days of start and another before t_2 days of actual completion of work.)

Then, total time required in completion of whole work

$= \left(\dfrac{\text{100\% – Work done (in \%) in } t_1 \text{ days by the person who leaves after the work starts + Work done (in \%) in } t_2 \text{ days by the person who leaves before completion of work}}{\text{Work done (in \%) by persons together (excluding one who leaves after the work starts) per day}} \right)$

Or

Time taken by the persons (excluding who leaves after the work starts) together to finish whole work

$\times \left(1 - \dfrac{t_1}{\text{Time taken by the person (who leaves after work starts) to complete the whole work}} + \dfrac{t_2}{\text{Time taken by the person (who leaves before completion of work) to complete whole work}} \right)$

Alternatively II

Let the whole work is finished in x days. It implies that A works for x days, B for $(x - 1)$ days and C for 2 days.

Obviously, $\dfrac{x}{10} + \dfrac{x-1}{20} + \dfrac{2}{25} = 1$

$\Rightarrow \quad 10x + 5x - 5 + 8 = 100$

$\Rightarrow \quad 15x = 97 \Rightarrow x = 6.47$ days

Problem 40. *A*, *B*, *C* and *D* individually can finish a piece of work in 10 days, 20 days, 25 days and 30 days, respectively. They together started the work but *C* left the work after 2 days of start and *B* left before 1 day of actual completion of work. In how many days will the work be finished?

Sol. $\begin{pmatrix} A \Rightarrow 10\% \text{ of work per day} \\ B \Rightarrow 5\% \text{ of work per day} \\ C \Rightarrow 4\% \text{ of work per day} \\ D \Rightarrow 3.33\% \text{ of work per day} \end{pmatrix}$ hence, 8% by C in 2 days.

Now, total time required to complete the whole work

$= \dfrac{(100 - 8 + 5)\%}{(10 + 5 + 3.33)\%} = 5.29$ days

Alternatively I

Total time required to complete the whole work

$= \dfrac{10 \times 20 \times 30}{10 \cdot 20 + 20 \cdot 30 + 30 \cdot 10}\left(1 - \dfrac{2}{25} + \dfrac{1}{20}\right)$

$= \dfrac{60}{11} \times \dfrac{97}{100} = 5.29$ days

Alternatively II

Let the whole work is finished in x days. It implies that A works for x days, B for $(x - 1)$ days, C for 2 days and D for x days.

Obviously, $\dfrac{x}{10} + \dfrac{x-1}{20} + \dfrac{2}{25} + \dfrac{x}{30} = 1$

$\Rightarrow \quad 30x + 15x - 15 + 24 + 10x = 300$

$\Rightarrow \quad 55x = 291 \Rightarrow x = 5.29$ days

The concepts associated with case-III to case-VI are depicted in the following chart for quick revision with the help of Fraction-Method (here in after referred to as 'F-M') as well as Percentage-Method (here in after referred to as 'P-M').

$\begin{pmatrix} A \Rightarrow x \text{ days} \\ B \Rightarrow y \text{ days} \end{pmatrix}$ They started together to work. Then, time required for completion of whole work in the following circumstances will be

Circumstances	Approaches	Models
If *A* leaves after *t* days	F-M	$y\left(1 - \dfrac{t}{x}\right)$
	P-M (all figures in %)	$\left(\dfrac{100 - A\text{'s } t \text{ days' work}}{B\text{'s 1 day's work}}\right)$
If *B* leaves after *t* days	F-M	$x\left(1 - \dfrac{t}{y}\right)$
	P-M	$\left(\dfrac{100 - B\text{'s } t \text{ days' work}}{A\text{'s 1 day's work}}\right)$
If *A* leaves **before** *t* days **of actual** completion of work	F-M	$\dfrac{xy}{x+y}\left(1 + \dfrac{t}{x}\right)$
	P-M	$\left[\dfrac{100 + A\text{'s } t \text{ days' work}}{(A+B)\text{'s 1 day's work}}\right]$
If *B* leaves **before** *t* days **of actual** completion of work	F-M	$\dfrac{xy}{x+y}\left(1 + \dfrac{t}{y}\right)$
	P-M	$\left(\dfrac{100 + B\text{'s } t \text{ days' work}}{(A+B)\text{'s 1 day's work}}\right)$
If *A* leaves **before** *t* days **of scheduled** completion of work	F-M	$\dfrac{xy}{x+y} + \dfrac{ty}{x}$
	P-M	$\left[\left(\dfrac{100}{(A+B)\text{'s 1 day's work}} - t \text{ days}\right) + \left(\dfrac{(A+B)\text{'s } t \text{ days' work}}{B\text{'s 1 day's work}}\right)\right]$
If *B* leaves **before** *t* days **of scheduled** completion of work	F-M	$\dfrac{xy}{x+y} + \dfrac{tx}{y}$
	P-M	$\left[\left(\dfrac{100}{(A+B)\text{'s 1 day's work}} - t \text{ days}\right) + \left(\dfrac{(A+B)\text{'s } t \text{ days' work}}{A\text{'s 1 day's work}}\right)\right]$

$\begin{pmatrix} A \Rightarrow x \text{ days} \\ B \Rightarrow y \text{ days} \\ C \Rightarrow z \text{ days} \end{pmatrix}$ They started together to work. Then, time required for completion of whole work in the following circumstances will be

Condition	Method	Expression
If *A* leaves after *t* days	F-M	$\frac{yz}{y+z}\left(1-\frac{t}{x}\right)$
	P-M	$\left(\frac{100 - A\text{'s } t \text{ days' work}}{(B+C)\text{'s 1 day's work}}\right)$
If *B* leaves after *t* days	F-M	$\frac{xz}{x+z}\left(1-\frac{t}{y}\right)$
	P-M	$\left\{\frac{100 - B\text{'s } t \text{ days' work}}{(A+C)\text{'s 1 day's work}}\right\}$
If *C* leaves after *t* days	F-M	$\frac{xy}{x+y}\left(1-\frac{t}{z}\right)$
	P-M	$\left(\frac{100 - C\text{'s } t \text{ days' work}}{(A+B)\text{'s 1 day's work}}\right)$
If *A* leaves after t_1 days and *B* **after** t_2 days	F-M	$z\left(1-\frac{t_1}{x}-\frac{t_2}{y}\right)$
	P-M	$\left(\frac{100 - A\text{'s } t_1 \text{ days' work} - B\text{'s } t_2 \text{ days' work}}{C\text{'s 1 day's work}}\right)$
If *B* leaves **after** t_1 days and *C* **after** t_2 days	F-M	$x\left(1-\frac{t_1}{y}-\frac{t_2}{z}\right)$
	P-M	$\left(\frac{100 - B\text{'s } t_1 \text{ days' work} - C\text{'s } t_2 \text{ days' work}}{A\text{'s 1 day's work}}\right)$
If *C* leaves **after** t_1 days and *A* **after** t_2 days	F-M	$y\left(1-\frac{t_1}{z}-\frac{t_2}{x}\right)$
	P-M	$\left(\frac{100 - C\text{'s } t_1 \text{ days' work} - A\text{'s } t_2 \text{ days' work}}{B\text{'s 1 day's work}}\right)$
If *A* leaves **before** *t* days of **actual** completion of work	F-M	$\frac{xyz}{xy+yz+zx}\left(1+\frac{t}{x}\right)$
	P-M	$\left(\frac{100 + A\text{'s } t \text{ days' work}}{(A+B+C)\text{'s 1 day's work}}\right)$
If *B* leaves **before** *t* days of **actual** completion of work	F-M	$\frac{xyz}{xy+yz+zx}\left(1+\frac{t}{y}\right)$
	P-M	$\left(\frac{100 + B\text{'s } t \text{ days' work}}{(A+B+C)\text{'s 1 day's work}}\right)$
If *C* leaves **before** *t* days of **actual** completion of work	F-M	$\frac{xyz}{xy+yz+zx}\left(1+\frac{t}{z}\right)$
	P-M	$\left(\frac{100 + C\text{'s } t \text{ days' work}}{(A+B+C)\text{'s 1 day's work}}\right)$
If *A* leaves **before** t_1 days and *B* **before** t_2 days of **actual** completion of work	F-M	$\frac{xyz}{xy+yz+zx}\left(1+\frac{t_1}{x}+\frac{t_2}{y}\right)$
	P-M	$\left(\frac{100 + A\text{'s } t_1 \text{ days' work} + B\text{'s } t_2 \text{ days' work}}{(A+B+C)\text{'s 1 day's work}}\right)$
If *B* leaves **before** t_1 days and *C* **before** t_2 days of **actual** completion of work	F-M	$\frac{xyz}{xy+yz+zx}\left(1+\frac{t_1}{y}+\frac{t_2}{z}\right)$
	P-M	$\left(\frac{100 + B\text{'s } t_1 \text{ days' work} + C\text{'s } t_2 \text{ days' work}}{(A+B+C)\text{'s 1 day's work}}\right)$
If *C* leaves **before** t_1 days and *A* **before** t_2 days of **actual** completion of work	F-M	$\frac{xyz}{xy+yz+zx}\left(1+\frac{t_1}{z}+\frac{t_2}{x}\right)$
	P-M	$\left(\frac{100 + C\text{'s } t_1 \text{ days' work} + A\text{'s } t_2 \text{ days' work}}{(A+B+C)\text{'s 1 day's work}}\right)$
If *A* leaves **before** *t* days **of scheduled** completion of work	F-M	$\left(\frac{xyz}{xy+yz+zx}-t\right)+t\left(\frac{xy+yz+zx}{x(y+z)}\right)$
	P-M	$\left(\frac{100}{(A+B+C)\text{'s 1 day's work}} - t \text{ days}\right) + \frac{t \times (A+B+C)\text{'s 1 day's work}}{(B+C)\text{'s 1 day's work}}$
If *B* leaves **before** *t* days **of scheduled** completion of work	F-M	$\left(\frac{xyz}{xy+yz+zx}-t\right)+t\left(\frac{xy+yz+zx}{y(z+x)}\right)$
	P-M	$\left(\frac{100}{(A+B+C)\text{'s 1 day's work}} - t \text{ days}\right) + \frac{t \times (A+B+C)\text{'s 1 day's work}}{(C+A)\text{'s 1 day's work}}$
If *C* leaves **before** *t* days **of scheduled** completion of work	F-M	$\left(\frac{xyz}{xy+yz+zx}-t\right)+t\left(\frac{xy+yz+zx}{z(x+y)}\right)$
	P-M	$\left(\frac{100}{(A+B+C)\text{'s 1 day's work}} - t \text{ days}\right) + \frac{t \times (A+B+C)\text{'s 1 day's work}}{(A+B)\text{'s 1 day's work}}$
If *A* leaves **after** t_1 days of start and *B* **before** t_2 days of completion of work	F-M	$\frac{yz}{y+z}\left(1-\frac{t_1}{x}+\frac{t_2}{y}\right)$
	P-M	$\left(\frac{100 - A\text{'s } t_1 \text{ days' work} + B\text{'s } t_2 \text{ days' work}}{(B+C)\text{'s 1 day's work}}\right)$
If *B* leaves **after** t_1 days of start and *C* **before** t_2 days of completion of work	F-M	$\frac{zx}{z+x}\left(1-\frac{t_1}{y}+\frac{t_2}{z}\right)$
	P-M	$\left(\frac{100 - B\text{'s } t_1 \text{ days' work} + C\text{'s } t_2 \text{ days' work}}{(C+A)\text{'s 1 day's work}}\right)$
If *C* leaves **after** t_1 days of start and *A* **before** t_2 days of completion of work	F-M	$\frac{xy}{x+y}\left(1-\frac{t_1}{z}+\frac{t_2}{x}\right)$
	P-M	$\left(\frac{100 - C\text{'s } t_1 \text{ days' work} + A\text{'s } t_2 \text{ days' work}}{(A+B)\text{'s 1 day's work}}\right)$

Please observe the pattern in the above model. The same pattern is applicable in case of more than three persons.

Note

(i) We observe that '–' stands for after and '+' for before. Further, we have not included the figure(s) associated with the person(s) leaving the work after certain period of start of work before the bracket in 'Fraction-Method'. However, in case of leaving the work before completion of work, we have included the respective figure(s).

(ii) The observation surfaced above is also recognised in 'Percentage-Method'.

(iii) Aspirants are advised to peruse the above chart meticulously before solving the questions.

Case VII When certain person(s) join(s) after start of the work

Problem 41. A and B alone can complete a work in 10 days and 20 days, respectively. A started to work and after 2 days B joined him. In how many days was the work completed?

Sol. $\left(\begin{array}{l} A \Rightarrow 10\% \text{ of work per day} \\ B \Rightarrow 5\% \text{ of work per day} \end{array}\right)$ hence, 15% by $(A + B)$ per day.

Now, work done by A in 2 days $= 2 \times 10\% = 20\%$.

Obviously, the remaining work i.e. 80% will be done by both together.

Then, time taken by both working together to complete 80% of the work $= (80\% \div 15\%)$ i.e. 5.33 days.

Hence, total time taken in completion of the work

$$= 2 + 5.33 = 7.33 \text{ days.}$$

Observation :

From the above, total time required $= 2 + (80 \div 15)$

$$= 2 + \frac{80}{15} = \frac{110}{15} = \frac{100 + (2 \times 5)}{15}$$

Hence, total time required

$$= \frac{100\% + B\text{'s 2 day's work in percentage}}{\text{work done by } (A+B)\text{'s per day in percentage}}$$

Generic Version :

Case-I

$\left(\begin{array}{l} A \Rightarrow x \text{ days} \\ B \Rightarrow y \text{ days} \end{array}\right)$ and (One starts and another joins him/her after t days.)

Then, time required in completion of whole work

$$= \left(\frac{100\% + \text{work done (in \%) in t days by the person who joined later}}{\text{work done (in \%) by them together per day}}\right)$$

Case-II $\left(\begin{array}{l} A \Rightarrow x \text{ days} \\ B \Rightarrow y \text{ days} \\ C \Rightarrow z \text{ days} \end{array}\right)$ and (Any two of them started together and one joins them after t days.)

On the above line, time required in completion of whole work

$$= \left(\frac{100\% + \text{work done (in \%) in } t \text{ days by the person who joined later}}{\text{work done (in \%) by them together per day}}\right)$$

Case-III $\left(\begin{array}{l} A \Rightarrow x \text{ days} \\ B \Rightarrow y \text{ days} \\ C \Rightarrow z \text{ days} \end{array}\right)$ and

(Anyone alone starts and second-one joins him/her after t_1 days and third-one after t_2 days from start of the work.)

On the above line, time required in completion of whole work

$$= \left(\frac{100\% + \text{work done (in \%) in } t_1 \text{ days by the person who joined after } t_1 \text{ days} + \text{ work done (in \%) in } t_2 \text{ days by the person who joined after } t_2 \text{ days}}{\text{work done (in \%) by them together per day}}\right)$$

Note *The above pattern works also in case of more than 3 persons.*

Alternatively

Let the time required for completion of whole work be x days.

It implies that A works for x days and B for $(x - 2)$ days.

Accordingly, A's x day's work $= \frac{x}{10}$

and B's $(x - 2)$ day's work $= \frac{x-2}{20}$

Obviously, $\frac{x}{10} + \frac{x-2}{20} = 1 \Rightarrow \frac{20x + 10(x-2)}{10 \times 20} = 1$

$$\Rightarrow \frac{x(20+10) - 10 \times 2}{10 \times 20} = 1$$

$$\Rightarrow x(20 + 10) = (10 \times 20) + (10 \times 2)$$

$$\Rightarrow x = \frac{10 \times 20}{20 + 10} + \frac{10 \times 2}{20 + 10} \Rightarrow x = \frac{10 \times 20}{10 + 20}\left(1 + \frac{2}{20}\right)$$

Generic Version

Case I $\left(\begin{array}{l} A \Rightarrow x \text{ days} \\ B \Rightarrow y \text{ days} \end{array}\right)$ and (One starts and another joins him/her after t days.)

Then, time required in completion of whole work

$$= \frac{xy}{x+y}\left(1 + \frac{t}{\text{time taken by the person to complete whole work who joins later}}\right)$$

i.e.

Time taken by both together to complete whole work

$$\times \left(1 + \frac{t}{\text{Time taken by the person to complete whole work who joins later}}\right)$$

Case II $\left(\begin{array}{l} A \Rightarrow x \text{ days} \\ B \Rightarrow y \text{ days} \\ C \Rightarrow z \text{ days} \end{array}\right)$ and (Any two of them started together and one joins them after t days.)

On the above line, time required in completion of the whole work $= \frac{xyz}{xy + yz + zx}$

$$\times \left(1 + \frac{t}{\text{Time taken by the person to complete whole work who joins later}}\right)$$

i.e. Time taken by them together to complete the whole work

$$\times \left(1 + \frac{t}{\text{Time taken by the person to complete whole work who joins later}}\right)$$

Case III $\left(\begin{array}{l} A \Rightarrow x \text{ days} \\ B \Rightarrow y \text{ days} \\ C \Rightarrow z \text{ days} \end{array}\right)$ and

(Anyone alone starts and second one joins him/her after t_1 days and third one after t_2 days from start of the work.)

On the above line, time required in completion of whole work

$$= \frac{xyz}{xy + yz + zx}\left(1 + \frac{t_1}{\text{time taken by the person who joins after } t_1 \text{ days}} + \frac{t_2}{\text{time taken by the person who joins after } t_2 \text{ days}}\right)$$

i.e. Time taken by them together to complete the whole work

$$\times\left(1 + \frac{t_1}{\text{time taken by the person who joins after } t_1 \text{ days}} + \frac{t_2}{\text{time taken by the person who joins after } t_2 \text{ days}}\right)$$

Note *The above pattern works also in case of more than 3 persons.*

Here, time required in completion of whole work

$$= \frac{10 \times 20}{10 + 20}\left(1 + \frac{2}{20}\right) = 7.33 \text{ days.}$$

Problem 42. **A, B and C alone can complete a work in 10 days, 20 days and 25 days, respectively. A and B started to work together and after 3 days C joined them. In how many days was the work completed?**

Sol. $\begin{pmatrix} A \Rightarrow 10\% \text{ of work per day} \\ B \Rightarrow 5\% \text{ of work per day} \\ C \Rightarrow 4\% \text{ of work per day} \end{pmatrix}$ hence, 19% by $(A + B + C)$ per day.

Now, work done by $(A + B)$ in 3 days

$= 3 \times (10 + 5)\%$

$= 45\%$

Obviously, the remaining work, i.e. 55% will be done by all three together.

Then, time taken by all three working together to complete 55% of the work = (55% ÷ 19%), i.e. 2.9 days.

Hence, total time taken in completion of the work

$= 3 + 2.9 = 5.9$ days.

Alternatively I

$$\text{Required time} = \left(\frac{100\% + \text{Work done (in \%) in } t \text{ days by the person who joins later}}{\text{Work done (in \%) by them together per day}}\right)$$

$$= \left(\frac{100\% + (3 \times 4\%)}{19\%}\right) = 5.9 \text{ days}$$

Alternatively II

Required time

= Time taken by them together to complete the whole work

$$\times\left(1 + \frac{t}{\text{Time taken by the person to complete whole work who joins later}}\right)$$

$$= \frac{10 \times 20 \times 25}{10 \cdot 20 + 20 \cdot 25 + 25 \cdot 10}\left(1 + \frac{3}{25}\right)$$

= 5.9 days

Problem 43. **A, B and C alone can complete a work in 10 days, 20 days and 25 days, respectively. A started to work and B and C joined him after 2 days and 3 days respectively from start of the work. In how many days was the work completed?**

Sol. $\begin{pmatrix} A \Rightarrow 10\% \text{ of work per day} \\ B \Rightarrow 5\% \text{ of work per day} \\ C \Rightarrow 4\% \text{ of work per day} \end{pmatrix}$ hence, 19% by $(A + B + C)$ per day.

Work done in first 3 days = A's 3 day's work + B's 1 day's work

$= (3 \times 10)\% + (1 \times 5)\% = 35\%$

Obviously, the remaining work i.e. 65% will be done by all three together.

Then, time taken by all three working together to complete 65% of the work = (65% ÷ 19%) i.e. 3.4 days.

Hence, total time taken in completion of the work

$= 3 + 3.42 = 6.42$ days.

Alternatively I

Required time

$$= \left(\frac{100\% + \text{Work done (in \%) in } t_1 \text{ days by the person who joined after } t_1 \text{ days} + \text{Work done (in \%) in } t_2 \text{ days by the person who joined after } t_2 \text{ days}}{\text{Work done (in \%) by them together per day}}\right)$$

$$= \left(\frac{100\% + (2 \times 5\%) + (3 \times 4\%)}{19\%}\right) = 6.42 \text{ days}$$

Alternatively II

Required time

= Time taken by them together to complete the whole work

$$\times\left(1 + \frac{t_1}{\text{Time taken by the person who joins after } t_1 \text{ days}} + \frac{t_2}{\text{time taken by the person who joins after } t_2 \text{ days}}\right)$$

$$= \frac{10 \times 20 \times 25}{10 \cdot 20 + 20 \cdot 25 + 25 \cdot 10}\left(1 + \frac{2}{20} + \frac{3}{25}\right) = 6.42 \text{ days}$$

Observation

The patterns for case-IV and case-VII are similar. The reason is that the contextual meaning of following questions is same:

(i) A and B alone can complete a work in 9 days and 18 days, respectively. They worked together, however 3 days before the completion of the work A left. In how many days was the work completed? ...... case-IV

(ii) A and B alone can complete a work in 9 days and 18 days, respectively. B alone started the work and after 3 days A joined him. In how many days was the work completed? ...... case-VII

In both questions, B worked for x days and A for $(x - 3)$ days assuming total time required in completion of whole work was x days.

Concept of Negative Work

In this case, one or more work for development/growth, while other try to negate/go against the same. Take one example for better conception of the negative work.

e.g. 'Shivansh can write a 100 pages notebook in 40 days while Saransh can erase his writings completely in 50 days. In how many days will the notebook be written if they together started to work?

Obviously, $\begin{pmatrix}\text{Shivansh} \Rightarrow 2.5\% \text{ of work per day (positive)}\\ \text{Saransh} \Rightarrow 2\% \text{ of work per day (negative)}\end{pmatrix}$

Hence, 0.5% of notebook is written per day if they work together. Then, time required to write 100% of the notebook = 200 days.

Note *This conception is generally applicable in the topic 'Pipes and Cisterns' where inlet pipes work for filling the cistern whereas outlet pipes/leakages go against it i.e. empty the cistern. We will discuss there with a lot of questions.*

Concept of Efficiency

We have already seen the relation among amount of work, time, efficiency and work-force in beginning of the chapter. Now, the following relation in context of efficiency are reproduced here.

(i) Efficiency $\propto \dfrac{1}{\text{time}}$

Or

Time $\propto \dfrac{1}{\text{efficiency}}$ (when amount of work is constant)

(ii) Efficiency $\propto$ amount of work

Or

Amount of work $\propto$ efficiency (when time is constant)

(iii) Efficiency $\propto \dfrac{1}{\text{work - force}}$

Or

Work-force $\propto \dfrac{1}{\text{efficiency}}$ (when amount of work is constant)

We will discuss the concept of efficiency at Individual-level as well as Group-level.

A. Individual-Level

Under this case comparison among efficiencies of workers at individual level is made, sometimes, their efficiencies are given directly in ratio. Take one example for building of the conception in this regard.

If working efficiencies of workers A, B and C are in ratio $x:y:z$ and they together finish a piece of work in D days, then :

Ratio of time taken by them to finish whole work individually $= \dfrac{1}{x}:\dfrac{1}{y}:\dfrac{1}{z}$

The logic behind it is that efficiency is indirectly proportional to time when amount of work is constant.

Further, time taken by A to finish whole work alone

$$= D \times \frac{(x+y+z)}{x}$$

Time taken by B to finish whole work alone

$$= D \times \frac{(x+y+z)}{y}$$

and time taken by C to finish whole work alone

$$= D \times \frac{(x+y+z)}{z}$$

Derivation

From above, time taken by them are $\dfrac{k}{x}$ days, $\dfrac{k}{y}$ days and $\dfrac{k}{z}$ days, respectively.

Now, their 1 day's work $= \dfrac{x}{k}, \dfrac{y}{k}$ and $\dfrac{z}{k}$ respectively.

As given, their 1 day's work together $= \dfrac{1}{D}$

Obviously, $\dfrac{x}{k} + \dfrac{y}{k} + \dfrac{z}{k} = \dfrac{1}{D}$

$\Rightarrow \quad k = D(x+y+z)$

$\Rightarrow \quad \dfrac{k}{x} = \dfrac{D(x+y+z)}{x}, \dfrac{k}{y} = \dfrac{D(x+y+z)}{y}$

and $\quad \dfrac{k}{z} = \dfrac{D(x+y+z)}{z}$

Note
(i) The above pattern works irrespective of number of workers.
(ii) The problems based on efficiency can be solved on the above concepts.
(iii) Efficiency can be worked out dividing the amount of work by time taken.

Problem 44. Babu and Asha can do a job together in 7 days. Asha is $1\dfrac{3}{4}$ times as efficient as Babu. The same job can be done by Asha alone in?

[SSC (CGL-PT)-03, SSC (CGL-T-I)-17, MAT-18]

Sol. Given, ratio of efficiencies of Babu and Asha $= 1 : \dfrac{7}{4} = 4:7$

Then, ratio of time taken by them individually = 7 : 4

($\because$ Time $\propto \dfrac{1}{\text{Efficiency}}$, when amount of work is constant)

Let Babu and Asha take $7x$ days and $4x$ days to finish the work alone.

Then, Babu's 1 day's work $= \dfrac{1}{7x}$ and Asha's 1 day's work $= \dfrac{1}{4x}$

Further, their 1 day's work together $= \dfrac{1}{7}$

Obviously, $\quad \dfrac{1}{7x} + \dfrac{1}{4x} = \dfrac{1}{7}$

$\Rightarrow \quad \frac{11}{28x} = \frac{1}{7} \Rightarrow x = \frac{11}{4}$

Then, time taken by Asha to finish the whole work alone

$= 4x = 4 \times \frac{11}{4} = 11$ days.

Alternatively

Given, ratio of efficiencies of Babu and Asha = $x : y$ (let)

$= 1 : \frac{7}{4} = 4 : 7$

Now, time taken by Asha to finish the work alone $= D \times \frac{(x + y)}{y}$

(where, D is time taken by them together to finish the work.)

$= 7 \times \frac{(4 + 7)}{7} = 11$ days

Problem 45. A man and a woman working together can do a certain work in 18 days. Their skills in doing the work are in ratio 3 : 2. How many days will the woman take to finish the work alone?

Sol. The required time $= 18 \times \frac{(3 + 2)}{2} = 45$ days

Note *The other methods have been skipped here and will be skipped deliberately for other questions too owing to space constraints.*

Problem 46. A does half as much work as B in one-sixth of the time. If together they take 10 days to complete a work, how much time shall B alone take to do it?

Sol. Ratio of work done by A and $B = \frac{1}{2} : 1$

and ratio of time taken by them $= \frac{1}{6} : 1$

$\therefore$ Ratio of their efficiencies

$= \left(\frac{1}{2} \div \frac{1}{6}\right) : (1 \div 1) = 3 : 1$

Now, time taken by B alone to do the work

$= 10 \times \frac{(3 + 1)}{1} = 40$ days.

Problem 47. A is three times more efficient worker than B and is, therefore, able to complete a work in 60 days earlier. The number of days that A and B together will take to complete the work is? ***[SSC (CPO-SI)-05, RRB (Bhopal)-07]***

Sol. Ratio of efficiencies of A and $B = 3 : 1$

Then, ratio of time taken by them = 1 : 3

$(\because \text{Time} \propto \frac{1}{\text{Efficiency}}$, when amount of work is constant)

Further, given, $(3 - 1)$ units = 60 days

Then, 1 unit = 30 days

Hence, time taken by A to finish a work = 30 days.

Now, time taken by A to finish a work $= D \times \frac{(3 + 1)}{3}$

where, D is time taken by them together to finish the work.

$\Rightarrow \quad 30 = D \times \frac{4}{3} \Rightarrow D = 22.5$ days

Problem 48. The ratio of time taken by A and B together and C to finish a piece of work is 1 : 3. What is the time taken by C alone to finish the work if A, B and C together take 12 days to complete the work?

Sol. Ratio of efficiencies of $(A + B)$ and $C = 3 : 1$

Here, $(A + B)$ may be considered as single work-force.

Then, time taken by C alone to finish the work

$= 12 \times \frac{(3 + 1)}{1}$

= 48 days

Problem 49. To complete a work, A takes 50% more time than B. If together they take 18 days to complete the work, how much time shall B take to do it?

Sol. Ratio of time taken by A and B

= 150% : 100% = 3 : 2

Then, ratio of their efficiencies = 2 : 3

Now, time taken by B alone to do the work

$= 18 \times \frac{(2 + 3)}{3} = 30$ days

Problem 50. A is twice as good a workman as B and B is twice as good a workman as C. If A and B can together finish a piece of work in 4 days, then C can do it by himself in?

Sol. Ratio of efficiencies of A, B and C

$= 2 : 1 : \frac{1}{2} = 4 : 2 : 1$

Then, time taken by them

$= \frac{1}{4} : \frac{1}{2} : \frac{1}{1} = 1 : 2 : 4$

Given, time taken by $(A + B)$ = 4 days

Then, time taken by $A = 4 \times \frac{(4 + 2)}{4} = 6$ days

Clearly, 1 unit = 6 days

Then, 4 units = 24 days

Hence, time taken by C alone to do the work = 24 days

Problem 51. A is 50% more efficient than B. C does half the work done by A and B together. If C alone does the work in 20 days, then A, B and C together can do the work in? ***[SSC (CGL-T-I)-11]***

Sol. Ratio of efficiencies of A, B and C

$= 150\% : 100\% : \frac{(150 + 100)\%}{2}$

$(\because$ Efficiency $\propto$ Amount of work, when time is constant)

= 6 : 4 : 5

Now, time taken by C alone to do the work

$= D \times \frac{6 + 4 + 5}{5}$

(where, D is time taken by them together to finish the work.)

$\Rightarrow \quad 20 = 3D$

$\Rightarrow \quad D = 6\frac{2}{3}$ days

Problem 52. A does 20% less work than B. If A can complete a piece of work in $7\frac{1}{2}$ h, then B can do it in?

Sol. Ratio of work done by A and B = 80% : 100% = 4 : 5

Then, ratio of their efficiencies = 4 : 5

($\because$ Efficiency $\propto$ Amount of work, when time is constant)

$\therefore$ Ratio of time taken by them = 5 : 4

($\because$ Time $\propto \frac{1}{\text{Efficiency}}$, when amount of work is constant)

As given, 5 units = 7.5 h

Then, 4 units = 6 h

Hence, time taken by B alone to do the work = 6 h

Problem 53. Efficiency of Shivansh is 40% more than Alka and Alka takes 35 days to complete a piece of work alone. Shivansh started to do a work alone and then Alka joined him 7 days before the completion of work. For how many days Shivansh worked alone?

Sol. Ratio of efficiencies of Shivansh and Alka

= 140% : 100% = 7 : 5

Then, ratio of time taken by them = 5 : 7

($\because$ Time $\propto \frac{1}{\text{Efficiency}}$, when amount of work is constant)

As given, 7 units = 35 days

$\therefore$ 5 units = 25 days

Now, $\left(\begin{array}{l}\text{Shivansh} \Rightarrow 25 \text{ days} \Rightarrow 4\% \text{ of work per day} \\ \text{Alka} \Rightarrow 35 \text{ days} \Rightarrow 2.86\% \text{ of work per day}\end{array}\right)$

hence, 6.86% by them together per day.

Then, work done by them together in last 7 days

= 7 × 6.86 = 48%

It implies that the remaining work, i.e. 52% has been done by Shivansh alone.

$\therefore$ Number of days Shivansh worked alone

= 52% ÷ 4% = 13 days

Problem 54. Ram starts working on a job and works on it for 12 days and complete 40% of the work. To help him complete the work, he employs Ravi and together they work for another 12 days and the work gets completed. How much more efficient is Ram than Ravi? ***[CAT-10]***

Sol. It is clear that Ram did 80% of the work in 24 days and Ravi did 20% of the work in 12 days. It may be said that Ravi can do 40% of the work in 24 days.

Hence, Ram is 100% more efficient than Ravi as 80 is 100% more than 40. Needless to say that the efficiency is directly proportional to the work done if time is constant.

Problem 55. The time taken by A and B to complete a piece of work are in ratio 2 : 3 and the efficiencies of B and C are in ratio 3 : 4. B takes 5 days more than A if they work alone. All together started to work, but A leaves the work after 1 day. In how many days will the work be finished?

Sol. Ratio of time taken by A and B = 2 : 3

Then, ratio of efficiencies of A and B = 3 : 2

As given, efficiencies of B and C = 3 : 4

Hence, ratio of efficiencies of A, B and C

= (3 × 3) : (2 × 3) : (2 × 4) = 9 : 6 : 8

(Recall the relevant concept in chapter "Ratio, Proportion and Variation")

Then, ratio of time taken by them

$= \frac{1}{9} : \frac{1}{6} : \frac{1}{8}$

($\because$ Time $\propto \frac{1}{\text{Efficiency}}$, when amount of work is constant)

= 8 : 12 : 9

Further, (12 − 8) units = 5 days

$\therefore$ 8 units = 10 days, 12 units = 15 days and 9 units = 11.25 days

Hence, time taken by A, B and C respectively = 10, 12 and 11.25 days

i.e. $\left(\begin{array}{l}A \Rightarrow 10\% \text{ of work per day} \\ B \Rightarrow 6.66\% \text{ of work per day} \\ C \Rightarrow 8.88\% \text{ of work per day}\end{array}\right)$

Now, required time $= \frac{(100-10)\%}{(6.66+8.88)\%} = 5.8$ days

Problem 56. B is 20% more efficient than A. B started the work and did it for x days and then B is replaced by A. A completed the remaining work in $(x+8)$ days. Ratio of work done by A and B is 3 : 2. In how many days can A and B working together complete the whole work? ***[SBI (PO)-17]***

Sol. Ratio of efficiencies of B and A

= 120 : 100 = 6 : 5

Let B and A do 6 units and 5 units work per day respectively.

Then, work done by B in x days = $6x$ units

and work done by A in $(x + 8)$ days

= $5(x + 8)$ units

As given, $6x : 5(x + 8) = 2 : 3$

$\Rightarrow \quad 18x = 10x + 80 \Rightarrow x = 10$ days

Total work done by them = $6x + 5(x + 8)$ units = 150 units

Then, required number of days

$= \frac{\text{total work to be done}}{\text{work done by them per day}}$

$= \frac{150 \text{ units}}{(6+5) \text{ units}}$

= 13.64 days

Alternatively I

Ratio of efficiencies of B and A

= 120 : 100 = 6 : 5

Then, ratio of time taken by B and A to finish the equal amount of work = 5 : 6

$\therefore$ Ratio of time taken by B and A to finish the work in ratio 2 : 3 = 5 · 2 : 6 · 3 = 5 : 9

As per question, $x : (x + 8) = 5 : 9$

$\Rightarrow \quad 9x = 5x + 40 \Rightarrow x = 10$

Clearly, $\begin{pmatrix} B \Rightarrow 4\% \text{ of work per day as he finishes 40\% of work in 10 days} \\ A \Rightarrow 3.33\% \text{ of work per day as he finishes 60\% of work in } (10 + 8) \text{ days} \end{pmatrix}$

Hence, 7.33% of work is done by $(A + B)$ per day.

Then, required number of day

$$= 100\% \div 7.33\% = 13.64 \text{ days}$$

Alternatively II

Ratio of efficiencies of B and A

$$= 120 : 100 = 6 : 5$$

Then, ratio of time taken by B and A to finish the equal amount of work = 5 : 6

If B had done the work equal to what A did, then time taken by

$$B = 1.5x \ (\because 3 \text{ is } 1.5 \text{ times } 2)$$

Obviously, $1.5x : (x + 8) = 5 : 6$

$\Rightarrow \quad 9x = 5x + 40 \Rightarrow x = 10$

B can finish 40% of work in 10 days, then he will take 25 days to finish 100% of the work.

Now, time taken by B to finish the whole work $= D \times \frac{(6 + 5)}{6}$

(where, D is time taken by them together to finish the work.)

$\Rightarrow \quad 25 = \frac{11D}{6} \Rightarrow D = 13.64$ days

Problem 57. Ganu and Manu together finish a piece of work in 12 days. If Ganu had worked twice efficiently as he did and Manu thrice efficiently as he actually did, they together would have finished the work in 5 days. In how many days can Manu finish the work alone? *[SSC (CGL-T-I)-17, MAT-19, CAT-19]*

Sol. In first case, work done by $(G + M)$ per day = 8.33%

$\Rightarrow$ Work done by G per day = 8.33% – work done by M per day

In second case, work done by $(G + M)$ per day = 20%

$\Rightarrow$ 2 (work done by G per day in first case) + 3 (work done by M per day in first case) = 20%

$\Rightarrow$ 2 (8.33% – work done by M per day in first case) + 3 (work done by M per day in first case) = 20%

$\Rightarrow$ (work done by M per day in first case) = 3.33%

Hence, M can do the work alone in (100% ÷ 3.33%) i.e. 30 days.

Problem 58. Ratio of efficiency of A and C is 2 : 1. A and B together can finish a piece of work in 20 days while B and C together can do it in 25 days. How long will B take to finish the work alone?

Sol. Work done by $(A + B)$ per day = 5%

and work done by $(B + C)$ per day = 4%

Hence, difference between amount of work done by A and C per day = 1% and this difference is on account of difference in their efficiencies. Also given, the efficiency of A is twice that of C.

Thus, work done by C daily = 1%

Then, work done by B per day

$$= (4 - 1)\% = 3\%$$

$\therefore$ Time taken by B alone to finish the work

$$= (100\% \div 3\%) = 33\frac{1}{3} \text{ days}$$

Problem 59. A person can complete a job in 120 days. He works alone on Day 1. On Day 2, he is joined by another person who also can complete the job in exactly 120 days. On Day 3, they are joined by another person of equal efficiency. Like this, everyday a new person with the same efficiency joins the work. How many days are required to complete the job? *[CAT-17]*

Sol. Clearly, each and every worker can do 0.83% of the work per day.

Work done on Day 1 = 0.83%

Work done on Day 2 = 0.83 + 0.83 = 1.66%

Work done on Day 3 = 3 × 0.83 = 2.50%

....

Hence, the sequence is 0.83% + 1.66% + 2.50% + 3.33% + … and its sum must be 100%.

Let the work is completed in n days.

Then, $0.83 (1 + 2 + 3 + \ldots + n) = 100 \Rightarrow \frac{n(n + 1)}{2} = 120$

$\Rightarrow \quad n = 15$ days

B. Group-Level

Type I When a work is done by a group of workers and all workers have equal efficiencies

Take one example

Problem 60. 8 men can do a work in 12 days. After 6 days of work, 4 more men were engaged to finish the work. In how many days would the remaining work be completed?

Sol. It is clear that $\frac{1}{2}$ the work will be completed by 8 men in 6 days.

Then, time taken by 12 men to finish the remaining $\frac{1}{2}$ of work

$$= \frac{\text{multiplication of figures appeared in the sentence completed in all aspects}}{\text{multiplication of figures appeared in sentence in which any item is missing}}$$

[with reciprocal ratio of amount of work]

$$= \frac{1/2 \times 8 \times 6}{1/2 \times 12} \text{ days} = 4 \text{ days}$$

A lot of such types of questions have been discussed in detail in topic 'Proportion' under 'Application of the concepts associated with Proportion' in chapter 'Ratio, Proportion and Variation'.

Type II When a work is done by different groups and all workers in the same group have equal efficiencies.

Case I **Concept of OR/AND**

Build your conception through examples.

Problem 61. 3 men OR 5 women can do a work in 12 days. How long will 6 men AND 5 women take to finish the work?

Sol. Given, efficiency of 3 men = efficiency of 5 women

Say simply, 3 men = 5 women

$\therefore \quad$ 6 men = 10 women

Now, 6 men + 5 women = 10 women + 5 women = 15 women

∵ 5 women can do a work in 12 days.

∴ 1 woman can do a work in = 12 × 5 i.e. 60 days.

∴ 15 women can do a work in = $\frac{60}{15}$ i.e. 4 days.

(**Note** If you wish, you may convert the efficiency of women in terms of that of men on the above pattern)

Alternatively I

∵ 3 men can do a work in 12 days.

Hence, 6 men can do 16.66% of work per day.

Also given, 5 women can do a work in 12 days. It implies that 5 women can do 8.33% of the work per day.

Obviously, (6 men + 5 women)'s 1 day's work

= (16.66 + 8.33)% = 25%.

Then, time taken by (6 men + 5 women) to complete the work

= (100% ÷ 25%) days = 4 days.

Alternatively II

Given that the efficiency of 3 men is equal to that of 5 women. We may treat either of both groups as one unit of work-force.

Given, **3 men (i.e. one unit)** or **5 women (i.e. one unit)** ⇒ 12 days

Then, **6 men (two units)** + **5 women (one unit)** ⇒ ?

We also know that, size of work-force is indirectly proportional to time taken to do the work if amount of work is constant. It is seen from the above that the size of work-force has increased from one unit to three units, i.e. three times. Then, time taken by 3 units of work-force will be $\frac{1}{3}$ of time taken by one unit.

Hence, required time = $\frac{1}{3}$ of 12 = 4 days.

In other words,

∵ 1 unit of work-force can do a work in 12 days.

∴ (2 + 1) units of work-force will take $\frac{12}{(2+1)}$ i.e. 4 days to do the work.

In brief 3 men **or** 5 women ⇒ 12 days

↓ × 2 ↓ × 1

6 men + 5 women ⇒ ?

Then, required time = $\frac{12}{(2+1)}$ = 4 days.

Remarks *The aspirants should stress on the 'Alternatively II' so as to arrive at answer quickly. The other approaches have been left in next questions deliberately on account of space constraints.*

Problem 62. 20 men or 24 women can complete a piece of work in 20 days. If 30 men and 12 women undertake to complete the work, the work will be completed in *[SSC (Investigator)-10, CDS-19]*

Sol. 20 men **or** 24 women ⇒ 20 days

↓ × 1.5 ↓ × 0.5

30 men + 12 women ⇒ ?

Then, required time = $\frac{20}{(1.5+0.5)}$ = 10 days.

Note *Sometimes you will feel uncomfortable to find the decimal figures, then please take approx. value keeping in mind the values of given options)*

Problem 63. If 1 man or 2 women or 3 boys can complete a piece of work in 88 days, then 1 man, 1 woman and 1 boy together will complete in?

Sol. 1 man **or** 2 women **or** 3 boys ⇒ 88 days

↓ × 1 ↓ × 0.5 ↓ × 0.33

1 man + 1 woman + 1 boy ⇒ ?

Then, required time = $\frac{88}{(1+0.5+0.33)}$ = 48 days

Problem 64. 5 men can do a piece of work in 6 days while 10 women can do it in 5 days. In how many days can 5 women and 3 men do it?

Sol. Number of men required to finish the work in 1 day = 5 × 6 = 30 and Number of women required to finish the work in 1 day = 10 × 5 = 50

Hence, efficiency of 30 men = efficiency of 50 women

Now, 30 men **or** 50 women ⇒ 1 day

↓ × 0.1 ↓ × 0.1

3 men + 5 women ⇒ ?

Then, required time = $\frac{1}{(0.1+0.1)}$ = 5 days

Problem 65. If 10 men or 20 boys can make 260 mats in 20 days, then how many mats will be made by 8 men and 4 boys in 20 days?

Sol. 10 men **or** 20 boys ⇒ 260 mats

↓ × 0.8 ↓ × 0.2

8 men + 4 boys ⇒ ?

Then, required time = 260 × (0.8 + 0.2)

(∵ size of work-force is directly proportional to work done)

= 260 mats

Problem 66. Either 8 men or 17 women can paint a house in 35 days. The number of days required to paint three such houses by 12 men and 34 women working at half of their efficiencies is

Sol. 8 men **or** 17 women ⇒ 35 days

↓ × 1.5 ↓ × 2

12 men + 34 women = ?

Then, number of days required to paint 3 houses at same efficiencies = $\frac{35}{(1.5+2)}$ × 3 = 30 days

In case of half rate of efficiencies, required time = 30 × 2 = 60 days.

Case II Concept of AND/AND

Problem 67. 6 men AND 8 women can do a work in 10 days, then 3 men AND 4 women can do the same work in? *[SSC (10+2- LDC/DEO)-II]*

Sol. 6 men + 8 women ⇒ 10 days

3 men + 4 women ⇒ ? days

Clearly, the latter group is half the former group in size. Therefore, the required time will be twice the given time i.e. 20 days.

Problem 68. 2 men and 3 women can do a piece of work in 15 days while 3 men and 2 women can do the same work in 10 days. Then, 3 men and 3 women can do the work in?

Sol. $\left(\begin{array}{l}\text{2 men + 3 women} \Rightarrow \text{15 days} \Rightarrow \text{6.66\% of work per day}\\ \text{3 men + 2 women} \Rightarrow \text{10 days} \Rightarrow \text{10\% of work per day}\end{array}\right)$

If both the groups work together, number of days taken by them to finish the work = [100 ÷ (6.66 + 10)] = 6 days

Now, 5 men + 5 women ⇒ 6 days

$\therefore$ 1 man + 1 woman ⇒ 6×5

$\therefore$ 3 men + 3 women ⇒ $\frac{6 \times 5}{3} = 10$ days

Alternatively

As detailed below the next question.

Problem 69. 2 men and 3 women can complete a piece of work in 10 days. 3 men and 2 women can do the same work in 8 days. Then, 2 men and 1 woman can finish in:

[SSC (10+2- LDC/DEO)- 11, CLAT-17, SNAP-11]

Sol. Given, $\left(\begin{array}{l}2M + 3W \Rightarrow \text{10 days} \Rightarrow \text{10\% of work per day}\\ 3M + 2W \Rightarrow \text{8 days} \Rightarrow \text{12.5\% of work per day}\end{array}\right)$

Hence, 5M + 5W ⇒ 22.5% of the work per day.

$\therefore$ 2M + 2W ⇒ $\frac{22.5 \times 2}{5}$ i.e. 9% of the work per day

From above, 2M + 3W ⇒ 10% of the work per day

Hence, 1W ⇒ (10 – 9) i.e. 1% of the work per day

Then, 2M ⇒ [10 – (3 × 1)]% i.e. 7% of the work per day

Now, work done by (2M + 1W) per day = (7 + 1)% = 8%

$\therefore$ Time taken by (2 men + 1 woman) to finish the work = (100% ÷ 8%) days = 12.5 days

Alternatively I

$\because$ 2 men and 3 women can complete a piece of work in 10 days.

Then, to finish the work in 1 day, 10 (2 men + 3 women) are required.

Or, to finish the work in 1 day, 8 (3 men + 2 women) are required.

In both cases, sizes of work-forces would obviously be equal.

Hence, 10 (2 men + 3 women) = 8 (3 men + 2 women)

⇒ $30W - 16W = 24M - 20M$

⇒ $2M = 7W \Rightarrow 1M = \frac{7}{2}W$

Now, $2M + 3W = 7W + 3W = 10W$

and $2M + 1W = 7W + 1W = 8W$

$\because$ (2M + 3W) i.e. 10W can do a piece of work in 10 days.

$\therefore$ (2M + 1W) i.e. 8W can do the same work in $\frac{10 \times 10}{8}$ days i.e. 12.5 days.

Alternatively II

$\left(\begin{array}{ll}2M + 3W \Rightarrow \text{10 days} & \text{......(i)}\\ 3M + 2W \Rightarrow \text{8 days} & \text{......(ii)}\\ 2M + 1W \Rightarrow \text{? days} & \text{......(iii)}\end{array}\right)$

Positive difference between (10 × 2M) and (8 × 3M) = Positive difference between (10 × 3W) and (8 × 2W)

⇒ $4M = 14W$

⇒ $2M = 7W$, this relation should be maintained when we take any two equations out of above three. But, it is the fact that Eq. (iii) is incomplete, which can be completed by putting the values of options in place of ?. Let options be 12, 12.5, ...

(a) 12 ⇒ Given, 2M + 3W ⇒ 10 days ...(i)

and 2M + 1W ⇒ 12 days ...(ii)

Now, positive difference between (10 × 2M) and (12 × 2M) = Positive difference between (10 × 3W) and (12 × 1W)

⇒ 4M = 18W ⇒ 2M = 9W, which is not same as 2M = 7W.

Hence, option (a) is not correct.

(b) 12.5 ⇒ Given, 2M + 3W ⇒ 10 days ...(i)

and 2M + 1W ⇒ 12.5 days ...(iii)

Now, positive difference between (10 × 2M) and (12.5 × 2M) = Positive difference between (10 × 3W) and (12.5 × 1W)

⇒ 5M = 17.5W ⇒ 2M = 7W, which is the same as relation established from Eqs. (i) and (ii).

Hence, option (b) is correct.

Note *The above processes seem to be very lengthy, but all these can be done mentally without pen and paper. Thus, aspirants must have command over this technique.*

Problem 70. 4 men and 6 women can complete a work in 8 days, while 3 men and 7 women can complete it in 10 days. In how many days will 10 women complete it?

[SSC (CGL-PT)-04, SSC (CGL-T-II)- 10, 12, CLAT-18]

Sol. $\left(\begin{array}{ll}4M + 6W \Rightarrow \text{8 days} & \text{......(i)}\\ 3M + 7W \Rightarrow \text{10 days} & \text{......(ii)}\\ 0M + 10W \Rightarrow \text{? days} & \text{......(iii)}\end{array}\right)$

Now, proceed through options (let these be 30, 40, 15, 20) on the pattern as envisaged in 'Alternatively II' meant for preceding question. Answer is 40 days.

Problem 71. 5 men and 2 women working together can do four times as much work per hour as a man and a woman together. The work done by a man and a woman should be in the ratio:

Sol. Given, $\left(\begin{array}{l}5M + 2W \Rightarrow 1\,h\\ 1M + 1W \Rightarrow 4\,h\end{array}\right)$ to finish a work.

Now, positive difference between (1 × 5M) and (4 × 1M) = positive difference between (1 × 2W) and (4 × 1W)

⇒ M = 2W ⇒ M : W = 2 : 1

Since ratio of their efficiency is 2 : 1, the work done by them will also be in ratio 2 : 1.

Problem 72. 2 men and 3 women together or 4 men together can complete a piece of work in 20 days. 3 men and 3 women will complete the same work in :

Sol. 2 M + 3 W ⇒ 20 days ⇒ 5% of work per day

Then, 4 M + 6 W ⇒ 10% of work per day

Further, 4M ⇒ 20 days ⇒ 5% of work per day (i.e. 3M ⇒ 3.75%)

Hence, 6W ⇒ (10 – 5)% of the work per day.

$\therefore$ 3W ⇒ 2.5% of the work per day

Now, work done by (3M + 3W) per day = (3.75 + 2.5)% per day

Then, time taken by (3M + 3W) to complete the work = 100% ÷ 6.25% = 16 days.

Alternatively I

Given, $2M + 3W = 4M \Rightarrow 2M = 3W$

Then, $3M + 3W = 3M + 2M = 5M$

$\because$ 4M can complete a piece of work in 20 days.

$\therefore$ 5M can complete the same in $\frac{20 \times 4}{5}$ i.e. 16 days.

Alternatively II

From above, relation established be 2M = 3W. This relation can also be established through

$$\begin{pmatrix} 2M + 3W \Rightarrow 20 \text{ days} & \dots(i) \\ 4M + 0W \Rightarrow 20 \text{ days} & \dots(ii) \end{pmatrix}$$

Now, proceed through options. Let options be 12, 16,...

(a) $12 \Rightarrow$ Given, $2M + 3W \Rightarrow 20$ days ...(i)

and $3M + 3W \Rightarrow 12$ days ...(iii)

Now, positive difference between $(20 \times 2M)$ and $(12 \times 3M)$

= positive difference between $(20 \times 3W)$ and $(12 \times 3W)$

$\Rightarrow 4M = 24W \Rightarrow 1M = 6W$, which is not same as 2M = 3W. Hence, option (a) is not correct.

(b) $16 \Rightarrow$ Given, $2M + 3W \Rightarrow 20$ days ...(i)

and $3M + 3W \Rightarrow 16$ days ...(iii)

Now, positive difference between $(20 \times 2M)$ and $(16 \times 3M)$

= positive difference between $(20 \times 3W)$ and $(16 \times 3W)$

$\Rightarrow 8M = 12W \Rightarrow 2M = 3W$, which is the same to the relation established through Eq. (i) and Eq. (ii).

Hence, option (b) is correct.

Problem 73. 2 men, 1 woman and 4 children can finish a piece of work in $6\frac{2}{3}$ days. 1 man and 2 children together can finish in $16\frac{2}{3}$ days, 2 men and 5 women can do in $4\frac{8}{23}$ days. The time taken by 3 men and 4 children is

Sol.
$$\begin{bmatrix} 2M + 1W + 4C \Rightarrow 15\% \text{ of work per day} & \dots(i) \\ 1M + 2C \Rightarrow 6\% \text{ of work per day} & \dots(ii) \\ 2M + 5W \Rightarrow 23\% \text{ of work per day} & \dots(iii) \\ 3M + 4C \Rightarrow ? \text{ days} & \end{bmatrix}$$

Eq. (ii) may be written as $2M + 4C \Rightarrow 12\%$ of work per day

From Eqs. (i) and (ii), work done by 1W per day

$= (15 - 12)\% = 3\%$

Now, from Eq. (iii), work done by 2M per day

$= (23 - 5 \cdot 3)\% = 8\%$

$\therefore$ Work done by 3M per day = 12%

and from Eq. (i), work done by 4C per day

$= (15 - 3 - 8) = 4\%$

Thus, work done by (3M + 4C) per day

$= (12 + 4)\% = 16\%$

$\therefore$ Time required $= 100\% \div 16\% = 6.25$ days.

Problem 74. 4 men and 10 women were put on a work. They completed 1/3 of the work in 4 days. After this, 2 men and 2 women were increased. They completed 2/9 more of the work in 2 days. If the remaining work is to be completed in 3 days, then how many more women must be increased? ***[MAT-06]***

Sol. 4 men and 10 women can complete $\frac{1}{3}$ of work in 4 days. Then, to finish the whole work in 1 day, 12 (4 men + 10 women) are required.

Further, 6 men and 12 women can complete $\frac{2}{9}$ of work in 2 days. Then, to finish the whole work in 1 day, 9 (6 men + 12 women) are required.

In both cases, sizes of work-forces would obviously be equal.

Hence, 12 (4 men + 10 women) = 9 (6 men + 12 women)

$\Rightarrow \quad 48M + 120W = 54M + 108W$

$\Rightarrow \quad 6M = 12W \Rightarrow 1M = 2W$

Now, $\quad 4M + 10W = 8W + 10W = 18W$

$\because$ To finish a piece of work in 12 days, (4M + 10W), i.e. 18W are required.

$\therefore$ To finish $\left(1 - \frac{1}{3} - \frac{2}{9}\right)$ of work in 3 days, required number of women $= \frac{4 \times 18 \times 12}{9 \times 3} = 32$

Since (6 men + 12 women), i.e. $(6 \times 2 + 12)$, women have already been engaged, only (32 – 24) i.e. 8 women are required.

Alternatively

$4M + 10W \Rightarrow 12$ days ...(i) $\left[\because \frac{1}{3} \Rightarrow 4 \;\therefore\; 1 \Rightarrow 12\right]$

$6M + 12W \Rightarrow 9$ days ...(ii) $\left[\because \frac{2}{9} \Rightarrow 2 \;\therefore\; 1 \Rightarrow 9\right]$

$6M + ?W \Rightarrow 6.75$ days ...(iii)

$\left[\because \left(1 - \frac{1}{3} - \frac{2}{9}\right) \Rightarrow 3 \therefore 1 \Rightarrow 6.75\right]$

Positive difference between $(12 \times 4M)$ and $(9 \times 6M)$

= Positive difference between $(12 \times 10W)$ and $(9 \times 12W)$

$\Rightarrow \quad 6M = 12W$

$\Rightarrow \quad 1M = 2W,$

this relation should be maintained when we take any two equations out of above three. But, it is the fact that Eq. (iii) is incomplete, which can be completed by putting the values of options in place of ?. Let options be 32, 8...

(a) $32 \Rightarrow$ Given, $4M + 10W \Rightarrow 12$ days ...(i)

and $6M + (12 + 32)W \Rightarrow 6.75$ days ...(iii)

Now, positive difference between $(12 \times 4M)$ and $(6.75 \times 6M)$

= positive difference between $(12 \times 10W)$ and $(6.75 \times 44W)$

$\Rightarrow 7.5M = 177W$, which is not same as 1M = 2W. Hence, option (a) is incorrect.

(b) $8 \Rightarrow$ Given, $4M + 10W \Rightarrow 12$ days ...(i)

and $6M + (12 + 8)W \Rightarrow 6.75$ days ...(iii)

Now, positive difference between $(12 \times 4M)$ and $(6.75 \times 6M)$

= positive difference between $(12 \times 10W)$ and $(6.75 \times 20W)$

$\Rightarrow 7.5M = 15W \Rightarrow 1M = 2W$, which is same as the relation established through Eqs. (i) and (ii). Hence, option (b) is correct.

Problem 75. The time taken by 2 men to finish a work is $2\frac{2}{5}$ days lesser than what 4 women take. By how much percentage is the efficiency of a man greater than that of a woman if 3 men and 2 women together can finish the work in 2 days?

Sol. $3M + 2W \Rightarrow 50\%$ of work per day

i.e. $6M + 4W \Rightarrow 100\%$ of work per day.

Let one man does $x\%$ of the work per day.

Then, work done by $4W = (100 - 6x)\%$ per day

Further, work done by $2M = 2x\%$ per day.

As per question,

Time taken by 4W – time taken by 2W $= 2\frac{2}{5}$

$$\Rightarrow \frac{100\%}{(100-6x)\%} - \frac{100\%}{2x\%} = 2\frac{2}{5}$$

$$\Rightarrow \frac{50}{50-3x} - \frac{50}{x} = \frac{12}{5}$$

$$\Rightarrow (50x - 2500 + 150x)5 = 12(50x - 3x^2)$$

$$\Rightarrow -12500 + 1000x = 600x - 36x^2$$

$$\Rightarrow 36x^2 + 400x - 12500 = 0$$

$$\Rightarrow 9x^2 + 100x - 3125 = 0$$

$$\Rightarrow 9x^2 + 225x - 125x - 3125 = 0$$

$$\Rightarrow 9x(x+25) - 125(x+25) = 0$$

$\Rightarrow x = -25$ and $\frac{125}{9} \Rightarrow x = \frac{125}{9}$ as value of x cannot be negative.

Hence, work done by 1M per day $= \frac{125}{9}\%$

Now, work done by 1W per day

$$= \left(100 - 6 \times \frac{125}{9}\right)\% \div 4 = 4\frac{1}{6}\%$$

Hence, ratio of efficiency (1M : 1W)

$$= \frac{125}{9}\% : 4\frac{1}{6}\% = 10 : 3$$

Clearly, a man is $\left(\frac{(10-3)}{3} \times 100\right)\%$, i.e. 233.33% more efficient than a woman.

Alternatively I

Given, $3M + 2M \Rightarrow 2$ days

Then, $\frac{2}{3}(3M + 2W) \Rightarrow 2 \times \frac{3}{2}$ days

Hence, $2M + \frac{4}{3}W \Rightarrow 3$ days

Let $\frac{4}{3}W$ can do the work in x days.

Then, number of days required for 2M to finish the work $= \frac{3x}{x-3}$

Since $\frac{4}{3}W$ can do the work in x days, 4W can finish the same in $\frac{x}{3}$ days.

As per question, $\frac{x}{3} - \frac{3x}{x-3} = \frac{12}{5}$

$$\Rightarrow 5x^2 - 60x = 36x - 108$$

$$\Rightarrow 5x^2 - 96x + 108 = 0$$

$$\Rightarrow 5x^2 - 90x - 6x + 108 = 0$$

$$\Rightarrow 5x(x-18) - 6(x-18) = 0$$

$$\Rightarrow (5x-6)(x-18) = 0 \Rightarrow x = \frac{6}{5} \text{ and } 18$$

$\Rightarrow x = 18$ as $\frac{6}{5}$ is not possible.

Now, time taken by $\frac{4}{3}W = 18$ days

$\therefore$ Time taken by 1 W = 24 days

and time taken by $2M = \frac{3x}{x-3} = \frac{3 \times 18}{18-3} = \frac{18}{5}$

$\therefore$ Time taken by $1M = \frac{36}{5}$

Hence, ratio of efficiency (Man : Woman) $\frac{1}{36/5} : \frac{1}{24} = \frac{5}{36} : \frac{1}{24}$

$= 10 : 3$

Clearly, a man is $\left[\frac{(10-3)}{3} \times 100\right]\%$, i.e. 233.33% more efficient than a woman.

Alternatively II

The best way is to proceed through options. Let options be 300%, 223.33%, ...

(a) 300% $\Rightarrow$ It means efficiency of a man is 4 times that of a woman.

Then, $3M + 2W = 3 \times 4 + 2 = 14W$

$\because$ 14W can do the work in 2 days.

$\therefore$ 4W can do the same work in 7 days

$\therefore$ 8W ($\because 2M = 2 \times 4 = 8W$ can do in $3\frac{1}{2}$ days)

But, $7 - 3\frac{1}{2} \neq 2\frac{2}{5}$

Hence, option (a) is incorrect.

(b) 233.33% $\Rightarrow$ It means efficiency of a man is 3.33 times that of a woman.

Then, $3M + 2W = 3 \times 3.33 + 2 = 12W$

$\because$ 12W can do the work in 2 days.

$\therefore$ 4W can do the same work in 6 days.

$\therefore$ 6.66W ($\because 2M = 2 \times 3.33 = 6.66$ W) can do in $\frac{12 \times 2}{6.66}$

$= 3\frac{3}{5}$ days

Also, $6 - 3\frac{3}{5} = 2\frac{2}{5}$

Hence, option (b) is correct.

Miscellaneous

Problem 76. If 8 men and 12 boys can finish a piece of work in 12 days, in what time will 40 men and 45 boys finish another piece of work 3 times as great supposing 16 men can do as much work in 8 h as 12 boys do in 24 h?

[SSC (Asstt. Gr.-Mains)-94, SSC (CGL-Mains)-99, 03]

Sol. Given, 16 men can do as much work in 8 h as 12 boys do in 24 h.

Then, to finish the work in 1 h, either (16×8) men or (12×24) boys are required.

Hence, (16×8) men $= (12 \times 24)$ boys

$\Rightarrow 4M = 9B$

Now, (8M + 12B) i.e. (18B + 12B) can finish a work in 12 days.

Then, (40M + 45B) i.e. (90B + 45B) can finish 3 times the work in $\frac{12 \times 30 \times 3}{135}$ i.e. 8 days.

Problem 77. A man, a woman and a boy can complete a work in 20 days, 30 days and 60 days respectively. How many boys must assist 2 men and 8 women so as to complete the work in 2 days?

Sol. Work done by a man per day = 5% of the work

Work done by a woman per day = 3.33% of the work

Work done by a boy per day = 1.66% of the work

Obviously, amount of the work to be done by boys in 2 days

$= 100\% - (2 \times 2 \times 5\%) - (8 \times 2 \times 3.33\%) = 26.66\%$

Hence, number of boys required

$= 26.66\% \div (2 \times 1.66\%) = 8$

Alternatively I

Work done by (2M + 8W) in 2 days $= 2\left(\frac{2}{20} + \frac{8}{30}\right) = \frac{11}{15}$

Remaining work i.e. 4/15 is to be done by boys in 2 days.

Now, work done by 1 boy in 2 days $= \frac{2}{60} = \frac{1}{30}$

Then, number of boys required $= \left(\frac{4}{15} \div \frac{1}{30}\right) = 8$

Alternatively II

A man, a woman and a boy can complete a work in 20 days, 30 days and 60 days respectively. It implies that

20M = 30W = 60B i.e. 2M = 3W = 6B.

$2M + 8W + xB$

$= 6B + 16B + xB = (22 + x)B$

(imagining number of boys in 2nd group be x.)

Given, 1B can do a piece of work in 60 days

Then,? B can do that work in 2 days

Obviously, number of boys $= \frac{1 \times 60}{2} = 30$

∴ Number of boys to assist 22 boys (i.e. 2M + 8W)

$= 30 - 22 = 8$

Problem 78. 24 men can complete a work in 16 days, 32 women can complete the same work in 24 days. 16 men and 16 women started working and worked for 12 days. How many more men are to be added to complete the remaining work in 2 days?

Sol. Clearly, 16 men and 16 women worked for (12 + 2) days.

Now, work done by 16 men in 14 days $= \frac{14 \times 16}{24 \times 16} = \frac{7}{12}$

(∵ 24 men do a work in 16 days)

and work done by 16 women in 14 days $= \frac{14 \times 16}{32 \times 24} = \frac{7}{24}$

(∵ 32 women do a work in 24 days)

∴ Remaining work $= 1 - \frac{7}{12} - \frac{7}{24} = \frac{1}{8}$, which will be done by the additional men in 2 days.

As given, a piece of work is done in 16 days by 24 men.

Then, $\frac{1}{8}$ of work will done in 2 days by ? men

Clearly, ? $= \frac{16 \times 24 \times 1}{2 \times 8} = 24$ men

Hence, 24 men must be added to complete the remaining work in 2 days.

Alternatively I

To finish the work in one day either (24 × 16) men or (32 × 24) women are required.

Hence, (24 × 16) men = (32 × 24) women

⇒ 1M = 2W

Then, 16M + 16W = 16M + 8M

= 24M

Given, 24 men in 16 days do a piece of work

∴ 24 men in 14 days will do $\frac{14 \times 24}{24 \times 16} = \frac{7}{8}$

Clearly, the remaining work i.e. $\frac{1}{8}$ is done by additional men in 2 days.

As given, a piece of work is done in 16 days by 24 men.

Then, $\frac{1}{8}$ of work will done in 2 days by ? men

Clearly, ? $= \frac{16 \times 24 \times 1}{2 \times 8} = 24$ men

Hence, 24 men must be added to complete the remaining work in 2 days.

Alternatively II

To finish the work in one day either (24 × 16) men or (32 × 24) women are required.

Hence, (24 × 16) men = (32 × 24) women

⇒ 1M = 2W

Then, 16M + 16W = 16M + 8M

= 24M

Work done by 24M in 12 days $= \frac{12}{16} = \frac{3}{4}$

(∵ 24 men do a piece of work in 16 days)

Now, 24 men does 1 unit of work in 16 days

Then, ? men will do $\frac{1}{4}$ of work in 2 days

$$? = \frac{24 \times \frac{1}{4} \times 16}{2 \times 1}$$

= 48 men

Hence, number of additional men required = 48 − 24 = 24.

Problem 79. 12 men can complete a piece of work in 4 days, while 15 women can complete the same work in 4 days, 6 men start working on job and after working for 2 days, all of them stopped working. How many women should be put on the job to complete the remaining work, if it is to be completed in 3 days?

Sol. Work done by 6 men in 2 days $= \frac{6 \times 2}{12 \times 4} = \frac{1}{4}$

Then, remaining work $= 1 - \frac{1}{4} = \frac{3}{4}$

Now, 15 women do 1 unit of work in 4 days

Then, ? women will do $\frac{3}{4}$ of work in 3 days

Clearly, ? $= \frac{15 \times \frac{3}{4} \times 4}{3 \times 1} = 15$ women

Note Try to solve with other approaches.

Problem 80. A group of women in a society decided to execute interior and exterior decoration of the society in a week's time. Since 11 women dropped out every day from the second day, the entire decoration was completed on 12th day. How many women participated in the beginning? (answer to the nearest integer)

[IIFT-19]

Sol. Let w women participated in the beginning and each of them does x% work daily.

Then, work is to be done by w women in 7 days = 7xw%

As per question, number of women for first day = w

Number of women for 2nd day = w − 11

Number of women for 3rd day = w − (2 × 11)

....

....

Number of women for 12th day = w − (11 × 11)

∴ Number of women engaged

$$= 12w - 11(0 + 1 + 2 + \ldots + 11)$$

$$= 12w - 11 \times \frac{11 \times 12}{2}$$

$$= 12w - 726$$

Then, work done by (12w − 726) = (12w − 726)x%

Clearly, $7xw\% = (12w - 726)x\%$

$\Rightarrow$ $5w = 726$

$\Rightarrow$ $w = 145.2$

∴ Required answer = 145

Problem 81. A task can be completed by 3 men working for a day, or by 8 women working for $\frac{3}{4}$th of a day, or by 18 children working for half a day. If 3 women and 3 children have already been hired for a day, how many men should be hired and for how many days for the completion of the task?

Sol. ∵ In $\frac{3}{4}$ days, 8 women finish 1 unit of work

∴ In 1 day, 3 women will do $\frac{3}{8 \times \frac{3}{4}}$, i.e. $\frac{1}{2}$ units of work.

Further, in one day 3 children will do $\frac{3 \times 1}{18 \times \frac{1}{2}}$ i.e. 1/3 units of work.

Remaining work which is to be done by men $= 1 - \frac{1}{2} - \frac{1}{3} = \frac{1}{6}$

As given, to finish a piece of work in 1 day, 3 men are required.

∴ To finish 1/6 of the work in 1 day, $\frac{3 \times \frac{1}{6}}{1 \times 1}$ i.e. $\frac{1}{2}$ men required.

Since, the number of men cannot be in fraction, it may be said that 1 man is required for $\frac{1}{2}$ days or 2 men are required for $\frac{1}{4}$ days.

Note *Try to solve with other approaches.*

WORKOUT

1. *A*, *B* and *C* can do a piece of work individually in 8 days, 12 days and 15 days respectively. *A* and *B* start working together but *A* quits after working for 2 days. After this, *C* joins and works till completion of the work. In how many days, will the work be completed? ***[CDS-14]***

(a) $3\frac{8}{9}$ days (b) $5\frac{8}{9}$ days

(c) $5\frac{2}{3}$ days (d) $6\frac{1}{18}$ days

2. Ganu, Kanu and Manu together completed a piece of work in certain days. Ganu and Kanu together executed 72% of the work while Kanu and Manu together 63% of the work. Then, which of the following statement is true? ***[SSC (CGL-T-I)-17]***

(a) Kanu is less efficient than Ganu.

(b) Kanu is the most efficient.

(c) Kanu is less efficient than Manu

(d) None of the above

3. *A*, *B* and *C* working together completed a job in 10 days. However, *C* only worked for the first three days when 37/100 of the job was done. Also, the work done by *A* in 5 days is equal to the work done by *B* in 4 days. How many days would be required by the fastest worker to complete the entire work? ***[MAT-13]***

(a) 25 days (b) 20 days

(c) 30 days (d) 40 days

4. *A* and *B* can do a piece of work in 21 and 24 days, respectively. They start the work together and after some days *A* leaves the work and *B* completes the remaining work in 9 days. After how many days did *A* leave? ***[SNAP-11]***

(a) 5 (b) 7

(c) 8 (d) 6

5. A person can do a job as fast as his two sons working together. If one son does the job in 6 days and the other in 12 days, how many days does it take the father to do the job?

(a) 9 days (b) 6 days

(c) 4 days (d) 3 days

6. *A* can complete a piece of work in 18 days, *B* in 20 days and *C* in 30 days. *B* and *C* together start the work and are forced to leave after 2 days. The time taken by *A* alone to complete the remaining work is?

(a) 20 days (b) 15 days

(c) 25 days (d) 10 days

7. *C* is twice efficient as *A*. *B* takes thrice as many days as *C*. A takes 12 days to finish the work alone. If they work in pairs (i.e. *AB*, *BC*, *CA*) starting with *AB* on the first day, *BC* on the second day and *AC* on the third day and so on, then how many days are required to finish the work?

(a) $6\frac{1}{5}$ days (b) 4.5 days

(c) $5\frac{1}{9}$ days (d) 8 days

8. *A* and *B* can finish a job in 10 days while *B* and *C* can do it in 18 days. *A* started the job worked for 5 days, then *B* worked for 10 days and the remaining job was finished by *C* in 15 days. In how many days could *C* alone have finished the whole job?

(a) 30 days (b) 15 days

(c) 45 days (d) 24 days

9. *A* is twice as good a workman as *B* and together can finish a piece of work in 14 days. The number of days taken by *A* alone to finish the work is ***[LIC (AAO)-10]***

(a) 11 (b) 21

(c) 28 (d) 42

10. *A* and *B* can do a piece of work in 10 h. *B* and *C* can do it in 15 h, while *A* and *C* take 12 h to complete the work. *B* independently can complete the work in ***[CDS-14, 09, SSC (CGL-T-II)-20]***

(a) 12 h (b) 16 h

(c) 20 h (d) 24 h

11. 10 hectare field is reaped by 2 men, 3 women and 4 children together in 10 days. If working capabilities of a man, a woman and a child are in the ratio 5 : 4 : 2, then a 16 hectare field will be reaped by 6 men, 4 women and 7 children in

(a) 5 days (b) 6 days

(c) 7 days (d) 8 days

12. *A* can do a piece of work in 40 days. He starts working, but having some other engagements he drops out after 5 days. Thereafter, *B* complete this work in 21 days. How many days would *A* and *B* take to complete this work working together?

(a) 16 days (b) 15 days

(c) 17 days (d) 11 days

13. Manoj and Ajita can do a job alone in 10 days and 12 days, respectively. Manoj starts the work and after 6 days Ajita also join to finish the work together. For how many days did Ajita actually work on the job?

(a) $2\frac{3}{11}$ (b) $2\frac{1}{11}$

(c) $2\frac{2}{11}$ (d) $3\frac{1}{12}$

14. Two typists undertake to do a job. The second typist begin working one hour after the first. Three hours after the first typist has begun working, there is still 9/20 of the work to be done. When the assignment is completed, it turns out that each typist has done half the work. How many hours would it take each one to do the whole job individually? ***[CAT-95]***

(a) 12h, 8h (b) 5h, 4h

(c) 8h, 5.6h (d) 10h, 8h

15. *A* is thrice as efficient as *B* and hence completes a work in 40 days less than the number of days taken by *B*. What will be the number of days taken by both of them when working together? ***[CDS-14, MAT-10, 99]***

(a) 22.5 days (b) 15 days

(c) 20 days (d) 18 days

16. 3 men and 4 boys can complete a piece of work in 12 days. 4 men and 3 boys can do the same work in 10 days. Then, 2 men and 3 boys can finish the work in number of days is : *[SSC (CGL-T-I)-12, SNAP-11]*

(a) $17\frac{1}{2}$ (b) $5\frac{5}{11}$
(c) 8 (d) 22

17. A and B completed a work together in 5 days. Had A worked at twice the speed and B at half the speed, it would have taken them 4 days to complete the job. How much time would it take for A alone to do the work? *[MAT-14]*

(a) 20 days (b) 10 days
(c) 15 days (d) 25 days

18. Ram completes 60% of a task in 15 days and then takes the help of Rahim and Rachel. Rahim is 50% as efficient as Ram is and Rachel is 50% as efficient as Rahim is. In how many more days will they complete the work? *[SNAP-11]*

(a) 121/3 (b) 51/3
(c) 55/3 (d) None of these

19. 6 men and 8 boys can do a piece of work in 10 days while 26 men and 48 boys can do the same in 2 days, what is the time taken by 15 men and 20 boys in doing the same type of work?

(a) 5 days (b) 4 days
(c) 6 days (d) 7 days

20. A alone would take 8h more to complete the job than if both A and B worked together. If B worked alone, he took $4\frac{1}{2}$ h more to complete the job than if A and B worked together. What time would they take if both A and B worked together? *[MAT-10, 08, SSC (CGL-T-II)-12]*

(a) 8 days (b) 5 days
(c) 2 days (d) 6 days

21. Madhu takes twice as much time as Uma to complete a work and Ramesh does it in the same time as Madhu and Uma together. If all three working together can finish the work in 6 days, then the time taken by Madhu to finish the work is :

(a) 12 days (b) 14 days
(c) 36 days (d) 40 days

22. Work done by A in one day is half of the work done by B in one day. Work done by B is half of the work done by C in one day. If C alone can complete the work in 7 days, in how many days can A, B and C together complete the work? *[SBI (PO)-08]*

(a) 28 (b) 14
(c) 4 (d) 21
(e) None of these

23. X can do a piece of work in 25 days. Y is 25% more efficient than X. The number of days taken by Y is :

(a) 15 days (b) 20 days
(c) 21 days (d) 30 days

24. A can do a piece of work in x days and B can do the same work in $3x$ days. To finish the work together they take 12 days. What is the value of x? *[CDS-14]*

(a) 6 days (b) 10 days
(c) 12 days (d) 16 days

25. B can do a piece of work in 6h, B and C together can do it in 4h and A, B and C together in $2\frac{2}{3}$ h. In how many hours can A and B together do the same piece of work? *[MAT-12]*

(a) 11 h (b) $6\frac{1}{7}$ h
(c) $2\frac{3}{7}$ h (d) $3\frac{3}{7}$ h

26. A can complete a piece of work in 36 days, B in 54 days and C in 72 days. All the three began the work together but A left 8 days before the completion of the work and B, 12 days before the completion of work. Only C worked upto the end. In how many days was the work completed?

(a) 24 (b) 25
(c) 27 (d) 30

27. Two men alone or three women alone can complete a piece of work in 4 days. In how many days can one woman and one man together complete the same piece of work?

(a) 6 days
(b) 24/5 days
(c) 12/1.75 days
(d) Cannot be determined
(e) None of the above

28. A can do a piece of work in 4 days and B can complete the same work in 12 days. What is the number of days required to do the same work together?

(a) 2 days (b) 3 days
(c) 4 days (d) 5 days

29. X can do a work in 16 days. In how many days will the work be completed by Y, if the efficiency of Y is 60% more than that of X? *[CDS-14, SSC (CGL-T-II)-10, 11]*

(a) 10 days (b) 12 days
(c) 25 days (d) 30 days

30. The ratio of efficiency of A is to C is 5 : 3. The ratio of number of days taken by B is to C is 2 : 3. A takes 6 days less than C, when A and C complete the work individually. B and C started the work and left after 2 days. The number of days taken by A to finish the remaining work is : *[MAT-13]*

(a) 4.5 days (b) 5 days
(c) 6 days (d) 9.33 days

31. A can complete a piece of work in 2 days, and B in 3 days. With the help of C they together completed it in one day. C alone could complete the work in?

(a) 4 days (b) 5 days
(c) 6 days (d) 7 days

32. If one man or two women or three boys can do a piece of work in 55 days, then one man, one woman and one boy will do it in how many days?

(a) 20 days (b) 30 days
(c) 40 days (d) 50 days

33. A and B can complete a task in 30 days when working together. After A and B have been working together for 11 days, B is called away and A completes the remaining task in the next 28 days. Had A been working alone, the number of days taken by him to complete the task would have been :

(a) $33\frac{3}{19}$ days (b) $19\frac{6}{25}$ days
(c) $44\frac{4}{19}$ days (d) $33\frac{7}{19}$ days

34. A can do a piece of work in 30 days. B and C together can do the same work in 20 days. A and B worked together and completed

2/3 of the work in 12 days. How many days will be required by C to complete the remaining work?

(a) 28 (b) 12
(c) 4 (d) 21

35. B is twice efficient as A and A can do a piece of work in 15 days. A started the work and after few days B joined him. They completed the work in 11 days, from the starting. For how many days did they work together? ***[SNAP-11]***

(a) 1 day (b) 2 days
(c) 6 days (d) 5 days

36. X can complete a job in 12 days. If X and Y work together, they can complete the job in $6\frac{2}{3}$ days. Y alone can complete the job in ***[CDS-14]***

(a) 10 days (b) 12 days
(c) 15 days (d) 18 days

37. X can do 1/4 of a work in 10 days. Y can do 40% of the work in 40 days and Z can do 1/3 of the work in 13 days. Who will complete the work first?

(a) X
(b) Y
(c) Z
(d) Both (a) and (c)

38. 8 men can complete a piece of work in 20 days. 8 women can complete the same work in 32 days. In how many days will 5 men and 8 women together complete the same work?

(a) 16 days (b) 12 days
(c) 14 days (d) 10 days
(e) None of these

39. A and B together can complete a piece of work in 12 days and B and C together in 15 days. If A is twice as good a workman as C, then in how many days will B alone complete the same work?

(a) 30 (b) 25
(c) 24 (d) 20

40. If 3 men and 4 boys can do a piece of work in 8 days, then 6 men and 8 boys can do the same work in ***[CDS-14]***

(a) 2 days (b) 4 days
(c) 6 days (d) 16 days

41. A and B can separately do a piece of work in 20 days and 15 days, respectively. They worked together for 6 days, after which B was replaced by C. If the work was finished in the next 4 days, then the number of days in which C alone could do the work will be :

(a) 60 days (b) 40 days
(c) 35 days (d) 30 days

42. A, B and C can do a job working alone in 50 days, 75 days and 20 days, respectively. They all work together for 4 days, then C quits. How many days will A and B take to finish the rest of the job? ***[SSC (CGL-T-II)-17]***

(a) 20 (b) 30
(c) 18 (d) 40

43. A and B complete a work in 12 days, B and C in 8 days and C and A in 16 days. A left after working for 3 days. In how many days more will B and C finish the remaining work? ***[CLAT-17]***

(a) $7\frac{3}{4}$ (b) $3\frac{3}{4}$
(c) $4\frac{3}{4}$ (d) $6\frac{5}{6}$

44. A and B can do a piece of work in 8 days. B and C can complete the same work in 12 days. If A, B and C can complete the same work in 6 days, in how many days can A and C complete the same work?

(a) 8 days (b) 10 days
(c) 12 days (d) 16 days

45. 8 children and 12 men complete a certain piece of work in 9 days. If each child takes twice the time taken by a man to finish the work, 12 men will finish the work in? ***[MAT-09, 08]***

(a) 15 days (b) 12 days
(c) 14 days (d) 8 days

46. Sunil completes a work in 4 days, whereas Dinesh completes the work in 6 days. Ramesh works 1.5 times as fast as Sunil. The three together can complete the work in ***[SSC (CGL-T-II)-13]***

(a) $1\frac{5}{12}$ days (b) $1\frac{5}{7}$ days
(c) $1\frac{3}{8}$ days (d) $1\frac{5}{19}$ days

47. A can finish a work in 8 days and B can do it in 12 days. After A had worked for 3 days, B also joins A to finish the remaining work. In how many days will the remaining work be finished?

(a) 2 days (b) 3 days
(c) 4 days (d) 5 days

48. Efficiency of Asha is 25% more than Usha and Usha takes 25 days to complete a piece of work. Asha started the work alone and then Usha joined her 5 days before actual completion of the work. For how many days Asha worked alone? ***[MAT-13]***

(a) 13 days (b) 11 days
(c) 10 days (d) 15 days

49. A, B and C can do a piece of work in 30 days, 20 days and 10 days, respectively. A is assisted by B on one day and by C on the next day, alternately. How long would the work take to finish? ***[SSC (CGL-T-II)-11]***

(a) $9\frac{3}{8}$ days (b) $4\frac{3}{8}$ days
(c) $8\frac{4}{13}$ days (d) $3\frac{9}{13}$ days

50. A, B and C can do a work in 8 days, 16 days and 24 days, respectively. They all begin together. A continues to work till it is finished. C left after 2 days and B one day before its completion. In what time is the work finished?

(a) 7 days (b) 5 days
(c) 6 days
(d) Cannot be determined

51. A can do a work in 18 days, B in 9 days and C in 6 days. A and B start working together and after 2 days C joins them. What is the total number of days taken to finish the work?

(a) 4.33 days (b) 4.0 days
(c) 4.66 days (d) 5.33 days

52. Anu can complete a work in 10 days. Manu is 25% more efficient than Anu and Sonu is 60% more efficient than Manu. Working together, how long would they take to finish the job?

(a) $2\frac{6}{17}$ days (b) $5\frac{6}{7}$ days
(c) $3\frac{5}{8}$ days (d) $4\frac{5}{8}$ days

53. 12 men can complete a work in ten days. 20 women can complete the same work in twelve days. 8 men and 4 women started working and after 9 days 10 more women joined them. How many days will they now take to complete the remaining work? ***[IIFT-12]***

(a) 2 (b) 5
(c) 8 (d) 10

54. 10 men can finish a piece of work in 10 days, whereas it takes 12 women to finish it in 10 days. If 15 men and 6 women undertake the work, how many days will they take to complete it?

(a) 4.5 days (b) 4 days
(c) 5 days (d) 6 days

55. *A* can do a piece of work in 20 days and *B* in 30 days. They work together for 7 days and then both leave the work. Then *C* alone finishes the remaining work in 10 days. In how many days will *C* finish the full work? ***[SSC (CGL-T-II)-13]***

(a) 25 (b) 30
(c) 24 (d) 20

56. Working together, Asha and Sudha can complete an assigned task in 20 days. However, if Asha worked alone and complete half the work and then Sudha takes over the task and completes the second half of the task, the task will be completed in 45 days. How long will Asha take to complete the task if she worked alone? Assume that Sudha is more efficient that Asha.

(a) 25 days
(b) 65 days
(c) 60 days
(d) 30 days

57. *A* can complete a piece of work in 4 days. *B* takes double the time taken by *A*, *C* takes double that of *B* and *D* takes double that of *C* to complete the same task. They are paired in groups of two each. One pair takes two-third the time needed by the second pair to complete the work. Which is the first pair? ***[CAT-01]***

(a) *A*, *B* (b) *A*, *C*
(c) *B* , *C* (d) *A*, *D*

58. *A* is 1.5 times efficient than *B* therefore takes 8 days less than *B* to complete a work. If *A* and *B* work on alternate days and *A* works on first day, then in how many days the work will be completed? ***[SSC (CGL-T-I)-17]***

(a) 17 (b) 19
(c) 20 (d) 21

59. If 5 men can do a piece of work in 10 days and 12 women can do the same work in 15 days, the number of days required to complete the work by 5 men and 6 women is : ***[CDS-17]***

(a) $7\frac{1}{2}$ days (b) 8 days
(c) $9\frac{1}{2}$ days (d) 12 days

60. When they work alone, *B* needs 25% more time to finish a job than *A* does. They two finish the job in 13 days in the following manner: *A* works alone till half the job is done, then *A* and *B* work together for four days, and finally *B* works alone to complete the remaining 5% of the job. In how many days can *B* alone finish the entire job? ***[CAT-18]***

(a) 20 (b) 16
(c) 22 (d) 18

61. Ramesh and Ganesh can together complete a work in 16 days. After seven days of working together, Ramesh got sick and his efficiency fell by 30%. As a result, they completed the work in 17 days instead of 16 days. If Ganesh had worked alone after Ramesh got sick, in how many days would he have completed the remaining work? ***[CAT-18]***

(a) 13.5 (b) 11
(c) 12 (d) 14.5

62. Humans and robots can both perform a job but at different efficiencies. Fifteen humans and five robots working together take thirty days to finish the job, whereas five humans and fifteen robots working together take sixty days to finish it. How many days will fifteen humans working together (without any robot) take to finish it? ***[CAT-18]***

(a) 36 (b) 32
(c) 45 (d) 40

63. To complete a certain task, *X* is 40% more efficient than *Y* and *Z* is 40% less efficient than *Y*. Working together, they can complete the task in 21 days. *Y* and *Z* together worked for 35 days. The remaining work will be completed by *A* alone in ***[SSC (CGL-T-I)-19]***

(a) 8 days (b) 4 days
(c) 6 days (d) 5 days

64. There are two groups of work-force. One has 4 men which can complete the task 8 days before the time taken by another group comprising 3 women to complete a task. The ratio of their efficiencies is given below.

(A) 7 : 5 (B) 8 : 3
(C) 6 : 5 (D) 5 : 6

Which of the following ratio can give the value of time in positive number? ***[IBPS (PO-Mains)-18]***

(a) Only A (b) A and B only
(c) A, B and C only (d) Only D

65. *A* and *B*, working together, can complete a work in *d* days. Working alone, *A* takes (8 + *d*) days and *B* takes (18 + *d*) days to complete the same work. *A* works for 4 days. The remaining work will be completed by *B* alone, in ***[SSC (CGL-T-I)-19]***

(a) 24 days (b) 16 days
(c) 18 days (d) 20 days

66. Three men and eight machines can finish a job in half the time taken by three machines and eight men to finish the same job. If two machines can finish the job in 13 days, then how many men can finish the job in 13 days? ***[CAT-19]***

(a) 17 (b) 19
(c) 13 (d) 21

67. Anil alone can do a job in 20 days while Sunil alone can do it in 40 days. Anil starts the job, and after 3 days, Sunil joins him. Again after a few more days, Bimal joins them and they together finish the job. If Bimal has done 10% of the job, then in how many days was the job done? ***[CAT-19]***

(a) 14 (b) 13
(c) 15 (d) 12

Answer Key

1. (b)	**2.** (a)	**3.** (b)	**4.** (b)	**5.** (c)	**6.** (b)	**7.** (c)	**8.** (c)	**9.** (b)	**10.** (d)
11. (d)	**12.** (b)	**13.** (c)	**14.** (d)	**15.** (b)	**16.** (a)	**17.** (b)	**18.** (d)	**19.** (b)	**20.** (d)
21. (c)	**22.** (c)	**23.** (b)	**24.** (d)	**25.** (d)	**26.** (a)	**27.** (b)	**28.** (b)	**29.** (a)	**30.** (c)
31. (c)	**32.** (b)	**33.** (c)	**34.** (b)	**35.** (b)	**36.** (c)	**37.** (c)	**38.** (a)	**39.** (d)	**40.** (b)
41. (b)	**42.** (a)	**43.** (c)	**44.** (a)	**45.** (b)	**46.** (d)	**47.** (b)	**48.** (b)	**49.** (a)	**50.** (b)
51. (b)	**52.** (a)	**53.** (a)	**54.** (c)	**55.** (c)	**56.** (c)	**57.** (d)	**58.** (b)	**59.** (a)	**60.** (a)
61. (a)	**62.** (b)	**63.** (d)	**64.** (c)	**65.** (a)	**66.** (c)	**67.** (b)			

Hints and SOLUTIONS

1. $\left(\begin{array}{l} A \Rightarrow 12.5\% \text{ of work per day} \\ B \Rightarrow 8.33\% \text{ of work per day} \\ C \Rightarrow 6.66\% \text{ of work per day} \end{array}\right)$ hence,

41.66% in 2 days by $(A + B)$.

Then, remaining work i.e. 58.33% will be done by $(B + C)$ and they will take $\dfrac{58.33\%}{(8.33 + 6.66)\%}$ i.e. 3.89 days

Thus, total time taken = 2 + 3.89 = 5.89 days.

2. Given, (Ganu + Kanu) ⇒ 72% and (Kanu + Manu) ⇒ 63%

On addition, (Ganu + Kanu + Manu) + Kanu ⇒ 135%

Clearly, (Ganu+ Kanu + Manu) ⇒ 100%

Hence, contribution of Kanu in the work done = 35%

Then, contribution of Ganu = (72 – 35)% = 37%

and contribution of Manu = (63 – 35)% = 28%.

Hence, option (a) is correct.

3. Clearly, $(A + B)$ finished 63/100 (i.e. 63%) of the job in (10 – 3) days. Hence, 9% of the job per day by $(A + B)$.

Further, efficiencies of A and $B = 4:5$ ($\because$ work done by A in 5 days = work done by B in 4 days)

It implies that A does 4% of the job daily and B does 5% daily.

In 3 days, work done by

$(A + B) = 3 \times 9\% = 27\%$

Hence, work done by C alone in 3 days

= (37 – 27)% = 10%

It means, C does 3.33% of the job per day.

Obviously, B is the fastest worker. Then, time taken by B to complete the entire work = (100% ÷ 5%) = 20 days.

4. Clearly, $\left(\begin{array}{l} A \Rightarrow 4.76\% \text{ of work per day} \\ B \Rightarrow 4.16\% \text{ of work per day} \end{array}\right)$

Work done by B in 9 days

$= 9 \times 4.16\% = 37.5\%$

It implies that (100 – 37.5)% i.e. 62.5% of the work has been done by them together in beginning.

Now, time taken by them together to finish 62.5% of the work

= [62.5% ÷ (4.76 + 4.16)%]

= 7 days.

Hence, A left the work after 7 days.

5. $\left(\begin{array}{l} \text{1st son} \Rightarrow 16.66\% \text{ of work per day} \\ \text{2nd son} \Rightarrow 8.33\% \text{ of work per day} \end{array}\right)$

hence, 25% by both sons per day.

Obviously, the father will do 25% of the work per day.

Then, time taken by the father to do the job = (100% ÷ 25%) = 4 days.

6. (*b*) $\left(\begin{array}{l} A \Rightarrow 5.55\% \text{ of work per day} \\ B \Rightarrow 5\% \text{ of work per day} \\ C \Rightarrow 3.33\% \text{ of work per day} \end{array}\right)$

hence, 16.66% in 2 days by $(B + C)$.

$\therefore$ Time taken by A to complete the remaining work = (83.33% ÷ 5.55%) days = 15 days.

7. $\left(\begin{array}{l} A \Rightarrow 12 \text{ days} \Rightarrow 8.33\% \text{ of work per day} \\ C \Rightarrow 16.66\% \text{ of work per day} \\ \quad (\because C \text{ is twice efficient as } A) \\ B \Rightarrow 5.55\% \text{ of work per day} \\ \quad (\because C \text{ is thrice efficient as } B) \end{array}\right)$

Hence,

$\left\{\begin{array}{l} (A + B) \Rightarrow 13.88\% \text{ of work per day} \\ (B + C) \Rightarrow 22.21\% \text{ of work per day} \\ (C + A) \Rightarrow 25\% \text{ of work per day} \end{array}\right\}$

In first 3 days work done by $(A + B)$, $(B + C)$ and $(C + A)$ = 61.09%

On 4th day, $(A + B)$ will work.

$\therefore$ In 4 days, work done

= (61.09 + 13.88)%

= 74.97%

On 5th day, $(B + C)$ will work.

$\therefore$ In 5 days, work done

= (74.97 + 22.21)% = 97.18%

On 6th day, $(C + A)$ will work. Then, time required by $(C + A)$ to do the remaining work i.e. 2.82%

= (2.82% ÷ 25%) = 0.11

Hence, total time required to finish the work = 5 + 0.11 = 5.11 days or $5\frac{1}{9}$ days

8. $\left\{\begin{array}{l} (A + B) \Rightarrow 10\% \text{ of work per day} \\ (B + C) \Rightarrow 5.55\% \text{ of work per day} \end{array}\right\}$

Given, A's 5 day's work + B's 10 day's work + C's 15 day's work = 100%

⇒ $(A + B)$'s 5 day's work + $(B + C)$'s 5 day's work + C's 10 day's work = 100%

⇒ (5 × 10%) + (5 × 5.55%) + C's 10 day's work = 100%

⇒ C's 10 day's work = 22.25%

$\therefore$ Time taken by C to do 100 % of the work = 45 days

9. Ratio of efficiencies of A and $B = 2:1$

Now, time taken by A alone to finish the work $= 14 \times \dfrac{(2+1)}{2} = 21$ days

10. $\left.\begin{cases}(A+B) \Rightarrow 10\% \text{ of work per hr}\\ (B+C) \Rightarrow 6.66\% \text{ of work per hr}\\ (C+A) \Rightarrow 8.33\% \text{ of work per hr}\end{cases}\right\}$

Hence, 25% by $2(A + B + C)$ per hr i.e. 12.5% by $(A + B + C)$ per hr.

Now, work done by B independently

$= (12.5 - 8.33)\%$ i.e. 4.16% per hr.

Then, time taken by B independently to complete the work

$= (100\% \div 4.16\%) = 24$ hr.

11. Ratio of efficiency of a man, a woman and a child = 5 : 4 : 2

Then, ratio of their numbers to do the same work $= \frac{1}{5} : \frac{1}{4} : \frac{1}{2} = 4 : 5 : 10$

Hence, 4M = 5W = 10C

Now, both the team/group must be converted either in terms of men or in terms of women or in terms of children.

Suppose we convert both teams in terms of children.

From above, 4M = 5W = 10C

i.e. 1M = 2.5C and 1W = 2C

Then, 1st team = 2M + 3W + 4C

= 5C + 6C + 4C = 15C

and 2nd team = 6M + 4W + 7C

= 15C + 8C + 7C = 30C

Given, 15C can reap 10 hectare field in 10 days

Then, 30C can reap 16 hectare in ? Days

Clearly,

$$? = \frac{\text{Multiplication of figures appeared in the sentence completed in all aspects}}{\text{Multiplication of figures appeared in sentence in which any item is missing}}$$

(with reciprocal ratio of amount of work)

$$= \frac{15 \times 16 \times 10}{30 \times 10} = 8 \text{ days}$$

Alternatively

Capabilities of 1st team

$= 2 \cdot 5 + 3 \cdot 4 + 4 \cdot 2 = 30$ units

and Capabilities of 2nd team $= 6 \cdot 5 + 4 \cdot 4 + 7 \cdot 2 = 60$ units

Now, 30 units can reap 10 hectare in 10 days

Then, 60C units can reap 16 hectare in ? Days

Clearly, $? = \frac{30 \times 16 \times 10}{60 \times 10} = 8$ days

12. Work done by A in 5 days

$= 5 \times 2.5\% = 12.5\%$

Obviously, the remaining work i.e. 87.5% has been done by B alone in 21 days. Then, his 1 day's work = 4.16%.

Further,

$\left.\begin{cases}A \Rightarrow 2.5\% \text{ of work per day}\\ B \Rightarrow 4.16\% \text{ of work per day}\end{cases}\right\}$ hence,

6.66% by $(A + B)$ per day

Then, time taken by both together to do the work = (100% ÷ 6.66%)

= 15 days

13. $\left.\begin{cases}\text{Manoj} \Rightarrow 10\% \text{ of work per day}\\ \text{Ajita} \Rightarrow 8.33\% \text{ of work per day}\end{cases}\right\}$

hence, 18.33% by (Manoj + Ajita) per day.

Now, work done by Manoj in 6 days

$= 6 \times 10 = 60\%$

It is obvious that the remaining work i.e. 40% has been done by them together.

Then, time taken by them together

= (40% ÷ 18.33%) = 2.18 days

Hence, Ajita worked for only 2.18 days.

Alternatively

Time required in completion of whole work

$$= \frac{100\% + \text{Work done (in \%) in } t \text{ days by the person who joined later}}{\text{work done (in \%) by them together per day}}$$

$$= \frac{100\% + (6 \times 8.33\%)}{18.33\%} = 8.18 \text{ days}$$

Hence, the required time

$= 8.18 - 6 = 2.18 = 2\frac{2}{11}$ days

14. Work done in 3 h from the beginning = 55% (The first typist works for 3 h and the second for 2 h).

Let the first typist and the second typist respectively complete x% and y% of the work per hour.

Then, $3x + 2y = 55$ ...(i)

Further, $\frac{50}{x} - \frac{50}{y} = 1$h {∵ Difference between time taken by them to complete 50% – 50% is 1 h}

$\Rightarrow$ $50y - 50x = xy$ ...(ii)

Solving Eqs. (i) and (ii), we get

$x = 10\%$ and $y = 12.5\%$

Hence, the first typist can complete the whole job in 10 h and the second one can do the whole work in 8 h

∴ Option (d) is correct.

Alternatively

The best way is to proceed through options.

(a) per hour work done by the first and the second typists respectively

= 8.33% and 12.5%.

But, $3(8.33) + 2(12.5) \neq 55$

(b) $3(20) + 2(25) \neq 55$

(c) $3(12.5) + 2(17.85) \neq 55$

(d) $3(10) + 2(12.5) = 55$

Hence, option (d) is correct.

15. Ratio of efficiencies of A and B = 3 : 1

Then, ratio of time taken by A and B = 1 : 3 {∵ time $\propto \frac{1}{\text{efficiency}}$, when amount of work is constant}

As given, (3 – 1) units = 40 days

∴ 1 unit = 20 days

Hence, time taken by A alone to complete the work = 20 days

Further, time taken by A alone

$= D \times \frac{(3+1)}{3}$

(Where D is time taken by them together to finish the work)

$\Rightarrow$ $20 = D \times \frac{4}{3} \Rightarrow D = 15$ days

16. 3M + 4B ⇒ 12 days ...(i)

4M + 3B ⇒ 10 days ...(ii)

2M + 3B ⇒ ? days ...(iii)

Positive difference between (12 × 3M) and (10 × 4M) = Positive difference between (12 × 4B) and (10 × 3B)

$\Rightarrow$ 4M = 18B

⇒ 2M = 9B, this relation should be maintained when we take any two equations out of above three. But, it is the fact that Eq. (iii) is incomplete, which can be completed by putting the values of options in place of?

(a) $17\frac{1}{2}$ ⇒ Given, 3M + 4B

⇒ 12 days ...(i)

and 2M + 3B ⇒ $17\frac{1}{2}$ days ...(iii)

Now, positive difference between (12 × 3M) and $\left(17\frac{1}{2} \times 2M\right)$ = Positive difference between (12 × 4B) and $\left(17\frac{1}{2} \times 3B\right)$

$\Rightarrow$ M = $4\frac{1}{2}$B ⇒ 2M = 9B, which is the same as relation established from Eqs. (i) and (ii). Hence, option (a) is correct.

17. Work done by $(A + B) = 20\%$ per day

Let the contribution of A per day $= x\%$

$\therefore$ Contribution of B per day

$= (20 - x)\%$

Now, work done by $(A + B)$ per day at altered speed $= \left\{2x + \frac{20-x}{2}\right\}\%$

As per question,

$\left\{2x + \frac{20-x}{2}\right\}\% = 25\%$

$\{\because (A + B) \Rightarrow 4$ days $\Rightarrow 25\%$ of work per day$\}$

$\Rightarrow \quad 4x + 20 - x = 50$

$\Rightarrow \quad x = 10\%$

Then, time taken by A alone to do the work $= (100\% \div 10\%) = 10$ days

Alternatively,

The best way is to proceed through options.

(a) 20 days : $A \Rightarrow 5\%$ daily and $(A + B) \Rightarrow 20\%$ daily, then $B \Rightarrow 15\%$ daily.

If $A \Rightarrow 10\%$ and $B \Rightarrow 7.5\%$, the time taken i.e $(100\% \div 17.5\%) \neq 4$

(b) 10 days : $A \Rightarrow 10\%$ daily and $(A + B) \Rightarrow 20\%$ daily, then $B \Rightarrow 10\%$ daily. If $A \Rightarrow 20\%$ and $B \Rightarrow 5\%$, the time taken i.e. $(100\% \div 25\%) = 4$

Hence, option (b) is correct.

18. Ratio of efficiencies of Ram, Rahim and Rachel = 100% : 50% : 25% = 4 : 2 : 1

Time taken by Ram to complete 40% of work $= D \times \frac{(4+2+1)}{4}$

(where, D is the time taken by them together to finish 40% of work)

$\Rightarrow \quad 10 = D \times \frac{7}{4}$

$\Rightarrow \quad D = 5\frac{5}{7}$ days.

Hence, option (d) is correct.

19. 6M + 8B ⇒ 10 days

<u>15M + 20B ⇒ ? days</u>

Clearly, the latter group is 2.5 times the former group in size. Therefore, the required time will be (10 ÷ 2.5) i.e. 4 days

20. Required time

$= \sqrt{\text{Product of time exceeding time taken by both together}}$

$= \sqrt{8 \times 4\frac{1}{2}} = 6$ days.

21. Obviously, Uma is twice as efficient as Madhu.

Now, ratio of efficiencies of Madhu, Uma and Ramesh = 1 : 2 : (1 + 2)

Now, time taken by Madhu to finish the work $= 6 \times \frac{1+2+3}{1} = 36$ days

22. Ratio of efficiencies of A, B and C

$= 1 : 2 : 4$

Time taken by C to complete the work $= D \times \frac{(1+2+4)}{4}$

(where, D is time taken by them together to finish the work)

$\Rightarrow \quad 7 = D \times \frac{7}{4} \Rightarrow D = 4$ days

23. Ratio of efficiencies of X and Y

$= 100\% : 125\% = 4 : 5$

Then, ratio of time taken by X and $Y = 5 : 4$

As given, 5 units = 25 days

$\therefore$ 4 units = 20 days

24. Clearly, ratio of efficiencies of A and $B = 3 : 1$

Then, time taken by A to do the work

$= 12 \times \frac{(3+1)}{3} = 16$ days

25. $\left\{\begin{array}{l} B \Rightarrow 16.66\% \text{ of work per day} \\ (B + C) \Rightarrow 25\% \text{ of work per day} \\ (A + B + C) \Rightarrow 37.5\% \text{ of work per day} \end{array}\right\}$

Hence, $C \Rightarrow (25 - 16.66)\%$ i.e. 8.33% per day. Then, $(A + B)$

$\Rightarrow (37.5 - 8.33)\%$ i.e. is 29.16% per day

$\therefore$ Time taken by $(A + B)$ to do the work $= (100\% \div 29.16\%)$

$= 3.42$ days or $3\frac{3}{7}$ days

26. $\left\{\begin{array}{l} A \Rightarrow 2.77\% \text{ of work per day} \\ B \Rightarrow 1.85\% \text{ of work per day} \\ C \Rightarrow 1.38\% \text{ of work per day} \end{array}\right\}$

Now, required time

$= \frac{(100\% + 8 \times 2.77\% + 12 \times 1.85\%)}{(2.77 + 1.85 + 1.38)\%}$

= 24 days

27. 2M or 3W ⇒ 4 days

↓ × 0.5   ↓ × 0.33

1M   1W ⇒ ? days

Then, required time

$= \frac{4}{(0.5 + 0.33)} = \frac{4}{0.83}$

$= \frac{4 \times 6}{0.83 \times 6} = \frac{24}{5}$ days

28. $\left\{\begin{array}{l} A \Rightarrow 25\% \text{ of work per day} \\ B \Rightarrow 8.33\% \text{ of work per day} \end{array}\right\}$ hence, 33.33% by $(A + B)$ per day.

Then, time taken by them together

$= (100\% \div 33.33\%) = 3$ days

29. Ratio of efficiencies of X and $Y = 100\% : 160\% = 5 : 8$

Then, ratio of time taken by X and $Y = 8 : 5$

As given, 8 units = 16 days, then 5 units = 10 days

Hence, time taken by Y to complete the work = 10 days

30. Ratio of time taken by A and $C = 3 : 5$

$\{\because$ ratio of efficiency $= 5 : 3\}$

As given, (5 − 3) units = 6 days

Then, 3 units = 9 days and 5 units

= 15 days.

Hence, time taken by A and C = 9 and 15 days (respectively)

Further, ratio of time taken by B and $C = 2 : 3$

Clearly, 3 units = 15 days, then 2 units = 10 days

Thus, time taken by B = 10 days

Now, $\left\{\begin{array}{l} A \Rightarrow 9 \text{ days} \Rightarrow 11.11\% \text{ of work per day} \\ B \Rightarrow 10 \text{ days} \Rightarrow 10\% \text{ of work per day} \\ C \Rightarrow 15 \text{ days} \Rightarrow 6.66\% \text{ of work per day} \end{array}\right\}$

Then, work done by $(B + C)$ in 2 days

$= 2(10 + 6.66)\% = 33.33\%$

The time taken by A to finish remaining work

$= (66.66\% \div 11.11\%) = 6$ days

31. Given,

$\left\{\begin{array}{l} A \Rightarrow 50\% \text{ of work per day} \\ B \Rightarrow 33.33\% \text{ of work per day} \\ A + B + C \Rightarrow 100\% \text{ of work per day} \end{array}\right\}$

Hence, $C \Rightarrow (100 - 50 - 33.33)\%$ i.e. 16.66% of the work per day

$\therefore$ Time taken by C to complete the whole work

$= (100\% \div 16.66\%) = 6$ days

32.

1 man or 2 women or 3 boys ⇒ 55 days

↓ × 1   ↓ × 1/2   ↓ × 1/3

1 man + 1 women + 1 boy ⇒ ?

Then, required time

$$= \frac{55}{\left(1 + \frac{1}{2} + \frac{1}{3}\right)} = 30 \text{ days}$$

33. Clearly, work done by $(A + B)$ per day = 3.33%

Then, work done by $(A + B)$ in 11 days = 11 × 3.33% = 36.66%

As given, the remaining work i.e. 63.33% is done by A in 28 days.

Then, time taken by A to finish 100% of work = 44.21 days or $44\frac{4}{19}$

34. Clearly, $(A + B)$ can complete the whole work in 18 days.

Now,

$$\left[\begin{array}{l} A \Rightarrow 30 \text{ days} \Rightarrow 3.33\% \text{ of work per day} \\ (A+B) \Rightarrow 18 \text{ days} \Rightarrow 5.55\% \text{ of work per day} \\ (B+C) \Rightarrow 20 \text{ days} \Rightarrow 5\% \text{ of work per day} \end{array}\right]$$

Hence, $B \Rightarrow (5.55 - 3.33)\%$ i.e. 2.22% of work per day

$\therefore$ $C \Rightarrow (5 - 2.22)\%$ i.e. 2.78% of the work per day.

Then, time taken by C to complete the remaining 33.33% of the work = (33.33% ÷ 2.78%) = 12 days

35. Clearly,

$$\left\{\begin{array}{l} A \Rightarrow 6.66\% \text{ of work per day} \\ B \Rightarrow 13.33\% \text{ of work per day} \\ (\because B \text{ is twice efficient as } A) \end{array}\right\}$$

Work done by A in 11 days

= 11 × 6.66% = 73.33%.

It is obvious that the remaining work i.e. 26.66% has been done by B alone.

Then, time taken by B to finish 26.66% of the work

= (26.66% ÷ 13.33%) = 2 days

Hence, they together worked for 2 days only.

36.
$$\left\{\begin{array}{l} X \Rightarrow 8.33\% \text{ of work per day} \\ (X+Y) \Rightarrow 15\% \text{ of work per day} \end{array}\right\}$$

Hence, 6.66% by Y alone per day

$\therefore$ Time taken by Y alone to complete the job = (100% ÷ 6.66%) = 15 days.

37. Time taken by X to do the work

= 10 × 4 = 40 days

and time taken by Y to do 100% of the work = 40 × 2.5 = 100 days

and time taken by Z to do the work

= 13 × 3 = 39 days

Thus, Z will complete the work first.

38. $\because$ 8 men ⇒ 20 days, then 5 men ⇒ 32 days.

Now, 5M or 8W ⇒ 32 days

↓ ×1 ↓ ×1

5M + 8W ⇒ ?

Then, required time $= \frac{32}{(1+1)} = 16 \text{ days}$

39. Clearly,

$$\left[\begin{array}{l} (A+B) \Rightarrow 8.33\% \text{ of work per day} \\ \therefore A \Rightarrow (8.33 - b)\% \text{ per day} \\ (B+C) \Rightarrow 6.66\% \text{ of work per day} \\ \therefore C \Rightarrow (6.66 - b)\% \text{ per day} \end{array}\right]$$

(where, work done by B per day = b%)

As per question, the efficiency of A is twice that of C.

Hence, $(8.33 - b)\% = 2(6.66 - b)\%$

$\Rightarrow b = 5$

Then, time taken by B alone to complete the same work

= (100% ÷ 5%) = 20 days

40. 3M + 4B ⇒ 8 days

<u>6M + 8W ⇒ ? days</u>

Clearly, the latter group is two times the former group in size. Therefore, the required time will be half the given time i.e. 4 days.

41.
$$\left\{\begin{array}{l} A \Rightarrow 5\% \text{ of work per day} \\ B \Rightarrow 6.66\% \text{ of work per day} \end{array}\right\}$$

It is clear that A worked for 10 days, B for 6 days and C for 4 days to finish the work.

Hence, (10 × 5%) + (6 × 6.66%) + work done by C in 4 days

= 100%

⇒ Work done by C in 4 days

= 10% of the work

Then, time taken by C to do 100% of the work = 40 days

42. Given,
$$\left\{\begin{array}{l} A \Rightarrow 2\% \text{ of work per day} \\ B \Rightarrow 1.33\% \text{ of work per day} \\ C \Rightarrow 5\% \text{ of work per day} \end{array}\right\}$$

Hence, 8.33% by all together per day.

Now, work done by all together in 4 days = 8.33% × 4 = 33.33%

Then, time taken by $(A + B)$ to finish the remaining work i.e. 66.66%

= 66.66% ÷ (2 + 1.33)% = 20 days.

43. Given,

$$\left\{\begin{array}{l} A+B \Rightarrow 8.33\% \text{ of work per day} \\ B+C \Rightarrow 12.5\% \text{ of work per day} \\ C+A \Rightarrow 6.25\% \text{ of work per day} \end{array}\right\}$$

Hence, $2(A + B + C)$ do 27.08% of the work per day.

Then, $(A + B + C)$ will do 13.54% of the work per day.

$\therefore$ Work done by A per day

= 13.54 – 12.5 = 1.04%

Then, work done by A for 3 days

= 1.04 × 3 = 3.12%

The remaining work i.e. 96.88% have been done by B and C together since beginning of the work and they took (96.88% ÷ 12.5%) days i.e. $7\frac{3}{4}$ days

$\therefore$ Required answer

$$= 7\frac{3}{4} - 3 = 4\frac{3}{4} \text{ days.}$$

44.
$$\left\{\begin{array}{l} A+B \Rightarrow 12.5\% \text{ of work per day} \\ B+C \Rightarrow 8.33\% \text{ of work per day} \\ C+A \Rightarrow ?\% \text{ of work per day} \end{array}\right\}$$

Also, $(A + B + C) \Rightarrow$ 16.66% of work per day

Now, work done by C = (16.66 – 12.5)% i.e. 4.16% per day

and work done by A = (16.66 – 8.33)% i.e. 8.33% per day

Then, work done by $(A + C)$ per day

= (4.16 + 8.33)% = 12.5%

$\therefore$ Time taken by $(A + C)$ to complete the work = (100% ÷ 12.5%) = 8 days.

45. Given, 1 man = 2 children,

$\therefore$ 8 children = 4 men

Now, 4 men + 12 men ⇒ 9 days

Then, 12 men ⇒ ?

$\therefore$ Required time $= \frac{9 \times 16}{12} = 12$ days.

Alternatively

8 C + 12 M ⇒ 9 ...(i)

0 C + 12 M ⇒ ? ...(ii)

Given, 1 man = 2 children. This relation must be maintained from the above equations. But, it is the fact that Eq. (ii) is incomplete, which can be completed by putting the values of options in place of ?.

(a) 15 ⇒ Given,

8C + 12M ⇒ 9 ...(i)

<u>0C + 12M ⇒ ?(15) ...(ii)</u>

Now, positive difference between (9 × 8C) and (15 × 0C) = Positive difference between (9 × 12M) and (15 × 12M)

⇒ 72C = 72M ⇒ 1M = 1C, which is not same as 1M = 2C.

(b) 12 ⇒ Given, 8C + 12M ⇒ 9 ...(i)

<u>0C + 12M ⇒ ?(12) ...(ii)</u>

Now, positive difference between
(9× 8C) and (12× 0C)
= positive difference between
(9×12M) and (12×12M)
⇒ 72C = 36M ⇒ 1M = 2C, which is same as given in question.
Hence, option (b) is correct.

46. Ratio of efficiencies of Sunil and Dinesh = 3 : 2
{∵ they take time in ratio 2 : 3}
Further, ratio of efficiencies of Ramesh and Sunil = 15 : 1 = 4.5 : 3
Hence, ratio of efficiencies of Sunil, Ramesh and Dinesh = 3 : 4.5 : 2
= 6 : 9 : 4
Now, time taken by Sunil to complete the work $= D \times \frac{(6+9+4)}{6}$
(where, D is time taken by them together to finish the work)
$\Rightarrow \quad 4 = D \times \frac{19}{6} \Rightarrow D = 1\frac{5}{19}$ days.

47. $\left\{\begin{matrix} A \Rightarrow 12.5\% \text{ of work per day} \\ B \Rightarrow 8.33\% \text{ of work per day} \end{matrix}\right\}$ hence,
37.5% by A in 3 days.
Now, time taken by $(A + B)$ to finish the remaining 62.5% of work
= {62.5% ÷ (12.5 + 8.33)%} days
= 3 days

48. Ratio of efficiencies of Asha and Usha = 125% : 100% = 5 : 4
Then, ratio of time taken by Asha and Usha = 4 : 5
As given, 5 units = 25 days, then 4 units = 20 days
Now, $\left\{\begin{matrix} \text{Usha} \Rightarrow 25 \text{ days} \Rightarrow 4\% \text{ of work per day} \\ \text{Asha} \Rightarrow 20 \text{ days} \Rightarrow 5\% \text{ of work per day} \end{matrix}\right\}$
Hence, work done by $(A + B)$ in last 5 days = 5(4 + 5)% = 45%
Clearly, the remaining work i.e., 55% has been done by Asha alone in (55% ÷ 5%) i.e. 11 days.

49. $\left\{\begin{matrix} A \Rightarrow 3.33\% \text{ of work per day} \\ B \Rightarrow 5\% \text{ of work per day} \\ C \Rightarrow 10\% \text{ of work per day} \end{matrix}\right\}$, then
$\left\{\begin{matrix} (A+B) \Rightarrow 8.33\% \text{ of work/day} \\ (A+C) \Rightarrow 13.33\% \text{ of work/day} \end{matrix}\right\}$
In first 2 days work done by $(A + B)$ and $(A + C)$
= (8.33 + 13.33)% = 21.66%
Then, in 8 days, work done by them = 86.66%
On 9th day, $(A + B)$ will work.
∴ In 9 days, work done
= (86.66 + 8.33)% = 95%
On 10th day, $(A + C)$ will work.
Then, time required by $(A + C)$ to do the remaining work i.e. 5%
= (5% ÷ 13.33%) = 0.37
Hence, total time required to finish the work $= 9 + 0.37 = 9\frac{3}{8}$ days

50. $\left\{\begin{matrix} A \Rightarrow 12.5\% \text{ of work per day} \\ B \Rightarrow 6.25\% \text{ of work per day} \\ C \Rightarrow 4.16\% \text{ of work per day} \end{matrix}\right\}$
Now, required time
$= \left(\frac{100\% - 2 \times 4.16\% + 6.25\%}{(12.5 + 6.25)\%}\right)$
= 5.22 days
Hence, option (b) seems to be correct.

51. (*b*) Clearly,
$\left\{\begin{matrix} A \Rightarrow 5.55\% \text{ of work per day} \\ B \Rightarrow 11.11\% \text{ of work per day} \\ C \Rightarrow 16.66\% \text{ of work per day} \end{matrix}\right\}$
Required time
$= \left\{\frac{100\% + (2 \times 16.66\%)}{(5.55 + 11.11 + 16.66\%)}\right\} = 4$ days

52. (*a*) Ratio of efficiencies of Anu, Manu and Sonu = 100 : 125 : (160% of 125)
= 4 : 5 : 8
Further, time taken by Anu alone
$= D \times \frac{4+5+8}{4}$
(Where, D is time taken by them together to finish the work)
$\Rightarrow \quad 10 = D \times \frac{17}{4} \Rightarrow D = 2\frac{6}{17}$ days.

53. Clearly, 12 × 10 men = 20 × 12 women
⇒ 1 man = 2 women
Then, 8 men + 4 women = 20 women
Work done by 20 women in 9 days
$= 9/12 = \frac{3}{4}$
∴ Remaining work $= 1 - \frac{3}{4} = \frac{1}{4}$
∵ 20 women do 1 unit of work in 12 days
Then, (20 + 10) women will do $\frac{1}{4}$ of work in $\frac{20 \times 12 \times 1}{30 \times 4} = 2$ days

54. 10M or 12W ⇒ 10 days
↓ × 1.5 ↓ × 0.5
15M 6W ⇒ ? days
Then, required time
$= \frac{10}{1.5 + 0.5} = 5$ days

55. (*c*) $\left\{\begin{matrix} A \Rightarrow 5\% \text{ of work per day} \\ B \Rightarrow 3.33\% \text{ of work per day} \end{matrix}\right\}$
hence, 8.33% by $(A + B)$ per day
Then, work done by $(A + B)$ in 7 days = 8.33% × 7 = 58.33%
Further, C does the remaining work i.e. (100 − 58.33)% in 10 days.
∴ C's per day work = 4.16%
Then, time taken by C to finish the full work = (100% ÷ 4.16%) = 24 days

56. (*c*) Clearly, (Asha + Sudha) ⇒ 5% of work per day
Let Asha does x% per day. Then Sudha will do $(5 - x)$% per day.
As per question,
time taken by Asha to do 50% of work + time taken by Sudha to do 50% of work = 45 days
$\Rightarrow \quad \frac{50\%}{x\%} + \frac{50\%}{(5-x)\%} = 45$
$\Rightarrow \quad (5x - x^2)9 = 50$
$\Rightarrow \quad 9x^2 - 45x + 50 = 0$
$\Rightarrow \quad x = \frac{-b \pm \sqrt{b^2 - 4ac}}{2a}$
$= \frac{-(-45) \pm \sqrt{(-45)^2 - 4 \cdot 9 \cdot 50}}{2 \cdot 9}$
$= \frac{45 \pm 15}{2 \cdot 9} = 3.33$ and 1.66
If we take x = 3.33%, Sudha's 1's day work = (5 − 3.33)% = 1.66%, which is contrary to the fact that Sudha is more efficient than Asha. Thus, we should take x = 1.66%. Then, time taken by Asha to complete the task alone = (100% ÷ 1.66%) = 60 days.

Alternatively

(a) 25 ⇒ Asha does 4% per day, then Sudha will do (5 − 4)% i.e. 1% per day. Here, Sudha appears to be less efficient than Asha. Hence, this option is incorrect.

(b) 65 ⇒ Asha does 1.54% per day, then Sudha will do 3.46% per day. Now, time taken by Asha to complete 50% of work + time taken by Sudha to complete 50% of work = 65/2 + (50% ÷ 3.46%) ≠ 45.
Hence, this option is incorrect.

(c) 60 ⇒ Asha does 1.66% per day, then Sudha will do 3.33% per day. Now, time taken by Asha to complete 50% of work + time taken by Sudha to complete 50% of work = 60/2 + (50% ÷ 3.33%) = 45.

Hence, this option is correct.

57. Ratio of efficiencies of A and $B = 2:1$

Ratio of efficiencies of B and $C = 2:1$

Ratio of efficiencies of D and $C = 2:1$

∴ Ratio of efficiencies of A, B, C and $D = 8:4:2:1$

{Detailed under chapter 'Ratio, Proportion and Variation'}

We know, efficiency is directly proportion to the work done if time is constant. Hence, the efficiency of one pair, as per question, must be 1.2 times that of the second pair as first pair takes 2/3 the time taken by the second.

Clearly, (8 + 1) is 1.5 times (4 + 2). Hence, efficiency of pair (A, D) is 1.5 times that of pair (B, C).

∴ Option (d) is correct.

58. Ratio of efficiencies of A and $B = 15:1$

∴ Ratio of time taken by A and B

$= 1:1.5$

As given, (1.5 − 1) units = 8 days

∴ 1 unit = 16 days and 1.5 units = 24 days

It means A alone completes the work in 16 days while B alone in 24 days.

Now,

$\left.\begin{array}{l} A \Rightarrow 6.25\% \text{ of work per day} \\ B \Rightarrow 4.17\% \text{ of work per day} \end{array}\right\}$ hence,

10.42% in 2 days by them.

Then, in 18 (i.e. 2 × 9) days

$\Rightarrow 10.42 \times 9 \approx 93.75\%$

Clearly, 6.25% of the work remains left after 18 days.

Since A starts the work, A will work on 19th day and finish all the remaining work.

Thus, total time taken

$= 18 + 1 = 19$ days

{**Note**: No need to calculate decimal value exactly. Observe the options.}

59. Time taken by 6 women to do the work = 30 days

$\left.\begin{array}{l} 5 \text{ men} \Rightarrow 10\% \text{ of work per day} \\ 6 \text{ women} \Rightarrow 3.33\% \text{ of work per day} \end{array}\right\}$

Hence, 13.33% by ($5M + 6W$) per day

∴ Required time

$= 100\% \div 13.33\% = 7\frac{1}{2}$ days

60. Ratio of time taken by A and B $= 100:125 = 4:5$

Then, ratio of efficiencies of A and B

$= 5:4$

[∵ Efficiency $\propto \frac{1}{\text{time}}$, when amount of work is constant]

Let B does b% of work per day.

Then, A will do $\frac{5b}{4}$% of work per day.

Thus, they together can do 2.25b% of work per day.

As given, A and B together worked for four days only and finished (100 − 50 − 5)% of the work.

$\Rightarrow 4 \times 2.25b\% = 45\% \Rightarrow b = 5\%$

Then, time taken by B alone to finish the entire job = (100% ÷ 5%) = 20 days

61. They together can do 6.25% of work per day. Then, work done by them together in first 7 days = 43.75%

The remaining work i.e. 56.25% is done by them together in next 10 days. It means, they do together only 5.625% of work per day and obviously it is due to reduction in efficiency of Ramesh. It may be construed that Ramesh can do (6.25 − 5.625)% of work per day with 30% of his efficiency. Then, with his full efficiency (i.e. 100%) Ramesh can do 2.08% of work per day.

∴ Work done by Ganesh per day

= (6.25 − 2.08)% = 4.17%

Thus, time taken by Ganesh to complete the remaining work, if he works alone

= 56.25% ÷ 4.17% = 13.5 days

{**Note** The whole process can be done mentally without pen and paper, if you have command over concept of percentage and fundamentals}

62. Given,

$\left\{\begin{array}{l} 15H + 5R \Rightarrow 30 \text{ days} \Rightarrow 3.33\% \text{ of work per day} \\ 5H + 15R \Rightarrow 60 \text{ days} \Rightarrow 1.67\% \text{ of work per day} \end{array}\right.$

Hence, $20H + 20R \Rightarrow 5\%$ of the work per day

∴ $5H + 5R \Rightarrow 1.25$ of the work per day

From above, $15H + 5R \Rightarrow 3.33\%$ of work per day

Hence, $10H \Rightarrow (3.33 - 1.25)$ i.e. 2.08% of the work per day

Then, $15H \Rightarrow 3.12\%$ of the work per day

∴ Time taken by $15H$ to finish the work = (100% ÷ 3.12%) = 32 days

63. Ratio of efficiency of X and

$(Y + Z) = 140:(100 + 60) = 7:8$

Time taken by X to complete the task

$= 21 \times \frac{(7+8)}{7} = 45$ days

Time taken by $(Y + Z)$ to complete the task $= 21 \times \frac{(7+8)}{8}$ days

Then, work done by $(Y + Z)$ in 35 days

$= \frac{35 \times 8}{21 \times 15} = \frac{8}{9}$

Now, time taken by X to complete the remaining work 1/9

$= 45 \times \frac{1}{9} = 5$ days

64. Since the first group takes lesser time in comparison to the second group, the efficiency of the former is more than that of the latter. Hence, 1st term of the given ratio must be greater than the second term. First three given ratios depict the same. Thus, option (c) is correct.

65. $d = \sqrt{8 \times 18} = 12$

Hence, A and B, working alone, can complete a work in (8 + 12) and (18 + 12) days respectively.

Clearly,

$\left\{\begin{array}{l} A \Rightarrow 5\% \text{ of work per day} \\ B \Rightarrow 3.33\% \text{ of work per day} \end{array}\right\}$

Work done by A in 4 days = 20%.

Then the time taken by B to complete the remaining 80% of work

= (80% ÷ 3.33%) = 24 days

66. Given that, (3 Men + 8 Machines) takes half the time taken by (3 Machines + 8 Men). It implies that efficiency of the former is twice that of the latter. Hence, (3 Men + 8 Machines) = 2 (3 Machines + 8 Men)

⇒ 13 Men = 2 Machines

Since 2 Machines take 13 days to finish the job, 13 Men will take the same time i.e. 13 days as efficiency of 13 Men is equal to that of 2 Machines.

67. $\left\{\begin{array}{l} \text{Anil} \Rightarrow 5\% \text{ of work per day} \\ \text{Sunil} \Rightarrow 2.5\% \text{ of work per day} \end{array}\right\}$

Hence, 7.5% by both per day.

Work done by Anil and Sunil together

= 100% − (3 × 5)% − 10% = 75%.

Then, time taken by Anil and Sunil together to finish 75% of work

= 75 ÷ 7.5 = 10 days.

∴ Required time

= 10 + 3 = 13 days

24

Work and Wages

Learning OBJECTIVES

After completing this chapter, aspirants will be able to :

- Understand the relations among wages, time, amount of work, work-force and efficiency;
- Apply the above relations in diverse cases;
- Realise the difference between traditional approach and new approach;
- Understand the concept of wages both at individual level as well group level;
- Realise that they can solve one question in many ways;
- Understand the philosophy of the division of this chapter in various parts;
- Analyse the trend of the questions being asked in various competitive examinations;
- Solve the questions in moments;
- Grip on concepts helping to solve the questions without pen and paper;
- Learn new concepts/fundamentals;
- Apply the pragmatic approaches in lieu of academic approaches; and
- Ponder on the alternative methods.

Introduction

Questions from this chapter are also asked in diverse competitive exams like CAT, MAT, SSC, CDS, Bank-PO, Railway, etc. It has importance in our daily life as well. There are so many alternatives to solve the questions; however, our stress will be upon the solution with the help of concept of ratio. No doubt, it is the most effective method.

Wages of workers depend upon their efficiencies or amount of work done by them. If they work together for equal number of days/time, we take their efficiencies for consideration otherwise amount of work is taken into consideration. It is the fact that

(i) Efficiency α Wages

or

Wages α Efficiencies {when, time is constant}

Inside

(ii) Amount of work α wages

or

Wages α Amount of Work

{when, time is constant or not}

(iii) size of work-force α Wages

or

Wages α size of work-force

{when time and efficiency at individual level are constant}

(iv) Time $\alpha \frac{1}{\text{wages}}$

or

Wages $\alpha \frac{1}{\text{time}}$ {when, amount of work is constant}

From the above, it is seen that wages are directly proportional to efficiency as well as work done and vice-versa. Also, wages are directly proportional to size of work-force under certain circumstances while wages are indirectly proportional to time. It will not be out to mention that if you command over the concepts envisaged in the chapter 'Time and Work', you will take only few seconds to solve the questions based on this topic as this chapter is technically an extension of the chapter 'Time and Work'.

Illustrations

Case I At Individual Level

Problem 1. Suman can do a work in 3 days. Sumati can do the same work in 2 days. Both of them finish the work together and get ₹ 150. What is the share of Suman?

Sol. Here, both work for the same time to finish the work. Hence, time is constant.

Now, ratio of time taken by Suman and Sumati = 3 : 2

Then, ratio of efficiencies of Suman and Sumati = 2 : 3

{∵ Efficiency $\alpha \frac{1}{\text{time}}$, when amount of work is constant}

Further, Wages α Efficiencies {when time is constant}

It is clear that both worked for same period.

Hence, the wages will be divided between Suman and Sumati in ratio 2 : 3.

Now, (2 + 3) units ⇒ ₹ 150

∴ 2 units ⇒ ₹ 60

Thus, share of Suman = ₹ 60

Alternatively

Work done by Suman per day = 33.33%

and work done by Sumati per day = 50%

Now, ratio of work done by Suman and Sumati in one day

= 33.33% : 50% = 2 : 3

Since wages α amount of work, the wages will be divided between Suman and Sumati in ratio 2 : 3.

Now, (2 + 3) units ⇒ ₹ 150

∴ 2 units ⇒ ₹ 60

Thus, share of Suman = ₹ 60

Problem 2. A alone can do a piece of work in 6 days and B alone in 8 days. A and B undertook to do it for ₹ 3200. With the help of C, they completed the work in 3 days. How much is to be paid to C?

Sol. Work done by A in 3 days = $\frac{3}{6} = \frac{1}{2}$

and work done by B in 3 days = $\frac{3}{8}$

∴ Work done by C in 3 days = $1 - \frac{1}{2} - \frac{3}{8} = \frac{1}{8}$

Since Wages α Amount of work, the wages will be divided among A, B and C in ratio $\frac{1}{2} : \frac{3}{8} : \frac{1}{8}$ i.e. 4 : 3 : 1

Now, (4 + 3 + 1) units ⇒ ₹ 3200

∴ 1 unit ⇒ ₹ 400

Thus, share of C = ₹ 400

Alternatively I

Work done by A in 3 days = $\frac{3}{6} = \frac{1}{2}$

and work done by B in 3 days = $\frac{3}{8}$

∴ Work done by C in 3 days = $1 - \frac{1}{2} - \frac{3}{8} = \frac{1}{8}$

Obviously, their wages will be as per their contributions towards the work.

Then, C's share = ₹ 3200 × $\frac{1}{8}$ = ₹ 400

Alternatively II

Work done by A per day = 16.66%

and work done by B per day = 12.5%

and work done by (A + B + C) per day = 33.33%

∴ Work done by C per day

= {33.33% – 16.66% –12.5%} = 4.16%

Now, ratio of work done by A, B and C

= 16.66% : 12.5% : 4.16% = 4 : 3 : 1

Clearly, (4 + 3 + 1) units ⇒ ₹ 3200

∴ 1 unit ⇒ ₹ 400

Hence, share of C = ₹ 400

Problem 3. Two persons A and B undertake to do a piece of work for ₹ 4800. A could do it alone in 5 days and B could do it alone in 8 days. With the help of C and D they finished it in 3 days. If the work done by C be twice that of D, determine the share of D.

Sol. Work done by A in 3 days = $\frac{3}{5}$

and work done by B in 3 days = $\frac{3}{8}$

∴ Work done by (C + D) in 3 days = $1 - \frac{3}{5} - \frac{3}{8} = \frac{1}{40}$

Obviously, their wages will be as per their contributions towards the work.

Then, (C + D)'s share = ₹ 4800 × $\frac{1}{40}$ = ₹ 120

As given, since work done by C is twice that of D, the ratio of wages of C and D = 2 : 1

Clearly, $(2 + 1)$ units $\Rightarrow 120$

$\therefore$ 1 unit $\Rightarrow$ ₹ 40

Alternatively I

Work done by A per day = 20%

and work done by B per day = 12.5%

and work done by $(A + B + C + D)$ per day = 33.33%

$\therefore$ Work done by $(C + D)$ per day

$= \{33.33\% - 20\% - 12.5\%\} = 0.83\%$

Now, ratio of work done by A, B and $(C + D)$

$= 20\% : 12.5\% : 0.83\% = 24 : 15 : 1$

Clearly, $(24 + 15 + 1)$ units $\Rightarrow$ ₹ 4800

$\therefore$ 1 unit $\Rightarrow$ ₹ 120

Hence, share of $(C + D)$ = ₹ 120

As given, work done by C is twice that of D, then the ratio of wages of C and $D = 2 : 1$.

Clearly, $(2 + 1)$ units $\Rightarrow 120$

$\therefore$ 1 unit $\Rightarrow$ ₹ 40

Problem 4. **A alone can complete a work in 16 days and B alone in 12 days. Starting with A, they work on alternate days. If the wages of the work is ₹ 480, what is the share of B?**

Sol. $\left.\begin{cases} A \Rightarrow 6\,.\,25\% \text{ of work per day} \\ B \Rightarrow 8.33\% \text{ of work per day} \end{cases}\right\}$ hence, 14.58% in 2 days by them

In 12 (i.e. 2×6) days $\Rightarrow 14.58 \times 6 = 87.50\%$

Clearly, 12.5% of the work remains left after 12 days.

Since A starts the work, A will work on 13th day.

Then, remaining work = $(12.5 - 6.25)\% = 6.25\%$, which is to be done by B on 14th day. Obviously, B will take $\frac{6.25\%}{8.33\%}$ i.e. $\frac{3}{4}$ days

Thus, total time taken $= 12 + 1 + \frac{3}{4} = 13\frac{3}{4}$ days

Clearly, B worked for $6\frac{3}{4}$ days and accordingly work done by

him $= \frac{6\frac{3}{4}}{12}$

Hence, share of $B = \frac{6\frac{3}{4}}{12} \times$ ₹ $480 =$ ₹ 270

Alternatively

Work done by them in 2 days

$= \frac{1}{16} + \frac{1}{12} = \frac{7}{48}$

Now, we should multiply $\frac{7}{48}$ by a suitable number so as to get the product closer or equal to 1 but not more than 1.

Here, $\frac{7}{48} \times 6$ i.e. $\frac{7}{8}$ of the work is done by them in (6×2) i.e. 12 days.

Now, romaining work after 12 days $= 1 - \frac{7}{8} = \frac{1}{8}$

Clearly, A will work on 13th day as A starts the work and the work remaining after 13 days $= \frac{1}{8} - \frac{1}{16} = \frac{1}{16}$, which will be done by B on 14th day.

Then, time required by B to finish $\frac{1}{16}$ of the work

$= 12 \times \frac{1}{16} = \frac{3}{4}$ days

Thus, total time taken $= 12 + 1 + \frac{3}{4} = 13\frac{3}{4}$ days

Clearly, B worked for $6\frac{3}{4}$ days and accordingly work done by

him $= \frac{6\frac{3}{4}}{12}$

Hence, share of $B = \frac{6\frac{3}{4}}{12} \times$ ₹ $480 =$ ₹ 270

Problem 5. **A, B and C completed a work costing ₹ 1800. A work for 6 days, B for 4 days and C for 9 days. If their daily wages are in the ratio 5 : 6 : 4, how much amount will be received by A?** ***[SSC (CGL-PT)-07]***

Sol. 3. Obviously, ratio of their wages

$= (6 \times 5) : (4 \times 6) : (9 \times 4) = 5 : 4 : 6$

Now, $(5 + 4 + 6)$ units $\Rightarrow$ ₹ 1800

$\therefore$ 5 units $\Rightarrow$ ₹ 600

Hence, A's share = ₹ 600

Problem 6. **A, B and C are employed to do a piece of work for ₹ 529. A and B together are supposed to do 19/23 of the work and B and C together 8/23 of the work. What amount should A be paid?** ***[CBI-97]***

Sol. Given, B and C together are supposed to do 8/23 of the work

$\therefore$ A is supposed to do the remaining work i.e. $1 - 8/23$, hence 15/23.

Then, A's wages $= 529 \times \frac{15}{23} =$ ₹ 345

Problem 7. **Pritam and Ananya finish a piece of work and are paid ₹ 46 and ₹ 69 respectively. How many days must they have taken to finish the work together, if Pritam takes 15 days to finish the work alone.**

Sol. Ratio of efficiency of Pritam and Ananya = Ratio of their wages $= 46 : 69 = 2 : 3$

Now, time taken by Pritam to complete the work $= D \times \frac{(2 + 3)}{2}$

(Where, D is time taken by them together to finish the work. Recall the concept of Efficiency, elaborated in 'Time and Work')

$\Rightarrow \quad 15 = D \times \frac{5}{2} \Rightarrow D = 6$ days.

Problem 8. **To complete a work, A takes 50% more time than B. If together they take 18 days to complete the work and earn ₹ 1800, what is the share of B?**

Sol. Ratio of time taken by A and $B = 150\% : 100\% = 3 : 2$

Then, ratio of their efficiencies = 2 : 3

We know wages are proportional to efficiencies when time is constant, then share of B = ₹ 1080

$\{\because (2 + 3)$ units $\Rightarrow$ ₹ 1800 $\therefore$ 3 units $\Rightarrow$ ₹ 1080$\}$

Problem 9. A is twice as good a workman as B and B is twice as good a workman as C. A and B can together finish a piece of work in 4 days. They all together complete the work, what is the share of C if he/she works for only 1 day and the wages for the whole work is fixed as ₹ 9600.

Sol. Ratio of efficiencies of A, B and $C = 2:1:1/2 = 4:2:1$

Then, time taken by them $= \frac{1}{4} : \frac{1}{2} : \frac{1}{1} = 1:2:4$

Given, time taken by $(A + B) = 4$ days

Then, time taken by $A = 4 \times \frac{(4+2)}{4} = 6$ days

Clearly, 1 unit = 6 days

Then, 4 units = 24 days

Hence, time taken by C alone to do the work = 24 days

Now, work done by C in 1 day = 1/24

∴ Share of C = ₹ 9600 × 1/24 = ₹ 400

Problem 10. A, B and C can independently do a work in 15 days, 20 days and 30 days respectively. They work together for some time after which C leaves. A total of ₹ 18000 is paid for the work and B gets ₹ 6000 more than C. For how many days did A work?

Sol. Ratio of efficiency of A, B and C respectively

$= \frac{1}{15} : \frac{1}{20} : \frac{1}{30} = 4:3:2$

Let C receives ₹ x.

Then, amount received by $B = x + 6000$

and amount received by $A = \frac{4}{3}(x + 6000)$

{∵ they work for equal number of days but ratio of their efficiency is 4 : 3}

Then, $x + (x + 6000) + \frac{4}{3}(x + 6000) = 18000$

⇒ $10x = 12000 \Rightarrow x = 1200$

Hence, wages received by A, B and C

= ₹ 9600, ₹ 7200, ₹ 1200

= 8 : 6 : 1 = ratio of work done

∴ Work done by A = 53.33%

{∵ (8 + 6 + 1) units = 100%, ∴ 8 units = 53.33%}

Since A does 6.66% of work per day, required number of days

= 53.33% ÷ 6.66% = 8 days.

Case II At Group Level

Problem 11. If 5 men or 7 women can earn ₹ 5250 per day, how much would 7 men and 13 women earn per day? **[SSC (CGL-T-I)-10]**

Sol. Given, 5 Men = 7 Women

∴ 7 men $= \frac{49}{5}$ Women

∵ 7 women can earn daily ₹ 5250

∴ $\left(\frac{49}{5} + 13\right)$ women can earn ₹ $\frac{5250}{7} \times \left(\frac{49}{5} + 13\right)$

= 750 × 22.8 = ₹ 17100

Alternatively

5M or W ⇒ 5250

↓ × 1.4 ↓ × 1.86

7M 13W ⇒ ?

Then, required earnings = ₹ 5250 × (1.4 + 1.86) = ₹ 17100

{∵ wages are directly proportional to work-forces}

Problem 12. 2 men and 1 woman together can complete a piece of work in 14 days while 4 women and 2 men together can do it in 8 days. If a man gets ₹ 600 per day, how much should a woman get per day?

Sol. 2M + 1W ⇒ 14 days ...Eq. (i)

2M + 4W ⇒ 8 days ...Eq. (ii)

Now, positive difference between (14 × 2M) and (8 × 2M) = Positive diference between (14 × 1W) and (8 × 4W)

⇒ 12M = 18W ⇒ 2M = 3W

or

To complete the work in one day, either 14(2M + 1W) or 8(4W + 2M) are required.

Hence, 28M + 14W = 32W + 16M

⇒ 12M = 18W ⇒ 2M = 3W

∴ Ratio of efficiency of one man and one woman

= 3 : 2 = ratio of wages of one man and one woman

As given, 3 units ⇒ ₹ 600

∴ 2 units ⇒ ₹ 400

Thus, one woman should get ₹ 400 per day.

Problem 13. If 4 men and 6 boys earn ₹ 1600 in 5 days and 3 men and 7 boys earn ₹ 1740 in 6 days, in what time will 7 men and 6 boys earn ₹ 3760?

[SSC (CGL-Mains)-2000, 01, SSC (CGL-PT)-07]

Sol. Earnings of 4 men and 6 boys in one day = (1600 ÷ 5) = ₹ 320

and earnings of 3 men and 7 boys in one day

= (1740 ÷ 6) = ₹ 290

Let each man earns ₹ M per day and each boy earns ₹ B per day.

Then, 4M + 6B = 320 ...Eq. (i)

and 3M + 7B = 290 ...Eq. (ii)

Now, Eq. (i) × 3 ⇒ 12M + 18B = 960

and Eq. (ii) × 4 ⇒ 12M + 28B = 1160

Eqn (i) – Eq. (ii) ⇒ –10B = –200

∴ B = 20

Putting the value of B in Eq. (i), we get M = 50

Now, per day earnings of 7 men and 6 boys

= (7 × 50) + (6 × 20) = ₹ 470

Then, number of days required = (3760 ÷ 470) = 8 days.

Note *More questions based on the concept of efficiency (already discussed in detail in chapter 'Time and Work') may be asked in examinations associating with wages. However, it is not desirable to reproduce such questions here only with incorporating wages aspects. Certain questions where in concept of ratio or proportion is substantially required, the same have been given under Chapter 'Ratio, Proportion and Variation'.*

WORKOUT

1. A sum of ₹ 25 was paid for a work which A can do in 32 days, B in 20 days, B and C in 12 days and D in 24 days. How much did C receive if all four work together?
   (a) ₹ 15/3 (b) ₹ 14/3
   (c) ₹ 13/3 (d) ₹ 16/3

2. 2 men and 4 boys can do a piece of work in 10 days, while 4 men and 5 boys can do it in 6 days. Men and boys are paid wages according to their output. If the daily wage of a man is ₹ 40, then the ratio of daily wages of a man and a boy will be : ***[SSC (CGL-T-II)-11]***
   (a) 5 : 3 (b) 5 : 2
   (c) 7 : 4 (d) 7 : 3

3. 2 men and 1 woman can complete a piece of work in 14 days. While 4 women and 2 men can do the same work in 8 days. If a man gets ₹ 90 per day. What should be the wages per day of a woman? ***[CDS-14]***
   (a) ₹ 48 (b) ₹ 60
   (c) ₹ 72 (d) ₹ 135

4. Two men A and B working together complete a piece of work which it would have taken them 30 and 40 days respectively to complete it they worked separately. If they received a payment of ₹ 2100, B's share is :
   (a) ₹ 900 (b) ₹ 800
   (c) ₹ 1200 (d) ₹ 1300

5. Two men undertake to do a piece of work for ₹ 600. One alone could do it in 6 days and the other in 8 days. With the assistance of a boy, they finish it in 3 days. The boy's share should be : ***[MAT-18, 11, 07, SSC (SO-Audit)-97]***
   (a) ₹ 75 (b) ₹ 225
   (c) ₹ 300 (d) ₹ 100

6. A can finish a work in 15 days. B in 20 days and C in 25 days. All these worked together and earned ₹ 4700. The share of C is ***[CDS-14]***
   (a) ₹ 1200 (b) ₹ 1500
   (c) ₹ 1800 (d) ₹ 2000

7. If 2 men or 3 women can earn ₹ 1200 per day, how much would 6 men 12 women earn per day?
   (a) ₹ 3600 (b) ₹ 8400
   (c) ₹ 3000 (d) ₹ 6800

8. A can do a piece of work in 10 days, while B alone can do it in 15 days. They work together for 5 days and the rest of the work is done by C in 2 days. If they get ₹ 450 for the whole work, how should they divide the money?
   (a) ₹ 225, 150, 75
   (b) ₹ 250, 100, 100
   (c) ₹ 200, 150, 100
   (d) ₹ 175, 175, 100

9. Three persons undertake to complete a piece of work for ₹ 1200. The first person can complete the work in 8 days, second person in 12 days and third person in 16 days. They complete the work with the help of fourth person in 3 days. What does the fourth person get? ***[SSC (CGL-T-II)- 13]***
   (a) ₹ 180 (b) ₹ 200
   (c) ₹ 225 (d) ₹ 250

10. X completed a job in 2 days and Y completed it in 3 days and Z takes 4 days to complete it. If they work together and get ₹ 3900 for the job, then how much amount does Y get?
    (a) ₹ 1800 (b) ₹ 1200
    (c) ₹ 900 (d) ₹ 800

11. One man can do as much work in one day as a woman can do in 2 days. A child does one-third the work in a day as a woman. If an estate owner hires 39 pairs of hands- men, women and children in ratio 6 : 5 : 2 and pays them in all ₹ 1113 at the end of the day's work, what must the daily wages of a child be, if the wages are proportional to the amount of work done? ***[CAT-94]***
    (a) ₹ 14 (b) ₹ 5 (c) ₹ 20 (d) ₹ 7

12. Ganu and Manu together finish a piece of work in 12 days. If Ganu had worked twice efficiently as he did and Manu thrice efficiently as he actually did, they together would have finished the work in 5 days. What is ratio of wages of Ganu and Manu considering their former efficiencies.
    (a) 3 : 2 (b) 2 : 3
    (c) 12 : 5 (d) 5 : 12

13. John gets ₹ 57 per hour of regular work and ₹ 114/h of overtime work. He works altogether 172 h and his income from overtime hours is 15% of his income from regular houre. Then, for how many hours did he work overtime? ***[CAT-19]***
    (a) 10 (b) 12
    (c) 16.5 (d) 15

Answer key

1. (d)	2. (b)	3. (b)	4. (a)	5. (a)	6. (a)	7. (b)	8. (a)	9. (c)	10. (b)
11. (d)	12. (a)	13. (b)							

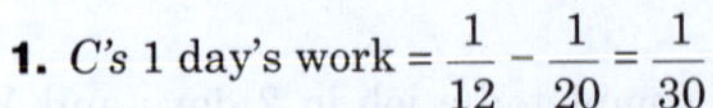

1. C's 1 day's work $= \frac{1}{12} - \frac{1}{20} = \frac{1}{30}$

Now, ratio of efficiencies of A, B, C and D

$= \frac{1}{32} : \frac{1}{20} : \frac{1}{30} : \frac{1}{24}$

$= 15 : 24 : 16 : 20$

Clearly, $(15 + 24 + 16 + 20)$ units $\Rightarrow$ ₹ 25

$\therefore$ 16 units $\Rightarrow$ ₹ 16/3

2. $2M + 4B \Rightarrow 10$ days .... Eq. (i)

$4M + 5B \Rightarrow 6$ days .... Eq. (ii)

Now, positive difference between $(10 \times 2M)$ and $(6 \times 4M)$ = Positive difference between $(10 \times 4B)$ and $(6 \times 5B)$

$\Rightarrow \quad 4M = 10B \Rightarrow 2M = 5B$

$\Rightarrow \quad \frac{M}{B} = \frac{5}{2}$

Hence, ratio of efficiencies of a man and a boy = 5 : 2

Then, obviously, ratio of their daily wages will be 5 : 2.

3. $2M + 1W \Rightarrow 14$ days ...Eq. (i)

$2M + 4W \Rightarrow 8$ days ...Eq. (ii)

Now, positive difference between $(14 \times 2M)$ and $(8 \times 2M)$ = Positive difference between $(14 \times 1W)$ and $(8 \times 4W)$

$\Rightarrow \; 12M = 18W \Rightarrow 2M = 3W \Rightarrow \frac{M}{W} = \frac{3}{2}$

Hence, ratio of efficiencies of a man and a woman = 3 : 2

As given 3 units $\Rightarrow$ ₹ 90

$\therefore$ 2 units $\Rightarrow$ ₹ 60

4. Ratio of time taken by A and B

$= 30 : 40 = 3 : 4$

$\therefore$ Ratio of efficiencies of A and B

$= 4 : 3$

Then, ratio of wages of A and $B = 4 : 3$

Clearly, $(3 + 4)$ units $\Rightarrow$ ₹ 2100

$\therefore$ 3 units $\Rightarrow$ ₹ 900

5. Work done by 1st man in 3 days

$= \frac{3}{6} = \frac{1}{2}$

and work done by 2nd man in 3 days

$= \frac{3}{8}$

$\therefore$ Work done by the boy in 3 days

$= 1 - \frac{1}{2} - \frac{3}{8} = \frac{1}{8}$

Obviously, their wages will be as per their contributions towards the work.

Then, the boy's share

$=$ ₹ $600 \times \frac{1}{8} =$ ₹ 75

6. Since wages are indirectly proportional to their time, when amount of work is constant, ratio of wages of A, B and C

$= \frac{1}{15} : \frac{1}{20} : \frac{1}{25} = 20 : 15 : 12$

Clearly, $(20 + 15 + 12)$ units $\Rightarrow$ ₹ 4700

$\therefore$ 12 units $\Rightarrow$ ₹ 1200

7. 2M or 3W $\Rightarrow$ 1200

$\downarrow \times 3 \qquad \downarrow \times 4$

6M $\qquad$ 12W $\Rightarrow$?

Since wages are directly proportional to work-forces, the required earnings

$=$ ₹ $1200 \times (3 + 4) =$ ₹ 8400

8. Work done by A in 5 days $= \frac{5}{10} = \frac{1}{2}$

and work done by B in 5 days

$= \frac{5}{15} = \frac{1}{3}$

$\therefore$ Work done by C in 2 days

$= 1 - \frac{1}{2} - \frac{1}{3} = \frac{1}{6}$

Obviously, their wages will be as per their contributions towards the work.

Then, A's share $=$ ₹ $450 \times \frac{1}{2} =$ ₹ 225

B's share $=$ ₹ $450 \times \frac{1}{3} =$ ₹ 150

C's share $=$ ₹ $450 \times \frac{1}{6} =$ ₹ 75

9. Work done by the fourth person

$= \left\{1 - \frac{3}{8} - \frac{3}{12} - \frac{3}{16}\right\} = \frac{3}{16}$

Then, fourth person's share

$=$ ₹ $1200 \times \frac{3}{16} =$ ₹ 225

10. Since wages are indirectly proportional to their time, when amount of work is constant, ratio of wages of X, Y and Z

$= \frac{1}{2} : \frac{1}{3} : \frac{1}{4} = 6 : 4 : 3$

Clearly, $(6 + 4 + 3)$ units $\Rightarrow$ ₹ 3900

$\therefore$ 4 units $\Rightarrow$ ₹ 1200

11. Ratio of efficiency of Man, Woman and Child $= 2 : 1 : \frac{1}{3} = 6 : 3 : 1$

Ratio of number of men, women and children = 6 : 5 : 2

$\therefore$ Ratio of amount of wages

$= (6 \times 6) : (3 \times 5) : (1 \times 2)$

$= 36 : 15 : 2$

Then, amount received by all children = 42

$\{\because (36 + 15 + 2)$ units $=$ ₹ 1113

$\therefore$ 2 units $= 42\}$

Number of children = 6

$\{\because (6 + 5 + 2)$ units $= 39$

$\therefore$ 2 units $= 6\}$

$\therefore$ Daily wages of a child $= 42 \div 6 =$ ₹ 7

12. In first case, work done by (G + M) per day = 8.33%

$\Rightarrow$ Work done by G per day

= 8.33% – Work done by M per day

In second case, work done by (G + M) per day = 20%

$\Rightarrow$ 2(work done by G per day in first case) + 3(work done by M per day in first case) = 20%

$\Rightarrow$ 2(8.33% – Work done by M per day in first case) + 3(work done by M per day in first case) = 20%

$\Rightarrow$ (work done by M per day in first case) = 3.33%

Further, work done by G per day

= 8.33% – 3.33% = 5%

Hence, ratio of efficiencies of G and M

= 5 : 3.33% = 3 : 2

Then, ratio of wages of G and M

= 3 : 2 as wages are directly proportional to efficiencies.

13. Given, income from regular work : income from overtime work

= 100 : 15 = 20 : 3

Also, wages per hour for regular work : wages per hour for overtime work = 57 : 114 = 1 : 2

It implies that ratio of working hours for regular work to that for overtime work $= (20 \div 1) : (3 \div 2) = 40 : 3$

$\left(\because \text{Working hours} = \frac{\text{income from work}}{\text{wages per hour}}\right)$

As given, $(40 + 3)$ units $\Rightarrow$ 172 h

$\therefore$ 3 units = 12 h

25 Pipes and Cisterns

Learning OBJECTIVES

After completing this chapter, aspirants will be able to :

- Observe that there is no difference between the chapter 'Time and Work' and 'Pipes and Cisterns' on conceptual level;
- Recognise how percentage method works to solve the problems quickly;
- Comprehend the concept of negative work and their utility;
- Apply the concept of efficiency in this chapter;
- Realise that percentage method is easier than the fraction method;
- Understand the philosophy of the division of this chapter in various parts;
- Analyse the trend of the questions being asked in various competitive examinations;
- Recognise the types of question which can be easily solved through the given options;
- Solve the questions in moments;
- Grip on concepts helping to solve the questions without pen and paper;
- Learn new concepts/fundamentals;
- Apply the pragmatic approaches in lieu of academic approaches; and
- Ponder on the alternative methods.

Introduction

This chapter is critical for the aspirants as questions are asked in diverse competitive exams like CAT, MAT, SSC, CDS, Bank- PO, Railway, etc. It has also importance in our daily life. Indeed, this chapter is an extension of the chapter 'Time and Work' that is why the concepts associated with this chapter are the same as applicable in case of chapter 'Time and Work', hence no difference between both topics on conceptual level. No doubt, if you are better in 'Time and Work' you will obviously be better in this topic.

Inlet pipes fill the cistern/tank whereas outlet pipes empty it. Where both types of pipes work, the 'concept of negative work' is applied.

Inside

Problems based on basic concepts (other than concept of efficiency) detailed in the chapter 'Time and Work'.

A cistern is fitted with 3 inlet pipes namely A, B and C and 2 outlet pipes (D and E). Pipes A, B and C can fill the cistern individually in 15, 20 and 25 min respectively, when outlet pipes don't function, whereas pipes D and E can empty the fully filled cistern individually in 25 and 40 min respectively, when inlet pipes don't function. Now, answer the following questions (1-20).

Note *Solved with only Percentage method due to space constraints.*

It is clear that

$$\left\{\begin{array}{l} A \text{ can fill} \Rightarrow 6.66\% \text{ of the cistern per min} \\ B \text{ can fill} \Rightarrow 5\% \text{ of the cistern per min} \\ C \text{ can fill} \Rightarrow 4\% \text{ of the cistern per min} \end{array}\right\}$$

Further,

$$\left\{\begin{array}{l} D \text{ can empty} \Rightarrow 4\% \text{ of the filled cistern per min} \\ E \text{ can empty} \Rightarrow 2.5\% \text{ of the filled cistern per min} \end{array}\right\}$$

Problem 1. **If only two pipes A and B are opened simultaneously, the time taken to fill the cistern is?**

Sol. Part filled by $(A + B) = (6.66 + 5)\%$ of the cistern per min

Then, time required to fill the cistern $= \dfrac{100\%}{(6.66 + 5)\%} = 8.58$ min.

Problem 2. **If only all three inlet pipes are opened simultaneously, the time taken to fill the cistern is :**

Sol. Part filled by $(A + B + C) = (6.66 + 5 + 4)\%$ of the cistern per min

Then, time required to fill the cistern

$$= \frac{100\%}{(6.66 + 5 + 4)\%} = 6.38 \text{ min}$$

Problem 3. **If only pipes B and D are opened simultaneously, the time taken to fill the cistern is :**

Sol. Part filled by pipe B when pipe D is open $= (5 - 4)\%$ of the cistern per min

Then, time required to fill the cistern $= \dfrac{100\%}{(5 - 4)\%} = 100$ min

Problem 4. **If pipes B and C are opened simultaneously and pipe E is opened after 5 min, the time taken to fill the cistern is :**

Sol. Part filled by $(B + C)$ upto 5 min. $= 5(5 + 4)\% = 45\%$. Obviously, for remaining 55%, all the three pipes work.

Then, time taken by them to fill 55% of cistern

$$= \frac{55\%}{(5 + 4 - 2.5)\%} = 8.5 \text{ min (approx.)}$$

Hence, total time taken $= 5 + 8.5 = 13.5$ min (approx.)

Problem 5. **If pipes A, B and C are opened simultaneously and A is closed after 3 min, then time taken to fill the cistern is :**

Sol. Part filled by A in 3 min $= 3 \times 6.66\%$. Obviously, the remaining part is to be filled by $(B + C)$ from starting to ending.

Then, required time to fill the cistern

$$= \frac{(100 - 3 \times 6.66)\%}{(5 + 4)\%} = 8.89 \text{ min}$$

Problem 6. **Pipe A is opened at 9:00 AM, B is opened at 9:02 AM and E is opened at 9:05 AM. At what time will the cistern be filled?**

Sol. Part filled by $(A + B)$ from 9:00 AM to 9:05 AM

$$= (5 \times 6.66)\% + (3 \times 5\%)$$
$$= 48.33\%$$

Clearly, since 9:05 AM, all the three pipes work. Then time taken by them to fill the remaining part

$$= \frac{(100 - 48.33)\%}{(6.66 + 5 - 2.5)\%} = 5.6 \text{ min}$$

Hence, the cistern will be filled at 5.6 min beyond 9:05 AM i.e. at 10.6 min past 9 O'clock.

Problem 7. **Pipes A and B are opened simultaneously and A is closed 3 min before the cistern is filled. The time taken to fill the cistern is :** ***[IAS(Pre.)-02]***

Sol. Required time taken $= \dfrac{(100 + 3 \times 6.66)\%}{(6.66 + 5)\%} = 10.3$ min (approx.)

[Recall the concept 'Starting together some persons leave before certain time of actual completion of the work']

Problem 8. **If B and C are opened for one minute each alternately (B is opened first), the cistern will be full in :**

Sol. In 2 min, part of cistern filled $= (5 + 4)\% = 9\%$

Then, in 22 min $\Rightarrow$ 99%

At 23rd min pipe B will function and will take 1/5 min to fill the remaining 1% of the cistern.

Hence, required time

$$= 22 + \frac{1}{5} = 22\frac{1}{5} \text{ min}$$

Problem 9. **If B is open for all time and C and E are opened for one min each alternately (started with E), the cistern will be full in :**

[SSC (CGL-PT)-99, SSC (SO-Commercial) -05]

Sol. In 2 min, part of cistern filled

$$= (5 - 2.5)\% + (5 + 4)\%$$
$$= 11.5\%$$

Then, in 16 min $= 92\%$

In 17 min $= 92\% + (5 - 2.5)\%$
$= 94.5\%$

The remaining part i.e. 5.5% will be filled by $(B + C)$ and they will take $\dfrac{5.5\%}{(5 + 4)\%}$ i.e. 0.6 min

Hence, the cistern will be full in 17.6 min.

Problem 10. If A, B and C are filling three different chemical solution X, Y and Z respectively in the cistern, then what is the proportion of the chemical solution Z in the liquid in the cistern after 2 min? **[MAT-10]**

Sol. Part filled by $(A + B + C)$ in 2 min = 2(6.66 + 5 + 4)% = 31.33%

Then, proportion of solution $Z = \frac{(2 \times 4)\%}{31.33\%} = \frac{8 \times 3}{94} = \frac{12}{47}$

Problem 11. If the cistern is 5/6th full and C and E are opened simultaneously, the time taken to fill the cistern is :

Sol. Cistern is 1/6th empty i.e. 16.66%

Then, required time = $\frac{16.66\%}{(4 - 2.5)\%}$ = 11.1 min

Problem 12. If B, C and F(an outlet pipe) are opened simultaneously, cistern is full in $12\frac{1}{2}$ min. In how much time can F empty the full cistern alone?

Sol. Part of cistern filled by $(B + C)$ in 1 min = (5 + 4)% = 9%

and part of cistern filled by $(B + C + F)$ in 1 min = 8%

[∵ (100 ÷ 12.5) = 8]

Hence, pipe F is emptying at the rate of 1% per min.

Then, time taken by F to empty the full cistern

= (100% ÷ 1%) = 100 min

Problem 13. Pipe B is opened to fill the cistern, but due to a leak, it took 25 min to fill. The leak can drain all the water of the cistern in :

Sol. Part filled by B = 5% of the cistern per min

Part filled by B due to leak = 4% [∵ 100 ÷ 25 = 4]

It means the leak is draining at the rate of 1% per minute.

Then, time taken to drain all the water by the leak

= (100% ÷ 1%) = 100 min.

Problem 14. How many minutes will it take to fill the cistern from empty state if B is used for half the time and B and C together for the other half? **[DMRC-03]**

Sol. Let the required time be x minutes.

Then, $(5\%) \times \frac{x}{2} + (5 + 4)\% \times \frac{x}{2} = 100\%$

$\Rightarrow \frac{x}{2} = \frac{100}{14} \Rightarrow x = 14\frac{2}{7}$ min

Problem 15. Pipes B, C and G (an outlet) working together can fill the cistern in $12\frac{1}{2}$ min. Pipe G can empty 4 gallons per min. What is the capacity of the cistern? **[Bank(PO)-01, SSC(CPO)-07]**

Sol. Part of cistern filled by $(B + C)$ in 1 min = (5 + 4)% = 9%

and part of cistern filled by $(B + C + F)$ in 1 min = 8%

[∵ (100 ÷ 12.5) = 8]

It means pipe F is emptying at the rate of 1% per min.

As given, 1% of the cistern = 4 gallons

∴ 100% of the cistern = 4 × 100 = 400 gallons

Hence, capacity of the cistern = 400 gallons

Problem 16. Both the pipes B and C are opened simultaneously. The cistern will be filled in just $17\frac{1}{2}$ min, if the pipe B is turned off after : **[SSC (CGL-PT)- 04, SSC (SO-Commerical)-06, CLAT-17]**

Sol. Let the pipe B be closed after x minutes. It implies that B worked for only x minutes and C worked for 17.5 min.

Obviously, $(x \times 5\%) + (17.5 \times 4\%) = 100\%$

$\Rightarrow 5x = 30 \Rightarrow x = 6$ min

Note *Such types of questions must be solved by options so as to save time in examination hall.*

Problem 17. If pipes D and E are opened simultaneously, how much time is needed to empty the full cistern?

Sol. Part emptied in one minute = (4 + 2.5)% = 6.5%

Then, required time = (100% ÷ 6.5%) = 15.38 min

Problem 18. Pipe B is opened and after half the cistern is filled, three more similar pipes are opened. Total time taken to fill the cistern is :

Sol. Time taken by B to fill 50% of the cistern = (50% ÷ 5%) = 10 min

Then, time taken by (B + three more pipes) to fill the remaining 50% = 50% ÷ (5% × 4) = 2.5 min

Hence, total time taken = 10 + 2.5 = 12.5 min

Problem 19. If diameter of pipe D is made double, then time taken by D to empty the completely filled cistern with water is :

Sol. It is obvious that when diameter of the pipe becomes double, the volume of the cylinder (assuming pipe be a cylinder) will be 4 times. It implies that the volume of water will be 4 times the volume of water earlier drained in the same time. Then, the time taken by pipe D with increased diameter will be $\frac{1}{4}$th of the time taken earlier.

Hence, required time = (25 ÷ 4) = 6.25 min

[**Note**: It may be said that the area of cross section of the tap i.e. that of circle has been increased to 4 times the earlier.]

Problem 20. Pipes B and C are opened simultaneously in order to fill the cistern. However, a leak in the cistern caused a delay of $3\frac{8}{9}$ min in filling the cistern. Time taken by leak to empty a filled cistern is : **[SSC (CGL-T-I)-11, IIFT-19, SSC (CGL-T-II)-20]**

Sol. Ideal time taken by $(B + C)$ to fill the cistern

= {100% ÷ (5 + 4)%} = $11\frac{1}{9}$ min

Time taken by $(B + C)$ to fill the cistern due to leak

= $11\frac{1}{9} + 3\frac{8}{9}$ = 15 min i.e. filling rate per minute

= 6.66%

Hence, the leak is emptying {(5 + 4) – 6.66}% per min.

Then, time taken by the leak to empty a filled cistern

= (100% ÷ 2.33%) = 42.9 min

Problem 21. 8 taps are fitted to a water tank. Some of them are inlet taps to fill the tank and the remaining are outlet taps used to empty tank. Each inlet tap can fill the tank in 12 h and each outlet tap can empty it in 36 h. On opening all the taps, the tank is filled in 3 h. Find the number of inlet taps.

Sol. On opening all the taps, the tank is filled in 3 h, it implies that they are filling 33.33% per min.

Let number of inlet taps be x.

Then, number of outlet taps = $8 - x$

Now, part filled by inlet taps in 1 min – part emptied by outlet taps in 1 min = 33.33%

$\Rightarrow \quad (x \times 8.33\%) - (8 - x) \times 2.77\% = 33.33\%$

$\Rightarrow \quad 11.1x = 55.50 \Rightarrow x = 5$

Alternatively

The best way is to proceed through options. Let options be 4, 5, ...

On opening all the taps, the tank is filled in 3 h, it implies that they are filling 33.33% per min.

(a) $4 \times 8.33\% - (8 - 4) \times 2.77 \neq 33.33\%$

(b) $5 \times 8.33\% - (8 - 5) \times 2.77\% = 33.33\%$

Hence, option (b) is correct.

Problem 22. A tank having capacity of 75 L, is fitted with an inlet and an outlet pipes. The tank is filled in $7\frac{1}{2}$ min if both the pipes are made operational simultaneously. The tank never gets filled up if the outflow rate is made 2.5 times the original. Which of the following is the possible initial outflow rate?

(a) 10 L/min (b) 6 L/min
(c) 8 L/min (d) 7 L/min

Sol. Effective filling rate per minute = $75 \div 7.5 = 10$ L. It implies that the inflow rate (I) exceeds to the outflow rate (P) by 10 L/min. Further, the tank will never get filled up only when the inflow rate is either equal to or less than increased outflow rate.

Hence, $I - P = 10$ ...(i)

and $I - 2.5P = 0$ or any negative value ...(ii)

If $I - 2.5P = 0$, value of $P = 6.66$ and if $I - 2.5P$ = negative value, value of $P > 6.66$.

Considering the value of options, option (d) is correct.

Problem 23. Three pipes A, B and C can fill a tank in 6 h. After working it together for 2 h, C is closed and A and B can fill the remaining part in 7 h. The number of hours taken by C alone to fill the tank is:

[LIC (AAO)-03, SSC (CPO)-09]

Sol. Part filled by all three together in one hour = 16.66%

$\therefore$ Part filled by all three together in 2 hrs = 33.33%

Remaining part i.e. 66.66% has been filled by $(A + B)$ in 7 hr.

Then, part filled by both per minute

$= (66.66\% \div 7) = 9.52\%$

Hence, part filled by C alone per minute

$= (16.66 - 9.52)\% = 7.14\%$

Then, time taken by C alone to fill the tank

$= (100\% \div 7.14\%) = 14$ h.

Problem 24. A tank can be filled with water by two pipes A and B together in 36 min. If the pipe B was stopped after 30 min, the tank is filled in 40 min. The pipe B can alone fill the tank in?

Sol. Part filled by both in 30 minutes = $\frac{30}{36}$ = 83.33%

Hence, the remaining part i.e. 16.66% has been filled by A alone in (40 – 30) min. i.e. 1.66% per min.

Now, part filled by both in 1 min. = $100\% \div 36 = 2.77\%$

Then, part filled by B alone in 1 min

$= (2.77 - 1.66)\% = 1.11\%$

$\therefore$ Time taken by B alone to fill the tank

$= (100\% \div 1.11\%) = 90$ min

Problem 25. A tank has a leak which would empty the completely filled tank in 10 h. If the tank is full of water and a tap is opened which admits 4 L of water per min in the tank, the leak takes 15 h to empty the tank. How many litres of water does the tank hold?

Sol. Part emptied by the leak per h = 10%

and part emptied by the leak per h while a tap is opened = 6.66%

Hence, the tap is filling water at (10 – 6.66)% i.e. 3.33% per h.

Then, time taken by the tap to fill a empty tank

$= (100\% \div 3.33\%) = 30$ h

$\therefore$ Capacity of the tank

$= 30 \times 240$ [$\because$ tap admits 240 L per h.]

$= 7200$ L

Problem 26. Three pipes P, Q and R can separately fill a cistern in 4, 8 and 12 h respectively. Another pipe S can empty the completely filled cistern in 10 h. Which of the following arrangements will fill the cistern in less than others?

(a) Q alone is open
(b) P and S are open
(c) P, R and S are open
(d) P, Q and S are open

Sol. Option (a) filling rate per h = 12.5%

(b) filling rate per h = (25 – 10)% = 15%

(c) filling rate per h = (25 + 8.33 – 10) = 23.33%

(d) filling rate per h = (25 + 12.5 – 10) = 27.5%

Hence, option (d) will take least time to fill the tank than others.

Problem 27. A pipe can empty a tank in 40 min. A second pipe with diameter twice as that of the first is also attached with the tank to empty it. The two together can empty the tank in

Sol. Since area of cross section of the second pipe i.e. area of a circle is 4 times area of cross section of the first pipe, the volume of water to be drained by second pipe will be 4 times the volume of water drained by the first pipe in the same time. Thus, the time taken by the second pipe will be $\frac{1}{4}$th of the time taken by first pipe to empty the tank.

Hence, time taken by second pipe to empty the tank

$= (40 \div 4) = 10$ min

Now, part emptied by them together in 1 min.

$= (2.5 + 10\%) = 12.5\%$

$\therefore$ Time taken to empty the tank $= (100\% \div 12.5\%) = 8$ min

Alternatively

Since the capacity to empty has increased to 5 times the earlier capacity, the required time $= \frac{40}{5} = 8$ min

Problem 28. **A tank has two taps- one can fill it in 10 min individually and another can empty it in 25 min alone. How many more outlet taps having same efficiency, can be fitted with the tank so as to avoid overflowing?**

Sol. Part filled by inlet tap = 10%/min

and part emptied by outlet tap = 4%/min

Hence, a total number of 2.5 outlet taps (i.e. 3) is required to manage the overflow.

$\therefore$ Additional number of outlet taps = 3 − 1 = 2

Problem 29. **Two pipes A and B together can fill a cistern in 4 h. Had they been opened separately, then B would have taken 6 h more than A to fill the cistern. How much time will be taken by A to fill the cistern separately?** ***[NABARD-01]***

Sol. Let A takes x h. Then, B will take $(x + 6)$ h to fill the cistern.

Obviously, $\frac{1}{x} + \frac{1}{x+6} = \frac{1}{4} \Rightarrow x^2 + 6x = 8x + 24$

$\Rightarrow \quad x^2 - 2x - 24 = 0 \Rightarrow x^2 - 6x + 4x - 24 = 0$

$\Rightarrow \quad x(x-6) + 4(x-6) = 0$

$\Rightarrow \quad (x-6)(x+4) = 0$

$\Rightarrow x = 6$ and -4. Time cannot be negative in this context.

$\therefore \quad x = 6$ h.

Alternatively

The best way is to proceed through options. Let options be 8, 6, ...

(a) 8 $\Rightarrow A$ in 8 h (i.e. 12.5% per hr) and B in 14 h (i.e. 7.14% per hour) but 12.5% + 7.14% ≠ 25%

[$\because (A + B) \Rightarrow$ 4 h i.e. 25% per hour]

(b) 6 $\Rightarrow A$ in 6 h (i.e. 16.66% per hour) and B in 12 h (i.e. 8.33% per hour) and 16.66% + 8.33% = 25%

Hence, option (b) is correct.

Problem 30. **Four two-way pipes A, B, C and D can either fill an empty tank or drain the full tank in 4, 10, 12 and 20 min respectively. All four pipes were opened simultaneously when the tank is empty. Under which of the following conditions the tank would be half filled after 30 min?**

(a) Pipe A drained and pipes B, C and D filled

(b) Pipe A filled and pipes B, C and D drained

(c) Pipes A and D drained and pipes B and C filled

(d) Pipes A and D filled and pipes B and C drained

(e) None of the above ***[XAT-17]***

Sol. Part filled/emptied by pipes A, B, C and D per min respectively

= 25%, 10%, 8.33% and 5%.

Now, proceed through options.

(a) part filled in 30 min

$= -(25 \times 30)\% + \{(10 + 8.33 + 5) \times 30\}\%$

$\neq +50\%$

(b) part filled in 30 min.

$= +(25 \times 30)\% - \{(10 + 8.33 + 5) \times 30\}\%$

$= +50\%$

Hence, option (b) is correct.

Problem 31. **A tank is fitted with pipes, some filling it and the rest draining it. All filling pipes fill at the same rate, and all the draining pipes drain at the same rate. The empty tank gets completely filled in 6 h when 6 fillings and 5 draining pipes are on, but this time becomes 60 h when 5 fillings and 6 draining pipes are on. In how many hours will the empty tank get completely filled when one draining and two filling pipes are on?** ***[CAT-18]***

Sol. Let one filling pipe fills x% of tank per hour and one draining pipe empties y% of the tank per hour.

As given, $6x - 5y = 16.66\%$ {$\because$ 100% of tanks is filled in 6 hrs)

and $5x - 6y = 1.66\%$ {$\because$ 100% of tank is filled in 60 hrs}

On adding both Equations we get, $11x - 11y = 18.33\%$

$\Rightarrow \quad x - y = 1.66$

$\therefore \quad 4x - 4y = 6.66$

Now, $\quad 6x - 5y = 16.66\%$ (from the above)

$\Rightarrow \quad 4x - 4y + 2x - y = 16.66\%$

$\Rightarrow \quad 2x - y = (16.66 - 6.66)\% = 10\%$

$\therefore$ Time taken by one draining pipe and two filling pipes to fill the empty tank completely

$= 100\% \div 10\% = 10$ h

Problems based on Application of Concept of Efficiency as Detailed in Chapter 'Time and Work'

Problem 32. **One pipe can fill a tank three times as fast as another pipe. If together the two pipes can fill the tank in 36 min, the slower pipe alone will be able to fill the tank in**

Sol. Given, ratio of efficiencies of faster and slower pipe

= 3 : 1

Now, time taken by slower pipe to fill the tank alone

$= 36 \times \frac{(3+1)}{1} = 144$ min

Problem 33. **A tank is filled in 5 h by three pipes A, B and C together. The pipe C is twice as fast as B and B is twice as fast as A. How much time will pipe A alone take to fill the tank?**

Sol. Ratio of efficiencies of pipes A, B and C = 1 : 2 : 4

$\therefore$ Time taken by pipe A to fill the tank alone

$= 5 \times \frac{(1+2+4)}{1} = 35$ h

Problem 34. A tank of capacity 8000 m^3 is fitted with an inlet and an outlet taps. The efficiency of outlet tap is 25% more than that of inlet tap, that is why, the inlet tap takes 16 min more time to fill the tank than that required to empty by the outlet tap. What is the inflow rate ?

Sol. Ratio of efficiency of inlet to outlet = 100% : 125% = 4 : 5

Then, ratio of time taken by them (to fill and to empty) = 5 : 4

As given, (5 – 4) units = 16 min

$\Rightarrow$ 5 units = 80 min

Hence, the inlet tap can fill the empty tank in 80 min.

$\therefore$ Inflow rate = $(8000 \div 80) = 100\ m^3$ / min

Problem 35. A swimming pool has 3 drain pipes. The first two pipes A and B, operating simultaneously, can empty the pool in half the time that C, the 3rd pipe, alone takes to empty it. Pipe A, working alone, takes half the time taken by pipe B. Together they take 6 h 40 min to empty the pool. Time taken by pipe A to empty the pool, in hours, is *[SSC (CGL-T-II)-11]*

Sol. Given, ratio of efficiencies of A to B = 2 : 1

and ratio of efficiencies of (A + B) to C = 2 : 1

$\left[\because \text{Efficiency} \propto \frac{1}{\text{time}}, \text{ when amount of work is constant}\right]$

Then, ratio of efficiencies of A to B to C $= 2:1:\left(\frac{2+1}{2}\right)$

[$\because$ (2 + 1) units $\Rightarrow$ 2 $\therefore$ 1 units 3/2]

$= 4:2:3$

$\therefore$ Time taken by A to empty the pool

$= 6\text{ h } 40\text{ min.} \times \left(\frac{4+2+3}{4}\right) = 15\text{ h}$

Problem 36. A tank is emptied everyday at a fixed time point. Immediately thereafter, either pump A or pump B or both start working until the tank is full. On Monday, A alone completed filling the tank at 8 pm. On Tuesday, B alone completed filling the tank at 6 pm. On Wednesday, A alone worked till 5 pm, and then B worked alone from 5 pm to 7 pm, to fill the tank. At what time was the tank filled on Thursday if both pumps were used simultaneously all along? *[CAT-18]*

Sol. Clearly, B takes 2 h less than A to complete filling of the tank as A completes filling the tank at 8 pm whereas B does at 6 pm.

Further, it is given that, on Wednesday, A stopped to work 3 h before the tank is filled. It means the part of the tank, which were to be filled by pump A in 3 h, was filled by pump B in only 2 h (i.e. 5 pm to 7 pm). It implies that ratio of efficiencies of A to that of B is 2 : 3.

$\therefore$ Ratio of time taken by A and B = 3 : 2.

$\Rightarrow \quad t:(t-2) = 3:2$ (let A fills the tank in t h)

$\Rightarrow \quad t = 6\text{h}.$

$\therefore$ Time when tank is emptied every day

= 6 h before 8 pm = 2 pm

Since A and B individually takes 6 h and 4 h respectively to fill the tank, the time taken by them working together to fill the tank = 2 h 24 min.

Thus, on Thursday, the tank will be filled by them at 2 h 24 min beyond 2 pm i.e. at 4 : 24 pm.

Alternatively

Ratio of efficiencies of pumps A and B = 2 : 3

Then, time taken by A alone to fill the tank $= D \times \frac{(2+3)}{2}$

[Where, D is time taken by them together to fill the tank]

$\Rightarrow \quad 6\text{ hr} = D \times \frac{5}{2} \Rightarrow D = 2\text{ h } 24\text{ min}$

Thus, on Thursday, the tank will be filled by them at 2 h 24 min beyond 2 pm i.e. at 4 : 24 pm.

Problem 37. A tank has two-way taps with different efficiencies. When both are used to fill, it takes 1/3rd of the time taken to fill the tank when one is used to fill while another is used to empty. Find the ratio of efficiency of the tap used to fill to that of the tap used to empty.

Sol. Let the required ratio of efficiency be x : y.

Effective efficiency when both taps are used to fill = (x + y)

and Effective efficiency when one is used to fill and another to empty = (x – y)

As given, time taken in case of efficiency $(x + y) = \frac{1}{3}$ of time taken in case of efficiency (x – y)

$\Rightarrow \quad \frac{1}{3}(x+y) = (x-y)$

$\left[\because \text{time} \propto \frac{1}{\text{efficiency}}, \text{ when amount of work is constant}\right]$

$\Rightarrow \quad x + y = 3x - 3y$

$\Rightarrow \quad 2x - 4y = 0$

$\Rightarrow \quad 2x = 4y$

$\Rightarrow \quad \frac{x}{y} = \frac{4}{2}$

Hence, x : y = 2 : 1

Alternatively

Then best way is to proceed through options. Let options be 3 : 1, 2 : 1, ...

(a) $3:1 \Rightarrow \frac{1}{3}(3+1) \neq (3-1)$

(b) $2:1 \Rightarrow \frac{1}{3}(2+1) = (2-1)$

Hence, option (b) is correct.

Note

(i) All the questions under this chapter may be solved through different methods as have been depicted in the chapter 'Time and Work'. The logic is that the there is no difference between both chapters on conceptual level. The other methods have deliberately been left on account of space constraints.

(ii) I hope that aspirants are now able to solve any sorts of questions based on 'Pipes and Cisterns' with the help of concepts/patterns envisaged in chapter 'Time and Work'.

WORKOUT

1. Two pipes A and B can fill a tank in 20 h and 30 h respectively. Both the pipes are opened to fill the tank but when the tank is one-third full, a leak develops in the tank through which one-fourth water supplied by both pipes goes out. What is the total time taken to fill the tank?

(a) $14\frac{2}{3}$ h (b) 15 h

(c) $12\frac{1}{2}$ h (d) $9\frac{1}{2}$ h

2. Pipe A can fill a tank in 3 h but there is a leakage also, due to which it takes 3.5 h for the tank to be filled. How much time will the leakage take in emptying the tank, if the tank is filled initially? ***[CDS-14, CDS-17, MAT-14, 18, 19]***

(a) 21 h (b) 20 h

(c) 18 h (d) 10.5 h

3. A cistern can be filled by two pipes filling separately in 12 and 16 min respectively. Both pipes are opened together for a certain time but being clogged, only 7/8 of full quantity water flows through the former and only 5/6 through the latter pipe. The obstructions, however, being suddenly removed, the cistern is filled in 3 min from that moment. How long was it before the full flow began? ***[MAT-09, 06]***

(a) 4.5 min (b) 2.5 min

(c) 3.5 min (d) 5.5 min

4. The ratio of the radii of the taps T_1 and T_2 is 2 : 1. Water is flowing through them with the same velocity. What is ratio of the time required to completely fill two identical drums kept under T_1 and T_2?

(a) 2 : 1 (b) 1: 2

(c) 4 : 1 (d) 1: 4

5. A tank of 4800 m^3 capacity is full of water. The discharging capacity of the pump is 10 m^3/min. higher than its filling capacity. As a result the pump needs 16 min. less to discharge the fuel than to fill up the tank. Find the filling capacity of the pump? ***[MAT-19, 14, 13, 11, 08, CAT-10]***

(a) 25 m^3/min

(b) 50 m^3/min

(c) 55 m^3/min

(d) None of the above

6. Four taps can individually fill a cistern of water in 1h, 2h, 3h and 6h respectively. If all the four taps are opened simultaneously, the cistern can be filled in how many minutes?

(a) 20 (b) 30

(c) 35 (d) 40

7. Six pipes are fitted to a water tank. Some of these are inlet pipes and others outlet pipes. Each inlet pipe can fill the tank in 9 h and each outlet pipe can empty it in 6 h. On opening all the pipes, an empty tank is filled in 9 h. The number of inlet pipes is: ***[MAT-13, 08, IIFT-15]***

(a) 4 (b) 3

(c) 2 (d) 5

8. A tap can fill a tank in 16 min and another can empty it in 8 min. If the tank is already ½ full and both the tapes are opened together, will the tank be filled or emptied? How long will it take before the tank is either filled or emptied completely as the case may be?

(a) Emptied, 16 min

(b) Filled, 8 min

(c) Emptied, 8 min

(d) Filled, 12 min

9. Pipe A can fill a tank in 10 min and pipe B can empty it in 15 min. If both the pipes are opened in an empty tank, the time taken to make it full is : ***[CDS-14]***

(a) 20 min (b) 25 min

(c) 30 min (d) None of these

10. A tank of capacity 25 L has an inlet and an outlet tap. If both are opened simultaneously, the tank is filled in 5 min and if outlet flow rate is doubled and taps opened, the tank never gets filled up. Which of the following may be outflow rate? ***[MAT-14, 19]***

(a) 3 L/min (b) 4 L/min

(c) 5 L/min (d) None of these

11. One fill pipe A is 3 times faster than second fill pipe B and takes 32 min less than the fill pipe B. When will the cistern be full if both pipes are opened together?

(a) 12 min

(b) 24 min

(c) 30 min

(d) Data inadequate

12. There are two taps A and B to fill up a water tank. The tank can be filled in 40 min. if both taps are on. The same tank can be filled in 60 min. if tap A alone is on. How much time will tap B alone take, to fill up the same tank?

(a) 64 min (b) 80 min

(c) 96 min (d) 120 min

13. Two taps can fill a tub in 5 min and 7 min respectively. A pipe can empty it in 3 min. If all the three are kept open simultaneously, when will the tub be full? ***[CDS-09, MAT-11]***

(a) 60 min

(b) 85 min

(c) 90 min

(d) 105 min

14. Pipe A fills a water tank in 24 min. Pipe B fills the same tank 7 times as fast as pipe A. If both the pipes are kept open, in how many minutes will the empty tank be filled? ***[SSC(SAS)-10]***

(a) 2 (b) 2.5

(c) 3 (d) 3.5

15. A, B and C are three pipes attached to a cistern. A and B can fill it in 20 min and 30 min respectively, while C can empty it in 15 min. If A, B and C be kept open successively for 1 min each, how soon will the cistern be filled? ***[MAT-12, 08]***

(a) 180 min (b) 60 min

(c) 157 min (d) 155 min

16. Two taps A and B can fill a cistern in 12 min and 15 min respectively.They are opened together but after a few minutes, A is turned off and the rest of the cistern is filled by B in 5 min. After how many minutes was A turned off?

(a) 4 min (b) 7 min
(c) 6 min (d) None of these

17. Two pipes A and B can fill a tank in 30 min and 40 min respectively. If both the pipes are opened simultaneously, after how much time should B be closed so that the tank gets filled in 20 min.?

(a) 12 min (b) $13\frac{1}{3}$ min
(c) $11\frac{1}{3}$ min (d) $15\frac{1}{2}$ min

18. Two pipes can fill a tank in 8 h and 12 h respectively, whereas an escape pipe can empty it in 6 h. If the three pipes are opened at 1 pm, 2 pm, and 3 pm respectively, at what time will the tank be filled? ***[MAT-11, SSC(CGL-T-II)-11, 12]***

(a) 8 am (b) 7 am
(c) 5 am (d) 7:30 am

19. In what time would a cistern be filled by three pipes whose diameters are 2 cm, 3 cm and 4 cm running together, when the largest alone can fill it in 58 min.? The amount of water flowing in each pipe is proportional to the square of its diameter.

(a) 30 min (b) 32 min
(c) 35 min (d) 48 min

20. Pipe A can fill an empty tank in 30 h while pipe B can fill it in 45 h. Pipe A and B are opened and closed alternately, i.e. First pipe A is opened, then B, again A and B and so on for 1 h each time without any time lapse. In how many hours will the tank be filled when it was empty, initially?

(a) 36 h (b) 54 h
(c) 48 h (d) 60 h

21. There are three pipes, A, B and C, opening into a tank. Each pipe can be used to fill or empty the tank at the same respective rate. The ratio of the rates of the three pipes at which they either fill or empty the tank is $2:3:4$. Pipes A and B working together, take 3 h to fill the tank when both are used as inlet pipes. In the first, second and third hours of operation, the pipes, A, B and C respectively are used as emptying pipes, while the other two pipes are used as filling pipes. What is the fraction of the tank to be filled at the end of the three hours?

(a) 1/5 (b) 3/5
(c) 2/5 (d) 4/5

22. A water tank has M inlet pipes and N outlet pipes. An inlet pipe can fill the tank in 8 h while an outlet pipe can empty the full tank in 12 h. If all pipes are left open simultaneously, it takes 6 h to fill the empty tank. What is the relationship between M and N? ***[XAT-16]***

(a) $M:N = 1:1$
(b) $M:N = 2:1$
(c) $M:N = 2:3$
(d) $M:N = 3:2$
(e) None of the above

23. Two taps can separately fill a cistern in 10 min. and 15 min. respectively. If these two pipes and a waste pipe are kept open simultaneously, the cistern gets filled in 18 min. The waste pipe can empty the full cistern in? ***[MAT-12, 11]***

(a) 7 min (b) 13 min
(c) 23 min (d) 9 min

24. Two pipes A and B can fill a cistern in 12 min. and 15 min. respectively, while a third pipe C can empty it in 6 min. Both A and B pipes are opened together for 5 min. in the beginning and then the third pipe C is also opened. In what time will the cistern be emptied? ***[MAT-12, 11, 04]***

(a) 30 min
(b) 33 min
(c) 37.5 min
(d) 45 min

25. Capacity of tap Y is 60% more than that of X. If both the taps are opened simultaneously, they take 40 h to fill the tank. The time taken by Y alone to fill the tank is ***[IIFT-13]***

(a) 60 h (b) 65 h
(c) 70 h (d) 75 h

26. There is a leak in the bottom of a cistern. Before the leak, it could be filled in $4\frac{1}{2}$ h. It now takes $\frac{1}{2}$ h longer. It the cistern is full, in how much time would the leakage empty the full cistern? ***[MAT-10, 09]***

(a) 23 h (b) 35 h
(c) 52 h (d) 45 h

27. A water tank has three taps A, B and C. A fills four buckets in 24 min, B fills 8 buckets in 1 h and C fills 2 buckets in 20 min. If all the taps are opened together a full tank is emptied in 2 h. If a bucket can hold 5 L of water, what is the capacity of the tank in litres. ***[CAT-94]***

(a) 60 L (b) 120 L
(c) 180 L (d) 240 L

28. To fill a certain tank pipes A, B and C take 20 min., 15 min. and 12 min. respectively. If the three pipes are opened every alternate minute, how long it take to fill the tank?

(a) 5 min (b) 10 min
(c) 12 min (d) 15 min

29. A bath can be filled by the cold water and hot water pipes in 10 min and 15 min respectively. A person leaves the bathroom after turning on both pipes simultaneously and returns at the moment when the bath should be full. Finding, however, that the waste pipe has been open, he then closes it. In exactly four minutes more the bath is full. In how much time would the waste pipe empty the full bath, if it alone is opened? ***[MAT-12, 08]***

(a) 9 min (b) 10 min
(c) 12 min (d) None of these

30. A leak in the bottom of a tank can empty the full tank in 6h. An inlet pipe fills water at the rate of 4 L per min. When the tank is full, the inlet pipe is opened and due to the leak the tank is emptied in 8h. The capacity of the tank is?

(a) 9600 L
(b) 5760 L
(c) 2880 L
(d) None of the above

31. Two pipes A and B can fill a water tank in 20 and 24 min respectively and third pipe C can empty at the rate of 3 gallons per min. If A, B and C are opened together filled the tank in 15 min, the capacity in gallons of the tank is? ***[MAT-11, SSC (CGL-T-II)-10]***

(a) 120 (b) 60
(c) 150 (d) 180

32. Two pipes P and Q can fill the cistern in 3 and 6 min respectively, while an empty pipe R can empty the cistern in 4 min. All the three pipes are opened together and after 2 min pipe R is closed. Find when the tank will be full.

(a) 3 min (b) 6 min
(c) 5 min (d) 8 min

33. Two filling pipes A and B can fill a tank in 30 h and 20 h respectively. Pipe B alone is kept open for half the time and both pipes are kept open for the remaining time. In how many hours, will the tank be completely full?

(a) 25 h (b) 40 h
(c) 15 h (d) 28 h

34. A tank is filled by three pipes with uniform flow. The first two pipes operating simultaneously fill the tank in the same time during which the tank is filled by the third pipe alone. The second pipe fills the tank 5 h faster than the first pipe and 4 h slower than the third pipe. The time required by the first pipe is :

(a) 10 h (b) 15 h
(c) 6 h (d) 30 h

35. Two pipes can fill a cistern in 15 min and 18 min respectively. Both the pipes are operating together, but 3 min. before the cistern is full, one of the pipes is closed. The cistern will be filled now in :

(a) $9\frac{1}{7}$ min (b) $3\frac{3}{11}$ min
(c) $7\frac{3}{11}$ min (d) None of these

36. Three taps A, B and C can fill a tank in 12, 15 and 20 h respectively. If A is open all the time and B and C are open for one hour each alternately, the tank will be filled in

(a) 6 h (b) 7 h
(c) 5 h (d) None of these

37. Three pipes are connected to an inverted cone, with its base at the top. Two inlet pipes, A and B, are connected to the top of the cone and can fill the empty cone individually in 8 h and 12 h respectively. The outlet pipe C, connected to the bottom, can empty a filled cone in 4 h.

When the cone is completely filled with water, all three pipes are opened. Two of the three pipes remain opon for 20 h continuously and the third pipe remains open for a lesser time. As a result, the height of the water inside the cone comes down to 50%.

Which of the following options would be possible? ***[XAT-15]***

(a) Pipe A was open for 19 h
(b) Pipe A was open for 19 h 30 min
(c) Pipe B was open for 19 h 30min
(d) Pipe C was open for19 h 50 min
(e) The situation is not possible

38. Pipes A and B can fill a tank in 10 h and 40 h respectively. C is an outlet pipe attached to the tank. If all the three pipes are opened simultaneously, it takes 80 min more time than what A and B together take to fill the tank. A and B are kept opened for 7 h and then closed and C was opened. C will now empty the tank in : ***[SSC (CGL-T-I)-19]***

(a) 42 h (b) 45.5 h
(c) 49 h (d) 38.5 h

39. A water tank has inlets of two types A and B. All inlets of type A when open, bring in water at the same rate. All inlets of type B, when open, bring in water at the same rate. The empty tank is completely filled in 30 min if 10 inlets of type A and 45 inlets of type B are open, and in 1 h if 8 inlets of type A and 18 inlets of type B are open. In how many minutes will the empty tank get completely filled if 7 inlets of type A and 27 inlets of type B are open? ***[CAT-18]***

(a) 56 min (b) 48 min
(c) 30 min (d) None of these

Answer key

1. (a)	2. (a)	3. (a)	4. (d)	5. (b)	6. (b)	7. (a)	8. (c)	9. (c)	10. (c)
11. (a)	12. (d)	13. (d)	14. (c)	15. (a)	16. (d)	17. (b)	18. (b)	19. (b)	20. (a)
21. (c)	22. (e)	23. (d)	24. (d)	25. (b)	26. (d)	27. (d)	28. (d)	29. (a)	30. (b)
31. (a)	32. (a)	33. (c)	34. (b)	35. (d)	36. (b)	37. (c)	38. (c)	39. (b)	

Hints and SOLUTIONS

1. $\left.\begin{array}{l} A\text{'s filling rate} = 5\% \text{ of tank per hour} \\ B\text{'s filling rate} = 3.33\% \text{ of the tank per hour} \\ \text{Leakage's emptying rate} = \frac{1}{4}(5 + 3.33)\% \\ \text{i.e. } 2.08\% \text{ of tank per hour} \end{array}\right\}$

Time taken to fill one-third of the tank

$= \{33.33\% \div (5 + 3.33)\%\} = 4 \text{ hrs}$

Clearly, after the tank is 33.33% full, both pipes and leakage work.

Then, time taken to fill remaining 66.66% of the tank

$= \{66.66\% \div (5 + 3.33 - 2.08\%)\} = 10.66 \text{ h}$

Hence, total time taken = 4 + 10.66 = 14.66 h

Alternatively

$\left\{\begin{array}{l} A\text{'s filling rate} = 5\% \text{ of tank per hour} \\ B\text{'s filling rate} = 3.33\% \text{ of tank per hour} \end{array}\right\}$

Then, time taken to fill the tank (assuming no leakage develops)

$= \{100\% \div (5 + 3.33)\%\} = 12 \text{ h}$

Now, time taken to fill 2/3 of the tank (assuming no leakage)

$= \frac{2}{3} \times 12 = 8 \text{ h}$

It is given that due to leakage, $\frac{1}{4}$ of the water supplied by both goes out after the tank is one-third full. It implies that the filling rate has reduced by 25%, then the time taken will be $\left(\frac{25}{100-25} \times 100\right)$% i.e. 33.33% more than earlier time taken.

Hence, due to leakage, time taken to fill $\frac{2}{3}$ of the tank

$= 133.33\% \text{ of } 8 \text{ h} = 10.66 \text{ h}$

Then, total time taken to fill the tank = time taken to fill initially $\frac{1}{3}$ of the tank + 10.66 h

$= \frac{1}{3} \times 12 + 10.66 = 14.66 \text{ h}$

2. Part filled by A = 33.33% of the cistern per h

Part filled by A due to leak = 28.57% $[\because 100 \div 3.5 = 28.57]$

It means the leakage is drawing at the rate of (33.33 – 28.57)% i.e. 4.76% per hour.

Then, time taken to empty the tank

$= (100\% \div 4.76)\% = 21 \text{ h}$

[**Note** No need to get the figure in exact decimal. Only approx. values are required to arrive at option]

3. $\left\{\begin{array}{l} \text{1st pipe's filling rate} = 8.33\% \text{ of the cistern per minute} \\ \text{2nd pipe's filling rate} = 6.25\% \text{ of the cistern per minute} \end{array}\right\}$

Part filled in last 3 min = 3 (8.33 + 6.25)% = 43.74%

Before obstructions are removed, filling rate of 1st pipe = $\frac{7}{8}$ of 8.33% per min and filling rate of 2nd pipe = $\frac{5}{6}$ of 6.25% per min.

Then, time taken by them to fill remaining part i.e. (100 – 43.74)% at their lower filling rates

$= \frac{(100 - 43.74)\%}{\left(\frac{7}{8} \times 8.33 + \frac{5}{6} \times 6.25\right)\%} = \frac{56.26}{7.29 + 5.20}$

$= 4.5 \text{ min}$

Hence, after 4.5 min both pipes were in full flow.

4. Ratio of areas of cross sections of taps T_1 and $T_2 = \pi(2)^2 : \pi(1)^2 = 4:1$

It is obvious that the time taken by taps will be inversely proportion to the areas of respective cross-section.

Hence, required ratio = 1 : 4

5. Let the filling capacity be x m^3 /min.

Then, discharging capacity = $(x + 10)$ m^3 /min

As per question, $\frac{4800}{x} - \frac{4800}{(x+10)} = 16$

$\Rightarrow x^2 + 10x = 300x + 3000 - 300x$

$\Rightarrow x^2 + 10x - 3000 = 0$

$\Rightarrow x^2 - 50x + 60x - 3000 = 0$

$\Rightarrow x(x - 50) + 60(x - 50) = 0$

$\Rightarrow (x - 50)(x + 60) = 0$

$\Rightarrow x = 50 \text{ and } -60$

Since the filling capacity cannot be negative, $x = 50$ m^3 /min

Alternatively

The best option is to proceed through options.

(a) $25 \Rightarrow \frac{4800}{25} - \frac{4800}{(25+10)} \neq 16$

(b) $50 \Rightarrow \frac{4800}{50} - \frac{4800}{(50+10)} = 16$

Hence, option (b) is correct.

6. $\left.\begin{array}{l} \text{1st} = 100\% \text{ of tank per hour} \\ \text{2nd} = 50\% \text{ of tank per hour} \\ \text{3rd} = 33.33\% \text{ of tank per hour} \\ \text{4th} = 16.66\% \text{ of tank per hour} \end{array}\right\}$ hence, 200% by all four per hour.

Then, time taken by them working together

$= (100\% \div 200\%) = \frac{1}{2} \text{ h} = 30 \text{ min}$

7. On opening all the pipes, the tank is filled in 9 h, it implies that they are filling 11.11% per minute.

Let number of inlet pipes be x.

Then, number of outlet pipes = $6 - x$

Now, part filled by inlet pipes in 1 min – part emptied by outlet pipes in 1 min = 11.11%

$\Rightarrow \quad (x \times 11.11\%) - (6 - x) \times 16.66\% = 11.11\%$

$\Rightarrow \quad 27.77x = 111.11 \Rightarrow x = 4$

Alternatively You may proceed through options.

8. $\left\{\begin{array}{l}\text{Filling rate} = 6.25\% \text{ of the tank per minute.}\\ \text{Emptying rate} = 12.5\% \text{ of the tank per minute.}\end{array}\right\}$

Since filling rate is less than emptying rate, the tank will be emptied if both pipes are opened together.

The required time to empty $\frac{1}{2}$ of the tank

$= \{50\% \div (12.5 - 6.25)\%\} = 8$ min

Hence, option (c) is correct.

9. $\left\{\begin{array}{l}A\text{'s filling rate} = 10\% \text{ of the tank per minute.}\\ B\text{'s emptying rate} = 6.66\% \text{ of the tank per minute.}\end{array}\right\}$

Hence, time taken to make the tank full

$= \{100\% \div (10 - 6.66)\%\} = 30$ min

10. Effective filling rate per minute = 25 ÷ 5 = 5 L. It implies that the inflow rate (I) exceeds to the outflow rate (P) by 5 L/min. Further, the tank will never get filled only when the inflow rate is either equal to or less than increased outflow rate.

Hence, $\quad I - P = 5 \quad$ ...(i)

and $\quad I - 2P = 0$ or any negative value $\quad$ ...(ii)

If $I - 2P = 0$, value of $P = 5$ and if $I - 2P$ = negative value, value of $P > 5$.

Considering the value of options, option (c) is correct.

11. Ratio of efficiencies of A and $B = 3:1$

Then, ratio of time taken by A and B to fill the cistern = 1: 3

As given, (3 – 1) units = 32, then 1 unit = 16 min.

Now, time taken by A alone $= D \times \frac{(3+1)}{3}$

[Where, D is time taken by them together to fill the cistern]

$\Rightarrow \quad 16 = D \times \frac{4}{3} \Rightarrow D = 12$ min

12. $\left\{\begin{array}{l}(A + B) = 2.5\% \text{ of tank per min.}\\ A = 1.66\% \text{ of tank per min.}\end{array}\right\}$ hence, 0.84% by B per min.

Time taken by B alone to fill the tank

$= (100\% \div 0.84\%) = 120$ min

13. $\left\{\begin{array}{l}\text{1st tap's filling rate} = 20\% \text{ of the tub per minute}\\ \text{2nd tap's filling rate} = 14.28\% \text{ of the tub per minute}\\ \text{3rd tap's emptying rate} = 33.33\% \text{ of the tub per minute}\end{array}\right\}$

Hence, time taken to make the tank full

$= \{100\% \div (20 + 14.28 - 33.33)\%\} = 105$ min

14. Ratio of efficiencies of pipe A and $B = 1:7$

Now, time taken by A alone $= D \times \frac{(1+7)}{1}$

(Where, D is taken by them together to fill the cistern)

$\Rightarrow \quad 24 = D \times \frac{8}{1} \Rightarrow D = 3$ min

15. $\left\{\begin{array}{l}A\text{'s filling rate} = 5\% \text{ of the cistern per minute.}\\ B\text{'s filling rate} = 3.33\% \text{ of the cistern per minute.}\\ C\text{'s filling rate} = 6.66\% \text{ of the cistern per minute.}\end{array}\right\}$

In first three minutes part filled when all three are kept open successively = (5 + 3.33 – 6.66) = 1.66%

Then, time required to fill 100% of the cistern

$= 3 \times (100\% \div 1.66\%) = 3 \times 60 = 180$ min

16. $\left\{\begin{array}{l}A\text{'s filling rate} = 8.33\% \text{ of the cistern per minute}\\ B\text{'s filling rate} = 6.66\% \text{ of the cistern per minute}\end{array}\right\}$

Part filled by B alone after A is turned off

$= 5 \times 6.66\% = 33.33\%$

It implies that 66.66% has initially been filled by both together and time taken by them to fill 66.66% of the cistern = $\{66.66\% \div (8.33 + 6.66)\%\} = 4.44$ min

It means A is turned of after 4.44 min.

17. $\left\{\begin{array}{l}A\text{'s filling rate} = 3.33\% \text{ of tank per minute}\\ B\text{'s filling rate} = 2.5\% \text{ of tank per minute}\end{array}\right\}$

Let the pipe B be closed after x minutes. It implies that B worked for only x min. and A worked for 20 min.

Obviously, $(x \times 2.5\%) + (20 \times 3.33\%) = 100\%$

$\Rightarrow \quad 2.5x = 33.33$

$\Rightarrow \quad x = 13\frac{1}{3}$ min

18. $\left\{\begin{array}{l}\text{1st inlet pipe's filling rate} = 12.5\% \text{ of}\\ \text{the tank per hour}\\ \text{2nd inlet pipe's filling rate} = 8.33\% \text{ of}\\ \text{the tank per hour}\\ \text{Escape pipe's emptying rate} = 16.66\% \text{ of}\\ \text{the tank per hour}\end{array}\right\}$

Part filled by both inlet pipes from 1 pm to 3 pm

$= (2 \times 12.5\%) + (1 \times 8.33\%) = 33.33\%$

Clearly, since 3 pm, all the three pipes work. Then time taken by them to fill the remaining part

$= \frac{(100 - 33.33)\%}{(12.5 + 8.33 - 16.66)\%} = 16$ h

Hence, the cistern will be filled at 16 h beyond 3 pm i.e. at 7 am next day.

19. Ratio of filling rates of three pipes $= 2^2 : 3^2 : 4^2 = 4:9:16$

Then, ratio of time taken by them $= \frac{1}{4} : \frac{1}{9} : \frac{1}{16} = 36:16:9$

As given, 9 units = 58 min. then, 36 units

= 232 min and 16 units = 103.111 min

Hence, $\left\{\begin{array}{l}\text{1st pipe's filling rate} \Rightarrow 232 \text{ min}\\ \Rightarrow 0.43\% \text{ of the cistern per minute}\\ \text{2nd pipe's filling rate} \Rightarrow 103.11 \text{ min}\\ \Rightarrow 0.97\% \text{ of the cistern per min}\\ \text{3rd pipe's filling rate} \Rightarrow 58 \text{ min}\\ \Rightarrow 1.72\% \text{ of the cistern per min}\end{array}\right\}$

Now, time taken by them together to fill the cistern

$= \{100\% \div (0.43 + 0.97 + 1.72)\%\} = 32$ min

20. $\left\{\begin{array}{l} A\text{'s filling rate} = 3.33\% \text{ of the tank per hour} \\ B\text{'s filling rate} = 2.22\% \text{ of the tank per hour}\end{array}\right\}$

In 2 h, part of cistern filled = (3.33 + 2.22)% = 5.55%

Then, in 36 h ⇒ 100%

Hence, required time = 36 h

21. Ratio of efficiencies of A, B and $C = 2:3:4$

Also, $(A+B) \Rightarrow 3$ h ⇒ 33.33% per hour.

It means, (2 + 3) units ⇒ 33.33

∴ 2 units = 13.33%, 3 units = 20% and 4 units = 26.66%

Hence, A's, B's and C's contribution per hour = 13.33%, 20% and 26.66 %

As given, A is used as emptying pipe, B and C are used as filling pipe.

Then, in 3 h, part filed

$= (-13.33\% + 20\% + 26.66\%) + (13.33\% - 20\% + 26.66\%) + (13.33\% + 20\% - 26.66\%)$

$= 33.33\% + 20\% + 6.66\% = 60\%$

Hence, 40% i.e. $\frac{2}{5}$ of the tank is to be filled at the end of three hours.

22. $\left\{\begin{array}{l} \text{One Inlet pipe's filling rate} \\ = 12.5\% \text{ of the tank per hour} \\ \text{One Outlet pipe's emptying rate} \\ = 8.33\% \text{ of the tank per hour}\end{array}\right\}$

As given, $M \times 12.5\% - N \times 8.33\% = 16.66\%$

[∵ they together fill in 6 h]

$$\Rightarrow \quad M = \frac{16.66 + 8.33N}{12.5}$$

If $N = 1$, $M = 2$. Further, if $N = 4$, $M = 4$ and so on. Hence, there is no definite relation between M and N.

∴ Option (e) is correct.

23. $\left\{\begin{array}{l} \text{1st tap's filling rate} \\ = 10\% \text{ of the cistern per minute} \\ \text{2nd tap's filling rate} \\ = 6.66\% \text{ of the cistern per minute} \\ \text{(1st + 2nd + waste) pipes' filling rate} \\ = 5.55\% \text{ of the cisten per minute}\end{array}\right\}$

Hence, waste pipe's emptying rate

$= (10 + 6.66 - 5.55)\% = 11.11\%$

Then, time taken by waste pipe to empty the full cistern

$= (100\% \div 11.11\%) = 9$ min

24.

$\left\{\begin{array}{l} A\text{'s filling rate} = 8.33\% \text{ of the cistern per minute} \\ B\text{'s filling rate} = 6.66\% \text{ of the cistern per minute} \\ C\text{'s emptying rate} = 16.66\% \text{ of the cistern per minnute}\end{array}\right\}$

Part filled by $(A+B)$ in first 5 min

$= 5(8.33 + 6.66)\% = 75\%$

Clearly, 75% of the cistern will be emptied at the rate of {16.66 – (8.33 + 6.66)}% per minute.

Then, time required = (75% ÷ 1.66%) = 45 min

25. Ratio of efficiencies of X and Y = 100% : 160% = 5 : 8

Now, time taken by Y alone $= 40 \times \frac{(5+8)}{8} = 65$ h

26. $\left\{\begin{array}{l} \text{Inlet's filling rate} = 22.22\% \text{ of the cistern per hour} \\ \text{(Inlet + leak)'s filling rate} \\ = 20\% \text{ of the cistern per hour [∵ they take 5 hr]}\end{array}\right\}$

Hence, part emptied by the leakage per hour

$= (22.22 - 20)\% = 2.22\%$

Then, time taken by leakage to empty the full cistern

$= (100\% \div 2.22\%) = 45$ h

27. Clearly, A fills a bucket in 6 min, B fills in 7.5 min and C fills in 10 min.

Hence, $\left\{\begin{array}{l} A\text{'s filling rate} = 16.66\% \text{ of the bucket per minute} \\ B\text{'s filling rate} = 13.33\% \text{ of the bucket per minute} \\ C\text{'s filling rate} = 10\% \text{ of the bucket per minute}\end{array}\right\}$

Part of bucket filled by them together = 40% of the bucket per minute.

Then, time taken by them to fill a bucket = 2.5 min

∴ They all together can fill 48 buckets in 2 h. It means the capacity of the tank is equal to the capacity of 48 buckets. Further, capacity of 48 buckets = 48 × 5 = 240 L

28. $\left\{\begin{array}{l} A\text{'s filling rate} = 5\% \text{ of the tank per minute} \\ B\text{'s filling rate} = 6.66\% \text{ of the tank per minute} \\ C\text{'s filling rate} = 8.33\% \text{ of the tank per minute}\end{array}\right\}$

Clearly, in 3 min part filled by all three pipe = 20%

Then, time taken by them to fill 100% of the tank

$= 3 \times 5 = 15$ min

29. $\left\{\begin{array}{l} \text{Cold water pipe's filling rate} = 10\% \text{ of the bath per minute} \\ \text{Hot water pipe's filling rate} = 6.66\% \text{ of the bath per minute}\end{array}\right\}$

Obviously, part filled by cold and hot water pipes in 4 min

= part emptied by waste pipe in 6 min.

⇒ 4 (10 + 6.66)% of the bath is emptied by waste pipe in 6 min

∴ Time taken by waste pipe to empty 100% of the bath

= 9 min

30. $\left\{\begin{array}{l} \text{Leakage's emptying rate} = 16.66\% \text{ of tank per hour} \\ \text{(Inlet + leak)'s emptying rate} = 12.50\% \text{ of tank per hour}\end{array}\right\}$

Hence, part filled by inlet pipe per hour

$= (16.66\% - 12.50)\% = 4.16\%$

As given, 4.16% of the tank = 4 × 60L [∵ in hour = 4 × 60L]

∴ 100% of the tank = 4 × 60 × 24 = 5760 L

31. $\left\{\begin{array}{l} A\text{'s filling rate} = 5\% \text{ of the tank per minute} \\ B\text{'s filling rate} = 4.16\% \text{ of the tank per minute} \\ (A+B+C)\text{'s filling rate} = 6.66\% \text{ of the tank per minute}\end{array}\right\}$

Hence, C's emptying rate per min

$= (5 + 4.16 - 6.66)\% = 2.5\%$

As given, 2.5% of the tank = 3 gallons

∴ 100% of the tank = 3 × 40 = 120 gallons

32. $\left\{\begin{array}{l}P\text{'s filling rate} = 33.33\% \text{ of the cistern per minute}\\ Q\text{'s filling rate} = 16.66\% \text{ of the cistern per minute}\\ R\text{'s emptying rate} = 25\% \text{ of the cistern per minute}\end{array}\right\}$

Part filled in first 2 min = 2 (33.33 + 16.66 – 25)% = 50%

As given, after 2 min from beginning, only pipes P and Q works.

Then, time taken by $(P + Q)$ to fill remaining 50% of the cistern = {50% ÷ (33.33 + 16.66)%} = 1 min

Hence, total time taken to fill the cistern = 2 + 1 = 3 min

33. $\left\{\begin{array}{l}A\text{'s filling rate} = 3.33\% \text{ of tank per hour}\\ B\text{'s filling rate} = 5\% \text{ of tank per hour}\end{array}\right\}$

Let the tank be completely full in x h.

It is clear that pipe B is open for x h and pipe A is for $\frac{x}{2}$ h.

Hence, $(x \times 5\%) + \left(\frac{x}{2} \times 3.33\%\right) = 100\%$

$\Rightarrow \quad x \times 13.33\% = 200\%$

$\Rightarrow \quad x = 15\text{ h}$

34. The best way is proceed through options.

(a) 10 h ⇒ then 2nd and 3rd pipe will take 5 h and 1 h respectively.

Hence, 1st ⇒ 10%, 2nd ⇒ 20% and

3rd ⇒ 100%. Further, (10 + 20)% ≠ 100%.

(b) 15 h ⇒ then 2nd and 3rd pipe will take 10 h and 6 h respectively. Hence, 1st ⇒ 6.66%, 2nd ⇒ 10% and 3rd ⇒ 16.66%.

Further, (6.66 + 10)% = 16.66%

Hence, option(b) is correct.

Alternatively

Let 1st pipe can fill the tank in x h.

Then, time taken by 2nd pipe to fill the tank = $(x - 5)$ h and time taken by 3rd pipe to fill the tank = $(x - 9)$ h

Now, time taken by first two pipes to fill the tank

$$= \frac{x(x-5)}{x+(x-5)} = (x - 9), \text{ as given}$$

$\Rightarrow \quad x^2 - 5x = 2x^2 - 23x + 45$

$\Rightarrow \quad x^2 - 18x + 45 = 0$

$\Rightarrow \quad x^2 - 15x - 3x + 45 = 0$

$\Rightarrow \quad x(x - 15) - 3(x - 15) = 0$

$\Rightarrow \quad (x - 3)(x - 15) = 0$

$\Rightarrow \quad x = 3$ and 15

$\Rightarrow \quad x = 3$ is not possible as value of time taken by 2nd and 3rd pipes will be negative at $x = 3$.

Hence, $x = 15$ h

35. $\left\{\begin{array}{l}A\text{'s filling rate} = 6.66\% \text{ of the cistern per minute}\\ B\text{'s filling rate} = 5.55\% \text{ of the cister per minute}\end{array}\right\}$

In case 1st pipe is closed, the required time

$$= \frac{100 + (3 \times 6.66)}{(6.66 + 5.55)\%} = 9.82 \text{ min}$$

In case 2nd pipe is close, required time

$$= \frac{100 + (3 \times 5.55)}{(6.66 + 5.55)\%} = 9.55 \text{ min.}$$

Hence, option (d) is correct.

36. $\left\{\begin{array}{l}A\text{'s filling rate} \Rightarrow 8.33\% \text{ of the cistern per hour}\\ B\text{'s filling rate} \Rightarrow 6.66\% \text{ of the cister per hour}\\ C\text{'s filling rate} \Rightarrow 5\% \text{ of the cister per hour}\end{array}\right\}$

Now, in first 2 h part filled

$$= (8.33 + 6.66)\% + (8.33 + 5)\% = 28.33\%$$

∴ In 6 h part filled = 85%

In 7th h, only A and B will be together. Then, time taken by $(A + B)$ to fill the remaining 15% = 1 h

Hence, total time taken = 6 + 1 = 7 h

37. $\left\{\begin{array}{l}A\text{'s filling rate} = 12.5\% \text{ of the cone per hour}\\ B\text{'s filling rate} = 8.33\% \text{ of the cone per hour}\\ C\text{'s emptying rate} = 25\% \text{ of the cone per hour}\end{array}\right\}$

It is obviously that when the height of water comes down to 50%, the volume of water in cone will only be 12.5% of the total capacity of the cone. It implies that when all pipes are opened, 87.5% of water of the cone has been drained.

[Recall, the properties regarding frustum of a cone in chapter 'Cylinder and Cone' : The radius of the cone cut will be half of the radius of the parent cone if the height of the cone cut is half of the height of the parent cone. Obviously, the volume of the cone cut will be 1/8 of the volume of the parent cone.]

Now, proceed through options :

(a) 20 (8.33 – 25)% + 19 × (12.5%) ≠ 87.5%

(b) 20 (8.33 – 25)% + 19.5 × (12.5%) ≠ 87.5%

(c) 20 (12.5 – 25)% + 19.5 × (8.33%) = 87.5%

Hence, option (c) is correct.

38. $\left\{\begin{array}{l}\text{Pipe } A\text{'s filling rate} = 10\% \text{ of the tank per hour}\\ \text{Pipe } B\text{'s filling rate} = 2.5\% \text{ of the tank per hour}\end{array}\right\}$

Time taken by $(A + B)$ to fill the tank = (100% ÷ 12.5%) = 8 h.

Clearly, part filled by $(A + B)$ in 80 min. (i.e. 1.33 h) is emptied by pipe C in 8 h 80 min. (i.e. 9.33 h)

Then, part filled by $(A + B)$ in 7 h, will be emptied by pipe C in $\frac{9.33}{1.33} \times 7$ h i.e. 49 h

39. Ratio of time taken by (10 type-A inlet + 45 type-B inlets) and (8 type-A inlets + 18 type-B inlets)

= 30 min : 60 min = 1 : 2

∴ Ratio of their efficiency = 2 : 1

Thus, (10 type-A inlets + 45 type-3 inlets)

= 2 (8 type-A inlets + 18 type-B inlets)

⇒ 1 type-A inlet = 1.5 type-B inlet

→ $1A = 1.5B$ (in short)

Now, $10A + 45B = (10 \times 1.5)B + 45B = 60B$

and $7A + 27B = (7 \times 1.5)B + 27B = 37.5B$

Since 60 B take 30 min to fill the tank, 37.5B will take 48 min.

Unit 4

Ratio and Its Extensive Applications

• Ratio, Proportion and Variation • Problems based on Ages • Partnership • Alligation /Mixture

26

Ratio, Proportion and Variation

Learning OBJECTIVES

After completing this chapter, aspirants will be able to

- Understand the concepts associated with ratio, proportion and variation and relation between them;
- Observe the diverse sorts of ratio with illustrations;
- Know critical properties of ratio and proportion and their applications directly or indirectly;
- Learn how value of a ratio derived from given ratios is computed;
- Observe the trends of alternation in ratio of numbers when numbers are changed either all or one or two;
- Realise the difference between approaches to know the trends of alternation when numbers are changed by same number and by different number;
- Observe the trend of questions based on number;
- Learn how to apply the concept of ratio while solving the problems based on income, expenditures and savings in a magical way;
- Understand the concept of value of different denominations whose number are given in a certain ratio;
- Observe the concept to alternation in ratio when the terms are changed by percentage;
- Know how to divide a sum into different parts given in ratio;
- Understand how the concept of ratio is useful in solving the problems based on wages-determination;
- Understand the different terms associated with the proportion;
- Observe the applicability of concept of proportion as substitute to unitary method with special reference to work and time, camp/garrison and provisions therein, etc.;
- Understand the real meaning of variations and their utility with illustrations;

Inside

- *Brief Introduction of Ratio, Proportion and Variation With Their Critical Properties*
- *Sorts of Ratio and Concepts of Ratio Derived From Given Ratios*
- *Problems Based on Ratio : Case-Wise viz. (i) Change in Ratio in Case a Number is Added to/Subtracted From Terms of Ratio; (ii) Changes in Ratio of Numbers When They All/One are (is) Either Increased/Decreased by Same/Different Number(s); (iii) Concepts of Income, Expenditures, Savings, Total Value of Different Denominations, Distribution of Amount in Terms of Ratio; etc.*
- *Problems Based on Concepts Associated With Proportion and Variation*
- ....
- .... *& Many More*
- *Work Out with Solutions*

- Understand the philosophy of the division of this chapter in various parts;
- Analyse the trend of the questions being asked in various competitive examinations;
- Recognise the types of question which can be easily solved through the given options;
- Solve the questions in moments;
- Grip on concepts helping to solve the questions without pen and paper;
- Learn many new concepts/fundamentals;
- Apply the pragmatic approaches in lieu of academic approaches; and
- Ponder on the alternative methods.

Introduction

The concepts associated with ratio, proportion and variation are substantially applicable in comparisons of two or more quantities and changes in their magnitudes **e.g.** comparison of income, savings, expenditures, ages, weights, heights, area, volume, perimeters, density, etc. In addition, questions related to data interpretation also seeks the concepts of ratio. It is also fact that the concepts associated with ratio is critical in solving the maximum number of arithmetic problems.

It is imperative to mention that the central concepts associated with the chapters namely 'Time and Distance', Races and Games, Problems based on Ages/Train, Partnership, Alligation/Mixture, etc. are based on conception of Ratio. Hence, we should have no hesitation to say that these chapters are broadly extension of conception of Ratio and its applications.

Further, it is to note that questions based on these concepts are regularly asked in various competitive examination like CAT, MAT, SSC, CDS, Bank-PO, Railway, etc, each and every year either directly or indirectly. Moreover, it has also importance in our daily life. That is why aspirants must have command over these concepts.

Ratio

Ratio is relation between two quantities of same kind which shows how many times one contains the other. In other words, the ratio of two quantities a and b in the same units, is the fraction a/b and we write it as $a : b$. Here, 1st term i.e. a is called antecedent and the 2nd term i.e. b is known as consequent.

e.g. The ratio 2 : 3 may be written as 2/3. Needless to say that the ratio connotes comparison among two or more than two things of the same kind, thus there may be a ratio between ₹ 25 and ₹ 50, but there cann't be the ratio between ₹ 25 and 50 mangoes.

Critical properties of Ratio

(i) The multiplication or division of each term of a ratio by the same non-zero number doesn't affect the ratio.

Hence, $a : b = ma : mb = \frac{a}{m} : \frac{b}{m}$

e.g. $2:3 = (2\times 5):(3\times 5) = (2\times 7):(3\times 7)$

$\Rightarrow \quad 2:3 = 10:15 = 14:21$

Also $4:6 = \frac{4}{2}:\frac{6}{2}$ (i.e. $2:3$) $= \frac{4}{3}:\frac{6}{3}\left(\frac{4}{3}:2\right)$

(ii) The ratio of two fractions can be expressed as a ratio of two integers.

$\Rightarrow \frac{a}{b}:\frac{c}{d} = \frac{ad}{bc} = ad : bc$

e.g. $\frac{2}{3}:\frac{4}{5} = (2\times 5):(4\times 3) = 5:6$

(iii) If either or both the terms of a ratio are a surd quantity, the ratio will never be into integral numbers, unless the surd quantities are equal.

$\Rightarrow \sqrt{3}:\sqrt{5}$ cannot be expressed as a ratio of integers.

(iv) If $\frac{a}{b} = \frac{c}{d} = \frac{e}{f} = ---- = k$, then $k = \frac{a+c+e+---}{b+d+f+---}$

(v) If $\frac{a}{b}, \frac{c}{d}, \frac{e}{f}, \frac{g}{h}$ --- are unequal fractions, then the ratio $\frac{(a+c+e+g+---)}{(b+d+f+h+---)}$ lies between the lowest and highest ratios.

e.g. Given ratios are $\frac{2}{3}, \frac{3}{4}, \frac{1}{3}$. Then, the ratio $\frac{2+3+1}{3+4+3}$ i.e. $\frac{3}{5}$ lies between the lowest ratio (i.e. 1/3) and the highest ratio (i.e. 3/4).

(vi) Equality of a ratio may be maintained by adding different numbers (in same ratio as given) to both numerator (antecedent) and denominator (consequent) of the given ratio.

i.e. $\frac{a}{b} = \frac{a+c}{b+d}$ $\left[\text{when, } \frac{c}{d} = \frac{a}{b}\right]$

e.g. $\frac{15}{45} = \frac{15+7}{45+21}$ $\left[\because \frac{7}{21} = \frac{15}{45}\right]$

Also, $\frac{a}{b} < \frac{a+c}{b+d}$ $\left[\text{when, } \frac{a}{b} < \frac{c}{d}\right]$

e.g. $\frac{15}{45} < \frac{15+5}{45+10}$ $\left[\because \frac{15}{45} < \frac{5}{10}\right]$

Further, $\frac{a}{b} > \frac{a+c}{b+d}$ $\left[\text{when, } \frac{a}{b} > \frac{c}{d}\right]$

e.g. $\frac{15}{45} > \frac{15+1}{45+5}$ $\left[\because \frac{15}{45} > \frac{1}{5}\right]$

(vii) $\dfrac{a+k}{b+k} > \dfrac{a}{b}$ [when, $1 > \dfrac{a}{b}$ and k is positive number]

$\dfrac{a+k}{b+k} < \dfrac{a}{b}$ [when, $1 < \dfrac{a}{b}$ and k is positive number]

$\dfrac{a-k}{b-k} > \dfrac{a}{b}$ [when, $1 < \dfrac{a}{b}$ and k is positive number]

$\dfrac{a-k}{b-k} < \dfrac{a}{b}$ [when, $1 > \dfrac{a}{b}$ and k is positive number]

(viii) If two equations each having three unknowns are given as $a_1x + b_1y + c_1z = 0$

and $a_2x + b_2y + c_2z = 0$

Then, $x : y : z = (b_1c_2 - b_2c_1) : (c_1a_2 - c_2a_1) : (a_1b_2 - a_2b_1)$

Note *The value of x, y, z cannot be determined unless a 3rd equation is given.*

e.g. Given, $2x + 4y - 16z = 0$

and $x - 3y + 7z = 0$

Then, $x : y : z = [(4 \times 7) - (-3 \times -16)]$

$: [(-16 \times 1) - (7 \times 2)] : [2 \times (-3) - (1 \times 4)]$

$= (-20) : (-30) : (-10) = 2 : 3 : 1$

In case of two equations each having two unknowns as

$a_1x + b_1y + c_1 = 0$

and $a_2x + b_2y + c_2 = 0$

Then, $x : y = (b_1c_2 - b_2c_1) : (c_1a_2 - c_2a_1)$

e.g. $7x - 12y + 44 = 0$ ...(i)

and $-x + 2y - 8 = 0$ ...(ii)

Then, $x : y = [(-12 \times -8) - (2 \times 44)]$

$: [44 \times (-1) - (-8 \times 7)]$

= [illegible] : 12 = 2 : 3

Sorts of Ratio

(i) Duplicate ratio of $(a : b)$ is $(a^2 : b^2)$.

(ii) Sub-duplicate ratio of $(a : b)$ is $(\sqrt{a} : \sqrt{b})$.

(iii) Triplicate ratio of $(a : b)$ is $(a^3 : b^3)$.

(iv) Sub-triplicate ratio of $(a : b)$ is $(\sqrt[3]{a} : \sqrt[3]{b})$ or $\left(a^{\frac{1}{3}} : b^{\frac{1}{3}}\right)$.

(v) Compounded ratio of $(a : b)$, $(c : d)$ and $(e : f)$ is (ace : bdf).

(vi) Inverse/reciprocal ratio of $(a : b)$ is (1/a : 1/b) or $(b : a)$.

Inverse/reciprocal ratio of $(a : b : c :$ ------) is $\left(\dfrac{1}{a} : \dfrac{1}{b} : \dfrac{1}{c} : \text{------}\right)$

(vii) Ratio of equality : when antecedent (A) = consequent (C). **e.g.** 4 : 4

(viii) Ratio of greater inequality : when $A > C$

e.g. 8 : 5

(ix) Ratio of less inequality : when $A < C$

e.g. 4 : 7

Concept of ratios derived from given ratios

Let $a : b = 3 : 2$ and $b : c = 3 : 4$, then you may find the values of the following ratios.

(i) $a : b : c$ (ii) $a : c$

Clearly, when a is 3 units, b is 2 units [$\because a : b = 3 : 2$], Further, given that when b is 3 units, c is 4 units [$\because b : c = 3 : 4$], Here, we can compare a to c only when value of b is same in both cases.

a (3) b (2)

b (3) c (4)

Now, in order to make values of b equal, we may multiply $b(2)$ by 3 and $b(3)$ by 2. But, it is to note that when $b(2)$ is multiplied by 3, a must be multiplied by 3 so as to maintain the earlier value of ratio $a : b$. Similarly, when we multiply $b(3)$ by 2, we must multiply c (4) by 2. Now, on multiplication, we have

a (9) b (6)

b (6) c (8)

Here, it is clear that, when a is 9, b is 6 and when b is 6, c is 8. Thus, we may construe that when a is 9, b is 6 and c is 8.

Hence, $a : b : c = 9 : 6 : 8$

The above result may be obtained through the following operations on the above principles/ concepts.

$a : b = 3 : 2$

$b : c = 3 : 4$

$a : b : c = (3\times3) : (2\times3) : (2\times4)$

$= 9 : 6 : 8$

Generic Version

$a : b = x : y$

$b : c = m : n$

$a : b : c = xm : ym : yn$

Hence, $a : c = xm : yn$

(ii) $a : c = ?$

From the above, $a : c = (3 \times 3) : (2 \times 4) = 9 : 8$

Alternatively

$\dfrac{a}{c} = \dfrac{a}{b} \times \dfrac{b}{c} = \dfrac{3}{2} \times \dfrac{3}{4} \Rightarrow \dfrac{a}{c} = \dfrac{9}{8}$

$\therefore \quad a : c = 9 : 8$

If $a : b = 2 : 3$, $b : c = 5 : 6$, $c : d = 7 : 5$, then what about $a : b : c : d$?

$a : b = 2 : 3$

$b : c = 5 : 6$

$a : b : c = (2\times5) : (3\times5) : (3\times6)$ ...Case-I

$c : d = 7 : 8$ as given ...Case-II

Now, we should equalise values of c appear in both cases so as to find $a : b : c : d$ and it may be done by multiplying values of c in Case I by 7 and value of c in Case II by (3×6).

Further, other associated ratios will also require the multiplication accordingly

Thus, $a:b:c = (2\times 5)\times 7 : (3\times 5)\times 7 : (3\times 6)\times 7$

and $c:d = 7\times(3\times 6) : 8\times(3\times 6)$

Hence, $a:b:c:d = (2\times 5\times 7):(3\times 5\times 7):(3\times 6\times 7):(8\times 3\times 6) = 70:105:126:144$

In brief,

If $a:b:c = x:y:z$

and $c:d = m:n$

Then, $a:b:c:d = xm:ym:zm:zn$

In direct way,

$a:b = 2:3$

$b:c = 5:6$

$c:d = 7:8$

$a:b:c:d = (2\times5\times7):(3\times5\times7):(3\times6\times7):(3\times6\times8) = 70:105:126:144$

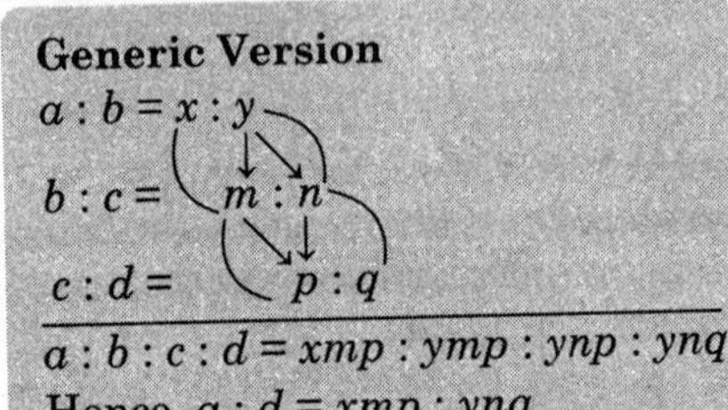

Here, $a:d$ may also be obtained either on above line or on the multiplication of given ratios i.e.

$$\frac{a}{d} = \frac{a}{b}\times\frac{b}{c}\times\frac{c}{d}$$

On the same analogy, if

$a:b:c:d = w:x:y:z$

$d:e = m:n$

$a:b:c:d:e = wm:xm:ym:zm:zn$

Or

$a:b = x:y$

$b:c = m:n$

$c:d = p:q$

$d:e = r:t$

$a:b:c:d:e = xmpr:ympr:ynpr:ynqr:ynqt$

Also, $\frac{a}{e} = \frac{a}{b}\times\frac{b}{c}\times\frac{c}{d}\times\frac{d}{e}$

Further,

If $a:b:c:d = w:x:y:z$

and $d:e:f = l:m:n$

Then, $a:b:c:d:e:f = wl:xl:yl:zl:zm:zn$

Problems Based on Ratio

Case I **Computation of value of ratio derived from given ratios**

Problem 1 If $a:b = \frac{2}{9}:\frac{1}{3}$, $b:c = \frac{2}{7}:\frac{5}{14}$ and $d:c = \frac{7}{10}:\frac{3}{5}$, then $a:b:c:d = ?$

Sol. Given, $a:b = \frac{2}{9}:\frac{1}{3}$

$b:c = \frac{2}{7}:\frac{5}{14}$

$c:d = \frac{3}{5}:\frac{7}{10}$

$$\therefore a:b:c:d = \left(\frac{2}{9}\times\frac{2}{7}\times\frac{3}{5}\right):\left(\frac{1}{3}\times\frac{2}{7}\times\frac{3}{5}\right):\left(\frac{1}{3}\times\frac{5}{14}\times\frac{3}{5}\right):\left(\frac{1}{3}\times\frac{5}{14}\times\frac{7}{10}\right)$$

$$= \frac{4}{105}:\frac{2}{35}:\frac{1}{14}:\frac{1}{12}$$

$$= \frac{4}{10}\times 420:\frac{2}{35}\times 420:\frac{1}{14}\times 420:\frac{1}{12}\times 420$$

[we multiply by LCM of denominators]

$= 16:24:30:35$

Note *You may convert fractions into integers before processing.*

e.g. $\Rightarrow a:b = \frac{2}{9}:\frac{1}{3} = \frac{2}{9}\times 9:\frac{1}{3}\times 9 = 2:3$ *and so on.*

Problem 2 If $A:B = 3:2$ and $B:C = 3:4$, then $A:C$ is equal to

Sol.

$A:B = 3:2$

$B:C = 3:4$

$A:B:C = 9:6:8$

Hence, $A:C = 9:8$

Alternatively

$$\frac{A}{C} = \frac{A}{B}\times\frac{B}{C} = \frac{3}{2}\times\frac{3}{4} = \frac{9}{8} \Rightarrow A:C = 9:8$$

Problem 3 If $W_1:W_2 = 2:3$ and $W_1:W_3 = 1:2$, then $W_2:W_3 = ?$ **[SSC (CGL-T-I)-10]**

Sol. $\frac{W_2}{W_3} = \frac{W_2}{W_1}\times\frac{W_1}{W_3} = \frac{3}{2}\times\frac{1}{2}$ $\left[\because \frac{W_1}{W_2} = \frac{2}{3}, \therefore \frac{W_2}{W_1} = \frac{3}{2}\right]$

$\Rightarrow W_2:W_3 = 3:4$

Problem 4 If $A:B = 3:4$ and $B:C = 6:5$, then $A:(A+C)$ is equal to **[SSC (CPO)-10]**

Sol.

$A:B = 3:4$

$B:C = 6:5$

$A:B:C = 18:24:20$

$= 9:12:10$

Now, $A:(A+C) = 9:(9+10) = 9:19$

Alternatively

$$\frac{A}{C} = \frac{A}{B} \times \frac{B}{C} = \frac{3}{4} \times \frac{6}{5} = \frac{9}{10}$$

$$\Rightarrow \quad \frac{C}{A} = \frac{10}{9}$$

$$\Rightarrow \quad \frac{C}{A} + 1 = \frac{10}{9} + 1$$

$$\Rightarrow \quad \frac{A+C}{A} = \frac{19}{9}$$

$$\therefore \quad A:(A+C) = 9:19$$

Problem 5 If $x = \frac{1}{3}y$ and $y = \frac{1}{2}z$, then $x:y:z = ?$

Sol. Given, $x = \frac{1}{3}y$ and $y = \frac{1}{2}z$

Hence, $x:y = 1:3$ and $y:z = 1:2$

Now, $x:y = 1:3$

and $y:z = \quad 1:2$

$\therefore$ $x:y:z = 1:3:6$

Problem 6 If $a:b = 5:7$ and $c:d = 2a:3b$, then $ac:bd = ?$ ***[SSC (CGL-PT)-05]***

Sol. Given, $\frac{a}{b} = \frac{5}{7}$ and $\frac{c}{d} = \frac{2a}{3b}$

$$\Rightarrow \quad \frac{a}{b} \times \frac{c}{d} = \frac{5}{7} \times \frac{2a}{3b}$$

$$\Rightarrow \quad \frac{ac}{bd} = \frac{10}{21} \times \frac{a}{b} = \frac{10}{21} \times \frac{5}{7} = \frac{50}{147}$$

$$\Rightarrow \quad ac:bd = 50:147$$

Problem 7 If $a:(b+c) = 1:3$ and $c:(a+b) = 5:7$, then $b:(a+c) = ?$

Sol. Given, $\frac{a}{b+c} = \frac{1}{3} \Rightarrow \frac{b+c}{a} = \frac{3}{1}$

$$\Rightarrow \quad \frac{b+c}{a} + 1 = \frac{3}{1} + 1$$

$$\Rightarrow \quad \frac{a+b+c}{a} = \frac{4}{1} \quad \text{...(i)}$$

Also given,

$$\frac{(a+b)}{c} = \frac{7}{5}$$

$$\Rightarrow \quad \frac{a+b+c}{c} = \frac{12}{5} \quad \text{...(ii)}$$

Further, Eq. (i) may be written as

$$\frac{a+b+c}{a} = \frac{4 \times 3}{1 \times 3} = \frac{12}{3} \quad \text{...(iii)}$$

On comparing Eqs. (ii) and (iii), we get $a = 3$ and $c = 5$

$\therefore$ $b = 12 - 3 - 5 = 4$

Now, $b:(a+c) = 4:(3+5) = 4:8 = 1:2$

Problem 8 If $a:b = 2:3$, $b:c = 4:5$ and $c:d = 3:4$, then $a:c = ?$

Sol. $a:b = 2:3$

and $b:c = \quad 4:5$

$\therefore$ $a:b:c = 8:12:15$

Thus, $a:c = 8:15$

Problem 9 If $a:b:c:d = 2:3:4:5$ and $d:e:f = 2:3:4$, then $a:b:e = ?$

Sol. $a:b:c:d = 2:3:4:5$

$d:e:f = \quad 2:3:4$

$\therefore a:b:c:d:e:f = 4:6:8:10:15:20$

Thus, $a:b:e = 4:6:15$

Problem 10 If $A:B:C = 2:3:4$, then $\frac{A}{B}:\frac{B}{C}:\frac{C}{A}$ is equal to

Sol. $\Rightarrow \frac{A}{B}:\frac{B}{C}:\frac{C}{A} = \frac{2}{3}:\frac{3}{4}:\frac{4}{2} = \frac{2}{3}:\frac{3}{4}:2$

$= \frac{2}{3} \times 12 : \frac{3}{4} \times 12 : 2 \times 12$ [multiplying by LCM of denominators]

$= 8:9:24$

Problem 11 If $2A = 3B = 4C$, then $A:B:C = ?$

Sol. Given, $2A = 3B \Rightarrow \frac{A}{B} = \frac{3}{2} \Rightarrow A:B = 3:2$

Further, $3B = 4C \Rightarrow \frac{B}{C} = \frac{4}{3} \Rightarrow B:C = 4:3$

Now, $A:B = 3:2$

and $B:C = \quad 4:3$

$\therefore$ $A:B:C = 12:8:6 = 6:4:3$

Alternatively

Given, $2A = 3B = 4C$

$$\Rightarrow \quad \frac{2A}{12} = \frac{3B}{12} = \frac{4C}{12} \Rightarrow \frac{A}{6} = \frac{B}{4} = \frac{C}{3}$$

$$\Rightarrow \quad A:B:C = 6:4:3$$

Problem 12 If $A:B = 1:2$, $B:C = 3:4$, $C:D = 6:9$ and $D:E = 12:16$, then $A:B:C:D:E$ is equal to

Sol.

$A:B = 1:2$

$B:C = \quad 3:4$

$C:D = \quad\quad 2:3$

$D:E = \quad\quad\quad 3:4$

$A:B:C:D:E = 18:36:48:72:96$

$= 3:6:8:12:16$

Problem 13 If $A:B = 4:9$ and $A:C = 2:3$, then $(A+B):(A+C)$ is ? ***[SSC (10 + 2) LDC/DEO-11]***

Sol. $\frac{B}{C} = \frac{B}{A} \times \frac{A}{C} = \frac{9}{4} \times \frac{2}{3} = \frac{3}{2}$

Now, $A:B = 4:9$

$B:C = \quad 3:2$

$\therefore$ $A:B:C = 12:27:18 = 4:9:6$

Then, $(A+B):(A+C) = (4+9):(4+6) = 13:10$

Alternatively

$A:B = 4:9$

$A:C = 2:3 = 2 \times 2 : 3 \times 2 = 4:6$

$\therefore A:B:C = 4:9:6$

Now, $(A+B):(A+C) = (4+9):(4+6) = 13:10$

Problem 14 If $\frac{2}{3}$ of $A = 75\%$ of $B = 0.6$ of C, then $A:B:C$ is? **[SSC (CGL-PT)-08]**

Sol. Given, $\frac{2}{3}A = B \times \frac{75}{100} = \frac{6}{10} \times C = k$ [suppose]

Now, $\frac{2}{3}A = k \quad \therefore A = \frac{3k}{2}$

$B \times \frac{75}{100} = k \quad \therefore B = \frac{4k}{3}$

$C \times \frac{6}{10} = k \quad \therefore C = \frac{5k}{3}$

$\therefore \quad A:B:C = \frac{3k}{2} : \frac{4k}{3} : \frac{5k}{3} = \frac{3k}{2} \times 6 : \frac{4k}{3} \times 6 : \frac{5k}{3} \times 6$

[we multiply by LCM of denominators to bring into integers]

$= 9k : 8k : 10k = 9:8:10$

Alternatively

$\frac{2}{3}A = B \times \frac{75}{100} \Rightarrow \frac{A}{B} = \frac{3 \times 75}{2 \times 100} = \frac{9}{8}$

and $B \times \frac{75}{100} = C \times \frac{6}{10} \Rightarrow \frac{B}{C} = \frac{6 \times 100}{10 \times 75} = \frac{4}{5}$

Now, $A:B = 9:8$

$B:C = 4:5$

$\therefore A:B:C = 9 \times 4 : 8 \times 4 : 8 \times 5 = 9:8:10$

Case II When individual value of antecedent and consequent may be used in finding the ratio of expressions.

Problem 15 If $x:y = 3:2$, then the ratio $(2x^2 + 3y^2):(3x^2 - 2y^2)$ is equal to **[SSC (CGL-PT)-05]**

Sol. $\frac{x}{y} = \frac{3}{2} = k$ [let]

$\therefore \quad \frac{x}{3} = \frac{y}{2} = k$

$\Rightarrow \quad x = 3k$ and $y = 2k$

Now, $(2x^2 + 3y^2):(3x^2 - 2y^2)$

$= [2 \cdot (3k)^2 + 3(2k)^2] : [3(3k)^2 - 2(2k)^2]$

$= 30k^2 : 19k^2 = 30:19$

Alternatively

Clearly, $x = 3$ and $y = 2$ [Technically, $x = 3$ and $y = 2$ is wrong, however we will assume to find the answer of all questions covered under this type so as to save the time]

$\therefore \quad (2x^2 + 3y^2):(3x^2 - 2y^2)$

$= (2 \cdot 3^2 + 3 \cdot 2^2):(3 \cdot 3^2 - 2 \cdot 2^2)$

$= 30:19$

Problem 16 If $a:b = c:d = e:f = 1:2$, then $(pa + dc + re) : (pb + dd + rf) = ?$

(a) $p:(d + r)$ (b) $(p + d):r$

(c) 2 : 3 (d) 1: 2

Sol. Given, $\frac{a}{b} = \frac{c}{d} = \frac{e}{f} = \frac{1}{2}$

i.e. $a = c = e = 1$ and $b = d = f = 2$ [technically, it is not true, however, we assume so to find the answer]

$\therefore \quad \frac{pa + dc + re}{pb + dd + rf} = \frac{(p \times 1) + (d \times 1) + (r \times 1)}{(p \times 2) + (d \times 2) + (r \times 2)}$

$= \frac{p + d + r}{2(p + d + r)} = \frac{1}{2}$

Problem 17 If $p:q:r = 1:2:4$, then $\sqrt{5p^2 + q^2 + r^2}$ is equal to

Sol. We may assume, $p = 1, q = 2$ and $r = 4$

$\therefore \quad \sqrt{[5 \times (1^2)] + (2^2) + (4^2)} = \sqrt{25} = 5 = 5 \cdot 1 = 5p.$

Case III Changes in ratio when a number is either added to or subtracted from both terms of ratio and that number is to be found.

Problem 18 What number should be added to or subtracted from each term of the ratio 17: 24 so that it becomes equal to 1: 2 ? **[SSC (CGL-T-1)-11]**

(a) 5 is subtracted

(b) 10 is added

(c) 7 is added

(d) 10 is subtracted

Sol. Let the number x be added.

Then, $\frac{17 + x}{24 + x} = \frac{1}{2} \Rightarrow 34 + 2x = 24 + x$

$\therefore \quad x = -10$

Hence, 10 should be subtracted.

Alternatively

Proceed through options.

(a) $\frac{17 - 5}{24 - 5} \neq \frac{1}{2}$ (b) $\frac{17 + 10}{24 + 10} \neq \frac{1}{2}$

(c) $\frac{17 + 7}{24 + 7} \neq \frac{1}{2}$ (d) $\frac{17 - 10}{24 - 10} = \frac{1}{2}$

Hence, correct option is (d).

Problem 19 The smallest integer, which when subtracted from both the terms of 6 : 7 give a ratio less than 16 : 21, is

(a) 5 (b) 4

(c) 3 (d) 2

Sol. Proceed through options

(a) $\frac{6 - 5}{7 - 5}$ is not less than $\frac{16}{21}$

(b) $\frac{6 - 4}{7 - 4}$ is not less than $\frac{16}{21}$

(c) $\frac{6 - 3}{7 - 3}$ is less than $\frac{16}{21}$

Hence, option (c) is correct.

Case IV **Changes in ratio of numbers when they all are either increased by the same number or decreased by the same number.**

Problem 20 The ratio between two numbers is 2 : 3. If each number is increased by 4, the ratio between them becomes 5 : 7. The difference between the numbers is? **[SSC (CGL-T-I)-II]**

Sol. Let the numbers be $2x$ and $3x$.

As given, $\frac{2x+4}{3x+4} = \frac{5}{7}$

$\Rightarrow \quad 15x + 20 = 14x + 28$

$\Rightarrow \quad x = 8$

$\therefore$ Required difference $= 3x - 2x$

$= x = 8.$

Alternatively I

Initial ratio = 2 : 3

Changed ratio = 5 : 7

Here, differences between antecedent-antecedent and consequent-consequent are not equal. Thus, we should try to make differences equal by multiplying both the terms of one ratio by a suitable number so that the terms of changed ratio are greater than respective terms of initial ratio if the original numbers are increased. In case the original numbers are decreased, the terms of the changed ratio must be lesser than respective terms of initial ratio. Initial ratio $= 2 : 3 = 2 \times 2 : 3 \times 2 = 4 : 6$ [No changed in value of ratio as both the terms have been multiplied by the same number]

Now, initial ratio = 4 : 6

and changed ratio = 5 : 7

Here, difference between antecedents = (5 – 4) = 1

and difference between consequents = (7 – 6) = 1

Clearly, differences are equal and terms of changed ratio are greater than respective terms of initial ratio.

It may be construed that difference of 1 unit is due to increment in each number by 4 i.e 1 unit $\Rightarrow$ 4

$\therefore$ Difference between numbers i.e. (6 – 4) units $= 2 \times 4 = 8.$

Alternatively II

Initial ratio = 2 : 3

Changed ratio = 5 : 7

Increment = 4 4

[Positive difference between outcomes obtained on cross multiplication in respect of upper two lines] units = [Positive difference between outcomes obtained on cross multiplication in respect of lower two lines]

i.e. $[(3 \times 5) - (2 \times 7)]$ units $= [(7 \times 4) - (5 \times 4)]$

$\Rightarrow$ 1 unit = 8

$\therefore$ (3 – 2) unit = 8

Remark

(i) If the both numbers are decreased by the same number, the same concept is applicable.

(ii) It is important to note that the unit-value [as mentioned in Alternatively II] is applicable for computation of terms given in original/initial ratio, not for terms as envisaged in the changed ratio. For illustration 1 unit = 8 (from the above). Then, increased 1st number = 5 units = 5 × 8 = 40, it is wrong.

To find the increased/changed 1st number, compute the original 1st number first. Here, original 1st number = 2 units $= 2 \times 8 = 16$ and then add 4. Hence, increased 1st number $= 16 + 4 = 20$, not 40. However, the unit- value [as mentioned in Alternatively I] is applicable for computation of terms given in original/initial ratio as well as the changed ratio.

Observation The above concept will play a vital role in solving questions based on ages and income-expenditures mentally without using pen and paper.

Problem 21 Two numbers are in the ratio 3 : 5. If each number is increased by 10, the ratio becomes 5 : 7. The smaller number is **[SSC (Investigator)-10]**

Sol. Let the number be $3x$ and $5x$.

As given, $\frac{3x+10}{5x+10} = \frac{5}{7} \Rightarrow 21x + 70 = 25x + 50$

$\Rightarrow \quad 4x = 20 \Rightarrow x = 5$

$\therefore$ Smaller number $= 3x = 3 \times 5 = 15$

Alternatively I Initial ratio = 3 : 5

Changed ratio = 5 : 7

Since the difference between antecedents and that between consequents are already equal, no need to enhance the terms of a ratio. Each difference is 2.

As construed, 2 units = 10

$\therefore$ 3 units = 15 [$\because$ smaller number = 3 units]

Hence, smaller number is 15.

Alternatively II

Initial ratio = 3 : 5

Changed ratio = 5 : 7

Increment = 10 10

Now, [Positive difference between outcomes obtained on cross multiplication in respect of upper two lines] units = [Positive difference between outcomes obtained on cross multiplication in respect of lower two lines]

i.e. $[(5 \times 5) - (3 \times 7)]$ units $= [(7 \times 10) - (5 \times 10)] \Rightarrow 4$ units = 20

$\therefore$ 3 units = 15

Hence, smaller number is 15.

Problem 22 The scores of Amal and Bimal in an examination are in the ratio 11 : 14. After an appeal, their scores increase by the same amount and their new scores are in the ratio 47 : 56. The ratio of Bimal's new score to that of his original score is **[CAT-18]**

Sol. Initial ratio = 11 : 14

Changed ratio = 47 : 56

Increment = x x

Now, $[(47 \times 14) - (56 \times 11)]$ units

$= [(56 \times x) - (47 \times x)] \Rightarrow 42$ units $= 9x$

$\therefore$ 14 units $= 3x$ [i.e. Bimal's original score]

Then, Bimal's new score $= 3x + x = 4x$

Now, required ratio $= 4x : 3x = 4 : 3$

Note *Try with other methods too.*

Problem 23 Two numbers are in the ratio 3 : 5. If 9 is subtracted from each, the new numbers are in the ratio 12 : 23. The smaller number is *[SSC (CGL-PT)-03, CDS-11, JMET-11]*

Sol. Here, numbers are decreased, thus, the terms of changed ratio must be lesser than respective terms of initial ratio even after multiplying the terms of one ratio by a suitable number.

Initial ratio $\Rightarrow 3 : 5 = 3 \times \frac{11}{2} : 5 \times \frac{11}{2} = 16\frac{1}{2} : 27\frac{1}{2}$

and changed ratio = $12 : 23$

Here, difference between respective terms = $4\frac{1}{2}$

As construed, $4\frac{1}{2}$ units = 9

$\therefore 16\frac{1}{2}$ units $= 9 \times \frac{2}{9} \times \frac{33}{2} = 33$ [$\because$ smaller number is $16\frac{1}{2}$ units]

Observation If you are feeling not comfortable with this method, adopt the other method given below

Alternatively

Initial ratio = 3 : 5

Changed ratio = 12 : 23

Decrease = 9 9

Now, $[(23 \times 3) - (12 \times 5)] = [(23 \times 9) - (12 \times 9)] \Rightarrow 9$ units = 99

$\therefore$ 3 units = 33

Hence, smaller number is 33.

Problem 24 The students in three classes are in the ratio 2 : 3 : 5. If 40 students are increased in each class, the ratio changes to 4 : 5 : 7. Originally, the total number of students was *[SSC (CGL-PT)-02, SSC (CGL-T-II-11]*

Sol. Let the students be $2x$, $3x$ and $5x$.

As given, $\frac{2x + 40}{3x + 40} = \frac{4}{5} \Rightarrow 10x + 200 = 12x + 160$

$\Rightarrow \quad 2x = 40 \Rightarrow x = 20$

Hence, total number of students

$= 2x + 3x + 5x = 10x = 10 \times 20 = 200$

Alternatively I

Initial ratio = 2 : 3 ; Changed ratio = 4 : 5

Since difference between antecedents and that between consequents are already equal, no need to enhance the terms of a ratio. Each difference is 2. As construed, 2 units = 40

$\therefore \quad (2 + 3 + 5)$ units = 200

Hence, total number of students = 200

Alternatively II

Initial ratio = 3 : 5 [you are free to take any two terms]

Changed ratio = 5 : 7

Increment = 40 40

Now, $[(5 \times 5) - (3 \times 7)]$ units $= [(7 \times 40) - (5 \times 40)]$

$\Rightarrow$ 4 units = 80

$\therefore$ $(2 + 3 + 5)$ units = 200

Hence, total number of students = 200

Case IVA Change in Ratio of Numbers when they all are Either Increased by Different Numbers or Decreased by Different Numbers.

We adopt the same process as envisaged in Type IV.

Problem 25 There are two numbers in ratio 4 : 5. When 1st number is increased by 4 and 2nd is increased by 11, new ratio becomes 2 : 3. Find the sum of the original numbers.

Sol. Let the original numbers be $4x$ and $5x$.

Then, $\frac{4x + 4}{5x + 11} = \frac{2}{3} \Rightarrow 12x + 12 = 10x + 22$

$\Rightarrow \quad 2x = 10 \Rightarrow x = 5$

$\therefore$ Their sum $= 4x + 5x = 9x = 9 \times 5 = 45$

Alternatively

Initial ratio = 4 : 5

Changed ratio = 2 : 3

Increment = 4 11

Now, $[(4 \times 3) - (5 \times 2)]$ units $= (11 \times 2) - (4 \times 3) \Rightarrow 2$ units = 10

$\therefore \quad (4 + 5)$ units = 45

Note *Try with other methods too.*

Problem 26 In a regiment, the ratio of number of officers and soldiers was 3 : 7.8 officers and 28 soldiers proceeded on leave and the ratio became 1 : 2. Find the number of soldiers before proceeding on leave. *[SSC (CGL-T-II)-17]*

Sol. Let the number of officers and soldiers before the battle be $3x$ and $7x$.

As given, $\frac{3x - 8}{7x - 28} = \frac{1}{2} \Rightarrow 6x - 16 = 7x - 28 \Rightarrow x = 12$

$\therefore \quad 7x = 84$

Alternatively

Initial ratio = 3 : 7

Changed ratio = 1 : 2

Decrease = 8 28

Now, $[(7 \times 1) - (3 \times 2)]$ unit

$= [(28 \times 1) - (8 \times 2)]$

$\therefore \quad 7$ units $= 12 \times 7 = 84$

Note *Try with other methods too.*

Problem 27 The ratio of the number of boys to the number of girls in a school of 640 students, is 5 : 3. If 30 more girls are admitted in the school, then how many more boys should be admitted so that the ratio of boys to that of the girls, becomes 14 : 9. *[SSC (CGL-T-1)-19]*

Sol. Initial ratio = 5 : 3

Changed ratio = 14 : 9

Increment = x 30

[let x more boys are admitted]

Now, $[(5\times 9)-(14\times 3)]$ units
$=(30\times 14)-(9\times x)$

$\Rightarrow$ 3 units $= 420 - 9x$

$\therefore$ $(5+3)$ units $= 8(140-3x) = 640$ [given]

$\Rightarrow$ $3x = 60 \Rightarrow x = 20$

Note *Try with other methods too.*

Problem 28 Three numbers are in the ratio 2 : 3 : 4. Their ratio is changed as 5 : 6 : 10, when 1st number is decreased by 5, 2nd by 15 and 3rd by 10. Find the smallest number.

Sol. We may take any two terms of the ratio.

Initial ratio = 2 : 3
Changed ratio = 5 : 6
Decrease = 5   15

Now, $[(5\times 3)-(6\times 2)]$ units
$=[(15\times 5)-(5\times 6)]$

$\Rightarrow$ 3 units = 45

$\therefore$ 2 units = 30

Hence, the smallest number is 30.

Problem 29 In an examination, Rama's score was $\frac{1}{12}$th of the sum of the scores of Mohan and Anjali. After a review, the score of each of them increased by 6. The revised scores of Anjali, Mohan and Rama were in the ratio 11 : 10 : 3. Then Anjali's score exceeded Rama's score by ***[CAT -19]***

Sol. Initial ratio = 1 : 12 [Rama's : (Mohan + Anjali)'s]
Changed ratio = 3 : (10+11)
Increment = 6   (6+6)

Now, $[(12\times 3)-(21\times 1)]$ units $=(21\times 6)-(12\times 3)$

$\Rightarrow$ 15 units = 90

$\therefore$ 1 unit = 6 i.e. Rama's initial score

$\therefore$ 12 units = 72 i.e. (Mohan + Anjali)'s initial score

Hence, Rama's new score = 6 + 6 = 12

and (Mohan + Anjali)'s new score = 72 + 6 + 6 = 84

Now, Anjali's new score = 44
[$\because$ (10 + 11) units = 84, $\therefore$ 11 units = 44]

$\therefore$ Required answer = 44 − 12 = 32

Note *Try with other methods too.*

Case IVB **Change in ratio of numbers when one is increased and another is decreased or *vice versa* either by the same number or by different numbers.**

Note *We adopt the same process but we add the outcomes obtained on cross multiplication in respect of lower two lines instead of difference between them as was done in earlier cases.*

Problem 30 Two numbers are in ratio 5 : 7. 1st is decreased by 5 and 2nd is increased by 3. In doing so, the new ratio becomes 5 : 9. Find the difference between original numbers.

Sol. Initial ratio = 5 : 7
Changed ratio = 5 : 9
Decrease and increment = 5   3

Now, [Positive difference between outcomes obtained on cross multiplication in respect of upper two lines] units = [Sum of outcomes obtained on cross multiplication in respect of lower two lines]

i.e. $[(5\times 9)-(5\times 7)]$ units $=[(5\times 3)+(5\times 9)]$

$\Rightarrow$ 10 units = 60

$\therefore$ (7 − 5) units = 12

Hence, difference between original numbers = 12

Problem 31 The ratio of the number of ducklings to that of ducks in a pond is 3 : 5. After few months, the new ratio becomes 3 : 8 as 63 ducklings become ducks. Find the initial number of ducks.

Sol. Initial ratio [ducklings to ducks] = 3 : 5
Changed ratio = 3 : 8
Decrease and increment = 63   63

Now, $[(8\times 3)-(5\times 3)]$ units
$=[(63\times 3)+(63\times 8)]$

$\Rightarrow$ 9 units = 693

$\therefore$ 5 units = 77 × 5 = 385

Hence, initial number of ducks = 385

Problem 32 Raju and Lalitha originally had marbles in the ratio 4: 9. Then Lalitha gave some of her marbles to Raju. As a result, the ratio of the number of marbles with Raju to that with Lalitha became 5 : 6. What fraction of her original number of marbles was given by Lalitha to Raju? ***[CAT-18]***

Sol. Initial ratio = 4 : 9
Changed ratio = 5 : 6
Increment and Decrease = x x

[let Lalitha gives x marbles to Raju]

Now, $[(9\times 5)-(4\times 6)]$ units
$=(6\times x)+(5\times x)$

$\Rightarrow$ 21 units $= 11x$

$\therefore$ 9 units $=\frac{33x}{7}$ i.e. Lalitha's initial marbles

Then, required answer $= x \div \frac{33x}{7}$

$=\frac{7}{33}$

Note *Try with other methods too.*

Problem 33 There are three numbers in ratio 2 : 3 : 5.1st number is increased by 4, 2nd is decreased by 8 and 3rd is increased by 2. In doing so, new ratio becomes 10 : 8 : 21. Find the 3rd number.

Sol. We may take any two terms of the ratios [say, 1st and 2nd term]

Initial ratio = 2 : 3

Changed ratio = 10 : 8

Increase and Decrease = 4 8

Now, [(10 × 3) − (2 × 8)] units = [(10 × 8) + (8 × 4)]

⇒ 14 units = 112

∴ 5 units = 40

Hence, the 3rd number = 40.

Alternatively

When we take 1st and 3rd terms of the ratio, then

Initial ratio = 2 : 5

Changed ratio = 10 : 21

Increase and Increase = 4 2

Now, [(10 × 5) − (21 × 2)] units = (21 × 4) − (10 × 2)

⇒ 8 units = 64

∴ 5 units = 40

Note *Here, the concept associated with Type IVA is applicable as 3rd line indicates increase and increase.*

Extract of the concepts associated with Type IV, IVA & IVB

(i) We arrange the given figures in three lines in same sequences as shown in above questions i.e. Initial Ratio as 1st line, Changed Ratio as 2nd line and alternation as 3rd line.

(ii) If third line indicates **increase** and **increase** or **decrease and decrease**, then [Positive difference between outcomes obtained on cross multiplication in respect of upper two lines] units = [Positive difference between outcomes obtained on cross multiplication in respect of lower two lines]

(iii) If third line indicates **increase and decrease** or **decrease and increase**, then [Positive difference between outcomes obtained on cross multiplication in respect of upper two lines] units = [Sum of outcomes obtained on cross multiplication in respect of lower two lines]

(iv) Value of unit(s) as referred at point (ii) and (iii) will be used for computation of terms as envisaged only in initial ratio. It means, unit-value is not applicable for the changed ratio.

Case V **Change in ratio of two numbers when one number is altered.**

Problem 34 Two numbers are in ratio 3 : 5. If 1st number is increased by 4, what is the new ratio?

Sol. Let the numbers be $3x$ and $5x$.

Then, new ratio = $(3x + 4) : 5x$

Conclusion We cannot find the new ratio in terms of integers unless relation between two numbers are given.

Problem 35 Two numbers are in ratio 3 : 4 and their sum is 42. 2nd number has been increased resulting in the ratio becomes 2 : 3. By how much the 2nd number has been increased.

Sol. Given, (3 + 4) units = 42

∴ 3 units = 18 [i. e. 1st number]

and 4 units = 24 [i.e. 2nd number]

As given, $\frac{18}{24 + x} = \frac{2}{3}$

⇒ $48 + 2x = 54 \Rightarrow x = 3$

Alternatively I

Given that only 2nd number has been increased, it means the 1st number remains unchanged. Further, initial ratio is 3 : 4 and new ratio = 2 : 3.

Since 1st number remains unchanged, we make the 1st term of both ratios equal by multiplying with a suitable number(s).

Now, initial ratio = 3 : 4 = 6 : 8

and new ratio = 2 : 3 = 6 : 9

Clearly, the 1st term of both ratio are equal and difference between 2nd terms is (9 − 8) unit and obviously, this is on account of increment in 2nd number. Hence, 2nd number has been increased by 1 unit.

As given, (6 + 8) units = 42

∴ 1 unit = 3

Thus, 2nd number has been increased by 3.

Alternatively II

Initial ratio = 3 : 4

Changed ratio = 2 : 3

Increment = 0 x [Let 2nd number is increased by x]

Now, [(3 × 3) − (2 × 4)] unit = $(2 \times x) - (3 \times 0)$

⇒ 1 unit = $2x$

∴ (3 + 4) units = $14x$ = 42 [given] ⇒ $x = 3$

Problem 36 The ratio of the numbers of boys and girls of a school with 504 students is 13 :11. What will be the new ratio if 12 more girls are admitted? *[SSC (CGL-PT)-02, CDS-17]*

Sol. Given, (13 + 11) units = 504

$\therefore 13 \text{ units} = \frac{504}{(13 + 11)} \times 13 = 273$ [i.e. number of boys]

and 11 units = 504 − 273 = 231 [i.e. number of girls]

Now, new ratio = 273 : (231 + 12) = 273 : 243 = 91 : 81

Alternatively I

$\because$ 504 students = (13 + 11) units

$\therefore$ 12 students = 12/21 units

Now, new ratio $= 13 : \left(11 + \frac{12}{21}\right) = 13 \times 21 : 243$

$= 3 \times 91 : 3 \times 81 = 91 : 81$

Alternatively II

Proceed through options. Let options be 91: 81, 81 : 91 etc.

(a) Given ratio = 13 : 11 = 91 : 77 [we equalised the first terms of ratios as number of boys remains unchanged]

New ratio = 91 : 81

Hence, difference between 2nd terms = 81 − 77 = 4 units

i.e. Number of girls has been increased by 4 units.

$\because$ (91 + 77) units = 504

$\therefore$ 4 units = 12 [Already given in question]

Hence, correct option is (a).

Problem 37 The ratio of the number of ladies to that of gents at a party was 3 : 2. When 20 more gents joined the party, the ratio was reversed. The number of ladies present at the party was

Sol. Let the number of ladies and gents be $3x$ and $2x$ respectively.

Then, $\frac{3x}{2x + 20} = \frac{2}{3}$

$\Rightarrow 9x = 4x + 40$

$\therefore x = 8$

$\therefore$ Number of ladies $= 3x = 3 \times 8 = 24$

Alternatively I

Initial ratio [ladies to gents] = 3 : 2 = 6 : 4

New ratio = 2 : 3 = 6 : 9 [we equalised the 1st terms of both ratios as number of ladies remains unchanged]

Here, difference between 2nd terms = (9 − 4) = 5

i.e number of gents has been increased by 5 units, but given that number of gents has been increased by 20.

Thus, 5 units = 20

$\therefore$ 6 units = 24 [$\because$ number of ladies : number of gents = 6 : 4]

Hence, number of ladies at party was 24.

Alternatively II

Initial ratio = 3 : 2 [Ladies to Gents]

Changed ratio = 2 : 3

Increment = 0 20

Now, $[(3 \times 3) - (2 \times 2)]$ units $= (20 \times 2) - (3 \times 0)$

$\Rightarrow$ 5 units = 40

$\therefore$ 3 units = 24

Problem 38 The numerator and denominator of a fraction are in the ratio 2 : 3. If 6 is subtracted from the numerator, the result is a fraction that has a value 2/3 of the original fraction. The numerator of the original fraction is

Sol. Initial ratio [numerator to denominator] = 2 : 3 = 6 : 9

New ratio $= \frac{2}{3} \times \frac{2}{3} = 4 : 9$

Since denominator has been unchanged, we equalised the second terms of both ratio.

Here, difference between 1st terms = 6 − 4 = 2 units

But given, 2 units = 6

$\therefore$ 6 units = 18

Hence, numerator = 18

Alternatively

Initial ratio = 2 : 3

Changed ratio = 4 : 9 [$\because$ new fraction = 2/3 of 2/3 = 4/9]

Decrease = 6 0

Now, $[(9 \times 2) - (4 \times 3)]$ units

$= (9 \times 6) - (4 \times 0)$

$\Rightarrow$ 6 units = 54

$\therefore$ 2 units = 18

Problem 39 The number of boys in a school was 30 more than the number of girls. Subsequently, a few more girls joined the same school. Consequently, the ratio of boys and girls became 3 : 5. Find the minimum number of girls, who joined subsequently. *[XAT-18]*

Sol. Initial ratio $= (x+30) : x$ [let number of girls be x initially]

Changed ratio = 3 : 5

Increment = 0 G [let G girls join later]

Now, $[5(x + 30) - (3 \times x)]$ units

$= (3 \times G) - (5 \times 0)$

$\Rightarrow (2x + 150)$ units $= 3G$

$\Rightarrow G = 50 + \frac{2}{3}x$

Minimum possible value of x to get value of $\frac{2}{3}x$ as an integer = 3

Hence, $G = 50 + \frac{2}{3} \times 3 = 52$

Note *Try with other methods too.*

Observation You will be, I hope, able to solve the above all questions without pen and paper. The above concept is very useful in chapter 'Alligation' also.

Case V A Change in ratio of three numbers when one or two numbers are altered.

Problem 40 There are three numbers in ratio 2 : 3 : 4. If 1st two numbers are increased by 6, the new ratio becomes 11:15:16. Find the difference between 1st and 3rd number.

Sol. Let the numbers be $2x$, $3x$ and $4x$.

But given that $(2x + 6) : (3x + 6) : 4x = 11 : 15 : 16$

$\Rightarrow \frac{2x + 6}{3x + 6} = \frac{11}{15} \Rightarrow 30x + 90 = 33x + 66$

$\Rightarrow 3x = 24$

$\therefore x = 8$

$\therefore$ Difference between 1st and 3rd number

$= 4x - 2x = 2x = 2 \times 8 = 16$

Alternatively

Initial ratio = 2 : 3 :4

Changed ratio = 11 : 15 : 16

Given that 3rd number is constant,thus, we should equalise 3rd term of both ratios.

Initial ratio = 2 : 3 : 4 = 8 : 12 : 16

Changed ratio = 11 : 15 : 16

Clearly, increase in 1st term of initial ratio = (11 – 8) = 3 units and increase in 2nd term of initial ratio = (15 – 12) = 3 units

Further given, 3 units = 6

$\therefore$ 1 unit = 2

Then, difference between 1st and 3rd number = (16 – 8) units

= 8 × 2 = 16

Problem 41 The numbers of mangoes in three baskets are in the ratio 2 : 3 : 5. When the number of mangoes in 2nd basket is decreased by 4, new ratio becomes 6 : 7 : 15. What is the number of mangoes in 3rd basket ?

Sol. It is clear that number of mangoes in 1st and 3rd basket remain unchanged. Thus, we should make the 1st terms of both ratios equal and also 3rd terms of both ratios equal by multiplying with a suitable number.

Initial ratio = 2 : 3 : 5 = 6 : 9 : 15

Changed ratio = 6 : 7 : 15

Clearly, decrease in 2nd term of initial ratio = 9 – 7 = 2 units

As given, 2 units = 4

$\therefore$ 15 units = 30

Hence, number of mangoes in 3rd basket = 30

Case VI Number-related problems.

Problem 42 The ratio of two numbers is 2 : 5. Find the 2nd number if 1st number is 16.

Sol. Let the number be $2x$ and $5x$.

Given, $2x = 16$

$\therefore$ $x = 8$

Hence, 2nd number = $5x = 5 \times 8 = 40$

Alternatively

$\because$ 2 units = 16

$\therefore$ 5 units = 40

Problem 43 If 78 is divided into three parts which are in the ratio $1 : \frac{1}{3} : \frac{1}{6}$, the middle part is

Sol. Given ratio $= 1 : \frac{1}{3} : \frac{1}{3} = 1 \times 6 : \frac{1}{3} \times 6 : \frac{1}{6} \times 6 = 6 : 2 : 1$

Clearly, (6 + 2 + 1) units = 78

$\therefore$ 2 units $= \frac{78}{(6+2+1)} \times 2 = 17\frac{1}{3}$

It may be constructed from the above that, Middle part

$= \frac{\text{term of middle part in ratio}}{\text{sum of all terms}}$

× the number which is to be divided

Or

Desired part $= \frac{\text{respective term}}{\text{sum of all terms}} \times$ the number to be divided.

On the above line, middle part

$$= \frac{\frac{1}{3}}{\left(1 + \frac{1}{3} + \frac{1}{6}\right)} \times 78 = \frac{2}{9} \times 78 = 17\frac{1}{3}$$

Problem 44 The ratio of two numbers is 3 : 8 and their difference is 115. The smaller of the two numbers is

Sol. Let the numbers be $3x$ and $8x$.

Given, $8x - 3x = 115$

$\therefore$ $x = 23$

Hence, smaller number = $3x = 3 \times 23 = 69$

Alternatively

Given, (8 – 3) units = 115

$\therefore$ 3 units = 69

Problem 45 Four numbers are in the ratio 1 : 2 : 3 : 4. Their sum is 16. The sum of the 1st and 4th number is equal to

Sol. Clearly, (1 + 2 + 3 + 4) units = 16

$\therefore$ (1 + 4) units = 8

Problem 46 Two whole numbers, whose sum is 144, can't be in the ratio.

(a) 5 : 7 (b) 3 : 5

(c) 3 : 4 (d) 4 : 5

Sol. Since the numbers are whole, the sum must exactly be divisible by the sum of terms of the ratio.

Hence, correct option is (c) as (3 + 4) can't divide 144.

Problem 47 The product of two positive integers is 1575 and their ratio is 9 : 7. The smaller integers is

Sol. Let the integers be $9x$ and $7x$.

Given, $9x \times 7x = 1575$

$\Rightarrow$ $x^2 = 25$ $\therefore x = 5$

Hence, smaller integer = $7x = 7 \times 5 = 35$

Alternatively

9 unit × 7 unit = 1575 = 1 (unit)^2 = 25

$\because$ 1 unit = $\sqrt{25} = 5$

$\therefore$ 7 units = 7 × 5 = 35

Problem 48 Two numbers are in the ratio 17 : 45. $\frac{1}{3}$rd of the smaller is less than $\frac{1}{5}$ of the bigger by 15. The smaller number is

Sol. Let the numbers be $17x$ and $45x$.

As given, $\left(17x \times \frac{1}{3}\right) = \left(45x \times \frac{1}{5}\right) - 15$

$\Rightarrow$ $9x - \frac{17x}{3} = 15 \Rightarrow \frac{10x}{3} = 15$

$\therefore$ $x = 4.5$

Hence, smaller number = 17 × 4.5 = 76.5

Alternatively

Given ratio = 17 : 45

Difference between $\frac{1}{3}$ of 17 and $\frac{1}{5}$ of 45 = $3\frac{1}{3}$

As given, $3\frac{1}{3}$ units = 15

∴ 17 units = 76.5

Problem 49 There is a ratio of 5 : 4 between two numbers. If 40% of the 1st is 12, then 50% of the 2nd number is

Sol. Let the numbers be $5x$ and $4x$.

Given that 40% of $5x = 12 \Rightarrow 2x = 12$

∴ $x = 6$

Thus, 50% of $4x = 2x = 2 \times 6 = 12$

Alternatively

$5 \times 40\% \Rightarrow 12$

Or, $2 \Rightarrow 12$

∴ 50% of $4 \Rightarrow 12$

Problem 50 Two numbers are such that the square of one is 224 less than 8 times the square of the other. If the numbers are in the ratio of 3 : 4, then their values are *[SSC (FCI-12 Asstt)-12]*

(a) 12,16 (b) 6,8
(c) 9,12 (d) 12,9

Sol. Let the numbers be $3x$ and $4x$.

Given, $(3x)^2 = 8(4x)^2 - 224$

$\Rightarrow 9x^2 = 128x^2 - 224$

$\Rightarrow 119x^2 = 224$

∴ $x = \sqrt{\frac{224}{119}}$

Clearly, we are not getting the value of x as positive integer, but the options are as positive integers. Thus, it may be possible that

$(4x)^2 = 8(3x)^2 - 224$

$\Rightarrow 16x^2 = 72x^2 - 224 \Rightarrow 56x^2 = 224$

$\Rightarrow x^2 = 4$

∴ $x = 2$

Now, numbers are 3×2 and 4×2 i.e. 6 and 8.

Alternatively

Proceed through options.

(a) $(12)^2 \neq 8(16)^2 - 224$

and also $16^2 \neq 8(12)^2 - 224$

(b) $6^2 \neq 8(8)^2 - 224$, but $8^2 = 8(6)^2 - 224$

Hence, option (b) is correct.

Problem 51 Three numbers are in the ratio of 3 : 2 : 5 and the sum of their squares is 1862. The smallest of these numbers is

Sol. Let the numbers be $3x$, $2x$ and $5x$.

Given, $(3x)^2 + (2x)^2 + (5x)^2 = 1862$

$\Rightarrow 9x^2 + 4x^2 + 25x^2 = 1862$

$\Rightarrow 38x^2 = 1862 \Rightarrow x^2 = 49$

∴ $x = 7$

Hence, the smallest number = $2x = 2 \times 7 = 14$

Alternatively

$[(3)^2 + (2)^2 + (5)^2]\,\text{unit}^2 = 1862$

$\Rightarrow \text{Unit}^2 = 49$

∴ 1 unit = 7

∴ 2 units = 14

Problem 52 Of the three numbers, the ratio of the 1st and the 2nd is 8 : 9 and that of the 2nd and 3rd is 3 : 4. If the product of the 1st and 3rd numbers is 2400, then the 2nd number is :

Sol. Given, 1st : 2nd = 8 : 9

and 2nd : 3rd = 3 : 4

∴ 1st: 2nd : 3rd = $(8 \times 3) : (9 \times 3) : (9 \times 4) = 8 : 9 : 12$

As given, $(8 \times 12)\,\text{unit}^2 = 2400 \Rightarrow \text{unit}^2 = 25$

∴ 1 unit = 5

∴ 9 units = $5 \times 9 = 45$

Hence, 2nd number = 45

Problem 53 Two numbers are in the ratio 2 : 3. If 2 is subtracted from the 1st and 2 is added to the 2nd, the ratio becomes 1: 2. The sum of the numbers is *[SSC (CGL-PT)-05]*

Sol. Let the numbers be $2x$ and $3x$.

As given, $(2x - 2) : (3x + 2) = 1 : 2$

$\Rightarrow 4x - 4 = 3x + 2 \Rightarrow x = 6$

∴ Sum of numbers = $2x + 3x = 5x = 5 \times 6 = 30$

Alternatively I

Initial ratio = 2 : 3

Changed ratio = 1 : 2

Decrease and Increment = 2 2

Now, $[(2 \times 2) - (3 \times 1)]$ unit = $(2 \times 2) + (1 \times 2) \Rightarrow$ 1 unit = 6

∴ (2 + 3) units = 30

Alternatively II

You may proceed through options. Let options be 28, 30, ...

(a) 28 ⇒ If sum is 28, numbers will be 11.2 and 16.8, then

$\frac{11.2 - 2}{16.8 + 2} \neq \frac{1}{2}$

(b) 30 ⇒ If sum is 30, then number will be 12 and 18, then

$\frac{12 - 2}{18 + 2} = \frac{1}{2}$

Hence, option (b) is correct.

Problem 54 If the sum of two quantities is equal to three times their difference, then the ratio of the two quantities is : *[SSC (CISF-ASI-10)]*

Sol. Let quantities be x and y.

Given, $x + y = 3(x - y)$

$\Rightarrow x + y = 3x - 3y$

$\Rightarrow -2x = -4y$

$\Rightarrow x = 2y \Rightarrow x : y = 2 : 1$

Alternatively

Proceed through options. Let options be 1 : 3, 3 : 1, 2 :1 and 2 : 3

(a) 1 : 3 ⇒ (1 + 3) ≠ 3 (3 − 1)

(b) 3 : 1 ⇒ (3 + 1) ≠ 3 (3 − 1)

(c) 2 : 1 ⇒ (2 + 1) = 3 (2 − 1)

Hence, option (c) is correct.

Problem 55 From each of the two given unequal numbers, half the smaller number is subtracted. Then, of the resulting numbers, the larger one is five times the smaller one. Then, the ratio of the larger to smaller one is *[SSC (10 + 2), LDC/DEO-11]*

Sol. Let numbers be x and y (where, $x > y$)

As given, $x - \frac{y}{2} = 5\left(y - \frac{y}{2}\right)$

$\Rightarrow \quad x = \frac{5y}{2} + \frac{y}{2} \Rightarrow x = 3y$

$\therefore \quad x : y = 3 : 1$

Alternatively

Proceed through options. Let options be 2 : 1, 3 : 2, 3 : 1 and 1 : 4.

(a) $2 - \frac{1}{2} \neq 5\left(1 - \frac{1}{2}\right)$ (b) $3 - \frac{2}{2} \neq 5\left(2 - \frac{2}{2}\right)$

(c) $3 - \frac{1}{2} = 5\left(1 - \frac{1}{2}\right)$

Hence, correct option is (c).

Problem 56 The sum of 5% of a number and 4% of the other is $\frac{2}{3}$ of the sum of 6% of the 1st number and 8% of the 2nd. The ratio of the 1st number to the 2nd is

Sol. Let the first number be x and second be y.

Then, $(x \times 5\%) + (y \times 4\%) = \frac{2}{3}\ [(x \times 6\%) + (y \times 8\%)]$

$\Rightarrow (x \times 5\%) + (y \times 4\%) = (x \times 4\%) + \left(y \times \frac{16}{3}\%\right)$

$\Rightarrow \quad (x \times 1\%) = \left(y \times \frac{4}{3}\%\right)$

$\Rightarrow \quad x \times 3\% = y \times 4\% \Rightarrow x : y = 4 : 3$

Alternatively

You may proceed through options. Let options be 2 : 3, 4 : 3, 3 : 4 and 3 : 2

(a) $(2 \times 5\%) + (3 \times 4\%) \neq \frac{2}{3}\ [(2 \times 6\%) + (3 \times 8\%)]$

(b) $(4 \times 5\%) + (3 \times 4\%) = \frac{2}{3}\ [(4 \times 6\%) + (3 \times 8\%)]$

Hence, option (b) is correct.

Problem 57 Two numbers are in the ratio $P : Q$. When 1 is added to both numbers, the ratio becomes $m : n$. Further, when 2 is added to both the numbers, it becomes 1 : 2. Find the sum of P and Q.

Sol. Clearly, when the original numerator and denominator each is increased by (1 + 2) the fraction becomes 1/2. Let fraction be $\frac{1}{x}$.

Hence, $\frac{1 + 3}{x + 3} = \frac{1}{2}$

$\Rightarrow \quad 8 = x + 3 \Rightarrow x = 5$

$\therefore$ Original fraction $= \frac{1}{5}$

Thus, $P = 1$ and $Q = 5$.

Then, $P + Q = 1 + 5 = 6$

If we assume the original fraction be $\frac{2}{x}$, then $\frac{2 + 3}{x + 3} = \frac{1}{2}$

Hence, $x = 7$

$\therefore$ Original fraction $= \frac{2}{7}$

Then, $P + Q = 2 + 7 = 9$

It is clear from the above that sum of P and Q will have infinite values. Therefore, we may proceed through options.

Let options be 9, 5,...

(a) $P + Q = 9$, then possible values of $P : Q = 1 : 8$ or $2 : 7$ or $3 : 6$ or $4 : 5$ but not like $8 : 1$ or $7 : 2$ as after addition of 3 to each term, the 1st term must be lesser than the 2nd term because the final ratio is 1 : 2.

Now, $1 : 8 \Rightarrow (1 + 3) : (8 + 3) \neq 1 : 2$

$2 : 7 \Rightarrow (2 + 3) : (7 + 3) = 1 : 2$

Hence, $P : Q = 2 : 7$. Then, sum of P and $Q = 9$

Problem 58 The ratio of a two-digit natural number to a number formed by reversing its digits is 4 : 7. The number of such pairs is *[CSAT-19]*

Sol. Let the unit digit be y and tens digit be x.

Then, given number $= 10x + y$

As given, $10x + y : 10y + x = 4 : 7$

$\Rightarrow \quad 66x = 33y$

$\Rightarrow \quad 2x = y$

The possible values of x and y may be

x	1	2	3	4
y	2	4	6	8

Hence, number of such pairs = 4

[Pairs of numbers (12, 21), (24, 42), (36, 63) and (48, 84)]

Case VII Concept of Income, Expenditures and Savings in terms of ratio.

Problem 59 The ratio of the income to the expenditure of a family is 10 : 7. If the family's expenses are ₹ 10500, then savings of the family is

Sol. Let income and expenditure of the family are $10x$ and $7x$ respectively.

Then, savings $= 10x - 7x = 3x$

As given, $7x = 10500$

$\therefore \quad 3x = 4500$

Hence, savings = ₹ 4500

Alternatively

Given, 7 units ⇒ ₹ 10500

∴ (10 − 7) units ⇒ ₹ 4500

Problem 60 Total salary of three persons A, B and C is ₹ 144000. They spend 80%, 85% and 75% respectively. If their savings are in the ratio 8 : 9 : 20, find C's salary. ***[CLAT-09]***

Sol. Given, (100 – 80)% of salary of A = 8 units

$\Rightarrow$ 100% of salary of A = 40 units

Similarly, 15% of the salary of B = 9 units

$\therefore$ 100% = 60 units

and 25% of salary of C = 20 units,

$\therefore$ 100% = 80 units

As given, (40 + 60 + 80) units = ₹ 144000

$\Rightarrow$ 80 units = ₹ 64000

Hence, C's salary = ₹ 64000

Problem 61 A, B and C received an amount of ₹ 8400 and distributed among themselves in the ratio 6 : 8 :7 respectively. If they save in the ratio 3 : 2 : 4 respectively and B saves ₹ 400, then what is the ratio of the expenditures of A, B and C respectively? ***[SSC (CGL-T-I)-17]***

Sol. Given, in context of income, (6 + 8 + 7) units = 8400

$\Rightarrow$ 1 unit = 400

Hence, A's income = 6 × 400 = 2400, B's income = 8 × 400 = 3200 and C's income = 7 × 400 = 2800

Further, in context of savings, 2 units = 400 $\Rightarrow$ 1 unit = 200

Hence, A's savings = 3 × 200 = 600, B's savings = 400 and C's savings = 4 × 200 = 800

Now, ratio of their expenditures

= (2400 – 600) : (3200 – 400) : (2800 – 800)

= 1800 : 2800 : 2000 = 9 : 14 : 10

Alternatively

Given, in context of income, (6 + 8 + 7) units = 8400

$\Rightarrow$ 1 unit = 400

Further, in context of savings, 2 units = 400 $\Rightarrow$ 1 unit – 200

Since unit value in context of income i.e. 400 is twice the unit value in context of savings i.e. 200, we should multiply the income ratio by (400 ÷ 200). Thereafter, subtract the saving ratio from the income ratio so as to obtain the expenditure ratio

Hence, required ratio

= (2 · 6 – 3) : (2 · 8 – 2) : (2 · 7 – 4) = 9 : 14 : 10

Problem 62 The income of A is 60% less than that of B, and the expenditure of A is equal to 60% of B's expenditure. If A's income is equal to 70% of B's expenditure, then what is the ratio of the savings of A and B? ***[SSC (CGL-T-I)-01]***

Sol. Ratio of income of A and B = 40 : 100 = 2 : 5

Ratio of expenditures of A and B = 3 : 5

Given, A's income is 70% of B's expenditure. It implies that (70% of 5) units should be A's income.

Thus, ratio of income of A and B = 3.5 : 8.75

[$\because$ 2 $\Rightarrow$ 3.5, $\therefore$ 5 $\Rightarrow$ 8.75]

and ratio of expenditures of A and B = 3 : 5

$\therefore$ Ratio of their savings = (3.5 – 3) : (8.75 – 5) = 2 : 15

Note *Try with other methods too.*

Problem 63 The annual incomes of A and B are in the ratio 4 : 3 and the ratio of their expenditures is 3 : 2. If each of them saves ₹ 600 in the year, the annual income of A is : ***[SSC(CGL-PT)-02, CPO-05, 06, CDS-07, IAS (Pre)-2000]***

Sol. Let annual incomes of A and B be ₹ $4x$ and ₹ $3x$ respectively.

$\therefore$ A's Annual expenditure = A's annual income – A's annual savings

= $4x - 600$

Similarly, annual expenditure of $B = 3x - 600$

As given, $\frac{4x - 600}{3x - 600} = \frac{3}{2}$

$\Rightarrow$ $8x - 1200 = 9x - 1800 \Rightarrow x = 600$

$\therefore$ Annual income of $A = 4x = 4 \times 600 =$ ₹ 2400

Alternatively I

Recall the concept of changes in ratio of numbers when they all are either increased by the same number or decreased by the same number. Now, the question reads as There are two numbers in the ratio 4 : 3 and it becomes 3 : 2 when both are subtracted by 600 each. Find the value of 1st number. Here,

Initial ratio [incomes] = 4 : 3

Changed ratio [expenditures] = 3 : 2

Since difference between antecedents and that between consequents are already equal, no need to enhance the terms of a ratio. Each difference is 1.

As construed, 1 unit = 600

$\therefore$ 4 units = 2400

Hence, annual income of A = ₹ 2400

Alternatively II

Recall the concept of changes in ratio of numbers when they all are either increased by the same number or decreased by the same number. Now, the question reads as There are two numbers in the ratio 4 : 3 and it becomes 3 : 2 when both are subtracted by 600 each. Find the value of 1st number. Here,

Initial ratio [incomes] = 4 : 3

Changed ratio [expenditures] = 3 : 2

Decrease (savings) = 600 600

Now, [(3 × 3) – (4 × 2)] unit (s) = [(600 × 3) – (600 × 2)]

$\Rightarrow$ 1 unit = 600

$\therefore$ 4 units = 2400

Hence, annual income of A = ₹ 2400

Problem 64 A and B have monthly incomes in the ratio 5 : 6 and monthly expenditures in the ratio 3 : 4. If they save ₹ 1800 and ₹ 1600 respectively, find the monthly income of B.

Sol. Let income of A and B be ₹ $5x$ and $6x$ respectively.

Then, $\frac{5x - 1800}{6x - 1600} = \frac{3}{4}$

$\Rightarrow$ $20x - 7200 = 18x - 4800$

$\Rightarrow$ $2x = 2400$

$\therefore$ $6x = 7200$

Hence, monthly income of B = ₹ 7200

Alternatively I

Ratio of incomes ⇒ 5 : 6 = 30 : 36

Ratio of expenditures ⇒ 3 : 4 = 21 : 28

∴ Ratio of savings ⇒ (30 – 21) : (36 – 28) = 9 : 8

Also, ratio of savings as given = 1800 : 1600 i.e. 9 : 8

It means, 1 unit = 200

∴ 36 units = 7200

Hence, monthly income of B = ₹ 7200.

Note *We write the ratio of incomes and expenditures in such a way that the difference between antecedents i.e. (30 – 21) is equal to 1st term of ratio of savings (i.e. 9) and also difference between consequents i.e. (36 – 28) is equal to 2nd term of the ratio of savings(i.e. 8).*

Alternatively II

Recall the concept of "Change in ratio of numbers when they all are either increased by different numbers or decreased by different numbers".

Ratio of incomes = 5 : 6

Ratio of expenditure = 3 : 4

Savings in ₹ = 1800 1600

Now, [(5 × 4) – (6 × 3)] unit(s)

= [(1800 × 4) – (1600 × 3)]

⇒ 2 units = 2400

∴ 6 units = 7200

Hence, monthly income of B = ₹ 7200

Problem 65 **A and B have monthly incomes in the ratio 5 : 6 and their monthly expenditures are ₹ 4200 and ₹ 5600. If they save in ratio 9 : 8. Find the monthly income of A.**

Sol. Let monthly incomes of A and B be ₹ $5x$ and ₹ $6x$ respectively.

∴ A's monthly savings = A's monthly income - A's monthly expenditure = $5x - 4200$

Similarly, monthly savings of $B = 6x - 5600$

As given, $\frac{5x - 4200}{6x - 5600} = \frac{9}{8}$

⇒ $40x - 33600 = 54x - 50400$

⇒ $14x = 16800$

⇒ $x = 1200$

∴ Monthly income of $A = 5x = 5 \times 1200$ = ₹ 6000

Alternatively

Ratio of incomes = 5 : 6

Expenditures in ₹ = 4200 5600

Ratio of savings = 9 : 8

Now, [(6 × 9) – (8 × 5)] units

= (5600 × 9) – (4200 × 8)

⇒ 14 units = 16800 ⇒ 5 units = 6000

Hence, monthly income of A = ₹ 6000

Problem 66 **The monthly incomes of A and B are ₹ 6000 and ₹ 7200 respectively and their expenditures are in ratio 3 : 4. Find the savings of A if their savings are in ratio 9 : 8.**

Sol. Let expenditures of A and B be ₹ $3x$ and $4x$ respectively. Then, their savings = $(6000 - 3x)$ and $(7200 - 4x)$

As given, $\frac{(6000 - 3x)}{(7200 - 4x)} = \frac{9}{8}$

⇒ $48000 - 24x = 64800 - 36x$

⇒ $12x = 16800$

⇒ $x = 1400$

∴ Monthly savings of $A = (6000 - 3x)$

= [6000 – (3 × 1400)] = ₹ 1800

Alternatively

Incomes = 600 7200

Expenditures in ₹ = 3 : 4

Ratio of savings = 9 : 8

Now, [(4 × 9) – (8 × 3)] units

= (7200 × 9) – (6000 × 8)

⇒ 12 units = 16800

⇒ 3 units = 4200

Hence, monthly savings of A = 6000 – 4200 = ₹ 1800

List of the above three questions/cases

Case I	Case II	Case III
Incomes = 5 : 6 Exp. = 3 : 4 Saving = 1800 1600	Incomes = 5 : 6 Exp. = 4200 5600 Savings = 9 : 8	Incomes = 6000 7200 Exp. = 3 : 4 Savings = 9 : 8
[(5 × 4) – (6 × 3)] unit (s) = [(1800 × 4) – (1600 × 3)]	[(6 × 9) – (8 × 5)] units = (5600 × 9) – (4200 × 8)	[(4 × 9) – (8 × 3)] units = (7200 × 9) – (6000 × 8)

Generic version

[Positive difference between outcomes obtained on cross multiplication of terms of ratios] units = [Positive difference between outcomes obtained on cross multiplication of values in ₹ with terms of ratio written lower].	[Positive difference between outcomes obtained on cross multiplication of terms of ratios] units = [Positive difference between outcomes obtained on cross multiplication of values in ₹ with terms of ratio written lower].	[Positive difference between outcomes obtained on cross multiplication of terms of ratios] units = [Positive difference between outcomes obtained on cross multiplication of values in ₹ with terms of ratio written lower].
The computation of incomes will be based on unit-value. It means unit-value is used for computation of terms as envisaged only in ratio written upper.	The computation of incomes will be based on unit-value. It means unit-value is used for computation of terms as envisaged only in ratio written upper.	The computation of expenditures will be based on unit-value. It means unit-value is used for computation of terms as envisaged only in ratio written upper.

Problem 67 Between two consecutive years my incomes are in the ratio of 2 : 3 and expenses in the ratio 5 : 9. If my income in the 2nd year is ₹ 45000 and my expenses in the 1st year is ₹ 25000, my total savings for the two years is

Sol. Incomes [1st yr and 2nd yr] ⇒ 2 : 3

Expenditure ⇒ 5 : 9

Given, 3 units = 45000

∴ 2 units = 30000

Hence, incomes are ₹ 30000 and ₹ 45000 for 1st and 2nd yr respectively.

Further, given that 5 units = 25000

∴ 9 units = 45000

Hence, expenses are ₹ 25000 and ₹ 45000 respectively for 1st and 2nd yr.

Thus, total savings =

(30000 – 25000) + (45000 – 45000) = ₹ 5000

Problem 68 The incomes of A, B and C are in the ratio 7 : 9 : 12 and their spending are in the ratio 8 : 9 :15. If A saves $\frac{1}{4}$th of his income, then the savings of A, B and C are in the ratio of **[SSC (CGL-T-I) -11, CDS-21]**

Sol. Let their incomes be $7x$, $9x$ and $12x$ respectively and their expenditure be $8y$, $9y$ and $15y$ respectively.

Then, A's savings $= 7x - 8y = 7x \times \frac{1}{4} \Rightarrow 21x = 32y$

$\therefore \quad y = \frac{21x}{32}$

Now B's savings $= 9x - 9y = 9x - 9 \times \frac{21x}{32} = \frac{99x}{32} \quad \left[\because y = \frac{21x}{32}\right]$

and C's savings $= 12x - 15y = 12x - 15 \times \frac{21x}{32} = \frac{69x}{32}$

$\therefore$ Ratio of their savings $= \frac{7x}{4} : \frac{99x}{32} : \frac{69x}{32}$

$= \frac{7x}{4} \times 32 : \frac{99x}{32} \times 32 : \frac{69x}{32} \times 32 = 56 : 99 : 69$

Alternatively I

Ratio of incomes = 7 : 9 : 12

Ratio of expenditures = 8 : 9 : 15

Now, proceed through options. Let options be 69 : 56 : 48, 56 : 99 : 69 ....

(a) 69 : 56 : 48 ⇒ savings of A = 69,

∴ His income = 69 × 4 = 276 $\quad [\because \text{savings} = \frac{1}{4} \text{ of income}]$

Then, his expenditure = 276 – 69 = 207 ⇒ 8 units [given]

Further, income of $B = \frac{276 \times 9}{7}$

$[\because 7 \text{ units} = 276 \therefore 9 \text{ units} = \frac{276 \times 9}{7}] = 355$ (approx.)

and expenditure of $B = \frac{207 \times 9}{7}$

$[\because 8 \text{ units} = 207 \therefore 9 \text{ units} = \frac{207 \times 9}{7}] = 233$ (approx.)

Hence, B's savings = 355 – 233 ≠ 56

∴ Option (a) is incorrect.

(b) 56 : 99 : 69 ⇒ Savings of A = 56

∴ His income = 56 × 4 = 224

Then, his expenditures = 224 – 56 = 168 ⇒ 8 units (given)

Hence, income of $B = \frac{224 \times 9}{7} = 288$

$[\because 7 \text{ units} = 224, \therefore 9 \text{ units} = \frac{224 \times 9}{7}]$

and expenditure of $B = \frac{168 \times 9}{8} = 189$

$[\because 8 \text{ units} = 168, \therefore 9 \text{ units} = \frac{168 \times 9}{8}]$

Then, savings of B = 288 – 189 = 99 [as given in option]

Similarly, savings of

$C = \frac{224 \times 12}{7} - \frac{168 \times 15}{8} = 384 - 315 = 69$ [as given in option]

Hence, option(b) is correct.

Alternatively II

Clearly, 75% of A's income = A's expenditure = 8 units

$\therefore 100\%$ of A's income $= 8 \times \frac{100}{75} = 10.66$

Hence, 7 units = 10.66

Then, 9 units = 13.70 and 12 units = 18.27

Now, ratio of their savings = (10.66 – 8) : (13.70 – 9) : (18.27 – 15)

= 2.66 : 4.70 : 3.27 = 56 : 99 : 69

Problem 69 The income of A and B are in the ratio 5 : 3. The expenses of A, B and C are in the ratio 8 : 5 : 2. If C spends ₹ 2000 and B saves ₹ 700, then A saves

Sol. Ratio of their expenditure = 8 : 5 : 2

Their expenditure = 8000, 5000, 2000 [as given 2 units = 2000]

Their savings = x, 700, y

Now, $\frac{8000 + x}{5000 + 700} = \frac{5}{3}$

[∵ income of A and B are in the ratio 5 : 3]

⇒ 24000 + $3x$ = 25000 + 3500 ⇒ $3x$ = 4500

∴ x = 1500

Hence, savings of A = ₹ 1500

Alternatively

You may proceed through options. Let options be ₹ 1500, ₹ 1000 ....

Their expenditures ⇒ 8000, 5000, 2000

Their savings ⇒ ?, 700,

Now, which number can replace "?" satisfying that

(8000 + ?) : (5000 + 700) = 5 : 3

From option (a) (8000 + 1500) : (5000 + 700) = 5 : 3

Hence, option (a) is correct.

Problem 70 Ratio of the earnings of A and B is 4: 7. If the earnings of A increase by 50% and those of B decrease by 25%, the new ratio of their earnings becomes 8 : 7. What are A's earnings? **[Bank -PO-02]**

Sol. Let earnings of A and B be ₹ $4x$ and $7x$ respectively.

Then, $\frac{4x \times 150\%}{7x \times 75\%} = \frac{8}{7}$

Now, we can't find the value of x from the above. It means the given data is inadequate.

Problem 71 A and B have their income in ratio 3 : 4 and the expenditure of B is equal to the savings of A. What is the ratio of sum of their savings to that of their income if sum of their expenditures is ₹ 9000.

Sol. Let expenditures of B be ₹ x.

Income = 3 : 4

Expenditures = $(9000 - x)$, x

Savings = x, ?

Clearly, income of

$A = (9000 - x) + x = 9000 = 3$ units [as given]

$\therefore$ 4 units = 12000

i.e. Income of B = 12000

$\therefore$ Savings of $B = 12000 - x$

Hence, sum of their savings = $x + (12000 - x) =$ ₹ 12000

Now, required ratio = 12000 : (9000 + 12000) = 4 : 7

Problem 72 The salaries of A, B and C together amount to ₹ 6060. If they spend 80%, 85% and 75% of their respective incomes, their savings are as 5 : 6 : 9. Find the salary of B.

Sol. Let salaries of A, B and C be ₹ x, y and z respectively.

Savings of A = 20% of x

Savings of B = 15% of y

Savings of C = 25% of z

But given that, 20% of x : 15% of y : 25% of z = 5 : 6 : 9

$\Rightarrow$ 100% of x : 100% of y : 100% of z = $5 \times 5 : 6 \times \frac{100}{15} : 9 \times 4$

$\Rightarrow \quad x : y : z = 25 : 40 : 36$

Clearly, (25 + 40 + 36) units = 6060

$\therefore$ 40 units = 2400

Hence, salary of B = ₹ 2400

Alternatively

Savings of A = 20% of his income = 5 units [as given]

Then, 100% of his income = 25 units

Also, savings of B = 15% of his income = 6 units [as given]

Then, 100% of his income = 40 units

Further, savings of C = 25% of his income = 9 units [as given]

Then, 100% of his income = 36 units

Clearly, (25 + 40 + 36) units = 6060

$\therefore$ 40 units = 2400. Hence, salary of B = ₹ 2400

Case VIII Concept of the total value of different denominations whose numbers are in a ratio.

Problem 73 ₹ 180 contained in a box consists of one rupee, 50 paise and 25 paise coins in the ratio 2 : 3 : 4. What is the number of 50 paise coins?

Sol. Let the number of one rupee, 50 paise and 25 paise coins be $2x$, $3x$ and $4x$ respectively.

Then, value of $2x$ coins of one rupee = ₹ $2x$

Value of $3x$ coins of 50 paise = ₹ $\frac{3x}{2}$

and value of $4x$ coins of 25 paise in = ₹ $\frac{4x}{4}$

As given, $2x + \frac{3x}{2} + \frac{4x}{4} = 180$

$\Rightarrow \frac{9x}{2} = 180 \Rightarrow x = 40$

Hence, number of 50 paise coins = $3x = 3 \times 40 = 120$

Alternatively

Their number $\Rightarrow 2 : 3 : 4$

Their values in ₹ $\Rightarrow \frac{2}{1} : \frac{3}{2} : \frac{4}{4} \Rightarrow 2 : \frac{3}{2} : 1$

As given, $\left(2 + \frac{3}{2} + 1\right)$ units = 180

$\therefore$ 1 unit = 40

Hence, number of 50 paise coin

= 3 units = 3 × 40 = 120

Problem 74 In a bag, there are coins of 25p, 10p and 5p in the ratio 1 : 2 : 3. Find the total amount therein, if value of 5p coins is ₹ 15.

Sol. Clearly, there are 300 coins of 5 paise

[$\because$ ₹ 1 = 20 coins of 5p, $\therefore$ ₹ 15 = 300 coins of 5p]

It implies, 3 units $\Rightarrow$ 300

$\therefore$ 2 units $\Rightarrow$ 200 and 1 unit $\Rightarrow$ 100

Hence, there are 100 coins of 25p and 200 coins of 10p.

$\therefore$ Total amount in bag

$= \left(100 \times \frac{1}{4}\right) + \left(200 \times \frac{1}{10}\right) + \left(300 \times \frac{1}{20}\right)$

= 25 + 20 + 15 = ₹ 60

Problem 75 A box has 210 coins of denominations one-rupee and fifty paise only. The ratio of their respective values is 13 : 11. The number of one-rupee coins is **[SSC (CGL-PT)-08]**

Sol. Ratio of the number of coins of one rupee and 50p

$= (13 \times 1) : (11 \times 2) = 13 : 22$

[$\because$ ₹ 1 = 1 coin of one-rupee = 2 coins of 50p]

As given, (13 + 22) units = 210 $\Rightarrow$ 13 units = 78

Hence, number of one-rupee coins = 78

Problem 76 ₹ 21 contained in a box consists of one-rupee, 50 paise and 25 paise coins. Find the minimum number of one-rupee coin if the ratio of number of coins of 50 paise and 25 paise is 1 : 4.

Sol. Let number of coins of 50 paise and 25 paise be x and $4x$ respectively.

Also, number of coins of one-rupee be y.

Now, sum of their values in ₹ = 21

$\Rightarrow \quad \frac{x}{2} + \frac{4x}{4} + \frac{y}{1} = 21 \Rightarrow y = 21 - \frac{3x}{2}$

It is very clear that maximum value of x in the above equation will yield minimum value of y. Also, keep in mind that x and y are integers as x, $4x$ and y are number of coins.

$y = 21 - \frac{3 \times 12}{2}$

[putting $x = 12$, the maximum possible value of x.]

$y = 3$

Hence, minimum number of one-rupee coin = 3

Problem 77 Jyoti, Rahul and Vishal have coins in ratio 7 : 6 : 5 and coins are of 25 paise, 50 paise and ₹ 1 denominations. 25 paise coins with them are in the ratio 10 : 7 : 3. 50 paise coins in ratio 8 : 11 : 7 and ₹ 1 coin in the ratio 7 : 5 : 12. Find the amount with Vishal if there are total 52 coins of 50 paise with them.

Sol. They have 50 paise coins in the ratio 8 : 11 : 7 and total number of 50 paise coins = 52

Hence, Jyoti has ⇒ $52 \times \dfrac{8}{(8+11+7)} = 16$ coins of 50 paise

Rahul has ⇒ $52 \times \dfrac{11}{26} = 22$ coins of 50 paise

and Vishal has ⇒ $52 \times \dfrac{7}{26} = 14$ coins of 50 paise

Let number of 25 paise coins with them are $10x$, $7x$ and $3x$ respectively and that of ₹ 1 coins are $7y$, $5y$ and $12y$.

Name	25 Paise Coins	50 Paise Coins	₹ 1 Coins	Ratio of number of all types of coins (as given)
Jyoti	$10x$	16	$7y$	7
Rahul	$7x$	22	$5y$	6
Vishal	$3x$	14	$12y$	5

Now, $\dfrac{10x + 16 + 7y}{7x + 22 + 5y} = \dfrac{7}{6}$

⇒ $60x + 96 + 42y = 49x + 154 + 35y$

⇒ $11x + 7y = 58$ ...(i)

Further, $\dfrac{7x + 22 + 5y}{3x + 14 + 12y} = \dfrac{6}{5}$

⇒ $35x + 110 + 25y = 18x + 84 + 72y$

⇒ $17x - 47y = -26$ ...(ii)

From Eqs. (i) and (ii), we have

$x = 4$ and $y = 2$

It means Vishal has (3×4) coins of 25 paise, 14 coins of 50 paise and (12×2) coins of ₹ 1.

∴ Total amount with Vishal $= \left(12 \times \dfrac{1}{4}\right) + \left(14 \times \dfrac{1}{2}\right) + (24 \times 1)$

$=$ ₹ 34

Case IX **Distribution of an amount in parts.**

Problem 78 ₹ 180 are to be divided among 66 persons (men and women). The ratio of the total amount of money received by men and women is 5 : 4. But the ratio of the money received by each man and woman is 3 : 2. The number of men is

Sol. Total amount received by men = 100

[∵ (5 + 4) units = 180 ∴ 5 units = 100]

and total amount received by women = 80 [180 – 100]

Let each man receives 3x and each woman receives $2x$.

Then, number of men

$= \dfrac{\text{total amount received by all men}}{\text{amount received by each man}} = \dfrac{100}{3x}$

and number of women $= \dfrac{80}{2x}$

But given, $\dfrac{100}{3x} + \dfrac{80}{2x} = 66$

⇒ $\dfrac{200 + 240}{6x} = 66$

⇒ $440 = 6x \times 66$

∴ $x = \dfrac{440}{6 \times 66} = \dfrac{10}{9}$

Hence, each man receives ₹ $3x$ i.e. $3 \times \dfrac{10}{9} =$ ₹ $\dfrac{10}{3}$

∴ Number of men $= \dfrac{100}{\frac{10}{3}} = 100 \times \dfrac{3}{10} = 30$

Alternatively

Let number of men be x.

Then, number of women $= 66 - x$

According to question,

$\dfrac{\text{number of men} \times \text{amount of each man}}{\text{number of women} \times \text{amount of each woman}} = \dfrac{5}{4}$

⇒ $\dfrac{x \times 3}{(66 - x) \times 2} = \dfrac{5}{4}$

⇒ $12x = 660 - 10x$

⇒ $22x = 660$

∴ $x = 30$

Problem 79 A sum of ₹ 370 is to be divided among A, B and C such that $\dfrac{A\text{'s share}}{B\text{'s share}} = \dfrac{B\text{'s share}}{C\text{'s share}} = \dfrac{3}{4}$. A's share in ₹ is **[SSC (So-Commercial-03), MAT-01]**

Sol. Given, $A : B = 3 : 4$

$B : C = 3 : 4$

∴ $A : B : C = 9 : 12 : 16$

Clearly, (9 + 12 + 16) units = 370

∴ 9 units = 90

Hence, A's share = ₹ 90

Problem 80 ₹ 8000 is to be distributed among A, B, C and D in such a way that ratio of shares of B and C be 1: 4, that of shares of C and D be 2 : 3. What is the share of C, if A and B together get $\dfrac{1}{3}$rd of what C and D together get ?

Sol. $B : C = 1 : 4$

$C : D = 2 : 3$

∴ $B : C : D = 2 : 8 : 12 = 1 : 4 : 6$

Also given, $(A + B) : (C + D) = 1 : 3$

$(? + 1) : (4 + 6) = 1 : 3$

Clearly, $? = \dfrac{7}{3}$

Now, $\left(\dfrac{7}{3} + 1 + 4 + 6\right)$ units = 8000

∴ 4 units = 2400

i.e, C's share = ₹ 2400

Problem 81 ₹ 395 are divided among A, B and C in such a manner that B gets 25% more than A and 20% more than C. The share of A will be? *[SSC (CGL-PT) -07]*

Sol. Let A gets ₹ 100. Then B's share = 125

$\therefore \quad A : B = 4 : 5$

Similarly, $B : C = \quad 6 : 5$

[if C's share be 100, B's share will be 120]

$\therefore A : B : C = (4 \times 6) : (5 \times 6) : (5 \times 5)$

$= 24 : 30 : 25$

Clearly, $(24 + 30 + 25)$ units = 395

$\therefore$ 24 units = 120

Hence, A's share = ₹ 120

Problem 82 An amount of ₹ 2760 is divided among A, B and C such that if their shares be reduced by ₹ 10, ₹ 20 and ₹ 30 respectively, the remainders shall be in the ratio 2 : 3 : 4. Then, C's share was

Sol. Remainder = [2760 − (10 + 20 + 30)] = 2700

As given, (2 + 3 + 4) units = 2700

$\therefore$ 4 units = 1200

$\therefore$ C's share = 1200 + 30 = 1230

Problem 83 A sum of ₹ 160 is divided among A, B and C in such a way that A gets ₹ 16 more than what B gets and B gets ₹ 18 more than what C gets. The ratio of their shares is

Sol. Let C gets ₹ x.

Then, B gets $(x + 18)$ and A gets $(x + 18 + 16)$

As given, $x + (x + 18) + (x + 18 + 16) = 160$

$\Rightarrow \quad 3x = 108$

$\therefore \quad x = 36$

Now, A's share $= x + 18 + 16$

$= 36 + 18 + 16 = 70$

B's share $= x + 18 = 36 + 18 = 54$

and C's share $= x = 36$

Hence, their ratio = 70 : 54 : 36 = 35 : 27 : 18

Alternatively

You may proceed through options.

Let options be 36 : 26 : 18, 35 : 27 : 18, ...

(a) 36 : 26 : 18 ⇒ (36 + 26 + 18) units = 160

⇒ 1 unit = 2

Now, their shares are ₹ (36 × 2), ₹ (26 × 2) and ₹ (18 × 2) which are not satisfying that A gets ₹ 16 more than what B gets and B gets ₹ 18 more than what C gets.

Hence, option (a) is not correct.

(b) 35 : 27 : 18 ⇒ (35 + 27 + 18) units = 160

⇒ 1 unit = 2

Now, their shares are ₹ (35 × 2), ₹ (27 × 2) and ₹ (18 × 2) which are satisfying that

70 − 54 = 16 and 54 − 36 = 18

Hence, option(b) is correct.

Problem 84 Divide ₹ 52000 among 3 professors, 5 teachers and 7 clerks such that 2 clerks get as much as 1 teacher and 3 teachers as much as 2 professors. Find the share of a teacher.

Sol. Let share of each professor be ₹ x.

$\therefore$ Share of each teacher $= \frac{2x}{3}$

[∵ 3 teachers get what 2 professors get]

and share of each clerk $= \frac{2x}{3 \times 2} = \frac{x}{3}$

[∵ 2 clerks get what 1 teacher gets]

As per question,

$$(3 \times x) + \left(5 \times \frac{2x}{3}\right) + \left(7 \times \frac{x}{3}\right) = 52000$$

$$\Rightarrow \quad \frac{9x + 10x + 7x}{3} = 52000 \Rightarrow x = 6000$$

$\therefore$ Share of each teacher $= 6000 \times \frac{2}{3} = 4000$

Alternatively

Share of each clerk (C) : Share of each teacher (T) = 1 : 2

Share of each teacher (T) : Share of each professor (P) = 2 : 3

$C : T : P = 2 : 4 : 6 = 1 : 2 : 3$

As given, [(1 × 7) + (2 × 5) + (3 × 3)] units = 52000

$\therefore$ 1 unit = 2000

$\therefore$ 2 units = 4000

Hence, share of each teacher is ₹ 4000.

Problem 85 A man distributes some money in ₹ among his four sons in the ratio $\frac{1}{2} : \frac{1}{3} : \frac{1}{4} : \frac{1}{6}$, Find the minimum value of the money in ₹ if each son gets the amount in integer.

Sol. Given ratio $= \frac{1}{2} : \frac{1}{3} : \frac{1}{4} : \frac{1}{6}$

$= \frac{1}{2} \times 12 : \frac{1}{3} \times 12 : \frac{1}{4} \times 12 : \frac{1}{6} \times 12 = 6 : 4 : 3 : 2$

Minimum value of money to be distributed

= 6 + 4 + 3 + 2 = ₹ 15

Problem 86 Distribution of ₹ 2100 among A, B and C is made in such a way that A should receive $\frac{2}{5}$th part of amount received to B and C together and B should receive $\frac{3}{7}$th part of amount received to A and C together. Find the C's share.

Sol. Given, $\quad A = \frac{2}{5}(B + C)$

$\therefore \quad B + C = \frac{5}{2}A$

Further, $\quad B = \frac{3}{7}(A + C)$

$\Rightarrow \quad 7B = 3A + 3C$

$\Rightarrow \quad 7B = 3 \times \frac{2}{5}(B + C) + 3C \quad [\because A = \frac{2}{5}(B + C)]$

$\Rightarrow \quad 35B = 6B + 21C \Rightarrow 29B = 21C$

$\therefore \quad B : C = 21 : 29$

Also, $A + B + C = 2100$

$\Rightarrow \quad A + \frac{5}{2}A = 2100 \quad [\because B + C = \frac{5}{2}A]$

$\therefore \quad A = 600$

Now, share of $(B + C) = 2100 - 600 = 1500$

$\therefore$ Share of $C = 1500 \times \frac{29}{(21 + 29)} = 870 \quad [\because B : C = 21 : 29]$

Problem 87 **A man divided two sums of money among 4 sons R, S, T and U the 1st in the ratio 4 : 3 : 2 : 1 and the 2nd in the ratio 5 : 6 : 7 : 8. If the 2nd sum of money is twice the 1st sum, which son receives the largest part?**

Sol. Let shares of R, S, T and U in the first sum be $4x$, $3x$, $2x$ and x respectively.

Hence, 1st sum $= 4x + 3x + 2x + x = 10x$

Then, 2nd sum $= 20x$

Now, share of R in 2nd sum

$$= 20x \times \frac{5}{(5 + 6 + 7 + 8)} = \frac{50x}{13}$$

Share of S in 2nd sum $= 20x \times \frac{6}{26} = \frac{60x}{13}$

Share of T in 2nd sum $= 20x \times \frac{7}{26} = \frac{70x}{13}$

Share of U in 2nd sum $= 20x \times \frac{8}{26} = \frac{80x}{13}$

Further, share of R in both sums $= 4x + \frac{50x}{13} = \frac{102x}{13}$

Share of S in both sums $= 3x + \frac{60x}{13} = \frac{99x}{13}$

Share of T in both sums $= 2x + \frac{70x}{13} = \frac{96x}{13}$

Share of U in both sums $= x + \frac{80x}{13} = \frac{93x}{13}$

Clearly, son R is getting largest part.

Alternatively I

1st sum is divided into 4 : 3 : 2 :1

2nd sum is divided into 5 : 6 : 7 : 8

But given that, 2nd sum is twice the 1st sum. Had the 1st sum been twice their respective shares would have been doubled.

$\therefore$ 2 × [1st sum] may be divided into 8 : 6 : 4 : 2 and 2nd sum is divided into 5 : 6 : 7 : 8

Clearly, R is getting (8 + 5) units.

S is getting (6 + 6) units.

T is getting (4 + 7) units.

U is getting (2 + 8) units.

Hence, R is getting the largest part.

Alternatively II

Let first sum be ₹ 100. Then, their shares are ₹ 40, ₹ 30, ₹ 20 and ₹ 10 respectively.

Obviously, 2nd sum will be of ₹ 200. Then, their shares are ₹ 38.5, ₹ 46.2, ₹ 54 and ₹ 61.6 respectively [approx, values taken].

Obviously, R is getting the largest part.

Case X **When numbers given in ratio are altered by percentage.**

Problem 88 **The salaries of A, B and C are in the ratio 1 : 3 : 4. If the salaries are increased by 5%, 10% and 15% respectively, then the increased salaries will be in the ratio** ***[SSC(CGL-PT)-08, CLAT-11]***

Sol. Let the initial salaries of A, B and C be ₹ x, $3x$ and $4x$ respectively.

Then, ratio of increased salaries

$$= x \times \frac{105}{100} : 3x \times \frac{110}{100} : 4x \times \frac{115}{100}$$

$$= 105 : 330 : 460 = 21 : 66 : 92$$

Alternatively

Required ratio $= (1 \times 105) : (3 \times 110) : (4 \times 115)$

$= 21 : 66 : 92$

Problem 89 **The salaries of Ramesh, Ganesh and Rajesh were in the ratio 6 : 5 : 7 in 2010, and in the ratio 3 : 4 : 3 in 2015. If Ramesh's salary increased by 25% during 2010-2015, then the percentage increase in Rajesh's salary during this period is closest to** ***[CAT-19]***

Sol. Ratio of salaries of Ramesh, Ganesh and Rajesh in 2010 = 6 : 5 : 7

Changed ratio in 2015 = 3 : 4 : 3

Given that Ramesh's salary is increased by 25%. It means his salary increases from 6 to 7.5. Thus, we may say that 3 units = 7.5.

Rajesh's increased salary is also 3 units i.e. 7.5.

Clearly, Rajesh's salary increases from 7 to 7.5 i.e. 7% [approx.] increase.

Note *Try with other methods too.*

Problem 90 **Two numbers are respectively 20% and 50% more than a 3rd number. The ratio of the two numbers is**

Sol. Let the 3rd number be 100.

Then 1st number = 120 and 2nd number = 150

Now, their ratio = 120 : 150 = 4 : 5

Miscellaneous

Problem 91 **If $\frac{a}{b + c} = \frac{b}{c + a} = \frac{c}{a + b}$, then $\frac{2a}{b + c} = ?$**

Sol. Recall the property of ratio, if $\frac{a}{b} = \frac{c}{d} = \frac{e}{f} = \ldots\ldots = k$,

then $k = \frac{a + c + e \ldots\ldots}{b + d + f \ldots\ldots}$

Here, let $\frac{a}{b + c} = \frac{b}{c + a} = \frac{c}{a + b} = k$

$\Rightarrow \quad \frac{a + b + c}{(b + c) + (c + a) + (a + b)} = k$

$\Rightarrow \quad \frac{a + b + c}{2(a + b + c)} = k \Rightarrow k = \frac{1}{2}$

Then, $\frac{2a}{b+c} = 2 \times \frac{1}{2} = 1$

Alternatively

Given, $\frac{a}{b+c} = \frac{b}{c+a}$

$\Rightarrow \frac{a+b+c}{a-b-c} = \frac{b+c+a}{b-c-a}$ [∵ by Componendo and Dividendo]

$\Rightarrow a-b-c = b-c-a$

$\Rightarrow 2a = 2b \Rightarrow a = b$

Further, $\frac{b}{c+a} = \frac{c}{a+b}$

$\Rightarrow \frac{b+c+a}{b-c-a} = \frac{c+a+b}{c-a-b}$ [∵ by Componendo and Dividendo]

$\Rightarrow b-c-a = c-a-b$

$\Rightarrow 2b = 2c \Rightarrow b = c$

From both outcomes, $a = b = c$

Then, $\frac{a}{b+c} = \frac{b}{c+a} = \frac{c}{a+b} = \frac{a}{a+a} = \frac{1}{2}$

Now, $\frac{2a}{b+c} = 2 \times \frac{1}{2} = 1.$

Problem 92 **In an examination rates of passed students and failed students is 3 : 1. If there had been 40 fewer candidates and 10 more failed, the ratio would have been 5 : 2. Find the total number of candidates.**

[SSC (CGL-T-II)-12, 17]

Sol. Let number of passed students be $3x$ and number of failed students be x.

∴ Total number of students $= 4x$

Now, new number of failed students $= x + 10$

∴ Number of passed students $= (4x - 40) - (x + 10) = 3x - 50$

As given, $\frac{3x - 50}{x + 10} = \frac{5}{2} \Rightarrow 6x - 100 = 5x + 50 \Rightarrow x = 150$

Hence, total number of students $= 4x = 4 \times 150 = 600$

Alternatively

You may proceed through options. Let options be 400, 600, ...

(a) 400 ⇒ Passed(300) + Failed (100)

New number of students = 400 − 40 = 360

and number of failed = 110

∴ Passed students = 360 − 110 = 250

Now, 250 : 110 ≠ 5 : 2

(b) 600 ⇒ Passed (450) + Failed (150)

New number of students = 600 − 40 = 560 and Failed = 160

∴ Passed = 560 − 160 = 400

Now, 400 : 160 = 5 : 2

Hence, correct option is (b).

Problem 93 **A stick is broken into two parts in such a way that the ratio of both parts is equal to that of the initial length of the stick to larger broken parts. Find the ratio of initial length to smaller broken part.**

Sol. Let both parts be a and b $[a > b]$.

Then, $\frac{a}{b} = \frac{a+b}{a}$ [as given]

$\Rightarrow a^2 - b^2 - ab = 0$

$\Rightarrow a^2 - a - 1 = 0$ [If $b = 1$ (assume)]

$\therefore a = \frac{-(-1) \pm \sqrt{(-1)^2 - 4 \cdot 1(-1)}}{2 \cdot 1}$

$= \frac{1 + \sqrt{5}}{2}$

[∵ negative value is not possible]

$\therefore a : b = \frac{1+\sqrt{5}}{2} : 1 = (1 + \sqrt{5}) : 2$

Then, $\frac{a+b}{b} = \frac{a}{b} + 1 = \frac{1+\sqrt{5}}{2 \cdot 1} + 1$

$= \frac{(3+\sqrt{5})}{2} = (3 + \sqrt{5}) : 2$

Problem 94 **In an organisation, ratio of number of managers, supervisors and clerks is 2 : 5 : 8 and wages of one manager, one supervisor and one clerk are in ratio 8 : 5 : 2. If there are 16 managers there and total daily wages of the employees are ₹ 912. Find daily wages of one supervisor.**

Sol. Given, 2 units = 16 = Number of managers

∴ 5 units = 40 = Number of supervisors

and 8 units = 64 = Number of clerks

Let daily wages of one manager, one supervisor and one clerk be $8x$, $5x$ and $2x$ respectively.

∴ Their total daily wages

$= (16 \times 8x) + (40 \times 5x) + (64 \times 2x) = 912$

$\Rightarrow 456x = 912$

$\therefore x = 2$

∴ Daily wages of one supervisor

$= 5x = 5 \times 2 =$ ₹ 10

Alternatively

Ratio of their number = 2 : 5 : 8

Ratio of wages post- wise [per person] = 8 : 5 : 2

Ratio of their wages [post-wise] = 16 : 25 : 16

But given, (16 + 25 + 16) units = 912

∴ 1 unit = 16

∴ 25 units = 400 [total wages of all 40 supervisors]

∴ Daily wages of one supervisor

$= \frac{400}{40} =$ ₹ 10

Problem 95 **The salary of employees of an organization is increased in ratio 7 : 8 but their number is reduced in the ratio 5 : 4. Then, in what ratio expenditures on salary is increased or decreased ?**

[SSC (CGL T-II) 17, CLAT-14]

Sol.

	Earlier	Later
Number of employees ⇒	5	4
Salary of each employee ⇒	7	8
Total expenditures on salary =	35	32

It means expenditure on salary is decreased in the ratio = 35 : 32

Problem 96 In the above question, find the increase or decrease % if earlier expenditures on salary was ₹ 7000.

Sol. Decrease % = $\left(\frac{35-32}{35}\times 100\right)\% = 8\frac{4}{7}\%$

[No need to take ₹ 7000 in consideration for solution]

Problem 97 One year ago, the ratio between Mahesh's and Suresh's salaries was 3 : 5. The ratio of their individual salaries of last year and present year are 2 : 3 and 4 : 5 respectively. If their total salaries for the present year are ₹ 43000, what is the present salary of Mahesh? ***[MAT-11]***

Sol.

	Mahesh's Salary		Suresh's Salary
Initial ratio	3	:	5
Present ratio	$3\times\frac{3}{2}$	:	$5\times\frac{5}{4}$
	$\frac{9}{2}$	:	$\frac{25}{4}$
	18	:	25

Given, (18 + 25) units = 43000

∴ 18 units = 18000

Hence, salary of Mahesh at present = ₹ 18000.

Problem 98 Two candles of the same height are lighted at the same time. The 1st is consumed in 4 h and the 2nd in 3 h. Assuming that each candle burns at the constant rate, in how many hours after being lighted, was the 1st candle twice the height of the 2nd?

Sol. Let after x h the height of 1st candle is twice the height of the 2nd. Suppose their original heights be 1 unit each.

Now, consumed part of 1st candle in x h = $\frac{x}{4}$

and consumed part of 2nd candle in x h = $\frac{x}{3}$

Given, $\left(1-\frac{x}{4}\right):\left(1-\frac{x}{3}\right) = 2:1$

$\Rightarrow \quad 1-\frac{x}{4} = 2-\frac{2x}{3} \Rightarrow \frac{5x}{12} = 1$

$\therefore \quad x = \frac{12}{5} = 2$ h 24 min

Problem 99 The ratio for land to water for the whole earth is 1 : 2 and is 2 : 3 in the northern hemisphere. What is the ratio of land to water in the southern hemisphere?

Sol. (Land + water) on whole earth = $x + 2x = 3x$

(Land + Water) in northern hemisphere = $\frac{3x}{2}$

Land in northern hemisphere = $\frac{3x}{2}\times\frac{2}{(2+3)} = \frac{3x}{5}$

Water in northern hemisphere = $\frac{3x}{2}\times\frac{3}{(2+3)} = \frac{9x}{10}$

Thus, land in southern hemisphere = $x - \frac{3x}{5} = \frac{2x}{5}$

and water in southern hemisphere = $2x - \frac{9x}{10} = \frac{11x}{10}$

∴ Their ratio = $\frac{2x}{5}:\frac{11x}{10}$

= 4 : 11

Alternatively

Let whole the earth be 300 units

[Land ⇒ 100 and Water ⇒ 200]

Then, (Land + Water) in Northern hemisphere = 150 units and (Land + Water) in Southern hemisphere = 150 units

Now, land in northern hemisphere = $150\times\frac{2}{(2+3)} = 60$

Then, water in northern hemisphere = 150 – 60 = 90

Now, land in southern hemisphere = 100 – 60 = 40

and water in southern hemisphere = 200 – 90 = 110

∴ Required ratio = 40 : 110 = 4 : 11

Problem 100 The ratio of the marks obtained by a candidate in Physics and Chemistry is 3 : 4, in Mathematics and Biology is 2 : 3, in Physics and Mathematics is 3 : 5 and in Biology and Home science is 3 : 2. In which subject(s) his marks is greater than 72%, if his overall marks is 49% and each subject has equal maximum marks.

Sol. Let maximum marks for each subject be 100.

Given, $P:C = 3:4$

$M:B = 2:3$

$P:M = 3:5$

$B:H = 3:2$

Hence, $P:C:M = 3:4:5$ $\quad [\because P:C = 3:4$ and $P:M = 3:5]$

and $M:B:H = 2:3:2$ $\quad [\because M:B = 2:3$ and $B:H = 3:2]$

$\therefore \quad P:C:M:B:H = 6:8:10:15:10$

Now, $\frac{6x+8x+10x+15x+10x}{5} = 49$

$\Rightarrow \quad x = 5$

Thus, Marks in Physics = 6 × 5 = 30

Marks in Chemistry = 8 × 5 = 40

Marks in Mathematics = 10 × 5 = 50

Marks in Biology = 15 × 5 = 75

Marks in Home science = 10 × 5 = 50

Clearly, his marks in Biology is greater than 72%.

WORKOUT

1. $(x+y):(x-y)=3:5$ and $xy=$ positive imply that **[CDS-14]**
(a) x and y are both positive
(b) x and y are both negative
(c) one of them is positive and one of them is negative
(d) no real solutions for x and y exist

2. If $A:B=2:3$, $B:C=5:7$ and $C:D=3:10$, then what is $A:D$ equal to ?
(a) 1 : 7 (b) 2 : 7 (c) 1 : 5 (d) 5 : 1

3. Three numbers *A*, *B* and *C* are in the ratio 12 : 15 : 25. If sum of these numbers is 312, ratio between the difference of *B* and *A* and the difference of *C* and *B* is **[MAT-10]**
(a) 3 : 7 (b) 10 : 3 (c) 3 : 10 (d) 5 : 1

4. If $P=\frac{x^2-36}{x^2-49}$ and $Q=\frac{x+6}{x+7}$, then the value of $\frac{P}{Q}$ is
(a) $\frac{x-6}{x-7}$ (b) $\frac{x-6}{x+7}$
(c) $\frac{x-7}{x+6}$ (d) $\frac{x+6}{x-7}$

5. Ratio of incomes of *A*, *B* and *C* last year was 3 : 4 : 5. The ratios of their individual incomes of last year and this year are 4 : 5, 2 : 3 and 3 : 4 respectively. If the sum of their present incomes is ₹ 78800, find the present income of *B*.
(a) 18000 (b) 28800
(c) 3200 (d) 44000

6. A bag contains 50 paise, ₹ 1 and ₹ 2 coins in the ratio 2 : 3 : 4. If the total amount is ₹ 240, what is the total number of coins? **[CDS-08, CLAT-17]**
(a) 90 (b) 150 (c) 180 (d) 200

7. If $x:y=1:3$, $y:z=5:k$, $z:t=2:5$ and $t:x=3:4$, then what is the value of k? **[CDS-09]**
(a) 1/2 (b) 1/3 (c) 2 (d) 3

8. If the ratio of *x* to *y* is 25 times the ratio of *y* to *x*, then what is the ratio of *x* to *y*?
(a) 1:5 (b) 5 : 1
(c) 25 : 1 (d) 1 : 25

9. Men, women and children are employed to do a work in the proportion 1 : 2 : 3 and their wages are in the proportion 6 : 3 : 2. When 50 men are employed, total wages of all amount to ₹ 4500. What is the weekly wages paid to a man, a woman and a child, in rupees? **[MAT-12]**
(a) 210, 105, 80 (b) 210, 105, 70
(c) 210, 105, 90 (d) 200, 105, 70

10. One man can do as much work in one day as a woman can do in 2 days. A child does one third the work in a day as a woman. If an estate-owner hires 39 pairs of hands, men, women and children in the ratio 6 : 5 : 2 and pays them in all ₹ 1113 at the end of the day's work. What must the daily wages of a child be, if the wages are proportional to the amount of work done? **[CAT-94]**
(a) ₹ 14 (b) ₹ 5
(c) ₹ 20 (d) ₹ 7

11. A certain amount of money has been divided between two persons *P* and *Q* in the ratio 3 : 5. But it was divided in the ratio 2 : 3 and thereby *Q* losses ₹ 10. What was the amount?
(a) ₹ 250 (b) ₹ 300
(c) ₹ 350 (d) ₹ 400

12. Two numbers are in ratio $P:Q$. When 1 is added to both the numerator and denominator, the ratio gets changed to R/S. Again, when 1 is added to both numerator and the denominator, it becomes 1/2. Find the sum of *P* and *Q*. **[MAT-14, 19]**
(a) 3 (b) 4
(c) 5 (d) 6

13. The wages of labourers in a factory has increased in ratio 22 : 25 and their number decreased in the ratio 3 : 2. What was the original wage bill of the factory if the present bill is ₹ 5000 ? **[CDS-10, SSC (CGL-T-II)-12]**
(a) ₹ 4000 (b) ₹ 6000
(c) ₹ 8000 (d) None of these

14. A sum of money is divided among *A*, *B*, *C* and *D* in the ratio 5 : 8 : 9 : 11. If the share of *B* is ₹ 2475 more than share of *A*, then what is the total amount of money of *A* and *C* together? **[UBI (PO)-08, SBI (PO) -08, Canara Bank (PO)-09, Allahabad Bank (PO)-10]**
(a) ₹ 9900 (b) ₹ 11550
(c) ₹ 10725 (d) ₹ 9075
(e) None of these

15. If *p*% of ₹ *x* is equal to *t* times *q*% of ₹ *y*, then what is the ratio of *x* to *y*?
(a) $pt:q$ (b) $p:qt$
(c) $qt:p$ (d) $q:pt$

16. The cost of manufacture of an article is made up of four components *A*, *B*, *C* and *D* which have a ratio of 3 : 4 : 5 : 6 respectively. If there are respective changes in the cost of + 10%, – 20%, –30% and +40%, then what would be the percentage change in the cost? **[MAT-14, 19]**
(a) 2.22% (b) 1.80%
(c) 3.28% (d) 0.95%

17. A sum of ₹ 36.90 is made of 180 coins which are either 10 paise coins or 25 paise coins. The number of 10 paise coins is **[LIC (AAO)-09]**
(a) 54 (b) 60
(c) 80 (d) 120

18. The ratio of *A* to *B* is $x:8$ and the ratio of *B* to *C* is $12:z$. If the ratio of *A* to *C* is 2 : 1, then what is the ratio of $x:z$?
(a) 2 : 3 (b) 3 : 2
(c) 4 : 3 (d) 3 : 4

19. If the ratio of boys to girls in a class is *B* and the ratio of girls to boys is *G*, then $3(B+G)$ is
(a) equal to 3 (b) less than 3
(c) more than 3 (d) less than 1/3

20. If $a:b=2:5$, then the value of $(2a+3b):(7a+5b)$ is
(a) 19/39 (b) 99/13
(c) 31/19 (d) 19/31

21. A sum of ₹ 1400 is divided amongst A, B, C and D such that A's share : B's share = B's share : C's share = C's share : D's share = 3/4. How much is C's share ? **[IIFT-12]**

(a) ₹ 72 (b) ₹ 288
(c) ₹ 216 (d) ₹ 384

22. If $P:Q=\frac{3}{5}:\frac{5}{7}$ and $Q:R=\frac{3}{4}:\frac{2}{5}$, then what is $P:Q:R$ equal to ? **[CDS-11, 09]**

(a) $\frac{3}{5}:\frac{5}{7}:\frac{2}{5}$ (b) $\frac{9}{20}:\frac{15}{28}:\frac{2}{7}$
(c) $\frac{3}{5}:\frac{3}{4}:\frac{2}{5}$ (d) $\frac{3}{5}:\frac{5}{7}:\frac{3}{4}$

23. In a demonstration against the demonetization, participants outnumbered the police by 9 : 1. The police arrested 135 participants averaging 5 for every 3 policemen. How many participants were there in the demonstration ?

(a) 1215 (b) 665
(c) 405 (d) None of these

24. If $a+b+c=1$ and $ab+bc+ca=1/3$, then $a:b:c=?$ **[SSC (CGL-T-II)-10]**

(a) 1 : 2 : 2 (b) 1 : 1 : 1
(c) 2 : 1 : 2 (d) 1 : 2 : 1

25. In a certain school, the ratio of boys to girls is 7 : 5. If there are 2400 students in the school, then how many girls are there ?

(a) 500 (b) 700 (c) 800 (d) 1000

26. Two numbers are in the ratio 2 : 3. If 9 is added to each number, they will be in the ratio 3 : 4. What is the product of the two numbers? **[CDS-13]**

(a) 360 (b) 480 (c) 486 (d) 512

27. ₹ 5783 is divided among Ramesh, Shiv and Bhuwan in such a way that if ₹ 28, ₹ 37 and ₹ 18 be deducted from their respective shares, they have money in the ratio 4 : 6 : 9. Find Ramesh's share. **[MAT-14, 10, 08]**

(a) ₹ 1256 (b) ₹ 1084
(c) ₹ 1456 (d) ₹ 1228

28. 80 is divided into three parts such that the 1st part is 24 and the ratio of the 2nd to the 3rd is 4 : 3. The 2nd part is : **[SSC (SAS)-10]**

(a) 40 (b) 36
(c) 32 (d) 24

29. If $(a-b):(a+b)=1:5$, then what is $(a^2-b^2):(a^2+b^2)$ equal to

(a) 2 : 3 (b) 3 : 2
(c) 5 : 13 (d) 13 : 15

30. If $a:b=b:c$, then the ratio of $a^4:b^4$ is equal to

(a) $ac:b^2$ (b) $a^2:c^2$
(c) $c^2:a^2$ (d) $b^2:ac$

31. In a class, the number of boys is more than the number of girls by 12% of the total students. What is the ratio of number of boys to that of girls ? **[CDS- 10]**

(a) 11 : 14 (b) 14 : 11
(c) 28 : 25 (d) 25 : 28

32. At the start of a seminar, the ratio of the number of male participants to the number of female participants was 3 : 1. During the tea break, 16 male participants left and 6 more female participants registered. The ratio of the male to the female participants became 2 : 1. The total number of participants at the start of the seminar was

(a) 112 (b) 48 (c) 54 (d) 72

33. There is a ratio of 5 : 4 between two numbers. If 40% of the 1st is 12, then 50% of the 2nd number is **[SSC (CGL-T-II) -11]**

(a) 12 (b) 24 (c) 18 (d) 20

34. A bag contains ₹ 112 in the form of ₹ 1, 50 paise and 10 paise coins in the ratio 3 : 8 : 10. What is the number of 50 paise coins? **[CDS-11, 09, MAT-03, 01, 99 SSC (SAS)-10, CAT-98]**

(a) 112 (b) 108
(c) 96 (d) 84

35. ₹ 1066 are divided among A, B, C and D such that $A:B=3:4$, $B:C=5:6$ and $C:D=7:5$, who will get the maximum?

(a) A (b) B (c) C (d) D

36. Shivransh is fond of shopping. He took nearly ₹ 15 with him in the form of one rupee notes and 20 paise coins. When he came back, he had as many one rupee notes as he originally had 20 paise coins and as many 20 paise coins as he originally had one rupee notes. The total amount was also reduced in ratio 3 : 1. How much did he spend ?

(a) ₹ 4.28 (b) ₹ 9.30
(c) ₹ 9.60 (d) ₹ 10.20

37. If $x:y=7:5$, then what is the value of $(5x-2y):(3x+2y)$?

(a) 5/4 (b) 6/5
(c) 25/31 (d) 31/42

38. If ₹ 8400 is divided among A, B and C in the ratio 1/5 : 1/6 : 1/10, what is the share of A ? **[CDS-11, 10, 08]**

(a) ₹ 1800 (b) ₹ 3000
(c) ₹ 3600 (d) ₹ 4200

39. The 1st, 2nd and 3rd class fares between two stations were 10 : 8 : 3 and the number of 1st, 2nd and 3rd class, passengers between the two stations in a day was 3 : 4 : 10. The sale of tickets to passengers running between two stations on that day was ₹ 8050. How much was realised by the sale of 2nd class tickets ? **[MAT-11, 08, 07]**

(a) ₹ 3000 (b) ₹ 2800
(c) ₹ 4500 (d) ₹ 3500

40. The number of employees in Obelix Menhin Co. is a prime number and is less than 300. The ratio of the number of employees who are graduates and above to that of employees who are not, can possibly be **[CAT - 06]**

(a) 101: 88 (b) 87 : 100
(c) 110 : 111 (d) 85 : 98
(e) 97 : 84

41. Salaries of A, B and C were in the ratio 3 : 5 : 7 respectively. If their salaries were increased by 50%, 60% and 50% respectively, what will be the new ratio of their respective salaries ?

(a) 4 : 5 : 7 (b) 3 : 6 : 7
(c) 4 : 15 : 18 (d) 9 : 16 : 21

42. If $\frac{y}{x-z}=\frac{y+x}{z}=\frac{x}{y}$, then find $x:y:z$.

(a) 1 : 2 : 3 (b) 3 : 2 : 1
(c) 4 : 2 : 3 (d) 2 : 4 : 7

43. A and B earn in the ratio 2 : 1. They spend in the ratio 5 : 3 and save in the ratio of 4 : 1. Find the monthly income of A if the total monthly savings of both A and B is ₹ 5000. **[SSC (CGL-Mains)-2000, SSC (CGL-T-II)-12]**

(a) ₹ 14000 (b) ₹ 7000
(c) ₹ 6000 (d) ₹ 1500

44. Ms. Sharma's expenditure and savings are in the ratio of 3 : 2. Her income increases by 10%. Her expenditure also increases by 12%. How much percent does her savings increase ?

(a) 7% (b) 6%
(c) 13% (d) 11%

45. The incomes of Chandra and Kim are in the ratio 9 : 4 and their expenditures are in the ratio 7 : 3. If each saves ₹ 2000, then Chandra's expenditure is *[MAT-09, 02, SSC (CGL-T-II)-11]*

(a) ₹ 60000
(b) ₹ 80000
(c) ₹ 70000
(d) None of these

46. Ramsukh Bhai sells Rasgulla (a favourite Indian sweets) at ₹ 15/kg. A Rasgulla is made up of flour and sugar in the ratio of 5 : 3. The ratio of price of sugar and flour is 7 : 3 [per kg]. Thus, he earns $66\frac{2}{3}$% profit. What is the cost price of sugar ? *[SNAP-11]*

(a) ₹ 10/kg (b) ₹ 9/kg
(c) ₹ 18/kg (d) ₹ 14/kg

47. A contractor employed 25 labourers on a job. He was paid ₹ 275 for the work. After retaining 20% of this sum, he distributed the remaining amount amongst the labourers. If the number of men to women labourers was in the ratio 2 : 3 and their wages in the ratio 5 : 4, what wages did a woman labourer get ?

(a) ₹ 10 (b) ₹ 8
(c) ₹ 12 (d) ₹ 15

48. Ram got twice as many marks in English as in Science. His total marks in English, Science and Maths are 180. If the ratio of his marks in English and Maths is 2 : 3, what are his marks in Science? *[SSC (CGL-T-II)-13]*

(a) 30 (b) 60
(c) 72 (d) 90

49. The number that must be added to each of the numbers 8, 21, 13 and 31 to make the ratio of first two numbers equal to the ratio of last two numbers is

(a) 7 (b) 5
(c) 9 (d) None of these

50. The sum of money is to be divided amongst A, B and C in the respective ratio of 3 : 4 : 5 and another sum of money is to be divided between E and F equally, If F got ₹ 1050 less than A, how much amount did B receive?

(a) ₹ 750
(b) ₹ 2000
(c) ₹ 1500
(d) Cannot be determined

51. Using only 2, 5, 10, 25 and 50 paise coins, what will be the minimum number of coins required to pay exactly 78 paise, 69 paise and ₹ 1.01 to three different persons? *[CAT-03]*

(a) 19 (b) 20
(c) 17 (d) 18

52. The prime cost of an article is three times the value of the raw material used. The cost of raw material increases in the ratio 5 : 12 and manufacturing expenses in the ratio 4 : 5. The article, which originally cost ₹ 6 will have new cost

(a) ₹ 10 (b) ₹ 17
(c) ` 20.50 (d) None of these

53. From each of two given numbers, half the smaller number is subtracted. Of the resulting numbers the larger one is three times as large as the smaller. What is the ratio of two numbers ? *[CAT-93]*

(a) 3 : 1 (b) 2 : 1
(c) 3 : 2 (d) None of these

54. A person distributes his pens among four friends A, B, C and D in the ratio $\frac{1}{3}:\frac{1}{4}:\frac{1}{5}:\frac{1}{6}$. What is the minimum number of pens that the person should have?

(a) 65 (b) 55
(c) 23 (d) 57

55. The ratio of number of cans of orange, pineapple and mixed fruit juices kept in a store is 8 : 9 : 15. If the store sells 25%, 33.33% and 20% orange, pineapple and mixed fruit juices cans respectively, then what is the ratio of number of cans of these juices in the remaining stock? *[SSC (CGL-T-I)-17]*

(a) 1 : 1 : 2 (b) 1 : 2 : 2
(c) 2 : 1 : 2 (d) 3 : 1 : 2

56. There are a total of 43800 students in 4 schools of a city. Half the number of students of the 1st school, $\frac{2}{3}$rd of the 2nd, $\frac{3}{4}$rd of the 3rd and $\frac{4}{5}$th of the 4th are all equal. What is the ratio of the number of students of A to D, if A, B, C and D be the 1st, 2nd, 3rd and 4th schools respectively? *[MAT-12]*

(a) 8 : 5 (b) 1 : 3
(c) 2 : 3 (d) 7 : 9

57. Two companies A and B quote for a tender. On the tender opening day, A realises that the two quotes are in the ratio 7 : 4 and hence decreases its price during negotiations to make ₹ 1 lakh lower than B's quoted price. B then realises that the final quotes of the two were in the ratio 3 : 4. By how much did A decrease its price in order to win the bid?

(a) ₹ 7 lakhs (b) ₹ 4 lakhs
(c) ₹ 9 1akhs (d) None of these

58. A stall sells popcorn and chips in packets of three sizes : large, super and jumbo. The numbers of large, super and jumbo packets in its stock are in the ratio 7 : 17 : 16 for popcorn and 6 : 15 : 14 for chips. If the total number of popcorn packets in its stock is the same as that of chips packets, then numbers of jumbo popcorn packets and jumbo chips packets are in the ratio *[CAT-17]*

(a) 8 : 7 (b) 4 : 3 (c) 6 : 5 (d) 1 : 1

59. Three numbers are in the ratio 2 : 3 : 4. If each number is increased by 5, the new numbers are in the ratio 5 : 7 : 9. Find the original first number. *[MAT-17]*

(a) 20 (b) 40 (c) 50 (d) 70

60. A man who recently died left a sum of ₹ 390000 to be divided among his wife, five sons and four daughters. He directed that each son should receive 3 times as much as each daughter receives and each daughter should receive twice as much as their mother receives. What was the wife's share ? *[CDS -19]*

(a) ₹ 14000 (b) ₹ 12000
(c) ₹ 10000 (d) ₹ 9000

Answer Key

1. (d)	*2.* (a)	*3.* (c)	*4.* (a)	*5.* (b)	*6.* (c)	*7.* (a)	*8.* (b)	*9.* (b)	*10.* (d)
11. (d)	*12.* (c)	*13.* (d)	*14.* (b)	*15.* (c)	*16.* (a)	*17.* (a)	*18.* (c)	*19.* (c)	*20.* (a)
21. (d)	*22.* (b)	*23.* (d)	*24.* (b)	*25.* (d)	*26.* (c)	*27.* (d)	*28.* (c)	*29.* (c)	*30.* (b)
31. (b)	*32.* (a)	*33.* (a)	*34.* (a)	*35.* (c)	*36.* (c)	*37.* (c)	*38.* (d)	*39.* (b)	*40.* (e)
41. (d)	*42.* (c)	*43.* (a)	*44.* (a)	*45.* (c)	*46.* (d)	*47.* (b)	*48.* (a)	*49.* (b)	*50.* (d)
51. (a)	*52.* (d)	*53.* (b)	*54.* (d)	*55.* (a)	*56.* (a)	*57.* (b)	*58.* (d)	*59.* (a)	*60.* (c)

Hints and SOLUTIONS

1. Given, $\frac{(x+y)}{(x-y)} = \frac{3}{5}$

$\Rightarrow \quad \frac{(x+y)+(x-y)}{(x+y)-(x-y)} = \frac{3+5}{3-5}$

[By componendo and Dividendo]

$\Rightarrow \quad \frac{2x}{2y} = \frac{-4}{1}$

$\Rightarrow \quad \frac{x}{y} = -4 \Rightarrow x = -4y$

But given, xy = positive

$\Rightarrow \quad -4y \times y$ = positive

$\Rightarrow \quad -4y^2$ = positive

[it is not possible]

Hence, no real solutions for x and y exist.

2. $A : D = \frac{2}{3} \times \frac{5}{7} \times \frac{3}{10} = 1 : 7$

3. Required ratio

$= (15-12) : (25-15) = 3 : 10$

4. (*a*) P/Q

$= \frac{x^2-36}{x^2-49} \div \frac{x+6}{x+7} = \left(\frac{x+6}{x+7} \times \frac{x-6}{x-7}\right)$

$\div \frac{x+6}{x+7} = \frac{x-6}{x-7}$

5.

	A's	B's,	C's income
Initial ratio ⇒	3 :	4 :	5
Present ratio ⇒	$3 \times \frac{5}{4}$:	$4 \times \frac{3}{2}$:	$5 \times \frac{4}{3}$
=	$\frac{15}{4}$:	6 :	$\frac{20}{3}$

$= \frac{15}{4} \times 12 : 6 \times 12 : \frac{20}{3} \times 12$

$= 45 : 72 : 80$

Given, (45 + 72 + 80) units = 78800

$\therefore$ 72 units = 18000

Hence, present income of B = ₹ 28800

6. Ratio of their number = 2 : 3 : 4

Ratio of their values in ₹ $= \frac{2}{2} : \frac{3}{1} : 4 \cdot 2$

$= 1 : 3 : 8$

As given, (1 + 3 + 8) units = 240

$\therefore$ 1 unit = 20

Hence, total number coins

= (2 + 3 + 4) units

= 9 × 20 = 180

7. Clearly, $x : t = \frac{1}{3} \times \frac{5}{k} \times \frac{2}{5}$

$= 2 : 3k = 4 : 3$ [given]

$\Rightarrow \quad 12k = 6 \Rightarrow k = 1/2$

8. Given, $\frac{x}{y} = 25\frac{y}{x} \Rightarrow 25y^2 = x^2$

$\Rightarrow \quad x^2 : y^2 - 25 : 1$

$\Rightarrow \quad x : y = 5 : 1$

9. Since only option (b) is in the ratio 6 : 3 : 2, the correct option is (b). However, explanation is being given for better understanding.

Total wages

= [(1 × 6) + (2 × 3) + (3 × 2)] units

= ₹ 4500

⇒ [6 + 6 + 6] units = 4500

⇒ 6 units = 1500

Obviously, wages of 50 men = 6 units = ₹ 1500

∴ Weekly wages of one man

= (1500 ÷ 50) × 7 = ₹ 210

Then, weekly wages of one woman

= 105

[∵ ratio of wages of man and woman = 6 : 3]

and weekly wages of one child = 70

[ratio of wages of woman and child = 3 : 2]

10. Since one man can do as much work in one day as a woman can do in 2 days and a child does $\frac{1}{3}$rd the work in a day as a woman, ratio of their wages = 2 : 1 : 1/3 = 6 : 3 : 1

Now, their total wages

= [(6 × 6) + (5 × 3) + (2 × 1)] units

= ₹ 1113

⇒ [36 + 15 + 2] units = 1113

⇒ 2 units = 42

Total wages of total children = ₹ 42

Further, number of children

$= 39 \times \frac{2}{(6+5+2)} = 6$

∴ Daily wages of a child

= 42 ÷ 6 = ₹ 7

11. Initial ratio = 3 : 5, then share of Q = 5/8 of the amount

New ratio = 2 : 3, then share of Q = 3/5 of the amount

As given, 5/8 of the amount − 3/5 of the amount = ₹ 10

⇒ 1/40 of the amount = ₹ 10

∴ Amount to be divided

= 10 × 40 = ₹ 400

12. Clearly, when the original numerator and denominator each is increased by (1 + 1) the fraction becomes 1/2.

Let fraction be $\frac{1}{x}$.

Hence, $\frac{1+2}{x+2} = \frac{1}{2}$

$\Rightarrow \quad 6 = x + 2 \Rightarrow x = 4$

∴ Original fraction $= \frac{1}{4}$

Thus, $P = 1$ and $Q = 4$.

Then. $P + Q = 1 + 4 = 5$

If we assume that the original fraction be $\frac{2}{x}$, then $\frac{2+2}{x+2}=\frac{1}{2}$

$\Rightarrow \quad x = 6$

$\therefore$ Original fraction $=\frac{2}{6}$

No doubt, there may be infinite number of fractions like 3/8, 5/12, etc. Thus, we ought to keep the given options in mind while solving the problem.

13.

	Earlier	Later
Number of employees ⇒	3	2
Wages of each employee ⇒	22	25
Total expenditures on wages =	66	50

It means, expenditures on wages is decreased in the ratio = 66 : 50

As given, 50 units = ₹ 5000

$\therefore$ Original/earlier wage bill = ₹ 6600

14. Given, (8 – 5) units = ₹ 2475

$\Rightarrow$ (5 + 9) units = ₹ 11550

15. Given, $p\%$ of $x = t \times q\%$ of y

$\Rightarrow \quad x : y = qt : p$

16. Change in the cost of the article

$= [+(3\times 10\%) - (4\times 20\%) - (5\times 30\%) + (6\times 40\%)]$ units = + 0.4 units

Original cost of the article

$= 3 + 4 + 5 + 6$

$= 18$ units

Hence, percentage change in cost

$=\left(\frac{0.4}{18}\times 100\right)\%$

$= 2.22\%$

Obviously, the cost will increase by 2.22%.

17. (*a*) Let number of 10 paise coins be x.

Then number of 25 paise coins

$= 180 - x$

Value of coins in term of ₹

$=\frac{x}{10}+\frac{180-x}{4}=$ ₹ 36.90

$\Rightarrow \frac{3x}{20} = 8.10$

$\Rightarrow \quad x = 54$

18. (*c*) $A : B = x : 8$

$B : C = 12 : z$

Then, $A : C = 12x : 8z = 2 : 1$ [given]

$\Rightarrow \quad x : z = 16 : 12 = 4 : 3$

19. (*c*) Boys : Girls = $b : g$ [let] = B [given]

and Girls : Boys = $g : b = G$ [given]

Now,

$$3(B+G) = 3\left(\frac{b}{g}+\frac{g}{b}\right) = 3\left(\frac{b}{g}+\frac{1}{\frac{b}{g}}\right)$$

For any value of $\frac{b}{g}$, $3(B+G)$ will have value more than 3.

20. Required ratio = $(2\cdot 2 + 3\cdot 5)$ units : $(7\cdot 2 + 5\cdot 5)$ units = 19/39

21. Given, $A : B = 3 : 4$

$B : C = 3 : 4$

$C : D = 3 : 4$

$\therefore A : B : C : D$

$= (3\times 3\times 3) : (4\times 3\times 3) : (4\times 4\times 3) : (4\times 4\times 4)$

$= 27 : 36 : 48 : 64$

Now, C's share

$= 1400 \times \frac{48}{(27+36+48+64)} =$ ₹ 384

22. Given, $P : Q = \frac{3}{5} : \frac{5}{7}$

and $Q : R = \frac{3}{4} : \frac{2}{5}$

$P : Q : R = \left(\frac{3}{5}\times\frac{3}{4}\right) : \left(\frac{5}{7}\times\frac{3}{4}\right) : \left(\frac{5}{7}\times\frac{2}{5}\right)$

$= \frac{9}{20} : \frac{15}{28} : \frac{2}{7}$

23. Number of police involved

$= 135 \times \frac{3}{5} = 81$

Now, 1 unit = 81

$\therefore$ 9 units = 81 × 9 = 729

Hence, number of participants = 729.

24. Given, $a + b + c = 1$

$\Rightarrow \quad (a+b+c)^2 = 1^2$

[$\because$ squaring on both sides]

$\Rightarrow a^2 + b^2 + c^2 + 2(ab + bc + ca) = 1$

$\Rightarrow a^2 + b^2 + c^2 + \left(2\times\frac{1}{3}\right) = 1$

$\Rightarrow \quad a^2 + b^2 + c^2 = \frac{1}{3}$ ...(i)

Further, $ab + bc + ca = \frac{1}{3}$ ...(ii)

From Eqs. (i) and (ii), $a^2 = ab$, $b^2 = bc$, $c^2 = ca$

Hence, $a = b$, $b = c$, $c = a$

$\therefore \ a : b : c = 1 : 1 : 1$

Alternatively

You may proceed through options also.

25. Given, (7 + 5) units = 2400

$\therefore$ 5 units = 1000

26. Here, difference between antecedents and that between consequents are equal and each difference is 1.

As construed, 1 unit = 9

$\therefore$ 2 units = 18 and 3 units = 27

Hence, product of numbers

= 18 × 27 = 486

27. Remainder

= [5783 – (28 + 37 + 18)] = 5700

As given, (4 + 6 + 9) units = 5700

$\therefore$ 4 units = 1200

$\therefore$ Ramesh's share = 1200 + 28 = 1228

28. Clearly, 2nd part + 3rd part

= 80 – 24 = 56

As given, (4 + 3) units = 56

$\therefore$ 4 units = 32

29. Given, $\frac{(a-b)}{(a+b)} = \frac{1}{5}$

$\Rightarrow \frac{(a-b)+(a+b)}{(a-b)-(a+b)} = \frac{1+5}{1-5}$

[By Componendo and Dividendo]

$\Rightarrow \frac{2a}{-2b} = \frac{-3}{2} \Rightarrow \frac{a}{b} = \frac{3}{2}$

$\Rightarrow \frac{a^2}{b^2} = \frac{9}{4}$

$\Rightarrow \frac{a^2+b^2}{a^2-b^2} = \frac{9+4}{9-4}$

[By Componendo and Dividendo]

$\therefore \ (a^2 - b^2) : (a^2 + b^2) = 5 : 13$

Alternatively

$a - b = 1$ and $a + b = 5$, then $a = 3$ and $b = 2$. It is not technically true, however, the same has been done for finding correct answer only. Now, $(a^2 - b^2) : (a^2 + b^2)$

$= (3^2 - 2^2) : (3^2 + 2^2) = 5 : 13$

30. Given, $a : b = b : c$

$\Rightarrow \quad b^2 = ac$

Now, $a^4 : b^4 = (a^2)^2 : (b^2)^2$

$= a^2 \cdot a^2 : ac \cdot ac$

$= a^2 : c^2$

31. Let total number of students = 100

Then, 12% of 100 = 12

Now, desired ratio

= Number of boys : Number of girls

$=\frac{(100-12)}{2} + 12 : \frac{(100-12)}{2}$

= 56 : 44 = 14 : 11

32. Initial ratio

[Male to Female] = 3 : 1

Changed ratio = 2 : 1

Decrease and increase = 16 6

Now, $[(3 \times 1) - (2 \times 1)]$ unit

$= (16 \times 1) + (2 \times 6)$

$\Rightarrow$ 1 unit = 28

$\therefore$ [3 + 1] units = 112

33. Given, 40% of 5 units = 12

Then, 50% of 4 units = 12

34. Ratio of their number = 3 : 8 : 10

Ratio of their values in ₹ $= \frac{3}{1} : \frac{8}{2} : \frac{10}{10}$

$= 3 : 4 : 1$

As given, (3 + 4 + 1) units = 112

$\therefore$ 1 unit = 14

Hence, number of 50 paise coin

= 8 units

$= 8 \times 14 = 112$

35. Given, $A : B = 3 : 4$

$B : C = 5 : 6$

$C : D = 7 : 5$

$\therefore A : B : C : D$

$= (3 \times 5 \times 7) : (4 \times 5 \times 7) : (4 \times 6 \times 7) : (4 \times 6 \times 5)$

$= 105 : 140 : 168 : 120$

Clearly, C gets maximum share.

36. Let he had x one-rupee coins and y 20-paise coins in the beginning.

As per quostion,

$[1x + 0.2y] = 3[0.2x + y]$

$\Rightarrow 0.4x = 2.8y \Rightarrow x = 7y$

Since he spent 2/3 of the money and he had nearly ₹ 15 in beginning, the expenditure is near to ₹ 10. So, proceed with trial and error method accordingly.

If $y = 1$, then $x = 7$. Hence, total money he spent

$= \frac{2}{3}[1x + 0.2y]$, which is not near to ₹ 10.

If $y = 2$, then $x = 14$. Hence, total money he spent

$= \frac{2}{3}[(1 \times 14) + (0.2 \times 2)] =$ ₹ 9.6.

Thus, option(c) is correct.

37. $(5x - 2y) : (3x + 2y)$

$= (5 \cdot 7 - 2 \cdot 5) : (3 \cdot 7 + 2 \cdot 5)$

$= 25 : 31$

$= 25/31$

38. Given ratio

$= \frac{1}{5} : \frac{1}{6} : \frac{1}{10}$

$= \frac{1}{5} \times 30 : \frac{1}{6} \times 30 : \frac{1}{10} \times 30$

$= 6 : 5 : 3$

Now, (6 + 5 + 3) units = ₹ 8400

$\therefore$ 6 units = ₹ 4200

39. Total revenue

$= [(10 \times 3) + (8 \times 4) + (3 \times 10)]$ units

= ₹ 8050

$\Rightarrow$ [30 + 32 + 30] units = 8050

$\Rightarrow$ 32 units = ₹ 2800

Hence, ₹ 2800 was realised by the sale of second class tickets.

40. (101 + 88), (87 + 100), (110 + 111) and (85 + 98) all are non prime numbers except (97 + 84). Hence, option(e) is correct.

41. Required ratio = (150% of 3) : (160% of 5) : (150% of 7)

$= 45 : 80 : 105 = 9 : 16 : 21$

42. Given, $\frac{y}{x - z} = \frac{y + x}{z}$

$\Rightarrow yz = xy + x^2 - yz - zx$

$\Rightarrow 2yz = xy + x^2 - zx$ ...(i)

Further, $\frac{y}{x - z} = \frac{x}{y}$,

$\Rightarrow y^2 = x^2 - zx$ ...(ii)

From Eqs.(i) and (ii),

$2yz = xy + y^2$

$\Rightarrow 2z = x + y$ and option(c) satisfies this equation.

43. Clearly, their savings are ₹ 4000 and ₹ 1000. [$\because$ ₹ 5000 are divided in 4 : 1]

Now, ratio of incomes = 2 : 1

Ratio of expenditures = 5 : 3

Saving in ₹ = 4000 1000

Then, $[(2 \times 3) - (1 \times 5)]$ unit(s)

$= [(4000 \times 3) - (1000 \times 5)]$

$\Rightarrow$ 1 unit = 7000

$\therefore$ 2 units = 14000

Hence, monthly income of A

= ₹ 14000

44. Income = Expenditure + savings

= (3 + 2) = 5 units

Now, increment in savings = [(110% of 5) – (112% of 3)] units – 2 units = 0.14 units

Increase % $= \left(\frac{0.14}{2} \times 100\right)\% = 7\%$

45. Ratio of incomes = 9 : 4

Ratio of expenditures = 7 : 3

Saving in ₹ = 2000 2000

Now, $[(7 \times 4) - (9 \times 3)]$ unit(s)

$= [(2000 \times 7) - (2000 \times 3)]$

$\Rightarrow$ 1 unit = ₹ 8000

$\therefore$ 9 units = 72000 = Chandra's income

Hence, Chandra's expenditure

= 72000 – 2000

= ₹ 70000

46. Cost price of Rasgulla per kg

$= \left(\frac{15}{166\frac{2}{3}} \times 100\right) = 9$

Clearly, $[(5 \times 3) + (3 \times 7)]$ units

= ₹ 9(5 + 3)

$\Rightarrow$ 7 units = ₹ 14

Hence, cost of price of sugar per kg

= ₹ 14

47. Total wages paid

$= [(2 \times 5) + (3 \times 4)]$ units

= 80% of 275

$\Rightarrow$ (3×4) units = ₹ 120

Hence, total women are getting ₹ 120 and number of women labourers

$= 25 \times \frac{3}{5} = 15$

Then, wages per woman

$= 120 \div 15 =$ ₹ 8

48. Science : English = 1 : 2

and English : Maths = 2 : 3

Hence, Science : English : Maths

= 1 : 2 : 3

Clearly, (1 + 2 + 3) units = 180

$\therefore$ 1 unit = 30

49. The best way is to proceed through options. Correct option is (b). However, explanation is being given for better understanding

$\Rightarrow \frac{8 + x}{21 + x} = \frac{13 + x}{31 + x}$

$\Rightarrow x = 5$

50. Since the sum is not known, we cannot determine the amount to be received by B.

51. Since we require minimum number of coins, we must have coins of possible greatest denominations.

78 paise ⇒
50 + 10 + 10 + 2 + 2 + 2 + 2

69 paise ⇒ 50 + 10 + 5 + 2 + 2

₹ 1.01 ⇒ 50 + 25 + 10 + 10 + 2 + 2 + 2

Hence, required number of coins
= 7 + 5 + 7 = 19

52.

	Manufacturing cost	Raw material cost	Total
Initial prime Cost	₹ 4	₹ 2	₹ 6
Increased prime cost	₹ $4 \times \frac{5}{4}$	₹ $2 \times \frac{12}{5}$	₹ 9.80

53. Let the numbers be a and b $(a > b)$.

Then, $\frac{a-b}{2} : \frac{b-b}{2} = 3:1$

$\Rightarrow \quad 2a - b = 3b$

$\Rightarrow \quad a : b = 2:1$

Alternatively

The best way is to proceed through options.

(a) 3 : 1 ⇒ (3 – 0.5) ≠ 3 (1 – 0.5)

(b) 2 : 1 ⇒ (2 – 0.5) = 3 (1 – 0.5)

Hence, option(b) is correct.

54. Given ratio $= \frac{1}{3} : \frac{1}{4} : \frac{1}{5} : \frac{1}{6}$

$= \frac{1}{3} \times 60 : \frac{1}{4} \times 60 : \frac{1}{5} \times 60 : \frac{1}{6} \times 60$

$= 20 : 15 : 12 : 10$

Minimum number of pens that he should have

$= 20 + 15 + 12 + 10 = 57$

55. Required ratio = (75% of 8) : (66.66% of 9) : (80% of 15)

$= 6 : 6 : 12 = 1 : 1 : 2$

56. Given, $\frac{A}{2} = \frac{2B}{3} = \frac{3C}{4} = \frac{4D}{5}$

$\Rightarrow \quad \frac{A}{2} = \frac{4D}{5} \quad \Rightarrow \quad \frac{A}{D} = \frac{8}{5}$

$\Rightarrow \quad A : D = 8 : 5$

57. Final ratio = 3 : 4

Clearly, A's quotation is 1 unit less than B's quotation.

Hence, 1 unit = 1 lakh

Then, (7 – 3) units = 4 lakhs.

58. Let popcorn packets in large, super and jumbo sizes be $7x$, $17x$ and $16x$ respectively and chips packets be $6y$, $15y$ and $14y$ respectively.

As given, $7x + 17x + 16x$

$= 6y + 15y + 14y$

$\Rightarrow \quad 40x = 35y$

$\Rightarrow \quad 8x = 7y$

Now, number of jumbo popcorn packets

$= 16x = 2 \cdot 8x = 2 \cdot 7y = 14y$

and number of jumbo chips packets

$= 14y$

Then, ratio $= 14y : 14y = 1:1$

59. We may take any two terms of the ratio (say, 1st and 2nd).

Then,

Initial ratio = 2 : 3

Changed ratio = 5 : 7

Increase = 5   5

Now, [(5 × 3) – (7 × 2)] units

= (7 × 5) – (5 × 5)

⇒ 1 unit = 10

∴ 2 units = 20

Hence, the first original number is 20.

60. Ratio of share of wife, one son and one daughter = 1/2 : 3 : 1

As per question, (1/2 + 3 · 5 + 1 · 4) units = ₹ 390000

⇒ 1/2 units = ₹ 10000

Proportion

The equality of two ratios is called proportion. If $a : b = c : d$, we write $a : b :: c : d$ and we say that a, b, c and d are in proportion. We may also say that a, b, c and d are proportionals.

$a : b : : c : d$ is read as "a is to b as c is to d".

Here, the terms a and d are called the 'extremes' (extreme terms) while the terms b and c are called the 'means' (middle terms).

Critical Properties of Proportion

(A) If four quantities are in proportion (say, $a : b : : c : d$), then the product of the extremes is equal to the product of means.

$\Rightarrow \quad \dfrac{a}{b} = \dfrac{c}{d}$

$\Rightarrow \quad ad = bc$

Here, d is called the 4th proportional to a, b, c. 'a' is called 1st proportional, 'b' is 2nd proportional and 'c' is 3rd proportional.

e.g. The fourth proportional to 5, 8,15 is? ***[RRB-02, SSC (CGL-T-1-17)]***

$\Rightarrow \quad 5 : 8 : : 15 : ?$

$\Rightarrow \quad 5 \times ? = 8 \times 15 \Rightarrow ? = \dfrac{8 \times 15}{5} = 24$

(B) Three quantities are said to be in **continued proportion** if $a : b : : b : c$.

$\Rightarrow \quad \dfrac{a}{b} = \dfrac{b}{c}$

$\Rightarrow \quad b^2 = ac$

$\therefore \quad \boldsymbol{b = \sqrt{ac}}$

Further, $\quad a : c = a \cdot a : c \cdot a$

$\Rightarrow \quad \boldsymbol{a : c = a^2 : b^2}$

Here, b is said to be a mean proportional between a and c. c is said to be a 3rd proportional to a and b.

e.g. The mean proportional between $(3 + \sqrt{2})$ and $(12 - \sqrt{32})$ is

[SSC (CPO)-06]

Mean proportional $= \sqrt{(3 + \sqrt{2}) \times (12 - \sqrt{32})}$

$= \sqrt{(3 + \sqrt{2}) \times 4(3 - \sqrt{2})} = 2\sqrt{9 - 2} = 2\sqrt{7}$

e.g. The 3rd proportional to 0.8 and 0.2 is

$\Rightarrow \quad \dfrac{0.8}{0.2} = \dfrac{0.2}{x}$

$\Rightarrow \quad x = \dfrac{0.2 \times 0.2}{0.8} = 0.05$

Note *Other example of continued proportion*

$\Rightarrow \quad a : b = b : c = c : d = d : e$ *and so on.*

(C) Many other proportions may be derived from $a : b : : c : d$.

(i) If $\dfrac{a}{b} = \dfrac{c}{d}$, then $\dfrac{b}{a} = \dfrac{d}{c} \rightarrow$ **[Invertendo]**

(ii) If $\dfrac{a}{b} = \dfrac{c}{d}$, then $\dfrac{a}{c} = \dfrac{b}{d} \rightarrow$ **[Alternando]**

(iii) If $\dfrac{a}{b} = \dfrac{c}{d}$, then $\dfrac{a}{b} + 1 = \dfrac{c}{d} + 1 \Rightarrow \dfrac{a + b}{b} = \dfrac{c + d}{d}$ **[Componendo]**

(iv) If $\dfrac{a}{b} = \dfrac{c}{d}$, then $\dfrac{a}{b} - 1 = \dfrac{c}{d} - 1$

$\Rightarrow \dfrac{a - b}{b} = \dfrac{c - d}{d} \rightarrow$ **[Dividendo]**

(v) If $\dfrac{a}{b} = \dfrac{c}{d}$, then $\dfrac{a + b}{a - b} = \dfrac{c + d}{c - d} \rightarrow$

[Componendo and Dividendo]

e.g. If $(a + b) : (a - b) = 18 : 1$, then the value of $a^2 : b^2 = ?$

$\Rightarrow \quad \dfrac{(a + b)}{(a - b)} = \dfrac{13}{1}$

$\Rightarrow \quad \dfrac{a + b + a - b}{a + b - a + b} = \dfrac{13 + 1}{13 - 1}$

[by componendo and dividendo]

$\Rightarrow \quad \dfrac{2a}{2b} = \dfrac{14}{12}$

$\Rightarrow \quad \dfrac{a}{b} = \dfrac{7}{6}$

Now, $a^2 : b^2 = 7^2 : 6^2 = 49 : 36$

Problems based on Proportion

Problem 1 If b is the mean proportional of a and c, then $(a - b)^3 : (b - c)^3$ equals

Sol. Given, $a : b : : b : c$

$\Rightarrow \quad \dfrac{a}{b} = \dfrac{b}{c} = k$ [suppose]

$\Rightarrow \quad a = bk$ and $b = ck$

Now, $\dfrac{(a - b)^3}{(b - c)^3} = \dfrac{(bk - b)^3}{(ck - c)^3} = \dfrac{b^3 (k - 1)^3}{c^3 (k - 1)^3} = b^3 : c^3$

We keep options given in mind to observe their forms/terms while solving the question.

Problem 2 If $a : b = b : c$, then $a^4 : b^4$ is equal to

Sol. Given, $a : b = b : c$

$\Rightarrow \quad \dfrac{a}{b} = \dfrac{b}{c} = k$ [suppose]

$\Rightarrow \quad a = bk$ and $b = ck$

Then, $a^4 = b^4k^4$ and $b^4 = c^4k^4$

Now, $\dfrac{a^4}{b^4} = \dfrac{b^4k^4}{c^4k^4} = \dfrac{b^4}{c^4} = \dfrac{(b^2)^2}{(c^2)^2}$

$= \dfrac{(ac)^2}{(c^2) \cdot (c^2)} = a^2 : c^2$

We keep options given in mind to observe their forms/terms while solving the question.

Alternatively

Given, $\frac{a}{b}=\frac{b}{c}$

$\therefore \quad ac = b^2$

Now, $\frac{a^4}{b^4}=\frac{(a)^2(a)^2}{(b)^2(b)^2}=\frac{(a^2)\cdot(a^2)}{(ac)^2}=a^2:c^2$

Problem 3 If $a:b=c:d$, then $\frac{ma+nc}{mb+nd}$ is not equal to **[SSC (CGL-PT)-07]**

(a) $\frac{a}{b}$ (b) $\frac{c}{d}$ (c) $\frac{a+c}{b+d}$ (d) $\frac{c-a}{b-d}$

Sol. Given, $\frac{a}{b}=\frac{c}{d}=k$ [suppose]

$\Rightarrow \quad a = bk$ and $c = dk$

Now, $\frac{ma+nc}{mb+nd}=\frac{m\cdot bk+n\cdot dk}{mb+nd}=\frac{k(mb+nd)}{(mb+nd)}=k$

i.e. $\frac{ma+nc}{mb+nd}=k=\frac{a}{b}=\frac{c}{d}$

Further, $\frac{a}{b}=\frac{c}{d}\Rightarrow\frac{a}{c}=\frac{b}{d}$

$\Rightarrow \quad \frac{a+c}{c}=\frac{b+d}{d}$ [by componendo]

$\Rightarrow \quad \frac{a+c}{b+d}=\frac{c}{d}$

Hence, $\frac{ma+nc}{mb+nd}=\frac{a}{b}=\frac{c}{d}=\frac{a+c}{b+d}$

Alternatively I

Given, $\frac{a}{b}=\frac{c}{d}=k$ [suppose]

$\Rightarrow \quad a = bk$ and $c = dk$

Now, $\frac{ma+nc}{mb+nd}=\frac{m\cdot bk+ndk}{mb+nd}=\frac{k(mb+nd)}{(mb+nd)}=k$

i.e. $\frac{ma+nc}{mb+nd}=k=\frac{a}{b}=\frac{c}{d}$

or, $\frac{ma+nc}{mb+nd}=\frac{a}{b}=\frac{c}{d}=\frac{a+c}{b+d}$

As we know that equality of a ratio may be maintained by adding different numbers (in same ratio as given) to both numerator (antecedent) and denominator (consequent) of the given ratio i.e.

$\frac{a}{b}=\frac{a+c}{b+d}$ [when $\frac{c}{d}=\frac{a}{b}$]

Alternatively II

Given, $\frac{a}{b}=\frac{c}{d}=k$ [suppose]

$\Rightarrow \quad \frac{ma}{mb}=\frac{nc}{nd}=k$

$\Rightarrow \quad \frac{ma+nc}{mb+nd}=k$ [recall the properties of ratio]

$\Rightarrow \quad \frac{ma+nc}{mb+nd}=\frac{a}{b}=\frac{c}{d}=\frac{a+c}{b+d}$

[applying the same property of ratio]

Hence, correct answer is (d).

Problem 4 The 1st, 3rd and 4th terms of a proportion are 5,15 and 60 respectively, find the 2nd term.

Sol. Given, $5:x::15:60$

$\Rightarrow \quad \frac{5}{x}=\frac{15}{60}$

$\therefore \quad x=\frac{5\times 60}{15}=20$

Problem 5 When a particular number is subtracted from each of 7, 9, 11 and 15, the resulting numbers are in proportion. The number to be subtracted is **[SSC (CPO)-07, SSC (CGLT-I)-17]**

Sol. Let the required number be x.

Then, $(7-x):(9-x)::(11-x):(15-x)$

$\Rightarrow \quad 105-7x-15x+x^2=99-9x-11x+x^2$

$\Rightarrow \quad -2x=-6$

$\therefore \quad x=3$

Alternatively

You may proceed through options. Let options be 2, 3,1, 5.

(a) $2\Rightarrow\frac{7-2}{9-2}\neq\frac{11-2}{15-2}$ (b) $3\Rightarrow\frac{7-3}{9-3}=\frac{11-3}{15-3}$

Hence, correct option is (b).

Problem 6 A fraction which bears the same ratio to 1/27 that 3/11 does to 5/9, is equal to

Sol. Given, $x:\frac{1}{27}::\frac{3}{11}:\frac{5}{9}$

$\Rightarrow \quad x\times\frac{5}{9}=\frac{1}{27}\times\frac{3}{11}\Rightarrow x=\frac{1}{55}$

Problem 7 If a, b, c, d and e are in continued proportion, then a/e is equal to **[CDS-13]**

(a) a^3/b^3 (b) a^4/b^4
(c) b^3/a^3 (d) b^4/a^4

Sol. Since a, b, c, d and e are in continued proportion,

$\frac{a}{b}=\frac{b}{c}=\frac{c}{d}=\frac{d}{e}$

$\Rightarrow \quad \frac{a}{b}=\frac{b}{c}=\frac{c}{d}=\frac{d}{e}=k$ [let]

$\Rightarrow \quad a=bk, b=ck, c=dk$ and $d=ek$

$\Rightarrow \quad a=bk=ckk=dkkk=ekkkk=ek^4$

Now, $a/e=\frac{ek^4}{e}=k^4=\left(\frac{a}{b}\right)^4=a^4/b^4$

Application of the Concepts Associated with Proportion

We may apply the concept of direct proportion and indirect proportion to solve the questions, which are generally solved through "unitary method". Basically, two quantities are said to be directly proportional, if one increases, then another increases and in case one decreases, another also decreases.

e.g.

(a) Cost is directly proportional to number of articles **(More articles, More cost and Less articles, Less cost).**

(b) Work done is directly proportional to number of men/workforces **(More men/workforces, More work and Less men/workforces, Less work).**

(c) Wages to paid is directly proportional to number of labourers/workforces **(More labourers/workforces, More wages and Less labourers/workforces, Less wages)** and so many such examples.

It must be kept in mind that other items/aspects, which may be related to above examples, have been considered constant. On the other hand, when one quantity increases, another decreases and if one decreases, another increases, then both are said to be "indirectly proportional" or inversely proportional. **e.g.**

(a) Number of men is indirectly proportional to the time taken to finish a piece of work **(More men, Less time and Less men, More time).**

(b) Speed is indirectly proportional to the time taken to cover a certain distance **(More speed, Less time and Less speed, More time).**

(c) Number of soldiers is indirectly proportional to time/number of days in which foods are to be consumed **(More soldiers, Less days and Less soldiers, More days)** and so many such examples.

[Here, also other items/aspects have been kept constant.]

Now, build your concept through some Illustrations

If cost of 4 cows is ₹ 252, what is the cost of 3 cows?

$\because$ Cost of 4 cows = ₹ 252

$\therefore$ Cost of 1 cow = $\frac{252}{4}$

$\therefore$ Cost of 3 cows = $\frac{252}{4} \times 3 =$ ₹ 189

Note *The above method of solution is called Unitary Method.*

Alternatively

Let the required cost be ₹ x.

Obviously, Less cows, Less cost [direct proportion]

$\Rightarrow \quad 4 : 3 :: 252 : x$

$\Rightarrow \quad 4 \times x = 3 \times 252$

$\Rightarrow \quad x = \frac{3 \times 252}{4} = 189$

Problem 8 **40 men can finish a piece of work in 10 days, in how many days can 16 men finish it?**

Sol. **Unitary Method**

$\because$ 40 men takes 10 days to finish a piece of work.

$\therefore$ 1 man will take (40×10) days to finish the work.

$\therefore$ 16 men will finish in $\frac{40 \times 10}{16}$ days i.e. 25 days.

Alternatively

Let the required number of days be x.

Clearly, Less men, More time (days) [indirect proportion]

$\Rightarrow \quad 40 : 16 :: x : 10$

[Mind that we have written x : 10 instead of 10 : x]

$\Rightarrow \quad 40 \times 10 = 16 \times x$

$\therefore \quad x = \frac{40 \times 10}{16} = 25$ days

Problem 9 **39 persons can finish a piece of work in 12 days working 5 h a day. In how many days will 30 persons working 6 h a day, complete twice the work.**

Sol. **Unitary Method**

$\because$ 39 persons working 5 h per day can finish one unit of work in 12 days.

$\therefore$ 1 person working 5 h per day can finish one unit of work in (12×39) days

$\therefore$ 1 person working 1 h per day can finish one unit work in $(12 \times 39 \times 5)$ days

$\therefore$ 30 persons working 1 h per day can finish one unit work in $\frac{12 \times 39 \times 5}{30}$ days

$\therefore$ 30 persons working 6 h per day can finish one unit work in $\frac{12 \times 39 \times 5}{30 \times 6}$ days

$\therefore$ 30 persons working 6 h per day can finish two units work in $\frac{12 \times 39 \times 5 \times 2}{30 \times 6} = 26$ days

Alternatively

Let the required number of days be x.

Also, Less persons, More days [indirect proportion]

and More working h per day, Less day [indirect proportion]

and More work, More days [direct proportion]

Observation We compared each and every item with the term to be found out considering all the remaining items constant.

$$\text{Now, } \left(\begin{array}{ll} \text{Persons} & 30 : 39 \\ \text{Working h/day} & 6 : 5 \\ \text{Work-unit} & 1 : 2 \end{array} \right) :: 12 : x$$

Note *We have written two upper ratios in reverse order on account of relation of indirect proportion.*

Then, $30 \times 6 \times 1 \times x = 39 \times 5 \times 2 \times 12$

$\therefore \quad x = \frac{39 \times 5 \times 2 \times 12}{30 \times 6 \times 1} = 26$ *days*

In all the above 3 questions, we compared each and every item with the term to be found out considering all the remaining items constant.

From the above, the required answer $= \frac{39 \times 5 \times 12 \times 2}{30 \times 6 \times 1} = 26$

In other words,

Required answer

$$= \frac{\text{Multiplication of figures appeared in sentence completed in all aspects}}{\text{Multiplication of figures appeared in sentence in which any item is missing}}$$

[with reciprocal ratio of amount of work]

It is to note that in this solution 1 comes for a piece of work and 2 for twice the work. Further, we have taken amount of work in reciprocal ratio as it is directly proportional to number of days (More work, More days). As regards other items i.e. persons, working hours, they have relation of indirectly proportion with number of days (Less persons, More days and More working hours/day, Less days.)

The items, generally come in questions, are as below

Working Force	Amount of Work / Wages / Consumption	Working/ Consumption rate	Duration of Work
Persons, Pipes, Machines and Others	Quantity of liquid/solid, Units, Length, Area, Volume, Production of articles, Value in currency and others	Minutes or Hours/ per day, Consumption of food at the rate of — gram per day and others	Hours, Days, Month, Years and others

It is reiterated that you have to write the amount of work/wages/consumption only in reciprocal ratio to find the missing item. If you have to find the amount of work, then it will be reciprocal of

$$\frac{\text{Multiplication of figures appeared in sentence completed in all aspects}}{\text{Multiplication of figures appeared in sentence in which any item is missing}}$$

[with reciprocal ratio of amount of work, assuming that amount of work to be obtained is 1.]

Note *If amount of work is same, you may leave this figure. Now, persue the following four illustrations for better understanding of the above concepts.*

Problem 10 15 men working 8 h per day complete a piece of work in 20 days. In how many days will 10 men finish $\frac{1}{2}$ the work by working 6 h per day.

Sol. Required number of days =

$$\frac{\text{Multiplication of figures appeared in sentence completed in all aspects}}{\text{Multiplication of figures appeared in sentence in which any item is missing}}$$

[with reciprocal ratio of amount of work]

$$= \frac{15 \times 8 \times \frac{1}{2} \times 20}{10 \times 6 \times 1}$$

$$= \frac{15 \times 8 \times 1 \times 20}{10 \times 6 \times 2} = 20 \text{ days}$$

Problem 11 If 80 persons can finish a work within 16 days by working 6 h a day, the number of hours a day, should 64 persons work to finish that very job within 15 days is ***[SSC (10 + 2-LDC/DEO-12)]***

Sol. Required working-hours a day =

$$\frac{\text{Multiplication of figures appeared in sentence completed in all aspects}}{\text{Multiplication of figures appeared in sentence in which any item is missing}}$$

[with reciprocal ratio of amount of work]

$$= \frac{80 \times \mathbf{1} \times 16 \times 6}{64 \times 15 \times \mathbf{1}} = 8 \text{ h}$$

[No need to write the amount of work if they are same, however, here is mentioned only for understanding-purpose.]

Problem 12 12 men working 6 h a day complete a piece of work in 10 days. To complete thrice the work in 9 days, working 8 h a day, the number of men required is

Sol. Required number of men

$$= \frac{\text{Multiplication of figures appeared in the sentence completed in all aspects}}{\text{Multiplication of figures appeared in sentence in which any item is missing}}$$

[with reciprocal ratio of amount of work]

$$= \frac{12 \times 6 \times \mathbf{3} \times 10}{9 \times 8 \times \mathbf{1}} = 30$$

Problem 13 4 mat-weavers weave 4 mats in 4 days. At the same rate, how many mats would be woven by 8 mat-weavers in 8 days?

Sol. Number of mats (i.e. amount of work)

= **Reciprocal of**

$$\frac{\text{Multiplication of figures appeared in sentence completed in all aspects}}{\text{Multiplication of figures appeared in sentence in which any item is missing}}$$

[with reciprocal ratio of amount of work, assuming number of mats (i.e. amount of work) be 1]

$$= \text{Reciprocal of } \frac{4 \times \mathbf{1} \times 4}{8 \times \mathbf{4} \times 8}$$

$$= \text{Reciprocal of } \frac{1}{16} = 16$$

I hope that you would now be capable to replace the unitary/ proportion method meant for these types of questions by using this amazing method. However, more illustrations are given type-wise.

***Type* I Direct application of**

$$\frac{\text{Multiplication of figures appeared in the sentence completed in all aspects}}{\text{Multiplication of figures appeared in sentence in which any item is missing}}$$
[with reciprocal ratio of amount of work]

Problem 14 If the expenditure of gas on burning 6 burners for 6 h a day for 8 days is ₹ 450, then how many burners can be used for 10 days at 5 h a day for ₹ 625?

Sol. Required number of burners $= \frac{6 \times 6 \times 8 \times \mathbf{625}}{10 \times 5 \times \mathbf{450}} = 8$

Problem 15 If 10 men or 20 boys can make 260 mats in 20 days, then how many mats will be made by 8 men and 4 boys in 20 days?

Sol. Obviously, the working force should be in same unit either in men or in boys.

Given, 10 men = 20 boys

∴ 8 men = 16 boys.

Thus, 8 men + 4 boys = 16 + 4 = 20 boys

Now, the question is as below.

20 boys 260 mats 20 days

20 boys ? 20 days

Clearly, answer is 260 mats.

Problem 16 If 10 burners, each burning 8 h a day consume 120 L of diesel, then how much diesel will be required for 15 burners, each burning 12 h a day, provided consumption by each burner of the former type is 50% more than that by each of the latter type ?

Sol. Required amount of diesel if both types of burners consume the equal amount of diesel in the same period

$= \text{Reciprocal of}\left(\frac{10 \times 1 \times 8}{15 \times 120 \times \mathbf{12}}\right) = \text{Reciprocal of } \frac{1}{270} = 270 \text{ L}$

Now, 150% of the consumption by burners of the latter type = 270 L

Then, 100% of the consumption by burners of the latter type = 180 L

Problem 17 If 10 men can do a piece of work in 12 days, the time taken by 12 men to do the same piece of work will be ***[SSC (CPO-SI)-08]***

Sol. The time taken $= \frac{10 \times 12}{12} = 10$ days

[no need to write the work- amount as they are same]

***Type* II Application of**

$$\frac{\text{Multiplication of figures appeared in sentence completed in all aspects}}{\text{Multiplication of figures appeared in sentence in which any item is missing}}$$
[with reciprocal ratio of amount of work]

after modification /alteration in some figures.

Problem 18 A certain number of men can do a work in 60 days. If there were eight more men, it would be completed in 10 days less. How many men were there in beginning ?

Sol. Let there were x men in beginning.

x men can do a work in 60 days. Then, number of days taken by $(x + 8)$ men to complete the work $= \frac{x \times 60}{(x + 8)}$

As given, $(60 - 10) = \frac{x \times 60}{(x + 8)} \Rightarrow 50x + 400 = 60x \Rightarrow 10x = 400$

∴ $x = 40$

Alternatively $\frac{x}{(x + 8)} = \frac{50}{60}$ $\left[\because \text{work force } \alpha \frac{1}{\text{time}}\right]$

$\Rightarrow 60x = 50x + 400$

∴ $x = 40$

Problem 19 A contractor undertook to do a job in 10 days and engaged so me labourers. 5 of them proceeded on leave from the 2nd day and the remaining part of job was finished in 12 days. Find the initial number of labourers.

Sol. Let there were x labourers in beginning.

Clearly, all x labourers (x) worked for 1st day and completed 1/10 of the work.

Now, x labourers can complete the whole work in 10 days, then number of days required by $(x - 5)$ labourers to complete the remaining work $\left(\text{i.e. } 1 - \frac{1}{10} = \frac{9}{10}\right) = \frac{x \times \frac{\mathbf{9}}{\mathbf{10}} \times 10}{(x - 5) \times 1} = 12$ [given]

$\Rightarrow 9x = 12x - 60$

∴ $x = 20$

Alternatively

Since 10% of work has been done by x labourers in one day, the time taken by x labourers to complete 90% of the work = 9 days

Hence, $\frac{x}{(x - 5)} = \frac{12}{9}$ $\left[\because \text{work force } \alpha \frac{1}{\text{time}}\right]$

$\Rightarrow 9x = 12x - 60 \Rightarrow x = 20$

Problem 20 A contractor undertakes to make a road in 40 days and employs 25 men. After 24 days, he finds that only 1/3rd of the road is made. How many extra men should he employ so that he is able to complete the work 4 days earlier ?

Sol. 25 men makes 1/3 of the road in 24 days, then number of men required to complete the remaining work [i.e. 1 – 1/3 = 2/3] in remaining time [i.e. 40 – 24 – 4 = 12 days]

$$= \frac{\text{Multiplication of figures appeared in the sentence completed in all aspects}}{\text{Multiplication of figures appeared in sentence in which any item is missing}}$$
[with reciprocal ratio of amount of work]

$$= \frac{25 \times \left(\frac{2}{3}\right) \times 24}{\left(\frac{1}{3}\right) \times 12} = 100$$

Thus, extra number of men to be employed = 100 – 25 = 75

Problem 21 A road of 5 km length will be constructed in 100 days. So, 280 workers were employed. But after 80 days it was found that only $3\frac{1}{2}$ km road was completed. Now how many more people were needed to finish the work in specified time ? *[SSC (CPO-SI)-09, CAT-20]*

Sol. 280 workers finish $3\frac{1}{2}$ km road in 80 days. The number of workers required to finish the remaining road $\left(i.e.\ 5 - 3\frac{1}{2} = 1\frac{1}{2}\text{km}\right)$ in the remaining days [*i.e.*100 – 80 = 20]

$$= \frac{\text{Multiplication of figures appeared in the sentence completed in all aspects}}{\text{Multiplication of figures appeared in sentence in which any item is missing}}$$

[with reciprocal ratio of amount of work]

$$= \frac{280 \times 1\frac{1}{2} \times 80}{3\frac{1}{2} \times 20} = 480$$

$\therefore$ Extra workers to be employed = 480 – 280 = 200

Problem 22 A contractor estimates that he will finish the road in 80 days by employing 60 men. After 40 days he finds that the only 40% work has been completed. Now, how many more days are required in order to complete the work ? *[CDS-21]*

Sol. 60 men complete 40% work in 40 days, then number of days required to complete the remaining work (i.e. 60%) by them

$= 1\frac{1}{2}$ times of 40 = 60 days.

$\therefore$ Extra number of days required = (40 + 60) – 80
= 20 days.

Alternatively

40% work by 60 men in 40 days.

Then, number of days required by 60 men to finish 100% work

$$= \frac{100 \times \mathbf{60} \times 40}{\mathbf{40} \times 60} = 100$$

$\therefore$ Extra number of days required = 100 – 80
= 20 days

Problem 23 A job can be completed by 12 men in 12 days. How many extra days will be needed to complete the job if 6 men leave after working for 6 days ?

Sol. 12 men completed $\frac{1}{2}$ of work in 6 days, then (12 – 6) men will finish the remaining work $\left(i.e.\ 1 - \frac{1}{2} = \frac{1}{2}\right)$ in $\frac{12 \times \frac{1}{2} \times 6}{6 \times \frac{1}{2}}$ days

[i.e. 12 days]

$\therefore$ Number of extra days = (6 + 12) – 12
= 6 days

Problem 24 A contractor deployed 60 men to finish a piece of work in 50 days. If he had deployed 20 men more after 30 days, the work would have been finished 2 days before the stipulated time. Suppose he had not deployed additional work force, for how many days the completion of the work would have been delayed ?

Sol. Clearly, after 30 days, (60 + 20) men finished the remaining work in (50 – 2 – 30) days.

Hence, 60 men can do the same remaining work in $\left(\frac{80 \times 18}{60}\right)$

i.e. in 24 days.

$\therefore$ Total time taken by 60 men in case no additional men is deployed = 30 + 24 = 54 days, which is 4 days more than the scheduled time.

Note *To get the work completed in 50 days with the help of 60 men is only estimation of the contractor, the number of days may differ.*

Problem 25 25 workers can finish a piece of work in 40 days. After how many days should 10 workers leave the job so that the work is completed in 50 days.

Sol. Let 10 workers leave the work after x days from starting.

$\therefore$ Work done in x days $= \frac{x}{40}$

$\therefore$ Remaining work $= 1 - \frac{x}{40} = \frac{40 - x}{40}$

Now, 25 workers can finish a piece of work in 40 days, then number of days required by (25 – 10) workers to finish the remaining work $\left(\frac{40 - x}{40}\right)$

$$= \frac{25 \times \left(\frac{\mathbf{40 - x}}{\mathbf{40}}\right) \times 40}{15 \times 1} = 50 - x$$

$\Rightarrow \quad 1000 - 25x = 750 - 15x$

$\Rightarrow \quad -10x = -250$

$\therefore \quad x = 25$

Hence, 10 workers should leave the work after 25 days.

Alternatively

You may also proceed through options.

Problem 26 30 men can complete a job in 40 days. However, after 24 days some men out of the assigned 30 left the job. The remaining people took another 40 days to complete the job. The number of men who left the job is *[CDS-17]*

Sol. Work done in 24 days $= \frac{24}{30} = \frac{4}{5}$

$\therefore$ Remaining work $= 1 - \frac{4}{5} = \frac{1}{5}$

Let x men out of 30 left the job.

30 men can complete the remaining job in (40 – 24) days. Then, number of days taken by $(30 - x)$ men to complete the remaining job $= \frac{30 \times 16}{(30 - x)}$

As given, $40 = \dfrac{30 \times 16}{(30 - x)} \Rightarrow 30 - x = 12 \Rightarrow x = 18$

Alternatively

$$\frac{30}{(30 - x)} = \frac{40}{(40 - 24)} \qquad \left[\because \text{work force } \alpha \frac{1}{\text{time}}\right]$$

$\Rightarrow \quad 480 = 1200 - 40x$

$\Rightarrow \quad x = 18$

***Type* III Problems based on camp/garrison and provisions therefore, etc. wherein quantity has been kept fixed.**

Problem 27 **A fort had provision of food for 80 soldiers for 150 days. After 10 days, 16 soldiers left the fort. The number of days for which the remaining food will last, is**
[SSC (CGL-PT)-01, CLAT-12]

Sol. Clearly, 80 soldiers had provisions for (150 – 10) i.e. 140 days.

$\therefore$ (80 – 16) i.e. 64 soldiers would have provisions for $\dfrac{140 \times 80}{64}$ i.e for 175 days. [Here, provisions are same, so not mentioned]

Alternatively

$$\frac{80}{80 - 16} = \frac{?}{(150 - 10)}$$

$$\left[\therefore \text{number of soldiers } \alpha \frac{1}{\text{Provision}}\right]$$

$\therefore \quad ? = \dfrac{80 \times 140}{64} = 175$ days

Problem 28 **A garrison of 1200 men had provisions for 90 days at the rate of 500 grams per capita per day. After 15 days, a reinforcement of certain number of men arrives resulting in provisions will last 60 days more if each of them is provided provisions at 20% less than the earlier provisions per day. Find the number of men in the reinforcement ?**

Sol. Clearly, 1200 men had provisions for 75 days if they were given 500 g each per day. Then, number of men for whom the same provisions would last for 60 days at the rate of (80% of 500) i.e 400 g per capita per day =

$$\frac{\text{Multiplication of figures appeared in sentence completed in all aspects}}{\text{Multiplication of figures appeared in sentence in which any item is missing}}$$

[with reciprocal ratio of provisions (i.e. quantity)]

$$= \frac{1200 \times 75 \times 500}{60 \times 400} = 1875$$

[here, provisions are same, so not mentioned]

Number of men in the reinforcement = 1875 – 1200 = 675

Problem 29 **A garrison of 200 men has provisions for 35 days. After how many days should a reinforcement of 50 men join them so that the provisions lasts in next 8 days.**

Sol. Let after x days, the reinforcement joins.

Now, it is clear that 200 men have the provisions for $(35 - x)$ days.

Then, number of days in which the provisions will be consumed by (200 + 50) men

$$= \frac{200 \times (35 - x)}{250} = 80 \text{ [given]}$$

$\Rightarrow \quad 140 - 4x = 40 \Rightarrow 4x = 100$

$\therefore \quad x = 25$ days

Problem 30 **A camp of certain number of soldiers had provisions for a certain number of days. If 2/3 of the soldiers had departed from the camp after 20 days, the provision would have lasted as long as before. For how many days was the provisions in the beginning?**

Sol. Let number of soldiers was x and had provisions for y days.

Now, it is clear that x men have the provisions for $(y - 20)$ days.

$\therefore \left(x - \dfrac{2x}{3}\right)$ will have provisions for $\dfrac{x \times (y - 20)}{\left(x - \dfrac{2}{3}x\right)}$ days

But given, $\dfrac{x \times (y - 20)}{\left(x - \dfrac{2}{3}x\right)} = y$

$\Rightarrow \quad \dfrac{x \times (y - 20)}{\dfrac{1}{3}x} = y$

$\Rightarrow \quad y - 20 = \dfrac{y}{3}$

$\dfrac{2y}{3} = 20$

$y = 30$ days

Remarks

[illegible]tions may be crafted on the same line as was d[illegible] under type II, however, the approach for the solving the questions would be the same.

WORKOUT

1. If 10 persons can dig 8 ft. trench in 12 days, then how many days will 8 persons take to dig 6 ft. trench? *[CDS-14]*

(a) 10 days (b) 10.25 days
(c) 11 days (d) 11.25 days

2. In a proportion, the multiplication of 1st and 4th terms is 40 and that of 2nd and 3rd terms is $2.5x$, then the value of x is

(a) 16 (b) 36
(c) 75 (d) 90

3. The fourth proportional to 7,11,14 is *[CDS-11, SSC (CGL-T-II)-17]*

(a) 16 (b) 18
(c) 20 (d) 22

4. If 7 spiders make 7 webs in 7 days, then 1 spider will make 1 web in how many days? *[Railway-03]*

(a) 1 (b) 7/2
(c) 7 (d) 49

5. 24 men each of equal efficiency, can complete a job in $12\frac{1}{2}$ days. They started to work together but every day of the work, one man leaves. Find the minimum number of days required to complete the job.

(a) 15 (b) 24
(c) 25 (d) 27

6. What number must be added to each of 4, 10, 12 and 24 so that the resulting numbers are in proportion?

(a) 6 (b) 4
(c) 10 (d) 8

7. In a camp, there is a meal for 120 men or 200 children. If 150 children have taken the meal, how many men will be catered with the remaining meal ?

(a) 30 (b) 20
(c) 50 (d) 40

8. Running at the same constant rate, 6 identical machines can produce a total of 270 bottles per min. At this rate, how many bottles could 10 such machines produce in 4 min.?

(a) 648 (b) 1800
(c) 2700 (d) 10800

9. What is the mean proportional between $(15 + \sqrt{200})$ and $(27 - \sqrt{648})$ is *[CDS -12]*

(a) $3\sqrt{5}$ (b) $5\sqrt{3}$ (c) $14\sqrt{7}$ (d) 4

10. In a company XYZ Ltd. a certain number of engineers can develop a design in 40 days. If there were 5 more engineers, it could be finished in 10 days less. How many engineers were there in the beginning ? *[MAT-14, 19]*

(a) 18 (b) 20 (c) 25 (d) 15

11. Find two numbers such that their mean proportion is 6 and 3rd proportional 20.25.

(a) 4 and 9 (b) 2 and 18
(c) 3 and 12 (d) 10 and 6

12. Two coal loading machines, each working 12 h per day for 8 days handle 9000 tonne of coal with an efficiency 90%, while 3 other coal loading machines at an efficiency of 80% are set to handle 12000 tonne of coal in 6 days. How many hours per day should each machine work ? *[MAT-13]*

(a) 18 h (b) 16 h (c) 20 h (d) 14 h

13. The 3rd proportional of 12 and 18 is *[SSC (CGL-T-III)-13]*

(a) 3 (b) 6 (c) 27 (d) 144

14. 40 persons can finish a piece of work in 20 days working 5 h a day. In how many days will 50 persons working 6 h a day, complete thrice the work.

(a) 18 (b) 20 (c) 25 (d) 40

15. The mean proportion between the numbers 8 and 18 is *[RRB (ASM)/Ajmer-01]*

(a) 12 (b) 13
(c) 32/9 (d) None of these

16. Four numbers are in proportion. The sum of the squares of the four numbers is 50 and the sum of the mean is 5. The ratio of the 1st two terms is 1 : 3. What is the average of the four numbers ?

(a) 1 (b) 3
(c) 5 (d) 6

17. If x is the mean proportional between 12.8 and 64.8 and y is the 3rd proportional to 38.4 and 57.6, then $2x$: y is *[SSC (CGL-T-I)-19]*

(a) 2 : 3 (b) 1 : 2
(c) 3 : 4 (d) 4 : 5

18. An engineering firm is allotted the work of building a railway bridge in 100 days. The engineers estimate that the work requires 200 men working for 100 days to complete the project. However, due to meager fund they are able to employee only 100 persons for first 40 days. After 40 days, total funds are released. How many men need to be employed now to complete the project in time ? *[MAT-18]*

(a) 315 (b) 243
(c) 267 (d) 298

19. A contract is to be completed in 75 days and 187 men are to work 15 h per day. After 65 days, 3/5 of the work is completed. How many additional men may be employed, so that the work may be completed in time, each man now working 17 h per day ? *[SSC (CGL-T-I)-19]*

(a) 514 (b) 532
(c) 528 (d) 495

Answer Keys

1. (d)	2. (a)	3. (d)	4. (c)	5. (b)	6. (b)	7. (a)	8. (b)	9. (a)	10. (d)
11. (a)	12. (b)	13. (c)	14. (d)	15. (a)	16. (b)	17. (a)	18. (c)	19. (c)	

Hints and SOLUTIONS

1. Required number of days =

$$\frac{\text{Multiplication of figures appeared in sentence completed in all aspects}}{\text{Multiplication of figures appeared in sentence in which any item is missing [with reciprocal ratio of amount of work]}}$$

$$= \frac{10 \times 6 \times 12}{8 \times 8}$$

$= 11.25$ days

2. We know, product of the extremes is equal to the product of means.

$\Rightarrow \quad 40 = 2.5x$

$\Rightarrow \quad x = 16$

3. 7 : 11 :: 14 : ?

$\Rightarrow 7 \times ? = 11 \times 14$

$\therefore \quad ? = 22$

4. Required number of days

$$= \frac{7 \times 1 \times 7}{1 \times 7} = 7$$

5. Work done by 1 man in 1 day

$$= \frac{1}{24 \times 12\frac{1}{2}} = \frac{1}{300}$$

Now, work done by 24 men for 1st day $= \frac{1}{300} \times 24$

Work done by 23 men for 2nd day

$$= \frac{1}{300} \times 23$$

Work done by 22 men for 3rd day

$$= \frac{1}{300} \times 22$$

and so on.

$\therefore$ Total work done in 24 days

$$= \frac{1}{300}(24 + 23 + 22 + \ldots\ldots + 1)$$

$$= \frac{1}{300} \times \left(\frac{24 \times 25}{2}\right) = 1$$

Thus, in 24 days the job will be completed.

6. Let the number x be added.

Then, $\frac{4 + x}{10 + x} = \frac{12 + x}{24 + x}$

$\Rightarrow 96 + 28x + x^2 = 120 + 22x + x^2$

$\Rightarrow \quad 6x = 24 \Rightarrow x = 4$

Alternatively

The best way is to proceed through options.

(a) $\Rightarrow \frac{4 + 6}{10 + 6} \neq \frac{12 + 6}{24 + 6}$

(b) $\Rightarrow \frac{4 + 4}{10 + 4} = \frac{12 + 4}{24 + 4}$

Hence, correct option is (b).

7. Clearly, $\frac{3}{4}$ of the meal has been consumed. Now, the question may be read as: '3/4 of meal is consumed by 150 children, then how many children can be catered with 1/4 of meal?

Required number of children

$$= \frac{150 \times 1/4}{3/4} = 50$$

$\because$ 200 children = 120 men

$\therefore$ 50 children = 30 men

8. Number of bottles produced per min.

$$= \text{reciprocal of } \frac{6 \times 1}{10 \times 270}$$

$= 450$

Then, number of bottles produced in 4 min.

$= 4 \times 450 = 1800$

9. Mean proportional

$$= \sqrt{(15 + \sqrt{200}) \times (27 - \sqrt{648})}$$

$$= \sqrt{5(3 + \sqrt{8}) \times 9(3 - \sqrt{8})}$$

$$= \sqrt{45\,(3^2 - (\sqrt{8})^2)}$$

$$= \sqrt{45} = 3\sqrt{5}$$

10. Let there were x men in beginning.

$$\frac{x}{(x + 5)} = \frac{(40 - 10)}{40} \qquad \left[\because \text{work force } \alpha \frac{1}{\text{time}}\right]$$

$\Rightarrow \quad 40x = 30x + 150$

$\therefore \quad x = 15$

11. Let the numbers be x and y.

Then, case I $\Rightarrow x : 6 :: 6 : y$

$\Rightarrow \quad xy = 36 \Rightarrow y = \frac{36}{x}$

and case II $\Rightarrow x : y :: y : 20.25$

$\Rightarrow \quad y^2 = 20.25x$

$$\Rightarrow \quad \left(\frac{36}{x}\right)^2 = 20.25x$$

$$\Rightarrow \quad x^3 = \frac{36 \times 36}{20.25} = 64$$

$\Rightarrow \quad x = 4$

Then, $\quad y = \frac{36}{4} = 9$

Hence, two numbers are 4 and 9.

12. Required time =

$$\frac{\text{Multiplication of figures appeared in sentence completed in all aspects}}{\text{Multiplication of figures appeared in sentence in which any item is missing [with reciprocal ratio of amount of work]}}$$

$$= \frac{2 \times 12 \times 12000 \times 90 \times 8}{3 \times 80 \times 9000 \times 6}$$

$= 16$ h

13. As given, 12 : 18 :: 18 : ?

$\Rightarrow \quad 18 \times 18 = 12 \times ?$

$\Rightarrow \quad ? = 27$

14. Required number of days

$$= \frac{40 \times 20 \times 5 \times 3}{50 \times 6 \times 1}$$

$= 40$ days

15. Mean proportional $= \sqrt{8 \times 18}$

$= 12$

16. Given, $a : b = c : d$

$= 1 : 3$

$\Rightarrow a = k, b = 3k$ and $c = m$ and $d = 3m$, where k and m are constants.

Now, $\quad a^2 + b^2 + c^2 + d^2 = 50$

$\Rightarrow (k)^2 + (3k)^2 + (m)^2 + (3m)^2 = 50$

$\Rightarrow \quad (k)^2 + (m)^2 = 5$

$$\Rightarrow \quad \left(\frac{5 - m}{3}\right)^2 + (m)^2 = 5$$

$$\left[\because b + c = 3k + m = 5 \text{ (given)} \quad \therefore k = \frac{5 - m}{3}\right]$$

$\Rightarrow \quad 25 + m^2 - 10m + 9m^2 = 45$

$\Rightarrow \quad 10m^2 - 10m - 20 = 0$

$\Rightarrow \quad m^2 - m - 2 = 0$

$\Rightarrow \quad m = 2$, then $k = \frac{5 - 2}{3} = 1$

Hence, $a = 1, b = 3, c = 2$

and $d = 6$

Then, their average

$= (1+3+2+6) \div 4 = 3$

17. $2x : y = 2 \times \sqrt{12.8 \times 64.8} : \dfrac{57.6 \times 57.6}{38.4}$

$= 57.6 : \dfrac{57.6 \times 57.6}{38.4}$

$= 1 : \dfrac{3}{2} = 2 : 3$

18. Clearly, 100 men have completed 20% of the work in 40 days. Then, number of men required to complete the remaining 80% work in remaining time [i.e. 60 days] =

$$= \frac{\text{Multiplication of figures appeared in the sentence completed in all aspects}}{\text{Multiplication of figures appeared in sentence in which any item is missing}}$$

[with reciprocal ratio of amount of work]

$$= \frac{100 \times 80\% \times 40}{20\% \times 60} \approx 267$$

19. 187 men, working 15 h per day, completes 3/5 of the work in 65 days, then, number of men, working 17 h per day, required to complete the remaining work [i.e. 1 – 3/5 = 2/5] in remaining time [i.e. 75 – 65 = 10 days]

$$= \frac{\text{Multiplication of figures appeared in sentence completed in all aspects}}{\text{Multiplication of figures appeared in sentence in which any item is missing}}$$

[with reciprocal ratio of amount of work]

$$= \frac{187 \times 15 \times \left(\frac{2}{5}\right) \times 65}{\left(\frac{3}{5}\right) \times 17 \times 10} = 715$$

Thus, additional number of men to be employed

$= 715 - 187 = 528$

Variation

Variation shows how one quantity changes when the other quantity is changed. It is denoted by 'α'. If one quantity (x) is increased/decreased, the other (y) is also increased/decreased proportionately, then it is said that 'x is directly proportional to y' (written as $x \propto y$). Sometimes we omit 'directly' and simply say "x is proportional to y". If $x \propto y$, then $x = ky$ [where k is a 'constant of proportionality'.]

If one quantity (x) is decreased/increased, the other quantity (y) is increased/decreased, then it is said that "x is inversely proportional to y" and written as $x \propto \frac{1}{y}$.

If $x \propto \frac{1}{y}$, then $x = k.\frac{1}{y}$ [where k is a 'constant of proportionality'.]

Note *$x \propto y$ is also read as "x varies directly as y" or "x varies as y" and $x \propto \frac{1}{y}$ is also read as "x varies inversely as y."*

Sometimes, a quantity varies jointly i.e. directly proportional to any quantity and inversely proportional to another quantity.

e.g. $x \propto y$ and $x \propto \frac{1}{z}$

On combination, $x \propto \frac{y}{z} \Rightarrow x = k\frac{y}{z}$

In addition, other relations may be as

(a) $x \propto yz \Rightarrow x = kyz$ (b) $x \propto \frac{1}{yz} \Rightarrow x = \frac{k}{yz}$, etc.

Some critical results/relations

(i) If $A \propto B$ and $B \propto C$, then $A \propto C$.

(ii) If $A \propto C$ and $B \propto C$, then $(A+B) \propto C$ and also, $(A-B) \propto C$.

(iii) If $A \propto B$ and $A \propto C$, then $A \propto (B-C)$ and also, $A \propto (B+C)$.

(iv) If $A \propto B$ and $C \propto D$, then $AC \propto BD$.

(v) If $A \propto BC$, then $B \propto \frac{A}{C}$ and also, $C \propto \frac{A}{B}$.

(vi) If $A \propto B$, then $A^n \propto B^n$ (where, n is any number).

(vii) If $A \propto B$, then $nA \propto nB$ (where n is any quantity, constant or variable).

Problems Based on Variation

Problem 1 A is proportional to B and $A = 3$ when $B = 5$. Find the value of B, if $A = 12$.

Sol. Given, $A \propto B \Rightarrow A = kB$ [where, k is a constant]

$\Rightarrow 3 = k \cdot 5 \quad \therefore \quad k = \frac{3}{5}$

Further, $A = kB \Rightarrow 12 = \frac{3}{5} \times B$

$\therefore \quad B = 20$

Alternatively

Given, $A \propto B \Rightarrow 3 \propto 5$ [case I]

and $12 \propto ?$ [case II]

Clearly, 12 is 4 times 3.

$\therefore$? will also be 4 times 5 i.e. ? = 20

Problem 2 A varies as B directly and as C inversely. B is 15 when A is 15 and C is 2. Find the value of B when A is 9 and C is 4.

Sol. Given that $A \propto B$ and $A \propto \frac{1}{C}$

$\Rightarrow A \propto \frac{B}{C} \Rightarrow A = k \cdot \frac{B}{C}$ [where k is a constant]

$\Rightarrow 15 = k \cdot \frac{10}{2} \Rightarrow 15 = 5k$

$\therefore \quad k = 3$

Further, $A = k \cdot \frac{B}{C} \Rightarrow 9 = 3 \times \frac{B}{4} \Rightarrow B = 12$

Alternatively

Given, $A \propto \frac{B}{C} \Rightarrow 15 \propto \frac{10}{2}$ [case I] and $9 \,\alpha\, \frac{?}{4}$ [case II]

Clearly, 15 is 3 times $\frac{10}{2}$.

$\therefore$ 9 will also be 3 times $\frac{?}{4}$ i.e. $\frac{?}{4} = 3 \Rightarrow ? = 12$

Problem 3 **Two numbers a and b are inversely proportional to each other. If a increases by 100%, then b decreases by** *[XAT-19, CLAT-18]*

Sol. Given, $a \,\alpha\, \frac{1}{b} \Rightarrow a = k\frac{1}{b}$ [where k is constant]

$\Rightarrow \quad k = ab$

It implies that for inversely proportional numbers, the product of the numbers is always constant.

$\Rightarrow \quad k = 2a \times \frac{b}{2}$

It is evident from the above that a becomes $2a$ when it is increased by 100% and b becomes $b/2$ when it is decreased by 50%. Hence, required answer = 50%

Alternatively

Since for inversely proportional numbers, the product of the numbers is always constant, recall the concept of product constancy envisaged in chapter 'Percentage'. Here, a is increased by 100% then b will decrease by $\left(\frac{100}{100+100} \times 100\right)\%$

i.e. by 50%.

Problem 4 **The period of the pendulum is directly proportional to the square root of the length of the string. The period is 51 sec when the string is of 9 cm. If the period is 85 sec, find the length of the string.**

Sol. Given, period (T) is directly proportional to the square root of length (L).

$\Rightarrow \quad T \,\alpha\, \sqrt{L} \Rightarrow T = k\sqrt{L}$, where k is a constant.

$\Rightarrow \quad k = \frac{51}{\sqrt{9}} = 17$

Now, $\quad T = k\sqrt{L} \Rightarrow 85 = 17\sqrt{L} \Rightarrow 5 = \sqrt{L}$

$\therefore \quad L = 25$ cm.

Alternatively

Given, $T \,\alpha\, \sqrt{L} \Rightarrow 51 \,\alpha\, \sqrt{9} \Rightarrow 51 \,\alpha\, 3$

$\therefore \quad 85 \,\alpha\, 5$ (unitary method)

$\Rightarrow \quad 85 \,\alpha\, \sqrt{25}$

Hence, $L = 25$ cm.

Problem 5 **The resistance of a wire is directly proportional to its length and inversely proportional to the square of its radius. Two wires of the same material have the same resistance and their radii are in the ratio 9 : 8. If the length of first wire is 162 cm, find the length of other wire.** *[CAT-09, SSC (CGL-Mains)-01]*

Sol. As given, resistance (R) α length(L) and $R \,\alpha\, \frac{1}{[\text{radius}(r)]^2}$

On combination, $R \,\alpha\, L/r^2 \Rightarrow R = k\frac{L}{r^2}$, where k is a constant.

$\Rightarrow \quad R = k \times \frac{162}{(9x)^2}$ [let their radii be $9x$ and $8x$]

$\Rightarrow \quad k = \frac{Rx^2}{2}$

Now, $\quad R = k\frac{L}{r^2} \Rightarrow R = \frac{Rx^2}{2} \times \frac{?}{(8x)^2}$

$\therefore \quad ? = 2 \times 64 = 128$ cm.

Problem 6 **m varies inversely as $n^2 + 1$ and is equal to 36, when $n = 4$. Find the value of n, when $m = 102$.**

Sol. Given, $\quad m \,\alpha\, \frac{1}{n^2+1} \Rightarrow m = k \times \frac{1}{n^2+1}$, where k is a constant.

$\Rightarrow \quad 36 = k \times \frac{1}{(n^2+1)} \Rightarrow 36 = k \times \frac{1}{(4^2+1)}$

$\therefore \quad k = 36 \times 17$

Further, $\quad 102 = k \times \frac{1}{(n^2+1)} \Rightarrow 102 = \frac{36 \times 17}{(n^2+1)} \Rightarrow n^2 + 1 = 6$

$\Rightarrow \quad n^2 = 5 \Rightarrow n = 2.236$

Alternatively

As given, $\quad m \,\alpha\, \frac{1}{n^2+1}$

$\Rightarrow \quad 36 \,\alpha\, \frac{1}{4^2+1} \Rightarrow 36 \,\alpha\, \frac{1}{17}$

$\therefore \quad 102 \,\alpha\, \frac{1}{6} \Rightarrow \frac{1}{n^2+1} = \frac{1}{6} \Rightarrow 6 = n^2 + 1$

$\therefore \quad n = 2.236$

Problem 7 **Cost of a diamond varies directly as the squares of its weight. A diamond is broken into four pieces with their weights in the ratio 1 : 2 : 3 : 4. What is the price of the original diamond if the loss in the total value of diamond amounts to ₹ 70000.** *[CAT-96, CDS-17, SSC (CGL-T-II)-17]*

Sol. Let the weights of four pieces be x, $2x$, $3x$ and $4x$.

$\therefore$ Total weight $= x + 2x + 3x + 4x = 10x$

As per question, Price(P) α [weight(w)]^2 $\Rightarrow p = kw^2$

$\therefore$ Price of original diamond $= k(10x)^2 = k100x^2$

Now, price of all four parts on the above line

$= k(x)^2 + k(2x)^2 + k(3x)^2 + k(4x)^2 = k30x^2$

Clearly, loss in value $= k100x^2 - k30x^2 = k70x^2$

But given that, $k70x^2 = 70000$

$\therefore \quad kx^2 = 1000$

Then, price of original diamond

$= k100x^2 = 1000 \times 100 = 100000$

Alternatively

Total weight $= 1 + 2 + 3 + 4 = 10$

Difference $= (10)^2 - (1^2 + 2^2 + 3^2 + 4^2) = 100 - 30 = 70$

Obviously, 70 ⇒ ₹ 70,000

$\therefore \quad$ 100 ⇒ 100000

Hence, price of original diamond = ₹ 100000

Problem 8 The expenses of a boys' institute are partly constant and partly vary as the number of boys. The expenses were ₹ 10000 for 150 boys and ₹ 8400 for 120 boys. What will the expenses be when there are 330 boys? *[MAT-14]*

Sol. Let the fixed expenditures be ₹ k and per boy expenses be ₹ x.

Then, $10000 = k + 150x$ ...(i)

and $8400 = k + 120x$ ...(ii)

or $40000 = 4k + 600x$ ...(i)

and $42000 = 5k + 600x$ ...(ii)

On solving Eqs. (i) and (ii), we have k = ₹ 2000 and x = ₹ 53.33

Now, expenditures on 330 boys = 2000 + (330 × 53.33) = 19600

Alternatively

Decrease in expenditures on account of decrease in number of boys from 150 to 120 = 10000 − 8400 = ₹ 1600

Hence, expenditure per boy = 1600 ÷ (150 − 120) = ₹ 53.33

Then, fixed expenditure = 10000 − (150 × 53.33) = ₹ 2000

∴ Expenditure on 330 boys = 2000 + (330 × 53.33) = ₹ 19600

Problem 9 Cost of a tree is jointly varies as its height and its age. Its cost is ₹ 600 when its height is 2 m and the age is 3 yrs . How old is the tree if its cost is ₹ 800 and height is 2.5 m.

Sol. Given, cost α height × age

$$600 = k \times 2 \times 3$$

where k is a constant.

$\therefore \quad k = 100$

Now, $800 = 100 \times 2.5 \times \text{age}$

$\therefore \quad \text{Age} = \dfrac{800}{100 \times 2.5} = 3.2$ yrs

Alternatively

As given, $600 \,\alpha\, 2 \times 3$

$\Rightarrow \quad 600 \,\alpha\, 6$

$\therefore \quad 800 \,\alpha\, 8$

As given, $8 = 2.5 \times ?$

$\therefore \quad ? = 3.2$ yrs

WORKOUT

1. The height of a tree varies as the square root of its age (between 5 to 17 yr). When the age of the tree is 9 yr, its height is 4 ft. What will be the height of the tree at the age of 16 yr ? *[CDS -14]*

(a) 5 ft 4 inch (b) 5 ft 5 inch
(c) 4 ft 4 inch (d) 4 ft 5 inch

2. The electricity bill of a certain establishment is partly fixed and partly varies with the number of units of electricity consumed. When in a certain month 540 units are consumed, the bill is ₹ 1800. In another month 620 units are consumed and the bill is ₹ 2040. In yet another month 500 units are consumed. The bill for that month would be : *[MAT-13]*

(a) ₹ 1680 (b) ₹ 1560
(c) ₹ 1840 (d) ₹ 1950

3. x varies inversely as the square of y in such a way that if $x = 1$, then $y = 6$. If $y = 3$, then what is the value of x ?

(a) 1/3 (b) 1/2
(c) 2 (d) 4

4. The ratio of the rate of flow of water in pipes varies inversely as the square of the radius of the pipes. What is the ratio of the rates of flow in 2 pipes of diameter 2 cm and 4 cm ?

(a) 1 : 2 (b) 2 : 1
(c) 1 : 8 (d) 4 : 1

5. If x varies as the mth power of y, y varies as the nth power of z and x varies as the pth power of z, then which one of the following is correct ?

(a) $p = m + n$
(b) $p = m - n$
(c) $p = mn$
(d) None of the above

6. x and y vary inversely with each other. When x is 12, y is 9. The pair which is not a possible pair of corresponding values of x and y is :

(a) 9 and 12 (b) 18 and 6
(c) 24 and 18 (d) 36 and 3

7. Suppose x and y are inversely proportional and positive. If x increases by 10%, then y decreases by *[LIC (AAO)-11]*

(a) 10% (b) $\dfrac{10}{11}$%
(c) $9\dfrac{1}{11}$% (d) $\dfrac{1}{11}$%

8. The duration of the journey from your home to the college in the local train varies directly as the distance and inversely as the velocity. The velocity varies directly as the square root of the diesel used per km, and inversely as the number of carriages in the train. If in a journey of 70 km in 45 min. with 15 carriages, 10 L of diesel is required, then the diesel that will be consumed in a journey of 50 km in half an hour with 18 carriages is *[IIFT-13]*

(a) 2.9 L (b) 11.8 L
(c) 15.7 L (d) None of these

9. If a quantity y varies as the sum of three quantities of which the 1st varies as x, the 2nd varies as $-x + x^2$, the 3rd varies as $x^3 - x^2$, then what is y equal to?

(a) kx^3 , where k is a constant
(b) $kx + lx^2 + mx^3$, where k, l m are constants
(c) kx^2, where k is a constant
(d) kx, where k is a constant

10. A locomotive engine without a train can run at 24 km/h. The speed diminishes by a quantity which varies as the square root of the number of wagons attached.

With 4 wagons, the speed is 20 km/h. Find the greatest number of wagons that the engine can move. *[SSC (Asstt-Mains)-95, IIFT-12]*

(a) 140 (b) 151
(c) 96 (d) 143

11. Let y is equal to the sum of two quantities of which one varies directly as x and the other inversely as x. If $y = 6$ when $x = 4$ and $y = 10/3$, when $x = 3$, then what is the relation between x and y?

(a) $y = x + \frac{4}{x}$ (b) $y = -2x + \frac{4}{x}$
(c) $y = -2x + \frac{8}{x}$ (d) $y = 2x - \frac{8}{x}$

12. x varies directly as y and inversely as square of z. When $y = 4$ and $z = 14$, $x = 10$. If $y = 16$ and $z = 7$, then what is value of x?

(a) 180 (b) 160
(c) 154 (d) 140

13. A quantity Q is obtained by adding three quantities. The 1st is a constant, the 2nd varies directly with the square root of y and the 3rd varies directly with the cube root of y. If $y = 1$, $Q = 60$, when $y = 64$, $Q = 230$ and when $y = 729$, $Q = 660$, then find the constant. *[CAT-14]*

(a) 10 (b) 20
(c) 30 (d) 40

Answer Key

1. (a)	2. (a)	3. (d)	4. (d)	5. (c)	6. (c)	7. (c)	8. (b)	9. (b)	10. (d)
11. (d)	12. (b)	13. (c)							

Hints and SOLUTIONS

1. Given, height(h) α $\sqrt{\text{age }(a)}$

$\Rightarrow h = k\sqrt{a}$, where k is a constant.

$\Rightarrow 4 = k\sqrt{9} \Rightarrow k = 4/3$

In case age is 16 yrs, $h = k\sqrt{16}$

$\Rightarrow h = 4/3 \times 4 = 16/3 = 5$ ft 4 inch.

Alternatively

Given, $h \alpha \sqrt{a}$

$\Rightarrow$ $4 \alpha \sqrt{9}$

$\Rightarrow$ $4 \alpha 3$

$\therefore$ $? \alpha \sqrt{16}$

$\Rightarrow$ $16/3 \alpha \sqrt{16}$ [unitary method]

Hence, $h = 16/3$ ft = 5 ft 4 inch.

2. Let the fixed bill be ₹ k and per unit bill be ₹ x.

Then, $1800 = k + 540x$ ...(i)

and $2040 = k + 620x$ ...(ii)

On solving Eqs. (i) and (ii), we have $k =$ ₹ 180 and $x =$ ₹ 3

Now, bill for 500 units
= 180 + (500 × 3) = ₹ 1680

Alternatively

Increase in bill-amount on account of increase in number of units from 540 to 620 = 2040 − 1800 = ₹ 240

Hence, expenditure per unit
= 240 ÷ (620 − 540) = ₹ 3

Then, fixed bill
= 1800 − (540 × 3) = ₹ 180

∴ Bill for 500 units
= 180 + (500 × 3) = ₹ 1680

3. Given, $x \alpha \frac{1}{y^2}$

$\Rightarrow x = k \cdot \frac{1}{y^2}$, where k is a constant.

$\Rightarrow 1 = k \cdot \frac{1}{6^2} \Rightarrow k = 36$

Further, $x = k \cdot \frac{1}{y^2}$

$= 36 \times \frac{1}{3^2} = 4$

4. Radii of both pipes = 1 cm and 2 cm

Now, required ratio $= \frac{1}{1^2} : \frac{1}{2^2} = 4:1$

5. Given, $x \alpha z^p$ and $y \alpha z^n$

Further, $x \alpha y^m$

$\Rightarrow z^p \alpha (z^n)^m \Rightarrow p = nm$

6. Given, $x \alpha \frac{1}{y} \Rightarrow x = k \cdot \frac{1}{y}$

$\Rightarrow 12 = k \frac{1}{9}$

$\Rightarrow k = 108$

Clearly, $9 \times 12 = 18 \times 6 = 36 \times 3 = 108$, but $24 \times 18 \neq 108$

Hence, correct option is (c).

7. Given, $x \alpha \frac{1}{y} \Rightarrow x = k \cdot \frac{1}{y} \Rightarrow y = \frac{k}{x}$

In case x is increased by 10%,

$y = k \cdot \frac{1}{110\% \text{ of } x}$

$\Rightarrow y = k \cdot \frac{100}{110x} = \frac{10k}{11x}$

Decrease in value of y

$= \frac{k}{x} - \frac{10k}{11x} = \frac{k}{11x}$

∴ Decrease percent

$= \left(\frac{\frac{k}{11x}}{\frac{k}{x}} \times 100\right)\% = 9\frac{1}{11}\%$

Alternatively

Decrease percent

$= \left(\frac{10}{100+10} \times 100\right)\% = 9\frac{1}{11}\%$

[recall the concept of product constancy as enumerated under chapter 'Percentage'.]

8. Given, Time(T) α Distance(D) and

$T \alpha \frac{1}{\text{velocity }(v)}$, hence, $T \alpha \frac{D}{V}$

Further, $V \alpha \sqrt{\frac{F \text{ (diesel used)}}{D}}$

and $V \propto \dfrac{1}{\text{Number of carriages}(N)}$,

hence, $V \propto \dfrac{\sqrt{F/D}}{N}$

On combining we get, $T \propto \dfrac{D}{\dfrac{\sqrt{F/D}}{N}}$

$\Rightarrow T = k \cdot \dfrac{DN}{\sqrt{F/D}}$

$\Rightarrow T = \dfrac{k \cdot D\sqrt{DN}}{\sqrt{F}}$

$\Rightarrow 45 = \dfrac{k \cdot 70\sqrt{70} \times 15}{\sqrt{10}}$

$\Rightarrow k = \dfrac{3}{70\sqrt{7}}$

Further, $T = \dfrac{k \cdot D\sqrt{DN}}{\sqrt{F}}$

$\Rightarrow 30 = \dfrac{\dfrac{3}{70\sqrt{7}} \cdot 50\sqrt{50} \times 18}{\sqrt{F}}$

$\Rightarrow \sqrt{F} = \dfrac{3 \times 50\sqrt{50} \times 18}{70\sqrt{7} \times 30}$

$\Rightarrow \quad F = 11.8$ L

9. 1st term $\propto x \Rightarrow$ 1st term $= gx$, where g is a constant.

2nd term $\propto (-x + x^2)$

$\Rightarrow$ 2nd term $= h(-x + x^2)$, where h is a constant.

and 3rd term $\propto (x^3 - x^2)$

$\Rightarrow$ 3rd term $= t\,(x^3 - x^2)$, where t is a constant

Now, $y \propto gx + h(-x + x^2) + t(x^3 - x^2)$

$\Rightarrow y = u\,[gx + h(-x + x^2)$
$+ t(x^3 - x^2)]$, where u is a constant.

$\Rightarrow y = u\,[x(g - h) + x^2(h - t) + x^3 \cdot t]$

$\Rightarrow y = x(g - h)u + x^2(h - t)u + x^3 \cdot tu$

$\Rightarrow y = kx + lx^2 + mx^3$ where k, l, m are constants and $k = (g - h)u$, $l = \{h - t\}u$ and $m = tu$

10 Given, decrease in speed $\propto \sqrt{\text{Number of wagons attached}}$

$\Rightarrow \quad 4 \propto \sqrt{4} \Rightarrow 4 = k2$, where k is constant.

$\therefore \quad k = 2$

Now, assuming that the initial speed is diminished by 24. It means the speed becomes zero.

Then, $24 \propto \sqrt{\text{Number of wagons attached}}$

$\Rightarrow 24 \propto k \cdot \sqrt{w}$

$\Rightarrow \quad 24 = k\sqrt{w}$

$\Rightarrow \quad 24 = 2 \cdot \sqrt{w}$

$\Rightarrow \quad w = 144$

Obviously, the speed is zero when 144 wagons are attached. Thus, in order to move the train, maximum number of wagons to be attached should be (144 – 1) *i.e.* 143.

11. 1st quantity $\propto x \Rightarrow$ 1st quantity $= mx$, where m is a constant and 2nd quantity $\propto \dfrac{1}{x} \Rightarrow$ 2nd quantity $= \dfrac{n}{x}$, where n is a constant.

As given, $y = mx + \dfrac{n}{x}$

$\Rightarrow \quad 6 = m \cdot 4 + \dfrac{n}{x} \quad [\because y = 6 \text{ and } x = 4]$

$\Rightarrow \quad 16m + n = 24$ ...(i)

Further, $\dfrac{10}{3} = m \cdot 3 + \dfrac{n}{3}$

$\left[\because y = \dfrac{10}{3} \text{ and } x = 3\right]$

$\Rightarrow \quad 9m + n = 10$ ...(ii)

From Eqs. (i) and (ii), we have $m = 2$ and $n = -8$

Now, $y = mx + \dfrac{n}{x}$ [from the above]

$\Rightarrow \quad y = 2x + \dfrac{(-8)}{x}$

$\Rightarrow \quad y = 2x - \dfrac{8}{x}$

12. Clearly, $x \propto \dfrac{y}{z^2}$

$\Rightarrow \quad x = k\dfrac{y}{z^2}$, where k is a constant.

$\Rightarrow 10 = k \cdot \dfrac{4}{(14)^2} \Rightarrow k = 490$

Further, value of x

[when $y = 16$ and $z = 7$]

$= 490 \times \dfrac{16}{(7)^2} = 160$

13. Given, $Q = a + m + n$

$Q = a + b\sqrt{y} + c\sqrt[3]{y}$, where, a, b and c are constant.

Then, $60 = a + b\sqrt{1} + c\sqrt[3]{1}$

$\Rightarrow \quad 60 = a + b + c$

$\Rightarrow \quad b + c = 60 - a$ ...(i)

Also, $230 = a + b\sqrt{64} + c\sqrt[3]{64}$

$\Rightarrow \quad 230 = a + 8b + 4c$

$\Rightarrow 8b + 4c = 230 - a$

$\Rightarrow \quad 2b + c = \dfrac{230 - a}{4}$ ...(ii)

Further, $660 = a + b\sqrt{729} + c\sqrt[3]{729}$

$\Rightarrow \quad 660 = a + 27b + 9c$

$\Rightarrow \quad 3b + c = \dfrac{660 - a}{9}$ ...(iii)

From Eqs. (ii) – (i), we get

$b = \dfrac{230 - a}{4} - (60 - a)$

and from Eqs. (iii) - (ii), we get

$b = \dfrac{660 - a}{9} - \dfrac{230 - a}{4}$

Hence, $\dfrac{230 - a}{4} - (60 - a)$

$= \dfrac{660 - a}{9} - \dfrac{230 - a}{4}$

$\Rightarrow \quad \dfrac{230 - a}{2} = \dfrac{660 - a + 540 - 9a}{9}$

$\Rightarrow 2070 - 9a = 2400 - 20a$

$\Rightarrow \quad 11a = 330 \Rightarrow a = 30$

27 Problems Based on Ages

Learning OBJECTIVES

After completing this chapter, aspirants will be able to

- Understand the application of concepts associated with ratio to solve the problems based on ages;
- Observe the diverse sorts of categories of the questions;
- Learn how to solve the maximum questions with the concepts associated with 'Ratio';
- Understand the philosophy of the division of this chapter in various parts;
- Analyse the trend of the questions being asked in various competitive examinations;
- Recognise the types of question which can be easily solved through the given options;
- Solve the questions in moments;
- Grip on concepts helping to solve the questions without pen and paper;
- Learn many new concepts/ fundamentals;
- Apply the pragmatic approaches in lieu of academic approaches; and
- Ponder on the alternative methods.

Introduction

'Problems based on Ages', asked in the competitive examinations substantially adheres to two concepts namely, 'average' and 'ratio'. It is mentioned that we have already discussed the problems on ages seeking concept on 'Average' in detail in the chapter 'Average'.

Thus, it will be appropriate to stress on the problems on 'ages' seeking concepts associated with 'ratio'. In other words, we will extend the concept of 'ratio' to solve such problems. Although types of such problems are too many, we will try to bring them under one umbrella, i.e. under concept related to 'Ratio' for better understanding as well as to avoid confusion/dilemma as to which approach(es) is/are suitable for a certain type of question(s).

Inside

- *Introduction including nature of questions to be discussed*
- *Determination of ratio of ages with reference to a certain time interval and vice-versa*
- *Alternation in ratio of ages on account of a time interval [same direction : either (ago-ago) or (hence-hence)]*
- *Alternation in ratio of ages on account of a time interval [opposite direction : either (ago-hence) or (hence-ago)]*
- *Cases wherein ages are not given in the ratio*
- ....
- .... *& Many More*
- *Work Out with Solutions*

Further, it is germane to indicate that the questions like the ratio of ages of A and B is $3:4$, that of ages of B and C is $3:5$ and that of ages of C and D is $5:6$, what is the ratio between ages of A and D? should have been under this chapter, as it appears. But indeed, this question is similar as '$A:B=3:4, B:C=3:5$ and $C:D=5:6$, then $A:D=?$' Apart from this one, there are many questions which are based on the concepts of ratio-proportion and should be dealt with there accordingly. Thus, aspirants should not assume these types of questions under this chapter and indeed no such types of questions have been discussed here to avoid duplicacy of the questions.

Determination of ratio of ages with reference to a certain time interval and *vice-versa*

Problem 1 Harsha is 40 yrs old and Ritha is 60 yrs old. How many years ago was the ratio of their ages 3 : 5 ?

Sol. Let x yrs ago, ratio of their ages was 3 : 5.

As per question, $\frac{40-x}{60-x}=\frac{3}{5}$

$\Rightarrow \quad 200-5x=180-3x \Rightarrow 2x=20$

$\therefore \quad x=10$ yrs

Alternatively

You may proceed through options. Let options be 20, 10, 37 and 5.

(a) $20 \Rightarrow \frac{40-20}{60-20} \neq \frac{3}{5}$ (b) $10 \Rightarrow \frac{40-10}{60-10}=\frac{3}{5}$

Hence, option (b) is correct.

Problem 2 The ratio between the present ages of P and Q is 5 : 7. If the difference between Q's present age and P's age after 6 yrs is 2, what is the total of P's and Q's present ages ? *[Bank PO-99]*

Sol. Let present ages of P and Q be $5x$ and $7x$ respectively.

Given, $7x-(5x+6)=2 \Rightarrow 2x=8 \Rightarrow x=4$

Then, total of their present ages $=5x+7x=12x=12\times4=48$ yrs

Alternatively I

Obviously, difference between their present ages $=(6+2)$ $=8$ yrs

Now, $(7-5)$ units $=8$ yrs

$\therefore \quad (7+5)$ units $=48$ yrs

Alternatively II

You may proceed through options. Let options be 60, 48, ...

(a) $60 \Rightarrow (7+5)$ units $=60$ yrs, then $(7-5)$ units $\neq 8$

(b) $48 \Rightarrow (7+5)$ units $=48$ yrs, then $(7-5)$ units $=8$

Hence, option (b) is correct.

Problem 3 At present the ratio between the ages of Arun and Deepak is 4 : 3. After 6 yr, Arun's age will be 26 yrs, what is the age of Deepak at present ? *[RRB-03]*

Sol. Given that Arun's age after 6 yrs = 26. It implies that his present age is 20 yrs.

Now, 4 units = 20, then 3 units = 15.

Hence, the age of Deepak at present = 15 yrs.

Problem 4 The present ages of three persons are in proportion 4 : 7 : 9. Eight years ago, the sum of their ages was 56. Find the age of the person who is youngest.

Sol. $\because$ Sum of their ages, 8 yrs ago = 56 yrs

$\therefore$ Sum of their present ages $=56+(3\times8)$ $=80$

As given, $(4+7+9)$ units $=80$

$\therefore \quad 4$ units $=16$

Hence, age of the youngest person = 16 yrs

Problem 5 Rajan got married 8 yrs ago. His present age is 6/5 times his age at the time of his marriage. Rajan's sister was 10 yrs younger to him at time of his marriage. Find the age of Rajan's sister. *[IAS(Pre.)-03]*

Sol. Let Rajan's age was x yrs at the time of his marriage.

Then, his present age $=\frac{6x}{5}$

As given, $\frac{6x}{5}-x=8$

$\Rightarrow \quad x=40$

Then, age of his sister at the time of his marriage $=40-10$ $=30$ yrs

$\therefore$ His sister's present age $=30+8=38$ yrs

Alternatively

Ratio of present age of Rajan to his age at time of marriage $=6:5$

As given, $(6-5)$ unit $=8$ yrs

6 units = 48 yrs (i.e. his present age)

Then, present age of his sister = 48 − 10 = 38 yrs.

Alternation in ratio of ages on account of a time interval [same direction: either (ago-ago) or (hence-hence)]

Problem 6 The present ages of A and B are in the ratio 4 : 5 and after 5 yrs they will be in the ratio 5 : 6. Find the present age of A. *[SSC(CGL-PT)-08]*

Sol. Let the present ages of A and B be $4x$ and $5x$ respectively.

After 5 yrs, A's age $=4x+5$

and after 5 yrs B's age $=5x+5$

Given, $\frac{4x+5}{5x+5}=\frac{5}{6}$

$\Rightarrow \quad 24x+30=25x+25$

$\Rightarrow \quad x=5$

$\therefore$ Present age of $A=4x=4\times5$ $=20$ yrs

Alternatively I

Recall the concept of "Change in ratio of numbers when they all are either increased by same/different number(s) or all decreased by same/different numbers" as envisaged in chapter "Ratio, Proportion and Variation" and read this question as "There are two numbers in ratio 4 : 5. If each number is increased by 5, the ratio becomes 5 : 6. Find the first number".

Initial ratio [ratio of present ages] = 4 : 5

Changed ratio [ratio of ages after 5 yrs] = 5 : 6

Here, difference between antecedents i.e. [5 – 4] and difference between consequents i.e. [6 – 5] are equal. In case they are not equal, we make the differences equal by multiplying both the terms of one ratio by a suitable number so that the terms of changed ratio are greater than respective terms of initial ratio if the original numbers are increased. In case the original numbers are decreased, the terms of the changed ratio must be lesser than respective terms of initial ratio [details in the chapter "Ratio, Proportion and Variation"].

It may be construed that difference of 1 unit is due to increment in each number by 5, i.e. 1 unit = 5

∴ 4 units = 5 × 4 = 20

Hence, present age of A = 20 yrs.

Alternatively II

Initial ratio [i.e. ratio of present ages] = 4 : 5

Changed ratio [i.e. ratio of ages after 5 yrs] = 5 : 6

Increment [i.e. time interval] = 5 5

[Positive difference between outcomes obtained on cross multiplication in respect of upper two lines] units = [Positive difference between outcomes obtained on cross multiplication in respect of lower two lines]

i.e. [(5 × 5) – (6 × 4)] units = [(6 × 5) – (5 × 5)]

⇒ 1 unit = 5 yrs

∴ 4 units = 20 yrs

Hence, present age of A = 20 yrs.

Problem 7 **The ratio of present ages of two brothers is 1 : 2 and 5 yrs back the ratio was 1 : 3. What will be the ratio of their ages after 5 yrs?**

Sol. Initial ratio [i.e. ratio of present ages] = 1 : 2

Changed ratio [i.e. ratio of ages 5 yrs ago] = 1 : 3

Decrease [i.e. time interval] = 5 5

[Positive difference between outcomes obtained on cross multiplication in respect of upper two lines] units = [Positive difference between outcomes obtained on cross multiplication in respect of lower two lines]

i.e. [(3 × 1) – (2 × 1)] unit = [(5 × 3) – (5 × 1)]

⇒ 1 unit = 10 yrs

∴ 2 units = 20 yrs

Hence, their present ages are 10 yrs and 20 yrs, respectively.

Now, new ratio after 5 yrs = (10 + 5) : (20 + 5) = 3 : 5

Problem 8 **Four years ago, the ratio of the ages of *A* and *B* was 2 : 3 and after four years, it will become 5 : 7. Find their present ages.**

[SSC(CGL-PT-)-02, SSC(CGL-T-II)-10, CDS-17, CLAT-17]

Sol. Initial ratio [i.e. ratio of ages, 4 yrs ago] = 2 : 3

Changed ratio [i.e. ratio of ages, 4 yrs hence] = 5 : 7

Time interval = 8 8

Now, [(5 × 3) – (7 × 2)] unit = [(8 × 7) – (8 × 5)]

⇒ 1 unit = 16 yrs

∴ 2 units = 32 and 3 units = 48

Now, their present ages = (32 + 4) yrs and (48 + 4) yrs, i.e. 36 yrs and 52 yrs.

Alternatively

You may proceed through options.

Let options be 36 and 40 yrs, 36 and 52 yrs, ...

(a) 36 and 40 ⇒ 4 yrs ago, ratio = 32 : 36 = 8 : 9 ≠ 2 : 3

(b) 36 and 52 ⇒ 4 yrs ago, ratio = 32 : 48 = 2 : 3 and after 4 yrs, ratio = 40 : 56 = 5 : 7

Hence, option (b) is correct.

Problem 9 **My grandfather was 8 times older than me 16 yrs ago. He would be 3 times of my age, 8 yrs from now. Eight years ago, what was the ratio of my age to that of my grandfather ?**

Sol. Initial ratio [i.e. ratio of ages of grandfather and mine 16 yrs back from now] = 8 : 1

Changed ratio [i.e. ratio of ages after 8 yrs from now] = 3 : 1

Time interval [i.e. 16 + 8] = 24 24

Now, [(8 × 1) – (1 × 3)] units = [(24 × 3) – (24 × 1)]

⇒ 5 units = 48

∴ 1 unit = $\frac{48}{5}$

and 8 units = $\frac{384}{5}$

Hence, ages of grandfather and mine 16 yrs back, were $\frac{48}{5}$ yrs and $\frac{384}{5}$ yrs respectively.

Now, ratio of their ages, only 8 yrs ago = $\left(\frac{48}{5} + 8\right) : \left(\frac{384}{5} + 8\right)$

= 88 : 424

= 11 : 53

Problem 10 **The ratio of present ages of two sisters is 5 : 6 and after 5 yrs and 10 yrs respectively, the ratio will be 3 : 4 . What will be the difference between their ages after 7 yrs ?**

Sol. Initial ratio [i.e. ratio of present ages] = 5 : 6

Changed ratio [i.e. after 5 yrs sand 10 yrs] = 3 : 4

Time interval = 5 10

Now, [(5 × 4) – (6 × 3)] units = (10 × 3) – (5 × 4)

⇒ 2 units = 10 yrs

∴ (6 – 5) unit = 5 yrs

Hence, difference between their present ages = 5 yrs and the difference will be same even after 7 yrs.

Problem 11 **The ratio of the present ages of a son and his father is 1 : 5 and that of his mother and father is 4 : 5. After 2 yrs the ratio of the ages of the son to that of his mother becomes 3 : 10. What is the present age of the father.** ***[BSRB, Patna (Bank PO)-01]***

Sol. Son : Father = 1 : 5

Father : Mother = 5 : 4

∴ Ratio of ages of son, his father and his mother = 1 : 5 : 4

Initial ratio [i.e. ratio of present ages (son to mother)] = 1 : 4

Changed ratio [ratio of ages, 2 yrs hence] = 3 : 10

Time interval = 2 2

Now, $[(4\times 3)-(1\times 10)]$ units = $[(10\times 2)-(3\times 2)]$

⇒ 2 units = 14 yrs

∴ 1 unit = 7 yrs

∴ 5 units = 35 yrs

Hence, present age of the father = 35 yrs.

Problem 12 **The ratio of the present age of a man and his wife is 5 : 4 and the ratio, 5 yrs hence, would be 11 : 9. If the ratio of their ages at the time of marriage was 7 : 5, then how many years ago were they married ?**

Sol. Initial ratio [i.e. ratio of present ages (man to his wife)] = 5 : 4

Changed ratio [ratio of ages, 5 yrs hence] = 11 : 9

Time interval = 5 5

Now, $[(5\times 9)-(11\times 4)]$ units = $[(11\times 5)-(9\times 5)]$

⇒ 1 unit = 10 yrs

∴ 5 units = 50 yrs and 4 units = 40 yrs

Hence, present ages of man and his wife are 50 yrs and 40 yrs, respectively.

Let x yrs ago, they were married.

Then, $$\frac{50-x}{40-x}=\frac{7}{5}$$

⇒ $250-5x=280-7x$

⇒ $x=15$ yrs

Alternatively I

From above, present ages of man and his wife are 50 yrs and 40 yrs respectively.

Initial ratio [i.e. ratio of present ages (man to his wife)] = 5 : 4

Changed ratio [Ratio of ages, x yrs ago] = 7 : 5

Time interval = x x

Then, $[(7\times 4)-(5\times 5)]$ units = $[(7\times x)-(5\times x)]$

⇒ 3 units = $2x$

⇒ 5 units = $10x/3$

Hence, present age of man = 50 = $10x/3$

⇒ $x=15$ yrs

Alternatively II

From above, present ages of man and his wife are 50 yrs and 40 yrs, respectively.

Now, you may proceed through options. Let options be 10, 15, 20 and 18.

(a) $10\Rightarrow \frac{50-10}{40-10}\neq\frac{7}{5}$ (b) $15\Rightarrow \frac{50-15}{40-15}=\frac{7}{5}$

Hence, option (b) is correct.

Problem 13 **Father is aged three times more than his son Yajju. After 8 yrs, he would be two and half times of Yajju's age. After further 8 yrs, how many times would he be of Yajju's age ?**

Sol. Initial ratio [i.e. ratio of present ages (father to son)] = 4 : 1

Changed ratio [ratio of ages, 8 yrs hence] = 5 : 2

Time interval = 8 8

Then, $[(2\times 4)-(5\times 1)]$ units = $[(8\times 5)-(8\times 2)]$

⇒ 3 units = 24 yrs

∴ 1 unit = 8 yrs and 4 units = 32 yrs

Hence, present ages of man and son are 32 yrs and 8 yrs, respectively.

Now, new ratio after further (8 + 8) yrs

= $(32+8+8):(8+8+8)=2:1$

It implies that father would be 2 times of Yajju's age.

Problem 14 **The ratio of ages of A and B is 5 : 3 and that of ages of C and D is 7 : 9. Four years hence, the ratio of ages of A and C will be 11 : 15. Find the ratio of ages of B and D, 8 yrs hence, provided their present ages are is an arithmetic progression.**

Sol. Let present ages of A and B be $5x$ and $3x$ respectively and those of C and D be $7y$ and $9y$, respectively.

Given that $5x$, $3x$, $7y$ and $9y$ are in an arithmetic progression.

∴ $5x-3x=9y-7y\Rightarrow x=y$

Given, ratio of ages of A and C = $5x:7y$

Then, $$\frac{5x+4}{7y+4}=\frac{11}{15}\Rightarrow 75x+60=77y+44$$

⇒ $75x-77y=-16\Rightarrow 75x-77x=-16$ $[\because x=y]$

⇒ $x=8$

Hence, present ages of A, B, C and D are 40 yrs, 24 yrs, 56 yrs and 72 yrs. Then, ratio of ages of B and D, 8 yrs hence

= $(24+8):(72+8)=32:80=2:5$

Alternatively

Since their ages are in arithmetic progression and the difference between ages of A and B is same as difference between ages of C and D, it is obvious that the ratio of their ages is 5 : 3 : 7 : 9.

Initial ratio [i.e. ratio of present ages (A to C)] = 5 : 7

Changed ratio [ratio of ages, 4 yrs hence] = 11 : 15

Time interval = 4 4

Then, $[(11\times 7)-(15\times 5)]$ units = $[(15\times 4)-(11\times 4)]$

⇒ 2 units = 16

⇒ 3 units = 24 and 9 units = 72

Now, required ratio = $(24+8):(72+8)=2:5$

Problem 15 Arun's present age in years is 40% of Barun's. In another few years, Arun's age will be half of Barun's. By what percentage will Barun's age increase during this period ? ***[CAT-17]***

Sol. Initial ratio [i.e. ratio of present ages (Arun to Barun)] = 2 : 5

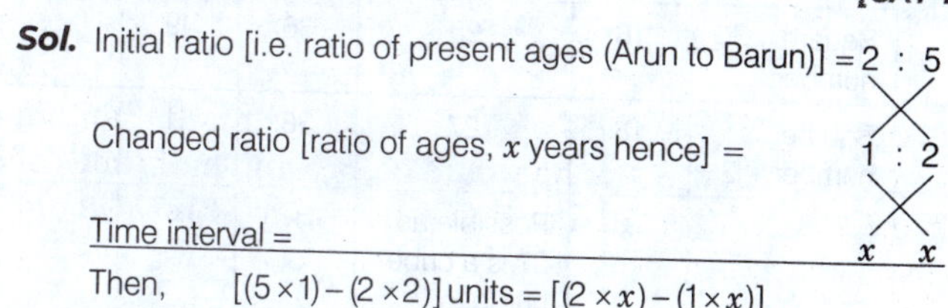

Changed ratio [ratio of ages, x years hence] = 1 : 2

Time interval = x x

Then, $[(5\times1)-(2\times2)]$ units $=[(2\times x)-(1\times x)]$

$\Rightarrow$ 1 unit $= x$

$\Rightarrow$ 5 units $= 5x$

Hence, Barun's present age $= 5x$

Then, his age after x yrs $= 6x$

It means Barun's age increases by 20% in x yrs.

Hence, required answer is 20%.

Alternation in ratio of ages on account of a time interval [Opposite directions either(ago-hence) or (hence-ago)]

Recall the concept of "Change in ratio of numbers when one is increased and another is decreased or *vice-versa* either by the same number or by different numbers" envisaged in chapter "Ratio, Proportion and Variation".

Problem 16 The ratio between the present ages of M and N is 5 : 3, respectively. The ratio between M's age 4 yrs ago and N's age after 4 yrs is 1 : 1. What is the ratio between M's age after 4 yrs and N's age 4 yrs ago ? ***[SIDBI (Officers' grade) Exam. 2000]***

Sol. Initial ratio [i.e. ratio of present ages (M to N)] = 5 : 3

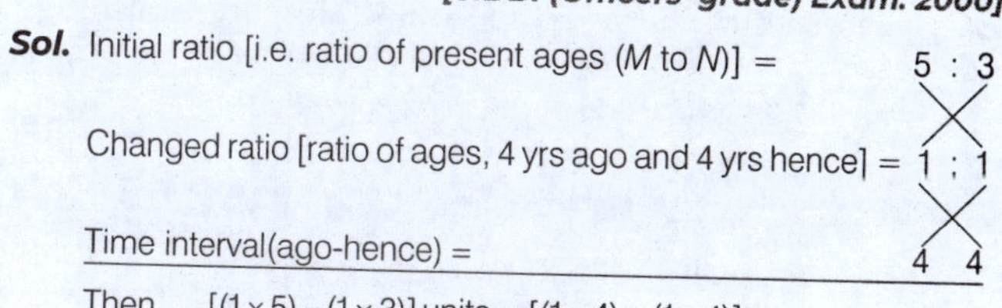

Changed ratio [ratio of ages, 4 yrs ago and 4 yrs hence] = 1 : 1

Time interval(ago-hence) = 4 4

Then, $[(1\times5)-(1\times3)]$ units $=[(1\times4)+(1\times4)]$

$\Rightarrow$ 2 units = 8

$\Rightarrow$ 3 units = 12 and 5 units = 20

Now, required ratio $=(20+4):(12-4)=3:1$

Alternatively

Let the present ages of M and N be $5x$ and $3x$ respectively.

As given, $\dfrac{5x-4}{3x+4}=\dfrac{1}{1}$

$\Rightarrow$ $5x-4=3x+4 \Rightarrow x=4$

Hence, present ages of M and N are (5×4) and (3×4), respectively.

Now, required ratio $=(20+4):(12-4)=3:1$

Problem 17 The ratio of ages of Ram and Shyam is 3 : 5. The ratio of Ram's age, 4 yrs hence to Shyam's age 5 yrs ago is 8 : 11. Find the ratio between Ram's age 4 yrs ago and Shyam's age 5 yrs hence.

Sol. Initial ratio [i.e. ratio of present ages of Ram and Shyam] = 3 : 5

Changed ratio [ratio of ages, 4 yrs hence and 5 yrs ago] = 8 : 11

Time interval [hence-ago] = 4 5

Then, $[(8\times5)-(11\times3)]$ units $=[(8\times5)+(11\times4)]$

$\Rightarrow$ 7 units = 84 yrs

$\Rightarrow$ 3 units = 36 yrs and 5 units = 60 yrs

Then, required ratio $=(36-4):(60+5)=32:65$

Ages are not given in the ratio

Problem 18 The present age of a father is 3 yrs more than three times the age of his son. 3 yrs hence, father's age will be 10 yrs more than twice the age of the son. Find the father's present age.

Sol. Let father's present age be f yrs and son's be x yrs.

Given, $f=3x+3$ ...(i)

After 3 yrs, father's age $=f+3$ and son's age $=x+3$

But given, $(f+3)=2(x+3)+10$

$\Rightarrow$ $f=2x+13$ ...(ii)

From Eqs. (i) and (ii), $3x+3=2x+13\Rightarrow x=10$

Hence, father's present age $=(3\times10)+3=33$ yrs

Alternatively

Process through options. Let options be 39, 33, 45, 40.

(a) 39 $\Rightarrow$ Son's present age $=\dfrac{39-3}{3}=12$

After 3 yrs, father's age $=39+3=42$ yrs and son's age $=10+3=13$ yrs

Here, $42\neq(2\times13)+10$

Hence, option (a) is not correct.

(b) 33 $\Rightarrow$ Son's present age $=\dfrac{33-3}{3}=10$

After 3 yrs, father's age $=33+3=36$ yrs and son's age $=10+3=13$ yrs

Here, $36=(2\times13)+10$

Hence, option (b) is correct.

Problem 19 A person was asked to state his age in years. His reply was, "take my age 3 years hence, multiply it by 3 and then subtract three times my age 3 years ago and you will know how old I am." What was the age of the person ?

Sol. Let the present age of the person be x yrs.

As per question, $x=3(x+3)-3(x-3)$

$=3x+9-3x+9=18$ yrs.

Problem 20 A father said to his son, "I was as old as you are present at the time of your birth". If the father's age is 38 yrs now, the son's age five year back was :

Sol. Obviously, father's age is twice the age of his son.

Hence, present age of son $=38/2=19$ yrs

Then, son's age 5 yrs back $=19-5=14$ yrs

Problem 21 Ayesha's father was 38 yrs of age when she was born while her mother was 36 yrs old when her brother 4 yrs younger to her was born. What is the difference between the ages of her parents? ***[Hotel Mgmt 2002]***

Sol. Father's age when Ayesha was born = 38 yrs

Father's age when Ayesha's brother was born= 38 + 4 = 42 yrs

Mother's age when Ayesha's brother was born = 36 yrs

Hence, difference between the ages of her parents

$= 42 - 36 = 6$ yrs

Problem 22 Priya's age was cube of an integral number (different from 1) 4 yrs ago and square of an integral number after 4 yrs. How long should she wait so that her age becomes square of a number in the previous year and cube of a number in the next year? ***[CDS 2019]***

(a) 7 yrs (b) 14 yrs

(c) 12 yrs (d) 21 yrs

Sol. Clearly, a cube number, after 8 yr, turns into a square number. Now, explore that number

Cube number	8	27	64	125	.....
Cube number + 8	16	35	72	133	.....
	Possible as 16 is a square number.	×	×	×	.....

Hence, Priya's present age = 16 – 4 = 12 yrs.

Further, as per 2nd stipulation, a square number, after 2 yrs, turns into a cube number. Now, explore that number. Since her present age is 12, we should try with a square number more than 12.

Square number	16	25	36	49	.....
Square number + 2	18	27	38	51	.....
	×	Possible as 27 is a cube number.	×	×	.....

It may be construed from the above that at the age of 26 yrs, she is satisfying the given stipulation as one year back her age would be 25 and the next year it will be 27.

Hence, she should wait for (26 – 12) i.e. 14 yrs.

Alternatively

Let the present age be x yrs.

Then, $x - 4 = n^3$ and $x + 4 = k^2$

Since n is different from 1, on putting $n = 2$,

we get $x - 4 = 2^3 = 8 \Rightarrow x = 12$

$\Rightarrow x + 4 = 16$, a square number, consistent with the given condition.

∴ Her present age = 12 yrs

Now, proceed through given options.

(a) 7 yrs $\Rightarrow 12 + 7 = 19$, then, $(19 - 1) \neq$ a square number and $(19 + 1) \neq$ a cube number.

(b) 14 yrs $\Rightarrow 12 + 14 = 26$, then $(26 - 1)$ = a square and $(26 + 1)$ = a cube number.

Thus, option (b) is correct.

WORKOUT

1. Age of X is six times that of Y. After 4 yrs, X is four times elder to Y. What is the present age of Y?
(a) 4 yrs (b) 5 yrs **[CDS-12]**
(c) 6 yrs (d) 7 yrs

2. If 1 is added to the age of the elder sister, then the ratio of the ages of two sisters becomes 0.5 : 1, but if 2 is subtracted from the age of the younger one, the ratio becomes 1 : 3. Find the age of the two sisters.
(a) 8 and 5 yrs (b) 9 and 5 yrs
(c) 11 and 6 yrs (d) 8 and 6 yrs

3. Ravindra and Rekha got married 10 yrs ago, their ages were in the ratio 5 : 4. Today, Ravindra's age is $\frac{1}{6}$ th more than Rekha's age.
After marriage, they had 6 children including a triplet and twins. The age of the triplets, twins and the 6th child is in the ratio of 3 : 2 : 1. What is the largest possible value of the present total age of the family? **[IIFT-14]**
(a) 79 (b) 93 (c) 101 (d) 107

4. The age of a mother, before 2 yrs, was eight times the age of her daughter. After 1 yr mother's age will be five times the daughter's age. After how many years from now, the mother's age will become three times the daughter's age? **[CDS-07, (CGL-T-I)-17]**
(a) 10 yrs (b) 8 yrs
(c) 6 yrs (d) 12 yrs

5. Three years ago, X was twice as old as Y. If the difference between their ages at present is 3 yrs, how old is X at present ?
(a) 18 yrs (b) 12 yrs
(c) 9 yrs (d) 8 yrs

6. The ratio between the ages of A and B is 2 : 5. After 8 yrs, their ages will be in the ratio 1 : 2. What is the difference between their present ages ?
[CDS-10, MAT-08, BOB (PO)-07, Andhra Bank (Specialist) -02, UBI(PO)- 08, SBI(PO)-08, BOB(AO)-08, BOM(PO)-09, OBC(PO)-09]
(a) 20 yrs (b) 22 yrs
(c) 24 yrs (d) 25 yrs

7. Ratio of Ashok's age to Pradeep's age is 4 : 3. Ashok will be 26 yrs old after 6 yrs. How old is Pradeep now?
(a) 18 yrs (b) 21 yrs
(c) 15 yrs (d) 24 yrs

8. 10 yrs ago, Ram was 5 times as old as Shyam but 20 yrs later from now he will be only twice as old as Shyam. How many year old is Shyam?
[CDS-08, RRB (ASM)/Bhubneshwar-02]
(a) 20 yrs (b) 30 yrs
(c) 40 yrs (d) 50 yrs

9. The ratio of the ages of a father and a son at present is 5 : 2. 4 yrs hence, the ratio of the ages of the son and his mother will be 1 : 2. What is the ratio of the present ages of the father and the mother?
(a) 3 : 4
(b) 5 : 4
(c) 4 : 3
(d) Cannot be determined
(e) None of the above

10. If the ages of P and R are added to twice the age of Q, the total becomes 59. If the ages of Q and R are added to thrice the age of P, the total becomes 68 and if the age of P is added to thrice the age of Q and thrice the age of R, the total becomes 108. What is the age of P? **[MAT-13]**
(a) 19 yrs (b) 15 yrs
(c) 17 yrs (d) 12 yrs

11. The age of a man is three times the sum of the ages of his two sons. Five year hence, his age will be double of the sum of the ages of his sons. The father's present age is ? **[IAS(Pre.)-02]**
(a) 45 yrs
(b) 29 yrs
(c) 49 yrs
(d) Cannot be determined
(e) None of the above

12. 6 yrs hence, a father's age will be three times his son's age and 3 yrs ago father was nine times as old as his son. What is the present age of the father?
(a) 48 yrs (b) 42 yrs
(c) 36 yrs (d) 30 yrs

13. The ratio of ages of the father and mother was 11 : 10 when their son was born. The ratio of ages of his father and mother will be 19 : 18 when the son will be twice his present age. What is the ratio of present ages of father and mother ? **[SSC(CGL-T-II)-17]**
(a) 15 : 14 (b) 14 : 13
(c) 16 : 15 (d) 17 : 16

14. Farah got married 8 yrs ago. Today, her age is $1\frac{2}{7}$ times her age at the time of her marriage. At present her daughter's age is $\frac{1}{6}$th of her age. What was her daughter's age 3 yrs ago?
[PNB(MT)-09, CDS-17]
(a) 6 yrs
(b) 4 yrs
(c) 3 yrs
(d) Cannot be determined

15. The sum of reciprocals of the ages of two brothers is five times the difference of the reciprocals of their ages. If the ratio of the product of their ages to the sum of their ages is 14.4 : 1, find their ages.
(a) 24 and 20 yrs
(b) 36 and 24 yrs
(c) 18 and 15 yrs
(d) 12 and 9 yrs

16. Radha's present age is 3 yrs less than twice her age 12 yrs ago. Also, the ratio of Raj's present age to Radha's present age is 4 : 9. What will be Raj's age after 5 yrs ?
(a) 12 yrs
(b) 7 yrs
(c) 21 yrs
(d) Cannot be determined
(e) None of the above

17. The sum of the age of a father and the age of son is 75 yrs. If the product of their ages before 5 yrs was 750, then what is the present age of the father ?
(a) 60 yrs
(b) 55 yrs
(c) 52 yrs
(d) 50 yrs

18. If 6 yrs are subtracted from the present age of Shyam and the remainder is divided by 18, then the present age of his grandson Anup is obtained. If Anup is 2 yrs younger to Mahesh whose age is 5 yrs, then what is the age of Shyam ?
(a) 48 yrs (b) 60 yrs
(c) 84 yrs (d) 96 yrs

19. 9 yrs later, age of B will be equal to the present age of A. Sum of A's age 3 yrs later and B's 4 yrs ago is 76. If C is half of the present age of B, then what will be C's age(in yrs) after 10 yrs ? *[SSC(CGL-T-I)-17]*
(a) 31 yrs (b) 27 yrs
(c) 32 yrs (d) 36 yrs

20. The present ages of Amit and his father are in ratio of 2 : 5. 4 yrs hence, the ratio of their ages will become 5:11, what was the father's age five years ago? *[Andhra Bank(PO)-09, Andhra Bank(MA)-09]*
(a) 40 yrs
(b) 45 yrs
(c) 30 yrs
(d) 35 yrs
(e) None of the above

21. The ratio of present ages of Ram and Rakesh is 6 : 11. 4 yrs ago, the ratio of their ages was 1 : 2. What will be Rakesh's age after 5 years?
(a) 45 yrs
(b) 29 yrs
(c) 49 yrs
(d) Cannot be determined
(e) None of the above

22. The age of the father is 30 yrs more than the son's age. 10 yrs hence, the father's age will become three times the son's age that time. What is the son's present age in years ? *[SBI (Associates)-11]*
(a) 8
(b) 7
(c) 5
(d) cannot be determined
(e) None of the abvoe

23. The ratio of present age of Manisha and Deepali is 5 : X. Manisha is 9 yrs younger than Parineeta. Parineeta's age after 9 yrs will be 33 yrs. The difference between Deepali's and Manisha's age is the same as the present age of Parineeta. What should come in place of X ? *[IBPS(CWE-PO/MT)-11]*
(a) 23
(b) 39
(c) 15
(d) Cannot be determined
(e) None of the above

24. Ravi is 12 yrs younger than Surya. Ravi's age is 40% of the sum of his and Surya's age. What will be Surya's age 9 yrs hence? *[SSC(CGL-T-II)-17]*
(a) 24 yrs (b) 36 yrs
(c) 33 yrs (d) 45 yrs

25. Hari's family consisted of his younger brother (Chari), younger sister (Gouri) and their father and mother. When Chari was born, the sum of the ages of Hari, his father and mother was 70 yrs. The sum of the ages of four family members, at the time of Gouri's birth, was twice the sum of ages of Hari's father and mother at the time of Hari's birth, if Chari is 4 yrs older than Gouri, then find the difference in age between Hari and Chari. *[XAT-17]*
(a) 5 yrs (b) 6 yrs
(c) 7 yrs (d) 8 yrs
(e) 9 yrs

26. Radhika has two daughters by name Rinku and Sindhu. The ratio of the age of Radhika and Rinku is 3 : 1 and that of Rinku and Sindhu is 8 : 5. Given that Rinku is 6 yrs elder to Sindhu, find the ratio of their ages after 12 yrs. *[SBI(PO)-17]*
(a) 29 : 14 : 41 (b) 30 : 14 : 11
(c) 29 : 15 : 12 (d) 30: 14: 10
(e) 30 : 13 : 11

27. Akbar will turn 50 when his son Jehangir turns 18. What will be Akbar's age when it will exactly 5 times that of Jehangir ? *[CLAT-11]*
(a) 48 yrs (b) 44 yrs
(c) 40 yrs (d) 36 yrs

28. In 2002, Meenu's age was $\frac{1}{3}$rd of the age of Meera, whereas in 2010, Meenu's age was half the age of Meera. What is Meenu's year of birth? *[CSAT-19]*
(a) 1992 (b) 1994
(c) 1996 (d) 1997

29. The ratio of the ages of A and B, 8 yrs ago, was 2 : 3. four years ago, the ratio of their ages was 5 : 7. What will be the ratio of their ages 8 yrs from now? *[SSC (CGL-T-I)-19]*
(a) 4 : 5 (b) 3 : 4
(c) 7 : 8 (d) 5 : 6

30. Ena was born 4 yrs after parents marriage. Her mother is 3 yrs younger than her father and 24 yrs older than Ena, who is 13 yrs old. At what age did Ena's father get married? *[CSAT-19]*
(a) 22 yrs (b) 23 yrs
(c) 24 yrs (d) 25 yrs

31. Four years ago, the sum of ages of B and C was 36 yrs and after 4 yrs from now the sum of ages of A and C will be 24 yrs more than that of present ages of A and B. What is the present age of B ? *[SBI(PO)-Pre.-19]*
(a) 14 yrs (b) 13 yrs
(c) 24 yrs (d) 18 yrs

Answer Key

1. (c)	2. (b)	3. (d)	4. (b)	5. (c)	6. (c)	7. (c)	8. (a)	9. (d)	10. (d)
11. (a)	12. (d)	13. (a)	14. (c)	15. (b)	16. (e)	17. (b)	18. (b)	19. (b)	20. (d)
21. (c)	22. (c)	23. (e)	24. (d)	25. (e)	26. (b)	27. (c)	28. (b)	29. (a)	30. (b)
31. (a)									

Hints and SOLUTIONS

1. Initial ratio [X to Y] = 6 : 1

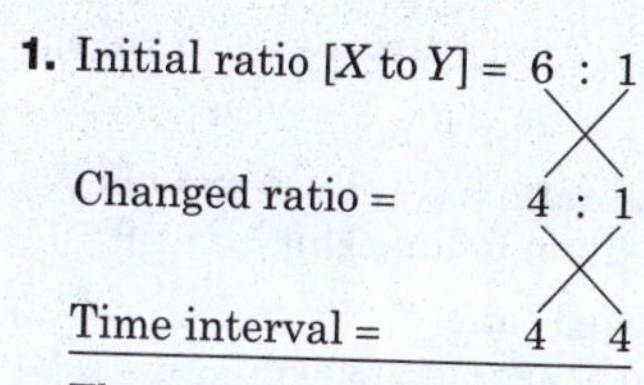

Changed ratio = 4 : 1

Time interval = 4 4

Then, $[(6 \times 1) - (4 \times 1)]$ units

$= [(4 \times 4) - (1 \times 4)]$

$\Rightarrow$ 2 units = 12 yrs

$\Rightarrow$ 1 unit = 6 yrs

2. The best way is to proceed through options.

(a) $5 : (8 + 1) \neq 0.5 : 1$

(b) $5 : (9 + 1) = 0.5 : 1$

and $(5 - 2) : 9 = 1 : 3$

Hence, option (b) is correct.

Remarks The details of these types of questions have been given in 'Number-System'

3. Initial ratio [i.e. ratio of present ages (Ravindra to Rekha)]

= 7 : 6

Changed ratio [ratio of ages, 10 yrs ago] 5 : 4

Time interval = 10 10

Then, $[(6 \times 5) - (7 \times 4)]$ unit(s)

$= [(10 \times 5) - (10 \times 4)]$

$\rightarrow$ 2 units = 10

$\therefore$ [7 + 6] units = 65

Hence, sum of the present ages of Ravindra and Rekha = 65 yrs.

Since the age of the triplets, twins and the sixth child is in the ratio of 3 : 2 : 1, the minimum sum of their ages = $[3 \times (1 + 1 + 1)] + [2 \times (1 + 1)] + (1 \times 1)] = 14$ yrs. This sum may be 14, 28, 42 and so on. i.e. multiple of 14. Hence, maximum possible value of present total age of the family = 65 + 42 = 107 as given in option.

4. Initial ratio [mother to daughter, 2 yrs ago]

= 8 : 1

Changed ratio [1 yr hence] = 5 : 1

Time interval [2 + 1] = 3 3

Then, $[(8 \times 1) - (5 \times 1)]$ unit(s)

$= [(5 \times 3) - (1 \times 3)]$

$\Rightarrow$ 3 units = 12 yrs

$\Rightarrow$ 1 unit = 4 yrs and 8 units = 32 yrs

Hence, present ages of mother and daughter are [4 + 2] yrs and [32 + 2] yrs respectively.

Now, initial ratio [mother to daughter, present age] 17 : 3

Changed ratio [1 yr hence] = 3 : 1

Time interval [let, x yr hence] = x x

Then, $[(17 \times 1) - (3 \times 3)]$ unit(s)

$= [(x \times 3) - (x \times 1)]$

$\Rightarrow$ 8 units = $2x$

$\Rightarrow$ 3 units = $\frac{3x}{4}$

$\Rightarrow$ 6 year = $\frac{3x}{4}$

[$\because$ 3 units = present age of daughter = 6 yrs]

$\Rightarrow$ $x = 8$ yrs

Alternatively

From above, present ages of mother and daughter are 6 yrs and 34 yrs, respectively.

Now, you may proceed through options.

(a) $10 \Rightarrow \frac{6 + 10}{34 + 10} \neq \frac{1}{3}$

(b) $8 \Rightarrow \frac{6 + 8}{34 + 8} = \frac{1}{3}$

Hence, option (b) is correct.

5. Let present ages of X and Y be x yrs and $(x - 3)$ yrs respectively.

Then, $(x - 3) = 2\ (x - 3 - 3)$

$\Rightarrow$ $x = 9$ yrs

6. Initial ratio [X to Y] = 2 : 5

Changed ratio = 1 : 2

Time interval = 8 8

Then, $[(5 \times 1) - (2 \times 2)]$ unit(s)

$= [(8 \times 2) - (8 \times 1)]$

$\Rightarrow$ 1 unit = 8 yrs

$\Rightarrow$ (5 − 2) units = 24 yrs

7. The present age of Ashok

= 26 − 6 = 20 yrs

As given, 4 units = 20 yrs

Then, 3 units = 15 yrs

8. Initial ratio [Ram to Shyam, 10 yrs ago] = 5 : 1

Changed ratio [20 yrs hence] = 2 : 1

Time interval [10 + 20] = 30 30

Then, $[(5 \times 1) - (2 \times 1)]$ unit(s)

$= [(30 \times 2) - (30 \times 1)]$

$\Rightarrow$ 3 units = 30 yrs

$\Rightarrow$ 1 unit = 10 yrs

Hence, present age of Shyam

= 10 + 10 = 20 yrs.

9. Let the present ages of the father and the son be $5x$ yrs and $2x$ yrs respectively and the age of the mother be y yrs, 4 yrs ago.

As given, $(2x - 4) : y = 1 : 2$

$\Rightarrow$ $y = 4x - 8$

Now, required ratio = $5x : (y + 4)$

$= 5x : (4x - 8 + 4)$

Hence, we cannot determine the required ratio.

10. Let ages of P, Q and R be x, y and z yrs respectively.

As given,

(i) $x + 2y + z = 59$

(ii) $3x + y + z = 68$

(iii) $x + 3y + 3z = 108$

$\Rightarrow$ $3y + 3z = 108 - x$

$\Rightarrow$ $y + z = 36 - x/3$

From Eqs. (ii), we get $3x + y + z = 68$

$\Rightarrow$ $3x + 36 - x/3 = 68$

$\Rightarrow$ $8x/3 = 32 \Rightarrow x = 12$

Hence, age of P = 12 yrs

11. Initial ratio [i.e. ratio of present ages (father to sons together)]

= 3 : 1

Changed ratio [ratio of ages, 5 yrs hence] = 2 : 1

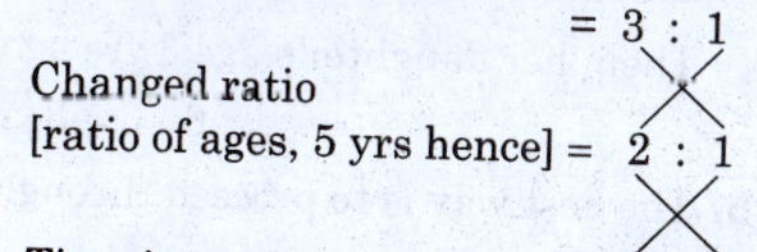

Time interval = 5 (5 + 5)

Since ages of both sons have been taken together, $(5+5)$ has been considered above.

Then, $[(3\times1)-(2\times1)]$ unit(s)

$= [(2\times10)-(1\times5)]$

$\Rightarrow$ 1 unit = 15 yrs

$\Rightarrow$ 3 units = 45 yrs

Hence, the father's present age = 45 yrs.

12. Initial ratio [father to son, 3 yrs ago] = 9 : 1

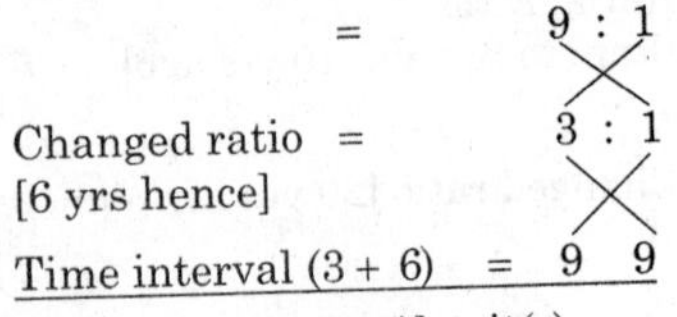

Changed ratio [6 yrs hence] = 3 : 1

Time interval (3 + 6) = 9 9

Then, $[(9\times1)-(3\times1)]$ unit(s)

$= [(9\times3)-(9\times1)]$

$\Rightarrow$ 6 units = 18 yrs

$\Rightarrow$ 1 unit = 3 yrs

$\Rightarrow$ 9 units = 27 yrs

Hence, age of the father 3 yrs ago = 27 yrs.

Then, present age of father = 27 + 3 = 30 yrs

13. Let the present age of son be x yrs.

Initial ratio [(father to mother, at time of birth of their son i.e. x yrs ago from now] = 11 : 10

Changed ratio [at time when son becomes twice his present age] = 19 : 18

Time interval = $2x$ $2x$

Now, $[(18\times11)-(10\times19)]$ units

$= [(19\times2x)-(18\times2x)]$

$\Rightarrow$ 8 units = $2x$

$\therefore$ 11 units $= \frac{11x}{4}$ and 10 units $= \frac{5x}{2}$

Now, ratio of their present ages

$= \left(\frac{11x}{4}+x\right):\left(\frac{5x}{2}+x\right) = 15:14$

14. Ratio of her present age to her age at time of marriage = 9 : 7

As given, $(9-7)$ units = 8 yrs

$\Rightarrow$ 9 units = 36 yrs

Now, present age of her daughter = $36 \div 6 = 6$

Then, her daughter's age 3 yrs ago = 6 − 3 = 3 yrs

15. The best way is to proceed through options.

(a) $\frac{1}{24}+\frac{1}{20} \neq 5\left(\frac{1}{20}-\frac{1}{24}\right)$

(b) $\frac{1}{24}+\frac{1}{36} = 5\left(\frac{1}{24}-\frac{1}{36}\right)$

and $(24\times36):(24+36) = 14.4:1$

Hence, option (b) is correct.

16. Let present age of Radha be x yrs.

Then, $x = 2(x-12)-3$

$\Rightarrow$ $x = 2x-27$

$\Rightarrow$ $x = 27$ yrs

Or directly, her present age

$= (12\times2)+3 = 27$ yrs

As given, 9 units = 27 yrs

$\therefore$ 4 units = 12 yrs

Hence, Raj's present age = 12 yrs

Then, Raj's age after 5 yrs

= 12 + 5 = 17 yrs.

17. Let the age of the father be x yrs at present.

Then, age of son $= 75-x$

As given, $(x-5)\times(75-x-5) = 750$

$\Rightarrow$ $-x^2+70x+5x-350 = 750$

$\Rightarrow$ $-x^2+75x-1100 = 0$

$\Rightarrow$ $-x^2+55x+20x-1100 = 0$

$\Rightarrow$ $-x(x-55)+20(x-55) = 0$

$\Rightarrow$ $x = 20$ and 55

The present age of father cannot be 20 yrs as it will yield the age of son as $(75-20)$ and the age of father cannot be lower than that of his son.

Hence, the present age of the father = 55 yrs.

Alternatively

The best way is to proceed through options to save the time.

18. Anup's present age = 5 − 2 = 3 yrs

Then, Shyam's present age

$= (18\times3)+6 = 60$ yrs.

19. Clearly, at present, B is 9 yrs younger than A.

Further, sum of their present ages

$= 76+(4-3) = 77$

Then, B's present age $= \frac{77-9}{2}$

= 34 yrs

$\therefore$ C's age after 10 yrs

$= (34\div2)+10 = 27$ yrs.

20. Initial ratio [i.e. ratio of present ages (son to father)] = 2 : 5

Changed ratio [ratio of ages, 4 yrs hence] = 5 : 11

Time interval = 4 4

Then, $[(5\times5)-(2\times11)]$ units

$= [(11\times4)-(5\times4)]$

$\Rightarrow$ 3 units = 24 yrs

$\therefore$ 5 units = 40 yrs

Hence, Amit's father's age 5 yrs ago

= 40 − 5 = 35 yrs

21. Initial ratio [i.e. ratio of present ages (Ram to Rakesh)] = 6 : 11

Changed ratio [ratio of ages, 4 yrs ago] = 1 : 2

Time interval = 4 4

Then, $[(6\times2)-(1\times11)]$ unit(s)

$= [(2\times4)-(1\times4)]$

$\Rightarrow$ 1 unit = 4 yrs

$\therefore$ 11 units = 44 yrs

Hence, Rakesh's age after 5 yrs

= 44 + 5 = 49 yrs

22. Let father's age be x yrs at present.

Then, son's present age $= x-30$

As given, $(x+10) = 3(x-30+10)$

$\Rightarrow$ $2x = 70$

$\Rightarrow$ $x = 35$

Hence, present age of the son

= 35 − 30 = 5 yrs

23. Present age of Parineeta

= 33 − 9 = 24 yrs

Then, the present age of Manisha

= 24 − 9 = 15 yrs

As given, $(X-5)$ units = 24 yrs

$\Rightarrow$ $(X-5)\times3 = 24$

[$\because$ 5 units = Manisha's age = 15, $\therefore$ 1 unit = 3]

$\Rightarrow$ $X-5 = 8$

$\Rightarrow$ $X = 13$

Alternatively

Present age of Parineeta

= 33 − 9 = 24 yrs

Then, the present age of Manisha

= 24 − 9 = 15 yrs

$\therefore$ Present age of Deepali

= 24 + 15 = 39 yrs

Now, ratio of present ages of Manisha and Deepali = 15 : 39 = 5 : 13

Hence, 13 will come in place of X.

24. Since Ravi's age is 40% of the sum of his and Surya's age, Surya's age will be 60% of the sum of his and Ravi's age.

It implies that $(60-40)\%$ of the sum of their ages = 12 yrs

$\therefore$ 60% of the sum of their ages

$= 36$ yrs

Then, Surya's age after 9 yrs

$= 36 + 9 = 45$ yrs

25. Let ages of the father and mother at time of birth of Hari be F and M respectively.

$\therefore$ Sum of ages of family members at time of birth of Chari $= (F + x) + (M + x) + x$, if Hari is x yrs older than Chari.

As given, $(F + x) + (M + x) + x = 70$

$\Rightarrow \quad F + M + 3x = 70$

Further, sum of ages of family members at time of birth of Gouri

$= 70 + (4 \cdot 4) = 86$

As per question, $86 = 2(F + M)$

$\Rightarrow \quad F + M = 43$

Then, $\quad 3x = 70 - 43$

$[\because F + M + 3x = 70$, from the above$]$

$\Rightarrow \quad x = 9$ yrs

26. Given, Radhika : Rinku = 3 : 1

Rinku : Sindhu = 8 : 5

Then, Radhika : Rinku : Sindhu

$= 3 \cdot 8 : 1 \cdot 8 : 1 \cdot 5$

$= 24 : 8 : 5$

$\because \ (8 - 5)$ units $= 6$

$\Rightarrow \quad 1$ unit $= 2$

Then, ratio of their ages after 12 yrs

$= (24 \cdot 2 + 12) : (8 \cdot 2 + 12) : (5 \cdot 2 + 12)$

$= 30 : 14 : 11$

Alternatively

Proceed through options.

(a) 29 : 14 : 41 $\Rightarrow$ Ratio of ages 12 yrs ago i.e. present ages = 17 : 2 : 29.
Here, 17 : 2 = 3 : 1

(b) 30 : 14 : 11 = 60 : 28 : 22 [we doubled as we cannot subtract 12 from 11] $\Rightarrow$ Ratio of present ages

= 48 : 16 : 10.

Here, 48 : 16 = 3 : 1 and 16 : 10 = 8 : 5 as given in the question. Hence, option(b) is correct.

27. Initial ratio

[Akbar to Jehangir, present age] 25 : 9

Changed ratio

[x yr ago as question suggests] = 5 : 1

Time interval $[x, x]$ $= x \quad x$

Now, $[(9 \times 5) - (25 \times 1)]$ units

$= [(5 \times x) - (1 \times x)]$

$\Rightarrow$ 20 units $= 4x$

$\therefore$ 25 units $= 5x = 50$ [given]

$\Rightarrow \quad x = 10$ yrs

Hence, required age of Akbar

$= 50 - 10 = 40$ yrs.

28. Initial ratio

[Meenu to Meera, in 2002] = 1 : 3

Changed ratio

[i.e. ratio of ages, 8 yrs hence] = 1 : 2

Time interval $= 8 \quad 8$

Now, $[(1 \times 3) - (1 \times 2)]$ unit

$= [(8 \times 2) - (8 \times 1)]$

$\Rightarrow$ 1 unit = 8 yrs

Hence, Meenu' age was 8 yrs in 2002.

Then, her year of birth $= 2002 - 8$

$= 1994$

29. Initial ratio

[i.e. ratio of ages, 8 yrs ago] = 2 : 3

Changed ratio

[i.e. ratio of ages, 4 yrs ago] = 5 : 7

Time interval $= 4 \quad 4$

Now, $[(5 \times 3) - (7 \times 2)]$ unit

$= [(4 \times 7) - (4 \times 5)]$

$\Rightarrow$ 1 unit = 8 yrs

$\therefore$ 2 units = 16 and 3 units = 24 yrs

It means A's and B's ages, 8 yrs ago, were 16 and 24 respectively.

Now, ratio of their ages 8 yrs hence

$= (16 + 16)$ yrs : $(24 + 16)$ yrs $= 4 : 5$

30. Age of Ena's mother at present $= 13 + 24 = 37$ yrs and age of Ena's father at present $= 37 + 3 = 40$ yrs

Since age of Ena is 13 yrs now, her father got married at the age of $(40 - 13 - 4)$ i.e. 23 yrs.

31. Present age of

$[B + C] = 36 + 4 + 4 = 44$

Present age of $[A + B] = x$ [let]

Then, age of $[A + C]$ after 4 yrs

$= x + 24$

$\therefore$ Present age of

$[A + C] = x + 24 - 4 - 4 = x + 16$

From the above, present age of $[2A + 2B + 2C] = 44 + x + x + 16$

$\Rightarrow$ Present age of $[A + B + C] = 30 + x$

$\Rightarrow$ Present age of

$A = 30 + x - 44 = x - 14$

Since the present age of $[A + B] = x$, present age of $B = x - (x - 14) = 14$ yrs

28

Partnership

Learning OBJECTIVES

After completing this chapter, aspirants will be able to :

- Understand the application of concepts associated with ratio to solve the problems based on 'Partnership';
- Recognise the General Rules of Partnership which is applicable for all of the questions;
- Observe the diverse sorts of categories of the questions;
- Learn how to solve the maximum questions with the concepts associated with 'Ratio';
- Understand the philosophy of the division of this chapter in various parts;
- Analyse the trend of the questions being asked in various competitive examinations;
- Recognise the types of question which can be easily solved through the given options;
- Solve the questions in moments;
- Grip on concepts helping to solve the questions without pen and paper;
- Learn many new concepts/fundamentals;
- Apply the pragmatic approaches in lieu of academic approaches; and
- Ponder on the alternative methods.

Inside

- *Brief Introduction*
- *Diverse Cases :*
  (i) When Different Investments Work for Equal Periods
  (ii) When Investments are Equal But Working for Different Time Periods
  (iii) When Different Investments Work for Different Time Periods
  (iv) When Initial Investment(s) is/are Changed Later
  (v) When the Partners get Allowances/Salaries/Interest etc. in Addition to Their Shares in Profit Depending Upon Their Investments
- ....
- .... *& Many More*
- *Work Out with Solutions*

Introduction

A partnership is the agreement between two or more persons who join to carry on a trade or business. They share the profits and liabilities of the business. In a certain agreements, all partners share profits and liabilities equally whereas in other arrangements some partners have limited liability. It is not necessary for every partner to involve in management and day to day operations of the venture. The partner whose involvement in the business is limited to providing capital only, not in day to day operations, is known as Sleeping or Dormant Partner.

General Rules of Partnership

Profits/losses are shared in the ratio of products of the investments and respective time periods. Suppose A invests ₹x for t_1 months and B invests ₹y for t_2 months, then

$$\frac{A\text{'s share in profit / loss}}{B\text{'s share in profit / loss}} = \frac{x \times t_1}{y \times t_2}$$

It may also be construed from the above that

(a) In case of same period of time, the profit/loss will be shared in the ratio of the investments made by the partners i.e.

$$\left[\frac{A\text{'s share in profit / loss}}{B\text{'s share in profit / loss}} = \frac{x \times \cancel{t_1}}{y \times \cancel{t_2}}\right]$$

(b) In case of equal investments by the partners, the profit/loss will be shared in the ratio of the time periods i.e.

$$\left[\frac{A\text{'s share in profit / loss}}{B\text{'s share in profit / loss}} = \frac{\cancel{x} \times t_1}{\cancel{y} \times t_2}\right]$$

(c) Ratio of shares in profit = Ratio of products of capitals and respective time periods.

∴ Ratio of time periods = Ratio of quotients obtained by dividing profits by respective capitals.

It also holds good that ratio of time periods = ratio of quotients obtained by dividing ratios of profits by respective ratios of capitals.

Or

Ratio of capitals invested = Ratio of quotients obtained by dividing profits by respective time periods.

It also holds good that ratio of capital invested = ratio of quotients obtained by dividing ratios of profits by respective time periods.

Note

(i) The above conceptions work for more than two partners too.

(ii) Time periods mentioned above means the period for which capital(s) work(s).

Since this chapter is based on the concepts associated with the 'Ratio and Proportion', it is advisable to go through those concepts before entering in this chapter.

Illustrations

When different amount of investments work for equal periods

Problem 1 X and Y started a business investing ₹ 85000 and ₹ 15000, respectively. In what ratio the profit earned after 2 yrs be divided between X and Y, respectively ?

Sol. Since both have invested for the same period, profit earned will be divided in the ratio of their investments.

∴ Required ratio = 85000 : 15000 = 17 : 3

Problem 2 A, B and C started a business investing ₹ 50000, ₹ 65000 and ₹ 90000 respectively. Find the share of each if an annual profit is of ₹ 8200.

Sol. Since, all of three have invested for the same period, the profit will be shared in the ratio of their investments.

∴ Ratio of shares of A, B and C
= 50000 : 65000 : 90000 = 10 : 13 : 18

Given, (10 + 13 + 18) units = ₹ 8200

∴ 10 units = 2000 i.e. share of A

and 13 units = 2600 i.e. share of B

and 18 units = 3600 i.e. share of C

Problem 3 A, B and C invest ₹ 2000, ₹ 4000 and ₹ 5000 respectively in a business. At the end of the year, a loss of 25% of the total initial investment is incurred. Find the share of loss of B.

Sol. Ratio of shares in loss = 2000 : 4000 : 5000 = 2 : 4 : 5

and total loss = 25% of (2000 + 4000 + 5000) = ₹2750

As given, (2 + 4 + 5) units = 2750

∴ 4 units = ₹1000

Problem 4 Gangadhar, Ramesh and Sudesh invest the money in a business in the ratio 4 : 5 : 6, respectively. If share of Sudesh on total profit is ₹ 16500, then find the total profit.

Sol. Clearly, time period is same. Thus, profit will be divided in same ratio as ratio of their investments i.e. 4 : 5 : 6.

Further given, 6 units = ₹ 16500

∴ (4 + 5 + 6) units = ₹ 41250

Problem 5 A, B and C entered into a partnership. A invested ₹ 2560 and ₹ 2000. At the end of the year, they gained ₹1105, out of which A got ₹ 320. C's capital was

[SSC(SO-Audit)-06]

Sol. Their capitals work for the same period.

Ratio of shares of A and B in profit = 2560 : 2000 = 32 : 25

But given, 32 units = 320

∴ 25 units = 250 i.e. B's share in profit

Obviously, C's share in profit = 1105 − (320 + 250) = 535

Further, 320 ⇒ 2560

∴ 535 ⇒ 4280

Hence, C's capital = ₹4280

When Investments are Equal but Working for Different Time Periods

Problem 6 A, B and C enter into a partnership and they invest ₹ 30000 each. A leaves the business after 4 months, B leaves after 6 months and C after 7 months. Find the share of B in the profit if total profits in ₹ 3400.

Sol. Since investments are equal, the profit will be shared in the ratio of time periods.

∴ Ratio of shares of A, B and C in profit = 4 : 6 : 7

As given, (4 + 6 + 7) units = ₹3400

∴ 6 units = ₹ 1200 i.e. share of B

When Different Investments Work for Different Time Periods

Problem 7 A, B and C enter into a partnership. A contributes ₹ 320000 for 4 months. B contributes ₹ 510000 for 3 months and C contributes ₹ 270000 for 5 months. If the total profit be ₹ 124800, then A's share in the profit is *[SSC(SO-Audit)-06, MAT-11]*

Sol. Ratio of shares of A, B and C in profit $= (320000 \times 4) : (510000 \times 3) : (270000 \times 5) = (32 \times 4) : (51 \times 3) : (27 \times 5)$
$= 128 : 153 : 135$

Now, $(128 + 153 + 135)$ units $= 124800$

$\therefore$ 128 units = ₹ 38400, i.e. A's share in profit.

Problem 8 Sunetra invested an amount of ₹ 50000 to start a software business. After 6 months, Nikhil joined her with an amount of ₹ 80000. At the end of 3 yr, they earned a profit of ₹ 24500. What is Sunetra's share in the profit ? *[SBI Associates (PO)-02]*

Sol. Ratio of shares in profit

$= (50000 \times 3) : (80000 \times 2\frac{1}{2}) = 3 : 4$

Now, $(3 + 4)$ units = ₹ 24500

$\therefore$ 3 units = 10500 i.e. share of Sunetra

Problem 9 A started a business with a capital of ₹ 100000. 1 yr later, B joined him a capital of ₹ 200000. At the end of 3 yrs from the start of the business, the profit earned was ₹ 84000. The share of B in the profit exceeded the share of A by *[SSC(CGL-PT)-10]*

Sol. The ratio of the shares of A and B in profit

$= (100000 \times 3) : (200000 \times 2) = 3 : 4$

As given, $(3 + 4)$ units $= 84000$

$\therefore$ $(4 - 3)$ units $= 12000$

Hence, share of B in the profit exceeds the share of A by ₹12000.

Problem 10 A, B and C rent a pasture. A puts in 10 oxen for 7 months, B puts in 12 oxen for 5 months and C puts in 15 oxen for 3 months for grazing. If the rent of the pasture is ₹ 175, how much must C pay as his share of rent?

Sol. Clearly, ratio of shares of A, B and C in rent

$= (10 \times 7) : (12 \times 5) : (15 \times 3) = 14 : 12 : 9$

Now, $(14 + 12 + 9)$ units $= 175$

$\therefore$ 9 units = ₹45 i.e. share of C in rent

Problem 11 A starts business with ₹ 3500 and after 5 months, B joins with A as his partner. After a year the profit is divided in the ratio 2: 3. What is B's contribution in the capital?

Sol. Let B's contribution towards capital be ₹x.

Then, $\frac{\text{share of A in profit}}{\text{share of B in profit}} = \frac{3500 \times 12}{x \times 7} = \frac{2}{3}$ [given]

$\Rightarrow \quad 14x = 36 \times 3500$

$\Rightarrow \quad x = 9000$

Alternatively

Ratio of their time periods (A to B) = 12 months : (12 – 5) months = 12 : 7

We know, ratio of capital invested = ratio of quotients obtained by dividing ratios of profits by respective time periods.

$\Rightarrow$ Ratio of their capitals $= \frac{2}{12} : \frac{3}{7} = 14 : 36 = 7 : 18$

As given, 7 units = ₹3500, then 18 units = ₹ 9000

Problem 12 A began business with ₹ 45000 and was joined afterwards by B with ₹ 54000. After how many months did B join, if the profits at the end of the year were divided in the ratio 2 : 1 ?

Sol. Let B joins after x months.

Then, $\frac{\text{share of A in profit}}{\text{share of B in profit}} = \frac{45000 \times 12}{54000 \times (12 - x)} = \frac{2}{1}$ [given]

$\Rightarrow \quad 45000 \times 12 = 108000 \times (12 - x)$

$\Rightarrow \quad 12 - x = 5$

$\Rightarrow \quad x = 7$ months

Alternatively I

Ratio of investments = 5 : 6

Ratio of shares in profit = 2 : 1

A invests for 12 months $\Rightarrow 5 \times 12 = 60$

Given, 2 units = 60

$\therefore$ 1 unit = 30

Further, number of months $= \frac{30}{6} = 5$

Hence, B's capital works for 5 months. It implies that B joins after (12 – 5) i.e. 7 months.

Alternatively II

We know, ratio of time periods = ratio of quotients obtained by dividing ratios of profits by respective ratios of capitals.

$\Rightarrow$ Ratio of time periods $= \frac{2}{5} : \frac{1}{6} = 12 : 5$

As given, 12 units = 12 months, then 5 units = 5 months

Hence, B's capital works for 5 months. It implies that B joins after [12 – 5], i.e. 7 months.

Problem 13 The ratio of investments of two partners is 11 : 12 and the ratio of their annual profits is 2 : 3. If A made investment for 8 months, then for how much time did B invest his money?

Sol. Let B invested his money for x months.

Then, $\frac{\text{share of A in profit}}{\text{share of B in profit}} = \frac{11 \times 8}{12 \times x} = \frac{2}{3}$ [given]

$\Rightarrow \quad 11 \times 8 \times 3 = 2 \times 12 \times x$

$\therefore \quad x = 11$ months

Alternativaly

We know, ratio of time periods = ratio of quotients obtained by dividing ratios of profits by respective ratios of capitals.

$\Rightarrow$ Ratio of time periods $= \frac{2}{11} : \frac{3}{12} = 24 : 33 = 8 : 11$

As given, 8 units = 8 months,

then 11 units = 11 months

Hence, B invested his money for 11 months.

Problem 14 A, B and C enter into a partnership and their capital contributions are in ratio $2:3:4$, respectively. At the end of the business term they share profits in the ratio $3:6:10$. Find the ratio of time-periods for which they invest their capitals. *[MAT-11]*

Sol. We know, ratio of time periods = ratio of quotients obtained by dividing ratios of profits by respective ratios of capitals.

$$= \frac{3}{2}:\frac{6}{3}:\frac{10}{4} = 18:24:30$$

$$= 3:4:5$$

Problem 15 Three partners shared the profit in a business in the ratio $5:7:8$. They had partnered for 14 months, 8 months and 7 months, respectively. What was the ratio of their investments? *[Hotel Mgmt-98]*

Sol. We know, ratio of capital invested = ratio of quotients obtained by dividing ratios of profits by respective time periods.

$\Rightarrow$ Ratio of their capitals $= \frac{5}{14}:\frac{7}{8}:\frac{8}{7} = 20:49:64$

When Initial Investment(s) is/are Changed Later

Problem 16 A, B and C started a business by investing ₹ 40500, ₹ 45000 and ₹ 60000, respectively. After 6 months, C withdrew ₹ 15000 while A invested ₹ 4500 more. In annual profit of ₹ 56100, the share of C will exceed that of A by *[SSC(CGL-PT)-08]*

Sol. Since profits are to be divided in ratio of their capitals worked for the business, ratio of the shares of A, B and C in profit

$$= [(40500\times 6)+(40500+4500)\times 6]:[45000\times 12]:[(60000\times 6)+(60000-15000)\times 6]$$

$$= [405+450]:[450\times 2]:[600+450]$$

$$= 855:900:1050 = 57:60:70$$

Now, $(57+60+70)$ units = ₹56100

$\therefore$ $(70-57)$ units = ₹3900

Hence, the share of C will exceed the share of A by ₹3900.

Problem 17 A, B and C started a business with investments in the ratio $5:6:8$ respectively. After 1 yr, C withdrew 50% of his capital and A increased his capital by 60% of his investment. After 2 yrs, in what ratio should the earned profit be distributed among A, B and C, respectively? *[Syndicate Bank(PO)-04]*

Sol. Ratio of shares of profits of A, B and C

$$= [(5\times 1)+(5\times 160\%)\times 1]:[6\times 2]:[(8\times 1)+(8\times 50\%)\times 1]$$

$$= 13:12:12$$

Problem 18 Mr. Nilesh Agrawal opened a workshop investing ₹ 40000. He invested additional amount of ₹ 10000 every year. After 2 yrs, his brother Suresh joined him with an amount of ₹ 85000. Thereafter, Suresh did not invest any additional amount. On completion of 4 yrs from the opening of workshop they earned an amount of ₹ 195000. What will be Nilesh's share in the earning? *[SBI(PO)-01]*

Sol. Ratio of shares of Nilesh and Suresh in earnings

$$= [(40000\times 1)+(50000\times 1)+(60000\times 1)+(70000\times 1)]:[85000\times 2]$$

$$= 220000:170000 = 22:17$$

Now, $(22+17)$ units = ₹195000

$\therefore$ 22 units = 110000 i.e. share of Nilesh in earnings.

When the Partners get Allowances/ Salaries/ Interest etc. in Addition to Their Shares in Profit Depending Upon Their Investments

Problem 19 A and B entered into a partnership by investing ₹ 10000 and ₹ 4000 respectively with the condition that B will receive ₹ 100 per month for managing the business. Annual profit will be divided in the ratio of their investments after paying 5% interest on the capitals. Find the total share of B in an annual profit of ₹ 4000.

Sol. Annual profit = ₹4000

Amount received by B for managing business for 1 yr

$= 100\times 12 =$ ₹1200

Interest on the investments of A and B

$= 5\%$ of $(10000+4000) =$ ₹700

Hence, amount of profit to be divided

$= 4000-1200-700 =$ ₹2100

Ratio of capital of A and $B = 10000:4000 = 5:2$

Now, $(5+2)$ units = 2100

$\therefore$ 2 units = ₹600 i.e. share of B

Then, total shares of $B = 1200+600 =$ ₹1800

Problem 20 A and B enter into a partnership. A invests total capital of ₹ 45000 on the condition that the profits are to be equally divided and B will pay to A interest on the half the capital at 10% per annum, but B gets ₹ 120 per month out of the yearly profit for working for the firm. Find their total yearly profit when B's share is equal to half of A's share.

Sol. Let the yearly profit be ₹ x.

Amount to be given to B for working for the firm in 1 yr

$= 120\times 12 =$ ₹1440

$\therefore$ Amount to be divided between them equally $= (x-1440)$

Hence, A's share $= \frac{x-1440}{2}$

and B's share $= \frac{x-1440}{2}$

Further, B will give 10% interest on ₹ 22500 from his own share i.e. ₹ 2250.

Now, as per question, B's share $= \frac{1}{2}\times A$'share

$$\Rightarrow \left[\left(\frac{x-1440}{2}\right)+1440-2250\right] = \frac{1}{2}\left[\left(\frac{x-1440}{2}\right)+2250\right]$$

$$\Rightarrow \frac{x-1440}{4} = 3375-1440$$

$$\Rightarrow x = (1935\times 4)+1440$$

$$= ₹\ 9180$$

Problem 21 **A partnership business has a capital ₹ 40000, of which one-half is contributed by A and the other half contributed equally by B and C. They appoint D as manager and agree to pay him 10% of the gross profit, B and C are also paid 5% each thereof, before the distribution of profit in the proportion to their capitals. Find the share of C out of a total profit of ₹ 12600.**

Sol. Gross profit = 12500

∴ Net profit to be distributed among A, B and C according to their capitals

= [100% – 10%(for D) – 5%(for B) – 5%(for C)] of 12600

= 80% of 12600 = 10080

Ratio of capitals of A, B and C = 20000 : 10000 : 10000

= 2 : 1 : 1

∴ C's share in net profit = $10800 \times \frac{1}{4}$ = ₹ 2520

Hence, C's total share = 2520 + (5% of 12600)

= 2520 + 630 = ₹ 3150

Problem 22 **X, Y and Z start a web-based venture together. X invests ₹ 2.5 lakh, Y invests ₹ 3.5 lakh and Z invests ₹ 4 lakh. In the first year, the venture makes a profit of ₹ 2 lakh. A part of the profit is shared between Y and Z in the ratio 2 : 3 and the remaining profit is divided among X, Y and Z in ratio of their initial investments. The amount that Z receives is four times the amount that X receives. How much amount does Y receive?** ***[XAT-20]***

Sol. Let ₹a is first distributed among Y and Z in ratio 2 : 3.

Then, amount received by $Z = \frac{3a}{5}$

Further, share of Z in the remaining $(200000 - a)$

$= \frac{8}{20}(200000 - a)$ and share of X in the remaining

$(200000 - a) = \frac{5}{20}(200000 - a)$

As given, $\frac{3a}{5} + \frac{8}{20}(200000 - a) = 4 \times \frac{5}{20}(200000 - a)$

$\Rightarrow \quad \frac{3a}{5} = \frac{3}{5}(200000 - a)$

$\Rightarrow \quad \frac{6a}{5} = 120000$

$\Rightarrow \quad a = 100000$

Now, share of $Y = \left(\frac{2}{5} + \frac{7}{20}\right)$ of 100000 = ₹ 75000

Alternatively

Ratio of initial investments of X, Y and Z = 5 : 7 : 8

Certain amount out of ₹2 lakh is to be divided among X, Y, Z in the ratio 5 : 7 : 8.

As given, the overall share of Z is 4 times the share of X, it implies that 12 units more is required over the given 8 units. Then, (8 + 12) will be four times the 5 units. These 12 units are equal to 3 units of another part of ₹2 lakh. If 3 units of the latter is equal to 12 units of the former, then 2 units of the latter will be equal to 8 units of the former.

Clearly, the latter part is containing two shares equal to 12 units and 8 units.

Since (5 + 7 + 8) units = (12 + 8) units, the ratio of the former part and the latter part = 1 : 1. It implies that one lakh out of the profit 2 lakh was first distributed among Y and Z in ratio 2 : 3 and the remaining amount i.e. one lakh was divided among X, Y and Z in ratio 5 : 7 : 8.

Now, the amount received by Y = (₹ 40000 + 35000) = ₹ 75000

Problem 23 **A, B and C invest ₹ 12000, ₹ 15000 and ₹ 18000 respectively, in a business. A gets 25% of the annual profit for managing the business and the remaining profit is divided among A, B and C in ratio of their investments. If at the end of the year A gets ₹ 100 less than what B and C get together, find the profit earned during the year.** ***[SSC(CGL-Mains)-01]***

Sol. Let the annual profit be ₹ x.

Amount paid to A for management = 25% of $x = \frac{x}{4}$

Now, remaining profit $\left(x - \frac{x}{4}\right)$ will be divided among them in ratio 4 : 5 : 6 as their capitals are in ratio 4 : 5 : 6.

Now, (4 + 5 + 6) units = ₹$\frac{3x}{4}$

$\Rightarrow \quad$ 4 units = $\frac{x}{5}$ and (5 + 6) units = $\frac{11x}{20}$

As given, difference between shares of A and $(B + C)$ = 100

$\Rightarrow \quad \frac{11x}{20} - \left(\frac{x}{4} + \frac{x}{5}\right) = 100$

$\Rightarrow \quad \frac{11x}{20} - \frac{9x}{20} = 100$

$\Rightarrow \quad x = 1000$

Alternatively I

Since 25% of the profit is to be paid to A, 75% of the profit will be left for distribution among A, B and C in ratio 4 : 5 : 6.

Obviously, their shares will be 20%, 25% and 30%, respectively.

In actual, A gets (20% + 25%) i.e. 45% of the profit.

As per question, [(25 + 30) – 45]% of the profit = ₹100

∴ 100% of the profit = ₹ 1000

Alternatively II

You may proceed through options.

Let options be ₹2250, 1000, ...

The ratio of shares in profit = 4 : 5 : 6

(a) 2250 $\Rightarrow$ $(B + C)$'s shares – A's share = $\left[\frac{5+6}{(4+5+6)}\right] \times 75\%$ of 2250 – 4(4 + 5 + 6) × 75% of 2250 + 25% of 2250 ≠ 100

(b) 1000 $\Rightarrow$ $(B + C)$'s shares – A's share = $\left[\frac{5+6}{(4+5+6)}\right] \times 75\%$ of 1000 – 4(4 + 5 + 6) × 75% of 1000 + 25% of 1000 = 100

Hence, option (b) is correct.

WORKOUT

1. In a business, A and C invested amounts in the ratio 2 : 1, whereas the ratio between amounts invested by A and B was 3 : 2. If ₹ 157300 was their profit, how much amount did B receive ? *[MAT-14, 19]*

(a) ₹ 24200 (b) ₹ 36300
(c) ₹ 48400 (d) ₹ 72600

2. A and B enter into a partnership with ₹ 50000 and ₹ 60000 respectively. C joins them after x months contributing ₹ 70000 and B leaves x months before the end of the year. If they share the profit in the ratio 20 : 18 : 21, then find the value of x ?

(a) 3 (b) 9
(c) 6 (d) 8

3. A and B enter into partnership with capitals in the ratio 5 : 6. At the end of 8 months, A withdraws his capital. They received profits in the ratio 5 : 9. B invested the capital for *[SSC(CGL-T-II) -10]*

(a) 6 months (b) 8 months
(c) 10 months (d) 12 months

4. Arvind began a business with ₹ 550 and was joined afterwards by Brij with ₹ 330. When did Brij join, if the profits at the end of the year were divided in the ratio 10 : 3.

(a) After 4 months
(b) After 6 months
(c) After 4.5 months
(d) None of the above

5. A and B started a partnership business investing in the ratio 3 : 8. C joined them after 4 months with an amount equal to 3/4th of B. What was their profit (in ₹)at the end of the year if C got ₹ 24000 as his share? *[SSC(CGL-T-II)-17]*

(a) 120000 (b) 150000
(c) 90000 (d) 180000

6. A, B and C are partners. A receives 9/10 of the profit and B and C share the remaining profit equally. A's income is increased by ₹ 270 when the profit rises from 12% to 15%. Find the capital invested by B and C each. *[MAT-12]*

(a) ₹5000 (b) ₹1000
(c) ₹500 (d) ₹1500

7. A and B started a business by investing ₹ 350000 and ₹ 140000, respectively. A gets 20% of the yearly profit for managing the business. Thereafter, the profit is divided in the ratio of capital. If A receives totally ₹ 38000 more than B at the end of a year, then the profit is *[SSC(CGL-T-II) -11]*

(a) ₹ 28000 (b) ₹ 2800000
(c) ₹ 105000 (d) ₹ 70000

8. Three partners invested capital in the ratio 2 : 7 : 9. The time period for which each of them invested was in the ratio of the reciprocals of the amount invested. Find the share of the partner who brought in the highest capital, if profit is ₹ 1080.

(a) ₹120 (b) ₹360
(c) ₹540 (d) ₹420

9. Meenakshi started a business investing ₹ 12000. After 5 months, Abhay joined her with a capital of ₹ 9000. At the end of the year the total profit was ₹ 4416. What is the share of profits of Meenakshi ? *[BOB(AO)-08, OBC(PO)-08, BOM(PO)-09, CLAT-14]*

(a) ₹1344 (b) ₹3364
(c) ₹3072 (d) ₹2568
(e) None of these

10. Sita and Gita enter into a partnership, Sita contributes ₹ 5000 while Gita contributes ₹ 4000. After 1 month, Gita withdraws $\frac{1}{4}$ part of her contribution and after 3 months from the starting, Sita puts ₹ 2000 more. When Gita withdraws her money Rita also joins them with ₹ 7000. If at the end of 1 yr, there is a profit of ₹ 1218, what will be the share of Rita in the profit ?

(a) ₹844.37 (b) ₹488.47
(c) ₹588.47 (d) None of these

11. Two businessmen A and B invest in a business in the ratio 5 : 8. They decided to reinvest 30% of the profit they earned back into the business. The remaining profit they distributed amongst themselves. If A's share of the profit was ₹ 87500, then how much profit (in ₹) did the business make? *[SSC(CGL-T-II)-17]*

(a) 227000 (b) 250000
(c) 375000 (d) 325000

12. Two partners invested ₹ 1250 and ₹ 850 respectively in a business. Both the partners shared 60% of the profit and distributed 40% as the interest on their capitals. If one partner received ₹ 30 more than the other, the total profit is

(a) ₹262.50 (b) ₹622.50
(c) ₹220.50 (d) ₹226.50

13. A, B and C are three partners. They altogether invested ₹ 14000 in business. At the end of the year, A got ₹ 337.50, B got ₹ 1125 and C got ₹ 637.50 as profit. The difference between the investments of B and A was *[MAT-10, 09]*

(a) ₹2200 (b) ₹3200
(c) ₹42000 (d) ₹5250

14. Anu is working partner and Bimla is a sleeping partner in a business. Anu puts in ₹ 5000 and Bimla puts in ₹ 6000. Anu receives 12.5% of the profit for managing the business and the rest is divided in proportion to their capitals. What does each get out of a profit of ₹ 880 ?

(a) ₹400 and ₹480 (b) ₹460 and ₹420
(c) ₹450 and ₹430 (d) ₹470 and ₹410

15. X and Y entered into partnership with ₹ 700 and ₹ 600, respectively. After 3 months, X withdrew 2/7 of his stock but after 3 months, he puts back 3/5 of what he had withdrawn. The profit at the end of the year is ₹ 726. How much of this should X receive ? *[MAT-12]*

(a) ₹336 (b) ₹366
(c) ₹633 (d) ₹663

16. *A, B* and *C* started a business in partnership with investments of ₹ 12000, ₹ 26000 and ₹ 32000, respectively. After 4 months, *A* leaves. After 6 months, *B* leaves and *A* joins with an amount equal to his earlier investment. After 10 months, *C* leaves and *B* joins with his prior investment. At the end of the year they earn a profit of ₹ 53622. Find the share of *B* in the annual profit.

[SIDBI Bank Officer-14]

(a) ₹ 17221 (b) ₹ 17212
(c) ₹ 17222 (d) ₹ 18212
(e) None of these

17. *X* and *Y* are partners in a business. *X* contributed 1/3 of the capital for 9 months and *Y* received 2/5 of the profits. For how long was *Y*'s money used in the business?

(a) 4 months
(b) 3 months
(c) 2 months
(d) 5 months

18. *A* started a business. After 4 months from the start of the business, *B* and *C* joined. The respective ratio between the investments of *A, B* and *C* was 4 : 6 : 5. If *A*'s share in annual profit was ₹ 250 more than *C*'s share, what was the total annual profit earned ?

[IBPS(CWE-V)-15, MAT-18]

(a) ₹ 3740 (b) ₹ 3910
(c) ₹ 4250 (d) ₹ 4350
(e) None of these

Answer key

1. (c)	2. (a)	3. (d)	4. (b)	5. (c)	6. (c)	7. (d)	8. (b)	9. (c)	10. (b)
11. (d)	12. (a)	13. (d)	14. (b)	15. (b)	16. (b)	17. (b)	18. (c)		

Hints and SOLUTIONS

1. Given, $A : C = 2 : 1 = 6 : 3$
and $A : B = 3 : 2 = 6 : 4$
Hence, $A : B : C = 6 : 4 : 3$
Since time period is same, the profit will be distributed in the ratio 6 : 4 : 3.
Now, (6 + 4 + 3) units = ₹157300
$\therefore$ 4 units = ₹48400

2. Ratio of shares of A, B and C in profit
$= (50000 \times 12) : 60000\,(12 - x) : 70000(12 - x)$
$\Rightarrow 60 : (72 - 6x) : (84 - 7x) = 20 : 18 : 21$
$\Rightarrow 60 : (72 - 6x) = 20 : 18$
$\Rightarrow 10 : (12 - x) = 10 : 9$
$\Rightarrow 12 - x = 9 \Rightarrow x = 3$

3. Ratio of time periods = Ratio of quotients obtained by dividing ratios of profits by respective ratios of capitals $= \frac{5}{5} : \frac{9}{6} = 2 : 3$
As given, 2 units = 8 months, then 3 units = 12 months

4. Ratio of their capitals = 550 : 330 = 5 : 3
and ratio of their profits = 10 : 3
Now, ratio of time periods = ratio of quotients obtained by dividing ratios of profits by respective ratios of capitals
$\Rightarrow$ Ratio of time periods $= \frac{10}{5} : \frac{3}{3} = 2 : 1$
As given, 2 units = 12 months, then 1 unit = 6 months
Hence, B joins to Arvind after (12 – 6) i.e. 6 months.

5. Ratio of shares of A, B and C in profit
$= (3 \times 12) : (8 \times 12) : \left[\left(\frac{3}{4} \times 8\right) \times (12 - 4)\right]$
$= 3 : 8 : 4$
As given, 4 units = 24000
$\therefore$ (3 + 8 + 4) units = ₹ 90000

6. Ratio of profits $(A : B : C) = 90 : 5 : 5 = 18 : 1 : 1$ = Ratio of their capitals
Given that when profit rises by (15 – 12)%, A's income increases by ₹270.
Obviously, 3% of 18 units = ₹270
$\therefore$ 100% of 18 units = ₹ 9000
Then, 1 unit = ₹500
Hence, B and C invested ₹500 each.

7. Ratio of shares of A and B in profit = 350000 : 140000 = 5 : 2
Obviously, (5 + 2) units = 80% of the profit
$\Rightarrow$ (5 – 2) units $= \frac{240}{7}$% of the profit
As per question, $\frac{240}{7}$% of the profit + 20% of the profit = ₹ 38000
$\Rightarrow \frac{380}{7}$% of the profit = ₹ 38000
$\Rightarrow$ 100% of the profit = ₹ 70000

8. Ratio of time periods $= \frac{1}{2} : \frac{1}{7} : \frac{1}{9} = 63 : 18 : 14$
Now, ratio of shares in profit
= ratio of products of capitals and respective time periods
$= (2 \times 63) : (7 \times 18) : (9 \times 14)$
$= 9 : 9 : 9$
$= 1 : 1 : 1$
Share of each partner in profit
$= 1080 \div 3$ = ₹360

9. Ratio of shares of Meenakshi and Abhay in profit
$= (12000 \times 12) : (9000 \times 7)$
$= 48 : 21$
Now, (48 + 21) units = ₹4416
$\therefore$ 48 units = ₹3072

10. (*b*) Ratio of shares of Sita, Gita and Rita $= (5000 \times 3) + (7000 \times 9) : (4000 \times 1) + (3000 \times 11) : (7000 \times 11)$
$= 78000 : 37000 : 77000$
$= 78 : 37 : 77$
Now, (78 + 37 + 77) units = ₹1218
$\therefore$ 77 units = ₹ 488.47

11. Ratio of shares of A and B in profits left after reinvestment = 5 : 8
As given, 5 units = 87500
$\therefore$ (5 + 8) units = 227500
Obviously, 70% of the total profit = ₹ 227500
$\therefore$ 100% of the total profit = ₹ 325000

12. Ratio of their capitals = 1250 : 850 = 25 : 17
= Ratio of their shares in profit
Now, (25 + 17) units = 60% of the profit
$\therefore$ (25 – 17) units $= \frac{80}{7}$% of the profit
= ₹ 30 [given]
$\Rightarrow$ 100% of the profit $= 30 \times \frac{7}{80} \times 100$
= ₹262.50

13. Since time period is same, the ratio of their capitals = ratio of their shares in profit.
Ratio of capitals = 337.50 : 1125 : 637.50 = 9 : 30 : 17
Now, (9 + 30 + 17) units = ₹ 14000
$\therefore$ (30 – 9) units = ₹ 5250

14. Amount received by Anu for management = 12.5% of 880 = 110
Now, ratio of their shares in profit
= 5000 : 6000
= 5 : 6
Then, (5 + 6) units = 880 – 110 = ₹ 770
$\therefore$ 5 units = 350 and 6 units = ₹ 420
Hence, Anu and Bimla are getting ₹ 460 and ₹420 respectively.

15. Ratio of shares of X and Y in profit
$= (700 \times 3) + \left(700 \times \frac{5}{7} \times 3\right) + \left\{\left(700 \times \frac{5}{7}\right) + \left(700 \times \frac{2}{7} \times \frac{3}{5}\right)\right\} \times 6 : (600 \times 12)$
$= 7320 : 7200$
Now, (7320 + 7200) units = ₹ 726
$\therefore$ 7320 units = ₹ 366

16. Clearly, A's capital worked for 10 months, B's capital for 8 months and C's capital for 10 months.
Then, ratio of shares of A, B and C in annual profit
$= (12000 \times 10) : (26000 \times 8) : (32000 \times 10)$
$= 15 : 26 : 40$
Note You may use ratio of their capitals instead of taking actual figure
$\because$ (15 + 26 + 40) units = ₹53622
26 units = ₹17212

17. We know, ratio of time periods = ratio of quotients obtained by dividing ratios of profits by respective ratios of capitals.
$= \frac{3/5}{1/3} : \frac{2/5}{2/3} = 3 : 1$
As given, 3 units = 9 months, then 1 unit = 3 months.

18. Ratio of shares of A, B and C in the annual profit
$= (4 \times 12) : (6 \times 8) : (5 \times 8)$
$= 6 : 6 : 5$
As given, (6 – 5) unit = ₹ 250
$\therefore$ (6 + 6 + 5) units = ₹ 4250

29

Alligation/ Mixture

Learning OBJECTIVES

After completing this chapter, aspirants will be able to :

- Understand the derivation of Alligation-Rule;
- Identify the diverse sorts of questions covered under different chapters/ topics which can easily be solved by Alligation-Rule;
- Apply the concepts associated with 'Ratio' to solve the problems based on Mixture;
- Recognise the fundamental approaches which is applicable for the questions;
- Feel comfortable to choose the appropriate approaches out of so many approaches as per his/her choice;
- Comprehend the solutions of complicated questions easily as usages of appropriate words have been made while solving the problems;
- Observe the diverse categories of the questions;
- Learn how to solve the maximum questions with the concepts associated with 'Ratio';
- Understand the philosophy of the division of this chapter in various parts;
- Analyse the trend of the questions being asked in various competitive examinations;
- Recognise the types of question which can be easily solved through the given options;
- Solve the questions in moments;
- Grip on concepts helping to solve the questions without pen and paper;
- Learn many new concepts/fundamentals;
- Apply the pragmatic approaches in lieu of academic approaches; and
- Ponder on the alternative methods.

Inside

- *Brief Introduction*
- ***Concept of** Mixture of Two or More Ingredients, Mixing of Ingredients and Mixture of Mixtures*
- *Changes in Ratio when Ingredient(s) Change(s)*
- *Adulteration in Ingredient to make Profit*
- ***Replacement of a Part of** Ingredient by Another Ingredient, Mixture by One of its Ingredients and Ingredient by a Mixture*
- *Application of Alligation Rule in Chapters-Profit & Loss, SI, Time and Distance and Average*
- *Miscellaneous*
- ....
- .... *& Many More*
- *Work Out with Solutions*

Introduction

This chapter is substantially based on combination of application of Alligation-Rule and application of concepts associated with 'Ratio'. No doubt, this chapter plays a vital role in securing marks/ranks in the competitive examinations like CAT, MAT, SSC, CDS, NDA, Bank, LIC, PSU, Railway, etc. I hope, aspirants will realise themselves totally different after perusal of this chapter.

Before discussing the topic in length, peruse the following questions with solutions for building the concepts

(a) Two classes of students having average marks 28 and 42 respectively are mixed; resulting in overall average becomes 36. Find the ratio of students of the classes.

(b) In what ratio must a grocer mix two varieties of pulses costing ₹ 15 and ₹ 20 per kg respectively so as to get a mixture worth ₹ 16.50 per kg? ***[RRB-03, SSC (CGL-T-II)-17, MAT-19]***

(c) The ratio in which two sugar solutions of the concentrations 15% and 40% are to be mixed to get a solution of concentration 30% is? ***[SSC (CGL-PT)-08, SSC (CGL-T-II)-17]***

(d) The ratio of the quantities of sugar, in which sugar costing ₹ 20 per kg and ₹ 15 per kg should be mixed so that there will be neither loss nor gain on selling the mixed sugar at the rate of ₹ 16 per kg is? ***[SSC 10+2 (DEO)-08]***

(e) A shopkeeper has some quantity of tea. He sells a part of it at 20% profit and the rest at 6% loss. If his overall profit is 13%, find the ratio between both parts. ***[RRB-05]***

(f) A money-lender lends his money in two parts, 1st part at 5% per annum and 2nd at 8% per annum simple interest. At the end of the year he found that he overall got 6% per annum. Find the ratio of the money invested two at 5% and 8% per annum respectively.

(g) A car travels at 40 km/h for a certain time and further travels at 60 km/h for a certain time. Find the ratio between both the time periods, if the average speed of the car in 45 km/h.

(h) Some amount is to be distributed among students in such a way that each boy gets ₹ 5 and each girl gets ₹ 12. Find the ratio of number of girls to that of boys, if each student gets average ₹ 10.

(i) The average daily wages of staff consisting of managers and labourers of a factory is ₹ 150. The average wages of managers is ₹ 200 while that of labourers is ₹ 140. Find the ratio of number of managers to that of labourers.

(j) Gold and copper are as heavy as water by 19 and 9 times respectively. The ratio in which these two metals be mixed so that the mixture is 17 times as heavy as water is ***[CLAT-17]***

Sol. (a) Let the required ratio be $x : y$.

As per question, $\frac{(x \times 28) + (y \times 42)}{(x + y)} = 36$

[Recall the concept of weighted average]

$\Rightarrow \quad 28x + 42y = 36x + 36y$

$\Rightarrow \quad (36x - 28x) = (42y - 36y)$

$\Rightarrow \quad \frac{x}{y} = \frac{(42 - 36)}{(36 - 28)}$

$\Rightarrow \quad x : y = (42 - 36) : (36 - 28)$

$\Rightarrow \quad x : y = 6 : 8 = 3 : 4$

The above resultant $[x : y = (42 - 36) : (36 - 28)]$ may be obtained by the following method.

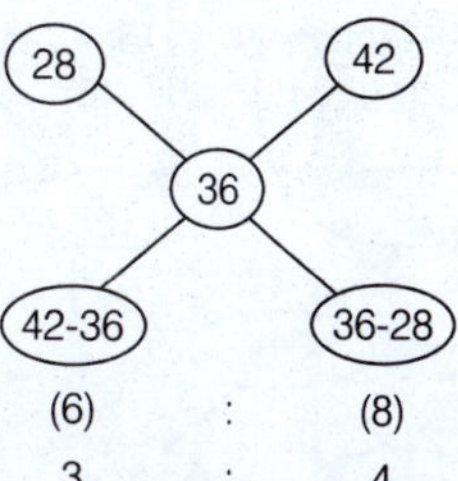

Hence, ratio of students of the classes = 3 : 4

It is clear from the above that there is no need to follow the detailed process as shown with reference to weighted average method. The cross method mentioned above will play a vital role in solving the questions. This cross method is popularly known as 'Alligation Method', which is, indeed, a substitute of weighted average method. No doubt, the problems envisaged in this chapter can be solved by both methods, however, we will stress on application of Alligation method so as to save time in examination hall.

It is also important to note that the figure placed in middle will lie between both figures written in 1st line. Here, 36 is between 28 and 42. It is to be kept in mind that the middle figure is subtracted from greater figure of the upper line while smaller figure of the upper line is subtracted from the middle figure.

(b)

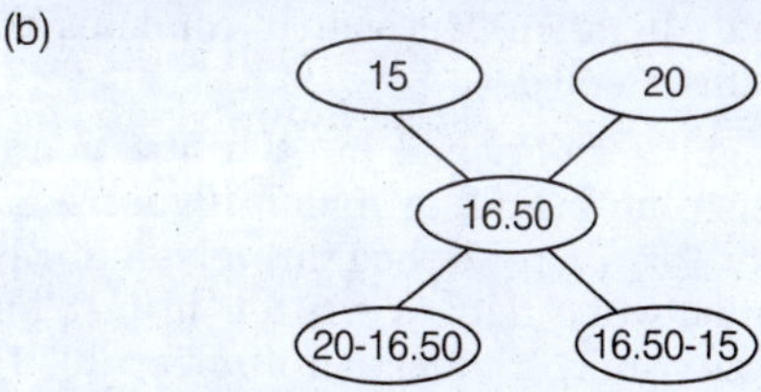

Now, required ratio = 3.50 : 1.50 = 7 : 3

Hence, the grocer must mix two varieties of pulses costing ₹ 15 and ₹ 20 per kg respectively in ratio 7 : 3, so as to get a mixture worth ₹ 16-50 per kg.

(c)

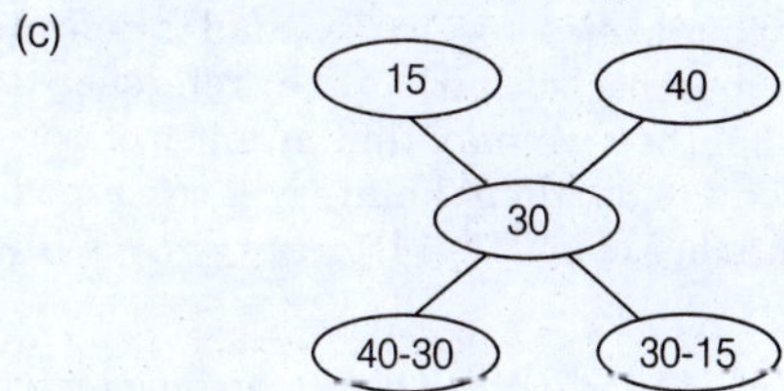

The required ratio = 10 : 15 = 2 : 3

(d) On the above line, the required ratio
$= (16 - 15) : (20 - 16) = 1 : 4$

(e)

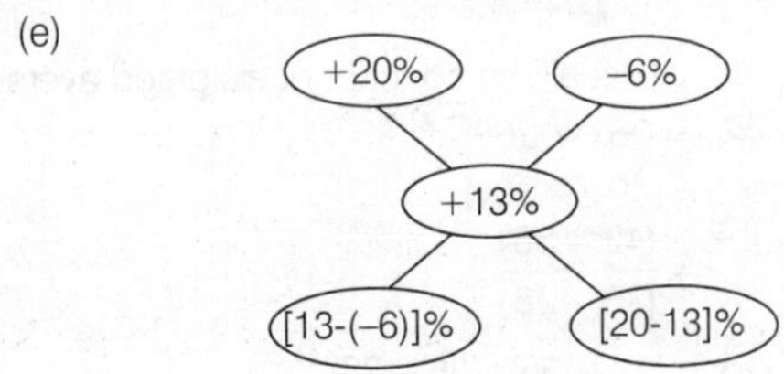

The ratio of 1st part to 2nd part = 19 : 7

(f) On the above line, the required ratio
$= (8 - 6) : (6 - 5) = 2 : 1$

(g) On the above line, the required ratio
$= (60 - 45) : (45 - 40) = 3 : 1$

(h)

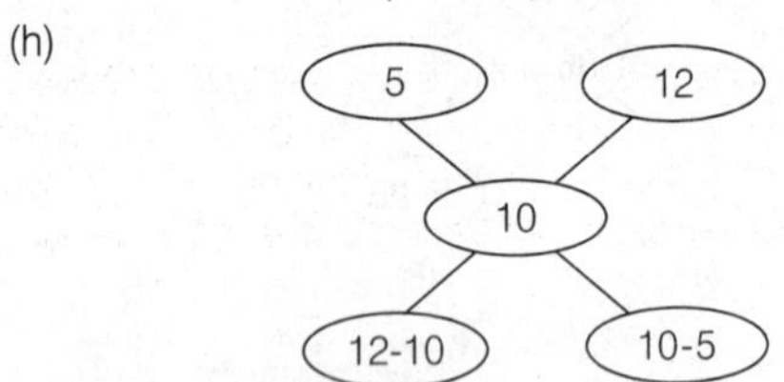

Hence, number of boys : number of girls = 2 : 5
Then, number of girls : number of boys = 5 : 2

(i) On the above line, the required ratio
$= (150 - 140) : (200 - 150) = 1 : 5$

(j) On the above line, the required ratio
$= (17 - 9) : (19 - 17) = 4 : 1$

Observation

(i) The solutions of all nine questions are similar as Alligation method has been adopted in all cases. No need to mention that questions are from diverse sectors like Average, Profit and Loss, Simple Interest, Time and Distance, Wages, etc. but solutions are made with only Alligation Method. It connotes the significance of this method.

(ii) All the three figures mentioned in each question or written in upper line including middle figure while solving through Alligation Method, are given in rate unit/average unit, while figures written in last line represent quantity unit. In question number (b), 15, 20 and 16.50 represent rate unit whereas (20 -16.50) and (16.50 -15) represent ratio of quantities.

(iii) The percentage values may be considered as average values.

(iv) Terms of ratio of quantities may correspond to definite sizes. Suppose number of girls with reference to question (h) is 35, then we may find number of boys. 5 units correspond to 35, then 2 units will correspond to 14. Hence, number of boys = 14.

Note *Many problems have been worked out in this chapter with the help of Alligation Method. However, many questions covered under diverse topics have been solved by Alligation Method in those particular chapters, hence, those questions have been left here deliberately.*

Concept of the Mixture of more than two Ingredients/Items

We have discussed the concept of mixture of two ingredients intensively. Now, we will discuss the concept of the mixture of more than two ingredients. Here, ingredients have very wide meanings which includes all the items which are to be mixed/merged. To build the fundamentals, take one example

"Find the proportion in which three types of tea at ₹ 10, ₹ 12 and ₹ 20 per kg may be mixed so as to obtain a mixture worth ₹ 15 per kg."

(a) 1 : 2 : 4 **(b) 5 : 5 : 8**
(c) 3 : 4 : 6 **(d) 2 : 4 : 7**

Let x, y and z kg of all three types of tea respectively are mixed. Then, the price of a mixture of $(x + y + z)$ kg $= 10x + 12y + 20z$. Further, it is given that overall price of the mixture is ₹ 15/per kg.

Hence, $10x + 12y + 20z = 15(x + y + z)$

$\Rightarrow$ $8z = 5x + 5y$

If $x = 1$ and $y = 1$, then $z = 1.25$.

Thus, required ratio $= 1 : 1 : 1.25 = 4 : 4 : 5$

In case $x = 4$ and $y = 5$, then $z = 45/8$.

Thus, required ratio $= 32 : 40 : 45$

Obviously, there will be infinite number of combinations. Thus, in such types of questions it is desirable to proceed through options.

(a) $1 : 2 : 4 \Rightarrow$ If $x = 1$ and $y = 2$, then $z = 15/8$ but the option shows $z = 4$. Hence, option (a) is not correct.

(b) $5 : 5 : 8 \Rightarrow$ If $x = 5$ and $y = 5$, then $z = 8$ and the option also shows $z = 8$.
Hence, option (b) is correct.

Remark *You may also proceed through options in case of mixing four or more ingredients/items.*

Mixing of Ingredients/items

Problem 1 In what proportion must a grocer mix wheat at ₹ 2.04 per kg and ₹ 2.88 per kg so as to make a mixture worth ₹ 2.52 per kg ?

Sol.

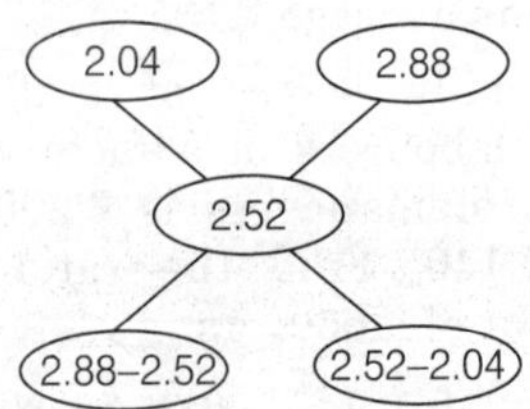

The required ratio = 0.36 : 0.48 = 3 : 4

Alternatively

You may proceed through options also, but it will take some more time. Let options be 2 : 3, 3 : 4, ...,

(a) $2:3 \Rightarrow \frac{(2\times 2.04)+(3\times 2.88)}{(2+3)} \neq 2.52$

(b) $3:4 \Rightarrow \frac{(3\times 2.04)+(4\times 2.88)}{(3+4)} = 2.52$

Hence, option (b) is correct.

Problem 2 The cost of type 1 rice is ₹ 15 per kg and type 2 rice is ₹ 20 per kg. If both type 1 and type 2 are mixed in the ratio 2 : 3, then price per kg of the mixed varieties of rice is *[MBA-02]*

Sol.

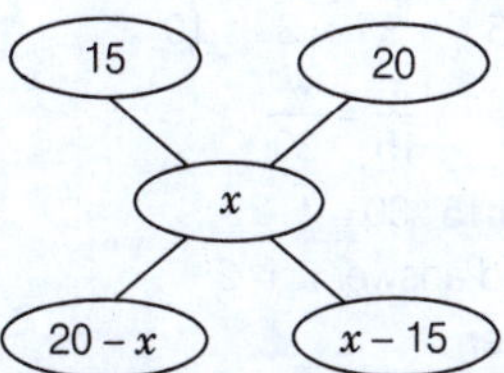

The required ratio $= (20-x):(x-15) = 2:3$ [as given]

$\Rightarrow \quad 2x - 30 = 60 - 3x \Rightarrow x = 18$

Hence, price of the mixture is ₹ 18 per kg.

Alternatively

Price of the mixture $= \frac{(2\times 15)+(3\times 20)}{(2+3)} =$ ₹ 18 per kg

[Recall weighted average concept]

Problem 3 How many kilograms of sugar costing ₹ 10 per kg must be mixed with 45 kg of sugar costing ₹ 8 per kg so as to obtain the mixture costing ₹ 9.10 per kg.

Sol.

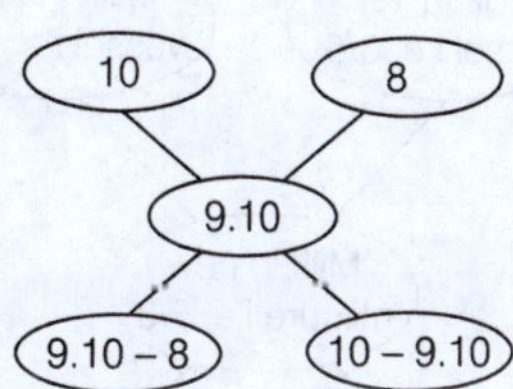

The required ratio = 1 .10 : 0 .90 = 11 : 9

As given, 9 units = 45 kg

$\therefore$ 11 units = 55 kg

Alternatively

Price of the mixture $= \frac{(10\times x)+(8\times 45)}{(x+45)} = 9.10$

$\Rightarrow \quad 10x + 360 = 9.10x + 409.5$

$\Rightarrow \quad 0.9x = 49.5 \Rightarrow x = 55$ kg

Problem 4 5 kg of rice at ₹ 8 per kg is mixed with 3 kg of costlier rice to get a mixture costing ₹ 10.50 per kg. Find the price of the costlier rice.

Sol.

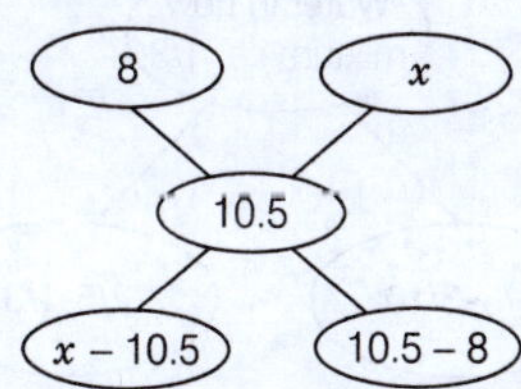

The required ratio $= (x-10.5):2.5 = 5:3$ [as given]

$\Rightarrow \quad 12.5 = 3x - 31.5 \Rightarrow 3x = 44 \Rightarrow x = 14.66$

Hence, the cost of costlier rice = ₹ 14.66 per kg.

Alternatively

Price of the mixture $= \frac{(8\times 5)+(x\times 3)}{(5+3)} = 10.5$ [given]

$\Rightarrow \quad 40 + 3x = 84 \Rightarrow x = 14.66$

Problem 5 Tea worth ₹ 126 per kg and ₹ 135 per kg are mixed with a 3rd variety in the ratio 1 : 1 : 2. If the mixture is worth ₹ 153 per kg, the price of the 3rd variety per kg will be

Sol. Price of mixture containing only 1st and 2nd varieties of tea

$= \frac{(126\times 1)+(135\times 1)}{(1+1)} =$ ₹ 130.50

Now, it may be assumed that there are only two varieties of tea worth ₹ 130.50 per kg and ₹ x per kg and both are to be mixed in the ratio (1 + 1) : 2 i.e. 1 : 1 to obtain the new mixture worth ₹ 153 per kg. Here, cost of 3rd variety of tea is ₹ x per kg.

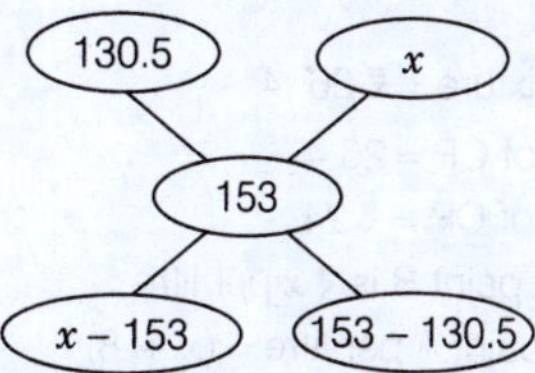

The required ratio $= (x-153):22.5 = 2:2$ [i.e. 1 : 1]

$\Rightarrow \quad x - 153 = 22.5 \Rightarrow x = 175.5.$

Alternatively I

Let the price of 3rd variety of tea be ₹ x per kg.

Then, $\frac{(126\times 1)+(135\times 1)+(x\times 2)}{(1+1+2)} = 153$

$\Rightarrow \quad 261 + 2x = 612 \Rightarrow x =$ ₹ 175.5

Alternatively II

You may proceed through options. Let options be ₹ 169.50, ₹ 175.5, ...

(a) $169.5 \Rightarrow \frac{(126\times 1)+(135\times 1)+(169.5\times 2)}{(1+1+2)} \neq 153$

(b) $175.5 \Rightarrow \frac{(126\times 1)+(135\times 1)+(175.5\times 2)}{(1+1+2)} = 153$

Hence, option (b) is correct.

Problem 6 The ratio, in which tea costing ₹ 192 per kg is to be mixed with tea costing ₹ 150 per kg so that the mixed tea, when sold for ₹ 194.40 per kg, gives a profit of 20 %, is *[SSC (CGL-PT)-08, SSC (T-I)-17]*

Sol. Selling price of the mixture at a profit of 20% = ₹ 194.40 per kg

$\therefore$ Cost price of the mixture $= 194.40 \times \frac{100}{120} =$ ₹ 162 per kg

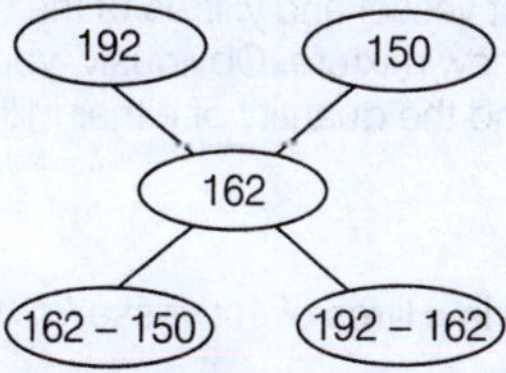

The required ratio = 12 : 30 = 2 : 5

Alternatively I

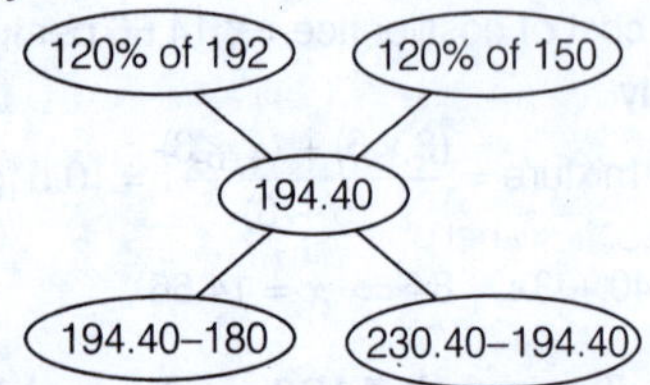

The required ratio = 14.4 : 36 = 2 : 5

Alternatively II

You may proceed through options also.

Problem 7 ***A*** **trader sells 10 L of a mixture of paints** ***A*** **and** ***B*****, where the amount of** ***B*** **in the mixture does not exceed that of** ***A*****. The cost of paint** ***A*** **per litre is ₹ 8 more than that of paint** ***B*****. If the trader sells the entire mixture for ₹ 264 and makes a profit of 10%, then the highest possible cost of paint** ***B*****, in ₹ per litre, is :**

[CAT-18]

Sol. SP of 1 L of mixture = ₹ 26 .4

$\because$ 110% of CP = 26.4

$\therefore$ 100% of CP = ₹ 24

Let the cost of paint B is ₹ x per litre.

Then, cost of paint A per litre = $(x + 8)$

Then,

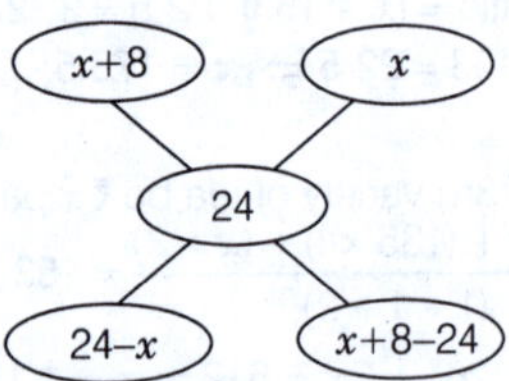

$\therefore$ Ratio of amount of paint A to that of B = $(24 - x) : (x - 16)$

Since the amount of B does not exceed that of A, maximum possible ratio will be 1 : 1.

Hence, $(24 - x) : (x - 16) = 1 : 1$

$\Rightarrow \quad x = 20$

Mixture of Mixtures

Case A Computation of ratio of quantities taken from different mixtures so as to form a new mixture with desired ratio of ingredients/items.

Problem 8 Two vessels contain milk and water in the ratio 3 : 2 and 7 : 3. Find the ratio in which the contents of the two vessels have to be mixed to get a new mixture in which the ratio of milk and water is 2 : 1.

[SSC (CGL-T-II)-11, SSC (CGL-T-II)-12]

Sol. Let x litres of 1st vessel and y litres of the second vessel be mixed to get a new mixture. Obviously, you have to find value of $x : y$. Now, find the quantity of either milk or water in all three mixtures.

In case of milk

Quantity of milk in x litres of 1st vessel/mixture $= \frac{3x}{5}$

and quantity of milk in y litres of 2nd vessel/mixture $= \frac{7y}{10}$

and quantity of milk in $(x + y)$ litres of new mixture $= \left(\frac{3x}{5} + \frac{7y}{10}\right)$

But, it has been given that the new mixture of $(x + y)$ contains milk and water in ratio 2 : 1. It means, quantity of milk in new mixture $= \frac{2(x + y)}{3}$

Obviously, $\left(\frac{3x}{5} + \frac{7y}{10}\right) = \frac{2(x + y)}{3}$

$\Rightarrow \quad \frac{3x}{5} - \frac{2x}{3} = \frac{2y}{3} - \frac{7y}{10}$

$\Rightarrow \quad \frac{x}{15} = \frac{y}{30}$

$\Rightarrow \quad x : y = 15 : 30 = 1 : 2$

Hence, required answer = 1 : 2

In case of water

On the above line, $\left(\frac{2x}{5} + \frac{3y}{10}\right) = \frac{1(x + y)}{3}$

$\Rightarrow \quad \frac{2x}{5} - \frac{x}{3} = \frac{y}{3} - \frac{3y}{10}$

$\Rightarrow \quad \frac{x}{15} = \frac{y}{30}$

$\Rightarrow \quad x : y = 15 : 30 = 1 : 2$

Hence, required answer = 1 : 2

Remarks *You are free to proceed with quantity of either milk or water.*

Alternatively I

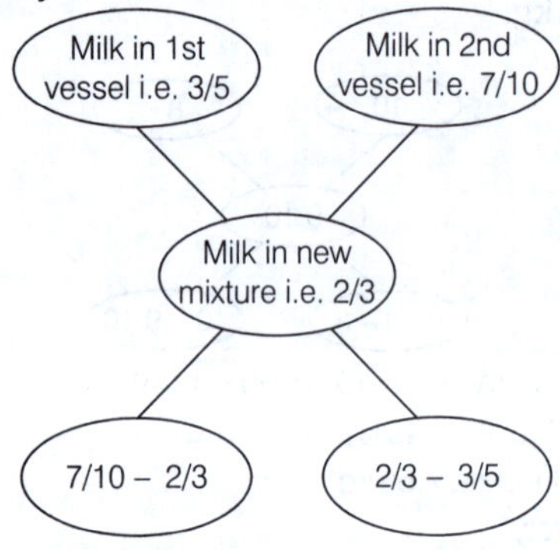

The required ratio $= \frac{1}{30} : \frac{1}{15} = 1 : 2$

or

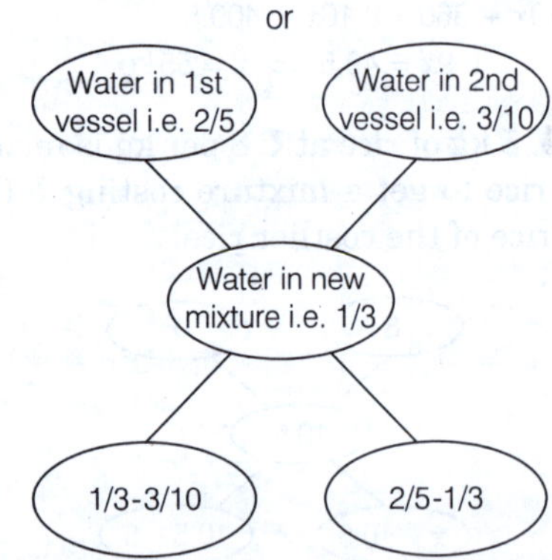

The required ratio = $\frac{1}{30}:\frac{1}{15} = 1:2$

Alternatively II

1st mixture (milk : water) = $a : b$

New mixture (milk : water) = $e : f$

2nd mixture (milk : water) = $c : d$

Then, required ratio = $(a + b)$ [positive difference between cf and de] : $(c + d)$ [positive difference between be and af]

$= (a + b)(cf - de) : (c + d)(be - af)$

In this question,

1st mixture (milk : water) = 3 : 2

New mixture (milk : water) = 2 : 1

2nd mixture (milk : water) = 7 : 3

Then, required ratio

$= (3 + 2)(7 \cdot 1 - 2 \cdot 3) : (7 + 3)(2 \cdot 2 - 3 \cdot 1)$

$= 5 : 10 = 1 : 2$

Alternatively III

Percentage of milk in 1st vessel/mixture = 60%

Percentage of milk in 2nd vessel/mixture = 70%

Percentage of milk in mixture = 66.66%

Now,

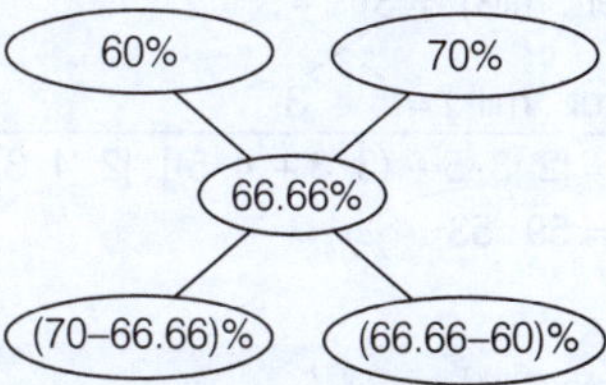

The required ratio = 3.33 : 6.66 = 1 : 2

Note *You may proceed through percentage of water in mixtures also.*

Problem 9 **Two types of alloys posses gold and silver in the ratio of 7 : 22 and 21: 37. In what ratio should these alloys be mixed so as to have a new alloy in which gold and silver would exist in the ratio 25 : 62 ?** ***[SSC (10 + 2/DEO-LDC)-08]***

Sol. The above format as envisaged in Alternatively II will always be maintained.

1st alloy (gold : silver) = 7 : 22

New alloy (gold : silver) = 25 : 62

2nd alloy (gold : silver) = 21 : 37

Then, required ratio

$= (7 + 22)(62 \cdot 21 - 25 \cdot 37) : (21 + 37)(22 \cdot 25 - 7 \cdot 62)$

$= 29 \times 377 : 58 \times 116$

$= 13 : 8$

[You may also proceed through percentage method as elaborated above.]

Problem 10 **A person stole wine from a bottle that contains 45% of spirit and replaced with equal amount of wine having 30% spirit. As a result, the bottle contains only 35% spirit. How much of the wine did he steal ?**

Sol. 1st mixture (spirit : other) = 9 : 11

New mixture (spirit : other) = 7 : 13

2nd mixture (spirit : other) = 3 : 7

Then, ratio of quantity taken from 1st mixture to that from 2nd mixture

$= (9 + 11)(7 \cdot 7 - 13 \cdot 3) : (3 + 7)(13 \cdot 9 - 7 \cdot 11)$

$= 20 \times 10 : 10 \times 40 = 1 : 2$

Clearly, the new mixture contains 1 part of the original mixture having 45% spirit and 2 part of 2nd mixture having 30% spirit. It implies that 2/3 part of the original mixture has been stolen.

Alternatively

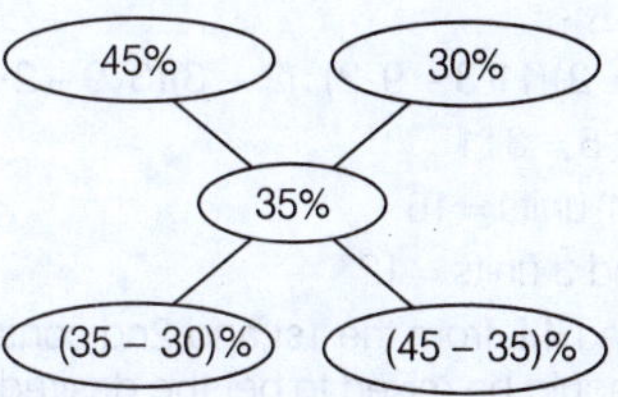

The required ratio = 5 : 10 = 1 : 2

Clearly, the new mixture contains 1 part of the original mixture having 45% spirit and 2 part of 2nd mixture having 30% spirit. It implies that 2/3 part of the original mixture has been stolen.

Problem 11 **A mixture contains wine and water in the ratio 3 : 2 and another mixture contains them in the ratio 4 : 5. How many litres of the latter must be mixed with 3 L of the former so that the resulting mixture may contain equal quantities of wine and water?** ***[SSC(SAS)-10]***

Sol. 1st mixture (wine : water) = 3 : 2

New mixture (spirit: other) = 1 : 1

2nd mixture (spirit: other) = 4 : 5

Then, ratio of quantity taken from 1st mixture to that from 2nd mixture $= (3 + 2)(5 \cdot 1 - 4 \cdot 1) : (4 + 5)(3 \cdot 1 - 2 \cdot 1)$

$= 5 : 9$

As given, 5 units = 3 L

$\therefore$ 9 units = $5\frac{2}{5}$ L.

Alternatively

Considering percentage of wine in mixtures, apply Alligation Rule. You may take percentage of water, if you wish.

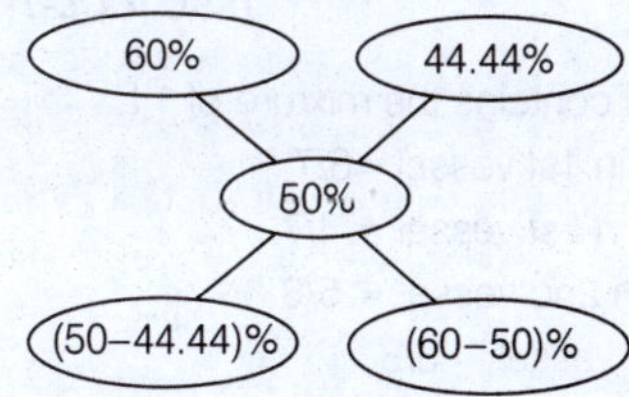

Then, ratio of quantity taken from 1st mixture to that from 2nd mixture = 5.56 : 10 = 5 : 9

As given, 5 units = 3 L

$\therefore$ 9 units = $5\frac{2}{5}$ L

Problem 12 **There are two containers of milk, first contains 40% water and second contains 60% water. How much milk should be mixed from each of the containers so as to get 16 L of milk containing water to milk in ratio 9 : 11.** *[SSC(CGL-T-II)-12, IIFT-17]*

Sol. 1st mixture (milk : water) = 3 : 2

New mixture (milk : water) = 11 : 9

2nd mixture (milk : water) = 2 : 3

Then, ratio of quantity taken from 1st mixture to that from 2nd mixture

$= (3+2)(11\cdot 3 - 9\cdot 2) : (2+3)(3\cdot 9 - 2\cdot 11)$

$= 15 : 5 = 3 : 1$

As given, (3 + 1) units = 16

$\therefore$ 1 unit = 4 and 3 units = 12

Hence, 12 L and 4 L from the 1st and 2nd containers respectively should be mixed to get the desired ratio.

Alternatively

Taking percentage of water in mixture, apply Alligation Rule.

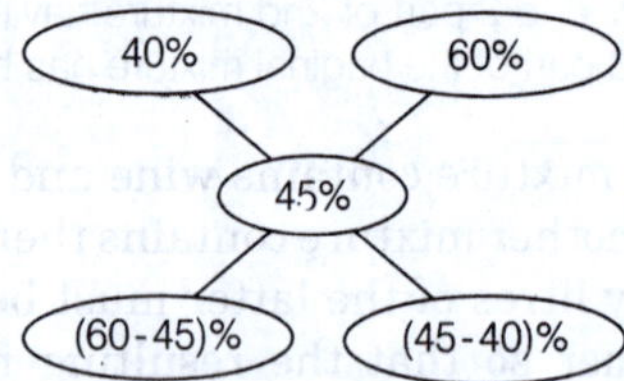

The required ratio = 15 : 5 = 3 : 1

As given, (3 + 1) units = 16

$\therefore$ 1 unit = 4 and 3 units = 12

Hence, 12 L and 4 L from the 1st and 2nd containers respectively should be mixed to get the desired ratio.

Case B **Computation of ratio of ingredients/items in a new mixture which is formed by taking whole/part/ equal quantities from the given different mixtures.**

Problem 13 **Two equal vessels are filled with the mixtures of water and milk in the ratio of 3 : 4 and 5 : 3 respectively. If the mixtures are poured into a 3rd vessel, the ratio of water and milk in the 3rd vessel will be** *[SSC (CGL-T-I-) 11, MAT-12]*

Sol. Let each vessel contains the mixture of 1 L

Then, water in 1st vessel = 3/7

and milk in 1st vessel = 4/7

Further, water in 2nd vessel = 5/8

and milk in 2nd vessel = 3/8

$\therefore$ Ratio of water and milk in the 3rd vessel

$$= \left(\frac{3}{7}+\frac{5}{8}\right) : \left(\frac{4}{7}+\frac{3}{8}\right) = 59 : 53$$

Note *The similar process will be made in case of three vessels/mixtures.*

Alternatively I

1st mixture (water : milk) = 3 : 4 = (3 : 4) × (5 + 3) = 24 : 32

2nd mixture (water : milk) = 5 : 3 = (5 : 3) × (3 + 4) = 35 : 21

$\therefore$ Required ratio (water: milk) = (24 + 35) : (32 + 21) = 59 : 53

Here, we have multiplied the each term of ratio of 1st mixture by sum of terms of ratio of 2nd mixture and *vice-versa* so as to get the quantities of both mixtures equal i.e. (24 + 32) = (35 + 21) = 56.

In other way, since LCM of (3 + 4) and (5 + 3) is 56, we multiply ratio of respective mixtures by [LCM ÷ sum of terms of that ratio].

Here, 3 : 4 = (3 : 4) × [56 ÷ (3 + 4)]

and 5 : 3 = (5 : 3) × [56 ÷ (5 + 3)].

This is very useful for three or more mixtures.

Alternatively II

1st mixture (water : milk) = a : b

2nd mixture (water : milk) = c : d

Required ratio = $[2ac + (ad + bc)] : [2bd + (ad + bc)]$

Here,

1st mixture (water : milk) = 3 : 4

2nd mixture (water : milk) = 5 : 3

$\therefore$ Required ratio = $[2\cdot 3\cdot 5 + (3\cdot 3 + 4\cdot 5)] : [2\cdot 4\cdot 3 + (3\cdot 3 + 4\cdot 5)]$

$= 59 : 53$

Alternatively III

1st mixture (water : milk) = a : b

2nd mixture (water : milk) = c : d

Required ratio = $[(a+b)c + (c+d)a] : [(a+b)d + (c+d)b]$

Here,

1st mixture (water : milk) = 3 : 4

2nd mixture (water : milk) = 5 : 3

$\therefore$ Required ratio = $[(3+4)5 + (5+3)3] : [(3+4)3 + (5+3)4]$

$= 59 : 53$

Alternatively IV

1st mixture (water : milk) = 3 : 4

New mixture (water : milk) = x : y [let]

2nd mixture (water : milk) = 5 : 3

Then, ratio of the quantities taken from both vessels

$= (3+4)(5\cdot y - 3\cdot x) : (5+3)(4\cdot x - 3\cdot y)$

$= 35y - 21x : 32x - 24y$

$= 1 : 1$

[$\because$ equal quantities have been taken from both the vessels]

$\Rightarrow \quad 59y = 53x$

$\Rightarrow \quad x : y = 59 : 53$

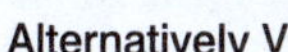

Alternatively V

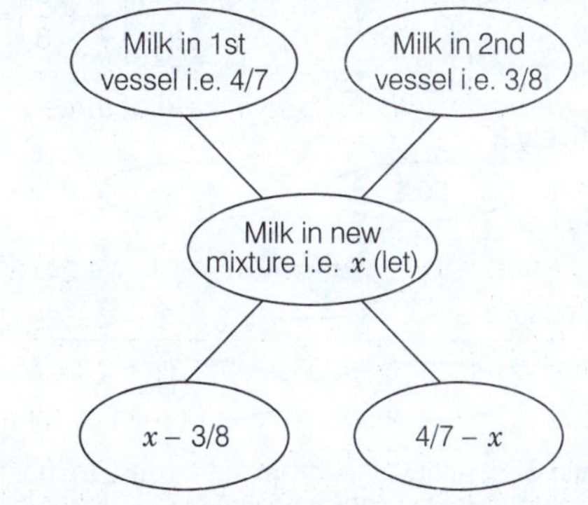

$\Rightarrow \left(x - \frac{3}{8}\right) : \left(\frac{4}{7} - x\right) = 1 : 1$ [as given]

$\Rightarrow \quad x - \frac{3}{8} = \frac{4}{7} - x \Rightarrow x = \frac{53}{112}$

Hence, quantity of water in 3rd vessel $= \frac{53}{112}$

Then, quantity of milk in 3rd vessel $= 1 - \frac{53}{112} = \frac{59}{112}$

$\therefore$ Ratio of milk and water in 3rd vessel $= \frac{59}{112} : \frac{53}{112} = 59 : 53$

Problem 14 **The ratio of spirit and water in two mixtures of 20 L and 36 L is 3 : 7 and 7 : 5 respectively. Both the mixtures are mixed together. Now, what is the ratio of the spirit and water in the new mixture?**

[SSC(10 + 2-DEO/LDC)-11, CDS-17]

Sol. Quantity of spirit in 1st mixture $= 20 \times \frac{3}{10} = 6$

Quantity of water in 1st mixture $= 20 \times \frac{7}{10} = 14$

and quantity of spirit in 2nd mixture $= 36 \times \frac{7}{12} = 21$

Quantity of water in 2nd mixture $= 36 \times \frac{5}{12} = 15$

Hence, required ratio $= (6 + 21) : (14 + 15) = 27 : 29$

Alternatively I

1st mixture (spirit : water)

$= 3 : 7 = (3 : 7) \times [(7 + 5) \times \mathbf{5}] = 180 : 420$

2nd mixture (spirit : water)

$= 7 : 5 = (7 : 5) \times [(3 + 7) \times \mathbf{9}] = 630 : 450$

$\therefore$ Required ratio (spirit : water) $= (180 + 630) : (420 + 450)$

$= 810 : 870 = 27 : 29$

Note *If we multiply 1st ratio by sum of terms of 2nd ratio and vice-versa, it means, we are taking equal quantity from both mixtures. Per contra, we have to take quantities in ratio 20 : 36 i.e. 5 : 9, therefore, we multiplied them by 5 and 9 respectively.*

Alternatively II

1st mixture (spirit: water) $= 3 : 7$

New mixture (spirit: water) $= x : y$ [let]

2nd mixture (spirit: water) $= 7 : 5$

Then, required ratio $= (3 + 7)(7 \cdot y - 5 \cdot x) : (7 + 5)(7 \cdot x - 3 \cdot y)$

$= 70y - 50x : 84x - 36y = 20 : 36$

[$\because$ quantities have been taken from both the vessels in ratio 20 : 36]

$\Rightarrow \quad 20(84x - 36y) = 36(70y - 50x)$

$\Rightarrow \quad 14x - 6y = 21y - 15x$

$\Rightarrow \quad 29x = 27y$

$\Rightarrow \quad x : y = 27 : 29$

Alternatively III

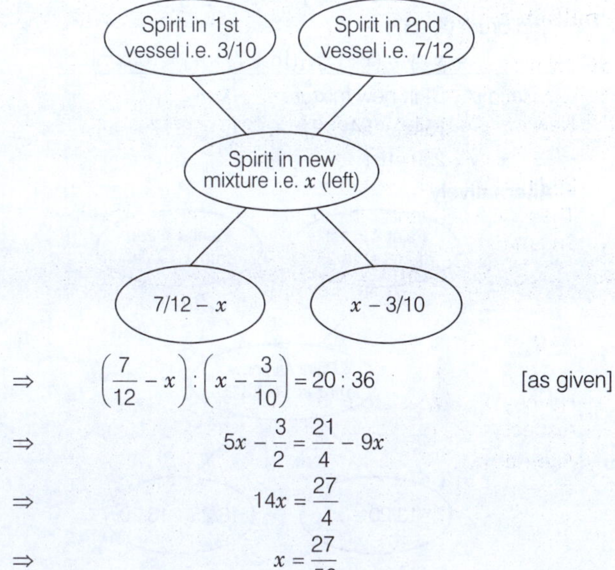

$\Rightarrow \quad \left(\frac{7}{12} - x\right) : \left(x - \frac{3}{10}\right) = 20 : 36$ [as given]

$\Rightarrow \quad 5x - \frac{3}{2} = \frac{21}{4} - 9x$

$\Rightarrow \quad 14x = \frac{27}{4}$

$\Rightarrow \quad x = \frac{27}{56}$

Hence, quantity of spirit in new mixture $= \frac{27}{56}$

Then, quantity of water in new mixture $= 1 - \frac{27}{56} = \frac{29}{56}$

$\therefore$ Ratio of spirit and water in new mixture $= \frac{27}{56} : \frac{29}{56} = 27 : 29$

Alternatively IV

1st mixture (spirit : water) $= a : b$

2nd mixture (spirit : water) $= c : d$

Required ratio $= \left\{(a + b)c + (c + d)a.\frac{m}{n}\right\}$

$: \left\{(a + b)d + (c + d)b.\frac{m}{n}\right\}$

[where, $m : n$ = ratio of quantity taken from 1st mixture to that from 2nd mixture]

Here,

1st mixture (spirit : water) $= 3 : 7$

2nd mixture (spirit : water) $= 7 : 5$

$\therefore$ Required ratio $= \left[(3 + 7)7 + (7 + 5) \times 3 \cdot \frac{20}{36}\right]$

$: \left[(3 + 7)5 + (7 + 5) . 7 \times \frac{20}{36}\right]$

$= (70 + 20) : \left(50 + \frac{140}{3}\right)$

$= 270 : 290 = 27 : 29$

Problem 15 There are two drums, each containing a mixture of paints A and B. In drum 1, A and B are in the ratio 18 : 7. The mixtures from drums 1 and 2 are mixed in the ratio 3 : 4 and in this final mixture, A and B are in the ratio 13 : 7. In drum 2, then A and B were in the ratio. *[CAT-18]*

Sol. 1st drum $(A : B) = 18 : 7 = (18 : 7) \times [(x + y) \times \mathbf{3}]$

$= 54(x + y) : 21(x + y)$

2nd drum $(A : B)$

$= x : y = (x : y) \times [(18 + 7) \times \mathbf{4}] = 100x : 100y$

$\therefore$ Ratio $(A : B)$ in new mixture

$= 154x + 54y : 21x + 121y = 13 : 7$

$\Rightarrow \quad x : y = 239 : 161$

Alternatively

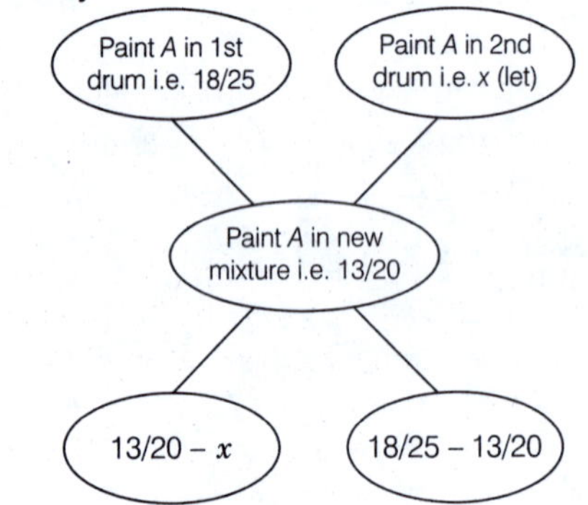

$\Rightarrow \quad \left(\frac{13}{20} - x\right) : \left(\frac{18}{25} - \frac{13}{20}\right) = 3 : 4$ [as given]

$\Rightarrow \quad x = \frac{239}{400}$

Hence, quantity of paint A in drum 2 $= \frac{239}{400}$

$\therefore$ Quantity of paint A in drum 2 $= 1 - \frac{239}{400} = \frac{161}{400}$

$\therefore$ Ratio of A and B in drum 2 $= \frac{239}{400} : \frac{161}{400} = 239 : 161$

Note *Try also with other approaches (supra).*

Problem 16 A man mixes 30 L of a mixture of milk and water with 50 L of another mixture of milk and water. Find the percentage of water in the final mixture, if 1st mixture contains 20% water and 2nd contains 40% water.

Sol. Quantity of water in 1st mixture = 20% of 30 = 6 L

and quantity of water in 2nd mixture = 40% of 50 = 20 L

Total quantity of water in new mixture = 6 + 20 = 26 L

$\therefore$ Percentage of water in new mixture $= \left(\frac{26}{30 + 50} \times 100\right)\%$

$= 32.5\%.$

Alternatively I

Ist mixture (milk : water) $= 4 : 1 = (4 : 1) \times [(3 + 2)3] = 60 : 15$

2nd mixture (milk : water) $= 3 : 2 = (3 : 2) \times [(4 + 1)5] = 75 : 50$

$\therefore$ Required ratio (milk : water) $= (60 + 75) : (15 + 50)$

$= 135 : 65 = 27 : 13$

Then, percentage of water in new mixture

$= \left(\frac{13}{27 + 13} \times 100\right)\% = 32.5\%$

Alternatively II

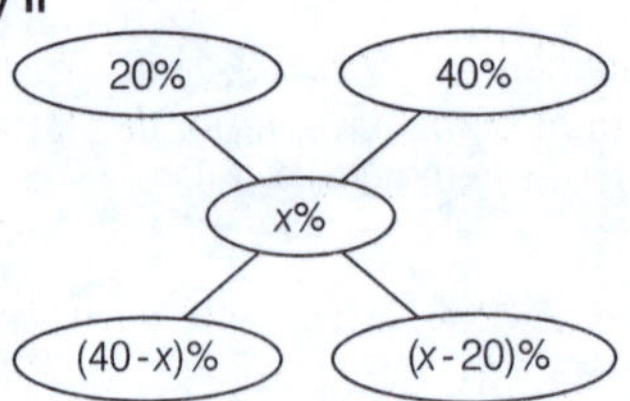

Then, ratio of quantity taken from 1st mixture to that from 2nd mixture $= (40 - x) : (x - 20) = 30 : 50$ [as given]

$\Rightarrow \quad 3x - 60 = 200 - 5x$

$\Rightarrow \quad 8x = 260$

$\Rightarrow \quad x = 32.5\%$

Alternatively III

1st mixture (milk : water) = 4 : 1

2nd mixture (milk : water) = 3 : 2

Here, $m : n$ = ratio of quantity taken from 1st mixture to that from 2nd mixture = 30 : 50 = 3 : 5

$\therefore$ Ratio of milk to water in new mixture

$= \left[(4 + 1)3 + (3 + 2) \cdot 4 \cdot \frac{3}{5}\right] : \left[(4 + 1)2 + (3 + 2) \cdot 1 \cdot \frac{3}{5}\right] = 27 : 13$

Then, percentage of water in new mixture

$= \left(\frac{13}{27 + 13} \times 100\right)\% = 32.5\%$

Problem 17 Three containers A, B and C are having mixtures of milk and water in the ratio 1 : 5, 3 : 5 and 5 : 7 respectively. If the capacities of the containers are in the ratio 5 : 4 : 5, find the ratio of milk to water, if the mixtures of all the three containers are mixed together? *[MAT-19, 14, 05]*

Sol. Let capacities of containers A, B and C be 5, 4 and 5 L respectively.

Then, total quantity of milk

$= \left(5 \times \frac{1}{6}\right) + \left(4 \times \frac{3}{8}\right) + \left(5 \times \frac{5}{12}\right) = \frac{53}{12}$ L

and quantity of water $= (5 + 4 + 5) - \frac{53}{12} = \frac{115}{12}$ L

Hence, required ratio $= \frac{53}{12} : \frac{115}{12} = 53 : 115$

Alternatively

LCM of (1 + 5), (3 + 5) and (5 + 7) = 24

1st mixture of **5** units (milk : water)

$= 1 : 5 = (1 : 5) \times \mathbf{5} \times 4$ $[\because 24 \div (1 + 5) = 4]$

$= 20 : 100$

2nd mixture of **4** units (milk : water)

$= 3 : 5 = (3 : 5) \times \mathbf{4} \times 3$ $[\because 24 \div (3 + 5) = 3]$

$= 36 : 60$

3rd mixture of **5** units (milk : water)

$= 5:7 = (5:7) \times \mathbf{5} \times 2 \qquad [\because 24 \div (5+7) = 2]$

$= 50:70$

∴ Required ratio (water : milk)

$= (20 + 36 + 50) : (100 + 60 + 70)$

$= 106 : 230 = 53 : 115$

Note *If the containers are of equal capacities of or they have equal amount of mixtures, multiplication by units of mixtures is not required at all.*

Problem 18 A 20% Ethanol solution is mixed with another Ethanol solution, say, S of unknown concentration in the proportion 1 : 3. This mixture is then mixed with an equal volume of 20% Ethanol solution. If the resultant mixture is a 31.25% Ethanol, then the unknown concentration of S is ***[CAT-18]***

Sol. Let the volume of 1st and 2nd solution be 100 and 300.

Quantity of Ethanol after both solutions are mixed

$= 20 + S\%$ of $300 = 20 + 3S$

Now, this solution is mixed with equal volume i.e. 400 of 3rd solution having 20% Ethanol.

Then, quantity of Ethanol in the resultant mixture

$= 20 + 3S + 20\%$ of $400 = 31.25\%$ of $(400 + 400)$

$\Rightarrow \quad 3S = 150 \Rightarrow S = 50\%$

Alternatively

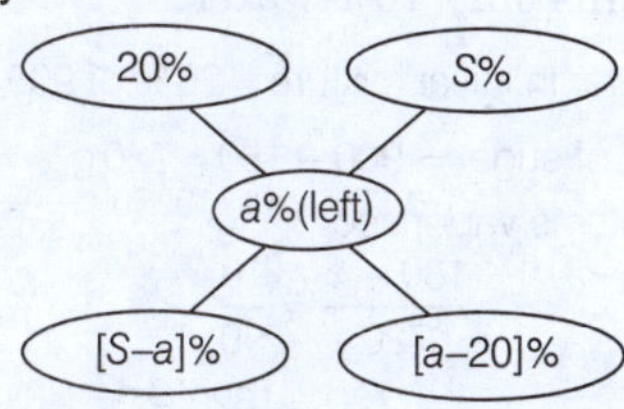

As given, $(S - a) : (a - 20) = 1 : 3$

$\Rightarrow \quad a = 5 + \dfrac{3S}{4}$

Further,

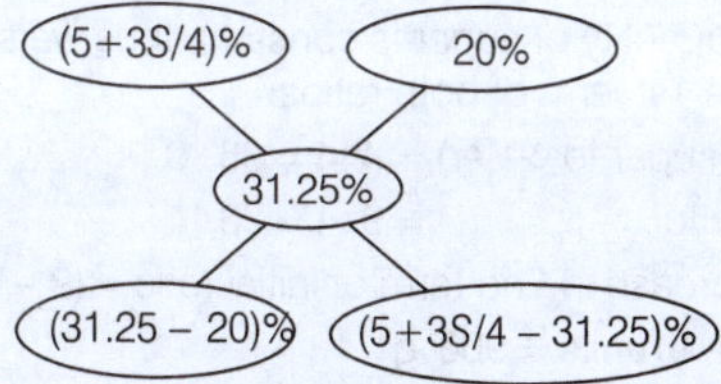

$\Rightarrow \quad 11.25\% : \left(\dfrac{3S}{4} - 26.25\right)\% = 1:1 \quad [\because$ equal volume is taken]

$\Rightarrow \quad S = 50\%$

Try also with other methods (supra).

Change in ratio of ingredients in a mixture when quantity of one or two ingredient(s)/ item(s) is/are increased or decreased.

Problem 19 The ratio of the quantities of an acid and water in a mixture is 1 : 3. If 5 L of acid is further added to the mixture, the new ratio becomes 1 : 2. The quantity of new mixture (in litres) is

Sol. Let initial quantities of acid and water be x litres and $3x$ litres respectively.

As per question, $\dfrac{x+5}{3x} = \dfrac{1}{2}$

$\Rightarrow \quad 2x + 10 = 3x \Rightarrow x = 10$

Hence, quantity of new mixture $= (x + 3x) + 5$

$= (4 \times 10) + 5 = 45$ L

Alternatively I

Recall the concepts of 'Change in ratio of two numbers when one number is altered' and 'Change in ratio of three numbers when one or two numbers are altered', elaborated in chapter 'Ratio, Proportion and Variation'.

Initial ratio (acid to water) = 1 : 3

Changed ratio = 1 : 2

Given that quantity of water is constant. Thus, we should equalise 2nd term of both ratios.

Initial ratio = 1 : 3 = 2 : 6

Changed ratio = 1 : 2 = 3 : 6

Clearly, increase in 1st term of initial ratio = (3 – 2) = 1 unit

Further given that quantity of acid has been increased by 5 L. It means, 1 unit = 5 L

∴ (3 + 6) units = 45 L

Hence, quantity of new mixture = 45 L

Alternatively II

You may proceed through options. Let options be 32, 45, 40, 42.

(a) 32 ⇒ quantity of initial mixture = 32 – 5 = 27

Then, quantities of acid and water = 6.75 L and 20.25 L respectively.

Now, $(6.75 + 5) : 20.25 \neq 1:2$

(b) 45 ⇒ quantity of initial mixture = 45 – 5 = 40

Then, quantities of acid and water = 10 L and 30 L respectively.

Now, (10 + 5) : 30 = 1 : 2

Hence, option (b) is correct.

Problem 20 In 40 L mixture of milk and water the ratio of milk to water is 7 : 1. In order to make the ratio of milk and water 3 : 1, the quantity of water (in litres) that should be added to the mixture will be ***[SSC(CGL-PT)-08, MAT-09, SSC(CGL-T-I)-17]***

Sol. Quantity of milk $= 40 \times \dfrac{7}{(7+1)} = 35$

and quantity of water = 40 – 35 = 5

Let x litres water is added. Then, $\dfrac{35}{5+x} = \dfrac{3}{1}$

$\Rightarrow \quad 15 + 3x = 35$

$\therefore \quad x = 6\dfrac{2}{3}$ L

Alternatively I

Initial ratio (milk to water) = 7 : 1

Changed ratio = 3 : 1

Given that quantity of milk is constant, thus, we should equalise Ist term of both ratios.

Initial ratio = 7 : 1 = 21 : 3

Changed ratio = 3 : 1 = 21 : 7

Clearly, increase in 2nd term of initial ratio = (7 – 3) = 4 units

As given,

(21 + 3) units = 40 L

$\therefore$ 4 units = $6\frac{2}{3}$ L

Alternatively II

You may proceed through options. Let options be 6, $6\frac{2}{3}$, ...

(a) 6 $\Rightarrow$ quantity of milk and water in initial mixture = 35 and 5
Now, 35 : (5 + 6) $\neq$ 3 : 1

(b) $6\frac{2}{3} \Rightarrow$ quantity of milk and water in initial mixture = 35 and 5

Now, $35 : \left(5 + 6\frac{2}{3}\right) = 3 : 1$

Hence, option(b) is correct.

Problem 21 **An alloy contains Copper, Zinc and Nickel in the ratio 5 : 3 : 2. The quantity of Nickel in kg must be added to 100 kg of this alloy to have the new ratio 5 : 3 : 3 is**

Sol. As given, (5 + 3 + 2) units = 100 kg

$\therefore$ 5 units = 50 kg, 3 units = 30 kg and 2 units = 20 kg

Hence, quantity of Copper = 50 kg, that of Zinc = 30 kg and that of Nickel = 20 kg.

Let x kg Nickel be added.

Then, $\frac{\text{quantity of Zinc in new alloy}}{\text{quanitity of Nickel in new alloy}} = \frac{3}{3}$

$\Rightarrow \frac{30}{20 + x} = \frac{3}{3}$

$\Rightarrow 20 + x = 30$

$\Rightarrow x = 10$ kg.

Alternatively I

Initial ratio (Copper : Zinc : Nickel) = 5 : 3 : 2

Changed ratio (Copper : Zinc : Nickel) = 5 : 3 : 3

Clearly, increase in 3rd term of initial ratio = (3 – 2) = 1 unit

It means quantity of Nickel has been increased by 1 unit.

Further, (5 + 3 + 2) units = 100 kg

$\therefore$ 1 unit = 10 kg.

Alternatively II

You may proceed through options. Let options be 20,10 ....

(a) 20 $\Rightarrow$ quantity of Copper = 50 kg, that of Zinc = 30 kg and that of Nickel = 20 kg in initial mixture.
Now, 50 : 30 : (20 + 20) $\neq$ 5 : 3 : 3

(b) 10 $\Rightarrow$ quantity of Copper = 50 kg, that of Zinc = 30 kg and that of Nickel = 20 kg in initial mixture.
Now, 50 : 30 : (20 + 10) = 5 : 3 : 3

Hence, option (b) is correct.

Problem 22 **What quantity of water should be added to 3 L of 10% solution of salt, so that it becomes a 5% salt solution ?** ***[ECGC of India Ltd. (PO)-01, CDS-21]***

Sol. Quantities of water and salt = 2.7 L and 0.3 L

Let x litres water be added.

Then, 5% of (3 + x) = 0.3

$\Rightarrow 3 + x = 6 \Rightarrow x = 3$ L

Alternatively I

Initial ratio (salt to water) = 1 : 9

Changed ratio = 1 : 19

Given that quantity of salt is constant and 1st term of both ratios are already equal.

Clearly, increase in 2nd term of initial ratio = (19 – 9) = 10 units

Given, (1 + 9) units = 3 L

$\therefore$ 10 units = 3 L

Hence, 3 L water should be added.

Alternatively II

The given solution has 10% salt and water to be added will have 0% salt concentration. Now,

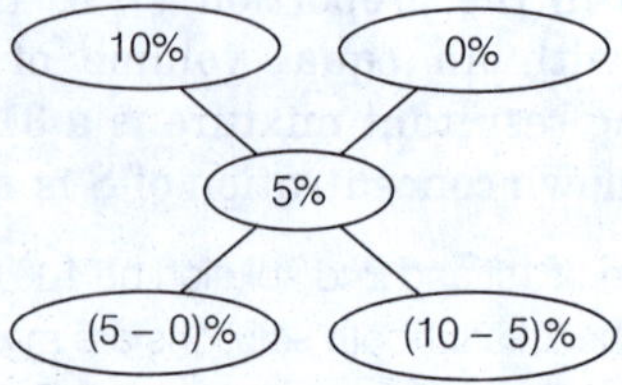

Hence, ratio of given solution to water(to be added) = 1 : 1

It implies that quantity of water = Quantity of given solution = 3 L

Problem 23 **In 900 g mixture of sugar and water, water is 20%. How much water must be vaporised so that the mixture contains only 10% water.**

Sol. Quantity of water in initial mixture = 20 % of 900 = 180 g

Then, quantity of sugar = 900 – 180 = 720 g

Let x gram water is vaporised.

Then, $\frac{180 - x}{720} = \frac{10}{90}$

$\Rightarrow 720 = 1620 - 9x$

$\Rightarrow x = 100$ g.

Alternatively

Initial ratio (sugar to water) = 4 : 1

Changed ratio = 9 : 1

Given that quantity of sugar is constant, thus we should equalise the 1st term of both ratios.

Initial ratio (sugar to water) = 4 : 1 = 36 : 9

Changed ratio = 9 : 1 = 36 : 4

Clearly, decrease in 2nd term of initial ratio = (9 – 4) = 5 units

Given, (36 + 9) units = 900 g

$\therefore$ 5 units = 100 g

Hence, 100 g water should be vaporised.

Adulteration in Ingredient to make a Profit

Problem 24 **In what ratio must water be mixed with milk to gain $16\frac{2}{3}$% on selling the mixture at cost price?** ***[LIC(AAO)-03]***

Sol. It is very clear that water appears to be milk when water is added to milk and the mixture so obtained is sold at the rate of milk i.e. profit is due to mixing of water. Hence, to earn a profit of $16\frac{2}{3}$%, $16\frac{2}{3}$ units (say gram) must be added to 100 g of milk.

$\therefore$ Required ratio $= 16\frac{2}{3} : 100 = 1 : 6$

Alternatively

Let CP of 1 L milk be ₹ 100

Then, SP of 1 L milk = ₹ 100 at a profit of $\frac{50}{3}\%$

[$\because$ milk is sold at cost price]

$\therefore$ CP of 1 L of mixture

$$= 100 \times \frac{100}{\left(100 + \frac{50}{3}\right)}$$

$$= \frac{600}{7} = 85\frac{5}{7}$$

Now,

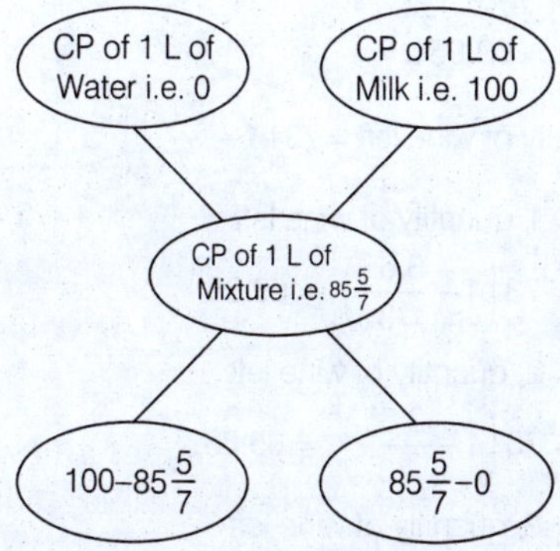

$\therefore$ Required ratio $= 14\frac{2}{7} : 85\frac{2}{7} = 1 : 6$

Problem 25 **A person has a chemical of ₹ 40 per L. In what ratio should water be mixed in that chemical so that on selling the mixture at ₹ 30 per L he may earn a profit of 20%.**

Sol. Profit% if mixture is sold at ₹ 40 per L [i.e. at cost price] = 60%

[$\because$ ₹ 30 $\Rightarrow$ 120%, $\therefore$ ₹ 40 $\Rightarrow$ 160%]

Hence, to earn a profit of 60%, 60 units water must be added in the 100 units of the chemical.

$\therefore$ Required ratio = 60 : 100 = 3 : 5

Hence, 3 units of water must be added to 5 units of chemical.

Alternatively

CP of the mixture $= 30 \times \frac{100}{120} =$ ₹ 25

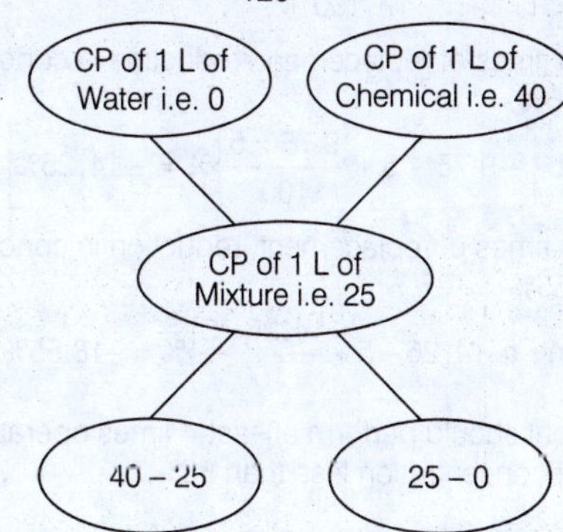

$\therefore$ Required ratio = 15 : 25 = 3 : 5

Problem 26 **In what ratio must water costing ₹ 8 per L be mixed with milk costing ₹ 44 per L so as to earn 10% profit on selling the mixture at cost of milk.**

Sol. Cost of mixture $= 44 \times \frac{100}{110} = 40$

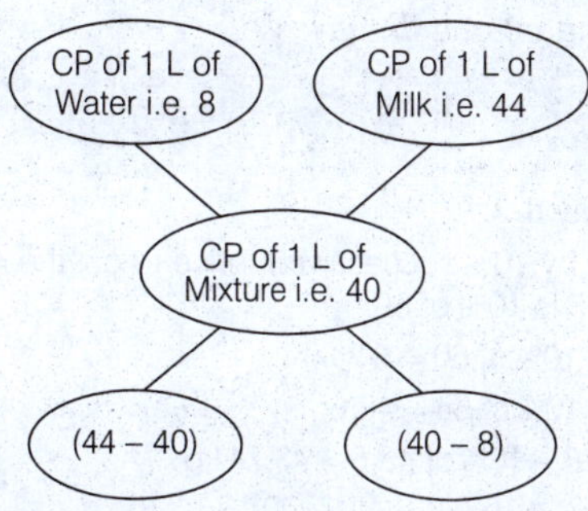

$\therefore$ Required ratio = 4 : 32 = 1 : 8

Replacement of a Part of Ingredient by Another Ingredient

Problem 27 **A container contains 60 kg of milk. From this container 6 kg of milk was taken out and replaced by water. This process was repeated further two times. The amount of milk left in the container is ?**

[SSC (10 + 2-LDC/DEO)-12, SSC(CGL-T-II)-17]

Sol. Out of 60 kg of milk, 6 kg is taken out and replaced by water.

Then, ratio of milk to water = (60 – 6) : 6 = 54 : 6 = 9 : 1

Again, 6 kg of the mixture having milk and water in the ratio 9 : 1 is taken out. It means, $\left(6 \times \frac{9}{(9+1)}\right)$ i.e. 5.4 kg milk and $\left(6 \times \frac{1}{(9+1)}\right)$ i.e. 0.6 kg water are taken out. But 6 kg water is added too.

$\therefore$ Ratio of milk to water = (54 – 5.4) : (6 + 5.4) = 48.6 : 11.4

= 8.1 : 1.9 = 81 : 19

Further, 6 kg of mixture having milk and water in ratio 81 : 19 is taken out. It means $\left(6 \times \frac{81}{(81+19)}\right)$ i.e. 4.86 kg milk and $\left(6 \times \frac{19}{(81+19)}\right)$ i.e. 1.14 kg water are taken out. But, 6 kg water is also added.

Now, amount of milk left in the container = 48.6 – 4.86

= 43.74 kg

Alternatively I

The amount of milk left in the container $= 60\left(1 - \frac{6}{60}\right)^3$

$$= 60\left(\frac{9}{10}\right)^3 = 60 \times \frac{729}{1000}$$

= 43.74 kg

Generic version :

milk left = initial quantity of milk

$$\times \left(1 - \frac{\text{Quantity of milk or mixture drawn}}{\text{initial quantity of milk}}\right)^n,$$

where n = number of operations

Note *If A kg milk is drawn 1st time and replaced by A kg of water. At 2nd time, B kg of mixture is drawn and replaced by equal quantity of water. Further, 3rd times C kg of mixture is drawn and replaced by C kg of water and so on. Then, amount of milk left =*
Initial amount of milk (say, x)
$$\times\left(1-\frac{A}{x}\right)\left(1-\frac{B}{x}\right)\left(1-\frac{C}{x}\right)\left(1-\frac{D}{x}\right)\ldots\ldots$$

Alternatively II

Reduce 60 by 10% successively thrice to get the required answer as 6 is 10% of 60.

$\Rightarrow$ 60 – 10% of 60 = 54

Then, 54 – 10% of 54 = 48.6

Further, 48.6 – 10% of 48.6 = 43.74 kg.

Or

Recall the concept of successive discounts.

$$-10-10+\frac{10\times 10}{100}=-19\%$$

Then, $-19-10+\frac{19\times 10}{100}=27.1\%$

Hence, milk left = (100 – 27.1)% of 60 = $60\times\frac{72.9}{100}=43.74$ kg.

Problem 28 There is 81 L pure milk in a container. 1/3rd of milk is replaced by water in the container. Again 1/3rd of mixture is extracted and equal amount of water is added. What is the ratio of milk to water in the new mixture. *[SSC (SO-Commercial Audit)-05]*

Sol. Milk left in container after 2 operations

$$=81\left(1-\frac{1}{3}\right)^2=81\times\frac{2}{3}\times\frac{2}{3}=36\text{ L}$$

$\therefore$ Ratio of milk to water = 36 : (81 – 36) = 36 : 45 = 4 : 5

Problem 29 A vessel contains 40 L of milk. 4 L of milk were taken out of the vessel and replaced by water. Further, 8 L of mixture were withdrawn and again replaced by water. How much milk is now left in the vessel?

Sol. Amount of milk left = $40\times\left(1-\frac{4}{40}\right)\left(1-\frac{8}{40}\right)$

$$=40\times\frac{9}{10}\times\frac{4}{5}=28.8\text{ L}$$

Problem 30 8 L are drawn from a cask full of wine and is then filled with water. This operation is performed three more times. The ratio of the quantity of wine now left in cask to that of water is 16 : 65. How much wine did the cask hold originally ? *[NIFT-03]*

Sol. Let the quantity of wine in the cask originally be x litres.

Then, quantity of wine left in cask after 4 operations

$$=x\left(1-\frac{8}{x}\right)^4$$

Now, $\frac{\text{quantiy of wine left}}{\text{quantity of wine originally}}=\frac{16}{(16+65)}$

$$\Rightarrow \frac{x\left(1-\frac{8}{x}\right)^4}{x}=\frac{16}{81}$$

$$\Rightarrow \left(1-\frac{8}{x}\right)^4=\left(\frac{2}{3}\right)^4$$

$$\Rightarrow 1-\frac{8}{x}=\frac{2}{3}\Rightarrow\frac{8}{x}=\frac{1}{3}$$

$$\therefore x=24\text{ L}$$

Problem 31 Gopal has 73 L of wine in a drum. He replaces 3.65 L of it with water and keeps doing so till the time the concentration of wine is less than 85%. The minimum number of operations that Gopal has to perform is *[MAT-13]*

Sol. 85% of 73 L = 62.05 L

Now, quantity of wine left = $73\left(1-\frac{3.65}{73}\right)^n$

In case, $n = 1$, quantity of wine left

$$=73\left(1-\frac{3.65}{73}\right)^1=69.35$$

In case, $n = 2$, quantity of wine left

$$=73\left(1-\frac{3.65}{73}\right)^2=65.88$$

In case, $n = 3$, quantity of wine left

$$=73\left(1-\frac{3.65}{73}\right)^3=62.59$$

In case, $n = 4$, quantity of wine left

$$=73\left(1-\frac{3.65}{73}\right)^4=59.46$$

Obviously, 59.46 is less than 85% of 73. Hence, Gopal should perform atleast 4 times operations to get wine having concentration less than 85%.

Alternatively

Reduce 73 by 5% successively so that the drum contains about 85% wine-concentration as 3.65 is 5% of 73.

If one time, replacement is made, reduction in concentration of wine = 5%

In case of 2 times of replacement, reduction in concentration of wine = 9.75%

$$\left[\text{Explanation}\left\{-5-5+\frac{5\times 5}{100}\right\}\%=-9.75\%\right]$$

In case of 3 times of replacement, reduction in concentration of wine = 14.26%

$$\left[\text{Explanation}\left\{-9.75-5+\frac{9.75\times 5}{100}\right\}\%=-14.26\%\right]$$

In case of 4 times of replacement, reduction in concentration of wine = 18.55%

$$\left[\text{Explanation}:\left\{-14.26-5+\frac{14.26\times 5}{100}\right\}\%=-18.55\%\right]$$

Hence, Gopal should perform atleast 4 times operations to get wine having concentration less than 85%.

Note *Explanations are based on concept of the successive discounts.*

Replacement of a Part of Mixture by one of its Ingredients

Problem 32 A Can contains a mixture of 175 ml water and 700 ml alcohol. Gopal takes out 10% of the mixture and substitutes it by water of the same amount. The process is repeated once again. The percentage of water in the mixture is now **[CAT-18]**

Sol. Here, 10% of the mixture = 10% of (175 + 700) = 87.5 ml.

87.5 ml of the mixture having alcohol and water in the ratio 4 : 1 is taken out . It means, $\left(87.5 \times \frac{4}{4+1}\right)$ i.e. 70 ml alcohol and $\left(87.5 \times \frac{1}{4+1}\right)$ i.e. 17.5 ml water are taken out.

But 87.5 ml water is added too.

$\therefore$ Ratio of alcohol to water = (700 – 70) : (175 + 70)

= 630 : 245 = 18 : 7

Further, 87.5 ml of mixture having alcohol and water in ratio 18 : 7 is taken out. It means $\left(87.5 \times \frac{18}{18+7}\right)$ i.e. 63 ml alcohol and $\left(87.5 \times \frac{7}{18+7}\right)$ i.e. 24.5 ml water are taken out.

But, 87.5 ml water is also added.

Now, amount of alcohol left in the Can = 630 – 63 = 567 ml

Then, amount of water = 875 – 567 = 308 ml

$\therefore$ Required percentage = $\left(\frac{308}{875} \times 100\right)\%$ = 35.2%

Alternatively

At every process, the amount of alcohol decreases by 10%. Initially, amount of alcohol was 700 ml, but it remains 90% i.e. 630 after 1st process. After 2nd process, it further decreases by 10% i.e. remains 81%. It means amount of alcohol remained after 2nd process

$$= \text{Initial amount of alcohol} \times \left(1 - \frac{\text{Quantity of mixture drawn}}{\text{Initial quantity of mixture}}\right)^2$$

In other words, the method/concept for 'Replacement of a part of ingredient by another ingredient' works in this case also.

Now, alcohol left $= 700 \times \left(1 - \frac{10\%}{100\%}\right)^2 = 567$ ml

Then, amount of water = 875 – 567 = 308 ml

$\therefore$ Required percentage = $\left(\frac{308}{875} \times 100\right)\%$ = 35.2%

Problem 33 A Can contains a mixture of two liquids A and B in the ratio 7 : 5, where 9 L of mixture are drained off and the Can is filled with B, the ratio of A and B becomes 7 : 9. How many litres of liquid A was contained by the Can initially?

Sol. Let quantities of liquid A and B were $7x$ and $5x$ respectively.

9 L from the mixture were drained off, it means $\left(9 \times \frac{7}{7+5}\right)$

i. e. 5.25 L of liquid A and (9 – 5.25) i.e. 3.75 L of liquid B were drained off. Further, 9 L of liquid B have been added too.

Now, after filling with liquid B,

$$\frac{\text{Quantity of liquid } A}{\text{Quantity of liquid } B} = \frac{7}{9}$$

$$\Rightarrow \quad \frac{7x - 5.25}{(5x - 3.75) + 9} = \frac{7}{9}$$

$$\Rightarrow \quad 63x - 47.25 = 35x + 36.75 \Rightarrow 28x = 84$$

$$\therefore \quad x = 3$$

Hence, quantity of liquid A in Can initially $= 7x = 7 \times 3 = 21$ L.

Alternatively I

Quantity of liquid A drained off $= \left(9 \times \frac{7}{7+5}\right) = 5.25$ L

It is obvious that decrease in quantity of liquid A = Increase in quantity of B = 5.25 L

As given, initial ratio (liquid A to B) = 7 : 5 = 28 : 20

and changed ratio (liquid A to B) = 7 : 9 = 21 : 27

Here, decrease in quantity of A i.e. (28 – 21) = increase in quantity of B i.e. (27 – 20) = 7 units

Hence, 7 units = 5.25

$\therefore$ 28 units = 21 L = Quantity of A in Can initially.

Alternatively II

Recall the concept of "Change in ratio of numbers when one is increased and another is decreased or *vice-versa* either by the same number or by different numbers" envisaged in chapter 'Ratio, Proportion and Variation'.

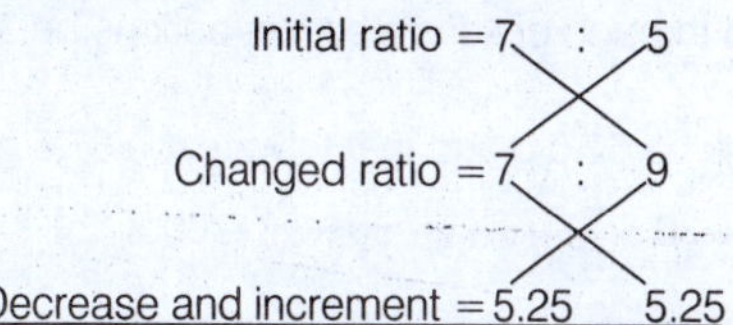

Now, [(7 × 9) – (5 × 7)] units = [(9 × 5.25) + (7 × 5.25)]

$\Rightarrow$ 28 units = 84

$\therefore$ 7 units = 21

Hence, quantity of A in Can initially = 21 L

Problem 34 A Can contains 5 parts of milk and 3 parts of water. How much of the mixture should be withdrawn and replaced by water so that the Can may have equal amount of milk and water? **[SSC(CGL-T-I)-17, CLAT-17]**

Sol. Let quantities of milk and water are $5x$ and $3x$ respectively. On withdrawing y litres from the mixture, $\left(y \times \frac{5}{8}\right)$ litres of milk and $\frac{3y}{8}$ liters of water would be withdrawn. But, given that y litres water is also added.

Now, quantity of milk $= 5x - \frac{5y}{8}$

and quantity of water $= \left(3x - \frac{3y}{8} + y\right)$

As given, $\left(5x - \frac{5y}{8}\right) : \left(3x - \frac{3y}{8} + y\right) = 1 : 1$

$$\Rightarrow \quad \left(5x - \frac{5y}{8}\right) = \left(3x - \frac{3y}{8} + y\right)$$

$\Rightarrow \quad 2x = \dfrac{5y}{4}$

$\Rightarrow \quad \dfrac{x}{y} = \dfrac{5}{8} \Rightarrow \dfrac{y}{x} = \dfrac{8}{5}$

$\Rightarrow \quad \dfrac{y}{8x} = \dfrac{1}{5}$

Hence, $\frac{1}{5}$ of the mixture must be drawn and replaced with water.

Alternatively

Let x litres be drawn from the mixture.

$\therefore$ Amount of milk drawn $= x \times \dfrac{5}{8} = \dfrac{5x}{8}$

Now, Initial ratio = 5 : 3

Changed ratio = 1 : 1

Decrease and increment $= \dfrac{5x}{8} \quad \dfrac{5x}{8}$

Now, $[(5 \times 1) - (3 \times 1)]$ units $= \left[\left(1 \times \dfrac{5x}{8}\right) + \left(1 \times \dfrac{5x}{8}\right)\right]$

$\Rightarrow \quad 2$ units $= \dfrac{10x}{8} \Rightarrow (5 + 3)$ units $= 5x$

Clearly, x litres are drawn from a mixture of $5x$ litres.

Hence, $\frac{1}{5}$ of the mixture must be drawn and replaced with water.

Alternatively II

You may proceed through options also. Let options be 30%, 20%...

Let quantity of milk be 62.5 g and that to water be 37.5 g.

(a) $30\% \Rightarrow (62.5 - 30 \times \frac{5}{8}) : (37.5 - 30 \times \frac{3}{8}) + 30 \neq 1:1$

(b) $20\% \Rightarrow (62.5 - 20 \times \frac{5}{8}) : (37.5 - 20 \times \frac{3}{8}) + 20 = 50 : 50 = 1:1$

Hence, option (b) is correct.

Problem 35 **Half of the volume of milk and water mixture of ratio 7 : 5 is converted into a mixture of ratio 3 : 1 by substitution (or replacement) method. The mixture of ratio 7 : 5 was formed from the mixture 7 : 3 by adding the water in it. If 240 L milk is required in the replacement method, what is the total amount of water added to prepare the mixture of 7 : 5 ?** ***[CAT-13]***

Sol. Clearly, 240 L from the mixture containing milk and water in ratio 7 : 5 be drawn and replaced by milk.

$\therefore$ Quantity of water drawn $= 240 \times \dfrac{5}{7+5} = 100$

Now,

Initial ratio (Milk to water) 7 : 5

Changed ratio (Milk to water) 3 : 1

Increase and decrease = 100 100

Now, $[(5 \times 3) - (7 \times 1)]$ units $= [(3 \times 100) + (1 \times 100)]$

$\Rightarrow \quad 8$ units $= 400$

$\Rightarrow \quad 1$ unit $= 50$ L

Clearly, ratio 7 : 3 becomes 7 : 5 when water is added.

Initial ratio (Milk to water) = 7 : 3

Changed ratio (Milk to water) = 7 : 5

Here, quantity of milk is constant, and quantity of water increases by $(5 - 3)$ i.e. 2 units.

Since 1 unit = 50 L, 2 units = 100 L

Now, if we talk about whole volume of mixture, the required answer $= 2 \times 100 = 200$ L

Replacement of a part of Ingredient by a Mixture

Problem 36 **A vessel contains 100 L of Nitric acid. 10 L from this vessel is taken out and replaced with a mixture containing Nitric acid and water in ratio 3 : 2. This process is repeated two times more, how much amount of Nitric acid will be left in the vessel ?**

Sol. Clearly, 10% of Nitric acid is taken out 1st time and $\left(10 \times \dfrac{3}{5}\right)$ i.e. 6 L of Nitric acid is added.

Hence, amount of Nitric acid in the vessel after 1st operation

$= 90\%$ of $100 + 6 = 96$ L

Further, amount of Nitric acid in the vessel after 2nd operation 90% of $96 + 6 = 92.4$ L

Now, amount of Nitric acid in the vessel after 3rd operation

$= 90\%$ of $92.4 + 6$

$= 89.16$ L

Note *we can continue the above pattern as per number of operations*

Problem 37 **There are two containers, the 1st contains 500 ml of alcohol while the 2nd contains 500 ml of water. Three cups of alcohol from the 1st container is taken out and is mixed well in the 2nd container. Then, three cups of this mixture is taken out and is mixed in the 1st container. Let A denote the proportion of water in the 1st container and B denote the proportion of alcohol in the 2nd container. Then,** ***[CAT-98]***

(a) $A > B$

(b) $A < B$

(c) $A = B$

(d) Cannot be determined

Sol. Let capacity of each cup be 100 ml.

Then, quantity of alcohol in the 1st container after withdrawal of three cups of alcohol = 500 − 300 = 200 ml

Then, ratio of alcohol and water in the 2nd container 300 : 500 = 3 : 5. It implies that concentration of alcohol in the 2nd container = 37.5%.

Quantity of alcohol and water in 300 ml of the mixture which is to be mixed in the 1st container = 112.5 ml and 187.5 as we have divided 300 ml in ratio 3 : 5.

Now, ratio of alcohol and water in the 1st container

$= (200 + 112.5) : 187.5 = 5 : 3$

It implies that concentration of water in the 1st container

$= 37.5\%$

Hence, option(c) is correct.

Application of Alligation-Rule in chapter 'Profit and Loss'

Suitable illustrations have been incorporated in chapter 'Profit and Loss'.

Application of Alligation-Rule in chapter 'Simple Interest'

Suitable illustrations have been incorporated in chapter 'Simple Interest'.

Application of Alligation-Rule in chapter 'Time and Distance'

Suitable illustrations have been incorporated in chapter 'Time and Distance'.

Application of Alligation-Rule in chapter 'Average'

Suitable illustrations have been incorporated in chapter 'Average'.

Miscellaneous

Problem 38 Three vessels contain water. The 1/3rd of water of the 1st vessel is poured into 2nd, and then 1/4th of water now in the 2nd vessel is poured into the 3rd and finally 1/10th of water now in 3rd vessel is poured into the 1st. Then, each vessel contains 9 L. How much water was there originally in each vessel?

Sol. The best way is to proceed through options. Let options be (a) 18, 8 and 4 L (b) 12, 8 and 7 L, ...

(a) $\frac{1}{3}$ of water in 1st vessel $= \frac{1}{3} \times 18 = 6$

$\therefore$ Water in 2nd vessel $= 8 + 6 = 14$

Now, $\frac{1}{4}$th of 2nd vessel $= \frac{14}{4} = 3.5$

$\therefore$ Water in 3rd vessel $= 4 + 3.5 = 7.5$

Further, $\frac{1}{10}$ of water in 3rd vessel $= \frac{7.5}{10} = 0.75$

$\therefore$ Water in 1st vessel $= (18 - 6) + 0.75 \neq 9$

(b) $\frac{1}{3}$ of water in 1st vessel $= \frac{1}{3} \times 12 = 4$

$\therefore$ Water in 2nd vessel $= 8 + 4 = 12$

Now, $\frac{1}{4}$th of 2nd vessel $= 12 \times \frac{1}{4} = 3$

$\therefore$ Water in 3rd vessel $= 7 + 3 = 10$

Further, $\frac{1}{10}$ of water in 3rd vessel $= \frac{10}{10} = 1$

$\therefore$ Water in 1st vessel $= (12 - 4) + 1 = 9$

and water in 2nd vessel $= 12 - 3 = 9$

and water in 3rd vessel $= 10 - 1 = 9$

Hence, option (b) is correct.

WORKOUT

1. A milkman bought 15 kg of milk and mixed 3 kg of water in it. If the price per kg of the mixture becomes ₹ 22, what is cost price of the milk per kg ? *[CDS-12]*

(a) ₹ 28 (b) ₹ 26.40
(c) ₹ 24 (d) ₹ 22.60

2. In a mixture of 80 L, the ratio of milk and water is 3 : 1. If the ratio of milk and water is to be 2 : 3, then how much amount of water is to be further added ?

(a) 70 L (b) 80 L (c) 100 L (d) 140 L

3. In what ratio should tea at ₹ 350 per kg be mixed with tea at ₹ 400 per kg in order that the mixture be worth ₹ 380 per kg ?

(a) 3 : 2 (b) 2 : 3 (c) 1 : 2 (d) 2 : 1

4. What will be the ratio of petrol and kerosene in the final solution formed by mixing petrol and kerosene that are present in three vessels in the ratio 4 : 1, 5 : 2 and 6 : 1 respectively, *[MAT-14, 19]*

(a) 166: 22 (b) 83 : 22
(c) 83 : 44 (d) None of these

5. A bottle is full of dettol. 1/3rd of it is taken out and then an equal amount of water is poured into the bottle to fill it. This operation is done four times. Find the final ratio of dettol and water in the bottle.

(a) 13 : 55 (b) 20 : 74
(c) 16 : 65 (d) 10 : 48

6. A milk vendor sells 10 L of milk from a Can containing 40 L of pure milk to the 1st customer. He then adds 10 L of water to the milk Can. He again sells 10 L of mixture to the 2nd customer and then he adds 10 L of water to the Can. Again he sells 10 L of mixture to the 3rd customer and then adds 10 L of water to the Can and so on. What amount of pure milk will the 5th customer receive ? *[IIFT-14]*

(a) 510/128 L (b) 505/128 L
(c) 410/128 L (d) 405/128 L

7. A bottle contains $\frac{3}{4}$ of milk and the rest water. How much of the mixture must be taken away and replaced by an equal quantity of water so that the mixture has half milk and half water ? *[MAT- 08, 07, SSC(CGL-T-II)-12]*

(a) 25% (b) 33.33%
(c) 45% (d) 50%

8. Tea costing ₹ 136 a kilogram is mixed with tea costing ₹ 141 a kilogram in the ratio 2:3. The cost of 1 kg of the mixture is *[SSC (SAS)-10]*

(a) ₹ 138 (b) ₹ 138.50
(c) ₹ 139 (d) ₹ 139.50

9. To x litres of an x% solution of acid, y litres of water is added to get $(x-10)$ % solution of acid. If $x > 20$, then value of y is *[LIC (AAO)-12]*

(a) $\frac{x^2}{100}$ (b) $\frac{10x^2}{x-10}$
(c) $\frac{10x}{x+10}$ (d) $\frac{10x}{x-10}$

10. A chemist has 10 L of a solution that is 10% Nitric acid by volume. He wants to dilute the solution to 4% strength by adding water. How many litres of water must be added ?

(a) 15 (b) 20 (c) 18 (d) 25

11. Two liquids A and B are in the ratio 5 : 1 in container 1 and 1 : 3 in container 2. In what ratio should the contents of the two containers be mixed so as to obtain a mixture of A and B in the ratio 1 : 1? *[CAT-96]*

(a) 2 : 3 (b) 4 : 3
(c) 3 : 2 (d) 3 : 4

12. A jar full of milk contains 40% water. A part of this milk is replaced by another containing 19% water and now the percentage of water is found to be 26%. The quantity of milk replaced is *[MAT-13]*

(a) 2/3 (b) 1/3 (c) 3/7 (d) 4/7

13. X and Y are the two alloys which were made by mixing Zinc and Copper in the ratio 6 : 9 and 7 : 11 respectively. If 40 g of alloy X and 60 g of alloy Y are melted and mixed to form another alloy Z, what is the ratio of Zinc and Copper in the new alloy Z? *[IIFT-14, SSC (CGL-T-II)-20]*

(a) 6 : 9 (b) 59 : 91
(c) 5 : 9 (d) 59 : 90

14. A jar contains 10 red marbles and 30 green ones. How many red marbles must be added to the jar so that 60% of the marbles will be red ? *[SSC (CGL-T-II)-12]*

(a) 25 (b) 30 (c) 35 (d) 40

15. Vijay purchased two different kinds of alcohol. In the 1st mixture, the ratio of alcohol to water is 3 : 4 and in the 2nd mixture it is 5 : 6. If he mixes, the two given mixtures and makes a 3rd mixture of 18 L in which the ratio of alcohol to water is 4 : 5, the quantity of the 1st mixture [whose ratio is 3 : 4] that is required to make 18 L of the 3rd kind of mixture is

(a) 6 (b) 7 (c) 8 (d) 9

16. There were two different copper alloys of total weight of 50 kg. The 1st contains 40% less copper than the 2nd. Determine the percentage of copper in the 1st and the 2nd alloys, if it is known that there were 6 kg in the 1st alloy and 12 kg in the 2nd.

(a) 20, 60 (b) 30, 70 (c) 25, 75 (d) 60, 40

17. A manufacturer has 200 L of acid solution which has 15% acid content. How many litres of acid solution with 30% acid content may be added so that the acid content in the resulting mixture will be more than 20 % but less than 25% ? *[XAT-10]*

(a) More than 100 L but less than 300 L
(b) More than 120 L but less than 400 L
(c) More than 100 L but less than 400 L
(d) More than 120 L but less than 300 L
(e) None of the above

18. A litre of water weighs a kilogram and a litre of another liquid weighs 1350 g. A mixture of two weighs 1250 g/L. The volume of water and the liquid respectively in a litre of mixture is *[MAT-13, CAT-19]*

(a) 2/7, 5/7 (b) 5/7, 2/7
(c) $\frac{3}{4}, \frac{1}{4}$ (d) $\frac{1}{4}, \frac{3}{4}$

19. Two qualities of tea are mixed in the ratio 4 : 1 and the mixture is sold at ₹ 72/kg for a profit of 12.5%. If the tea of the 2nd quality costs ₹ 3.25 more per kg than the tea of 1st quality, what is the cost per kg of the tea of the 1st quality ?

(a) ₹ 63.35 (b) ₹ 23.65
(c) ₹ 70.62 (d) ₹ 73.54

20. A container has 30 L of water. If 3 L of water is replaced by 3 L of spirit and this operation is repeated twice, what will be the quantity of water in the new mixture ? *[LIC (AAO)-13, MAT-19]*

(a) 24 L (b) 23 L (c) 24.3 L (d) 23.3 L
(e) None of these

21. A person has a chemical of ₹ 25 /L. In what ratio should water be mixed in that chemical so that after selling the mixture at ₹ 20/L, he may get a profit of 25%?

(a) 12 : 15 (b) 16 : 9
(c) 13 : 16 (d) 19 : 22

22. In a mixture of milk and water the proportion of water by weight was 75%. If in 60 g of mixture 15 g water was added, what would be the percentage of water ? [Weight in g] *[IOB (PO)-09]*

(a) 75% (b) 88%
(c) 90% (d) 100%
(e) None of the above

23. The concentration of petrol in three different mixtures [of petrol and kerosene] is 1/2, 3/5 and 4/5 respectively. If 2L, 3L and 1L are taken from these three different vessels and mixed, what is the ratio of petrol and kerosene in the new mixture ?

(a) 4 : 5 (b) 3 : 2 (c) 3 : 5 (d) 2 : 3

24. There are two alloys P and Q made up of Silver, Copper and Aluminium. Alloy P contains 45% Silver and rest Aluminium. Alloy Q contains 30% Silver, 35% Copper and rest Aluminium. Alloy P and Q are mixed in the ratio 1 : 4.5. The approximate percentages of Silver and Copper in the newly formed alloy is *[IIFT-15]*

(a) 33% and 29% (b) 29% and 26%
(c) 35% and 30% (d) None of these

25. Gopal sells fruit juice mixture using orange juice and pineapple juice. Gopal prepares this mixture by drawing out a jug of orange juice from a 10 L container filled with orange juice, and replacing it with pineapple juice. If Gopal draws out another jug of the resultant mixture and replaces it with pineapple juice, the container will have equal volumes of orange juice and pineapple juice. The volume of the jug in litres, is *[XAT-12]*

(a) 2 (b) > 2 and ≤ 2.5
(c) 2.5 (d) > 2.5 and ≤ 3
(e) ≥ 3

26. Two casks of 48 L and 42 L are filled with mixtures of milk and water, the proportions in the two casks being respectively 13 : 7 and 18 : 17. If the contents of the two casks be mixed and 20 L of water be added to the whole, what will be the proportion of milk and water in the resulting mixture ? **[MAT-12, MAT-11]**

(a) 5 :12 (b) 7 : 13
(c) 12 : 13 (d) 8 : 15

27. *A* and *B* are two alloys of Gold and Copper prepared by mixing metals in the ratio 7 : 2 and 7 : 11 respectively. If equal quantities of the alloys are melted to form a third alloy *C*, the ratio of Gold and Copper in *C* will be **[MAT-11, SSC (CGL-T-II)-12, LIC(AAO)-12]**

(a) 7 : 5 (b) 5 : 9
(c) 5 : 7 (d) 9 : 5

28. Product *M* is produced by mixing chemical *X* and chemical *Y* in the ratio of 5 : 4. Chemical *X* is prepared by mixing two raw materials A and B in the ratio of 1 : 3. Chemical *Y* is prepared by mixing raw materials B and C in the ratio of 2 : 1. Then the final mixture is prepared by mixing 864 units of product *M* with water. If the concentration of the raw material B in the final mixture is 50%, how much water had been added to product *M*? **[XAT-15]**

(a) 328 units (b) 368 units
(c) 392 units (d) 616 units
(e) Nono of these

29. Two vessels contain spirit of 0.5 and 0.75 concentrations. If 2 L from the 1st vessel and 3 L from the 2nd vessel are mixed, then what will be the ratio of the spirit and the water in the resultant solution ?

(a) 15 : 17 (b) 7 : 17
(c) 13 : 7 (d) 17 : 15

30. Two alloys of Aluminium have different percentages of Aluminium in them. The 1st one weighs 8 kg and the 2nd one weighs 16 kg. One piece each of equal weight was cut off from both the alloys and 1st piece was alloyed with the 2nd alloy and the 2nd piece alloyed with the 1st one. As a result, the percentage of Aluminium became the same in the resulting two new alloys. What was the weight of each cut-off piece ? **[IIFT-13]**

(a) 3.33 kg (b) 4.67 kg
(c) 5.33 kg (d) None of these

31. A vessel is filled to its capacity with pure milk. 9 L are withdrawn from the vessel and replaced with an equal amount of water. 9 L of the mixture is again withdrawn and then replaced with an equal amount of water. After these changes, the vessel contains 17.1 L of milk less than it did initially. Find the capacity of the vessel. **[CAT-14]**

(a) 120 L (b) 150 L
(c) 90 L (d) 75 L

32. A milkman mixes 20 L of water with 80L of milk. After selling 1/4th of this mixture, he adds water to replenish the quantity that he has sold. What is the current proportion of water to milk ?

(a) 2 : 3 (b) 1 : 2
(c) 1 : 3 (d) 3 : 4

33. In a 40 L mixture, Acetic acid and Sodium acetate are in the ratio 3 : 1, find the amount of Sodium acetate solution to be added to make the ratio 2 : 3. **[SBI (PO)-17]**

(a) 40 L (b) 35 L (c) 30 L (d) 20 L
(e) 15 L

34. 60 kg of an alloy *X* is mixed with 100 kg of an alloy *Y*. If alloy *X* has lead and tin in the ratio of 3 : 2 and alloy *Y* has Tin and Copper in the ratio of 1: 4, then the amount of Tin in the new alloy is **[CLAT-16]**

(a) 53 kg (b) 80 kg
(c) 44kg (d) 36 kg

35. Two types of steel is available. Type *A* contains 5% of Nickel and type *B* contains 40% of Nickel. How much (in tones) of each type be mixed to obtain 140 tons of steel containing 30% of Nickel.

(a) 40,100 (b) 80, 60
(c) 75, 65 (d) 90, 50

36. Two types of tea, *A* and *B* are mixed and then sold at ₹ 40/kg. The profit is 10%, if *A* and *B* are mixed in the ratio 3 : 2 and 5% if this ratio is 2 : 3. The cost price per kg of *A* and *B* are in the ratio **[CAT-18]**

(a) 18 : 25 (b) 19 : 24
(c) 21 : 25 (d) 17 : 25

37. The strength of a salt solution is *p*% if 100 ml of the solution contains *p* grams of salt. If three salt solutions *A*, *B*, *C* are mixed in the proportion 1 : 2 : 3, then the resulting solution has strength 20%. If instead the proportion is 3 : 2 : 1, then the resulting solution has strength 30%. A 4th solution *D*, is produced by mixing *B* and *C* in the ratio 2 : 7. The ratio of the strength of *D* to that of *A* is **[CAT-18]**

(a) 2 : 5 (b) 1 : 3
(c) 1 : 4 (d) 3 : 10

38. A pharmaceutical company produces two chemicals *X* and *Y*, such that *X* consists of 5% salt *A* and 10% salt *B* and *Y* consists of 10% salt *A* and 6% salt *B*. For producing the chemicals *X* and *Y*, the company requires at least 7 g of salt *A* and at least 7 g of salt *B*. If chemical *X* costs ₹ 10.50 per g and chemical *Y* costs ₹ 7.80 per g, what is the minimum cost at which the company can meet the requirement by using a combination of both types of chemicals ? **[IIFT-17]**

(a) ₹ 810
(b) ₹ 850
(c) ₹ 537
(d) None of the above

39. The strength of a salt solution is *p*% if 100 ml of the solution contains *p* grams of salt. Each of three vessels *A*, *B*, *C* contains 500 ml of salt solution of strengths 10%, 22% and 32% respectively. Now 100 ml of the solution in vessel *A* is transferred to vessel *B*. Then, 100 ml of the solution in vessel *B* is transferred to vessel *C*. Finally, 100 ml of the solution in vessel *C* is transferred to vessel *A*. The strength, in percentage, of the resulting solution in vessel *A* is **[CAT-19]**

(a) 12 (b) 14
(c) 13 (d) 15

Answer Key

1. (b)	2. (a)	3. (b)	4. (b)	5. (c)	6. (d)	7. (b)	8. (c)	9. (d)	10. (a)
11. (d)	12. (a)	13. (b)	14. (c)	15. (b)	16. (a)	17. (c)	18. (a)	19. (a)	20. (c)
21. (b)	22. (e)	23. (b)	24. (a)	25. (d)	26. (c)	27. (a)	28. (b)	29. (c)	30. (c)
31. (c)	32. (a)	33. (b)	34. (c)	35. (a)	36. (b)	37. (b)	38. (a)	39. (b)	

Hints and SOLUTIONS

1.

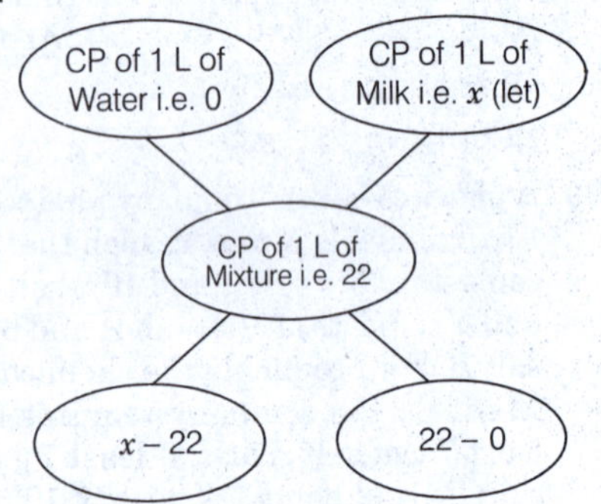

Ratio of quantities of water and milk to be mixed together

$= (x - 22) : 22 = 3 : 15$

$\Rightarrow \quad 5x - 110 = 22$

$\Rightarrow \quad x =$ ₹ 26.40/kg

2. Initial ratio (milk to water) $= 3 : 1$

Changed ratio $= 2 : 3$

Given that quantity of milk is constant, thus, we should equalise 1st term of both ratios.

Initial ratio $= 3 : 1 = 6 : 2$

Changed ratio $= 2 : 3 = 6 : 9$

Clearly, increase in 2nd term of initial ratio $= (9 - 2) = 7$ units

$\because \quad (6 + 2)$ units $= 80$ L

$\therefore \quad 7$ units $= 70$ L

Hence, quantity of water to be added $= 70$ L

3.

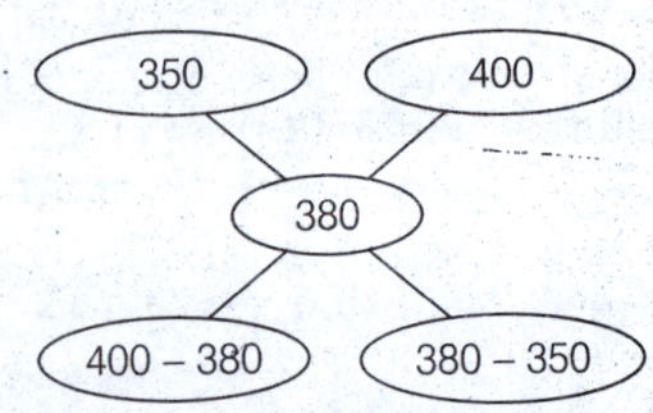

$\therefore$ Required ratio $= 20 : 30 = 2 : 3$

4. LCM of $(4 + 1)$, $(5 + 2)$ and $(6 + 1) = 35$

1st mixture [petrol : kerosene]

$= 4 : 1 = (4 : 1) \times 7 = 28 : 7$

$[\because 35 \div (4 + 1) = 7]$

2nd mixture [petrol : kerosene]

$= 5 : 2 = (5 : 2) \times 5 = 25 : 10$

$[\because 35 \div (5 + 2) = 5]$

3rd mixture [petrol : kerosene]

$= 6 : 1 = (6 : 1) \times 5 = 30 : 5$

$[\because 35 \div (6 + 1) = 5]$

$\therefore$ Required ratio [petrol : kerosene]

$= (28 + 25 + 30) : (7 + 10 + 5)$

$= 83 : 22$

5. Let quantity of the dettol initially be 100 L.

Now, amount of dettol after 4 operations $= 100\left(1 - \frac{1}{3}\right)^4$

$= 100 \times \frac{16}{81}$

$= \frac{1600}{81}$

Hence, ratio of dettol to water

$= \frac{1600}{81} : \left(100 - \frac{1600}{81}\right)$

$= 1600 : 6500$

$= 16 : 65$

6. Amount of milk left after 4 operations [i.e. number of drawing]

$= 40\left(1 - \frac{10}{40}\right)^4$

$= 40 \times \frac{81}{256} = \frac{405}{32}$ L

It is obvious that 5th customer will receive only $\frac{1}{4}$ of $\frac{405}{32}$ L as 10 L is $\frac{1}{4}$ of 40 L.

Hence, he will receive $\frac{1}{4}$ of $\frac{405}{32}$ L

i.e. $\frac{405}{128}$ L

7. Given, initial ratio of milk to water

$= \frac{3}{4} : \frac{1}{4} = 3 : 1$

Let x litres be drawn from the mixture.

$\therefore$ Amount of milk drawn $= x \times \frac{3}{4} = \frac{3x}{4}$

Now,

Initial ratio [milk to water] $= 3 : 1$

Changed ratio [milk to water] $= 1 : 1$

Decrease and Increment $= \frac{3x}{4} \quad \frac{3x}{4}$

Now, $[(3 \times 1) - (1 \times 1)]$ units

$= \left\{\left(1 \times \frac{3x}{4}\right) + \left(1 \times \frac{3x}{4}\right)\right\}$

$\Rightarrow 2$ units $= \frac{3x}{2} \Rightarrow (3 + 1)$ units $= 3x$

Clearly, x litres are drawn from a mixture of $3x$ litres.

Hence, 33.33% of mixture must be drawn and replaced with water.

Alternatively

The best way is to proceed through options.

Let quantity of milk be 75 g and that to water be 25 g.

(a) 25% $\Rightarrow \left(75 - 25 \times \frac{3}{4}\right) : \left(25 - 25 \times \frac{1}{4}\right) + 25 \neq 1 : 1$

(b) 33.33% $\Rightarrow$

$(75 - 33.33) \times \frac{3}{4} : \left(25 - 33.33 \times \frac{1}{4}\right) + 33.33$

$= 50 : 50 = 1 : 1$

Hence, option (b) is correct.

8.

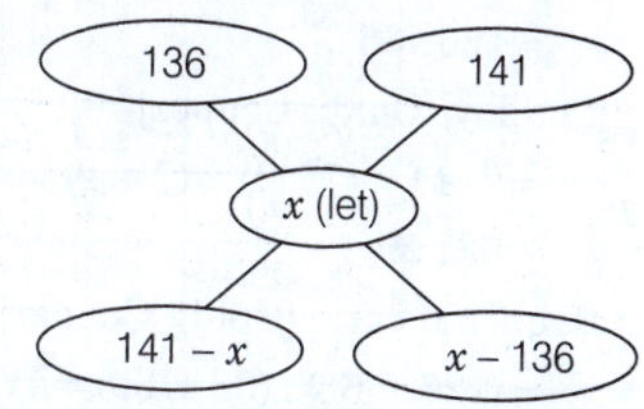

Ratio of quantities of both types of tea

$= (141 - x) : (x - 136) = 2 : 3$

$\Rightarrow 2x - 272 = 423 - 3x$

$\Rightarrow \quad 5x = 695$

$\Rightarrow \quad x = ₹\,139$

9. In fact, water has 0% solution of acid. Then,

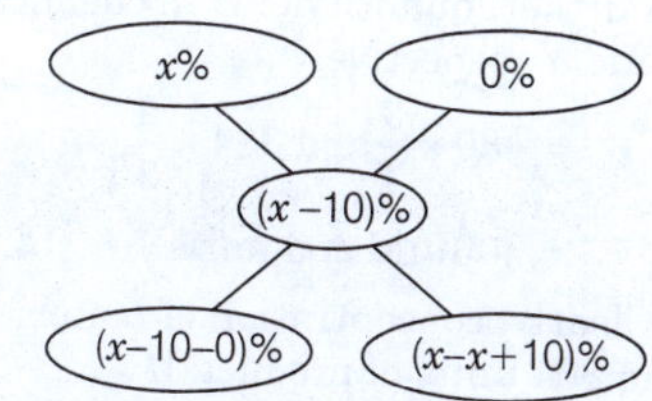

The ratio of quantity of solution to that of water

$= (x - 10) : 10$

As given, $(x - 10)$ units $= x$ litres

$\therefore 10 \text{ units} = \dfrac{10x}{x - 10}$ = quantity of water

$= y$ [given]

10. Intial ratio (nitric acid : water)

$= 1 : 9$

Changed ratio $= 1 : 24$

Clearly, the quantity of Nitric acid remains unchanged and here it is also equal. Here, increment in quantity of water

$= (24 - 9) = 15$ units

Given, $(1 + 9)$ units $= 10$ L

$\therefore \quad 15$ units $= 15$ L

11. 1st mixture $(A : B) = (5 : 1)$

New mixture $(A : B) = 1 : 1$

2nd mixture $(A : B) = 1 : 3$

Then, required ratio

$= (5 + 1)(3 \cdot 1 - 1 \cdot 1) : (1 + 3)(5 \cdot 1 - 1 \cdot 1)$

$= 12 : 16 = 3 : 4$

12.

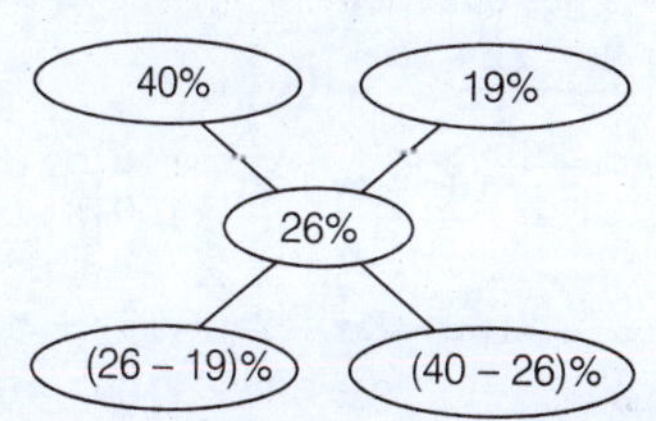

$\therefore$ Ratio of quantities of milk taken from both jars for mixing

$= 7 : 14 = 1 : 2$

It implies that only one part of initial jar is required out of (1 + 2) parts of milk to get the desired mixture having 26% water. Hence, 2/3 part of initial jar has been replaced.

13. 1st mixture (Zinc : Copper) $= 6 : 9$

2nd mixture (Zinc : Copper) $= 7 : 11$

$\therefore$ Required ratio

$= \left[(6 + 9)7 + (7 + 11) \times 6 \times \dfrac{40}{60}\right] :$

$\left[(6 + 9)11 + (7 + 11) \cdot 9 \cdot \dfrac{40}{60}\right]$

$= (105 + 72) : (165 + 108)$

$= 177 : 273 = 59 : 91$

14. Initial ratio (red : green) $= 1 : 3$

Changed ratio $= 3 : 2$

Given that number of green marbles is constant, thus we should equalise 2nd term of both ratios.

Intial ratio $= 1 : 3 = 2 : 6$

Changed ratio $= 3 : 2 = 9 : 6$

Clearly, increase in 1st term of initial ratio $= (9 - 2) = 7$ units

As given, $(2 + 6)$ units $= 40$

$\therefore \quad 7$ units $= 35$

15. 1st mixture (alcohol : water) $= 3 : 4$

New mixture (alcohol : water) $= 4 : 5$

2nd mixture (alcohol : water) $= 5 : 6$

Then, ratio of quantity taken from 1st mixture to that from 2nd mixture

$= (3 + 4)(5 \cdot 5 - 4 \cdot 6) : (5 + 6)(4 \cdot 4 - 3 \cdot 5)$

$= 7 : 11$

As given, $(7 + 11)$ units $= 18$ L

$\therefore \quad 7$ units $= 7$ L

16. Let percentage of Copper in 2nd alloy be x%. Then, percentage of Copper in 1st alloy be $(x - 40)$%

As given, $(x - 40)$% of 1st alloy $= 6$

$\Rightarrow$ 100% of 1st alloy $= \dfrac{600}{(x - 40)}$

= Further, x % of 2nd alloy $= 12$

$\Rightarrow \quad$ 100% of 2nd alloy $= \dfrac{1200}{x}$

Now, $\dfrac{600}{(x - 40)} + \dfrac{1200}{x} = 50$

$\Rightarrow \quad x^2 - 40x = 12x + 24x - 960$

$\Rightarrow \quad x^2 - 76x + 960 = 0$

$\Rightarrow \quad x^2 - 60x - 16x + 960 = 0$

$\Rightarrow \quad x(x - 60) - 16(x - 60) = 0$

$\Rightarrow \quad x = 16$ and 60

But the value of $(x - 40)$ will be negative, if we take $x = 16$.

Thus, $x = 16$ is not possible.

Hence, percentage of Copper in the 2nd alloy = 60%

and percentage of Copper in the 1st alloy = 20%

Alternatively

The best away is to proceed through options. Let options be 30% and70%, 20% and 60%,.........

(a) 30% = 6

$\therefore$ 100% = 20 kg and 70% = 12

$\therefore$ 100% = 17.14 kg

But, $20 + 17.14 \neq 50$

(b) 20% = 6

$\therefore$ 100% = 30 kg and 60% = 12

$\therefore$ 100% = 20 kg and 30 + 20 = 50

Hence, option (b) is correct.

17. Case I Let the resulting mixture has 20% acid content Then,

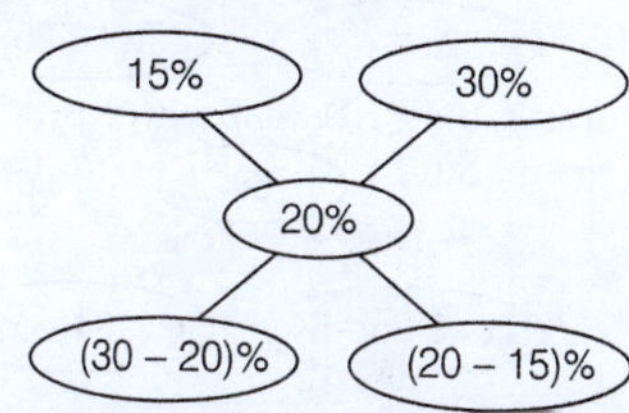

The ratio of quantities of both types of solutions $= 10 : 5 = 2 : 1$

As given, 2 units = 200 L

$\therefore \quad$ 1 unit = 100 L

Case II Let the resulting mixture has 25% acid content. Then,

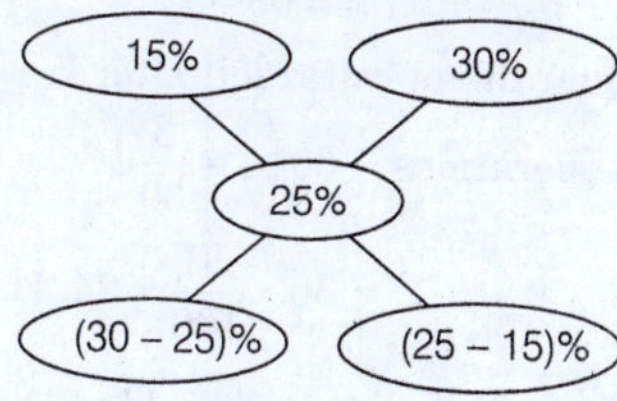

The ratio of quantities of both types of solutions $= 5 : 10 = 1 : 2$

As given, 1 unit = 200 L

$\therefore \quad$ 2 unit = 400 L

From case I and II, it is construed that the quantity of acid solution having 30% acid content must be between 100 L and 400 L.

Hence, option (c) is correct.

18.

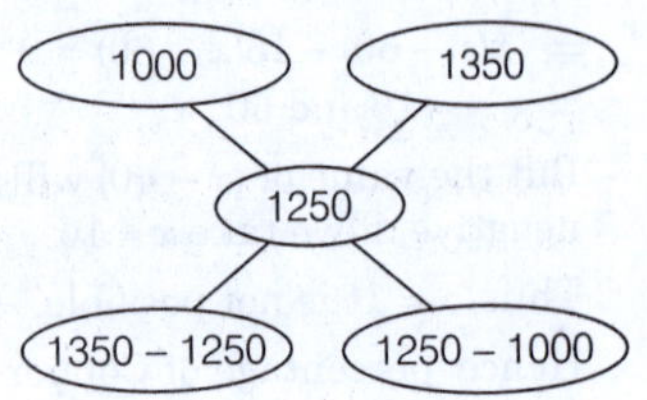

Ratio of quantities taken from both [i.e. water and another liquid]

$= 100 : 250 = 2 : 5$

Hence, volume of water $= \frac{2}{7}$

and volume of another liquid $= \frac{5}{7}$

$\therefore$ Optoin (a) is correct.

19. Cost price of the mixture

$= 72 \times \frac{100}{(100 + 12.5)} = 64$ per kg

Let cost of tea of the 1st quality $= x$/kg

Then, cost of tea of 2nd quality $= ₹ (x + 3.25)$/kg

Now,

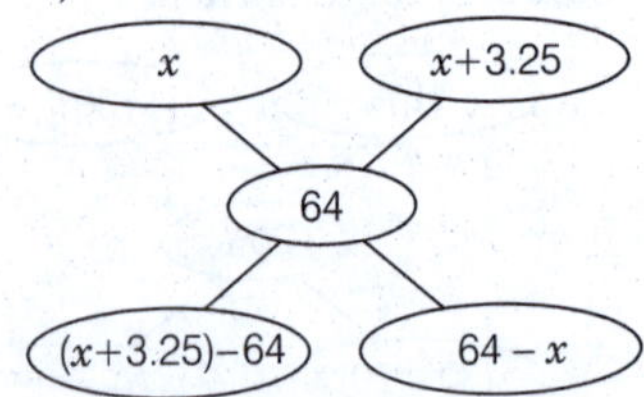

Ratio of quantities taken from both qualities of tea

$= (x + 3.25) - 64 : (64 - x)$

$= 4 : 1$ [given]

$\Rightarrow 256 - 4x = (x + 3.25) - 64$

$\Rightarrow 316.75 = 5x$

$\therefore x = 63.35$

20. Quantity of water left after 2 operations $= 30\left(1 - \frac{3}{30}\right)^2$

$= 30 \times \frac{81}{100} = 24.3$ L

21. Profit % if mixture is sold at the rate of ₹ 25/L = 56.25%

[$\because$ ₹ 20 $\Rightarrow$ 125% $\therefore$ ₹ 25 $\Rightarrow$ 156.25%]

Hence, to earn a profit of 56.25 %, 56.25 units water must be added in the 100 units of the chemical as water has no cost.

$\therefore$ Required ratio $= 100 : 56.25 = 16 : 9$

22. Quantity of water in the mixture of 60 g = 75% of 60 = 45 g.

Now, quantity of new mixture

$= 60 + 15 = 75$ g

Then, % of water

$= \left(\frac{45 + 15}{75} \times 100\right)\% = 80\%$

Hence, option (e) is correct.

23. Quantity of petrol in new mixture

$= \left(2 \times \frac{1}{2}\right) + \left(3 \times \frac{3}{5}\right) + \left(1 \times \frac{4}{5}\right)$

$= 3.6$ L

Then, quantity of kerosene in new mixture $= (2 + 3 + 1) - 3.6 = 2.4$ L

Hence, required ratio $= 3.6 : 2.4 = 3 : 2$

Try with other methods too.

24. Let the quantity of alloy P and Q be 100 kg and 450 kg respectively in newly formed alloy.

Then, quantity of Silver in new alloy

$= 45\%$ of $100 + 30\%$ of 450

$= 45 + 135 = 180$ kg

Further, quantity of Copper in new alloy $= 35\%$ of $450 = 157.5$ kg

Now, required percentages

$= \left(\frac{180}{100 + 450} \times 100\right)\%$

and $\left(\frac{157.5}{100 + 450} \times 100\right)\%$

$= 33\%$ and 29% [approx.]

25. Let volume of the jug be x litres.

Now, amount of orange juice after 2 operations [i.e. number of drawing]

$= 10\left(1 - \frac{x}{10}\right)^2 = 5$ [as given]

$\Rightarrow \frac{5}{10} \times 100 = (10 - x)^2$

$\Rightarrow 50 = (10 - x)^2$

$\Rightarrow 7.07 = 10 - x$

$\Rightarrow x = 2.93$

Hence, option (d) is correct.

26. Quantity of milk in new mixture

$= \left(48 \times \frac{13}{20}\right) + \left(42 \times \frac{18}{35}\right) = 52.8$ L

Then, quantity of water in new mixture

$= (48 + 42) - 52.8 = 37.2$ L

Hence, required ratio

$= 52.8 : (37.2 + 20) = 12 : 13$

Try with other methods too.

27. 1st alloy [Gold : Copper]

$= 7 : 2 = (7 : 2) \times (7 + 11)$

$= 126 : 36$

2nd alloy [Gold : Copper]

$= 7 : 11 = (7 : 11) \times (7 + 2)$

$= 63 : 99$

$\therefore$ Required ratio [Gold : Copper]

$= (126 + 63) : (36 + 99) = 7 : 5$

28. Given, quantity of product $M = 864$ units

Then, quantity of chemical

$X = 864 \times \frac{5}{9} = 480$ units

and quantity of chemical

$Y = 864 - 480 = 384$ units

Further, quantity of B in chemical X and Y respectively

$= 480 \times \frac{3}{4}$ and $384 \times \frac{2}{3}$

$= 360$ and 256 units

Clearly, concentration of material B in 864 units of product M

$= 360 + 256 = 616$ units.

But the concentration of material B in final mixture is 50%.

Now, let the quantity of water in final mixture be x units.

Then, 50% of $(864 + x) = 616$

$\Rightarrow 864 + x = 1232$

$\Rightarrow x = 1232 - 864 = 368$ units

29. LCM of $(1 + 1)$ and $(3 + 1) = 4$

1st vessel (spirit : water)

$= 1 : 1 = (1 : 1) \times 2 \times 2 = 4 : 4$

2nd vessel (spirit : water)

$= 3 : 1 = (3 : 1) \times 1 \times 3$

$= 9 : 3$

$\therefore$ Required ratio (spirit : water)

$= (4 + 9) : (4 + 3)$

$= 13 : 7$

30. Let the initial amount of Aluminium in the 1st and the 2nd alloys be $8x$ kg and $16y$ kg respectively and a piece of z kg is cut-off from each alloy.

Then, amount of Aluminium in z kg of the 1st alloy $= zx$

and amount of Aluminium in z kg of the 2nd alloy $= zy$

As per question,

$\left(\frac{8x - zx + zy}{8} \times 100\right)\%$

$= \left(\frac{16y - zy + zx}{16} \times 100\right)\%$

$\Rightarrow 16x - 2zx + 2zy = 16y - zy + zx$

$\Rightarrow 3zy - 3zx = 16y - 16x$

$\Rightarrow \quad 3z(y-x) = 16(y-x)$

$\Rightarrow \quad z = 16/3 = 5.33\,\text{kg}$

Alternatively

You may proceed through options.

31. Let the quantity of pure milk in the vessel originally be x litres.

Then, quantity of pure milk left after 2 operations $= x\left(1-\dfrac{9}{x}\right)^2$

$= x - 17.1$

$\Rightarrow x + \dfrac{81}{x} - 18 = x - 17.1$

$\Rightarrow \quad x = \dfrac{81}{0.9} = 90\,\text{L}$

Alternatively

Proceed through options.

32. Initial ratio (water to milk) = 1 : 4

$\therefore$ Quantity of milk sold

$= 25 \times \dfrac{4}{4+1} = 20$

[$\because$ 1/4 of the mixture = 25 L]

Then, quantity of milk left

$= 80 - 20 = 60\,\text{L}$

Now, required ratio (water to milk)

$= (100-60) : 60 = 2 : 3$

33. Initial ratio [Acetic acid to Sodium acetate] = 3 : 1

Changed ratio [Acetic acid to Sodium acetate] = 2 : 3

Given that quantity of Acetic acid is constant ,thus, we should equalize 1st term of both ratios.

Initial ratio = 3 : 1 = 6 : 2

Changed ratio = 2 : 3 = 6 : 9

Clearly, increase in 2nd term of initial ratio = (9 – 2) = 7 units

As given,

(6 + 2) units = 40 L

$\therefore$ 7 units = 35 L

34. Amount of Tin in new alloy

$= \left(60 \times \dfrac{2}{3+2}\right) + \left(100 \times \dfrac{1}{1+4}\right)$

= 44 kg

35.

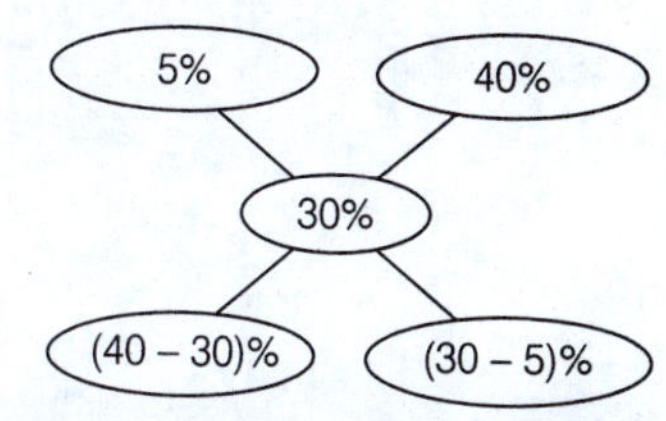

Ratio of quantities of type A and B to be mixed = 10 : 25 = 2 : 5

As given, (2 + 5) units = 140 tons

$\therefore$ 2 units = 40 tons

and 5 units = 100 tons

36. Let cost price of type A tea is ₹ a per kg and that of type B is ₹ b per kg and the both the mixtures are of 5 kg each.

Now, cost price of 1st mixture

$= 3a + 2b$

and cost price of 2nd mixture

$= 2a + 3b$

As given, 110% of $(3a + 2b)$

$= (3+2) \times 40$

and 105% of $(2a + 3b) = (2+3) \times 40$

Thus,

110% of $(3a+2b)$ = 105% of $(2a+3b)$

$\Rightarrow \quad 66a + 44b = 42a + 63b$

$\Rightarrow \quad 24a = 19b$

$\Rightarrow \quad a : b = 19 : 24$

37. Let three salt solutions A, B and C are of 100 ml, 200 ml and 300 ml respectively and their respective strength per 100 ml be a%, b% and c%.

Now, strength of 1st mixture

$= a + 2b + 3c$ = 20% of (100 + 200 + 300)

$\Rightarrow \quad a + 2b + 3c = 120 \quad \ldots\text{(i)}$

In case of reversed proportion,

$3a + 2b + c$ = 30% of (300 + 200 + 100)

$\Rightarrow \quad 3a + 2b + c = 180 \quad \ldots\text{(ii)}$

On dividing Eqs. (i) by (ii), we get

$\dfrac{a+2b+3c}{3a+2b+c} = \dfrac{2}{3}$

$\Rightarrow \quad 3a + 6b + 9c = 6a + 4b + 2c$

$\Rightarrow \quad 2b + 7c = 3a$

$\Rightarrow \quad \dfrac{2b+7c}{a} = 3$

Let 200 ml of solution B and 700 ml of solution C be mixed to produce solution D having d% strength.

Then, $2b + 7c = d\% \times (200 + 700)$

$\therefore \quad d = \dfrac{2b+7c}{9}$

Then, required ratio

$= \dfrac{2b+7c}{9} : a = \dfrac{2b+7c}{a} : 9$

$= 3 : 9 = 1 : 3$

38. Let the company produces m gram of chemical X and n gram of chemical Y.

Then, requirement of salt A

= 5% of m + 10% of $n \geq 7$ [as given]

and requirement of salt B

= 10% of m + 6% of $n \geq 7$ [as given]

For the minimum cost,

5% of m + 10% of n = 7 ...(i)

and 10% of m + 6% of n = 7 ...(ii)

$\Rightarrow$ 5% of m + 10% of n

= 10% of m + 6% of n

$\Rightarrow$ 5% of m = 4% of n

Putting this value in Eq. (i), we get 14% of n = 7 and 17.5% of m = 7

$\Rightarrow \quad m = 40\,\text{g}$ and $n = 50\,\text{g}$

Now, cost of

$m + n = 40 \times 10.50 + 50 \times 7.80 =$ ₹ 810

39. Quantity of salt in vessel B when 100 ml of vessel A is transferred to $B = 10\,\text{g} + (22 \times 5)\,\text{g} = 120\,\text{g}$ in 600 ml i.e. 20 g in 100 ml.

Quantity of salt in vessel C when 100 ml of vessel B is transferred to $C = 20\,\text{g} + (32 \times 5)\,\text{g} = 180\,\text{g}$ in 600 ml i.e. 30 g in 100 ml.

Finally, quantity of salt in vessel A when 100 ml of vessel C is transferred to $A = 30\,\text{g} + (10 \times 4)\,\text{g}$ = 70 g in 500 ml i.e. 14 g in 100 ml.

Hence, strength of resulting solution of A is 14%.

Unit 5

Time and Distance

- Concept of Time, Speed and Distance and Their Applications
- Concept of Average Speed • Relative Speed : Same Direction and Opposite Directions • To and Fro Motion in a Straight Line
- Races and Games • Concept of Circular Motion
- Concepts Associated with Problems Based on Trains • Boat and Stream
- Escalator • Clock

Time and Distance

This unit is substantially based on applications of the concepts of Percentage and the concepts of Ratio. You may say that it is the combination of unit-III and unit-IV in terms of the concepts associated with them. However, different approaches have been adopted in different sectors so as to solve the problems in moments. For the sake of convenience, this unit has divided into ten chapters with different titles.

Learning OBJECTIVES

Upon completion of this chapter, aspirants will be able to

- Understand the concepts associated with Time, Speed and Distance and their applications in solving the problems based on motion in straight line, relative motion, circular motion, trains, boat and stream, races etc;
- Observe the relation between different units of speed;
- Understand the relation between distance and speed, speed and time and time and distance and the effect on one aspect when another alters;
- Comprehend the concept of average speed and solve the questions based thereon;
- Recognise concept of relative speed : same direction and opposite direction with illustrations;
- Solve the questions based on the relative speed;
- Understand the concept of To and Fro motion in a straight line;
- Understand the different terms associated with the races and games;
- Solve the questions based on races and games;
- Be well versed with the concept of circular motion, same direction and opposite direction;
- Solve the questions based on Trains where the length of the train plays a critical role in determination of time as well as speed;
- Understand the conception of upstream speed and downstream speed and their applications in case of Boat and Stream;
- Know the facts related to clock and how to determine angle between hour hand and minute hand;
- Understand the concept of coinciding/ overtaking in case a clock gains/loses, concept of too fast and too slow and that of strokes and intervals;
- Ascertain the relation between actual time and the time reflected in mirror;
- Understand the philosophy of the division of this chapter in various parts;
- Analyse the trend of the questions being asked in various competitive examinations;
- Recognise the types of questions which can be easily solved through the given options;
- Solve the questions in moments;
- Grip on concepts helping to solve the questions without pen and paper;
- Learn many new concepts/fundamentals;
- Apply the pragmatic approaches in lieu of academic approaches; and
- Ponder on the alternative methods.

30

Concept of Time, Speed and Distance and their Applications

Introduction

This chapter is very critical as a lot of questions are asked in diverse competitive exams like CAT, MAT, SSC, CDS, Bank-PO, Railway, etc. It has also importance in our daily life. Concepts of percentage have been applied in this topic to the greater extent in addition of the application of the concept of Ratio.

Concept of Time, Speed and Distance (TSD)

The concepts associated with TSD is substantially applied in solving the problems based on motion in straight line, Relative motion, Circular motion, Trains, Boat and stream, Races, etc. What time you take to solve such types of problems depends upon your basic understanding of the concepts.

We know that movement occurs when a body changes its position with respect to any external stationary point and this displacement is known as **distance** covered by that body. **Time** is the duration over which movement occurs and **speed** is the rate at which distance is covered. Generally, **speed** is uniform unless it is specified.

In other words, when a body moves from point P to Q at a distance of D, then it requires some time (say 'T') to cover this distance (D) with a particular speed (S).

Hence, Distance = Time × Speed

$$\Rightarrow \quad D = TS \Rightarrow T = \frac{D}{S} \Rightarrow \quad S = \frac{D}{T}$$

It is also evident that when speed (S) increases so as to cover a certain distance (D), time(T) will decrease. It implies that T is inversely proportional to S, when D remains constant. Hence,

Inside

- *Concept of Time, Speed and Distance (TSD) and their applications*
- *Relation between Time and Speed, when Distance is constant*
- *Relation between Speed and Distance, when Time is constant*
- *Relation between Distance and Time, when Speed is constant*
- *Conversion of Speeds in different units*
- *Illustrations based on the above relation*
- *Illustrations based on relation among Distance, Revolutions and Circumference of a round object*
- ....
- .... *& Many More*
- *Work Out with Solutions*

(i) $T \propto \frac{1}{S}$ or $S \propto \frac{1}{T}$, when D is constant.

(ii) If time T_1 is taken at the speed of S_1 and time T_2 is taken at the speed of S_2 to cover a distance, then $\frac{S_1}{S_2} = \frac{T_2}{T_1}$, if distance is constant.

(iii) In terms of ratio, $S_1 : S_2 = T_2 : T_1$, when distance is constant.

Also, in case the speed is increased, distance will increase in a constant time. It means speed is directly proportional to distance, if time is kept constant. Hence,

(i) $S \propto D$ or $D \propto S$, when T is constant.

(ii) If a distance D_1 is covered at the speed of S_1 and a distance of D_2 at the speed of S_2, then $\frac{S_1}{S_2} = \frac{D_1}{D_2}$, when time is constant.

(iii) In terms of ratio, $S_1 : S_2 = D_1 : D_2$, when time is constant.

Further, when speed is constant, more time is required to cover more distance. It implies that time is directly proportional to distance if speed is kept constant. Hence,

(i) $D \propto T$ or $T \propto D$, when S is constant.

(ii) If T_1 is the time taken to cover a distance of D_1 and time T_2 is taken to cover a distance of D_2, then, $\frac{T_1}{T_2} = \frac{D_1}{D_2}$, if speed is constant.

(iii) In terms of ratio, $T_1 : T_2 = D_1 : D_2$, when speed is constant.

Aspirants are supposed to understand the above mentioned proportionality very well as its application is very wide while solving the problems. Before arriving at the application of the concept mentioned above, how to make conversion of speed in different units is indispensable.

Now, convert a speed of x km/h in m/sec. Speed of x km/h means a body moves x km in 1 h. In other words, in (60×60) sec, a body moves $(x \times 1000)$ m.

Then in 1 sec, it will move a distance of $= \frac{x \times 1000}{60 \times 60} = \frac{5x}{18}$ m

Hence, the speed is $\frac{5x}{18}$ m/sec.

Clearly,

km/h ....$\left\{\times \frac{5}{18}\right\}$........⇒ m/sec

or

m/sec ....$\left\{\times \frac{18}{5}\right\}$........⇒ km/h

Speed of x km/h may be interpreted that a body moves x km in 60 min.

In other words, in 60 min, a body moves $(x \times 1000)$ m.

Then, in 1 min, it will move a distance of

$$\frac{x \times 1000}{60} = \frac{50x}{3} \text{ m}$$

Hence, the speed is $\frac{50x}{3}$ m/min.

Clearly,

km/h ....$\left\{\times \frac{50}{3}\right\}$........⇒ m/min

or

m/min ....$\left\{\times \frac{3}{50}\right\}$........⇒ km/h

Further, if the speed of a body is x m/min, then convert the speed in m/sec.

Given that, in 60 sec, a body moves x m.

Then, in 1 sec, it will move a distance of

$$\frac{x \times 1}{60} \text{ m. i.e. } \frac{x}{60} \text{ m}$$

Hence, the speed is $\frac{x}{60}$ m/sec.

Clearly,

m/min ....$\left\{\times \frac{1}{60}\right\}$........⇒ m/sec

or

m/sec ....[× 60] .... ⇒ m/min

Note *It is important to note that all units are made uniform so as to solve the questions.*

Now, take some examples to understand the concepts mentioned above in practical way.

(i) Convert 36 km/h in m/sec

$$\Rightarrow 36 \text{ km/h} = 36 \times \frac{5}{18}$$
$$= 10 \text{ m/sec}$$

(ii) Convert 25 m/sec in km/h

$$\Rightarrow 25 \text{ m/sec} = 25 \times \frac{18}{5}$$
$$= 90 \text{ km/h}$$

(iii) Convert 27 km/h in m/min

$$\Rightarrow 27 \text{ km/h} = 27 \times \frac{50}{3} = 450 \text{ m/min}$$

(iv) Convert 100m/min in km/h

$$\Rightarrow 100 \text{ m/min} = 100 \times \frac{3}{50} = 6 \text{ km/h}$$

(v) Convert 90m/min in m/sec

$$\Rightarrow 90 \text{ m/min} = 90 \times \frac{1}{60} = 1.5 \text{ m/sec}$$

(vi) Covert 3 m/sec in m/min

$$\Rightarrow 3 \text{ m/sec} = 3 \times 60 = 180 \text{ m/min}$$

A. Illustrations based on $T \alpha \frac{1}{S}$ or $S \alpha \frac{1}{T}$, when D is Constant

Problem 1 If a man reduces his speed to $\frac{2}{3}$, he takes 1 h more in walking a certain distance. The time (in hours) to cover the distance with his normal speed is ? **[SSC (CGL-PT) 2000, CDS-21]**

Sol. Since distance is constant, $T \alpha \frac{1}{S}$, This means whatever happens to speed, the reciprocal of that will happen to time.

Here, the speed becomes $\frac{2}{3}$, then the time will obviously become $\frac{3}{2}$ (reciprocal of $\frac{2}{3}$).

Extra time taken $= \left(\frac{3}{2} - 1\right)$ of usual time taken $= \frac{1}{2}$ of usual time taken.

Given, $\frac{1}{2}$ of usual time taken = 1 h

Then, usual time taken = 2 h

Alternatively I

Recall the concept 'Product Constancy' envisaged in the chapter 'Percentage'.

We know, Time × Speed = Distance and in this question, distance has been kept constant.

Since speed has been reduced by 33.33%, the time will increase by $\left\{\frac{33.33}{100 - 33.33} \times 100\right\}$% i.e. by 50%.

But, given that he takes 1 h more to cover the distance. It implies that 50% of the usual time taken = 1 h

Then, 100% of the usual time taken = 2 h

Hence, usual time taken = 2 h

Alternatively II

We may frame the generic version of the question :

If a man reduces/increases his speed to $\frac{x}{y}$, he takes t hours more/less in walking a certain distance.

The time (in hours) to cover the distance with his normal speed will be as below.

Let the initial speed of the man is S km/h and distance is D km.

Then, normal time taken $= \frac{D}{S}$ h.

Assume, later his speed reduced to $\frac{x}{y}$.

It means his speed becomes $\frac{xS}{y}$ km/h.

Then, time taken to cover the same distance $= D \div \frac{xS}{y} = \frac{Dy}{xS}$ h

As given, $\frac{Dy}{xS} - \frac{D}{S} = t \Rightarrow \frac{D}{S}\left(\frac{y}{x} - 1\right) = t$

$\Rightarrow \frac{D}{S} = \frac{x}{(y - x)} \times t$

Hence, if a man reduces/increases his speed to $\frac{x}{y}$, he takes t hours more/less in walking a certain distance.

Then, usual time taken time (in hours) to cover the distance with his normal speed will be

$$\frac{x}{\text{positive difference between } x \,\&\, y} \times \text{time taken extra/less}$$

Here, the usual time taken $= \left\{\frac{2}{(3-2)} \times 1\right\} = 2$ h

Alternatively III

Let the initial speed of the man is 30 km/h and the distance to be covered is 90 km. [you are free to assume any figures instead of 30 and 90]

Then, initial time taken = 90 ÷ 30 = 3 h

Time taken after reduction in his speed $= 90 \div \left(30 \times \frac{2}{3}\right) = 4.5$ h

Clearly, in case the difference between both time taken is (4.5 – 3) h, the usual time taken is 3 h.

Then, if difference between them is 1 h, the usual time taken = 2 h.

Problem 2 A and B started at the same time from the same place for a certain destination. B walking at $\frac{5}{6}$ of A's speed reached the destination 1 h 15 min after A. B reaches the destination in : **[SSC (CGL-PT) 08]**

Sol. Extra time taken by $B = \left(\frac{6}{5} - 1\right)$ of time taken by A

$= \frac{1}{5}$ of time taken by A

Given, $\frac{1}{5}$ of usual time taken by A = 1 h 15 min

Then, time taken by A = 6 h 15 min

∴ Time taken by B = 6h 15 min + 1 h 15 min = 7 h 30 min

Alternatively I

Since the speed of B is 16.66% less than that of A, time taken by B will be $\left(\frac{16.66}{100 - 16.66} \times 100\right)$% i.e. 20% more than that by A.

But, given that B takes 1 h 15 min more than time taken by A.

It implies that 20% of the time taken by A = 1 h 15 min

Then, 100% of the time taken by A = 6 h 15 min

Hence, time taken by B = 6 h 15 min + 1 h 15 min = 7 h 30 min

Alternatively II

Assume if A had walked at 5/6 of his usual speed, he would have reached the destination 1 h 15 min late.

Now, time taken by $A = \left\{\frac{5}{(6-5)} \times 1 \text{ h } 15 \text{ min}\right\} = 6$ h 15 min

Then, required time = 6 h 15 min + 1 h 15 min = 7 h 30 min

Note *You are free to adopt other methods too as explained for the previous question.*

Problem 3 Walking at $\frac{6}{5}$ of its usual speed, a man is 20 min. before. Find its usual time to cover the journey.

Sol. His latter speed = $\frac{6}{5}$ of the usual speed

Then, latter time taken = $\frac{5}{6}$ of the usual time taken

As given, (usual time taken − $\frac{5}{6}$ of usual time taken) = 20 min

⇒ 1/6 of usual time taken = 20 min

⇒ Usual time taken = 120 min = 2 h

Alternatively I

Since speed has increased by 20%, the time will decrease by $\left\{\frac{20}{(100+20)} \times 100\right\}$% i.e. by 16.66%. But, given that he takes 20 min less to cover the distance.

It implies that 16.66% of his usual time taken = 20 min

Then, 100% of the usual time taken = 120 min

Hence, usual time taken = 2 h

Alternatively II

Time taken by that man = $\left[\frac{6}{(6-5)} \times 20 \text{ min.}\right] = 120$ min

= 2 h

Problem 4 **A car meets with an accident and moves at $\frac{6}{7}$ of its normal speed, resulting in 15 min late. Find the usual time taken for the journey ahead of the accident-point.**

Sol. Since speed has decreased by 14.28%, the time will increase by $\left\{\frac{14.28}{(100-14.28)} \times 100\right\}$% i.e. by 16.66%.

But, given that it takes 15 min more than the usual time taken.

It implies that 16.66% of its usual time taken = 15 min

Then, 100% of the usual time taken = 90 min

Hence, usual time taken = 1.5 h

Alternatively

Usual time taken by car = $\left\{\frac{6}{(7-6)} \times 15 \text{ min}\right\} = 90$ min = 1.5 h

Try yourself with other approaches mentioned earlier.

Problem 5 **A cycle increases its normal speed by 20% and reaches its destination $\frac{1}{2}$ h early. What is the actual time taken?**

Sol. Since the speed increases by 20%, time will decrease by $\left\{\frac{20}{(100+20)} \times 100\right\}$% i.e. 16.66%

As given, 16.66% of the actual time taken = $\frac{1}{2}$ h

Then, 100% of the actual time taken = 3 h

Alternatively

Increase in speed = 20%.

It means the speed has become 6/5 of the normal speed.

Time taken by that cycle = $\left\{\frac{6}{(6-5)} \times 0.5 \text{ h}\right\} = 3$ h

Try yourself with other approaches mentioned earlier.

Problem 6 **A train running at $\frac{7}{11}$ of its own speed reached a place in 22 h. How much time could be saved if the train runs at its own speed ?** ***[MAT-10]***

Sol. Since the speed becomes 7/11, actual time will be 11/7 of the usual time taken as distance is constant.

As given, 11/7 of the usual time taken = 22 h

Then, usual time taken = 14 h

Hence, saving in time = 22 − 14 = 8 h

Problem 7 **Radha and Hema are neighbours and study in the same school. Both of them use bicycle to go to the school. Radha's speed is 8 km/h whereas Hema's speed is 10 km/h. Hema takes 9 min less than Radha to reach the school. How far is the school from the locality of Radha and Hema ?** ***[CDS-19, CAT-20 (Partly)]***

Sol. Since speed of Hema is 25% more than that of Radha, Hema will take $\left\{\frac{25}{(100+25)} \times 100\right\}$% i.e. 20% less time than Radha. But, given that she takes 9 min less than Radha.

Hence, 20% of the time taken by Radha = 9 min

∴ Time taken by Radha to reach school = 45 min

Now, distance covered by Radha in 45 min at 8 km/h = 6 km

Try also with other approaches.

Problem 8 **Every day a person walks at a constant speed, V_1 for 30 min. On a particular day, after walking for 10 min at V_1, he rested for 5 min. He finished the remaining distance of his regular walk at a constant speed, V_2, in another 30 min. On that day, find the ratio of V_2 and his average speed [i.e. total distance covered/ total time taken including resting time].** ***[XAT-18]***

Sol. It is evident from the above that the distance [i.e. remaining after 10 min walk] to be covered at V_1m / min in 20 min, is covered by him in 30 min at V_2m/min. [you are free to assume any unit for speed like km/h, m/sec. etc.]

Since distance is constant,

$V_1 : V_2$ = time taken at the rate of

V_2 : time taken at the rate of $V_1 = 30 : 20 = 3 : 2 \Rightarrow V_2 = \frac{2}{3}V_1$

Clearly, walking distance = 30 V_1 m

Now, required ratio = $V_2 : \frac{30V_1}{(10+5+30)} = \frac{2}{3}V_1 : \frac{2}{3}V_1 = 1 : 1$

Alternatively It is clear that last 2/3 rd of the walking distance has been covered at the rate of V_2 in 30 min. It implies that V_2 is such that 1/3rd of the walking distance can be covered in 15 min at the rate of V_2. Further, the first 1/3rd of the walking distance has been covered in 15 min (including resting time). Now, it may be construed from the above that average speed is equal to V_2. Hence, their ratio = 1 : 1.

Problem 9 **A student goes to school at the rate of $2\frac{1}{2}$ km/h and reaches 6 min late. If he travels at the speed of 3 km/h, he is 10 min early. The distance (in km) between the school and his house is :** ***[SSC(CGL-PT)-11]***

Sol. Let the student takes T_1 and T_2 minutes at the speed of S_1 and S_2 respectively.

We know that, $\frac{S_1}{S_2} = \frac{T_2}{T_1}$, where D is constant.

$$\Rightarrow \frac{2\frac{1}{2}}{3} = \frac{T_2}{(T_2 + 16)}$$

[∵ $T_1 = T_2 + 16$ as the time-interval between T_1 and $T_2 = 10 + 6 = 16$ min]

$$\Rightarrow 2\frac{1}{2}T_2 + 40 = 3T_2$$

$$\Rightarrow T_2 = 80 \text{ min} = 1.33 \text{ h}$$

Now, distance $= S_2 \times T_2 = 3 \times 1.33 = 4$ km

Alternatively I

Since the speed increases from $2\frac{1}{2}$ to 3 i.e. by 20%, time will decrease by $\left\{\frac{20}{(100+20)} \times 100\right\}$% i.e. by 16.66%.

Obviously, 16.66% of [time taken at the rate of 2.5km/h] = (10 + 6) min

Then, 100% of the initial time taken = 96 min

Hence, distance $= 2\frac{1}{2} \times \frac{96}{60} = 4$ km

Or

New time [i.e. at rate of 3 km/h] = 96 – 16 = 80 min

Then, distance $= 3 \times \frac{80}{60} = 4$ km

Alternatively II

If the speed alters from x to y and the time interval is T, then the distance will be $\frac{xy}{\text{difference between } x \text{ and } y} \times T$. Here, unit of the time must be same as envisaged in the speed.

$$\text{Required answer} = \frac{2\frac{1}{2} \times 3}{\left(3 - 2\frac{1}{2}\right)} \times \frac{(10+6)}{60} = \frac{7\frac{1}{2}}{\frac{1}{2}} \times \frac{16}{60} = 4 \text{ km}$$

Problem 10 **If a man cycles at 10 km/h, then he arrives at a certain place at 1 : 00 PM. If he cycles at 15 km/h, he will arrive at the same place at 11 : 00 AM. At what speed must he cycle to get there at noon?** *[CAT-04]*

Sol. Clearly, his usual reaching time be 12 noon. Then, this question reads 'A cyclist goes to a place at the rate of 10 km/h and reaches 1 h late. If he travels at the speed of 15 km/h, he is 1 h early. What is his usual speed ?"

Let the cyclist takes T_1 and T_2 minutes at the speed of S_1 and S_2 respectively.

Now, $\frac{S_1}{S_2} = \frac{T_2}{T_1}$, where D is constant.

$$\Rightarrow \frac{10}{15} = \frac{T_2}{T_2 + 2}$$

[∵ $T_1 = T_2 + 2$ as the time - interval between T_1 and $T_2 = 1 + 1 = 2$h]

$$\Rightarrow T_2 = 4 \text{ h}$$

Now, distance $= S_2 \times T_2 = 4 \times 15 = 60$ km

and usual time taken = 4 + 1 = 5 h

Then, usual speed = 60 ÷ 5 = 12 km/h

Alternatively

Since the speed increases from 10 to 15 i.e. by 50%, time will decrease by $\left[\frac{50}{(100+50)} \times 100\right]$% i.e. by 33.33%

Obviously, 33.33% of time taken at the rate of 10km/hr = (1 + 1) h

Then, 100% of the time taken at the rate of 10 km/h = 6 h

Hence, distance = 10 × 6 = 60 km

Now, usual time taken = 6 – 1 = 5 h

Then, usual speed = 60 ÷ 5 = 12 km/h

Problem 11 **A man covered a certain distance at some speed. Had he moved 3 km/h faster, he would have taken 40 min less. If he had moved 2 km/h slower, he would have taken 40 min more. The distance (in km) is :**

Sol. Let the normal speed be x km/h.

In 1st case, speed becomes $\frac{(x+3)}{x}$ of normal speed, then time will be $\frac{x}{(x+3)}$ of usual time taken.

Then, $\left\{1 - \frac{x}{x+3}\right\}$ of the usual time taken $= \frac{40}{60}$

$\Rightarrow \left\{\frac{3}{x+3}\right\}$ of the usual time taken $= \frac{40}{60}$

$$\Rightarrow \text{Usual time taken} = \frac{2(x+3)}{9}$$

In 2nd case, the speed becomes $\frac{(x-2)}{x}$ of normal speed, then time taken $= \frac{x}{(x-2)}$ of usual time taken.

Then, $\left\{\frac{x}{(x-2)} - 1\right\}$ of usual time taken $= \frac{40}{60}$

$\Rightarrow \left\{\frac{2}{x-2}\right\}$ of usual time taken $= \frac{2}{3}$

$$\Rightarrow \text{Usual time taken} = \frac{2(x-2)}{6}$$

But usual time taken in both cases will be equal.

$$\Rightarrow \frac{2(x+3)}{9} = \frac{2(x-2)}{6}$$

$$\Rightarrow 2x + 6 = 3x - 6$$

$$\Rightarrow x = 12$$

Hence, normal speed = 12 km/h

Then, usual time taken $= \frac{2(x+3)}{9} = \frac{2(12+3)}{9} = \frac{10}{3}$ h

Thus, distance = normal speed × usual time taken $= 12 \times \frac{10}{3} = 40$ km

Alternatively I

Speed to be increased by (say, x) ...... Saving in time (say, T_1)

Speed to be reduced by (say, y) ...... Excess of time (say, T_2)

Then, normal speed $= \frac{xy(T_1 + T_2)}{\text{difference between } xT_2 \text{ and } yT_1}$

Here, the normal speed $= \frac{3 \times 2(40+40)}{(3 \times 40) - (2 \times 40)} = \frac{6 \times 80}{40} = 12$ km/h

Now, the speed increases by 3 km, it means increment in speed is 25%. Then, time will decrease by 20%.

Further, 20% of the usual time taken = 40 min

Then, usual time taken = 200 min

Now, distance = usual speed × usual time taken $= 12 \times \frac{200}{60}$

$= 40$ km

Alternatively II

Case I Speed(*S*) Time taken(*T*)

(+3) (−2/3)

Do cross-multiplication and write as below :

$$+3T - \frac{2}{3}S = \left|(+3) \times \left(\frac{-2}{3}\right)\right|$$

$$\Rightarrow \quad +3T - \frac{2}{3}S = 2 \quad \text{... (i)}$$

Case II Speed(*S*) Time taken(*T*)

(−2) (+2/3)

Do cross-multiplication and write as below :

$$-2T + \frac{2}{3}S = \left|(-2) \times \left(\frac{+2}{3}\right)\right|$$

$$\Rightarrow \quad -2T + \frac{2}{3}S = \frac{4}{3} \quad \text{... (ii)}$$

On solving Eqs. (i) and (ii), we get $T = \frac{10}{3}$ and $S = 12$

Then distance $= 12 \times \frac{10}{3} = 40$ km

Under the conditions given in the question, we may find the usual speed, increased speed, reduced speed, distance, usual time, increased time, reduced time. ***[IBPS (PO-Mains)-18]***

Problem 12 **Yesterday Julli was 2 h late than her normal time to leave for her office. Thus, she increased her speed by 6 km/h and reached at usual time. Find the increased speed, if she used to go 72 km.**

Sol. Let the normal speed be x km/h.

Then, $\frac{72}{x} - \frac{72}{(x+6)} = 2$

$\therefore \quad x = 12$

Hence, increased speed $= (x + 6) = 12 + 6 = 18$ km/h

Alternatively I

Let S_1 be the normal speed and T_1 is the time taken at this speed. Further, S_2 be the altered speed and T_2 is the time taken at this speed.

Then, $\frac{S_1}{S_2} = \frac{T_2}{T_1}$, when D is constant.

$$\Rightarrow \quad \frac{S_1 + S_2}{S_1 - S_2} = \frac{T_2 + T_1}{T_2 - T_1} \quad \text{[by componendo and dividendo]}$$

$$\Rightarrow \quad \frac{S_1 - S_2}{T_2 - T_1} = \frac{S_1 + S_2}{T_2 + T_1}$$

$$\Rightarrow \text{Distance} \times \frac{S_1 - S_2}{T_2 - T_1} = \text{Distance} \times \frac{S_1 + S_2}{T_2 + T_1}$$

$$\Rightarrow \text{Distance} \times \frac{S_1 - S_2}{T_2 - T_1} = \text{Distance} \times \frac{\frac{D}{T_1} + \frac{D}{T_2}}{T_2 + T_1}$$

$$\Rightarrow \text{Distance} \times \frac{S_1 - S_2}{T_2 - T_1} = \text{Distance} \times \frac{D\left\{\frac{T_2 + T_1}{T_1 T_2}\right\}}{T_2 + T_1}$$

$$\Rightarrow \quad D \times \frac{S_1 - S_2}{T_2 - T_1} = D \times D\left\{\frac{1}{T_1 T_2}\right\}$$

$$\Rightarrow \quad D \times \frac{S_1 - S_2}{T_2 - T_1} = \frac{D}{T_1} \times \frac{D}{T_2}$$

$$\Rightarrow D \times \frac{\text{Alternation in speed}}{\text{Alternation in time}} = S_1 \times S_2$$

$$S_1 \times S_2 = D \times \frac{\text{Alternation in speed}}{\text{Alternation in time}}$$

$$\Rightarrow \quad S_1 \times S_2 = D \times \frac{\Delta S}{\Delta T}$$

Similarly, $T_1 \times T_2 = D \times \frac{\Delta T}{\Delta S}$

For the above question, $S_1 \times S_2 = 72 \times \frac{6}{2} = 216$

$\Rightarrow \quad S_1 \times (S_1 + 6) = 216$

$\Rightarrow \quad S_1 = 12$

Then, $\quad S_2 = 12 + 6 = 18$ km/h

Alternatively II

Factorise 72 and select the correct combination.

Factorisation will be as under.

Time × Speed = Distance

1 × 72 = 72

2 × 36 = 72

3 × 24 = 72

4 × 18 = 72

6 × 12 = 72

8 × 9 = 72

Clearly, combinations written in bold satisfy the conditions envisaged in the question. Hence, correct answer = 18 km/h.

Problem 13 **A car covers a certain distance in $4\frac{1}{2}$ h. If the speed is increased by 5 km/h, it would take $\frac{1}{2}$ h less to cover the same distance. Find the slower speed of the car.**

Sol. We know that $\frac{S_1}{S_2} = \frac{T_2}{T_1}$, where D is constant.

$$\Rightarrow \quad \frac{x}{x+5} = \frac{4}{4\frac{1}{2}} \Rightarrow 4\frac{1}{2}x = 4x + 20$$

$$\Rightarrow \quad \frac{1}{2}x = 20 \Rightarrow x = 40$$

Hence, slower speed of the car = 40 km/h

Alternatively

Due to increment in speed, an additional distance of (5 × 4) km can be covered with the increased speed in 4 h. Obviously, this additional distance can be covered by the car at initial speed in $\frac{1}{2}$ h.

Hence, initial speed $= 20 \div \frac{1}{2} = 40$ km/h

Problem 14 With a uniform speed, a car covers the distance in 8 h. Had the speed been increased by 4 km/h the same could have been covered in $7\frac{1}{2}$ h. What is the distance covered? *[Indian Bank (PO)-08]*

Sol. We know that $\frac{S_1}{S_2} = \frac{T_2}{T_1}$, where D is constant.

$\Rightarrow \frac{7\frac{1}{2}}{8} = \frac{x}{x+4} \Rightarrow 8x = 7\frac{1}{2}x + 30$

$\Rightarrow \frac{1}{2}x = 30 \Rightarrow x = 60$

Hence, initial speed = 60 km/h

Then, required distance = 60 × 8 = 480 km

Alternatively Due to increment in speed, an additional distance of (4 × 7.5) km can be covered with the increased speed in 7.5 h. Obviously, this additional distance can be covered by the car at initial speed in $\frac{1}{2}$ h.

Hence, initial speed = $30 \div \frac{1}{2} = 60$ km/h

Then, required distance = 60 × 8 = 480 km

Problem 15 A car had to go from Patna to Lucknow, but it met accident at 80 km away from Patna resulting in it covered the remaining distance with the speed of $\frac{4}{5}$ of its former speed and took $1\frac{1}{2}$ h more than usual time to reach Lucknow. If the accident had occurred 50 km further, it would have reached Lucknow only 1 h late. What is the distance between Patna and Lucknow?

Sol. •<.........80 km ....>•< ....50 km .... >•.........•

Patna(P) (M) (N) (L) Lucknow

Suppose M is the place/point where accident occurs.

After accident, speed decreases by 20%, then time will increase by $\left(\frac{20}{80} \times 100\right)$% i.e. 25%.

As given, 25% of the usual time taken to cover the distance beyond the accident point = $1\frac{1}{2}$ h

Hence, usual time taken to cover the distance beyond the accident point = 6 h

i.e. normal time taken by car to reach from point M to L = 6 h

Further, in case an accident occurs at point N, which is 50 km far away from M, then 25% of the usual time taken to cover the distance beyond the accident point (N) = 1 h

Then, usual time taken to cover the distance beyond the accident point (N) = 4 h

i.e. normal time taken by car to reach from point N to L = 4 h

From the above, usual time taken from N to Lucknow = 4 h and usual time taken from M to Lucknow = 6 h. It means, usual time taken from M to N (i.e. 50 km) = 2 h

Then, usual speed = $\frac{50}{2}$ = 25 km/h

Now, distance from point M to Lucknow

= usual time taken × usual speed = 6 × 25 = 150 km

Hence, total distance from Patna to Lucknow = 80 + 150

= 230 km

Problem 16 A runs twice as fast as B and B runs thrice as fast as C. The distance covered by C in 72 min, will be covered by A in ?

Sol. Clearly ratio of speeds of A, B and C = 6 : 3 : 1

Then, ratio of time taken by A, B and $C = \frac{1}{6} : \frac{1}{3} : \frac{1}{1}$

$[\because T \propto \frac{1}{S}$, when D is constant$]$

$= \frac{1}{6} \times 6 : \frac{1}{3} \times 6 : \frac{1}{1} \times 6 = 1 : 2 : 6$

Given, 6 units = 72 min, then 1 unit = 12 min

Hence, time taken by A = 12 min

Problem 17 Two cars travel the same distance starting at 10:00 AM and 11:00 AM, respectively, on the same day. They reach their common destination at the same point of time. If the 1st car travelled for at least 6 h, then the highest possible value of the percentage by which the speed of the 2nd car could exceed that of the first car is : *[CAT-19]*

Sol. Since the 1st car travels for at least 6 h, the 2nd car will move for at least (6 – 1) h. Then ratio of their speeds [1st car's to 2nd car's] = 5 : 6. Clearly, speed of 2nd car is 20% more than that of 1st car.

Now, we have to check as to whether 20% is the highest possible value. We know that when two unequal numbers are increased gradually by the same number [in order to maintain same difference], the increase in the greater number [in terms of percentage], with comparison to smaller increased number, is gradually decreased. In other words, 6 is 20% greater than 5, but (6 + 1) will be greater than (5 + 1) by less than 20% and so on.

Hence, required answer = 20%

B. Illustrations Based on the Concept "Time α Distance, when Speed is Constant"

Problem 18 A man moves at a speed of 30 km/h for $2\frac{1}{2}$ h and another man moves at the same speed for 4 h. What is the ratio of distance covered by them?

Sol. We know that Distance α Time, when Speed is constant.

Then, ratio of distance covered by them = ratio of time taken by them = $2\frac{1}{2} : 4 = 5 : 8$

Problem 19 A covers some distance in 50 days when he rests 9 h a day. In how many days will he cover the double distance by resting twice as before? *[SSC(Asst-Mains-95, LIC(AAO)-12)]*

Sol. We know that $\frac{T_1}{T_2} = \frac{D_1}{D_2}$, if speed is constant

$\Rightarrow \quad \dfrac{50 \text{ days} \times (24-9)\text{ h}}{x \text{ days} \times (24-9\times2)\text{ h}} = \dfrac{D_1}{D_2}$

$\Rightarrow \quad \dfrac{50\times15}{x\times6} = \dfrac{D_1}{2D_1}$

$\Rightarrow \quad x = 250$

Hence, time required = 250 days

Problem 20 **A train approaches a tunnel AB. Inside the tunnel a cat located at a point that is 3/8 of the distance AB measured from the entrance A. When the train whistles, the cat runs. If the cat moves to the entrance of the tunnel 'A', the train catches the cat exactly at the entrance. If the cat moves to the exit 'B', the train catches the cat at exactly the exit. The speed of the train is greater than the speed of the cat by what order?** *[CAT-02]*

Sol. Let speed of the cat be 1m/sec. and the length of the tunnel AB be 8 m.

Then, time taken by the cat to reach at the entrance point 'A' = 3 sec. as it is at 3 m from the entrance. Further, it will take 5 sec. to reach at the exit.

Obviously, time taken by the train to reach at the entrance = 3 sec and time taken by the train to reach at the exit = 5 sec.

Let the train is d m before the entrance point 'A', when it whistles.

It implies that the train covers d m in 3 sec. and $(d + 8)$ m in 5 sec.

Since speed of the train is constant, $\dfrac{T_1}{T_2} = \dfrac{D_1}{D_2}$

$\Rightarrow \quad \dfrac{3}{5} = \dfrac{d}{d+8}$

$\Rightarrow \quad d = 12$ m

$\therefore$ Speed of the train = 12 ÷ 3 = 4 m/sec.

Hence, ratio of speeds of the train and that of the cat = 4 : 1

C. Illustrations Based on the Concept "Speed α Distance, when time is Constant"

Problem 21 **A man can reach a certain place in 30 h. If he reduces his speed by $\dfrac{1}{15}$th, he goes 10 km less in that time. Find his speed.**

Sol. Let the speed be x km/h.

Then, distance = $30 \times x = 30x$

Later, speed becomes $\dfrac{14x}{15}$

Then, distance = $\dfrac{14x}{15} \times 30 = 28x$

As per question, $30x - 28x = 10$

$\Rightarrow \quad x = 5$

Hence, speed of the car = 5 km/h

Alternatively I

We know, $\dfrac{S_1}{S_2} = \dfrac{D_1}{D_2}$, when time is constant.

$\Rightarrow \quad \dfrac{\text{Initial speed}}{\text{Reduced speed}} = \dfrac{\text{Initial distance}}{\text{Initial distance - 10}}$

$\Rightarrow \quad \dfrac{1}{\frac{14}{15}} = \dfrac{d}{d-10}$

$\Rightarrow \quad 15d - 150 = 14d \Rightarrow d = 150$

Now, Speed = $\dfrac{D}{T} = \dfrac{150}{30} = 5$ km/h

Alternatively II

We know that, Speed α Distance, when time is constant. If speed is reduced by 1/15, the distance will also be reduced by 1/15.

Hence, $\dfrac{1}{15}$ of the distance = 10 km

$\therefore$ Distance = 150 km

Then, speed = $\dfrac{150}{30} = 5$ km/h

Problem 22 **If a person walks at 14 km/h instead of 10 km/h, he would have walked 20 km more. The actual distance travelled by him is :** *[RRB-2000]*

Sol. Obviously, the speed increases by 40%. Then, the distance will also increase by 40%.

As given, 40% of the distance = 20 km

Then, the distance = 50 km

Alternatively

We know, $\dfrac{S_1}{S_2} = \dfrac{D_1}{D_2}$, when time is constant.

$\Rightarrow \quad \dfrac{10}{14} = \dfrac{x}{x+20}$

$\Rightarrow \quad 14x = 10x + 200$

$\Rightarrow \quad x = 50$

Hence, distance = 50 km

Problem 23 **Two buses move simultaneously from point M and N towards each other at the speed of 30 km/h and 35 km/h respectively. Find the meeting point if distance between M and N is 260 km.**

Sol. Here, both move simultaneously, it means time is constant.

Then, $\dfrac{S_1}{S_2} = \dfrac{D_1}{D_2}$

$\Rightarrow \quad \dfrac{30}{35} = \dfrac{D_1}{D_2} \Rightarrow D_1 : D_2 = 6 : 7$

Further, both buses meet at a point after moving towards each other, it means, the sum of distance travelled by 1st bus (D_1) and distance travelled by 2nd bus (D_2) is 260 km.

Hence, $D_1 + D_2 = 260$ km.

Clearly, (6 + 7) units = 260 km

Then, 6 units = 120 km

Hence, their meeting point will be at either 120 km from point M or (260 – 120) km from point N.

Problem 24 **Two buses move simultaneously from point M and N (180 km apart) towards each other. They meet at 50 km from N. Find the ratio of their speeds.**

Sol. Here, both buses move simultaneously, it means time is constant.

Then, $\frac{S_1}{S_2} = \frac{D_1}{D_2}$

$\Rightarrow \frac{S_1}{S_2} = \frac{\text{Distance travelled from point } M}{\text{Distance travelled from point } N}$

$= \frac{(180-50)}{50}$

$\Rightarrow S_1 : S_2 = 13 : 5$

Problem 25 A starts from a place P to go to a place Q. At the same time B starts from Q to P. If after meeting each other A and B took 4 and 9 h more respectively to reach their destination, the ratio of their speeds is: **[SSC(CGL-PT)-04, CAT-20]**

Sol. Since they take equal time to arrive at meeting point, we may say

$\frac{\text{Speed of } A}{\text{Speed of } B} = \frac{\text{distance covered by } A \text{ before meeting}}{\text{distance covered by } B \text{ before meeting}}$

$\Rightarrow \frac{S_1}{S_2} = \frac{\text{distance covered by } A \text{ before meeting}}{\text{distance covered by } B \text{ before meeting}}$

$= \frac{\text{distance covered by } B \text{ in 9 h after meeting}}{\text{distance covered by } A \text{ in 4 h after meeting}}$

$\Rightarrow \frac{S_1}{S_2} = \frac{9S_2}{4S_1}$

$\Rightarrow \frac{S_1}{S_2} = \sqrt{\frac{9}{4}} = \frac{3}{2}$

Hence, $\frac{\text{speed of } A}{\text{speed of } B} = \sqrt{\frac{\text{time taken by } B \text{ after meeting}}{\text{time taken by } A \text{ after meeting}}}$

Also, if they take x hours to meet each other and after meeting they take y and z hours respectively to reach the destination, then $x = \sqrt{yz}$.

Problem 26 With reference to the above question, how long will the faster car take to cover the whole distance?

Sol. Speed of faster car : Speed of slower car = 3 : 2

$\Rightarrow$ Time taken by faster car : Time taken by slower car = 2 : 3

Further, the distance covered by slower in 9 h, will be covered by the faster car in $\left(9 \times \frac{2}{3}\right)$ i.e. 6 h. In fact, this distance is the same which has been covered by the faster before the meeting.

Hence, total time taken by the faster car

= time taken before meeting + time taken after meeting

= 6 + 4 = 10 h

Problem 27 Points A, P, Q and B lie on the same line such that P, Q and B are, respectively, 100 km, 200 km and 300 km away from A. Cars 1 and 2 leave A at the same time and move towards B. Simultaneously, car 3 leaves B and moves towards A. Car 3 meets car 1 at Q, and car 2 at P. If each car is moving in uniform speed then the ratio of the speed of car 2 to that of car 1 is: **[CAT-18]**

Sol. A ……100……P……100…… Q……100………B

Here, AP = PQ = QB = 100 km

Speed of car 1 : Speed of car 3 = distance covered by car 1 : that by car 3 [$\because$ time is constant]

$\Rightarrow$ Speed of car 1 : Speed of car 3

= (100 + 100) : 100 = 2 : 1

= 4 : 2

Similarly, Speed of car 2 : Speed of car 3

= 100 : (100 + 100) = 1 : 2

Clearly, Speed of car 2 : Speed of car 1 = 1 : 4

Problem 28 Renu and Swati start at the same time to meet each other. Renu goes uphill from a location A and she moves at an average speed of 4 km/h. Swati goes downhill from a location B to meet Renu. Swati crosses a cafe at C, which is 2 km from A, after 10 min. Renu and Swati meet at point D which divides the distance A and B in the ratio 2:3. What is the distance between A and B? **[MAT-18]**

Sol. Since time is constant, $\frac{S_1}{S_2} = \frac{D_1}{D_2}$

$\Rightarrow \frac{\text{Speed of Renu}}{\text{Speed of Swati}} = \frac{\text{Distance covered by Renu}}{\text{Distance covered by Swati}} = \frac{2}{3}$

$\Rightarrow \frac{4}{\text{Speed of Swati}} = \frac{2}{3}$

$\Rightarrow$ Speed of Swati = 6 km/h

Now, distance covered by Swati in 10 min = 1 km. It implies that the distance between location B and the Cafe is 1 km.

Further, the distance between the Cafe and location A is 2 km, as given.

Then, the distance between A and B = 1 + 2 = 3 km

Problem 29 John jogs on track A at 6 km/h and Mary jogs on track B at 7.5 km/h. The total length of tracks A and B is 325 m. While John makes 9 rounds of track A, Mary makes 5 rounds of track B. In how many seconds will Mary make one round of track A? **[CAT-19]**

Sol. Let the length of track A is x m.

Then, length of track B = 325 − x

Since time is constant, $\frac{S_1}{S_2} = \frac{D_1}{D_2}$

$\Rightarrow \frac{\text{Speed of John}}{\text{Speed of Mary}} = \frac{\text{Distance covered by John}}{\text{Distance covered by Mary}}$

$\Rightarrow \frac{6}{7.5} = \frac{9x}{5(325-x)}$

$\Rightarrow x = 100$ m

Marry's speed = 7.5 km/h = $7.5 \times \frac{5}{18}$ m/sec

Then, time taken by Mary to make one round of track A [i.e. time to cover 100 m] = $100 \div \left(7.5 \times \frac{5}{18}\right)$ = 48 sec

D. Illustrations Based on Relation Among Distance, Revolutions and Circumference of a Round Object

The concepts associated with Time, Speed and Distance is also applicable in case the wheel(s) cover certain distance. Distance can be computed by multiplying the number of revolution of wheel with its circumference. Thus

Number of Revolutions (say, R) × Circumference (say, C) = Distance (say, D). Further,

(i) $R \alpha \frac{1}{C}$ or $C \alpha \frac{1}{R}$, when D is constant.

and $R_1 : R_2 = C_2 : C_1$, when D is constant.

(ii) $C \alpha D$ or $D \alpha C$, when R is constant.

and $C_1 : C_2 = D_1 : D_2$, when R is constant.

(iii) $R \alpha D$ or $D \alpha R$, when C is constant.

and $R_1 : R_2 = D_1 : D_2$, when C is constant.

Note *Since circumference α radius, we may replace circumference by radius, if required in the questions. Further, concept of 'Product Constancy' also works here.*

Problem 30 The speed of a car is 60 km/h and each of wheels of the car is rotating at the rate of 300 revolutions per minute. What is the circumference of the wheel ?

Sol. Speed of the car $= 60 \times \frac{50}{3} = 1000$ m/min

As given, the wheel makes 300 revolutions per minute. It implies that it covers 1000 m by making 300 revolutions. We know that in one revolution, the wheel covers a distance equal to its circumference.

Thus, circumference of the wheel $= 1000 \div 300 = 3.33$ m

Or, simply say, $R \times C = D$

$\Rightarrow \quad 300 \times C = 1000 \Rightarrow C = 3.33$ m

Problem 31 The wheels of three cars A, B and C have circumferences 80 cm, 100 cm and 120 cm, respectively. To reach their destinations they make revolutions in ratio 2 : 3 : 4. If difference between distances covered by car A and C, is 24 km. The distance covered by car B is :

Sol. Ratio of distances covered by them

$= 80 \times 2 : 100 \times 3 : 120 \times 4 \quad [\because R \times C = D]$

$= 8 : 15 : 24$

As given, $(24 - 8)$ units = 24 km

$\therefore$ 15 units = 22.5 km

Problem 32 The wheels of bicycles A and B have radii 30 cm and 40 cm, respectively. While travelling a certain distance, each wheel of A required 5000 more revolutions than each wheel of B. If bicycle B travelled this distance in 45 min, then its speed, in km/h was : ***[CAT-19]***

Sol. We know, $R_1 : R_2 = C_2 : C_1$, when D is constant. Since circumference is directly proportional to radius, we may say that $R_1 : R_2 =$ radius of 2nd wheel: radius of 1st wheel.

Hence, ratio of number of revolutions made by each wheel of A and that made by each wheel of B to cover the constant distance = 40 : 30 = 4 : 3

As given, $(4 - 3)$ unit $= 5000 \Rightarrow$ 3 units = 15000 revolutions

Now, distance covered by B in 15000 revolutions

$= 15000 \times 2\pi \times 40 \text{ cm} = 12\pi$ km

Then, speed of $B = \dfrac{12\pi}{\frac{3}{4}\text{h}} = 16\pi$ km/h

WORKOUT

1. A man rides $\frac{1}{3}$ rd of the distance from A to B at the rate of x km/h and the remain at the rate of $2y$ km/h. If he had travelled at a uniform rate of $6z$ km/h, then he could have ridden from A to B and back again in the same time. Which one of the following is correct ? *[CDS-14]*

(a) $z = x + y$ (b) $3z = x + y$
(c) $\frac{1}{z} = \frac{1}{x} + \frac{1}{y}$ (d) $\frac{1}{2z} = \frac{1}{x} + \frac{1}{y}$

2. A car covers a distance of 715 km at a constant speed. If the speed of the car had been 10 km/h more, then it would have taken 2 h less to cover the same distance. What is the original speed of the car ?

(a) 55 km/h (b) 50 km/h
(c) 45 km/h (d) 65 km/h

3. Two cars an Alto and a Swift start at the same time in opposite directions from two distinct points P and Q. Starting from P, the Alto reaches Q in 6 h 20 min and the Swift starting from Q reaches P in 4 h 45 min. What is the speed of the Swift, if the speed of the Alto is 60 km/h. *[JMET-11]*

(a) 110 km/h (b) 100 km/h
(c) 90 km/h (d) 80 km/h

4. A man and a woman 81 miles apart from each other, start travelling towards each other at the same time. If the man covers 5 miles/h to the woman's 4 miles/h, how far will the woman have travelled when they meet? *[SNAP-11]*

(a) 27 (b) 36
(c) 45 (d) None of these

5. A bike travels a distance of 200 km at a constant speed. If the speed of the bike is increased by 5 km/h, the journey would have taken 2 h less. What is the speed of the bike? *[CDS-13, MAT-10, 09]*

(a) 30 km/h (b) 25 km/h
(c) 20 km/h (d) 15 km/h

6. A scooterist completes a certain journey in 10 h. He covers half the distance at 30 km/h and the rest at 70 km/h. What is the total distance of the journey ?

(a) 210 km (b) 400 km
(c) 420 km (d) 500 km

7. A certain distance is covered by a cyclist at a certain speed. If a jogger covers half the distance in double the time, the ratio of the speed of the jogger to that of the cyclist is *[SSC(CGL-T-II)-13]*

(a) 1 : 4 (b) 4 : 1 (c) 1 : 2 (d) 2 : 1

8. I have to reach a certain place at a certain time and I find that I shall be 15 min. late, if I walk at 4 km an hour and 10 min too soon if I walk at 6 km an hour. How far have I to walk?

(a) 25 km (b) 5 km
(c) 10 km (d) None of these

9. The ratio between the speeds of Ramesh and Suresh is 3 : 4. If Ramesh takes 30 min more than Suresh to cover a distance, then the actual times taken by Ramesh and Suresh respectively are:

(a) 2 h and 1.5 h (b) 2.5 h and 1.5 h
(c) 2 h and 1 h (d) 2.5 h and 1 h

10. Sameer drove at the speed of 45 km/h from home to a resort. Returning over the same route, he got stuck in traffic and took an hour longer. Also, he could drive only at the speed of 40 km/h. How many kilometers did he drive each way ? *[BOB(PO)-08]*

(a) 250 km (b) 300 km
(c) 310 km (d) 275 km
(e) None of these

11. A person travels a certain distance at 3 km/h and reaches 15 min. late. If he travels at 4 km/h, he reaches 15 min earlier. The distance he has to travel is *[CDS-13, 08, SSC(SAS)-10, SSC(CGL-T-II)-10, SSC(CGL-T-II)-13, CAT-17]*

(a) 4.5 km (b) 6 km
(c) 7.2 km (d) 12 km

12. It was a rainy morning in Delhi when Rohit drove his mother to a dentist in his Maruti Alto. They started at 8:30 AM from home and Rohit maintained the speed of the vehicle at 30 km/h. However, while returning from the doctor's chamber, rain intensified and the vehicle could not move due to sever water logging. With no other alternative, Rohit kept the vehicle outside the doctor's chamber and returned home along with his mother in a rickshaw at a speed of 12 km/h. They reached home at 1 : 30 PM. If they stayed at the doctor's chamber for the dental check-up for 48 min, the distance of the doctor's chamber from Rohit's house is

(a) 15 km (b) 30 km *[IIFT-13]*
(c) 36 km (d) 45 km

13. A father and his son start at a point A with speeds of 12 km/h and 18 km/h respectively and reach another point B. If his son starts 60 min after his father at A and reaches B, 60 min before his father, what is the distance between A and B ? *[CDS-10]*

(a) 90 km (b) 72 km
(c) 36 km (d) None of these

14. P and Q are 27 km away. Two trains with speeds of 24 km/h and 18 km/h respectively start simultaneously from P and Q and travel in the same direction. They meet at a point R beyond Q. Distance QR is *[SSC(CGL-T-II)-11]*

(a) 126 km (b) 81 km
(c) 48 km (d) 36 km

15. Two trains starting at the same time from two stations, 200 km apart going in opposite directions, cross each other at a distance of 110 km from one of them. What is the ratio of their speeds ?

(a) 11 : 20 (b) 9 : 20
(c) 11 : 9 (d) 19 : 20

16. A train running between two stations A and B arrives at its destination 10 min late when its speed is 50 km/h and 50 min late when its speed is 30 km/h. What is the distance between the stations A and B?

(a) 40 km (b) 50 km
(c) 60 km (d) 70 km

17. The average speed of a car is 1.8 times the average speed of a bus. A tractor covers 575 km in 23 h. How much distance will the car cover in 4 h, if the speed of the bus is twice the speed of the tractor ? ***[Corporation Bank (PO)-11, Allahabad Bank (PO)-11]***

(a) 340 km (b) 480 km
(c) 360 km (d) 450 km
(e) None of these

18. Excluding stoppages, the speed of a bus is 54 km/h and including stoppages, it is 45 km/h. For how many minutes does the bus stop per hour ? ***[MAT-08, NIFT-02, PNB(MT)-09, IAS(Pre)-07, SSC(CGL-T-1)-17, SNAP-18]***

(a) 12 (b) 10 (c) 9 (d) 20

19. A man can walk uphill at the rate of 2.5 km/h and downhill at the rate of 3.25 km/h. If the total time required to walk a certain distance up the hill and return to the starting position is 4 h 36 min, what is the distance he walked up the hill ? ***[CDS-11, 07, LIC (AAO)-11]***

(a) 4.5 km (b) 6.5 km
(c) 7.2km (d) 12 km

20. The difference between the time taken by two cars to travel a distance of 350 km is 2 h 20 min. If the difference between their speeds is 5 km/h, then the speed of the faster car is : ***[CAT-09]***

(a) 30 km/h (b) 35 km/h
(c) 40 km/h (d) 45 km/h

21. A car starts running with the initial speed of 40 km/h, with its speed increasing every hour by 5 km/h, how many hours will it take to cover a distance of 385 km? ***[MAT-07, SSC(CGL-T-I)-17]***

(a) 9 h (b) 9.5 h
(c) 8.5 h (d) 7 h

22. A car travels a distance of 170 km in 2 h partly at a speed of 100 km/h and partly at 50 km/h. The distance travelled at speed of 50 km/h is :

(a) 50 km (b) 40 km
(c) 30 km (d) 60 km

23. The speeds of three cars are in the ratio 2:3:4. The ratio between the times taken by these cars to travel the same distance is ***[MAT-09]***

(a) 4 : 3 : 2 (b) 2 : 3 : 4
(c) 4 : 3 : 6 (d) 6 : 4 : 3

24. Two cars A and B start simultaneously from a certain place at the speed of 30 km/h and 45 km/h respectively. The car B reaches the destination 2 h earlier than A. What is the distance between the starting point and destination ?

(a) 90 km (b) 180 km
(c) 270 km (d) 360 km

25. A man cycles with a speed of 10 km/h and reaches his office at 1 PM. However, when he cycles with a speed of 15 km/h, he reaches his office at 11 AM. At what speed should he cycle so that he reaches his office at 12 noon ? ***[CDS-13, CAT-04]***

(a) 12.5 km/h (b) 12 km/h
(c) 13 km/h (d) 13.5 km/h

26. A man has to cover a distance of 6 km in 45 min. If he covers one-half of the distance in 2/3rd time, what should be his speed to cover the remaining distance in the remaining time ? ***[MAT-12, 11]***

(a) 12 km/h (b) 16 km/h
(c) 3 km/h (d) 8 km/h

27. A man travelled a total distance of 3990 km, part of it by air, part by water and the rest by land. The time he spent in travelling by air, water and land was in ratio 1 : 16 : 2 respectively and the average speed of each mode of travel was in the ratio 20 : 1 : 3 respectively. If his overall average speed was 42 km/h, find the distance covered by water.

(a) 1520 km (b) 2000 km
(c) 1330 km (d) 1975 km

28. Two trains A and B start from stations X and Y towards Y and X respectively. After passing each other, they take 4 hrs 48 min and 3 h 20 min to reach Y and X respectively. If train A is moving at 45 km/h, then the speed of the train B is? ***[SSC(CGL-T-II)-11]***

(a) 60 km/h (b) 64.8 km/h
(c) 54 km/h (d) 37.5 km/h

29. A man starts cycling from A to B and at the same time another man starts cycling from B to A along the same path. They completed their journeys in $1\frac{2}{3}$ h and $2\frac{2}{5}$ h respectively. At what speed has the second man cycled, if the first cycles at 16 km/h. ***[MAT-12]***

(a) 16.5 km/h (b) 18.5 km/h
(c) 11.11 km/h (d) 13.33 km/h

30. Anand travelled 300 km by train and 200 km by taxi. It took him 5 h 30 min. However, if he travels 260 km by train and 240 km by taxi, he takes 6 min more. The speed of the train is : ***[MAT-11, 01]***

(a) 100 km/h (b) 120 km/h
(c) 80 km/h (d) 110 km/h

31. A man driving at 3/4 of his original speed reaches his destination 20 min later than the usual time. Then, the usual time is ***[SSC(CGL-T-II)-12, SNAP-11]***

(a) 45 min (b) 60 min
(c) 75 min (d) 120 min

32. A man walks half of the journey at 4 km/h by cycle, does 1/3rd of journey at 12 km/h and rides the remaining journey in a horse cart at 9 km/h, thus completing the whole journey in 6 h 12 min. The length of the journey is:

(a) 40 km (b) 153 km
(c) 36 km (d) 28 km

33. A man in a train notices that he can count 21 telephone posts in 1 min. If they are known to be 50 m apart, then at what speed is the train travelling ?

(a) 57 km/h (b) 60 km/h
(c) 55 km/h (d) 63 km/h

34. A train, an hour after starting, meets with an accident which detains it for 30 min, after which it proceeds at 3/4th of its former rate and arrives 3.5 h late. Had the accident happened 90 km further along the line, it would have arrived only 3 h late. Find the length of the trip.

(a) 150 km (b) 200 km
(c) 330 km (d) 600 km

35. An express train travels 299 km between two cities. During the 1st 111 km of the trip, the train travelled through mountain terrain. The train travelled 10 km/h slower through mountainous terrain than through level terrain. If the total time to travel between two cities was 7 h, what is the speed of the train on level terrain ? ***[MAT-II]***

(a) 56 km/h (b) 55 km/h
(c) 47 km/h (d) 88 km/h

36. Prof. Mandal walks to the market and comes back in an auto. It takes him 90 min. to make the round trip. If he takes an auto both ways it takes him 30 min. On a Sunday, he decides to walk both ways. How long would it take him ? *[XAT-13]*

(a) 100 min (b) 120 min
(c) 140 min (d) 150 min
(e) None of these

37. A man covers a certain distance on a toy train. If the train moved 4 km/h faster, it would take 30 min less. If it moved 2 km/h slower, it would have taken 20 min more. Find the distance. *[MAT-10, 08, 06]*

(a) 60 km (b) 45 km
(c) 30 km (d) 20 km

38. The distance between two points (A and B) is 110 km. X starts running from point A at a speed of 60 km/h and Y starts running from point B at a speed of 40 km/h at the same time. They meet at a point C, somewhere on the line AB, what is the ratio of AC to BC ? *[CDS-14]*

(a) 3 : 2 (b) 2 : 3
(c) 3 : 4 (d) 4 : 3

39. A train after travelling 150 km meets with an accident and then proceeds with 3/5 of its former speed and arrives at its destination 8 h late. Had the accident occurred 360 km further, it would have reached the destination 4 h late. What is the total distance travelled by the train ?

(a) 840 km (b) 960 km
(c) 870 km (d) 1100 km

40. Ram, Shyam and Hari went out for a 100 km journey. Ram and Hari started the journey in Ram's car at the rate of 25 km/h, while Shyam walked at 5 km/h. After sometime, Hari got off and started walking at the rate of 5 km/h and Ram went back to pick up Shyam. All three reached the destination simultaneously. What was the number of hours required for the trip ? *[CAT-16]*

(a) 10 h (b) 12 h
(c) 7h (d) 8 h

41. A motor cyclist goes from Mumbai to Pune, a distance of 192 km, at an average speed of 32 km/h. Another man starts from Mumbai by car, 2.5 h after the 1st and reaches Pune half an hour earlier. What is the ratio of the speeds of the motor cycle and the car ?

(a) 10 : 27 (b) 1 : 3
(c) 1 : 2 (d) 5 : 4

42. Alok walks to a viewpoint and returns to the starting point by his car and thus takes a total time of 6 h 45 min. He would have gained 2 h by driving both ways. How long would it have taken for him to walk both ways ? *[MAT-09,08, BOM(PO)-10, IAS(Pre)-03-07]*

(a) 7 h 45 min (b) 8 h 45 min
(c) 5 h 30 min (d) None of these

43. A bus started its journey from Ramgarh and reached Devgarh in 44 min at its average speed of 50 km/h. If the average speed of the bus is increased by 5 km/h, how much time will it take to cover the same distance ? *[Corporation Bank (PO)-09]*

(a) 40 min (b) 38 min
(c) 36 min (d) 31 min
(e) 49 min

44. Deepa rides her bike at an average speed of 30 km/h and reaches her destination in 6 h. Hema covers the same distance in 4 h. If Deepa increases her average speed by 10 km/h and Hema increases her average speed by 5 km/h, what would be the difference in their time taken to reach the destination ? *[Syndicate Bank (PO)-10]*

(a) 54 min (b) 40 min
(c) 45 min (d) 1 h
(e) None of these

45. The average speed of a train is 20% less on the return journey than on the onward journey. The train halts for half an hour at the destination station before starting on the return journey. If the total time taken for the to and fro journey is 23 h, covering a distance of 1000 km, the speed of the train on the return journey is:

(a) 60 km/h (b) 40 km/h
(c) 50 km/h (d) 55 km/h

46. Mohan and Puran are running towards each other, each one from his own house. Mohan can reach Puran's house in 25 min of running, which is half the time it takes Puran to reach Mohan's house. If the two started to run at the same time, how much more time will it take Puran to reach the middle than Mohan? *[MAT-10]*

(a) 35 min (b) 25 min
(c) 12.5 min (d) 50 min

47. The average speed of a train in the onward journey is 25% more than that of the return journey. The train halts for one hour on reaching the destination. The total time taken for the complete to and fro journey is 17 h covering a distance of 800 km.
The speed of the train in the onward journey is *[IAS(Pre)-99]*

(a) 45 km/h (b) 47.06 km/h
(c) 50 km/h (d) 56.25 km/h

48. A train can travel 20% faster than a car. Both start from the point A at the same time and reach point B, 75 km away from A at the same time. On the way, however the train lost about 12.5 min while stopping at the stations. The speed of the car is :

(a) 60 km/h (b) 50 km/h
(c) 55 km/h (d) 65 km/h

49. A train covers a distance between station A and station B in 45 min. If the speed of the train is reduced by 5 km/h, then the same distance is covered in 48 min. The distance between stations A and B is : *[SSC(CGL-T-II)-11, CLAT-16]*

(a) 60 km (b) 64 km
(c) 80 km (d) 55 km

50. An express train travelled at an average speed of 100 km/h, stopping for 3 min after 75 km. A local train travelled at a speed of 50 km/h, stopping for 1 min after every 25 km. If the trains began travelling at the same time, how many kilometers did the local train travel in the time it took the express train to travel 600 km ?

(a) 900 km (b) 307.5 km
(c) 1200 km (d) 1000 km

51. Train A takes 1 h more than B to travel a distance of 720 km. Due to engine trouble speed of train B falls by a 3rd, so it takes 3 h more than train A to complete the same journey. What is the speed of train A (in km/hr) ? **[SSC(CGL-T-II)-17]**

(a) 80 (b) 90
(c) 60 (d) 70

52. Two buses, one moving towards North and the other towards East, leave the same place at the same time. The speed of one of them is greater than that of the other by 5 km/h. At the end of 2 h, they are at a distance of 50 km from each other. The speed of the bus going slower is : **[MAT-08]**

(a) 15 km/h (b) 12 km/h
(c) 10 km/h (d) 20 km/h

53. A passenger train departs from Delhi at 6 : 00 PM for Mumbai. At 9 : 00 PM, an express train, whose average speed exceeds that of the passenger train by 15 km/h leaves Mumbai for Delhi. Two trains meet each other mid-route. At what time do they meet, given that the distance between the cities is 1080 km ? **[CDS-17]**

(a) 4 PM (b) 2 PM
(c) 12 mid-night (d) 6 PM

54. Rampur is 100 km from Sitapur. At 3:00 pm, Bharat Express leaves Rampur for Sitapur and travels at a constant speed of 30 km/h. 1 h later, Laxman Mail leaves Sitapur for Rampur and travels at a constant speed of 40 km/h. Each train makes one stop only at a station 10 km from its starting point and remains there for 15 min. Which train is nearer to Rampur when they meet?

(a) both are equidistant
(b) Laxman Mail
(c) Bharat Express
(d) None of the above

55. Pradeep could either walk or drive to office. The time taken to walk to the office is 8 times the driving time. One day, his wife took the car making him walk to office. After walking 1 km, he reached a temple when his wife called to say that he can now take the car. Pradeep figured that continuing to walk to the office will take as long as walking back home and then driving to the office. Calculate the distance between the temple and the office. **[XAT-16]**

(a) 1 (b) 7/3
(c) 9/7 (d) 16/7
(e) 16/9

56. Train A leaves Ludhiana for Delhi at 11 : 00 AM, running at the speed of 60 km/h . Train B leaves Ludhiana for Delhi by the same route at 2 : 00 PM on the same day, running at the speed of 72 km/h. At what time will the two trains meet each other ?

(a) 5 : 00 AM on the next day
(b) 2 : 00 AM on the next day
(c) 5 : 00 PM on the next day
(d) None of the above

57. Train A running at 60 km/h leaves Mumbai for Delhi at 6 : 00 PM. Train B running at 90 km/h also leaves Mumbai for Delhi at 9 : 00 PM. Train C leaves Delhi for Mumbai at 9 : 00 PM. If all the trains meet at the same time between Mumbai and Delhi, then what is the speed of train C if distance between Delhi and Mumbai is 1260 km ?

(a) 60 km/h (b) 90 km/h
(c) 120 km/h (d) 135 km/h

58. The distance from A to B is 60 km. Partha and Narayan start from A at the same time and move towards B. Partha takes 4 h more than Narayan to reach B. Moreover, Partha reaches the mid-point of A and B, 2 h before Narayan reaches B. The speed of Partha, in km per hour, is : **[CAT-18]**

(a) 4 (b) 6 (c) 5 (d) 3

59. Train T leaves station X for station Y at 3 : 00 PM. Train S, travelling at three quarters of the speed of T, leaves Y for X at 4 pm. The two trains pass each other at a station Z, where the distance between X and Z is 3/5th of that between X and Y. How many hours does train T take for its journey from X to Y? **[CAT-18]**

(a) 15 (b) 10
(c) 8 (d) 12

60. Joseph diametrically crosses a semi-circle playground and takes 48 sec less than if he crosses the playground along the semicircular path. If he walks 50 m in 1 min the diameter of playground is **[IIFT-18]**

(a) 54 m (b) 70 m
(c) 35 m (d) 85 m

61. Two persons started simultaneously from a point, one to the West direction and another to the North. The distance between them is 100 km after 2 h. What is the speed of the faster person, if the speed of one of them was 75% of the speed of the other. **[MAT-19]**

(a) 30 km/h (b) 40 km/h
(c) 42 km/h (d) 48 km/h

62. Land Cruiser Prado, the latest SUV from Toyota Motors, consumes diesel at the rate of $\frac{1}{400}\left(\frac{1000}{x}+x\right)$ L/km when travelling at the speed of x km/h. The diesel costs ₹ 65/L and the driver is paid ₹ 50/h. Find the steady speed that will minimise the total cost of a 1000 km trip. **[IIFT-18]**

(a) 33 km/h (b) 36 km/h
(c) 39 km/h (d) 52km/h

63. On can use three different transports which move at 10, 20 and 30 km/h, respectively. To reach from A to B, Amal took each mode of transport 1/3 of his total journey time, while Bimal took each mode of transport 1/3 of the total distance. The percentage by which Bimal's travel time exceeds Amal's travel time is nearest to: **[CAT-19]**

(a) 21 (b) 22 (c) 20 (d) 19

64. A cyclist leaves A at 10 : 00 AM and reaches B at 11 : 00 AM. Starting from 10 : 01 AM, every minute a motor cycle leaves A and moves towards B. Forty-five such motor cycles reach B by 11 : 00 AM. All motor cycles have the same speed. If the cyclist had doubled his speed, how many motor cycles would have reached B by the time the cyclist reached B ? **[CAT-19]**

(a) 23 (b) 20 (c) 15 (d) 22

Answer Keys

1. (c)	2. (a)	3. (d)	4. (b)	5. (c)	6. (c)	7. (a)	8. (b)	9. (a)	10. (e)
11. (b)	12. (c)	13. (b)	14. (b)	15. (c)	16. (b)	17. (c)	18. (b)	19. (b)	20. (a)
21. (d)	22. (c)	23. (d)	24. (b)	25. (b)	26. (a)	27. (a)	28. (c)	29. (c)	30. (a)
31. (b)	32. (c)	33. (b)	34. (d)	35. (c)	36. (d)	37. (a)	38. (a)	39. (c)	40. (d)
41. (c)	42. (b)	43. (a)	44. (a)	45. (b)	46. (c)	47. (d)	48. (a)	49. (a)	50. (b)
51. (a)	52. (a)	53. (d)	54. (a)	55. (c)	56. (a)	57. (c)	58. (c)	59. (a)	60. (b)
61. (b)	62. (b)	63. (b)	64. (c)						

Hints and SOLUTIONS

1. Let the distance be $3d$ units.

Time taken to cover a distance d at the rate of x km/h $= \frac{d}{x}$

Time taken to cover a distance $2d$ at the rate of $2y$ km/h $= \frac{2d}{2y} = \frac{d}{y}$

and time taken to cover a distance $6d$ at the rate of $6z$ km/h

$= \frac{6d}{6z} = \frac{d}{z}$

As per question, $\frac{d}{z} = \frac{d}{x} + \frac{d}{y}$

$\Rightarrow \quad \frac{1}{z} = \frac{1}{x} + \frac{1}{y}$

2. We know, $S_1 \times S_2 = D \times \frac{\Delta S}{\Delta T}$

$S_1 \times S_2 = 715 \times \frac{10}{2}$

$\Rightarrow \quad S_1 \times (S_1 + 10) = 3575$

$\therefore \quad S_1 = 55$ km/h

Alternatively

The best way is to proceed through options.

3. Since time is constant, ratio of speed of Alto to that to Swift

= 4 h 45 min : 6 h 20 min

$\Rightarrow$ speed of Alto : speed of Swift

= 285 : 380

= 3 : 4

$\Rightarrow$ 60 : speed of Swift = 3 : 4

$\Rightarrow$ Speed of Swift = 80 km/h

4. Ratio of speed of man to that of woman = 5 : 4

Then, ratio of distance covered by man to that by woman

= 5 : 4 [$\because$ time is constant]

Hence, distance covered by the woman $= 81 \times \frac{4}{(5+4)} = 36$ miles

Alternatively

When they together cover 9 miles, woman covers 4 miles. Then, if they together cover 81 miles, woman will cover 36 miles.

5. Let the initial speed be x km/h. Then, time taken $= \frac{200}{x}$ h

We know that $\frac{S_1}{S_2} = \frac{T_2}{T_1}$, when D is constant.

$\Rightarrow \quad \frac{x}{x+5} = \frac{\frac{200}{x} - 2}{\frac{200}{x}}$

$\Rightarrow 200 = 200 - 2x + \frac{1000}{x} - 10$

$\Rightarrow \quad 2x^2 + 10x - 1000 = 0$

$\Rightarrow \quad x^2 + 5x - 500 = 0$

$\Rightarrow x^2 + 25x - 20x - 500 = 0$

$\Rightarrow x(x+25) - 20(x+25) = 0$

$\Rightarrow \quad x = 20$ km/h

Alternatively

The best way is to proceed through options.

6. Since the distance is constant[half-half], ratio of time taken by him = 70 : 30 = 7 : 3

[$\because$ time is inversely proportional to speed]

Now, time taken to cover 1st half of the distance at the rate of 30 km/h

$= 10 \times \frac{7}{(7+3)} = 7$ h

Hence, half of the distance

$= 30 \times 7 = 210$ km

$\therefore$ Total distance $= 2 \times 210 = 420$ km

7. Let distance be d km and time taken by the cyclist be t h. Then, ratio of speeds $= \left(\frac{d}{2} \div 2t\right) : (d \div t)$

$= \frac{d}{4t} : \frac{d}{t} = 1:4$

8. If the speed alters from x to y and the time interval is T, then the distance will be

$\frac{xy}{\text{difference between } x \text{ and } y} \times T$

$\therefore$ Required answer

$= \frac{4 \times 6}{(6-4)} \times \frac{(15+10)}{60} = 5$ km

9. Ratio of speeds of Ramesh and Suresh = 3 : 4

Then, ratio of time taken by Ramesh and Suresh = 4 : 3

[$\because$ distance is constant]

As given, (4 – 3) unit = 30 min.

$\therefore$ 4 units = 120 min and 3 units = 90 min

Hence, option(a) is correct.

10. We know, if the speed alters from x to y and the time interval is T, then the distance will be

$\frac{xy}{\text{difference between } x \text{ and } y} \times T.$

Here, unit of the time must be same as envisaged in the speed.

$\therefore$ Required answer

$= \frac{45 \times 40}{(45-40)} \times 1 = 360$ km

11. It is clear from the question that when he increases his speed by 33.33%, he saves (15 + 15) min to cover the same distance. In case speed increases by 33.33%, time will reduce by $\left(\frac{33.33}{133.33} \times 100\right)$% i.e. by 25%.

Hence, 25% of the usual time taken
$= 30$ min

Then, 100% of the usual time taken
$= 2$ h

Now, distance $= 3 \times 2 = 6$ km

12. Total time taken
$= (1:30 \text{ PM} - 8:30 \text{ AM}) = 5$ h

Actual time taken in journey
$= 5$ h $- 48$ min
$= 4$ h 12 min

Since the distance is constant for both modes of journey, ratio of time taken by car to that by rickshaw
$= 12:30 = 2:5$

Then, time taken by car
$= 4 \text{ h } 12 \text{ min} \times \frac{2}{(2+5)} = 72$ min

Hence, distance covered by car in 72 min $= 30 \times 1.2 = 36$ km

13. Since the speed of son is 50% more than that of his father, time taken by son will be

$\left(\frac{50}{100+50} \times 100\right)$% i.e. 33.33% less than that by his father.

But, given that he takes 2 h less than time taken by his father. It implies that 33.33% of the time taken by his father = 2 h.

Then, 100% of the time taken by his father = 6 h

Now, distance $= 12 \times 6 = 72$ km

14. Here, time is constant.

Then, $\frac{S_1}{S_2} = \frac{D_1}{D_2}$

$\Rightarrow \frac{24}{18} = \frac{27 + QR}{QR}$

$\Rightarrow 4QR = 81 + 3QR$

$\Rightarrow QR = 81$ km

15. Since time is constant, distance α speed.

Now, ratio of distances covered by them $= 110:(200 - 110) = 11:9$

$\therefore$ Ratio of their speeds $= 11:9$

16. We know, if the speed alters from x to y and the time interval is T, then the distance will be

$\frac{xy}{\text{difference between } x \text{ and } y} \times T.$

Here unit of the time must be same as envisaged in the speed.

$\therefore$ Required answer
$= \frac{50 \times 30}{(50-30)} \times \frac{(50-10)}{60} = 50$ km

17. Speed of the tractor
$= 575 \div 23 = 25$ km/h

Now, ratio of speeds of car, bus and tractor $= 1.8:1:\frac{1}{2} = 3.6:2:1$

Clearly, 1 unit = 25 km/h then, 3.6 units = 90 km/h

$\therefore$ Distance covered by the car in 4 h
$= 90 \times 4 = 360$ km

18. Clearly, due to stoppages, the bus covers 9 km less in 1 h. Now, time taken to cover 9 km at the rate of 54 km/h $= \frac{1}{6}$ h $= 10$ min

Alternatively

Let the distance to be covered be 270 km. You are free to take any figure instead of 270.

Without stoppage, time taken to cover 270 km $= 270 \div 54 = 5$ h

and with stoppage, time taken to cover 270 km $= 270 \div 45 = 6$ h

Hence, at every 6 h, the bus stops for 1 h.

$\therefore$ Per hour it stops for 10 min.

19. Ratio of speeds $= 2.5:3.25 = 10:13$

Then, ratio of time taken $= 13:10$

Hence, time taken to walk up the hill
$= 4 \text{ h } 36 \text{ min} \times \frac{13}{(13+10)} = 2 \text{ h } 36$ min
$= 2\frac{3}{5}$ h

$\therefore$ Required distance $= 2.5 \times 2\frac{3}{5}$
$= 6.5$ km

20. Let the speed of the faster car be x km/h.

Then, $\frac{350}{x-5} - \frac{350}{x} = 2\frac{1}{3}$

$\Rightarrow \frac{50x - 50x + 250}{x(x-5)} = \frac{1}{3}$

$\Rightarrow x^2 - 5x - 750 = 0$

$\Rightarrow (x + 25)(x - 30) = 0 \Rightarrow x = 30$

Alternatively

The best way is to proceed through option.

21. We know, sum of n terms in Arithmetic Progression

$= \frac{n}{2}[2a + (n-1)d]$ .... detailed in chapter 'Progression'.

$385 = \frac{n}{2}[(2 \times 40) + (n-1)5]$

$\Rightarrow 770 = 80n + 5n^2 - 5n$

$\Rightarrow 5n^2 + 75n - 770 = 0$

$\Rightarrow n^2 + 15n - 154 = 0$

$\Rightarrow n^2 + 22n - 7n - 154 = 0$

$\Rightarrow n(n + 22) - 7(n + 22) = 0$

$\Rightarrow (n - 7)(n + 22) = 0$

$\Rightarrow n = 7$

$\therefore$ Required answer = 7 h

22. Let the distance travelled at the rate of 50 km/h be x.

Then, $\frac{x}{50} + \frac{170 - x}{100} = 2$

$\Rightarrow 2x + 170 - x = 200$

$\Rightarrow x = 30$ km

23. Since distance is constant,

time α $\frac{1}{\text{speed}}$

Hence, the ratio of times taken
$= \frac{1}{2}:\frac{1}{3}:\frac{1}{4}$
$= \frac{1}{2} \times 12 : \frac{1}{3} \times 12 : \frac{1}{4} \times 12 = 6:4:3$

24. We know that $\frac{S_1}{S_2} = \frac{T_2}{T_1}$, when D is constant.

$\Rightarrow \frac{30}{45} = \frac{t-2}{t} = 30t = 45t - 90$

$\Rightarrow t = 6$ h

Now distance $= 30 \times 6$ or $45 \times (6-2)$ i.e. 180 km

Alternatively

The best way is to proceed through options.

25. It is clear from the question that when he increases his speed by 50%, he saves 2 h to cover the same distance.

In case speed increases by 50%, time will reduce by $\left(\frac{50}{150} \times 100\right)$% i.e. 33.33%

Hence, 33.33% of the usual time taken = 2 h

Then, 100% of the usual time taken
= 6 h

Now, distance $= 10 \times 6 = 60$ km

and desired time to be taken
$= 6 - 1 = 5\,h$

$\therefore$ Required speed $= \frac{60}{5} = 12\,km/h$

26. It is given that 3 km has been covered in $45 \times \frac{2}{3}$ min i.e. in $\frac{1}{2}$ h.

Then, remaining time $= \frac{3}{4} - \frac{1}{2} = \frac{1}{4}$ h

Now, required speed

$= (6-3) \div \frac{1}{4} = 12\,km/h$

27. Ratio of speeds in air, water and land $= 20:1:3$ and time taken in air, water and land $= 1:16:2$

Hence, distance covered by different modes

$= (20 \times 1):(1 \times 16):(3 \times 2) = 20:16:6$
$= 10:8:3$

Since the total distance covered by him is 3990 km, the distance covered by water $= 3990 \times \frac{8}{(10+8+3)}$

$= 1520\,km$

28. (c) We know, $\frac{\text{Speed of 1st train}}{\text{Speed of 2nd train}} = \sqrt{\frac{\text{time taken by 2nd train after meeting}}{\text{time taken by 1st train after meeting}}}$

$\Rightarrow \frac{45}{\text{Speed of 2nd train}} = \sqrt{\frac{3h\ 20\ min}{4h\ 48\ min}}$

$\Rightarrow \frac{45}{\text{Speed of 2nd train}} = \sqrt{\frac{200\ min}{288\ min}}$

$= \sqrt{\frac{25}{36}} = \frac{5}{6}$

$\Rightarrow$ Speed of 2nd train $= 54\,km/h$

29. We know that $\frac{S_1}{S_2} = \frac{T_2}{T_1}$, when D is constant.

$\Rightarrow \frac{16}{S_2} = \frac{2\frac{2}{5}}{1\frac{2}{3}}$

$\Rightarrow S_2 = 16 \times \frac{5}{3} \times \frac{5}{12} = 11.11\,km/h$

30. Let the train covers 300 km in t hours.

Then, speed of the train $= 300/t$ km/h

and speed of the taxi $= 200 \div (5.5 - t)$

Further, time taken by train to cover 260 km $= \frac{t}{300} \times 260 = \frac{13t}{15}$

and time taken by taxi to cover 240 km

$= 240 \div \frac{200}{(5.5-t)} = \frac{6\,(5.5-t)}{5}$

As given, $\frac{13t}{15} + \frac{6(5.5-t)}{5}$

$= 5\,h\ 30\ min + 6\,min$

$\Rightarrow 13t + 99 - 18t = \frac{28}{5} \times 15$

$\Rightarrow \quad -5t = -15$

$\therefore \quad t = 3\,h$

Hence, speed of the train

$= 300 \div 3 = 100\,km/h$

31. Here, the usual time taken

$= \left\{\frac{3}{(4-3)} \times 20\right\} = 60\,min$

32. Let the length of the journey be x km.

Then, time taken to cover half of the journey at the rate of 4 km/h

$= \frac{x}{2 \times 4} = \frac{x}{8}$

Time taken to cover $\frac{1}{3}$ of the journey at the rate of 12 km/h $= \frac{x}{3 \times 12} = \frac{x}{36}$

and time taken to cover $\frac{1}{6}$ of the journey at the rate of 9 km/h

$= \frac{x}{6 \times 9} = \frac{x}{54}$

But given, $\frac{x}{8} + \frac{x}{36} + \frac{x}{54} = 6\frac{12}{60}$

$\Rightarrow \quad x = 36\,km$

33. Distance between 1st pole to 21st pole $= (21-1) \times 50 = 1000$ m. Obviously, the speed of the train is 1000 m/min. i.e. 60 km/h.

34. After accident, speed decreases by 25%, then time will increase by $\left(\frac{25}{75} \times 100\right)$% i.e. by 33.33%.

As given, 33.33% of the usual time taken to cover the distance beyond the 1st accident point $= (3.5 - 0.5)$ h

Hence, usual time taken to cover the distance beyond the 1st accident point = 9 h

Further, in case an accident occurs at another point which is 90 km far from 1st accident point, then 33.33% of the usual time taken to cover the distance beyond latter accident point $= (3 - 0.5)$ h

Then, usual time taken to cover the distance beyond the 2nd accident point = 7.5 h

From the above, usual time taken from 1st accident point to destination = 9 h and usual time taken from 2nd accident point to destination = 7.5 h.

It means usual time taken from 1st accident point to 2nd accident point (i.e. 90 km) $= (9 - 7.5)$ h

Then, usual speed $= \frac{90}{1.5} = 60$ km/h

Now, distance from 1st accident point to destination $= 60 \times 9 = 540\,km$

Hence, total distance $= 540 + (60 \times 1)$

$= 600\,km$

35. Let the speed of the train on level terrain be x km/h.

Then, $\frac{111}{x-10} + \frac{299-111}{x} = 7$

$\Rightarrow 7x^2 - 70x = 111x + 188x - 1880$

$\Rightarrow 7x^2 - 369x + 1880 = 0$

$\Rightarrow \quad x = 47\,km/h$

Alternatively

The best way is to proceed through options.

36. Time taken by the auto one way

$= \frac{30}{2} = 15\,min$

Then, time taken by walk one way

$= 90 - 15 = 75\,min$

$\therefore$ Time taken by walk both ways

$= 2 \times 75 = 150\,min$

37. ***Case I*** Speed(S) Time taken(T)

(+4) (−1/2)

Do cross-multiplication and write as below :

$+4T - \frac{1}{2}S = \left|(+4) \times \left(-\frac{1}{2}\right)\right|$

$\Rightarrow \quad +4T - \frac{1}{2}S = 2$ ... (i)

Case II Speed(S) Time taken(T)

(−2) (+1/3)

Do cross-multiplication and write as below

$-2T + \frac{1}{3}S = \left|(-2) \times \left(+\frac{1}{3}\right)\right|$

$\Rightarrow -2T + \frac{1}{3}S = \frac{2}{3}$ ...(ii)

On solving Eqs. (i) and (ii), we get $T = 3$ and $S = 20$

Then, distance $= 20 \times 3 = 60\,km$

38. Since time is constant,

$S_1 : S_2 = D_1 : D_2$

$\Rightarrow D_1 : D_2 = 60:40 = 3:2$

39. After accident, speed decreases by 40%, then time will increase by $\left(\frac{40}{60} \times 100\right)\%$ i.e. 66.67%.

As given, 66.67% of the usual time taken to cover the distance beyond the 1st accident point = 8 h.

Hence, usual time taken to cover the distance beyond the 1st accident point = 12 h

Further, in case an accident occurs at another point which is 360 km far from 1st accident point, then 66.67% of the usual time taken to cover the distance beyond latter accident point = 4 h.

Then, usual time taken to cover the distance beyond the 2nd accident point = 6 h

From the above, usual time taken from 1st accident point to destination = 12 h and usual time taken from 2nd accident point to destination = 6 h. It means usual time taken from 1st accident point to 2nd accident point (i.e. 360 km) = (12 – 6) h.

Then, usual speed $= \frac{360}{6} = 60$ km/h

Now, distance from 1st accident point to destination $= 60 \times 12$
$= 720$ km

Hence, total distance
$= 150 + 720 = 870$ km

40. Let Shyam walked x km before he was picked up by Ram and Hari covered y km by the car.

Now, time taken by Shyam to reach the destination $= \frac{x}{5} + \frac{100-x}{25}$

Time taken by Hari to reach the destination $= \frac{y}{25} + \frac{100-y}{5}$

Time taken by Ram to reach the destination
$= \frac{y}{25} + \frac{y-x}{25} + \frac{100-x}{25}$

As given,
$$\frac{x}{5} + \frac{100-x}{5} = \frac{y}{25} + \frac{100-y}{5}$$
$$= \frac{y}{25} + \frac{y-x}{25} + \frac{100-x}{25}$$
$$\Rightarrow \frac{100+4x}{25} = \frac{500-4y}{25}$$
$$= \frac{100+2y-2x}{25}$$
$$\Rightarrow 100+4x = 500-4y$$
$$= 100+2y-2x$$
$$\Rightarrow 100+4x = 500-4y$$
and $100+4x = 100+2y-2x$
$$\Rightarrow 4(x+y) = 400 \text{ and } 6x = 2y$$
$$\Rightarrow x+y = 100 \text{ and } x:y = 1:3$$
Hence, $x = 25$ km, $y = 75$ km

Now, time taken by Shyam to reach the destination
$$= \frac{x}{5} + \frac{100-x}{25} = \frac{25}{5} + \frac{100-25}{25} = 8\text{ h}$$

41. Time taken by the motor cycle
$= 192 \div 32 = 6$ h

Ratio of time taken by motor cycle and car
$= 6 : (6 - 2.5 - 0.5) = 2 : 1$

$\therefore$ Ratio of speeds of motor cycle and car
$= 1 : 2$ [$\because$ distance is constant]

42. Time taken by the car for one way
$= \frac{6\text{h } 45\text{ min} - 2}{2} = 2\text{h } 22.5$ min

Then, time taken by walking for one way = 6 h 45 min – 2 h 22.5 min
= 4 h 22.5 min

$\therefore$ Time taken by walking for both ways
$= 2 \times$ (4h 22.5 min)
= 8 h 45 min

43. Since speed has been increased by 10%, the time taken will be $\left(\frac{10}{110} \times 100\right)$ % i.e. 9.09% less than the earlier time taken.

Hence, latter time taken = 90.9% of 44
= 40 min

44. (*a*) Distance covered by Deepa
$= 30 \times 6 = 180$ km

Then, speed of Hema
$= 180 \div 4 = 45$ km/h

Now, speeds of Deepa and Hema are (30 + 10) km/h and (45 + 5) km/h respectively.

$\therefore$ Difference between times taken by them
$= (180 \div 40) - (180 \div 50) = 4\frac{1}{2}\text{ h} - 3\frac{3}{5}\text{ h}$
= 54 min

45. Total time taken for the to and fro journey = 23 – 0.5 = 22.5 h

Ratio of speeds during onward and return journey = 100 : 80 = 5 : 4

Then, ratio of time taken for onward and return journey = 4 : 5

Hence, time taken for the return journey $= 22.5 \times \frac{5}{(4+5)} = 12.5$ h

$\therefore$ Speed of the train for the return journey $= 500 \div 12.5 = 40$ km/h

46. Time taken by Mohan to reach Puran's house = 25 min

Then, time taken by Mohan to reach the middle = 12.5 min

And time taken by Puran to reach the middle $= 2 \times 12.5 = 25$ min as Puran takes time twice the time taken by Mohan in order to cover the same distance.

Hence, extra time taken by Puran
= 25 – 12.5 = 12.5 min

47. The distance for onward journey and return journey each = 400 km and total effective journey time
= 17 – 1 = 16 h

Ratio of speeds during onward and return journeys = 125 : 100

Then, ratio of time taken
= 100 : 125 = 4 : 5

The time taken in onward journey
$= 16 \times \frac{4}{9} = \frac{64}{9}$ h

Hence, speed $= 400 \div \frac{64}{9} = 56.25$ km/h

48. Since, speed of the train is 20% more than that of a car, train will take $\left(\frac{20}{120} \times 100\right)$ % i.e. 16.66% less time than the time taken by the car. However, both reach at the same time due to stoppage of train for 12.5 min. It implies that this stoppage-time has been adjusted against the reduction in time taken by train due to its greater speed. Hence, 16.66% of the time taken by car = 12.5 min. Then, 100% of the time taken by car = 75 min

Clearly, the car takes 75 min to move 75 km.

Thus, its speed is 60 km/h.

49. Here, distance is constant.

Then, $\frac{S_1}{S_2} = \frac{T_2}{T_1}$

$$\Rightarrow \frac{S_1}{(S_1 - 5)} = \frac{48}{45} \Rightarrow 15S_1 = 16S_1 - 80$$
$$\Rightarrow S_1 = 80\text{ km/h}$$

Hence, required distance
$= 80 \times \frac{3}{4} = 60$ km

50. Time taken to cover 600 km by the express train = (600 ÷ 100) h + [(600 ÷ 75) – 1] × 3 min = 6 h 21 min

The local train travels 50 km in 1 h 2 min as it stops 1 min for every 25 km. Then, distance travelled by local train in
6 h 12 min. $= 50 \times 6 = 300$ km

Further, in remaining 9 min, it will cover $\left(\frac{50}{60} \times 9\right)$ i.e 7.5 km

Hence, total distance covered by the local train $= 300 + 7.5 = 307.5$ km

51. Factorise 720

1 × 720	6 × 120	15 × 48
2 × 360	<u>8 × 90</u>	16 × 45
3 × 240	<u>9 × 80</u>	18 × 40
4 × 180	10 × 72	20 × 36
5 × 144	<u>12 × 60</u>	24 × 30

Clearly, speeds of A and B are 80 km/h and 90 km/h respectively as $(9 - 8) = 1$provided other conditon is also satisfied. Now, check :

When 90 is reduced by 1/3rd, it becomes 60 and train B will take 12 h to cover 720 km and it is true that 12 h is 3 h more than 9 h. Hence, option (a) is correct.

52. Let the speed of the slower bus be x km/h.

Then, speed of the faster bus
$= (x + 5)$ km/h

The distances covered by them in 2 h are $2x$ and $(2x + 10)$ km respectively.

Since they are moving in North and East directions, the length of the hypotenuse will be 50 km.

Hence, $(2x)^2 + (2x + 10)^2 = (50)^2$

$\Rightarrow \quad 8x^2 + 40x - 2400 = 0$

$\Rightarrow \quad x^2 + 5x - 300 = 0$

$\Rightarrow x^2 + 20x - 15x - 300 = 0$

$\Rightarrow x(x + 20) - 15(x + 20) = 0$

$\Rightarrow \quad (x + 20)(x - 15) = 0$

$\Rightarrow \quad x = 15$ km/h

53. Let both the trains meet after x h since 6 : 00 PM and speed of the passenger train be y km/h.

Since distances covered be them are constant,

$$\frac{\text{Speed of passenger train}}{\text{Speed of express train}} = \frac{\text{Time taken by express train}}{\text{Time taken by passenger train}}$$

$$\Rightarrow \frac{540 \div x}{(540 \div x) + 15} = \frac{x - 3}{x}$$

$$\Rightarrow 540 + 15x - \frac{1620}{x} - 45 = 540$$

$$\Rightarrow \quad x - \frac{108}{x} - 3 = 0$$

$$\Rightarrow \quad x^2 - 3x - 108 = 0$$

$$\Rightarrow \quad x = 12$$

Hence, both the trains will meet each other 12 h beyond 6 : 00 PM i.e. at 6 : AM morning.

54. Wherever both trains meet, they will be equidistant from a given point.

55. Ratio of walking time and driving time $= 8 : 1$

$\therefore$ Ratio of speed of walking and driving $= 1 : 8$

Let walking speed and driving speed be y km/h and $8y$ km/h and the distance between the temple and the office be x km.

Then, distance between home and the office $= (1 + x)$ km

Now, time taken by walking from temple to office $= \frac{x}{y}$

and time taken by walking from temple to home $= \frac{1}{y}$

and time taken by driving from home to office $= \frac{1 + x}{8y}$

As per question, $\frac{x}{y} = \frac{1}{y} + \frac{1 + x}{8y}$

$$\Rightarrow \quad \frac{x}{y} = \frac{9 + x}{8y}$$

$$\Rightarrow \quad x = \frac{9}{7} \text{ km}$$

56. The distances travelled by the both trains are the same but the time taken by the faster train is 3 h less than that by the slower.

Further, we know that

$$\frac{S_1}{S_2} = \frac{T_2}{T_1}$$

when D is constant

$$\Rightarrow \frac{\text{Speed of train } B}{\text{Speed of train } A} = \frac{\text{Time taken by train } A}{\text{Time taken by train } B}$$

$$\Rightarrow \frac{72}{60} = \frac{\text{Time taken by train } A (\text{say } T)}{T - 3}$$

$\Rightarrow T = 18$ h

Hence, at 18 h beyond 11 am i.e. at 5 am on the next day, they will meet.

57. The distances travelled by the trains A and B are the same but the time taken by the train B is 3 hrs less than that by A. Further, we know that $\frac{S_1}{S_2} = \frac{T_2}{T_1}$, when D is contant

$$\Rightarrow \frac{\text{Speed of train } B}{\text{Speed of train } A} = \frac{\text{Time taken by train } A}{\text{Time taken by train } B}$$

$$\Rightarrow \frac{90}{60} = \frac{\text{Time taken by train } A \text{ (say } T)}{T - 3}$$

$\Rightarrow T = 9$ hrs

Hence, time taken by train B to overtake train $A = 9 - 3 = 6$ h

Since train B and C start at the same time, train C will also take 6 h to meet with train B/train A.

The distance covered by train B in 6 h $= 90 \times 6 = 540$ km

Obviously, the remaining distance i.e. $(1260 - 540)$ km is covered by the train C in 6 h.

Hence, speed of train C

$= (1260 - 540) \div 6 = 120$ km/h

58. Let Narayan takes t hours to reach at B. Then, Partha will take $(t + 4)$ hours.

Clearly, time taken by Partha to reach at mid-point of A and $B = \frac{t + 4}{2}$

As given, $\frac{t + 4}{2} = t - 2 \Rightarrow t = 8$ h

Now, speed of partha

$$= \frac{60}{8 + 4} = 5 \text{ km/h}$$

59. X ...... (60%)......Z......(40%)......Y

[Here, $XZ = 60$ % of XY and $ZY = 40\%$ of XY]

Ratio of time taken by train T to travel 60% : Ratio of time taken by train S to travel 40%

$$= \frac{60\%}{\text{speed of Train } T} : \frac{40\%}{75\% \text{ of speed of Train } T} = 9 : 8$$

Now, $(9 - 8)$ unit $= 1$ h

[$\because$ train S starts to move 1 h later]

$\therefore \quad$ 9 units $= 9$ h

It implies that train T takes 9 h to cover 60% of the total distance. Then, it will obviously take 15 h to move 100% of total distance i.e. XY.

60. Length of the semi-circular path $= \pi r$

and length of the diameter $= 2r$

It may be construed from the question that distance $(\pi r - 2r)$ can be covered by Joseph in 48 sec. at the rate of 50 m/min.

Hence, $(\pi r - 2r) = 50 \times \frac{48}{60} = 40$

$\Rightarrow \quad 2r\left(\frac{22}{14} - 1\right) = 40$

$\Rightarrow \quad 2r = 70\,\text{m}$

61. Let the speed of the faster person be x km/h.

Then, speed of the slower $= \frac{3x}{4}$

Obviously, the distances covered by the faster and the slower person in 2 h will be $2x$ and $\frac{3x}{2}$.

As given, $(2x)^2 + \left(\frac{3x}{2}\right)^2 = (100)^2$

[$\because$ the diagonal distance between both is 100 km]

$\Rightarrow \frac{25x^2}{4} = 10000 \Rightarrow x = 40\,\text{km/h}$

62. Diesel consumed

$= 1000 \times \frac{1}{400}\left(\frac{1000}{x} + x\right)$

Then, cost of diesel

$= 65 \times 1000 \times \frac{1}{400}\left(\frac{1000}{x} + x\right)$

$= \frac{162500}{x} + \frac{325x}{2}$

Cost of driver $= \frac{1000}{x} \times 50$

Now, total cost [let, c]

$= \frac{162500}{x} + \frac{325x}{2} + \frac{50000}{x}$

$= \frac{212500}{x} + \frac{325x}{2}$

$\frac{dc}{dx} = \frac{-212500}{x^2} + \frac{325}{2}$

[Differentiating with respect to x]

$\Rightarrow \frac{-212500}{x^2} + \frac{325}{2} = 0$

$\Rightarrow \quad x^2 = \frac{17000}{13}$

$\Rightarrow \quad x = 36.16$

Thus, option (b) is correct.

Alternatively

We may proceed through options by substituting the values from the options in the expression.

$\frac{212500}{x} + \frac{325x}{2}$.

At $x = 36$, its values is minimum.

63. Let the total distance from A to B is 60 km [i.e. their LCM, you are free to assume any value]

Then, time taken by Bimal

$= \frac{20}{10} + \frac{20}{20} + \frac{20}{30} = 3\frac{2}{3}\,\text{h}$

Let Amal's journey time be x h. Then, distance covered by Amal in x h

$= \frac{x}{3} \times 10 + \frac{x}{3} \times 20 + \frac{x}{3} \times 30 = 60$

$\Rightarrow \quad x = 3\,\text{h}$

Percentage increase

$= \left(\frac{\frac{2}{3}}{3} \times 100\right)\% = 22.22\% \approx 22\%$

64. Since every minute, starting from 10 : 01 AM, one motor cycle leaves A, 45th motor cycle will leave A at 10 : 45 AM and will reach B at 11 : 00 AM.

Needless to mention that the rest all 44 will reach before 11 : 00 AM. It implies that each motor cycle takes 15 min, to cover the distance from A to B. In case the speed of cyclist is doubled, the cyclist will reach B at 10 : 30 AM instead of 11 : 00 AM.

Since each motor cycle needs 15 min. to reach B from A, the last motor cycle should leave A at 10 : 15 AM to reach B by 10 : 30 AM.

Further, since each motor cycle leaves A at interval of one minute, starting from 10 : 01 AM, and the last one leaves A at 10 : 15 AM, the number of such motor cycles is 15.

31

Average Speed

Introduction

As we know, speed $= \dfrac{\text{distance}}{\text{time}}$, then it may be said that

average speed $= \dfrac{\text{total distance}}{\text{total time}}$

Computation of average speed when the same/equal distance is covered at the different speeds

Now take an example to build the concepts

Problem 1 A man covers a certain distance at 30 km/h and returns the same at 40 km/h. Find out his average speed.

Sol. Let the distance be d km.

Time taken to cover d km at the rate of 30 km/h $= \dfrac{d}{30}$ h

Time taken to cover d km at the rate of 40 km/h $= \dfrac{d}{40}$ h

Now, average speed $= \dfrac{\text{total distance}}{\text{total time}} = \dfrac{d + d}{\dfrac{d}{30} + \dfrac{d}{40}} = \dfrac{2d \times 30 \times 40}{d(30 + 40)}$

$$= \frac{2 \times 30 \times 40}{(30 + 40)} = \frac{2 \times \text{1st speed} \times \text{2nd speed}}{\text{1st speed} + \text{2nd speed}} = \frac{240}{7} = 34\frac{2}{7} \text{ km/h}$$

Generic version

If a certain distance is covered at a speed of S_1 km/h and further, the same/equal distance is covered at a speed of S_2 km/h, then the average speed will be equal to $\dfrac{2 \times S_1 \times S_2}{(S_1 + S_2)}$

Similarly, if a certain distance is covered at a speed of s_1 km/h and further, the same/equal distance is covered at a speed of s_2 km/h and again the same/equal distance is covered at s_3 km/h, then the average will be equal to

$$\frac{3s_1s_2s_3}{(s_1s_2) + (s_2s_3) + (s_3s_1)}$$

Alternatively I

Assume any distance which is exactly divisible by both 30 and 40 [say, LCM or its multiple]. Let the distance be 120 km.

Now, time taken at the rate of 30 km/h = 120 ÷ 30 = 4 h

and time taken at the rate of 40 km/h = 120 ÷ 40 = 3 h

Inside

- *Introduction*
- *Computation of average speed when the same/equal distance is covered at the different speeds*
- *Computation of average speed when a distance is covered at different speeds in parts*
- ....
- .... *& Many More*
- *Work Out with Solutions*

Then, average speed $= \frac{120+120}{4+3} = \frac{240}{7} = 34\frac{2}{7}$ km/h

Alternatively II

Ratio of speeds = 30 : 40 = 3 : 4

Then, ratio of time taken = 4 : 3 [∵ distance is constant]

Now, average speed $= \frac{(30\times 4)+(40\times 3)}{4+3} = \frac{240}{7} = 34\frac{2}{7}$ km/h

Alternatively III

Ratio of speeds = 30 : 40 = 3 : 4

and difference between the speeds = 10 km/h

Clearly, 10 km will be covered at the both speeds in ratio 3 : 4.

Hence, distance covered at the rate of 30 km/h

$= 10 \times \frac{3}{(3+4)} = 4\frac{2}{7}$

and distance covered at the rate of 40 km/h $= 10 \times \frac{4}{(3+4)} = 5\frac{5}{7}$

Now, average speed will be either

$\left(30 + 4\frac{2}{7}\right)$ or $\left(40 - 5\frac{5}{7}\right)$ i.e. $34\frac{2}{7}$ km/h

Problem 2 A man covers half of its journey at 6 km/h and the remaining half at 3 km/h. His average speed is what? ***[SSC(CGL-PT)-07]***

Sol. Average speed $= \frac{2\times s_1 \times s_2}{(s_1+s_2)} = \frac{2\times 6\times 3}{(6+3)} = 4$ km/h

Alternatively

Ratio of speeds = 6 : 3 = 2 : 1

Then, ratio of time taken = 1 : 2 [∵ distance is constant]

Now, average speed $= \frac{(6\times 1)+(3\times 2)}{1+2} = \frac{12}{3} = 4$ km/h

Computation of average speed when a distance is covered at different speeds in parts

Problem 3 A car covers 1/3rd of its journey at 45 km/h and the remaining distance at 60 km/h. What is its average speed for the whole journey?

Sol. Let the distance be 180 km. You are free to assume any figure.

Now, time to cover 1/3rd of 180 at the rate of 45 km/h = 4/3 h

and time to cover 2/3rd of 180 at the rate of 60 km/h = 2 h

Then, average speed $= \frac{\text{total distance}}{\text{total time}} = \frac{180}{\frac{4}{3}+2} = \frac{180\times 3}{10}$

= 54 km/h

Alternatively I

Whole journey may be covered at $\left(45 \times \frac{3}{1}\right)$ km/h in the same time. Further, the whole journey may be covered at $\left(60 \times \frac{3}{2}\right)$ km/h in the same given time. In both cases, there is no alternation in the original time. However, it is important to note that in actual, whole journey has been covered one time only.

Thus, average speed will be $\frac{s_1 \times s_2}{(s_1+s_2)}$ in lieu of $\frac{2\times s_1\times s_2}{(s_1+s_2)}$.

Hence, average speed $= \frac{\left(45\times\frac{3}{1}\right)\times\left(60\times\frac{3}{2}\right)}{\left(45\times\frac{3}{1}\right)+\left(60\times\frac{3}{2}\right)} = \frac{135\times 90}{135+90}$

= 54 km/h

Generic version

If *a/b* of the whole journey is covered at a speed of *x* km/h and the remaining *c/d* of the journey is covered at the speed of *y* km/h, then the average speed will be $\frac{\left(x\times\frac{b}{a}\right)\times\left(y\times\frac{d}{c}\right)}{\left(x\times\frac{b}{a}\right)+\left(y\times\frac{d}{c}\right)}$

Similarly, in case of three parts, average speed will be $\frac{s_1s_2s_3}{(s_1s_2)+(s_2s_3)+(s_3s_1)}$, where s_1, s_2 and s_3 are adjusted speeds on the above pattern.

Alternatively II

Here, distance is not constant. Thus, it should be brought to the same as below

Given that $\frac{1}{3}$ of the whole journey is covered at 45 km/h.

We may say that whole journey may be covered at $\left(45 \times \frac{3}{1}\right)$ km/h in the same time. Similarly, the whole journey may be covered at $\left(60 \times \frac{3}{2}\right)$ km/h in given time. In both cases, there is no alternation in the original time.

Ratio of speeds $= \left(45\times\frac{3}{1}\right) : \left(60\times\frac{3}{2}\right) = 3:2$

∴ Ratio of time taken = 2 : 3

and their speeds are 45 and 60

Then, average speed $= \frac{(2\times 45)+(60\times 3)}{2+3} = \frac{270}{5} = 54$ km/h

Problem 4 A man covers $\frac{1}{4}$th of journey at 20 km/h, $\frac{1}{3}$rd at 30 km/h and remaining at 25 km/h. Find out his average speed.

Sol. The average speed will be $= \frac{s_1s_2s_3}{(s_1s_2)+(s_2s_3)+(s_3s_1)}$

Where, $s_1 = 20\times\frac{4}{1} = 80$, $s_2 = 30\times\frac{3}{1} = 90$ and $s_3 = 25\times\frac{12}{5} = 60$

∴ Average speed $= \frac{80\times 90\times 60}{(80\times 90)+(90\times 60)+(60\times 80)} = 24.82$ km/h

Alternatively

Ratio of speeds for whole journey without alternation in time.

$\Rightarrow 20\times\frac{4}{1} : 30\times\frac{3}{1} : 25\times\frac{12}{5}$ $\left[\because 1-\frac{1}{3}-\frac{1}{4}=\frac{5}{12}\right]$

$\Rightarrow 80:90:60 \Rightarrow 8:9:6$

Then, ratio of time taken $= \frac{1}{8}:\frac{1}{9}:\frac{1}{6} = 9:8:12$

Their speed are 20, 30 and 25

∴ Average speed $= \frac{(20\times 9)+(30\times 8)+(25\times 12)}{9+8+12}$

$= \frac{720}{29} = 24.82$ km/h

WORKOUT

1. A car travels along the four sides of a square at speeds $v, 2v, 3v$ and $4v$ respectively. If u is the average speed of the car in its travel around the square, then which one of the following is correct ?

(a) $u = 2.25v$ (b) $u = 3v$
(c) $v < u < 2v$ (d) $3v < u < 4v$

2. A train covers a distance of $193\frac{1}{3}$ km in $4\frac{1}{4}$ h with one stoppage of 10 min, two of 5 min and one of 3 min on the way. Find the average speed of the train.

(a) 50 km/h (b) 55 km/h
(c) 30 km/h (d) 60 km/h

3. Pratibha covers a distance of 24 km at the speed of 8 km/h and a distance of 18 km at the speed of 9 km/h. Further, she covers a distance of 12 km at the speed of 3 km/h. What is her average speed in covering the whole distance ? ***[NABARD (PO)-08]***

(a) 8 km/h (b) 5.5 km/h
(c) 3 km/h (d) 6 km/h
(e) None of these

4. A car covers the 1st 39 km of its journey in 45 min and the remaining 25 km in 35 min. What is the average speed of the car ? ***[P&B Bank (PO)-11, Andhra Bank (PO)-11]***

(a) 40 km/h (b) 64 km/h
(c) 49 km/h (d) 48 km/h
(e) None of these

5. A person travelled from one place to another at an average speed of 40 km/h and back to the original place at an average speed of 50 km. What is the average speed in km/h during the entire round-trip ? ***[IAS (Pre)-95, SSC (SAS)-10, SSC (CGL-T-11)-13]***

(a) 4.5 km/h (b) $20\sqrt{5}$ km/h
(c) $\frac{400}{9}$ km/h
(d) Impossible to find out unless the distance between the two places is known

6. If a body covers a distance at the rate of x km/h and another equal distance at the rate of y km/h, then the average speed (in km/h) is : ***[CDS-13]***

(a) $\frac{x+y}{2}$ (b) $\frac{2xy}{x+y}$
(c) $\frac{x+y}{xy}$ (d) $\sqrt{xy}$

7. An aeroplane travels distance 2500 km, 1200 km and 500 km at the rate of 500 km/h, 400 km/h and 250 km/h respectively. The average speed is :

(a) 420 km/h (b) 405 km/h
(c) 410 km/h (d) 575 km/h

8. A car driver covers a distance between two cities at a speed of 60 km/h and on the return his speed is 40 km/h. He goes again from 1st to the 2nd city at twice that original speed and returns at half the original speed. Find his average speed for the entire journey. ***[LIC (AAO)-07]***

(a) 55 km/h
(b) 50 km/h
(c) 48 km/h
(d) 40 km/h
(e) None of the above

9. A chartered bus carrying office employees travels everyday in two shifts-morning and evening. In the evening, the bus travels at an average speed which is 50% greater than the morning average speed, but takes 50% more time than the amount of time it takes in the morning. The average speed of the chartered bus for the entire journey is greater/less than its average speed in the morning by : ***[IIFT-15]***

(a) 18% less
(b) 30% greater
(c) 37.5% greater
(d) 50% less

10. A boat goes upstream at a speed of 21 km/h and comes back the same distance at 28 km/h. What is the average speed (in km/h) for the total journey ? ***[SSC (CGL-T-II)-17]***

(a) 24.5 km/h
(b) 24 km/h
(c) 25 km/h
(d) 25.4 km/h

11. Shyam went from Delhi to Shimla via Chandigarh by car. The distance from Delhi to Chandigarh is $\frac{3}{4}$ times the distance from Chandigarh to Shimla. The average speed from Delhi to Chandigarh was half as much as that from Chandigarh to Shimla. If the average speed for the entire journey was 49 km/h, what was the average speed from Chandigarh to Shimla ? ***[CAT-94]***

(a) 39.2 km/h
(b) 63 km/h
(c) 42 km/h
(d) None of the above

12. A runner covers his onward journey from A to B at 10 km/h and during the return journey from B to A he covers the same distance at 8 km/h. If he finishes the onward and return journey in 4.5 h, then the total distance covered by him during the entire journey is : ***[MAT-18]***

(a) 35 km (b) 40 km
(c) 45 km (d) 50 km

Answer Keys

1. (c)	2. (a)	3. (d)	4. (d)	5. (c)	6. (b)	7. (a)	8. (d)	9. (b)	10. (b)
11. (d)	12. (b)								

Hints and SOLUTIONS

1. Since the distances are equal,

time $\alpha \dfrac{1}{\text{speed}}$

Hence, ratio of time taken to cover sides of the square

$= \dfrac{1}{v} : \dfrac{1}{2v} : \dfrac{1}{3v} : \dfrac{1}{4v} = 12 : 6 : 4 : 3$

Their speed are v, $2v$, $3v$ and $4v$

Then, average speed

$= \dfrac{(v \cdot 12) + (2v \cdot 6) + (3v \cdot 4) + (4v \cdot 3)}{12 + 6 + 4 + 3}$

$\Rightarrow \quad u = 1.92v$

Hence, option (c) is correct.

2. Total stoppage time

$= (10 \times 1) + (5 \times 2) + (3 \times 1)$

$= 23$ min.

Actual running time

$= 4\dfrac{1}{4}$ h $- \dfrac{23}{60}$ h $= \dfrac{232}{60}$ h

Then, average speed

$= 193\dfrac{1}{3} \div \dfrac{232}{60} = 50$ km/h

3. Total distance covered

$= 24 + 18 + 12 = 54$ km

and total time taken

$= (24 \div 8) + (18 \div 9) + (12 \div 3) = 9$ h

Then, average speed $= 54 \div 9 = 6$ km/h

4. Average speed $= \dfrac{\text{total distance}}{\text{total time}}$

$= \dfrac{(39 + 25)}{(45 + 35)} = 0.8$ km/min

$= 0.8 \times 60 = 48$ km/h

5. Average speed

$= \dfrac{2 \times 40 \times 50}{(40 + 50)} = \dfrac{400}{9}$ km/h

6. We know, if a certain distance is covered at a speed of S_1 km/h and further, the same/equal distance is covered at a speed of S_2 km/h, then the average speed will be equal to $\dfrac{2 S_1 S_2}{(S_1 + S_2)}$.

Hence, option (b) is correct.

7. Total distance covered

$= 2500 + 1200 + 500 = 4200$ km

and total time taken

$= (2500 \div 500) + (1200 \div 400) + (500 \div 250) = 10$ h

Then, average speed $= 4200 \div 10$

$= 420$ km/h

8. For the 1st trip [i.e. to and fro], average speed

$= \dfrac{2 \times 60 \times 40}{(60 + 40)} = 48$

and for 2nd trip [i.e. to and fro], average speed

$= \dfrac{2 \times 120 \times 20}{(120 + 20)} = \dfrac{240}{7}$

Hence, the average speed

$= \dfrac{2 \times 48 \times \dfrac{240}{7}}{\left(48 + \dfrac{240}{7}\right)} = \dfrac{2 \times 48 \times 240 \times 7}{7 \times 576}$

$= 40$ km/h

9. Let average speed in the morning be 100 km/h and the distance the bus travels in morning shift be 100 km.

Hence, time taken = 1 h

Then, average speed in the evening

= 150 km/h

and time taken by it in the morning

= 1.5 h

Then, distance covered in evening

$= 1.5 \times 150 = 225$ km

Now, average speed for the entire journey

$= \dfrac{100 + 225}{1 + 1.5} = 130$ km/h,

which is 30% greater than the average speed in the morning.

Alternatively

Increase% in the distance in the evening

$= \left(+ 50 + 50 + \dfrac{50 \times 50}{250} \right)\% = 125\%$

Let average speed in the morning be 100 km/h and the distance the bus travels in morning shift be 100 km.

Hence, time taken = 1 h

Now, average speed for the entire journey

$= \dfrac{100 + (225\% \text{ of } 100)}{1 + (150\% \text{ of } 1)}$

$= 130$ km/h,

which is 30% greater than the average speed in the morning.

10. Average speed

$= \dfrac{2 \times 21 \times 28}{(21 + 28)} = 24$ km/h

11. Ratio of average speed for Delhi to Chandigarh and that for Chandigarh to Shimla = 1 : 2

$\therefore$ Ratio of time taken = 2 : 1, if both the stretches have equal length.

But given that ratio of distance between Delhi and Chandigarh and that between Chandigarh to Shimla = 3 : 4

Then, ratio of actual time taken

$= (2 \times 3) : (1 \times 4) = 3 : 2$

Now, average speed for the whole journey

$= \dfrac{(1 \times 3) + (2 \times 2)}{3 + 2} = \dfrac{7}{5}$

As given, $\dfrac{7}{5}$ units = 49 km/h

$\therefore$ 2 units = 70 km/h

Hence, average speed from Chandigarh to Shimla = 70 km/h

$\therefore$ Option (d) is correct.

12. Average speed $= \dfrac{2 \times 10 \times 8}{(10 + 8)}$

$= \dfrac{80}{9}$ km/h

Then, total distance covered in 4.5 h

$= \dfrac{80}{9} \times 4.5$

$= 40$ km

32

Relative Speed: Same Direction and Opposite Directions

Introduction

Relative speed is the speed of one body with respect to that of other body. The relative speed may be computed in following two cases.

(a) **When two bodies (say, A and B) move in the same direction and don't return in reverse directions.**

In this case, the relative speed will be the difference between their speed i.e. $(S_A - S_B)$ or $(S_B - S_A)$ whichever is positive and the faster body will take a lead to the slower by a distance equal to product of relative speed and the time, if both start to move simultaneously from the same point. In other words, the distance between them will be equal to (relative speed × time).

Further, if a slower body is d units ahead of the point from where the faster body is to move then this gap of distance d will be bridged in time of $\left(\frac{d}{\text{relative speed}}\right)$ units. This concept may be applied in the problems based on overtaking like in cases of police-thief, dog-horse, tortoise-rabbit, hearing of sound of bomb-blast, running etc.

The relative speed will be the difference between S_A and S_B even after the meeting/crossing/overtaking point.

In nutshell, aspirants have to adhere to principles of time, distance and speed assuming speed is a relative speed.

Note *If three bodies start to move from the same point at speeds x, y, z km/h (where, $x < y < z$) respectively and the 2nd body starts m h after 1st body, then the 3rd body must start $\left(m \times \frac{y}{z} \times \frac{z-x}{y-x}\right)$ h after the 1st body so that the 2nd and the 3rd bodies overtake the 1st body at the same time.*

(b) **When two bodies (say, A and B) start to move in the opposite directions from two fixed points.**

In this case, the relative speed will be $(S_A + S_B)$ till and beyond the point of meeting/crossing. Further, the distance between both bodies at any point

Inside

of time will be covered at the speed of $(S_A + S_B)$ and also they will start separating from each other at the same relative speed. Here, principles of time, distance and speed is also applicable assuming speed is a relative speed.

Note *If A and B start to move from two fixed point towards each other simultaneously and they meet each other after x h and after meeting A and B take y and z h more respectively to reach each other's starting point, then*

(i) $\frac{\text{Speed of } A}{\text{Speed of } B} = \sqrt{\frac{\text{time taken by } B \text{ after meeting}}{\text{time taken by } A \text{ after meeting}}} = \sqrt{\frac{z}{y}}$

[Also, discussed in detail in the beginning under 'Problems based on the concept :Speed α Distance, when time is constant']

(ii) $x = \sqrt{yz}$ *[asked in CAT-14]*

Problems Based on Relative Speed in the same Direction

Problem 1 Two cars A and B start simultaneously from the same point at the speed of 20 km/h and 25 km/h respectively in the same direction. Then answer the following.

(a) What will be the distance between them after 4 h?

(b) In what time will the distance between them be 40 km?

Sol. Relative speed $= 25 - 20 = 5$ km/h

(a) Distance between them after 4 h = Relative speed × Time
$= 5 \times 4 = 20$ km

(b) Required time = Distance ÷ Relative speed $= 40 \div 5 = 8$ h

Problem 2 Taxi A starts to move with speed of 50 km/h from point R to point N, 560 km apart, after 3 h from starting of taxi B from the same point R with the speed of 30 km/h. When will taxi A overtake taxi B and at what distance from point N?

Sol. Since taxi B is 3 h ahead of taxi A, taxi B is (30 × 3) km ahead of taxi A. This distance i.e. 90 km is to bridged by the relative speed. Here, relative speed $= 50 - 30 = 20$ km/h

Then, time taken by A to overtake $B = 90 \div 20 = 4.5$ h

Further, distance travelled by A to meet $B = 50 \times 4.5 = 225$ km

Hence, at 225 km from the point R, A meets B.

It may be said that A meets B at a distance of (560 – 225) km from the point N as R and N are 560 km apart.

∴ Required answer $= 560 - 225 = 335$ km

Problem 3 Bus X and Y start from point M to point N, 300 km apart, at the interval of 2.5 h with the speed of 20 km/h and 30 km/h respectively. Answer the followings if bus X starts first.

(i) After how many hours will X and Y be separated by 25 km before the meeting?

(ii) After how many hours will Y be at 25 km ahead of X from the meeting point?

(iii) At what distance from point N will Y be at 30km ahead of X after meeting?

Sol. (i) Before moving of Y, the difference between X and Y is (2.5×20) i.e. 50 km, which is to be reduced by 25 km so as to make a separation of 25 km between them. Obviously, this 25 km will be covered at the relative speed.

∴ Required time $= 25 \div (30 - 20) = 2.5$ h

Hence, after 2.5h from starting time of Y, both will be separated by 25 km.

Or

After (2.5 + 2.5) h from starting time of X, both will be separated by 25 km.

(ii) Even after meeting point, Y will lead X by 10km/h but Y will have to lead 25km.

Thus, time taken to go 25 km ahead of X after meeting point
$= 25 \div 10 = 2.5$ h

Earlier, time taken by Y to meet $X = (2.5 \times 20) \div (30 - 20) = 5$ h

∴ Required answer $= 5 + 2.5 = 7.5$ h

Hence, after 7.5 h from starting time of Y, Y will be ahead of X by 25 km.

In other words, after (7.5 + 2.5) h from starting time of X, Y will be ahead of X by 25 km.

(iii) On the above line, time taken by Y to go 30 km ahead of X after meeting point $= 30 \div 10 = 3$ h.

Hence, after (5 + 3) h from starting time of Y, Y will be ahead of X by 30 km.

It means, distance covered by Y from point $M = 30 \times 8$
$= 240$ km

Then, required answer $= 300 - 240 = 60$ km

Problem 4 A thief steals a car at 2 : 30 PM and drives at 60 km/h. The theft is discovered at 3 : 00 PM and the owner sets off in another car at 75km/hr. When will he overtake the thief? ***[RRB-Bhubneshwar (ASM) -02]***

Sol. Distance travelled by the thief upto 3:00 pm $= (60 \times 1/2) = 30$ km. It means, the thief is 30 km ahead of the owner, which is to be bridged at the relative speed. Here, relative speed = $(75 - 60) = 15$ km/h.

Thus, time taken by the owner to overtake the thief
$= 30 \div 15 = 2$ h

Hence, the owner will overtake the thief at (3 : 00 + 2 : 00) PM.

Alternatively

The distances travelled by the owner and the thief are the same but the time taken by the owner is 1/2 h less than that by the thief. Further, we know that $\frac{S_1}{S_2} = \frac{T_2}{T_1}$, when D is contant

$$\Rightarrow \frac{\text{Speed of owner}}{\text{Speed of thief}} = \frac{\text{Time taken by thief}}{\text{Time taken by owner}}$$

$$\Rightarrow \frac{75}{60} = \frac{\text{Time taken by thief}(T)}{T - \frac{1}{2}}$$

$$\Rightarrow T = 2.5 \text{ h}$$

Hence, at (2 : 30 + 2 : 30) PM, the owner overtakes the thief.

Or

Time taken by the owner to overtake the thief $= 2.5 - 1/2 = 2$ h.

Hence, at (3 : 00 + 2 : 00) PM, the owner overtakes the thief.

Problem 5 Two men together start a journey in the same direction. They travel 9 km and 15 km per day respectively. After travelling for 6 days the man travelling 9 km per day doubles his speed and both of them finish the distance in the same time. Find the number of days taken by them to reach the destination?
[RRB-Gorakhpur (ASM-02)]

Sol. In 6 days, the faster man is $[6 \times (15-9)]$ i.e. 36 km ahead of the slower.

After 6 days, the relative speed $= (9 \times 2) - 15 = 3$ km/day.

Thus, time taken $= 36 \div 3 = 12$ days beyond the 6 days.

Thus, total time $= 12 + 6 = 18$ days

Alternatively

Let total days be x.

Distance travelled by the faster man $= 15 \times x = 15x$ km.

and distance travelled by the slower man

$$= (9 \times 6) + 18(x-6) = (18x - 54) \text{ km}$$

But, the distance is same for both. Then, $15x = (18x - 54)$

$\Rightarrow \quad 3x = 54 \Rightarrow x = 18$ days

Problem 6 Two cars start to move simultaneously from the same point and in the same direction at the speed of 30 km/h and 40 km/h respectively. A 3rd car, $1\frac{1}{2}$ h later, starts from the same point and overtakes the two. What is the speed of the 3rd car if difference between time taken by it to overtake them is $3\frac{3}{4}$ h.

Sol. Let the speed of 3rd car be x km/h.

Distance travelled by1st car in 1.5h $= 30 \times 1.5 = 45$ km

and distance travelled by 2nd car in 1.5 h $= 40 \times 1.5 = 60$ km

Now, relative speed of 3rd car with reference to 1st car $= (x-30)$ km/h and that with reference to 2nd car $= (x - 40)$ km/h.

Further, time taken by 3rd car to overtake 1st car $= \frac{45}{(x-30)}$

and time taken by 3rd car to overtake 2nd car $= \frac{60}{(x-40)}$

Obviously, the 3rd car will overtake 1st car first then the 2nd car.

Hence, $\quad \frac{45}{(x-30)} + 3\frac{3}{4} = \frac{60}{(x-40)}$

$$\Rightarrow \quad \frac{60}{(x-40)} - \frac{45}{(x-30)} = \frac{15}{4}$$

$$\Rightarrow \quad \frac{4}{(x-40)} - \frac{3}{(x-30)} = \frac{1}{4}$$

$$\Rightarrow \quad 4(4x - 120 - 3x + 120) = x^2 - 70x + 1200$$

$$\Rightarrow \quad 4x = x^2 - 70x + 1200$$

$$\Rightarrow \quad x^2 - 74x + 1200 = 0$$

$$\Rightarrow \quad x^2 - 50x - 24x + 1200 = 0$$

$$\Rightarrow \quad x(x-50) - 24(x-50) = 0$$

$$\Rightarrow \quad x = 50 \text{ or } 24$$

The speed of 3rd car must be more than speed of 1st car as well as 2nd car to overtake them.

Hence, speed of the 3rd car = 50 km/h

Alternatively I

From the above, $\frac{45}{(x-30)} + 3\frac{3}{4} = \frac{60}{(x-40)}$

Now, we may put the values of options in place of x so as to check as to which option is satisfying the equation.

Alternatively II

Let the speed of the 3rd car be x km/h and till overtake, 1st car moves for T h.

Now, distance is constant with respect to overtaking of 1st car by 3rd one. Then, $\frac{S_1}{S_2} = \frac{T_2}{T_1}$

$$\Rightarrow \quad \frac{30}{x} = \frac{T-1.5}{T}$$

$$\Rightarrow \quad 30T = xT - 1.5x \Rightarrow T(30-x) = -1.5x$$

$$\Rightarrow \quad T = \frac{-1.5x}{30-x}$$

Further, with respect to overtaking of 2nd car by 3rd one.

$$\frac{40}{x} = \frac{T + 3.75 - 1.5}{T + 3.75}$$

$$\Rightarrow \quad 40T + 150 = xT + 2.25x$$

$$\Rightarrow \quad T(40-x) = 2.25x - 150$$

$$\Rightarrow \quad T = \frac{2.25x - 150}{40-x}$$

Obviously, value of T in both cases will be equal.

Hence,

$$\Rightarrow \quad \frac{-1.5x}{30-x} = \frac{2.25x - 150}{40-x}$$

$$\Rightarrow \quad \frac{-x}{30-x} = \frac{1.5x - 100}{40-x}$$

$$\Rightarrow \quad -40x + x^2 = 45x - 3000 - 1.5x^2 + 100x$$

$$\Rightarrow \quad 2.5x^2 - 185x + 3000 = 0$$

$$\Rightarrow \quad x^2 - 74x + 1200 = 0$$

$$\Rightarrow \quad x^2 - 50x - 24x + 1200 = 0$$

$$\Rightarrow \quad x(x-50) - 24(x-50) = 0$$

$$\Rightarrow \quad x = 50 \text{ or } 24$$

The speed of 3rd car must be more than speed of 1st car as well as 2nd car to overtake them.

Hence, speed of the 3rd car = 50km/h

Problem 7 Yesterday, Shivansh and Kranti left Patna for Jehanabad. Shivansh left at 10 : 00 AM and Kranti left 2 h later with the speed 10km higher than that of Shivansh. Find the speed of Kranti, if the distance between them was 20 km at 6 : 00 PM of the same day.

Sol. Let the speed of Shivansh be x km/h.

Then, distance covered by Shivansh $= 8 \times x = 8x$

and distance covered by Kranti $= 6(x + 10) = 6x + 60$

There may be two cases.

(a) When Shivansh is 20 km ahead of Kranti

(b) When Kranti is 20 km ahead of Shivansh

Case (a) $8x - (6x + 60) = 20$

$\Rightarrow \quad x = 40$, then speed of Kranti = 50 km/h

Case (b) $(6x + 60) - 8x = 20$

$\Rightarrow \quad x = 20$, then speed of Kranti = 30 km/h

Problem 8 **An army bomb squad man set a fuse for blasting a rock to take place after one minute. He ran away from the site at the speed of 13 m/s. Sound travels at the speed of 325 m/sec. Upto what distance could the army man run, before he heard the sound of blast ?**

Sol. Before blasting, distance covered by the man $= 60 \times 13 = 780$ m

On blast, sound of the blasting will overtake the man, then,

relative speed $= 325 - 13 = 312$ m/s

Further, time taken by sound of blasting to overtake

$= 780/312$ sec

Hence, distance covered by the sound $= 325 \times \frac{780}{312} = 812.5$ m

$\therefore$ Required answer = 812.5 m

Alternatively

Since distance is constant, $\frac{S_1}{S_2} = \frac{T_2}{T_1}$, where S_1 and S_2 are speeds of the man and the sound of blasting. T_1 and T_2 are time taken by them respectively.

Then, $\frac{13}{325} = \frac{T_1 - 60}{T_1}$

[$\because$ 1 min. = 60 sec, and speeds are given in term of per sec.]

$\Rightarrow \frac{1}{25} = \frac{T_1 - 60}{T_1}$

$\Rightarrow T_1 = 25T_1 - 1500$

$\Rightarrow T_1 = 62.5$ sec

Hence, required distance $= 62.5 \times 13 = 812.5$ m

Problem 9 **A cat is 40 of its own leaps ahead of a dog and the dog takes 3 leaps for every 4 leaps of cat. If the dog and cat cover 2.5 m and 1.5 m/leap respectively, in how many leaps will dog catch the cat?**

Sol. Initially, cat is (40×1.5) i.e. 60 m ahead of the dog.

Now, distance covered by dog in 3 leaps $= 3 \times 2.5 = 7.5$ m

and distance covered by cat in 4 leaps $= 4 \times 1.5 = 6$ m

Then, relative speed $= (7.5 - 6) = 1.5$ m per 3 leaps of the dog.

$\because$ Dog bridges a gap of 1.5 m in its own 3 leaps.

$\therefore$ It will bridge a distance of 60 m in $\left(\frac{3}{1.5} \times 60\right)$ leaps

i.e. 120 leaps

Hence, required answer = 120 leaps

Problem 10 **Three cars *A*, *B* and *C* started from a point at 5 : 00 PM, 6 : 00 PM and 7 : 00 PM respectively and travelled at uniform speeds of 60 km/h, 80 km/h and *x* km/h respectively in the same direction. If all the three met at another point at the same instant during their journey, then what is the value of *x* ?** ***[CDS-19]***

Sol. By 7 : 00 PM, distance covered by $A = 2 \times 60 = 120$ km

By 7 : 00 PM distance covered by $B = 1 \times 80 = 80$ km

Hence, A is 40 km ahead of B by 7 : 00 PM, which is to be bridged at the relative speed.

Since relative speed $= 80 - 60 = 20$ km/h, time taken by B to overtake $A = 40 \div 20 = 2$ h

Thus, at (7 : 00 + 2 : 00) PM, A and B will meet together at the distance of (4×60) km from the starting point.

Since C has also to meet A and B at 9 : 00 PM, C must travel 240 km starting at 7 : 00 PM i.e. in 2 h.

$\therefore$ Speed of $C = 240 \div 2 = 120$ km/h

Problems Based on Relative Speed in the Opposite Directions

Problem 11 **Two auto rickshaws *R* and *T* start to move simultaneously at 10 : 00 AM towards each other form the points *A* and *B* respectively at the speed of 40 km/h and 50 km/h respectively. If points *A* and *B* are at a distance of 900 km, then at what time will they meet and at what distance from point *R*. Also, find ratio of distances covered by *R* and *T* respectively.**

Sol. *R*.......40 km/h... → ←...50 km/h...... *T*

(*A*) •<.......................•(*P*)...>•(*B*)

←...............................900 km...................................→

[Let *P* be the point where the both rickshaws meet]

(a) Relative speed $= 40 + 50 = 90$ km/h

Then, time required to cover 900 km $= 900 \div 90 = 10$ h

Hence, after 10 h they will meet i.e. at 8 : 00 PM of the same day.

(b) The distance covered by $R = 40 \times 10 = 400$ km

Hence, at the distance of 400 km from point *A*, they meet.

Alternatively

We have already discussed under concept "Speed α Distance, when time is constant" that if they move simultaneously, time taken by them to meet each other will be same.

Then, $\frac{S_1}{S_2} = \frac{D_1}{D_2}$

$\Rightarrow \frac{\text{Speed of } R}{\text{Speed of } T} = \frac{\text{Distance covered by } R}{\text{Distance covered by } T}$

$\Rightarrow \frac{\text{Distance covered by } R}{\text{Distance covered by } T} = \frac{40}{50}$

$\Rightarrow \frac{AP}{PB} = \frac{4}{5}$

$\Rightarrow AP : PB = 4 : 5$

But, given, $AP + PB = 900$

$\because (4 + 5)$ units $= 900$

$\therefore$ 4 units $= 400$

Hence, $AP = 400$ km

(c) From the above,

$\frac{\text{Distance covered by } R}{\text{Distance covered by } T} = \frac{40}{50}$

Hence, required answer = 4 : 5

Problem 12 **Two towns *A* and *B* are 500 km apart. A train starts at 8 : 00 AM from *A* towards *B* at a speed of 70 km/h. At 10 : 00 AM, another train starts from *B* towards *A* at a speed of 110 km/h. When will the two trains meet?** ***[SSC (CGL-T-II)-19, CAT-20]***

Sol. Let they meet after *x* h since 8 : 00 AM.

Then, $[70 \times x] + [110 \times (x - 2)] = 500$

$\Rightarrow \quad 180x = 720$

$\Rightarrow \quad x = 4$ h

Hence, they will meet after 4 h beyond 8 : 00 AM i.e. at 12 : 00 noon.

Alternatively

Distance travelled by 1st train by 10 : 00 AM = $70 \times 2 = 140$ km.

The distance between both trains at exactly 10 : 00 AM

$= 500 - 140 = 360$ km

Then, time required to meet after 10 : 00 AM = $360 \div$ (relative speed) = $360 \div (70 + 110) = 2$ h

Hence, required answer = 2 h beyond 10 : 00 AM = 12 noon

Problem 13 With reference to the above question, answer the followings

(i) At what time will they be separated by 30 km before they meet?

(ii) At what time will they be separated by 90 km after the meeting?

(iii) What is the ratio of time taken by the slower train and the faster train to reach their destination after meeting?

Sol. (i) From the above solution, both the trains are meeting at 12 noon. But there should be a distance of 30 km between them before the meeting.

Time taken by them to cover 30 km = $30 \div (110 + 70) = 1/6$ h = 10 min. It implies that at 11 : 50 AM, they will be at a distance of 30 km before the meeting. No need to mention that if they are allowed to move 10 min more beyond 11 : 50 AM, they will meet to each other.

(ii) They meet at 12 noon. Further, to make a distance of 90 km between them after the meeting, they will take $[90 \div (110 + 70)]$ i.e. 1/2 h. Hence, at 12 : 30 PM they will be at the distance of 90 km after meeting.

(iii) Since they are meeting at 12 noon, distance travelled by the slower train from 8 : 00 AM to 12 noon = $70 \times 4 = 280$ km and distance travelled by the faster train from 10 : 00 AM to 12 : 00 noon = $110 \times 2 = 220$ km

In order to reach their destination after they meet, the slower train and the faster train will cover 220 km and 280 km respectively.

Then, ratio of time taken = $(220 \div 70) : (280 \div 110) = 121 : 98$

Alternatively

Time taken by the slower train for the whole journey = $(500 \div 70)$ h and time taken by the faster train for whole journey = $(500 \div 110)$ h. Till meeting they have moved for 4 h and 2 h respectively.

Then, the required ratio

$= [(500 \div 70) - 4] : [(500 \div 110) - 2]$

$= 121 : 98$

Problem 14 Two runners A and B start running at 12 km/h and 16 km/h towards each other. They meet after 1 h and 30 min. How far (in km) were they from each other when they started? ***[SSC (CGL-T-II)-17]***

Sol. Relative speed = 12 + 16 = 28 km/h

Then, required answer = the distance covered by them in 1.5 h

$= 28 \times 1.5 = 42$ km

Problem 15 Two trains start from stations A and B and travel towards each other at speeds of 50 km/h and 60 km/h respectively. At the time of their meeting, the 2nd train has travelled 120 km more than the 1st. The distance between A and B is : ***[SSC (CPO-SI)-05, MAT-10]***

Sol. Since time is constant, $S_1 : S_2 = D_1 : D_2$

$\Rightarrow \quad D_1 : D_2 = 50 : 60 = 5 : 6$

[no need to mention that distance D_1 is covered by 1st train and D_2 is covered by 2nd train]

As given, $(6 - 5)$ unit = 120 km

Then, $(6 + 5)$ units = 1320 km

Alternatively

Suppose they meet after x h. Then, $(60 \times x) - (50 \times x) = 120$

$\Rightarrow \quad 10x = 120 \Rightarrow x = 12$ h

Then, total distance = $60x + 50x = 110x = 110 \times 12 = 1320$ km

Problem 16 Two cars P and Q start at the same time from A and B which are 120 km apart. If the two cars travel in opposite directions, they meet after 1 h and if they travel in the same direction (from A towards B), then P meets Q after 6 h. What is the speed of car P? ***[SBI (PO)-2000, SSC (CGL-T-II)-20]***

Sol. Let speeds of cars P and Q be x km/h and y km/h respectively.

In case of opposite directions, relative speed = $(x + y)$ km/h

Then, $\quad \dfrac{120}{(x+y)} = 1 \Rightarrow x + y = 120 \quad$ ...(i)

Further, when they move in same direction, then

(distance travelled by P in 6 h) − (distance travelled by Q in 6 h) = 120 km

$\Rightarrow \quad 6x - 6y = 120$

$\Rightarrow \quad x - y = 20 \quad$ ...(ii)

On solving both Eqs. we get, $x = 70$

Hence, speed of the car P = 70 km/h.

Problem 17 On a long stretch of East-West road, A and B are two points such that B is 350 km West of A. One car starts from A and another from B at the same time. If they move towards each other, then they meet after 1 h. If they both move towards East, they meet in 7 h. The difference between their speeds, in km/h, is : ***[CAT-18]***

Sol. ....→ ....→

B.......................350 km...A

It is evident from the question that when they both travel in same direction, they meet in 7 h. Hence, distance travelled by B − distance travelled by A = 350

$\Rightarrow$ (7 × speed of B) − (7 × speed of A) = 350

$\Rightarrow$ difference in their speeds = 50 km/h

Problem 18 Distance between two stations X and Y is 220 km. Trains P and Q leave station X at 8 : 00 AM and 9 : 51 AM respectively at the speed of 25 km/h and 20 km/h respectively for journey towards Y. A train R leaves station Y at 11 : 30 AM at a speed of 30 km/h for journey towards X. When will P be at equal distance from Q and R?

Sol. Suppose after T h beyond 8 : 00 AM, P will be at equal distance from Q and R.

Distance travelled by P from $X = 25 \times T = 25T$

Then, distance travelled by P from $Y = 220 - 25T$

and distance travelled by Q from $X = 20(T - 1\text{h } 51 \text{ min})$

$$= 20T - \left(20 \times \frac{37}{20}\right) = 20T - 37$$

Also, distance travelled by R from $Y = 30 \times (T - 3.5 \text{ h})$

$$= 30T - 105$$

As per question, distance between P and Q is equal to distance between P and R.

$$\Rightarrow \quad [25T] - [20T - 37] = [220 - 25T] - [30T - 105]$$

$$\Rightarrow \quad 5T + 37 = 325 - 55T \Rightarrow 60T = 288$$

$$\Rightarrow \quad T = 4\frac{4}{5} \text{ h} = 4 \text{ h } 48 \text{ min.}$$

Required answer = 4 h 48 min. beyond 8 : 00 AM = 12 : 48 PM

Problem 19 **Two men set out at the same time to walk towards each other from two points A and B, 72 km apart. The 1st man walks at the rate of 4 km/h. The 2nd man walks 2 km in 1st h, 2.5 km in 2nd h, 3km in 3rd h and so on. Find the time after which two men will meet?** ***[SSC (CGL-Mains)-99, CAT-98]***

Sol. Let the required time be x h.

Then, distance travelled by the 1st man = $4x$

and distance travelled by 2nd man = $2 + 2.5 + 3 + \dots\dots$ x terms

$$= \frac{x}{2}\left\{(2 \times 2) + (x - 1) \times \frac{1}{2}\right\}$$

[details in chapter 'progression']

As given, $4x + \frac{x}{2}\left\{(2 \times 2) + (x - 1) \times \frac{1}{2}\right\} = 72$

$$\Rightarrow \quad 6x + \frac{x^2}{4} - \frac{x}{4} = 72$$

$$\Rightarrow \quad x^2 + 23x - 288 = 0$$

$$\Rightarrow \quad x^2 + 32x - 9x - 288 = 0$$

$$\Rightarrow \quad x(x + 32) - 9(x + 32) = 0$$

$$\Rightarrow \quad (x - 9)(x + 32) = 0$$

$$\Rightarrow \quad x = 9 \text{ or } -32$$

Time can't be negative. Hence, correct answer is 9 h.

Alternatively

The best way is to proceed from the options. Let options be 5 h, 9 h, ...

(a) 5 h $\Rightarrow (5 \times 4) + (2 + 2.5 + 3 + 3.5 + 4) \neq 72$

(b) 9 h $\Rightarrow (9 \times 4) + (2 + 2.5 + 3 + 3.5 + 4 + 4.5 + 5 + 5.5 + 6) = 72$

Hence, option (b) is correct.

Problem 20 **Two trains start out towards each other from points 650 km apart. If they start out at the same time, they will meet in 10 h, but if one of them starts out 4 h 20 min. after the other, they will pass each other 8 h following the departure of the latter. Determine the average speed of each other.**

Sol. Relative speed = $650 \div 10 = 65$ km/h

In 2nd case, let the 2nd train starts after 4 h 20 min from starting of 1st train.

Then, distance covered by them in 8 h = $65 \times 8 = 520$ km

Now, difference = $650 - 520 = 130$ km

It is understood that 1st train travels 130 km in 4 h 20 min.

Thus, speed of 1st train = $130 \div 4.33 = 30$ km/h

Then, speed of 2nd train = $65 - 30 = 35$ km/h

Alternatively

The distance travelled by them in $(10 - 8)$ h

= distance travelled by 1st train in 4 hrs 20 min.

$\Rightarrow \left(\frac{650}{10} \times 2\right)$ km = distance travelled by 1st train in 4 h 20 min.

$\Rightarrow$ distance travelled in 1h by 1st train = $130 \div 4\frac{1}{3} = 30$ km

Thus, speed of 1st train = 30 km/h

Then, speed of 2nd train = $(650 \div 10) - 30 = 35$ km/h

Problem 21 **On a particular day a person starts walking from a place X at 2 : 00 AM and reaches place Y at 5 : 00 AM. A 2nd person starts walking from a place Y at 4 : 00 AM and reaches place X at 9:00 AM on the same day. At what time do they cross each other ?**

Sol. Clearly, 1st person takes 3 h to cover a distance from X to Y and 2nd person covers the same distance in 5 h.

Then, ratio of their speeds = 5 : 3 as $S \propto \frac{1}{T}$, when D is constant.

Before B starts (i.e. at 4 : 00 AM), A travels for 2 h. It implies that A can cover the remaining distance only in 1 h as total time taken by A is 3 h. But, the remaining distance equal to the distance A can cover in 1 h, is to be covered by both at the rate of relative speed since 4 : 00 AM.

A, at the speed of 5 units can cover the remaining distance in 1 h.

Then, $(A + B)$ at the relative speed of $(5 + 3)$ units can cover the same remaining distance in 5/8 h.

Hence, required answer = 5/8 h beyond 4 : 00 AM = $4\frac{5}{8}$ AM

Problem 22 **A train X departs from station A at 11 : 00 AM for station B, which is 180 km away. Another train Y departs from station B at 11 : 00 AM for the station A. Train X travels at an average speed of 70 km/h and does not stop anywhere until it arrives at station B. Train Y travels at an average speed of 50 km/h, but has to stop for 15 min at station C, which is 60 km away from station B in route to station A. Ignoring the lengths of the trains, what is the distance, to the nearest (in km) from station A to the point where the trains cross each other?** ***[CAT -01]***

Sol. The time when the train Y starts from the station C = [{60 ÷ 50} h + 15 min.)] beyond 11 : 00 AM = 12 : 27 PM

The distance covered by the train X in 1 h 27 min.

$$= 70 + 31.5 = 101.5 \text{ km}$$

Now, the distance between the train X and Y at 12 : 27 PM = $180 - (60 + 101.5) = 18.5$ km, which will be covered by them in ratio 7 : 5 as their speeds are in ratio 7 : 5.

Hence, distance of the meeting point from station A

$$= 101.5 + \left\{18.5 \times \frac{7}{5}\right\} = 112.29 \approx 112 \text{ km}$$

WORKOUT

1. Two men starting from the same place walk at the rate of 5km/h and 5.5 km/h respectively. What time will they take to be 8.5 km apart, if they walk in the same direction? *[MAT-11]*

(a) 16 h
(b) 8 h 30 min
(c) 4 h 15 min
(d) 17 h

2. Points A and B are 70 km apart on a highway. One car starts from A and the another one from B at the same time. If they travel in the same direction, they meet in 7 h. But if they travel towards each other, they meet in 1 h. The speeds of the two cars are *[MAT-01, 99, SSC (CGL-T-II)-10]*

(a) 45 km/h and 25 km/h
(b) 70 km/h and 10 km/h
(c) 40 km/h and 30 km/h
(d) 60 km/h and 10 km/h

3. Navjeevan express from Ahmedabad to Chennai leaves Ahmedabad at 6 : 30 AM and travels at 50 km/h towards Baroda situated 100 km away. At 7 : 00 AM, Howrah-Ahmedabad express leaves Baroda towards Ahmedabad and travels at 40 km/h. At 7 : 30 AM, Mr. Shah, the traffic controller at Baroda realises that both the trains are running on the same track. How much time does he have to avert a head-on collision between the two trains ? *[CAT-99]*

(a) 15 min (b) 20 min
(c) 25 min (d) 30 min

4. A bus is moving with a speed of 30 km/h ahead of a car with a speed of 50 km/h. How many km apart are they if it takes 15 min for the car to catch up with the bus ? *[IAS (Pre)-02]*

(a) 5 km (b) 7.5 km
(c) 12.5 km (d) 15 km

5. A man starts running from point P at 11 : 00 AM with a speed of 10 km/h. He runs for 2 h and take 1 h rest. He continues this till he is caught by another man who starts at 2 : 00 PM from point P and runs non-stop at a speed of 15 km/h towards the 1st man. At what time (in PM) will the 1st man be caught ? *[SSC(CGL-T-I)-17]*

(a) 6 : 20 PM (b) 4 : 40 PM
(c) 6 : 00 PM (d) 5 : 30 PM

6. Two boats travelling at 5 km/h and 10 km/h head directly towards each other. They begin at a distance of 20 km from each other. How far apart are they (in km) 1 min. before collide ? *[CAT-04]*

(a) $\frac{1}{12}$ km (b) $\frac{1}{6}$ km
(c) $\frac{1}{4}$ km (d) $\frac{1}{3}$ km

7. A thief running at 8 km/h is chased by a policeman whose speed is 10 km/h. If the thief is 100 m ahead of the policeman, then the time required for the policeman to catch the thief will be *[IAS (Pre)-95]*

(a) 2 min (b) 6 min
(c) 10 min (d) 3 min

8. X express which goes from Hyderabad to Chennai, leaves Hyderabad at 5: 30 AM and travels at a constant speed of 50 km/h towards Nalgonda which is 100 km away. At 6 : 00 AM, Y express leaves from Nalgonda for Hyderabad at a constant speed of 40 km/h. At 6 : 30 AM Mr. Shah, the Control Officer realizes that both the trains are on the same track. How much time does Mr. Shah have to avert the accident?

(a) 20 min (b) 30 min
(c) 25 min (d) 15 min

9. Arun, Barun and Kiranmala start from the same place and travel in the same direction at speeds of 30 km/h, 40 km/h and 60 km/h respectively. Barun starts 2 h after Arun. If Barun and Kiranmala overtake Arun at the same instant, how many hours after Arun did Kiranmala start ? *[CAT-06]*

(a) 3 h (b) 3.5 h
(c) 4 h (d) 4.5 h

10. Two cars A and B are travelling on the same road towards each other. If car A is travelling at a speed of 120 km/h and car B is travelling 15% slower than A, how much time will it take the cars to meet, if the initial distance between the two is 668.4 km and car A started to drive $1\frac{1}{2}$ h before car B started ?

(a) 2 h (b) 1 h 30 min
(c) 2 h 12 min (d) 3 h 15 min

Directions (Q. Nos. 11-13) Q started to move from point B towards point A exactly an hour after P started from A in the opposite direction. Q's speed was twice that of P. When P had covered $\frac{1}{6}$th of the distance between the points A and B, Q has also covered the same distance. Then answer the following.

11. The point where P and Q would meet, is

(a) Closer to A
(b) Exactly between A and B
(c) Closer to B
(d) P and Q will not meet at all

12. How many hours would P take to reach B?

(a) 2 h (b) 5 h
(c) 6 h (d) 12 h

13. How many more hours would P (compared to Q) take to complete his journey ? *[CAT-93]*

(a) 4 h (b) 5 h
(c) 6 h (d) 7 h

14. The distance between two stations A and B is 450 km. A train starts form A and moves towards B at an average speed of 15 km/h. Another train starts from B, 20 min earlier than the train A and moves towards A at an average speed of 20 km/h. How far from A will the two trains meet ? *[MAT-09, SSC (SAS)-10]*

(a) 190 km (b) 320 km
(c) 180 km (d) 260 km

15. A ship leaves on a long voyage. When it is 18 miles from the shore, a sea plane, whose speed is ten times that of the ship, is sent to deliver mail. How far from the shore does the sea plane catchup with the ship ? *[CAT-95]*

(a) 24 miles
(b) 25 miles
(c) 22 miles
(d) 20 miles

16. A train starts from Delhi at 6 : 00 AM and reaches Ambala Cantt. at 10:00 AM. The other train starts from Ambala Cantt. at 8 : 00 AM and reaches Delhi at 11 : 30 AM. If the distance between Delhi and Ambala Cantt. is 200 km, then at what time did the two trains meet each other ?

(a) 8 :56 AM (b) 8 : 46 AM
(c) 7 : 56 AM (d) 8 : 30 AM

17. Two stations A and B are 50 km apart. Trains start from A and B at the same time and move in the same direction along the parallel tracks. If the average speed of train from B is 90 km/h and the train from A crosses Station B after 25 min., after how much time will the train from A cross the train from B? *[MAT-18]*

(a) after 50 min (b) after 1 h 15 min
(c) after 1 h 40 min (d) after 2 h

Answer Keys

1. (d)	2. (c)	3. (b)	4. (a)	5. (b)	6. (c)	7. (d)	8. (a)	9. (c)	10. (c)
11. (a)	12. (d)	13. (c)	14. (a)	15. (d)	16. (a)	17. (c)			

Hints and SOLUTIONS

1. Relative speed for the same direction $= 5.5 - 5 = 0.5$ km/h

It means they are making a gap of 0.5 km/h.

Then, required time $= 8.5 \div 0.5 = 17$ h

2. Relative speed in case of same direction $= 70/7 = 10$ km/h and relative speed in case of opposite directions $= 70/1 = 70$ km/h

Clearly, options (c) satisfies the above conditions.

3. The distance between both the moving trains at 7 : 30 AM $= 100 - (50 \times 1) - (40 \times 1/2) = 30$ km

Relative speed $= 50 + 40 = 90$ km/h

Now, time taken by them to meet each other $= (30 \div 90)$ h $= 20$ min.

Hence, Mr. Shah has only 20 min. to avert a head-on collision between the two trains.

4. Relative speed $= 50 - 30 = 20$ km/h

$$= 20 \times \frac{1}{60} = \frac{1}{3} \text{ km/min}$$

Then, required distance

$$= \frac{1}{3} \times 15 = 5 \text{ km}$$

5. Distance covered by the 1st man from 11 : 00 AM to 2 : 00 PM

$= 10 \times 2 = 20$ km.

Hence, the 1st man is 20 km ahead of the 2nd man before the 2nd man starts to run.

At 4 : 00 PM the 1st man will cover a total of $(20 + 20)$ km and he will take rest from 4 : 00 PM to 5 : 00 PM.

On the other hand, the 2nd man will cover 30 km till 4 : 00 PM and continue to run. In next 40 min. he will cover 10 km more.

Hence, the 2nd man will catch the 1st man after 2 h 40 min since 2 : 00 PM i.e. at 4 : 40 PM.

6. Relative speed $= 5 + 10 = 15$ km/

$= 1/4$ km/min

Obviously, they will be at the distance 1 min before collide what will be covered in 1 min. at the rate of relative speed. It means, required answer is1/4 km.

7. Relative speed $= 10 - 8 = 2$ km/h

The time required to cover 100 m at the rate of 2 km/h = 3 min

[Distance covered by the thief beyond 100 m before he is overtaken

= distance covered by the thief in 3 min

$= \left\{8 \times \frac{50}{3} \times 3\right\} = 400$ m]

8. Relative speed for the opposite directions $= 50 + 40 = 90$ km/h

The distance covered by X express upto 6 : 00 AM = 25 km

Hence, time required for meeting of both trains $= (100 - 25) \div 90 = 5/6$ h

$= 50$ min

It means they will meet at 50 min beyond 6 : 00 AM i.e. at 6 : 50 AM.

∴ Control Officer has only 20 min. to avert the accident as he notices at only 6 : 30 AM.

9. Time taken by Barun to overtake Arun $= (2 \times 30) \div (40 - 30) = 6$ h

It implies that Arun moves for 8 h till Barun and Kiranmala both overtake him.

Since Kiranmala also overtakes Arun, the distance covered by Arun will be equal to the distance covered by Kiranmala i.e. (8×30) km.

Clearly, (8×30) km will be covered by Kiranmala in $(240 \div 60)$ i.e. 4 h.

Hence, Kiranmala starts $(8 - 4)$ h after Arun does.

Alternatively

We know, if three bodies start to move from the same point at speeds x, y, z km/h (where, $x < y < z$) respectively and the 2nd body starts m h after 1st body, then the 3rd body must start $\left(m \times \frac{y}{z} \times \frac{z-x}{y-x}\right)$ h after the 1st body so that the 2nd and the 3rd bodies overtake the 1st body at the same time.

Hence, required answer

$$= 2 \times \frac{40}{60} \times \frac{60-30}{40-30}$$

$= 4$ h

10. Distance covered by car A in 1.5 h

$= 120 \times 1.5 = 180$ km.

Speed of the car B

$= 85\%$ of $120 = 102$ km/h

Relative speed

$= 120 + 102 = 222$ km/h

Time required for their meeting

$= (668.4 - 180) \div 222 = 2.2$h

$= 2$h 12 min

11. It is given that at certain point of time, P and Q have covered 1/6 of the distance between A and B from the points A and B respectively. Now, it can be interpreted that both P and Q have to move simultaneously beyond 1/6 of the distance between A and B at the speeds of 1 : 2 to cover the remaining distance between them in order to meet each other. Since the speed of P is lesser than that of Q, P will cover lesser distance than Q. It implies that their meeting point will be closer to A.

12. Let P takes x h to cover 1/6 of the distance between A and B, then Q will take$(x-1)$ h to cover the same distance. Given that ratio of speeds of P and Q be 1 : 2.

Since distance is constant,

$$\frac{x}{x-1} = \frac{2}{1}$$

$\Rightarrow \quad x = 2x - 2$

$\Rightarrow \quad x = 2$

∵ 1/6 of the distance between A and B is covered by P in 2 h.

∴ Whole distance will be covered by P in 12 h.

13. From the above, time taken by P to cover 1/6 of the distance between A and B is 2 h.

Then, time taken by Q to cover the same distance

$= 2 - 1 = 1$h

∴ Difference between time taken by them to cover the whole distance between A and B

$= (2 \times 6) - (1 \times 6) = 6$h

14. Distance covered by train B in 20 min. at the rate of 15 km/h = 5 km

Now, relative speed = 15 + 20

$= 35$ km/h

Then, time required for their meeting

$= (450 - 5) \div 35 = 12.71$h

The distance of the meeting point from station $A = 15 \times 12.71 = 190$ km

15. Let the speed of ship be x miles/h.

Then, speed of the sea plane

$= 10x$ miles/h

Relative speed = $10x - x = 9x$ miles/h

∴ Time taken by the sea plane to catch the ship

$= 18 \div 9x = \frac{2}{x}$ h

Then, distance covered by the sea plane in $\frac{2}{x}$ h

$= 10x \times \frac{2}{x} = 20$ miles

16. Speed of the train running from Delhi to Ambala Cantt.

$= 200 \div 4 = 50$ km/h

and speed of the train running from Ambala Cantt. to Delhi

$= 200 \div 3.5 = 400/7$ km/h

Now, relative speed

$= 50 + 400/7$ km/h

The distance covered by the train running from Delhi to Ambala Cantt. till 8 : 00 AM

$= 2 \times 50 = 100$ km

Then, time required for their meeting

$= (200 - 100) \div (50 + 400/7)$

$= 14/15$ h $= 56$ min

Hence, they will meet at 56 min. beyond 8 : 00 AM i.e. at 8 : 56 AM.

17. Speed of train started from Station $A = 120$ km/h

[∵ it covers 50 km in 25 min.]

Now, relative speed

$= 120 - 90 = 30$ km/h

Then, time taken by the faster train to overtake the slower

$= 50/30$ h $= 1$h 40 min

33

To and Fro Motion in a Straight Line

Concept of To and Fro Motion in a Straight Line

Case A **When two bodies start moving in the same direction.**

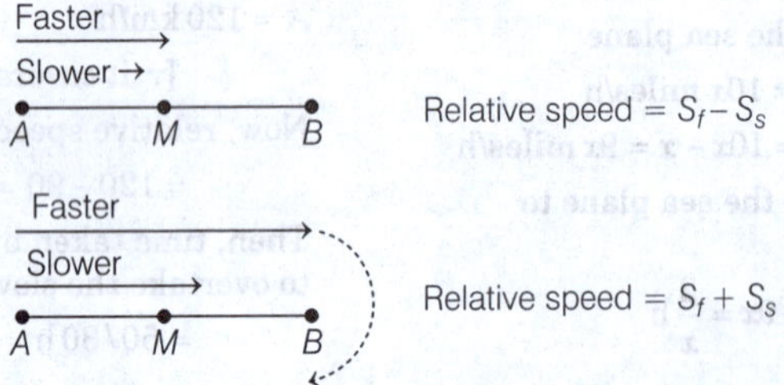

In this case the faster body will reach the ending point before the slower reaches there and will meet with the slower while returning on its way. Then, the relative speed will be $(S_f - S_s)$ till the faster reaches the ending point. Here, S_f is the speed of the faster body whereas the speed of the slower body is S_s. However, the relative speed will be $(S_f + S_s)$ since when the faster starts to reverse.

In lucid way, when they move in same direction, their relative speed will be $(S_f - S_s)$ and when they move in opposite direction, their relative speed will be $(S_f + S_s)$.

When both start simultaneously, the time taken by them individually to meet each other will be same. In case they start to move at an interval of time, time will be adjusted as has been done under topic 'Relative speed.'

So far as distance covered in this case is concerned, for every meeting the total distance covered by them together will be $2D$, where D is the distance between two given ending points. To illustrate, for nth meeting, total distance covered by them together will be $(n \times 2D)$ and the respective coverage of distance will be in ratio of individual speeds.

Hence, $\dfrac{\text{distance covered by the faster body}}{\text{distance covered by the slower body}} = \dfrac{\text{speed of the faster body}}{\text{speed to the slower body}}$,

when time is constant.

Inside

- Concept of To and Fro Motion in a Straight Line
  - (i) When Two Bodies Start Moving in Same Direction
  - (ii) When Two Bodies Start Moving Towards Each Other
- ....
- .... & Many More
- Work Out with Solutions

Case B **When two bodies start moving towards each other**

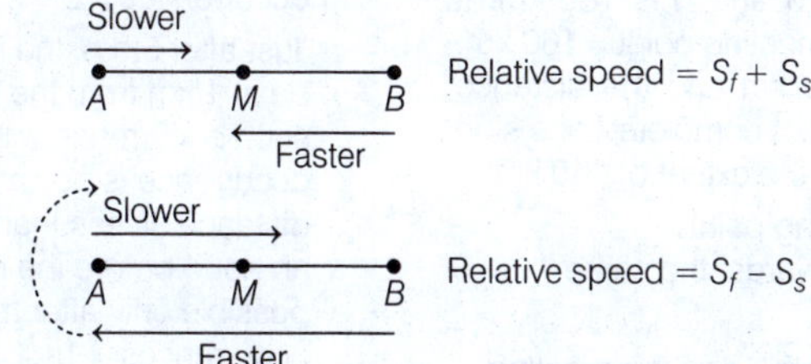

When they move in opposite directions, their relative speed will be $(S_f + S_s)$ and when they move in same direction, their relative speed will be $(S_f - S_s)$.

Further, when both start simultaneously, the time taken by them individually to meet each other will be same. In case they start to move at an interval of time, time will be adjusted as has been done under topic 'Relative speed'.

So far as distance covered in this case is concerned, for 1st meeting, the total distance covered by them together will be D, where D is the distance between two given ending points. For 2nd meeting onwards, a distance of $2D$ is covered for every meeting. To illustrate, for nth meeting, total distance covered by them together will be $[D + (n-1) \times 2D]$ i.e. $[2n - 1]\, D$ and the respective coverage of distance will be in ratio of individual speeds.

Hence, $\dfrac{\text{distance covered by the faster body}}{\text{distance covered by the slower body}} = \dfrac{\text{speed of the faster body}}{\text{speed of the slower body}}$, when time is constant.

Illustrations

Problem 1 The speeds of two objects namely X and Y are 40 km/h and 50 km/h respectively and they have to move between points M and N, 180 km apart. Now, prepare a chart comparing meeting-wise the distance covered by them together, time taken by them, distance covered by them individually, location of the meeting point from both points if (i) they start to move simultaneously from point M in the same direction, (ii) they start to move simultaneously from point M and N respectively.

Distance in km and time in h

1	2		3		4		5		6		7	
Sequence of Meeting	**Distance covered by them together**		**Time Required**		**Distance covered by X**		**Distance covered by Y**		**Distance of meeting point from M**		**Distance of meeting point from N**	
	Same direction (SD)	**Opposite direction (OD)**	**SD**	**OD**	**SD**	**OD**	**SD**	**OD**	**SD**	**OD**	**SD**	**OD**
1st	360	180	4	2	160	80	200	100	160	80	20	100
2nd	720	540	8	6	320	240	400	300	40	120	140	60
3rd	1080	900	12	10	480	400	600	500	120	40	60	140
4th	1440	1260	16	14	640	560	800	700	80	160	100	20

Sol. **Column 2** For same direction ⇒ For n th meeting, total distance covered by them together = $(n \times 2D)$

and for opposite directions ⇒ For n th meeting, total distance covered by them together = $(2n - 1)\, D$

Column 3 Cumulative time (applicable in both cases) = $\dfrac{\text{Cumulative distance}}{\text{Relative speed}} = \dfrac{\text{Distance covered by them togehter}}{\text{Sum of their speeds}}$

Column 4 and 5 Respective coverage of distance will be in ratio of their speed as time is constant because they have started to move simultaneously.

e.g. In case of 1st meeting in the same direction, total distance covered by them together is 360 km. Then, distances covered by them individually will be in ratio 4 : 5. It means distance covered by $X = 360 \times \dfrac{4}{4+5} = 160$ km and that by $Y = 360 - 160 = 200$ km.

The same approach is applicable in case of opposite directions.

Column 6 and 7 In case of same direction, X covered 160 km till 1st meeting and the distance between M and N is 180 km. It means the distance between M and the meeting point = 160 km. Further, till 2nd meeting, X covered 320 km and the distance between M and N is 180 km. It means X has completed one side from M to N and returned from point N to the extent of 140 km.

Then, distance between M and the meeting point $= 180 - 140 = 40$ km. For 3rd meeting onwards, the same approach is applicable.

So far as computation of distance between N and the meeting point is concerned, the same approach is applicable also. Alternatively, you may subtract the distance between M and meeting point from the 180 km to get the distance between N and the meeting point.

Problem 2 A hare and a tortoise run between points O and P located exactly 6 km from each other on a straight line. They start together at O, go straight to P and then return to O along the same line. They run at constant speeds of 12 km/h and 1 km/h respectively. Since the tortoise is slower than the hare, the hare shuttles between O and P until the tortoise goes once to P and returns to O. During this run, how many times are the hare and the tortoise separated by an exact distance of 1 km from each other? ***[XAT-20]***

(a) 24 (b) 42
(c) 22 (d) 48
(e) 40

Sol. Clearly, the tortoise will take 12 h to move from points O to point P and back to point O. The hare will move for 12 h to and fro.

Hare ...... → 12 km/h

Tortoise .... → 1 km/h

O..P

Both started to move simultaneously from point O. It is obvious that after $5\frac{5}{11}$ min, the hare will be 1 km ahead of the tortoise.

Such occurrence will happen once in 1st half an hour. After half an hour, the hare will reach at point P and will continue to move towards point O and they will be separated by 1 km twice i.e. before the meeting and after the meeting. After meeting both are separated by 1 km and the hare is at the starting point O. Hence, in 1st one hour, number of such occurrences $= 1 + 2 = 3$

Further, in next half an hour, number of such occurrences = 1 as the hare is now at point O and such occurrence will happen only when the hare will be 1 km ahead of the tortoise.

Such occurrence before crossing the tortoise has already been taken in consideration earlier. Hence, number of such occurrences during 2nd h = 3 as same as 1st h.

Just after 2 h, the hare will be at point O and tortoise will be at 2km from the point O. Then, there will be two such occurrences will happen i.e. before crossing and after crossing in next half an hour. After half an hour the hare will move towards point O and two such occurrences will happen i.e. before meeting and after meeting.

Hence, in 3rd hour, number of such occurrences $= 2 + 2 = 4$

Similarly, in 4th and 5th hour each, number of such occurrences $= 2 + 2 = 4$

Just after 5th h, the hare will be at point O and the tortoise will be at 1 km from the point P. Then, in next half an hour, only one such occurrence will happen i.e. before the crossing, no such occurrence is possible after crossing as the tortoise will be at a distance of less than 1 km from point P. Further, in the next half an hour, before the meeting such occurrence is not possible, possible only after meeting. Hence in 6th h number of such occurrences $= 1 + 1 = 2$

Just after 6th h, the hare will be at point O and the tortoise at point P. In next half an hour, only one such occurrence is possible before the meeting as after half an hour the tortoise will be within 1 km from the point P. Further, in next half an hour, only one occurrence is possible after crossing.

Hence, in 7th h, number of such occurrences $= 1 + 1 = 2$

On the above trends, number of such occurrences in 8th h, 9th h, 10th h, 11th h and 12th h will be 4, 4, 4, 4 and 2 respectively.

Thus, number of such occurrences

$$= 3 + 3 + 4 + 4 + 4 + 2 + 2 + 4 + 4 + 4 + 4 + 2 = 40$$

Alternatively

Since they start to move simultaneously from point O, the event of overtaking cannot happen during the 1st h. Term 'Overtaking' refers to the situation when they meet while moving in the same direction. For the opposite direction, term 'meeting' has been used for sake of convenience only.

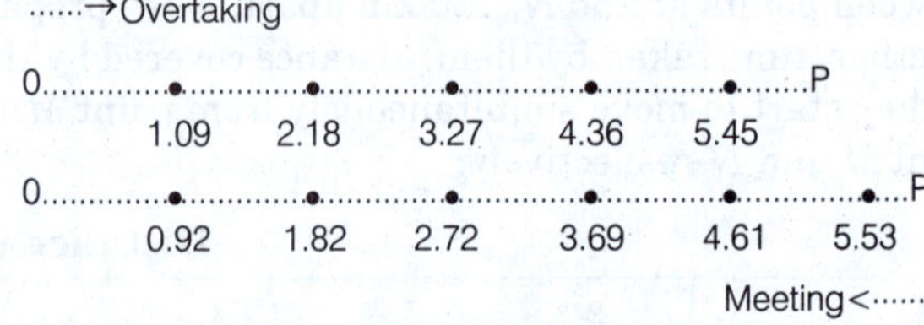

[distances have been shown from the point O and the above diagram is for first 6 h]

1st overtaking will occur at the distance of $\left\{1 + \frac{1}{(12-1)}\right\}$ km from point O and 2nd on its next multiple and so on (depicted above).

1st meeting will occur at the distance of $\left\{\frac{2 \times 6}{(12+1)}\right\}$ km from point O and further on its multiples (depicted above).

Clearly, with reference to first 4 overtakings, two occurrences of being seperated by 1 km are possible with each overtaking. However, only one such occurrence is possible with reference to 5th overtaking i.e. only before the overtaking. Hence, total of 9 occurrences.

Further, with reference to first 5 meetings, (5×2) occurrences are possible. Even after the meeting at a distance of 0.92 from point O, such occurrence is possible as after few moments tortoise will be at a distance of 1 km and the hare will at point O. However, with reference to 6th meeting, only one such occurrence is possible i.e.after meeting. Before 6th meeting it is not possible. Hence, total of 11 occurrences.

Just after 6 h, tortoise will be at point P and the hare at point O. The similar situations will happen again in next 6 h. Meeting will happen first.

Thus, number of such occurrences $= 2(9 + 11) = 40$

WORKOUT

1. Two bodies X and Y start to move simultaneously towards each other from the points M and N respectively and they meet at a point, lying at a distance of $0.6D$ from point M. Find the point of their 7th meeting if D is the distance between points M and N
   (a) At a distance of $0.8D$ from point N
   (b) At a distance of $0.8D$ from point M
   (c) At a distance of $0.2D$ from point N
   (d) None of the above

2. Two men A and B start walking simultaneously from P to Q, a distance of 21 km at the speed of 3 km/h and 4 km/h respectively. B reaching Q returns immediately and meet A at R. Find the distance from P to R.
   (a) 20 km (b) 18 km
   (c) 16 km (d) 24 km

3. Two persons P and Q start at the same time from city A for city B, 60 km away. P travels 4 km/h slower than Q. Q reaches city B and at once turns back meeting P, 12 km from city B. What is the speed of P? **[CDS-08, FMS-11]**
   (a) 8 km/h (b) 12 km/h
   (c) 16 km/h (d) 20 km/h

Directions (Q.Nos. 4-5) Ram and Shyam run a race between points A and B, 5 km apart. Ram starts at 9 : 00 AM from A at a speed of 5 km/h, reaches B and returns to A at the same speed. Shyam starts at 9 : 45 AM from A at a speed of 10 km/h, reaches B and comes back to A at the same speed.

4. At what time do Ram and Shyam first meet each other ?
   (a) 10 : 00 AM (b) 10 : 10 AM
   (c) 10 : 20 AM (d) 10 : 30 AM

5. At what time does Shyam overtake Ram ? **[CAT-05]**
   (a) 10 : 20 AM
   (b) 10 : 30 AM
   (c) 10 : 40 AM
   (d) 10 : 50 AM

6. Hari and Ravi started a race from opposite ends of the pool. After a minute and a half, they passed each other in the centre of the pool. If they lost no time in turning and maintained their respective speeds, how many minutes after starting did they pass each other the 2nd time? **[FMS-11]**
   (a) 3 min (b) 4.5 min
   (c) 6 min (d) 7.5 min

7. A car A starts from a point P towards another point Q. Another car B starts (also from P) 1h after the 1st car and overtakes it after covering 30% of the distance PQ. After that, the cars continue. On reaching Q, car B reverses and meets car A after covering $23\frac{1}{3}$ of the distance QP.
   Find the time taken by car B to cover the distance PQ (in h).
   (a) 3 (b) 4
   (c) 5 (d) $3\frac{1}{3}$

8. A and B walk from X to Y, a distance of 27 km at 5 km/h and 7 km/h respectively. B reaches Y and immediately turns back meeting A at Z. What is the distance from X to Z ? **[CAT-94]**
   (a) 25 km (b) 22.5 km
   (c) 24 km (d) 20 km

Answer Keys

1. (a)	2. (b)	3. (a)	4. (b)	5. (b)	6. (b)	7. (d)	8. (b)		

Hints and SOLUTIONS

1. They meet at distance of $0.6D$ from point M, it implies that the distances covered by them individually will be in ratio $0.6D : 0.4D$ i.e. $3 : 2$.

Since time is constant, their speeds be in ratio $3 : 2$.

Now, total distance covered by them together upto 7th meeting

$$= [(2 \times 7) - 1]\, D = 13D$$

Now, distance covered by X

$$= 13D \times \frac{3}{(3+2)} = 7.8\, D$$

This indicates that X goes from M to N, back, again M to N, back, again M to N and back, further, M to N and back to the extent of $0.8D$ only from point N.

It means, the meeting point is at a distance of $0.8D$ from the point N or at a distance of $0.2\, D$ from the point M.

2. In 1st meeting, the total distance to be covered by them togethe

$$= 2D = 2 \times 21 = 42\,\text{km}.$$

Since time is constant, ratio of distances covered by them individually will be equal to the ratio of their speeds i.e. $3 : 4$.

Now, distance covered by A

$$= 42 \times \frac{3}{(3+4)} = 18$$

Hence, distance between P and meeting point $R = 18\,\text{km}$.

3. Obviously, Q travels a distance of $(60 + 12)$ km whereas P travel only $(60 - 12)$ km in the equal time.

Since time is constant, $\frac{S_1}{S_2} = \frac{D_1}{D_2}$

$$\Rightarrow \quad \frac{\text{speed of } P}{\text{speed of } Q} = \frac{48}{72}$$

$$\Rightarrow \quad \frac{\text{speed of } P}{\text{speed of } P + 4} = \frac{48}{72}$$

$$\Rightarrow \quad \text{Speed of } P = 8\,\text{km/h}$$

4. Distance covered by Ram till 9 : 45 AM

$$= 5 \times \frac{3}{4} = 3.75 \text{ km}$$

For the 1st meeting, they have to move (2×5) km together. Out of 10 km, a distance of 3.75 km has been covered by Ram before Shyam starts to move. Now, they together have to move only $(10 - 3.75)$ km to meet each other.

$\therefore$ Time taken by them together to move 6.25 km

$$= \frac{6.25}{(5+10)} \text{ h} = 25 \text{ min}$$

Hence, they will meet at 25 min beyond 9 : 45 AM

i.e. at 10 : 10 AM.

5. Distance covered by Ram till 9 : 45 AM $= 5 \times \frac{3}{4} = 3.75\,\text{km}$

Obviously, this gap will be bridged by the relative speed. Here, relative speed will be $(10 - 5)$ km/h.

Hence, time taken by Shyam to overtake Ram

$$= 3.75 \div (10 - 5) = 45 \text{ min}$$

Hence, required time = 45 min beyond 9 : 45 AM

i.e. 10 : 30 AM.

6. Obviously, they together covered a distance equal to length of the pool in 1.5 min in order to meet each other. For 2nd meeting, they together have to move further a distance equal to two times the length of the pool. Accordingly, they together will take 3 min to cover two times the length of the pool.

Hence, the required time

$$= 1.5 + 3 = 4.5\,\text{min}$$

7. Let the distance between P and Q be 100 km.

Then distance covered by cars A and B each till 1st meeting = 30 km

Further, distance covered by car A for the 2nd meeting (after 1st meeting)

$$= 100 - 30 - 23.33 = 46.66 \text{ km}$$

and distance covered by car B for the 2nd meeting (after 1st meeting)

$$= 70 + 23.33 = 93.33\,\text{km}$$

It may be interpreted that from the point of 1st meeting they move simultaneously and the distance covered by car B is twice the distance covered by car A.

Hence. ratio of speeds of car A and B $= 1 : 2$

Also, for the 1st meeting both the cars cover the same distance but at different speeds and car A takes 1h more than car B.

We know,

$$\frac{\text{speed of car } A}{\text{speed of car } B} = \frac{\text{time taken by } B}{\text{time taken by } A},$$

when distance is constant

$$\Rightarrow \quad \frac{1}{2} = \frac{x \text{ (let)}}{(x+1)}$$

$$\Rightarrow \quad x = 1\,\text{h}$$

$\because$ Car B takes 1h to cover 30% of the distance between P and Q

$\therefore$ Time taken by car B to cover 100% of the distance $PQ = 3\frac{1}{3}$ h

8. Total distance covered by them for the meeting at point $z = 2 \times 27 = 54\,\text{km}$

Since ratio of speeds of A and B be $5 : 7$, the ratio of distance covered by them $= 5 : 7$

Hence, distance covered by A

$$= 54 \times \frac{5}{5+7} = 22.5\,\text{km}$$

Since A is still on the way from X to Y, the distance from X to $Z = 22.5\,\text{km}$.

34

Races and Games

Introduction

This topic is also based on the principles of Time, Distance and Speed. Before building the fundamentals in this topic, it is imperative to know certain terminology related to Races and Games.

Races It means a contest of speed over a certain distance and it may be in case of running, driving, riding, sailing, rowing, etc.

Race course The ground, track, path or place where race is held.

Starting point The point from which the race begins.

Winning or Goal The point set to bound a race.

Winner The person who reaches the winning post.

Dead-heat race The race in which all the participants reach the winning post exactly at the same time. It means, no one is winner.

Startup or headstart When a participant/runner (say, A) allows another (say, B) to stay ahead of some distance (say, x metres) before starting the race, then it is said that A gives x metres startup to B. In other words, B has startup of x metres. In this case, B runs x metres less than A does in the same time. Further, A beats B by x metres also refers to B runs x metres less than A does.

A may give B a start of T seconds (or any unit of time) instead of a distance and in this case, B will be ($T \times$ speed of B) metres (or any unit of distance) ahead of A.

Further, when it is called A beats B by T seconds it means A has reached at winning post and B will reach at that post after T sec, if he continues to run. In other words, B takes T sec more than A does to reach at the winning post. It also implies that the difference between distances covered by them may be covered by B in T sec.

Games A game of 100 means the person among the contestants who scores 100 points first is the winner. In a game of 100, A can give B 25 points– it means A scores 100 points whereas B scores only 75 points. *Vice versa,* if B scores 100 points while A scores only 75, then it is said that 'B can give A 25 points' in a game of 100.

Important to note In race, time is constant, then ratio of speeds of participants will be equal to the ratio of distances covered by them.

[Races on the circular path will be discussed under the next topic "Circular Motion"]

Inside

- *Introduction*
- *Terminology Related to Races and Games*
- *Illustrations*
- ....
- .... *& Many More*
- *Work Out with Solutions*

Illustrations

Problem 1 **The speeds of Ram and Shyam are in ratio 5 : 8 and Ram loses the race by 210 m. Find the length of the race course?**

Sol. In race, time is constant. Then, ratio of speeds of Ram and Shyam = ratio of distances covered by them.

$\Rightarrow$ Ratio of distances covered by Ram and Shyam = 5 : 8

As given, (8 – 5) units = 210 m

Then, 8 units = 560 m

Hence, the length of the race course = 560 m

Alternatively

When Shyam covers 8 m, Ram covers 5 m. It means Ram loses the race by 3 m.

$\because$ Ram loses 3 m when the race is of 8 m

$\therefore$ Ram will lose 210 m when the race will be of 560 m

Problem 2 **Abhay beats Bunty by 35 m or 7 sec in a km race, then find Abhay's time over the course.**

Sol. Obviously, Bunty can run 35 m in 7 sec.

Then, 1000 m (i.e. 1 km) in $\left(\frac{7}{35}\times 1000\right) = 200$ sec

Hence, Abhay's time over the course = 200 – 7 = 193 sec

Problem 3 **In a 100 m race, Kunal defeats Bimal by 5 sec. If the speed of Kunal is 18 km/h then the speed of Bimal is what?** *[SSC (CGL-T-I)-10]*

Sol. Let the speed of Bimal be x km/h. Then, ratio of speeds of Kunal and Bimal = ratio of distances covered by them

$$\Rightarrow \quad 18 : x = 100 : \left\{100 - \frac{25x}{18}\right\}$$

[$\because x$ km/h $= \frac{5x}{18}$ m/s and distance covered by Bimal in 5 sec $= \frac{25x}{18}$ m]

$$\Rightarrow \quad 1800 - 25x = 100x \Rightarrow x = 14.4 \text{ km/h}$$

Alternatively I

Let the speed of Bimal be x km/h.

Then, time taken by Kunal to cover 100 m (i.e. 0.1 km) $= \frac{0.1}{18}$ h

and time taken by Bimal to cover 0.1 km $= \frac{0.1}{x}$ h

As given, $\frac{0.1}{x} - \frac{0.1}{18} = \frac{5}{60\times 60}$

$$\Rightarrow \quad 0.1\left\{\frac{1}{x} - \frac{1}{18}\right\} = \frac{1}{720} \Rightarrow \frac{1}{x} = \frac{1}{72} + \frac{1}{18}$$

$$\Rightarrow \quad \frac{1}{x} = \frac{5}{72} \Rightarrow x = 14.4 \text{ km/h}$$

Alternatively II

Speed of Kunal $= 18\times\frac{5}{18}$ m/sec = 5 m/sec

Hence, time taken by Kunal in this race = 100 ÷ 5 = 20 sec

Now, $\frac{S_1}{S_2} = \frac{T_2}{T_1}$, as distance is constant

$$\Rightarrow \quad \frac{\text{Speed of Kunal}}{\text{Speed of Bimal}} = \frac{\text{Time taken by Bimal}}{\text{Time taken by Kunal}}$$

$$\Rightarrow \quad \frac{18}{\text{Speed of Bimal}} = \frac{(20+5)}{20}$$

$\Rightarrow$ Speed of Bimal = 18 × 0.8 = 14.4 km/h

Problem 4 **In a race of 800 m, A can beat B by 40 m. In a race of 500 m, B can beat C by 5 m. In a race of 200 m, A will beat C by what?** *[SSC (CPO)-07, MAT-07, 08, 09]*

Sol. Ratio of speeds of A and B = Ratio of distances covered by them

$\Rightarrow$ Ratio of speeds of A and B = 800 : 760 = 20 : 19

Similarly, ratio of speeds of B and C = 500 : 495 = 100 : 99

$\therefore$ Ratio of speeds of A, B and C = (20 × 100) : (19 × 100) : (19 × 99)

[how to find ratio, has been discussed in chapter 'Ratio, Proportion and variation' in detail]

$\Rightarrow$ Ratio of speeds of A, B and C = 2000 : 1900 : 1881

Therefore, in a race of 200 m, A will cover 200 m, B will cover 190 m and C will cover 188.1 m.

Hence, A will beat C by (200 – 188.1) m in a race of 200 m.

Hence, required answer = 11.9 m

Alternatively I

As given, A = 800 m, then B = 760 m. Hence, $A : B$ = 20 : 19

Further, B = 500, then C = 495 m. Hence, $B : C$ = 100 : 99

Now, $A : C = (A : B)\times(B : C)$ = (20 × 100) : (19 × 99) = 200 : 188.1

Obviously, A will beat C by (200 – 188.1) m in a race of 200 m.

Hence, required answer = 11.9m

Alternatively II

As given, A = 800 m, then B = 40 less than A

$\therefore$ When A = 200, then B = 10 less than A. (i.e. B = 190)

Further, B = 500, then C = 5 less than B

$\therefore$ When B = 190, then C = 1.9 less than B

Obviously, when A = 200, C will be (10 + 1.9) less than A.

Hence, required answer = 11.9 m

Problem 5 **In a race of 1 km, A gives B a start of 100 m and still wins by 20 sec. But if A gives B a start of 25 sec, B wins by 50 m. The time taken by A to run 1 km is what?**

Sol. Let the speeds of A and B be x m/sec and y m/sec respectively.

Then, ratio of their speeds = ratio of distances covered by them

$$\Rightarrow \quad x : y = 1000 : \{1000 - 100 - 20y\} = 50 : (45 - y)$$

For 2nd case,

Ratio of their speeds = Ratio of distances covered by them

$$\Rightarrow \quad x : y = 950 : \{1000 - 25y\} = 38 : (40 - y)$$

Obviously, $x : y = 50 : (45 - y) = 38 : (40 - y)$

$$\Rightarrow \quad 2000 - 50y = 1710 - 38y \Rightarrow y = \frac{290}{12} \text{ m/sec}$$

Further, $x : y = 50 : (45 - y)$

$$\Rightarrow x\left\{45 - \frac{290}{12}\right\} = 50\times\frac{290}{12}$$

$$\Rightarrow \quad x = 50\times\frac{290}{12}\times\frac{12}{250} = 58 \text{ m/sec}$$

Then, time taken by A to run 1000 m = 1000 ÷ 58 $= \frac{500}{29}$ sec

Alternatively

Let the time taken by A to run 1000 m be x sec.

Case I B covers (1000 – 100) m in $(x+20)$ sec.

Then, B will cover 1000 m in $\frac{10}{9}\times(x+20)$ sec.

Case II

A runs (1000 – 50) m and will take $\left(\frac{x}{1000}\times 950\right)$ i.e. $\frac{19x}{20}$ sec.

This time is 25 sec less than the time taken by B to run 1000 m.

Hence, $\frac{19x}{20}=\left\{\frac{10}{9}\times(x+20)\right\}-25$

$\Rightarrow \quad \frac{19x}{20}-\frac{10x}{9}=\frac{200}{9}-25$

$\Rightarrow \quad \frac{-29x}{20\times 9}=\frac{-25}{9}\Rightarrow 29x=500\Rightarrow x=\frac{500}{29}$ sec

Problem 6 **In a km race, A beats B by 30 sec and B beats C by 15 sec. If A beats C by 180 m, the time taken by A to run 1 km is ?**

Sol. Obviously, A beats C by (30 + 15) sec or 180 m.

$\because$ To cover 180 m, C takes 45 sec.

Then, to cover 1000 m C will take 250 sec.

Hence, time taken by $A = 250 - 45 = 205$ sec.

Problem 7 **Arun can run a km in 5 min 40 sec and Bhukhan in 6 min. How many metres' start should Arun give Bhukhan in a km race in order to end up the race in dead-heat?**

Sol. Obviously, Arun can given Bhukhan a start of (6 min – 5 min 40 sec) i.e. 20 sec.

Further, distance which may be travelled by Bhukhan in 20 sec

$=\left(\frac{1000}{340}\times 20\right)=58.82$ m

Hence, required answer = 58.82 m

Problem 8 **In a km race A gives B a startup of 5 sec and still beats B by 10 sec. What is the speed of A if ratio of speeds of A and B be 5 : 2 ?**

Sol. Clearly, B will take $(5+10)$ sec. more than A does to cover 1000 m.

As given, ratio of their speeds = 5 : 2

Then, ratio of time taken to cover 1000 m = 2 : 5

$\Rightarrow$ Time taken by A : Time taken by B = 2 : 5

$\Rightarrow \quad T:(T+15)=2:5\Rightarrow 5T=2T+30\Rightarrow T=10$ sec

Hence, speed of $A=1000\div 10=100$ m/s

Alternatively

Ratio of time taken to cover 1000 m = 2 : 5

But, B takes 15 sec more than A do.

Hence, $(5-2)$ units = 15 sec

Then, 2 Units = 10 sec

$\therefore$ Time taken by A to cover 1000 m = 10 sec

Then, speed of $A=1000\div 10=100$ m/s

Problem 9 **In a race of 100 m, A can give B 20 m and C 30 m while B can give C 2.5 sec. What is the time taken by A to cover 100 m.**

Sol. Ratio of speeds of A and B = 100 : 80 = 10 : 8

and ratio of speeds of A and C = 100 : 70 = 10 : 7

Then, ratio of speeds of A, B and C = 10 : 8 : 7

Hence, in a race of 100 m, if B covers 100 m, C will cover 87.5 m.

But, given that B can give C 2.5 sec.

It means, C can cover (100 – 87.5) m in 2.5 sec. Then, C can cover 100 m in 20 sec.

As given, ratio of speeds of A and C = 10 : 7

Then, ratio of time taken by them = 7 : 10 if distance is constant.

[we have assumed C moves 100 m, for which time taken is 20 sec.]

Obviously, 10 units = 20 sec,

$\therefore$ 7 units = 14 sec.

Hence, time taken by A to cover 100 m = 14 sec.

Alternatively

As given, A = 100 m, then B = 80 m

Further, A = 100 m, then C = 70 m.

Clearly, in a race of 80 m, B gives C (80 – 70) m

Then, in a race of 100 m, B will give C 12.5 m.

Also given that B gives C 2.5 sec.

It means, C covers 12.5 m in 2.5 sec.

Then, time taken by C to cover 100 m = 20 sec.

Further, time taken by B to cover 100 m = 20 – 2.5 = 17.5 sec as B gives C 2.5 sec. over the course.

Now, 100 m are covered by B in 17.5 sec.

Then, 80 m will be covered by B in 14 sec.

Since A gives B 20 m, the time taken by A to cover 100 m will be equal to the time taken by B to cover 80 m.

Hence, time taken by A to cover 100 m = 14 sec

Problem 10 **In a km race, A beats B by 200 m and B gives a start of 300 m to C. How many metres should A give C so as to end up the race (between A and C) in dead-heat ?**

Sol. Ratio of speeds of A and B = 1000 : 800 = 5 : 4

and ratio of speeds of B and C = 1000 : 700 = 10 : 7

Hence, ratio of speeds of A, B and C

$=(5\times 10):(4\times 10):(4\times 7)=25:20:14$

Then, in a race of 1000 m, when A moves 1000 m, C will move 560 m. It implies that A should give C a start up of (1000 – 560) m i.e. 440 m to end the race in dead-heat.

Alternatively

As given, $A \Rightarrow$ 1000 m, then $B \Rightarrow$ 800 m.

Hence, $A : B$ = 10 : 8

Further, $B \Rightarrow$ 1000 m, then $C \Rightarrow$ 700 m.

Hence, $B : C$ = 10 : 7

Now, $A:C=(A:B)\times(B:C)=(10\times 10):(8\times 7)=100:56$

Clearly, A beats C by (100 – 56) m in a race of 100 m.

Then, in a race of 1000 m, A will beat C by 440 m.

It implies that A should give C a start up of 440 m to end the race in dead-heat.

Problem 11 **In a game of 90 points, A can give B 10 points and C 15 points. Then, how many points can B give C in a game of 64 ?**

Sol. $A : B$ = 90 : 80 = 9 : 8

$A : C$ = 90 : 75 = 6 : 5

Then, $B : C = (B : A) \times (A : C)$

$= (8 : 9) \times (6 : 5) = 48 : 45 = 16 : 15$

Clearly, in a game of 16, *B* gives *C* 1 point.

Then, in a game of 64, *B* will give *C* 4 points.

Alternatively

As given, $A \Rightarrow 90$, then $B \Rightarrow 80$

Further, $A = 90$, then $C = 75$

Clearly, in a game of 80, *B* gives *C* (80 – 75) points.

Then, in a game of 64, *B* will give *C* 4 points.

Problem 12 **In a game of certain points, *A* can give *B* 50 points and *C* 62 points. *B* can give *C* 16 points. How many points make the game ?**

Sol. Let the game is of *x* points.

Then, $A : B = x : (x - 50)$

$A : C = x : (x - 62)$

Then, $B : C = (B : A) \times (A : C) = (x - 50)x : x(x - 62)$

$= (x - 50) : (x - 62)$

But given, $B : C = x : (x - 16)$

Hence, $(x - 50) : (x - 62) = x : (x - 16)$

$\Rightarrow \quad [x^2 - 66x + 800] = x^2 - 62x$

$\Rightarrow \quad 4x = 800 \Rightarrow x = 200$

Alternatively

Let the game is of *x* points.

Then, $A : B = x : (x - 50)$

$A : C = x : (x - 62)$

Clearly, in a game of $(x - 50)$ points, *B* gives *C* $[(x - 50) - (x - 62)]$ i.e. 12 points.

Then, in a game of *x*, *B* will give *C* $\left(\frac{12}{x - 50} \times x\right)$ points.

But given, $\left(\frac{12}{x - 50} \times x\right) = 16$

$\Rightarrow \quad 3x = 4x - 200 \Rightarrow x = 200$

Problem 13 **In a 100 m race, if *A* gives *B* a start of 20 m, then *A* wins the race by 5 sec. Alternatively, if *A* gives *B* a start of 40 m the race ends in a dead heat. How long does *A* take to run 200 m ?** ***[SNAP-11]***

Sol. Let the speeds of *A* and *B* be *x* m/s and *y* m/s respectively.

Then, ratio of their speeds = ratio of distances covered by them

$\Rightarrow x : y = 100 : [100 - 20 - 5y]$

For 2nd case, $x : y = 100 : 60$

Obviously, $[100 - 20 - 5y] = 60$

$\Rightarrow \quad 80 - 5y = 60$

$\Rightarrow \quad y = 4$ m/s

From above, $x : y = 5 : 3$

$\Rightarrow \quad x : 4 = 5 : 3$

$\Rightarrow \quad x = 6.67$ m/s.

Then, time taken by *A* to run 200 m $= 200 \div 6.67 = 30$ sec

Alternatively I

From the 2nd case, ratio of the their speeds = 100 : 60 = 5 : 3

Let the speed of *B* be *x* m/s.

For the 1st case, ratio of speeds $= 100 : \{100 - 20 - (5 \times x)\}$

$\Rightarrow \quad 5 : 3 = 20 : (16 - x)$

$\Rightarrow \quad 60 = 80 - 5x \Rightarrow x = 4$ m/s

From above, speed of *A* : speed of *B* = 5 : 3

$\Rightarrow$ speed of *A* : 4 = 5 : 3 $\Rightarrow$ speed of *A* = 6.67 m/s

Then, time taken by *A* to run 200 m $= 200 \div 6.67$

$= 30$ sec

Alternatively II

In 2nd case, A runs 100 m whereas *B* runs only 60 m. In 1st case, *A* runs 100 m while *B* reaches upto 80 m as *A* gives him a start of 20 m. Also, *A* wins the race by 5 sec. It implies that (100 – 80) m distance can be covered by *B* in 5 sec. Hence, speed of *B* = 4 m/s.

From the above, speed of *A* : speed of *B* = 100 : 60 = 5 : 3

$\Rightarrow$ speed of *A* : 4 = 5 : 3

$\Rightarrow$ speed of *A* = 6.67 m/s

Then, time taken by *A* to run 200 m $= 200 \div 6.67 = 30$ sec

Problem 14 **X runs 1.5 times as fast as Y. If X gives Y a start of 300 m, how far must X run before he catches up with Y ?** ***[MAT-09]***

Sol. Let X runs *D* m before he catches up with *Y*: Since time is constant, ratio of speeds of X and Y = ratio of distances covered by them

$\Rightarrow 1.5 : 1 = D : (D - 300)$

$\Rightarrow \quad D = 1.5D - 450 \Rightarrow D = 900$ m

Alternatively

Since the speed of X is 1.5 times that of *Y*, *X* will cover 1.5 times the distance covered by *Y*.

But given that *A* covers 300 m more than *Y* does.

Hence, 0.5 times of the distance covered by *Y* = 300 m

Then, distance covered by *Y* = 600 m

Hence, distance covered by *X* = 600 + 300 = 900 m

Problem 15 ***A* and *B* run a 5 km race on a round course of 400 m. If their speeds be in the ratio 5 : 4, how often does the winner pass the other ?**

Sol. Obviously, when *A* covers 500 m, *B* will cover 400 m as ratio of their speeds is 5 : 4. It implies that *A* is taking a lead of 100 m per 500 m. Hence, a lead of 400 m will be made by travelling a distance of 2000 m.

Hence, number of pass = 5 km $\div$ 2000 m = 2.5 times

Problem 16 **In a race of three horses, the 1st beats the 2nd by 11 m and the 3rd by 90 m. If the 2nd beats the 3rd by 80 m, what was the length, in metres, of the racecourse ?** ***[CAT-19, 20]***

Sol. Let the length of race-course be *x* m.

As given, when *A* covers *x* m, then *B* and *C* cover $(x - 11)$ m and $(x - 90)$ m respectively.

Further, when *B* covers *x* m then *C* covers only $(x - 80)$ m.

$\therefore$ When *B* will cover $(x - 11)$ m, *C* will cover

$$\left\{\frac{x - 80}{x} \times (x - 11)\right\} \text{m.}$$

But, from above, $\left\{\frac{x - 80}{x} \times (x - 11)\right\} = (x - 90)$

$\Rightarrow x^2 - 90x = x^2 - 91x + 880 \Rightarrow x = 880$ m

Try with other approaches mentioned earlier.

WORKOUT

1. A and B run a km and A wins by 25 sec. A and C run a km and A wins by 275 m. When B and C run the same distance, B wins by 30 sec. The time taken by A to run a km is what ? ***[SSC (CGL-PT)-05]***
(a) 2 min 20 sec (b) 2 min 25 sec.
(c) 2 min 30 sec. (d) 4 min 20 sec.

2. In a 200 m race, A beats B by 20 m, while in a 100 m race, B beats C by 5 m. Assuming that the speeds of A, B and C remain the same in various races, by how many metres will A beat C in 1 km race ? ***[SSC (CGL-Mains)-04]***
(a) 145 m (b) 155 m
(c) 125 m (d) 325 m

3. A can give 40 m start to B and 70 m to C in a race of 1 km. How many metres start can B give to C in race of 1 km ? ***[SSC (CPO)-08, SSC (SO-Audit)-06, MAT-08, 09, CSAT-19]***
(a) 30 m (b) 31.25 m
(c) 31.75 m (d) 32 m

4. In a 800 m race, A defeated B by 15 sec. If A's speed was 8 km/h, the speed of B was ?
(a) $7\frac{17}{25}$ km/h (b) $8\frac{17}{25}$ km/h
(c) $\frac{16}{27}$ km/h (d) $\frac{27}{2516}$ km/h

5. In a game of certain points, A can give B 25 points and C 31 points. B can give C 8 points. How many points make the game ?
(a) 90 (b) 200
(c) 100 (d) 165

6. The speeds of A and B are in ratio 7 : 10 and A loses the race by 120 m. What is the length of the race course ?
(a) 300 m (b) 100 m
(c) 400 m (d) 280 m

7. In 1 km race, A beats B by 45 m or 9 sec. What is the speed of A in m/s.
(a) 6.66 m/s (b) 5 m/s
(c) 6.25 m/s (d) 5.26 m/s

8. In a game of 180 points, A can give B 20 points and C 30 points. Then, how many points can B give C in game of 80 ?
(a) 50 (b) 10
(c) 5 (d) 6

9. In 1 km race A gives B a startup of 10 sec and still beats B by 20 sec. What is the speed of B if ratio of speeds of A and B be 10 : 7 ?
(a) 10 m/s (b) 15 m/s
(c) 20 m/s (d) 25 m/s

10. Karan and Arjun run a 100 m race, where Karan beats Arjun by 10 m. To do a favour to Arjun, Karan starts 10 m behind the starting line in a 2nd 100 m race. They both run at their earlier speeds. Which of the following is true in connection with the 2nd race? ***[CAT-04]***
(a) Karan and Arjun reach the finishing line simultaneously
(b) Arjun beats Karan by 1 m
(c) Arjun beats Karan by 11 m
(d) Karan Beats Arjun by 1 m

11. In a 1 km race, A can beat B by 30 m, while in a 500 m race B can beat C by 25 m. By how many metres will A beat C in a 100 m race ? ***[SSC (SAS)-10]***
(a) 7.85 m (b) 7.25 m
(c) 7.15 m (d) 7.03 m

12. Ram and Shyam run a race of 2000 m. First, Ram gives Shyam a start of 200 m and beats him by 30 sec. Next, Ram gives Shyam a start of 3 min and is beaten by 1000 m. Find the time in minutes in which Ram and Shyam can run the race separately ? ***[MAT-14, 19]***
(a) 8 min, 10 min (b) 4 min, 5 min
(c) 5 min, 9 min (d) 6 min, 9 min

13. Ganu runs 1.6 times as fast as Manu and Ganu gives a startup of 36 m in a race. What should be length of the race course so as to end up the race in dead-heat?
(a) 100 m (b) 76 m
(c) 96 m (d) 66 m

14. In a 100 m race, A runs at a speed of 5/3 m/s. If A gives a start of 4 m to B and still beats him by 12 sec, what is the speed of B ? ***[CDS-17]***
(a) 5/4 m/s (b) 7/5 m/s
(c) 4/3 m/s (d) 6/5 m/s

15. In a race of 200 m run, A beats S by 20 m and N by 40 m. If S and N are running a race of 100 m with exactly same speed as before, then by how many metres will S beat N ? ***[CAT-95]***
(a) 11.11 m (b) 10 m
(c) 12 m (d) 25 m

Answer Keys

1. (b)	2. (a)	3. (b)	4. (a)	5. (c)	6. (c)	7. (d)	8. (c)	9. (a)	10. (d)
11. (a)	12. (b)	13. (c)	14. (c)	15. (a)					

Hints and SOLUTIONS

1. A beats B by 25 sec and B beats C by 30 sec. It implies that A beats C by (25 + 30) sec.
Also, given that A beats C by 275 m.
$\because$ To cover 275 m, C takes 55 sec.
Then, to cover 1000 m, C will take 200 sec. Hence, time taken by $A = 200 - 55 = 145$ sec = 2 min 25 sec.

2. As given, $A \Rightarrow 200$ m, then $B \Rightarrow 180$ m.
Hence, $A : B = 10 : 9$
Further, $B \Rightarrow 100$, then $C \Rightarrow 95$ m.
Hence, $B : C = 20 : 19$
Now, $A : C = (A : B) \times (B : C)$
$= (10 \times 20) : (9 \times 19) = 200 : 171$
Clearly, A beats C by $(200 - 171)$ m in a race of 200 m.
Then, A will beat C in race of 1000 m by 145 m.

3. As given, $A \Rightarrow 1000$ m,
then $B \Rightarrow 960$ m.
Further, $A \Rightarrow 1000$ m, then $C \Rightarrow 930$ m.
Clearly, in a race of 960 m, B gives $(960 - 930)$ m starts to C.
Then, in race of 1000 m, B will give $\left(\frac{30}{960} \times 1000\right)$ m start to C.
Hence, required answer = 31.25 m.

4. Speed of $A = 8 \times \frac{5}{18} = \frac{20}{9}$ m/s
Hence, time taken by A in this race
$= 800 \div \frac{20}{9} = 360$ sec
Now, $\frac{S_1}{S_2} = \frac{T_2}{T_1}$, as distance is constant.
$\Rightarrow \frac{\text{Speed of } A}{\text{Speed of } B} = \frac{\text{Time taken by } B}{\text{Time taken by } A}$
$\Rightarrow \frac{8}{\text{Speed of } B} = \frac{(360 + 15)}{360}$
$\Rightarrow$ Speed of $B = \frac{8 \times 360}{375} = \frac{192}{25} = 7\frac{17}{25}$ km/h

5. Let the game is of x points.
Then, $A : B = x : (x - 25)$
$A : C = x : (x - 31)$
Clearly, in a game of $(x - 25)$ points, B gives C $[(x - 25) - (x - 31)]$ i.e. 6 points.
Then, in a game of x, B will give $C\left(\frac{6}{x - 25} \times x\right)$ points.
But given, $\left(\frac{6}{x - 25} \times x\right) = 8$
$\Rightarrow \quad 3x = 4x - 100$
$\Rightarrow \quad x = 100$

6. When B covers 10 m, A covers 7 m. It means A loses the race by 3 m.
$\because A$ loses 3 m when the race is of 10 m.
$\therefore A$ will lose 120 m when the race will be of 400 m.

7. Obviously, B can move 45 metres in 9 sec. Then, 1000 m (i.e. 1 km) in $\left(\frac{9}{45} \times 1000\right)$ i.e. in 200 sec.
Hence, A's time over the course
$= 200 - 9 = 191$ s
Now, speed of $A = 1000 \div 191$
$= 5.26$ m/s

8. As given, $A \Rightarrow 180$, then $B \Rightarrow 160$.
Further, $A \Rightarrow 180$, then $C \Rightarrow 150$.
Clearly, in a game of 160, B gives C $(160 - 150)$ points.
Then, in a game of 80, B will give C 5 points.

9. Ratio of their speeds = 10 : 7
Then, ratio of time taken to cover 1000 m = 7 : 10
But, B takes (10 + 20) sec. more than A do. Hence, (10 − 7) units = 30sec.
Then, 10 units = 100 sec.
$\therefore$ Time taken by B to cover 1000 m
= 100 sec.
Then, speed of $B = 1000 \div 100$
= 10 m/sec

10. Since time is constant, ratio of speeds of Karan and Arjun = ratio of distances covered by them = 100 : 90
= 10 : 9
In 2nd race, Karan is to run 110 m and Arjun is to run only 100 m.
Obviously, when Karan runs 110 m, Arjun will run only 99 m as ratio of their speeds is 10 : 9.
Hence, Karan beats Arjun by 1 m.

11. As given, $A \Rightarrow 1000$ m, then $B \Rightarrow 970$ m
Hence, $A : B = 100 : 97$
Further, $B \Rightarrow 500$ m, then
$C \Rightarrow 475$ m.
Hence, $B : C = 20 : 19$
Now, $A : C = (A : B) \times (B : C)$
$= (20 \times 100) : (19 \times 97)$
$= 2000 : 1843$
Obviously, A will beat C by (2000 − 1843) m in a race of 2000 m.
Then, in a race of 100 m, A will beat B by 7.85 m.

12. Let the speeds of Ram and Shyam be x m/s and y m/s respectively.
Then, ratio of their speeds = ratio of distances covered by them
$\Rightarrow x : y = 2000 : \{2000 - 200 - 30y\}$
$= 200 : (180 - 3y)$
For 2nd case,
Ratio of their speeds = Ratio of distances covered by them
$\Rightarrow x : y = (2000 - 1000) : [2000 - (60 \times 3)y]$
$= 50 : (100 - 9y)$
Obviously, $x : y = 200 : (180 - 3y)$
$= 50 : (100 - 9y)$
$\Rightarrow 9000 - 150y = 20000 - 1800y$
$\Rightarrow \quad y = \frac{20}{3}$ m/s
Further, $x : y = 50 : (100 - 9y)$
$\Rightarrow \quad x\left\{100 - 9 \times \frac{20}{3}\right\} = 50 \times \frac{20}{3}$
$\Rightarrow \quad 40x = 50 \times \frac{20}{3}$
$\Rightarrow \quad x = \frac{50 \times 20}{40 \times 3} = \frac{25}{3}$ m/s
Then, time taken by Ram to run 2000 m
$= 2000 \div \frac{25}{3} = 240$ sec. = 4 min.
and time taken by Shyam to run 2000 m $= 2000 \div \frac{20}{3} = 300$ sec = 5 min

13. Let the distance of winning post from the starting point be D m.
Since time is constant, ratio of their speeds = ratio of distances covered by them
$\Rightarrow 1.6 : 1 = D : (D - 36)$
$\Rightarrow \quad D = 1.6D - 36 \times 1.6$
$\Rightarrow \quad D = 96$ m

14. Time taken by A to run 100 m = 100 ÷ 5/3 = 60 sec.
Then, time taken by B to run (100 − 4) m = 60 + 12 = 72 sec.
$\therefore$ Speed of $B = 96 \div 72 = \frac{4}{3}$ m/s

15. In a certain time if A runs 200 m, S and N run 180 m and 160 m respectively.
$\because S$ runs 180 m then N runs 160 m
$\therefore S$ runs 100 m then N will run 88.88 m
Hence, S will beat N by (100 − 88.88) i.e. by 11.11 m.

35 Circular Motion

Inside

- *Introduction*
- *Concept of Circular Motion : Relative Speed in Same/ Opposite Direction(s)*
- *First Meeting at Any Point on Circular Path in Case of Same/Opposite Direction(s) When Two or More bodies move from the Same Point*
- *First Meeting at the Starting Point on Circular Path in Case of Same/Opposite Direction(s) When Two or More Bodies move from the Same Point*
- *Illustrations*
- ....
- .... *& Many More*
- *Work Out with Solutions*

Introduction

When two or more bodies move around a circular track, it is said that they are in circular motion. Here, relative speed of two bodies will be as below

(a) In case of same direction : $(S_1 - S_2)$ or $(S_2 - S_1)$, whichever is positive.

(b) In case of opposite direction : $(S_1 + S_2)$

I. Cases of Circular Motion in Same Direction

Suppose two bodies are moving around a circular track/path from the same point in the same direction (clockwise or anti-clockwise), then the faster body will take a lead of a distance for every unit time and when this distance becomes equal to length of the circular track, it is said that faster body overlaps the slower or slower body has been overlapped by the faster body. And, this event is called overlapping. Overlapping means, in simple words, their meeting.

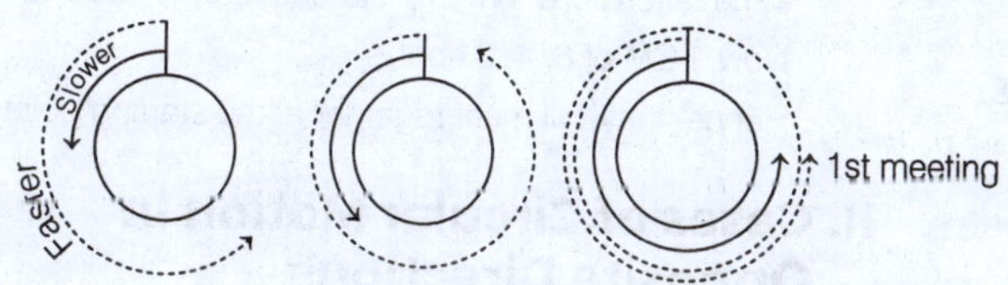

• 1st meeting at any point on the circular path

Case A **Two bodies moving in the same direction from the same point**

From the above descriptions, it is clear that the faster body goes ahead of a distance for every unit time at the rate of relative speed and obviously, the 1st meeting will occur when the length/circumference of the circular path will be covered with the relative speed.

Hence, time taken by them for the 1st meeting

$$= \frac{\text{Length or circumference of the circular path}}{\text{Relative speed}}$$

Now, take one question for illustration

Problem 1 Two bodies start to move simultaneously from the same point in the same direction on a circular path of length 24 km at the speeds of 4 km/h and 10 km/h respectively. Answer the followings :

(a) When will they meet further for the 1st time?

(b) What is the distance between starting point and the meeting point? *[LIC (AAO)-13 (Part (a)]* *[SSC (CGL-T-II)-19, CAT-20 (Partly)]*

Sol. It may be construed from the question that the faster body will be ahead of the slower body by 6 km (i.e. relative speed) for every 1 h and it has to bridge the length of circular path so as to meet with each other. Then,

$$\text{Time required} = \frac{\text{Length of the circular path}}{\text{Relative speed}} = \frac{24}{6} = 4\text{ h}$$

Note *They will be meeting at interval of every 4 h.*

Further, the distance between starting point and the meeting point = distance covered by faster body − the multiple of length of path just below the distance covered by the faster body

Or

distance covered by slower body − the multiple of length of path just below the distance covered by the slower body

Hence, required answer = $(4 \times 10) - 24 = 16$ km

Or $(4 \times 4) - (0 \times 24) = 16$ km

Alternatively

The above facts reveal that the distance between the starting point and the meeting point is the remainder arrived on dividing distance covered (either by faster body or slower body) by length of the path.

Hence, required answer = the remainder in case of $(4 \times 10) \div 24$ = 16 km

Remark *The distance between starting point and the meeting point i.e. 16 km has been measured in the same direction (clockwise/anticlockwise) in which they started to move. The distance between starting point and the meeting point may be (24–16) km with reference to opposite direction. Hence, the distance between starting point and the meeting point is 16 km from one direction and 8 km from another direction.*

Case B **More than two bodies moving in the same direction from the same point.**

In this case, all the bodies will meet again at the time equal to LCM of different time taken by the fastest body to overlap the bodies moving at slower speeds. e.g.

Problem 2 Three bodies A, B and C start to run simultaneously on a circular track of length 24 km from the same point in the same direction at the speeds of 6 km/h, 10 km/h and 18 km/h respectively. When will they meet again ?

Sol. Here, C is the fastest body.

Now, time taken by C to overlap $A = 24 \div (18 - 6) = 2$ h

and time taken by C to overlap $B = 24 \div (18 - 10) = 3$ h

Further, LCM of 2 and 3 = 6

Hence, they will meet together after 6 h.

Note *They will be meeting at interval of every 6 h.*

• 1st Meeting at the Starting Point

Two or more than two bodies moving in the same direction from the same point

They have to meet at the starting point and it is the fact that each body will take different time to complete one round. In order to ensure they all together meet at starting point, we should find a least number which is common multiple for all different pieces of time taken by them individually.

In other words, the LCM of different time taken to complete one full round by them individually will be the time required for their next meeting.

It is important to note that if 1st meeting occurs in T units of time, then subsequent meetings will be occurred at the interval of T units of time i.e. 2nd meeting will be in $2T$ units of time, 3rd meeting be in $3T$ units of time and so on.

e.g.

Problem 3 Four bodies A, B, C and D start to move on a circular path of length 36 km at the speeds of 6 km/h, 9 km/h, 12 km/h and 18 km/h respectively from the same time in the same direction. How long will they take to meet together at the starting point ?

Sol. Time taken by A to complete one round = $36 \div 6 = 6$ h

Time taken by B to complete one round = $36 \div 9 = 4$ h

Time taken by C to complete one round = $36 \div 12 = 3$ h

Time taken by D to complete one round = $36 \div 18 = 2$ h

Now, LCM of 6, 4, 3 and 2 = 12

Hence, they will meet together at the starting point after 12 h.

II. Cases of Circular Motion in Opposite Directions

• 1st Meeting at Any Point on the Circular Path

Case A **Two bodies moving in the opposite directions from the same point.**

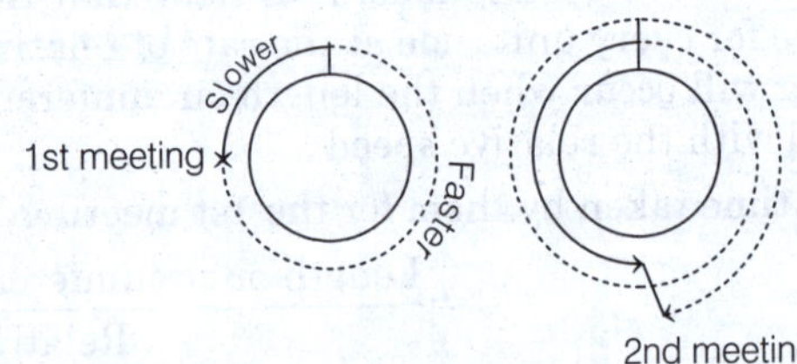

In case of two bodies moving on the circular path in the opposite directions from the same point, the whole length of the path will be covered by them together at the rate of their relative speed. Here, relative speed will be $(S_1 + S_2)$.

Hence, the time required for 1st meeting

$$= \frac{\text{Length of the circular path}}{\text{Relative speed}}$$

Further, it is important to note that if 1st meeting occurs in T units of time, then subsequent meetings will be occurred at the interval of T units of time i.e. 2nd meeting will be in $2T$ units of time, 3rd meeting be in $3T$ units of time and so on.
e.g.

Problem 4 Two bodies start to move simultaneously from the same point in the opposite direction on a circular path of length 24 km at the speeds of 4 km/h and 8 km/h respectively. When will they meet further for the 1st time? *[MAT-03, 02, 01]*

Sol. Time required $= \dfrac{\text{Length of the circular path}}{\text{Relative speed}}$

$= \dfrac{24}{(4+8)} = 2$ h

Case B **More than two bodies moving in the opposite directions from the same point.**

If we take an example of three bodies, then two will move in one direction and one will move in opposite direction. In case of four bodies, either one or two or three will move in one direction and the remaining will move in opposite direction. You may stress on the different combinations.
e.g.

Problem 5 Three bodies A, B and C start to move simultaneously from a point on a circular path of 24 km. A and B move in the same direction at the speed of 4 km/h and 16 km/h respectively and the speed of C moving in direction opposite to that of A and B, is 8 km/h. What time will they require to meet together?

Sol. Here, A and B are moving in the same direction, it means they will meet 1st time after $\left\{\dfrac{24}{(16-4)}\right\}$ i.e. 2 h. Further, they will be meeting at interval of 2 h. Obviously, C will meet them either at 2 h or at one of its other multiples as all three have to meet together. Let options be (a) 3 h (b) 4 h (c) 2 h (d) 5 h

Option (c) 2 h $\Rightarrow$ Distance moved by A in 2 h $= 4 \times 2 = 8$ km

and distance moved by C in 2 h $= 8 \times 2$
$= 16$ km

Since A and C are moving in opposite directions and the sum of distance covered by them is equal to length of the path, it will be true that they are meeting together. No need to mention that B will also be meeting with A at the interval of 2 h, as discussed above.

Hence, all three will meet together 1st time after 2 h.

[It is imperative to mention that at interval of 2 h, all three will be meeting together]

Alternatively

Time required for meeting of A and C, which are moving in opposite direction $= 24 \div (4+8) = 2$ h

Further, time required for meeting of C and B, which are moving in opposite directions $= 24 \div (16+8) = 1$ h

Hence, time required for meeting of all three together
$=$ LCM of 2 and 1 $= 2$ h

• 1st Meeting at Starting Point on the Circular Path

Two or more than two bodies moving in the opposite direction from the same point.

They have to meet at starting point and it is the fact that each body will take different time to complete one round. In order to ensure they all together meet at starting point, we should find a least number which is common multiple for all different pieces of time taken by them individually. In other words, the LCM of different time taken to complete one full round by them individually will be the time required for their next meeting.

It is also important to note that if 1st meeting occurs in T units of time, then subsequent meetings will be occurred at the interval of T units of time i.e. 2nd meeting will be in $2T$ units of time, 3rd meeting will be in $3T$ units of time and so on.

e.g.

Problem 6 Four bodies A, B, C and D start to move simultaneously on a circular path of length 36 km at the speeds of 6 km/h, 9 km/h, 12 km/h and 18 km/h. respectively. A and B moves in clockwise direction while C and D moves in anti-clockwise direction. How long will they take to meet together at the starting point ?

Sol. Time taken by A to complete one round $= 36 \div 6 = 6$ h

Time taken by B to complete one round $= 36 \div 9 = 4$ h

Time taken by C to complete one round $= 36 \div 12 = 3$ h

Time taken by D to complete one round $= 36 \div 18 = 2$ h

Now, LCM of 6, 4, 3 and 2 = 12

Hence, they will meet together at the starting point after 12 h.

Illustrations

Problem 7 Four bodies A, B, C and D start to move simultaneously around a circle of length 80 km, A and B are moving at 4 km/h and 20 km/h respectively in the same direction while C and D are moving at 8 km/h and 48 km/h respectively in opposite of the direction of A and B. After what time will they all of four meet together?

Sol. Time required for meeting of A and $B = 80 \div (20-4) = 5$ h

and time required for meeting of C and $D = 80 \div (48-8) = 2$ h

It is obvious that they all of four will meet at time equal to one of multiples of LCM of 5 and 2. Hence, required answer is either 10 or one of its other multiples. Let the options be

(a) 5 h (b) 10 h (c) 15 h (d) 20 h

Option (b) 10 h $\Rightarrow$ In 10 h, distance covered by A
$= 10 \times 4 = 40$ km

Further, in 10 h, C will cover a distance of 80 km.

It is clear that both A and C are at different distance, hence, no meeting with each other.

Option (d) 20 h $\Rightarrow$ In 20 h, distance covered by $A = 20 \times 4 = 80$ km. It means A is at starting point.

Further, in 20 h, C will cover a distance of 160 km. It means C is also at starting point.

Both being at starting point, indicate that they are meeting with each other. As regards B and D, B will be meeting with A and D with C at above mentioned interval of time.

Hence, required answer = 20 h.

It will not be out of context to say that they all of four are meeting at starting point.

Alternatively

Time required for meeting of A and C, which are moving in opposite direction $= 80 \div (4 + 8) = 80/12$ h.

Time required for meeting of B and C, which are moving in opposite directions $= 80 \div (20 + 8) = 80/28$ h

Time required for meeting of A and D, which are moving in opposite directions $= 80 \div (4 + 48) = 80/52$ h

Time required for meeting of B and D, which are moving in opposite directions $= 80 \div (20 + 48) = 80/68$ h

Hence, time required for meeting of all four together = LCM of 80/12, 80/28, 80/52 and 80/68 = 80/4 = 20 h.

Problem 8 There is a race of 2.1 km between Alakh and Rahul in a round course of 200 m. Alakh overlaps Rahul at the middle of the 4th round. By how many km will Alakh beat Rahul in the race?

Sol. Since, they are running in the same direction, Alakh must ahead of Rahul by one round to overlap Rahul. Given that Alakh overlaps Rahul at the middle of 4th round. It means Alakh runs 3.5 rounds whereas Rahul runs (3.5 – 1) rounds.

Now, Alakh runs (3.5×200) m whereas Rahul runs (2.5×200) m

$\therefore$ When Alakh runs 2100 m, Rahul will run 1500 m.

Hence, Alakh beats Rahul by (2100 – 1500) m i.e. by 0.6 km.

Alternatively

Ratio of speeds of Alakh and Rahul = Ratio of distances covered by them

$\Rightarrow$ Speed of Alakh : Speed of Rahul = 3.5 rounds : 2.5 rounds = 7 : 5

$\because$ While Alakh runs 7 km, Rahul runs 5 km.

$\therefore$ When Alakh runs 2.1 km, Rahul will run 1.5 km.

Hence, Alakh beats Rahul by (2100 – 1500) m i.e. by 0.6 km.

Problem 9 A can go round a circular path 8 times in 40 min. If the diameter of the circle is increased to 10 times the original diameter, the times required by A to go round the new path once travelling at the same speed as before is? *[SSC (CGL-PT)-2000]*

Sol. Time taken in one round $= 40 \div 8 = 5$ min.

When diameter increases to 10 times, the length of the path will also increase to 10 times. We know that, Time is directly propotional to Distance, when speed is constant.

Hence, time required $= 5 \times 10 = 50$ min

Problem 10 A and B run together on a circular track in the same direction from the same point. A can make one full round in 30 min, whereas B can in 50 min. How many times will A pass B till they together arrive at the starting point?

Sol. Let the length of the track be 150 m.

Then, speed of $A = 150 \div 30 = 5$ m/min

and speed of $B = 150 \div 50 = 3$ m/min

Time taken to meet 1st time at any point of track

$= 150 \div (5 - 3) = 75$ min.

Further, time required for their meeting at starting point = LCM of 30 and 50 = 150 min.

Hence, number of times that A will pass $B = 150 \div 75 = 2$ times.

It may also be interpreted that at 2 points (including the starting point) of the circular track they meet each other.

Generic version

Number of points (including starting point) at the circular path at which they meet

$$= \frac{\text{time taken to meet at the starting point 1st time}}{\text{time taken to meet anywhere 1st time}}$$

Problem 11 A, B and C start together from the same place to walk around a circular path of length 12 km. A walks at the rate of 4 km/h, B 3 km/h and C 1.5 km/h. They will meet together at the starting place at the end of *[SSC (CGL-PT) 08]*

Sol. Time taken by A to complete one round $= 12 \div 4 = 3$ h

Time taken by B to complete one round $= 12 \div 3 = 4$ h

Time taken by C to complete one round $= 12 \div 1.5 = 8$ h

Now, LCM of 3, 4 and 8 = 24

Hence, they will meet together at the starting point after 24 h.

Problem 12 Two men move from a fixed point along a circle of circumference 8 km. When they move in same direction, meets at interval of 4 h and when they move in opposite direction, meets each other at interval of 1h. Find the speed of the faster man?

Sol. Let the speed of the faster man be S_f and that of slower be S_s

Time taken to meet in same direction $= \dfrac{8}{S_f - S_s} = 4$ [given]

$\Rightarrow \quad S_f - S_s = 2$ ...(i)

Further, time taken to meet in opposite direction $= \dfrac{8}{S_f + S_s} = 1$

$\Rightarrow \quad S_f + S_s = 8$ ...(ii)

On adding both Eqs. (i) and (ii)

$2S_f = (2 + 8)$

$\Rightarrow \quad S_f = 5$ km/h

Alternatively

If relative speeds in context of both direction (i.e. same and opposite) are known, then speed of the faster body is the average of both relative speeds. For computation of speed of slower body, we simply subtract the speed of faster body from the relative speed meant for opposite direction.

Here, relative speeds are $(8 \div 4)$ km/h and $(8 \div 1)$ km/h

Then, speed of faster man = average of 2 and 8 = 5 km/h.

Problem 13 Yajat and Shivansh start to run simultaneously on the circular track of 800 m at the speeds of 60 m/sec and 40 m/sec respectively. What is the ratio of the time taken by them for the 2nd meeting moving in the same direction and that for the 3rd

meeting moving in the opposite direction assuming that they are positioned diametrically opposite to each other.

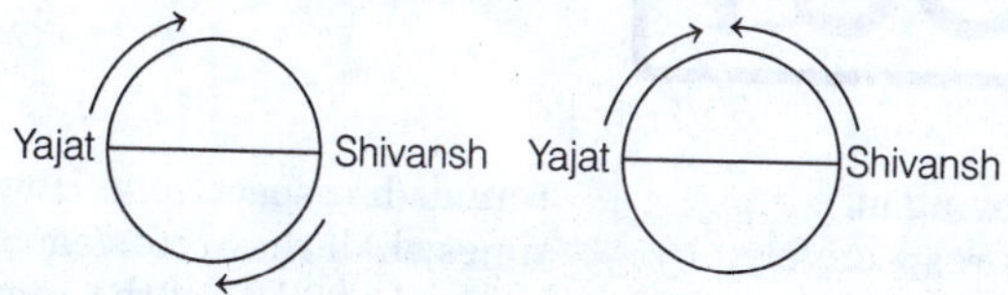

Sol. Relative speed for the same direction $= 60 - 40 = 20$ m/s

Here, Shivansh is $(800 \div 2)$ m ahead of the Yajat.

Then, time required for the 1st meeting $= 400 \div 20 = 20$ s

and time required for the 2nd meeting after the 1st meeting

$= 800 \div 20 = 40$ sec

Further, relative speed for the opposite directions

$= 60 + 40 = 100$ m/sec

It is clear that they have to travel only 400 m for the 1st meeting, further 800 m for the 2nd meeting and again 800 m for the 3rd meeting. Hence, total distance to be moved for 3 rd meeting

$= 2000$ m

Now, time taken for 3rd meeting $= 2000 \div 100 = 20$ sec

Hence, the required ratio $= 60 : 20 = 3 : 1$

➘ Problem 14 ***A*** **and** ***B*** **walk around a circular track. They start at 8 : 00 AM from the same point in the opposite directions.** ***A*** **and** ***B*** **walk at a speed of 2 rounds/h and 3 rounds/h respectively. How many times shall they cross each other before 9 : 30 AM ?** ***[MAT-13]***

Sol. Let one round be of x m.

Then, speed of $A = 2x$ m/h and speed of $B = 3x$ m/h

Now, relative speed $= 2x + 3x = 5x$ m/h

Now, time taken in 1st meeting

= length of track ÷ relative speed

$= x \div 5x = 1/5$ h $= 12$ min.

Time from 8 : 00 AM to 9 : 30 AM = 90 min.

Thus, number of meetings $= 90 \div 12 = 7.5$

It implies that they will cross each other 7 times before 9 : 30 AM.

Alternatively

Relative speed = 2 rounds/h + 3 rounds/h

= 5 rounds/h

Then, time taken to cover one round $= 1/5$ h $= 12$ min.

Time from 8 : 00 AM to 9 : 30 AM = 90 min.

Thus, number of meetings $= 90 \div 12 = 7.5$

It implies that they will cross each other 7 times before 9 : 30 AM.

➘ Problem 15 ***A*** **moves after** ***B*** **moves 300 m from the same point on a circular path having circumference of 2500 m but in the opposite directions. They meet each other when** ***A*** **moves 1200 m. What will be distance between them at the time when one reaches 1st to starting point ?**

Sol. A moves 1200 m, while B moves (2500 – 1200 – 300) i.e. 1000 m in the same time.

In order to reach to starting point, A has still to move (2500 – 1200) m i.e. 1300 m and B has still to move 1200 m.

$\because A$ moves 1200 m in a certain time while B moves 1000 m in the same time.

$\therefore$ When A moves 1300 m in a certain time, B will move 1083.33 m.

But, B is to move 1200 m to reach the destination. It implies that B is (1200 – 1083.33) m before the starting points and A reaches 1st to starting point.

Hence, the distance between them when A reaches to starting point = 1200 – 1083.33 = 116.67 m

➘ Problem 16 There are two concentric circular tracks of radii 100 m and 102 m respectively. ***A*** **runs on the inner track and goes once round the track in 1 min. 30 sec while** ***B*** **runs on the outer track in 1 min. 32 sec. Who runs faster ?**

Sol. Speed of $A = \dfrac{\text{Circumference of inner track}}{\text{Time taken}}$

$= \dfrac{2\pi 100}{90} = 2.22\pi$ m/s

and speed of $B = \dfrac{\text{circumference of inner track}}{\text{time taken}}$

$= \dfrac{2\pi 102}{92}$

$= 2.22\pi$ m/s

Hence, both A and B are equal in speed.

➘ Problem 17 ***A*** **walks around a circular field at the rate of 1 round/h while** ***B*** **runs around it at the rate of 6 rounds/h. They start in the same direction from the same point at 7 : 30 AM. They shall first cross each other at :** ***[IAS (Pre)-03]***

Sol. Relative speed = 6 rounds/h –1 rounds/h = 5 rounds/h

Then, time taken to cover one round $= 1/5$ h $= 12$ min.

Hence, they will meet further at 12 min. beyond 7:30 AM i.e. at 7 : 42 AM.

➘ Problem 18 Two ants ***A*** **and** ***B*** **start from a point** ***P*** **on a circle at the same time, with** ***A*** **moving clock-wise and** ***B*** **moving anti-clockwise. They meet for the 1st time at 10 : 00 AM when** ***A*** **has covered 60% of the track. If** ***A*** **returns to** ***P*** **at 10 : 12 AM, then** ***B*** **returns to** ***P*** **at** ***[CAT- 19]***

Sol. Speed of A : Speed of B = 60% : 40% = 3 : 2 [$\because$ time is constant]

A covers remaining 40% of the circumference in 12 min. to reach at point P. Then, B will take $\left\{12 \times \dfrac{3}{2}\right\}$ i.e. 18 min. to cover 40% of the circumference.

But, B has to travel 60% of the circumference to reach at point P.

Thus, it will take $\left\{18 \times \dfrac{3}{2}\right\}$ i.e. 27 min.

Hence, required time = 10 : 27 AM.

WORKOUT

1. There is a race of 27 km in a round course of 250 m. *A* overlaps *B* at the middle of the 5th round. By what distance will *A* beat *B*?

(a) 6 km (b) 8 km
(c) 17 km (d) 11.5 km

2. There are two concentric circular tracks of radii 50 m and 51 m respectively. *A* runs on the inner track and goes once round the track in 50 sec. while *B* runs on the outer track in 52 sec. Who runs faster?

(a) Both *A* and *B* are equal
(b) *A*
(c) *B*
(d) None of the above

3. *A* can go round a circular path 12 times in 84 min. If the radius of the circle is increaseo to 5 times the original radius, the time required by *A* to go round the new path once travelling at the same speed as before is?

(a) 30 min (b) 40 min
(c) 45 min (d) 35 min

4. Four bodies *A*, *B*, *C* and *D* start to move simultaneously around a circle of length 120 km, *A* and *B* are moving at 6 km/h and 30 km/h respectively in the same direction while *C* and *D* are moving at 12 km/h and 72 km/h respectively in opposite of the direction of *A* and *B*. After what time will they all of four meet together?

(a) 7.5 h (b) 20 h
(c) 22.5 h (d) 30 h

5. Two boys move from a fixed point along a circle of circumference 18 km. When they move in same direction, meet at interval of 3 h and when they move in opposite direction, meet each other at interval of 2 h. Find the speed of the slower boy ?

(a) 5 km/h (b) 1.5 km/h
(c) 3.5 km/h (d) 2.5 km/h

6. *A*, *B* and *C* are running on a circular track of 120 m at a speed of 5m/s, 8m/s and 10 m/s respectively. *A*, *B* and *C* all three are moving in the same direction. When will all the three meet again at the starting point?

[MAT-13, MAT-08]

(a) After 55 sec (b) After 2 min
(c) After 5 min (d) After 40 sec

7. *M* and *N* walk around a circular track. They start at 10 : 00 AM from the same point in the opposite directions. *M* and *N* walk at a speed of 3 rounds/h and 7 rounds/h respectively. How many times shall they cross each other before 1 : 00 PM of the same day?

(a) 16 times (b) 30 times
(c) 17 times (d) 48 times

8. *A* walks around a circular field at the rate of 2 round/h while *B* runs around it at the rate of 12 rounds/h. They start in the same direction from the same point at 6 : 00 AM. They shall first cross each other at

(a) 8 : 12 AM (b) 6 : 10 AM
(c) 6 : 48 AM (d) 6 : 06 AM

9. Three friends *A*, *B* and *C* start running around a circular stadium and complete a single round in 24 sec, 36 sec and 30 sec. respectively. After how many minutes will they meet again at the starting point?

[IDBI Bank (PO)-09]

(a) 12 min (b) 6 min
(c) 8 min (d) 15 min
(e) 18 min

10. There is a race of 27 km in a round course of 250 m. How many times will *A* overlap *B* if their speeds are in ratio 8 : 7 ?

(a) 13 times (b) 11 times
(c) 10 times (d) 17 times

Answer Keys

1. (a)	2. (b)	3. (d)	4. (b)	5. (b)	6. (b)	7. (b)	8. (d)	9. (b)	10. (a)

Hints and SOLUTIONS

1. Since they are running in the same direction, A must ahead of B by one round to overlap B. Given that A overlaps B at the middle of 5th rounds. It means A runs 4.5 rounds whereas B runs $(4.5-1)$ rounds.

Further, ratio of speeds of A and B = ratio of distances covered by them

$\therefore$ Speed of A : Speed of B = 4.5 rounds : 3.5 round = 9 : 7

$\because$ While A runs 9 km, B runs 7 km.

$\because$ While A runs 27 km, B will run 21 km.

Hence, A wins the race by $(27-21)$ km i.e. by 6 km.

2. Speed of A

$$= \frac{\text{Circumference of inner track}}{\text{Time taken}}$$

$$= \frac{2\pi \times 50}{50} = 2\pi \text{ m/s}$$

and speed of B

$$= \frac{\text{Circumference of outer track}}{\text{Time taken}}$$

$$= \frac{2\pi \times 51}{52} = 1.96\ \pi \text{ m/s}$$

Hence, A is faster then B.

3. Time taken in one round $= 84 \div 12 = 7$ min.

When the radius increases to 5 times, the length of the path will also increase to 5 times. We know that Time is directly propotional to Distance, when speed is constant.

Hence, time required $= 7 \times 5 = 35$ min

4. Time required for meeting of A and C, which are moving in opposite directions

$= 120 \div (6+12) = 120/18$ h

Time required for meeting of B and C, which are moving in opposite directions

$= 120 \div (30+12) = 120/42$ h

Time required for meeting of A and D, which are moving in opposite directions $= 120 \div (6+72)$

$= 120/78$ h

Time required for meeting of B and D, which are moving in opposite directions

$= 120 \div (30+72) = 120/102$ h

Hence, time required for meeting of all four together

= LCM of 120/18, 120/42, 120/78 and 120/102

$= 120/6 = 20$ h.

5. Here, relative speeds are $(18 \div 3)$ km/h and $(18 \div 2)$ km/h.

Then, speed of faster boy = average of 6 and 9 = 7.5 km/h.

Now, speed of slower boy = relative speed meant for opposite direction $-7.5 = 9 - 7.5 = 15$ km/h

6. Time taken by A to complete one round $= 120 \div 5 = 24$ sec.

Time taken by B to complete one round $= 120 \div 8 = 15$ sec.

Time taken by C to complete one round $= 120 \div 10 = 12$ sec.

Now, LCM of 24, 15 and 12 = 120

Hence, they will meet together at the starting point after 120 sec.

i.e. after 2 min.

7. Relative speed = 3 rounds/h + 7 rounds/ h = 10 rounds/h

Then, time taken to cover one round

$= 1/10$ h = 6 min.

Time from 10 AM to 1.00 PM = 180 min

Thus, number for meetings

$= 180 \div 6 = 30$

It implies that they will cross each other 30 times.

8. Relative speed = 12 rounds/h − 2 rounds/h = 10 rounds/h

Then, time taken to cover one rounds

$= 1/10$ h = 6 min.

Hence, they will meet further at 6 min. in beyond 6 : 00 AM i.e. at 6 : 06 AM.

9. Now, LCM of 24, 36 and 30 = 360 sec.

Hence, they will meet together at the starting point after 6 min.

10. Since their speeds are in ratio 8 : 7, A runs 8 rounds, when B runs 7 rounds. It impliop that A overlaps B when A runs 8 rounds.

The distance covered in 8 rounds by

$A = 250 \text{ m} \times 8 = 2$ km

Since 2 km is included 13 times in 27 km, A overlaps B 13 times.

36

Problems Based on Trains

Introduction

We have already discussed the questions where trains move and in those questions length(s) of the train(s) was/were ignored. There was no role of its length to affect the time, speed or distance. Here, the length of the train will be considered in respect of aspects of the problems.

Assume that you are standing at a platform and a train is to pass/cross you. The crossing process starts when the engine of the train comes in line with you and ends when last vehicle of the train comes in line with you. The diagram depicted below shows the process.

• (you)

.... train.........>

.... train.........>

Clearly, in order to cross you, the train is to move a distance equal to its length. Let train is moving at S_T m/sec and takes T sec. to cross you, then the distance covered by it in T sec. $= (S \times T)$ m. This distance is equal to the length of the train. In other words, we may say that **Length of the train = Speed of the train × Time taken to cross the man**. It is reiterated that the principles of Time, Speed and Distance, will be applicable here also. Now, we will consider all the possible cases.

Case A **When the train crosses a stationary object having no considerable length such as pole, man, tree, house, etc.**

From the above descriptions, Length of the train (L) = Speed of the train (S_T) × Time taken to cross the object (T) i.e.

$$L = S_T \times T$$

Case B **When the train crosses a stationary object having considerable length such as bridge, platform, tunnel, etc.**

= = = (l) = = = (platform, tunnel, etc)

.... train.... (L) ....>

•.... train.... (L) ....>

Obviously, the train is to move a distance of ($L + l$) to cross the platform having length (l). We know, distance = speed × time.

Hence, (length of the train + length of the stationary object) = Speed of the train (S_T) × Time taken to cross the object (T) i.e. $(L + l) = S_T \times T$

Inside

- Introduction
- Application of Relation Among Time, Speed and Distance in Following Cases :

  (i) When the Train Crosses a Stationary Object With/ Without considerable Length

  (ii) When the Train Crosses a Running Object With/Without Considerable Length in Same/Opposite Direction(s)
- Derivation of a Single Model for All Types of Questions
- Illustrations
- ....
- .... & Many More
- Work Out with Solutions

Case C When the train crosses a running object having no considerable length such as running man, running car, man sitting in other running train, etc.

(i) When both are running in the same direction.

On the same line as explained under Case A, Length of the train (L) = Relative speed ($S_T - S_o$) × Time taken to cross the object (T), i.e.

$L = (S_T - S_o) \times T$, where S_o is the speed of running object.

(ii) When both are running in the opposite directions,

$$L = (S_T + S_o) \times T$$

Case D When the train crosses a running object having considerable length such as running other trains, etc.

(i) When both are running in the same direction.

On the same line as explained under Case-B, (Length of the running train + length of the running object) = Relative speed × Time taken to cross the object (T)

$(L + l) = (S_T - S_o) \times T$, where S_o is the speed of running object.

(ii) When both are running in opposite directions.

$$(L + l) = (S_T + S_o) \times T$$

Observation A single model i.e. $(L + l) = (S_T \pm S_o) \times T$ may be suitable for all the cases mentioned above, you have only to apply it considering the cases. The applications are as below

Cases	Train length	Object length	Relative speed	Time taken to cross	Model
	(L)	(l or 0)	($S_T \pm S_o$)	T	$(L + l) = (S_T \pm S_o) \times T$
A	L	0	($S_T \pm 0$)	T	$L = S_T \times T$
B	L	l	($S_T \pm 0$)	T	$L + l = S_T \times T$
C(i)	L	0	($S_T - S_O$)	T	$L = (S_T - S_O) \times T$
C(ii)	L	0	($S_T + S_O$)	T	$L = (S_T + S_O) \times T$
D(i)	L	l	($S_T - S_O$)	T	$L + l = (S_T - S_O) \times T$
D(ii)	L	l	($S_T + S_O$)	T	$L + l = (S_T + S_O) \times T$

Illustrations

Problem 1 A train is 125 m long. If the train takes 30 sec to cross a tree by the railway line, then the speed of the train is **[SSC (CGL-PT)-02, Vijaya Bank (PO)-08]**

Sol. Clearly, the train moves a distance equal to its length in order to cross the tree.

Now, distance travelled by the train = speed of the train × time taken to cross the tree

$\Rightarrow$ 125 m = Speed of the train × 30 sec

$\Rightarrow$ Speed of the train = 125/30 m/sec.

$$= \left\{\frac{125}{30} \times \frac{18}{5}\right\} \text{km/h}$$

$$= 15 \text{ km/h}$$

Alternatively

We know, $(L + l) = (S_T \pm S_O) \times T$

$\Rightarrow$ $(125 + 0) = (S_T \pm 0) \times 30$

$\Rightarrow$ $S_T = 125/30$ m/sec.

$$= \left\{\frac{125}{30} \times \frac{18}{5}\right\} \text{km/h}$$

$$= 15 \text{ km/h}$$

Problem 2 A train travelling at the speed of 60 km/h crosses a platform in 20 sec. What is the length of the train? **[IOB-PO-08]**

Sol. We know, $(L + l) = (S_T \pm S_O) \times T$

$\Rightarrow$ $(L + l) = (16.67 \pm 0) \times 20$

We can not find the length of the train unless the length of platform is known. Hence, given data is inadequate.

Problem 3. A train passes a platform 90 m long in 30 sec. and a man standing on the platform in 15 sec. What is the speed of the train? **[SSC (CPO)-07, SSC(CGL-T-II)-19]**

Sol. The train passes a man standing on the platform in 15 sec, it implies that it covers a distance equal to its length in 15 sec. It is also given that it covers a distance equal to sum of its length and length of the platform in 30 sec. It means the train covers a distance equal to length of the platform i.e. 90 m in (30 – 15) sec. Hence, its speed = 90 ÷ 15 = 6 m/sec.

Alternatively I

Since speed of the train is constant, $\frac{T_1}{T_2} = \frac{D_1}{D_2}$

Let length of the train be L m.

Then, $\frac{15}{30} = \frac{L}{L + 90} \Rightarrow L = 90$ m

The train crosses the man in 15 sec, it means it covers a distance of its length i.e. 90 m in 15 sec

Hence, speed = 90 ÷ 15 = 6 m/sec

Alternatively II

We know, $(L + l) = (S_T \pm S_O) \times T$

Case I $L + 90 = (S_T \pm 0) \times 30 = 30S_T$ ...(i)

Case II $L + 0 = (S_T \pm 0) \times 15 = 15S_T$ ...(ii)

Subtracting Eq. (ii) from Eq. (i), we get, $15S_T = 90$

$\therefore$ $S_T = 6$ m/sec

Problem 4 A train passes two bridges of lengths 800 m and 400 m in 100 sec and 60 sec respectively. The length of the train is ?

Sol. In 100 sec. the train covers a distance of (Length of train + 800) and in 60 sec. the train covers a distance of (Length of train + 400)

Hence, in 40 sec. the train covers 400 m.

Then, in 60 sec. it will cover 600 m.

But, given that is covers a distance of (Length of train + 400) in 60 sec.

Thus, (Length of train + 400) = 600

$\therefore$ Length of the train = 600 − 400 = 200 m

Alternatively I

Let the length of the train be L m.

Since speed of the train is constant,

$$\frac{T_1}{T_2} = \frac{D_1}{D_2}$$

Then, $\frac{100}{60} = \frac{L+800}{L+400}$

$\Rightarrow 5L + 2000 = 3L + 2400$

$\Rightarrow 2L = 400$

$\therefore L = 200$ m

Alternatively II

We know, $(L + l) = (S_T \pm S_O) \times T$

Case I : $L + 800 = (S_T \pm 0) \times 100 = 100S_T$ ...(i)

Case II : $L + 400 = (S_T \pm 0) \times 60 = 60S_T$ ...(ii)

On Eqs. (i) - (ii) we get,

$40\,S_T = 400$

$\therefore S_T = 10$ m/sec

Now, in 100 sec it will cover (100 × 10) m. But as per question, it covers $(L + 800)$ m in 100 sec.

Hence, $L + 800 = (100 \times 10) \Rightarrow L = 200$ m

Or

Put the value of S_T in Eqs. (i), then $L + 800 = (100 \times 10)$

$\Rightarrow L = 200$ m

Problem 5 **A train passes two persons walking in the same direction at a speed of 3 km/h and 5 km/h in 10 sec. and 11 sec. respectively. What is the speed of the train?** ***[SSC (CGL-T-II)-19, CAT-20 (Partly)]***

Sol. Let the speed of the train be x km/h.

Then, relative speed with reference to 1st man = $(x - 3)$ km/h

and relative speed with reference to 2nd man = $(x - 5)$ km/h

Now, time taken in 1st case = $\frac{\text{Length of the train}}{(x-3)} = 10$

$\Rightarrow$ Length of the train = $10(x - 3)$

Further, time taken in 2nd case = $\frac{\text{Length of the train}}{(x-5)} = 11$

$\Rightarrow$ Length of the train = $11(x - 5)$

Hence, $10(x - 3) = 11(x - 5)$ as both are the lengths of train.

$\Rightarrow 11x - 10x = 55 - 30$

$\Rightarrow x = 25$ km/h

Alternatively I

Let the speed of the train be x km/h.

Then, relative speed with reference to 1st man = $(x - 3)$ km/h

and relative speed with reference to 2nd man = $(x - 5)$ km/h

In both cases, the train covers a distance of its length in order to cross the men.

Hence, $\frac{S_1}{S_2} = \frac{T_2}{T_1}$ as distance is constant here,

$\Rightarrow \frac{(x-3)}{(x-5)} = \frac{11}{10}$

$\Rightarrow 10(x - 3) = 11(x - 5)$

$\Rightarrow 11x - 10x = 55 - 30$

$\Rightarrow x = 25$ km/h

Alternatively II

We know, $(L + l) = (S_T \pm S_O) \times T$

Case I $(L + 0) = (S_T - 3) \times 10$

Case II $(L + 0) = (S_T - 5) \times 11$

Hence, $(S_T - 3) \times 10 = (S_T - 5) \times 11$

$\Rightarrow S_T = 25$ km/h

Problem 6 **Two trains 105 m and 90 m long run at the speeds of 45 km/h and 72 km/h respectively in opposite direction on parallel track. What is the time taken to cross each other ?** ***[SSC (CGL-PT)-07]***

Sol. Relative speed = 45 + 72 = 117 km/h

$= \left(117 \times \frac{5}{18}\right)$ m/sec

Total distance to be covered = 105 + 90 = 195 m

Then, time taken = $195 \div \left(117 \times \frac{5}{18}\right) = 6$ sec

Alternatively

We know, $(L + l) = (S_T \pm S_O) \times T$

$\Rightarrow (105 + 90)\text{ m} = (45 + 72) \times \frac{5}{18} \times T$

[$\because$ we have to maintain uniformity in units always]

$\Rightarrow T = 6$ sec

Problem 7 **A train 110 m long passes a man running at 6 km/h in the direction opposite to that of train in 6 sec. The speed of the train is** ***[RRB (ASM)-02]***

Sol. Let the speed of the train be x km/h.

Then, relative speed = $(x + 6)$ km/h

$= (x + 6) \times \frac{5}{18}$ m/sec

Time required to cross the man = $110 \div \left\{(x + 6) \times \frac{5}{18}\right\}$

$\Rightarrow 6 = \frac{22 \times 18}{(x+6)}$

$\Rightarrow x + 6 = 66 \Rightarrow x = 60$ km/h

Alternatively I

We know, $(L + l) = (S_T \pm S_O) \times T$

$\Rightarrow (110 + 0) = (S_T + 6) \times \frac{5}{18} \times 6$

$\Rightarrow 330 = 5S_T + 30$

$\Rightarrow 5\,S_T = 300$

$\therefore S_T = 60$ km/h

Problem 8 **Two trains, running at 68 km/h and 40 km/h in opposite directions, cross each other in 15 sec. If they run in same direction, faster train passes a man sitting in slower train in 36 sec. Find the length of the slower train.**

Sol. Relative speed in opposite direction

$= (68 + 40) \times \frac{5}{18}$ m/sec = 30 m/sec

Obviously, they cover a distance of sum of their lengths in 15 sec in order to cross each other.

Hence, sum of their lengths = 30 × 15 = 450 m

It is obvious that the speed of the man sitting in slower train is equal to the speed of slower train.

Then, relative speed $= (68-40) \times \frac{5}{18}$ m/sec

When the faster train passes a man sitting in slower train in 36 sec, it means a distance equal to length of faster train is covered at relative speed in 36 sec.

Now, distance travelled in 36 sec $= (68-40) \times \frac{5}{18} \times 36 = 280$ m

Hence, length of faster train = 280 m

and length of slower train = 450 − 280 = 170 m

Alternatively

We know, $(L + l) = (S_T \pm S_O) \times T$

Case I : $(L + l) = (68 + 40) \times \frac{5}{18} \times 15$

$\Rightarrow \quad (L + l) = 450$ m

Case II : $(L + l) = (68 - 40) \times \frac{5}{18} \times 36 \Rightarrow L = 280$ m

Then, $l = 450 - 280 = 170$ m

Problem 9 **A train having length of 300 m crosses a man travelling in another train having length of 200 m in the same direction, in 30 sec. However, it requires 20 sec to cross the same man if the train are running in the opposite directions. Find the speed of the train having length of 300 m.**

Sol. Let the speeds of train having length of 300 m and the man be S_T m/sec and S_O m/sec respectively,

Now, relative speed in the opposite directions

$$= S_T + S_O = 300/20$$

and relative speed in the same direction $= S_T - S_O = 300/30$

On addition, $2S_T = 25$

$\Rightarrow \quad S_T = 12.5$ m/sec

Alternatively I

We know, $(L + l) = (S_T \pm S_O) \times T$

Case I $(300 + 0) = (S_T - S_O) \times 30$

$\Rightarrow \quad S_T - S_O = 10 \quad$ ...(i)

Case II $(300 + 0) = (S_T + S_O) \times 20$

$\Rightarrow \quad S_T + S_O = 15 \quad$ ...(ii)

On solving both Eqs. we get, $S_T = 12.5$ m/sec

Hence, speed of the train having length of 300 m = 12.5 m/sec

Alternatively II

If relative speeds in context of both directions, (i.e. same and opposite) are known, then speed of the faster body is the average of both relative speeds. For computation of speed of slower body, we simply subtract the speed of faster body from the relative speed meant for opposite direction.

Here, relative speeds are $(300 \div 30)$ m/sec and $(300 \div 20)$ m/sec.

Then, speed of faster body (train)

= average of 10 and 15 = 12.5 m/sec

Problem 10 **A man standing on platform finds that a train takes 3 sec to pass him and another train of same length moving in the opposite direction, takes 4 sec. The time taken by the trains to pass each other will be :**

Sol. Let speeds of the 1st and 2nd trains be x m/sec and y m/sec. Also, let each train has length of L m.

Then, time taken by 1st train $= L \div x = 3$

$\Rightarrow \quad L = 3x$

Further, time taken by 2nd train $= L \div y = 4$

$\Rightarrow \quad L = 4y$

From the above, $3x = 4y$

$\Rightarrow \quad x = \frac{4}{3}y$

When a train passes another, relative speed $= (x + y)$ m/sec.

And total distance to be covered $= L + L = 2L$

Then, time taken to cross each other $= 2L \div (x + y)$

$$= (4y + 4y) \div \left(\frac{4}{3}y + y\right)$$

$$= \frac{24}{7} = 3\frac{3}{7} \text{ sec}$$

Alternatively I

We know, $(L + l) = (S_T \pm S_O) \times T$

Case I $(L + 0) = (S_{T1} \pm 0) \times 3$

Case II $(L + 0) = (S_{T2} \pm 0) \times 4$

Hence, $3S_{T1} = 4S_{T2}$

Further steps on the above line.

Alternatively II

Both the trains travel the same distance equal to each of their lengths,.

Given, ratio of time taken = 3 : 4

Then, ratio of their speeds = 4 : 3 $\left[\because S \alpha \frac{1}{T}, \text{ when } D \text{ is constant}\right]$

Total distance travelled $= (3 \times 4) + (4 \times 3) = 24$ units

and relative speed in opposite directions $= (4 + 3) = 7$

Then, time taken $= \frac{24}{7} = 3\frac{3}{7}$ sec

Hence, time required to cross each other $= 3\frac{3}{7}$ sec

Problem 11 **A moving train crosses a man standing on a platform and a bridge 300 m long in 10 sec and 25 sec respectively. What will be the time taken by the train to cross a platform 200 m long.**

Sol. Let the length of the train be L m.

As given, L m is covered by the train in 10 sec.

and $(L + 300)$ m is covered by it in 25 sec.

It means it covers 300 m in (25 − 10) sec.

Then, it will cover 200 m in 10 sec.

Obviously, it will cover $(L + 200)$ m in (10 + 10) sec.

Hence, required time taken = 20 sec

Alternatively I

Since speed of the train is constant, $\frac{T_1}{T_2} = \frac{D_1}{D_2}$

$\Rightarrow \quad \frac{10}{25} = \frac{L}{L + 300}$

$\Rightarrow \quad 15L = 3000$

$\Rightarrow \quad L = 200$ m

As given, it moves 200 m in 10 sec.

Then, it will move (200 + 200) in 20 sec.
Hence, required time taken = 20 sec

Alternatively II

We know, $(L + l) = (S_T \pm S_O) \times T$

Case I $(L + 0) = (S_T \pm 0) \times 10$ ...(i)

Case II $(L + 300) = (S_T \pm 0) \times 25$ ...(ii)

On solving, $S_T = 20$ and $L = 200$

Further, $(L + l) = (S_T \pm S_O) \times T$

$\Rightarrow (200 + 200) = (20 \pm 0) \times T$

$\Rightarrow T = 20$ sec

Problem 12 A train travelling at 48 km/h crosses another train having half its length and travelling in opposite direction at 42 km/h, in 12 sec. It also passes a railway platform in 45 sec. The length of the railway platform is *[MAT-10, 08, SSC (CGL-PT)-08]*

Sol. Let the length of the 1st train be L m.

Relative speed = 48 + 42 = 90 km/h

$= \left(90 \times \frac{5}{18}\right)$ m/sec

= 25 m/sec

Distance covered in 12 sec = 25 × 12 = 300 m

$\Rightarrow (L + L/2) = 300$ m

$\Rightarrow L = 200$ m

Further, distance covered by the 1st train in 45 sec

$= 45 \times \left(48 \times \frac{5}{18}\right) = 600$ m

$\Rightarrow$ (L + length of platform) = 600

$\Rightarrow$ (200 + length of platform) = 600

$\Rightarrow$ The length of platform = 600 − 200 = 400 m

Alternatively I

Let the 1st train takes T sec. to cover sum of both trains' lengths.

Here, it is assumed that sum of both trains' length i.e. 1.5 times the length of 1st train is constant.

Then, $\frac{S_1}{S_2} = \frac{T_2}{T_1} \Rightarrow \frac{(48 + 42)}{48} = \frac{T}{12}$

$\Rightarrow T = 22.5$ sec

∵ To cover 1.5 times of its length, the 1st train takes 22.5 sec.

∴ To cover length of itself, it will take 15 sec.

But, 1st train takes 45 sec, to cover (length of itself + length of platform). It implies that it covers the length of platform in (45 − 15) sec.

Now, distance covered in 30 sec $= 30 \times \left(48 \times \frac{5}{18}\right) = 400$ m

Hence, length of the platform = 400 m

Alternatively II

We know, $(L + l) = (S_T \pm S_O) \times T$

Case I $(L + L/2) = (48 + 42) \times \frac{5}{18} \times 12$

$\Rightarrow L = 200$ m

Case II $(L + \text{length of platform}) = (48 \pm 0) \times \frac{5}{18} \times 45$

$\Rightarrow$ (200 + length of platform) = 600

$\Rightarrow$ Length of the platform = 600 − 200 = 400 m

Problem 13 A train crosses two men running at 8 m/sec. and 15 m/sec in 8 sec and 6 sec respectively. Find the speed and length of the train provided both men are running in opposite to the direction of the train.

Sol. Let the speed of the train be x m/sec.

Then, relative speed with reference to 1st man = $(x + 8)$ m/sec

and relative speed with reference to 2nd man = $(x + 15)$ m/sec

Now, time taken in 1st case $= \frac{\text{Length of the train}}{(x + 8)} = 8$

$\Rightarrow$ Length of the train = $8(x + 8)$

Further, time taken in 2nd case $= \frac{\text{Length of the train}}{(x + 15)} = 6$

$\Rightarrow$ Length of the train = $6(x + 15)$

Hence, $8(x + 8) = 6(x + 15)$ as both are the lengths of train.

$\Rightarrow 8x - 6x = 90 - 64$

$\Rightarrow x = 13$ m/sec

Now, length of the train = $8(x + 8)$

$= 8(13 + 8) = 168$ m

Alternatively I

Let the speed of the train be x m/sec.

Then, relative speed with reference to 1st man = $(x + 8)$ m/sec.

and relative speed with reference to 2nd man = $(x + 15)$ m/sec.

In both cases, the train covers a distance equal to its length in order to cross the men.

Hence, $\frac{S_1}{S_2} = \frac{T_2}{T_1}$ as distance is constant here.

$\Rightarrow \frac{(x + 8)}{(x + 15)} = \frac{6}{8}$

$\Rightarrow 8(x + 8) = 6(x + 15)$

$\Rightarrow 8x - 6x = 90 - 64$

$\Rightarrow x = 13$ m/sec

Distance covered in 1st case = $8(x + 8) = 8(13 + 8) = 168$

Hence, the length of the train = 168 m.

Alternatively II

We know, $(L + l) = (S_T \pm S_O) \times T$

Case I $(L + 0) = (S_T + 8) \times 8$

Case II $(L + 0) = (S_T + 15) \times 6$

Hence, $8(x + 8) = 6(x + 15)$

$\Rightarrow 8x - 6x = 90 - 64$

$\Rightarrow x = 13$ m/sec

Further, $L = (S_T + 8) \times 8 = (13 + 8) \times 8$

= 168 m

Problem 14 The ratio of speeds of two trains running in opposite directions is 2 : 3. What is the time taken by them to cross each other if they can cross a pole in 4 sec. individually.

Sol. Given, ratio of their speeds = 2 : 3

and ratio of time taken = 4 : 4

Total distance travelled = (2 × 4) + (3 × 4) = 20 units

and relative speed in opposite directions = (2 + 3) = 5

Then, time taken $= \frac{20}{5} = 4$ sec

Hence, time required to cross each other = 4 sec

Alternatively I

We know, $(L + l) = (S_T \pm S_O) \times T$

Case I $(L_1 + 0) = (S_{T1} \pm 0) \times 4$

$\Rightarrow \quad L_1 = 4S_{T1}$

Case II $(L_2 + 0) = (S_{T2} \pm 0) \times 4$

$\Rightarrow \quad L_2 = 4S_{T2}$

Case III $(L_1 + L_2) = (S_{T1} + S_{T2}) \times T$

$\Rightarrow (4S_{T1} + 4S_{T2}) = (S_{T1} + S_{T2}) \times T$

$\Rightarrow \quad T = 4$ sec

Alternatively II

Required time will be 4 sec irrespective of their speeds as they take same time to cross a pole. It may be understood with the help of the following diagram.

| (pole)
..........train.........>
<..........train..........

[Before crossing]

| (pole)
..........train.........>
<..........train..........

[After crossing]

Problem 15 **The length of faster train is 100 m less than that of slower train. They take 20 sec. to cross each other running in opposite directions. However, they require 1 min 40 sec to cross each other while running in the same direction. Find out the speed of faster train if the speed of slower is 40 km/h. Find their lengths also.**

Sol. Let the speed of faster train be S_T m/sec.

Since they travel sum of their lengths in each case, $\frac{S_1}{S_2} = \frac{T_2}{T_1}$

$\Rightarrow \quad \frac{S_T + 40}{S_T - 40} = \frac{100}{20}$

$\Rightarrow \quad 5S_T - 200 = S_T + 40$

$\Rightarrow \quad S_T = 60$ km/h

Now, distance travelled in 20 sec at the rate of (60 + 40) km/h

$= 20 \times 100 \times \frac{5}{18} = 555.56$ m

Hence, sum of their lengths = 555.56 m

Since length of faster train is 100 m less than that of slower train, length of faster train = (555.56 – 100) ÷ 2 = 227.78

and length of slower train = 227.78 + 100 = 327.78 m

Alternatively

We know, $(L + l) = (S_T \pm S_O) \times T$

Case I $(L + L + 100) = (S_T + 40) \times \frac{5}{18} \times 20$

$\Rightarrow \quad 18L + 900 = 50S_T + 2000$

$\Rightarrow \quad 9L - 25S_T = 550$ ...(i)

Case II $(L + L + 100) = (S_T - 40) \times \frac{5}{18} \times 100$

$\Rightarrow \quad 18L + 900 = 250S_T - 10000$

$\Rightarrow \quad 9L - 125S_T = -5450$ ...(ii)

On Eqs. (i) - (ii) we get, $100\,S_T = 6000$

$\Rightarrow \quad S_T = 60$ km/h

Putting the value of S_T in Eq. (i) we get

$L = 227.78$ m

Then, length of slower train = 227.78 + 100 = 327.78 m

Problem 16 **Two trains *A* and *B* are running in the same direction and a passenger sitting in train *B* notices that he crosses the train *A* in 60 sec. but the loco-pilot of train *A* notices that train *B* crosses his train in 90 sec. What is ratio of their length if ratio of their speeds is 2 : 3 ?**

Sol. The passenger of train *B* notices that he crosses the train *A* in 60 sec, it means a distance equal to length of train *A* has been covered at the rate of relative speed. Further, the loco-pilot of train *A* notices that train *B* crosses his train in 90 sec. it means a distance equal to sum of both trains has been covered at the rate of relative speed.

Let the speeds of trains *A* and *B* be $2x$ m/sec. and $3x$ m/sec.

Then, in 1st case, length of train *A*

$= (3x - 2x) \times 60 = 60x$

In 2nd case, sum of their lengths $= (3x - 2x) \times 90 = 90x$

Hence, length of train $B = 90x - 60x = 30x$

$\therefore$ Required ratio $= 60x : 30x = 2 : 1$

Alternatively

Let the speeds of trains *A* and *B* be $2x$ m/sec. and $3x$ m/sec.

We know, $(L + l) = (S_T \pm S_O) \times T$

Case I $(L_A + 0) = (3x - 2x) \times 60$

$\Rightarrow \quad L_A = 60x$

Case II $(L_A + L_B) = (3x - 2x) \times 90$

$\Rightarrow \quad L_A + L_B = 90x$

Hence, $\quad L_B = 90x - 60x = 30x$

Then, required ratio $= 60x : 30x = 2 : 1$

Problem 17 **Two trains running in opposite directions cross a man standing on the platform in 27 sec and 17 sec respectively and they cross each other in 23 sec The ratio of their speeds is :** ***[Hotel Mgmt-97]***

Sol. Let the speeds be x m/sec and y m/sec.

Then, length of 1st train $= 27 \times x = 27x$ m

and length of 2nd train $= 17 \times y = 17y$ m

Now, relative speed $= (x + y)$ m/sec

Then, time required to cross each other

= sum of their lengths ÷ relative speed

$= (27x + 17y) \div (x + y) = 23$ [given]

$\Rightarrow (27x + 17y) = 23x + 23y$

$\Rightarrow \quad 4x = 6y$

$\Rightarrow \quad x : y = 3 : 2$

Alternatively I

Proceed through options. Let options be 1 : 3, 3 : 2, 3 : 4, None of these

(a) Ratio of time taken = 27 : 17

Ratio of speeds = 1 : 3 [1st option]

Then, distance travelled $= (27 \times 1) + (17 \times 3) = 78$

$\therefore$ Time taken to cross each other $= 78 \div (1 + 3) \neq 23$

(b) Ratio of time taken = 27 : 17

Ratio of speeds = 3 : 2 [2nd option]

Then, distance travelled

$$= (27 \times 3) + (17 \times 2) = 115$$

$\therefore$ Time taken to cross each other

$$= 115 \div (3+2) = 23$$

Hence, option (b) is correct.

Alternatively II

We know, $(L + l) = (S_T \pm S_O) \times T$

Case I $(L_1 + 0) = (S_{T1} \pm 0) \times 27$

$\Rightarrow \quad L_1 = 27 S_{T1}$

Case II $(L_2 + 0) = (S_{T2} \pm 0) \times 17$

$\Rightarrow \quad L_2 = 17 S_{T2}$

Case III $(L_1 + L_2) = (S_{T1} + S_{T2}) \times 23$

$\Rightarrow (27S_{T1} + 17S_{T2}) = (S_{T1} + S_{T2}) \times 23$

$\Rightarrow \quad 4S_{T1} = 6S_{T2}$

$\Rightarrow \quad S_{T1} : S_{T2} = 3 : 2$

Problem 18 Two trains are 2 km apart and their lengths are 200 m and 300 m. They are approaching towards each other with a speed of 20 m/sec and 30 m/sec respectively. After how much time will they cross each other? ***[MAT-06]***

Sol. Relative speed = 20 + 30 = 50 m/sec

Total distance to be covered

$$= 200 + 300 + 2000 = 2500 \text{ m}$$

Then, time taken = 2500 ÷ 50 = 50 sec

Alternatively

We know, $(L + l) = (S_T \pm S_O) \times T$

$\Rightarrow \quad (200 + 300) = (20 + 30) \times T$

$\Rightarrow \quad T = 10 \text{ sec}$

Hence, they will take 10 sec to cross each other and time has been counted since both engines meet. But they are 2 km apart which is also to be covered at the rate of relative speed.

To cover 2000 m, they will take (2000 ÷ 50) i.e. 40 sec

Thus, total required time = 40 + 10 = 50 sec

Problem 19 Two trains X and Y running in the same direction at the speed of 50 m/sec and 40 m/sec cross each other in 40 sec. While travelling in opposite directions, Y takes 3.25 sec more to cross the meeting point than what X does. What is the ratio of their lengths?

Sol. Relative speed in the same direction = 50 − 40 = 10 m/sec.

Then, sum of their length = 40 × 10 = 400 m

Further, time taken by train X to cross completely the meeting point = its own length (say, L_x) ÷ 50

Further, time taken by train Y to cross completely the meeting point = its own length (say, L_y) ÷ 40

As per question, $\frac{L_x}{50} + 3.25 = \frac{L_y}{40}$

$\Rightarrow \quad 4L_x - 5L_y = -650$ ...(i)

and from the above, $L_x + L_y = 400$ ...(ii)

Adding Eqs. (i) and (ii) × 5 we have $9L_x = 1350$

$\therefore \quad L_x = 150$

Then, $\quad L_y = 400 - 150$

$= 250$

Hence, required ratio = 150 : 250

= 3 : 5

Alternatively

Let the lengths of trains X and Y be L_x and L_y respectively and time taken by them to cross the meeting point while travelling in opposite directions are T_x and T_y respectively.

We know, $(L + l) = (S_T \pm S_O) \times T$

Case I $(L_x + L_y) = (50 - 40) \times 40 = 400$

Case II $(L_x + 0) = (50 \pm 0) \times T_x$

$\Rightarrow \quad T_x = 0.02\, L_x$

Case III $(L_y + 0) = (40 \pm 0) \times T_y$

$\Rightarrow \quad T_y = 0.025 L_y$

As per question, $T_x + 3.25 = T_y$

$\Rightarrow \quad 0.02L_x + 3.25 = 0.025L_y$

$\Rightarrow \quad 20L_x - 25L_y = -3250$ ...(i)

From the above,

$L_x + L_y = 400$ ...(ii)

On solving, we have $L_x = 150$ and $L_y = 250$

Hence, required ratio = 150 : 250 = 3 : 5

Observation Each of problems under 'Illustration' has been solved also with the help of one model, $(L + l) = (S_T \pm S_O) \times T$

Problem 20 Two trains A and B were moving in opposite directions, their speeds being in the ratio 5 : 3. The front end of A crossed the rear end of B 46 sec after the front ends of the trains had crossed each other. It took another 69 sec for the rear ends of the trains to cross each other. The ratio of length of train A to that of train B is ***[CAT-21]***

Sol. Clearly, the length of train B is covered by the relative speed in 46 sec and the length of train A is covered in 69 sec with the same relative speed.

Since the distance is directly proportional to the time, when speed is constant, the ratio of length of train A to that of train B be 69 : 46 i.e. 3 : 2.

WORKOUT

1. A train takes 10 sec to cross a pole and 20 sec to cross a platform of length 200 m. What is the length of the train? *[CDS-14, 12, SSC (SAS)-10]*
(a) 50 m (b) 100 m
(c) 150 m (d) 200 m

2. A train 100 m long running at 54 km/h takes 20 sec to pass a bridge. The length of the bridge is : *[CLAT-08]*
(a) 50 m (b) 150 m
(c) 200 m (d) 620 m

3. Two trains are running at 40 km/h and 20 km/h respectively in the same direction. The fast train completely passes a man sitting in the slow train in 5 sec. The length of the fast train is : *[SSC (CGL-T-II)-13, IAS (Pre)-98]*
(a) $23\frac{2}{9}$ m (b) 27 m
(c) $27\frac{7}{9}$ m (d) 23 m

4. A train takes 9 sec to cross a pole. If the speed of the train is 48 km/h, the length of the train is*[CDS-14,13]*
(a) 150 m (b) 120 m
(c) 90 m (d) 80 m

5. A train 280 m long is moving at a speed of 60 km/h. What is the time taken by the train to cross a platform 220 m long? *[CDS-12, CLAT-12]*
(a) 45 sec (b) 40 sec
(c) 35 sec (d) 30 sec

6. Two trains each 200 m long move towards each other on parallel lines with velocities 20 km/h and 30 km/h respectively. What is the time that elapses when they 1st meet until they have cleared each other ? *[CDS-11, 10 MAT-07]*
(a) 20 sec (b) 24.8 sec
(c) 28.8 sec (d) 30 sec

7. Two trains of equal length are running on parallel lines in the same direction at 46 km/h and 36 km/h. The faster train passes the slower train in 36 sec. The length of each train is :
(a) 50 m (b) 80 m (c) 72 m (d) 82 m

8. Two trains travel in the same direction at 50 km/h and 32 km/h respectively. A man in the slower train observes that 15 sec elapse before the faster train completely passes him. What is the length of the faster train ? *[CDS-19, SSC (CGL-T-I)-17]*
(a) 75 m (b) 125 m
(c) 150 m (d) 208.33 m

9. A train of length 100 m crosses another train of length 150 m, running on a parallel track in the opposite direction in 9 sec. If the speed of the train having length 150 m is 40 km/h, then what is the speed (in km/h) of the other train? *[SSC (CGL-T-I)-17]*
(a) 30 (b) 48
(c) 50 (d) 60

10. A train of length 150 m takes 10 sec to cross another train 100 m long coming from the opposite direction. If the speed of 1st train is 30 km/h. What is the speed of 2nd train ?
(a) 72 km/h (b) 60 km/h
(c) 54 km/h (d) 48 km/h

11. A train 110 m long is running with a speed of 60 km/h. What is the time in which it will pass a man who starts from the engine running at the speed of 6 km/h in the direction opposite to that of the train ? *[CDS-08, MAT-18]*
(a) 5 sec (b) 6 sec
(c) 10 sec (d) 15 sec

12. Two trains 130 m and 110 m long are going in the same direction. The faster train takes 1 min. to pass the other completely. If they are moving in opposite directions, they pass each other completely in 3 sec. What is the speed of the faster train? *[MAT-13, 08]*
(a) 42 m/sec (b) 38 m/sec
(c) 46 m/sec (d) 50 m/sec

13. A train overtakes two persons walking along a railway track. The first one walks at 4.5 km/h and the other one walks at 5.4 km/h. The train needs 8.4 sec and 8.5 sec respectively to overtake them. What is the speed of the train if both the persons are walking in the same direction as the train? *[MAT-10, LIC(AAO)-09]*
(a) 66 km/h (b) 72 km/h
(c) 78 km/h (d) 81 km/h

14. It takes 15 sec for a train travelling at 60 km/h to cross entirely another train half its length and travelling in opposite direction at 48 km/h. It also passes a bridge in 51 sec. The length of the bridge is *[IIFT-12]*
(a) 550 m (b) 450 m
(c) 500 m (d) 600 m

15. Train A crosses a pole in 20 sec and train B crosses the same pole in 1 min. The length of train *A* is half the length of train *B*. What is the ratio of the speed of train *A* to that of train *B* ? *[CB(PO)-11]*
(a) 3 : 2 (b) 3 : 4
(c) 4 : 3
(d) Cannot be determined
(e) None of the above

16. Ramu sees a train passing over a 1 km long bridge. The length of the train is half that of the bridge. If the train clears the bridge in 2 min, the speed of the train is
(a) 50 km/h (b) 43 km/h
(c) 45 km/h (d) None of these

17. A man sitting in a train travelling at the rate of 50 km/h observes that it takes 9 sec for a goods train travelling in the opposite direction to pass him. If the goods train is 187.5 m long, find its speed :
(a) 25 km/h (b) 40 km/h
(c) 35 km/h (d) 36 km/h

18. A 222 m long train crosses a pole in 6 sec. The same train crosses a man running in the same direction in 10 sec. What will be the speed of man ? *[LIC (AAO)-13]*
(a) 15 m/sec (b) 17 m/sec
(c) 18 m/sec (d) 14 m/sec
(e) None of these

19. A 200 m long train crosses a platform double its length in 36 sec. What is the speed of the train in km/h? **[SBI (PO)-08, CDS-21]**
(a) 66 km/h (b) 72 km/h
(c) 78 km/h (d) 60 km/h

20. A train passes a station platform in 36 sec and a man standing on the platform in 20 sec. If the speed of the train is 54 km/h, find the length of the platform ? **[MAT-08, SNAP-11]**
(a) 225 m (b) 235 m
(c) 230 m (d) 240 m

21. A train consists of 12 bogies, each bogie is 15 m long. The train crosses a telegraph post in 18 sec. Due to some problem, two bogies were detached. The train now crosses a telegraph post in :
(a) 20 sec (b) 12 sec
(c) 18 sec (d) 15 sec

22. A train 75 m long overtook a person who was walking at the rate of 6 km/h and passes him in 7.5 sec. Subsequently, it overtook a 2nd person and passes him in 6.75 sec. At what rate was the 2nd person travelling ?
(a) 4 km/h (b) 1 km/h
(c) 2 km/h (d) 5 km/h

23. A 180 m long train crosses another 270 m long train running in the opposite direction in 10.8 sec. If the speed of the 1st train is 60 km/h, what is the speed of the 2nd train in km/h? **[Allahabad Bank (PO)-10]**
(a) 80
(b) 90
(c) 150
(d) Cannot be determined
(e) None of the above

24. A train with 90 km/h crosses a bridge in 36 sec. Another train 100 m shorter crosses the same bridge at 45 km/h. What is the time taken by the 2nd train to crosses the bridge?
(a) 61 sec (b) 63 sec
(c) 62 sec (d) 64 sec

25. Train A crosses a stationary train B in 50 sec and a pole in 20 sec with the same speed. The length of the train A is 240 m. What is the length of the stationary train B ? **[CBI (PO)-10]**
(a) 360 m (b) 260 m
(c) 300 m
(d) Cannot be determined
(e) None of the above

26. A train 100 m long passes a bridge at the rate of 72 km/h in 25 sec. The length of the bridge is
(a) 150 m (b) 400 m
(c) 300 m (d) 200 m

27. A 225 m long train is running at a speed of 30 km/h. How much time does it take to cross a man running at 3 km/h in the same direction ? **[CDS-17]**
(a) 40 sec (b) 30 sec
(c) 25 sec (d) 15 sec

28. A train X travelling at 60 km/h overtakes another train Y, 225 m long and completely passes it in 72 sec. If the trains had been going in opposite directions, they would have passed each other in 18 sec. The length (in m) of X and the speed (in km/h) of Y are, respectively : **[SSC (CGL-T-1)-19]**
(a) 245, 54 (b) 245, 45
(c) 255, 36 (d) 255, 40

Answer Keys

1. (d)	2. (c)	3. (c)	4. (b)	5. (d)	6. (c)	7. (a)	8. (a)	9. (d)	10. (b)
11. (b)	12. (a)	13. (d)	14. (a)	15. (a)	16. (c)	17. (a)	18. (e)	19. (d)	20. (d)
21. (d)	22. (c)	23. (b)	24. (d)	25. (a)	26. (b)	27. (b)	28. (c)		

Hints and SOLUTIONS

1. The train takes 10 sec to move a distance equal to its own length and 20 sec to move a distance equal to its own length plus a platform 200 m long. It means it takes (20 − 10) sec to move 200 m (i.e length of platform). Also, given that, it takes 10 sec to move a distance equal to its own length. Hence, length of the train = 200 m.

2. We know, $(L + l) = (S_T \pm S_O) \times T$
$\Rightarrow (100 + l) = (54 \pm 0) \times \frac{5}{18} \times 20$
$\Rightarrow l = 200\text{ m}$

3. We know, $(L + l) = (S_T \pm S_O) \times T$
$\Rightarrow (L + 0) = (40 - 20) \times \frac{5}{18} \times 5$
$\Rightarrow L = \frac{250}{9} = 27\frac{7}{9}\text{ m}$

4. We know, $(L + l) = (S_T \pm S_O) \times T$
$\Rightarrow (L + 0) = (48 \pm 0) \times \frac{5}{18} \times 9$
$\Rightarrow L = 120\text{ m}$

5. We know, $(L + l) = (S_T \pm S_O) \times T$
$\Rightarrow (280 + 220) = (60 \pm 0) \times \frac{5}{18} \times T$
$\Rightarrow T = 30\text{ sec}$

6. We know, $(L + l) = (S_T \pm S_O) \times T$
$(200 + 200) = (20 + 30) \times \frac{5}{18} \times T$
$\Rightarrow T = 28.8\text{ sec}$

7. We know, $(L + l) = (S_T \pm S_O) \times T$
$\Rightarrow (L + L) = (46 - 36) \times \frac{5}{18} \times 36$
$\Rightarrow L = 50\text{ m}$

8. We know, $(L + l) = (S_T \pm S_O) \times T$
$\Rightarrow (L + 0) = (50 - 32) \times \frac{5}{18} \times 15$
$\Rightarrow L = 75\text{ m}$

9. We know, $(L+l)=(S_T \pm S_O)\times T$

$\Rightarrow \quad (100+150)=(S_T+40)\times\frac{5}{18}\times 9$

$\Rightarrow \quad S_T = 60\,\text{km/h}$

10. We know, $(L+l)=(S_T \pm S_O)\times T$

$\Rightarrow \quad (150+100)=(30+S_O)\times\frac{5}{18}\times 10$

$\Rightarrow \quad 2250 = 750+25S_O$

$\Rightarrow \quad S_O = 60\,\text{km/h}$

11. We know, $(L+l)=(S_T \pm S_O)\times T$

$\Rightarrow \quad (110+0)=(60+6)\times\frac{5}{18}\times T$

$\Rightarrow \quad T = 6\,\text{sec}$

12. We know, $(L+l)=(S_T \pm S_O)\times T$

Case I $(130+110)=(S_T-S_O)\times 60$

[$\because$ 1 min = 60 sec]

$\Rightarrow \quad S_T - S_O = 4 \quad$ ...(i)

Case II $(130+110)=(S_T+S_O)\times 3$

$\Rightarrow \quad S_T + S_O = 80 \quad$ ...(ii)

On Adding Eqs. (i) + (ii) we have

$2S_T = 84$

$\therefore \quad S_T = 42\,\text{m/sec}$

13. We know, $(L+l)=(S_T \pm S_O)\times T$

Case I $(L+0)=(S_T-4.5)\times\frac{5}{18}\times 8.4$

Case II $(L+0)=(S_T-5.4)\times\frac{5}{18}\times 8.5$

Hence, $(S_T-4.5)\times 8.4 = (S_T-5.4)\times 8.5$

$\Rightarrow 8.5\,S_T - 8.4S_T = -(8.4\times 4.5)+(8.5\times 5.4)$

$\Rightarrow \quad 0.1S_T = 8.1$

$\therefore \quad S_T = 81\,\text{km/h}$

14. We know,

$(L+l)=(S_T \pm S_O)\times T$

Case I $(L+L/2)=(60+48)\times\frac{5}{18}\times 15$

$\Rightarrow \quad L = 300\,\text{m}$

Case II $(300+l)=(60\pm 0)\times\frac{5}{18}\times 51$

$\Rightarrow \quad (300+l)=850$

$\Rightarrow \quad l = 550\,\text{m}$

15. Clearly, train A moves a distance equal to its own length (say, d m) in 20 sec. Then, its speed $= d/20$ m/sec

Further, speed of the train B

$= 2d/60$ m/sec

Then, required ratio $= d/20 : 2d/60$

$= 3:2$

16. We know, $(L+l)=(S_T \pm S_O)\times T$

$\Rightarrow \quad (500+1000)=(S_T\pm 0)\times\frac{5}{18}\times 120$

$\Rightarrow \quad S_T = 45\,\text{km/h}$

17. We know, $(L+l)=(S_T \pm S_O)\times T$

$\Rightarrow \quad (187.5+0)=(S_T+50)\times\frac{5}{18}\times 9$

$\Rightarrow \quad 375 = 5S_T + 250$

$\therefore \quad S_T = 25\,\text{km/h}$

18. We know,

$(L+l)=(S_T \pm S_O)\times T$

$\Rightarrow \quad (222+0)=(S_T\pm 0)\times 6$

$\Rightarrow \quad S_T = 37\,\text{m/sec}$

Further, $(L+l)=(S_T \pm S_O)\times T$

$\Rightarrow \quad (222+0)=(37-S_O)\times 10$

$\Rightarrow \quad S_O = 14.8\,\text{m/sec}$

19. Obviously, the train covers a distance of (200 + 400) m in 36 sec. Then, its speed

$= 600 \div 36 = 16.66$ m/sec

$= 16.66\times\frac{18}{5} = 60\,\text{km/h}$

20. We know, $(L+l)=(S_T \pm S_O)\times T$

Case I

$(L+l)=(54+0)\times\frac{5}{18}\times 36 = 540$

Case II

$(L+0)=(54+0)\times\frac{5}{18}\times 20 = 300$

Hence, $l = 540-300 = 240\,\text{m}$

21. We know, $\quad (L+l)=(S_T \pm S_O)\times T$

Case I $(12\times 15+0)=(S_T\pm 0)\times 18$

$\Rightarrow \quad S_T = 10\,\text{m/sec}$

Case II $(10\times 15+0)=(10+0)\times T$

$\Rightarrow \quad T = 15\,\text{sec}$

22. We know, $(L+l)=(S_T \pm S_O)\times T$

Case I $(75+0)=(S_T-6)\times\frac{5}{18}\times 7.5$

$\Rightarrow \quad (S_T-6)=36$

$\therefore \quad S_T = 42\,\text{km/h}$

Case II

$(75+0)=(42-S_O)\times\frac{5}{18}\times 6.75$

$\Rightarrow \quad (42-S_O)=40$

$\therefore \quad S_O = 2\,\text{km/h}$

23. We know, $(L+l)=(S_T \pm S_O)\times T$

$\Rightarrow \quad (180+270)=(60+S_O)\times\frac{5}{18}\times 10.8$

$\Rightarrow \quad S_O = 90\,\text{km/h}$

24. We know, $(L+l)=(S_T \pm S_O)\times T$

***Case* I** $\quad (L+l)=(90\pm 0)\times\frac{5}{18}\times 36$

$\Rightarrow \quad L+l = 900\,\text{m}$

***Case* II**

$(L-100+l)=(45\pm 0)\times\frac{5}{18}\times T$

$\Rightarrow \quad (900-100)=\frac{225}{18}\times T$

$\Rightarrow \quad T = 64\,\text{sec}$

25. The train takes 20 sec to move a distance equal to its own length in order to cross a pole and 50 sec to move a distance equal to its own length plus length of a stationary train. It means it takes (50 − 20) sec to move the length of the stationary train.

$\because$ In 20 sec train A moves a distance equal to its own length i.e. 240 m

$\because$ In 30sec, it will move 360 m.

Hence, the length of the stationary train = 360 m

26. We know, $(L+l)=(S_T \pm S_O)\times T$

$\Rightarrow \quad (100+l)=(72\pm 0)\times\frac{5}{18}\times 25$

$\Rightarrow \quad l = 400\,\text{m}$

27. We know, $\quad (L+l)=(S_T \pm S_O)\times T$

$\Rightarrow \quad (225+0)=(30-3)\times\frac{5}{18}\times T$

$\Rightarrow \quad T = 30\,\text{sec}$

28. We know, $(L+l)=(S_T \pm S_O)\times T$

***Case* I** $L+225=(60-S_O)\times\frac{5}{18}\times 72$

***Case* II** $L+225=(60+S_O)\times\frac{5}{18}\times 18$

Clearly, $(60-S_O)\times\frac{5}{18}\times 72$

$=(60+S_O)\times\frac{5}{18}\times 18$

$\Rightarrow \quad 240-4S_O = 60+S_O$

$\Rightarrow \quad S_O = 36\,\text{km/h}$

Substituting the value of S_O in Equation

$L+225=(60-S_O)\times\frac{5}{18}\times 72$

we get, $\quad L = 255\,\text{m}$

Hence, option (c) is correct.

37 Boat and Stream

Introduction

When a boat moves along the direction of current (stream) of the river/sea, then the relative speed of that boat is the sum of the speed of boat in still water (say, x) and speed of current (say, y). It is called downstream speed. Hence, Downstream speed $= x + y$

Further, when boat moves against the direction of stream, the relative speed of the boat is the difference between x and y and it is called upstream speed. Hence, Upstream speed $= x - y$

It is important to note that

(i) x is always greater than y otherwise a boat can't move ahead.

(ii) Downstream speed (DS) = Speed of boat in still water + Speed of water current

and Upstream speed (US) = Speed of boat in still water – Speed of water current.

Then, the followings may be derived through solving both equations,

(a) Speed of boat in still water $= \dfrac{\text{downstream speed} + \text{upstream speed}}{2}$

$\Rightarrow$ Speed of boat in still water $= \dfrac{DS + US}{2}$

(b) Speed of water current $= \dfrac{DS - US}{2}$

(iii) The concepts associated with Time, Speed and Distance is also applicable in this topic. e.g. In case of downstream, Distance = Downstream speed (DS) × Time taken for travelling downstream (DT) and in case of upstream, Distance = Upstream speed (US) × time taken for travelleing upstream (UT).

(iv) If the distances covered in case of upstream and in case of downstream are equal, $DS \times DT = US \times UT$ or $\dfrac{DS}{US} = \dfrac{UT}{DT}$

i.e. $\dfrac{\text{Downstream speed } (DS)}{\text{Upstream speed } (US)}$

$= \dfrac{\text{Time taken for travelling upstream } (UT)}{\text{Time taken for travelling downstream } (DT)}$

Inside

- *Introduction*
- *Concept of Upstream Speed (US) and Downstream Speed (DS) With Critical Results*
- *Computation of Speed of Current/Boat When Either US and DS are Known or Distance and (UT+DT) are Known*
- *Comparison Between UT and DT When DS/US is Known*
- *When Different Distances are Covered in Different Modes*
- *Working Out DS or US When DT : UT is known and vice-versa*
- *Concept of Two Boats Proceeding Towards Each Other*
- ....
- .... *& Many More*
- *Work Out with Solutions*

[Alternatively, it may be said that in case of constant distance, Speed $\alpha \frac{1}{\text{Time}}$]. You are supposed to apply this conception in process of solving the questions.

(v) If the UT is $\frac{m}{n}$ times DT, speed of the boat in still water will be equal to $\left[\text{speed of current} \times \frac{(m+n)}{(m-n)}\right]$.

In other way, speed of the current

$$= \left[\text{speed of the boat in still water} \times \frac{(m-n)}{(m+n)}\right]$$

Now, build your concepts through the question (infra).

A boat takes 4 h to travel from P to Q downstream and back from Q to P upstream. If the speed of the boat in still water and that of the current are 8 km/h and 2 km/h respectively, then what is the distance between P and Q ? *[CLAT -17]*

Ratio of downstream speed (DS) to upstream speed (US)

$$= (8+2):(8-2) = 10:6 = 5:3$$

Since the distances covered upstream and downstream are equal,

$$\frac{\text{Downstream speed } (DS)}{\text{Upstream speed } (US)} = \frac{\text{Time taken for travelling upstream } (UT)}{\text{Time taken for travelling downstream } (DT)}$$

$\Rightarrow \quad UT : DT = 5 : 3$

Then, $UT = \text{total time taken} \times \frac{5}{(5+3)}$

$$= 4 \times \frac{5}{(5+3)} = 2.5 \text{ h}$$

Then, distance between P and Q

$$= US \times UT = 6 \times 2.5 = 15 \text{ km}$$

Alternatively

$$DT = 4 \times \frac{3}{(5+3)} = 1.5 \text{h}$$

Then, distance between P and Q

$$= DS \times DT = 10 \times 1.5 \text{ km} = 15 \text{ km}$$

With reference to the above question, if total time taken had been 6 hours, what would have been distance between P and Q ?

If total time taken is 6 h, $DT = 6 \times \frac{3}{(3+5)} = \frac{9}{4}$ h

Then, distance $= DS \times DT = (8+2) \times \frac{9}{4} = 22.5$ km

Alternatively

$\left\{\begin{array}{l} \because 4 \text{ h} = 15 \text{ km} \\ \therefore 6 \text{ h} = 22.5 \text{ km} \end{array}\right\}$ as distance is directly proportional to time, when speed is constant.

Illustrations

A. Computation of Speed of Current/Boat When Downstream Speed and Upstream Speed are Known.

Problem 1 **In 1h, a boat goes 11km along the stream and 5 km against the stream. The speed of the boat in still water is what?**

Sol. Let speed of the boat in still water be x km/h and speed of stream be y km/h.

Now, downstream speed $(DS) = x + y = 11$ ...(i)

and upstream speed (US) $x - y = 5$ ...(ii)

Adding both equations we have $2x = 16$

$\therefore \quad x = 8$ km/h

Alternatively

Speed of boat in still water $= \frac{DS + US}{2} = \frac{11+5}{2} = 8$ km/h

B. Working Out DS or US When Ratio Between DT and UT is Known and *Vice-Versa*.

Problem 2 **A boat running upstream takes 8 h 48 min. to cover a certain distance, while it takes 4 h to cover the same distance running downstream. What is the ratio between the speed of the boat and speed of the water current ?** *[Bank PO-03]*

Sol. We know, $\frac{DS}{US} = \frac{UT}{DT}$

$$\Rightarrow \quad \frac{DS}{US} = \frac{8\frac{4}{5}}{4} = \frac{11}{5}$$

Now, proceed through options. Let the option be 2 : 1, 3 : 2, 8 : 3, Can't dotormincd and None of these.

Obviously, option (c) is correct as $(8+3) = 11$ and $(8-3) = 5$

Problem 3 **A boat takes 9 h to travel a distance upstream and 3 h to travel the same distance downstream. If its speed in still water is 4 km/h. What is the velocity of the stream?** *[Delhi Metro- 2002]*

Sol. Let velocity of the stream be x km/h.

We know, $\frac{DS}{US} = \frac{UT}{DT}$

$$\Rightarrow \quad \frac{DS}{US} = \frac{9}{3} = \frac{3}{1} \Rightarrow \frac{4+x}{4-x} = \frac{3}{1}$$

$\therefore \quad x = 2$ km/h

[you may also put the values of given options in place of x in order to ensure as to which option yields the result as 3.]

Alternatively

We know, if the UT is $\frac{m}{n}$ times DT, speed of the current

$$= \text{speed of the boat in still water} \times \frac{(m-n)}{m+n}$$

Here, UT is $\frac{\mathbf{3}}{\mathbf{1}}$ times DT.

Then, speed of the current $= 4 \times \frac{(3-1)}{(3+1)} = 2$ km/h

Problem 4 The speed of Motor-boat is that of the current of water as 36 : 5. The boat goes along with the current in 5 h 10 min. It will come back in ? *[SSC (CGL-PT)-07, 17]*

Sol. We know, $\frac{DS}{US}=\frac{UT}{DT} \Rightarrow \frac{36+5}{36-5}=\frac{UT}{DT}$

$\Rightarrow \frac{41}{31}=\frac{UT}{5\text{ h }10\text{ min}}$

$\Rightarrow UT=\left(\frac{41\times 5\text{ h }10\text{ min}}{31}\right)=410\text{ min}=6\text{ h }50\text{ min}$

C. Computation of Distance When Sum of *DT* and *UT* is Known in Addition to Speed of Current/Boat

Problem 5 A boat takes 6 h to travel from M to N downstream and back from *N* to *M* upstream. If the speed of the boat in still water is 4 km/h. What is the distance between the two places ?

Sol. We know, $\frac{DS}{US}=\frac{UT}{DT}$

$\Rightarrow \frac{4+\text{speed of stream}}{4-\text{speed of stream}}=\frac{UT}{DT}$

Here, $UT+DT=6$ h, but from the given data the relation between *UT* and *DT* cannot be established, so we can't determine the value of the speed of stream. Hence, distance cannot be determined with the given data.

Alternatively

Distance $= DS\times DT$ or $US\times UT$

From the given data, neither *DT* nor *UT* can be determined. Hence, given data is inadequate to compute the distance.

D. Concept of Comparison Between *UT* and *DT* While *DS*/*US* is Known For the Same Distance.

Problem 6 A boat takes 90 min less to travel 36 miles downstream than to travel the same distance upstream. If the speed of the boat in still water is 10 miles/h, the speed of the stream is ?

Sol. Let the speed of the current be x miles/h

Then, $DT=36\div(10+x)$ and $UT=36\div(10-x)$

But given, $\frac{36}{(10-x)}-\frac{36}{(10+x)}=1.5$

$\Rightarrow 150-1.5x^2=72x$

$\Rightarrow x^2+48x-100=0$

$\Rightarrow x^2+50x-2x-100=0$

$\Rightarrow x(x+50)-2(x+50)=0$

$\Rightarrow x=-50$ and 2

Since speed of the current cannot be negative, the required answer is 2 miles/h.

Alternatively

The best way is to proceed through options. Let options be 2.5 miles/h, 2 miles/h......

(a) $2.5 \Rightarrow DT=36\div(10+2.5)=2.88$ h

and $UT=36\div(10-2.5)=4.8$ h

Here, *DT* is not 90 min. less than *UT*.

(b) $2 \Rightarrow DT=36\div(10+2)=3$ h

and $UT=36\div(10-2)=4.5$ h

Here, *DT* is 90 min. less than *UT*. Hence, option (b) is correct.

E. When Different Distances are Covered by a Boat in Different Modes (Upstream and Downstream)

Problem 7 A boat takes 9 h to move 44 km downstream and 25 km upstream. Also, it may take 7 h to move 55 km downstream and 10 km upstream. Find out the speed of the boat in still water and velocity of current ?

Sol. If you are asked to answer both parts of the question, the best way is to proceed through options. But this approach doesn't work when only one part is asked in the examination. Now develop your concept. Let the time taken to move 44 km downstream be *T* h.

	Down-stream distance	*DT*	Upstream distance	*UT*	Total time
1st Case	44 →	T	25→	$(9-T)$	9
2nd Case	55 →	$\frac{T}{44}\times 55$ i.e. $\frac{5T}{4}$	10→	$\frac{(9-T)}{25}\times 10$ i.e. $\frac{2(9-T)}{5}$	7

From the above chart, $=\frac{5T}{4}+\frac{2(9-T)}{5}=7$

$\Rightarrow 25T+72-8T=140$

$\Rightarrow T=4$ h

Now, $DS=44\div 4=11$ km/h

and $US=25\div(9-4)=5$ km/h

Then, speed of the boat in still water $=\frac{DS+US}{2}$

$=\frac{11+5}{2}=8$ km/h

$\therefore$ Speed of the stream $=11-8=3$ km/h

F. Computation of Speed of Current/Boat When Distance and Sum of *DT* and *UT* are Known

Problem 8 The speed of a boat in still water is 11 km/h. It can go 12 km upstream and return downstream to the engine point in 2 h 45 min. What is the speed of the stream ? *[CDS-10, MAT-08, 01, 99, SSC (CGL-T-II)-12]*

Sol. Let the speed of the stream be x km/h.

Then, $DT=\frac{12}{11+x}$ and $UT=\frac{12}{(11-x)}$

But, given $\frac{12}{(11+x)}+\frac{12}{(11-x)}=\frac{11}{4}$

$\Rightarrow 12\times 22\times 4=11(121-x^2)$

$\Rightarrow x^2=25$

$\therefore x=5$

Alternatively

You may proceed through options. Let the options be 4 km/h, 5 km/h, ...

(a) 4 km/h $\Rightarrow \frac{12}{(11+4)} + \frac{12}{(11-4)} \neq \frac{11}{4}$

(b) 5 km/h $\Rightarrow \frac{12}{(11+5)} + \frac{12}{(11-5)} = \frac{11}{4}$

Hence, option (b) is correct.

G. Comparison Between Two Boats With Respect to Their *UTs* and *DTs*

Problem 9 A faster boat takes half the time taken by a slower boat while they go downstream. When they go upstream, the faster boat takes $\frac{1}{4}$ of the time taken by the slower. Find the speed of the slower boat in still water provided the speed of the slower is 9 km/h less than that of the faster.

Sol. Let the speed of the faster and the slower boat in still water be f km/h and m km/h respectively. Also, the speed of the current be y km/h.

Given, $f = (m+9)$ km/h

Now, $\frac{DS \text{ of the faster}}{DS \text{ of the slower}} = \frac{2}{1}$

$\Rightarrow \frac{f+y}{m+y} = \frac{2}{1}$

$\Rightarrow \frac{m+9+y}{m+y} = \frac{2}{1}$

$\Rightarrow m+9+y = 2m+2y$

$\Rightarrow m+y = 9$

$\Rightarrow y = 9-m$

Further, $\frac{US \text{ of the faster}}{US \text{ of the slower}} = \frac{4}{1}$

$\Rightarrow \frac{f-y}{m-y} = \frac{4}{1}$

$\Rightarrow m+9-y = 4m-4y$

$\Rightarrow 3m-3y = 9$

$\Rightarrow m-y = 3$

$\Rightarrow m-(9-m) = 3$ $[\because y = 9-m]$

$\Rightarrow 2m = 12$

$\therefore m = 6$ km/h

H. Concept of Two Boats Proceeding Towards Each Other

Problem 10 Two boats A and B start towards each other from two places, 108 km apart. Speeds of the boats A and B in still water are 12 km/h and 15 km/h respectively. If A proceeds down and B up the stream, they will meet after ? ***[SSC (CGL-PT)-08]***

Sol. Let the speed of current be x km/h and the both boats meet after T h.

Now, DS of $A = (12+x)$ and US of $B = (15-x)$

As given, $(12+x)T + (15-x)T = 108$

$\Rightarrow 12T + xT + 15T - xT = 108 \Rightarrow T = 4$ h

Alternatively

In case of downstream, stream enhances the speed of boat by its speed and in case of upstream it reduces the speed of boat by its speed. It is also fact that speed of the current remains same in both cases. It means effect of stream is negated while the both proceed in opposite directions.

Thus, relative speed = 15 + 12 = 27 km/h

$\therefore$ Time taken to meet $= \frac{108}{27} = 4$ h

WORKOUT

1. The speed of a boat in still water is 15 km/h and the rate of current is 3 km/h. The distance travelled downstream (in km) in 12 min. is *[LIC (AAO)-09]*

(a) 3.6 (b) 2.4
(c) 1.2 (d) 1.8

2. Speed of a speed-boat when moving in the direction perpendicular to the direction of the current is 16 km/h, speed of the current is 3 km/h. So, the speed of the boat against the current will be

(a) 22 km/h (b) 9.5 km/h
(c) 10 km/h (d) None of these

3. A motor boat takes 2 h to travel a distance of 9 km down the current and it takes 6 h to travel the same distance against the current. What is the speed of the boat in still water in km/h ? *[CDS-10, MAT-08]*

(a) 3 (b) 2
(c) 1.5 (d) 1

4. A boat travels upstream from B to A and downstream from A to B in 3 h. If the speed of the boat in still water is 9 km/h and the speed of the current is 3 km/h, what is the distance between A and B? *[MAT-13, 11, 09, 08]*

(a) 6 km (b) 4 km
(c) 8 km (d) 12 km

5. A boat which has a speed of 5 km/h in still water crosses a river of width 1 km along the shortest possible path in 15 min. The velocity of the river water in km/h is : *[IAS (PT)-97]*

(a) 3.6 (b) 2.4
(c) 1.2 (d) 1

6. The speed of a boat in still water is 6 km/h and the speed of the stream is 1.5 km/h. A man rows to a place at a distance of 22.5 km and comes back to the starting point. Find the total time taken by him.

(a) 8 h (b) 4 h
(c) 7 h (d) 2 h

7. A person can row a boat d km upstream and the same distance downstream in $5\frac{1}{4}$ h. Also, he can row the boat $2d$ km upstream in 7 h. He will row the same distance downstream in *[LIC (AAO)-12]*

(a) $3\frac{1}{2}$ h (b) $3\frac{1}{4}$ h
(c) $4\frac{1}{4}$ h (d) 4 h

8. A man who can swim 48 m/min. in still water swims 200 m against the current and 200 m with the current. If the difference between those two times is 10 min, what is the speed of the current ? *[MAT-12, 08]*

(a) 30 m/min (b) 31 m/min
(c) 29 m/min (d) 32 m/min

9. A man can row against the current $\frac{3}{4}$ th of a km in 15 min and returns the same distance in 10 min. The ratio of his speed to that of the current is : *[SSC (CGL-T-II)-10]*

(a) 3 : 5 (b) 5 : 3
(c) 1 : 5 (d) 5 : 1

10. A man rowed against a stream flowing 1.5 km/h to a certain point and then turned back, stopping 2 km short of the place from where he originally started. If the whole time occupied in rowing be 2 h 10 min and his uniform speed in still water be 4.5 km/h, the man went up the stream a distance of *[MAT-10, 09]*

(a) 4 km (b) 7 km
(c) 8 km (d) 5 km

11. The time taken for covering 'X' km downstream is equal to that for covering '$x-18$' km upstream. Upstream speed is 6 km less than downstream speed. If the speed of the boat in still water is 15 km/h, what is value of X ? *[SBI (PO)-17]*

(a) 51 (b) 52
(c) 53 (d) 54
(e) 55

12. A motorboat covers a certain distance downstream in 6 h but takes 8 h to return upstream to the starting point. If the speed of the stream be 6 km/h, find the speed of the motor boat in still water.

(a) 41 km/h (b) 43 km/h
(c) 42 km/h (d) 44 km/h

13. A man can row 40 km upstream and 55 km downstream in 13 h. Also, he can row 30 km upstream and 44 km downstream in 10 h. Find the speed of the man in still water. *[MAT-11, 09, SSC (CGL-T-II)-11, SSC (CGL-T-II)-12, SSC (CGL-Mains) -2000, SSC (CGL-T-I)-17, 19]*

(a) 5 km/h (b) 2 km/h
(c) 4 km/h (d) None of these

14. A boat covers a certain distance downstream in 1 h while it comes back in 1.5 h. If the speed of the stream be 3 km/h, what is the speed of the boat in still water ?

(a) 11 km/h (b) 10 km/h
(c) 16 km/h (d) None of these

15. A boat, while going downstream in a river covered a distance of 50 miles at an average speed of 60 miles/h. While returning because of the water resistance, it took 1 h 15 min to cover the same distance. What was the average speed during the whole journey ?

(a) 40 miles/h (b) 48 miles/h
(c) 50 miles/h (d) 55 miles/h

16. A boatman goes 2 km against the current of the stream in 1 h and goes 1 km along the current in 10 min. How long will he take to go 5 km in stationary water ? *[MAT-10, 09, 07]*

(a) 1.5 h (b) 1.25 h
(c) 1 h (d) 40 min

17. A man rows down a river 18 km in 4 h with the stream and returns in 10 h. Consider the following statements

1. The speed of the man against the stream is 1.8 km/h.

2. The speed of the man in still water is 3.15 km/h.
3. The speed of the stream is 1.35 km/h

Which of the above statements are correct? *[CDS -17]*

(a) 1 and 2 only (b) 2 and 3 only
(c) 1 and 3 only (d) 1, 2 and 3

18. A man swimming in a stream which flows 1.5 km/h finds that in a given time he can swim twice as far with the stream as he can against it. At what rate does he swim?

(a) 4.5 km/h (b) 5.5 km/h
(c) 7.5 km/h (d) None of these

19. The time taken by a boat to travel 13 km downstream is the same as time taken by it to travel 7 km upstream. If the speed of the stream is 3 km/h, then how much time (in h) will it take to travel a distance of 44.8 km in still water? *[SSC (CGL-T-I)-19]*

(a) $4\frac{12}{25}$ (b) $5\frac{3}{5}$
(c) $5\frac{2}{5}$ (d) $4\frac{13}{25}$

20. The ratio of the time taken by a boat to cover x km downstream to the time taken by it to cover $(x-70)$ km upstream is 2 : 3 and the ratio of the speed of a boat in still water to the speed of current is 8 : 3. What is the value of x in km? *[SBI (PO-Pre.)-19]*

(a) 220 (b) 170
(c) 90 (d) 110
(e) None of these

21. A man travels by a motor boat down a river to his office and back. With the speed of the river unchanged, if he doubles the speed of his motor boat, then his total travel time gets reduced by 75%. The ratio of the original speed of the motor boat to the speed of the river is *[CAT-17]*

(a) $\sqrt{6}:\sqrt{2}$ (b) $\sqrt{7}:2$
(c) $2\sqrt{5}:3$ (d) 3 : 2

22. A boat went down the river for a distance of 20 km. It then turned back and returned to its starting point, having travelled a total for 7 h. On its return trip, at a distance of 12 km from the starting point, it encountered a log, which had passed the starting point at the moment at which the boat had started downstream. The downstream speed of the boat is : *[MAT-17]*

(a) 7 km/h (b) 10 km/h
(c) 13 km/h (d) 16 km/h

23. Two ports A and B are situated along the same side of a river, with A being located up-stream, 10 km from B. The speed of stream is 4 km/h. A boat, whose speed in still water is 10 km/h starts from B at 2 : 00 PM, to reach A. At A it takes 15 min for offloading, and then it starts its return journey. Another boat, whose speed in still water is 12 km/h is to be sent from B to A. At what time this 2nd boat should start from B, so that is reaches A exactly when the 1st boat has finished offloading? *[MAT-18]*

(a) 2 : 25 PM
(b) 2 : 40 PM
(c) 2 : 55 PM
(d) 3 : 05 PM

Answer Keys

1. (a)	2. (c)	3. (a)	4. (d)	5. (d)	6. (a)	7. (a)	8. (d)	9. (d)	10. (d)
11. (d)	12. (c)	13. (d)	14. (d)	15. (b)	16. (b)	17. (d)	18. (a)	19. (a)	20. (a)
21. (b)	22. (b)	23. (b)							

Hints and SOLUTIONS

1. Downstream speed $= 15 + 3 = 18\,\text{km/h}$

Then, distance covered in 12 min $= \left(\frac{18}{60} \times 12\right) = 3.6$ km

2. Speed of the speed-boat in still water $= 16 - 3 = 13\,\text{km/h}$

Then, speed of the speed-boat against the current

$= 13 - 3 = 10\,\text{km/h}$

3. Speed of boat in still water $= \frac{DS + US}{2} = \frac{(9 \div 2) + (9 \div 6)}{2}$

$= 3$ km/h

4. DS : $US = (9 + 3) : (9 - 3) = 2 : 1$

Then, $DT : UT = 1 : 2$

Clearly, DT is 1 h as $DT + UT = 3$h

Hence, the distance between A and B

$= DS \times DT = 12 \times 1 = 12$ km

Alternatively, You may proceed through options.

5. The speed of the boat taking into consideration the resistance of the velocity of the river water = 4 km/ h as the boat takes 15 min to cross 1 km.

But the speed of the boat in still water = 5 km/h

Hence, velocity of the river water $= 5 - 4 = 1$ km/h

6. Total time taken $= \frac{22.5}{(6 + 1.5)} + \frac{22.5}{(6 - 1.5)} = 3 + 5 = 8$ h

7. Given, the person can row $2d$ km upstream in 7 h.

Then, he can row d km upstream in $3\frac{1}{2}$ h.

Obviously, he can row d km downstream in $\left(5\frac{1}{4} - 3\frac{1}{2}\right)$ h

Then, he can row $2d$ km downstream in $2 \times \left(5\frac{1}{4} - 3\frac{1}{2}\right)$

i.e. in $3\frac{1}{2}$ h

8. Let the speed of the stream be x km/h.

Then, $DT = \frac{200}{(48 + x)}$ and $UT = \frac{200}{(48 - x)}$

But, given $\frac{200}{(48 - x)} - \frac{200}{(48 + x)} = 10$

$\Rightarrow\ 200 \times 2x = 10(2304 - x^2)$

$\Rightarrow\ x^2 + 40x - 2304 = 0$

$\Rightarrow\ x^2 + 72x - 32x - 2304 = 0$

$\Rightarrow\ x(x + 72) - 32(x + 72) = 0$

$\Rightarrow\ (x - 32)(x + 72) = 0 \Rightarrow x = 32$

Alternatively

The best way is to proceed through options.

9. US $= \frac{3}{4} \times 4 = 3$ km/h and $DS = \frac{3}{4} \times 6 = 4.5$ km/h

Speed of the man $= \frac{DS + US}{2} = \frac{4.5 + 3}{2}$ km/h

and speed of the current $= \frac{4.5 - 3}{2}$ km/h

Hence, required ratio $= \frac{4.5 + 3}{2} : \frac{4.5 - 3}{2}$

$= 7.5 : 1.5 = 5 : 1$

10. Let the distance covered downstream be d km.

As given, $\frac{d}{(4.5 - 1.5)} + \frac{(d - 2)}{(4.5 + 1.5)} = 2$ h 10 min

$\Rightarrow\ 2d + d - 2 = 13 \Rightarrow d = 5$ km

11. Let the downstream speed be d km/h.

Then, upstream speed $= (d - 6)$ km/h

Then, speed of the boat in the still water

$= \frac{d + (d - 6)}{2} = d - 3$

$\Rightarrow\ d - 3 = 15 \Rightarrow d = 18$

As given, $\frac{x}{\text{downstream speed}} = \frac{x - 18}{\text{upstream speed}}$

$\Rightarrow\ \frac{x}{18} = \frac{x - 18}{12} \Rightarrow x = 54$ km

12. We know, $\frac{DS}{US} = \frac{UT}{DT} \Rightarrow \frac{\text{Speed of the boat} + 6}{\text{Speed of the boat} - 6} = \frac{8}{6}$

$\Rightarrow$ Speed of the boat = 42 km/h

Alternatively

We know, if the UT is $\frac{m}{n}$ times DT, speed of the boat in still water will be equal to $\left[\text{speed of current} \times \frac{(m + n)}{(m - n)}\right]$.

Hence, speed of the boat in still water

$= 6 \times \frac{(8 + 6)}{(8 - 6)} = 42$ km/h

13. Let the time taken to moves 55 km downstream be T hrs.

	Down-stream distance	DT	Upstream distance	UT	Total time
1st Case	55 →	T	40 →	$(13 - T)$	13
2nd Case	44 →	$\frac{T}{55} \times 44$ i.e. $\frac{4T}{5}$	30 →	$\frac{(13 - T)}{40} \times 30$ i.e. $\frac{3(13 - T)}{4}$	10

From the above chart, $\Rightarrow \frac{4T}{5} + \frac{3(13 - T)}{4} = 10$

$\Rightarrow\ 16T + 195 - 15T = 200 \Rightarrow T = 5$ h

Now, $DS = 55 \div 5 = 11$ km/h

and $US = 40 \div (13 - 5) = 5$ km/h

Then, speed of the boat in still water

$= \frac{DS + US}{2} = \frac{11 + 5}{2} = 8$ km/h

14. We know, $\frac{DS}{US} = \frac{UT}{DT}$

$\Rightarrow \frac{\text{Speed of the boat} + 3}{\text{Speed of the boat} - 3} = \frac{1.5}{1}$

$\Rightarrow$ Speed of the boat = 15 km/h

15. Time taken for going downstream = (50 ÷ 60) h = 50 min and time taken for going upstream = 75 min.

Now, average speed = $\frac{\text{total distance covered}}{\text{total time taken}}$

$= \frac{50+50}{(50+75)} = 0.8$ miles/min

Now, 0.8 miles/min. = 0.8 × 60 = 48 miles/h

16. US = (2 ÷ 1) = 2 km/h and $DS = \left(1 \div \frac{1}{6}\right) = 6$ km/h

Then, speed of the boat in stationary water

$= \frac{2+6}{2} = 4$ km/h

Now, time taken to go 5 km = (5 ÷ 4) = 1.25 h

17. Downstream speed = 18 ÷ 4 = 4.5 km/h

and Upstream speed = 18 ÷ 10 = 1.8 km/h

Then, speed of the man in still water = $\frac{4.5+1.8}{2} = 3.15$ km/h

and speed of the stream = $\frac{4.5-1.8}{2} = 1.35$ km/h

Hence, option (d) is correct.

18. Clearly, the *DS* is two times the *US*.

[(speed of man in still water + 1.5) = 2 (speed of man in still water − 1.5)]

$\Rightarrow$ Speed of the man in still water = 1.5 + 3 = 4.5 km/h

19. Since time is constant, *DS* : *US* = 13 : 7

Let the speed of boat in still water be *x* km/h.

Then, $(x+3) : (x-3) = 13 : 7$

$\Rightarrow x = 10$ km/h

∴ Time taken to cover 44.8 km = 44.8 ÷ 10

$= 4.48 \text{ h} = 4\frac{12}{25}$ h

20. As given, $\frac{x}{(8+3)\text{ units}} : \frac{x-70}{(8-3)\text{ units}} = 2:3$

$\Rightarrow \frac{2(x-70)}{5} = \frac{3x}{11}$

$\Rightarrow x = 220$ km

21. Let the speed of the motor boat in the still water and that of river be *x* km/h and *y* km/h respectively.

Then, time taken to go office and back

$= \frac{1}{x+y} + \frac{1}{x-y}$, assuming he travels 1 km to go his office.

Now, $\frac{1}{2x+y} + \frac{1}{2x-y} = 25\%$ of $\left[\frac{1}{x+y} + \frac{1}{x-y}\right]$

$\Rightarrow \frac{4x}{4x^2 - y^2} = \frac{2x}{4(x^2 - y^2)}$

$\Rightarrow 8x^2 - 8y^2 = 4x^2 - y^2$

$\Rightarrow 4x^2 = 7y^2$

$\Rightarrow x : y = \sqrt{7} : 2$

Hence, option (b) is correct.

22. Let the speed of the boat in still water and speed of the current be *x* km/h and *y* km/h respectively.

Then, time taken by the boat to meet the log

$= \frac{20}{(x+y)} + \frac{20-12}{(x-y)}$

and time taken by the log to meet the boat = $\frac{12}{y}$

[∵ speed of the log = speed of the current]

Since time is constant,

$\frac{20}{(x+y)} + \frac{8}{(x-y)} = \frac{12}{y}$

$\Rightarrow \frac{20y}{(x+y)} + \frac{8y}{(x-y)} = 12$

$\Rightarrow 12(x^2 - y^2) = 20xy - 20y^2 + 8xy + 8y^2$

$\Rightarrow 12x^2 - 12y^2 = 28xy - 12y^2$

$\Rightarrow 12x^2 = 28xy$

$\Rightarrow 12x = 28y$

$\Rightarrow 3x = 7y$

$\Rightarrow x = 7y/3$

Further, as given, $\frac{20}{(x+y)} + \frac{20}{(x-y)} = 7$

$\Rightarrow \frac{20}{(7y/3+y)} + \frac{20}{(7y/3-y)} = 7$

[putting the value of *x*]

$\Rightarrow \frac{60}{10y} + \frac{60}{4y} = 7$

$\Rightarrow 120 + 300 = 140y$

$\Rightarrow y = 3$

Then, $x = \frac{7}{3} \times 3 = 7$

Thus, downstream speed of the boat = 7 + 3 = 10 km/h

23. Upstream speed of the slower boat = 10 − 4 = 6 km/h

Time taken by the slower boat to reach the point *B* = 10/6 h = 1 h 40 min

Hence, it will reach the point *B* at 3 : 40 PM and will finish the offloading at 3 : 55 PM

Now, upstream speed of the faster boat = 12 − 4 = 8 km/h

Then, time taken by the faster boat to reach the point *B* = 10/8 h = 75 min

Clearly, the faster boat should start from *B* 75 min before 3 : 55 PM i.e. at 2 : 40 PM.

38

Escalator

Introduction

The concepts applicable in case of Escalator are similar to those meant for 'Boat and Stream'. If we talk about the escalator, it is a moving stair which goes up or down continuously, generally at constant speed and the number of visible steps/ stairs is always same in both cases.

It is noticeable that when a person and the escalator move in the same direction, effective speed is the sum of their speeds while in case of opposite direction, the effective speed is the difference between their speeds. Recall, the concept of downstream and upstream speed in chapter 'Boat and stream'. Now, take an example to observe the similarity between the concepts.

"A man can walk up a moving 'up' escalator in 30 sec. The same man can walk down this moving 'up' escalator in 90 sec. Assume that his walking speed is same upwards and downwards. How much time he takes to walk up the escalator, when it is not moving?" ***[CAT-95]***

Let the man's own speed be x m/s and the speed of the escalator be y m/s.

Here, upwards speed or say, effective speed

$= \text{man's own speed} + \text{speed of the escalator} = (x + y)$ m/s

and downwards speed or say, effective speed

$= \text{man's own speed} - \text{speed of the escalator}$

$= (x - y)$ m/s

Now, if the length of the lift is 90 m, then $(x + y)$

$= (90 \div 30) = 3$ m/s and $(x - y) = (90 \div 90) = 1$ m/s

On solving, we get $x = 2$ m/s

Then, time required to walk up the escalator when it is not moving

$= (90 \div 2) = 45$ s

[other approaches for such types of problems have been detailed under Illustration]

Keep the following Points in Mind

(a) If a person and the escalator move in the same direction upward or downward, the number of steps climbed/descended by the person of his own, is always less than the total number of visible steps of the escalator because the escalator also pushes the person forward.

(b) If they are moving in opposite direction, the number of steps climbed/descended by the person of his own, is always greater than the total number of visible steps of the escalator because the escalator also pulls the person to the backward.

Inside

- Introduction
- Critical Facts to be Kept in Mind
- Cases Wherein the Directions of Movement of Person(s) and Escalator are Same
- Cases Wherein the Directions of Movement of Persons and Escalator are Opposite to Each Other
- ....
- .... & Many More
- Work Out with Solutions

(c) If two persons A and B walk up a moving 'up' escalator of n visible steps and take x and y steps of their own to reach on the top of the escalator, then ratio of the time taken by A and B to reach the top of escalator $= (n - x) : (n - y)$. Clearly, it is a case of same direction.

[Explanation: Let A's speed be m steps/sec, then time taken by A to move x steps $= \frac{x}{m}$ sec. In the same time, the escalator has moved $(n - x)$ steps. Then, time taken by the escalator to move $(n - y)$ steps $= \frac{x \times (n - y)}{m \times (n - x)}$ sec

But, in $\frac{x \times (n - y)}{m \times (n - x)}$ sec. B also reaches on the top, when the escalator pushes B forward by $(n - y)$ steps. Hence, required ratio

$$= \frac{x}{m} : \frac{x \times (n - y)}{m \times (n - x)} = (n - x) : (n - y)]$$

(d) If two persons A and B walk up a moving 'down' escalator of n visible steps and take x and y steps of their own to reach on the top of the escalator, then ratio of the time taken by A and B to reach the top of the escalator $= (x - n) : (y - n)$. Clearly, it is a case of opposite direction.

(e) Let A walk up a moving 'up' escalator of n visible steps and take x steps of his own to reach the top of the escalator whereas B runs against this escalator and reaches its bottom after having taken y steps of his own. Then, ratio of time taken by A and B to reach the top and the bottom of the escalator respectively $= (n - x) : (y - n)$. It is a case of combination of the same and opposite direction.

(f) The faster body takes more number of steps on his/her own than the slower body to reach the other end of the escalator, in case they are going towards the moving direction of the escalator. On the contrary, the faster body takes lesser number of steps than the slower body in case they are going opposite to moving direction of the escalator.

Illustrations

Case I When the directions of movement of person(s) and escalator are same

Problem 1 A and B walk up a moving 'up' escalator. A and B take 48 and 40 steps of their own respectively to reach the top of the escalator. What is the number of visible steps in the escalator if walking speeds of A and B are in ratio 3 : 2 ?

Sol. Let the speeds of B and the escalator be $2x$ and y steps/sec. respectively. Then, speed of $A = 3x$ steps/sec

Now, time taken by A to reach at the top of the escalator = time taken by A to take 48 steps of his own

$$= \frac{48}{3x} = \frac{16}{x} \text{ sec}$$

∴ Total number of visible steps = number of steps taken by A of his own + number of steps by which escalator pushes him forward in $\frac{16}{x}$ sec.

$$= 48 + \left(\frac{16}{x} \times y\right)$$

Similarly, in case of B, total number of visible steps

$$= 40 + \left(\frac{40}{2x} \times y\right)$$

Hence, $48 + \left(\frac{16}{x} \times y\right) = 40 + \left(\frac{40}{2x} \times y\right)$

$\Rightarrow \quad 8 = \frac{4y}{x}$

$\Rightarrow \quad \frac{y}{x} = 2$

∴ Total number of visible steps of the escalator

$$= \text{either } 48 + \left(\frac{16}{x} \times y\right)$$

Or

$$40 + \left(\frac{40}{2x} \times y\right) = 48 + (16 \times 2) \text{ or } 40 + (20 \times 2) = 80$$

Alternatively I

Steps taken by A and B of their own = 48 40

Ratio of their speed = 3 2

∴ Ratio of time taken by them to reach at top of the escalator $= (48 \div 3) : (40 \div 2) = 4 : 5$

[or you may say, $48 \times 2 : 40 \times 3$, ratio of product on cross multiplication]

Now, number of steps by which escalator pushed A forward $= 4y$

and number of steps by which escalator pushed B forward $= 5y$

[let speed of the escalator be y steps/sec]

Clearly, total number of visible steps

$= 48 + 4y = 40 + 5y$

$\Rightarrow \quad y = 8$

Then, total number of visible steps of the escalator $= 48 + 4 \times 8 = 80$

Alternatively II

Steps taken by A and B of their own = 48 40

Ratio of their speed = 3 2

∴ Ratio of time taken to reach at the top $= 48 \times 2 : 40 \times 3 = 4 : 5$

[You may also say, $(48 \div 3) : (40 \div 2)$]

Or, ratio of time taken by the escalator to push them forward by certain number of steps = 4 : 5. Obviously, the escalator has to push B forward by greater number of steps in comparison to A as B has taken fewer number of steps of his own compared to A. In fact, B has taken only $(48 - 40)$ steps fewer than A.

Hence, $(5 - 4)$ units $\Rightarrow (48 - 40)$ steps

∴ 4 units $\Rightarrow$ 32 steps

It implies that the escalator has pushed A forward by 32 steps. Then, total number of visible steps $= 48 + 32 = 80$

Alternatively, since 5 units correspond to 40, required number of steps $= 40 + 40 = 80$

Alternatively III

Let total number of visible steps of the escalator be x.

Then, ratio of time taken by A and B to reach at the top of the escalator $= (x - 48) : (x - 40)$

Hence, $(x - 48) : (x - 40) = 4 : 5$

$\Rightarrow \quad 4x - 160 = 5x - 240$

$\Rightarrow \quad x = 80$

Problem 2 **Shyama and Vyom walk up an escalator (moving stairway). The escalator moves at a constant speed. Shyama takes 3 steps for every 2 steps of Vyom's steps. Shyama gets to the top of the escalator after having taken 25 steps, while Vyom (because his slower pace lets the escalator do a little more of the work) takes only 20 steps to reach the top. If the escalator were turned off, how many steps would they have to take to walk up?** ***[CAT-01]***

Sol. Steps taken by Shyama and Vyom = 25 20

Ratio of their speed = 3 2

$\therefore$ Ratio of time taken to reach at the top $= 25 \times 2 : 20 \times 3 = 5 : 6$

Now, $(6 - 5)$ units $\Rightarrow (25 - 20)$ steps

$\therefore \quad 5$ units $\Rightarrow 25$ steps

$\therefore$ Total number of visible steps in the escalator $= 25 + 25 = 50$ steps

[Try with other approaches mentioned for pervious problem]

Problem 3 **Saransh uses the escalator to move from Metro platform to customer care centre. If he runs up 9 steps of an escalator, he reaches at the top of escalator in 41 sec. Had he run up 5 steps, he would have reached the top in 45 sec. What is the time taken by the escalator to bring Saransh at the top if he does not run up any step?**

Sol. Let number of visible steps in the escalator be n.

Clearly, $(n - 9)$ steps are taken by the escalator itself and takes 41 sec. and also to take $(n - 5)$ steps, it takes 45 sec.

Hence, $\dfrac{n-9}{n-5} = \dfrac{41}{45}$ [$\because$ speed of escalator is constant]

$\Rightarrow 45n - 405 = 41n - 205 \Rightarrow n = 50$

$\because$ Time taken to move $(50 - 9)$ steps = 41 sec

$\therefore$ Time taken to move 50 steps = 50 sec

Alternatively I

Let the speed of the escalator be x steps/sec

Then, total number of steps $= 9 + 41x = 5 + 45x$

$\Rightarrow \quad x = 1$

Then, total number of steps $= 9 + (41 \times 1) = 50$

$\therefore$ Required time $= 50 \times 1 = 50$ sec

Alternatively II

Clearly, the escalator takes $(45 - 41)$ sec to cover $(9 - 5)$ steps, hence, 1 sec. per step.

Then, number of steps in the escalator $= 9 + (41 \div 1) = 50$

or $\quad 5 + (45 \div 1) = 50$

Then, required time $= 50 \times 1 = 50$ sec

Case II **When the directions of movement of person(s) and escalator are opposite to each other.**

Problem 4 **Kranti walks up a moving 'up' escalator and takes 20 steps to reach the top. Had he walked down this moving 'up' escalator at a speed 3 times the earlier, he would have taken 36 steps to reach the bottom. What is the number of visible steps in the escalator?**

Sol. Let the speeds of Kranti and the escalator be x steps/sec and y steps/sec respectively.

Now, time taken by Kranti to reach at the top of the escalator = time taken by Kranti to take 20 steps of his own $= \dfrac{20}{x}$ sec.

$\therefore$ Total number of visible steps = number of steps taken by Kranti of his own + number of steps by which escalator pushes him forward in $\dfrac{20}{x}$ sec $= 20 + \left(\dfrac{20}{x} \times y\right)$

Similarly, time taken by Kranti to reach at the bottom of the upward escalator = time taken by Kranti to take 36 steps of his own $= \dfrac{36}{3x} = \dfrac{12}{x}$ sec

$\therefore$ Total number of visible steps = number of steps taken by Kranti of his own − number of steps by which escalator pulls him back in $\dfrac{12}{x}$ sec $= 36 - \left(\dfrac{12}{x} \times y\right)$

Hence, total number of visible steps

$$= 20 + \left(\frac{20}{x} \times y\right) = 36 - \left(\frac{12}{x} \times y\right)$$

$\Rightarrow \quad \dfrac{32y}{x} = 16 \Rightarrow \dfrac{y}{x} = \dfrac{1}{2}$

$\therefore$ Total number of visible steps in the escalator

$$= 20 + \left(\frac{20}{x} \times y\right) = 20 + \left(20 \times \frac{1}{2}\right) = 30$$

Alternatively I

Steps taken in upward and downward movement = 20 36

Ratio of their speed = 1 3

$\therefore$ Ratio of time taken to reach at the top and bottom $= 20 \times 3 : 36 \times 1 = 5 : 3$

Or, ratio of time taken by the escalator to push him forward and pull him back by certain number of steps = 5 : 3. Obviously, in $(5 + 3)$ units of time the escalator covers only $(36 - 20)$ steps.

Now, $(5 + 3)$ units $\Rightarrow (36 - 20)$ steps.

$\therefore$ 5 units $\Rightarrow$ 10 steps and 3 units = 6 steps

$\therefore$ Total number of visible steps in the escalator $= 20 + 10$ or $36 - 6 = 30$ steps

Alternatively II

From the above, ratio of time taken to reach at the top and bottom = 5 : 3

Let total number of visible steps of the escalator be x.

Then, ratio of time taken by him to reach at the top and the bottom of the escalator $= (x - 20) : (36 - x)$

Hence, $(x - 20) : (36 - x) = 5 : 3$

$\Rightarrow \quad 180 - 5x = 3x - 60 \Rightarrow x = 30$

Problem 5 Swati can walk up a moving 'up' escalator in 20 sec and walk down this moving 'up' escalator in 60 sec Assume that her walking speed is same upwards and downwards. How much time she take to walk up the escalator, when it is not moving ?

Sol. Let the Swati's own speed be x m/s and the speed of the escalator be y m/s.

Here, upwards speed or say, effective speed = Swati's own speed + speed of the escalator = $(x + y)$ m/s

and downwards speed or say, effective speed = Swati's own speed − speed of the escalator = $(x - y)$ m/s

Now, if the length of the lift is 60 m,

then $(x + y) = (60 \div 20) = 3$ m/s

and $(x - y) = (60 \div 60) = 1$ m/s

On solving, we get $x = 2$ m/s

Then, time required to walk up the escalator when it is not moving = $(60 \div 2) = 30$ s

Alternatively I

Let she takes x steps while going up. Then, number of steps she taken while coming back = $3x$ as ratio of time taken is 3 : 1.

Steps taken in upward and downward movement = x $3x$

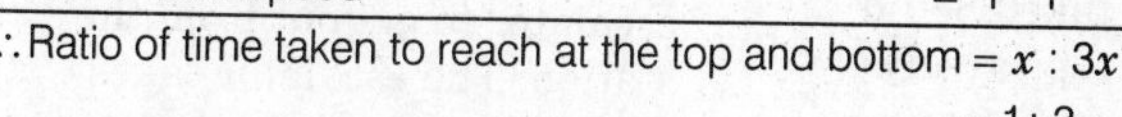

Ratio of their speed = 1 1

$\therefore$ Ratio of time taken to reach at the top and bottom = $x : 3x$

$= 1 : 3$

Now, (1 + 3) units $\Rightarrow (3x - x)$ steps

$\therefore$ 1 unit $\Rightarrow x/2$ steps

$\therefore$ Total number of visible steps in the escalator

$$= x + x/2 = \frac{3x}{2}$$

$\because$ Time taken to cover x steps is 30 sec

$\therefore$ Time taken to cover $\frac{3x}{2}$ steps = 45 sec

Or

$\because$ Time taken to cover $3x$ steps is 90 sec

$\therefore$ Time taken to cover $\frac{3x}{2}$ steps = 45 sec

Alternatively II

Let speeds of Swati and the escalator be x steps/sec and y steps/sec respectively.

Then, her upward speed = $(x + y)$ steps/sec and downward speed = $(x - y)$ steps/sec

Then, time taken by her in downward movement, when escalator not working and in upward movement respectively

$$= \frac{1}{x - y}, \frac{1}{x}$$

and $\frac{1}{x + y} = 90, \frac{1}{x}$ and 30

Recall the concept of Harmonic mean, HM of two quantities a and $b = \frac{2ab}{a + b}$

Hence, $\frac{1}{x} = \frac{2 \times 90 \times 30}{90 + 30} = 45$

$\therefore$ Required time taken = 45 sec

Problem 6 Julli is going up a moving 'up' escalator and takes 40 steps of her own to reach the top while Pintu is going down this escalator and takes 100 steps of his own to reach the other end. If they start to move simultaneously, then, after how many steps will Julli meet Pintu provided speeds of Julli and Pintu are in ratio 4 : 5.

Sol. Steps taken by Julli and Pintu of their own = 40 100

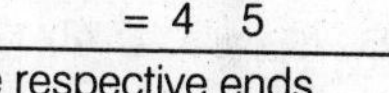

Ratio of their speed = 4 5

$\therefore$ Ratio of time taken to reach at the respective ends

$= 40 \times 5 : 100 \times 4 = 1 : 2$

Hence, (1 + 2) units $\Rightarrow$ (100 − 40) steps

$\therefore$ 1 unit $\Rightarrow$ 20 steps

Hence, total number of steps in the escalator = 40 + 20 = 60

Let speeds of Julli and escalator be 40 steps/sec and 20 steps /sec respectively. Then, speed of Pintu = 50 steps/sec as ratio of their speeds is 4 : 5.

Then, effective speed of Julli = (40 + 20) = 60 steps/sec

and effective speed of Pintu = (50 − 20) = 30 steps/sec

Hence, relative speed = 60 + 30 = 90 steps/sec

Then, time taken to meet = 60 ÷ 90 = 2/3 sec

$\therefore$ Steps moved by Julli till meeting time = 60 × 2 / 3 = 40 steps

Alternatively

$\because$ Ratio of time taken to reach at the respective ends = 1 : 2

$\therefore$ Ratio of effective speeds of Julli and Pintu = 2 : 1

Then, till meeting number of steps moved by Julli = number of steps in the escalator i.e. $60 \times \frac{2}{2 + 1} = 40$ steps, which also contains the steps covered by the escalator.

WORKOUT

1. Sajal is going up a moving 'up' escalator and takes 40 steps to reach the top while Swati is going down this escalator. Then, how many steps are there in the escalator if the speeds of Sajal and Swati are in ratio 2 : 3 and they take same time to reach their respective ends?

(a) 50 (b) 60
(c) 80 (d) 56

2. With reference to above question, what is the difference between number of steps taken by them of their own till they meet each other?

(a) 20 (b) 10
(c) 05 (d) 12

3. Ashok uses an escalator to move from ground floor to the 1st floor of a building. If he runs up 7 steps of an escalator, he reaches at the top of escalator in 16.5 sec. Had he run up 4 steps, he would have reached the top in 18 sec. What is the time (in sec.) taken by the escalator to bring Ashok at the top if he does not run up any step ?

(a) 20 (b) 30
(c) 10 (d) 05

4. Shishupal and Shivansh walk up an escalator (moving stairway). Shishupal takes 5 steps for every 3 steps of Shivansh's steps. Shishupal gets to the top of the escalator after having taken 40 steps, while Shivansh takes only 30 steps to reach the top. If the escalator were turned off, how many steps would they have to take to walk up?

(a) 50 (b) 60
(c) 80 (d) 56

5. Ganu and Chandrabhushan walk up a moving 'up' escalator. Ganu and Chandrabhushan take 28 and 24 steps of their own respectively to reach the top of the escalator. What is the number of visible steps in the escalator if walking speeds of Ganu and Chandrabhushan are in ratio 4 : 3?

(a) 50
(b) 60
(c) 80
(d) 56

6. Ritish can go up a moving 'up' escalator by taking 80 steps of his own to reach the top while Ganu can go down this escalator by taking 140 steps of his own to reach the other end. If Ganu starts to move after Ritish moves 9 steps including number of steps pushed by the escalator, then after how many steps will Ritish meet Ganu provided speeds of Ritish and Ganu are in ratio 2: 7

(a) 26 (b) 36
(c) 46 (d) 56

7. You travel by Delhi Metro every day from Botanical Graden, Noida to Hauz Khas, Delhi. At Hauz Khas Metro Station, you use an escalator to get out from the station. The escalator takes 80 sec. One day, escalator was not working and you walk up the escalator in 50 sec. How many minutes does it approximately take you to walk down the working escalator?

(a) 1.5 min (b) 2.2 min
(c) 2.8 min (d) 2.6 min

Answer Keys

1. (a)	2. (b)	3. (a)	4. (c)	5. (d)	6. (c)	7. (b)			

Hints and SOLUTIONS

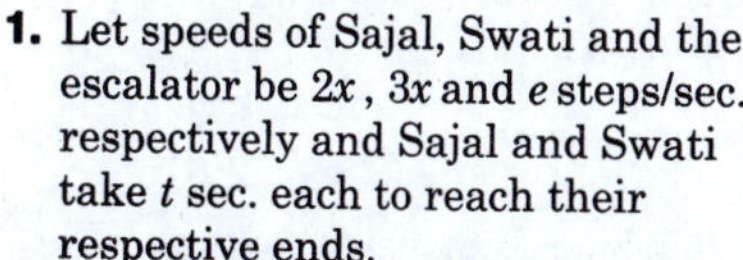

1. Let speeds of Sajal, Swati and the escalator be $2x$, $3x$ and e steps/sec. respectively and Sajal and Swati take t sec. each to reach their respective ends.

Then, total number of steps in the escalator

$$= (2x + e)t = (3x - e)t \Rightarrow x = 2e$$

$$\Rightarrow \quad 2x = 4e$$

Or

Since they have taken equal time, their effective speed will be equal.

Hence, $(2x + e) = (3x - e) \Rightarrow 2x = 4e$

It implies that the ratio of speeds of Sajal and Escalator = 4 : 1

Since time is constant for all three, 4 units $\Rightarrow$ 40 steps

$\therefore$ (1+ 4) units $\Rightarrow$ 50

$\therefore$ Total number of steps in the escalator = 50

Alternatively I

Ratio of speeds of Sajal and Swati = 2 : 3

Then, steps to be covered by them will be in ratio 2 : 3 as time is constant.

As given, 2 units $\Rightarrow$ 40 steps

$\therefore$ 3 units $\Rightarrow$ 60 steps

Now, ratio of time required by Sajal and Swati to reach their respective ends = $(n - 40) : (60 - n)$, if there are n steps in escalator.

$\Rightarrow$ $(n - 40) : (60 - n) = 1 : 1 \Rightarrow n = 50$

Alternatively II

Steps covered by Sajal and Swati = 40 60

Ratio of their speeds = 2 3

$\therefore$ Ratio of time taken to reach at the top and bottom = $40 \times 3 : 60 \times 2 = 1 : 1$

$\because$ (1+ 1) units $\Rightarrow$ (60 − 40) steps

$\therefore$ 1 unit $\Rightarrow$ 10 steps

Hence, number of steps in the escalator = 40 + 10 or 60 − 10 = 50

2. Since their effective speeds are equal, they will meet at the mid of the escalator i.e. at the distance of 25 steps from either end. Further, there is a difference of (60 − 40) steps to reach the respective ends. Then, difference between number of steps taken by them to reach the mid point of the escalator = 1/2 of (60 − 40) = 10 steps

Note Sajal and Swati will take 20 and 30 steps of their own respectively to meet at the mid point of the length of the escalator.

3. Let the speed of the escalator be x steps/sec.

Then, total number of steps

$$= 7 + 16.5x = 4 + 18x \Rightarrow x = 2$$

Then, total number of steps

$$= 7 + (16.5 \times 2) = 40$$

$\therefore$ Required time = $40 \times 1/2 = 20$ sec

4. Steps taken by Shishupal and Shivansh = 40 30

Ratio of their speeds = 5 3

$\therefore$ Ratio of time taken to reach at the top = $40 \times 3 : 30 \times 5 = 4 : 5$

Now, (5 − 4) units $\Rightarrow$ (40 − 30) steps

$\therefore$ 4 units $\Rightarrow$ 40 steps

$\therefore$ Total number of visible steps in the escalator = 40 + 40 = 80 steps

5. Steps taken by Ganu and Chandrabhushan of their own = 28 24

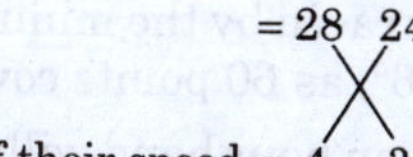

Ratio of their speed = 4 3

$\therefore$ Ratio of time taken by them to reach at top of the escalator = $(28 \div 4) : (24 \div 3) = 7 : 8$

Hence, (8 − 7) unit $\Rightarrow$ (28 − 24) steps

$\therefore$ 7 units $\Rightarrow$ 28 steps

Then, number of steps in the escalator = 28 + 28 = 56

6. Steps taken by Ritish and Ganu of their own = 80 140

Ratio of their speeds = 2 7

$\therefore$ Ratio of time taken to reach at the respective ends = 2 : 1

Hence, (2 + 1) units $\Rightarrow$ (140 − 80) steps

$\therefore$ 2 units $\Rightarrow$ 40 steps

Hence, total number of steps in the escalator = 80 + 40 = 120 steps

When Ritish takes 80 steps, the escalator pushed him forward by 40 steps. Thus, ratio of speeds of Ritish and escalator = 2 : 1

Let speeds of Ritish and escalator be $2x$ steps/sec and x steps/sec.

Then, speed of Ganu = $7x$

Now, effective speed of Ritish = $(2x + x) = 3x$ steps/sec and effective speed of Ganu = $(7x - x)$

$$= 6x \text{ steps/sec}$$

When Ganu starts to move, there is a gap of 111 steps between them, which is to be bridged by them in ratio $3x : 6x$ i.e. 1 : 2

$\therefore$ Steps covered by Ritish beyond his initial 9 steps = $111 \times \frac{1}{1+2} = 37$ steps

Hence, Ritish will meet Ganu after (9 + 37) = 46 steps.

Alternatively

From the above, ratio of time taken to reach at the respective ends by Ritish and Ganu = 2 : 1

$\therefore$ Ratio of effective speeds of Ritish and Ganu = 1: 2

$\therefore$ Steps covered by Ritish beyond his initial 9 steps = $111 \times \frac{1}{1+2} = 37$ steps

Hence, Ritish will meet Ganu after (9 + 37) = 46 steps.

7. Perusal of the question reveals that you are not walking up a moving 'up' escalator. In other words, you are standing on a moving 'up' escalator. ie. you, per se, are not taking any step.

Let number of steps in escalator be x

Then, speed of escalator

$$= \frac{x}{80} \text{ steps/sec}$$

and your speed = $\frac{x}{50}$ steps/sec

Your effective speed while walking down the working escalator

$$= \left(\frac{x}{50} - \frac{x}{80}\right) \text{ steps/sec}$$

$$\therefore \text{Required time} = \frac{x}{\frac{x}{50} - \frac{x}{80}} = \frac{400}{3} \text{ sec}$$

$$\approx 2.2 \text{ min}$$

Alternatively

Let number of steps in escalator be 400 [i.e. LCM of 50 and 80].

Then, speed of escalator will be 400/80 [i.e 5 steps/sec] and your speed will be 400/50 (i.e. 8 steps/sec).

Then, required time

$$= \frac{400}{8-5} \text{ sec} \approx 2.2 \text{ min}$$

39

Clock

Introduction

This chapter is critical for the aspirants appearing in UPSC, SSC, Bank-PO, MAT, Railway, Armed Forces-exams etc. Unique approaches have been adopted to solve the problems.

Facts

(i) The face (or dial) of a clock/watch is a circle whose circumference is divided into 60 min spaces. 1 min space may be referred as the distance between any two consecutive points.

(ii) In every hour, the minute hand aka long hand covers 60 min spaces i.e. 60 points whereas an hour hand (i.e. short hand) covers only 5 points. It means the minute hand gains 55 min (i.e. 55 points) on the hour hand per hour.

(iii) Angle traced by the hour hand in 12 h = angle traced by the minute hand in 1 h = 360°. It implies that each point carries 6°, as 60 points cover 360°.

(iv) If a minute hand moves 60 min. (or 60 points), an hour hand will move only 5 points ($= 5 \times 6° = 30°$). It implies that an hour hand moves $\left(\frac{1}{2}\right)^{\circ}$ every minute.

A. Measurement of Angles Between Hands of a Clock

Case I When the hour hand is ahead of the minute hand

Suppose you have to determine the angle between the hour hand and the minute hand at 6:15.

It may be assumed that the hour hand is at figure 6 exactly. Indeed, it will be ahead of 6. Then, number of points between the hour hand and the minute hand $= (6 \times 5 - 15) = 15$ points

$\therefore$ Angle between them $= 15 \times 6° = 90°$ [$\because$ each point carries 6°]

It is also fact that an hour hand moves $\left(\frac{1}{2}\right)^{\circ}$ every minute. Hence, instead of being at figure 6 exactly, the hour hand will move $\left(\frac{15}{2}\right)^{\circ}$ in 15 min.

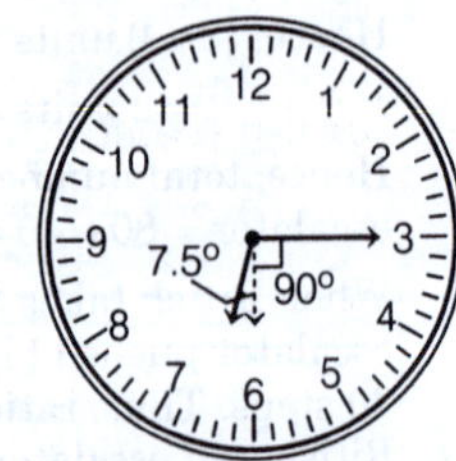

Thus, the required angle $= 90° + \left(\frac{15}{2}\right)^{\circ} = 97.5°$

Clearly, the minor angle between them is 97.5° and the major angle is (360° − 97.5°) i.e. 262.5°.

Inside

- *Introduction*
- *Critical Facts*
- *Measurement of Angles Between Hands of a Clock*
- *Determination of Time in Case Hands are in a Particular Position*
- *Number of Specific Events in a Particular Period*
- *Concept of Coinciding/ Overtaking, too Fast/Too Slow and Strokes and Intervals*
- *Reflection of Time in Mirror*
- ....
- .... *& Many More*
- *Work Out with Solutions*

The above approach may be elaborated as under

Given time $= \dfrac{6:15}{H\ M}$

Then, the required angle $= (H \cdot 5 - M)\,6 + \dfrac{M}{2}$

$= 30H - 6M + \dfrac{M}{2} = \left(30H - \dfrac{11M}{2}\right)$

***Case* II When the minute hand is ahead of the hour hand**

Suppose you have to determine the angle between hands at 3:35.

Let the hour hand is at figure 3, but it will be ahead of 3 in real. Then, number of points between the hour hand and the minute hand $= (35 - 3 \times 5) = 20$ points

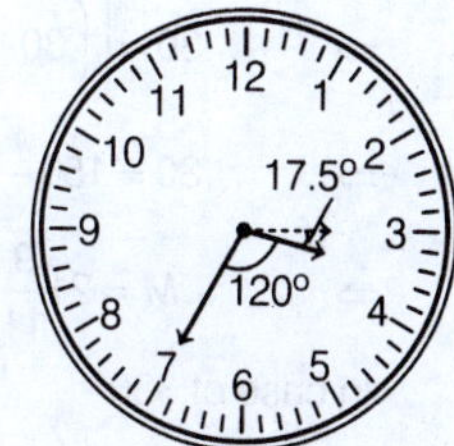

$\therefore$ Angle between them

$= 20 \times 6° = 120°$

[$\because$ each point carries 6°]

It is also fact that an hour hand moves $\left(\dfrac{1}{2}\right)^\circ$ every minute.

Hence, instead of being at figure 3 exactly, the hour hand will move $\left(\dfrac{35}{2}\right)^\circ$ in 35 min.

Thus, required angle $= 120° - \left(\dfrac{35}{2}\right)^\circ = 102.5°$

[Clearly, the minor angle is 102.5° and major angle is (360° − 102.5°) i.e. 257.5°.]

The above approach may be written as under

Given time $= \dfrac{3:35}{H\ M}$

Then, the required angle $= (M - H \cdot 5)6 - \dfrac{M}{2}$

$= 6M - 30H - \dfrac{M}{2} = -\left(30H - \dfrac{11M}{2}\right)$

Conclusion It may be construed from the above results that the angle between the hour hand and the minute hand is $\left|\left(30H - \dfrac{11M}{2}\right)\right|$.

Problem 1 Find the angle between the hour hand and the minute hand of a clock at 9.06.

Sol. The required angle $= \left|\left(30H - \dfrac{11M}{2}\right)\right|$

$= \left|30 \times 9 - \dfrac{11 \times 6}{2}\right| = 237°$

Hence, major angle = 237° and minor angle = 360°−237° = 123°

Problem 2 At 2:40, the hour hand and the minute hand of a clock form an angle of ?

Sol. The required angle $= \left|\left(30H - \dfrac{11M}{2}\right)\right|$

$= \left|30 \times 2 - \dfrac{11 \times 40}{2}\right| = 160°$

Hence, the minor angle is 160° and the major angle is 200°.

B. Determining Time When Hands are in a Particular Position.

Problem 3 At what time, in minutes, between 3 o' clock and 4 o' clock, both the hands will coincide each other ? ***[RRB-02, CDS-21]***

(a) $5\dfrac{1}{11}$ min (b) $12\dfrac{4}{11}$ min

(c) $13\dfrac{4}{11}$ min (d) $16\dfrac{4}{11}$ min

Sol. Needless to say that the angle between both the hands is 0° when they coincide each other.

Further, given time = 3 : M

Then, the angle between both hands $= \left|\left(30H - \dfrac{11M}{2}\right)\right|$

$\Rightarrow \quad 0 = \left|\left(30 \cdot 3 - \dfrac{11 \cdot M}{2}\right)\right|$

$\Rightarrow \quad \dfrac{11 \cdot M}{2} = 90 \Rightarrow M = 16\dfrac{4}{11}$

Hence, at $16\dfrac{4}{11}$ min. past 3 o' clock, both the hands will coincide.

Problem 4 At what time between 7 o' clock and 8 o' clock will the hands of a clock be in the same straight line but, not together ? ***[LIC (AAO)-03 IIFT-19]***

(a) 5 min past 7 (b) $5\dfrac{2}{11}$ min past 7

(c) $5\dfrac{3}{11}$ min past 7 (d) $5\dfrac{5}{11}$ min past 7

Sol. Clearly, the angle between both hands = 180°

$\Rightarrow \quad 180 = \left|\left(30H - \dfrac{11M}{2}\right)\right|$

$\Rightarrow \quad 180 = \left|\left(30 \cdot 7 - \dfrac{11M}{2}\right)\right|$ [$\because$ given time is 7 : M]

$\Rightarrow \quad 180 = 30 \times 7 - \dfrac{11M}{2}$ and $180 = -30 \cdot 7 + \dfrac{11M}{2}$

$\Rightarrow \quad M = 5\dfrac{5}{11}$ and $70\dfrac{10}{11}$

Hence, answer is $5\dfrac{5}{11}$ min. past 7 o' clock. If we consider the value of M as $70\dfrac{10}{11}$, the time will be $70\dfrac{10}{11}$ min. past 7 o' clock i.e. $10\dfrac{10}{11}$ min. past 8 o' clock. But we have to determine the time between 7 o' clock and 8 o' clock. Therefore, option (d) is correct.

Problem 5 **At what time between 5 o' clock and 6 o' clock will the hands of a clock be at right angle?**

Sol. Clearly, the angle between both hands = either 90° or 270°

In case of 90°

$$\Rightarrow \quad 90 = \left|\left(30H - \frac{11M}{2}\right)\right|$$

$$\Rightarrow \quad 90 = \left|\left(30\cdot 5 - \frac{11M}{2}\right)\right| \quad [\because \text{given time is } 5 : M]$$

$$\Rightarrow \quad 90 = 30\cdot 5 - \frac{11M}{2}$$

$$\text{and} \quad 90 = -30\cdot 5 + \frac{11M}{2}$$

$$M = 10\frac{10}{11} \text{ and } 43\frac{7}{11}$$

In case of 270°

$$\Rightarrow \quad 270 = \left|\left(30H - \frac{11M}{2}\right)\right|$$

$$\Rightarrow \quad 270 = \left|\left(30\cdot 5 - \frac{11M}{2}\right)\right| \quad [\because \text{given time is } 5 : M]$$

$$\Rightarrow \quad 270 = 30\cdot 5 - \frac{11M}{2} \text{ and } 270 = -30\cdot 5 + \frac{11M}{2}$$

$$\Rightarrow \quad M = -21\frac{9}{11} \text{ and } 76\frac{4}{11}$$

Obviously, the value of minutes can neither be negative nor be more than 60. Hence, at $10\frac{10}{11}$ min. past 5 o' clock and at $43\frac{7}{11}$ min. past 5 o' clock, both the hands will be at right angle.

Problem 6 **At what time between 12 o' clock and 1 o' clock both the hands be at right angle?**

Sol. Clearly, the angle between both hands = either 90° or 270°

In case of 90°

$$\Rightarrow \quad 90 = \left|\left(30H - \frac{11M}{2}\right)\right|$$

$$\Rightarrow \quad 90 = \left|\left(30\cdot 12 - \frac{11M}{2}\right)\right|$$

$[\because$ given time is 12 : M]

$$\Rightarrow \quad 90 = 360 - \frac{11M}{2}$$

$$\text{and} \quad 90 = -360 + \frac{11M}{2}$$

$$\Rightarrow \quad M = 49\frac{1}{11} \text{ and } 81\frac{9}{11}$$

In case of 270°

$$\Rightarrow \quad 270 = \left|\left(30H - \frac{11M}{2}\right)\right|$$

$$\Rightarrow \quad 270 = \left|\left(30\cdot 12 - \frac{11M}{2}\right)\right|$$

$[\because$ given time is 12 : M]

$$\Rightarrow \quad 270 = 360 - \frac{11M}{2}$$

$$\text{and} \quad 270 = -360 + \frac{11M}{2}$$

$$\Rightarrow \quad M = 16\frac{4}{11} \text{ and } 114\frac{6}{11}$$

Obviously, the value of minutes can neither be negative nor be more than 60. Hence, at $49\frac{1}{11}$ min. past 12 o' clock and at $16\frac{4}{11}$ min. past 12 o' clock, both the hands will be at right angle.

Problem 7 **At what time between 5 o' clock and 6 o' clock are the hour hand and the minute hand of a clock 5 min apart?**

Sol. Here, 5 min = 5 points = 5 × 6° = 30°

Clearly, the angle between both hands = either 30° or 330°

In case of 30°

$$\Rightarrow \quad 30 = \left|\left(30H - \frac{11M}{2}\right)\right|$$

$$\Rightarrow \quad 30 = \left|\left(30\cdot 5 - \frac{11M}{2}\right)\right| \quad [\because \text{given time is } 5 : \text{M}]$$

$$\Rightarrow \quad 30 = 150 - \frac{11M}{2} \text{ and } 30 = -150 + \frac{11M}{2}$$

$$\Rightarrow \quad M = 21\frac{9}{11} \text{ and } 32\frac{8}{11}$$

In case of 330°

$$\Rightarrow \quad 330 = \left|\left(30H - \frac{11M}{2}\right)\right|$$

$$\Rightarrow \quad 330 = \left|\left(30\cdot 5 - \frac{11M}{2}\right)\right| \quad [\because \text{given time is } 5 : M]$$

$$\Rightarrow \quad 330 = 150 - \frac{11M}{2} \text{ and } 330 = -150 + \frac{11M}{2}$$

$$\Rightarrow \quad M = -32\frac{8}{11} \text{ and } 87\frac{3}{11}$$

Obviously, the value of minutes can neither be negative nor be more than 60. Hence, at $32\frac{8}{11}$ min past 5 o' clock and at $21\frac{9}{11}$ min past 5 o' clock both the hands will be 5 min apart

Problem 8 **At what time between 5 o' clock and 6 o' clock clock will both hands of a clock be equidistant from the figure 6?**

Sol. Let both hands be at distance of x points from the figure 6.

Now, angle between both hands

$$= \left|\left(30\cdot 5 - \frac{11(30+x)}{2}\right)\right|$$

$[\because$ time is 5 : $(30 + x)]$

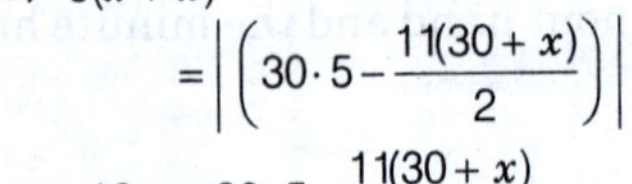

$$\Rightarrow 6(x + x) = \left|\left(30\cdot 5 - \frac{11(30+x)}{2}\right)\right|$$

$$\Rightarrow \quad 12x = 30\cdot 5 - \frac{11(30+x)}{2}$$

$$\text{and } 12x = -30\cdot 5 + \frac{11(30+x)}{2}$$

$$\Rightarrow \quad 24x = 300 - 330 - 11x$$

$$\text{and } 24x = -300 + 330 + 11x$$

$$\Rightarrow \quad x = \frac{-6}{7} \text{ and } 2\frac{4}{13}$$

Fig. 6

Hence, the required time is $5 : \left(30 + 2\frac{4}{13}\right)$ i.e. $32\frac{4}{13}$ min. past 5 o' clock.

Alternatively

Let both hands be at distance of x points from the figure 6.

Clearly, the minute hand has moved $(30+x)$ min. (or points).

We know that when the minute hand moves 60 points, the hour hand moves only 5 points.

It means the hour hand will move only $\left\{\frac{5}{60}\times(30+x)\right\}$ points from the figure 5 when the minute hand moves $(30+x)$ points.

Obviously, $\left\{\frac{5}{60}\times(30+x)\right\}$ points + x points = 5 points

$$\Rightarrow \quad 150+5x+60x=300$$

$$\Rightarrow \quad 65x=150$$

$$\Rightarrow \quad x=\frac{150}{65}=2\frac{4}{13}$$

It means, at $\left(30+2\frac{4}{13}\right)$ min past 5 o' clock, both the hands will be equidistant from the figure 6.

Generic Version

At what time between $(H-1)$ and H o' clock, will both the hands of a clock be equidistant from the latter figure.

Clearly, number of points upto figure $H=5H$

Let both hands be at distance of x points from the figure H.

Obviously, the hour hand will move $\frac{5}{60}(5H+x)$ points, when the minute hand moves $(5H+x)$ points.

Now, $\quad \frac{5}{60}(5H+x)+x=5$

$$\Rightarrow \quad 5H+x+12x=60$$

$$\Rightarrow \quad x=\frac{60-5H}{13}$$

$\therefore$ Required time $=5H+\left(\frac{60-5H}{13}\right)$ min past $(H-1)$ o' clock.

i.e. at $\left\{\frac{60}{13}(H+1)\right\}$ min past $(H-1)$ o' clock, both hands will be equidistant from the figure H.

Problem 9 **Shivansh went out between 2 o' clock and 3 o' clock and returning home between 6 o' clock and 7 o' clock he noticed that the position of the hour hand and the minute hand have been mutually changed. What is the time of his arrival at home ?**

Sol. Clearly, the hour hand is between 2 and 3 and the minute hand is between 6 and 7 while he departures. Further, the hour hand is between 6 and 7 and the minute hand is between 2 and 3 when he returns.

Let at m min past 2 o' clock he departs.

Then, angle between both hands $=\left|\left(30\cdot 2-\frac{11m}{2}\right)\right|$

Also, the hour hand will move $\frac{5m}{60}$ points from figure 2 when the minute hand moves m min. The reason is that when the minute hand moves 60 points, the hour hand moves only 5 points.

At time of his returning, the minute hand is at distance of $\left\{(2\times 5)+\frac{5m}{60}\right\}$ points.

Then, angle between hands $=\left|\left\{30\cdot 6-\frac{11\left(2\times 5+\frac{5m}{60}\right)}{2}\right\}\right|$

Since the positions of hands are interchanged, there will be no effect on angles between them.

$$\therefore \quad \left|\left(30\cdot 2-\frac{11m}{2}\right)\right|=\left|\left\{30\cdot 6-\frac{11\left(2\times 5+\frac{5m}{60}\right)}{2}\right\}\right|$$

$$\Rightarrow \quad \left(30\cdot 2-\frac{11m}{2}\right)=\left\{30\cdot 6-\frac{11\left(2\times 5+\frac{5m}{60}\right)}{2}\right\}$$

Or

$$\left(30\cdot 2-\frac{11m}{2}\right)=-\left\{30\cdot 6-\frac{11\left(2\times 5+\frac{5m}{60}\right)}{2}\right\}$$

$$\Rightarrow \quad 120-11m=360-110-\frac{55m}{60}$$

$$\textit{Or} \quad 120-11m=-360+110+\frac{55m}{60}$$

$$\Rightarrow \quad m=-\frac{1560}{121} \text{ or } \frac{4440}{143}$$

$$\Rightarrow \quad m=\frac{4440}{143} \quad [\because m \text{ cannot be negative}]$$

Hence, his returning time $=\left\{(2\times 5)+\frac{5m}{60}\right\}$ min past 6 o' clock.

$$=\left\{10+\frac{5\cdot\frac{4440}{143}}{60}\right\} \text{min past 6 o' clock}$$

$$=12\frac{84}{143} \text{ min past 6 o' clock}$$

Alternatively I Departure time $=2:m$

Then, number of points between hands $=m-\left(10+\frac{5m}{60}\right)$

[$\because$ when minute hand moves 60 points, hour hand moves only 5 points]

Arrival time $=6:\left(10+\frac{5m}{60}\right)$

Then, number of points between hands

$$=\left\{30+\frac{5}{60}\left(10+\frac{5m}{60}\right)\right\}-\left(10+\frac{5m}{60}\right)$$

Obviously, number of points between them will remain the same.

$$\therefore m-\left(10+\frac{5m}{60}\right)=\left\{30+\frac{5}{60}\left(10+\frac{5m}{60}\right)\right\}-\left(10+\frac{5m}{60}\right)$$

$$\Rightarrow \quad m=\frac{4440}{143}$$

Now, his returning time $=6:\left(10+\frac{5m}{60}\right)$

$$\Rightarrow \quad 6:\left\{10+\frac{5\cdot\frac{4440}{143}}{60}\right\}=6:12\frac{84}{143}$$

Alternatively II

You may proceed through options also.

Let options be $6 : 12\frac{84}{143}, 6 : 14\frac{84}{143}$, ...

(a) $6 : 12\frac{84}{143} \Rightarrow$ number of points between hands

$$= \left\{6 \cdot 5 + \frac{12\frac{84}{143}}{12}\right\} - 12\frac{84}{143} = 18.4 \text{ (approx.)}$$

During departure time, the hour hand will be exactly at $2\frac{84}{143}$ points from figure 2. Accordingly, minute hand will move $2\frac{84}{143} \times 12$ points.

Then, number of points between them

$$= 2\frac{84}{143} \times 12 - 12\frac{84}{143} = 18.4 \text{ (approx).}$$

Hence, option (a) is correct as number of points are same in both cases.

[Note : You may also proceed through angle approach]

C. Number of Specific Events in a Particular Period

(a) In every hour, both the hands coincide once. **e.g.** at $1 : 5\frac{5}{11}$, at $2 : 10\frac{10}{11}$, at $3 : 16\frac{4}{11}$ ......, at 11 : 60, at 12 : 00.

It is noticeable that both 11 : 60 and 12 : 00 are same. It implies that in every 12 h, both the hands coincide 11 times only.

(b) In every hour, both the hands be in the same straight line (but not together) once. **e.g.** at $1 : 38\frac{2}{11}$, at $2 : 43\frac{7}{11}$, at $3 : 49\frac{1}{11}$ ......, at 5 : 60, at 6 : 00 , ... at $12 : 32\frac{8}{11}$.

Since, 5 : 60 and 6 : 00 are same, such events occur only 11 times in every 12 h.

(c) In every hour, both the hands be at right angles twice. **e.g.** at $1 : 21\frac{9}{11}$ and $1 : 54\frac{6}{11}$, at $2 : 27\frac{3}{11}$, at 2 : 60, at 3 : 00, at $3 : 32\frac{8}{11}$, ..., at $8 : 27\frac{3}{11}$, at 8 : 60, ... at 9 : 00, at $9 : 32\frac{8}{11}$ and so on. Here, 2 : 60, 3 : 00 are same.

Also, 8 : 60 and 9 : 00 are same. It implies that in every 12 h both the hands are at right angles 22 times.

Problem 10 How many times are the hands of a clock at right angle in a day?

Sol. $\because$ In 12 h $\rightarrow$ 22 times

$\therefore$ In 24 h $\rightarrow$ 44 times

Problem 11 How many times in a day, are the hands of a clock in straight line but opposite in direction? ***[RRB-03]***

Sol. $\because$ In 12 h $\rightarrow$ 11 times

$\therefore$ In 24 h $\rightarrow$ 22 times

D. Concept of Coinciding/Overtaking in Case a Clock Gains/Loses

Take one example to understand this concept comprehensively.

The minute hand of a clock overtakes the hour hand at intervals of 62 min of the correct time. How much a day does the clock gain or lose ?

Or

If the hands of a clock coincide every 62 min, how much a day does the clock gain or lose ?

Recall the time when both the hands coincide. The time be $1 : 5\frac{5}{11}, 2 : 10\frac{10}{11}, 3 : 16\frac{4}{11}$ .... and so on.

It is observed that both hands coincide at interval of every $65\frac{5}{11}$ min.

In other words, after every $65\frac{5}{11}$ min the hands will meet or will be together or will overtake.

As per question, they meet only after 62 min. It implies that the clock is gaining $\left(65\frac{5}{11} - 62\right)$ i.e. $3\frac{5}{11}$ min in every 62 min.

$$\therefore \text{In 24 h, time gained} = \frac{3\frac{5}{11}}{62} \times 24 \times 60 \text{ min}$$

$$= 80\frac{80}{341} \text{ min}$$

Generic version

Gain or Loss/min. = $\left\{\dfrac{65\frac{5}{11} - \text{time taken for overtaking (in min).}}{\text{time taken for overtaking (in min).}}\right\}$

Note *If resultant is '+', clock is gaining and in case of '–', clock is losing.*

Problem 12 The minute hand of a clock overtakes the hour hand at intervals of 70 min of the correct time. How much a day the clock gain or lose ?

Sol. Gain or loss / min. $= \left(\dfrac{65\frac{5}{11} - 70}{70}\right)$ min

Then, gain or loss in one day $= \dfrac{-4\frac{6}{11}}{70} \times 24 \times 60$

$$= -93\frac{39}{77} \text{ min}$$

'–' indicates that the clock is losing.

Problem 13 A clock is gaining 16 min a day. At what intervals do both the hands coincide?

Sol. We know,

$$\text{Gain or loss / min} = \left\{\frac{65\frac{5}{11} - \text{time taken for coinciding}}{\text{time taken for coinciding}}\right\}\text{min}$$

$$\Rightarrow \quad (+)\frac{16}{24\times 60} = \left(\frac{65\frac{5}{11} - x}{x}\right)$$

[let they coincide at interval of x min.]

$$\Rightarrow \quad 16x = 24\times 60\times \frac{720}{11} - 1440x$$

$$\Rightarrow \quad x = \frac{24\times 60\times 720}{1456\times 11}$$

$$= 64.74 \text{ min (approx).}$$

E. Concept of Too Fast/Too Slow

If a clock indicates 4 : 20, when the correct time is 4, it is said to be 20 min. too fast. On the contrary, if it indicates 3 : 40, when the correct time is 4, it is said to be 20 min too slow.

Problem 14 A clock is set right at 10:00 AM and it gains 10 min in 12 h. What will be the correct time when it indicates 10 : 30 PM on the following day?

[SSC (CGL-T-I)-11, CSAT-19]

Sol. Time from 10:00 AM on a day to 10 : 30 PM on the following day = $36\frac{1}{2}$ h.

Time run by the faulty clock	Time run by the accurate clock
12 h 10 min	12 h
$12\frac{10}{60}$ h = $\frac{73}{6}$ h	12 h
1 h	$12\times\frac{6}{73}$
$36\frac{1}{2}$ h	$12\times\frac{6}{73}\times\frac{73}{2} = 36$ h

Hence, the correct time is 36 h beyond 10 : 00 AM i.e. 10:00 PM on the following day.

Alternatively

It may also be construed from the above that the correct time is $\left(36\frac{1}{2} - 36\right)$ i.e. $\frac{1}{2}$ h less than the incorrect time. Then, the correct time = 30 min. less than 10 : 30 PM = 10:00 PM on the following day.

Problem 15 A clock is set right at 8 : 30 AM and it loses 20 min in 24 h. What will be the true time if this clock indicates 7 : 00 PM on 5th day.

Sol. Time from 8 : 30 AM of a day to 7 : 00 PM on 5th day = $106\frac{1}{2}$ h

Time run by the faulty clock	Time run by the accurate clock
23 h 40 min	24 h
$23\frac{40}{60}$ h = $\frac{71}{6}$ h	24 h
1 h	$24\times\frac{3}{71}$
$106\frac{1}{2}$ h	$24\times\frac{3}{71}\times\frac{213}{2} = 108$ h

Hence, the correct time is 108 h beyond 8 : 30 AM i.e. 8 : 30 PM on the 5th day.

Alternatively

Since the correct time is $(108 - 106\frac{1}{2})$ i.e. $1\frac{1}{2}$ h greater than the incorrect time, the correct time will be $1\frac{1}{2}$ h beyond the 7 : 00 PM on the 5th day. Hence, the correct time is 8 : 30 PM on the 5th day.

Problem 16 A watch which gains uniformly, is 2 min slow at noon on Monday and is 4 min 48 sec fast at 2:00 PM on the following Monday. When was it correct?

Sol. Time from noon on Monday to 2:00 PM on the following Monday = 7 days 2 h = 170 h

Clearly, the clock gains $\left(2 + 4\frac{48}{60}\right)$ min i.e. $\frac{34}{5}$ min in 170 h

Then, it will gain 2 min in $170\times\frac{5}{34}\times 2$ h i.e. 50 h

It implies that the watch is correct at 50 h beyond noon on Monday i.e. at 2:00 PM on Wednesday.

Problem 17 A clock gains 5 min every h. What will be the angle traversed by the second hand in 1 min?

[SSC (CGL)-T-1-11]

Sol. The clock gains 5 min every hour i.e. 5 sec every min. It implies that the second hand will move 65 points instead of 60 points in a minute.

Then, the angle traversed by it = 65 × 6 = 390°

Problem 18 A clock goes slow from midnight by 5 min at the end of the Ist h, by 10 min at the end of the 2nd h, by 15 min at the end of the 3rd h and so on. What will be the time by this clock after 6 h?

Sol. Clearly, the clock will go 30 min slow in 6 h as the sequence is 5 min (1st h), 10 min (2nd h), 15 min (3rd h), 20 min (4th h), 25 min (5th h), 30 min (6th h), ...

Hence, time shown by clock after 6 h = 6 : 00 − 00 : 30
= 5 : 30 AM

Problem 19 Kranti observed that at 9:00 PM on Saturday, one clock is 1 min slow and another is 2 min fast whereas he had noticed earlier that at 5:00 PM on preceding Tuesday the 1st was 3 min fast and the 2nd was 4 min. slow. At what time and day did they agree and what time were shown by them at that moment?

Sol. The time from 5:00 PM on Tuesday to 9:00 PM on Saturday = 100 h

In 100 h, the 1st clock loses 3 + 1 i.e. 4 min. and in 100 h the 2nd clock gains (4 + 2) i.e. 6 min. in the same period.

It implies that the 2nd clock gains (4 + 6) i.e. 10 min. over the 1st clock in 100 h. It is also fact that the 2nd was (3 + 4) i.e. 7 min behind the 1st one on 5:00 PM on Tuesday.

$\because$ 2nd clock gains 10 minutes in 100 h

$\therefore$ It will gain 7 minutes in 70 hours.

Then, 70 hours after 5:00 PM on Tuesday = 3:00 PM on Friday.

Hence, both the clocks agreed at 3:00 PM on Friday.

Further, it is fact that the 1st clock loses 4 min in 100 h. Then it will lose 2 min 48 sec in 70 h.

It means it is still 12 sec fast as it was 3 min fast on 5:00 PM on Tuesday.

Hence, the time both clocks indicated when they agreed = 12 sec past 3:00 PM on Friday.

Alternatively

Time gained by the 2nd clock in 70 h = 4 min 12 sec as it gains 6 min in 100 h. Since the 2nd clock was only 4 min slow at 5:00 PM on Tuesday, it is now 12 sec fast at 3:00 PM on Friday.

Hence, the time both clocks indicated when they agreed = 12 sec past 3:00 PM on Friday.

F. Concept of Strokes and Intervals

In old days, a certain types of clocks were used to mark every hour by striking a number of times equal to the hour i.e. at 1 o' clock 1 stroke, at 2 o' clock-2 strokes, at 3 o' clock-3 strokes, at 4 o' clock-4 strokes and so on.

At 6 o' clock: 1st stroke + short interval + 2nd stroke + short interval + 3rd stroke + short interval + 4th stroke + short interval + 5th stroke + short interval + 6th stroke.

Clearly, there would be 6 strokes and 5 intervals at 6 o' clock.

Similarly, at 9 o' clock: 9 strokes + 8 intervals

At 10 o' clock : 10 strokes + 9 intervals.

Problem 20 **If a certain clock takes 18 sec to strike 5 o' clock, how many sec will it take to strike 12 o' clock provided the time required for a stroke is exactly equal to the time-interval between any two consecutive strokes.**

Sol. Clearly, at 5 o' clock, 5 strokes + 4 intervals i.e. 9 spans

[$\because$ duration of sound of each stroke (bong) = time interval between any two consecutive strokes]

Similarly, at 12 o' clock, 12 strokes + 11 intervals i.e. 23 spans

Given, 9 spans = 18 sec

$\therefore$ 23 spans = 46 sec

Hence, the required time = 46 sec

Problem 21 **If a clock takes 7 sec to strike 7 o' clock, how long will it take to strike 11 o' clock provided duration of bong is negligible.**

Sol. At 7 o' clock, 7 strokes + 6 intervals i.e. 6 spans

[$\because$ duration of bong is negligible]

At 11 o' clock, 11 strokes + 10 intervals i.e. 10 spans

Given, 6 spans = 7 sec

$\therefore$ 10 spans $= \frac{7}{6} \times 10$

$= 11\frac{2}{3}$ sec

Problem 22 **Two clocks start to strike 10 o' clock simultaneously, one takes 23 sec and another does 17.25 sec. What is the difference between time taken by the 1st to strike 4th stroke and that taken by the 2nd to to strike 6th stroke. Assume, the duration of the sound of each stroke is $\frac{1}{4}$ th of the time-interval between any two consecutive strokes.**

Sol. At 10 o' clock, 10 strokes + 9 intervals i.e. $\left(\frac{10}{4} + 9\right)$ spans

[$\because$ duration of bong $= \frac{1}{4}$ of interval]

For the 1st clock,

$\because$ $\left(\frac{10}{4} + 9\right)$ spans = 23 sec

$\therefore$ 1 span = 2 sec

Now, time taken by the 1st clock to strike 4th stroke

$= 2 \times \left(\frac{4}{4} + 3\right)$

= 8 sec

For the 2nd clock,

$\because$ $\left(\frac{10}{4} + 9\right)$ spans = 17.25 sec

$\therefore$ 1 span = 1.5 sec

Now, time taken by the 2nd clock to strike 6th stroke

$= 1.5 \times \left(\frac{6}{4} + 5\right) = 9.75$ sec

Hence, required difference = 9.75 − 8 = 1.75 sec

G. Reflection of Time in Mirror

Case I **When a mirror is kept in front/left side/right side/above/below the clock**

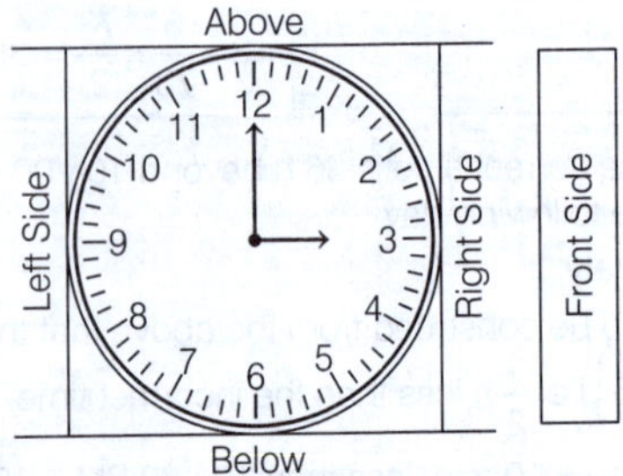

Time reflected in mirror = 12 : 00 − times shown by the clock

Or

Time shown by the clock = 12 : 00 − time reflected in the mirror,

Problem 23 A mirror is placed in front of the clock showing time 5 : 20. What is the time reflected in the mirror ?

Sol. Time reflected in the mirror = 12 : 00 – 5 : 20
= 6 : 40

Problem 24 A mirror is kept below a clock. The time reflected in mirror is 10 : 50. What is the actual time ?

Sol. Actual time or time shown by the clock
= 12 : 00 – 10 : 50
= 1 : 10

Case II **When a clock is kept upside-down in front of a mirror.**

Problem 25 A clock with only dots marking 3, 6, 9 and 12 position, has been kept upside-down in front of a mirror. A person reads the time in the reflection of the clock at 4 : 50. What is the actual time? **[SSC (Matric-level)-02]**

Sol.

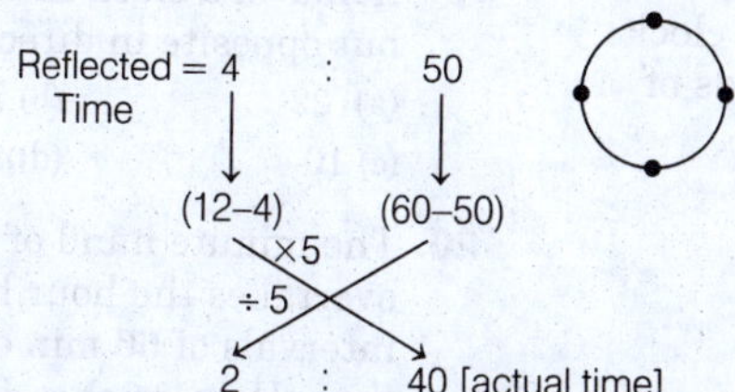

Hence, the actual time or time shown by the clock = 2 : 40

Generic version

If the time reflected in the mirror is $H : M$, then the time shown by clock or Actual time = $\left(\frac{60-M}{5}\right)$: $(12 - H)5$. This may be depicted as below

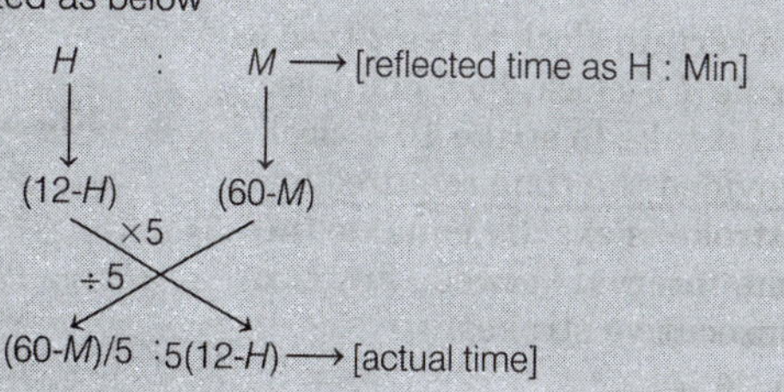

Note *The same concept is applicable to determine the reflected time if actual time is given.*

Problem 26 A clock with only dots marking 3, 6, 9 and 12 positions, has been kept upside – down in front of a mirror. If the actual time is 12 : 30, what is the reflected time?

Sol.

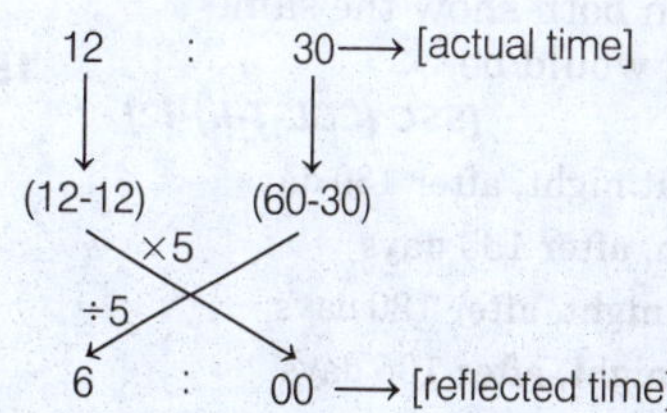

Hence, the time reflected in the mirror = 6 : 00

WORKOUT

1. An accurate clock shows 8 o' clock in the morning. Through how many degrees will the hour hand rotate when the clock shows 2 o' clock in the afternoon? **[IAS (Pre.)-2000]**

(a) 144° (b) 150° (c) 160° (d) 180°

2. At what angle the hands of a clock are inclined at 15 min past 5 ? **[LIC (AAO)-03]**

(a) $58\frac{1}{2}°$ (b) 64°
(c) $67\frac{1}{2}°$ (d) $72\frac{1}{2}°$

3. The reflex angle between the hands of a clock at 10:25 is : **[SCRA-96]**

(a) 180° (b) $192\frac{1}{2}°$
(c) 195° (d) $197\frac{1}{2}°$

4. Find the angle between hour hand and the minute hand at 12 : 20.

(a) 144° (b) 110°
(c) 160° (d) 180°

5. At what time between 12 o' clock and 1 o' clock will the hands of a clock be in straight line but opposite in direction.

(a) $32\frac{8}{11}$ min past 12
(b) $38\frac{2}{11}$ min past 12
(c) $36\frac{8}{11}$ min past 12
(d) $5\frac{5}{11}$ min past 12

6. If a clock started at noon, then the angle turned by hour hand at 3:45 PM is **[SSC (CGL-T-I)-15]**

(a) $117\frac{1}{2}°$ (b) $104\frac{1}{2}°$
(c) $97\frac{1}{2}°$ (d) $112\frac{1}{2}°$

7. A clock goes fast by 1 min during the 1st h, by 2 min at the end of the 2nd h, by 4 min at the end of 3rd h, by 8 min at the end of 4th h and so on. At the end of which hour, will it be fast by just over 60 min? **[SSC (Matirc PT)-02]**

(a) 5th (b) 6th
(c) 7th (d) 8th

8. Two watches are set right at 8:00 AM. One gains 10 sec and another loses 20 sec in a day. After how much time will the 1st watch have gained 4 min on the another and what time will they show at that moment ?

(a) 8 days, 8:01:20 AM, 7:57:20 AM
(b) 8 days, 8:01:30 AM, 7:57:20 AM
(c) 8 days, 8:01:20 AM, 7:57:40 AM
(d) 8 days, 8:01:25 AM, 7:57:40 AM

9. If a certain clock takes 26 sec to strike 7 o' clock, how many sec will it take to strike 10 o' clock provided the time required for a stroke is exactly equal to the time-interval between any two consecutive strokes.
(a) 36 sec
(b) 33.33 sec
(c) 38 sec
(d) None of the above

10. A walk clock gains 2 min in 12 h while a table clock loses 2 min every 36 h. Both are set right at 12 noon on Tuesday. The correct time when both show the same time next would be *[SSC (CGL-T-II)-12]*
(a) 12:30 at night, after 130 days
(b) 12 noon, after 135 days
(c) 1:30 at night, after 130 days
(d) 12 midnight, after 135 days

11. If a clock strikes appropriate number of times at each hour, how many times will it strike a day ? *[SSC (CGL-T-II)-13]*
(a) 300 (b) 156
(c) 68 (d) 78

12. The angle between the minute hand and hour hand of a clock when the time is 7 : 20 is equal to: *[SSC (CGL-T-I)-15, CMAT-19]*
(a) 45° (b) 90°
(c) 100° (d) 120°

13. A clock with only dots marking 3, 6, 9 and 12 position, has been kept upside-down in front of a mirror. If the actual time is 7 : 30, what is the reflected time ?
(a) 7 : 25 (b) 6 : 25
(c) 4 : 30 (d) 6 : 00

14. At what time between 2 o' clock and 3 o' clock, both the hands will coincide each other ?
(a) $2:10\frac{10}{11}$ (b) $2:12\frac{4}{11}$
(c) $2:13\frac{4}{11}$ (d) $2:16\frac{4}{11}$

15. At what time between 1 o' clock and 2 o' clock will the hands of a clock be at right angle ?
(a) $1:21\frac{9}{11}$ and $1:54\frac{6}{11}$
(b) $1:41\frac{9}{11}$ and $1:24\frac{6}{11}$
(c) $1:21\frac{6}{11}$ and $1:54\frac{9}{11}$
(d) $1:22\frac{8}{11}$ and $1:54\frac{6}{11}$

16. At what time between 3 o' clock and 4 o' clock will both hands of a clock be equidistant from the figure 4.
(a) $3:21\frac{9}{11}$ (b) $3:24\frac{9}{11}$
(c) $3:23\frac{1}{13}$ (d) $3:2\frac{8}{11}$

17. A mirror is kept in front of a clock. The time reflected in mirror is 8 : 40. What is the actual time ?
(a) 4 : 20 (b) 4 : 40 (c) 4 : 45 (d) 3 : 20

18. A clock is set right 8 : 30 AM and it loses 10 min in 12 h. What will be the true time if this clock indicates 8 : 00 PM on next day?
(a) 8 : 30 PM, next day
(b) 7 : 30 PM, next day
(c) 8 : 00 PM, next day
(d) 8 : 40 PM, next day

19. How many times in 12 h, are the hands of a clock in straight line but opposite in direction?
(a) 22 (b) 24
(c) 10 (d) 11

20. The minute hand of a clock overtakes the hour hand at intervals of 66 min of the correct time. How much a day the clock gain or lose?
(a) $11\frac{109}{121}$ min gain (b) $11\frac{109}{121}$ min loss
(c) $12\frac{13}{121}$ min gain (d) $12\frac{13}{121}$ min loss

Answer Keys

1. (d)	2. (c)	3. (d)	4. (b)	5. (a)	6. (d)	7. (c)	8. (a)	9. (c)	10. (b)
11. (b)	12. (c)	13. (b)	14. (a)	15. (a)	16. (c)	17. (d)	18. (a)	19. (d)	20. (b)

Hints and SOLUTIONS

1. Clearly, the hour hand moves 30 points

$\therefore$ Angle traced by it $= 30 \times 6° = 180°$

2. Required angle $= \left|\left(30H - \frac{11M}{2}\right)\right|$

$= \left|30 \times 5 - \frac{11 \times 15}{2}\right| = 67\frac{1}{2}°$

3. Angle between hands

$= \left|\left(30H - \frac{11M}{2}\right)\right|$

$= \left|30 \times 10 - \frac{11 \times 25}{2}\right| = 162\frac{1}{2}°$

Hence, the reflex angle

$= 360° - 162\frac{1}{2}° = 197\frac{1}{2}°$

4. Required angle $= \left|\left(30H - \frac{11M}{2}\right)\right|$

$= \left|30 \times 12 - \frac{11 \times 20}{2}\right| = 250°$

Hence, major angle between both hands = 250°

and minor angle between them

$= 360° - 250° = 110°$

5. Clearly, the angle between both hands = 180°

$\Rightarrow \quad 180 = \left|\left(30H - \frac{11M}{2}\right)\right|$

$\Rightarrow \quad 180 = \left|\left(30 \times 12 - \frac{11M}{2}\right)\right|$

[$\because$ given time is 12 : M]

$\Rightarrow \quad 180 = 30 \times 12 - \frac{11M}{2}$

and $\quad 180 = -30 \times 12 + \frac{11M}{2}$

$\Rightarrow \quad M = 32\frac{8}{11}$ and $M = 98\frac{2}{11}$

Hence, answer is $32\frac{8}{11}$ min past 12 o' clock. If we consider the value of M as $98\frac{2}{11}$, the time will be $98\frac{2}{11}$ min past 12 o' clock i.e. $38\frac{2}{11}$ min past 1 o' clock. But we have to determine the time between 12 and 1 o' clock. Therefore, option (a) is correct.

6. The angle turned by the hour hand in one hour = 30°

[$\because$ in 12 h the hour hand makes 360°]

$\therefore$ Angle turned by the hour hand in $3\frac{3}{4}$ h $= 30° \times \frac{15}{4} = 112\frac{1}{2}°$.

7. Given sequence be 1 min (1st h), 2 min (2nd h), 4 min (3rd h), 8 min (4th h), 16 min (5th h), 32 min (6th h), 64 min (7th h)........

Clearly, answer is 7th h.

8. Clearly, 1st watch is gaining (10 + 20) i.e. 30 sec over the 2nd watch in a day.

$\therefore$ It will gain 4 min in 8 days.

[Answer of 1st part]

Further, in 8 days the 1st watch will gain 10×8 i.e. 80 sec and the 2nd watch will lose 20 × 8 i.e. 160 sec.

Now, time after 8 days from 8:00 AM on a day = 8:00 AM at the end of 8 days = correct time.

Then, 1st watch will show 80 sec beyond 8:00 AM i.e. 8:01: 20 AM and 2nd watch will show 160 sec. less than 8:00 AM i.e. 7:57:20 AM

[Answer of 2nd part]

Hence, option (a) is correct.

9. Clearly at 7 o' clock, 7 strokes +6 intervals i.e. 13 spans

[$\because$ duration of sound of each stroke (bong) = time interval between any two consecutive strokes]

Similarly, at 10 o' clock, 10 strokes + 9 intervals i.e. 19 spans

Given, 13 spans = 26 sec

$\therefore \quad$ 19 spans = 38 sec

Hence, the required time = 38 sec

10. As given, the wall clock gains 6 min in 36 h whereas the table clock loses 2 min in 36 h. Hence, the wall clock is gaining 8 min over the table clock every 36 h.

It is obvious that both clocks will show the same time only when the wall clock gains either 12 h or 24 h (or multiple of 12) over the table clock.

Now, the required time when the wall clock gains 12 h over the table clock $= \frac{36}{8} \times 12 \times 60$ h

[$\because$ 8 min are gained in 36 h]

= 3240 h = 135 days

Then, 135 days after 12 noon on Tuesday = 12 noon after 135 days

Hence, option (b) is correct.

Note *After 135 days, the wall clock will show 9:00 PM whereas the table clock will show 9:00 AM, hence the same time. But, the correct time be 12 noon.*

11. In 1st 12 h, number of strokes = 1+ 2+ 3+..............+ 12 = 78

[$\because$ clock strikes 1 time at 1 o' clock, 2 times at 2 o' clock, 3 times at 3 o' clock and so on]

In next 12 h, number of strokes = 78

Hence, total number of strokes

= 78+ 78 = 156

12. Angle between hands

$= \left|\left(30H - \frac{11M}{2}\right)\right| = \left|30 \cdot 7 - \frac{11 \cdot 20}{2}\right| = 100°$

13. Reflected time $= \frac{(60 - M)}{5} : (12 - H)\,5$

$= \frac{(60 - 30)}{5} : (12 - 7)5$

$= 6 : 25$

14. Needless to say that the angle between both the hands is 0° when they coincide each other.

Further, given time = 2 : M

Then, the angle between both hands

$= \left|\left(30H - \frac{11M}{2}\right)\right|$

$\Rightarrow \quad 0 = \left|30 \cdot 2 - \frac{11 \cdot M}{2}\right|$

$\Rightarrow \quad \frac{11 \cdot M}{2} = 60 \Rightarrow M = 10\frac{10}{11}$

Hence, at $10\frac{10}{11}$ min. past 2 o' clock, both the hands will coincide.

15. Clearly, the angle between both hands = either 90° or 270°

In case of 90°

$\Rightarrow \quad 90 = \left|\left(30H - \frac{11M}{2}\right)\right|$

$\Rightarrow \quad 90 = \left|\left(30 \times 1 - \frac{11M}{2}\right)\right|$

[$\because$ given time is 1 : M]

$\Rightarrow \quad 90 = 30 - \frac{11M}{2}$ and

$$90 = -30 + \frac{11M}{2}$$

$\Rightarrow \quad M = -10\frac{10}{11}$ and $21\frac{9}{11}$

In case of 270°

$$\Rightarrow 270 = \left|\left(30H - \frac{11M}{2}\right)\right|$$

$$\Rightarrow 270 = \left|\left(30 \times 1 - \frac{11M}{2}\right)\right|$$

[∵ given time is 1 : M]

$$\Rightarrow 270 = 30 - \frac{11M}{2}$$

and $\quad 270 = -30 + \frac{11M}{2}$

$\Rightarrow \quad M = -43\frac{7}{11}$ and $54\frac{6}{11}$

Obviously, the value of minutes cannot be negative. Hence, at $21\frac{9}{11}$ min past 1 o' clock and at $54\frac{6}{11}$ min past 1 o' clock both the hands will at right angle.

16. We know, at $\left\{\frac{60}{13}(H+1)\right\}$ min past $(H-1)$ o' clock, both hands be equidistant from the figure H.

Hence, required time $= \left\{\frac{60}{13}(4+1)\right\}$ min past 3 o' clock $= 23\frac{1}{13}$ min past 3 o' clock.

17. Actual time or time shown by the clock = 12:00 − 8:40 = 3:20

18. Time from 8:30 AM of day to 8:00 PM of the next day $= 35\frac{1}{2}$ h

Time run by the faulty clock	Time run by the accurate clock
11 h 50 min	12 h
$11\frac{50}{60}$ h $= \frac{71}{6}$ h	12 h
1 h	$12 \times \frac{6}{71}$
$35\frac{1}{2}$ h	$12 \times \frac{6}{71} \times 35\frac{1}{2} = 36$ h

Since the correct time is $\left(36 - 35\frac{1}{2}\right)$ i.e $\frac{1}{2}$ h greater than the incorrect time, the correct time will be $\frac{1}{2}$ h beyond 8 : 00 PM of the next day. Hence, the correct time is 8 : 30 PM on the next day.

19. We know, the hands of a clock are in straight line but opposite in direction 11 times in 12 h.

20. Gain or loss / min $= \left(\frac{65\frac{5}{11} - 66}{66}\right)$ min

Then, gain or loss in one day

$$= \left(\frac{65\frac{5}{11} - 66}{66}\right) \times 24 \times 60$$

$= -11\frac{109}{121}$ min '−' indicates that the clock is losing.

Unit 6

Mensuration

A. Two Dimensional (Plane) Figures

- Rectangles and Square
- Triangle
- Quadrilateral
- Circle

B. Three Dimensional Figures

- Cuboid and Cube
- Cylinder and Cone
- Sphere and Hemisphere
- Prism and Pyramid

Mensuration

Mensuration is a branch of mathematics that deals with measuring, calculating and estimating length of lines, areas of surfaces and volumes of solids. Its knowledge is of immense use to the surveyors, architects and engineers. For sake of convenience this section has been divided into two segments namely, plane figures and solid figures.

Plane figures refer to two-dimensional figures i.e. the figures having only two components namely length and breadth and these occupy a certain surface. Rectangle, Square, Triangle, Quadrilateral, Circle are termed as two-dimensional figures whereas Cube, Cuboid, Cylinder, Cone, Sphere, Hemisphere, Prism, Pyramid etc. are three-dimensional figu res as they have three dimensions viz. length, breadth and height so as to occupy a certain space. It is worth mentioning that Geometry represent shapes and its properties whereas the Mensuration represents calculation of dimensions of the shapes as well as the surface/space occupied by them.

In my opinion this section is the easiest section to secure the certain marks/ranks in the competitive exams such as SSC, Banks, Railway, LIC etc. as it requires no much complexity in logical thinking. However, it is advisable that the aspirants must study the topic 'Mensuration' just after 'Geometry' for better understanding in totality.

It is also germane to mention that different units are used to measure the same length, same area, same volume etc. as and when required.

Table I

1 kilometre (km)	= 10 hectometre (hm)
1 hm	= 10 decametre (dam)
1 dam	= 10 metre (m)
1 m	= 10 decimetre (dm)
1 dm	= 10 centimetre (cm)
1 cm	= 10 millimetre (mm)
Hence, 1 km = 10 hm = 100 dam = 1000 m = 10000 dm = 100000 cm = 1000000 mm.	

You may replace word 'metre' by gram or litre when required.
For example :
1 kilogram = 10 hectogram and 1 Kilolitre
= 10 hectoletre and so on.

Further, 1 quintal = 100 kg and 10 quintals = 1 tonne

Table II

2.54 cm	= 1 inch (in)
12 inch	= 1 foot (ft)
3 feet	= 1 yard (yd)
220 yards	= 1 furling
8 furlong	= 1 mile
0.6214 mile	= 1 km or 5 miles = 8 kms (approx.)

Table III

100 sq m	= 1 are (a)
100 ares	= 1 hectare (ha)
100 hectares	= 1 square km
Hence, 1 hectare = 10000 square metre	

Table IV

Weight = Volume × Density

Structures

A. Two-Dimensional (Plane) Figures

I. Rectangle and Square
II. Triangle
III. Quadrilateral
IV. Circle

B. Three-Dimensional Figures

V. Cuboid and Cube
VI. Cylinder and Cone
VII. Sphere and Hemisphere
VIII. Prism and Pyramid

40

Rectangle and Square

LEARNING OBJECTIVE

After completing this chapter, aspirants will be able to :

- Understand the application of the rectangle and square related formulae in different cases;
- Understand the application of combination of rectangle/square related formulae in diverse cases;
- Observe how perimeter area and diagonal are co-related;
- Compare the rectangles and squares with reference to different aspects.
- Understand the philosophy of the division of this chapter in different parts;
- Analyse the trend of the questions being asked in various competitive examinations;
- Solve the questions in moments;
- Grip the concepts helping to solve the questions without pen and paper;
- Apply many concepts envisaged in the chapter 'PERCENTAGE' like 'effect on product in case of changes in factors', 'Product Constancy', etc;
- Apply many concepts of 'Ratio' in this chapter;
- Learn many new concepts/fundamentals;
- Apply the pragmatic approaches in lieu of academic approaches; and
- Ponder on the alternative methods.

Note *Illustrations based on geometric properties of a rectangle / square have been given in 'GEOMETRY' Section.*

Inside

- *Conceptual Observation of all Terms Associated With the Rectangle/Square and Some Peculiar Cases.*
- *Concept of the Perimeter, Diagonal, Area, Inner and Outer Path and their applications*
- *Interrelation Among Area, Perimeter, Diagonal and Problems Thereon*
- *Alteration in dimensions affecting the area*
- *Concept of Paving / Covering / Gardening, etc.*
- *Formation of a regular polygon when a rectangle/ square is cut on its four vertices*
- *Problems Based on Combination of Rectangle and Square*
- *..............................*
- *.............................. & Many More*
- *Work Out with Solutions*

Conceptual Observation of Terms Associated With the Rectangle/Square and Some Peculiar Cases

Perimeter Suppose both the rectangle and the square are made of a wire. The length of wire required to make the rectangle is termed as perimeter of the rectangle and in case of the square, the same is termed as perimeter of the square.

Hence,

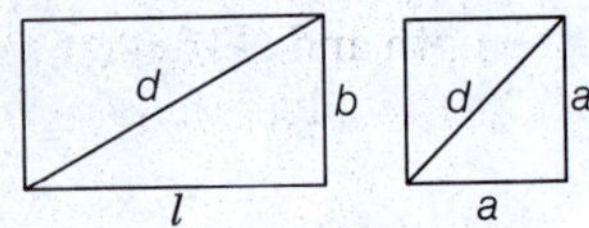

Perimeter of the rectangle $= l + b + l + b = 2(l + b)$

$= 2$ (length + breadth)

and perimeter of the square $= a + a + a + a = 4a = 4\times$ side

Area Assume a line having length of 1 unit is drawn on a plane paper and the another line of the same length is again drawn touching the earlier. Further, 3rd, 4th, 5th, ... line of the same length are drawn touching the just earlier till the shape thus formed has a breadth of b units. Then, the area covered by this shape is called 'area of the rectangle' and in case of square it is termed as 'area of the square' where length and breadth of the wire are of same length.

Hence, area of the rectangle = length × breadth

$= l \times b$

and area of the square = side × side = $(\text{side})^2 = (a)^2$

Diagonal A line of maximum length which can be drawn inside the rectangle is called its diagonal. Obviously, the diagonal is acting as hypotenuse for both the right angled Δs. As per Pythagoras Theorem, length of the diagonal will be as under :

Diagonal of the rectangle $= \sqrt{l^2 + b^2}$

and diagonal of the square $= \sqrt{a^2 + a^2} = a\sqrt{2}$

Each of the rectangle and the square will have two diagonals.

Concept of the inner path Assume a path x metre wide runs around the inside of a rectangular field l metre long and b metre wide.

Here, area of the field including path $= lb$

Clearly, area of the field without path $= (l - 2x)(b - 2x)$

$= lb - 2x(l + b) + 4x^2$

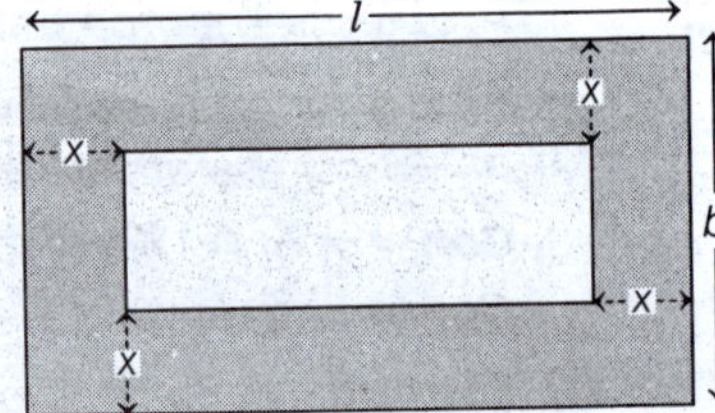

Now, area of the path (i.e. shaded portion)

$= lb - [lb - 2x(l + b) + 4x^2]$

$= 2x(l + b - 2x)$

In case of square field, area of the path $= 2x(2a - 2x) = 4x(a - x)$, where a is the side of that square field.

(This concept is also applicable when a table-cover measuring x cm × y cm is hanging all around the table by z cm.)

When two straight inner paths intersect each other perpendicularly

Let the path be x unit wide.

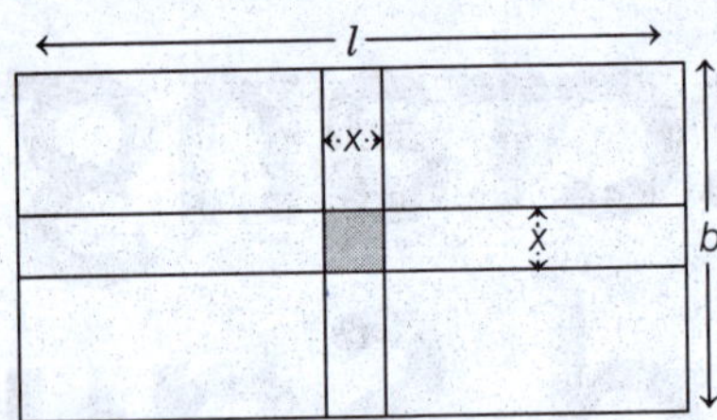

Then, area of the path parallel to length $= lx$ and the area of the path parallel to breadth $= bx$. Clearly, the area of a square (shaded region) is common in both the areas computed above.

It means actual area of the path $= lx + bx - x^2 = x(l + b - x)$

In case of square, the area of the path will be $x(2a - x)$, where a is the side of the square.

Concept of the outer path Assume a rectangular field l metre long and b metre wide is surrounded by a path x metre wide on its outside.

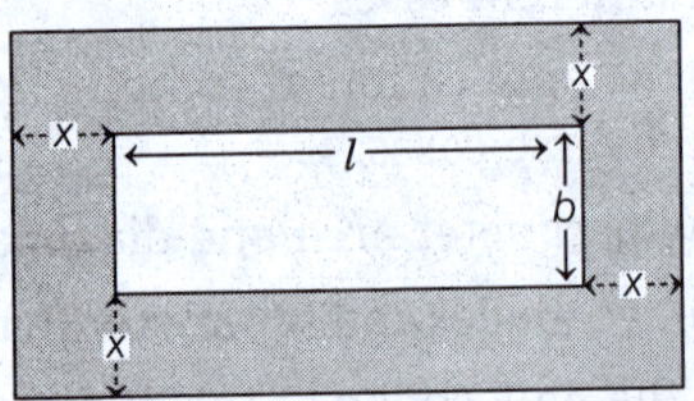

Here, area of the field excluding path $= lb$

Clearly, area of the field with path $= (l + 2x)(b + 2x)$

$= lb + 2x(l + b) + 4x^2$

Now, area of the path (i.e. shaded portion)

$= [lb + 2x(l + b) + 4x^2] - lb$

$= 2x(l + b + 2x)$

In case of square field, area of the path

$= 2x(2a + 2x) = 4x(a + x)$

where, a is the side of that square field.

Concept of the Perimeter and its Applications

- **In case of Rectangle**

1. When length and breadth both are made n times, the perimeter will become n times.
2. When length and breadth both are increased/decreased by n units, the perimeter will increase/decrease by $4n$ units.

3. When length and breadth both are increased by $x\%$, the perimeter will increase by $x\%$.
4. If length and breadth are in ratio $a : b$ and the former is increased or decreased by $x\%$ and the later is increased or decreased by $y\%$, the effect on the original perimeter $= \frac{a(\pm x) + b(\pm y)}{a + b}\%$.

If the resultant is positive, perimeter will increase and in case of negative, it will decrease. We have to append '+' for increase and '−' for decrease.

- **In case of Square**

1. When the side is made n times, the perimeter will become n times.
2. When the side is increased /decreased by n units, the perimeter will increase/decrease by $4n$ units.
3. If side of the square is increased / decreased by $x\%$, the perimeter will increase/decrease by $x\%$.

Problem 1 A rectangle of certain dimensions is chopped off from one corner of a larger rectangle as shown. As shown, $AB = 8$ cm and $BC = 4$ cm. The perimeter of the figure $ABCPQRA$ (in cm) is

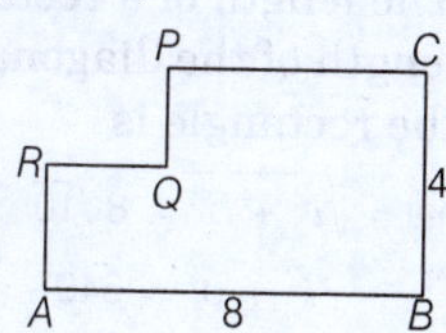

Sol. We know that the perimeter is the sum of length of all lines.

Here, perimeter

$= RA + AB + BC + CP + PQ + QR$

$= (RA + PQ) + AB + BC + (CP + QR)$

$= BC + AB + BC + AB$

$= 2(AB + BC)$

$= 2(8 + 4) = 24$ cm

Alternatively

We can lift RQ to join with PC and move PQ towards RA to join with AR, so that it takes a form of a rectangle. Now, perimeter of the rectangle $= 2(8 + 4) = 24$ cm.

Problem 2 The length and breadth of a rectangular field are in ratio 3 : 2. If the perimeter of the field is 80 m, its breadth (in m) is **[SSC (CGL-PT)-99, SNAP-18]**

Sol Perimeter $= 2(l + b)$

$\Rightarrow \quad 80 = 2(3 + 2)$ units

$\Rightarrow \quad 80 = 10$ units

$\Rightarrow$ 1 unit = 8 m = 2 units

= 16 m

Hence, breadth = 16 m

Alternatively

You may solve assuming length as $3x$ and breadth as $2x$.

Then, $\quad 80 = 2(3x + 2x)$

$\Rightarrow \quad 2x = 16$ m

Problem 3 If the length and perimeter of a rectangle are in the ratio 5 :16, then its length and breadth will be in the ratio : **[SSC (CPO-SI)-08]**

Sol. Perimeter $= 2(l + b) \Rightarrow 16 = 2(5 + b) \Rightarrow b = 3$

Hence, $l : b = 5 : 3$

Problem 4 A rectangular field 80 m long and 40 m wide is to be fenced with wire by erecting poles at a distance of 4 m. How many poles will be required ?

Sol. Obviously, the length of the fencing wire = perimeter of the field $= 2(80 + 40) = 240$ m

Then, number of poles $= \frac{240}{4} = 60$

Problem 5 The length of a rectangular plot is 20 m more than its breadth. If the cost of fencing the plot at the rate of ₹ 26.50/m is ₹ 5300, what is the length (in m) of the plot ? **[Bank (PO)-99]**

Sol. Clearly, length of the wire $= \frac{5300}{26.50} = 200$ m

Hence, perimeter of the plot = 200 m

$\Rightarrow \quad 200 = 2(l + l - 20) \quad [\because l = b + 20]$

$\Rightarrow \quad l = 60$ m

Alternatively

$200 = 2(l + b) \Rightarrow (l + b) = 100$

Since length is 20 m more than breadth, length will be $\frac{100 - 20}{2} + 20$ i.e. 60 m

Problem 6 The perimeter of a rectangle is 24 m. How many rectangles with integer sides are possible ?

Sol. Given, $2(l + b) = 24$

$\Rightarrow \quad (l + b) = 12$

Combinations may be as under.

Combinations	A	B	C	D	E	F	G	H
l	11	10	9	8	7	6	5	...
b	1	2	3	4	5	6	7	...

In combination (F), $l = b$ i.e. a square. It means no such combination is possible for a rectangle. Further, in combination (G), $l < b$, this combination is also not possible. Combination (G) onwards are not possible for the same reason.

Hence, only 5 integral values of length are possible (11, 10, 9, 8 and 7).

Problem 7 If the length and breadth of a rectangle are increased by 5 units each, what will be effect on its perimeter ?

Sol. Original perimeter $= 2(l + b)$

Changed perimeter $= 2\{(l + 5) + (b + 5)\} = 2(l + b) + 20$

Hence, the original perimeter will increase by 20 units.

Alternatively Observing the properties of the perimeter of the rectangle, we may arrive at the correct answer directly as $(4n = 4 \times 5 = 20$ units).

Problem 8 In a rectangle, length is three times its breadth. If the length and the breadth of the rectangle are increased by 30% and 100% respectively, then its perimeter increases by : **[CDS-19]**

Sol. Let breadth = 100, then length = 300

Now, perimeter = 2(300 + 100) = 800

Increased length = 130% of 300 = 390

and increased breadth = 200% of 100 = 200

$\therefore$ New perimeter = 2(390 + 200) = 1180

Then, percentage increase in perimeter

$$= \left(\frac{1180 - 800}{800} \times 100\right)\% = 47.5\%$$

Alternatively

Ratio of length to breadth = 3 : 1, as given.

Then, effect on the original perimeter

$$= \frac{3(+30) + 1(+100)}{3+1}\% = +47.5\%$$

Since the outcome is positive, the original perimeter will increase by 47.5%.

Application of the Concept of Diagonal

Problem 9 A horse 3 m long is at one corner of a square field of side $13\sqrt{2}$ m. Find the speed of the horse if it takes 10 sec. to reach to diagonally opposite corner.

Sol. Diagonal = $\sqrt{2} \times \text{side} = \sqrt{2} \times 13\sqrt{2} = 26$ m

Since the horse is 3 m long itself, it has to cover only (26 – 3) m to reach the destination.

$\therefore$ Speed $= \frac{23}{10} = 2.3$ m/s

Problem 10 The diagonal of a rectangle is twice its smaller side. The ratio of the length to the breadth of the rectangle is :

Sol. Given, $\sqrt{l^2 + b^2} = 2b$

$\Rightarrow l^2 + b^2 = 4b^2$ (squaring both sides)

$\Rightarrow l^2 = 3b^2 \Rightarrow \frac{l^2}{b^2} = \frac{3}{1} \Rightarrow l : b = \sqrt{3} : 1$

Problem 11 What is the length of side of the adjacent square ?

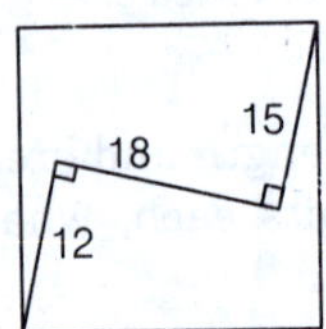

Sol. Clearly, ΔDOE and ΔBOF are similar triangles (details of concept of similarity of Δs are under Geometry Section.)

Then, $\frac{DE}{BF} = \frac{EO}{OF} \Rightarrow \frac{12}{15} = \frac{EO}{18 - EO}$

$\Rightarrow EO = 8$, then $OF = 18 - 8 = 10$

Now, hypotenuse $DO = \sqrt{12^2 + 8^2}$

and $BO = \sqrt{15^2 + 10^2}$

$\therefore BD = \sqrt{12^2 + 8^2} + \sqrt{15^2 + 10^2}$

$= 9\sqrt{13}$

$\Rightarrow \sqrt{2} \times \text{side} = 9\sqrt{13}$

$\Rightarrow \text{side} \approx 22.94$

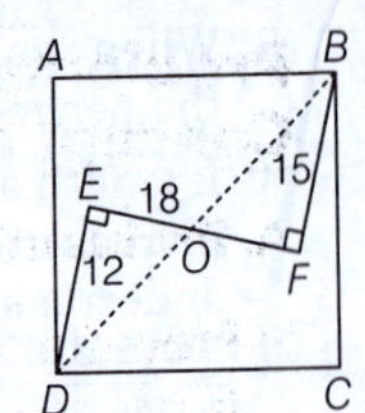

Problems Based on Relation Between Perimeter and Diagonal

In case of the square, we can obtain its perimeter by multiplying its diagonal with $2\sqrt{2}$.

Hence, perimeter = diagonal × $2\sqrt{2}$.

Problem 12 If the length of the diagonal of a square is 20 cm, then its perimeter must be

Sol. Given, diagonal = 20

$\Rightarrow \sqrt{2} \times \text{side} = 20 \Rightarrow \text{side} = 10\sqrt{2}$

Now, perimeter = 4 × side = $4 \times 10\sqrt{2} = 40\sqrt{2}$ cm

Alternatively

Perimeter = Diagonal × $2\sqrt{2} = 20 \times 2\sqrt{2} = 40\sqrt{2}$ cm

Problem 13 The length of a rectangle is three times its width. If the length of the diagonal is $8\sqrt{10}$ cm, then the perimeter of the rectangle is

Sol. Given, diagonal = $\sqrt{l^2 + b^2} = 8\sqrt{10}$

$\Rightarrow l^2 + b^2 = 640$ (squaring both sides)

$\Rightarrow (3b)^2 + b^2 = 640$ ($\because l = 3b$, given)

$\Rightarrow 10b^2 = 640 \Rightarrow b = 8$ cm

$\therefore$ Length = 3 × 8 = 24 cm

Now, perimeter = 2(24 + 8) = 64 cm

Problem 14 A man walked diagonally across a square lot. Approximately, what was the percent saved by not walking along the edges?

Sol. Distance covered by walking along the edges = $a + a = 2a$

and distance covered by walking diagonally = $a\sqrt{2}$

Saving = $2a - a\sqrt{2} = a(2 - \sqrt{2})$

$$\% \text{ saving} = \left(\frac{\text{saving}}{\text{earlier distance}} \times 100\right)\%$$

$$= \left\{\frac{a(2 - \sqrt{2})}{2a} \times 100\right\}\% = 29.30\%$$

Problem 15 The perimeter of a rectangle, one of whose sides is 10 cm and diagonal 26 cm, is **[UPPSC (Rev. Insp.)-15]**

Sol. Given, diagonal = $\sqrt{l^2 + b^2} = 26$

$\Rightarrow \sqrt{l^2 + (10)^2} = 26$

$\Rightarrow l^2 + (10)^2 = 676$

$\Rightarrow l^2 = 576$

$\therefore l = 24$

Then, perimeter = $2(l + b) = 2(24 + 10) = 68$ cm

Problems based on the Concept of the Area

A. Problems Based on the Concept of Inner Path

Problem 16 A path of uniform width runs around the inside of a rectangular field 38 m long and 32 m wide. If the path occupies 600 m^2, then the width of the path is : *[SSC (CGL-PT)-07]*

Sol. Area of the rectangular field = $38 \times 32 = 1216\ m^2$

Clearly, area of the field without path = $(38 - 2x)(32 - 2x)$

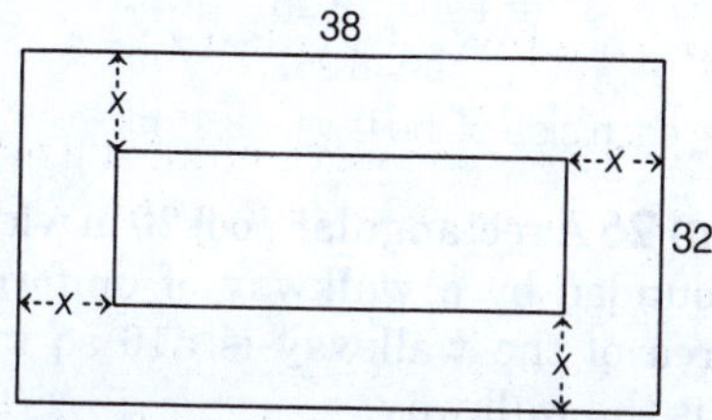

Now, area of the path = $1216 - [(38 - 2x)(32 - 2x)]$

$\Rightarrow \quad 600 = 1216 - 1216 + 140x - 4x^2$

$\Rightarrow \quad 4x^2 - 140x + 600 = 0$

$\Rightarrow \quad x^2 - 35x + 150 = 0$

$\Rightarrow x^2 - 30x - 5x + 150 = 0$

$\Rightarrow x(x - 30) - 5(x - 30) = 0$

$\Rightarrow \quad x = 5 \text{ or } 30$

But, $x = 30$ is not possible as a path 30 m wide inside the field is not possible.

Alternatively I

We know that area of the inner path = $2x(l + b - 2x)$

$\Rightarrow \quad 600 = 2x(38 + 32 - 2x)$

$\Rightarrow \quad 4x^2 - 140x + 600 = 0$

$\Rightarrow \quad x\ 5 \text{ or } 30$

But, $x = 30$ is not possible as a path 30 m wide inside the field is not possible.

Alternatively II

We may proceed through the given options on the above line.

Problem 17 A field is 30 m long and 25 m broad. It has a 3 m wide pavement all around it inside. Find the area of the pavement and cost of constructing the pavement at the rate of ₹ 2 /sq m.

Sol. Area of the pavement (say, inner path)

$= 2x \times (l + b - 2x)$

$= 2 \times 3[30 + 25 - (2 \times 3)] = 294\ m^2$

Now, cost of constructing the pavement = 294×2 = ₹ 588

Problem 18 A room is 15 ft long and 12 ft broad. A mat has to be placed on the floor of this room leaving 1.5 ft space from the walls. What will be the cost of mat at the rate of ₹ 3.50 /sq ft? *[RRB-02]*

Sol. Area of the space left (say, area of path 1.5 ft wide)

$= 2 \times 1.5\,[15 + 12 - (2 \times 1.5)] = 72$ sq ft

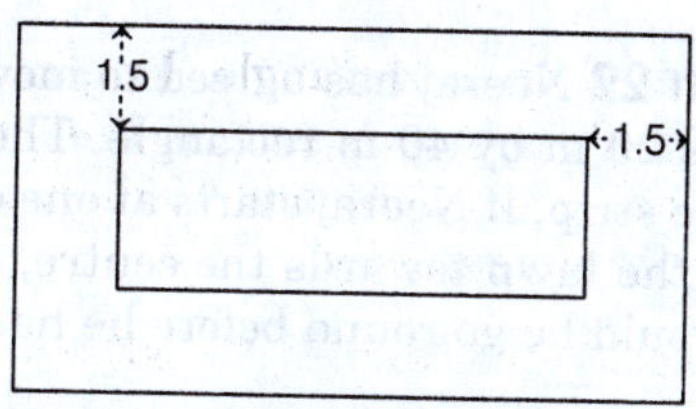

$\therefore$ Area of the floor to be covered by the mat

$= (15 \times 12) - 72 = 108$ sq ft

Now, cost = 108×3.50 = ₹ 378

Alternatively

Area of the floor to be covered by mat

$= (15 - 2 \times 1.5)(12 - 2 \times 1.5) = 108$ sq ft

Then, cost = 108×3.50 = ₹ 378

Problem 19 A park is 60 m long and 40 m broad. Two pathways 2 m wide each run through the centre of the park, one is parallel to its length and other is parallel to its breadth. What is the area of the path?

Sol. Observing the concept given in beginning of this chapter under title 'When two straight inner paths intersect each other perpendicularly' the area of the path will be $x\,(l + b - x)$, i.e. $2(60 + 40 - 2) = 196\ m^2$.

Problem 20 A rectangular park 60 m long and 40 m wide has two concrete crossroads running in the middle of the park and rest of the park has been used as a lawn. If area of the lawn is 2109 m^2, then what is the width of the road ? *[MAT-97, SNAP-11]*

Sol. Area of the path = $x \times (l + b - x) = x(60 + 40 - x) = 100x - x^2$

$\therefore$ Area of the lawn = $(60 \times 40) - (100x - x^2)$

$\Rightarrow \quad 2109 = 2400 - 100x + x^2 \Rightarrow x^2 - 100x + 291 = 0$

$\Rightarrow (x - 97)(x - 3) = 0 \Rightarrow x = 97 \text{ or } 3$

Since the width of the inner path cannot be greater than length/breadth of the park, the width of the park is 3 m.

Problem 21 Certain number of people is to sit in a rectangular hall measuring 100 m and 50 m in rows perpendicular to its length leaving a 2 m wide path all around the hall inside. Find the number of people if there is a distance of 2 m from people to people and also 2 m from row to row.

Sol. The length of the rectangular area in which people are to be sit

$= 100 - (2 + 2)$

$= 96$ m

$\therefore$ Number of rows

$= (96 \div 2) + 1$

$= 49$

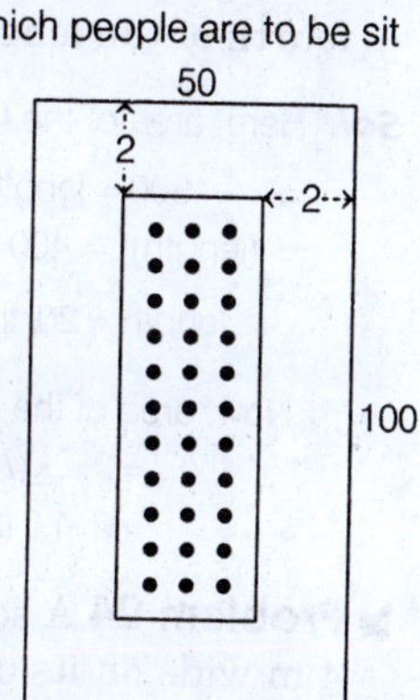

Further, the breadth of that area

$= 50 - (2 + 2)$

$= 46$ m

$\therefore$ Number of people in one row

$= (46 \div 2) + 1 = 24$

Hence, number of total people

$= 49 \times 24 = 1176$

Problem 22 Neeraj has agreed to mow the farm lawn, which is 20 m by 40 m rectangle. The mower mows a 1 m wide strip. If Neeraj starts at one corner and mows around the lawn towards the centre, about how many times would he go round before he has mowed half the lawn ? *[CAT-02]*

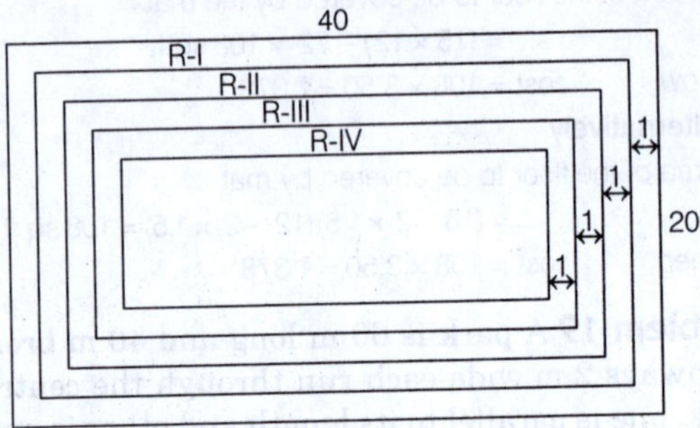

Sol. We know that area of the inner path $= 2x \times (l + b - 2x)$

Clearly, area mowed by Neeraj in 1st round
= area of inner path 1 m wide
$= 2 \cdot 1(40 + 20 - 2 \cdot 1) = 116$

Further, area mowed by Neeraj in 2nd round
$= 2 \cdot 1(38 + 18 - 2 \cdot 1) = 108$

Area mowed by Neeraj in 3rd round
$= 2 \cdot 1(36 + 16 - 2 \cdot 1) = 100$

Area mowed by Neeraj in 4th round
$= 2 \cdot 1(34 + 14 - 2 \cdot 1) = 92$

Hence, area covered by him in 4 rounds
$= 116 + 108 + 100 + 92 = 416$

which is little more than half the area of the lawn
($\because$ area of lawn $= 40 \times 20 = 800$)

Obviously, he would have mowed 16 sq m in $\frac{16}{92}$, i.e. 0.17 of the 4th round.

Thus, required number of rounds $= 4 - 0.17 = 3.83$

B. Problems Based on the Concept of Outer Path

Problem 23 What will be the area of 1.5 m wide garden developed around all the four sides of a rectangular field having area equal to 300 m^2 and breadth equal to three-fourth of the length ? *[IOB-PO-08]*

Sol. Here, area of the field = 300 m^2

$\Rightarrow \quad 300 = \text{length} \times (\text{length} \times 3/4)$ [$\because$ breadth = length $\times$ 3/4]

$\Rightarrow (\text{length})^2 = 400$

$\therefore$ length = 20, then breadth $= 20 \times \frac{3}{4} = 15$

Now, area of the path (say, garden)
$= 2x \times (l + b + 2x)$
$= 2 \cdot 1.5 (20 + 15 + 2 \cdot 1.5) = 114 \text{ m}^2$

Problem 24 A square field is surrounded by a path 4 m wide on its outside. The area of the path is 384 m^2. Find the area of that field.

Sol. Area of the outer path $= 4x \times (a + x)$

$\Rightarrow \quad 384 = 4 \cdot 4(a + 4)$

$\Rightarrow \quad (a + 4) = 24 \Rightarrow a = 20$

Then, area of the field $= (20)^2 = 400 \text{ m}^2$

Problem 25 A square lawn has a path 4 m wide around it. If the area of the path is 196 m^2, then each side of the lawn is : *[SSC (CGL-T-I)-14]*

Sol. Area of the outer path $= 4x \times (a + x)$

$196 = 4 \cdot 4(a + 4)$

$\Rightarrow \quad (a + 4) = 12.25$

$\Rightarrow \quad a = 8.25$

Hence, each side of the lawn = 8.25 m

Problem 26 A rectangular pool 20 m wide and 60 m long is surrounded by a walkway of uniform width. If the total area of the walkway is 516 sq m, how wide (in metre) is the walkway? *[CAT-01]*

Sol. Let width of the walkway be x metre.

Then, area of the walkway $= 2x \times (l + b + 2x)$

$\Rightarrow \quad 516 = 2 \cdot x \times (60 + 20 + 2x)$

$\Rightarrow \quad 129 = 40x + x^2$

$\Rightarrow \quad x^2 + 40x - 129 = 0$

$\Rightarrow \quad x^2 + 43x - 3x - 129 = 0$

$\Rightarrow \quad x(x + 43) - 3(x + 43) = 0$

$\Rightarrow \quad x = -43 \text{ and } 3$

Hence, width of the walkway is 3m (negative value not possible).

Problem 27 A rectangular garden has a path 4 m wide around it. The area of the path is reduced by $\frac{5}{33}$ of the earlier and the rectangle becomes a square when the length of the garden is reduced by 5 m. Find the area of the original garden.

Sol. Let l and b be length and breadth of the garden respectively. In 1st case, the area of the path $= 2x \times (l + b + 2x)$, where x is width of the path.

$= 2 \cdot 4(l + b + 2 \cdot 4)$
$= 8\{l + (l - 5) + 2 \cdot 4\} = 16l + 24$

[When the length is reduced by 5 m, the rectangle becomes a square. It implies that its length is greater than breadth by 5 m.]

In 2nd case, the area of the path $= 4x \times (a + x)$, where $a = (l - 5)$
$= 4 \cdot 4(l - 5 + 4) = 16l - 16$

As given, $16l - 16 = (16l + 24) \times \frac{28}{33}$

$\Rightarrow \quad 2l - 2 = (2l + 3) \times \frac{28}{33}$

$\Rightarrow \quad 66l - 66 = 56l + 84$

$\Rightarrow \quad 10l = 150 \Rightarrow l = 15$

Then, $\quad b = 15 - 5 = 10$

Hence, area of the original garden $= 15 \times 10 = 150 \text{ m}^2$

Alternatively

Since the length is reduced by 5 m, the area of the path will reduce by $(5 \cdot 4 + 5 \cdot 4)$ i.e. by 40 m^2.

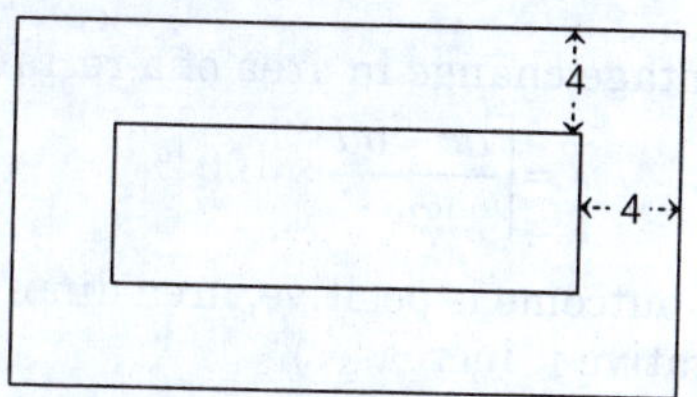

But, it is given that the reduction in area of the path is equal to $\frac{5}{33}$ times the original area of the path. i.e. $\frac{5}{33} \Rightarrow 40$

$\therefore 1 = 264$ (this is original area of the path)

Now, original area of the path $= 2x \times (l + b + 2x)$

$\Rightarrow \quad 264 = 2 \cdot 4(l + b + 2 \cdot 4)$

$\Rightarrow \quad 33 = l + (l - 5) + 8$

$\Rightarrow \quad l = 15\,\text{m}$, then $b = 10$

Hence, area of the original lawn $= 15 \times 10 = 150\ \text{m}^2$

C. Problems Based on the Area (Miscellaneous)

Problem 28 A rectangular room has been partitioned into two equal square rooms by making a partition 10 m long. What is the area of the rectangular room?

Sol. Clearly, area of one square $= (10)^2 = 100\,\text{m}^2$

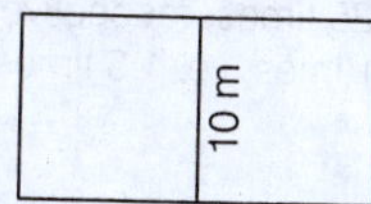

Then, area of the rectangle $= 2 \times 100 = 200\ \text{m}^2$

Problem 29 The difference between the areas of two squares drawn on two line segments of different lengths is $32\ \text{cm}^2$. Find the length of the greater line segment if one is longer than the other by 2 cm.

Sol. Given, $(a + 2)^2 - a^2 = 32$

$\Rightarrow a^2 + 4a + 4 - a^2 = 32$

$\Rightarrow \quad 4a = 28 \Rightarrow a = 7$

Hence, length of the greater line segment $= 7 + 2 = 9\,\text{cm}$.

Alternatively

You may proceed through options. Let options be 7, 9, ...

(a) $7 \Rightarrow 7^2 - (7-2)^2 \neq 32$

(b) $9 \Rightarrow 9^2 - (9-2)^2 = 32$, hence option (b) is correct.

Problem 30 In the given figure, there are four squares. The second biggest square has been made by joining mid-points of the biggest square. The third biggest has been made by joining mid-points of second biggest and so on. Find the area of smallest square if the side of the biggest square is 16 cm.

Sol. Given, $AB = BC = CD = DA =$ side of the biggest square $= 16\,\text{cm}$

$\therefore \quad AM = AP = 8\,\text{cm}$

Then, $PM = \sqrt{(AM)^2 + (AP)^2}$

$= \sqrt{8^2 + 8^2} = 8\sqrt{2}$

Now, $PM = MN = NO = OP = 8\sqrt{2}$ cm

$\therefore \quad PQ = PT = 8\sqrt{2} \div 2 = 4\sqrt{2}$

Then, $QT = \sqrt{(PQ)^2 + (PT)^2}$

$= \sqrt{(4\sqrt{2})^2 + (4\sqrt{2})^2} = 8$

Now, $QT = QR = RS = ST = 8\,\text{cm}$

$\therefore \quad QU = QV = 8 \div 2 = 4$

Then, $UV = \sqrt{(QU)^2 + (QV)^2} = \sqrt{4^2 + 4^2} = 4\sqrt{2}$

Since UV is the side of the smallest square, its area $= (4\sqrt{2})^2 = 32\ \text{cm}^2$

If side of the biggest square $= a$, then it is clear from the above that

side of the 2nd biggest square $= \frac{a}{(\sqrt{2})^{2-1}}$,

side of the 3rd biggest square $= \frac{a}{(\sqrt{2})^{3-1}}$

and side of the nth biggest square $= \frac{a}{(\sqrt{2})^{n-1}}$

Alternatively

Side of the nth (here, $n = 4$) biggest square

$= \frac{\text{side of the biggest square}}{(\sqrt{2})^{n-1}} = \frac{16}{(\sqrt{2})^{4-1}} = 4\sqrt{2}$

$\therefore$ Its area $= (4\sqrt{2})^2 = 32\ \text{cm}^2$

Note *Aspirants should stress on this result so as to arrive at correct answer in moments.*

Problem 31 l and b are length and breadth of a rectangle and their ratio is equal to ratio $(l + b) : l$, then which of the following is correct? ***[SSC (CGL-T-I)-11, CAT-15]***

(a) area $= (l + b)^2$ (b) $l : b = l : (b + l)$

(c) area $= (l + b)(l - b)$ (d) None of these

Sol. Given, $l : b = (l + b) : l$

Hence, option (b) is not correct.

$\Rightarrow \quad \frac{l}{b} = \frac{l + b}{l}$

$\Rightarrow \quad l^2 = lb + b^2$

$\Rightarrow \quad l^2 - b^2 = lb$

$\Rightarrow \quad lb = (l + b)(l - b)$

Thus, option (c) is correct.

Problem 32 A big rectangular plot of area $4320\ \text{m}^2$ is divided into 3 square-shaped smaller plots by fencing parallel to the smaller side of the plot. However, some area of land was still left as a square could not be formed. So, 3 more square-shaped plots were formed by fencing parallel to the longer side of the original plot such that no area of the plot was left surplus. What are the dimensions of the original plot? ***[IAS (Pre.) 05]***

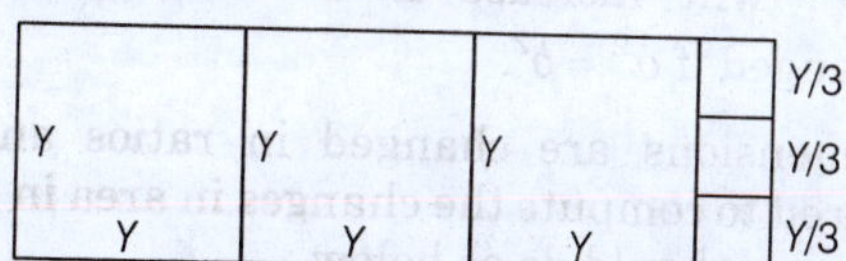

Sol. Let the length and breadth of the original plot be x metre and y metre. Clearly, area of the original plot

$$= y^2 + y^2 + y^2 + \left(\frac{y}{3}\right)^2 + \left(\frac{y}{3}\right)^2 + \left(\frac{y}{3}\right)^2 = 4320$$

$\Rightarrow$ $30y^2 = 4320 \times 9 \Rightarrow y^2 = 144 \times 9$

$\Rightarrow$ $y = 12 \times 3 = 36$ m

Then, $x = 4320 \div 36 = 120$ m

Thus, the original plot was 120 m long and 36 m broad.

Problem 33 A rectangular sheet of paper, when halved by folding it at the mid-point of its longer side, results in a rectangle, whose longer and shorter sides are in the same proportion as the longer and shorter sides of the original rectangle. If the shorter side of the original rectangle is 2, what is the area of the smaller rectangle? ***[CAT-04]***

Sol. Let the length of the original rectangular sheet of paper be x.

Then, $2 : \frac{x}{2} = x : 2$ [$\because \frac{x}{2} : 2 = x : 2$ is not possible]

$\Rightarrow$ $x = 2\sqrt{2}$

$\therefore$ Area of the smaller rectangle $= 2 \times \frac{2\sqrt{2}}{2} = 2\sqrt{2}$

Effect of Alteration in Dimensions on Area

- **If alternation is reflected in terms of ... times**

1. If length and breadth of a rectangle are made x and y times respectively, the area will become xy times the earlier.
2. If the side of a square is made n times, its area will become n^2 times.

- **If alternation is reflected in terms of percentage**

3. Recall the 'concept of effect on product in case of changes in factors' detailed in chapter percentage, if dimensions of a rectangle/square are changed by percentage. It should be kept in mind that an area is a product of two factors *viz.* length and breadth or side and side as the case may be.

- **If alternation is reflected in terms of ratio**

4. If the dimensions of a rectangle are changed in ratios (say, length in ratio $a : b$ and breadth in ratio $c : d$), then the area will change in ratio $ac : bd$.
   (a) If $ac > bd$, the area diminishes.
   (b) If $ac < bd$, the area is increases.
   (c) If $ac = bd$, no changes in the area.
5. If the sides of a square is changed in ratio $a : b$, its area will change in ratio $a^2 : b^2$. The area will decrease if $a^2 > b^2$, will increase if $a^2 < b^2$ and will remain unchanged if $a^2 = b^2$.
6. If dimensions are changed in ratios and you are required to compute the changes in area in percentage term, you should do as below :

Percentage change in area of a rectangle

$$= \left[\frac{ac - bd}{ac} \times 100\right]\%$$

(if the outcome is positive, area diminishes and in case of negative it increases)

Similarly, percentage change in area of a square

$$= \left[\frac{a^2 - b^2}{a^2} \times 100\right]\%$$

(if the outcome is positive, area diminishes and in case of negative, it increases).

Now, we will observe some illustrations.

Problem 34 The length and breadth of a rectangle are increased by 20% and 25%, respectively. The increase in the area of the resulting rectangle will be : ***[SSC (10+2, LDC/DEO)-10, SSC (CGL-T-I)-19]***

Sol. Effect on area $= \left[+20 + 25 + \frac{20 \times 25}{100}\right]\% = +50\%$

'+' indicates that the area will increase by 50%.

Alternatively I

Clearly, length and breadth of the rectangle have been made to 1.2 times and 1.25 times, respectively. Then, its area will become (1.2×1.25) times, i.e. 1.5 times. It means the area will increase by 50%.

Alternatively II

Length has been changed in ratio 100 : 120 i.e. 5 : 6 and breadth has been changed in ratio 100 : 125, i.e. 4 : 5.

Now, percentage change in area

$$= \left[\frac{ac - bd}{ac} \times 100\right]\% = \left[\frac{(5 \times 4) - (6 \times 5)}{(5 \times 4)} \times 100\right]\%$$

$$= -50\%$$

'–' indicates that the area has been increased and the increment is 50%.

Problem 35 If the length of a rectangle is increased by 10% and its breadth is decreased by 10%, the change in its area will be : ***[SSC (CGL-Tier-I)-10, SSC (CGL-Tier-II)-10]***

Sol. Effect on area $= \left[+10 - 10 - \frac{10 \times 10}{100}\right]\% = -1\%$

'–' indicates that the area will decrease by 1%.

Alternatively I

Clearly, length and breadth of the rectangle have been made to 1.1 times and 0.9 times, respectively. Then, its area will become (1.1×0.9) times, i.e. 0.99 times. It means the area will decrease by 1%.

Alternatively II

Length has been changed in ratio 100 : 110 i.e. 10 : 11 and breadth has been changed in ratio 100 : 90 i.e. 10 : 9.

Now, percentage change in area

$$= \left[\frac{(10 \times 10) - (11 \times 9)}{(10 \times 10)} \times 100\right]\% = +1\%$$

'+' indicates that the area has been decreased and the decrease is 1%.

Problem 36 If each side of a square is increased by 10%, what is the percentage of increase of area

[SSC (SAS)-10, MAT-18]

Sol. Effect on area $= \left[+10+10+\frac{10\times10}{100}\right]\% = +21\%$

Hence, the area will increase by 21%.

Alternatively I

Clearly, each side of the square has been made to 1.1 times. Then, its area will become $(1.1)^2$ times, i.e. 1.21 times. It means the area will increase by 21%.

Alternatively II

Each side has been changed in ratio 100 : 110 i.e. 10 : 11.

Now, percentage change in area

$$= \left[\frac{a^2-b^2}{a^2}\times100\right]\% = \left[\frac{(10)^2-(11)^2}{(10)^2}\times100\right]\% = -21\%$$

'–' indicates that the area has been increased and the increment is 21%.

It may be construed from the discussion made through the above three problems that all the techniques are correlated and aspirants are supposed to be well versed with all these techniques, so that they apply them appropriately observing the nature of questions.

Problem 37 If length of a rectangle is increased in the ratio 6 : 7 and its breadth is diminished in the ratio 5 : 4, then its area will be diminished in the ratio :

Sol. Clearly, length is increased by $\left(\frac{1}{6}\times100\right)\%$ i.e. by 16.66% and breadth is diminished by $\left(\frac{1}{5}\times100\right)\%$ i.e. by 20%

Now, effect on the area $= \left[+16.66-20-\frac{16.66\times20}{100}\right]\%$

$= -6.66\%$

It means the area will be diminished by 6.66%, i.e. by $\frac{1}{15}$. In other words, the area will be diminished in the ratio 15 : 14.

Alternatively

Observing the facts (supra), the required ratio will be $(6\times5):(7\times4)$ i.e. 15 : 14

Problem 38 The length of a rectangle is decreased by r% and the breadth is increased by $(r+5)$%. Find the value of r, if the area of the rectangle is unaltered.

[SCMHRD-02]

Sol. Effect on area $= \left[-r+(r+5)-\frac{r\times(r+5)}{100}\right]\%$

$\Rightarrow 0\% = \left[\frac{500-r(r+5)}{100}\right]\% \Rightarrow 500-r^2-5r=0$

$\Rightarrow r^2+5r-500=0 \Rightarrow r^2+25r-20r-500=0$

$\Rightarrow r(r+25)-20(r+25)=0 \Rightarrow r=20$ and -25

Problem 39 In measuring the sides of a rectangle, there is an excess of 5% on one side and 2% deficit on the other. Then, the error percentage in the area is :

Sol. Error % $= \left[+5-2-\frac{5\times2}{100}\right]\% = +2.9\%$

Problem 40 The length of a rectangle is increased by 20%, by what percentage should its breadth be reduced so as to keep the area constant ?

Sol. Recall the 'concept of Product Constancy' detailed in chapter 'percentage'. Area i.e. a product of the length and the breadth is to be kept constant.

Now, reduction in breadth $= \left[\frac{20}{(100+20)}\times100\right]\% = 16.66\%$

Alternatively

Effect on area $= \left[+20-r-\frac{20\times r}{100}\right]\%$

(let reduction in breadth be r %.)

$\Rightarrow \left[+20-r-\frac{20\times r}{100}\right]\% = 0\%$

[$\because$ no effect in area as it is to be kept constant]

$\Rightarrow -120r = -2000$

$\therefore r = 16.66\%$

Problem 41 The length of a room floor exceeds its breadth by 20 m. The area of the floor remains unaltered when the length is decreased by 10 m but the breadth is increased by 5 m. The area of the floor, (in sq m) is :

Sol. Since there is no alternation in the area, original area will be equal to the area after changes in the dimensions.

Hence, $(l-10)(b+5)=(b+20)b$ $\quad(\because l=b+20)$

$\Rightarrow (b+20-10)(b+5)=(b+20)b$

$\Rightarrow (b+10)(b+5)=b^2+20b$

$\Rightarrow b^2+15b+50-b^2-20b=0 \Rightarrow b=10$

Then, $l=10+20=30$

Thus, area $=30\times10=300\text{ m}^2$.

Alternatively

Proceed through options. Let option be 300, 400, ...

$300=30\times10$ or 20×15, both conditions are satisfied.

Hence, option (a) is correct.

Problem 42 When the length of a rectangular plot is increased by 5 m and the breadth is decreased by 2 m, the area remains unchanged. If the length be decreased by 5 m and the breadth is increased by $3\frac{1}{3}$ m, the area also remains unchanged. Find the original area of the rectangular plot.

Sol. Let the length and breadth be l and b, respectively.

Then, original area $= lb$

and New area $=(l+5)(b-2)=lb$ (Given)

$\Rightarrow lb+5b-2l-10=lb$

$\Rightarrow 2l-5b=-10 \quad \ldots(i)$

Further, the other new area

$=(l-5)\left(b+3\frac{1}{3}\right)=lb$ {$\because$ no change in area}

$\Rightarrow \frac{10}{3}l-5b=\frac{50}{3} \quad \ldots(ii)$

On subtracting Eq. (ii) from Eq. (i), we get

$$\frac{4l}{3} = \frac{80}{3} \Rightarrow l = 20$$

Putting the value of l in Eq. (i), we get $b = 10$

Hence, original area $= 20 \times 10 = 200\,m^2$

Alternatively

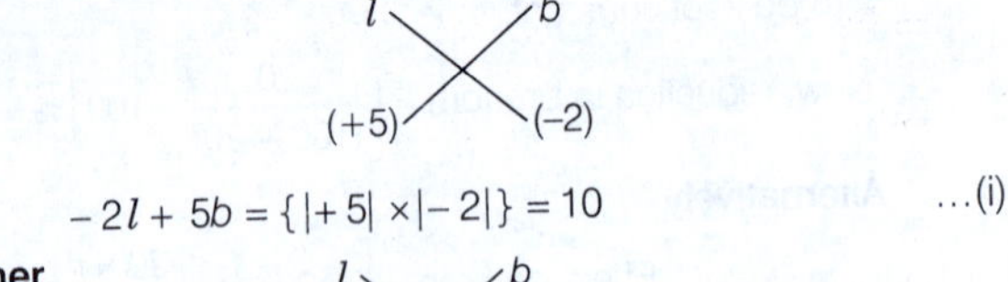

$$-2l + 5b = \{|+5| \times |-2|\} = 10 \quad \ldots(i)$$

Further,

l b

(–5) (+10/3)

$$\frac{10}{3}l - 5b = \left\{|-5| \times \left|+\frac{10}{3}\right|\right\} = \frac{50}{3} \quad \ldots(ii)$$

On solving both Eqs. (i) and (ii), we get $l = 20$ and $b = 10$

Hence, area $= 20 \times 10 = 200\,m^2$

Problem 43 The area of a rectangle gets reduced by $9\ m^2$ if its length is reduced by 5 m and breadth is increased by 3 m. If we increase the length by 2 m and breadth by 3 m, the area is increased by $75\ m^2$. The length of the rectangle is : *[LIC (AAO)-10]*

Sol. Let the length and breadth be l and b, respectively. Then, original area $= lb$.

New area $= (l - 5)(b + 3) = lb - 9$ (Given)

$\Rightarrow \quad lb - 5b + 3l - 15 = lb - 9$

$\Rightarrow \quad 3l - 5b = 6 \quad \ldots(i)$

Further, the other new area

$= (l + 2)(b + 3) = lb + 75$

$\Rightarrow \quad 3l + 2b = 69 \quad \ldots(ii)$

On subtracting Eq. (ii) from Eq. (i), we get

$-7b = -63$

$\Rightarrow \quad b = 9$

Putting the value of b in Eq. (i), we get $l = 17$

Alternatively

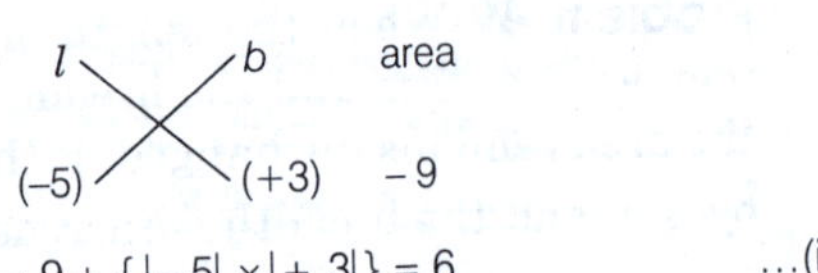

$$3l - 5b = -9 + \{|-5| \times |+3|\} = 6 \quad \ldots(i)$$

Further,

l b area

(+2) (+3) +75

$$3l + 2b = 75 - \{|+2| \times |+3|\} = 69 \quad \ldots(ii)$$

On solving both Eqs. (i) and (ii) we get $l = 17$ m

Aspirant should keep in mind that if length and breadth have opposite signs (i.e. one increases and another decreases), we add the product of the mode figures with the change occurred in the area. In case the length and breadth have the same sign (either both increase or decrease) we subtract the product of the mode figures from the change occurred in the area. In both cases, changes in area must be written with signs.

Problems Based on the Relation Between Area and Perimeter

1. **It holds good that area of square > area of rectangle > area of triangle under the given equal perimeters and this facts are based on the principles of 'more sides more area in a convex polygon'.**
2. **The perimeter of a rectangle is always greater than that of a square if they have equal areas.**
3. **If the dimensions (presented in ratio, say $l : b$) of a rectangle and its area are given, then its perimeter will be $2(l + b) \times \sqrt{\frac{\text{area}}{lb}}$.**

Problem 44 If the perimeter of a square and a rectangle are the same and their areas are P and Q respectively, then which of the following is true? *[SSC (FCI-Asstt. Gr.-III)-12]*

(a) $P < Q$ (b) $P \le Q$

(c) $P > Q$ (d) $P = Q$

Sol. We already discussed in beginning that the area of a square is greater than that of a rectangle provided they have equal perimeters

Alternatively

You may assume a figure for both perimeter and then compute the areas accordingly. You will realise that the area of the square is greater than that of the rectangle.

Problem 45 The area of a rectangular field is $4500\ m^2$. If its length and breadth are in the ratio 9 : 5, then its perimeter is : *[SSC (SAS)-10, CDS-13]*

Sol. Let length and breadth be 9x and 5x, respectively.

Now, area $= 4500$ (given)

$\Rightarrow \quad 9x \times 5x = 4500$

$\Rightarrow \quad x^2 = 100$

$\Rightarrow \quad x = 10$

Hence, length $= 9 \times 10 = 90$ m and breadth $= 5 \times 10 = 50$ m

Then, perimeter $= 2(90 + 50) = 280$ m

Alternatively

$$\text{Perimeter} = 2(9 + 5) \times \sqrt{\frac{4500}{9 \times 5}} = 280\text{ m}$$

Problem 46 If the perimeter of a rectangle is 10 cm and the area is $4\ cm^2$, then its length is : *[CDS-17]*

Sol. Given, $lb = 4$

and $2(l + b) = 10$

$\Rightarrow \quad l + b = 5 \quad \ldots(i)$

Now, $(l - b)^2 = (l + b)^2 - 4lb$

$= 5^2 - 4 \cdot 4 = 9$

$\Rightarrow \quad l - b = 3 \quad \ldots(ii)$

On adding both Eqs. (i) and (ii), we get

$2l = 8 \Rightarrow l = 4$ cm

Problem 47 The perimeters of two squares are 40 cm and 24 cm. The perimeter of a third square, whose area is equal to the difference of the areas of these squares is :

Sol. Side of 1st square $= \frac{40}{4} = 10$, then its area $= (10)^2 = 100\text{ cm}^2$

and side of 2nd square $= \frac{24}{4} = 6$, then its area $= (6)^2 = 36\text{ cm}^2$

Hence, area of the 3rd square $= 100 - 36 = 64$

Then, its perimeter $= 4 \times \sqrt{64} = 32$ cm

Problem 48 The perimeter of the top of a rectangular table is 28 m, whereas its area is 48 m². What is the length of its diagonal ?

Sol. Given, perimeter $= 2(l + b) = 28$

$\Rightarrow \quad (l + b) = 14$

Also given, $\quad lb = 48$

Now, $\quad l^2 + b^2 = (l + b)^2 - 2\,lb = (14)^2 - 2 \times 48 = 100$

$\Rightarrow \quad \sqrt{l^2 + b^2} = \sqrt{100} = 10$

Hence, diagonal = 10 m

Problem 49 The area of a rectangular field is 1500 sq m and the ratio of the length and the breadth is 5 : 3. Find the cost of the wire required to fence around the field at the rate of ₹ 55/m of length of the wire.

Sol. Given, $lb = 1500$

$\Rightarrow \quad 5x \times 3x = 1500$ (Let l be $5x$ and b be $3x$.)

$\Rightarrow \quad x^2 = 100$

$\therefore \quad x = 10$

Hence, length = 50 m and breadth = 30 m

Now, perimeter $= 2(50 + 30) = 160$ m

$\therefore$ Cost $= 160 \times 55 =$ ₹ 8800

Alternatively

Perimeter $= 2(5 + 3) \times \sqrt{\frac{1500}{5 \times 3}}$ m $= 160$ m

Then, cost $= 160 \times 55 =$ ₹ 8800

Problem 50 The area of a rectangle is less than that of a square by 100 m² but their perimeters are 160 m each. The length of the rectangle is :

Sol. Side of the square $= \frac{160}{4} = 40$, then its area

$= (40)^2 = 1600\text{ m}^2$

Then, area of the rectangle

$(lb) = 1600 - 100 = 1500\text{ m}^2$

Also given, perimeter of the rectangle = 160

$\Rightarrow \quad 2(l + b) = 160$

$\Rightarrow \quad (l + b) = 80 \quad$ ...(i)

Now, $\quad (l - b)^2 = (l + b)^2 - 4\,lb$

$= (80)^2 - 4 \times 1500$

$= 400$

$\Rightarrow \quad l - b = \sqrt{400} = 20 \quad$ ...(ii)

Solving the Eqs. (i) and (ii), we get $l = 20$ m

Problem 51 The ratio between the length and breadth of a rectangular park is 3 : 2. If a man cycling along the boundary of the park at speed of 12 km/h, completes one round in 8 min, then the area of the park (in m²) is :

Sol. Clearly, distance covered in 8 min = perimeter of the park

$\Rightarrow$ Perimeter of the park $= \frac{8 \times 12000}{60} = 1600$ m

Now, $\quad 2(l + b) = 1600$

$\Rightarrow \quad 2(3 + 2)$ units $= 1600$ m

$\Rightarrow \quad 1$ unit $= 160$ m

Hence, $\quad$ length $= 3$ units $= 3 \times 160 = 480$ m

and $\quad$ breadth $= 2$ units $= 2 \times 160 = 320$ m

$\therefore \quad$ Area $= 480 \times 320 = 153600\text{ m}^2$

Problem 52 A rectangular paper, when folded into two congruent parts, had a perimeter of 34 cm for each part folded along one set of sides and the same is 38 cm when folded along the other set of sides. What is the area of the paper ?

Sol. When the paper is folded along the breadth, the perimeter be

$= l + \frac{b}{2} + l + \frac{b}{2} = 2l + b$

When the paper is folded along the length, the perimeter be

$\frac{l}{2} + b + \frac{l}{2} + b = l + 2b$

As given, $\quad 2l + b = 38 \quad$ ...(i)

$l + 2b = 34 \quad$ ...(ii)

Eq. (i) – Eq. (ii),

$-3b = -30 \Rightarrow b = 10$

Putting the value of b in Eq. (i) or Eq. (ii), we get $l = 14$

Thus, area $= 14 \times 10 = 140\text{ cm}^2$

Problem 53 *A* took 15 sec to cross a rectangular field diagonally walking at the rate of 52 m/min and *B* took the same time to cross the same field along its sides walking at the rate of 68 m/min. The area of the field is :

Sol. Clearly, diagonal of the field = distance covered in 15 sec at the rate of 52 m/s.

$\Rightarrow \quad \sqrt{l^2 + b^2} = \frac{52 \times 15}{60} = 13$ m

$\Rightarrow \quad l^2 + b^2 = 169$

Further, sum of the length and the breadth = distance covered in 15 s at the rate of 68 m/s

$\Rightarrow \quad l + b = \frac{68 \times 15}{60} = 17$ m

Now, $\quad (l + b)^2 = (l)^2 + (b)^2 + 2\,lb$

$\Rightarrow \quad (17)^2 = 169 + 2\,lb$

$\Rightarrow \quad lb = 60\text{ m}^2$

Hence, area of the field $= 60\text{ m}^2$

Problem 54 A square and a rectangle have equal areas. If their perimeters are x and y respectively, then

(a) $x = y$ (b) $x < y$
(c) $x > y$ (d) None of these

Sol. Observing the facts (supra) under this section, it is stated that the perimeter of a rectangle is always greater than that of a square if they have equal areas.

Alternatively

You may assume a figure for both areas and then compute the perimeters accordingly. You will realise that the perimeter of the rectangle is greater than that of the square.

Problem 55 The areas of a square and a rectangle are equal. The length of the rectangle is greater than the length of any side of the square by 5 cm and breadth is less by 3 cm. Find the perimeter of the rectangle.

Sol. Let each side of the square is 'a'.

Then its area $= a^2$

Further, area of the rectangle $= (a+5)(a-3) = a^2$ (as given)

$\Rightarrow \quad a^2 + 2a - 15 = a^2$

$\Rightarrow \quad a = 7.5$

Hence, perimeter of the rectangle

$= 2[(7.5+5)+(7.5-3)] = 34$ cm

Problem 56 l and b are the length and breadth of a rectangle respectively. If its area is numerically equal to its perimeter, then what is its perimeter provided values of l and b are in integer ?

Sol. Given, $\quad lb = 2(l+b)$

$\Rightarrow \quad lb = 2l + 2b$

$\Rightarrow \quad l(b-2) = 2b$

$\therefore \quad l = \dfrac{2b}{b-2} = \dfrac{2(b-2)+4}{b-2} = 2 + \dfrac{4}{b-2}$

Since, l is an integer, 4 must be divisible exactly by $(b-2)$ and it is possible only when $b = 3$ or 4 or 6

Now,

1. If $b = 3$, $l = 2 + \dfrac{4}{3-2} = 6$, it is possible as $l > b$.

2. If $b = 4$, $l = 2 + \dfrac{4}{4-2} = 4$, it is not possible as $l = b$.

3. If $b = 6$, $l = 2 + \dfrac{4}{6-2} = 3$, it is not possible as $l < b$.

Hence, perimeter of the rectangle $= 2(6+3) = 18$ units

Problem 57 Shri Bhagwan made a fence around his square field by using 13 poles for each side, each at distance of 5 m. Find the area of the field.

Sol. Between 13 poles, total length $= (13-1)\times 5 = 60$ m

Hence, each side of the field is 60 m.

Then, its area $= (60)^2 = 3600\ \text{m}^2$

Problem 58 The area of a rectangle and the square of its perimeter are in the ratio 1 : 25. Then, the lengths of the shorter and longer sides of the rectangle are in the ratio : ***[CAT-18]***

Sol. Let length and breadth be l and b, respectively.

As given, $lb : [2(l+b)]^2 = 1:25$

$\Rightarrow \quad 25lb = 4l^2 + 4b^2 + 8lb$

$\Rightarrow \quad l^2 + b^2 = \dfrac{17}{4} lb$

$\Rightarrow \quad \dfrac{l}{b} + \dfrac{1}{\frac{l}{b}} = \dfrac{17}{4} \Rightarrow \dfrac{l}{b} + \dfrac{1}{\frac{l}{b}} = 4 + \dfrac{1}{4}$

$\Rightarrow \quad \dfrac{l}{b} = 4 \Rightarrow b : l = 1 : 4$

Alternatively

The best way is to proceed through options.

Let options be 2 : 3, 1 : 4, ...:

Option (a) $\Rightarrow 3\times 2 : [2(3+2)]^2 \neq 1:25$

Option (b) $\Rightarrow 1\times 4 : [2(1+4)]^2 = 1:25$

Hence, option (b) is correct.

Problems Based on the Relation Between Area and Diagonal

1. **If the diagonal of a square is made n times, its area will become n^2 times. But, this property does not hold good in case of rectangle. Similarly, when the area is made n times, its diagonal will become $\sqrt{n}$ times (applicable only in case of a square).**
2. **If ratio of diagonals of two squares is $a : b$, the ratio of their areas will be $a^2 : b^2$.**
3. **Area of the square $= \dfrac{d^2}{2}$, where $d =$ its diagonal.**

Problem 59 The ratio of the area of a square to that of the square drawn on its diagonal is : ***[SSC (CGL-PT)-05, CDS-21]***

Sol. Area of the first square $= a^2$

and area of the latter square

$= (\sqrt{2}\,a)^2 = 2a^2$

Hence, ratio $= a^2 : 2a^2 = 1:2$

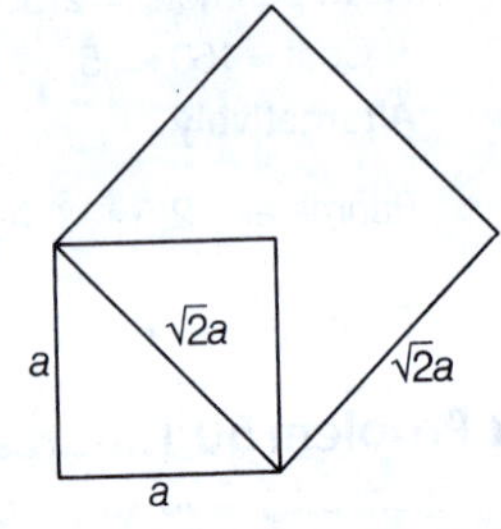

Alternatively

Ratio of their sides = side : $\sqrt{2}$ side

$= 1 : \sqrt{2}$

Then, ratio of their areas

$= (1)^2 : (\sqrt{2})^2 = 1:2$

Problem 60 The diagonal of a square is $4\sqrt{2}$ cm. The diagonal of another square whose area is double that of the first square is :

Sol. Diagonal of the first square $= \sqrt{2}\,a$ {let a be its side}

$\Rightarrow \quad 4\sqrt{2} = \sqrt{2}\,a \Rightarrow a = 4$

Hence, its area $= 16\ \text{cm}^2$

Now, area of the second square $= 2\times 16 = 32\ \text{cm}^2$

$\therefore \quad$ Its side $= \sqrt{32} = 4\sqrt{2}$

Then, its diagonal $= \sqrt{2}\times 4\sqrt{2} = 8$ cm

Alternatively

Since the area of the second square is double that of the first square, it may be assumed for the sake of convenience that the area of the first has been made 2 times. If area is made 2 times, its diagonal will become $\sqrt{2}$ times.

Hence, its diagonal $= \sqrt{2}\times 4\sqrt{2} = 8$ cm

Problem 61 If the measurement of a diagonal and the area of a rectangle are 25 cm and 168 cm^2 respectively, then what is the length of the rectangle ?

[SSC (CGL-Tier-II)-10]

Sol. Given, $lb = 168$ and $\sqrt{l^2 + b^2} = 25$

i.e. $l^2 + b^2 = 625$

Now, $(l + b)^2 = l^2 + b^2 + 2lb = 625 + (2 \times 168) = 961$

$\Rightarrow l + b = 31$ ...(i)

Further, $(l - b)^2 = l^2 + b^2 - 2lb = 625 - (2 \times 168) = 289$

$\Rightarrow l - b = 17$ ...(ii)

Solving Eq. (i) and Eq. (ii), we get $l = 24$ cm

Alternatively

You may proceed through options. Let options be 31, 24,......

(a) If $l = 31$, then $b = \frac{168}{31}$ but $\sqrt{(31)^2 + \left(\frac{168}{31}\right)^2} \neq 25$

(b) If $l = 24$, then $b = \frac{168}{24} = 7$ and $\sqrt{(24)^2 + (7)^2} = 25$

Hence, option (b) is correct.

Concept of Paving / Covering / Gardening etc.

Problem 62 The floor of a corridor is 100 m long and 3 m wide. Cost of covering the floor with carpet 50 cm wide at the rate of ₹ 15/m is :

Sol. Obviously, area of the floor = area of the carpet

$\Rightarrow 100 \times 3 = \text{length} \times \frac{1}{2} \Rightarrow \text{length} = 600$ m

Then, cost = 600 × 15 = ₹ 9000

Problem 63 Find the number of tiles with which the floor measuring 22.5 m × 16 m is to be paved, if each tile is of size 2.5 m × 2 m.

Sol. Clearly, number of tiles $= \frac{\text{area of the floor}}{\text{area of one tile}} = \frac{22.5 \times 16}{2.5 \times 2} = 72$

Problem 64 Find the least number of square tiles for flooring a hall measuring 28 m × 20 m.

Sol. We need the least number of tiles, it means tiles must be of maximum size. Further, the HCF of the length and the breadth will be the greatest number which can divide the length (28) and breadth (20) exactly. It means the tile having each side equals to that greatest number, will be of maximum size.

Hence, each side of the largest possible tile

= HCF of 28 and 20 = 4 m

$\therefore$ Number of tiles required $= \frac{\text{Area of the hall}}{\text{Area of one tile}} = \frac{28 \times 20}{4 \times 4} = 35$

Problem 65 A playground is in the shape of a rectangle. A sum of ₹ 1000 was spent to make the ground usable at the rate of 25 paise per m^2. The breadth of the ground is 50 m. If the length of the ground is increased by 20 m, what will be the expenditure in rupees at the same rate per m^2?

[SSC (CGL-PT)-07]

Sol. Clearly, area of the playground $= \frac{1000}{0.25} = 4000\ m^2$

Now, $\text{length} = \frac{\text{area}}{\text{breadth}} = \frac{4000}{50} = 80$ m

Then, area of the playground = (80 + 20) × 50 = 5000

$\because$ Expenditure for an area 4000 m² is ₹ 1000.

$\therefore$ Expenditure for an area 5000 is ₹ 1250.

Problem 66 The central portion of a rectangular park 50 m × 30 m, is laid with flower bed leaving a path paved with tiles all around the flower bed. The total cost of laying the flower bed and paving the tiles is ₹ 3800. Find the breadth of the path, if laying the flower bed and paving the tiles cost ₹ 3 and ₹ 2 per m^2 respectively.

Sol. Let the width of the path paved with tiles be x metre.

Then, area of this path (say, inner path)

$= 2x \times (l + b - 2x)$

$= 2x \times (50 + 30 - 2x)$

$= 160x - 4x^2$

$\therefore$ Cost for this area

$= 2\,(160x - 4x^2)$

$= 320x - 8x^2$

Further, area of the flower bed

= (50 × 30) – area of the path

$= 1500 - (160x - 4x^2) = 1500 - 160x + 4x^2$

$\therefore$ Cost for this area $= 3 \times [1500 - 160x + 4x^2]$

$= 4500 - 480x + 12x^2$

But given, total cost = ₹ 3800

$\Rightarrow [320x - 8x^2] + [4500 - 480x + 12x^2] = 3800$

$\Rightarrow 4x^2 - 160x + 700 = 0 \Rightarrow x^2 - 40x + 175 = 0$

$\Rightarrow x^2 - 35x - 5x + 75 = 0$

$\Rightarrow x(x - 35) - 5(x - 35) = 0$

$\Rightarrow x = 35$ and 5

Here, $x = 35$ is not possible as the breadth is 30 m only.

Hence, width of the path is 5 m.

Alternatively

Area of the field = 50 × 30 = 1500 m^2

Combined cost per $m^2 = \frac{3800}{1500} = 2.533$

Now, by rule of allegation (details in topic 'Allegation')

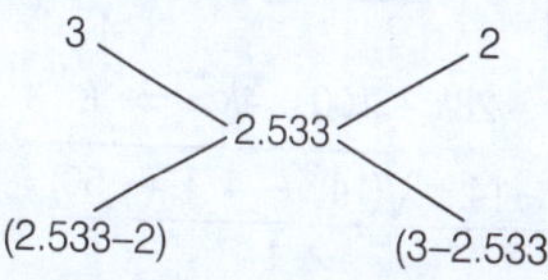

$\Rightarrow$ 0.533 : 0.466 = 8 : 7

It means area of the flower bed is 800 m² and that of the path is 700 m^2.

Now, area of the path $= 2x \times (50 + 30 - 2x)$

$\Rightarrow 700 = 160x - 4x^2 \Rightarrow 4x^2 - 160x + 700 = 0$

$\Rightarrow x^2 - 40x + 175 = 0 \Rightarrow x^2 - 35x - 5x + 175 = 0$

$\Rightarrow x(x-35)-5(x-35)=0 \Rightarrow x=35$ and 5.

Here, $x=35$ is not possible as the breadth is 30 m only.

Hence, width of the path is 5 m.

Problem 67 What will be the cost of gardening 1 m broad boundary around a rectangular plot having perimeter of 340 m at the rate of ₹ 10 per m^2?

Sol. Area to be gardened $= 340 \times 1 = 340 \text{ m}^2$

Then, cost $= 340 \times 10 =$ ₹ 3400

Problem 68 The cost of carpeting a room is ₹ 120. If the width had been 4 m less, the cost of the carpet would have been ₹ 20 less. The width of the room is :

Sol. Let the cost of carpeting is ₹ 1 per sq m.

Then, area $= 120 \text{ m}^2$.

Now, length $= \dfrac{120}{\text{breadth } (b)}$

If breadth is reduced by 4 m, the cost is only ₹ 100. It implies that the area is 100 m^2.

Now, length × breadth = 100

$$\Rightarrow \frac{120}{b} \times (b-4) = 100 \Rightarrow 120b - 480 = 100b$$

$$\therefore \quad b = 24 \text{ m}$$

When a rectangle/square is cut on its four vertices so as to form a regular polygon

Problem 69 Length and breadth of a rectangle are 8 cm and 6 cm, respectively. The rectangle is cut on its four vertices such that the resulting figure is a regular octagon. What is the side (in cm) of the octagon? **[SSC (CGL-T-II)-17]**

Sol. Clearly, each side of the octagon be $\sqrt{x^2+y^2}$.

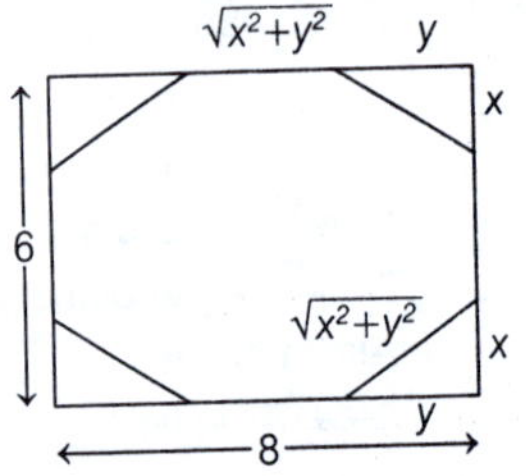

Then, $\sqrt{x^2+y^2} + x + x = 6$

$$\Rightarrow \quad x = \frac{6-k}{2}$$

where, $k = \sqrt{x^2+y^2}$

and $\sqrt{x^2+y^2} + y + y = 8$

$$\Rightarrow \quad y = \frac{8-k}{2}$$

Further, $x^2 + y^2 = k^2$

$$\Rightarrow \left(\frac{6-k}{2}\right)^2 + \left(\frac{8-k}{2}\right)^2 = k^2$$

$$\Rightarrow \quad 2k^2 - 28k + 100 = 4k^2 \Rightarrow k^2 + 14k - 50 = 0$$

$$\Rightarrow \quad k = \frac{-14 \pm \sqrt{(14)^2 - 4\cdot 1\cdot(-50)}}{2\cdot 1}$$

$$= -7 \pm \sqrt{99} = 3\sqrt{11} - 7 \{\because -3\sqrt{11} - 7 \text{ is not possible}\}$$

Problem 70 $PQRS$ is a square whose side is 16 cm. What is the value of the side (in cm) if the largest regular octagon that can be cut from the given square? **[SSC (CGL-T-II)-17, CAT-01]**

Sol. Clearly, each side of the octagon is $\sqrt{2}x$

Then, $x + \sqrt{2}x + x = 16$

$$\Rightarrow \quad x(2+\sqrt{2}) = 16$$

$$\Rightarrow \quad x = \frac{16}{2+\sqrt{2}}$$

Then, $\sqrt{2}x = \dfrac{16\sqrt{2}}{2+\sqrt{2}} = \dfrac{16}{\sqrt{2}+1}$

$$= \frac{16(\sqrt{2}-1)}{2-1} = 16\sqrt{2} - 16$$

Generic version Side of the largest regular octagon that can be cut from a square of side $x = x(\sqrt{2}-1)$. **Asked in CDS-19**

Problems based on Combination of Rectangle and Square

Problem 71 A rectangle plank $\sqrt{10}$ m wide, is placed symmetrically along the diagonal of a square of side 10 m as shown in the figure. The area of the plank is : **[IIFT-17]**

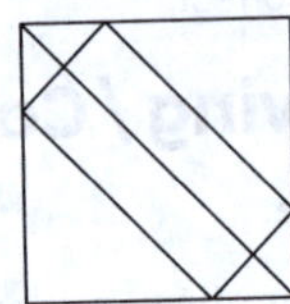

Sol. Given, $HE = GF = \sqrt{10}$

Since plank is placed symmetrically,

$AH = AE$.

Clearly, HAE is a right-angled triangle,

therefore $HE^2 = AH^2 + AE^2$

$$\Rightarrow \quad (\sqrt{10})^2 = 2 \times AE^2 \Rightarrow AE = \sqrt{5}$$

Then, $EB = 10 - \sqrt{5}$

Now, in right-angled ΔBEF,

$$EF^2 = EB^2 + BF^2$$

$$\Rightarrow \quad EF^2 = 2(10-\sqrt{5})^2 \qquad \{\because EB = BF\}$$

$$\Rightarrow \quad EF = \sqrt{2}(10-\sqrt{5})$$

Then, area of the plank $= HE \times EF = \sqrt{10} \times \sqrt{2}\,(10-\sqrt{5})$

$$= 10(\sqrt{20}-1) \text{ m}^2$$

Problem 72 Twelve equal squares are placed to fit in a rectangle of diagonal 5 cm. There are three rows containing four squares each. No gaps are left between adjacent squares. What is the area of each square? **[CSAT-18]**

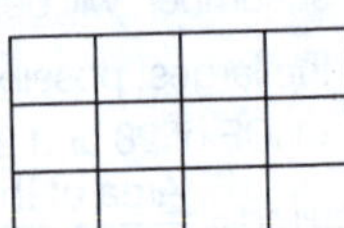

Sol. Clearly, ratio of length to breadth of rectangle = 4 : 3

Let length and breadth be $4x$ and $3x$ cm.

As given, $\sqrt{l^2+b^2} = 5 \Rightarrow (4x)^2 + (3x)^2 = 25 \Rightarrow x = 1$

Hence, length = 4 cm and breadth = 3 cm

Obviously, side of each square = 1 cm

$\therefore$ Area of each square $= 1^2 = 1 \text{ cm}^2$

WORKOUT

1. Rahim plans to draw a square *JKLM* with a point *O* on the side *JK* but is not successful. Why is Rahim unable to draw the square? **[CAT-07]**

(a) The length of *OM* is twice that of *OL*.

(b) The length of *OM* is 4 cm.

2. The adjoining figure contains three squares with areas of 100, 16 and 49 lying side by side as shown. By how much should the area of the middle square be reduced in order that the total length *PQ* of the resulting three squares is 19 ?

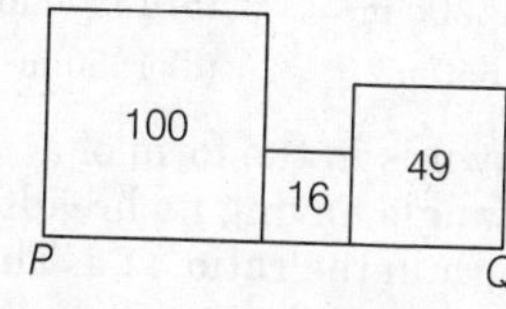

(a) 12 (b) 4
(c) 2 (d) $\sqrt{2}$

3. The length of a rectangle is increased to $\frac{3}{2}$ times of its length and breadth is reduced to $\frac{1}{3}$rd of its breadth. What is the change in area?

(a) $\frac{1}{3}$ (b) $\frac{1}{4}$
(c) $\frac{1}{2}$ (d) No change

4. A rectangular plot of lawn has dimensions *x* and *y* and is surrounded by a gravel pathway of width 2 m. What is the total area of the pathway ? **[IAS (Pre.)-97]**

(a) $2x + 2y + 4$
(b) $2x + 2y + 8$
(c) $4x + 4y + 8$
(d) $4x + 4y + 16$

5. The area of a rectangle, whose one side is *a* is $2a^2$. What is the area of a square having one of the diagonals of the rectangle as side ? **[CDS-12]**

(a) $2a^2$ (b) $3a^2$
(c) $4a^2$ (d) $5a^2$

6. The length of a rectangular field is double its width. Inside the field, there is a square-shaped pond 8 m long. If the area of the pond is $\frac{1}{8}$ of the area of the field, what is the length of the field ?

(a) 32 m (b) 16 m
(c) 64 m (d) 20 m

7. A person wishes to make a 100 sq m rectangular garden. Since he has only 30 m barbed wire for fencing, he fences only three sides letting the house wall act as the fourth side. The width of the garden is :

(a) 10 m (b) 5 m
(c) 50 m (d) 100 m

8. If a lawn 30 m long and 16 m wide is surrounded by a path 2 m wide, then what is the area of the path ? **[CDS-11, CDS-19]**

(a) 200 m^2 (b) 280 m^2
(c) 300 m^2 (d) 320 m^2

9. A square is of area 200 sq m. A new square is formed in such a way that the length of its diagonal is $\sqrt{2}$ times of the diagonal of the given square. Then, area of the new square formed is : **[SSC (CGL-Tier-II)-12]**

(a) $200\sqrt{2}\ m^2$ (b) $400\sqrt{2}\ m^2$
(c) $400\ m^2$ (d) $800\ m^2$

10. In measuring the sides of a rectangular plot, there is an excess of 5% on one side and 6% deficit on the other. Then the error percentage in the area calculated of the plot is : **[MAT-10]**

(a) 1% (b) 1.3%
(c) 1.5% (d) 3%

11. A rectangular area of 6 sq m is to be painted on a 3 m × 4 m board leaving a border of uniform width on all sides. What should be the width of the border ? **[CDS-10]**

(a) 0.25 m (b) 0.5 m
(c) 1 m (d) 3 m

12. Twenty nine times the area of a square is one square metre less than six times the area of the second square and nine times the side of it exceeds the perimeter of other square by one metre. The difference in sides of these squares is :

(a) 5 m (b) $\frac{54}{11}$ m
(c) 11 m (d) 6 m

13. How many tiles, each 4 decimeter square, will be required to cover the floor of a room 8 m long and 6 m broad ? **[SSC (CGL-Tier-II)-13]**

(a) 200 (b) 260
(c) 280 (d) 300

14. A gardener increased the area of his rectangular garden by increasing its length by 40% and decreasing its width by 20%. The area of the new garden : **[IAS (Pre.)-14]**

(a) has increased by 20%
(b) has increased by 12%
(c) has increased by 8%
(d) is exactly the same as the old area

15. The length of a rectangular blackboard is 8 m more than its breadth. If its length is increased by 7 m and its breadth is decreased by 4 m, its area remains unchanged. The length and breadth of the rectangular blackboard is : **[MAT-09]**

(a) 24 m, 16 m (b) 20 m, 24 m
(c) 28 m, 20 m (d) 28 m, 16 m

16. If the diagonals of two squares are in the ratio of 2 : 5, their areas will be in the ratio of : **[SSC (SO-Comm.)-03]**

(a) $\sqrt{2} : \sqrt{5}$ (b) 2 : 5
(c) 4 : 25 (d) 4 : 5

17. If the length of the diagonal of a square and that of the side of another square are both 10 cm, the ratio of the area of the first square to that of the second is :

(a) 1 : 2 (b) 1 : 3
(c) 1 : 4 (d) 2 : 3

18. A rectangular field has dimensions 25 m by 15 m. Two mutually perpendicular passages of 2 m width have been made in its central part and grass has been grown in rest of the field. Find the area under the grass. ***[MBA-07, IIFT-13]***

(a) 295 m^2 (b) 299 m^2
(c) 300 m^2 (d) 375 m^2
(e) None of these

19. On increasing each side of a square by 50%, the ratio of the area of new formed square and the given square will be : ***[SSC (CGL-Tier-II)-14]***

(a) 9 : 5 (b) 9 : 7
(c) 9 : 3.5 (d) 9 : 4

20. A rectangular farm has to be fenced on one long side, one short side and the diagonal. If the cost of fencing is ₹ 100 / m, the area of the farm is 1200 sq m and the short side is 30 m long. How much would the job cost ?

(a) ₹ 14000 (b) ₹ 12000
(c) ₹ 7000 (d) ₹ 15000

21. A landowner increased the length and breadth of a rectangular plot by 10% and 20% respectively. Find the percentage change in the cost of the plot.

(a) 35% (b) 33%
(c) 22.22% (d) None of these

22. The area of a square is 1024 cm^2. What is the ratio of the length to the breadth of a rectangle whose length is twice the side of the square and breadth is 12 cm less than the side of this square ?

(a) 5 : 18 (b) 16 : 7
(c) 14 : 5 (d) 32 : 5
(e) None of these

23. The length of the diagonal of a square is 'a' cm. Which of the following represents the area of the square (in sq cm)?

(a) $2a$ (b) $\frac{a}{\sqrt{2}}$
(c) $\frac{a^2}{2}$ (d) $\frac{a^2}{4}$

24. A garden is 24 m long and 14 m wide. There is a path 1 m wide outside the garden along its sides. If the path is to be constructed with square marble tiles 20 cm × 20 cm, find the number of tiles required to cover the path.

(a) 1800 (b) 2000
(c) 200 (d) 2150

25. If the ratio of the areas of two squares is 25 : 36, then ratio of their perimeters is : ***[CLAT-12]***

(a) 5 : 6 (b) 6 : 5
(c) 36 : 25 (d) 25 : 36

26. The length and breadth of a playground are 36 m and 21 m respectively. Poles are required to be fixed all along the boundary at a distance 3 m apart. The number of poles required will be : ***[MAT-08]***

(a) 39 (b) 38
(c) 37 (d) 40

27. In a rectangle, the difference between the sum of the adjacent sides and the diagonal is half the length of longer side. What is the ratio of the shorter to the longer side ? ***[CAT-97]***

(a) $\sqrt{3} : 2$ (b) $1 : \sqrt{3}$
(c) 2 : 5 (d) 3 : 4

28. Four sheets 50 cm × 5 cm are arranged without overlapping to form a square having side 55 cm. What is the area of inner square so formed ?

(a) 2500 cm^2
(b) 2025 cm^2
(c) 1600 cm^2
(d) None of the above

29. What would be the cost of building a fence around a square plot with area equal to 361 sq ft, if the price per foot of building the fence is ₹ 62. ***[OBC (PO)-09]***

(a) ₹ 4026
(b) ₹ 4712
(c) ₹ 3948
(d) Cannot be determined
(e) None of the above

30. Instead of walking along two adjacent sides of a rectangular field, a boy took a short-cut along the diagonal of the field and saved a distance equal to 1/2 of the longer side. The ratio of the shorter side of the rectangle to the longer side is : ***[FMS-11]***

(a) $\frac{1}{2}$ (b) $\frac{2}{3}$
(c) $\frac{1}{4}$ (d) $\frac{3}{4}$

31. A rectangular lawn 80 m by 60 m has two roads each 10 m wide running in the middle of it, one parallel to the length and the other parallel to the breadth. Find the cost of gravelling them at ₹ 30 per sq m.

(a) ₹ 39000 (b) ₹ 3900
(c) ₹ 3600 (d) ₹ 36000

32. The length of a rectangular floor is twice its breadth. If ₹ 256 is required to paint the floor at the rate of ₹ 2 / m^2, then what would be the length of the floor ? ***[Corporation Bank-PO-09]***

(a) 16 m (b) 8 m
(c) 12 m (d) 32 m
(e) 20 m

33. The perimeter of a rectangular field is 480 m and the ratio between the length and the breadth is 5 : 3. The area is :

(a) 15500 m^2 (b) 13500 m^2
(c) 1550 m^2 (d) 1350 m^2

34. A lawn is in the form of a rectangle having its breadth and length in the ratio 3 : 4. The area of the lawn is $\frac{1}{12}$ hectare. The breadth of the lawn is : ***[SSC (CGL-Tier-II)-13]***

(a) 25 m (b) 50 m
(c) 75 m (d) 100 m

35. The area (in m^2) of the square which has the same perimeter as a rectangle whose length is 48 m and is 3 times its breadth is :

(a) 1000 (b) 1024
(c) 1042 (d) 1600

36. The difference between the length and breadth of a rectangle is 23 m. If the perimeter is 206 m, then area is :

(a) 1520 m^2 (b) 2420 m^2
(c) 2480 m^2 (d) 2520 m^2

37. When the length of a rectangle field is increased by 3 m and the breadth is decreased by 3 m, the area is decreased by 18 m^2. If length and breadth each is increased by 3 m, the area is increased by 60 m^2. Find the length of the field.

(a) 10 m (b) 18 m
(c) 12 m (d) 32 m

38. The side of a square exceeds the side of another square by 3 cm and the sum of the areas of two squares is 549 cm^2. The perimeters of the two squares (in cm) are : *[UPPSC (Rev. Insp.)-15]*

(a) 40, 52 (b) 50 , 62
(c) 60 , 72 (d) 70, 82

39. Rectangular tiles each of size 70 cm × 30 cm must be laid horizontally on a rectangular floor of size 110 cm × 130 cm, such that the tiles do not overlap. A tile can be placed in any orientation so long as its edges are parallel to the edges of the floor. No tile should overshoot any edge of the floor. The maximum number of tiles that can be accommodated on the floor is : *[CAT-05]*

(a) 4 (b) 5 (c) 6 (d) 7

40. A farmer has decided to build a wire fence along one straight side of his property. For this, he planned to place several fence-posts at 6 m intervals, with posts fixed at both ends of the side. After he bought the posts and wire, he found that the number of posts he had bought was 5 less than required. However, he discovered that the number of posts he had bought would be just sufficient if he spaced them 8 m apart. What is the length of the side of his property and how many posts did he buy ? *[CAT-2000]*

(a) 100 m, 15 (b) 100 m, 16
(c) 120 m, 15 (d) 120 m, 16

41. The adjoining figure shows a set of concentric squares. If the diagonal of the innermost square is 2 unit and if the distance between the corresponding corners of any two successive squares is 1 unit, then find the difference between the areas of the eighth and the seventh squares, counting from the innermost square. *[CAT-1997]*

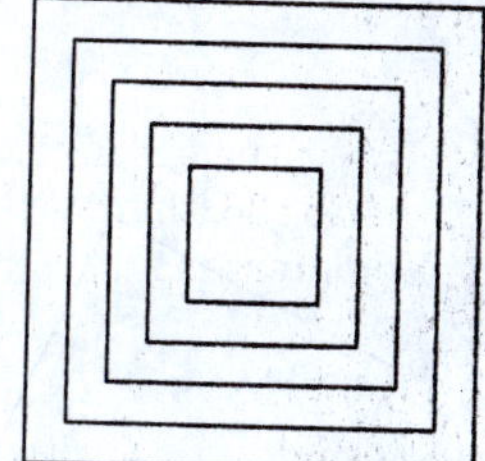

(a) $10\sqrt{2}$ sq unit (b) 30 sq unit
(c) $35\sqrt{2}$ sq unit (d) None of these

42. In the adjoining figure, $AC + AB = 5AD$ and $AC - AD = 8$. Then, the area of the rectangle $ABCD$ is : *[CAT-95]*

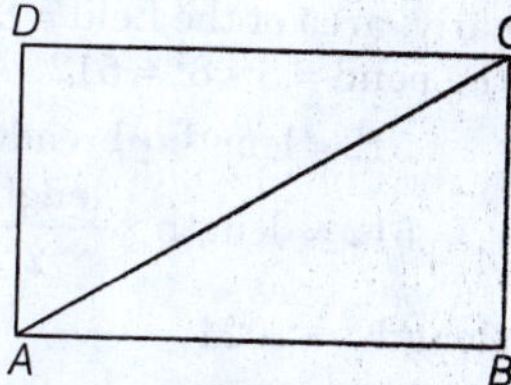

(a) 36 (b) 50 (c) 60
(d) cannot be answered

43. Akhtar plans to cover a rectangular floor of dimensions 9.5 m and 11.5 m using tiles. Two types of square shaped tiles are available in the market. A tile with side 1 m cost ₹ 100 and a tile with side 0.5 m costs ₹ 30. The tiles can be cut if required. What will be the minimum cost of covering the entire floor with tiles ? *[XAT-16]*

(a) 10930 (b) 10900
(c) 11000 (d) 10950
(e) 10430

44. In an engineering college there is a rectangular garden of dimensions 34 m by 21 m. Two mutually perpendicular walking corridors of 4 m width have been made in the central part and flowers have been grown in the rest of the garden. The area under the flowers is : *[IIFT-13]*

(a) 320 sq m (b) 400 sq m
(c) 510 sq m (d) 630 sq m

45. The central park of the city is 40 m long and 30 m wide. The mayor wants to construct two roads of equal width in the park such that the roads intersect each other at right angles and the diagonals of the park are also the diagonals of the small rectangle formed at the intersection of the two roads. Further, the mayor wants that the area of the two roads to be equal to the remaining area of the park. What should be the width of the roads ? *[XAT-13]*

(a) 10 m (b) 12.5 m
(c) 14 m (d) 15 m
(e) 16 m

46. If each of the dimensions of a rectangle is increased by 200%, the area is increased by : *[CDS-17]*

(a) 300% (b) 400%
(c) 600% (d) 800%

47. A square field has its area equal to 324 m^2. The perimeter of the field is : *[CLAT-14]*

(a) 36 m (b) 72 m (c) 18 m (d) 65 m

48. The area of a square is 225 sq cm which is equal to the area of a rectangle. The length of the rectangle is 16 cm more than the breadth of the rectangle. What is the respective ratio between the side of the square and the breadth of the rectangle ? *[CLAT-12]*

(a) 3 : 5 (b) 5 : 3
(c) 5 : 4 (d) 4 : 5

49. The diagonal of a square A is $(a + b)$ units. What is the area (in square units) of the square drawn on the diagonal of square B whose area is twice the area of A ?

(a) $4(a + b)^2$ (b) $(a + b)^2$
(c) $8(a + b)^2$ (d) $2(a + b)^2$

Answer Sheet

1. (a)	2. (a)	3. (c)	4. (d)	5. (d)	6. (a)	7. (b)	8. (a)	9. (c)	10. (b)
11. (b)	12. (d)	13. (d)	14. (b)	15. (c)	16. (c)	17. (a)	18. (b)	19. (d)	20. (b)
21. (d)	22. (e)	23. (c)	24. (b)	25. (a)	26. (b)	27. (d)	28. (b)	29. (b)	30. (d)
31. (a)	32. (a)	33. (b)	34. (a)	35. (b)	36. (d)	37. (a)	38. (c)	39. (c)	40. (d)
41. (b)	42. (c)	43. (a)	44. (c)	45. (a)	46. (d)	47. (b)	48. (b)	49. (d)	

Hints and SOLUTIONS

1. Draw a square $JKLM$ and point O on side JK and draw a line joining points O and M.

It is clear that OM will have maximum value if it is diagonal of the square and will be equal to $\sqrt{2}$ times of any one side.

It means, when OM is maximum,

$$OM = \sqrt{2} \times OL.$$

Thus, $OM \neq 2 \times OL$ if O lies on JK. That is why, Rahim is unable to draw the square. Statement (b) offers no additional or relevant information. Hence, option (a) is correct.

2. Length of

$$PQ = \sqrt{100} + \sqrt{16} + \sqrt{49} = 21$$

But we have to make 21 as 19 by reducing the area of the middle square. If we replace 16 by 4 we can get 19 instead of 21. Thus, reduction in area of the middle square

$$= 16 - 4 = 12$$

3. Clearly, length has increased by 50% and breadth has reduced by 66.66%.

Now, effect on the area

$$= \left[+ 50 - 66.66 - \frac{50 \times 66.66}{100} \right] \%$$

$$= - 50\%$$

'–' indicates that the area will decrease by 50%, i.e. $\frac{1}{2}$.

4. (*d*) We have already derived the formula of the area of the outer path as $2x\,(l + b + 2x)$.

Here, area $= 2 \cdot 2\,(x + y + 2 \cdot 2)$

$$= 4x + 4y + 16$$

5. Another side of the rectangle (say, length)

$$= \frac{\text{area}}{\text{one side}}$$

$$= \frac{2a^2}{a} = 2a \text{ unit}$$

Now, its diagonal

$$= \sqrt{l^2 + b^2}$$

$$= \sqrt{(2a)^2 + a^2}$$

$$= a\sqrt{5} \text{ unit}$$

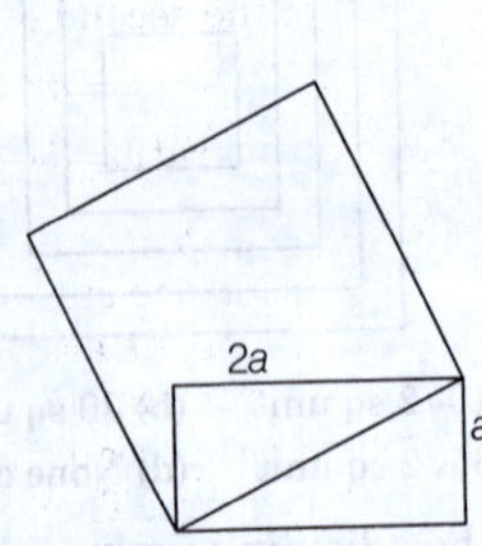

Since this diagonal is a side of the square, the area of the square will be $(a\sqrt{5})^2$, i.e. $5a^2$ sq units.

6. Clearly, area of the field $= 8 \times$ area of the pond $= 8 \times 8^2 = 512$

$$\Rightarrow \quad 512 = \text{length} \times \text{breadth}$$

$$\Rightarrow \quad 512 = \text{length} \times \frac{\text{length}}{2}$$

$$\Rightarrow \text{length}^2 = 1024$$

$$\Rightarrow \quad \text{length} = 32 \text{ m}$$

7. Given, sum of three sides = 30, it means there are two cases

Case (i), $l + b + l = 30$

$$\Rightarrow \quad l = 15 - \frac{b}{2}$$

Also given, area = 100

$$\Rightarrow \quad lb = 100$$

$$\Rightarrow \quad \left(15 - \frac{b}{2}\right) b = 100$$

$$\Rightarrow 30b - b^2 - 200 = 0$$

$$\Rightarrow b^2 - 30b + 200 = 0$$

$$\Rightarrow b^2 - 20b - 10b + 200 = 0$$

$$\Rightarrow b(b - 20) - 10(b - 20) = 0$$

$$\Rightarrow b = 10 \text{ and } 20 \text{ then, } l = 10 \text{ and } 5$$

If $b = 10$, the garden will become a square and if $b = 20$, l will be 5, which is not possible. Hence, we should now analyse the second case.

Case (ii), $b + l + b = 30$

$$\Rightarrow \quad l = 30 - 2b$$

Now, $lb = 100$ (as given)

$$\Rightarrow \quad (30 - 2b) \times b = 100$$

$$\Rightarrow \quad 2b^2 - 30b + 100 = 0$$

$$\Rightarrow \quad b^2 - 15b + 50 = 0$$

$$\Rightarrow b^2 - 10b - 5b + 50 = 0$$

$$\Rightarrow b(b - 10) - 5(b - 10) = 0$$

$$\Rightarrow b = 10 \text{ or } 5 \text{ then, } l = 10 \text{ or } 20$$

If $b = 10$, the garden will become a square and if $b = 5$, l will be 20.

Hence the garden is measuring 20 m × 5 m.

Alternatively

Proceed through options

(a) $b = 10$, then l will be 10, not a rectangular field.

(b) $b = 5$, then l will be 20, forms a rectangular shape. Also, $5 + 20 + 5 = 30$

Hence, option (b) is correct.

8. We know, an area of the outer path

$$= 2x\,(l + b + 2x)$$

$$= 2 \cdot 2(30 + 16 + 2 \cdot 2) = 200 \text{ m}^2$$

{Details may be observed in the beginning of this chapter}

9. Given, ratio of their diagonals $= 1 : \sqrt{2}$

Then, ratio of their areas

$$= (1)^2 : (\sqrt{2})^2 = 1 : 2$$

But it is given that area of 1st square is 200 m², then area of the new square will be 2×200 i.e. 400 m².

10. Error %

$$= \left[+ 5 - 6 - \frac{5 \times 6}{100} \right] \% = - 1.3\%$$

11. We know, area of the inner path (say, border) $= 2x\,(l + b - 2x)$

$$\Rightarrow \quad (12 - 6) = 2x\,(4 + 3 - 2x)$$

$$\Rightarrow \quad 4x^2 - 14x + 6 = 0$$

$$\Rightarrow \quad 2x^2 - 7x + 3 = 0$$

$$\Rightarrow \quad 2x^2 - 6x - x + 3 = 0$$

$$\Rightarrow 2x(x - 3) - 1(x - 3) = 0$$

$$\Rightarrow \quad x = 3 \text{ or } \frac{1}{2}$$

If the breadth of the board is 3 m, the width of the border cannot be 3 m. Hence, the width of the border is 0.5 m.

Alternatively

Proceed through the options so as to test which option suffices that the area of the border is 6 sq m.

(a) $0.25 \Rightarrow$

$$2 \cdot (0.25)\,[4 + 3 - 2 \cdot (0.25)] \neq 6$$

(b) $0.5 \Rightarrow$

$$2 \cdot (0.5)\,[4 + 3 - 2 \cdot (0.5)] = 6$$

Hence, correct option is (b).

12. Let their sides be x metre and y metre, respectively. As given, $9x-1=4y$

$\Rightarrow \quad y=\frac{9x-1}{4}$

Also given, $29x^2+1=6y^2$

$\Rightarrow \quad 29x^2+1=6\left(\frac{9x-1}{4}\right)^2$

$\Rightarrow \quad 232x^2+8=243x^2+3-54x$

$\Rightarrow \quad 11x^2-54x-5=0$

$\Rightarrow \quad 11x^2-55x+x-5=0$

$\Rightarrow 11x(x-5)+1(x-5)=0$

$\Rightarrow \quad x=5$ and $\frac{-1}{11}$

But the side of a square cannot be negative. Hence, $x=5$.

Then, $y=\frac{9\times5-1}{4}=11$

Now, $y-x=11-5=6$ m

13. Number of tiles $=\frac{\text{Area of the floor}}{\text{Area of one tile}}$

$=\frac{80\times60}{4\times4}=300$

[$\because$ 1 m = 10 decimeter]

14. Effect on area

$=\left[+40-20-\frac{40\times20}{100}\right]\%=+12\%$

Hence, the area will increase by 12%.

15. Since there is no alternation in the area, original area will be equal to the area after changes in the dimensions.

Hence, $(l+7)(b-4)=(b+8)b$

$[\because l=b+8]$

$\Rightarrow \quad (b+8+7)(b-4)=(b+8)b$

$\Rightarrow \quad (b+15)(b-4)=b^2+8b$

$\Rightarrow b^2+11b-60-b^2-8b=0$

$\Rightarrow \quad 3b=60\Rightarrow b=20$

Then, $\quad l=20+8=28$

Alternatively

Proceed through options :

(a) $24\times16\neq(24+7)\times(16-4)$

(b) is not possible as length does not exceed breadth by 8 m.

(c) $28\times20=(28+7)\times(20-4)$, hence, option (c) is correct.

16. We know that if ratio of diagonals of two squares is $a:b$, the ratio of their areas will be $a^2:b^2$. Thus, the required ratio

$=2^2:5^2=4:25$

17. Given, diagonal of the first square $=10$

$\Rightarrow \quad \sqrt{2}a=10\Rightarrow a=5\sqrt{2}$,

then area $=(5\sqrt{2})^2=50$

and area of the second square

$=a^2=(10)^2=100$

Now, required ratio $=50:100=1:2$

Alternatively I

We know when the side of a square is made n times, its area becomes n^2 times. Here, the diagonal of the first square has been treated as a side of the second square. Then, desired ratio $=1:(\sqrt{2})^2=1:2$

Alternatively II

Clearly, ratio of the sides (first square : second square)

$=\frac{\text{given diagonal}}{\sqrt{2}}$: given side

$=1:\sqrt{2}$

Then, ratio of their squares

$=(1)^2:(\sqrt{2})^2=1:2$

18. We know, area of the path $=x(l+b-x)$, details given in the beginning.

$=2(25+15-2)=76\text{ m}^2$

Thus, area under the grass

$=(25\times15)-76=299\text{ m}^2$

19. Given, ratio of their sides (new : original) $=15:1$

Then, ratio of their areas

$=(1.5)^2:(1)^2=2.25:1=9:4$

20. Length of the farm

$=\frac{1200}{30}=40$ m

and diagonal $=\sqrt{l^2+b^2}$

$=\sqrt{(40)^2+(30)^2}=50$ m

Total length to be fenced

$=40+30+50=120$ m

Then, cost $=120\times100=$ ₹ 12000

21. Effect on the cost

$=\left[+10+20+\frac{10\times20}{100}\right]\%$

$=+32\%$

Hence, the cost will increase by 32%.

22. Side of the square $=\sqrt{1024}$

$=32$ cm

Now, the required ratio

$=(2\times32):(32-12)$

$=64:20=16:5$

23. Side of the square

$=\frac{\text{diagonal}}{\sqrt{2}}=\frac{a}{\sqrt{2}}$

Then, its area $=\left(\frac{a}{\sqrt{2}}\right)^2=\frac{a^2}{2}\text{ cm}^2$

24. Number of tiles

$=\frac{\text{Area of the outer path}}{\text{Area of one tile}}$

$=\frac{2x(l+b+2x)}{0.20\times0.20}$

$=\frac{2\cdot1(24+14+2\cdot1)}{0.20\times0.20}=2000$

25. Required ratio

$=4\sqrt{25}:4\sqrt{36}=5:6$

26. Perimeter of the playground

$=2(36+21)=114$ m

Then, number of poles required

$=\frac{114}{3}=38$

27. Let the length and breadth be l and b, respectively.

Then, diagonal $=\sqrt{l^2+b^2}$

As given, $l+b-\sqrt{l^2+b^2}=\frac{1}{2}$

$\Rightarrow \quad \sqrt{l^2+b^2}=\frac{1}{2}+b$

$\Rightarrow \quad l^2+b^2=\frac{l^2}{4}+b^2+lb$

$\Rightarrow \quad \frac{3l^2}{4}=lb$

$\Rightarrow \quad b:l=3:4$

28. Area of the inner square

$=(55-2\times5)\times(55-2\times5)$

$=2025\text{ cm}^2$

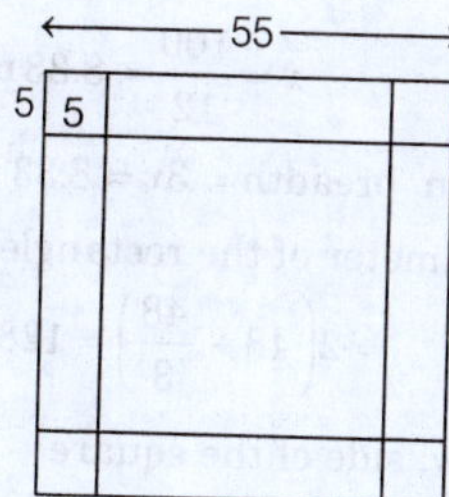

29. Side of the square $=\sqrt{361}=19$ ft

Cost = Perimeter of the square × Rate

$=(4\times19)\times62=$ ₹ 4712

30. Let the length and the breadth be l and b, respectively.

By given conditions,

$l+b-\frac{1}{2}l=\sqrt{l^2+b^2}$

$\Rightarrow \quad b + \frac{1}{2}l = \sqrt{l^2 + b^2}$

Squaring both sides, we get

$b^2 + \frac{1}{4}l^2 + bl = l^2 + b^2$

$\Rightarrow \quad bl = \frac{3}{4}l^2 \Rightarrow b : l = \frac{3}{4}$

31. We know, area of this type of path $= x(l + b - x)$, details given in beginning.

$= 10(80 + 60 - 10) = 1300\text{ m}^2$

Now, cost of gravelling $= 1300 \times 30$

$= ₹\,39000$

32. Area of the floor $= \frac{256}{2} = 128\text{ m}^2$

$\Rightarrow$ length $\times \frac{\text{length}}{2} = 128$

($\because$ length $= 2 \times$ breadth, as given)

$\Rightarrow$ length $\times$ length $= 256$

$\Rightarrow \quad$ length $= 16$ m

33. Perimeter $= 2(5 + 3)$ units

$\Rightarrow$ 480 m

$\therefore$ 1 unit $\Rightarrow$ 30 m

Hence, length $= 5$ units

$= 5 \times 30 = 150$ m

and breadth $= 3$ units $= 3 \times 30$

$= 90$ m

Now, area $= 150 \times 90 = 13500\text{ m}^2$

34. Let the breadth and length be $3x$ m and $4x$ m, respectively.

Given, $3x \times 4x = \frac{1}{12} \times 10000\text{ m}^2$

$\{\because 1 \text{ hectare} = 10000\text{ m}^2\}$

$\Rightarrow \quad x^2 = \frac{10000}{12 \times 12}$

$\Rightarrow \quad x = \frac{100}{12} = 8.33$ m

Then, breadth $= 3x = 8.33 \times 3 = 25$ m

35. Perimeter of the rectangle

$= 2\left(48 + \frac{48}{3}\right) = 128$

Now, side of the square

$= \frac{\text{perimeter}}{4} = \frac{128}{4} = 32$

Then, its area $= (32)^2 = 1024\text{ m}^2$

36. Perimeter of the rectangle

$= 2(l + b) = 206$

$\Rightarrow (l + b) = 103$

$\Rightarrow \quad (l + l - 23) = 103 \quad \{\because l = b + 23\}$

$\Rightarrow l = 63$ and $b = 63 - 23 = 40$

Now, area $= 63 \times 40 = 2520\text{ m}^2$

37.

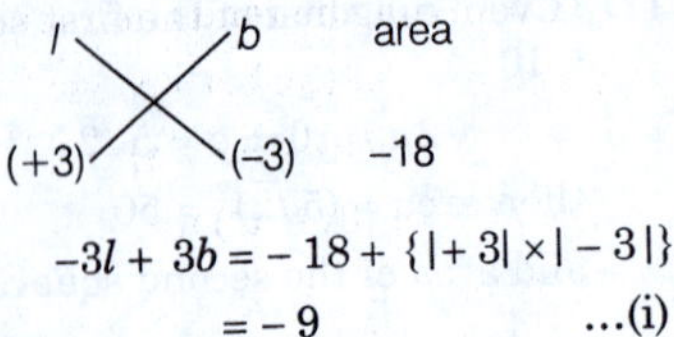

$-3l + 3b = -18 + \{|+3| \times |-3|\}$

$= -9 \quad$ ...(i)

Further,

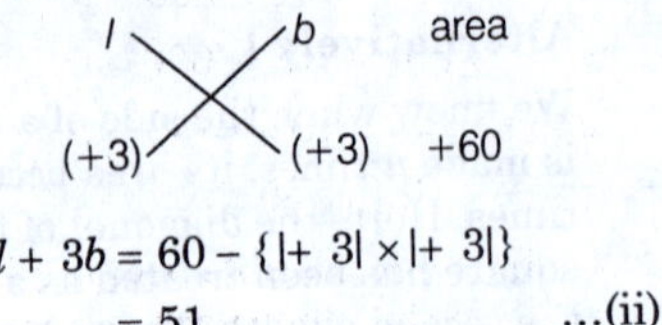

$3l + 3b = 60 - \{|+3| \times |+3|\}$

$= 51 \quad$ ...(ii)

On solving both equations, we get $b = 7$ m and $l = 10$ m

38. Clearly, their sides are a and $(a + 3)$.

Given, $\quad (a)^2 + (a + 3)^2 = 549$

$\Rightarrow \quad a^2 + a^2 + 6a + 9 = 549$

$\Rightarrow \quad 2a^2 + 6a - 540 = 0$

$\Rightarrow \quad a^2 + 3a - 270 = 0$

$\Rightarrow a^2 + 18a - 15a - 270 = 0$

$\Rightarrow a(a + 18) - 15(a + 18) = 0$

$\Rightarrow \quad a = 15$ and -18

$\Rightarrow \quad a = 15$

[$\because$ sides cannot be negative]

Hence, perimeters will be (4×15) and $\{4 \times (15 + 3)\}$ i.e. 60 and 72

39. The maximum number of tiles can be arranged in the above manner.

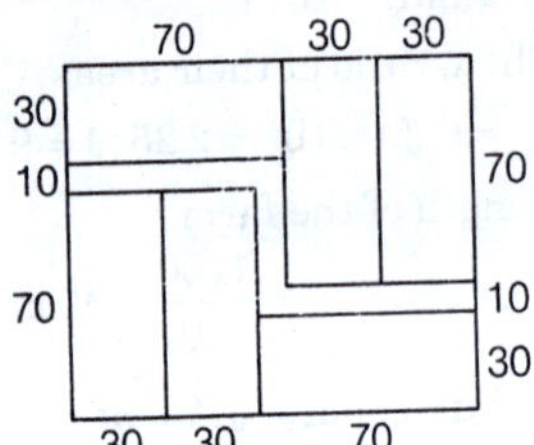

Hence, option (c) is correct.

40. Since the farmer covers entire length placing posts at a distance of either 6 m or 8 m, the length of the property must be divisible by both of 6 and 8. LCM of 6 and 8 = 24.

Number of posts if it is placed at 6 m intervals $= (24 \div 6) + 1$

$= 5$

and number of posts if it is placed at 8 m intervals $= (24 \div 8) + 1$

$= 4$

$\because$ There is a difference of 1 post when length of the side be 24 m

$\therefore$ There will be a difference of 5 posts when length be 120 m.

Hence, length of the side of the property = 120 m and number of posts he bought

$= (120 \div 8) + 1 = 16$

Or, number of posts he bought

$= \{(120 \div 6) + 1\} - 5 = 16$

41. Diagonal of innermost square = 2 (given)

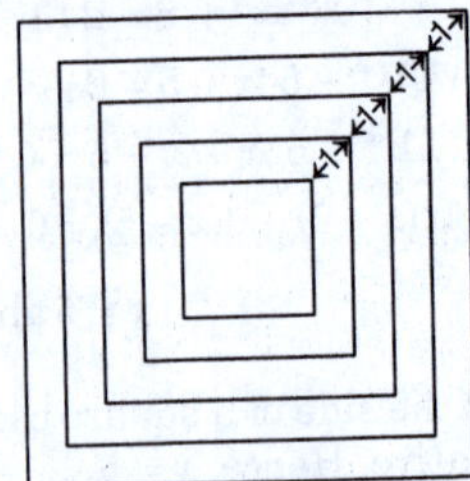

Then, diagonal of the square just bigger than the innermost

$= 2 + 1 + 1 = 4$

Diagonal of the square just bigger than the previous i.e. 3rd square, counting from the innermost square

$= 4 + 2 = 6$

$\therefore$ Diagonal of the 7th square and 8th square respectively = 14 and 16

We know, area of the square $= \frac{d^2}{2}$

where, d = its diagonal.

Then, required difference

$= \frac{(16)^2}{2} - \frac{(14)^2}{2}$

$= 128 - 98 = 30$ sq unit

42. Given, $AC + AB = 5AD$

$\Rightarrow \quad 8 + AD + AB = 5AD$

$[\because AC - AD = 8 \;\therefore AC = 8 + AD]$

$\Rightarrow \quad AB = 4(AD - 2)$

Now, $\quad AC^2 = AB^2 + BC^2$

$= AB^2 + AD^2$

$[\because BC = AD]$

$\Rightarrow (8 + AD)^2 = \{4(AD - 2)\}^2 + AD^2$

$\Rightarrow (8 + AD)^2 - AD^2$

$= 16AD^2 + 64 - 64AD$

$\Rightarrow 8 \times (8 + 2AD)$

$= 16AD^2 + 64 - 64AD$

$\Rightarrow \quad 80AD = 16AD^2$

$\Rightarrow \quad AD = 5$

Then, $\quad AB = 4(5 - 2) = 12$

$\therefore$ Area of rectangle $ABCD$

$= AB \times AD$

$= 12 \times 5 = 60$ sq unit

43. Area of the floor

$= 9.5 \times 11.5 = 109.25$ sq m

$= 109 + 0.25$

109 tiles of size 1 m × 1 m are required to cover 109 sq m and 1 tile of size 0.5 m × 0.5 m is required to cover 0.25 sq m.

Then, total cost

$= (109 \times 100) + (1 \times 30) = 10930$

{We used maximum number of bigger tiles as it is cheaper than the smaller tiles. The reason is that area to be covered by 4 smaller tiles is equal to the area to be covered by one bigger tile but cost of 4 smaller tiles is greater than 1 big tile.}

44. Observing the concept given in beginning of this chapter under title 'When two straight inner paths intersect each other perpendicularly' the area of the path will be $x(l + b - x)$'

i.e. $4(34 + 21 - 4) = 204\,m^2$.

$\therefore$ Area under the flowers

$= (34 \times 21) - 204 = 510$ sq m

45. Let width of the roads be x metre.

Then, area of the both roads

$= x(40 + 30 - x)$

$= (70x - x^2)\,m^2$

As given,

$(70x - x^2) = (40 \times 30) - (70x - x^2)$

$\Rightarrow \quad x^2 - 70x + 600 = 0$

$\Rightarrow x^2 - 60x - 10x + 600 = 0$

$\Rightarrow x(x - 60) - 10(x - 60) = 0$

$\Rightarrow \quad x = 10$ and 60

$x = 60$ is not possible because the width of the inner path cannot be greater than any side of the park. Hence, width of the roads = 10 m.

Note Intersection of two roads will obviously form a rectangle with two adjacent sides equal i.e. a square.

46. Effect on area

$= \left\{+ 200 + 200 + \frac{200 \times 200}{100}\right\}\%$

$= +800\%$

Hence, area will increase by 800%.

47. Perimeter = 4 × Side

$= 4 \times \sqrt{324} = 72$ m

48. Let breadth of the rectangle be b.

Then, $(b + 16)b = 225$

$\Rightarrow \quad (b + 16)b = (9 + 16)9$

$\Rightarrow \quad b = 9$

Further, side of the square

$= \sqrt{225} = 15$

Hence, required ratio $= 15 : 9 = 5 : 3$

49. Area of square $A = \frac{(a + b)^2}{2}$

Then, area of square B

$= 2 \times \frac{(a + b)^2}{2} = (a + b)^2$

$\therefore$ Side of square B

$= \sqrt{(a + b)^2} = (a + b)$

Now, diagonal of square B

$= \sqrt{2}\,(a + b)$

= side of the third square

$\therefore$ Area of 3rd square

$= \{\sqrt{2}(a + b)\}^2 = 2(a + b)^2$

41

Triangle

Learning OBJECTIVES

After completing this chapter, aspirants will be able to :

- Understand the significance of this chapter in competitive exams;
- Know the types of triangles;
- Be familiar with fundamental properties of triangle (with reference to Mensuration only);
- Compute the area of various triangles in different ways;
- Determine the effect on area and perimeter in case alternation in dimension(s) is made;
- Recognise the critical Pythagorean triplets with their properties;
- Compare between square and equilateral triangle in diverse cases;
- Observe how the concept of 'Product Constancy, works in this chapter;
- Apply the concept of concept of 'effect on product in case of changes in factors';
- Understand the philosophy of the division of this chapter in various parts;
- Realise themselves competent to solve the questions in ways other than the conventional way;
- Recognise the types of question which can be easily solved through the given options;
- Analyse the trend of the questions being asked in various competitive examinations;
- Solve the questions in moments;
- Grip on concepts helping to solve the questions without pen and paper.
- Learn many new concepts/fundamentals; and
- Apply the pragmatic approaches in lieu of academic approaches.

Concepts of Median, Inradius, Circumradius, Congruency, Similarity, Angle-bisector, Geometrical properties of triangle, etc. and problems based thereon have been detailed in 'Geometry Section'. Here, aspects related to Mensuration only have been considered.

Inside

Introduction

A triangle is a polygon having three sides. There are three interior angles $\angle A$, $\angle B$ and $\angle C$ and three sides AB (i.e. c), AC (i.e. b) and BC (i.e. a).

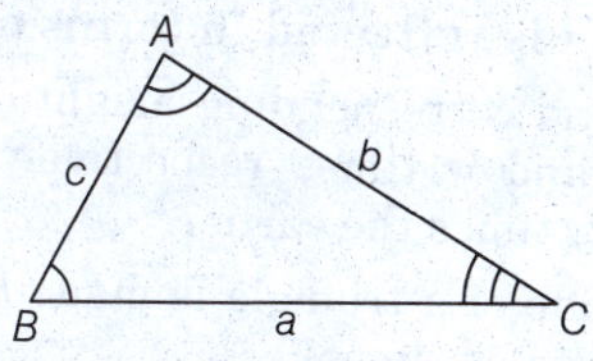

Types of Triangles

A. Based on Interior Angles

(i) **Acute angle triangle :** Triangle with all three angles acute ($0° < \theta < 90°$).

(ii) **Right angled triangle :** Triangle with one of the angles equal to 90°.

(iii) **Obtuse angle triangle :** Triangle with one of angles obtuse ($90° < \theta < 180°$).

B. Based on Length of Sides

(i) **Equilateral triangle :** Triangle with all sides equal.

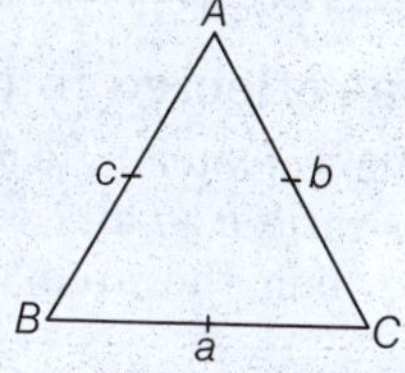

Here, $AB = BC = AC$

i.e. $c = a = b$

(ii) **Isosceles triangle :** Triangle with two of its sides equal.

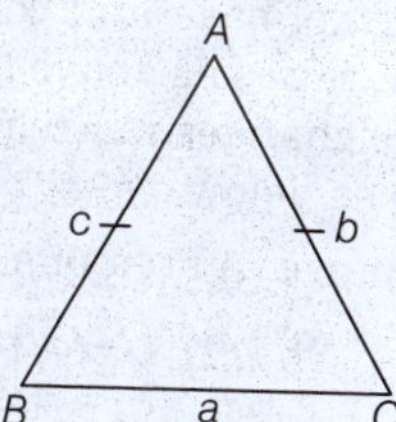

Here, $AB = AC$ i.e. $c = b$

(iii) **Scalene triangle :** Triangle with none of the sides equal to any other side.

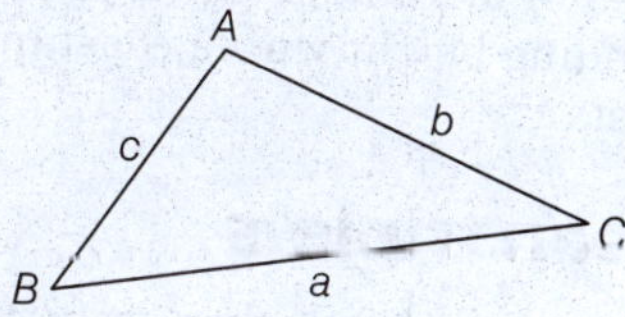

Here, $AB \neq BC \neq AC$ i.e. $c \neq a \neq b$

Fundamental Properties of Triangle (with reference to Mensuration only)

A. Perimeter Related

1. Perimeter of a triangle = sum of lengths of all sides
2. If each side of a triangle is made n times, the perimeter will become n times.
3. If each side of a triangle is increased or decreased by n units the perimeter will increase/decrease by $3n$ units.
4. If areas of the different types of triangles are same, then perimeter of an equilateral triangle will be minimum.
5. The perimeter of the triangle formed by joining the mid-points of the sides of a triangle is half the perimeter of the triangle.

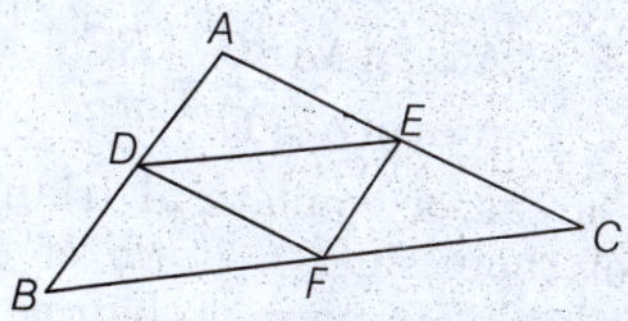

Here, perimeter of $\Delta DEF = \frac{1}{2} \times$ perimeter of ΔABC

6. If mid-points of all sides of a triangle are joined with each other, the triangles formed, 4 in number, are congruent to each other. Here, ΔADE, ΔBDF, ΔEFC and ΔDEF all are congruent.

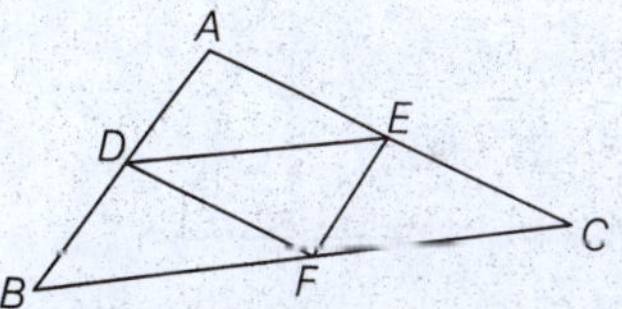

Further, each of the four small triangles is similar to the original triangle. Hence, if ΔABC is an equilateral triangle, all the four triangles will also be equilateral. In case of isosceles, all four will be isosceles and in case of scalene, all will be scalene.

B. Area Related

1. Area of square > area of rectangle > area of triangle under the given equal perimeters.
2. For the given equal perimeter, an equilateral triangle has maximum area in comparison to other types of triangles.
3. For the given equal hypotenuse, an isosceles right triangle has maximum area in comparison to other right-angled triangles.
4. Ratio of areas of triangles = ratio of their heights [if their bases are equal or same]

Here, $\frac{\text{Area of } \Delta ABC}{\text{Area of } \Delta EFG} = \frac{AD}{GH}$ $[\because BC = EF]$

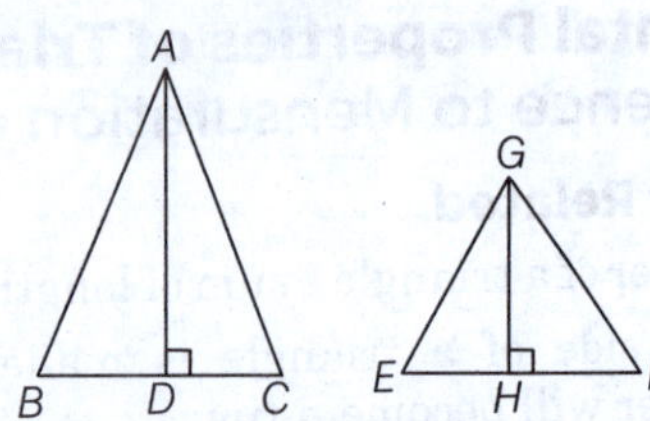

5. Ratio of areas of triangles = ratio of their bases [if their heights are equal or same]

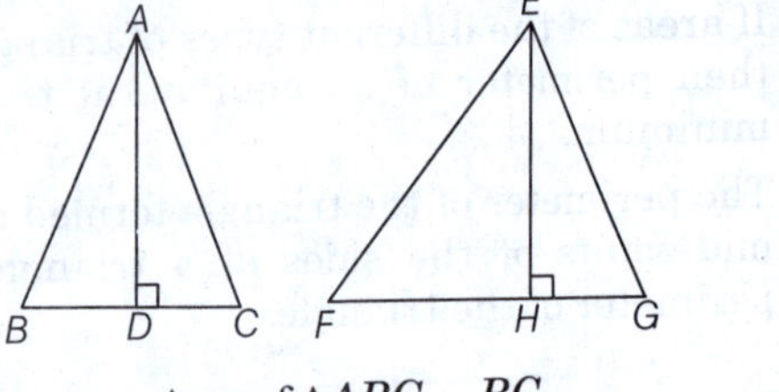

Here, $\frac{\text{Area of } \Delta ABC}{\text{Area of } \Delta EFG} = \frac{BC}{FG}$ $[\because AD = EH]$

6. Ratio of areas of equilateral triangles = Ratio of square of their sides = Ratio of square of their heights = Ratio of square of their perimeters.

Computation of Area of Triangles

(i) Area $= \frac{1}{2} \times$ Base $\times$ Height

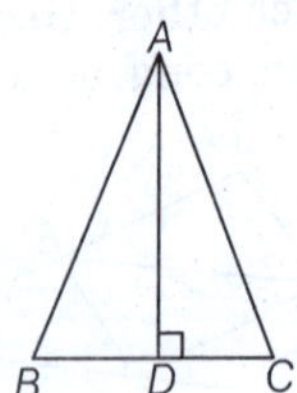

Here, Height refers to a perpendicular distance between the base and the vertex opposite to it.

Area $= \frac{1}{2} \times BC \times AD$

(ii) Area $= \sqrt{s(s-a)(s-b)(s-c)}$, where, $s = \frac{a+b+c}{2}$

It is important to note that each of the above approaches/formulae is applicable for all types of triangles.

(iii) Area, Height and Perimeter of an equilateral triangle with side 'a' are respectively $\frac{\sqrt{3}}{4} a^2, \frac{\sqrt{3}\, a}{2}$ and $3a$.

(iv) Area of an equilateral triangle $= \frac{(\text{Perimeter})^2}{12\sqrt{3}}$

(v) Area of an isosceles triangle $= \frac{b}{4}\sqrt{4a^2 - b^2}$, where a is the length of each of equal sides and b is the length of the third side.

(vi) The maximum area of an isosceles triangle under given perimeter P, is $\frac{P^2}{12\sqrt{3}}$.

(vii) Effect on area due to alteration in dimension(s) :

- **If alternation is reflected in terms of ... times**

1. If base and corresponding height of a triangle are made x and y times respectively, the area will become xy times the earlier.
2. If each side of a triangle is made n times, its area will become n^2 times.

****

- **If alternation is reflected in terms of percentage**

3. Recall the 'concept of effect on product in case of changes in factors' detailed in chapter percentage, if dimensions of a triangle are changed by percentage. It should be kept in mind that an area is a product of two factors *viz* base and height or side and side with a constant term [*viz.* Area $= \frac{1}{2} \times$ product of any two sides $\times$ sine of the included angle (discussed in Geometry section) or Area $= \frac{1}{2} \times$ base $\times$ height]

****

- **If alternation is reflected in terms of ratio**

4. If the dimensions of a triangle are changed in ratios (say, base in ratio $a : b$ and corresponding height in ratio $c : d$), then the area will change in ratio $ac : bd$.
   (a) If $ac > bd$, the area diminishes.
   (b) If $ac < bd$, the area increases.
   (c) If $ac = bd$, no changes in the area.

****

5. If dimensions (i.e. base and corresponding height) are changed in ratios as above and you are required to compute the changes in area in percentage term, you should do as below.

Percentage change in area of a triangle

$$= \left(\frac{ac - bd}{ac} \times 100\right)\%$$

[if the outcome is positive, area diminishes and in case of negative it increases]

(viii) We can apply the concept of 'Product Constancy' when area of a triangle (i.e. a product of base and corresponding height with a constant term) is to be kept constant.

Pythagorean triplets

A set of three numbers which satisfy the Pythagoras theorem is known as Pythagorean triplet.

$$3^2 + 4^2 = 5^2$$

[Here, (3, 4, 5) is termed as Pythagorean triplet.]

Examples of such triplets

(5, 12, 13), (7, 24, 25), (8, 15, 17), (9, 40, 41), (11, 60, 61), (12, 35, 37), (16, 63, 65), (20, 21, 29), (28, 45, 53), (33, 56, 65), etc.

No doubt, these triplets are very useful as many questions are based directly on them. It needs to be mentioned that (3, 4, 5) is the only triplet , which has **consecutive integers.** (asked in **CDS-19**).

When a Pythagorean triplet is either multiplied or divided by a real positive number, other triplets come out. For example :

$$\begin{array}{ccc} 5 & 12 & 13 \Rightarrow \text{a triplet} \\ \downarrow \times 1.2 & \downarrow \times 1.2 & \downarrow \times 1.2 \\ 6 & 14.4 & 15.6 \Rightarrow \text{another triplet} \end{array}$$

$$\begin{array}{ccc} 5 & 12 & 13 \Rightarrow \text{a triplet} \\ \downarrow \div 2 & \downarrow \div 2 & \downarrow \div 2 \\ 2.5 & 6 & 6.5 \Rightarrow \text{another triplet} \end{array}$$

Comparison between square and equilateral triangle in diverse cases

S. No.	Cases	Area of square: Area of equilateral triangle
(i)	a square and an equilateral triangle are drawn on the same base	$4:\sqrt{3}$
(ii)	When an equilateral triangle is drawn on the diagonal of a square	$2:\sqrt{3}$
(iii)	Perimeters are equal	$3\sqrt{3}:4$
(iv)	When a square is inscribed in an equilateral triangle	$12:(12+7\sqrt{3})$ (approximately 1 : 2)
		Perimeter of square : Perimeter of equilateral triangle
(v)	Sides are equal	4 : 3
(vi)	Areas are equal	$4:3\sqrt{3}$

Illustrations

(Illustrations based on Median, Inradius. Circumradius, Congruency, Similarity, Angle-bisector, Geometrical properties of triangle, etc. have been detailed in 'Geometry Section')

Problems Based on Triangle Irrespective of its Types

Problem 1 If every side of a triangle is doubled, the area of the new triangle is K times the area of the old one. K is equal to :

[CBI-1997, CDS-19]

Sol. Clearly, every side has increased by 100%.

Recall 'concept of effect on product in case of changes in factors' detailed in chapter percentage. Area of a triangle is the product of side and side with a constant term.

Hence, effect on area

$$= \left[+100 + 100 + \frac{100 \times 100}{100}\right]\%$$

$= 300\%$

It implies that the new area will be 300% greater than the old one, hence, 4 times.

$\therefore \quad K = 4$

Alternatively

We know, if each side of a triangle is made n times, its area will become n^2 times. Here, $n = 2$

$\therefore \quad K = 2^2 = 4$

Problem 2 If the height of a triangle is decreased by 40% and its base is increased by 40%, what will be effect on its area ?

[SBI (PO)-2000]

Sol. Effect on area $= \left[-40 + 40 - \frac{40 \times 40}{100}\right]\%$

$= -16\%$

'–' indicates that the area will decrease.

Hence, area will decrease by 16%.

Problem 3 What will be the ratio between the area of a rectangle and the area of a triangle with one of the sides of the rectangle as base and a vertex on the opposite side of the rectangle ?

Sol.

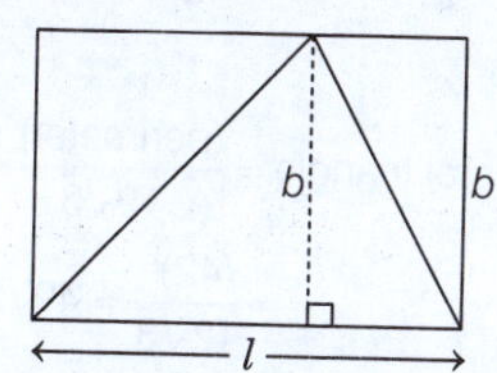

Required ratio $= lb : \frac{1}{2} \times l \times b = 2:1$

Problem 4 There is a triangular building (ABC) located in the heart of Jaipur, the Pink City. The length of the one wall (BC) in the East direction is 597 feet. If the length of South wall (AB) is perfect cube, the length of South West wall (AC) is a power of three and the length of wall in South-West (AC) is thrice the length of side AB, determine the perimeter of this triangular building.

(a) 1507 feet (b) 1513 feet
(c) 1569 feet (d) 1613 feet

Sol. Let, $AB = a^3$ and $AC = 3^n$

As given, $AC = 3AB$

$\Rightarrow \quad 3^n = 3 \times a^3$

$\Rightarrow \quad a^3 = 3^{n-1}$

Now, perimeter $= AB + BC + AC$

$= a^3 + 597 + 3^n$

$= 3^{n-1} + 597 + 3^n$

$= 597 + 3^{n-1}(1+3)$

$= 597 + 3^{n-1} \cdot 4$

$\Rightarrow$ Perimeter $- 597 = 3^{n-1} \cdot 4$

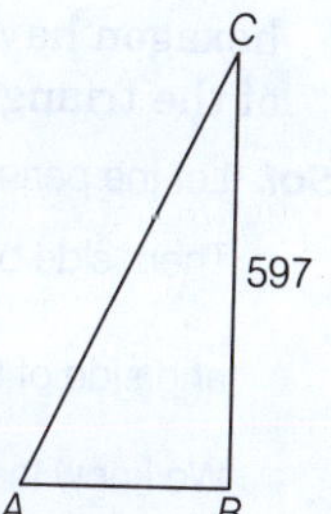

It implies that (perimeter – 597) must be divisible by both 3 and 4.

Now, proceed through options.

(a) 1507 ⇒ (1507 – 597) is neither divisible by 3 nor by 4

(b) 1513 ⇒ (1513 – 597) is divisible by 4 but not by 3

(c) 1569 ⇒ (1569 – 597) is divisible by both 3 and 4

(d) 1613 ⇒ (1569 – 597) is divisible by 4 but not by 3

Hence, option (c) is correct.

Problem 5 If the altitude of a triangle is increased by 10% while its area remains same, its corresponding base will have to be decreased by how much percentage ? *[SSC (So-commercial)-07]*

Sol. Since area is product of base and corresponding height with a constant term, concept of 'Product Constancy' works.

Hence, required percentage $= \left[\frac{10}{100+10} \times 100\right]\% = 9.09\%$

Problems Based on Equilateral Triangle

Problem 6 If the perimeter of an equilateral triangle is 42 cm, its area, in cm^2, will be : *[SSC (CGL-T-I)-14]*

Sol. Given, perimeter = 42 cm

Then, each side = (42 ÷ 3) = 14 cm

Now, area $= \frac{\sqrt{3}}{4}(\text{side})^2 = \frac{\sqrt{3}}{4} \times 14 \times 14 = 49\sqrt{3}$ cm^2

Alternatively

Area of an equilateral triangle $= \frac{(\text{perimeter})^2}{12\sqrt{3}}$

$= \frac{(42)^2}{12\sqrt{3}} = 49\sqrt{3}$ cm^2

Problem 7 The area of an equilateral triangle is $4\sqrt{3}$ cm^2. Its perimeter is : *[SSC (Asstt. Grade-III)-12, SSC (CGL-T-I)-19]*

Sol. Given, area $= 4\sqrt{3}$

$\Rightarrow \frac{\sqrt{3}}{4}(\text{side})^2 = 4\sqrt{3}$

$\Rightarrow (\text{side})^2 = 16 \Rightarrow$ side = 4 cm

Then, perimeter = 3 × side

= 3 × 4 = 12 cm

Alternatively

Given, area $= \frac{(\text{perimeter})^2}{12\sqrt{3}} = 4\sqrt{3}$

perimeter $= \sqrt{144} = 12$ cm

Problem 8 An equilateral triangle and a regular hexagon have the same perimeter. The ratio of the area of the triangle to that of the hexagon is :

Sol. Let the perimeter be a cm.

Then, side of the equilateral triangle $= \frac{a}{3}$

and side of the hexagon $= \frac{a}{6}$

We know that, a regular hexagon contains 6 equilateral triangles with the side equal to side of the hexagon as depicted.

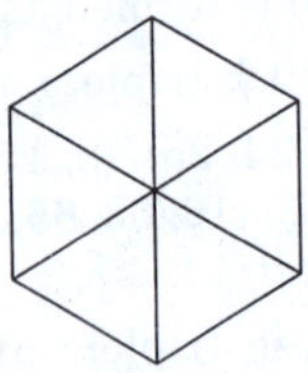

Then, required ratio $= \frac{\sqrt{3}}{4}\left(\frac{a}{3}\right)^2 : 6 \times \frac{\sqrt{3}}{4}\left(\frac{a}{6}\right)^2 = 2:3$

Alternatively

Since ratio of side of the equilateral triangle and side of the equilateral triangle envisaged in the hexagon is 2 : 1, ratio of their areas will be in ratio 4 : 1. But, there are 6 equilateral triangles in the hexagon, hence, the required ratio

= 4 : (1 × 6) = 2 : 3

Problem 9 Lengths of the perpendiculars from a point in the interior of an equilateral triangle on its sides are 3 cm, 4 cm and 5 cm. Area of the triangle is : *[SSC (So-commercial)-07, SSC (DEO)-09]*

Sol. Clearly, area of the equilateral ΔABC

= area of ΔAOB + area of ΔBOC + area of ΔAOC

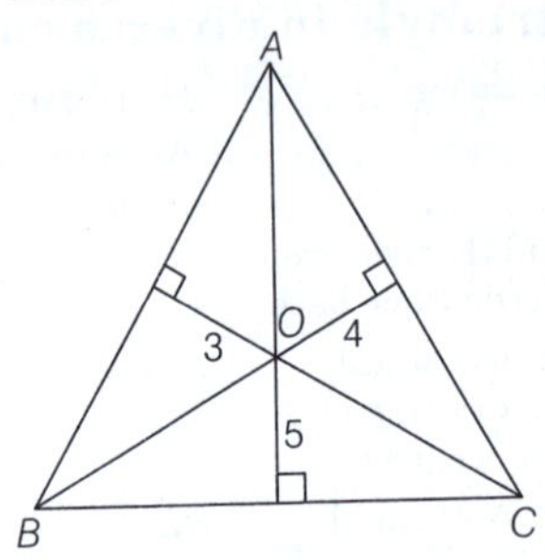

$\Rightarrow \frac{\sqrt{3}}{4}a^2 = \left(\frac{1}{2} \times a \times 5\right) + \left(\frac{1}{2} \times a \times 3\right) + \left(\frac{1}{2} \times a \times 4\right)$

[let the side of ΔABC be a]

$\Rightarrow \frac{\sqrt{3}}{4}a^2 = \frac{1}{2}a(5+3+4)$

$\Rightarrow \frac{\sqrt{3}}{4}a = 6$

$\Rightarrow a = 8\sqrt{3}$

Now, area of $\Delta ABC = \frac{\sqrt{3}}{4}(8\sqrt{3})^2 = 48\sqrt{3}$ cm^2

Problem 10 If the numerical value of the perimeter of an equilateral triangle is $\sqrt{3}$ times the area of it, then the length of each side of the triangle is : *[SSC (Asstt. Grade-II)-12]*

Sol. Given, perimeter of triangle $= \sqrt{3} \times$ area of triangle

$\Rightarrow 3 \times \text{side} = \sqrt{3} \times \frac{\sqrt{3}}{4} \times \text{side} \times \text{side} \Rightarrow$ side = 4 unit

Alternatively

Given, perimeter of triangle $= \sqrt{3} \times$ area of triangle

⇒ Given, perimeter of triangle $= \sqrt{3} \times \frac{(\text{perimeter})^2}{12\sqrt{3}}$

⇒ Perimeter = 12

∴ Side = 12 ÷ 3 = 4 unit

Problem 11 The areas of two equilateral triangles are in the ratio 25 : 36. Their altitudes will be in the ratio

Sol. We know, ratio of areas of equilateral triangles

= Ratio of square of their altitudes

$\Rightarrow$ 25 : 36 = ratio of square of their altitudes.

$\Rightarrow$ Ratio of their altitudes $= \sqrt{25} : \sqrt{36} = 5 : 6$

Problem 12 If an equilateral triangle of area X and a square of area Y have the same perimeter, then what is the relation between X and Y. **[CDS-03]**

Sol. We know, Area of square > area of rectangle > area of triangle under the given equal perimeters.

Hence, $Y > X$.

Problem 13 If the side of an equilateral triangles is decreased by 20%, its area is decreased by what per cent ?

Sol. Recall the concept of 'Effect on product in case of changes in factors'. The area of an equilateral triangle is the product of side and side with a constant term.

Then, effect on area $= \left[-20 - 20 + \dfrac{20 \times 20}{100}\right]\% = -36\%$

'–' indicates that the area will decrease.

Hence, area will decrease by 36%.

Problem 14 The perimeter of an equilateral triangle and a rectangle are same and each side of the triangle is equal to the length of the rectangle. Find the ratio of areas of the triangle and the rectangle.

Sol. As given, perimeter of triangle = perimeter of rectangle

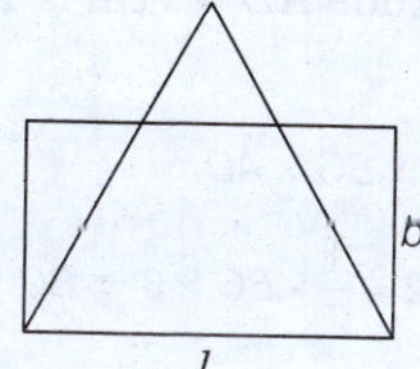

$$\Rightarrow \quad 3l = 2(l + b) \Rightarrow b = \frac{l}{2}$$

Then, required ratio $= \dfrac{\sqrt{3}}{4} l^2 : l \cdot \dfrac{l}{2} = \sqrt{3} : 2$.

Problem 15 Consider an equilateral triangle of a side of unit length. A new equilateral triangle is formed by joining the mid-points of one, then a third equilateral triangle is formed by joining the mid-points of second. The process is continued. The perimeter of all triangles, thus formed is : **[CDS-14, SNAP-11]**

Sol. Sum of perimeters of all triangles

$= 3(1) + 3(0.5) + 3(0.25) + 3(0.125) + 3(0.0625) + \ldots$ upto ∞ terms

$$= 3\left[1 + \frac{1}{2} + \frac{1}{4} + \frac{1}{8} + \frac{1}{16} + \ldots \text{upto } \infty \text{ terms})\right] = 3\left[\frac{1}{1 - \frac{1}{2}}\right]$$

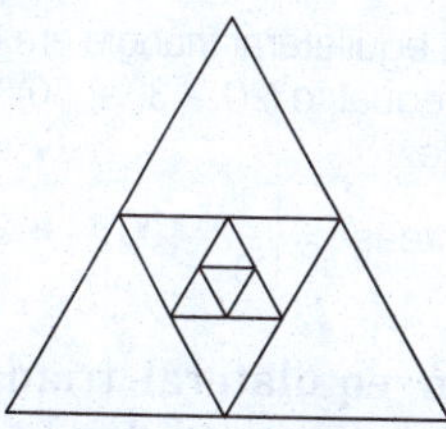

[$\because$ sum to infinity of a geometric progression with initial term a and common ratio $r = \dfrac{a}{1-r}$, where $-1 < r < 1$]

= 6 units

Problem 16 With reference to above, what is the sum of the areas of infinitely many such triangles if the largest triangle is with side 24 cm. **[CAT-18]**

Sol. Clearly, sum of areas

$$= \frac{\sqrt{3}}{4}(24)^2 + \frac{\sqrt{3}}{4}(12)^2 + \frac{\sqrt{3}}{4}(6)^2 + \ldots\ldots \text{ upto } \infty \text{ terms}$$

$$= \frac{\sqrt{3}}{4}[(24)^2 + (12)^2 + (6)^2 + \ldots\ldots \infty]$$

$$= \frac{\sqrt{3}}{4}\left[\frac{(24)^2}{1 - \frac{1}{4}}\right] = 192\sqrt{3} \text{ cm}^2$$

Problem 17 The side of an equilateral triangle is 10 cm long. By drawing parallels to all its sides, the distance between any two parallel lines being the same, the triangle is divided into smaller equilateral triangle, each of which has sides of length 1 cm. How many such small triangles are formed ?

Sol. Suppose, the bigger equilateral triangle is with side 4 cm. Then, number of smaller triangles with side 1 cm will be computed as below :

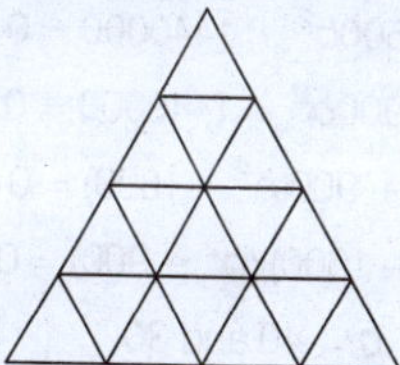

1st row : number of triangles = 1

2nd row : number of triangles = 3

3rd row : number of triangles = 5

4th row : number of triangles = 7

Hence, number of smaller triangles $= 1 + 3 + 5 + 7 = 4^2 = 16$

[Recall sum of first n odd numbers $= n^2$]

Obviously, number of smaller triangles in case the bigger equilateral triangle is with 10 cm $= 10^2 = 100$

Problem 18 If the sides of an equilateral triangle are increased by 20%, 30% and 50% respectively to form a new triangle, the percentage increase in the perimeter of the equilateral triangle is :

Sol. Since sides of an equilateral triangle are equal, total increase in perimeter will be equal to (20 + 30 + 50)% of side of the equilateral triangle

$\therefore$ Percentage increase $= \left[\frac{1}{3} \times 100\right]\% = 33.33\%$

Problem 19 An equilateral triangle is cut from its corners so as to form a regular hexagon. What is the ratio of area of the equilateral triangle to that of hexagon? **[SSC (CGL-T-II)-11, CAT-19]**

Sol. Clearly, there were 9 small equilateral triangles of equal areas in the original triangle before it is cut from vertices.

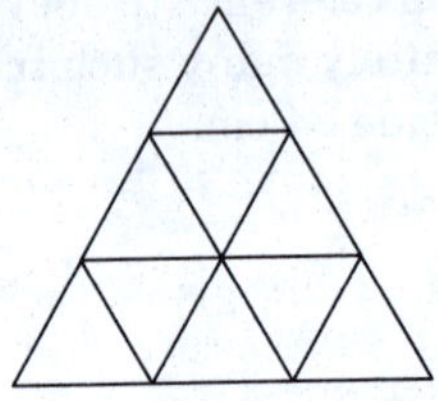

Hence, the required ratio $= \frac{\text{Area of 9 triangles}}{\text{Area of 6 triangles}} = 3:2.$

Problems Based on Isosceles Triangle

Problem 20 The area of an isosceles triangle is 300 cm^2 and the length of one of the equal sides is 25 cm. Find the third side.

Sol. Area of an isosceles triangle $= \frac{b}{4}\sqrt{4a^2 - b^2}$, where a is the length of each of equal sides and b is the length of the third side.

$\Rightarrow \quad 300 = \frac{b}{4}\sqrt{4a^2 - b^2}$

$\Rightarrow \quad 1200 = b\sqrt{4 \cdot (25)^2 - b^2}$

$\Rightarrow \quad 1200 \times 1200 = b^2(2500 - b^2)$ [squaring both sides]

$\Rightarrow \quad b^4 - 2500b^2 + 1440000 = 0$

$\Rightarrow b^4 - 1600b^2 - 900b^2 + 1440000 = 0$

$\Rightarrow \quad b^2(b^2 - 1600) - 900(b^2 - 1600) = 0$

$\Rightarrow \quad (b^2 - 1600)(b^2 - 900) = 0$

$\Rightarrow \quad b = 40$ and 30

Hence, length of the side other than equal sides = either 30 cm or 40 cm.

Alternatively

You may proceed through options.

Let options be 20, 40.

(a) $20 \Rightarrow \text{area} = \frac{20}{4}\sqrt{4(25)^2 - 20^2} \neq 300$

(b) $40 \Rightarrow \text{area} = \frac{40}{4}\sqrt{4(25)^2 - 40^2} = 300$

Hence, option(b) is correct.

Problem 21 The area of a square is 500 m^2 and four equal isosceles triangles of maximum possible area are cut off from its corners. What is the area of the resulting square?

Sol.

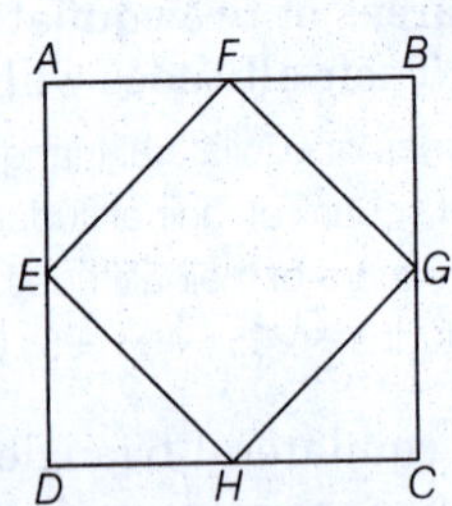

Clearly, the corners of the resulting square will be at the mid-points of sides of the original square so as to make the areas of isosceles triangles equal and maximum.

Side $\quad AB = \sqrt{500} = 10\sqrt{5}$ cm

$\therefore \quad AF = 5\sqrt{5}$ cm

Also, $\quad AE = 5\sqrt{5}$ cm

Now, $\quad EF = \sqrt{(5\sqrt{5})^2 + (5\sqrt{5})^2} = \sqrt{250}$

Then, area of the new square $EFGH$

$= (\sqrt{250})^2 = 250$ m^2

Alternatively

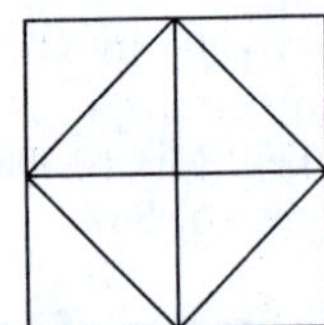

Clearly, there are 8 equal triangles having total area of 500 m^2

Then, area of new square containing 4 triangles = 250 m^2.

Problem 22 The area of an isosceles ΔABC with $AB = AC$ and altitude $AD = 3$ cm is 12 sq cm. What is its perimeter? **[CDS-13]**

Sol. Area of $\Delta ABC = \frac{1}{2} \times BC \times AD$

$\Rightarrow \quad 12 = \frac{1}{2} \times BC \times 3 \Rightarrow BC = 8$ cm

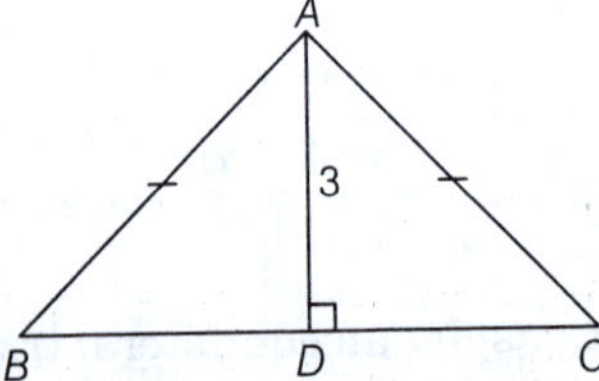

We know, in an isosceles triangle, the altitude having the side other than congruent sides as its base has the mid-point of that side as its foot.

Then, $\quad BD = DC = 4$ cm

Now, in right-angled ΔADC, hypotenuse $AC = 5$ as (3, 4, 5) is a Pythagorean triplet. Then, perimeter of $\Delta ABC = 5 + 8 + 5 = 18$ cm

Problem 23 If an isosceles right-angled triangle has area 1 sq unit, then what is its perimeter?

Sol. Area of $\Delta ABC = \frac{b}{4}\sqrt{4a^2 - b^2}$, where a is the length of each of equal sides and b is the length of the third side.

$\Rightarrow \quad 1 = \frac{b}{4}\sqrt{4a^2 - b^2} \Rightarrow 4 = b\sqrt{4a^2 - b^2}$

$\Rightarrow \quad 4 = \sqrt{2}\,a\,\sqrt{4a^2 - 2a^2}$

$[\because b^2 = a^2 + a^2 \Rightarrow b = \sqrt{2}\,a]$

$\Rightarrow \quad 16 = 2a^2 \times 2a^2$ [squaring both sides]

$\Rightarrow \quad a = \sqrt[4]{4} = \sqrt{2}$

Then, $b = \sqrt{2} \times \sqrt{2} = 2$

Now, perimeter $= a + a + b$

$= \sqrt{2} + \sqrt{2} + 2 = 2(1 + \sqrt{2})$ units

Note *The alternative method adhering to the Geometric concept has deliberately been left and will be discussed there.*

Problem 24 **A lawn is in the form of an isosceles triangle. The cost of turfing it came to ₹ 1200 at ₹ 4 per m². If the base be 40 m long, find the length of equal side.** ***[MAT-09]***

Sol. Clearly, area of the triangular lawn = 1200 ÷ 4 = 300

$\Rightarrow \quad \frac{b}{4}\sqrt{4a^2 - b^2} = 300$

[where a is the length of each of equal sides and b is the length of the third side]

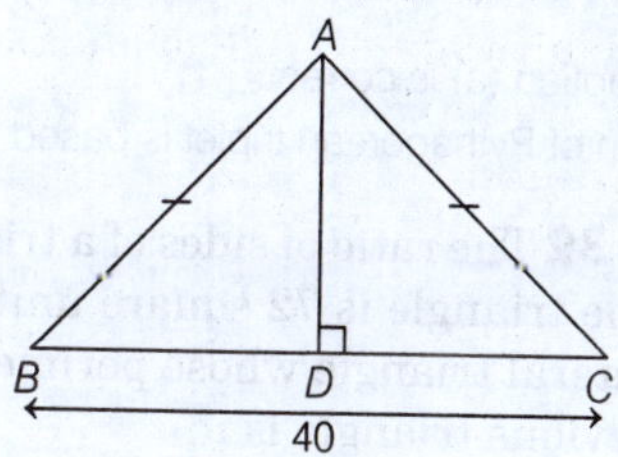

$\Rightarrow \quad \frac{40}{4}\sqrt{4a^2 - 40^2} = 300$

$\Rightarrow \quad 4a^2 - 40^2 = 900 \Rightarrow a = 25\text{ m}$

Alternatively

Area of $\Delta ABC = \frac{1}{2} \times BC \times AD$

$\Rightarrow \quad 300 = \frac{1}{2} \times 40 \times AD \Rightarrow AD = 15$

Now, in right-angled ΔADC,

$AC = \sqrt{AD^2 + DC^2} = \sqrt{15^2 + 20^2}$

[we know, in an isosceles triangle, the altitude having the side other than congruent sides as its base has the mid-point of that side as its foot]

$\Rightarrow \quad AC = 25\text{ m}$

Problem 25 **The altitude drawn to the base of an isosceles triangle is 8 cm and its perimeter is 64 cm. The area (in cm²) of the triangle is :** ***[SSC (10+2, LDC/DEO)-13]***

Sol. Let each of equal sides be a and the base be b.

Then, perimeter $= 2a + b = 64$

$\Rightarrow \quad 2a = 64 - b$

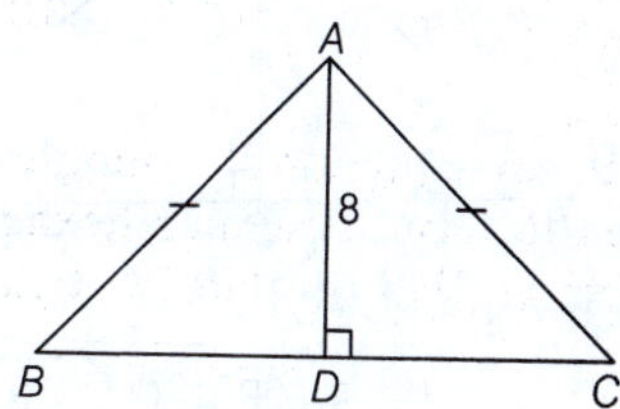

Now, area of the triangle $= \frac{b}{4}\sqrt{4a^2 - b^2}$

$\Rightarrow \quad \frac{1}{2} \times b \times 8 = \frac{b}{4}\sqrt{(64 - b)^2 - b^2}$

$\Rightarrow \quad 4b = \frac{b}{4}\sqrt{4096 - 128b}$

$\Rightarrow \quad 16 = \sqrt{4096 - 128b}$

$\Rightarrow \quad 256 = 4096 - 128b$ [squaring both sides]

$\Rightarrow \quad 128b = 3840$

$\Rightarrow \quad b = 30$

Hence, area $= \frac{1}{2} \times 30 \times 8 = 120\text{ cm}^2$

Problems Based on Scalene Triangle

Problem 26 **What is area of the triangle whose sides are 9 cm, 10 cm and 11 cm ?**

Sol. Here, semi-perimeter $(S) = \frac{\text{sum of all sides}}{2}$

$= \frac{9 + 10 + 11}{2} = 15\text{ cm}$

Then, area $= \sqrt{s(s-a)(s-b)(s-c)}$

$= \sqrt{15(15-9)(15-10)(15-11)}$

$= \sqrt{15 \times 6 \times 5 \times 4}$

$= \sqrt{3 \times 5 \times 3 \times 2 \times 5 \times 2 \times 2}$

$= 30\sqrt{2}\text{ cm}^2$

Problem 27 **The sides of a triangle are in the ratio 2 : 3 : 4. The perimeter of the triangle is 18 cm. The area (in cm²) of the triangle is :** ***[SSC (CGL-T-I)-11]***

Sol. Given, (2 + 3 + 4) units = 18 cm

∴ 2 units = 4 cm, 3 units = 6 cm and 4 units = 8 cm

Hence, sides are 4 cm, 6 cm and 8 cm

Now, semi-perimeter = 9 cm

Then, area of the triangle $= \sqrt{9(9-4)(9-6)(9-8)}$

$= \sqrt{9 \times 5 \times 3 \times 1} = 11.62\text{ cm}^2$

Problem 28 **Euclid has a triangle in mind. Its longest side has length 20 and another of its side has length 10. Its area is 80. What is the exact length of its third side ?** ***[CAT-01]***

Sol. Area of $\Delta ABC = \frac{1}{2} \times BC \times AD$

$\Rightarrow \quad 80 = \frac{1}{2} \times 20 \times AD \Rightarrow AD = 8$

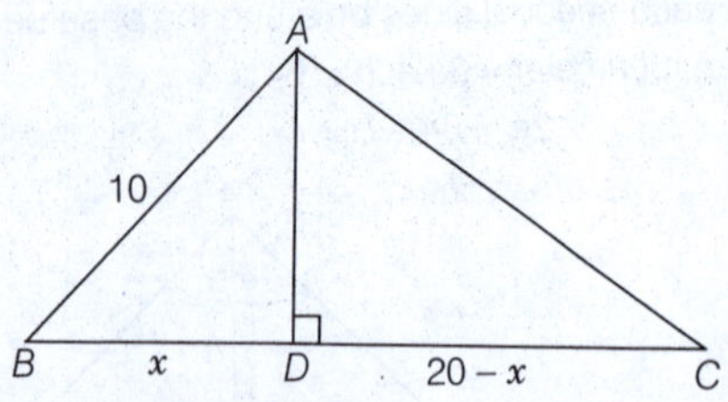

Now in right-angled ΔABD,

$$BD = \sqrt{AB^2 - AD^2} = \sqrt{10^2 - 8^2} = 6$$

$\therefore \quad CD = 20 - 6 = 14$

Now, in right-angled ΔACD,

$$AC = \sqrt{AD^2 + CD^2} = \sqrt{8^2 + 14^2} = \sqrt{260}$$

Problem Based on Right-angled Triangle

Problem 29 The base and altitude of a right-angled triangle are 12 cm and 5 cm respectively. The perpendicular distance of its hypotenuse from the opposite vertex is :

Sol. Clearly, hypotenuse $AC = \sqrt{5^2 + 12^2} = 13$ cm

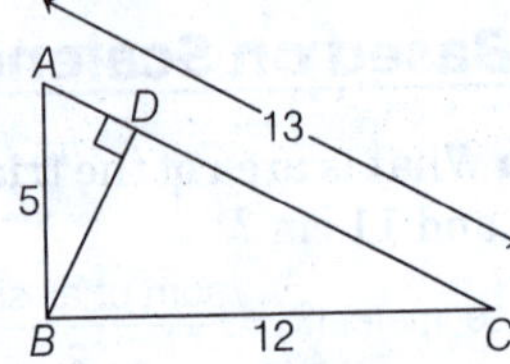

You may write directly as $AC = 13$ cm because (5, 12, 13) is a Pythagorean triplet.

Now, area of the triangle $= \frac{1}{2} \times BC \times AB = \frac{1}{2} \times AC \times BD$

$\Rightarrow \quad \frac{1}{2} \times 12 \times 5 = \frac{1}{2} \times 13 \times BD \Rightarrow BD = \frac{60}{13} = 4.61$ cm

Problem 30 The perimeter of a triangle is 30 cm and its area is 30 cm^2. If the largest side measures 13 cm, what is the length of the smallest side to the triangle ?

Sol. Sum of smaller sides = 30 − 13 = 17 cm

Let the smallest side be x cm. Then, the other side will be $(17 - x)$ cm. Also, semi-perimeter $(S) = \frac{30}{2} = 15$ cm

Now, area of the triangle $= \sqrt{15(15-13)(15-x)(15-17+x)}$

$\Rightarrow \quad 30 = \sqrt{30(15-x)(15-17+x)}$

$\Rightarrow \quad 900 = 30(-x^2 + 17x - 30)$

[squaring both sides]

$\Rightarrow \quad x^2 - 17x + 60 = 0 \Rightarrow x^2 - 12x - 5x + 60 = 0$

$\Rightarrow \quad x(x-12) - 5(x-12) = 0 \Rightarrow x = 5$ or 12

Then, $\quad (17 - x) = 12$ or 5

It implies that the smallest side is 5 cm.

Alternatively

Given, the largest side is 13 cm. This side may be hypotenuse if the triangle is right-angled.

Recall, (5, 12 , 13) is a Pythagorean triplet.

It is also seen that perimeter = 5 + 12 + 13 = 30

and $\quad$ area $= \frac{1}{2} \times 5 \times 12 = 30$

Hence, the given triangle is a right-angled triangle with sides 5 cm, 12 cm and 13 cm.

$\therefore$ The smallest side is 5 cm.

Problem 31 If the perimeter of a right angled triangle is 56 cm and area of the triangle is 84 cm^2, then the length of the hypotenuse is (in cm)

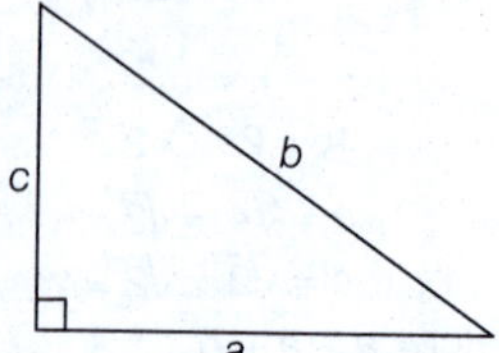

$$\text{Area} = \frac{1}{2}ac \Rightarrow 84 = \frac{1}{2}ac \Rightarrow ac = 168$$

Further, $b^2 = a^2 + c^2$

$\Rightarrow \quad b^2 = (a + c)^2 - 2ac$

$\Rightarrow \quad b^2 = (56 - b)^2 - 2 \cdot 168 \quad [\because a + b + c = 56]$

$\Rightarrow \quad b^2 = 3136 - 112b + b^2 - 2 \cdot 168$

$\Rightarrow \quad 112b = 2800 \Rightarrow b = 25$ cm

Hence, $\quad$ hypotenuse = 25 cm

Alternatively

Proceed through options. Let, options be 25, 50, 7 and 24.

(a) 25 $\Rightarrow$ (7, 24, 25) is a Pythagorean triplet and perimeter $= 7 + 24 + 25 = 56$ cm and also area $= \frac{1}{2} \times 7 \times 24 = 84$

Hence, option (a) is correct.

[selection of Pythagorean triplet is based on assumption only]

Problem 32 The ratio of sides of a triangle is 3 : 4 : 5 and area of the triangle is 72 square unit. Then, the area of an equilateral triangle whose perimeter is same as that of the previous triangle is :

Sol. Since (3, 4, 5) is a Pythagorean triplet, the given triangle is a right-angled triangle.

Let, sides be $3x$, $4x$ and $5x$.

Then, $\quad$ area $= \frac{1}{2} \times 3x \times 4x = 72 \Rightarrow x = \sqrt{12}$

Now, perimeter of the right-angled triangle = perimeter of an equilateral triangle

$\Rightarrow (3x + 4x + 5x)$ = perimeter of an equilateral triangle

$\Rightarrow 12\sqrt{12}$ = perimeter of an equilateral triangle

$\therefore$ Area of an equilateral triangle $= \frac{(\text{Perimeter})^2}{12\sqrt{3}}$

$$= \frac{(12\sqrt{12})^2}{12\sqrt{3}}$$

$$= 48\sqrt{3} \text{ sq unit}$$

Problem 33 The distance between feet of two trees 12 m and 27 m high, be 20 m. Then, what is the distance between their tops ?

Sol. Distance $AE = \sqrt{AB^2 + BE^2} = \sqrt{(AC - BC)^2 + 20^2}$

$= \sqrt{(27 - 12)^2 + 20^2} = 25$ m

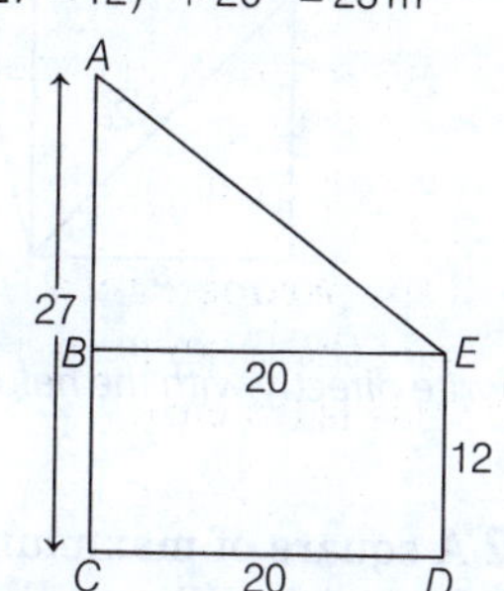

↘ Problem 34 If the measures of the sides of a triangle are $(x^2 - 1)$, $(x^2 + 1)$ and $2x$, the triangle would be :

[SSC (CGL-T-I)-14]

Sol. Assume any value for x, say 2.

Then, sides are: 3, 5 and 4.

Since (3, 4, 5) is a Pythagorean triplet, the given triangle is a right-angled triangle.

↘ Problem 35 ABC is a right-angled triangle, B being the right angle. Mid-points of BC and AB are respectively B' and A'. Area of $\Delta A'B'C$ is :

Sol. Area of $\Delta ABC = \frac{1}{2} \times BC \times AB$

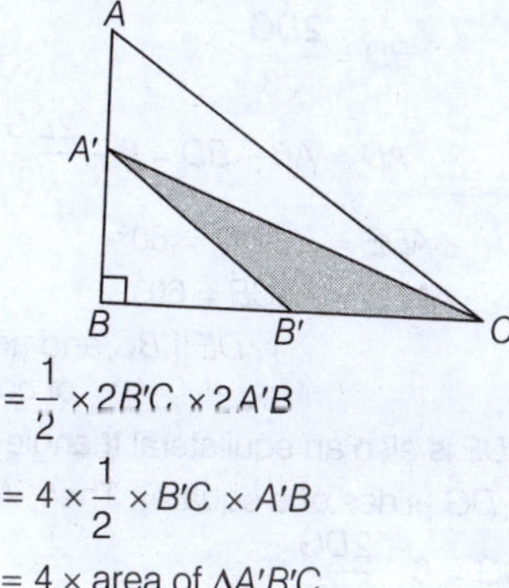

$= \frac{1}{2} \times 2B'C \times 2A'B$

$= 4 \times \frac{1}{2} \times B'C \times A'B$

$= 4 \times$ area of $\Delta A'B'C$

Hence, area of $\Delta A'B'C$ would be one-fourth of area of ΔABC.

Alternatively

You may take any Pythagorean triplet into consideration for comparing the areas of both triangles.

↘ Problem 36 The perimeter of a right-angled triangle is 60 cm. Its hypotenuse is 26 cm. What is its area ?

[CDS-21]

Sol. Clearly, base + height = 60 − 26 = 34 cm

Now, $(\text{hypotenuse})^2 = (\text{base})^2 + (\text{height})^2$

$\Rightarrow \quad (26)^2 = (x)^2 + (34 - x)^2$ [let base be x cm]

$\Rightarrow \quad x^2 - 34x + 240 = 0$

$\Rightarrow \quad x^2 - 24x - 10x + 240 = 0$

$\Rightarrow \quad x(x - 24) - 10(x - 24) = 0$

$\Rightarrow \quad (x - 10)(x - 24) = 0 \Rightarrow x = 10$ or 24

Then, $34 - x = 24$ or 10

In both cases, area $= \frac{1}{2} \times 10 \times 24 = 120 \text{ m}^2$

Alternatively

Recall a Pythagorean triplet (5, 12, 13).

Obviously, (10, 24, 26) will be a Pythagorean triplet.

Here, 26 is the hypotenuse and perimeter

$= 10 + 24 + 26 = 60$

Hence, 10 and 24 are base and height or height and base.

Then, area $= \frac{1}{2} \times 10 \times 24 = 120 \text{ m}^2$

[selection of Pythagorean triplet is based on assumption only]

↘ Problem 37 If the perimeter of a right-angled isosceles triangle is $(4\sqrt{2} + 4)$ cm, the length of the hypotenuse is :

Sol. Let $AB = BC = x$ cm.

Then, $AC = \sqrt{x^2 + x^2}$

$= \sqrt{2}\, x$ cm

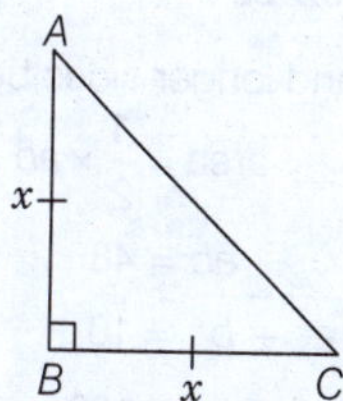

As given, $x + x + \sqrt{2}\, x = 4\sqrt{2} + 4$

$\Rightarrow \quad x(2 + \sqrt{2}) = 2\sqrt{2}(2 + \sqrt{2})$

$\Rightarrow \quad x = 2\sqrt{2}$

$\therefore$ hypotenuse $= \sqrt{2}x$

$= \sqrt{2} \times 2\sqrt{2} = 4$ cm

↘ Problem 38 In a triangular field having sides 30 m, 72m and 78 m, the length of the altitude to the side measuring 72 m is :

Sol. Semi-perimeter $(S) = \frac{30 + 72 + 78}{2} = 90$

Then, area of the triangle $= \sqrt{90(90 - 30)(90 - 72)(90 - 78)}$
$= 1080$

Now, area of the triangle $= \frac{1}{2} \times$ side $\times$ altitude to this side

$\Rightarrow \quad 1080 = \frac{1}{2} \times 72 \times$ altitude

$\Rightarrow \quad$ altitude $= 30$ m

Alternatively

We may take a Pythagorean triplet on assumption basis as (5, 12, 13)

Obviously, (30, 72, 78) will also be a Pythagorean triplet.

Hence, the given triangle is a right-angled triangle and perpendicular sides are 30 and 72 and hypotenuse is 78.

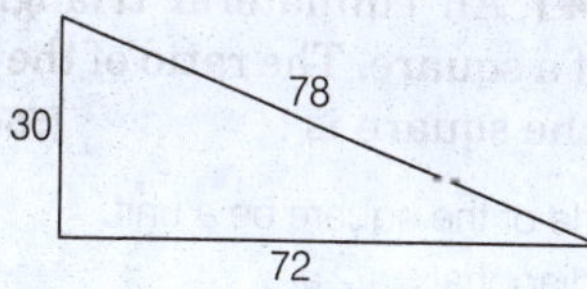

It implies that the length of the altitude to side measuring 72 m = 30 m

Problem 39 What is the area of the greatest right-angled triangle whose hypotenuse is m cm ?

Sol. We know, for the given equal hypotenuse, an isosceles right triangle has maximum area in comparison to other right-angled triangles.

Hence, base = perpendicular = x (let)

Then, $x^2 + x^2 = m^2 \Rightarrow x = \frac{m}{\sqrt{2}}$

Now, area $= \frac{1}{2} \times \text{base} \times \text{perpendicular}$

$$= \frac{1}{2} \times \frac{m}{\sqrt{2}} \times \frac{m}{\sqrt{2}} = \frac{m^2}{4}$$

Problem 40 The hypotenuse of a right-angled triangle is 10 cm and its area is 24 cm². If the shorter side is halved and the longer side is doubled, the new hypotenuse becomes : **[CDS-19]**

Sol. Let the shorter and longer sides be a and b (in cm) respectively.

As given, area $= \frac{1}{2} \times ab = 24$

$\Rightarrow ab = 48$

Further, $\sqrt{a^2 + b^2} = 10$

$\Rightarrow a^2 + b^2 = 100$

$\Rightarrow (a + b)^2 - 2ab = 100$

$\Rightarrow (a + b)^2 - 2 \cdot 48 = 100$

$\Rightarrow a + b = 14$ ...(i)

Similarly, $(a - b)^2 + 2ab = 100$

$\Rightarrow a - b = 2$ ...(ii)

Solving both Eqs. (i) and (ii) we get, $a = 6$ and $b = 8$.

Now, changed shorter side = 3 and changed longer side = 16.

Then, new hypotenuse $= \sqrt{3^2 + 16^2} = \sqrt{265}$.

Alternatively

We know, (3, 4, 5) is a Pythagorean triplet.

Therefore, the triplet obtained by multiplying it with certain number (say, 2) will also be Pythagorean triplet.

Hence, (6, 8, 10) is a Pythagorean triplet.

Further, area $= \frac{1}{2} \times 6 \times 8 = 24$.

It implies that sides of the triangle are 6 cm, 8 cm and 10 cm.

Now, changed shorter side = 3 and changed longer side = 16.

Then, new hypotenuse $= \sqrt{3^2 + 16^2} = \sqrt{265}$

Problems Based on Combination of Shapes Discussed so far

Problem 41 An equilateral triangle is drawn on the diagonal of a square. The ratio of the area of the triangle to that of the square is : **[SSC (Asstt. Grade-III)-12]**

Sol. Let the side of the square be a unit.

Then, its diagonal $= \sqrt{2}\,a$

Now, the required ratio $= \frac{\sqrt{3}}{4}(\sqrt{2}\,a)^2 : a^2 = \sqrt{3} : 2$

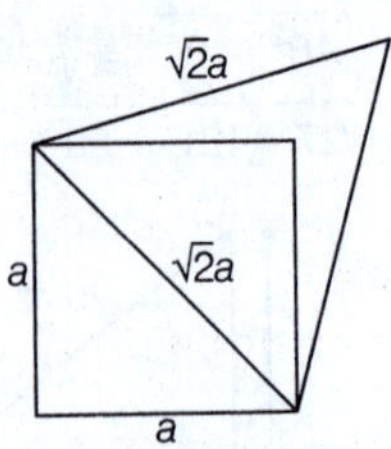

Note *You may write directly with the help of the facts given in the beginning.*

Problem 42 A square of maximum area is inscribed in an equilateral triangle. What is the area of the square if the side of the triangle is 6 cm.

Sol.

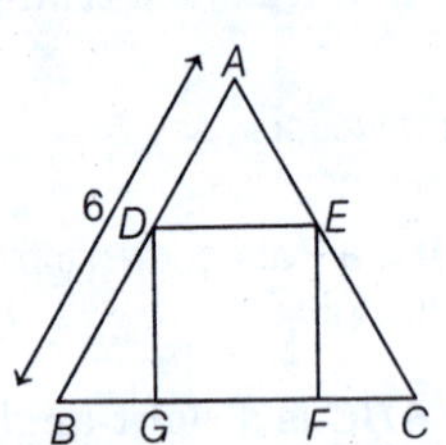

$\sin B = \frac{DG}{BD}$

$\Rightarrow \sin 60° = \frac{DG}{BD}$

$\Rightarrow BD = \frac{2DG}{\sqrt{3}}$

Then, $AD = AB - BD = 6 - \frac{2DG}{\sqrt{3}}$

Further, $\angle ADE = \angle ABC = 60°$

and $\angle AED = \angle ACB = 60°$

[$\because DE \parallel BC$ and and each pair is a pair of corresponding angles.]

Hence, ΔADE is also an equilateral triangle i.e. $AD = DE = AE$.

Also, $DE = DG$ [sides of a square]. Then, $AD = DG$

Now, $DG = 6 - \frac{2DG}{\sqrt{3}}$

$\Rightarrow \sqrt{3}DG = 6\sqrt{3} - 2DG$

$\Rightarrow DG = \frac{6\sqrt{3}}{2 + \sqrt{3}} = \frac{6\sqrt{3}(2 - \sqrt{3})}{(2 + \sqrt{3})(2 - \sqrt{3})}$

$= \frac{12\sqrt{3} - 18}{4 - 3} = 12\sqrt{3} - 18$

$\therefore$ Area of square $DEFG = (12\sqrt{3} - 18)^2 = 7.75 \text{ cm}^2$

Alternatively

Area of the triangle $= \frac{\sqrt{3}}{4}(6)^2 = 9\sqrt{3}$

We know, ratio of area of square to that of an equilateral triangle is $12 : (12 + 7\sqrt{3})$ when a square is inscribed in the equilateral triangle.

$\Rightarrow$ Area of square : $9\sqrt{3} = 12 : (12 + 7\sqrt{3})$

$\therefore$ Area of square $= \frac{12 \times 9\sqrt{3}}{12 + 7\sqrt{3}} = \frac{108\sqrt{3}}{12 + 7\sqrt{3}} = 7.75 \text{ cm}^2$

Problem 43 The sides of a triangle are 6 cm, 8 cm and 10 cm. The area of the greatest square that can be inscribed in it is :

Sol. Since (6, 8, 10) is Pythagorean triplet, the given ΔABC is a right-angled triangle.

Let the side of the square be x.

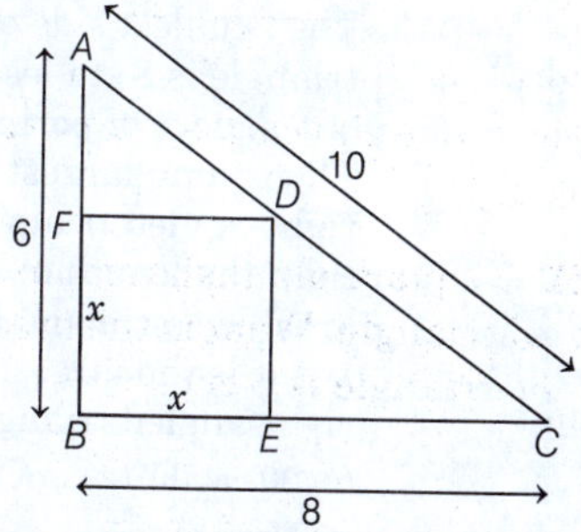

Then, area of ΔAFD + area of ΔDEC + Area of square $FDEB$ = area of ΔABC.

$$\Rightarrow \left(\frac{1}{2} \times FD \times AF\right) + \left(\frac{1}{2} \times DE \times EC\right) + x^2 = \frac{1}{2} \times 8 \times 6$$

$$\Rightarrow \left(\frac{1}{2} \times x \times (6 - x)\right) + \left(\frac{1}{2} \times x \times (8 - x)\right) + x^2 = 24$$

$$\Rightarrow \quad 6x - x^2 + 8x - x^2 + 2x^2 = 48$$

$$\Rightarrow \quad x = 48 \div 14 = \frac{24}{7} \text{ cm}$$

$$\therefore \text{ Area of the square} = \left(\frac{24}{7}\right)^2 = 11.76 \text{ cm}^2$$

Note *The alternative method adhering to the Geometric concept has intentionally been left and will be discussed in Geometry section.*

Problem 44 There is a triangle inside a square such that one side of the square is its base and its height is a part of another side of the square. Find the area of the square if the area of the triangle is 150 cm² and the side of the square exceed the height of the triangle by 5 cm.

Sol. Clearly, $AE = 5$ cm

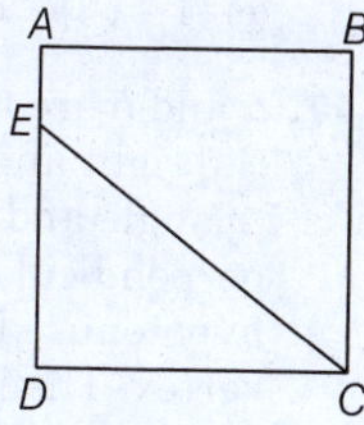

Given, area of $\Delta EDC = 150$

$$\Rightarrow \frac{1}{2} \times CD \times ED = 150$$

$$\Rightarrow \frac{1}{2} \times CD \times (CD - 5) = 150 \qquad [\because AD = CD]$$

$$\Rightarrow CD^2 - 5CD - 300 = 0$$

$$\Rightarrow CD^2 - 20CD + 15CD - 300 = 0$$

$$\Rightarrow CD(CD - 20) + 15(CD - 20) = 0$$

$$\Rightarrow CD = -15 \text{ or } 20$$

$$\Rightarrow CD = 20$$

[$\because$ side of a square cannot be negative]

Then, area of the square $= 20^2 = 400$ cm²

Problem 45 In the adjacent picture, *EFGH, ABCD* are squares, and *ABE, BCF, CDG, DAH* are equilateral triangles. What is the ratio of the area of the square *EFGH* to that of *ABCD* ? **[XAT-19]**

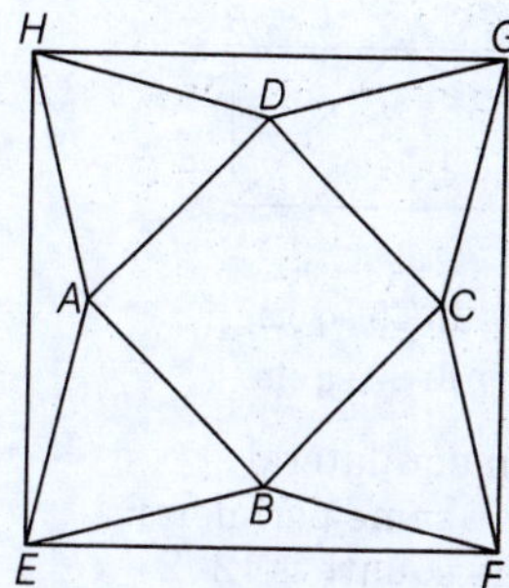

Sol. Let sides of squares $ABCD$ and $EFGH$ be a and b respectively. Clearly, height of ΔAHD + height of ΔBCF + length of dotted line inside square $ABCD$ = diagonal of square $EFGH$.

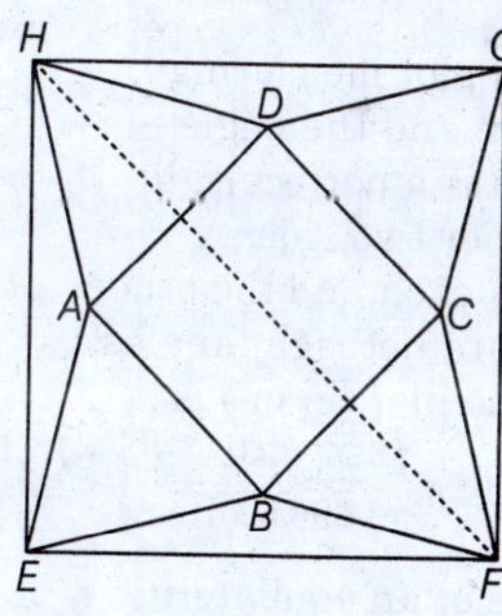

$$\Rightarrow \frac{\sqrt{3}}{2}a + \frac{\sqrt{3}}{2}a + a = \sqrt{2}\, b$$

$$\Rightarrow (\sqrt{3} + 1)a = \sqrt{2}\, b$$

$$\Rightarrow a : b = \sqrt{2} : (\sqrt{3} + 1)$$

$$\Rightarrow b^2 : a^2 = (4 + 2\sqrt{3}) : 2$$

$$= (\sqrt{3} + 2)$$

WORKOUT

1. The hypotenuse of a right-angled isosceles triangle is 5 cm. Its area will be : ***[SSC (SO-05, CDS-09)]***

(a) 8 sq cm (b) 6.25 sq cm
(c) 12.5 sq cm (d) 14 sq cm

2. In the figure given below, the area of rectangle $ABCD$ is 100 sq cm, O is any point on AB and $CD = 20$ cm. Then, the area of ΔCOD is ***[CDS-13]***

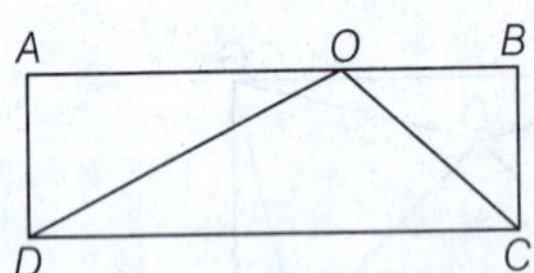

(a) 40 sq cm (b) 45 sq cm
(c) 50 sq cm (d) 80 sq cm

3. A square and an equilateral triangle have the same perimeter. If diagonal of the square is $12\sqrt{2}$ cm, then the area of the triangle is :

(a) $24\sqrt{3}$ cm² (b) $24\sqrt{2}$ cm²
(c) $64\sqrt{3}$ cm² (d) $32\sqrt{3}$ cm²

4. Two sides of a plot measuring 32 m and 24 m and the angle between them is a perfect right angle. The other two sides measure 25 m each and the other three angles are not right angles. The area of the plot (in m²) is : ***[SSC (CGL-T-II)-14]***

(a) 768 (b) 534 (c) 696.5 (d) 684

5. If the altitude of an equilateral triangle is $\sqrt{3}$ cm, then what is its perimeter ? ***[CDS-12]***

(a) 3 cm (b) $3\sqrt{3}$ cm
(c) 6 cm (d) $6\sqrt{3}$ cm

6. The sides of a triangle are 5, 12 and 13 unit. A rectangle is constructed, which is equal in area to the triangle, has a width of 10 unit. Then, the perimeter of the rectangle is : ***[CAT-95]***

(a) 30 unit (b) 36 unit
(c) 13 unit (d) None of these

7. A wire of length 60 m is used to enclose a field which is in the shape of a right-angled triangle. The area of the field is 150 m². Then, the largest side of the field has a length of :

(a) 25 m (b) 20 m
(c) 30 m (d) 15 m

8. The lengths of two sides of a right-angled triangle which contain the right angle are a and b respectively. Three squares are drawn on the three sides of the triangle on the outer side. What is the total area of the triangle and three squares ?

(a) $2(a^2 + b^2) + 0.5\,ab$
(b) $2(a^2 + b^2) + ab$
(c) $2(a^2 + b^2) + 2.5\,ab$
(d) $2.5(a^2 + b^2)$

9. In an isosceles triangle, the length of each equal side is twice the length of the third side. The ratio of areas of the isosceles triangle and an equilateral triangle with same perimeter is : ***[SSC (CGL-T-II)-16]***

(a) $36\sqrt{5} : 100$ (b) $32\sqrt{5} : 100$
(c) $\sqrt{5} : 100$ (d) $42\sqrt{5} : 100$

10. The ratio of sides of a triangle is $3:4:5$ and area of the triangle is 72 sq unit. Then, the area of an equilateral triangle, whose perimeter is same as that of the previous triangle, is : ***[SSC (10+2, LDC/DEO)-12]***

(a) $32\sqrt{3}$ sq unit (b) $48\sqrt{3}$ sq unit
(c) 96 sq unit (d) $60\sqrt{3}$ sq unit

11. The perimeter of a triangular field is 240 m. If two of its sides are 78 m and 50 m, then what is the length of the perpendicular on the side of length 50 m from the opposite vertex ?

(a) 43 m (b) 52.2 m
(c) 67.2 m (d) 70 m

12. If a and b are the lengths of the sides of a right triangle whose hypotenuse is 10 and whose area is 20, then the value of $(a + b)^2$ is : ***[SSC (CGL-T-II)-15]***

(a) 140 (b) 180
(c) 120 (d) 160

13. The smallest side of a right-angled triangle is 8 cm less than the side of a square of perimeter 56 cm. The second largest side of the right-angled triangle is 4 cm less than the length of the rectangle of area 96 sq cm and breadth 8 cm. What is the largest side of the right-angled triangle ? ***[IOB (PO)-11]***

(a) 20 cm (b) 12 cm (c) 10 cm (d) 15 cm
(e) None of these

14. A piece of wire 78 cm is bent in the form of an isosceles triangle. If the ratio of one of the equal sides to the base is 5 : 3, then what is the length of the base ? ***[CDS-11]***

(a) 16 cm (b) 18 cm
(c) 20 cm (d) 30 cm

15. The area of an isosceles triangle is 4 sq. unit. If the length of the third side is 2 unit, the length of each equal side is :

(a) 4 units (b) $2\sqrt{3}$ units
(c) $\sqrt{17}$ units (d) $3\sqrt{2}$ units

16. The difference between the area of a square and that of an equilateral triangle on the same base is $\frac{1}{4}$ cm². What is the length of side of triangle?

(a) $(4 - \sqrt{3})^{\frac{1}{2}}$ cm (b) $(4 + \sqrt{3})^{\frac{1}{2}}$ cm
(c) $(4 - \sqrt{3})^{\frac{-1}{2}}$ cm (d) $(4 + \sqrt{3})^{\frac{-1}{2}}$ cm

17. a and b are two sides adjacent to the right angle of a right-angled triangle and p is the perpendicular drawn to the hypotenuse from the opposite vertex. Then, p^2 is equal to : ***[SSC (CGL-T-II)-14]***

(a) $a^2 + b^2$ (b) $a^2 - b^2$
(c) $\frac{1}{a^2} + \frac{1}{b^2}$ (d) $\frac{a^2b^2}{a^2 + b^2}$

18. Point M is the mid-point of side AB of a square $ABCD$ and point N is lying on side BC such that BN is $\frac{1}{3}$rd of BC. What is the area of ΔBMN if $BD = 60\sqrt{2}$ cm.

(a) 270 cm² (b) 450 cm²
(c) 300 cm² (d) 180 cm²

19. The base of a triangle is 15 cm and height is 12 cm. The height of another triangle of double the area having the base 20 cm is :
(a) 9 cm (b) 18 cm
(c) 8 cm (d) 12.5 cm

20. Given that the ratio of altitudes of two triangles is 4 : 5, ratio of their areas is 3 : 2. The ratio of their corresponding bases is *[SSC (CGL-T-II)-15]*
(a) 8 : 15 (b) 15 : 8
(c) 5 : 8 (d) 8 : 5

21. What will the effect on the area of an isosceles triangle if all the sides are doubled in length ? *[ECGX of India Ltd. (PO)-01]*
(a) Increase to four times the area of original triangle
(b) Increase to two times the area of original triangle
(c) Increase by 150%
(d) Increase by 200%
(e) None of the above

22. The hypotenuse of a right-angled triangle is 39 cm and the difference of other two sides is 21 cm. Then, the area of the triangle is : *[SSC (10+2, LDC/DEO)-15]*
(a) 270 cm² (b) 450 cm²
(c) 540 cm² (d) 180 cm²

23. The area of a right-angled triangle is $\frac{2}{3}$rd of the area of a rectangle. The base of the triangle is 80% of the breadth of the rectangle. If the perimeter of the rectangle is 200 cm, the height of the triangle is :
(a) 30 cm (b) 20 cm
(c) 15 cm (d) Data inadequate

24. If the sides of an equilateral triangle be increased by 1 m, its area is increased by $\sqrt{3}$ sq m. The length of any of its sides is : *[SSC (10+2, LDC/DEO)-14]*
(a) 2 m (b) 2.5 m (c) 1.5 m (d) $\sqrt{3}$ m

25. A piece of paper is in the shape of a right-angled triangle and is cut along a line that is parallel to the hypotenuse, leaving a smaller triangle. If the area of the original triangle was 40 cm², then what is the area (in cm²) of the smaller triangle formed after the cut provided there was a 40 % reduction in the length of the hypotenuse of the triangle due to cutting.
(a) 15 (b) 22.5 (c) 14.4 (d) 18.6

26. The perimeter of a right-angled triangle is k times the shortest side. If the ratio of the other side to hypotenuse is 4 : 5, then what is the value of k ? *[CDS-19]*
(a) 2 (b) 3 (c) 4 (d) 5

27. An umbrella is made by stitching 10 triangular pieces of cloth of two different colours each piece measuring 30 cm, 60 cm, 60 cm. How much cloth of each colour is required for umbrella ? *[SNAP-18]*
(a) $1015\sqrt{45}$ (b) $1125\sqrt{15}$
(c) $1220\sqrt{45}$ (d) $1235\sqrt{15}$

28. The perimeter of a triangle is equal to that of a rectangle having length and breadth in ratio 4 : 3. The length of the rectangle is 80% of side of a square. What is the perimeter of the triangle if the difference between perimeter of the square and rectangle is 30 cm ? *[SBI (PO-Pre)-19]*
(a) 70 cm (b) 100 cm
(c) 120 cm (d) 85 cm
(e) None of these

Answer Key

1. (b)	2. (c)	3. (c)	4. (d)	5. (c)	6. (d)	7. (a)	8. (a)	9. (a)	10. (b)
11. (c)	12. (b)	13. (c)	14. (b)	15. (c)	16. (c)	17. (d)	18. (c)	19. (b)	20. (b)
21. (a)	22. (a)	23. (d)	24. (c)	25. (c)	26. (c)	27. (b)	28. (a)		

Hints and SOLUTIONS

1. Clearly, $5^2 = x^2 + x^2 \Rightarrow x^2 = 12.5$

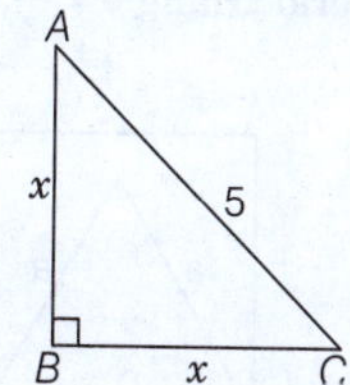

Then, area of $\Delta ABC = \frac{1}{2} \times x \times x$

$= \frac{1}{2}x^2 = \frac{1}{2} \times 12.5 = 6.25\,\text{cm}^2$

2. Area of rectangle = $CD \times AD$

$\Rightarrow \quad 100 = 20 \times AD$

$\Rightarrow \quad AD = 5\,\text{cm}$

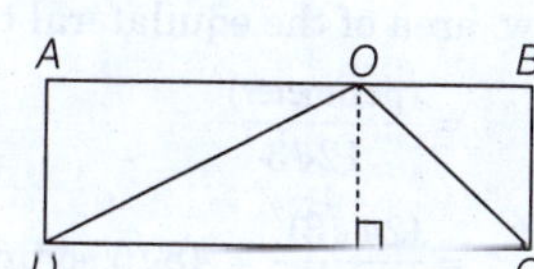

Now, area of ΔODC

$= \frac{1}{2} \times CD \times OE$

$= \frac{1}{2} \times 20 \times AD$

$= \frac{1}{2} \times 20 \times 5 = 50$ sq cm

3. Diagonal of the square

$= \sqrt{2} \times$ side

$\Rightarrow \quad 12\sqrt{2} = \sqrt{2} \times$ side

$\Rightarrow \quad$ Side = 12 cm

Now, perimeter of the triangle

= Perimeter of the square

$= 4 \times 12$

Then, area of the triangle

$= \frac{(\text{perimeter})^2}{12\sqrt{3}} = \frac{(48)^2}{12\sqrt{3}} = 64\sqrt{3}\text{ cm}^2$

4. Area of the right-angled ΔBCD

$= \frac{1}{2} \times 32 \times 24 = 384$

and $BD = \sqrt{32^2 + 24^2} = 40$

Now, area of isosceles ΔABD

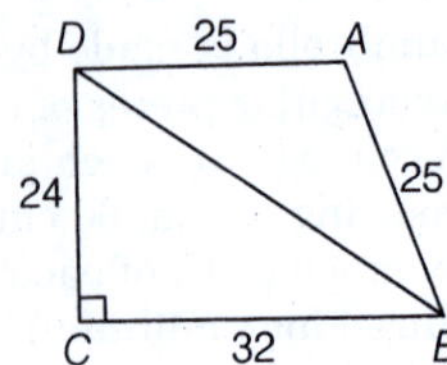

$= \frac{b}{4}\sqrt{4a^2 - b^2}$, where a is the length of each of equal sides and b is the length of the 3rd side.

$= \frac{40}{4}\sqrt{4(25)^2 - 40^2} = 300$

Hence, area of the plot

$= 384 + 300 = 684\text{ m}^2$

5. Altitude of an equilateral triangle

$= \frac{\sqrt{3}}{2} \times \text{side}$

$\Rightarrow \quad \sqrt{3} = \frac{\sqrt{3}}{2} \times \text{side}$

$\Rightarrow \quad \text{side} = 2\text{ cm}$

Then, its perimeter $= 3 \times 2 = 6\text{ cm}$

6. Since (5, 12, 13) is a Pythagorean triplet, the given triangle is a right-angled triangle.

Now, area of rectangle

= area of triangle

$= \frac{1}{2} \times 5 \times 12$

$\Rightarrow$ length $\times$ breadth = 30

$\Rightarrow$ length $\times$ 10 = 30

$\Rightarrow$ length = 3

Now, perimeter of the rectangle

$= 2(3 + 10) = 26$ unit

7. Proceed through options.

(a) 25 $\Rightarrow$ (7, 24, 25) is a Pythagorean triplet and perimeter

$= 7 + 24 + 25 \neq 60$

Another triplet be (15, 20, 25).

Then, perimeter = 60

and area $= \frac{1}{2} \times 15 \times 20 = 150$

Hence, option(a) is correct.

[selection of Pythagorean triplet is based on assumption only]

8. Hypotenuse $AB = \sqrt{a^2 + b^2}$

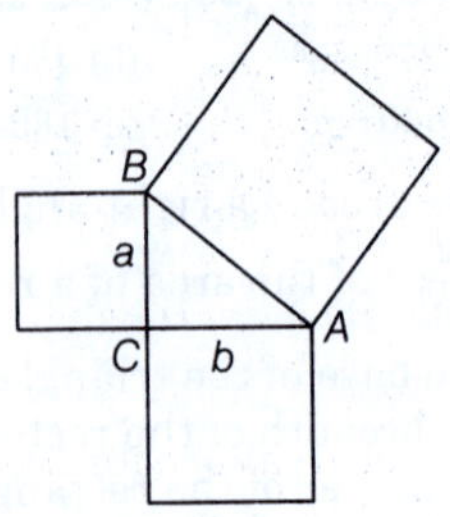

Now, required area

$= \frac{1}{2} \times ab + a^2 + b^2 + (\sqrt{a^2 + b^2})^2$

$= 2(a^2 + b^2) + 0.5\,ab$

9. Area of isosceles triangle

$= \frac{b}{4}\sqrt{4a^2 - b^2}$, where a is the length of each of equal sides and b is the length of the 3rd side.

$= \frac{b}{4}\sqrt{4(2b)^2 - b^2} \quad [\because a = 2b]$

$= \frac{\sqrt{15}\,b^2}{4}$

Now, perimeter of the equilateral triangle

$= a + a + b$

$= 2b + 2b + b = 5b$

Then, its area $= \frac{(5b)^2}{12\sqrt{3}} = \frac{25b^2}{12\sqrt{3}}$

$\therefore$ Required ratio $= \frac{\sqrt{15}\,b^2}{4} : \frac{25\,b^2}{12\sqrt{3}}$

$= 9\sqrt{5} : 25$

$= 36\sqrt{5} : 100$

10. Since (3, 4, 5) is a Pythagorean triplet, the former triangle is a right-angled triangle.

Let sides be $3x$, $4x$ and $5x$.

Then, its area $= \frac{1}{2} \times 3x \times 4x$

$\Rightarrow \quad 72 = 6x^2$

$\Rightarrow \quad x = \sqrt{12} = 2\sqrt{3}$

$\therefore$ Its perimeter $= 12x$

$= 12 \times 2\sqrt{3}$

$= 24\sqrt{3}$ = perimeter of the equilateral triangle,

Now, area of the equilateral triangle

$= \frac{(\text{perimeter})^2}{12\sqrt{3}}$

$= \frac{(24\sqrt{3})^2}{12\sqrt{3}} = 48\sqrt{3}$ sq unit

11. Length of 3rd side

$= 240 - (78 + 50) = 112$

Semi-perimeter $(S) = 240 \div 2 = 120$

Then, area of the triangle

$= \sqrt{(120(120 - 112)(120 - 78)(120 - 50)}$

$\Rightarrow \quad \frac{1}{2} \times \text{side} \times \text{altitude to this side}$

$= 1680$

$\Rightarrow \quad \frac{1}{2} \times 50 \times \text{altitude} = 1680$

$\Rightarrow \quad \text{altitude} = 67.2\text{ m}$

12. Area of the triangle $= \frac{1}{2} \times a \times b$

$\Rightarrow \quad 20 = \frac{1}{2} \times a \times b \Rightarrow ab = 40$

Further, hypotenuse $= \sqrt{a^2 + b^2}$

$\Rightarrow \quad 10 = \sqrt{a^2 + b^2}$

$\Rightarrow \quad 100 = a^2 + b^2$

Now, $(a + b)^2 = a^2 + b^2 + 2ab$

$= 100 + 2 \cdot 40 = 180$

13. Side of the square $= 56 \div 4 = 14$

and length of the rectangle

$= 96 \div 8 = 12$

Then, the smallest and 2nd largest side of the right-angled triangle are respectively (14 – 8) cm and (12 – 4) cm i.e. 6 and 8

$\therefore$ The largest side i.e. hypotenuse

$= \sqrt{6^2 + 8^2} = 10\text{ cm}$

14. Clearly, sides are in ratio 5 : 5 : 3

Now, (5 + 5 + 3) units = 78 cm

$\therefore$ 3 units (i.e. base) = 18 cm

15. Area of an isosceles triangle

$= \frac{b}{4}\sqrt{4a^2 - b^2}$, where a is the length of each of equal sides and b is the length of the 3rd side.

$\Rightarrow \quad 4 = \frac{2}{4}\sqrt{4a^2 - 2^2} \Rightarrow 8 = \sqrt{4a^2 - 2^2}$

$\Rightarrow \quad 64 = 4a^2 - 4$

[squaring both sides]

$\Rightarrow \quad 16 = a^2 - 1 \Rightarrow a = \sqrt{17}$

16. As given, area of square – area of equilateral triangle $= \frac{1}{4}$

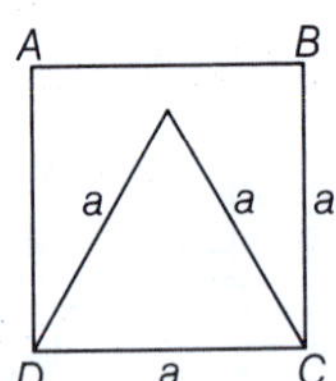

$\Rightarrow \quad a^2 - \frac{\sqrt{3}}{4}a^2 = \frac{1}{4}$

$\Rightarrow \quad 4a^2 - \sqrt{3}a^2 = 1$

$\Rightarrow \quad a^2(4 - \sqrt{3}) = 1$

$\Rightarrow \quad a^2 = \dfrac{1}{(4 - \sqrt{3})}$

$\Rightarrow \quad a = (4 - \sqrt{3})^{\frac{-1}{2}}$ cm

17. Clearly, $AB = \sqrt{a^2 + b^2}$

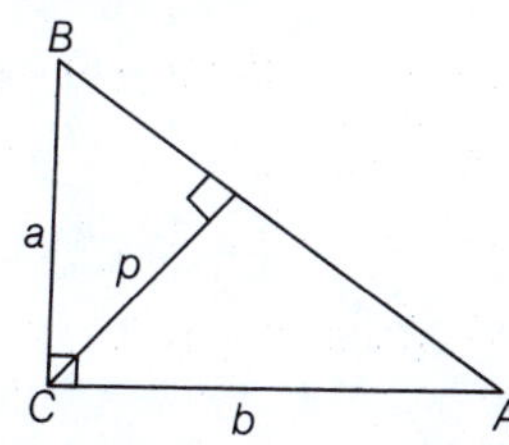

Area of $\Delta ABC = \dfrac{1}{2}ab$

$= \dfrac{1}{2} \times AB \times p$

$\Rightarrow \quad \dfrac{1}{2}ab = \dfrac{1}{2} \times \sqrt{a^2 + b^2} \times p$

$\Rightarrow \quad ab = \sqrt{a^2 + b^2} \times p$

$\Rightarrow \quad a^2b^2 = (a^2 + b^2)p^2$ [squaring both sides]

$\Rightarrow \quad p^2 = \dfrac{a^2b^2}{a^2 + b^2}$

18. Diagonal $BD = \sqrt{2} \times$ side

$\Rightarrow \quad 60\sqrt{2} = \sqrt{2} \times$ side

$\Rightarrow \quad$ Side = 60 cm

Now, area of ΔBMN

$= \dfrac{1}{2} \times BM \times BN$

$= \dfrac{1}{2} \times \dfrac{60}{2} \times \dfrac{60}{3} = 300 \text{ cm}^2$

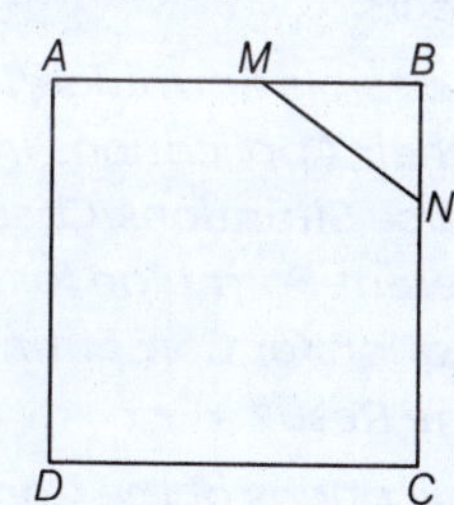

19. Area of the 1st triangle

$= \dfrac{1}{2} \times 15 \times 12 = 90$

Now, area of the 2nd triangle

$= \dfrac{1}{2} \times 20 \times$ Height

$\Rightarrow \; 2 \times 90 = 10 \times$ Height

$\Rightarrow$ Height = 18 cm

20. Ratio of areas of both triangles

$= 3 : 2$

$\Rightarrow \; \dfrac{1}{2} \times \text{base}_1 \times \text{height}_1 : \dfrac{1}{2} \times \text{base}_2 \times \text{height}_2$

$= 3 : 2$

$\Rightarrow \text{base}_1 \times 4 : \text{base}_2 \times 5 = 3 : 2$

$\Rightarrow \quad \text{base}_1 : \text{base}_2 = 15 : 8$

21. Effect on area

$= \left[+ 100 + 100 + \dfrac{100 \times 100}{100}\right]\%$

$= 300\%$

It implies that the new area will be 300% greater than the old one, hence, 4 times.

22. Recall a Pythagorean triplet (5, 12, 13). Obviously, (15, 36, 39) will also be a Pythagorean triplet.

Since the difference between 36 and 15 is 21, the sides are 15 cm, 36 cm and 39 cm.

Then, area of the triangle $= \dfrac{1}{2} \times 15 \times 36 = 270 \text{ cm}^2$

23. Given, 2(length + breadth) = 200

$\Rightarrow$ length + breadth = 100 cm

Now, we are unable to compute breadth resulting in we cannot find the area or height of the triangle. Hence, option (d) is correct.

24. Let, side of the triangle be x m.

Then, $\dfrac{\sqrt{3}}{4}[(x + 1)^2 - x^2] = \sqrt{3}$

$\Rightarrow \quad 2x + 1 = 4$

$\Rightarrow \quad x = 1.5$ m

25. If the hypotenuse is reduced by 40%, the perpendicular sides will also be reduced by 40%.

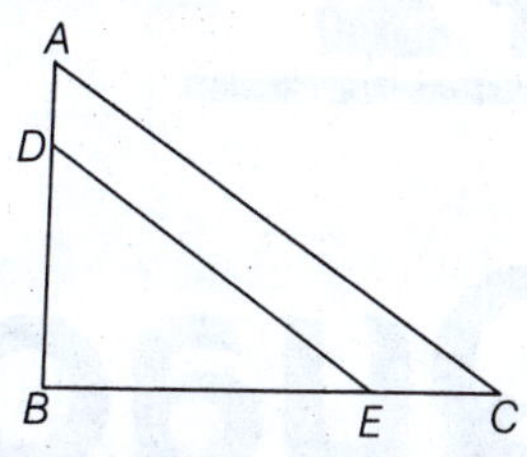

[Explanation ΔABC is similar to ΔBDE, $\therefore \dfrac{AB}{BD} = \dfrac{BC}{BE} = \dfrac{AC}{DE} = \dfrac{100}{60}$]

Since the area of a right-angled triangle is a product of perpendicular sides with a constant term, effect on area

$= \left[- 40 - 40 + \dfrac{40 \times 40}{100}\right]\% = - 64\%$

It implies that the area of the newly formed triangle will be 64% less than that of the original triangle.

$\therefore$ Required answer

$= 36\% \text{ of } 40 = 14.4 \text{ cm}^2$

26. Clearly, the sides of the triangle are in ratio 3 : 4 : 5 as (3, 4, 5) is a Pythagorean triplet.

As given, $(3 + 4 + 5) = 3k$

$\therefore \quad k = 4$

27. Area of 10 isosceles triangles

$= 10 \times \dfrac{30}{4}\sqrt{4 \times (60)^2 - 30^2} = 2250\sqrt{15}$

$\therefore$ Area of each colour

$= 2250\sqrt{15} \div 2 = 1125\sqrt{15}$

28. Length of the rectangle is 80% of the side of the square, it implies that side of the square is 5 units.

As given,

(4×5) units $- 2(4 + 3)$ units = 30 cm

$\Rightarrow \quad$ 1 unit = 5 cm

$\Rightarrow 2(4 + 3)$ units = 70 cm

Hence, perimeter of the rectangle

= 70 cm = perimeter of triangle

42

Quadrilateral (Parallelogram, Rhombus and Trapezium)

LEARNING OBJECTIVE

After completing this chapter, aspirants will be able to :

- Understand the quadrilateral and its various types with their properties;
- Understand the application of different quadrilateral related formulae in different cases;
- Understand the application of combination of quadrilateral related formulae in diverse cases;
- Compare different types of quadrilateral with reference to different aspects;
- Understand the philosophy of the division of this chapter in different parts;
- Analyse the trend of the questions being asked in various competitive examinations;
- Solve the questions in moments;
- Grip the concepts helping to solve the questions without pen and paper;
- Apply many concepts envisaged in the chapter 'PERCENTAGE' like 'effect on product in case of changes in factors', etc.;
- Apply many concepts of 'Ratio' in this chapter;
- Learn many new concepts/fundamentals;
- Apply the pragmatic approaches in lieu of academic approaches; and
- Ponder on the alternative methods.

Note *Geometric properties of quadrilaterals and relevant concepts have been given in 'GEOMETRY' section.*

Inside

- *Quadrilateral and its Various Types With Their Properties*
- *Comparison Among Types of Quadrilaterals Through a Chart with Reference to Properties Related to Various Aspects*
- *All Possible Formulae/Results for their Applications in Diverse Situations/Cases Even Different Formulae for Same Situation for Computing the Same Result*
- *Applications of the Concepts Related to Parallelogram*
- *Rhombus and Trapezium*
- *Problems Based on Combination of Shapes Discussed So Far*
- ....
- .... *& Many More*
- *Work Out with Solutions*

Quadrilateral

A quadrilateral is a closed plane figure bounded by four straight lines.

Perimeter $= AB + BC + CD + DA$

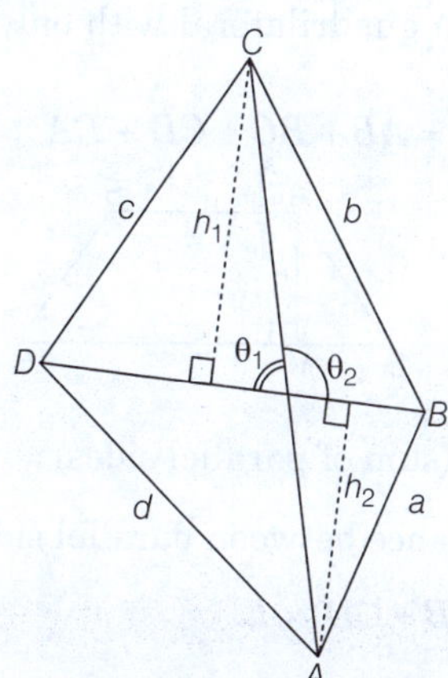

Area $= \frac{1}{2} \times$ one of diagonals $\times$ sum of perpendiculars drawn on it from the both opposite vertices

$$= \frac{1}{2} \times BD \times (h_1 + h_2)$$

OR

Area $= \frac{1}{2} \times$ product of diagonals $\times$ sine of angle between them

$$= \frac{1}{2} \times AC \times BD \times \sin\theta_1$$

or $\frac{1}{2} \times AC \times BD \times \sin\theta_2$

Diagonal $BD = \sqrt{a^2 + d^2 - 2ad\cos A}$

or $\sqrt{b^2 + c^2 - 2bc\cos C}$

Diagonal $AC = \sqrt{c^2 + d^2 - cd\cos D}$

or $\sqrt{a^2 + b^2 - 2ab\cos B}$

Types of Quadrilaterals Parallelogram, Rectangle, Rhombus, Square and Trapezium. Their properties are as under

Side related properties	Parallelogram	Rectangle	Rhombus	Square
Opposite sides are parallel	✓	✓	✓	✓
Opposite sides are equal	✓	✓	✓	✓
All sides are equal	✗	✗	✓	✓
Any line joining the mid-points of opposite sides bisects the area	✓	✓	✓	✓

Diagonal related properties	Parallelogram	Rectangle	Rhombus	Square
Diagonals bisect each other	✓	✓	✓	✓
Diagonals bisect each other at 90°	✗	✗	✓	✓
Diagonals are equal	✗	✓	✗	✓
Diagonals form four triangles of equal area	✓	✓	✓	✓
Each diagonal divides the figure into two congruent triangles	✓	✓	✓	✓
Length of diagonal $= \sqrt{a^2 + b^2 - 2ab\cos B}$ where, B is included angle between two adjacent sides a and b.	✓	✓	✓	✓
Area related properties				
Area = Base × Height	✓	✓	✓	✓
Area $= \frac{1}{2} \times$ product of diagonals × sine of angle between them	✓	✓	✓	✓
Area $= \frac{1}{2} \times$ one diagonal × sum of perpendiculars on it from the both opposite vertices	✓	✓	✓	✓
Area = Product of any two adjacent sides × sine of included angle	✓	✓	✓	✓
The properties with reference to various aspects have been given in details in Geometry Section.				

Parallelogram

A parallelogram is a quadrilateral with opposite sides parallel. Parallelogram also includes Rhomboid, Rectangle, Rhombus and Square.

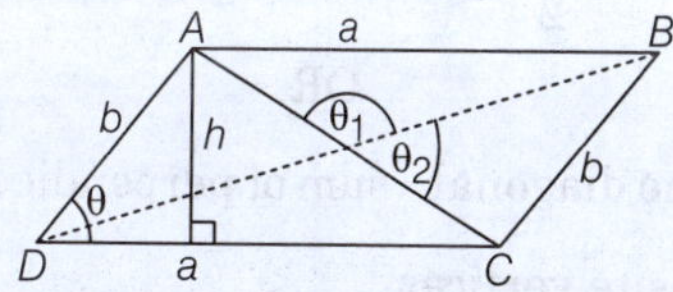

Perimeter $= 2(a + b)$, where a and b are any two adjacent sides.

Area = Base × Height $= a \times h$

OR

Area $= \frac{1}{2} \times$ product of diagonals $\times$ sine of angle between them $= \frac{1}{2} \times AC \times BD \times \sin\theta_1$ or $\frac{1}{2} \times AC \times BD \times \sin\theta_2$

OR

Area $= \frac{1}{2} \times$ one diagonal $\times$ sum of perpendiculars on it from the both opposite vertices.

OR

Area = Product of any two adjacent sides × sine of included angle = $AD \times DC \times \sin\theta$

Diagonal $BD = \sqrt{a^2 + b^2 - 2ab\cos A}$

or $\sqrt{a^2 + b^2 - 2ab\cos C}$

Diagonal $AC = \sqrt{a^2 + b^2 - 2ab\cos D}$

or $\sqrt{a^2 + b^2 - 2ab\cos B}$

Also, $2(a^2 + b^2) = (d_1)^2 + (d_2)^2$, where a and b are adjacent sides and d_1 and d_2 are diagonals.

Rhombus

A parallelogram having all the sides equal is a rhombus.

Perimeter $= 4a$, where a is length of each side.

Area = Base × Height $= a \times h$

OR

Area $= \frac{1}{2} \times$ product of diagonals × sine of angle between them

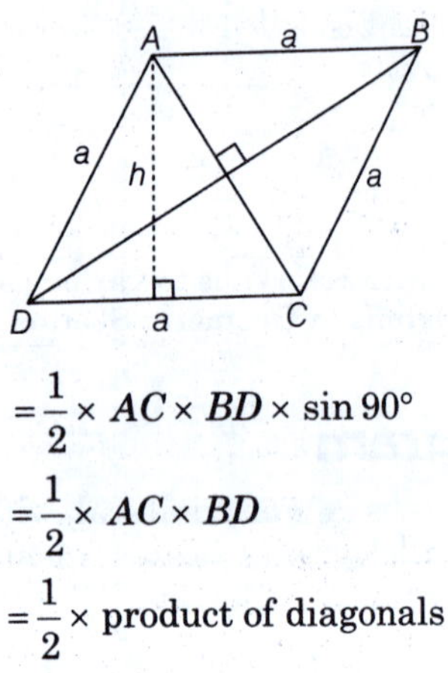

$= \frac{1}{2} \times AC \times BD \times \sin 90°$

$= \frac{1}{2} \times AC \times BD$

$= \frac{1}{2} \times$ product of diagonals

OR

Area $= \frac{1}{2} \times$ one diagonal × sum of perpendiculars on it from the both opposite vertices.

OR

Area = Product of any two adjacent sides × sine of included angle $= AD \times DC \times \sin D$

or $BC \times DC \times \sin C$

or $AD \times AB \times \sin A$

or $AB \times BC \times \sin B$

OR

Area $= 2ar$, where, r = inradius

Diagonal $BD = \sqrt{a^2 + a^2 - 2a^2\cos C}$

$= \sqrt{2}\, a\sqrt{1 - \cos C}$ or $\sqrt{2}\, a\sqrt{1 - \cos A}$

Diagonal $AC = \sqrt{2}\, a\sqrt{1 - \cos D}$ or $\sqrt{2}\, a\sqrt{1 - \cos B}$

Also, $4a^2 = (d_1)^2 + (d_2)^2$, where a is each side of the rhombus and d_1 and d_2 are its diagonals.

Trapezium

A trapezium is a quadrilateral with only one pair of sides parallel.

Perimeter $= AB + BC + CD + DA$

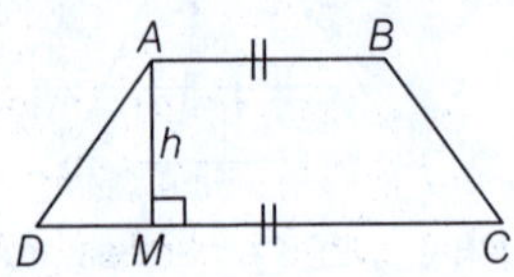

Area $= \frac{1}{2} \times$ (sum of parallel sides) × perpendicular distance between parallel sides

$= \frac{1}{2}(AB + CD) \times h$

It is noted that applications of many formulae mentioned above have been discussed in **Geometry Section** and in order to avoid repetitions applications of only mensuration related formulae have been discussed here.

Applications of the concepts Related to Parallelogram

Problem 1 The perimeter of a parallelogram is 32 cm. What is the length of its sides if its area is maximum ?

Sol. The area of the parallelogram is maximum when it is a rhombus.

$\therefore$ Its side $= \frac{32}{4} = 8$ cm

Problem 2 Sides of a parallelogram are in the ratio 5 : 4. Its area is 1000 sq unit. Altitude on the greater side is 20 units. Altitude on the smaller side is

Sol.

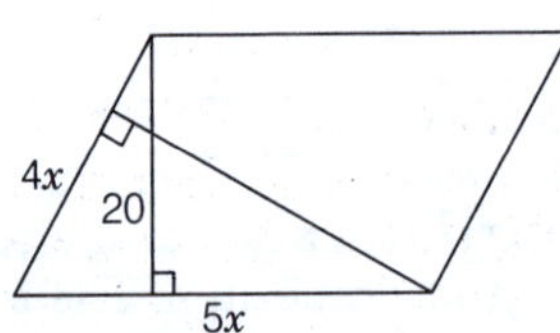

Let the sides be $5x$ and $4x$.

Then, area $= 5x \times 20 = 1000$ (given)

$\Rightarrow x = 10$

$\therefore$ Smaller side $= 4x = 4 \times 10 = 40$ units

Now, area of the parallelogram = 40 × altitude on the smaller side

$\Rightarrow$ 1000 = 40 × altitude on the smaller side.

$\therefore$ Altitude on the smaller side $= \frac{1000}{40} = 25$ units

Problem 3 A parallelogram has sides 15 cm and 7 cm long. The length of one of diagonals is 20 cm. The area of the parallelogram is ***[SSC (CGL-T-II)-14]***

Sol. Area of parallelogram $ABCD$ = Area of ΔABD + Area of ΔBCD

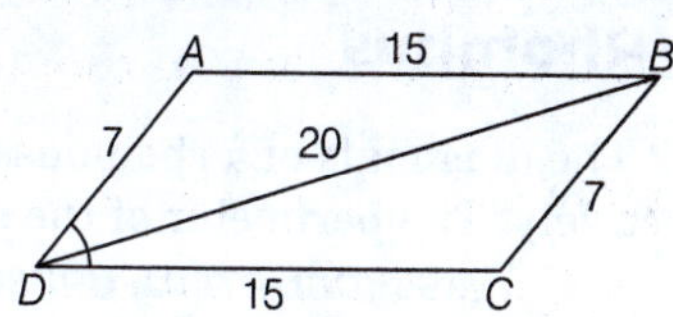

Now, area of $\Delta ABD = \sqrt{s(s-a)(s-b)(s-c)}$,

$$\left(s = \frac{15 + 7 + 20}{2} = 21\right)$$

$$= \sqrt{21(21-15)(21-7)(21-20)}$$

$$= \sqrt{21 \times 6 \times 14 \times 1} = 42$$

Since each diagonal divides a parallelogram into two congruent triangles, area of ΔABD will be equal to the area of ΔBCD.

Thus, area of parallelogram = 42 + 42 = 84 cm²

Problem 4 In a parallelogram $ABCD$ of area 72 sq cm, the sides CD and AD have lengths 9 cm and 16 cm respectively. Let P be a point on CD such that AP is perpendicular to CD. Then, the area, in sq cm, of ΔAPD is **[CAT-18]**

Sol. Area of parallelogram $ABCD = CD \times AP = 72$

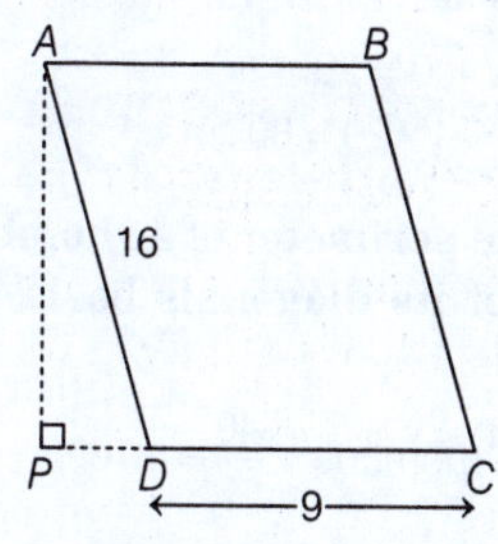

$\Rightarrow \quad 9 \times AP = 72$

$\therefore \quad AP = 8$

Then, in right angled ΔAPD,

$$PD = \sqrt{16^2 - 8^2} = 8\sqrt{3}$$

Now, area of $\Delta APD = \frac{1}{2} \times 8\sqrt{3} \times 8 = 32\sqrt{3}$

Problem 5 A parallelogram $ABCD$ has sides $AB = 24$ cm, $AD = 16$ cm. The distance between the sides AB and DC is 10 cm. Find the distance between sides AD and BC. **[SSC (10+2)- LDC/DEO-11]**

Sol. Area of the parallelogram = Base × Height

$$= 24 \times 10 = 240 \text{ cm}^2$$

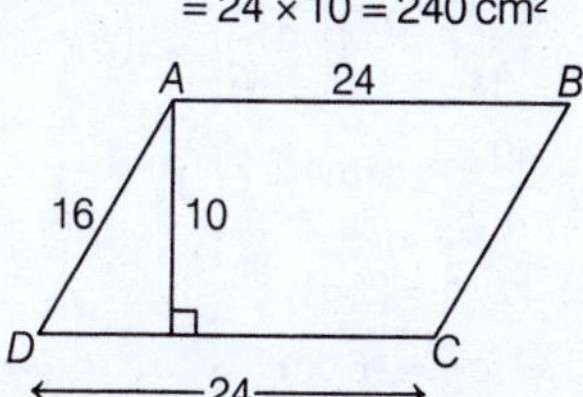

Also, area of the parallelogram = base AD × height (i.e. distance between sides AD and BC)

$\Rightarrow$ 240 = 16 × distance between sides AD and BC

$\Rightarrow$ distance between sides AD and BC = 240 ÷ 16 = 15 cm

Directions (Q. Nos. 6-8) *A piece of land is in the form of a parallelogram and the perimeter of the land is 86 m. The length of one side exceeds the other by 13 m and one of the diagonals is 41 m.*

Problem 6 What is the area of the parallelogram?

Problem 7 What is the shortest height of the parallelogram?

Problem 8 Consider the following statements

(i) The difference between the diagonals of the parallelogram is more than 20 m.

(ii) The difference between the heights of the parallelogram is more than 10 m.

Which of the above statements is/are correct ?

(a) Only (i) (b) Only (ii)
(c) Both (i) and (ii) (d) Neither (i) nor (ii) **(CDS-14)**

Solutions

6. Let one side of parallelogram be x m.

Then, other side will be $(x + 13)$ m.

Now, perimeter $= 2[x + (x + 13)] = 86$ [given]

$\Rightarrow \quad x = 15$

It means, shorter side is 15 m and longer is (15 + 13) m.

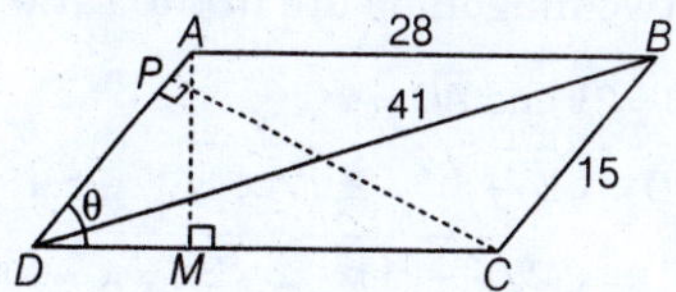

Now, area of $\Delta ABD = \sqrt{s(s-a)(s-b)(s-c)}$,

$$\left\{s = \frac{15 + 28 + 41}{2} = 42\right\}$$

$$= \sqrt{42(42-15)(42-28)(42-41)}$$

$$= \sqrt{42 \times 27 \times 14 \times 1} = 126$$

Since each diagonal divides a parallelogram into two congruent triangles, area of ΔABD will be equal to the area of ΔBCD.

Thus, area of parallelogram = 126 + 126 = 252 m²

7. Area of parallelogram = Base × Height

$\Rightarrow \quad 252 = 28 \times \text{Height}$

[∵ longer side will have shorter height and shorter side will have longer height]

$\Rightarrow \quad \text{Shorter height} = \frac{252}{28} = 9 \text{ m}$

8. Shorter height (AM) = 9 m [calculated above]

Now, longer height (CP) $= \frac{252}{15} = 16.8$ m

Hence, the difference between heights (i.e. 9 m and 16.8 m) of the parallelogram is not more than 10 m.

Further, $2(a^2 + b^2) = (d_1)^2 + (d_2)^2$, where a and b are adjacent sides and d_1 and d_2 are diagonals.

$\Rightarrow \quad 2[(15)^2 + (28)^2] - (41)^2 + (d_2)^2$

$\Rightarrow \quad (d_2)^2 = 337 \Rightarrow d_2 = 18.36$

Thus, the difference between diagonals (i.e. 41 m and 18.36 m) of the parallelogram is more than 20 m.

$\therefore$ Option (a) is correct.

Problem 9 $ABCD$ is a parallelogram P, Q, R and S are points on sides AB, BC, CD and DA respectively such that $AP = DR$. If the area of the parallelogram $ABCD$ is 16 cm², then the area of the quadrilateral $PQRS$ is

Sol. Area of the quadrilateral = Area of ΔPRS + Area of ΔPQR

$$= \left\{\frac{1}{2} \times PR \times AP\right\} + \left\{\frac{1}{2} \times PR \times PB\right\}$$

$$= \frac{1}{2} PR(AP + PB) = \frac{1}{2} \times PR \times AB$$

$$= \frac{1}{2} \times AD \times AB$$

$$= \frac{1}{2} \times \text{area of parallelogram } ABCD$$

$$= \frac{1}{2} \times 16 = 8\,\text{cm}^2$$

Problem 10 $ABCD$ is a parallelogram, where $AB : AD = 2 : 1$. One of the angles of the parallelogram is 60°. The two diagonals are in the ratio **[CDS-15]**

Sol. Let side $AB = 2x$ and $AD = x$

$$\text{Diagonal } BD = \sqrt{a^2 + b^2 - 2\,ab\cos A}$$

$$= \sqrt{(2x)^2 + (x)^2 - 2 \times 2x \times x \times \cos 60^\circ}$$

$$= \sqrt{3}x$$

$$\text{Diagonal } AC = \sqrt{a^2 + b^2 - 2ab\cos D}$$

$$= \sqrt{(2x)^2 + (x)^2 - 2 \times 2x \times x \times \cos 120^\circ}$$

$$= \sqrt{7}\,x$$

Now, required ratio = $\sqrt{7} : \sqrt{3}$

Problem 11 If the base and the altitude of a parallelogram are increased by 10% and 20% respectively, what will be effect on its area ?

Sol. Here, area is the product of the base and the height. Now apply the concept of 'effect on product in case of changes in factors'.

$$\text{Thus, effect on area} = \left[+ 10 + 20 + \frac{10 \times 20}{100}\right]\%$$

$$= +\, 32\%$$

'+' indicates that area will increase by 32%.

Applications of the Concepts Related to Rhombus

Problem 12 The diagonals of a rhombus are 32 cm and 24 cm respectively. The perimeter of the rhombus is **[SSC (CGL-PT)-02, UPPSC (Rev. Insp.)-15]**

Sol. We know that diagonals of a rhombus bisect each other at 90°.

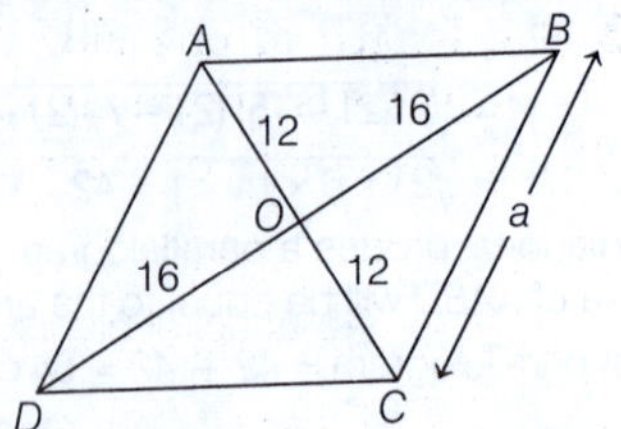

Hence, $(AD)^2 = (AO)^2 + (OD)^2 = (12)^2 + (16)^2 = 400$

$\Rightarrow \quad AD = \sqrt{400} = 20$

Then, perimeter = $4 \times 20 = 80$ cm

Alternatively

$$4a^2 = (d_1)^2 + (d_2)^2 = (24)^2 + (32)^2 = 1600$$

$\Rightarrow \quad a^2 = 400$

$\therefore \quad a = \sqrt{400} = 20$

Then, perimeter = $4 \times 20 = 80$ cm

Problem 13 The perimeter of a rhombus is 40 cm. If the length of one of its diagonals be 12 cm, the length of other diagonal is

Sol. Each side of the rhombus = $\frac{40}{4} = 10$ cm

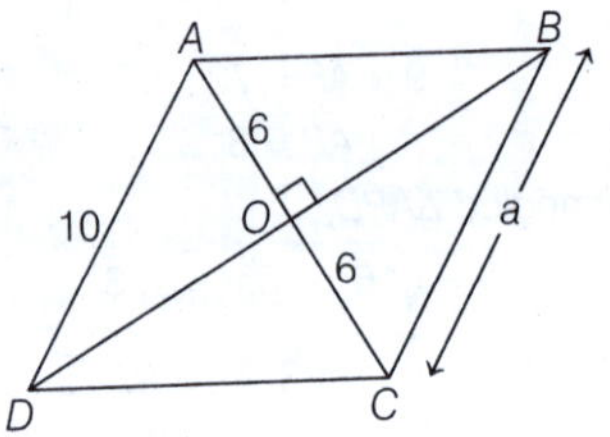

Clearly, $(OD)^2 = (AD)^2 - (AO)^2$

$= (10)^2 - (6)^2 = 64$

$\Rightarrow \quad OD = \sqrt{64} = 8$

Since diagonals bisect each other, OB will be equal to OD.

$\therefore \quad BD = 8 + 8 = 16$ cm

Alternatively

$$4a^2 = (d_1)^2 + (d_2)^2$$

$\Rightarrow \quad 4\left(\frac{40}{4}\right)^2 = (12)^2 + (d_2)^2$

$\Rightarrow \quad (d_2)^2 = 256$

$\therefore \quad d_2 = 16$ cm

Problem 14 The area of a rhombus is 150 cm². The length of one of its diagonal is 10 cm. The length of the other diagonal is

Sol. Area of the rhombus $= \frac{1}{2} \times$ product of diagonals

$\Rightarrow \quad 150 = \frac{1}{2} \times 10 \times$ 2nd diagonal

$\Rightarrow$ 2nd diagonal = 30 cm

Problem 15 The length of one side of a rhombus is 6.5 cm and its altitude is 10 cm. If the length of one diagonal be 26 cm, the length of other diagonal will be

Sol. Area of the rhombus = Base × Height

$= \frac{1}{2} \times$ product of diagonals

$\Rightarrow \quad 6.5 \times 10 = \frac{1}{2} \times 26 \times$ 2nd diagonal

$\Rightarrow$ 2nd diagonal = 5 cm

Problem 16 The perimeter of a rhombus is 100 cm. If one of its diagonals is 14 cm, then the area of the rhombus is

Sol. We know, $4a^2 = (d_1)^2 + (d_2)^2$

$\Rightarrow \quad 4\left(\frac{100}{4}\right)^2 = (14)^2 + (d_2)^2$

$\Rightarrow \quad (d_2)^2 = 2304$

$\therefore \quad d_2 = 48$ cm

Now, area of the rhombus $= \frac{1}{2} \times$ Product of diagonals

$= \frac{1}{2} \times 14 \times 48 = 336 \text{ cm}^2$

Problem 17 If the diagonals of a rhombus are 8 cm and 6 cm, then the square of its side is **[SSC (CGL-T-II)-11]**

Sol. We know, $4a^2 = (d_1)^2 + (d_2)^2$

$= (8)^2 + (6)^2 = 100$

$\Rightarrow \quad a^2 = 25 \text{ cm}^2$

Problem 18 The two diagonals of a rhombus are of lengths 55 cm and 48 cm. If p is the perpendicular height of the rhombus, then which one of the following is correct ? **[CDS-11]**

(a) 36 cm < p < 37 cm (b) 35 cm < p < 36 cm

(c) 34 cm < p < 35 cm (d) 33 cm < p < 34 cm

Sol. We know, $4a^2 = (d_1)^2 + (d_2)^2$

$= (55)^2 + (48)^2 = 5329$

$\Rightarrow \quad a^2 = 1332.25$

$\therefore \quad a = \sqrt{1332.25}$

$= 36.5$

It means each side of the rhombus = 36.5 cm

Now, area of the rhombus = Base × Height

$= \frac{1}{2} \times$ product of diagonals

$\Rightarrow \quad 36.5 \times p = \frac{1}{2} \times 55 \times 48$

$\Rightarrow \quad p = 36.16$ cm

Hence, correct option is (a).

Problem 19 The area of a rhombus is 2016 cm² and its side is 65 cm. The lengths of the diagonals (in cm) are **[MAT-03, CAT-21]**

Sol. Let $AO = x$ and $OD = y$

$\therefore \quad x^2 + y^2 = (65)^2 = 4225 \quad \ldots(i)$

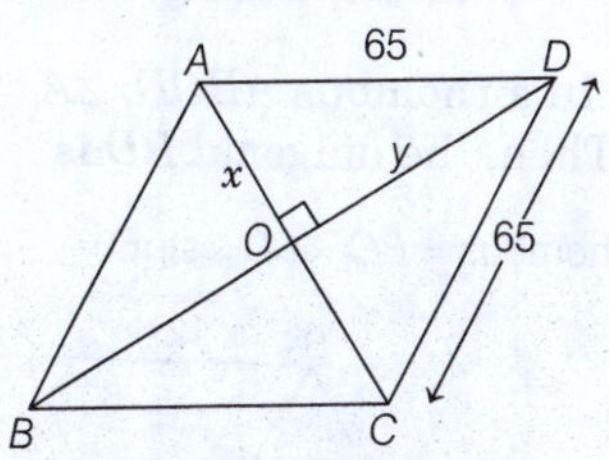

Further, area of the rhombus $= \frac{1}{2} \times 2x \times 2y$

$\Rightarrow \quad 2016 = 2xy$

$\Rightarrow \quad xy = 1008 \quad \ldots(ii)$

Now, $\quad (x - y)^2 = x^2 + y^2 - 2xy$

$\Rightarrow \quad (x - y)^2 = 4225 - 2 \times 1008 = 2209$

$\Rightarrow \quad x - y = 47 \quad \ldots(iii)$

And $\quad (x + y)^2 = (x - y)^2 + 4xy$

$\Rightarrow \quad (x + y)^2 = (47)^2 + 4 \times 1008$

$\Rightarrow \quad x + y = \sqrt{6241} = 79 \quad \ldots(iv)$

From Eqs. (iii) and (iv), we get $x = 63$ and $y = 16$.

Hence, diagonals are (2 × 63) and (2 × 16) i.e. 126 and 32

Alternatively

We know, $\quad 4a^2 = (d_1)^2 + (d_2)^2$

$\Rightarrow \quad 4 \times (65)^2 = (d_1)^2 + (d_2)^2$

Now, proceed through the options. Let options be (a) 125, 35 (b) 126, 32 (c) ………

(a) 125, 35 $\Rightarrow 4 \times (65)^2 \neq (125)^2 + (35)^2$

(b) 126, 32 $\Rightarrow 4 \times (65)^2 = (126)^2 + (32)^2$

Hence, option (b) is correct.

Problem 20 The measure of each of two opposite angles of a rhombus is 60° and the measure of one of its sides is 10 cm. The length of its smaller diagonal is

Sol. Area of the rhombus $= BC \times CD \times \sin 60°$

$= 10 \times 10 \times \frac{\sqrt{3}}{2} = 50\sqrt{3}$

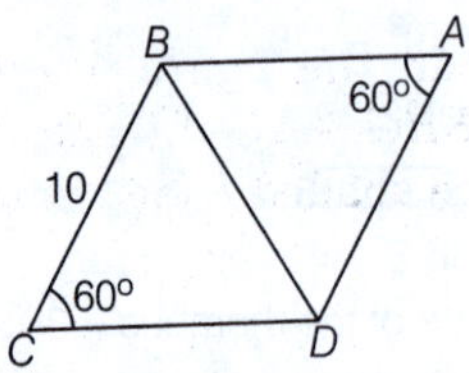

Since the diagonal divides the rhombus in two congruent triangles, area of $\Delta BCD = 50\sqrt{3} \div 2 = 25\sqrt{3}$

Now, proceed through options, let option be 10 cm, $10\sqrt{3}$ cm ...

(a) 10 cm $\Rightarrow$ If $BD = 10$ cm, then it is an equilateral triangle and its area will be $\frac{\sqrt{3}}{4} \times (10)^2 = 25\sqrt{3}$

Hence, option (a) is correct.

Alternatively

$$\angle BAD = \angle BCD = 60^\circ$$

Then, $\angle ABC = \angle ADC = 120^\circ$

$\therefore$ $\angle CBD = \angle CDB = 60^\circ$

[$\because$ diagonals bisect vertex angles]

Hence, ΔBCD is an equilateral triangle. It means $BD = 10$ cm

Problem 21 In a rhombus $ABCD$, $\angle A = 60^\circ$ and $AB = 12$ cm. Then, the diagonal BD is *[SSC (CGL-T-I)-15]*

Sol. Area of the rhombus $= BC \times CD \times \sin 60^\circ$

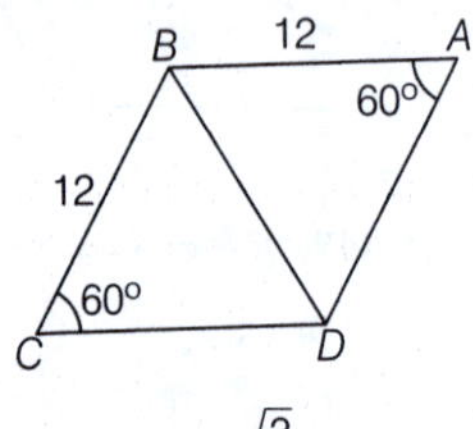

$$= 12 \times 12 \times \frac{\sqrt{3}}{2} = 72\sqrt{3}$$

Since the diagonal divides the rhombus in two congruent triangles, area of $\Delta BCD = 72\sqrt{3} \div 2 = 36\sqrt{3}$

Now, proceed through options. Let options be 12 cm, $2\sqrt{3}$ cm, ...

(a) 12 cm $\Rightarrow$ If $BD = 12$ cm, then it is an equilateral triangle and its area will be $\frac{\sqrt{3}}{4} \times (12)^2 = 36\sqrt{3}$

Hence, option (a) is correct.

Alternatively

$\angle BAD = \angle BCD = 60^\circ$ [given]

Then, $\angle ABC = \angle ADC = 120^\circ$

$\therefore$ $\angle CBD = \angle CDB = 60^\circ$

[$\because$ diagonals bisect vertex angles]

Hence, $\angle BCD$ is an equilateral triangle. It means $BD = 12$ cm

Problem 22 If the diagonals of a rhombus of side 15 cm are in ratio 3 : 4, find the area of the rhombus. *[XAT-18]*

Sol. We know, $4a^2 = (d_1)^2 + (d_2)^2$

$\Rightarrow$ $4 \times (15)^2 = (3x)^2 + (4x)^2$

[let diagonal be $3x$ and $4x$ respectively]

$\Rightarrow$ $x = 6$

Hence, diagonals are $(3 \cdot 6)$ and $(4 \cdot 6)$ i.e.18 and 24.

Now, area of the rhombus $= \frac{1}{2} \times 18 \times 24 = 216 \text{cm}^2$

Applications of the Concepts Related to Trapezium

Problem 23 The ratio of the length of the parallel sides of a trapezium is 3 : 2 .The shortest distance between them is 15 cm. If the area of the trapezium is 450 cm², the sum of the lengths of the parallel sides are

Sol. Let the parallel sides be $3x$ and $2x$.

Now, area of the trapezium $= \frac{1}{2} \times$ (sum of parallel sides) $\times$ altitude

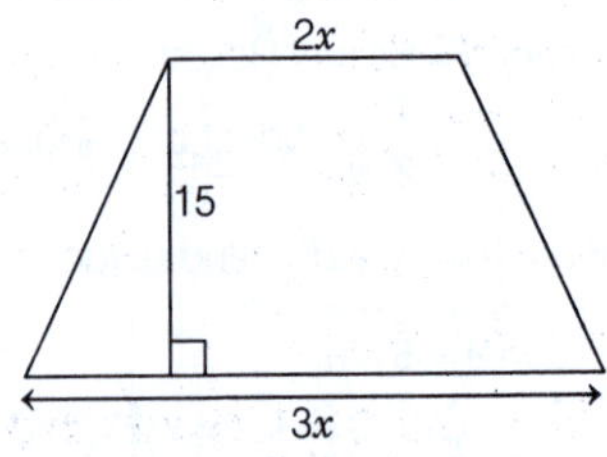

$\Rightarrow$ $450 = \frac{1}{2} \times (3x + 2x) \times 15$

$\Rightarrow$ $5x = 60$

$\therefore$ $x = 12$

Then, sum of the parallel sides

$$= 3x + 2x = 5x$$
$$= 5 \times 12 = 60 \text{ m}$$

Alternatively

Area of the trapezium $= \frac{1}{2} \times$ (sum of parallel sides) $\times$ altitude

$\Rightarrow$ $450 = \frac{1}{2} \times$ (sum of parallel sides) $\times 15$

$\therefore$ Sum of parallel sides = 60 cm

Problem 24 The parallel sides of a field, which is in the shape of a trapezium, are 20m and 41m and the remaining two sides are 10 m and 17 m. Find the cost of levelling the field at the rate of ₹ 30 per sq m. *[SSC (CGL-Mains)-01, CLAT-18, SNAP-18]*

Sol. Clearly, $AB = EF = 20$ m

$\therefore$ $DE + FC = 41 - 20 = 21$ cm

Let DE be x m, then FC will be $(21 - x)$ m.

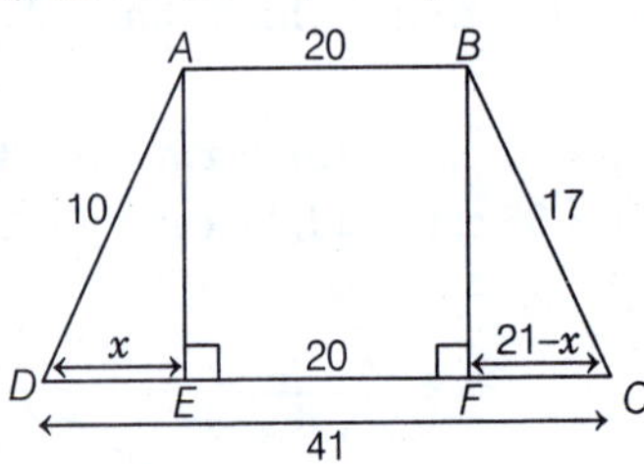

In ΔADE, $(AD)^2 = DE)^2 + (AE)^2$

$\Rightarrow$ $(10)^2 = x^2 + (AE)^2$

$\Rightarrow$ $(AE)^2 = 100 - x^2$

In ΔBFC, $(BC)^2 = (BF)^2 + (FC)^2$

$\Rightarrow$ $(17)^2 = (BF)^2 + (21 - x)^2$

$\Rightarrow \quad (BF)^2 = 289 - 441 + 42x - x^2$

$\Rightarrow \quad (BF)^2 = -152 + 42x - x^2$

But, $\quad AE = BF$

$\Rightarrow \quad (AE)^2 = (BF)^2$

$\Rightarrow \quad 100 - x^2 = -152 + 42x - x^2$

$\Rightarrow \quad 42x = 252$

$\therefore \quad x = 6$

Then, $\quad AE = \sqrt{(10)^2 - 6^2} = 8\text{ m}$

Now, area of the field $= \frac{1}{2} \times$ (sum of parallel sides) $\times$ altitude

$= \frac{1}{2} \times (AB + DC) \times AE$

$= \frac{1}{2} \times (20 + 41) \times 8 = 244\text{ m}^2$

Hence, total cost of levelling the field

$= 244 \times 30 =$ ₹ 7320

Problems Based on Combination of Shapes Discussed so Far

Problem 25 A triangle and a parallelogram are constructed on the same base such that their areas are equal. If the altitude of the parallelogram is 100 m, then the altitude of the triangle is

Sol. Area of parallelogram = Base $(DC) \times$ Altitude (AE)

$= DC \times 100$

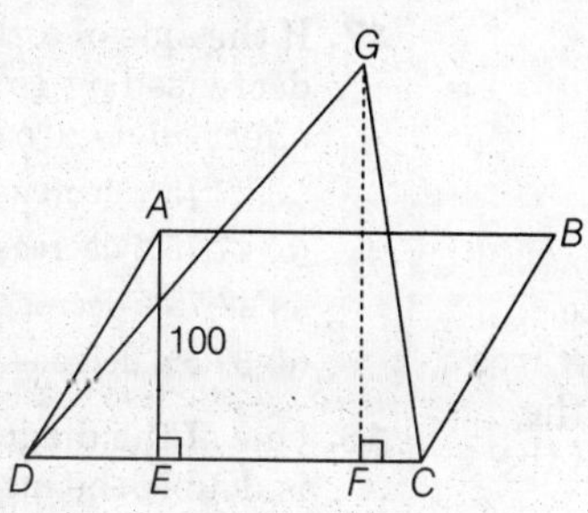

Now, area of $\Delta DGC = \frac{1}{2} \times$ base $(DC) \times$ altitude (GF)

$\Rightarrow \quad DC \times 100 = \frac{1}{2} \times DC \times GF$

$\therefore \quad GF = 200\text{ m}$

Problem 26 In the figure, there are two rectangles $ABCD$ and $DEBG$, each of length 7 cm and width 3 cm. The area of the shaded region in cm^2, is approximately *[LIC (AAO)-11]*

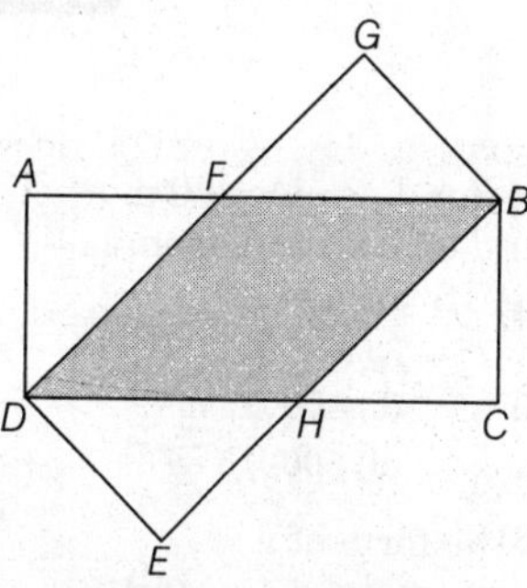

Sol. Let $BF = x$ cm, then $AF = (7 - x)$ cm

Now, $BF = DH = x$ and $AF = HC = (7 - x)$

Area of the parallelogram $FBHD$ = base $\times$ height

$\Rightarrow DH \times BC = BH \times DE$ [we have taken both bases]

$\Rightarrow \quad x \times 3 = \sqrt{(BC)^2 + (HC)^2} \times 3 \Rightarrow x^2 = (3)^2 + (7 - x)^2$

$\Rightarrow \quad x^2 = 9 + 49 - 14x + x^2 \Rightarrow 14x = 58$

$\Rightarrow \quad x = \frac{58}{14} = 4.14$

Then, area of the parallelogram (i.e. shaded region)

$= DH \times BC = 4.14 \times 3 = 12.42$ cm

Problem 27 What is the ratio of area of rhombus to that of the square if they have same base and the angle between adjacent sides of the rhombus is 45°?

Sol. Required ratio $= \frac{\text{Area of the rhombus}}{\text{Area of the square}}$

$= \frac{\text{Product of adjacent sides} \times \text{sine of included angle}}{(\text{side})^2}$

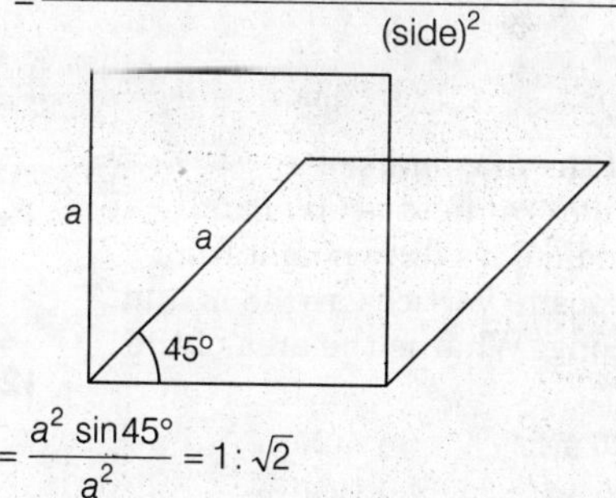

$= \frac{a^2 \sin 45°}{a^2} = 1 : \sqrt{2}$

WORKOUT

1. A parallelogram has sides 60 m and 40 m and one of its diagonals is 80 m long. Its area is **[SSC (CGL-T-I)-14]**
(a) $600\sqrt{15}$ m² (b) $400\sqrt{15}$ m²
(c) $450\sqrt{15}$ m² (d) $500\sqrt{15}$ m²

2. A wall is of the form of a trapezium with height 4 m and parallel sides being 3 m and 5 m. What is the cost of painting the wall, if the rate of painting is ₹ 25 per sq m ? **[CDS-11]**
(a) ₹ 240 (b) ₹ 400
(c) ₹ 480 (d) ₹ 800

3. The diagonals of a rhombus are 24 cm and 10 cm. Its area is **[MAT-08, CDS-17, CLAT-13]**
(a) 240 cm² (b) 312 cm²
(c) 130 cm² (d) 120 cm²

4. The area of a field in the shape of a trapezium measures 1440 m². The perpendicular distance between its parallel sides is 24 m. If the ratio of the parallel sides is 5 : 3, the length of the longer parallel side is
(a) 75 m (b) 45 m
(c) 120 m (d) 60 m

5. One of the diagonals of a quadrilateral field is 60 m and perpendiculars drawn on it from the opposite vertices are 48 m and 24 m long. What is the area of the field ?
(a) 2400 m² (b) 3120 m²
(c) 2160 m² (d) 1280 m²

6. The diagonals of a rhombus are 12 cm and 16 cm respectively. The length of one side is **[SSC (CGL-T-II)-13, CDS-17]**
(a) 8 cm (b) 6 cm
(c) 10 cm (d) 12 cm

7. The area of a rhombus is 256 sq cm and one of its diagonal is twice the other in length. Then, length of its longer diagonal is **[SSC (CGL-T-II)-14]**
(a) 32 cm (b) 48 cm
(c) 16 cm (d) 24 cm

8. The lengths of two parallel sides of a trapezium are 28 cm and 40 cm. If the length of each of its other two sides be 12 cm, then area of the trapezium (in cm²) is **[SSC (CGL-T-II)-14]**
(a) $204\sqrt{3}$ cm² (b) $504\sqrt{3}$ cm²
(c) $408\sqrt{3}$ cm² (d) $312\sqrt{5}$ cm²

9. The area of a trapezium is 480 cm². If the difference between its parallel sides is 12 cm and the distance between them is 16 cm, the parallel sides (in cm) are **[UPPSC (Rev. Insp.)-15]**
(a) 12, 24 (b) 16, 28
(c) 20, 32 (d) 24, 36

10. The measure of each of two opposite angles of a rhombus is 60° and the measure of one of its sides is 10 cm. The length of its smaller diagonal is
(a) 12 cm (b) $10\sqrt{3}$ cm
(c) $10\sqrt{2}$ cm (d) $\frac{5}{2}\sqrt{2}$ cm

11. The ratio of the length of the parallel sides of a trapezium is 5 : 3 . The shortest distance between them is 16 cm. If the area of the trapezium is 640 cm², the difference between lengths of the parallel sides is
(a) 15 cm (b) 20 cm
(c) 45 cm (d) 60 cm

12. The length of one side of a rhombus is 8 cm and its altitude is 12 cm. If the length of one diagonal be 24 cm, the length of other diagonal will be
(a) 6 cm (b) 8 cm
(c) 12 cm (d) 24 cm

13. The perimeter of a rhombus is 40 cm and the measure of an angle is 60°, then its area is
(a) 100 cm² (b) $100\sqrt{3}$ cm²
(c) $50\sqrt{3}$ cm² (d) $160\sqrt{3}$ cm²

14. The two diagonals of a rhombus are of lengths 40 cm and 30 cm. If h is the perpendicular height of the rhombus, then value of h is
(a) 30 cm (b) 32 cm
(c) 34 cm (d) 24 cm

15. If diagonals of a parallelogram $ABCD$ intersect at point O in such a way that area of $\Delta OAB = k \times$ Area of the parallelogram $ABCD$. Then, the value of k is **[UPPSC (Rev. Insp.)-15]**
(a) $\frac{1}{2}$ (b) 1
(c) $\frac{1}{3}$ (d) $\frac{1}{4}$

16. Which of the following is true?
(a) All sides of a parallelogram are equal
(b) Diagonals of a parallelogram are equal
(c) Any line joining the mid-points of opposite sides of a rhombus bisects the area
(d) None of the above

17. If the side of a rhombus is decreased by 15%, what will be effect on its area?
(a) 27.75% decrease
(b) 47.75% decrease
(c) 22.75% decrease
(d) 37.5% decrease

18. One of the diagonal of a rhombus is double the other diagonal. Its area is 25 sq cm. The sum of the diagonal is
(a) 15 cm (b) 10 cm
(c) 6.5 cm (d) 26 cm

19. The cross-section of a waterway is trapezium in shape. The waterway is 30 m wide at the top and 18 m wide at the bottom. What is the depth of the waterway if the area of the cross- section is 960 sq m ?
(a) 75 m (b) 45 m
(c) 40 m (d) 60 m

20. In a quadrilateral $ABCD$, $AB = 10$ cm, $AD = 12$ cm and $\angle A = 60°$. Find the length of diagonal BD.
(a) 12.36 cm (b) 11.13 cm
(c) 15.26 cm (d) 28.10 cm

21. Two sides of a plot measure 32 m and 24 m and the angle between them is a perfect right angle. The other two sides measure 25 m each and the other three angles are not right angles. What is the area of the plot (in m²) ? *[CAT-01]*

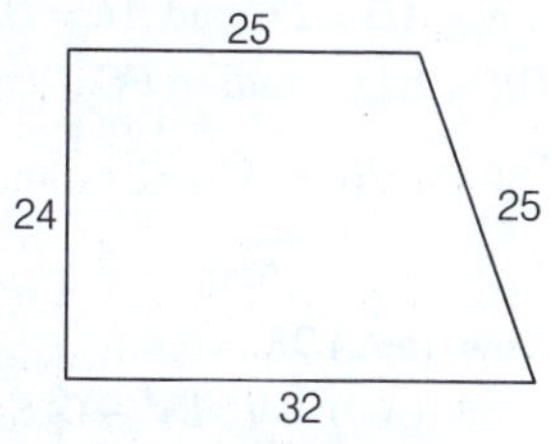

(a) 768 (b) 534
(c) 696.5 (d) 684

22. $ABCD$ is a parallelogram in which $AB = 7$ cm, $BC = 9$ cm and $AC = 8$ cm. What is the length (in cm) of other diagonal ? *[SSC (CGL-T-I)-17]*

(a) 14 (b) $14\sqrt{2}$
(c) 7 (d) $7\sqrt{2}$

23. Smaller diagonal of a rhombus is equal to length of its sides. If length of each side is 6 cm, then what is the area (in cm²) of an equilateral triangle whose side is equal to the bigger diagonal of the rhombus ? *[SSC (CGL-T-I)-17]*

(a) $18\sqrt{3}$
(b) $27\sqrt{3}$
(c) $32\sqrt{3}$
(d) $36\sqrt{3}$

24. If length of each side of a rhombus $PQRS$ is 8 cm and $\angle PQR = 120°$, then what is the length (in cm) of QS ? *[SSC (CGL-T-I)-17]*

(a) $4\sqrt{5}$ (b) 6 (c) 8 (d) 12

25. One of the diagonals of a rhombus is 70% of the other diagonal. What is the ratio of area of rhombus to the square of the length of the larger diagonal ? *[SSC (CGL-T-I)-17]*

(a) 3 : 10 (b) 3 : 20
(c) 7 : 20 (d) 7 : 10

26. AB, CD and EF are three parallel lines, in that order. Let d_1 and d_2 be the distances from CD to AB and EF respectively. d_1 and d_2 are integers, where $d_1 : d_2 = 2 : 1$. P is a point on AB, Q and S are points on CD and R is a point on EF. If the area of the quadrilateral $PQRS$ is 30 sq units, what is the value of QR when value of SR is the least ? *[XAT-17]*

(a) slightly greater than 20 units
(b) slightly greater than 10 units
(c) slightly less than 20 units
(d) slightly less than 10 units
(e) 10 units

27. $ABCD$ is a rectangle. P, Q and R are the mid-points of BC, CD and DA. The point S lies on the line QR in such a way that $SR : QS = 1 : 3$. The ratio of areas of ΔAPS and rectangle $ABCD$ is *[XAT-17]*

(a) 36 : 128 (b) 39 : 128
(c) 44 :128 (d) 48 : 128
(e) 64 : 128

28. A square piece of paper is folded three times along its diagonal to get an isosceles triangle whose equal sides are 10 cm. What is the area of the unfolded original piece of paper ? *[XAT-16]*

(a) 400 cm²
(b) 800 cm²
(c) $800\sqrt{2}$ cm²
(d) 1600 cm²
(e) Insufficient data to answer

Directions (Q. Nos. 29-31) Consider the following for the next three questions: $ABCD$ is a quadrilateral with $AB = 9$ cm, $BC = 40$ cm, $CD = 28$ cm, $DA = 15$ cm and angle ABC is a right- angle.

29. What is the area of quadrilateral $ABCD$?

(a) 300 cm² (b) 306 cm²
(c) 312 cm² (d) 316 cm²

30. What is the difference between perimeter of ΔABC and perimeter of ΔADC ?

(a) 4 cm (b) 5 cm
(c) 6 cm (d) 7 cm

31. What is the area of ΔADC? *[CDS-19]*

(a) 126 cm² (b) 124 cm²
(c) 122 cm² (d) 120 cm²

Answer Key

1. (a)	2. (b)	3. (d)	4. (a)	5. (c)	6. (c)	7. (a)	8. (a)	9. (d)	10. (a)
11. (b)	12. (b)	13. (c)	14. (d)	15. (d)	16. (c)	17. (a)	18. (a)	19. (c)	20. (b)
21. (d)	22. (a)	23. (b)	24. (c)	25. (c)	26. (a)	27. (a)	28. (a)	29. (b)	30. (c)
31. (a)									

Hints and SOLUTIONS

1. Area of parallelogram $ABCD$

$= \text{Area of } \Delta ABD + \text{Area of } \Delta BCD$

Now, area of ΔABD

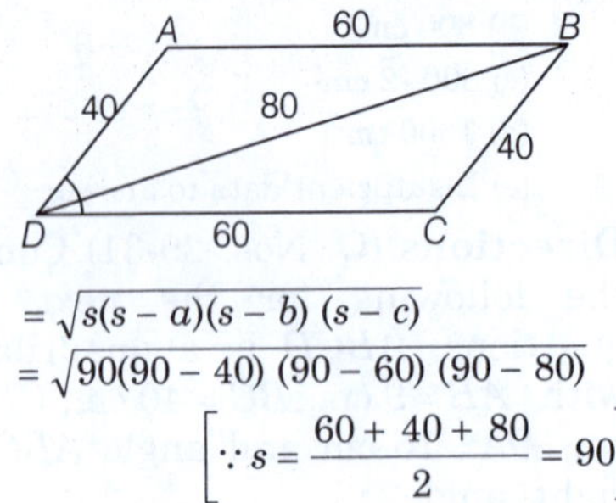

$= \sqrt{s(s-a)(s-b)(s-c)}$

$= \sqrt{90(90-40)(90-60)(90-80)}$

$\left[\because s = \dfrac{60+40+80}{2} = 90\right]$

$= \sqrt{90 \times 50 \times 30 \times 10}$

$= 300\sqrt{15}\ \text{m}^2$

Since each diagonal divides a parallelogram into two congruent triangles, area of ΔABD will be equal to the area of ΔBCD.

Thus, area of parallelogram

$= 300\sqrt{15} + 300\sqrt{15}$

$= 600\sqrt{15}\ \text{m}^2$

2. Area of the trapezium $= \dfrac{1}{2} \times$ (sum of parallel sides) $\times$ Altitude

$= \dfrac{1}{2} \times (3+5) \times 4 = 16\ \text{m}^2$

Now, cost of painting $= 16 \times 25$

$= ₹\ 400$

3. Area of the rhombus

$= \dfrac{1}{2} \times$ product of diagonals

$= \dfrac{1}{2} \times 24 \times 10 = 120\ \text{cm}^2$

4. Area of the trapezium $= \dfrac{1}{2} \times$ (sum of parallel sides) $\times$ Altitude

$\Rightarrow 1440 = \dfrac{1}{2} \times$ (sum of parallel sides) $\times 24$

$\Rightarrow$ Sum of parallel sides $= 120$

Then, longer side $= 120 \times \dfrac{5}{(5+3)}$

$= 75\ \text{m}$

5. Area of the field $= \dfrac{1}{2} \times$ one of the diagonals $\times$ (sum of length of its offsets)

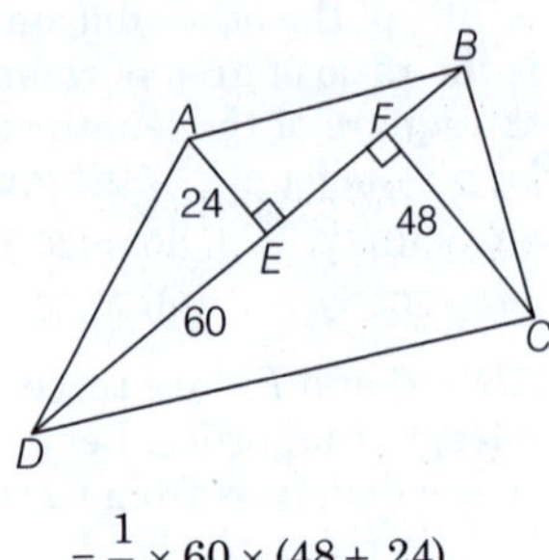

$= \dfrac{1}{2} \times 60 \times (48 + 24)$

$= 2160\ \text{m}^2$

6. We know, $4a^2 = (d_1)^2 + (d_2)^2$

$= (12)^2 + (16)^2 = 400$

$\Rightarrow a^2 = 100$

$\Rightarrow a = 10\ \text{cm}$

Alternatively

Then, hypotenuse of the $\Delta COD = \sqrt{6^2 + 8^2} = 10$ i.e. each side.

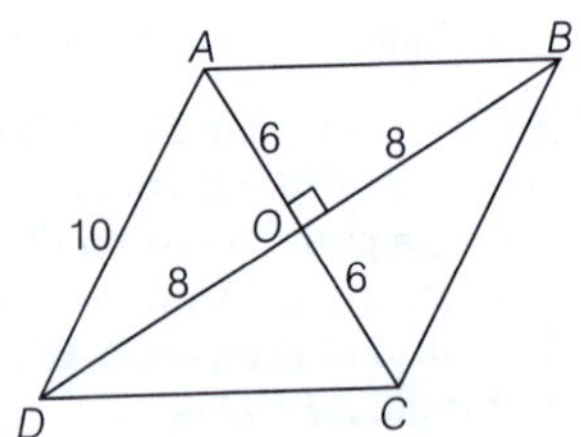

7. Let diagonals be d and $2d$.

Then, area of the rhombus

$= \dfrac{1}{2} \times$ product of diagonals

$\Rightarrow 256 = \dfrac{1}{2} \times d \times 2d$

$\Rightarrow d = 16$ cm

Hence, longer diagonal $= 2 \times 16$

$= 32\ \text{cm}$

8. Clearly, $EF = 28\ \text{cm}$,

Then, $DE + FC = 40 - 28$

$= 12\ \text{cm}$

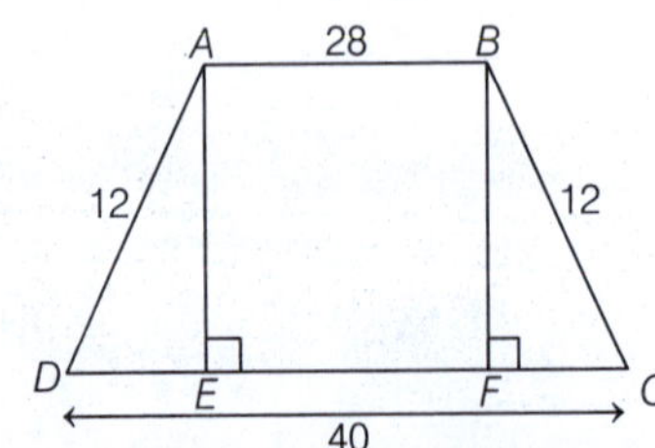

Since $AD = BC$ and $AE = BF$, DE will be equal to FC.

Hence, $DE = FC = \dfrac{12}{2}$

$= 6\ \text{cm}$

Now, in ΔADE,

$(AD)^2 = (DE)^2 + (AE)^2$

$\Rightarrow (12)^2 = (6)^2 + (AE)^2$

$\Rightarrow AE = 6\sqrt{3}$ cm

Then, area of trapezium $= \dfrac{1}{2} \times$ (sum of parallel sides) $\times$ Altitude

$= \dfrac{1}{2} \times (28 + 40) \times 6\sqrt{3}$

$= 204\sqrt{3}\ \text{cm}^2$

9. Area of the trapezium $= \dfrac{1}{2} \times$ (sum of parallel sides) $\times$ Altitude

$\Rightarrow 480 = \dfrac{1}{2} \times$ (sum of parallel sides) $\times 16$

$\Rightarrow$ Sum of parallel sides $= 60$

Since the difference between parallel sides is 12 cm, the sides will be $\dfrac{60-12}{2}$ and $\dfrac{60-12}{2} + 12$ i.e. 24 cm and 36 cm.

10. Area of the rhombus

$= BC \times CD \times \sin 60°$

$= 12 \times 12 \times \dfrac{\sqrt{3}}{2}$

$= 72\sqrt{3}$

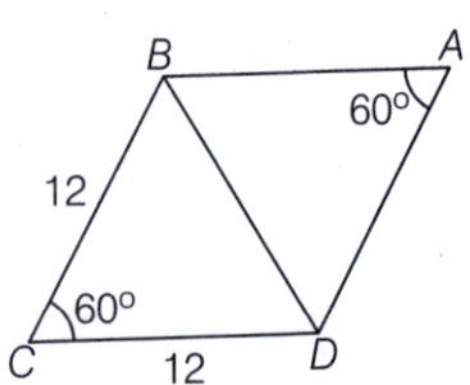

Since the diagonal divides the rhombus in two congruent triangle, area of $\Delta BCD = 72\sqrt{3} \div 2 = 36\sqrt{3}$

Now, proceed through option,

(a) 12 cm $\Rightarrow$ If $BD = 12$ cm, then it is an equilateral triangle and its area will be $\dfrac{\sqrt{3}}{4} \times (12)^2 = 36\sqrt{3}$

Hence, option (a) is correct.

Alternatively

$\angle BAD = \angle BCD = 60°$

Then, $\angle ABC = \angle ADC = 120°$

$\therefore \quad \angle CBD = \angle CDB = 60°$

[$\because$ diagonals bisect vertex angles]

Hence, $\angle BCD$ is an equilateral triangle. It means $BD = 12$ cm.

11. Area of the trapezium

$= \frac{1}{2} \times$ (sum of parallel sides) $\times$ Altitude

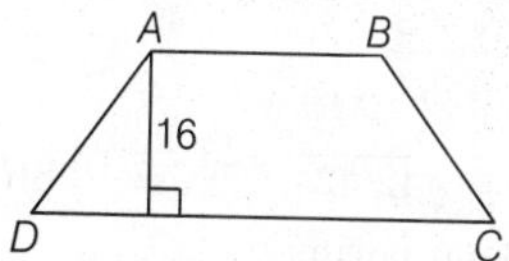

$\Rightarrow 640 = \frac{1}{2} \times$ (sum of parallel sides) $\times 16$

$\Rightarrow$ sum of parallel sides $= 80$ cm

Given, $(5 + 3)$ units $= 80$ cm

Then, $(5 - 3)$ units $= 20$ cm

12. Area of the rhombus = Base $\times$ Height

$= \frac{1}{2} \times$ product of diagonals

$\Rightarrow 8 \times 12 = \frac{1}{2} \times 24 \times$ 2nd diagonal

$\Rightarrow$ 2nd diagonal $= 8$ cm

13. Each side $= \frac{40}{4} = 10$ cm

Area = Product of any two adjacent sides $\times$ sine of included angle

$= 10 \times 10 \times \sin 60°$

$= 10 \times 10 \times \frac{\sqrt{3}}{2}$

$= 50\sqrt{3}$ cm^2

14. We know, $4a^2 = (d_1)^2 + (d_2)^2$

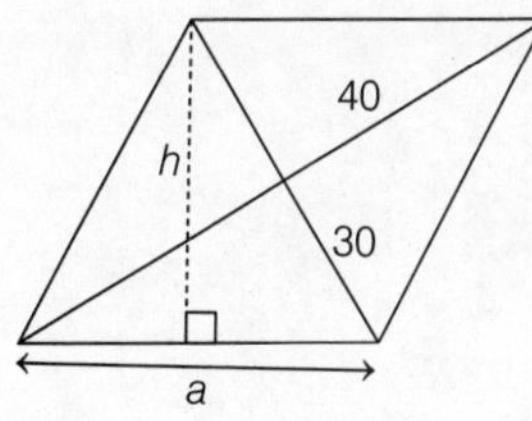

$= (40)^2 + (30)^2 = 2500$

$\Rightarrow \quad a^2 = 625$

$\therefore \quad a = \sqrt{625} = 25$

It means each side of the rhombus $= 25$ cm

Now, area of the rhombus

= base $\times$ height

$= \frac{1}{2} \times$ product of diagonals

$\Rightarrow \quad 25 \times h = \frac{1}{2} \times 40 \times 30$

$\Rightarrow \quad h = 24$ cm

15. Since diagonals of a parallelogram form four triangles of equal area, the value of k will obviously be $\frac{1}{4}$.

(Observe diagonal related properties).

16. Observe the properties of quadrilaterals as discussed.

17. Here, the area is a product of two adjacent sides with a constant value of sine of included angle. Also, both adjacent sides are equal in case of rhombus.

Thus, effect on area

$= \left[-15 - 15 + \frac{15 \times 15}{100}\right]\%$

$= -27.75\%$

'–' indicates that area will decrease by 27.75%.

18. Area $= \frac{1}{2} \times$ product of diagonals

$\Rightarrow \quad 25 = \frac{1}{2} \times d \times 2d$

[let diagonals be d and $2d$]

$\Rightarrow \quad d = 5$

Now, sum of diagonals

$= d + 2d = 3d$

$= 3 \times 5 = 15$ cm

19. Area of the trapezium $= \frac{1}{2} \times$ (sum of parallel sides) $\times$ height

$\Rightarrow \quad 960 = \frac{1}{2} \times (30 + 18) \times$ depth

$\Rightarrow$ Depth $= 40$ m

20. Diagonal

$BD = \sqrt{(AB)^2 + (AD)^2 - 2AB \cdot AD \cdot \cos A}$

$= \sqrt{(10)^2 + (12)^2 - 2 \times 10 \times 12 \times \cos 60°}$

$= \sqrt{(10)^2 + (12)^2 - 2 \times 10 \times 12 \times \frac{1}{2}}$

$= \sqrt{124}$

$= 11.13$ cm

21. $AC = \sqrt{24^2 + 32^2} = 40$

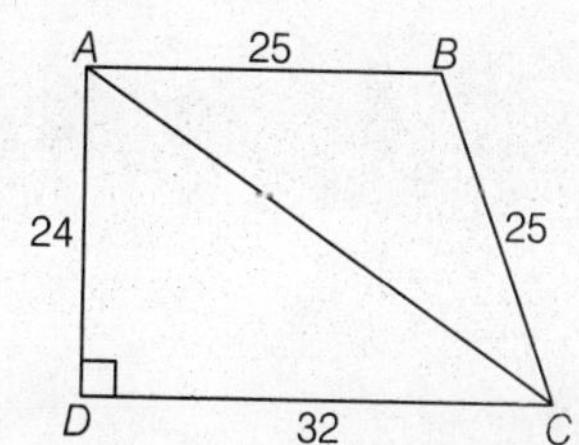

Then, area of right-angled ΔADC

$= \frac{1}{2} \times 32 \times 24 = 384$

We have, area of the isosceles Δ

$= \frac{b}{4}\sqrt{4a^2 - b^2}$, where a is the length of each of equal sides and b is the length of the third side.

Hence, area of isosceles

$\Delta ACB = \frac{40}{4}\sqrt{4 \cdot 25^2 - 40^2} = 300$

$\therefore$ Area of quadrilateral $ABCD$

$= 384 + 300 = 684$ m^2

22. We have, $2(a^2 + b^2) = (d_1)^2 + (d_2)^2$, where a and b are adjacent sides and d_1 and d_2 are diagonals.

$\Rightarrow \quad 2(7^2 + 9^2) = (8)^2 + (d_2)^2$

$\Rightarrow \quad d_2 = 14$

23. We have, $4a^2 = (d_1)^2 + (d_2)^2$, where a is each side of the rhombus and d_1 and d_2 are its diagonals.

$\Rightarrow \quad 4 \cdot 6^2 = (6)^2 + (d_2)^2$

$\Rightarrow \quad d_2 = 6\sqrt{3}$ cm

Now, area of the given equilateral triangle

$= \frac{\sqrt{3}}{4} \times (6\sqrt{3})^2 = 27\sqrt{3}$ cm^2

24. Given, $\angle PQR = 120°$, then

$\angle SPQ = 180° - 120°$

$= 60°$ [$\because PS \parallel QR$]

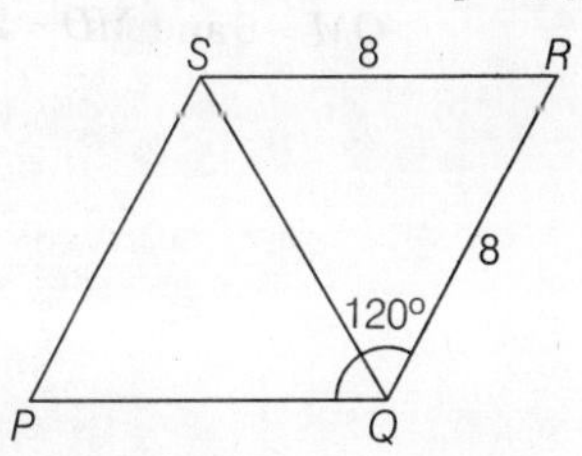

Diagonal $QS = \sqrt{2} \times 8\sqrt{1 - \cos 60°}$

$= \sqrt{2} \times 8 \times \frac{1}{\sqrt{2}} = 8$

25. Let the larger diagonal of the rhombus be x.

Then, required ratio

$= \left(\frac{1}{2} \times x \times 0.7x\right) : x^2$

$= 7 : 20$

26.

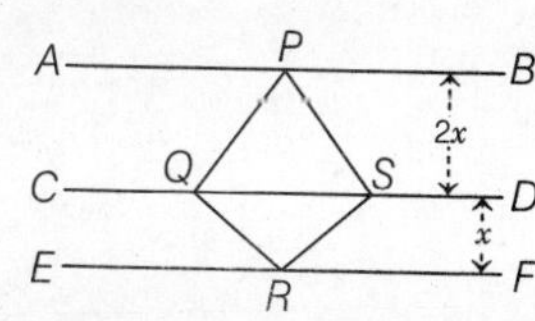

Area of quadrilateral $PQRS$ = area of ΔPQS + area of ΔQRS

$\Rightarrow \quad 30 = \frac{1}{2} \times QS \times 2x + \frac{1}{2} \times QS \times x$

$\Rightarrow \quad 60 = 3x \times QS$

$\Rightarrow QS \times x = 20$

Clearly, the least value of

$SR = x = 1$ unit

Then, $QS \times 1 = 20$

$\Rightarrow \quad QS = 20$

Now, $\quad QR = \sqrt{SR^2 + QS^2}$

$= \sqrt{1^2 + 20^2}$

Hence, the value of QR will be slightly greater than 20 units.

27. Let $AD = BC = 8$ units and $CD = AB = 16$ units.

Then, $AR = RD = 4$ and $DQ = QC = 8$

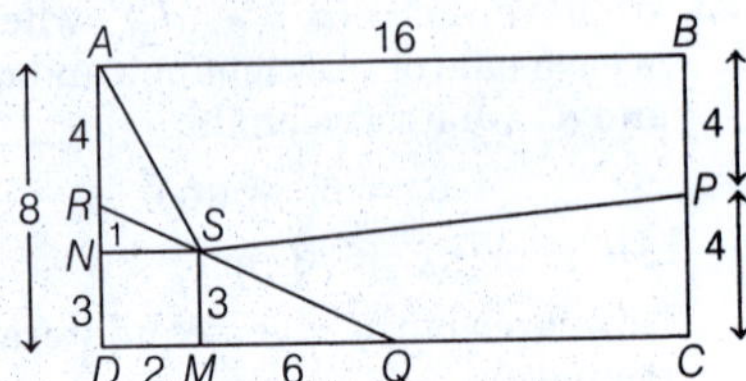

Since ΔRNS is similar to ΔRDQ,

$\frac{RS}{SQ} = \frac{RN}{ND} = \frac{1}{3}$

Similarly, $\frac{QS}{SR} = \frac{QM}{MD} = \frac{3}{1}$

$\therefore \quad QM = 6$ and $MD = 2$

Now, area of rectangle $ABCD$

$= 8 \times 16 = 128$ sq units

Area of ΔAPS = area ($\square ABCD$) − area (ΔABP) − area (trapezium $ASMD$) − area (trapezium $SPCM$)

$= 128 - \left(\frac{1}{2} \times 4 \times 16\right) - \frac{1}{2}(3 + 8)$

$\times 2 - \frac{1}{2}(3 + 4) \times 14 = 36$

$\therefore$ Required ratio = 36 : 128

28. It is clear that when we fold a square along its diagonal, the area of the resulting figure (i.e. isosceles triangle) will be half of the area of the square.

Let area of the square be x cm^2.

Then, area of the isosceles triangle after first folding $= \frac{x}{2}$

Area of the isosceles triangle after second folding $= \frac{x}{4}$

and area of the isosceles triangle after third folding $= \frac{x}{8}$

Hence, area of the square is 8 times that of the isosceles triangle obtained after three times folding.

Now, area of the isosceles

triangle $= \frac{1}{2} \times 10 \times 10 = 50$

[$\because$ the resulting isosceles triangle is a right isosceles triangle]

$\therefore$ Area of the square $= 8 \times 50$

$= 400$ cm^2

29. Since (9, 40, 41) is a Pythagorean triplet, $AC = 41$

[Practice to memorise certain triplets to save time in exam.]

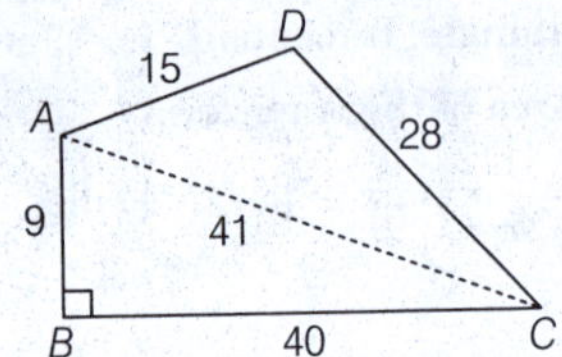

Area of ΔADC,

$= \sqrt{42(42 - 15)(42 - 28)(42 - 41)}$

$\because$ Semi-perimeter

$\left[\because \text{Semi-perimeter} = \frac{15 + 28 + 41}{2} = 42\right]$

$= 126$

Then, area of Quadrilateral $ABCD$

= area of ΔABC + area of ΔADC

$= \left(\frac{1}{2} \times 9 \times 40\right) + 126$

$= 306$ cm^2

30. Difference between perimeter of ΔABC and perimeter of triangle ΔADC

$= (9 + 40 + 41) - (15 + 28 + 41)$

$= 6$ cm

31. From above,

Area of $\Delta ADC = 126$ cm^2

43

Circle

Learning OBJECTIVES

After completing this chapter, aspirants will be able to :

- Understand the application of the circle related formulae in different cases;
- Understand the application of combination of circle related formulae in diverse cases;
- Understand the philosophy of the division of this chapter in 5 parts;
- Analyse the trend of the questions being asked in various competitive examinations;
- Solve the questions in moments;
- Grip the concepts helping to solve the questions without pen and paper;
- Apply many concepts envisaged in the chapter 'PERCENTAGE' like 'effect on product in case of changes in factors', Product Constancy, etc.;
- Apply many concepts of 'Ratio' in this chapter;
- Learn many new concepts/fundamentals;
- Apply the pragmatic approaches in lieu of academic approaches;
- Ponder on the alternative methods; and
- Compare the circles with reference to different ratios.

Inside

- *Application of the Concept of Circumference.*
- *Changes in Circumference(s) When Radius(dii)/Diameter(s) Alter.*
- *Application of the Concept of Area.*
- *Alteration in Area.*
- *Relation Between Circumference and Area.*
- *Comparisons Between/Among Circles With Reference to Different Ratios.*
- *Case of Concentric Circles.*
- *Application of the Concept of the Sector and Segment of the Circle.*
- *Combinations of Two or More Figures.*
- *Counting of Number of Triangles Having Vertices on Circumference of a Circle.*
- ....
- .... *& Many More*
- *Work Out with Solutions*

Note *Illustrations based on geometric properties of a circle/ tangents/secant have been given in 'GEOMETRY' section.*

Application of the Concept of Circumference

(i) The circumference of a circle is the distance around the circle. It may be termed as its perimeter and measured by the product of its diameter (d) and π i.e. Circumference of the circle = πd.

Since diameter is twice the radius, circumference will be equal to $2\pi r$.

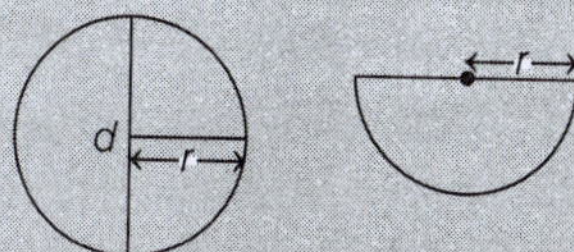

Obviously, the perimeter of the semi-circle

$$= \frac{2\pi r}{2} + 2r = r(\pi + 2) = \pi r + d = d\left(\frac{\pi}{2} + 1\right).$$

Clearly, the circumference/perimeter of a circle/ semi-circle is directly proportional to its radius or diameter. It implies that if radius or diameter is increased to n times, the circumference/perimeter will also be increased to n times and *vice-versa*. Now, we will observe the applications of the above concepts in following diverse problems.

(ii) The perimeter of any polygon circumscribed about a circle is greater than the circumference of the circle and the circumference of the circle is greater than the perimeter of any polygon inscribed in the circle.

Problem 1 The diameter of a Toy wheel is 14 cm. What is the distance travelled by it in 15 revolutions?
[SSC (CGL-T-II)-13]

Sol. Here, $r = \frac{14}{2} = 7$ cm

Clearly, the distance travelled by the wheel in one revolution
= circumference of the wheel.

$\therefore$ Distance travelled in 15 revolutions $= 15 \times 2\pi r$

$= 15 \times 2 \times \frac{22}{7} \times 7$

$= 660$ cm

Problem 2 The perimeter of a square is equal to twice the perimeter of a rectangle of length 8 cm and breadth 7 cm. What is the circumference of a semi-circle whose diameter is equal to the side of the square ?
[Punjab & Sind Bank (PO)-10]

Sol. Perimeter of the square = 2 × perimeter of the rectangle

$\Rightarrow \quad 4 \times \text{side} = 2 \times 2(8+7)$

$\Rightarrow \quad \text{side} = 15$

As given, diameter of the semi-circle
= side of the square i.e. 15 cm

$\therefore$ Radius of the semi-circle $= \frac{15}{2} = 7.5$

Now, circumference of the semi-circle $= \pi r + 2r$

$= \left(\frac{22}{7} \times 7.5\right) + 15 = 23.57 + 15$

$= 38.57$ cm

Problem 3 *A* can go round a circular path 8 times in 40 min. If the diameter of the circle is increased to 10 times the original diameter, the time required by *A* to go round the new path once travelling at the same speed as before is :

Sol. We know that circumference $= 2\pi r$ i.e. $2\pi(d/2)$. It means if radius/diameter is increased to n times, the circumference will also increase to n times and *vice-versa*.

Since the diameter is increased to 10 times, the circumference will also increase to 10 times. It means the length of the circular path will increase to 10 times.

Earlier, *A* takes 5 min. But, now he/she will take 10 times the time taken earlier i.e. 50 min.

Problem 4 A piece of wire is when bent to form a circle, will have a radius of 84 cm. If the wire is bent to form a square, the length of a side of the square is :
[SSC CGL-T-(II)-14]

Sol. Since the wire is the same, circumference of the circle will be equal to the perimeter of the square.

$\Rightarrow \quad 2\pi r = 4 \times \text{side}$

$\Rightarrow \quad \frac{2 \times 22 \times 84}{7 \times 4} = \text{side}$

$\therefore \quad$ Side of the square $= 132$ cm

Problem 5 A path of uniform width surrounds a circular park. The difference of internal and external circumferences of this circular path is 132 m. Its width is :
[SSC (CGL-PT) 2000, MAT-07,08, SSC (CGL-T-II)-19]

Sol. Clearly, the difference between internal and external circumferences $= 2\pi R - 2\pi r = 2\pi(R - r)$

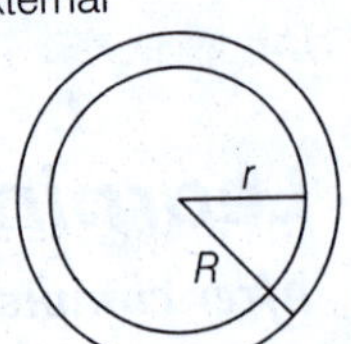

As given, $2\pi(R - r) = 132$

$\therefore \quad R - r = 132 \times \frac{7}{2 \times 22} = 21$

Here, $(R - r)$ is the width of the path.

Problem 6 Find the diameter of a wheel that makes 113 revolutions to go 2 km 26 decameters.

Sol. Distance covered in one revolution
= circumference of the wheel

$\Rightarrow \quad \frac{2 \text{ km } 26 \text{ decameters}}{113} = 2\pi r$

$\Rightarrow \quad \frac{(2000 + 260)}{113} m = 2\pi r$

{$\because$ 1 km = 1000 m and 1 decameter = 10 m}

$\Rightarrow \quad \frac{2260 \times 7}{113 \times 22} = 2r \Rightarrow d = 6\frac{4}{11}$ m

Problem 7 Four equal sized maximum circular plates are cut off from a square paper sheet of area 784 sq cm. The circumference of each plate is :
[CDS-15, SSC (CGL-PT)-03]

Sol. Side of the square $= \sqrt{784} = 28$ cm

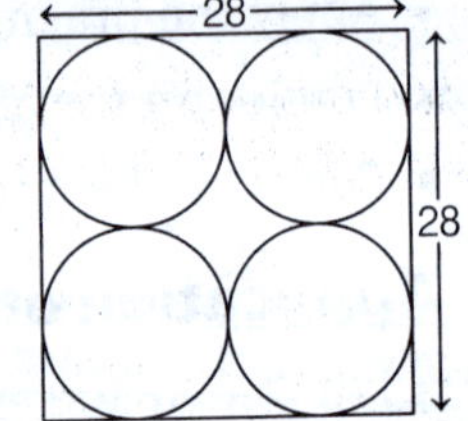

Clearly, diameter of each plate

$= \frac{28}{2} = 14$ cm

$\therefore$ Circumference of each plate

$= 2\pi \times \frac{14}{2} = 2 \times \frac{22}{7} \times 7 = 44$ cm

Problem 8 The ratio of the outer and the inner perimeters of a circular path is 23:22. If the path is 5 m wide, the diameter of the inner circle is :

Sol. Given, $\frac{2\pi R}{2\pi r} = \frac{23}{22}$

$\Rightarrow \quad R : r = 23 : 22$

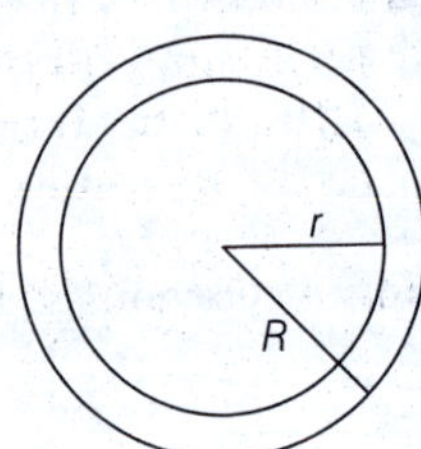

But, there is difference between R and r is 5 m as given.

Hence, (23 - 22) unit = 5 m

$\therefore \quad$ 22 units = 110 m

Then, $\quad$ diameter = 2r

= 2 × 110

= 220 m

Alternatively

From above, $\frac{R}{r} = \frac{23}{22}$

$\Rightarrow \quad \frac{r+5}{r} = \frac{23}{22} \quad [\because R - r = 5]$

$\Rightarrow \quad 22r + 110 = 23r$

$\Rightarrow \quad r = 110$

$\therefore \quad d = 220 \text{ m}$

Problem 9 When the circumference of a Toy Ballon is increased from 20 cm to 25 cm, its radius is increased by :

Sol. Given, $2\pi r = 20$

$\therefore \quad r = \frac{10}{\pi}$

Further, $2\pi R = 25$ [after increasing]

$\therefore \quad R = 12.5/\pi$

Thus, radius is increased by $\frac{12.5}{\pi} - \frac{10}{\pi}$ i.e. by $\frac{5}{2\pi}$ cm.

Problem 10 If the difference between the circumference and diameter of a circle is 30 cm, then the radius of the circle must be :

Sol. As given, $2\pi r - 2r = 30$

$\Rightarrow \quad 2r(\pi - 1) = 30$

$\Rightarrow \quad r = \frac{30}{2\left(\frac{22}{7} - 1\right)} = \frac{30 \times 7}{2 \times 15} = 7 \text{ cm}$

Problem 11 There are two circles of different radii. The area of a square is 196 sq cm, whose side is half the radius of the larger circle. The radius of the smaller circle is three-seventh that of the larger circle. What is the circumference of the smaller circle ?

[Syndicate Bank (PO)-10]

Sol. Radius of the larger circle $(R) = 2 \times$ side of the square

$\Rightarrow \quad R = 2 \times \sqrt{\text{area of the square}}$

$= 2 \times \sqrt{196} = 28 \text{ cm}$

Then, radius of the smaller circle (r)

$= 28 \times \frac{3}{7} = 12$

$\therefore$ Circumference of the smaller circle

$= 2\pi r = 2\pi \times 12 = 24\pi \text{ cm}$

Problem 12 A wire when bent in the form of an equilateral triangle encloses a region having area of $121\sqrt{3}\text{ cm}^2$. If the same wire is rebent into the form of a circle, its radius will be :

[SSC (CGL-PT) 08]

Sol. Area of an equilateral triangle $= \frac{\sqrt{3}}{4}(\text{side})^2 = 121\sqrt{3}$

$\Rightarrow \quad (\text{side})^2 = 484 \Rightarrow \text{side} = 22$

$\therefore$ Its perimeter $= 22 \times 3 = 66$ cm

Obviously, the perimeter of the triangle will be equal to circumference of the circle as the same wire has been rebent.

i.e. $\quad 2\pi r = 66$

$\Rightarrow \quad 2 \times \frac{22}{7} \times r = 66$

$\therefore \quad r = \frac{66 \times 7}{2 \times 22} = 10.5 \text{ cm}$

Problem 13 The length of the minute hand of a clock is 7 cm, then the distance covered by its tip in 2 days is :

Sol. In 2 days, the minute hand will complete 48 rounds. Then, length moved by it = 48 × (circumference of the circle with radius 7 cm) $= 48 \times 2\pi \times 7 = 2112$ cm

Problem 14 A bucket is raised from a well by means of a rope which is round a wheel of diameter 75 cm. If the bucket ascends in 1 min 12 sec with a uniform speed of 1.3 m/s, find the number of complete revolutions made by the wheel in raising the bucket.

Sol. Radius of the wheel $= \frac{75}{2} = 37.5$ cm

Then, its circumference $= 2\pi \times 37.5 = 75\pi$

Now, time taken to pull the bucket = 72 sec

$\therefore$ Distance covered by the bucket $= 72 \times 1.3 \text{ m} = 72 \times 130$ cm

Obviously, this distance has been travelling by means of revolutions of the wheel and the fact that the wheel pulls the bucket by 75π cm in one revolution.

$\therefore$ Number of revolutions $= \frac{72 \times 130}{75\pi} = \frac{72 \times 130 \times 7}{75 \times 22} = 39.70$

Hence, number of complete revolutions = 39

Problem 15 Wheels of diameters 7 cm and 14 cm start rolling simultaneously from X and Y, which are 1980 cm apart, towards each other in opposite directions. Both of them make the same number of revolutions per second. If both of them meet after 10 sec, the speed of the smaller wheel (in cm/sec) is :

Sol. Both of the wheels have made the same number of revolutions in 10 sec, however, the ratio of distances covered by them will be 1 : 2 as ratio of their diameters or radii is 1 : 2.

Now, sum of distances covered by them = 1980 cm.

$\therefore$ Distance covered by the smaller wheel in 10 seconds

$= 1980 \times \frac{1}{3} = 660$ cm

$\therefore$ Distance covered by the smaller wheel in 1 sec

$= \frac{660}{10} = 66$ cm

Hence, the speed of the smaller wheel = 66 cm/sec

Problem 16 The front wheels of a wagon are 2π feet in circumference and the rear wheels are 3π feet in circumference. When the front wheels have made 10 more revolutions than the rear wheels, how many feet has the wagon travelled ?

Sol. Ratio of circumferences (front to rear)

$= 2\pi : 3\pi$

$= 2 : 3$

$\therefore$ Ratio of revolutions made by them to cover the same distance = 3 : 2

As given, (3 units − 2 units) = 10

$\therefore$ 1 unit = 10

Hence, 3 units = 30 and 2 units = 20

It means the front wheels have made 30 revolutions and rear have made 20 revolutions.

Then, the distance covered by the front wheels or by the wagon

$= 30 \times 2\pi = 60\pi$ feet

Alternatively

The distance covered by the rear wheels or by the wagon

$= 30 \times 2\pi = 60\,\pi$ feet

Changes in Circumference(s) when Radius (dii)/Diameter(s) Alter

Since the radius/diameter is directly proportional to the circumference, the followings hold good :

(i) If radius/ diameter of a circle is made n times, its circumference will also become n times and *vice-versa.*

(ii) If radius/diameter of a circle is increased/ decreased by x%, its circumference will also increase/decrease by x% and *vice-versa.*

Problem 17 If the radius of a circle is increased by 75%, then its circumference will increase by :

[CDS-03, SSC (CGL-T-I)-14]

Sol. The circumference will increase by 75% as per the above 2nd observation.

Application of the Concept of Area

- Area of a circle $= \pi r^2$

  And, Area of a semicircle $= \dfrac{\pi r^2}{2}$

- The area of the circle is always greater than that of square/rectangle/triangle under given circumference/ perimeter of equal length.
- It also holds good that the area of square > area of rectangle > area of triangle, under the given equal perimeters. (asked in CAT-03)
- The above facts are based on the principles of 'more sides more area in a convex polygon'.

Problem 18 The perimeters of a circle, a square and an equilateral triangle are same and their areas are C, S and T respectively. Which of the following statement is true ? *[SSC(CGL-T-I)-14]*

(a) C = S = T (b) C > S > T

(c) C < S < T (d) S < C < T

Sol. The second option is the correct answer adhering the above principle.

Problem 19 The area of the largest circle that can be drawn inside a square of side 28 cm is :

Sol. Clearly, the diameter of the circle = 28 cm

$\therefore$ Radius (r) = 14 cm

Now, area $= \pi r^2 = \dfrac{22}{7} \times 14 \times 14$

$= 616 \text{ cm}^2.$

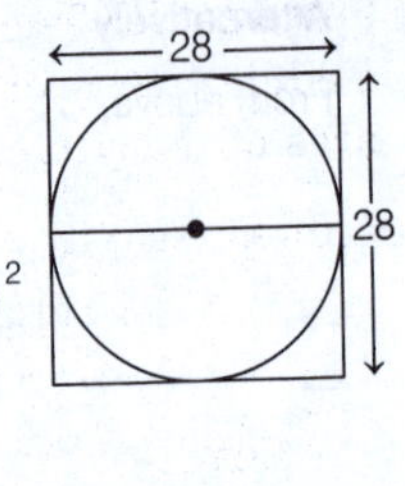

Alternatively

Area of the circle $= \dfrac{\pi}{4} \times$ (side of a square)^2

$= \dfrac{22}{7 \times 4} \times 28 \times 28$

$= 616 \text{ cm}^2$

Problem 20 Between a square of perimeter 44 cm and a circle of circumference 44 cm, which figure has larger area and by how much? *[SSC (CGL-T-I)-14]*

Sol. Here, side of the square $= \dfrac{44}{4} = 11$ cm

$\therefore$ Its area $= (11)^2 = 121 \text{ cm}^2$

Further, $2\pi r = 44$ (Given)

$\Rightarrow \quad r = 44 \times \dfrac{7}{2 \times 22} = 7$ cm

$\therefore$ Area of the circle $= \dfrac{22}{7} \times (7)^2$

$= 154 \text{ cm}^2$

Hence, area of the circle is greater than that of the square by 33 cm^2.

Problem 21 If a wire is bent into the shape of a square, the area of the square is 81 sq cm. When the wire is bent into a semi-circular shape, the area of the semi-circle is : *[SSC (CGL-T-II)-10]*

Sol. Side of the square $= \sqrt{\text{area}} = \sqrt{81}$

$= 9$ cm

$\therefore$ Perimeter $= 9 \times 4 = 36$ cm

Since the length of the wire is same, the perimeter of the semi-circle will also be 36 cm.

$\Rightarrow \quad (\pi r + 2r) = 36$

$\Rightarrow \quad r = \dfrac{36}{\left(\dfrac{22}{7} + 2\right)} = 36 \times \dfrac{7}{36} = 7$

Now, area $= \dfrac{\pi}{2} r^2 = \dfrac{22}{7 \times 2} \times 7 \times 7$

$= 77 \text{ cm}^2$

Problem 22 A circular wire of diameter 42 cm is folded in the shape of a rectangle whose sides are in the ratio 6 : 5. Find the area enclosed by the rectangle.

Sol. Since the length of the wire is same, the circumference of the circle will be equal to the perimeter of the rectangle.

$\therefore$ Perimeter of rectangle $= 2\pi r$

$\Rightarrow \quad 2(l + b) = 2 \times \dfrac{22}{7} \times 21$ [Given, $d = 42$ cm $\therefore r = 21$ cm]

$\Rightarrow \quad (l + b) = 66$

$\Rightarrow \quad 6x + 5x = 66$

[let l be $6x$ and b be $5x$ as their ratio is 6 : 5]

$\Rightarrow \quad x = 6$

$\therefore \quad l = 6 \times 6 = 36$ and $b = 5 \times 6 = 30$

Thus, area $= 36 \times 30 = 1080 \text{ cm}^2$

Problem 23 The side of a square is half the diameter of a circle. The area of the square is 1225 cm^2. What is the area of the circle ? **[Corporation Bank (PO)-11]**

Sol. Given, diameter of the circle = $2x$ side of the square

$$\Rightarrow \quad 2 \times \text{radius} = 2 \times \sqrt{\text{area of the square}}$$

$$\Rightarrow \quad \text{radius} = \sqrt{1225} = 35 \text{ cm}$$

$$\therefore \text{Area of the circle} = \frac{22}{7} \times 35 \times 35 = 3850 \text{ cm}^2$$

Problem 24 The area of a rectangle is equal to the area of a circle whose radius is 14 cm. If the breadth of the rectangle is 22 cm, what is its length ? **[Andhra Bank (Marketing Associate) 09]**

Sol. Given, area of the rectangle = area of the circle

$$\Rightarrow \text{length} \times \text{breadth} = \pi r^2$$

$$\Rightarrow \quad \text{length} = \frac{\pi r^2}{\text{breadth}} = \frac{22 \times 14 \times 14}{7 \times 22} = 28 \text{ cm}$$

Problem 25 If the perimeter of a square is equal to the radius of a circle whose area is 39424 sq cm, what is the area of the square ? **[IDBI (PO)-09]**

Sol. Given, $\pi r^2 = 39424$

$$\Rightarrow \quad r^2 = \frac{39424 \times 7}{22} = 12544$$

$$\therefore \quad r = 112$$

Now, perimeter of the square = r (as given)

$$\Rightarrow 4 \times \text{side} = 112$$

$$\therefore \quad \text{side} = \frac{112}{4} = 28$$

Then, area of the square = $(28)^2 = 784 \text{ cm}^2$

Problem 26 The area of a circle whose radius is 6 cm is trisected by two concentric circles. The radius of the smallest circle is :

Sol. The area of biggest circle = $\pi \, (6)^2$

$$\therefore \text{The area of smallest circle} = \frac{\pi}{3} (6)^2$$

Now, $\frac{\pi}{3}(6)^2 = \pi \times r^2 \Rightarrow r^2 = 12$

$$\therefore \quad r = 2\sqrt{3} \text{ cm}$$

Problem 27 The ratio of the areas of the incircle and circumcircle of a square is : **[SSC (SO-Commer. Audit)-06]**

Sol. In case of incircle,

diameter (d) = side of the square

$$\therefore \quad r = \frac{\text{side}}{2}$$

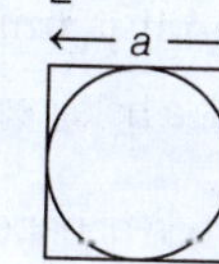

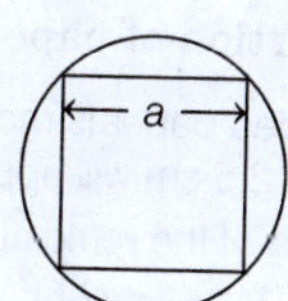

In case of Circumcircle, diameter (D) = diagonal of the square

$$\therefore \quad R = \frac{(\sqrt{2} \times \text{side})}{2}$$

Now, the required ratio $= \frac{\pi r^2}{\pi R^2} = \frac{\left(\frac{\text{side}}{2}\right)^2}{\left(\frac{\sqrt{2} \times \text{side}}{2}\right)^2} = 1:2$

Problem 28 The area of the largest circle that can be drawn inside a rectangle with sides 18 cm by 14 cm is : **[SSC(CGL-PT)-07, MAT-08, SSC (CGL-T-I)-13]**

Sol. Obviously, diameter = 14

$$\therefore \quad \text{Area} = \pi \times \left(\frac{14}{2}\right)^2 = \frac{22}{7} \times 7 \times 7 = 154 \text{ cm}^2$$

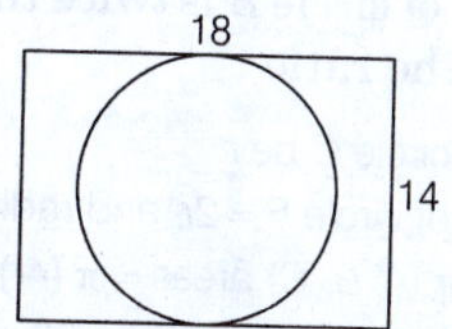

Problem 29 If the difference between areas of the circumcircle and the incircle of an equilateral triangle is 44 cm^2. Then, the area of the triangle is :

Sol. Circumradius of the equilateral triangle $= \frac{\text{side}}{\sqrt{3}}$

$$\therefore \text{Area of circumcircle} = \pi \left(\frac{\text{side}}{\sqrt{3}}\right)^2 = \pi \, \frac{(\text{side})^2}{3}$$

Further, inradius $= \frac{1}{2\sqrt{3}} \times \text{side}$

$$\therefore \quad \text{Area of incircle} = \pi \, \frac{(\text{side})^2}{12}$$

As given, $\pi \frac{(\text{side})^2}{3} - \pi \frac{(\text{side})^2}{12} = 44$

$$\Rightarrow \quad \pi \, (\text{side})^2 = 44 \times 4$$

$$\Rightarrow \quad (\text{side})^2 = 44 \times 4 \times \frac{7}{22} = 56$$

Now, the area of the equilateral triangle $= \frac{\sqrt{3}}{4} \times (\text{side})^2$

$$= \frac{\sqrt{3}}{4} \times 56 = 14\sqrt{3} \text{ cm}^2$$

Alternatively

In equilateral triangle, circumradius (R) : inradius (r) = 2 : 1

i.e. $R = 2r$

As given, $\pi R^2 - \pi r^2 = 44$

$$\Rightarrow \quad \pi (2r)^2 - \pi r^2 = 44 \Rightarrow \pi r^2 = \frac{44}{3}$$

$$\therefore \quad r = \sqrt{\frac{44 \times 7}{3 \times 22}} = \sqrt{\frac{14}{3}}$$

Further, inradius $= \frac{1}{2\sqrt{3}} \times \text{side} \Rightarrow \sqrt{\frac{14}{3}} = \frac{1}{2\sqrt{3}} \times \text{side}$

$$\Rightarrow \frac{14}{3} = \frac{1}{12} \times (\text{side})^2 \Rightarrow (\text{side})^2 = \frac{14 \times 12}{3} = 56$$

Now, Area of the triangle $= \frac{\sqrt{3}}{4} \times (\text{side})^2$

$$= \frac{\sqrt{3}}{4} \times 56 = 14\sqrt{3} \text{ cm}^2$$

Problem 30 **A circle and a square have equal areas. The ratio of a side of the square and the radius of the circle is:** *[SSC (10+2, LDC/DEO)-10, SSC (CGL-T-II)-14]*

Sol. As given, $\pi r^2 = a^2 \Rightarrow \frac{a^2}{r^2} = \frac{\pi}{1}$

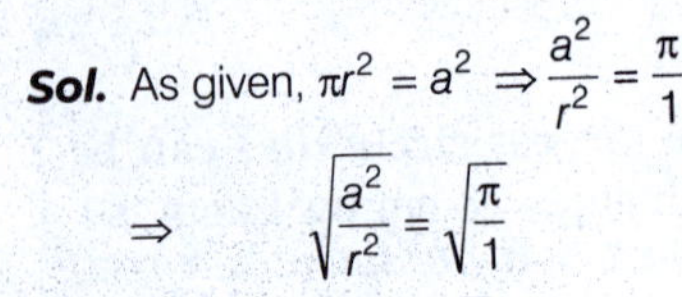

$\Rightarrow \quad \sqrt{\frac{a^2}{r^2}} = \sqrt{\frac{\pi}{1}}$

$\Rightarrow \quad a : r = \sqrt{\pi} : 1$

Problem 31 **The radius of circle *A* is twice that of circle *B* and the radius of circle *B* is twice that of circle *C*. Their areas will be in the ratio :**

Sol. Let the radius of circle *C* be *r*.

Then, the radius of circle $B = 2r$ and radius of circle $A = 4r$

Now, ratio of their (*A, B, C*) areas $= \pi (4r)^2 : \pi (2r)^2 : \pi r^2$

$= 16r^2 : 4r^2 : 1r^2 = 16 : 4 : 1$

Alternatively

Ratio of radii $= 4 : 2 : 1$

$\therefore$ Ratio of areas $= (4)^2 : (2)^2 : (1)^2 = 16 : 4 : 1$

Problem 32 **The radius of a circle is 20% more than the height of a *right-angled* triangle. The base of triangle is 36 cm. If the area of triangle and circle be equal, what will be the area of circle ?**

Sol. Let the height of the triangle be *h* cm.

Then, the radius of the circle $= h \times \frac{120}{100} = \frac{6h}{5}$

$\therefore \quad$ Its area $= \pi \left(\frac{6h}{5}\right)^2$

$\Rightarrow \frac{36}{25} \times (h)^2 \times \pi = \frac{1}{2} \times 36 \times h \quad [\because \text{area of } \Delta = \frac{1}{2} \times \text{base} \times \text{height}]$

$\Rightarrow \quad \frac{h}{25}\pi = \frac{1}{2} \Rightarrow h = 25 \times \frac{7}{44}$

Now, area of the triangle $= \frac{1}{2} \times \text{base} \times \text{height}$

$= \frac{1}{2} \times 36 \times 25 \times \frac{7}{44} = 71.59$

Therefore, area of circle $= 72 \text{ cm}^2$ (approx.)

Problem 33 **The radius of a circle is 20 cm. Three more concentric circles are drawn inside it in such a manner that it is divided into 4 equal parts. Find the radius of smallest circle.**

Sol. Area of biggest circle $= \pi (20)^2$

$= 400\pi$

$\therefore$ Area of the smallest circle

$400\pi/4 = 100\pi$

i.e. $\quad \pi r^2 = 100\pi$

$\therefore \quad r = 10$ cm

Problem 34 **Find the ratio of area of a square inscribed in a semi-circle of radius *r* to the area of another square inscribed in the entire circle of radius *r*.**

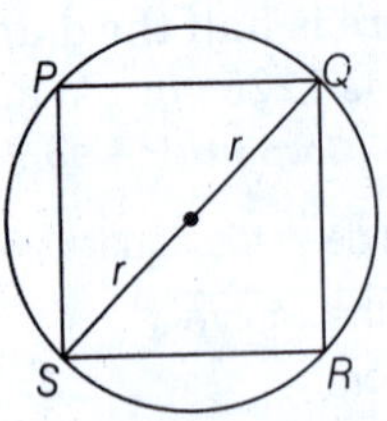

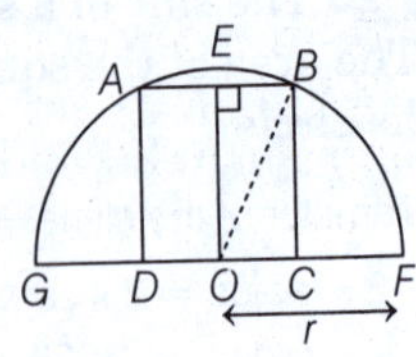

Sol. As given, radius $= r$

$\therefore \quad OB = OF = \text{radius}(r)$

Let the side of the square be a.

Now, in right triangle OEB, $(OB)^2 = (EO)^2 + (EB)^2$

$\Rightarrow \quad r^2 = a^2 + \left(\frac{1}{2}a\right)^2 \Rightarrow r^2 = a^2 + \frac{a^2}{4}$

$\Rightarrow \quad \frac{5a^2}{4} = r^2 \Rightarrow a^2 = \frac{4r^2}{5}$

Now, diagonal of the square *PQRS* = diameter of the circle

i.e. $\quad \sqrt{2} \times \text{side} = 2r \Rightarrow \text{side} = \sqrt{2}\, r$

$\therefore$ Area of the square $= (\text{side})^2 = (\sqrt{2}r)^2 = 2r^2$

Hence, required ratio $= \frac{4r^2}{5} : 2r^2 = 2 : 5$

Problem 35 **If the area of a square inscribed in a circle is 15 sq cm, then the area of the square inscribed in a semi-circle of the same circle in cm^2, is :** *[LIC (AAO)-11]*

Sol. From the above result,

Area of a square inscribed in a circle : Area of the square inscribed in a semi-circle = 5 : 2

$\Rightarrow \quad$ 15 : required area = 5 : 2

$\therefore \quad$ Required area $= 6 \text{ cm}^2$

Problem 36 **A circular swimming pool with a diameter of 28 ft has a deck of uniform width built around it. If the area of the deck is 60 π sq ft, find its width.** *[MAT-07, MAT- 10]*

Sol. Area of the swimming pool $= \pi r^2 = \pi \times \frac{28}{2} \times \frac{28}{2} = 196\pi$

Then, Area of the pool with the deck $= 196\pi + 60\pi$

$= 256\pi$

$\Rightarrow \quad \pi R^2 = 256$

$\therefore \quad R = 16$

Now, Outer radius (*R*) – Inner radius (*r*)

$= 16 - 14 = 2$ ft i.e. width of the deck.

Problem 37 **A circular paper of radius 11 cm is folded along its diameter and then again folded so as to make a quadrant. Thereafter, a sector having radius 2.5 cm is cut, as shown in the figure below. Find the area of the remaining portion of paper when it is unfolded.**

Sol. When the folded paper is reopened, a hole in shape of circle having radius 2.5 cm will appear.

Thus, the area of the remaining portion of paper

$= \pi (11)^2 - \pi (2.5)^2$

$= 121\pi - 6.25\pi$

$= 360.64 \text{ cm}^2$

Problem 38 A cow is tethered in the middle of a field with a 14 feet long rope. If the cow grazes 100 sq ft/day, then approximately what time (in days) will be taken by the cow to graze the whole field ?

Sol. Area to be grazed by the cow $= \pi r^2$

$= \frac{22}{7} \times 14 \times 14 = 616$ sq ft

14

Then, number of days required to graze the field

$= \frac{616}{100} = 6.16$

$= 6$ days (approx.)

Alteration in Area

- Recall the concept "How the product is changed when relevant factors are changed under topic "effect on product in case of changes in factors" under chapter percentage. Here, the area of the circle is product of radius and radius with a constant term 'π'.
- If radius or diameter of a circle is made n times, its circumference will become n times and *vice-versa*. The logic is that the circumference is directly proportional to radius and diameter.
- If radius or diameter or circumference is made n times, the area will become n^2 times as the area of the circle is directly proportional to the square of its radius.

Problem 39 If the radius of a circle is increased by 50%, its area is increased by :

[MAT- 08, SSC (CGL-T-II)-14. UPPSC(Rev.Insp)-15, SSC (CGL-T-I)-19]

Sol. Effect on area $= \left[+ 50 + 50 + \frac{50 \times 50}{100}\right]\%$

$= + 125\%$

Hence, the area will increase by 125%.

Alternatively

Since the radius has been made to 1.5 times the earlier, the new area will become $(1.5)^2$ times the original area i.e. 2.25 times. It implies that the original area will increase by 125%. The logic behind this concept is that the area of the circle is directly proportional to square of its radius.

Problem 40 If the circumference of a circle is increased by 50% then the area will increase by :

Sol. For an increase in circumference by 50%, the radius should also be increased by 50%.

Now, effect on area $= \left[+ 50 + 50 + \frac{50 \times 50}{100}\right]\%$

$= + 125\%$

Hence, the area will increase by 125%.

Alternatively

Since the radius has been made to 1.5 times the earlier, the new area will become $(1.5)^2$ times the original area i.e. 2.25 times, It implies that the original area will increase by 125%.

Problem 41 If the circumference of a circle increases from 4π to 8π, what change occurs in its area?

Sol. The circumference has been increased by 100% ($4\pi \rightarrow 8\pi$). It implies that the radius has also been increased by 100%.

Now, effect on area $= \left[+ 100 + 100 + \frac{100 \times 100}{100}\right]\% = + 300\%$

Hence, the original area will increase by 300% i.e. it quadruples.

Alternatively,

Since the radius has been made doubled, the new area will be quadrupled.

Problem 42 The area of a circle is increased by 22 cm^2 when its radius is increased by 1 cm. The original radius of the circle is :

[SSC (CGL-T-II)-12]

Sol. Given, $\pi (r + 1)^2 - \pi r^2 = 22 \Rightarrow \pi (r^2 + 2r + 1 - r^2) = 22$

$\Rightarrow \quad 2r + 1 = 22 \times \frac{7}{22} = 7 \Rightarrow 2r = 6$

$\therefore \quad r = 3$ cm

Alternatively

You may proceed through the given options.

Relation Between Circumference and Area

- **If we divide the area of a circle by $\frac{r}{2}$, we get its circumference.**
- **Under Given Equal Perimeter/Circumference**
  (i) Area of the circle : Area of the equilateral triangle $= 21\sqrt{3} : 22$
  (ii) Area of the circle : Area of the square $= 14 : 11$
  and
- **Under given equal areas**
  (i) Circumference of the circle : Perimeter of the equilateral triangle $= \sqrt{\pi} : \sqrt[4]{27}$
  (ii) Circumference of the circle: Perimeter of the square $= \sqrt{\pi} : 2$

  Aspirants are advised to keep these facts in mind while solving the problems based on the circle.

Problem 43 The area of a circle 38.5 sq cm. Its circumference (in cm) is :

[BOI (PO)-08, Andhra Bank (PO)-09, SSC (CGL-T-II)-19]

Sol. Given, Area of the circle $= 38.5 \Rightarrow \pi r^2 = 38.5$

$\Rightarrow \quad r^2 = 38.5 \times \frac{7}{22} = 12.25$

$\therefore \quad r = 3.5$

Now, its circumference $= 2\pi r = 2 \times \frac{22}{7} \times 3.5 = 22$ cm

Alternatively

Circumference $= \frac{\text{area}}{r/2} = \frac{38.5}{3.5/2} = 22$ cm

Problem 44 If the circumference and area of a circle are numerically equal, then the diameter is equal to :

Sol. Given, $2\pi r = \pi r^2 \Rightarrow r = 2$

$\therefore$ Diameter $= 2 \times 2 = 4$

Problem 45 Perimeter of an equilateral triangle is equal to the circumference of a circle. The ratio of their areas is : *[SSC (CGL-T-II)-14]*

Sol. Let the perimeter of the equilateral triangle be 30 cm. (any figure with or without unit can be assumed).

Then, its area $= \frac{\sqrt{3}}{4} \times (10)^2 = 25\sqrt{3}$

As per question, $2\pi r = 30 \Rightarrow r = \frac{30}{2\pi}$

$\therefore$ Area of the circle $= \pi r^2 = \pi \times \frac{30}{2\pi} \times \frac{30}{2\pi} = \frac{225}{\pi}$

Then, required ratio $= 25\sqrt{3} : \frac{225}{\pi} = \sqrt{3} : \frac{9 \times 7}{22}$

$= 22\sqrt{3} : 9 \times 7 = 22 : 21\sqrt{3}$

Alternatively I

Given, $3s = 2\pi r$ [here, s = side of the equilateral triangle]

$\Rightarrow 9s^2 = 4 \times \frac{22}{7} \times \pi r^2$

$\Rightarrow \frac{\sqrt{3}}{4} s^2 \times 3\sqrt{3} \times 4 = \frac{88}{7} \times \pi r^2$

$\Rightarrow$ Area of triangle $\times 3\sqrt{3} \times 4 = \frac{88}{7} \times$ Area of the circle

$\Rightarrow$ Area of triangle : Area of the circle

$= \frac{88}{7 \times 12\sqrt{3}} = 22 : 21\sqrt{3}$

Alternatively II

If we remember the fact (given in shaded portion), we can directly arrive at the correct option.

Problem 46 If the areas of a circle and a square are equal, then the ratio of their perimeters is :

Sol. Given, $\pi r^2 = a^2$ [a = side of the square and, r = radius]

$\Rightarrow r^2 : a^2 = 1 : \pi \Rightarrow r : a = 1 : \sqrt{\pi}$

Then, $2\pi r : 4a = 2\pi (1) : 4\sqrt{\pi} = \pi : 2\sqrt{\pi} = \sqrt{\pi} : 2$

Alternatively

If we remember the fact (given in shaded portion), we can directly arrive at the correct option.

Problem 47 The perimeter (in m) of a semi-circle is numerically equal to its area. The length of its diameter is :

Sol. Given, $(\pi r + 2r) = \frac{\pi}{2} r^2 \Rightarrow r(\pi + 2) = \frac{\pi}{2} r^2$

$\Rightarrow (\pi + 2) = \frac{\pi r}{2} \Rightarrow r = \frac{2(\pi + 2)}{\pi} = 2 + \frac{4}{\pi}$

$\Rightarrow 2r = 2\left(2 + \frac{4}{\pi}\right) = 4 + \frac{8 \times 7}{22} = 6\frac{6}{11}$ m

Problem 48 The area of the ring between two concentric circles, whose circumferences are 88 cm and 132 cm, is : *[Indian Bank (PO)-10, Corporation Bank (PO)-11, CDS-19]*

Sol. Given, Circumference of the outer circle = 132

$\Rightarrow 2\pi R = 132$

$\therefore R = 21$

Further, circumference of the inner circle $= 88$

$\Rightarrow 2\pi r = 88$

$\therefore r = 14$

Now, area of the ring [shaded region]

$= \pi R^2 - \pi r^2$

$= \pi(R + r)(R - r) = \frac{22}{7} \times (21 + 14)(21 - 14)$

$= 770 \text{ cm}^2$

Problem 49 The sum of the circumference of a circle and the perimeter of a square is equal to 272 cm. The diameter of the circle is 56 cm. What is the sum of the areas of the circle and the square? *[Allahabad Bank (PO)-11]*

Sol. The radius of the circle $(r) = \frac{56}{2} = 28$ cm

$\therefore$ Its circumference $= 2 \times \frac{22}{7} \times 28 = 176$ cm

Then, perimeter of the square $= 272 - 176 = 96$ cm

$\therefore$ Side of the square $= \frac{96}{4} = 24$ cm.

Now, sum of the areas of the circle and the square

$= \pi r^2 + (\text{side})^2 = \left(\frac{22}{7} \times 28 \times 28\right) + (24)^2 = 3040 \text{cm}^2$

Comparisons between/among circles with reference to different ratios

	If ratio of radii is ⇩	Then ⇨	Ratio of diameters	Ratio of circumferences	Ratio of areas
In case of two circles	$a : b$	⇨	$a : b$	$a : b$	$a^2 : b^2$
	$2 : 3$ (let)	⇨	$2 : 3$	$2 : 3$	$2^2 : 3^2$ i.e. $4 : 9$
In case of three circles or more	$a : b : c$	⇨	$a : b : c$	$a : b : c$	$a^2 : b^2 : c^2$
	$2 : 3 : 5$ (let)	⇨	$2 : 3 : 5$	$2 : 3 : 5$	$2^2 : 3^2 : 5^2$ i.e. $4 : 9 : 25$

	If ratio of areas is ⇩	Then ⇨	Ratio of radii	Ratio of diameters	Ratio of circumferences
In case of two circles	$x : y$	⇨	$\sqrt{x} : \sqrt{y}$	$\sqrt{x} : \sqrt{y}$	$\sqrt{x} : \sqrt{y}$
	$1 : 9$ (let)	⇨	$\sqrt{1} : \sqrt{9}$ i.e. $1 : 3$	$\sqrt{1} : \sqrt{9}$ i.e. $1 : 3$	$\sqrt{1} : \sqrt{9}$ i.e. $1 : 3$
In case of three circles or more	$x : y : z$	⇨	$\sqrt{x} : \sqrt{y} : \sqrt{z}$	$\sqrt{x} : \sqrt{y} : \sqrt{z}$	$\sqrt{x} : \sqrt{y} : \sqrt{z}$
	$1 : 4 : 25$ (let)	⇨	$\sqrt{1} : \sqrt{4} : \sqrt{25}$ i.e. $1 : 2 : 5$	$\sqrt{1} : \sqrt{4} : \sqrt{25}$ i.e. $1 : 2 : 5$	$\sqrt{1} : \sqrt{4} : \sqrt{25}$ i.e. $1 : 2 : 5$

Note *The logic for the above contents are:*

(i) The diameter/circumference is directly proportional to radius.

(ii) The area is directly proportional to square of radius/diameter/circumference.

Problem 50 If the ratio of area of two circles is 4:9, then ratio of their circumference will be : **[RRB-03]**

Sol. The required ratio $= \sqrt{4} : \sqrt{9} = 2 : 3$

Problem 51 The diameter of two circular coins are in the ratio of 1 : 3. The smaller coin is made to roll around the bigger coin till it returns to the position from where the process of rolling started. How many times the smaller coin rolled around the bigger coin ? **[IAS (Pre)-10]**

Sol. Obviously, their circumferences will be in ratio 1 : 3.

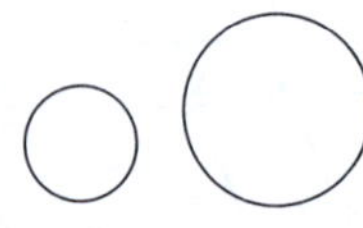

$\therefore$ Number of rounds $= \dfrac{3}{1} = 3$

Cases of Concentric Circles

- Area of the space (say, circular path) enclosed between the circumferences of two concentric circles $= \pi(R^2 - r^2)$, where R and r are radii of the outer and inner circles respectively.

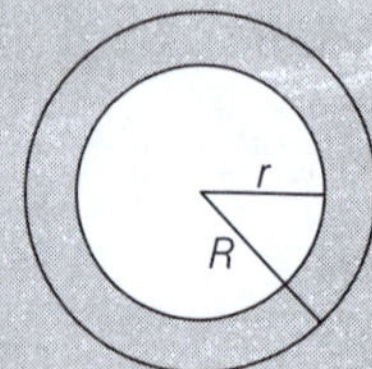

- If circumferences of concentric circles are in ratio $a : b : c : \ldots$ the ratio of their radii will be $a : b : c : \ldots$ (Needless to say that it is applicable for non-concentric circles also, described already through a chart).

Problem 52 A circular road is to be constructed around a circular garden which costs ₹ 15840 at the rate ₹ 4/sq m. What is the width of the circular road provided the inner radius is 58 m.

Sol. Area of the circular road = 15840/4 = 3960 m^2

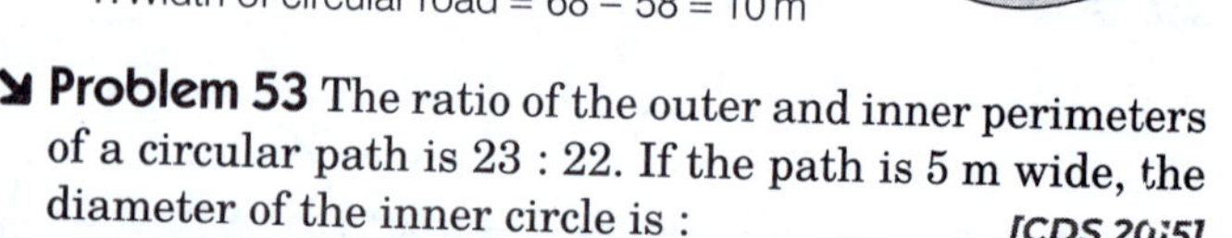

$\Rightarrow \quad \pi R^2 - \pi r^2 = 3960$

$\Rightarrow \quad \pi(R^2 - 58^2) = 3960$

$\Rightarrow \quad R^2 - 3364 = 3960 \times \dfrac{7}{22}$

$\Rightarrow \quad R^2 = 1260 + 3364 = 4624$

$\Rightarrow \quad R = 68$

$\therefore$ Width of circular road = 68 – 58 = 10 m

Problem 53 The ratio of the outer and inner perimeters of a circular path is 23 : 22. If the path is 5 m wide, the diameter of the inner circle is : **[CDS 2015]**

Sol. Let outer and inner perimeters be $23x$ and $22x$.

Then, their radii $= \dfrac{23x}{2\pi}$ and $\dfrac{22x}{2\pi}$

As given, $\dfrac{23x}{2\pi} - \dfrac{22x}{2\pi} = 5$

$\Rightarrow \quad x = 10\pi$

Now, radius of inner circle $= \dfrac{22x}{2\pi} = \dfrac{22 \times 10\pi}{2\pi}$

$= 110$ m

Then, its diameter $= 2 \times 110$

$= 220$ m

Alternatively

We know, ratio of circumferences of concentric circles = ratio of their radii

$\Rightarrow$ 23 : 22 = ratio of radii of outer and inner circles

But, given that width of the path is 5 m, it means the difference between their radii is 5 m.

Hence, (23 – 22) unit = 5 m

$\therefore$ 22 units = 110 m i.e. radius of the inner circle.

Thus, its diameter $= 2 \times 110 = 220$ m

Application of the Concept of the Sector of the Circle

(i) Area of the sector of a circle $= \dfrac{\pi r^2}{360°}\theta$

[where, θ is the angle of the sector]

(ii) Length of the arc (i.e. AMB) of the sector $= \dfrac{2\pi r}{360°}\theta$

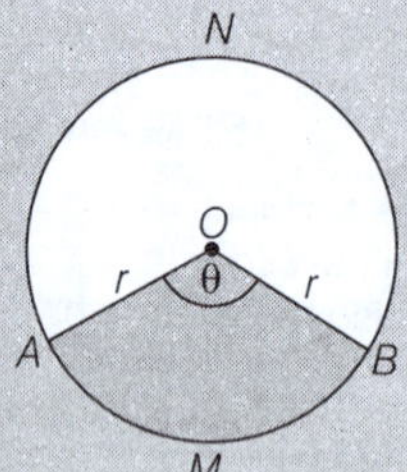

(iii) Area of sector of a circle may be in terms of arc as under :

$$\text{Area} = \frac{\pi r^2}{360°}\theta = \frac{1}{2} \times \frac{2\pi r}{360°}\theta \times r$$

$$= \frac{1}{2} \times \text{length of arc} \times \text{radius}$$

(iv) Perimeter of the Sector = length of the arc + $2r$

$$= \frac{2\pi r}{360°}\theta + 2r$$

It is observed from the above that the area of the Sector depends upon the value of θ provided the area of the circle is known. Further, if the area of the Sector is $1/n$ of that of the circle, the concerned arc will also be $1/n$ of the circumference of the circle as the Sector is a part of the circle in terms of both area and circumference with variable values of θ. Also, in case θ is equal to 360° the Sector becomes a circle and the arc turns into the circumference as well. It means **area of Circle: area of Sector = circumference of Circle: arc of Sector = 360° : θ.**

Problem 54 The area of a sector of a circle of radius 5 cm, formed by an arc of length 3.5 cm is : *[CDS-21]*

Sol. Area of the circle $= \pi r^2 = \pi(5)^2 = 25\pi \text{ cm}^2$

Circumference $= 2\pi r = 2 \times \frac{22}{7} \times 5$

$= \frac{220}{7}$ cm

$\because \frac{220}{7}$ cm covers 25π cm^2

$\therefore$ 3.5 cm will cover $\frac{25 \times 22 \times 7 \times 3.5}{7 \times 220}$

i.e. 8.75 cm^2

Alternatively I

Area of the sector $= \frac{1}{2} \times$ length of an arc $\times$ radius

$= \frac{1}{2} \times 3.5 \times 5 = 8.75 \text{ cm}^2$

Alternatively II

Area of circle : Area of sector = Circumference of circle : Arc of sector

$\Rightarrow \quad \pi(5)^2$: Area of sector $= 2\pi \cdot 5 : 3.5$

$\Rightarrow \quad$ Area of sector $= \frac{\pi(5)^2 \times 3.5}{2\pi \times 5} = 8.75 \text{ cm}^2$

Problem 55 The ratio of the area of a sector of a circle to the area of the circle is 1 : 4. If the area of the circle is 154 cm^2, the perimeter of the sector is : *[SSC (CGL-T-II) 2010]*

Sol. Area of the circle $= \pi r^2 = 154$ (Given) $\Rightarrow r = 7$ cm

Since the area of the sector is one–fourth of that of the circle, the concerned arc will also be one-fourth of the circumference of the circle.

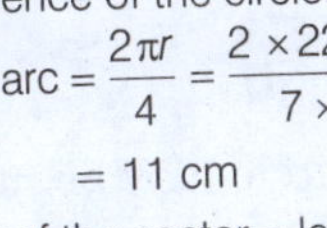

$\therefore$ Length of the arc $= \frac{2\pi r}{4} = \frac{2 \times 22 \times 7}{7 \times 4}$

$= 11$ cm

Now, Perimeter of the sector = length of an arc $+ 2r = 11 + (2 \times 7) = 25$ cm

Alternatively I

Area of the circle $= \pi r^2 = 154 \Rightarrow r = 7$ cm

Now, Area of the sector $= \frac{1}{2} \times$ length of the arc $\times$ radius

$\Rightarrow \quad \frac{154}{4} = \frac{1}{2} \times$ length of the arc $\times 7$

$\Rightarrow$ Length of the arc $= \frac{154 \times 2}{4 \times 7} = 11$ cm

$\therefore$ The perimeter of the sector $= 11 + (2 \times 7) = 25$ cm

Alternatively II

Given, area of the circle $= \pi r^2 = 154$ (given)

$\therefore \quad r = 7$ cm

Now, Area of circle : Area of sector = Circumference of circle : Arc of sector

$\Rightarrow \quad 4 : 1 = 2\pi r$: arc of sector

$\Rightarrow \quad 4 : 1 = (2 \times \pi \times 7)$: arc of sector

$\Rightarrow$ Arc of sector = 11 cm

$\therefore$ The perimeter of the sector $= 11 + (2 \times 7) = 25$ cm

Problem 56 At each corner of a triangular field of sides 26 m, 28 m and 30 m, a cow is tethered by a rope of length 7 m. The area (in m^2) ungrazed by the cow is : *[SSC(CGL-T-II)-10, SSC (CGL-T-I)-11]*

Sol. Area of the sector at Angle $A = \frac{\pi r^2 \angle A}{360°}$

and, Area of the sector at Angle $B = \frac{\pi r^2 \angle B}{360°}$

and, Area of sector at Angle $C = \frac{\pi r^2 \angle C}{360°}$

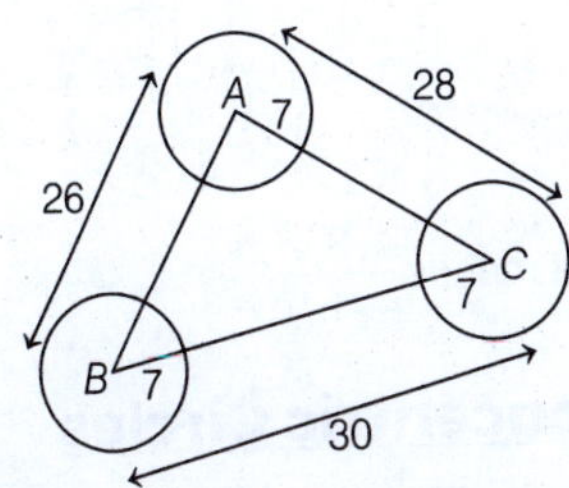

Now, area grazed by the cow

$= \frac{\pi r^2 \angle A}{360°} + \frac{\pi r^2 \angle B}{360°} + \frac{\pi r^2 \angle C}{360°}$

$= \frac{\pi r^2}{360°}(\angle A + \angle B + \angle C)$

$= \frac{\pi r^2}{360°} \times 180° = \frac{\pi r^2}{2}$

Obviously, here $r = 7$ m

$\therefore$ Grazed area $= \frac{\pi(7)^2}{2} = \frac{22 \times 7 \times 7}{7 \times 2} = 77 \text{ m}^2$

[you may compute the grazed area directly using the result $\frac{\pi r^2}{2}$ in such type of questions]

Now, Area of $\Delta ABC = \sqrt{s(s-a)(s-b)(s-c)}$

$\left[\text{here, } s = \frac{a+b+c}{2} = \frac{26+28+30}{2} = 42\right]$

$= \sqrt{42(42-26)(42-28)(42-30)}$

$= \sqrt{42 \times 16 \times 14 \times 12}$

$= \sqrt{14 \times 3 \times 4 \times 4 \times 14 \times 3 \times 4}$

$= 14 \times 4 \times 3 \times 2 = 336 \text{ m}^2$

$\therefore$ Area not grazed $= (336 - 77) \text{ m}^2 = 259 \text{ m}^2$

Problem 57 The circumference of a circle is 11 cm and the angle of a sector of the circle is 60°. The area of the sector is :

Sol. Given, $\quad 2\pi r = 11 \Rightarrow 2 \times \frac{22}{7} \times r = 11$

$\therefore \quad r = \frac{11 \times 7}{2 \times 22} = \frac{7}{4}$

Then, area of sector $= \frac{\pi r^2}{360°}\theta = \frac{\pi r^2}{360°} \times 60°$

$= \frac{22 \times 7 \times 7 \times 1}{7 \times 4 \times 4 \times 6}$

$= \frac{11 \times 7}{48} = 1\frac{29}{48} \text{ cm}^2$

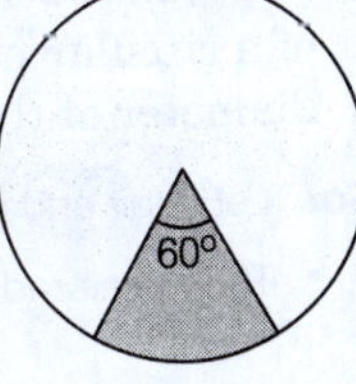

Alternatively

Given, $2\pi r = 11 \Rightarrow r = \frac{7}{4}$

Now, Area of circle : Area of sector = 360° : θ

$\Rightarrow \quad \pi r^2$: Area of sector = 360° : 60°

$\Rightarrow \quad \pi\left(\frac{7}{4}\right)^2$: Area of sector = 6 : 1

Area of the sector $= \pi\left(\frac{7}{4}\right)^2 / 6 = 1\frac{29}{48}\text{ cm}^2$

Problem 58 **A cow is tied on the corner of a rectangular field of size 30 m × 20 m by a 14 m long rope. The area of the region, that it can graze is :**

Sol. Area of the sector (grazed area) $= \frac{\pi r^2}{360°}\theta$

$$= \frac{22 \times 14 \times 14 \times 90}{7 \times 360} = 154\text{ m}^2$$

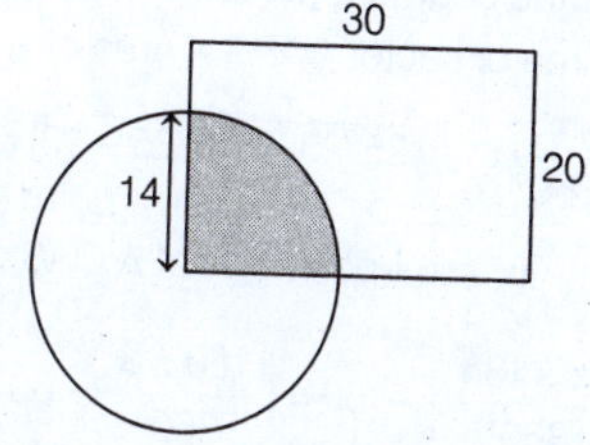

Problem 59. **A 72° sector of a circle has an area 7 cm². What is the length of the arc of the sector ?**

Sol. ∵ 72° describes an area of 7 cm²

∴ 360° will describe an area of 35 cm²

Hence, area of the circle = 35 cm² $\Rightarrow \pi r^2 = 35$

$\Rightarrow \quad \frac{22}{7} \times r^2 = 35$

$\Rightarrow \quad r^2 = \frac{35 \times 7}{22}$

$\Rightarrow \quad r = \frac{7\sqrt{5}}{\sqrt{22}}$

Now, 360° is describing the circumference $2\pi r$ cm

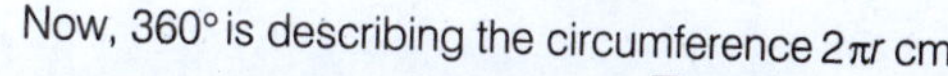

∴ 72° will describe $= \frac{2 \times 22 \times 7\sqrt{5}}{7 \times \sqrt{22} \times 5} = \frac{2\sqrt{22}}{\sqrt{5}} = 4.18$ cm

Hence, the length of the arc AB = 4.18 cm

Problem 60 **A pendulum swings through an angle of 45° and describes an arc 11 cm in length. Find the length of the pendulum.**

Sol. Here, 45° is describing an arc of length 11 cm

∴ 360° will describe an arc of length

11 × 8 i.e. 88 cm

Obviously, $2\pi r$ = 88 cm

$\Rightarrow \quad r = \frac{88 \times 7}{2 \times 22} = 14$

Hence, the length of the pendulum

= 14 cm

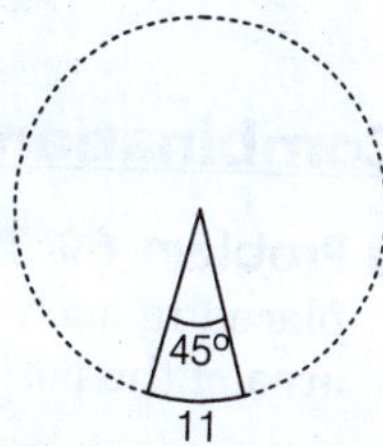

On the same concept :

Problem 61 **A horse is tied to a post by a rope. If the horse moves along a circular path always keeping the rope stretched and describes 88 m when it has traced out 72° at the centre, the length of the rope is :**

[SSC (CGL-T-I)-13]

Sol. Answer is 70 m.

Problem 62 **The length of minute hand of a clock is 7 cm. Find the area swept by this hand in 3 m (in cm²).**

Sol. Clearly, Area of the circle $= \frac{22}{7} \times (7)^2 = 154\text{ cm}^2$

∵ Area swept by the minute hand in 60 min. = 154 cm²

∴ Area swept by the minute hand in 3 min $= \frac{154}{20}\text{ cm}^2 = 7.7\text{ cm}^2$

Problem 63 **Sectors of two concentric circes of radii 7 cm and 10.5 cm are shown. Find the area of the shaded region, if the angle is 60°.**

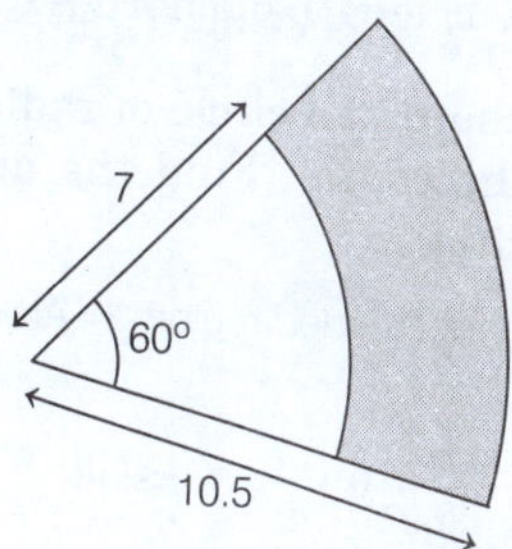

Sol. Area of shaded region $= \pi R^2 \frac{\theta}{360°} - \pi r^2 \frac{\theta}{360°}$

$$= \pi (10.5)^2 \times \frac{60°}{360°} - \pi (7)^2 \times \frac{60°}{360°}$$

$$= \frac{1}{6}\pi (10.5 + 7)(10.5 - 7)$$

$$= \frac{1}{6} \times \frac{22}{7} \times 17.5 \times 3.5 = 32.08\text{ cm}^2$$

Problem 64 **The minute hand of the clock is 10 cm long. What is the area on the face of clock described by the minute hand between 9 am and 9.35 am ?**

Sol. Area of the circle $= \frac{22}{7} \times (10)^2 = \frac{2200}{7}\text{ cm}^2$

∵ Area swept by the minute hand in 60 min $= \frac{2200}{7}\text{ cm}^2$

∴ Area swept by the minute hand in 35 min

$$= \frac{2200 \times 35}{7 \times 60}$$

$$= 183.33\text{ cm}^2$$

Application of the Concept of the Segment of the Circle

- A line joining any two points of a circle divides the circle into two regions. The bigger one is called major segment and the smaller one is called minor segment.

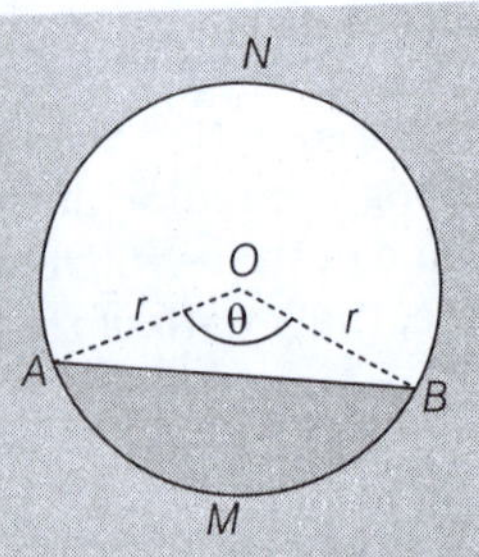

- In other words, the sector region formed by two radii - triangle region formed by them = segment of the circle.
  Area of the segment AMB = area of sector OAMB – area of ΔOAB

$$= \frac{\pi r^2}{360°}\theta - \frac{1}{2}r^2 \sin\theta$$

- Perimeter of segment AMB = length of arc AMB + length of line joining the points A and B i.e. length of chord AB Further, length of chord $AB = 2r\sin(\theta/2)$

Problem 65 A chord of a circle of radius 7 cm makes a right angle at the centre. Find the area of the minor segment of the circle.

Sol. Area of the segment (shaded region) = Area of sector $OACB$ – Area of $\Delta\ OAB$

$$= \frac{\pi r^2}{360°}\theta - \frac{1}{2}\times r^2 \times \sin\theta$$

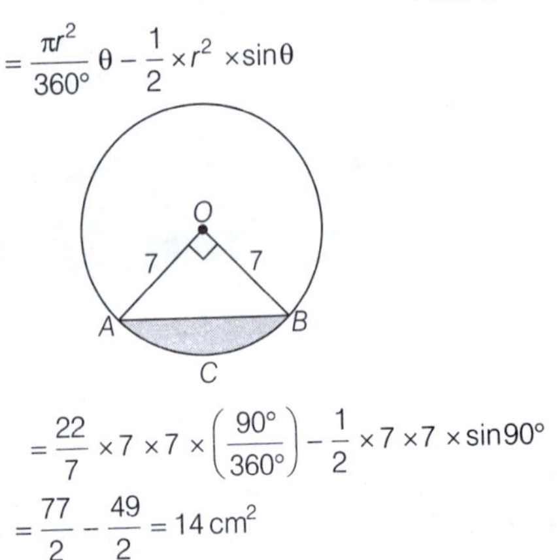

$$= \frac{22}{7}\times 7\times 7\times\left(\frac{90°}{360°}\right) - \frac{1}{2}\times 7\times 7\times\sin 90°$$

$$= \frac{77}{2} - \frac{49}{2} = 14\,\text{cm}^2$$

Problem 66 A chord 12 cm long is drawn in a circle whose radius is $6\sqrt{2}$ cm. Calculate the area of the major segment of the circle.

Sol. Here, $(AB)^2 = (OA)^2 + (OB)^2 \quad \{\because (12)^2 = (6\sqrt{2})^2 + (6\sqrt{2})^2\}$

It means ΔOAB is a right-angled triangle

$\therefore \quad \angle AOB = 90°$

Now, Area of minor segment = Area of sector $OACB$ – Area of ΔOAB

$$= \pi (6\sqrt{2})^2 \times \frac{90°}{360°} - \frac{1}{2}\times 6\sqrt{2}\times 6\sqrt{2}$$

$$= \frac{22}{7}\times 72\times\frac{1}{4} - \frac{1}{2}\times 72$$

$= 56.57 - 36$

$= 20.57\ \text{cm}^2$

Thus, the area of the major segment

= Area of the Circle – Area of the minor segment

$$= \pi (6\sqrt{2})^2 - 20.57 = \frac{22}{7}\times 72 - 20.57$$

$$= 226.28 - 20.57 = 205.71\,\text{cm}^2$$

Problem 67 A chord AB of length 24 cm is drawn in a circle of radius 13 cm. find the area of the shaded portion APB. **[IIFT-17]**

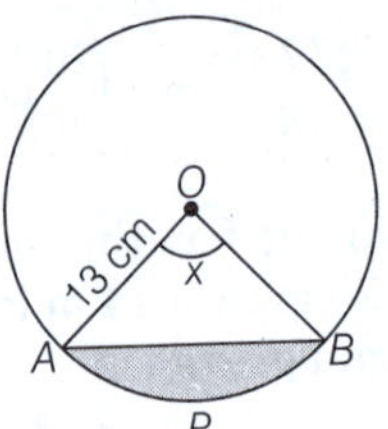

Sol. Given, $\angle AOB = x$

and sides of ΔOAB = 13,13 and 24

Now, area of the segment (shaded region)

= Area of sector $OAPB$ – Area of $\Delta\ OAB$

$$= \frac{\pi r^2}{360°}x - \sqrt{25(25-13)(25-13)(25-24)}$$

$$\left[\because \text{semi-perimeter of } \Delta\ OAB = \frac{13+13+24}{2} = 25\right]$$

$$= \frac{\pi\times 13^2}{360°}x - 60 = \left(\frac{169\pi x}{360°} - 60\right)\text{cm}^2$$

Problem 68 In a circle with centre O and radius 1 cm, an arc AB makes an angle 60° at O. Let R be the region bounded by the radii OA, OB and the arc AB. If C and D are two points on OA and OB respectively, such that $OC = OD$ and the area of ΔOCD Is half that of R, then the length of OC is : **[CAT-18]**

Sol. Area of region $R = \pi\times 1^2\times\frac{60°}{360°} = \frac{\pi}{6}$

Now, area of $\Delta OCD = \frac{1}{2}\times OC\times OD\times\sin 60°$

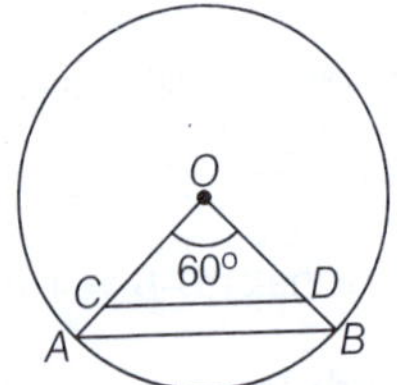

$$\Rightarrow \quad \left(\frac{\pi}{6}\div 2\right) = \frac{1}{2}\times OC\times OC\times\frac{\sqrt{3}}{2}$$

$$\Rightarrow \quad OC = \sqrt{\frac{\pi}{3\sqrt{3}}}$$

Combinations of Two or more Figures

Problem 69 Three circles of radius 3.5 cm each are placed in such a way that each touches other two. The area of the portion enclosed by the circles is :

Sol. Clearly, ΔABC is an equilateral triangle having each side of 7 cm.

$\therefore$ Its area $= \frac{\sqrt{3}}{4} \times (7)^2 = \frac{49\sqrt{3}}{4} = 21.217 \text{ cm}^2$

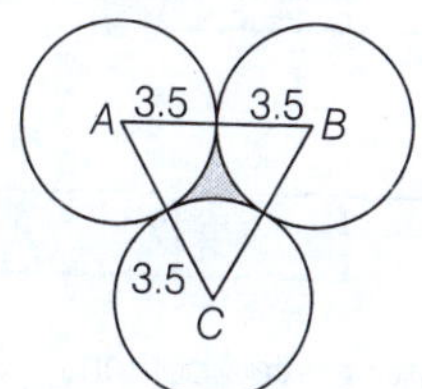

Further, there are three sectors at angles A, B and C whose areas in totality

$$= \pi r^2 \times \frac{(\angle A + \angle B + \angle C)}{360^\circ}$$

$$= \pi (3.5)^2 \times \frac{180^\circ}{360^\circ} = \frac{22}{7} \times 3.5 \times 3.5 \times \frac{1}{2} = 19.25 \text{ cm}^2$$

$\therefore$ Area of shaded region $= 21.217 - 19.25 = 1.967 \text{ cm}^2$

Problem 70 The area of the shaded region in the figure given below is :

Sol. Clearly, the radius of the circle $= a$

$\therefore$ Area of upper semi-cricle $= \frac{\pi a^2}{2}$

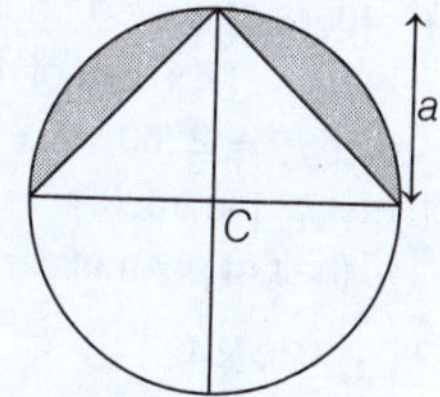

Now, area of two triangles in upper semi-circle

$$= 2 \times \frac{1}{2} \times a \times a = a^2$$

$\therefore$ Area of shaded region $= \frac{\pi a^2}{2} - a^2 = a^2 \left(\frac{\pi}{2} - 1 \right)$ sq units.

Problem 71 A circle is inscribed in an equilateral triangle of side 8 cm. The area of the portion between the triangle and the circle is :

[SSC (SO-Commer. Audit)- 07, SSC (CGL-T-II)-13]

Sol.

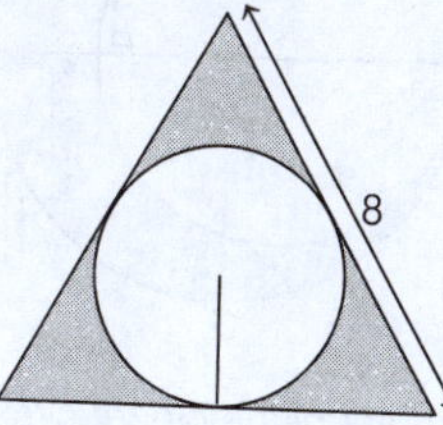

Area of the traingle $= \frac{\sqrt{3}}{4} \times 8^2 = 16\sqrt{3} = 27.71$

Its inradius $= \frac{1}{2\sqrt{3}} \times \text{side} = \frac{1}{2\sqrt{3}} \times 8 = \frac{4}{\sqrt{3}}$

$\therefore$ Area of the incircle $= \pi \left(\frac{4}{\sqrt{3}} \right)^2 = \frac{16}{3} \pi = \frac{16}{3} \times \frac{22}{7} = 16.76$

Thus, the area of the shaded region

$$= 27.71 - 16.76 = 10.95 \text{ cm}^2$$

Problem 72 Let $ABCD$ be a rectangle inscribed in a circle of radius 13 cm. Which one of the following pairs can represent, in centimetre, the possible length and breadth of $ABCD$?

[CAT-18]

(a) 24, 10
(b) 25,9
(c) 24, 12
(d) 25, 10

Sol. We know, (5, 12, 13) is a Pythagorean triplet. Therefore, the triplet obtained by multiplying it with certain number (say, 2) will also be Pythagorean triplet.

Hence, (10, 24, 26) is a Pythagorean triplet. Thus, option (a) is correct.

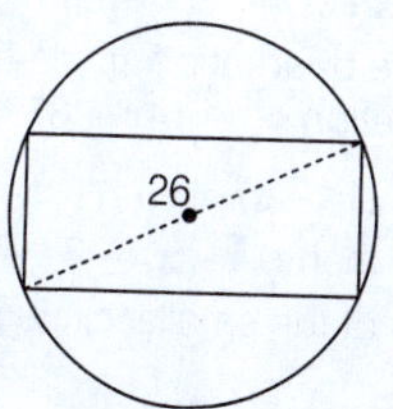

Problem 73 The four equal circles of radius 4 cm drawn on the four corners of a square touch each other externally. Then, the area of the portion between the square and the four sectors is (in sq cm) :

[SSC (10+2 LDC/DEO)-11]

Sol. Side of the square $= 4 + 4 = 8$ cm

$\therefore$ Its area $= (8)^2 = 64 \text{ cm}^2$

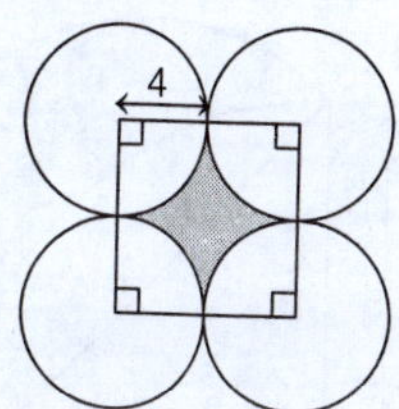

Now, area of all four sectors

$$= \pi r^2 \left(\frac{90^\circ + 90^\circ + 90^\circ + 90^\circ}{360^\circ} \right)$$

$$= \pi r^2 = \pi 4^2$$

$$= 16\pi$$

$\therefore$ Area of the shaded portion $= 64 - 16\pi = 16(4 - \pi) \text{ cm}^2$

Problem 74 If the four equal circles of radius 3 cm touch each other externally, then the area of the region bounded by the four circles is (in sq cm) :

Sol. On the above pattern, the area of the shaded portion

$$= (3 + 3)^2 - \pi r^2 \left(\frac{90^\circ + 90^\circ + 90^\circ + 90^\circ}{360^\circ} \right)$$

$$= 36 - \pi (3)^2 = 9(4 - \pi) \text{ cm}^2$$

Problem 75 Two circles touch externally. The sum of their areas is 130π sq cm and the distance between their centres is 14 cm. Find the radius of the smaller one.

[SSC (CGL-T-I)-15, SSC (CGL-Mains)-03]

Sol. Let the radius of the smaller circle be x.

Then, the radius of the bigger circle $= 14 - x$

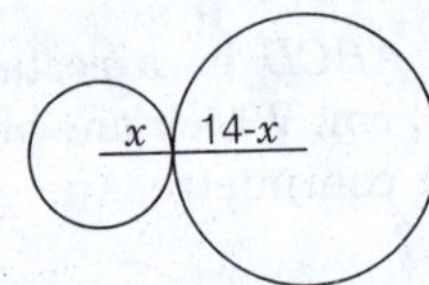

As given, $\pi x^2 + \pi(14 - x)^2 = 130\pi$

$\Rightarrow x^2 + 196 + x^2 - 28x = 130$

$\Rightarrow 2x^2 - 28x + 196 - 130 = 0$

$\Rightarrow x^2 - 14x + 33 = 0$

$\Rightarrow x^2 - 11x - 3x + 33 = 0$

$\Rightarrow x(x - 11) - 3(x - 11) = 0 \Rightarrow x = 3$ or 11

[you may apply the quadratic formula to find the value of x from the quadratic equation $x^2 - 14x + 33 = 0$]

When, $x = 3$, then $(14 - x) = 11$

and, when $x = 11$, then $(14 - x) = 3$

Hence, the radius of the smaller circle is 3 cm.

Alternatively

We should proceed through the options so as to save the time in the examination hall.

Assume, 1st option is 3 cm, then the radius of bigger be circle 11 cm. Now, ensure as to whether $\pi(3)^2 + \pi(11)^2$ is equal to 130π or not.

Problem 76 **In the given figure, $ABCMA$ is a quadrant of a circle of radius 28 cm. A semi-circle is drawn with AC as diameter. Find the area of the shaded portion.**

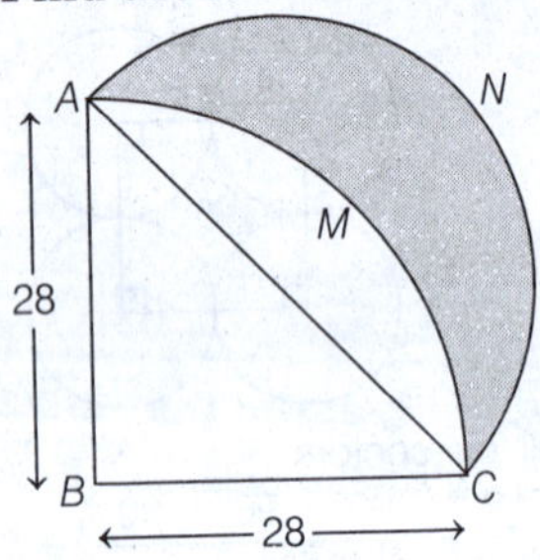

Sol. Since $ABCMA$ is a quadrant of a circle of radius 28 cm, $AB = BC = 28$ cm and $\angle ABC = 90°$

Then, $AC = \sqrt{AB^2 + BC^2} = \sqrt{28^2 + 28^2} = 28\sqrt{2}$ cm

$\therefore$ Radius of the semi-circle ANC with AC as diameter

$= \frac{28\sqrt{2}}{2} = 14\sqrt{2}$

Now, Required area = Area of semicircle ANC − (Area of quadrant $ABCMA$ − Area of ΔABC)

$$= \frac{\pi(14\sqrt{2})^2}{2} - \left[\pi(28)^2 \times \frac{90°}{360°} - \frac{1}{2} \times 28 \times 28\right]$$

$$= \frac{22}{7} \times 196 - \left[\frac{22}{7} \times 28 \times 28 \times \frac{1}{4} - 392\right]$$

$$= 616 - [616 - 392] = 392 \text{ cm}^2$$

Problem 77 **The inner perimeter of a running track, as shown in the figure, is 340 m. The length of each straight portion is 60 m, and the curved portions are semi-circles. If the track is 7 m wide, find the area of the track. Also find the outer perimeter of the track.**

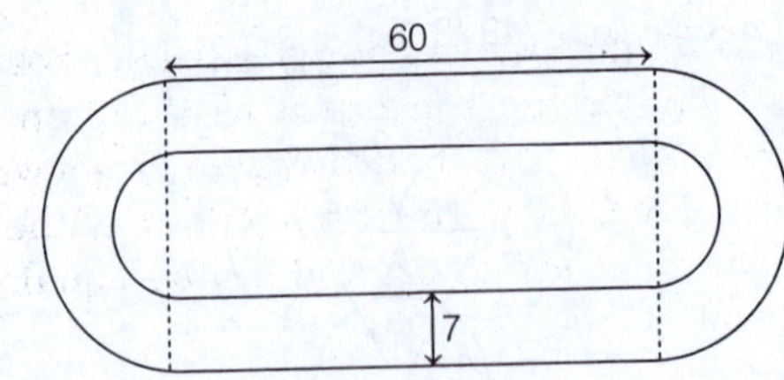

Sol. Length of both inner curved portions $= 340 - (2 \times 60) = 220$ m

$\therefore$ Length of each inner curved portion $= \frac{220}{2} = 110$ m

Obviously, 110 m is the half of the perimeter of a circle.

$\Rightarrow \frac{2\pi r}{2} = 110$ [here, r = radius of inner curved part]

$\Rightarrow r = 110 \times \frac{7}{22} = 35$ m

Then, radius of outer curved part $= (35 + 7) = 42$ m

Now, the area of curved track at both ends

$$= 2\left[\frac{\pi(42)^2}{2} - \frac{\pi(35)^2}{2}\right] = 2\left[\frac{\pi}{2} \times 77 \times 7\right]$$

$$= 2\left[\frac{22}{7 \times 2} \times 77 \times 7\right] = 2 \times 847 = 1694 \text{ m}^2$$

$\therefore$ Area of the track = [area of curved track at both ends] − [2 × area of rectangle (60 m × 7 m)]

$= 1694 + 2(60 \times 7) = 1694 + 840 = 2534 \text{ m}^2$

Further, outer perimeter of the track

$= 2 \times 60 + 2$ (half of perimeter of circle of radius 42 m)

$= 120 + 2\left(\frac{1}{2} \times 2\pi \times 42\right)$

$= 120 + 2 \times \frac{22}{7} \times 42 = 120 + 264 = 384$ m

Problem 78 **In the given figure, O is the centre of the bigger circle and AC is its diameter. Another circle with AB as diameter is drawn. If $AC = 54$ cm and $BC = 10$ cm, find the area of the shaded region.**

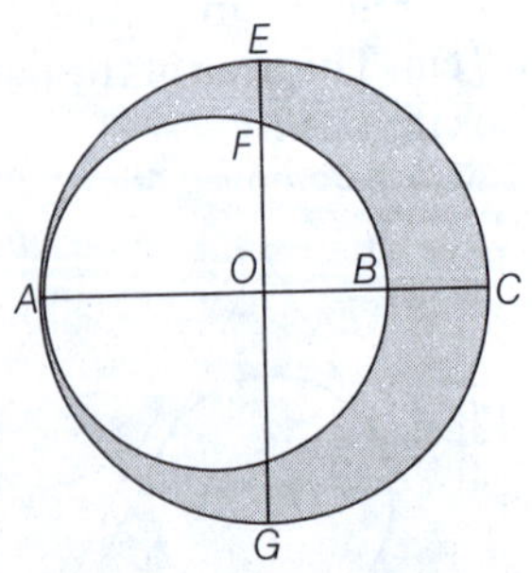

Sol. Given $AC = 54$ (Diameter of the bigger circle)

$\therefore$ $AO = OC$ = radius of bigger circle = 27

Now, $AB = AC - BC = 54 - 10 = 44$

Since AB is the diameter of the smaller circle, its radius

$= \frac{44}{2} = 22$ cm

Now, area of shaded region

= area of bigger circle − area of smaller circle

$= \pi(27)^2 - \pi(22)^2 = \frac{22}{7} \times 49 \times 5 = 770 \text{ cm}^2$

Problem 79 $PQRS$ is a diameter of a circle whose radius is 21 cm. Semi-circles are drawn with PQ and QS as diameters to create the shaded region as shown in the given figure. Find the perimeter and area of the shaded region, if the lengths PQ, QR and RS are equal.

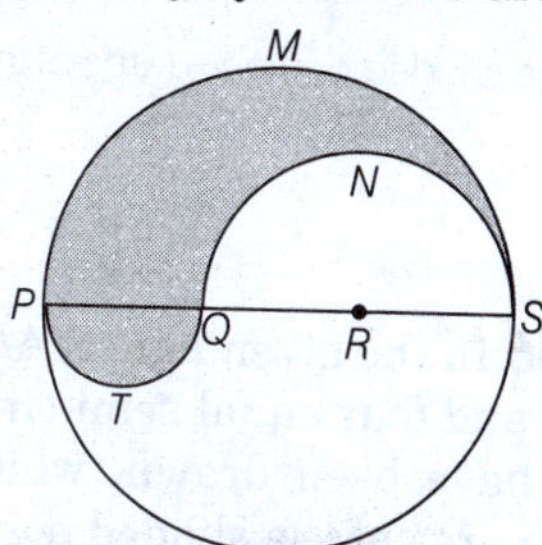

Sol. As given, $PQ = QR = RS = \frac{21 \times 2}{3} = 14$ cm

Now, area of the shaded region = Area of semi-circle PMS − Area of semi-circle QNS + Area of semi-circle PTQ.

$$= \frac{1}{2}\pi(21)^2 - \frac{1}{2}\pi(14)^2 + \frac{1}{2}\pi(7)^2$$

$$= \frac{1}{2} \times \frac{22}{7}(441 - 196 + 49) = \frac{1}{2} \times \frac{22}{7} \times 294$$

$$= 462 \text{ cm}^2$$

Further, perimeter of the shaded region

= Arc PTQ + Arc QNS + Arc PMS

$= \pi(7) + \pi(14) + \pi(21) = 42\pi = 42 \times 3.14 = 131.88$ cm

Problem 80 In the given figure, $OMNRO$ is a quadrant of a circle of radius 7 cm with centre O. What is the area of the shaded portion?

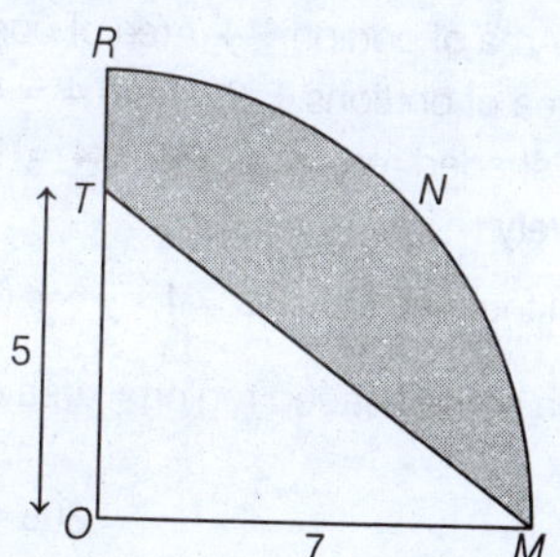

Sol. Area of the quadrant $= \frac{1}{4} \times \pi \times (7)^2 = \frac{1}{4} \times \frac{22}{7} \times 7 \times 7 = 38.5 \text{ cm}^2$

and Area of $\Delta TOM = \frac{1}{2} \times 7 \times 5 = 17.5 \text{ cm}^2$

$\therefore$ Area of shaded region = 38.5 − 17.5

= 21 cm^2

Problem 81 A brick, 5 cm thick(high), is placed against a wheel to act for a stop. The horizontal distance of the face of the brick stopping the wheel from the point where the wheel touches the ground is 15 cm. What is the radius of the wheel?

Sol. Let the radius of the wheel be r, i.e. $OB = r = OD$

As given, the height of the brick = DC = 5 cm

$\therefore$ AB = 5 cm

Now, in ΔAOD, $(OD)^2 = (OA)^2 + (AD)^2$

$\Rightarrow (OD)^2 = (OB - AB)^2 + (BC)^2$ [$\because BC = AD$]

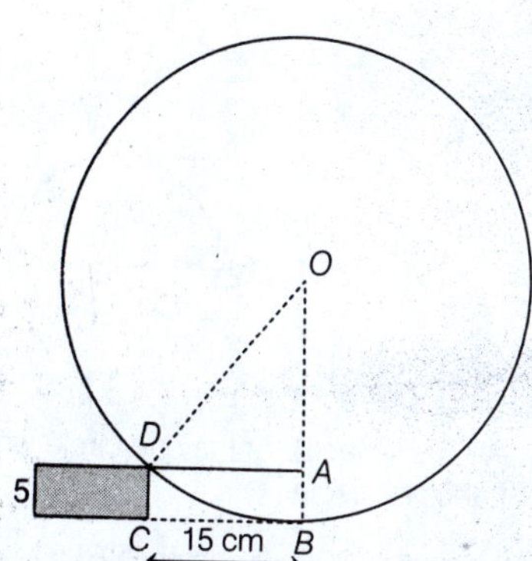

$\Rightarrow r^2 = (r-5)^2 + (15)^2$

$\Rightarrow r^2 = r^2 - 10r + 25 + 225 \Rightarrow 10r = 250$

$\therefore r = 25$ cm

Problem 82 In the given figure, A, B and C are centres of respective circles. The radii of the larger circles are 8 cm each. Find the circumference of the smaller circle, if AC =14 cm.

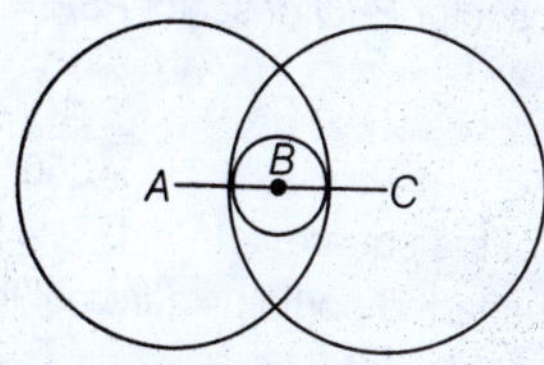

Sol. Clearly, radius of 1st larger circle + radius of 2nd larger circle − diameter of smaller circle (d) = AC

$\Rightarrow 8 + 8 - d = 14 \Rightarrow d = 2$ cm

Then, radius of the smaller circle $= \frac{2}{2} = 1$ cm

$\therefore$ Its circumference $= 2\pi r = 2\pi$ cm

Problem 83 A rectangular field is surrounded by four semi-circular flower-beds. If the length and the breadth of the field are 6 m and 4 m respectively, find the cost of raising flower-beds at the rate of ₹ 8/m^2.

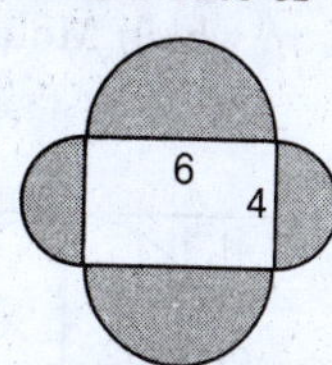

Sol. Two opposite semi-circles will become one complete circle when both are combined i.e. two different complete circles will be formed with diameters 6 m and 4 m.

$\therefore$ Area of flower-beds $= \pi\left(\frac{6}{2}\right)^2 + \pi\left(\frac{4}{2}\right)^2 = 13 \times \frac{22}{7} \text{ m}^2$

Then, total cost at the rate of ₹ 8/m^2

$= 13 \times \frac{22 \times 8}{7} =$ ₹ 326.85

Problem 84 $PQRS$ is a square of side 1 unit and Q, S are the centres of the two circles. The area of the shaded portion is : **[IAS (Pre.)-95]**

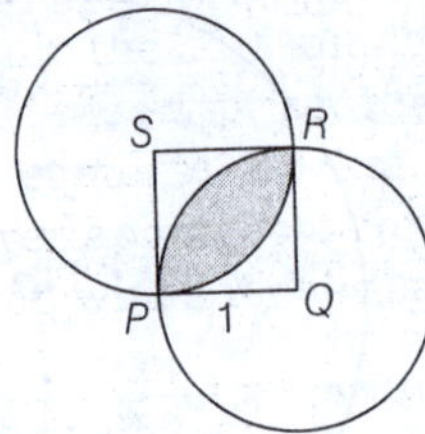

Sol. Clearly, PRQ is a sector of the one circle whereas PRS is the sector of another circle.

Area of the sector PRQ or sector $PSR = \frac{\pi r^2}{360°}\theta = \frac{\pi r^2}{360°} \times 90°$

$= \frac{\pi 1^2}{360°} \times 90° = \frac{\pi}{4}$

Now, area of the ΔRQP or $\Delta RSP = \frac{1}{2} \times 1 \times 1 = \frac{1}{2}$

$\therefore$ Area of the shaded portion $= 2\left(\frac{\pi}{4} - \frac{1}{2}\right) = \frac{\pi}{2} - 1$

Alternatively

Area of the sector PRQ or sector $PSR = \frac{\pi r^2}{360°}\theta$

$= \frac{\pi r^2}{360°} \times 90° = \frac{\pi 1^2}{360°} \times 90° = \frac{\pi}{4}$

Now, area of the square $= 1^2 = 1$

Area of the shaded portion = (area of the sector PRQ and sector PSR) − area of the square $= \left\{\frac{\pi}{4} + \frac{\pi}{4}\right\} - 1 = \frac{\pi}{2} - 1$

Problem 85 Mohanlal, a prosperous farmer, has a square land of side 2 km. For the current season, he decides to have some fun. He marks two distinct points on one of the diagonals of the land. Using these points as centres, he constructs two circles. Each of these circles falls completely within the land and touches at least two sides of the land. To his surprise, the radii of both the circles are exactly equal to 2/3 km. Mohanlal plants potatoes on the overlapping portion of these circles. Calculate the area on which Mohanlal planted potatoes (in sq km). **[XAT-20]**

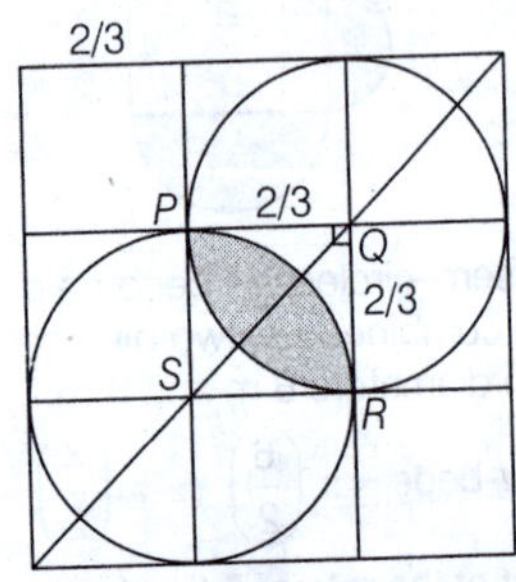

Sol. Area of the sector PRQ or sector $PSR = \frac{\pi r^2}{360°}\theta$

$= \frac{\pi r^2}{360°} \times 90° = \frac{\pi\left(\frac{2}{3}\right)^2}{360°} \times 90° = \frac{\pi}{9}$

Now, area of the square $PQRS = \left(\frac{2}{3}\right)^2 = \frac{4}{9}$

$\therefore$ Area of the shaded portion = (area of the sector PRQ and sector PSR) − area of the square

$= \left\{\frac{\pi}{9} + \frac{\pi}{9}\right\} - \frac{4}{9} = \frac{2(\pi - 2)}{9}$ km^2

Problem 86 In the given figure, $ABCD$ is a square with side 14 cm and four equal semi-circles with its sides as diameters have been drawn, which meet at point O. What is the area of the shaded portion ?

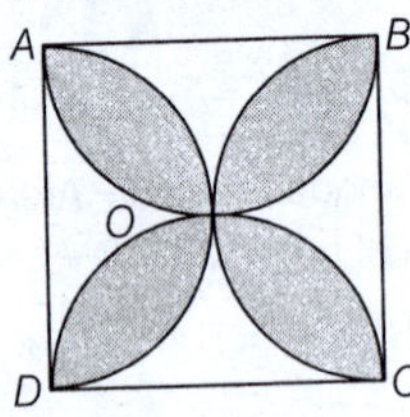

Sol. Area of the square $= (14)^2 = 196$

Area of semi-circle DOC

$= \frac{1}{2} \times \pi (7)^2 = 77$

Then, area of semi-circle DOC + area of semi-circle AOB

$= 77 + 77 = 154$

$\therefore$ Area of portion 2 + Area of portion 4

$= 196 - 154 = 42$

Similarly, Area of portion 1 + Area of portion 3 = 42

Hence, area of portions 1, 2, 3 and 4 = 42 + 42 = 84

$\therefore$ Area of shaded region = 196 − 84 = 112 cm^2

Alternatively

Area of all four semi-circles $= 4 \times \frac{1}{2} \times \pi (7)^2 = 308$

Then, area of the shaded region = area of all four semi-circles − area of the square

$= 308 - 196 = 112$ cm^2

Problem 87 The adjacent figure consists of circles and squares. What is the area of the smallest circle if the circumference of the largest circle be $10\sqrt{2}\,\pi$ cm.

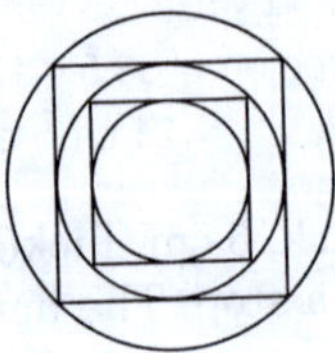

Sol. Diameter of the largest circle = diagonal of the bigger square

$\Rightarrow$ $10\sqrt{2}\,\pi \div \pi = \sqrt{2} \times$ side of the bigger square

$\Rightarrow$ Side of the bigger square = 10 cm

Further, side of the bigger square

= diameter of the 2nd largest circle

$= 10 =$ diagonal of the smaller square

Then, side of the smaller square $= 10 \div \sqrt{2} = 5\sqrt{2}$ cm

$=$ diameter of the smallest circle

$\therefore$ Radius of the smallest circle $= 5\sqrt{2} \div 2 = \dfrac{5}{\sqrt{2}}$ cm

Then, its area $= \pi\left(\dfrac{5}{\sqrt{2}}\right)^2 = 39.29\ \text{cm}^2$

Problem 88 **In the given figure, a circle has been inscribed in a square and also four quadrants have been drawn at its vertices. What is the area of the shaded portion?**

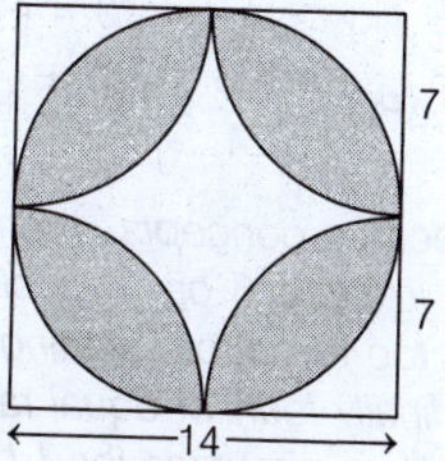

Sol. Area of the square $= (14)^2 = 196$

and area of the circle $= \pi(14 \div 2)^2 = 154$

$\therefore$ Area of the square outside the circle $= 196 - 154 = 42$

Further, area of all four sectors $= 4 \times \dfrac{\pi r^2}{360^\circ} \times 90^\circ$

$= \pi(7)^2 = 154$

Then, area of the shaded portion $=$ area of all four sectors $-$ area of square outside the circle

$= 154 - 42 = 112\ \text{cm}^2$

Problem 89 **Three equal circles have been cut from a rectangle as shown in the figure. Find the area of the rectangle if each circle has radius of 2 cm.**

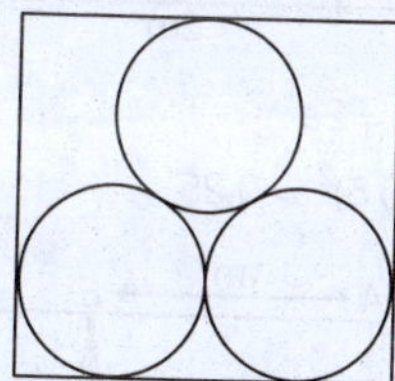

Sol. Length of the rectangle $= 4 + 4 = 8$ cm

Clearly, $OM = \sqrt{OB^2 - MB^2} = \sqrt{4^2 - 2^2} = 2\sqrt{3}$ cm

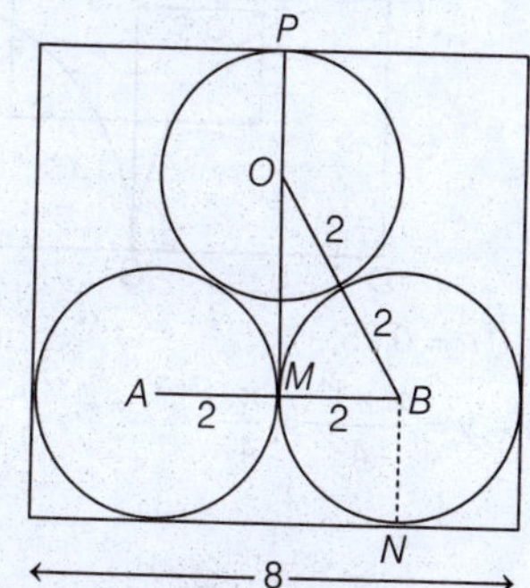

Then, breadth of the rectangle $= NB + MO + OP$

$= 2 + 2\sqrt{3} + 2 = 2(2 + \sqrt{3}) = 7.46$

Then, area of the rectangle $= 8 \times 7.46 = 59.68\ \text{cm}^2$

Problem 90 **What is the circumference of a circle of maximum possible area, inscribed in a quadrant of a circle of radius 1 cm?**

Sol. Draw a bisector of $\angle PQR$.

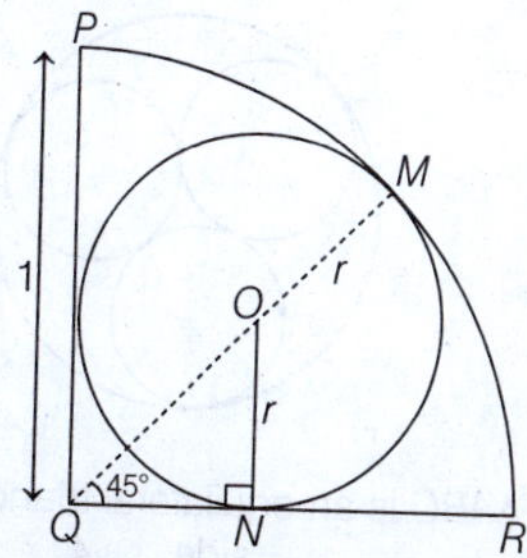

Now, in ΔQON, $\angle OQN = 45^\circ$ and $\angle ONQ = 90^\circ$

[$\because$ the tangent at any point of a circle is perpendicular to the radius through the point of contact]

In right-angled triangle OQN, $\sin \angle OQN = \dfrac{ON}{OQ}$

$\Rightarrow \quad \dfrac{1}{\sqrt{2}} = \dfrac{r}{OQ} \Rightarrow OQ = r\sqrt{2}$

Here, $PQ = QR = QM =$ radius of the circle whose quadrant is given $= 1$

$\Rightarrow \quad OQ + OM = 1$

$\Rightarrow \quad r\sqrt{2} + r = 1$

$\Rightarrow \quad r = \dfrac{1}{\sqrt{2} + 1} = \dfrac{1(\sqrt{2} - 1)}{(\sqrt{2} + 1)(\sqrt{2} - 1)} = \sqrt{2} - 1$

Now, circumference $= 2\pi r = 2\pi(\sqrt{2} - 1) = 0.82\pi$ cm

Problem 91 **Three circles of diameter 10 cm each, are bound together by a rubber band, as shown in the figure. The length of the rubber band is :** ***[SSC (CPO-S)I-11]***

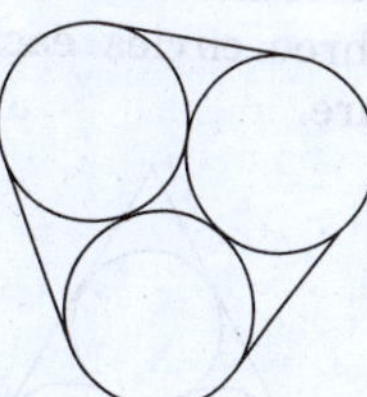

Sol. Length of the rubber Band $= PQ + MN + RS + (QM + NR + SP)$

$= AC + BC + AB + 22\pi r$

[$\because PQ = AC$, $MN = BC$ and $RS = AB$]

$= 10 + 10 + 10 + (2\pi \times 5) = 30 + 10\pi$

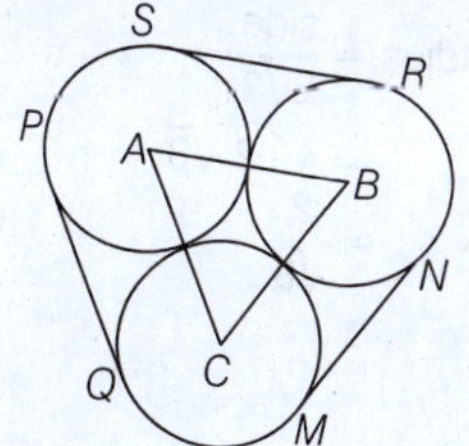

It may be construed from the above that the length of the rubber-band in such cases
$= (3 \times \text{diameter}) + \text{diameter} \times \pi$
$= \text{diameter } (3 + \pi)$ unit.

Problem 92 What is the circumference of a cicle circumscribing three circles each of radius 2 cm, as shown in the figure.

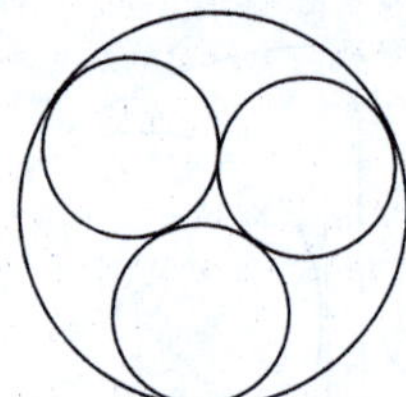

Sol. Obviously, ΔABC is an equilateral triangle of side (2 + 2) cm

Then, its circumradius $= \frac{\text{side}}{\sqrt{3}} = \frac{4}{\sqrt{3}}$ cm

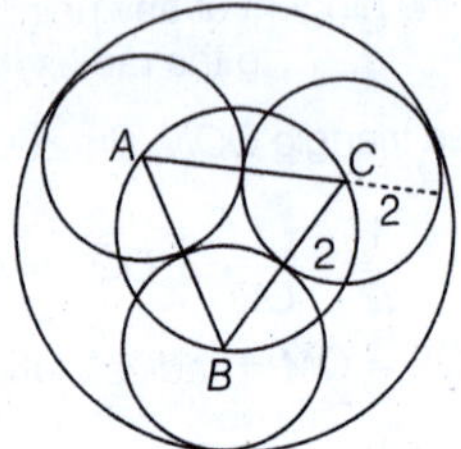

$\therefore$ Radius of the largest circle $= \left(\frac{4}{\sqrt{3}} + 2\right)$ cm

Now, circumference of the largest circle

$= 2\pi \times \left(\frac{4}{\sqrt{3}} + 2\right)$

$= \left(\frac{8}{\sqrt{3}} + 4\right) \pi$ cm

Problem 93 What is the area of an equilateral triangle circumscribing three circles each of radius 2 cm, as shown in the figure.

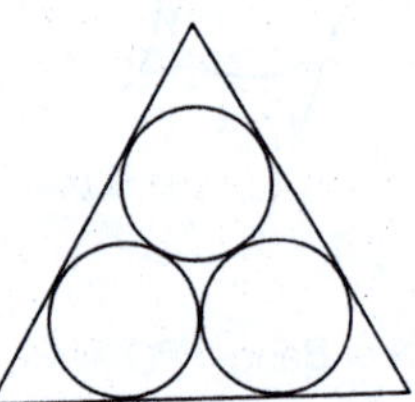

Sol. Clearly, side of the smaller equilateral $\Delta ABC = 2 + 2 = 4$ cm

Then, its inradius $= \frac{\text{side}}{2\sqrt{3}}$

$= 4 \div 2\sqrt{3}$

$= \frac{2}{\sqrt{3}}$

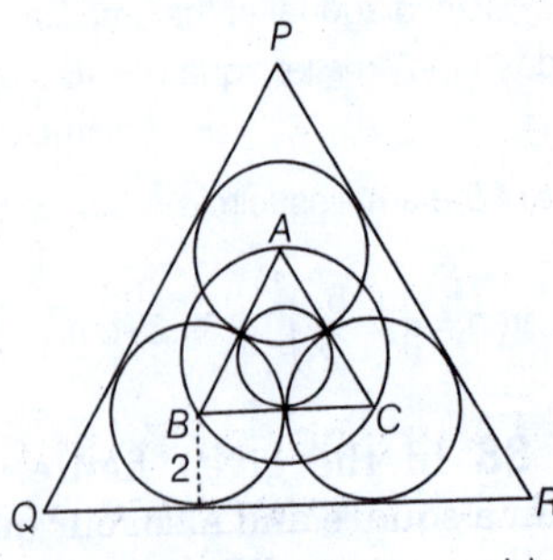

$\therefore$ Inradius of equilateral $\Delta PQR = \frac{2}{\sqrt{3}} + 2 \Rightarrow \frac{\text{side}}{2\sqrt{3}} = \frac{2}{\sqrt{3}} + 2$

$\Rightarrow$ side $= 4 + 4\sqrt{3} = 4(\sqrt{3} + 1)$

Now, Area of $\Delta PQR = \frac{\sqrt{3}}{4} \times \{4(\sqrt{3} + 1)\}^2 = (16\sqrt{3} + 24)\text{cm}^2$

Note *The respective concepts mentioned above are also applicable in case 6 or 10 circles (placed in ascending order from top to bottom starting with one, then two, then three and finally four) of equal radii are either bound by a rubber-band or circumscribed by larger circle/equilateral triangle.*

Problem 94 A square of length 1 m is inside a square of length 2 m and four quarter circles are joined as shown in the figure. The value of $y - x$ is given by :

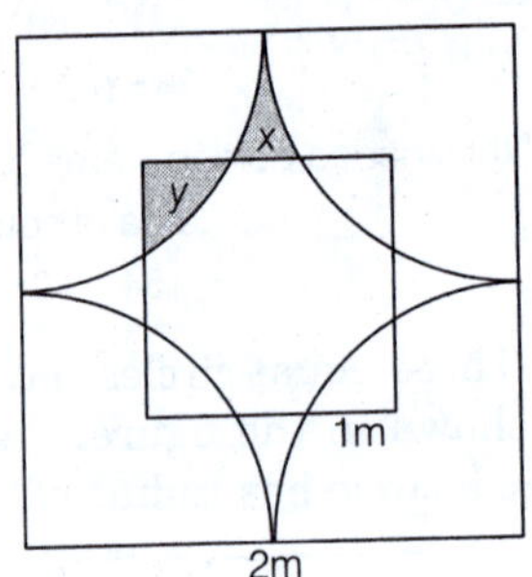

[IIFT-19]

Sol. Clearly, $x = a + a = 2a$

Further, $y + n = (0.5)^2 = 0.25$

A ← 1m → E B
m
1m
a a
y
0.5
n
a
H F
0.5
D G C

$\Rightarrow$ $n = 0.25 - y$

and $m + y = \frac{\pi \times 1^2}{4} = \frac{\pi}{4}$

Now, $m + y + n + a + a = 1^2$

$\Rightarrow \quad m + y + n + x = 1$

$\Rightarrow \quad (m + y) + n + x = 1$

$\Rightarrow \quad \frac{\pi}{4} + 0.25 - y + x = 1 \Rightarrow x - y = 0.75 - \frac{\pi}{4}$

$\Rightarrow \quad y - x = \frac{\pi - 3}{4}$

Counting of Number of Triangles having Vertices on Circumference of a Circle

Problem 95 There are 12 equally spaced points lying on the circumference of a circle. What is the maximum number of (i) equilateral triangles (ii) isosceles triangles (iii) right-angled triangles and (iv) scalene triangles, those can be drawn by taking sets of three points as the vertices ?

Sol. (i) **For equilateral triangles** It needs to be mentioned that the each chord i.e. each side of the triangles must be equal in length so as to form equilateral triangles and it is possible only when all the three vertices are equidistant. Adhering to this stipulation, equilateral triangles may be formed by joining the following points :

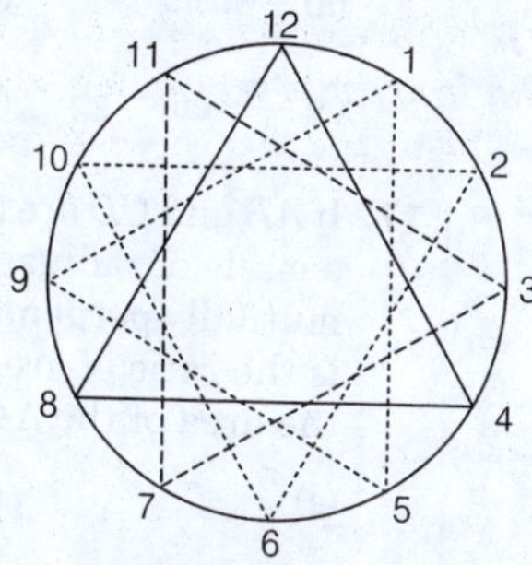

(1, 5, 9), (2, 6, 10), (3, 7, 11), (4, 8, 12), **(5, 9, 1)**, (6, 10, 2), ... (12, 4, 8)

It is observed that from 5th triplet onwards the repetition starts. Hence, number of distinct equilateral triangles = 4

(ii) **For Isosceles triangles** In order to form isosceles triangles, two chords i.e., two sides of a triangle must be equal in length and it is possible only when two points are equidistant from the third point. Under this stipulation, isosceles triangles may be formed by joining the following points :

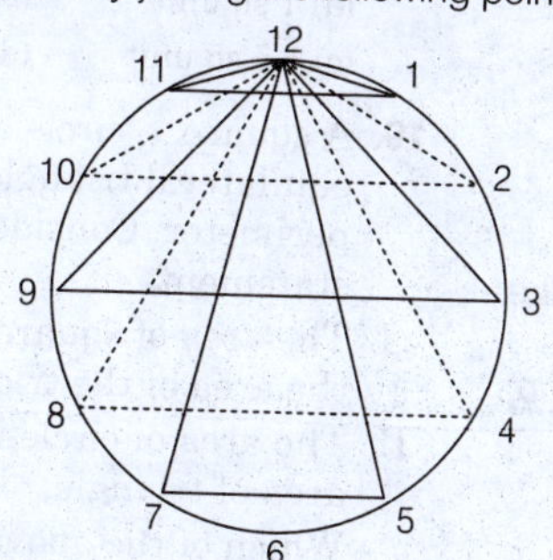

I. (12, 11, 1), (12, 10, 2), (12, 9, 3), **(12, 8, 4)**, (12, 7, 5) – shown in the above diagram.

II. (1, 12, 2), (1, 11, 3), (1, 10, 4), **(1, 9, 5)**, (1, 8, 6)

III. (2, 1, 3), (2,.....), (2,.........), ...

....

XII. (11, 10, 12), (11, 9, 1), (11, 8, 2), **(11, 7, 3)**, (11, 6, 4)

It is observed that each row contains 5 triangles and one of them is an equilateral triangle (shown in bold) and in fact there will be 12 rows.

Hence, number of isosceles triangles = (12 × 5) – 12 = 48

(iii) **For right-angled triangles** There will be 6 diameters, one of them is shown in the diagram. Since an angle in a semi-circle is a right angle, with each diameter, 10 right-angled triangles can be formed. 5 such triangles of one side of the diameter is shown in picture. Hence, total number of right-angled triangles = 6 × 10 = 60

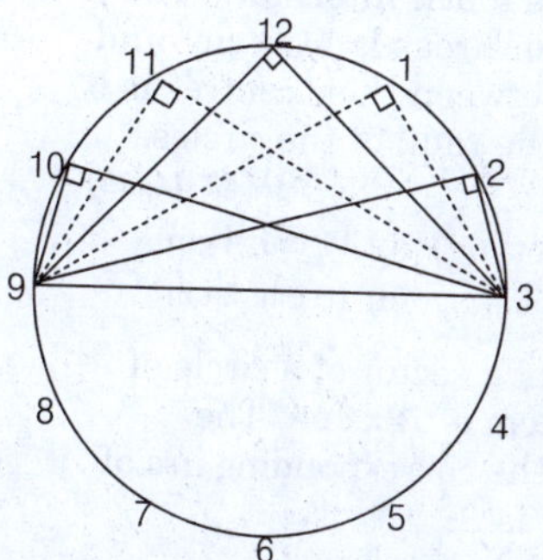

(iv) **For scalene triangles** Total number of triangles that can be formed by taking 3 points out of given 12 points = $^{12}C_3$. Hence, Total number of scalene triangles = $^{12}C_3$ – number of equilateral triangles – number of isosceles triangles

Generic version

(i) **Number of equilateral triangles =**

$$\frac{\text{Number of points equally spaced on circumference (let, } n)}{3} = \frac{n}{3}$$

[if n is not exactly divisible by 3, no equilateral triangle can be formed]

(ii) **Number of isosceles triangles** $= \frac{n^2}{2} - 2n$

[This pattern is applicable if n is even number. In case n is odd, number of isosceles triangles $= n\left(\frac{n-1}{2}\right)$]

(iii) **Number of right-angled triangles** $= n\left(\frac{n}{2} - 1\right)$

[if n is not an even number, no right-angled triangle can be formed as no diameter can be drawn by joining any two given points].

Problem 96 There are 24 equally spaced points lying on the circumference of a circle. What is the maximum number of equilateral triangles that can be drawn by taking sets of three points as the vertices ? ***[CSAT-18]***

Sol. Number of equilateral triangles $= \frac{24}{3} = 8$

WORKOUT

1. How many circular plates of diameter d be taken out of a square plate of side $2d$ with minimum loss of material ? **[CDS-14]**
(a) 8 (b) 6
(c) 4 (d) 2

2. Two circles touch internally. The sum of their areas is 116π cm² and distance between their centres is 6 cm. Find the radii of the circles. **[MAT-14, 19]**
(a) 10 cm, 4 cm (b) 11 cm, 4 cm
(c) 9 cm, 5 cm (d) 10 cm, 5 cm

3. The area of a sector of a circle of radius 36 cm is 72π cm². The length of the corresponding arc of the sector is :
(a) π cm (b) 2π cm
(c) 3π cm (d) 4π cm

4. The diameter of a circle is 8 cm less than its semi-circumference. Its radius is : **[SSC (SAS)-10]**
(a) 3.5 cm (b) 7 cm
(c) 10.5 cm (d) 14 cm

5. Four horses are tied on the four corners of a square field of 14 m length so that each horse can just touch the other two horses. They were able to graze in the area accessible to them for 11 days. For how many days is the ungrazed area sufficient for them ?
(a) 3 days (b) 4 days
(c) 5 days (d) 2 days

6. A wire when bent into the form of a square encloses an area of 484 sq cm. What will be the enclosed area when the same wire is bent into the form of a circle ?
(a) 462 cm² (b) 539 cm²
(c) 616 cm² (d) 693 cm²

7. The circumference of a circle is 14 cm greater than its diameter, find the diameter.
(a) 12.28 cm (b) 6.54 cm
(c) 12.48 cm (d) 8.96 cm

8. The circumference of a circular ground is 88 m. A strip of land, 3 m wide, inside and along circumference of the ground is to be leveled. What is the budgeted expenditure if the leveling cost is ₹ 7 per sq m.
(a) ₹ 1050 (b) ₹ 1125
(c) ₹ 1325 (d) ₹ 1650

9. What is the area of the larger segment of a circle formed by a chord of length 5 cm subtending an angle of 90° at the centre ? **[CDS-14]**
(a) $\frac{25}{4}\left(\frac{\pi}{2}+1\right)$ cm² (b) $\frac{25}{4}\left(\frac{\pi}{2}-1\right)$ cm²
(c) $\frac{25}{4}\left(\frac{3\pi}{2}+1\right)$ cm² (d) None of these

10. Seven semi-circular areas are removed from the rectangle $ABCD$ as shown in the figure, in which $AB = 2$ cm and $AD = 0.5$ cm. The radius of each semi-circle, r, s, t, u and v is half of that of semi-circle p or q. What is the area of the remaining portion?

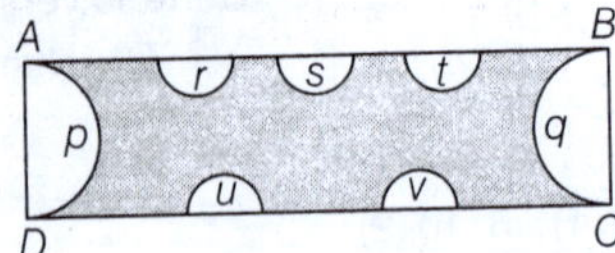

(a) $\frac{128-14\pi}{128}$ cm² (b) $\frac{128-12\pi}{128}$ cm²
(c) $\frac{128-13\pi}{128}$ cm² (d) None of these

11. Semi-circular lawns are attached to both the edges of a rectangular field measuring 42 m × 35 m. The area of the total field (in sq. m) is :
(a) 3818.5 (b) 8318
(c) 5813 (d) 1358

12. If a circle and a semi-circle have the same radius of 14 cm, then the ratio of their perimeters is : **[SSC (SAS)-10]**
(a) 3 : 1 (b) 2 : 1
(c) 10 : 9 (d) 11 : 9

13. From a circular piece of cardboard of radius 3 cm, two sectors of 40° each have been cut off. The area of the remaining portion is : **[CDS-15]**
(a) 11 cm² (b) 22 cm²
(c) 33 cm² (d) 44 cm²

14. If the circumference of a circle is reduced by 50%, its area will be reduced by ?
(a) 12.5% (b) 25%
(c) 50% (d) 75%

15. The areas of two circles are in the ratio 1 : 2. If the two circles are bent in the form of squares, then what is the ratio of their areas ? **[CDS-11]**
(a) 1 : 2 (b) 1 : 3
(c) 1: $\sqrt{2}$ (d) 1 : 4

16. The area of an equilateral triangle inscribed in a circle is $4\sqrt{3}$ cm². The area of the circle is :
(a) $\frac{16}{3}\pi$ cm² (b) $\frac{22}{3}\pi$ cm²
(c) $\frac{28}{3}\pi$ cm² (d) $\frac{32}{3}\pi$ cm²

17. If AB and CD are two diameters of a circle of radius r and they are mutually perpendicular, then what is the ratio of area of the circle to the area of the ΔACD ? **[CDS-14]**
(a) $\frac{\pi}{2}$ (b) π
(c) $\frac{\pi}{4}$ (d) 2π

18. A square $ABCD$ is inscribed in a circle of unit radius. Semi-circles are described on each side as a diameter. The area of the region bounded by the four semi-circles and the circle is : **[SSC (CGL-T-I)-13]**
(a) 1 sq unit (b) 2 sq unit
(c) 1.5 sq unit (d) 2.5 sq unit

19. A square, a circle and an equilateral triangle have same perimeter. Consider the following statements :
I. The area of square is greater than the area of the triangle.
II. The area of circle is less than the area of triangle.
Which of the above statement is/are correct?
(a) Only I
(b) Only II
(c) Both I and II
(d) Neither I nor II

20. The circumference of a circle is 100 cm, the side of a square inscribed in the circle is :

(a) $\frac{100\sqrt{2}}{\pi}$ (b) $\frac{50\sqrt{2}}{\pi}$

(c) $100\sqrt{2}$ (d) $50\sqrt{2}$

21. At the centre of a city's municipal park there is a large circular pool. A fish is released in the water at the edge of the pool. The fish swims North for 300 feet before it hits the edge of the pool. It then turns East and swims for 400 feet before hitting the edge again. What is the area of the pool (in sq feet)? ***[XAT-13]***

(a) 62500 (b) $125000\,\pi$

(c) $250000\,\pi$ (d) $500000\,\pi$

(e) Cannot be determined

22. A square is inscribed in a circle of diameter 2a and another square is circumscribing circle. The difference between the areas of outer and inner squares is :

(a) a^2 (b) $2a^2$ ***[CDS-14]***

(c) $3a^2$ (d) $4a^2$

23. P, Q, S and R are points on the circumference of a circle of radius r, such that PQR is an equilateral triangle and PS is a diameter of the circle. What is the perimeter of the quadrilateral $PQSR$? ***[CAT-05]***

(a) $2r(1+\sqrt{3})$ (b) $2r(2+\sqrt{3})$

(c) $r(1+\sqrt{5})$ (d) $2r+\sqrt{3}$

24. If the circumference of a circle is equal to the perimeter of square, then which one of the following correct ? ***[CDS-10, 12]***

(a) Area of circle = Area of square

(b) Area of circle ≥ Area of square

(c) Area of circle >Area of square

(d) Area of circle < Area of square

25. A man by walking diametrically across a circular grass plot finds that it has taken him 45 sec less than if he had kept to path round the outside. If he walks at 90 m/min, what is the circumference of the grass plot ? ***[MAT-13]***

(a) 251.23 m (b) 371.25 m

(c) 421.32 m (d) 125.12 m

26. $PQRS$ is a square. SR is a tangent (at point S) to the circle with centre O and $TR = OS$. Then, the ratio of area of the circle to the area of the square is : ***[CAT-95]***

(a) $\frac{\pi}{3}$ (b) $\frac{11}{7}$

(c) $\frac{3}{\pi}$ (d) $\frac{7}{11}$

27. A rectangle of maximum area is drawn inside a circle of diameter 5 cm. What is the maximum area of such a rectangle ? ***[CDS-14]***

(a) 25 cm² (b) 12.5 cm²

(c) 12 cm² (d) None of these

28. Consider a circle C of radius 6 cm with centre at O. What is the difference in the area of the circle C and the area of the sector of C subtending an angle of 80°at O ?

(a) 26π cm² (b) 16π cm²

(c) 28π cm² (d) 30π cm²

Directions (Q. Nos. 29-30) *Study the given diagram to answer the questions.* ***[IBPS (CWE)-13]***

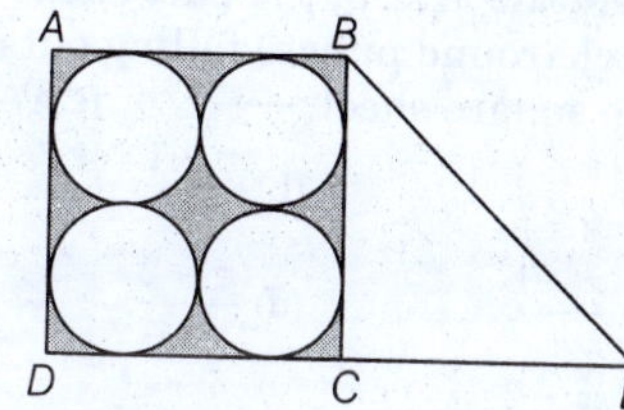

29. If the diameter of each circle is 14 cm and $DC = CE$, the area of ΔBDE is :

(a) 784 cm² (b) 748 cm²

(c) 874 cm² (d) 441 cm²

(e) None of these

30. The area of the shaded region of square $ABCD$ is :

(a) 186 cm² (b) 168 cm²

(c) 188 cm² (d) 441 cm²

(e) None of these

31. Two small circular parks of diameters 16 m, 12 m are to be replaced by a bigger circular park. What would be the radius of this new park, if the new park occupied the same space as the two small parks?

(a) 10 m (b) 20 m

(c) 15 m (d) 25 m

32. If the circumferences of two circles are in the ratio 2 : 3, then what is the ratio of their areas ?

(a) 2 : 3 (b) 4 : 9

(c) 1 : 3 (d) 8 : 27

33. Rekha drew a circle of radius 2 cm on a graph paper of grid 1 cm × 1 cm. She then calculated the area of the circle by adding up only the number of full unit-squares that fell within the perimeter of the circle. If the value that Rekha obtained was d sq. cm less than the correct value, then find the maximum possible value of d.

(a) 6.28 (b) 7.28

(c) 7.56 (d) 8.56

34. The area of the largest triangle that can be inscribed in a semi-circle of radius r is : ***[SSC (SO)-01, CDS-15]***

(a) r^2 (b) $2r^2$

(c) r^3 (d) $2r^3$

35. What is the perimeter of a square of maximum possible area, inscribed in a quadrant of a circle of radius 1 cm ?

(a) $2\sqrt{2}$ cm (b) $8\sqrt{2}$ cm

(c) 8 cm (d) None of these

36. The figure shows the rectangle $ABCD$ with a semi-circle and a circle inscribed inside it as shown. What is the ratio of the area of the circle to that of the semi-circle ? ***[CAT-96]***

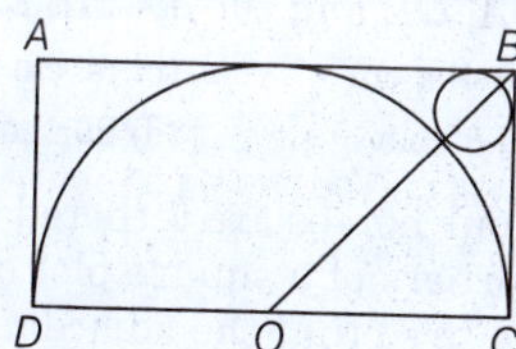

(a) $(\sqrt{2}-1)^2 : 1$

(b) $2(\sqrt{2}-1)^2 : 1$

(c) $(\sqrt{2}-1)^2 : 2$

(d) None of the above

37. What is the area of a circle whose area is equal to that of a triangle with sides 7 cm, 24 cm and 25 cm?

(a) 80 cm² (b) 84 cm² ***[CDS-13]***

(c) 88 cm² (d) 90 cm²

38. If a wire of length 36 cm is bent in the form of a semi-circle, what is the radius of the semi-circle ?

(a) 9 cm (b) 8 cm

(c) 7 cm (d) 6 cm

39. Four identical coins are placed in a square. For each coin, as shown in the figure and the ratio of area to circumference is same as the ratio of circumference to area. Then, find the area of the square that is not covered by the coins. *[CAT-98]*

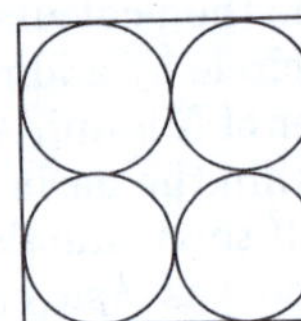

(a) $16(\pi - 1)$ (b) $16(8 - \pi)$
(c) $16(4 - \pi)$ (d) $16\left(4 - \frac{\pi}{2}\right)$

40. In a circular field, there is a rectangular tank of length 130 m and breadth 110 m. If the area of the land portion of the field is 20350 m^2, then the radius of the field is : *[IIFT-12]*

(a) 85 m (b) 95 m
(c) 105 m (d) 115 m

41. A circular water fountain 6.6 m in diameter is surrounded by a path of width 1.5 m. The area of the path(in sq m) is :

(a) $13.62\ \pi$ (b) $13.15\ \pi$
(c) $12.15\ \pi$ (d) None of these

42. The arc AB of the circle with centre O and radius 10 cm has length 16 cm. What is the area of the sector bounded by the radii OA, OB and the arc AB? *[CDS-13]*

(a) 40π cm^2 (b) 40 cm^2
(c) 80 cm^2 (d) 20π cm^2

43. Four horses are tethered at four corners of a square plot of side 14 m so that the adjacent horses can just reach one another. There is a small circular pond of area 20 m^2 at the centre. Find the ungrazed area. *[CAT-02]*

(a) 22 m^2 (b) 42 m^2
(c) 84 m^2 (d) 168 m^2

44. The area enclosed between the circumferences of two concentric circles is 16π sq cm and their radii are in the ratio 5 : 3. What is the area of the outer circle ?

(a) 9π cm^2
(b) 16π cm^2
(c) 25π cm^2
(d) 36π cm^2

Directions (Q. Nos. 45 - 46) *Answer the questions based on the following information.*

A punching machine is used to punch a circular hole of diameter 2 units from a square sheet of aluminium of width 2 unit, as shown below. The hole is punched such that the circular hole touches one corner P of the square sheet and the diameter of the hole originating at P is in the line with a diagonal of the square.

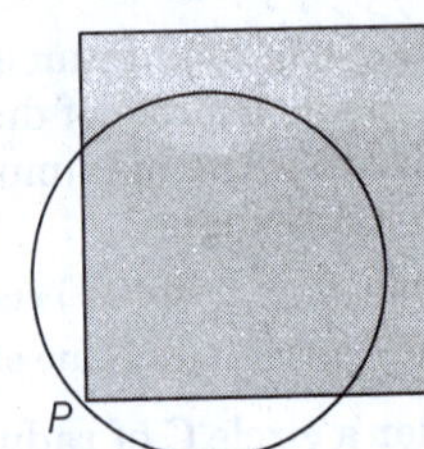

45. The proportion of the sheet area that remains after punching is :

(a) $\frac{\pi + 2}{8}$ (b) $\frac{6 - \pi}{8}$
(c) $\frac{4 - \pi}{4}$ (d) $\frac{\pi - 2}{4}$
(e) $\frac{14 - 3\pi}{6}$

46. Find the area of the part of the circle(round punch) falling outside the square sheet. *[CAT-06]*

(a) $\frac{\pi}{4}$ (b) $\frac{\pi - 1}{2}$
(c) $\frac{\pi - 1}{4}$ (d) $\frac{\pi - 2}{4}$
(e) $\frac{\pi - 2}{2}$

Directions (Q. Nos. 47-48) *Answer the questions based on the following information.* *[CAT-98]*

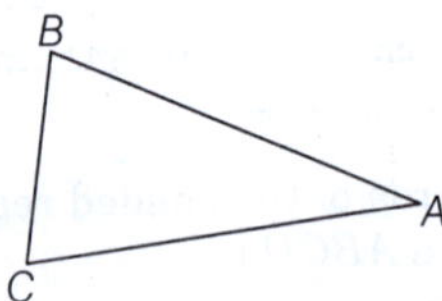

A cow is tethered at point A by a rope. Neither the rope nor the cow is allowed to enter the ΔABC. $\angle BAC = 30°$, $AB = AC = 10$ m.

47. What is the area that can be grazed by the cow, if the length of the rope is 8 m ?

(a) $\frac{134\pi}{3}$ m^2 (b) 121π m^2
(c) 132π m^2 (d) $\frac{176\pi}{3}$ m^2

48. What is the area that can be grazed by the cow, if the length of the rope is 12 m ?

(a) $\frac{133\pi}{6}$ m^2 (b) 121π m^2
(c) 132π m^2 (d) $\frac{176\pi}{3}$ m^2

49. The radius of a circle is increased so that its circumference increases by 15%. The area of the circle will increase by : *[CDS-17]*

(a) 31.25% (b) 32.25%
(c) 33.25% (d) 34.25%

50. In the given figure below, D is the diameter of each circle. What is the diameter of the shaded circle ? *[CDS-17]*

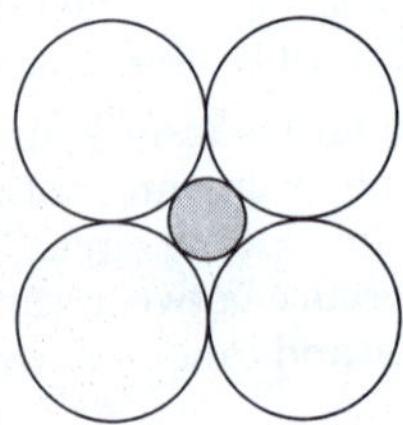

(a) $D(\sqrt{2} - 1)$ (b) $D(\sqrt{2} + 1)$
(c) $D(\sqrt{2} + 2)$ (d) $D(2 - \sqrt{2})$

51. There are 8 equidistant points A, B, C, D, E, F, G and H (in same order) on a circle. What is the value of $\angle FDH$ (in degrees) ? *[SSC (CGL-T-II)-17]*

(a) 22.5 (b) 45
(c) 30 (d) 42.5

52. A square is inscribed in a quarter circle in such a way that two of its adjacent vertices on the radius are equidistant from the centre and other two vertices lie on the circumference. If the side of square is $\sqrt{\frac{5}{2}}$ cm, then what is the radius (in cm) of the circle ? *[SSC (CGL-T-I)-17]*

(a) 2 (b) 2.5 (c) 5 (d) 10

53. From a rectangle $ABCD$ of area 768 sq cm, a semi-circle part with diameter AB and area 72π sq cm is removed. The perimeter of the leftover portion, in cm, is : *[CAT-18]*

(a) $80 + 16\pi$ (b) $86 + 8\pi$
(c) $82 + 24\pi$ (d) $88 + 12\pi$

54. Considering two opposite vertices of a square of side 'a' as centres, two circular arcs are drawn with the square joining the other two vertices, thus forming two sectors. What is the common area in these two sectors ?

(a) $a^2\left(\pi + \frac{1}{2}\right)$ (b) $a^2\left(\pi - \frac{1}{2}\right)$

(c) $a^2\left(\frac{\pi}{2} - 1\right)$ (d) $a^2\left(\frac{\pi}{2} + 1\right)$

55. A 12 m long wire is cut into two pieces, one of which is bent into a circle and the other into a square enclosing the circle. What is the radius of the circle ? ***[CDS-19]***

(a) $\frac{12}{\pi + 4}$ (b) $\frac{6}{\pi + 4}$

(c) $\frac{3}{\pi + 4}$ (d) $\frac{6}{\pi + 2\sqrt{2}}$

56. In the given diagram, there are four semi-circular arcs and a shaded region. The diameter of largest semi-circle is 28 cm and of the smallest is 7 cm. The area of shaded region is : ***[SNAP-18]***

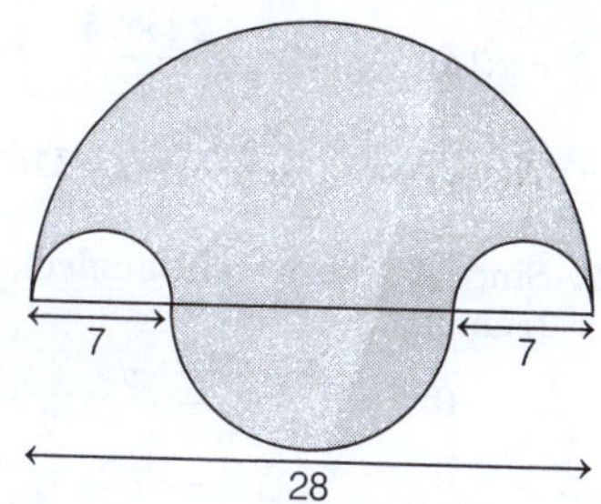

(a) 98.75π cm^2 (b) 120.5π cm^2

(c) 105.5π cm^2 (d) 110.25π cm^2

Answer Key

1. (c)	2. (a)	3. (d)	4. (b)	5. (a)	6. (c)	7. (b)	8. (d)	9. (c)	10. (c)
11. (a)	12. (d)	13. (b)	14. (d)	15. (a)	16. (a)	17. (b)	18. (b)	19. (a)	20. (b)
21. (a)	22. (b)	23. (a)	24. (c)	25. (b)	26. (a)	27. (c)	28. (c)	29. (a)	30. (b)
31. (a)	32. (b)	33. (d)	34. (a)	35. (a)	36. (d)	37. (b)	38. (c)	39. (c)	40. (c)
41. (c)	42. (c)	43. (a)	44. (c)	45. (b)	46. (e)	47. (d)	48. (c)	49. (b)	50. (a)
51. (b)	52. (b)	53. (d)	54. (c)	55. (b)	56. (d)				

Hints and SOLUTIONS

1. It is clear from the figure that 4 circular plates of diameter d can be made of a square plate of side $2d$ with minimum loss of material.

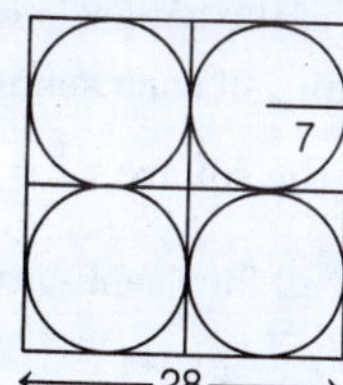

2. Let the radius of the smaller circle be r.

Then, the radius of the bigger $= r + 6$

As given, $\pi r^2 + \pi(r + 6)^2 = 116\pi$

$\Rightarrow \quad r^2 + 36 + r^2 + 12r = 116$

$\Rightarrow \quad 2r^2 + 12r - 80 = 0$

$\Rightarrow \quad r^2 + 6r - 40 = 0$

$\Rightarrow \quad r^2 + 10r - 4r - 40 = 0$

$\Rightarrow \quad r(r + 10) - 4(r + 10) = 0$

$\Rightarrow r = 4$ or -10

$\therefore r = 4$ cm.

[you may apply the quadratic formula to find the value of x from the quadratic equation $r^2 + 6r - 40 = 0$]

Hence, the radii of the circles are 4 cm and (4 + 6) cm.

Alternatively

We should proceed through the options so as to save the time in the examination hall. 1st option is 10 cm, 4 cm, then difference between radii is 6 cm .... 1st condition satisfied

Now, $\pi(10)^2 + \pi(4)^2 = 116\pi$ ...... 2nd condition is also satisfied.

Hence, correct option is (a).

3. We know, Area of circle : area of sector = circumference of circle : arc of sector

$\Rightarrow \pi(36)^2 : 72\pi = 2\pi \cdot 36 :$ arc of sector

$\Rightarrow 18 = 72\pi :$ arc of sector

$\therefore$ arc of sector $= 4\pi$ cm

4. Given, $\frac{2\pi r}{2} - 2r = 8$

$\Rightarrow r(\pi - 2) = 8 \Rightarrow r = 7$ cm

5. Here, radius for each sector = 7 cm

$\therefore$ Area grazed by them in 11 days

$= 4 \times \frac{\pi}{4} \times 7 \times 7 = 154$ cm^2

Thus, ungrazed area

$= (14)^2 - 154 = 42$

$\therefore$ Required number of days

$= 11 \times \frac{42}{154} = 3$ days

6. Side of the square $= \sqrt{\text{area}}$

$= \sqrt{484} = 22$ cm

$\therefore$ Perimeter $= 22 \times 4 = 88$ cm

Since the length of the wire is same, the circumference of the circle will also be 88 cm.

$\Rightarrow \quad 2\pi r = 88 \Rightarrow r = 14$

Now, area $= \pi r^2 = \frac{22}{7} \times 14 \times 14$

$= 616$ cm^2

7. Clearly, $2\pi r - 2r = 14$

$\Rightarrow \quad 2r = 6.54$ cm

8. Here, outer radius = 14 $[\because 2\pi r = 88]$

$\therefore$ Inner radius $= 14 - 3 = 11$

Area of the ground to the leveled

$= \pi(14)^2 - \pi(11)^2 = \frac{22 \times 75}{7}$

Then, cost $= \frac{22 \times 75}{7} \times 7 = ₹1650$

9. Since AOB is a right-angled triangle,

$$(5)^2 = r^2 + r^2 \Rightarrow r^2 = \frac{25}{2}$$

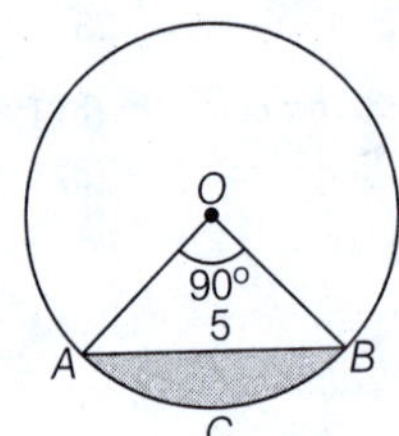

Area of the minor segment (i.e. shaded region)

$=$ Area of sector $OACB -$ Area of ΔOAB

$= \frac{\pi r^2}{360°}\theta - \frac{1}{2} \times r^2 \times \sin\theta$

$= \pi \times \frac{25}{2} \times \left(\frac{90°}{360°}\right) - \frac{1}{2} \times \frac{25}{2} \times \sin 90°$

$= \frac{25}{8}(\pi - 2)$

Now, area of larger segment = area of circle − area of minor segment

$= \pi r^2 - \left[\frac{25}{8}(\pi - 2)\right]$

$= \pi \times \frac{25}{2} - \left[\frac{25}{8}(\pi - 2)\right]$

$= \frac{25}{2}\pi - \frac{25}{8}\pi + \frac{25}{4}$

$= \frac{25}{4}\left(2\pi - \frac{1}{2}\pi + 1\right) = \frac{25}{4}\left(\frac{3\pi}{2} + 1\right) \text{cm}^2$

10. Clearly, radius of semi-circle p or

$$q = \frac{0.5}{2} = 0.25$$

Then, radius of each semi-circle, r, s, t, u and $v = \frac{0.25}{2} = 0.125$

Sum of areas of both bigger semi-circles p and $q = 2 \times \frac{\pi R^2}{2}$

$= 2 \times \frac{\pi(0.25)^2}{2} = \frac{\pi}{16}$

Further, sum of all 5 smaller semi-circles $= 5 \times \frac{\pi r^2}{2}$

$= 5 \times \frac{\pi(0.125)^2}{2} = \frac{5\pi}{128}$

Now, area of the remaining portion

= area of rectangle − area of all semi-circles

$= (2 \times 0.5) - \frac{\pi}{16} - \frac{5\pi}{128}$

$= \frac{(128 - 13\pi)}{128} \text{cm}^2$

11. Area of the rectangular field

$= 42 \times 35 = 1470$

Sum of areas of the both semi-circles attached to length of the field

$= \pi\left(\frac{42}{2}\right)^2 = 441\pi$

and sum of area of the both semi-circles attached to breadth of the field $= \pi\left(\frac{35}{2}\right)^2 = 306.25\pi$

Thus, total area

$= 1470 + 441\pi + 306.25\pi$

$= 3818.5 \text{ m}^2$

12. $2\pi r : (\pi r + 2r) = 2 \times \frac{22}{7} \times 14 : \left[\frac{22}{7} \times 14 + (2 \times 14)\right]$

$= 88 : 72 = 11 : 9$

13. Area of the circle $= \pi(3)^2$, described by 360°.

Then, (40° + 40°) will describe $2\pi \text{ cm}^2$.

Hence, area of remaining portion

$= 9\pi - 2\pi = 7\pi = 22 \text{ cm}^2$

14. Effect on area

$= \left\{-50 - 50 + \frac{50 \times 50}{100}\right\}\%$

$= -75\%$.

Hence, the area will decrease by 75%.

Alternatively

Since the radius has been made to 0.5 times the earlier, the new area will become $(0.5)^2$ times the original area i.e. 0.25 times. It implies that the original area will decrease by 75%. The logic behind this concept is that the area of the circle is directly proportional to square of its radius.

15. Ratio of circumferences of circles

$= \sqrt{1} : \sqrt{2} = 1 : \sqrt{2}$

It means ratio of perimeters of both squares $= 1 : \sqrt{2}$

Then, ratio of areas of both squares

$= (1)^2 : (\sqrt{2})^2 = 1 : 2$

16. Area of the equilateral triangle

$= \frac{\sqrt{3}}{4}(\text{side})^2 = 4\sqrt{3}$

$\Rightarrow (\text{side})^2 = 16$

$\Rightarrow \quad \text{side} = 4$

Now, the circumradius of an equilateral triangle $= \frac{\text{side}}{\sqrt{3}} = \frac{4}{\sqrt{3}}$

$\therefore$ Area of the circumcircle

$= \pi \times \left(\frac{4}{\sqrt{3}}\right)^2 = \frac{16}{3}\pi \text{ cm}^2$

17. Required ratio $= \frac{\text{Area of circle}}{\text{Area of } \Delta ACD}$

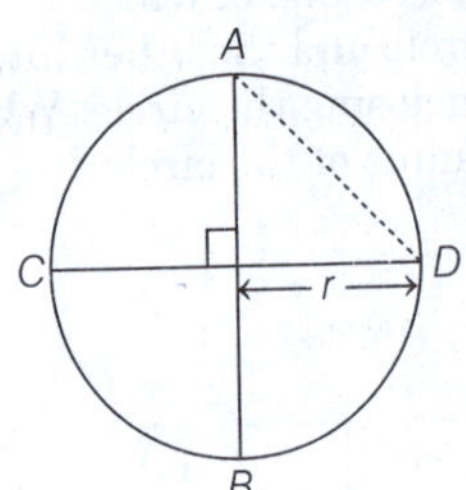

$= \frac{\pi r^2}{\frac{1}{2} \times 2r \times r} = \pi$

18. Given, radius = 1 unit

$\therefore$ Area of the circle $= \pi$ sq unit

Clearly, diagonal of the square

$= 1 \times 2 = 2$

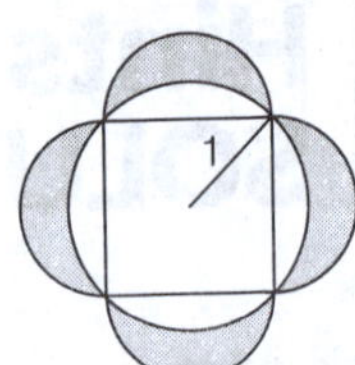

Then, side of the square $= \frac{2}{\sqrt{2}} = \sqrt{2}$

$\therefore$ Area of the square $= (\sqrt{2})^2 = 2$

Now, radius of each semi-circle

= side of the square $\div 2 = \frac{1}{\sqrt{2}}$

$\therefore$ area of all four semi-circles

$= 4 \times \frac{\pi\left(\frac{1}{\sqrt{2}}\right)^2}{2} = \pi$

Now, area of the shaded region = area of the four semi-circles − (area of the circle − area of the square)

$= \pi - (\pi - 2) = 2$ sq unit

19. It has been described in the shaded portion under title 'Application of the concept of Area' that the area of a circle is always greater than that of square/ rectangle/triangle under given circumference/ perimeter of equal length and also area of square > area of rectangle > area of triangle.

Hence, option (a) is correct.

20. Given, $2\pi r = 100$

$\Rightarrow \quad 2r = \frac{100}{\pi}$

Here, the diameter $(2r)$ is diagonal of the square.

$\therefore \sqrt{2} \times (\text{side}) = \frac{100}{\pi} \Rightarrow \text{side} = \frac{50\sqrt{2}}{\pi}$

21. We know, an angle in a semi-circle is a right angle.

Since $\angle CAB$ is a right angle, CB must be the diameter of the circle.

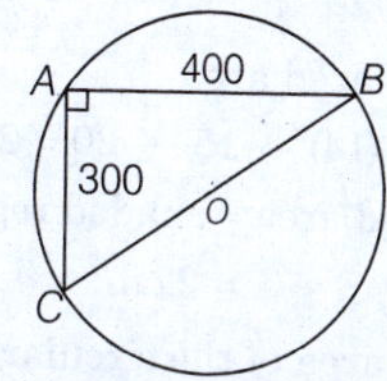

Clearly, $CB = \sqrt{300^2 + 400^2} = 500$

Then, radius $OC = 250$ feet

Now, area of the pool $= \pi(250)^2$

$= 62500\pi$ sq feet

22. Clearly, diameter of circle = diagonal of square inscribed in circle.

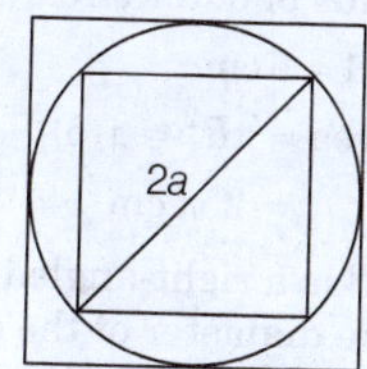

$\Rightarrow \quad 2a = \sqrt{2} \times \text{side}$

$\Rightarrow \quad \text{side} = a\sqrt{2}$

and diameter of circle = side of the square circumscribing the circle = $2a$

Now, difference between both squares' areas $= (2a)^2 - (a\sqrt{2})^2 = 2a^2$.

23. Height of ΔPQR

$= PN = PO + OM$

$= r + \frac{r}{2} = \frac{3r}{2}$

[where, r is the radius of the circle and circumradius = 2· inradius]

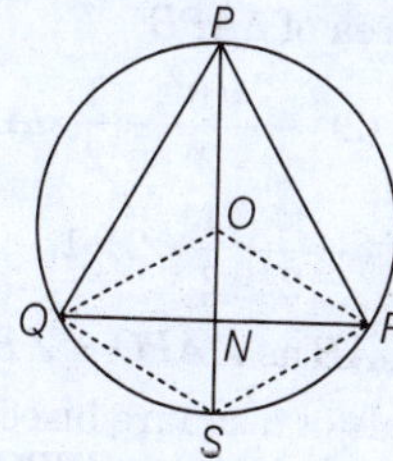

$\Rightarrow \frac{\sqrt{3}}{2} \times \text{side of the } \Delta PQR = \frac{3r}{2}$

$\Rightarrow$ Side of the $\Delta PQR = r\sqrt{3}$

Further, in ΔONQ and ΔRNS,

$ON = NS$,

$QN = NR$ and $\angle ONQ = \angle RNS$

$\therefore OQ = SR = r$

Similarly, $OR = QS = r$

Now, perimeter of the quadrilateral

$PQSR = PQ + QS + SR + RP$

$= r\sqrt{3} + r + r + r\sqrt{3}$

$= 2r + 2r\sqrt{3}$

$= 2r(1 + \sqrt{3})$

24. It has been described in the shaded portion under title 'Application of the concept of Area' that the area of a circle is always greater than that of square/rectangle/triangle under given circumference/ perimeter of equal length. Hence, option (c) is correct.

25. Distance covered by man in 45 sec $= 90 \times \frac{45}{60} = 67.5$ m

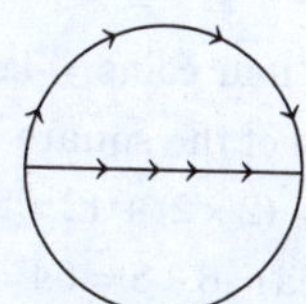

If man walks diametrically instead of walking through semi-circular path, he saves 45 sec. It implies that the difference between length of diameter and length of semi-circular path is the same as he can cover the distance in 45 sec i.e. 67.5 m.

i.e. πr (length of semi-circular path) $- 2r$ (diameter) $= 67.5$

$\Rightarrow \quad r(\pi - 2) = 67.5$

$\Rightarrow \quad r = 59.06$ m

Now, circumference of the grass plot

$= 2\pi r = 2 \times \frac{22}{7} \times 59.21$

$= 371.25$m

26. Let the radius of the circle be R.

Then, its area $= \pi R^2$

Further,

$OR = TR + OT = OS + OT$

$= R + R = 2R$

and $SR = \sqrt{(OR)^2 - (OS)^2}$

$= \sqrt{(2R)^2 - (R)^2} = R\sqrt{3}$

Now, required ratio

$= \pi(R)^2 : (R\sqrt{3})^2$

$= \pi : 3$

$\therefore$ Option (a) is correct.

27. Here, diameter of the circle = diagonal of the rectangle

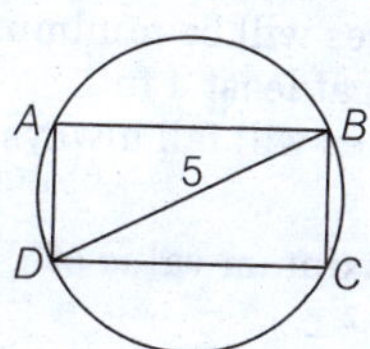

$\Rightarrow \quad 5 = \sqrt{l^2 + b^2}$

$\Rightarrow \quad l^2 + b^2 = 25$

Since they form Pythagoras triplet, $l = 4$ and $b = 3$

Thus, area of the rectangle

$= 4 \times 3 = 12 \text{ cm}^2$

28. The required answer

$= \pi r^2 - \frac{\pi r^2}{360°}\theta$

$= \pi r^2\left(1 - \frac{80°}{360°}\right)$

$= \pi(6)^2 \times \frac{7}{9} = 28\pi \text{ cm}^2$

29. Given, diameter of each circle

$= 14$ cm

Then, $\quad BC = DC = 28$ cm

Further, $DE = 2DC$

$[\because DC = CE \text{ (given)}]$

$\Rightarrow \quad DE = 2 \times 28 = 56$ cm

Now, area of $\Delta BDE = \frac{1}{2} \times DE \times BC$

$= \frac{1}{2} \times 56 \times 28$

$= 784 \text{ cm}^2$

30. Area of square

$ABCD = (28)^2 = 784$

and area of each circle

$= \pi \cdot (7)^2 = 49\pi$

$\therefore$ Area of all four circles

$= 4 \times 49\pi = 196\pi = 616$

Then, area of the shaded portion

$= 784 - 616 = 168 \text{ cm}^2$

31. Given, radii = 8 m and 6 m

Let the radius of new park be R m

Then, area of the new park

$= \pi(8)^2 + \pi(6)^2$

$\Rightarrow \quad \pi R^2 = 100\pi$

$\therefore \quad R = 10$ m

32. Observe the facts given in chart under title 'Comparisons between/among circles with reference to different ratios'.

The required ratio $= (2)^2 : (3)^2 = 4 : 9$

33. The value of d will be maximum when the number of full unit-squares will be minimum. Obviously, at least 4 full unit-squares will fall always within the circle.

Hence, maximum value of

$$d = \pi r^2 - 4$$
$$= \pi \cdot (2)^2 - 4 = 8.56 \text{ sq cm}$$

34. Area of the largest triangle

$$= \frac{1}{2} \times \text{base} \times \text{height}$$
$$= \frac{1}{2} \times 2r \times r = r^2$$

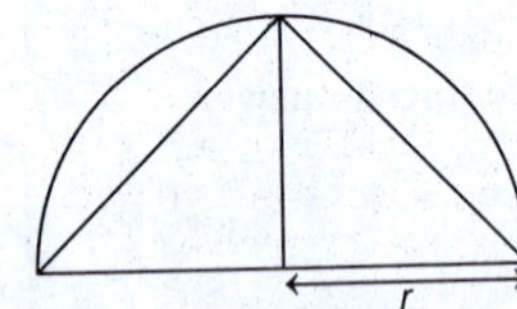

35. Given, $AB = 1$ cm

Clearly, diagonal of square $MNOB$

$= BN = AB = 1$ cm

Then, side of the square $= \frac{1}{\sqrt{2}}$

Now, its perimeter $= 4 \times \frac{1}{\sqrt{2}}$

$= 2\sqrt{2}$ cm

36. Clearly, $OC = OE = OK$ = radius of the semi-circle (say, R) $= BC$.

Then, $OB = \sqrt{OC^2 + BC^2}$

$= \sqrt{R^2 + R^2} = R\sqrt{2}$

Similarly, $FG = FH = FE$ = radius of the circle (say, r) $= BG$

Then, $FB = r\sqrt{2}$

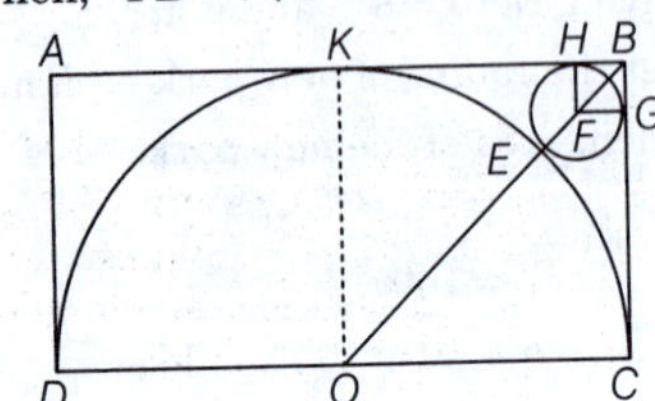

Now, $EB = OB - OE$

$\Rightarrow FB + EF = R\sqrt{2} - R$

$\Rightarrow r\sqrt{2} + r = R\sqrt{2} - R$

$\Rightarrow r : R = (\sqrt{2} - 1) : (\sqrt{2} + 1)$

Then, required ratio $= \pi r^2 : \frac{1}{2}\pi R^2$

$= \pi(\sqrt{2} - 1)^2 : \frac{1}{2}\pi(\sqrt{2} + 1)^2$

$= 2(\sqrt{2} - 1)^2 : (\sqrt{2} + 1)^2$

Hence, option (d) is correct.

37. The given triangle is a right angled triangle as the set of figures i.e. 7, 24 and 25, is a Pythagoras triplet.

Thus, area of the circle

$= \frac{1}{2} \times 24 \times 7 = 84 \text{ cm}^2$

38. Obviously, perimeter of the semi-circle = 36 cm

$\Rightarrow \quad r(\pi + 2) = 36$

$\Rightarrow \quad r = 36 \div (\pi + 2) = 7$ cm

39. Let the radius of each coin be r units.

Then, $\frac{\pi r^2}{2\pi r} = \frac{2\pi r}{\pi r^2}$

$\Rightarrow \quad r^2 = 4 \Rightarrow r = 2$

$\therefore$ Area of all four coins $= 4\pi r^2 = 16\pi$

Further, side of the square

$= (2 \times 2) + (2 \times 2) = 8$

Then, its area $= 8 \times 8 = 64$

Now, the required area $= 64 - 16\pi$

$= 16(4 - \pi)$

40. Clearly, area of the field

$= (130 \times 110) + 20350$

$= 34650 \text{ m}^2$

$\Rightarrow \quad \pi r^2 = 34650$

$\Rightarrow \quad r = 105$ m

Hence, option (c) is correct.

41. Area of the path = area of outer circle − area of inner circle

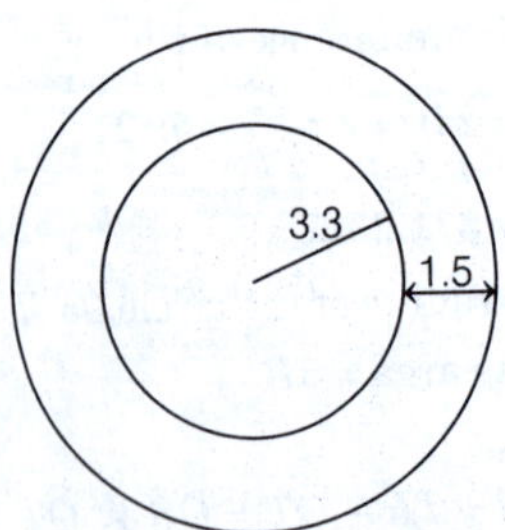

$\Rightarrow \pi(3.3 + 1.5)^2 - \pi(3.3)^2 = 12.15\,\pi \text{ m}^2$

42. Area of the sector $= \frac{1}{2} \times$ length of arc $\times$ radius

$= \frac{1}{2} \times 16 \times 10 = 80 \text{ cm}^2$

43. Here, radius for each sector

$= 14 \div 2 = 7$ m

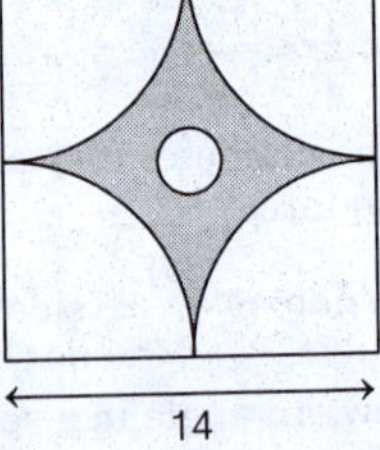

$\therefore$ Area grazed by them

$= 4 \times \frac{\pi}{4} \times 7 \times 7 = 154 \text{ m}^2$

Thus, ungrazed area

$= (14)^2 - 154 - 20 = 22$

$\therefore$ Ungrazed area = shaded region

$= 22 \text{ m}^2$

44. We know, area of the circular path

$= \pi(R^2 - r^2)$

$\Rightarrow 16\pi = \pi[(5x)^2 - (3x)^2]$

[let radii be $5x$ and $3x$]

$\Rightarrow \quad 16 = (5x + 3x)(5x - 3x)$

$\Rightarrow 16x^2 = 16$

$\therefore \quad x = 1$

Hence, radius of outer circle

$= 5 \times 1 = 5$ cm

Then, its area $= \pi R^2 = \pi(5)^2$

$= 25\pi \text{ cm}^2$

45. Since ΔAPB is a right-angled triangle, the diameter of the circle circumscribing this triangle will be hypotenuse of the triangle.

Hence, $AB = 2$ units

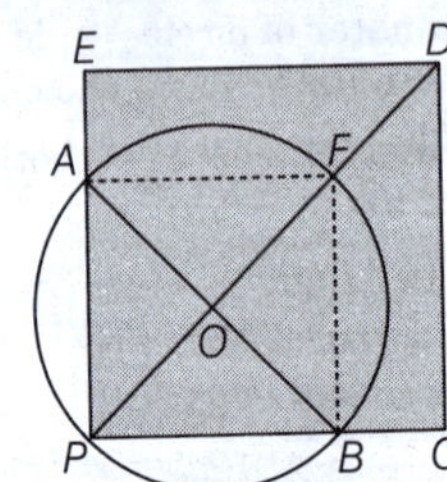

The area remains after punching = area of square − area of semi-circle AFB − area of ΔAPB

$= (2)^2 - \frac{\pi(1)^2}{2} - \frac{1}{2} \times AB \times OP$

$= 4 - \frac{\pi}{2} - \frac{1}{2} \times 2 \times 1$

[$\because OP \perp AB$ as $\angle APO = \angle BPO = 45°$ (diagonals of a square bisect vertices angles) and $\angle PAO = \angle PBO = 45°$ (angles opposite to equal sides are equal)]

$= \frac{6 - \pi}{2}$

$\therefore$ Required proportion

$$= \frac{\frac{6-\pi}{2}}{4} = \frac{6-\pi}{8}$$

46. Area of the part of the circle(round punch) falling outside the square sheet

= area of semi-circle APB – area of ΔAPB

$$= \frac{\pi(1)^2}{2} - \frac{1}{2} \times 2 \times 1 = \frac{\pi - 2}{2}$$

47. Required area = area of the circle with radius 8 m – area of the sector of the same circle with angle 30°

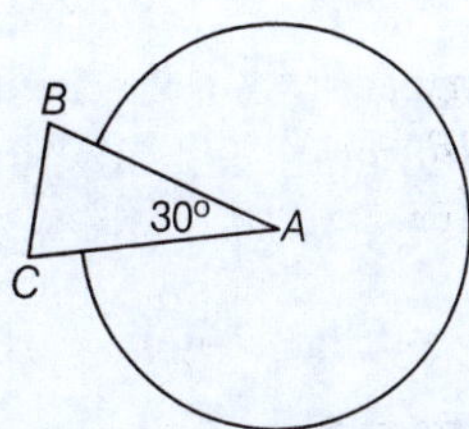

$$= 64\pi - \frac{30°}{360°} \times \pi \times 8 \times 8 = \frac{176\pi}{3} \text{ m}^2$$

48. On the above line, required area

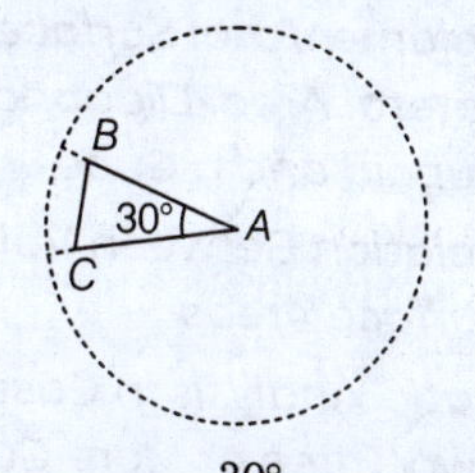

$$= 144\pi - \frac{30°}{360°} \times \pi \times 12 \times 12$$

$$= 132\pi \text{ m}^2$$

49. Effect on the area

$$= \left[+ 15 + 15 + \frac{15 \times 15}{100} \right] \%$$

$$= + 32.25\%$$

Hence, area of the circle will increase by 32.25%.

50. Clearly, $AO = OE = D$

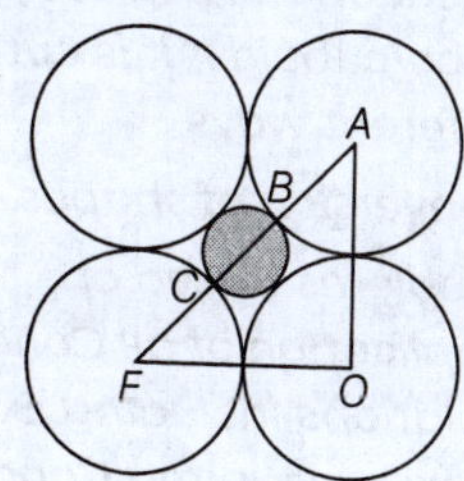

Then, $AE = \sqrt{D^2 + D^2} = D\sqrt{2}$

Now, $BC = AE - AB - CE$

$$= D\sqrt{2} - \frac{D}{2} - \frac{D}{2} = D(\sqrt{2} - 1)$$

Hence, diameter of the shaded circle $= D(\sqrt{2} - 1)$

51. Since all the given points are equidistant, line HD is the diameter of the circle.

Clearly, $\angle HFD = 90°$

{$\because$ an angle in a semi-circle is a right angle}

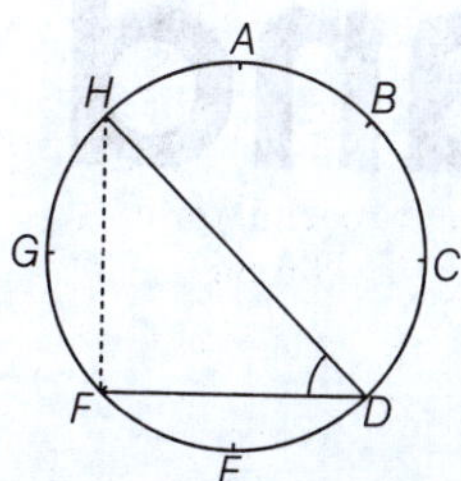

Now, in ΔHFD, $\angle FDH = \angle FHD$

[$\because FH = FD$]

and $\angle HFD = 90°$

Then, $90° + 2\angle FDH = 180°$

$\Rightarrow \quad \angle FDH = 45°$

52. Clearly, $EF = m\sqrt{2}$

[$\because \Delta EOF$ is a right-angled triangle]

$\Rightarrow$ side of the square $= m\sqrt{2}$

$$\Rightarrow \sqrt{\frac{5}{2}} \text{ cm} = m\sqrt{2} \Rightarrow m = \frac{\sqrt{5}}{2}$$

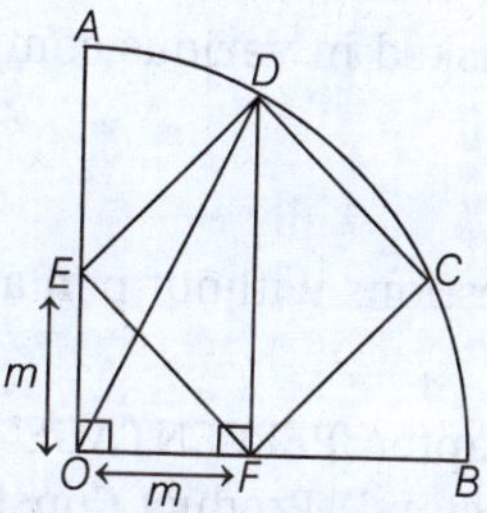

Further, diagonal of the square

$= EF\sqrt{2} = m\sqrt{2} \times \sqrt{2} = 2\text{ m}$

$\Rightarrow \quad DF = 2\text{ m}$

Now, $OD = \sqrt{DF^2 + OF^2}$

$= \sqrt{(2m)^2 + m^2} = m\sqrt{5}$

$$\Rightarrow \text{radius} = m\sqrt{5} = \frac{\sqrt{5}}{2} \times \sqrt{5} = 2.5$$

53. Area of semi-circle drawn with diameter $AB = \frac{1}{2}\pi\left(\frac{AB}{2}\right)^2$

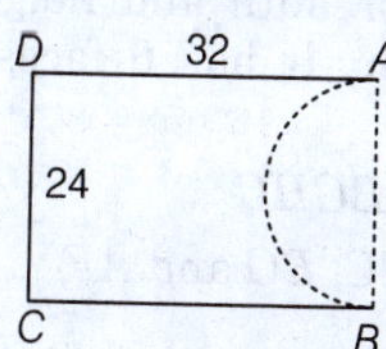

$$\Rightarrow \quad 72\pi = \frac{1}{2}\pi\left(\frac{AB}{2}\right)^2$$

$\Rightarrow \quad AB = 24$

Then, $24 \times AD = 768 \Rightarrow AD = 32$

Now, perimeter of the left over part

$= AD + DC + CB + \text{arc } AB$

$= 32 + 24 + 32 + \pi \cdot 12$

$= 88 + 12\pi$

54. Area of shaded portion = area of sector $ABFD$ + area of sector $CBED$ – area of square $ABCD$

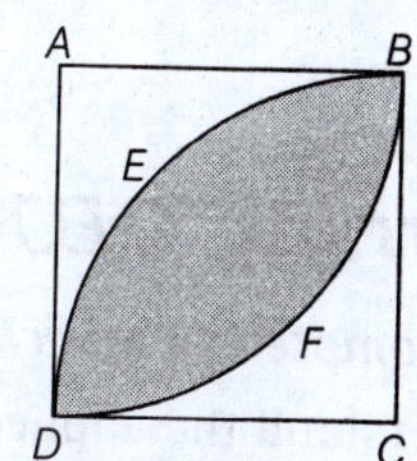

$$= \frac{\pi a^2 \times 90°}{360°} + \frac{\pi a^2 \times 90°}{360°} - a^2$$

$$= a^2\left(\frac{\pi}{2} - 1\right)$$

55. Clearly, diameter of the circle = side of the square

$\Rightarrow \quad 2r = a$

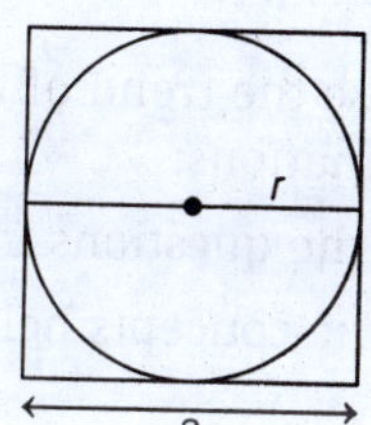

Since circle and square have been made by bending the wire, circumference of the circle + perimeter of the square = 12

$\Rightarrow \quad 2\pi r + 4a = 12$

$\Rightarrow \quad 2\pi r + 4(2r) = 12$

$\Rightarrow \quad r(\pi + 4) = 6$

$$\Rightarrow \quad r = \frac{6}{\pi + 4}$$

56. Area of shaded region = area of largest semi-circle having diameter 28 cm – area of two smallest semi-circles each having diameter 7 cm + area of the shaded semi-circle, shown in lower part, having diameter (28 – 7 – 7) cm

$$= \frac{1}{2}\pi(14)^2 - 2 \times \frac{1}{2}\pi(3.5)^2 + \frac{1}{2}\pi(7)^2$$

$= 98\pi - 12.25\pi + 24.5\pi$

$= 110.25\pi \text{ cm}^2$

44

Cuboid and Cube

Learning OBJECTIVES

After completing this chapter, aspirants will be able to :

- Understand the application of the cuboid and cube related formulae in different cases;
- Understand the application of combination of cuboid and cube related formulae in diverse cases;
- Compare the cuboids and cubes with reference to different ratios;
- Learn how the surface area is related with volume;
- Understand the philosophy of the divisions of this chapter in various parts;
- Analyse the trend of the questions being asked in various competitive examinations;
- Solve the questions in moments;
- Grip the concepts helping to solve the questions without pen and paper;
- Apply many concepts envisaged in the chapter 'PERCENTAGE' like 'effect on product in case of changes in factors', 'Product Constancy' etc.;
- Apply many concepts of 'Ratio' in this chapter;
- Learn many new concepts/fundamentals;
- Apply the pragmatic approaches in lieu of academic approaches; and
- Ponder on the alternative methods.

Inside

- *Understanding the Shapes With Derivation of Formulae of Volume, Total Surface Areas, Lateral Area, Diagonal of a Cuboid and a Cube*
- *Relation Between Volume and Surface Areas*
- *Ratio-Analysis in Case of Two/Three or More Cubes*
- *Application of the Concept of the Volume in Diverse Cases*
- *Alteration in Dimensions Affecting Volume/Surface Area*
- *Concept of Volume and Surface Area When Faces' Thickness is Considered*
- *Effect on the surface area of a cube/cuboid if it is cut in different ways*
- *Conversion of shapes*
- *Problems Based on Application of all Concepts/ Relations Including Based on Maxima/Minima Concepts*
- *..............................*
- *.............................. & Many More*
- *Work Out with Solutions*

CUBOID

Cuboid is a three dimensional solid like a room, matchbox, a tile, a book etc. which can be defined by virtue of its length, breadth and height. Length, breadth and height are not equal to each other. It has 6 faces, 12 edges, 8 vertices (corners) and 4 diagonals,

Faces *ABCD*, *EFGH*, *DEHC*, *AFGB*, *AFED* and *BGHC*

Edges *AB*, *BC*, *CD*, *DA*, *FG*, *GH*, *HE*, *EF*, *ED*, *HC*, *BG* and *AF*

Vertices *A*, *B*, *C*, *D*, *E*, *F*,*G* and *H*

Diagonals *DG*, *CF*, *BE* and *AH*

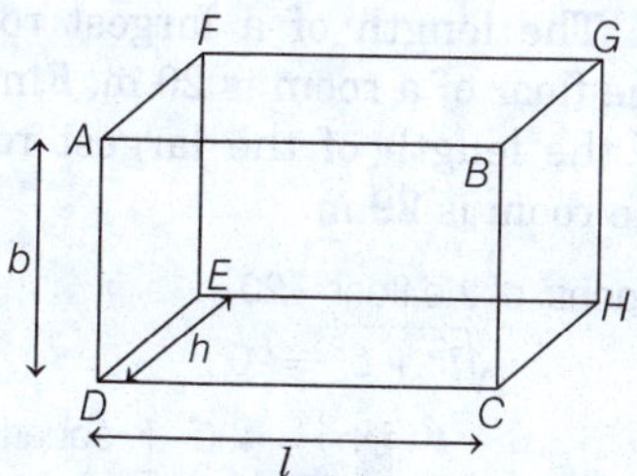

Volume Assume, a rectangular sheet $ABCD$ is being made thick by h units. Then, material/sheet required will be equal to the product of area of base and height. This material is called volume of the cuboid.

Thus, Volume = Area of base $ABCD \times$ Height(h)

$$= lb \times h = lbh \text{ cu. units}$$

The volume of a cuboid may be referred as the space occupied by it. The air existing in the room is its volume. Water filled up in the cuboidal pot is the volume of the pot. The material/bricks to construct a wall is the volume of the wall. It means volume is basically the capacity to hold something. This conception is applicable in all solid shapes like cylinder, cone, sphere, prism, pyramid, etc.

Diagonal Diagonal of a cuboid is a line joining one vertex of the base and the vertex just above its diagonally opposite vertex. If you have to place a rod of maximum length in a room, the length will be equal to diagonal of the cuboid/room.

In this figure, BD is a diagonal of rectangle $ABCD$,

i.e. $$BD = \sqrt{l^2 + b^2}$$

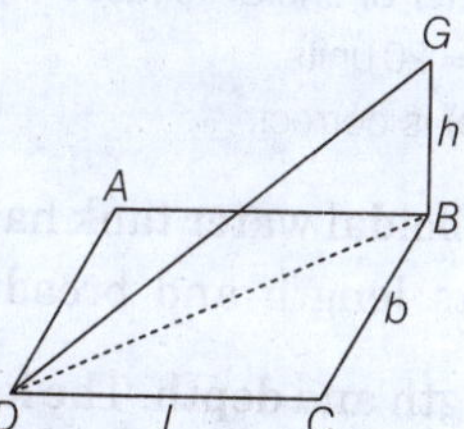

Now, in the right angled ΔBGD

$$(DG)^2 = (BD)^2 + (BG)^2 = (\sqrt{l^2 + b^2})^2 + h^2$$

$$\therefore \quad DG = \sqrt{l^2 + b^2 + h^2}$$

Hence, diagonal of a cuboid $= \sqrt{l^2 + b^2 + h^2}$ unit

Needless to say that all four diagonals of a cuboid are equal in length.

Total surface area There are six faces and the sum of areas of these faces is called total surface area of a cuboid.

It means total surface area = Area of $ABCD$ + Area of $EFGH$ + Area of $DEHC$ + Area of $AFGB$ + Area of $AFED$ + Area of $BGHC$

$$= lb + lb + lh + lh + bh + bh$$
$$= 2(lb + bh + hl) \text{ sq. units.}$$

Lateral surface area Suppose a wire has been bent to form a rectangular base $ABCD$. Another wire of the same length i.e. $2(l + b)$ is being placed on the previous. Further, 3rd wire and many more are being placed above those wires to raise by h units. The areas of faces, thus formed, is called lateral surface area. Hence, lateral surface area = perimeter of base × height. This concept is also applicable in case of cylinder and prism.

Thus, lateral surface area of the cuboid $= 2(l + b)h$

Alternatively

Lateral surface area

$$= \text{Total surface area} - (\text{base area} + \text{top area})$$
$$= 2(lb + bh + hl) - (lb + lb)$$
$$= 2(bh + hl)$$
$$= 2(l + b)h \text{ sq. units.}$$

Lateral surface area may be termed as areas of all four walls in case of a room.

Relation Between Volume and Surface Areas

If the areas of three adjacent faces of a cuboid are x, y, z square units respectively and the volume be v cu. units, then $v^2 = xyz$.

Reason $\quad v = lbh$

$\Rightarrow \quad v^2 = l^2b^2h^2$

$\Rightarrow \quad v^2 = lb \times bh \times hl = xyz$

i.e. (Volume)^2 = Product of areas of all three adjacent faces.

CUBE

Cube is a special type of the cuboid which has all its edges equal. i.e. length = breadth = height

Volume $= lbh = a \cdot a \cdot a \quad$ [let $l = b = h = a$ (side)]

$$= a^3 \text{ cu. units}$$

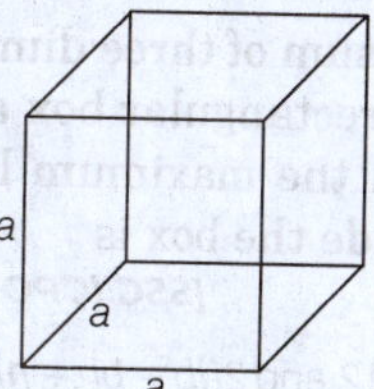

Diagonal $= \sqrt{l^2 + b^2 + h^2}$

$$= \sqrt{a^2 + a^2 + a^2}$$
$$= a\sqrt{3} \text{ units}$$

All four diagonals are equal in length.

Total Surface Area $= 2(lb + bh + hl)$

$$= 2(a \cdot a + a \cdot a + a \cdot a)$$
$$= 6a^2 \text{ sq. units}$$

Note *If the numbers representing volume and surface area of a cube are equal, the length of edge will always be 6 units. [explanation $a^3 = 6a^2 \Rightarrow a = 6$].*

Lateral surface area = Perimeter of the base × Height
$= 4a \times a = 4a^2$ sq. units

Note *8 equilateral triangles can be formed by joining any three vertices of a cube.*

Ratio Analysis in Case of Two/Three or More Cubes

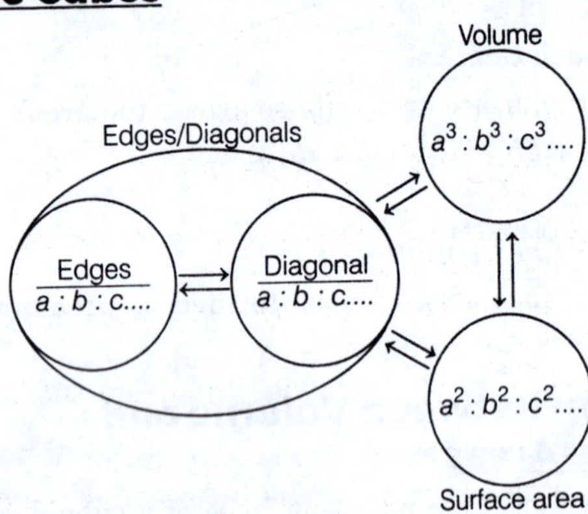

Problems based on the Diagonal and its Applications

Problem 1 Find the length of the largest rod that can be placed in a room 16 m long, 12 m broad and $10\frac{2}{3}$ m high.

[SSC (CGL-PT)-99,CDS-13,CDS-10, MAT-18]

Sol. We know that the diagonal has the maximum length in a cuboid.

$\therefore$ Diagonal $= \sqrt{l^2 + b^2 + h^2}$

$$= \sqrt{(16)^2 + (12)^2 + \left(10\frac{2}{3}\right)^2}$$

$$= \sqrt{\left(256 + 144 + \frac{1024}{9}\right)} = \sqrt{\frac{4624}{9}} = \frac{68}{9} = 22\frac{2}{3}\text{ m}$$

Problem 2 If the sum of three dimensions and the total surface area of a rectangular box are 12 cm and 94 cm^2 respectively, then the maximum length of a stick that can be placed inside the box is

[SSC (CPO-SI)-11, SSC (CGL-T-II)-19]

Sol. Given, $l + b + h = 12$ and $2(lb + bh + hl) = 94$

We know, $(l + b + h)^2 = l^2 + b^2 + h^2 + 2(lb + bh + hl)$

$\Rightarrow \quad (12)^2 = l^2 + b^2 + h^2 + 94$

$\Rightarrow \quad l^2 + b^2 + h^2 = 144 - 94 = 50$

$\Rightarrow \quad$ Diagonal $= \sqrt{50} = 5\sqrt{2}$ cm

Problem 3 The surface area of a cuboid is 21 cm^2 and sum of the length of all its edges is 24 cm. Length of each diagonal of the cuboid (in cm) is ***[LIC(AAO)-12]***

Sol. Given, sum of all edges = 24 cm

$\Rightarrow \quad 4(l + b + h) = 24 \Rightarrow l + b + h = 6$

Also given, $2(lb + bh + hl) = 21$

On the above pattern, diagonal $= \sqrt{15}$ cm

Problem 4 The length of a largest rod that can be placed on the floor of a room is 20 m. Find the height of the room if the length of the largest rod that can be placed in the room is 29 m.

Sol. Clearly, diagonal of the floor = 20

$\Rightarrow \quad \sqrt{l^2 + b^2} = 20$

$\Rightarrow \quad l^2 + b^2 = 400 \quad [\because$ squaring on both sides$]$

Further, diagonal of the room = 29

$\Rightarrow \quad \sqrt{l^2 + b^2 + h^2} = 29$

$\Rightarrow \quad l^2 + b^2 + h^2 = 841$

$\Rightarrow \quad 400 + h^2 = 841$

$\Rightarrow \quad h^2 = 441$

$\therefore \quad h = \sqrt{441} = 21$ m

Application of the Concept of the Volume in Diverse Cases

Problem 5 In which one of the following pairs does the 1st number represent the perimeter of one face of a cube and the 2nd number represent the volume of the cube? ***[CDS-08]***

(a) (16, 32) (b) (20, 125)
(c) (9, 27) (d) (10,100)

Sol. The volume refers to three dimensions and in case of the cube the volume will be cube of any number. Thus, options (a) and (d) can be eliminated. Now consider option (b).

Side of the cube $= \sqrt[3]{125} = 5$ units

Then, the perimeter of one of its faces = 4 × side

$= 4 \times 5 = 20$ units.

Hence, option (b) is correct.

Problem 6 A cuboidal water tank has 216 L of water. Its depth is $\frac{1}{3}$ of its length and breadth is $\frac{1}{2}$ of $\frac{1}{3}$ of the difference of length and depth. The length of the tank is

Sol. Given, Volume/capacity of the tank = 216 L

$\Rightarrow \quad l \times b \times h = 216 \times 1000 \text{ cm}^3 \quad [\because 1 \text{ L} = 1000 \text{ cm}^3]$

$\Rightarrow \quad [l] \times \left[\frac{1}{2} \times \frac{1}{3}(l - h)\right] \times \left[\frac{1}{3}l\right] = 216000$

$\Rightarrow \quad l \times \frac{1}{6}\left(l - \frac{l}{3}\right) \times \frac{l}{3} = 216000$

$\Rightarrow \quad l^3 = 216 \times 27 \times 1000$

$\Rightarrow \quad l = \sqrt[3]{216 \times 27 \times 1000} = 6 \times 3 \times 10$

$= 180 \text{ cm} = 18 \text{ dm} \quad [\because 1 \text{ dm} = 10 \text{ cm}]$

Problem 7 The volume (in cm^3) of rain water that can be collected from 1.5 hec. of ground in a rainfall of 5 cm is : ***[SSC (CGL-T-I)-11]***

Sol. Volume = Area of the ground × Desired height

$= (1.5 \times 10000 \text{ m}^2) \times \left(\frac{5}{100}\text{m}\right) = 750 \text{ m}^3$

$= 750 \times (100)^3 \text{ cm}^3.$

Problem 8 If the volume of two cubes are in ratio 27 : 1, the ratio of their edges is :

Sol. Observe the concept of "ratio analysis in case of two/three or more cubes". The required ratio = $\sqrt[3]{27}:\sqrt[3]{1} = 3:1$

Problem 9 A river 3 m deep and 40 m wide is flowing at the rate of 2 km/h. How much water (in L) will fall into the sea in a min ? **[SSC(CGL-T-I)-11]**

Sol. Volume of water flowed in an h = $3\text{m} \times 40\text{m} \times (2 \times 1000)$ m

$\therefore$ Volume of the water flowed in a min. $= \frac{3 \times 40 \times 2 \times 1000}{60}$

$= 4000\text{m}^3 = 4000 \times (100)^3 \text{cm}^3 = 4000000000 \text{ cm}^3$

$= 4000000$ L $\quad [\because 1 \text{ L} = 1000 \text{ cm}^3]$

Problem 10 The water in a rectangular reservoir having a base 80 m by 60 m is 6.5 m deep. In what time can the water be emptied by a pipe of which the cross section is a square of side 20 cm and the water runs through the pipe at the rate of 15 km/h ?

Sol. Clearly, volume of water running per hour

$= 0.2 \text{ m} \times 0.2 \text{ m} \times 15000 \text{ m} = 600 \text{ m}^3$

$\therefore$ Time required $= \frac{80 \times 60 \times 6.5}{600} = 52$ h

Problem 11 A rectangular tank measuring 5.5 m by 4 m by 2.5 m is dig in the middle of a field 35 m by 15.4 m. The earth dug out is spread evenly on the remaining portion of the field. How much is the level of field raised ?

Sol. The volume of the earth dug out = $5.5 \times 4 \times 2.5 = 55 \text{m}^3$

The area to be raised by spreading the earth dug out

$= (35 \times 15.4) - (5.5 \times 4) = 517 \text{ m}^2$

Let us assume that the level of the field be raised by h m on spreading the earth over it.

Then, $517 \times h = 55$

$\therefore \quad h = \frac{55}{517} \text{ m} = \frac{55 \times 100}{517} \text{ cm} = 10.64 \text{ cm}$

Problem 12 The surface of water in a swimming pool is a rectangle 32 m long and 9.5 m wide and the depth of water increases uniformly from 1.5 m at one end to 4.5 m at other end. What is the volume of water in the pool ? **[SSC(Asstt. Gr-mains) 96, XAT-20]**

Sol. Assume, face $DCFE$ as top/bottom face, then AD will act as the depth of the swimming pool.

Now, area of the surface $DCFE = \frac{1}{2} \times$

(Sum of parallel sides) $\times$ Height $\quad [\because DCFE$ is a trapezium$]$

$= \frac{1}{2} \times (1.5 + 4.5) \times 32 = 96 \text{ m}^2$

$\therefore$ Volume of the water in the swimming pool

= Area of the base $\times$ Height $= 96 \times 9.5 = 912 \text{ m}^3$

Alternatively

Point E and H may be joined.

Then, volume of cuboid with height 1.5 m

$= 32 \times 9.5 \times 1.5 = 456 \text{m}^3$

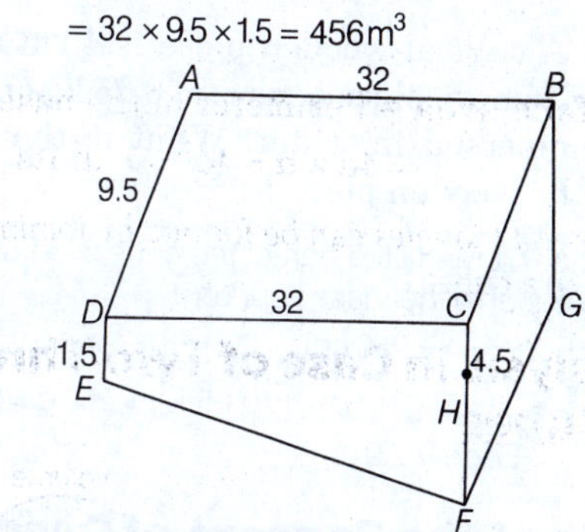

Now, ΔEHF is a right-angled triangle

Then, its area $= \frac{1}{2} \times 32 \times (4.5 - 1.5) = 48 \text{m}^2$

$\therefore$ Volume of water in triangular area = Area of triangle $\times$ corresponding height $= 48 \times 9.5 = 456 \text{ m}^3$

Hence, volume of water in the pool $= 456 + 456 = 912 \text{ m}^3$

Problem 13 $\frac{2}{3}$ rd of a cuboidal aquarium is filled with water and then tilted on its side until the water level coincides with one edge on the bottom and one edge on the top. During the tilting process, 6 L of water is poured out. What is the volume of the aquarium?

Sol. Let the volume of the aquarium be x L.

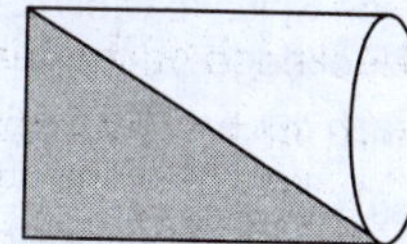

The volume of the water before tilting $= \frac{2x}{3}$

and volume of water after tilting $= \frac{x}{2}$

But given, $\frac{2x}{3} - \frac{x}{2} = 6 \Rightarrow x = 36$ L

Problem 14 A tank measuring 28 cm by 25 cm contains water to a depth of 7 cm. An iron cube of edge 11cm is placed inside. How much water must be added in order to just cover the cube.

Sol. Obviously, Volume of existing water + Required volume of water + Volume of the cube = Area of the base $\times$ Height

$\Rightarrow (28 \times 25 \times 7)$ + Required volume of water $+ (11)^3 = 28 \times 25 \times 11$

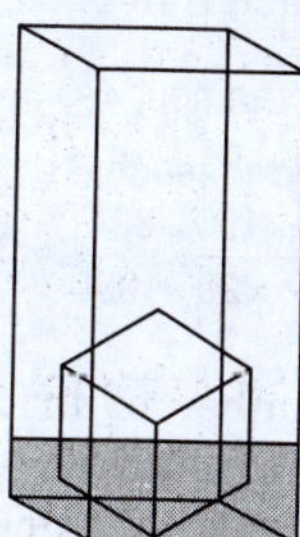

$\Rightarrow 4400 +$ Required volume of water $+ 1331 = 7700$

$\therefore \quad$ Required volume of water $= 1469 \text{ cm}^3 = 1.47$ L

Problem 15 A cubical wooden piece is 4 cm above the water level. When we place a mass of 0.25 kg on it, it gets completely immersed in water. What is the length of each side of the wooden piece ?

Sol. Obviously, the volume of the part of the cube seen outside the water will be equal to the mass kept on it.

$\Rightarrow$ Edge × Edge × 4 = 250 cm^3

[$\because$ 0.25 kg = 250 g = 250 cu. cm]

$\Rightarrow$ Edge = 7.9 cm

Application of the Concept of Dividing One Volume by Another Volume

Problem 16 How many cubes, each of edge 3 cm, can be cut from a cube of edge 15 cm?

Sol. Number of small cubes $= \dfrac{\text{Volume of bigger cube}}{\text{Volume of smaller cube}}$

$$= \frac{(15)^3}{(3)^3} = 125$$

Problem 17 A cuboidal block of 6 cm × 9 cm × 12 cm is cut up into exact number of equal cubes. The least possible number of cubes will be

Sol. Since we require least number of cubes, we will find the greatest side of the cube which may divide the edges of the cuboid exactly i.e. HCF of edges of the cuboid.

It means side of the each cube = HCF of 6, 9 and 12 = 3

Now, number of cubes $= \dfrac{\text{Volume of the cuboid}}{\text{Volume of one cube}}$

$$= \frac{6 \times 9 \times 12}{3 \times 3 \times 3} = 24$$

Problem 18 A soap cake is of size 8 cm × 5 cm × 4 cm. The number of such soap cakes that can be packed in a box measuring 56 cm × 35 cm × 28 cm is

Sol. Required number of soap cakes $= \dfrac{\text{Volume of the box}}{\text{Volume of one soap cake}}$

$$= \frac{56 \times 35 \times 28}{8 \times 5 \times 4} = 343$$

Problem 19 Water flows into a tank which is 200 m long and 150 m wide, through a pipe of cross-section 0.3 m × 0.2 m at 20 km/h. Then, the time (in hours) for the water level in the tank to reach 8 m is : *[SSC(CGL-T-I)-11]*

Sol. Volume of water flows in an hour = $0.3 \times 0.2 \times (20 \times 1000)$ m^3

Volume of water required to reach 8 m = $200 \times 150 \times 8$

$\therefore$ Time taken $= \dfrac{200 \times 150 \times 8}{0.3 \times 0.2 \times 20 \times 1000} = 200$ h

Problem 20 The number of bricks, each measuring 25 cm × 12.5 cm × 7.5 cm, required to construct a wall 6 m long, 5 m high and 0.5 m thick, while the mortar occupies 5% of the volume of the wall, is : *[MAT-08]*

Sol. Volume of each brick = $(25 \times 12.5 \times 7.5)$ cm^3

Now, volume of wall excluding mortar

= 95% of $(600 \times 500 \times 50)$ cm^3

$\therefore$ Number of bricks $= \dfrac{\frac{95}{100} \times 600 \times 500 \times 50}{25 \times 12.5 \times 7.5} = 6080$

Problem 21 The internal dimensions of a wooden box are 80 cm, 60 cm and 40 cm. Find the number of cubes each of side 9 cm which can be placed in the box with faces parallel to the sides of the box.

Sol. Since length is 80 cm, only 8 cubes can be placed along the length $\left[\dfrac{80}{9} = 8\dfrac{8}{9}\right]$ and 6 cubes can be placed along the breadth $\left[\dfrac{60}{9} = 6\dfrac{6}{9}\right]$ and 4 cubes can be placed along the height $\left[\dfrac{40}{9} = 4\dfrac{4}{9}\right]$.

Thus, number of cubes that can be placed in the box

$= 8 \times 6 \times 4 = 192$

Alternatively

Actual length used = Multiple of 9 just below the given length

= 72

Similarly, actual breadth and actual height used are 54 cm and 36 cm respectively.

Hence, volume = $(72 \times 54 \times 36)$ cm^3

and volume of each cube = $(9)^3$ cm^3

Thus, number of cubes $= \dfrac{72 \times 54 \times 36}{(9)^3} = 192$

Problem Based on Relation Between Diagonal and Volume

Problem 22 If diagonal of a cube is $\sqrt{12}$ cm, then its volume in cubic cm is *[SSC (CGL-PT)-99, SSC (CGL-T-II)-10]*

Sol. Diagonal of the cube = $\sqrt{3} \cdot$ Side

$\Rightarrow$ $\sqrt{3} \cdot \text{Side} = \sqrt{12}$

$\therefore$ Side $= \dfrac{\sqrt{12}}{\sqrt{3}} = 2$

Then, volume = $(\text{side})^3 = (2)^3 = 8$ cm^3

Problem 23 Three cubes with sides in the ratio 3 : 4 : 5 are melted to form a single cube whose diagonal is $12\sqrt{3}$ cm. The sides of the cubes are

Sol. Let sides be $3x$, $4x$ and $5x$.

Sum of volumes of all three small cubes = Volume of new cube

$\Rightarrow$ $(3x)^3 + (4x)^3 + (5x)^3$ = Volume of new cube

$\Rightarrow$ $216x^3 = (\text{side})^3 \Rightarrow$ Side

But given, diagonal of the new cube = $12\sqrt{3}$

$\Rightarrow$ $\sqrt{3} \times \text{side} = 12\sqrt{3}$

$\Rightarrow$ $\sqrt{3} \times 6x = 12\sqrt{3}$

$x = 2$

Hence, sides are 6 cm, 8 cm and 10 cm.

Problem 24 The diagonals of the three faces of a cuboid are p, q and r respectively. What is the volume of the cuboid ?

Sol. Given, $p^2 = l^2 + b^2, q^2 = b^2 + h^2$ and $r^2 = h^2 + l^2$

On adding all three we get, $2(l^2 + b^2 + h^2) = p^2 + q^2 + r^2$

$$\Rightarrow \quad (l^2 + b^2 + h^2) = \frac{1}{2}(p^2 + q^2 + r^2)$$

$$\Rightarrow \quad (p^2 + h^2) = \frac{1}{2}(p^2 + q^2 + r^2) \; [\because l^2 + b^2 = p^2, \text{given}]$$

$$\Rightarrow \quad h^2 = \frac{1}{2}(q^2 + r^2 - p^2)$$

$$\therefore \quad h = \sqrt{\frac{1}{2}(q^2 + r^2 - p^2)}$$

Similarly, $l = \sqrt{\frac{1}{2}(p^2 + r^2 - q^2)}$

and $b = \sqrt{\frac{1}{2}(p^2 + q^2 - r^2)}$

Now, volume = lbh

$$= \frac{1}{2\sqrt{2}}\sqrt{(q^2 + r^2 - p^2)(p^2 + r^2 - q^2)(p^2 + q^2 - r^2)}$$

Problems Based on Relationship Between Volume and Surface Area

Problem 25 The volume of a cube is numerically equal to sum of its edges. What is the total surface area in sq. units ? *[MAT-99, CDS-12]*

Sol. Given, $(\text{side})^3 = 12 \cdot \text{side} \Rightarrow (\text{side})^2 = 12$

Then, total surface area = $6 \cdot (\text{side})^2 = 6 \cdot 12 = 72$ sq. units

Problem 26 The volume of a cuboid is twice the volume of a cube. If the dimensions of the cuboid are 9 cm, 8 cm and 6 cm, the total surface area of the cube is : *[SSC (CGL-PT)-02, MAT-08]*

Sol. Given, volume of the cube = $\frac{1}{2} \times$ volume of the cuboid

$$\Rightarrow \quad \text{Volume of the cube} = \frac{1}{2} \times 9 \times 8 \times 6 = 216$$

$$\Rightarrow \quad \text{Side of the cube} = \sqrt[3]{216} = 6 \text{ cm}$$

Now, total surface area of the cube = $6 \cdot (\text{side})^2 = 6 \cdot (6)^2 = 216 \text{ cm}^2$

Problem 27 The edges of a cuboid are in the ratio 1 : 2 : 3 and its surface area is 88 cm^2. The volume of the cuboid is : *[SSC (CGL-PT)-99, SSC (CGL-T-I)-19]*

Sol. Let $l = 3x$, $b = 2x$ and $h = x$

Now, surface area = 88 (given)

$$\Rightarrow \quad 2(lb + bh + hl) = 88$$

$$\Rightarrow (3x \times 2x) + (2x \times x) + (x \times 3x) = 44$$

$$\Rightarrow \quad 11x^2 = 44$$

$$\Rightarrow \quad x = 2$$

Hence, $l = 3 \times 2 = 6$ cm, $b = 4$ cm and $h = 2$ cm

$\therefore$ Volume = $lbh = 6 \times 4 \times 2 = 48 \text{ cm}^3$

Problem 28 A cube of side 1 m is reduced 3 times in the ratio 2 : 1. The area of one face of the reduced cube to that of the original cube is in the ratio

Sol. Clearly, the length of each side of the cube after three times reduction = 1/8 m

Obviously, the required ratio will be the ratio of squares of their edges i.e. $\left(\frac{1}{8}\right)^2 : 1^2 = 1 : 64$

Problem 29 The areas of three adjacent faces of a cuboid are x, y, z sq. units respectively. If the volume of the cuboid be v cu. units, then the correct relation between v, x, y, z is

(a) $v^2 = xyz$ (b) $v^2 = x^3y^3z^3$

(c) $v^3 = x^2y^2z^2$ (d) $v^3 = xyz$

Sol. (a) Observe the details envisaged under heading 'Relation between volume and surface areas' given in beginning of this chapter.

Problem 30 The areas of three consecutive faces of a cuboid are 12 cm^2, 20 cm^2 and 15 cm^2, then the volume (in cm^3) of the cuboid is *[SSC (CGL-T-I)-11]*

Sol. We know, $(\text{volume})^2$ = product of areas of all three adjacent faces

$$\therefore \text{Volume} = \sqrt{12 \times 20 \times 15}$$

$$= \sqrt{3 \times 4 \times 4 \times 5 \times 3 \times 5} = 60 \text{ cm}^3$$

Problem 31 A hall is 15 m long and 12 m broad. If the sum of the areas of the floor and the ceiling is equal to the sum of the areas of the four walls, the volume of the hall in cubic metres is

Sol. As given, Area of the floor + Area of the ceiling = Lateral area

$$\Rightarrow \quad lb + lb = 2(l + b) \times h$$

$$\Rightarrow \quad 2\,lb = 2\,(15 + 12) \times h$$

$$\Rightarrow \quad 15 \times 12 = 27 \times h$$

$$\Rightarrow \quad h = \frac{20}{3}$$

$$\therefore \text{Volume} = lbh = 15 \times 12 \times \frac{20}{3} = 1200 \text{ m}^3$$

Problem 32 If the volumes of two cubes are in the ratio 27 : 64, then the ratio of their total surface areas is

Sol. Observe the details of 'Ratio analysis in case of two/three or more cubes' given in beginning.

Given, ratio of volumes = $27 : 64 = (3)^3 : (4)^3$

$\therefore$ Ratio of total surface areas = $(3)^2 : (4)^2 = 9 : 16$

Problem 33 The inner surface area of walls and the height of a room are 300 m^2 and 5 m respectively. Find its breadth if it accommodates 200 people, each requires 5 m^3 space.

Sol. Clearly, volume of the room = $200 \times 5 = 1000 \text{ m}^3$ as one person requires 5 m^3.

Also given, lateral area = 300 m^2

$\Rightarrow \quad 2(l + b) \times h = 300$

$\Rightarrow \quad 2(l + b) \times 5 = 300$

$\Rightarrow \quad l + b = 30$

Option (a) is correct as the remaining options are not possible. The reason is that the sum of length and breadth is less than the value of the breadth given in other options.

Problem 34 The length, breadth and height of a rectangular parallelepiped are in ratio 6 : 3 : 1. If the surface area of a cube is equal to the surface area of this parallelepiped, then what is the ratio of the volume of the cube to the volume of the parallelepiped ?

Sol. Let the side of a cube be a.

Given, surface area of the cube

= surface area of the parallelepiped

$\Rightarrow \quad 6a^2 = 2(lb + bh + hl)$

$= [2(6x \cdot 3x + 3x \cdot x + x \cdot 6x)]$

$\Rightarrow \quad 6a^2 = 54x^2$

$\Rightarrow \quad a = 3x$

Now, required ratio = (side)^3 : lbh

$= (a)^3 : (6x \cdot 3x \cdot x) = (3x)^3 : (6x \cdot 3x \cdot x) = 3 : 2$

Alteration in Dimensions Affecting Volume and Surface areas

- **If alteration is reflected in terms of times**

1. If length, breadth and height of a cuboid are made x, y and z times respectively, the volume will become xyz times the earlier.
2. If each side/edge of a cube is made x times, its volume will become x^3 times the earlier.
3. If each side/edge of a cube is made x times, its total surface area will become x^2 times the earlier. Also, if each dimension of a cuboid is made x times, its total surface area will become x^2 times the earlier.

- **If alteration is reflected in terms of percentage**

4. Recall the **'concept of effect on product in case of changes in factors'** detailed in chapter percentage if dimensions of a cuboid/cube are changed by percentage. It should be kept in mind that the volume refers to three dimensions and the area does two. In case of volume of a cuboid, length is multiplied by breadth and further by height, whereas in case of surface area, any two out of three dimensions are multiplied. Needless to say that in case of cube, all the three dimensions are equal. The 'concept of effect on product in case of changes in factors' will not be applicable in case of total surface area/lateral area of a cuboid as they are a sum of certain products.

- **If alteration is reflected in terms of ratio**

5. If the dimensions of a cuboid are changed in ratios (say, length in ratio $a : b$, breadth in ratio $c : d$ and height in ratio $e : f$), then its volume will be changed in ratio $ace : bdf$.
   (a) If $ace > bdf$, the volume diminishes.
   (b) If $ace < bdf$, the volume increases.
   (c) If $ace = bdf$, no changes in the volume.
6. If the side/edge of a cube is changed in ratio $a : b$, its volume will be changed in ratio $a^3 : b^3$. The volume will decrease if $a^3 > b^3$, will increase if $a^3 < b^3$ and will remain unchanged if $a^3 = b^3$.
7. If dimensions are changed in ratios and you are required to compute the changes in volume in percentage term, you should do as below.

   Percentage change in volume of a cuboid

   $$= \left(\frac{ace - bdf}{ace} \times 100\right)\%$$

   [If the outcome is positive, volume diminishes and in case of negative it increases].

   Similarly, percentage change in volume of a cube

   $$= \left(\frac{a^3 - b^3}{a^3} \times 100\right)\%$$

   [if the outcome is positive, volume diminishes and in case of negative, it increases].
8. If the side/edge of a cube is changed in ratio $a : b$, its surface area will be changed in ratio $a^2 : b^2$. The surface area will decrease if $a^2 > b^2$, will increase if $a^2 < b^2$ and will remain unchanged if $a^2 = b^2$. If we intend to compute the changes in surface area in percentage term, the required percentage change will be $\left(\frac{a^2 - b^2}{a^2} \times 100\right)\%$ [if the outcome is positive, surface area diminishes and in case of negative, it increases].
9. The observations given at point (8) will also be applicable for the surface area of a cuboid if its each dimension is changed in ratio $a : b$.

   Now, we will observe some illustrations as below.

Problem 35 The length, breadth and height of a cuboid are in the ratio 1 : 2 : 3. If they are increased by 100%, 200% and 200% respectively, then compared to the original volume the increase in the volume of the cuboid will be

Sol. Length = 1 unit $\Rightarrow$ after increasing it becomes 2 units

Breadth = 2 units $\Rightarrow$ after increasing it becomes 6 units

Height = 3 units $\Rightarrow$ after increasing it becomes 9 units

Original volume = $1 \times 2 \times 3 = 6$ cu. units and it is increased to $2 \times 6 \times 9$ i.e. 108 cu. units.

i.e. increase in volume = $108 - 6 = 102$ (unit)^3

It means the increase in volume is 17(= 102 / 6) times as compared to the original volume.

Alternatively I

Recall the concept of "Effect on product in case of changes in factors" detailed in chapter 'percentage'.

Effect on taking any two figures (say 100 and 200)

$$\Rightarrow \left\{+100+200+\frac{100\times200}{100}\right\}\% = +500\%$$

Now, final effect $= \left\{+500+200+\frac{500\times200}{100}\right\}\% = +1700\%$

It means, in compared to the original volume the increase in the volume of the cuboid will be 17 times.

Alternatively II

We know that if length, breadth and height of a cuboid are made x, y and z times respectively, the volume will become xyz times the earlier. Here, length, breadth and height have become twice, thrice and thrice respectively the earlier. Then, volume will become $(2\times3\times3)$ times i.e. 18 times.

Hence, increase in volume is 17 times the original volume.

Alternatively III

Clearly, length, breadth and height have been changed in ratio 1 : 2, 1 : 3 and 1 : 3 respectively. Then, change in volume in ratio $= (1\times1\times1) : (2\times3\times3)$ i.e. 1 : 18.

Hence, increase in volume is 17 times the original volume.

Problem 36 If each edge of a cube is increased by 20%, then the percentage increase in its surface area is

[SSC (CGL-T-II)-13]

Sol. Since the surface area is concerned with two dimensional aspects, effect on the surface area

$$= \left\{+20+20+\frac{20\times20}{100}\right\}\% = +\ 44\%$$

Hence, the percentage increase = 44%

Alternatively I

If each side/edge of a cube is made x times, its surface area will become x^2 times the earlier (observe the facts given above). Since each edge is made to 1.2 times, the surface area will become $(1\cdot2)^2$ times the earlier i.e. 1.44 times.

Hence, increase in percentage is 44%.

Alternatively II

It may be said that each edge of the cube has increased in ratio 100 : 120 i.e. 5 : 6. Also, we know that if the side/edge of a cube is changed in ratio $a : b$, its surface area is changed in ratio $a^2 : b^2$ i.e. $5^2 : 6^2$.

Now, percentage change in surface area $= \left\{\frac{a^2-b^2}{a^2}\times100\right\}\%$

$$= \left\{\frac{25-36}{25}\times100\right\}\%$$

$$= -44\%$$

'–' indicates that surface area has increased and the increment is 44%.

Problem 37 If each edge of a cube is changed in ratio 4 : 5, then the percentage increase in its surface area is

Sol. It may be said that each edge of the cube has been increased from 4 units to 5 units i.e. increment of 25%.

Now, effect on the surface area

$$= \left\{+25+25+\frac{25\times25}{100}\right\}\% = 56.25\%$$

Hence, the percentage increase in surface area

= 56.25%.

Alternatively I

If the side/edge of a cube is changed in ratio $a : b$, its surface area will be changed in ratio $a^2 : b^2$ i.e. $4^2 : 5^2$.

Now, percentage change in surface area $= \left\{\frac{a^2-b^2}{a^2}\times100\right\}\%$

$$= \left\{\frac{16-25}{16}\times100\right\}\% = -56\cdot25\%$$

'–' indicates that surface area has been increased by 56.25%.

Alternatively II

Each edge of the cube has been increased from 4 units to 5 units. It implies that it has become 1.25 times the earlier. We know that if each side/edge of a cube is made x times, its total surface area will become x^2 times the earlier.

Here, the surface area will become $(1.25)^2$ times i.e. 1.5625 times. Hence, percentage increase in surface area = 56.25%.

Note *It may be construed from the discussion made through the above three problems that all the techniques are correlated and aspirants are supposed to be well versed with all these techniques so that they apply them appropriately observing the nature of questions.*

Problem 38 If the dimensions of a cuboid have been changed in ratio 2 : 3, 3 : 7 and 5 : 3 respectively, the percentage change in its volume is

Sol. Clearly, 1st dimension is increased from 2 units to 3 units i.e. increment of 50%. 2nd dimension is increased by 133.33% and 3rd has been reduced by 40%.

Then, effect on taking any two figures (say 50 and 40)

$$\Rightarrow \left(+50-40-\frac{50\times40}{100}\right)\% = -10\%$$

Then, final effect $= \left(-10+133.33-\frac{10\times133.33}{100}\right)\% = +110\%$

Hence, the new volume will increase by 110%.

Alternatively

Percentage change in volume $= \left\{\frac{ace-bdf}{ace}\times100\right\}\%$

$$= \left\{\frac{(2\times3\times5)-(3\times7\times3)}{(2\times3\times5)}\times100\right\}\%$$

$$= -110\%$$

'–' indicates that volume has increased and the increment is 110%.

Concept of Volume/Surface Area When Thickness of Faces is Taken Into Consideration

It is stressed that in case of cuboidal/cubical wooden box or other such items, the thickness of the wood/other metals etc. is taken into consideration while computing volumes. The volume of the wood/metal/material used will be equal to the difference between their external and internal volumes. The illustrations given below will emphasise manifestly on applications of these conceptions. Further, the computation of the surface areas considering the thickness of the faces varies on case to case basis.

Problem 39 A wooden box measures 20 cm by 12 cm by 10 cm. Thickness of wood is 1 cm. Volume of wood to make the box (in cu. cm) is :

Sol. External measurements of the box are given as

$l = 20, b = 12$ and $h = 10$

$\therefore$ External volume $= 20 \times 12 \times 10 = 2400 \text{ cm}^3$

Also, thickness of wood = 1 cm (as given), then

internal length $= 20 - (1 + 1) = 18$ cm

internal breadth $= 12 - (1 + 1) = 10$ cm

and internal height $= 10 - (1 + 1) = 8$ cm

Thus, internal volume $= 18 \times 10 \times 8 = 1440 \text{ cm}^3$

Hence, volume of the wood $= 2400 - 1440 = 960 \text{ cm}^3$

Problem 40 The external dimensions of a wooden box closed at both ends are 50 cm × 40 cm × 30 cm. Thickness of wood is 2.5 cm. Find the weight of 1 cu. cm of wood, if the weight of the empty box is 2062.5 kg.

Sol. Here, external volume $= 50 \times 40 \times 30 = 60000 \text{ cm}^3$

Then, internal volume

$= [50 - (2 \times 2.5)] \times [40 - (2 \times 2.5)] \times [30 - (2 \times 2.5)]$

$= 45 \times 35 \times 25 = 39375 \text{ cm}^3$

$\therefore$ Volume of the wood $= 60000 - 39375 = 20625 \text{ cm}^3$

Now, total weight of the wood/box $= 20625 \times$ weight of 1 cm^3

$\Rightarrow \quad 2062.5 = 20625 \times$ weight of 1 cm^3

$\Rightarrow$ Weight of $1 \text{ cm}^3 = \frac{2062.5}{20625} = \frac{1}{10}$ kg $= 100$ g

Problem 41 What is the total surface area of a one-side open cubical box of outer side of length 5 cm and thickness 0.5 cm?

Sol. Surface area of the outer cubical box $= 5 \cdot (\text{side})^2$

$= 5 \cdot (5)^2 = 125 \text{ cm}^2$

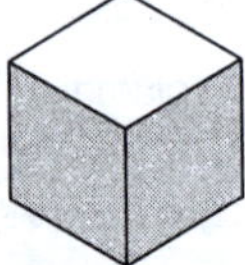

and surface area of inner cubical box

= Lateral area (inner) + Area of the bottom (inner)

$= 4[(5 - 0.5) \times 4] + 4 \cdot 4$

$= 72 + 16 = 88 \text{ cm}^2$

and surface area of upper portion

$= (4 \times 4 \times 0.5) + (4 \times 0.5 \times 0.5) = 9 \text{ cm}^2$

Thus, total surface area $= 125 + 88 + 9 = 222 \text{ cm}^2$

Problem 42 The length, breadth and height of a wooden box with a lid are 10 cm, 9 cm and 7 cm respectively. The total inner surface of the closed box is 262 cm². The thickness of the wood (in cm.) is : *[SSC (CGL-T-II)-14]*

Sol. Let the thickness of the wood is $\frac{a}{2}$ cm.

Then, inner surface area $= 2[\{(10 - a) \times (9 - a)\} + \{(9 - a) \times (7 - a)\} + \{(7 - a) \times (10 - a)\}] = 262$

$\Rightarrow 2[90 - 19a + a^2 + 63 - 16a + a^2 + 70 - 17a + a^2] = 262$

$\Rightarrow \quad 223 - 52a + 3a^2 = 131$

$\Rightarrow \quad 3a^2 - 52a + 92 = 0$

$\Rightarrow \quad 3a^2 - 46a - 6a + 92 = 0$

$\Rightarrow \quad a(3a - 46) - 2(3a - 46) = 0$

$\Rightarrow \quad a = 2$ and $\frac{46}{3}$

$\therefore \quad \frac{a}{2} = 1$ cm and $\frac{23}{3}$ cm

Application of the Concept of Surface Areas (Total/Lateral)

Problem 43 A cistern 6 m long and 4 m wide, contains water up to a depth of 1 m 25 cm. The total area of the wet surface is : *[SSC (CGL-PT)-04, CDS-11]*

Sol. Clearly, the wet surface area = Lateral surface area upto the height of 1.25 m + Area of the bottom of the cistern

$= [2(l + b) \times h] + (l \times b)$

$= [2(6 + 4) \times 1.25] + (6 \times 4)$

$= 25 + 24 = 49 \text{ m}^2$

Problem 44 The perimeter of the floor of a room is 18 m. What is the area of the walls of the room, if the height of the room is 3 m ?

Sol. Area of the walls = Lateral surface area

= Perimeter of base × Height

$= 18 \times 3 = 54 \text{ m}^2$

Problem 45 The floor of a room is of size 4 m × 3 m and its height is 3 m. The walls and ceiling of the room require painting. The area to be painted is : *[SSC CGL-T-I)-11]*

Sol. The area to be painted = Lateral area + Area of the roof or (total surface area – area of the floor)

$= [2(4 + 3) \times 3] + [4 \times 3] = 54 \text{ m}^2$

Problem 46 If the length of the diagonal of a cube is $8\sqrt{3}$ cm, then its surface area is : *[FCI(Asstt. Gr-III) 10-12]*

Sol. Given, diagonal $= 8\sqrt{3}$

$\Rightarrow \quad \sqrt{3} \times \text{side} = 8\sqrt{3}$

$\Rightarrow \quad \text{Side} = 8$

Now, its surface area $= 6 \times (\text{side})^2 = 6 \cdot (8)^2 = 384 \text{ cm}^2$

Problem 47 The area of the four walls of a room is 660 m² and its length is twice its breadth. If the height of the room is 11 m, then area of its floor (in m²) is :

Sol. Given, lateral area = 660

$\Rightarrow \quad 2(l + b) \times h = 660$

$\Rightarrow \quad 2(2b + b) \times 11 = 660 \quad [\because l = 2b]$

$\Rightarrow \quad 3b = 30$

$\therefore \quad b = 10$

Then, $\quad l = 20$

$\therefore$ Area $= lb = 20 \times 10 = 200 \text{ m}^2$

Problem 48 The four walls of a room can be fully covered by 70 square wall papers of 2 m × 2 m size. The length of the room is 18 m and its breadth is twice that of its height. If the cost of carpeting is ₹ 20 per sq. m, what will be the total expenditure in carpeting the room ?

Sol. Given, lateral surface area = 70 (2 ×2) i.e. areas of wall papers.

$\Rightarrow \quad 2(l+b)\times h = 280$

$\Rightarrow \quad 2(18+2h)\times h = 280$

$\Rightarrow \quad 36h + 4h^2 = 280$

$\Rightarrow \quad h^2 + 9h - 70 = 0$

$\Rightarrow \quad h^2 + 14h - 5h - 70 = 0$

$\Rightarrow \quad h = 5 \text{ or } -14$

Negative height is not possible, $\therefore b = 10$

Hence, the area which is to be carpeted = area of the floor = lb

$= 18\times 10 = 180\text{ m}^2$

$\therefore$ Total expenditure at ₹ 20 per m^2 = 180 × 20 = ₹ 3600

Problem 49 The height of a room is 40% of its semi-perimeter. It costs ₹ 260 to paper the walls of the room with paper 50 cm wide at the rate of ₹ 2 per m allowing an area of 15 m^2 for doors and windows. The height of the room is

Sol. Let the length, breadth and height of the room be l, b and h (all be in m), respectively.

As given, $\quad h = 0.4(l+b)$

Now, area of the walls $= 2(l+b)h = 2(l+b)\times 0.4(l+b)$

$= 0.8\,(l+b)^2$

Clearly, the area of the paper $= 0.8\,(l+b)^2 - 15$

$\Rightarrow \quad \text{Length}\times\text{Breadth} = 0.8\,(l+b)^2 - 15$

$\Rightarrow \quad (260\div 2)\times\frac{1}{2} = 0.8\,(l+b)^2 - 15$

$\Rightarrow \quad l+b = 10$

Then, $\quad h = 0.4\times 10 = 4\text{ m}$

Problem 50 The sum of length, breadth and height of a cuboid is 40 cm and its diagonal is $17\sqrt{2}$ cm. Find the surface area.

Sol. Given, diagonal of the cuboid = $17\sqrt{2}$ cm

$\Rightarrow \quad \sqrt{l^2+b^2+h^2} = 17\sqrt{2}$

$\Rightarrow \quad l^2+b^2+h^2 = 578$ [squaring both sides]

Further, $\quad l+b+h = 40$ [given]

$\Rightarrow \quad (l+b+h)^2 = (40)^2$ [squaring both sides]

$\Rightarrow l^2+b^2+h^2+2(lb+bh+hl) = 1600$

$\Rightarrow \quad 578 + 2(lb+bh+hl) = 1600$

$\Rightarrow$ Total surface area $= 1600 - 578 = 1022\text{ cm}^2$

Problem 51 If the ratio of diagonals of three cubes is 2 : 3 : 4, what is the ratio of their surface areas :

Sol. Observe the concept of "ratio analysis in case of two/three or more cubes" detailed in beginning.

Now, required ratio $= (2)^2 : (3)^2 : (4)^2 = 4 : 9 : 16$

Problem 52 The dimensions of a rectangular box are in the ratio 1 : 2 : 4 and the difference between the costs of covering it with the cloth and sheet at the rate of ₹ 20 and ₹ 20.5 per sq. m respectively is ₹ 126. Find its longest side.

Sol. Since there is difference of ₹ 0.5 per m^2, the total surface area will be $\frac{126}{0.5} = 252\text{ m}^2$.

Now, $2[(1a\times 2a)+(2a\times 4a)+(4a\times 1a)] = 252\text{ m}^2$

$\Rightarrow \quad 28a^2 = 252$

$\Rightarrow \quad a^2 = 9$

$\therefore \quad a = 3$

Hence, longest side = 4 × 3 = 12 m

Effect on the Surface Area (Total/Lateral) of a Cube/Cuboid if it is Cut in Different Ways

Problem 53 A cube of edge 5 cm is cut into cubes each of edge of 1 cm. The ratio of the total surface area of one of the small cubes to that of the large cube is equal to

Sol. Required ratio $= \frac{6\times(1)^2}{6\times(5)^2} = 1:25$

Alternatively

Observing the 'ratio analysis' given in beginning of this chapter, the required ratio will be ratio of squares of their edges i.e. 1 : 25.

Problem 54 A cuboid of size 50 cm × 40 cm × 30 cm is cut into 8 identical parts by 3 cuts. What is the total surface area (in cm^2) of all the 8 parts ? ***[SSC (CGL-T-II)-17]***

Sol. In order to obtain 8 identical parts, we will cut the cuboid with two perpendicular cuts through the centre, then will make a horizontal cut at half the height of the cuboid. Obviously, the length, breadth and height of the each part obtained by the cutting of the cuboid will become half of the original length, breadth and height of the original cuboid.

Now, total surface area of all the 8 parts

$= 8\times 2[25\cdot 20 + 20\cdot 15 + 15\cdot 25]\text{ cm}^2$

$= 18800\text{ cm}^2$

Problem 55 A cuboid having total surface area 10000 cm^2 is cut into 8 identical parts by 3 cuts. What is the total surface area (in cm^2) of all the 8 parts?

Sol. Obviously, the length, breadth and height of the each part obtained by the cutting of the cuboid will become half of the original length, breadth and height of the original cuboid.

Further, we know that if each side/edge of a cube is made x times, its total surface area will become x^2 times the earlier.

Hence, the total surface area of each part will be $\left(\frac{1}{2}\right)^2$ times of 10000 i.e. 2500.

Hence, total surface area of all 8 parts

$= 8\times 2500 = 20000\text{ cm}^2$

Problem 56 **A solid cube has side 8 cm. It is cut along diagonals of the top face to get 4 equal parts. What is the total surface area (in cm^2) of each part ?** ***[SSC (CGL-T-II)-17]***

Sol. Diagonal DF = Diagonal $CE = 8\sqrt{2}$ cm

Then, $OD = OC = 4\sqrt{2}$ cm

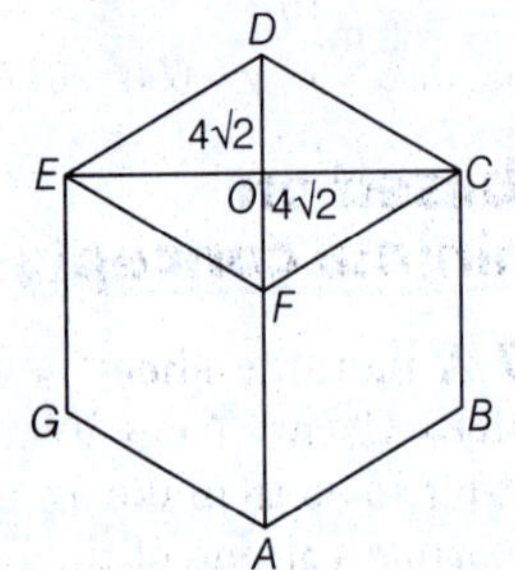

[$\because$ diagonals of a square bisect each other]

We know, area of an isosceles triangle $= \frac{b}{4}\sqrt{4a^2 - b^2}$

[where, a = length of each of equal sides and b = length of the third side]

$\therefore$ Area of an isosceles $\Delta DOC = \frac{8}{4}\sqrt{4(4\sqrt{2})^2 - 8^2} = 16$

Now, total surface area of each part $= 2(16) + 2(4\sqrt{2} \times 8) + 8 \cdot 8$
$= 96 + 64\sqrt{2}$ cm^2

Problem 57 **Identical cubes of largest possible size are cut from a solid cuboid of size 65 cm × 26 cm × 3.9 cm. What is the total surface area (in cm^2) of all the small cubes taken together ?** ***[SSC(CGL-T-II)-17]***

Sol. Largest side of each small cube = HCF of 65, 26 and 3.9 = 1.3

Then, total number of cubes $= \frac{\text{Volume of the cuboid}}{\text{Volume of one cube}}$

$$= \frac{65 \times 26 \times 3.9}{1.3 \times 1.3 \times 1.3} = 3000$$

$\therefore$ Total surface area of all 3000 cubes $= 3000 \times 6\,(1.3)^2$
$= 30420$ cm^2

Conversion of Shapes

Problem 58 **$\frac{1}{2}$ m cubic gold sheet is extending by hammering so as to cover an area of 1 hec. The thickness of the sheet (in cm) is :** ***[LIC(AAO)-09]***

Sol. We know, 1 hec. = 10000 m^2

Clearly, volume of the gold sheet = Area of 1 hec. × Height

$\Rightarrow$ $1/2\ \text{m}^3 = 10000\text{m}^2 \times h$

$\Rightarrow$ $h = 0.00005\ \text{m} = 0.005$ cm

Problem 59 **The weight of a cu. metre of a certain metal is 480 kg. It is melted and then rolled into a square bar 4 m long. Now an exact cube is cut from it. Find the weight of the cube.**

Sol. Volume of the metal cube = 1 m^3

Let the height and breadth of new bar be x m each.

Then, volume of the bar $= 4 \times x \times x$

Clearly, $1\ \text{m}^3 = 4 \times x \times x$

[$\because$ original metal cube is melted]

$\Rightarrow$ $x^2 = \frac{1}{4} \Rightarrow x = \frac{1}{2}$

It means a cube of side $\frac{1}{2}$ m can be cut from this new bar.

Then, its volume $= \left(\frac{1}{2}\right)^3 = \frac{1}{8}\text{m}^3$

As given, weight of 1 m^3 metal = 480 kg

$\therefore$ Weight of $\frac{1}{8}$ m^3 metal $= 480 \times \frac{1}{8} = 60$ kg

Problem 60 **A metal bar of dimensions 19 cm × 4 cm × 2 cm and a metal cube of side 4 cm were melted together and re-casted into a new single cube. Find the length of the edge of new single cube.**

Sol. Volume of the metal bar $= 19 \times 4 \times 2 = 152$ cm^3

and volume of the metal cube $= (4)^3 = 64$ cm^3

$\therefore$ Volume of both together $= 152 + 64 = 216$ cm^3 and the fact that this becomes the volume of the new re-casted single cube i.e. $(\text{edge})^3 = 216$

$\therefore$ Edge of the new single cube $= \sqrt[3]{216} = 6$ cm

Problem 61 **A metallic sheet is of rectangular shape with dimensions 48 m × 36 m. From each of its corners, a square is cut off so as to make an open box. If the length of the square is 8 m, the volume of the box (in m^3) is :**

Sol. Clearly, length of the open box $= [48 - (8 + 8)] = 32$ m

its breadth $= [36 - (8 + 8)] = 20$ m

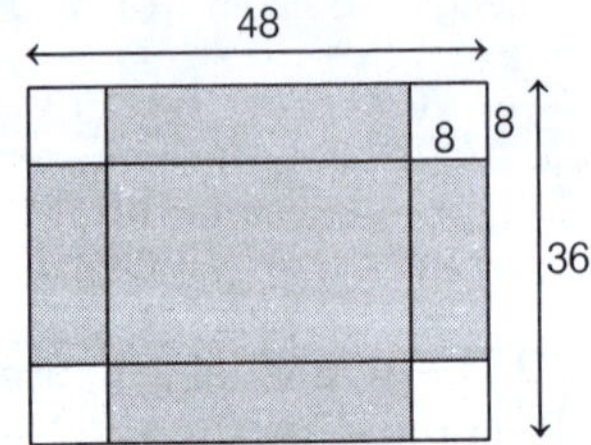

and its height = 8 m

$\therefore$ Volume of the box $= 32 \times 20 \times 8$
$= 5120\text{m}^3$

Problem 62 **An iron cube of side 10 cm is hammered into a rectangular sheet of thickness 0.5 cm. If the sides of the sheet are in the ratio 1 : 5, the sides are :** ***[Hotel Mgmnt-99]***

Sol. Clearly, volume of the cube = volume of the sheet

$\Rightarrow$ $(10)^3 = x \times 5x \times 0.5$

[let l and b be $5x$ and x respectively]

$\Rightarrow$ $5x^2 = 2000$

$\Rightarrow$ $x = 20$

Hence, sides are 20 cm and 100 cm.

Alternatively

You may proceed through options to solve quickly.

Problem 63 A metallic cube of edge 2.5 cm is melted and re-casted into the form of a cuboid of base 1.25 cm × 0.25 cm. Find the increase in the surface area.

Sol. Obviously, the volume of the metallic cube = volume of the cuboid

$\Rightarrow \quad (2.5)^3 = 1.25 \times 0.25 \times \text{Height}$

$\Rightarrow \quad \text{Height} = 50 \text{ cm}$

Now, surface area of the metallic cube

$= 6 \times (\text{side})^2 = 6 \times (2.5)^2 = 37.5 \text{ cm}^2$

and surface area of the cuboid $= 2(lb + bh + hl)$

$= 2[(1.25 \times 0.25) + (0.25 \times 50) + (50 \times 1.25)]$

$= 2[75.3125] = 150.625 \text{ cm}^2$

Hence, increase in surface area

$= 150.625 - 37.5 = 113.125 \text{ cm}^2$

Problem 64 64 identical cubes are cut from a large cube and all are arranged in the row so as to form a long cuboid. Find the ratio of surface area of larger cube to that of the cuboid. *[MAT-14]*

Sol. Let each side of the smaller cube be 1 cm.

Then, volume of the bigger cube $= 64 \times (1)^3 = 64$

Side of the bigger cube $= \sqrt[3]{64} = 4 \text{ cm}$

$\therefore$ Its surface area $= 6(4)^2 = 96 \text{ cm}^2$

Now, length, breadth and height of the newly formed cuboid will be 64 cm, 1 cm and 1 cm respectively as all smaller cubes have been arranged in a row.

Then, its surface area $= 2[(64 \times 1) + (1 \times 1) + (1 \times 64)]$

$= 2(129) = 258 \text{ cm}^2$

$\therefore \quad \text{Required ratio} = \dfrac{96}{258}$

$= 48 : 129$

Problem 65 Some identical cubes of side 1 cm are arranged to form a big cube of side 4 cm. Then, four corner cubes of top most layer are removed. Find the change percentage in surface areas of the former and latter cubes.

Sol. If four corner cubes are removed, there will be no effect on its surface area as the area of visible faces of removed cubes is equal to that of visible faces of other cubes adjacent to the cubes removed.

Problem 66 A solid metallic cube of edge 4 cm is melted and recast into solid cubes of edge 1 cm. If x is the surface area of the melted cube and y is the total surface area of all the cubes recast, then what is $x : y$?

Sol. Number of small cubes $= \dfrac{\text{Volume of the original cube}}{\text{Volume of the one recast cube}}$

$= \dfrac{(4)^3}{(1)^3} = 64$

Now, required ratio $= x : y = 6(4)^2 : 64 \times 6(1)^2 = 1 : 4$

Problems based on maxima/minima concepts

Problem 67 A metallic sheet is of rectangular shape with dimensions 12 cm × 5 cm. From each of its corner, a square is cut-off so as to make an open box. What is the maximum possible volume of the box.

Sol. Let length of the square be x cm.

Clearly, length of the open box $= [12 - (x + x)] = (12 - 2x)$ cm

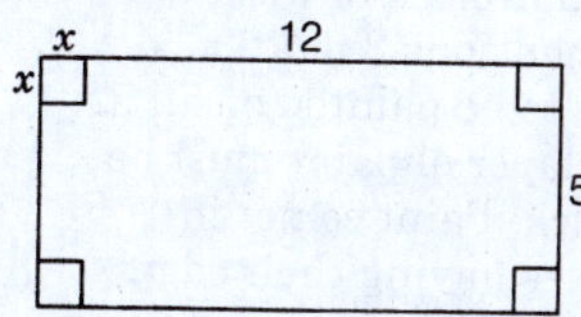

and its breadth $= [5 - (x + x)] = (5 - 2x)$ cm

and its height $= x$ m

$\therefore$ Volume of the open box $(v) = (12 - 2x) \times (5 - 2x) \times x$

$= 4x^3 - 34x^2 + 60x$

$\Rightarrow \quad \dfrac{dv}{dx} = 12x^2 - 68x + 60$ [First derivative]

$\Rightarrow \quad 12x^2 - 68x + 60 = 0$

$\Rightarrow \quad x = 1.09 \text{ and } 4.57$

$\Rightarrow \quad x = 1.09$

[$\because$ at $x = 4.57$, the value of $(5 - 2x)$ is negative. The length can not be negative]

Since $\dfrac{d^2v}{dx^2}$ i.e. $24x - 68$ is negative at $x = 1.09$, the volume is maximum at the same value of x.

Hence, maximum volume of the open box $= 4x^3 - 34x^2 + 60x$

$= 4(1.09)^3 - 34(1.09)^2 + 60(1.09) = 30.18 \text{ cm}^3$

WORKOUT

1. The volume of a rectangular block of stone is 10368 dm^2, its dimensions are in the ratio of 3 : 2 : 1. If its entire surface is polished at 2 paise per dm^2, then what is the total cost? *[MAT-11]*

(a) ₹ 31.68 (b) ₹ 31.50
(c) ₹ 63 (d) ₹ 63.36

2. The perimeter of one face of a cube is 20 cm. Its volume will be: *[SSC (CGL-T-I)-15]*

(a) 125 cm^3 (b) 400 cm^3
(c) 625 cm^3 (d) 100 cm^3

3. A rectangular block of wood having dimensions 3 m × 2 m × 1.75 m has to be painted on all its faces. The layer of paint must be 0.1 mm thick. Paint comes in cubical boxes having their edges equal to 10 cm. The minimum number of boxes of paint to be purchased is : *[CDS-15]*

(a) 5 (b) 4 (c) 3 (d) 2

4. The length of an edge of a hollow cube open at one face is $\sqrt{3}$ m. What is the length of the largest pole that it can accommodate?

(a) $\sqrt{3}$ m (b) $3\sqrt{3}$ m
(c) 3 m (d) $1/\sqrt{3}$ m

5. A cistern of dimensions 2.4 m × 2m × 1.5 m takes 2 h 30 min. to get filled with water. The rate at which water flows into the cistern is :

(a) 2.88 cu. m/h (b) 800 cu. m/min.
(c) 800 cu. m/sec. (d) 80 cu. m/min.

6. The diagonal of three faces of a cuboid are 13, $\sqrt{281}$ and 20 linear units. Then the total surface area of the cuboid is :

(a) 650 sq. units (b) 658 sq. units
(e) 664 sq. units (d) 672 sq. units

7. The length, breadth and height of a room are l, b and h respectively. The perimeter of the ceiling expressed as a percentage of the total area of the four walls is : *[IAS (Pre)-97]*

(a) 100 h (b) h
(c) $\frac{100}{h}$ (d) $\frac{h}{100}$

8. A cistern open at the top, is to be lined with sheet lead which weights 27 kg/m^3. The cistern is 4.5 m long and 3 m wide and holds 50 m^3. The weight of the lead required is : *[MAT-13]*

(a) 1764.60 kg (b) 1864.5 kg
(c) 1660.62 kg (d) 1860.62 kg

9. A cube has each edge 2 cm and a cuboid is 1 cm long, 2 cm wide and 3 cm high. The paint in a certain container is sufficient to paint an area of 54 sq. cm. Which of the following is correct?

(a) Both cube and cuboid can be painted
(b) Only cube can be painted
(c) Only cuboid can be painted
(d) Neither cube nor cuboid can be painted

10. The length of the diagonal of a cube is 6 cm. The volume of the cube (in cm^3) is *[SSC (CGL-T-II)-10]*

(a) $18\sqrt{3}$ (b) $24\sqrt{3}$ (c) $28\sqrt{3}$ (d) $30\sqrt{3}$

11. A school room is to be built to accommodate 70 children so as to allow 2.2 m^2 of floor and 11 m^3 of space for each child. If the room be 14 m long, what must be its breadth and height (in m)?

(a) 12, 5.5 (b) 13,6
(c) 11,5 (d) 11,4

12. If the total surface area of a cube is 6 sq. units, then what is the volume of the cube? *[CDS-13]*

(a) 1 cu. unit (b) 2 cu. units
(c) 4 cu. units (d) 6 cu. units

13. The volume of a cube is V. The total length of its edges is : *[MAT-99, 01]*

(a) $6v^{\frac{2}{3}}$ (b) $12v^{\frac{2}{3}}$
(c) $12v^{\frac{1}{3}}$ (d) $8\sqrt{v}$

14. The dimensions of a rectangle box are in the ratio 1:2:4 and the difference between the costs of covering in with the cloth and sheet at the rate of ₹ 20 and ₹ 20.5 per m^2 respectively is ₹ 126. Find the dimensions of the box.

(a) 3 m, 6 m, 12 m (b) 6m, 12 m, 24 m
(c) 1 m, 2 m, 4 m (d) 3 m, 9 m, 12 m

15. The paint in certain container is sufficient to paint an area equal to 5.875 m^2. How many bricks of dimensions 12.5 cm ×10 cm × 7.5 cm can be painted out of this container ?

(a) 225 (b) 180
(c) 150 (d) 100

16. A square tin sheet of side 12 inches is converted into a box with open top in the following steps-the sheet is placed horizontally. Then, equal sized squares, each of side x inches, are cut from the four corners of the sheet. Finally, the four resulting sides are bent vertically upwards in the shape of a box. If x is an integer, then what value of x maximises the volume of the box ? *[CAT-03]*

(a) 3 (b) 1
(c) 2 (d) 4

17. What are the dimensions (length, breadth, height respectively) of a cuboid with volume 720 cu. cm, surface area 484 sq. cm and the area of the base 72 sq. cm? *[CDS-12]*

(a) 12,6,10 cm (b) 9,8,10 cm
(c) 18, 4, 10 cm (d) 30,2,12 cm

18. The sum of length, breadth and height of a room is 19 m. The length of the diagonal is 11 m. The cost of painting the total surface area of the room at the rate of ₹ 10 per m^2 is : *[MAT-03, CDS-19]*

(a) ₹ 240 (b) ₹ 2400
(c) ₹ 420 (d) ₹ 4200

19. The percentage increase in the surface area of a cube when each side is doubled, is *[SSC (CGL-T-I)-15]*

(a) 300% (b) 50%
(c) 200% (d) 150%

20. If three cubic biscuits having edges 0.3 m, 0.4 m and 0.5 m respectively are melted and formed into a single cubic biscuit, then what is the total surface area of the cubic biscuit ?

(a) 2.16 m^2 (b) 2.46 m^2
(c) 36.16 m^2 (d) 4.16 m^2

21. A solid cube with an edge of 10 cm is melted to form two equal cubes. The ratio of the edge of the smaller cube to the bigger cube is :

(a) $\left(\frac{1}{3}\right)^{\frac{1}{3}}$ (b) $\frac{1}{2}$ (c) $\left(\frac{1}{2}\right)^{\frac{1}{3}}$ (d) $\left(\frac{1}{4}\right)^{\frac{1}{3}}$

22. The dimensions of a field are 15 m by 12 m. A pit 8 m long, 2.5 m wide and 2 m deep is dug in one corner of the field and the earth removed is evenly spread over the remaining area of the field. The level of the field is raised by : *[CDS-14]*

(a) 15 cm (b) 20 cm
(c) 25 cm (d) 200/9 cm

23. A wooden box (open at the top) of thickness 0.5 cm, length 21 cm, width 11 cm and height 6 cm is painted on the inside. The expenses of painting are ₹ 70. What is the rate of painting per cm^2 ? *[CAT-96]*

(a) ₹ 0.7 (b) ₹ 0.5
(c) ₹ 0.1 (d) ₹ 0.2

24. The length of a room is double the breadth. The cost of colouring the ceiling at ₹ 25 per sq m is ₹ 5000 and the cost of painting the four walls at ₹ 240 per sq m is ₹ 64800. Find the height of the room.

(a) 4.5 m (b) 4 m
(c) 3.5 m (d) 5 m

25. A hospital room is to accommodate 56 patients. It should be done in such a way that every patient gets 2.2 m^2 of floor and 8.8 m^3 of space. If the length of the room is 14 m, then breadth and the height of the room are respectively *[CDS-13]*

(a) 8.8 m, 4 m (b) 8.4 m, 4.2 m
(c) 8 m, 4 m (d) 7.8 m, 4.2 m

26. A plot of land in the form of a rectangle has a dimension 240 m × 180 m. A drainlet 10 m wide is dug all around it (on the outside) and the earth dug out is spread over the plot, increasing its surface level by 25 cm. The depth of the drainlet is :

(a) 1.225 m (b) 1.229 m
(c) 1.227 m (d) 1.223 m

27. For a plot of land of 100 m × 80 m, the length to be raised by spreading the earth from stack of a rectangular base 10 m × 8m and vertical section being a trapezium of height 2 m. The top of the stack is 8 m × 5 m. How many cm can the level raised ? *[CDS-14]*

(a) 3 cm (b) 2.5 cm
(c) 2 cm (d) 1.5 cm

28. The dimensions of a room are 10 m × 7m × 5 m. There are 2 doors and 3 windows in the room. The dimensions of the doors are 1 m × 3 m. One window is of size 2 m × 1.5 m and the other two windows are of size 1 m × 1.5 m. The cost (in ₹) of painting the walls at ₹ 3 per m^2 is : *[MAT-13]*

(a) 584 (b) 474
(c) 604 (d) 342

29. How many small cubes, each of 96 cm^2 surface area, can be formed from the material obtained by melting a larger cube of 384 cm^2 surface area ?

(a) 8 (b) 5
(c) 800 (d) 8000

30. The volume of a cuboid is 1120 cm^3 and its height is 5 cm while the length and the breadth of the cuboid are in the ratio 8 : 7. The length of this cuboid exceeds the breadth by :

(a) 2 cm (b) 4 cm
(c) 7 cm (d) 5 cm

31. 1 m^3 piece of copper is melted and recast into a square cross section bar 36 m long. An exact cube is cut off from this bar. If 1 m^3 of copper cost ₹ 108, then the cost of the cube is : *[MAT-11]*

(a) 50 p (b) 75 p
(c) 25 p (d) ₹ 1

32. 2 cm of rain has fallen on a square km of land. Assuming that 50% of raindrops could have been collected and contained in a pool having a 100 m × 10 m base, by what level would the water level in the pool have increased ? *[MAT-07, 08, SSC (CGL-T-II)-11]*

(a) 15 m (b) 20 m
(c) 10 m (d) 25 m

33. Water has been poured into an empty rectangular tank at the rate of 8 cu. ft/min. for 2.5 min. The length of the tank is 3 ft. and the width is $\frac{1}{2}$ of the length. How deep is the water in the tank ?

(a) 4 ft.
(b) 3.86 ft.
(c) 3.23 ft.
(d) 4.44 ft.

34. A 4 cm cube is cut into 1 cm cubes. Find the percentage increase in surface area. *[MAT-09, SNAP-12]*

(a) 200% (b) 100%
(c) 400% (d) 300%

35. A solid metallic cube is melted to form five solid cubes whose volumes are in ratio 1 : 1 : 8 : 27 : 27. The percentage by which the sum of the surface areas of these five cubes exceeds the surface area of the original cube is nearest to : *[CAT-17]*

(a) 60 (b) 50
(c) 20 (d) 10

36. A hollow cube is formed by joining six identical squares. A rectangular cello tape of length 4 cm and breadth 0.5 cm is used for joining each pair of edges. What is the total area (in sq. cm) of cello tape used ? *[CDS-17]*

(a) 12 (b) 24
(c) 36 (d) 48

37. A rectangular piece of cardboard 18 cm × 24 cm is made into an open box by cutting a square of 5 cm side from each corner and build up the side. Find the volume of the box (in cm^3) :

(a) 432
(b) 560
(c) 216
(d) None of the above

38. If the rectangular faces of a brick have their diagonals in the ratio $3:2\sqrt{3}:\sqrt{15}$, then the ratio of the length of the shortest edge of the brick to that of its longest edge is : *[CAT-19]*

(a) $\sqrt{3}:2$ (b) $2:\sqrt{5}$
(c) $1:\sqrt{3}$ (d) $\sqrt{2}:\sqrt{3}$

Answer Key

1. (d)	2. (a)	3. (c)	4. (b)	5. (a)	6. (c)	7. (c)	8. (b)	9. (a)	10. (b)
11. (c)	12. (a)	13. (c)	14. (a)	15. (d)	16. (c)	17. (b)	18. (b)	19. (a)	20. (a)
21. (c)	22. (c)	23. (c)	24. (a)	25. (a)	26. (c)	27. (d)	28. (b)	29. (a)	30. (a)
31. (a)	32. (c)	33. (d)	34. (d)	35. (b)	36. (b)	37. (b)	38. (c)		

Hints and SOLUTIONS

1. Given, $3x \times 2x \times 1x = 10368$

$\Rightarrow \quad x^3 = 1728$

$\therefore \quad x = 12\,\text{dm}$

It means length = 36 dm, breadth = 24 dm and height = 12 dm

Now, total surface area

$= 2[(36\times24)+(24\times12)+(12\times36)]$

$= 3168\,\text{dm}^2$

Then, total cost $= 3168\times0.02$

$= ₹\,63.36$

2. Perimeter of one face of cube = 20 cm

$\therefore$ Side of the cube = 20/4 = 5 cm

Then, its volume $= (5)^3 = 125\,\text{cm}^3$

3. Total surface area

$= 2[(3\times2)+(2\times1.75)+(1.75\times3)] = 29.5$

Thus, volume of paint required = total surface area × thickness of paint $= 29.5\times0.0001\,\text{m}^3$

Volume of paint comes in one box

$= (0.1)^3$

$\therefore$ Number of boxes required

$= \dfrac{29.5\times0.0001}{(0.1)^3} = 2.95$

It means that minimum 3 boxes of paint are required.

4. Obviously, a largest pole can be accommodated along the diagonal of the cube.

Hence, length of the largest pole

$= \sqrt{3}\cdot\text{side} = 3\sqrt{3}\,\text{m}$

5. Clearly, the cistern takes 2.5 h to get filled with $(2.4\times2\times1.5)\,\text{m}^3$ water.

Hence, the rate of flow

$= \dfrac{2.4\times2\times1.5}{2.5}$

$= 2.88\,\text{m}^3/\text{h}$

6. Given, $\sqrt{l^2+b^2} = 13$,

$\sqrt{b^2+h^2} = \sqrt{281}$ and $\sqrt{h^2+l^2} = 20$

$\Rightarrow \quad l^2+b^2 = 169,$

$b^2+h^2 = 281$

and $\quad h^2+l^2 = 400$

On adding all three Eqs. we get,

$2(l^2+b^2+h^2) = 850$

$\Rightarrow \quad l^2+b^2+h^2 = 425$

Now, $(l^2+b^2+h^2)-(l^2+b^2)$

$= 425-169 = 256$

$\Rightarrow h^2 = 256 \Rightarrow h = 16$

Similarly, $(l^2+b^2+h^2)-(b^2+h^2)$

$= 425-281 = 144$

$\Rightarrow \quad l^2 = 144 \Rightarrow l = 12$

and $(l^2+b^2+h^2)-(h^2+l^2)$

$= 425-400 = 25$

$\Rightarrow \quad b^2 = 25 \Rightarrow b = 5$

Now, total surface area

$= 2(lb+bh+hl)$

$= 2[(12\times5)+(5\times16)+(16\times12)]$

$= 664$ sq. units.

Alternatively

(Hit and trial method) 13 is a figure of a Pythagoras triplet (5, 12, 13) then it may be assumed as two dimensions (say, l and b) are 12 and 5. Further, 20 is also a figure of a Pythagoras triplet (12, 16, 20), then it may be construed that 16 is also a dimension. Now, we may check the correctness of our assumption

$\sqrt{(12)^2+(5)^2} = 13,$

$\sqrt{(16)^2+(5)^2} = \sqrt{281}$

and $\quad \sqrt{(16)^2+(12)^2} = 20.$

Since assumptions are correct, dimensions are 12, 5 and 16. Now, total surface area $= 2(lb+bh+hl)$

$= 2[(12\times5)+(5\times16)+(16\times12)]$

$= 664$ sq units

7. Required answer

$= \left(\dfrac{\text{Perimeter of the ceiling}}{\text{Lateral surface area i.e. areas of four walls}}\times100\right)\%$

$= \left\{\dfrac{2(l+b)}{2h(l+b)}\times100\right\}\% = \dfrac{100}{h}\%$

8. Height of the cistern

$= \dfrac{\text{Volume}}{\text{Area of the base}} = \dfrac{50}{4.5\times3} = \dfrac{100}{27}$

Now, area to be lined with the lead sheet

= lateral area + area of the base

$= \left[2(4.5+3)\times\dfrac{100}{27}\right]+(4.5\times3)$

$= \dfrac{1864.5}{27}\,\text{m}^2$

Now, weight of the required lead

$= \dfrac{1864.5}{27}\times27$

$= 1864.5\,\text{kg}$

9. Surface area of the cube $= 24\,\text{cm}^2$ and surface area of the cuboid $= 22\,\text{cm}^2$. Since sum of surface areas of both is less than $54\,\text{cm}^2$, both can be painted.

10. Given, diagonal of the cube $= \sqrt{3}$ side = 6

$\Rightarrow \quad$ Side $= 6/\sqrt{3} = 2\sqrt{3}$

Then, its volume $= (2\sqrt{3})^3 = 24\sqrt{3}\,\text{cm}^3$

11. Area of the floor = Length × Breadth

$\Rightarrow \quad 70\times2.2 = 14\times\text{Breadth}$

$\Rightarrow \quad$ Breadth = 11 m

Also, volume = length × breadth × height

$\Rightarrow (70 \times 11)\ m^3 = 14 \times 11 \times \text{height}$

$\therefore \quad \text{Height} = 5\ m$

12. Given, $6a^2 = 6 \Rightarrow a = 1$

Then, Volume $= a^3 = 1$ cu. unit.

13. Given, volume$(v) = (\text{side})^3$

$\Rightarrow \text{side} = (v)^{\frac{1}{3}}$

Since there are 12 edges in a cube, their total length $= 12v^{\frac{1}{3}}$.

14. Let the dimensions be x, $2x$ and $4x$. Then,

Total surface area $= 2(lb + bh + hl)$

$= 2\,(4x \cdot 2x + 2x \cdot x + x \cdot 4x)$

$= 28x^2$ sq. m

Given that the difference between the costs of covering in with the cloth and sheet is ₹ 126.

Then,

$(28x^2 \times 20.5) - (28x^2 \times 20) = 126$

$\Rightarrow \quad 14x^2 = 126 \Rightarrow x = 3$

Hence, dimensions are 3 m, 6 m and 12 m.

Alternatively

The difference of ₹ 0.5 is for 1 sq m. Then, difference of ₹ 126 will be for 252 sq. m. Here, covering the box means covering its total surface area. It means total surface area = 252 m². Now, you have to ensure that which option will fetch 252 m² as total surface area.

15. Number of bricks painted

$= \dfrac{\text{Area which can be painted}}{\text{Surface area of one brick}}$

$= \dfrac{5.875 \times 100 \times 100\ cm^2}{2(12.5 \times 10 + 10 \times 7.5 + 7.5 \times 12.5)cm^2}$

$= \dfrac{5.875 \times 100 \times 100}{587.5} = 100$

16. Clearly, the volume of the resulting box $= (12 - 2x)^2 \times x$

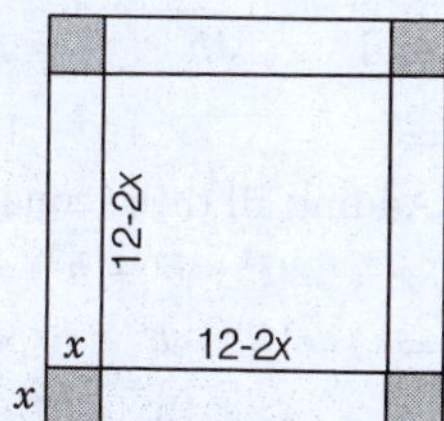

Now, proceed through options.

(a) At $x = 3$, volume of the resulting box $= (12 - 2 \cdot 3)^2 \times 3 = 108$

(b) At $x = 1$, volume of the resulting box $= (12 - 2 \cdot 1)^2 \times 1 = 100$

(c) At $x = 2$, volume of the resulting box $= (12 - 2 \cdot 2)^2 \times 2 = 128$

(d) At $x = 4$, volume of the resulting box $= (12 - 2 \cdot 4)^2 \times 4 = 64$

Hence, option (c) is correct.

17. Height $= \dfrac{\text{Volume}}{\text{Base area}} = \dfrac{720}{72} = 10,$

it means answer is except (d).

Now, proceed through the remaining options to check which option is yielding the surface area 484 sq. cm.

Option (b) is correct.

18. Given, diagonal

$= \sqrt{l^2 + b^2 + h^2} = 11$

$\Rightarrow l^2 + b^2 + h^2 = 121$

$\Rightarrow (l + b + h)^2 - 2(lb + bh + hl) = 121$

$\Rightarrow (19)^2 - 2\,(lb + bh + hl) = 121$

$[\because \text{given}, l + b + h = 19]$

$\Rightarrow 2\,(lb + bh + hl) = 240$

Hence, total surface area = 240 m²

$\therefore$ Total cost $= 240 \times 10$

= ₹ 2400

19. If each side/edge of a cube is made x times, its surface area will become x^2 times the earlier. Here, each edge is made 2 times, then surface area will become $(2)^2$ times the earlier i.e. 4 times. Hence, increase in percentage is 300%.

Alternatively I

Here, each side has been increased by 100%. Since the surface area is concerned with two dimensional aspects, effect on the surface area

$= \left\{+100 + 100 + \dfrac{100 \times 100}{100}\right\}$

$= +300\%$

Hence, the percentage increase

$= 300\%$

Alternatively II

It may be said that each edge of the cube has increased in ratio 100 : 200 i.e. 1 : 2. Also, we know that if the side/edge of a cube is changed in ratio $a : b$, its surface area is changed in ratio $a^2 : b^2$ i.e. $1^2 : 2^2$

Now, percentage change in surface area $= \left\{\dfrac{a^2 - b^2}{a^2} \times 100\right\}\%$

$= \left\{\dfrac{1 - 4}{1} \times 100\right\}\% = -300\%$

'–' indicates that surface area has been increased and the increment is 300%.

20. Sum of volumes of all three cubic biscuits

$= (0.3)^3 + (0.4)^3 + (0.5)^3 = 0.216$

Hence, volume of the newly formed biscuit = 0.216

Then, its edge $= \sqrt[3]{0.216} = 0.6$

Now, its surface area $= 6(0.6)^2$

$= 216\ m^2$

21. Volume of each smaller cube = 1/2 of $(10)^3 = 500\ cm^3$

Edge of each cube $= \sqrt[3]{500} = 5\sqrt[3]{4}$

$\therefore$ Required ratio $= 5\sqrt[3]{4} : 10 = \sqrt[3]{4} : 2$

$= \sqrt[3]{4} : \sqrt[3]{8} = \sqrt[3]{4/8} = \left(\dfrac{1}{2}\right)^{\frac{1}{3}}$

22. The volume of the earth dug out

$= 8 \times 2.5 \times 2 = 40\ m^3$

The area to be raised by spreading the earth dug out

$= (15 \times 12) - (8 \times 2.5) = 160\ m^2$

Let us assume that the level of the field be raised by h m on spreading the earth over it.

Then, $160 \times h = 40$

$\therefore h = \dfrac{40}{160}\ m = \dfrac{40 \times 100}{160}\ cm = 25\ cm$

23. Clearly, internal length

$= 21 - (0.5 + 0.5) = 20\ cm$

Internal width $= 11 - (0.5 + 0.5)$

$= 10\ cm$

and height of the box $= 6 - 0.5$

$= 5.5\ cm$

The area to be painted

$= 2 \times 5.5\,(20 + 10) + (20 \times 10) = 530$

Then, rate of painting per cm²

$= (70 \div 530) =$ ₹ 0.13

Hence, option (c) is correct.

24. Area of the ceiling $= lb$

$\Rightarrow \quad 5000/25 = 200\ m^2$

$\Rightarrow 2b \times b = 200 \quad [\because l = 2b, \text{given}]$

$\Rightarrow \quad b = 10m$

$\therefore \quad l = 20\ m$

Also given, area of all four walls i.e. lateral surface area

$= 2h(l + b) = 64800/240 = 270\ m^2$

$\Rightarrow 2h(20 + 10) = 270$

$\Rightarrow \quad h = 4.5\ m$

25. Area to be occupied by 56 patients $= 2.2 \times 56 = 123.2$

Then, breadth of the floor = area/length $= 123.2/14 = 8.8\ m$

Space (say, volume) to be required by all the patients $= 8.8 \times 56$

Then, height = volume / floor area

$= 8.8 \times 56 / 123.2 = 4$ m

26. Let the depth of the drainlet be d m.

Then, volume of the earth dug out

$= (240 \times 10 \times d) + (180 \times 10 \times d) + (240 \times 10 \times d) + (180 \times 10 \times d) + 4(10 \times 10 \times d)$

$= 8800d$ m^3

Now, volume of the earth dug out = volume of the earth spread over the plot

$\Rightarrow \quad 8800d = 240 \times 180 \times 0.25$

$\Rightarrow \quad d = 1.227$ m

Alternatively

Recall the concept of outer path elaborated in chapter 'Rectangle and Square'.

Area of the outer path 10 m wide (say, drainlet) $= 2x(l + b + 2x)$

$= 2 \cdot 10(240 + 180 + 2 \cdot 10) = 8800$ m^2

Then, volume of the earth dug out

= area of the drainlet × depth(d)

$= 8800d$ m^3

Now, volume of the earth dug out = volume of the earth spread over the plot

$\Rightarrow 8800d = 240 \times 180 \times 0.25$

$\Rightarrow \quad d = 1.227$ m

27. Given that the stack is in the shape having vertical cross section of trapezium.

Then, volume of the earth in the stack = area of cross section of trapezium × height

= 1/2 (Sum of parallel sides) × Height × Height

$= 1/2 \times (10 + 5) \times 2 \times 8 = 120$

Let us assume that the level of the field be raised by h m on spreading the earth over it.

Then, $(100 \times 80) \times h = 120$

$\therefore \quad h = \dfrac{120}{100 \times 80} = 0.015$ m $= 1.5$ cm

28. Area of all four walls i.e. lateral surface area $= 2h(l + b)$

$= 2 \times 5(10 + 7) = 170$

Area occupied by the windows

$= (2 \times 1.5) + 2(1 \times 1.5) = 6$

Area occupied by doors $= 2(1 \times 3) = 6$

Thus, area to be painted

$= 170 - 6 - 6 = 158$ m^2

Now, cost of painting $= 158 \times 3$

= ₹ 474

29. For each smaller cube,

$6 \cdot (\text{Side})^2 = 96$

$\therefore$ Side = 4, then volume = 64

For bigger cube, $6 \cdot (\text{side})^2 = 384$

$\therefore$ Side = 8, then volume = 512

Hence, number of small cubes

$= \dfrac{512}{64} = 8$

30. Let length and breadth be $8x$ and $7x$ respectively.

Volume of cuboid $= l \times b \times h$

$\Rightarrow 1120 = 8x \times 7x \times 5 \Rightarrow x^2 = 4$

$\therefore \quad x = 2$

Now, difference between l and b

$= 8x - 7x \Rightarrow x = 2$ cm

31. Volume of the square cross section bar $= 36 \times (a \times a)$, where 'a' is dimension of square section of bar in metres

$\Rightarrow \quad 36 \times (a \times a) = 1$

$\Rightarrow \quad a = 1/6$ m

From this bar, we can cut a cube of maximum edge of 1/6 m

Then, its volume $= \left(\dfrac{1}{6}\right)^3 = \dfrac{1}{216}$ m^3

Since 1 m^3 of copper cost ₹ 108, the cost of $\dfrac{1}{216}$ m^3 will be ₹ $\dfrac{1}{2}$ i.e. 50 p.

32. Obviously, volume of water collected = volume of water in the pool

$\Rightarrow$ 50% of (1000 m × 1000 m × 0.02 m)

= 100 m × 10 m × Height

$\Rightarrow$ Height = 10 m

33. Volume of the tank = Volume of water

$\Rightarrow 3 \times \dfrac{3}{2} \times \text{Depth} = 8 \times 2.5$

$\Rightarrow \quad$ Depth = 4.44 ft

34. Number of small cubes

$= \dfrac{\text{Volume of bigger cube}}{\text{Volume of smaller cube}} = \dfrac{4^3}{1^3} = 64$

Surface area of bigger cube

$= 6 \times 4^2 = 96$ m^2

and surface area of all smaller cubes

$= 64 \times (6 \times 1^2) = 384$ m^2

Increase% $= \dfrac{384 - 96}{96} \times 100$

$= 300\%$

35. Let volumes of five small cubes be $x^3, x^3, 8x^3, 27x^3$ and $27x^3$.

Then, their sides be $x, x, 2x, 3x$ and $3x$.

Now, sum of surface areas of all five cubes $= 6[x^2 + x^2 + 4x^2 + 9x^2 + 9x^2]$

$= 144x^2$

Obviously, volume of the original cube $= x^3 + x^3 + 8x^3 + 27x^3 + 27x^3$

$= 64x^3$

$\therefore$ Side of the original cube $= 4x$

Then, its surface area $= 6(4x)^2 = 96x^2$

Now, the required percentage

$= \left\{\dfrac{144x^2 - 96x^2}{96x^2} \times 100\right\}\% = 50\%$

36. Since a cube has 12 edges, the total area of cello tape used

$= 12(4 \times 0.5) = 24$ cm^2

37. Clearly, length of the open box

$= [24 - (5 + 5)] = 14$ cm

its breadth $= [18 - (5 + 5)] = 8$ cm

and its height = 5 cm

$\therefore$ Volume of the box $= 14 \times 8 \times 5$

$= 560$ cm^3

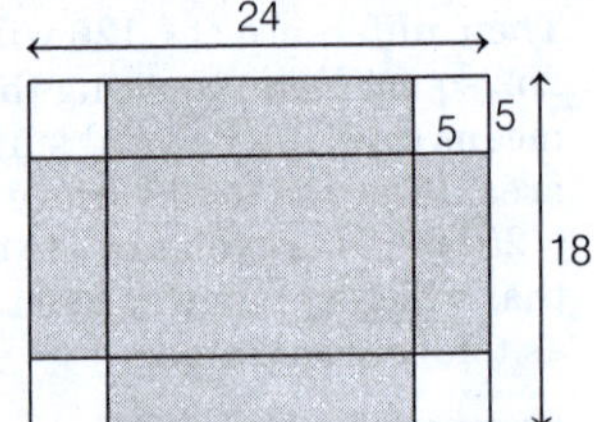

38. Let sides of brick be l, b and h and given diagonals are $3x$, $\sqrt{3}x$ and $\sqrt{15}$ x.

As given, $\sqrt{l^2 + b^2} = 3x$

$\Rightarrow \quad l^2 + b^2 = 9x^2 \quad$ ...(i)

and $\quad \sqrt{b^2 + h^2} = 2\sqrt{3}x$

$\Rightarrow \quad b^2 + h^2 = 12x^2 \quad$ ...(ii)

and $\quad \sqrt{h^2 + l^2} = \sqrt{15}x$

$\Rightarrow \quad h^2 + l^2 = 15x^2 \quad$ ...(iii)

Adding all three equations we get,

$2(l^2 + b^2 + h^2) = 36x^2$

$\Rightarrow \quad l^2 + b^2 + h^2 = 18x^2$

$\Rightarrow \quad 9x^2 + h^2 = 18x^2$

$\Rightarrow \quad h^2 = 9x^2 \Rightarrow h = 3x$

On the above line, $l = \sqrt{6}x$ and $b = \sqrt{3}x$

Then, required ratio (shortest to longest) $= \sqrt{3}x : 3x = 1 : \sqrt{3}$

45

Cylinder and Cone

Learning OBJECTIVES

After completing this chapter, aspirants will be able to :

- Understand the application of the cylinder and cone related formulae in different cases;
- Understand the application of combination of cylinder and cone related formulae in diverse cases;
- Compare the cylinder and cones with reference to different ratios;
- Understand the properties regarding frustum of a cone and ratio analysis in case of cylinders/cones;
- Understand the philosophy of the division of this chapter in various parts;
- Analyse the trend of the questions being asked in various competitive examinations;
- Solve the questions in moments;
- Grip the concepts helping to solve the questions without pen and paper;
- Apply many concepts envisaged in the chapter 'PERCENTAGE' like 'effect on product in case of changes in factors', 'Product constancy' etc.;
- Apply many concepts of 'Ratio' in this chapter;
- Learn many new concepts/fundamentals;
- Apply the pragmatic approaches in lieu of academic approaches; and
- Ponder on the alternative methods.

Inside

- *Derivation of Diverse Formulae Including for Hollow Cylinder*
- *Properties Regarding Frustum of a Cone*
- *Ratio-Analysis in Case of Two/Three or More Cylinders/Cones*
- *Concept of Volume/Area When Thickness of Layer/Surface is taken into*
- *Alteration in Dimensions Affecting Volume and Areas*
- *Comparison Between Cylinder and Cone w.r.t Diverse Aspects*
- *Conversions and Combination of Shapes (Cuboid, Cube, Cylinder and Cone)*
- *Problems Based on Above All Aspects*
- *Maximum/Minimum Value of Volume/Area*
- *Concept of Water/Liquid Flow*
- ....
- .... *& Many More*
- *Work Out with Solutions*

CYLINDER

Cylinder is a three dimensional shape bounded by a curved surface and two parallel and congruent circles at the ends. This curved surface is formed by all the line segments joining corresponding points of these circles. Here, we are discussing the right circular cylinder.

Volume

Assume a circular sheet (say, base) is being made thick by h units. Then, material/sheet required will be equal to the product of the area of base and the height.

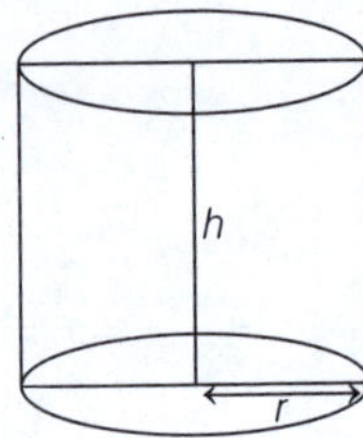

Thus, volume = Area of the circle (i.e. base area) × height (h)

$$= \pi r^2 h$$

(What volume actually refers to? May be seen in chapter 'Cuboid and Cube")

Curved Surface Area

Suppose a wire has been bent to form a circular base. Another and many more wires are being placed on the former and so on to reach upwards till h units, then the area of faces, thus formed, is termed as curved surface area, just like lateral surface area of a cube/cuboid.

Hence, curved surface area

$$= \text{Length of the wire} \times \text{height } (h)$$
$$= \text{circumference of a circle} \times h = 2\pi rh$$

Total Surface Area

It is sum of areas of both top and bottom base circles and curved surface area i.e. $\pi r^2 + \pi r^2 + 2\pi rh$

$$= 2\pi r(r + h) = \pi d(r + h),$$

where d = diameter of the base/top circle.

Concept Of Hollow Cylinder

Solids like iron pipes, tube well are in shape of hollow cylinders and the ***thickness of the pipe*** will be equal to the difference between its external and internal radii i.e. R-r, if external and internal radii be R and r respectively.

Volume of the Material Used

Obviously, the volume of metal/material used will be equal to the difference between their external and internal volumes. Hence, volume of the material used

$$= \pi h \times [(\text{exterior radius})^2 - (\text{interior radius})^2]$$
$$= \pi h (R^2 - r^2)$$

Total Surface Area of Pipe

Clearly, it will be equal to sum of exterior curved area, interior curved area and area of both top and bottom base rings

$$= 2\pi Rh + 2\pi rh + 2(\pi R^2 - \pi r^2)$$
$$= 2\pi Rh + 2\pi rh + 2\pi(R^2 - r^2)$$
$$= 2\pi h(R + r) + 2\pi(R + r)(R - r) = 2\pi(R + r)(h + R - r)$$

CONE

A cone is a solid formed by rotating a right-angled triangle around one of its perpendicular sides. Its dimensions are defined by the radius of the circle (base), the height and the slant height.

Volume

It is the fact that the space covered by the cone is 1/3 of the space covered by a cylinder provided both have equal base and equal height.

Hence, volume of a cone $= \frac{1}{3}\pi r^2 h$

Curved/Lateral Surface Area

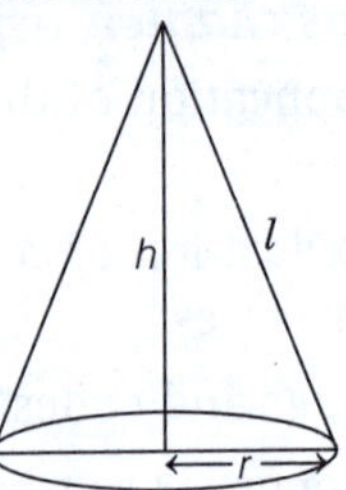

Curved surface area $= \pi rl$

[where, l = slant height $= \sqrt{r^2 + h^2}$]

Total Surface Area

Total surface area = area of the circle (base) + curved surface area

$$= \pi r^2 + \pi rl = \pi r(r + l)$$

Frustum of a Cone

A frustum of a cone is the piece left after cutting off the top portion of that cone with a plane parallel to the base. Here, $BCDE$ is a frustum of a cone.

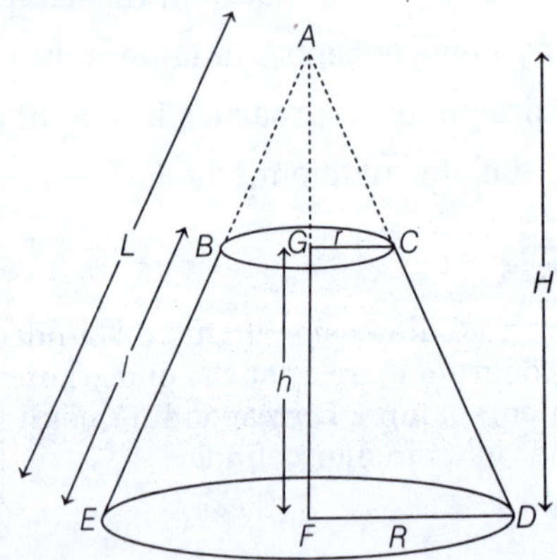

Volume

Clearly, volume of figure $BCDE$

= volume of original cone AED – volume of smaller cone ABC

$$= \frac{1}{3}\pi R^2 H - \frac{1}{3}\pi r^2 (H-h)$$

$$= \frac{1}{3}\pi h \left\{ R^2 \frac{H}{h} - r^2 \left(\frac{H}{h} - 1 \right) \right\}$$

Since, ΔAGC and ΔAFD are similar, $\frac{H-h}{H} = \frac{r}{R}$

$$\Rightarrow \quad 1 - \frac{h}{H} = \frac{r}{R} \Rightarrow \frac{H}{h} = \frac{R}{R-r}$$

Now, the volume of the frustum of the cone

$$= \frac{1}{3}\pi h \left\{ R^2 \frac{R}{R-r} - r^2 \left(\frac{R}{R-r} - 1 \right) \right\}$$

$$= \frac{1}{3}\pi h \left\{ \frac{R^3}{R-r} - \frac{r^3}{R-r} \right\}$$

$$= \frac{1}{3}\pi h (R^2 + r^2 + Rr)$$

$$= \frac{1}{3} h (\pi R^2 + \pi r^2 + \pi Rr)$$

$$= \frac{1}{3} h \left(\pi R^2 + \pi r^2 + \sqrt{\pi R^2 \times \pi r^2} \right)$$

$= \frac{1}{3} h$ [area of lower base [say, A_1] + area of upper base [say A_2] + $A_1 . A_2$

$$= \frac{1}{3} h \left[A_1 + A_2 + \sqrt{A_1 . A_2} \right]$$

Curved/Lateral/Slant Surface Area

Curved surface area of figure $BCDE$

= Curved area of cone AED–Curved area of cone ABC

$= \pi Rl - \pi r [L - l]$ [where, slant height $L = AE = AD$]

$$= \pi l \left[\frac{RL}{l} - r \left(\frac{L}{l} - 1 \right) \right]$$

$$= \pi l \left[\frac{R \times R}{(R-r)} - r \left\{ \frac{R}{(R-r)} - 1 \right\} \right]$$

$\left[\because \Delta AGC \sim \Delta AFD \therefore \frac{L}{l} = \frac{R}{R-r} \text{ just like } \frac{H}{h} = \frac{R}{R-r} \right]$

$$= \pi l \left[\frac{R^2}{(R-r)} - \frac{r^2}{(R-r)} \right] = \pi l \left(\frac{(R+r)(R-r)}{(R-r)} \right) = \pi l (R+r)$$

$$= \frac{1}{2} l (2\pi R + 2\pi r)$$

$= \frac{1}{2} \times$ Slant height [circumference of the lower base (say C_1) + circumference of the top base (say, C_2)]

$$= \frac{1}{2} l (C_1 + C_2)$$

Total Surface Area

Total surface area = Curved area + bases areas

$$= \pi l (R+r) + \pi R^2 + \pi r^2$$

$$= \pi [R^2 + r^2 + l(R+r)]$$

Slant Height

Slant height (l) of the frustum of a cone $= \sqrt{h^2 + (R-r)^2}$

Properties related to Frustum of a cone

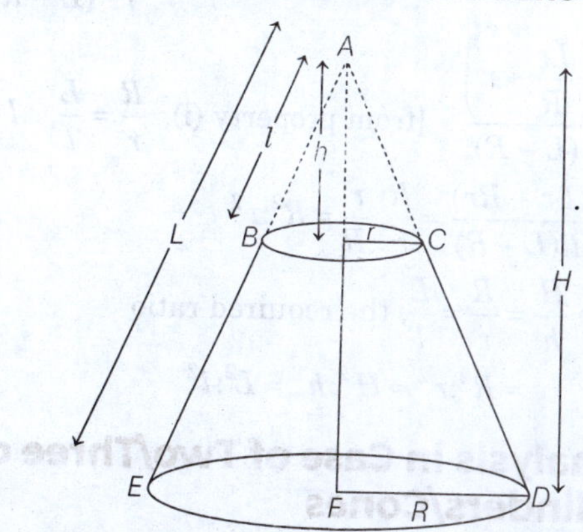

(i) $\frac{\boldsymbol{H}}{\boldsymbol{h}} = \frac{\boldsymbol{R}}{\boldsymbol{r}} = \frac{\boldsymbol{L}}{\boldsymbol{l}}$ $\quad (\because \Delta AED \sim \Delta ABC)$

i.e. $\frac{\text{Height of the parent cone } (H)}{\text{Height of the cone cut } (h)}$

$$= \frac{\text{Radius of the parent cone } (R)}{\text{Radius of the cone cut } (r)}$$

$$= \frac{\text{Slant height of parent cone } (L)}{\text{Slant height of cone cut } (l)}$$

(ii) Volume of the parent cone (*AED*) : Volume of the cone cut (*ABC*)

$$= \boldsymbol{R^3 : r^3 = H^3 : h^3 = L^3 : l^3}$$

Explanation Volume of the parent cone (AED) : Volume of the cone cut (ABC) $= \frac{1}{3}\pi R^2 H : \frac{1}{3}\pi r^2 h$

$$= R^2 H : r^2 h = \frac{R^2}{r^2} : \frac{h}{H}$$

$= \frac{R^2}{r^2} : \frac{r}{R}$ $\quad$ [from property (i), $\frac{H}{h} = \frac{R}{r}$]

$= R^3 : r^3$

Since $\frac{H}{h} = \frac{R}{r} = \frac{L}{l}$, the required ratio $= R^3 : r^3 = H^3 : h^3$ $= L^3 : l^3$

(iii) Curved area of parent cone (AED) : Curved area of cone cut (*ABC*) $= \boldsymbol{R^2 : r^2 = H^2 : h^2 = L^2 : l^2}$

Explanation $\pi RL : \pi rl = RL : rl = \frac{R}{r} : \frac{l}{L} = \frac{R}{r} : \frac{r}{R}$

[from property (i), $\frac{l}{L} = \frac{r}{R}$]

$= R^2 : r^2$

Since $\frac{H}{h} = \frac{R}{r} = \frac{L}{l}$,

the required ratio $= R^2 : r^2 = H^2 : h^2 = L^2 : l^2$

(iv) Total surface area of parent cone (*AED*) : Total surface area of cone cut (*ABC*) $= R^2 : r^2 = H^2 : h^2 = L^2 : l^2$

Explanation $\pi R(L+R) : \pi r(l+r) = \frac{R}{r} : \frac{(l+r)}{(L+R)}$

$= \frac{R}{r} : \frac{\left(\frac{Lr}{R} + r\right)}{(L+R)}$ [from property (i), $\frac{R}{r} = \frac{L}{l}, \therefore l = \frac{Lr}{R}$]

$= \frac{R}{r} : \frac{(Lr + Rr)}{R(L+R)} = \frac{R}{r} : \frac{r}{R} = R^2 : r^2$

Since $\frac{H}{h} = \frac{R}{r} = \frac{L}{l}$, the required ratio

$= R^2 : r^2 = H^2 : h^2 = L^2 : l^2$

Ratio-Analysis in Case of Two/Three or More Cylinders/Cones

(i) When heights are same

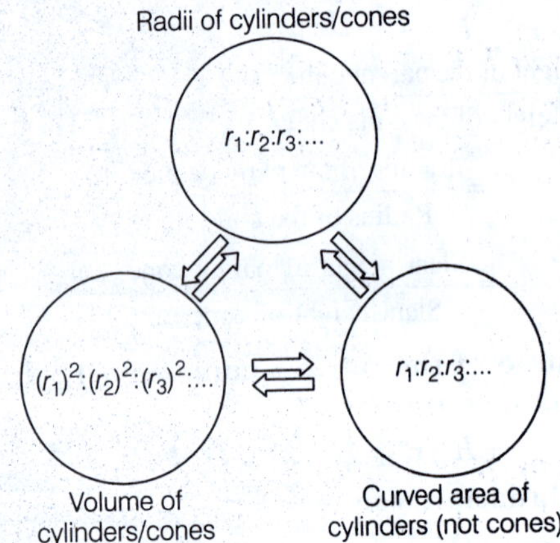

(ii) When radii are same

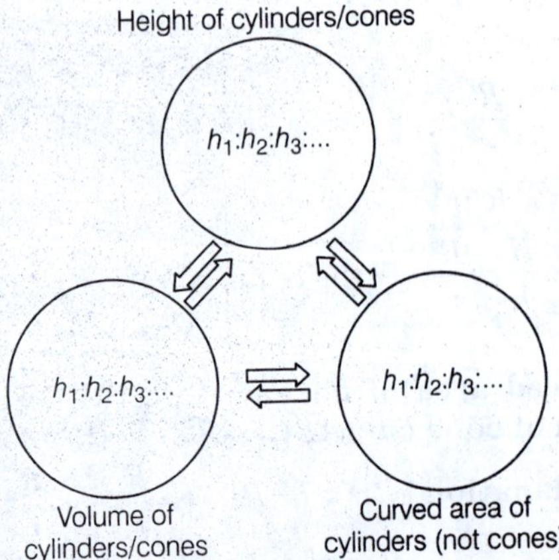

(iii) When neither heights are same nor radii are same

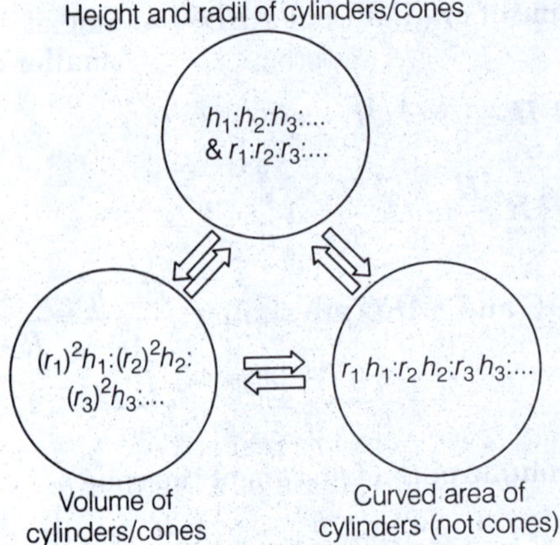

(iv) When curved areas of cylinders are same

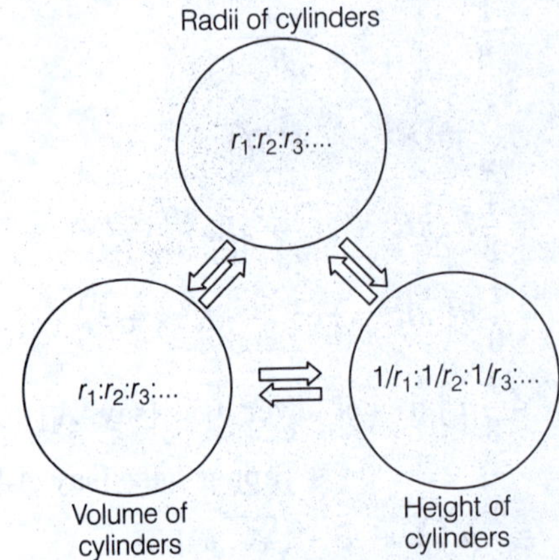

(v) When volumes of cylinders/cones are same

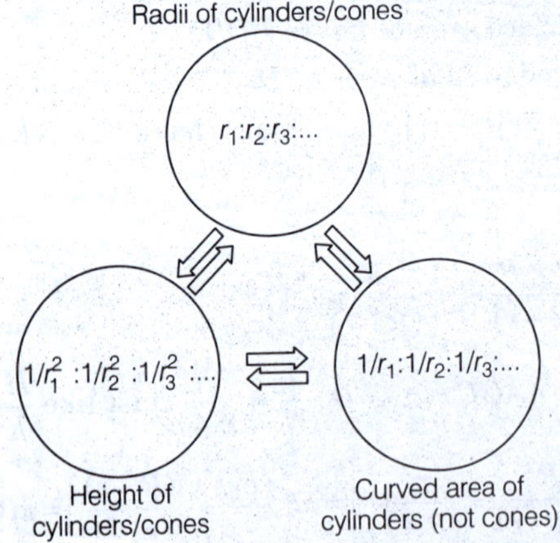

[In the above context, also ratio of radii = ratio of diameters]

Applications of the Concepts of Volume or /and Surface Area

When we multiply the curved surface area of a cylinder by $\frac{r}{2}$, we get its volume.

Problem 1 The curved surface area of a cylindrical pillar is 264 m^2 and its volume is 924 m^3. What is the ratio of its diameter to its height?

Sol. From the above observation,

Curved surface area $\times \frac{r}{2}$ = Volume

$\Rightarrow \quad \frac{r}{2} = \frac{924}{264}, \quad \therefore r = 7$

Further, $\quad 2\pi rh = 264$

$\Rightarrow \quad 2 \times \frac{22}{7} \times 7 \times h = 264, \quad \therefore h = 6$

Thus, required ratio = $(7 \times 2) : 6 = 7 : 3$

Problem 2 If h, c, v are respectively the height, curved surface area and volume of a right circular cone, then the value of $3\pi vh^3 - c^2h^2 + 9v^2$ is :

[SSC (CGL T-II)-13, SNAP-18]

Sol. Given expression

$= 3\pi\left(\frac{1}{3}\pi r^2 h\right) \times h^3 - \left[\pi r\sqrt{(h)^2 + (r)^2}\right]^2 \times h^2 + 9\left(\frac{1}{3}\pi r^2 h\right)^2$

$= \pi^2 r^2 h^4 - \pi^2 r^2 h^2 (h^2 + r^2) + \pi^2 r^4 h^2 = 0$

Problem 3 If x is the curved surface area and y is the volume of a right circular cylinder, then which one of the following is correct?

(a) Only the ratio of the height to radius of the cylinder is independent of x

(b) Only the ratio of the height to radius of the cylinder is independent of y

(c) Either (a) or (b)

(d) Neither (a) nor (b)

Sol. Given, $\frac{2\pi rh}{\pi r^2 h} = \frac{x}{y} \Rightarrow r = \frac{2y}{x}$

Also, $\quad h = \frac{x}{2\pi r} \quad [\because 2\pi rh = x]$

Now, $\quad \frac{r}{h} = \frac{2y}{x} \times \frac{2\pi r}{x} = \frac{2y}{x} \times \frac{2\pi \times \frac{2y}{x}}{x} = \frac{8\pi y^2}{x^3}$

$\Rightarrow \quad \frac{h}{r} = \frac{x^3}{8\pi y^2}$

Hence, the ratio mentioned in the question is not independent of x or y.

Problem 4 A right circular cylinder of height 16 cm is covered by a rectangular tin foil of size 16 cm × 22 cm. The volume of the cylinder is

Sol. Curved surface area $= 2\pi rh = 16 \times 22$ (as given)

$\Rightarrow \quad 2\pi r \times 16 = 16 \times 22 \Rightarrow r = 3.5$

Now, volume $= \pi r^2 h = \frac{22}{7} \times (3.5)^2 \times 16 = 616 \text{ cm}^3$

Problem 5 If a right circular cone of height 24 cm has a volume of 1232 cm^3, then the area of its curved surface is :

[SSC (CGL-PT)-03, SSC (CGL-T-II)-11]

Sol. Given, volume $= \frac{1}{3}\pi r^2 h = 1232$

$\Rightarrow \quad \frac{1}{3}\pi r^2 \times 24 = 1232$

$\Rightarrow \quad r^2 = \frac{1232 \times 3 \times 7}{22 \times 24} = 49$

$\Rightarrow \quad r = 7 \text{ cm}$

Now, the slant height $(l) = \sqrt{7^2 + 24^2} = 25$

$\therefore$ Curved surface area $= \pi rl = \frac{22}{7} \times 7 \times 25$

$= 550 \text{ cm}^2$

Problem 6 The height of a conical tank is 60 cm and the diameter of its base is 64 cm. The cost of painting it from outside at the rate of ₹ 35 per sq. m is :

[SSC (CGL-PT)-99, CDS-13]

Sol. Given, $h = 60$ cm, $r = \frac{64}{2} = 32$ cm

$\therefore \quad l = \sqrt{(60)^2 + (32)^2} = 68$

Needless to say that only curved area of the tank will be painted.

$\therefore \quad$ Cost = curved area × rate

$= \left(\frac{22}{7} \times \frac{32 \times 68}{100 \times 100}\right) \times 35 \quad [\because 1 \text{ m}^2 = 100 \times 100 \text{ cm}^2]$

= ₹ 23.94

Problem 7 A cylindrical vessel of base radius 14 cm is filled with water to some height. If a rectangular solid of dimensions 22 cm × 7 cm × 5 cm is immersed in it, what is the rise in water level ?

[CDS-09]

Sol. Let water rise be h cm.

Then, volume of the water risen in vessel = volume of the solid immersed

$\Rightarrow \quad \pi(14)^2 h = 22 \times 7 \times 5$

$\Rightarrow \quad h = 1.25 \text{ cm}$

Problem 8 The base diameter of a right circular cylinder is 3 cm. There is a section making an angle of 30° with the cross-section. What is its area ?

Sol. In ΔABO, $\cos 30° = AB/AO = 1.5/AO$

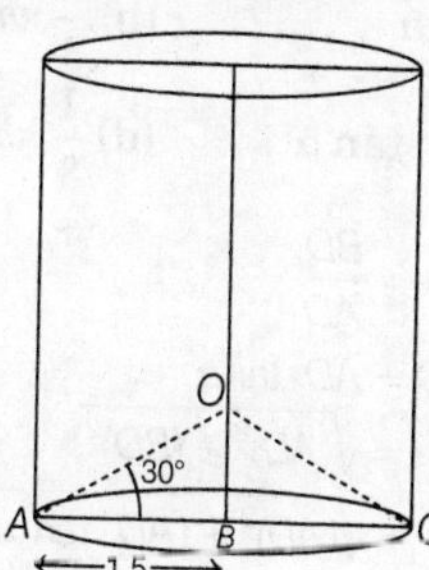

$\Rightarrow \quad \sqrt{3}/2 = 1.5/AO \Rightarrow AO = \sqrt{3}$

Now, curved area of the cone $AOC = \pi rl$

$= \pi \times 1.5 \times \sqrt{3}$

$= \frac{3\sqrt{3}\pi}{2}$ sq. cm

Problem 9 A well 20 m in diameter is dug 14 m deep and the earth taken out is spread all around it to a width of 5 m to form an embankment. The height of the embankment is : *[SSC (CGL-T-II)-20]*

Sol. When a well is dug, a shape of cylinder will appear.

Thus, the volume of the earth taken out = $\pi r^2 h$

$$= \pi \times \left(\frac{20}{2}\right)^2 \times 14 = \pi(10)^2 \times 14 \text{ m}^3$$

The earth taken out is spread around the well to a width of 5 m, it means the area of the base of embankment × its height = volume of the earth spread

$$\Rightarrow \quad \pi[(10+5)^2 - 10^2] \times h = \pi(10)^2 \times 14$$

$$\Rightarrow \quad 125h = 1400$$

$$\therefore \quad h = 11.2 \text{ m}$$

Problem 10 If the area of the base of a cone is 770 cm^2 and the area of the curved surface is 814 cm^2, then its volume (in cm^3) is

Sol. Given, $\pi r^2 = 770 \Rightarrow \frac{22}{7} r^2 = 770$

$$\therefore \quad r = 7\sqrt{5}$$

Further, $\pi r l = 814 \Rightarrow \frac{22}{7} \times 7\sqrt{5} \times l = 814$

$$\Rightarrow \quad l = \frac{37}{\sqrt{5}}$$

Now, $l = \sqrt{r^2 + h^2} \Rightarrow \frac{37}{\sqrt{5}} = \sqrt{(7\sqrt{5})^2 + h^2}$

$$\Rightarrow \quad \frac{1369}{5} = 245 + h^2 \quad \text{[on squaring both sides]}$$

$$\Rightarrow \quad 5h^2 = 1369 - 1225 = 144$$

$$\therefore \quad h = \frac{22}{\sqrt{5}}$$

Thus, its volume $= \frac{1}{3} \times \frac{22}{7} \times (7\sqrt{5})^2 \times \frac{12}{\sqrt{5}}$

$$= 616\sqrt{5} \text{ cm}^3$$

Problem 11 If s denotes the area of the curved surface of a right circular cone of height h and semi vertical angle α then s equals

(a) $\pi h^2 \tan^2 \alpha$ (b) $\frac{1}{3} \pi h^2 \tan^2 \alpha$

(c) $\pi h^2 \sec \alpha \cdot \tan \alpha$ (d) $\frac{1}{3} \pi h^2 \cdot \sec \alpha \cdot \tan \alpha$

Sol. Clearly, $\tan \alpha = \frac{BD}{AD}$

$$\Rightarrow \quad BD = AD \cdot \tan \alpha$$

Now, $l = \sqrt{(AD)^2 + (BD)^2}$

$$= \sqrt{(AD)^2 + (AD \cdot \tan\alpha)^2}$$

$$= AD \cdot \sec \alpha = h \cdot \sec \alpha$$

$\therefore$ Curved area $= \pi r l = \pi \cdot BD \cdot h \cdot \sec \alpha$

$$= \pi \cdot (AD \cdot \tan\alpha) \cdot h \sec \alpha$$

$$= \pi \cdot h \tan\alpha \cdot h \sec \alpha$$

$$= \pi h^2 \cdot \sec \alpha \cdot \tan \alpha$$

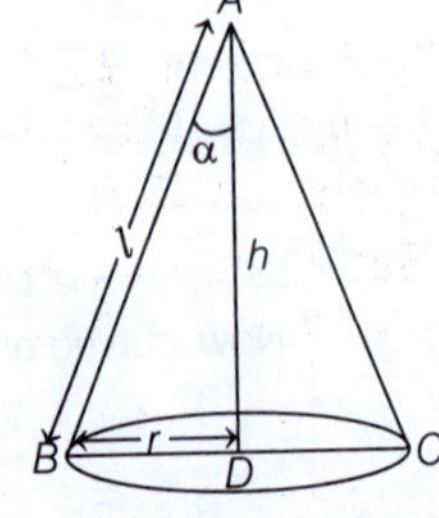

Problem 12 The curved surface area of a cylinder is 1000 sq. cm. A wire of diameter 5 mm is wound around it so as to cover it completely. What is the length of the wire used ?

Sol. Given, length of wire used in one round = circumference of the base of the cylinder = $2\pi r$

And number of rounds $= \frac{\text{height of the cylinder}}{\text{diameter of the wire}} = \frac{h}{0.5}$

Then, length of the wire used

= Length of wire used in one round × number of rounds

$$= 2\pi r \times \frac{h}{0.5} = \frac{2\pi r h}{0.5} = \frac{1000}{0.5}$$

[given, curved area $(2\pi r h) = 1000$]

$$= 2000 \text{ cm} = 20 \text{ m}$$

Problem 13 Certain number of cubes of side 1.4 cm are dropped into a cylindrical beaker containing some water and they are fully submerged. The diameter of the beaker is 14 cm. Find how many cubes have been dropped in it if the water rises by 2.8 cm.

Sol. Obviously, number of cubes $= \frac{\text{Volume of water raised}}{\text{Volume of each cube}} = \frac{\pi (r)^2 h}{(\text{side})^3}$

$$= \frac{\pi\left(\frac{14}{2}\right)^2 \times 2.8}{(1.4)^3} = \frac{\frac{22}{7} \times \left(\frac{14}{2}\right)^2 \times 2.8}{(1.4)^3}$$

$$= 157.14$$

Hence, number of cubes dropped in the beaker = 157

Problem 14 A right circular cone and a right circular cylinder have the same base and their heights in the ratio 2:3. The ratio of their volumes will be *[SSC (CGL-T-II)-14]*

Sol. Required ratio $= \frac{1}{3} \pi r^2 (2h) : \pi r^2 (3h) = 2 : 9$

Problem 15 From a right circular cylinder of radius 10 cm and height 21 cm, a right circular cone of same base is removed. If the volume of the remaining portion is 4400 cm^3, then the height of the removed cone is

Sol. Volume of the cylinder $= \frac{22}{7} \times (10)^2 \times 21 = 6600$

$\therefore$ Volume of the removed cone = 6600 − 4400 = 2200

$$\Rightarrow \quad \frac{1}{3} \pi (10)^2 \times h = 2200 \Rightarrow \frac{1}{3} \times \frac{22}{7} \times 100 \times h = 2200$$

$$\therefore \quad h = 21 \text{ cm}$$

Problem 16 If the height of a cylinder is 4 times its circumference, the volume of the cylinder in terms of its circumference c, is : *[SSC (CGL-T-II)-14]*

Sol. Given, $h = 4c$

Then, its volume $= \pi r^2 h = \pi r^2 4c$

$$= \frac{2\pi r \times 2\pi r \times 4c}{4\pi} = \frac{c \times c \times c}{\pi} = \frac{c^3}{\pi}$$

Problem 17 A right circular cylinder and a cone have equal base radius and equal heights. If their curved surfaces are in the ratio 8 : 5, then the radius of the base to the height are in the ratio

Sol. Given, $\frac{2\pi rh}{\pi rl} = \frac{8}{5} \Rightarrow \frac{h}{l} = \frac{4}{5}$

We know,

$$l^2 = h^2 + r^2 \Rightarrow 5^2 = 4^2 + r^2 \Rightarrow r = 3$$

Hence, required ratio = 3 : 4

Problem 18 The height of a solid right circular cylinder is 6 m and three times the sum of the areas of its two end faces is twice the area of its curved surface. The radius of its base (in metres) is :

Sol. Given, $3(\pi r^2 + \pi r^2) = 2\,(2\pi rh) \Rightarrow 6\pi r^2 = 4\pi rh$

$\Rightarrow \quad 3r = 2h \Rightarrow 3r = 2 \times 6 \quad \therefore r = 4\text{ m}$

Problem 19 The radius of a cylinder is 10 cm and height is 4 cm. The number of centimeters that may be added either to the radius or to the height to get the same increase in the volume of the cylinder is

Sol. Volume $= \pi r^2 h = \pi(10)^2 \times 4$

As per question, $\pi(10 + x)^2 \times 4 = \pi(10)^2 \times (4 + x)$

[let x cm are increased]

$\Rightarrow \quad 100 + 20x + x^2 = 100 + 25x$

$\Rightarrow \quad x^2 = 5x$

$\Rightarrow \quad x = 5$

Problem 20 The sum of radius and height of a cone is 7.5 cm. What is the maximum volume of the cone ?

Sol. Volume of the cone $= \frac{1}{3}\pi r^2 h = \left(\frac{1}{3}\pi\right) hr^2$

We know that xy^2 has maximum value if $x : y = 1 : 2$ provided $x + y = a$ constant number.

Obviously, $h = 2.5$ and $r = 5$

$\therefore$ Maximum volume $= \frac{1}{3}\pi \times 2.5 \times (5)^2 = 20.8\pi \text{ cm}^3$

Problem 21 The area of the curved surface and the area of the base of a right circular cylinder are a cm^2 and b cm^2 respectively. The height of the cylinder is

Sol. Area of the base $= \pi r^2 = b$ (given)

$\therefore \quad r = \sqrt{\frac{b}{\pi}}$

Also given, $2\pi rh = a \Rightarrow 2\pi \times \sqrt{\frac{b}{\pi}} \times h = a$

$\Rightarrow \quad h = \frac{a}{\sqrt{b} \times 2\sqrt{\pi}} = \frac{a}{2\sqrt{\pi b}}$

Problem 22 Increasing the radius of the base of cylinder by 6 units increase the volume by y cubic units. Increasing the altitude of the cylinder by 6 units also increase the volume by y cubic units. If the original altitude is 2 units, find the original radius.

Sol. Original volume $= \pi r^2 h$

Volume if radius is increased by 6 units $= \pi(r + 6)^2 h$

and volume if altitude is increased by 6 units $= \pi r^2 (h + 6)$

In both cases, the original volume is increased by y cubic units, it means,

$$\pi(r + 6)^2 h = \pi r^2 (h + 6)$$

$\Rightarrow \quad \pi(r + 6)^2 \times 2 = \pi r^2 (2 + 6)$

$\Rightarrow \quad r^2 + 12r + 36 = 4r^2 \Rightarrow -3r^2 + 12r + 36 = 0$

$\Rightarrow \quad -r^2 + 4r + 12 = 0 \Rightarrow -r^2 + 6r - 2r + 12 = 0$

$\Rightarrow \quad -r(r - 6) - 2(r - 6) = 0$

$\Rightarrow \quad r = 6 \text{ or } r = -2$

$\therefore$ Radius = 6 cm. [$\because -2$ cannot be value of the radius]

Alternatively

We should proceed through the options given in such cases.

Problem 23 The total surface area of the solid cylinder is 1628 cm^2 and sum of the radius of the base and its height is 37 cm. Find its volume.

Sol. Given, total surface area = 1628

$\Rightarrow \quad 2\pi r(r + h) = 1628$

$\Rightarrow \quad 2\pi r(37) = 1628 \quad [\because r + h = 37, \text{ given}]$

$\Rightarrow \quad r = \frac{1628 \times 7}{2 \times 22 \times 37} = 7$

Hence, $h = 37 - r = 37 - 7 = 30$ cm

Now, its volume $= \pi r^2 h = \frac{22}{7} \times 7 \times 7 \times 30 = 4620 \text{ cm}^3$

Problem 24 A cylinder is filled to $\frac{4}{5}$th of volume. It is then tilted so that the level of water coincides with one edge of its bottom and top edge of the opposite side. In the process, 30 cc of the water is spilled. What is the volume of the cylinder ?

Sol. Given, $\frac{4}{5}\pi r^2 h - 30 = \frac{1}{2}\pi r^2 h$

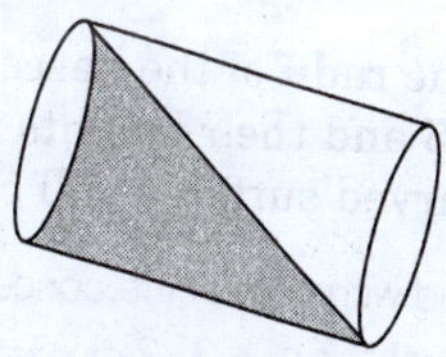

$\Rightarrow \quad \pi r^2 h\left(\frac{4}{5} - \frac{1}{2}\right) = 30$

$\therefore \quad \pi r^2 h = 100 \text{ cc.}$

Problems based on Ratio Analysis in Case of two/three or more Cylinders/Cones

Problem 25 Two right circular cylinders of equal volume have their heights in the ratio 1 : 2. The ratio of their radii is

Sol. Let their height be h and $2h$.

Then, $\pi(r_1)^2 h = \pi(r_2)^2 2h$

$\Rightarrow \quad \frac{(r_1)^2}{(r_2)^2} = \frac{2h}{h} \Rightarrow \frac{r_1}{r_2} = \sqrt{\frac{2h}{h}} = \sqrt{2} : 1$

Alternatively

Proceed through concept of ratio analysis detailed in beginning of this chapter.

If volumes are same, then ratio of heights $= \frac{1}{(r_1)^2} : \frac{1}{(r_2)^2}$

$$\Rightarrow \quad 1:2 = \frac{1}{(r_1)^2} : \frac{1}{(r_2)^2} \Rightarrow \frac{1}{2} = \frac{(r_2)^2}{(r_1)^2}$$

$$\Rightarrow \quad \frac{(r_1)^2}{(r_2)^2} = \frac{2}{1} \Rightarrow r_1 : r_2 = \sqrt{2} : 1$$

Problem 26 The base radii of two cylinders are in the ratio 2 : 3 and their heights are in ratio 5 : 3. The ratio of their volumes is : *[SSC (CGL-PT)-02, SSC (CGL-T-I)-19]*

Sol. You may proceed directly through the concept of ratio analysis detailed in beginning of this chapter.

The required ratio $= 2^2 \cdot 5 : 3^2 \cdot 3 = 20 : 27$

Problem 27 The ratio of the volumes of two cones is 2 : 3 and the ratio of radii of their base is 1 : 2. The ratio of their height is

Sol. Here, neither radii nor heights are same. Now, you can proceed with help of the concept of ratio analysis.

Ratio of volumes $= (r_1)^2 h_1 : (r_2)^2 h_2$

$$\Rightarrow \quad 2:3 = 1^2 h_1 : 2^2 h_2$$

$$\Rightarrow \quad \frac{2}{3} = \frac{h_1}{4h_2} \Rightarrow h_1 : h_2 = 8 : 3$$

Problem 28 If the height of a given cone be doubled and radius of the base remains the same, the ratio of the volume of the given cone to that of the second cone will be

Sol. Given, ratio of their heights $= h_1 : h_2 = 1:2$

Thus, ratio of their volumes $= h_1 : h_2 = 1:2$

Problem 29 The radii of the bases of two cylinders are in the ratio 3 : 5 and their heights in the ratio 2: 3. The ratio of their curved surfaces will be *[SSC (CPO-SI)-07]*

Sol. You can proceed with help of the concept of ratio analysis.

Required ratio $= r_1 h_1 : r_2 h_2 = 3 \times 2 : 5 \times 3 = 2 : 5$

Problem 30 A cylinder has '*r*' as the base radius and *h* as the height. The radius of base of another cylinder, having double the volume but the same height as that of the 1st cylinder must be equal to : *[FCI (Asst Gr.-III)-12]*

Sol. Given, volumes' ratio = 1: 2 and ratio of heights = 1 : 1

We know, ratio of volumes $= (r_1)^2 h_1 : (r_2)^2 h_2$

$$\Rightarrow \quad 1:2 = (r)^2 h : (r_2)^2 h$$

$$\Rightarrow \quad \frac{1}{2} = \frac{r^2 h}{(r_2)^2 h}$$

$$\Rightarrow \quad (r_2)^2 = 2r^2 \therefore r_2 = \sqrt{2r^2} = r\sqrt{2}$$

Problem 31 If the curved surface area of a cone is thrice that of another cone and slant height of the 2nd cone is thrice that of the former. Find the ratio of areas of their bases *[MAT-14, 19]*

Sol. $\frac{\text{Curved area of the 1st cone}}{\text{Curved area of the 2nd cone}} = \frac{\pi Rl}{\pi r(3l)} \Rightarrow \frac{3}{1} = \frac{\pi Rl}{3\pi rl} \Rightarrow R = 9r$

Now, required ratio $= \frac{\text{Area of base of 1st cone}}{\text{Area of base of 2nd cone}}$

$$= \frac{\pi R^2}{\pi r^2} = \frac{\pi (9r)^2}{\pi r^2} = 81 : 1$$

Problem 32 The radii of two cylinders are in the ratio 2 : 3 and their curved surface areas are in the ratio 5:3. What is the ratio of their volumes? *[CDS-11]*

Sol. We know that when we multiply the curved surface area of a cylinder by *r*/2, we get its volume.

Thus, the required ratio $= 5 \times \frac{2}{2} : 3 \times \frac{3}{2} = 10 : 9$

Concept of Volume/Area When Thickness of Layer/Surface is Taken Into Consideration

It has earlier been stated that in case of hollow cylindrical pipes/items, the thickness of the pipe will be equal to the difference between its external and internal radii and the volume of the metal/material used will be equal to the difference between their external and internal volumes.

Hence, volume of the material used

$$= \pi h \times [(\text{exterior radius})^2 - (\text{interior radius})^2]$$

$$= \pi h (R^2 - r^2)$$

Further, total surface area $= 2\pi(R + r)(h + R - r)$

[formulae have already been derived in the beginning]

The illustrations given below will emphasize manifestly the applications of these conceptions.

Problem 33 The internal radius and thickness of a hollow metallic pipe are 24 cm and 1 cm respectively. It is melted and recast into a solid cylinder of equal length. The diameter of the solid cylinder will be : *[FCI (Asst Gr.-III)-12]*

Sol. The internal volume $= \pi r^2 h = \pi(24)^2 \times h = 576\pi h$

and external volume $= \pi(25)^2 h = 625\pi h$

[$\because$ external radius = 24 + 1]

$\therefore$ Volume of the metal used $= 625\pi h - 576\pi h = 49\pi h$

Obviously, volume of the metal used = volume of the solid cylinder

$$\Rightarrow \quad 49\pi h = \pi(r)^2 h \Rightarrow r = 7$$

$\therefore$ Its diameter $= 7 \times 2 = 14$ cm

Note *The volume of the metal may be computed directly as* $\pi h \times [(\text{exterior radius})^2 - (\text{interior radius})^2]$

Problem 34 **A metal pipe is 21 cm long and its exterior diameter is 8 cm. If the thickness of the pipe is 1 cm and the metal weighs 8 gm/cm^3, the weight of the pipe (in kg) is :** ***[SSC (10+2, LDC/DEO)-11, CDS-10]***

Sol. Obviously, exterior radius = 4 cm

and interior radius = 4 − 1 = 3 cm

$\therefore$ Volume of the metal $= \pi h(4^2 - 3^2)$

$= \frac{22}{7} \times 21 \times 7 = 462 \text{ cm}^3$

Now, its weight = 462 × 8g = 3.696 kg

Problem 35 **The volume of the metal of a cylindrical pipe is 748 cm^3. The length of the pipe is 14 cm and its external radius is 9 cm. Its thickness is :** ***[SSC (CGL-PT)-08, SSC (CGL-T-II)-14]***

Sol. Volume of the metal $= \pi h[(9)^2 - (r)^2]$ (here, r is internal radius)

$\Rightarrow 748 = \frac{22}{7} \times 14[81 - r^2]$

$\Rightarrow r^2 = 81 - 17 = 64$

$\therefore r = 8$

Then, thickness of the pipe = exterior radius − interior radius

= 9 − 8 = 1 cm

Problem 36 **The difference between the outside and inside surfaces of a cylindrical metallic pipe 14 cm long is 44 cm^2. If the pipe is made of 99 cm^3 of metal, find the outer and inner radii of the pipe.**

Sol. Difference between the outside and inside surfaces

$= 2\pi Rh - 2\pi rh$

[R and r be outer and inner radii respectively]

$\Rightarrow 44 = 2\pi \times 14 \times [R - r]$

$\Rightarrow R - r = \frac{1}{2}$ cm ...(i)

Further, volume of the metal $= \pi h(R^2 - r^2)$

$\Rightarrow 99 = \frac{22}{7} \times 14(R^2 - r^2)$

$\Rightarrow R^2 - r^2 = \frac{99}{44}$

$\Rightarrow (R - r)(R + r) = \frac{99}{44} \Rightarrow \frac{1}{2}(R + r) = \frac{9}{4}$

$\Rightarrow R + r = \frac{9}{2}$ ...(ii)

On solving Eqs. (i) and (ii), we get $R = 2.5$ cm and $r = 2$ cm

Problem 37 **A hollow metallic pipe is 28 cm long and its exterior and interior radii are 5 cm and 4 cm respectively. Find the cost of polishing the pipe all over at ₹ 1 per cm^2.**

Sol. The interior radius (r) = 4 cm

and the exterior radius (R) = 5 cm

Now, exterior curved area $= 2\pi Rh$

and Interior curved area $= 2\pi rh$

Area of one base ring $= (\pi R^2 - \pi r^2)$

$\therefore$ Area of both base rings $= 2\pi(R^2 - r^2)$

Then, total surface area (the area to be polished) = exterior curved area + interior curved area + area of both base rings

$= 2\pi Rh + 2\pi rh + 2\pi(R^2 - r^2)$

$= 2\pi h(R + r) + 2\pi(R^2 - r^2)$

$= 2\pi(R + r)(h + R - r)$

$= 2 \times \frac{22}{7}(5 + 4)(28 + 5 - 4) = 1640.57 \text{ cm}^2$

$\therefore$ Cost of polishing = 1640.57 × 1 = ₹ 1640.57

Alternatively

We may write directly as below :

Total surface area $= 2\pi(R + r)(h + R - r)$ [already discussed]

$= 2 \times \frac{22}{7}(5 + 4)(28 + 5 - 4) = 1640.57 \text{ cm}^2$

$\therefore$ Cost of polishing = 1640.57 × 1 = ₹ 1640.57

Alteration in Dimensions Affecting Volume and Surface areas

- **If alteration is reflected in terms of ...... times.**

1. **If radius and height of a cylinder/cone are made x and y times respectively, the volume will become x^2y times the earlier.**
2. **If radius and height of a cylinder are made x and y times respectively, the curved surface area will become xy times the earlier.**

- **If alteration is reflected in terms of Percentage**

3. Recall the '**concept of effect on product in case of changes in factors**' detailed in chapter percentage. It should be kept in mind that the volume refers to three dimensions and the area does two. In case of volume, radius is multiplied by radius and further by the height whereas in case of lateral area radius is multiplied by either height or slant height as the case may be. The 'concept of effect on product in case of changes in factors' will not be applicable in case of total surface area of a cylinder/cone as they are a sum of certain products.

- **If alteration is reflected in terms of ratio**

4. If the dimensions of a cylinder/cone are changed in ratios (say, radius in ratio $a : b$ and height in ratio $c : d$), then its volume will be changed in ratio $a^2c : b^2d$
   (a) If $a^2c > b^2d$, the volume diminishes.
   (b) If $a^2c < b^2d$, the volume increases.
   (c) If $a^2c = b^2d$, no changes in the volume.
5. If dimensions are changed in ratios and you are required to compute the changes in volume in percentage term, then the percentage change in volume of a cylinder/cone

$$= \left\{\frac{a^2c - b^2d}{a^2c} \times 100\right\}\%$$

[If the outcome is positive, volume diminishes and in case of negative, it increases]

****

6. If the dimensions of a cylinder are changed in ratios (say, radius in ratio $a:b$ and height in ratio $c:d$), then its curved surface area will be changed in ratio $ac:bd$.
   (a) If $ac > bd$, curved surface area diminishes.
   (b) If $ac < bd$, curved surface area increases.
   (c) If $ac = bd$, no changes in curved surface area.
7. If dimensions of a cylinder are changed in ratios and you are required to compute the changes in curved surface area in percentage term, then
   Percentage change in curved surface area
   $$= \left(\frac{ac - bd}{ac} \times 100\right)\%$$
   [If the outcome is positive, area diminishes and in case of negative it increases]
   Now, we will observe some illustrations as below.

Problem 38 If the radius of a cylinder is decreased by 50% and the height is increased by 50% to form a new cylinder, the volume will decrease by how much percentage?

Sol. Obviously, three figures are −50%, −50% and +50% as radius is repeated twice in case of volume. New, effect on taking any two figures (say, −50 and +50)

$$\Rightarrow \left(-50 + 50 - \frac{50 \times 50}{100}\right)\% = -25\%$$

$$\therefore \text{Final effect} = \left(-25 - 50 + \frac{25 \times 50}{100}\right)\% = -62.5\%$$

Hence, the volume of new cylinder will decrease by 62.5%.

Alternatively I

If radius and height of a cylinder are made x and y times respectively, the volume will become x^2y times the earlier (observe the facts given above). Here, radius is made to 0.5 times and height is made to 1.5 times the earlier. Then, the volume will become $[(0.5)^2 \times 1.5]$ times the earlier i.e. 0.375 times. Hence, decline in volume in percentage = 62.5%

Alternatively II

It may be said that radius has changed in ratio 100 : 50 i.e. 2 : 1 (say, $a:b$) and height in ratio 100 : 150 i.e. 2 : 3 (say, $c:d$) Then, percentage change in volume of the cylinder

$$= \left(\frac{a^2c - b^2d}{a^2c} \times 100\right)\%$$

$$= \left\{\frac{(2^2 \times 2) - (1^2 \times 3)}{(2^2 \times 2)} \times 100\right\}\% = \left\{\frac{8-3}{8} \times 100\right\}\%$$

$$= +62.5\%$$

'+' indicates that volume has decreased and the decline is 62.5%.

Problem 39 If the height of a cylinder is increased by 15% and radius of its base is decreased by 10%, then by what per cent will its curved surface area change?

Sol. Effect on the curved surface area $= \left(+15 - 10 - \frac{15 \times 10}{100}\right)\%$

$$= +3.5\%$$

It means the curved surface will increase by 3.5%

Alternatively I

If radius and height of a cylinder are made x and y times respectively, the curved surface area will become xy times the earlier (observe the facts given above). Here, radius is made to 0.9 times and height is made to 1.15 times the earlier. Then, the curved surface area will become $[0.9 \times 1.15]$ times the earlier i.e. 1.035 times. Hence, increase in curved area in percentage = 3.5%.

Alternatively II

It may be said that radius has changed in ratio 100 : 115 i.e. 20 : 23 (say, $a:b$) and height in ratio 100 : 90 i.e. 10 : 9 (say, $c:d$) Then, percentage change in curved surface area of the cylinder

$$= \left\{\frac{ac - bd}{ac} \times 100\right\}\% = \left\{\frac{(20 \times 10) - (23 \times 9)}{(20 \times 10)} \times 100\right\}\% = -3.5\%$$

∴ '−' indicates that curved area has increased and the increment is 3.5%.

Problem 40 The height of a circular cylinder is increased to six times and the base area is decreased to one-ninth of its value. The factor by which the lateral surface of the cylinder increases is : ***[SSC (CGL-T-II)-12]***

Sol. The lateral surface area $= 2\pi rh$

Now, height (h) becomes $6h$

and radius becomes $\frac{1}{3}r$ [∵ base area (πr^2) becomes $\frac{1}{9}\pi r^2$]

∴ Latest lateral surface area $= 2\pi \times \frac{1}{3} r \times 6h = 4\pi rh$

Hence, latest lateral area will become twice.

Alternatively I

$$h = +500\%, r = -66.66\%$$

Total effect on lateral surface area

$$= \left(+500 - 66.66 - \frac{500 \times 66.66}{100}\right)\%$$

$$= +433.33 - 333.33 = +100\%$$

The latest surface area will increase by 100% i.e. becomes twice as compared to earlier.

Alternatively II

Clearly, height has been made to 6 times and radius to $\frac{1}{3}$ times the earlier. Then, lateral surface of the cylinder will become $\left[6 \times \frac{1}{3}\right]$ times the earlier i.e. 2 times.

Alternatively III

It may be said that height has changed in ratio 1 : 6 (say, $a:b$) and radius in ratio 3 : 1 (say, $c:d$) Then, percentage change in lateral surface area of the cylinder $= \left\{\frac{ac - bd}{ac} \times 100\right\}\%$

$$= \left\{\frac{(1 \times 3) - (6 \times 1)}{(1 \times 3)} \times 100\right\}\% = -100\%$$

∴ '−' indicates that curved area has increased and the increment is 100%. It means this area has become twice the earlier.

It may be construed from the discussion made through the above three problems that all the techniques are correlated and aspirants are supposed to be well versed with all these techniques so that they apply them appropriately observing the nature of questions.

Problem 41 **The radius of the base of a right circular cone is doubled to keep the volume fixed, the height of the cone will be :**

Sol. Original volume $= \frac{1}{3} \times \pi r^2 h$

New volume $= \frac{1}{3} \times \pi (2r)^2 H$

But, $\frac{1}{3} \times \pi r^2 h = \frac{1}{3} \times \pi 4r^2 H$ [$\because$ volumes have to be kept same]

$\Rightarrow \quad H = \frac{h}{4}$

Hence, height of the cone will be $\frac{1}{4}$ of the previous height.

Alternatively

Here, r is increased by 100%.

i.e. its effect $= \left\{+100 + 100 + \frac{100 \times 100}{100}\right\}\% = +300\%$

Now, recall the concept of "Product constancy" detailed in chapter 'percentage'.

$\Rightarrow \quad \left\{\frac{300}{(100+300)} \times 100\right\}\% = \left\{\frac{300}{400} \times 100\right\}\% = 75\%$

It means height must be reduced by 75% to keep the volume fixed. i.e. height of the cone will be $\frac{1}{4}$ of the previous height.

Problem 42 **Height of a cylindrical jar is decreased by 36%. By what per cent must the radius be increased so that there is no change in its volume ?** *[LIC (AAO)-12]*

Sol. Let each of height and radius be 100 cm, then volume of the cylinder $= \pi(100)^2 \times 100 = 1000000\pi$

Assume, radius is increased by x% to keep the volume constant, then $\pi(100 + x)^2 \times (100 - 36) = 1000000\pi$

$\Rightarrow \quad 100 + x = \sqrt{\frac{1000000}{64}} = 125 \Rightarrow x = 25\%$

Alternatively I

Let the radius is increased by r%.

Then, effect on the base area $= \left\{+r + r + \frac{r \times r}{100}\right\}\% = \left\{2r + \frac{r^2}{100}\right\}\%$

Now, volume is to be kept constant i.e. 0% increment and the fact that volume is a product of two variable factors *viz.* r^2 and h with a constant term.

Hence, $\left[+\left\{2r + \frac{r^2}{100}\right\} - 36 - \frac{\left\{2r + \frac{r^2}{100}\right\} \times 36}{100}\right]\% = 0\%$

$\Rightarrow \quad +\left[2r + \frac{r^2}{100}\right]\left[1 - \frac{9}{25}\right] = 36$

$\Rightarrow \quad 200r + r^2 - 5625 = 0$

$\Rightarrow \quad r = 25\%$

Alternatively II

Volume is the product of two variable factors *viz.* r^2 and h, with a constant term. In order to keep the volume unchanged, when h is decreased by 36%, the r^2 must be increased by $\left\{\frac{36}{(100-36)} \times 100\right\}\%$ i.e. by 56.25%

Now, proceed through the options and check as to what value of r can yield 56.25% when it is put in the format : $\left\{+r + r + \frac{r \times r}{100}\right\}\%$.

Obviously, $r = 25$ will yield 56.25%.

Problem 43 **The radius of the base of a right circular cone is doubled keeping its height fixed. The volume of the cone will be**

Sol. Only radius has been increased by 100%.

$\therefore$ Effect on volume $= \left\{+100 + 100 + \frac{100 \times 100}{100}\right\}\% = +300\%$

[$\because r$ is repeated twice as factor in case of volume]

Hence, new volume will increase by 300%

i.e. it will become 4 times the previous volume.

Problem 44 **If the area of the base of a cone is increased by 100%, then the volume is increased by :** *[SSC (CGL-T-II)-14]*

Sol. Since the area of the base is multiplied by the height in order to get the volume, the volume will have the same effect as the area has. It means the volume is also increased by 100%.

Comparison Between Cylinder and Cone w.r.t Diverse Aspects

(i) Volume of cylinder : volume of cone = 3 : 1
(if they have equal radii and equal heights)

(ii) Height of cylinder : Height of cone = 1 : 3
(if they have equal radii and equal volumes)

(iii) Radius of cylinder : Radius of cone $= 1 : \sqrt{3}$
(if they have equal heights and equal volumes)

(iv) Height of cylinder : Height of cone $= r^2 : 3R^2$
(if they have equal volumes and ratio of their radii is $R : r$)

(v) Radius of cylinder : Radius of cone $= \sqrt{h} : \sqrt{3H}$
(if they have equal volumes and ratio of their heights is $H : h$)

Remarks Observe the pattern of the above ratios and it will categorically help you solve the problems without paper and pen as I hope.

Problem 45 **If C_1 is a right circular cone with base radius r_1 cm and height h_1 cm and C_2 is a right circular cylinder with base radius r_2 cm and height h_2 cm and if $r_1 : r_2 = 1 : n$ (where n is a positive integer) and their volumes are equal, then which one of the following is correct ?**

(a) $h_1 = 3nh_2$ (b) $h_1 = 3n^2 h_2$
(c) $h_1 = 3h_2$ (d) $h_1 = n^2 h_2$

Sol. Let r_1 be k and r_2 be nk.

Given that their volumes are same, then $\frac{1}{3} \times \pi k^2 h_1 = \pi(nk)^2 h_2$

$\Rightarrow \quad h_1 = 3n^2 h_2$

Alternatively

Observing the facts given above the ratio of heights ($h_2 : h_1$) will $(1)^2 : 3(n)^2$. Then, $h_1 = 3n^2 h_2$.

Conversions of Shapes (Cuboid, Cube, Cylinder, Cone)

(i) If a solid cylinder is converted into a right circular solid cone or vice-versa, with the same base, the height of the cone will be thrice that of the cylinder as the volume of the cone is 1/3rd of that of the cylinder. Needless to say that in case of same base radii will be same.

(ii) If a solid cylinder is converted into a right circular solid cone or vice-versa, with the same height, the radius of the cone will be $\sqrt{3}$ times that of the cylinder as the volume of the cone is 1/3rd of that of the cylinder.

Problem 46 A conical vessel whose internal radius is 12 cm and height 50 cm is full of liquid. the contents are emptied into a cylindrical vessel with radius (internal) 10 cm. The height to which the liquid rises in the cylindrical vessel is:

Sol. Here, the liquid has changed its shape, but the fact is that the volume will remain the same.

$\therefore$ Volume of liquid in the cone = Volume of liquid in the cylinder

$\Rightarrow \quad \frac{1}{3}\pi(12)^2 \times 50 = \pi(10)^2 \times h$

$\Rightarrow \quad h = 24$ cm

Problem 47 The edge of an ice cube is 14 cm. The volume of the largest cylindrical ice cube that can be formed out of it is : **[IBPS-13]**

Sol. Obviously, each of the height and the diameter of the circular base of the largest possible cylindrical ice cube = edge of the ice cube = 14 cm

Then, its volume = $\pi r^2 h = \frac{22}{7} \times \left(\frac{14}{2}\right)^2 \times 14 = 2156$ cu cm.

Problem 48 A cone of height 7cm and base radius 1 cm is carved from a cuboidal block of wood 10 cm × 5cm × 2cm. The percentage wood wasted in the process is

Sol. Volume of the cone = $\frac{1}{3} \times \pi(1)^2 \times 7 = \frac{22}{3}$ cm^3

Volume of the cuboid = $10 \times 5 \times 2 = 100$ cm^3

Then, wastage% = $\frac{\left(100 - \frac{22}{3}\right) \times 100}{100}\% = 92\frac{2}{3}\%$

Problem 49 A rectangular paper sheet of dimensions 22 cm × 12 cm is folded in the form of a cylinder along its length. What will be the volume of this cylinder? **[SSC (SO Comm. Audit)-05, SSC (SAS)-2010, MAT-18]**

Sol. Obviously, the circumference of the base of the cylinder is 22 cm and it height is 12 cm.

Hence, $\quad 2\pi r = 22$ and $h = 12$

$\Rightarrow \quad 2 \times \frac{22}{7} \times r = 22 \Rightarrow r = 3.5$

Now, volume = $\pi r^2 h = \frac{22}{7} \times (3.5)^2 \times 12 = 462$ cm^3

Problem 50 A right triangle with sides 3 cm, 4 cm and 5 cm is rotated about the side 3 cm to form a cone. The volume of the cone so formed is

Sol. Clearly, $r = 4$ cm and $h = 3$ cm

$\therefore$ Volume = $\frac{1}{3} \times \pi r^2 h = \frac{1}{3} \times \pi \times 4 \times 4 \times 3 = 16\pi$ cm^3

Problem 51 A solid metallic cone is melted and recast into a solid cylinder of same base as that of the cone. If the height of the cylinder is 7 cm the height of the cone is

Sol. Clearly, Volume of the cone = Volume of the cylinder

$\Rightarrow \quad \frac{1}{3}\pi r^2 h = \pi r^2 \times 7 \quad \therefore h = 21$ cm

Alternatively

Since, the volume of the cone is 1/3rd of that of the cylinder, the height of the cone will be thrice that of the cylinder provided bases are same. This fact is also given above.

Problem 52 The rain water from a roof 22 m ×20 m drains into a cylindrical vessel having a diameter of 2 m and height 3.5 m. If the vessel is just full, then the rainfall in cm is

Sol. Volume of water on roof = volume of water in the vessel

$\Rightarrow \quad 22 \times 20 \times h = \frac{22}{7} \times \left(\frac{2}{2}\right)^2 \times 3.5$

$\Rightarrow \quad h = \frac{22 \times 1 \times 3.5}{22 \times 20 \times 7} = 0.025$ m $= 0.025 \times 100 = 2.5$ cm

Problem 53 A semi-circular sheet of metal of diameter 28 cm is bent into an open conical cap. The depth of the cup is approximately : **[SSC (10+2 LDC/DEO)-11]**

Sol. If point A and B are joined , an open cone will be formed. Then, OB or OA will act as slant height l.

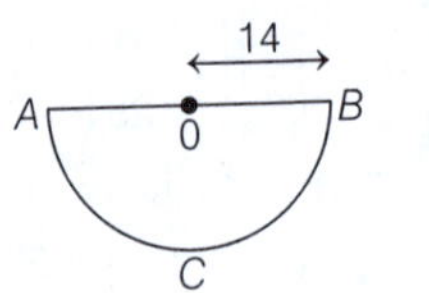

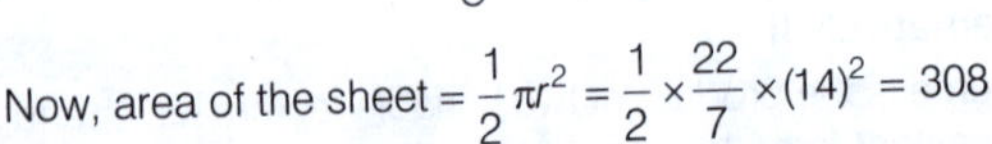

Now, area of the sheet = $\frac{1}{2}\pi r^2 = \frac{1}{2} \times \frac{22}{7} \times (14)^2 = 308$

Obviously, area of the sheet = curved area of the cone

$$\Rightarrow \quad 308 = \frac{22}{7} \times r \times 14 \qquad (\because l = 14)$$

$$\therefore \quad r = 7$$

Further, $h^2 = l^2 - r^2 = (14)^2 - (7)^2 = 147$

$$\therefore \quad h = \sqrt{147} = 12.12 \text{ i.e. } 12 \text{ cm approx.}$$

Problem 54 From a right angled triangular sheet having perpendicular sides 3 cm and 4 cm, a cone of maximum size is formed with vertex as right angle vertex of the sheet. Find the curved surface area of that cone.

Sol. Area of $\Delta ABC = \frac{1}{2} \times AC \times BD = \frac{1}{2} \times 3 \times 4$

$$\Rightarrow \quad \frac{1}{2} \times 5 \times BD = \frac{1}{2} \times 3 \times 4$$

$$\therefore \quad BD = \frac{3 \times 4}{5} = 2.4 \text{ cm.}$$

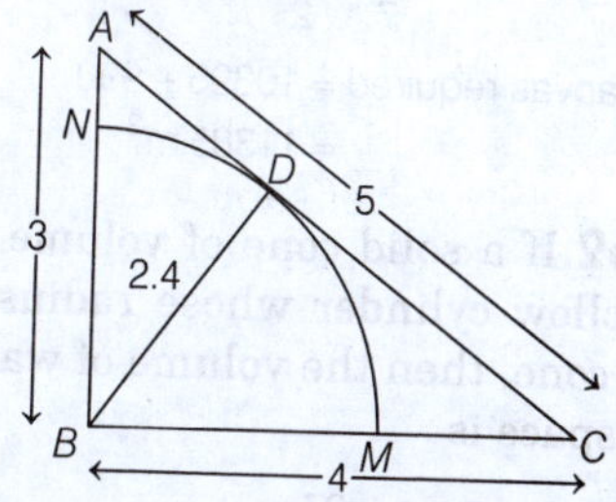

Obviously, BD will act as slant height of the cone and arc NDM will work as circumference of the base of the cone.

Now, circumference of the base = length of the arc NDM

$$\Rightarrow \quad 2\pi r = 2\pi \times 2.4 \times \frac{90°}{360°}$$

$$\Rightarrow \quad r = 0.6 \text{ cm}$$

Then, curved surface area $= \pi r l = \frac{22}{7} \times 0.6 \times 2.4 = 4.52 \text{ cm}^2$

Problem 55 The radius of a metallic cylinder is 3 cm and its height is 5 cm. It is melted and moulded into small cones, each of height 1 cm and base radius 1 mm. The number of such cones formed, is:

[FCI (Asst. Gr.-III)-12, SSC (CGL-Tier-II)-13]

Sol. Number of small cubes $= \frac{\text{Volume of the cylinder}}{\text{Volume of each cone}}$

$$= \frac{\pi(3)^2 \times 5}{\frac{1}{3}\pi(0.1)^2 \times 1} = 13500$$

Problem 56 A gold ingot in the shape of a cylinder is melted and the resulting molten metal molded into a few identical conical ingots. If the height of each cone is half the height of the original cylinder and the area of the circular base of each cone is 1/5th that of the circular base of the cylinder, then how many conical ingots can be made?

[XAT-19]

Sol. Number of conical ingots $= \frac{\text{Volume of the cylinder}}{\text{Volume of each cone}}$

$$= \frac{\pi r^2 \times h}{\frac{1}{3} \times \left(\frac{1}{5}\pi r^2\right) \times \frac{h}{2}} = 30$$

Alternatively

If a cylinder and a cone have same height and same radius, volume of the cylinder is thrice that of the cone. If height of the cone is half that of cylinder, then volume of the cylinder will be (3×2) times that of cone. Further, if the base area of cone is 1/5 of that of cylinder, then volume of cylinder will be (6×5) times that of cone.

Hence, number of conical ingots = 30. Try yourself to solve these types of questions on this line.

Problem 57 An open conical cap is formed by joining any two sides of an equilateral triangle of side x cm. Find the perpendicular height of this cap.

Sol. Clearly, the circumference of the base = x

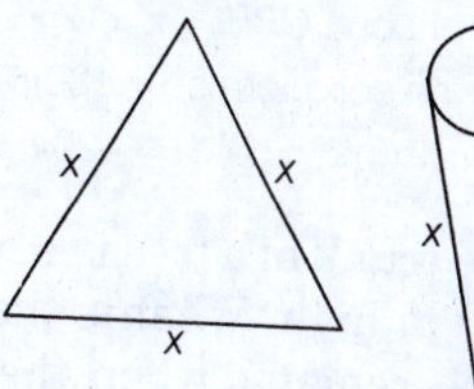

$$\Rightarrow \quad 2\pi r = x \quad \therefore r = \frac{x}{2\pi}$$

Now, slant height $(l) = \sqrt{h^2 + r^2}$

$$\Rightarrow l^2 = h^2 + r^2 \Rightarrow x^2 = h^2 + \left(\frac{x}{2\pi}\right)^2$$

$$\Rightarrow \quad h^2 = x^2 - \frac{x^2}{4(\pi)^2}$$

$$\therefore \quad h = \frac{x}{2\pi}\sqrt{4(\pi)^2 - 1}$$

Problem 58 A right angled triangle, whose sides are 15 cm and 20 cm, is rotated about its hypotenuse. Find the volume and surface area of the solid so formed.

[MAT-14]

Sol. Hypotenuse $AC = \sqrt{20^2 + 15^2} = 25$

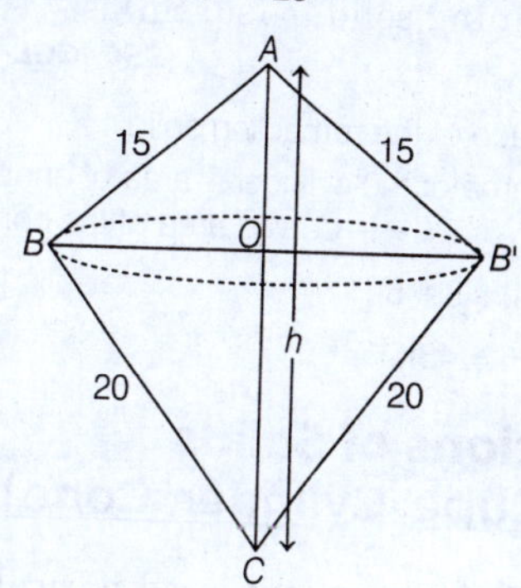

Area of right angled $\Delta ABC = \frac{1}{2} \times 15 \times 20$

$$\Rightarrow \frac{1}{2} \times AC \times BO = \frac{1}{2} \times 15 \times 20$$

$\Rightarrow \quad \frac{1}{2}\times 25\times BO = \frac{1}{2}\times 15\times 20$

$\therefore \quad BO = \frac{15\times 20}{25} = 12$ cm

Volume of cone $ABB' = \frac{1}{3}\pi(12)^2 \times AO \quad \{\because BO = \text{radius}\}$

Volume of cone $CBB' = \frac{1}{3}\pi(12)^2 \times OC \quad [\text{Also, radius} = BO]$

$\therefore$ Volume of the solid formed $= \frac{1}{3}\pi(12)^2 AO + \frac{1}{3}\pi(12)^2 \times OC$

$= \frac{1}{3}\pi(12)^2 [AO + OC]$

$= \frac{1}{3}\pi \times 144 \times AC$

$= \frac{1}{3}\pi \times 144 \times 25 = 3771 \text{ cm}^3$ (Approx).

Now, curved area of cone $ABB' = \pi \times 12 \times 15 = 180\pi$

And curved area of cone $CBB' = \pi \times 12 \times 20 = 240\pi$

$\therefore$ Curved area of the solid formed $= 180\pi + 240\pi$

$= 420\pi = 1320 \text{ cm}^2$

Problem 59 The trunk of a tree is a right cylinder 1.5 m in radius and 10 m high. What is the volume (in m^3) of the timber which remains when the trunk is trimmed just enough to reduce it to a rectangular parallelopiped on a square base?

Sol. The diagonal of maximum possible square carved from the base of the trunk = diameter of the base of the trunk

$\Rightarrow \quad \sqrt{2} \times \text{side} = 1.5 \times 2$

$\therefore \quad \text{side} = \frac{3}{\sqrt{2}}$

Now, area of the base of the rectangular parallelopiped

$= \text{Area of the square} = \left(\frac{3}{\sqrt{2}}\right)^2 = \frac{9}{2}$

$\therefore$ Its volume = base area × height $= \frac{9}{2} \times 10 = 45 \text{ m}^3$

Problem 60 From a solid right circular cylinder of length 4 cm and diameter 6 cm, a conical cavity of the same height and base is hollowed out. The whole surface of the remaining solid in (sq. cm) is

[SSC (CGL-Tier-II)-14, CDS-17]

Sol. Whole surface of the remaining solid

= Curved area of the cylinder + area of one base of the cylinder + curved area of the cone $= 2\pi rh + \pi r^2 + \pi rl$

$= \pi r[(2\times 4) + 3 + 5)] \quad [\because l = \sqrt{4^2 + 3^2} = 5]$

$= \pi \times 3 \times 16 = 48\pi$

Combinations of Solids (Cuboid, Cube, Cylinder, Cone)

Problem 61 A circus tent is cylindrical upto a height of 3 m and conical above it. If the diameter is 105 m and the slant height of the conical part is 63 m, then the total area of the canvas required to make the tent is

[SSC (10+2, LDC/DEO)-11, MAT-12]

Sol. Curved surface area of conical part $= \pi rl$

$= \frac{22}{7} \times \left(\frac{105}{2}\right) \times 63 = 10395$

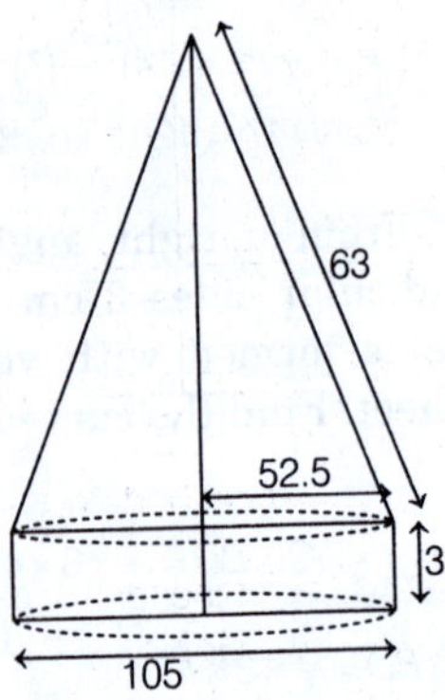

And curved surface area of cylindrical part $= 2\pi rh$

$= 2 \times \frac{22}{7}\left(\frac{105}{2}\right) \times 3 = 990$

$\therefore$ Area of canvas required $= 10395 + 990$

$= 11385 \text{ m}^2$

Problem 62 If a solid cone of volume 27π cm^3 is kept inside a hollow cylinder whose radius and height are that of the cone, then the volume of water needed to fill the empty space is

Sol. Given, volume of cone $= 27\pi$

$\therefore$ Volume of the cylinder = 3 × volume of the cone

$= 3 \times 27\pi = 81\pi$

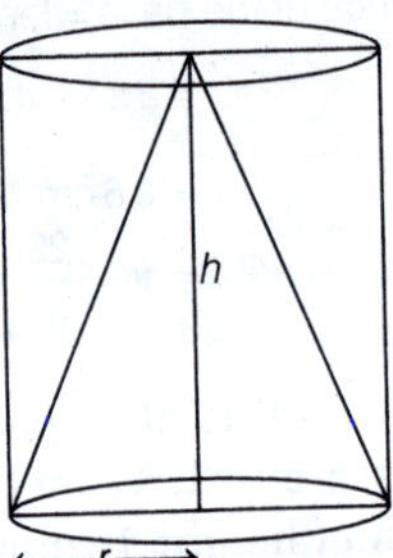

Thus, volume of water needed to fill the empty space

$= 81\pi - 27\pi = 54\pi \text{ cm}^3$

Problem 63 A circus tent is cylindrical up to a height of 7 m and conical above it. The diameter of the cylindrical portion is 10 m and the total height of the tent from the ground to the vertex of the conical part is 19 m. Find the cost of canvas required to build the tent at 7.70 per m^2.

[SSC (Mains)-2000, CDS-13]

Sol. Curved area of the cylindrical portion

$= 2 \times \frac{22}{7} \times 5 \times 7 = 220$

Slant height of the conical part

$= \sqrt{(\text{height})^2 + (\text{radius})^2}$

$= \sqrt{(19-7)^2 + (5)^2} = 13$

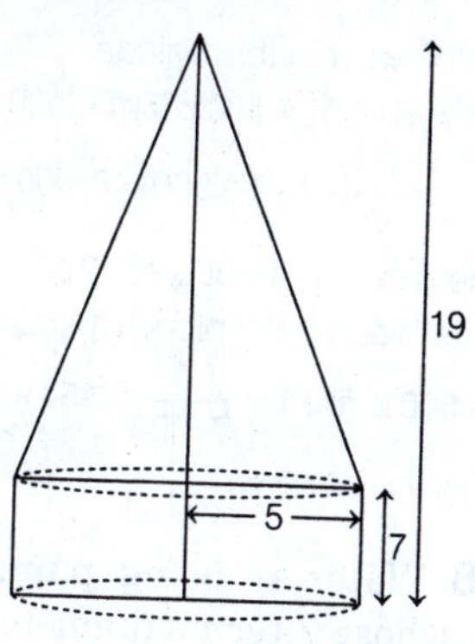

$\therefore$ Curved area of the conical portion $= \frac{22}{7} \times 5 \times 13 = 204.28$

Hence, total lateral area of the tent

$= 220 + 204.28 = 424.28 \text{ m}^2$

$\therefore$ Cost $= 424.28 \times 7.70 = 3266.95 = 3267$ (approx.)

Maximum/Minimum Value of Volume/Area

Problem 64 What is the maximum volume of a cylinder which can be cut from a cone of height h and radius r?

Sol. If we assume $DECB$ as a frustum of a cone ABC, then

$$\frac{\text{Height of the parent cone}}{\text{Height of the cone cut}} = \frac{\text{Radius of the parent cone}}{\text{Radius of the cone cut}}$$

$$\Rightarrow \quad \frac{h}{AK} = \frac{r}{KE}$$

$$\Rightarrow \quad \frac{h}{h - \text{height of cylinder}} = \frac{r}{x}$$

[let radius of cylinder be x]

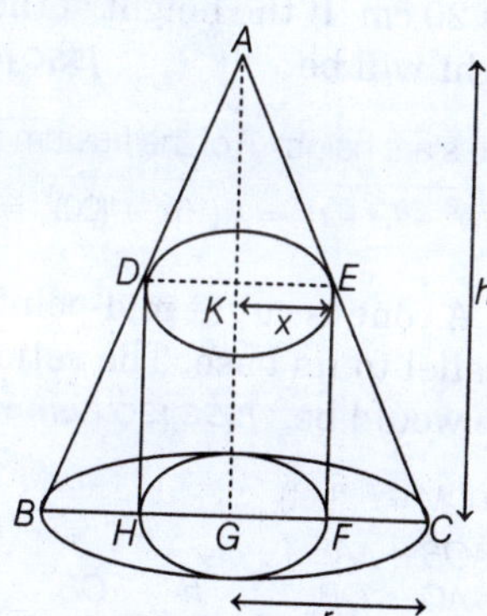

$$\Rightarrow \quad \text{Height of cylinder} = h - \frac{hx}{r}$$

Now, volume of the cylinder $(V) = \pi x^2 \left(h - \frac{hx}{r}\right)$

$$\Rightarrow \quad V = \pi x^2 h - \frac{\pi h x^3}{r}$$

$$\Rightarrow \quad \frac{dV}{dx} = 2\pi x h - \frac{3\pi h x^2}{r}$$

[differentiating with respect to x]

$$\Rightarrow \quad 2\pi x h - \frac{3\pi h x^2}{r} = 0$$

$$\Rightarrow \quad 2\pi x h = \frac{3\pi h x^2}{r}$$

$$\Rightarrow \quad x = 2r/3$$

Since $\frac{d^2v}{dx^2}$ i.e. $2\pi h - \frac{6\pi h x}{r}$ is negative at $x = \frac{2r}{3}$, the volume is maximum at the same value of x.

Hence, maximum volume of the cylinder $= \pi x^2 \left(h - \frac{hx}{r}\right)$

$$= \pi \left(\frac{2r}{3}\right)^2 \left(h - h\frac{\frac{2r}{3}}{r}\right) = \frac{4}{27}\pi r^2 h$$

Observation The height and radius of a cylinder of maximum possible volume, inscribed in a cone of radius r and height h, are $h/3$ and $\frac{2r}{3}$ respectively. The volume is $\frac{4}{9}$ times the volume of the cone.

Problem 65 What is the maximum volume of a cube which can be cut from a cone of height h and radius r.

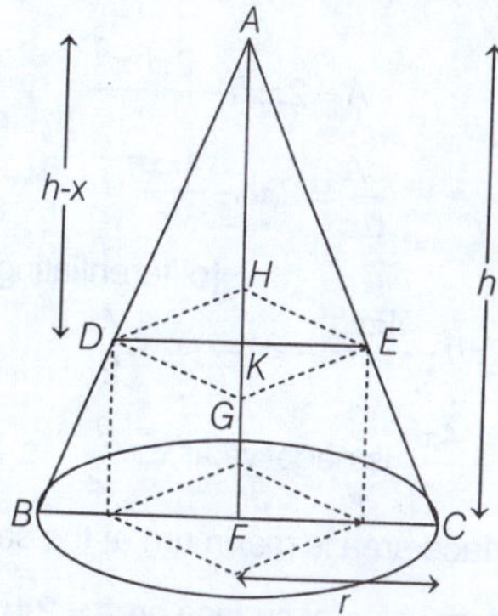

Sol. Let each side of cube be x.

Then, diagonal $DE = \sqrt{2}x$ as $DHEG$ is a square.

Since ΔADE and ΔABC are similar,

$$\frac{AK}{AF} = \frac{DE}{BC}$$

$$\Rightarrow \quad \frac{h - x}{h} = \frac{\sqrt{2}x}{2r}$$

$$\Rightarrow \quad 1 - \frac{x}{h} - \frac{x}{\sqrt{2}\,r} = 0$$

$$\Rightarrow \quad 1 = x\left(\frac{1}{h} + \frac{1}{\sqrt{2}\,r}\right)$$

$$\Rightarrow \quad x = \frac{\sqrt{2}rh}{\sqrt{2}\,r + h}$$

Then, maximum volume of the cube $= \left(\frac{\sqrt{2}\,rh}{\sqrt{2}\,r + h}\right)^3$

Problem 66 What is the maximum curved surcace area of a cylinder inscribed in a cone of radius r and height h?

Sol. If we assume $DECB$ as a frustum of a cone ABC, then

$$\frac{\text{Height of the parent cone}}{\text{Height of the cone cut}} = \frac{\text{Radius of the parent cone}}{\text{Radius of the cone cut}}$$

$$\Rightarrow \quad \frac{h}{AK} = \frac{r}{KE}$$

$$\Rightarrow \quad \frac{h}{h - \text{height of cylinder}} = \frac{r}{x}$$

[let radius of cylinder be x]

$$\Rightarrow \quad \text{Height of cylinder} = h - \frac{hx}{r}$$

Now, curved surface area of the cylinder $(A) = 2\pi x\left(h - \frac{hx}{r}\right)$

$$\Rightarrow \quad A = 2\pi xh - \frac{2\pi hx^2}{r}$$

$$\Rightarrow \quad \frac{dA}{dx} = 2\pi h - \frac{4\pi xh}{r}$$

[differentiating with respect to x]

$$\Rightarrow \quad 2\pi h - \frac{4\pi xh}{r} = 0 \Rightarrow x = \frac{r}{2}$$

Since $\frac{d^2A}{dx^2}$ i.e. $-\frac{4\pi h}{r}$ is negative at $x = \frac{r}{2}$,

the curved surface area is maximum at the same value of x.

Hence, maximum value of surface area $= 2\pi x\left(h - \frac{hx}{r}\right)$

$$= 2\pi\frac{r}{2}\left(h - \frac{h \times \frac{r}{2}}{r}\right) = \frac{\pi rh}{2}$$

Observation The height and radius of a cylinder of maximum curved surface area, inscribed in a cone of radius r and height h, are $h/2$ and $\frac{r}{2}$ respectively.

Concept of Water/Liquid Flow

Problem 67 Water flows through a cylindrical pipe, whose radius is 7cm, at 5 m/sec. The time, it takes to fill an empty water tank, with height 1.54 m and area of the base $(3 \times 5)\ m^2$ is :

Sol. It is obvious that water filled in the cylinder (figure as drawn) will flow in 5 sec.

Now, volume of water in this cylinder

$$= \pi \times (7)^2 \times 500 \quad [\because 5\ m = 500\ cm]$$

$$= \frac{22}{7} \times 7 \times 7 \times 500 = 77000\ cm^3$$

Volume of the tank $= 154 \times 300 \times 500\ cm^3$

$\because 77000\ cm^3$ is filled by the pipe in 1 sec.

$\because 154 \times 300 \times 500$ is filled in $\frac{1}{77000} \times 154 \times 300 \times 500$ sec.

$$= 300\ \text{sec} = 5\ \text{min}$$

Problem 68 Water is being pumped out through a circular pipe whose internal diameter is 7 cm. If the flow of water is 12 cm/sec, how many litres of water is being pumped out in 1 h ? ***[SSC (CPO-SI)-09]***

Sol. Volume of water pumped out in 1 sec $= \frac{22}{7} \times \left(\frac{7}{2}\right)^2 \times 12\ cm^3$

$\therefore$ Volume of water pumped out in 3600 sec $= 1663200\ cm^3$

$$= 1663.2\ L$$

Problem 69 It is required to design a circular pipe such that water flowing through it at a speed of 7 m/min fills a tank of capacity 440 cubic m in 10 min. The inner radius of the pipe should be ? ***[MAT-10]***

Sol. Volume of water being pumped in 1 min. $= \pi \times (r)^2 \times 7$

$\therefore$ In 10 min., volume $= 10 \times \pi \times (r^2) \times 7 = 440$ (given)

$$\Rightarrow \quad r = \sqrt{2}\ m$$

Problems based on "Frustum of a Cone"

Problem 70 The difference in radii of the circular ends of a bucket is 20 cm. If the height of the bucket is 15 cm, its slant height will be ***[SSC (CGL-T-I)-14, CDS-21]***

Sol. We know, the slant height (l) of the frustum of a cone

$$= \sqrt{h^2 + (R - r)^2} = \sqrt{(15)^2 + (20)^2} = 25\ cm$$

Problem 71 A cone is cut at mid-point of its height by a frustum parallel to its base. The ratio between the two parts of cone would be ***[SSC (SO Comm. Audit)-05, CDS-14]***

Sol. Given, $AO = OM = h$ (Let)

Clearly, $\Delta AOB \sim \Delta AEM$

$$\therefore \quad \frac{AO}{AM} = \frac{OB}{EM} \Rightarrow \frac{h}{h+h} = \frac{OB}{EM} \Rightarrow EM = 2OB$$

Now, required ratio $= \frac{\text{Volume (cone } ABC)}{\text{Volume (frustum } BCDE)}$

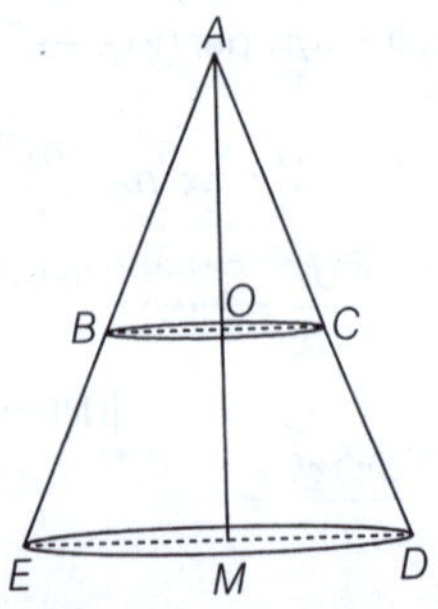

$$= \frac{\frac{1}{3}\pi(OB)^2 h}{\frac{1}{3}\pi h\,[(EM)^2 + (OB)^2 + (EM \times OB)]}$$

$$= \frac{\frac{1}{3}\pi(OB)^2 h}{\frac{1}{3}\pi h\,[(2OB)^2 + (OB)^2 + 2OB \times OB]}$$

$$= \frac{(OB)^2}{7(OB)^2} = 1:7$$

Alternatively I

Volume (cone ABC) $= \frac{1}{3}\pi(OB)^2 h$

and Volume (cone AED) $= \frac{1}{3}\pi(EM)^2 \times AM$

$\therefore$ Required ratio $= \dfrac{\frac{1}{3}\pi(OB)^2 h}{\frac{1}{3}\pi 8(OB)^2 h - \frac{1}{3}\pi(OB)^2 h} = 1:7$

Alternatively II

Observe the properties regarding frustum of a cone given in beginning of this chapter.

Volume (parent cone) : Volume (cone cut)

$$= H^3 : h^3 = 1^3 : \left(\frac{1}{2}\right)^3 = 8:1$$

$\therefore$ Volume (cone cut) : Volume {frustum of the cone}

$$= 1 : (8-1) = 1:7$$

[Aspirants are advised to adopt this alternative to solve the questions quickly]

Problem 72 The height and the radius of the base of a right circular cone are 12 cm and 6 cm respectively. The radius of the circular cross-section of the cone cut by a plane parallel to its base at a distance of 3 cm from the base is

[SSC (CGL-PT)-08]

Sol. Clearly, $\Delta AOC \sim \Delta AMD$

$\Rightarrow \frac{AO}{AM} = \frac{OC}{MD} \Rightarrow \frac{9}{12} = \frac{OC}{6}$

$\therefore$ $OC = 4.5$

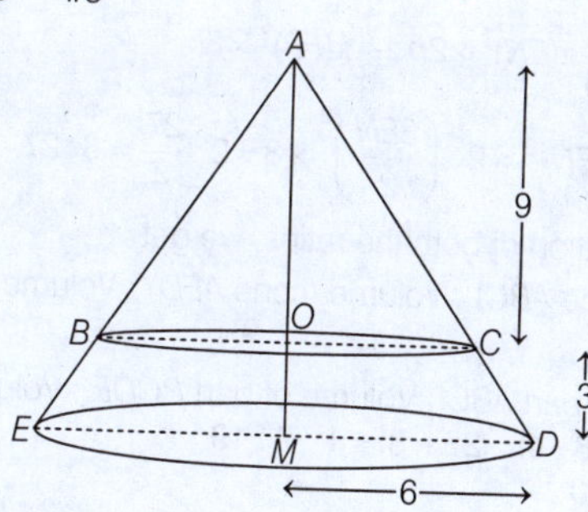

Alternatively

$$\frac{\text{Height of the parent cone } (H)}{\text{Height of the cone cut } (h)} = \frac{\text{Radius of the parent cone } (R)}{\text{Radius of the cone cut } (r)}$$

[one of properties regarding frustum of a cone]

$\Rightarrow \frac{12}{9} = \frac{6}{r}$

$\therefore$ $r = 4.5$ cm

Problem 73 The height of the cone is 30 cm. A small cone is cut off at the top by a plane parallel to its base. If its volume is $\frac{1}{27}$ of the volume of the cone, at what height above the base, is the section made ?

Sol. Given, $AM = 30$ cm, Let AO be x

Clearly, $\Delta AOC \sim \Delta AMD$

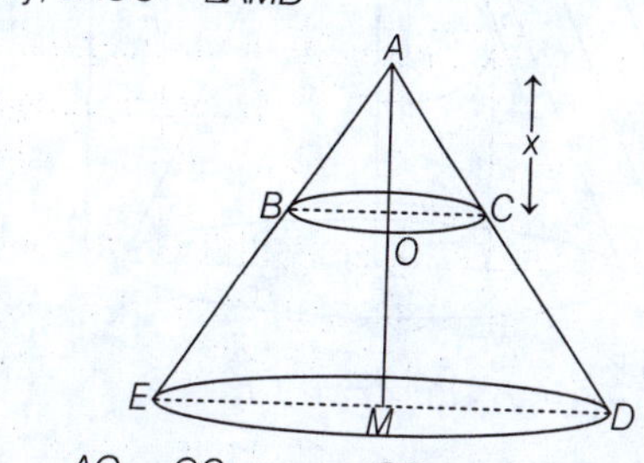

$\therefore \frac{AO}{AM} = \frac{OC}{MD} \Rightarrow \frac{x}{30} = \frac{OC}{MD}$

$\Rightarrow MD = 30 \times \frac{OC}{x}$

Now, $\frac{\text{Volume (small cone)}}{\text{Volume (big cone)}} = \frac{1}{27}$

$\Rightarrow \dfrac{\frac{1}{3}\pi(OC)^2 \times x}{\frac{1}{3}\pi(MD)^2 \times 30} = \frac{1}{27}$

$\Rightarrow \dfrac{(OC)^2 \times x}{\left(\frac{30 \times OC}{x}\right)^2 \times 30} = \frac{1}{27}$

$\Rightarrow \frac{x \times x^2}{900 \times 30} = \frac{1}{27}$

$\Rightarrow x^3 = 1000 \Rightarrow x = 10$

$\therefore$ $OM = 30 - 10 = 20$ cm

Hence, at height of 20 cm from the base, the section is made.

Alternatively

Volume (parent cone) : Volume (cone cut)

$= H^3 : h^3 = 27:1$ (as given)

$\Rightarrow (30)^3 : h^3 = 27:1$

$\Rightarrow h^3 = 1000$

$\therefore$ $h = 10$ cm

Clearly, the section is made at height of 20 cm (= 30 − 10) from the base of the parent cone.

Problem 74 The height of a right circular cone and the radius of its circular base are respectively 9 cm and 3 cm. The cone is cut by a plane parallel to its base so as to divide it into two parts. The volume of the frustum (i.e. the lower part) of the cone is 44 cubic cm. The radius of the upper circular surface of the frustum is :

[SSC (CGL-T-I)-12]

Sol. Here, $\frac{H}{h} = \frac{R}{r} \Rightarrow \frac{9}{h} = \frac{3}{r}$

$\Rightarrow h = 3r$

Now, volume of the frustum

$= \frac{1}{3}\pi \times \text{height of frustum} \times (R^2 + r^2 + R \cdot r)$

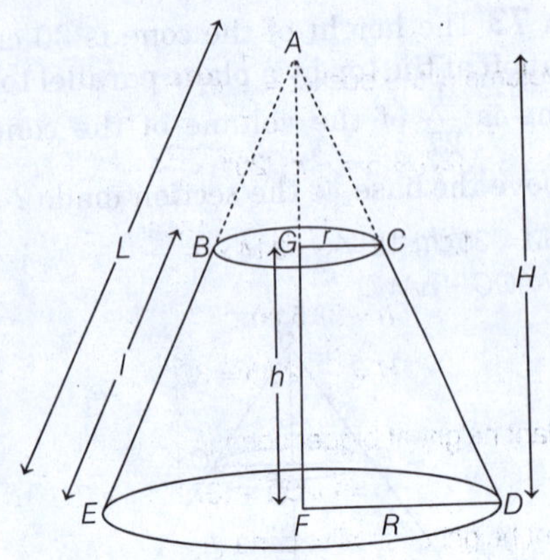

$$\Rightarrow \quad 44 = \frac{1}{3} \times \frac{22}{7} \times (9 - 3r) \times (3^2 + r^2 + 3 \cdot r)$$

$$\Rightarrow \quad 14 = (3 - r) \times (3^2 + r^2 + 3 \cdot r)$$

$$\Rightarrow \quad 14 = 3^3 - r^3 \Rightarrow r = \sqrt[3]{13}$$

Alternatively

We know, $\dfrac{\text{Volume (parent cone)}}{\text{Volume (cone cute)}} = \dfrac{R^3}{r^3}$

$$\Rightarrow \quad \frac{\text{Volume (parent cone)}}{\text{Volume (parent cone) - Volume (cone cute)}} = \frac{R^3}{R^3 - r^3}$$

$$\Rightarrow \quad \frac{\frac{1}{3}\pi 3^2 \times 9}{\text{Volume of the frustum}} = \frac{3^3}{3^3 - r^3}$$

$$\Rightarrow \quad \frac{27\pi}{44} = \frac{3^3}{3^3 - r^3} \Rightarrow r^3 = 27 - 14 \therefore r^3 = \sqrt[3]{13}$$

Problem 75 If the radii of the circular ends of a truncated conical bucket which is 45 cm high, be 28 cm and 7 cm, then the capacity of the bucket in cubic centimeter is ***[SSC (10+2 LDC/DEO)-11, CDS-11]***

Sol. Given, height of the bucket $(h) = 45$ cm, $r = 7$ and $R = 28$ cm

Now, Volume/Capacity of the bucket

$$= \frac{1}{3}\pi h(R^2 + r^2 + Rr)$$

$$= \frac{1}{3} \times \frac{22}{7} \times 45[(28)^2 + (7)^2 + (28 \times 7)]$$

$$= 48510 \text{ cm}^3$$

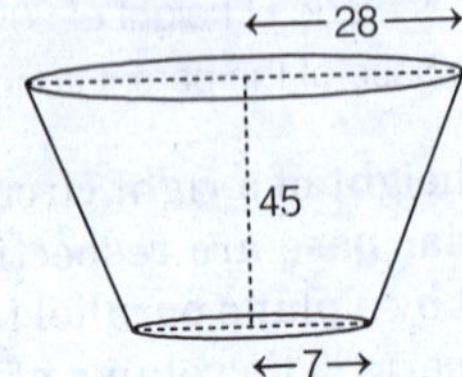

Alternatively

We know, $\dfrac{\text{Height of the parent cone } (H)}{\text{Height of the cone cut } (h)} = \dfrac{\text{Radius of the parent cone } (R)}{\text{Radius of the cone cut } (r)}$

$$\Rightarrow \quad \frac{H}{h} = \frac{28}{7} \Rightarrow H = 4h$$

Given, $\quad H - h = 45 \Rightarrow 4h - h = 45$

$\therefore \quad h = 15$

Clearly, $H = 60$ cm, and $h = 15$ cm

$\therefore$ Volume of the bucket $= \frac{1}{3}\pi R^2 H - \frac{1}{3}\pi r^2 h$

$$= \frac{1}{3} \times \frac{22}{7}[(28^2 \times 60) - (7^2 \times 15)] = 48510 \text{ cm}^3$$

Problem 76 A right circular cone is cut by two planes parallel to the base such that the height is divided into three equal parts. Compare the volumes of the 3 parts of the cone.

Sol. Given, $AM = MN = NP = h$ (let)

Since $\quad \Delta ABM \sim \Delta AFP, \dfrac{AM}{AP} = \dfrac{BM}{FP}$

$$\Rightarrow \quad \frac{h}{3h} = \frac{BM}{FP} \Rightarrow FP = 3BM$$

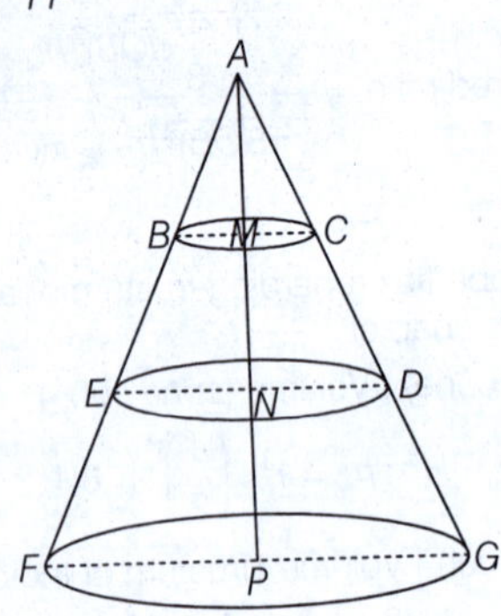

$\therefore$ Volume (cone ABC) : Volume (cone AFG)

$$= \frac{1}{3}\pi(BM)^2 \times h : \frac{1}{3}\pi(FP)^2 \times 3h$$

$$= (BM)^2 : (3BM)^2 \times 3 = 1:27$$

Further, $\quad \Delta AEN \sim \Delta AFP$

$$\therefore \quad \frac{AN}{AP} = \frac{EN}{FP}$$

$$\Rightarrow \quad \frac{2h}{3h} = \frac{EN}{FP}$$

$$\Rightarrow \quad FP = \frac{3EN}{2}$$

$\therefore$ Volume (cone AED) : Volume (cone AFG)

$$= \frac{1}{3}\pi(EN)^2 \times 2h : \frac{1}{3}\pi(FP)^2 \times 3h$$

$$= (EN)^2 \times 2 : \left(\frac{3EN}{2}\right)^2 \times 3 = 2 : \frac{27}{4} = 8:27$$

On combination of both the ratios, we get

Volume (cone ABC) : Volume (cone AED) : Volume (cone AFG) $= 1:8:27$

$\therefore$ Volume of part ABC : Volume of part $BCDE$: Volume of Part $EDGF = 1:(8-1):(27-8) = 1:7:19$

Alternatively

Volume (cone ABC) : Volume (cone AFG) $= h^3 : (3h)^3 = 1^3 : 3^3 = 1:27$

Volume (cone AED) : Volume (cone AFG) $= (2h)^3 : (3h)^3 = 8:27$

On combination, volume (cone ABC) : Volume (cone AED) : Volume (cone AFG) $= 1:8:27$

$\therefore$ Volume of cone ABC : Volume of part $BCDE$: Volume of part $EDGF = 1:(8-1):(27-8) = 1:7:19$

Problem 77 A bucket contains 200 cc of liquid. A solid ball is dropped in the bucket resulting in the rise of liquid level to 1.3 times of its original level. If the radius of the base of the bucket is 3 cm and the radius of the surface of the liquid level is 1 cm more than the radius of the base of the bucket before the ball is dropped. Find the volume of the solid metal ball. **[IIFT-18]**

Sol. Clearly, $\frac{\text{Height of cone } AED}{\text{Height of cone } ABC} = \frac{\text{radius of base of cone } AED}{\text{radius of base of cone } ABC} = \frac{4}{3}$

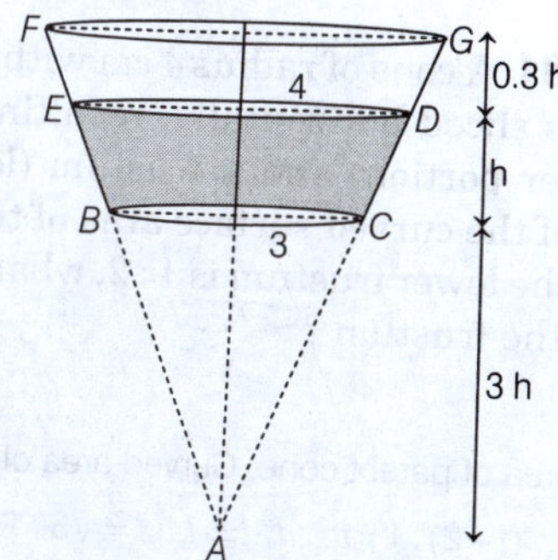

Let height of cone ABC be $3h$, then height of cone $AED = 4h$

We know, $\frac{\text{Volume (parent cone)}}{\text{Volume (cone cut)}} = \frac{R^3}{r^3} = \frac{H^3}{h^3}$

Volume (cone ADE) : Volume (cone ABC)

$= 4^3 : 3^3 = 64 : 27$

Similarly, Volume (cone AFG) : Volume (cone ABC)

$= (4.3h)^3 : (3h)^3 = 79.5 : 27$

Hence, Volume (cone ADE) : Volume (cone AFG) = 64 : 79.5

Then, volume of the metallic ball = 79.5 − 64 = 15.5 units.

Now, volume of liquid before the ball is dropped = 64 − 27 = 37 units = 200 cc(given)

$\Rightarrow$ 15.5 units = 83.7 cc

Problem 78 A bucket is in the form of a frustum of a cone with a capacity of 12308.8 cm^3 of water. The radii of the top and bottom circular ends are 20 cm and 12 cm respectively. Find the area of the metal sheet used in its making. (Given $\pi = 3.14$)

Sol. We know, $\frac{H}{h} = \frac{R}{r} \Rightarrow \frac{H}{h} = \frac{20}{12}$

$\Rightarrow \quad H = \frac{5}{3}h$

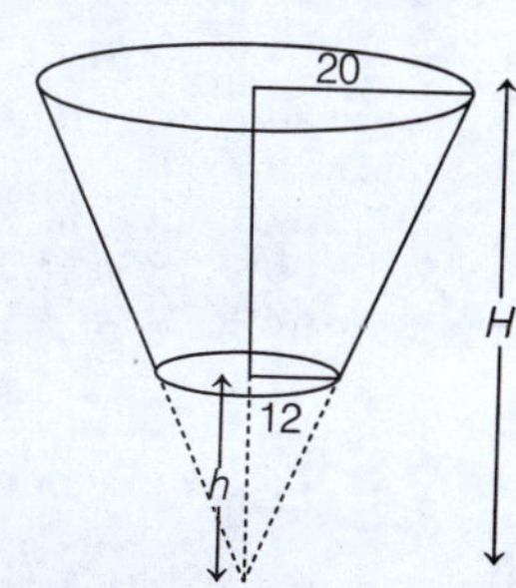

Now, volume of the bucket $= \frac{1}{3}\pi R^2 H - \frac{1}{3}\pi r^2 h$

$\Rightarrow \quad 12308.8 = \frac{1}{3}\pi\left[(20)^2 \times \frac{5}{3}h - (12)^2 h\right]$

$\Rightarrow \quad 12308.8 = \frac{1}{3} \times 3.14 \times \frac{1568h}{3}$

$\Rightarrow \quad h = 22.5$ cm

Then, $\quad H = \frac{5}{3} \times 22.5 = 37.5$ cm

Now, slant height of bigger cone (l)

$= \sqrt{(20)^2 + (37.5)^2} = 42.5$ cm

and slant height of smaller cone (l)

$= \sqrt{(12)^2 + (22.5)^2} = 25.5$ cm

Then, curved area of the bucket $= \pi RL - \pi rl$

$= \pi[20 \times 42.5 - 12 \times 25.5] = 3.14 \times 544 = 1708.16$ cm^2

Now, area of metal sheet used = curved area + area of the bottom $= 1708.16 + \pi(12)^2$

$= 1708.16 + (3.14 \times 144) = 2160.32$ cm^2

Problem 79 A hollow cone is cut by a plane parallel to the base and the upper portion is removed. If the curved surface of the remainder is 8/9 of the curved surface of the whole cone, find the ratio of the line segments into which the altitude of the cone is divided by the plane. **[MAT-10]**

Sol. Given, $\frac{1}{9}\pi RL = \pi rl \Rightarrow \frac{R}{r} \times \frac{L}{l} = \frac{9}{1}$

$\Rightarrow \quad \left(\frac{H}{h} \times \frac{H}{h}\right) = \frac{9}{1} \qquad \left[\because \frac{H}{h} = \frac{R}{r} = \frac{L}{l}\right]$

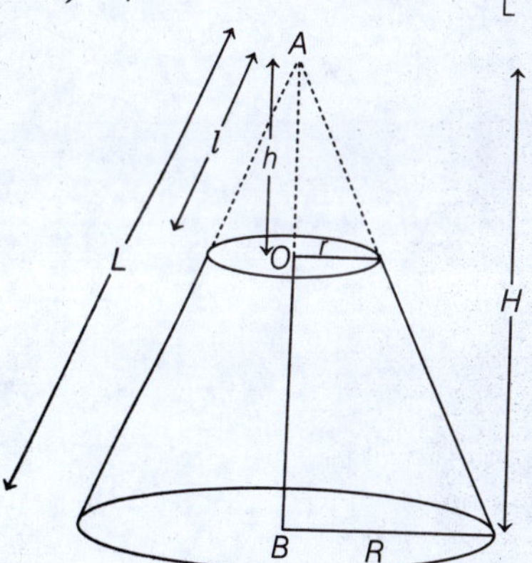

$\Rightarrow \quad H^2 : h^2 = 9 : 1$

$\Rightarrow \quad H : h = 3 : 1$

$\Rightarrow \quad h : (H - h) = 1 : (3 - 1) = 1 : 2$

Hence, $AO : OB = 1 : 2$, the required ratio.

Problem 80 The radii of the circular ends of solid frustum of a cone are 33 cm and 27 cm and its slant height is 10 cm. Find its capacity.

Sol. We know, $\frac{R}{r} = \frac{L}{l}$

$\Rightarrow \quad \frac{33}{27} = \frac{L - l + l}{l}$

$$\Rightarrow \quad \frac{11}{9} = \left(\frac{L-l}{l}\right) + 1$$

$$\Rightarrow \quad \frac{11}{9} = \frac{10}{l} + 1 \Rightarrow \frac{2}{9} = \frac{10}{l} \Rightarrow l = 45$$

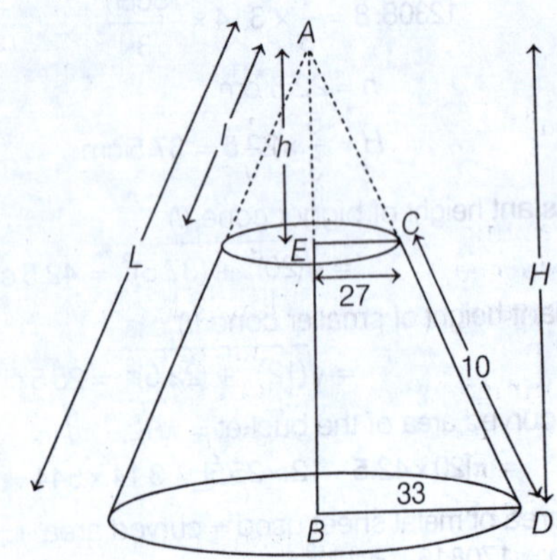

Now, $h = \sqrt{l^2 - r^2} = \sqrt{(45)^2 - (27)^2} = 36$

We know, $\frac{H}{h} = \frac{R}{r}$

$$\Rightarrow \quad \frac{H}{36} = \frac{33}{27} \Rightarrow H = 44$$

$\therefore$ Capacity of the frustum $= \frac{1}{3}\pi R^2 H - \frac{1}{3}\pi r^2 h$

$$= \frac{1}{3}\pi[(33)^2 \times 44 - (27)^2 \times 36]$$

$$= \frac{1}{3}\pi \times 21672 = 22704 \text{ cm}^3$$

Alternatively

Given, slant height = 10, $R = 32, r = 27$

We know, (slant height)$^2 = h^2 + (R - r)^2$

$$\Rightarrow \quad (10)^2 = h^2 + (33 - 27)^2 \Rightarrow h = 8$$

Now, volume of the frustum $= \frac{1}{3}\pi h(R^2 + r^2 + R \cdot r)$

$$= \frac{1}{3} \times \frac{22}{7} \times 8 \times [(33)^2 + (27)^2 + 33 \cdot 27]$$

$$= 22704 \text{ cm}^3$$

Problem 81 A cone of radius 4 cm with a slant height of 12 cm was sliced horizontally, resulting into a smaller cone (upper portion) and a frustum (lower portion). If the ratio of the curved surface area of the upper smaller cone and the lower frustum is 1 : 2, what will be the slant height of the frustum ? ***[XAT-18]***

Sol. We know,

Curved area of parent cone: Curved area of cone cut $= L^2 : l^2$

$$\Rightarrow \quad (1+2) : 1 = L^2 : l^2 \Rightarrow L : l = \sqrt{3} : 1$$

As given, $\sqrt{3}$ units = 12 cm

$$\therefore \quad 1 \text{ unit} = 12 / \sqrt{3} = 4\sqrt{3}$$

Then, slant height of the frustum $= (12 - 4\sqrt{3})$ cm

WORKOUT

1. A conical iron piece having diameter 28 cm and height 30 cm is totally immersed into the water of a cylindrical vessel, resulting in the rise of water level by 6.4 cm. The diameter, in cm, of the vessel is **[SSC (CGL-T-I) 15]**
(a) 35 (b) 32
(c) 3.5 (d) $\frac{35}{2}$

2. Two cans have the same height equal to 21 cm. One can is cylindrical, the diameter of whose base is 10 cm. The other can has square base of side 10 cm. What is the differences in their capacities (in cm^3)? **[MAT-10]**
(a) 350 (b) 450
(c) 250 (d) 300

3. From a solid wooden right circular cylinder, a right circular cone whose radius and height are same as the radius and height of the cylinder, respectively is carved out. What is the ratio of the volume of the utilized wood to that of the wasted wood ?
(a) 1 : 2 (b) 2 : 1
(c) 2 : 3 (d) 1 : 3

4. A solid metal cylinder of 10 cm height and 14 cm diameter is melted and recast into two cones in the proportion of 3 : 4 (volume), keeping the height 10 cm. What would be the percentage change in the flat surface area before and after ? **[XAT-15]**
(a) 9% (b) 16%
(c) 25% (d) 50%

5. A reservoir is in the shape of a frustum of a right circular cone. It is 8 m across at the top and 4 m across the bottom. It is 6 m deep. Find the area of its curved surface ?
(a) 118.4 m^2 (b) 162.3 m^2
(c) 452 m^2 (d) 181.6 m^2

6. The height of a cylinder is 15 cm. The lateral surface area is 660 sq cm. Its volume is **[CDS-13]**
(a) 1155 cm^3 (b) 1215 cm^3
(c) 1230 cm^3 (d) 2310 cm^3

7. A square pond has 2 m sides and is 1 m deep. If it is to be enlarged, the depth remaining the same, into a circular pond with the diagonal of the square as diameter as shown in the figure, then what would be the volume of earth to be removed ? **[IAS (Pre)-98]**

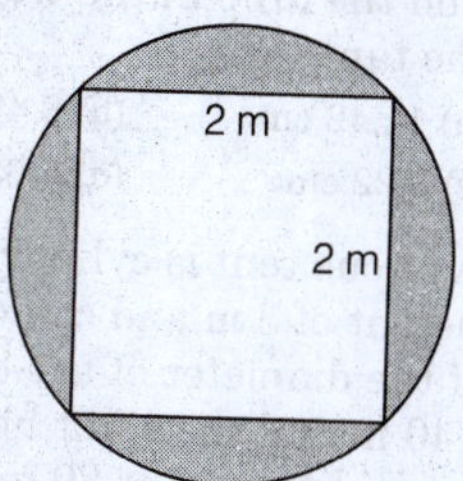

(a) $(2\pi - 4)$ m^3 (b) $(4\pi - 4)$ m^3
(c) $(4\pi - 2)$ m^3 (d) $(2\pi - 2)$ m^3

8. A right triangle has hypotenuse 15 cm and legs in the ratio 3 : 4. It is made to revolve about its hypotenuse. What is the volume of the double cone so formed ? ($\pi = 3.14$)
(a) 815 cm^3 (b) 1813 cm^3
(c) 2310 cm^3 (d) 666 cm^3

9. A conical vessel of base radius 2 cm and height 3 cm is filled with kerosene. This liquid leaks through a hole in the bottom and collects in a cylindrical jar of radius 2 cm. The kerosene level in the jar is :
(a) π cm (b) 1.5 cm
(c) 1 cm (d) 3 cm

10. What will be the cost to plaster the inner surface of a well 14 m deep and 4 m in diameter at the rate of ₹ 25/m^2 ? **[CDS-12]**
(a) ₹ 4000 (b) ₹ 4200
(c) ₹ 4400 (d) ₹ 5400

11. What part of a ditch, 48 m long, 16.5 m broad and 4 m deep can be filled by the earth got by digging a cylindrical tunnel of diameter 4 m and length 56 m ? **[SSC (CGL-PT)-07]**
(a) $\frac{1}{9}$ (b) $\frac{2}{9}$
(c) $\frac{7}{9}$ (d) $\frac{8}{9}$

12. What is the whole surface area of a cone of base radius 7 cm and height 24 cm ? **[CDS-13]**
(a) 654 m^2 (b) 704 m^2
(c) 724 m^2 (d) 964 m^2

13. A solid cylinder and a solid cone have equal base and equal height. If the radius and the height be in the ratio of 4 : 3, the ratio of the total surface area of the cylinder to that of the cone is in the ratio of
(a) 10 : 9 (b) 11 : 9
(c) 12 : 9 (d) 14 : 9

14. The radius of a cone is r cm and its height is h cm. The change in volume when the height is decreased by x cm is the same as the change in volume when the radius is decreased by x cm. Find the relation between x, r and h. **[CAT-14]**
(a) $x = \frac{2rh - r^2}{h}$ (b) $x = \frac{2rh + r^2}{h}$
(c) $x = \frac{r^2 - 2rh}{h}$ (d) $x = 2r + r^2$

15. What is the height of a solid cylinder of radius 5 cm and total surface area 660 cm^2 ?
(a) 10 cm (b) 12 cm
(c) 15 cm (d) 16 cm

16. The volume of the largest cylinder formed, when a rectangular sheet of paper of size 22 cm × 15 cm is rolled along its larger side, is **[SSC (SAS)-10]**
(a) 288.75 cm^3
(b) 577.5 cm^3
(c) 866.25 cm^3
(d) 115 cm^3

17. Consider a right circular cone of base radius 4 cm and height 10 cm. A cylinder is to be placed inside the cone with one of the flat surfaces resting on the base of the cone. Find the largest possible

total surface area (in sq. cm) of the cylinder. *[CAT-08]*

(a) $\frac{100\pi}{3}$ (b) $\frac{80\pi}{3}$

(c) $\frac{120\pi}{7}$ (d) $\frac{130\pi}{7}$

(e) $\frac{110\pi}{7}$

18. If radius of a right circular cone is increased by $p\%$ without increasing its height, then what is the percentage increase in the volume of the cone ? *[CDS-17]*

(a) P^2 (b) $2P^2$

(c) $\frac{P^2}{100}$ (d) $P\left(2+\frac{P}{100}\right)$

19. The volume of a right circular cone of height 3 cm and slant height 5 cm is

(a) 49.3 cm^3 (b) 50.3 cm^3

(c) 52 cm^3 (d) 53 cm^3

20. The volume of a right circular cylinder 9 cm high with diameter of the base 60 cm is equal to that of a right circular cone 54 cm high. What is the diameter of the base of the cone ?

(a) $30\sqrt{2}$ cm (b) $40\sqrt{2}$ cm

(c) $50\sqrt{2}$ cm (d) $60\sqrt{2}$ cm

21. 1496 cm^3 of metal is used to cast a pipe of length 28 cm. If the internal radius of the pipe is 8 cm, the outer radius of the pipe is

(a) 7 cm (b) 10 cm

(c) 9 cm (d) 12 cm

22. If the lateral surface of a cylinder is 94.2 cm^2 and its height is 5 cm, then find radius of its base. *[NICL (Asst.)-15]*

(a) 5 cm (b) 7 cm

(c) 3 cm (d) 4 cm

23. A cylinder having base of circumference 60 cm is rolling without sliding at a rate of 5 round per second. How much distance will the cylinder roll in 5 sec ?

(a) 15 m (b) 1.5 m

(c) 30 m (d) 3 m

24. The radius of the base of a conical tent is 5 m. If the tent is 12 m high, then the area of the canvas required in making the tent is : *[MAT-13]*

(a) 300π m^2 (b) 60π m^2

(c) 90π m^2 (d) None of these

25. Each of the height and radius of the base of a right circular cone is increased by 100%. The volume will be increased by : *[SSC (SAS)-10]*

(a) 700% (b) 500%

(c) 300% (d) 100%

26. The outer and inner diameters of a circular pipe are 6 cm and 4 cm respectively. If the length is 10 cm, then what is the total surface area in sq cm ?

(a) 35π (b) 110π

(c) 150π (d) None of these

27. A cylindrical tank of diameter 35 cm is full of water. If 11 L water is taken out from the tank, find the drop in the water level in the tank.

(a) 11.42 cm (b) 8.42 cm

(c) 3.22 cm (d) 5.83 cm

28. A circus tent is cylindrical to a height of 3 m and conical above it. If the diameter of the base is 140 m and the slant height of the conical portion is 80 m, the length of canvas 2 m wide required to make the tent is : *[MAT-11]*

(a) 8960 m (b) 9960 m

(c) 9460 m (d) 9860 m

29. The material of a solid cone is converted into the shape of a solid cylinder of equal radius. If the height of the cylinder is 5 cm, what is the height of the cone ?

(a) 15 cm (b) 20 cm

(c) 25 cm (d) 30 cm

30. The length of canvas, 75 cm wide required to build a conical tent of height 14 m and the floor area 346.5 m^2 is : *[SSC (CGL-T-I)-15]*

(a) 490 m (b) 770 m

(c) 860 m (d) 665 m

31. A right circular cylinder has a radius of 6 and a height of 24. A rectangular solid with a square base and a height of 20, is placed in the cylinder such that each of the corners of the solid is tangent to the cylinder wall. If water is poured into the cylinder such that it reaches the rim, then the volume of water is ? *[IIFT-14]*

(a) $288(\pi - 5)$

(b) $288(2\pi - 3)$

(c) $288(3\pi - 5)$

(d) None of the above

32. Diameter of the base of a water filled inverted right circular cone is 26 cm. A cylindrical pipe, 5 mm in radius, is attached to the surface of the cone at a point. The perpendicular distance between the point and the base (the top) is 15 cm. The distance from the edge of the base to the point is 17 cm, along the surface. If water flows at the rate of 10m/min through the pipe, how much time would elapse before water stops coming out of the pipe ? *[XAT-14]*

(a) < 4.5 min

(b) ≥ 4.5 min but < 4.8 min

(c) ≥ 4.8 min but < 5 min

(d) ≥ 5 min but < 5.2 min

(e) ≥ 5.2 min

33. Let C be a right circular cone. It is given that the two ends of a frustum of C are of radii 3 cm and 6 cm and the height of the frustum is 9 cm. What is the slant height of the given frustum ?

(a) $3\sqrt{10}$ cm (b) $6\sqrt{10}$ cm

(c) 12 cm (d) 15 cm

34. What is the height of the cone? (Ref. above question)

(a) 9 cm (b) 12 cm (c) 13.5 cm (d) 18 cm

35. What is the total surface area of the given frustum? (Ref-above question)

(a) $9\pi(2\sqrt{10}+5)$ cm^2

(b) $9\pi(3\sqrt{10}+5)$ cm^2

(c) $9\pi(3\sqrt{10}+4)$ cm^2

(d) $27\pi(\sqrt{10}+1)$ cm^2

Directions (Q. Nos. 36-38) *Answer the questions based on the following information.*

Consider a cylinder of height h cm and radius $r = 2/\pi$ as shown in the figure (not drawn to scale). A string of a certain length when wound on its cylindrical surface starting at point A and ending at point B, gives a maximum of n turns (in other words, the string's length is the minimum length required to wind n turns).

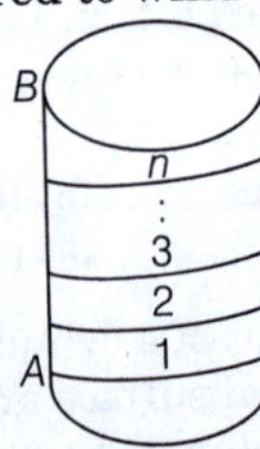

36. What is the vertical spacing in cm between two consecutive turns ?

(a) h/n
(b) $h/\sqrt{n}$
(c) h/n^2
(d) cannot be determined with the given information

37. The same string, when wound on the exterior four walls of a cube of side n cm, starting at point C and ending at point D, can give exactly one turn (see figure, not to scale). The length of the string, in cm, is

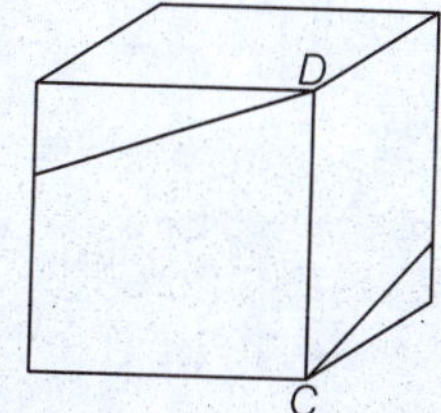

(a) $\sqrt{2}\ n$ (b) $\sqrt{17}\ n$
(c) n (d) $\sqrt{13}\ n$

38. In the setup of the previous two questions, how is h related to n? **[CAT-03]**

(a) $h = \sqrt{2}\ n$ (b) $h = \sqrt{17}\ n$
(c) $h = n$ (d) $h = \sqrt{13}\ n$

39. A cylindrical bucket of height 36 cm and radius 21 cm is filled with sand. The bucket is emptied on the ground and conical heap of sand is formed, the height of the conical is 12 cm, the radius of the heap at the base is :

(a) 63 cm (b) 53 cm
(c) 56 cm (d) 66 cm

40. The ratio of radii of two cylinders is 2 : 3 and their heights are in ratio 5 : 4. The ratio of their curved surface area is : **[SSC (CGL T-II)-12]**

(a) 5 : 6 (b) 3 : 4
(c) 4 : 5 (d) 2 : 3

41. Water flows through a cylindrical pipe of internal diameter 7 cm, at 5 m/s. The time, in minutes, the pipe would take to fill an empty rectangular tank 4 m × 3 m × 2.31m is : **[CDS-15]**

(a) 28 (b) 24
(c) 20 (d) 12

42. The ratio of volume of two cylinders is 7:3 and the ratio of their heights is 7:9. If the area of the base of the 2nd cylinder is 154 cm^2, then what will be the radius (in cm) of the 1st cylinder ? **[SSC (CGL T-I)-17]**

(a) $6\sqrt{2}$ (b) $6\sqrt{3}$
(c) $7\sqrt{2}$ (d) $7\sqrt{3}$

43. A right circular cone of height h is cut by a plane parallel to the base at a distance $h/3$ from the base, then the volumes of the resulting cone and frustum are in the ratio : **[CAT-95]**

(a) 1 : 3 (b) 8 : 19
(c) 1 : 4 (d) 1 : 7

44. The slant height of a conical mountain is 2.5 km and the area of its base is 1.54 sq km. The height of the mountain is :

(a) 2.2 km (b) 2.4 km
(c) 3 km (d) 3.11 km

45. Three identical cones with base radius r are placed on their bases so that each is touching the other two. The radius of the circle drawn through their vertices is : **[CAT-93]**

(a) smaller than r
(b) equal to r
(c) larger than r
(d) depends on the height of the cones

46. The ratio of curved surface area of a right circular cylinder to the total area of its bases is 2:1. If the total surface area of cylinder is 23100 cm^2, then what is the volume (in cm^3) of cylinder ? **[SSC (CGL T-II)-17]**

(a) 247200 (b) 269500
(c) 312500 (d) 341800

47. The radii of the frustum of a right circular cone are in the ratio 2 : 1. What is the ratio of the volume of the frustum of the cone to that of the whole cone ? **[CDS-17]**

(a) 1 : 8 (b) 3 : 4
(c) 7 : 8 (d) 1 : 4

48. A slab of ice 8 inches in length, 11 inches in breadth and 2 inches thick was melted and resolidified in the form of a rod of 8 inches diameter. The length of such a rod, in inches, is nearest to : **[CAT-93]**

(a) 3 (b) 3.5
(c) 4 (d) 4.5

49. A right circular cylinder has height as 18 cm and radius as 7 cm. The cylinder is cut in three equal parts (by 2 cuts parallel to base). What is the percentage increase in total surface area ? **[SSC (CGL T-II)-17]**

(a) 62 (b) 56
(c) 48 (d) 52

50. An iron pipe 20 cm long has exterior diameter equal to 25 cm. If the thickness of the pipe is 1 cm, the whole surface area of the pipe is :

(a) 3068 cm^2
(b) 3268 cm^2
(c) 3168 cm^2
(d) 3368 cm^2

51. A solid cylinder of height 9 m has its curved surface area equal to 1/3rd of the total surface area. What is the radius of the base ?

(a) 9 m (b) 18 m
(c) 27 m (d) 30 m

52. The ratio of curved surface area and volume of a cylinder is 1 : 7. The ratio of total surface area and volume is 187 : 770. What is the respective ratio of its base radius and height ? **[SSC (CGL T-II)-17]**

(a) 5 : 8 (b) 4 : 9
(c) 3 : 7 (d) 7 : 10

53. A conical cavity is drilled in a circular cylinder of 15 cm height and 16 cm base diameter. The height and the base diameter of the cone are same as those of the cylinder. Determine the total surface area of the remaining solid.

(a) 440π cm^2 (b) 215π cm^2
(c) 542π cm^2 (d) 376π cm^2

54. If height of a right circular cone is increased by 200% and the radius of the base is reduced by 50%, then the volume of the cone : **[CDS-15]**

(a) remains unaltered
(b) decreases by 25%
(c) increases by 25%
(d) increases by 50%

55. If the radius of the cylinder is decreased by 20%, then by how much per cent the height must be increased so that the volume of the cylinder remains the same. **[SSC (CGL T-I)-17]**

(a) 20 (b) 36.25
(c) 56.25 (d) 65

56. Ratio between heights of two cylinders are in the ratio 3 : 5 and their volumes are in ratio 27 : 80. Find the ratio between their radii. **[SBI (PO)-17]**

(a) 1 : 2 (b) 2 : 3
(c) 3 : 4 (d) 4 : 5
(e) NT

57. Two rectangular sheets of sizes $2\pi \times 4\pi$ and $\pi \times 5\pi$ are available. A hollow right circular cylinder can be formed by joining a pair of parallel sides of any sheet. What is the maximum possible volume of the circular cylinder that can be formed this way? **[CDS-17]**

(a) $4\pi^2$ (b) $8\pi^2$
(c) $1.25\pi^2$ (d) $6.25\pi^2$

58. The vertical angle of a right circular cone is $\pi/2$ and the slant height is $\sqrt{2}\,r$ cm. What is the volume of the cone in cubic cm ? **[CDS-17]**

(a) πr^3 (b) $9\pi r^3$
(c) $3\pi r^3$ (d) $\pi r^3/3$

59. The radius of base of solid cone is 9 cm and its height is 21 cm. It is cut into 3 parts by two cuts, which are parallel to its base. The cuts are at height of 7 cm and 14 cm from the base respectively. What is the ratio of curved surface areas of top, middle and bottom parts respectively ? **[SSC (CGL T-II)-17]**

(a) 1 : 4 : 8 (b) 1 : 3 : 5
(c) 1 : 3 : 9 (d) 1 : 6 : 12

Directions (Q. Nos. 60-63) *Answer the following questions if a right cylinder is inscribed in a right cone of radius 10 cm and height 14 cm.*

60. What is the radius of cylider if its volume is maximum ?

(a) 6.6 cm (b) 4.6 cm
(c) 7 cm (d) 12 cm

61. What is the maximum volume of the cylinder?

(a) 207.4π cm^3
(b) 66π cm^3
(c) 70π cm^3
(d) 66.66π cm^3

62. What is the height of the cylinder at which its curved surface area is maximum ?

(a) 6.6 cm (b) 4.6 cm
(c) 7 cm (d) 12 cm

63. What is the maximum curved surface area of the cylinder ?

(a) 140π cm^2 (b) 70π cm^2
(c) 142π cm^2 (d) 35π cm^2

64. A right circular cone, of height 12 ft, stands on its base which has diameter 8 ft. The tip of the cone is cut off with a plane which is parallel to the base and 9 ft from the base. With $\pi = \frac{22}{7}$, the volume in cube ft, of the remaining part of the cone is : **[CAT-18]**

(a) 198 (b) 156
(c) 216 (d) None of these

65. A metallic solid is made up of a solid cylindrical base with a solid cone on its top. The radius of the base of the cone is 5 cm and the ratio of the height of the cylinder and the cone is 3 : 2. A cylindrical hole is drilled through the solid with height equal to 2/3rd of the height of solid. What should be the radius (in cm) of the hole so that the volume of the hole is 1/3rd of the volume of the metallic solid after drilling ? **[IIFT-18]**

(a) $\sqrt{\frac{45}{8}}$
(b) $\sqrt{\frac{65}{8}}$
(c) $\sqrt{\frac{55}{8}}$
(d) $\sqrt{\frac{35}{8}}$

66. Consider a cuboidal underground tunnel of length 500 m whose cross-section is given in the figure. If 1m^3 of concrete costs ₹ 1000, find the amount of money needed to build the tunnel. **[IIFT-19]**

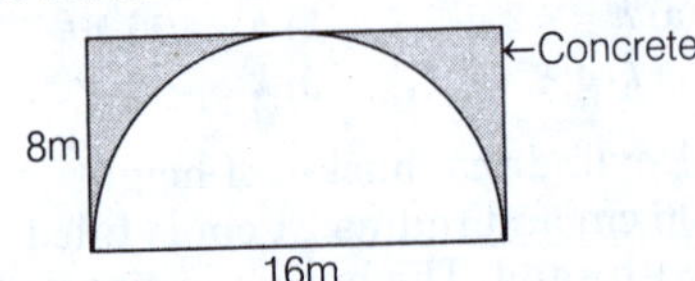

(a) ₹ $8(4-\pi)10^6$
(b) ₹ $64(4-\pi)10^6$
(c) ₹ $16(4-\pi)10^6$
(d) ₹ $32(4-\pi)10^6$

Answer Key

1. (a)	2. (b)	3. (a)	4. (d)	5. (a)	6. (d)	7. (a)	8. (a)	9. (c)	10. (c)
11. (b)	12. (b)	13. (d)	14. (a)	15. (d)	16. (b)	17. (a)	18. (d)	19. (b)	20. (a)
21. (c)	22. (c)	23. (a)	24. (d)	25. (a)	26. (b)	27. (a)	28. (c)	29. (a)	30. (b)
31. (c)	32. (d)	33. (a)	34. (d)	35. (b)	36. (a)	37. (b)	38. (c)	39. (a)	40. (a)
41. (b)	42. (d)	43. (b)	44. (b)	45. (c)	46. (b)	47. (c)	48. (b)	49. (b)	50. (c)
51. (b)	52. (d)	53. (a)	54. (b)	55. (c)	56. (c)	57. (b)	58. (d)	59. (b)	60. (a)
61. (a)	62. (c)	63. (b)	64. (a)	65. (c)	66. (c)				

Hints and SOLUTIONS

1. Volume of the iron piece $= \frac{1}{3}\pi r^2 h$

$= \frac{1}{3}\pi(14)^2 \times 30 = 6160$

Obviously, volume of the water risen in vessel = volume of the solid immersed

$\Rightarrow \quad \pi R^2 \times 6.4 = 6160$

$\Rightarrow \quad R^2 = 306.25$

$\Rightarrow \quad R = 17.5\text{ cm}$

Hence, diameter of the vessel

$= 2 \times 17.5 = 35\text{ cm}$

2. Required difference, $= lbh - \pi r^2 h$

$= (10 \times 10 \times 21) - \left\{\frac{22}{7} \times 5 \times 5 \times 21\right\} = 450$

3. We know that ratio of volume of cone to that of cylinder in case of same radius and height = 1 : 3

$\therefore$ Volume of cone (i.e. utilized wood) : wasted wood = 1 : (3 − 1) = 1 : 2

4. Flat surface area of the cylinder

$= \pi R^2 + \pi R^2 = 2\pi(7)^2 = 98\pi$

Volume of the cylinder

$= \pi R^2 h = \pi \times 7 \times 7 \times 10$

$= 490\pi$

Then, volume of smaller cone

$= 490\pi \times \frac{3}{7} = 210\pi$

$\Rightarrow \quad \frac{1}{3}\pi r^2 \times 10 = 210\pi$

$\Rightarrow \quad \pi r^2 = 63\pi$

and volume of bigger cone

$= 490\pi - 210\pi$

$= 280\pi$

$\Rightarrow \frac{1}{3}\pi r^2 \times 10 = 280\pi$

$\Rightarrow \quad \pi r^2 = 84\pi$

Hence, total flat surface area of two cones = $63\pi + 84\pi = 147\pi$

Then, required percentage change

$= \left\{\frac{147\pi - 98\pi}{98\pi} \times 100\right\}\%$

$= 50\%$

5. $R = 4$, $r = 2$ and $h = 6$

Now, (slant height)$^2 = h^2 + (R - r)^2$

$\Rightarrow \quad l = \sqrt{40}$

Then, curved surface area

$= \pi l(R + r) = 118.4\text{ m}^2$

6. Given, $2\pi rh = 660$

$\Rightarrow \quad 2\pi \times r \times 15 = 660$

$\Rightarrow \quad r = 7\text{ cm}$

Then, volume of the cylinder

$= \text{lateral surface area} \times \frac{r}{2}$

$= 660 \times \frac{7}{2} = 2310\text{ cm}^3$

7. Clearly, the diameter of the circular pond = $2\sqrt{2}$

$\therefore$ Its radius = $\sqrt{2}$

Now, volume of earth to be removed = volume of the circular pound (i.e. cylinder) − volume of square pond (i.e. cuboid)

$\Rightarrow \quad \pi \times (\sqrt{2})^2 \times 1 - 2 \times 2 \times 1$

$= (2\pi - 4)\text{ m}^3$

8. Hypotenuse AC

$= \sqrt{(3x)^2 + (4x)^2} = 15$

$\Rightarrow \quad 5x = 15$

$\therefore \quad x = 3$

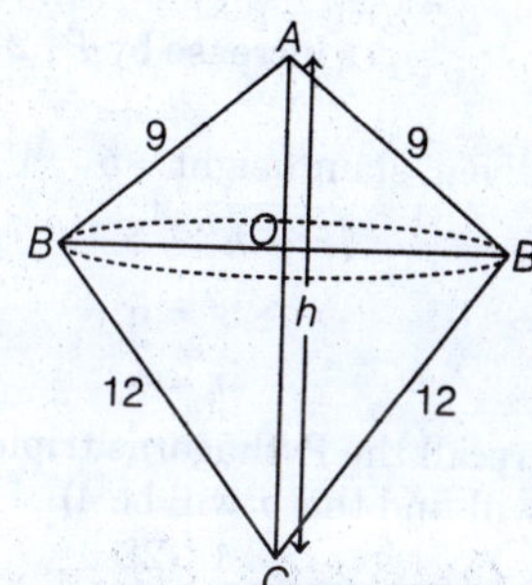

Hence, perpendicular sides are (3 × 3) and (4 × 3) i.e. 9 and 12 cm

Now, area of right angled ΔABC

$= \frac{1}{2} \times 9 \times 12$

$\Rightarrow \quad \frac{1}{2} \times AC \times BO = \frac{1}{2} \times 9 \times 12$

$\Rightarrow \quad \frac{1}{2} \times 15 \times BO = \frac{1}{2} \times 9 \times 12$

$\therefore \quad BO = \frac{9 \times 12}{15} = 7.2\text{ cm}$

Volume of cone ABB'

$= \frac{1}{3}\pi(7.2)^2 \times AO$

[BO = radius]

Volume of cone $CBB' = \frac{1}{3}\pi(7.2)^2 \times OC$

[also, radius = BO]

$\therefore$ Volume of the solid formed

$= \frac{1}{3}\pi(7.2)^2 \times AO + \frac{1}{3}\pi(7.2)^2 \times OC$

$= \frac{1}{3}\pi((7.2)^2[AO + OC]$

$= \frac{1}{3}\pi \times 51.84 \times AC$

$= \frac{1}{3}\pi \times 51.84 \times 15$

$= 814.62 = 815\text{ cm}^3$ (approx.)

9. Let the level be h cm.

Then, $\frac{1}{3}\pi \times 2^2 \times 3 = \pi \times 2^2 \times h$

$\Rightarrow \quad h = 1\text{ cm}$

10. Lateral area of the well $= 2\pi rh$

$= 2\pi \times \frac{d}{2} \times 14$

[here, d is diameter]

$= 2\pi \times \frac{4}{2} \times 14 = 176\text{ m}^2$

Then, total cost = 176 × 25 = ₹ 4400

11. Volume of ditch $= 48 \times 16.5 \times 4\text{ m}^3$

and volume of the earth dug out

$= \pi r^2 h = \frac{22}{7} \times \frac{4}{2} \times \frac{4}{2} \times 56\text{ m}^3$

Now, part of the ditch filled

$= \frac{\frac{22}{7} \times \frac{4}{2} \times \frac{4}{2} \times 56}{48 \times 16.5 \times 4} = \frac{2}{9}$

12. Total surface area of a cone

$= \pi r(r + l)$

$= \frac{22}{7} \times 7 \times (7 + 25)$

$[\because l = \sqrt{7^2 + (24)^2} = 25]$

$= 704\text{ m}^2$

13. Required ratio $= \frac{2\pi r(h + r)}{\pi r(l + r)}$

$= \frac{2\pi r(3 + 4)}{\pi r(5 + 4)}$

$[\because l = \sqrt{3^2 + 4^2}]$

$= 14 : 9$

14. Decrease in volume in case height is decreased by x cm

$= \frac{1}{3}\pi r^2 h - \frac{1}{3}\pi r^2 (h - x) = \frac{1}{3}\pi r^2 x$

Further, decrease in volume in case radius is decreased by x cm

$$= \frac{1}{3}\pi r^2 h - \frac{1}{3}\pi(r-x)^2 h$$

$$= \frac{1}{3}\pi(2xrh - x^2 h)$$

According to the question,

$$\frac{1}{3}\pi r^2 x = \frac{1}{3}\pi(2xrh - x^2 h)$$

$$\Rightarrow \quad r^2 = 2rh - xh$$

$$\Rightarrow \quad x = \frac{2rh - r^2}{h}$$

15. Total surface area $= 2\pi r(r+h)$

$= 660$ (given)

$$\Rightarrow \quad 2\pi \cdot 5(5+h) = 660$$

$$\Rightarrow \quad 5 + h = 21$$

$$\therefore \quad h = 16\,\text{cm}$$

16. Obviously, height of the cylinder will be width of the paper sheet and circumference of the base of cylinder will be equal to larger side of the sheet (i.e. 22 cm).

$$\Rightarrow \quad 2\pi r = 22$$

$$\therefore \quad r = 3.5\,\text{cm}$$

Now, volume $= \pi r^2 h$

$$= \frac{22}{7} \times 3.5 \times 3.5 \times 15$$

$$= 577.50\,\text{cm}^3$$

17. If we assume $DECB$ as a frustum of a cone ABC, then

$$\frac{\text{Height of the parent cone } (H)}{\text{Height of the cone cut } (h)} = \frac{\text{Radius of the parent cone } (R)}{\text{Radius of the cone cut } (r)}$$

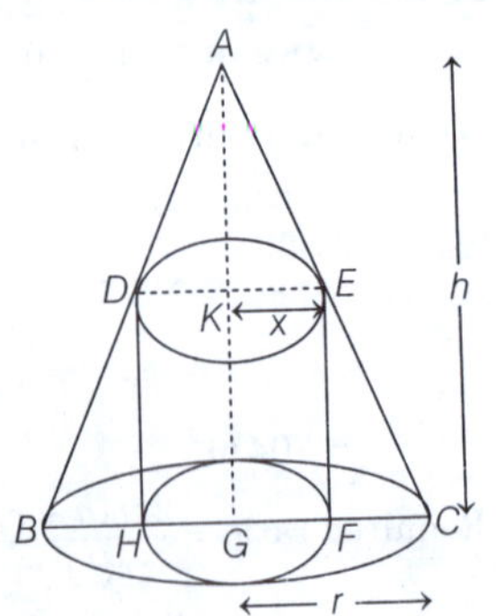

$$\Rightarrow \frac{10}{AK} = \frac{4}{KE}$$

$$\Rightarrow \frac{10}{10 - \text{height of cylinder}} = \frac{4}{x}$$

[let radius of cylinder be x]

$$\Rightarrow \text{Height of cylinder} = 10 - \frac{5x}{2}$$

Now, total surface area of the cylinder $= 2\pi r(r+h)$

$$= 2\pi x\left\{x + 10 - \frac{5x}{2}\right\}$$

$$= 2\pi x\left\{10 - \frac{3x}{2}\right\}$$

$$= 2\pi\left\{10x - \frac{3x^2}{2}\right\}$$

We are aware that for x be real, the maximum value of $ax^2 + bx + c$ is $\frac{4ac - b^2}{4a}$, if $a < 0$.

Hence, maximum value of $-\frac{3x^2}{2} + 10x$

$$= \frac{4 \times \frac{-3}{2} \times 0 - 10^2}{4 \times \frac{-3}{2}} = \frac{50}{3}$$

Then, maximum value of total surface area of the cylinder

$$= 2\pi \times \frac{50}{3} = \frac{100\pi}{3}$$

18. Total effect on volume

$$= \left(+P + P + \frac{P \times P}{100}\right)\%$$

$$= +P\left(2 + \frac{P}{100}\right)\%$$

+ indicates that the volume will increase by $P\left(2 + \frac{P}{100}\right)\%$.

19. Given, slant height = 5

$$\Rightarrow \quad \sqrt{r^2 + h^2} = 5$$

$$\Rightarrow \quad r^2 + 3^2 = 5^2$$

$$\Rightarrow \quad r = 4$$

(recall the Pythagoras triplets. You will find that r will be 4)

Now, volume $= \frac{1}{3} \times \frac{22}{7} \times 4^2 \times 3$

$= 50.3\,\text{cm}^3$ (approx.)

20. Here, $\frac{1}{3}\pi \times r^2 \times 54 = \pi \times (30)^2 \times 9$

$$\Rightarrow \quad r = 15\sqrt{2}$$

$$\therefore \quad d = 30\sqrt{2}\,\text{cm}$$

21. Volume of metal in hollow pipe

$$= \pi h(R^2 - r^2)$$

Here, h = length,

R = external radius

and r = internal radius

$$\Rightarrow \quad 1496 = \pi \times 28(R^2 - 8^2)$$

$$\Rightarrow (R^2 - 8^2) = 17$$

$$\Rightarrow \quad R^2 = 64 + 17 = 81$$

$$\therefore \quad R = \sqrt{81} = 9\,\text{cm}$$

22. Given, $2\pi rh = 94.2$

$$\Rightarrow 2 \times 3.14 \times r \times 5 = 94.2$$

$$\Rightarrow \quad r = 3\,\text{cm}$$

23. Clearly, distance covered by cylinder in one round = 60 cm.

Then, distance covered in 5 sec

$= (5 \times 5)$ rounds $\times$ 60 cm

$= 1500$ cm

$= 15$ m

24. Area of canvas required = curved area $= \pi rl$

$$= \pi \times 5 \times 13$$

$$[\because l = \sqrt{r^2 + h^2} = \sqrt{5^2 + (12)^2} = 13]$$

$$= 65\pi\,\text{m}^2$$

25. Here, three figures are +100%, + 100% and +100%

Effect on taking two figures

$$= \left\{+100 + 100 + \frac{100 \times 100}{100}\right\}\%$$

$$= +300\%$$

Then, final effect

$$= \left\{+300 + 100 + \frac{300 \times 100}{100}\right\}\%$$

$$= +700\%$$

Hence, volume will increase by 700%.

Alternatively

If radius and height of a cylinder/cone are made x and y times respectively, the volume will become $x^2 y$ times the earlier. Here, each of the radius and height has been made to 2 times the earlier, then, the volume will become $(2^2 \times 2)$ times the earlier. It means it will increase by 700%. For other methods, see the relevant illustrations.

26. Given, exterior radius $= 6/2 = 3$ cm, interior radius $= 4/2 = 2$ cm and height or length = 10 cm.

Now, Total surface area

$$= 2\pi(R + r)(h + R - r)$$

$$= 2\pi(3 + 2)(10 + 3 - 2)$$

$$= 110\pi\,\text{cm}^2$$

27. Clearly, 11000 cm^3

$$= \pi \times \left(\frac{35}{2}\right)^2 \times h \Rightarrow \quad h = 11.42$$

Hence, drop in water level = 11.42 cm

28. Surface area of the tent = Curved area of cylindrical portion + curved area of conical portion

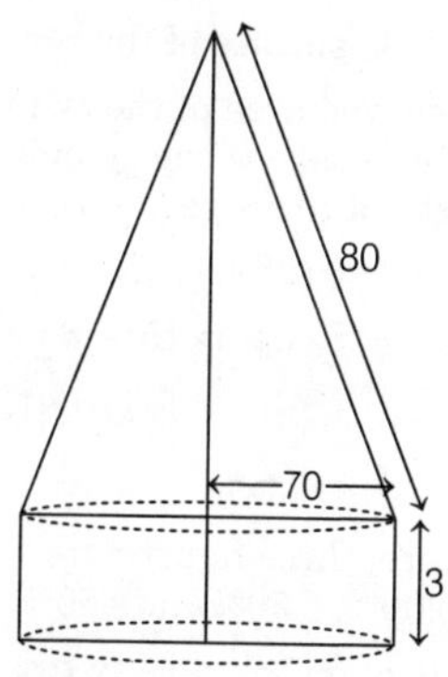

$= 2\pi rh + \pi rl$

$= (2\pi \times 70 \times 3) + \pi \times 70 \times 80$

$= \pi \times 70(6 + 80) = 18920\text{ m}^2$

Then, length of canvas = 18920/2

= 9460 m

29. Given, $\frac{1}{3}\pi r^2 h = \pi r^2 \times 5$

$\Rightarrow \quad h = 15\text{ cm}$

Alternatively

We know that if a solid cylinder is converted into a right circular solid cone or *vice-versa*, with the same base, the height of the cone will be thrice that of the cylinder. Hence, height of the cone = 3 × 5 = 15 cm.

30. Given, $\pi r^2 = 346.5 \Rightarrow r^2 = 110.25$

$\Rightarrow \quad r = 10.5\text{ cm}$

Area of canvas required

= curved area

$= \pi rl$

$= \pi \times 10.5 \times 17.5$

$[\because l = \sqrt{r^2 + h^2} = \sqrt{(10.5)^2 + (14)^2} = 17.5]$

$= 577.5\text{ cm}^2$

Now, length of the canvas

= 577.5/0.75 = 770 m

31. Volume of the cylinder

$= \pi r^2 h = \pi \times 6^2 \times 24$

$= 864\pi$

Obviously, diameter of the cylinder = diagonal of the square base of the cuboid

$\Rightarrow$ side of the square $\times \sqrt{2} = 12$

$\Rightarrow$ side of the square $= 6\sqrt{2}$

Now, required volume of water

$= 864\pi$ – volume of the cuboid

$= 864\pi - (6\sqrt{2} \times 6\sqrt{2} \times 20)$

$= 864\pi - 1440 = 288(3\pi - 5)$

32. Clearly, $FB = \sqrt{17^2 - 15^2} = 8\text{ cm}$

Then, $CF = (26/2 - 8) = 5\text{ cm} = DE$

Hence, radii of bases of a frustum of a cone are 5 cm and 13 cm and its height is 15 cm.

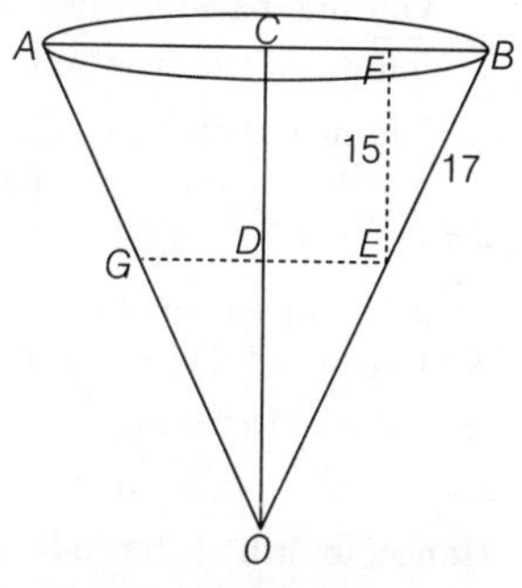

Then, volume of the figure *ABEG*

$= \frac{1}{3}\pi h(R^2 + r^2 + Rr)$

$= \frac{1}{3} \times \pi \times 15(13^2 + 5^2 + 13 \times 5)$

$= 1295\pi\text{ cm}^3$

Now, volume of water being flowing in 1 min $= \pi \times (0.5)^2 \times (10 \times 100)$

$= 250\pi\text{ cm}^3$

$\therefore$ Required time $= 1295\pi \div 250\pi$

= 5.18 min

33. Slant height (l) of the frustum of a cone

$= \sqrt{h^2 + (R - r)^2}$

$= \sqrt{(9)^2 + (6 - 3)^2}$

$= 3\sqrt{10}\text{ cm}$

34. $\frac{\text{Height of the parent cone } (H)}{\text{Height of the cone cut } (h)} = \frac{\text{Radius of the parent cone } (R)}{\text{Radius of the cone cut } (r)}$

[one of properties regarding frustum of a cone]

$\Rightarrow \frac{\text{height of frustum} + \text{height of cone cut } (h)}{h} = \frac{6}{3}$

$\Rightarrow \quad 3(9 + h) = 6h$

$\Rightarrow \quad h = 9$

Hence, height of the parent cone

= 9 + 9 = 18 cm

35. Total surface area

$= \pi[R^2 + r^2 + l(R + r)]$

$= \pi[6^2 + 3^2 + 3\sqrt{10}\,(6 + 3)]$

$= 9\pi(3\sqrt{10} + 5)\text{ cm}^2$

36. Vertical spacing between two consecutive turns

$= \frac{\text{Height of the cylinder}}{\text{Number of turns}}$

$= \frac{h}{n}$

37. If we arrange all the four walls of the cube in a sequence, the following image will be emerged.

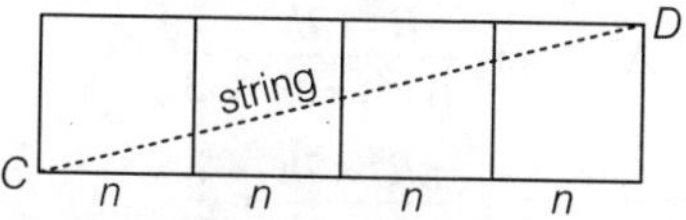

Clearly, length of the string

$= \sqrt{n^2 + (4n)^2} = \sqrt{17}\,n$

38. Clearly, length of each turn

$= 2\pi r = 2\pi.\frac{2}{\pi} = 4$

Then, length needed for n turns

= 4n cm

Hence,

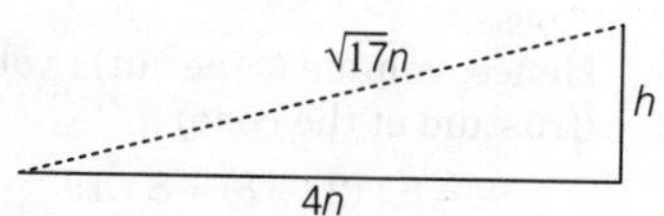

Then, $h^2 + (4n)^2 = (\sqrt{17}n)^2$

$\Rightarrow \quad h^2 = n^2$

$\Rightarrow \quad h = n$

39. Clearly, volume of sand in bucket = volume of the heap

$\Rightarrow \pi(21)^2 \times 36 = \frac{1}{3}\pi \times r^2 \times 12$

$\Rightarrow \quad r^2 = (21)^2 \times 9$

$\therefore \quad r = 21 \times 3 = 63\text{ cm}$

40. You may proceed directly through the concept of ratio analysis detailed in beginning of this chapter.

The required ratio $= (2 \times 5) : (3 \times 4)$

$= 5 : 6$

41. It is obvious that water filled in the cylinder (as below figure) will flow in 5 sec.

3.5 cm

5 m

Now, volume of water in this cylinder $= \pi \times (3.5)^2 \times 500$

[∵ 5 m = 500 cm]

$= \frac{22}{7} \times 3.5 \times 3.5 \times 500$

$= 19250\text{ cm}^3$

$\because 19250\text{ cm}^3$ is filled by the pipe in 1 sec.

$\therefore 400 \times 300 \times 231\text{ cm}^3$ is filled in $\frac{1}{19250} \times 400 \times 300 \times 231$ sec.

i.e. 1440 sec. (= 24 min)

42. Let radii of the 1st and 2nd cylinder be R and r respectively and their heights be $7h$ and $9h$ respectively.

As given, $\frac{\pi R^2 \times 7h}{(\pi r^2) \times 9h} = \frac{7}{3}$

$\Rightarrow \frac{\pi R^2 \times 7h}{154 \times 9h} = \frac{7}{3}$

$\Rightarrow \pi R^2 = 154 \times 3$

$\Rightarrow R = 7\sqrt{3}$ cm

43. Observe the properties regarding frustum of a cone given in beginning of this chapter.

Volume (parent cone) : Volume (cone cut)

$= H^3 : h^3 = 1^3 : \left(\frac{2}{3}\right)^3 = 27:8$

Hence, volume (cone cut) : volume (frustum of the cone)

$= 8 : (27 - 8) = 8 : 19$

44. Given, $\pi r^2 = 154 \Rightarrow r = 0.7$ km

Now, slant height $= \sqrt{r^2 + h^2}$

$\Rightarrow 2.5 = \sqrt{(0.7)^2 + h^2}$

$\Rightarrow 6.25 = 0.49 + h^2$

$\Rightarrow h^2 = 5.76 \Rightarrow h = 2.4$ km

45. It is obvious that the triangle formed with all three vertices of the cones will be an equilateral triangle with side $2r$ and it will be inscribed in the circle drawn through all three vertices of cones.

Then, circumradius $= \frac{\text{side}}{\sqrt{3}} = \frac{2r}{\sqrt{3}}$

$= 1.15r$

46. Total area of bases of cylinder

$= 23100 \times \frac{1}{(2+1)} = 7700$

$\Rightarrow 2\pi r^2 = 7700$

$\Rightarrow r^2 = 1225$

$\Rightarrow r = 35$

Now, volume of cylinder

$=$ curved surface area $\times \frac{r}{2}$

$= (23100 - 7700) \times \frac{35}{2} = 269500$

47. We know that

$\frac{\text{Volume (parent cone)}}{\text{Volume (cone cut)}} = \frac{R^3}{r^3}$

[where, R and r be radii of parent cone and the cone cut respectively]

$\Rightarrow \frac{\text{Volume (parent cone)}}{\text{Volume (parent cone)} - \text{Volume (cone cut)}}$

$= \frac{R^3}{R^3 - r^3}$

$\Rightarrow \frac{\text{Volume (parent cone)}}{\text{Volume of the frustum}} = \frac{2^3}{2^3 - 1^3}$

$\Rightarrow$ Volume of frustum : Volume of parent cone

$= 7 : 8$

48. Clearly, volume of slab of ice = volume of rod (cylindrical)

$\Rightarrow 8 \times 11 \times 2 = \pi(4)^2 \times h$

$\Rightarrow h = 3.5$

Hence, length of the rod = 3.5 inches

49. Total surface area of the cylinder before cutting $= 2\pi r(r + h)$

$= 2\pi \times 7(7 + 8) = 350\pi$

When the cylinder is cut in three equal parts, there will be no change in curved surface area but the total surface area will increase by $(\pi r^2 + \pi r^2 + \pi r^2 + \pi r^2)$ as four more bases will be emerged.

Then, increase in total surface area

$= 4\pi r^2 = 4\pi \times 49 = 196\pi$

Hence, percentage increase in total surface area $= \left(\frac{196\pi}{350\pi} \times 100\right)\% = 56\%$

50. Here, $h = 20$ cm, $R = 12.5$ cm, $r = 12.5 - 1 = 11.5$ cm

Now, total surface area

$= 2\pi(R + r)(h + R - r)$

$= 2 \times \frac{22}{7} \times (12.5 + 11.5)(20 + 12.5 - 11.5)$

$= 3168 \text{ cm}^2$

51. Given, $2\pi rh = \frac{1}{3} \times 2\pi r(h + r)$

$\Rightarrow 3h = h + r$

$\Rightarrow r = 2h = 2 \times 9 = 18$ cm

52. Given, curved surface area : volume $= 1 : 7$

We know, volume = curved surface area $\times \frac{r}{2}$

$\Rightarrow 7 = 1 \times \frac{r}{2} \Rightarrow r = 14$

Further, $2\pi r(r + h) : \pi r^2 h = 187 : 770$

$\Rightarrow 2(r + h) : rh = 187 : 770$

$\Rightarrow \frac{2}{h} + \frac{2}{r} = \frac{187}{770}$

$\Rightarrow \frac{2}{h} + \frac{2}{14} = \frac{187}{770}$

$\Rightarrow h = 20$

Hence, the required ratio

$= 14:20 = 7 : 10$

53. Whole surface of the remaining solid

= curved area of the cylinder + area of one base of the cylinder + curved area of the cone $= 2\pi rh + \pi r^2 + \pi rl$

$= \pi r(2h + r + l)$

$= \pi \times 8[(2 \times 15) + 8 + 17]$

$[\because l = \sqrt{(15)^2 + 8^2} = 17]$

$= 440\pi \text{ cm}^2$

54. Here, three figures are +200%, – 50% and –50%

Effect on taking any two figures (say, +200, – 50)

$= \left\{+200 - 50 - \frac{200 \times 50}{100}\right\}\% = +50\%$

Then, final effect

$= \left\{+50 - 50 - \frac{50 \times 50}{100}\right\}\%$

$= -25\%$

Hence, volume will decrease by 25%.

Alternatively

If radius and height of a cylinder/cone are made x and y times respectively, the volume will become x^2y times the earlier. Here, radius has been made to 0.5 times the earlier and height has been made to thrice the earlier. Then, the volume will become $(0.5^2 \times 3)$ i.e. 0.75 times the earlier. It means it will be decreased by 25%. For other methods, see the relevant illustrations.

55. Effect on the value of

$r^2 = \left\{-20 - 20 + \frac{20 \times 20}{100}\right\}\%$

$= -36\%$

We know that volume is the product of r^2 and h with a constant term. When r^2 is decreased by 36%, the h must be increased by $\left\{\frac{36}{(100 - 36)} \times 100\right\}\%$ i.e. by 56.25%.

56. Ratio of volumes $= (r_1)^2 h_1 : (r_2)^2 h_2$

$\Rightarrow 27:80 = (r_1)^2 \times 3 : (r_2)^2 \times 5$

$\Rightarrow (r_1)^2 : (r_2)^2 = 9 : 16$

$\Rightarrow r_1 : r_2 = 3 : 4$

57. Consider 1st sheet

(i) If it is folded along length, then circumference of the base of the cylinder $= 4\pi$ and height of the cylinder thus formed $= 2\pi$

$\Rightarrow 2\pi r = 4\pi \Rightarrow r = 2$

Then, volume of the cylinder

$= \pi \times 2^2 \times 2\pi = 8\pi^2$

(ii) If it is folded along breadth, then circumference of the base of the cylinder = 2π and height of the cylinder thus formed = 4π

$\Rightarrow 2\pi r = 2\pi \Rightarrow r = 1$

Then, volume of the cylinder

$= \pi 1^1 \times 4\pi = 4\pi^2$

(b) Consider 2nd sheet

(i) If it folded along length, then circumference of the base of the cylinder = 5π and height of the cylinder thus formed = π

$\Rightarrow \quad 2\pi r = 5\pi$

$\Rightarrow \quad r = 2.5$

Then, volume of the cylinder

$= \pi(2.5)^2 \times \pi = 6.25\pi^2$

(ii) If it is folded along breadth, then circumference of the base of the cylinder = π and height of the cylinder thus formed = 5π

$\Rightarrow \quad 2\pi r = \pi$

$\Rightarrow \quad r = 0.5$

Then, volume of the cylinder

$= \pi(0.5)^2 \times 5\pi = 1.25\pi^2$

58. Given, slant height AC

$= \sqrt{2}\, r \text{ cm} = AB$

Since $\angle BAC = 90°$, Hypotenuse

$BC = \sqrt{(\sqrt{2}r)^2 + (\sqrt{2}r)^2} = 2r$

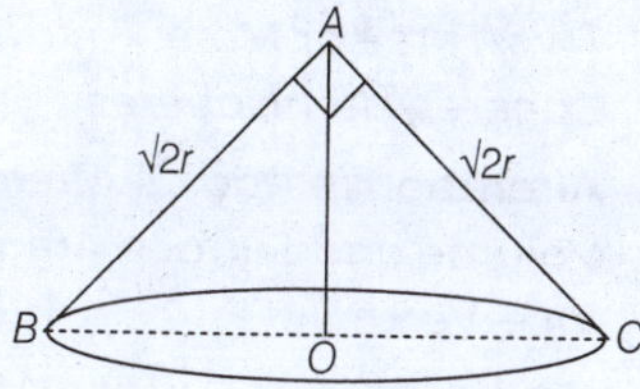

Then, radius $OC = BC \div 2$

$= 2r \div 2 = r$

Now, $AO = \sqrt{(AC)^2 - (OC)^2}$

$= \sqrt{(\sqrt{2}r)^2 - (r)^2} = r$

$\therefore$ Volume of the cube $= \frac{1}{3}\pi r^2 h$

$= \frac{1}{3}\pi r^2 \times r = \frac{\pi r^3}{3}$

59. Curved area (cone ABC) : Curved area (cone AFG)

$= h^2 : (3h)^2 = 1 : 9$

Curved area (cone AED) : Curved area (cone AFG) = $(2h)^2 : (3h)^2$

$= 4 : 9$

On combination, Curved area (cone ABC) : Curved area (cone AED) : Curved area (cone AFG) = 1 : 4 : 9

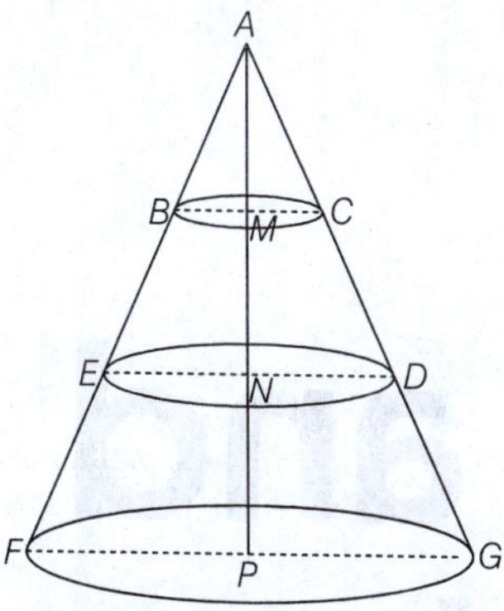

$\therefore$ Curved area of cone ABC : Curved area of part $BCDE$: Curved area of part $EDGF$ = 1: (4 − 1) : (9 − 4)

$= 1 : 3 : 5$

60. We know that the height and radius of a cylinder of maximum possible volume, inscribed in a cone of radius r and height h, are $h/3$ and $\frac{2r}{3}$ respectively.

Hence, the required radius $= \frac{2}{3} \times 10$

$= 6.66$ cm

61. We know that the maximum volume of a cylinder inscribed in a cone, is $\frac{4}{9}$ times the volume of the cone.

Hence, required answer

$= \frac{4}{9} \times \frac{1}{3}\pi r^2 h = \frac{4}{27} \times \pi \times (10)^2 \times 14$

$= 207.4\pi \text{ cm}^3$

Alternatively

Radius of cylinder $= \frac{2}{3} \times 10$ and

Height of the cylinder $= \frac{14}{3}$

Then, its volume $= \pi\left(\frac{20}{3}\right)^2 \times \frac{14}{3}$

$= 207.4\pi \text{ cm}^3$

62. We know that the height and the radius of a cylinder of maximum curved surface area, inscibed in a cone of radius r and height h, are $h/2$ and $\frac{r}{2}$ respectively.

Hence, required answer $= \frac{14}{2} = 7$ cm

63. In case of maximum curved surface area, height of the cylinder

$= \frac{14}{2} = 7$ cm

and radius of the cylinder

$= \frac{10}{2} = 5$ cm

Then, maximum curved surface area of the cylinder $= 2\pi \cdot 5 \cdot 7 = 70\pi \text{ cm}^2$

64. We know, Volume of the parent cone : Volume of the cone cut

$= H^3 : h^3 = 12^3 : (12-9)^3 = 64 : 1$

Now, volume of the remaining part of the cone $= \frac{(64-1)}{64} \times \frac{1}{3}\pi R^2 H$

$= \frac{(64-1)}{64} \times \frac{1}{3} \times \frac{22}{7} \times 4^2 \times 12 = 198 \text{ ft}^3$

65. Let heights of the solid cylinder and the solid cone be $3h$ and $2h$ respectively. Then, height of the hole = 2/3 of $(3h + 2h) = \frac{10h}{3}$

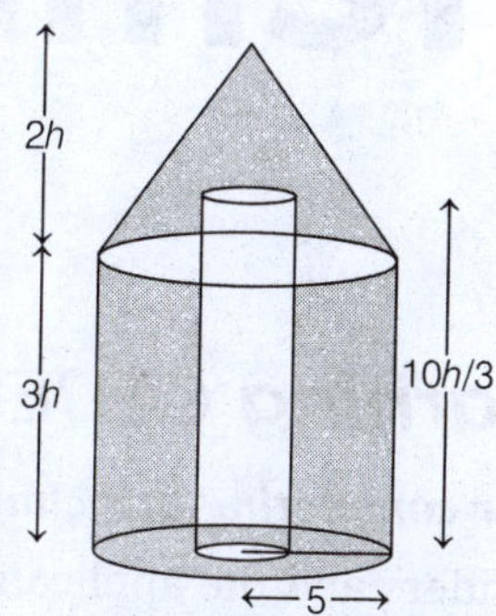

Volume of the solid before a hole is drilled $= \pi \times 5^2 \times 3h + \frac{1}{3}\pi \times 5^2 \times 2h$

$= 25\pi \times \frac{11h}{3}$

Let the radius of the hole be r cm.

Then, volume of the hole $= \pi r^2 \times \frac{10h}{3}$

As given, $\pi r^2 \times \frac{10h}{3}$

$= \frac{1}{3}\left[\left(25\pi \times \frac{11h}{3}\right) - \left(\pi r^2 \times \frac{10h}{3}\right)\right]$

$\Rightarrow 40\pi r^2 = 275\pi$

$\Rightarrow \quad r^2 = \frac{275}{40} \Rightarrow r = \sqrt{\frac{55}{8}}$

66. Clearly, volume of the concrete = volume of the tunnel (cuboidal) − volume of the cross section

= (length × breadth × height)

$-\frac{1}{2}(\pi r^2 h)$

[here, h is basically length of the tunnel]

$= (500 \times 16 \times 8) - \frac{1}{2} \times \pi\left(\frac{16}{2}\right)^2 \times 500$

$= 16000(4 - \pi) \text{ m}^3$

Then, amount of money needed to build the tunnel $= 16000(4 - \pi) \times 1000$

$= ₹\, 16(4 - \pi)10^6$

46

Sphere and Hemisphere

Learning OBJECTIVES

After completing this chapter, aspirants will be able to :

- Understand the application of the sphere and hemisphere related formulae in different cases;
- Understand the application of combination of sphere and hemisphere related formulae in diverse cases;
- Learn how the surface area is related with volume;
- Distinguish between concepts on solid and hollow spheres/hemispheres;
- Analyse different shapes with reference to diverse aspects;
- Compare the sphere and hemisphere with reference to different ratios;
- Understand the philosophy of the division of this chapter in various parts;
- Analyse the trend of the questions being asked in various competitive examinations;
- Solve the questions in moments;
- Grip the concepts helping to solve the questions without pen and paper;
- Apply many concepts envisaged in the chapter 'PERCENTAGE' like 'effect on product in case of changes in factors', 'Product Constancy' etc.;
- Apply many concepts of 'Ratio' in this chapter;
- Learn many new concepts/fundamentals;
- Apply the pragmatic approaches in lieu of academic approaches; and
- Ponder on the alternative methods.

Inside

- *Understanding Shapes of Sphere and Hemisphere with Formulae of Volume, Curved Area and Total Surface Areas*
- *Concept of the Hollow Sphere and the Hollow Hemisphere*
- *Ratio-Analysis in Case of Two/Three or More Spheres/Hemispheres*
- *Alteration in Radius Affecting Volume and Surface Area and Vice-Versa*
- *Comparison of Surface Areas of Various Shapes*
- *Ratio Analysis in Case of Different Shapes with Reference to Diverse Aspects*
- *Conversions and Combination of Shapes (Cubiod, Cube, Cylinder, Cone, Sphere and Hemisphere)*
- *Concept of Maxima/Minima Value of Volume/Area*
- *Problems Based on Above all Aspects*
- ....
- .... *& Many More*
- *Work Out with Solutions*

SPHERE

Sphere is a 3-dimensional object shaped like a football, a cricket ball, etc. In sphere, every point on the surface is at the same distance from the centre.

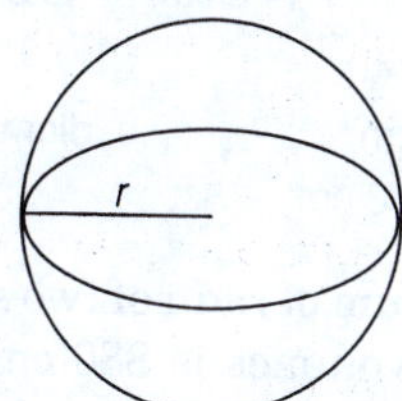

Volume of the sphere $= \frac{4}{3}\pi r^3$, where r = radius

Surface area of the sphere $= 4\pi r^2$

HEMISPHERE

When a plane through the centre of a sphere cuts it into two equal parts, each part is said to have the shape of a hemisphere.

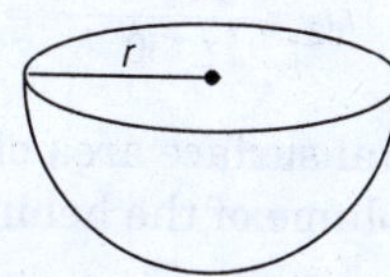

Volume of the hemisphere

$$= \frac{\text{Volume of the sphere}}{2} = \frac{\frac{4}{3}\pi r^3}{2} = \frac{2}{3}\pi r^3$$

Curved surface area of the hemisphere

$$= \frac{\text{Curved area of the sphere}}{2} = \frac{4\pi r^2}{2} = 2\pi r^2$$

Total surface area of the hemisphere = Curved surface area + Area of the base $= 2\pi r^2 + \pi r^2 = 3\pi r^2$

Concept of the hollow sphere and the hollow hemisphere

Hollow Sphere

Imagine a hollow spherical ball like tennis ball, whose external radius and internal radius are 'R' and 'r' respectively. Then, thickness of the ball wall will be equal to the difference between its external and internal radii. The volume of the rubber used to make the tennis ball will be equal to the difference between the external volume and the internal volume, i.e.

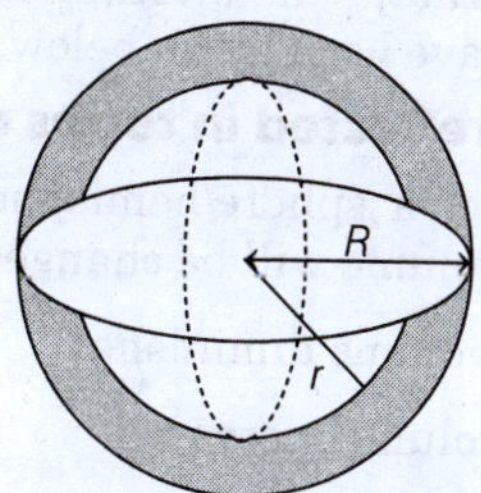

Volume of rubber to make hollow tennis ball

$$= \frac{4}{3}\pi R^3 - \frac{4}{3}\pi r^3$$

$$= \frac{4}{3}\pi[R^3 - r^3]$$

Hollow Hemisphere

In case of hollow hemispherical items like bowl made of metal, etc. having external and internal radii 'R' and 'r' respectively, the volume of the metal used **to make the bowl** is $\frac{2}{3}\pi R^3 - \frac{2}{3}\pi r^3$

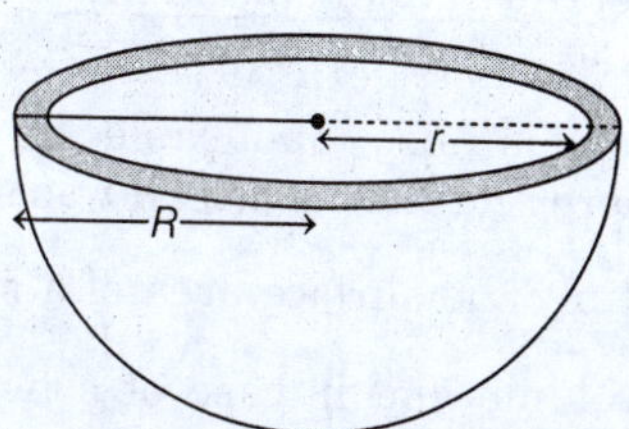

i.e. $\frac{2}{3}\pi\,[R^3 - r^3]$

The thickness of bowl wall = Difference between R and r.

Total surface area of the bowl = External curved area + internal curved area + area of the base ring (i.e. shaded region)

$$= 2\pi R^2 + 2\pi r^2 + \pi(R^2 - r^2)$$

$$= \pi(3R^2 + r^2)$$

The illustrations given under title 'concept of volume/area when thickness of layer/wall is taken into consideration' will emphasize manifestly on applications of these conceptions.

Ratio-analysis in case of two/three or more spheres/hemispheres

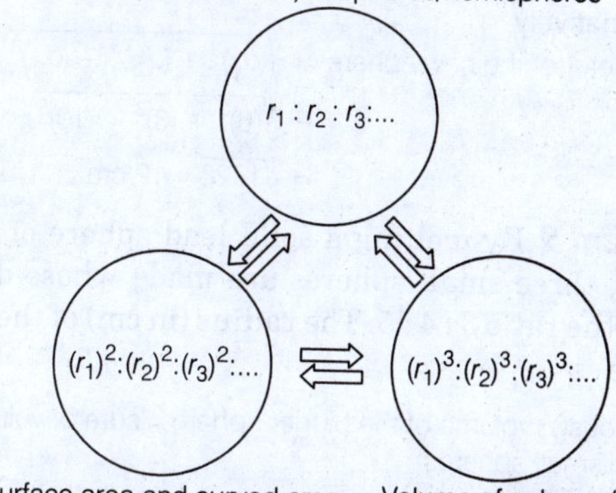

Illustrations

Application of the Concepts of Volume or/and Surface Area of the Sphere/Hemisphere

(i) If two or more solid spheres with radii r_1, r_2, r_3, ...are melted to form a new solid sphere, then radius of the newly formed sphere will be

$$\sqrt[3]{(r_1)^3 + (r_2)^3 + (r_3)^3 + \ldots\ldots}$$

This pattern also holds good if we replace radius r by diameter(d) i.e. diameter (D) of newly formed sphere $= \sqrt[3]{(d_1)^3 + (d_2)^3 + (d_3)^3 + \ldots\ldots}$ Further, this conception is also applicable in case of the hemisphere and the cube(read r as side).

(ii) We multiply the surface area of a sphere by $\frac{r}{3}$ to get its volume and in case of a hemisphere, we multiply its curved area by $\frac{r}{3}$ for getting its volume.

Problem 1 Three solid metallic spheres of diameters 6 cm, 8 cm and 10 cm are melted and recast into a new solid sphere. The diameter of the new sphere is **[SSC(CGL-PT)- 02, DMRC-03]**

Sol. Sum of volumes of all three spheres

$$= \frac{4}{3}\pi\left(\frac{6}{2}\right)^3 + \frac{4}{3}\pi\left(\frac{8}{2}\right)^3 + \frac{4}{3}\pi\left(\frac{10}{2}\right)^3$$

$$= \frac{4}{3}\pi[27 + 64 + 125] = \frac{4}{3}\pi(216)$$

This sum will form a new sphere, thus volume of the new sphere i.e. $\frac{4}{3}\pi R^3 = \frac{4}{3}\pi(216)$

[where, R is the radius of newly formed sphere]

$\Rightarrow \quad R = \sqrt[3]{216} = 6$ cm

Hence, its diameter $= 6 \times 2 = 12$ cm.

Alternatively

Diameter of the new sphere $= \sqrt[3]{(d_1)^3 + (d_2)^3 + (d_3)^3}$

$$= \sqrt[3]{(6)^3 + (8)^3 + (10)^3}$$

$$= \sqrt[3]{1728} = 12 \text{ cm}$$

Problem 2 By melting a solid lead sphere of diameter 12 cm, three small spheres are made whose diameters are in the ratio 3 : 4 : 5. The radius (in cm) of the smallest sphere is

Sol. Obviously, volume of the bigger sphere = sum of volumes of all three small spheres

$$\Rightarrow \frac{4}{3}\pi\left(\frac{12}{2}\right)^3 = \frac{4}{3}\pi\left(\frac{3x}{2}\right)^3 + \frac{4}{3}\pi\left(\frac{4x}{2}\right)^3 + \frac{4}{3}\pi\left(\frac{5x}{2}\right)^3$$

[let diameters be $3x$, $4x$ and $5x$ respectively]

$$\Rightarrow \quad 216 = \frac{216x^3}{8} \Rightarrow x^3 = 8 \Rightarrow x = \sqrt[3]{8} = 2$$

$\therefore$ Radius of the smallest sphere $= \frac{3x}{2} = \frac{3 \times 2}{2} = 3$ cm

Alternatively

Diameter of the sphere which has been melted

$$= \sqrt[3]{(d_1)^3 + (d_2)^3 + (d_3)^3}$$

$$= \sqrt[3]{(3 \text{ units})^3 + (4 \text{ units})^3 + (5 \text{ units})^3} = 6 \text{ units}$$

Given, 6 units = 12 cm

$\therefore$ 3 units = 6 cm [i.e. diameter of smallest sphere]

Hence, its radius = 3 cm

Problem 3 The sum of radii of two spheres is 10 cm and the sum of their volumes is 880 cm^3. What will be the product of their radii ? **[SSC(SO-Audit)-05]**

Sol. Given, $r_1 + r_2 = 10$ and $\frac{4}{3}\pi(r_1)^3 + \frac{4}{3}\pi(r_2)^3 = 880$

$$\Rightarrow \quad \frac{4}{3}\pi[(r_1)^3 + (r_2)^3] = 880$$

$$\Rightarrow \quad (r_1)^3 + (r_2)^3 = 210$$

Now, $\quad (r_1 + r_2)^3 = (r_1)^3 + (r_2)^3 + 3r_1r_2(r_1 + r_2)$

$$\Rightarrow \quad (10)^3 = 210 + 3r_1r_2(10)$$

$$\Rightarrow \quad r_1r_2 = \frac{1000 - 210}{30} = 26\frac{1}{3}$$

Problem 4 The total surface area of a solid hemisphere is $108\pi \text{ cm}^2$. The volume of the hemisphere is

Sol. Given, $3\pi r^2 = 108\pi \Rightarrow r^2 = 36 \Rightarrow r = 6$

$$\therefore \quad \text{Volume} = \frac{2}{3}\pi r^3 = \frac{2}{3}\pi(6)^3$$

$$= 144\pi \text{ cm}^3$$

Alteration in Radius Affecting Volume and Surface Area and *Vice-Versa*

- **If alteration is reflected in terms of ...... times**

1. If radius of a sphere/hemisphere is made x times, its volume will become x^3 times the earlier.
2. If radius of a sphere/hemisphere is made x times, its total/curved surface area will become x^2 times the earlier.

- **If alteration is reflected in terms of percentage**

3. Recall the concept 'Effect on product in case of changes in factors' detailed in chapter 'PERCENTAGE' when alteration in dimensions is occurred in percentage term. In case of volume of a sphere/hemisphere, variable factor radius is multiplied by radius and further by radius whereas in case of total surface area/ curved area, radius is multiplied by radius once. Illustrations have been given below.

- **If alteration is reflected in terms of ratio**

4. If the radius of a sphere/hemisphere is changed in ratio $a : b$, its volume will be changed in ratio $a^3 : b^3$.

   (a) If $a^3 > b^3$, volume diminishes.

   (b) If $a^3 < b^3$, volume increases.

   (c) If $a^3 = b^3$, no changes in the volume.

[Percentage change in volume $=\left(\dfrac{a^3-b^3}{a^3}\times 100\right)\%$, if the outcome is positive, volume diminishes and in case of negative, it increases].

5. If radius of a sphere/hemisphere is changed in ratio $a:b$, its total/curved surface area will be changed in ratio $a^2:b^2$. The surface area will decrease if $a^2>b^2$, will increase if $a^2<b^2$ and will remain unchanged, if $a^2=b^2$.

   [Percentage change in total/curved surface area $=\left(\dfrac{a^2-b^2}{a^2}\times 100\right)\%$, if the outcome is positive, surface area diminishes and in case of negative, it increases].

Problem 5 If the radius of a sphere is doubled, its volume becomes :

Sol. Obviously, the radius has been increased by 100%.

Here, three figures are = + 100%, + 100% and + 100%

Effect on taking any two figures

$$=\left(+100+100+\frac{100\times 100}{100}\right)\% = +300\%$$

$$\text{Now, final effect} = \left(+300+100+\frac{300\times 100}{100}\right)\% = +700\%$$

Hence, the original volume will increase by 700% i.e. it will become 8 times.

Alternatively 1

Since, the radius is made to 2 times, the volume will become $(2)^3$ times.

Alternatively II

It is known that if the radius of a sphere/hemisphere is changed in ratio $a:b$, its volume will be changed in ratio $a^3:b^3$. Here, radius is changed in ratio 1 : 2, then volume will change in 1 : 8 i.e. the original volume will become 8 times.

Problem 6 If the radius of a sphere be doubled, the area of its surface will become :

Sol. Obviously, two figures are = + 100% and + 100%

$$\therefore \quad \text{Effect} = \left(+100+100+\frac{100\times 100}{100}\right)\% = +300\%$$

It means the surface area will increase by 300% i.e. it will become 4 times.

Alternatively I

Since the radius is made to 2 times, the surface area will become $(2)^2$ times.

Alternatively II

We know that if radius of a sphere/hemisphere is changed in ratio $a:b$, its total/curved surface area will be changed in ratio $a^2:b^2$. Here, radius is changed in ratio 1 : 2, then area of surface will change in ratio 1 : 4. It means area of surface will become 4 times.

It may be construed from the discussion made through the above two problems that all the techniques are correlated and aspirants are supposed to be well versed with all these techniques so that they apply them appropriately observing the nature of questions.

Problem 7 If the radius of a sphere is increased by 2 cm, its surface area is increased by 352 cm^2. The radius of sphere before change is :

[SSC(CGL-PT)-99, CBI-03, SSC(CGL-T-II)-10]

Sol. Clearly, $4\pi(r+2)^2-4\pi r^2=352$

[here, r = radius before the change]

$$\Rightarrow \quad r^2+4r+4-r^2 = 352\times\frac{7}{22\times 4}$$

$$\Rightarrow \quad 4r = 28-4$$

$$\Rightarrow \quad r = 6\text{ cm}$$

Alternatively

You may proceed through the options. Let options be 3, 4, 6, ...

(a) $4\pi(3+2)^2-4\pi(3)^2$ is not equal to 352

(b) $4\pi(4+2)^2-4\pi(4)^2$ is not equal to 352

(c) $4\pi(6+2)^2-4\pi(6)^2$ is equal to 352

Thus, (c) is correct option.

Problem 8 The surface area of a sphere is 616 cm^2. If its radius is changed so that the area gets reduced by 75%, then the radius becomes ***[CDS-11]***

Sol. Given, new surface area = 25% of 616

$$\Rightarrow \quad 4\pi r^2 = 154 \Rightarrow r^2 = 12.25 \Rightarrow r = 3.5\text{ cm}$$

Comparison of Surface Areas of Various Shapes

Under given equal volume, the solid with the maximum number of sides will have minimum surface area. Conversely, the solid with least number of sides will have maximum surface area. It is remarkable that a sphere has infinite number of sides with least possible length. However, in case of cuboid and cube under given equal volume, the cuboid will have maximum surface area despite the fact that they have equal number of sides.

Illustration If the volumes of a sphere, a cube, a octahedron are equal, which of these will have minimum surface area

Obviously, the sphere will have minimum surface area as it has infinite number of sides with least possible length.

Problems Based on Ratio Analysis in Case of Two/Three or more Spheres/Hemispheres

The details of ratio analysis have already been texted in the beginning of this chapter

Problem 9 The diameter of two hollow spheres made from the same metal sheet are 21 cm and 17.5 cm respectively. The ratio of the areas of metal sheets required for making the two spheres is :

Sol. Given, ratio of their diameters = 21 : 17.5

$$\therefore \text{Ratio of their surface area} = (21)^2:(17.5)^2 = 9:6.25$$

$$= 36:25$$

Problem 10 If the radii of two spheres are in the ratio 1 : 4, then their surface areas are in the ratio : *[SSC(CPO-SI)-07]*

Sol. The required ratio = $(1)^2 : (4)^2 = 1:16$

Problem 11 The radii of two spheres are in the ratio 3 : 2. Their volumes will be in ratio :

Sol. The required ratio = $(3)^3 : (2)^3 = 27 : 8$

Problem 12 The volumes of two spheres are in the ratio 8 : 27. The ratio of their surface areas is :

Sol. Ratio of radii = $\sqrt[3]{8} : \sqrt[3]{27} = 2 : 3$

$\therefore$ Ratio of surface areas = $(2)^2 : (3)^2 = 4 : 9$

Problem 13 The surface areas of two spheres are in the ratio 4 : 9, their volumes will be in the ratio *[SSC(10+2, LDC/DEO)-12, SSC(CGL-T-1I)-13, SSC(CGL-T-II)-14]*

Sol. Ratio of radii = $\sqrt{4} : \sqrt{9} = 2 : 3$

$\therefore$ Ratio of volumes = $(2)^3 : (3)^3 = 8 : 27$

Ratio Analysis in Case of Different Shapes With Reference to Diverse Aspects

- **Ratio analysis in case of Hemisphere and Sphere**

(i) If radii of the hemisphere and the sphere are in ratio $a : b$, the ratio of their volumes will be $a^3 : 2b^3$.

(ii) If volumes of the hemisphere and the sphere are in ratio $a : b$, the ratio of their ratii is $\sqrt[3]{a} : \sqrt[3]{\frac{b}{2}}$.

(iii) If volumes of the hemisphere and the sphere are same, ratio of their radii is $\sqrt[3]{2} : 1$.

(iv) If volumes of the hemisphere and the sphere are same, ratio of their curved areas is $1 : \sqrt[3]{2}$.

- **Ratio analysis in case of Cone and Sphere**

(i) If a cone and a sphere have the same radius and the same volume, the height of the cone will be 4 times the radius.

(ii) If a cone and sphere have the same radius and the height of the cone is equal to the radius, the volume of the sphere will be 4 times that of the cone.

(iii) If a cone and a sphere have same volume and the height of the cone is equal to the radius of the sphere, then radius of the cone will be twice that of the sphere.

(iv) If a cone and a sphere have equal radius and equal volume, the ratio of their curved areas will be $\sqrt{17} : 4$.

- **Ratio analysis in case of Cylinder and Sphere**

If a cylinder and a sphere have equal radius and equal volume, the ratio of their curved areas will be 2 : 3 and the ratio of their whole surface areas will be 7 : 6.

- **Ratio analysis in case of Cube and Sphere**

(i) If an edge of a cube and the radius of a sphere are equal, the ratio of their volumes will be 21 : 88. Conversely, if the volumes of a cube and a sphere are equal, the ratio of edge to radius will be $\sqrt[3]{88} : \sqrt[3]{21}$.

(ii) If an edge of a cube and the radius of a sphere are equal, the ratio of their total surface areas will be 21 : 44. Conversely, in case of equal total surface areas, the ratio of edge to radius will be $\sqrt[2]{44} : \sqrt[2]{21}$.

(iii) If the total surface areas of a cube and a sphere are equal, the ratio of their volumes will be $\sqrt{\pi} : \sqrt{6}$.

- **Ratio analysis in case of Cone, Hemisphere/ Sphere and Cylinder**

Table I

Ratio of the volumes of a cylinder, a cone and a hemisphere /sphere	3 : 1 : 2	Provided they have equal radius and equal height
Ratio of the heights of a cylinder, a cone and a hemisphere /sphere	2 : 6 :3	Provided they have equal radius and equal volume
Ratio of the radii of a cylinder, a cone and a hemisphere /sphere	$\sqrt{2} : \sqrt{6} : \sqrt{3}$	Provided they have equal heights and equal volume

Table II

Ratio of the curved surface areas of a cylinder, a cone and a hemisphere	$\sqrt{2} : 1 : \sqrt{2}$	Provided they have equal radius and equal height
Ratio of the curved surface areas of a cylinder, a cone and a sphere	$4 : \sqrt{5} : 4$	
Ratio of the whole surface areas of a cylinder, a cone and a hemisphere	$4 : (\sqrt{2} + 1) : 3$	
Ratio of the whole surface areas of a cylinder, a cone and a sphere	$6 : (\sqrt{5} + 1) : 4$	

Note *Height of the hemisphere means its radius and height of the sphere means its diameter - applicable for both tables.*

Problem 14 Each of the measure of the radius of base of a cone and that of a sphere is 8 cm. Also, the volumes of these two solids are equal. The slant height of the cone is :

Sol. Given, volume of the sphere = volume of the cone

$$\Rightarrow \quad \frac{4}{3}\pi(8)^3 = \frac{1}{3}\pi(8)^2 \times h \Rightarrow h = 32$$

$\therefore$ Slant height of the cone

$$= \sqrt{8^2 + (32)^2} = \sqrt{64(1+16)} = 8\sqrt{17} \text{ cm}$$

Alternatively

From the above observation, the height of the cone will be 4 times the radius i.e. 32 cm.

Now, slant height of the cone

$$= \sqrt{8^2 + (32)^2} = \sqrt{64(1+16)} = 8\sqrt{17} \text{ cm}$$

Problem 15 A hemisphere and a cone have equal bases. If their heights are also equal, the ratio of their curved surfaces will be :

[SSC(CGL-PT)-02, 05, SSC(CGL- T-I)-11]

Sol. Required ratio $= \dfrac{\text{Curved surface area of the hemisphere}}{\text{Curved surface area of the cone}}$

$$= \frac{2\pi r^2}{\pi r l} = \frac{2r}{l} = \frac{2r}{\sqrt{r^2+h^2}} = \frac{2r}{\sqrt{r^2+r^2}}$$

[given, height of the cone = height of the hemisphere i.e. $h = r$]

$$= \frac{2r}{\sqrt{2}\, r} = \sqrt{2} : 1$$

Alternatively

Adhering the observation (supra), you may arrive directly at the correct answer.

Problem 16 The total surface area of a cube and a sphere are equal. What will be the ratio between their volumes ?

[SSC(SO-CommAudit)-05, SSC(10 + 2), (LDC/DEO)-10, SSC(MTS-NT)-11, MAT-03]

Sol. Given, $6(\text{side})^2 = 4\pi r^2$

$$\Rightarrow \left(\frac{\text{side}}{r}\right)^2 = \frac{4\pi}{6} = \frac{2\pi}{3}$$

$$\Rightarrow \frac{\text{side}}{r} = \frac{\sqrt{2\pi}}{\sqrt{3}}$$

$$\Rightarrow \frac{(\text{side})^3}{r^3} = \frac{2\sqrt{2}.\pi\sqrt{\pi}}{3\sqrt{3}}$$

$$\Rightarrow \frac{(\text{side})^3}{\frac{4}{3}\pi r^3} = \frac{2\sqrt{2}.\pi\sqrt{\pi}}{3\sqrt{3} \times \frac{4}{3}\pi}$$

$$\Rightarrow \frac{\text{Volume of the cube}}{\text{Volume of the sphere}} = \frac{2\sqrt{2}\sqrt{\pi}}{4\sqrt{3}} = \frac{\sqrt{\pi}}{\sqrt{6}} = \sqrt{\pi} : \sqrt{6}$$

Alternatively

Adhering to the observation texted above, you may arrive directly at the correct option.

Problem 17 The volumes of a sphere and a right circular cylinder having the same radius are equal. The ratio of the diameter of the sphere to the height of the cylinder is :

Sol. Given, $\frac{4}{3}\pi r^3 = \pi r^2 h \Rightarrow 4r = 3h$

$$\Rightarrow 2r = \frac{3}{2}h \Rightarrow \frac{2r}{h} = \frac{3}{2}$$

$$\Rightarrow \frac{\text{Diameter of the sphere}}{\text{Height of the cylinder}} = \frac{3}{2} = 3:2$$

Alternatively

Adhering the observation texted in the above box, you may arrive directly at the correct answer.

Problem 18 A cone, a hemisphere and a cylinder stand on equal bases and have the same height. The ratio of their respective volumes is :

[SSC(SO- Comm. Audit)-07, SSC(10 + 2 LDC/DEO)-11, FCI(AssttGr.III)-12, SSC(CGL- T-I)-14]

Sol. Required ratio $= \frac{1}{3}\pi r^2 h : \frac{2}{3}\pi r^3 : \pi r^2 h = \pi r^2 h : 2\pi r^3 : 3\pi r^2 h$

$$= 1h : 2r : 3h = 1h : 2h : 3h$$

[height of the hemisphere is its radius too, i.e. $h = r$]

$$= 1 : 2 : 3$$

Alternatively

Adhering the observation (supra), you may arrive at the correct answer directly.

Problem 19 A sphere and a cylinder have equal volume and equal radius. The ratio of the curved surface area of the cylinder to that of the sphere is :

Sol. Given, $\frac{4}{3}\pi r^3 = \pi r^2 h$

$$\Rightarrow 4r = 3h \Rightarrow r = \frac{3}{4}h$$

Now, the required ratio $= \dfrac{\text{curved surface area of the cylinder}}{\text{surface area of the sphere}}$

$$= \frac{2\pi r h}{4\pi r^2} = \frac{h}{2r} = \frac{h}{2 \times \frac{3}{4}h} = 2:3$$

Alternatively

Adhering the observation (supra), you may arrive at the correct answer directly.

Problem 20 The respective heights and volumes of a hemisphere and a right circular cylinder are equal, then the ratio of their radii is :

Sol. Given, $\frac{2}{3}\pi r^3 = \pi R^2 h$

$$\Rightarrow \frac{2}{3}\pi r^2 \times r = \pi R^2 h$$

$$\Rightarrow \frac{2}{3}\pi r^2 \times h = \pi R^2 h$$

$$\Rightarrow \frac{r^2}{R^2} = \frac{3}{2} \Rightarrow r : R = \sqrt{3} : \sqrt{2}$$

Alternatively

Adhering the observation (supra), you may arrive at the correct answer directly.

Problem 21 The ratio of the volume of a cube and a solid sphere is 363 : 49. The ratio of an edge of the cube and the radius of the sphere is :

Sol. Given, $\dfrac{(\text{side})^3}{\frac{4}{3}\pi r^3} = \dfrac{363}{49}$

$$\Rightarrow \frac{(\text{side})^3}{r^3} = \frac{363}{49} \times \frac{4}{3} \times \frac{22}{7} = \frac{11 \times 11 \times 11 \times 8}{7 \times 7 \times 7}$$

$$\Rightarrow \frac{\text{side}}{r} = \frac{11 \times 2}{7} = \frac{22}{7}$$

$$\Rightarrow \text{Edge : Radius} = 22 : 7$$

Problem 22 A sphere and a hemisphere have the same volume. The ratio of their curved surface areas is :

Sol. Given, $\frac{4}{3}\pi(R)^3 = \frac{2}{3}\pi r^3$

[let R and r be radius of the sphere and the hemisphere respectively]

$\Rightarrow \quad 2R^3 = r^3 \Rightarrow \frac{R^3}{r^3} = \frac{1}{2}$

$\Rightarrow \quad R : r = \sqrt[3]{1} : \sqrt[3]{2} = 1 : \sqrt[3]{2}$

Now, the required ratio

$$= \frac{\text{Curved surface area of the sphere}}{\text{Curved surface area of the hemisphere}} = \frac{4\pi R^2}{2\pi r^2}$$

$$= 2 \times \left(\frac{1}{\sqrt[3]{2}}\right)^2 = 2 \times \frac{1}{2^{\frac{2}{3}}} = \frac{2^{\frac{1}{3}}}{1} = 2^{\frac{1}{3}} : 1$$

Alternatively

Adhering the observation (supra), you may arrive directly at the correct answer.

Problem 23 The base of a right circular cone has the same radius a as that of a sphere. Both the sphere and the cone have the same volume. Height of the cone is :

Sol. Given, $\frac{1}{3}\pi r^2 h = \frac{4}{3}\pi r^3$

$\Rightarrow \quad h = 4r \Rightarrow h = 4a \quad [\because r = a]$

Alternatively

The ratio of the heights of a cone and a sphere is 6 : 3 provided they have equal radius and equal volume

[Reference – observation given above].

$\Rightarrow$ Height of the cone : Height of the sphere = 6 : 3

$\Rightarrow$ Height of the cone : $2a$ = 6 : 3

$\therefore$ Height of the cone = $4a$

Problem 24 A cone and a hemisphere stand on equal base and have the same height. The ratio of their whole surfaces is **[SSC(CGL-T-I)-12]**

Sol. Required ratio $= \frac{\text{Whole surface area of the cone}}{\text{Whole surface area of hemisphere}}$

$$= \frac{\pi r l + \pi r^2}{3\pi r^2}$$

$$= \frac{\pi r(l + r)}{3\pi r^2} = \frac{l + r}{3r} = \frac{\sqrt{h^2 + r^2} + r}{3r}$$

[$\because$ height the cone (h) = height of the hemisphere = r]

$$= \frac{\sqrt{r^2 + r^2} + r}{3r}$$

$$= \frac{\sqrt{2}.r + r}{3r} = \frac{(\sqrt{2} + 1)r}{3r}$$

$$= (\sqrt{2} + 1) : 3$$

Alternatively

Adhering the observation (supra), you may arrive at the correct answer directly.

Problem 25 Let the largest possible right circular cone and largest possible sphere be fitted into two cubes of same length. If C and S denote the volume of cone and volume of sphere respectively, then which one of the following is correct? **[CDS-12]**

(a) $C = 2S$ (b) $S = 2C$

(c) $C = S$ (d) $C = 3S$

Sol. Clearly, diameter of the base of the cone = side of the cube(a) = height of the cone

Hence, radius of the cone $= \frac{a}{2}$ and its height = a

Then, volume of the cone $= \frac{1}{3}\pi\left(\frac{a}{2}\right)^2 \times a = \frac{\pi a^3}{12}$

Further, diameter of the sphere = a

$\therefore$ Its radius $= \frac{a}{2}$

Then, its volume $(S) = \frac{4}{3}\pi\left(\frac{a}{2}\right)^3 = \frac{\pi a^3}{6}$

It is clear that $S = 2C$

Alternatively

If a cone and a sphere have equal radius and equal height, the ratio of their volumes = 1 : 2 (supra). It means, $S = 2C$.

Conversions of Shapes (Cuboid, Cube, Cylinder, Cone, Sphere and Hemisphere)

Problem 26 A solid metallic cone of height 10 cm, radius of base 20 cm is melted to make spherical balls each of 4 cm diameter. How many such balls can be made?

Sol. Number of spherical balls $= \frac{\text{Volume of the cone}}{\text{Volume of each ball}}$

$$= \frac{\frac{1}{3}\pi(20)^2 \times 10}{\frac{4}{3}\pi\left(\frac{4}{3}\right)^3} = 125$$

Problem 27 The diameter of the iron ball used for the shot-put game is 14 cm. It is melted and then a solid cylinder of height $2\frac{1}{3}$ cm is made. What will be the diameter of the base of the cylinder ?

Sol. Obviously, the volume of the iron ball = the volume of the cylinder

$\Rightarrow \quad \frac{4}{3}\pi\left(\frac{14}{2}\right)^3 = \pi r^2 \times 2\frac{1}{3}$

$\Rightarrow \quad r^2 = \frac{4 \times 14 \times 14 \times 14 \times 3}{3 \times 2 \times 2 \times 2 \times 7}$

$= 196$

$\Rightarrow \quad r = 14$ cm

Hence, its diameter $= 14 \times 2 = 28$ cm

Problem 28 A cone of height 15 cm and base diameter 30 cm is carved out of a wooden sphere of radius 15 cm. The percentage of wasted wood is :

[SSC (CPQ-SI)-04, FCI(Asstt Gr-II)-12]

Sol. Volume of wooden sphere $= \frac{4}{3}\pi(15)^3$

and volume of the cone $= \frac{1}{3}\pi\left(\frac{30}{2}\right)^2 \times 15$

Here, it is very clear that the volume of the cone is $\frac{1}{4}$ of the volume of the sphere i.e. 75% of wood has been wasted.

Problem 29 12 spheres of the same size are made by melting a solid cylinder of 16 cm diameter and 2 cm height. The diameter of each sphere is :

Sol. Obviously, volume of the cylinder = volume of all 12 spheres

$\Rightarrow \quad \pi\left(\frac{16}{2}\right)^2 \times 2 = 12 \times \frac{4}{3}\pi r^3$

$\Rightarrow \quad 64 \times 2 = 16 \times r^3$

$\Rightarrow \quad 8 = r^3 \Rightarrow r = 2$ cm

Hence, diameter $= 2 \times 2 = 4$ cm

Problem 30 The total surface area of a metallic hemisphere is 1848 cm^2. The hemisphere is melted to form a solid right circular cone. If the radius of the base of the cone is the same as the radius of the hemisphere, its height is :

[SSC(SO-Comm. Audit)-07]

Sol. Given, $3\pi r^2 = 1848$

$\Rightarrow \quad r^2 = \frac{1848 \times 7}{3 \times 22} = 196 \Rightarrow r = 14$ cm

Now, volume of the hemisphere = volume of the cone

$\Rightarrow \quad \frac{2}{3}\pi r^3 = \frac{1}{3}\pi r^2 h$

$\Rightarrow \quad 2r = h \Rightarrow h = 2 \times 14 = 28$ cm

Problem 31 Two solid right cones of equal heights and of radii r_1 and r_2 are melted and made to form a solid sphere of radius R. Then, the height of the cone is :

(a) $\frac{4R^2}{r_1^2 + r_2^2}$ (b) $\frac{4R}{r_1 + r_2}$

(c) $\frac{4R^3}{r_1^2 + r_2^2}$ (d) $\frac{R^2}{r_1^2 + r_2^2}$

Sol. Obviously, volume of both the cones = volume of the sphere

$\Rightarrow \quad \frac{1}{3}\pi(r_1)^2 h + \frac{1}{3}\pi(r_2)^2 h = \frac{4}{3}\pi(R)^3$

$\Rightarrow \quad \frac{1}{3}\pi h[r_1^2 + r_2^2] = \frac{4}{3}\pi(R)^3 \Rightarrow h = \frac{4R^3}{r_1^2 + r_2^2}$

Problem 32 A cylindrical rod of iron whose height is eight times its radius is melted and cast into spherical balls each of half the radius of the cylinder. The number of such spherical balls is :

Sol. Volume of the cylinder = $n \times$ volume of the spherical balls

[let number of balls be n]

$\Rightarrow \quad \pi r^2 h = n \times \frac{4}{3}\pi\,(\text{radius})^3$

$\Rightarrow \quad \pi r^2 h = n \times \frac{4}{3}\pi\left(\frac{r}{2}\right)^3$

$\Rightarrow \quad \pi r^2 \times 8r = n \times \frac{4}{3}\pi \times \frac{r^3}{8}$ [$\because$ height of the cylinder, $h = 8r$]

$\Rightarrow \quad n = \frac{r^2 \times 8r \times 24}{4r^3} = 48$

Problem 33 A hemispherical cup of radius 4 cm is filled to the brim with coffee. The coffee is then poured into a vertical cone of radius 8 cm and height 16 cm. The percentage of the volume of the cone that remains empty is :

[SSC(10+2, LDC/DEO)-11]

Sol. $\frac{\text{Volume of the hemisphere}}{\text{Volume of the cone}} = \frac{\frac{2}{3}\pi(4)^3}{\frac{1}{3}\pi(8)^2 \times 16} = \frac{1}{8}$

It means $\frac{7}{8}$ part of the cone remains empty.

i.e. $\frac{7}{8} \times 100\% = 87.5\%$

Problem 34 The radius of a solid metallic sphere is 1.5 cm. It is melted and drawn into a wire of diameter 0.2 cm. Find the length of the wire.

Sol. Obviously, volume of the sphere = volume of the wire(cylinder)

$\Rightarrow \quad \frac{4}{3}\pi\,(1.5)^3 = \pi\left(\frac{0.2}{2}\right)^2 \times h$

$\Rightarrow \quad h = \frac{4 \times 1.5 \times 1.5 \times 1.5}{3 \times 0.1 \times 0.1} = 450 \text{ cm} = 4.5 \text{ m}$

Problem 35 A hemispherical bowl is filled to the brim with a beverage. The contents of the bowl are transferred into a cylindrical vessel whose radius is 50% more than its height. If the diameter is same for both the bowl and the cylinder, the volume of the beverage in the cylindrical vessel is :

[IAS(Pre.)-99]

Sol. Let the radius for both the bowl and the cylinder be r as their diameter are same.

Now, volume of the bowl $= \frac{2}{3}\pi r^3$

and volume of the cylindrical vessel $= \pi r^2 h$

$= \pi r^2 \times \frac{r \times (100 - 33.33)}{100}$

[$\because$ r is 50% greater than h, it means h is $\left(\frac{50}{150} \times 100\right)$% i.e. 33.33% less than r]

$= \pi r^3 \times \frac{2}{3} = \frac{2}{3}\pi r^3$, exactly equal to the volume of the bowl.

It means volume of the beverage in the cylindrical vessel is 100%.

Problem 36 A metallic sphere of radius 3 cm is melted to form a hollow cylindrical vessel of uniform thickness. If the height and internal radius of the vessel are $5\frac{1}{7}$ cm and 3 cm respectively, then what is the thickness of the cylinder ?

Sol. Volume of the sphere = Volume of metal used to form a hollow cylinder

$$\Rightarrow \frac{4}{3}\pi(3)^3 = \pi h(R^2 - r^2)$$

$$\Rightarrow \frac{4}{3}\pi \times 27 = \pi \times \frac{36}{7}[R^2 - (3)^2]$$

$$\Rightarrow 7 = R^2 - 9 \Rightarrow R^2 = 16 \Rightarrow R = 4$$

Then, thickness $= R - r = 4 - 3 = 1$ cm

Problem 37 A cylindrical container is filled with ice cream. Its diameter is 12 cm and height is 15 cm. The whole ice cream is distributed among 10 children in equal cones having hemispherical tops. If the height of the conical portion is twice the diameter of its base, the diameter of the ice cream cone is : *[MAT-11]*

Sol. Let the radius of the cone base be r. Then, its height $= 4r$

Obviously, volume of the cylinder = 10 × [volume of the cone + volume of hemisphere]

$$\Rightarrow \pi R^2 h = 10\left[\frac{1}{3}\pi r^2 h + \frac{2}{3}\pi r^3\right]$$

$$\Rightarrow \pi(6)^2 \times 15 = 10\pi\left[\frac{1}{3}r^2 4r + \frac{2}{3}r^3\right]$$

$$\Rightarrow 36 \times 15 = 10 \times 2r^3$$

$$\Rightarrow r^3 = 27 \quad \therefore r = 3\text{ cm}$$

Combination of Solids (Cuboid, Cube, Cylinder, Cone, Sphere and Hemisphere)

Problem 38 A sphere of radius 2 cm is put into water contained in a cylinder of base radius 4 cm. If the sphere is completely immersed in the water, the water level in the cylinder rises by :

Sol. Clearly, volume of cylinder up to h cm

= volume of the sphere + volume of the existing water

$$\Rightarrow \pi(4)^2 \times 4 = \frac{4}{3}\pi(2)^3 + \text{volume of existing water}$$

[∵ height of the cylinder (h) = diameter of the sphere]

$$\Rightarrow 64\pi - \frac{32}{3}\pi = \pi(4)^2 \times \text{level(height) before putting the sphere}$$

$$\Rightarrow \frac{160}{3} = 16 \times \text{height of water before putting the sphere}$$

⇒ ∴ Height of water before putting the sphere

$$= \frac{10}{3} = 3\frac{1}{3}$$

$$\therefore \text{Rise in the water level} = 4 - 3\frac{1}{3} = \frac{2}{3}\text{ cm}$$

Alternatively

Volume of sphere = Volume of water displaced

$$\Rightarrow \frac{4}{3}\pi(2)^3 = \pi(4)^2 \times h$$

$$\Rightarrow \frac{4}{3}\pi(2)^3 = \pi(4)^2 \times \text{rise in water level}$$

$$\Rightarrow \text{Rise in water level} = \frac{2}{3}\text{ cm}$$

Porоblem 39 Half of a large cylindrical tank open at the top is filled with water and identical heavy spherical balls are to be dropped into the tank without spilling water out. If the radius and the height of the tank are equal and each is four times the radius of a ball, then what is the maximum number of balls that can be dropped ?

Sol. Sum of the volumes of all balls = Volume of displaced water

⇒ Number of balls (n) × Volume of each ball = Volume of water displaced

$$\Rightarrow n \times \frac{4}{3}\pi r^3 = \pi(4r)^2 \times \text{rise in water level}$$

[∵ radius of the cylinder(tank) is 4 times that of the sphere(ball)]

$$\Rightarrow n \times \frac{4}{3}\pi r^3 = \pi(4r)^2 \times \frac{\text{height}}{2}$$

[∵ half of tank was already filled]

$$\Rightarrow n \times \frac{4}{3}\pi r^3 = \pi(4r)^2 \times \frac{4r}{2}$$

[given, height of the tank = its radius]

$$\Rightarrow n = 24$$

Problem 40 A cylindrical tub of radius 12 cm contains water up to a depth of 20 cm. A spherical iron ball is dropped into the tub and thus the level of water is raised by 675 cm. The radius of the ball is : *[MAT-13]*

Sol. Volume of spherical ball = Volume of water displaced

$$\Rightarrow \frac{4}{3}\pi r^3 = \pi(12)^2 \times h$$

$$\Rightarrow \frac{4}{3}\pi r^3 = \pi(12)^2 \times \text{rise in water level}$$

$$\Rightarrow \frac{4}{3}\pi r^3 = \pi(12)^2 \times 6.75$$

$$\Rightarrow r^3 = 729$$

$$\therefore r = 9\text{ cm}$$

Problem 41 Marbles of diameter 1.4 cm are dropped into a cylindrical beaker containing some water and are fully submerged. The diameter of the beaker is 7 cm. Find how many marbles have been dropped in it, if the water rises by 5.6 cm. *[SSC (CGL-T-1)-11]*

Sol. Volume of marbles = Volume of the water displaced

$$\Rightarrow n \times \frac{4}{3}\pi\left(\frac{1.4}{2}\right)^3 = \pi\left(\frac{7}{2}\right)^2 \times 5.6$$

[let n marbles are dropped]

$$\Rightarrow n \times \frac{4}{3}\pi (0.7)^3 = \pi(3.5)^2 \times 5.6$$

$$\Rightarrow n = \frac{(3.5)^2 \times 5.6 \times 3}{4 \times (0.7)^3} = 150$$

Problem 42 A cylindrical can whose base is horizontal and is of internal radius 3.5 cm contains sufficient water so that when a solid sphere is placed inside, water just covers the sphere. The sphere fits the can exactly. The depth of water in the can before the sphere was put is

Sol. Obviously, volume of sphere = volume of water displaced

$\Rightarrow \quad \frac{4}{3}\pi(r)^3 = \pi r^2 \times$ rise in water-level

$\Rightarrow \quad \frac{4}{3}\pi(3.5)^3 = \pi(3.5)^2 \times$ rise in water-level

$\therefore$ Rise in water-level $= \frac{14}{3}$ cm

Then, depth of the water before the sphere was put = height after putting the sphere - rise in water level after putting the sphere

$= 7$ [$\because$ water just covers the sphere after the sphere is put inside] $- \frac{14}{3} = \frac{7}{3}$ cm

Problem 43 A solid wooden toy is in the shape of a right circular cone mounted on a hemisphere. If the radius of the hemisphere is 4.2 cm and the total height of the toy is 10.2 cm. Find the volume of the wooden toy (nearly)? ***[FCI (Asstt Gr.III)-12, CDS-21]***

Sol. Radius of the cone = Radius of the hemisphere = 4.2 cm and height of the cone = 10.2 – (4.2) = 6 cm

Now, volume of the wooden toy

= volume of the cone + volume of the hemisphere

$= \frac{1}{3}\pi(4.2)^2 \times 6 + \frac{2}{3}\pi(4.2)^3$

$= \frac{1}{3}\pi \times (4.2)^2[6 + 8.4]$

$= \frac{1}{3} \times \frac{22}{7} \times 4.2 \times 4.2 \times 14.4$

$= 266.1 = 266 \text{ cm}^3$ (approx.)

Problem 44 A sphere has a diameter of $500\sqrt{3}$. A biggest cube is fitted in it. Now a biggest sphere is fitted within this cube. Again a biggest cube is fitted within this smaller sphere. Determine the ratio of volume of bigger cube to the volume of smaller cube?

Sol. Clearly, the diagonal of the bigger of cube $= 500\sqrt{3}$

$\Rightarrow \quad \sqrt{3} \times \text{side} = 500\sqrt{3}$

$\Rightarrow \quad \text{side} = 500$

$\therefore$ Volume of the bigger cube $= (500)^3 \text{ cm}^3$

Now, side of the bigger cube = diameter of the smaller sphere

$\Rightarrow$ 500 = diameter of the smaller sphere

But, diameter of the smaller sphere = diagonal of the smaller cube

$\Rightarrow$ $500 = \sqrt{3} \times$ side of the smaller cube

$\therefore$ Side of the smaller cube $= \frac{500}{\sqrt{3}}$

$\therefore$ Volume of the smaller cube $= \left(\frac{500}{\sqrt{3}}\right)^3 = \frac{(500)^3}{3\sqrt{3}}$

Hence, required ratio $= \frac{\text{volume of bigger cube}}{\text{volume of smaller cube}}$

$= \frac{(500)^3}{\frac{(500)^3}{3\sqrt{3}}} = 3\sqrt{3} : 1$

Problem 45 A solid is in the form of a right circular cylinder with hemispherical ends. The total length of the solid is 35 cm. The diameter of the cylinder is $\frac{1}{4}$ of its height. Find the volume and total surface area of the solid.

Sol. Clearly, height of the cylinder $= 35 - (r + r) = 35 - 2r$

$\therefore$ Its diameter $= \frac{1}{4}(35 - 2r)$

$\Rightarrow \quad 2r = \frac{1}{4}(35 - 2r)$

$\Rightarrow \quad 8r = 35 - 2r$

$\Rightarrow \quad r = 3.5$ cm

Hence, $\quad h = 35 - 2r = 28$ cm

Now, volume of the cylinder $= \pi(3.5)^2 \times 28$

and curved area of the cylinder $= 2\pi(3.5) \times 28$

Further, volumes of hemispheres at both ends

$= 2 \times \frac{2}{3}\pi(3.5)^3$

$= \frac{4}{3}\pi(3.5)^3$

and curved areas of the hemispheres at both ends

$= 2 \times 2\pi(3.5)^2 = 4\pi(3.5)^2$

Thus, volume of the solid = volume of the cylinder + volumes of hemispheres at both ends

$= [\pi(3.5)^2 \times 28] + [\frac{4}{3}\pi(3.5)^3]$

$= \pi(3.5)^2\left[28 + \left(\frac{4}{3} \times 3.5\right)\right]$

$= \frac{22}{7} \times 3.5 \times 3.5 \times \frac{98}{3}$

$= 1257.66 \text{ cm}^3$

Now, total surface area of the solid = curved area of the cylinder + curved areas of the hemispheres at both ends

$= [2\pi(3.5) \times 28] + [4\pi(3.5)^2]$

$= 2\pi(3.5)[28 + (2 \times 3.5)]$

$= 2 \times \frac{22}{7} \times 3.5 \times 35 = 770 \text{ cm}^2$

Problem 46 A toy is in the shape of a right circular cylinder with a hemisphere on one end and a cone on the other. The radius and the height of the cylindrical part are 5 cm and 13 cm respectively. The radii of the hemispherical and conical parts are the same as that of the cylindrical part. Find the surface area of the toy, if the total height of the toy is 30 cm. ***[CDS-14]***

Sol. Curved surface area of the cylindrical part $= 2\pi rh$

$= 2\pi \times 5 \times 13 = 130\pi$

Surface area of the hemispherical part

$= 2\pi r^2 = 2\pi(5)^2 = 50\pi$

and curved area of the conical part

$= \pi rl = \pi \times 5 \times \sqrt{h^2 + r^2}$

$= \pi \times 5 \times \sqrt{12^2 + 5^2} = 65\pi \quad [\because h = 30 - 13 - 5 = 12]$

$\therefore$ Total surface area of the toy

$= 130\,\pi + 50\pi + 65\pi = 245\pi$

$= 770\text{ cm}^2$

Problem 47 10 circular plates each of thickness 3 cm, each are placed one above the other and a hemisphere of radius 6 cm is placed on the top just to cover the cylindrical solid. What is the volume of the solid so formed ?

Sol. Placing one plate above the other will form a cylinder with height 30 (= 3 × 10) cm. The radius of this cylinder will be 6 cm as a hemisphere with radius 6 cm is placed on the top of the cylinder just to cover the cylinder.

Now, volume of the solid = volume of the cylinder + volume of the hemisphere

$= \pi r^2 h + \frac{2}{3}\pi r^3 = \pi r^2 (h + \frac{2}{3} r)$

$= \pi 6^2 (30 + \frac{2}{3} \cdot 6)$

$= 1224\pi \text{ cm}^3$

Concept of Maximum/Minimum Value of Volume/Area

Problem 48 What is the maximum volume of a right circular cylinder which can be inscribed in a sphere of radius *r*.

Sol. Clearly, $R^2 = r^2 - \left(\frac{h}{2}\right)^2$, where R and h respectively are radius and height of the cylinder.

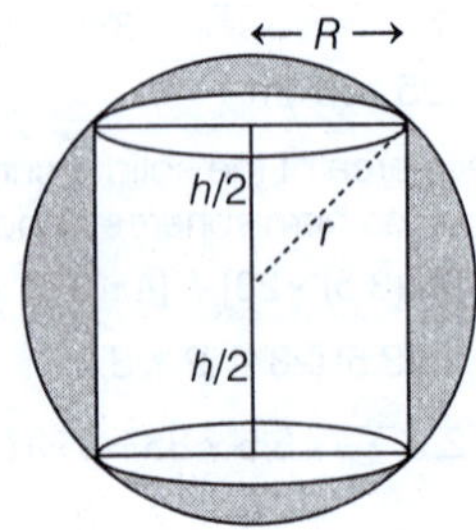

Now, volume of the cylinder $(V) = \pi R^2 h = \pi h\left\{r^2 - \left(\frac{h}{2}\right)^2\right\}$

$\Rightarrow \quad V = \pi\left\{r^2 h - \frac{h^3}{4}\right\}$

$\Rightarrow \quad \frac{dV}{dh} = \pi\left\{r^2 - \frac{3h^2}{4}\right\} = 0$

[differentiating with respect to h]

$\Rightarrow \quad r^2 - \frac{3h^2}{4} = 0$

$\Rightarrow \quad h = \frac{2r}{\sqrt{3}}$

Since $\frac{d^2 v}{dh^2}$ i.e. $\frac{-6\pi h}{4}$ is negative at $h = \frac{2r}{\sqrt{3}}$, the volume is maximum at the same value of h.

Hence, at $h = \frac{2r}{\sqrt{3}}$, the cylinder has maximum volume.

Then, $\quad R^2 = r^2 - \left(\frac{h}{2}\right)^2$

$= r^2 - \left(\frac{\frac{2r}{\sqrt{3}}}{2}\right)^2 = \frac{2r^2}{3}$

$\therefore$ Maximum volume of the cylinder

$= \pi R^2 h = \pi \times \frac{2r^2}{3} \times \frac{2r}{\sqrt{3}} = \frac{4\pi r^3}{3\sqrt{3}}$

Problem 49 What is the volume of the largest cone that can be inscribed in a sphere of radius 'a' ?

Sol. Clearly, $r^2 = a^2 - x^2$, where r is radius of the cone.

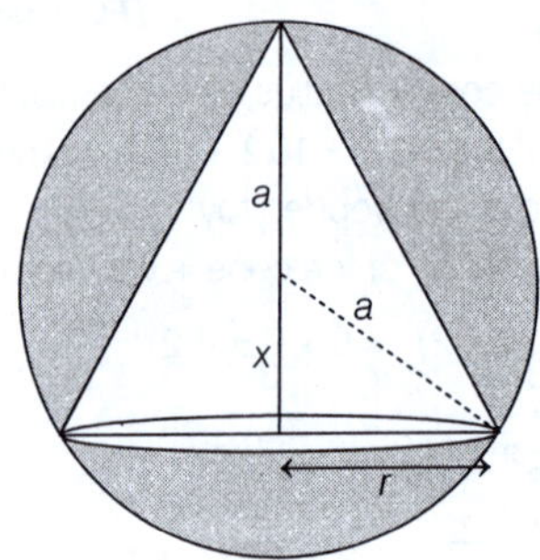

Now, volume of the cone (V)

$= \frac{1}{3}\pi r^2 h = \frac{1}{3}\pi[a^2 - x^2][a + x]$

$\Rightarrow \quad V = \frac{1}{3}\pi[a^3 + a^2 x - ax^2 - x^3]$

$\Rightarrow \quad \frac{dV}{dx} = \frac{1}{3}\pi[-3x^2 - 2ax + a^2]$

[$\because$ differentiating with respect to x]

$\Rightarrow \quad \frac{1}{3}\pi\{-3x^2 - 2ax + a^2\} = 0$

$\Rightarrow \quad -3x^2 - 2ax + a^2 = 0$

$\Rightarrow \quad x = \frac{a}{3}$ and $-a$

$\Rightarrow x = \frac{a}{3}$ as the value of x cannot be negative.

Since $\frac{d^2 v}{dx^2}$ i.e. $\frac{1}{3}\pi(-6x - 2a)$ is negative at $x = \frac{a}{3}$, the volume is maximum at the same value of x.

$\therefore$ Maximum volume of the cone $= \frac{1}{3}\pi[a^2 - x^2][a + x]$

$= \frac{1}{3}\pi\left[a^2 - \left(\frac{a}{3}\right)^2\right]\left[a + \frac{a}{3}\right] = \frac{32\pi a^3}{81}$

Problem 50 What is the minimum volume of a cone circumscribed about a sphere of radius r?

Sol. Clearly, $AB = AC = \sqrt{h^2 + a^2}$

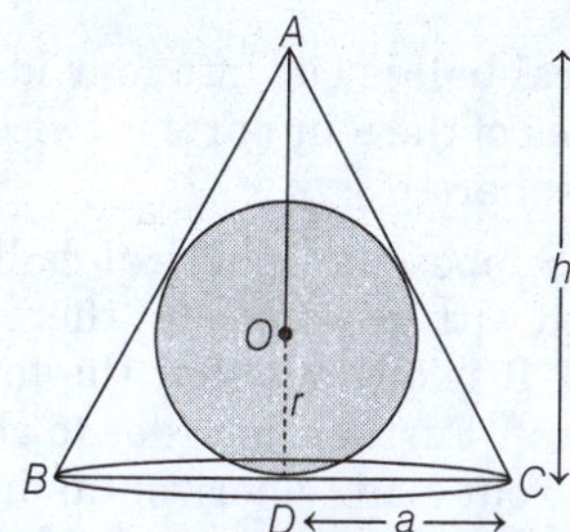

Further, area of ΔABC = inradius × semi-perimeter

$$\Rightarrow \frac{1}{2} \times 2a \times h = r \times \frac{\sqrt{h^2 + a^2} + \sqrt{h^2 + a^2} + 2a}{2}$$

$$\Rightarrow ah = r \times (\sqrt{h^2 + a^2} + a)$$

$$\Rightarrow (ah - ar)^2 = r^2h^2 + r^2a^2$$

$$\Rightarrow a^2h^2 + r^2a^2 - 2a^2hr = r^2h^2 + r^2a^2$$

$$\Rightarrow a^2 = \frac{r^2h^2}{h^2 - 2hr} = \frac{r^2h}{h - 2r}$$

Now, volume of the cone $(V) = \frac{1}{3}\pi a^2 h$

$$\Rightarrow V = \frac{1}{3}\pi \times \frac{r^2h}{h - 2r} \times h$$

$$\Rightarrow V = \frac{\pi r^2}{3}\left(\frac{h^2}{h - 2r}\right)$$

$$\Rightarrow \frac{dV}{dh} = \frac{\pi r^2}{3}\left\{\frac{(h - 2r)2h - h^2 \cdot 1}{(h - 2r)^2}\right\}$$

$$\Rightarrow \frac{\pi r^2}{3}\left[\frac{(h - 2r)2h - h^2 \cdot 1}{(h - 2r)^2}\right] = 0$$

$$\Rightarrow (h - 2r)2h - h^2 \cdot 1 = 0 \Rightarrow h = 4r$$

Since $\frac{d^2v}{dh^2}$ is positive at $h = 4r$, the volume of the cone is minimum at the same value of h.

Now, minimum volume of the cone $= \frac{\pi r^2}{3}\left(\frac{h^2}{h - 2r}\right)$

$$= \frac{\pi r^2}{3}\left\{\frac{(4r)^2}{4r - 2r}\right\} = \frac{8\pi r^3}{3}$$

Observations

(i) The height and radius of a cylinder of maximum possible volume, inscribed in a sphere of radius r, are $\frac{2r}{\sqrt{3}}$ and $\sqrt{\frac{2}{3}}\,r$ respectively. The volume is $\frac{1}{\sqrt{3}}$ times the volume of the sphere.

(ii) The height and radius of a cone of maximum possible volume, inscribed in a sphere of radius r, are $\frac{4r}{3}$ and $\frac{2\sqrt{2}\,r}{3}$ respectively. The volume is $\frac{8}{27}$ times the volume of the sphere.

(iii) The height and radius of a cone of minimum possible volume, circumscribed about a sphere of radius r, are $4r$ and $\sqrt{2}\,r$ respectively. The volume of the cone is twice that of the sphere.

Problem 51 A sphere of maximum volume is cut from a solid hemisphere of radius r. Find the ratio of the volume of the hemisphere to that of the sphere.

Sol. Required ratio $= \frac{\text{Volume of the hemisphere}}{\text{Volume of the sphere}} = \frac{\frac{2}{3}\pi r^3}{\frac{4}{3}\pi\left(\frac{r}{2}\right)^3}$

[Obviously, radius of the sphere is half of the radius of the hemisphere]

$= 4:1$

Problem 52 The volume of the greatest sphere that can be cut off from a cylindrical log of wood of base radius 1 cm and height 5 cm is :

Sol. Obviously, a sphere of maximum radius 1 cm may be cut off.

$\therefore$ Its volume $= \frac{4}{3}\pi(1)^3 = \frac{4}{3}\pi$ cm^3

Problem 53 The ratio of the volume of a cube to that of a sphere, which will fit exactly inside the cube is :

[SSC(CGL-PT)-05, 08, SSC(SAS)-10]

Sol. Required ratio $= \frac{\text{Volume of the cube}}{\text{Volume of the sphere}} = \frac{(x)^3}{\frac{4}{3}\pi\left(\frac{x}{2}\right)^3}$

[let each side of the cube be x, then radius of the sphere $= \frac{x}{2}$]

$= 6 : \pi$

Problem 54 The ratio of the surface area of a sphere and the curved surface area of the cylinder circumscribing the sphere is :

Sol. Here, diameter of sphere = diameter of cylinder = height of the cylinder.

$\therefore$ Required ratio $= \frac{4\pi r^2}{2\pi rh} = \frac{2r}{h} = \frac{2r}{d}$ [d is diameter]

$= \frac{2r}{2r} = 1:1$

Problem 55 A cube of maximum volume is cut from a solid hemisphere of radius r. What is the volume of the cube ?

Sol. Clearly, OP is the radius of the hemisphere.

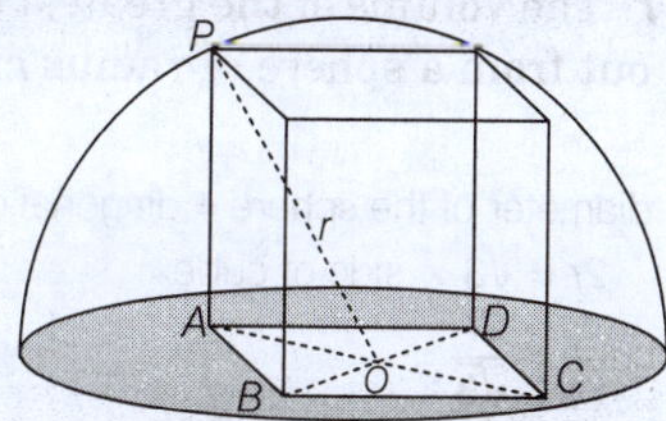

Let edge of the cube be x.

Then, diagonal of square $ABCD = AC = \sqrt{2}x$

$\therefore \quad AO = \sqrt{2}x \div 2 = \frac{x}{\sqrt{2}}$ cm

Now, $\quad AP^2 = OP^2 - AO^2 = r^2 - \left(\frac{x}{\sqrt{2}}\right)^2$

$\Rightarrow \quad x^2 = r^2 - \frac{x^2}{2} \Rightarrow x = \sqrt{\frac{2}{3}}\,r$

Thus, volume of the cube $= \left(\sqrt{\frac{2}{3}}\,r\right)^3 = \frac{2\sqrt{2}}{3\sqrt{3}}r^3$ cu unit.

Problem 56 **A conical vessel of radius 6 cm and height 8 cm is completely filled with water. A sphere is lowered into the water and its size is such that when it touches the sides, it is just immersed. The fraction of water that overflows is :** *[MAT-12]*

[This question can be read as : A greatest sphere is fitted in a conical vessel whose radius and height are 6 cm and 8 cm respectively. What is ratio of their volumes ?]

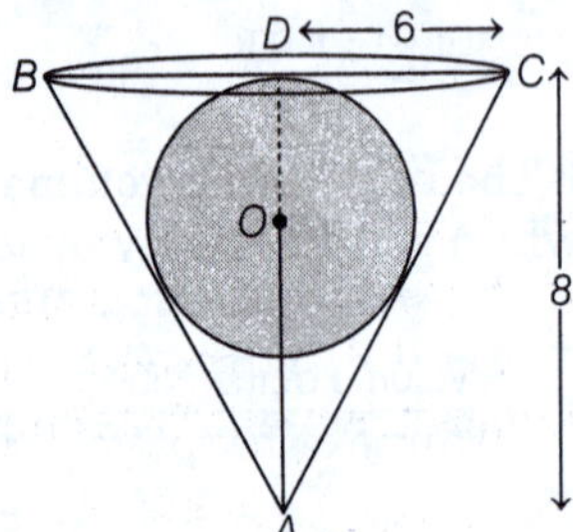

Sol. In right-angled $\Delta ADC \Rightarrow AC = \sqrt{8^2 + 6^2} = 10$

Hence, $AB = AC = 10$ and $BC = 12$

We know, area of $\Delta ABC = \frac{1}{2} \times BC \times AD$

$\Rightarrow$ Inradius $\times$ semi-perimeter $= \frac{1}{2} \times 12 \times 8$

$\Rightarrow \quad OD \times \frac{10 + 10 + 12}{2} = \frac{1}{2} \times 12 \times 8 \Rightarrow OD = 3$ cm

Obviously, volume of water overflowed = volume of sphere

$= \frac{4}{3}\pi(3)^3$

Then, required fraction $= \frac{\text{volume of water overflowed}}{\text{volume of cone}}$

$= \frac{\frac{4}{3}\pi(3)^3}{\frac{1}{3}\pi(6)^2 \times 8} = \frac{3}{8}$

Problem 57 **The volume of the greatest cube which can be chiseled out from a sphere of radius r is :** *[RRB-02, CDS-11]*

Sol. Obviously, diameter of the sphere = diagonal of the cube

$\Rightarrow \quad 2r = \sqrt{3} \times$ side of cube

$\Rightarrow$ Side of cube $= \frac{2r}{\sqrt{3}}$

$\therefore$ Volume of the cube $= \left(\frac{2r}{\sqrt{3}}\right)^3 = \frac{8r^3}{3\sqrt{3}}$

Concept of Total Surface Area When a Sphere is Cut Into Identical Parts

Facts

When a spherical ball is cut into four identical parts, the total surface area of the four parts is twice the surface area of the original sphere.

Explanation Suppose a spherical ball made of white metal is painted with green colour, thereafter it is cut into 4 similar parts. It is obvious that the total green surface areas will be $4\pi r^2$. Further, in order to get 4 similar parts the ball must be cut a cross horizontal and vertical. When we cut horizontally, we get two circular white areas i.e. $2\pi r^2$. Similarly, we get two circular white areas while cutting vertically i.e. $2\pi r^2$

$\therefore$ Total white surface areas $= 2\pi r^2 + 2\pi r^2 = 4\pi r^2$

Hence, total surface area of the four parts = total white surface areas + total green surface areas

$= 4\pi r^2 + 4\pi r^2 = 8\pi r^2$

Hence, **If a sphere of radius r is cut into 4 identical parts by 2 cuts, then the total surface area of each part $= 2\pi r^2$.**

Also,

(i) **If a sphere of radius r is cut into 8 identical parts by 3 cuts, then the total surface area of each part $= 1.25\pi r^2$.**

(ii) **If a sphere of radius r is cut into 2 identical parts (i.e. two hemispheres) by a cut, then the total surface area of each part $= 3\pi r^2$.**

Problem 58 **A spherical ball made of white metal is painted with red colour, thereafter it is cut into 4 similar parts. What is the ratio of the painted area to the non-painted area ?**

Sol. Painted surface area $= 4\pi r^2$

In order to get 4 similar parts the ball must be cut a cross horizontal and vertical. When we cut horizontally, we get two circular white areas i.e. $2\pi r^2$. Similarly, we get two circular white areas while cutting vertically i.e. $2\pi r^2$.

$\therefore$ Sum of white surface areas $= 2\pi r^2 + 2\pi r^2$

$= 4\pi r^2$

Hence, required ratio $= 4\pi r^2 : 4\pi r^2 = 1:1$

Alternatively

You may directly arrive at the final conclusion observing the facts (supra).

Problem 59 **It takes 2 L to paint the surface of a solid sphere. If this solid sphere is sliced into 4 identical pieces, how many litres will be required to paint all the surfaces of these 4 pieces ?** *[CAT-18]*

Sol. We know, when a spherical ball is cut into four identical parts, the total surface area of the four parts is twice the surface area of the original sphere. Hence, 4 L paint is required.

Problem 60 A sphere of radius 21 cm is cut into 8 identical parts by 3 cuts. What will be the total surface area (in cm^2) of each part ? ***[SSC(CGL-T-II)-17]***

Sol. In order to obtain 8 identical parts, we will cut the sphere with two perpendicular cuts through the centre, then will make a plane cut at half of the height of the sphere.

Now, total surface area of each part

$= [4\pi r^2 \div 8] + 3[\pi r^2 \div 4]$

$= \frac{5}{4}\pi r^2 = \frac{5}{4} \times \frac{22}{7} \times (21)^2$

$= 1732.5 \text{ cm}^2$

Alternatively

We know that if a sphere of radius r is cut into 8 identical parts by 3 cuts, then the total surface area of each part $= 1.25\pi r^2$

Hence, total surface area of each part $= 1.25 \times \frac{22}{7} \times (21)^2$

$= 1732.5 \text{ cm}^2$

Concept of Volume/Surface Area When Thickness of Layer/Wall is Taken Into Consideration

Problem 61 Find the volume of the metal used to make a hollow sphere of internal and external diameter 8 cm and 10 cm respectively.

Sol. Here, internal radius $(r) = \frac{8}{2} = 4$ cm

and external radius $(R) = \frac{10}{2} = 5$ cm

Now, volume of metal used $= \frac{4}{3}\pi R^3 - \frac{4}{3}\pi r^3$

$= \frac{4}{3}\pi(5^3 - 4^3)$

$= \frac{244}{3}\pi \text{ cm}^3$

Problem 62 The internal and external radius of a hollow hemispherical vessel are 10 cm and 12 cm respectively. Find the cost of painting the vessel all over at ₹ 0.2/ cm^2.

Sol. Internal curved area $= 2\pi(10)^2 = 200\pi$

External curved area $= 2\pi(12)^2 = 288\pi$

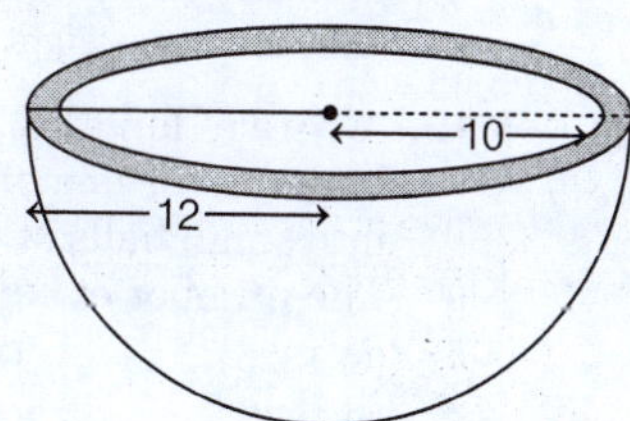

Area of the top plane (shaded) surface (say base ring)

$= \pi(12^2 - 10^2) = 44\pi$

Hence, area to be painted $= 200\pi + 288\pi + 44\pi = 532\pi$

$= 532 \times \frac{22}{7} = 1672 \text{ cm}^2$

$\therefore$ Cost $= 1672 \times 0.2 = ₹ 334.4$

Problem 63 The total surface area of a hollow hemispherical bowl having external radius 24 cm, is 6688 cm^2. Find the volume of the liquid which may be filled in.

Sol. Let internal radius be r. Then, total surface area

$= 2\pi(24)^2 + 2\pi r^2 + \pi(24^2 - r^2)$

$\Rightarrow \quad 6688 = 3\pi(24)^2 + \pi r^2$

$\Rightarrow \quad 6688 = \frac{22}{7}[(3 \times 576) + r^2]$

$\Rightarrow \quad 1728 + r^2 = 2128$

$\Rightarrow \quad r^2 = 400$

$\Rightarrow \quad r = 20$ cm

Now, volume of the liquid = internal volume of the bowl

$= \frac{2}{3}\pi(20)^3$

$= \frac{2}{3} \times \frac{22}{7} \times (20)^3$

$= 16762 \text{ cm}^3$ (approx.)

Problem 64 A hollow sphere of internal and exterior diameter 4 cm and 8 cm respectively is melted into a cone of base diameter 8 cm. The height of the cone is

Sol. Given, internal radius $= \frac{4}{2} = 2$ cm

and external radius $= \frac{8}{2} = 4$ cm

$\therefore$ Volume of the metal used in sphere $= \frac{4}{3}\pi(4)^3 - \frac{4}{3}\pi(2)^3$

$= \frac{4}{3}\pi \times 56$

Obviously, volume of the metal used = volume of the cone

$\Rightarrow \quad \frac{4}{3}\pi \times 56 = \frac{1}{3}\pi(r)^2 \times h$

$\Rightarrow \quad \frac{4}{3}\pi \times 56 = \frac{1}{3}\pi\left(\frac{8}{2}\right)^2 \times h$

$\Rightarrow \quad h = \frac{56 \times 4}{16} = 14$ cm

Problem 65 A hemisphere is made of a sheet of a metal 1 cm thick. If the outer radius is 5 cm, what is the weight of the hemisphere? (1 cm^3 of the metal weighs 9 g)

Sol. Given, external radius = 5 cm

Then, internal radius = 5 − 1 = 4 cm

Now, volume of the metal $= \frac{2}{3}\pi(5^3 - 4^3) = \frac{122}{3}\pi \text{ cm}^3$

$\therefore$ Weight of the metal used $= \frac{122}{3}\pi \times 9\text{g} = 366$ g

WORKOUT

1. A cone is inscribed in a hemisphere such that their bases are common. If C is the volume of the cone and H that of the hemisphere, then what is the value of $C : H$?
(a) 3 : 4 (b) 1 : 2
(c) 2 : 3 (d) 4 : 5

2. A hemispherical bowl is 176 cm round the brim. Supposing it to be half full, how many persons may be served from it in hemispherical glasses 4 cm in diameter at the top ?
(a) 1372 (b) 1272
(c) 1172 (d) 1472

3. The volumes of a right circular cylinder and a sphere are equal. The radius of the cylinder and the diameter of the sphere are equal. The ratio of height and radius of the cylinder is : *[SSC (CGL-T-I)-14]*
(a) 1 : 3 (b) 3 :1
(c) 1 : 6 (d) 6 : 1

4. If 64 identical small spheres are made out of big sphere of diameter 8 cm, then what is surface area of each small sphere ? *[CDS-14]*
(a) π cm^2 (b) 2π cm^2
(c) 4π cm^2 (d) 8π cm^2

5. Three spherical balls of radii 1 cm, 2 cm and 3 cm are melted to form a single spherical ball. In the process, the loss of material is 25%. The radius of the new ball is : *[SSC (CGL-T-II)-10]*
(a) 6 cm (b) 5 cm
(c) 3 cm (d) 2 cm

6. A cylindrical rod of iron whose radius is 1/4th of its height is melted and cast into spherical balls of the same radius as that of the cylinder. What is the number of spherical balls ?
(a) 3 (b) 2
(c) 4 (d) 5

7. The diameters of internal and external surfaces of a hollow spherical shell are 6 cm and 10 cm respectively. If it is melted and made of solid cylinder of length $\frac{8}{3}$ cm, then diameter (in cm) of the cylinder is : *[SSC (CGL-T-I)-14]*
(a) 16 (b) 7
(c) 10 (d) 14

8. A cylindrical vessel of radius 4 cm contains water. A solid sphere of radius 3 cm is lowered into the water until it is completely immersed. The water level in the vessel will rise by : *[CDS-15]*
(a) 1.5 cm (b) 2 cm
(c) 2.25 cm (d) 4.5 cm

9. A hemispherical bowl of thickness 1 cm and external diameter 10 cm is to be painted all over. What is the cost of painting at the rate of ₹ 0.70 /cm^2 ? *[MAT-11]*
(a) ₹ 200 (b) ₹ 400
(c) ₹ 800 (d) ₹ 100

10. The total number of spherical bullets , each of diameter 5 dm, that can be made by utilizing the maximum of a rectangular block of lead $11\text{m} \times 10\text{m} \times 5\text{m}$ is [assume that $\pi = 3$] *[SSC(CGL-T-II)-13]*
(a) equal to 8800
(b) less than 8800
(c) equal to 8400
(d) greater than 9000

11. The volume of the material of a hemispherical shell with outer and inner radii 9 cm and 7 cm respectively, is approximately
(a) 808 cm^3 (b) 800 cm^3
(c) 816 cm^3 (d) 824 cm^3

12. A right circular cylinder is circumscribing a hemisphere such that their bases are common. The ratio of their volumes is
(a) 1 : 3 (b) 1 : 2
(c) 2 : 3 (d) 3 : 4

13. If the ratio of the diameters of two spheres is 3 : 5, then what is the ratio of their surface areas ? *[CDS-12]*
(a) 9 : 25 (b) 9 : 10
(c) 3 : 5 (d) 27 : 125

14. The heights of a cone, cylinder and a hemisphere are equal. If their radii are in the ratio 2:3:1, then the ratio of their volumes is : *[SSC(CGL-T-II)-12]*
(a) 4 : 27 : 2 (b) 2 : 9 : 2
(c) 4 : 9 : 1 (d) 8: 27 : 1

15. A right circular metal cone is 8 cm high and the radius is 2 cm. It is melted and recast into a sphere. What is the radius of the sphere ?
(a) 3 cm (b) 2 cm
(c) 4 cm (d) 5 cm

16. A large solid metallic cylinder whose radius and height are equal to each other is to be melted and 48 identical solid balls are to be recast from the liquid metal, so formed. What is the ratio of the radius of a ball to the radius of the cylinder ? *[CDS-12]*
(a) 1 : 16 (b) 1 : 12
(c) 1 : 8 (d) 1 : 4

17. Assume that a drop of water is spherical and its diameter is 1/10th of a cm. A conical glass has a height equal to the diameter of its rim. If 32000 drops of water fill the glass completely, then the height of the glass, in cm, is : *[SSC(CCGL-T-II)-13]*
(a) 1 (b) 2 (c) 3 (d) 4

18. If the diameter of a sphere is doubled, then how does its surface area change ?
(a) it increases two times
(b) it increases three times
(c) it increases four times
(d) it increases eight times

19. A wire of length 22 cm and 0.2 cm in diameter is melted and recast into small balls of diameter 0.1 cm. The number of balls made is
(a) 1225 (b) 1350
(c) 1320 (d) 1280

20. What is the number of wax balls, each of radius 1 cm that can be molded out of a sphere of radius 8 cm ?
(a) 256 (b) 512 (c) 768 (d) 1024

21. A cylindrical vessel of diameter 24 cm contains some water. If two spheres of radii 6 cm each are lowered into the water until they are completely immersed, then water level (in cm) in the vessel will rise by **[SSC(CGL-T-II)-14]**

(a) 4 (b) 9 (c) 12 (d) 16

22. A sphere is inscribed in a cubical box such that the sphere is tangent to all faces of the box. What is the ratio of the volume of the cubical box to the volume of the sphere ?

(a) $6\pi : 1$ (b) $36 : \pi$
(c) $4\pi : 3$ (d) $6 : \pi$

23. A solid is hemispherical at the bottom and conical above. If the surface areas of the two parts are equal, then the ratio of radius and height of its conical part, is

(a) 1 : 3 (b) 1 : 1
(c) $\sqrt{3} : 1$ (d) $1 : \sqrt{3}$

24. A cylindrical vessel of height 10 cm has base radius 60 cm. If d is the diameter of a spherical vessel of equal volume, then what is d equal to ? **[CDS-11]**

(a) 30 cm (b) 60 cm
(c) 90 cm (d) 120 cm

25. The volume of a cone is equal to that of a sphere. If the diameter of base of cone is equal to the diameter of the sphere, then what is the ratio of height of cone to the diameter of the sphere ?

(a) 2 :1 (b) 1 : 2 (c) 3 :1 (d) 4 :1

26. A toy is in the shape of a hemisphere surmounted by a cone. If radius of base of the cone is 3 cm and its height is 4 cm, the total surface area of the toy is

(a) 33π cm^2 (b) 42π cm^2
(c) 66π cm^2 (d) 56π cm^2

27. From a cylindrical log whose height is equal to its diameter, the greatest possible sphere has been taken out. What is the fraction of the original log which is cut away ?

(a) $\frac{1}{2}$ (b) $\frac{1}{3}$ (c) $\frac{1}{4}$ (d) $\frac{2}{3}$

28. A hemispherical bowl of internal radius 9 cm contains a liquid. This liquid is to be filled into cylindrical shaped small bottles of diameter 3 cm and height 4 cm. How many bottles will be needed to empty the bowl? **[LIC(AAO)-09]**

(a) 27 (b) 35 (c) 45 (d) 54

29. The diameter of the Moon is approximately 1/4th of that of the Earth. What is the ratio of the volume of the Moon to that of Earth ?

(a) 1 :16 (b) 1 : 32 (c) 1 : 48 (d) 1 : 64

30. A cylindrical container of height 14 m and base diameter 12 m contains oil. This oil is to be transferred to one cylindrical can, one conical can and a spherical can. The base radius of all the containers is same. The height of the conical can is 6 m. While pouring some oil is dropped and hence only $\frac{3}{4}$ th of cylindrical can could be filled. How much oil is dropped ? **[MAT-10]**

(a) 54π m^3 (b) 36π m^3
(c) 46π m^3 (d) 50π m^3

31. If a sphere of radius r is divided into four identical parts, then the total surface area of the four parts is

(a) $4\pi r^2$ sq unit (b) $2\pi r^2$ sq unit
(c) $8\pi r^2$ sq unit (d) $3\pi r^2$ sq unit

32. A spherical ball of radius 1 cm is dropped into a conical vessel of radius 3 cm and slant height 6 cm. The volume of water (in cu cm) that can just immerse the ball is **[SSC(CGL-T-II)-14]**

(a) $\frac{5}{3}\pi$ (b) 3π
(c) $\frac{\pi}{3}$ (d) $\frac{4}{3}\pi$

33. The surface area of a spherical dome-shaped roof of a cylindrical water tank shown in the figure is **[IAS(Pre)-99]**

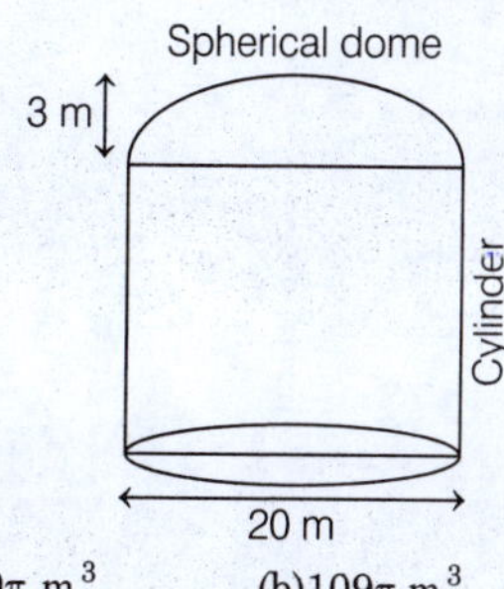

(a) 60π m^3 (b) 109π m^3
(c) 120π m^3 (d) 300π m^3

34. The diameter of hollow cone is equal to the diameter of a spherical ball. If the ball is placed at the base of the cone, what portion of the ball will be outside the cone ? **[CAT-93]**

(a) 50% (b) less than 50%
(c) more than 50% (d) 100%

35. The radius of a sphere is equal to the radius of the base of a right circular cone, and the volume of the sphere is double the volume of the cone. The ratio of the height of the cone to the radius of its base is : **[CDS-15]**

(a) 2 : 1 (b) 1 : 2
(c) 3 : 1 (d) 4 : 1

36. If the volume of a sphere is numerically equal to its surface area, then its diameter is : **[SSC(CGL-T-I)-15]**

(a) 4 cm (b) 2 cm
(c) 3 cm (d) 6 cm

37. If a right circular cylinder of height 14 is inscribed in a sphere of radius 8, then the volume of the cylinder is : **[IIFT-15]**

(a) 110 (b) 220 (c) 440 (d) 660

38. If the radius of a sphere is increased by 10%, then the volume will be increased by : **[SSC (CPO)-04, SSC (CGL-T-II)-20]**

(a) 33.1% (b) 30%
(c) 50% (d) 10%

39. A solid sphere of radius 12 inches is melted and cast into a right circular cone whose base diameter is $\sqrt{2}$ times its slant height. If the radius of the sphere and the cone are the same, how many such cones can be made and how much material is left out ? **[CAT-12]**

(a) 4 and 1 inch^3
(b) 3 and 12 inch^3
(c) 4 and 0 inch^3
(d) 3 and 6 inch^3

40. Two identical hemispheres of maximum possible size are cut from a solid cube of side 14 cm. The bases of hemispheres are part of the two opposite faces of cube. What is the total volume (in cm^3) of the remaining part of the cube ? **[SSC(CGL-T-II)-17]**

(a) 1556.33 (b) 898.5
(c) 1467.33 (d) 1306.67

41. A solid sphere is melted and recast into a right circular cone with a base radius equal to the radius of the sphere. What is the ratio of the height and radius of the cone so formed ? ***[SNAP-11]***

(a) 4 : 3 (b) 2 : 3
(c) 3 : 4 (d) None of these

42. If three metallic spheres of radii 6 cm, 8 cm and 10 cm are melted to form a single sphere, then the diameter of the new sphere will be :

(a) 12 cm (b) 24 cm
(c) 30 cm (d) 36 cm

43. Three solid spheres of radius 3 cm, 4 cm and 5 cm are melted and recast into a solid sphere. What will be the percentage decrease in the surface area ?

(a) 12 (b) 14
(c) 16 (d) 28

44. If the surface area of a sphere is reduced to 1/9th of the area, its radius reduces to : ***[CDS-17]***

(a) One-fourth (b) One-third
(c) One-fifth (d) One-ninth

45. Radius of base of hollow cone is 8 cm and its height is 15 cm. A sphere of largest radius is put inside the cone. What is the ratio of radius of the cone to the radius of sphere ? ***[SSC(CGL-T-II)-17]***

(a) 5 : 3 (b) 4 : 1
(c) 2 : 1 (d) 7 : 3

46. What is the largest sphere that can be carved out of a cube of edge 3 cm ?

(a) 9π cm^3 (b) 6π cm^3
(c) 4.5π cm^3 (d) 3π cm^3

47. If the volume of a sphere is divided by its surface area, we obtain 27 cm. The radius of the sphere is ***[SSC(CGL-T)-17]***

(a) 9 cm (b) 81 cm
(c) 27 cm (d)24 cm

48. A sphere and a cube have same surface area. The ratio of square of their volumes is

(a) $6 : \pi$ (b) $5 : \pi$
(c) 3 : 5 (d) 1 : 1

Direction *Consider the following for the next three items*

A cube is inscribed in a sphere. A right circular cylinder is within the cube touching all the vertical faces. A right circular cone is inside the cylinder. Their heights are same and the diameter of the cone is equal to that of the cylinder.

49. What is the ratio of the volume of the sphere to that of cone ?

(a) 7 : 2 (b) $5\sqrt{3} : 1$
(c) $3\sqrt{3} : 1$ (d) $6\sqrt{3} : 1$

50. What is the ratio of the volume of the cube to that of the cylinder ?

(a) 4: 3 (b) 21 : 16
(c) 14 : 11 (d) 45 : 32

51. Condiser the following statements

I. The surface area of the sphere is $\sqrt{5}$ times the curved surface area of the cone.

II. The surface area of the cube is equal to the curved surface area of the cylinder.

Which of the above statements is/are correct ? ***[CDS-19]***

(a) Only I (b) Only II
(c) Both I and II (d) Neither I nor II

52. The volumes of spheres A and B are in the ratio 125 : 64. If the sum of radii of A and B is 36 cm, then the surface area (in cm^2) of A is : ***[SSC(CGL-T-I)-19]***

(a) 512π (b) 800π
(c) 1600π (d) 1024π

53. A ball of 4 cm diameter is kept on top of a hollow cylinder standing vertically. The height of the cylinder is 3 cm, while its volume is 9π cm^3. Then, the vertical distance, in cm, of the topmost point of the ball from the base of the cylinder is : ***[CAT-17]***

(a) 4 (b) 6
(c) 8 (d) 10

Answer Key

1. (b)	2. (a)	3. (c)	4. (c)	5. (c)	6. (a)	7. (d)	8. (c)	9. (a)	10. (a)
11. (a)	12. (c)	13. (a)	14. (a)	15. (b)	16. (d)	17. (d)	18. (c)	19. (c)	20. (b)
21. (a)	22. (d)	23. (d)	24. (b)	25. (a)	26. (a)	27. (d)	28. (d)	29. (d)	30. (b)
31. (c)	32. (a)	33. (b)	34. (c)	35. (a)	36. (d)	37. (d)	38. (a)	39. (c)	40. (d)
41. (d)	42. (a)	43. (d)	44. (b)	45. (a)	46. (c)	47. (b)	48. (a)	49. (d)	50. (c)
51. (d)	52. (c)	53. (b)							

Hints and SOLUTIONS

1. Here, the cone and the hemisphere will have equal radius and equal height, then ratio of their volumes will be 1 : 2. [reference : Ratio analysis in case of different shapes with reference to diverse aspects].

2. Given, $2\pi r = 176 \Rightarrow r = 28\,\text{cm}$

Number of persons

$$= \frac{\text{Half of the volume of the bowl}}{\text{Volume of each glass}}$$

$$= \frac{\frac{2}{3}\pi(28)^3 \times \frac{1}{2}}{\frac{2}{3}\pi(2)^3} = 1372$$

3. Let radii of cylinder and sphere be r and R respectively.

Given, $\pi r^2 h = \frac{4}{3}\pi R^3$

$$\Rightarrow \pi r^2 h = \frac{4}{3}\pi\left(\frac{2R}{2}\right)^3$$

$$\Rightarrow \pi r^2 h = \frac{4}{3}\pi\left(\frac{r}{2}\right)^3 \quad [\text{given, } r = 2R]$$

$$\Rightarrow 24r^2 h = 4r^3$$

$$\Rightarrow \frac{h}{r} = \frac{1}{6} = 1 : 6$$

4. (c) Number of small spheres

$$= \frac{\text{Volume of the bigger sphere}}{\text{Volume of each smaller sphere}}$$

$$\Rightarrow 64 = \frac{\frac{4}{3}\pi R^3}{\frac{4}{3}\pi r^3} \Rightarrow 64 = \frac{R^3}{r^3}$$

$$\Rightarrow 64 = \frac{4^3}{r^3} \Rightarrow r = 1\,\text{cm}$$

Then, surface area of each small sphere $= 4\pi r^2 = 4\pi 1^2 = 4\pi\ \text{cm}^2$

Alternatively

Since the ratio of the volume of the bigger sphere to that of smaller is 64 :1, the ratio of their radii will be $\sqrt[3]{64} : \sqrt[3]{1} = 4 : 1$

Further, it is given that radius of the bigger is 4 cm, then radius of the smaller one be 1 cm.

Now, surface area of each small sphere $= 4\pi r^2 = 4\pi 1^2 = 4\pi\ \text{cm}^2$

5. Volume of new ball = 75% of [sum of volumes of all three balls]

$$\Rightarrow \frac{4}{3}\pi R^3 = 75\% \text{ of } \frac{4}{3}\pi[1^3 + 2^3 + 3^3]$$

$$\Rightarrow \frac{4}{3}\pi R^3 = 36\pi$$

$$\Rightarrow R^3 = 27$$

$$\therefore R = 3\,\text{cm}$$

6. Number of balls

$$= \frac{\text{Volume of the cylinder}}{\text{Volume of each ball}} = \frac{\pi r^2 h}{\frac{4}{3}\pi r^3}$$

$$= \frac{\pi r^2 \times 4r}{\frac{4}{3}\pi r^3}$$

[$\because h = 4r$, given in the question]

$= 3$

7. Volume of the material used to make shell = Volume of the cylinder

$$\Rightarrow \frac{4}{3}\pi[R^3 - r^3] = \pi r^2 h$$

$$\Rightarrow \frac{4}{3}\pi[5^3 - 3^3] = \pi r^2 \times \frac{8}{3}$$

$$\Rightarrow \frac{98 \times 4}{3} = r^2 \times \frac{8}{3}$$

$$\Rightarrow r^2 = 49 \Rightarrow r = 7$$

$\therefore$ Diameter = 14 cm

8. Obviously, volume of water risen in cylindrical vessel = volume of sphere

$$\Rightarrow \pi R^2 h = \frac{4}{3}\pi r^3$$

[let rise in water level be h cm]

$$\Rightarrow (4)^2 h = \frac{4}{3}\pi(3)^3$$

$$\Rightarrow 4h = 9$$

$$\Rightarrow h = 2.25\,\text{cm}$$

9. Here, external radius = 5 cm and internal radius = 5 – 1 = 4 cm.

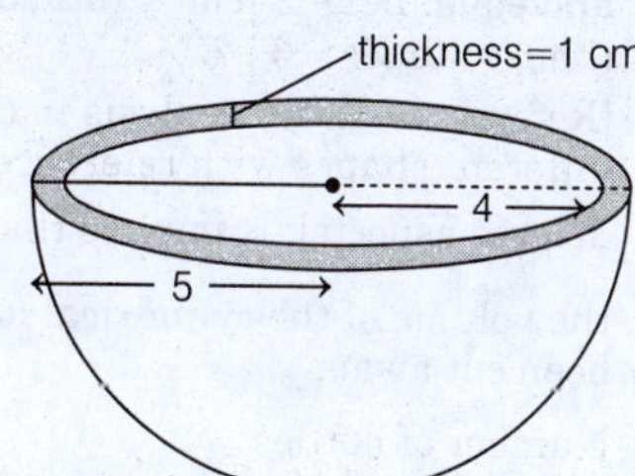

Now, internal curved area

$$= 2\pi(4)^2 = 32\pi$$

and external curved area

$$= 2\pi(5)^2 = 50\pi$$

Area of the top plane (shaded) surface (say base ring)

$$= \pi(5^2 - 4^2) = 9\pi$$

Hence, area to be painted

$$= 32\pi + 50\pi + 9\pi = 91\pi$$

$$= 91 \times \frac{22}{7} = 286\,\text{cm}^2$$

$\therefore$ Cost $= 286 \times 0.70$

$= ₹\ 200.20$ (approx. 200)

10. Number of bullets

$$= \frac{\text{Volume of the block}}{\text{Volume of each spherical bullet}}$$

$$= \frac{110 \times 100 \times 50}{\frac{4}{3}\pi(2.5)^3} = \frac{110 \times 100 \times 50}{\frac{4}{3} \times 3 \times (2.5)^3}$$

$= 8800$

11. Volume of the material

$$= \frac{2}{3}\pi R^3 - \frac{2}{3}\pi r^3 = \frac{2}{3}\pi(R^3 - r^3)$$

$$= \frac{2}{3}\pi(9^3 - 7^3) = 808\,\text{cm}^3 \text{ (approx.)}$$

12. In case of equal radius and equal height, the ratio of volume of the hemisphere to that of the cylinder = 2 : 3

13. Observing the 'ratio analysis in case of two/three or more spheres /hemispheres' the required ratio will be $(3)^2 : (5)^2$ i.e. 9 : 25

14. Since height of a hemisphere is its radius itself, the height of the hemisphere will be 1 unit and the heights of the cone and cylinder will also be 1 unit as per question.

Now, required ratio

$$= \frac{1}{3}\pi 2^2 h : \pi 3^2 h : \frac{2}{3}\pi 1^3$$

$$= \frac{1}{3}\pi 2^2 \times 1 : 3^2\pi \times 1 : \frac{2}{3}\pi 1^3$$

$= 4 : 27 : 2$

15. Obviously, volume of the cone = volume of the sphere

$$\Rightarrow \frac{1}{3}\pi \times 2^2 \times 8 = \frac{4}{3}\pi r^3$$

$$\Rightarrow r = 2\,\text{cm}$$

16. Number of balls

$$= \frac{\text{Volume of the cylinder}}{\text{Volume of each ball}} = \frac{\pi R^2 h}{\frac{4}{3}\pi r^3}$$

$\Rightarrow \quad 48 = \dfrac{\pi R^2 h}{\frac{4}{3}\pi r^3}$

$\Rightarrow \quad 64 = \dfrac{R^2 \times R}{R^3}$

[$\because R = h$, given in the question]

$\Rightarrow \quad 64 = \dfrac{R^3}{r^3}$

$\Rightarrow \quad \dfrac{r^3}{R^3} = \dfrac{1}{64} \Rightarrow r : R = 1 : 4$

17. Let the height of the cone be h, then its radius $= \dfrac{h}{2}$

Now, volume of cone

$= 32000 \times$ volume of each drop of water

$\Rightarrow \dfrac{1}{3}\pi\left(\dfrac{h}{2}\right)^2 \times h = 32000 \times \dfrac{4}{3}\pi\left(\dfrac{0.1}{2}\right)^3$

$\Rightarrow \quad h^3 = 64$

$\therefore \quad h = 4\,\text{cm}$

18. Since the diameter is made to 2 times, the surface area will become $(2)^2$ times.

19. Number of small balls

$= \dfrac{\text{Volume of the cylinderical wire}}{\text{Volume of each spherical ball}}$

$= \dfrac{\pi(0.1)^2 \times 22}{\frac{4}{3}\pi(0.05)^3} = 1320$

20. Number of balls

$= \dfrac{\text{Volume of the bigger sphere}}{\text{Volume of each ball}}$

$= \dfrac{\frac{4}{3}\pi(8)^3}{\frac{4}{3}\pi(1)^3} = 512$

[$\because$ You can directly divide the $(8)^3$ by $(1)^3$ so as to get number of balls in such types of questions]

21. Volume of both spheres = Volume of water displaced

$\Rightarrow 2 \times \dfrac{4}{3} \times \pi 6^3 = \pi r^2 \times$ rise in water level

$\Rightarrow 2 \times \dfrac{4}{3}\pi 6^3 = \pi(12)^2 \times$ rise in water level

$\Rightarrow$ Rise in water level = 4 cm

22. Obviously, the radius of the sphere = half of side of the cube

Then, required ratio

$= \dfrac{\text{volume of the cube}}{\text{volume of the sphere}}$

$= \dfrac{(\text{side})^3}{\frac{4}{3}\pi\left(\frac{\text{side}}{2}\right)^3} = 6 : \pi$

23. Given, surface area of cone = curved surface area of hemisphere

$\Rightarrow \pi r l = 2\pi r^2$ [$\because$ radii of both shapes will be same i.e. r]

$\Rightarrow \quad \sqrt{r^2 + h^2} = 2r$

$\Rightarrow \quad r^2 + h^2 = 4r^2$

$\Rightarrow \quad 3r^2 = h^2$

$\Rightarrow \quad \dfrac{r}{h} = \sqrt{\dfrac{1}{3}} = 1 : \sqrt{3}$

24. Given, volume of the cylindrical vessel = volume of the sphere

$\Rightarrow \quad \pi r^2 h = \dfrac{4}{3}\pi\left(\dfrac{d}{2}\right)^3$

$\Rightarrow \quad (60)^2 \times 10 = \dfrac{4}{3} \times \left(\dfrac{d}{2}\right)^3$

$\Rightarrow \quad d^3 = 60 \times 60 \times 60$

$\therefore \quad d = 60\,\text{cm}$

25. Here, a cone and a sphere have equal radius and equal volume, then ratio of their heights will be 6 : 3 i.e. 2 : 1.

[Reference : Ratio analysis in case of different shapes with reference to diverse aspects]. Needless to say that the height of the sphere is also its diameter.

26. Radius of the cone = Radius of the hemisphere = 3 cm

Slant height of the cone (l)

$= \sqrt{r^2 + h^2} = \sqrt{3^2 + 4^2} = 5\,\text{cm}$

Now, total surface area of the toy = surface area of the toy + surface area of the cone

$= 2\pi(3)^2 + \pi 3 \times 5 = 33\pi\,\text{cm}^2$

27. It is obvious that the cylinder and the sphere will have equal radius and equal height. Then, ratio of their volumes = 3 : 2

[Reference : Ratio analysis in case of different shapes with reference to diverse aspects]. It implies that $\dfrac{2}{3}$ of the volume of the cylindrical log has been cut away.

28. Number of bottles

$= \dfrac{\text{Volume of the bowl}}{\text{Volume of each bottle}}$

$= \dfrac{\frac{2}{3}\pi 9^3}{\pi(1.5)^2 \times 4} = 54$

29. Obviously, the Moon and the Earth are spheres and ratio of their radii is 1 : 4. Then, ratio of their volumes will be $(1)^3 : (4)^3$

[Reference : Ratio analysis in case of two/three or more spheres/ hemispheres].

30. Volume of the oil

$= \pi r^2 h = \pi(6)^2 \times 14 = 504\pi$

Volume of the spherical can

$= \dfrac{4}{3}\pi r^3 = \dfrac{4}{3}\pi(6)^3 = 288\pi$

Volume of the conical can

$= \dfrac{1}{3}\pi r^2 h = \dfrac{1}{3}\pi(6)^2 \times 6 = 72\pi$

Thus, capacity/ volume of cylindrical can $= 504 - 288 - 72 = 144\pi\,\text{m}^3$. But the fact that $\dfrac{1}{4}$th of this capacity has been dropped i.e. $144 \times \dfrac{1}{4} = 36\pi\,\text{m}^3$.

31. We know that when a spherical ball is cut into four identical parts, the total surface area of the four parts is twice the surface area of the original sphere i.e. $8\pi r^2$ square unit.

32. Height of the cone = BD

$= \sqrt{l^2 - r^2} = \sqrt{6^2 - 3^2}$

$= 3\sqrt{3}$

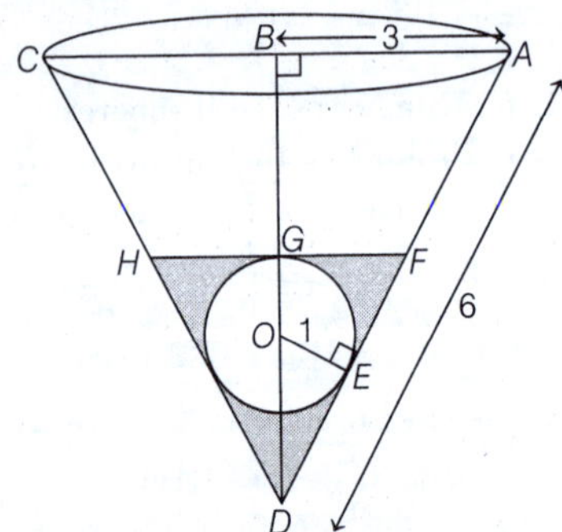

Clearly, ΔABD is similar to ΔOED

[$\because \angle ADB = \angle ODE$ (common)]

and $\angle ABD = \angle OED = 90°$

$\therefore \quad \dfrac{AD}{OD} = \dfrac{AB}{OE} \Rightarrow \dfrac{6}{OD} = \dfrac{3}{1}$

$\Rightarrow \quad OD = 2$

Then, $DG = OD + OG = 2 + 1 = 3$

Further, ΔABD is similar to ΔFGD.

Then, $\dfrac{AB}{GF} = \dfrac{BD}{DG}$

$\Rightarrow \quad \dfrac{3}{GF} = \dfrac{3\sqrt{3}}{3}$

$\Rightarrow \quad GF = \sqrt{3}$

Now, volume of cone $DFH = \frac{1}{3}\pi r^2 h$

$$= \frac{1}{3}\pi (GF)^2 \times DG$$

$$= \frac{1}{3}\pi(\sqrt{3})^2 \times 3 = 3\pi$$

$\therefore$ Volume of the water

$$3\pi - \frac{4}{3}\pi 1^3 = \frac{5}{3}\pi \text{ cm}^3$$

33. We know, $AO \times OB = CO \times OD$

[Refer to chapter circle under Geometry]

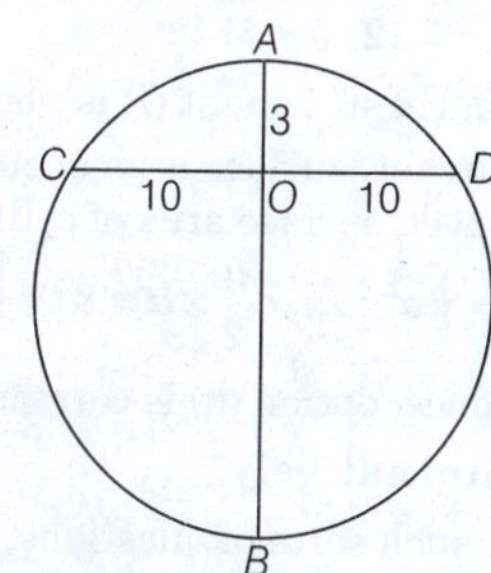

$\Rightarrow$ $3 \times OB = 10 \times 10$

$\Rightarrow$ $3(2R - 3) = 10 \times 10$

[where, R is radius of curvature]

$\Rightarrow$ $R = \frac{109}{6}$

Now, surface area of the roof

$$= 2\pi Rh$$

$$= 2\pi \times \frac{109}{6} \times 3$$

$$= 109\pi$$

Alternatively

Surface area of the roof $= \pi(r^2 + h^2)$

$$= \pi(10^2 + 3^2)$$

$$= 109\pi$$

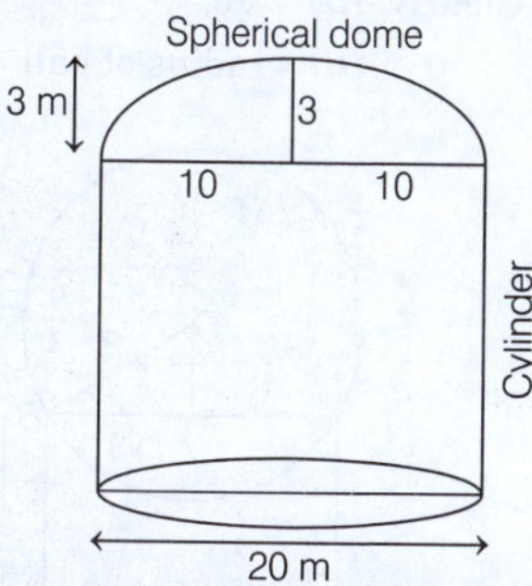

34. Since diameters of both cone and spherical ball are equal, the ball cannot be placed in the cone till the ball's equator. Only some portion of the ball, as shown in the figure, can be placed inside the cone. Hence, more than 50% of the portion of the ball will be outside the cone.

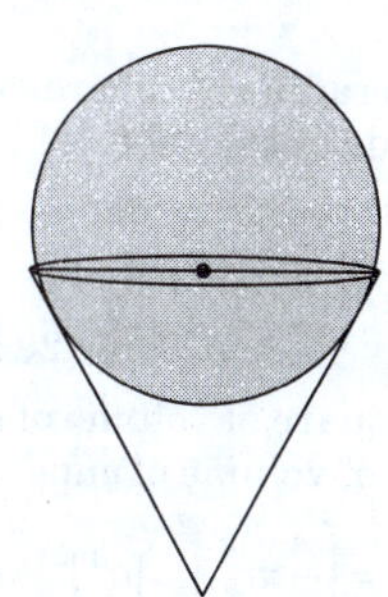

35. Given, volume of sphere

$$= 2 \times \text{volume of cone}$$

$\Rightarrow$ $\frac{4}{3}\pi r^3 = 2 \times \frac{1}{3}\pi r^2 h$

$\Rightarrow$ $h : r = 2 : 1$

36. Given, $\frac{4}{3}\pi r^3 = 4\pi r^2 \Rightarrow r = 3$

Hence, diameter = 6 cm

37. Clearly, $DC = \sqrt{16^2 - 14^2} = 2\sqrt{15}$

$\therefore$ Radius of the cylinder $= \sqrt{15}$

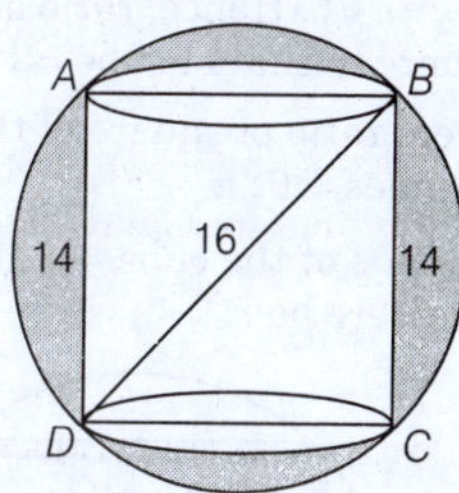

Now, volume of the cylinder $= \pi r^2 h$

$$= \frac{22}{7} \times (\sqrt{15})^2 \times 14 = 660$$

38. Here, radius has been increased by 10%.

It means, three figures are

$$= +10\% + 10\% \text{ and } + 10\%$$

Effect on taking any two figures

$$= \left(+10 + 10 + \frac{10 \times 10}{100}\right)\% = +21\%$$

Now, final effect

$$= \left(+21 + 10 + \frac{21 \times 10}{100}\right)\%$$

$$= +33.1\%$$

Hence, the original volume will increase by 33.1%.

Alternatively I

Since the radius is made to 1.10 times, the volume will become $(1.10)^3$ i.e. 1.331 times the earlier. It means it will increase by 33.1%.

Alternatively II

It is known that if the radius of a sphere/hemisphere is changed in ratio $a : b$, then percentage change in volume $= \left(\frac{a^3 - b^3}{a^3} \times 100\right)\%$.

Here, radius is changed in ratio 100 : 110 i.e. 10 :11 (say, $a : b$), then percentage change

$$= \left(\frac{1000 - 1331}{1000} \times 100\right)\%$$

$$= -33.1\%$$

Outcome is negative, it means volume has been increased by 33.1%.

39. Base diameter of the cone

$$= 12 \times 2 = 24 \text{ inches}$$

Then, its slant height

$$= 24 \div \sqrt{2} = 12\sqrt{2} \text{ inches}$$

Now, height of the cone

$$= \sqrt{(12\sqrt{2})^2 - (12)^2} = 12$$

$\therefore$ Volume of one cone $= \frac{1}{3}\pi(12)^2 \times 12$

and volume of sphere $= \frac{4}{3}\pi(12)^3$

$\therefore$ Number of cones

$$= \frac{4}{3}\pi(12)^3 \div \frac{1}{3}\pi(12)^2 \times 12 = 4$$

40. Clearly, diameter of each hemisphere = 14 cm, hence radius of each hemisphere = 7 cm

Now, volume of both hemispheres

$$= 2 \times \frac{2}{3}\pi r^3$$

$$= 2 \times \frac{2}{3} \times \frac{22}{7} \times (7)^3$$

$$= 1437.33$$

Volume of cube $= (14)^3 = 2744$

$\therefore$ Required volume

$$= 2744 - 1437.33 = 1306.67$$

41. Refer to 'Ratio analysis in case of Cone and Sphere. If a cone and a sphere have the same radius and the same volume, the height of the cone will be 4 times the radius.

Hence, the required ratio = 4 : 1

42. We know that if two or more solid spheres with radii $r_1, r_2, r_3, \ldots$ are melted to form a new solid sphere, then radius (R) of the newly formed sphere will be

$$\sqrt[3]{(r_1)^3 + (r_2)^3 + (r_3)^3 + \ldots}$$

Here, radius of new formed sphere

$$= \sqrt[3]{(6)^3 + (8)^3 + (10)^3}$$

$$= \sqrt[3]{1728} = \sqrt[3]{12 \times 12 \times 12}$$

$$= 12 \text{ cm}$$

43. Sum of surface area of all three smaller spheres

$= 4\pi(3^2 + 4^2 + 5^2) = 200\pi$

Now, radius of the bigger sphere

$= \sqrt[3]{(3)^3 + (4)^3 + (5)^3} = 6$

Then, its surface area

$= 4\pi 6^2 = 144\pi$

$\therefore$ Percentage decrease

$= \left(\frac{200\pi - 144\pi}{200\pi} \times 100\right)\% = 28\%$

44. We know, if radius of a sphere/hemisphere is made x times, its total/curved surface area will become x^2 times the earlier.

As given, $x^2 = \frac{1}{9} \Rightarrow x = \frac{1}{3}$

45. In right-angled ΔADC,

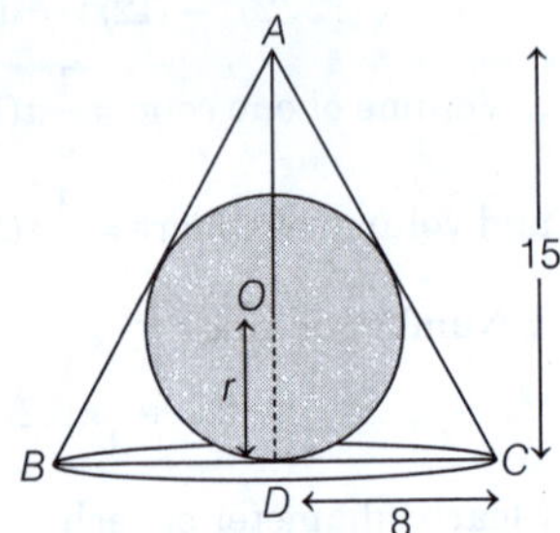

$AC = \sqrt{8^2 + 15^2} = 17$

Hence, $AB = AC = 17$

and $BC = 16$

We know, area of ΔABC

$= \frac{1}{2} \times BC \times AD$

$\Rightarrow$ inradius $\times$ semi-perimeter

$= \frac{1}{2} \times 16 \times 15$

$\Rightarrow OD \times \frac{17 + 17 + 16}{2} = \frac{1}{2} \times 16 \times 15$

$\Rightarrow \quad OD = \frac{24}{5}$

Now, required ratio

$= DC : OD$

$= 8 : \frac{24}{5} = 5 : 3$

46. It is obvious that the diameter of the sphere = side of the cube = 3

Hence, its radius = 1.5 cm

Now, its volume

$= \frac{4}{3}\pi (1.5)^3 = 4.5\pi \text{ cm}^3$

47. As given, $\frac{4}{3}\pi r^3 \div 4\pi r^2 = 27 \text{cm}$

$\Rightarrow \quad r = 81 \text{cm}$

48. Let r is radius of sphere and a be each side of the cube.

Given, $\quad 4\pi r^2 = 6a^2$

$\Rightarrow \quad r^2 : a^2 = 6 : 4\pi$

$\Rightarrow \quad r : a = \sqrt{6} : 2\sqrt{\pi}$

Now, square of volume of sphere : square of volume of cube

$= \left[\frac{4}{3}\pi r^3\right]^2 : [a^3]^2 = \frac{16}{9}\pi^2 r^6 : a^6$

$= \frac{16}{9}\pi^2 [r^2]^3 : [a^2]^3$

$= \frac{16}{9}\pi^2 [6]^3 : [4\pi]^3$

$= 16 \times 24\pi^2 : 64\pi^3$

$= 24 : 4\pi = 6 : \pi$

Alternatively

It is known from the observation (supra) that if the total surface areas of a cube and a sphere are equal, the ratio of their volumes will be $\sqrt{\pi} : \sqrt{6}$. Hence, ratio of volume of sphere to that of cube $= \sqrt{6} : \sqrt{\pi}$.

Then, ratio of square of their volumes $= 6 : \pi$

49. Let side of the cube $= a$ and radius of sphere be r.

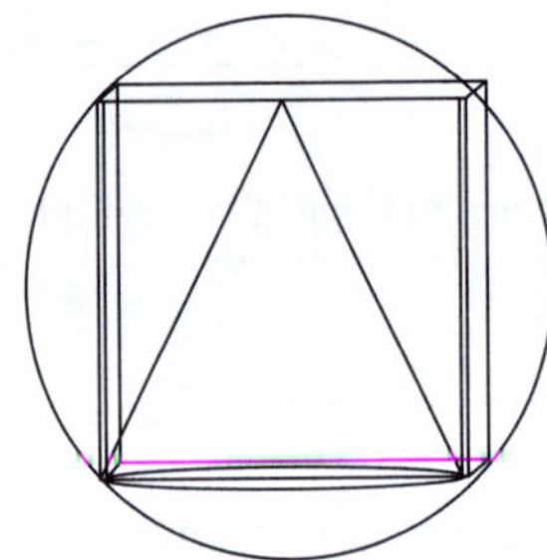

Clearly, diagonal of cube = diameter of sphere

$\Rightarrow \quad \sqrt{3}a = 2r \Rightarrow a = 2r/\sqrt{3}$

Further, diameter of cone = height of cone = side of the cube $= a$

Now, required ratio

$= \frac{4}{3}\pi r^3 : \frac{1}{3}\pi\left(\frac{a}{2}\right)^2 \times a$

$= 4r^3 : \left(\frac{2r}{2\sqrt{3}}\right)^2 \times \frac{2r}{3} = 6\sqrt{3} : 1$

50. Side of cube = Diameter of cylinder $= a$

Now, volume of the cube : volume of the cylinder

$= a^3 : \pi\left(\frac{a}{2}\right)^2 \times a = 4 : \pi$

$= 4 : \frac{22}{7} = 14 : 11$

51. Radius of cone $= \frac{a}{2}$ and height of cone $= a$

$\therefore$ Slant height of cone $= \sqrt{a^2 + \left(\frac{a}{2}\right)^2}$

Surface area of sphere : Curved surface area of cone

$= 4\pi r^2 : \pi \times \frac{a}{2} \times \sqrt{a^2 + \left(\frac{a}{2}\right)^2}$

$= 4\pi\left(\frac{\sqrt{3}}{2}a\right)^2 : \pi \times \frac{a}{2} \times \frac{\sqrt{5}}{2}a$

$= 12 : 5 \neq 5 : 1$

Hence, statement (I) is not correct.

Further, surface area of cube : curved surface area of cylinder

$= 6a^2 : 2\pi \times \frac{a}{2} \times a = 6 : \pi \neq 1 : 1$

Hence, option (d) is correct.

Alternatively

In such sorts of questions, we should assume the value of side or radius, as the case may be, in number, not in terms of variables like a, b, x, y, z etc. so as to solve them in moments.

52. Observing ratio analysis, given in beginning of this chapter, ratio of their radii (A to B)

$= \sqrt[3]{125} : \sqrt[3]{64} = 5:4$

As given, (5 + 4) units = 36

$\Rightarrow$ 5 units = 20 cm i.e. radius of sphere A.

Now, surface area of sphere A

$= 4\pi(20)^2 = 1600\pi \text{ cm}^2$

53. Volume of the cylinder $= \pi r^2 h$

$\Rightarrow \quad 9\pi = \pi r^2 \times 3 \Rightarrow r = \sqrt{3}$

Clearly, $AB = \sqrt{3}$,

OB = radius of ball = 2

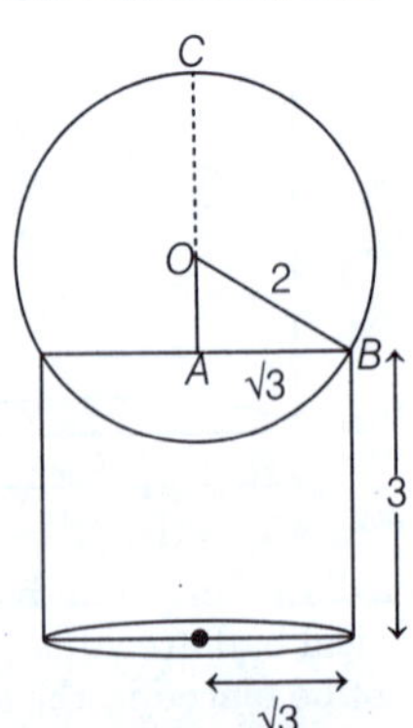

Then, $OA = \sqrt{2^2 - (\sqrt{3})^2} = 1$

Now, required distance = height of the cylinder + AO + OC

$= 3 + 1 + 2 = 6 \text{cm}$

47

Prism and Pyramid

Learning OBJECTIVES

After completing this chapter, aspirants will be able to :

- Understand certain types of prism and pyramid with their specific properties;
- Learn derivation of formulae as to prism and pyramid;
- Understand the frustum of a regular pyramid in totality;
- Understand the application of different formulae in different cases;
- Know the concept of maxima/minima value of volume of prism/pyramid;
- Understand the philosophy of the division of this chapter in different parts;
- Analyse the trend of the questions being asked in various competitive examinations;
- Solve the questions in moments;
- Grip the concepts helping to solve the questions without pen and paper;
- Apply many concepts envisaged in the chapter 'PERCENTAGE' like 'effect on product in case of changes in factors', etc.;
- Apply many concepts of 'Ratio' in this chapter;
- Learn many new concepts/fundamentals;
- Apply the pragmatic approaches in lieu of academic approaches; and
- Ponder on the alternative methods.

Inside

- *Introduction: Certain Types of Prism and Pyramid*
- *Derivation of Formulae for Volume, Surface Area of Prism/Pyramid*
- *Frustum of a Regular Pyramid With Significant Properties*
- *Application of the Concept of Volume/Surface Areas in Diverse Cases*
- *Relation Between Volume and Surface Areas and its Applications*
- *Alteration in Dimensions Affecting Volume and Surface Areas*
- *Conversion of Shapes (Cuboid, Cube, Cylinder, Cone, Sphere, Hemisphere, Prism & Pyramid)*
- *Problems Based on Above All Aspects Including Maximum/ Minimum Value of Volume*
- *..............................*
- *.............................. & Many More*
- *Work Out with Solutions*

PRISM

A prism is a solid with two faces parallel and congruent (these faces are called bases), other faces are parallelograms. The prism can have any polygon as its bases.

Prisms are named on the basis of shapes of the base like triangular prism, rectangular prism, etc. If the base is regular polygon, it is called regular prism and in case of irregular polygon as base, it is known as irregular prism.

Right prism is a prism that has two bases, one directly above the other and that has lateral faces as rectangles. It means, in a right prism, the edges of the lateral faces are perpendicular to the bases. If lateral faces are parallelograms, it is an oblique prism. A right rectangular prism is also called a cuboid. We will discuss only right prism.

Volume Just like cube/cuboid/cylinder, we compute volume of the prism by multiplying area of the base with perpendicular distance between the two base faces (i.e. height).

Hence, **volume of the right prism = base area × perpendicular height (*h*).**

Further, if the base is a regular polygon having n sides each of length a, volume of the right prism will be

$$\frac{na^2}{4}\cot\left(\frac{180°}{n}\right)\times h$$

Needless to say that the right prism having regular polygon as its base is called right regular prism and literally $\frac{na^2}{4}\cot\left(\frac{180°}{n}\right)\times h$ is the volume of a right regular prism.

Lateral Surface Area Same as cube/cuboid/ cylinder,

Lateral surface area of the right prism = perimeter of the base × height

Total Surface Area = Lateral surface area + 2 (area of the base)

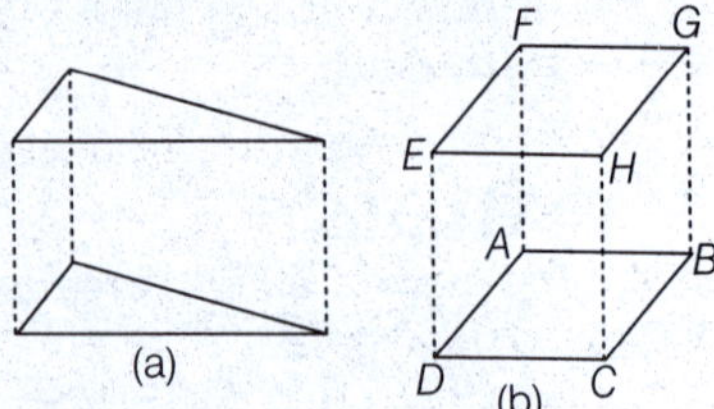

Fig.(a) Right triangular prism

Fig.(b) Right square prism : If *ABCD* are base, then *BG, CH, DE* and *AF* may be called as lateral edges

PYRAMID

A pyramid is a 3-dimensional figure made up of a base and triangular faces that meet at a point above the base. This point is called apex of the pyramid. A pyramid with n-sided base will have $(n+1)$ vertices, $(n+1)$ faces and $2n$ edges. Pyramids are named on the basis of shapes of the base like triangular pyramid, square pyramid, etc.

The base can be any polygon. If the base of the pyramid is regular polygon, the pyramid is known as regular pyramid and in case of irregular polygon as base, it is an irregular pyramid.

If the apex is directly over the centre of the base, it is known as right pyramid, otherwise, oblique pyramid.

It is clear from the above that "right regular pyramid" means a right pyramid whose base is a regular polygon. In a right regular pyramid, all the triangular faces are congruent and isosceles. We will discuss here only right pyramid.

Volume Volume of a right pyramid is $\frac{1}{3}$rd of the volume of the right prism, if both have equal base and equal height.

Hence, the volume of the right pyramid $=\frac{1}{3}\times$ area of the base × perpendicular height (from apex to base).

If the base of a right pyramid is a regular polygon having n sides each of length a, volume of the right pyramid will be

$$\frac{1}{3}\times\frac{na^2}{4}\cot\left(\frac{180°}{n}\right)\times h \text{ or } \frac{1}{12}\times na^2\times\cot\left(\frac{180°}{n}\right)\times h$$

It may be construed from above that the 1st method of computing volume is meant for both right pyramid and right regular pyramid. But, the latter method is restricted for only right regular pyramid.

Lateral/Slant Surface Area

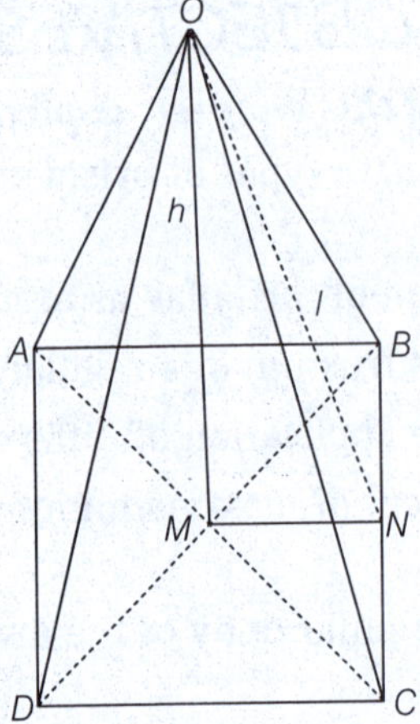

Lateral/ Slant surface area = Area of ΔOCB + Area of ΔOAB + Area of ΔOAD + Area of ΔODC

$$=\left[\frac{1}{2}\times BC\times ON\right]+\left[\frac{1}{2}\times AB\times ON\right]+\left[\frac{1}{2}\times AD\times ON\right]+\left[\frac{1}{2}\times CD\times ON\right]$$

[here, ON = slant height, all the lateral faces will have slant heights of equal length in case of right regular pyramid. *ABCD* is a regular polygon.]

$$=\frac{1}{2}\times ON\times(BC+AB+AD+CD)$$

$$=\frac{1}{2}\times ON\times \text{perimeter of the base}$$

$$=\frac{1}{2}\times \text{perimeter of the base}\times\text{slant height}$$

Clearly, slant height $(l)=\sqrt{h^2+(MN)^2}$

$$=\sqrt{h^2+(\text{Inradius of the base})^2}$$

This formula is applicable only if the base of a right pyramid is a regular polygon i.e. applicable only in case of right regular pyramid. In case the base is an irregular polygon, the slant heights are of different lengths and we compute lateral / slant surface area after finding relevant slant heights individually.

It is important to note that in the given figure OA, OB, OC and OD are four slant edges, not slant heights.

Total Surface Area

Total surface area = Area of the lateral surfaces + Area of the base

FRUSTUM OF A RIGHT REGULAR PYRAMID

A pyramid left after cutting a portion at the top by a plane parallel to the base is known as a frustum of a pyramid. The frustum of a right regular pyramid is a portion of a right regular pyramid included between the base of the pyramid and a section parallel to this base. Recall the concept of the frustum of a cone. The formulae meant for the frustum of a right regular pyramid follow the same pattern as applicable in case of frustum of a cone. Due to space constraint their derivation is not being given. Needless to say that the derivation process is almost the same. You may compare the formulae.

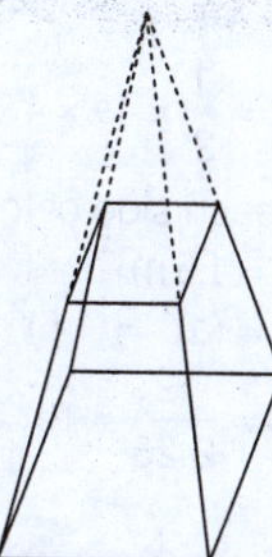

A frustum of a regular pyramid with n-sided base will have $2n$ vertices, $(n+2)$ faces and $3n$ edges(edges include base edges and lateral edges).

Volume

Recall, volume of the frustum of a cone

$$= \frac{h}{3}\,[\text{area of lower base } (A_1) + \text{area of upper base } (A_2) + \sqrt{A_1 \cdot A_2}]$$

Similarly, volume of the frustum of a right regular pyramid is also equal to $\frac{h}{3}$ [area of lower base (A_1) + area of upper base $(A_2) + \sqrt{A_1 \cdot A_2}$]

[where, h is perpendicular distance between both the parallel bases]

If a_1 and a_2 are sides of bases (n-sided) of a frustum of a right regular pyramid, then volume will be equal to

$$\frac{nh}{12}\,[(a_1)^2 + a_1 a_2 + (a_2)^2]\cot\left(\frac{180°}{n}\right)$$

Curved/Lateral/Slant Surface Area

We know, Curved/lateral/slant surface area of the frustum of a cone $= \frac{1}{2} \times$ slant height $\times$ circumference of the lower base (C_1) + circumference of the top base (C_2)]

$$= \frac{1}{2}\,l\,(C_1 + C_2)$$

Similarly, lateral/slant surface area of the frustum of a right regular pyramid

$$= \frac{1}{2} \times \text{Slant height [perimeter of the lower base } (P_1) + \text{perimeter of the top base } (P_2)]$$

$$= \frac{1}{2}\,l\,(P_1 + P_2)$$

Total Surface Area

In case of the frustum of a cone = curved area + bases areas and in case of the frustum of a right regular pyramid = curved area + bases areas.

It is observed from the above that there is no theoretical difference between the frustum of a cone and the frustum of a right regular pyramid.

Properties regarding frustum of a right regular pyramid

1. The slant height of a frustum of a right regular pyramid is the altitude of the face and faces are equal isosceles trapezoids.
2. The lateral edges of a frustum of a right regular pyramid are equal.
3. The bases of a frustum of a right regular pyramid are similar regular polygons.
4. (lower base edge − upper base edge)^2 + 4 (slant height)^2 = 4(lateral edge)^2

   Since, the parent right pyramid and the pyramid cut are similar, they will hold the following properties (point 5 to 8)
5. $$\frac{\text{height of the parent right regualr pyramid } (H)}{\text{height of the right regular pyramid cut } (h)} = \frac{\text{length of the base edge of the parent pyramid } (B)}{\text{length of the base edge of the pyramid cut } (b)} = \frac{\text{length of the lateral edge of the parent pyramid } (E)}{\text{length of the lateral edge of the pyramid cut } (e)} = \frac{\text{slant height of the parent pyramid } (L)}{\text{slant height of the pyramid cut } (l)}$$
6. Volume of the parent pyramid : Volume of the pyramid cut
   $= H^3 : h^3 = B^3 : b^3 = E^3 : e^3 = L^3 : l^3$
7. Lateral surface area of the parent pyramid : Lateral surface area of the pyramid cut
   $= H^2 : h^2 = B^2 : b^2 = E^2 : e^2 = L^2 : l^2$
8. Base area of the parent pyramid : Base area of the pyramid cut
   $= H^2 : h^2 = B^2 : b^2 = E^2 : e^2 = L^2 : l^2$

Application of the Concept of Volume in Diverse Cases

Problem 1 Area of the base of a pyramid is 57 cm^2 and height is 10 cm, then its volume in cm^3, is

Sol. Volume of the pyramid $= \frac{1}{3} \times$ area of the base $\times$ height

$$= \frac{1}{3} \times 57 \times 10 = 190 \text{ cm}^3$$

Problem 2 There is a pyramid on a base which is a regular hexagon of side $2a$ cm. If every slant edge of this pyramid is of length $\frac{5a}{2}$ cm, then the volume of this pyramid is

Sol. We know that there are six equilateral triangles in a regular hexagon and each side of the triangle is equal to that of the regular hexagon. It means, $OC = BC = 2a$

Further, MC (slant edge) $= \frac{5a}{2}$ (given)

$$\therefore \text{Perpendicular height} = \sqrt{(MC)^2 - (OC)^2} = \sqrt{\left(\frac{5a}{2}\right)^2 - (2a)^2}$$

$$= \sqrt{\frac{25a^2}{4} - 4a^2} = \frac{3a}{2} \text{ cm}$$

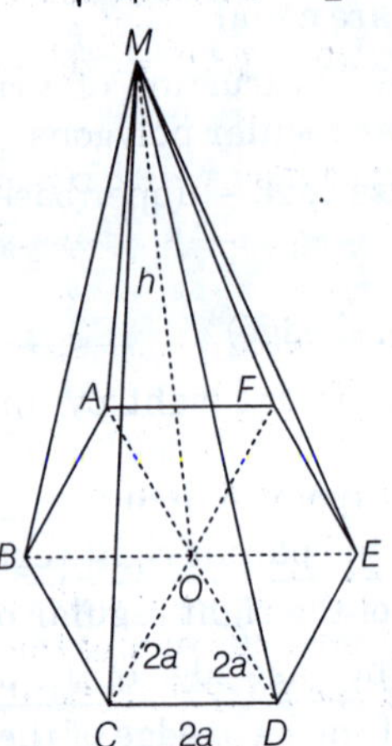

Now, volume of the pyramid

$$= \frac{1}{3} \times \text{area of the base} \times \text{perpendicular height}$$

$$= \frac{1}{3} \times [6 \times \text{area of the equilateral triangle}] \times \frac{3a}{2}$$

$$= 2 \times \frac{\sqrt{3}}{4} \times (2a)^2 \times \frac{3a}{2} = 3\sqrt{3}\, a^3 \text{cm}^3$$

Problem 3 The base of a right pyramid is a square of side 40 cm long. If the volume of the pyramid is 8000 cm^3, then its height is

Sol. Volume of the pyramid $= \frac{1}{3} \times$ area of the base $\times$ height

$$\Rightarrow \quad 8000 = \frac{1}{3} \times (40)^2 \times \text{height}$$

$$\Rightarrow \quad \text{height} = \frac{800 \times 3}{40 \times 40} = 15 \text{ cm}$$

Problem 4 The base of a right prism is a trapezium. The length of the parallel sides are 8 cm and 14 cm and the distance between the parallel sides is 8 cm. If the volume of the prism is 1056 cm^3, then the height of the prism is :

Sol. Given, volume of the right prism = 1056

$\Rightarrow$ Area of the base (trapezium) $\times$ height = 1056

$$\Rightarrow \quad \frac{1}{2}(8 + 14) \times 8 \times \text{height} = 1056$$

$$\Rightarrow \quad \text{height} = \frac{1056}{88} = 12 \text{ cm}$$

Problem 5 If the slant height of a right pyramid with square base is 4 m and the total slant surface of the pyramid is 12 m^2, the ratio of total slant surface and area of the base is : ***[SSC(10 + 2, LDC/DEO)-12]***

Sol. Total slant surface area $= \frac{1}{2} \times$ (perimeter of the base) $\times$ slant height

$$\Rightarrow \quad 12 = \frac{1}{2} \times 4a \times 4$$

[let each side of the square base be a metre]

$$\Rightarrow \quad a = 1.5 \text{ m}$$

$$\therefore \quad \text{Area of the base} = (a)^2 = (1.5)^2 = 2.25 \text{ m}^2$$

Then, required ratio $= \frac{12}{2.25} = 16 : 3$

Problem 6 The base of a right pyramid is an equilateral triangle of side 4 cm. The height of the pyramid is half of its slant height. Its volume is : ***[FCI (Mains)-12]***

Sol. Obviously, OM = inradius of the equilateral ΔABC

$$= \frac{\text{side}}{2\sqrt{3}} = \frac{4}{2\sqrt{3}} = \frac{2}{\sqrt{3}}$$

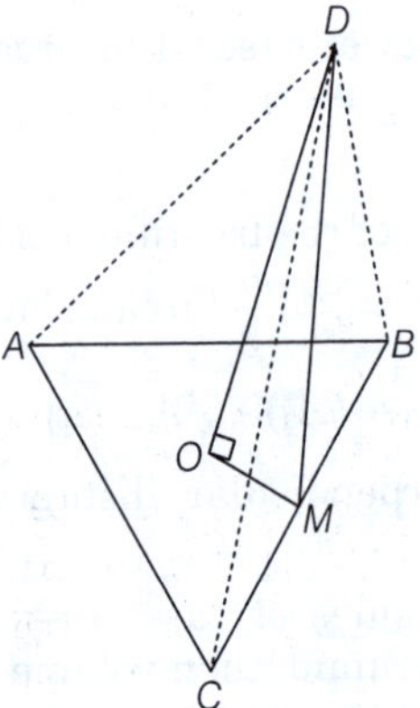

Now, [slant height $(DM)^2$] = [perpendicular height $(DO)^2$] + $(OM)^2$

$$\Rightarrow \quad (2h)^2 = h^2 + \left(\frac{2}{\sqrt{3}}\right)^2 \quad \text{[let perpendicular height be } h \text{ cm]}$$

$$\Rightarrow \quad 3h^2 = \left(\frac{2}{\sqrt{3}}\right)^2$$

$$\Rightarrow \quad h = \frac{2}{\sqrt{3}} \times \frac{1}{\sqrt{3}} = \frac{2}{3}$$

Then, volume of the pyramid $= \frac{1}{3} \times$ area of the base $\times$ height

$$= \frac{1}{3} \times \left[\frac{\sqrt{3}}{4} \times 4^2\right] \times \frac{2}{3}$$

$$= \frac{1}{3} \times 4\sqrt{3} \times \frac{2}{3} = \frac{8\sqrt{3}}{9} \text{ cm}^3$$

Problem 7 The base of a right pyramid is a rectangle of sides 40 cm and 30 cm. The height of the pyramid is 45 cm, find its volume.

Sol. Volume $= \frac{1}{3} \times$ area of the base $\times$ height

$$= \frac{1}{3} \times 40 \times 30 \times 45 = 18000 \text{ cm}^3$$

Problem 8 The base of a right pyramid is an equilateral triangle of side 4 cm each. Each slant edge is 5 cm long. The volume of the pyramid is :

[SSC(CGL-T-I)-14, SSC (CGL-T-II)-20]

Sol. Given, OC = slant edge = 5 cm

Now, CM = circumradius of the equilateral ΔABC

$$= \frac{\text{side}}{\sqrt{3}} = \frac{4}{\sqrt{3}} \text{ cm}$$

[In case of equilateral triangle, an incentre is also circumcentre]

$\therefore$ Perpendicular height (OM)

$$= \sqrt{(OC)^2 - (CM)^2} = \sqrt{(5)^2 - \left(\frac{4}{\sqrt{3}}\right)^2}$$

$$= \sqrt{25 - \frac{16}{3}} = \sqrt{\frac{59}{3}}$$

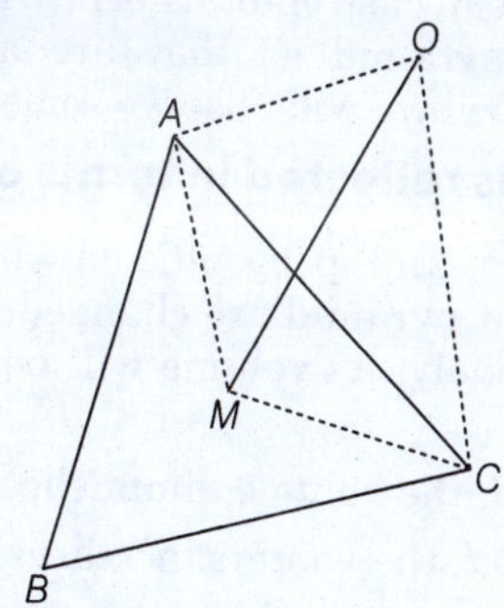

Then, Volume $= \frac{1}{3} \times$ area of the base $\times$ perpendicular height

$$= \frac{1}{3} \times \left[\frac{\sqrt{3}}{4} \times 4^2\right] \times \sqrt{\frac{59}{3}}$$

$$= \frac{1}{3} \times \frac{\sqrt{3}}{4} \times 4 \times 4 \times \frac{\sqrt{59}}{\sqrt{3}} = \frac{4\sqrt{59}}{3} \text{ cm}^3$$

Application of the Concept of Surface Areas

Problem 9 The base of a solid right prism is a triangle whose sides are 9 cm, 12 cm and 15 cm. The height of the prism is 5 cm, then the total surface area of the prism is :

[SSC (CGL-T-II)-20]

Sol. Total surface area = lateral surface area + 2 (area of the base)

$$= (\text{perimeter of the base}) \times \text{height} + 2\left[\frac{1}{2} \times 9 \times 12\right]$$

[$\because$ base of the prism is a right angled triangle as $(15)^2 = (9)^2 + (12)^2$]

$$= (9 + 12 + 15) \times 5 + [2 \times 54]$$

$$= 180 + 108 = 288 \text{ cm}^2$$

Problem 10 Find the total surface area of a right square pyramid with a perpendicular height of 16 cm and base edge of 24 cm.

Sol. Given, OM = perpendicular height = 16 cm

Clearly, $MN = \frac{24}{2} = 12$ cm

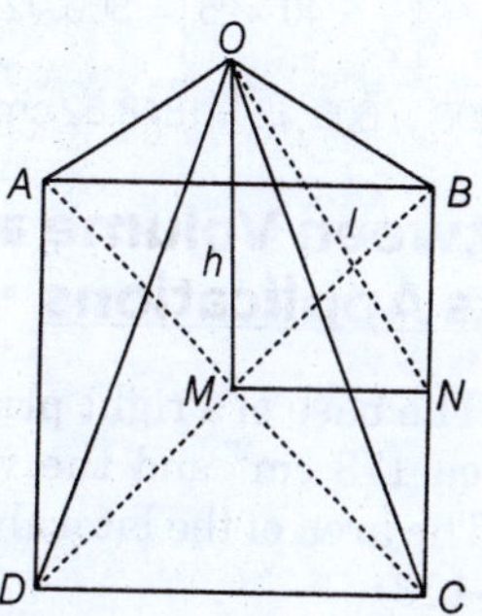

Now, $ON^2 = (OM)^2 + (MN)^2 = (16)^2 + (12)^2 = 400$

$\therefore$ $ON = 20$ [ON is slant height]

Now, lateral surface area of the pyramid

$$= \frac{1}{2} (\text{Perimeter of the base}) \times \text{slant height}$$

$$= \frac{1}{2}(4 \times 24) \times 20 = 960 \text{ cm}^2$$

Further, area of the base $= (\text{side})^2 = (24)^2 = 576 \text{ cm}^2$

$\therefore$ Total surface area $= 960 + 576 = 1536 \text{ cm}^2$

Problem 11 The base of a right pyramid is a rectangle measuring 40 cm × 30 cm and the perpendicular height is 20 cm. Find the lateral surface area.

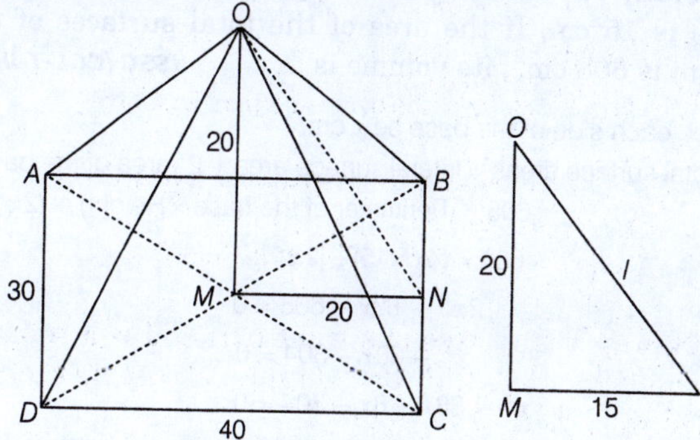

Sol. Since the base is a non- regular polygon, all the slant heights will not be equal in length. Thus, the formula meant for lateral surface area (i.e. $\frac{1}{2} \times$ perimeter of the base $\times$ slant height) will not be applicable here. We will compute the same finding slant height in each case.

Given, OM = perpendicular height = 20

and $AB = 40$ cm, $BC = 30$ cm, $MN = \frac{40}{2} = 20$ cm

Now, lateral surface area = area of ΔOBC + area of $\Delta\ OAB$ + area of ΔOAD + area of ΔOCD

$$= \left[\frac{1}{2} \times BC \times ON\right] + \left[\frac{1}{2} \times AB \times \text{relevant slant height } (l)\right] + \left[\frac{1}{2} \times AD \times ON\right] + \left[\frac{1}{2} \times CD \times l\right]$$

[here, $ON = \sqrt{OM^2 + (MN)^2} = \sqrt{(20)^2 + (20)^2} = 20\sqrt{2}$

and $l = \sqrt{(20)^2 + (15)^2} = 25$]

$$= \left[\frac{1}{2} \times 30 \times 20\sqrt{2}\right] + \left[\frac{1}{2} \times 40 \times 25\right] + \left[\frac{1}{2} \times 30 \times 20\sqrt{2}\right] + \left[\frac{1}{2} \times 40 \times 25\right] = 300\sqrt{2} + 500 + 300\sqrt{2} + 500$$

$$= 1000 + 600\sqrt{2} = 1848.52 \text{ cm}^2$$

Relation Between Volume and Surface Areas and its Applications

Problem 12 The base of a right prism is an equilateral triangle of area 173 cm^2 and the volume of the prism is 10380 cm^3. The area of the lateral surface of the prism is (given $\sqrt{3} = 1.73$)

Sol. Given, area of the base (equilateral triangle) = 173

$\Rightarrow \quad \frac{\sqrt{3}}{4}(\text{side})^2 = 173$

$\Rightarrow \quad (\text{side})^2 = 400 \Rightarrow \text{side} = 20$

Perimeter of the base = $3 \times 20 = 60$ cm

Now, Volume = Area of the base × Height

$\Rightarrow \quad 10380 = 173 \times \text{height}$

$\Rightarrow \quad \text{height} = \frac{10380}{173} = 60$ cm

$\therefore$ Area of the lateral surface = Perimeter of the base × Height $= 60 \times 60 = 3600$ cm^2

Problem 13 The height of a right prism with a square base is 15 cm. If the area of the total surfaces of the prism is 608 cm^2, its volume is : ***[SSC (CGL-T-II)-11]***

Sol. Let, each side of the base be x cm.

Total surface areas = lateral surface area + 2 (area of the base)

$\Rightarrow \quad 608 = (\text{Perimeter of the base} \times \text{Height}) + (2x^2)$

$\Rightarrow \quad 608 = (4x \times 15) + [2x^2]$

$\Rightarrow \quad 2x^2 + 60x - 608 = 0$

$\Rightarrow \quad x^2 + 30x - 304 = 0$

$\Rightarrow \quad x^2 + 38x - 8x - 304 = 0$

$\Rightarrow \quad x(x + 38) - 8(x + 30) = 0$

$\Rightarrow \quad (x - 8)(x + 38) = 0$

$\Rightarrow \quad x = 8$ and -38 (not possible)

It means each side of the square = 8 cm

$\therefore$ Volume of the prism = Area of the base × Height $= (8)^2 \times 15 = 64 \times 15 = 960$ cm^3

Alteration in dimensions affecting volume and surface areas

• If alteration is reflected in terms of ...... times

1. If base area and perpendicular height of a right prism/ a right pyramid are made x and y times respectively, the volume will become xy times the earlier.

2. If each side of base of right regular prism/right regular pyramid is made x times and perpendicular height is made y times, its volume will become x^2y times the earlier.

3. If perimeter of a right prism is made to x times and perpendicular height is made to y times, its lateral surface area will be xy times the earlier. In case of a right regular pyramid if its perimeter is made to x times and the slant height is made to y times, its lateral surface area becomes xy times the earlier.

• If alteration is reflected in terms of percentage

4. Recall the 'concept of effect on product in case of changes in factors' detailed in chapter percentage. It should be kept in mind that the volume refers to three dimensions and the area does two. In case of volume of a right prism/ a right pyramid base-area is multiplied by the perpendicular height with/without a constant figure, whereas in case of lateral surface area, perimeter of the base is either multiplied by perpendicular height or slant height with/without a constant figure as the case may be. The 'concept of effect on product in case of changes in factors' will not be applicable in case of total surface area / lateral area of a prism/pyramid as they are a sum of certain products. Now, we will observe some illustrations.

• If alteration is reflected in terms of ratio

5. If base area and perpendicular height of a right prism/a right pyramid are changed in ratios $a : b$ and $c : d$ respectively, its volume will be changed in ratio $ac : bd$,

(a) If $ac > bd$, the volume diminishes.

(b) If $ac < bd$, the volume increases.

(c) If $ac = bd$, no changes in the volume.

[Percentage change in volume $= \left[\frac{ac - bd}{ac} \times 100\right]\%$, if the outcomes is positive, volume diminishes and in case of negative it increases.]

6. If each side of base of right regular prism/right regular pyramid is changed in ratio $a : b$ and perpendicular height is changed in ratio $c : d$, then its volume will be changed in ratio $a^2c : b^2d$

(a) If $a^2c > b^2d$, the volume diminishes.

(b) If $a^2c < b^2d$, the volume increases.

(c) If $a^2c = b^2d$, no changes in the volume.

[Percentage change in volume $= \left\{\dfrac{a^2c - b^2d}{a^2c} \times 100\right\}\%$, if the outcome is positive, volume diminishes and in case of negative it increases.]

7. If perimeter of a right prism/ a right regular pyramid is changed in ratio $a:b$ and perpendicular height (slant height in case of right regular pyramid) is changed in ratio $c:d$, its lateral surface area will be changed in ratio $ac:bd$.
   (a) If $ac > bd$, lateral area diminishes.
   (b) If $ac < bd$, lateral area increases.
   (c) If $ac = bd$, no changes in lateral area.

   [Percentage change in volume $= \left(\dfrac{ac - bd}{ac} \times 100\right)\%$, if the outcome is positive, lateral area diminishes and in case of negative it increases.]

   Now, we will observe some illustrations as below.

Problem 14 The area of base of a right prism remains constant but the lateral edges are doubled, find the effect on its volume?

Sol. Volume = Area of the base × Height
= Area of the base × Lateral edge [∵ height = lateral edge]
Obviously, if lateral edges are doubled, its volume will be doubled.

Alternatively

Since out of three variables attached with the volume only one variable is changed, the volume will change by the same degree. Suppose this right prism is rectangular and either length or breadth of the base is increased by 20%, the volume will increase only by 20% as only one variable has been changed.

Problem 15 The length and breadth of the base of a right rectangular prism are reduced by 50% each and the height is doubled, then the lateral surface area will increase by :

Sol. If the length and the breadth both are reduced by 50%, the perimeter of the base will reduce by 50%. It is fact that lateral surface is the product of two variables *viz.* perimeter and height.

$\therefore$ Effect on lateral surface area $= \left[-50 + 100 - \dfrac{50 \times 100}{100}\right]\%$
$= 0\%$.

It means the original lateral surface area will remain the same.

Alternatively I

Clearly, perimeter has become 0.5 times the earlier and height has become 2 times. Then, lateral surface area will become $[0.5 \times 2]$ times the earlier. It means the original lateral surface area will remain the same.

Alternatively II

Here, perimeter has been changed in ratio 2 : 1 and height in ratio 1 : 2. Then, the lateral surface area will change in ratio $(2 \times 1):(1 \times 2)$, i.e. 1 : 1. It means the original lateral surface area will remain the same.

Alternatively III

You may assess the effect on lateral surface area taking certain values of all three variables.

Problem 16 The base of a right pyramid is a rectangle, whose sides are increased by 50% and 20%. What is the effect on its volume if perpendicular height is reduced to 40% ?

Sol. Effect on area of the base
$= \left[+50 + 20 + \dfrac{50 \times 20}{100}\right]\% = +80\%$

Now, effect on the volume
$= \left[+80 - 60 - \dfrac{80 \times 60}{100}\right]\% = -28\%$

Hence, the volume will decrease by 28%.

Alternatively I

Clearly, base-area has become 1.8 times the earlier and height has become 0.4 times. Then, volume will become $[1.8 \times 0.4]$ times i.e. 0.72 times the earlier. It means the original volume will reduce by 28%.

Alternatively II

Here, base-area has been changed in ratio 100 : 180 i.e. 5 : 9 (say, $a:b$) and height in ratio 100 : 40 i.e. 5 : 2 (say, $c:d$). Then, Percentage change in volume
$= \left[\dfrac{ac - bd}{ac} \times 100\right]\% = \left[\dfrac{25 - 18}{25} \times 100\right]\% = 28\%$

'+' indicates that original volume will reduce and the reduction is 28%.

Alternatively III

You may assess the effect on volume taking certain values of all three variables.

It may be construed from the discussion made through the above three problems that all the techniques are correlated and aspirants are supposed to be well versed with all these techniques so that they apply them appropriately observing the nature of questions.

Problems Based on Maximum/ Minimum Value of Volume

Problem 17 What is the maximum volume of a right pyramid with a square base that can be cut from a sphere of radius r ?

Sol. Clearly, $AF = FC = r$

Let FO be x.

Then, length from centre of the base to the corner of the base i.e.

$OC(\text{say}, d) = \sqrt{r^2 - x^2}$

$\Rightarrow \quad d^2 = r^2 - x^2$

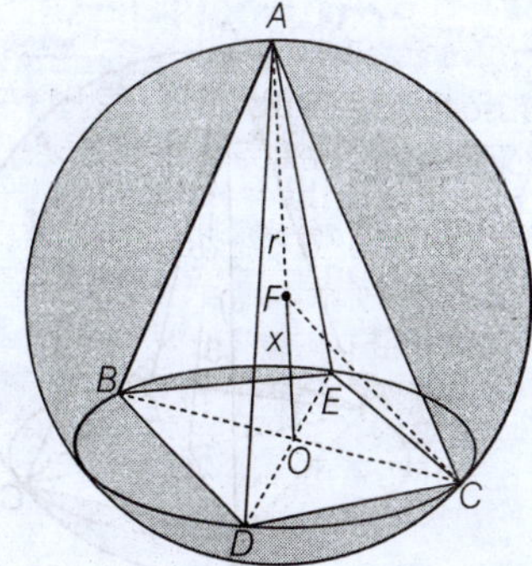

Now, area of the base $= \frac{1}{2} \times n \times \sin\left(\frac{360°}{n}\right) d^2$

where n is number of sides of the regular polygon.

$\Rightarrow$ Area of the base $= \frac{1}{2} \times 4 \times \sin\left(\frac{360°}{4}\right) \times (r^2 - x^2)$

$= 2(r^2 - x^2)$

Then, volume of the pyramid $= \frac{1}{3} \times$ area of base $\times$ height

$\Rightarrow \quad V = \frac{1}{3} \times 2(r^2 - x^2) \times (r + x)$

$\Rightarrow \quad V = \frac{2}{3}(r^2 - x^2) \times (r + x)$

$\Rightarrow \quad V = \frac{2}{3}(r^3 + r^2x - rx^2 - x^3)$

$\Rightarrow \quad \frac{dV}{dx} = \frac{2}{3}[- 3x^2 - 2rx + r^2]$

[$\because$ on differentiating with respect to x]

$\Rightarrow \quad \frac{2}{3}[-3x^2 - 2rx + r^2] = 0$

$\Rightarrow \quad - 3x^2 - 2rx + r^2 = 0 \Rightarrow x = \frac{r}{3}$ and $- r$

$\Rightarrow x = \frac{r}{3}$ as the value of x cannot be negative.

Since $\frac{d^2V}{dx^2}$ i.e $\frac{2}{3}(- 6x - 2r)$ is negative at $x = \frac{r}{3}$, the volume is maximum at the same value of x.

$\therefore$ Maximum volume of the pyramid $= \frac{2}{3}(r^2 - x^2) \times (r + x)$

$$= \frac{2}{3}\left[r^2 - \left(\frac{r}{3}\right)^2\right]\left(r + \frac{r}{3}\right) = \frac{64r^3}{81}$$

Observation Altitude and volume of the largest right pyramid with a square base that can be inscribed in a sphere of radius r, are $\frac{4r}{3}$ and $\frac{64r^3}{81}$ respectively. The area of square base $= \frac{16r^2}{9}$.

Problem 18 What is the maximum volume of a right pyramid of height 20 cm that can be cut from a sphere of radius 12 cm if the base of pyramid is square?

Sol. Given, radius of sphere = 12 cm

and height of pyramid = 20 cm

Hence, $AF = 12$ and $FO = 20 - 12 = 8$

Then, length from centre to the corner of the base i.e. OC (say, d) $= \sqrt{12^2 - 8^2} \Rightarrow d^2 = 80$

Now, area of the base $= \frac{1}{2} \times n \times \sin\left(\frac{360°}{n}\right) d^2$, where n is number of sides of the regular polygon.

$\Rightarrow$ Area of the base $= \frac{1}{2} \times 4 \times \sin\left(\frac{360°}{4}\right) \times 80 = 160$

Then, volume of the pyramid $= \frac{1}{3} \times$ area of base $\times$ height

$= \frac{1}{3} \times 80 \times 20 = 533.33 \text{ cm}^3$

With reference to the above question, what is the maximum volume if the base of the pyramid is hexagon.

Area of the base $= \frac{1}{2} \times n \times \sin\left(\frac{360°}{n}\right) d^2$

$= \frac{1}{2} \times 6 \times \sin\left(\frac{360°}{6}\right) \times 80 = 120\sqrt{3}$

Then, volume of the pyramid $= \frac{1}{3} \times 120\sqrt{3} \times 20 = 800\sqrt{3} \text{ cm}^3$

Problems Based on the Frustum of a Right Regular Pyramid

Problem 19 If the areas of the both ends of a truncated right regular pyramidal bucket which is 15 cm high, be 16cm^2 and 25 cm^2, then the capacity of the bucket in cubic centimeter is :

Sol. Given, height of the bucket (h) = 15 cm,

$A_1 = 25$ and $A_2 = 16$

Then, volume of the bucket

$= \frac{h}{3}$ [area of lower base (A_1) + area of upper base (A_2) $+ \sqrt{A_1 \cdot A_2}$]

$= \frac{1}{3} \times 15[25 + 16 + \sqrt{25 \cdot 16}] = 305 \text{ cm}^3$

Problem 20 A bucket is in the form of a frustum of a right regular pyramid with slant height of 10 cm. The areas of the top and bottom square ends are 144 sq cm and 64 sq cm respectively. Find the area of the metal sheet used in its making.

Sol. Slant surface area of the bucket

$= \frac{1}{2} \times$ slant height [perimeter of the lower base (P_1) + perimeter of the top base (P_2)]

$= \frac{1}{2} \times 10[4 \cdot \sqrt{64} + 4 \cdot \sqrt{44}] = 400 \text{ cm}^2$

Now, area of metal sheet used = slant area + area of the bottom

$= 400 + 64 = 464 \text{ cm}^2$

Problem 21 A right regular pyramid is cut at mid point of its height by a frustum parallel to its base. The ratio between the two parts of pyramid would be :

Sol. Observe the properties regarding frustum of a regular pyramid given in beginning of this chapter.

Volume (parent pyramid) : Volume (pyramid cut)

$$= H^3 : h^3 = 1^3 : \left[\frac{1}{2}\right]^3 = 8:1$$

$\therefore$ Volume (pyramid cut) : Volume (frustum of the pyramid)

$$= 1:(8-1) = 1:7$$

Problem 22 The height of a right regular pyramid is 30 cm. A small pyramid is cut off at the top by a plane parallel to its base. If its volume is $\frac{1}{27}$ of the volume of the original pyramid, at what height above the base, is the section made ?

Sol. Volume (parent pyramid) : Volume (pyramid cut)

$$= H^3 : h^3 = 27:1 \text{ (as given)}$$

$\Rightarrow \quad (30)^3 : h^3 = 27:1 \Rightarrow h^3 = 1000$

$\therefore \quad h = 10$

Clearly, the section is made at height of 20 cm [$\because$ 30 − 10 = 20] from the base of the parent pyramid.

Problem 23 A right regular pyramid is cut by two planes parallel to the base such that the height is divided into three equal parts. What is the ratio of the volume of the top, middle and bottom part?

[SSC(CGL-T-II)-17]

Sol. Volume (pyramid $ABCNM$) : Volume (pyramid $AFGYX$)

$$= h^3 : (3h)^3$$

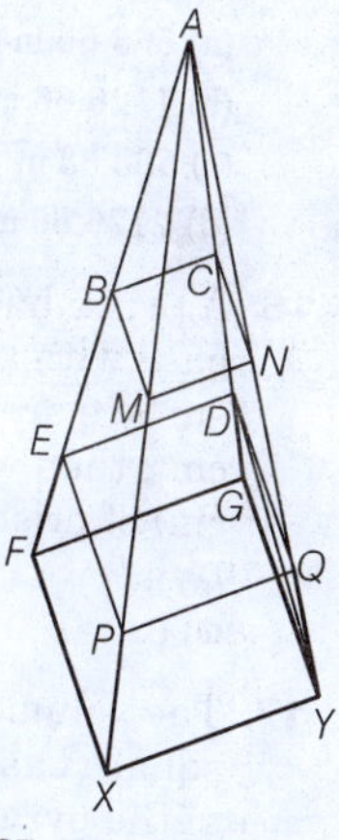

$$= 1^3 : 3^3 = 1:27$$

Volume (pyramid $AEDQP$) : Volume (pyramid $AFGYX$)

$$= (2h)^3 : (3h)^3 = 8:27$$

On combination, Volume (pyramid $ABCNM$) : Volume (pyramid $AEDQP$) : Volume (pyramid $AFGYX$) = 1 : 8 : 27

$\therefore$ Volume of pyramid $ABCNM$: Volume of middle part : Volume of lower part = 1 : (8 − 1) : (27 − 8) = 1 : 7 : 19

Problem 24 A right regular pyramid is cut by a plane parallel to the base and the upper portion is removed. If the lateral surface of the remainder is 8/9 of the lateral surface of the whole pyramid, find the ratio of the line segments into which the altitude of the pyramid is divided by the plane.

Sol. We know, lateral surface area of the parent pyramid : lateral surface area of the pyramid cut = $H^2 : h^2$

[where, H is height of the parent pyramid and h is that of pyramid cut]

$\Rightarrow \quad 9:1 = H^2 : h^2$

$\Rightarrow \quad H : h = 3:1$

$\Rightarrow \quad h:(H-h) = 1:(3-1) = 1:2$

Hence, required ratio = 1 : 2

Problem 25 A frustum of a right square pyramid has lower base edge of 5 cm and its slant height and the lateral height are 1.5 cm and 2 cm respectively. Find the perimeter of the upper end.

Sol. We know,

(lower base edge − upper base edge)^2 + 4 (slant height)^2 = 4(lateral edge)^2

$\Rightarrow$ (5 − upper base edge)^2 + 4 $(1.5)^2$ = $4(2)^2$

$\Rightarrow \quad$ (5 − upper base edge)^2 = 7

$\Rightarrow \quad$ upper base edge = $5 - \sqrt{7}$

Then, perimeter of the upper end

$$= 4(5-\sqrt{7}) = 9.41 \text{ cm}$$

Conversions of Shapes (Cuboid, Cube, Cylinder, Cone, Sphere, Hemisphere, Prism and Pyramid)

Problem 26 The base of a right solid prism of height 12 cm is a trapezium. The length of the parallel sides are 8 cm and 14 cm and the distance between the parallel sides is 8 cm. This prism is transformed into a right pyramid whose base is a square of side 18 cm. What is the height of the pyramid ?

Sol. Volume of the right prism

= Area of the base (trapezium) × height

$$= \frac{1}{2}(8+14) \times 8 \times 12 = 1056$$

Clearly, volume of the pyramid = volume of the prism

$\Rightarrow \frac{1}{3} \times$ area of the base × height = 1056

$\Rightarrow \quad \frac{1}{3} \times (18 \times 18) \times \text{height} = 1056$

$\Rightarrow \quad$ Height of the pyramid = 9.77 cm

Problem 27 A regular pyramid has a square base. The height of the pyramid is 22 cm and side of its base is 14 cm. Volume of pyramid is equal to the volume of a sphere. What is the radius(in cm) of the sphere ?

[SSC(CGL- T-II)-17]

Sol. Volume of the pyramid = Volume of the sphere

$\Rightarrow \frac{1}{3} \times$ area of the base × height $= \frac{4}{3}\pi r^3$

$\Rightarrow \quad \frac{1}{3} \times 14 \times 14 \times 22 = \frac{4}{3} \times \frac{22}{7} \times r^3$

$\Rightarrow \quad r = 7$ cm

WORKOUT

1. Let A be a pyramid on a square base and B be a cube. If a, b and c denote the number of edges, number of faces and number of corners respectively, then the result $a = b + c$ is true for : ***[CDS-13]***

(a) Only A (b) Only B
(c) Both A and B (d) Neither A or B

2. If the heights and the areas of the base of a right-circular cone and a pyramid with square base are the same, then they have

(a) same volume and the same surface area
(b) same surface area but different volumes
(c) same volume but different surface areas
(d) different volumes and different surface areas

3. Find the total surface area of a square pyramid with a perpendicular height of 8 cm and base edge of 12 cm. ***[SSC (CGL-T-II)-19]***

(a) 144 cm^2 (b) 544 cm^2
(c) 384 cm^2 (d) 182 cm^2

4. Base of a right pyramid is a square, length of diagonal of the base is $24\sqrt{2}$ m. If the volume of the pyramid is 1728 cu m, its height is : ***[SSC(CGL-T-II)-12]***

(a) 7 m (b) 8 m
(c) 9 m (d) 10 m

5. Base of a right prism is an equilateral triangle of side 6 cm. If the volume of the prism is $108\sqrt{3}$ cu cm, its height is : ***[SSC(CGL-T-II)-12]***

(a) 9 cm (b) 10 cm
(c) 11 cm (d) 12 cm

6. Base of a right pyramid is a square whose area is 324 sq m. If the volume of the pyramid is 1296 cu m, then area (in sq m) of the slant surface is : ***[SSC(CGL-T-II)-14]***

(a) 1080 (b) 432 (c) 360 (d) 540

7. A right pyramid 6 m high has a square base of which diagonal is $\sqrt{1152}$ m. Volume of the pyramid is : ***[SSC(CGL-T-I)-13]***

(a) 144 m^3 (b) 288 m^3
(c) 576 m^3 (d) 1152 m^3

8. The area of a right prism remains constant but the lateral edges are made three times. Find the effect on its volume.

(a) Volume becomes 3 times
(b) Volume becomes 9 times
(c) Volume becomes 3/2 times
(d) No change in volume

9. A hole is to be dug in a wooden block in the form of a frustum of a pyramid. The top end of the hole is a square of side 6 m and the bottom end is the square of side 4 m. What is the volume of the wood to be removed if the depth of the hole is 3 m ?

(a) $52 + 2\sqrt{6}$ m^3 (b) $42 + 2\sqrt{3}$ m^3
(c) $42 + 2\sqrt{6}$ m^3 (d) $52 + \sqrt{12}$ m^3

10. The height of a right regular pyramid is 18 cm. A small pyramid is cut off at the top by a plane parallel to its base. If its volume is $\frac{1}{8}$ of the volume of the original pyramid, at what height above the base, is the section made?

(a) 6 cm (b) 9 cm
(c) 10 cm (d) 20 cm

11. The base of a right pyramid is a rectangle, whose sides are increased by 50% and 40%. What is the effect on its volume if perpendicular height is reduced by 50%.

(a) 5% increase (b) 10% decrease
(c) No change (d) We cannot say

12. If the area of the base, height and volume of a right prism be $(3\sqrt{3}/2)P^2$ cm^2, $100\sqrt{3}$ cm and 7200 cm^3 respectively, then the value of P will be : ***[SSC(CGL-T-I)-15]***

(a) $\frac{2}{\sqrt{3}}$ (b) $\frac{3}{2}$
(c) $\sqrt{3}$ (d) 4

13. The areas of the ends of a frustum of a pyramid are P and Q where $P < Q$ and H is thickness. What is the volume of the frustum? ***[CDS-12]***

(a) $3H[P + Q + \sqrt{PQ}]$
(b) $H[P + Q + \sqrt{PQ}]$
(c) $\frac{H}{3}[P + Q + \sqrt{PQ}]$
(d) $\frac{H}{3}[P + Q - \sqrt{PQ}]$

14. The base of a right pyramid is a rectangle measuring 40 cm × 32 cm. Its slant height to the longer side of the base is 20 cm. The volume of the pyramid is :

(a) 5760 cm^3 (b) 5276 cm^3
(c) 2650 cm^3 (d) 5120 cm^3

15. The lower base edge and upper base edge of a hollow frustum of a right square pyramid are 16 m and 12 m respectively. This frustum is to be filled by water upto height of 5 m. What is the volume of water if the perpendicular distance between upper and lower bases is 10 m ?

(a) 563.66 m^3
(b) 1126.66 m^3
(c) 563.33 m^3
(d) 2126.66 m^3

16. A prism has a regular hexagonal base with side 6 cm. If the total surface area of prism is $216\sqrt{3}$ cm^2, then what is the height (in cm) of prism ? ***[SSC(CGL-T-II)-17]***

(a) $3\sqrt{3}$ (b) $6\sqrt{3}$
(c) 6 (d) 3

17. The volume of a pyramid with a square base is 200 cm^3. The height of the pyramid is 13 cm. What will be the length of the slant height (i.e. the distance between the apex and any other vertex), rounded to the nearest integer ? ***[XAT-17]***

(a) 12 cm (b) 13 cm
(c) 14 cm (d) 15 cm
(e) 17 cm

18. The base of a regular pyramid is a square and each of the other four sides is an equilateral triangle, length of each side being 20 cm. The vertical height of the pyramid (in cm) is : ***[CAT-19]***

(a) $8\sqrt{3}$ (b) $5\sqrt{5}$
(c) 12 (d) $10\sqrt{2}$

Answer Key

1. (d)	2. (c)	3. (c)	4. (c)	5. (d)	6. (d)	7. (d)	8. (a)	9. (a)	10. (b)
11. (a)	12. (d)	13. (c)	14. (d)	15. (b)	16. (a)	17. (c)	18. (d)		

Hints and SOLUTIONS

1.

	Pyramid with square base i.e. $n = 4$, detailed in the beginning.	Cube
Number of edges (a)	$2n = 2 \times 4 = 8$	12
Number of faces (b)	$n + 1 = 4 + 1 = 5$	6
Number of corners (c)	$n + 1 = 4 + 1 = 5$	8

Clearly, $a = b + c$ is neither true for the cube nor for the pyramid.

2. Since volumes of the both are computed through the same formula i.e. $\frac{1}{3} \times$ base-area $\times$ perpendicular height (h), their volumes will be same. However, their surface areas will differ as their slant heights may not be equal and the formula for computation is also different.

3. Given OM = perpendicular height $= 8\,\text{cm}$

Clearly, $MN = \frac{12}{2} = 6\,\text{cm}$

$\therefore$ [slant height $(ON)]^2$

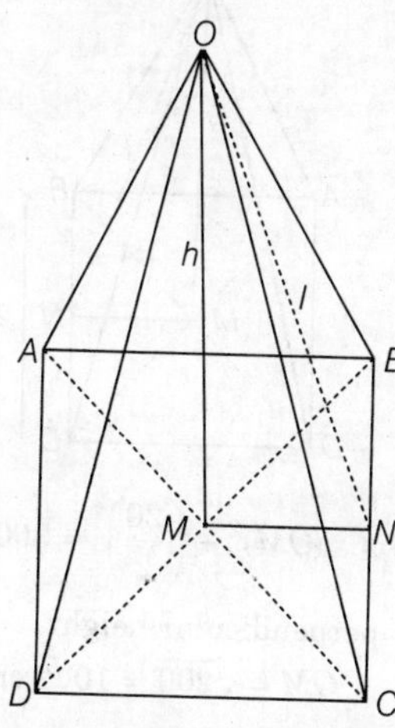

$= (OM)^2 + (MN)^2 = (8)^2 + (6)^2 = 100$

$\therefore \; ON = 10$

Now, lateral surface area of the pyramid

$= \frac{1}{2}$ (perimeter of the base) $\times$ slant height

$= \frac{1}{2}(4 \times 12) \times 10 = 240\,\text{cm}^2$

Further, area of the base

$= (\text{side})^2 = (12)^2 = 144\,\text{cm}^2$

$\therefore$ Total surface area

$= 240 + 144 = 384\,\text{cm}^2$

4. Volume of the pyramid

$= \frac{1}{3} \times$ base area $\times$ perpendicular height (h)

$\Rightarrow \; 1728 = \frac{1}{3} \times (24)^2 \times h$

$\left[\because \text{side of the square} = \frac{\text{diagonal}}{\sqrt{2}} = \frac{24\sqrt{2}}{\sqrt{2}} = 24\right]$

$\Rightarrow \; h = 9\,\text{m}$

5. Volume of the prism = base area $\times$ height

$\Rightarrow 108\sqrt{3} = \frac{\sqrt{3}}{4} \times 6^2 \times \text{height}$

$\Rightarrow$ height $= 12\,\text{cm}$

6. Volume of the pyramid

$= \frac{1}{3} \times$ base area $\times$ perpendicular height (OM)

$\Rightarrow 1296 = \frac{1}{3} \times 324 \times$ perpendicular height (OM)

$\Rightarrow \qquad OM = 12\,\text{m}$

Clearly, $MN = \frac{\text{side of the square}}{2}$

$= \frac{18}{2} = 9\,\text{m}$

$\therefore$ [slant height $(ON)]^2$

$= (OM)^2 + (MN)^2$

$= (12)^2 + (9)^2 = 225$

$\therefore \; ON = 15$

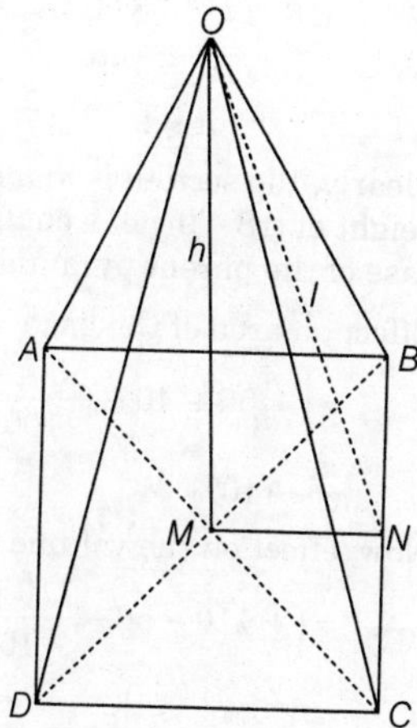

Now, lateral/slant surface area of the pyramid

$= \frac{1}{2}$ (perimeter of the base) $\times$ slant height

$= \frac{1}{2}(4 \times 18) \times 15 = 540\,\text{m}^2$

7. Volume of the pyramid

$= \frac{1}{3} \times$ base area $\times$ height

$= \frac{1}{3} \times (24)^2 \times 6$

$\left[\because \text{side of the square} = \frac{\text{diagonal}}{\sqrt{2}} = \frac{\sqrt{1152}}{\sqrt{2}} = 24\right]$

$= 1152\,\text{m}^3$

8. Volume = area of the base $\times$ height

= area of the base $\times$ lateral edge

[$\because$ height = lateral edge]

Obviously, if lateral edges are made three times, its volume will become three times.

Alternatively

Since out of three variables attached with the volume of the prism only one variable is changed, the volume will change by the same degree.

9. Volume of the frustum of a right regular pyramid

$= \frac{h}{3}$ [area of lower base (A_1) + area of upper base (A_2) + $\sqrt{A_1 \cdot A_2}$]

$= \frac{3}{3}[(4)^2 + (6)^2 + \sqrt{4 \cdot 6}]$

$= 52 + 2\sqrt{6}\ m^3$

10. Volume (parent pyramid) : Volume (pyramid cut)

$= H^3 : h^3 = 8:1$ (as given)

$\Rightarrow (18)^3 : h^3 = 8:1$

$\Rightarrow h^3 = 729$

$\therefore h = 9$

Clearly, the section is made at height of (18 – 9) i.e. 9 cm from the base of the parent pyramid.

11. Effect on area of the base

$= \left(+50 + 40 + \frac{50 \times 40}{100}\right)\%$

$= +110\%$

Now, effect on the volume

$= \left(+110 - 50 - \frac{110 \times 50}{100}\right)\%$

$= +5\%$

Hence, the volume will increase by 5%.

Alternatively I

Clearly, base area has become 2.1 times the earlier and height has become 0.5 times. Then, volume will become $[2.1 \times 0.5]$ times i.e. 1.05 times the earlier. It means the original volume will increase by 5%.

Alternatively II

Here, base-area has been changed in ratio 100 : 210 i.e. 10 : 21$(a : b)$ and height in ratio 100 : 50 i.e. 2 : 1 $(c : d)$.

Then, Percentage change in volume

$= \left(\frac{ac - bd}{ac} \times 100\right)\%$

$= \left(\frac{20 - 21}{20} \times 100\right)\% = -5\%$

'–' indicates that original volume will increase and the increment is 5%.

Alternatively III

You may assess the effect on volume taking certain values of all three variables.

12. Volume = area of the base × height

$\Rightarrow 7200 = \left(\frac{3\sqrt{3}}{2}\right)P^2 \times 100\sqrt{3}$

$\Rightarrow 7200 = 450\,P^2$

$\Rightarrow P = 4$

13. Clearly, volume of the frustum of a right regular pyramid

$= \frac{H}{3}$ [area of lower base (Q) + area of upper base (P) + $\sqrt{PQ}$]

$= \frac{H}{3}[P + Q + \sqrt{PQ}]$

14. Here, $BC = 40$ cm and $AB = 32$ cm

As given, slant height $ON = 20$

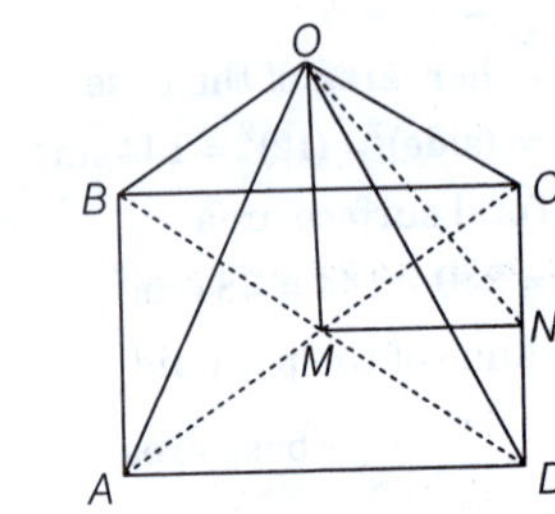

$\Rightarrow \sqrt{(OM)^2 + (MN)^2} = 20$

$\Rightarrow (OM)^2 + \left(\frac{32}{2}\right)^2 = 400$

$\Rightarrow$ perpendicular height $OM = 12$

Now, volume of the pyramid

$= \frac{1}{3} \times$ area of the base × height

$= \frac{1}{3} \times (40 \times 32) \times 12 = 5120\ cm^3$

15. Median $EF = \frac{1}{2}(12 + 16) = 14$

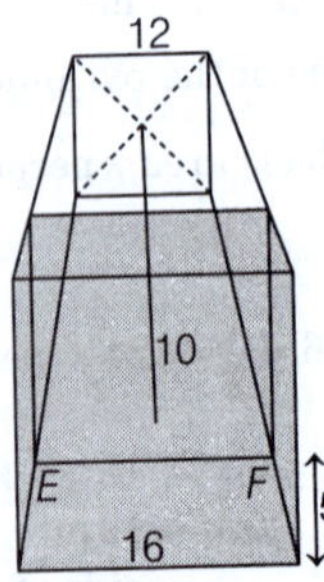

Then, volume of water

$= \frac{h}{3}$ [area of lower base (A_1) + area of upper base (A_2) + $\sqrt{A_1 \cdot A_2}$]

$= \frac{5}{3}[16^2 + 14^2 + \sqrt{16^2 \cdot 14^2}]$

$\frac{5}{3}[676] = 1126.66\ m^3$

16. Area of the hexagonal base

$= 6 \times \frac{\sqrt{3}}{4} \times (6)^2$

$= 54\sqrt{3}$

Then, total surface area of prism = lateral surface area + $(2 \times 54\sqrt{3})$

$\Rightarrow 216\sqrt{3} =$ perimeter of base × height + $108\sqrt{3}$

$\Rightarrow 216\sqrt{3} = (6 \times 6) \times \text{height} + 108\sqrt{3}$

$\Rightarrow$ height $= 3\sqrt{3}$ cm

17. Volume of the pyramid $= \frac{1}{3} \times$ base area × perpendicular height

$\Rightarrow 200 = \frac{1}{3} \times \text{base area} \times 13$

$\Rightarrow$ base area = 46.15

$\Rightarrow$ Side of the base

$= \sqrt{46.15} = 6.8$

Now, the distance between the apex and any other vertex i.e. lateral height =

$= \sqrt{\left(\frac{1}{2}\text{ of the diagonal of square base}\right)^2 + 13^2}$

$= \sqrt{\left(\frac{1}{2} \times 6.8\sqrt{2}\right)^2 + 13^2} \approx 14\ cm$

18. Slant height ON = Height of equilateral ΔOCB

$\Rightarrow \sqrt{(OM)^2 + (MN)^2} = \frac{\sqrt{3}}{2} \times 20$

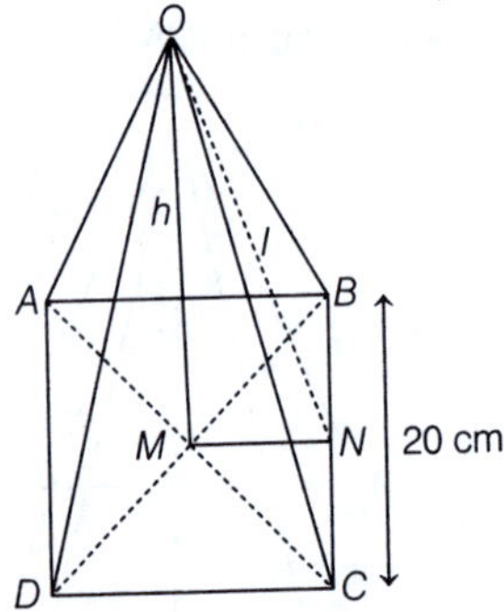

$\Rightarrow (OM)^2 + \left(\frac{20}{2}\right)^2 = 300$

$\Rightarrow$ perpendicular height

$OM = \sqrt{200} = 10\sqrt{2}$ cm

Unit 7

Data Interpretation

Data Interpretation

Data Interpretation (DI) is a part of our daily life. It is basically the process of making sense of the given numerical data and data may be related to observations, experiments, etc.

Interpreting data is a critical skill which helps us comprehend the relevant graphs, charts, tables, etc. in diverse aspects. No doubt, this section plays a pivotal role in securing marks in the competitive exams like CAT, MAT, Bank PO, SSC, CDS, Railway, etc.

As regards competitive exams, your job is to find the data from the given format(s) {graph, chart, table, etc.} which is/are relevant for the given question and thereafter to interpret the same to arrive at answer as soon as possible. Hence, DI examines how quickly you can take decision and needless to mention that the govt. organizations deploy the personnel who are quick in decision-making.

Certain questions require your approximation ability as you are supposed to find the approximate value instead of accurate value so as to arrive at correct option. Furthermore, sometimes you are supposed to use two or more formats simultaneously.

It is worth mentioning that the aspirants must have command over techniques related to basic calculations, comparison of fractions, fundamental concept of percentage including percentage-conversion chart, product constancy, ratio and proportion, SI, Cl, Profit and loss, average (including weighted average), etc. Thus, you are advised to revise them before solving the DI questions. At the last, while solving the questions, read the footnotes meticulously.

48

Data Interpretation

Learning OBJECTIVES

After completing this Section, aspirants will be able to :

- Understand the significance of this Section in competitive exams and in our daily life as well;
- Interpret the data in actual sense or as per demand of the questions;
- Be familiar with fundamental techniques useful for Data Interpretation;
- Solve all types of DI being asked in the competitive examinations as per latest patterns;
- Recognise the types of question which can be easily solved through the given options;
- Analyse the trend of the questions being asked in various competitive examinations;
- Solve the questions in moments;
- Grip on concepts helping to solve the questions without pen and paper;
- Learn many new concepts/fundamentals;
- Apply the pragmatic approaches in lieu of academic approaches; and
- Ponder on the alternative methods.

Inside

- *Introduction*
- *Techniques meant for comparison of values*
- *Techniques meant for computing percentage increase or decrease*
- *Percentage related other critical techniques*
- *Profit and loss related techniques*
- *Compound Interest-related techniques*
- *Ratio-related techniques*
- *Average-related techniques*
- *Illustrations Based on Data-table, Bar graphs, Line graphs, Pie charts, Funnel charts, Radar charts, Ternary diagrams, Combination of different graphs/tables/lines, etc*
- ....
- .... *& Many More*
- *Work Out with Solutions*

Introduction

The DI questions being asked in the competitive examinations may broadly be categorised as below for the sake of convenience.

(a) DI based on data table
(b) DI based on bar graphs (bar/multiple/subdivided)
(c) DI based on ternary diagrams
(d) DI based on line graphs
(e) DI based on pie-charts
(f) DI based on radar diagram
(g) DI based on the combination of the above graphs

Before you start to solve the DI questions, you must observe the following fundamental techniques, already discussed in detail in diverse chapters.

I. Techniques Meant for Comparison of Values

1. Comparison among fractions

We divide the numerators by respective differences between numerator and denominator, i.e. $\frac{N}{D-N}$ not $\frac{N}{N-D}$

e.g. Which is least among $\frac{8}{25}, \frac{7}{23}, \frac{11}{23}, \frac{14}{53}$?

$\Rightarrow \frac{8}{(25-8)}, \frac{7}{(23-7)}, \frac{11}{(23-11)}, \frac{14}{(53-14)}$

$\Rightarrow \frac{8}{17}, \frac{7}{16}, \frac{11}{12}, \frac{14}{39}$

However, it is difficult to find their values in decimals. Thus, you are advised to repeat the process of division.

Then, $\frac{8}{9}, \frac{7}{9}, \frac{11}{1}, \frac{14}{25}$

If you are still facing the difficulties to find their values, you may repeat the process further. But, the third fraction will converge into negative figure i.e. $\left(\frac{11}{1-11}\right)$ on further processing. Thus, we should stop the process.

$\Rightarrow$ 0.9, 0.7, 11, 0.56

Since 0.56 is the smallest, the corresponding fraction i.e. $\frac{14}{53}$ is the smallest fraction.

It may be constructed from the above that whenever any fraction converts into improper fraction, we should stop the further processing. In improper fraction, numerator is greater than denominator.

It is important to note that among $\frac{8x}{25}, \frac{7x}{23}, \frac{11x}{23}, \frac{14x}{53}; \frac{14x}{53}$ is the smallest and also among $\frac{8}{25x}, \frac{7}{23x}, \frac{11}{23x}, \frac{14}{53x}; \frac{14}{53x}$ is the smallest for $x > 0$.

2. Comparison among expressions in form of x% of y

***Case* I When values of y are same for all expressions.**

e.g. which has least value : 42% of 1241, 57% of 1241 and 15% of 1241.

It is obvious that more percentage term, more value and less percentage term less value in case of same base number. Hence, 15% of 1241 is the least.

It also holds good that more base number, more value and less base number, less value if the percentage terms are same.

***Case* II When values of y are not same for all expressions.**

e.g. Arrange in ascending order : 23% of 2250, 26% of 1801, 51% of 1129.

There are many ways to compare the given expressions

First Way

$\because$ 10% of 2250 = 225 $\quad\therefore$ 20% of 2250 = 450

$\because$ 1% of 2250 = 22.5 $\quad\therefore$ 3% of 2250 = 67.5

Hence, 23% of 2250 = 450 + 67.5 = **517.5**

Also, 10% of 1801 = 180.1

$\therefore$ 20% of 1801 = 360.2 and 1% of 1801 = 18.01

$\therefore$ 6% of 1801 = 108.06

Hence, 26% of 1801 = 360.2 + 108.06 = **468.26**

Further, 50% of 1129 = 564.5

and 1% of 1129 = 11.29

Thus, 51% of 1129 = 564.5+11.29 = **575.79**

Then, the ascending order is 26% of 1801, 23% of 2250 and 51% of 1129.

Second way

We know that, a% of b = b% of a

Hence, given expression is 23 × 2250%, 26 × 1801% and 51 × 1129%.

$\Rightarrow$ 23 × 22.5, 26 × 18.01 and 51 × 11.29

$\Rightarrow$ 517.5, 468.26 and 575.79

Then, the ascending order is 26% of 1801, 23% of 2250 and 51% of 1129.

Third way

We may make the base numbers same as 2250 or what you wish to select from the given base numbers. 26% of 1801 ≈ 20.8% of 2250

[We have changed 1801 to 2250, hence approx. 25% increase. Then, 26 will decrease by $\left(\frac{25}{125} \times 100\right)$%, i.e. by 20%. Thus, 26 becomes (26 − 5.2). ... Recall the **concept of Product Constancy]**

Now, compare : 23% of 2250, 20.8% of 2250 and 25.5% of 2250.

Further, 51% of 1129 ≈ 25.5% of 2250

[We have changed 1129 to 2250, hence approx. 100% increase. Then, 51 will decrease by $\left(\frac{100}{200} \times 100\right)$%, i.e. by 50%. Thus, 51 becomes 25.5.]

Now conpare : 23% of 2250, 20.8% of 2250 and 25.5% of 2250.

Clearly, the ascending order is 26% of 1801, 23% of 2250 and 51% of 1129.

The other main advantages of this technique is that we will be able to

(a) Compute the difference between given expressions in terms of percentage. *e.g.* difference between 23% of 2250 and 26% of 1801 = (23 − 20.8)% of 2250 = 2.2% of 2250.

Now, compute the absolute value of difference

$\because$ 1% of 2250 = 22.50, then 2% of 2250 = 45

Also, 0.1% of 2250 = 2.250, then 0.2% of 2250 = 4.5

Hence, required difference = 45 + 4.5 = 49.5

(b) Compute the percentage by which one expression is greater or lesser than others. *e.g.* (23% of 2250) is approximately 10.5% greater than (26% of 1801) as 23 is approximately 10.5% greater than 20.8. Similarly, we may do for other expressions.

3. Comparison among expressions in form of x% of y% of z.

Compare 50% of 1% of 20855, 40% of 5% of 22647 and 27% of 4% of 25050.

Clearly, given expressions are 0.5% of 20855, 2% of 22647 and 1.08% of 25050.

Now, the expressions become in form of x% of y. You may compute the value of all expressions in many ways as explained above.

II. Techniques Meant for Computing Percentage Increase or Decrease

4. When a number changes

What is percentage increase or decrease when 397 changes to 478 ?

Required percentage increase

$$=\left(\frac{478-397}{397}\times 100\right)\%$$

$$=\left(\frac{81\times 2.5}{397\times 2.5}\times 100\right)\%$$

(We made the denominator in form of 1000 for easy computation, you are free to make the denominator in another form like 100, 500, 2000, 2500, etc., as per your convenience)

$$\approx\left(\frac{81\times 2.5}{1000}\times 100\right)\%\approx 20.25\%$$

It is also fact that the actual percentage is little more than 20.25% as the denominator is less than 1000.

5. When an expression in form of (x% of y) changes in the same form

Case I **When values of x and y for one expression are totally different from those for another expression.**

What is percentage increase or decrease when 37% of 3057 changes to 28% of 3672 ?

Or

By what percent is 28% of 3672 greater or lesser than 37% of 3057?

Clearly, the base expression is 37% of 3057 as we have to compare 28% of 3672 with reference to 37% of 3057, in other words, we have to view 28% of 3672 as compared to 37% of 3057.

Thus, we must convert 3672 into 3057.

Now, 28% of 3672 = 33.6% of 3057.

[We have changed 3672 to 3057, hence approximately 1/6, i.e. 16.66% decrease. Then, 28 will increase by $\left(\frac{16.66}{83.33}\times 100\right)$% i.e. by 20%. Thus, 28 becomes (28 + 5.6). ...... Recall the concept of Product Constancy]

Obviously, 33.6% of 3057 is approximately 9% lesser than 37 % of 3057 as 33.6 is approximately 9% lesser than 37.

Now, alter the question

By what percent is 37% of 3057 greater or lesser than 28% of 3672 ?

Here, the base number is 28% of 3672, so convert 3057 to 3672.

Now, 37% of 3057 ≈ 31% of 3672 following the concept of product constancy.

Clearly, 31% of 3672 is approximately 10% greater than 28% of 3672 as 31 is approximately 10% greater than 28.

Case II **When value of either x or y is common for given expressions.**

By how much percent is 48% of 1906 greater or lesser than 48% of 1466 ?

Clearly, base expression is 48% of 1466, i.e. 1466% of 48 because we have to view 48% of 1906 as compared to 48% of 1466.

Also, 48% of 1906 = 1906% of 48

Obviously, 1906% of 48 is approximately 30% greater than 1466% of 48 as 1906 is approximately 30% more than 1466.

Take another example:

76% of 24538 is how much percentage lesser or greater than 56.8% of 24538 ?

Since 76 is approximately 1/3 i.e. 33.33% greater than 56.8, 76% of 24538 is approximately 33.33% greater than 56.8% of 24538.

Now, consider 56.8 % of 24538 is how much percentage lesser or greater than 76% of 24538 ?

Since 56.8 is approximately 25% lesser than 76, 56.8% of 24538 is approximately 25% lesser than 76% of 24538.

III. Percentage Related other Critical Techniques

6. (a% of b) will increase (or decrease) by m% if b is inc eased (or decreased) by m%.

It may be construed that (a% of b) will change to (a% of c) if b is changed to c and the changing rate for b to c is as same as that for (a% of b) to (a% of c).

e.g. The profit and the investment of a company in year 1990 were 19% and ₹ 236.78 lakh respectively. Next year, the investment increased to ₹ 339.27 lakh.

What was the profit in rupees on the investment in 1991 if the profit had increased at the same rate as was for the investment?

The required profit = 19% of 339.27 ≈ ₹ 64.5 lakh

($\because$ 20% of 339.27 = 67. 854 and 1% of 339.27 = 3.3927, then 19% of 339.27 = 67.854 − 3.3927 = 64.46)

Note *No need to solve conventionally as below:*

Percentage increase in investment

$$=\left(\frac{339.27-236.78}{236.78}\times 100\right)$$

$$=43.28\%$$

Now, profit = 143.28% of 19% of 236.78 ≈ ₹ 64.5 lakh

With reference to above question, what was the profit in rupees on the investment in 1991 if the profit had increased at one third of the rate as was for the investment.

The required profit = 19% of $\left(236.78+\frac{339.27-236.78}{3}\right)$

{Recall the concept 'How to insert n arithmetic mean between two numbers a and b' where first arithmetic mean $= a+\left(\frac{b-a}{n+1}\right)$} = 19% of 270.94

= ₹ 51.48 lakh

7. Effect on the Product of two Factors

(a) When one factor is increased by x% and another by y%, the effect on their product $=\left(+x+y+\frac{xy}{100}\right)\%$

(b) When one factor is increased by x% and another is decreased by y%, the effect on their product $=\left(+x-y-\frac{xy}{100}\right)\%$

(c) When one factor is decreased by x% and another is increased by y%, the effect on their product $=\left(-x+y-\frac{xy}{100}\right)\%$

(d) When one factor is decreased by x% and another by y%, the effect on their product $=\left(-x-y+\frac{xy}{100}\right)\%$

Note

(i) *If outcome is positive, the product increases and in case of negative outcome, the product decreases.*

(ii) *If one factor is increased or decreased, the product will be changed by the same magnitude.*

e.g. Sales of a company increases by 10% if it reduces the cost by 20%, what is effect on revenue?

Effect on revenue $=\left(+10-20-\frac{10\times 20}{100}\right)\% = -12\%$

Hence, the revenue will decrease by 12%.

IV. Profit and Loss Related Techniques

8. If profit or loss is calculated on investment, then income will be (100 + profit)% or (100 − loss)% of the investment, if income = investment + profit or income = investment − loss

e.g. If the income of a firm in 2015 and that in 2016 were equal and the amount invested in 2015 was ₹ 64 lakh, what was the amount invested in 2016 provided percent profit earned were 45% and 60% respectively in year 2015 and 2016.

Note *Income = Investment + Profit earned*

Obviously, 145% of amount invested in 2015 = 160% of the amount invested in 2016

⇒ amount invested in 2015 : that in 2016 = 160 : 145

= 32 : 29

As given, 32 units = ₹ 64 lakh

∴ 29 units = ₹ 58 lakh

Other forms may be like : revenue = cost + profit, expenditure = cost + taxes, etc.

V. Compound Interest-related Techniques

9. Compound Interest (CI) for 2 yrs

= (Cumulative SI).$\boxed{r^2}$ % of the principal money

CI for 3 yrs = (Cumulative SI).$\boxed{3r^2}\boxed{r^3}$ % of the principal money

CI for 4 yrs = (Cumulative SI).$\boxed{6r^2}\boxed{4r^3}\boxed{r^4}$% of the principal money

Note *The box will contain only two digits and in case of more than two digits, the remaining figure(s) will be carried over left hand side. It is also important to note that the box will always have two digits.*

e.g. CI for 2 yr at the rate of 6% per annum = $(2\times 6).\boxed{6^2}$% i.e. 12.36% of the principal money.

Then, amount = 112.36% of principal money.

CI for 3 yrs at the rate of 6% per annum = 19.1016% of the principal money and amount = 119.1016% of the principal money

CI for 4 yrs at the rate of 6% per annum

= 26.247696% of the principal money

and amount = 126.247696% of the principal money

CI for 5 yrs $=\left(+\text{CI for 4 yrs}+6+\frac{\text{CI for 4 yrs}\times 6}{100}\right)\%$

CI for 6 yrs $=\left(+\text{CI for 5 yrs}+6+\frac{\text{CI for 5 yrs}\times 6}{100}\right)\%$

and so on.

VI. Ratio-Related Techniques

10. The multiplication or division of each term of a ratio by the same non-zero number does not affect the ratio.

Hence, $a : b = ma : mb = \frac{a}{m} : \frac{b}{m}$

e.g. A firm earns a profit of 34% on its investment. The profits were ₹ 4.9 lakh and 6.3 lakh respectively in years 2016 and 2017. What is the ratio of investments made in 2016 and 2017?

In both years, the investments will be (100 ÷ 34) times the given profits. Then, the ratio of investments will be same as ratio of profits as the multiplication or division of each term of a ratio by the same non-zero number doesn't affect the ratio. Hence, required ratio = 4.9 : 6.3 = 7 : 9

VII. Average-Related Techniques

11. Average = average (assumed)

$$+ \frac{\text{Sum of values obtained by subtrating the assumed average from the given each value}}{\text{Number of given values}}$$

e.g. A firm earns profits of ₹ 1132, ₹ 1249, ₹ 956 and ₹ 746 in four consecutive weeks, what is its average weekly profit?

Let assumed average be ₹ 1000 (we should assume the figure nearer to expected average or a given value for our convenience, however free to assume any figure)

Then, average $= 1000 + \frac{132 + 249 - 44 - 254}{4}$

$= 1000 + 20.75 = 1020.75$

12. Application of concept of Weighted average

Suppose there are x groups having n_1, n_2, n_3, n_4, ..., n_x elements with averages a_1, a_2, a_3, a_4, ... a_x, respectively. Then,

Weighted Average

$$= \frac{n_1 a_1 + n_2 a_2 + n_3 a_3 + n_4 a_4 + \ldots\ldots + n_x a_x}{n_1 + n_2 + n_3 + n_4 + \ldots\ldots + n_x}$$

Note *If we take ratio of $n_1, n_2, n_3, n_4, \ldots\ldots, n_x$, instead of their real values, there will be no effect on the value of weighted average*

e.g. A company produces three items I, II and III in ratio 2 : 3 : 5 and earns 25%, 30% and 40% profits there on respectively. What is its profit if cost of the total production is ₹ 758 lakh ?

Percentage profit on items I, II and III together

$$= \frac{(2 \times 25\%) + (3 \times 30\%) + (5 \times 40\%)}{2 + 3 + 5} = 34\%$$

Then, total profit earned for items I, II and III together

= 34% of 758 lakh

= ₹ 257.72 lakh

It is reiterated that the techniques discussed above are critical for solving the DI questions. These techniques are fundamentals in nature. Solving DI questions more and more by these techniques will make you winner in this section. Now, observe the diverse types of DI questions.

Illustrations

A. Problems Based on Data Table

Directions (Problems 1-5) Study the table carefully to answer the questions that follow. ***[Corporation Bank (PO)-11]***

Number of girls and boys (in hundreds) in six different years in five different schools.

School →	A		B		C		D		E	
Years ↓	Boys	Girls	Boys	Girls	Boys	Girls	Boys	Girls	Boys	Girls
2005	3.3	3.6	5.2	3.1	5.5	4.5	2.4	1.4	6.5	6.6
2006	6.6	4.2	4.9	2.2	6.9	3.3	4.4	2.3	5.5	3.6
2007	9.3	6.9	4.7	4.2	5.8	4.9	6.4	3.3	2.7	2.4
2008	5.4	9.6	6.3	5.4	6.6	5.2	5.3	5.4	5.4	5.7
2009	8.4	12.9	7.5	5.9	8.7	6.6	12.1	5.2	6.8	6.5
2010	12.3	14.4	9.8	4.4	11.7	4.2	12.2	9.4	10.8	12.7

Problem 1 What is the approximate percentage decrease in the number of boys in school D in the year 2008 as compared to that in the previous year ?

(a) 17 (b)12 (c) 9 (d) 5 (e) 23

Sol. (a) Number of boys in 2008 = 5.3 and number of boys in 2007 = 6.4

Clearly, base value is 6.4 as we have to compare the figure in 2008 with the figure in 2007, i.e. by how much percent is figure in 2008 less or more than that in 2007.

Now, the required decrease percentage $= \left(\frac{6.4 - 5.3}{6.4} \times 100\right)\%$

$= \left(\frac{1.1}{6.4} \times 100\right)\% = 17.19\% \approx 17\%$

Alternatively

We may write $\frac{1.1}{6.4}$ as $\frac{1}{6}$ as approximate value. Now, $\frac{1}{6} = 16.66\%$

(Observe 'Percentage Conversion Table' as given under chapter 'Percentage' for quick answer)

Obviously, 16.66% ≈ 17%

Problem 2 The number of girls in school B in the year 2009 is approximately what percent of the total number of students (both boys and girls) in school E in the year 2006 ?

(a) 46 (b) 52 (c) 70 (d) 58 (e) 65

Sol. (e) Number of girls in school B in 2009 = 5.9

Number of students in school E in 2006 = 5.5 + 3.6 = 9.1

Then, 5.9 is what percent of 9.1?

$\therefore$ Required answer $= \left(\frac{5.9}{9.1} \times 100\right)\%$

$= 64.84\% \approx 65\%$

Alternatively

Since the question seeks approximate value, we may write $\frac{5.9}{9.1}$ as $\frac{6}{9}$ i.e. $\frac{2}{3}$.

Now, $\frac{2}{3} = 66.66\%$ (Observe 'Percentage Conversion Table' as given under chapter 'Percentage' for quick answer)

Since options are 46, 52, 70, 58 and 65, the correct answer is (e).

Problem 3 What is the average number of girls in school A in all the years taken together ?

(a) 760 (b) 800 (c) 860 (d) 600

(e) None of these

Sol. (c) Let the average be 9.6.

Then, the required average

$= \text{Assumed average} + \frac{-6 - 5.4 - 2.7 + 0 + 3.3 + 4.8}{6} = 9.6 - 1$

$= 8.6$ hundreds $= 860$

Alternatively

If you feel uncomfortable, you may write the given decimal number in actual sense i.e. 360, 420, 690, 960, 1290, 1440.

Let the average be 800.

Then, required average

$= 800 + \frac{-440 - 380 - 110 + 160 + 490 + 640}{6}$

$= 800 + 60 = 860$

Problem 4 What is the ratio of the number of boys in school C in the year 2009 to the number of girls in school A in the year 2009 ?

(a) 29 : 41 (b) 36 : 11

(c) 29 : 43 (d) 36 : 13

(e) None of these

Sol. (c) Number of boys in school C in 2009 = 8.7

Number of girls in school A in 2009 = 12.9

Then, the required ratio = 8.7 : 12.9 = 29 : 43

Problem 5 In which year is the total number of students (both girls and boys together) the third highest in school E ?

(a) 2006 (b) 2007 (c) 2008 (d) 2005

(e) 2010

Sol. (d) Total number of students in school E in 2005

= 6.5 + 6.6 = 12.1

Total number of students in school E in 2006 = 5.5 + 3.6 = 9.1

Total number of students in school E in 2007 = 2.7 + 2.4 = 5.1

Total number of students in school E in 2008 = 5.4 + 5.7

= 11.1

Total number of students in school E in 2009

= 6.8 + 6.5 = 13.3

Total number of students in school E in 2010

= 10.8 + 12.7 = 23.5

Here, the third highest figure = 12.1

Hence, the required answer = 2005

Directions (Problems 6-11) *Study the table carefully to answer the questions that follow.* ***[SBI(Associates-PO)-11]***

Number of candidates found eligible and the number of candidates short listed for interview for a recent recruitment process for six posts from different states.

Post	I		II		III		IV		V		VI	
State	E	S	E	S	E	S	E	S	E	S	E	S
A	2500	65	7200	240	5200	76	3600	200	4600	110	5400	380
B	3200	220	8500	420	8400	190	6200	320	5800	180	6200	430
C	2800	280	4500	350	7600	160	8200	440	7300	310	3700	250
D	2400	85	4800	200	2600	55	7500	350	3900	160	4800	360
E	3000	120	5600	280	3800	75	6800	280	6100	260	7800	520
F	4800	325	6400	320	4400	220	4700	180	4900	220	8800	640
G	6500	550	7000	140	6000	325	5500	220	8100	410	2700	200

[E- Eligible S- Short listed]

Problem 6 From state B, which post had the highest percentage of candidates short listed ?

(a) V (b) IV

(c) VI (d) II

(e) None of these

Sol. (c) Percentage of short listed candidates will be as under :

$\left(\frac{220}{3200} \times 100\right)\%$ ... for post I, $\left(\frac{420}{8500} \times 100\right)\%$ ...for post II,

$\left(\frac{190}{8400} \times 100\right)\%$ ... for post III, $\left(\frac{320}{6200} \times 100\right)\%$ ...for post IV,

$\left(\frac{180}{5800} \times 100\right)\%$ ... for post V and $\left(\frac{430}{6200} \times 100\right)\%$ ...for post VI.

Note *The above steps are not required to be written. We may, directly write as below for comparison*

$\frac{22}{32}, \frac{42}{85}, \frac{19}{84}, \frac{32}{62}, \frac{18}{58}, \frac{43}{62}$

$\Rightarrow$ 0.68, 0.49, 0.22, 0.51, 0.31,0.69

Hence, required answer is VI.

Alternatively

$\frac{22}{32}, \frac{42}{85}, \frac{19}{84}, \frac{32}{62}, \frac{18}{58}, \frac{43}{62}$

$\Rightarrow \frac{22}{32-22}, \frac{42}{85-42}, \frac{19}{84-19}, \frac{32}{62-32}, \frac{18}{58-18}, \frac{43}{62-43}$

$\Rightarrow \frac{22}{10}, \frac{42}{43}, \frac{19}{65}, \frac{32}{30}, \frac{18}{40}, \frac{43}{19}$ (please refer to technique meant for 'Comparison of Fractions')

$\Rightarrow$ 2.2, less than 1, less than 1, less than 1, 2.26

Hence, correct option is (c).

Problem 7 What is the average number of candidates (approximately) found eligible for post III from all states?

(a) 6700 (b) 6200
(c) 4200 (d) 4500
(e) 5500

Sol. (e) Required average = 5000 (assumed)

$$+\frac{200+3400+2600-2400-1200-600+1000}{7}$$

$$=5000+\frac{3000}{7}\approx 5500$$

Problem 8 What is the overall percentage (rounded off to one digit after decimal) of candidates short listed over the total number of candidates eligible for post I from all the states together?

(a) 9.5% (b)12.5%
(c) 37.2% (d) 6.5%
(e) None of these

Sol. (d) Total short listed candidates

$= 65 + 220 + 280 + 85 + 120 + 325 + 550$
$= 1645$

Total eligible candidates

$= 2500 + 3200 + 2800 + 2400 + 3000 + 4800 + 6500$
$= 25200$

Now, the question is '1645 is what percent of 25200 ?'

$\because$ 10% of 25200 = 2520
$\therefore$ 5% of 25200 = 1260
$\therefore$ 1% of 25200 = 252
i.e. 6% of 25200 = 1260 + 252 = 1512

Further, 0.5% of 25200 = 126

Hence, 6.5% of 25200 = 1512 + 126
= 1638 ≈ 1645

Hence, option (d) is correct.

Alternatively

Required percentage $=\left(\frac{1645}{25200}\times 100\right)\%$

$$=\left(\frac{1645\times 4}{25200\times 4}\times 100\right)\%$$

$$=\left(\frac{6580}{100000\text{ (approximately)}}\times 100\right)\%$$

$\approx 6.5\%$

Problem 9 What is the ratio of the total number of candidates short listed for all the posts together from state E to that from state G ?

(a) 307:309 (b) 73: 79
(c) 6:5 (d) 9:7
(e) None of these

Sol. (a) Required ratio = (120 + 280 + 75 + 280 + 260 + 520) : (550 + 140 +325 + 220 + 410 + 200) = 307 : 369

Problem 10 The total number of candidates found eligible for post I from all states together is approximately what percent of the total number of candidates found eligible for post VI from all states together ?

(a) 45% (b) 50% (c) 60% (d) 55%
(e) 65%

Sol. (e) Total number of candidates found eligible for post I = 25200

Total number of candidates found eligible for post VI = 39400

Then, required percentage

$$=\left(\frac{25200}{39400}\times 100\right)\%$$

$$\approx\left(\frac{25000}{40000}\times 100\right)\%\text{ i.e. }62.5\%$$

Since we have increased the denominator and decreased the numerator, the actual percentage will be greater than 62.5%. Hence, 65% (approximately) as options suggest.

Problem 11 Which state had the lowest percentage of candidates shortlisted with respect to candidate eligible for post IV ?

(a) G (b) F (c) E (d) C
(e) None of these

Sol. (b) Percentage of short listed candidates will be as under :

$\left(\frac{200}{3600}\times 100\right)\%$ ... for state A, $\left(\frac{320}{6200}\times 100\right)\%$ ...for state B,

$\left(\frac{440}{8200}\times 100\right)\%$ ... for state C, $\left(\frac{350}{7500}\times 100\right)\%$ ...for state D,

$\left(\frac{280}{6800}\times 100\right)\%$...for state E, $\left(\frac{180}{4700}\times 100\right)\%$...for state F,

and $\left(\frac{220}{5500}\times 100\right)\%$..... for state G.

Note *The above steps are not required to be written. We may, directly write as below for comparison)*

$$\frac{200}{36},\frac{320}{62},\frac{440}{82},\frac{350}{75},\frac{280}{68},\frac{180}{47},\frac{220}{55}$$

$$\Rightarrow \frac{20}{36},\frac{32}{62},\frac{44}{82},\frac{35}{75},\frac{28}{68},\frac{18}{47},\frac{22}{55}$$

$$\Rightarrow \frac{20}{16},\frac{32}{30},\frac{44}{38},\frac{35}{40},\frac{28}{40},\frac{18}{29},\frac{22}{33}$$

(Please refer to technique meant for 'Comparison of Fractions')

$\Rightarrow$ more than 1, more than 1, more than 1, $\frac{28}{40},\frac{18}{29},\frac{22}{33}$

Since we require lowest value, we have to compare only $\frac{28}{40},\frac{18}{29},\frac{22}{33}$ as each of these three has value lower than 1.

$$\Rightarrow \frac{28}{40},\frac{18}{29},\frac{22}{33}$$

$$\Rightarrow \frac{28}{12},\frac{18}{11},\frac{22}{11}$$

$\Rightarrow$ more than 2, between 1 and 2, 2.

Hence, state F has lowest percentage.

Directions (Problems 12-15) *Study the table given below to answer these questions.* **[CAT-16]**

The following table shows the profits (in ₹ lakh) earned by 5 shops A, B, C, D and E during 2007-2012.

Shops	Years					
	2007	2008	2009	2010	2011	2012
A	4	6	8	7	5	10
B	3	5	6	4	7	8
C	9	11	8	7	10	12
D	5	4	6	3	7	3
E	6	7	5	5	4	8

Income = Profit + Investment

Problem 12 By what percent is the profit earned by A in 2012 more than that in 2008 ?

(a) 40% (b) 66.6% (c) 266% (d) 160%

Sol. (b) Required percentage $= \left(\frac{10-6}{6} \times 100\right)\%$

$= 66.6\%$

Problem 13 Consider the percentage change in profit for A, B, D and E for the pair of years 2007 and 2008, 2007 and 2008, 2007 and 2011 and 2007 and 2011, respectively. Which shop has the maximum percentage change in profits for the years mentioned ?

(a) A (b) B (c)D (d)E

Sol. (b) Percentage change in profit for A and B for the pair of years 2007 and 2008

$= \left(\frac{6-4}{4} \times 100\right)\%$ and $\left(\frac{5-3}{3} \times 100\right)\%$

= 50% and 66.6% (respectively.)

Percentage change in profit for D and E for the pair of years 2007 and 2011 $= \left(\frac{7-5}{5} \times 100\right)\%$ and $\left(\frac{4-6}{6} \times 100\right)\%$

= 40% and – 33.3%, respectively.

Clearly, Shop B has maximum percentage change in profits.

Problem 14 What should be the income of D in 2009 so that its profit percent is 30%?

(a) ₹ 14 lakh (b) ₹ 20 lakh
(c) ₹ 26 lakh (d) ₹ 11 lakh

Sol. (c) Clearly, 30% of the investment = 6 lakh

∴ 100% of the investment = 20 lakh

Then, income = profit + investment
= 6 + 20 = 26 lakh

Alternatively,

Recall the concept of computation of amount if the interest and the principal are given. Here, income, investment and profit may be termed as amount, principal and interest respectively.

∵ 30% of investment = 6 lakh

∴ 130% of investment = 26 lakh

Hence, income = 26 lakh

Problem 15 To earn a profit of 20% by each of C and E, what is the ratio of investment made by them in 2008 ?

(a) 7 : 11 (b) 1: 5
(c) 5 : 1 (d) 11 : 7

Sol. (d) In case of C, 20% of the investment = 11 lakh

∴ 100% of the investment = 55 lakh

In case of E, 20% of the investment = 7 lakh

∴ 100% of the Investment = 35 lakh

Then, required ratio = 55 : 35 = 11 : 7

Alternatively

In both cases, the investments will be 5 times the given profits. Then, the ratio of investments will be same as ratio of profits as the multiplication or division of each term of a ratio by the same non zero-number does not affect the ratio.

Directions (Problems 16-19) *Study the table given below to answer these questions.*

The following table shows the quantities of minerals exported from India to six different countries A, B, C, D, E and F (in 000 tonne) in the year 2010.

Country → Item↓	A	B	C	D	E	F
Iron ore	18	12	25	11	32	5
Coal	12	38	22	6	29	42
Silver	14	18	26	3	48	23
Copper	23	12	20	2	2	16
Gold	2	4	8	1	3	10

Problem 16 If the second largest importing country increases its off-take of all the minerals, taken together, from India and is at par with the largest importer, its share of the total exports of minerals from India would be how much percentage ?

(a) 23.4% (b) 66.6%
(c) 26.6% (d) 16.2%

Sol. (a) Quantity imported by country
A = 18 + 12 + 14 + 23 + 2 = 69

Quantity imported by country
B = 12 + 38 + 18 + 12 + 4 = 84

Quantity imported by country
C = 25 + 22 + 26 + 20 + 8 = 101

Quantity imported by country
D = 11 + 6 + 3 + 2 + 1 = 23

Quantity imported by country
E = 32 + 29 + 48 + 2 + 3 = 114

Quantity imported by country
F = 5 + 42 + 23 + 16 + 10 = 96

Clearly, second largest importing country, i.e. C is at par with country E when it increases its off-take.

Then, required percentage

$= \left(\frac{114}{69 + 84 + 101 + 23 + 114 + 96} \times 100\right)\%$

$$= \left(\frac{114}{487} \times 100\right)\% = \left(\frac{114 \times 2}{487 \times 2} \times 100\right)\%$$

$$\approx \left(\frac{228}{1000} \times 100\right)\% \text{ i.e. } 22.8\%$$

Hence, option (a) is correct.

Problem 17 If India were to ban the export of gold altogether, by what approximat percentage should the export of the rest of the minerals increase so that the total quantity of minerals exported remains the same ?

(a) 4% (b) 6.1% (c)2.6% (d) 1.6%

Sol. (b) Quantity of exports excluding gold = 487 (from above) − (2 + 4 + 8 + 1 + 3 + 10)

= 487 − 28 = 459

Then, percentage increase in the export of the remaining minerals

$$= \left(\frac{28}{459} \times 100\right)\% = \left(\frac{28 \times 2}{459 \times 2} \times 100\right)\%$$

$$\approx \left(\frac{56}{900} \times 100\right)\% \text{ i.e. } 6.2\%$$

Hence, option (b) is correct.

Problem 18 18% of country E's total requirement of iron ore is being met by imports from India. Next year, its total requirement of iron ore is expected to fall by 22% and only 12% of the revised requirement will be imported from India. Assuming all other exports to country E by India remain at the same levels, what would be its total percentage drop in quantity of exports to E ?

(a) 24% (b) 13.4% (c) 26.6% (d) 19%

Sol. (b) In 2011, the total requirement of iron ore = 78% of the requirement of iron ore in year 2010.

Quantity of iron ore to be imported from India in 2011 = 12% of 78% of requirement of iron ore in year 2010

= 9.36% of the requirement of iron ore in year 2010

Then, decrease in imports from India = 18% − 9.36% = 8.64%

$\because$ 18% = 32 (as given in table)

$\therefore$ $8.64\% = \frac{32}{18} \times 8.64$

Now, total percentage drop in exports to E

$$= \left(\frac{\frac{32}{18} \times 8.64}{32 + 29 + 48 + 2 + 3} \times 100\right)\% = 13.4\%$$

Problem 19 If international demand for coal from India rises by 12%, that of iron ore by 8% and copper by 15%, by how much percent will total exports from India will rise, assuming export of gold and silver remain at the same level ?

(a) 4.8% (b) 10.6% (c) 7.6% (d) 2.4%

Sol. (c) In 2010, quantity of coal exported by India

= 12 + 38 + 22 + 6 + 29 + 42 = 149

Iron ore exported by India = 18 + 12 + 25 + 11 + 32 + 5 = 103

Copper exported by India = 23 + 12 + 20 + 2 + 2 + 16 = 75

Then, overall rise in percentage

$$= \left(\frac{12\% \text{ of } 149 + 8\% \text{ of } 103 + 15\% \text{ of } 75}{487} \times 100\right)\%$$

$$= \left(\frac{17.88 + 8.24 + 11.25}{487} \times 100\right)\% = \left(\frac{37.37}{487} \times 100\right)\%$$

$$= \left(\frac{37.37 \times 2}{487 \times 2} \times 100\right)\% \approx \left(\frac{74.74}{1000} \times 100\right)\%,$$

i.e. 7.4%

Hence, option (c) is correct.

Directions (Problems 20-23) *Answer the following questions based on the information given below* **[CAT-07]**

The following table shows the break-up of actual costs incurred by a company in last five years (year 2002 to year 2006) to produce a particular product.

Year	2002	2003	2004	2005	2006
Volume of production and sale (units)	1000	900	1100	1200	1200
					Costs (₹)
Material	50000	45100	55200	59900	60000
Labour	20000	18000	22100	24150	24000
Consumables	2000	2200	1800	1600	1400
Rent of building	1000	1000	1100	1100	1200
Rates and taxes	400	400	400	400	400
Repair and maintenance expenses	800	820	780	790	800
Operating cost of machines	30000	27000	33500	36020	36000
Selling and marketing expenses	5750	5800	5800	5750	5800

The production capacity of the company is 2000 units. The selling price for the year 2006 was ₹ 125 per unit. Some costs change almost in direct proportion to the change in volume of production, while others do not follow any obvious pattern of change with respect to the volume of production and hence are considered fixed. Using the information provided for the year 2006 as the basis for projecting the figures for the year 2007, answer the following questions.

Problem 20 What is the approximate cost per unit in rupees, if the company produces and sells 1400 units in the year 2007 ?

(a) ₹ 104 (b) ₹ 107
(c) ₹ 110 (d) ₹ 115
(e) ₹ 116

Sol. (b) **Variable cost** for production of 1200 units in year 2006 = Material cost + Labour cost + Operating cost of machines

= 60000 + 24000 + 36000 = 120000

Hence, per unit variable cost = ₹ 100

Fixed cost in year 2006 for production and sale upto 2000 units
= Consumables + Rent of building
+ Rates and taxes + Repair and maintenance expenses
+ Selling and marketing expenses = 1400 + 1200
+ 400 + 800 + 5800 = 9600

Now, cost per unit in 2007
= variable cost per unit in 2006 + (fixed cost ÷ 1400)
= ₹ 100 + ₹ (9600 ÷ 1400) = ₹ 106.86 ≈ ₹ 107

Hence, option (b) is correct.

Problem 21 What is the minimum number of units that the company needs to produce and sell to avoid any loss?

(a) 313 (b) 350 (c) 384 (d) 747
(e) 928

Sol. (c) Let minimum number of units to produced be x.

Then, cost of x units = $(100 \times x) + 9600$

and selling price (SP) of x units = $125x$

(∵ SP per unit was ₹ 125 in 2006)

As per question, $(100 \times x) + 9600 = 125x$

$\Rightarrow \quad x = 384$

Problem 22 If the company reduces the price by 5%, it can produce and sell as many units as it desires. How many units the company should produce to maximise its profit?

(a) 1400 (b) 1600 (c) 1800 (d) 1900
(e) 2000

Sol. (e) Since maximum production capacity of the company is 2000 units, the fixed cost per unit will be minimum at level of production of 2000 units. Further, (selling price – variable cost) per unit will be same at all levels of productions. It implies that the profit per unit is maximum at lowest fixed cost. Hence, the company should produce maximum number of units as per production capacity i.e. 2000 units to maximise its profit.

Problem 23 Given that the company cannot sell more than 1700 units, and it will have to reduce the price by ₹ 5 for all units, if it wants to sell more than 1400 units, what is the maximum profit (in rupees), that the company can earn?

(a) 25400 (b) 24400
(c) 31400 (d) 32900
(e) 32000

Sol. (a) Profit in case of sale of 1400 units

$= (1400 \times 125) - [(1400 \times 100) + 9600)]$

[∵ variable cost per unit is ₹ 100 as computed earlier]

= ₹ 25400

Profit in case of sale of $(1400 + x)$ units

$= [(1400 + x) \times 120] - [(1400 + x) \times 100) + 9600]$

$= 18400 + 20x$

The maximum value of $x = 1700 - 1400 = 300$

Then, profit at $x = 300$ will be $18400 + (20 \times 300)$ i.e. 24400.

Hence, the company can earn the maximum profit of ₹ 25400 by selling 1400 units.

Hence, option (a) is correct.

Directions (Problems 24-26) *Answer the following questions based on the information given below*

The proportion of male students and the proportion of vegetarian students in a school are given below. The school has a total of 800 students, 80% of whom are in the Secondary Section and rest equally divided between class 11 and class 12.

	Male(M)	Vegetarian (V)
Class 12	0.60	
Class 11	0.55	0.50
Secondary Section		0.55
Total	0.475	0.53

Problem 24 What is the percentage of vegetarian students in class 12?

(a) 40 (b) 45 (c) 50 (d) 55
(e) 60

Sol. (a) The following table can be formed with the help of given data

	Male (M)	Female (F)	Vegetarian (V)	Non-vegetarian (NV)	Total
Class 12	80 × 0.6 = 48 ④	80 − 48 = 32 ⑤	424 − (40 + 352) = 32 ⑰	80 − 32 = 48	80 ③
Class 11	80 × 0.55 = 44 ⑥	80 − 44 = 36 ⑦	80 × 0.5 = 40 ⑧	80 − 40 = 40 ⑨	(800 − 640) ÷ 2 = 80 ②
Secondary Section	380 − (48 + 44) = 288 ⑪	640 − 288 = 352 ⑫	640 × 0.55 = 352 ⑬	640 − 352 = 288 ⑭	800 × 0.8 = 640 ①
Total	800 × 0.475 = 380 ⑩	800 − 380 = 420 ⑮	800 × .53 = 424 ⑯	800 − 424 = 376 ⑱	800

Numbers in dark circle indicate the steps in order starting with ①

Now, percentage of vegetarian students in class 12

$= \left(\frac{32}{80} \times 100\right)\% = 40\%$

Problem 25 In class 12, 25% of the vegetarians are male. What is the difference between the number of female vegetarians and male non-vegetarians?

(a) less than 8 (b) 10
(c) 12 (d) 14
(e) 16

Sol. (e) Number of male vegetarians in class 12 = 25% of total number of vegetarians = 25% of 32 = 8

Then, number of male non-vegetarians

= 48 − 8 = 40 and number of female vegetarians

= 75% of 32 = 24

Now, required difference = 40 − 24 = 16

Problem 26 What is the percentage of male students in the secondary section?

(a) 40 (b) 45 (c) 50 (d) 55
(e) 60

Sol. (b) Required percentage $= \left(\frac{288}{640} \times 100\right)\% = 45\%$

Directions (Problems 27-30) : *Refer to the following tables A, B and C to answer the questions given below.*

Table-A Manpower for a building project in Narhi

Project Phase	Managers	Engineers	Skilled workers	Unskilled workers	Super-visors
Design	1	3	-	-	-
Foundation	1	2	5	23	2
Construction	1	2	12	43	4
Finishing (Internal)	1	3	15	31	6
Painting	1	0	5	23	3

Table-B Cost per month for select category of workers (in ₹)

Manpower	Cost (In ₹)
Managers	10000
Engineers	8000
Supervisors	5000
Skilled workers	3000
Unskilled workers	1800

Table - C Estimated time and the actual time required for each phase of the projects (in months)

	Design	Foundation	Construc-tion	Finishing (Internal)	Painting
Actual	5	3	8	4	2
Estimated	3	2	6	3	2

Problem 27 What was the phase in which there was the highest absolute increase between the estimated and actual cost ?

(a) Design (b) Foundation
(c) Construction (d) Cannot be determined

Sol. (c) Perusal of Table-A reveals that comparatively greater number of manpower is required in each of foundation, construction and finishing (internal) phases. The difference between actual time and estimated time in foundation phase is one month, that in construction phase is 2 months and that in finishing (internal) is one month.

No doubt, manpower costs for 2 months in construction phase > manpower costs for one month in design phase or manpower costs for one month in finishing (internal) phase.

Hence, in construction phase there was the highest absolute increase between the estimated and actual cost.

Note *No need to compute difference between actual cost and estimated cost phase-wise)*

Problem 28 For which phase of the project was there the highest percentage increase in cost (as compared to the estimated cost) ?

(a) Design (b) Foundation
(c) Construction (d) Cannot be determined

Sol. (a) **For design phase**, percentage increase in time

$$= \left(\frac{5-3}{3} \times 100\right)\% = 66.66\%$$

Since time is directly proportional to the cost, percentage increase in cost = 66.66%

For foundation phase, percentage increase in cost

$$= \left(\frac{3-2}{2} \times 100\right)\% = 50\%$$

For construction phase, percentage increase in cost

$$= \left(\frac{8-6}{6} \times 100\right)\% = 33.33\%$$

Hence, option (a) is correct.

Problem 29 What is the value of the percentage overflow in costs during construction ? (approximately)

(a) 33.33% (b) 50%
(c) 66.66% (d) None of these

Sol. (a) On the above concept, percentage overflow in costs during construction $= \left(\frac{8-6}{6} \times 100\right)\% = 33.33\%$

Problem 30 If the internal finishing and painting can be started simultaneously at the end of construction, then what is the overflow of the project as a percentage of the estimated total time of completion of the project?

(a) 33.33% (b) 27.27%
(c) 42.85% (d) Cannot be determined

Sol. (c) If the internal finishing and painting can be started simultaneously at the end of construction, then total time taken in completion of the project = 5 + 3 + 8 + 4 = 20 months

(∵ both the internal finishing and painting will be done in next 4 days after end of construction)

Estimated time for completion of the project = 3 + 2 + 6 + 3 = 14 months

Then, percentage overflow in time

$$= \left(\frac{20-14}{14} \times 100\right)\% = 42.85\%$$

Directions (Problems 31-34) *There are only four brands of entry level smartphones called Azra, Bysi, Cxqi and Dipq in a country. Details about their market share, unit selling price and profitability (defined as the profit as a percentage of the revenue) for the year 2016 are given in the table below.*

Brand	Market share (%)	Unit selling price (₹)	Profitability (%)
Azra	40	15000	10
Bysi	25	20000	30
Cxqi	15	30000	40
Dipq	20	25000	30

In 2017, sales volume of entry level smartphones grew by 40% as compared to that in 2016. Cxqi offered a 40% discount on its unit selling price in 2017, which resulted in a 15% increase in its market share. Each of the other three brands lost 5% market share. However, the profitability of

Cxqi came down to half of its value in 2016. The unit selling prices of the other three brands and their profitability values remained the same in 2017 as they were in 2016. *[CAT-18]*

Problem 31 The brand that had the highest revenue in 2016 is :

(a) Dipq (b) Bysi (c) Cxqi (d) Azra

Problem 32 The brand that had the highest profit in 2016 is :

(a) Dipq (b) Bysi (c) Cxqi (d) Azra

Problem 33 The brand that had the highest profit in 2017 is :

(a) Dipq (b) Bysi (c) Cxqi (d) Azra

Problem 34 The complete list of brands whose profits went up in 2017 from 2016 is :

(a) Azra, Bysi, Cxqi (b) Bysi, Cxqi, Dipq
(c) Cxqi, Azra, Dipq (d) Azra, Bysi, Dipq

The given data may be summerised as below :

Brand	Market share (%)		Unit selling price (₹)		Profitability (%)	
	2016	2017	2016	2017	2016	2017
Azra	40	35	15k	15k	10	10
Bysi	25	20	20k	20k	30	30
Cxqi	15	30	30k	18k	40	20
Dipq	20	15	25k	25k	30	30

Sol.

31. (d) Since we have to compare only in term of revenue, no need to calculate exactly. We may compare with the help of concept of ratio.

The ratio of revenues from Azra, Bysi, Cxqi and Dipq in 2016

$= 40 \times 15 : 25 \times 20 : 15 \times 30 : 20 \times 25$

$= 12 : 10 : 9 : 10$

Hence, brand Azra had the highest revenue in year 2016.

32. (c) Ratio of profits of Azra, Bysi, Cxqi and Dipq in 2016

$= 12 \times 1 : 10 \times 3 : 9 \times 4 : 10 \times 3$

(as profitability ratio is 1 : 3 : 4 : 3)

Clearly, Cxqi had the highest profit in year 2016.

33. (b) Ratio of profits of Azra, Bysi, Cxqi and Dipq in 2017

$= 35 \times 15 \times 1 : 20 \times 20 \times 3 : 30 \times 18 \times 2 : 15 \times 25 \times 3$

$= 105 : 240 : 216 : 225 = 35 : 80 : 72 : 75$

Hence, Bysi had the highest profit in 2017.

Note *No need to utilise the figures related to increase in sales volume as it will be cancelled out while computing ratio)*

34. (d) Recall the concept associated with 'effect on product in case factor(s) is/are altered' given in chapter 'Percentage'.

In case of Azra, the market share is decreased from 40 to 35 i.e. by 12.5%. Then, effect on revenue if market share is decreased by 12.5% and sales volume is increased by 40%

$= \left(-12.5 + 40 - \frac{12.5 \times 40}{100}\right)\%$

$= +22.5\%$

It implies that the profit will increase by 22.5% if the profitability percentage remains unchanged. It is also to note that we need not to find the exact effect, only realisation of positive or negative outcome will suffice the requirement of the question.

In case of Bysi, the market share is decreased by 20% and sales volume is increased by 40%. Then, effect on the revenue or profit $= \left(-20 + 40 - \frac{20 \times 40}{100}\right)\% = +12\%$ (no need to find exact value, however done for your clarity) if profitability percentage remains unchanged.

In case of Cxqi, the market share is increased by 100% and sales volume is increased by 40%. Then, effect on the revenue $= \left(+100 + 40 + \frac{100 \times 40}{100}\right)\% = +180\%$ provided the selling price remains unchanged.

If selling price is decreased by 40%, effect on revenue $= \left(+180 - 40 - \frac{180 \times 40}{100}\right)\% = 68\%$ if profitability remains unchanged. Further, if profitability is decreased by 50% then effect on the profit $= \left(+68 - 50 - \frac{68 \times 50}{100}\right)\% = -16\%$

Hence, the profit of brand Cxqi had gone down. Thus, option (d) is correct.

Remark

conventional method has deliberately been left. The whole process can be done mentally in moments without pen and paper. Thus, the aspirant should attempt such questions on the above line}

B. Problems Based on Bar Graphs (Simple bar/Multiple/Subdivided)

Directions (Problems 35-38) Answer the following questions based on the information given below.

The bar chart below shows the revenue received, in million US Dollars (USD), from subscribers to a particular Internet Service. The data covers the period 2003 to 2007 for the United States (US) and Europe. The bar chart also shows the estimated revenues from subscription to this service for the period 2008 to 2010.

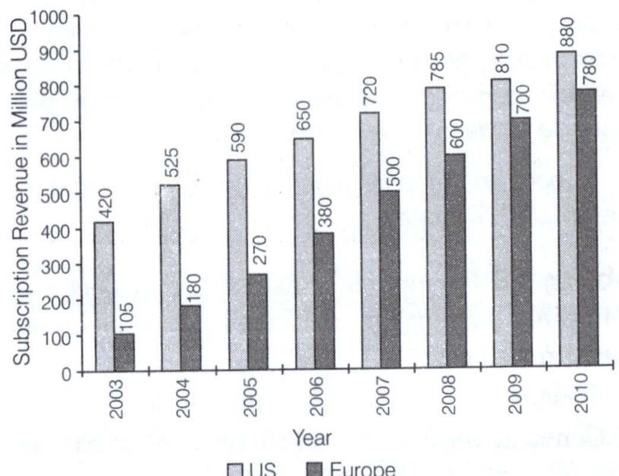

Problem 35 The difference between the estimated subscription in Europe in 2008 and what it would have been if it were computed using the percentage growth rate of 2007 (over 2006), is closest to :

(a) 50 (b) 80 (c) 20 (d) 10
(e) Zero

Sol. (a) Estimated subscription in 2008 = 600

Growth rate of 2007 over 2006 $= \left(\frac{500-380}{380} \times 100\right)\% \approx 31.6\%$

Then, computed subscription in 2008 using the aforesaid growth rate = 131.6% of 500 = 658

Hence, required difference = 658 − 600 = 58

Hence, option (a) is correct.

Problem 36 In 2003, 60% of subscribers in Europe were men. Given that women subscribers increase at the rate of 10% per annum and men at the rate of 5% per annum, what is the approximate percentage growth of subscribers between 2003 and 2010 in Europe ? The subscription prices are volatile and may change each year.

(a) 62 (b) 15 (c) 78 (d) 84
(e) 50

Sol. (a) Let total subscribers in 2003 in Europe be 100.

∴ Number of men subscriber = 60 and that of women subscribers = 40

Now, we have to find the percentage increase in number of men subscribers over 7 yrs.

Percentage increase in number of men subscribers over 4 yrs.

$= (5 \times 4). \boxed{6 \cdot 5^2}\boxed{4 \cdot 5^3}\boxed{5^4}\% = 21.550625\%$

(Recall the concept of computation of compound interest over 4 yrs)

Percentage increase over 5 yrs $= 21.5 + 5 + \frac{21.5 \times 5}{100} \approx 27.5\%$

Percentage increase over 6 yrs $= 27.5 + 5 + \frac{27.5 \times 5}{100} \approx 33.8\%$

Percentage increase over 7 yrs $= 33.8 + 5 + \frac{33.8 \times 5}{100} \approx 40.5\%$

∴ Number of men subscribers in 2010 = 140.5% of 60 ≈ 84

Similarly, percentage increase in number of women subscribers over 7 yrs ≈ 95%

∴ Number of women subscribers in 2010 = 195% of 40 = 78

Hence, number of subscribers in 2010 = 84 + 78 = 162, whereas that in 2003 was 100 as assumed.

∴ Percentage growth in subscribers between 2003 and 2010 = 62%(approximately)

Problem 37 Consider the annual percent change in the gap between subscription revenues in the US and Europe. What is the year in which the absolute value of this change is the highest ?

(a) 03-04 (b) 05-06
(c) 06-07 (d) 08-09
(e) 09-10

Sol. (d) Gap in subscription revenues in US and Europe in 2003
= 420 − 105 = 315

Gap in subscription revenues in US and Europe in 2004 = 345
Gap in subscription revenues in US and Europe in 2005 = 320
Gap in subscription revenues in US and Europe in 2006 = 270
Gap in subscription revenues in US and Europe in 2007 = 220
Gap in subscription revenues in US and Europe in 2008 = 185
Gap in subscription revenues in US and Europe in 2009 = 110
Gap in subscription revenues in US and Europe in 2010 = 100

Now, absolute percentage change in 2003-04

$= \left|\left(\frac{345-315}{315} \times 100\right)\%\right| \approx 9.5\%$

Absolute percentage change in 2005-06

$= \left|\left(\frac{270-320}{320} \times 100\right)\%\right| \approx 15.6\%$

Now, percentage change in 2006-07

$= \left|\left(\frac{220-270}{270} \times 100\right)\%\right| \approx 18.5\%$

Absolute percentage change in 2008-09

$= \left|\left(\frac{110-185}{185} \times 100\right)\%\right| \approx 40.5\%$

Absolute percentage change in 2009-10

$= \left|\left(\frac{100-110}{110} \times 100\right)\%\right| \approx 9\%$

Hence, correct option is (d).

Alternatively,

We can compare $\frac{30}{315}, \frac{50}{320}, \frac{50}{270}, \frac{75}{185}$ and $\frac{10}{110}$

$\Rightarrow \frac{2}{21}, \frac{5}{32}, \frac{5}{27}, \frac{15}{37}$ and $\frac{1}{11}$

$\Rightarrow \frac{2}{17}, \frac{5}{22}, \frac{5}{17}, \frac{15}{7}$ and $\frac{1}{9}$ (Recall the technique for comparison of fraction, we have subtracted 2 times the numerators from the respective denominators.)

⇒ less than 1, less than 1, less than 1, more than 2 and less than 1

Hence, option (d) is correct.

Problem 38 While the subscription in Europe has been growing steadily towards that of US, the growth rate in Europe seems to be declining. Which of the following is closest to the percent change in growth rate of 2007 (over 2006) relative to the growth rate of 2005 (over 2004) ?

(a) 17 (b) 20 (c) 35 (d) 60
(e) 100

Sol. (c) Growth rate of 2007 over 2006

$= \left(\frac{500-380}{380} \times 100\right)\% = 31.5\%$

Growth rate of 2005 over 2004

$= \left(\frac{270-180}{180} \times 100\right)\% = 50\%$

Then, percent change in growth rate of 2007(over 2006) relative to the growth rate of 2005 (over 2004)

$= \left(\frac{31.5-50}{50} \times 100\right)\% = 37\% \approx 35\%$

Hence, option (c) is correct.

Directions (Problems 39 - 43) *Study the graphs which show the seats won and the percentage of valid votes polled for different political parties in Gujrat over the years.* ***[MAT-06]***

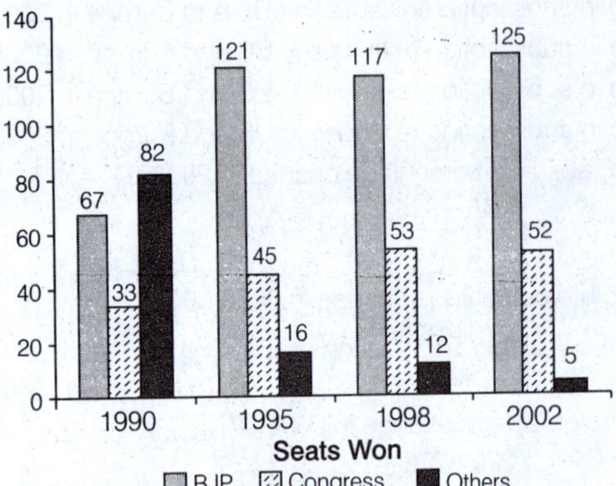

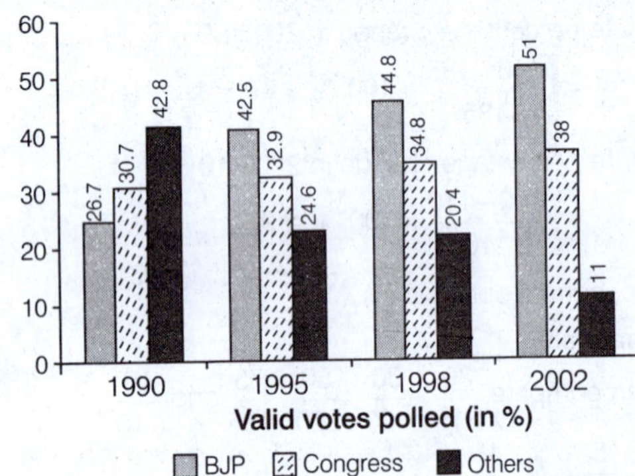

Problem 39 If the total number of valid votes in 2002 Gujrat Election was 5 crore, then find the average number of votes for winning one seat for other political parties.

(a) 11 lakh
(b) 1.10 lakh
(c) 1.10 crore
(d) Data is inadequate

Sol. (d) We cannot find the average number of votes for winning one seat for other political parties under given data as there is no relation between the valid votes polled and number of seats won. Hence, option (d) is correct.

Problem 40 In which of the following years was the number of seats won by BJP maximum with respect to the previous given year ?

(a) 1998 (b) 1995
(c) 2002 (d) 1995 and 2002

Sol. (b) Percentage increase in seats won in 1995 (over 1990)

$$= \left(\frac{121-67}{67} \times 100\right)\% \approx 80\%$$

Percentage increase in seats won in 1998 (over 1995)

= negative figure

Percentage increase in seats won in 2002 (over 1998)

$$= \left(\frac{125-117}{117} \times 100\right)\% \approx 7\%$$

Hence, option (b) is correct.

Alternatively,

No need to calculate like above, we may arrive at answer by only observing the graph.

Problem 41 In 1998, if 2.24 crore people votes were valid for BJP whereas in 1990 there were 1.228 crore people votes valid for Congress, then by what percent was the number of valid votes less in 1990 with respect to that in 1998 ?

(a) 20% (b) 30%
(c) 24% (d) 25%

Sol. (a) Clearly, 44.8% of the total valid votes polled in 1998 = 2.24 crore

$\Rightarrow$ 100% of the total valid votes polled $= \frac{2.24}{44.8} \times 100 = 5$ crore

Similarly, 30.7% of the total valid votes polled in 1990 = 1.228 crore.

$\Rightarrow$ 100% of the total valid votes polled

$$\frac{1.228}{30.7} \times 100 = 4 \text{ crore}$$

Obviously, 4 crore is 20% less than 5 crore. Hence, option (a) is correct.

Alternatively

As given, 44.8% of the total valid votes polled in 1998 = 2.24 crore and 30.7% of the total valid votes polled in 1990 = 1.228 crore i.e. 55% (approximately) of total valid votes polled in 1990 = 2.2 crore (It can be done mentally without paper and pen (1.2 $\rightarrow$ 30, then 2.2 $\rightarrow$ 55)). Then, ratio of total votes polled in 1998 to that in 1990 = 55 : 44

Clearly, 44 is 20% less than 55.

Problem 42 In which of the following years did the BJP secure more than $66\frac{2}{3}\%$ of the total seats?

(a) 1990 (b) 1995
(c) 1998 (d) 2002

Sol. (d) $66\frac{2}{3}\%$ of total seats = two-third of 182

$$= 121.33 \approx 121$$

Clearly, in year 2002, BJP secured more than 121 seats, i.e. 125.

Hence, option (d) is correct.

Problem 43 In which of the following years was the difference in the number of valid votes for any two political parties maximum ?

(a) 1990 (b) 1995
(c) 1998 (d) Cannot be determined

Sol. (d) In a particular year, we are able to find the difference between valid votes for any two political parties in term of percentage of total valid votes. However, we cannot determine the difference in number as number of total valid votes has not been provided.

Directions (Problems 44 - 47) *Answer the questions based on the data presented in the figure below :* **[CAT-2000]**

FEI for a country in a year, is the ratio (expressed as a percentage) of its foreign equity inflows to its GDP. The figure given displays the FEIs for select asian countries for 1997 and 1998.

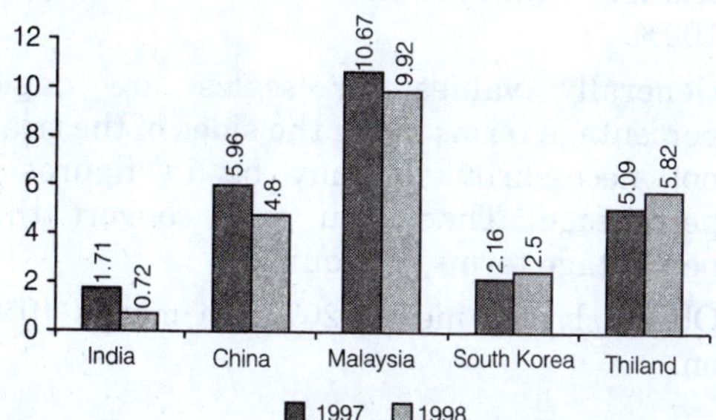

Problem 44 **In country with the largest change in FEI in 1998 relative to its FEI in 1997, is :**

(a) India (b) China
(c) Malaysia (d) Thailand

Sol. (a) Percentage change in FEI of India in 1998 relative to 1997

$$= \left(\frac{0.72 - 1.71}{1.71} \times 100\right)\% \approx -58\%$$

Percentage change in FEI of China in 1998 relative to 1997

$$= \left(\frac{4.8 - 5.96}{5.96} \times 100\right)\% \approx -20\%$$

Percentage change in FEI of Malaysia in 1998 relative to 1997

$$= \left(\frac{9.92 - 10.67}{10.67} \times 100\right)\% \approx -7\%$$

Percentage change in FEI of South Korea in 1998 relative to 1997

$$= \left(\frac{2.5 - 2.16}{2.16} \times 100\right)\% \approx 15\%$$

Percentage change in FEI of Thailand in 1998 relative to 1997

$$= \left(\frac{5.82 - 5.09}{5.09} \times 100\right)\% \approx 14\%$$

Clearly, the largest change appears in India. Hence, option (a) is correct.

Alternatively

$$\text{Change in FEI in India} = \frac{0.72 - 1.71}{1.71} = \frac{-0.99}{1.71}$$

$$\text{Change in FEI in China} = \frac{4.8 - 5.96}{5.96} = \frac{-1.16}{5.96}$$

$$\text{Change in FEI in Malaysia} = \frac{9.92 - 10.67}{10.67} = \frac{-0.75}{10.67}$$

$$\text{Change in FEI in South Korea} = \frac{2.5 - 2.16}{2.16} = \frac{0.34}{2.16}$$

$$\text{Change in FEI in Thailand} = \frac{5.82 - 5.09}{5.09} = \frac{0.73}{5.09}$$

Now, compare (taking absolute values)

$$\frac{0.99}{1.71}, \frac{1.16}{5.96}, \frac{0.75}{10.67}, \frac{0.34}{2.16}, \frac{0.73}{5.09}$$

$$\Rightarrow \quad \frac{99}{72}, \frac{116}{480}, \frac{75}{982}, \frac{34}{182}, \frac{73}{436}$$

(Please refer to technique meant for 'Comparison of Fractions') more than 1, less than 1, less than 1, less than 1, less than 1

Clearly, the largest change appears in India. Hence, option (a) is correct.

Problem 45 **Based on the data provided, it can be concluded that :**

(a) Absolute value of Foreign Equity Inflows in 1998 was higher than that in 1997 for both Thailand and South Korea.

(b) Absolute value of Foreign Equity Inflows was higher in 1998 for Thailand and lower for China than the corresponding values in 1997

(c) Absolute value of Foreign Equity Inflows was lower in 1998 for both India and China than the corresponding values in 1997

(d) None of the above can be inferred

Sol. (d) Since values of GDP are not given, none of the above mentioned conclusions can be inferred.

Problem 46 **It is known that China's GDP in 1998 was 7% higher than its value in 1997, while India's GDP grew by 2% during the same period. The GDP of South Korea, on the other hand, fell by 5%. Which of the following statement(s) is/are true ?**

I. Foreign Equity Inflows to China were higher in 1998 than in 1997.

II. Foreign Equity Inflows to China were lower in 1998 than in 1997.

III. Foreign Equity Inflows to India were higher in 1998 than in 1997.

IV. Foreign Equity Inflows to South Korea decreased in 1998 relative to 1997.

V. Foreign Equity Inflows to South Korea increased in 1998 relative to 1997.

(a) I, III and IV (b) II,. III and IV
(c) I, III and V (d) II and V

Sol. (d) Let China's GDP be ₹ 100 in year 1997.

Then, in 1998, it will be ₹ 107.

Clearly, 4.80% of 107 < 5.96% of 100.

(Alternatively, 4.80% of 7 $\not>$ (5.96 – 4.80))

Hence, Statement I is wrong but Statement II is correct.

Similarly, in respect of India, 0.72% of 2 $\not>$ (1.71 – 0.72), hence, Statement III is wrong.

In case of South Korea, 2.5% of 5 $\not>$ (2.50 – 2.16).

Hence Statement IV is wrong and Statement V is correct.

∴ option (d) is correct.

Problem 47 **China's Foreign Equity Inflows in 1998 were 10 times that into India. It can be concluded that :**

(a) China's GDP in 1998 was 40% higher than that of India

(b) China's GDP in 1998 was 70% higher than that of India

(c) China's GDP in 1998 was 50% higher than that of India

(d) No inference can be drawn about relative magnitudes of China's and India's GDPs

Sol. (c) As given, 4.80% of GDP of China = 10 × 0.72% of GDP of India

$\Rightarrow$ GDP of China: GDP of India = 3 : 2

Hence, China's GDP in 1998 was 50% higher than that of India.

∴ Option (c) is correct.

C. Problems Based on Ternary Diagrams

How to get information from a ternary diagram :

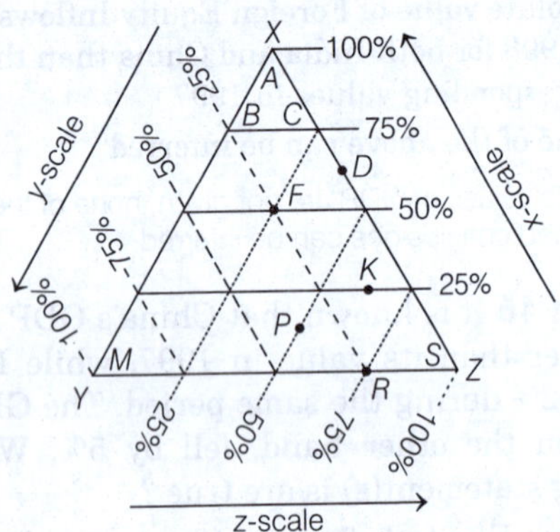

(i) Any three variables are represented by a ternary diagram. Here, these variables are X, Y and Z. These variables are represented by vertices of the triangle and vertices denote 100% and the sides opposite to them denote 0%. *e.g.* vertex A denotes 100% value with reference to variable X whereas the base opposite to vertex A i.e. MS denotes 0% value with reference to variable X.

Also, value of any point lying on base MS will be 0% with reference to variable X.

Further, the lines between vertex A and the base MS, parallel to MS and each other, have values in decreasing order if we read from vertex to base. No doubt, values of all points lying on a line parallel to base MS will be same with reference to variable X. The same concept is also applicable for vertices B and C.

(ii) It is also construed from the above that different lines between a vertex and the side opposite to it, which are parallel to that side represent values of all points lying on these parallel lines only with reference to the variable represented by that vertex. Hence, lines (—) represent the values of all points lying thereon with reference to variable X, lines (------) represent the value of points lying thereon with reference to variable Y and lines (...) does with reference to Z.

Note *In actual questions, all the lines are of same type.*

(iii) For sake of convenience, one end of all three types of lines (including vertices) are extended outside of the triangle towards corresponding scales to depict the values.

(iv) Sum of the values of any point lying inside or on the triangle with reference to given three variables is 100%.

(v) Generally values for scales are depicted in percentage terms along the sides of the triangle but not necessarily. It may be in figures without percentage. Then, you may convert them into percentage terms, if required.

Obviously, 0.2 means 20%, 0.4 means 40% and so on.

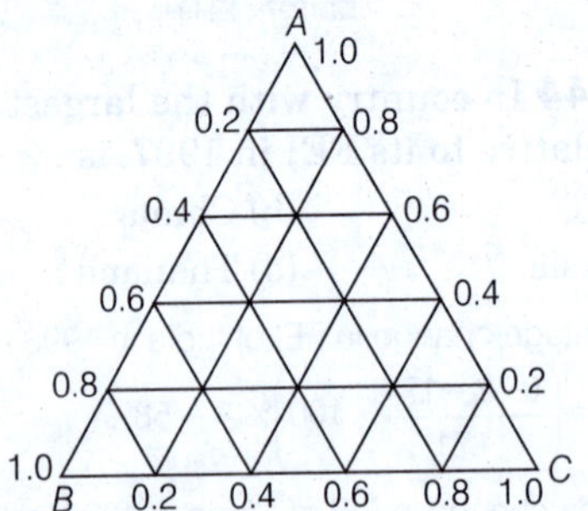

(vi) In common practice, the x-scale, y-scale and z-scale are not written, but values are depicted. Thus, ignore the scales written above along sides of the triangle while understanding 'how to read'.

Now, try to read values of different points lying inside or on the triangle with reference to given three variables X, Y and Z.

Values of point A with reference to variables X, Y, Z, respectively = (100%, 0%, 0%)

[∵ A is lying on intersection of sides SA and MA and values of all points lying on side SA are 0 with reference to Y and also values of all points lying on side MA are 0 with reference to Z.]

Values (co-ordinates) of point B = (75%, 25%, 0%)

[∵ B is lying on intersection of all three types of lines. B lies on side MA, hence its value is 0 with reference to Z. It also lies on line BC, parallel to MS, hence its value is 75% with reference to X. Since, it lies also on line BR, parallel to SA, its value is 25% with reference to Y.]

Note *Since the sum of the values of any point lying inside or on the triangle with reference to given three variables is 100%, we may find the third value subtracting sum of two known values from 100%*

Values (co-ordinates) of point C = (75%, 0%, 25%)

Values (co-ordinates) of point D = (62.5%, 0%, 37.5%)

Values (co-ordinates) of point F = (50%, 25%, 25%)

Values (co-ordinates) of point K = (25%, 12.5%, 62.5%)

Values (co-ordinates) of point P =(12.5%, 37.5%, 50%)

Please try yourself for other points.

Problem 48 What are the values of the points M and R with reference to A, B and C respectively ?

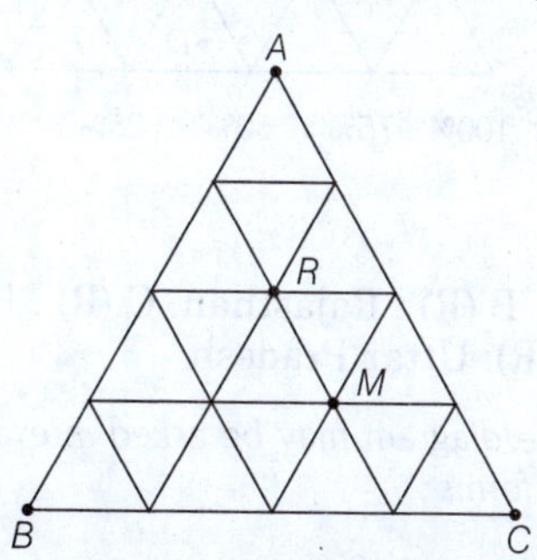

Sol. Since there are four intervals along the sides of the triangle, each interval will have value of 25%. Now, the diagram will as follow :

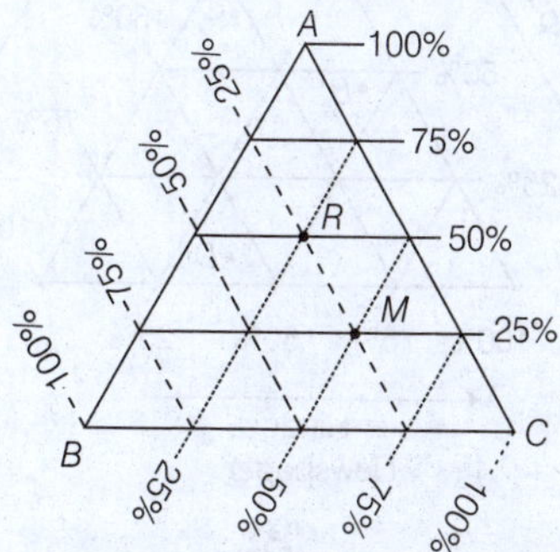

Values of R with reference to A, B and C respectively = (50%, 25%, 25%) and values of M with reference to A, B and C respectively = (25%, 25%, 50%)

Directions (Problems 49-53) *A survey was conducted in five cities viz. Pune, Kanpur, Raisen, Surat and Trivandrum for the percentage of people using T (trains), B (buses), C (cars) as modes of transport. Number of persons surveyed in the cities Pune, Kanpur, Raisen, Surat and Trivandrum are 2000, 4000, 6000, 3000 and 8000, respectively. Refer to the data in the triangular bar diagram to answer the questions that follow :*

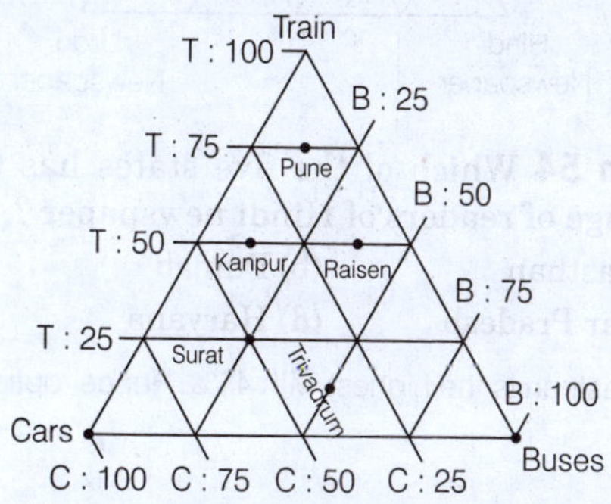

Problem 49 The city where the least number of persons uses buses is :

(a) Surat (b) Raisen

(c) Kanpur (d) Pune

Sol. (d)

Cities (person surveyed)	Pune (2000)	Kanpur (4000)	Raisen (6000)	Surat (3000)	Trivand rum (8000)
Percentage of persons using buses	12.5%	12.5%	37.5%	25%	50%

Now, we have to find least values among (12.5% of 2000), (12.5% of 4000), (37.5% of 6000) and (25% of 3000).

Clearly, (12.5% of 2000) < (12.5% of 4000) as 4000 is greater than 2000.

Further, others have greater percentage term as well as greater base number as compared to Pune.

Hence, (12.5% of 2000) is the least among the above.

∴ Option (d) is correct.

Alternatively-I

Compare (12.5% of 2000), (12.5% of 4000), (37.5% of 6000) and (25% of 3000)

$\Rightarrow 1 \times 2000$, 1×4000, 3×6000 and 2×3000

[∵ 12.5% = 1 (as assumed)]

Clearly, 1×2000 is the least.

Hence, option (d) is correct.

Alternatively-II

Obviously, the city(ies) located farthest from the vertex representing buses, will have the least percentage of persons using buses.

Hence, Pune and Kanpur will have least and equal percentage of persons using buses. Further, number of persons surveyed in Pune is least among those in all cities. Therefore, Pune will have least number of persons using buses.

Problem 50 The average number of persons using trains for transportation in Pune, Kanpur, Raisen and Trivandrum is :

(a) 1880

(b) 1750

(c) 1950

(d) None of the above

Sol. (d)

Cities (person surveyed)	Pune (2000)	Kanpur (4000)	Raisen (6000)	Surat (3000)	Trivan drum (8000)
Percentage of persons using trains	75%	50%	50%	25%	12.5%
Number of persons using trains	1500	2000	3000	750	1000

Now, average of 1500, 2000, 3000 and 1000 = 1875

Hence, option (d) is correct. (No need to try with another approach as we are much familiar with percentage terms)

Problem 51 The mode of transport used by the least number of persons in all the given cities.

(a) trains (b) buses
(c) cars (d) cars and buses

Sol. (c)

Cities (person surveyed)	Pune (2000)	Kanpur (4000)	Raisen (6000)	Surat (3000)	Trivan drum (8000)
Percentage of persons using trains	75%	50%	50%	25%	12.5%
Percentage of persons using buses	12.5%	12.5%	37.5%	25%	50%
Percentage of persons using cars	12.5%	37.5%	12.5%	50%	37.5%

Ratio of person surveyed in Pune, Kanpur, Raisen, Surat and Trivandrum =1 : 2 : 3 : 1.5 : 4

Now, number of persons using trains = {1 (75) + 2 (50) + 3 (50) + 1.5 (25) + 4 (12.5)}% of a certain number (here, 2000, but no need to mention as we have to compare only).

= 412.5% of a certain number

Similarly, number of persons using buses = 387.5%

and number of persons using cars = 350%

Hence, option (c) is correct.

Problem 52 Among the given five cities, the cities where less than 30% of the people use cars in transport are :

(a) Kanpur and Trivandrum
(b) Pune, Kanpur and Raisen
(c) Pune and Raisen
(d) Pune, Kanpur and Surat

Sol. (c) In Pune and Raisen, only 12.5% people are using cars. Hence, option (c) is correct.

Problem 53 Which of the following statement is not true ?

(a) 50% of the people use trains for transport in the cities Kanpur and Raisen.
(b) In city Trivandrum, more than 50% of the people use cars for transport.
(c) More percentage of people use buses for transport in the city Surat than in the city Pune.
(d) In city Raisen, there are more percentage of people using trains for transport than buses.

Sol. (b) In Trivandrum, only 37.5% people use cars, hence correct answer is option (b).

Directions (Problems 54 - 57) *These questions are based on the figure below. The diagram given here describes the percentage of readers in five states who purchase newspapers in English, Hindi and Urdu languages. Assuming the readers do not buy newspaper in any other language.* **[MAT-08]**

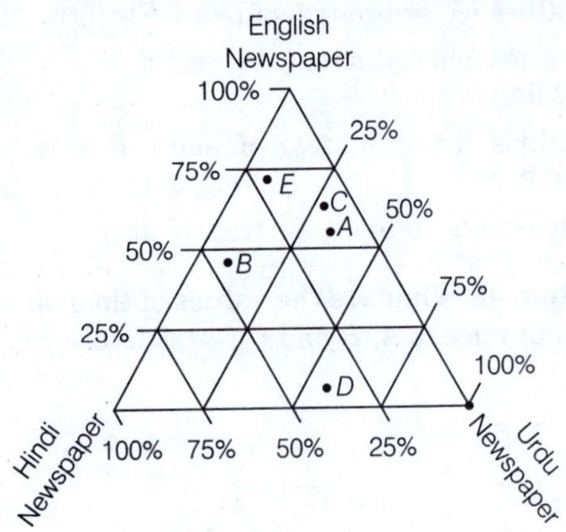

A (R) : Punjab, B (R) : Rajasthan, C (R) : Haryana, D (R) : Delhi NCR, E(R): Uttar Pradesh

Note *The above diagram may be asked in examinations in the following forms :*

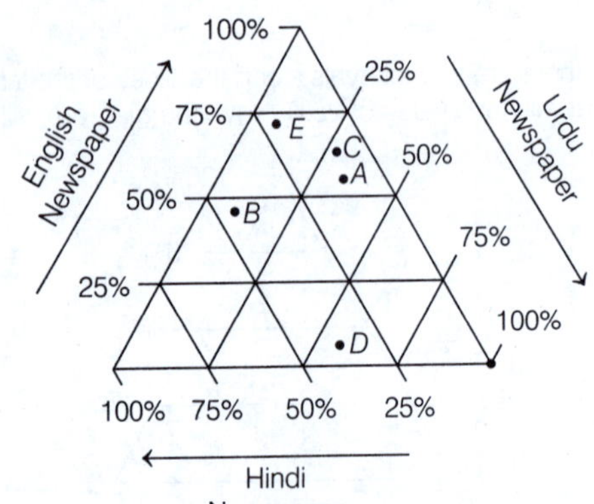

Or

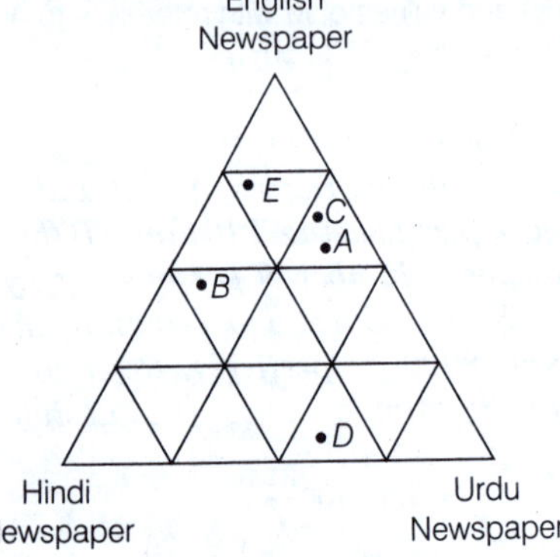

Problem 54 Which of the five states has the highest percentage of readers of Hindi newspaper ?

(a) Rajasthan (b) Punjab
(c) Uttar Pradesh (d) Haryana

Sol. (a) Rajasthan is the highest with 45%. Hence, option (a) is correct.

Problem 55 From the data provided above we can infer that :

(a) Each reader can read newspapers in at least two languages
(b) Each reader can read newspapers in only one of the languages
(c) Each reader can read newspapers in the three languages discussed
(d) Each reader can read newspapers in more than one languages

Sol. (b) Clearly, option (b) is correct. For example in state of Rajasthan, 45% people read Hindi newspaper, 45% read English newspaper and the rest 10% read Urdu newspaper. No one reads newspaper in more than one language. For every state, sum of newspaper readers of all three languages is 100%.

Problem 56 If UP has 2500000 newspaper readers and Rajasthan has 1500000, how many more readers buy English newspapers in UP than in Rajasthan ?

(a) 1075000 (b) 1800000 (c) 1200000 (d) 300000

Sol. (a) Number of English newspaper readers in UP

= 70% of 2500000 = 1750000

and number of English newspaper readers in Rajasthan = 45% of 1500000 = 675000

∴ Required difference = 1750000 − 675000 = 1075000

Problem 57 About what percent of readers in Rajasthan read Urdu newspapers?

(a) 10% (b) 40% (c) 50% (d) 60%

Sol. (a) Clearly, percentage of Urdu newspaper readers in Rajasthan = 10%

D. Problems Based on Line Graphs

Directions (Problems 58-65) *Study the given graph, depicting percent profit earned by two companies producing electronic goods over the years, carefully to answer these questions.* $\% \text{ profit} = \frac{\text{Profit earned}}{\text{Total investment}} \times 100$

Profit earned = Total income − Total investment in the year

[SBI (Associates-PO)-11]

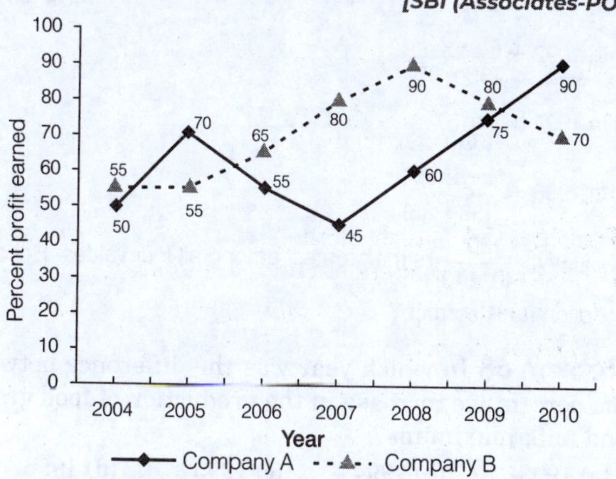

Problem 58 If the profit earned in 2006 by Company B was ₹ 812500. What was the total income of the company in that year ?

(a) ₹ 1250000 (b) ₹ 2062500
(c) ₹ 1650000 (d) ₹ 1825000
(e) None of these

Sol. (b) Clearly, total income = Total investment + Profit earned

∵ 65% of the total investment = ₹ 812500

∴ (100 + 65)% of the total investment = ₹ 2062500

Hence, total income = ₹ 2062500

Problem 59 If amount invested by the two companies in 2005 was equal, what was the ratio of the total income of the Company A to that of B in 2005 ?

(a) 31 : 33 (b) 33 : 31
(c) 34 : 31 (d) 14 : 11
(e) None of these

Sol. (e) Required ratio = 170% of investment of Company A: 155% of investment of Company B

= 170 : 155 (∵ investments of both companies are equal)

= 34 : 31

Problem 60 If the total amount invested by the two companies in 2009 was ₹ 27 lakh, while the amount invested by Company B was 50% of the amount invested by Company A, what was the total profit earned by the two companies together ?

(a) ₹ 21.15 lakh
(b) ₹ 20.70 lakh
(c) ₹ 18.70 lakh
(d) ₹ 20.15 lakh
(e) None of the above

Sol. (b)

	Company A	Company B
Investments	100(let)	50
Profit earned	75% of 100 i.e. 75	80% of 50 i.e. 40

As given, (100 + 50) units = ₹ 27 lakh

∴ (75 + 40) units = ₹ 20.70 lakh

Problem 61 If the income of Company A in 2007 and that in 2008 were equal and the amount invested in 2007 was ₹ 12 lakh, what was the amount invested in 2008 ?

(a) ₹ 1087500 (b) ₹ 1085700
(c) ₹ 1245000 (d) ₹ 1285000
(e) None of these

Sol. (a) Obviously, 145% of amount invested in 2007 = 160% of the amount invested in 2008

⇒ Amount invested in 2007 : that in 2008 = 160 : 145 = 32:29

As given, 32 units = ₹ 12 lakh

∴ 29 units = ₹ 1087500

Hence, option (a) is correct.

Problem 62 If the amount of profit earned by Company A in 2006 was ₹ 10.15 lakh, what was the total investment ?

(a) ₹ 13.8 lakh (b) ₹ 15.4 lakh
(c) ₹ 14.9 lakh (d) ₹ 14.2 lakh
(e) None of these

Sol. (e) ∵ 55% of the total investment = ₹ 10.15 lakh
∴ 100% of the total investment = ₹ 18.45 lakh
Hence, option (e) is correct.

Problem 63 If the amount invested by Company B in 2004 is ₹ 12 lakh and the income of 2004 is equal to the investment in 2005, what is the amount of profit earned in 2005 by Company B ?

(a) ₹ 6.6 lakh (b) ₹ 10.23 lakh
(c) ₹ 18.6 lakh (d) ₹ 9.6 lakh
(e) None of these

Sol. (b) Clearly, income in 2004 = 155% of the total investment in 2004 = Total investment in 2005 = 155% of ₹ 12 lakh
Then, profit in 2005 = 55% of 155% of 12 lakh = 10.23 lakh

Problem 64 If the investment of Company A in 2007 and 2008 were equal, what is the difference between the profits earned in the two years if the income in 2008 was ₹ 24 lakh ?

(a) ₹ 2.25 lakh
(b) ₹ 3.6 lakh
(c) ₹ 1.8 lakh
(d) ₹ 2.6 lakh
(e) None of the above

Sol. (a) Income in 2008 = 160% of the total investment in 2008
= ₹ 24 lakh
⇒ 100% total investment in 2008
= ₹ 15 lakh
= total investment in 2007
Now, required difference = (60 – 45)% of ₹ 15 lakh
= ₹ 2.25 lakh

Problem 65 If each of the Companies A and B invested ₹ 25 lakh in 2010, what was the average profit earned by the two companies ?

(a) ₹ 18 lakh (b) ₹ 22.5 lakh
(c) ₹ 17.5 lakh (d) ₹ 20 lakh
(e) None of the above

Sol. (d) Average profit earned
$= \left(\frac{90+70}{2}\right)$% of ₹ 25 lakh
= ₹ 20 lakh

Directions (Problems 66-71) *Answer the questions based on the following graph. The graph shows the quantity of milk and food grains consumed annually along with female and male population (in millions)*

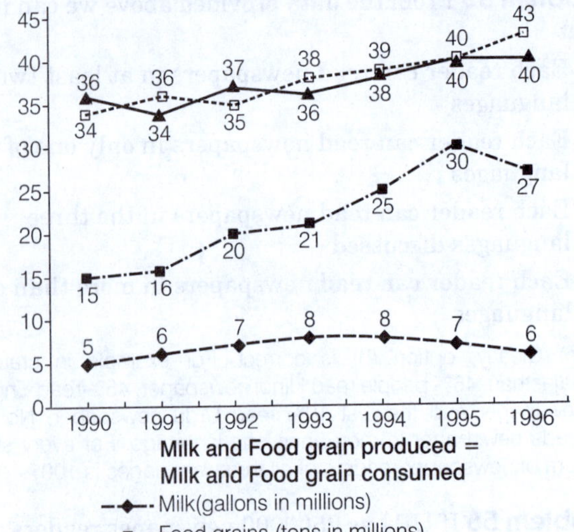

Problem 66 When was the per capita production of milk least ?

(a) 1990 (b) 1992 (c) 1994 (d) 1996

Sol. (a) In 1990, 1992, 1994 and 1996, per capita production of milk were: $\frac{5}{(34+36)}, \frac{7}{(35+37)}, \frac{8}{(38+39)}$ and $\frac{6}{(40+43)}$, respectively.

Hence, $\frac{5}{70}, \frac{7}{72}, \frac{8}{77}$ and $\frac{6}{83}$

$\Rightarrow \frac{5}{25}, \frac{7}{9}, \frac{8}{5}$ and $\frac{6}{29}$ (please refer to technique meant for 'Comparison of Fractions'. We have subtracted 9 times numerator from the denominator and wrote as new denominator)

Clearly, $\frac{5}{25}$ is the least among all four values. Hence, option (a) is correct.

Problem 67 When was the per capita consumption of food grains maximum ?

(a) 1995 (b) 1992 (c) 1994 (d) 1993

Sol. (a) In 1995, per capita production of food grains = $\frac{30}{(40+40)}$

In 1992 ⇒ $\frac{20}{(35+37)}$

In 1994 ⇒ $\frac{25}{(38+39)}$

In 1993 ⇒ $\frac{21}{(38+36)}$

Clearly, $\frac{30}{(40+40)}$ is the greatest among all four values. Hence, option (a) is correct.

Problem 68 In which year was the difference between the percentage increase in the production of food grains and milk maximum ?

(a) 1993 (b) 1994 (c) 1995 (d) 1996

Sol. (c)

Year	Percentage increase in production of food grains	Percentage increase in production of milk
1993	$\left(\frac{21-20}{20}\times 100\right)\% = 5\%$	$\left(\frac{8-7}{7}\times 100\right)\% = 14.28\%$
1994	$\left(\frac{25-21}{21}\times 100\right)\% = 19.04\%$	$\left(\frac{8-8}{8}\times 100\right)\% = 0\%$
1995	$\left(\frac{30-25}{25}\times 100\right)\% = 20\%$	$\left(\frac{7-8}{8}\times 100\right)\% = -12.5\%$
1996	$\left(\frac{27-30}{30}\times 100\right)\% = -10\%$	$\left(\frac{6-7}{7}\times 100\right)\% = -14.28\%$

Clearly, maximum difference appears in the year 1995.

Problem 69 If one gallon milk contains 120 g of a particular nutrient and one tonne of food grains contains 80 g of the same nutrient, in which year was the availability of this nutrient maximum ?

(a) 1993 (b) 1994 (c) 1995 (d) 1996

Sol. (c)

Year	Availability of nutrient (ratio of nutrient in milk and food grains = 3 : 2)
1993	$(8\times 3) + (21\times 2) = 66$
1994	$(8\times 3) + (25\times 2) = 74$
1995	$(7\times 3) + (30\times 2) = 81$
1996	$(6\times 3) + (27\times 2) = 72$

Clearly, availability of nutrient is maximum in year 1995.

Problem 70 Referring to the above question, in which year was the per capita consumption of this nutrient highest ?

(a) 1993 (b) 1994 (c) 1995 (d) 1996

Sol. (c)

Year	Per capita consumption of the particular nutrient
1993	$\frac{66}{36+38} = \frac{66}{74}$ = less than 1
1994	$\frac{74}{38+39} = \frac{74}{77}$ = less than 1
1995	$\frac{81}{40+40} = \frac{81}{80}$ = more than 1
1996	$\frac{72}{40+43} = \frac{72}{83}$ = less than 1

Clearly, consumption of nutrient per capita is highest in year 1995.

E. Problems Based on Pie-charts

Directions (Problems 71 - 75) *The following pie-chart shows the monthly expenditure of a family on various items. If the family spends ₹ 825 on clothing, answer these questions.* **[SSC(CGL-T-II)-16]**

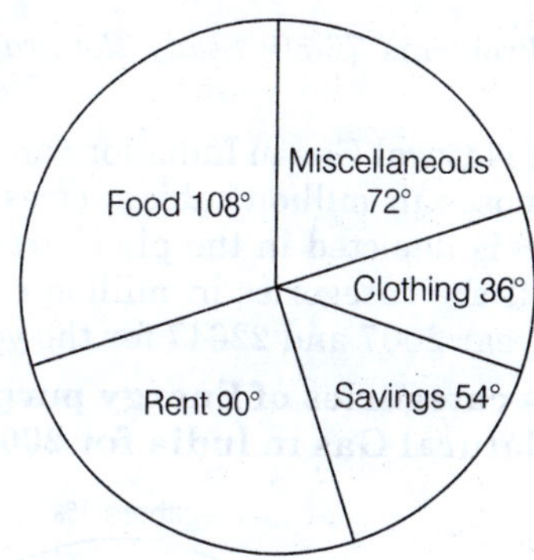

Problem 71 What is the total monthly income of the family?

(a) ₹ 8025 (b) ₹ 8250
(c) ₹ 8520 (d) ₹ 8052

Sol. (b) As given, 36° = ₹ 825

∴ 360° = ₹ 8250 (∵ 360° indicates total monthly income)

Problem 72 What percent of the total income does the family save?

(a) 15% (b) 50%
(c) 20% (d) 25%

Sol. (a) Option (a) is correct as 54° is 15% of 360°.

Problem 73 What is the ratio of expenses on food and miscellaneous ?

(a) 3 : 4 (b) 2 : 3
(c) 3 : 2 (d) 2 : 5

Sol. (c) Required ratio = 108° : 72°
= 3 : 2

Problem 74 What is the average of expenses on clothing and rent ?

(a) ₹ 1443.75 (b) ₹ 1344.57
(c) ₹ 1574.34 (d) ₹ 1734.45

Sol. (a) Average in terms of degree = $\frac{90°+36°}{2} = 63°$

∵ 36° = ₹ 825
∴ 18° = ₹ 412.5
∴ 9° = ₹ 206.25

63° = ₹ 1443.75 (on addition)

Problem 75 The ratio of average of expenses on food, clothing and miscellaneous items to the average of expenses on savings and rent is :

(a) 1 : 1 (b) 1 : 3
(c) 3 : 2 (d) 2 : 1

Sol. (a) Required ratio = $\frac{108°+36°+72°}{3} : \frac{54°+90°}{2}$
= 72 : 72 = 1 : 1

Directions (Problems 76-80) *Study the graphs to answer these questions.* ***[MAT-2011]***

The off-take of Natural Gas in India for various categories of energy purposes in million cubic metres for the years 2007 and 2008 is depicted in the pie charts given below. The total of all the categories in million cubic metres is 20855 for the year 2007 and 22647 for the year 2008.

Various categories of Energy purposes of Natural Gas in India for 2007

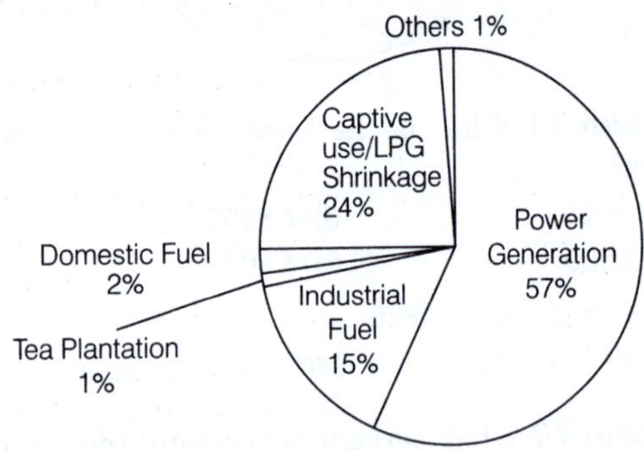

Various categories of Energy purposes of Natural Gas in India for 2008

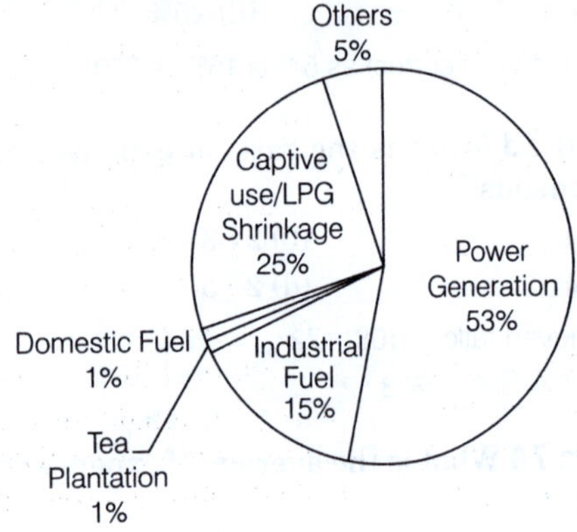

Problem 76 The category of captive use/LPG Shrinkage usage in million cubic metres in 2008 is approximately equal to the sum of which of the categories in 2007 ?

(i) Tea Plantation
(ii) Domestic Fuel
(iii) Captive use/LPG Shrinkage
(iv) Industrial Fuel

(a) (i), (ii) and (iii) only (b) (i) and (iv) only
(c) (i), (ii) and (iv) only (d) All of (i), (ii), (iii) and (iv)

Sol. (a) Clearly, 25% of 22 647 ≈ 27% of 20855 (do mentally without pen and paper recalling the concept envisaged under chapter Percentage')

Also, 27% covers each of (captive use/LPG Shrinkage usage + Domestic Fuel + Tea Plantation) and (captive use/LPG Shrinkage usage + Domestic Fuel + Others).

Hence, option (a) is correct.

Alternatively

You may solved conventionally.

Problem 77 The Natural Gas was wasted by 50% in 2007 and 40% in 2008 of the Others' category in both the years. What is the percentage increase or decrease in wastage of Natural Gas from 2007 to 2008 ?

(a) 300% increase (b) 405% increase
(c) 405% decrease (d) 300% decrease

Sol. (a) Clearly, we have to find percentage increase or decrease when (50% of 1% of 20855) changes to (40% of 5% of 22647).
Also, 5% of 22647 ≈ 5.4% of 20855 (do mentally without pen and paper)

Now, increase = (40% of 5.4% of 20855) – (50% of 1% of 20855)
= (2.1 – 0.5)% of 20855 = 1.6% of 20855

Obviously, 1.6 is approx. 300% of 0.5.

Hence, option (a) is correct.

Alternatively,

You may solve conventionally.

Increase = (40% of 5% of 22647) – (50% of 1% of 20855)
= 2% of 22647 – 0.5% of 20855
= 452.94 – 104.27 = 348.67

Then, percentage increase

$$= \left(\frac{348.67}{104.27} \times 100\right)\% \approx 300\%$$

Problem 78 What is the percentage increase in usage of Natural Gas as industrial Fuel in 2008 compared to 2007?

(a) 8.6% (b) 4.71%
(c) 6.8% (d) 7.98%

Sol. (a) Clearly, we have to find percentage increase or decrease when (15% of 20855) changes to (15% of 22647)
Also, 15% of 22647 ≈ 16.3% of 20855 (do mentally without pen and paper)

Now, increase = 1.3% of 20855

Obviously, 1.3 is 8.6% of 15.

Hence, option (a) is correct.

Alternatively,

You may solve conventionally.

Increase = (15% of 22647) – (15% of 20855)
= 15% of (22647 – 20855) ≈ 270

Then, percentage increase =

$$= \left(\frac{270}{15\% \text{ of } 20855} \times 100\right)\% = \left(\frac{270}{3130} \times 100\right)\% = 8.6\%$$

Problem 79 The two categories which had shown an increase by more than 10% in usage in 2008 as compared to 2007 are :

(a) Captive use/LPG Shrinkage and Others
(b) Power Generation and Industrial Fuel
(c) Domestic Fuel and Industrial Fuel
(d) Domestic Fuel and Tea Plantation

Sol. (a) Proceed through options. (25 + 5)% of 22647 is greater than (24 + 1)% of 20855 by more than 10%.

Since, we arrived now at correct option, no need to proceed further.

Problem 80 The ratio of percentage change of Power Generation to the percentage change in Captive use/LPG Shrinkage in the year 2008 compared to 2007 is

(a) 7.39 : 100 (b) 9.37 : 1000
(c) 7.39 : 1000 (d) 9.37 : 100

Sol. (a) Clearly, we have to find percentage change in respect of Power Generation when (57 % of 20855) changes to (53% of 22647)

Also, 53% of 22647 ≈ 57.5 % of 20855

Obviously, (57.5 – 57) is 0.9 (approx.) % of 57.

Also, find the percentage change in respect of Captive use/LPG Shrinkage when (24 % of 20855) changes to (25% of 22647)

Further, 25% of 22647 ≈ 27% of 20855

Obviously, (27– 24) is 12.5% of 24.

Now, required ratio $= \frac{0.9}{12.5} \approx 7.39 : 100$

Alternatively You may solve conventionally.

Directions (Problems 81 and 82) *A town divided into 4 zones Z_1, Z_2, Z_3 and Z_4, had certain number of registered (valid and invalid) voters. The pie-chart given below shows the registered valid voters distributed in these four zones in degree. The ratio between values of degrees of Z_2 and Z_4 is 3: 7.*

Valid Voters

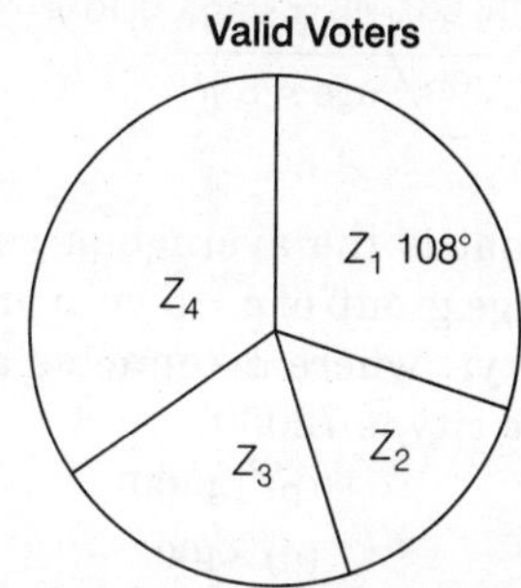

Note

1. *Number of valid voters in Z_3 is 1/3 more than the difference between number of valid voters in Z_1 and that in Z_2.*
2. *Difference between number of valid voters in Z_3 and Z_4, is 480 ($Z_4 > Z_3$).*
3. *20% of the total registered voters in the town were invalid.*

[IBPS(PO-Mains)-19]

Problem 81 What is the central angle corresponding to number of valid voters in Z_3 ?

Sol. Clearly, 108° will correspond to $\left(\frac{108°}{360°} \times 100\right)$% i.e. 30%

Then, Z_2, Z_3 and Z_4 together will equal to 70% of the total registered voters assuming zones Z_2 , Z_3 and Z_4 have Z_2, Z_3 and Z_4 valid voters, respectively.

Further, $Z_2 : Z_4 = 3 : 7$

$\Rightarrow$ $Z_4 = 233.33\%$ of Z_2

Now, $Z_2 + Z_3 + Z_4 = 70\%$

$\Rightarrow$ $Z_2 + 133.33\%$ of $(Z_1 - Z_2) + 233.33\%$ of $Z_2 = 70\%$

$\Rightarrow$ $Z_2 + 133.33\%$ of $(30\% - Z_2) + 233.33\%$ of $Z_2 = 70\%$

$\Rightarrow$ $Z_2 + 40\% - 133.33\%$ of $Z_2 + 233.33\%$ of $Z_2 = 70\%$

$\Rightarrow$ $2Z_2 = 30\%$

$\Rightarrow$ $Z_2 = 15\%$

Then, $Z_3 = 133.33\%$ of $(Z_1 - Z_2) = 133.33\%$ of $(30\% - 15\%) = 20\%$, corresponding to 72°.

Alternatively,

The best way is to proceed through options. Let options be 108° and 72°.

(a) 108° ⇒ Clearly, $Z_2 + Z_4 = 360° - 108° - 108° = 144$, then $Z_2 = \left(\frac{3}{3+7} \times 144°\right) = 43.2°$ but $108° \neq 133.33\%$ of $(108° - 43.2°)$

(b) 72° ⇒ Clearly, $Z_2 + Z_4 = 360° - 108° - 72° = 180°$, then $Z_2 = \left(\frac{3}{3+7} \times 180°\right) = 54°$ but $108° = 133.33\%$ of $(108° - 54°)$

Hence, option (b) is correct.

Problem 82 Number of valid voters in Z_3 is approximate, what percent less than number of valid voters in Z_4 ?

Sol. From above, $Z_1 = 30\%$, $Z_2 = 15\%$ and $Z_3 = 20\%$, then

$Z_4 = (100 - 30 - 15 - 20)\% = 35\%$

Clearly, Z_3 is $\left(\frac{35-20}{35} \times 100\right)$% i.e. 42.8% less than Z_4.

Directions (Problems 83-86) *Study the pie-charts to answer these questions.*

Percentage distribution of workforce : Region-wise (Total = 68700)

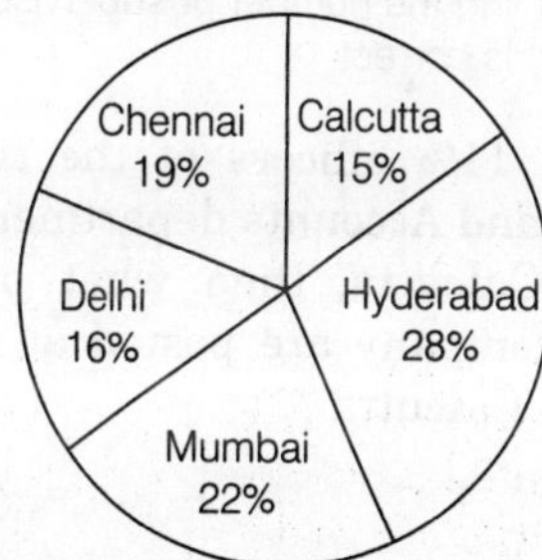

Percentage distribution of workforce : Category-wise (Total = 68700)

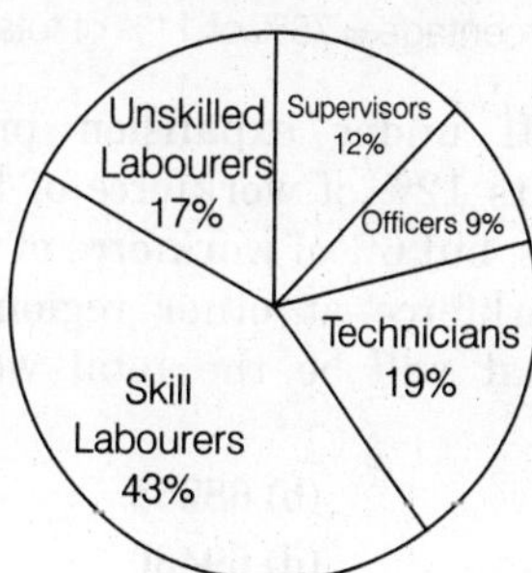

Percentage distribution of workforce : Department-wise (total = 68700)

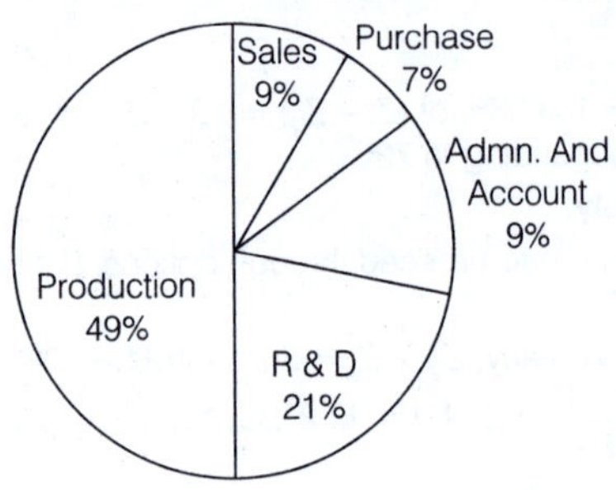

Problem 83 If 22% of the Production department persons are posted at Hyderabad region, then what percent of Hyderabad workforce are in Production department ?

(a) 38.5 (b) 78 (c) 68 (d) 22

Sol. (a) Clearly, we have to find, (22% of 49% of 68700) is what percent of (28% of 68700)

Now, required percentage $= \left(\frac{49 \times 0.22}{28} \times 100\right)\% = 38.5\%$

($\because$ bases are same)

Problem 84 How many supervisors are posted in Calcutta region?

(a) 1237 (b) 985
(c) 1144 (d) Data insufficient

Sol. (d) Since percentage of supervisors posted in Calcutta is not given, we cannot find the number of supervisors posted there. Hence, option (d) is correct.

Problem 85 If 11% officers of the company are in Administration and Accounts department of which 75% are posted at Calcutta, then what percent of total officers of the company are posted at Administration and Accounts in Calcutta ?

(a) 6
(b) 8.25
(c) 3.75
(d) Data insufficient

Sol. (b) Required percentage = 75% of 11% of total officers = 8.25%

Problem 86 If under expansion programme, the company recruits 12% of workforce of Hyderabad and Mumbai regions, but 6% of workforce of Calcutta region retires and workforce at other regions remains the same, then what will be the total workforce of the company ?

(a) 65196 (b) 68238
(c) 72204 (d) 69430

Sol. (c) Percentage of newly recruited persons = 12% of (28 + 22)% of total workforce of the company = 6% of total workforce of the company

and percentage of retired persons = 6% of 15% of total workforce of the company = 0.9% of total workforce of the company

Hence, total workforce of the company

= (100 + (6 − 0.9))% of 68700

= 105.1% of 68700 ≈ 72204

F. Problems Based on Funnel-Charts

In certain examinations, the data interpretation based on funnel-chart is also asked. Basically, it is only a way of presentation in a cone shaped diagram. There is no difference on level of arithmetic conceptions. Consider the following questions based on funnel-charts.

Directions (Problems: 87 - 91) *Study the funnel-chart given below to answer these questions. The chart shows the distribution of five different age groups of a particular city in percentage.* **[SBI (PO-Mains)-19]**

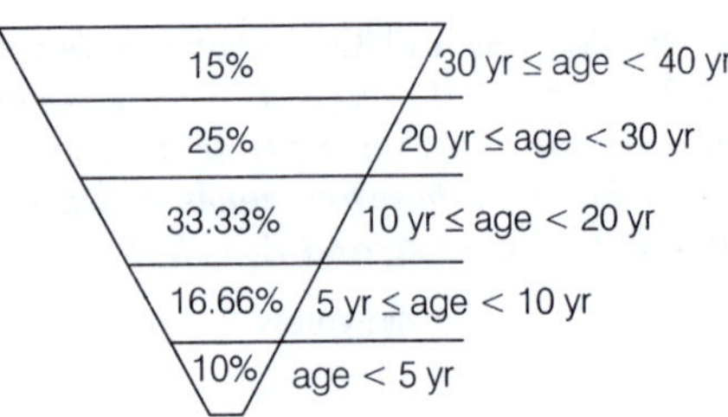

Problem 87 What is the average of population of the city who are in age group of $a < 5$ yr, 5 yr $\leq a < 10$ yr and 10 yr $\leq a < 20$ yr, where a connotes age, if the total population of the city is 72000.

(a) 11400 (b) 14400
(c) 14300 (d) 8400
(e) None of these

Sol. (b) Average of all three mentioned groups

= (10% + 16.66% + 33.33%) ÷ 3

= 20%

Then, required number

= 20% of 72000 = 14400

Problem 88 If the ratio of males to females in age groups 10 yr $\leq a <$ 20 yr and 20 yr $\leq a <$ 30 yr are 1 : 3 and 3 : 2 respectively, then number of males in age group 10 yr $\leq a <$ 20 yr is how much percent more or less than the number of females in age group 20 yr $\leq a <$ 30 ?

(a) 16.66% (b) 8.33%
(c) 22.78% (d) 33.33%
(e) None of these

Sol. (b) Number of males in age group 10 yr $\leq a <$ 20

= 25% of 33.33% = 8.33%

and number of females in age group 20 yr $\leq a <$ 30

= 40% of 25% = 10%

Clearly, 8.33 is 16.66% lesser than 10.

Problem 89 The current populations of the city is 50000 and it is increasing at the rate of 10% per annum. What will be the sum of population of age groups 5 yr $\le a <$ 10 yr and 10 yr $\le a <$ 20 yr after 3 yr.

(a) 36930 (b) 28500
(c) 33275 (d) 36430
(e) None of these

Sol. (c) Sum of current population of both age groups (16.66 + 33.33)% of 50000 = 25000

Clearly, it will increase by 33.1%.

[∵ increase for 3 yr = (3 × 10). $\boxed{3 \times 10^2}\boxed{10^3}$ %]

Then, sum of their population after 3 yr = 133.1% of 25000 = 33275

Problem 90 What is the total population of the persons whose ages are 15 yr, 25 yr and 35 yr provided they are 60%, 80% and $33\frac{1}{3}$% respectively of their respective age groups? (Assume that the total population of the city is 140000.)

(a) 62000 (b) 74400
(c) 61300 (d) 63000
(e) None of these

Sol. (d) Number of the persons having ages 15 yr, 25 yr and 35 yr = (60% of $33\frac{1}{3}$% + 80% of 25% + $33\frac{1}{3}$% or 15%) of 140000

= 45% of 140000 = 63000

Problem 91 Find the number of persons in age group 5 yr $\le a <$ 10 yr if the difference between number of persons in group 10 yr $\le a <$ 20 yr and 20 yr $\le a <$ 30 yr group is 1000.

(a) 4000 (b) 4400
(c) 2000 (d) 3000
(e) None of these

Sol. (c) ∵ (33.33 − 25) % of the population of city = 1000

∴ 16.66% of the population of city = 2000

G. Problems Based on Radar-Charts

Observe the following data table

Region	%distribution of workforce	Wages index
Chennai	19%	28
Kolkata	15%	25
Hyderabad	28%	30
Mumbai	22%	32
Delhi	16%	24

The above data may be represented through radar-chart as under:

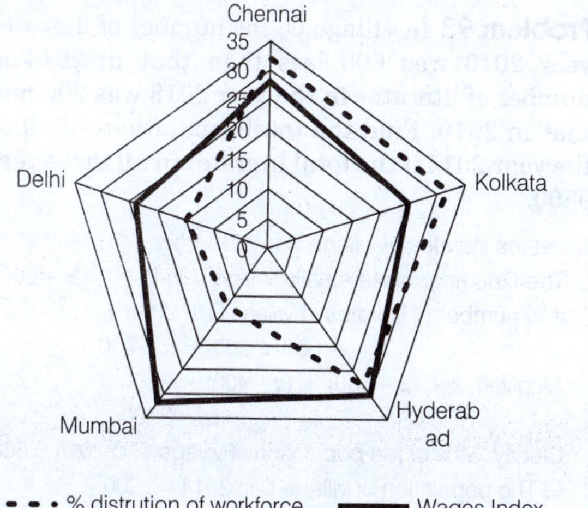

% distrution of workforce Wages Index

It is evident from the above that there is no conceptual difference between radar-charts and other charts/graphs, only ways of presentations are different. Now, take few examples for better understanding:

Directions (Problems 92 and 93) *Study the following graph carefully and answer the given questions. The graph shows the percentage of literates in three different villages in three different years.* ***[IBPS (PO-Mains)-18]***

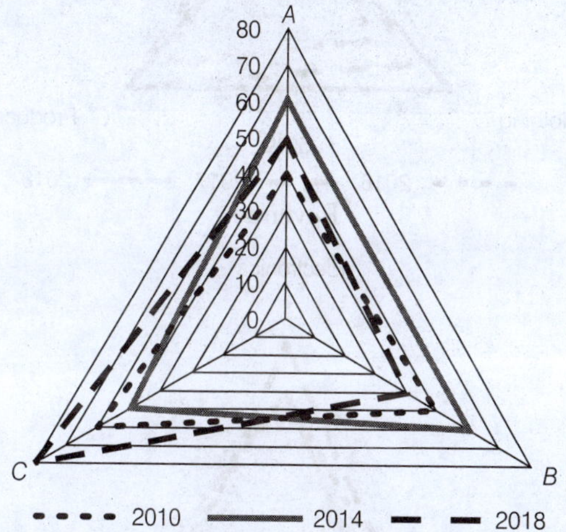

2010 2014 2018

Problem 92 In village B, total populations in the years 2010, 2014 and 2018 were in ratio 2 : 3 : 5 and the average number of total literates in those three years was 1600. Find the number of literates of village B in the year 2010.

Sol. Ratio of total population in three years = 2 : 3 : 5

Number of literates in three years = 50% of 2, 60% of 3 and 40% of 5 = 1, 1.8, 2

Average of 1, 1.8 and 2 = 1.6

As given, 1.6 units = 1600

Then, 3 units = 3000

Problem 93 In village C, the number of literates in the year 2010 was 600 less than that in 2014 and the number of literates in the year 2018 was 200 more than that in 2010. Find the total populations of village C in the year 2014 if the total literates in all three years were 9800.

Sol. Let the literates of village C in 2014 be x.

Then, number of literates in village C in 2010 = $x - 600$

and number of literates in village C in 2018

$= x - 600 + 200 = x - 400$

As given, $x + (x - 600) + (x - 400) = 9800$

$\Rightarrow \qquad x = 3600$

Clearly, 50% of the population of village C in 2014 = 3600

$\Rightarrow$ The population of village C in 2014 = 7200

Directions (Problems 94 - 97)

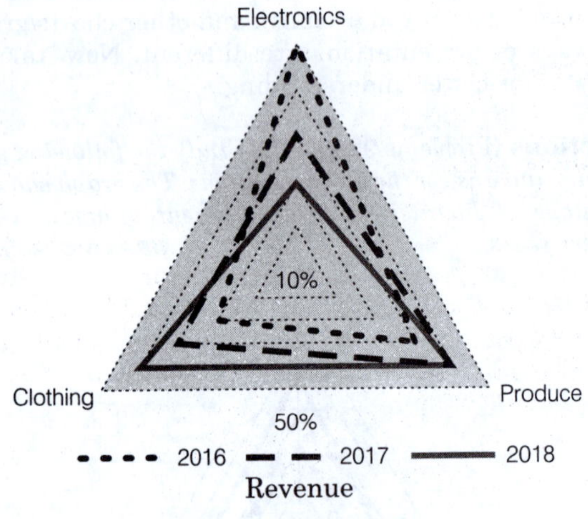

Revenue

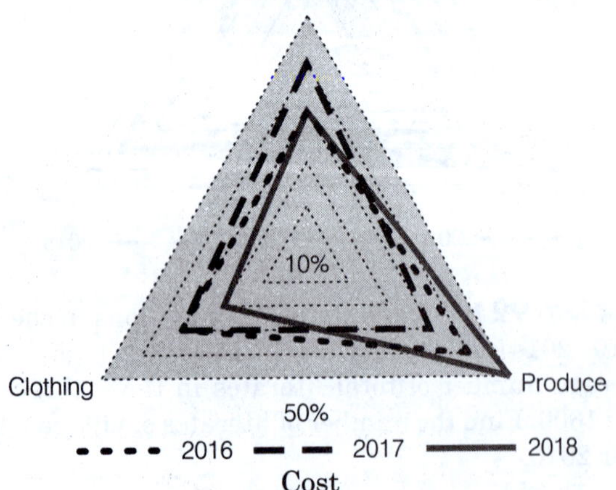

Cost

A large store has only three departments-Clothing, Produce and Electronics. The above figure shows the percentages of revenue and cost from the three departments for the years 2016, 2017 and 2018. The dotted lines(thin) depict percentage levels. *e.g.* In 2016, 50% of store's revenue came from its Electronics department while 40% of its costs were incurred in the Produce department.

In this setup, profit is computed as (revenue − cost) and percentage profit as $\left(\frac{\text{profit}}{\text{cost}} \times 100\right)\%$. It is known that:

1. The percentage profit for the store in 2016 was 100%.
2. The store's revenue doubled from 2016 to 2017, and its cost doubled from 2016 to 2018.
3. There was no profit from the Electronics department in 2017.
4. In 2018, the revenue from the Clothing department was the same as the cost incurred in the Produce department. ***[CAT-19]***

Problem 94 What was the percentage profit of the store in 2018 ?

Problem 95 What was the ratio of revenue generated from the Produce department in 2017 to that in 2018 ?

(a) 16 : 9 (b) 9 : 16
(c) 8 : 5 (d) 4 : 3

Problem 96 What percentage of the total profit for the store in 2016 was from the Electronics department ?

Problem 97 What was the approximate difference in profit percentages of the store in 2017 and 2018 ?

(a) 33.3 (b) 8.3 (c) 15.5 (d) 25.0

Sol. (Problem 94-97) : The given data is as under :

	2016		2017		2018	
	Cost	Revenue	Cost	Revenue	Cost	Revenue
Electronics	30%	50%	40%	30%	30%	20%
Produce	40%	30%	30%	40%	50%	40%
Clothing	30%	20%	30%	30%	20%	40%
	100 (let)	200	300	400	200	250
As per given information		Revenue is at 100% Profit.	30% of Revenue = 40% of cost as there was no profit in Electronics, in 2017	(∵ doubled from 2016)	Cost doubled from 2016	50% of cost = 40% of revenue as given

94. Profit percentage in 2018 = $\left(\frac{250-200}{200} \times 100\right)\% = 25\%$

95. (c) Required ratio = 40% of 400 : 40% of 250 = 8 : 5

96. Profit from Electronics department in 2016 = 50% of 200 − 30% of 100 = 70

Obviously, 70 is 70% of the total profit of 100.

97. (b) Required difference = $\left(\frac{400-300}{300} \times 100\right)\%$

$-\left(\frac{250-200}{200} \times 100\right)\% = 8.33\%$

H. Problems Based on Combination of different Graphs/Tables

Directions (Problems 98-101) *Study the given graphs to answer these questions.*

Production of wheat indifferent countries

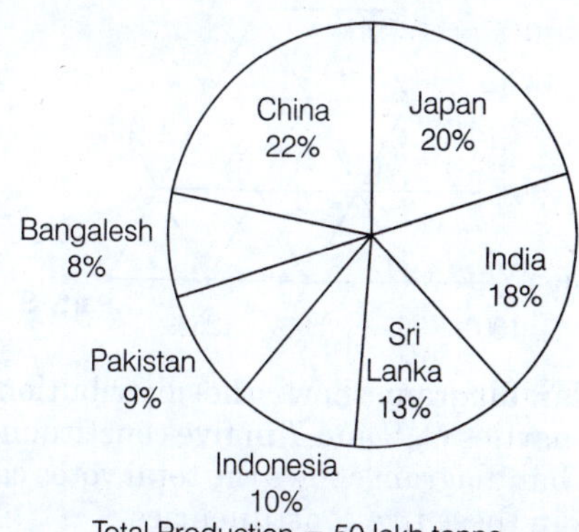

Production : Scientific Vs Conventional Methods in Percentage

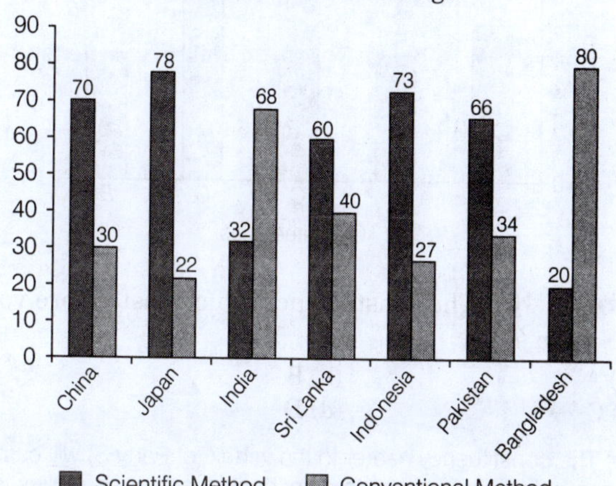

Problem 98 What is the difference between the production by scientific method and conventional method in Indonesia ?

(a) 2.3 lakh tonnes (b) 2.8 lakh tonnes
(c) 2.5 lakh tonnes (d) 3 lakh tonnes

Sol. (a) Required difference = (73 – 27)% of 10% of 50 lakh tonnes = 2.3 lakh tonnes

Problem 99 What is average production of wheat by scientific method for all the countries ?

(a) 3.1 lakh tonnes
(b) 4.24 lakh tonnes
(c) 2.6 lakh tonnes
(d) 3.07 lakh tonnes

Sol. (b) Total production by scientific method for all the countries (in lakh tonnes)

$$= \{(0.7 \times 22) + (0.78 \times 20) + (0.32 \times 18) + (0.6 \times 13) + (0.73 \times 10) + (0.66 \times 9) + (0.2 \times 8)\}\% \text{ of } 50$$

$$= (15.4 + 15.6 + 5.76 + 7.8 + 7.3 + 5.94 + 1.6)\% \text{ of } 50$$

$$= 59.4\% \text{ of } 50 = 29.7$$

Then, required average $= \frac{29.7}{7} = 4.24$ lakh tonnes

Problem 100 What is the ratio of production by conventional method in Pakistan to that by scientific method in Japan ?

(a) 9 : 40 (b) 99 : 260
(c) 51 : 260 (d) 48 : 77

Sol. (c) Required ratio= 34% of 9% of 50 : 78% of 20% of 50

$$= 34 \times 9 : 78 \times 20 = 51 : 260$$

Problem 101 The production of wheat in Sri Lanka by conventional method is approximately, how many times the production in India by scientific method ?

(a) 1.2 (b) 1.5
(c) 0.9 (d) None of these

Sol. (c) Required answer $= \frac{40\% \text{ of } 13\% \text{ of } 50}{32\% \text{ of } 18\% \text{ of } 50} = 0.9$

Directions (Problems 102 -106) *Answer the questions based on the following information.* ***[CAT-96]***

The first table gives the percentage of students in MBA class, who sought employment in the areas of finance, marketing and software. The second table gives the average starting salaries of the students per month (₹ in thousands) in these areas. The third graph gives the number of students who passed out in each year.

	Finance	Marketing	Software	Others
1992	22	36	19	23
1993	17	48	23	12
1994	23	43	21	13
1995	19	37	16	28
1996	32	32	20	16

	Finance	Marketing	Software
1992	5450	5170	5290
1993	6380	6390	6440
1994	7550	7630	7050
1995	8920	8960	7760
1996	9810	10220	8640

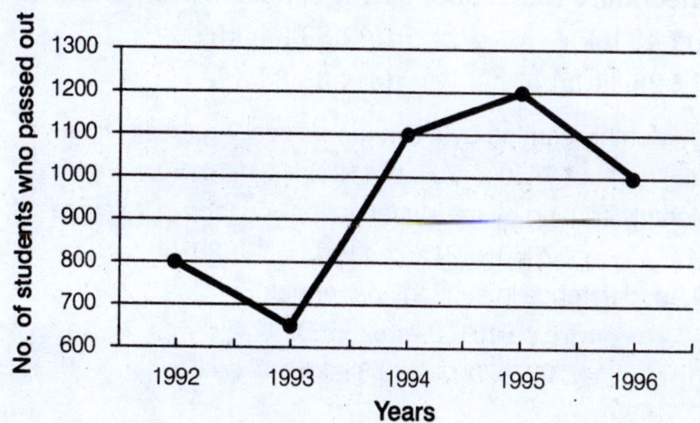

Problem 102 The number of students who get jobs in finance is less than the students getting marketing jobs in the 5 yrs, by :

(a) 826 (b) 650 (c) 750 (d) 548

Sol. (c)

Year	Number of students passed out or employed	Number of students getting jobs in marketing – Number of students getting jobs in finance
1992	800	(36 – 22)% of 800 = 112
1993	650	(48 – 17)% of 650 = 201.5
1994	1100	(43 – 23)% of 1100 = 220
1995	1200	(37 – 19)% of 1200 = 216
1996	1000	(32 – 32)% of 1000 = 00

Hence, required answer = 112 + 201.5 + 220 + 216 + 00 = 749.5 ≈ 750

Problem 103 What is the percentage increase in the average salary of finance from 1992 to 1996?

(a) 60% (b) 32%
(c) 96% (d) 80%

Sol. (d) Percentage increase $= \left(\frac{9810-5450}{5450} \times 100\right)\% = 80\%$

Problem 104 The average annual rate at which the initial salary offered in software increases is :

(a) 21% (b) 33% (c) 15.9% (d) 65%

Sol. (c) Required answer $= \frac{1}{4}\left(\frac{8640-5290}{5290} \times 100\right)\% \approx 15.9\%$

Problem 105 What is the average monthly salary offered to a management graduate in 1993 ?

(a) ₹ 6403 (b) ₹ 6330
(c) ₹ 6333 (d) Cannot be determined

Sol. (a) Average of 6380, 6390 and 6440

$= 6400 \text{ (assumed)} + \frac{-20-10+40}{3}$

$= 6400 + 3.33$

$\approx$ ₹ 6403

Problem 106 In 1994, students seeking jobs in finance earned more than those opting for software(per annum)

(a) ₹ 43 lakh (b) ₹ 33.8 lakh
(c) ₹ 28.41 lakh (d) ₹ 38.8 lakh

Sol. (b) Monthly earning by the students seeking jobs in finance

= ₹ 7550 × (23% of 1100) = ₹ 1910150

Monthly earning by the students seeking jobs in software

= ₹ 7050 × (21% of 1100) = ₹ 1628550

Then, difference in earnings per annum

= 12 (1910150 – 1628550)

= ₹ 3379200 ≈ ₹ 33.8 lakh

Directions (Problems 107-109) *These questions are based on the triangular diagram and bar diagram given below. Study them carefully and answer.*

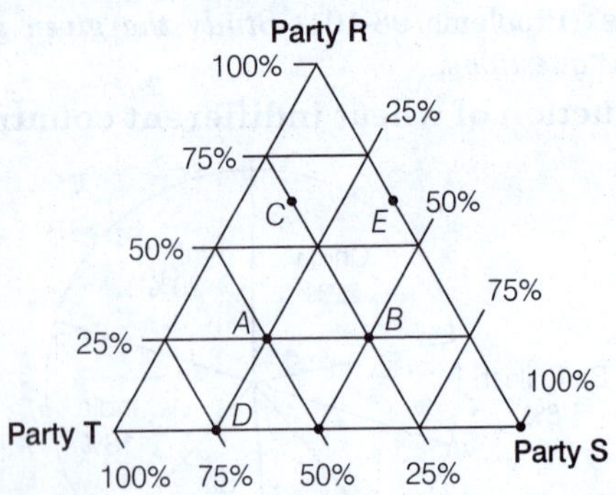

The triangular diagram shows the distribution of votes among three parties *R*, *S* and *T* in five constituencies in an election. The bar diagram shows the total votes cast for the three parties in these five constituencies.

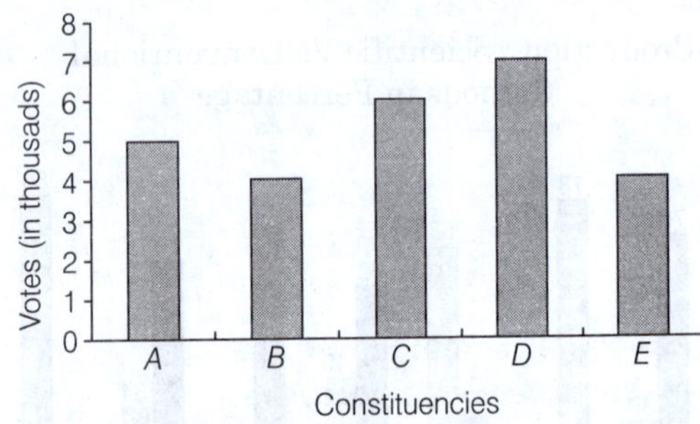

Problem 107 The constituency which casts more votes for *S* than *T* is :

(a) A (b) B
(c) C (d) D

Sol. (b) The constituency nearer to the vertex (of Party S) will cast more votes as compared to other constituencies. It implies that votes casted for Party S by constituency *B* > that by other than *B*. Therefore, we should proceed to check from constituency *B* for the convenience as it is nearest to the vertex (of Party S).

Clearly, constituency B castes 50% for Party S and does 25% for Party T.

Hence, option (b) is correct.

Alternatively,

We may depict the information in the tabular form for our convenience as below:

Parties	Votes casted by constituencies				
	A (5000)	*B* (4000)	*C* (6000)	*D* (7000)	*E* (4000)
R	25%	25%	62.5%	0%	62.5%
S	25%	50%	12.5%	25%	37.5%
T	50%	25%	25%	75%	0%

Clearly, option (b) is correct as 50% of a number is greater than 25% of that number.

Problem 108 Which constituencies cast the same number of votes for Party *R* ?

(a) *A, B* (b) *B, E*
(c) *B, C* (d) None of thsee

Sol. (d) Proceed through options

(a) The votes casted by *A* and *B* are equal in terms of percentage However, the number of votes casted by them shall not be equal as the total votes casted by them(i.e. base numbers) are different.

(b) The votes casted by *B* and *E* are different in terms of percentage. Therefore, the number of votes casted by them shall not be equal as the total votes casted by them (i.e. base numbers) are same.

(c) The votes casted by *B* and *C* are different in terms of percentage. Also, their bases are different. Of course, greater percentage of greater number cannot be equal to lower percentage of lower number.

Hence, option (d) is correct.

Alternatively,

Votes casted by *A* = 25% of 5000 = 1250
Votes casted by *B* = 25% of 4000 = 1000
Votes casted by *C* = 62.5% of 6000 = 3750
Votes casted by *D* = 0% of 7000 = 00
Votes casted by *E* = 62.5% of 4000 = 2500

Clearly, no two constituencies cast the same number of votes. Hence, option (d) is correct.

Problem 109 The total number of votes secured by *T* in the five constituencies together is

(a) 2000–4000 (b) 8000–10000
(c) 4000–6000 (d) 10000–12000

Sol. (d) Required number of votes = 50% of 5000 + 25% of (4000 + 6000) + 75% of 7000 + 0% of 4000
= 25% of 10000 + 25% of 10000 + 52.5% of 10000
= 102.5% of 10000
= 10000 – 12000 = 10250
i.e. between 10000 and 12000
Hence, option (d) is correct.

Directions (Problems 110 - 112) The break-up of the students in a university by subject major is given in the polar pie-chart. The bar chart shows the number of students who major in Physics by geographic location **[XAT-19]**

Percentage of students by Major

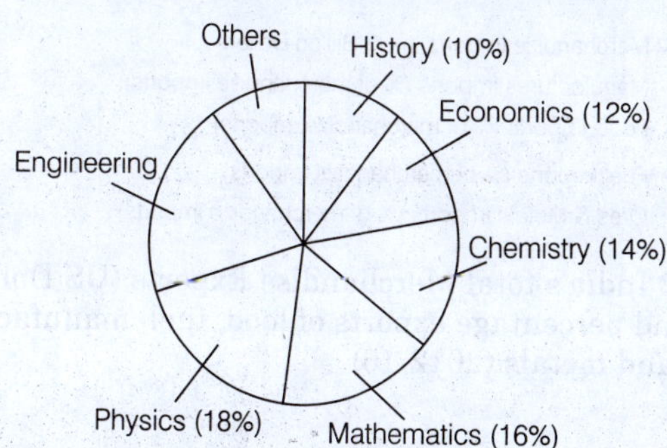

Location of Physics major

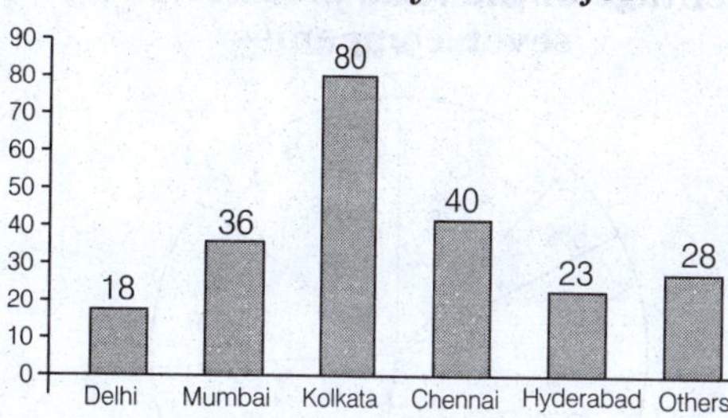

Problem 110 How many students major in Chemistry ?

(a) 170 (b) 175
(c) 190 (d) 180
(e) 200

Sol. (b) Number of students major in Physics
= 18 + 36 + 80 + 40 + 23 + 28
= 225
As given, 18% of total students by major = 225
∴ 14% of total students by major = 175

Problem 111 If the proportion of Physics majors who are from Delhi is the same as the proportion of Engineering majors who are from Delhi, how many Engineering majors are from Delhi ?

(a) 24 (b) 18
(c) 20 (d)26
(e) 22

Sol. (c) As given, 18% of the total students by major = 225
∴ 20% of the total students by major = 250 (visualised from the pie-chart that share of Engineering majors is 20%, also corroborated by AP series)

Now, $\frac{18}{225} = \frac{\text{Number of Engineering majors from Delhi}}{250}$

⇒ Number of Engineering majors from Delhi = 20

Problem 112 12% of all students are from Chennai. What is the largest possible percentage of Economics students that can be from Chennai, rounded off to the nearest integer ?

(a) 75% (b) 73%
(c) 77% (d) 69%
(e) 71%

Sol. (b) From the above, 18% of total students by major = 225
∴ 12% of total students by major = 150 i.e. from Chennai
It is clear from the bar chart that 40 Physics students are from Chennai. Then, maximum possible number of Economics students from Chennai = 150 – 40 = 110
Clearly, 110 is equal to approximately 73% of 150.

Directions (Problems 113 - 116) *Seven companies A, B, C, D, E, F and G are engaged in production of two items I and II. The comparative data about production of these items by the seven companies is given in the following graph and table. Study the information carefully to answer these questions.* **[MAT-09]**

Percentage of the total production by the seven companies

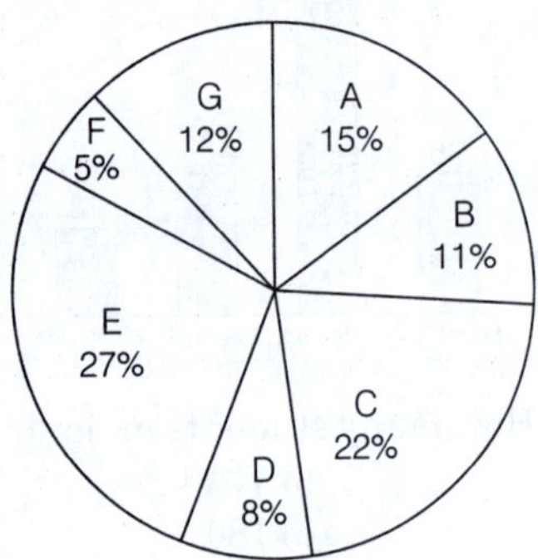

Cost of the total production (both items together) by seven comparies = ₹ 25 cr

Ratio of production between items I and II and the percent profit earned for the two items

Company	Ratio of production		Percentage profit earned	
	Item I	Item II	Item I	Item II
A	2	3	25	20
B	3	2	32	35
C	4	1	20	22
D	3	5	15	25
E	5	3	28	30
F	1	4	35	25
G	1	2	30	24

Problem 113 What is the total profit earned by Company G for items I and II together ?

(a) ₹ 78 lakh (b) ₹ 1.62 crore
(c) ₹ 16.2 lakh (d) None of these

Sol. (a) Recall the concept of weighted average. Percentage profit on items I and II together

$$= \frac{(1 \times 30\%) + (2 \times 24\%)}{1 + 2} = 26\%$$

Then, total profit earned by Company G for items I and II together = 26% of 12% of 25 crore = 0.78 crore

Problem 114 What is the total of the profit earned by Company B on production of item I and the profit earned by Company A on the production of item II ?

(a) ₹ 97.8 lakh (b) ₹ 9.78 crore
(c) ₹ 52.8 lakh (d) ₹ 30.65 lakh

Sol. (a) Profit earned by Company B on item I

= 32% of 60% of 11% of 25 crore [∵ 3 : 2 = 60% : 40%]

and profit earned by Company A on item II

= 20% of 60% of 15% of 25 crore [∵ 2 : 3 = 40% : 60%]

Then, total profit = 60% of (0.88 + 0.75)

= 0.978 crore = 97.8 lakh

Problem 115 What is the total cost of the production of item I by companies A and C together (in ₹ crore) ?

(a) 9.25
(b) 5.9
(c) 4.1625
(d) None of the above

Sol. (b) Required total cost (in ₹ crore)

= 40% of 15% of 25 + 80% of 22% of 25

= 40% of (15 + 44)% of 25

[∵ 80% of 22% of 25 = 40% of 44% of 25]

= 5.9

Problem 116 Cost production of item I by Company F is what percent of the cost of production of item II by Company D ?

(a) 16% (b) 33.33%
(c) 20% (d)12.5%

Sol. (c) Clearly, we have to find, (20% of 5% of 25) is what percent of (62.5% of 8% of 25)?

Then, required percentage $= \left(\frac{20 \times 5}{62.5 \times 8} \times 100\right)\%$

$= 20\%$

Directions (Problems 117 - 121) *Exhibit 1 as under provides the data of India's Merchandize Imports(Billion US Dollar) on left axis and Percentage of Food, Fuel, Manufactures and Ores and Metals Imports of India's on the right axis. Similarly, Exhibit 2 provides data of India's Merchandize Exports(Billion US Dollar) on left axis and Percentage exports of Food, Fuel, Manufactures and Ores and Metals on the right axis.*

Exhibit 1 India's Total Merchandize Imports (US Dollar in Billion) and Percentage Imports of Food, Fuel, Manufactures and Ores and Metals (2012-16).

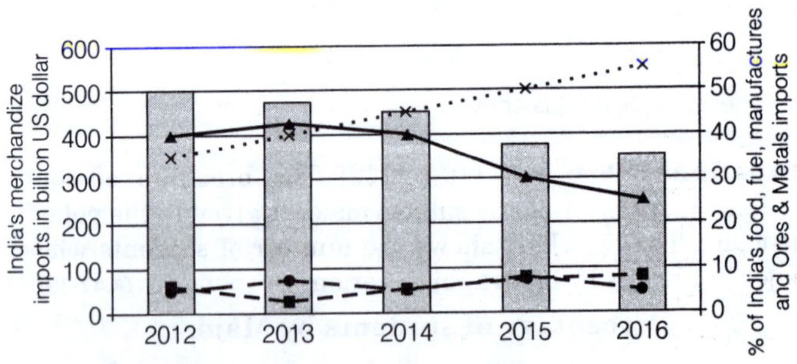

Exhibit 2 India's total Merchandise Exports (US Dollar in Billion) and percentage exports of food, fuel, manufactures and ores and metals (2012-16)

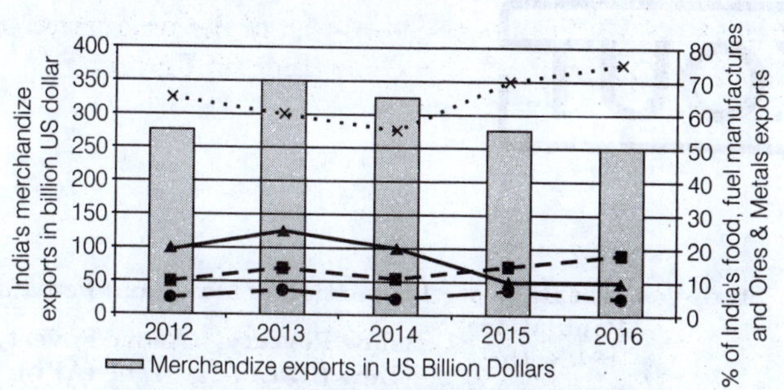

Attempt the questions in the context of the information provided as under:

a. Trade balance = imports minus export

b. Trade deficit = If imports are more than exports

c. Trade surplus = If exports are more than imports

[IIFT-17]

Problem 117 What shall be approximate manufactures exports of India in the year 2016 based on average exports for the period (2012 -2016)?

(a) 221 billion US dollar
(b) 209 billion US dollar
(c) 239 billion US dollar
(d) 236 billion US dollar

Sol. (a) Average exports for (2012 -16) = Average of 275, 350, 325, 275, 250

$$= \text{Assumed average (300)} + \frac{-25 + 50 + 25 - 25 - 50}{5} = 295$$

Manufactures exports in 2016 = 75% of 295 ≈ 221 billion US dollars

Problem 118 What is the proportion of positive and negative manufactures trade balance in the period (2012-16) ?

(a) 3 : 2 (b) 2 : 3
(c) 1 : 4 (d) None of these

Sol. (b) For year 2012 :

Manufactures trade balance = Imports – Exports

= 35% of 500 – 65% of 275 = –3.75

For year 2013 :

Manufactures trade balance

= 40% of 475 – 60% of 350 = – 20

For year 2014 :

Manufactures trade balance

= 45% of 450 – 55% of 325

= 23.75

For year 2015 :

Manufactures trade balance

= 50% of 375 – 70% of 275 = – 5

For year 2016,

Manufactures trade balance

= 55% of 350 – 75% of 250 = 5

Now, required ratio = Number of positive trade balance : Number of negative trade balances = 2 : 3

Problem 119 In which year, trade deficit of fuel has been the second lowest ?

(a) 2015 (b) 2014
(c) 2013 (d) 2016

Sol. (a) For the year 2013,

trade deficit of fuel = 42% of 475 – 25% of 350 = 112

For the year 2014,

trade deficit of fuel = 40% of 450 – 20% of 325 = 115

For the year 2015,

trade deficit of fuel = 30% of 375 – 10% of 275 = 85

For the year 2016,

trade deficit of fuel = 25% of 350 – 10% of 250 = 62.5

Clearly, trade deficit of fuels was second lowest in year 2015.

Problem 120 Assuming India's import and exports grow by 10% in 2017 and sectoral share of fuel in both imports and exports grow by 5% basis, what shall be the approximate deficit of Fuel trade in the year 2017 ?

(a) 88 billion fuel trade deficit
(b) 74 billion fuel trade deficit
(c) 85 billion fuel trade deficit
(d) 78 billion fuel trade deficit

Sol. (b) Fuel trade deficit in 2017 = (25 + 5)% of 110% of 350 – (10 + 5)% of 110% of 250 ≈ 74 billion US dollar

Problem 121 Assuming India's absolute trade deficit grows by 54 billion US dollar and exports become 324 billion US dollar in the year 2017. What shall be India's absolute imports of food and fuel in 2017 if sectoral composition of food, fuel, manufactures Ores and metals remain same as that of 2016 ?

(a) 36 billion US dollar and 119.5 billion US dollar
(b) 38 billion US dollar and 117.5 billion US dollar
(c) 34 billion US dollar and 118.5 billion US dollar
(d) 38 billion US dollar and 116.5 bilion US dollar

Sol. (a) India's trade deficit in 2017 = 54 + (350 – 250) = 154 and India's exports in 2017 = 324

Then, India's imports in 2017 = 324 + 154 = 478

Now, food imports in 2017 = 7.5% of 478

≈ 36 and fuel imports in 2017 = 25% of 478 = 119.5

Hence, option (a) is correct.

WORKOUT

Directions (Q. Nos. 1- 4) *Answer these questions based on the data provided in the table below.* ***[CAT-2000]***

Factory Sector by type of ownership

Sector	Factories	Employment	Fixed capital	Gross output	Value added
Public	7	27.7	43.2	25.8	30.8
Central Govt.	1	10.5	17.5	12.7	14.1
State Local Govt.	5.2	16.2	24.3	11.6	14.9
Central and State/Local Govt.	0.8	1.0	1.4	1.5	1.8
Joint	1.8	5.1	6.8	8.4	8.1
Wholly Private	90.3	64.6	46.8	63.8	58.7
Others	0.9	2.6	3.2	2.0	2.4
Total	100	100	100	100	100

All figures in the table are in percent of the total for the corresponding column.

1. Suppose the average employment level is 60 per factory. The average employment in 'Wholly Private, sector is approximately :
(a) 43 (b) 50
(c) 47 (d) 54

2. Among the firms in different sectors, value added per employee is highest in :
(a) Central Govt
(b) Central and State/Local Govt.
(c) Joint Sector
(d) Wholly Private

3. Capital productivity is defined as the gross output value per rupee of fixed capital. The three sectors with the higher capital productivity, arranged in descending order are :
(a) Joint, Wholly Private, Central and State/Local Govt.
(b) Wholly Private, Joint, Central and State/Local Govt.
(c) Wholly Private, Central and State/Local Govt. Joint
(d) Joint, Wholly Private, Central Govt.

4. The total value added in all sectors is estimated at ₹ 140000 crore. Suppose the number of firms in the Joint sector is 2700. The average value added per factory (in ₹ crore) in the Central Govt. is :
(a) 141
(b) 14.1
(c) 131
(d) 13.1

Directions (Q. Nos. 5-8) *The following table gives the percentage distribution of population of five states, P, Q R, S and T on the basis of poverty line and also on the basis of sex.*

State	Percentage of Population below the Poverty Line	Proportion of Males and Females	
		Below Poverty Line (BPL)	Above Poverty Line (APL)
		M : F	M : F
P	35	5 : 6	6 : 7
Q	25	3 : 5	4 : 5
R	24	1 : 2	2 : 3
S	19	3 : 2	4 : 3
T	15	5 : 3	3 : 2

5. If the male population above poverty line for State R is 1.9 million, then the total population of State R is
(a) 4.5 million (b) 4.85 million
(c) 5.35 million (d) 6.25 million
(e) 7.6 million

6. What will be the male population above poverty line for State P if the female population below poverty line for State P is 2.1 million ?
(a) 2.1 million (b) 2.3 million
(c) 2.7 million (d) 3.3 million
(e) 3.4 million

7. What will be the number of females above the poverty line in the State S if it is known that the population of State S is 7 million ?
(a) 3 million (b) 2.43 million (c) 1.33 million (d) 5.7 million
(e) 1.61 million

8. If the population of males below poverty line for State Q is 2.4 million and that for State T is 6 million, then the total populations of States Q and T are in the ratio :
(a) 1 : 3 (b) 2 : 5 (c) 3 : 7 (d) 4 : 9
(e) 5 :12

Directions (Q.Nos. 9-12) *Study the following tables carefully and answer the questions given below.* ***[SBI (PO)-10]***

Number of candidates appeared in competitive examination from five centres over the years

Centre → / Year ↓	Mumbai	Delhi	Kolkata	Hyderabad	Chennai
2001	35145	65139	45192	51124	37346
2002	17264	58248	52314	50248	48932
2003	24800	63309	56469	52368	51406
2004	28316	70316	71253	54196	52315
2005	36503	69294	69632	58360	55492
2006	29129	59216	64178	48230	57365
2007	32438	61345	56304	49178	58492

Approximate percentage of candidates qualified in the competitive examination from five centres over the years

Mumbai	Delhi	Kolkata	Hyderabad	Chennai
12	24	18	17	9
10	28	12	21	12
15	21	23	25	10
11	27	19	24	8
13	23	16	23	13
14	20	21	19	11
16	19	24	20	14

9. In which of the following years was the difference in number of candidates appeared from Mumbai over the previous year the minimum ?

(a) 2004 (b) 2006
(c) 2007 (d) 2002
(e) None of these

10. In which of the following years was the number of candidates qualified from Chennai the maximum among the given years ?

(a) 2005 (b) 2006
(c) 2007 (d) 2003
(e) None of these

11. Approximately what was the total number of candidates qualified from Delhi in 2002 and 2006 together ?

(a) 27250 (b) 25230
(c) 30150 (d) 28150
(e) 26250

12. Approximately what was the difference between the number of candidates qualified from Hyderabad in 2001 and that in 2002 ?

(a) 1680 (b) 2440
(c) 1450 (d) 2060
(e) 1860

Directions (Q. Nos. 13-16) *Study the bar graph below to answer these questions.* **[MAT-09]**

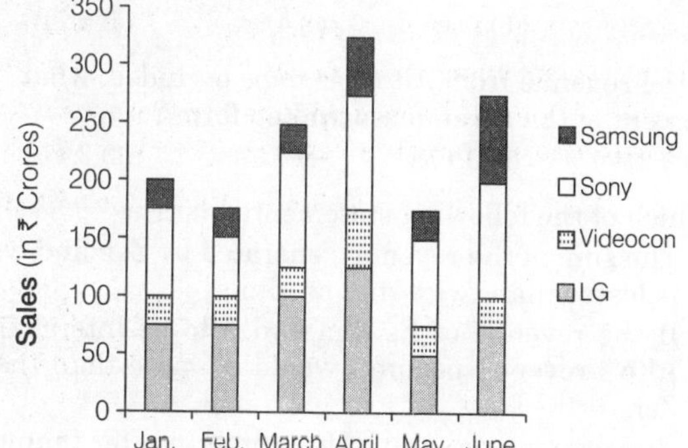

13. In which of the following month(s) is Sony's profit the highest, if it makes 25% profit on its sales turnover on each month ?

(a) January and May (b) May
(c) February, May and June (d) March, April and June

14. In which of the given months is the sales turnover of LG as a percentage of the combined sales turnover of Videocon and Samsung the least ?

(a) March (b) June
(c) January (d) May

15. In how many months was the sales turnover of Sony more than one-third of the total sales turnover of the four companies ?

(a) Three months (b) One month
(c) Five months (d) None of these

16. In how many months from February to June, did the sales turnover of LG as a percentage of total sales turnover of the four companies increase over that of its immediate preceding month ?

(a) Three months (b) One month
(c) Two months (d) None of these

Directions (Q. Nos. 17-23) *Answer the questions based on the following information.* **[CAT-98]**

In the following chart, the price of logs shown is per cubic metre and that of plywood and Sawn timber is per tonne.

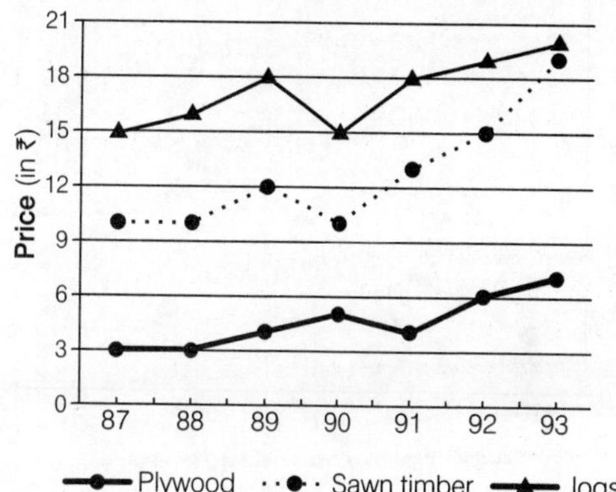

17. What is the maximum percentage increase in price per cubic metre or per tonne over the previous year ?

(a) 33.33% (b) 85%
(c) 50% (d) Cannot be determined

18. Which product shows the maximum percentage increase in price over the period ?

(a) Sawn timber (b) Plywood
(c) Logs (d) Cannot be determined

19. If 1 m^3 = 750 kg for Sawn timber, find in which year was the difference in prices of Sawn timber and Logs the least ?

(a) 1989 (b) 1990 (c) 1991 (d) 1992

20. If one cubic metre = 700 kg for Plywood and 800 kg for Sawn timber, find in which year was the difference in the prices of Plywood and Sawn timber (per cubic metre) the maximum ?

(a) 1989 (b) 1990 (c) 1991 (d) 1992

21. If the volumes of sales of Plywood, Sawn timber and Logs were 40%, 30% and 30% respectively, then what was the average realisation in 1993 per cubic metre of sales ? (Weight of one cubic metre of Sawn timber and Plywood both = 800 kg)

(a) 18 (b) 15
(d) 13 (c) 16

22. In the previous question, if in 1994, prices increased by 5%, 1% and 10% while the volume sales break–up was 40%, 30% and 30% for Plywood, Sawn timber and Logs respectively, then what was the average realisation ?

(a) 18.95 (b) 16.45
(c) 13.55 (d) 10.25

Directions (Q. Nos. 23 - 26) *Answer the questions based on the following information.*

Ghosh Babu has a manufacturing unit. The following graph gives the cost for various number of units.

Given that Profit = Revenue – Variable cost – Fixed cost

The fixed cost remains constant up to 34 units after which additional investment is to be done in fixed assets. In any case, production cannot exceed 50 units.

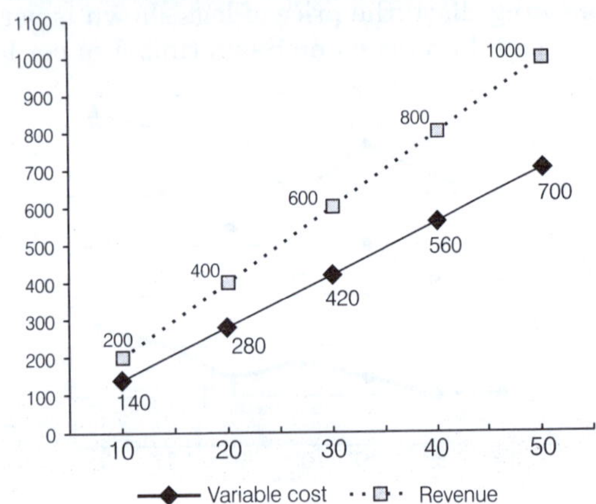

Note

I. Fixed cost upto 34 units is ₹ 70 and above 34, it is ₹ 100.

23. What is the minimum number of units that need to be produced to make sure that there was no loss ?

(a) 5 (b) 12
(c) 20 (d) Cannot be determined

24. How many units should be manufactured such that profit was at least ₹ 50 ?

(a) 20 (b) 34
(c) 45 (d) 30

25. If at the most 40 units can be manufactured, then what is the number of units that can be manufactured to maximize profit per unit ?

(a) 40 (b) 34 (c) 35 (d) 25

26. If the fixed cost of production goes up by ₹ 40, then what is the minimum number of unit that need to be manufactured to make sure that there is no loss ?

(a) 10 (b) 19 (c) 15 (d) 20

Directions (Q. Nos. 27-31) *Answer the questions based on the following information.*

Break-up of total revenue

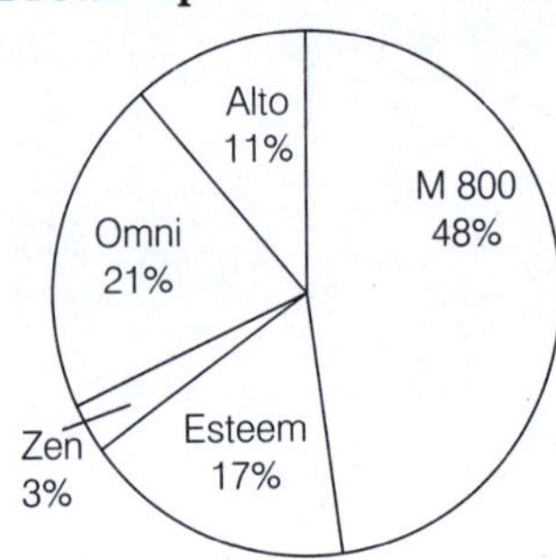

Break-up of Total Sales

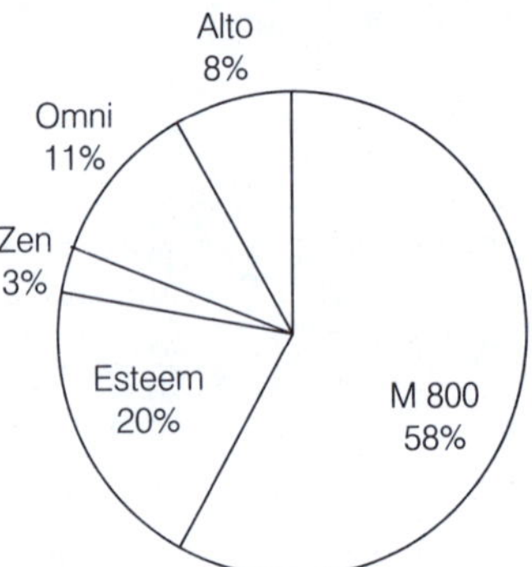

The pie–charts show the break–up of the total revenue and total sales volume for a company, for the year ended 2003. (Total revenue = ₹ 1139200 million and total sales volume = 653610 unit)

27. What is the revenue per Esteem sold, approximately ?

(a) ₹ 1481000 (b) ₹ 1532000
(c) ₹ 1348000 (d) ₹ 1326000

28. The highest revenue per vehicle comes from :

(a) M 800 (b) Omni (c) Esteem (d) Zen

29. If volume of sales of Omni was increased by 25% and this sales increase was entirely at the cost of M 800, then what is the percentage decrease in M 800 volume sales ?

(a) 1.3% (b) 2.4% (c) 4.3% (d) 4.74%

30. If the revenue from Alto was to be excluded, what percent of the revenue would Zen form ?

(a) 3.27% (b) 3.17% (c) 3.37% (d) 3.47%

31. Which of the following statement(s) is true ?

(i) The sum of the revenue generated by Zen and Omni is lesser than twice that of Alto.

(ii) If the revenue of Esteem and Alto be interchanged, Alto's revenue per unit would be more than that for Zen.

(iii) The revenue per unit of Esteem is greater than that for M 800.

(a) Only (i) (b) Only (ii)
(c) Only (iii) (d) Both (ii) and (iii)

Directions (Q. Nos. 32-35) *These questions are based on the table and pie charts. The table gives the demand for gas in Million Metric Standard Cubic Metres Per Day (MMSCMPD).*

Region	Year (1999–2000)	Year (2004–2005)
North	49.48	64.81
West	37.00	78.13
South	39.90	88.31
East	20.16	52.65
Total	146.54	283.90

Sector-wise share : 1999-2000

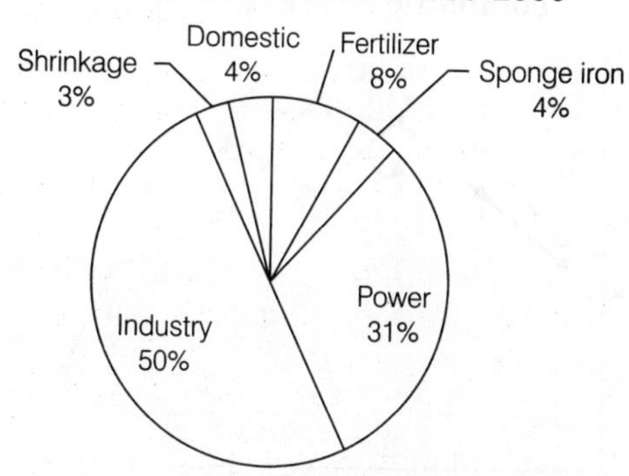

Sector-wise share : 2004-2005

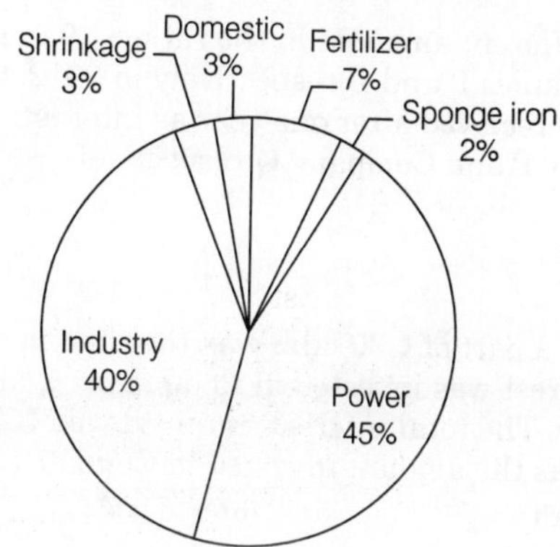

32. What is the percentage growth in the all India demand for Industry sector between 1999-2000 and 2004-2005 ?

(a) 53% (b) 55%
(c) 54% (d) 52%

33. Had the all India fertilizer sector demand shown the same growth rate as the total all India demand between 1999 and 2004. What would be share of fertilizer sector in total demand in 2004-2005 ?

(a) 22.71 (b) 24.21
(c) 23.81 (d) 21.91

34. By what percentage did the southern demand grow between 1999 and 2004 ?

(a) 120.21% (b) 121.32%
(c) 125.61% (d) 134.1%

35. If the all India sector-wise demand distribution is exactly the same in South, what is the industry sector demand from North, East and West combined in 2004-2005 ?

(a) 71.56 (b) 77.456 (c) 78.236 (d) 81.236

Directions (Q. Nos. 36-40) *Study the graphs to answer these questions :*

Total sales volume = 1230 kg

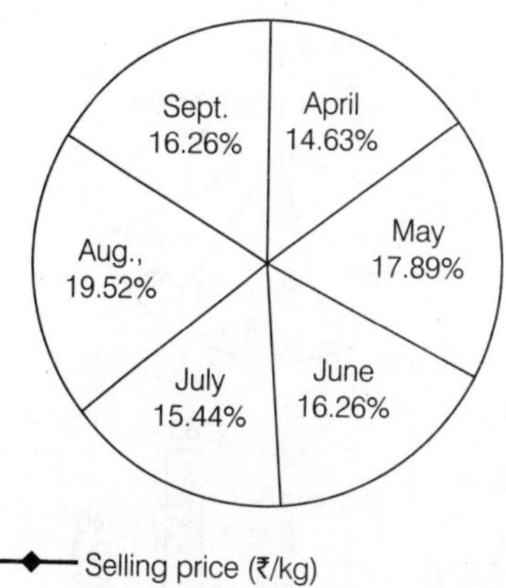

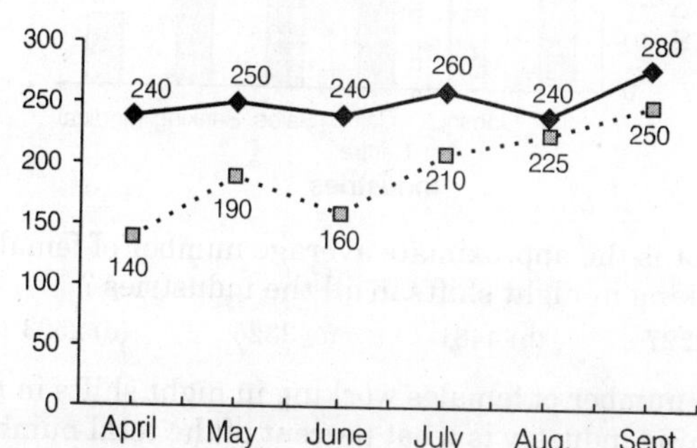

36. In which month was the cost/sales ratio the lowest ?

(a) April (b) July (c) August (d) June

37. Maximum profit is in the month of :

(a) April (b) September (c) August (d) June

38. In which month, the cost as percentage of the previous month cost escalated the most ?

(a) May (b) September (c) August (d) July

39. What was the decrease/increase in profits from July to August ?

(a) ₹ 2400 less (b) ₹ 7000 less
(c) ₹ 5900 less (d) ₹ 3400 less

40. If the cost price growth from September to October is equal to that from August to September, what would be the total cost per kg in October ?

(a) ₹ 290.87 (b) ₹ 281.58
(c) ₹ 61.85 (d) ₹ 277.78

Directions (Q. Nos. 41-45) *Study the following graphs to answer these questions :* *[MAT-13]*

The following pie–chart gives the percentage of adults from various industries working in night shifts. The bar graph shows the percentage of females from these industries working in the night shifts. Total number of people = 40250

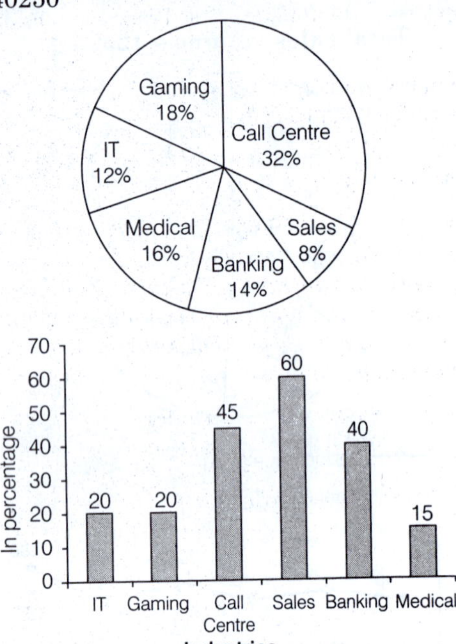

41. What is the approximate average number of females working in night shifts in all the industries ?
(a) 2227 (b) 4481 (c) 3326 (d) 2523

42. The number of females working in night shifts in the gaming industry is what percent of the total number of people working in the night shifts from all the industries taken together ?
(a) 5.6 (b) 3.6
(c) 3.2 (d) 4.4

43. What is the difference in percentage of males and females working in night shifts in these industries ?
(a) 33.2 (b) 23.8
(c) 33.6 (d) None of these

44. If the number of females working in night shifts in banking were 3206, while the number of females in rest of industries remained unchanged, the new percentage of females working in night shifts will be : (assume that total number of people working in night shifts remains constant)
(a) 35.56% (b) 33.72%
(c) 23.82% (d) Cannot be determined

45. Percentage of males working in night shifts is lowest in which of the following industries ?
(a) Sales (b) IT and Medical
(c) Banking (d) Gaming

Directions (Q. Nos. 46-50) *Study the following graph to answer these questions :*

Three different finance companies declare fixed annual rate of interest on the amounts invested with them by investors. The rate of interest offered by these companies may differ from year-to-year depending on the variation in the economy of the country and the bank's rate of interest. The annual rate of interest offered by the three companies P, Q and R over the years are shown by the line graph provided below:

Annual rate of interest offered by three finance companies over the years

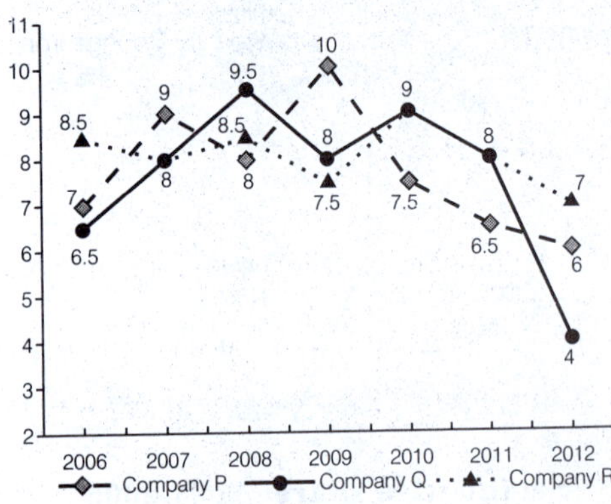

46. If two different amounts in the ratio 8 : 9 are invested in Companies P and Q respectively in 2012, then the amounts received after one year as interests from Company P and Company Q respectively, are in the ratio :
(a) 2 : 3 (b) 3 : 4
(c) 6 : 7 (d) 4 : 3

47. In 2010, a part of ₹ 30 lakh was Invested in company P and the rest was invested in Companies Q and R for one year. The total interest, received was ₹ 2.43 lakh. What was the amount invested in Company P ?
(a) ₹ 7 lakh (b) ₹ 18 lakh
(c) ₹ 9 lakh (d) Cannot be determined

48. An investor invested a sum of ₹ 15 lakh in Company P in 2008. The total amount received after one year was re-invested in the same Company for one more year. The total appreciation received by the investor on his investment was :
(a) ₹ 282000 (b) ₹ 296200
(c) ₹ 216000 (d) ₹ 225600

49. In 2010, if ₹ 30 lakh were invested in companies P, Q and R for one year in the ratio 2 : 1 : 4, then what was the total interest earned in that year ?
(a) ₹ 2.67 lakh (b) ₹ 2.57 lakh
(c) ₹ 3.12 lakh (d) ₹ 2.48 lakh

50. The average annual rate of interest offered for the given period is highest for :
(a) Company P (b) Company Q
(c) Company R (d) Cannot be determined

Directions (Q. Nos. 51 - 53) *Consider the following for the next three items.*

In a certain town of population size 100000 three types of newspapers (I, II and III) are available. The percentage of the people in the town who read these papers are as follows:

Newspaper	Proportion of readers
I	10%
II	30%
III	5%
Both I and II	8%
Both II and III	4%
Both I and III	2%
All of three (I, II and III)	1%

[CDS-19]

51. What is the number of people who read only one newspaper ?

(a) 20000 (b) 25000 (c) 30000 (d) 35000

52. What is the number of people who read at least two newspapers ?

(a) 12000 (b) 13000 (c) 14000 (d) 15000

53. What is the number of people who do not read any of these three newspapers ?

(a) 62000 (b) 64000 (c) 66000 (d) 68000

Directions (Q. Nos. 54-58) *Study the table carefully to answer the questions that follow :* *[IBPS–CWE (PO/MT)–11]*

Percentage of marks obtained by different students in different subjects of MBA.

Students	Subjects (Maximum Marks)					
	Strategic Management (150)	Brand Management (100)	Compensation Management (150)	Consumer Behaviour (125)	Service Marketing (75)	Training and Development (50)
Anushka	66	75	88	56	56	90
Archit	82	76	84	96	92	88
Arpan	76	66	78	88	72	70
Garvita	90	88	96	76	84	86
Gunit	64	70	68	72	68	74
Pranita	48	56	50	64	64	58

54. How many marks did Anushka get in all the subjects together ?

(a) 369 (b) 463 (c) 558 (d) 496
(e) None of these

55. The marks obtained by Garvita in Brand management is what percent of the marks obtained by Archit in the same subject ? (Rounded off to two digits after decimal)

(a) 86.36 (b) 101.71
(c) 115.79 (d) 133.33
(e) None of these

56. What is the average marks obtained by all students together in Compensation Management ?

(a) 116 (b) 120 (c) 123 (d) 131
(e) None of these

57. Who has scored highest total marks in all the subjects together ?

(a) Archit (b) Gunit (c) Pranita (d) Garvita
(e) Arpan

58. How many students have scored the highest marks in more than one subject ?

(a) 3 (b) 2 (c) 1 (d) 0
(e) None of these

Directions (Q. Nos. 59-63) *The given table shows the number(in percent) of employees working in different departments of an organization. The table also shows the ratio of males and females and the ratio of employees living in City Z and employees living in City Y. The total number of employees in the organization is 80000.* *[SSC (CGL-T-II)-17]*

Department	Number of employees	Gender	City
		M : F	Z : Y
A	10%	7 : 3	1 : 9
B	22%	13 : 9	3 : 19
C	12%	1 : 2	5 : 1
D	20%	3 : 2	1 : 3
E	36%	8 : 1	5 : 13

59. How many employees of department A and C together are living in City Z ?

(a) 8800 (b) 9000 (c) 8200 (d) 9200

60. Male employees of department E is what percent of the employees living in City Z from department A ?

(a) 1600 (b) 4200 (c) 3200 (d) 2400

61. What is the ratio of male employees working in department B and department D together to female employees working in department A and E together ?

(a) 7 : 9 (b) 25 : 7 (c) 23 : 9 (d) 13 : 8

62. What are the total number of employees in department A and department E together ?

(a) 36800 (b) 17600 (c) 46400 (d) 29400

63. On an average how many residents of City Y are working in each department ?

(a) 9130 (b) 10940 (c) 12420 (d) 11360

Directions (Q. Nos. 64-68) *Study the data- table carefully and answer the given questions.*

Position	No. of applications received	No. of duplicate applicants	Average no. of duplicate applications received from duplicate applicant
A	1040	63	4
B	880		6
C	600	28	–
D	—	48	–
E	420	—	–

Note

I. A duplicate applicant is an applicant who has submitted additional (duplicate) applications after submitting their original application.

II. All application forms (original and duplicate) received from duplicate applicants were rejected, remaining all applications were accepted.

III. None of the applicants applied for more than one post.

[IBPS(PO–Mains)–19]

64. For position A, if the respective ratio of number of accepted applications from males to that of rejected applications from males is 5 : 3 and if the respective ratio of number of accepted applications from females to that of rejected applications from females is 5 :1, then what is number of rejected applications from males ?

(a) 210 (b) 255
(c) 270 (d) 240
(e) 285

65. If the average number of accepted applications for positions A and B is 659, what is the number of duplicate applicants for Position B ?

(a) 41 (b) 43
(c) 39 (d) 37
(e) 45

66. For Position D, if the ratio of accepted and rejected applications is 4 : 1, which of the following can be true?

(a) Number of applications received (all originals + all duplicates) can be 240.

(b) Number of applications accepted (all originals + all duplicates) can be 786.

(c) The least number of applications were received for position D is a possibility.

(d) None of the above

67. For Position E, the number of accepted applications from the males, number of accepted applications from the females and the number of rejected applications(males and females) are X, X + Y and X + 2Y, respectively. Then, which of the following is true ? Considering that the average number of duplicate applications received from duplicate applicant is non–zero integer.

(a) Number of accepted applications from the males can be 131.

(b) Number of accepted applications from the males can be 141.

(c) Number of accepted applications from the males can be 139.

(d) None of the above

68. For Position C, the number of accepted applications from males is between 150 and 200 and that from females is between 130 and 180. Which of the following can be a possible value(s) of the average number of duplicate applications submitted by the duplicate applicants for Position C ?

(I) 5 (II) 7 (III) 9 (IV) 11 (V) 13

(a) Only I (b) Only II
(c) II and III (d) All of I, II, III, IV and V
(e) None of these

Directions (Q. Nos. 69-72) *Five vendors are being considered for a service. The evaluation committee evaluated each vendor on six aspects–Cost, Customer Service, Features, Quality, Reach and Reliability. Each of these evaluations are on a scale of 0 (worst) to 100 (perfect). The evaluation scores on these aspects are shown in the radar chart. For example, Vendor 1 obtains a score of 52 on Reliability, Vendor 2 obtains a score of 45 on Features and Vendor 3 obtains a score of 90 on Cost.*

[CAT–19]

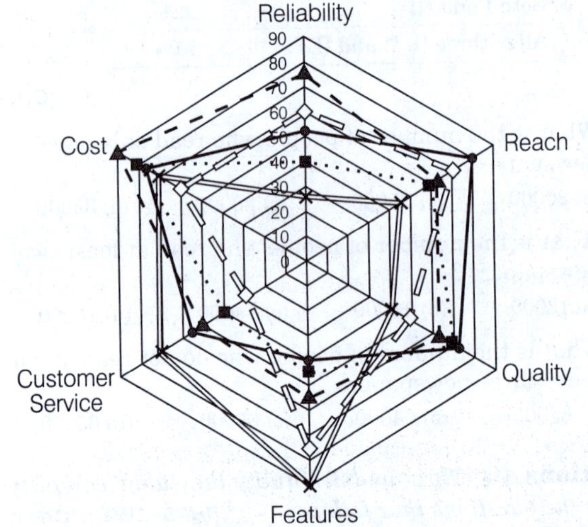

69. On which aspect is the median score of the five vendors the least ?

(a) Customer Service (b) Reliability
(c) Cost (d) Quality

70. A vendor's final score is the average of their scores on all six aspects. Which vendor has the highest final score ?

(a) Vendor 1 (b) Vendor 2
(c) Vendor 3 (d) Vendor 4

71. List of all the vendors who are among the top two scorers on the maximum number of aspects is :

(a) Vendor 1,5 (b) Vendor 2, 3, 4
(c) Vendor 2,5 (d) Vendor 1,2

72. List of all the vendors who are among the top three vendors on all six aspects is :

(a) Vendor 1

(b) Vendor 3

(c) Vendor 1, 3

(d) None of the vendors

Directions (Q. Nos. 73 - 76) *The analyst at a retail giant has modelled the consumer spending and consumer sentiment to prepare a strategy on product offers in the upcoming festive season. The data given in the figure are from last 10 yrs on three parameters namely food spending (Food-spend), style spending (Style-Spend) and sentiment index (Sentiment-Index). Study the figure below and answer the following questions.* **[IIFT-19]**

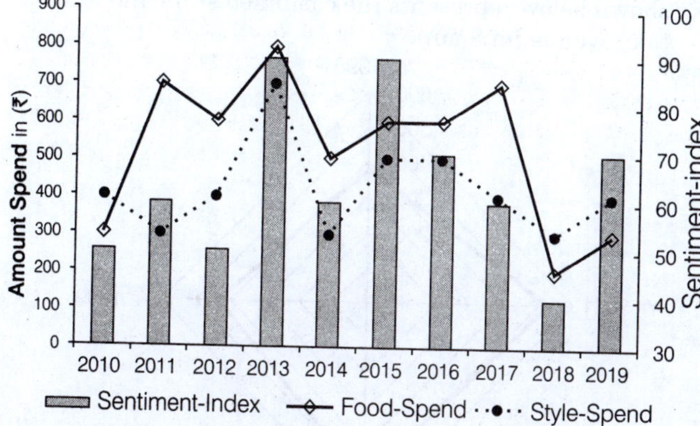

73. From the year mentioned above, in which year total of Food-Spend and Style-Spend as a proportion of Sentiment-Index is lowest ?

(a) 2011 (b) 2013
(c) 2016 (d) 2017

74. From the year mentioned above, in which year Food-Spend as a proportion of Style-Spend is lowest ?

(a) 2012 (b) 2013
(c) 2015 (d) 2016

75. In which year, the highest annual percentage change in Style-Spend is recorded ?

(a) 2011 (b) 2013
(c) 2014 (d) 2015

76. In which year, ratio of annual change in Style-Spend to annual change in Sentiment-Index is highest?

(a) 2011 (b) 2012 (b) 2014 (d) 2017

Directions (Q. Nos. 77-80) *To compare the rainfall data, India Meteorological Department (IMD) calculated the Long Period Average(LPA) of rainfall during period June–August for each of the 16 states. The figure given below shows the actual rainfall (measured in mm) during June-August, 2019 and the percentage deviations from LPA of respective states in 2018. Each state along with its actual rain fall is presented in the figure.* **[CAT-19]**

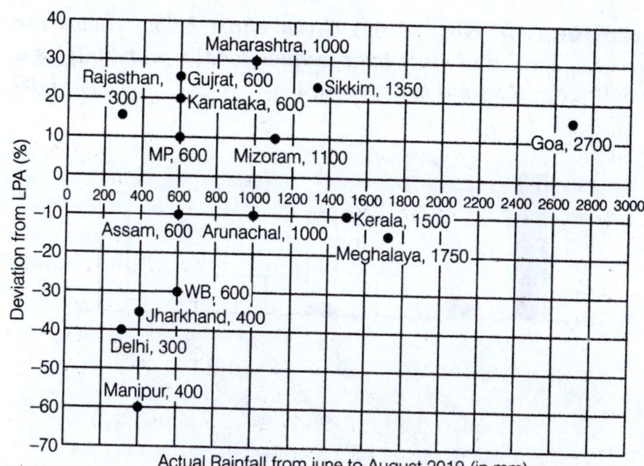

77. If a 'Heavy Monsoon State' is defined as a state with actual rainfall from June–August, 2019 of 900 mm or more, then approximately what percentage of 'Heavy Monsoon State' have a negative deviation from respective LPAs in 2019 ?

(a) 75.00 (b) 42.86
(c) 57.14 (d) 14.29

78. If a 'Low Monsoon State' is defined as a state with actual rainfall from June–August, 2019 of 750 mm or less, then what is the median 'deviation from LPA' (as defined in the Y–axis of the figure) of 'Low Monsoon State'?

(a) –10% (b) –30%
(c) –20% (d) 10%

79. What is the average rainfall of all states that have actual rainfall of 600 mm or less in 2019 and have a negative deviation from LPA?

(a) 500 mm (b) 460 mm
(c) 367 mm (d) 450 mm

80. The LPA of a state for a year is defined as the average rainfall in the preceding 10 yrs considering the period of June–August. *For example*, LPA in 2018 is the average rainfall during 2009–2018 and LPA in 2019 is the average rainfall during 2010–2019. It is observed that the actual rainfall in Gujrat in 2019 is 20% more than the rainfall in 2009. The LPA of Gujrat in 2019 is closest to :

(a) 525 mm
(b) 490 mm
(c) 475 mm
(d) 505 mm

Directions (Q. Nos. 81-82) *Given chart below shows the sources of local and state tax revenues for the year 2001 as a percentage of the total tax revenues.* **[CAT-15]**

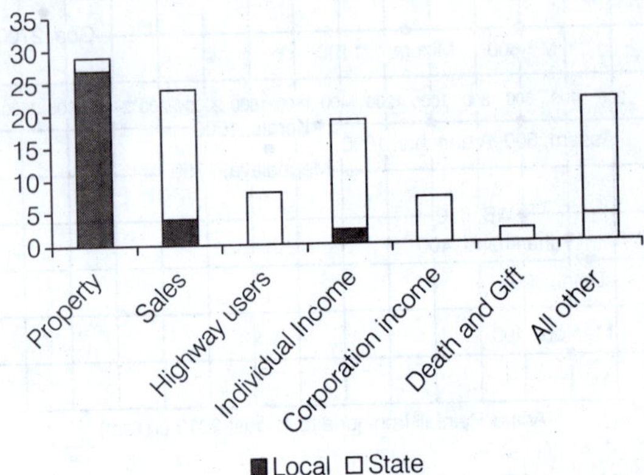

81. Find the difference between the tax revenue generated from the individual income at local and state levels in 2001 in \$ mn.

(a) 64.5

(b) 78.75

(c) 40.25

(d) None of the above

82. If revenue distribution in 2000 was same as in 2001 and tax from highway users was collected for the first time, then what was the percentage increase in tax revenue from source other than highway users in 2000 over the previous year ?

(a) 8.5% (b) 10%

(c) 16.5% (d) 4.5%

With reference to the above question, the diagram shown below represents the combined state and local tax revenue (in \$ mn).

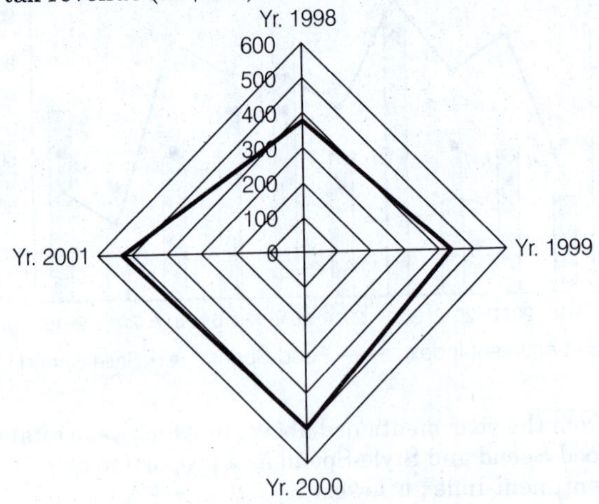

Answer Sheet

1. (a)	2. (b)	3. (b)	4. (d)	5. (d)	6. (d)	7. (b)	8. (b)	9. (c)	10. (c)
11. (d)	12. (e)	13. (d)	14. (b)	15. (d)	16. (b)	17. (c)	18. (b)	19. (b)	20. (d)
21. (d)	22. (c)	23. (b)	24. (a)	25. (b)	26. (b)	27. (a)	28. (b)	29. (d)	30. (c)
31. (d)	32. (b)	33. (a)	34. (b)	35. (c)	36. (a)	37. (a)	38. (a)	39. (c)	40. (d)
41. (a)	42. (b)	43. (c)	44. (a)	45. (a)	46. (d)	47. (b)	48. (a)	49. (b)	50. (c)
51. (a)	52. (a)	53. (d)	54. (b)	55. (c)	56. (a)	57. (d)	58. (b)	59. (a)	60. (c)
61. (b)	62. (a)	63. (d)	64. (b)	65. (a)	66. (b)	67. (c)	68. (c)	69. (a)	70. (c)
71. (a)	72. (b)	73. (c)	74. (b)	75. (b)	76. (c)	77. (b)	78. (a)	79. (b)	80. (b)
81. (b)	82. (d)								

Hints and SOLUTIONS

1. Total employments in 'Wholly Private' sector = 64.6 × 60

$\therefore$ Average employment $= \dfrac{64.6 \times 60}{90.3} \approx 43$

2. Value added per employee in Public, Central Govt., State Local Govt., Central and State/Local Govt, Joint, Wholly Private and Others are Respectively : $\dfrac{30.8}{27.7}, \dfrac{14.1}{10.5}, \dfrac{14.9}{16.2}, \dfrac{1.8}{1}, \dfrac{8.1}{5.1}, \dfrac{58.7}{64.6}, \dfrac{2.4}{2.6}$

$\Rightarrow$ 1.1, 1.3, less than 1, 1.8, 1.6, less than 1, less than 1

Hence, option (b) is correct.

3. Capital productivity in Central Govt., Wholly Private, Joint, Central and State/Local Govt. are

$\dfrac{12.7}{17.5}, \dfrac{63.8}{46.8}, \dfrac{8.4}{6.8}, \dfrac{1.5}{1.4}$, respectively.

$\Rightarrow$ 0.7, 1.4, 1.3, 1.07

Hence, option (b) is correct.

4. Value added in Central Govt. = 14.1% of 140000

As given, 1.8% of total factories = 2700

$\therefore$ 1% of total factories = 1500

Then, value added per factory in Central Govt.

$= \dfrac{14.1\% \text{ of } 140000}{1500} = \dfrac{14.1 \times 14}{15} = 13.1$

5. APL in State R = 76% of the total population

As given, (2 + 3) units = 76% of the total population

$\Rightarrow 2 \text{ units} = \dfrac{2}{5} \times 76\%$ of the total population

$\Rightarrow 1.9 \text{ million} = \dfrac{2}{5} \times 76\%$ of the total population

$\Rightarrow$ 16% of the total population = 1 million

$\Rightarrow$ 100% of the total population = 6.25 million

Hence, option (d) is correct.

6. As given, in BPL, M : F = 5 : 6

Hence, 6 units = 2.1

$\therefore$ (5 + 6) units = 3.85

Clearly, 35% of the population = 3.85

$\therefore$ 65% of the population i.e. population (APL) $= \dfrac{3.85}{35} \times 65$

Now, $(6 + 7) \text{ units} = \dfrac{3.85}{35} \times 65$

$\therefore 6 \text{ units} = \dfrac{3.85}{35} \times 65 \times \dfrac{6}{13} = 3.3$

Hence, option (d) is correct.

7. Population (APL) = (100 − 19)% of 7

Clearly, (4 + 3) units = 81% of 7

$\therefore \quad 3 \text{ units} = 81\% \text{ of } 7 \times \dfrac{3}{7} = 2.43$

Hence, option (b) is correct.

8. For State Q In BPL, M : F = 3 : 5 Clearly, 3 units = 2.4

$\therefore$ (3 + 5) units = 0.8 × 8

$\Rightarrow$ 25% of total population = 0.8 × 8

$\Rightarrow$ 100% of total population = 0.8 × 8 × 4

For State T In BPL, M : F = 5 : 3

As given, 5 units = 6

$\therefore$ (5 + 3) units = 1.2 × 8

$\Rightarrow$ 15% of total population = 1.2 × 8

$\Rightarrow$ 100% of total population = 1.2 × 8 × 6.66

Now, required ratio = 0.8 × 8 × 4 : 1.2 × 8 × 6.66 = 2 : 5

Alternatively

Sum of units are same in both cases. But, value of one unit of T is 1.5 times that of State Q ($\because$ 1.2 = 1.5 × 0.8).

Then, required ratio = 4 × 1 : 6.66 × 1.5 = 2 : 5

9. Looking at the table carefully we get the desired difference is minimum either in case of 2004 or 2007.

In case of 2004, desired difference

= 28316 − 24800 = 3516

In case of 2007, desired difference

= 32438 − 29129 = 3309

Hence, option (c) is correct.

10.

Year	Qualified from Chennai
2001	9% of 37346 = 3361.14
2002	12% of 48932 = 5871.84
2003	10% of 51406 = 5140.60
2004	8% of 52315 = 4185.20
2005	13% of 55492 = 7213.96
2006	11% of 57365 = 6310.15
2007	14% of 58492 = 8188.88

Clearly, option (c) is correct.

Alternatively,

Since both the number of candidates appeared and the percentage of qualified candidates in 2007 are highest in comparison to other years, the number of qualified candidate will be maximum in 2007 among the given years.

11. The required number of candidates

= 28% of 58248 + 20% of 59216

= 14% of 116496 + 11843.20

= 16309.44 + 11843.20 ≈ 28150

Alternatively,

The required number = 20% of (58248 + 59216) + 8% of 58248 ≈ 28150

12. The required number of candidates

= 21% of 50248 − 17% of 51124

= 10552.08 − 8691.08

= 1861 ≈ 1860

13. Clearly, more turnover, more profit as profit percentage is fixed on turnover.

Since sales in each of March, April and June months is 100 crore (maximum), option (d) is correct.

14. In March, required percentage

$$= \left(\frac{100}{25+25} \times 100\right)\% = 200\%$$

In June, required percentage

$$= \left(\frac{75}{25+75} \times 100\right)\% = 75\%$$

In January, required percentage

$$= \left(\frac{75}{25+25} \times 100\right)\% = 150\%$$

In May, required percentage

$$= \left(\frac{50}{25+25} \times 100\right)\% = 100\%$$

Hence, option (b) is correct.

15. Required fraction in January $= \frac{75}{200} = \frac{3}{8} > \frac{1}{3}$

Required fraction in February $= \frac{50}{175} = \frac{2}{7} < \frac{1}{3}$

Required fraction in March $= \frac{100}{250} = \frac{2}{5} > \frac{1}{3}$

Required fraction in April $= \frac{100}{325} = \frac{4}{13} < \frac{1}{3}$

Required fraction in May $= \frac{75}{175} = \frac{3}{7} > \frac{1}{3}$

Required fraction in June $= \frac{100}{275} = \frac{4}{11} > \frac{1}{3}$

Hence, option (d) is correct.

16. Sales turnover of LG as percentage of total sales turnover in January.

$$= \left(\frac{75}{200} \times 100\right)\% = 37.5\%$$

Aforesaid percentage in February.

$$= \left(\frac{75}{175} \times 100\right)\% = 42.85\%$$

Aforesaid percentage in March

$$= \left(\frac{100}{250} \times 100\right)\% = 40\%$$

Aforesaid percentage in April

$$= \left(\frac{125}{325} \times 100\right)\% = 38.46\%$$

Aforesaid percentage in May

$$= \left(\frac{50}{175} \times 100\right)\% = 28.57\%$$

Aforesaid percentage in June

$$= \left(\frac{75}{275} \times 100\right)\% = 27.27\%$$

Clearly, in month of February the aforesaid percentage is greater than that in January.

Hence, option (b) is correct.

Alternatively,

Compare

$$\frac{75}{200}, \frac{75}{175}, \frac{100}{250}, \frac{125}{325}, \frac{50}{175}, \frac{75}{275}$$

$$\Rightarrow \frac{3}{8}, \frac{3}{7}, \frac{2}{5}, \frac{5}{13}, \frac{2}{7}, \frac{3}{11}$$

Clearly, $\frac{3}{8} < \frac{3}{7} > \frac{2}{5} > \frac{5}{13} > \frac{2}{7} > \frac{3}{11}$

($\because$ 3.7 < 3.8, 3.5 > 2.7 and so on (recall the method for 'Comparison of Fractions')

17. It is clear from the table that the price of Plywood has increased by 50% (from 1991 to 1992), which is maximum among other percentage increases over the previous year in respect of all three products.

18. Percentage increase in price of Sawn timber over the period = 90% ($\because 10 \to 19$)

Percentage increase in price of Plywood timber over the period = 133.33% ($\because 3 \to 7$)

Percentage increase in price of Logs timber over the period = 33.33% ($\because 15 \to 20$)

Hence, option (b) is correct.

19. Clearly, weight of 1 m^3 Log :1 tonne

= 750 :1000 = 3 : 4

Year	Price of Sawn timber in ₹ (per tonne)	Price of Sawn timber Price in ₹ (per cubic metre)	Price of Logs in ₹ (per cubic metre)	Difference in Prices in ₹ (per cubic metre)
1989	12	$12 \times \frac{3}{4} = 9$	18	9
1990	10	$10 \times 0.75 = 7.5$	15	7.5
1991	13	$13 \times 0.75 = 9.75$	18	8.25
1992	15	$15 \times 0.75 = 11.25$	19	7.75

Clearly, the difference is least in year 1990.

20.

Year	Price of Plywood in ₹ (per tonne)	Price of Plywood in ₹ (Per cubic metre)	Price of Sawn timber in ₹ (per tonne)	Price of Sawn timber in ₹ (per cubic metre)	Difference in Prices in ₹ (per cubic metre)
1989	4	$4 \times 0.7 = 2.8$	12	$12 \times 0.8 = 9.6$	6.8
1990	5	$5 \times 0.7 = 3.5$	10	$10 \times 0.8 = 8$	4.5
1991	4	$4 \times 0.7 = 2.8$	13	$13 \times 0.8 = 10.4$	7.6
1992	6	$6 \times 0.7 = 4.2$	15	$15 \times 0.8 = 12$	7.8

Clearly, the required difference is maximum in year 1992.

21. Ratio of volumes of sales of Plywood, Sawn timber and Logs = 4 : 3 : 3

Then, average realisation in 1993 per cubic metre of sales

$$= \frac{4(7 \times 0.8) + 3(19 \times 0.8) + 3(20)}{4 + 3 + 3}$$

$$= 12.8 \approx ₹ 13$$

22. Average realisation in 1994 per cubic metre of sales

$$= \frac{\{4(7 \times 0.8) \times 1.05\} + \{3 \ (19 \times 0.8) \times 1.01\} + \{3(20) \times 1.1\}}{4 + 3 + 3}$$

$$= ₹ 13.55$$

23. Let minimum number of units produced be x, where $x \le 34$.

Then, Revenue – (fixed cost + variable cost) ≥ 0

$$\Rightarrow 20x - 70 - 14x \ge 0$$

($\because$ revenue and variable cost per unit are ₹ 20 and 14 respectively as seen from the graph)

$\Rightarrow$ $6x \ge 70$

$\Rightarrow$ $x \ge 1166$

$\Rightarrow$ $x \ge 12$

($\because$ number of units cannot be in fraction)

Hence, option (b) is correct.

(**Note** If $x > 34$, then fixed cost be ₹ 100. Further, $20x - 100 - 14x \ge 0 \Rightarrow x \ge 17$. Obviously, $x \ge 17$ is not possible as it contradicts $x > 34$)

Alternatively, You may proceed through options.

24. Let number of units produced be x, where $x \le 34$.

Then, $20x - 70 - 14x \ge 50$

$\Rightarrow x \ge 20$

Hence, option(a) is correct.

{**Note** If x > 34, then fixed cost be ₹ 100. Further, 20x – 100 – 14x ≥ 50 ⇒ x ≥ 25. Obviously, x ≥ 25 is not possible as it contradicts x > 34}

Alternatively, You may proceed through options.

25. Profit per unit in case of production of 40 units

$$= \frac{800 - 100 - 560}{40} = ₹ 3.5$$

Profit per unit in case of production of 34 units

$$= \frac{34 \cdot 20 - 70 - 34 \cdot 14}{34} = ₹ 3.94$$

Profit per unit in case of production of 35 units

$$= \frac{35 \cdot 20 - 100 - 35 \cdot 14}{35} = ₹ 3.14$$

Profit per unit in case of production of 25 units

$$= \frac{25 \cdot 20 - 70 - 25 \cdot 14}{25} = ₹ 3.20$$

Hence, option (b) is correct.

26. Let minimum number of units produced be x, where $x \le 34$.

Then, revenue – (fixed cost + variable cost) ≥ 0

$\Rightarrow$ $20x - (70 + 40) - 14x \ge 0$

($\because$ revenue and variable cost per unit are ₹ 20 and 14 respectively as seen from the graph)

$\Rightarrow$ $6x \ge 110$

$\Rightarrow$ $x \ge 18.33$

$\Rightarrow$ $x \ge 19$

($\because$ number of units cannot be in fraction)

Hence, option (b) is correct.

(**Note** If $x > 34$, then fixed cost be ₹ 100 + 40. Further, $20x - 140 - 14x \ge 0 \Rightarrow x \ge 24$. Obviously, $x \ge 24$ is not possible as it contradicts $x > 34$)

Alternatively, You may proceed through options.

27. Net revenue per Esteem $= \dfrac{17\% \text{ of } 1139200 \text{ millions}}{20\% \text{ of } 653610}$

$$= 1.48 \text{ millions} \approx ₹ 1481000$$

28. Proceed through options:

(a) M 800 ⇒ Revenue per vehicle (in millions)

$$= \frac{48\% \text{ of } 1139200}{58\% \text{ of } 653610} = \frac{48}{58} \times \left(\frac{1139200}{653610}\right)$$

$$= (\text{less than } 1) \times x, \left[\text{Assume} \left(\frac{1139200}{653610}\right) = x\right]$$

(b) Omni ⇒ Revenue per vehicle (in millions) $= \dfrac{21}{11} \times x$

$$= (\text{more than } 1) \times x$$

(c) Esteem ⇒ Revenue per vehicle (in millions)

$$= \frac{17}{20} \times x$$

$$= (\text{less than } 1) \times x$$

(d) Zen ⇒ Revenue per vehicle (in millions) $= \dfrac{3}{3} \times x = x$

Hence, option (b) is correct.

29. Clearly, we have to find (25% of 11% of 653610) is what percent of (58% of 653610) as increase in volume of sales of Omni is equal to decrease in volume of sales of M 800.

⇒ (2.75% of 653610) is what percent of (58% of 653610)

Obviously, required percentage $= \left(\dfrac{2.75}{58} \times 100\right)\%$

$$= 4.74\%$$

30. Obviously, exclusion of revenue generated by Alto (i.e. 11%) will augment the percentage of revenue generated by each of the remaining vehicles in the ratio of existing revenue generated. Hence, new contribution of Zen towards revenue in percentage = previous contribution + share of Zen in 11% as per ratio

$$= 3\% + \frac{11\%}{(100 - 11)} \times 3$$

$$= 3\% + 0.37\%$$

$$= 3.37\%$$

31. Statement (i) is false as (21 + 3)% is not lesser than (2 × 11%).

After interchange, Alto's revenue per unit

$$= \frac{17\% \text{ of } 1139200 \text{ millions}}{8\% \text{ of } 653610}$$

$$= \frac{17}{8} \times \left(\frac{1139200}{653610}\right) = \frac{17}{8} \times x, \left(\text{assuming } \frac{1139200}{653610} \text{ as } x\right)$$

Zen's revenue per unit $= \dfrac{3}{3} \times x$

Since $\frac{17}{8} > \frac{3}{3}$, statement (ii) is true.

Further, revenue per unit of Esteem

$= \frac{17x}{20}$ and revenue per unit of M 800 $= \frac{48x}{58}$.

Since $\frac{17}{20} > \frac{48}{58}$ $\left(\because \frac{17}{20-17} > \frac{48}{58-48}\right)$,

statement (iii) is true.

Hence, option (d) is correct.

32. All India demand in 1999-2000 = 50% of 146.54

and demand in 2004-2005 = 40% of 283.90 = 80% of 141.95 $\approx$ 77.5% of 146.54 (Recall of concept of 'Product constancy' detailed in chapter 'Percentage'. 141.95 is increased by approximately 3.2%, then 80 will decrease by 3.1%. Hence, 80 will become approximately 77.5)

Clearly, growth = 27.5% of 146.54

Then, growth% $= \left(\frac{27.5}{50} \times 100\right)\% = 55\%$

Hence, option (b) is correct.

33. Share of Fertilizer sector in demand of gas in 1999-2000 = 8% of 146.54

All India demands for gas (in MMSCMPD) in 1999-2000 = 146.54, which changes to 283.90 in 2004-2005, hence certain percentage growth in demand (say x%).

Then, new share of fertilizer in all India demand in 2004-2005 = $(100 + x)$% of 8% of 146.54 = 8% of {$(100 + x)$% of 146.54} = 8% of 283.90 = 22.71 MMSCMPD

34. As given, 39.90 increases to 88.31

$\therefore$ 100 will increase to approximately 88.31 × 2. 5

i.e. 176.62 + 44.15 = 220.77

But the actual answer will be little more than 220.71 as 100 is little more than 2.5 times of 39.90.

Hence, required percentage = little more than (220.77 – 100)% = little more than 120.77

Hence, option (b) is correct.

35. Required Industry sector demand in 2004-2005 = 40% of (283.90 – 88.31) = 78.23

36. Cost/sales ratio in months of April, June, July and August are $\frac{140}{240}, \frac{160}{240}, \frac{210}{260}$ and $\frac{225}{240}$, respectively.

The lowest among $\frac{140}{240}, \frac{160}{240}$ and $\frac{225}{240}$ is $\frac{140}{240}$.

Now, compare $\frac{210}{260}$ and $\frac{140}{240}$ in order to find the lower.

These are $\frac{21}{26}$ and $\frac{14}{24} \Rightarrow \frac{21}{5}$ and $\frac{14}{10}$ (Refer to technique meant for 'Comparison of Fractions')

Clearly, 1.4 is less than 4.2. Hence, option (a) is correct.

37. Profit/kg in month of April = ₹ (240 – 140) = ₹ 100

Then, total profit in April = ₹ 100 × (14.63% of 1230)

Similarly, profits in months of June, August and September are (240 – 160) {16.26% of 1230}, (240 – 225) {19.52% of 1230} and (280 – 250) {16.26% of 1230}, respectively.

Now, compare : 100 × 14.63, 80 × 16.26, 15 × 19.52 and 30 × 16.26

$\Rightarrow$ 1463, 1300.8, 292.8 (30 × 16.26 has been left as 80 × 16.26 > 30 × 16.26 at the first instance)

Clearly, 1300.8 is the maximum. Hence, option (a) is correct.

Alternatively, You may apply the 'concept of product constancy' to compare the above.

38. Escalations in months of May, July, August and September are 50 over 140, 50 over 160, 15 over 210 and 25 over 225.

Obviously, percentage cost– escalation will be corresponding to $\frac{5}{14}, \frac{5}{16}, \frac{1}{14}$ and $\frac{1}{9}$.

Clearly, $\frac{5}{14} > \frac{5}{16}$ ($\because$ denominator of the former is lesser than that of the latter)

Further, $\frac{5}{14} > \frac{1}{14}$ ($\because$ 5 > 1) and $\frac{5}{14} > \frac{1}{9}$ ($\because$ 9 × 5 > 1 × 14)

Hence, $\frac{5}{14}$ is the greatest.

Hence, option (a) is correct.

Alternatively Corresponding values are $\frac{5}{14}, \frac{5}{16}, \frac{1}{14}$ and $\frac{1}{9}$, respectively i.e. 0.35, 0.31, 0.07, 0.11. Obviously, 0.35 is the greatest.

Note *You may also take other approaches for comparison of fractions into consideration as detailed in chapter 'Comparison of Fractions'.*

39. In month of July, profit = (260 – 210) {15.44% of 1230}
= 772% of 1230

In month of August, profit = (240 – 225) {19.52% of 1230} = 292.8% of 1230

Hence, decrease in profit

= 772% of 1230 – 292.8% of 1230

= 479.2% of 1230 = 500% of 1230 – 20.8% of 1230

= 6150 – (246 + 9.84) $\approx$ ₹ 5900

Hence, option (c) is correct.

40. Percentage growth in cost/kg from August to September $= \left(\frac{250-225}{225} \times 100\right)\% = 11.11\%$

Now, cost/kg in month of October = 111.11% of 250 = 250 + 27.78 = ₹ 277.78

41. Required average number

$$= \left(\frac{\begin{array}{c}20\% \text{ of } 12 + 20\% \text{ of } 18 + 45\% \text{ of } 32\\ + 60\% \text{ of } 8 + 40\% \text{ of } 14 + 15\% \text{ of } 16\end{array}}{6}\right)\% \text{ of } 40250$$

= (0.4 + 0.6 + 2.4 + 0.8 + 0.93 + 0.4)% of 40250

= 5.53% of 40250 $\approx$ 2227

($\because$ 10% of 40250 = 4025

$\therefore$ 5% of 40250 = 2012.5

and $\because$ 1% of 40250 = 402.50

$\therefore$ $0.5\% = 201.25.$

Further, 0.01% of $40250 = 4.025$

$\therefore$ Its $0.03\% = 12.075$, then 5.53% of 40250

$= 2012.5 + 201.25 + 12.075 = 2225.825)$

Note *No need to proceed further after we obtain 5% of 40250 = 2012.5 as options suggest. Here, done for your clarity only.*

42. Required percentage = 20% of 18% of total people working in night shifts in all industries = 3.6% of total people working in night shifts in all industries

Hence, option (b) is correct.

43. From the previous question, average number of females = 5.53% of total people

Then, total number of females = $6 \times 5.53\%$ i.e. 33.18% of total people

Now, number of males = (100 – 33.18)% of total people

$\therefore$ Required percentage = (100 – 33.18)% – 33.18% = 33.6%

44. From the above question, total number of females = $6 \times$ 5.53% i.e. 33.18% of total people

Number of females in banking industry = 40% of 14% of 40250 = 5.6% of 40250

As given, number of females in banking industry = 3206, which is approximately 8% of 40250.

Hence, new percentage of females

$= \{33.18 + (8 - 5.6)\}\% = 35.5\%$

45. Number of males working in Sales

= (100 – 60)% of 8% of total working people = 3.2% of total people working

Number of males working in IT = (100 – 20)% of 12% of total working people = 9.6% of total people working

Number of males working in Medical

= (100 – 15)% of 16% of total working people

= 13.6% of total people working

Hence, number of males in IT and Medical = (9.6 + 13.6)% i.e. 23.2% of total working people

Number of males working in Banking

= (100 – 40)% of 14% of total people working = 8.4% of total working people

Number of males working in Gaming

= (100 – 20)% of 18% of total people working = 14.4% of total people working

Clearly, option (a) is correct.

46. Required ratio = 6% of 8 : 4% of 9 = 4 : 3

47. Let ₹ x lakh be invested in Company P.

Then, 7.5% of $x + 9\%$ of $(30 - x) = 2.43$

($\because$ rates of interest for Q and R are same in 2010)

$\Rightarrow \quad 0.075x - 0.09x = 2.43 - 2.70$

$\Rightarrow \quad x = 18$

Hence, amount of investments in companies P, Q and R are 18 lakh, 6 lakh and 6 lakh, respectively. The amount invested in Q and R may be different but their sum must be ₹ 12 lakh like (2, 10), (3, 9), etc.

Hence, option (b) is correct.

48. Rates of interest in 2008 and 2009 were 8% and 10% per annum.

Then, overall interest in two years $= \left(+\ 8 + 10 + \frac{8 \times 10}{100}\right)\%$

[recall the concept of 'Successive Increas/Decrease/ combination of both' envisaged in chapter 'Percentage'.]

$= 18.8\%$

Then, total appreciation = 18.8% of 15 lakh

= ₹ 282000

49. Overall interest $= \dfrac{(2 \times 7.5\%) + (1 \times 9\%) + (4 \times 9\%)}{2 + 1 + 4}$

$= \dfrac{60}{7}\%$

(recall the concept of 'Weighted Average')

Then, total interest $= \dfrac{60}{7}\%$ of 30 lakh = ₹ 2.57 lakh

50. For Company P, average annual rate of interest

$= 7 \text{ (assumed)} + \dfrac{0 + 2 + 1 + 3 + 0.5 - 0.5 - 1}{7}$

$= 7 + 0.7 = 7.7\%$

For Company Q, average annual rate of interest

$= 6 \text{ (assumed)} + \dfrac{0.5 + 2 + 3.5 + 2 + 3 + 2 - 2}{7}$

$= 6 + 1.5 = 7.5\%$

For Company R, average annual rate of interest

$= 8 \text{ (assumed)} + \dfrac{0.5 + 0 + 0.5 - 0.5 + 1 + 0 - 1}{7}$

$= 8 + 0.07 = 8.07\%$

Hence, Company R has the highest average annual rate of interest.

Solution (Q.Nos. 51 - 53)

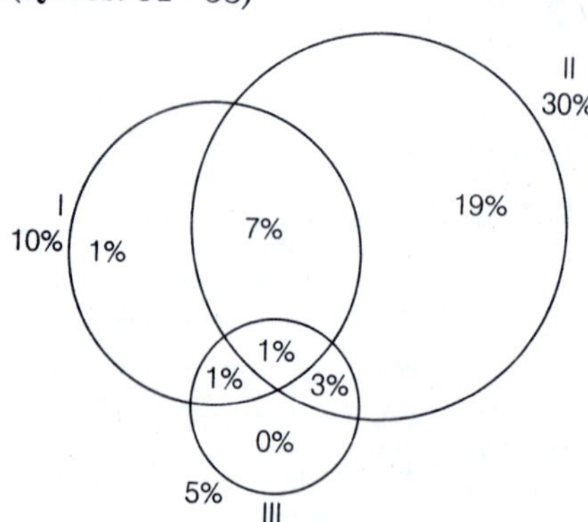

51. Number of people reading only one newspaper = (1 + 19 + 0)% of 100000 = 20000

52. Number of people reading at least two newspapers = Number of people reading two newspapers + Number of people reading three newspapers = {(7 + 3 + 1)% + 1%} of 100000 = 12000

53. Number of people reading at least one newspaper = Number of people reading one newspaper + Number of people reading two newspapers + Number of people reading three newspapers = 20000 + 12000 = 32000

Then, number of people not reading any newspaper = 100000 – 32000 = 68000

54. Total number of marks

$= (66\% \text{ of } 150) + (75\% \text{ of } 100) + (88\% \text{ of } 150) + (56\% \text{ of } 125) + (56\% \text{ of } 75) + (90\% \text{ of } 50)$

$= 99 + 75 + 132 + 70 + 42 + 45 = 463$

Alternatively,

Compare the maximum marks with base 100 in each subject. Strategic management : Since 150 is 1.5 times of 100, her marks will be 1.5 times of 66 i.e. 99. In other subjects, same approach will be adopted. In doing so, we will be able to compute the required answer orally.

Now her marks $= (66 \times 1.5) + (75 \times 1) + (88 \times 1.5) + (56 \times 1.25) + (56 \times 0.75) + (90 \times 0.5) = 99 + 75 + 132 + 70 + 42 + 45 = 463$

55. Clearly, 88 is what percent of 76 ?

Then, the required percentage $= \left(\frac{88}{76} \times 100\right)\% = 115.79\%$

Alternatively,

76 is 100% of 76 and (88 – 76) is 15.79% of 76.

Hence, required percentage is (100 + 15.79) i.e. 115.79%.

56. Average percentage

$= 80 \text{ (assumed)} + \frac{+\ 8 + 4 - 2 + 16 - 12 - 30}{6}$

$= 80 + (-2.67) = 77.33\%$

$\therefore$ Average marks = 77.33% of 150

$= 77.33 + \frac{77.33}{2} = 116$

57. Archit' marks $= (82 \times 1.5) + (76 \times 1) + (84 \times 1.5) + (96 \times 1.25) + (92 \times 0.75) + (88 \times 0.5) = 123 + 76 + 126 + 120 + 69 + 44 = 558$

Similarly, Gunit's marks = 446

Pranita's marks = 360

Garvita's marks = 568

Arpan's marks = 496

Hence, option (d) is correct.

Alternatively,

If we look at the table meticulously, it appears that either Archit or Garvita may have highest scores. Hence, we need to compute marks of only Archit and Garvita.

58. Archit (in Consumer behaviour and Service marketing) and Garvita(in Strategic management, Brand management and Compensation management) have scored the highest marks. Hence, option (b) is correct.

Solution (Q.Nos. 59-63) *The given table may be interpreted in terms of percent of total number of employees as below :*

Department	Number of employees	Gender	City
		M and F	Z and Y
A	10%	7% and 3%	1% and 9%
B	22%	13% and 9%	3% and 19%
C	12%	4% and 8%	10% and 2%
D	20%	12% and 8%	5% and 15%
E	36%	32% and 4%	10% and 26%

59. Required answer = (1 + 10)% of 80000 = 8800

60. 32% of 80000 is 3200% of 1% of 80000.

61. Required ratio = (13 + 12)% : (3 + 4)% = 25 : 7

62. Required number of employees = (10 + 36)% of 80000 = 36800

63. Number of residents living in city Y

= (9 + 19 + 2 + 15 + 26)% of 80000 = 56800

Then, average number of residents of City Y working in each department $= 56800 \div 5 = 11360$

64. Total accepted applications for Position A $= 1040 - (63 \times 5)$

$= 1040 - 315 = 725$

($\because$ Number of rejected applications from each duplicate applicant = 1 his/her original application + 4 his/her duplicate applications)

Let $5x$ applications from the males and 5y applications from the females have been accepted. Then, $3x$ applications of males and y applications from the females will be rejected.

Now, $5x + 5y = 725$ ...(i)

and $3x + y = 315$ ... (ii)

Solving the Eqs. (i) and (ii), we get $x = 85$

Then, $3x = 3 \times 85 = 255$

65. Total number of accepted applications for positions A and B $= 2 \times 659 = 1318$

Then, number of rejected applications for positions A and B = (1040 + 880) – 1318 = 602

As given, number of duplicate applicants for position A is 63. Then, the number of rejected applications for position B $= 602 - (63 \times 5) = 287$

Now, number of duplicate applicants for position B $= 287 \div (6 + 1) = 41$

66. As given, ratio of number of accepted applications and rejected applications = 4 : 1.

Now, consider the given statements:

(a) If number of applications received (all originals + all duplicates) be 240. Then, accepted applications = 80% of 240 = 192 and rejected applications = 48

As given, number of duplicate applicants is 48. Then, it is obvious that atleast 48 (1 + 1) i.e. 96 applications must be rejected. Hence, statement (a) is not possible.

(b) If number of applications accepted (all originals + all duplicates) be 768. Then, number of rejected applications = 25% of 768 = 192 which is possible as 192 is divisible by 48 (1 + 1) and by 48 (1 + 3). No need to mention that in this case the average number of duplicate applications received from the duplicate applicants will either be 1 or 3.

(c) Minimum number of rejected application must be 48 (1 + 1) = 96, since the ratio of number of accepted applications and rejected applications = 4 : 1, the minimum number of applications received for the position D $= 96 \times 5 = 480$. Further, number of applications received for position E is 420. Hence, statement (c) is not true.

67. Given, number of accepted applications from the males = X

Number of accepted applications from the females = X + Y

Number of rejected applications (males and females)

= X + 2Y

Then, total number of applications received

= X + (X + Y) + (X + 2 Y)

= 3 X + 3 Y = 3 (X + Y) = 420 (given)

⇒ (X + Y) = 140

Hence, number of accepted applications from the females = 140

Now, proceed through the given statements.

(a) If number of accepted applications from the males is 131, then number of duplicate applications (males + females) = 420 – 140 – 131 = 149.

If number of duplicate applications be 149, it must be divisible by (1 + 1) or (1+ 2) or (1 + 3) or so on.... assuming that the average number of duplicate applications received from duplicate applicants be 1, 2, 3........... But 149 is not divisible by them as it is a prime number. Further, since it is divisible by (1 + 148), average number of duplicate applications received from the duplicate applicants will be 148. It implies that number of duplicate applicants is only one, if it is true, then concept of average will be defeated. Hence, statement (a) is not possible.

(b) If number of accepted applications from the males is 141, then number of duplicate applications (males + females) = 420 – 140 – 141 = 139. It is also a prime number. Thus, statement (b) is not possible.

(c) If number of accepted applications from the males is 139, then number of duplicate applications (males + females) = 420 – 140 – 139 = 141. It is not a prime number and divisible by 3, 47. It may be possible that average number of applications received from the duplicate applicants is either 2 or 46. Hence, statement (c) is true.

68. Minimum number of accepted applications = 151 + 131 = 282 and maximum number of accepted applications = 199 + 179 = 378

It implies that maximum number of rejected applications = 600 – 282 = 318 and minimum number of rejected applications = 600 – 378 = 222

Now, consider the given possible values.

(I) 5⇒ Number of rejected applications = 28 (5 + 1) = 168, not possible as minimum number of rejected applications is 222.

(II) 7⇒ Number of rejected applications = 28 (7 + 1) = 224, possible.

(III) 9⇒ Number of rejected applications = 28 (9 + 1) = 280, possible.

(IV) 11⇒ Number of rejected applications = 28 (11 + 1) = 336, possible.

(V) 13⇒ Obviously, number of rejected applications will be higher than 336, thus, not possible.

Hence, option (c) is correct.

69.

	Vendor 1	Vendor 2	Vendor 3	Vendor 4	Vendor 5	
Customer Service	55	40	50	70	28	Median of observations 28, 40, **50**, 55, 70 (arranged in ascending order)= 50
Reliability	52	40	75	26	60	Median of observations 26, 40, **52**, 60, 75 = 52
Cost	76	80	90	70	60	Median of observations 60, 70, **76**, 80, 90 = 76
Quality	72	70	62	40	48	Median of observations 40, 48, **62**, 70, 72 = 62

Hence, option (a) is correct

Alterantively,

Observe the radar chart meticulously. You have to compare the scores lying on respective aspects as middle figure. In other words, compare the scores lying on third (either from top or bottom) data series irrespective of vendors. *For examples*, with reference to 'Customer Service'. 50 is lying on third line/data series, 52 is lying on third line with respect to Reliability, 76 is lying on third line in case of cost and 62 is lying on third line with regard to quality. Since 50 is least, option(a) is correct.

70.

	Vendor 1	Vendor 2	Vendor 3	Vendor 4
Customer Service	55	40	50	70
Reliability	52	40	75	26
Cost	76	80	90	70
Quality	72	70	62	40
Reach	80	58	62	45
Features	40	45	55	90
Let assumed average be 70. Then, sum of deviations of scores from 70⇒	– 15 –18 + 6 + 2 + 10 – 30 = – 45	– 30 – 30 + 10 + 0 – 12 –25 = –87	– 20 + 5 + 20 – 8 – 8 – 15 = –26	0 – 44 + 0 – 30 – 25 + 20 = –79

Hence, Vendor 3 has the highest final score as it has minimum deviations.

71.

	Top two scorers		
Customer Service	Vendor-4, 1	Vendor-4⇒ 2 times	Hence, option (a) is correct.
Reliability	Vendor-3, 5	Vendor-1⇒ 3 times	
Cost	Vendor-3, 2	Vendor 3⇒ 2 times	
Quality	Vendor-1, 2	Vendor-5⇒ 3 times	
Reach	Vendor-1, 5		
Features	Vendor-4, 5	Vendor-2⇒ 2 times	

72.

	Top three Vendors
Customer Service	Vendor-4, 1, 3
Reliability	Vendor-3, 5, 1
Cost	Vendor-3, 2, 1
Quality	Vendor-1, 2, 3
Reach	Vendor-1, 5, 3
Features	Vendor-4, 5, 3

Vendor-4 $\Rightarrow$ 2 times
Vendor-1 $\Rightarrow$ 5 times
Vendor-3 $\Rightarrow$ 6 times
Vendor-5 $\Rightarrow$ 3 times
Vendor-2 $\Rightarrow$ 2 times

Hence, option (b) is correct.

73. For 2011, $\frac{700+400}{60} = 18.3$

For 2013, $\frac{800+700}{90} = 16.66$

For 2016, $\frac{600+500}{70} = 15.7$

For 2017, $\frac{700+400}{60} = 18.3$

Hence, option (c) is correct

74. For 2012, $\frac{600}{400} = 1.5$

For 2013, $\frac{800}{700} = 1.1$

For 2015, $\frac{600}{500} = 1.2$

For 2016, $\frac{600}{500} = 1.2$

Hence, option (b) is correct.

75. For 2011, (400 – 300) is 25% of 400

For 2013, (700 – 400) is 75% of 400

For 2014, (700 – 300) is 57.14% of 700

For 2015, (500 – 300) is 66.66% of 300

Hence, option (b) is correct.

76. For 2011, $\frac{|300-400|}{|60-50|} = 10$

For 2012, $\frac{|400-300|}{|50-60|} = 10$

For 2014, $\frac{|300-700|}{|60-90|} = 13.33$

For 2017, $\frac{|400-500|}{|60-70|} = 10$

Hence, option (c) is correct.

77. 'Heavy Monsoon States' are Arunachal, Kerala, Meghalaya, Mizoram, Goa, Sikkim, Maharashtra in which rainfall is 900 mm or more. Out of these 7 states first three states have negative LPA. Obviously, 3 is 42.86% of 7.

78. Deviation from LPA of 'Low Monsoon States' are –60% (Manipur), –40% (Delhi), –35% (Jharkhand), –30%(WB), –10% (Assam), + 10% (MP), + 15% (Rajasthan), + 20% (Karnataka) and + 25% (Gujrat). Clearly, median of the given observations is – 10%.

79. States, with actual rainfall of 600 mm or less, having negative deviations from LPA are Manipur, Delhi, Jharkhand, WB and Assam.

Average of respective actual rainfall = average of 400, 300, 400, 600 and 600 = 460 mm

80. As given, 120% of rainfall in Gujrat in 2009 = rainfall in Gujrat in 2019 = 600 mm

$\therefore$ Rainfall in 2009 = 500 mm

Further, rainfall in 2019 is 25% more than the LPA in 2018 as it is given that the deviation from LPA is + 25%. It implies that 600 mm is 25% more than the average rainfall during 2009-2018. Hence, the average rainfall during 2009-2018 is 480 mm. Then, sum of rainfall during 2010-2018 = (480 ×10) – 500

Now, LPA in 2019 = average rainfall during 2010-2019

= {(sum of rainfall during 2010–2018) + rainfall during 2019} ÷ 10

= {(480 × 10) – 500 + 600} ÷ 10 = 490 mm

81. Required difference = (17–2)% of 525

= $ 78.75 mn.

82. As given, tax revenue from the highway user is 8%. Then, tax revenue from the sources other than highway users in yr 2000 = (100 – 8)% of 500 = $ 460 mn.

As shown, the combined tax revenue for year 1999 is $ 440 mn. Clearly, 460 is 20 more than 440 i.e. an increase of 4.5%.

Unit 8

Trigonometry

- Trigonometric Ratios of Acute Angles and Allied Angles
- Trigonometric Ratios of Compound and Multiple Angles
- Height and Distance

Trigonometry

Introduction

The word trigonometry is derived from the greek words 'tris' (meaning - three), gon (meaning - sides) and metron (measure). It means trigonometry is the study of relationships between the sides and angles of a triangle. Needless to say that this Section is very important for the competitive examinations like SSC, MBA, CDS, IIFT, NIFT, FMS, etc. In the geometry section we have studied about the angles, their types and measurements. Now we will discus about the ratios of the sides of a right angled triangle with respect to its angles, called 'trigonometric ratios of the angle'. The application of these ratios will be observed under topic 'Height and Distance'. For the sake of convenience, we may discuss the above into three parts as below.

Part I : Trigonometric ratios of acute angles and allied angles

Part II : Trigonometric ratios of compound and Multiple Angles

Part III : Height and Distance

Learning OBJECTIVES

After completing this chapter, aspirants will be able to :

- Understand the practical meaning of the trigonometric ratios, concepts of their signs and their range and locate the quadrant of the given angles;
- Understand the concept of trigonometric ratios of acute angles, complem- entary angles, allied angles, compound angles and multiple angles;
- Recognise the trigonometric identities and their use in solving the problems;
- Establish the relation among the trigonometric ratios;
- Solve the problems based on trigonometric equations/expressions;
- Solve the problems based on elimination of trigonometric ratios;
- Find the maximum and minimum value of a trigonometric expression;
- Simplify the trigonometric expressions based on acute angles, compleme- ntary angles, allied angles, compound angles and multiple angles;
- Be acquainted with the concept of trigonometric ratios of angles of triangle/quadrilateral;
- Be aware of Addition Formulae, Difference Formulae and Transformation formulae;
- Apply the concept of trigonometric ratio in the practical problems under 'Height and Distance';
- Understand the philosophy of the division of this section in different parts;
- Analyse the trend of the questions being asked in various competitive examinations;
- Solve the questions in moments;
- Grip the concepts helping to solve the questions without pen and paper;
- Learn many new concepts/fundamentals; and
- Apply the pragmatic approaches in lieu of academic approaches.
- Ponder on the alternative methods.

49

Part-I

Trigonometric Ratios of Acute Angles and Allied Angles

A. Trigonometric Ratios of Acute Angles

We know that, an acute angle is between 0° and 90°. Let ABC be a right angle triangle.

Here, $\angle B = 90°$ and $\angle A$ and $\angle C$ are acute angles.

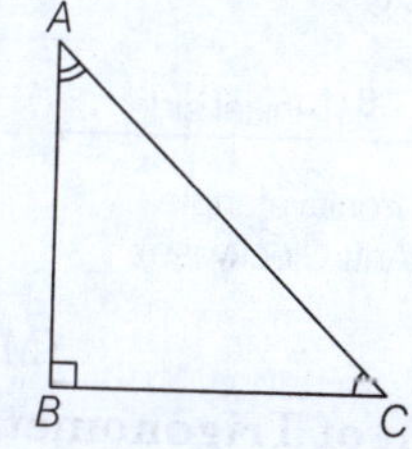

For $\angle C$

- AB is the perpendicular(P) [i.e. side opposite to $\angle C$]
- BC is the base(B) [i.e. a smaller side making $\angle C$]
- AC is the hypotenuse(H) [side opposite to the right angle]

For $\angle A$

- BC is the perpendicular [i.e. side opposite to $\angle A$]
- AB is the base [i.e. a smaller side making $\angle A$]
- AC is the hypotenuse [side opposite to the right angle]

Observations Base and perpendicular are defined with reference to a particular angle. The same side is a base for one angle and a perpendicular for another angle.

It is also noticed that only six ratios are possible with the help of all three sides and these ratios are called **'trigonometric ratios'**. All six trigonometric ratios are as under with reference to $\angle C$.

(I) $\dfrac{AB}{AC} = \dfrac{P}{H}$ = sine of $\angle C$ (in brief, **sin C**)

Inside

- Trigonometric Ratios (T-R) of Acute Angles/ Complementary Angles
- Trigonometric Identities
- T-R of Allied Angles Including their Conversion w.r.t. Quadrants
- Range of T-R and Trends of Changes in Values of T-R
- Illustrations Based on Above Aspect Including Maximum and Minimum Value of T-R/ Expression Under Diverse Stipulations
- ....
- .... & Many More
- Work Out with Solutions

(II) $\dfrac{AC}{AB} = \dfrac{H}{P}$ = cosecant of $\angle C$ (in brief, **cosec C**)

(III) $\dfrac{BC}{AC} = \dfrac{B}{H}$ = cosine of $\angle C$ (in brief, **cos C**)

(IV) $\dfrac{AC}{BC} = \dfrac{H}{B}$ = secant of $\angle C$ (in brief, **sec C**)

(V) $\dfrac{AB}{BC} = \dfrac{P}{B}$ = tangent of $\angle C$ (in brief, **tan C**)

(VI) $\dfrac{BC}{AB} = \dfrac{B}{P}$ = cotangent of $\angle C$ (in brief, **cot C**)

Remarks

(A) Sin C is not a product of 'sin' and 'C'. Sin separated from C has no meaning. Similar interpretations follow for other ratios, it may be construed from the above that

(B) cosec C, sec C and cot C are the reciprocals of ratios sin C, cos C, and tan C respectively. It means that sin C × cosec C = 1, cos C × sec C = 1 and tan C × cot C = 1

(C) $\tan C = \dfrac{AB}{BC} = \dfrac{AB}{AC} \div \dfrac{BC}{AC} = \sin C \div \cos C \Rightarrow \tan C = \dfrac{\sin C}{\cos C}$

(D) All six are ratios, hence, have no any units.

[You may try to define the trigonometric ratios with reference to $\angle A$ of the triangle ABC right-angled at B. Further, it is to mention that the values of the trigonometric ratios of an angle don't vary with the lengths of the sides of a triangle, if the angle remains the same]

Trigonometric Ratios of Complementary Angles

Recall that two angles are said to be complementary if their sum is 90°.

In given $\Delta\ ABC$, right angled at B, $\angle A + \angle C = 90°$ i.e. $\angle A$ and $\angle C$ are said to be complementary.

Let $\angle C = \theta$, then $\angle A = (90 - \theta)$.

Then,

$$\sin\theta = \frac{\text{perpendicular}}{\text{hypotenuse}} = \frac{AB}{AC}$$

and $$\cos(90° - \theta) = \frac{\text{base}}{\text{hypotenuse}} = \frac{AB}{AC}$$

Hence, $\sin\theta = \cos(90° - \theta)$

Further, $$\cos\theta = \frac{\text{base}}{\text{hypotenuse}} = \frac{BC}{AC}$$

and $$\sin(90° - \theta) = \frac{BC}{AC}$$

Hence, $\cos\theta = \sin(90° - \theta)$

Similarly, $\tan\theta = \cot(90° - \theta)$ and $\cot(90° - \theta) = \tan\theta$ and $\sec\theta = \text{cosec}(90° - \theta)$ and $\text{cosec}\,\theta = \sec(90° - \theta)$

Remarks The above all hold good when θ is lying between 0° and 90°.

Trigonometric Identities

An equation involving trigonometric ratios of an angle is called a trigonometric identity if it is true for all real values of angle(s) involved.

(i) $\sin^2\theta + \cos^2\theta = ?$

$$\Rightarrow \frac{p^2}{h^2} + \frac{b^2}{h^2} = \frac{p^2 + b^2}{h^2} = \frac{h^2}{h^2} = 1 \quad (\because p^2 + b^2 = h^2)$$

Hence, $\mathbf{\sin^2\theta + \cos^2\theta = 1}$

(ii) $$1 + \tan^2\theta = 1 + \frac{p^2}{b^2} = \frac{b^2 + p^2}{b^2} = \frac{h^2}{b^2} = \sec^2\theta$$

Hence, $\mathbf{1 + \tan^2\theta = \sec^2\theta}$

(iii) On the above line, $\mathbf{1 + \cot^2\theta = \text{cosec}^2\theta}$

B. Trigonometric Ratios of Allied Angles

Before discussing the trigonometric ratios of allied angle(viz, $-\theta$, $90° \pm \theta$, $180° \pm \theta$, $360° \pm \theta$, etc, we should keep eyes on two important aspects).

I. Sense of an angle

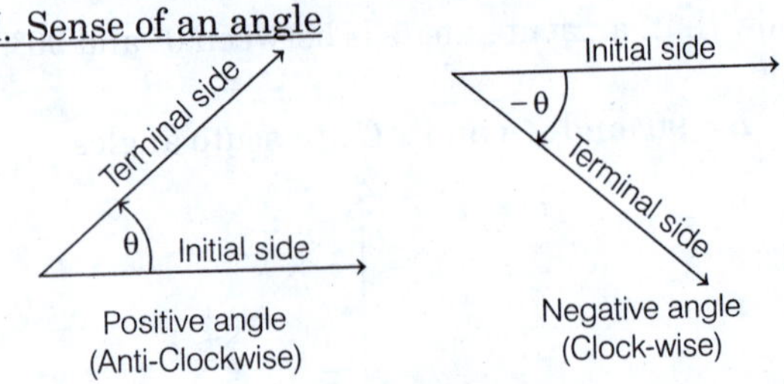

(Mind the directions of rotation)

II. Sign of Trigonometric Ratios

(90° < A < 180°) *2nd Quadrant* Only sin and cosec ⇒ '+' A = 180° − θ	**0° < angle (A) < 90°)** *1st Quadrant* All six T-ratios ⇒ '+' A = θ
(180° < A < 270°) *3rd Quadrant* Only tan and cot ⇒ '+' A = 180° + θ	**(270° < A < 360°)** *4th Quadrant* Only cos and sec ⇒ '+' A = 360° − θ

Coding for memorizing the above

All -I (Qt)

Students-II ('S' for sin and cosec)

To-III ('T' for tan and cot)

Consider-IV ('C' for cos and sec)

It is worth mentioning that when 180° is written as (180° − 0°), it lies in 2nd quadrant and it lies in 3rd quadrant if it is expressed as (180° + 0°).

Now, we may have a look on the trigonometric ratios of allied angles

	1st Quadrant		II Quadrant	III Quadrant	IV Quadrant	
	θ, $(0^\circ < \theta < 90^\circ)$	$(n \times 360^\circ + \theta)$	$(180^\circ - \theta)$	$(180^\circ + \theta)$	$(n \times 360^\circ - \theta)$	$-\theta$
sin	sin	sin	sin	– sin	– sin	– sin
cosec	cosec	cosec	cosec	– cosec	– cosec	– cosec
cos	cos	cos	– cos	– cos	cos	cos
sec	sec	sec	– sec	– sec	sec	sec
tan	tan	tan	– tan	tan	– tan	– tan
cot	cot	cot	– cot	cot	– cot	– cot
	All '+'		**Only sin and cosec '+'**	**Only tan and cot '+'**	**Only cos and sec '+'**	

From the above chart and with the help of trigonometric ratios of complementary angles, we may find the trigonometric ratios of $(90^\circ + \theta)$, $(270^\circ \pm \theta)$. Needless to say that it has already been discussed for $(90^\circ - \theta)$.

For $(90^\circ + \theta)$

(i) $\sin(90^\circ + \theta) = \sin[180^\circ - (90^\circ - \theta)]$

$= \sin(90^\circ - \theta)$ $\quad [\because \sin(180^\circ - A) = \sin A]$

$= \cos\theta$ $\quad [\because \sin(90^\circ - \theta) = \cos\theta]$

(ii) $\cos(90^\circ + \theta) = \cos[180^\circ - (90^\circ - \theta)]$

$= -\cos(90^\circ - \theta)$ $\quad [\because \cos(180^\circ - A) = -\cos A]$

$= -\sin\theta$ $\quad [\because \cos(90^\circ - \theta) = \sin\theta]$

On the same line, we may find for the remaining trigonometric ratios as below

$\text{cosec}(90^\circ + \theta) = \sec\theta$

$\sec(90^\circ + \theta) = -\text{cosec}\,\theta$

$\tan(90^\circ + \theta) = -\cot\theta$

$\cot(90^\circ + \theta) = -\tan\theta$

For $(270^\circ + \theta)$

$\sin(270^\circ + \theta) = \sin[360 - (90^\circ - \theta)]$

$= -\sin(90^\circ - \theta)$ $\quad [\because \sin(360 - A) = -\sin A]$

$= -\cos\theta$ $\quad [\because \sin(90 - \theta) = \cos\theta]$

For $(270^\circ - \theta)$

Now, $\sin(270^\circ - \theta) = \sin[180^\circ + (90^\circ - \theta)]$

$= -\sin(90^\circ - \theta)$

$[\because \sin(180 + A) = -\sin A]$

$= -\cos\theta$

Hence, $\sin(270^\circ - \theta) = -\cos\theta$

We may find values for the remaining trigonometric ratio on the above line. In brief, the table in the above context is given here.

	1st Quadrant	2nd Quadrant	3rd Quadrant	4th Quadrant
	$(90^\circ - \theta)$	$(90^\circ + \theta)$	$(270^\circ - \theta)$	$(270^\circ + \theta)$
sin	$\cos\theta$	$\cos\theta$	$-\cos\theta$	$-\cos\theta$
cosec	$\sec\theta$	$\sec\theta$	$-\sec\theta$	$-\sec\theta$
cos	$\sin\theta$	$-\sin\theta$	$-\sin\theta$	$\sin\theta$
sec	$\text{cosec}\,\theta$	$-\text{cosec}\,\theta$	$-\text{cosec}\,\theta$	$\text{cosec}\,\theta$
tan	$\cot\theta$	$-\cot\theta$	$\cot\theta$	$\cot\theta$
cot	$\tan\theta$	$-\tan\theta$	$\tan\theta$	$\tan\theta$

Remarks *In case of $(90^\circ \pm \theta)$ and $(270^\circ \pm \theta)$, we change sin $\to$ cos and vice-versa, sec $\to$ cosec and vice-versa, tan $\to$ cot and vice-versa with an appropriate sign according to specific quadrant. In my opinion, no need to stress on trigonometric ratios of $(90^\circ + \theta)$ and $(270^\circ \pm \theta)$ as we can derive the same from $(180^\circ \pm \theta)$ and $(360^\circ - \theta)$ when required.*

C. Values of Trignometric Ratios w.r.t Some Specific Angles

	0°	30° $\left(\frac{\pi}{6}\right)$	45° $\left(\frac{\pi}{4}\right)$	60° $\left(\frac{\pi}{3}\right)$	90° $\left(\frac{\pi}{2}\right)$	180° (π)	270° $\left(\frac{3\pi}{2}\right)$	360° (2π)
sin	0	$\frac{1}{2}$	$\frac{1}{\sqrt{2}}$	$\frac{\sqrt{3}}{2}$	1	0	–1	0
cosec (i.e. $\frac{1}{\sin}$)	not defined (ND)	2	$\sqrt{2}$	$\frac{2}{\sqrt{3}}$	1	ND	–1	ND
cos	1	$\frac{\sqrt{3}}{2}$	$\frac{1}{\sqrt{2}}$	$\frac{1}{2}$	0	–1	0	1
sec (i.e. $\frac{1}{\cos}$)	1	$\frac{2}{\sqrt{3}}$	$\sqrt{2}$	2	ND	–1	ND	1
tan (i.e. $\frac{\sin}{\cos}$)	1	$\frac{1}{\sqrt{3}}$	1	$\sqrt{3}$	ND	0	ND	0
cot	ND	$\sqrt{3}$	1	$\frac{1}{\sqrt{3}}$	0	ND	0	ND

Observations

(I) It is clear that value of $\tan\theta$ increases faster than $\sin\theta$ as θ increases.

(II) It is clear that value of $\cos\theta$ decreases $(1 \to 0)$ from 0° to 90° and value of $\sin\theta$ increases $(0 \to 1)$ from 0° to 90° but at 45° their values are same. Hence, value of $\cos\theta$ will be greater than that of $\sin\theta$ provided $0^\circ \le \theta < 45^\circ$. Further, in case of $45^\circ < \theta \le 90^\circ$, the value of $\cos\theta$ will be lesser than that of $\sin\theta$.

(III) No need to memorize the values for 180°, 270° and 360° as you are now capable to derive these values from the values of t-ratios for 90°/0°.

e.g. $\sin 180° = ?$

1st method $\sin 180° = \sin(180° - 0°)$

$= \sin 0°$ [$\because (180° - 0°)$ is lying in 2nd quadrant]

$= 0$

2nd Method $\sin 180° = ?$

$\Rightarrow \sin(180° + 0°) = -\sin 0°$

[$\because (180° + 0°)$ is lying is 3rd quadrant]

$= 0$

3rd method $\sin 180° = \sin(90° + 90°)$

$= \cos 90°$ $[\because \sin(90° + \theta) = \cos\theta]$

$= 0$

Now, $\sin 270° = ?$

1st method $\sin 270° = \sin(360° - 90°)$

$= -\sin 90° = -1$ $[\because \sin(360 - \theta) = -\sin\theta]$

2nd method $\sin 270° = \sin(270° - 0°) = -\cos 0° = -1$

3rd method $\sin 270° = \sin(270° + 0°) = -\cos 0° = -1$

Furthermore, $\sin 360° = ?$

1st method $\sin 360° = \sin(360° + 0°) = \sin 0° = 0$

2nd method $\sin 360° = \sin(360° - 0°) = -\sin 0° = 0$

You may find the value of all the remaining T-ratios for 180°, 270° and 360° on the above line.

D. Range of Trigonometric Ratios

Here, $h = \sqrt{x^2 + h^2}$, $\therefore |x| \le h$ and $|y| \le h$

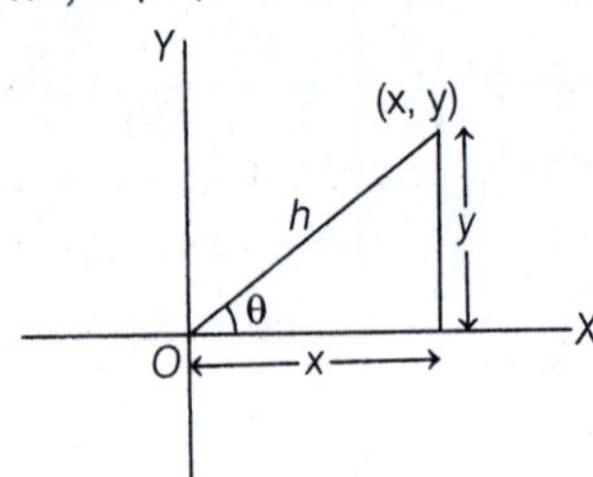

Now, $|\sin\theta| = \left|\frac{y}{h}\right| = \frac{|y|}{|h|} = \frac{|y|}{h} \le 1$

$\Rightarrow$ Hence, $\boxed{-1 \le \sin\theta \le 1}$

- $|\operatorname{cosec}\theta| = \left|\frac{h}{y}\right| = \frac{h}{|y|} \ge 1$

$\Rightarrow$ Hence, $\boxed{\operatorname{cosec}\theta \le -1 \text{ or } \operatorname{cosec}\theta \ge 1}$

- $|\cos\theta| = \left|\frac{x}{h}\right| = \frac{|x|}{h} \le 1$

$\Rightarrow$ Hence, $\boxed{-1 \le \cos\theta \le 1}$

- $|\sec\theta| = \left|\frac{h}{x}\right| = \frac{h}{|x|} \ge 1$

$\Rightarrow$ Hence, $\boxed{\sec\theta \le -1 \text{ or } \sec\theta \ge 1}$

- $|\tan\theta| = \left|\frac{y}{x}\right| = \frac{|y|}{|x|}$ = any positive number.

$\Rightarrow$ Hence, $\boxed{-\infty < \tan\theta < \infty}$

- $|\cot\theta| = \left|\frac{x}{y}\right| = \frac{|x|}{|y|}$ = any positive number

$\Rightarrow$ Hence, $\boxed{-\infty < \cot\theta < \infty}$

In brief

(i) $-1 \le \sin\theta$ or $\cos\theta \le 1 \Rightarrow 0 \le \sin^2\theta$ or $\cos^2\theta \le 1$

(ii) $-1 \ge \operatorname{cosec}\theta$ or $\sec\theta$ **or** $\operatorname{cosec}\theta$ or $\sec\theta \ge 1$

$\Rightarrow \sec^2\theta$ or $\operatorname{cosec}^2\theta \ge 1$

(iii) $-\infty < \tan\theta$ or $\cot\theta < \infty$

Observe the diagram

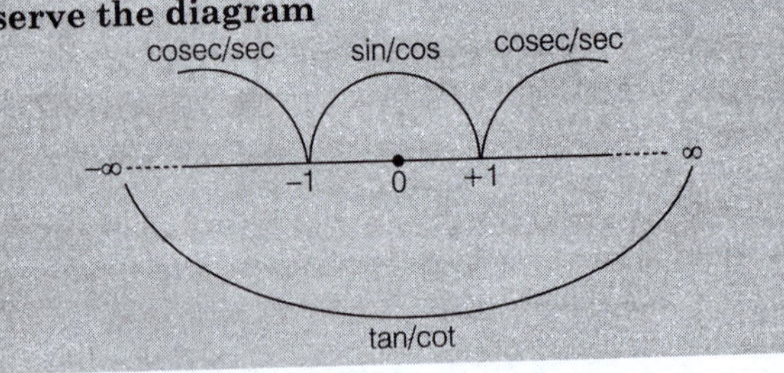

Critical facts

I. The maximum, and minimum value of $\sin\theta + \cos\theta$ are $+\sqrt{2}$ and $-\sqrt{2}$ respectively. Details have been given under illustrations titled as "Problems based on range of the values of T-ratios or Problems based on Maximum/minimum value of T-ratios/ expressions". Thus, the range of $\sin\theta + \cos\theta$ is $-\sqrt{2}$ to $+\sqrt{2}$.

II. $(\sin\theta + \operatorname{cosec}\theta) \ge 2$, $(\cos\theta + \sec\theta) \ge 2$ and $(\tan\theta + \cot\theta) \ge 2$ provided trigonometric ratios have positive values.

[We know that Arithmetic Means (AM) ≥ Geometric mean (GM) in case of positive numbers. Then, $\frac{(\sin\theta + \operatorname{cosec}\theta)}{2} \ge (\sin\theta \cdot \operatorname{cosec}\theta)^{1/2}$

$\Rightarrow \frac{(\sin\theta + \operatorname{cosec}\theta)}{2} \ge (1)^{1/2} \Rightarrow (\sin\theta + \operatorname{cosec}\theta) \ge 2$.

The concepts are same for the remaining two facts.

III. On the same relation between AM and GM, we may conclude that $(\sin^2\theta + \operatorname{cosec}^2\theta) \ge 2$, $(\cos^2\theta + \sec^2\theta) \ge 2$ and $(\tan^2\theta + \cot^2\theta) \ge 2$ irrespective of nature of value (positive/negative) of trigonometric ratios.

E. Trend of Changes in values of Trigonometric Ratios-Quadrant-wise

Quadrant	I	II	III	IV
sin	Increases from 0 to 1 $(0 \to 1)$	Decreases from 1 to 0 $(1 \to 0)$	Decreases from 0 to −1 $(0 \to -1)$	Increases from −1 to 0 $(-1 \to 0)$
cosec	$\infty \to 1$	$1 \to \infty$	$-\infty \to -1$	$-1 \to -\infty$
cos	$1 \to 0$	$0 \to -1$	$-1 \to 0$	$0 \to 1$
sec	$1 \to \infty$	$-\infty \to -1$	$-1 \to -\infty$	$\infty \to 1$
tan	$0 \to \infty$	$-\infty \to 0$	$0 \to \infty$	$-\infty \to 0$
cot	$\infty \to 0$	$0 \to -\infty$	$\infty \to 0$	$0 \to -\infty$

Illustrations

I. Problem based on Location (quadrant) of Angles and Sign of Trigonometric Ratios

- If the given angle is positive and greater then 360°, divide it by 360. The quadrant of the remainder will be the quadrant of the given angle.
- If the given angle is negative and greater than – 360°, add it with 360° and the quadrant of the resultant will be the quadrant of the given angle.
- If the given angle is negative and less than – 360°, divide it (ignoring negative sign) by 360 and subtract the remainder from 360°. The quadrant of the resultant will be the quadrant of the given angle.

Problem 1. Find the location of θ measuring 460°.

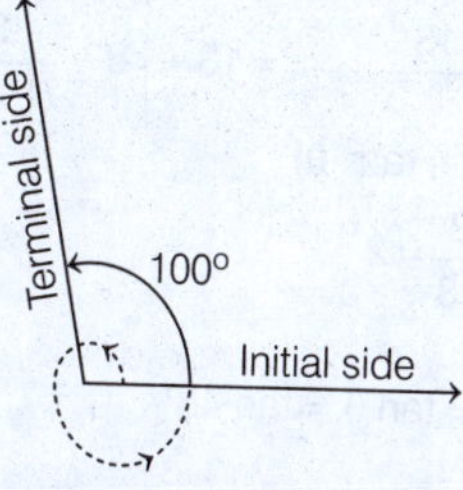

Clearly, $460° = (1 \times 360°) + 100°$

It means the terminal side completes one rotation in anti-clockwise direction and further moves 100°. Clearly, 100° lies in 2nd quadrant.

Alternatively

Find the remainder dividing the given angle by 360° and the quadrant of the remainder will be required quadrant.

Problem 2. Find the quadrant for θ measuring 180°.

Sol. If we assume $\theta = (180° - 0°)$, it is in 2nd quadrant. In case we assume $\theta = (180° + 0°)$, it is in 3rd quadrant.

Problem 3. Find the location of θ measuring –1100°.

Sol. Since measurement of the given angle is negative, it is clear that the terminal side moves in clock-wise direction.

Now, $-1100 = (-3 \times 360°) - 20°$. It means, the terminal side completes three rotation and further moves 20° in the same direction i.e. clock-wise.

Now, if we intend to measure the angle between initial side and terminal side with reference to anti-clockwise direction, the angle will be $(-20 + 360)° = 340°$ i.e. in 4th quadrant.

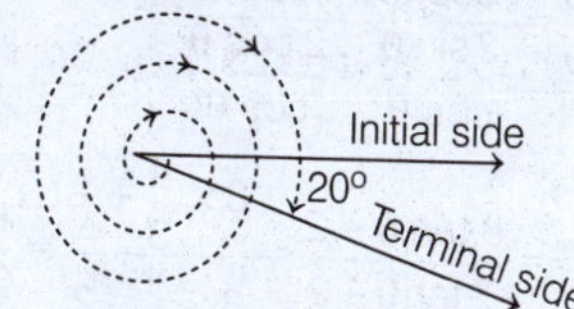

Alternatively

Find the remainder dividing the given angle (ignore negative sign while dividing) by 360 and subtract the remainder so obtained, from 360°. The quadrant of the resultant will be the required quadrant. We will get remainder 20° in case 1100° is divided by 360. Now, 360° – 20° = 340° i.e. 4th quadrant.

Problem 4. Find the sign of $\tan\theta$, if $\theta = 170°$

$90°<\theta<180°$ II Quadrant	$0°<\theta<90°$ I Quadrant
$180°<\theta<270°$ III Quadrant	$270°<\theta<360°$ IV Quadrant

Clearly θ is lying in 2nd quadrant in which $\tan\theta$ and $\cot\theta$ are 'negative'.

Problem 5. Find the sign of $\sin\theta$, if $\theta = -1000°$.

Sol. We get remainder 280 if 1000° is divided by 360. Now, subtract 280 from 360° i.e. 80° and it is clear that 80° is in 1st quadrant. Since, sign of sine is positive in the 1st quadrant, sign of sin – 1000° will be positive.

II. Computation of value of a Trigonometric Ratio (T-ratio)/ Expression when the Value of one T-ratio is given

Problem 6. If $\sin\theta = \dfrac{3}{5}$, then $\cos\theta = ?$

Sol. Given, $\sin\theta = \dfrac{3}{5} \Rightarrow \dfrac{p}{h} = \dfrac{3}{5} = k$ (let)

$\Rightarrow \quad \dfrac{p}{3} = \dfrac{h}{5} = k$

$\Rightarrow \quad p = 3k$ and $h = 5k$

We know, $b^2 = h^2 - p^2 = (5k)^2 - (3k)^2 = 16k^2$

$\therefore \quad b = \sqrt{16k^2} = \pm 4k$

Now, $\cos\theta = \dfrac{b}{h} = \dfrac{\pm 4k}{5k} = \pm\dfrac{4}{5}$

Since $\sin\theta = +\dfrac{3}{5}$, θ is either in 1st quadrant or in 2nd quadrant as sine is positive in 1st and 2nd quadrants. If θ is in 1st quadrant, $\cos\theta = +\dfrac{4}{5}$ and in cos θ is in 2nd quadrant, $\cos\theta$ will be negative, i.e. $-\dfrac{4}{5}$ as $\cos\theta$ is positive in 1st quadrant and negative in 2nd quadrant.

Alternatively I

Given, $\sin\theta = \dfrac{3}{5}$, it means, $p = 3, h = 5$ then, $b = \sqrt{5^2 - 3^2} = \pm 4$

Now, $\cos\theta = \dfrac{b}{h} = \pm\dfrac{4}{5}$

Alternatively II

We know, $\sin^2\theta + \cos^2\theta = 1$

$\Rightarrow \quad \cos^2\theta = 1 - \left(\dfrac{3}{5}\right)^2 = \dfrac{16}{25}$

$\therefore \quad \cos\theta = \pm\dfrac{4}{5}$

Problem 7. Find the value of $\sin\theta + \cos\theta$ if $\tan\theta = \dfrac{5}{12}$ and θ is in 3rd quadrant.

Sol. Given, $\tan\theta = \dfrac{5}{12}$ i.e. $p = 5, b = 12$

$\therefore \quad h = \sqrt{5^2 + 12^2} = \pm 13$

$\therefore \quad \sin\theta = \pm\frac{5}{13}$ and $\cos\theta = \pm\frac{12}{13}$

Since θ is in 3rd quadrant, sine and cosine will be negative. It means, $\sin\theta = -\frac{5}{13}$ and $\cos\theta = -\frac{12}{13}$

Now, $\sin\theta + \cos\theta = \left(\frac{-5}{13}\right) + \left(\frac{-12}{13}\right) = -\frac{17}{13}$

Note *Generally, we assume θ as a positive acute angle (i.e. in 1st quadrant) unless it is specified.*

Problem 8. If $\tan\theta = \frac{4}{3}$, then $\sqrt{\frac{1-\sin\theta}{1+\sin\theta}} = ?$

(a) ± 3
(b) $-\frac{1}{2}$
(c) $\frac{1}{4}$
(d) ± 1

Sol. (a) Given, $p = 4, b = 3$

$\therefore \quad h = 5$

Then, $\sin\theta = \frac{4}{5}$ [let θ be in 1st quadrant, as tangent is '+' in 1st quadrant]

Now, $\sqrt{\frac{1-\sin\theta}{1+\sin\theta}} = \sqrt{\frac{1-\frac{4}{5}}{1+\frac{4}{5}}} = \sqrt{\frac{1}{9}} = \pm\frac{1}{3}$. But $\pm\frac{1}{3}$ is not in option.

Now, assume θ be in 3rd quadrant as tangent is '+' in 3rd quadrant. Then, $\sin\theta = -\frac{4}{5}$

(∵ sine is negative in 3rd quadrant)

Now, $\sqrt{\frac{1-\sin\theta}{1+\sin\theta}} = \sqrt{\frac{1-\left(\frac{-4}{5}\right)}{1+\left(\frac{-4}{5}\right)}} = \sqrt{\frac{9}{1}} = \pm 3$

Hence, option (a) is correct.

Problem 9. If $\sin\theta = -\frac{7}{25}$, then $\sec\theta = ?$ assuming that θ is not in 3rd quadrant.

Sol. We know that sine is '–' in both III and IV quadrants.

Since sin θ has '–' value, θ is obviously in 4th quadrant as θ is not in 3rd quadrant as given.

Also, secant is '+' is 4th quadrant.

Now, given that $p = -7, h = 25$

$\therefore \quad b = \pm 24$

Then, $\sec\theta = \frac{25}{+24}$ or $\frac{25}{-24}$

Since θ is in 4th quadrant, $\sec\theta = \frac{25}{24}$.

Problem 10. If $\tan\theta = 1$, then the value of $\frac{8\sin\theta + 5\cos\theta}{\sin^3\theta - 2\cos^3\theta + 7\cos\theta}$ is

Sol. Given, $\tan\theta = 1 = \frac{p}{b}$

$\therefore \quad h = \sqrt{1^2 + 1^2} = \sqrt{2}$, then $\sin\theta = \frac{1}{\sqrt{2}}$ and $\cos\theta = \frac{1}{\sqrt{2}}$

Now, $\frac{8\times\frac{1}{\sqrt{2}} + 5\times\frac{1}{\sqrt{2}}}{\left(\frac{1}{\sqrt{2}}\right)^3 - 2\left(\frac{1}{\sqrt{2}}\right)^3 + 7\times\frac{1}{\sqrt{2}}}$

$$= \frac{\frac{13}{\sqrt{2}}}{\frac{1}{2\sqrt{2}} - \frac{2}{2\sqrt{2}} + \frac{7}{\sqrt{2}}} = \frac{\frac{13}{\sqrt{2}}}{\frac{13}{2\sqrt{2}}} = 2$$

Alternatively I

$$\frac{\frac{8\sin\theta}{\cos\theta} + \frac{5\cos\theta}{\cos\theta}}{\frac{\sin^3\theta}{\cos\theta} - \frac{2\cos^3\theta}{\cos\theta} + \frac{7\cos\theta}{\cos\theta}} = \frac{8\tan\theta + 5}{\tan\theta \times \sin^2\theta - 2\cos^2\theta + 7}$$

$$= \frac{(8\times 1) + 5}{1(1-\cos^2\theta) - 2\cos^2\theta + 7} = \frac{13}{1 - \frac{3}{\sec^2\theta} + 7}$$

$$= \frac{13}{8 - \frac{3}{(1+\tan^2\theta)}} = 13 \div \left[8 - \frac{3}{(1+1)}\right]$$

$$= 13 \times \frac{2}{13} = 2$$

Alternatively II

Given, $\tan\theta = 1 \Rightarrow \tan\theta = \tan 45^\circ$

$\therefore \quad \theta = 45^\circ$

Now,

$$\frac{8\sin 45^\circ + 5\cos 45^\circ}{(\sin 45^\circ)^3 - 2(\cos 45^\circ)^3 + 7\cos 45^\circ}$$

$$= \frac{8\times\frac{1}{\sqrt{2}} + 5\times\frac{1}{\sqrt{2}}}{\left(\frac{1}{\sqrt{2}}\right)^3 - 2\left(\frac{1}{\sqrt{2}}\right)^3 + 7\times\frac{1}{\sqrt{2}}}$$

$$= \frac{13\times\frac{1}{\sqrt{2}}}{\frac{1}{2\sqrt{2}} - 2\times\frac{1}{2\sqrt{2}} + 7\times\frac{1}{\sqrt{2}}} = \frac{\frac{13}{\sqrt{2}}}{\frac{13}{2\sqrt{2}}}$$

$$= \frac{13}{\sqrt{2}} \times \frac{2\sqrt{2}}{13} = 2$$

Problem 11. If $\tan\theta = \frac{4}{3}$, then the value of $\frac{3\sin\theta + 2\cos\theta}{3\sin\theta - 2\cos\theta}$ is :

[SSC (10+2 LDC/DEO)-11, SSC (CGL-T-I)-19]

Sol. $\frac{3\sin\theta + 2\cos\theta}{3\sin\theta - 2\cos\theta} = \frac{\frac{3\sin\theta}{\cos\theta} + \frac{2\cos\theta}{\cos\theta}}{\frac{3\sin\theta}{\cos\theta} - \frac{2\cos\theta}{\cos\theta}}$

$$= \frac{3\tan\theta + 2}{3\tan\theta - 2} = \frac{3\times\frac{4}{3} + 2}{3\times\frac{4}{3} - 2} = \frac{6}{2} = 3$$

Alternatively

Let $\frac{3\sin\theta + 2\cos\theta}{3\sin\theta - 2\cos\theta} = x$

$$\Rightarrow \frac{(3\sin\theta + 2\cos\theta) + (3\sin\theta - 2\cos\theta)}{(3\sin\theta + 2\cos\theta) - (3\sin\theta - 2\cos\theta)} = \frac{x+1}{x-1}$$

[by rule of Componendo and Dividendo]

$\Rightarrow \quad \dfrac{6\sin\theta}{4\cos\theta} = \dfrac{x+1}{x-1} \Rightarrow \dfrac{3}{2} \times \tan\theta = \dfrac{x+1}{x-1}$

$\Rightarrow \quad \dfrac{3}{2} \times \dfrac{4}{3} = \dfrac{x+1}{x-1} \Rightarrow x + 1 = 2x - 2$

$\Rightarrow \quad -x = -3$

$\therefore \quad x = 3$

III. Problem based Trigonometric Identity

You are not supposed to stress on the quadrant of an angle in this section as the Trigonometric identity is true for all possible values of the angle involved.

Type A

Computation of value of a trigonometric expression/ ratio when the value of other equation is given

Problem 12. If $\sec^2\theta + \tan^2\theta = \dfrac{7}{12}$, then $\sec^4\theta - \tan^4 = ?$ *[FCI-Asstt. Gr.-III-12, SSC (CGL-T-II)-19]*

Sol. $\sec^4\theta - \tan^4\theta = (\sec^2\theta)^2 - (\tan^2\theta)^2$

$= (\sec^2\theta - \tan^2\theta)(\sec^2\theta + \tan^2\theta)$

$= (1 + \tan^2\theta - \tan^2\theta) \times \dfrac{7}{12} = \dfrac{7}{12}$

$[\because \sec^2\theta = 1 + \tan^2\theta]$

Problem 13. On the same line, if $\mathrm{cosec}^2\theta + \cot^2 = 7 + 2\sqrt{3}$, then $\mathrm{cosec}^4\theta - \cot^4\theta = 7 + 2\sqrt{3}$ i.e. the same value.

Problem 14. If $\dfrac{\tan\theta + \cot\theta}{\tan\theta - \cot\theta} = 2, (0^\circ \le \theta \le 90^\circ)$, then the value of $\sin\theta$ is *[SSC (CPO-SI)-11]*

Sol. $\dfrac{\dfrac{\sin\theta}{\cos\theta} + \dfrac{\cos\theta}{\sin\theta}}{\dfrac{\sin\theta}{\cos\theta} - \dfrac{\cos\theta}{\sin\theta}} = \dfrac{\sin^2\theta + \cos^2\theta}{\sin^2\theta - \cos^2\theta} = \dfrac{1}{\sin^2\theta - \cos^2\theta}$

$= \dfrac{1}{\sin^2\theta - (1 - \sin^2\theta)} = 2$ (given)

$\Rightarrow \quad \dfrac{1}{2\sin^2\theta - 1} = 2 \Rightarrow 4\sin^2\theta = 3$

$\therefore \quad \sin\theta = \dfrac{\sqrt{3}}{2}$

Alternatively

Given, $\dfrac{\tan\theta + \dfrac{1}{\tan\theta}}{\tan\theta - \dfrac{1}{\tan\theta}} = 2 \Rightarrow \dfrac{\tan^2\theta + 1}{\tan^2\theta - 1} = 2$

$\Rightarrow \quad \tan^2\theta + 1 = 2\tan^2\theta - 2 \Rightarrow \tan^2\theta = 3$

$\Rightarrow \quad \tan\theta = \sqrt{3} \Rightarrow \tan\theta = \tan 60^\circ$

$\Rightarrow \quad \theta = 60^\circ$

Then, $\sin\theta = \sin 60^\circ = \dfrac{\sqrt{3}}{2}$

[Another way : $\tan\theta = \dfrac{\sqrt{3}}{1}$, $p = \sqrt{3}$ and $b = 1$, then $h = 2$, Now, $\sin\theta = p/h = \dfrac{\sqrt{3}}{2}$]

Problem 15. If $\cos^4\theta - \sin^4\theta = \dfrac{2}{3}$, than the value of $2\cos^2\theta - 1$ is

Sol. Given, $\cos^4\theta - \sin^4\theta = \dfrac{2}{3} \Rightarrow (\cos^2\theta)^2 - (\sin^2\theta)^2 = \dfrac{2}{3}$

$\Rightarrow (\cos^2\theta + \sin^2\theta)(\cos^2\theta - \sin^2\theta) = \dfrac{2}{3}$

$\Rightarrow [\cos^2\theta - (1 - \cos^2\theta)] = \dfrac{2}{3} \Rightarrow 2\cos^2\theta - 1 = \dfrac{2}{3}$

Problem 16. If θ be a positive acute angle satisfying $\cos^2\theta + \cos^4\theta = 1$, then the value of $\tan^2\theta + \tan^4\theta$ is

Sol. Given, $\cos^2\theta + \cos^4\theta = 1$

$\Rightarrow \quad \cos^4\theta = 1 - \cos^2\theta \Rightarrow \cos^4\theta = \sin^2\theta$

$\Rightarrow \quad \cos^2\theta = \dfrac{\sin^2\theta}{\cos^2\theta} = \tan^2\theta \Rightarrow \cos\theta = \tan\theta$

Replace $\cos\theta$ by $\tan\theta$ in the given equation $\cos^2\theta + \cos^4\theta = 1$.

Then, $\tan^2\theta + \tan^4\theta = 1$.

Alternatively

$\cos^2\theta + \cos^4\theta = 1$

$\Rightarrow \quad \dfrac{\cos^2\theta}{\cos^2\theta} + \dfrac{\cos^4\theta}{\cos^2\theta} = \dfrac{1}{\cos^2\theta}$

[dividing by $\cos^2\theta$ both sides]

$\Rightarrow \quad 1 + \cos^2\theta = \sec^2\theta$

$\Rightarrow \quad 1 + \cos^2\theta = 1 + \tan^2\theta$

$\Rightarrow \quad \cos\theta = \tan\theta$

Replace $\cos\theta$ by $\tan\theta$ in the given equation $\cos^2\theta + \cos^4\theta = 1$

Then, $\tan^2\theta + \tan^4\theta = 1$.

Problem 17. If θ be an acute angle and $7\sin^2\theta + 3\cos^2\theta = 4$, then the value of $\tan\theta$ is *[SSC (10 +2, LDC/DEO)-11, CDS-17]*

Sol. Given, $7\sin^2\theta + 3\cos^2\theta = 4$

$\Rightarrow \quad 4\sin^2\theta + 3(\sin^2 + \cos^2\theta) = 4$

$\Rightarrow \quad 4\sin^2\theta + 3 = 4$

$\Rightarrow \quad 4\sin^2\theta = 1$

$\Rightarrow \quad \sin^2\theta = \dfrac{1}{4}$

$\therefore \quad \sin\theta = \sqrt{\dfrac{1}{4}} = \pm\dfrac{1}{2}$

Since θ is an acute angle, $\sin\theta = \dfrac{1}{2}$

Now, $\quad p = 1, h = 2$

$\therefore \quad b = \sqrt{3}$

Then, $\tan\theta = \dfrac{1}{\sqrt{3}}$.

Alternatively I

$7\sin^2\theta + 3\cos^2\theta = 4$

$\Rightarrow \quad \dfrac{7\sin^2\theta}{\cos^2\theta} + \dfrac{3\cos^2\theta}{\cos^2\theta} = \dfrac{4}{\cos^2\theta}$

$\Rightarrow \quad 7\tan^2\theta + 3 = 4\sec^2\theta$

$\Rightarrow \quad 3\tan^2\theta + 3 = 4(\sec^2\theta - \tan^2\theta)$

$\Rightarrow \quad 3\tan^2\theta = 1$

$\Rightarrow \quad \tan\theta = \sqrt{\dfrac{1}{3}}$

$$\Rightarrow \quad \tan\theta = \pm\frac{1}{\sqrt{3}}$$

Since θ is an acute angle, $\tan\theta = \frac{1}{\sqrt{3}}$.

Alternatively II

$$7\sin^2\theta + 3\cos^2\theta = 4$$

$$\Rightarrow \quad 7\sin^2\theta + 3\cos^2\theta = 4(\sin^2\theta + \cos^2\theta)$$

$$\Rightarrow \quad 3\sin^2\theta = \cos^2\theta$$

$$\Rightarrow \quad \frac{\sin^2\theta}{\cos^2\theta} = \frac{1}{3}$$

$$\Rightarrow \quad \tan^2\theta = \frac{1}{3}$$

$$\Rightarrow \quad \tan\theta = \pm\frac{1}{\sqrt{3}}$$

Since, θ is an acute angle, $\tan\theta = \frac{1}{\sqrt{3}}$.

Problem 18. If $\frac{\sin\theta + \cos\theta}{\sin\theta - \cos\theta} = \frac{5}{8}$, the value of $\frac{\tan^2\theta + 1}{\tan^2\theta - 1}$ is

Sol. $\Rightarrow \dfrac{\frac{\sin\theta}{\cos\theta} + \frac{\cos\theta}{\cos\theta}}{\frac{\sin\theta}{\cos\theta} - \frac{\cos\theta}{\cos\theta}} = \frac{\tan\theta + 1}{\tan\theta - 1} = \frac{5}{4}$ (given)

$$\Rightarrow \quad \frac{(\tan\theta + 1) + (\tan\theta - 1)}{(\tan\theta - 1) - (\tan\theta - 1)} = \frac{5+4}{5-4}$$

[by rule of Componendo and Dividendo]

$$\Rightarrow \quad \frac{2\tan\theta}{2} = 9$$

$$\Rightarrow \quad \tan\theta = 9$$

$$\Rightarrow \quad \tan^2\theta = 81$$

Now, $\frac{\tan^2\theta + 1}{\tan^2\theta - 1} = \frac{81+1}{81-1} = \frac{41}{40}$

Alternatively

Given, $\frac{\sin\theta + \cos\theta}{\sin\theta - \cos\theta} = \frac{5}{4}$

$$\Rightarrow \quad 4\sin\theta + 4\cos\theta = 5\sin\theta - 5\cos\theta$$

$$\Rightarrow \quad 9\cos\theta = \sin\theta$$

$$\Rightarrow \quad \frac{\sin\theta}{\cos\theta} = 9$$

$$\Rightarrow \quad \tan\theta = 9$$

$$\Rightarrow \quad \tan^2\theta = 81$$

Now, $\frac{\tan^2\theta + 1}{\tan^2\theta - 1} = \frac{81+1}{81-1} = \frac{41}{40}$

Problem 19. If $\frac{\sec\theta - \tan\theta}{\sec\theta + \tan\theta} = \frac{3}{5}$, then the value of $\frac{\operatorname{cosec}\theta + \cot\theta}{\operatorname{cosec}\theta - \cot\theta}$ is : ***[SSC (CGL-T-I)-19, SSC (CGL-T-II)-20]***

Sol. Given, $\frac{\sec\theta - \tan\theta}{\sec\theta + \tan\theta} = \frac{3}{5}$

$$\Rightarrow \quad 5\sec\theta - 5\tan\theta = 3\sec\theta + 3\tan\theta$$

$$\Rightarrow \quad 2\sec\theta = 8\tan\theta$$

$$\Rightarrow \quad \frac{\sec\theta}{\tan\theta} = 4$$

$\Rightarrow \operatorname{cosec}\theta = 4$ [you may arrive at this conclusion by applying rule of Componendo and Dividendo too]

Then, $\cot\theta = \sqrt{\operatorname{cosec}^2\theta - 1} = \sqrt{15}$

Now, $\frac{\operatorname{cosec}\theta + \cot\theta}{\operatorname{cosec}\theta - \cot\theta} = \frac{(\operatorname{cosec}\theta + \cot\theta)}{(\operatorname{cosec}\theta - \cot\theta)} \times \frac{(\operatorname{cosec}\theta + \cot\theta)}{(\operatorname{cosec}\theta + \cot\theta)}$

$$= (\operatorname{cosec}\theta + \cot\theta)^2 = (4 + \sqrt{15})^2$$

$$= 31 + 8\sqrt{15}$$

Problem 20. If $\cos^2\theta - \sin^2\theta = \tan^2 A$, then $\cos^2 A - \sin^2 A = ?$

(a) $\tan^2(\theta + A)$
(b) $\tan^2 A$
(c) $\tan^2\theta$
(d) None of these

Sol. $\cos^2 A - \sin^2 A = \frac{\cos^2 A - \sin^2 A}{\cos^2 A + \sin^2 A} = \dfrac{\frac{\cos^2 A - \sin^2 A}{\cos^2 A}}{\frac{\cos^2 A + \sin^2 A}{\cos^2 A}}$

$$= \frac{1 - \tan^2 A}{1 + \tan^2 A} = \frac{1 - (\cos^2\theta - \sin^2\theta)}{1 + (\cos^2\theta - \sin^2\theta)}$$

$$= \frac{(1 - \cos^2\theta) + \sin^2\theta}{(1 - \sin^2\theta) + \cos^2\theta} = \frac{2\sin^2\theta}{2\cos^2\theta} = \tan^2\theta$$

Problem 21. If $\sin^2\theta + 3\cos\theta - 2 = 0$, then $\cos^3\theta + \sec^3\theta = ?$

Sol. Given, $\sin^2\theta + 3\cos\theta - 2 = 0$

$$\Rightarrow \quad (1 - \cos^2\theta) + 3\cos\theta - 2 = 0$$

$$\Rightarrow \quad \cos^2\theta + 1 = 3\cos\theta$$

$$\Rightarrow \quad \frac{\cos^2\theta}{\cos\theta} + \frac{1}{\cos\theta} = 3\frac{\cos\theta}{\cos\theta}$$

[dividing both sides by $\cos\theta$]

$$\Rightarrow \quad \cos\theta + \frac{1}{\cos\theta} = 3$$

$$\Rightarrow \quad \cos\theta + \sec\theta = 3$$

On cubing both sides, $(\cos\theta + \sec\theta)^3 = (3)^3$

$$\Rightarrow \quad \cos^3\theta + \sec^3\theta + 3\cdot\cos\theta\cdot\sec\theta(\cos\theta + \sec\theta) = 27$$

$$\Rightarrow \quad \cos^3\theta + \sec^3\theta + 3 \times 1 \times 3 = 27$$

$$\therefore \quad \cos^3\theta + \sec^3\theta = 27 - 9 = 18$$

Problem 22. If $\sec A + \tan A = 4$, then the value of $\sin A$ is

Sol. Given, $\sec A + \tan A = 4$ ...(i)

$$\Rightarrow \quad (\sec A - \tan A)(\sec A + \tan A) = 4(\sec A - \tan A)$$

$$\Rightarrow \quad \sec^2 A - \tan^2 A = 4(\sec A - \tan A)$$

$\therefore \quad \sec A - \tan A = \frac{1}{4}$ ...(ii)

On adding Eqs. (i) and (ii),

$$2\sec A = 4 + \frac{1}{4} \Rightarrow \sec A = \frac{17}{8}$$

Hence, $h = 17, b = 8$

$\therefore \quad p = 15$

Then, $\sin A = \frac{15}{17}$

Problem 23. If $\frac{\cos^4\alpha}{\cos^2\beta}+\frac{\sin^4\alpha}{\sin^2\beta}=1$ then, $\frac{\cos^4\beta}{\cos^2\alpha}+\frac{\sin^4\beta}{\sin^2\alpha}=?$

Sol. $\frac{\cos^4\alpha}{\cos^2\beta}+\frac{\sin^4\alpha}{\sin^2\beta}=\sin^2\alpha+\cos^2\alpha$

$\Rightarrow \frac{\cos^4\alpha}{\cos^2\beta}-\cos^2\alpha=\sin^2\alpha-\frac{\sin^4\alpha}{\sin^2\beta}$

$\Rightarrow \cos^2\alpha\left(\frac{\cos^2\alpha}{\cos^2\beta}-1\right)=\sin^2\alpha\left(1-\frac{\sin^2\alpha}{\sin^2\beta}\right)$

$\Rightarrow \cos^2\alpha\left(\frac{\cos^2\alpha-\cos^2\beta}{\cos^2\beta}\right)=\sin^2\alpha\left(\frac{\sin^2\beta-\sin^2\alpha}{\sin^2\beta}\right)$

$\Rightarrow \frac{\sin^2\beta(\cos^2\alpha-\cos^2\beta)}{\sin^2\alpha}=\frac{\cos^2\beta(\sin^2\beta-\sin^2\alpha)}{\cos^2\alpha}$

$\Rightarrow \frac{\sin^2\beta(1-\sin^2\alpha-1+\sin^2\beta)}{\sin^2\alpha}$

$=\frac{\cos^2\beta(1-\cos^2\beta-1+\cos^2\alpha)}{\cos^2\alpha}$

$\Rightarrow \frac{\sin^2\beta(\sin^2\beta-\sin^2\alpha)}{\sin^2\alpha}=\frac{\cos^2\beta(\cos^2\alpha-\cos^2\beta)}{\cos^2\alpha}$

$\Rightarrow \frac{\sin^4\beta-\sin^2\beta\cdot\sin^2\alpha}{\sin^2\alpha}=\frac{\cos^2\beta\cdot\cos^2\alpha-\cos^4\beta}{\cos^2\alpha}$

$\Rightarrow \frac{\sin^4\beta}{\sin^2\alpha}-\sin^2\beta=\left(\cos^2\beta-\frac{\cos^4\beta}{\cos^2\alpha}\right)$

$\Rightarrow \frac{\cos^4\beta}{\cos^2\alpha}+\frac{\sin^4\beta}{\sin^2\alpha}=\cos^2\beta+\sin^2\beta=1$

Problem 24. If $\frac{\sin\theta+\cos\theta}{\sin\theta-\cos\theta}=3$, then the value of $\sin^4\theta-\cos^4\theta$ is

Sol. $\frac{\sin\theta+\cos\theta}{\sin\theta-\cos\theta}=3 \Rightarrow \frac{\frac{\sin\theta}{\cos\theta}+\frac{\cos\theta}{\cos\theta}}{\frac{\sin\theta}{\cos\theta}-\frac{\cos\theta}{\cos\theta}}=3$

$\Rightarrow \frac{\tan\theta+1}{\tan\theta-1}=\frac{3}{1}$

$\Rightarrow \frac{(\tan\theta+1)+(\tan\theta-1)}{(\tan\theta+1)-(\tan\theta-1)}=\frac{3+1}{3-1}$

[by rule of componendo and dividendo]

$\Rightarrow \frac{2\tan\theta}{2}=2 \Rightarrow \tan\theta=2$

Hence, $p=2$ and $b=1$, then $h=\sqrt{2^2+1^2}=\sqrt{5}$

Now, $\sin^4\theta-\cos^4\theta=(\sin^2\theta)^2-(\cos^2\theta)^2$

$=(\sin^2\theta+\cos^2\theta)(\sin^2\theta-\cos^2\theta)$

$=1\times(\sin^2\theta-\cos^2\theta)$

$=\sin^2\theta-\cos^2\theta=\left(\frac{2}{\sqrt5}\right)^2-\left(\frac{1}{\sqrt5}\right)^2=\frac{3}{5}$

Alternatively

Given, $\frac{\sin\theta+\cos\theta}{\sin\theta-\cos\theta}=3$

$\Rightarrow \frac{(\sin\theta+\cos\theta)+(\sin\theta-\cos\theta)}{(\sin\theta-\cos\theta)-(\sin\theta-\cos\theta)}=\frac{3+1}{3-1}$

[by rule of componendo and dividendo]

$\Rightarrow \frac{2\sin\theta}{2\cos\theta}=2 \Rightarrow \tan\theta=2$

Hence, $p=2$ and $b=1$, then $h=\sqrt{2^2+1^2}=\sqrt5$

Now, $\sin^4\theta-\cos^4\theta=(\sin^2\theta)^2-(\cos^2\theta)^2$

$=(\sin^2\theta+\cos^2\theta)(\sin^2\theta-\cos^2\theta)$

$=1\times(\sin^2\theta-\cos^2\theta)$

$=\sin^2\theta-\cos^2\theta=\left(\frac{2}{\sqrt5}\right)^2-\left(\frac{1}{\sqrt5}\right)^2=\frac{3}{5}$

Problem 25. If $\tan\theta-\cot\theta=a$ and $\cos\theta+\sin\theta=b$, then $(a^2+4)(b^2-1)^2=?$

Sol. Given, $\tan\theta-\cot\theta=a$

$\Rightarrow (\tan\theta-\cot\theta)^2=a^2$ [squaring both sides]

$\Rightarrow (\tan\theta+\cot\theta)^2-4\tan\theta\cdot\cot\theta=a^2$

$\Rightarrow (\tan\theta+\cot\theta)^2=a^2+4$

Further given, $\cos\theta+\sin\theta=b$

$\Rightarrow (\cos\theta+\sin\theta)^2=(b)^2$ [squaring both sides]

$\Rightarrow \cos^2\theta+\sin^2\theta+2\cos\theta\cdot\sin\theta=b^2$

$\Rightarrow 2\cos\theta\cdot\sin\theta=b^2-1$

Now, $(a^2+4)(b^2-1)^2$

$=[(\tan\theta+\cot\theta)^2][2\cos\theta\cdot\sin\theta]^2$

$=[(\tan\theta+\cot\theta)(2\cos\theta\cdot\sin\theta)]^2$

$=[(\tan\theta\cdot2\cos\theta\cdot\sin\theta)+\cot\theta(2\cos\theta\cdot\sin\theta)]^2$

$=[2\sin^2\theta+2\cos^2\theta]^2$

$=[2(\sin^2\theta+\cos^2\theta)]^2=4$

Problem 26. If $\sec\theta-\tan\theta=x$, then the value of $\cos\theta$ (in terms of x) is :

Sol. Given, $\sec\theta-\tan\theta=x$ ...(i)

$\Rightarrow (\sec\theta-\tan\theta)(\sec\theta+\tan\theta)=x(\sec\theta+\tan\theta)$

$\Rightarrow \sec^2\theta-\tan^2\theta=x(\sec\theta+\tan\theta)$

$\therefore \sec\theta+\tan\theta=\frac{1}{x}$ ...(ii)

On adding Eqs. (i) and (ii) we get,

$2\sec\theta=x+\frac{1}{x}=\frac{x^2+1}{x} \Rightarrow 2\times\frac{1}{\cos\theta}=\frac{x^2+1}{x}$

$\therefore \cos\theta=\frac{2x}{1+x^2}$

Problem 27. If $5\sec\theta-3\tan\theta=5$, then what is the value of $5\tan\theta-3\sec\theta$? ***[SSC (CGL-T-I) 2017]***

Sol. Given, $5\sec\theta-3\tan\theta=5$

$\Rightarrow 5\sec\theta-5=3\tan\theta \Rightarrow 5(\sec\theta-1)=3\tan\theta$

$\Rightarrow 5(\sec\theta-1)(\sec\theta+1)=3\tan\theta(\sec\theta+1)$

$\Rightarrow 5\tan^2\theta=3\tan\theta(\sec\theta+1)$

$\Rightarrow 5\tan\theta=3(\sec\theta+1) \Rightarrow 5\tan\theta-3\sec\theta=3$

Problem 28. If $2\sin\theta+15\cos^2\theta=7, 0^\circ<\theta<90^\circ$, then $\tan\theta+\cos\theta+\sec\theta=?$ ***[SSC (CGL-T-I)-19]***

Sol. Given, $2\sin\theta+15\cos^2\theta=7$

$\Rightarrow 2\sin\theta+15(1-\sin^2\theta)=7$

$\Rightarrow \quad 15\sin^2\theta - 2\sin\theta - 8 = 0$

$\Rightarrow \quad 15\sin^2\theta - 12\sin\theta + 10\sin\theta - 8 = 0$

$\Rightarrow \quad 3\sin\theta(5\sin\theta - 4) + 2(5\sin\theta - 4) = 0$

$\Rightarrow \quad \sin\theta = \frac{-2}{3}, \frac{4}{5} \Rightarrow \sin\theta = \frac{4}{5}$

[$\because$ the value of $\sin\theta$ can not be negative if $0° < \theta < 90°$]

Then, $\tan\theta = \frac{4}{3}, \cos\theta = \frac{3}{5}$ and $\sec\theta = \frac{5}{3}$

Now, $\tan\theta + \cos\theta + \sec\theta = \frac{4}{3} + \frac{3}{5} + \frac{5}{3} = 3\frac{3}{5}$

Type B

Finding value of a trigonometric expression whose terms are same as of the given equation but with opposite sign and interchanged coefficients.

If $m\cos\theta - n\sin\theta = r$, then the value of the expression having the same terms with mutually interchanged coefficients and opposite sign (i.e. $n\cos\theta + m\sin\theta$) will be $\sqrt{(m^2+n^2)-r^2}$.

Similarly, if $m\cos\theta + n\sin\theta = r$, then

$$n\cos\theta - m\sin\theta = \sqrt{(m^2+n^2)-r^2}.$$

Problem 29. If $2\cos\theta - \sin\theta = \frac{1}{\sqrt{2}}$ $(0° < \theta < 90°)$, then the value of $(2\sin\theta + \cos\theta)$ is

Sol. $2\cos\theta - \sin\theta = \frac{1}{\sqrt{2}}$ (given)

and $\quad 2\sin\theta + \cos\theta = x$ (let)

On squaring both equations and adding,

$(\sin^2\theta + 4\cos^2\theta - 2\sin\theta 2\cos\theta) +$

$(4\sin^2\theta + \cos^2\theta + 2\cdot 2\sin\theta\cdot\cos\theta) = \frac{1}{2} + x^2$

$\Rightarrow \quad 4(\sin^2\theta + \cos^2\theta) + (\sin^2\theta + \cos^2\theta) = \frac{1}{2} + x^2$

$\Rightarrow \quad 5 - \frac{1}{2} = x^2 \quad \therefore x = \sqrt{\frac{9}{2}} = \pm\frac{3}{\sqrt{2}}$

Since θ is in 1st quadrant, both sine and cosine will have '+' value and their sum will also have '+' value.

$\therefore \quad x = +\frac{3}{\sqrt{2}}$

Alternatively

We know that if $m\cos\theta - n\sin\theta = r$, then

$n\cos\theta + m\sin\theta = \sqrt{(m^2+n^2)-r^2}$

Here, $2\cos\theta - 1\sin\theta = \frac{1}{\sqrt{2}}$

Then, $1\cos\theta + 2\sin\theta = \sqrt{(2^2+1^2) - \left(\frac{1}{\sqrt{2}}\right)^2} = \sqrt{\frac{9}{2}} = \pm\frac{3}{\sqrt{2}}$

Since θ is in 1st quadrant, both sine and cosine will have '+' value and their sum will also have '+' value. Hence, the required answer is $+\frac{3}{\sqrt{2}}$.

Problem 30. If $\sin\theta - \cos\theta = \frac{7}{13}$ and $0 < \theta < 90°$, then the value of $(\sin\theta + \cos\theta)$ is

Sol. Given, $1\sin\theta - 1\cos\theta = \frac{7}{13}$

$\Rightarrow \quad 1\sin\theta + 1\cos\theta = \sqrt{(1^2+1^2) - \left(\frac{7}{13}\right)^2}$

$\Rightarrow \quad \sin\theta + \cos\theta = \pm\frac{17}{13}$

Since, θ is in1st quadrant, both sine and cosine will have '+' value and their sum will also have '+' value.

Hence, $\sin\theta + \cos\theta = +\frac{17}{13}$

Problem 31. If $\cos\theta + \sin\theta = \sqrt{2}\cos\theta$, then $\cos\theta - \sin\theta = ?$ *[SSC (CGL-T-I) 2013]*

Sol. On the above line,

$\cos\theta - \sin\theta = \sqrt{(1^2+1^2) - (\sqrt{2}\cos\theta)^2}$

$= \sqrt{2 - 2\cos^2\theta} = \sqrt{2(1-\cos^2\theta)} = \pm\sqrt{2}\sin\theta$

Problem 32. If $\sin\theta + \cos\theta = 1$, then $\sin\theta - \cos\theta = ?$

Sol. On the above line, $\sin\theta - \cos\theta = \sqrt{(1^2+1^2)-(1)^2} = \pm 1$.

Type C

Finding the value of a trigonometric expression.

Problem 33. The simplified value of $(\sec A - \cos A)^2 + (\text{cosec}\, A - \sin A)^2 - (\cot A - \tan A)^2$ is :

Sol. $(\sec A - \cos A)^2 + (\text{cosec}\, A - \sin A)^2 - (\cot A - \tan A)^2$

$= \sec^2 A + \cos^2 A - 2\sec A\cdot\cos A$
$+ \text{cosec}^2 A + \sin^2 A - 2\,\text{cosec}\, A\cdot\sin A$
$- [\cot^2 A + \tan^2 A - 2\cot A\cdot\tan A]$

$= \sec^2 A + \cos^2 A - 2 + \text{cosec}^2 A$
$+ \sin^2 A - 2 - \cot^2 A - \tan^2 A + 2$

$= (\sec^2 A - \tan^2 A) + (\text{cosec}^2 A - \cot^2 A)$
$+ (\cos^2 A + \sin^2 A) - 2$

$= 1 + 1 + 1 - 2 = 1$

Alternatively

Take any value for angle A, say, 30°.

Then, value of expression $= (\sec 30° - \cos 30°)^2 + (\text{cosec}\, 30° - \sin 30°)^2 - (\cot 30° - \tan 30°)^2$

$= \left(\frac{2}{\sqrt{3}} - \frac{\sqrt{3}}{2}\right)^2 + \left(2 - \frac{1}{2}\right)^2 - \left(\sqrt{3} - \frac{1}{\sqrt{3}}\right)^2 = 1$

Remarks

Aspirants are advised to adhere the alternative method to solve the questions quickly. You have to assume the angles like 30°/45°/60°as the values of some T-ratio for 0° or 90° are not defined. However, you are free to assume any angle (s) in addition to 30°/45°/60° for which T-ratios have defined value.

Problem 34. The simplified value of $(\sec x\cdot\sec y + \tan x\cdot\tan y)^2 - (\sec x\cdot\tan y + \tan x\cdot\sec y)^2$ is :

(a) -1 (b) 0 (c) $\sec^2 x$ (d) 1

Sol. (d) $\sec^2 x \cdot \sec^2 y + \tan^2 x \cdot \tan^2 y$

$+ 2 \cdot \sec x \cdot \sec y \cdot \tan x \cdot \tan y \; \sec^2 x \cdot \tan^2 y - \tan^2 x \cdot \sec^2 y$

$- 2 \sec x \cdot \tan y \cdot \tan x \cdot \sec y$

$= \sec^2 x (\sec^2 y - \tan^2 y) + \tan^2 x (\tan^2 y - \sec^2 y)$

$= (\sec^2 x \times 1) - \tan^2 x (\sec^2 y - \tan^2 y)$

$= \sec^2 x - \tan^2 x = 1$

Alternatively

Let $x = 0°$ and $y = 60°$

(you are free to take any angles for which the given *T*-ratios have defined values.)

Now, expression will be as $(\sec 0° \cdot \sec 60° + \tan 0° \cdot \tan 60°)^2$

$- (\sec 0° \cdot \tan 60° + \tan 0° \cdot \sec 60°)^2$

$= (1 \cdot 2 + 0 \cdot \sqrt{3})^2 - (1 \cdot \sqrt{3} + 0 \cdot 2)^2 = 4 - 3 = 1$

Problem 35. The expression $3 \sec^2 \theta \tan^2 \theta + \tan^6 \theta - \sec^6 \theta$ is equal to **[SSC (CGL-T-I)-19]**

Sol. Let $\theta = 45°$ (you are free to take any angles for which the given *T*-ratios have defined values.)

Now the expression $= 3 \sec^2 45° \cdot \tan^2 45° + \tan^6 45° - \sec^6 45°$

$= 3(2)(1) + 1 - 8 = -1$

Problem 36. The numerical value of $\frac{5}{\sec^2 \theta} + \frac{2}{1 + \cot^2 \theta} + 3 \sin^2 \theta$ is **[SSC (10+2 LDC/DEO)-12]**

Sol. $5 \times \frac{1}{\sec^2 \theta} + \frac{2}{1 + \cot^2 \theta} + 3 \sin^2 \theta$

$= 5 \cos^2 \theta + \frac{2}{\operatorname{cosec}^2 \theta} + 3 \sin^2 \theta$

$= 5 \cos^2 \theta + 2 \sin^2 \theta + 3 \sin^2 \theta$

$= 5 (\cos^2 \theta + \sin^2 \theta) = 5$

Alternatively

Let θ be 30°, then $\frac{5}{\sec^2 30°} + \frac{2}{1 + \cot^2 30°} + 3 \sin^2 30°$

$= \frac{5}{\left(\frac{2}{\sqrt{3}}\right)^2} + \frac{2}{1 + (\sqrt{3})^2} + 3 \cdot \left(\frac{1}{2}\right)^2$

$= \frac{15}{4} + \frac{2}{4} + \frac{3}{4} = \frac{20}{4} = 5$

Problem 37. The numerical value of $\left(\frac{1}{\cos \theta} + \frac{1}{\cot \theta}\right)\left(\frac{1}{\cos \theta} - \frac{1}{\cot \theta}\right)$ is

Sol. $\left(\frac{1}{\cos \theta} + \frac{\sin \theta}{\cos \theta}\right)\left(\frac{1}{\cos \theta} - \frac{\sin \theta}{\cos \theta}\right) = \left(\frac{1 + \sin \theta}{\cos \theta}\right) \times \left(\frac{1 - \sin \theta}{\cos \theta}\right)$

$= \left(\frac{1 - \sin^2 \theta}{\cos^2 \theta}\right) = \frac{\cos^2 \theta}{\cos^2 \theta} = 1$

Alternatively

Let $\theta = 60°$. Then $\left(\frac{1}{\frac{1}{2}} + \frac{1}{\frac{1}{\sqrt{3}}}\right)\left(\frac{1}{\frac{1}{2}} - \frac{1}{\frac{1}{\sqrt{3}}}\right) = \left(\frac{1}{\frac{1}{2}}\right)^2 - \left(\frac{1}{\frac{1}{\sqrt{3}}}\right)^2$

$= 4 - 3 = 1$

Problem 38. The value of $(2 \cos^2 \theta - 1)\left(\frac{1 + \tan \theta}{1 - \tan \theta} + \frac{1 - \tan \theta}{1 + \tan \theta}\right)$ is **[FCI-Asstt. Gr.-III-12]**

Sol. $(2 \cos^2 \theta - 1)\left[\frac{(1 + \tan \theta)^2 + (1 - \tan \theta)^2}{(1 - \tan \theta)(1 + \tan \theta)}\right]$

$= (2 \cos^2 \theta - 1)\left[\frac{2 + 2 \tan^2 \theta}{1 - \tan^2 \theta}\right] = (2 \cos^2 \theta - 1)\left[\frac{2(1 + \tan^2 \theta)}{1 - \tan^2 \theta}\right]$

$= (2 \cos^2 \theta - 1)\left(\frac{2 \sec^2 \theta}{1 - \frac{\sin^2 \theta}{\cos^2 \theta}}\right) = (2 \cos^2 \theta - 1)\left(\frac{2 \sec^2 \theta \cdot \cos^2 \theta}{\cos^2 \theta - \sin^2 \theta}\right)$

$= (2 \cos^2 \theta - 1) \frac{2 \times (1)^2}{\cos^2 \theta - (1 - \cos^2 \theta)}$

$= (2 \cos^2 \theta - 1) \times \frac{2}{(2 \cos^2 \theta - 1)} = 2$

Alternatively

Let θ be 0°.

Then, Given expression

$= (2 \cos^2 0° - 1)\left(\frac{1 + \tan 0°}{1 - \tan 0°} + \frac{1 - \tan 0°}{1 + \tan 0°}\right)$

$= (2 \cdot 1^2 - 1)\left(\frac{1}{1} + \frac{1}{1}\right) = 1 \times 2 = 2$

Problem 39. The value of $(1 - \sin \theta)(1 + \sin \theta)(1 + \tan^2 \theta) + 1$ is :

Sol. $(1^2 - \sin^2 \theta) \cdot \sec^2 \theta + 1$

$= (\cos^2 \theta \times \sec^2 \theta) + 1$

$= \left(\cos^2 \theta \times \frac{1}{\cos^2 \theta}\right) + 1 = 1 + 1 = 2$

Alternatively

Let θ be 0°.

Then, $(1 - 0)(1 + 0)(1 + 0^2) = 1$

Problem 40. The value of $2 (\sin^6 \theta + \cos^6 \theta) - 3 (\sin^4 \theta + \cos^4 \theta)$ is :

Sol. $2 [(\sin^2 \theta)^3 + (\cos^2 \theta)^3] - 3 [(\sin^2 \theta)^2 + (\cos^2 \theta)^2]$

$= 2 [(\sin^2 \theta + \cos^2 \theta) (\sin^4 \theta + \cos^4 \theta - \sin^2 \theta \cdot \cos^2 \theta)$

$- 3 [(\sin^2 \theta + \cos^2 \theta)^2 - 2 \sin^2 \theta \cdot \cos^2 \theta]$

$= 2 \sin^4 \theta + 2 \cos^4 \theta - 2 \sin^2 \theta \cdot \cos^2 \theta - 3 + 6 \sin^2 \theta \cdot \cos^2 \theta$

$[\because \sin^2 \theta + \cos^2 \theta = 1]$

$= 2[(\sin^2 \theta)^2 + (\cos^2 \theta)^2] + 4 \sin^2 \theta \cdot \cos^2 \theta - 3$

$= 2 - 4 \sin^2 \theta \cdot \cos^2 \theta + 4 \sin^2 \theta \cdot \cos^2 \theta - 3$

$= 2 - 3 = -1$

Alternatively

Let $\theta = 0°$.

Then, $2(0^6 + 1^6) - 3(0^4 + 1^4) = 2 - 3 = -1$

Problem 41. The value of $\sqrt{\frac{1 + \sin \theta}{1 - \sin \theta}} + \sqrt{\frac{1 - \sin \theta}{1 + \sin \theta}} = ?$

(a) $2 \sec \theta$ (b) $2 \sin \theta$

(c) $1 + \cos \theta$ (d) 0

Sol. (a) $\sqrt{\frac{1+\sin\theta}{1-\sin\theta}\times\frac{1+\sin\theta}{1+\sin\theta}}+\sqrt{\frac{1-\sin\theta}{1+\sin\theta}\times\frac{1-\sin\theta}{1-\sin\theta}}$

$$=\sqrt{\frac{(1+\sin\theta)^2}{1^2-\sin^2\theta}}+\sqrt{\frac{(1-\sin\theta)^2}{1^2-\sin^2\theta}}=\frac{1+\sin\theta}{\cos\theta}+\frac{1-\sin\theta}{\cos\theta}$$

$$=\frac{1}{\cos\theta}+\frac{\sin\theta}{\cos\theta}+\frac{1}{\cos\theta}-\frac{\sin\theta}{\cos\theta}=\frac{2}{\cos\theta}=2\sec\theta$$

Alternatively

Let θ be 30°. Then, $\sqrt{\frac{1+\sin\theta}{1-\sin\theta}}+\sqrt{\frac{1-\sin\theta}{1+\sin\theta}}=\sqrt{\frac{1+\frac{1}{2}}{1-\frac{1}{2}}}+\sqrt{\frac{1-\frac{1}{2}}{1+\frac{1}{2}}}$

$$=\sqrt{3}+\frac{1}{\sqrt{3}}=\frac{4}{\sqrt{3}}=2\cdot\frac{2}{\sqrt{3}}=2\sec 30^\circ=2\sec\theta$$

Problem 42. The value of $\frac{\tan\theta+\sec\theta-1}{\tan\theta-\sec\theta+1}-(\sec\theta+\tan\theta)$ is

Sol. $\frac{\tan\theta+\sec\theta-1}{\tan\theta-\sec\theta+1}-(\sec\theta+\tan\theta)$

$$=\frac{\tan\theta+\sec\theta-(\sec^2\theta-\tan^2\theta)}{\tan\theta-\sec\theta+1}-(\sec\theta+\tan\theta)$$

$$=\frac{(\tan\theta+\sec\theta)-(\sec\theta+\tan\theta)(\sec\theta-\tan\theta)}{\tan\theta-\sec\theta+1}-(\sec\theta+\tan\theta)$$

$$=\frac{(\tan\theta+\sec\theta)[1-(\sec\theta-\tan\theta)]}{\tan\theta-\sec\theta+1}-(\sec\theta+\tan\theta)$$

$$=\frac{(\tan\theta+\sec\theta)(\tan\theta-\sec\theta+1)}{(\tan\theta-\sec\theta+1)}-(\sec\theta+\tan\theta)$$

$$=(\tan\theta+\sec\theta)-(\sec\theta+\tan\theta)=0$$

Alternatively

Let θ be 45°. Then, $\frac{\tan\theta+\sec\theta-1}{\tan\theta-\sec\theta+1}-(\sec\theta+\tan\theta)$

$$=\frac{1+\sqrt{2}-1}{1-\sqrt{2}+1}-(\sqrt{2}+1)=\frac{\sqrt{2}}{\sqrt{2}(\sqrt{2}-1)}-(\sqrt{2}+1)$$

$$=\frac{(\sqrt{2}+1)}{(\sqrt{2}+1)(\sqrt{2}-1)}-(\sqrt{2}+1)=(\sqrt{2}+1)-(\sqrt{2}+1)=0$$

Problem 43. $(1+\cot A-\operatorname{cosec} A)(1+\tan A+\sec A)$ is equal to.

Sol. $\left(1+\frac{\cos A}{\sin A}-\frac{1}{\sin A}\right)\left(1+\frac{\sin A}{\cos A}+\frac{1}{\cos A}\right)$

$$=\frac{(\sin A+\cos A-1)}{\sin A}\times\frac{(\cos A+\sin A+1)}{\cos A}$$

$$=\frac{(\sin A+\cos A)^2-(1)^2}{\sin A\cos A}$$

$$=\frac{(\sin^2 A+\cos^2 A)+2\sin A\cos A-1}{\sin A\cos A}=2$$

Alternatively

Let $A=45^\circ$, then $(1+\cot A-\operatorname{cosec} A)(1+\tan A+\sec A)$

$$=(1+1-\sqrt{2})(1+1+\sqrt{2})$$

$$=(2-\sqrt{2})(2+\sqrt{2})=(2)^2-(\sqrt{2})^2$$

$$=4-2=2$$

Problem 44. The value of $\sin^6\theta+\sin^4\theta\cdot\cos^2\theta-\sin^2\theta\cdot\cos^4\theta-\cos^6\theta$ is

(a) $\sin^2\theta-\cos^3\theta$ (b) $\sin^3\theta+\cos^3\theta$

(c) $\sin^4\theta+\cos^4\theta$ (d) $\sin^2\theta-\cos^2\theta$

Sol. (d) $\sin^4\theta(\sin^2\theta+\cos^2\theta)-\cos^4\theta(\sin^2\theta+\cos^2\theta)$

$$=\sin^4\theta-\cos^4\theta$$

$$=(\sin^2\theta)^2-(\cos^2\theta)^2$$

$$=(\sin^2\theta+\cos^2\theta)(\sin^2\theta-\cos^2\theta)$$

$$=\sin^2\theta-\cos^2\theta$$

Alternatively

Let $\theta=30^\circ$, then $\sin^6\theta+\sin^4\theta\cdot\cos^2\theta-\sin^2\theta\cdot\cos^4\theta-\cos^6\theta$

$$=\left(\frac{1}{2}\right)^6+\left(\frac{1}{2}\right)^4\cdot\left(\frac{\sqrt{3}}{2}\right)^2-\left(\frac{1}{2}\right)^2\cdot\left(\frac{\sqrt{3}}{2}\right)^4-\left(\frac{\sqrt{3}}{2}\right)^6=-\frac{1}{2}$$

Also, $\sin^2 30^\circ-\cos^2 30^\circ=-\frac{1}{2}$

Hence, option (d) is correct.

Problem 45. The value of $\sin^6\alpha+\cos^6\alpha+3\sin^2\alpha\cdot\cos^2\alpha$ is :

Sol. $(\sin^2\alpha)^3+(\cos^2\alpha)^3+3\cdot\sin^2\alpha\cdot\cos^2\alpha(\sin^2\alpha+\cos^2\alpha)$

$$=(\sin^2\alpha+\cos^2\alpha)^3=(1)^3=1$$

Alternatively

Let α be 0°. Then, $\sin^6\alpha+\cos^6\alpha+3\sin^2\alpha\cdot\cos^2\alpha$

$$=0^6+1^6+3\cdot 0^2\cdot 1^2=1$$

IV. Problems Seeking Specific Approaches

Problem 46. If θ be acute and $\tan\theta+\cot\theta=2$, then the value of $\tan^5\theta+\cot^{10}\theta$ is.

Sol. Given, $\tan\theta+\cot\theta=2\Rightarrow\tan\theta+\frac{1}{\tan\theta}=2$

Obviously, $\tan\theta=1$

Now, $\tan^5\theta+\frac{1}{\tan^{10}\theta}=(1)^5+\frac{1}{(1)^{10}}=2$

Alternatively I

$$\tan\theta+\cot\theta=2$$

$$\Rightarrow\quad\tan 45^\circ+\cot 45^\circ=2$$

$$\therefore\quad\theta=45^\circ$$

Now, $\tan^5 45^\circ+\cot^{10} 45^\circ=1+1=2$

Alternatively II

Given, $\tan\theta+\cot\theta=2$

$$\Rightarrow\quad\tan\theta+\frac{1}{\tan\theta}=2,$$

$$\Rightarrow\quad\tan^2\theta-2\tan\theta+1=0$$

$$\Rightarrow\quad(\tan\theta-1)^2=0\Rightarrow\tan\theta=1$$

Now, $\tan^5\theta+\cot^{10}\theta=1^5+\frac{1}{1^{10}}=2$

Problem 47. If $\sin\theta+\operatorname{cosec}\theta=2$, then value of $\sin^{100}\theta+\operatorname{cosec}^{100}\theta$ is *[SSC (10+2, LDC/DEO)-11]*

Sol. Given, $\sin\theta+\frac{1}{\sin\theta}=2$

Clearly, $\sin\theta = 1$

$\therefore \quad \sin^{100}\theta + \dfrac{1}{\sin^{100}\theta} = 1^{100} + \dfrac{1}{1^{100}} = 2$

Alternatively I

Given, $\sin\theta + \text{cosec}\,\theta = 2$

$\Rightarrow \sin 90° + \text{cosec}\, 90° = 1 + 1 \Rightarrow \theta = 90°.$

Now, $\sin^{100} 90° + \text{cosec}^{100} 90° = 1 + 1 = 2$

Alternatively II

Given, $\sin\theta + \text{cosec}\,\theta = 2 \Rightarrow \sin\theta + \dfrac{1}{\sin\theta} = 2$

$\Rightarrow \sin^2\theta - 2\sin\theta + 1 = 0$

$\Rightarrow (\sin\theta - 1)^2 = 0 \Rightarrow \sin\theta = 1$

Now, $\sin^{100}\theta + \dfrac{1}{\sin^{100}\theta} = 1^{100} + \dfrac{1}{1^{100}} = 2$

Problem 48. If $\cos\theta + \sec\theta = 2$ then, $\cos^6\theta + \sec^6\theta = ?$

Sol. On the above line, $\cos^6\theta + \sec^6\theta$ will also be equal to 2.

Problem 49. If $\sin\theta + \text{cosec}\,\theta = -2$, then the value of $\sin^8\theta + \text{cosec}^5\theta$ is

Sol. Given, $\sin\theta + \dfrac{1}{\sin\theta} = -2$

Clearly, $\sin\theta = -1$

Now, $\sin^8\theta + \text{cosec}^5\theta = \sin^8\theta + \dfrac{1}{\sin^5\theta} = (-1)^8 + \dfrac{1}{(-1)^5} = 0$

Problem 50. If $\cos\theta + \sec\theta = 1$, then the value of $\cos^{16}\theta + \cos^{13}\theta + \cos^5\theta + \cos^2\theta + 3$ is

Sol. Given, $\cos\theta + \dfrac{1}{\cos\theta} = 1$

$\Rightarrow \cos^2\theta - \cos\theta + 1 = 0$

Now, $\cos^{16}\theta + \cos^{13}\theta + \cos^5\theta + \cos^2\theta + 3$

$= \cos^{13}\theta\,(\cos^3\theta + 1) + \cos^2\theta(\cos^3\theta + 1) + 3$

$= \cos^{13}\theta\,(\cos\theta + 1)(\cos^2\theta + 1 - \cos\theta)$
$\quad + \cos^2\theta(\cos\theta + 1)(\cos^2\theta + 1 - \cos\theta) + 3$

$= [\cos^{13}\theta \times (\cos\theta + 1) \times 0]$
$\quad + [\cos^2\theta \times (\cos\theta + 1) \times 0] + 3$

$= 0 + 0 + 3 = 3$

> ***Observation*** It may be construed from the above that the sum of two same trigonometric ratios having difference of 3 between their powers, is 0 provided sum of the trigonometric ratio and its reciprocal is 1.
>
> **e.g.**
>
> (i) If $\sin\theta + \text{cosec}\,\theta = 1$, then $\sin^n\theta + \sin^{(n-3)}\theta = 0$
>
> (ii) If $\cos\theta + \sec\theta = 1$, then $\cos^n\theta + \cos^{(n-3)}\theta = 0$
>
> (iii) If $\tan\theta + \cot\theta = 1$, then $\tan^n\theta + \tan^{(n-3)}\theta = 0$

Problem 51. If $\cos^2\alpha + \cos^2\beta = 2$, then $\tan^3\alpha + \sin^5\beta = ?$

Sol. Given, $\cos^2\alpha + \cos^2\beta = 2$

$\Rightarrow \cos^2\alpha = 1$ and $\cos^2\beta = 1$

[$\because \cos\alpha$ and $\cos\beta$ may have maximum value of 1]

Clearly, $\alpha = 0°$ and $\beta = 0°$

Now, $\tan^3 0° + \sin^5 0° = 0 + 0 = 0$

Problem 52. If $0° < \theta < 90°$ and $\dfrac{\sin\theta}{\cos\theta} + \dfrac{\cos\theta}{\sin\theta} = 2$, then $\theta = ?$

Sol. Given, $\tan\theta + \cot\theta = 2$

$\Rightarrow \tan\theta + \dfrac{1}{\tan\theta} = 2$

Clearly, $\tan\theta = 1 \Rightarrow \tan\theta = \tan 45°$

$\therefore \theta = 45°$

V. Problem Based on Elimination of Trigonometric Ratios

Problem 53. If $a = m\sin\theta$ and $b = n\cos\theta$, then which of the following is true?

(a) $a^2 + b^2 = m^2 + n^2$
(b) $\dfrac{a^2}{m^2} + \dfrac{b^2}{n^2} = 1$
(c) $\dfrac{a^2 \times b^2}{n^2} = m^2$
(d) $\dfrac{m^2 + n^2}{mn} = ab$

Sol. (b) Given, $a = m\sin\theta$ and $b = n\cos\theta$

$\Rightarrow a^2 = m^2\sin^2\theta$ and $b^2 = n^2\cos^2\theta$

$\Rightarrow \sin^2\theta = \dfrac{a^2}{m^2}$ and $\cos^2\theta = \dfrac{b^2}{n^2}$

$\Rightarrow$ On adding, $\dfrac{a^2}{m^2} + \dfrac{b^2}{n^2} = \sin^2\theta + \cos^2\theta$

$\therefore \dfrac{a^2}{m^2} + \dfrac{b^2}{n^2} = 1$

Alternatively

Let θ be 45° (you are free to take any angle)

Then, $a = m\sin 45°$ and $b = n\cos 45°$

$\Rightarrow a = \dfrac{m}{\sqrt{2}}$ and $b = \dfrac{n}{\sqrt{2}} \Rightarrow a^2 = \dfrac{m^2}{2}$ and $b^2 = \dfrac{n^2}{2}$

$\Rightarrow \dfrac{a^2}{m^2} = \dfrac{1}{2}$ and $\dfrac{b^2}{n^2} = \dfrac{1}{2} \Rightarrow$ On adding, $\dfrac{a^2}{m^2} + \dfrac{b^2}{n^2} = 1$

Problem 54. If $\tan\theta - \sin\theta = a$ and $\tan\theta + \sin\theta = b$, then the value of $4\sqrt{ab}$ is

(a) $a^2 + b^2$
(b) $b^2 - a^2$
(c) $a^2 - b^2$
(d) $\dfrac{a+b}{ab}$

Sol. Given, $\tan\theta - \sin\theta = a$ and $\tan\theta + \sin\theta = b$

On multiplication, $\tan^2\theta - \sin^2\theta = ab$

$\Rightarrow \dfrac{\sin^2\theta}{\cos^2\theta} - \sin^2\theta = ab \Rightarrow \sin^2\theta\left(\dfrac{1}{\cos^2\theta} - 1\right) = ab$

$\Rightarrow \dfrac{\sin^2\theta(1 - \cos^2\theta)}{\cos^2\theta} = ab$

$\Rightarrow \tan^2\theta \cdot \sin^2\theta = ab$

$\Rightarrow \tan\theta \cdot \sin\theta = \sqrt{ab}$

$\Rightarrow 4\tan\theta \cdot \sin\theta = 4\sqrt{ab}$

$\Rightarrow (\tan\theta + \sin\theta)^2 - (\tan\theta - \sin\theta)^2 = 4\sqrt{ab}$

[$\because (x + y)^2 - (x - y)^2 = 4xy$]

$\Rightarrow b^2 - a^2 = 4\sqrt{ab}$

Alternatively

Let $\theta = 45°$,

Now, $\tan 45° - \sin 45° = a$, and $\tan 45° + \sin 45° = b$

$\Rightarrow \quad 1 - \frac{1}{\sqrt{2}} = a$ and $1 + \frac{1}{\sqrt{2}} = b$

$\Rightarrow \quad ab = (1)^2 - \left(\frac{1}{\sqrt{2}}\right)^2 = \frac{1}{2}$

$\Rightarrow \quad 4\sqrt{ab} = 4 \times \sqrt{\frac{1}{2}} = 2\sqrt{2}$

Now, ensure which option has the value of $2\sqrt{2}$.

Option (a) $a^2 + b^2 = \left(1 - \frac{1}{\sqrt{2}}\right)^2 + \left(1 + \frac{1}{\sqrt{2}}\right)^2$

$= 1 + \frac{1}{2} - 2 \cdot 1 \cdot \frac{1}{\sqrt{2}} + 1 + \frac{1}{2} + 2 \cdot 1 \cdot \frac{1}{\sqrt{2}} = 3$

Option (b) $b^2 - a^2 = \left(1 + \frac{1}{\sqrt{2}}\right)^2 - \left(1 - \frac{1}{\sqrt{2}}\right)^2$

$= 4 \cdot 1 \cdot \frac{1}{\sqrt{2}} = 2\sqrt{2}$

Hence, option (b) is correct.

Problem 55. If $\sin^2\theta + \operatorname{cosec}^2\theta = m^2$ and $\sin\theta - \operatorname{cosec}\theta = n$, then the value of $n^2 - m^2$ is :

Sol. Given, $\sin^2\theta + \operatorname{cosec}^2\theta = m^2$

$\Rightarrow \quad \sin^2\theta + \operatorname{cosec}^2\theta - 2 = m^2 - 2$

$\Rightarrow \quad \sin^2\theta + \operatorname{cosec}^2\theta - 2 \cdot \sin\theta \cdot \operatorname{cosec}\theta = m^2 - 2$

$\Rightarrow \quad (\sin\theta - \operatorname{cosec}\theta)^2 = m^2 - 2$

$\Rightarrow \quad n^2 - m^2 = -2$

Alternatively

Let $\theta = 45°$,

Then, $\sin^2 45 + \operatorname{cosec}^2 45° = m^2$

and $\sin 45° - \operatorname{cosec} 45° = n$

$\Rightarrow \quad \left(\frac{1}{\sqrt{2}}\right)^2 + (\sqrt{2})^2 = m^2$ and $\frac{1}{\sqrt{2}} - \sqrt{2} = n$

$\Rightarrow \quad \frac{1}{2} + 2 = m^2$

and $\quad n^2 = \frac{1}{2} + 2 - 2 \cdot \frac{1}{\sqrt{2}} \cdot \sqrt{2} = \frac{1}{2}$

Now, $\quad n^2 - m^2 = \frac{1}{2} - \left(\frac{1}{2} + 2\right) = \frac{1}{2} - \frac{1}{2} - 2 = -2$

VI. Problems Based on Range of the Values of *T*-ratios or Problems based on Maximum/Minimum Values of *T*-ratios/Expressions

Problem 56. If θ is an acute angle, which of the followings is true ?

(a) $\sin\theta + \cos\theta < 1$ (b) $\sin\theta > \tan\theta$

(c) $\tan\theta > \sin\theta$ (d) None of these

Sol. (c) Proceed through options

(a) $\sin\theta + \cos\theta = \frac{p}{h} + \frac{b}{h} = \frac{p+b}{h} > 1$

[$\because$ sum of two sides of a triangle is greater than its 3rd side]

Thus, $\sin\theta + \cos\theta > 1$

(b) $\sin\theta = \frac{p}{h}$ and $\tan\theta = \frac{p}{b}$

$\therefore \quad \tan\theta > \sin\theta$ [$\because h > b$ in a triangle]

Hence, option (c) is correct.

Problem 57. If $0° < \theta < 90°$, the value of $\sin\theta + \cos\theta$ is: *[SSC (CGL-T-II)-12]*

(a) equal to 1 (b) greater than 1

(c) less than 1 (d) equal to 2

Sol. $\sin\theta + \cos\theta = \frac{p}{h} + \frac{b}{h} = \frac{p+b}{h} > 1$

[$\because$ sum of two sides of a triangle is greater than its 3rd side]

Thus, $\sin\theta + \cos\theta > 1$

Problem 58. The equation $\sec^2\theta = \frac{4xy}{(x+y)^2}$ is only possible when :

(a) $x = y$ (b) $x < y$

(c) $x > y$ (d) None of these

Sol. (a) Since $\sec^2\theta \geq 1$, $\frac{4xy}{(x+y)^2} \geq 1$

$\Rightarrow 4xy \geq (x+y)^2 \Rightarrow 0 \geq (x+y)^2 - 4xy$

$\Rightarrow (x-y)^2 \leq 0$

$\Rightarrow$ It is possible only when $x = y$.

Problem 59. If for real values of x, $\cos\theta = x + \frac{1}{x}$, then which of the following is true?

(a) θ is an acute angle (b) θ is an right angle

(c) θ is an obtuse angle (d) No value of θ is possible

Sol. (d) Given, $\cos\theta = x + \frac{1}{x} \Rightarrow x^2 + x\cos\theta + 1 = 0$

Since x is real, $\cos^2\theta - 4 \cdot 1 \cdot 1 \geq 0$

$\Rightarrow \cos^2\theta \geq 4$, which is impossible as maximum value $\cos\theta$ is 1.

Hence, option (d) is correct.

Alternatively

Since $\quad$ AM $\geq$ GM, $\frac{x + \frac{1}{x}}{2} \geq \sqrt{x \cdot \frac{1}{x}} \Rightarrow x + \frac{1}{x} \geq 2$

It implies that $\cos\theta \geq 2$. It is not possible as the maximum value of $\cos\theta$ is 1. Hence, option (d) is correct.

Problem 60. Find the maximum value of $5\cos^2\theta + 7\sin^2\theta$.

Sol. $5\cos^2\theta + 7\sin^2\theta = 5(\cos^2 + \sin^2\theta) + 2\sin^2\theta = 5 + 2\sin^2\theta$

$\because$ Maximum value of $\sin\theta = +1$

$\therefore$ Maximum value of $\sin^2\theta = 1$

Hence, maximum value of $(5\cos^2\theta + 7\sin^2\theta) = 5 + 2 \cdot 1 = 7$.

Problem 61. The minimum value of $3\cos^2\theta + 2\sin^2\theta$ is *[SSC (CPO-SI)-11, SSC (CGL-T-I)-15, CDS-21]*

Sol. $2\sin^2\theta + 3\cos^2\theta = 2\sin^2\theta + 3(1 - \sin^2\theta)$

$= 3 - \sin^2\theta$

[$(3 - \sin^2\theta)$ takes minimum value if $\sin\theta$ will have maximum value (i.e., 1)]

$\therefore$ Required minimum value $= 3 - 1^2 = 2$

Alternatively

$2\sin^2\theta + 3\cos^2\theta = 2\sin^2\theta + 2\cos^2\theta + \cos^2\theta$

$= 2(\sin^2\theta + \cos^2\theta) + \cos^2\theta = 2 + \cos^2\theta$

[$(2 + \cos^2\theta)$ takes minimum value if $\cos^2\theta$ will have minimum value and the fact that minimum value of $\cos^2\theta$ is 0.]

$\therefore$ Required minimum value $= 2 + 0 = 2$

Observations from the above questions

- The maximum value of $m\sin^2\theta + n\cos^2\theta = m$ or n, whichever is greater.
- The minimum value of $m\sin^2\theta + n\cos^2\theta = m$ or n, whichever is lower.
- The maximum/minimum value of $x\sin^2\theta + x\cos^2\theta = x$.
- $m\sin^2\theta - n\cos^2\theta$ may be written as $m\sin^2\theta + (-n)\cos^2\theta$ for application of the above conceptions.

Problem 62. Find the minimum value of $6\cos^2\theta - 11\sin^2\theta$.

Sol. $6\cos^2\theta - 11\sin^2\theta = (+6)\cos^2\theta + (-11)\sin^2\theta$

$\therefore$ Required minimum value $= -11$ $\quad [\because (-11) < (+6)]$

Problem 63. Find the maximum value of the $6\sin^2\theta - 11\cos^2\theta$.

Sol. $6\sin^2\theta - 11\cos^2\theta = (+6)\sin^2\theta + (-11)\cos^2\theta$

Required maximum value $= +6$ as 6 is greater than (-11).

Problem 64. Find the maximum value of $-4\cos^2\theta - 7\sin^2\theta$.

Sol. $-4\cos^2\theta - 7\sin^2\theta = +(-4)\cos^2\theta + (-7)\sin^2\theta$

$\therefore$ Required maximum value $= -4$ $\quad [\because (-4) > (-7)]$

Problem 65. Find the maximum value of $\sin^6\theta + \cos^6\theta$.

Sol. $\sin^6\theta + \cos^6\theta = (\sin^2\theta)^3 + (\cos^2\theta)^3$

$= (\sin^2\theta + \cos^2\theta)(\sin^4\theta + \cos^4\theta - \sin^2\theta\cdot\cos^2\theta)$

$[\because a^3 + b^3 = (a + b)(a^2 - ab + b^2)]$

$= 1[(\sin^2\theta + \cos^2\theta)^2 - 2\sin^2\theta\cos^2\theta - \sin^2\theta\cdot\cos^2\theta]$

$= 1 - 3\sin^2\theta\cdot\cos^2\theta = 1 - \frac{3}{4}(2\sin\theta\cdot\cos\theta)^2$

$= 1 - \frac{3}{4}(\sin 2\theta)^2$

$\left[1 - \frac{3}{4}(\sin 2\theta)^2\right]$ will have maximum value if $(\sin 2\theta)^2$ takes the minimum value and the fact that minimum value of $(\sin 2\theta)^2$ is 0.

Hence, required maximum value $= 1 - \frac{3}{4}\times 0 = 1$

Problem 66. Find the minimum and maximum value of $5\sin\theta + 12\cos\theta$.

Sol. $5\sin\theta + 12\cos\theta = 13\left(\frac{5}{13}\sin\theta + \frac{12}{13}\cos\theta\right)$

$[\because 13 = \sqrt{5^2 + (12)^2}]$

$= 13(\sin\alpha\cdot\sin\theta + \cos\alpha\cdot\cos\theta)$

$\left[\text{assuming } \frac{5}{13} = \sin\alpha, \text{ then } \frac{12}{13} = \cos\alpha\right]$

$= 13\cdot\cos(\alpha - \theta)$

[$\because \cos(\alpha - \theta) = \cos\alpha\cdot\cos\theta + \sin\alpha\cdot\cos\theta$, detailed at topic compound angles.]

Now, maximum value of $5\sin\theta + 12\cos\theta = 13\times 1 = 13$

[$\because$ maximum value of $\cos(\alpha - \theta)$ is 1]

and minimum value of $5\sin\theta + 12\cos\theta = 13\times(-1) = -13$

[$\because$ minimum value of $\cos(\alpha - \theta)$ is -1]

Problem 67. Find the minimum and maximum value of $7\cos\theta - 24\sin\theta$.

Sol. $7\cos\theta - 24\sin\theta = 25\left(\frac{7}{25}\cos\theta - \frac{24}{25}\sin\theta\right)$ $[\because 25 = \sqrt{7^2 + (24)^2}]$

$= 25(\cos\alpha\cdot\cos\theta - \sin\alpha\cdot\sin\theta)$

$\left[\text{assuming } \frac{7}{25} = \cos\alpha, \text{ then } \frac{24}{25} = \sin\alpha\right]$

$= 25\cdot\cos(\alpha + \theta)$

Now, maximum value of $7\cos\theta - 24\sin\theta = 25\times 1 = 25$

[$\because$ maximum value of $\cos(\alpha + \theta)$ is 1]

and minimum value of $7\cos\theta - 24\sin\theta = 25(-1) = -25$

[$\because$ Minimum value of $\cos(\alpha + \theta)$ is (-1)]

Observations from the above last two questions

- The maximum value of $m\sin\theta \pm n\cos\theta = +\sqrt{m^2 + n^2}$
- The minimum value of $m\sin\theta \pm n\cos\theta = -\sqrt{m^2 + n^2}$

Problem 68. Find the minimum value of $\sin\theta + \cos\theta$.

Sol. $\sin\theta + \cos\theta = 1\cdot\sin\theta + 1\cdot\cos\theta$

Now, minimum value $= -\sqrt{1^2 + 1^2}$

$= -\sqrt{2}$

Problem 69. Find the minimum value of $-5\cos\theta - 8\sin\theta$.

Sol. Given expression $= (-5)\cos\theta - 8\sin\theta$

$\therefore$ Required minimum value $= -\sqrt{(-5)^2 + (8)^2}$

$= -\sqrt{89}$

Problem 70. Find the range of the value of expression $y = 3\cos\left(\theta + \frac{\pi}{3}\right) + 5\cos\theta + 7.5$.

(a) $3 \le y \le 13.5$ (b) $0.5 \le y \le 14.5$

(c) $-2.5 \le y < 12$ (d) $-3.5 \le y \le 12.5$

Sol. Given, expression

$= 3\cos\left(\theta + \frac{\pi}{3}\right) + 5\cos\theta + 7.5$

$= 3\left(\cos\theta\cdot\cos\frac{\pi}{3} - \sin\theta\cdot\sin\frac{\pi}{3}\right) + 5\cos\theta + 7.5$

$= 3\left(\cos\theta\cdot\frac{1}{2} - \sin\theta\cdot\frac{\sqrt{3}}{2}\right) + 5\cos\theta + 7.5$

$= \frac{13}{2}\cos\theta - \frac{3\sqrt{3}}{2}\sin\theta + 7.5$

$$= 7\left(\frac{13}{2\times7}\cos\theta - \frac{3\sqrt{3}}{2\times7}\sin\theta\right) + 7.5$$

$$\left[\because 7 = \sqrt{\left(\frac{13}{2}\right)^2 + \left(\frac{3\sqrt{3}}{2}\right)^2}\right]$$

$$= 7(\cos\alpha\cdot\cos\theta - \sin\alpha\cdot\sin\theta) + 7.5$$

$$\left[\text{Assuming } \frac{13}{14} = \cos\alpha, \text{ then } \frac{3\sqrt{3}}{14} = \sin\alpha\right]$$

$$= 7[\cos(\alpha+\theta)] + 7.5$$

Maximum value of expression, $y = (7\times1) + 7.5 = 14.5$

$\{\because$ maximum value of $\cos(\alpha+\theta)$ is 1$\}$

and minimum value of expression, $y = (7\times-1) + 7.5 = 0.5$

$\{\because$ minimum value of $\cos(\alpha+\theta)$ is $-1\}$

Hence, option (b) is correct.

Problem 71. The maximum value of $\sin\theta\cdot\cos\theta = ?$

Sol. $\sin\theta\cdot\cos\theta = \frac{1}{2}\times 2\sin\theta\cdot\cos\theta$

$= \frac{1}{2}(\sin 2\theta)$

Since the maximum value of $\sin(2\theta)$ is '+1', maximum value of expression will be $\frac{1}{2}\times(+1) = \frac{1}{2}$.

[minimum value of expression $= \frac{1}{2}\times(-1) = -\frac{1}{2}$ as minimum value of $\sin(2\theta)$ is -1]

Problem 72. What is the maximum value of $\sin^5\theta\cdot\cos^5\theta$?

Sol. $\sin^5\theta\cdot\cos^5\theta = (\sin\theta\cdot\cos\theta)^5$

$= \left(\frac{1}{2}\times 2\cdot\sin\theta\cdot\cos\theta\right)^5 = \left(\frac{1}{2}\right)^5\cdot(\sin 2\theta)^5$

Since maximum value of $\sin(2\theta)$ is $+1$, maximum value of

$\sin^5\theta\cdot\cos^5\theta = \left(\frac{1}{2}\right)^5\times(+1)^5 = \frac{1}{32}$

{Here, minimum value $= \left(\frac{1}{2}\right)^5\times(-1)^5 = -\frac{1}{32}$}

Problem 73. Find the maximum and minimum value of $\sin^6\theta\cdot\cos^6\theta$.

Sol. $\sin^6\theta\cdot\cos^6\theta = (\sin\theta\cdot\cos\theta)^6$

$= \left(\frac{1}{2}\cdot 2\sin\theta\cdot\cos\theta\right)^6 = \left(\frac{1}{2}\right)^6\cdot(\sin 2\theta)^6$

$\because$ Maximum value of $\sin(2\theta) = 1$

$\therefore$ Maximum value of

$$\sin^6\theta\cdot\cos^6\theta = \left(\frac{1}{2}\right)^6\times(1)^6 = \frac{1}{64}$$

Further, minimum value of $\sin 2\theta = -1$

$\therefore$ Minimum value of $\sin^6\theta\cdot\cos^6\theta = \left(\frac{1}{2}\right)^6\times(-1)^6 = \frac{1}{64}$, but it is not correct as $(-1)^6$ yields positive value. If we take the value of $\sin 2\theta$ as 0, the minimum value of

$\sin^6\theta\cdot\cos^6\theta = \left(\frac{1}{2}\right)^6\times(0)^6 = 0$, which is correct.

Observations from the above last three questions

- The maximum value of $(\sin\theta\cdot\cos\theta)^n = \left(\frac{1}{2}\right)^n$, when n = odd/even number.
- The minimum value of $(\sin\theta\cdot\cos\theta)^n = -\left(\frac{1}{2}\right)^n$, when n = odd number.
- The minimum value of $(\sin\theta\cdot\cos\theta)^n = 0$, when n = even number.

Problem 74. If $\sin^2\theta + \cos^4\theta = A$, for any value of θ, then the value of A is *[SSC (10+2 LDC/DEO)-11]*

(a) $1\le A\le 2$ (b) $\frac{3}{4}\le A\le 1$

(c) $\frac{13}{16}\le A\le 1$ (d) $\frac{3}{4}\le A\le\frac{13}{16}$

Sol. Given, $\sin^2\theta + \cos^4\theta = A$

$\Rightarrow 1-\cos^2\theta + \cos^4\theta = A$

$\Rightarrow 1-\cos^2\theta(1-\cos^2\theta) = A$

$\Rightarrow 1-\cos^2\theta\cdot\sin^2\theta = A$

$\Rightarrow$ Maximum value of $(\cos\theta\cdot\sin\theta)^2 = \left(\frac{1}{2}\right)^2 = \frac{1}{4}$

and minimum value of $(\cos\theta\cdot\sin\theta)^2 = 0$ [$\because$ 2 is an even number]

Hence, A will have maximum value when $(\cos\theta\cdot\sin\theta)^2$ takes minimum value,

$\therefore$ Maximum value of $A = 1-0 = 1$

and A will have minimum value when $(\cos\theta\cdot\sin\theta)^2$ takes maximum value.

$\therefore$ Minimum value of $A = 1-\frac{1}{4} = \frac{3}{4}$

Thus, option (b) is correct.

Alternatively

Given, $\sin^2 + \cos^4\theta = A$

Putting the value of θ as 0°, 30°, 45°, 60° and 90°.

We get minimum value of $A = \frac{3}{4}$ $\left[\because \sin^2 45° + \cos^2 45° = \frac{3}{4}\right]$

and maximum value of $A = 1$ $[\because \sin^2 0° + \cos^4 0° = 1]$

Hence, correct option is (b).

Problem 75. For what value of x, $\cos 3x$ has minimum value.

Sol. We know that the minimum value of $\cos\theta = -1$

$\therefore \cos 3x = -1$

$\Rightarrow \cos 3x = \cos(180\pm 0) \Rightarrow 3x = 180°$

$\therefore x = 60°$

Alternatively

You may proceed through the options. Let options be 0°, 45°, 60°, 90°.

(a) $0° \Rightarrow \cos 3\cdot 0° = \cos 0° = 1$

(b) $45° \Rightarrow \cos 3\cdot 45° = \cos 135° = \cos(180°-45°)$

$= -\cos 45° = -\frac{1}{\sqrt{2}}$

(c) $60° \Rightarrow \cos 3\cdot 60° = \cos 180°$

$= \cos(180°-0°) = -\cos 0° = -1$

(d) $90° \Rightarrow \cos 3\cdot 90 = \cos 270° = \cos(180° + 90°)$
$= -\cos 90° = 0$

Clearly, for $x = 60°$, $\cos 3x$ has minimum value.

VII. Problems Based on Numerical Values of T-ratios/Expressions with Acute Angles/Allied Angles.

- If the given angle is positive and greater than 360°, divide it by 360. The numerical value of *T*-ratio with an angle equal to the remainder will be the same as numerical value of the *T*-ratio with the given angle.
- If given angle is negative and greater than −360°, add it with 360°. The numerical value of *T*-ratio with an angle equal to the resultant will be the same as numerical value of the *T*-ratio with the given angle.
- If the given angle is negative and less than −360°, divide it (ignoring negative sign) by 360 and subtract the remainder from 360°. The numerical value of *T*-ratio with an angle equal to the resultant will be the same as numerical value of the *T*-ratio with the given angle.
- We should also mind the concept of sign of *T*-ratios and concept of *T*-ratios with the allied angles, concepts have already been discussed.

Problem 76. Find the numerical value of the $\tan 420°$.

Sol. As discussed, if the given angle is positive and greater than 360°, divide it by 360. The numerical value of *T*-ratio with an angle equal to the remainder will be the same as numerical value of the *T*-ratio with the given angle.

Here, remainder in case of $\frac{420°}{360} = 60°$

Then, $\tan 420° = \tan 60° = +\sqrt{3}$

[∵ 60° is in 1st quadrant in which tangent is positive]

Problem 77. Find the numerical value of the $\cos(-300°)$.

Sol. As discussed, if the given angle is negative and greater than −360°, add it with 360°. The numerical value of *T*-ratio with an angle equal to the resultant will be the same as numerical value of the *T*-ratio with the given angle.

Then, $\cos(-300°) = \cos[360° + (-300°)]$
$= \cos 60° = \frac{1}{2}$

Problem 78. Numerical value of $\cos(-2130°) = ?$

Sol. As discussed, if given angle is negative and less than −360°, divide it (ignoring negative sign) by 360 and subtract the remainder from 360°. The numerical value of *T*-ratio with an angle equal to the resultant will be the same as numerical value of the *T*-ratio with the given angle.

Here, remainder in case of $\frac{2130°}{360} = 330°$

Then, $\cos(-2130°) = \cos(360° - 330°)$
$= \cos 30° = \frac{\sqrt{3}}{2}$

Problem 79. Find the numerical value of $\sec(-675°)$.

Sol. Here, remainder in case of $\frac{675°}{360} = 315°$

Then, $\sec(-675°) = \sec(360° - 315°)$
$= \sec 45° = \sqrt{2}$

Problem 80. Find the numerical value of $\tan 510°$.

Here, remainder in case of $\frac{510°}{360} = 150°$

Then, $\tan 510° = \tan 150° = \tan(180° - 30°)$
$= -\tan 30°$

[∵ $(180° - \theta)$ is in 2nd quadrant in which tangent is negative]

$= \frac{-1}{\sqrt{3}}$

Problem 81. $\tan 225°\cdot\cot 405°° + \tan 765°\cdot\cot 675° = ?$

Sol. $\tan(180° + 45°)\cot(360° + 45°) + \tan(45°)\cdot\cot(315°)$

[∵ $\frac{765°}{360}$ yields 45° as remainder and $\frac{675°}{360}$ yields 315° as remainder]

$= \tan 45°\cdot\cot 45° + \tan 45°\cdot\cot(360° - 45°)$
$= \tan 45°\cdot\cot 45° + \tan 45°\cdot(-\cot 45°)$

[∵ (360° − 45°) lies in 4th quadrant in which cot is negative]

$= \tan 45°\cdot\cot 45° - \tan 45°\cdot\cot 45° = 0$

Problem 82. $\sin 660° + \cos 585° = ?$

Sol. $\sin 300° + \cos 225°$

[∵ $\frac{660°}{360}$ yields 300° as remainder and $\frac{585°}{360}$ yields 225° as remainder]

$= \sin(360° - 60°) + \cos(180° + 45°)$
$= -\sin 60° - \cos 45°$

[∵ (360° − 60°) lies in 4th quadrant in which sine is negative and (180 + 45°) lies in 3rd quadrant in which cos is negative]

$= -\frac{\sqrt{3}}{2} - \frac{1}{\sqrt{2}} = \frac{-\sqrt{3} - \sqrt{2}}{2} = \frac{-(\sqrt{2} + \sqrt{3})}{2}$

Problem 83. Find the value of $2\cos^2 60° + \sin^2 120° + \frac{1}{2}\cos 0° + \sin 90°$.

Sol. $2\cos^2 60° + \sin^2 120° + \frac{1}{2}\cos 0° + \sin 90°$

$= 2\cdot\left(\frac{1}{2}\right)^2 + \sin^2(180° - 60°) + \frac{1}{2}\cdot 1 + 1$

$= \frac{1}{2} + \sin^2 60° + \frac{1}{2} + 1$

$= 2 + \left(\frac{\sqrt{3}}{2}\right)^2 = 2\frac{3}{4}$

Problem 84.

$\sin 420°\cdot\cos 390° + \cos(-300°)\cdot\sin(-330°) = ?$

Sol. $\sin 420° = \sin 60°$ [∵ 420°/360 yields remainder 60°]

$\cos 390° = \cos 30°$ [∵ 390°/360 yields remainder 30°]

$\cos(-300°) = \cos 60°$ $[\because -300° + 360° = 60°]$

$\sin(-330°) = \sin 30°$ $[\because -330° + 360° = 30°]$

Now expression be $\sin 60° \cdot \cos 30° + \cos 60° \cdot \sin 30°$

$$= \left(\frac{\sqrt{3}}{2} \times \frac{\sqrt{3}}{2}\right) + \left(\frac{1}{2} \times \frac{1}{2}\right) = \frac{3}{4} + \frac{1}{4} = 1$$

Problem 85. $\sin(-1890°) + \cos(630°) = ?$

Sol. sin [360°– remainder in case of –(1890°)/360] + cos (remainder in case of 630°/360) = sin 270° + cos 270°

$= \sin(360° - 90°) + \cos(360° - 90°)$

$= -\sin 90° + \cos 90°$

[∵ (360° – θ) lies in 4th quadrant in which cosine is '+' and sine is '–']

$= -1 + 0 = -1$

Problem 86. $\tan^2 135° - 4\sin^2 120° + 2\cos^2 45° + \sec^2 180° = ?$

Sol. $\tan^2(180° - 45°) - 4\sin^2(180° - 60°) + 2\left(\frac{1}{\sqrt{2}}\right)^2 + \sec^2(180° - 0°)$

$= (-\tan 45°)^2 - 4(\sin 60°)^2 + 2 \cdot \frac{1}{2} + (-\sec 0°)^2$

$= 1 - 4 \cdot \left(\frac{\sqrt{3}}{2}\right)^2 + 1 + (-1)^2$

$= 1 - 3 + 1 + 1 = 0$

Problem 87. $\dfrac{\sin\frac{\pi}{6} + \cos\frac{\pi}{3} + \tan\frac{\pi}{4} + \cos\frac{\pi}{2}}{\sin\frac{5\pi}{6} - \cos\frac{2\pi}{3} - \tan\frac{3\pi}{4} + \tan\pi} = ?$

Sol. $? = \dfrac{\frac{1}{2} + \frac{1}{2} + 1 + 0}{\sin(180° - 30°) - \cos(180° - 60°) - \tan(180° - 45°) + \tan(180° - 0°)}$

$$= \frac{2}{\sin 30° - (-\cos 60°) - (-\tan 45°) + (-\tan 0°)}$$

$$= \frac{2}{\frac{1}{2} + \frac{1}{2} + 1 - 0} = \frac{2}{2} = 1$$

Problem 88. Find the value of $\tan x + \tan(-y) - \tan(180° - y)$, if $x = 45°$

Sol. $\tan x + \tan(360° - y) - (-\tan y)$

$= \tan x - \tan y - (-\tan y)$

$= \tan x - \tan y + \tan y$

$= \tan x = \tan 45° = 1$

Problem 89. $\sin\left[n\pi + (-1)^n \cdot \frac{\pi}{4}\right] = ?$, if n is a negative integer.

Sol. $\sin\left[n\pi + (-1)^n \cdot \frac{\pi}{4}\right]$

$= \sin\left(n\pi - \frac{\pi}{4}\right)$, when n is a negative odd integer.

$= -\left(-\sin\frac{\pi}{4}\right) = \sin\frac{\pi}{4} = \frac{1}{\sqrt{2}}$

Further, in case n is a negative even integer.

$\Rightarrow \sin\left[n\pi + (-1)^n \frac{\pi}{4}\right] = \sin\left[n\pi + \frac{\pi}{4}\right]$

$= -\left(-\sin\frac{\pi}{4}\right) = \sin\frac{\pi}{4} = \frac{1}{\sqrt{2}}$

Note *aspirants are advised to try either with odd or with even negative integer, not with both.*

Alternatively

Let $n = -1$ then, $\sin\left[n\pi + (-1)^n \cdot \frac{\pi}{4}\right]$

$= \sin(-180 - 45°)$

$= \sin(-225°) = -\sin 225°$

$= -\sin(180° + 45°)$

$= -(-\sin 45°) = \sin 45° = \frac{1}{\sqrt{2}}$

Problem 90. Find the value of $\sin(270° + \theta)(-\cos\theta) - \cos(270° - \theta) \cdot \sin\theta + \sec^2\theta$, where $\theta = 45°$.

Sol. $\sin(270° + \theta)(-\cos\theta) - \cos(270° - \theta) \cdot \sin\theta + \sec^2\theta$

$= (-\cos\theta)(-\cos\theta) - [(-\sin\theta)\sin\theta] + \sec^2\theta$

$= \cos^2\theta + \sin^2\theta + \sec^2\theta$

$= 1 + \sec^2 45° = 1 + (\sqrt{2})^2 = 3$

Alternatively I

$\sin(270° + \theta)(-\cos\theta) - \cos(270° - \theta)\sin\theta + \sec^2\theta$

$= (-\cos\theta) \cdot (-\cos\theta) - (-\sin\theta) \cdot \sin\theta + \sec^2\theta$

[∴ $\sin(270° + \theta) = -\cos\theta$ and $\cos(270° - \theta) = -\sin\theta$]

$= \cos^2\theta + \sin^2\theta + \sec^2 45°$

$= 1 + (\sqrt{2})^2 = 3$

Alternatively II

You may directly put the value of θ in given terms.

$\Rightarrow \sin(270° + 45°)(-\cos 45°) - \cos(270° - 45°) \cdot \sin 45° + \sec^2 45°$

$= \sin(360° - 45°)\left(-\frac{1}{\sqrt{2}}\right) - \cos(180° + 45°) \cdot \frac{1}{\sqrt{2}} + (\sqrt{2})^2$

$= (-\sin 45°)\left(-\frac{1}{\sqrt{2}}\right) - (-\cos 45°) \cdot \frac{1}{\sqrt{2}} + 2$

$= \frac{1}{2} + \frac{1}{2} + 2 = 3$

Problem 91. If $\sin\alpha + \cos\beta = 2\,(0° \le \beta < \alpha \le 90°)$, then, $\sin\left(\frac{2\alpha + \beta}{3}\right) = ?$ ***[SSC (10+2 LDC/DEO)-11]***

(a) $\sin\frac{\alpha}{2}$ (b) $\cos\frac{\alpha}{3}$

(c) $\sin\frac{\alpha}{3}$ (d) $\cos\frac{2\alpha}{3}$

Sol. (b) Given, $\sin\alpha + \cos\beta = 2$

$\Rightarrow \sin\alpha = 1$ and $\cos\beta = 1$

[∵ sine and cosine have maximum value of 1]

$\Rightarrow$ $\sin\alpha = \sin 90°$ and $\cos\beta = \cos 0°$

$\therefore$ $\alpha = 90°$ and $\beta = 0°$

Now, $\sin\left[\dfrac{(2\times 90)+0}{3}\right] = \sin 60°$

$= \dfrac{\sqrt{3}}{2} = \cos 30° = \cos\dfrac{\alpha}{3}$

Hence, option (b) is correct.

VIII. Determination of Angle from a given Trigonometric Equation

Problem 92. If $2(\cos^2\theta - \sin^2\theta) = 1$, θ is a positive acute angle, then the value of θ is *[FCI-Asst. Gr.-III-12]*

Sol. Given, $2[(1-\sin^2\theta) - \sin^2\theta] = 1$

$\Rightarrow$ $1 - 2\sin^2\theta = \dfrac{1}{2}$

$\Rightarrow$ $\sin^2\theta = \dfrac{1}{4}$

$\Rightarrow$ $\sin\theta = \dfrac{1}{2}$

$\Rightarrow$ $\sin\theta = \sin 30°$

$\therefore$ $\theta = 30°$

Problem 93. If $\sec^2\theta + \tan^2\theta = 7$, then the value of θ when $0° \le \theta \le 90°$, is :

Sol. Given, $(1+\tan^2\theta) + \tan^2\theta = 7$

$\Rightarrow$ $2\tan^2\theta = 6 \Rightarrow \tan^2\theta = 3$

$\Rightarrow$ $\tan\theta = \sqrt{3}$

$\Rightarrow$ $\tan\theta = \tan 60°$ ($\because \theta$ is not exceeding 90°)

$\Rightarrow$ $\theta = 60°$

Problem 94. If $\dfrac{\cos^2\theta}{\cot^2\theta - \cos^2\theta} = 3$ and $0° < \theta < 90°$, then the value of θ is

Sol. Given, $\dfrac{\cot^2\theta - \cos^2\theta}{\cos^2\theta} = \dfrac{1}{3}$

$\Rightarrow$ $\dfrac{\cos^2\theta}{\sin^2\theta\cdot\cos^2\theta} - \dfrac{\cos^2\theta}{\cos^2\theta} = \dfrac{1}{3}$

$\Rightarrow$ $\text{cosec}^2\theta = \dfrac{1}{3} + 1 = \dfrac{4}{5}$

$\Rightarrow$ $\text{cosec}\,\theta = \dfrac{2}{\sqrt{3}} = \text{cosec}\,60°$

$\therefore$ $\theta = 60°$

Problem 95. If $\tan\theta = \sqrt{2}$, the value of θ is

(a) less than 45° (b) equal to 45°

(c) between 45° and 60° (d) greater than 60°

Sol. (c) We know, $\tan 45° = 1$ and $\tan 60° = \sqrt{3}$

It means, value of θ lies between 45° and 60°.

Problem 96. If $\sin(A+B) = \dfrac{1}{2}$ and $\cos(A-B) = 1$, then $\dfrac{A}{B} = ?$

Sol. Given, $\sin(A+B) = \dfrac{1}{2}$ and $\cos(A-B) = 1$

$\Rightarrow$ $\sin(A+B) = \sin 30°$ and $\cos(A-B) = \cos 0°$

$\therefore$ $A + B = 30°$ and $A - B = 0°$

On solving both equations, $A = B = 15°$

$\therefore$ $\dfrac{A}{B} = 1$

Problem 97. If $\sin^2\theta = \dfrac{1}{4}$ and $0° < \theta < \pi$, then how many values of θ are possible :

Sol. Given, $\sin^2\theta = \dfrac{1}{4} \Rightarrow \sin\theta = \pm\dfrac{1}{2}$

For $+\dfrac{1}{2}$, $\theta = 30°$ or $(180° - 30°)$ i.e. 150°

For $-\dfrac{1}{2}$, $\theta = -30°$ or 330° or $(180° + 30°)$ i.e. 210°

Since $0° < \theta < \pi$, value of θ as $-30°$ or 330° or 210° is not possible.

Hence, two values of θ are possible.

Problem 98. If $2\sin^2\theta - 3\cos\theta = 0$, then $\theta = ?$, if $0° \le \theta \le 90°$.

Sol. Given, $2(1-\cos^2\theta) - 3\cos\theta = 0$

$\Rightarrow$ $2 - 2\cos^2\theta - 3\cos\theta = 0$

$\Rightarrow$ $2\cos^2\theta + 3\cos\theta - 2 = 0$

On solving quadratic equation (pattern $2x^2 + 3x - 2 = 0$) we get, $\cos\theta = -2$ and $\dfrac{1}{2}$

Since the value of $\cos\theta$ cannot be less than -1, $\cos\theta \ne -2$,

Hence, $\cos\theta = \dfrac{1}{2} = \cos 60° \Rightarrow \theta = 60°$

IX. Simplification Based on the Concept of Trigonometric Ratios of Complementary Angles

Type A

Problem 99. The value of $\tan 4°\cdot\tan 43°\cdot\tan 47°\cdot\tan 86°$ is : *[SSC (CPO-SI)-11]*

Sol. $\tan 4°\cdot\tan 43°\cdot\cot(90° - 47°)\cdot\cot(90° - 86°)$

$= (\tan 4°\cdot\cot 4°)\times(\tan 43°\cdot\cot 43°)$

$= 1\times 1 = 1$

Problem 100. The value of $\tan 1°\cdot\tan 2°\cdot\tan 3° \ldots \tan 89°$ is *[FCI-Asstt. Gr.-III-12, SSC (CGL-T-I)-14, CDS-09]*

Sol. $\tan 1°\cdot\tan 2°\cdot\tan 3° \ldots \tan 87°\cdot\tan 88°\cdot\tan 89°$

$= \tan 1°\cdot\tan 2°\cdot\tan 3° \ldots \tan 45° \ldots \cot(90° - 87°)\cdot\cot(90° - 88°)\cdot\cot(90° - 89°)$

$= \tan 1°\cdot\tan 2°\cdot\tan 3° \ldots \tan 45° \ldots \cot 3°\cdot\cot 2°\cdot\cot 1°$

$= (\tan 1°\cdot\cot 1°)(\tan 2°\cdot\cot 2°)(\tan 3°\cdot\cot 3°)(\ldots)(\ldots)\times\ldots\times\tan 45°$

$= 1\times 1\times\ldots\times 1\times 1\times 1 = 1$

Note *There will be 44 pairs of like $(\tan\theta_1\cdot\cot\theta_1)$, $(\tan\theta_2\cdot\cot\theta_2)$, $(\tan\theta_3\cdot\cot\theta_3)$ ... and so on and $\tan 45°$ remains left (i.e. not included in pair). Also, each pair has value of 1.*

Observations from the above

(a) $\tan 1° \cdot \tan 2° \cdot \tan 3° \dots \tan 89° = 1$

(b) $\cot 1° \cdot \cot 2° \cdot \cot 3° \dots \cot 89° = 1$

In other words, if $\theta_1, \theta_2, \theta_3 \dots, \theta_n$ are in AP series and sum of θ_1 and $\theta_n = 90°$, the following holds good

(i) $\tan \theta_1 \cdot \tan \theta_2 \cdot \tan \theta_3 \dots \tan \theta_n = 1$

(ii) $\cot \theta_1 \cdot \cot \theta_2 \cdot \cot \theta_3 \dots \cot \theta_n = 1$

Problem 101. $\tan 9° \cdot \tan 15° \cdot \tan 21° \dots \tan 81° = ?$

Sol. Since 9°, 15°, 21°, ... are in AP series and 9° + 81° = 90°, the value of the expression will be 1.

Problem 102. $\sin 1° \cdot \sin 2° \cdot \sin 3° \dots \cdot \sin 180° = ?$

Sol. $\sin 1° \cdot \sin 2° \cdot \sin 3° \dots \sin (180° - 0°)$

$\Rightarrow \sin 1° \cdot \sin 2° \cdot \sin 3° \dots \sin 0°$

$\Rightarrow \sin 1° \cdot \sin 2° \cdot \sin 3° \dots \times 0 = 0$

Problem 103. $\cos 1° \cdot \cos 2° \cdot \cos 3° \dots \cos 90° = ?$ *[CDS-10]*

Given expression $= \cos 1° \cdot \cos 2° \cdot \cos 3° \dots \times 0 = 0$

Type B

Problem 104. The value of $\sin^2 1° + \sin^2 2° + \sin^2 3° + \dots + \sin^2 89° = ?$

[SSC (CGL-T-I)-14]

Sol. $\sin^2 1° + \sin^2 2° + \sin^2 3° \dots + \sin^2 45° + \dots + \sin^2 87° + \sin^2 88° + \sin^2 89°$

$= (\sin^2 1° + \sin^2 89°) + (\sin^2 2° + \sin^2 88°) + \dots + \sin^2 45°$

$= [\sin^2 1 + \cos^2 (90° - 89°)] + [\sin^2 2° + \cos^2 (90° - 88°) + .. + (1/\sqrt{2})^2]$

$= [\sin^2 1° + \cos^2 1°] + [\sin^2 2° + \cos^2 2°] + \dots + \frac{1}{2}$

$= 1 + 1 + 1 + \dots$ (up to 44 terms) $+ \frac{1}{2}$

[∵ number of pairs like $(\sin^2 \theta + \cos^2 \theta) = 44$]

$= 44 + \frac{1}{2} = 44\frac{1}{2}$

Problem 105. The value of $\sin^2 1° + \sin^2 5° + \sin^2 9° + \dots + \sin^2 89°$ is

Sol. We know that $t_n = a + (n-1)d$

[$t_n = n^{th}$ term, a = 1st term, n = number of terms, d = common difference ...... recall the concepts of AP series]

$\Rightarrow 89 = 1 + (n-1)4 \quad \therefore n = 23$

Now, expression be $(\sin^2 1° + \sin^2 89°) + (\sin^2 5° + \sin^2 85°) + (\sin^2 9° + \sin^2 81°) + (\dots\dots) + \dots + \sin^2 45°$

$= 1 + 1 + 1 + \dots + 1$ (total 11 terms) $+ \left(\frac{1}{\sqrt{2}}\right)^2 = 11 + \frac{1}{2} = 11\frac{1}{2}$

Observation forms the above two questions

(a) $\sin^2 \theta_1 + \sin^2 \theta_2 + .. + \sin^2 \theta_n = \frac{\text{number of terms}}{2}$

(b) $\cos^2 \theta_1 + \cos^2 \theta_2 + \dots \cos^2 \theta_n = \frac{\text{number of terms}}{2}$

Subject to the condition that $\theta_1, \theta_2, \theta_3, \dots, \theta_n$ are in AP series and sum of θ_1 and $\theta_n = 90°$ [for both (a) and (b)].

Problem 106. $\sin^2 1° + \sin^2 3° + \sin^2 5° + \dots + \sin^2 87° + \sin^2 89° = ?$

Sol. Here, $t_n = a + (n-1)d$

$\Rightarrow 89 = 1 + (n-1) \cdot 2 \Rightarrow n = 45$

Alternatively

Number of term $= \frac{(89-1)}{2} + 1 = 45$

Since the given expression suffices both the conditions mentioned above, the required value of the expression will be half of the number of terms.

Hence, $\frac{45}{2}$ i.e., $22\frac{1}{2}$.

Problem 107.

$\sin^2 5° + \sin^2 10° + \sin^2 15° + \dots + \sin^2 85° = ?$

Sol. Clearly, both conditions are satisfied here.

∴ Required value of the expression

$= \frac{\text{number of terms}}{2} = \frac{17}{2} = 8\frac{1}{2}$

Problem 108. $\cos^2 6° + \cos^2 19° + \cos^2 32° + \dots + \cos^2 84° = ?$

Sol. Clearly, both conditions are satisfied here.

∴ Required value of the expression

$= \frac{\text{number of terms}}{2} = \frac{7}{2} = 3\frac{1}{2}$

Problem 109. $\cos^2 12° + \cos^2 18° + \cos^2 24° + \dots\dots + \cos^2 78° = ?$

Sol. Clearly, both conditions are satisfied here.

∴ required value of the expression

$= \frac{\text{number of terms}}{2} = \frac{12}{2} = 6$

Problem 110. The simplest value of $\sin^2 5° + \sin^2 10° + \sin^2 15° + \dots + \sin^2 85° + \sin^2 90°$ is

Sol. Expression is $[\sin^2 5° + \sin^2 10° + \sin^2 15° + \dots + \sin^2 85°] + \sin^2 90°$

Expression under bracket satisfies both conditions.

Then, its value $= \frac{\text{number of terms}}{2} = \frac{17}{2} = 8\frac{1}{2}$

∴ Value of the whole expression $= 8\frac{1}{2} + (1)^2 = 9\frac{1}{2}$

Type C

Problem 111. The value of $\cot 18° \left(\cot 72° \cdot \cos^2 22° + \frac{1}{\tan 72° \cdot \sec 268°}\right)$ is :

[SSC (10+2 LDC/DEO)-11]

Sol. $\Rightarrow \tan (90° - 18°)\left(\cot 72° \cdot \cos^2 22° + \frac{1}{\tan 72° \sec^2 68°}\right)$

$= \tan 72° \cdot \cot 72° \cdot \cos^2 22° + \tan 72° \cdot \frac{1}{\tan 72° \sec^2 68°}$

$= 1 \cdot \cos^2 22° + \cos^2 68°$

$= \cos^2 22° + \sin^2 (90 - 68°) = 1$

Problem 112. The value of $\dfrac{\cot 30^\circ - \cot 75^\circ}{\tan 15^\circ - \tan 60^\circ}$ is

Sol. $\Rightarrow \dfrac{\cot 30^\circ - \cot 75^\circ}{\tan 15^\circ - \tan 60^\circ} = \dfrac{\tan(90^\circ - 30^\circ) - \tan(90^\circ - 75^\circ)}{\tan 15^\circ - \tan 60^\circ}$

$= \dfrac{\tan 60^\circ - \tan 15^\circ}{\tan 15^\circ - \tan 60^\circ} = \dfrac{-1(\tan 15^\circ - \tan 60^\circ)}{(\tan 15^\circ - \tan 60^\circ)} = -1$

Problem 113. The value of $\dfrac{(\cos 9^\circ + \sin 81^\circ)(\sec 9^\circ + \operatorname{cosec} 81)}{2\sin^2 63^\circ + 1 + 2\sin^2 27^\circ}$ is :

[SSC (CGL-T-I)-19]

Sol. Given expression

$= \dfrac{[\cos 9^\circ + \cos(90 - 81)^\circ][\sec 9^\circ + \sec(90 - 81)^\circ]}{2\sin^2 63^\circ + 1 + 2\cos^2(90 - 27)^\circ}$

$= \dfrac{2\cos 9^\circ \times 2\sec 9^\circ}{2(\sin^2 63^\circ + \cos^2 63^\circ) + 1} = \dfrac{4}{3}$

Problem 114. If cosec $39^\circ = x$, the value of $\dfrac{1}{\operatorname{cosec}^2 51^\circ} + \sin^2 39^\circ + \tan^2 51^\circ - \dfrac{1}{\sin^2 51^\circ \cdot \sec^2 39^\circ}$ is:

[SSC (CPO-SIAS LIO)-11]

(a) $\sqrt{x^2 - 1}$ (b) $\sqrt{1 - x^2}$

(c) $x^2 - 1$ (d) $1 - x^2$

Sol. Given expression $= \sin^2 51^\circ + \sin^2 39^\circ + \tan^2 51^\circ - \dfrac{1}{\sin^2 51^\circ \cdot \sec^2 39^\circ}$

$= \sin^2 51^\circ + \cos^2(90^\circ - 39^\circ) + \cot^2(90^\circ - 51^\circ) - \dfrac{1}{\cos^2(90^\circ - 51^\circ)\cdot \sec^2 39^\circ}$

$= \sin^2 51^\circ + \cos^2 51^\circ + \cot^2 39^\circ - \dfrac{1}{\cos^2 39^\circ \times \dfrac{1}{\cos^2 39^\circ}}$

$= 1 + (\operatorname{cosec}^2 39^\circ - 1) - \dfrac{1}{1} = x^2 - 1$

Problem 115. The expression $\dfrac{\tan 57^\circ + \cot 37^\circ}{\tan 33^\circ + \cot 53^\circ}$ is equal to

[SSC (CGL-T-II)-12]

(a) $\tan 33^\circ \cdot \cot 57^\circ$ (b) $\tan 57^\circ \cdot \cot 37^\circ$

(c) $\tan 33^\circ \cdot \cot 53^\circ$ (d) $\tan 53^\circ \cdot \cot 37^\circ$

Sol. (b) $\dfrac{\tan 57^\circ + \cot 37^\circ}{\tan 33^\circ + \cot 53^\circ} = \dfrac{\cot(90^\circ - 57^\circ) + \tan(90^\circ - 37^\circ)}{\tan 33^\circ + \cot 53^\circ}$

$= \dfrac{\cot 33^\circ + \tan 53^\circ}{\tan 33^\circ + \cot 53^\circ} = \dfrac{\dfrac{1}{\tan 33^\circ} + \tan 53^\circ}{\tan 33^\circ + \dfrac{1}{\tan 53^\circ}}$

$= \dfrac{(1 + \tan 33^\circ \times \tan 53^\circ)}{\tan 33^\circ} \times \dfrac{\tan 53^\circ}{(\tan 33^\circ \cdot \tan 53^\circ + 1)}$

$= \dfrac{\tan 53^\circ}{\tan 33^\circ} = \tan 53^\circ \cdot \cot 33^\circ$

$= \cot(90^\circ - 53^\circ)\cdot \tan(90^\circ - 33^\circ) = \cot 37^\circ \cdot \tan 57^\circ$

Problem 116. If $\sin 17^\circ = \dfrac{x}{y}$, then the value of $\sec 17^\circ - \sin 73^\circ$ is :

[FCI-Asstt. Gr.III-12]

(a) $\dfrac{y^2 - x^2}{xy}$ (b) $\dfrac{x^2}{\sqrt{y^2 - x^2}}$

(c) $\dfrac{x^2}{y\sqrt{y^2 + x^2}}$ (d) $\dfrac{x^2}{y\sqrt{y^2 - x^2}}$

Sol. (d) $\sec 17^\circ - \sin 73^\circ = \sec 17^\circ - \cos(90^\circ - 73^\circ)$

$= \sec 17^\circ - \cos 17^\circ = \dfrac{1}{\cos 17^\circ} - \cos 17^\circ$

$= \dfrac{1 - \cos^2 17^\circ}{\cos 17^\circ} = \dfrac{\sin^2 17^\circ}{\cos 17^\circ} = \dfrac{\left(\dfrac{x}{y}\right)^2}{\sqrt{1 - \sin^2 17^\circ}}$

$= \dfrac{\left(\dfrac{x}{y}\right)^2}{\sqrt{1 - \left(\dfrac{x}{y}\right)^2}} = \dfrac{x^2}{y^2} \times \dfrac{y}{\sqrt{y^2 - x^2}} = \dfrac{x^2}{y\sqrt{y^2 - x^2}}.$

Problem 117. If $\tan 15^\circ = 2 - \sqrt{3}$, then the value of $\tan 15^\circ \cdot \cot 75^\circ + \tan 75^\circ \cdot \cot 15^\circ$ is :

Sol. $\Rightarrow \tan 15^\circ \cdot \cot 75^\circ + \tan 75^\circ \cdot \cot 15^\circ$

$= \tan 15^\circ \cdot \tan(90^\circ - 75^\circ) + \cot(90^\circ - 75^\circ)\cdot \cot 15^\circ$

$= \tan^2 15^\circ + \cot^2 15^\circ$

Given, $\tan 15^\circ = 2 - \sqrt{3}$, then $\cot 15^\circ = \dfrac{1}{2 - \sqrt{3}} = 2 + \sqrt{3}$

Now, $\tan^2 15^\circ + \cot^2 15^\circ = (2 - \sqrt{3})^2 + (2 + \sqrt{3})^2$

$= 7 - 4\sqrt{3} + 7 + 4\sqrt{3} = 14$

Problem 118. If $\angle A$ and $\angle B$ are complementary to each other, then the value of $\sec^2 A + \sec^2 B - \sec^2 A \cdot \sec^2 B$ is : ***[FCI-Asstt. Gr.-III-12]***

Sol. Given, $\angle A + \angle B = 90^\circ \Rightarrow \angle B = 90^\circ - \angle A$

Now, $\sec^2 A + \sec^2 B - \sec^2 A \cdot \sec^2 B$

$= \sec^2 A - \sec^2 B(\sec^2 A - 1)$

$= \sec^2 A - \sec^2 B \tan^2 A$

$= \sec^2 A - \sec^2(90^\circ - A)\cdot \tan^2 A$

$= \sec^2 A - \operatorname{cosec}^2 A \cdot \tan^2 A$

$= \sec^2 A - \operatorname{cosec}^2 A \times \dfrac{\sin^2 A}{\cos^2 A}$

$= \sec^2 A - \dfrac{1}{\cos^2 A} = \sec^2 A - \sec^2 A$

$= 0$

Type D

Problem 119. If θ is a positive acute angle and $\tan 2\theta \cdot \tan 3\theta = 1$, then the value of $\left(2\cos^2\dfrac{5\theta}{2} - 1\right)$ is :

[SSC (CGL-T-I)-12]

Sol. Given, $\tan 2\theta \cdot \tan 3\theta = 1$

$\Rightarrow \quad \tan 2\theta = \dfrac{1}{\tan 3\theta}$

$\Rightarrow \quad \tan 2\theta = \cot 3\theta$

$\Rightarrow \quad \tan 2\theta = \tan(90^\circ - 3\theta)$

$\Rightarrow \quad 2\theta = 90^\circ - 3\theta$

$\therefore \quad \theta = 18^\circ$

Now, $2\cos^2 \dfrac{5\cdot 18^\circ}{2} - 1 = 2\cos^2 45^\circ - 1$

$$= 2\left(\frac{1}{\sqrt{2}}\right)^2 - 1 = 0$$

Alternatively

Given, $\tan 2\theta \cdot \tan 3\theta = 1$

$\Rightarrow \quad \tan 2\theta = \dfrac{1}{\tan 3\theta}$

$\Rightarrow \quad \tan 2\theta = \cot 3\theta$

$\Rightarrow \quad \tan 2\theta = \tan(270^\circ - 3\theta)$

$\Rightarrow \quad 2\theta = 270 - 3\theta \Rightarrow \theta = 54^\circ$

Now, $2\cos^2 \dfrac{5\times 54^\circ}{2} - 1$

$$= 2\cos^2(180 - 45^\circ) - 1$$

$$= 2(-\cos 45^\circ)^2 - 1 = 2\times\left(\frac{-1}{\sqrt{2}}\right)^2 - 1 = 0$$

Problem 120. If $\tan(2\theta + 45^\circ) = \cot 3\theta$ where $(2\theta + 45^\circ)$ and 3θ are acute angles, then the values of θ is

[FCI-Asstt. Gr-III-12]

Sol. Given, $\tan(2\theta + 45^\circ) = \cot 3\theta$

$\Rightarrow \quad \tan(2\theta + 45^\circ) = \tan(90^\circ - 3\theta)$

$\Rightarrow \quad 2\theta + 45^\circ = 90^\circ - 3\theta \Rightarrow \theta = 9^\circ$

Problem 121. If $0^\circ < x < \pi$ and $\sec x = \operatorname{cosec} y$, then the value of $\sin(x + y)$ is

Sol. Given, $\sec x = \operatorname{cosec} y$

$\Rightarrow \quad \dfrac{1}{\cos x} = \dfrac{1}{\sin y} \Rightarrow \cos x = \sin y$

$\Rightarrow \quad \sin(90^\circ - x) = \sin y$

$\Rightarrow \quad 90^\circ - x = y \Rightarrow x + y = 90^\circ$

Now, $\sin(x + y) = \sin 90^\circ = 1$

Problem 122. Value of $\cot\theta \cdot \tan(90^\circ - \theta) - \sec(90^\circ - \theta)\operatorname{cosec}\theta + (\sin^2 25^\circ + \sin^2 65^\circ) + \sqrt{3}$ $[\tan 5^\circ \tan 15^\circ \tan 30^\circ \tan 75^\circ \tan 85^\circ]$ is :

Sol. $\cot\theta\cot\theta - \operatorname{cosec}\theta\operatorname{cosec}\theta + [\sin^2 25^\circ + \cos^2(90^\circ - 65^\circ + 3\tan 5^\circ \tan 85^\circ$ $(\tan 15\cdot\tan 75^\circ)\cdot\tan 30^\circ$

$= \cot^2\theta - \operatorname{cosec}^2\theta + 1 + \sqrt{3}(\tan 5^\circ\cdot\cot 5^\circ)(\tan 15^\circ\cdot\cot 15^\circ)\cdot\dfrac{1}{\sqrt{3}}$

$= \cot^2\theta - \operatorname{cosec}^2\theta + 1 + \sqrt{3}\left(1\cdot 1\cdot\dfrac{1}{\sqrt{3}}\right)$

$= \cot^2\theta - \operatorname{cosec}^2\theta + 1 + 1$

$= \operatorname{cosec}^2\theta - 1 - \operatorname{cosec}^2\theta + 2 = 1$

Problem 123. If $\sin\alpha\cdot\sec(30^\circ + \alpha) = 1\ (0^\circ < \alpha < 60^\circ)$, then the value of $\sin\alpha + \cos 2a$ is :

Sol. Given, $\sin\alpha\cdot\dfrac{1}{\cos(30^\circ + \alpha)} = 1$

$\Rightarrow \quad \sin\alpha = \cos(30^\circ + \alpha)$

$\Rightarrow \quad \cos(90^\circ - \alpha) = \cos(30^\circ + \alpha)$

$\Rightarrow \quad (90^\circ - \alpha) = (30^\circ + \alpha)$

$\Rightarrow \quad \alpha = 30^\circ$

Now, $\sin\alpha + \cos 2\alpha = \sin 30^\circ + \cos 60^\circ$

$$= \frac{1}{2} + \frac{1}{2} = 1$$

Problem 124. If $\cos\theta\cdot\operatorname{cosec} 23^\circ = 1$, the value of θ is

Sol. Given, $\cos\theta\cdot\dfrac{1}{\sin 23^\circ} = 1 \Rightarrow \cos\theta = \sin 23^\circ$

$\Rightarrow \quad \cos\theta = \cos(90^\circ - 23^\circ)$

$\Rightarrow \quad \theta = 67^\circ$

X. Simplification Based on the Concept of Trigonometric Ratios of Allied Angles

Problem 125. $\sin 112^\circ + \cos 74^\circ - \sin 68^\circ + \cos 106^\circ = ?$

Sol. Expression $= \sin(180^\circ - 68^\circ) + \cos(180^\circ - 106^\circ) - \sin 68^\circ + \cos 106^\circ$

$$= \sin 68^\circ - \cos 106^\circ - \sin 68^\circ + \cos 106^\circ = 0$$

Problem 126. $\sin^2 560^\circ + \sin^2 1330^\circ + \sin^2 390^\circ + \sin^2 300^\circ = ?$

Sol. $\sin^2$ (remainder if 560° is divided by 360) + $\sin^2$ (remainder if 1330° is divided by 360) + $\sin^2$ (remainder if 390° is divided by 360) + $\sin^2 300^\circ$

$= \sin^2 200^\circ + \sin^2 250^\circ + \sin^2 30^\circ + \sin^2 300^\circ$

$= \sin^2(180^\circ + 20^\circ) + \sin^2(270^\circ - 20^\circ) + \sin^2 30^\circ + \sin^2(270^\circ + 30^\circ)$

$= (-\sin 20^\circ)^2 + (-\cos 20^\circ)^2 + \sin^2 30^\circ + (-\cos 30^\circ)^2$

$= \sin^2 20^\circ + \cos^2 20^\circ + \sin^2 30^\circ + \cos^2 30^\circ = 1 + 1 = 2$

Problem 127. $\cos 30^\circ + \cos 70^\circ + \cos 110^\circ + \cos 150^\circ = ?$

Sol. Given, expression $= \cos 30^\circ + \cos 70^\circ + \cos(180^\circ - 70^\circ) + \cos(180^\circ - 30^\circ)$

$= \cos 30^\circ + \cos 70^\circ - \cos 70^\circ - \cos 30^\circ = 0$

XI. Problems based on the Concept of Trigonometric Ratios of Angles of Triangle/Quadrilateral

Problem 128. If A, B and C be the angles of a triangle, then of the following the incorrect relation is :

(a) $\sin\left(\dfrac{A+B}{2}\right) = \cos\dfrac{C}{2}$ (b) $\cos\left(\dfrac{A+B}{2}\right) = \sin\dfrac{C}{2}$

(c) $\tan\left(\dfrac{A+B}{2}\right) = \sec\dfrac{C}{2}$ (d) $\cot\left(\dfrac{A+B}{2}\right) = \tan\dfrac{C}{2}$

[FCI-Asstt. Gr.III-12]

Sol. We know, $A + B + C = 180^\circ$

$\Rightarrow \quad \dfrac{A+B+C}{2} = 90^\circ$

$\Rightarrow \quad \dfrac{A+B}{2} = 90^\circ - \dfrac{C}{2}$

$\Rightarrow \quad \sin\left(\dfrac{A+B}{2}\right) = \sin\left(90^\circ - \dfrac{C}{2}\right)$

$\Rightarrow \quad \sin\left(\dfrac{A+B}{2}\right) = \cos\dfrac{C}{2}$

Similarly, $\cos\left(\frac{A+B}{2}\right) = \sin\frac{C}{2}$, $\tan\left(\frac{A+B}{2}\right) = \cot\frac{C}{2}$

and $\cot\left(\frac{A+B}{2}\right) = \tan\frac{C}{2}$

Hence, option (c) is incorrect.

Problem 129. If A, B and C be the angles of a triangle, then find the value of $\cos B + \cos(A+C)$.

Sol. We have, $A+B+C = 180°$

$\Rightarrow B = 180° - (A+C)$

$\Rightarrow \cos B = \cos[180° - (A+C)]$

$\Rightarrow \cos B = -\cos(A+C)$

$\Rightarrow \cos B + \cos(A+C) = 0$

Problem 130. In a ΔABC, $\sin\frac{C}{2} - \cos\left(\frac{A+B}{2}\right) = ?$

Sol. We know, $A+B+C = 180°$

$\Rightarrow \frac{A+B+C}{2} = 90°$

$\Rightarrow \frac{A+B}{2} = 90° - \frac{C}{2}$

$\Rightarrow \cos\left(\frac{A+B}{2}\right) = \cos\left(90° - \frac{C}{2}\right)$

$\Rightarrow \cos\left(\frac{A+B}{2}\right) = \sin\frac{C}{2}$

$\Rightarrow \sin\frac{C}{2} - \cos\left(\frac{A+B}{2}\right) = 0.$

Problem 131. In a Δ ΔABC, $\cos\left(\frac{A-B+C}{2}\right) + \cos\left(\frac{A+B-C}{2}\right) + \cos\left(\frac{B+C-A}{2}\right) = ?$

(a) $\cos A + \cos B + 2\cos C$

(b) $\sin A + \sin B + \sin C$

(c) $\sin(A+B+C) - 3\sin A \cdot \sin B \cdot \sin C$

(d) $\sin 2(A+B+C) - 1$

Sol. We know, $A+B+C = 180°$

$\Rightarrow \frac{A}{2} + \frac{B}{2} + \frac{C}{2} = 90°$

Now, $\cos\left(\frac{A+C-B}{2}\right) + \cos\left(\frac{A+B-C}{2}\right) + \cos\left(\frac{B+C-A}{2}\right)$

$= \cos\left(90° - \frac{B}{2} - \frac{B}{2}\right) + \cos\left(90° - \frac{C}{2} - \frac{C}{2}\right)$

$+ \cos\left(90° - \frac{A}{2} - \frac{A}{2}\right)$

$= \cos(90° - B) + \cos(90° - C) + \cos(90° - A)$

$= \sin A + \sin B + \sin C$

Problem 132. In quadrilateral $ABCD$, $\sin(A+C) + \sin(B+D) = ?$

Sol. We have, $A+B+C+D = 360°$

$\Rightarrow A+C = 360° - (B+D)$

$\Rightarrow \sin(A+C) = \sin[360° - (B+D)]$

$\Rightarrow \sin(A+C) = -\sin(B+D)$

$\Rightarrow \sin(A+C) + \sin(B+D) = 0$

Problem 133. In a cyclic quadrilateral $ABCD$, $\cos A + \cos B + \cos C + \cos D = ?$

[CDS-11, SSC (10+2 DEO/LDC)-14]

Sol. We know that in a cyclic quadrilateral $ABCD$, sum of opposite angles is 180° i.e. $A+C = B+D = 180°$

Now, $\cos A + \cos B + \cos C + \cos D$

$= \cos A + \cos B + \cos(180° - A) + \cos(180° - B)$

$= \cos A + \cos B - \cos A - \cos B = 0$

Problem 134. In a quadrilateral $ABCD$, $\cos\left(\frac{A+D}{2}\right) + \cos\left(\frac{B+C}{2}\right) = ?$

Sol. We have, $A+B+C+D = 360°$

$\Rightarrow \frac{A+B+C+D}{2} = 180°$

$\Rightarrow \left(\frac{A+D}{2}\right) = 180° - \left(\frac{B+C}{2}\right)$

$\Rightarrow \cos\left(\frac{A+D}{2}\right) = \cos\left[180° - \left(\frac{B+C}{2}\right)\right]$

$\Rightarrow \cos\left(\frac{A+D}{2}\right) = -\cos\left(\frac{B+C}{2}\right)$

$\Rightarrow \cos\left(\frac{A+D}{2}\right) + \cos\left(\frac{B+C}{2}\right) = 0$

Problem 135. In a quadrilateral $ABCD$, $\sin\left(\frac{A+B}{2}\right) - \sin\left(\frac{C+D}{2}\right) = ?$

Sol. We have, $A+B+C+D = 360°$

$\Rightarrow \frac{A+B+C+D}{2} = 180°$

$\Rightarrow \left(\frac{A+B}{2}\right) = 180° - \left(\frac{C+D}{2}\right)$

$\Rightarrow \sin\left(\frac{A+B}{2}\right) = \sin\left[180° - \left(\frac{C+D}{2}\right)\right]$

$\Rightarrow \sin\left(\frac{A+B}{2}\right) = \sin\left(\frac{C+D}{2}\right)$

$\Rightarrow \sin\left(\frac{A+B}{2}\right) - \sin\left(\frac{C+D}{2}\right) = 0$

WORKOUT

1. What is $\dfrac{\sec x}{\cot x + \tan x}$ equal to? ***[CDS-14]***

(a) $\sin x$ (b) $\cos x$
(c) $\tan x$ (d) $\cot x$

2. What is the value of $\dfrac{\cos(90° + A)}{\sec(270° - A)} + \dfrac{\sin(270° + A)}{\operatorname{cosec}(630° - A)}$? ***[SSC (CGL-T-II)-17]***

(a) $3 \sec A$ (b) $\tan A \sec A$
(c) 0 (d) 1

3. If $\tan\theta + \cot = 2$, then what is $\sin\theta + \cos\theta$ equal to?

(a) $\dfrac{1}{\sqrt{3}}$ (b) $\dfrac{1}{2}$
(c) $\sqrt{2}$ (d) 1

4. What is the least value of $\sin^2\theta + \cos^2\theta + \tan^2\theta + \operatorname{cosec}^2\theta + \sec^2\theta + \cot^2\theta$? ***[SSC (CGL-T-I)-17]***

(a) 3 (b) 7 (c) 5 (d) 1

5. If $0° < \theta < \dfrac{\pi}{4}$, then what is value of $\sqrt{1 - 2\sin\theta\cos\theta}$?

(a) $\cos\theta - \sin\theta$ (b) $\sin\theta - \cos\theta$
(c) $\pm(\cos\theta - \sin\theta)$
(d) $\cos\theta \cdot \sin\theta$

6. What is $\dfrac{\sin x - \cos x + 1}{\sin x + \cos x - 1}$ equal to? ***[CDS-14]***

(a) $\dfrac{\sin x - 1}{\cos x}$ (b) $\dfrac{\sin x + 1}{\cos x}$
(c) $\dfrac{\sin x - 1}{\cos x + 1}$ (d) $\dfrac{\sin x + 1}{\cos x + 1}$

7. The least value of $4\sec^2\theta + 9\operatorname{cosec}^2\theta$ is ***[SSC (CGL-T-I)-12]***

(a) 21 (b) 19 (c) 25 (d) 7

8. What is $(1 + \cot x - \operatorname{cosec} x)(1 + \tan x + \sec x)$ equal to?

(a) $\sin x$ (b) $\cos x$
(c) 1 (d) 2

9. If $\sin 3\theta \sec 2\theta = 1$, then what is the value of $\left[3\tan^2\left(\dfrac{5\theta}{2}\right) - 1\right]$? ***[SSC (CGL-T-I)-17]***

(a) 0 (b) 1
(c) 2 (d) 3

10. If $\sin x + \operatorname{cosec} x = 2$, then what is $\sin^9 x + \operatorname{cosec}^9 x$ equal to?

(a) 512 (b) 2
(c) 18 (d) 1024

11. If $\sin x + \cos x = p$ and $\sin^3 x + \cos^3 x = q$, then what is $p^3 - 3p$ equal to? ***[CDS-14]***

(a) 0 (b) $-2q$
(c) $2q$ (d) $4q$

12. If $\sin^4\theta + \cos^4\theta = 2\cdot\sin^2\theta\cdot\cos^2\theta$, θ is an acute angle, then the value of $\tan\theta$ is ***[SSC (CAPF, Delhi Police)-15]***

(a) 1 (b) 2
(c) $\sqrt{2}$ (d) 0

13. The value of $\cos 25° - \sin 25°$ is

(a) positive but less than 1
(b) positive but greater than 1
(c) negative
(d) 0

14. If α and β are complementary angles, then what is $\sqrt{\cos\alpha \operatorname{cosec}\beta - \cos\alpha \sin\beta}$ equal to ***[CDS-14]***

(a) $\sec\beta$
(b) $\cos\alpha$
(c) $\sin\alpha$
(d) $-\tan\beta$

15. If $\tan(x + y)\cdot\tan(x - y) = 1$, then the value of $\tan\left(\dfrac{2x}{3}\right)$ is ***[SSC (CGL-T-I)-12]***

(a) $\dfrac{2}{\sqrt{3}}$ (b) $\dfrac{1}{\sqrt{3}}$
(c) 1 (d) $\sqrt{3}$

16. If $\sin\theta\cos\theta = \dfrac{1}{2}$, then what is $\sin^6\theta + \cos^6\theta$ equal to

(a) 1 (b) 2
(c) 3 (d) $\dfrac{1}{4}$

17. If $\sec\theta + \tan\theta = 2$, then the value of $\sec\theta$ is ***[CDS-2014,13]***

(a) $\dfrac{3}{2}$ (b) $\dfrac{5}{2}$
(c) $\dfrac{5}{4}$ (d) $\sqrt{2}$

18. The maximum value of $\sin^4\theta + \cos^4\theta$ is ***[SSC (CGL-T-I)-15]***

(a) $\dfrac{1}{3}$ (b) 1
(c) 2 (d) 3

19. What is $\operatorname{cosec}(75° + \theta) - \sec(15° - \theta) - \tan(55° + \theta) + \cot(35° - \theta)$ equal to

(a) 0 (b) -1
(c) 1 (d) $\dfrac{3}{2}$

20. If $5\cos\theta + 12\sin\theta = 13$, $0° < \theta < 90°$, then the value of $\sin\theta$ is ***[SSC (CGL-T-II)-15]***

(a) $\dfrac{5}{13}$ (b) $\dfrac{12}{13}$
(c) $\dfrac{6}{13}$ (d) $\dfrac{12}{13}$

21. If $\sin\theta + 2\cos\theta = 1$, where $0° < \theta < \dfrac{\pi}{2}$, what is $2\sin\theta - \cos\theta$ equal to

(a) 1 (b) -1
(c) 2 (d) $\dfrac{1}{2}$

22. If $\cos x + \sec x = 2$, then what $\cos^n x + \sec^n x$ equal to, where n is a positive integer ***[CDS-14]***

(a) 2 (b) $2^{n-2} - 1$
(c) 2^{n-1} (d) 2^n

23. If $\sec\alpha + \tan\alpha = 2$, then the value of $\sin\alpha$ is (assume that $0 < \alpha < 90°$) ***[SSC (CGL-T-I)-14]***

(a) 0.4 (b) 0.5
(c) 0.6 (d) 0.8

24. $\sin 25° \sin 35° \sec 65° \sec 55°$ equal to

(a) 1 (b) -1
(c) 2 (d) $\dfrac{1}{2}$

25. If $\tan 8\theta = \cot 2\theta$, where $0° < \theta < \dfrac{\pi}{2}$, then what is the value of $\tan 5\theta$?

(a) 1 (b) 0
(c) $\dfrac{1}{\sqrt{3}}$ (d) $\sqrt{3}$

26. Which of the following cannot be true ?

(a) $\sin\theta = -\frac{1}{5}$ (b) $\cos\theta = 1$

(c) $\sec\theta = \frac{1}{2}$ (d) $\tan\theta = 20$

27. $\cot(45^\circ + \theta)\cdot\cot(45^\circ - \theta) = ?$

(a) 0 (b) 1

(c) –1 (d) ∞

28. If $\sin^2\alpha + \sin^2\beta = 2$, then $\cos\left(\frac{\alpha+\beta}{2}\right) = ?$ *[SSC (CGL-T-II)-12]*

(a) 1 (b) –1

(c) 0 (d) 0.5

29. If $\sin\theta - \cos\theta = 0$, then what is $\sin^4\theta + \cos^4\theta$ equal to *[CDS-13]*

(a) 1 (b) –1

(c) 2 (d) $\frac{1}{2}$

30. Consider the following statements

I. $\tan\theta$ increases faster than $\sin\theta$ as θ increases.

II. The value of $\sin\theta + \cos\theta$ is always greater than 1.

Which of the statements given above is/are correct?

(a) only I

(b) only II

(c) both I and II

(d) neither I nor II

31. If θ be positive acute angle and $4\cos^2\theta - 4\cos\theta + 1 = 0$, then the value of $\tan(\theta - 15^\circ)$ is equal to : *[SSC (10+2, DEO/LDC)-14]*

(a) 0 (b) 1

(c) $\sqrt{3}$ (d) $\frac{1}{\sqrt{3}}$

32. What is $\frac{(\sin\theta + \cos\theta)(\tan\theta + \cot\theta)}{\sec\theta + \text{cosec}\,\theta}$ equal to ?

(a) 1 (b) $\sin\theta$

(c) 2 (d) $\cos\theta$

33. The value of $\sin^2 22^\circ + \sin^2 68^\circ + \cot^2 30^\circ$ is : *[SSC (CGL-T-I)-15]*

(a) 4 (b) $\frac{3}{4}$

(c) $\frac{5}{4}$ (d) 3

34. If $\sin\theta + \cos\theta = \sqrt{3}$, then what is $\tan\theta + \cot\theta$ equal to ?

(a) 1 (b) 2

(c) $\sqrt{2}$ (d) $\sqrt{3}$

35. If α, β and γ are acute angles such that $\sin\alpha = \frac{\sqrt{3}}{2}$, $\cos\beta = \frac{\sqrt{3}}{2}$ and $\tan\gamma = 1$, then what is $\alpha + \beta + \gamma$ equal to : *[CDS-13]*

(a) 105° (b) 120°

(c) 135° (d) 150°

36. If $2\sin\theta + \cos\theta = \frac{7}{3}$, then the value of $(\tan^2\theta - \sec^2\theta)$ is : *[SSC (CGL-T-I)-14]*

(a) 0 (b) –1 (c) $\frac{3}{7}$ (d) $\frac{7}{3}$

37 If $\cos A + \cos^2 A = 1$, then what is the value of $2(\sin^2 A + \sin^4 A)$? *[CDS-13, 09]*

(a) 4 (b) 1 (c) 2 (d) $\frac{1}{2}$

38. Consider the following statements for $0^\circ < \theta < 90^\circ$;

I. The value of $\sin\theta + \cos\theta$ is always greater than 1

II. The value of $\tan\theta + \cot\theta$ is always greater than 1

Which of the statements given above is/are correct? *[CDS-13]*

(a) only I (b) only II

(c) both I and II (d) neither I nor II

39. If θ is positive acute angle and $3(\sec^2\theta + \tan^2\theta) = 5$, then the value of $\cos 2\theta$ is : *[SSC (CGL-T-II)-15]*

(a) $\frac{1}{2}$ (b) 1

(c) $\frac{\sqrt{3}}{2}$ (d) $\frac{1}{\sqrt{2}}$

40. If $a^2 = \frac{1 + 2\sin\theta\cos\theta}{1 - 2\sin\theta\cos\theta}$, then what is the value of $\frac{a+1}{a-1}$?

(a) $\sec\theta$ (b) 1

(c) 0 (d) $\tan\theta$

41. The value of $\tan(-945^\circ)$ is :

(a) –2 (b) –3

(c) –1 (d) –4

42. If $\sin\theta + \cos\theta = m$ and $\sec\theta + \text{cosec}\,\theta = n$, then $n(m+1)(m-1)$ is equal to :

(a) m (b) n

(c) $2m$ (d) $2n$

43. If $a\sec\theta + b\tan\theta = 1$, $a^2\sec^2\theta - b^2\tan^2\theta = 5$, then $a^2b^2 + 4a^2 = ?$

(a) $9b^2$ (b) 9

(c) $\frac{9}{a^2}$ (d) $\frac{-2}{b}$

44. If $0^\circ < \theta < 90^\circ$, then all the trigonometric ratios can be obtained when *[CDS-12]*

(a) only $\sin\theta$ is given

(b) only $\cos\theta$ is given

(c) only $\tan\theta$ is given

(d) any one of the six ratios is given

45. If $\sin\theta + \sin^2\theta = 1$, then the value of $\cos^{12}\theta + 3\cos^{10}\theta + 3\cos^8\theta + \cos^6\theta - 1$ is : *[SSC (CGL-T-II)-14, 15]*

(a) 1 (b) 2

(c) 3 (d) 0

46. What is the value of $\frac{\sin\theta}{1+\cos\theta} + \frac{1+\cos\theta}{\sin\theta}$?

(a) $\sec\theta$ (b) $2\sec\theta$

(c) $\text{cosec}\,\theta$ (d) $2\,\text{cosec}\,\theta$

47. If x, y are positive acute angles, $x + y < 90^\circ$ and $\sin(2x - 20^\circ) = \cos(2y + 20^\circ)$, then the value of $\sec(x + y)$ is: *[SSC (CGL-T-I)-12]*

(a) $\sqrt{2}$ (b) 1

(c) $\frac{1}{\sqrt{2}}$ (d) 0

48. The expression $\sin^2 x + \cos^2 x - 1 = 0$ is satisfied by how many values of x? *[CDS-12]*

(a) only one value of x

(b) two values of x

(c) infinite value of x

(d) no values of x

49. Consider the following statements :

I. There is only one value of x in the first quadrant that satisfies $\sin x + \cos x = 2$

II. There is only one value of x in the first quadrant that satisfies $\sin x - \cos x = 0$

Which of the statements given above is/are correct?

(a) only I (b) only II

(c) both I and II (d) neither I nor II

50. If $3\sin x + 5\cos x = 5$, then what is the value of $(3\cos x - 5\sin x)$?

(a) 0 (b) 3

(c) 2 (d) 5

51. If $\text{cosec}\,A + \cot A = \frac{11}{2}$, then $\tan A = ?$

(a) $\frac{21}{22}$ (b) $\frac{15}{16}$ (c) $\frac{44}{117}$ (d) $\frac{117}{43}$

52. If $x = y\cos\frac{2\pi}{3} = z\cos\frac{4\pi}{3}$, then $xy + yz + zx = ?$

(a) 0 (b) 1
(c) –1 (d) 2

53. The minimum value of $\sin^2\theta + \cos^2\theta + \sec^2\theta + \text{cosec}^2\theta + \tan^2\theta + \cot^2\theta$ is : *[SSC (CGL-T-II)-11]*

(a) 1 (b) 3
(c) 5 (d) 7

54. If $\tan A = \frac{1-\cos B}{\sin B}$, then what is $\frac{2\tan A}{1-\tan^2 A}$ equal to? *[CDS-11]*

(a) $\tan B$ (b) $2\tan B$
(c) $\text{cosec } B$ (d) $2\,\text{cosec } B$

55. If $\sin\theta + \cos\theta = a$ and $\sec\theta + \text{cosec}\theta = b$, then the value of $b(a^2 - 1)$ is equal to? *[MAT-03]*

(a) $2a$ (b) $3a$
(c) 0 (d) $2ab$

56. If $\cos\theta \geq \frac{1}{2}$ in the 1st quadrant, then which one of the following is correct ?

(a) $\theta \geq \frac{\pi}{6}$ (b) $\theta \geq \frac{\pi}{3}$
(c) $\theta \leq \frac{\pi}{6}$ (d) $\theta \leq \frac{\pi}{3}$

57. If $\cos 1° = p$ and $\cos 89° = q$, then which one of the following is correct ?

(a) $p < q$
(b) p is close to 0 and q is close to 1
(c) $p = q$
(d) p is close to 1 and q is close to 0

58. If $\sin^2\theta = \frac{x^2 + y^2 + 1}{2x}$, then x must be :

(a) –2 (b) 1
(c) –3 (d) None of these

59 If $7\cos^2\theta + 3\sin^2\theta = 4$ and $0° < \theta < \frac{\pi}{2}$, then what is the value of $\tan\theta$?

(a) 3 (b) $\sqrt{7}$
(c) $\frac{7}{3}$ (d) $\sqrt{3}$

60. If $0° \leq \theta \leq \frac{\pi}{2}$ and $p = \sec^2\theta$, then which one of the following is correct ? *[CDS-10]*

(a) $p < 1$ (b) $p = 1$
(c) $p > 1$ (d) $p \geq 1$

61. $\frac{2\sin\theta\tan\theta(1-\tan\theta) + 2\sin\theta\sec^2\theta}{(1+\tan\theta)^2} = ?$

(a) $\frac{\sin\theta}{1+\tan\theta}$ (b) $\frac{2\sin\theta}{1+\tan\theta}$
(c) $\frac{2\sin\theta}{(1+\tan\theta)^2}$ (d) None of these

62 $\sin 15° + \cos 105° = ?$

(a) 0 (b) $2\sin 15°$
(c) $\cos 15° + \sin 15°$
(d) $\sin 15° - \cos 15°$

63 If $\sec\theta + \tan\theta = m$, then $\frac{m^2-1}{m^2+1} = ?$

(a) $\sin\theta$ (b) $\cos\theta$
(c) $\sec\theta$ (d) $\tan\theta$

64. If $\cos\theta + \cos^2\theta = 1$, then $\sin^2\theta + \sin^4\theta = ?$

(a) 0 (b) 2
(c) 1 (d) $\frac{1}{2}$

65. The value of $\text{cosec}^2\theta - 2 + \sin^2\theta$ is always : *[CDS-09]*

(a) less than zero (b) non-negative
(c) zero (d) 1

66. If $0 < \theta < \Phi < 90°$, then which one of the following is correct?

(a) $(\sin\theta + \cos\theta)^2 > 2$
(b) $(\sin^2\theta + \cos^2\Phi) \leq 2$
(c) $(\sin^2\theta + \cos^2\Phi) < 2$
(d) $(\sin^2\theta + \cos^2\Phi) > 2$

67. If $(\sin\alpha + \text{cosec}\,\alpha)^2 + (\cos\alpha + \sec\alpha)^2 = k + \tan^2\alpha + \cot^2\alpha$, then the value of k is : *[SSC (CGL-T-I)-14, SSC (CGL-T-II)-19]*

(a) 1 (b) 7 (c) 3 (d) 5

68. If $\sin\theta\cos\theta = \frac{1}{2}$, then what is the value of $\sin\theta - \cos\theta$?

(a) 0 (b) 1
(c) 2 (d) 4

69. $\tan^2 y\,\text{cosec}^2 x - 1 = \tan^2 y$, then which one of the following is correct ? *[CDS-09]*

(a) $x - y = 0$ (b) $x = 2y$
(c) $y = 2x$ (d) $x - y = 1°$

70. The value of $\frac{1}{\text{cosec}\theta - \cot\theta} - \frac{1}{\sin\theta}$ is : *[SSC (CGL-T-I)-13]*

(a) $\text{cosec}\,\theta$ (b) $\cot\theta$
(c) 1 (d) $\tan\theta$

71. If α is the angle of 1st quadrant such that $\text{cosec}^4\alpha = 17 + \cot^4\alpha$, then what is the value of $\sin\alpha$?

(a) $\frac{1}{3}$ (b) $\frac{1}{4}$
(c) $\frac{1}{9}$ (d) $\frac{1}{16}$

72. If $x + \frac{1}{x} = 2\cos\alpha$, then what is the value of $x^2 + \frac{1}{x^2}$?

(a) $4\cos^2\alpha$
(b) $2\cos^2\alpha - 2\sin^2\alpha$
(c) $4\cos^2\alpha - 1$
(d) $\cos^2\alpha - \sin^2\alpha$

73. $\tan A + \cot A = 4$, then $\tan^4 A + \cot^4 A$ is equal to

(a) 110 (b) 191
(c) 80 (d) 194

74. The value of $\sin 10° + \sin 20° + \sin 30° + \ldots + \sin 360°$ is

(a) 1 (b) 0
(c) –1 (d) None of these

75. If $\sin x = \cos y$ and angle x and y are acute, then what is the relation between x and y ?

(a) $x - y = \frac{\pi}{2}$ (b) $x + y = \frac{3\pi}{2}$
(c) $x + y = \frac{\pi}{2}$ (d) $x + y = \frac{\pi}{4}$

76. If $\sin(x - y) = \frac{1}{2}$ and $\cos(x + y) = \frac{1}{2}$, then what is the value of x? *[CDS-08]*

(a) 0° (b) 30°
(c) 25° (d) 45°

77. If $a(\tan\theta + \cot\theta) = 1$, $\sin\theta + \cos\theta = b$ with $0° < \theta < 90°$, the no the relation between a and b is. *[SSC (CGL-T-II)-14]*

(a) $b^2 = 2(a + 1)$ (b) $b^2 = 2(a - 1)$
(c) $2a = b^2 - 1$ (d) $2a = b^2 + 1$

78. If $\sin x - \cos x = 0$, then what is the value of $\sin^4 x + \cos^4 x$?

(a) 1 (b) $\frac{3}{4}$
(c) $\frac{1}{2}$ (d) $\frac{1}{4}$

79. If $0° \leq \theta \leq \frac{\pi}{2}$, $2y\cos\theta = x\sin\theta$ and $2x\sec\theta - y\,\text{cosec}\,\theta = 3$, then the value of $x^2 + 4y^2$ is :

(a) 4 (b) 1
(c) 2 (d) 3

80. If $\tan x = \frac{3}{4}$, where $0° < x < 90°$, then what is the value of $\sin x \cos x$? **[CDS-08]**

(a) $\frac{3}{5}$ (b) $\frac{4}{5}$

(c) $\frac{12}{25}$ (d) $\frac{13}{25}$

81. What is the value of $\frac{[\tan(90 - A) + \cot(90 - A)]^2}{2\sec^2(90 - 2A)}$? **[SSC (CGL-T-II)-17]**

(a) 0 (b) 1

(c) 2 (d) –1

82. Which one of the following is correct ?

(a) $\tan x > 1, 45° < x < 90°$

(b) $\sin x > \frac{1}{2}, 0° < x < 30°$

(c) $\cos x > \frac{1}{2}, 60° < x < 90°$

(d) $\sin x = \cos x$ for some value of x, $30° < x < 45°$

83. If $\tan\theta \cdot \tan 2\theta = 1$, then the value of $\sin^2 2\theta + \tan^2 2\theta$ is equal to : **[SSC (CGL-T-II)-12]**

(a) $\frac{5}{17}$ (b) $\frac{3}{19}$

(c) $\frac{7}{10}$ (d) $\frac{7}{13}$

84. If $p = \frac{2\sin\theta}{1 + \cos\theta + \sin\theta}$ and $q = \frac{\cos\theta}{1 + \sin\theta}$, then which of the following is true ?

(a) $q + p = 1$ (b) $\frac{p}{q} = 1$

(c) $q - p = 1$ (d) $pq = 1$

85. If $p = \sin^{10} x$, then which one of the following is correct for any value of x? **[CDS-07]**

(a) $0 \le p \le 1$ (b) $1 \le p \le 2$

(c) $p \ge 1$ (d) None of these

86. If $\cos\alpha + \sec\alpha = \sqrt{3}$, then the value of $\cos^3\alpha + \sec^3\alpha$ is : **[SSC (CGL-T-I)-14]**

(a) 0 (b) 1

(c) 2 (d) 4

87. Which of the following expression for $0° < \theta < 90°$ is/are independent of θ?

I. $\cos\theta(1 - \sin\theta)^{-1} + \cos\theta(1 + \sin\theta)^{-1}$

II. $\cos\theta(1 + \text{cosec}\,\theta)^{-1} + \cos\theta(\text{cosec}\,\theta - 1)^{-1}$

Select the correct answer using the codes given below.

(a) Only I (b) Only II

(c) Both I and II (d) Neither I and II

88. If $5\tan\theta = 4$, then the value of $\frac{5\sin\theta - 3\cos\theta}{5\sin\theta + 3\cos\theta}$ is : **[SSC (CGL-T-I)-12]**

(a) $\frac{1}{7}$ (b) $\frac{2}{7}$ (c) $\frac{5}{7}$ (d) $\frac{2}{5}$

89. Under which one of the following conditions is the trigonometrical identity $\frac{\sin x}{1 + \cos x} = \frac{1 - \cos x}{\sin x}$ true ?

(a) x is not a multiple of 360°

(b) x is not an odd multiple of 180°

(c) x is not a multiple of 180°

(d) none of the above

90. If $x = \text{cosec}\,\theta - \sin\theta$ and $y = \sec\theta - \cos\theta$, then the value of $x^2y^2(x^2 + y^2 + 3) = ?$ **[SSC (CGL-T-I)-12]**

(a) 0 (b) 1

(c) 2 (d) 3

91. If $\sin x + \sin^2 x = 1$, then the value of $\cos^{12} x + 3\cos^{10} x + 3\cos^8 x + \cos^6 x - 2$ is equal to :

(a) 0 (b) 1

(c) –1 (d) 2

92. $\cos 1° \cdot \cos 2° \cdot \cos 3 \dots \cos 179° = ?$

(a) 0 (b) 1

(c) 2 (d) $\frac{1}{2}$

93. If $\tan(A + B) = \sqrt{3}$ and $\tan A = 1$, then $\tan(A - B) = ?$ **[CDS-15]**

(a) 0 (b) 1

(c) $\sqrt{2}$ (d) $\frac{1}{\sqrt{3}}$

94. If $x = a\sin\theta + b\cos\theta$ and $y = a\cos\theta - b\sin\theta$, then $x^2 + y^2 = ?$ **[CDS-12, 17]**

(a) $a^2 + b^2$ (b) a^2b^2

(c) $a^2 - b^2$ (d) $\frac{a^2}{b^2}$

95. If $\sin\theta + \cos\theta = \sqrt{2}\cos(90° - \theta)$, then $\cot\theta = ?$

(a) $\sqrt{2}$ (b) $\sqrt{2} + 1$

(c) $\sqrt{2} - 1$ (d) 0

96. If $\sin\theta + \cos\theta = \sqrt{2}\sin(90° - \theta)$, then $\cot\theta = ?$ **[SSC (CGL-T-II)-15]**

(a) $\sqrt{2}$ (b) $\sqrt{2} + 1$

(c) $\sqrt{2} - 1$ (d) 0

97. If $\tan\alpha = n\tan\beta$ and $\sin\alpha = m\sin\beta$, then $\cos^2\alpha$ is : **[SSC (CGL-T-I)-13]**

(a) $\frac{m^2}{n^2}$ (b) $\frac{m^2 - 1}{n^2 - 1}$

(c) $\frac{m^2 + 1}{n^2 + 1}$ (d) $\frac{m^2}{n^2 + 1}$

98. If $\sin\theta_1 + \sin\theta_2 + \sin\theta_3 = 3$, then $\cos\theta_1 + \cos\theta_2 + \cos\theta_3 = ?$ **[CDS-17]**

(a) 0 (b) 1

(c) 3 (d) 2

99. What is the value of $\frac{(\sin x + \sin y)(\sin x - \sin y)}{(\cos x + \cos y)(\cos y - \cos x)}$? **[SSC (CGL-T-II)-17]**

(a) 0 (b) 1

(c) –1 (d) 2

100. If $\sec\theta(\cos\theta + \sin\theta) = \sqrt{2}$, then what is the value of $\frac{2\sin\theta}{\cos\theta - \sin\theta}$? **[SSC (CGL-T-II) 2017]**

(a) $3\sqrt{2}$ (b) $\frac{3}{\sqrt{2}}$ (c) $\frac{1}{\sqrt{2}}$ (d) $\sqrt{2}$

101. Suppose $0° < \theta < 90°$, then for every θ, $4\sin^2\theta + 1$ is greater than or equal to : **[CDS-19]**

(a) 2 (b) $4\sin\theta$

(c) $4\cos\theta$ (d) $4\tan\theta$

102. Consider the following for real numbers α, β, γ and δ?

I. $\sec\alpha = \frac{1}{4}$ II. $\tan\beta = 20$

III. $\text{cosec}\,\gamma = \frac{1}{2}$ IV. $\cos\delta = 2$

How many of the above statements are not possible. **[CDS-19]**

(a) one (b) two

(c) thee (d) four

103. If $x = 4\cos A + 5\sin A$ and $y = 4\sin A - 5\cos A$, then the value of $x^2 + y^2$ is : **[SSC (CGL-T-I)-19]**

(a) 25 (b) 16

(c) 0 (d) 41

104. The value of expression $\text{cosec}(85°+\theta) - \sec(5°-\theta) - \tan(55°+\theta) + \cot(35°-\theta)$ is : **[SSC (CGL-T-I)-19]**

(a) 1 (b) 3/2

(c) –1 (d) 0

Answer key

1. (a)	**2.** (d)	**3.** (c)	**4.** (b)	**5.** (c)	**6.** (b)	**7.** (a)	**8.** (d)	**9.** (c)	**10.** (b)
11. (b)	**12.** (a)	**13.** (a)	**14.** (c)	**15.** (b)	**16.** (d)	**17.** (c)	**18.** (b)	**19.** (a)	**20.** (b)
21. (c)	**22.** (a)	**23.** (c)	**24.** (a)	**25.** (a)	**26.** (c)	**27.** (b)	**28.** (c)	**29.** (d)	**30.** (a)
31. (b)	**32.** (a)	**33.** (a)	**34.** (a)	**35.** (c)	**36.** (b)	**37.** (c)	**38.** (b)	**39.** (a)	**40.** (d)
41. (c)	**42.** (c)	**43.** (a)	**44.** (d)	**45.** (d)	**46.** (d)	**47.** (a)	**48.** (c)	**49.** (b)	**50.** (b)
51. (c)	**52.** (a)	**53.** (d)	**54.** (a)	**55.** (a)	**56.** (d)	**57.** (d)	**58.** (b)	**59.** (d)	**60.** (d)
61. (b)	**62.** (a)	**63.** (a)	**64.** (c)	**65.** (b)	**66.** (c)	**67.** (b)	**68.** (a)	**69.** (a)	**70.** (b)
71. (a)	**72.** (b)	**73.** (d)	**74.** (b)	**75.** (c)	**76.** (d)	**77.** (c)	**78.** (c)	**79.** (a)	**80.** (c)
81. (c)	**82.** (a)	**83.** (d)	**84.** (a)	**85.** (a)	**86.** (a)	**87.** (d)	**88.** (a)	**89.** (c)	**90.** (b)
91. (c)	**92.** (a)	**93.** (d)	**94.** (a)	**95.** (c)	**96.** (b)	**97.** (b)	**98.** (a)	**99.** (b)	**100.** (d)
101. (b)	**102.** (c)	**103.** (d)	**104.** (d)						

Hints and SOLUTIONS

1. $\dfrac{\sec x}{\cot x+\tan x}=\dfrac{\dfrac{1}{\cos x}}{\dfrac{\cos x}{\sin x}+\dfrac{\sin x}{\cos x}}$

$=\dfrac{\dfrac{1}{\cos x}}{\dfrac{\sin^2 x+\cos^2 x}{\cos x\cdot\sin x}}$

$=\dfrac{\dfrac{1}{\cos x}}{\dfrac{1}{\cos x\cdot\sin x}}=\sin x$

Alternatively

Take a value of x, say 30° (you are free to assume any angles for which the given T-ratios have a defined value).

Then,

$\dfrac{\sec x}{\cot x+\tan x}=\dfrac{\sec 30^\circ}{\cot 30^\circ+\tan 30^\circ}$

$=\dfrac{\dfrac{2}{\sqrt{3}}}{\sqrt{3}+\dfrac{1}{\sqrt{3}}}$

$=\dfrac{1}{2}=\sin 30^\circ=\sin x.$

2. $\dfrac{\cos(90^\circ+A)}{\sec(270^\circ-A)}+\dfrac{\sin(270^\circ+A)}{\operatorname{cosec}(630^\circ-A)}$

$=\dfrac{-\sin A}{-\operatorname{cosec} A}+\dfrac{-\cos A}{\operatorname{cosec}[360^\circ+(270^\circ-A)]}$

$=\sin^2 A+\dfrac{-\cos A}{\operatorname{cosec}(270^\circ-A)}$

$=\sin^2 A+\dfrac{-\cos A}{-\sec A}$

$=\sin^2 A+\cos^2 A=1$

Alternatively

Put any value of A and solve the given expression. Then, match the resultant with values of the given options.

3. Given, $\dfrac{\sin\theta}{\cos\theta}+\dfrac{\cos\theta}{\sin\theta}=2$

$\Rightarrow\quad \dfrac{\sin^2\theta+\cos^2\theta}{\cos\theta\cdot\sin\theta}=2$

$\Rightarrow\quad \sin\theta\cdot\cos\theta=\dfrac{1}{2}$

Now, $(\sin\theta+\cos\theta)^2=\sin^2\theta+\cos^2\theta+2\sin\theta\cos\theta$

$=1+2\times\dfrac{1}{2}=2$

$\therefore\quad \sin\theta+\cos\theta=\sqrt{2}$

Alternatively I

Given, $\tan\theta+\dfrac{1}{\tan\theta}=2$

$\Rightarrow\quad \tan\theta=1$

$\Rightarrow\ \tan\theta=\tan 45^\circ\Rightarrow\theta=45^\circ$

Now, $\sin\theta+\cos\theta=\sin 45^\circ+\cos 45^\circ$

$=\dfrac{1}{\sqrt{2}}+\dfrac{1}{\sqrt{2}}=\sqrt{2}$

Alternatively II

Given, $\tan\theta+\dfrac{1}{\tan\theta}=2$

$\Rightarrow\quad \tan\theta=1$

$\Rightarrow\quad \dfrac{p}{b}=\dfrac{1}{1}\Rightarrow h=\sqrt{2}$

Now, $\sin\theta+\cos\theta=\dfrac{1}{\sqrt{2}}+\dfrac{1}{\sqrt{2}}=\sqrt{2}$

4. $(\sin^2\theta+\cos^2\theta)+\tan^2\theta+\operatorname{cosec}^2\theta+\sec^2\theta+\cot^2\theta$

$=1+\tan^2\theta+(1+\cot^2\theta)+(1+\tan^2\theta)+\cot^2\theta$

$=3+2(\tan^2\theta+\cot^2\theta)$

$=3+2\,(\geq 2)\quad [\because \tan^2\theta+\cot^2\theta\geq 2]$

Hence, least value of the expression

$=3+2\cdot 2=7$

Alternatively

Put any value of θ and solve the given expression. Then, match the resultant with given options.

5. $\sqrt{1-2\sin\theta\cos\theta}$

$=\sqrt{\sin^2\theta+\cos^2\theta-2\sin\theta\cos\theta}$

$[\because \sin^2\theta+\cos^2\theta=1]$

$=\sqrt{(\sin\theta-\cos\theta)^2}=\pm(\sin\theta-\cos\theta)$

$=\pm(\cos\theta-\sin\theta)$

Alternatively

Put any value of θ between 0° and $\dfrac{\pi}{4}$ and solve. Then, match the resultant with results of given options.

6. $\dfrac{\sin x-\cos x+1}{\sin x+\cos x-1}=\dfrac{(\sin x-\cos x)+1}{(\sin x+\cos x)-1}\times\dfrac{(\sin x+\cos x)+1}{(\sin x+\cos x)+1}$

$$= \frac{\sin^2 x - \cos^2 x + 2\sin x + 1}{\sin^2 x + \cos^2 x + 2\sin x \cos x - 1}$$

$$= \frac{\sin^2 x - (1 - \sin^2 x) + 2\sin x + 1}{1 + 2\sin x \cos x - 1}$$

$$= \frac{2\sin^2 x + 2\sin x}{2\sin x \cos x}$$

$$= \frac{2\sin x(\sin x + 1)}{2\sin x \cos x} = \frac{\sin x + 1}{\cos x}$$

Alternatively

Take a value of x, say 45° and solve (you are free to assume any angle for which the given T-ratios have a defined value). Then,

$$\frac{\sin x - \cos x + 1}{\sin x + \cos x - 1} = \frac{\sin 45° - \cos 45° + 1}{\sin 45° + \cos 45° - 1}$$

$$= \frac{\frac{1}{\sqrt{2}} - \frac{1}{\sqrt{2}} + 1}{\frac{1}{\sqrt{2}} + \frac{1}{\sqrt{2}} - 1} = \frac{1}{\sqrt{2} - 1}$$

$$= \frac{1}{\sqrt{2} - 1} \times \frac{\sqrt{2} + 1}{\sqrt{2} + 1} = \sqrt{2} + 1$$

Now, ensure as to which option gives result as $(\sqrt{2} + 1)$.

7. $4\sec^2\theta + 9\operatorname{cosec}^2\theta = 4(1 + \tan^2\theta) + 9(1 + \cot^2\theta)$

$= 13 + 4\tan^2\theta + 9\cot^2\theta$

$= 13 + 5\cot^2\theta + 4(\tan^2\theta + \cot^2\theta)$

$= 13 + 5\cot^2\theta + 4[(\tan\theta - \cot\theta)^2 + 2\tan\theta \cdot \cot\theta]$

$= 13 + 8 + 5\cot^2\theta + 4(\tan\theta - \cot\theta)^2$

$= 21 + 5\cot^2\theta + 4(\tan\theta - \cot\theta)^2$

Since $(\tan - \cot\theta)^2 \geq 0$ and also $\cot^2\theta \geq 0$, the minimum value of $21 + 5\cot^2\theta + 4(\tan\theta - \cot\theta)^2$

$= 21 + 0 + 0 = 21$

8. Take a value of x, say 45°. Then,

$(1 + \cot x - \operatorname{cosec} x)(1 + \tan x + \sec x)$

$= (1 + \cot 45° - \operatorname{cosec} 45°)(1 + \tan 45° + \sec 45°)$

$= (1 + 1 - \sqrt{2})(1 + 1 + \sqrt{2})$

$= 2^2 - (\sqrt{2})^2 = 2$

Alternatively,

Put any value of x and solve the given expression. Then, match the resultant with the given options.

9. Given, $\sin 3\theta = \frac{1}{\sec 2\theta} = \cos 2\theta$

$\Rightarrow \sin 3\theta = \sin(90° - 2\theta)$

$\Rightarrow 3\theta = 90° - 2\theta \Rightarrow \theta = 18°$

Now, $\left[3\tan^2\left(\frac{5\theta}{2}\right) - 1\right]$

$= \left[3\tan^2\left(\frac{5 \cdot 18}{2}\right) - 1\right] = 3 - 1 = 2$

10. Given, $\sin x + \frac{1}{\sin x} = 2$

Clearly, $\sin x = 1$

$\therefore \sin^9 x + \frac{1}{\sin^9 x} = 1^9 + \frac{1}{1^9} = 2$

11. Given, $p = \sin x + \cos x$

$\Rightarrow p^3 = (\sin x + \cos x)^3$

$\Rightarrow p^3 = \sin^3 x + \cos^3 x + 3\sin x \cdot \cos x(\sin x + \cos x)$

$\Rightarrow p^3 = q + 3\sin x \cdot \cos x\,(p)$ ...(i)

Further,

$p^2 = (\sin x + \cos x)^2$

$= 1 + 2\sin x \cdot \cos x$

$\Rightarrow \sin x \cdot \cos x = \frac{p^2 - 1}{2}$ ...(ii)

From Eqs. (i) and (ii),

$$p^3 = q + \left[3 \times \frac{p^2 - 1}{2}\right](p)$$

$\Rightarrow 2p^3 = 2q + 3p^3 - 3p$

$\Rightarrow p^3 - 3p = -2q$

Alternatively

Let x be 0°.

Then, $\sin 0° + \cos 0° = p \Rightarrow p = 1$

Further, $\sin^3 0° + \cos^3 0° = q \Rightarrow q = 1$

Now, $p^3 - 3p = 1 - 3 = -2$

$= -2 \times 1 = -2q$

12. Given,

$\sin^4\theta + \cos^4\theta = 2 \cdot \sin^2\theta \cdot \cos^2\theta$

$\Rightarrow \sin^4\theta + \cos^4\theta - 2 \cdot \sin^2\theta \cdot \cos^2\theta = 0$

$\Rightarrow (\sin^2\theta - \cos^2\theta)^2 = 0$

$\Rightarrow \sin^2\theta = \cos^2\theta$

$\Rightarrow \frac{\sin^2\theta}{\cos^2\theta} = 1$

$\Rightarrow \tan^2\theta = 1 \Rightarrow \tan\theta = 1$

13. Observe the chart titled as 'Trend of changes in values of Trigonometric ratios-Quadrant-wise". It is clear that values of $\cos\theta$ decrease $(1 \to 0)$ form 0° to 90° and value of $\sin\theta$ increase $(0 \to 1)$ from 0° to 90° but at 45° their values are same. Hence, value of $\cos\theta$ will be greater than that of $\sin\theta$ provided $0° \leq \theta < 45°$. Further, in case of $45° < \theta \leq 90°$, the value of $\cos\theta$ will be lesser than that of $\sin\theta$. Thus, the value of $(\cos 25° - \sin 25°)$ will always be positive but less than 1.

14. $\sqrt{\cos\alpha \operatorname{cosec}\beta - \cos\alpha \sin\beta}$

$= \sqrt{\cos(90° - \beta)\operatorname{cosec}\beta - \cos(90° - \beta)\sin\beta}$

given, $\alpha + \beta = 90°$

$= \sqrt{\sin\beta \operatorname{cosec}\beta - \sin\beta \sin\beta}$

$= \sqrt{1 - \sin^2\beta} = \sqrt{\cos^2\beta} = \cos\beta$

$= \cos(90° - \alpha) = \sin\alpha$

Alternatively

Assume the values of α and β and solve (you are free to assume any angles for which the given T-ratios have a defined value). Then, match the resultant with values of the given options.

15. Given, $\tan(x + y)\tan(x - y) = 1$

$\Rightarrow \tan(x + y) = \frac{1}{\tan(x - y)}$

$\Rightarrow \tan(x + y) = \cot(x - y)$

$\Rightarrow \tan(x + y) = \tan(90° - x + y)$

$\Rightarrow x + y = 90° - x + y$

$\Rightarrow 2x = 90°$

Now, $\tan\left(\frac{2x}{3}\right) = \tan\left(\frac{90°}{3}\right) = \frac{1}{\sqrt{3}}$

16. $\sin^6\theta + \cos^6\theta = (\sin^2\theta)^3 + (\cos^2\theta)^3$

$= (\sin^2\theta + \cos^2\theta)(\sin^4\theta + \cos^4\theta - \sin^2\theta \cdot \cos^2\theta)$

$= (\sin^4\theta + \cos^4\theta - \sin^2\theta \cdot \cos^2\theta)$

$= (\sin^2\theta + \cos^2\theta)^2 - 2\sin^2\theta\cos^2\theta - \sin^2\theta \cdot \cos^2\theta$

$= 1 - 3\sin^2\theta\cos^2\theta$

$= 1 - 3\left(\frac{1}{2}\right)^2 = \frac{1}{4}$

Alternatively

Clearly, $\theta = 45°$ as given that $\sin\theta\cos\theta = \frac{1}{2}$

Now, $\sin^6\theta + \cos^6\theta$

$= \sin^6 45° + \cos^6 45°$

$= \left(\frac{1}{\sqrt{2}}\right)^6 + \left(\frac{1}{\sqrt{2}}\right)^6 = \frac{1}{8} + \frac{1}{8} = \frac{1}{4}$

17. Given, $\sec\theta + \tan\theta = 2$ ...(i)

$\Rightarrow (\sec\theta - \tan\theta)(\sec\theta + \tan\theta) = 2(\sec\theta - \tan\theta)$

$\Rightarrow \sec^2\theta - \tan^2\theta = 2(\sec\theta - \tan\theta)$

$\Rightarrow \sec\theta - \tan\theta = \frac{1}{2}$ ...(ii)

On adding Eqs. (i) and (ii),

$2\sec\theta = 2 + \frac{1}{2}$

$\Rightarrow \sec\theta = \frac{5}{4}$

18. $\sin^4\theta + \cos^4\theta$

$= \sin^4\theta + \cos^4\theta + 2 \cdot \sin^2\theta \cdot \cos^2\theta - 2 \cdot \sin^2\theta \cdot \cos^2\theta$

$= (\sin^2\theta + \cos^2\theta)^2 - 2 \cdot \sin^2\theta \cdot \cos^2\theta$

$= 1 - \frac{4\sin^2\theta \cdot \cos^2\theta}{2}$

$= 1 - \frac{(2\sin\theta\cos\theta)^2}{2} = 1 - \frac{(\sin 2\theta)^2}{2}$

The value of $1 - \frac{(\sin 2\theta)^2}{2}$ will be maximum if $\frac{(\sin 2\theta)^2}{2}$ takes minimum value. Since the minimum value of $(\sin 2\theta)^2$ be 0, the required maximum value $= 1 - \frac{0}{2} = 1$

19. $\text{cosec}(75° + \theta) - \sec(15° - \theta) - \tan(55° + \theta) + \cot(35° - \theta)$

$= \sec[90° - (75° + \theta)] - \sec(15° - \theta) - \cot[90° - (55° + \theta)] + \cot(35° - \theta)$

$= \sec(15° - \theta) - \sec(15° - \theta) - \cot(35° - \theta) + \cot(35° - \theta) = 0$

Alternatively

Assume any value of θ as per your convenience and solve (you are free to assume any angles for which the given *T*-ratios have a defined value). Then, match the resultant with the given options.

20. Given, $5\cos\theta + 12\sin\theta = 13$

$\Rightarrow \frac{5}{13}\cos\theta + \frac{12}{13}\sin\theta = 1$

$\Rightarrow \frac{5}{13}\cos\theta + \frac{12}{13}\sin\theta = \sin^2\theta + \cos^2\theta$

$\Rightarrow \cos\theta = \frac{5}{13}$ and $\sin\theta = \frac{12}{13}$

21. $\sin\theta + 2\cos\theta = 1$ (given) and $2\sin\theta - \cos\theta = x$ (let)

On squaring both equation and adding.

$(\sin^2\theta + 4\cos^2\theta + 2\sin\theta \cdot 2\cos\theta) + (4\sin^2\theta + \cos^2\theta - 2\cdot 2\sin\theta\cos\theta) = 1 + x^2$

$\Rightarrow 4(\sin^2\theta + \cos^2\theta) + (\sin^2\theta + \cos^2\theta) = 1 + x^2$

$\Rightarrow 5 - 1 = x^2 \quad \therefore x = \pm 2$

Since θ is in 1st quadrant, both sine and cosine will have '+' value.

Hence, $2\sin\theta - \cos\theta = +2$

Alternatively

We know, if

$m\cos\theta + n\sin\theta = r$, then

$n\cos\theta - m\sin\theta = \sqrt{(m^2 + n^2) - r^2}$

Here, $2\cos\theta + 1\sin\theta = 1$

Then,

$1\cos\theta - 2\sin\theta = \sqrt{(2^2 + 1^2) - (1)^2}$

$= \sqrt{4} = \pm 2$

$\Rightarrow 2\sin\theta - \cos\theta = -(\pm 2) = \pm 2$

Hence, correct option is (c).

22. Given, $\cos x + \frac{1}{\cos x} = 2$

Clearly, $\cos x = 1$

$\therefore \cos^n x + \sec^n x = 1^n + \frac{1}{1^n} = 2$

Alternatively

Given, $\cos x + \sec x = 2$

$\Rightarrow \cos 90° + \sec 90° = 1 + 1 \Rightarrow x = 90°$

Now, $\cos^n x + \sec^n x = \cos^n 90° + \sec^n 90°$

$= 1^n + 1^n = 2$

23. Given, $\sec\alpha + \tan\alpha = 2$ ...(i)

$\Rightarrow (\sec\alpha + \tan\alpha)(\sec\alpha - \tan\alpha) = 2(\sec\alpha - \tan\alpha)$

$\Rightarrow \sec^2\alpha - \tan^2\alpha = 2(\sec\alpha - \tan\alpha)$

$\Rightarrow \sec\alpha - \tan\alpha = \frac{1}{2}$ ...(ii)

On adding Eqs. (i) and (ii), we get

$2\sec\alpha = \frac{5}{2}$

$\Rightarrow \sec\alpha = \frac{5}{4}$

Then, $\cos\alpha = \frac{4}{5}$

$\therefore \sin\alpha = \sqrt{1 - \left(\frac{4}{5}\right)^2} = 0.6$

24. $\sin 25° \sin 35° \sec 65° \sec 55°$

$= \sin 25° \cdot \sin 35° \cdot \frac{1}{\cos 65°} \cdot \frac{1}{\cos 55°}$

$= \sin 25° \cdot \sin 35° \cdot \frac{1}{\sin(90 - 65°)} \cdot \frac{1}{\sin(90° - 55°)}$

$= \sin 25° \cdot \sin 35° \cdot \frac{1}{\sin 25°} \cdot \frac{1}{\sin 35°} = 1$

25. Given, $\tan 8\theta = \cot 2\theta$

$\Rightarrow \tan 8\theta = \tan(90° - 2\theta)$

$\Rightarrow 8\theta = 90° - 2\theta \Rightarrow \theta = 9°$

Now, $\tan 5\theta = \tan 45° = 1$

26. The value of $\sec\theta$ can be $\frac{1}{2}$ as $-1 \geq \text{cosec}\,\theta$ or $\sec\theta$ or $\text{cosec}\,\theta$ or $\sec\theta \geq 1$ Hence, option (c) is correct.

27. Given expression

$\cot(45° + \theta) \cdot \cot(45° - \theta)$

$= \cot[90° - (45° - \theta)] - \cot(45° - \theta)]$

$= \tan(45° - \theta) \cdot \cot(45° - \theta) = 1$

Alternatively

Take any value of θ so that the value of *T*-ratios be defined. At $\theta = 15°$, the given expression $= \cot 60° \cdot \cot 30° = 1$

Hence, option (b) is correct.

28. Given, $\sin^2\alpha + \sin^2\beta = 2$

$\Rightarrow \sin^2\alpha + \sin^2\beta = 1 + 1$

$\Rightarrow \sin^2\alpha = \sin^2\beta = 1$

$[\because \sin\alpha \text{ or } \sin\beta \leq 1]$

Then, $\alpha = \beta = 90°$

Now, $\cos\left(\frac{90° + 90°}{2}\right) = \cos 90° = 0$

29. Given, $\sin\theta = \cos\theta$

$\Rightarrow \sin 45° = \cos 45°$

$\Rightarrow \theta = 45°$

Now, $\sin^4\theta + \cos^4\theta$

$= \sin^4 45° + \cos^4 45°$

$= \left(\frac{1}{\sqrt{2}}\right)^4 + \left(\frac{1}{\sqrt{2}}\right)^4 = \frac{1}{2}$

30. Observe the chart titled 'Trend of changes in values of Trigonometric ratios - Quadrant - wise". It is clear that value of $\tan\theta$ increases faster than $\sin\theta$ as θ increases. As regards second statement, the value of $\sin\theta + \cos\theta$ may also be equal to 1 or less than 1. [maximum and minimum value of $1\cdot\sin\theta + 1\cdot\cos\theta$ are $+\sqrt{1^2 + 1^2}$ and $-\sqrt{1^2 + 1^2}$ respectively].

Hence, option (a) is correct.

31. Given,

$4\cos^2\theta - 4\cos\theta + 1 = 0$

Then, $\cos\theta = \frac{4 \pm \sqrt{16 - 16}}{2\cdot 4} = \frac{1}{2}$

$\Rightarrow \cos\theta = \cos 60°$

$\Rightarrow \theta = 60°$

Then, $\tan(\theta - 15°)$

$= \tan(60° - 15°) = 1$

32. $\frac{(\sin\theta + \cos\theta)(\tan\theta + \cot\theta)}{\sec\theta + \text{cosec}\,\theta}$

$= \frac{(\sin\theta + \cos\theta)\left(\frac{\sin\theta}{\cos\theta} + \frac{\cos\theta}{\sin\theta}\right)}{\frac{1}{\cos\theta} + \frac{1}{\sin\theta}}$

$= \frac{(\sin\theta + \cos\theta)\left(\frac{\sin^2\theta + \cos^2\theta}{\cos\theta \cdot \sin\theta}\right)}{\frac{1}{\cos\theta} + \frac{1}{\sin\theta}}$

$= \frac{(\sin\theta + \cos\theta)}{(\sin\theta + \cos\theta)} = 1$

Alternatively

Let θ be 30° (you are free to take any angle for which the given *T*-ratios have a defined value.)

Then, $\frac{(\sin\theta + \cos\theta)(\tan\theta + \cot\theta)}{\sec\theta + \text{cosec}\,\theta}$

$$= \frac{\left(\frac{1}{2}+\frac{\sqrt{3}}{2}\right)\left(\frac{1}{\sqrt{3}}+\sqrt{3}\right)}{\frac{2}{\sqrt{3}}+2}$$

$$= \frac{\left(\frac{4+4\sqrt{3}}{2\sqrt{3}}\right)}{\frac{2}{\sqrt{3}}+2} = \frac{\left(\frac{2}{\sqrt{3}}+2\right)}{\frac{2}{\sqrt{3}}+2} = 1$$

33. $\sin^2 22^\circ + \sin^2 68^\circ + \cot^2 30^\circ$

$= \sin^2 22^\circ + \cos^2 (90^\circ - 22^\circ) + \cot^2 30^\circ$

$= 1 + (\sqrt{3})^2 = 4$

34. Given, $\sin\theta + \cos\theta = \sqrt{3}$

On squaring both sides,

$\sin^2\theta + \cos^2\theta + 2\sin\theta\cdot\cos\theta = 3$

$\Rightarrow 1 + 2\sin\theta\cdot\cos\theta = 3$

$\Rightarrow \quad \sin\theta\cdot\cos = 1$

Now, $\tan\theta + \cot\theta = \frac{\sin x}{\cos x} + \frac{\cos x}{\sin x}$

$= \frac{\sin^2 x + \cos^2 x}{\cos x\cdot\sin x}$

$= \frac{1}{\cos x\cdot\sin x} = \frac{1}{1} = 1$

35. (c) Given, $\sin\alpha = \frac{\sqrt{3}}{2} \Rightarrow \alpha = 60^\circ$

and $\cos\beta = \frac{\sqrt{3}}{2} \Rightarrow \beta = 30^\circ$

and $\tan\gamma = 1 \Rightarrow \gamma = 45^\circ$

Now, $\alpha + \beta + \gamma = 60^\circ + 30^\circ + 45^\circ$

$= 135^\circ$

36. $\tan^2\theta \quad \sec^2 0$

$= -(\sec^2\theta - \tan^2\theta) = -1$

[Here, no need of the given other value.]

37. Given, $\cos A + \cos^2 A = 1$

$\Rightarrow \cos A + \cos^2 A = \sin^2 A + \cos^2 A$

$\Rightarrow \quad \cos A = \sin^2 A$

Then, $\sin^4 A = \cos^2 A$

Now, $2(\sin^2 A + \sin^4 A)$

$= 2(\cos A + \cos^2 A)$

$= 2\cdot 1 = 2$

38. We have already discussed that the range of $\sin\theta + \cos\theta$ is $-\sqrt{2}$ to $+\sqrt{2}$. It means the value of $\sin\theta + \cos\theta$ is not always greater than 1. We also known that $(\tan\theta + \cot\theta) \geq 2$ provided these ratios have positive values and for $0^\circ < \theta < 90^\circ$, $\tan\theta$ and $\cot\theta$ both have positive values.

Hence, option (b) is correct.

39. Given, $3(\sec^2\theta + \tan^2\theta) = 5$

$\Rightarrow \quad 3(\sec^2\theta + \sec^2\theta - 1) = 5$

$\Rightarrow \quad 6\sec^2 = 8$

$\Rightarrow \quad \sec^2\theta = \frac{4}{3}$

$\Rightarrow \quad \cos\theta = \frac{\sqrt{3}}{2}$

$\Rightarrow \quad \cos\theta = \cos 30^\circ$

$\Rightarrow \quad \theta = 30^\circ$

Now, $\cos(2\cdot 30^\circ) = \frac{1}{2}$

40. Given, $a^2 = \frac{1 + 2\sin\theta\cos\theta}{1 - 2\sin\theta\cos\theta}$

$\Rightarrow a^2 = \frac{\sin^2\theta + \cos^2\theta + 2\sin\theta\cos\theta}{\sin^2\theta + \cos^2\theta - 2\sin\theta\cos\theta}$

$\Rightarrow a^2 = \frac{(\sin\theta + \cos\theta)^2}{(\sin\theta - \cos\theta)^2}$

$\Rightarrow \frac{a}{1} = \frac{\sin\theta + \cos\theta}{\sin\theta - \cos\theta}$

$\Rightarrow \frac{a+1}{a-1}$

$= \frac{(\sin\theta + \cos\theta) + (\sin\theta - \cos\theta)}{(\sin\theta + \cos\theta) - (\sin\theta - \cos\theta)}$

[applying compnendo and dividendo Rule]

$\Rightarrow \quad \frac{a+1}{a-1} = \frac{2\sin\theta}{2\cos\theta} = \tan\theta$

41. We know, if the given angle is negative and less than -360°, divide it (ignoring negative sign) by 360 and subtract the remainder from 360°. The quadrant of the resultant will be the quadrant of the given angle.

The remainder when 945° is divided by 360 = 225°

Now, $360^\circ - 225^\circ = 135^\circ$

Hence, $\tan(-945^\circ) = \tan(135^\circ)$

$= \tan(90^\circ + 45^\circ)$

$= -\cot 45^\circ = -1$

42. $n(m+1)(m-1) = n(m^2 - 1)$

$= \sec\theta + \operatorname{cosec}\theta\,[(\sin\theta + \cos\theta)^2 - 1]$

$= (\sec\theta + \operatorname{cosec}\theta)[2\sin\theta\cdot\cos\theta]$

$= 2\sin\theta\cdot\cos\theta\sec\theta + 2\sin\theta\cdot\cos\theta\cdot\operatorname{cosec}\theta$

$= 2\sin\theta + 2\cos\theta$

$= 2(\sin\theta + \cos\theta) = 2m$

43. $a^2\sec^2\theta - b^2\tan^2\theta = 5$

$\Rightarrow (a\sec\theta + b\tan\theta)(a\sec\theta - b\tan\theta) = 5$

$\Rightarrow (1)(a\sec\theta - b\tan\theta) = 5$

$\Rightarrow a\sec\theta - b\tan\theta = 5$

Now, solving two equations

i.e. $a\sec\theta + b\tan\theta = 1$ and

$a\sec\theta - b\tan\theta = 5$, we get,

$2a\sec\theta = 6$ and $2b\tan\theta = -4$

Hence, $a = 3\cos\theta$

and $\quad b = -2\cot\theta$

Then, $a^2b^2 + 4a^2 = a^2(b^2 + 4)$

$= (3\cos\theta)^2[(-2\cot\theta)^2 + 4]$

$= (3\cos\theta)^2[4(\cot^2\theta + 1)]$

$= 9\cos^2\theta \times 4\operatorname{cosec}^2\theta$

$= 9 \times 4\cot^2\theta$

$= 9[(-2\cot\theta)^2] = 9b^2$

44. We can get value of all the trigonometric ratios it anyone of the six ratios is given with the help of trigonometric identities or Pythagoras theorem.

45. Given, $\sin\theta + \sin^2\theta = 1$

$\Rightarrow \quad \sin\theta = 1 - \sin^2\theta = \cos^2\theta$

$\Rightarrow \quad \sin^2\theta = \cos^4\theta$

Now, $\cos^{12}\theta + 3\cos^{10}\theta + 3\cos^8\theta + \cos^6\theta - 1$

$= (\cos^4\theta + \cos^2\theta)^3 - 1$

$= (\sin^2\theta + \cos^2\theta)^3 - 1$

$= 1 - 1 = 0$

46. $\frac{\sin\theta}{1+\cos\theta} + \frac{1+\cos\theta}{\sin\theta}$

$= \frac{\sin^2\theta + \cos^2\theta + 1 + 2\cos\theta}{(1+\cos\theta)\sin\theta}$

$= \frac{2(1+\cos\theta)}{(1+\cos\theta)\sin\theta}$

$= 2\operatorname{cosec}\theta$

Alternatively

Assume any value of θ as per your convenience and solve (you are free to assume any angle for which the given T-ratios have a defined value). Therefore, match the resultant with the values of the given options.

47. Given, $\sin(2x - 20^\circ) = \cos(2y + 20^\circ)$

$\Rightarrow \sin(2x - 20^\circ) = \sin(90^\circ - 2y - 20^\circ)$

$\Rightarrow \quad 2x - 20^\circ = 90^\circ - 2y - 20^\circ$

$\Rightarrow \quad x + y = 45^\circ$

Then, $\sec(x+y) = \sec(45^\circ) = \sqrt{2}$

48. We have already discussed that an equation involving trigonometric ratios of an angle is called a trigonometric identity, if it is true for all real values of angle(s) involved. Obviously, $\sin^2 x + \cos^2 x - 1 = 0$ (i.e. $\sin^2 x + \cos^2 x = 1$) is a trigonometric identity.

49. We have already discussed that the maximum value of $\sin x + \cos x$ is $\sqrt{2}$. Hence, there is no any value of x in the Ist quadrant that satisfies $\sin x + \cos x = 2$.

Further, at angle of $\frac{\pi}{4}$, sin x and cos x are equal.

Hence there difference will be 0.

Thus, option (b) is correct.

50. We know that, if
$m\cos\theta + n\sin\theta = r$, then
$n\cos\theta - m\sin\theta = \sqrt{(m^2+n^2)-r^2}$.

Here, $5\cos x + 3\sin x = 5$,
then $3\cos x - 5\sin x$
$= \sqrt{(5^2+3^2)-5^2} = 3$

51. (c) Given, $\operatorname{cosec} A + \cot A = \frac{11}{2}$ ...(i)

$\Rightarrow (\operatorname{cosec} A + \cot A)(\operatorname{cosec} A - \cot A)$
$= \frac{11}{2} \times (\operatorname{cosec} A - \cot A)$

$\Rightarrow \quad 1 = \frac{11}{2} \times (\operatorname{cosec} A - \cot A)$

$\Rightarrow \quad (\operatorname{cosec} A - \cot A) = \frac{11}{2}$ ...(ii)

On subtracting Eq. (ii) from Eq. (i), we get

$2\cot A = \frac{117}{22}$

$\Rightarrow \quad \cot A = \frac{117}{44} \Rightarrow \tan A = \frac{44}{117}$

52. Given, $x = y\cos\frac{2\pi}{3} = z\cos\frac{4\pi}{3}$

$\Rightarrow \quad x = y\cos(90° + 30°)$
$= z\cos(270° - 30°)$

$\Rightarrow \quad x = y(-\sin 30°) = z(-\sin 30°)$

$\Rightarrow \quad x = \frac{y}{-2} = \frac{z}{-2} = k$ (let)

$\Rightarrow \quad x = k, y = -2k$ and $z = -2k$

Now, $xy + yz + zx = k(-2k)$
$+ (-2k)(-2k) + (-2k)k = 0$

53. $\sin^2\theta + \cos^2\theta + \sec^2\theta + \operatorname{cosec}^2\theta + \tan^2\theta + \cot^2\theta$
$= 1 + \sec^2\theta + \operatorname{cosec}^2\theta + \tan^2\theta + \cot^2\theta$
$= 1 + \tan^2\theta + \sec^2\theta + \operatorname{cosec}^2\theta + \cot^2\theta$
$= 1 + \tan^2\theta + 1 + \tan^2\theta + 1 + \cot^2\theta + \cot^2\theta$
$= 3 + 2(2\tan^2\theta + \cot^2\theta)$
$= 3 + 2[(\tan\theta - \cot\theta)^2 + 2]$

Since the minimum value of $(\tan\theta - \cot\theta)^2$ is 0, the minimum value of $3 + 2[(\tan\theta - \cot\theta)^2 + 2]$
$= 3 + 2(0 + 2) = 7$

54. $\frac{2\tan A}{1-\tan^2 A} = \frac{2 \times \frac{1-\cos B}{\sin B}}{1 - \left(\frac{1-\cos B}{\sin B}\right)^2}$

$= \frac{2(1-\cos B)\sin B}{(\sin^2 B - 1) - \cos^2 B + 2\cos B}$

$= \frac{2(1-\cos B)\sin B}{-\cos^2 B - \cos^2 B + 2\cos B}$

$= \frac{2(1-\cos B)\sin B}{-2\cos^2 B + 2\cos B}$

$= \frac{2(1-\cos B)\sin B}{2\cos B(1-\cos B)} = \tan B$

55. $b(a^2 - 1) = (\sec\theta + \operatorname{cosec}\theta)$
$[(\sin\theta + \cos\theta)^2 - 1]$
$= (\sec\theta + \operatorname{cosec}\theta)$
$[\sin^2\theta + \cos^2\theta + 2\sin\theta\cos\theta - 1]$
$= (\sec\theta + \operatorname{cosec}\theta)[2\sin\theta\cos\theta]$
$= 2\sin\theta + 2\cos\theta$
$= 2(\sin\theta + \cos\theta) = 2a$

56. We know that $\cos\frac{\pi}{3} = \frac{1}{2}$ and it is also fact that value of cosine decreases when the value of angle increases.

Hence, $\theta \le \frac{\pi}{3}$

57. We know that value of cosine decreases from 0° to 90°.

e.g. $\cos 0° = 1$ and $\cos 90° = 0$. If $\cos 1° = p$, it means p is close to 1 and if $\cos 89° = q$, it means q is close to 0.

Hence, option (d) is correct.

58. Given, $\sin^2\theta = \frac{x^2+y^2+1}{2x}$

$\Rightarrow \quad \frac{x^2+y^2+1}{2x} \le 1$

$[\because 0 \le \sin^2\theta$ or $\cos^2\theta \le 1]$

$\Rightarrow \quad x^2 + y^2 + 1 \le 2x$

$\Rightarrow \quad x^2 - 2x + 1 + y^2 \le 0$

$\Rightarrow \quad (x-1)^2 + y^2 \le 0$

This is possible only if $x = 1$ and $y = 0$.

Hence, option (b) is correct.

59. $7\cos^2\theta + 3\sin^2\theta = 4$

$\Rightarrow 4\cos^2\theta + 3\cos^2\theta + 3\sin^2\theta = 4$

$\Rightarrow 4\cos^2\theta + 3(\cos^2\theta + \sin^2\theta) = 4$

$\Rightarrow 4\cos^2\theta = 1$

$\Rightarrow \quad \cos\theta = \pm\frac{1}{2}$

Since θ is an acute angle, $\cos\theta = +\frac{1}{2}$.

Clearly, $\theta = 60°$.

Now, $\tan\theta = \tan 60° = \sqrt{3}$

60. We know that value of $\sec\theta$ increase ($1 \Rightarrow 2 \Rightarrow$ not defined, see value chart) when angle increases for $0° \le \theta \le \frac{\pi}{2}$.

Hence, $\sec^2\theta \ge 1$

Note It has already been mentioned earlier that $\sec^2\theta$ or $\operatorname{cosec}^2\theta \ge 1$

61. Given, expression
$\frac{2\sin\theta\tan\theta(1-\tan\theta) + 2\sin\theta\sec^2\theta}{(1+\tan\theta)^2}$

$= \frac{2\sin\theta}{(1+\tan\theta)^2}$
$[\tan\theta - \tan^2\theta + 1 + \tan^2\theta]$

$= \frac{2\sin\theta}{1+\tan\theta}$

Alternatively

Assume any value of θ as per your convenience and solve (you are free to assume any angles for which the given T-ratios have a defined value). Thereafter, match the resultant with the values of the given options.

62. $\sin 15° + \cos 105° = \sin 15° + \cos(90° + 15°)$
$= \sin 15° - \sin 15° = 0$

63. (a) $\frac{m^2-1}{m^2+1} = \frac{(\sec\theta + \tan\theta)^2 - 1}{(\sec\theta + \tan\theta)^2 + 1}$

$= \frac{\sec^2\theta + \tan^2\theta + 2\sec\theta\tan\theta - 1}{\sec^2\theta + \tan^2\theta + 2\sec\theta\tan\theta + 1}$

$= \frac{(\sec^2\theta - 1) + \tan^2\theta + 2\sec\theta\tan\theta}{\sec^2\theta + (\tan^2\theta + 1) + 2\sec\theta\tan\theta}$

$= \frac{2\tan^2\theta + 2\sec\theta\tan\theta}{2\sec^2\theta + 2\sec\theta\tan\theta}$

$= \frac{2\tan\theta(\tan\theta + \sec\theta)}{2\sec\theta(\sec\theta + \tan\theta)}$

$= \frac{\tan\theta}{\sec\theta} = \frac{\sin\theta}{\cos\theta} \times \cos\theta$

$= \sin\theta$

Alternatively

Let θ be 30° (You are free to take any angle provided the trigonometric ratios have defined values)

Then,

$m = \sec 30° + \tan 30° = \frac{2}{\sqrt{3}} + \frac{1}{\sqrt{3}} = \sqrt{3}$

Now, $\frac{m^2-1}{m^2+1} = \frac{(\sqrt{3})^2 - 1}{(\sqrt{3})^2 + 1} = \frac{1}{2} = \sin 30°$

Hence, option (a) is correct.

64. Given, $\cos\theta + \cos^2\theta = 1$

$\Rightarrow \quad \cos\theta = 1 - \cos^2\theta$

$\Rightarrow \quad \cos\theta = \sin^2\theta$

$\Rightarrow \quad \cos^2\theta = \sin^4\theta$

Now, $\sin^2\theta + \sin^4\theta = \cos\theta + \cos^2 = 1$

[On the above line, if $\sin\theta + \sin^2\theta = 1$, then $\cos^2\theta + \cos^4\theta = 1$]

65. $\operatorname{cosec}^2\theta - 2 + \sin^2\theta = \operatorname{cosec}^2\theta - 2\operatorname{cosec}\theta\sin\theta + \sin^2\theta$

$= (\text{cosec}\,\theta - \sin\theta)^2$

Obviously, value of $(\text{cosec}\,\theta - \sin\theta)^2$ will be non-negative.

Alternatively

We know that $(\sin^2\theta + \text{cosec}^2\theta) \geq 2$, irrespectivey of nature of value (positive/negative) of trigonometric ratios. Then, $\text{cosec}^2\theta - 2 + \sin^2\theta$ will obviously be non-negative.

66. We know that maximum value of $(\sin\theta + \cos\theta)$ is $+\sqrt{2}$. Then, its square cannot be greater than 2. Hence, option (a) is incorrect. Since θ and Φ are lying between 0° and 90°, the value of $\sin\theta$ and $\cos\Phi$ each will be between 0 and 1. It means, $\sin^2\theta < 1$ and $\cos^2\Phi < 1$ Hence, $(\sin^2\theta + \cos^2\Phi) < 2$.

67. Given,

$$(\sin\alpha + \text{cosec}\,\alpha)^2 + (\cos\alpha + \sec\alpha)^2 = k + \tan^2\alpha + \cot^2\alpha$$

$$\Rightarrow \sin^2\alpha + \text{cosec}^2\alpha + 2 + \cos^2\alpha + \sec^2\alpha + 2 = k + \tan^2\alpha + \cot^2\alpha$$

$$\Rightarrow (\sin^2\alpha + \cos^2\alpha) + (1 + \cot^2\alpha) + (1 + \tan^2\alpha) + 4 = k + \tan^2\alpha + \cot^2\alpha$$

$$\Rightarrow \quad 7 = k$$

68. $(\sin\theta - \cos\theta)^2 = \sin^2\theta + \cos^2\theta - 2\sin\theta\cos\theta = 1 - 2 \times \frac{1}{2} = 0$

Then, $\sin\theta - \cos\theta = 0$

69. Given, $\tan^2 y\,\text{cosec}^2 x - 1 = \tan^2 y$

$$\Rightarrow \quad \tan^2 y(\text{cosec}^2 x - 1) = 1$$

$$\Rightarrow \quad \tan^2 y \cot^2 x = 1$$

$$\Rightarrow \quad \tan^2 y = \frac{1}{\cot^2 x}$$

$$\Rightarrow \quad \tan^2 y = \tan^2 x$$

$$\therefore \quad y = x$$

Then, $x - y = 0$

70. $$\frac{1}{\text{cosec}\,\theta - \cot\theta} - \frac{1}{\sin\theta}$$

$$= \frac{1}{\frac{1}{\sin\theta} - \frac{\cos\theta}{\sin\theta}} - \frac{1}{\sin\theta}$$

$$= \frac{\sin\theta}{1 - \cos\theta} - \frac{1}{\sin\theta}$$

$$= \frac{\sin^2\theta - 1 + \cos\theta}{(1 - \cos\theta)\sin\theta} = \frac{-\cos^2\theta + \cos\theta}{(1 - \cos\theta)\sin\theta}$$

$$= \frac{(1 - \cos\theta)\cos\theta}{(1 - \cos\theta)\sin\theta} = \cot\theta$$

Alternatively

Let θ be 30°, then

$$\frac{1}{\text{cosec}\,30° - \cot 30°} - \frac{1}{\sin 30°}$$

$$= \frac{1}{2 - \sqrt{3}} - \frac{1}{\frac{1}{2}} = \frac{1}{2 - \sqrt{3}} - 2$$

$$= \frac{(2 + \sqrt{3})}{(2 + \sqrt{3})(2 - \sqrt{3})} - 2 = \sqrt{3}$$

$$= \cot 30° = \cot\theta$$

[You are free to assume any value for θ at which the given trigonometric ratios are defined]

71. Given, $\text{cosec}^4\alpha = 17 + \cot^4\alpha$

$$\Rightarrow \text{cosec}^4\alpha - \cot^4\alpha = 17$$

$$\Rightarrow (\text{cosec}^2\alpha - \cot^2\alpha)(\text{cosec}^2\alpha + \cot^2\alpha) = 17$$

$$\Rightarrow \quad 1 \cdot \left(\frac{1 + \cos^2\alpha}{\sin^2\alpha}\right) = 17$$

$$\Rightarrow \quad 1 + \cos^2\alpha = 17\sin^2\alpha$$

$$\Rightarrow \quad 1 + (1 - \sin^2\alpha) = 17\sin^2\alpha$$

$$\Rightarrow \quad 2 = 18\sin^2\alpha \Rightarrow \sin^2\alpha = \frac{1}{9}$$

$$\Rightarrow \quad \sin\alpha = \sqrt{\frac{1}{9}} = \pm\frac{1}{3}$$

Since α is an acute angle, $\sin\alpha = \frac{1}{3}$

72. Given, $x + \frac{1}{x} = 2\cos\alpha$

On squaring both sides,

$$x^2 + \frac{1}{x^2} + 2 = 4\cos^2\alpha$$

$$\Rightarrow \quad x^2 + \frac{1}{x^2} = 4\cos^2\alpha - 2$$

$$\Rightarrow \quad x^2 + \frac{1}{x^2} = 4\cos^2\alpha - 2(\cos^2\alpha + \sin^2\alpha)$$

$$= 2\cos^2\alpha - 2\sin^2\alpha$$

Alternatively

Let α be 0°, then $x + \frac{1}{x} = 2$ i.e. $x = 1$

Further, $x^2 + \frac{1}{x^2} = 2$

Now, match this resultant with results of given options considering α be 0°.

73. Given, $\tan A + \cot A = 4$

$$\Rightarrow (\tan A + \cot A)^2 = 16$$

$$\Rightarrow \tan^2 A + \cot^2 A + 2\tan A \cot A = 16$$

$$\Rightarrow \tan^2 A + \cot^2 A = 14$$

$$\Rightarrow (\tan^2 A + \cot^2 A)^2 = 196$$

$$\Rightarrow \tan^4 A + \cot^4 A + 2\tan^2 A \cdot \cot^2 A = 196$$

$$\Rightarrow \tan^4 A + \cot^4 A = 194$$

74. $\sin 190° = \sin(180° + 10°) = -\sin 10°$

$\sin 200° = \sin(180° + 20°) = -\sin 20°$

....

$\sin 270° = \sin(180° + 90°) = -\sin 90°$

$\sin 280° = \sin(180° + 100°) = -\sin 100°$

....

$\sin 360° = \sin(180° + 180°) = -\sin 180°$

Now, the expression be :

$\sin 10° + \sin 20° + \sin 30° + \ldots + \sin 180° - \sin 10° - \sin 20° - \sin 30° - \ldots\ldots - \sin 180° = 0$

75. We know that at angle of 45°, sine and cosine have equal value. Hence, $x = y = 45°$. Then, $x + y = \frac{\pi}{2}$.

76. Given, $\sin(x - y) = \frac{1}{2}$ and $\cos(x + y) = \frac{1}{2}$,

$$\Rightarrow \sin(x - y) = \sin 30°$$

and $\cos(x + y) = \cos 60°$

$$\Rightarrow x - y = 30° \text{ and } x + y = 60°$$

On solving both equation, $x = 45°$.

77. Given, $a(\tan\theta + \cot\theta) = 1$

$$\Rightarrow \quad a\left(\frac{\sin\theta}{\cos\theta} + \frac{\cos\theta}{\sin\theta}\right) = 1$$

$$\Rightarrow \quad a\left(\frac{\sin^2\theta + \cos^2\theta}{\cos\theta \cdot \sin\theta}\right) = 1$$

$$\Rightarrow \quad \sin\theta\cos\theta = a$$

Also given, $\sin\theta + \cos\theta = b$

On squaring both sides,

$$\sin^2\theta + \cos^2\theta + 2\sin\theta\cos\theta = b^2$$

$$\Rightarrow \quad 1 + 2a = b^2$$

78. Given, $\sin x - \cos x = 0$

$$\Rightarrow \sin x = \cos x \Rightarrow x = 45°$$

Now, $\sin^4 x + \cos^4 x$

$$= \sin^4 45° + \cos^4 45°$$

$$= \left(\frac{1}{\sqrt{2}}\right)^4 + \left(\frac{1}{\sqrt{2}}\right)^4 = \frac{1}{2}$$

79. Given, $2y\cos\theta = x\sin\theta$

$$\Rightarrow \quad \frac{2y}{\sin\theta} = \frac{x}{\cos\theta}$$

$$\Rightarrow \quad 2y\,\text{cosec}\,\theta = x\sec\theta$$

$$\Rightarrow \quad 2y\,\text{cosec}\,\theta - x\sec\theta = 0 \quad \ldots\text{(i)}$$

But also given,

$$-y\,\text{cosec}\,\theta + 2x\sec\theta = 3 \quad \ldots\text{(ii)}$$

On solving Eqs. (i) and (ii) we get

$x = 2\cos\theta$ and $y = \sin\theta$

Now, $x^2 + 4y^2 = (2\cos\theta)^2 + 4(\sin\theta)^2$
$= 4(\sin^2\theta + \cos^2\theta) = 4$

Alternatively
Let θ be 30° (you are free to assume any value of θ)
Then, $2y\cos 30° = x\sin 30°$

$\Rightarrow \sqrt{3}y = \frac{x}{2} \Rightarrow x = 2\sqrt{3}y$

Further, $2x\sec 30° - y\,\text{cosec}\,30° = 3$

$\Rightarrow \frac{4x}{\sqrt{3}} = 3 + 2y$

$\Rightarrow \frac{4\times 2\sqrt{3}y}{\sqrt{3}} = 3 + 2y \quad [\because x = 2\sqrt{3}y]$

$\Rightarrow 6y = 3 \Rightarrow y = \frac{1}{2}$ then,

$x = 2\sqrt{3}\times\frac{1}{2} = \sqrt{3}$

Now, $x^2 + 4y^2 = (\sqrt{3})^2 + 4\left(\frac{1}{2}\right)^2 = 4$

80. Given, $\tan x = \frac{3}{4}$, it means perpendicular (p) = 3 unit and base (b) = 4 units, then, hypotenuse (h) = 5

Now, $\sin x\cos x = \frac{3}{5}\times\frac{4}{5} = \frac{12}{25}$

81. Let $A = 30°$ [you are free to assume any value of θ such that the value of T-ratios must be defined]

Then, $\frac{[\tan(90° - A) + \cot(90° - A)]^2}{2\sec^2(90° - 2A)}$

$= \frac{[\tan 60° + \cot 60°]^2}{2\sec^2 30°}$

$= \frac{\left(\sqrt{3} + \frac{1}{\sqrt{3}}\right)^2}{2\left(\frac{2}{\sqrt{3}}\right)^2} = 2$

Alternatively
Assume any value of A as per your convenience and solve (you are free to assume any angle for which the given T-ratios have a defined values). Thereafter, match the resultant with the values of the given options.

82. Observing the trend depicted in chart "Values of trigonometric ratios w.r.t some specific angles", we may say that $\tan x > 1$, when $45° < x < 90°$.

83. Given, $\tan\theta\cdot\tan 2\theta = 1$

$\Rightarrow \tan\theta = \cot 2\theta$

$\Rightarrow \tan\theta = \tan(90° - 2\theta)$

$\Rightarrow \theta = 90° - 2\theta$

$\Rightarrow \theta = 30°$

Now, $\sin^2 2\theta + \tan^2 2\theta$
$= \sin^2 60° + \tan^2 60° = \frac{3}{4} + 3 = 3\frac{3}{4}$

84. Proceed through options.

(a) $q + p = \frac{2\sin\theta}{1+\cos\theta+\sin\theta} + \frac{\cos\theta}{1+\sin\theta}$

$= \frac{2\sin\theta + 2\sin^2\theta + \cos\theta + \cos^2\theta + \sin\theta\cdot\cos\theta}{1+\sin\theta+\cos\theta+\sin\theta\cdot\cos\theta + \sin\theta + \sin^2\theta}$

$= \frac{2\sin\theta + \sin^2\theta + \sin^2\theta + \cos\theta + \cos^2\theta + \sin\theta\cdot\cos\theta}{1 + 2\sin\theta + \cos\theta + \sin\theta\cdot\cos\theta + \sin^2\theta}$

$= \frac{2\sin\theta + \sin^2\theta + 1 + \cos\theta + \sin\theta\cdot\cos\theta}{1 + 2\sin\theta + \cos\theta + \sin\theta\cdot\cos\theta + \sin^2\theta} = 1$

Hence, option (a) is correct.

Alternatively
Let θ be 0°.

Then, $p = \frac{2\sin 0°}{1+\cos 0° + \sin 0°} = 0$

and $q = \frac{\cos 0°}{1+\sin 0°} = 1$

Hence, $p + q = 1$ and $q - p = 1$

At $\theta = 0°$, two options are correct. Hence, we should take another value of θ (say, 30°)

Then,

$p = \frac{2\sin 30°}{1+\cos 30° + \sin 30°} = \frac{2}{3+\sqrt{3}}$

and $q = \frac{\cos 30°}{1+\sin 30°} = \frac{1}{\sqrt{3}}$

Now, $p + q = 1$, but $q - p \neq 1$

Hence, option (a) is correct.

{**Note** you are free to take any value of θ for which the given T-ratios have defined values}

85. We know that $0 \leq \sin^2 x$ or $\cos^2 x \leq 1$.

It means, $[0]^5 \leq [\sin^2 x]^5 \leq [1]^5$

$\Rightarrow 0 \leq \sin^{10} x \leq 1 \Rightarrow 0 \leq p \leq 1$

86. $\cos^3\alpha + \sec^3\alpha = (\cos\alpha + \sec\alpha)^3 - 3\cos\alpha\cdot\sec\alpha(\cos\alpha + \sec\alpha)$
$= (\sqrt{3})^3 - 3\cdot 1\cdot(\sqrt{3}) = 0$

87. I $\Rightarrow \cos\theta(1 - \sin\theta)^{-1} + \cos\theta(1 + \sin\theta)^{-1}$

$= \frac{\cos\theta}{1-\sin\theta} + \frac{\cos\theta}{1+\sin\theta}$

$= \frac{\cos\theta + \sin\theta\cos\theta + \cos\theta - \sin\theta\cos\theta}{1 - \sin^2\theta}$

$= \frac{2\cos\theta}{\cos^2\theta} = 2\sec\theta$

II $\Rightarrow \cos\theta(1 + \text{cosec}\,\theta)^{-1} + \cos\theta(\text{cosec}\,\theta - 1)^{-1}$

$\frac{\cos\theta}{1+\text{cosec}\,\theta} + \frac{\cos\theta}{\text{cosec}\,\theta - 1}$

$= \frac{-\cos\theta + \text{cosec}\,\theta\cos\theta + \cos\theta + \text{cosec}\,\theta\cos\theta}{\text{cosec}^2\theta - 1}$

$= \frac{2\,\text{cosec}\,\theta\cos\theta}{\cot^2\theta} = \frac{2\cot\theta}{\cot^2\theta} = 2\tan\theta$

Clearly, Neither I nor II is independent of θ.

88. Given, $5\tan\theta = 4 \Rightarrow \tan\theta = \frac{4}{5}$

Now, $\frac{5\sin\theta - 3\cos\theta}{5\sin\theta + 3\cos\theta}$

$= \frac{\frac{5\sin\theta}{\cos\theta} - \frac{3\cos\theta}{\cos\theta}}{\frac{5\sin\theta}{\cos\theta} + \frac{3\cos\theta}{\cos\theta}}$

[dividing each trigonometric ratio by $\cos\theta$]

$= \frac{5\tan\theta - 3}{5\tan\theta + 3} = \frac{5\times\frac{4}{5} - 3}{5\times\frac{4}{5} + 3} = \frac{1}{7}$

89. $\frac{\sin x}{1+\cos x} = \frac{1-\cos x}{\sin x}$ is possible only when each term has a definite value. But, in case denominator(s) of any/both of the terms is/are 0, they will be undefined. However, at $x = 180°$, $1 + \cos x = 0$ and also, $\sin x = 0$.

Further, at 360°, $1 + \cos x = 2$ and $\sin x = 0$

As conclusion, we may say that $\frac{\sin x}{1+\cos x} = \frac{1-\cos x}{\sin x}$, provided x is not a multiple of 180°.

90. Let θ be 30°, then
$x = \text{cosec}\,30° - \sin 30° = 2 - \frac{1}{2} = \frac{3}{2}$

and $y = \sec 30° - \cos 30°$
$= \frac{2}{\sqrt{3}} - \frac{\sqrt{3}}{2} = \frac{4-3}{2\sqrt{3}} = \frac{1}{2\sqrt{3}}$

Now, $x^2y^2(x^2 + y^2 + 3)$

$= \left(\frac{3}{2}\right)^2\left(\frac{1}{2\sqrt{3}}\right)^2\left[\left(\frac{3}{2}\right)^2 + \left(\frac{1}{2\sqrt{3}}\right)^2 + 3\right]$

$= \frac{9}{4}\times\frac{1}{12}\left[\frac{9}{4} + \frac{1}{12} + 3\right] = \frac{3}{16}\times\frac{16}{3} = 1$

{you are free to take any value of θ as per your convenience. You may try through conventional method also.}

91. Given, $\sin x + \sin^2 x = 1$

$\Rightarrow \sin x = 1 - \sin^2 x = \cos^2 x$

Now, $\cos^{12} x + 3\cos^{10} x + 3\cos^8 x + \cos^6 x - 2$

$= \sin^6 x + 3\sin^5 x + 3\sin^4 x + \sin^3 x - 2$

$= (\sin^2 x + \sin x)^3 - 2$

$= (1)^3 - 2 = -1$

92. Given expression

$\cos 1° \cdot \cos 2° \cdot \cos 3° \ldots\ldots \cos 90° \ldots \cos 179°$

$= \cos 1° \cdot \cos 2° \cdot \cos 3° \ldots\ldots \times 0 \times \ldots \cos 179° = 0$

93. Given, $\tan(A+B) = \sqrt{3}$

$\Rightarrow \tan(A+B) = \tan 60°$

$\Rightarrow A + B = 60°$

Further, $\tan A = 1$

$\Rightarrow \tan A = \tan 45°$

$\Rightarrow A = 45°$

Clearly, $B = 60° - 45° = 15°$

Now, $\tan(A - B) = \tan(45° - 15°)$

$= \tan 30° = \frac{1}{\sqrt{3}}$

94. Given, $x = a\sin\theta + b\cos\theta$ and $y = a\cos\theta - b\sin\theta$

Squaring on both sides,

$\Rightarrow x^2 = a^2\sin^2\theta + b^2\cos^2\theta + 2ab\sin\theta\cos\theta$

and $y^2 = a^2\cos^2\theta + b^2\sin^2\theta - 2ab\sin\theta\cos\theta$

$\Rightarrow$ On adding,

$x^2 + y^2 = a^2(\sin^2\theta + \cos^2\theta) + b^2(\sin^2\theta + \cos^2\theta) = a^2 + b^2$

Alternatively

Let θ be 45° (you are free to take any angle)

Then, $x = a\sin 45° + b\cos 45°$

and $y = a\cos 45° - b\sin 45°$

$\Rightarrow x = \frac{(a+b)}{\sqrt{2}}$ and $y = \frac{(a-b)}{\sqrt{2}}$

$\Rightarrow x^2 = \frac{(a+b)^2}{2}$ and $y^2 = \frac{(a-b)^2}{2}$

$\Rightarrow$ On adding, $x^2 + y^2 = a^2 + b^2$

95. Given, $\sin\theta + \cos\theta = \sqrt{2}\cos(90° - \theta)$

$\Rightarrow \sin\theta + \cos\theta = \sqrt{2}\sin\theta$

$\Rightarrow \cos\theta = \sin\theta(\sqrt{2} - 1)$

$\Rightarrow \frac{\cos\theta}{\sin\theta} = (\sqrt{2} - 1) \Rightarrow \cot\theta = \sqrt{2} - 1$

96. Given,

$\sin\theta + \cos\theta = \sqrt{2}\sin(90° - \theta)$

$\Rightarrow \sin\theta + \cos\theta = \sqrt{2}\cos\theta$

$\Rightarrow \sin\theta = \cos\theta(\sqrt{2} - 1)$

$\Rightarrow \tan\theta = (\sqrt{2} - 1)$

$\Rightarrow \cot\theta = \frac{1}{(\sqrt{2} - 1)}$

$= \frac{1 \times (\sqrt{2} + 1)}{(\sqrt{2} - 1)(\sqrt{2} + 1)} = \sqrt{2} + 1$

97. Let α be 30° and β be 45° (you are free to assume the values of angles as per your convenience)

Then, $\tan\alpha = n\tan\beta$

$\Rightarrow \tan 30° = n\tan 45°$

$\Rightarrow n = \frac{1}{\sqrt{3}}$

Also, given, $\sin\alpha = m\sin\beta$

$\Rightarrow \sin 30° = m\sin 45°$

$\Rightarrow m = \frac{1}{\sqrt{2}}$

Now, proceed through options

(a) $\frac{m^2}{n^2} = \frac{\left(\frac{1}{\sqrt{2}}\right)^2}{\left(\frac{1}{\sqrt{3}}\right)^2} = \frac{3}{2} \neq \cos^2 30°$

(b) $\frac{m^2 - 1}{n^2 - 1} = \frac{\left(\frac{1}{\sqrt{2}}\right)^2 - 1}{\left(\frac{1}{\sqrt{3}}\right)^2 - 1} = \frac{3}{4}$

$= \cos^2 30° = \cos^2\alpha$

Hence, option (b) is correct.

Alternatively

Try yourself through conventional method.

98. Given, $\sin\theta_1 + \sin\theta_2 + \sin\theta_3 = 3$

Then, $\sin\theta_1 = \sin\theta_2 = \sin\theta_3 = 1$

[$\because$ maximum value of $\sin\theta$ be 1]

$\therefore \theta_1 = \theta_2 = \theta_3 = 90°$

Now, $\cos 90° + \cos 90° + \cos 90° = 0$

99. $\frac{(\sin x + \sin y)(\sin x - \sin y)}{(\cos x + \cos y)(\cos y - \cos x)}$

$= \frac{\sin^2 x - \sin^2 y}{\cos^2 y - \cos^2 x}$

$= \frac{\sin^2 x - \sin^2 y}{1 - \sin^2 y - 1 + \sin^2 x} = 1$

Alternatively

Assume any value of A as per your convenience and solve (you are free to assume any angles for which the given T-ratios have a defined value). Thereafter, match the resultant with the values of the given options.

100. Given, $\sec\theta(\cos\theta + \sin\theta) = \sqrt{2}$

$\Rightarrow \cos\theta + \sin\theta = \sqrt{2}\cos\theta$

$\Rightarrow \sin\theta = \cos\theta(\sqrt{2} - 1)$

Now,

$\frac{2\sin\theta}{\cos\theta - \sin\theta} = \frac{2\cos\theta(\sqrt{2} - 1)}{\cos\theta - \cos\theta(\sqrt{2} - 1)}$

$= \frac{2\cos\theta(\sqrt{2} - 1)}{\cos\theta(1 - \sqrt{2} + 1)} = \frac{2(\sqrt{2} - 1)}{\sqrt{2}(\sqrt{2} - 1)} = \sqrt{2}$

101. Since, AM $\geq$ GM,

$\frac{4\sin^2\theta + 1}{2} \geq \sqrt{4\sin^2\theta \times 1}$

$\Rightarrow 4\sin^2\theta + 1 \geq 4\sin\theta$

Hence, option (b) is correct.

102.

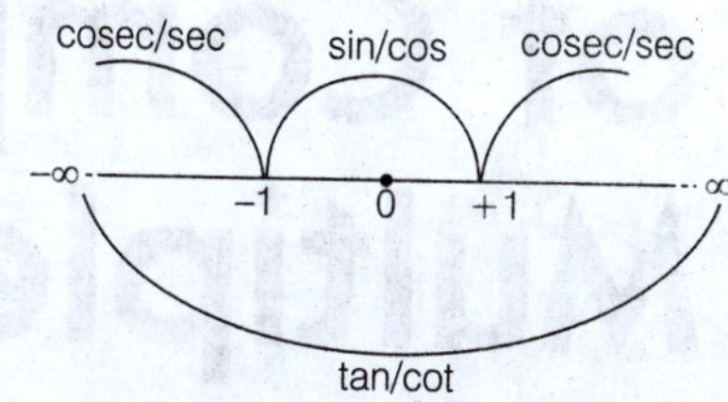

It is clear from the above that Statements I, III and IV are not possible. Thus, option (c) is correct.

103. Given, $x = 4\cos A + 5\sin A$ and $y = 4\sin A - 5\cos A$

Squaring on both sides,

$\Rightarrow x^2 = 16\cos^2 A + 25\sin^2 A + 2\cdot4\cdot5\cos A\sin A$

and $y^2 = 16\sin^2 A + 25\cos^2 A - 2\cdot4\cdot5\cos A\sin A$

On adding,

$x^2 + y^2 = 16(\sin^2 A + \cos^2 A) + 25(\sin^2 A + \cos^2 A) = 41$

Alternatively

You may compute the value of the given expression assuming any value of A. However, you have to keep in mind that the value of given T-ratios at A must be defined.

104. Given expression

$\text{cosec}(85° + \theta) - \sec(5° - \theta) - \tan(55° + \theta) + \cot(35° - \theta)$

$= \sec(90° - 85° - \theta) - \sec(5° - \theta) - \tan(55° + \theta) + \tan(90° - 35° + \theta)$

$= 0$

Alternatively

Let θ be 5° (you are free to assume any value of θ for which T-ratios have defined value).

Then, expression

$= \text{cosec}(85° + 5°) - \sec(5° - 5°) - \tan(55° + 5°) + \cot(35° - 5°)$

$= 1 - 1 - \sqrt{3} + \sqrt{3} = 0$

50

Part-II

Trigonometric Ratios of Compound and Multiple Angles

Box A

Addition Formulae

$$\sin(A+B) = \sin A\cdot\cos B + \cos A\cdot\sin B$$
$$\cos(A+B) = \cos A\cdot\cos B - \sin A\cdot\sin B$$
$$\tan(A+B) = \frac{\tan A + \tan B}{1 - \tan A\cdot\tan B}$$

Difference Formulae

$$\sin(A-B) = \sin A\cdot\cos B - \cos A\cdot\sin B$$
$$\cos(A-B) = \cos A\cdot\cos B + \sin A\cdot\sin B$$
$$\tan(A-B) = \frac{\tan A - \tan B}{1 + \tan A\cdot\tan B}$$

Box B

Now, $\cot(A-B) = \dfrac{\cos(A-B)}{\sin(A-B)} = \dfrac{\cos A\cdot\cos B + \sin A\cdot\sin B}{\sin A\cdot\cos B - \cos A\cdot\sin B}$

$$= \frac{\dfrac{\cos A\cos B}{\sin A\sin B} + 1}{\dfrac{\cos B}{\sin B} - \dfrac{\cos A}{\sin A}}$$

[dividing numerator and denominator by $\sin A\cdot\sin B$]

$$= \frac{\cot A\cdot\cot B + 1}{\cot B - \cot A}$$

Similarly, $\cot(A+B) = \dfrac{\cot A\cdot\cot B - 1}{\cot A + \cot B}$

Inside

- *Addition and Difference Formulae (Boxes-A and B)*
- *Transformation Formulae (Boxes-C and D)*
- *Formulae for Trigonometric Ratios of Multiple Angles (Boxes-E and F)*
- *Illustrations Based on Above Boxes*
- ....
- .... *& Many More*
- *Work Out with Solutions*

Box C

Formulae will be derived from the Box A.

$\sin(A+B)\cdot\sin(A-B)$

$= \{\sin A\cdot\cos B + \cos A\cdot\sin B\} \times \{\sin A\cdot\cos B - \cos A\cdot\sin B\}$

$= (\sin A\cdot\cos B)^2 - (\cos A\cdot\sin B)^2$

$= \sin^2 A(1-\sin^2 B) - (1-\sin^2 A)\cdot\sin^2 B$

$= \sin^2 A - \sin^2 A\cdot\sin^2 B - \sin^2 B + \sin^2 A\cdot\sin^2 B$

$= \sin^2 A - \sin^2 B$

$= (1-\cos^2 A) - (1-\cos^2 B)$

$= \cos^2 B - \cos^2 A$

Hence, $\mathbf{\sin(A+B)\cdot\sin(A-B) = \sin^2 A - \sin^2 B = \cos^2 B - \cos^2 A}$

On the same line,

$\mathbf{\cos(A+B)\cdot\cos(A-B) = \cos^2 A - \sin^2 B = \cos^2 B - \sin^2 A}$

Box D

(i) $\sin(A+B) + \sin(A-B)$

$= \sin A\cdot\cos B + \cos A\cdot\sin B + \sin A\cdot\cos B - \cos A\cdot\sin B$

$= 2\sin A\cdot\cos B$

If it is assumed that $A+B = C$ and $A-B = D$, then

$\sin C + \sin D = 2\sin\left(\frac{C+D}{2}\right)\cdot\cos\left(\frac{C-D}{2}\right)$

Similarly,

(ii) $\sin(A+B) - \sin(A-B) = 2\cos A\cdot\sin B$

Or, $\sin C - \sin D = 2\cos\left(\frac{C+D}{2}\right)\cdot\sin\left(\frac{C-D}{2}\right)$

(iii) $\cos(A+B) + \cos(A-B) = 2\cos A\cdot\cos B$

Or, $\cos C + \cos D = 2\cos\left(\frac{C+D}{2}\right)\cdot\cos\left(\frac{C-D}{2}\right)$

(iv) $\cos(A+B) - \cos(A-B) = -2\sin A\cdot\sin B$

Or, $\cos C - \cos D = -2\sin\left(\frac{C+D}{2}\right)\cdot\sin\left(\frac{C-D}{2}\right)$

Box E

We know, $\sin(A+B) = \sin A\cdot\cos B + \cos A\cdot\sin B$

Then, $\sin(A+A) = \sin A\cdot\cos A + \cos A\cdot\sin A$

$\Rightarrow \mathbf{\sin 2A = 2\sin A\cdot\cos A}$

$$= \frac{2\sin A\cdot\cos A}{\sin^2 A + \cos^2 A} = \frac{\frac{(2\sin A\cdot\cos A)}{\cos^2 A}}{\frac{(\sin^2 A + \cos^2 A)}{\cos^2 A}}$$

$$= \frac{2\tan A}{1+\tan^2 A}$$

Similarly,

(i) $\cos 2A = \cos^2 A - \sin^2 A = 2\cos^2 A - 1 = 1 - 2\sin^2 A = \dfrac{1-\tan^2 A}{1+\tan^2 A}$

(ii) $\tan 2A = \dfrac{2\tan A}{1-\tan^2 A}$

(iii) $\cot 2A = \dfrac{\cot^2 A - 1}{2\cot A}$

Box F

We know, $\sin(A+B) = \sin A\cdot\cos B + \cos A\cdot\sin B$

Then, $\sin(2A+A) = \sin 2A\cdot\cos A + \cos 2A\cdot\sin A$

$\Rightarrow \sin 3A = 2\sin A\cdot\cos A\cdot\cos A + (1-2\sin^2 A)\cdot\sin A$

$= 2\sin A(1-\sin^2 A) + \sin A - 2\sin^3 A$

$= 2\sin A - 2\sin^3 A + \sin A - 2\sin^3 A$

$= 3\sin A - 4\sin^3 A$

Hence, $\mathbf{\sin 3A = 3\sin A - 4\sin^3 A}$

Similarly,

(i) $\cos 3A = 4\cos^3 A - 3\cos A$

(ii) $\tan 3A = \dfrac{3\tan A - \tan^3 A}{1 - 3\tan^2 A}$

Illustrations

Problems based on Boxes A and B

Problem 1. Find the value of sin 75°.

Sol. $\sin 75° = \sin(45° + 30°)$

$= \sin 45°\cdot\cos 30° + \cos 45°\cdot\sin 30°$

$= \frac{1}{\sqrt{2}}\cdot\frac{\sqrt{3}}{2} + \frac{1}{\sqrt{2}}\cdot\frac{1}{2} = \frac{\sqrt{3}+1}{2\sqrt{2}}$

Problem 2. Find the value of cos 15°.

Sol. $\cos 15° = \cot(45° - 30°)$

$= \cos 45°\cdot\cos 30° + \sin 45°\cdot\sin 30°$

$= \frac{1}{\sqrt{2}}\cdot\frac{\sqrt{3}}{2} + \frac{1}{\sqrt{2}}\cdot\frac{1}{2} = \frac{\sqrt{3}+1}{2\sqrt{2}}$

[Remarks : $\cos 15° = \sin 75° = \frac{\sqrt{3}+1}{2\sqrt{2}}$]

Problem 3. Find the value of cot 105°.

Sol. $\cot 105° = \cot(60° + 45°)$

$= \frac{\cot 60°\cdot\cot 45° - 1}{\cot 60° + \cot 45°}$

$= \frac{\frac{1}{\sqrt{3}}\cdot 1 - 1}{\frac{1}{\sqrt{3}} + 1} = \frac{1-\sqrt{3}}{1+\sqrt{3}}$

Problem 4. In a ΔABC, the value of $\tan A + \tan B + \tan C = ?$

(a) $\tan A + \tan B \cdot \tan C$
(b) $\tan A \cdot \tan B \cdot \tan C$
(c) $\tan(A+B+C)$
(d) $\cot(A+B+C)$

Sol. We know, $A + B + C = 180°$

$$\Rightarrow A + B = 180° - C$$
$$\Rightarrow \tan(A+B) = \tan(180° - C)$$
$$\Rightarrow \frac{\tan A + \tan B}{1 - \tan A \cdot \tan B} = -\tan C$$
$$\Rightarrow \tan A + \tan B = -\tan C + \tan A \cdot \tan B \cdot \tan C$$
$$\Rightarrow \tan A + \tan B + \tan C = \tan A \cdot \tan B \cdot \tan C$$

Problem 5. If $\sin A = \frac{12}{13}$ and $\sin B = \frac{15}{17}$, then $\sin(A+B) = ?$, where A and B both are acute angles.

Sol. $\sin(A+B) = \sin A \cdot \cos B + \cos A \cdot \sin B$

$$\Rightarrow \sin(A+B) = \frac{12}{13}\sqrt{1-\left(\frac{15}{17}\right)^2} + \sqrt{1-\left(\frac{12}{13}\right)^2} \cdot \frac{15}{17}$$
$$= \frac{12}{13} \cdot \frac{8}{17} + \frac{5}{13} \cdot \frac{15}{17} = \frac{171}{221}$$

Problem 6. $\frac{\cos 9° + \sin 9°}{\cos 9° - \sin 9°} = ?$

(a) $\tan 54°$ (b) $\tan 36°$
(c) $\tan 18°$ (d) None of these

Sol. $$\frac{\cos 9° + \sin 9°}{\cos 9° - \sin 9°} = \frac{\frac{(\cos 9° + \sin 9°)}{\cos 9°}}{\frac{(\cos 9° - \sin 9°)}{\cos 9°}}$$
$$= \frac{1 + \tan 9°}{1 - \tan 9°}$$
$$= \frac{\tan 45° + \tan 9°}{1 - \tan 45° \cdot \tan 9°} = \tan(45° + 9°)$$
$$= \tan 54°$$

Problem 7. Find the value of $\tan 35° + \tan 10° + \tan 35° \cdot \tan 10°$.

Sol. $$\tan(35° + 10°) = \frac{\tan 35° + \tan 10°}{1 - \tan 35° \cdot \tan 10°}$$
$$\Rightarrow 1 = \frac{\tan 35° + \tan 10°}{1 - \tan 35° \cdot \tan 10°}$$
$$\Rightarrow \tan 35° + \tan 10° = 1 - \tan 35° \cdot \tan 10°$$
$$\Rightarrow \tan 35° + \tan 10° + \tan 35° \cdot \tan 10° = 1$$

Problem 8. If $\tan\alpha = \frac{m}{m+1}$ and $\tan\beta = \frac{1}{2m+1}$, then $\alpha + \beta = ?$

Sol. We know, $\tan(\alpha+\beta) = \frac{\tan\alpha + \tan\beta}{1 - \tan\alpha \cdot \tan\beta}$

$$\Rightarrow \tan(\alpha+\beta) = \frac{\frac{m}{m+1} + \frac{1}{2m+1}}{1 - \frac{m}{m+1} \cdot \frac{1}{2m+1}}$$
$$= \frac{2m^2 + 2m + 1}{2m^2 + 2m + 1} = 1$$
$$\Rightarrow \tan(\alpha+\beta) = \tan(45°)$$
$$\Rightarrow \alpha + \beta = \frac{\pi}{4}$$

Alternatively

Let $m = 1$, then $\tan\alpha = \frac{1}{1+1} = \frac{1}{2}$

and $\tan\beta = \frac{1}{2 \cdot 1 + 1} = \frac{1}{3}$

Now, $\tan(\alpha+\beta) = \frac{\frac{1}{2} + \frac{1}{3}}{1 - \frac{1}{2} \cdot \frac{1}{3}} = 1$

$$\Rightarrow \tan(\alpha+\beta) = \tan\left(\frac{\pi}{4}\right)$$
$$\Rightarrow (\alpha+\beta) = \frac{\pi}{4}$$

Problem 9. If $\tan A = -\frac{1}{3}$ and $\tan B = -\frac{1}{2}$, then the value of $(A+B)$ is

Sol. $$\tan(A+B) = \frac{\tan A + \tan B}{1 - \tan A \cdot \tan B}$$
$$= \frac{-\frac{1}{3} - \frac{1}{2}}{1 - \left(-\frac{1}{3}\right) \times \left(-\frac{1}{2}\right)}$$
$$= \frac{-5}{6} \times \frac{-6}{5} = 1$$

Problem 10. If $\frac{\cos 20° - \sin 20°}{\cos 20° + \sin 20°} = \tan A$, then $A = ?$

Sol. $$\frac{\frac{\cos 20° - \sin 20°}{\cos 20°}}{\frac{\cos 20° + \sin 20°}{\cos 20°}} = \tan A$$
$$\Rightarrow \frac{1 - \tan 20°}{1 + \tan 20°} = \tan A$$
$$\Rightarrow \frac{\tan 45° - \tan 20°}{1 + \tan 45° \cdot \tan 20°} = \tan A$$
$$\Rightarrow \tan(45° - 20°) = \tan A$$
$$\Rightarrow A = 25°$$

Problem 11. The value of $\cos 15° - \sin 15°$ is equal to :

Sol. $$\cos 15° - \sin 15° = \sqrt{2}\,[\sin 45° \cdot \cos 15° - \cos 45° \cdot \sin 15°]$$
$$= \sqrt{2}\,[\sin 45° \cdot \cos 15° - \cos 45° \cdot \sin 15°]$$
$$= \sqrt{2}\,[\sin(45° - 15°)] = \sqrt{2} \times \frac{1}{2} = \frac{1}{\sqrt{2}}$$

Problem 12. Find the value of $\sin\left(\theta - \frac{\pi}{4}\right) + \cos\left(\theta + \frac{\pi}{4}\right)$

Sol. Given, expression $= \sin\theta \cdot \cos 45^\circ - \cos\theta \cdot \sin 45^\circ + \cos\theta \cdot \cos 45^\circ - \sin\theta \cdot \sin 45^\circ$

$$= \sin\theta \cdot \frac{1}{\sqrt{2}} - \cos\theta \cdot \frac{1}{\sqrt{2}} + \cos\theta \cdot \frac{1}{\sqrt{2}} - \sin\theta \cdot \frac{1}{\sqrt{2}} = 0$$

Alternatively

Let θ be 90°, then $\sin(90^\circ - 45^\circ) + \cos(90^\circ + 45^\circ)$
$= \sin 45^\circ - \sin 45^\circ = 0$

Remarks *You may put any angle as per your convenience*

Problem 13. If $\sin A = \frac{1}{\sqrt{10}}$ and $\sin B = \frac{1}{\sqrt{5}}$, where A and B are positive acute angles, then $A + B$ is equal to

Sol. We know, $\sin(A+B) = \sin A \cdot \cos B + \cos A \cdot \sin B$

$$\Rightarrow \sin(A+B) = \frac{1}{\sqrt{10}} \cdot \sqrt{1 - \frac{1}{5}} + \sqrt{1 - \frac{1}{10}} \cdot \frac{1}{\sqrt{5}}$$

$$\Rightarrow \sin(A+B) = \frac{1}{\sqrt{50}} \cdot (2+3) = \frac{1}{\sqrt{2}}$$

$$\Rightarrow \sin(A+B) = \sin 45^\circ$$

$$\Rightarrow A + B = 45^\circ$$

Problem 14. If $A + B = 225^\circ$, then $\frac{\cot A}{1 + \cot A} \cdot \frac{\cot B}{1 + \cot B} = ?$

Sol. $$\frac{\cot A}{1+\cot A} \cdot \frac{\cot B}{1+\cot B} = \frac{\frac{1}{\tan A}}{1 + \frac{1}{\tan A}} \cdot \frac{\frac{1}{\tan B}}{1 + \frac{1}{\tan B}}$$

$$= \frac{1}{(1+\tan A)(1+\tan B)}$$

$$= \frac{1}{1 + \tan A + \tan B + \tan A \cdot \tan B}$$

We know, $\tan(A+B) = \frac{\tan A + \tan B}{1 - \tan A \cdot \tan B}$

$$\Rightarrow \tan 225^\circ = \frac{\tan A + \tan B}{1 - \tan A \cdot \tan B} \quad [\because A + B = 225^\circ \text{ (given)}]$$

$$\Rightarrow 1 = \frac{\tan A + \tan B}{1 - \tan A \cdot \tan B}$$

$$\Rightarrow \tan A + \tan B = 1 - \tan A \cdot \tan B$$

Now, given expression $= \frac{1}{1 + \tan A + \tan B + \tan A \cdot \tan B}$

$$= \frac{1}{1 + 1 - \tan A \cdot \tan B + \tan A \cdot \tan B} = \frac{1}{2}$$

Problem 15. If $\tan A = 2$, $\tan B = 3$, then $\angle C = ?$, provided $A + B + C = \pi$.

Sol. We know, $A + B + C = 180^\circ$

$$\Rightarrow A + B = 180^\circ - C$$

$$\Rightarrow \tan(A+B) = \tan(180^\circ - C)$$

$$\Rightarrow \frac{\tan A + \tan B}{1 - \tan A \cdot \tan B} = -\tan C$$

$$\Rightarrow \tan A + \tan B = -\tan C + \tan A \cdot \tan B \cdot \tan C$$

$$\Rightarrow \tan A + \tan B + \tan C = \tan A \cdot \tan B \cdot \tan C$$

$$\Rightarrow 2 + 3 + \tan C = 2 \cdot 3 \cdot \tan C$$

$$\Rightarrow 5 \tan C = 5$$

$$\therefore \tan C = 1$$

$$\Rightarrow \tan C = \tan 45^\circ$$

$$\Rightarrow \angle C = 45^\circ$$

Alternatively

We have, $\tan A + \tan B + \tan C = \tan A \cdot \tan B \cdot \tan C$, where A, B and C are angles of a triangle.

$$\Rightarrow 2 + 3 + \tan C = 2 \cdot 3 \cdot \tan C$$

$$\Rightarrow \tan C = 1$$

$$\Rightarrow \tan C = \tan 45^\circ \Rightarrow C = 45^\circ$$

Problem 16. If $A + B + C = 90^\circ$, then $\tan A \cdot \tan B + \tan B \cdot \tan C + \tan C \cdot \tan A = ?$

Sol. Given, $A + B = 90^\circ - C$

$$\Rightarrow \tan(A+B) = \tan(90^\circ - C)$$

$$\Rightarrow \frac{\tan A + \tan B}{1 - \tan A \cdot \tan B} = \cot C$$

$$\Rightarrow \frac{\tan A + \tan B}{1 - \tan A \cdot \tan B} = \frac{1}{\tan C}$$

$$\Rightarrow \tan C \cdot \tan A + \tan B \cdot \tan C = 1 - \tan A \cdot \tan B$$

$$\Rightarrow \tan A \cdot \tan B + \tan B \cdot \tan C + \tan C \cdot \tan A = 1$$

Alternatively

Let $A = 0^\circ$, $B = 30^\circ$ and $C = 60^\circ$

Then, $\tan 0^\circ \cdot \tan 30^\circ + \tan 30^\circ \cdot \tan 60^\circ + \tan 60^\circ \cdot \tan 0^\circ$

$$= 0 + \frac{1}{\sqrt{3}} \cdot \sqrt{3} + 0 = 1$$

Remarks *you may put any angle as per your convenience.*

Problem 17. $\tan 5x \cdot \tan 3x \cdot \tan 2x = ?$

(a) $\tan 5x - \tan 3x - \tan 2x$

(b) 0

(c) $\frac{\sin 5x - \sin 3x - \sin 2x}{\cos 5x - \cos 3x - \cos 2x}$

(d) None of the above

Sol. $\tan 5x = \tan(3x + 2x) = \frac{\tan 3x + \tan 2x}{1 - \tan 3x \cdot \tan 2x}$

$$\Rightarrow \tan 5x - \tan 5x \cdot \tan 3x \cdot \tan 2x = \tan 3x + \tan 2x$$

$$\Rightarrow \tan 5x \cdot \tan 3x \cdot \tan 2x = \tan 5x - \tan 3x - \tan 2x$$

Alternatively

You may assume the values of x and simplify the expression. Thereafter, match with the options. In case two or more options yield the same result as the resultant of the given expression, you should try with another value of the angle.

Remarks *In order to solve the questions quickly, aspirants must memorise the following results*

(i) $\tan A + \tan B + \tan C = \tan A \cdot \tan B \cdot \tan C$, *if* $A + B + C = \pi$

(ii) $\tan A \cdot \tan B + \tan B \cdot \tan C + \tan C \cdot \tan A = 1$, *if* $A + B + C = \frac{\pi}{2}$

(iii) $\tan X \cdot \tan Y \cdot \tan Z = \tan X - \tan Y - \tan Z$, *if* $X = Y + Z$.

Problem 18. If $\sqrt{3}\sin\theta + \cos\theta = 2$, then $\theta = ?(0° < \theta < 90°)$ **[CDS-10]**

Sol. We can divide all the terms by 2 as

$$\sqrt{(\text{coefficient of } \sin\theta \text{ i.e. } \sqrt{3})^2 + (\text{coefficient of } \cos\theta \text{ i.e. } 1)^2}$$

is equal to 2.

We compared the given terms with $m\sin\theta + n\cos\theta$.

$\Rightarrow \quad \frac{\sqrt{3}}{2}\cdot\sin\theta + \frac{1}{2}\cdot\cos\theta = \frac{2}{2}$

$\Rightarrow \quad \cos 30°\cdot\sin\theta + \sin 30°\cdot\cos\theta = 1$

$\Rightarrow \quad \sin(\theta + 30°) = \sin 90°$ or $\sin 450°$ or ...

$\Rightarrow \quad \theta = 60°$ or $420°$ or ...

Required angle $= 60°$ as given that $0° < \theta < 90°$.

Problem 19. If $\cos\theta - \sin\theta = \frac{1}{\sqrt{2}}$, then find the value of θ.

Sol. As given, $1\cdot\cos\theta - 1\cdot\sin\theta = \frac{1}{\sqrt{2}}$

$\Rightarrow \quad \frac{1}{\sqrt{2}}\cdot\cos\theta - \frac{1}{\sqrt{2}}\cdot\sin\theta = \frac{1}{\sqrt{2}\cdot\sqrt{2}}$

[by dividing all terms by $\sqrt{2}$ as $\sqrt{(\text{coefficient of } \sin\theta \text{ i.e.} -1)^2 + (\text{coefficient of } \cos\theta \text{ i.e. } 1)^2}$ is equal to $\sqrt{2}$].

We compared the given terms with $m\sin\theta + n\cos\theta$.

$\Rightarrow \quad \sin 45°\cdot\cos\theta - \cos 45°\cdot\sin\theta = \frac{1}{2}$

$\Rightarrow \quad \sin(45° - \theta) = \sin 30°$ or $\sin 150°$

$\Rightarrow \quad 45° - \theta = 30°$ and $45° - \theta = 150°$

$\Rightarrow \quad \theta = 15°$ or $-105°$

$\Rightarrow \quad \theta = 15°$ or $255°$

[$\because -105° = -105° + 360°$, recall the concept of direction of rotation]

Problem Based on Boxes C and D

Problem 20. Find the value of $\cot\left(\frac{\pi}{4}+\theta\right)\cdot\cot\left(\frac{\pi}{4}-\theta\right)$.

Sol. $\dfrac{\cos\left(\frac{\pi}{4}+\theta\right)}{\sin\left(\frac{\pi}{4}+\theta\right)} \times \dfrac{\cos\left(\frac{\pi}{4}-\theta\right)}{\sin\left(\frac{\pi}{4}-\theta\right)}$

$$= \frac{\cos^2\frac{\pi}{4} - \sin^2\theta}{\sin^2\frac{\pi}{4} - \sin^2\theta} = \frac{\left(\frac{1}{\sqrt{2}}\right)^2 - \sin^2\theta}{\left(\frac{1}{\sqrt{2}}\right)^2 - \sin^2\theta} = 1$$

Alternatively Let θ be 180°.

Then, $\cot\left(\frac{\pi}{4} + 180°\right)\cdot\cot\left(\frac{\pi}{4} - 180°\right)$

$= \cot\frac{\pi}{4}\cdot\cot(-135°) = \cot\frac{\pi}{4}\cdot(-\cot 135°)$

$= -\cot\frac{\pi}{4}\cdot\cot(180° - 45°)$

$= \cot\frac{\pi}{4}\cdot\cot 45° = 1\cdot 1 = 1.$

Remarks *you may put any angle as per your convenience*

Problem 21. Find the maximum value of $\cos A\cdot\sin B$, if $A + B = \frac{\pi}{2}$.

Sol. We have, $\sin(A + B) - \sin(A - B) = 2\cos A\cdot\sin B$

$\Rightarrow \quad \sin 90° - \sin(A - B) = 2\cos A\cdot\sin B$

$\Rightarrow \quad \frac{1 - \sin(A - B)}{2} = \cos A\cdot\sin B$

Obviously, $\cos A\cdot\sin B$ will take maximum value if value of $\sin(A - B)$ is minimum.

Hence, maximum value of $\cos A\cdot\sin B = \frac{1-(-1)}{2} = 1$

[$\because$ minimum value of $\sin(A - B)$ is -1]

Problem 22. Find the value of $\sin 20°\cdot\sin 40°\cdot\sin 60°\cdot\sin 80°$.

Sol. Expression : $\sin 20°\cdot\sin 40°\cdot\sin 60°\cdot\sin 80°$

$= \sin 60°\cdot\frac{1}{2}[2\sin 40°\sin 80°]\cdot\sin 20°$

$= \frac{\sqrt{3}}{2}\cdot\frac{1}{2}\cdot[\cos(40° - 80°) - \cos(40° + 80°)]\cdot\sin 20°$

$= \frac{\sqrt{3}}{4}\times[\cos 40° - \cos 120°]\cdot\sin 20°$

$= \frac{\sqrt{3}}{4}\times[\sin 20°\cdot\cos 40° - \cos 120°\cdot\sin 20°]$

$= \frac{\sqrt{3}}{4\times 2}\times[2\sin 20°\cdot\cos 40° - 2\cos 120°\cdot\sin 20°]$

$= \frac{\sqrt{3}}{8}\cdot[\sin(20° + 40°) + \sin(20° - 40°)$

$- 2\cos(180° - 60°)\cdot\sin 20°]$

$= \frac{\sqrt{3}}{8}\times\left[\frac{\sqrt{3}}{2} - \sin 20° - 2\cdot(-\cos 60°)\cdot\sin 20°\right]$

$= \frac{\sqrt{3}}{8}\times\left[\frac{\sqrt{3}}{2} - 0\right] = \frac{3}{16}$

Problem 23. If $\sin 5\theta + \sin 2\theta = 0$, then how many values of θ are possible provided $0° \le \theta \le 90°$.

Sol. Given, $\sin 5\theta + \sin 2\theta = 0$

$\Rightarrow \quad 2\sin\frac{5\theta + 2\theta}{2}\times\cos\frac{5\theta - 2\theta}{2} = 0$

$\Rightarrow \quad \sin\frac{7\theta}{2}\times\cos\frac{3\theta}{2} = 0$

$\therefore \quad \sin\frac{7\theta}{2} = 0$ or $\cos\frac{3\theta}{2} = 0$

When $\sin\frac{7\theta}{2} = 0$

$\Rightarrow \quad \sin\frac{7\theta}{2} = \sin 0°/\sin 180°/\sin 360°/\sin 540°/...$

$\Rightarrow \quad \frac{7\theta}{2} = 0°/180°/360°/540°/...$

$\Rightarrow \quad \theta = 0°/51\frac{3}{7}°/102\frac{6°}{7}/....$

When $\cos\frac{3\theta}{2} = 0$

$\Rightarrow \quad \cos\frac{3\theta}{2} = \cos 90°/\cos 270°/....$

$\Rightarrow \quad \frac{3\theta}{2} = 90°/270°/....$

$\Rightarrow \quad \theta = 60°/180°/...$

Since $0° \le \theta \le 90°$, $\theta = 0°, 51\frac{3}{7}°$ and $60°$.

Hence, there are three possible value of θ.

Problem 24. The value of $\cos 52° + \cos 68° + \cos 172°$ is :

Sol. $(\cos 52° + \cos 172°) + \cos 68°$

$= 2\cos 112° \cdot \cos 60° + \cos 68°$

$= \cos 112° + \cos 68° = 2\cos 90° \cdot \cos 22° = 0$ $[\because \cos 90° = 0]$

Problem 25. The value of $\sin 75° + \cos 75°$ is

Sol. $\sin 75° + \cos 75° = \sin 75° + \sin 15°$

$= 2\sin\left(\frac{75° + 15°}{2}\right) \cdot \cos\left(\frac{75° - 15°}{2}\right)$

$= 2\sin 45° \cdot \cos 30° = \sqrt{\frac{3}{2}}$

Problem 26. If $\sin A + \sin B = C$, $\cos A + \cos B = D$, then the value of $\sin(A + B) = ?$

(a) $\frac{CD}{C^2 + D^2}$ (b) $\frac{2CD}{C^2 + D^2}$

(c) $\frac{C^2 + D^2}{2CD}$ (d) CD

Sol. Given, $\frac{\sin A + \sin B}{\cos A + \cos B} = \frac{C}{D}$

$\Rightarrow \quad \frac{2\sin\left(\frac{A+B}{2}\right) \cdot \cos\left(\frac{A-B}{2}\right)}{2\cos\left(\frac{A+B}{2}\right) \cdot \cos\left(\frac{A-B}{2}\right)} = \frac{C}{D}$

$\Rightarrow \quad \tan\left(\frac{A+B}{2}\right) = \frac{C}{D}$

Now, $\sin(A + B) = \sin\left[2 \cdot \left(\frac{A+B}{2}\right)\right]$

$= \frac{2\tan\left(\frac{A+B}{2}\right)}{1 + \tan^2\left(\frac{A+B}{2}\right)} = \frac{2 \cdot \frac{C}{D}}{1 + \left(\frac{C}{D}\right)^2} = \frac{2CD}{C^2 + D^2}$

Problems Based on Boxes E and F

Problem 27. If $\sin\theta = \frac{8}{17}$ and $0° < \theta < 90°$, then find the value of $\sin 2\theta$ and $\tan 2\theta$.

Sol. We know, $\sin^2\theta + \cos^2\theta = 1$

$\Rightarrow \quad \cos^2\theta = 1 - \left(\frac{8}{17}\right)^2 = \frac{225}{289}$

$\therefore \quad \cos\theta = \pm\frac{15}{17}$

Since θ is an acute angle, $\cos\theta = +\frac{15}{17}$

Then, $\tan\theta = \frac{\sin\theta}{\cos\theta} = \frac{8}{15}$

Now, $\sin 2\theta = 2\sin\theta \cdot \cos\theta = 2 \cdot \frac{8}{17} \cdot \frac{15}{17} = \frac{240}{289}$

and $\tan 2\theta = \frac{2\tan A}{1 - \tan^2 A} = \frac{2 \cdot \frac{8}{15}}{1 - \left(\frac{8}{15}\right)^2} = 1\frac{79}{161}$

Problem 28. Find the value of $\cot\theta - \tan\theta - 2\cot 2\theta$.

Sol. $\frac{\cos\theta}{\sin\theta} - \frac{\sin\theta}{\cos\theta} - 2\cot 2\theta$

$= \frac{\cos^2\theta - \sin^2\theta}{\sin\theta \cdot \cos\theta} - 2\cot 2\theta$

$= \frac{\cos 2\theta}{\sin\theta \cdot \cos\theta} - 2\cot 2\theta$

$= \frac{2\cos 2\theta}{2\sin\theta \cdot \cos\theta} - 2\cot 2\theta$

$= \frac{2\cos 2\theta}{\sin 2\theta} - 2\cot 2\theta$

$= 2\cot 2\theta - 2\cot 2\theta = 0$

Alternatively

You may assume the values of θ and simplify the expression. Thereafter match with the options. In case two or more options yield the same result as the resultant of the given expression, you should try with another value of the angle.

Problem 29. Find the value of $\sin 22\frac{1}{2}°$.

Sol. We know, $\cos 2\theta = 1 - 2\sin^2\theta$

$\Rightarrow \cos\left(2 \times 22\frac{1}{2}°\right) = 1 - 2\sin^2 22\frac{1}{2}°$

$\Rightarrow \quad \cos 45° = 1 - 2\sin^2 22\frac{1}{2}°$

$\Rightarrow \quad 2\sin^2 22\frac{1}{2}° = 1 - \frac{1}{\sqrt{2}}$

$\Rightarrow \quad \sin^2 22\frac{1}{2}° = \frac{\sqrt{2} - 1}{2\sqrt{2}}$

$\Rightarrow \quad \sin 22\frac{1}{2}° = \sqrt{\frac{\sqrt{2} - 1}{2\sqrt{2}}}$

$= \sqrt{\frac{\sqrt{2}(\sqrt{2} - 1)}{\sqrt{2} \cdot 2\sqrt{2}}} = \frac{1}{2}\sqrt{2 - \sqrt{2}}$

Problem 30. Find the value of $\operatorname{cosec} 10° - \sqrt{3}\sec 10°$.

Sol. $\frac{1}{\sin 10°} - \frac{\sqrt{3}}{\cos 10°} = \frac{\cos 10° - \sqrt{3}\sin 10°}{\sin 10° \cdot \cos 10°}$

$= \frac{2\left[\frac{1}{2} \cdot \cos 10° - \frac{\sqrt{3}}{2}\sin 10°\right]}{\sin 10° \cdot \cos 10°}$

$= \frac{2[\cos 60° \cdot \cos 10° - \sin 60° \cdot \sin 10°]}{\sin 10° \cdot \cos 10°}$

$= \frac{2\cos(60 + 10)°}{\sin 10° \cos 10°} = \frac{2 \cdot 2\cos 70°}{2 \cdot \sin 10° \cos 10°} = \frac{4\cos 70°}{\sin 20°}$

$= \frac{4\sin(90° - 70°)}{\sin 20°} = 4$

Problem 31. $\dfrac{1}{\sin 10^\circ} - \dfrac{\sqrt{3}}{\cos 10^\circ} = ?$

$$\frac{1}{\sin 10^\circ} - \frac{\sqrt{3}}{\cos 10^\circ} = \frac{1\cos 10^{\circ\circ} - \sqrt{3}\sin 10^\circ}{\sin 10^\circ \cos 10^\circ}$$

$$= \frac{2\left(\frac{1}{2}\cos 10^\circ - \frac{\sqrt{3}}{2}\sin 10^\circ\right)}{\frac{2\sin 10^\circ \cos 10^\circ}{2}}$$

$$= \frac{2(\sin 30^\circ \cos 10^\circ - \cos 30^\circ \sin 10^\circ)}{\frac{2\sin 10^\circ \cos 10^\circ}{2}}$$

$$= \frac{2\sin(30^\circ - 10^\circ)}{\frac{\sin 2\cdot 10^\circ}{2}} = 4$$

Problem 32. Find the value of $\sin 3A$, if $\cos A = \dfrac{24}{25}$.

Sol. We know, $\sin^2\theta + \cos^2\theta = 1$

$$= \sin^2\theta = 1 - \left(\frac{24}{25}\right)^2 = \frac{49}{625}$$

$\therefore \quad \sin\theta = \pm\dfrac{7}{25}$

Since value of θ is not specified, we may assume it as an acute angle. Thus, $\sin\theta = \dfrac{7}{25}$

Now, $\sin 3A = 3\sin A - 4\sin^3 A$

$$\Rightarrow \quad \sin 3A = 3\times\frac{7}{25} - 4\times\left(\frac{7}{25}\right)^3 = \frac{11753}{15625}$$

Problem 33. If $\cos A = \dfrac{3}{4}$, then numerical value of

$$32\sin\frac{A}{2}\cdot\cos\frac{5A}{2} = ?$$

Sol. Given, $\cos A = \dfrac{3}{4}$

$\therefore \quad \sin A = \dfrac{\sqrt{7}}{4}$

Now, $32\sin\dfrac{A}{2}\cdot\cos\dfrac{5A}{2}$

$$= 16\left[2\sin\frac{3A-2A}{2}\cdot\cos\frac{3A+2A}{2}\right]$$

$$= 16[\sin 3A - \sin 2A]$$

$$= 16[3\sin A - 4\sin^3 A - 2\sin A\cos A]$$

$$= 16\sin A[3 - 4\sin^2 A - 2\cos A]$$

$$= 16\times\frac{\sqrt{7}}{4}\left[3 - 4\left(\frac{\sqrt{7}}{4}\right)^2 - 2\cdot\frac{3}{4}\right]$$

$$= -\sqrt{7}$$

Problem 34. If $90^\circ < A < 180^\circ$ and $\sin A = \dfrac{4}{5}$, then $\tan\dfrac{A}{2}$ is equal to

Sol. Given, $\sin A = \dfrac{4}{5}$

$\therefore$ $\tan A = \dfrac{-4}{3}$, as A is in 2nd quadrant in which tangent is negative.

$$\Rightarrow \quad \tan\left(2\cdot\frac{A}{2}\right) = \frac{-4}{3} \Rightarrow \frac{2\tan\left(\frac{A}{2}\right)}{1-\tan^2\left(\frac{A}{2}\right)} = \frac{-4}{3}$$

$\Rightarrow \quad 6x = -4 + 4x^2 \qquad$ [let $\tan\dfrac{A}{2}$ be x]

$\Rightarrow \quad 4x^2 - 6x - 4 = 0$

$\Rightarrow \quad 4x^2 - 8x + 2x - 4 = 0$

$\Rightarrow \quad 4x(x-2) + 2(x-2) = 0$

$\Rightarrow \quad x = -\dfrac{1}{2}$ and 2

Since $\dfrac{A}{2}$ is in 1st quadrant, the value of $\tan\dfrac{A}{2}$ cannot be negative. Hence, $\tan\dfrac{A}{2} = 2$.

Problem 35. $\dfrac{\cot^2 15^\circ - 1}{\cot^2 15^\circ + 1} = ?$

Sol. $$\frac{\cot^2 15^\circ - 1}{\cot^2 15^\circ + 1} = \frac{\frac{\cos^2 15^\circ}{\sin^2 15^\circ} - 1}{\frac{\cos^2 15^\circ}{\sin^2 15^\circ} + 1} = \frac{\cos^2 15^\circ - \sin^2 15^\circ}{\cos^2 15^\circ + \sin^2 15^\circ}$$

$$= \frac{\cos(2\cdot 15^\circ)}{1} = \frac{\sqrt{3}}{2}$$

Problem 36. If $\tan A = \dfrac{1-\cos B}{\sin B}$, find the value of $\tan 2A$ in terms of $\tan B$.

Sol. Given, $$\tan A = \frac{1-\cos B}{\sin B} = \frac{1-\cos\left(2\times\frac{B}{2}\right)}{\sin\left(2\times\frac{B}{2}\right)}$$

$$= \frac{1-\left[1-2\sin^2\left(\frac{B}{2}\right)\right]}{2\sin\left(\frac{B}{2}\right)\cos\left(\frac{B}{2}\right)} = \tan\left(\frac{B}{2}\right)$$

$\Rightarrow \quad \tan A = \tan\left(\dfrac{B}{2}\right)$

$\Rightarrow \quad \tan 2A = \tan B$

WORKOUT

1. If $\sin(A+B)=1$, where $0° < B < 45°$, then what is $\cos(A-B)$ equal to? ***[CDS-14]***

(a) $\sin 2B$ (b) $\sin B$
(c) $\cos 2B$ (d) $\cos B$

2. Find the value of $\tan 105°$.

(a) $2+\sqrt{3}$ (b) $-(2+\sqrt{3})$
(c) $\sqrt{3}$ (d) $\sqrt{3}-1$

3. $\dfrac{\sin(B-C)}{\cos B\cdot \cos C}+\dfrac{\sin(C-A)}{\cos C\cdot \cos A}+\dfrac{\sin(A-B)}{\cos A\cdot \cos B}=?$

(a) 0 (b) $\cos 2B$
(c) 1 (d) $\cos B$

4. If $\sin A=\sin B$ and $\cos A=\cos B$, then which of the following is true?

(a) $\sin\left(\dfrac{A-B}{2}\right)=0$
(b) $\sin\left(\dfrac{A+B}{2}\right)=0$
(c) $\cos\left(\dfrac{A-B}{2}\right)=0$
(d) $\sin(A+B)=0$

5. If $\sin\theta+\sin 5\theta=\sin 3\theta$ and $0°<\theta<\dfrac{\pi}{2}$, then what is the value of θ in degrees? ***[SSC (CGL-T-I)-17]***

(a) 30 (b) 45
(c) 60 (d) 75

6. If $\cos\theta=\dfrac{1}{2}\left(a+\dfrac{1}{a}\right)$, then the value of $\cos 3\theta$ is :

(a) $\dfrac{1}{8}\left(a^3+\dfrac{1}{a^3}\right)$
(b) $\dfrac{3}{2}\left(a+\dfrac{1}{a}\right)$
(c) $\dfrac{1}{2}\left(a^3+\dfrac{1}{a^3}\right)$
(d) $\dfrac{1}{3}\left(a^3+\dfrac{1}{a^3}\right)$

7. If $\cot A=\dfrac{\sin B}{1-\cos B}$, then what is the value of $\cot 2A$? ***[SSC (CGL-T-I)-17]***

(a) $\cot\left(\dfrac{B}{2}\right)$ (b) $\cot 2B$
(c) $\cot B$ (d) $\tan B$

8. $\dfrac{\sin\theta+\sin 2\theta+\sin 3\theta}{\cos\theta+\cos 2\theta+\cos 3\theta}=?$

(a) $\tan 6\theta$ (b) $\tan 2\theta$
(c) $\tan 3\theta$ (d) $\tan 4\theta$

9. If θ is an acute angle and $\sin\dfrac{\theta}{2}=\sqrt{\dfrac{x-1}{2x}}$, then $\tan\theta=?$

(a) x^2-1 (b) $\sqrt{x^2-1}$
(c) $\sqrt{x^2+1}$ (d) x^2+1

10. If $\sin\alpha=-\dfrac{3}{5}$, where $\pi<\alpha<\dfrac{3\pi}{2}$, then $\cos\dfrac{\alpha}{2}=?$

(a) $\dfrac{-1}{\sqrt{10}}$ (b) $\dfrac{1}{\sqrt{10}}$
(c) $\dfrac{3}{\sqrt{10}}$ (d) $\dfrac{-3}{\sqrt{10}}$

11. Find the value of $\dfrac{\cos 20°-\sin 20°}{\cos 20°+\sin 20°}$.

(a) $\tan 20°$ (b) $\cot 20°$
(c) $\tan 25°$ (d) $\cot 25°$

12. What is the value of $\dfrac{[\sin(y-z)+\sin(y+z)+2\sin y]}{[\sin(x-z)+\sin(x+z)+2\sin x]}$? ***[SSC (CGL-T-II)-17]***

(a) $\cos x\sin y$ (b) $\dfrac{\sin y}{\sin x}$
(c) $\sin z$ (d) $\sin x\tan y$

13. $\sin 50°-\sin 70°+\sin 10°=?$

(a) 0 (b) 1
(c) 2 (d) $\dfrac{1}{2}$

14. What is the simplified value of $\operatorname{cosec} 2A+\cot 2A$? ***[SSC (CGL-T-I)-17]***

(a) $\sec A$ (b) $\sec\left(\dfrac{A}{2}\right)$
(c) $\cot A$ (d) $\cot^2 A$

15. If $\tan\alpha=\dfrac{1}{7}$ and $\tan\beta=\dfrac{1}{3}$, then $\cos 2\alpha=?$

(a) $\sin 2\beta$ (b) $\sin 4\beta$
(c) $\sin 3\beta$ (d) None of these

16. $\sqrt{2}\sin 54°-\sin 9°=?$

(a) $\cos 9°$ (b) $\tan 36°$
(c) $\tan 9°$ (d) Undefined

17. If $\sin^8\theta+\cos^8\theta-1=0$, then what is the value of $\cos^2\theta\sin^2\theta$, if $\theta\neq 0$ or $\dfrac{\pi}{2}$. ***[SSC (CGL-T-I)-17]***

(a) 0 (b) 1
(c) 2 (d) -1

18. If $\cos x+\cos y+\cos\alpha=0$ and $\sin x+\sin y+\sin\alpha=0$, then $\cot\left(\dfrac{x+y}{2}\right)=?$

(a) $\sin\alpha$ (b) $\cos\alpha$
(c) $\sin\left(\dfrac{x+y}{2}\right)$ (d) $\cot\alpha$

19. $2\cos^2\theta-2\sin^2\theta=1$, then the value of θ is :

(a) 15° (b) 30°
(c) 45° (d) 60°

20. $\sin\theta\cdot\sin(60°-\theta)\cdot\sin(60°+\theta)=?$

(a) $\dfrac{1}{4}\sin 3\theta$ (b) $\dfrac{1}{2}\tan 2\theta$
(c) $\tan 3\theta$ (d) $\tan\theta$

21. What is the value of $\dfrac{1-\sin(90°-2A)}{1+\sin(90°+2A)}$? ***[SSC (CGL-T-II)-17]***

(a) $\cot^2 A$ (b) $\tan^2 A$
(c) $\sin A\cos A$ (d) $\sin^2 A\cos A$

22. If $a\tan\theta=b$, then $a\cos 2\theta+b\sin 2\theta=?$

(a) a (b) $-a$
(c) b (d) $-b$

23. $\dfrac{\sin^2 A-\sin^2 B}{\sin A\cdot\cos A-\sin B\cdot\cos B}=?$

(a) $\tan(A-B)$ (b) $\tan(A+B)$
(c) $\cot(A-B)$ (d) $\cot(A+B)$

24. $\dfrac{1+\tan^2(45°-\theta)}{1-\tan^2(45°-\theta)}=?$

(a) $\sin 2\theta$
(b) $\cos 2\theta$
(c) $\operatorname{cosec} 2\theta$
(d) 1

25. What is the value of $\dfrac{2\sin(45°+\theta)\sin(45°-\theta)}{\cos 2\theta}$? ***[SSC (CGL-T-II)-17]***

(a) 0 (b) $\cot 2\theta$
(c) $\tan 2\theta$ (d) 1

26. $2\sin A\cdot\cos^3 A-2\sin^3 A\cdot\cos A=?$

(a) $\sin 4A$ (b) $\frac{1}{2}\sin 4A$

(c) $\frac{1}{4}\sin 4A$ (d) None of these

27. If $\sin 2\theta+\sin 2\alpha=\frac{1}{2}$ and $\cos 2\theta+\cos 2\alpha=\frac{3}{2}$, then $\cos^2(\theta-\alpha)$ is equal to :

(a) $\frac{3}{8}$ (b) $\frac{3}{4}$

(c) $\frac{5}{8}$ (d) None of these

28. $1+\cot(45°+\theta)\times\cot(45°-\theta)=?$

(a) 0 (b) 2

(c) 1 (d) $\sqrt{3}$

29. $\cos 20°\cdot\cos 40°\cdot\cos 80°=?$

(a) $\frac{1}{2}$ (b) $\frac{1}{4}$

(c) $\frac{1}{8}$ (d) $\frac{1}{6}$

30. If $x+\frac{1}{x}=2\cos\alpha$, then $x^n+\frac{1}{x^n}=?$

(a) $2^n\cos\alpha$ (b) $2^n\cos n\alpha$

(c) $2i\sin n\alpha$ (d) $2\cos n\alpha$

31. What is the value of $\frac{\sin(90°-10\theta)-\cos(\pi-6\theta)}{\cos\left(\frac{\pi}{2}-10\theta\right)-\sin(\pi-6\theta)}$?

[SSC (CGL-T-II)-17]

(a) $\tan 2\theta$ (b) $\cot 2\theta$

(c) $\cot\theta$ (d) $\cot 3\theta$

32. If $0°<x<\frac{\pi}{4}$, then $\sec 2x-\tan 2x=?$

(a) $\tan\left(x-\frac{\pi}{4}\right)$ (b) $\tan\left(\frac{\pi}{4}-x\right)$

(c) $\tan\left(x+\frac{\pi}{4}\right)$ (d) $\tan^2\left(x-\frac{\pi}{4}\right)$

33. $\frac{\sin\theta+\sin 2\theta}{1+\cos\theta+\cos 2\theta}=?$

(a) $\cot\theta$ (b) $\tan\theta$

(c) $\frac{1}{2}\tan\theta$ (d) $\frac{1}{2}\cot\theta$

34. $\cot 70°+4\cos 70°=?$

(a) $\frac{1}{\sqrt{3}}$ (b) $2\sqrt{3}$

(c) $\frac{1}{2}$ (d) $\sqrt{3}$

35. Find the value of $\tan 33°+\tan 12°+\tan 33°\cdot\tan 12°$.

(a) 0 (b) 2

(c) 1 (d) $\sqrt{3}$

36. If $A=130°$ and $x=\sin A+\cos A$, then

(a) $x>0$ (b) $x<0$

(c) $x=0$ (d) $x\le 0$

37. What is the value of $\frac{\tan 5\theta+\tan 3\theta}{4\cos 4\theta(\tan 5\theta-\tan 3\theta)}$?

[SSC (CGL-T-II)-17]

(a) $\sin 2\theta$ (b) $\cot 2\theta$

(c) $\tan 4\theta$ (d) $\cos 2\theta$

38. What is the value of $\frac{\cos 3\theta+2\cos 5\theta+\cos 7\theta}{\cos\theta+2\cos 3\theta+\cos 5\theta}+\sin 2\theta\cdot\tan 3\theta$?

[SSC (CGL-T-II)-17]

(a) $\cos 2\theta$ (b) $\sin 2\theta$

(c) $\tan 2\theta$ (d) $\cot\theta\sin 2\theta$

39. $\cot x-\tan x=?$

(a) $\cot 2x$ (b) $2\cot^2 x$

(c) $2\cot 2x$ (d) $\cot^2 2x$

40. What is the value of $\frac{[\sin(x+y)-2\sin x+\sin(x-y)]}{[\cos(x-y)+\cos(x+y)-2\cos x]}\times\frac{[\sin 10x-\sin 8x]}{[\cos 10x+\cos 8x]}$?

[SSC (CGL-T-II)-17]

(a) 0 (b) $\tan^2 x$

(c) 1 (d) $2\tan x$

41. $\cos 15°=?$

(a) $\sqrt{\frac{1+\cos 30°}{2}}$

(b) $\sqrt{\frac{1-\cos 30°}{2}}$

(c) $\pm\sqrt{\frac{1+\cos 30°}{2}}$

(d) $\pm\sqrt{\frac{1-\cos 30°}{2}}$

42. If $A+B=45°$, then the value of $2(1+\tan A)(1+\tan B)$ is :

[SSC (CGL-T-I)-19]

(a) 0 (b) 2

(c) 4 (d) 1

Answer key

1. (a)	2. (b)	3. (a)	4. (a)	5. (a)	6. (c)	7. (c)	8. (b)	9. (b)	10. (a)
11. (c)	12. (b)	13. (a)	14. (c)	15. (b)	16. (a)	17. (c)	18. (d)	19. (b)	20. (a)
21. (b)	22. (a)	23. (b)	24. (c)	25. (d)	26. (b)	27. (c)	28. (b)	29. (c)	30. (d)
31. (b)	32. (b)	33. (b)	34. (d)	35. (c)	36. (a)	37. (d)	38. (a)	39. (c)	40. (b)
41. (a)	42. (c)								

Hints and SOLUTIONS

1. Given, $\sin (A+B)=1$

$\Rightarrow \quad \sin (A+B)=\sin 90^\circ$

$\Rightarrow \quad A+B=90^\circ$

$\Rightarrow \quad A=90^\circ-B$

Now,

$\cos (A-B)=\cos A\cdot\cos B+\sin A\cdot\sin B$

$=\cos (90^\circ-B)\cdot\cos B+\sin (90^\circ-B)\cdot\sin B$

$=\sin B\cdot\cos B+\cos B\cdot\sin B$

$=2\sin B\cdot\cos B=\sin 2B$

2. $\tan 105^\circ=\tan (60^\circ+45^\circ)$

$$=\frac{\tan 60^\circ+\tan 45^\circ}{1-\tan 60^\circ\cdot\tan 45^\circ}$$

$$=\frac{1+\sqrt{3}}{1-\sqrt{3}}=\frac{1+\sqrt{3}}{1-\sqrt{3}}\times\frac{1+\sqrt{3}}{1+\sqrt{3}}$$

$$=\frac{4+2\sqrt{3}}{1-3}=-(2+\sqrt{3})$$

3. $\dfrac{\sin (B-C)}{\cos B\cos C}$

$$=\frac{\sin B\cos C-\cos B\sin C}{\cos B\cos C}$$

$$=\frac{\sin B\cos C}{\cos B\cos C}-\frac{\cos B\sin C}{\cos B\cos C}$$

$=\tan B-\tan C$

Similarly, 2nd term i.e.

$$\frac{\sin (C-A)}{\cos C\cos A}=\tan C-\tan A$$

and 3rd term i.e.

$$\frac{\sin (A-B)}{\cos A\cdot\cos B}=\tan A-\tan B$$

Now, on adding,

$\tan B-\tan C+\tan C-\tan A+\tan A-\tan B=0$

Alternatively

You may assume the values of x, y and z and simplify the expression. Thereafter match with the options. In case two or more options yield the same result as the resultant of the given expression, you should try with another value of the angle.

4. Clearly, $\dfrac{\sin A}{\sin B}=1$ and $\dfrac{\cos A}{\cos B}=1$,

hence, $\quad \dfrac{\sin A}{\sin B}=\dfrac{\cos A}{\cos B}$

$\Rightarrow \sin A\cdot\cos B-\cos A\cdot\sin B=0$

$\Rightarrow \quad \sin (A-B)=0$

$\Rightarrow \quad \sin\left(\dfrac{A-B}{2}\right)=0$

5. Given, $\sin 5\theta+\sin\theta=\sin 3\theta$

$\Rightarrow \quad 2\sin 3\theta\cdot\cos 2\theta=\sin 3\theta$

$\Rightarrow \quad \cos 2\theta=\dfrac{1}{2}$

$\Rightarrow \quad \cos 2\theta=\cos 60^\circ$

$\Rightarrow \quad \theta=30^\circ$

6. $\cos 3\theta=4\cos^3\theta-3\cos\theta$

$=\cos\theta\,(4\cos^2\theta-3)$

$$=\frac{1}{2}\left(a+\frac{1}{a}\right)\left[4\times\frac{1}{4}\left(a+\frac{1}{a}\right)^2-3\right]$$

$$=\frac{1}{2}\left[\left(a+\frac{1}{a}\right)^3-3a\cdot\frac{1}{a}\left(a+\frac{1}{a}\right)\right]$$

$$=\frac{1}{2}\left(a^3+\frac{1}{a^3}\right)$$

7. Given, $\cot A=\dfrac{\sin B}{1-\cos B}$

$$\cot^2 A=\frac{\sin^2 B}{1+\cos^2 B-2\cos B}$$

Now, $\cot 2A=\dfrac{\cot^2 A-1}{2\cot A}$

$$=\frac{\dfrac{\sin^2 B}{1+\cos^2 B-2\cos B}-1}{\dfrac{2\sin B}{1-\cos B}}$$

$$=\frac{\dfrac{\sin^2 B-1-\cos^2 B+2\cos B}{1+\cos^2 B-2\cos B}}{\dfrac{2\sin B}{1-\cos B}}$$

$$=\frac{1-\cos^2 B-1-\cos^2 B+2\cos B}{2\sin B\,(1-\cos B)}$$

$$=\frac{1-\cos^2 B-1-\cos^2 B+2\cos B}{2\sin B\,(1-\cos B)}$$

$$=\frac{2\cos B\,(1-\cos B)}{2\sin B\,(1-\cos B)}=\cot B$$

8. $\dfrac{\sin\theta+\sin 2\theta+\sin 3\theta}{\cos\theta+\cos 2\theta+\cos 3\theta}$

$$=\frac{\sin 2\theta+\sin 3\theta+\sin\theta}{\cos 2\theta+\cos 3\theta+\cos\theta}$$

$$=\frac{\sin 2\theta+\sin (2\theta+\theta)+\sin (2\theta-\theta)}{\cos 2\theta+\cos (2\theta+\theta)+\cos (2\theta-\theta)}$$

$$=\frac{\sin 2\theta+2\sin 2\theta\cos\theta}{\cos 2\theta+2\cos 2\theta\cos\theta}$$

$$=\frac{\sin 2\theta\,(1+2\cos\theta)}{\cos 2\theta\,(1+2\cos\theta)}=\tan 2\theta$$

Alternatively

Let θ be 30°.

Then, $\dfrac{\sin\theta+\sin 2\theta+\sin 3\theta}{\cos\theta+\cos 2\theta+\cos 3\theta}$

$$=\frac{\sin 30^\circ+\sin 60^\circ+\sin 90^\circ}{\cos 30^\circ+\cos 60^\circ+\cos 90^\circ}=\sqrt{3}$$

Now, ensure which one of the options results into $\sqrt{3}$.

(**Remarks** you may put any angle as per your convenience)

9. Given, $\sin\dfrac{\theta}{2}=\sqrt{\dfrac{x-1}{2x}}$

Now, $\cos\dfrac{\theta}{2}=\sqrt{1-\sin^2\dfrac{\theta}{2}}=\sqrt{\dfrac{x+1}{2x}}$

$\therefore \quad \tan\dfrac{\theta}{2}=\sin\dfrac{\theta}{2}\div\cos\dfrac{\theta}{2}=\sqrt{\dfrac{x-1}{x+1}}$

We know, $\tan\left(2\cdot\dfrac{\theta}{2}\right)=\dfrac{2\tan\left(\dfrac{\theta}{2}\right)}{1-\tan^2\left(\dfrac{\theta}{2}\right)}$

$$\Rightarrow \tan\theta=\frac{2\times\sqrt{\dfrac{x-1}{x+1}}}{1-\left(\sqrt{\left(\dfrac{x-1}{x+1}\right)}\right)^2}$$

$$=\sqrt{x^2-1}$$

10. Given, $\sin\alpha=\dfrac{-3}{5}$

$\therefore \quad \cos\alpha=\dfrac{-4}{5}$

[$\because \alpha$ lies in 3rd quadrant]

We know, $\cos 2\alpha=2\cos^2\alpha-1$

$\Rightarrow \quad \cos\alpha=2\cos^2\left(\dfrac{\alpha}{2}\right)-1$

$\Rightarrow \cos^2\left(\dfrac{\alpha}{2}\right)=\dfrac{1+\cos\alpha}{2}$

$\Rightarrow \quad \cos\left(\dfrac{\alpha}{2}\right)=\pm\sqrt{\dfrac{1+\cos\alpha}{2}}$

$$=\pm\sqrt{\frac{1-\dfrac{4}{5}}{2}}=\frac{-1}{\sqrt{10}}$$

$\left[\because \dfrac{\alpha}{2}\text{ lies in 2nd quadrant}\right]$

11. $\dfrac{\cos 20^\circ-\sin 20^\circ}{\cos 20^\circ+\sin 20^\circ}$

$$= \frac{\frac{\cos 20° - \sin 20°}{\cos 20°}}{\frac{\cos 20° + \sin 20°}{\cos 20°}} = \frac{1 - \tan 20°}{1 + \tan 20°}$$

$$= \frac{\tan 45° - \tan 20°}{1 + \tan 45° \tan 20°}$$

$= \tan (45° - 20°) = \tan 25°$

12. Given expression $\frac{[\sin (y - z) + \sin (y + z) + 2 \sin y]}{[\sin (x - z) + \sin (x + z) + 2 \sin x]}$

$$= \frac{[2 \sin y \cdot \cos z + 2 \sin y]}{[2 \sin x \cdot \cos z + 2 \sin x]}$$

$[\because \sin (A + B) + \sin (A - B) = 2 \sin A \cdot \cos B]$

$$= \frac{2 \sin y (\cos z + 1)}{2 \sin x (\cos z + 1)} = \frac{\sin y}{\sin x}$$

Alternatively

You may assume the values of x, y and z and simplify the expression. Thereafter match with the options. In case two or more options yield the same result as the resultant of the given expression, you should try with another values of angle.

13. $2 \cos\left(\frac{50 + 70}{2}\right)° \cdot \sin\left(\frac{50 - 70}{2}\right)° + \sin 10°$

$= 2\cos 60° \cdot \sin (-10°) + \sin 10°$

$= -2 \cos 60° \cdot \sin 10° + \sin 10°$

$= -\sin 10° + \sin 10° = 0$

14. $\text{cosec } 2A + \cot 2A = \frac{1}{\sin 2A} + \frac{\cos 2A}{\sin 2A}$

$$= \frac{1 + \cos 2A}{\sin 2A} = \frac{1 + 2 \cos^2 A - 1}{2 \sin A \cdot \cos A}$$

$= \cot A$

Alternatively

You may assume the values of θ and simplify the expression. Thereafter match with the values of options. In case two or more options yield the same result as the resultant of the given expression, you should try with another value of the angle.

15. $\cos 2\alpha = \frac{1 - \tan^2 \alpha}{1 + \tan^2 \alpha} = \frac{1 - \left(\frac{1}{7}\right)^2}{1 + \left(\frac{1}{7}\right)^2} = \frac{24}{25}$

Further, $\tan \beta = \frac{1}{3}$

Then, $\sin 2\beta = \frac{2 \tan \beta}{1 + \tan^2 \beta}$

$$= \frac{2 \times \frac{1}{3}}{1 + \left(\frac{1}{3}\right)^2} = \frac{3}{5}$$

$$\cos 2\beta = \frac{1 - \tan^2 \beta}{1 + \tan^2 \beta} = \frac{1 - \left(\frac{1}{3}\right)^2}{1 + \left(\frac{1}{3}\right)^2} = \frac{4}{5}$$

$\therefore \sin 4\beta = 2 \sin 2\beta \cdot \cos 2\beta$

$$= 2 \cdot \frac{3}{5} \cdot \frac{4}{5} = \frac{24}{25}$$

Hence, $\cos 2\alpha = \sin 4\beta$

16. $\sqrt{2} \sin 54° - \sin 9°$

$= \sqrt{2} \sin (45° + 9°) \cdot \sin 9°$

We know,

$\sin (A + B) = \sin A \cdot \cos B + \cos A \cdot \sin B$

Then, $\sin (45° + 9°) = \sin 45° \cos 9° + \cos 45° \sin 9°$

$$= \frac{1}{\sqrt{2}} (\cos 9° + \sin 9°)$$

Now, $\sqrt{2} \sin (45° + 9°) - \sin 9°$

$$= \sqrt{2} \times \frac{1}{\sqrt{2}} (\cos 9° + \sin 9°) - \sin 9°$$

$= \cos 9°$

17. We know, $\sin^2 \theta + \cos^2 \theta = 1$

On squaring both sides,

$\sin^4 \theta + \cos^4 \theta + 2 \sin^2 \theta \cos^2 \theta = 1$

$\Rightarrow \sin^4 \theta + \cos^4 \theta = 1 - 2 \sin^2 \theta \cos^2 \theta$

$$\Rightarrow \sin^4 \theta + \cos^4 \theta = 1 - \frac{\sin^2 2\theta}{2}$$

On squaring both sides again,

$$\sin^8 \theta + \cos^8 \theta + 2 \sin^4 \theta \cos^4 \theta = 1 + \frac{\sin^4 2\theta}{4} - \sin^2 2\theta$$

$$\Rightarrow 1 + \frac{\sin^4 2\theta}{8} = 1 + \frac{\sin^4 2\theta}{4} - \sin^2 2\theta$$

$$\Rightarrow \sin^2 2\theta = \frac{\sin^4 2\theta}{8}$$

$\Rightarrow \sin^2 2\theta = 8$

$\Rightarrow (2 \sin \theta \cos \theta)^2 = 8$

$\Rightarrow \cos^2 \theta \sin^2 \theta = 2$

18. Given, $\cos x + \cos y + \cos \alpha = 0$

$\Rightarrow \cos x + \cos y = -\cos \alpha$

$$\Rightarrow 2 \cos\left(\frac{x + y}{2}\right) \cdot \cos\left(\frac{x - y}{2}\right) = -\cos \alpha \quad \text{...(i)}$$

Also given, $\sin x + \sin y + \sin \alpha = 0$

$$\Rightarrow 2 \sin\left(\frac{x + y}{2}\right) \cdot \cos\left(\frac{x - y}{2}\right) = -\sin \alpha \quad \text{...(ii)}$$

On dividing Eqs. (i) by (ii), we get

$$\cot\left(\frac{x + y}{2}\right) = \cot \alpha$$

19. Given, $2 \cos^2 \theta - 2 \sin^2 \theta = 1$

$\Rightarrow 2 (\cos^2 \theta - \sin^2 \theta) = 1$

$\Rightarrow 2 \times \cos 2\theta = 1 \Rightarrow \cos 2\theta = \frac{1}{2}$

$\Rightarrow \cos 2\theta = \cos 60°$

$\Rightarrow 2\theta = 60° \Rightarrow \theta = 30°$

Alternatively

You may proceed through options. You have to ascertain as to which option satisfies the given equation.

20. $\sin \theta \cdot \sin (60° - \theta) \cdot \sin (60° + \theta)$

$= \frac{1}{2} \sin \theta [2 \sin (60° + \theta) \cdot \sin (60° - \theta)]$

$= \frac{1}{2} \sin \theta [\cos (60° + \theta - 60° + \theta) - \cos (60° + \theta + 60° - \theta)]$

$[\because 2 \sin A \cdot \sin B = \cos (A - B) - \cos (A + B)]$

$= \frac{1}{2} \sin \theta [\cos 2\theta - \cos 120°]$

$= \frac{1}{2} \sin \theta \left[\cos 2\theta - \left(-\frac{1}{2}\right)\right]$

$= \frac{1}{4} \cdot 2 \sin \theta \left[\cos 2\theta - \left(-\frac{1}{2}\right)\right]$

$= \frac{1}{4} \cdot 2 \cos 2\theta \cdot \sin \theta + \frac{1}{4} \sin \theta$

$= \frac{1}{4} [\sin (2\theta + \theta) - \sin (2\theta - \theta)] + \frac{1}{4} \sin \theta$

$[\because \sin (A + B) - \sin (A - B) = 2 \cos A \sin B]$

$= \frac{1}{4} \{\sin 3\theta - \sin \theta\} + \frac{1}{4} \sin \theta$

$= \frac{1}{4} \sin 3\theta$

Alternatively

Let θ be 30°.

Then, $\sin \theta \sin (60° - \theta) \cdot \sin (60° + \theta)$

$= \sin 30° \sin 30° \sin 90° = \frac{1}{4}$

Now, ensure which one of the options results into $\frac{1}{4}$.

Remarks you may put any angle as per your convenience

21. $\frac{1 - \sin (90° - 2A)}{1 + \sin (90° + 2A)} = \frac{1 - \cos 2A}{1 + \cos 2A}$

$$= \frac{1 - (\cos^2 A - \sin^2 A)}{1 + \cos^2 A - \sin^2 A}$$

$$= \frac{1 - \cos^2 A + \sin^2 A}{1 + \cos^2 A - \sin^2 A} = \frac{2 \sin^2 A}{2 \cos^2 A}$$

$= \tan^2 A$

Alternatively

Let $A = 30°$ (you are free to assume any other angle as per your convenience ascertaining that the value of given trigonometric ratios as well as the expression have a defined value at that angle.)

Then,

$$\frac{1-\sin(90°-2A)}{1+\sin(90°+2A)} = \frac{1-\sin 30°}{1+\cos 60°} = \frac{1}{3}$$

Now proceed through options.

(a) $\cot^2 A \Rightarrow \cot^2 30° \neq \frac{1}{3}$

(b) $\tan^2 A \Rightarrow \tan^2 30° = \frac{1}{3}$

Hence, option (b) is correct.

Remarks In case two or more options yield the same result as the resultant of the given expression, you should try with another value of the angle

22. Given, $a \tan\theta = b$

$$\Rightarrow \tan\theta = \frac{b}{a}$$

Now, $a\cos 2\theta + b\sin 2\theta$

$$= a\left(\frac{1-\tan^2\theta}{1+\tan^2\theta}\right) + b\left(\frac{2\tan\theta}{1+\tan^2\theta}\right)$$

$$= a\left(\frac{1-\frac{b^2}{a^2}}{1+\frac{b^2}{a^2}}\right) + b\left(\frac{\frac{2b}{a}}{1+\frac{b^2}{a^2}}\right)$$

$$= a\left(\frac{a^2-b^2}{a^2+b^2}\right) + b\left(\frac{2ab}{a^2+b^2}\right)$$

$$= a\left(\frac{a^2+b^2}{a^2+b^2}\right) = a$$

23. $$\frac{\sin^2 A - \sin^2 B}{\sin A\cdot\cos A - \sin B\cdot\cos B}$$

$$= \frac{2(\sin^2 A - \sin^2 B)}{2\sin A\cdot\cos A - 2\sin B\cdot\cos B}$$

$$= \frac{2\sin(A+B)\cdot\sin(A-B)}{\sin 2A - \sin 2B}$$

$$= \frac{2\sin(A+B)\cdot\sin(A-B)}{2\cos(A+B)\cdot\sin(A-B)}$$

$= \tan(A+B)$

Alternatively

Let $A = 90°$ and $b = 30°$ [you are free to take any angle for which the given trigonometric-ratios as well as the given expression have defined values.]

Then,

$$\frac{\sin^2 90° - \sin^2 30°}{\sin 90°\cdot\cos 90° - \sin 30°\cos 30°}$$

$= -\sqrt{3}$

Now, proceed through options.

(a) $\tan(90°-30°) = \sqrt{3} \neq -\sqrt{3}$

(b) $\tan(90°+30°) = -\cot 30° = -\sqrt{3}$

Hence, option (b) is correct.

Remarks In case two or more options yield the same result as the resultant of the given expression, you should try with another value of the angle

24. We know, $\cos 2A = \frac{1-\tan^2 A}{1+\tan^2 A}$

then, $$\frac{1+\tan^2(45°-\theta)}{1-\tan^2(45°-\theta)}$$

$$= \frac{1}{\cos 2(45°-\theta)}$$

$$= \frac{1}{\cos(90°-2\theta)}$$

$$= \frac{1}{\sin 2\theta} = \text{cosec } 2\theta$$

Alternatively

Let θ be 15°, then

$$\frac{1+\tan^2(45°-\theta)}{1-\tan^2(45°-\theta)} = \frac{1+\tan^2(45°-15°)}{1-\tan^2(45°-15°)}$$

$$= \frac{1+\left(\frac{1}{\sqrt{3}}\right)^2}{1-\left(\frac{1}{\sqrt{3}}\right)^2} = 2$$

Now, ensure that the value of which option is equal to 2.

25. $$\frac{2\sin(45°+\theta)\sin(45°-\theta)}{\cos 2\theta}$$

$$= \frac{2\sin(45°+\theta)\cos[90°-(45°-\theta)]}{\cos 2\theta}$$

$$= \frac{2\sin(45°+\theta)\cos(45°+\theta)}{\cos 2\theta}$$

$$= \frac{\sin 2(45°+\theta)}{\cos 2\theta}$$

$$= \frac{\sin(90°+2\theta)}{\cos 2\theta}$$

$$= \frac{\cos 2\theta}{\cos 2\theta} = 1$$

Alternatively

You may assume the values of θ and simplify the expression. Thereafter match with the values of options. In case two or more options yield the same result as the resultant of the given expression, you should try with another value of the angle.

26. $2\sin A\cdot\cos^3 A - 2\sin^3 A\cdot\cos A$

$= 2\sin A\cos A(\cos^2 A - \sin^2 A)$

$= 2\sin A\cos A\cos 2A = \sin 2A\cos 2A$

$$= \frac{2\sin 2A\cos 2A}{2} = \frac{1}{2}\sin 4A$$

Alternatively

You may assume the values of A and simplify the expression. Thereafter match with the values of options. In case two or more options yield the same result as the resultant of the given expression, you should try with another value of the angle.

27. Given, $\sin 2\theta + \sin 2\alpha = \frac{1}{2}$

On squaring both sides,

$$\sin^2 2\theta + \sin^2 2\alpha + 2\cdot\sin 2\theta\cdot\sin 2\alpha = \frac{1}{4} \quad \text{...(i)}$$

Similarly, $$\cos^2 2\theta + \cos^2 2\alpha + 2\cdot\cos 2\theta\cdot\cos 2\alpha = \frac{9}{4} \quad \text{...(ii)}$$

On adding both Eqs. (i) + (ii)

$$[\sin^2 2\theta + \cos^2 2\theta] + [\sin^2 2\alpha + \cos^2 2\alpha] + 2[\cos 2\theta\cdot\cos 2\alpha + \sin 2\theta\cdot\sin 2\alpha] = \frac{5}{2}$$

$$\Rightarrow 1 + 1 + 2[\cos(2\theta - 2\alpha)] = \frac{5}{2}$$

$$\Rightarrow \cos(2\theta - 2\alpha) = \frac{1}{4}$$

$$\Rightarrow \cos 2(\theta - \alpha) = \frac{1}{4}$$

$$\Rightarrow 2\cos^2(\theta-\alpha) - 1 = \frac{1}{4}$$

$$\Rightarrow \cos^2(\theta - \alpha) = \frac{5}{8}$$

28. $1 + \cot(45°+\theta)\times\cot(45°-\theta)$

$$= 1 + \frac{\cos(45°+\theta)\cos(45°-\theta)}{\sin(45°+\theta)\sin(45°-\theta)}$$

$$= 1 + \frac{\cos^2 45° - \sin^2\theta}{\sin^2 45° - \sin^2\theta}$$

$$= 1 + \frac{\left(\frac{1}{\sqrt{2}}\right)^2 - \sin^2\theta}{\left(\frac{1}{\sqrt{2}}\right)^2 - \sin^2\theta} = 1 + 1 = 2$$

Alternatively,

Let θ be 180°.

Then, $\cot\left(\frac{\pi}{4} + 180°\right)\cdot\cot\left(\frac{\pi}{4} - 180°\right)$

$= \cot\frac{\pi}{4}\cdot\cot(-135°)$

$= \cot\frac{\pi}{4}\cdot(-\cot 135°)$

$= \cot\frac{\pi}{4}[-\cot(180° - 45°)]$

$= \cot\frac{\pi}{4}\cdot\cot 45° = 1\cdot 1 = 1$

Then, $1+\cot(45°+\theta)\times\cot(45°-\theta)$
$=1+1=2$

Remarks you may put any angle as per your convenience.

29. $\cos 20°\cdot\cos 40°\cdot\cos 80°$
$$=\frac{(2\sin 20°\cdot\cos 20°)\cdot\cos 40°\cdot\cos 80°}{2\sin 20°}$$
$$=\frac{\sin 40°\cdot\cos 40°\cdot\cos 80°}{2\sin 20°}$$
$$=\frac{(2\sin 40°\cdot\cos 40°)\cdot\cos 80°}{2\cdot 2\sin 20°}$$
$$=\frac{\sin 80°\cdot\cos 80°}{4\sin 20°}$$
$$=\frac{2\cdot\sin 80°\cdot\cos 80°}{2\cdot 4\sin 20°}$$
$$=\frac{\sin 160°}{8\sin 20°}=\frac{\sin(180°-20°)}{8\sin 20°}$$
$$=\frac{\sin 20°}{8\sin 20°}=\frac{1}{8}$$

30. Given, $x+\frac{1}{x}=2\cos\alpha$
$$\Rightarrow\quad x^2+\frac{1}{x^2}+2x\times\frac{1}{x}=4\cos^2\alpha$$
$$\Rightarrow\quad x^2+\frac{1}{x^2}=4\cos^2\alpha-2$$
$$\Rightarrow\quad x^2+\frac{1}{x^2}=2(2\cos^2\alpha-1)$$
$$\Rightarrow\quad x^2+\frac{1}{x^2}=2\cos 2\alpha$$
Similarly, $x^n+\frac{1}{x^n}=2\cos n\alpha$

31. $$\frac{\sin(90°-10\theta)-\cos(\pi-6\theta)}{\cos\left(\frac{\pi}{2}-10\theta\right)-\sin(\pi-6\theta)}$$
$$=\frac{\cos 10\theta+\cos 6\theta}{\sin 10\theta-\sin 6\theta}$$
$$=\frac{2\cos 8\theta\cdot\cos 2\theta}{2\cos 8\theta\cdot\sin 2\theta}=\cot 2\theta$$

Alternatively

Let $\theta=15°$ [you are free to assume any other angle as per your convenience ascertaining that the value of given trigonometric ratios as well as the expression have a defined value at that angle.]

Then, $$\frac{\sin(90°-10\theta)-\cos(\pi-6\theta)}{\cos\left(\frac{\pi}{2}-10\theta\right)-\sin(\pi-6\theta)}$$
$$=\frac{\sin(-60°)-\cos(90°)}{\cos(-60°)-\sin(90°)}$$
$$=\frac{-\frac{\sqrt{3}}{2}}{\frac{1}{2}-1}=\sqrt{3}=\cot(2\times 15°)=\cot 2\theta$$

Remarks In case two or more options yields the same result as the resultant of the given expression, you should try with another value of the angle.

32. $\sec 2x-\tan 2x=\frac{1-\sin 2x}{\cos 2x}$
$$=\frac{\sin^2 x+\cos^2 x-2\sin x\cos x}{\cos^2 x-\sin^2 x}$$
$$=\frac{(\cos x-\sin x)^2}{(\cos x-\sin x)(\cos x+\sin x)}$$
$$=\frac{(\cos x-\sin x)}{(\cos x+\sin x)}$$
$$=\frac{\frac{(\cos x-\sin x)}{\cos x}}{\frac{(\cos x+\sin x)}{\cos x}}=\frac{1-\tan x}{1+\tan x}$$
$$=\frac{\tan\frac{\pi}{4}-\tan x}{1+\tan\left(\frac{\pi}{4}\right)\tan x}=\tan\left(\frac{\pi}{4}-x\right)$$

Alternatively

You may assume the values of x and simplify the expression. Thereafter match with the values of options. In case two or more options yield the same result as the resultant of the given expression, you should try with another value of the angle.

33. $$\frac{\sin\theta+\sin 2\theta}{1+\cos\theta+\cos 2\theta}$$
$$=\frac{\sin\theta+2\sin\theta\cdot\cos\theta}{1+\cos\theta+2\cos^2\theta-1}$$
$$=\frac{\sin\theta(1+2\cos\theta)}{\cos\theta(1+2\cos\theta)}=\tan\theta$$

Alternatively

Let $\theta=30°$ [you are free to take any angle for which the given T-ratios as well as the expression have defined values.]

Then, $$\frac{\sin\theta+\sin 2\theta}{1+\cos\theta+\cos 2\theta}$$
$$=\frac{\sin 30°+\sin 60°}{1+\cos 30°+\cos 60°}$$
$$=\frac{1}{\sqrt{3}}=\tan 30°=\tan\theta$$

Remarks In case two or more options yield the same result as the resultant of the given expression, you should try with another value of the angle.

34. $\cot 70°+4\cos 70°$
$$=\frac{\cos 70°+4\sin 70°\cos 70°}{\sin 70°}$$
$$=\frac{\cos 70°+2\cdot 2\sin 70°\cos 70°}{\sin 70°}$$
$$=\frac{\cos 70°+2\sin 140°}{\sin 70°}$$
$$=\frac{\cos 70°+2\sin(180°-40°)}{\sin 70°}$$
$$=\frac{\cos 70°+2\sin 40°}{\sin 70°}$$
$$=\frac{(\sin 20°+\sin 40°)+\sin 40°}{\sin 70°}$$
$$=\frac{2\sin 30°\cdot\cos 10°+\sin 40°}{\sin 70°}$$
$$=\frac{\cos 10°+\sin 40°}{\sin 70°}$$
$$=\frac{\sin 80°+\sin 40°}{\sin 70°}$$
$$=\frac{2\sin 60°\cdot\cos 20°}{\sin 70°}$$
$$=\frac{\sqrt{3}\cdot\cos 20°}{\cos(90°-70°)}=\sqrt{3}$$

35. $\tan(33°+12°)=\frac{\tan 33°+\tan 12°}{1-\tan 33°\cdot\tan 12°}$
$$\Rightarrow\quad 1=\frac{\tan 33°+\tan 12°}{1-\tan 33°\cdot\tan 12°}$$
$\Rightarrow \tan 33°+\tan 12°$
$=1-\tan 33°\cdot\tan 12°$
$\Rightarrow \tan 33°+\tan 12°$
$+\tan 33°\cdot\tan 12°=1$

36. Given, $x=\sin 130°+\cos 130°$
$\Rightarrow\quad x=\sin(90°+40°)+\cos 130°$
$\Rightarrow\quad x=\cos 40°+\cos 130°$

We know, $\cos C+\cos D$
$$=2\cos\left(\frac{C+D}{2}\right)\cos\left(\frac{C-D}{2}\right)$$
$\therefore\quad x=2\cos 85°\cos 45°$
$=\sqrt{2}\cos 85°$

Hence, $x>0$ [$\because$ in 1st quarter, value of cosine lies between 1 and 0]

37. $$\frac{\tan 5\theta+\tan 3\theta}{4\cos 4\theta(\tan 5\theta-\tan 3\theta)}$$
$$=\frac{\frac{\sin 5\theta}{\cos 5\theta}+\frac{\sin 2\theta}{\cos 3\theta}}{4\cos 4\theta\left(\frac{\sin 5\theta}{\cos 5\theta}-\frac{\sin 3\theta}{\cos 3\theta}\right)}$$
$$=\frac{\sin 5\theta\cdot\cos 3\theta+\cos 5\theta\cdot\sin 3\theta}{4\cos 4\theta(\sin 5\theta\cdot\cos 3\theta-\cos 5\theta\cdot\sin 3\theta)}$$
$$=\frac{\sin(5\theta+3\theta)}{4\cos 4\theta\times\sin(5\theta-3\theta)}$$

$$= \frac{\sin (2 \cdot 4\theta)}{4 \cos 4\theta \times \sin 2\theta}$$

$$= \frac{2 \sin 4\theta \times \cos 4\theta}{4 \cos 4\theta \times \sin 2\theta} = \frac{2 \sin 2 \cdot 2\theta}{4 \times \sin 2\theta}$$

$$= \frac{2 \cdot 2 \sin 2\theta \times \cos 2\theta}{4 \times \sin 2\theta} = \cos 2\theta$$

Alternatively

Let $\theta = 60°$ [You are free to assume any other angle as per your convenience]

Then, $\dfrac{\tan 5\theta + \tan 3\theta}{4 \cos 4\theta (\tan 5\theta - \tan 3\theta)}$

$$= \frac{\tan 300° + \tan 180°}{4 \cos 240° (\tan 300° - \tan 180°)}$$

$= -0.5$

Now, proceed through options.

(a) $\sin 2\theta \Rightarrow \sin (2.60°) \neq -0.5$

(b) $\cot 2\theta \Rightarrow \cot (2.60°) \neq -0.5$

(c) $\tan 4\theta \Rightarrow \tan (4.60°) \neq -0.5$

(d) $\cos 2\theta \Rightarrow \cos (2.60°) = -0.5$

Remarks In case two or more options yield the same result as the resultant of the given expression, you should try with another value of the angle

38. $\dfrac{\cos 7\theta + \cos 3\theta + 2 \cos 5\theta}{\cos 5\theta + \cos\theta + 2 \cos 3\theta} + \sin 2\theta \tan 3\theta$

$$\Rightarrow \frac{2 \cos 5\theta \cdot \cos 2\theta + 2 \cos 5\theta}{2 \cos 3\theta \cdot \cos 2\theta + 2 \cos 3\theta} + \sin 2\theta \tan 3\theta$$

$$\Rightarrow \frac{2 \cos 5\theta (\cos 2\theta + 1)}{2 \cos 3\theta (\cos 2\theta + 1)} + \sin 2\theta \tan 3\theta$$

$$= \frac{\cos 5\theta}{\cos 3\theta} + \sin 2\theta \tan 3\theta$$

$$= \frac{\cos 5\theta + \sin 2\theta \tan 3\theta \cos 3\theta}{\cos 3\theta}$$

$$= \frac{\cos (2\theta + 3\theta) + \sin 2\theta \sin 3\theta}{\cos 3\theta}$$

$$= \frac{\cos 2\theta \cos 3\theta - \sin 2\theta \sin 3\theta + \sin 2\theta \sin 3\theta}{\cos 3\theta}$$

$= \cos 2\theta$

Alternatively

You may assume the values of θ and simplify the expression. Thereafter match with the options. In case two or more options yield the same result as the resultant of the given expression, you should try with another value of the angle.

39. $\cot x - \tan x = \cot x - \dfrac{1}{\cot x}$

$$= \frac{\cot^2 x - 1}{\cot x} = 2\left(\frac{\cot^2 x - 1}{2 \cot x}\right) = 2 \cot 2x$$

Alternatively

You may assume the values of x and simplify the expression. Thereafter match with the options. In case two or more option yield the same result as the resultant of the given expression, you should try with another value of the angle.

40. $\dfrac{[\sin (x + y) - 2 \sin x + \sin(x - y)]}{[\cos (x - y) + \cos (x + y) - 2 \cos x]}$

$$= \frac{[2 \sin x \cdot \cos y - 2 \sin x]}{[2 \cos x \cdot \cos y - 2 \cos x]}$$

$[\because \sin (A + B) + \sin (A - B) = 2 \sin A \cdot \cos B$ and $\cos (A + B) + \cos (A - B) = 2 \cos A \cdot \cos B]$

$= \tan x$

Further,

$$\frac{[\sin 10x - \sin 8x]}{[\cos 10x + \cos 8x]} = \frac{[2 \cos 9x \cdot \sin x]}{[2 \cos 9x \cdot \cos x]}$$

$= \tan x$

Hence, required answer

$= \tan x \cdot \tan x = \tan^2 x$

Alternatively

You may assume the values of x and y and simplify the expression. Thereafter match with the options. In case two or more options yield the same result as the resultant of the given expression, you should try with another value of the angle.

41. We know,

$$\cos 2\alpha = 2 \cos^2 \alpha - 1$$

$$\Rightarrow \quad \cos^2 \alpha = \frac{1 + \cos 2\alpha}{2}$$

$$\Rightarrow \quad \cos \alpha = \pm \sqrt{\frac{1 + \cos 2\alpha}{2}}$$

$$\Rightarrow \quad \cos 15° = \pm \sqrt{\frac{1 + \cos (2 \cdot 15°)}{2}}$$

$$\Rightarrow \quad \cos 15° = \pm \sqrt{\frac{1 + \cos 30°}{2}}$$

Since 15° is in the 1st quadrant (where cosine is positive),

$$\cos 15° = + \sqrt{\frac{1 + \cos 30°}{2}}$$

42. Given, $A + B = 45°$

$$\Rightarrow \quad \tan (A + B) = \tan 45°$$

$$\Rightarrow \quad \frac{\tan A + \tan B}{1 - \tan A \cdot \tan B} = 1$$

$\Rightarrow \tan A + \tan B = 1 - \tan A \cdot \tan B$

Now, $2 (1 + \tan A)(1 + \tan B)$

$= 2 [1 + \tan A + \tan B + \tan A \cdot \tan B]$

$= 2 [1 + (1 - \tan A \cdot \tan B) + \tan A \cdot \tan B]$

$= 4$

Part-III

51 Height and Distance

Introduction

Topic "Height and distance" is concerned with the applications of trigonometry which helps us to find the height of an object and distance of the object from a given point without measuring them physically. It is a vital topic from which questions are being asked in diverse competitive examinations. The following terminology needs your concentration.

Line of Sight A line joining an eye of an observer and the object/ point being viewed by him/her, is known as line of sight.

Angle of Elevation The angle of elevation of the object/point viewed is the angle subtended by the line of Sight drawn from the eye of the observer which is parallel to the earth in case the object/point is above the eye.

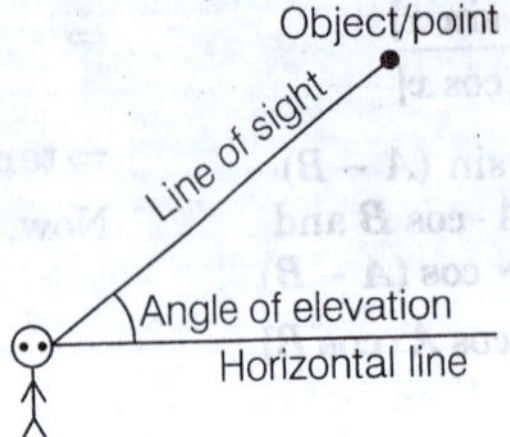

Angle of Depression The angle of depression of an object/point viewed is the angle subtended by the line of Sight with the line drawn from the eye of the observer, parallel to earth in case the object/point is below the eye.

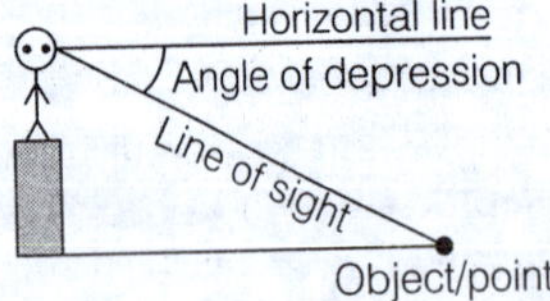

Note *Numerically an angle of elevation is equal to the angle of depression. Recall the concept of 'Alternate interior angles' in Geometry Section.*

Before solving the problems related to this topic we must know some fundamental concepts so as to save time in examination hall.

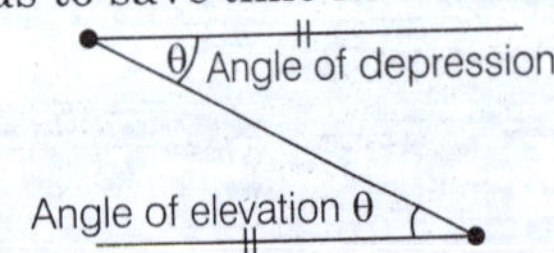

Inside

- *Introduction*
- *Terminology*
- *Fundamental Concepts Associated with Diverse Circumstances*
- *Illustrations Based on All the Above Concepts*
- ....
- .... *& Many More*
- *Work Out with Solutions*

Type A

When perpendiculars are same/equal for both given angles

Case I

Clearly, $\theta_2 > \theta_1$

Now, $\tan\theta_2 = \dfrac{AB}{BC}$ and $\tan\theta_1 = \dfrac{AB}{BD}$

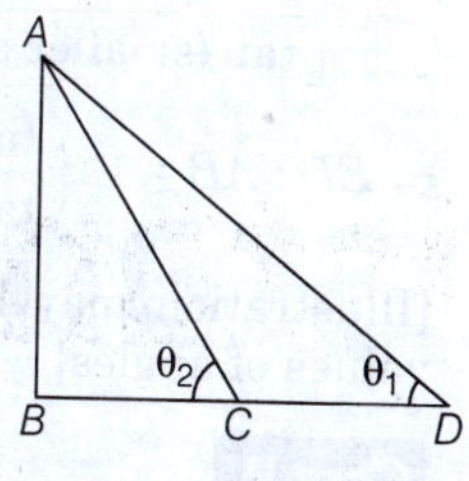

Then, $\dfrac{\tan\theta_2}{\tan\theta_1} = \dfrac{BD}{BC}$

$\Rightarrow \quad BC : BD = 1 : \left(\dfrac{\tan\theta_2}{\tan\theta_1}\right)$

Base for greater angle : Base for smaller angle

$$= 1 : \left\{\frac{\textbf{tan (greater angle)}}{\textbf{tan (smaller angle)}}\right\}$$

Illustrations

(i) Base for greater angle (60°) : Base for smaller angle (30°)

$$= 1 : \left[\frac{\tan 60^\circ}{\tan 30^\circ}\right]$$

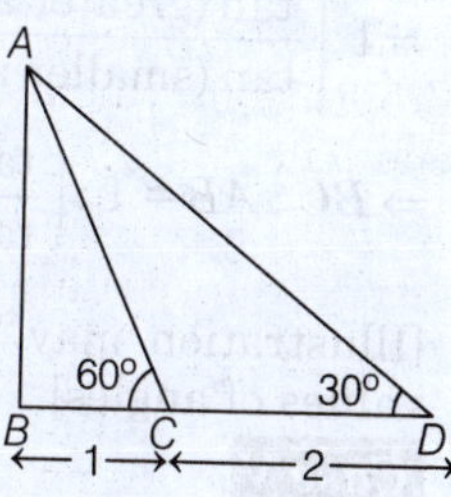

$$\Rightarrow BC : BD = 1 : \left[\frac{\tan 60^\circ}{\tan 30^\circ}\right]$$

$\boldsymbol{BC : BD = 1 : 3}$ **or** $\boldsymbol{BC : CD = 1 : 2}$

(ii) Base for greater angle (45°) : Base for smaller angle (30°) $= 1 : \left[\dfrac{\tan 45^\circ}{\tan 30^\circ}\right]$

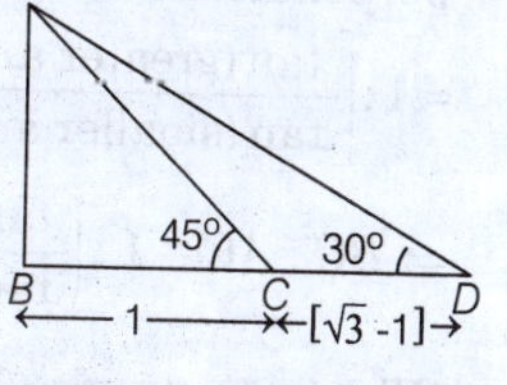

$\Rightarrow \quad BC : BD = 1 : \left\{\dfrac{\tan 45^\circ}{\tan 30^\circ}\right\}$

$\Rightarrow \quad \boldsymbol{BC : BD = 1 : \sqrt{3}}$

or $\quad \boldsymbol{BC : CD = 1 : (\sqrt{3} - 1)}$

(iii) Base for greater angle (60°) : Base for smaller angle (45°) $= 1 : \left[\dfrac{\tan 60^\circ}{\tan 45^\circ}\right]$

$\Rightarrow BC : BD = 1 : \left[\dfrac{\tan 60^\circ}{\tan 45^\circ}\right]$

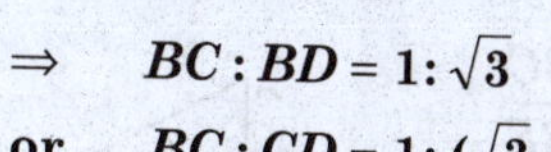

$\Rightarrow \quad \boldsymbol{BC : BD = 1 : \sqrt{3}}$

or $\quad \boldsymbol{BC : CD = 1 : (\sqrt{3} - 1)}$

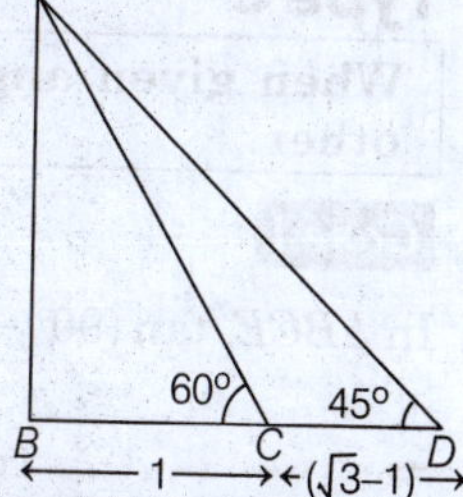

Observations You may write directly as, $\dfrac{\tan 60^\circ}{\tan 45^\circ} = \dfrac{\tan 45^\circ}{\tan 30^\circ} = \sqrt{3}$ and $\dfrac{\tan 60^\circ}{\tan 30^\circ} = 3$

Hint If difference is of 15° write $\sqrt{3}$ and in case of 30°, write 3.

Case II

Clearly, $\theta_2 > \theta_1$

Here, perpendiculars AB and ED are equal in length.

Then, base for greater angle : base for smaller angle

$$= 1 : \left[\frac{\tan(\text{greater angle})}{\tan(\text{smaller angle})}\right]$$

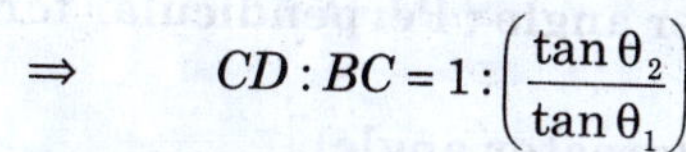

$\Rightarrow \quad CD : BC = 1 : \left(\dfrac{\tan\theta_2}{\tan\theta_1}\right)$

[Illustrations may be framed by taking different sets of values of angles]

Case III

Clearly, $\theta_2 > \theta_1$

Here, perpendiculars are same.

Then, base for greater angle: base for smaller angle

$$= 1 : \left[\frac{\tan(\text{greater angle})}{\tan(\text{smaller angle})}\right]$$

$\Rightarrow BD : DC = 1 : \left(\dfrac{\tan\theta_2}{\tan\theta_1}\right)$

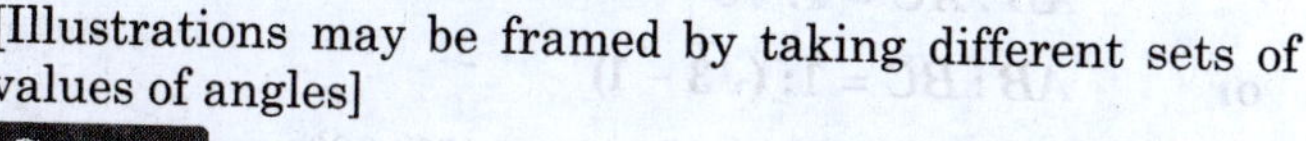

[Illustrations may be framed by taking different sets of values of angles]

Case IV

Clearly, $\theta_2 > \theta_1$

Here, ΔACE and ΔBDE have equal perpendiculars.

Then, base for greater angle: base for smaller angle

$$= 1 : \left[\frac{\tan(\text{greater angle})}{\tan(\text{smaller angle})}\right]$$

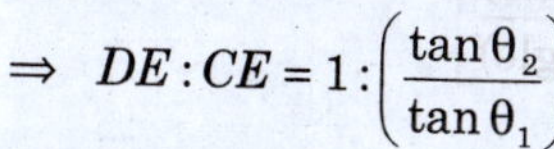

$\Rightarrow \quad DE : CE = 1 : \left(\dfrac{\tan\theta_2}{\tan\theta_1}\right)$

[Illustrations may be framed by taking different sets of values of angles]

Type B

When bases are same/equal for both given angles

Case I

Clearly, $\theta_2 > \theta_1$

Now, $\quad \tan\theta_2 = \dfrac{AC}{CD}$ and $\tan\theta_1 = \dfrac{AB}{BE}$

Then, $\quad \dfrac{\tan\theta_2}{\tan\theta_1} = \dfrac{AC}{AB} \Rightarrow AB : AC = 1 : \left(\dfrac{\tan\theta_2}{\tan\theta_1}\right)$

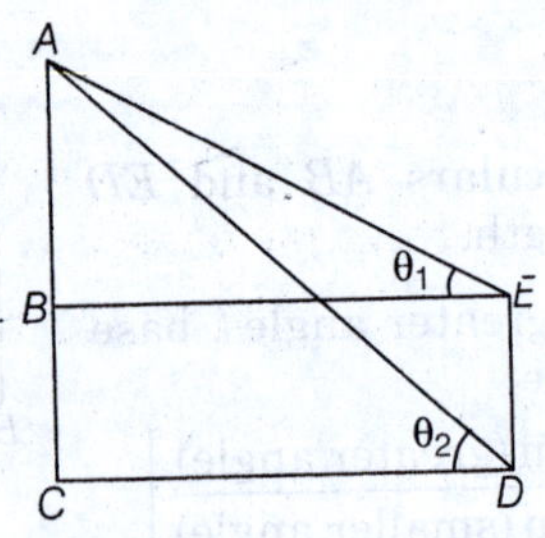

Perpendicular for smaller angle : Perpendicular for greater angle

$$= 1 : \left[\frac{\textbf{tan (greater angle)}}{\textbf{tan (smaller angle)}}\right]$$

Illustration

Perpendicular for smaller angle (30°): Perpendicular for greater angle (60°) $= 1 : \left[\frac{\tan 60^\circ}{\tan 30^\circ}\right]$

$\Rightarrow \quad AB : AC = 1 : 3$

$\Rightarrow \quad \boldsymbol{AB : AC = 1 : 3}$

or $\quad \boldsymbol{AB : BC = 1 : 2}$

Similarly,

$\quad AB : AC = 1 : \sqrt{3}$

or $\quad AB : BC = 1 : (\sqrt{3} - 1)$

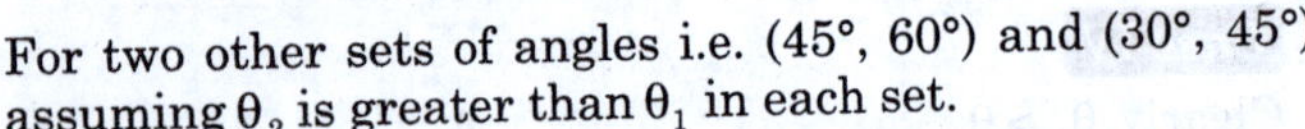
For two other sets of angles i.e. (45°, 60°) and (30°, 45°) assuming θ_2 is greater than θ_1 in each set.

Case II

Clearly, $\theta_2 > \theta_1$

Here, bases are same.

Then, perpendicular for smaller angle : perpendicular for greater angle

$$= 1 : \left[\frac{\tan(\text{greater angle})}{\tan(\text{smaller angle})}\right]$$

$$\Rightarrow CD : AB = 1 : \left(\frac{\tan\theta_2}{\tan\theta_1}\right)$$

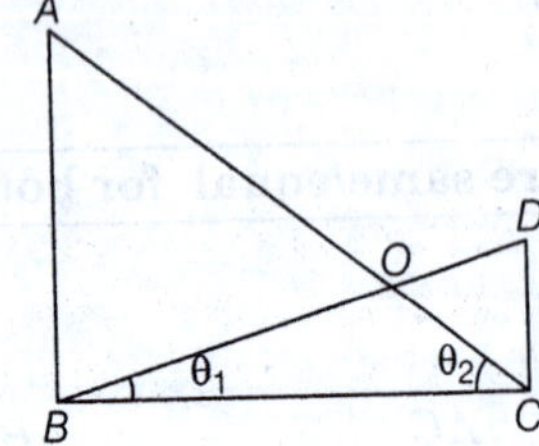

[Illustration may be framed by taking different sets of values of angles]

Case III

Clearly, $\theta_2 > \theta_1$

Here, bases are equal,

Then, perpendicular for smaller angle : perpendicular for greater angle

$$= 1 : \left[\frac{\tan(\text{greater angle})}{\tan(\text{smaller angle})}\right]$$

$$\Rightarrow ED : AB = 1 : \left(\frac{\tan\theta_2}{\tan\theta_1}\right)$$

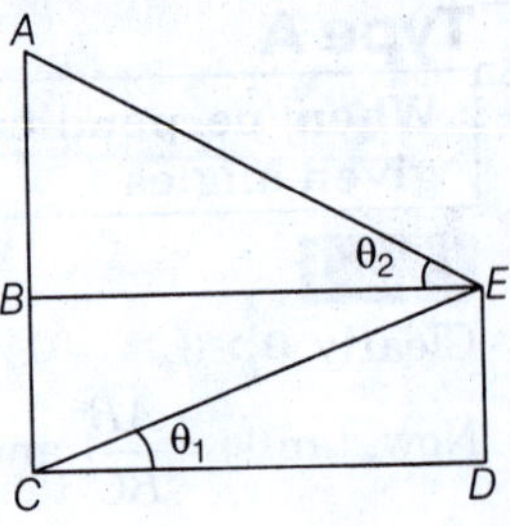

[Illustration may be framed by taking different sets of values of angles]

Case IV

Clearly, $\theta_2 > \theta_1$

Here, bases are same,

Then, perpendicular for smaller angle : perpendicular for greater angle

$$= 1 : \left[\frac{\tan(\text{greater angle})}{\tan(\text{smaller angle})}\right]$$

$$\Rightarrow BC : AB = 1 : \left(\frac{\tan\theta_2}{\tan\theta_1}\right)$$

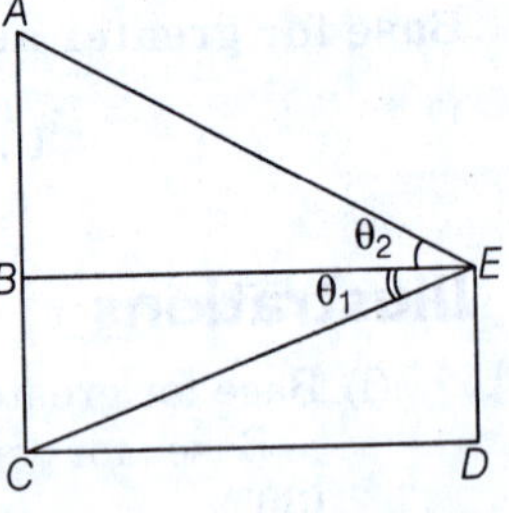

[Illustration may be framed by taking different sets of values of angles]

Case V

Clearly, $\theta_2 > \theta_1$

Here, bases are same, Then, perpendicular for smaller angle : perpendicular for greater angle

$$= 1 : \left[\frac{\tan(\text{greater angle})}{\tan(\text{smaller angle})}\right]$$

$$\Rightarrow BC : AC = 1 : \left(\frac{\tan\theta_2}{\tan\theta_1}\right)$$

[Illustrations may be framed by taking different sets of values of angles].

Type C

When given angles are complementary to each other.

Case I

In ΔBCE, $\tan(90^\circ - \theta) = \dfrac{BC}{BE}$

$\Rightarrow \quad \cot\theta = \dfrac{BC}{BE}$

In ΔABE, $\quad \cot\theta = \dfrac{BE}{AB}$

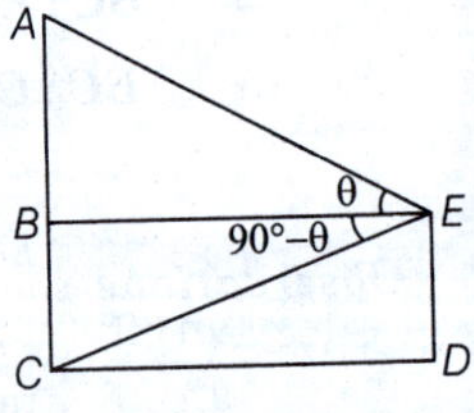

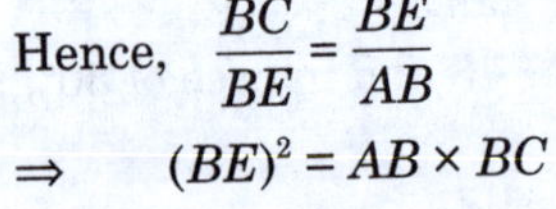
Hence, $\dfrac{BC}{BE} = \dfrac{BE}{AB}$

$\Rightarrow \quad (BE)^2 = AB \times BC$

$\Rightarrow (\text{common base})^2 = \text{product of perpendiculars}$

In other words, product of bases for both angles = product of perpendiculars for both angles.

Case II

In $\Delta ABC, \tan(90° - \theta) = \dfrac{AB}{BC}$

$\Rightarrow \quad \cot\theta = \dfrac{AB}{BC}$

In $\Delta CDE, \quad \cot\theta = \dfrac{CD}{ED}$

Hence, $\dfrac{AB}{BC} = \dfrac{CD}{ED} \Rightarrow BC \times CD = AB \times ED$

i.e. Product of bases for both angles = Product of perpendiculars for both angles.

Case III

In $\Delta ABC, \tan(90° - \theta) = \dfrac{AB}{BC}$

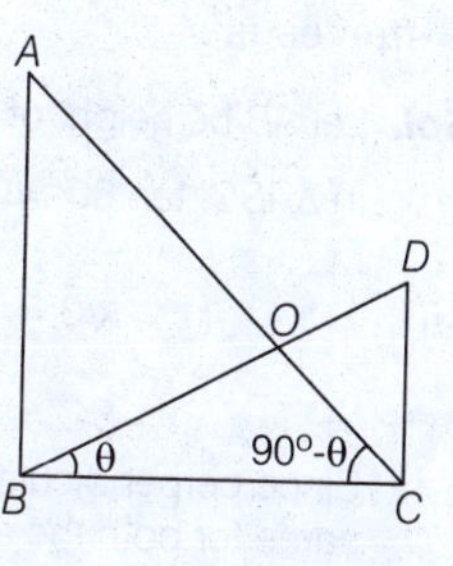

$\Rightarrow \quad \cot\theta = \dfrac{AB}{BC}$

In $\Delta BCD, \quad \cot\theta = \dfrac{BC}{CD}$

Hence, $\dfrac{AB}{BC} = \dfrac{BC}{CD}$

$\Rightarrow \quad BC \times BC = AB \times CD$

i.e. Product of bases for both angles

= Product of perpendiculars for both angles.

Remarks

θ_1 and θ_2 may be complementary to each other in some cases of Types A and B, where we may apply the concept of Type C.

Alternative Approach

This approach is basically filtered/tailored form of the concepts distussed before and also very effective.

Type A

When perpendiculars are same/equal for both given angles.

It is known that $\dfrac{\tan 60°}{\tan 45°} = \dfrac{\tan 45°}{\tan 30°} = \sqrt{3}$ and $\dfrac{\tan 60°}{\tan 30°} = 3.$

Further, in case of same/equal perpendiculars, base for greater angle : base for smaller angle

$= 1 : \left[\dfrac{\tan(\text{greater angle})}{\tan(\text{smaller angle})}\right]$

$\Rightarrow$ smaller base: greater base $= 1 : \left[\dfrac{\tan(\text{greater angle})}{\tan(\text{smaller angle})}\right].$

$\Rightarrow$ greater base = smaller base $\times \left[\dfrac{\tan(\text{greater angle})}{\tan(\text{smaller angle})}\right]$

Conclusion :

Greater base = Smaller base $\times\sqrt{3}$

[if difference of angles is 15°]

and Greater base = Smaller base × 3

[if difference of angles is 30°]

Note *Smaller angle will have greater base and vice-versa. These conclusions are not applicable if angles are not from a set of 30°, 45° and 60°.*

Type B

When bases are same/equal for both given angles

In such cases, perpendicular for smaller angle : perpendicular for greater angle $= 1 : \left[\dfrac{\tan(\text{greater angle})}{\tan(\text{smaller angle})}\right]$

$\Rightarrow$ smaller perpendicular : greater perpendicular

$= 1 : \left[\dfrac{\tan(\text{greater angle})}{\tan(\text{smaller angle})}\right].$

$\Rightarrow$ greater perpendicular

$= \text{smaller perpendicular} \times \left[\dfrac{\tan(\text{greater angle})}{\tan(\text{smaller angle})}\right]$

Conclusion :

Greater Perpendicular = Smaller Perpendicular $\times\sqrt{3}$

[if difference of angles is 15°]

and Greater Perpendicular = Smaller Perpendicular × 3

[if difference of angles is 30°]

Note *Greater angle will have greater perpendicular and the smaller angle will have the smaller. These conclusions are not applicable if angles are not from a set of 30°, 45° and 60°.*

Type C

When given angles are complementary to each other

Products of bases for both angles

= Product of perpendiculars for both angles.

For the above both types also, this result will hold good provided both angles are complementary to each other.

Illustrations

Problems Based on the Concepts Associated with the Cases Having Same/Equal Perpendiculars for Both Angles :

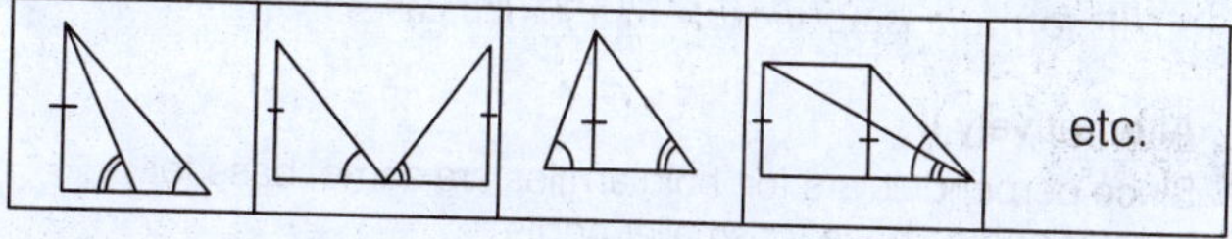

Base for greater angle : Base for smaller angle

$$= 1 : \left(\frac{\tan\theta_2}{\tan\theta_1}\right), (\theta_2 > \theta_1)$$

Problem 1. If angles of elevation of the sun changes from 30° to 45°, the length of the shadow of a pillar decreases by 20 m. The height of the pillar is :
[SSC(CPO-SI/ASI/IO)-11, RRB Chennai-01]

Sol. Let the height of pillar AB be h metres and the initial length of shadow (i.e. BD) be x metres. Then, the latter length of shadow $= (x - 20)$ m

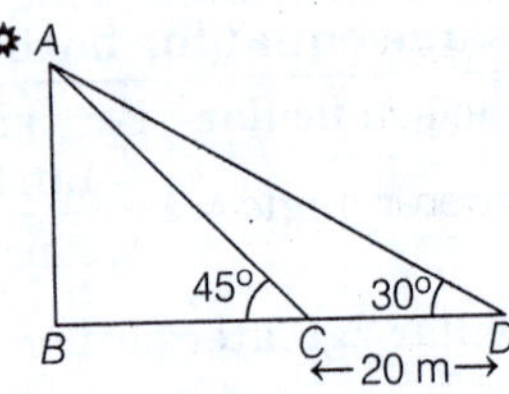

Now, in ΔABC, $\tan 45° = \frac{AB}{BC} \Rightarrow 1 = \frac{h}{x - 20} \Rightarrow h = (x - 20)$

$\therefore \quad x = h + 20$

Further, in ΔABD,

$$\tan 30° = \frac{AB}{BD} \Rightarrow \frac{1}{\sqrt{3}} = \frac{h}{x}$$

$\Rightarrow \quad \frac{1}{\sqrt{3}} = \frac{h}{h + 20} \qquad [\because x = h + 20]$

$\Rightarrow \quad \sqrt{3}h = h + 20 \Rightarrow (\sqrt{3}h - h) = 20$

$\Rightarrow \quad h(\sqrt{3} - 1) = 20$

$\Rightarrow \quad h = \frac{20}{\sqrt{3} - 1} = \frac{20}{\sqrt{3} - 1} \times \frac{\sqrt{3} + 1}{\sqrt{3} + 1}$

$$= \frac{20(\sqrt{3} + 1)}{(3 - 1)} = 10(\sqrt{3} + 1)$$

Alternatively I

In ΔABC, $\tan 45° = \frac{AB}{BC} \Rightarrow BC = \frac{AB}{\tan 45°}$

$\Rightarrow \quad BC = AB \times \cot 45°$

Further, in ΔABD, $\tan 30° = \frac{AB}{BD} \Rightarrow BD = AB \times \cot 30°$

Now, $\quad CD = BD - BC$

$\Rightarrow \quad 20 = AB \times \cot 30° - AB \times \cot 45°$

$\Rightarrow \quad 20 = AB(\cot 30° - \cot 45°)$

$\therefore \quad AB = \frac{20}{\cot 30° - \cot 45°} = \frac{20}{(\sqrt{3} - 1)} = 10(\sqrt{3} + 1)$

Remarks

Height of the pillar, $h = \frac{d}{\cot\theta_1 - \cot\theta_2}$

[Where, d = difference between lengths of shadow and $\theta_1 < \theta_2$*]*

This formula is applicable in all such cases.

Alternatively II

Since perpendiculars for both angles are same, base for greater angle : base for smaller angle

$$= 1 : \left[\frac{\tan \text{(greater angle)}}{\tan \text{(smaller angle)}}\right]$$

$\Rightarrow \quad BC : BD = 1 : \frac{\tan 45°}{\tan 30°} \Rightarrow BC : BD = 1 : \sqrt{3}$

Hence, $\quad BC : CD = 1 : (\sqrt{3} - 1)$

But given, $(\sqrt{3} - 1)$ units $\Rightarrow$ 20 m.

$\therefore \quad 1 \text{ unit} = \frac{20}{\sqrt{3} - 1} = 10(\sqrt{3} + 1)$ m i.e. BC

Now, in ΔABC, $\tan 45° = \frac{AB}{BC}$

$\Rightarrow \quad AB = BC = 10(\sqrt{3} + 1)$ m

Problem 2. A man standing at a point D is watching the top of a tower, which makes an angle of elevation of 30°. The man walks some distance towards the tower and then his angle of elevation of the top of the tower is 60°. If the height of the tower is 30 m, then the distance he moves is :
[SSC (10+2, LDC/DEO)-12, MAT-12]

Sol. Let AB be height of the tower.

In ΔABC, $\tan 60° = \frac{AB}{BC}$

$\Rightarrow \quad \sqrt{3} = \frac{30}{BC}$

$\Rightarrow \quad BC = 10\sqrt{3}$

Since perpendiculars are same for both the angles.
Base for greater angle :
Base for smaller angle

$$= 1 : \left[\frac{\tan \text{(greater angle)}}{\tan \text{(smaller angle)}}\right]$$

$\Rightarrow \quad BC : BD = 1 : 3 \qquad [\because$ difference between angles is 30°]

$\Rightarrow \quad BC : CD = 1 : (3 - 1) = 1 : 2$

Hence, $\quad CD = 2\,BC = 2 \times 10\sqrt{3} = 20\sqrt{3}$ m

Alternatively I

Since the both angles are complementary to each other,
Product of bases = Product of perpendiculars

$\Rightarrow \quad BC \times BD = AB \times AB$

$\Rightarrow \quad 10\sqrt{3} \times BD = 30 \times 30$

$\Rightarrow \quad BD = 30\sqrt{3}$

Now, $CD = BD - BC = 30\sqrt{3} - 10\sqrt{3} = 20\sqrt{3}$ m

Alternatively II

We know, $h = \frac{d}{\cot 30° - \cot 60°} \Rightarrow 30 = \frac{d}{\sqrt{3} - \frac{1}{\sqrt{3}}}$

$\Rightarrow \quad d = 30 \times \frac{2}{\sqrt{3}} = 20\sqrt{3}$ m

Alternatively III

Since perpendiculars are same, smaller angle will have greater base.

Now, greater base = smaller base × 3

$[\because$ difference of angles is 30°]

$\Rightarrow \quad BD = 3BC = 3 \times 10\sqrt{3} = 30\sqrt{3}$

$$\left[\because \tan 60° = \frac{AB}{BC} \Rightarrow \sqrt{3} = \frac{30}{BC} \Rightarrow BC = 10\sqrt{3}\right]$$

Now, $\quad CD = BD - BC = 30\sqrt{3} - 10\sqrt{3} = 20\sqrt{3}$ m

Problem 3. A man on a cliff observes a boat an angle of depression of 30°, which is approaching the shore to the point immediately beneath the observer with a uniform speed. 6 min later, the angle of depression of the boat is found to be 60°. Find the time taken by the boat to reach the shore.

Sol. Let AB be height of the cliff.

In ΔABC, $\tan 60^\circ = \dfrac{AB}{BC}$

$\Rightarrow \quad \sqrt{3} = \dfrac{AB}{BC}$

$\Rightarrow \quad AB = \sqrt{3}BC$

and in ΔABC, $\tan 30^\circ = \dfrac{AB}{BD}$

$\Rightarrow \quad \dfrac{1}{\sqrt{3}} = \dfrac{AB}{BD}$

$\Rightarrow \quad BD = AB \times \sqrt{3}$

$\Rightarrow \quad BD = \sqrt{3}BC \times \sqrt{3} \qquad (\because AB = \sqrt{3}BC)$

$\Rightarrow \quad BD = 3BC \Rightarrow BD : BC = 3 : 1$

$\Rightarrow \quad (BD - BC) : BC = (3-1) : 1$

$\Rightarrow \quad CD : BC = 2 : 1$

Since the boat takes 6 min to cover the distance of 2 units, it will take 3 min to cover a distance of 1 unit. Hence, the boat will reach the shore in next 3 min.

Alternatively I

Since perpendiculars are same for both the angles. Base for greater angle : Base for smaller angle

$= 1 : \left[\dfrac{\tan \text{(greater angle)}}{\tan \text{(smaller angle)}}\right]$

$\Rightarrow BC : BD = 1 : 3 \qquad [\because$ difference between angles is 30°]

$\Rightarrow BC : CD = 1 : (3-1) = 1 : 2$

Since the boat takes 6 min to cover the distance of 2 units, it will take 3 min to cover a distance of 1 unit. Hence, the boat will reach the shore in next 3 min.

Alternatively II

In ΔABC, $\tan 60^\circ = \dfrac{AB}{BC} \Rightarrow \sqrt{3} = \dfrac{AB}{BC}$

$\Rightarrow \quad AB = \sqrt{3}BC$

Since the both angles are complementary to each other, product of bases = product of perpendiculars

$\Rightarrow \quad BC \times BD = AB \times AB \Rightarrow BC \times BD = \sqrt{3}BC \times \sqrt{3}BC$

$\Rightarrow \quad BD = 3BC \Rightarrow BC : BD = 1 : 3$

$\Rightarrow \quad BC : CD = 1 : (3-1) = 1 : 2$

Since the boat takes 6 min to cover the distance of 2 units, it will take 3 min to cover a distance of 1 unit. Hence, the boat will reach the shore in next 3 min.

Alternatively III

Since perpendiculars are same, smaller angle will have greater base.

Now, greater base = smaller base × 3

$[\because$ difference of angles is 30°]

$\Rightarrow \quad BD = 3BC \rightarrow BC . BD = 1 : 3$

$\Rightarrow \quad BC : CD = 1 : (3-1) = 1 : 2$

Since the boat takes 6 min to cover the distance of 2 units, it will take 3 min to cover a distance of 1 unit. Hence, the boat will reach he shore in next 3 min.

Problem 4. At a point on a horizontal line through the base of monument, the angle of elevation of the top of the monument is found to be such that its tangent is 1/5. On walking 138 m towards the monument, the secant of the angle of the elevation is found to be $\dfrac{\sqrt{193}}{12}$. The height of the monument (in metres) is :

Sol. Let AB be height of the monument. Since perpendiculars are same for both the angles, Base for greater angle : Base for smaller angle $= 1 : \left[\dfrac{\tan \text{(greater angle)}}{\tan \text{(smaller angle)}}\right]$

$\Rightarrow BC : BD = 1 : \left(\dfrac{\tan \theta_2}{\tan \theta_1}\right)$

$= 1 : \left(\dfrac{\tan \theta_2}{1/5}\right)$

$\Rightarrow BC : BD = 1 : \dfrac{\sqrt{\left(\dfrac{\sqrt{193}}{12}\right)^2 - 1}}{\dfrac{1}{5}}$

$= 1 : \dfrac{7/12}{1/5} \qquad [\because \tan \theta_2 = \sqrt{\sec^2 \theta_2 - 1}]$

$\Rightarrow \quad BC : BD = 1 : \dfrac{35}{12} = 12 : 35 \Rightarrow BC : CD = 12 : (35 - 12)$

$\Rightarrow \quad BC : CD = 12 : 23$

Given, 23 units = 138 m

$\therefore$ 12 units = 72 m

Hence, $BC = 72$ m

Now, in ΔABC, $\tan \theta_2 = \dfrac{AB}{BC} \Rightarrow \dfrac{7}{12} = \dfrac{AB}{72} \Rightarrow AB = 42$ m

Alternatively

We know, $h = \dfrac{d}{\cot \theta_1 - \cot \theta_2}$

$\Rightarrow$ Height of monument $= \dfrac{138}{\dfrac{1}{\tan \theta_1} - \dfrac{1}{\tan \theta_2}}$

$= \dfrac{138}{\dfrac{1}{1/5} - \dfrac{1}{7/12}} = \dfrac{138}{\dfrac{5}{1} - \dfrac{12}{7}} = \dfrac{138 \times 7}{23} = 42$ m

Problem 5. The angles of elevation of the top of a tower from two points A and B lying on the horizontal through the foot of the tower are respectively 15° and 30°. If A and B are on the same side of the tower and AB = 48 m, then the height of the tower is : ***[FCI(Asstt. Gr.III)-12]***

Sol. Obviously, $BC : AC$

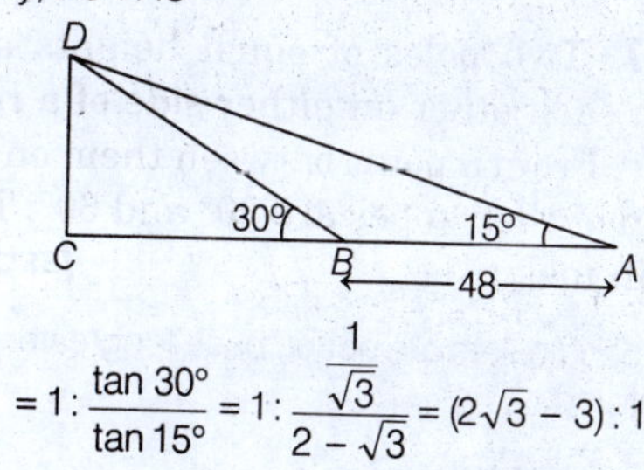

$= 1 : \dfrac{\tan 30^\circ}{\tan 15^\circ} = 1 : \dfrac{\frac{1}{\sqrt{3}}}{2 - \sqrt{3}} = (2\sqrt{3} - 3) : 1$

$\Rightarrow \quad BC : AB = (2\sqrt{3} - 3) : [1 - (2\sqrt{3} - 3)]$

$\Rightarrow \quad CB : BA = (2\sqrt{3} - 3) : (4 - 2\sqrt{3})$

As given, $4 - 2\sqrt{3} \Rightarrow 48$

$\therefore \quad 1 \Rightarrow \dfrac{48}{4 - 2\sqrt{3}}$

$\therefore \quad (2\sqrt{3} - 3) \Rightarrow \dfrac{48}{(4 - 2\sqrt{3})} \times (2\sqrt{3} - 3)$

$= \dfrac{24}{(2\sqrt{3} - 3)} \times \sqrt{3}\,(2\sqrt{3} - 3) = 24\sqrt{3}$

It means $CB = 24\sqrt{3}$ m

Now, in ΔDCB, $DC = CB \times \tan 30^\circ$

$= 24\sqrt{3} \times \dfrac{1}{\sqrt{3}} = 24$ m

Alternatively

We know that, height of the tower,

$$h = \frac{d}{\cot \theta_1 - \cot \theta_2}$$

$$h = \frac{48}{\cot 15^\circ - \cot 30^\circ}$$

$$= \frac{45}{(2 + \sqrt{3}) - \sqrt{3}} = 24 \text{ m}$$

Problem 6. The angle of elevation of the top of a tower from a point A due South of the tower is α and from a point B due East of the tower is β. If $AB = d$, then the height of the tower is

(a) $\dfrac{d}{\sqrt{\tan^2 \alpha - \tan^2 \beta}}$ (b) $\dfrac{d}{\sqrt{\tan^2 \alpha + \tan^2 \beta}}$

(c) $\dfrac{d}{\sqrt{\cot^2 \alpha + \cot^2 \beta}}$ (d) $\dfrac{d}{\sqrt{\cot^2 \alpha - \cot^2 \beta}}$

Sol. Here, OM is the height of the tower.

Clearly, $\tan \beta = \dfrac{OM}{OB}$

$\Rightarrow \quad OB = OM \times \cot \beta$

Also, $\quad \tan \alpha = \dfrac{OM}{OA}$

$\Rightarrow \quad OA = OM \times \cot \alpha$

Now, in right-angled ΔOAB,

$d = \sqrt{(OA)^2 + (OB)^2}$

$\Rightarrow \quad d = \sqrt{(OM \times \cot \alpha)^2 + (OM \times \cot \beta)^2}$

$\Rightarrow \quad OM = \dfrac{d}{\sqrt{\cot^2 \alpha + \cot^2 \beta}}$

Problem 7. Two poles of equal heights are standing opposite to each other on either side of a road which is 100 m wide. From a point between them on road, angles of elevation of their topes are 30° and 60°. The height of each pole in metre, is : ***[SSC (CGL-T-II)-20]***

Sol. Since perpendiculars are equal, base for greater angle : base for smaller angle

$= 1 : \dfrac{\tan 60^\circ}{\tan 30^\circ}$

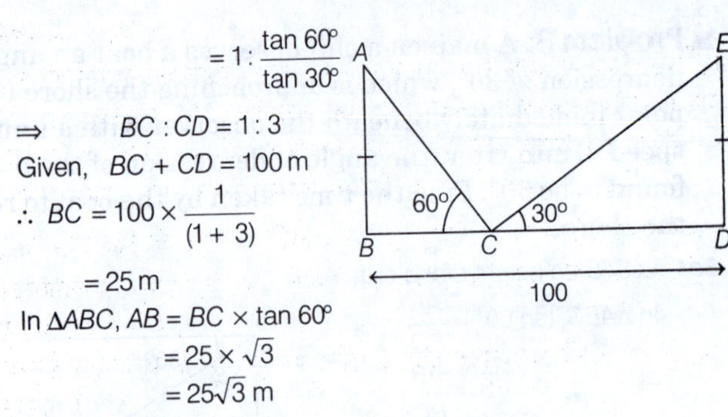

$\Rightarrow \quad BC : CD = 1 : 3$

Given, $BC + CD = 100$ m

$\therefore \; BC = 100 \times \dfrac{1}{(1 + 3)}$

$= 25$ m

In ΔABC, $AB = BC \times \tan 60^\circ$

$= 25 \times \sqrt{3}$

$= 25\sqrt{3}$ m

Alternatively

In ΔABC, $\tan 60^\circ = \dfrac{AB}{BC} \Rightarrow BC = \dfrac{AB}{\sqrt{3}}$

Since both angles are complementary to each other, product of bases = product of perpendiculars.

$\Rightarrow \quad BC \times (100 - BC) = AD \times ED$

$\Rightarrow \quad \dfrac{AB}{\sqrt{3}}\left(100 - \dfrac{AB}{\sqrt{3}}\right) = AB \times AB$

$\Rightarrow \quad \dfrac{100AB}{\sqrt{3}} - \dfrac{(AB)^2}{3} = (AB)^2 \Rightarrow \dfrac{100AB}{\sqrt{3}} = \dfrac{4(AB)^2}{3}$

$\Rightarrow \quad AB = \dfrac{300}{4\sqrt{3}} = 25\sqrt{3}$ m

Problem 8. The angles of depression of the banks on opposite sides of a river from a point on a bridge across the river, are 45° and 60° respectively. Find the width of the the river if the height of the bridge from the banks is $10\sqrt{3}$ m.

Sol. Let B and C be two points on the opposite banks. It means BC is the width of the river.

In ΔABC, $BD = \dfrac{AD}{\tan 45^\circ} = AD$

$= 10\sqrt{3}$ (as given)

Since perpendiculars are same, base for greater angle : base for smaller angle

$= 1 : \dfrac{\tan 60^\circ}{\tan 45^\circ}$

$\Rightarrow \quad CD : BD = 1 : \sqrt{3}$

$\Rightarrow \quad CD = \dfrac{BD}{\sqrt{3}} = \dfrac{10\sqrt{3}}{\sqrt{3}} = 10$

Now, $\quad BC = BD + CD = 10\sqrt{3} + 10$

$= 10(\sqrt{3} + 1)$ m

Alternatively

In ΔABD, $BD = \dfrac{AD}{\tan 45^\circ} = AD = 10\sqrt{3}$ (as given)

In ΔADC, $DC = \dfrac{AD}{\tan 60^\circ} = \dfrac{10\sqrt{3}}{\sqrt{3}} = 10$

Hence, $BC = 10\sqrt{3} + 10 = 10(\sqrt{3} + 1)$ m

Problem 9. An observer, 1 m tall, is standing at certain distance from the tower 31 m high. The angle of elevation of the top of the tower from his eyes changes from 30° to 60° when he walks towards the tower. What is the distance he walked towards the tower?

Sol. Here, EF and GD are two positions of the boy.

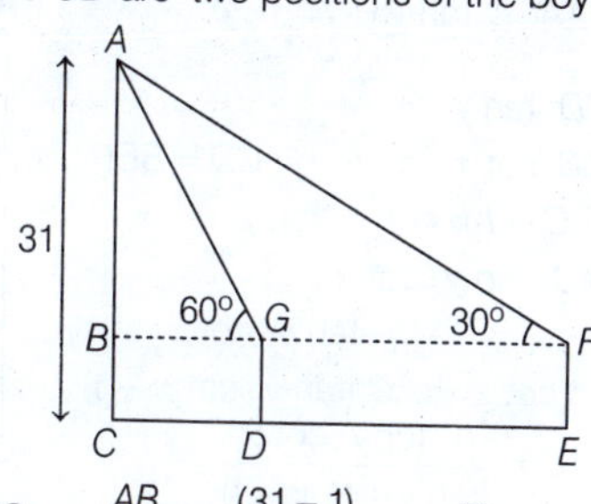

In ΔABG, $BG = \frac{AB}{\tan 60^\circ} = \frac{(31-1)}{\sqrt{3}} = 10\sqrt{3}$ m

Here, perpendiculars are same. Thus, greater base = smaller base × 3 [as difference between angles is 30° and angles are from a set of 30°, 45° and 60°.]

$\Rightarrow \quad BF = 3 \times BG$

$\Rightarrow \quad BG : GF = 1 : 2$

$\therefore \quad GF = 2 \times 10\sqrt{3} = 20\sqrt{3}$

Hence, the observer walked $20\sqrt{3}$ m towards the tower.

Alternatively I

We know, height of the tower, $h = \frac{d}{\cot \theta_1 - \cot \theta_2}$

$\Rightarrow \quad (31-1) = \frac{d}{\cot 30^\circ - \cot 60^\circ}$

$\Rightarrow \quad (31-1) = \frac{d}{\sqrt{3} - \frac{1}{\sqrt{3}}}$

$\Rightarrow \quad d = 20\sqrt{3}$ m

Alternatively II

In ΔABG, $BG = \frac{AB}{\tan 60^\circ} = \frac{(31-1)}{\sqrt{3}} = 10\sqrt{3}$ m

Since both angles are complementary, $BG \times BF = AB \times AB$

$\Rightarrow \quad 10\sqrt{3} \times BF = 30 \times 30$

$\Rightarrow \quad BF = 30\sqrt{3}$

Hence, $GF = 30\sqrt{3} - 10\sqrt{3} = 20\sqrt{3}$ m

Problem 10. The angle of elevation of an aeroplane from a point on the ground is 60°. After 15 sec flight, the elevation changes to 30°. If the aeroplane is flying at a height of $1500\sqrt{3}$ m , find the speed of the plane.

[SSC (Delhi Police-SI)-12, SSC (CGL-T-II)-17]

Sol. Let the plane moves from point A to B and accordingly, the angle of elevation changes.

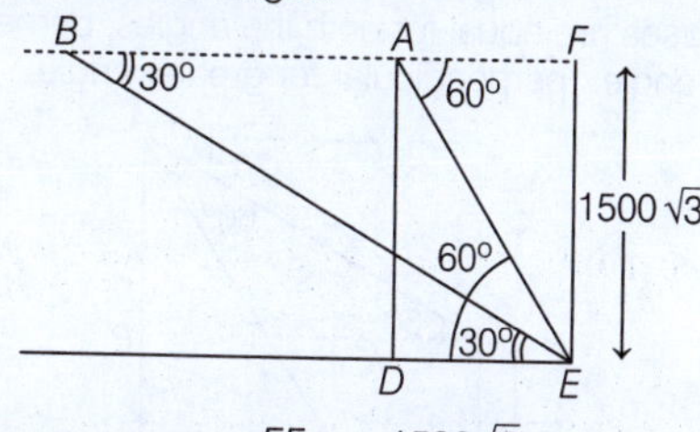

In ΔAFE, $AF = \frac{EF}{\tan 60^\circ} = \frac{1500\sqrt{3}}{\sqrt{3}} = 1500$

Here, perpendiculars are same. Thus, greater base = smaller base × 3 [as difference between angles is 30° and angles are form a set of 30°, 45° and 60°.]

$\Rightarrow \quad BF = 3 \times AF$

$\Rightarrow \quad BF : AF = 3 : 1$

$\Rightarrow \quad AF : AB = 1 : 2$

$\therefore \quad AB = 2 \times 1500 = 3000$ m

Now, speed $= \frac{3000}{15} = 200$ m/sec

Alternatively I

Height of the plane, $h = \frac{d}{\cot 30^\circ - \cot 60^\circ}$

$\Rightarrow \quad 1500\sqrt{3} = \frac{d}{\sqrt{3} - \frac{1}{\sqrt{3}}}$

$\Rightarrow \quad d = 1500\sqrt{3} \times \frac{2}{\sqrt{3}} = 3000$ m

Now, speed $= \frac{3000}{15} = 200$ m/sec.

Alternatively II

In ΔBDC, $CD = \frac{BD}{\tan 60^\circ}$

$= \frac{1500\sqrt{3}}{\sqrt{3}} = 1500$

Clearly, $CD : EC = 1 : 3$

$\Rightarrow \quad CD : ED = 1 : 2$

Hence, $ED = 1500 \times 2$

$= 3000$ m

Now, speed $= \frac{3000}{15} = 200$ m/sec.

Alternatively III

In ΔBDC, $CD = \frac{BD}{\tan 60^\circ} = \frac{1500\sqrt{3}}{\sqrt{3}} = 1500$

Since both angles are complementary,

$EC \times DC = AE \times BD$

$\Rightarrow \quad EC \times 1500 = 1500\sqrt{3} \times 1500\sqrt{3}$

$\Rightarrow \quad EC = 4500$

$\therefore \quad ED = AB = 4500 - 1500 = 3000$ m

Now, speed $= \frac{3000}{15} = 200$ m/sec

Problem 11. A tower standing on the horizontal plane subtends a certain angle at a point 160 m apart from the foot of the tower. On advancing 100 m towards it, the tower is found to subtend an angle twice as before, the height of the tower is :

[SSC(CGL-T-II)-13]

Sol. Here, $BD = 160$ m

Since perpendiculars are same for both the angles, base for greater angle : base for smaller angle

$= 1 : \left[\frac{\tan(\text{greater angle})}{\tan(\text{smaller angle})}\right]$

$\Rightarrow \quad BC : BD = 1 : \frac{\tan(2x)}{\tan x}$

$\Rightarrow \quad (160 - 100) : 160$

$= 1 : \left[\frac{2\tan x}{1 - \tan^2 x} \div \tan x\right]$

$\Rightarrow \quad \frac{3}{8} = \frac{1 - \tan^2 x}{2}$

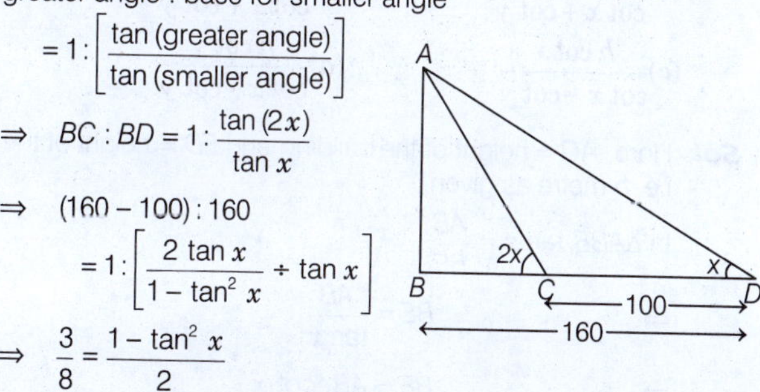

$\Rightarrow \quad 4\tan^2 x = 4 - 3 \Rightarrow \tan x = \frac{1}{2}$

Now, in ΔABD, $\tan x = \frac{AB}{BD}$

$\Rightarrow \quad \frac{1}{2} = \frac{AB}{160} \Rightarrow AB = 80\text{ m}$

Problem 12. The angle of elevation of the top of the tower observed from each of the three points A, B and C on the ground, forming a triangle is the same angle α. If R is the circumradius of the triangle ABC, then height of the tower is :

(a) $R\sin\alpha$ (b) $R\cos\alpha$

(c) $R\cot\alpha$ (d) $R\tan\alpha$

Sol. Here, OM is the height of the tower.

Given that the angle of elevation of the top of the tower observed from each of three points is the same. It implies that the foot of the tower is at circumcentre (i.e. at O) of the ΔABC.

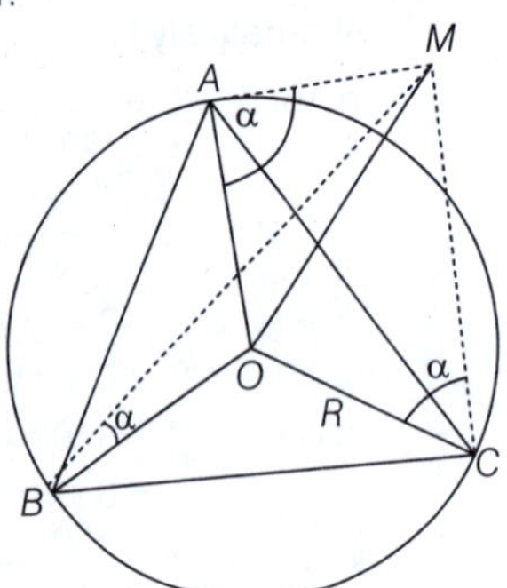

In ΔOMC, $\tan\alpha = \frac{OM}{OC}$

$\Rightarrow \quad \tan\alpha = \frac{OM}{R}$

$\Rightarrow \quad OM = R\tan\alpha$

Problems Based on the Concepts Associated with the cases having same/equal bases for both Angles

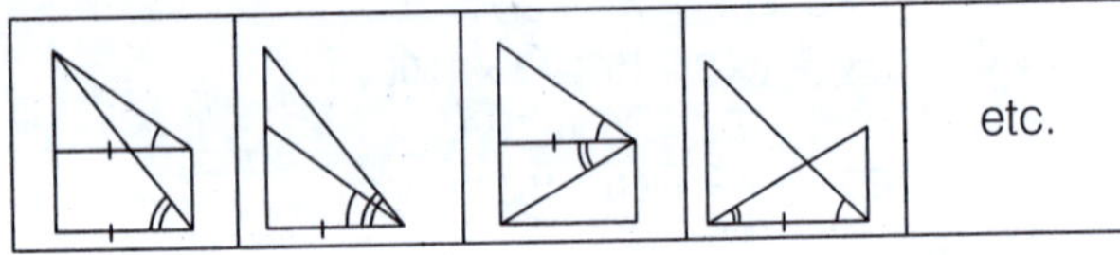

Perpendicular for smaller angle : Perpendicular for greater angle $= 1 : \left(\frac{\tan\theta_2}{\tan\theta_1}\right)$, $[\theta_2 > \theta_1]$

Problem 13. The angles of elevation of the top of the building from the top and bottom of a tree are x and y respectively. If the height of the tree is h metre, then, in metre the height of the building is :

(a) $\frac{h\cot x}{\cot x + \cot y}$ (b) $\frac{h\cot y}{\cot x + \cot y}$

(c) $\frac{h\cot x}{\cot x - \cot y}$ (d) $\frac{h\cot y}{\cot x - \cot y}$

Sol. Here, AC = height of the building and ED = height of the tree i.e. h metre as given.

In ΔABE, $\tan x = \frac{AB}{BE}$

$\Rightarrow \quad BE = \frac{AB}{\tan x}$

$\Rightarrow \quad BE = AB\cdot\cot x$

Further, in ΔACD, $\tan y = \frac{AC}{CD}$

$\Rightarrow AC = CD\cdot\tan y$

$\Rightarrow AC = AB\cdot\cot x\cdot\tan y \quad [\because CD = BE]$

$\Rightarrow AC = (AC - h)\cdot\cot x\cdot\tan y$

$\Rightarrow AC = AC\cdot\tan y\cdot\cot x - h\cdot\tan y\cdot\cot x$

$\Rightarrow h\cdot\tan y\cdot\cot x = AC(\tan y\cdot\cot x - 1)$

$$\Rightarrow \quad AC = \frac{h\cdot\tan y\cdot\cot x}{(\tan y\cdot\cot x - 1)} = \frac{h\cdot\frac{\cot x}{\cot y}}{\left(\frac{\cot x}{\cot y} - 1\right)} = \frac{h\cot x}{\cot x - \cot y}$$

Alternatively

Since bases are equal for both the angles, Perpendicular for smaller angle : Perpendicular for greater angle

$$= 1 : \left[\frac{\tan\text{(greater angle)}}{\tan\text{(smaller angle)}}\right]$$

$\Rightarrow \quad AB : AC = 1 : \left[\frac{\tan y}{\tan x}\right]$ [obviously, $y > x$]

$\Rightarrow \quad \frac{AC - h}{AC} = \frac{1\cdot\cot y}{\cot x}$

$\Rightarrow \quad AC\cdot\cot x - h\cot x = AC\cdot\cot y$

$\Rightarrow \quad AC(\cot x - \cot y) = h\cot x$

$\Rightarrow \quad AC = \frac{h\cot x}{\cot x - \cot y}$

Height of the taller object

$$= \frac{\text{height of the smaller object}\cdot\cot(\text{smaller angle})}{\cot(\text{smaller angle}) - \cot(\text{greater angle})}$$

[This formula for computing height of the taller/smaller object, may be applied in similar questions].

Problem 14. There are two vertical posts, one on each side of a road, just opposite to each other. One post is 108m high. From the top of this post, the angles of depression of the top and foot of the other post are 30° and 60° respectively. The height of the other post is

Sol. Since bases are equal for both the angles, perpendicular for smaller angle : perpendicular for greater angle

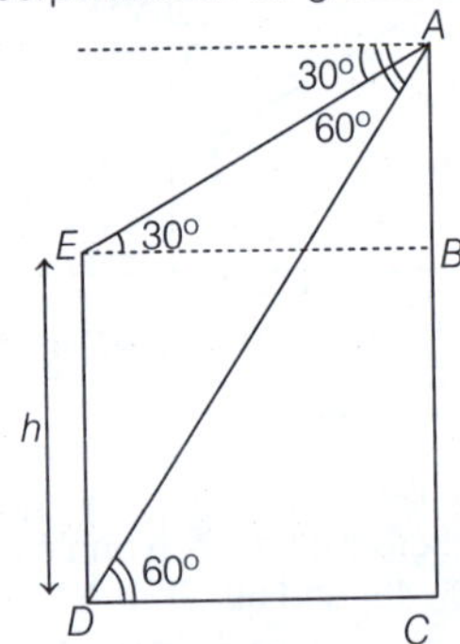

$$= 1 : \left[\frac{\tan \text{(greater angle)}}{\tan \text{(smaller angle)}}\right]$$

$\Rightarrow \quad AB : AC = 1 : 3$

$\Rightarrow \quad AB : 108 = 1 : 3$ (given, $AC = 108$ m)

$\Rightarrow \quad AB = 36$

Now, height of the other post $= AC - AB = 108 - 36 = 72$ m

Alternatively I

Let height of the smaller post be h metre. Then, height of the taller object

$$= \frac{\text{height of the smaller object} \cdot \cot \text{(smaller angle)}}{\cot \text{(smaller angle)} - \cot \text{(greater angle)}}$$

$$\Rightarrow \quad 108 = \frac{h \cdot \cot 30^\circ}{\cot 30^\circ - \cot 60^\circ} \quad \Rightarrow 108 = \frac{h \cdot \sqrt{3}}{\sqrt{3} - \frac{1}{\sqrt{3}}}$$

$$\Rightarrow \quad h = 108 \times \frac{2}{\sqrt{3}} \times \frac{1}{\sqrt{3}} = 72 \text{ m}$$

Alternatively II

In ΔADC, $\quad \tan 60^\circ = \frac{AC}{CD} \Rightarrow CD = \frac{108}{\sqrt{3}} = 36\sqrt{3}$ m

Since both the angles are complementary to each other, the product of bases for both angles will be equal to the product of perpendiculars for both angles.

$\Rightarrow \quad 36\sqrt{3} \times 36\sqrt{3} = 108 \times AB$

$\Rightarrow \quad AB = 36$

Then, $\quad BC = 108 - 36 = 72$ m

Alternatively III

Since bases are equal, greater perpendicular = smaller perpendicular × 3 [as difference between angles is 30° and angles are from a set of 30°, 45° and 60°.]

$\Rightarrow \quad 108 = 3 \times AB \Rightarrow AB = 36$

Then, $\quad BC = 108 - 36 = 72$ m

Problem 15. **The angles of elevation of the top of a building and top of the chimney on the roof of the building from a point on the ground are x and 45° respectively. The height of the building is h metre, then height of the chimney, in metre is :**

(a) $h \cot x + h$ (b) $h \tan x + h$

(c) $h \cot x - h$ (d) $h \tan x - h$

Sol. Here, BD = height of the building (h) and AD = height of the chimney

Since bases are same for both the angles, perpendicular for smaller angles : perpendicular for greater angle

$$= 1 : \left[\frac{\tan \text{(greater angle)}}{\tan \text{(smaller angle)}}\right]$$

$$\Rightarrow BD : AB = 1 : \frac{\tan 45^\circ}{\tan x}$$

$\Rightarrow AB = BD \cdot \cot x = h \cot x$

Now, $AD = AB - BD = h \cot x - h$

Alternatively

Obviously, $\angle BDC = 90^\circ - x$ and $\angle BAC = 45^\circ$

Since 45° is greater than x, $(90^\circ - x)$ will be greater than 45°.

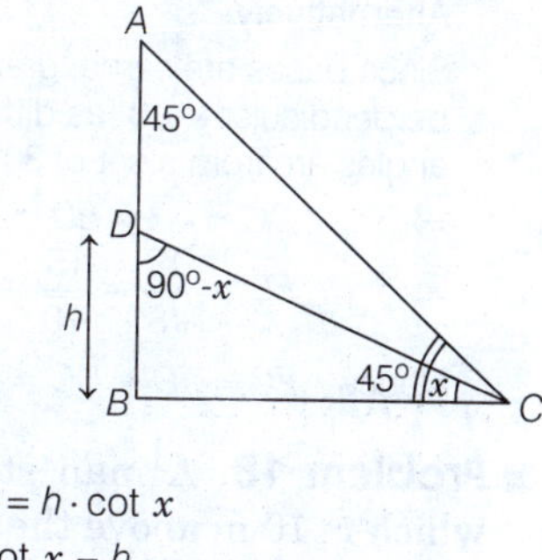

It is clear that perpendiculars are same for both the angles i.e. $\angle BAC$ and $\angle BDC$.

Then, base for greater angle : base for smaller angle

$$= 1 : \left[\frac{\tan \text{(greater angle)}}{\tan \text{(smaller angle)}}\right]$$

$$\Rightarrow \quad BD : AB = 1 : \frac{\tan (90^\circ - x)}{\tan 45^\circ}$$

$AB = BD \cdot \cot x \Rightarrow AB = h \cdot \cot x$

Then, $\quad AD = AB - BD = h \cdot \cot x - h$

Problem 16. **An aeroplane when flying at a height of 3125 m from the ground passes vertically below another plane at an instant when the angles of elevation of the two planes from the same point on the ground are 30° and 60° respectively. The distance between two planes at that instant is :** ***[SSC(10+2,LDC/DEO)-12, CDS-17]***

Sol. Here, bases are same for both angles,

$$\therefore \quad BC : AC = 1 : \frac{\tan 60^\circ}{\tan 30^\circ}$$

$\Rightarrow \quad BC : AC = 1 : 3$

$\Rightarrow \quad 3125 : AC = 1 : 3$

$\Rightarrow \quad AC = 9375$

Now, $AB = AC - BC$

$= 9375 - 3125 = 6250$

Hence, distance between both planes = 6250 m

Alternatively

In ΔBCE, $\tan 30^\circ = \frac{BC}{CD}$

$\Rightarrow CD = 3125\sqrt{3}$

Since both angles are complementary to each other, product of bases = product of perpendiculars.

$\Rightarrow \quad 3125\sqrt{3} \times 3125\sqrt{3} = 3125 \times AC \Rightarrow AC = 9375$

$\therefore \quad AB = 9375 - 3125 = 6250$ m

Problem 17. **From a window 15 m high above the ground, in a street, the angles of elevation and depression of the top and foot of the another house on the opposite side of the street are 30° and 45° respectively. The height of the opposite house is :**

Sol. Since bases are same for both the angles, perpendicular for smaller angles : perpendicular for greater angle

$$= 1 : \left\{\frac{\tan \text{(greater angle)}}{\tan \text{(smaller angle)}}\right\}$$

$\Rightarrow \quad ED : DC = 1 : \sqrt{3}$

$$\Rightarrow \quad ED = \frac{DC}{\sqrt{3}} = \frac{15}{\sqrt{3}} = 5\sqrt{3}$$

[$\because DC = AB = 15$ m]

Hence, $EC = ED + DC$

$= 5\sqrt{3} + 15$

$= 23.66$ m

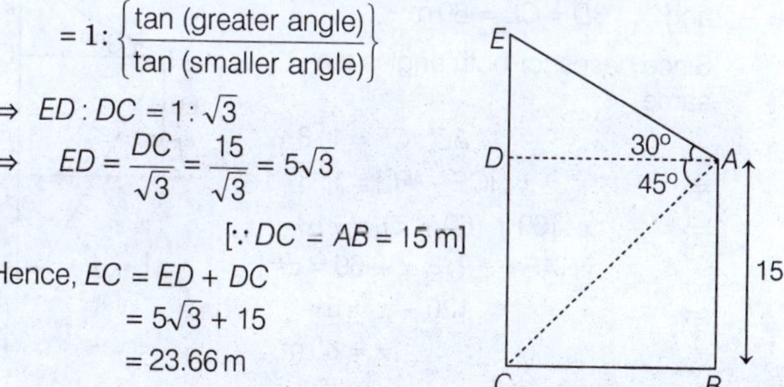

Alternatively

Since bases are same, greater perpendicular = smaller perpendicular × $\sqrt{3}$ [as difference between angles is 15° and angles are from a set of 30°, 45° and 60°.]

$$\Rightarrow \quad DC = \sqrt{3} \times ED$$

$$\Rightarrow \quad ED = \frac{DC}{\sqrt{3}} = \frac{15}{\sqrt{3}} = 5\sqrt{3} \qquad [\because DC = AB = 15\,\text{m}]$$

Hence, $EC = ED + DC = 5\sqrt{3} + 15 = 23.66\,\text{m}$

Problem 18. **A man standing on the deck of a ship, which is 10 m above the water level, observes the angle of elevation of the top of a hill as 60° and the angle of depression of the base of the hill as 30°. Find the distance of the hill from the ship and the height of the hill.**

Sol. Here, height of the hill = AC

and distance between the ship and the hill is EB.

Since bases are same, greater perpendicular = smaller perpendicular × 3 [as difference between angles is 30° and angles are from a set of 30°, 45° and 60°.]

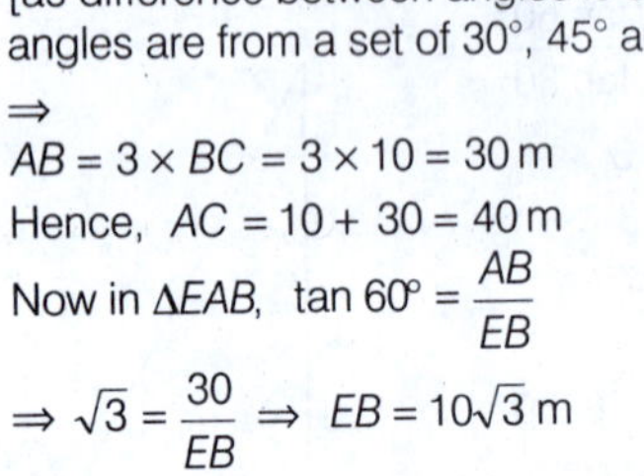

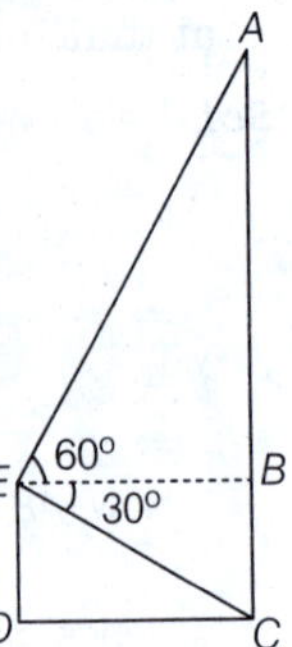

$$\Rightarrow$$

$$AB = 3 \times BC = 3 \times 10 = 30\,\text{m}$$

Hence, $AC = 10 + 30 = 40\,\text{m}$

Now in ΔEAB, $\tan 60° = \dfrac{AB}{EB}$

$$\Rightarrow \sqrt{3} = \frac{30}{EB} \Rightarrow EB = 10\sqrt{3}\,\text{m}$$

Alternatively

In ΔEAB, $\tan 60° = \dfrac{AB}{EB} \Rightarrow \sqrt{3} = \dfrac{30}{EB}$

$$\Rightarrow \quad EB = 10\sqrt{3}$$

Hence, distance of the hill from the ship = $10\sqrt{3}$ m.

Since both angles are complementary to each other, product of bases will be equal to product of perpendiculars.

$$\Rightarrow \quad 10\sqrt{3} \cdot 10\sqrt{3} = AB \cdot 10$$

$$\Rightarrow \quad AB = 30$$

Thus, $AC = 10 + 30 = 40\,\text{m}$

Problem 19. **The angle of elevation of a cloud from a point 30 m above the lake is 30° and the angle of depression of the reflection of the cloud in the lake is 60°. What is the height of the cloud ?**

Sol. Here, DE = the surface of the lake; point A is a point of the cloud and point F is a point of its reflection.

Clearly, $AE = EF$

and $BD = CE = 60\,\text{m}$

Since bases for both angles are same,

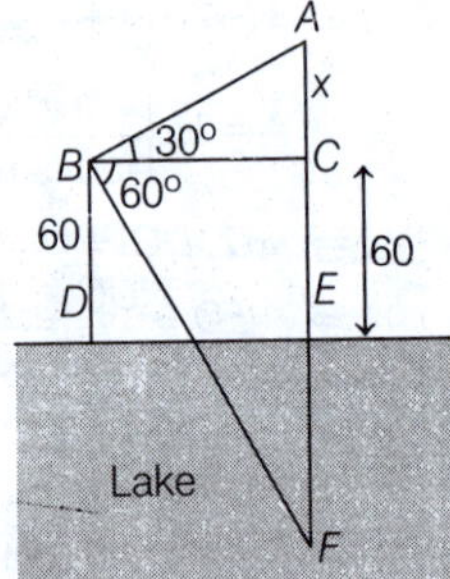

$$AC : CF = 1 : 3$$

$$\Rightarrow \quad x : (CE + FE) = 1 : 3$$

$$\Rightarrow \quad x : [60 + (60 + x)] = 1 : 3$$

$$[\because AE = EF \Rightarrow x + 60 = EF]$$

$$\Rightarrow \quad 120 + x = 3x$$

$$\therefore \quad x = 60\,\text{m}$$

$\therefore$ Height of the cloud = $AE = 60 + x = 60 + 60 = 120\,\text{m}$

Alternatively

In ΔABC, $\quad \tan 30° = \dfrac{AC}{BC}$

$$\therefore \quad BC = \sqrt{3}x$$

Since both angles are complementary to each other,

$$BC \times BC = AC \times CF$$

$$\Rightarrow \quad \sqrt{3}x \times \sqrt{3}x = x[(60 + (60 + x)]$$

$$\Rightarrow \quad 3x^2 = 120x + x^2 \Rightarrow 2x^2 = 120x$$

$$\therefore \quad x = 60$$

Hence, height of cloud = $AE = AC + CE = 60 + 60 = 120\,\text{m}$

Problem 20. **The angle of elevation of the top of the hill at the foot of a tower is 60° and the angle of elevation of the top of the tower from the foot of the hill is 30°. If the tower is 50 m high, find the height of the hill.**

Sol. Here, bases are same for both angles.

$$\therefore \quad CD : AB = 1 : \frac{\tan 60°}{\tan 30°}$$

$$\Rightarrow \quad 50 : AB = 1 : 3$$

$$\therefore \quad AB = 150\,\text{m}$$

Alternatively

In ΔBCD, $\tan 30° = \dfrac{CD}{BC}$

$$\Rightarrow \quad BC = 50\sqrt{3}$$

Here, both angles are complementary to each other.

$$\therefore \quad BC \times BC = AB \times CD$$

$$\Rightarrow \quad 50\sqrt{3} \times 50\sqrt{3} = AB \times 50$$

$$\Rightarrow \quad AB = 150\,\text{m}$$

Problem 21. **A vertical pole consists of two parts, the lower part being 1/3rd of the whole. At a point in the horizontal plane through the base of the pole and distance 20 m from it, the upper part of the pole subtends an angle whose tangent is 1/2. The possible heights of the pole are :**

(a) 20 m and $20\sqrt{3}$ m (b) 20m and 60m
(c) 16m and 48m (d) None of these

Sol. Let $\angle ACB$ be α.

Since bases are same for both the angles, perpendicular for smaller angle: perpendicular for greater angle

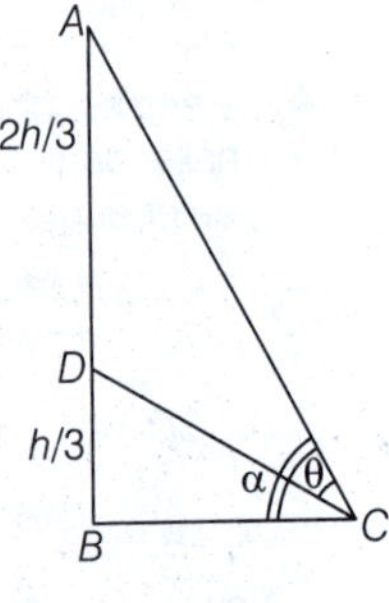

$$= 1 : \left[\frac{\tan\text{(greater angle)}}{\tan\text{(smaller angle)}}\right]$$

$$\Rightarrow \quad BD : AB = 1 : \frac{\tan\alpha}{\tan(\alpha - \theta)}$$

$$\Rightarrow \quad 1 : 3 = \tan(\alpha - \theta) : \tan\alpha$$

$$\Rightarrow \quad \tan\alpha = 3\tan(\alpha - \theta)$$

$$\Rightarrow \quad \tan\alpha = 3 \times \frac{\tan\alpha - \tan\theta}{1 + \tan\alpha \cdot \tan\theta}$$

$$= 3 \times \frac{\tan\alpha - \frac{1}{2}}{1 + \tan\alpha \cdot \frac{1}{2}} \qquad \left[\because \tan\theta = \frac{1}{2}, \text{given}\right]$$

$$\Rightarrow \quad x = 3 \times \frac{x - \frac{1}{2}}{1 + x \cdot \frac{1}{2}} \quad \text{[let } \tan\alpha \text{ be } x]$$

$$\Rightarrow \quad x + \frac{x^2}{2} = 3x - \frac{3}{2} \Rightarrow x^2 - 4x + 3 = 0$$

$$\Rightarrow \quad x^2 - 3x - x + 3 = 0$$

$$\Rightarrow \quad x(x-3) - 1(x-3) = 0$$

$\Rightarrow x = 1$ and 3

Hence, $\tan\alpha = 1$ and 3

Now, in ΔABC, $\tan\alpha = \frac{AB}{BC} \Rightarrow \tan\alpha = \frac{AB}{20}$

$$\Rightarrow \quad AB = 20 \cdot \tan\alpha = 20 \times 1$$

and $20 \times 3 = 20$ m and 60 m.

Problems based on the concepts associated with the cases having angles, complementary to each other

Please observe the concepts in

Problem No. 2,3,7,9,10,14,16,18,19,20.

Problems based on the concepts associated with the cases having neither equal perpendiculars nor equal bases

Problem 22. Consider a regular hexagon $ABCDEF$. Two towers are situated at B and C. The angle of elevation from A to the top of the tower at B is 30°, and the angle of elevation to the top of the tower at C is 45°. What is the ratio of the height of the towers at B and C? ***[CDS-19]***

Sol. Let each side of the hexagon be x unit.

Clearly, AC = sum of height of equilateral ΔCOB and ΔAOB

$\Rightarrow AC = \frac{\sqrt{3}}{2}x + \frac{\sqrt{3}}{2}x = \sqrt{3}x$

Now, in ΔAEC, $\tan 45° = \frac{EC}{AC}$

$\Rightarrow EC = AC = \sqrt{3}x$

In ΔABD,

$$\tan 30° = \frac{BD}{AB}$$

$$\Rightarrow \quad BD = \frac{AB}{\sqrt{3}} = \frac{x}{\sqrt{3}}$$

$\therefore$ Required ratio $= BD : EC$

$$= \frac{x}{\sqrt{3}} : \sqrt{3}x = 1 : 3$$

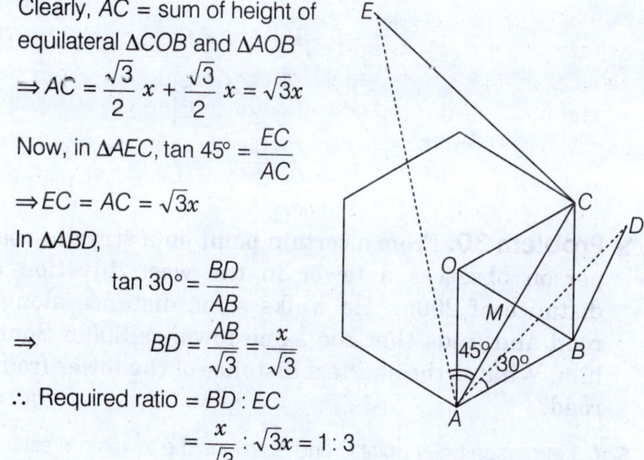

Problem 23. A man standing on the line joining the two poles finds that the top of the poles make an angle of elevation of 60° and 45° respectively. After walking for sometime towards the other pole, the angles change to 30° and 60° respectively. The ratio of the height of the poles is : ***[IIFT-19]***

Sol.

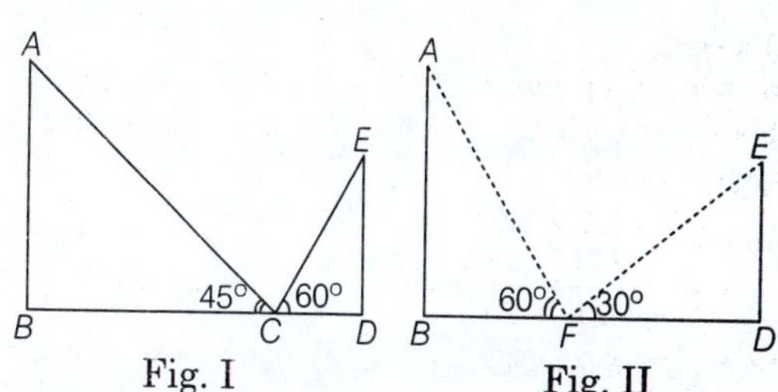

Fig. I

In ΔABC, $\tan 45° = \frac{AB}{BC}$

$$\Rightarrow \quad BC = AB$$

Also, in ΔCDE, $\tan 60° = \frac{ED}{CD} = \frac{ED}{\sqrt{3}}$

Fig. II

In ΔABF, $\quad \tan 60° = \frac{AB}{BF}$

$$\Rightarrow \quad BF = \frac{AB}{\sqrt{3}}$$

Also, in ΔFDE, $\tan 30° = \frac{ED}{FD}$

$$\Rightarrow \quad FD = ED\sqrt{3}$$

Clearly, $BC + CD = BF + FD$ [$\because$ distance between both poles is fixed]

$$\Rightarrow \quad AB + \frac{DE}{\sqrt{3}} = \frac{AB}{\sqrt{3}} + ED\sqrt{3}$$

$$\Rightarrow \quad AB\left(\frac{\sqrt{3}-1}{\sqrt{3}}\right) = ED\left(\frac{2}{\sqrt{3}}\right)$$

$$\Rightarrow \quad ED : AB = (\sqrt{3}-1) : 2$$

Cases having application of T-ratios other than tangent/cotangent with/ without tangent/ cotangent

Problem 24. One flies a kite with a thread 150 m long. If the thread of the kite makes an angle of 60° with the horizontal line, then the height of the kite from the ground (assuming the thread to be in a straight line) is : ***[FCI(Asstt. Gr.III)-12, MAT-16]***

Sol. Here, AC = length of the thread

Now, $\quad \sin 60° = \frac{AB}{AC}$

$$\Rightarrow \quad \frac{\sqrt{3}}{2} = \frac{AB}{150}$$

$$\Rightarrow \quad AB = 75\sqrt{3}$$

Hence, height of the kite from the ground $= 75\sqrt{3}$ m

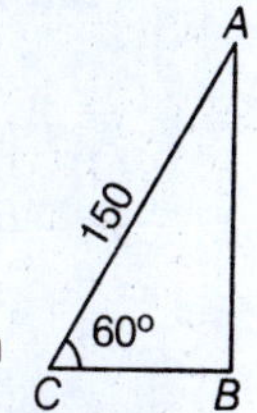

Problem 25. A telegraph post is bent at a point above the ground due to storm. Its top just meets the ground at a distance of $8\sqrt{3}$ m from its foot and makes an angle of 30°. The height of the post is :

Sol. Let AC is telegraph post which bends at point B.

In ΔBCD, $\tan 30^\circ = \dfrac{BC}{CD}$

$\Rightarrow \quad BC = 8\sqrt{3} \times \dfrac{1}{\sqrt{3}} = 8$

Further, $\sin 30^\circ = \dfrac{BC}{BD}$

$\Rightarrow \quad \dfrac{1}{2} = \dfrac{8}{BD}$

$\Rightarrow \quad BD = 16\text{ m}$

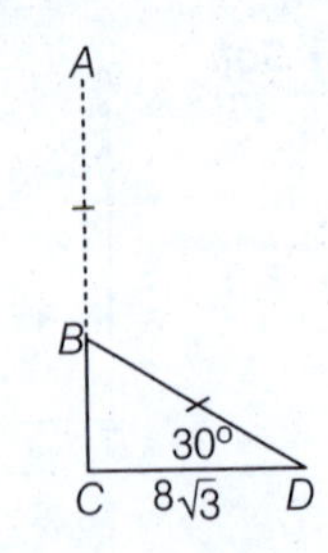

But, $AB = BD$ as part AB becomes CD when it is bent at point B.

Hence, $\quad AB = BD = 16$

Now, height of the post $= BC + AB = 8 + 16 = 24\text{ m}$

Problem 26. A vertical post 15 feet high is broken at a certain height and its upper part, not completely separated, meets the ground at an angle of 30°. Find the height at which the post is broken.

Sol. Let BC be x feet.

Then, $\sin 30^\circ = \dfrac{BC}{BD}$

$\Rightarrow \quad \dfrac{1}{2} = \dfrac{x}{AB}$

$[\because BD = AB \Rightarrow \text{same part}]$

$\Rightarrow \quad \dfrac{1}{2} = \dfrac{x}{(15 - x)}$

$\Rightarrow \quad 2x = 15 - x$

$\therefore \quad x = 5$

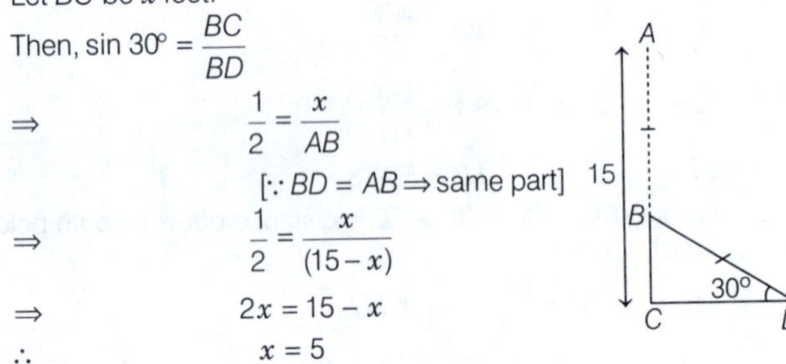

Hence, the post is broken at a height of 5 feet.

Problem 27. A boy standing on a horizontal plane finds a bird flying at a distance of 100 m from him at an elevation of 30°. A girl standing on the roof of a 20 m high building finds the angle of elevation of the same bird to be 45°. Boy and girl are on the opposite sides of the bird. Find the distance of the bird from the girl.

Sol. Here, GE is a building 20 m high; AB is the slant height of the bird;

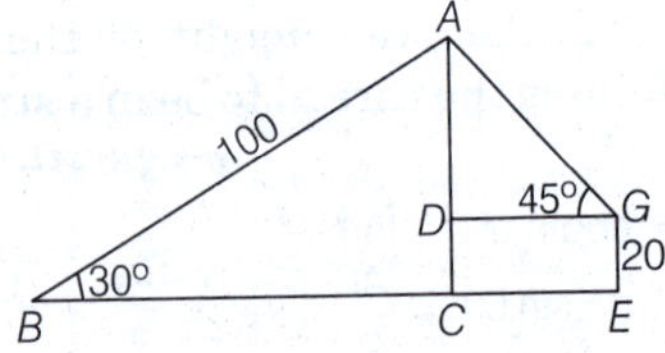

A, B and G are positions of bird, boy and girl respectively.

In ΔABC, $\sin 30^\circ = \dfrac{AC}{AB}$

$\Rightarrow \quad \dfrac{1}{2} = \dfrac{20 + AD}{100} \Rightarrow AD = 30\text{ m}$

Now, in ΔADG, $\sin 45^\circ = \dfrac{AD}{AG} \Rightarrow \dfrac{1}{\sqrt{2}} = \dfrac{30}{AG}$

$\therefore \quad AG = 30\sqrt{2}\text{ m}$

Problem 28. The angle of elevation of the top of the mountain from its foot is 45° which changes to 60° when a person climbs 300 m towards the mountain up with an incline of 30°. What is the height of the mountain?

Sol. In ΔDEC, $\quad \sin 30^\circ = \dfrac{DE}{DC}$

$\Rightarrow \quad \dfrac{1}{2} = \dfrac{DE}{300}$

$\therefore \quad DE = 150 = FB$

Further, $\cos 30^\circ = \dfrac{EC}{300}$

$\Rightarrow \quad \dfrac{\sqrt{3}}{2} = \dfrac{EC}{300}$

$\Rightarrow \quad EC = 150\sqrt{3}$

In ΔABC, $\tan 45^\circ = \dfrac{AB}{BC}$

$\Rightarrow \quad AB = BC$

Now, in ΔAFD,

$\tan 60^\circ = \dfrac{AF}{FD} \Rightarrow \sqrt{3} = \dfrac{AF}{BE} \Rightarrow BE = \dfrac{AF}{\sqrt{3}}$

$\because \quad AB = BC$

$\therefore \quad AF + FB = BE + EC \Rightarrow AF + 150 = \dfrac{AF}{\sqrt{3}} + 150\sqrt{3}$

$\Rightarrow \quad \dfrac{\sqrt{3}AF - AF}{\sqrt{3}} = 150\sqrt{3} - 150 \Rightarrow \dfrac{AF(\sqrt{3} - 1)}{\sqrt{3}} = 150(\sqrt{3} - 1)$

$\therefore \quad AF = 150\sqrt{3}$

Now, height of the mountain $= AF + FB$

$= 150\sqrt{3} + 150 = 150(\sqrt{3} + 1)\text{ m}$

Problem 29. A bridge across a river subtends an angle of 60° with river bank. Find the width of the river if the length of the bridge across the river is $70\sqrt{3}$ m.

Sol. In ΔABC, $\sin 60^\circ = \dfrac{AC}{70\sqrt{3}}$

$AC = \dfrac{\sqrt{3}}{2} \times 70\sqrt{3} = 105\text{ m}$

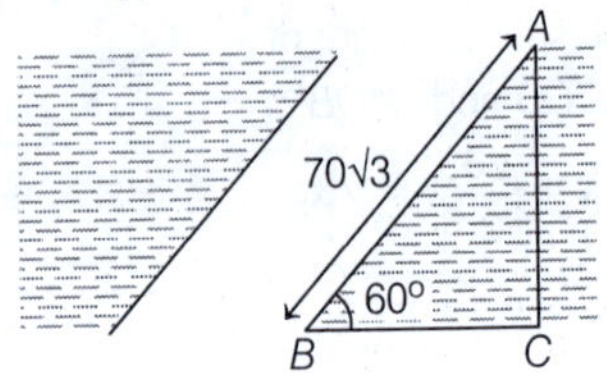

Problem 30. From a certain point on a straight road, a person observes a tower in the west direction at a distance of 200m. He walks some distance along the road and finds that the same tower is 300m South of him. What is the shortest distance of the tower from the road? **[CDS-14]**

Sol. Let person be at point C and observes the tower in West direction at point B. Hence, $BC = 200$ m. Now, he reaches at point A and observes that the same tower is in South at point B.

Clearly, BD will be the shortest distance of tower from the road and BD is a perpendicular distance. Let $\angle ABD = \theta$, then $\angle CBD = 90^\circ - \theta$, as angle between South and West is equal to 90°.

In ΔCBD, $\quad \cos(90^\circ - \theta) = \dfrac{BD}{BC}$

$\Rightarrow \quad \sin\theta = \dfrac{BD}{200} \quad \ldots(\text{i})$

In $\Delta ABD, \cos\theta = \dfrac{BD}{AB}$

$\Rightarrow \quad \cos\theta = \dfrac{BD}{300}$ ...(ii)

We know, $\sin^2\theta + \cos^2\theta = 1$

$\Rightarrow \left(\dfrac{BD}{200}\right)^2 + \left(\dfrac{BD}{300}\right)^2 = 1$ [from Eqs. (i) and (ii)]

$\Rightarrow BD^2\left(\dfrac{130000}{3600000000}\right) = 1$

$\Rightarrow BD = \dfrac{600}{\sqrt{13}}$

Alternatively

We know that $\left(\dfrac{1}{BD}\right)^2 = \left(\dfrac{1}{AB}\right)^2 + \left(\dfrac{1}{BC}\right)^2$

[recall the properties of a right angled triangle in Geometry section]

$\Rightarrow \left(\dfrac{1}{BD}\right)^2 = \left(\dfrac{1}{300}\right)^2 + \left(\dfrac{1}{200}\right)^2 \Rightarrow BD = \dfrac{600}{\sqrt{13}}$

➘ Problem 31. A round balloon of unit radius subtends an angle of 90° at the eye of an observer standing at a point, say A. What is the distance of the centre of the balloon from the point A?

Sol. Clearly, $OB = OC$ = radii of the balloon

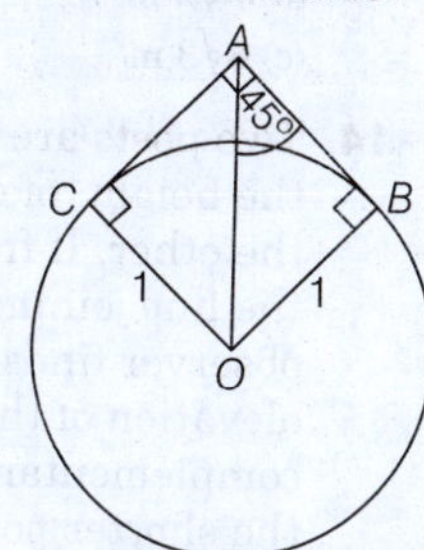

Now, in $\Delta OAB, \sin 45° = \dfrac{OB}{AO} \Rightarrow AO = \sqrt{2}\, OB = \sqrt{2} \times 1 = \sqrt{2}$

➘ Problem 32. A spherical balloon of radius r subtends angle of 60° at the eye of an observer. If the angle of elevation of its centre is 60° and h is the height of the centre of the balloon, then which one of the following is correct?

(a) $h = r$ (b) $h = \sqrt{2}\, r$

(c) $h = \sqrt{3}\, r$ (d) $h = 2r$

Sol.

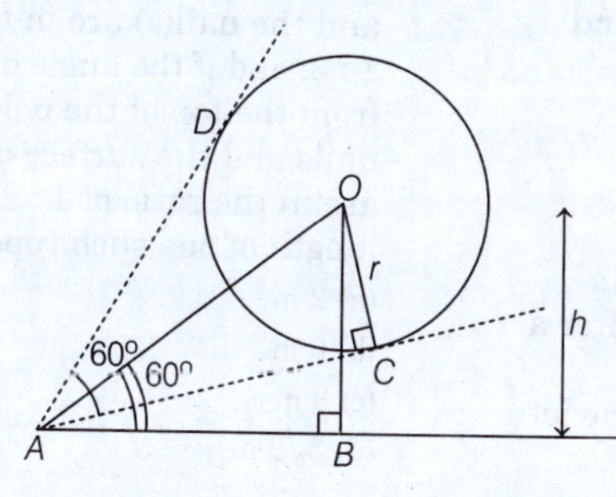

In $\Delta AOB, \sin 60° = \dfrac{OB}{AO} \Rightarrow AO = \dfrac{2OB}{\sqrt{3}} = \dfrac{2h}{\sqrt{3}}$ ...(i)

and in $\Delta AOC, \angle OAC = \dfrac{60°}{2}$, then $\sin\left(\dfrac{60°}{2}\right) = \dfrac{OC}{AO}$

$\Rightarrow \quad AO = 2OC = 2r$ ...(ii)

From Eqs. (i) and (ii),

$\dfrac{2h}{\sqrt{3}} = 2r \Rightarrow h = \sqrt{3}r$

➘ Problem 33. The length of the shadow of a pole inclined at 10° to the vertical towards the sun is 2.05 m, when the elevation of the sun is 38°. The length of the pole is :

(a) $\dfrac{2.05 \sin 38°}{\sin 42°}$ (b) $\dfrac{2.05 \sin 42°}{\sin 38°}$

(c) $\dfrac{2.05 \cos 38°}{\cos 42°}$ (d) None of these

Sol.

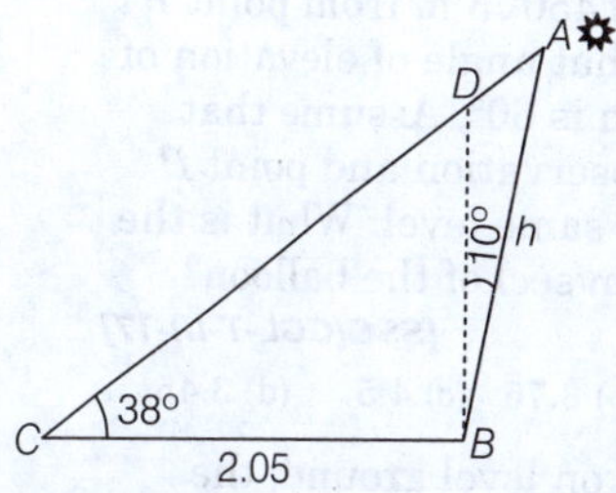

Clearly, $\angle CAB = 180° - 90° - 10° - 38° = 42°$

Now, as per sine-rule,

$\dfrac{a}{\sin A} = \dfrac{b}{\sin B} = \dfrac{c}{\sin C}$

$\Rightarrow \quad \dfrac{h}{\sin 38°} = \dfrac{2.05}{\sin 42°} = \dfrac{2.05 \sin 38°}{\sin 42°}$

➘ Problem 34. AB is a vertical pole resting at the end A on the level ground. P is a point on the level ground such that $AP = 3AB$. If C is the mid-point of AB and CB subtends an angle of β at P, the value of $\tan\beta$ is :

Sol. Let $\angle APC$ be α.

Then, $\tan\alpha = \dfrac{AC}{AP} = \dfrac{\frac{1}{2}AB}{3AB} = \dfrac{1}{6}$

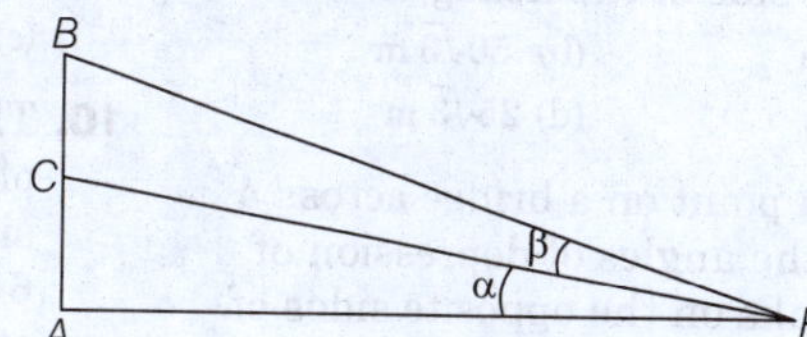

Further, $\tan(\alpha + \beta) = \dfrac{\tan\alpha + \tan\beta}{1 - \tan\alpha \cdot \tan\beta}$

$\Rightarrow \dfrac{AB}{AP} = \dfrac{\frac{1}{6} + \tan\beta}{1 - \frac{1}{6}\cdot\tan\beta} \Rightarrow \dfrac{AB}{3AB} = \dfrac{\frac{1}{6} + \tan\beta}{1 - \frac{1}{6}\cdot\tan\beta}$

$\Rightarrow 1 - \dfrac{1}{6}\cdot\tan\beta = \dfrac{1}{2} + 3\tan\beta \Rightarrow \dfrac{19}{6}\cdot\tan\beta = \dfrac{1}{2}$

$\therefore \quad \tan\beta = \dfrac{3}{19}$

WORKOUT

1. The angle of elevation and angle of depression both are measured with
(a) the vertical only
(b) the horizontal line only
(c) both horizontal and vertical
(d) None of the above

2. A balloon leaves from a point P rises at a uniform speed. After 6 min, an observer situated at a distance of $450\sqrt{3}$ m from point P observes that angle of elevation of the balloon is 60°. Assume that point of observation and point P are on the same level. What is the speed (in m/sec) of the balloon? *[SSC(CGL-T-II)-17]*
(a) 4.25 (b) 3.75 (c) 4.5 (d) 3.45

3. At a point on level ground, the angle of elevation of a vertical tower is found to be such that its tangent is $\frac{5}{12}$. On walking 192m towards the tower, the tangent of the angle of elevation is 3/4. The height of the tower is *[MAT-13]*
(a) 96 m (b) 150 m
(c) 180 m (d) 226 m

4. The angle of depression of vertex of a regular hexagon lying in a horizontal plane, from the top of the tower of height 75 m located at the centre of the regular hexagon is 60°. What is the length of each side of the hexagon?
(a) 75 m (b) $50\sqrt{3}$ m
(c) 25 m (d) $25\sqrt{3}$ m

5. From a point on a bridge across a river, the angles of depression of the banks on the opposite sides of the river are 45° and 60° respectively. Find the width of the river if the height of the bridge from banks is $20\sqrt{3}$ m. *[SSC (CGL-T-II)-20]*
(a) $20(\sqrt{3}+1)$ m (b) $20(\sqrt{3}-1)$ m
(c) 25 m (d) $25\sqrt{3}$ m

6. A person standing on the bank of a river finds that the angle of elevation of the top of the tower on the opposite bank is 45°. Which of the following statements is correct?
(a) Breadth of the river is twice the height of the tower
(b) Breadth of the river is half of the height of the tower
(c) Breadth of the river and the height of the tower are the same
(d) None of the above

7. The angle of elevation of the top of a unfinished pillar at a point 150 m from its base is 30°. If the angle of elevation at the same point is to be 45°, then the pillar has to be raised to a height of how many metres?
(a) 59.4 m (b) 61.4 m
(c) 62.4 m (d) 63.4 m

8. As observed from the top of a lighthouse, 100 m high above the sea level, the angle of depression of a ship, sailing directly towards it, changes from 30° to 60°. The distance travelled by the ship during the period of observation is : *[MAT-12]*
(a) 173.2 m (b) 115.5 m
(c) 57.7 m (d) None of these

9. A boat is being rowed away in still water, from a 210 m high cliff at the speed of 3 km/h. What is the approx. time taken for the angle of depression of the cliff at the boat to change from 60° to 45°. *[IIFT-14]*
(a) 1 min (b) 2 min
(c) 4 min (d) 5 min

10. The angles of elevation of the top of a tower from two points situated at distances 36 m and 64m from its base and in the same straight line with it are 58° and 32° respectively. What is the height of the tower?
(a) 48 m (b) 50 m
(c) 25 m (d) 24 m

11. The angle of a jet plane from a point on the ground is 60°. After a flight of 20 sec, the angle of elevation changes to 30°. If the jet plane is flying at a constant height of $1000\sqrt{3}$ m, find the speed of the jet plane.
(a) 100 m/s (b) 80 m/s
(c) 120 m/s (d) None of these

12. The shadow of a tower standing on a level plane is found to be 50 m longer when the sun's elevation is 30°, than when it is 60°. What is the height of the tower? *[CDS-14]*
(a) 25 m (a) 30 m
(c) $\frac{25}{\sqrt{3}}$ m (d) $25\sqrt{3}$ m

13. A vertical lamp post of height 9 m stands at the corner of a rectangular field. The angle of elevation of its top from the farthest corner is 30°, while from another corner it is 45°. The area of the field is :
(a) $9\sqrt{2}$ m^2 (b) $81\sqrt{2}$ m^2
(c) $8\sqrt{3}$ m^2 (d) $9\sqrt{3}$ m^2

14. Two posts are x metres apart and the height of one is double that of the other. If from the mid-point of the line joining their feet, an observer finds the angular elevation of their tops to be complementary, then the height of the shorter post is : *[SSC(CGL-T-II)-11]*
(a) $\frac{x}{2\sqrt{2}}$ m (b) $\frac{x}{4}$ m
(c) $x\sqrt{2}$ m (d) $\frac{x}{\sqrt{2}}$ m

15. The central pole of conical tent is 3/2 m high. The pole is supported by ropes tied to its top and nails on the ground. If, on the ground from the foot of the pole, the distances of the surface of the tent and the nail(s) are in the ratio of 1 : 3, and if the angle of depression from the top of the pole of the nails and the surface of the tent are in the ratio of 1 : 2, then the length of one such rope is :
(a) 2 m
(b) 6 m
(c) 3 m
(d) $3\sqrt{2}$ m

16. The angles of elevation of the top of the tower from two points which are at distances of 10m and 5m from the base of the tower and in the same straight line with it are complementary. The height of the tower is : *[CDS-12, SSC(CGL-T-II)-17]*

(a) 5 m (b) $\sqrt{50}$ m
(c) 15 m (d) $\sqrt{75}$ m

17. At the foot of the mountain, the angle of elevation of the summit at the top of the mountain is 45°. After ascending 100 m, at a slope of 30° up the mountain towards summit, the angle of elevation of the summit is 60°. Find the height of the summit. *[IIFT-18]*

(a) $50(\sqrt{3}+1)$ m (b) $50(\sqrt{5}+1)$ m
(c) $50\sqrt{3}$ m (d) $50(\sqrt{3}+2)$ m

18. An electrician has to repair an electric fault on a pole of height 4 m. He needs to reach a point 1.3 m below the top of the pole to undertake the repair work. The length of the ladder that he should use, which inclined at an angle of 60° to the horizontal would enable him to reach the required position is :

(a) $\frac{9\sqrt{3}}{10}$ m (b) $\frac{3\sqrt{3}}{10}$ m
(c) $\frac{3\sqrt{3}}{5}$ m (d) $\frac{9\sqrt{3}}{5}$ m

19. A ladder 20 m long is placed against a wall so that the foot of the ladder is 10 m from the wall. The angle of inclination of the ladder to the horizontal will be : *[CDS-12]*

(a) 30° (b) 45°
(c) 60° (d) 75°

20. An aeroplane when flying at a height of 5000 m from the ground passes vertically above another plane at an instant when the angles of elevation of the two planes from the same point on the ground are 60° and 45° respectively. The distance between two planes at that instant is : *[SSC(CGL-T-II)-11]*

(a) 4500 m
(b) $5000\left(1-\frac{1}{\sqrt{3}}\right)$ m
(c) $5000(\sqrt{3}-1)$ m
(d) $5000(3-\sqrt{3})$ m

21. A flagstaff of 5 m high stands on a building of 25 m height. At an observer at a height of 30 m, the flagstaff and the building subtend equal angles. The distance of the observer from the top of the flagstaff is: *[SSC(CGL-T-II)-12, IIFT-17]*

(a) $\frac{5\sqrt{3}}{2}$ (b) $5\sqrt{\frac{3}{2}}$
(c) $5\sqrt{\frac{2}{3}}$ (d) None of these

22. Two poles of heights 6 m and 11 m stand on a plane ground. If the distance between their feet is 12 m, what is the distance between their tops ?

(a) 13 m (b) $50\sqrt{2}$ m
(c) $5(\sqrt{3}-1)$ m (d) 17 m

23. If a flagstaff of 6 m high placed on the top of the tower throws a shadow of $2\sqrt{3}$ m along the ground, then the angle (in degrees) that the sun makes with the ground is :

(a) 60° (b) 80°
(c) 75° (d) None of these

24. The angles of elevation of the top of a building from a foot of a tower is 30° and the angle of elevation of the top of the tower from the foot of the building is 60°. If the height of the tower is 50 m, the height of the building is : *[MAT-11]*

(a) $\frac{50\sqrt{3}}{3}$ m (b) $50\sqrt{3}$ m
(c) 16.66 m (d) None of these

25. Let AB represent a building of height h metre with A being its top, B being its bottom. Let $A'B'$ represent a tower of height $(h+x)$ metre $(x>0)$ with A' being its top and B' being its bottom. Let $BB'=d$ metre. Let the angle of elevation of A' as seen from A be 45°. *[CDS-17]*

Consider the following statements:

Statement I $h+x>d$

Statement II The angle of depression of B as seen from A' is less than 45°.

Which one of the following is correct in respect of the above statements ?

(a) Both Statement I and Statement II are true and Statement II is correct explanation of Statement I
(b) Both Statement I and Statement II are true and Statement II is not the correct explanation of Statement I
(c) Statement I is true but Statement II is false
(d) Statement I is false but Statement II is true

26. $ABCD$ is a rectangular field. A vertical lamp post of height 12 m stands at the corner A. If the angle of elevation of its top from B is 60° and from C is 45°, then the area of the field is :

(a) $48\sqrt{2}$ m^2 (b) $48\sqrt{3}$ m^2
(c) 48 m^2 (d) $12\sqrt{2}$ m^2
(e) $12\sqrt{3}$ m^2

27. The angle of elevation of the top of an incomplete vertical pillar at a horizontal distance of 100 m from its base is 45°. If the angle of elevation of the top of complete pillar at the same point is to be 60°, then the height of the incomplete pillar is to increased by :

(a) 100 m (b) $50\sqrt{2}$ m
(c) $100(\sqrt{3}-1)$ m (d) $100(\sqrt{3}+1)$ m

28. From an aeroplane just over a straight road, the angles of depression of two consecutive kilometer stones situated at the opposite sides of the aeroplane were found to be 60° and 30° respectively. The height of the aeroplane (in km) from the road at that instant, is : *[SSC(CGL-T-II)-14, CDS-17]*

(a) $\frac{\sqrt{3}}{4}$ (b) $\frac{\sqrt{3}}{2}$
(c) $\frac{\sqrt{3}}{3}$ (d) $\sqrt{3}$

29. From the top of a cliff 200 m high, the angles of depression of the top and bottom of the tower are observed to be 30° and 45° respectively. What is the height of the tower ?

(a) 400 m
(b) $400\sqrt{3}$ m
(c) $400/\sqrt{3}$ m
(d) None of the above

30. A house subtends a right angle at the window of an opposite house and the angle of elevation of the window from the bottom of the 1st house is 60°. If the distance between the two houses be 6 m, then the height of the 1st house is:

(a) $6\sqrt{3}$ m (b) $8\sqrt{3}$ m
(c) $4\sqrt{3}$ m (d) None of these

31. From a lighthouse the angles of depression of two ships on opposite sides of the lighthouse are observed to be 30° and 45°. If the height of the lighthouse is h, what is the distance between the ships ?

(a) $\sqrt{3}h$ (b) $h\left(1+\frac{1}{\sqrt{3}}\right)$

(c) $h(\sqrt{3}-1)$ (d) $h(\sqrt{3}+1)$

32. A tower subtends angles α, 2α, 3α respectively at the points A, B and C, all lying on a horizontal line through the foot of the tower. Then, $\frac{AB}{BC}$ =?

(a) $\frac{\sin 3\alpha}{\sin 2\alpha}$ (b) $1+2\cos 2\alpha$

(c) $2+\cos 3\alpha$ (d) $\frac{\sin 2\alpha}{\sin \alpha}$

33. $ABCDEF$ is a regular polygon. Two poles at C and D are standing vertically and subtend angles of elevation 30° and 60° at A respectively. What is the ratio of the height of the pole at C to that of the pole at D? ***[CDS-17]***

(a) 1: 1

(b) $1 : 2\sqrt{3}$

(C) $2\sqrt{3} : 1$

(d) $2 : \sqrt{3}$

34. A boat stationed at the North of the lighthouse, is making an angle of 30° with the top of the lighthouse. Simultaneously, another boat, stationed at the East of the same lighthouse is making an angle 45° with the top of the lighthouse. What will be the shortest distance between these two boats? The height of the lighthouse is 300 ft. Assume both the boats are of negligible dimensions. ***[XAT-18]***

(a) 300 ft (b) $300\sqrt{3}$ ft

(c) $600\sqrt{3}$ ft (d) 600 ft

(e) None of these

Answer key

1. (b)	2. (b)	3. (c)	4. (d)	5. (a)	6. (c)	7. (d)	8. (b)	9. (b)	10. (a)
11. (a)	12. (d)	13. (b)	14. (a)	15. (c)	16. (b)	17. (a)	18. (d)	19. (c)	20. (b)
21. (b)	22. (a)	23. (a)	24. (c)	25. (c)	26. (a)	27. (c)	28. (a)	29. (d)	30. (b)
31. (d)	32. (b)	33. (b)	34. (d)						

Hints and SOLUTIONS

[Conventional methods for solution have deliberately been left due to space constraint but given in illustrations.]

1. The angle of elevation and angle of depression both are measured with the horizontal line only.

2. $\tan 60° = \dfrac{BP}{PA}$

$\Rightarrow \quad BP = \sqrt{3} \times 450\sqrt{3} = 1350$ m.

Then, speed of balloon per second = $1350 \div (6 \times 60) = 3.75$ m/sec.

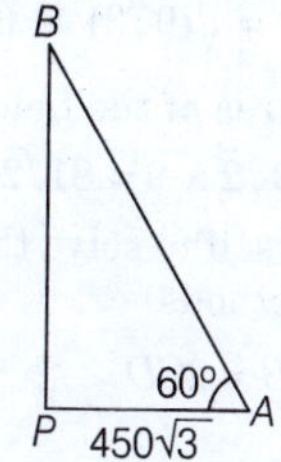

3. Since perpendiculars are same for both the angles, base for greater angle : base for smaller angle

$= 1 : \left[\dfrac{\tan \text{(greater angle)}}{\tan \text{(smaller angle)}}\right]$

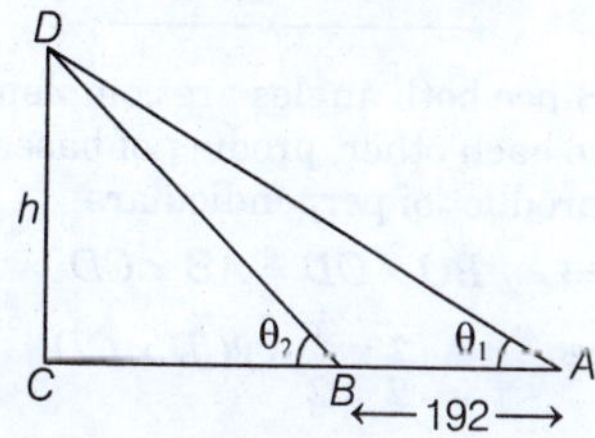

$\Rightarrow BC : AC = 1 : \dfrac{\frac{3}{4}}{\frac{5}{12}} = 1 : \dfrac{9}{5} = 5 : 9$

$\Rightarrow BC : AB = 5 : 4$

$\Rightarrow BC : 192 = 5 : 4$

$\therefore \quad BC = 240$

Now, in ΔBCD, $\tan \theta_2 = \dfrac{CD}{BC}$

$\Rightarrow \quad \dfrac{3}{4} = \dfrac{CD}{240} \Rightarrow CD = 180$ m

Try yourself to solve the question by other methods.

4. Here, OK = height of the tower, d = distance between O and E

In ΔEOK, $\tan 60° = \dfrac{OK}{OE} = \dfrac{75}{d}$

$\Rightarrow \sqrt{3} = \dfrac{75}{d}$

$\therefore \quad d = 25\sqrt{3}$

We know that all the six triangles of regular hexagon are equilateral triangles.

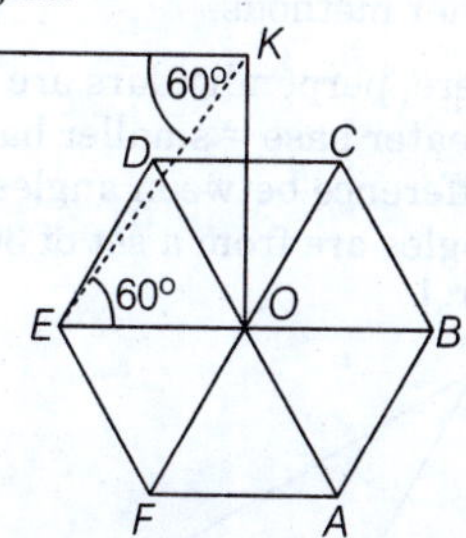

i.e. $OE = DE = OD$.

Then, $DE = OE = 25\sqrt{3} = d$

Hence, each side of the hexagon = $25\sqrt{3}$ m.

5. Let B and C be two points on the opposite banks. It means BC is the width of the river.

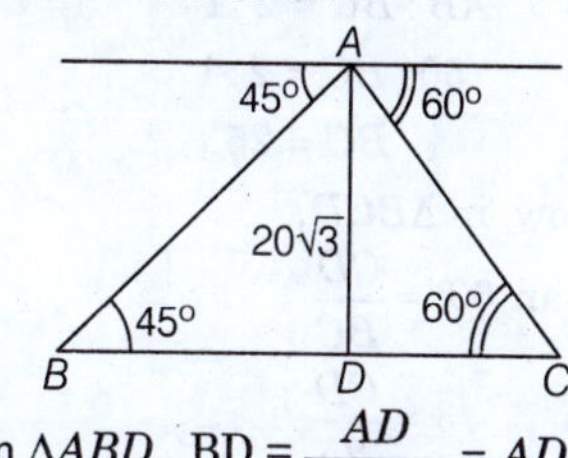

In ΔABD, BD $= \dfrac{AD}{\tan 45°} = AD$

$= 20\sqrt{3}$ (as given)

In ΔADC, $DC = \dfrac{AD}{\tan 60°}$

$= \dfrac{20\sqrt{3}}{\sqrt{3}} = 20$

Hence, $BC = 20\sqrt{3} + 20$

$= 20(\sqrt{3} + 1)$ m

Try yourself to solve the questions by other methods.

6. Here, AB is the breadth of the river and BC is the height of the tower.

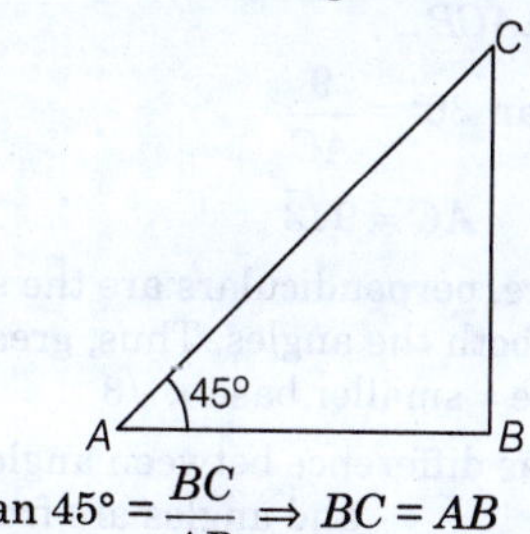

$\tan 45° = \dfrac{BC}{AB} \Rightarrow BC = AB$

Hence, option (c) is correct.

7. In ΔABC, $\tan 30° = \dfrac{BC}{AB}$

$\Rightarrow \quad BC = \dfrac{AB}{\sqrt{3}} = \dfrac{150}{\sqrt{3}} = 50\sqrt{3}$

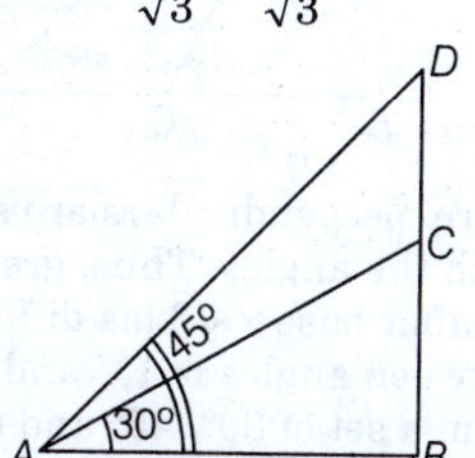

Since bases are same for both the angles, greater perpendicular = smaller perpendicular × $\sqrt{3}$ [as difference between angles is 15° and angles are from a set of 30°, 45° and 60°.]

$\Rightarrow \quad BD = BC \times \sqrt{3}$

$= 50\sqrt{3} \times \sqrt{3} = 150$

Now, $CD = 150 - 50\sqrt{3} = 63.4$ m (approx.)

Try yourself to solve the questions by other methods.

8. In ΔACD, $\tan 60° = \dfrac{AD}{CD} = \dfrac{100}{CD}$

$\Rightarrow \quad \sqrt{3} = \dfrac{100}{CD}$

$\therefore \quad CD = \dfrac{100}{\sqrt{3}}$

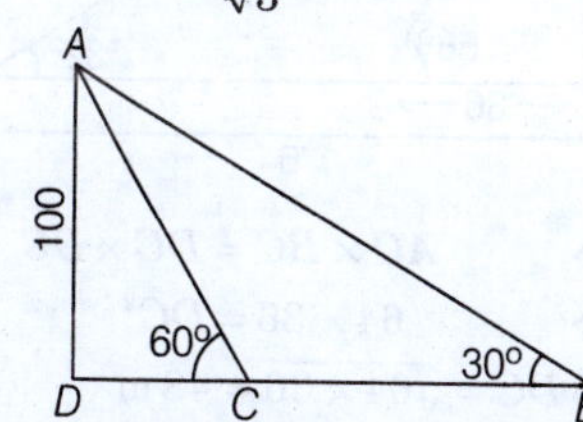

Since both angles are complimentary to each other, product of bases = product of perpendiculars.

$\Rightarrow \quad CD \times BD = AD \times AD$

$\Rightarrow \quad \dfrac{100}{\sqrt{3}} \times BD = 100 \times 100$

$\Rightarrow \quad BD = 100\sqrt{3}$

Now, $BC = BD - CD = 100\sqrt{3} - \dfrac{100}{\sqrt{3}}$

$= 100(\sqrt{3} - \dfrac{1}{\sqrt{3}}) = \dfrac{200}{\sqrt{3}} = 115.5$ m

Try yourself to solve the questions by other methods.

9. Here, AB = height of cliff = 210 m

In ΔABC, $\tan 60° = \dfrac{AB}{BC}$

$\Rightarrow \quad BC = \dfrac{210}{\sqrt{3}} = 70\sqrt{3}$

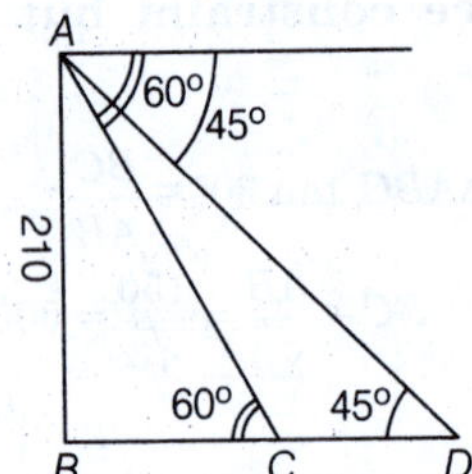

Here, perpendiculars are same for both the angles. Thus, greater base = smaller base × $\sqrt{3}$ [as difference between angles is 15° and angles are from a set of 30°, 45° and 60°.]

$\Rightarrow \quad BD = BC \times \sqrt{3}$

$\Rightarrow \quad 70\sqrt{3} \times \sqrt{3} = 210$

Now, $CD = BD - BC = 210 - 70\sqrt{3}$

$= 88.76$ m

To cover 88.76m at the speed of 3 km/h, the boat will take 1.77 min. i.e. approx. 2 min.

Try yourself of solve the question by other methods.

10. Since both angles are complementary to each other, product of bases = product of perpendiculars.

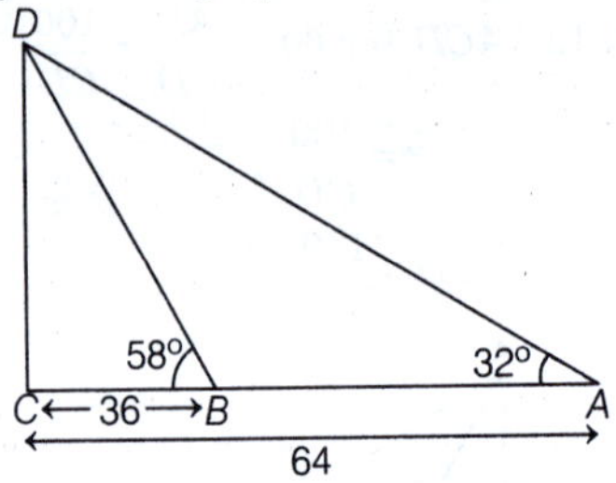

$\Rightarrow \quad AC \times BC = DC \times DC$

$\Rightarrow \quad 64 \times 36 = DC^2$

$\Rightarrow DC = \sqrt{64 \times 36} = 48$ m

Try yourself to solve the question by other methods.

11. In $\Delta BDC, CD = \dfrac{BD}{\tan 60°}$

$= \dfrac{1000\sqrt{3}}{\sqrt{3}} = 1000$

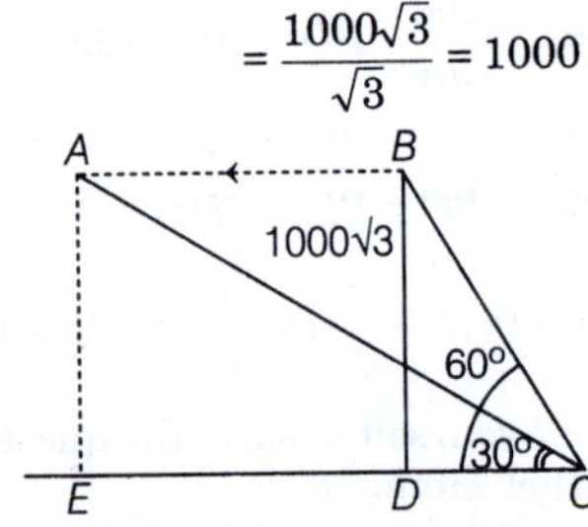

Since both angles are complementary, $EC \times DC = AE \times BD$

$\Rightarrow EC \times 1000 = 1000\sqrt{3} \times 1000\sqrt{3}$

$\Rightarrow EC = 3000$

$\therefore \; ED = AB = 3000 - 1000$

$= 2000$ m

Now, speed $= \dfrac{2000}{20} = 100$ m/sec.

Try yourself to solve the question by other methods.

12. Here, perpendiculars are same. Thus, greater base = smaller base × 3 [as difference between angles is 30° and angles are from a set of 30°, 45° and 60°.]

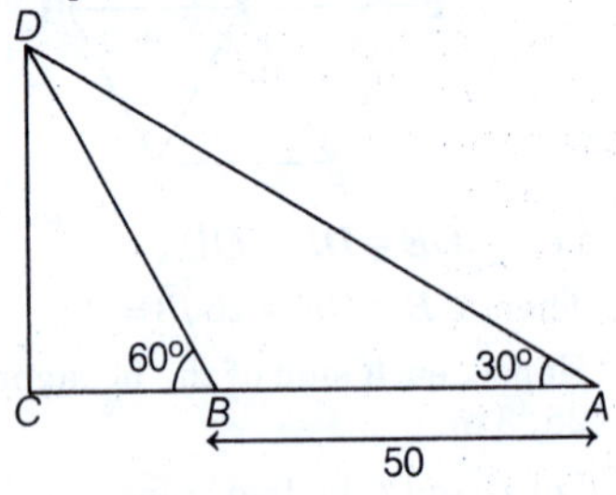

$\Rightarrow \quad AC : BC = 3 : 1$

$\Rightarrow \quad AB : BC = 2 : 1$

$\Rightarrow \quad 50 : BC = 2 : 1$

$\therefore \quad BC = 25$

Now, in ΔBCD,

$\tan 60° = \dfrac{CD}{BC}$

$\Rightarrow \quad \sqrt{3} = \dfrac{CD}{25}$

$\therefore \quad CD = 25\sqrt{3}$ m

Alternatively

Height of the tower,

$$h = \frac{d}{\cot \theta_1 - \cot \theta_2}$$

$$= \frac{50}{\cot 30° - \cot 60°}$$

$= 25\sqrt{3}$ m

Note Conventional method has deliberately been left due to space constraint, but given in illustrations.

13. In ΔACP,

$\tan 30° = \dfrac{9}{AC}$

$\Rightarrow \quad AC = 9\sqrt{3}$

Here, perpendiculars are the same for both the angles. Thus, greater base = smaller base × $\sqrt{3}$

[as difference between angles is 15° and angles are from a set of 30°, 45° and 60°.]

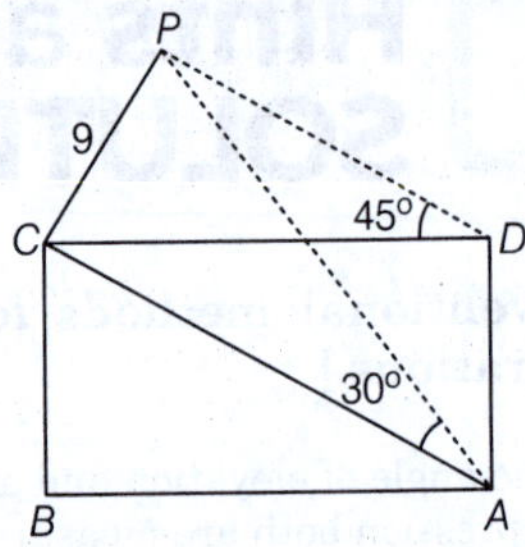

$\Rightarrow \quad AC = CD \times \sqrt{3}$

$\Rightarrow \quad 9\sqrt{3} = CD \times \sqrt{3}$

$\Rightarrow \quad CD = 9$

Now, in ΔACD,

$AD = \sqrt{(AC)^2 - (DC)^2}$

$= \sqrt{(9\sqrt{3})^2 - (9)^2} = 9\sqrt{2}$

Hence, area of the field = $AD \times CD$

$= 9\sqrt{2} \times 9 = 81\sqrt{2}$ m^2

Try yourself to solve the question by other methods

14. Here, $AB = 2CD$

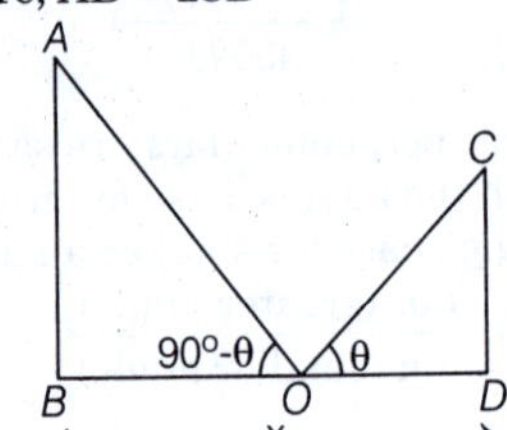

Since both angles are complementary to each other, product of bases = product of perpendiculars

$\Rightarrow \quad BO \times OD = AB \times CD$

$\Rightarrow \quad \dfrac{x}{2} \times \dfrac{x}{2} = 2CD \times CD$

$\Rightarrow \quad CD = \dfrac{x}{2\sqrt{2}}$ m

Try yourself to solve the question by other methods.

15. Let the surface of the tent be at d metres from the base of the central pole. Then, nails are at $3d$ metres from the base of the central pole as the given ratio is 1: 3.

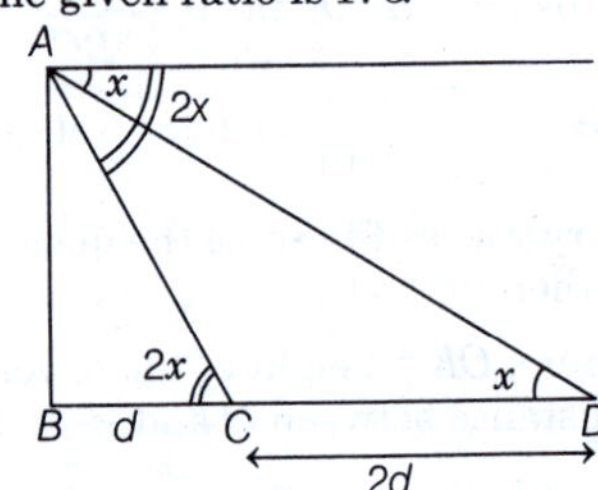

Since perpendiculars are same for both the angles, base for greater angles : base for smaller angle

$$= 1:\left[\frac{\tan \text{(greater angle)}}{\tan \text{(smaller angle)}}\right]$$

$$\Rightarrow BC : BD = 1 : \frac{\tan(2x)}{\tan x}$$

$$\Rightarrow \quad 1:3 = 1:\left[\frac{2\tan x}{1-\tan^2 x} \div \tan x\right]$$

$$\Rightarrow \quad \frac{1}{3} = \frac{1-\tan^2 x}{2}$$

$$\Rightarrow \quad 3\tan^2 x = 3 - 2$$

$$\Rightarrow \quad \tan x = \frac{1}{\sqrt{3}}$$

$$\Rightarrow \tan x = \tan 30^\circ \Rightarrow x = 30^\circ$$

Now, in ΔABD, $\sin x = \dfrac{AB}{AD}$

$$\Rightarrow \quad \sin 30^\circ = \frac{\frac{3}{2}}{AD} \Rightarrow \frac{1}{2} = \frac{\frac{3}{2}}{AD}$$

$$\Rightarrow \quad AD = 3\text{ m}$$

Hence, length of the rope = 3 m

16. Since, both angles are complementary to each other, product of bases = product of perpendiculars.

$$\Rightarrow \quad AC \times BC = DC \times DC$$

$$\Rightarrow \quad 5 \times 10 = DC^2 \Rightarrow DC = \sqrt{50}\text{ m}$$

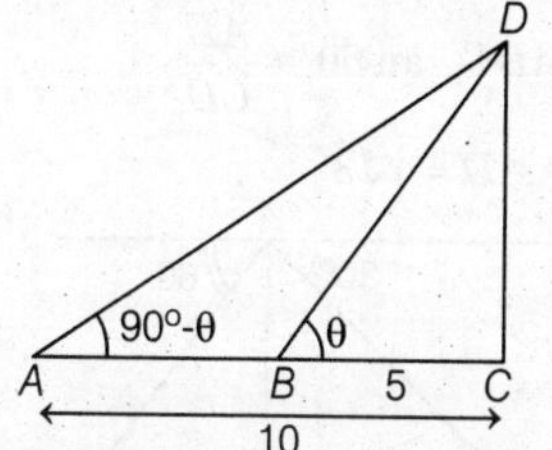

Try yourself to solve the question by other methods.

17. In ΔDEC, $\sin 30^\circ = \dfrac{DE}{DC}$

$$\Rightarrow \quad \frac{1}{2} = \frac{DE}{100}$$

$\therefore$ $DE = 50 = FB$

Further, $\cos 30^\circ = \dfrac{EC}{100}$

$$\Rightarrow \quad \frac{\sqrt{3}}{2} = \frac{EC}{100} \Rightarrow EC = 50\sqrt{3}$$

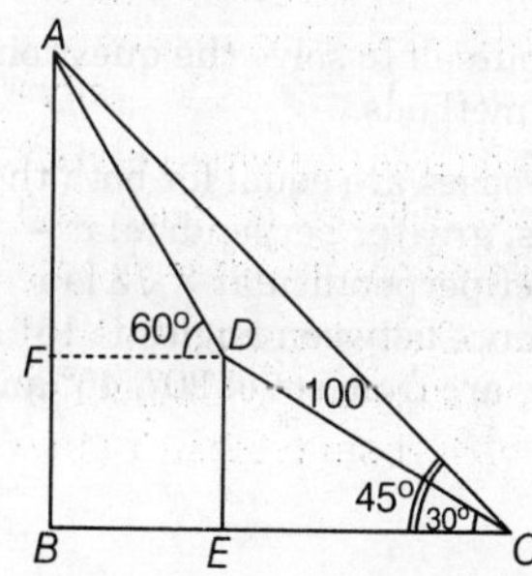

In ΔABC, $\tan 45^\circ = \dfrac{AB}{BC}$

$$\Rightarrow \quad AB = BC$$

Now, in ΔAFD, $\tan 60^\circ = \dfrac{AF}{FD}$

$$\Rightarrow \quad \sqrt{3} = \frac{AF}{BE} \Rightarrow BE = \frac{AF}{\sqrt{3}}$$

$\because$ $AB = BC$

$\therefore$ $AF + FB = BE + EC$

$$\Rightarrow \quad AF + 50 = \frac{AF}{\sqrt{3}} + 50\sqrt{3}$$

$$\Rightarrow \frac{\sqrt{3}AF - AF}{\sqrt{3}} = 50\sqrt{3} - 50$$

$$\Rightarrow \frac{AF(\sqrt{3}-1)}{\sqrt{3}} = 50(\sqrt{3}-1)$$

$$\Rightarrow \quad AF = 50\sqrt{3}$$

Now, height of the mountain

$= AF + FB$

$= 50\sqrt{3} + 50 = 50(\sqrt{3}+1)$ m

Try yourself to solve the question by other methods.

18. Here, length of the ladder = BD

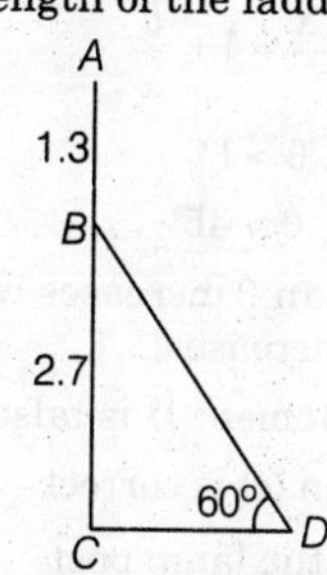

In ΔBCD, $\sin 60^\circ = \dfrac{BC}{BD}$

$$\Rightarrow \quad \frac{\sqrt{3}}{2} = \frac{(4-1.3)}{BD}$$

$$\Rightarrow \quad BD = \frac{2 \times 2.7}{\sqrt{3}} = \frac{9\sqrt{3}}{5}\text{ m}$$

Try yourself to solve the question by other method.

19. Clearly, $\cos\theta = \dfrac{BC}{AC} = \dfrac{10}{20} = \dfrac{1}{2}$

$$\Rightarrow \cos\theta = \cos 60^\circ \Rightarrow \theta = 60^\circ$$

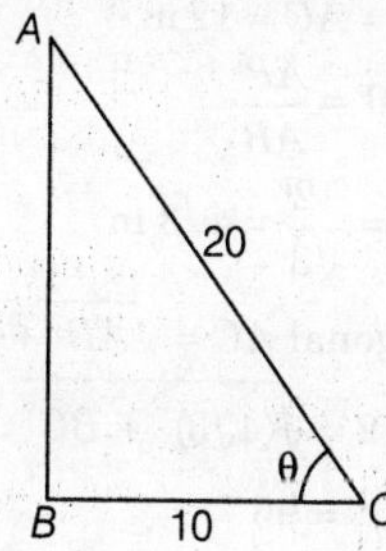

20. Since bases are same for both the angles, greater perpendicular = smaller perpendicular × 3 [as the difference between angles is 15° and angles are from a set of 30°, 45° and 60°.]

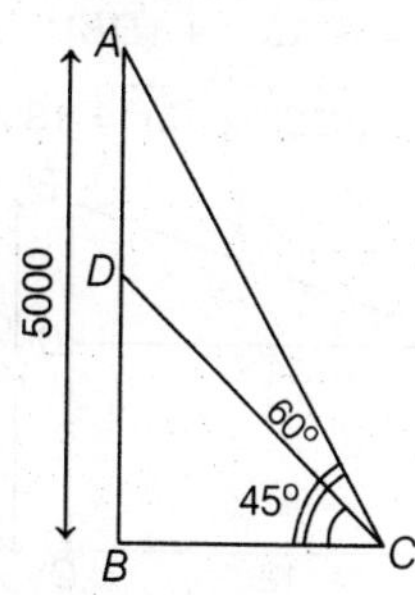

$$\Rightarrow \quad AB = BD \times \sqrt{3}$$

$$\Rightarrow \quad 5000 = BD \times \sqrt{3}$$

$$\therefore \quad BD = \frac{5000}{\sqrt{3}}$$

Now, $AD = AB - BD = 5000 - \dfrac{5000}{\sqrt{3}}$

$$= 5000\left(1 - \frac{1}{\sqrt{3}}\right)\text{ m}$$

Try yourself to solve the question by other methods.

21. Since, the bases for $\angle AEB$ and $\angle AEC$ are same, perpendicular for smaller angle : perpendicular for greater angle

$$= 1:\left[\frac{\tan \text{(greater angle)}}{\tan \text{(smaller angle)}}\right]$$

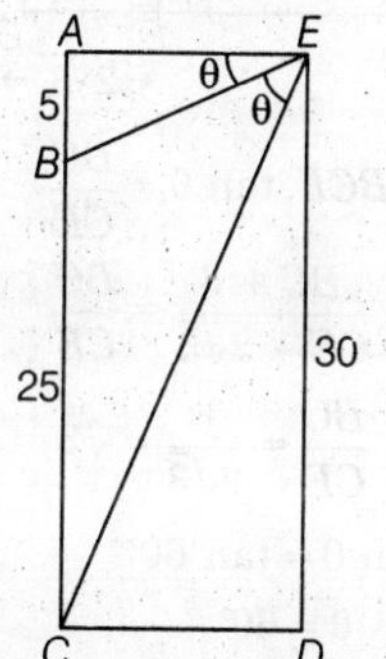

$$\Rightarrow \quad AB = AC = 1 : \frac{\tan 2\theta}{\tan\theta}$$

$$\Rightarrow \quad 5 : 30 = \tan\theta : \tan 2\theta$$

$$\Rightarrow \quad 1 : 6 = (1 - \tan^2\theta) : 2$$

$$\Rightarrow \quad \tan^2\theta = \frac{4}{6}$$

$$\Rightarrow \quad \tan\theta = \sqrt{\frac{2}{3}}$$

Now, in ΔABE, $\tan\theta = \dfrac{5}{AE}$

$$\Rightarrow \quad \sqrt{\frac{2}{3}} = \frac{5}{AE}$$

$$\Rightarrow \quad AE = 5\sqrt{\frac{3}{2}}\text{ m}$$

22. Given, $AB = 6$ m,
then $DE = 11 - 6 = 5$ m

Now, $AE = \sqrt{(AD)^2 + (DE)^2}$

$= \sqrt{(12)^2 + (5)^2} = 13$ m

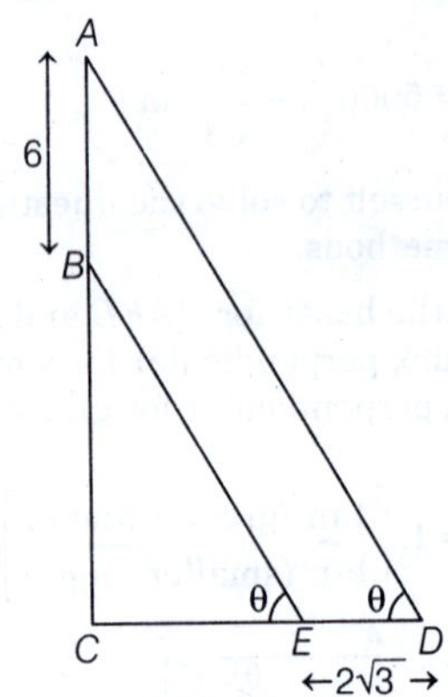

23. Obviously, ΔACD and ΔBCE are similar.

Hence, $\angle BEC = \angle ADC = \theta$

Now, in ΔACD, $\tan\theta = \dfrac{BC + 6}{CE + 2\sqrt{3}}$

and in ΔBCE, $\tan\theta = \dfrac{BC}{CE}$

Hence, $\dfrac{BC+6}{CE+2\sqrt{3}} = \dfrac{BC}{CE}$

$\Rightarrow \quad \dfrac{BC}{CE} = \dfrac{6}{2\sqrt{3}} = \sqrt{3}$

$\Rightarrow \quad \tan\theta = \tan 60^\circ$

$\Rightarrow \quad \theta = 60^\circ$

24. Here, height of the tower = AB and height of the building = CD

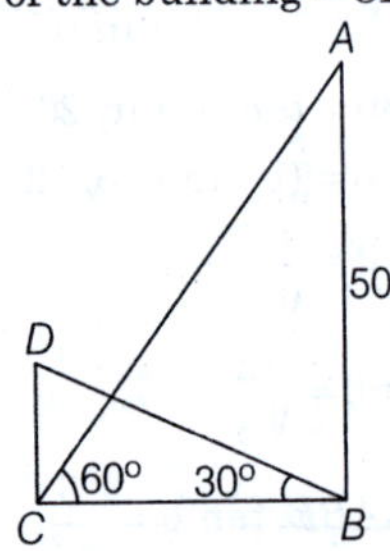

Bases are same for both angles, therefore,

$CD : AB = 1 : \dfrac{\tan 60^\circ}{\tan 30^\circ}$

$\Rightarrow \quad CD : 50 = 1 : 3 \Rightarrow CD = 16.66$ m

Try yourself to solve the question by other methods.

25. In $\Delta AA'M$, $\tan 45^\circ = \dfrac{A'M}{AM}$

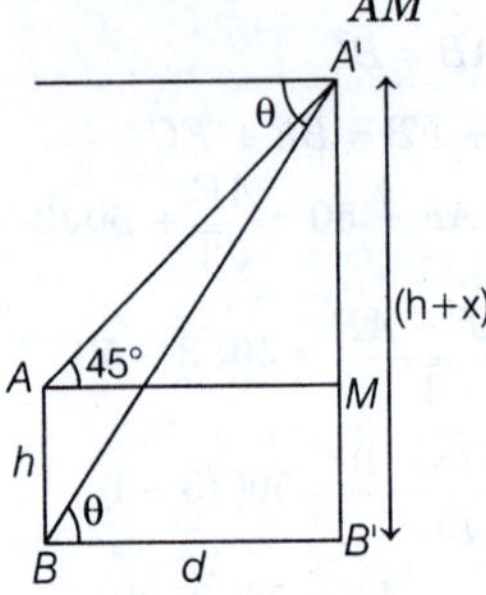

$\Rightarrow \quad AM = A'M = h + x - h = x$

Hence, $d = x$

Then, $h + x > d$

Further, in $\Delta A'B'B$,

$\tan\theta = \dfrac{A'B'}{d} = \dfrac{h+x}{d}$

$= \dfrac{h+x}{x} = 1 + \dfrac{h}{x}$

$\Rightarrow \quad \tan\theta > 1$

$\Rightarrow \quad \theta > 45^\circ$

[$\because$ value of $\tan\theta$ increases when the value of θ increases]

Hence, Statement II is false.

Thus, option (c) is correct.

26. Here, AP is the lamp post.

Clearly, $\tan 45^\circ = \dfrac{AP}{AC}$

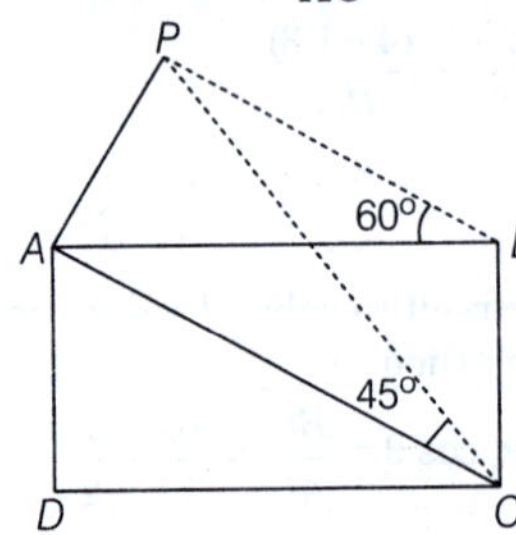

$\Rightarrow \quad AP = AC = 12$ m

and $\tan 60^\circ = \dfrac{AP}{AB}$

$\Rightarrow \quad AB = \dfrac{12}{\sqrt{3}} = 4\sqrt{3}$ m

Now, diagonal $AC = \sqrt{AB^2 + BC^2}$

$\Rightarrow \quad 12 = \sqrt{(4\sqrt{3})^2 + BC^2}$

$\Rightarrow \quad BC^2 = 96$

$\Rightarrow \quad BC = 4\sqrt{6}$ m

Then, area of rectangle $ABCD$ =

$AB \times BC = 4\sqrt{3} \times 4\sqrt{6} = 48\sqrt{2}$ m^2

27. In ΔABC, $\tan 45^\circ = \dfrac{BC}{AB}$

$\Rightarrow \quad BC = AB = 100$

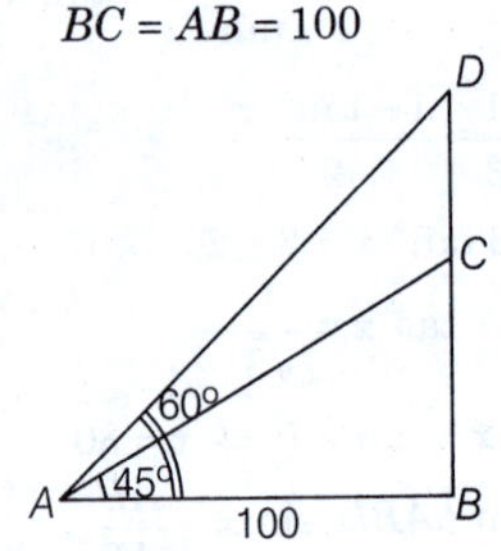

Since bases are the same for both the angles, greater perpendicular = smaller perpendicular × $\sqrt{3}$ [as difference between angles is 15° and angles are from set of 30°, 45° and 60°.]

$\Rightarrow \quad BD = BC \times \sqrt{3} = 100\sqrt{3}$ m

Now, $CD = 100\sqrt{3} - 100$

$= 100(\sqrt{3} - 1)$ m

Try yourself to solve the questions by other methods.

28. Let CD be x km.

In ΔADC, $\tan 60^\circ = \dfrac{AD}{CD}$

$\Rightarrow \quad AD = x\sqrt{3}$

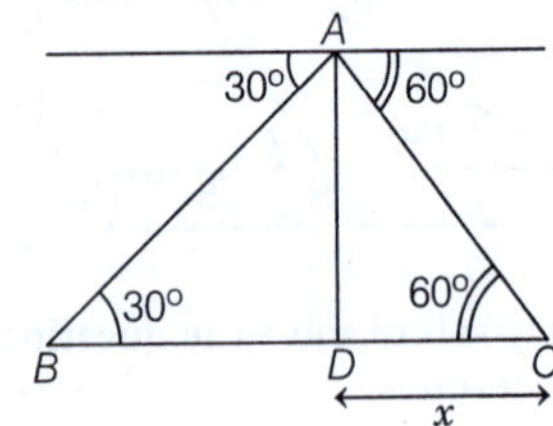

Since both angles are complementary to each other, product of of bases = product of perpendiculars.

$\Rightarrow x \times (1 - x) = AD \times AD$

$\Rightarrow x \times (1 - x) = x\sqrt{3} \times x\sqrt{3} \Rightarrow x = \dfrac{1}{4}$

Now, $AD = x\sqrt{3} = \dfrac{\sqrt{3}}{4}$ km

Try yourself to solve the questions by other methods.

29. Since bases are equal for both the angles, greater perpendicular = smaller perpendicular × $\sqrt{3}$ [as difference between angles is 15° and angles are from set of 30°, 45° and 60°.]

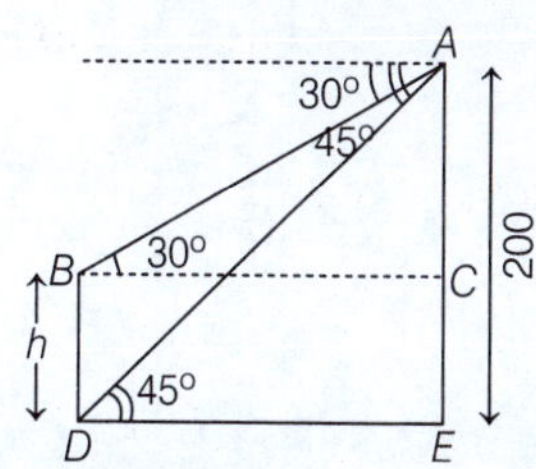

$\Rightarrow \quad AE = AC \times \sqrt{3} \Rightarrow 200 = AC \times \sqrt{3}$

$\Rightarrow \quad AC = \dfrac{200}{\sqrt{3}}$

Now, $EC = 200 - \dfrac{200}{\sqrt{3}}$

$= 200(1 - \dfrac{1}{\sqrt{3}})$ m

Try yourself to solve the question by other methods.

30. Since, $\angle AEB$ and $\angle BEC$ are complementary to each other, product of bases = product of perpendiculars

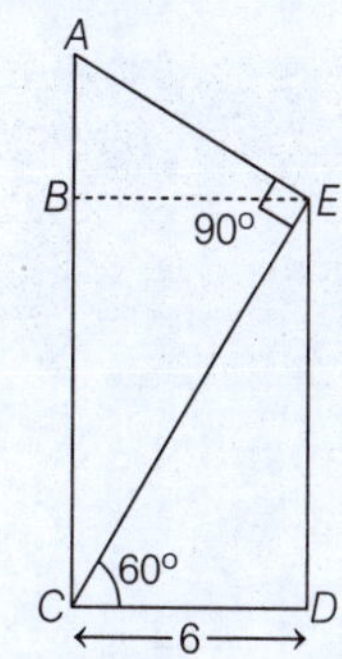

$\Rightarrow \quad BE \times BE = AB \times BC$

$\Rightarrow \quad 6 \times 6 = AB \times BC$

Now, in

$\Delta CDE, \tan 60° = \dfrac{ED}{6} \Rightarrow ED = 6\sqrt{3}$

$\therefore \ AB = 36 \div 6\sqrt{3} = 2\sqrt{3}$

Then, $AC = AB + BC$

$= 2\sqrt{3} + 6\sqrt{3} = 8\sqrt{3}$ m

31. In ΔCBD, $\tan 45° = \dfrac{h}{BD} \Rightarrow BD = h$

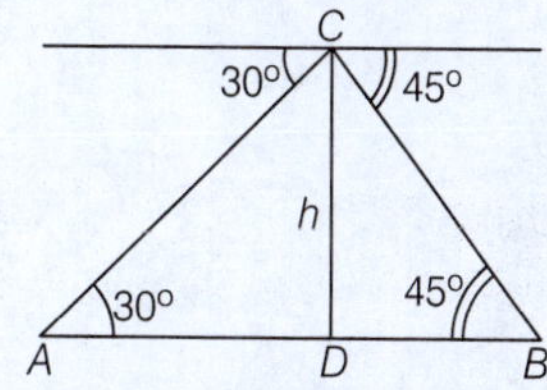

Here, perpendiculars are same for both the angles. Thus, greater base = smaller base $\times \sqrt{3}$ [as difference between angles is 15° and angles are from a set of 30°, 45° and 60°.]

$\Rightarrow \quad AD = BD \times \sqrt{3}$

$\Rightarrow \quad AD = \sqrt{3}h$

Hence, distance between both ships

$= h + \sqrt{3}h = h(\sqrt{3} + 1)$ m

Try yourself to solve the question by other methods.

32.

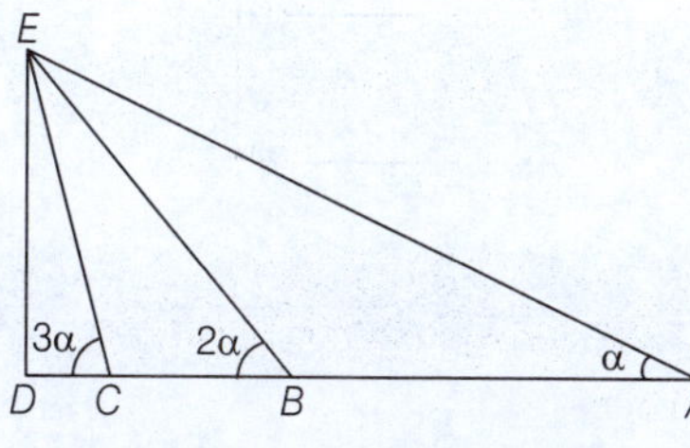

For ΔBCE, $3\alpha = \angle CEB + \angle CBE$

$\Rightarrow \quad 3\alpha = \angle CEB + 2\alpha$

$\Rightarrow \quad \angle CEB = \alpha$

For ΔABE, $2\alpha = \angle BEA + \angle BAE$

$\Rightarrow \quad 2\alpha = \angle BEA + \alpha$

$\Rightarrow \angle BEA = \alpha$ [$\because$ exterior angle = sum of opposite interior angles]

According to Sine-Rule (in ΔBCE),

$$\frac{BC}{\sin \angle CEB} = \frac{BE}{\sin \angle BCE}$$

$$\Rightarrow \quad \frac{BC}{\sin \alpha} = \frac{BE}{\sin (180° - 3\alpha)}$$

$$\Rightarrow \quad \frac{BC}{\sin \alpha} = \frac{AB}{\sin 3\alpha}$$

[$\because \angle EAB = \angle AEB$, $\therefore$ BE = AB]

$$\Rightarrow \quad \frac{AB}{BC} = \frac{\sin 3\alpha}{\sin \alpha} = \frac{3 \sin \alpha - 4 \sin^3 \alpha}{\sin \alpha}$$

$= 3 - 4 \sin^2 \alpha = 3 - 4(1 - \cos^2 \alpha)$

$= 4 \cos^2 \alpha - 2 + 1 = 2(2\cos^2 \alpha - 1) + 1$

$= 2 \cos 2\alpha + 1$

33. Let each side of the polygon be x unit.

Clearly, $AD = 2OD$ and $OD = CD = x$

Then, $\tan 60° = \dfrac{MD}{AD}$

$\Rightarrow \quad \sqrt{3} = \dfrac{MD}{2x}$

$\Rightarrow \quad MD = 2x\sqrt{3}$

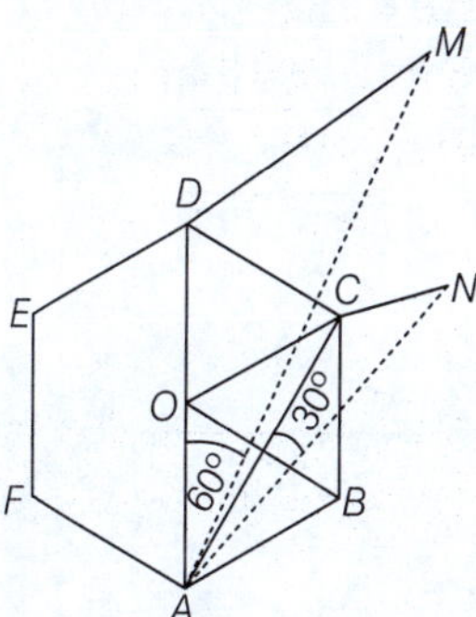

Further, AC is sum of heights of equilateral ΔCOB and ΔAOB.

$\therefore \quad AC = \dfrac{\sqrt{3}}{2}x + \dfrac{\sqrt{3}}{2}x = \sqrt{3}x$

Then, $\tan 30° = \dfrac{NC}{AC} \Rightarrow \dfrac{1}{\sqrt{3}} = \dfrac{NC}{\sqrt{3}x}$

$\Rightarrow \quad NC = x$

Hence, the required ratio

$= NC : MD = x : 2x\sqrt{3} = 1 : 2\sqrt{3}$

34. In ΔAOB, $\tan 45° = \dfrac{OB}{OA}$

$\Rightarrow \quad OA = OB = 300$

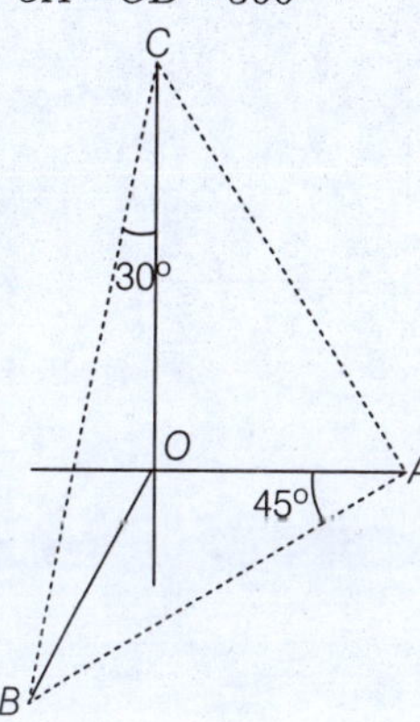

Here, perpendiculars are same for both the angles. Thus, greater base = smaller base $\times \sqrt{3}$ [as difference between angles is 15° and angles are from a set of 30°, 45° and 60°.]

$\Rightarrow \quad OC = OA \times \sqrt{3} = 300\sqrt{3}$

The shortest distance between points A and C = hypotenuse of right-angled

$\Delta AOC = \sqrt{(OC)^2 + (OA)^2}$

$= \sqrt{(300\sqrt{3})^2 + (300)^2}$

$= 600$ ft

Unit 9

The ABC of Algebra

- Algebraic Expression, Algebraic Identities and Remainder/ Factor Theorem
- Concept of Equations : Linear, Quadratic, Cubic and Others
- Concept of Inequality

The ABC of Algebra

Algebra is one of the core branches of Mathematics in which letters and symbols are used to represent numbers or quantities and they are processed as per rules of arithmetical operations and relationship. Algebraists are mathematicians who do research in algebra. Algebra plays a vital role in securing marks in diverse competitive examinations either directly or indirectly. The special characteristics of this section is that a lot of problems can be solved quickly through options. For the sake of convenience, this Section is divided into following parts :

Part-I Algebraic Expression, Algebraic Identities and Remainder/ Factor-Theorem

Part-II Concept of Equations: Linear, Quadratic, Cubic and others

Part-III Concept of Inequality

Learning OBJECTIVES (Part I)

After completing this chapter, aspirants will be able to :

- Understand the significance of this sub -section in diverse competitive exams;
- Comprehend the diverse types of algebraic expression on various basis with illustration;
- Find the techniques how to do mathematical operation with the algebraic expressions;
- Compute the minimum and maximum values of an algebraic expression;
- Know the Remainder-theorem and Factor-theorem and its utilities;
- Recognise various types of algebraic identities;
- Realise themselves competent to solve all types of problems based on algebraic identities;
- Be familiar with the diverse types of problems seeking specific approaches and the techniques how to apply logics in consonance with the given context;
- Understand the concept of perfect square and cube with illustrations;
- Observe the problems seeking concept of ratio;
- Understand the philosophy of division of 'Illustrations';
- Identify the problems which can easily be solved through options;
- Realise themselves competent to solve the questions in unique way;
- Analyse the trend of the questions being asked in various competitive examinations;
- Solve the questions in moments;
- Learn many new concepts/fundamentals;
- Apply the pragmatic approaches in lieu of academic approaches; and
- Ponder on the alternative methods.

52

Part-I

Algebraic Expression, Identities and Remainder/ Factor-Theorem

Introduction

A combination of constants and variables with some or all of the operations of +, − , × and ÷ is called algebraic expression. Each part of the algebraic expression separated either by + or − is known as 'terms' of the algebraic expression.

Example of algebraic expression $7x^2 - 3xy + c$

[In algebraic expression, as per convention, last letters of the English Alphabet (i.e. ...., *x, y, z*) are used to represent variables and beginning letters (i.e. *a, b, c*, ...) to represent constants]

Here, $7x^2$, $3xy$ and c ⇒ terms (3 in number)

+, − ⇒ operators

c ⇒ constant

x, y ⇒ variables

Note *x and y are called variables as they carry varying values. c is a constant as its value never changes in the expression. Variables are also known as 'unknowns'.*

Types of Algebraic Expressions

Monomial An algebraic expression which consists of only one term is called monomial.

e.g.

'x' is a monomial in one variable x.

'$7x^2$' is a monomial in one variable x.

'$-5x^2y$' is a monomial in two variables x and y.

'$\frac{-2y^2}{3xz}$' is a monomial in three variables x, y and z.

Polynomial

An algebraic expression which consists of two or more terms is called polynomial.

Inside

- *Introduction*
- *Algebraic Expressions, Types and Mathematical Operations*
- *Minimum and Maximum Value of an Algebraic Expression*
- *Remainder and Factor Theorem*
- *Algebraic Identities*
- *Illustrations Based on Above Aspects Including Those Seeking Specific Approaches*
- ....
- .... *& Many More*
- *Work Out with Solutions*

e.g.

'$7x^2 + 9x + 17$' is a polynomial of three terms in one variable x.

'$x + 7y$' is a polynomial of two terms in two variables x and y.

'$6x^2 + 3xy + 11y$' is a polynomial of three terms in two variables x and y.

Remarks

We can use words 'Binomial' and 'Trinomial' for the algebraic expression consisting of two and three terms respectively.

Polynomials in one variable (in detail)

Take a few examples :

(a) $8y^3 + 5y^2 + 4y + 5$ [read as 'a polynomial in y of degree 3' (take highest power)].

(b) $2x^2 + 5x - 6$ (read as 'a polynomial in x of degree 2')

It is important to note :

(i) The greatest power(or exponents) of the variable in the polynomial in one variable is known as the 'degree of the polynomial'.

(ii) General form of the polynomial in one variabile is $a_n x^n + a_{n-1} x^{n-1} + \ldots\ldots\ldots + a_2 x^2 + a_1 x + a_0$, where $a_n, a_{n-1}, \ldots\ldots\ldots\ldots a_2, a_1, a_0$ are real numbers and they are called coefficients of the polynomial and n is non-negative integer.

(iii) For more clarity, $\frac{1}{5}x^2 - \sqrt{3}\,x + \sqrt[4]{3}$ is a polynomial in x whereas $5x^2 - \sqrt{3x} + 5$ and $x + \frac{1}{x^2}$ are not polynomial because the variable x has power in fraction in first case and the power of variable x is not non-negative integer in the second case.

Polynomials in two or more variables (in detail)

Take a few examples :

(a) $8x^2 y^3 + 5y^2 - 4xy + 7$ (a polynomial in two variables x and y)

(b) $x^2y^3z - 3y^2z^2 - 4xy + 7xz - 9$ (a polynomial in three variables x, y and z)

It is important to note

(i) In case of polynomial in two or more variables, sum of the powers of the variables in each term is the degree of the term and the greatest among the degrees of all the terms, is the degree of the polynomial. The 1st example can be written as $8x^2y^3 + 5x^0y^2 - 4x^1y^1 + 7x^0y^0$. The degree of 1st term is $(2 + 3)$, that of 2nd is $(0 + 2)$. The 3rd and the 4th term have degrees $(1 + 1)$ and $(0 + 0)$, respectively. Then, degree of the polynomial is 5 as it is the highest among the degrees of all the terms.

(ii) A polynomial in two or more variables is in standard form when the term having the highest degree is the 1st term, the term having 2nd highest degree is the 2nd term and so on.

(iii) Power of each of variables must be non-negative integer.

Types of polynomials on basis of their degrees

(i) **Linear Polynomial** A polynomial of degree 1 is called a linear polynomial. It may be in one or more variables.

e.g.

(a) $3x + 5$ is a linear polynomial in x.

(b) $5x + 4y + 3$ is a linear polynomial in x and y.

(ii) **Quadratic Polynomial** A polynomial of degree 2 is called quadratic polynomial.

e.g.

(a) $2x^2 + 5x + 6$ is a quadratic polynomial in x.

(b) $2x^2 + xy + 6$ is a quadratic polynomial in x and y.

(iii) **Cubic Polynomial** A polynomial of degree 3 is called a cubic polynomial.

e.g.

(a) $2x^2y + 5xy + 3y + 6$ is a cubic polynomial in x and y.

(b) $xyz + xy + yz + z + 4$ is a cubic polynomial in x, y and z.

(iv) **Biquadratic Polynomial** A polynomial of degree 4 is called a biquadratic polynomial.

e.g.

(a) $2x^2yz + 5xyz + 3yz + 6y + 3$ is a biquadratic polynomial in x, y and z.

(b) $x^4 + 5x^2y + 3xy + 2y + 5$ is a biquadratic polynomial in x and y.

Mathematical Operations of Algebraic Expressions

(i) **Addition and Subtraction of Algebraic Expressions** Only like terms of an algebraic expressions are added or subtracted. Like terms refer to the terms having same variable raised to same power.

e.g. (a) $(3x + 4y) + (3y + 5) = ?$ Here, $4y$ and $3y$ are like terms as they have same variable with same power.

Then, $(3x + 4y) + (3y + 5) = 3x + (4y + 3y) + 5 = 3x + 7y + 5$

(b) $(3x^2 + 2y) - (8 + 2x^2 - 2x) = ?$

$= 3x^2 + 2y - 8 - 2x^2 + 2x$

$= (3x^2 - 2x^2) + 2y + 2x - 8$

$= x^2 + 2y + 2x - 8$

(ii) **Multiplication of Algebraic Expressions** In case of two expressions, each term of one expression is multiplied with each term of another keeping the sign in mind.

e.g. $(a + b) \times (a - b) = ?$

$= (a \times a) - (a \times b) + (b \times a) - (b \times b) = a^2 - b^2$

$$\text{Please mind}\begin{Bmatrix} + & \times & + & = & + \\ - & \times & - & = & + \\ + & \times & - & = & - \\ - & \times & + & = & - \end{Bmatrix}$$

In case of more than two expressions, we first get the product of any two expression as explained above. Thereafter each term of the product is multiplied with each term of the 3rd expression and so on until all the given expressions are used.

e.g. $(a + b) \times (a + b) \times (a + b) = ?$

$= (a^2 + ab + ab + b^2) \times (a + b)$

$= (a^2 + 2ab + b^2) \times (a + b)$

$= a^3 + a^2b + 2a^2b + 2ab^2 + ab^2 + b^3$

$= a^3 + 3a^2b + 3ab^2 + b^3$

$= a^3 + b^3 + 3ab(a + b)$

(iii) **Division of Algebraic Expressions** We divide each term of the dividend by the divisor to obtain the quotient keeping the sign in mind.

e.g. (a) $\dfrac{4x^3 + 6x^2 - 8x}{2x} = ?$

$$\frac{4x^3 + 6x^2 - 8x}{2x} = \frac{4x^3}{2x} + \frac{6x^2}{2x} - \frac{8x}{2x}$$

$$= 2x^2 + 3x - 4$$

(b) $\dfrac{32x^3\, y^3\, z^5}{8xy^2} = ?$

$$\frac{32x^3\, y^3\, z^5}{8xy^2} = 4x^2yz^5$$

(iv) **LCM and HCF of Algebraic Expressions** Discussed in the chapter 'LCM and HCF' in detail.

Maximum and Minimum Value of Algebraic Expressions

(i) For positive numbers a and b, if $(a + b)$ is constant, then the value of ab is maximum at $a = b$.

Similarly, if $(a + b + c + \ldots\ldots\ldots)$ is constant, then the value of abc ...... is maximum at $a = b = c = \ldots\ldots\ldots$

Note *If the question does not permit to take their values equal, we take their values as close as possible. It is applicable in all such types of cases including the followings :*

(ii) For positive numbers a and b, if the value of ab is constant, the value of $(a + b)$ is minimum at $a = b$.

Similarly, if the value of abc ..... is constant, the value of $(a + b + c + \ldots)$ is minimum at $a = b = c \ldots\ldots$

(iii) For real numbers a and b, if the value of $(a + b)$ is constant, $a^2 + b^2$ will have minimum value at $a = b$.

Similarly, if the value of $(a + b + c + \ldots)$. is constant, $a^2 + b^2 + c^2 + \ldots\ldots$ will have minimum value at $a = b = c = \ldots\ldots$

(iv) For positive numbers a, b and c, if the value of $(a + b + c)$ is constant, the value of $ab + bc + ca$ is maximum at $a = b = c$.

(v) For real positive numbers, Root-Mean Square ≥ Arithmetic Mean ≥ Geometric mean ≥ their Harmonic mean i.e.

$$\sqrt{\frac{x_1^2 + x_2^2 + x_3^2 + \ldots + x_n^2}{n}} \ge \frac{x_1 + x_2 + x_3 + \ldots + x_n}{n} \ge \sqrt[n]{x_1\, x_2\, x_3 \ldots\ldots x_n} \ge \frac{n}{\frac{1}{x_1} + \frac{1}{x_2} + \frac{1}{x_3} + \ldots\ldots + \frac{1}{x_n}}$$

It is important to note that RMS = AM = GM = HM, when $x_1 = x_2 = x_3 = \ldots \ldots = x_n$.

Hence, RMS > AM > GM > HM if all the given numbers are not equal.

e.g. (a) $\dfrac{a + b}{2} \ge \sqrt{ab} \ge \dfrac{2ab}{a + b}$ where $a, b > 0$.

(b) $\dfrac{a + b + c}{3} \ge \sqrt[3]{abc} \ge \dfrac{3abc}{ab + bc + ca}$, where $a, b, c > 0$.

Hence, the minimum value of the arithmetic mean of positive real numbers is either equal to the value of geometric mean or equal to harmonic mean of those numbers. Critical inequalities have been given in chapter 'Inequality'.

(vi) Concept of maximum and minimum value of a quadratic equation has been discussed under 'Quadratic equation'.

Remainder-Theorem

If a polynomial $f(x)$ is divided by $(x - a)$, the remainder is $f(a)$. **e.g.**

(a) Find the remainder if $3x^2 - 5x + 8$ is divided by $(x - 2)$.

Given, $f(x) = 3x^2 - 5x + 8$

and $(x - a) = x - 2$ i.e. $a = 2$

Then, $f(a) = 3a^2 - 5a + 8 = 3(2)^2 - (5 \times 2) + 8 = 10$

Hence, the required remainder = 10

Alternatively

Say, divisor $= 0 \Rightarrow x - 2 = 0 \Rightarrow x = 2$

Now, put the value of x in the given expression, the outcome will be the required remainder.

Hence, required remainder $= 3(2)^2 - 5(2) + 8 = 10$

(b) Find the remainder if $4x^3 - 5x^2 + 6x + 2$ is divided by $(2x - 1)$.

$$2x - 1 = 0 \Rightarrow x = \frac{1}{2}$$

$\therefore$ Remainder $= 4\left(\frac{1}{2}\right)^3 - 5\left(\frac{1}{2}\right)^2 + 6\left(\frac{1}{2}\right) + 2 = 4\frac{1}{4}$

Factor-Theorem

It is obvious that if the remainder is equal to zero, then the given expression is exactly divisible by the divisor. It means, the divisor is a factor of that expression.

Hence, if $f(a) = 0$, then $(x - a)$ would be a factor of $f(x)$. This is known as 'Factor- Theorem'. It may be said that factor theorem is a special case of remainder-theorem.

Points to be remembered

(i) **$x^n - a^n$ is exactly divisible by $(x - a)$, whether n is odd or even.** (Clarification $x - a = 0 \Rightarrow x = a$. Putting the value of $x = a$ in $x^n - a^n$, we get the remainder 0 whether n is odd or even)

(ii) **$x^n - a^n$ is exactly divisible by $(x + a)$, when n is even.**
(Clarification $x + a = 0 \Rightarrow x = -a$. Putting the value of $x = -a$ in $x^n - a^n$, we get the remainder 0 if n is even)

(iii) **$x^n + a^n$ is exactly divisible by $(x + a)$, when n is odd.**
(Clarification $x + a = 0 \Rightarrow x = -a$. Putting the value of $x = -a$ in $x^n + a^n$, we get the remainder 0 if n is odd)

(iv) **$x^n + a^n$ is never divisible by $(x - a)$.**
(Clarification $x - a = 0 \Rightarrow x = a$. Putting the value of $x = a$ in $x^n + a^n$, we can never get the remainder 0)

Algebraic Identities

The algebraic identities are the algebraic equations which are true for all values of the variables involved therein.

$(a + b)^2$ $= a^2 + b^2 + 2ab$	When b is substituted by $\frac{1}{a}$, then $\rightarrow$	$\left(a + \frac{1}{a}\right)^2 = a^2 + \frac{1}{a^2} + 2$
$(a - b)^2 = [a + (-b)]^2$ $= a^2 + (-b)^2 + 2 \cdot a \cdot (-b)$ $\Rightarrow (a - b)^2$ $= a^2 + b^2 - 2ab$	-do-	$\left(a - \frac{1}{a}\right)^2 = a^2 + \frac{1}{a^2} - 2$
$\because a^2 + b^2 - 2ab$ $= (a - b)^2$ $\Rightarrow a^2 + b^2 - 2ab + 4ab$ $= (a - b)^2 + 4ab$ $\Rightarrow a^2 + b^2 + 2ab$ $= (a - b)^2 + 4ab$ $\Rightarrow (a + b)^2$ $= (a - b)^2 + 4ab$	-do-	$\left(a + \frac{1}{a}\right)^2 = \left(a - \frac{1}{a}\right)^2 + 4$
$(a - b)^2$ $= (a + b)^2 - 4ab$	-do-	$\left(a - \frac{1}{a}\right)^2 = \left(a + \frac{1}{a}\right)^2 - 4$

$(a - b)^2 + (b - c)^2 + (c - a)^2 = 2(a^2 + b^2 + c^2 - ab - bc - ca)$

$(a + b)(a - b) = a^2 - b^2$

$(a + b + c)^2 = a^2 + b^2 + c^2 + 2ab + 2bc + 2ca$

$(a + b - c)^2 = [a + b + (-c)^2]$
$\Rightarrow (a + b - c)^2 = a^2 + b^2 + c^2 + 2ab - 2bc - 2ca$

$(a - b + c)^2 = a^2 + b^2 + c^2 - 2ab - 2bc + 2ac$

$(-a + b + c)^2 = a^2 + b^2 + c^2 - 2ab + 2bc - 2ac$

$(a + b)^3 = a^3 + b^3$ $+ 3ab(a + b)$	When b is substituted by $\frac{1}{a}$, then $\rightarrow$	$\left(a + \frac{1}{a}\right)^3$ $= a^3 + \frac{1}{a^3} + 3\left(a + \frac{1}{a}\right)$
$(a - b)^3 = a^3 - b^3$ $- 3ab(a - b)$	-do-	$\left(a - \frac{1}{a}\right)^3$ $= a^3 - \frac{1}{a^3} - 3\left(a - \frac{1}{a}\right)$

$(a + b)^3 + (a - b)^3 = 2a^3 + 6ab^2$

$\because (a + b)^3 = a^3 + b^3 + 3ab(a + b)$
$\Rightarrow a^3 + b^3 = (a + b)^3 - 3ab(a + b) = (a + b)[(a + b)^2 - 3ab]$
$\Rightarrow a^3 + b^3 = (a + b)(a^2 + b^2 - ab)$
Similarly, $a^3 - b^3 = (a - b)(a^2 + b^2 + ab)$

$(a + b + c)^3 = a^3 + b^3 + c^3 + 3(a + b)(b + c)(c + a)$

$a^3 + b^3 + c^3 - 3abc = (a + b + c)(a^2 + b^2 + c^2 - ab - bc - ca)$
$= (a + b + c)[(a + b + c)^2 - 3(ab + bc + ca)]$
$= \frac{1}{2}(a + b + c)[(a - b)^2 + (b - c)^2 + (c - a)^2]$

Note (i) If $a + b + c = 0$, then $a^3 + b^3 + c^3 = 3abc$
(ii) On the above analogy, $(a - b)^3 + (b - c)^3 + (c - a)^3 = 3(a - b)(b - c)(c - a)$ as $(a - b) + (b - c) + (c - a) = 0$

$a^4 - b^4 = (a - b)(a + b)(a^2 + b^2) = (a^2 - b^2)(a^2 + b^2)$

$a^5 - b^5 = (a - b)(a^4 + a^3 b + a^2 b^2 + ab^3 + b^4)$

$a^n - b^n = (a - b)(a^{n-1} + a^{n-2} b^1 + a^{n-3} b^2 + ... + a^1 b^{n-2} + b^{n-1})$

Note *Applications of the above algebraic identities/formulae have been discussed separately.*

Problems Based on Algebraic Identities

Type 1

Computation of value of algebraic expression when relation among different variables is known.

Problem 1 If $a^3 - b^3 - c^3 = 3abc$, then which of the following is true?

(a) $a = b = c$
(b) $a + b + c = 0$
(c) $a + c = b$
(d) $a = b + c$

Sol. We have,

$$a^3 + b^3 + c^3 - 3abc = (a + b + c)(a^2 + b^2 + c^2 - ab - bc - ca)$$

$$\Rightarrow a^3 + (-b)^3 + (-c)^3 - 3a(-b)(-c) = [(a + (-b) + (-c)] [a^2 + (-b)^2 + (-c)^2 - a(-b) - (-b)(-c) - (-c)a]$$

Substituting b by $-b$ and c by $-c$ so as to get

$a^3 - b^3 - c^3 - 3abc$

$\Rightarrow \quad a^3 - b^3 - c^3 - 3abc = [a - b - c][a^2 + b^2 + c^2 + ab - bc + ca]$

$\Rightarrow \quad 0 = [a - b - c][a^2 + b^2 + c^2 + ab - bc + ca]$

$\Rightarrow \quad [a - b - c] = 0 \Rightarrow a = b + c$

Alternatively

Given, $a^3 - b^3 - c^3 = 3abc$

$\Rightarrow a^3 + (-b)^3 + (-c)^3 = 3a(-b)(-c)$

We know that $a^3 + b^3 + c^3 = 3abc$, if $a + b + c = 0$

Hence, $a + (-b) + (-c) = 0$

$\Rightarrow \quad a = b + c$

Problem 2 If $ax + by = 6$, $bx - ay = 2$ and $x^2 + y^2 = 4$, then $(a^2 + b^2) = ?$

Sol. Given, $ax + by = 6$

$\Rightarrow \quad a^2x^2 + b^2y^2 + 2abxy = 36$ (squaring both sides)

Also, $bx - ay = 2$

$\Rightarrow \quad b^2x^2 + a^2y^2 - 2abxy = 4$ (squaring both sides)

On addition, $x^2(a^2 + b^2) + y^2(a^2 + b^2) = 40$

$\Rightarrow \quad (x^2 + y^2)(a^2 + b^2) = 40$

$\Rightarrow \quad 4(a^2 + b^2) = 40$

$\therefore \quad a^2 + b^2 = 10$

Problem 3 If $a^3 - b^3 = 56$ and $a - b = 2$, then $a^2 + b^2 = ?$

[SSC(CGL-T-I)-12, SSC (10+2-LDC/DEO)-12]

Sol. We have, $a^3 - b^3 = (a - b)(a^2 + b^2 + ab)$

$\Rightarrow \quad 56 = 2(a^2 + b^2 + ab) \Rightarrow a^2 + b^2 + ab = 28$

$\Rightarrow (a^2 + b^2 - 2ab) + 3ab = 28 \Rightarrow (a - b)^2 - 28 = -3ab$

$\Rightarrow \quad (2)^2 - 28 = -3ab \therefore ab = \frac{-24}{-3} = 8$

Now, $a^2 + b^2 = (a - b)^2 + 2ab = (2)^2 + 2 \cdot 8 = 4 + 16 = 20$

Problem 4 If $(a - b) = 3$, $(b - c) = 5$ and $(c - a) = 1$, then $\frac{a^3 + b^3 + 1 - 3ab}{a + b + c} = ?$

Sol. We have, $(a - b)^2 + (b - c)^2 + (c - a)^2$

$= 2(a^2 + b^2 + c^2 - ab - bc - ca)$

$\Rightarrow \quad 3^2 + 5^2 + 1^2 = 2(a^2 + b^2 + c^2 - ab - bc - ca)$

$\Rightarrow \quad (a^2 + b^2 + c^2 - ab - bc - ca) = 17.5$

Now, $\frac{a^3 + b^3 + c^3 - 3abc}{a + b + c}$

$= \frac{(a + b + c)(a^2 + b^2 + c^2 - ab - bc - ca)}{(a + b + c)}$

$= (a^2 + b^2 + c^2 - ab - bc - ca) = 17.5$

Problem 5 If $x^3 + y^3 = 35$ and $x + y = 5$, then $\left(\frac{1}{x} + \frac{1}{y}\right) = ?$

[SSC (CGL-T-1)-12, SSC 10+2-LDC/DEO-12]

Sol. We have, $x^3 + y^3 = (x + y)^3 - 3xy(x + y)$

$\Rightarrow \quad 35 = (5)^3 - 3xy \cdot 5$

$\Rightarrow \quad xy = \frac{90}{15} = 6$

Now, $\frac{1}{x} + \frac{1}{y} = \frac{x + y}{xy} = \frac{5}{6}$

Problem 6 If $x + y + z = 1$, $xy + yz + zx = -1$ and $xyz = -1$, then $x^3 + y^3 + z^3 = ?$

[SSC(FCI-Mains)-12, SSC (CGL-T-I)-19, SSC (CGL-T-II)-20]

Sol. We have, $(x + y + z)^2 = x^2 + y^2 + z^2 + 2(xy + yz + zx)$

$\Rightarrow \quad 1^2 = x^2 + y^2 + z^2 + 2(-1)$

$\Rightarrow x^2 + y^2 + z^2 = 1 + 2 = 3$

Now, $x^3 + y^3 + z^3 - 3xyz$

$= (x + y + z)[x^2 + y^2 + z^2 - xy - yz - zx]$

$\Rightarrow \quad x^3 + y^3 + z^3 - 3(-1) = 1[3 - (-1)]$

$\Rightarrow \quad x^3 + y^3 + z^3 = 4 - 3 = 1$

Problem 7 If $a^4 + b^4 = a^2b^2$, then $(a^6 + b^6) = ?$

Sol. Clearly, $a^6 + b^6 = (a^2)^3 + (b^2)^3$

$= (a^2 + b^2)(a^4 + b^4 - a^2b^2)$

$= (a^2 + b^2)(a^2b^2 - a^2b^2)$

$(\because a^4 + b^4 = a^2b^2)$

$= (a^2 + b^2) \times 0 = 0$

Problem 8 For $a > b$, if $a + b = 5$ and $ab = 6$, then the value of $a^2 - b^2 = ?$

[SSC (Investigator)-10]

Sol. We have, $(a - b)^2 = (a + b)^2 - 4ab$

$\Rightarrow \quad (a - b)^2 = (5)^2 - 4 \cdot 6 = 1$

$\Rightarrow \quad a - b = 1$

Now, $a^2 - b^2 = (a + b)(a - b) = 5 \times 1 = 5$

Problem 9 If $ax + by - 2 = 0$ and $axby = 1$, where $a \neq 0, b \neq 0$, then what is the value of $a^2x + b^2y$ equal to?

[CDS-14]

(a) $a + b$ (b) $2ab$

(c) $a^3 + b^3$ (d) $a^4 + b^4$

Sol. As given, $ax + by = 2$

$\Rightarrow \quad a^2x^2 + b^2y^2 + 2axby = 4$ (squaring both sides)

$\Rightarrow \quad a^2x^2 + b^2y^2 + 2 \cdot 1 = 4$

$\Rightarrow \quad a^2x^2 + b^2y^2 = 2$

$\Rightarrow \quad (ax)^2 + (by)^2 = 2$

Since values of $(ax)^2$ and $(by)^2$ are positive, $(ax)^2 = 1$ and $(by)^2 = 1$.

Then, $ax = 1$ and $by = 1$

$\Rightarrow \quad x = \frac{1}{a}$ and $y = \frac{1}{b}$

Now, $(a^2x + b^2y) = \left(a^2 \cdot \frac{1}{a} + b^2 \cdot \frac{1}{b}\right) = a + b$

Problem 10 If u, v and w are real numbers such that $u^3 - 8v^3 - 27w^3 = 18\,uvw$, then which one of the following is correct?

(a) $u - v + w = 0$ (b) $u = -v = -w$
(c) $u - 2v = 3w$ (c) $u + 2v = -3w$

Sol. Given, $u^3 - 8v^3 - 27w^3 = 18uvw$

$\Rightarrow \quad u^3 + (-2v)^3 + (-3w)^3 = 3u(-2v)(-3w)$

We have, if $a + b + c = 0$

then $a^3 + b^3 + c^3 = 3abc$

It implies that $u + (-2v) + (-3w) = 0$

$\Rightarrow \quad u - 2v = 3w$

Problem 11 If $x + y + z = 6$ and $xy + yz + zx = 11$, then what is the value of $x^3 + y^3 + z^3 - 3xyz$?

[CDS-08, SSC (CGL-T-I)-19]

Sol. $x^3 + y^3 + z^3 - 3xyz = (x + y + z)$

$[(x + y + z)^2 - 3(xy + yz + zx)]$

$= 6[(6)^2 - 3(11)] = 18$

In following types of problems, we may assume the value of variable(s) under the restriction given in the question and find the value of the expression so as to save the time in the examination hall. For illustrations :

Problem 12 If $x + y + z = 0$, then $\frac{x^2}{yz} + \frac{y^2}{xz} + \frac{z^2}{xy} = ?$

Sol. $\frac{x^2}{yz} + \frac{y^2}{xz} + \frac{z^2}{xy} = \frac{x^3 + y^3 + z^3}{xyz}$

$= \frac{(x^3 + y^3 + z^3 - 3xyz) + 3xyz}{xyz}$

$= \frac{(x + y + z)(x^2 + y^2 + z^2 - xy - yz - zx) + 3xyz}{xyz}$

$\frac{0 \times (x^2 + y^2 + z^2 - xy - yz - zx) + 3xyz}{xyz}$

$= \frac{0 + 3xyz}{xyz} = 3$

Alternatively I

$\frac{x^2}{yz} + \frac{y^2}{xz} + \frac{z^2}{xy} = \frac{x^3 + y^3 + z^3}{xyz} = \frac{3xyz}{xyz}$

$(\because a^3 + b^3 + c^3 = 3abc$, if $a + b + c = 0)$

$= 3$

Alternatively II

Let $x = 1$, $y = 2$ and $z = -3$ $(\because x + y + z = 0)$

(You are free to take any value of x, y, z satisfying $x + y + z = 0$. However, the value of denominator of any term at the assumed values must not be equal to 0, otherwise the term will be undefined.)

Now, $\frac{x^2}{yz} + \frac{y^2}{xz} + \frac{z^2}{xy} = \frac{1^2}{2(-3)} + \frac{2^2}{1(-3)} + \frac{(-3)^2}{1 \cdot 2} = 3$

Problem 13 If $a + b + c = 8$, then $(a - 4)^3 + (b - 3)^3 + (c - 1)^3 - 3(a - 4)(b - 3)(c - 1) = ?$

Sol. $(a - 4)^3 + (b - 3)^3 + (c - 1)^3 - 3(a - 4)(b - 3)(c - 1)$

$= (a - 4 + b - 3 + c - 1) \times [(a - 4)^2 + (b - 3)^2 + (c - 1)^2 -$
$(a - 4)(b - 3) - (b - 3)(c - 1) - (c - 1)(a - 4)]$

$= (a + b + c - 8) \times [(a - 4)^2 + (b - 3)^2 + (c - 1)^2 - (a - 4)$
$(b - 3 - b - 3c - 1 - c - 1a - 4]$

$= (8 - 8) \times [(a - 4)^2 + (b - 3)^2 + (c - 1)^2 - (a - 4)(b - 3) -$
$b - 3c - 1 - c - 1a - 4 = 0$

Alternatively

Let $a = 5$, $b = 2$ and $c = 1$. $(\because a + b + c = 8)$

(You are free to take any value of a, b, c satisfying $a + b + c = 8$)

Now, $(a - 4)^3 + (b - 3)^3 + (c - 1)^3 - 3(a - 4)(b - 3)(c - 1)$

$= (5 - 4)^3 + (2 - 3)^3 + (1 - 1)^3 - 3(5 - 4)(2 - 3)(1 - 1)$

$= 1 - 1 + 0 - 0 = 0$

Problem 14 If $a + b + 1 = 0$, then $a^3 + b^3 + 1 - 3ab = ?$

[SSC (CGL-T-I)-12]

Sol. Clearly, $a^3 + b^3 + 1 - 3ab \cdot 1 = a^3 + b^3 + 1^3 - 3ab \cdot 1$

$= (a + b + 1)(a^2 + b^2 + 1 - ab - b \cdot 1 - 1 \cdot a)$

$= 0 \times (a^2 + b^2 + 1 - ab - b \cdot 1 - 1 \cdot a) = 0$

Alternatively

Let $a = 1$ and $b = -2$. $(\because a + b + 1 = 0)$

(You are free to take any value of a, b satisfying $a + b + 1 = 0$)

Now, $a^3 + b^3 + 1 - 3ab = 1^3 + (-2)^3 + 1 - 3 \cdot 1 \cdot (-2)$

$= 1 - 8 + 1 + 6 = 0$

Problem 15 If $x + y + z = 2s$, then what is the value of $(s - x)^3 + (s - y)^3 + 3(s - x)(s - y)z$?

(a) x^3 (b) y^3
(c) z^3 (d) $-z^3$

Sol. As given, $x + y + z = 2s$

$\Rightarrow \quad x + y + z = s + s$

$\Rightarrow \quad (s - x) + (s - y) - z = 0$

$\Rightarrow \quad (s - x) + (s - y) + (-z) = 0$

We know, $a^3 + b^3 + c^3 = 3abc$,

if $a + b + c = 0$

Hence, $(s - x)^3 + (s - y)^3 + (-z)^3 = 3(s - x)(s - y)(-z)$

$\Rightarrow \quad (s - x)^3 + (s - y)^3 + 3(s - x)(s - y)(z) = z^3$

Alternatively

Let $x = 1$, $y = 2$, $z = 3$, then $2s = 6 \Rightarrow s = 3$

Now, $(s - x)^3 + (s - y)^3 + 3(s - x)(s - y)z$

$= (3 - 1)^3 + (3 - 2)^3 + 3(3 - 1)(3 - 2)3$

$= 8 + 1 + 18 = 27 = 3^3 = z^3$

Problem 16 If $a^{\frac{1}{3}} + b^{\frac{1}{3}} + c^{\frac{1}{3}} = 0$, then a relation among a, b, c is :

[SSC (10+2-LDC/DEO)-14]

(a) $a + b + c = 0$ (b) $(a + b + c)^3 = 27abc$
(c) $a + b + c = 27abc$ (d) $a^3 + b^3 + c^3 = 0$

Sol. We have, $a^3 + b^3 + c^3 = 3abc$, if $a + b + c = 0$

Then, $\left(a^{\frac{1}{3}}\right)^3 + \left(b^{\frac{1}{3}}\right)^3 + \left(c^{\frac{1}{3}}\right)^3$

$= 3a^{\frac{1}{3}} \cdot b^{\frac{1}{3}} \cdot c^{\frac{1}{3}}$ $\quad (\because a^{\frac{1}{3}} + b^{\frac{1}{3}} + c^{\frac{1}{3}} = 0)$

$\Rightarrow \quad a + b + c = 3a^{\frac{1}{3}} \cdot b^{\frac{1}{3}} \cdot c^{\frac{1}{3}}$

$\Rightarrow \quad (a + b + c)^3 = \left(3a^{\frac{1}{3}} \cdot b^{\frac{1}{3}} \cdot c^{\frac{1}{3}}\right)^3$ (cubing both sides)

$\Rightarrow \quad (a + b + c)^3 = 27abc$

Alternatively

Let $a = 1$, $b = 8$ and $c = -27$ as $a^{\frac{1}{3}} + b^{\frac{1}{3}} + c^{\frac{1}{3}} = 0.$

[You are free to assume other values of a, b, c like $(1, 1, -8)$]

Then, $a + b + c = 1 + 8 - 27 = -18$

Clearly, $-18 \neq 0$ and $-18 \neq 27\ [1 \cdot 8 \cdot (-27)]$

$\therefore$ Option (a) and (c) are eliminated.

Further, $(a + b + c)^3 = (-18)^3 = -(18 \times 18 \times 18)$

$= -(27 \cdot 27 \cdot 8) = 27\ [1 \cdot 8 \cdot (-27)] = 27abc$

Thus, option (b) is correct.

Problem 17 If $x + y + z = 0$, then what is the value of $\dfrac{1}{x^2 + y^2 - z^2} + \dfrac{1}{y^2 + z^2 - x^2} + \dfrac{1}{z^2 + x^2 - y^2} = ?$

Sol. Given, $x + y + z = 0 \Rightarrow x + y = -z$

$\Rightarrow x^2 + y^2 + 2xy = z^2$ (squaring both sides)

$\Rightarrow \quad x^2 + y^2 = z^2 - 2xy$

Similarly, $y^2 + z^2 = x^2 - 2yz$ and $z^2 + x^2 = y^2 - 2zx$

Now, $\dfrac{1}{x^2 + y^2 - z^2} + \dfrac{1}{y^2 + z^2 - x^2} + \dfrac{1}{z^2 + x^2 - y^2}$

$= \dfrac{1}{z^2 - 2xy - z^2} + \dfrac{1}{x^2 - 2yz - x^2} + \dfrac{1}{y^2 - 2zx - y^2}$

$= \dfrac{-z - x - y}{2xyz} = \dfrac{-(x + y + z)}{2xyz} = \dfrac{-0}{2xyz} = 0$

Alternatively

Let $\quad x = 1, y = 2$ and $z = -3$ $\quad (\because x + y + z = 0)$

(you are free to take any values of x, y, z satisfying $x + y + z = 0$ like $(1, 1, -2)$. However, the value of denominator of any term at the assumed values must not be equal to 0, otherwise the term will be undefined.)

Now, $\dfrac{1}{x^2 + y^2 - z^2} + \dfrac{1}{y^2 + z^2 - x^2} + \dfrac{1}{z^2 + x^2 - y^2}$

$= \dfrac{1}{1^2 + 2^2 - (-3)^2} + \dfrac{1}{2^2 + (-3)^2 - 1^2} + \dfrac{1}{(-3)^2 + 1^2 - 2^2}$

$= \dfrac{1}{-4} + \dfrac{1}{12} + \dfrac{1}{6} = 0$

Problem 18 If $x = a(b - c)$, $y = b(c - a)$ and $z = c(a - b)$, then $\left(\dfrac{x}{a}\right)^3 + \left(\dfrac{y}{b}\right)^3 + \left(\dfrac{z}{c}\right)^3 = ?$

(a) $\dfrac{xyz}{3abc}$ (b) $\dfrac{3xyz}{abc}$

(c) $\dfrac{xyz}{abc}$ (d) $3xyzabc$

Sol. Given, $\dfrac{x}{a} = b - c$, $\dfrac{y}{b} = c - a$ and $\dfrac{z}{c} = a - b$

Now, $\left(\dfrac{x}{a}\right)^3 + \left(\dfrac{y}{b}\right)^3 + \left(\dfrac{z}{c}\right)^3 = 3 \cdot \left(\dfrac{x}{a}\right)\left(\dfrac{y}{b}\right)\left(\dfrac{z}{c}\right) = \dfrac{3xyz}{abc}$

$\left(\because \dfrac{x}{a} + \dfrac{y}{b} + \dfrac{z}{c} = b - c + c - a + a - b = 0\right)$

Alternatively

Let $a = 1$, $b = 2$ and $c = 3$.

Then, $x = 1(2 - 3) = -1$, $y = 2(3 - 1) = 4$ and $z = 3(1 - 2) = -3$

Now, $\left(\dfrac{x}{a}\right)^3 + \left(\dfrac{y}{b}\right)^3 + \left(\dfrac{z}{c}\right)^3 = \left(\dfrac{-1}{1}\right)^3 + \left(\dfrac{4}{2}\right)^3 + \left(\dfrac{-3}{3}\right)^3$

$= -1 + 8 - 1 = 6$

$= \dfrac{3(-1)(4)(-3)}{1 \cdot 2 \cdot 3} = \dfrac{3xyz}{abc}$

Problem 19 If $ab + bc + ca = 0$, then what is the value of $a^3 + b^3 + c^3 - 3abc$?

(a) 0 (b) $(a + b + c)^3$

(c) 1 (d) $3abc$

Sol. We have, $a^3 + b^3 + c^3 - 3abc$

$= (a + b + c)\,[(a + b + c)^2 - 3(ab + bc + ca)]$

$= (a + b + c)\,[(a + b + c)^2 - 3 \times (0)] = (a + b + c)^3$

Alternatively

Let $a = -1$, $b = 2$ and $c = 2$. $[\because ab + bc + ca = 0]$

[you are free to take any value of a, b, c satisfying $ab + bc + ca = 0$ like $(-6, 10, 15)$. However, the value of denominator of any term (if exists) at the assumed values must not be equal to 0, otherwise the term will be undefined.]

Now, $a^3 + b^3 + c^3 - 3abc = (-1)^3 + 2^3 + 2^3 - 3(-1) \cdot 2 \cdot 2$

$= 27 = (-1 + 2 + 2)^3 = (a + b + c)^3$

Problem 20 If $bc + ab + ca = abc$, then the value of $\dfrac{b^3 + c^3}{bc(a - 1)(b^2 + c^2 - bc)} + \dfrac{a^3 + c^3}{ac(b - 1)(a^2 + c^2 - ac)} + \dfrac{a^3 + b^3}{ab(c - 1)(a^2 + b^2 - ab)}$ is

Sol. Given expression $\dfrac{(b + c)(b^2 + c^2 - bc)}{bc(a - 1)(b^2 + c^2 - bc)}$

$+ \dfrac{(a + c)(a^2 + c^2 - ac)}{ac(b - 1)(a^2 + c^2 - ac)} + \dfrac{(a + b)(a^2 + b^2 - ab)}{ab(c - 1)(a^2 + b^2 - ab)}$

$= \dfrac{b + c}{bc(a - 1)} + \dfrac{a + c}{ac(b - 1)} + \dfrac{a + b}{ab(c - 1)}$

$= \dfrac{b + c}{abc - bc} + \dfrac{a + c}{abc - ac} + \dfrac{a + b}{abc - ab}$

$= \dfrac{b + c}{ab + ac} + \dfrac{a + c}{ab + bc} + \dfrac{a + b}{bc + ca}$ $\quad [\because bc + ab + ca = abc]$

$= \dfrac{1}{a} + \dfrac{1}{b} + \dfrac{1}{c} = \dfrac{bc + ab + ca}{abc} = \dfrac{abc}{abc} = 1$

Alternatively

Let $a = 2$, $b = 4$ and $c = 4$ as $bc + ab + ca = abc$.

(you are free to assume other values of a, b, c like $(1, 2, -2)$, $(1, 3, -3)$, etc.)

Now, $\frac{b+c}{bc(a-1)}+\frac{a+c}{ac(b-1)}+\frac{a+b}{ab(c-1)}$

$=\frac{4+4}{4\cdot 4(2-1)}+\frac{2+4}{2\cdot 4(4-1)}+\frac{2+4}{2\cdot 4(4-1)}=\frac{1}{2}+\frac{1}{4}+\frac{1}{4}=1$

Problem 21 If $pqr = 1$, then the value of the expression $\frac{1}{1+p+q^{-1}}+\frac{1}{1+q+r^{-1}}+\frac{1}{1+r+p^{-1}}$ is equal to : **[CAT-02]**

(a) $p+q+r$ (b) $\frac{1}{p+q+r}$

(c) 1 (d) $p^{-1}+q^{-1}+r^{-1}$

Sol. 1st term $=\frac{1}{1+p+q^{-1}}=\frac{1}{1+p+\frac{1}{q}}=\frac{q}{q+pq+1}$

$=\frac{q}{q+\frac{1}{r}+1}$ $(\because pqr=1 \therefore pq=\frac{1}{r})$

$=\frac{qr}{qr+1+r}=\frac{qr}{\frac{1}{p}+1+r}=\frac{pqr}{1+p+pr}=\frac{1}{1+p+pr}$

2nd term $=\frac{1}{1+q+r^{-1}}=\frac{1}{1+q+\frac{1}{r}}=\frac{r}{r+qr+1}$

$=\frac{r}{r+\frac{1}{p}+1}$ $(\because pqr=1 \therefore qr=\frac{1}{p})$

$=\frac{pr}{1+p+pr}$

Third term $=\frac{1}{1+r+p^{-1}}=\frac{1}{1+r+\frac{1}{p}}=\frac{p}{1+p+pr}$

Now, sum of all three terms $=\frac{1}{1+p+pr}+\frac{pr}{1+p+pr}+\frac{p}{1+p+pr}=\frac{1+p+pr}{1+p+pr}=1$

Alternatively

Let $a=1$, $b=1$ and $c=1$ as $abc=1$.

(you are free to take any value of a, b, c satisfying $abc = 1$ like $\left(1, 2, \frac{1}{2}\right), \left(1, 3, \frac{1}{3}\right)$, etc. However, the value of denominator of any term at the assumed values must not be equal to 0, otherwise the term will be undefined.)

Now, value of the expression

$=\frac{1}{1+1+1^{-1}}+\frac{1}{1+1+1^{-1}}+\frac{1}{1+1+1^{-1}}=\frac{1}{3}+\frac{1}{3}+\frac{1}{3}=1$

Type 2

Computation of value of one-variable algebraic expression when value of either $x^r \pm \frac{1}{x^r}$ or $mx^r \pm \frac{1}{x^r}$ or $x^r \pm \frac{1}{mx^r}$ or $mx^r \pm \frac{1}{nx^r}$ is known directly or indirectly and *vice-versa*.

We take certain cases to comprehend the concepts

Case I

Suppose $\left(x+\frac{1}{x}\right)=a$ and we have to find out the values of $\left(x^2+\frac{1}{x^2}\right), \left(x^4+\frac{1}{x^4}\right)$ and $\left(x^8+\frac{1}{x^8}\right)$.

Given, $x+\frac{1}{x}=a$

$\Rightarrow \left(x+\frac{1}{x}\right)^2=(a)^2$ (squaring both sides)

$\Rightarrow x^2+\frac{1}{x^2}+2\cdot x\cdot\frac{1}{x}=a^2$

$\Rightarrow x^2+\frac{1}{x^2}=a^2-2=R$ (say)

Similarly, $x^4+\frac{1}{x^4}=R^2-2=R_1$ (say)

Also, $x^8+\frac{1}{x^8}=(R_1)^2-2$ and so on.

Case II

If $\left(x-\frac{1}{x}\right)=a$, then what are the values of $\left(x^2+\frac{1}{x^2}\right)$, $\left(x^4+\frac{1}{x^4}\right)$, and $\left(x^8+\frac{1}{x^8}\right)$?

Given, $\left(x-\frac{1}{x}\right)=a$

$\Rightarrow \left(x-\frac{1}{x}\right)^2=a^2$ (squaring both sides)

$\Rightarrow x^2+\frac{1}{x^2}-2\cdot x\cdot\frac{1}{x}=a^2$

$\Rightarrow x^2+\frac{1}{x^2}=a^2+2=R$ (say)

Now, on the pattern emerged in case (i),

$x^4+\frac{1}{x^4}=(R)^2-2=R_1$ (say)

Also, $x^8+\frac{1}{x^8}=(R_1)^2-2$ and so on.

Case III

If $x+\frac{1}{x}=a$, then what are values of $\left(x^3+\frac{1}{x^3}\right), \left(x^6+\frac{1}{x^6}\right)$ and $\left(x^9+\frac{1}{x^9}\right)$?

Given, $x+\frac{1}{x}=a \Rightarrow \left(x+\frac{1}{x}\right)^3=a^3$

(cubing both sides)

$\Rightarrow x^3+\frac{1}{x^3}+3\cdot x\cdot\frac{1}{x}\left(x+\frac{1}{x}\right)=a^3$

$\Rightarrow \quad x^3 + \frac{1}{x^3} + 3a = a^3$

$\Rightarrow \quad x^3 + \frac{1}{x^3} = a^3 - 3a = P$ (say)

Similarly, $\quad x^9 + \frac{1}{x^9} = P^3 - 3P = P_1$ (say)

Also, $\quad x^{27} + \frac{1}{x^{27}} = (P_1)^3 - 3P_1$ and so on.

From above, $\quad x^3 + \frac{1}{x^3} = P$

Then, $\quad x^6 + \frac{1}{x^6} + 2 = P^2$ (squaring both sides)

$\Rightarrow \quad x^6 + \frac{1}{x^6} = P^2 - 2$

Case IV

If $x - \frac{1}{x} = a$, then what are values of $\left(x^3 - \frac{1}{x^3}\right), \left(x^6 + \frac{1}{x^6}\right)$ and $\left(x^9 - \frac{1}{x^9}\right)$?

Given, $\quad x - \frac{1}{x} = a \Rightarrow \left(x - \frac{1}{x}\right)^3 = a^3$ (cubing both sides)

$\Rightarrow \quad x^3 - \frac{1}{x^3} - 3 \cdot x \cdot \frac{1}{x}\left(x - \frac{1}{x}\right) = a^3$

$\Rightarrow \quad x^3 - \frac{1}{x^3} - 3a = a^3$

$\Rightarrow \quad x^3 - \frac{1}{x^3} = a^3 + 3a = P$ (say)

Similarly, $\quad x^9 - \frac{1}{x^9} = P^3 + 3P = P_1$ (say)

Also, $\quad x^{27} - \frac{1}{x^{27}} = (P_1^3 + 3P_1)$ and so on.

From above, $x^3 - \frac{1}{x^3} = P$

Then, $\quad x^6 + \frac{1}{x^6} - 2 = P^2$ (squaring both sides)

$\Rightarrow \quad x^6 + \frac{1}{x^6} = P^2 + 2$

Case V

Let $mx + \frac{1}{x} = a$, then what are the values of (a) $\left(m^2 x^2 + \frac{1}{x^2}\right)$, (b) $\left(m^4 x^4 + \frac{1}{x^4}\right)$, (c) $\left(m^3 x^3 + \frac{1}{x^3}\right)$, (d) $\left(m^9 x^9 + \frac{1}{x^9}\right)$ and (e) $m^6 x^6 + \frac{1}{x^6}$

(a) Given, $mx + \frac{1}{x} = a$

$\Rightarrow \quad \left(mx + \frac{1}{x}\right)^2 = a^2$ (squaring both sides)

$\Rightarrow \quad m^2 x^2 + \frac{1}{x^2} + 2 \cdot mx \cdot \frac{1}{x} = a^2$

$\Rightarrow \quad m^2 x^2 + \frac{1}{x^2} = a^2 - 2m$

(b) Let $m^2 x^2 + \frac{1}{x^2} = a^2 - 2m = R$

$\Rightarrow \quad \left(m^2 x^2 + \frac{1}{x^2}\right)^2 = R^2$ (squaring both sides)

$\Rightarrow \quad m^4 x^4 + \frac{1}{x^4} + 2 \cdot m^2 x^2 \cdot \frac{1}{x^2} = R^2$

$\Rightarrow \quad m^4 x^4 + \frac{1}{x^4} = R^2 - 2m^2$

(c) Given, $mx + \frac{1}{x} = a$

$\Rightarrow \quad \left(mx + \frac{1}{x}\right)^3 = a^3$ (cubing both sides)

$\Rightarrow \quad m^3 x^3 + \frac{1}{x^3} + 3 \cdot mx \cdot \frac{1}{x}\left(mx + \frac{1}{x}\right) = a^3$

$\Rightarrow \quad m^3 x^3 + \frac{1}{x^3} + 3 \cdot mx \cdot \frac{1}{x} \cdot a = a^3$

$\Rightarrow \quad m^3 x^3 + \frac{1}{x^3} = a^3 - 3am$

(d) Let $m^3 x^3 + \frac{1}{x^3} = a^3 - 3am = P$

$\Rightarrow \quad \left(m^3 x^3 + \frac{1}{x^3}\right)^3 = P^3$ (cubing both sides)

$\Rightarrow \quad m^9 x^9 + \frac{1}{x^9} + 3 \cdot m^3 x^3 \cdot \frac{1}{x^3}(m^3 x^3 + \frac{1}{x^3}) = P^3$

$\Rightarrow \quad m^9 x^9 + \frac{1}{x^9} + 3 \cdot m^3 x^3 \cdot \frac{1}{x^3} \cdot P = P^3$

$\Rightarrow \quad m^9 x^9 + \frac{1}{x^9} = P^3 - 3P\,m^3$

(e) From above, $m^3 x^3 + \frac{1}{x^3} = P$

Then, $(m^3 x^3)^2 + \left(\frac{1}{x^3}\right)^2 = P^2 - 2m^3$

$\left[\because m^2 x^2 + \frac{1}{x^2} = a^2 - 2m \text{ if } mx + \frac{1}{x} = a\text{, refer to (a)}\right]$

$\Rightarrow \quad m^6 x^6 + \frac{1}{x^6} = p^2 - 2m^3$

Case VI

Let $x - \frac{1}{mx} = a$, then what are the values of (a) $(x^2 + \frac{1}{m^2 x^2}$ and (b) $\left(x^3 - \frac{1}{m^3 x^3}\right)$?

(a) Given, $x - \frac{1}{mx} = a$

$$\left(x-\frac{1}{mx}\right)^2=a^2 \quad \text{(squaring both sides)}$$

$$\Rightarrow x^2+\frac{1}{m^2x^2}-2\cdot x\cdot\frac{1}{mx}=a^2$$

$$\Rightarrow x^2+\frac{1}{m^2x^2}=a^2+\frac{2}{m}$$

(b) Given, $x-\frac{1}{mx}=a$

$$\Rightarrow \quad \left(x-\frac{1}{mx}\right)^3=a^3 \text{ (cubing both sides)}$$

$$\Rightarrow \quad x^3-\frac{1}{m^3x^3}-3\cdot x\cdot\frac{1}{mx}\left(x-\frac{1}{mx}\right)=a^3$$

$$\Rightarrow \quad x^3-\frac{1}{m^3x^3}=a^3+\frac{3a}{m}$$

Case VII

Let $mx-\frac{1}{nx}=a,$ then what is the value of $\left(m^3x^3-\frac{1}{n^3x^3}\right)$?

Given, $\quad mx-\frac{1}{nx}=a$

$$\Rightarrow \quad \left(mx-\frac{1}{nx}\right)^3=a^3 \text{ (cubing both sides)}$$

$$\Rightarrow \quad m^3x^3-\frac{1}{n^3x^3}-3\cdot mx\cdot\frac{1}{nx}\left(mx-\frac{1}{nx}\right)=a^3$$

$$\Rightarrow \quad m^3x^3-\frac{1}{m^3x^3}=a^3+\frac{3am}{n}$$

We may peruse the above results in the following chart to comprehend the trends.

If $x+\frac{1}{x}=a$ then ⇒	$R=x^2+\frac{1}{x^2}=a^2-2$	$R_1=x^4+\frac{1}{x^4}=R^2-2$
If $(m)x+\frac{1}{x}=a$ then ⇒	$R=m^2x^2+\frac{1}{x^2}=a^2-2(m)$	$R_1=m^4x^4+\frac{1}{x^4}=R^2-2(m^2)$
If $x+\frac{1}{(m)x}=a$ then ⇒	$R=x^2+\frac{1}{m^2x^2}=a^2-\frac{2}{(m)}$	$R_1=x^4+\frac{1}{m^4x^4}=R^2-\frac{2}{(m^2)}$
If $(m)x+\frac{1}{(n)x}=a$ then ⇒	$R=m^2x^2+\frac{1}{n^2x^2}=a^2-2\left(\frac{m}{n}\right)$	$R_1=m^4x^4+\frac{1}{n^4x^4}=R^2-2\left(\frac{m^2}{n^2}\right)$
If $x-\frac{1}{x}=a$ then ⇒	$R=x^2+\frac{1}{x^2}=a^2+2$	$R_1=x^4+\frac{1}{x^4}=R^2-2$
If $(m)x-\frac{1}{x}=a$ then ⇒	$R=m^2x^2+\frac{1}{x^2}=a^2+2(m)$	$R_1=m^4x^4+\frac{1}{x^4}=R^2-2(m^2)$
If $x-\frac{1}{(m)x}=a$ then ⇒	$R=x^2+\frac{1}{m^2x^2}=a^2+\frac{2}{(m)}$	$R_1=x^4+\frac{1}{m^4x^4}=R^2-\frac{2}{(m^2)}$
If $(m)x-\frac{1}{(n)x}=a$ then ⇒	$R=m^2x^2+\frac{1}{n^2x^2}=a^2+2\left(\frac{m}{n}\right)$	$R_1=m^4x^4+\frac{1}{n^4x^4}=R^2-2\left(\frac{m^2}{n^2}\right)$
If $x+\frac{1}{x}=a$ then ⇒	$P=x^3+\frac{1}{x^3}=a^3-3a$	$P_1=x^9+\frac{1}{x^9}=P^3-3P$
If $(m)x+\frac{1}{x}=a$ then ⇒	$P=m^3x^3+\frac{1}{x^3}=a^3-3a(m)$	$P_1=m^9x^9+\frac{1}{x^9}=P^3-3P(m^3)$
If $x+\frac{1}{mx}=a$ then ⇒	$P=x^3+\frac{1}{m^3x^3}=a^3-\frac{3a}{(m)}$	$P_1=x^9+\frac{1}{m^9x^9}=P^3-\frac{3P}{(m^3)}$
If $(m)x+\frac{1}{(n)x}=a$ then ⇒	$P=m^3x^3+\frac{1}{n^3x^3}=a^3-3a\left(\frac{m}{n}\right)$	$P_1=m^9x^9+\frac{1}{n^9x^9}=P^3-3P\left(\frac{m^3}{n^3}\right)$
If $x-\frac{1}{x}=a$ then ⇒	$Q=x^3-\frac{1}{x^3}=a^3+3a$	$Q_1=x^9-\frac{1}{x^9}=Q^3+3Q$
If $(m)x-\frac{1}{x}=a$ then ⇒	$P=m^3x^3-\frac{1}{x^3}=a^3+3a(m)$	$P_1=m^9x^9+\frac{1}{x^9}=P^3+3P(m^3)$
If $(x)-\frac{1}{(m)x}=a$ then ⇒	$P=x^3-\frac{1}{m^3x^3}=a^3+\frac{3a}{(m)}$	$P_1=x^9+\frac{1}{m^9x^9}=P^3+\frac{3P}{m^3}$
If $(m)x-\frac{1}{(n)x}=a$ then ⇒	$P=m^3x^3+\frac{1}{n^3x^3}=a^3+3a\left(\frac{m}{n}\right)$	$P_1=m^9x^9+\frac{1}{n^9x^9}=P^3+3P\left(\frac{m^3}{n^3}\right)$

Please mind the sign.

Type 2A

When value of an expression $mx^r\pm\frac{1}{nx^r}$, where $r=1$ is known directly or indirectly.

Problem 22 If $a+\frac{1}{a}=3$, then what is the value of $a^4+\frac{1}{a^4}$?

Sol. Given, $a+\frac{1}{a}=3\Rightarrow\left(a+\frac{1}{a}\right)^2=(3)^2$ (squaring both sides)

$$\Rightarrow \quad a^2+\frac{1}{a^2}+2\cdot a\cdot\frac{1}{a}=9$$

$$\Rightarrow \quad a^2+\frac{1}{a^2}=9-2=7$$

$\Rightarrow \quad \left(a^2 + \frac{1}{a^2}\right)^2 = (7)^2$ (squaring both sides)

$\Rightarrow \quad a^4 + \frac{1}{a^4} + 2 \cdot a^2 \cdot \frac{1}{a^2} = 49$

$\therefore \quad a^4 + \frac{1}{a^4} = 49 - 2 = 47$

Alternatively

$a^4 + \frac{1}{a^4} = (3^2 - 2)^2 - 2$ (from the above chart)

$= 47$

No doubt, this approach shall save the time in the examination hall.

Problem 23 If $2a - \frac{2}{a} + 3 = 0$, then what is the value of $\left(a^3 - \frac{1}{a^3} + 2\right)$?

Sol. Given, $2a - \frac{2}{a} + 3 = 0$

$\Rightarrow \quad 2\left(a - \frac{2}{2a}\right) + 3 = 0$

$\Rightarrow \quad a - \frac{1}{a} = \frac{-3}{2}$

$\Rightarrow \quad a^3 - \frac{1}{a^3} = \left(\frac{-3}{2}\right)^3 + 3\left(\frac{-3}{2}\right)$ (from the chart)

$\Rightarrow \quad a^3 - \frac{1}{a^3} + 2 = \frac{-27}{8} - \frac{9}{2} + 2 = \frac{-47}{8}$

Alternatively

From above, $a - \frac{1}{a} = \frac{-3}{2}$

Clearly, $a = -2$ (by hit and trial method)

Then, $a^3 - \frac{1}{a^3} + 2 = (-2)^3 - \frac{1}{(-2)^3} + 2$

$= -8 + \frac{1}{8} + 2$

$= \frac{-47}{8}$

Problem 24 If $a + \frac{1}{a+2} = 0$, then find the value of $(a+2)^3 + \frac{1}{(a+2)^3}$.

[SSC (CGL-T-I)-12]

Sol. Given, $a + \frac{1}{a+2} = 0$

$\Rightarrow (a+2) + \frac{1}{(a+2)} = 0 + 2$ (adding 2 to both sides)

$\Rightarrow (a+2)^3 + \frac{1}{(a+2)^3} = (2)^3 - 3 \times 2$ (from the chart)

$\Rightarrow \quad (a+2)^3 + \frac{1}{(a+2)^3} = 2$

Alternatively

Given, $a + \frac{1}{a+2} = 0$

Clearly, $a = -1$ (by hit and trial method)

Then, $(a+2)^3 + \frac{1}{(a+2)^3} = (-1+2)^3 + \frac{1}{(-1+2)^3}$

$= 1 + 1 = 2$

Problem 25 If $2x - \frac{1}{2x} = 6$, then find the value of $x^2 + \frac{1}{16x^2}$.

Sol. Given, $2x - \frac{1}{2x} = 6 \Rightarrow 2\left(x - \frac{1}{4x}\right) = 6 \Rightarrow x - \frac{1}{4x} = 3$

$\Rightarrow x^2 + \frac{1}{16x^2} = (3)^2 + \frac{2}{4} = \frac{19}{2}$ (from the chart)

Problem 26 If $5a + \frac{1}{3a} = 5$, then $9a^2 + \frac{1}{25a^2}$?

[SSC (CGL-T-I)-12, SSC (CGL-T-I)-19]

Sol. Given, $5a + \frac{1}{3a} = 5$

$\Rightarrow \quad \frac{5}{3}\left(3a + \frac{1}{5a}\right) = 5 \Rightarrow \left(3a + \frac{1}{5a}\right) = 3$

$\Rightarrow \quad 9a^2 + \frac{1}{25a^2} = (3)^2 - \frac{2 \cdot 3}{5} = \frac{39}{5}$ (from the chart)

Problem 27 If $x\left(3 - \frac{2}{x}\right) = \frac{3}{x}$, $x \neq 0$, then find the value of $x^2 + \frac{1}{x^2}$.

Sol. Given, $x\left(3 - \frac{2}{x}\right) = \frac{3}{x} \Rightarrow 3x - 2 = \frac{3}{x}$

$\Rightarrow \quad 3x - \frac{3}{x} = 2 \Rightarrow x - \frac{1}{x} = \frac{2}{3}$

$\Rightarrow \quad x^2 + \frac{1}{x^2} = \left(\frac{2}{3}\right)^2 + 2$

(from the chart)

$\Rightarrow \quad x^2 + \frac{1}{x^2} = 2\frac{4}{9}$

Problem 28 If $x + \frac{1}{x} = 3$, then find the value of $x^5 + \frac{1}{x^5}$.

[SSC (CGL-T-I)-12, SSC (10+2-LDC/DEO)-13]

Sol. Given, $x + \frac{1}{x} = 3 \Rightarrow x^2 + \frac{1}{x^2} = 3^2 - 2 = 7$

Further, $x^3 + \frac{1}{x^3} = (3)^3 - 3 \times 3 = 18$

Now, $\left(x^2 + \frac{1}{x^2}\right) \times \left(x^3 + \frac{1}{x^3}\right) = 7 \times 18$

$\Rightarrow \quad x^5 + \left(x + \frac{1}{x}\right) + \frac{1}{x^5} = 126$

$\Rightarrow \quad x^5 + 3 + \frac{1}{x^5} = 126 \Rightarrow x^5 + \frac{1}{x^5} = 126 - 3 = 123$

Problem 29 If $x + \frac{1}{x} = 5$, then find value of $\frac{x^4 + \frac{1}{x^2}}{x^2 - 3x + 1}$.

Sol. Given, $x + \frac{1}{x} = 5$

$\Rightarrow \quad x^2 - 5x + 1 = 0$

Now, $\dfrac{x^4+\frac{1}{x^2}}{x^2-3x+1}=\dfrac{x\left(x^3+\frac{1}{x^3}\right)}{(x^2-5x+1)+2x}=\dfrac{x\left(x^3+\frac{1}{x^3}\right)}{0+2x}$

$=\dfrac{1}{2}\left(x^3+\dfrac{1}{x^3}\right)$

Given, $x+\dfrac{1}{x}=5$

$\Rightarrow \quad x^3+\dfrac{1}{x^3}=(5)^3-3\times5=110$

Then, $\dfrac{1}{2}\left(x^3+\dfrac{1}{x^3}\right)=\dfrac{1}{2}\times110=55$

Problem 30 If $x+\dfrac{1}{x}=5$, then find the value of $\dfrac{x^4+3x^3+5x^2+3x+1}{x^4+1}$.

Sol. Given, $x+\dfrac{1}{x}=5$

$\Rightarrow \quad x^2+\dfrac{1}{x^2}=(5)^2-2=23$ (from the chart)

Now, $\dfrac{x^4+3x^3+5x^2+3x+1}{x^4+1}=\dfrac{(x^4+1)+(3x^3+3x)+5x^2}{x^4+1}$

$=\dfrac{x^2\left(x^2+\frac{1}{x^2}\right)+3x^2\left(x+\frac{1}{x}\right)+5x^2}{x^2\left(x^2+\frac{1}{x^2}\right)}$

$=\dfrac{\left(x^2+\frac{1}{x^2}\right)+3\left(x+\frac{1}{x}\right)+5}{\left(x^2+\frac{1}{x^2}\right)}=\dfrac{23+3\cdot5+5}{23}=\dfrac{43}{23}$

Changed Pattern

Problem 31 If $x+\dfrac{1}{x}=5$, then find the value of $\dfrac{2x}{3x^2-5x+3}$.

Sol. Given, $x+\dfrac{1}{x}=5\Rightarrow x^2+1=5x$

Now, $\dfrac{2x}{3x^2-5x+3}=\dfrac{2x}{3(x^2+1)-5x}=\dfrac{2x}{3(5x)-5x}=\dfrac{1}{5}$

Problem 32 If $\dfrac{x}{2x^2+5x+2}=\dfrac{1}{6}$, then find the value of $x+\dfrac{1}{x}$.

Sol. Given, $\dfrac{x}{2x^2+5x+2}=\dfrac{1}{6}$

$\Rightarrow \quad 2x^2+5x+2=6x \Rightarrow 2(x^2+1)=x$

$\Rightarrow \quad \dfrac{2(x^2+1)}{x}=1\Rightarrow x+\dfrac{1}{x}=\dfrac{1}{2}$

Problem 33 If $\dfrac{x^2-x+1}{x^2+x+1}=\dfrac{2}{3}$, then find the value of $\left(x+\dfrac{1}{x}\right)$. **[SSC (CISF-ASI)-10]**

Sol. Given, $\dfrac{x^2-x+1}{x^2+x+1}=\dfrac{2}{3}\Rightarrow\dfrac{\frac{x^2-x+1}{x}}{\frac{x^2+x+1}{x}}=\dfrac{2}{3}$

(dividing numerator and denominator by x)

$\Rightarrow \quad \dfrac{\left(x+\frac{1}{x}\right)-1}{\left(x+\frac{1}{x}\right)+1}=\dfrac{2}{3}\Rightarrow3\left(x+\dfrac{1}{x}\right)-3=2\left(x+\dfrac{1}{x}\right)+2$

$\Rightarrow \quad \left(x+\dfrac{1}{x}\right)=2+3=5$

Problem 34 If $30x^2-15x+1=0$, then what is the value of $25x^2+(36x^2)^{-1}$? **[SSC (CGL-T-I)-19]**

Sol. Given, $30x^2-15x+1=0 \quad\Rightarrow 30x^2+1=15x$

$\Rightarrow \quad 6x\left(5x+\dfrac{1}{6x}\right)=15x\Rightarrow\left(5x+\dfrac{1}{6x}\right)=\dfrac{5}{2}$

$\Rightarrow 25x^2+(36x^2)^{-1}+\dfrac{5}{3}=\dfrac{25}{4}$ (squaring both sides)

$\Rightarrow \quad 25x^2+(36x^2)^{-1}=\dfrac{55}{12}$

Type 2B

When value of an expression $mx^r\pm\dfrac{1}{nx^r}$, where $r>1$ is known directly or indirectly.

Problem 35 If $4b^2+\dfrac{1}{b^2}=2$, then the value of $8b^3+\dfrac{1}{b^3}$ is **[SSC (CPO)-06]**

Sol. In such types of problems, we initially find the value of an algebraic expression $mx^r\pm\dfrac{1}{nx^r}$, where $r=1$, interpreting the expression given with the numerical value. Thereafter, we compute the desired value accordingly.

Given, $(2b)^2+\dfrac{1}{b^2}=2$

Let, $2b+\dfrac{1}{b}=a$

Then, $(2b)^2+\dfrac{1}{b^2}=a^2-2\cdot2$ (from the chart)

$\Rightarrow \quad a^2-2\cdot2=2$ (given)

$\Rightarrow \quad a^2=6\Rightarrow a=\sqrt{6}$

Now, $(2b)^3+\dfrac{1}{b^3}=a^3-3a\cdot2$ (from the chart)

$=(\sqrt{6})^3-3\cdot\sqrt{6}\cdot2=0$

Alternatively

Given, $(2b)^2+\dfrac{1}{b^2}=2$

We may find the value of $2b+\dfrac{1}{b}$ directly interpreting the result envisaged in the chart. Hence, $2b+\dfrac{1}{b}=\sqrt{2\cdot2+2}=\sqrt{6}$.

Now, $(2b)^3+\dfrac{1}{b^3}=(\sqrt{6})^3-3\cdot\sqrt{6}\cdot2=0$

Problem 36 If $x^4 + \frac{1}{x^4} = 119$ and $x > 1$, then find the value of $x^3 - \frac{1}{x^3}$ is :

Sol. It is very clear that we can find the value of $x^3 - \frac{1}{x^3}$ only if the value of $x - \frac{1}{x}$ is known.

Let, $x - \frac{1}{x} = a$

Then, $x^2 + \frac{1}{x^2} = a^2 + 2$

Further, $x^4 + \frac{1}{x^4} = (a^2 + 2)^2 - 2 = 119$

$\Rightarrow (a^2 + 2)^2 = 121 \Rightarrow (a^2 + 2) = 11$

$\Rightarrow a^2 = 9 \Rightarrow a = 3$

Now, $x^3 - \frac{1}{x^3} = 3^3 + 3 \cdot 3 = 36$

Alternatively

Given, $x^4 + \frac{1}{x^4} = 119$

Then, $x^2 + \frac{1}{x^2} = \sqrt{119 + 2} = 11$

and $x - \frac{1}{x} = \sqrt{11 - 2} = 3$

Now, $x^3 - \frac{1}{x^3} = 3^3 + 3 \cdot 3 = 36$

Problem 37 If x is real, $x + \frac{1}{x} \neq 0$ and $x^3 + \frac{1}{x^3} = 0$, then the value of $\left(x + \frac{1}{x}\right)^4$ is : *[SSC (CGL-T-I)-12]*

Sol. $\left(x + \frac{1}{x}\right)^3 = x^3 + \frac{1}{x^3} + 3 \cdot x \cdot \frac{1}{x}\left(x + \frac{1}{x}\right)$

$\Rightarrow \left(x + \frac{1}{x}\right)^3 = 0 + 3 \cdot x \cdot \frac{1}{x}\left(x + \frac{1}{x}\right)$

$\Rightarrow \left(x + \frac{1}{x}\right)^2 = 3 \Rightarrow \left(x + \frac{1}{x}\right)^4 = 9$

Type 3

Computation of value of algebraic expression when numerical value of the variable(s) is/are known.

Type 3A

When one-variable expression is in form of $x^n \pm \frac{1}{x^n}$ or its series and value of the variable is known.

Problem 38 If $x = 3 + 2\sqrt{2}$, then find the value of $x^3 + \frac{1}{x^3}$.

Sol. Given, $x = 3 + 2\sqrt{2}$

$\Rightarrow \frac{1}{x} = \frac{1}{3 + 2\sqrt{2}} = \frac{1}{3 + 2\sqrt{2}} \times \frac{3 - 2\sqrt{2}}{3 - 2\sqrt{2}}$

$= \frac{3 - 2\sqrt{2}}{(3)^2 - (2\sqrt{2})^2} = \frac{3 - 2\sqrt{2}}{9 - 8} = 3 - 2\sqrt{2}$

Now, $x + \frac{1}{x} = 3 + 2\sqrt{2} + 3 - 2\sqrt{2} = 6$

$\therefore x^3 + \frac{1}{x^3} = (6)^3 - 3 \times 6 = 198$

(Recall the chart meant for computation of value of algebraic expression when relation between a variable (or its square, cube ....) and its reciprocal is given)

Problem 39 If $x = \frac{\sqrt{5} - 2}{\sqrt{5} + 2}$, then the value of $x^4 + x^{-4}$ is

(a) an integer

(b) an irrational number but not a surd

(c) a surd

(d) a rational number but not an integer

Sol. Given, $x = \frac{\sqrt{5} - 2}{\sqrt{5} + 2}$

In such forms of x, rationalise the denominator.

$\Rightarrow x = \frac{\sqrt{5} - 2}{\sqrt{5} + 2} \times \frac{\sqrt{5} - 2}{\sqrt{5} - 2}$

$\Rightarrow x = \frac{5 + 4 - 4\sqrt{5}}{1} = 9 - 4\sqrt{5}$

$\Rightarrow \frac{1}{x} = \frac{1}{9 - 4\sqrt{5}} = \frac{1}{9 - 4\sqrt{5}} \times \frac{9 + 4\sqrt{5}}{9 + 4\sqrt{5}} = 9 + 4\sqrt{5}$

Now, $x + \frac{1}{x} = 9 - 4\sqrt{5} + 9 + 4\sqrt{5} = 18$

$\Rightarrow x^2 + \frac{1}{x^2} = (18)^2 - 2 = 322$

(Recall the chart meant for computation of value of algebraic expression when relation between a variable (or its square, cube, ...) and its reciprocal is given)

$\Rightarrow x^4 + \frac{1}{x^4} = (322)^2 - 2 = 103682$

(recall the chart as mentioned above)

If implies that $x^4 + x^{-4}$ is an integer.

It may be construed from the above that

(i) If $x = p + \sqrt{q}$ (where, p is rational and $\sqrt{q}$ is a surd), then $\frac{1}{x}$ must be its conjugate i.e. $p - \sqrt{q}$ and *vice-versa*.

(ii) (a) If $x = \frac{1}{p + \sqrt{q}}$, then x may be written as $(p - \sqrt{q})$.

(b) If $x = \frac{1}{p - \sqrt{q}}$, then x may be written as $(p + \sqrt{q})$.

(iii) (a) If $x = p + \sqrt{q}$, then $x + \frac{1}{x} = 2p$

and $x - \frac{1}{x} = 2(+\sqrt{q}) = 2\sqrt{q}$

(b) If $x = p - \sqrt{q}$, then $x + \frac{1}{x} = 2p$

and $x - \frac{1}{x} = 2(-\sqrt{q}) = -2\sqrt{q}$

(iv) (a) If $x = \frac{p - \sqrt{q}}{p + \sqrt{q}}$, then $x + \frac{1}{x} = 2[p^2 + (\sqrt{q})^2]$

and $x - \frac{1}{x} = 4p(-\sqrt{q}) = -4p\sqrt{q}$

(b) If $x = \frac{p+\sqrt{q}}{p-\sqrt{q}}$, then $x + \frac{1}{x} = 2[p^2 + (\sqrt{q})^2]$

and $x - \frac{1}{x} = 4p(+\sqrt{q}) = +4p\sqrt{q}$

Note

(i) *All the above facts are true only if (first term)^2 – (second term)^2 = 1 [say, $p^2 - (\sqrt{q})^2 = 1$].*

(ii) *The above facts are also true if x is in form of (surd ± surd) like $\sqrt{p} \pm \sqrt{q}$, $m\sqrt{p} \pm \sqrt{q}$, etc. within the above restriction.*

(iii) *If (first term)^2 – (second term)$^2 \neq 1$, we divide the resultants by [(first term)^2 – (second term)^2] where first term is either less or greater than the second term. This is applicable for all cases except case (iii) (supra).*

Problem 40 If $x = 1 + \sqrt{3}$, then what is the value of $\frac{1}{x}$?

Sol. Clearly, $\frac{1}{x} = \frac{\text{Conjugate of } (1+\sqrt{3})}{(1)^2 - (\sqrt{3})^2}$ $[\because (1)^2 - (\sqrt{3})^2 \neq 1]$

$= \frac{(1-\sqrt{3})}{-2} = \frac{(\sqrt{3}-1)}{2}$

Problem 41 If $x = \frac{\sqrt{5}+\sqrt{2}}{\sqrt{5}-\sqrt{2}}$, then find the value of

(a) $x + \frac{1}{x}$ (b) $x - \frac{1}{x}$

Sol. (a) $x + \frac{1}{x} = \frac{2[(\sqrt{5})^2 + (\sqrt{2})^2]}{(\sqrt{5})^2 - (\sqrt{2})^2}$ $[\because (\sqrt{5})^2 - (\sqrt{2})^2 \neq 1]$

$= \frac{14}{3}$

(b) $x - \frac{1}{x} = \frac{4\sqrt{5}\sqrt{2}}{(\sqrt{5})^2 - (\sqrt{2})^2}$ $[\because (\sqrt{5})^2 - (\sqrt{2})^2 \neq 1]$

$= \frac{4\sqrt{10}}{3}$

Problem 42 If $x = 2 + \sqrt{3}$, then find the value of $\frac{x^6 + x^4 + x^2 + 1}{x^3}$.

Sol. $\frac{x^6 + x^4 + x^2 + 1}{x^3} = \frac{x^6}{x^3} + \frac{x^4}{x^3} + \frac{x^2}{x^3} + \frac{1}{x^3}$

$= x^3 + x + \frac{1}{x} + \frac{1}{x^3}$

$= \left(x^3 + \frac{1}{x^3}\right) + \left(x + \frac{1}{x}\right)$

Given, $x = 2 + \sqrt{3}$

Then, $x + \frac{1}{x} = 2 \cdot 2 = 4$ $[\because (2)^2 - (\sqrt{3})^2 = 1]$

and $x^3 + \frac{1}{x^3} = (4)^3 - 3 \times 4 = 52$

(Recall the chart meant for computation of value of algebraic expression when relation between a variable (or its square, cube, ...) and its reciprocal is given)

Now, $\left(x^3 + \frac{1}{x^3}\right) + \left(x + \frac{1}{x}\right) = 52 + 4 = 56$

Problem 43 If $x = 3 + 2\sqrt{2}$, then find the value of $\sqrt{x} - \frac{1}{\sqrt{x}}$.

[SSC (CGL-T-I)-12, SSC (CGL-T-II)-12, SSC (CGL-T-II)-19]

Sol. Let $\sqrt{x} - \frac{1}{\sqrt{x}} = a$

Then, $x + \frac{1}{x} = a^2 + 2$ (Recall the chart meant for computation of value of algebraic expression when relation between a variable (or its square, cube, -----) and its reciprocal is given)

As given, $x = 3 + 2\sqrt{2}$

Then, $x + \frac{1}{x} = 2 \cdot 3 = 6$ $[\because 3^2 - (2\sqrt{2})^2 = 1]$

$\Rightarrow a^2 + 2 = 6 \Rightarrow a^2 = 4 \therefore a = 2$

Conventionally

Given, $x = 3 + 2\sqrt{2} = (\sqrt{2}+1)^2 \Rightarrow \sqrt{x} = \sqrt{2} + 1$

and $\frac{1}{\sqrt{x}} = \sqrt{2} - 1$ $[\because (\sqrt{2})^2 - (1)^2 = 1]$

Then, $\sqrt{x} - \frac{1}{\sqrt{x}} = \sqrt{2} + 1 - \sqrt{2} + 1 = 2$

Problem 44 If $x = 2 + \sqrt{3}$, then find the value of $\sqrt{x} + \frac{1}{\sqrt{x}}$.

Sol. Let $\sqrt{x} + \frac{1}{\sqrt{x}} = a$

Then, $x + \frac{1}{x} = a^2 - 2$ (Recall the chart meant for computation of value of algebraic expression when relation between a variable (or its square, cube, ----) and its reciprocal is given)

As given, $x = 2 + \sqrt{3}$

Then, $x + \frac{1}{x} = 2 \cdot 2 = 4$ $[\because (2)^2 - (\sqrt{3})^2 = 1]$

$\Rightarrow a^2 - 2 = 4 \Rightarrow a^2 = 6 \therefore a = \sqrt{6}$

Problem 45 If $x = 1 + \sqrt{2} + \sqrt{3}$, then find the value of $x + \frac{1}{x-1}$?

[SSC (FCI-Mains)-12]

Sol. Given, $x = 1 + \sqrt{2} + \sqrt{3}$

$\Rightarrow x - 1 = \sqrt{3} + \sqrt{2}$

Then, $(x-1) + \frac{1}{(x-1)} = 2 \cdot \sqrt{3}$ $[\because (\sqrt{3})^2 - (\sqrt{2})^2 = 1]$

$\Rightarrow x + \frac{1}{(x-1)} = 1 + 2\sqrt{3}$

Problem 46 If $x = (\sqrt{2}+1)^{\frac{-1}{3}}$, then find the value of $x^3 - \frac{1}{x^3}$.

Sol. Given, $x = (\sqrt{2}+1)^{\frac{-1}{3}}$

$\Rightarrow x^3 = \frac{1}{\sqrt{2}+1} = \sqrt{2} - 1$ $[\because (\sqrt{2})^2 - (1)^2 = 1]$

$\Rightarrow x^3 - \frac{1}{x^3} = -2(1) = -2$

Type 3B

When one variable expression is not in form of either $x^n \pm \frac{1}{x^n}$ or its series and value of the variable is known.

Problem 47 If $x = 1 + \sqrt{2} + \sqrt{3}$, then find the value of $2x^4 - 8x^3 - 5x^2 + 26x - 28$.

[SSC (CGL-T-I)-12, SSC (CGL-T-II)-16]

Sol. Given, $x = 1 + \sqrt{2} + \sqrt{3} \Rightarrow x - 1 = \sqrt{2} + \sqrt{3}$

$\Rightarrow \quad (x-1)^2 = (\sqrt{2} + \sqrt{3})^2$ (squaring both sides)

$\Rightarrow \quad x^2 - 2x + 1 = 5 + 2\sqrt{6}$

$\Rightarrow \quad x^2 - 2x - 4 = 2\sqrt{6}$

Now, express the expression in terms of $x^2 - 2x - 4$ and then put its vlaue.

Given expression, $2x^4 - 8x^3 - 5x^2 + 26x - 28$

$= 2x^2(x^2 - 2x - 4) - 4x(x^2 - 2x - 4) - 5(x^2 - 2x - 4) - 48$

$= (x^2 - 2x - 4)(2x^2 - 4x - 5) - 48$

$= 2\sqrt{6}\,[2(x^2 - 2x - 4) + 3] - 48$

$= 2\sqrt{6}\,[2 \cdot 2\sqrt{6} + 3] - 48$

$= 48 + 6\sqrt{6} - 48 = 6\sqrt{6}$

Alternatively

Given, $\quad x = 1 + \sqrt{2} + \sqrt{3} \Rightarrow x - 1 = \sqrt{2} + \sqrt{3}$

$\Rightarrow \quad (x-1)^2 = (\sqrt{2} + \sqrt{3})^2$ (squaring both sides)

$\Rightarrow \quad x^2 - 2x + 1 = 5 + 2\sqrt{6} \Rightarrow x^2 - 2x - 4 = 2\sqrt{6}$

$\Rightarrow \quad x^4 + 4x^2 + 16 - 4x^3 + 16x - 8x^2 = 24$ (squaring both sides)

$\Rightarrow \quad x^4 - 4x^2 - 8 - 4x^3 + 16x = 0$

$\Rightarrow 2x^4 - 8x^3 - 8x^2 + 32x - 16 = 0$

$\Rightarrow 2x^4 - 8x^3 - 5x^2 + 26x - 28 - 3x^2 + 6x + 12 = 0$

$\Rightarrow 2x^4 - 8x^3 - 5x^2 + 26x - 28 - 3(x^2 - 2x - 4) = 0$

$\Rightarrow \quad 2x^4 - 8x^3 - 5x^2 + 26x - 28 = 3 \times 2\sqrt{6} = 6\sqrt{6}$

Problem 48 If $p = 999$, then find the value of $\sqrt[3]{p(p^2 + 3p + 3) + 1}$.

Sol. $\sqrt[3]{p(p^2 + 3p + 3) + 1} = \sqrt[3]{p^3 + 3p^2 \cdot 1 + 3p \cdot 1^2 + 1^3}$

$= \sqrt[3]{(p+1)^3} = p + 1 = 999 + 1 = 1000$

Type 3C

When expression has more than one variable and value of each variable is known.

Problem 49 If $x = 997$, $y = 998$ and $z = 999$, then what is the value of $x^2 + y^2 + z^2 - xy - yz - zx$?

Sol. We have, $(x - y)^2 + (y - z)^2 + (z - x)^2$

$= 2(x^2 + y^2 + z^2 - xy - yz - zx)$

$\Rightarrow (997 - 998)^2 + (998 - 999)^2 + (999 - 997)^2$

$= 2(x^2 + y^2 + z^2 - xy - yz - zx)$

$\Rightarrow \quad 1 + 1 + 4 = 2(x^2 + y^2 + z^2 - xy - yz - zx)$

$\Rightarrow \quad x^2 + y^2 + z^2 - xy - yz - zx = \frac{6}{2} = 3$

Problem 50 If $a = 299$, $b = 298$, $c = 297$, then the value of $2a^3 + 2b^3 + 2c^3 - 6abc$ is :

[SSC (CGL-T-II)-16, 20]

Sol. $2a^3 + 2b^3 + 2c^3 - 6abc = 2(a^3 + b^3 + c^3 - 3abc)$

$= 2 \cdot \frac{1}{2}(a + b + c)[(a - b)^2 + (b - c)^2 + (c - a)^2]$

$= (299 + 298 + 297)[(299 - 298)^2 + (298 - 297)^2 + (297 - 299)^2]$

$= 894 \times 6 = 5364$

Problem 51 If $x = \sqrt{a} + \frac{1}{\sqrt{a}}$, $y = \sqrt{a} - \frac{1}{\sqrt{a}}$, then $x^4 + y^4 - 2x^2y^2 = ?$

Sol. Clearly, $x^4 + y^4 - 2x^2y^2 = (x^2)^2 + (y^2)^2 - 2x^2y^2$

$= (x^2 - y^2)^2 = \left[\left(\sqrt{a} + \frac{1}{\sqrt{a}}\right)^2 - \left(\sqrt{a} - \frac{1}{\sqrt{a}}\right)^2\right]^2$

$= \left[a + \frac{1}{a} + 2 - a - \frac{1}{a} + 2\right]^2 = 16$

Alternativaly

Let, $a = 1$ (you are free to taken any value)

Then, $\quad x = \sqrt{1} + \frac{1}{\sqrt{1}} = 2$

and $\quad y = \sqrt{1} - \frac{1}{\sqrt{1}} = 0$

Now, $x^4 + y^4 - 2x^2y^2 = 2^4 + 0^4 - 2x^2.0 = 16$

Problem 52 If $x = 2 + \sqrt{3}$, $y = 2\sqrt{3}$, then find the value of $\frac{x^2 + y^2}{x^3 + y^3}$?

[SSC (CGL-T-I)-12]

Sol. $\frac{x^2 + y^2}{x^3 + y^3} = \frac{(x + y)^2 - 2xy}{(x + y)^3 - 3xy(x + y)}$

$= \frac{(2 + \sqrt{3} + 2 - \sqrt{3})^2 - 2(2 + \sqrt{3})(2 - \sqrt{3})}{(2 + \sqrt{3} + 2 - \sqrt{3})^3 - 3(2 + \sqrt{3})(2 - \sqrt{3})(2 + \sqrt{3} + 2 - \sqrt{3})}$

$= \frac{16 - 2}{64 - 12} = \frac{14}{52} = \frac{7}{26}$

Alternatively

Given, $x = 2 + \sqrt{3}$ and $y = 2 - \sqrt{3} = \frac{1}{x}$

Then, $x + \frac{1}{x} = 4$ and $x \cdot \frac{1}{x} = 1$

Now, $\quad \frac{x^2 + y^2}{x^3 + y^3} = \frac{(x + y)^2 - 2xy}{(x + y)^3 - 3xy(x + y)}$

$= \frac{\left(x + \frac{1}{x}\right)^2 - 2 \cdot x \cdot \frac{1}{x}}{\left(x + \frac{1}{x}\right)^3 - 3 \cdot x \cdot \frac{1}{x}\left(x + \frac{1}{x}\right)}$

$= \frac{(4)^2 - 2}{(4)^3 - 3(4)} = \frac{7}{26}$

Problem 53 If $a = \frac{2+\sqrt{3}}{2-\sqrt{3}}$ and $b = \frac{2-\sqrt{3}}{2+\sqrt{3}}$, then find the value of $a^2 + b^2 + ab$.

Sol. Clearly, $a^2 + b^2 + ab = (a^2 + b^2 + 2ab) - ab = (a+b)^2 - ab$

$$= \left(\frac{2+\sqrt{3}}{2-\sqrt{3}} + \frac{2-\sqrt{3}}{2+\sqrt{3}}\right)^2 - \left(\frac{2+\sqrt{3}}{2-\sqrt{3}} \times \frac{2-\sqrt{3}}{2+\sqrt{3}}\right)$$

$$= \left(\frac{14}{(2)^2 - (\sqrt{3})^2}\right)^2 - 1 = 196 - 1 = 195$$

Alternatively

Given, $a = \frac{2+\sqrt{3}}{2-\sqrt{3}}$ and $b = \frac{2-\sqrt{3}}{2+\sqrt{3}} = \frac{1}{a}$

Now, $a^2 + b^2 + ab = a^2 + \frac{1}{a^2} + a \cdot \frac{1}{a}$

$$= \left(a + \frac{1}{a}\right)^2 - 1$$

Also, $a + \frac{1}{a} = 2\,[(2)^2 + (\sqrt{3})^2]$ $\quad [\because (2)^2 - (\sqrt{3})^2 = 1]$

$\Rightarrow$ $a + \frac{1}{a} = 14$

Then, $\left(a + \frac{1}{a}\right)^2 - 1 = (14)^2 - 1 = 195$

Problem 54 If $a = \frac{\sqrt{3}-\sqrt{2}}{\sqrt{3}+\sqrt{2}}$ and $b = \frac{\sqrt{3}+\sqrt{2}}{\sqrt{3}-\sqrt{2}}$, then find the value of $\frac{a^2}{b} + \frac{b^2}{a}$. **[SSC (CGL-T-I)-12]**

Sol. $\frac{a^2}{b} + \frac{b^2}{a} = \frac{a^3 + b^3}{ab} = \frac{(a+b)^3 - 3ab\,(a+b)}{ab}$

Now, $a + b = \frac{\sqrt{3}-\sqrt{2}}{\sqrt{3}+\sqrt{2}} + \frac{\sqrt{3}+\sqrt{2}}{\sqrt{3}-\sqrt{2}}$

$$= \frac{3 + 2 - 2\sqrt{6} + 3 + 2 + 2\sqrt{6}}{3 - 2} = 10$$

and $ab = \frac{\sqrt{3}-\sqrt{2}}{\sqrt{3}+\sqrt{2}} \times \frac{\sqrt{3}+\sqrt{2}}{\sqrt{3}-\sqrt{2}} = 1$

Then, $\frac{(a+b)^3 - 3ab\,(a+b)}{ab} = \frac{(10)^3 - 3 \cdot 1 \cdot 10}{1} = 970$

Alternatively

Clearly, $b = \frac{1}{a}$

Also, $\frac{a^2}{b} + \frac{b^2}{a} = \frac{a^3 + b^3}{ab}$

$$= \frac{a^3 + \left(\frac{1}{a}\right)^3}{a \cdot \frac{1}{a}} = \left(a + \frac{1}{a}\right)^3 - 3 \cdot a \cdot \frac{1}{a}\left(a + \frac{1}{a}\right)$$

Now, $\left(a + \frac{1}{a}\right) = 2\,[(\sqrt{3})^2 + (\sqrt{2})^2]$ $\quad [\because (\sqrt{3})^2 - (\sqrt{2})^2 = 1]$

$\Rightarrow$ $\left(a + \frac{1}{a}\right) = 10$

Then, $\left(a + \frac{1}{a}\right)^3 - 3 \cdot a \cdot \frac{1}{a}\left(a + \frac{1}{a}\right) = (10)^3 - 3 \cdot a \cdot \frac{1}{a}(10) = 970$

Problem 55 If $x = \frac{1}{2+\sqrt{3}}$ and $y = \frac{1}{2-\sqrt{3}}$, then find the value of $\frac{1}{x+1} + \frac{1}{y+1}$.

Sol. $x = \frac{1}{2+\sqrt{3}} \times \frac{2-\sqrt{3}}{2-\sqrt{3}} = 2 - \sqrt{3}$

and $y = \frac{1}{2-\sqrt{3}} \times \frac{2+\sqrt{3}}{2+\sqrt{3}} = 2 + \sqrt{3}$

Now, $\frac{1}{x+1} + \frac{1}{y+1} = \frac{1}{2-\sqrt{3}+1} + \frac{1}{2+\sqrt{3}+1}$

$$= \frac{1}{3-\sqrt{3}} + \frac{1}{3+\sqrt{3}} = \frac{3+\sqrt{3}+3-\sqrt{3}}{(3)^2 - (\sqrt{3})^2} = \frac{6}{6} = 1$$

Alternatively

Clearly, $y = \frac{1}{x}$

Then, $\frac{1}{x+1} + \frac{1}{y+1} = \frac{1}{x+1} + \frac{1}{\frac{1}{x}+1} = \frac{1+x}{x+1} = 1$

Problem 56 Let $x = \frac{\sqrt{13}+\sqrt{11}}{\sqrt{13}-\sqrt{11}}$ and $y = \frac{1}{x}$, then the value of $3x^2 - 5xy + 3y^2$ is : **[SSC (CGL-T-II)-15]**

Sol. Let, $3x^2 - 5xy + 3y^2 = 3x^2 - 6xy + 3y^2 + xy$

$$= 3\,(x - y)^2 + xy$$

$$= 3\left(x - \frac{1}{x}\right)^2 + x \cdot \frac{1}{x} \qquad \left(\because y = \frac{1}{x}\right)$$

$$= 3\left(x - \frac{1}{x}\right)^2 + 1$$

Given, $x = \frac{\sqrt{13}+\sqrt{11}}{\sqrt{13}-\sqrt{11}}$

Then, $x - \frac{1}{x} = \frac{4\sqrt{13} \cdot \sqrt{11}}{2}$ $\quad [\because (\sqrt{13})^2 - (\sqrt{11})^2 = 2]$

$\Rightarrow$ $x - \frac{1}{x} = \frac{4\sqrt{13} \cdot \sqrt{11}}{2} = 2\sqrt{143}$

Now, $3\left(x - \frac{1}{x}\right)^2 + 1 = 3\,(2\sqrt{143})^2 + 1 = 1717$

Type 4

Simplification based on algebraic expression

Problem 57 What is $\frac{(x-y)^3 + (y-z)^3 + (z-x)^3}{3\,(x-y)\,(y-z)\,(z-x)}$ equal to : **[CDS-19]**

Sol. We have, $(x-y)^3 + (y-z)^3 + (z-x)^3$

$$= 3\,(x-y)\,(y-z)\,(z-x)$$

$\therefore$ Required answer = 1

Problem 58 The value of the expression $\frac{(a-b)^2}{(b-c)\,(c-a)} + \frac{(b-c)^2}{(a-b)\,(c-a)} + \frac{(c-a)^2}{(a-b)\,(b-c)}$ is **[SSC(10+2-LDC/DEO)-11, 13]**

Sol. $\frac{(a-b)^2}{(b-c)(c-a)}+\frac{(b-c)^2}{(a-b)(c-a)}+\frac{(c-a)^2}{(a-b)(b-c)}$

$=\frac{(a-b)^3+(b-c)^3+(c-a)^3}{(a-b)(b-c)(c-a)}$

$=\frac{3(a-b)(b-c)(c-a)}{(a-b)(b-c)(c-a)}=3$

Alternatively

We assume the values of variables involved in the expression and thereafter we simply. This approach is suitable for all types of the simplifications based on the algebraic expression. However, the value of denominator of any term at the assumed values must not be equal to 0, otherwise the term will be undefined

Let $a=1$, $b=2$ and $c=3$.

Then, $\frac{(a-b)^2}{(b-c)(c-a)}+\frac{(b-c)^2}{(a-b)(c-a)}+\frac{(c-a)^2}{(a-b)(b-c)}$

$=\frac{(1-2)^2}{(2-3)(3-1)}+\frac{(2-3)^2}{(1-2)(3-1)}+\frac{(3-1)^2}{(1-2)(2-3)}$

$=\frac{1}{-2}+\frac{1}{-2}+\frac{4}{1}=3$

Problem 59 Ramesh is trying to simplify expression $(p+q)^3-(p-q)^3-6q(p^2-q^2)$ and $q=1$. You helped him and the solution arrived was ------. **[SNAP-18]**

Sol. Given expression, $(p+q)^3-(p-q)^3-6q(p^2-q^2)$

$=(p+1)^3-(p-1)^3-6(p^2-1)$

$=p^3+1+3p^2+3p-(p^3-1-3p^2+3p)-6(p^2-1)=8$

Alternatively

Take any value of p, preferably smaller, as per your convenience.

Let p be 2.

Then, the given expression $=(2+1)^3-(2-1)^3-6(2^2-1)=8$

Problem 60 If $P=\frac{x^4-8x}{x^3-x^2-2x}$, $Q=\frac{x^2+2x+1}{x^2-4x-5}$ and $R=\frac{2x^2+4x+8}{x-5}$, then $(P\times Q)\div R$ is equal to : **[SSC (CGL-T-I)-19]**

Sol. $P=\frac{x(x^3-2^3)}{x(x^2-x-2)}=\frac{x(x-2)(x^2+2x+4)}{x(x-2)(x+1)}$

$=\frac{x^2+2x+4}{(x+1)}$

$Q=\frac{x^2+2x+1}{x^2-4x-5}=\frac{(x+1)(x+1)}{(x-5)(x+1)}=\frac{(x+1)}{(x-5)}$

Now, $(P\times Q)\div R=\left[\frac{(x^2+2x+4)}{(x+1)}\times\frac{(x+1)}{(x-5)}\right]\div\frac{2x^2+4x+8}{x-5}$

$=\frac{(x^2+2x+4)}{(x-5)}\times\frac{(x-5)}{2(x^2+2x+4)}=\frac{1}{2}$

Alternatively

We may assume the values of variables involved in the expression and get the required answer. However, the value of denominator of any term at the assumed values must not be equal to 0.

Problems Based on Other than Algebraic Identities, However Can be Solved by Assuming the Values of Given Variables.

Problem 61 If $x+y+z=0$, then what is the value of $(x+y)(y+z)(z+x)$?

(a) $-xyz$ (b) $x^2+y^2+z^3$

(c) $x^3+y^3+z^3+3xyz$ (d) xyz

Sol. Given, $x+y+z=0$

$\Rightarrow$ $x+y=-z$, $y+z=-x$ and $z+x=-y$

Now, $(x+y)(y+z)(z+x)=(-z)(-x)(-y)=-xyz$

Alternatively

Let, $x=1$, $y=2$ and $z=-3$ $[\because x+y+z=0]$

(you are free to take any value of x, y, z satisfying $x+y+z=0$)

Now, $(x+y)(y+z)(z+x)=(1+2)(2-3)(-3+1)$

$=6=-xyz$

Problem 62 If $\frac{a}{b}=\frac{c}{d}=\frac{e}{f}=3$, then $\frac{2a^2+3c^2+4e^2}{2b^2+3d^2+4f^2}$? **[SSC (CGL-T-I)-11]**

Sol. Given, $\frac{a}{b}=3\Rightarrow a=3b$

Also, $c=3d$ and $e=3f$

Now, $\frac{2(3b)^2+3(3d)^2+4(3f)^2}{2b^2+3d^2+4f^2}$

$=\frac{9(2b^2+3d^2+4f^2)}{2b^2+3d^2+4f^2}=9$

Alternatively

Let $a=3$, $b=1$, $c=6$, $d=2$, $e=9$ and $f=3$ $\left(\because\frac{a}{b}=\frac{c}{d}=\frac{e}{f}=3\right)$

(you are free to take any value of a, b, c, d, e and f)

Now, $\frac{2a^2+3c^2+4e^2}{2b^2+3d^2+4f^2}=\frac{2\cdot3^2+3\cdot6^2+4\cdot9^2}{2\cdot1^2+3\cdot2^2+4\cdot3^2}=9$

Problem 63 If $pq+qr+rp=0$, then what is the value of $\frac{p^2}{p^2-qr}+\frac{q^2}{q^2-rp}+\frac{r^2}{r^2-pq}$? **[CDS-09]**

Sol. Given, $pq+qr+rp=0$

$\therefore$ $qr=-pq-rp$, $rp=-pq-qr$ and $pq=-qr-rp$

Now, $\frac{p^2}{p^2-qr}+\frac{q^2}{q^2-rp}+\frac{r^2}{r^2-pq}$

$=\frac{p^2}{p^2-(-pq-rp)}+\frac{q^2}{q^2-(-pq-qr)}+\frac{r^2}{r^2-(-qr-rp)}$

$=\frac{p}{p+q+r}+\frac{q}{p+q+r}+\frac{r}{p+q+r}=\frac{p+q+r}{p+q+r}=1$

Alternatively

Let $p=-1$, $q=2$ and $r=2$ $(\because pq+qr+rp=0)$

(you are free to take any value of p, q, r satisfying $pq+qr+rp=0$ like $(-6, 10, 15)$. However, the value of denominator of any term at the assumed values must not be equal to 0, otherwise the term will be undefined).

Now, $\frac{p^2}{p^2 - qr} + \frac{q^2}{q^2 - rp} + \frac{r^2}{r^2 - pq}$

$= \frac{(-1)^2}{(-1)^2 - 2\cdot2} + \frac{2^2}{2^2 - 2(-1)} + \frac{2^2}{2^2 - (-1)2} = \frac{1}{-3} + \frac{4}{6} + \frac{4}{6} = 1$

Problems Seeking Specific Approaches

Type I

Computation of value of $x^a + \frac{1}{x^b}$, when value of $x^m + \frac{1}{x^n}$ is given.

Problem 64 If $x + \frac{1}{x} = 2$, then what is the value of $x^9 + \frac{1}{x^{27}}$?

Sol. Clearly, $x + \frac{1}{x} = 2$ is satisfied at $x = 1$

Then, $x^9 + \frac{1}{x^{27}} = 1^9 + \frac{1}{1^{27}} = 2$

Problem 65 If $x^{17} + \frac{1}{x^{23}} = -2$, then what is the value of $x^{20} + \frac{1}{x^{29}}$?

Sol. Given, $x^{17} + \frac{1}{x^{23}} = -2$

Clearly, $x = -1$

Then, $x^{20} + \frac{1}{x^{29}} = (-1)^{20} + \frac{1}{(-1)^{29}} = 1 - 1 = 0$

Problem 66 If $x + \frac{1}{x} = -2$, then $x^{2n+1} + \frac{1}{x^{2n+1}} = ?$ where n is a positive integer. *[SSC (CPO)-08]*

Sol. Given, $x + \frac{1}{x} = -2$

Clearly, $x = -1$

Now, $x^{2n+1} + \frac{1}{x^{2n+1}} = x^{\text{odd number}} + \frac{1}{x^{\text{odd number}}}$

(As, $2n + 1$ will always be odd)

$= (-1)^{\text{odd}} + \frac{1}{(-1)^{\text{odd}}} = -1 - 1 = -2$

Problem 67 If $x + \frac{1}{x} = 2$, then $x^2 + \frac{1}{x^3}$? *[SSC (CGL-T-I)-12]*

Sol. Given, $x + \frac{1}{x} = 2$

Clearly, $x = 1$

Then, $x^2 + \frac{1}{x^3} = 1^2 + \frac{1}{1^3} = 2$

Type 2

Problems based on 'Sum of square of any term cannot be negative'.

Problem 68

If $(x - 3)^2 + (y - 4)^2 + (z - 5)^2 = 0$, then what is the value of $x + y + z$? (If x, y and z all are real numbers)

Sol. Given, $(x - 3)^2 + (y - 4)^2 + (z - 5)^2 = 0$

Since square of any number cannot be negative, each of the given terms will be either positive or zero. If they are positive, their sum cannot be equal to zero. It implies that each of terms is equal to zero.

$\Rightarrow \quad (x - 3)^2 = 0, (y - 4)^2 = 0$ and $(z - 5)^2 = 0$

$\Rightarrow \quad x = 3, y = 4$ and $z = 5$

$\therefore \quad x + y + z = 3 + 4 + 5 = 12$

Problem 69 If $x^2 + y^2 + z^2 = xy + yz + zx \ (x \neq 0)$, then what is the value of $\frac{4x + 2y - 3z}{2x}$?

Sol. We know, $(a - b)^2 + (b - c)^2 + (c - a)^2$

$= 2(a^2 + b^2 + c^2 - ab - bc - ca)$

$\Rightarrow (x - y)^2 + (y - z)^2 + (z - x)^2$

$= 2(x^2 + y^2 + z^2 - xy - yz - zx)$

$\Rightarrow (x - y)^2 + (y - z)^2 + (z - x)^2$

$= 2(xy + yz + zx - xy - yz - zx)$

$(\because x^2 + y^2 + z^2 = xy + yz + zx$, as given)

$\Rightarrow (x - y)^2 + (y - z)^2 + (z - x)^2 = 0$

$\Rightarrow \quad (x - y)^2 = 0, (y - z)^2 = 0$ and $(z - x)^2 = 0$

$\Rightarrow \quad x = y, y = z$ and $z = x$

Observation If $x^2 + y^2 + z^2 = xy + yz + zx$, then $x = y = z$.

Now, $\frac{4x + 2y - 3z}{2x} = \frac{4x + 2x - 3x}{2x} = \frac{3}{2} = 1.5$

Alternatively

Given, $x^2 + y^2 + z^2 = xy + yz + zx$

Clearly, $x^2 = xy$, $y^2 = yz$ and $z^2 = zx$

$\Rightarrow \quad x = y, y = z$ and $z = x$

$\therefore \quad x = y = z$

Now, $\frac{4x + 2y - 3z}{2x} = \frac{4x + 2x - 3x}{2x} = \frac{3}{2} = 1.5$

Problem 70 If $x \neq 0$, $y \neq 0$ and $z \neq 0$ and $\frac{1}{x^2} + \frac{1}{y^2} + \frac{1}{z^2} = \frac{1}{xy} + \frac{1}{yz} + \frac{1}{zx}$, then what is the relation among x, y and z? *[SSC (CGL-T-I)-13]*

Sol. Given, $\frac{1}{x^2} + \frac{1}{y^2} + \frac{1}{z^2} = \frac{1}{xy} + \frac{1}{yz} + \frac{1}{zx}$

$\Rightarrow \quad x^2 y^2 z^2 \left(\frac{1}{x^2} + \frac{1}{y^2} + \frac{1}{z^2}\right)$

$= x^2 y^2 z^2 \left(\frac{1}{xy} + \frac{1}{yz} + \frac{1}{zx}\right)$

$\Rightarrow y^2 z^2 + x^2 z^2 + x^2 y^2 = xyz^2 + x^2 yz + y^2 zx$

$\Rightarrow x^2 y^2 + y^2 z^2 + z^2 x^2 = y^2 zx + z^2 xy + x^2 yz$

$\Rightarrow x^2y^2 + y^2z^2 + z^2x^2 - y^2zx - z^2xy - x^2yz = 0$

We have,

$(a-b)^2 + (b-c)^2 + (c-a)^2 = 2(a^2+b^2+c^2-ab-bc-ca)$

$\Rightarrow (xy-yz)^2 + (yz-zx)^2 + (zx-xy)^2$

$= 2(x^2y^2 + y^2z^2 + z^2x^2 - y^2zx - z^2xy - x^2yz)$

$\Rightarrow (xy-yz)^2 + (yz-zx)^2 + (zx-xy)^2 = 2 \times 0 = 0$

$\Rightarrow (xy-yz)^2 = 0, (yz-zx)^2 = 0$ and $(zx-xy)^2 = 0$

$\Rightarrow xy = yz, yz = zx$ and $zx = xy$

$\Rightarrow x = y = z$

Obervation If $\frac{1}{x^2} + \frac{1}{y^2} + \frac{1}{z^2} = \frac{1}{xy} + \frac{1}{yz} + \frac{1}{zx}$, then $x = y = z$.

Problem 71 If a, b, c are real and $a^2 + b^2 + c^2 = 2(a - b - c) - 3$, then the value of $2a - 3b + 4c$ is : **[SSC (10+2-LDC/DEO)-11, 13, SSC (CGL-T-I)-13]**

Sol. Given, $a^2 + b^2 + c^2 = 2(a-b-c) - 3$

$\Rightarrow (a^2 - 2a + 1) + (b^2 + 2b + 1) + (c^2 + 2c + 1) = 0$

$\Rightarrow (a-1)^2 + (b+1)^2 + (c+1)^2 = 0$

$\Rightarrow (a-1)^2 = 0, (b+1)^2 = 0, (c+1)^2 = 0$

$\Rightarrow a = 1, b = -1, c = -1$

Then, $2a - 3b + 4c = 2(1) - 3(-1) + 4(-1) = 1$

Problem 72 If $x^2 + y^2 + z^2 + 2 = 2(y - x)$, then find the value of $x^3 + y^3 + z^3$.

Sol. Given, $x^2 + y^2 + z^2 + 2 = 2y - 2x$

$\Rightarrow x^2 + 2x + y^2 - 2y + 2 + z^2 = 0$

$\Rightarrow (x^2 + 2x + 1) + (y^2 - 2y + 1) + z^2 = 0$

$\Rightarrow (x+1)^2 + (y-1)^2 + z^2 = 0$

$\Rightarrow x + 1 = 0, y - 1 = 0$, and $z^2 = 0$

$\Rightarrow x = -1, y - 1$ and $z = 0$

Now, $x^3 + y^3 + z^3$

$= (-1)^3 + (1)^3 + (0)^3 = -1 + 1 + 0 = 0$

Problem 73 If $x^2 + y^2 + 2x + 1 = 0$,then the value of $x^{31} + y^{35}$ is : **[SSC (CGL-T-II)-16]**

Sol. Given, $x^2 + y^2 + 2x + 1 = 0 \Rightarrow (x+1)^2 + y^2 = 0$

$\Rightarrow (x+1)^2 = 0$ and $y^2 = 0 \Rightarrow x = -1$ and $y = 0$

Then, $x^{31} + y^{35} = (-1)^{31} + (0)^{35} = -1$

Problem 74 If $u^2 + (u - 2v - 1)^2 = -4v(u + v)$, then what is the value of $u + 3v$? **[CAT-18]**

Sol. Given, $u^2 + (u - 2v - 1)^2 = -4v(u+v)$

$\Rightarrow u^2 + u^2 + 4v^2 + 1 - 4uv + 4v - 2u = -4uv - 4v^2$

$\Rightarrow 2u^2 - 2u + 8v^2 + 4v + 1 = 0$

$\Rightarrow 2\left[u^2 - 2\cdot u\cdot\frac{1}{2} + \left(\frac{1}{2}\right)^2\right] + 2\left[(2v)^2 + 2\cdot 2v\cdot\frac{1}{2} + \left(\frac{1}{2}\right)^2\right] = 0$

$\Rightarrow \left(u - \frac{1}{2}\right)^2 + \left(2v + \frac{1}{2}\right)^2 = 0$

$\Rightarrow \left(u - \frac{1}{2}\right)^2 = 0$ and $\left(2v + \frac{1}{2}\right)^2 = 0 \Rightarrow u = \frac{1}{2}$ and $v = -\frac{1}{4}$

Now, $u + 3v = \frac{1}{2} - \frac{3}{4} = -\frac{1}{4}$

Problem 75 If the value of $(a + b - 2)^2 + (b + c - 5)^2 + (c + a - 5)^2 = 0$, then the value of $\sqrt{(b+c)^a + (c+a)^b - 1}$ is : **[SSC (CGL-T-I)-19]**

Sol. Clearly, $(a + b - 2)^2 = 0, (b + c - 5)^2 = 0$ and $(c + a - 5)^2 = 0$

$\Rightarrow a + b = 2, b + c = 5$ and $c + a = 5$

$\Rightarrow 2(a + b + c) = 12 \Rightarrow (a + b + c) = 6$

Then, $a = 6 - 5 = 1$ and $b = 6 - 5 = 1$

Now, $\sqrt{(b+c)^a + (c+a)^b - 1} = \sqrt{(5)^1 + (5)^1 - 1} = \pm 3$

Type 3

Problems based on series wherein numerators are the different variables and respective denominators are (1± multiple of the variable).

Problem 76 If $\frac{a}{1-xa} + \frac{b}{1-xb} + \frac{c}{1-xc} + \ldots\ldots + n$th term $= m$, then the value of $\frac{1}{1-xa} + \frac{1}{1-xb} + \frac{1}{1-xc} + \ldots\ldots + n$th term = ?

Sol. Given, $\frac{a}{1-xa} + \frac{b}{1-xb} + \frac{c}{1-xc} + \ldots\ldots + n$th term $= m$

$\Rightarrow \frac{1}{x}\left(\frac{ax}{1-xa} + \frac{bx}{1-xb} + \frac{cx}{1-xc} + \ldots\ldots + n\text{th term}\right) = m$

$\Rightarrow \left(\frac{ax}{1-xa} + \frac{bx}{1-xb} + \frac{cx}{1-xc} + \ldots\ldots + n\text{th term}\right) = mx$

$\Rightarrow \left(\frac{ax}{1-xa} + 1\right) + \left(\frac{bx}{1-xb} + 1\right) + \left(\frac{cx}{1-xc} + 1\right) + \ldots + n$ th term $= mx + n$

$\Rightarrow \frac{1}{1-xa} + \frac{1}{1-xb} + \frac{1}{1-xc} + \ldots\ldots + n$ th term $= mx + n$

Observations

If $\frac{a}{1-xa} + \frac{b}{1-xb} + \frac{c}{1-xc} + \ldots\ldots + n$th term $= m$, then $\frac{1}{1-xa} + \frac{1}{1-xb} + \frac{1}{1-xc} + \ldots\ldots + n$th term = number of variables + ($m \times$ absolute value of coefficient of variable)

Similarly, if $\frac{a}{1+xa} + \frac{b}{1+xb} + \frac{c}{1+xc} + \ldots\ldots + n$th term $= m$, then $\frac{1}{1+xa} + \frac{1}{1+xb} + \frac{1}{1+xc} + \ldots\ldots + n$th term = number of varibles − ($m \times$ absolute value of coefficient of variable)

Problem 77 If $\frac{a}{1-a} + \frac{b}{1-b} + \frac{c}{1-c} = 1$, then the value of $\frac{1}{1-a} + \frac{1}{1-b} + \frac{1}{1-c}$ is : **[SSC (10+2-LDC/DEO)-13]**

Sol. Given, $\frac{a}{1-a} + \frac{b}{1-b} + \frac{c}{1-c} = 1$

$\Rightarrow \left(\frac{a}{1-a} + 1\right) + \left(\frac{b}{1-b} + 1\right) + \left(\frac{c}{1-c} + 1\right) = 1 + 3$

$$\Rightarrow \quad \frac{1}{1-a}+\frac{1}{1-b}+\frac{1}{1-c}=4$$

Alternatively

Required answer = number of variables + (m × absolute value of coefficient of variable) = 3 + 1 (1) = 4

Type 4

Computation of either sum of squares of variables or sum of their reciprocals when sum of squares of variables and their reciprocals together is twice the number of variables.

Problem 78 If $x^2+y^2+z^2+\frac{1}{x^2}+\frac{1}{y^2}+\frac{1}{z^2}=6$, then find the value of $x^2+y^2+z^2$.

Sol. Given, $x^2+y^2+z^2+\frac{1}{x^2}+\frac{1}{y^2}+\frac{1}{z^2}=6$

$$\Rightarrow \left(x^2+\frac{1}{x^2}-2\right)+\left(y^2+\frac{1}{y^2}-2\right)+\left(z^2+\frac{1}{z^2}-2\right)=0$$

$$\Rightarrow \left(x-\frac{1}{x}\right)^2+\left(y-\frac{1}{y}\right)^2+\left(z-\frac{1}{z}\right)^2=0$$

$$\Rightarrow \left(x-\frac{1}{x}\right)^2=0, \left(y-\frac{1}{y}\right)^2=0 \text{ and } \left(z-\frac{1}{z}\right)^2=0$$

$$\Rightarrow \quad x-\frac{1}{x}=0,\ y-\frac{1}{y}=0 \text{ and } z-\frac{1}{z}=0$$

$$\Rightarrow \quad x^2=1,\ y^2=1 \text{ and } z^2=1$$

Then, $x^2+y^2+z^2=1+1+1=3$

Hence, sum of squares of variables = sum of reciprocals of squares of variables = $\frac{1}{2}$ (sum of squares of variables + sum of reciprocals of squares of variables) provided (sum of squares of variables + sum of reciprocals of squares of variables) is twice the number of variables.

Problem 79 If $x^2+y^2+\frac{1}{x^2}+\frac{1}{y^2}=4$, then find the value of x^2+y^2 is.

Sol. Given, $x^2+y^2+\frac{1}{x^2}+\frac{1}{y^2}=4$

$$\Rightarrow \quad \left(x^2+\frac{1}{x^2}-2\right)+\left(y^2+\frac{1}{y^2}-2\right)=0$$

$$\Rightarrow \quad \left(x-\frac{1}{x}\right)^2+\left(y-\frac{1}{y}\right)^2=0$$

$$\Rightarrow \quad x-\frac{1}{x}=0 \text{ and } y-\frac{1}{y}=0 \Rightarrow x^2=1 \text{ and } y^2=1$$

Then, $x^2+y^2=1+1=2$

Alternatively

Since (sum of squares of variables + sum of reciprocals of squares of variables) is twice the number of variables, the sum of squares of variables = $\frac{1}{2}$ (sum of squares of variables + sum of reciprocals of squares of variables).

Hence, required answer = $\frac{1}{2}\times 4=2$

Type 5

Computation of sum of reciprocals of (variable + 1) when square of each of the variables given is equal to sum of other variables

Problem 80 If $x^2=y+z$, $y^2=z+x$ and $z^2=x+y$, then the value of $\frac{1}{x+1}+\frac{1}{y+1}+\frac{1}{z+1}$ is :

[SSC (CPO)-11, SSC (10+2)-DEO/LDC-11, SSC (CGL-T-I)-13, CDS-07]

Sol. Given, $x^2=y+z$

$$\Rightarrow \quad x+x^2=x+y+z \Rightarrow x(1+x)=x+y+z$$

$$\Rightarrow \quad \frac{1}{x+1}=\frac{x}{x+y+z}$$

Similarly, $\frac{1}{y+1}=\frac{y}{x+y+z}$ and $\frac{1}{z+1}=\frac{z}{x+y+z}$

Now, $\frac{x}{x+y+z}+\frac{y}{x+y+z}+\frac{z}{x+y+z}=\frac{x+y+z}{x+y+z}=1$

Problem 81 If $a^2=b+c+d$, $b^2=a+c+d$, $c^2=a+b+d$ and $d^2=a+b+c$, then find the value of $\frac{1}{a+1}+\frac{1}{b+1}+\frac{1}{c+1}+\frac{1}{d+1}$.

Sol. On the above pattern, $\frac{1}{a+1}+\frac{1}{b+1}+\frac{1}{c+1}+\frac{1}{d+1}=1$

Type 6

Computation of product of three variables when sum of first variable and reciprocal of second variable = sum of second variable and reciprocal of third variable = sum of third variable and reciprocal of first variable.

Problem 82 If $a+\frac{1}{b}=b+\frac{1}{c}=c+\frac{1}{a}$ $(a \neq b \neq c)$, then the value of abc is : *[SSC (CPO, Delhi Police-SI)-14]*

Given, $a+\frac{1}{b}=b+\frac{1}{c}$

$$\Rightarrow \quad a-b=\frac{1}{c}-\frac{1}{b}=\frac{b-c}{bc}$$

Also, $b+\frac{1}{c}=c+\frac{1}{a} \Rightarrow b-c=\frac{c-a}{ca}$

and $c+\frac{1}{a}=a+\frac{1}{b} \Rightarrow c-a=\frac{a-b}{ab}$

Multiplying side-wise, we get

$$(a-b)(b-c)(c-a)=\frac{b-c}{bc}\times\frac{c-a}{ca}\times\frac{a-b}{ab}$$

$$\Rightarrow \quad (abc)^2=1 \Rightarrow abc=\pm 1$$

Problem 83 What is the possible value of a, b and c in the above question?

Sol. From above, $abc=\pm 1$

Let $abc=+1$ and $a=1$

Then, $bc=1 \Rightarrow c=\frac{1}{b}$

As given, $a+\frac{1}{b}=b+\frac{1}{c}$

$\Rightarrow \quad 1+\frac{1}{b}=b+\frac{1}{\left(\frac{1}{b}\right)} \Rightarrow 1+\frac{1}{b}=2b$

$\Rightarrow \quad 2b^2-b-1=0 \Rightarrow b=-\frac{1}{2}$ and 1

$\therefore \quad b=-\frac{1}{2} \quad (\because a \neq b)$

Then, $\quad c=\frac{1}{-\left(\frac{1}{2}\right)}=-2$

Hence, values of a, b, c respectively $=\left(1,-\frac{1}{2},-2\right)$

Similarly, in case $abc=-1$, values of a, b, c, respectively

$=\left(-1,-\frac{1}{2},-2\right)$ [mind the sign]

Here, a lot of sets of values is possible.

Type 7

When sum of fractions whose numerator is conjugate to the denominator is known and the sum of fractions having numerator as either first term or second term of former numerator and the denominator as before, is required.

Problem 84 If $\frac{a^2-bc}{a^2+bc}+\frac{b^2-ca}{b^2+ca}+\frac{c^2-ab}{c^2+ab}=m$, then find the value of $\frac{a^2}{a^2+bc}+\frac{b^2}{b^2+ca}+\frac{c^2}{c^2+ab}$ in terms of m.

Sol. Given, $\frac{a^2-bc}{a^2+bc}+\frac{b^2-ca}{b^2+ca}+\frac{c^2-ab}{c^2+ab}=m$

$\Rightarrow \left(\frac{a^2-bc}{a^2+bc}+1\right)+\left(\frac{b^2-ca}{b^2+ca}+1\right)+\left(\frac{c^2-ab}{c^2+ab}+1\right)=m+3$

$\Rightarrow \frac{2a^2}{a^2+bc}+\frac{2b^2}{b^2+ca}+\frac{2c^2}{c^2+ab}=m+3$

$\Rightarrow \frac{a^2}{a^2+bc}+\frac{b^2}{b^2+ca}+\frac{c^2}{c^2+ab}=\frac{m+3}{2}$

Problem 85 If $\frac{a^2-bc}{a^2+bc}+\frac{b^2-ca}{b^2+ca}+\frac{c^2-ab}{c^2+ab}=1$, then the value of $\frac{a^2}{a^2+bc}+\frac{b^2}{b^2+ca}+\frac{c^2}{c^2+ab}$ is :

[SSC (CGL-T-II)-14]

Sol. On the above pattern, the required answer $=\frac{1+3}{2}=2$

Problem 86 If $\frac{a^4+bcd}{a^4-bcd}+\frac{b^4+cda}{b^4-cda}+\frac{c^4+dab}{c^4-dab}+\frac{d^4+abc}{d^4-abc}=m$, then find the value of $\frac{bcd}{a^4-bcd}+\frac{cda}{b^4-cda}+\frac{dab}{c^4-dab}+\frac{abc}{d^4-abc}$ in terms of m.

Sol. Given, $\frac{a^4+bcd}{a^4-bcd}+\frac{b^4+cda}{b^4-cda}+\frac{c^4+dab}{c^4-dab}+\frac{d^4+abc}{d^4-abc}=m$

$\Rightarrow \frac{a^4+bcd}{a^4-bcd}-1+\frac{b^4+cda}{b^4-cda}-1+\frac{c^4+dab}{c^4-dab}-1+\frac{d^4+abc}{d^4-abc}-1$

$=m-4$

$\Rightarrow \frac{2bcd}{a^4-bcd}+\frac{2cda}{b^4-cda}+\frac{2dab}{c^4-dab}+\frac{2abc}{d^4-abc}=m-4$

$\Rightarrow \frac{bcd}{a^4-bcd}+\frac{cda}{b^4-cda}+\frac{dab}{c^4-dab}+\frac{abc}{d^4-abc}=\frac{m-4}{2}$

Type 8

Value of the constant term when the sum of fractions in form of

$$\frac{\text{constant}-(\text{variable})^n}{\text{sum of remaining variables to the power } n}$$

is equal to number of terms.

Problem 87 If $\frac{k-a^2}{b^2+c^2+d^2}+\frac{k-b^2}{c^2+d^2+a^2}+\frac{k-c^2}{d^2+a^2+b^2}+\frac{k-d^2}{a^2+b^2+c^2}=4$, then find the value of k.

Sol. Given, $\frac{k-a^2}{b^2+c^2+d^2}+\frac{k-b^2}{c^2+d^2+a^2}+\frac{k-c^2}{d^2+a^2+b^2}+\frac{k-d^2}{a^2+b^2+c^2}=4$

$\Rightarrow \frac{k-a^2}{b^2+c^2+d^2}-1+\frac{k-b^2}{c^2+d^2+a^2}-1+\frac{k-c^2}{d^2+a^2+b^2}-1+\frac{k-d^2}{a^2+b^2+c^2}-1=0$

$\Rightarrow \frac{k-(a^2+b^2+c^2+d^2)}{b^2+c^2+d^2}+\frac{k-(a^2+b^2+c^2+d^2)}{c^2+d^2+a^2}+\frac{k-(a^2+b^2+c^2+d^2)}{d^2+a^2+b^2}+\frac{k-(a^2+b^2+c^2+d^2)}{a^2+b^2+c^2}=0$

$\Rightarrow k-(a^2+b^2+c^2+d^2)\left(\frac{1}{b^2+c^2+d^2}+\frac{1}{c^2+d^2+a^2}+\frac{1}{d^2+a^2+b^2}+\frac{1}{a^2+b^2+c^2}\right)=0$

$\Rightarrow \quad k-(a^2+b^2+c^2+d^2)=0$

$\Rightarrow \quad k=a^2+b^2+c^2+d^2$

Alternatively

Each term = 1

$\frac{k-a^2}{b^2+c^2+d^2}=1$

$\Rightarrow \quad k-a^2=b^2+c^2+d^2$

$\Rightarrow \quad k=a^2+b^2+c^2+d^2$

Problem 88 If $\frac{m-a^2}{b^2+c^2}+\frac{m-b^2}{c^2+a^2}+\frac{m-c^2}{a^2+b^2}=3$, then the value of m is :

[SSC (CGL-T-I)-15]

(a) $a^2+b^2+c^2$ (b) $a^2+b^2-c^2$
(c) a^2+b^2 (d) $a^2-b^2-c^2$

Sol. Since sum is equal to number of terms, $\frac{m-a^2}{b^2+c^2}=1$

$\Rightarrow \quad m-a^2=b^2+c^2$

$\Rightarrow \quad m=a^2+b^2+c^2$

Problem 89 If $\frac{a^3-p}{b^3+c^3}+\frac{b^3-p}{c^3+a^3}+\frac{c^3-p}{a^3+b^3}=-3$, then find the value of p.

Sol. Given, $\frac{a^3-p}{b^3+c^3}+\frac{b^3-p}{c^3+a^3}+\frac{c^3-p}{a^3+b^3}=-3$

$\Rightarrow \quad \frac{p-a^3}{b^3+c^3}+\frac{p-b^3}{c^3+a^3}+\frac{p-c^3}{a^3+b^3}=3$

$\Rightarrow \quad \frac{p-a^3}{b^3+c^3}=1$

$\Rightarrow \quad p=a^3+b^3+c^3$

Type 9

Computation of value of $a^2 \pm ab+b^2$ and ab, when values of $a^4+a^2b^2+b^4$ and its one factor are known.

It is clear that $a^4+a^2b^2+b^4=(a^2+ab+b^2)(a^2-ab+b^2)$. We can find the value of either (a^2+ab+b^2) or (a^2-ab+b^2) if values of $(a^4+a^2b^2+b^4)$ and one of the above two factors are known.

Let $a^4+a^2b^2+b^4=48$ and $(a^2+ab+b^2)=8$, then what are the values of (a^2-ab+b^2) and ab?

Clearly, value of $(a^2-ab+b^2)=48 \div 8=6$

Now, $\quad (a^2+ab+b^2)=8 \quad$ ...(i)

and $\quad (a^2-ab+b^2)=6 \quad$ ...(ii)

Subtracting Eq. (ii) from Eq. (i), we get

$2ab=2 \Rightarrow ab=1$

No doubt, we are now able to find values of a and b too as below.

From Eq. (i), $a^2+ab+b^2=8$

$\Rightarrow \quad a^2+ab+b^2+ab=8+ab$

$\Rightarrow \quad (a+b)^2=8+1$

$\Rightarrow \quad a+b=3$

Further, $\quad a^2-ab+b^2=6$

$\Rightarrow \quad a^2-ab+b^2-ab=6-ab$

$\Rightarrow \quad (a-b)^2=6-1 \Rightarrow a-b=\sqrt{5}$

Now, we can find the values of a and b as their sum and difference are known.

Problem 90 *If* $a^4+a^2b^2+b^4=32$ and $a^2-ab+b^2=8$, then what is the value of a^2+ab+b^2?

Sol. The required value $=32 \div 8=4$

Problem 91 If $x^4+x^2y^2+y^4=21$ and $x^2+xy+y^2=7$, then find the value of $\frac{1}{x^2}+\frac{1}{y^2}$. *[SSC (CGL-T-I)-19]*

Sol. Clearly, $\quad x^2-xy+y^2=21 \div 7=3 \quad$ ...(i)

and $\quad x^2+xy+y^2=7 \quad$ ...(ii)

Additing, Eqs. (i) and (ii), we get

$2(x^2+y^2)=10 \Rightarrow x^2+y^2=5$

Then, $xy=7-5=2$

Now, $\frac{1}{x^2}+\frac{1}{y^2}=\frac{x^2+y^2}{(xy)^2}=\frac{5}{(2)^2}=\frac{5}{4}$

Type 10

Value of $mx^n+mx^{n\pm 6}$ (where, $n \ge 6$) is always 0 if $x+\frac{1}{x}=\sqrt{3}$.

Problem 92 Let $x+\frac{1}{x}=\sqrt{3}$, then mx^n+mx^{n+6}?

Sol. Given, $x+\frac{1}{x}=\sqrt{3}$

$\Rightarrow \quad x^3+\frac{1}{x^3}=3\sqrt{3}-3\sqrt{3}=0 \quad$ (Recall the chart)

$\Rightarrow \quad x^6+1=0$

Now, $mx^n+mx^{n+6}=mx^n(x^6+1)$

$=mx^n \times 0=0$

(It may be treated as evidence of the above facts)

Problem 93 If $x+\frac{1}{x}=\sqrt{3}$, then find the value of $x^{18}+x^{12}+x^6+1$. *[SSC (CPO)-11, SSC (CGL-T-II)-16]*

Sol. Given expression, $x^{18}+x^{12}+x^6+1$

$=x^{12}(x^6+1)+1(x^6+1)$

$=(x^6+1)(x^{12}+1)$

As given, $\quad x+\frac{1}{x}=\sqrt{3}$

$\Rightarrow \quad x^3+\frac{1}{x^3}=3\sqrt{3}-3\sqrt{3}=0 \quad$ (Recall the chart)

$\Rightarrow \quad x^6+1=0$

Now, $\quad (x^6+1)(x^{12}+1)=0 \times (x^{12}+1)=0$

Alternatively I

Given, $\quad x+\frac{1}{x}=\sqrt{3}$

$\Rightarrow \quad x^3+\frac{1}{x^3}=3\sqrt{3}-3\sqrt{3}=0 \quad$ (Recall the chart)

$\Rightarrow \quad x^3=\frac{-1}{x^3}$

$\Rightarrow \quad x^6=-1$

Now, $x^{18}+x^{12}+x^6+1$

$=(x^6)^3+(x^6)^2+x^6+1$

$=(-1)^3+(-1)^2+(-1)+1=0$

Alternatively II

$x^{18} + x^{12} + x^6 + 1 = (x^{18} + x^{12}) + (x^6 + x^0)$

$= 0 + 0$

($\because$ difference between powers in each group is 6)

$= 0$

Problem 94 If $\left(x + \frac{1}{x}\right)^2 = 3$, then find the value of $x^{206} + x^{200} + x^{90} + x^{84} + x^{18} + x^{12} + x^6 + 1$.

Sol. Given, $\left(x + \frac{1}{x}\right)^2 = 3$

$\Rightarrow \quad x + \frac{1}{x} = \sqrt{3}$

$\Rightarrow \quad x^3 + \frac{1}{x^3} = 3\sqrt{3} - 3\sqrt{3} = 0$ (Recall the chart)

$\Rightarrow \quad x^3 = \frac{-1}{x^3} \Rightarrow x^6 + 1 = 0$

Now, $x^{206} + x^{200} + x^{90} + x^{84} + x^{18} + x^{12} + x^6 + 1$

$= x^{200}(x^6 + 1) + x^{84}(x^6 + 1) + x^{12}(x^6 + 1) + (x^6 + 1)$

$= x^{200}(0) + x^{84}(0) + x^{12}(0) + (0) = 0$

Alternatively

$x^{206} + x^{200} + x^{90} + x^{84} + x^{18} + x^{12} + x^6 + 1$

$= (x^{206} + x^{200}) + (x^{90} + x^{84}) + (x^{18} + x^{12}) + (x^6 + x^0)$

$= 0 + 0 + 0 + 0$

($\because$ difference between powers in each group is 6)

$= 0$

Type 11

When $x + \frac{1}{x} = \sqrt{3}$ and terms of the expression are in form of mx^n, where n is a multiple of 6.

Problem 95 If $x + \frac{1}{x} = \sqrt{3}$, then find the value of $x^{72} + x^{66} + x^{54} + x^{36} + x^{24} + x^6 + 1$. *[SSC (CGL-T-II)-12]*

Sol. Given, $x + \frac{1}{x} = \sqrt{3}$

$\Rightarrow \quad x^3 + \frac{1}{x^3} = 3\sqrt{3} - 3\sqrt{3} = 0$ (Recall the Chart)

$\Rightarrow \quad x^3 = \frac{-1}{x^3} \Rightarrow x^6 = -1$

Now, $x^{72} + x^{66} + x^{54} + x^{36} + x^{24} + x^6 + 1$

$= (x^6)^{12} + (x^6)^{11} + (x^6)^9 + (x^6)^6 + (x^6)^4 + x^6 + 1$

$= (-1)^{12} + (-1)^{11} + (-1)^9 + (-1)^6 + (-1)^4 + (-1) + 1$

$= 1 - 1 - 1 + 1 + 1 - 1 + 1 = 1$

Problem 96 If $x + \frac{1}{x} = \sqrt{3}$, then find the value of $x^{42} + x^{30} + x^{12} + 8$.

Sol. $x^{42} + x^{30} + x^{12} + 8 = (x^6)^7 + (x^6)^5 + (x^6)^2 + 8$

$= (-1)^7 + (-1)^5 + (-1)^2 + 8$

(simply putting the value of x^6 as -1)

$= -1 - 1 + 1 + 8 = 7$

Type 12

Value of $mx^n + mx^{n \pm 3}$ (where, $n \geq 3$) is always 0 if $x + \frac{1}{x} = 1$.

Problem 97 Let $x + \frac{1}{x} = 1$, then find the value of $mx^n + mx^{n+3}$.

Sol. Given, $x + \frac{1}{x} = 1 \Rightarrow x^2 + 1 = x$

$\Rightarrow \quad x^2 - x + 1 = 0$

Now, $mx^n + mx^{n+3} = mx^n(x^3 + 1^3)$

$= mx^n(x + 1)(x^2 - x \cdot 1 + 1^2)$

$= mx^n(x + 1) \times 0 = 0$

(It may be treated as evidence of the above facts)

Problem 98 If $x + \frac{1}{x} = 1$, then find the value of $x^9 + x^6 + 1$.

Sol. Given, $x + \frac{1}{x} = 1$

$\Rightarrow \quad x^2 + 1 = x \Rightarrow x^2 - x + 1 = 0$

Now, $x^9 + x^6 + 1 = x^6(x^3 + 1) + 1$

$= x^6(x^3 + 1) + 1$

$= x^6[(x + 1)(x^2 - x + 1)] + 1$

$= x^6[(x + 1) \times 0] + 1 = 0 + 1 = 1$

Alternatively

$x^9 + x^6 + 1 = (x^9 + x^6) + 1$

$= 0 + 1$ ($\because$ difference between powers in the group is 3)

$= 1$

Problem 99 If $\left(x + \frac{1}{x}\right)^2 = 1$, then find the value of $5x^{72} + 5x^{69} + x^{54} + x^{51} + 1$.

Sol. Given, $\left(x + \frac{1}{x}\right)^2 = 1 \Rightarrow x + \frac{1}{x} = 1$

Then, $5x^{72} + 5x^{69} + x^{54} + x^{51} + 1$

$= (5x^{72} + 5x^{69}) + (x^{54} + x^{51}) + 1$

$= 0 + 0 + 1$

($\because$ difference between powers in each group is 3)

$= 1$

Type 13

If $x^2 \pm mx + 1 = 0$, then $x + \frac{1}{x} =$ linear coefficient with opposite sign. Also, if $x^2 \pm mx - 1 = 0$, then $x - \frac{1}{x} =$ linear coefficient with opposite sign.

Problem 100 If $x^2 - 3x + 1 = 0$, then find the value of $x^2 + \frac{1}{x^2}$.

Sol. Given, $x^2 - 3x + 1 = 0$

$\Rightarrow \quad x\left(x-3+\frac{1}{x}\right)=0$

$\Rightarrow \quad \left(x-3+\frac{1}{x}\right)=0 \Rightarrow x+\frac{1}{x}=3$

$\Rightarrow \quad x^2+\frac{1}{x^2}=(3)^2-2=7$

Alternatively

$\Rightarrow$ Given, $x^2-3x+1=0$

$\Rightarrow \quad x+\frac{1}{x}=$ linear coefficient with opposite sign $=+3$

$\Rightarrow \quad x^2+\frac{1}{x^2}=(3)^2-2=7$

Problem 101 If $x^2+4x-1=0$, then find the value of $x^3-\frac{1}{x^3}$.

Sol. Given, $x^2+4x-1=0$

$\Rightarrow \quad x-\frac{1}{x}=$ linear coefficient with opposite sign $=-4$

$\Rightarrow \quad x^3-\frac{1}{x^3}=(-4)^3+3(-4)=-76$

Problem 102 If $x^2-2\sqrt{5}x+1=0$, then what is the value of $x^5+\frac{1}{x^5}$? **[SSC (CGL-T-I)-19]**

Sol. $x+\frac{1}{x}=$ linear coefficient with opposite sign $=2\sqrt{5}$

Then, $\quad x^2+\frac{1}{x^2}=(2\sqrt{5})^2-2=18$

and $x^3+\frac{1}{x^3}=(2\sqrt{5})^3-3(2\sqrt{5})=34\sqrt{5}$ (Recall the chart)

Now, $\left(x^2+\frac{1}{x^2}\right)\times\left(x^3+\frac{1}{x^3}\right)=18\times 34\sqrt{5}$

$\Rightarrow \quad x^5+\frac{1}{x^5}+x+\frac{1}{x}=612\sqrt{5}$

$\Rightarrow \quad x^5+\frac{1}{x^5}+2\sqrt{5}=612\sqrt{5}$

$\Rightarrow \quad x^5+\frac{1}{x^5}=610\sqrt{5}$

Type 14

If $x^2 \pm x+1=0$, then $x^3=$ linear coefficient.

Problem 103 If $x^2+x+1=0$, then find the value of x^3.

Sol. $x^3=x^3-1+1=x^3-1^3+1$

$=(x-1)(x^2+x+1)+1$

$=(x-1)0+1=1$

Alternatively

$x^3=$ linear coefficient $=1$

Problem 104 If $a^2+a+1=0$, then find the value of a^9. **[SSC (CGL-T-I)-13, SSC (CGL-T-II)-19]**

Sol. Given, $\quad a^2+a+1=0$

We have, $\quad a^3-1^3=(a-1)(a^2+a+1)$

$\Rightarrow \quad a^3-1^3=(a-1)\times 0$

$\Rightarrow \quad a^3=1^3$

$\Rightarrow \quad a=1$

Then, $\quad a^9=1^9=1$

Alternatively

$a^3=$ linear coefficient $=1$

Then, $\quad a^9=1^3=1$

Problem 105 If $x+\frac{1}{x}=1$, then find the value of $x^{4000}+\frac{1}{x^{4000}}$.

Sol. Given, $x+\frac{1}{x}=1$

$\Rightarrow x^2-x+1=0$

Then, $x^3=$ linear coefficient $=-1$

Now, $x^{4000}+\frac{1}{x^{4000}}=(x^3)^{1333}\cdot x+\frac{1}{(x^3)^{1333}\cdot x}$

$=(-1)^{1333}\cdot x+\frac{1}{(-1)^{1333}\cdot x}=-\left(x+\frac{1}{x}\right)=-1$

Problem 106 If $x^2+x+1=0$, then $x^{2018}+x^{2019}$ equals which of the following? **[XAT-19]**

(a) x (b) $x-1$

(c) $x+1$ (d) $-x$

(e) None of these

Sol. Clearly, $x^3=$ linear coefficient $=1$

Now, $x^{2018}+x^{2019}=x^{2018}(1+x)=(x^3)^{672}\cdot x^2(1+x)$

$=(1)^{672}\cdot x^2(-x^2)=-x^3\cdot x=-x$

Type 15

If $a^2+b^2+c^2=m$, $x^2+y^2+z^2=n$ and $ax+by+cz=\sqrt{mn}$, then $\frac{a+b+c}{x+y+z}=\sqrt{\frac{m}{n}}$.

Problem 107 If $a^2+b^2+c^2=16$, $x^2+y^2+z^2=25$ and $ax+by+cz=20$, then find the value of $\frac{a+b+c}{x+y+z}$. **[SSC (CGL-T-II)-16]**

Sol. $(a^2+b^2+c^2)\times(x^2+y^2+z^2)=16\times 25=(ax+by+cz)^2$

$\Rightarrow a^2y^2+a^2z^2+b^2x^2+b^2z^2+c^2x^2+c^2y^2$

$=2abxy+2bcyz+2cazx$

$\Rightarrow (a^2y^2-2abxy+b^2x^2)+(b^2z^2-2bcyz+c^2y^2)$

$+(a^2z^2+2cazx+c^2x^2)=0$

$\Rightarrow (ay-bx)^2+(bz-cy)^2+(az-cx)^2=0$

$\Rightarrow \quad (ay-bx)^2=0, (bz-cy)^2=0$ and $(az-cx)^2=0$

$\Rightarrow \quad ay=bx, bz=cy$ and $az=cx$

$\Rightarrow \quad \frac{a}{b}=\frac{x}{y}$ and $\frac{a}{c}=\frac{x}{z}$

$\Rightarrow a=kx, b=ky$ and $c=kz$, where k is a constant.

Also, $a^2 + b^2 + c^2 = 16$

$\Rightarrow (kx)^2 + (ky)^2 + (kz)^2 = 16$

$\Rightarrow k^2(x^2 + y^2 + z^2) = 16$

$\Rightarrow k^2 \times 25 = 16$

$\Rightarrow k = \sqrt{\frac{16}{25}} = \frac{4}{5}$

Now, $\frac{a+b+c}{x+y+z} = \frac{kx+ky+kz}{x+y+z} = k = \frac{4}{5}$

Alternatively

$\frac{a+b+c}{x+y+z} = \sqrt{\frac{16}{25}} = \frac{4}{5}$

Problem 108 Let a, b, x, y be real numbers such that $a^2 + b^2 = 25$, $x^2 + y^2 = 169$ and $ax + by = 65$. If $k = ay - bx$, then **[CAT-19]**

(a) $k = 0$ (b) $0 < k \le \frac{5}{13}$ (c) $k = \frac{5}{13}$ (d) $k > \frac{5}{13}$

Sol. $(a^2 + b^2)(x^2 + y^2) = 25 \times 169 = (65)^2 = (ax + by)^2$

$\Rightarrow a^2x^2 + a^2y^2 + b^2x^2 + b^2y^2 = a^2x^2 + b^2y^2 + 2abxy$

$\Rightarrow a^2y^2 + b^2x^2 = 2abxy$

$\Rightarrow a^2y^2 + b^2x^2 - 2abxy = 0$

$\Rightarrow (ay - bx)^2 = 0 \Rightarrow ay - bx = 0 = k$

Problem 109 If $a^2 + b^2 = 4$, $x^2 + y^2 = 16$ and $ax + by = 8$, then find the value of $\frac{x+y}{a+b}$.

Sol. Here, $8 = \sqrt{4 \times 16}$. Then, $\frac{a+b}{x+y} = \sqrt{\frac{4}{16}} = \frac{1}{2}$

$\therefore \frac{x+y}{a+b} = 2$

Type 16

When LHS is equal to RHS term-wise

Problem 110 If $47.2506 = 4A + \frac{7}{B} + 2C + \frac{5}{D} + 6E$, then find the value of $5A + 3B + 6C + D + 3E$.

Sol. Given, $4A + \frac{7}{B} + 2C + \frac{5}{D} + 6E = 47.2506$

$\Rightarrow 4 \cdot 10 + 7 \cdot 1 + 2 \cdot (0.1) + 5(0.01) + 6(0.0001) = 4A + \frac{7}{B} + 2C + \frac{5}{D} + 6E$

Hence, $A = 10$, $B = 1$, $C = 0.1$, $D = 100$ and $E = 0.0001$

Now, $5A + 3B + 6C + D + 3E = 5(10) + 3(1) + 6(0.1) + 100 + 3(0.0001) = 153.6003$

Problem 111 If $2x^2 + y^2 + 8z^2 - 2\sqrt{2}xy + 4\sqrt{2}yz - 8zx = (Ax + y + Bz)^2$, then find the value of $(A^2 + B^2 - AB)$. **[SSC (CGL-T-I)-19]**

Sol. $(Ax + y + Bz)^2 = A^2x^2 + y^2 + B^2z^2 + 2Axy + 2Byz + 2ABxz$

Given, $A^2x^2 + y^2 + B^2z^2 + 2Axy + 2Byz + 2BzAx$

$= (\pm\sqrt{2})^2x^2 + y^2 + (\pm 2\sqrt{2})^2z^2 + 2(-\sqrt{2})xy + 2(2\sqrt{2})yz + 2(2\sqrt{2})z(-\sqrt{2})x$

Comparing term-wise, we get $A = -\sqrt{2}$, $B = 2\sqrt{2}$

Now, $(A^2 + B^2 - AB) = (A - B)^2 + AB = (-\sqrt{2} - 2\sqrt{2})^2 + (-\sqrt{2} \cdot 2\sqrt{2}) = 18 - 4 = 14$

Problems Based on Maximum and Minimum Value of an Expression

(Concept of maximum and minimum value of a quadratic equation has been discussed under 'Quadratic Equation.')

Problem 112 For positive real numbers a, b, c, if $a + b + c = 21$, then what is the maximum value of abc?

Sol. We know, Arithmetic Mean (AM) ≥ Geometric Mean (GM)

$\Rightarrow \frac{a+b+c}{3} \ge \sqrt[3]{abc} \Rightarrow \frac{21}{3} \ge \sqrt[3]{abc} \Rightarrow abc \le 343$

Hence, maximum value of $abc = 343$

Alternation : We know, the value of abc is maximum at $a = b = c$ if $(a + b + c)$ is constant.

$\because a + b + c = 21$

$\therefore a = b = c = 7$

Hence, the maximum value of $abc = 7 \times 7 \times 7 = 343$

Problem 113 If a, b, c are positive real numbers and $abc = 64$, then what is the minimum value of $(a + b + c)$?

Sol. AM ≥ GM

$\Rightarrow \frac{a+b+c}{3} \ge \sqrt[3]{abc} \Rightarrow \frac{a+b+c}{3} \ge \sqrt[3]{64}$

$\Rightarrow a + b + c \ge 12$

Hence, minimum value of $(a + b + c)$ is 12.

Alternation. We know, if abc is constant, then $(a + b + c)$ is minimum at $a = b = c$.

$\because abc = 64$

$\therefore a = b = c = 4$

Then, minimum value of $(a + b + c) = 4 + 4 + 4 = 12$

Problem 114 If a and b are positive real numbers and $ab = 4$, then what is the minimum value of $(a + 2)(b + 2)$ is ?

Sol. $(a + 2)(b + 2) = ab + 2(a + b) + 4$

Obviously, $(a + b)$ will be minimum at $a = b$.

Hence, minimum value of $(a + b) = 2 + 2 = 4$.

Then, minimum value of $ab + 2(a + b) + 4 = 4 + 2(4) + 4 = 16$

Problem 115 If $x + y + z = 19$ and x, y, z be positive real numbers, then what is the maximum value of $(x - 5)(y + 4)(z - 3)$?

Sol. We know, the value of $(x - 5)(y + 4)(z - 3)$ will be maximum if $(x - 5) = (y + 4) = (z - 3)$

Now, $(x - 5) + (y + 4) + (z - 3) = x + y + z - 4 = 19 - 4 = 15$

$\therefore (x - 5) = (y + 4) = (z - 3) = 5$

Then, maximum value of $(x - 5)(y + 4)(z - 3) = 5 \times 5 \times 5 = 125$

Problem 116 Let x, y be two positive numbers such that $x + y = 1$. Then, find the minimum value of

$$\left(x+\frac{1}{x}\right)^2+\left(y+\frac{1}{y}\right)^2$$ **[CAT-01]**

Sol. $\left(x+\frac{1}{x}\right)^2+\left(y+\frac{1}{y}\right)^2 = x^2+y^2+\frac{1}{x^2}+\frac{1}{y^2}+4$

$$= x^2+y^2+\frac{x^2+y^2}{(xy)^2}+4$$

$$= [(x+y)^2-2xy]+\frac{[(x+y)^2-2xy]}{(xy)^2}+4$$

Obviously, $[(x+y)^2-2xy]$ will take minimum value if $2xy$ is maximum and we know that xy will have maximum value at $x = y$.

Hence, maximum value of $xy = \frac{1}{2}\times\frac{1}{2}=\frac{1}{4}$ $(\because x + y = 1)$

Then, the minimum value of

$[(x+y)^2-2xy] = (1)^2 - 2\cdot\frac{1}{4}=\frac{1}{2}$

Further, $\frac{[(x+y)^2-2xy]}{(xy)^2}$ will have minimum value if numerator is minimum and the denominator is maximum.

Then, minimum value of $\frac{[(x+y)^2-2xy]}{(xy)^2}=\frac{\left(\frac{1}{2}\right)}{\left(\frac{1}{4}\right)^2}=8$

Thus, minimum value of $[(x+y)^2-2xy]+\frac{[(x+y)^2-2xy]}{(xy)^2}+4$

$$=\frac{1}{2}+8+4=12.5$$

Alternatively

$$\left(x+\frac{1}{x}\right)^2+\left(y+\frac{1}{y}\right)^2 = x^2+y^2+\frac{1}{x^2}+\frac{1}{y^2}+4$$

$$= x^2+y^2+\frac{x^2+y^2}{(xy)^2}+4$$

We know, if the value of $(a + b)$ is constant, $a^2 + b^2$ will have minimum value at $a = b$.

Then, minimum value of $x^2+y^2=\left(\frac{1}{2}\right)^2+\left(\frac{1}{2}\right)^2$

$$=\frac{1}{2}$$

Further, minimum value of

$$\frac{x^2+y^2}{(xy)^2}=\frac{\text{minimum value of } x^2+y^2}{\text{maximum value of } (xy)^2}$$

$$=\frac{\left(\frac{1}{2}\right)}{\left(\frac{1}{2}\times\frac{1}{2}\right)^2}=8$$

Hence, minimum value of $x^2+y^2+\frac{x^2+y^2}{(xy)^2}+4$

$$=\frac{1}{2}+8+4=12.5$$

Problem 117 If x and y are positive real numbers and $xy = 8$, then find the minimum value of $2x + y$.

Sol. We know, for real positive numbers,

Arithmetic Mean ≥ Geometric Mean.

$\Rightarrow \frac{2x+y}{2}\ge\sqrt{2x\cdot y}\Rightarrow 2x+y\ge\sqrt{4\cdot 2x\cdot y}$

$\Rightarrow 2x+y\ge\sqrt{8\cdot 8}\Rightarrow 2x+y\ge 8$

Hence, minimum value of $2x + y = 8$

Problem 118 If a, b, c are positive and $a + b + c = 1$, then find the least value of $\frac{1}{a}+\frac{1}{b}+\frac{1}{c}$. **[SSC (CGL-T-I)-14]**

Sol. We know, for real positive numbers,

Arithmetic Mean ≥ Geometric Mean.

Then, $\frac{1}{3}\left(\frac{1}{a}+\frac{1}{b}+\frac{1}{c}\right)\ge\left(\frac{1}{a}\cdot\frac{1}{b}\cdot\frac{1}{c}\right)^{\frac{1}{3}}$

Also, $\frac{1}{3}(a+b+c)\ge(abc)^{\frac{1}{3}}$

On multiplying, we get

$$\frac{1}{3}\left(\frac{1}{a}+\frac{1}{b}+\frac{1}{c}\right)\times\frac{1}{3}(a+b+c)\ge\left(\frac{1}{a}\cdot\frac{1}{b}\cdot\frac{1}{c}\right)^{\frac{1}{3}}\times(abc)^{\frac{1}{3}}$$

$$\Rightarrow \left(\frac{1}{a}+\frac{1}{b}+\frac{1}{c}\right)\times(a+b+c)\ge 9$$

$$\Rightarrow \left(\frac{1}{a}+\frac{1}{b}+\frac{1}{c}\right)\times 1\ge 9 \quad (\because a+b+c=1)$$

Hence, minimum value of $\frac{1}{a}+\frac{1}{b}+\frac{1}{c}$ is 9.

Alternatively

We know, Arithmetic mean ≥ Harmonic mean

$$\Rightarrow \frac{1}{3}(a+b+c)\ge\frac{3}{\frac{1}{a}+\frac{1}{b}+\frac{1}{c}}$$

$$\Rightarrow \left(\frac{1}{a}+\frac{1}{b}+\frac{1}{c}\right)\times(a+b+c)\ge 9$$

(It is an important result)

$$\Rightarrow \left(\frac{1}{a}+\frac{1}{b}+\frac{1}{c}\right)\times 1\ge 9 \quad (\because a+b+c=1)$$

Hence, minimum value of $\frac{1}{a}+\frac{1}{b}+\frac{1}{c}$ is 9.

Problem 119 Consider the expression

$$\frac{(a^2+a+1)(b^2+b+1)(c^2+c+1)(d^2+d+1)(e^2+e+1)}{abcde}$$

where a, b, c, d and e are positive numbers. Find the minimum value of the expression. **[XAT-13, CAT-16]**

Sol. Given expression,

$$\frac{(a^2+a+1)}{a}\times\frac{(b^2+b+1)}{b}\times\frac{(c^2+c+1)}{c}\times\frac{(d^2+d+1)}{d}\times\frac{(e^2+e+1)}{e}$$

$$= \left(a + \frac{1}{a} + 1\right) \times \left(b + \frac{1}{b} + 1\right)\left(c + \frac{1}{c} + 1\right)\left(d + \frac{1}{d} + 1\right)\left(e + \frac{1}{e} + 1\right)$$

We know, AM ≥ GM

$$\Rightarrow \quad \frac{a + \frac{1}{a}}{2} \geq \sqrt{a \cdot \frac{1}{a}} \Rightarrow a + \frac{1}{a} \geq 2$$

$$\Rightarrow \quad a + \frac{1}{a} + 1 \geq 2 + 1$$

Hence, minimum value of $\left(a + \frac{1}{a} + 1\right)$ is 3. Then, minimum value of the expression is $3 \times 3 \times 3 \times 3 \times 3$, i.e. 243.

Problem 120 p, q and r are three non-negative integers such that $p + q + r = 10$. The maximum value of $pq + qr + rp + pqr$ is : **[CAT-16]**

(a) ≥ 40 and < 50 (b) ≥ 50 and < 60

(c) ≥ 60 and < 70 (d) ≥ 70 and < 80

Sol. Given, $p + q + r = 10$

$\Rightarrow \quad (p + 1) + (q + 1) + (r + 1) = 13$

Further $pq + qr + rp + pqr = (p + 1)(q + 1)(r + 1) - (p + q + r) - 1 = (p + 1)(q + 1)(r + 1) - 11$

If $[(p + 1) + (q + 1) + (r + 1)]$ is constant, $(p + 1)(q + 1)(r + 1)$ will have maximum value when they are equal or as close as possible.

Hence, $(p + 1) = 4$, $(q + 1) = 4$ and $(r + 1) = 5$

Then maximum value of $(p + 1)(q + 1)(r + 1) = 4 \times 4 \times 5 = 80$

∴ Maximum value of $pq + qr + rp + pqr = 80 - 11 = 69$

Thus, option (c) is correct.

Alternatively

We know, if $(p + q + r)$ is constant, then the value of $pq + qr + rp$ is maximum at $p = q = r$. Since, $p + q + r = 10$, $p = q = r = 3.33$, but as per question, all three are integers. Therefore, we ought to take their values in integers which must be as close as possible. Maximum possible values of a, b, c be 3, 3 and 4 (order does not matter).

Now, maximum value of $pq + qr + rp$ under the given restriction

$= 3 \cdot 3 + 3 \cdot 4 + 4 \cdot 3 = 33$

Further, maximum value of $pqr = 3 \times 3 \times 4 = 36$

Hence, maximum value of $pq + qr + rp + pqr = 33 + 36 = 69$.

Problem 121 Let w, x, y and z be four natural numbers such that their sum is $8m + 10$, where m is a natural number. Which of the following is necessarily true?

(a) The maximum possible value of $w^2 + x^2 + y^2 + z^2$ is $6m^2 + 40m + 26$.

(b) The maximum possible value of $w^2 + x^2 + y^2 + z^2$ is $16m^2 + 40m + 28$.

(c) The minimum possible value of $w^2 + x^2 + y^2 + z^2$ is $16m^2 + 40m + 28$.

(d) The minimum possible value of $w^2 + x^2 + y^2 + z^2$ is $16m^2 + 40m + 26$. **[CAT-14]**

Sol. We know, if $(w + x + y + z)$ is constant, then value of $w^2 + x^2 + y^2 + z^2$ is minimum at $w = x = y = z$.

Given, $w + x + y + z = 8m + 10$

Then, $w = x = y = z = 2m + 2.5$, but $2m + 2.5$ is not a natural number if m is a natural number. Thus, we should take their values in natural numbers as close as possible.

Such values are $2m + 2, 2m + 2, 2m + 3, 2m + 3$.

Now, minimum value of $w^2 + x^2 + y^2 + z^2$

$= (2m + 2)^2 + (2m + 2)^2 + (2m + 3)^2 + (2m + 3)^2$

$= 16m^2 + 40m + 26$

Problems Based on Remainder Theorem and Factor Theorem

(Since concepts of remainder and factor have also been discussed in arithmetical section, only algebraic examples have been discussed here.)

Problem 122 If $(x + 1)$ and $(x - 2)$ are factors of $x^3 + (a + 1)x^2 - (b - 2)x - 6$, then values of a and b are : **[SSC (CGL-T-I)-12, 13]**

Sol. Given that $(x + 1)$ is a factor of the given expression. It implies that the expression leaves the remainder 0 if it is divided by $(x + 1)$.

Now, $x + 1 = 0 \Rightarrow x = -1$

Then, remainder $0 = (-1)^3 + (a + 1)(-1)^2 - (b - 2)(-1) - 6$

$\Rightarrow \quad -1 + a + 1 + b - 2 - 6 = 0$

$\Rightarrow \quad a + b = 8$ ...(i)

Also, $(x - 2)$ is a factor of given expression, then $x - 2 = 0$

$\Rightarrow x = 2$

Then, remainder $0 = (2)^3 + (a + 1)(2)^2 - (b - 2)2 - 6$

$\Rightarrow \quad 8 + 4a + 4 - 2b + 4 - 6 = 0$

$\Rightarrow \quad 4a - 2b = -10$

$\Rightarrow \quad 2a - b = -5$ ...(ii)

Solving Eqs. (i) and (ii), we get

$a = 1$ and $b = 7$

Problem 123 If $(x^2 + 3x - 10)$ is a factor of $3x^4 + 2x^3 - ax^2 + bx - a + b - 4)$, then the closest approximate values of a and b are : **[IIFT-13]**

Sol. $x^2 + 3x - 10 = x^2 + 5x - 2x - 10$

$= x(x + 5) - 2(x + 5) = (x - 2)(x + 5)$

Obviously, factors of $(x^2 + 3x - 10)$ are also the factors of $(3x^4 + 2x^3 - ax^2 + bx - a + b - 4)$.

Now, remainder [in case $(x - 2)$ is a factor]

$= 3 \cdot 2^4 + 2 \cdot 2^3 - a \cdot 2^2 + b \cdot 2 - a + b - 4$

$\Rightarrow \quad 0 = 60 - 5a + 3b$

$\Rightarrow 5a - 3b = 60$ ...(i)

Remainder [in case $(x + 5)$ is a factor]

$= 3 \cdot (-5)^4 + 2 \cdot (-5)^3 - a \cdot (-5)^2 + b \cdot (-5) - a + b - 4$

$\Rightarrow \quad 0 = 1875 - 250 - 25a - 5b - a + b - 4$

$\Rightarrow \quad 26a + 4b = 1621$ ...(ii)

Solving Eqs. (i) and (ii), we get $a = 52$ and $b = 67$

Problem 124 If $x^3 + 5x^2 + 10k$ leaves remainder $-2x$ when divided by $x^2 + 2$, then what is the value of k ? **[CDS-12, 08]**

Sol. Clearly, $x^2 + 2 = 0 \Rightarrow x^2 = -2$

Now, remainder $= x^2 \cdot x + 5(-2) + 10k = -2x$ (given)

$\Rightarrow \quad (-2)\cdot x + 5(-2) + 10k = -2x$

$\Rightarrow \quad 10k = 10 \Rightarrow k = 1$

Problem 125 If the expression $px^3 + 3x^2 - 3$ and $2x^3 - 5x + p$, when divided by $(x-4)$, leave the same remainder, then what is the value of p?

Sol. $x - 4 = 0 \Rightarrow x = 4$

Remainder in first case $= p(4)^3 + 3(4)^2 - 3 = 64p + 45$

and remainder in second case $= 2(4)^3 - 5(4) + p = p + 108$.

As per question, $64p + 45 = p + 108$

$\Rightarrow \quad 63p = 63 \Rightarrow p = 1$

Problem 126 When $(x^3 - 2x^2 + px - q)$ is divided by $(x^2 - 2x - 3)$ the remainder is $(x - 6)$. What are the values of p and q, respectively?

Sol. $x^2 - 2x - 3 = 0 \Rightarrow x^2 = 2x + 3$

Then, remainder $= x^2 \cdot x - 2x^2 + px - q$

$= (2x+3)\cdot x - 2x^2 + px - q$

$= 3x + px - q$

$= x(3 + p) - q = x - 6$ (given)

$\Rightarrow 3 + p = 1$ and $q = 6$

$\Rightarrow \quad p = -2$ and $q = 6$

Problem 127 What should be subtracted from $27x^3 - 9x^2 - 6x - 5$ to make it exactly divisible by $(3x - 1)$?

Sol. Remainder if the expression is divided by $(3x - 1)$

$$= 27\left(\frac{1}{3}\right)^3 - 9\left(\frac{1}{3}\right)^2 - 6\left(\frac{1}{3}\right) - 5 \quad (\because 3x - 1 = 0 \Rightarrow x = \frac{1}{3})$$

$= 1 - 1 - 2 - 5 = -7$

Hence, if (–7) is subtracted from the expression, the resultant will be divisible by (3x – 1).

Problem 128 Given that the polynomial $(x^2 + ax + b)$ leaves the same remainder when divided by $(x - 1)$ or $(x + 1)$. What are the values of a and b respectively? **[CDS-19]**

Sol. Remainder in case the polynomial is divided by $(x - 1) = 1^2 + a\cdot 1 + b = a + b + 1$.

Further, the remainder in case the polynomial is divided by $(x + 1) = (-1)^2 + a\cdot(-1) + b = b - a + 1$

As given, $a + b + 1 = b - a + 1$

$\Rightarrow \quad 2a = 0 \Rightarrow a = 0$

Clearly, b can take any value which can be cancelled out.

Problems Based on Perfect Square/Cube

Problem 129 For what value of a is $x + \frac{1}{4}\sqrt{x} + a^2$ a perfect square ? **[SSC (CPO)-06]**

Sol. $x + \frac{1}{4}\sqrt{x} + a^2 = (\sqrt{x})^2 + 2\cdot\sqrt{x}\cdot\frac{1}{8} + a^2$

Clearly, at $a = \frac{1}{8}$, the expression will be $\left(\sqrt{x} + \frac{1}{8}\right)^2$, i.e. a perfect square.

Alternatively

Assume $x = 1$ (you may take another value except 0)

Then, $x + \frac{1}{4}\sqrt{x} + a^2 = 1 + \frac{1}{4}\sqrt{1} + a^2$

$= \frac{5}{4} + a^2$

Now, proceed through options. Let options be 1/18, 1/8, etc.

(a) $\frac{1}{18} \Rightarrow \frac{5}{4} + \left(\frac{1}{18}\right)^2 = \frac{406}{324}$, not a perfect square.

(b) $\frac{1}{8} \Rightarrow \frac{5}{4} + \left(\frac{1}{8}\right)^2 = \frac{81}{64}$, i.e. a perfect square.

Hence, option (b) is correct.

Problem 130 For what value of a and b $(a > 0, b < 0)$, $8x^3 - ax^2 + 54x + b$ is a perfect cube?

(a) $a = 18, b = -27$ (b) $a = 36, b = -27$

(c) $a = 12, b = -9$ (d) $a = 16, b = -9$

Sol. $8x^3 - ax^2 + 54x + b = (2x)^3 - ax^2 + 3\cdot 2x\cdot(3)^2 + b$,

apparently, it is a cube of $(2x - 3)$.

Then, $(2x - 3)^3 = (2x)^3 - 3\cdot(2x)^2\cdot 3 + 3\cdot 2x\cdot(3)^2 - 3^3$

On comparing, we get $a = 36$, and $b = -27$

Alternatively

You may proceed through options.

(a) $a = 18, b = -27$

$\Rightarrow 8x^3 - 18x^2 + 54x - 27 = (2x)^3 - 3\times\frac{3}{2}(2x)^2 + 54x - 27$

Clearly, it is not a cube of $\left(2x - \frac{3}{2}\right)$

(b) $a = 36, b = -27$

$\Rightarrow 8x^3 - 36x^2 + 54x - 27$

$= (2x)^3 - 3\cdot(2x)^2\cdot 3 + 3\cdot 2x\cdot(3)^2 - 3^3$

$= (2x - 3)^3$, a perfect cube.

Hence, option (b) is correct.

Problem 131 Find n, if $2^{200} - 2^{192}\cdot 31 + 2^n$ is a perfect square. **[CAT-16]**

Given expression, $2^{200} - 2^{192}\cdot 31 + 2^n$

$\Rightarrow \quad 2^{192}(2^8 - 31) + 2^n$

$\Rightarrow \quad 2^{192}\cdot 225 + 2^n$

$\Rightarrow \quad 2^{192}(225 + 2^{n-192})$

Clearly, $(225 + 2^{n-192})$ must be a perfect square as 2^{192} is a perfect square number.

Also, the perfect square number next to 225 is 256 and 2^{integer} cannot yield (256 – 225).

Further, a perfect square next to 256 is 289 and 2^6 yields (289 – 225).

Hence, $\quad 2^{n-192} = 2^6$

$\Rightarrow \quad n - 192 = 6 \Rightarrow n = 198$

Problems Seeking Concept of Ratio

Problem 132 If $\frac{a}{2}=\frac{b}{3}=\frac{c}{5}$, then find the value of $\frac{a+b+c}{c}$.

Sol. Let $\frac{a}{2}=\frac{b}{3}=\frac{c}{5}=k \Rightarrow a=2k, b=3k, c=5k$

Then, $\frac{a+b+c}{c}=\frac{2k+3k+5k}{5k}=2$

Alternatively

We know, $\frac{a}{e}=\frac{b}{f}=\frac{c}{g}=\frac{a+b+c}{e+f+g}$ (Recall the property of ratio)

$\Rightarrow \frac{c}{g}=\frac{a+b+c}{e+f+g} \Rightarrow \frac{c}{5}=\frac{a+b+c}{2+3+5}$

$\Rightarrow \frac{10}{5}=\frac{a+b+c}{c} \Rightarrow \frac{a+b+c}{c}=2$

Problem 133 If $a:b=c:d=1:6$, then what is the value of $\frac{a^2+c^2}{b^2+d^2}$? **[CDS-19]**

Sol. Let $a=x$, then $b=6x$

and $d=6y$ if $c=y$ (let)

Now, $\frac{a^2+c^2}{b^2+d^2}=\frac{x^2+y^2}{(6x)^2+(6y)^2}=\frac{1}{36}$

Alternatively

$\frac{a^2+c^2}{b^2+d^2}=\frac{1^2+1^2}{6^2+6^2}=\frac{1}{36}$

(Such types of questions have also been discussed under chapter 'Ratio, Proportion and Variation').

Miscellaneous

Problem 134 If $\frac{1}{x+1}+\frac{2}{y+2}+\frac{1009}{z+1009}=1$, then what is the value of $\frac{x}{x+1}+\frac{y}{y+2}+\frac{z}{z+1009}$? **[CDS-09]**

Sol. Given, $\frac{1}{x+1}+\frac{2}{y+2}+\frac{1009}{z+1009}=1$

$\Rightarrow \frac{1}{x+1}-1+\frac{2}{y+2}-1+\frac{1009}{z+1009}-1=1-3$

$\Rightarrow \frac{-x}{x+1}+\frac{-y}{y+2}+\frac{-z}{z+1009}=-2$

$\Rightarrow \frac{x}{x+1}+\frac{y}{y+2}+\frac{z}{z+1009}=2$

Problem 135 If $ab-b+1=0$ and $bc-c+1=0$, then what is the value of $(a-ac)$?

Sol. Given, $ab-b+1=0$

$\Rightarrow b(a-1)=-1$

$\Rightarrow b=\frac{-1}{a-1}$

Also given, $bc-c+1=0$

$\Rightarrow bc=c-1$

$\Rightarrow b=\frac{c-1}{c}$

Hence, $\frac{-1}{a-1}=\frac{c-1}{c}$

$\Rightarrow -c=ac-a-c+1$

$\Rightarrow a-ac=1$

Problem 136 If $\frac{5x-3}{x}+\frac{5y-3}{y}+\frac{5z-3}{z}=0$, then what is the value of $\frac{1}{x}+\frac{1}{y}+\frac{1}{z}$? **[SSC (CGL-T-I)-12]**

Sol. Given, $\frac{5x-3}{x}+\frac{5y-3}{y}+\frac{5z-3}{z}=0$

$\Rightarrow 5-\frac{3}{x}+5-\frac{3}{y}+5-\frac{3}{z}=0$

$\Rightarrow 15=3\left(\frac{1}{x}+\frac{1}{y}+\frac{1}{z}\right)$

$\therefore \frac{1}{x}+\frac{1}{y}+\frac{1}{z}=\frac{15}{3}=5$

WORKOUT

1. If $(a-1)^2+(b+2)^2+(c+1)^2=0$, then the value of $2a-3b+7c$ is :

(a) 12 (b) 3
(c) −11 (d) 1

2. What is the value of k which will make the expression $4k^2+12k+k$ a perfect square? *[CDS-12]*

(a) 5 (b) 7
(c) 8 (d) 9

3. The expression x^4-2x^2+k will be a perfect square when the value of k is :

(a) −1 (b) 1
(c) −2 (d) 2

4. Consider a function $f(x)=x^4+x^3+x^2+x+1$, where x is a positive integer greater than 1. What will be the remainder, if $f(x^5)$ is divided by $f(x)$? *[XAT-13]*

(a) 1
(b) 4
(c) 5
(d) A monomial in x
(e) A polynomial in x

5. If $\dfrac{2p}{p^2-2p+1}=\dfrac{1}{4}$, $p\neq 0$, then $p+\dfrac{1}{p}=?$

(a) 4 (b) 5
(c) 10 (d) 12

6. If $x=\dfrac{4ab}{a+b}$, then $\dfrac{x+2a}{x-2a}+\dfrac{x+2b}{x-2b}=?$ *[SSC (CGL-T-I)-12, MAT-03]*

(a) a (b) b
(c) 0 (d) 2

7. If $x=2+2^{\frac{2}{3}}+2^{\frac{1}{3}}$, then the value of x^3-6x^2+6x is : *[MAT-02, CDS-17]*

(a) 3 (b) 2
(c) 1 (d) None of these

8. If $x+\dfrac{1}{x}=\dfrac{\sqrt{3}+1}{2}$, then $x^4+\dfrac{1}{x^4}=?$ *[SSC (CGL-T-II)-17]*

(a) $\dfrac{4\sqrt{3}-1}{4}$ (b) $\dfrac{4\sqrt{3}+1}{2}$

(c) $\dfrac{-4\sqrt{3}-1}{4}$ (d) $\dfrac{-4\sqrt{3}-1}{2}$

9. If $x=\dfrac{2\sqrt{24}}{\sqrt{3}+\sqrt{2}}$, then $\dfrac{x+\sqrt{8}}{x-\sqrt{8}}+\dfrac{x+\sqrt{12}}{x-\sqrt{12}}=?$

(a) 1 (b) 2
(c) 0 (d) −2

10. Find the remainder when a^3-5a^2+7a-9 is divided by a^2+a-6. *[CAT-09]*

(a) $19a-31$ (b) $19a-38$
(c) $19a-49$ (d) $19a-45$

11. If $(x-2a)(x-5a)(x-8a)(x-11a)+ka^4$ is a perfect square, then the value of k is : *[SSC (FCI-Mains)-12]*

(a) 81 (b) 49
(c) 64 (d) 72

12. $x(y^2-z^2)+y(z^2-x^2)+z(x^2-y^2)$ is divisible by : *[CDS-13]*

(a) only $(y-z)$
(b) only $(z-x)$
(c) both $(y-z)$ and $(z-x)$
(d) neither $(y-z)$ nor $(z-x)$

13. If $\dfrac{4x-3}{x}+\dfrac{4y-3}{y}+\dfrac{4z-3}{z}=0$, then the value of $\dfrac{1}{x}+\dfrac{1}{y}+\dfrac{1}{z}$ is :

(a) 9 (b) 3
(c) 4 (d) 6

14. If $x+\dfrac{1}{x}=1$, then $x^{19}+x^{16}+x^{12}+x^9+3=?$

(a) 9 (b) 3
(c) 4 (d) 6

15. If $x=\sqrt{5}+2$, then the value of $\dfrac{2x^2-3x-2}{3x^2-4x-3}$ is equal to : *[SSC (CGL-T-I)-14]*

(a) 0.185 (b) 0.525
(c) 0.625 (d) 0.785

16. If $(x-3)(2x+1)=0$, then possible value of $2x+1$ are :

(a) only 0 (b) 0 and 3
(c) $-\dfrac{1}{2}$ and 3 (d) 0 and 7

17. If $a^2-4a-1=0$, then $a^2+\dfrac{1}{a^2}+3a-\dfrac{3}{a}$ is :

(a) 25 (b) 30
(c) 35 (d) 40

18. If $\dfrac{x}{(2x+y+z)}=\dfrac{y}{(x+2y+z)}=\dfrac{z}{(x+y+2z)}=a$, then find the value of a, if $x+y+z\neq 0$. *[MAT-05]*

(a) $\dfrac{1}{3}$ (b) $\dfrac{1}{4}$

(c) $\dfrac{1}{8}$ (d) $\dfrac{1}{2}$

19. If $a+a^2+a^3-1=0$, then $a^3+\dfrac{1}{a}=?$ *[SSC (CGL-T-II)-17]*

(a) 1 (b) 4
(c) 2 (d) 3

20. If $a+\dfrac{1}{b}=1$ and $b+\dfrac{1}{c}=1$, then $c+\dfrac{1}{a}$ is equal to :

(a) 0 (b) $\dfrac{1}{2}$
(c) 1 (d) 2

21. If one of the two factors of an expression which is the difference of two cubes, is $(x^4+x^2y+y^2)$, then what is the other factor?

(a) $x+y$
(b) $x-y$
(c) x^2+y
(d) x^2-y

22. If $x-\dfrac{1}{x}=4$, then $x+\dfrac{1}{x}$ is equal to :

(a) $4\sqrt{5}$ (b) $2\sqrt{5}$
(c) $5\sqrt{2}$ (d) $4\sqrt{2}$

23. If $3x+4y-2z+9=17$, $7x+2y+11z+8=23$ and $5x+9y+6z-4=18$, then what is the value of $x+y+z-34$? *[SSC (CGL-T-II)-17]*

(a) −28 (b) −14
(c) −31 (d) −45

24. If $\dfrac{a^2-ab+b^2}{a^2+ab+b^2}=\dfrac{1}{3}$, then find the value of $\dfrac{a}{b}$. *[CAT-09]*

(a) 1 (b) 2
(c) 3 (d) 4

25. x, y and z are real numbers. If $x^3 + y^3 + z^3 = 13$, $x + y + z = 1$ and $xyz = 1$, then what is the value of $xy + yz + zx$? *[SSC (CGL-T-II)-17]*

(a) -3 (b) -1

(c) 3 (d) 1

26. If $\frac{a+b}{c} = \frac{6}{5}$ and $\frac{b+c}{a} = \frac{9}{2}$, then what is the value of $\frac{a+c}{b}$? *[SSC (CGL-T-II)-17]*

(a) $\frac{9}{5}$ (b) $\frac{11}{7}$

(c) $\frac{7}{11}$ (d) $\frac{7}{4}$

27. $a^2 + b^2 = 2$ and $c^2 + d^2 = 1$, then $(ad - bc)^2 + (ac + bd)^2$ is :

(a) $\frac{4}{9}$ (b) $\frac{1}{2}$

(c) 1 (d) 2

28. If $x^2 + y^2 = 34$ and $x^4 - y^4 = 544$, then the values of x and y are : *[IAS (pre.)-01]*

(a) $\pm 4, \pm 3$

(b) $\pm 5, \pm 3$

(c) $\pm 3, \pm 5$

(d) $\pm 3, \pm 4$

29. If $x = 3 + 2\sqrt{2}$ and $xy = 1$, then the value of $\frac{x^2 + 3xy + y^2}{x^2 - 3xy + y^2}$ is :

(a) $\frac{30}{31}$ (b) $\frac{70}{31}$

(c) $\frac{35}{31}$ (d) $\frac{37}{31}$

30. $(a + 1)^4 - a^4$ is divisible by

(a) $-2a^2 + 2a - 1$ (b) $2a^3 - 2a - 1$

(c) $2a^3 - 2a + 1$ (d) $2a^2 + 2a + 1$

31. If $x^3 + y^3 + z^3 = 3(1 + xyz)$, $P = y + z - x$, $Q = z + x - y$ and $R = x + y - z$, then what is the value of $P^3 + Q^3 + R^3 - 3PQR$? *[SSC (CGL-T-II)-17]*

(a) 9 (b) 8

(c) 12 (d) 6

32. If $\frac{x}{b+c} = \frac{y}{c+a} = \frac{z}{a+b}$, then which of the following is true? *[SSC (10 + 2-LDC/DEO)-13]*

(a) $\frac{x-y}{b-a} = \frac{y-z}{c-b} = \frac{z-x}{a-c}$

(b) $\frac{x}{a} = \frac{y}{b} = \frac{z}{c}$

(c) $\frac{x-y}{c} = \frac{y-z}{b} = \frac{z-x}{c}$

(d) None of the above

33. If $x(x + y + z) = 9$, $y(x + y + z) = 16$ and $z(x + y + z) = 144$, then what is the value of x?

(a) $\frac{9}{5}$ (b) $\frac{9}{7}$

(c) $\frac{9}{13}$ (d) $\frac{16}{13}$

34. If the expression $\frac{x^2}{y^2} + tx + \frac{y^2}{4}$ is a perfect square, then the value of t is :

(a) ± 1 (b) ± 2

(c) 0 (d) ± 3

35. The least value of $3^x + 3^{-x}$ is : *[MAT-05]*

(a) 2 (b) $\frac{1}{2}$

(c) 0 (d) $\frac{2}{3}$

36. If $\frac{x-a^2}{b+c} + \frac{x-b^2}{c+a} + \frac{x-c^2}{a+b} = 4(a + b + c)$, then x is equal to : *[SSC (CGL-T-II)-13]*

(a) $(a + b + c)^2$

(b) $a^2 + b^2 + c^2$

(c) $ab + bc + ca$

(d) $a^2 + b^2 + c^2 - ab - bc - ca$

37. If x, y and z are three positive integers such that $x > y > z$. Which of the following is closest to the product xyz? *[CAT-16]*

(a) $(x - 1)yz$ (b) $(y - 1)zx$

(c) $(z - 1)xy$ (d) $(y + 1)zx$

38. If $\frac{b-c}{a} + \frac{a+c}{b} + \frac{a-b}{c} = 1$ and $a - b + c \neq 0$, then which one of the following relation is true? *[SSC (CGL-T-I)-14]*

(a) $\frac{1}{c} = \frac{1}{a} + \frac{1}{b}$ (b) $\frac{1}{a} = \frac{1}{b} + \frac{1}{c}$

(c) $\frac{1}{b} = \frac{1}{a} - \frac{1}{c}$ (d) $\frac{1}{b} = \frac{1}{a} + \frac{1}{c}$

39. If p, q, r and s are positive real numbers such that $p + q + r + s = 2$, then $m = (p + q)(r + s)$ satisfies the relation :

(a) $0 < m \leq 1$ (b) $1 \leq m \leq 2$

(c) $2 \leq m \leq 3$ (d) $3 \leq m \leq 4$

40. If $a + b = 1$, $c + d = 1$ and $a - b = \frac{d}{c}$, then $c^2 - d^2$ is : *[SSC (CGL-T-I)-13]*

(a) $\frac{a}{b}$ (b) $\frac{b}{a}$

(c) 1 (d) -1

41. When $x^{40} + 2$ is divided by $x^4 + 1$, what is the remainder?

(a) 1 (b) 2

(c) 3 (d) 4

42. If $a + b = 1$, then $a^4 + b^4 - a^3 - b^3 - 2a^2b^2 + ab$ is equal to : *[SSC (CGL-T-II)-16]*

(a) 4 (b) 2

(c) 1 (d) 0

43. If $x + y = 7$, then the value of $x^3 + y^3 + 21xy$ is :

(a) 243 (b) 143

(c) 343 (d) 443

44. $x^{21} + y^{21}$ is divisible by which one of the following ?

(a) $(x + y)$ (b) $(x - y)$

(c) $x^2 - y^2$ (d) None of these

45. If $x^2 + y^2 + 6x + 5 = 4(x - y)$, then $(x - y)$ is : *[SSC (CGL-T-II)-16]*

(a) 4 (b) -1

(c) 1 (d) 0

46. If $x^2 - 3x + 1 = 0$, then the value of $x^3 + \frac{1}{x^3}$ is equal to :

(a) 9 (b) 18

(c) 27 (d) 1

47. If $(x^4 + x^{-4}) = 322$, then what is one of the value of $(x - x^{-1})$?

(a) 18 (b) 16

(c) 8 (d) 4

48. If $\left(x - \frac{1}{x}\right)^2 = 3$, then the value of $\left(x^6 + \frac{1}{x^6}\right)$ equal to : *[SSC (CGL-T-II)-16]*

(a) 90 (b) 100

(c) 110 (d) 120

49. If $x^{\frac{1}{3}} + y^{\frac{1}{3}} = z^{\frac{1}{3}}$, then $(x + y - z)^3 + 27xyz$ is equal to :

(a) -1 (b) 1

(c) 0 (d) 27

50. If $x^4 + 2x^3 + ax^2 + bx + 9$ is a perfect square, where a and b are positive real numbers, then the value of a and b are : *[SSC (CGL-T-II)-16]*

(a) $a = 5, b = 6$ (b) $a = 6, b = 7$

(c) $a = 7, b = 6$ (d) $a = 7, b = 8$

51. If $x^2 - x + 1 = 0$, then $5x^3 - 4 = ?$

(a) -9 (b) 18

(c) 27 (d) 1

52. If $a + b + c + d = 4$, then find the value of $\frac{1}{(1-a)(1-b)(1-c)} + \frac{1}{(1-b)(1-c)(1-d)} + \frac{1}{(1-c)(1-d)(1-a)} + \frac{1}{(1-d)(1-a)(1-b)}$

[SSC (10+2-LDC/DEO)-14]

(a) 0 (b) 5 (c) 1 (d) 4

53. If $x = a^{\frac{1}{2}} + a^{-\frac{1}{2}}$, $y = a^{\frac{1}{2}} - a^{-\frac{1}{2}}$, then the value of $(x^4 - x^2y^2 - 1) + (y^4 - x^2y^2 + 1)$ is **[SSC (CGL-T-II)-15]**

(a) 16 (b) 13
(c) 12 (d) 14

54. If $a^2 + b^2 + c^2 = 2a - 2b - 2$, then the value of $3a - 2b + c$ is :

(a) 0 (b) 3
(c) 5 (d) 2

55. If $\frac{1}{x+y} = \frac{1}{x} + \frac{1}{y}$ $(x \neq 0, y \neq 0, x \neq y)$, then the value of $x^3 - y^3$ is :

(a) 0 (b) 1
(c) –1 (d) 2

56. If $2x = \sqrt{a} + \frac{1}{\sqrt{a}}$, $a > 0$, then the value of $\frac{\sqrt{x^2-1}}{x - \sqrt{x^2-1}}$ is :

[SSC (CGL-T-II)-14]

(a) $a + 1$ (b) $\frac{1}{2}(a + 1)$
(c) $\frac{1}{2}(a - 1)$ (d) $a - 1$

57. If $x = \frac{2\sqrt{6}}{\sqrt{3} + \sqrt{2}}$, then $\frac{x+\sqrt{2}}{x-\sqrt{2}} + \frac{x+\sqrt{3}}{x-\sqrt{3}} = ?$

(a) $\sqrt{2}$ (b) $\sqrt{3}$ (c) $\sqrt{6}$ (d) 2

58. If $\frac{a+b-c}{a+b} = \frac{b+c-a}{b+c} = \frac{c+a-b}{c+a}$ and $a + b + c \neq 0$, then

[SSC (CGL-T-II)-14]

(a) $a \neq b \neq c$ (b) $a = b \neq c$
(c) $a \neq b = c$ (d) $a = b = c$

59. If $(a^2 + b^2)^3 = (a^3 + b^3)^2$, then $\frac{a}{b} + \frac{b}{a}$ is :

(a) $\frac{1}{3}$ (b) $\frac{2}{3}$
(c) $\frac{-1}{3}$ (d) $\frac{-2}{3}$

60. If $x = \frac{a-b}{a+b}$, $y = \frac{b-c}{b+c}$, $z = \frac{c-a}{c+a}$, then $\frac{(1-x)(1-y)(1-z)}{(1+x)(1+y)(1+z)}$ is equal to : **[SSC (CGL-T-I)-15]**

(a) 1 (b) 0 (c) 2 (d) $\frac{1}{2}$

61. If $\frac{a}{b} + \frac{b}{a} = 1$, $a \neq 0$ and $b \neq 0$, then the value of $a^3 + b^3$ is :
[SSC (CGL-PT)-07, SSC (CGL-T-I)-13, SSC (FCI-Mains)-12, SSC (CPO)-11]

(a) 0 (b) 1 (c) –1 (d) 2

62. If $x^2 + 1 = 2x$, then the value of $\frac{x^4 + \frac{1}{x^2}}{x^2 - 3x + 1}$ is : **[SSC (Delhi Police-SI)-12]**

(a) 0 (b) 1 (c) 2 (d) –2

63. If $\frac{x}{x^2 - 2x + 1} = \frac{1}{3}$, then value of $x^3 + \frac{1}{x^3}$ is :

(a) 112 (b) $\frac{1}{27}$
(c) 28 (d) 110

64. If $x^2 + x = 5$, then the value of $(x+3)^3 + \frac{1}{(x+3)^3}$ is : **[SSC (CGL-T-I)-15]**

(a) 140 (b) 110
(c) 130 (d) 120

65. If $x + y + z = 6$ and $x^2 + y^2 + z^2 = 20$, then what is the value of $x^3 + y^3 + z^3 - 3xyz$?
[SSC (CGL-T-I)-13]

(a) 64 (b) 70
(c) 72 (d) 76

66. If $\frac{x^{24}+1}{x^{12}} = 7$, then the value of $\frac{x^{72}+1}{x^{36}}$ is :

(a) 433 (b) 322
(c) 343 (d) 432

67. If $(x - a)(x - b) = 1$ and $a - b + 5 = 0$, then the value of $(x-a)^3 - \frac{1}{(x-a)^3}$ is : **[SSC (CGL-T-II)-13]**

(a) –125 (b) 1
(c) 125 (d) 140

68. If $a + b = 10$ and $\sqrt{\frac{a}{b}} - 13 = -\sqrt{\frac{b}{a}} - 11$, then what is the value of $3ab + 4a^2 + 5b^2$?
[SSC (CGL-T-II)-17]

(a) 450 (b) 300
(c) 600 (d) 750

69. If $a + b - c = 0$, then the value of $2b^2c^2 + 2c^2a^2 + 2a^2b^2 - a^4 - b^4 - c^4$ is : **[SSC (CGL-T-I)-15]**

(a) 140 (b) 110
(c) 130 (d) 0

70. If $x = \sqrt[3]{2 + \sqrt{3}}$, then the value of $x^3 + \frac{1}{x^3}$ is :

(a) 8 (b) 9
(c) 2 (d) 4

71. If $a^2 + b^2 + c^2 = ab + bc + ca$, then the value of $\frac{a+c}{b}$ is :

(a) 0 (b) 2
(c) 1 (d) –1

72. If $x^3 + \frac{3}{x} = 4(a^3 + b^3)$ and $3x + \frac{1}{x^3} = 4(a^3 - b^3)$, then $a^2 - b^2$ is equal to : **[SSC (CGL-T-I)-13]**

(a) 4 (b) 0
(c) 1 (d) 2

73. If $\frac{p}{a} + \frac{q}{b} + \frac{r}{c} = 1$ and $\frac{a}{p} + \frac{b}{q} + \frac{c}{r} = 0$, where p, q, r and a, b, c are non-zero, then the value of $\frac{p^2}{a^2} + \frac{q^2}{b^2} + \frac{r^2}{c^2}$ is : **[SSC (CGL-T-I)-14]**

(a) –1 (b) 0
(c) 1 (d) 2

74. If $ab + bc + ca = 0$, then what is the value of $\frac{a^2}{a^2 - bc} + \frac{b^2}{b^2 - ca} + \frac{c^2}{c^2 - ab}$? **[CDS-17]**

(a) 3 (b) 0
(c) 1 (d) –1

75. If $a + \frac{1}{a} = \sqrt{3}$, then the value of $a^6 - \frac{1}{a^6} + 2$ is :

(a) 5 (b) 1
(c) 2 (d) $3\sqrt{3}$

76. If $\frac{xy}{x+y} = a$, $\frac{xz}{x+z} = b$ and $\frac{yz}{y+z} = c$, where a, b, c are all non-zero numbers, then x is equal to :

(a) $\frac{2abc}{ab + bc - ac}$ (b) $\frac{2abc}{ab + ac - bc}$
(c) $\frac{2abc}{ac + bc - ab}$ (d) $\frac{2abc}{ab + bc + ac}$

77. If $a = 2.234, b = 3.121$ and $c = -5.355$, then the value of $a^3 + b^3 + c^3 - 3abc$ is :

(a) −1 (b) 1
(c) 0 (d) 2

78. If $a^3 + 3a^2 + 9a = 1$, then $a^3 + \frac{3}{a} = ?$ **[SSC (CGL-T-II)-17]**

(a) 31 (b) 26
(c) 28 (d) 24

79. If $\frac{x}{a} = \frac{y}{b} = \frac{z}{c}$, then $xy + yz + zx$ is equal to :

(a) $\frac{(a+b+c)^2}{x^2+y^2+z^2}$

(b) $\frac{x^2(a+b+c)^2 - a^2(x^2+y^2+z^2)}{2a^2}$

(c) $\frac{ax+by+cz}{(a+b+c)^2}$

(d) None of the above

80. If $x = \sqrt{5} + \sqrt{3}$ and $y = \sqrt{5} - \sqrt{3}$, then the value of $(x^4 - y^4)$ is : **[SSC (CGL-T-II)-14]**

(a) $64\sqrt{15}$
(b) 16
(c) 544
(d) $32\sqrt{15}$

81. Let a, b, c and d be four integers such that $a + b + c + d = 4m + 1$, where m is a positive integer. Given m, which one of the following is necessarily true ? **[CAT-03]**

(a) The minimum possible value of $a^2 + b^2 + c^2 + d^2$ is $4m^2 - 2m + 1$.
(b) The minimum possible value of $a^2 + b^2 + c^2 + d^2$ is $4m^2 + 2m + 1$.
(c) The maximum possible value of $a^2 + b^2 + c^2 + d^2$ is $4m^2 - 2m + 1$.
(d) The maximum possible value of $a^2 + b^2 + c^2 + d^2$ is $4m^2 + 2m + 1$.

82. If a and b are integers of opposite signs such that $(a+3)^2 : b^2 = 9:1$ and $(a-1)^2 : (b-1)^2 = 4:1$, then the ratio $a^2 : b^2$ is : **[CAT-17]**

(a) 9 : 4 (b) 81 : 4
(c) 1 : 4 (d) 25 : 4

83. If a, b, c and d are integers such that $a + b + c + d = 30$, then the minimum value of $(a-b)^2 + (a-c)^2 + (a-d)^2$ is : **[CAT-17]**

(a) 0 (b) 1
(c) 2 (d) 4

84. If $\sqrt{\frac{x}{y}} = \frac{24}{5} + \sqrt{\frac{y}{x}}$ and $x + y = 26$, then what is the value of xy ? **[CDS-17]**

(a) 5 (b) 15
(c) 25 (d) 30

85. If $P = 7 + 4\sqrt{3}$ and $PQ = 1$, then what is the value of $\frac{1}{P^2} + \frac{1}{Q^2}$? **[SSC (CGL-T-II)-17]**

(a) 196 (b) 194
(c) 206 (d) 182

86. If x, y, z are distinct positive real numbers, then $\frac{x^2(y+z) + y^2(x+z) + z^2(x+y)}{xyz}$ would be : **[CAT-03]**

(a) greater than 4 (b) greater than 5
(c) greater than 6 (d) None of these

87. If $x = \frac{\sqrt{a+b} - \sqrt{a-b}}{\sqrt{a+b} + \sqrt{a-b}}$, then what is $bx^2 - 2ax + b$ equal to $(b \neq 0)$? **[CDS-17]**

(a) 0 (b) 1
(c) ab (d) $2ab$

88. If $x + y + z = 0$, then what is the value of $\frac{3y^2 + x^2 + z^2}{2y^2 - xz}$? **[SSC (CGL-T-II)-17]**

(a) 2 (b) $\frac{3}{2}$
(c) 1 (d) $\frac{5}{3}$

89. If $x + y + z = 22$ and $xy + yz + zx = 35$, then what is the value of $(x-y)^2 + (y-z)^2 + (z-x)^2$?

(a) 793 (b) 681
(c) 715 (d) 758

90. If x and y are real numbers, the least possible value of the expression $4(x-2)^2 + 4(y-3)^2 - 2(x-3)^2$ is : **[XAT-17]**

(a) −8 (b) −4
(c) −2 (d) 0
(e) 2

91. If $a - \frac{1}{a} = b$, $b - \frac{1}{b} = c$ and $c - \frac{1}{c} = a$, then what is the value of $\frac{1}{ab} + \frac{1}{bc} + \frac{1}{ca}$? **[SSC (CGL-T-II)-17]**

(a) − 1 (b) −3
(c) −6 (d) −9

92. If $x^2 + 3x + 1 = 0$, then the value of $x^6 + \frac{1}{x^6}$ is : **[SSC (CGL-T-I)-19]**

(a) 324 (b) 327
(c) 322 (d) 318

93. If $\frac{a}{b+c} = \frac{b}{c+a} = \frac{c}{a+b}$, then which one of the following statements is correct ? **[CDS-19]**

(a) Each fraction is equal to 1 or −1.
(b) Each fraction is equal to $\frac{1}{2}$ or 1.
(c) Each fraction is equal to $\frac{1}{2}$ or −1.
(d) Each fraction is equal to 1.2 only

94. If $12x^2 - 21x + 1 = 0$, then what is the value of $9x^2 + (16x^2)^{-1}$?

(a) $\frac{429}{8}$ (b) $\frac{465}{16}$
(c) $\frac{417}{16}$ (d) $\frac{453}{8}$

95. If $a^4 + \frac{1}{a^4} = 50$ and $a > 0$, then the value of $a^3 + \frac{1}{a^3}$ is : **[SSC (CGL-T-I)-19]**

(a) $\sqrt{2(1+\sqrt{13})}\,(-1+2\sqrt{13})$
(b) $\sqrt{2(1+\sqrt{13})}\,(-1-2\sqrt{13})$
(c) $\sqrt{2(1-\sqrt{13})}\,(-1+2\sqrt{13})$
(d) $\sqrt{2(1+\sqrt{13})} + (-1-2\sqrt{13})$

Answer Keys

1. (d)	2. (d)	3. (b)	4. (c)	5. (c)	6. (d)	7. (d)	8. (c)	9. (b)	10. (d)
11. (a)	12. (c)	13. (c)	14. (b)	15. (c)	16. (d)	17. (b)	18. (b)	19. (c)	20. (c)
21. (d)	22. (b)	23. (c)	24. (a)	25. (a)	26. (d)	27. (d)	28. (b)	29. (d)	30. (d)
31. (c)	32. (a)	33. (c)	34. (a)	35. (a)	36. (a)	37. (a)	38. (c)	39. (a)	40. (b)
41. (c)	42. (d)	43. (c)	44. (a)	45. (c)	46. (b)	47. (d)	48. (c)	49. (c)	50. (c)
51. (a)	52. (a)	53 (a)	54. (c)	55. (a)	56. (c)	57. (d)	58. (d)	59. (b)	60. (a)
61. (a)	62. (d)	63. (d)	64. (b)	65. (c)	66. (b)	67. (d)	68. (b)	69. (d)	70. (d)
71. (b)	72. (c)	73. (c)	74. (c)	75. (c)	76. (c)	77. (c)	78. (c)	79. (b)	80. (a)
81. (b)	82. (d)	83. (c)	84. (c)	85. (b)	86. (c)	87. (a)	88. (a)	89. (d)	90. (b)
91. (b)	92. (c)	93. (c)	94. (c)	95. (a)					

Hints and SOLUTIONS

1. Given,
$(a-1)^2+(b+2)^2+(c+1)^2=0$

Since the square of any number cannot be negative, each of the given terms will be either positive or zero. If they are positive, their sum cannot be equal to zero. It implies that each of terms is equal to zero.

$\Rightarrow \quad (a-1)^2=0, (b+2)^2=0$

and $\quad (c+1)^2=0$

$\Rightarrow \quad a=1, b=-2$ and $c=-1$

Then, $2a-3b+7c=2(1)-3(-2)+7(-1)=1$

2. Given, expression $4k^2+12k+k$
$=(2k)^2+2\cdot 2k\cdot 3+k$

At $k=(3)^2$, the expression will be $(2k+3)^2$ i.e. a perfect square.

Hence, option (d) is correct.

3. $x^4-2x^2+k=(x^2)^2-2.x^2.1+k$

Clearly, at $k=1$, the expression will be $(x^2-1)^2$, i.e. a perfect square.

4. Let $x=2$ as $x>1$, then

$f(2)=2^4+2^3+2^2+2+1=2^5-1$

and $f(2^5)=(2^5)^4+(2^5)^3+(2^5)^2+(2^5)+1$

Remainder in case $f(2^5)\div f(2)=$

Remainder $\left\{\frac{(2^5)^4}{2^5-1}\right\}$ + Rem. $\left\{\frac{(2^5)^3}{2^5-1}\right\}$

+ Rem. $\left\{\frac{(2^5)^2}{2^5-1}\right\}$ + Rem. $\left\{\frac{(2^5)}{2^5-1}\right\}$ +

Rem. $\left\{\frac{1}{2^5-1}\right\}$

$=1+1+1+1+1=5$

5. Given, $\frac{2p}{p^2-2p+1}=\frac{1}{4}$

$\Rightarrow \quad \frac{p^2-2p+1}{2p}=\frac{4}{1}$

$\Rightarrow \quad \frac{p^2-2p+1}{p}=8$

$\Rightarrow \quad p-2+\frac{1}{p}=8$

$\Rightarrow \quad p+\frac{1}{p}=8+2=10$

6. Let $a=1$ and $b=3$, then

$x=\frac{4\cdot 1\cdot 3}{1+3}=3$

Now, $\frac{x+2a}{x-2a}+\frac{x+2b}{x-2b}$

$=\frac{3+(2\times 1)}{3-(2\times 1)}+\frac{3+(2\times 3)}{3-(2\times 3)}$

$=\frac{5}{1}+\frac{9}{(-3)}$

$=5-3=2$

(It is reiterated that aspirants are free to assume any value of a and b)

7. Given, $x=2+2^{\frac{2}{3}}+2^{\frac{1}{3}}$

$\Rightarrow \quad x-2=2^{\frac{2}{3}}+2^{\frac{1}{3}}$

$\Rightarrow \quad x^3-6x^2+12x-8=4+2+3\cdot 2\left\{2^{\frac{2}{3}}+2^{\frac{1}{3}}\right\}$ [cubing both sides]

$\Rightarrow x^3-6x^2+12x-8=6+3\cdot 2x$

$\Rightarrow \quad x^3-6x^2+6x=14$

Hence, option (d) is correct.

8. $x^4+\frac{1}{x^4}=\left\{\left(\frac{\sqrt{3}+1}{2}\right)^2-2\right\}^2-2$

(from the relevant chart)

$=\left\{\frac{2\sqrt{3}-4}{4}\right\}^2-2=\frac{-16\sqrt{3}-4}{16}$

$=\frac{-4\sqrt{3}-1}{4}$

9. Given, $x=\frac{2\sqrt{24}}{\sqrt{3}+\sqrt{2}}$

$\Rightarrow \quad x=\frac{2\times\sqrt{8}\times\sqrt{3}}{\sqrt{3}+\sqrt{2}}$

$\Rightarrow \quad \frac{x}{\sqrt{8}}=\frac{2\sqrt{3}}{\sqrt{3}+\sqrt{2}}$

$\Rightarrow \frac{x+\sqrt{8}}{x-\sqrt{8}}=\frac{2\sqrt{3}+\sqrt{3}+\sqrt{2}}{2\sqrt{3}-\sqrt{3}-\sqrt{2}}$

(by Componendo and Dividendo)

$\Rightarrow \quad \frac{x+\sqrt{8}}{x-\sqrt{8}}=\frac{3\sqrt{3}+\sqrt{2}}{\sqrt{3}-\sqrt{2}}$

Further, $x=\frac{2\sqrt{24}}{\sqrt{3}+\sqrt{2}}$

$\Rightarrow \quad x=\frac{2\times\sqrt{12}\times\sqrt{2}}{\sqrt{3}+\sqrt{2}}$

$\Rightarrow \quad \frac{x}{\sqrt{12}}=\frac{2\sqrt{2}}{\sqrt{3}+\sqrt{2}}$

$\Rightarrow \frac{x+\sqrt{12}}{x-\sqrt{12}}=\frac{2\sqrt{2}+\sqrt{3}+\sqrt{2}}{2\sqrt{2}-\sqrt{3}-\sqrt{2}}$

(by Componendo and Dividendo)

$\Rightarrow \frac{x+\sqrt{12}}{x-\sqrt{12}}=\frac{3\sqrt{2}+\sqrt{3}}{\sqrt{2}-\sqrt{3}}$

Now, on adding $\frac{x+\sqrt{8}}{x-\sqrt{8}}+\frac{x+\sqrt{12}}{x-\sqrt{12}}=$

$\frac{3\sqrt{3}+\sqrt{2}}{\sqrt{3}-\sqrt{2}}+\frac{3\sqrt{2}+\sqrt{3}}{\sqrt{2}-\sqrt{3}}$

$=\frac{3\sqrt{3}+\sqrt{2}}{\sqrt{3}-\sqrt{2}}-\frac{3\sqrt{2}+\sqrt{3}}{\sqrt{3}-\sqrt{2}}$

$=\frac{3\sqrt{3}+\sqrt{2}-3\sqrt{2}-\sqrt{3}}{\sqrt{3}-\sqrt{2}}$

$=\frac{2\sqrt{3}-2\sqrt{2}}{\sqrt{3}-\sqrt{2}}=\frac{2(\sqrt{3}-\sqrt{2})}{(\sqrt{3}-\sqrt{2})}=2$

10. $a^2+a-6=0$

$\Rightarrow \quad a^2=6-a$

Now, remainder

$=(6-a)a-5(6-a)+7a-9$

$=6a-a^2-30+5a+7a-9$

$=6a-(6-a)-30+5a+7a-9$

$=19a-45$

11. Let the value of a be 1.

Then, $(x-2a)(x-11a)(x-5a)$

$(x-8a)+ka^4$

$=(x-2)(x-11)(x-5)(x-8)+k$

$=(x^2-13x+22)(x^2-13x+40)+k$

$\Rightarrow (m+22)(m+40)+k$

(Let $x^2-13x=m$)

$\Rightarrow \quad m^2+62m+880+k$

$\Rightarrow \quad m^2+2\cdot m\cdot 31+(880+k)$

For perfect square, $880+k=(31)^2$

$\therefore \quad k=81$

(Try yourself assuming suitable values of x and a)

12. If the given expression is divisible by $(y-z)$, then $y-z=0$

$\Rightarrow \quad y=z$

Putting $y=z$, we get

$x(z^2-z^2)+z(z^2-x^2)$

$+z(x^2-z^2)=0$

Thus, $(y-z)$ is a factor.

Further, if $(z-x)$ is a factor, then $z-x=0 \Rightarrow z=x$

Putting $z=x$, we get

$x(y^2-x^2)+y(x^2-x^2)+x(x^2-y^2)$

$=0$

Hence, option (c) is correct.

13. Given, $\frac{4x-3}{x}+\frac{4y-3}{y}+\frac{4z-3}{z}=0$

$\Rightarrow \frac{4x}{x}-\frac{3}{x}+\frac{4y}{y}-\frac{3}{y}+\frac{4z}{z}-\frac{3}{z}=0$

$\Rightarrow \quad 3\left(\frac{1}{x}+\frac{1}{y}+\frac{1}{z}\right)=12$

$\Rightarrow \quad \frac{1}{x}+\frac{1}{y}+\frac{1}{z}=4$

14. $x^{19}+x^{16}+x^{12}+x^9+3$

$=(x^{19}+x^{16})+(x^{12}+x^9)+3$

$=0+0+3$

($\because$ difference between powers in the group is 3)

$=3$

15. $\frac{2x^2-3x-2}{3x^2-4x-3}=\frac{2x-\frac{2}{x}-3}{3x-\frac{3}{x}-4}$

(dividing numerator and denominator by x)

$=\frac{2\left(x-\frac{1}{x}\right)-3}{3\left(x-\frac{1}{x}\right)-4}$

Given, $x=\sqrt{5}+2$

Then, $x-\frac{1}{x}=4 \quad [\because (\sqrt{5})^2-2^2=1]$

Now, $=\frac{2\left(x-\frac{1}{x}\right)-3}{3\left(x-\frac{1}{x}\right)-4}=\frac{2\cdot 4-3}{3\cdot 4-4}=\frac{5}{8}$

$=0.625$

16. Given, $(x-3)(2x+1)=0$

$\Rightarrow (x-3)=0$ and $(2x+1)=0$

$\Rightarrow \quad x=3$ and $(2x+1)=0$

At $x=3$, $2x+1=7$

Hence, possible value of $(2x+1)$ are 0 and 7.

17. Given, $a^2-4a-1=0$

$\Rightarrow \quad a^2-1=4a$

$\Rightarrow \quad \frac{a^2-1}{a}=\frac{4a}{a}$ (on dividing by a)

$\Rightarrow \quad a-\frac{1}{a}=4$

$\Rightarrow \quad a^2+\frac{1}{a^2}=(4)^2+2=18$

(recall the relevant chart)

Now, $a^2+\frac{1}{a^2}+3a-\frac{3}{a}=a^2+\frac{1}{a^2}$

$+3\left(a-\frac{1}{a}\right)$

$=18+3\cdot 4=30$

18. $\frac{x}{(2x+y+z)}=\frac{y}{(x+2y+z)}$

$=\frac{x}{(x+y+2z)}$

$=\frac{x+y+z}{(2x+y+z)+(x+2y+z)+(x+y+2z)}=a$

$\Rightarrow \quad a=\frac{1}{4}$

19. Given, $a+a^2+a^3-1=0$

$\Rightarrow \quad a^3=1-a-a^2$

$\Rightarrow \quad 1+a+a^2-\frac{1}{a}=0$

(dividing by a)

$\Rightarrow \quad \frac{1}{a}=1+a+a^2$

Now, $a^3+\frac{1}{a}=1-a-a^2+1+a+a^2$

$=2$

20. Given, $a+\frac{1}{b}=1$

$\Rightarrow \quad a=\frac{b-1}{b}$

$\Rightarrow \quad \frac{1}{a}=\frac{b}{b-1}$

Also, $\quad b+\frac{1}{c}=1$

$\Rightarrow \quad \frac{1}{c}=1-b$

$\Rightarrow \quad c=\frac{1}{1-b}$

Now, $c+\frac{1}{a}=\frac{1}{1-b}+\frac{b}{b-1}=\frac{b-1}{b-1}=1$

21. Clearly, $(x^2-y)(x^4+x^2y+y^2)$

$=(x^2)^3-y^3$

Hence, the other factor is (x^2-y).

22. Given, $x-\frac{1}{x}=4$

$\Rightarrow x^2+\frac{1}{x^2}=4^2+2=18$

(observe the relevant chart)

Let $x+\frac{1}{x}=a$

Then, $x^2+\frac{1}{x^2}=a^2-2$

(observe the relevant chart)

Obviously, $a^2-2=18$

$\Rightarrow \quad a=\sqrt{20}=2\sqrt{5}$

23. Adding all three equations 'we get

$15x+15y+15z=45$

$\Rightarrow \quad x+y+z=3$

$\therefore x+y+z-34=3-34=-31$

24. Given, $\frac{a^2-ab+b^2}{a^2+ab+b^2}=\frac{1}{3}$

$\Rightarrow \quad 3a^2-3ab+3b^2=a^2+ab+b^2$

$\Rightarrow \quad 2a^2-4ab+2b^2=0$

$\Rightarrow \quad a^2-2ab+b^2=0$

$\Rightarrow \quad (a-b)^2=0$

$\Rightarrow \quad a-b=0$

$\Rightarrow \quad a=b$

$\therefore \quad \frac{a}{b}=1$

25. We have,
$(x+y+z)^2 = x^2+y^2+z^2+2$
$(xy+yz+zx)$
$\Rightarrow 1^2 = x^2+y^2+z^2+2(xy+yz+zx)$
$\Rightarrow x^2+y^2+z^2 = 1-2(xy+yz+zx)$
Now,
$x^3+y^3+z^3-3xyz = (x+y+z)$
$[x^2+y^2+z^2-xy-yz-zx]$
$\Rightarrow 13-3(1) = 1[1-2(xy+yz+zx)$
$-(xy+yz+zx)]$
$\Rightarrow \quad 10 = 1-3(xy+yz+zx)$
$\Rightarrow xy+yz+zx = -3$

26. Given, $\frac{a+b}{c} = \frac{6}{5}$
$\Rightarrow \quad 5a+5b = 6c \quad \ldots(i)$
Also, $\frac{b+c}{a} = \frac{9}{2}$
$\Rightarrow \quad 2c = 9a-2b \quad \ldots(ii)$
From Eqs. (i), $6c = 5a+5b$
$\Rightarrow \quad 3\times 2c = 5a+5b$
$\Rightarrow \quad 3(9a-2b) = 5a+5b$
$\Rightarrow \quad 22a = 11b \Rightarrow b = 2a$
Putting the value of b in Eqs. (ii), we get $2c = 9a-4a = 5a$
$\Rightarrow \quad c = 2.5a$
Now, $\frac{a+c}{b} = \frac{a+2.5a}{2a} = \frac{7}{4}$

27. No need to solve through conventionally so as to save the time.
Let $a=1$, $b=1$ (as $a^2+b^2=2$), $c=1$ and $d=0$ (as $c^2+d^2=1$). [You are free to assume other values].
Then, $(ad-bc)^2+(ac+bd)^2$
$= (1\cdot 0-1\cdot 1)^2+(1\cdot 1+1\cdot 0)^2 = 2$

28. Given, $x^4-y^4 = 544$
$\Rightarrow \quad (x^2+y^2)(x^2-y^2) = 544$
$\Rightarrow \quad (x^2-y^2) = 544\div 34 = 16 \quad \ldots(i)$
and $\quad x^2+y^2 = 34 \quad \ldots(ii)$
On adding Eqs. (i) and (ii), we get
$2x^2 = 50$
$\Rightarrow \quad x^2 = 25 \Rightarrow x = \pm 5$
Putting the value of x in Eqs. (i), we get $25-y^2 = 16$
$\Rightarrow \quad y^2 = 9$
$\Rightarrow \quad y = \pm 3$
Hence, option (b) is correct.

29. Given, $x = 3+2\sqrt{2}$
They, $y = \frac{1}{3+2\sqrt{2}} = \frac{1}{x}$
Then, $x+\frac{1}{x} = 2\cdot 3 = 6$
and $x-\frac{1}{x} = 2\cdot 2\sqrt{2} = 4\sqrt{2}$
Now, $\frac{x^2+3xy+y^2}{x^2-3xy+y^2} = \frac{(x+y)^2+xy}{(x-y)^2-xy}$
$= \frac{\left(x+\frac{1}{x}\right)^2+xy}{\left(x+\frac{1}{x}\right)^2-xy} = \frac{(6)^2+1}{(4\sqrt{2})^2-1} = \frac{37}{31}$

30. $(a+1)^4-a^4 = [(a+1)^2]^2-(a^2)^2$
$= (a^2+2a+1+a^2)(a^2+2a+1-a^2)$
$= (2a^2+2a+1)(2a+1)$
Hence, $(a+1)^4-a^4$ is divisible by $(2a^2+2a+1)$

31. Given, $x^3+y^3+z^3 = 3(1+xyz)$
$\Rightarrow x^3+y^3+z^3-3xyz = 3$
Now, $P+Q+R = y+z-x+z$
$+x-y+x+y-z$
$= x+y+z$
We have, $P^3+Q^3+R^3-3PQR$
$= \frac{1}{2}(P+Q+R)[(P-Q)^2+(Q-R)^2$
$+(R-P)^2]$
$= \frac{1}{2}(x+y+z)[(2y-2x)^2$
$+(2z-2y)^2+(2x-2z)^2]$
$= 2(x+y+z)\times 2[$
$x^2+y^2+z^2-xy-yz-zx]$
$= 4[x^3+y^3+z^3-3xyz]$
$= 4\times 3 = 12$

32. Given, $\frac{x}{b+c} = \frac{y}{c+a} = \frac{z}{a+b} = k$ (let)
$\Rightarrow \frac{x}{b+c} = \frac{y}{c+a} = k$
$\Rightarrow x = bk+ck$ and $y = ck+ak$
Then, $\frac{x-y}{b-a} = \frac{bk+ck-ck-ak}{b-a} = k$
Similarly, $\frac{y-z}{c-b} = \frac{z-x}{a-c} = k$
Hence, option (a) is correct.

33. On adding all three equations, we get
$(x+y+z)(x+y+z) = 169$
$\Rightarrow \quad (x+y+z) = 13$
As given, $x(x+y+z) = 9$
$\Rightarrow \quad x\times 13 = 9$
$\Rightarrow \quad x = \frac{9}{13}$

34. $\frac{x^2}{y^2}+tx+\frac{y^2}{4} = \left(\frac{x}{y}\right)^2+tx+\left(\frac{y}{2}\right)^2$
Now,
$\left(\frac{x}{y}+\frac{y}{2}\right)^2 = \left(\frac{x}{y}\right)^2+2\cdot\frac{x}{y}\cdot\frac{y}{2}+\left(\frac{y}{2}\right)^2$
$= \left(\frac{x}{y}\right)^2+1x+\left(\frac{y}{2}\right)^2$
and
$\left(\frac{x}{y}-\frac{y}{2}\right)^2 = \left(\frac{x}{y}\right)^2-2\cdot\frac{x}{y}\cdot\frac{y}{2}+\left(\frac{y}{2}\right)^2$
$= \left(\frac{x}{y}\right)^2-1x+\left(\frac{y}{2}\right)^2$
Clearly, $t = \pm 1$

35. We know, AM $\geq$ GM
Then, $\frac{3^x+3^{-x}}{2} \geq \sqrt{3^x\times 3^{-x}}$
$\Rightarrow 3^x+3^{-x} \geq 2$
Hence, option (a) is correct.

36. Let $a=1$, $b=2$ and $c=3$
(You are free to assume other values)
Then, $\frac{x-1^2}{2+3}+\frac{x-2^2}{3+1}+\frac{x-3^2}{1+2} =$
$4(1+2+3)$
$\Rightarrow$
$\frac{12x-12+15x-60+20x-180}{60} = 24$
$\Rightarrow 47x-252 = 1440$
$\Rightarrow x = 36 = (1+2+3)^2$
$= (a+b+c)^2$
Hence, option (a) is correct.

37. Proceed through options.
(a) $(x-1)yz = xyz-yz$
(b) $(y-1)zx = xyz-zx$
(c) $(z-1)xy = xyz-xy$
(d) $(y+1)zx = xyz+zx$
Obviously, yz will be the smallest among zx, xy and yz as $x>y>z$.
Hence, option (a) is correct.

Alternatively

You may assume the values of x, y, z and then proceed through options.

38. Given, $\frac{b-c}{a}+\frac{a+c}{b}+\frac{a-b}{c} = 1$
$\Rightarrow \quad \frac{b-c}{a}+\frac{a-b}{c} = 1-\frac{a+c}{b}$
$\Rightarrow \frac{bc-c^2+a^2-ab}{ac} = \frac{b-a-c}{b}$
$\Rightarrow \frac{bc-ab+(a-c)(a+c)}{ac} = \frac{b-a-c}{b}$
$\Rightarrow \frac{b(c-a)+(a-c)(a+c)}{ac} = \frac{b-a-c}{b}$
$\Rightarrow \frac{(c-a)(b-a-c)}{ac} = \frac{b-a+c}{b}$
$\Rightarrow \quad \frac{(c-a)}{ac} = \frac{1}{b}$
$\Rightarrow \quad \frac{1}{a}-\frac{1}{c} = \frac{1}{b}$

39. Let $(p + q) = x$ and $(r + s) = y$.

Then, $x + y = 2$ and $m = xy$

We know, AM $\geq$ GM

$\Rightarrow \quad \dfrac{x + y}{2} \geq \sqrt{xy}$

$\Rightarrow \quad \dfrac{2}{2} \geq \sqrt{xy}$

$\Rightarrow 1 \geq xy$ (suqaring both sides)

Since p, q, r and s are positive real numbers, xy will be greater than 0.

Hence, $0 < m \leq 1$.

40. Given, $a - b = \dfrac{d}{c}$

$\Rightarrow \quad \dfrac{c}{d} = \dfrac{1}{a - b}$

$\Rightarrow \dfrac{c + d}{c - d} = \dfrac{1 + a - b}{1 - a + b}$ (by componendo and dividendo)

$\Rightarrow \quad \dfrac{1}{c - d} = \dfrac{a + b + a - b}{a + b - a + b}$

$\Rightarrow \quad \dfrac{1}{c - d} = \dfrac{a}{b}$

$\Rightarrow \quad c - d = \dfrac{b}{a}$

Now, $c^2 - d^2 = (c + d)(c - d)$

$= 1 \times \dfrac{b}{a} = \dfrac{b}{a}$

Alternatively

Given, $a - b = \dfrac{d}{c}$

$\Rightarrow \quad a - b = \dfrac{1 - c}{c}$

$\Rightarrow \quad a - b = \dfrac{1}{c} - 1$

Let $a = \dfrac{3}{4}$ and $b = \dfrac{1}{4}$ (as $a + b = 1$. You are free to assume other values), then $c = \dfrac{2}{3}$.

$\because \quad c + d = 1$

$\therefore \quad d = 1 - \dfrac{2}{3} = \dfrac{1}{3}$

Now, $c^2 - d^2 = (c + d)(c - d)$

$= 1 \times \left(\dfrac{2}{3} - \dfrac{1}{3}\right) = \dfrac{1}{3} = \dfrac{b}{a}$

41. $x^4 + 1 = 0 \Rightarrow x^4 = -1$

Given expression is $(x^4)^{10} + 2$.

$\therefore$ Remainder $= (-1)^{10} + 2 = 3$

42. Let $a = 1$ and $b = 0$ as $a + b = 1$.

Now, $a^4 + b^4 - a^3 - b^3 - 2a^2 b^2 + ab$

$= 1^4 + 0^4 - 1^3 - 0^3 - 2 \cdot 1^2 \cdot 0^2 + 1 \cdot 0$

$= 0$

Note *No need to solve conventionally.*

43. $x^3 + y^3 + 21xy = x^3 + y^3 + 3xy(x + y)$ $\quad (\because x + y = 7)$

$= (x + y)^3 = 7^3 = 343$

44. We know that $x^n + a^n$ is exactly divisible by $(x + a)$, when n is odd. Since n is odd here, the given expression will be divisible by $(x + y)$.

Also, $x^n + a^n$ is never divisible by $(x - a)$. It means $x^{21} + y^{21}$ will never be divisible by $(x - y)$. Obviously, $x^{21} + y^{21}$ will not be divisible by $(x + y)(x - y)$ i.e. by $x^2 - y^2$.

Hence, option (a) is correct.

Alternatively

Proceed through options.

(a) $\Rightarrow x + y = 0 \Rightarrow x = -y$, then remainder $= (-y)^{21} + y^{21} = 0$

(b) $\Rightarrow x - y = 0 \Rightarrow x = y$, then remainder $= y^{21} + y^{21}$

(c) $\Rightarrow x^2 - y^2 = 0 \Rightarrow x = y$, then remainder $= y^{21} + y^{21}$

Since first option yields the remainder as 0, $(x + y)$ is a factor of $x^{21} + y^{21}$.

45. Given, $x^2 + y^2 + 6x + 5 = 4(x - y)$

$\Rightarrow x^2 + y^2 + 6x - 4x + 5 + 4y = 0$

$\Rightarrow (x^2 + 2x + 1) + (y^2 + 2 \cdot 2y + 4) = 0$

$\Rightarrow (x + 1)^2 + (y + 2)^2 = 0$

$\Rightarrow (x + 1)^2 = 0$ and $(y + 2)^2 = 0$

$\Rightarrow \quad x = -1$ and $y = -2$

Then, $x - y = -1 + 2 = 1$

46. Given, $x^2 - 3x + 1 - 0$

$\Rightarrow \quad x^2 + 1 = 3x$

$\Rightarrow \quad \dfrac{x^2}{x} + \dfrac{1}{x} = \dfrac{3x}{x} \Rightarrow x + \dfrac{1}{x} = 3$

Then, $x^3 + \dfrac{1}{x^3} = 3^3 - 3 \cdot 3 = 18$

47. Let $x - \dfrac{1}{x} = a$

Then, $x^2 + \dfrac{1}{x^2} = a^2 + 2$

and $x^4 + \dfrac{1}{x^4} = (a^2 + 2)^2 - 2$

$= 322$

$\Rightarrow \quad a^2 + 2 = \sqrt{324} = 18$

$\Rightarrow \quad a^2 = 16$

$\Rightarrow \quad a = 4$

48. Given, $x - \dfrac{1}{x} = \sqrt{3}$

Then, $x^2 + \dfrac{1}{x^2} = (\sqrt{3})^2 + 2 = 5$

Now, $x^6 + \dfrac{1}{x^6} = 5^3 - 3 \cdot 5 = 110$

49. Let $x = 1$, $y = 8$ and $z = 27$ as $x^{\frac{1}{3}} + y^{\frac{1}{3}} = z^{\frac{1}{3}}$

Then, $(x + y + z)^3 + 27xyz$

$= (1 + 8 - 27)^3 + 27 \cdot 1 \cdot 8 \cdot 27$

$= 0$

50. $x^4 + 2x^3 + ax^2 + bx + 9$

$= (x^2)^2 + x^2 + 3^2 + 2x^2 \cdot x + bx + (a - 1)x^2$

Clearly, at $(a - 1) = 2 \cdot 3$ and $b = 2 \cdot 3$, the expression will be $(x^2 + x + 3)^2$, i.e. a perfect square.

Hence, $a - 1 = 6$ and $b = 6$

$\Rightarrow \quad a = 7$ and $b = 6$

51. $5x^3 - 4 = 5 \times$ (linear coefficient)

$-4 = 5(-1) - 4 = -9$

52. Given expression

$= \dfrac{(1 - d) + (1 - a) + (1 - b) + (1 - c)}{(1 - a)(1 - b)(1 - c)(1 - d)}$

$= \dfrac{4 - (a + b + c + d)}{(1 - a)(1 - b)(1 - c)(1 - d)}$

$= \dfrac{4 - 4}{(1 - a)(1 - b)(1 - c)(1 - d)} = 0$

53. $(x^4 - x^2 y^2 - 1) + (y^4 - x^2 y^2 + 1)$

$= x^2(x^2 - y^2) - 1 + y^2(y^2 - x^2) + 1$

Now, $x^2 - y^2 = (x - y)(x + y)$

$= 2a^{-\frac{1}{2}} \times 2a^{\frac{1}{2}} = 4$

$\therefore x^2(x^2 - y^2) - 1 + y^2(y^2 - x^2) + 1$

$= x^2(4) - 1 + y^2(-4) + 1$

$= 4(x^2 - y^2)$

$= 4 \cdot 4 = 16$

54. Given, $a^2 + b^2 + c^2 = 2a - 2b - 2$

$\Rightarrow a^2 - 2a + 1 + b^2 + 2b + 1 + c^2 = 0$

$\Rightarrow (a - 1)^2 + (b + 1)^2 + c^2 = 0$

$\Rightarrow (a - 1)^2 = 0$, $(b + 1)^2 = 0$ and $c^2 = 0$

$\Rightarrow a = 1$, $b = -1$ and $c = 0$

Then,

$3a - 2b + c = 3 \cdot 1 - 2 \cdot (-1) + 0 = 5$

55. Given, $\dfrac{1}{x + y} = \dfrac{1}{x} + \dfrac{1}{y}$

$\Rightarrow \quad \dfrac{1}{x + y} = \dfrac{y + x}{xy}$

$\Rightarrow (x + y)^2 = xy \Rightarrow x^2 + y^2 + xy = 0$

Now, $x^3 - y^3 = (x - y)(x^2 + y^2 + xy)$

$= (x - y) \times 0 = 0$

56. No need to solve conventionally.

Let $a = 4$, then $2x = \dfrac{5}{2}$

$\Rightarrow \quad x = \dfrac{5}{4}$

Now, $\dfrac{\sqrt{x^2-1}}{x-\sqrt{x^2-1}} = \dfrac{\sqrt{\left(\frac{5}{4}\right)^2-1}}{\frac{5}{4}-\sqrt{\left(\frac{5}{4}\right)^2-1}}$

$= \dfrac{\frac{3}{4}}{\frac{1}{2}} = \dfrac{3}{2} = \dfrac{1}{2}(4-1) = \dfrac{1}{2}(a-1)$

57. $x = \dfrac{2\times\sqrt{2}\times\sqrt{3}}{\sqrt{3}+\sqrt{2}}$

$\Rightarrow \dfrac{x}{\sqrt{2}} = \dfrac{2\sqrt{3}}{\sqrt{3}+\sqrt{2}}$

$\Rightarrow \dfrac{x+\sqrt{2}}{x-\sqrt{2}} = \dfrac{2\sqrt{3}+\sqrt{3}+\sqrt{2}}{2\sqrt{3}-\sqrt{3}-\sqrt{2}}$

$= \dfrac{3\sqrt{3}+\sqrt{2}}{\sqrt{3}-\sqrt{2}}$ ...(i)

(by Componendo and Dividendo)

Further, $x = \dfrac{2\times\sqrt{2}\times\sqrt{3}}{\sqrt{3}+\sqrt{2}}$

$\Rightarrow \dfrac{x}{\sqrt{3}} = \dfrac{2\sqrt{2}}{\sqrt{3}+\sqrt{2}}$

$\Rightarrow \dfrac{x+\sqrt{3}}{x-\sqrt{3}} = \dfrac{2\sqrt{2}+\sqrt{3}+\sqrt{2}}{2\sqrt{2}-\sqrt{3}-\sqrt{2}}$

$= \dfrac{3\sqrt{2}+\sqrt{3}}{\sqrt{2}-\sqrt{3}}$ ...(ii)

(by Componendo and Dividendo)

On adding Eqs. (i) and (ii), we get

$\dfrac{x+\sqrt{2}}{x-\sqrt{2}} + \dfrac{x+\sqrt{3}}{x-\sqrt{3}} = \dfrac{3\sqrt{3}+\sqrt{2}}{\sqrt{3}-\sqrt{2}} + \dfrac{3\sqrt{2}+\sqrt{3}}{-(\sqrt{3}-\sqrt{2})}$

$= \dfrac{3\sqrt{3}+\sqrt{2}-3\sqrt{2}-\sqrt{3}}{\sqrt{3}-\sqrt{2}}$

$= \dfrac{2\sqrt{3}-2\sqrt{2}}{\sqrt{3}-\sqrt{2}} = \dfrac{2(\sqrt{3}-\sqrt{2})}{(\sqrt{3}-\sqrt{2})} = 2$

58. Given, $\dfrac{a+b-c}{a+b} = \dfrac{b+c-a}{b+c}$

$\Rightarrow \dfrac{a+b-c+a+b}{a+b-c-a-b} = \dfrac{b+c-a+b+c}{b+c-a-b-c}$

$= \dfrac{c+a-b+c+a}{c+a-b-c-a}$

(by Componendo and Dividendo)

$\Rightarrow \dfrac{2a+2b-c}{-c} = \dfrac{2b+2c-a}{-a}$

$= \dfrac{2c+2a-b}{-b}$

$\Rightarrow \dfrac{2a+2b-c}{-c} - 3 = \dfrac{2b+2c-a}{-a} - 3$

$= \dfrac{2c+2a-b}{-b} - 3$

$\Rightarrow \dfrac{2(a+b+c)}{-c} = \dfrac{2(a+b+c)}{-a}$

$= \dfrac{2(a+b+c)}{-b}$

$\Rightarrow \dfrac{1}{c} = \dfrac{1}{a} = \dfrac{1}{b}$

$\Rightarrow a = b = c$

59. Given, $(a^2+b^2)^3 = (a^3+b^3)^2$

$\Rightarrow a^6 + b^6 + 3a^2b^2(a^2+b^2)$

$= a^6 + b^6 + 2a^3b^3$

$\Rightarrow 3a^2b^2(a^2+b^2) = 2a^3b^3$

$\Rightarrow (a^2+b^2) = \dfrac{2ab}{3}$

$\Rightarrow \dfrac{a^2+b^2}{ab} = \dfrac{2}{3}$

$\Rightarrow \dfrac{a}{b} + \dfrac{b}{a} = \dfrac{2}{3}$

60. Given, $x = \dfrac{a-b}{a+b} \Rightarrow \dfrac{1}{x} = \dfrac{a+b}{a-b}$

$\Rightarrow \dfrac{1+x}{1-x} = \dfrac{a+b+a-b}{a+b-a+b}$

(by Componendo and Dividendo)

$= \dfrac{1+x}{1-x} = \dfrac{a}{b}$

Similarly, $\dfrac{1+y}{1-y} = \dfrac{b}{c}$

and $\dfrac{1+z}{1-z} = \dfrac{c}{a}$

Then, $\dfrac{1+x}{1-x}\times\dfrac{1+y}{1-y}\times\dfrac{1+z}{1-z}$

$= \dfrac{a}{b}\times\dfrac{b}{c}\times\dfrac{c}{a} = 1$

$\Rightarrow \dfrac{(1-x)(1-y)(1-z)}{(1+x)(1+y)(1+z)} = 1$

Alternatively

Solve by assuming values of a, b and c.

61. Given, $\dfrac{a}{b} + \dfrac{b}{a} = 1$

$\Rightarrow \dfrac{a^2+b^2}{ab} = 1$

$\Rightarrow a^2 + b^2 - ab = 0$

Now, $a^3 + b^3 = (a+b)(a^2 - ab + b^2)$

$= (a+b)\times 0 = 0$

62. Given, $x^2 + 1 = 2x$

$\Rightarrow x^2 - 2x + 1 = 0$

$\Rightarrow (x-1)^2 = 0$

$\Rightarrow x - 1 = 0$

$\Rightarrow x = 1$

Now, $\dfrac{x^4 + \frac{1}{x^2}}{x^2 - 3x + 1} = \dfrac{1^4 + \frac{1}{1^2}}{1^2 - 3\cdot 1 + 1} = -2$

63. Given, $\dfrac{x}{x^2-2x+1} = \dfrac{1}{3}$

$\Rightarrow x^2 - 2x + 1 = 3x \Rightarrow x^2 + 1 = 5x$

$\Rightarrow x + \dfrac{1}{x} = 5$

$\Rightarrow x^3 + \dfrac{1}{x^3} = 5^3 - 3\cdot 5 = 110$

[from the chart]

64. Given, $x^2 + x = 5$

$\Rightarrow x + 1 = \dfrac{5}{x} \Rightarrow x + 3 = \dfrac{5}{x} + 2$

$\Rightarrow x + 3 + \dfrac{1}{x+3} = \dfrac{5}{x} + 2 + \dfrac{1}{x+3}$

$\Rightarrow x + 3 + \dfrac{1}{x+3} = \dfrac{5x+15+x}{x(x+3)} + 2$

$= \dfrac{5x+15+x}{x^2+3x} + 2$

$= \dfrac{5x+15+x}{5-x+3x} + 2$

$(\because x^2 + x = 5 \therefore x^2 = 5 - x)$

$= \dfrac{6x+15}{2x+5} + 2 = 3 + 2 = 5$

Then,

$(x+3)^3 + \dfrac{1}{(x+3)^3} = (5)^3 - 3\cdot 5 = 110$

65. Given, $x + y + z = 6$

$\Rightarrow x^2 + y^2 + z^2 + 2xy + 2yz + 2zx = 36$

$\Rightarrow 20 + 2(xy + yz + zx) = 36$

$\Rightarrow xy + yz + zx = 8$

Now, $x^3 + y^3 + z^3 - 3xyz$

$= (x+y+z)[x^2 + y^2 + z^2 - (xy + yz + zx)]$

$= 6(20 - 8) = 72$

66. Given, $\dfrac{x^{24}+1}{x^{12}} = 7$

$\Rightarrow x^{12} + \dfrac{1}{x^{12}} = 7$

Then, $(x^{12})^3 + \dfrac{1}{(x^{12})^3} = 7^3 - 3\cdot 7$

$\Rightarrow \dfrac{x^{72}+1}{x^{36}} = 322$

67. Given, $(x-a)(x-b) = 1$

$\Rightarrow (x-a) = \dfrac{1}{x-b}$

Now, $(x-a)^3 - \dfrac{1}{(x-a)^3}$

$= \left((x-a) - \dfrac{1}{(x-a)}\right)^3$

$+ 3(x-a)\dfrac{1}{(x-a)}\left((x-a) - \dfrac{1}{(x-a)}\right)$

$= [(x-a)-(x-b)]^3$

$+ 3[(x-a)-(x-b)]$

$[\because (x-a)(x-b) = 1]$

$= [-(a-b)]^3 + 3[-(a-b)]$

$[\because a - b + 5 = 0 \therefore a - b = -5]$

$= (5)^3 + 3(5) = 140$

68. Given, $\sqrt{\frac{a}{b}} - 13 = -\sqrt{\frac{b}{a}} - 11$

$\Rightarrow \sqrt{\frac{a}{b}} + \sqrt{\frac{b}{a}} = 2$

$\Rightarrow \frac{a}{b} + \frac{b}{a} + 2 = 4$

(squaring both sides)

$\Rightarrow \frac{a^2+b^2}{ab} = 2$

$\Rightarrow \frac{(a+b)^2 - 2ab}{ab} = 2$

$\Rightarrow \frac{(10)^2 - 2ab}{ab} = 2$

$\Rightarrow \frac{100}{ab} - 2 = 2$

$\Rightarrow ab = 25$

Hence, $a + b = 10$ and $ab = 25$

It implies, $a = b = 5$

$[\because (a-b)^2 = (a+b)^2 - 4ab = 100 - 4 \cdot 25 = 0 \therefore a - b = 0]$

Now, $3ab + 4a^2 + 5b^2$

$= 3 \cdot 5 \cdot 5 + 4 \cdot 5^2 + 5 \cdot 5^2 = 300$

69. Let $a = 1$, $b = 2$ and $c = 3$ as $a + b - c = 0$

[you are free to assume any value of a, b, c as per your convenience]

Now, $2b^2c^2 + 2c^2a^2 + 2a^2b^2 - a^4 - b^4 - c^4$

$= 2 \cdot 2^2 \cdot 3^2 + 2 \cdot 3^2 \cdot 1^2 + 2 \cdot 1^2 \cdot 2^2 - 1^4 - 2^4 - 3^4 = 0$

70. Given, $x = \sqrt[3]{2+\sqrt{3}}$

$\Rightarrow x^3 = 2 + \sqrt{3}$

$\Rightarrow \frac{1}{x^3} = 2 - \sqrt{3}$ $[\because (2)^2 - (\sqrt{3})^2 = 1]$

Then, $x^3 + \frac{1}{x^3} = 2 + \sqrt{3} + 2 - \sqrt{3}$

$= 4$

71. As given,

$a^2 + b^2 + c^2 = ab + bc + ca$

$\Rightarrow a = b = c$

Then, $\frac{a+c}{b} = \frac{b+b}{b} = 2$

72. Given, $x^3 + \frac{3}{x} = 4(a^3 + b^3)$ and

$3x + \frac{1}{x^3} = 4(a^3 - b^3)$

On adding, we get

$x^3 + \frac{3}{x} + 3x + \frac{1}{x^3} = 8a^3$

$\Rightarrow \left(x + \frac{1}{x}\right)^3 = (2a)^3$

$\Rightarrow x + \frac{1}{x} = 2a$

$\Rightarrow a = \frac{1}{2}\left(x + \frac{1}{x}\right)$

On subtracting, we get

$x^3 + \frac{3}{x} - 3x - \frac{1}{x^3} = 8b^3$

$\Rightarrow \left(x - \frac{1}{x}\right)^3 = (2b)^3$

$\Rightarrow x - \frac{1}{x} = 2b$

$\Rightarrow b = \frac{1}{2}\left(x - \frac{1}{x}\right)$

Now, $a^2 - b^2 = (a+b)(a-b)$

$= \left[\frac{1}{2}\left(x + \frac{1}{x}\right) + \frac{1}{2}\left(x - \frac{1}{x}\right)\right]\left[\frac{1}{2}\left(x + \frac{1}{x}\right) - \frac{1}{2}\left(x - \frac{1}{x}\right)\right]$

$= x \cdot \frac{1}{x} = 1$

73. Let $\frac{p}{a}, \frac{q}{b}$ and $\frac{r}{c}$ be x, y and z respectively.

As given, $x + y + z = 1$

and $\frac{1}{x} + \frac{1}{y} + \frac{1}{z} = 0$

$\Rightarrow \frac{xy + yz + zx}{xyz} = 0$

$\Rightarrow xy + yz + zx = 0$

Now, $\frac{p^2}{a^2} + \frac{q^2}{b^2} + \frac{r^2}{c^2} = x^2 + y^2 + z^2$

$= (x+y+z)^2 - 2(xy + yz + zx)$

$= (1)^2 - 2(0) = 1$

74. Let $a = -1$, $b = 2$ and $c = 2$

$(\because ab + bc + ca = 0)$

(you are free to take any value of a, b, c satisfying $ab + bc + ca = 0$, like (–6, 10, 15). However, the value of denominator of any term (if exists) at the assumed values must not be equal to 0, otherwise, the term will be undefined).

Now, $\frac{a^2}{a^2 - bc} + \frac{b^2}{b^2 - ca} + \frac{c^2}{c^2 - ab}$

$= \frac{(-1)^2}{(-1)^2 - 2 \cdot 2} + \frac{2^2}{2^2 - 2(-1)} + \frac{2^2}{2^2 - (-1)2} = 1$

75. Given, $a + \frac{1}{a} = \sqrt{3}$

$\Rightarrow a^3 + \frac{1}{a^3} = 3\sqrt{3} - 3\sqrt{3} = 0$

(recall the chart)

$\Rightarrow a^3 = \frac{-1}{a^3} \Rightarrow a^6 = -1$

Now, $a^6 - \frac{1}{a^6} + 2 = (-1) - \frac{1}{(-1)} + 2$

$= 2$

Alternatively

Given, $a + \frac{1}{a} = \sqrt{3}$

$\Rightarrow a^3 + \frac{1}{a^3} = 3\sqrt{3} - 3\sqrt{3} = 0$ (recall the chart)

Now, $a^6 - \frac{1}{a^6} + 2 = (a^3)^2 - \left(\frac{1}{a^3}\right)^2 + 2$

$= \left(a^3 + \frac{1}{a^3}\right)\left(a^3 - \frac{1}{a^3}\right) + 2$

$= 0\left(a^3 - \frac{1}{a^3}\right) + 2 = 2$

76. Given, $\frac{xy}{x+y} = a$, $\frac{xz}{x+z} = b$

and $\frac{yz}{y+z} = c$

$\Rightarrow \frac{x+y}{xy} = \frac{1}{a}$, $\frac{x+z}{xz} = \frac{1}{b}$

and $\frac{y+z}{yz} = \frac{1}{c}$

$\Rightarrow \frac{1}{y} + \frac{1}{x} = \frac{1}{a}$, $\frac{1}{z} + \frac{1}{x} = \frac{1}{b}$

and $\frac{1}{z} + \frac{1}{y} = \frac{1}{c}$

On adding, we get $2\left(\frac{1}{x} + \frac{1}{y} + \frac{1}{z}\right)$

$= \frac{1}{a} + \frac{1}{b} + \frac{1}{c}$

$\Rightarrow \frac{1}{x} + \frac{1}{y} + \frac{1}{z} = \frac{bc + ac + ab}{2abc}$

$\Rightarrow \frac{1}{x} = \frac{bc + ac + ab}{2abc} - \frac{1}{c}$

$\left(\because \frac{1}{y} + \frac{1}{z} = \frac{1}{c}\right)$

$\Rightarrow \frac{1}{x} = \frac{bc + ac + ab - 2ab}{2abc}$

$\Rightarrow x = \frac{2abc}{ac + bc - ab}$

77. $a^3 + b^3 + c^3 - 3abc = (a + b + c)(a^2 + b^2 + c^2 - ab - bc - ca)$

$= (2.234 + 3.121 - 5.355)(a^2 + b^2 + c^2 - ab - bc - ca)$

$= 0 \times (a^2 + b^2 + c^2 - ab - bc - ca) = 0$

78. Given, $a^3 + 3a^2 + 9a = 1$

$\Rightarrow \quad a^2 + 3a + 9 = \frac{1}{a}$

($\because$ on dividing by a)

$\Rightarrow 3(a^2 + 3a + 9) = \frac{3}{a}$

Now, $a^3 + \frac{3}{a} = a^3 + 3(a^2 + 3a + 9)$

$= (a^3 + 3a^2 + 9a) + 27$

$= 1 + 27 = 28$

79. Given, $\frac{x}{a} = \frac{y}{b} = \frac{z}{c} = k$ (let)

$\Rightarrow \quad x = ak,\ y = bk$ and $z = ck$

We have, $(x + y + z)^2$

$= x^2 + y^2 + z^2 + 2(xy + yz + zx)$

$\Rightarrow (ak + bk + ck)^2 = x^2 + y^2 + z^2 + 2(xy + yz + zx)$

$\Rightarrow k^2(a + b + c)^2 - x^2 + y^2 + z^2 = 2(xy + yz + zx)$

$\Rightarrow \frac{x^2}{a^2}(a + b + c)^2 - x^2 + y^2 + z^2$

$= 2(xy + yz + zx) \quad \left(\because k = \frac{x}{a}\right)$

$\Rightarrow \quad (xy + yz + zx)$

$= \frac{x^2(a + b + c)^2 - a^2(x^2 + y^2 + z^2)}{2a^2}$

80. Given, $x = \sqrt{5} + \sqrt{3}$

$\Rightarrow \quad x^2 = 5 + 3 + 2\sqrt{15} = 8 + 2\sqrt{15}$

and $\quad y^2 = 8 - 2\sqrt{15}$

Now, $(x^4 - y^4) = (x^2 - y^2)(x^2 + y^2)$

$= (8 + 2\sqrt{15} - 8 + 2\sqrt{15})(8 + 2\sqrt{15} + 8 - 2\sqrt{15})$

$= 4\sqrt{15}(16) = 64\sqrt{15}$

Note *You may also write $(x^2 - y^2)$ as $(x + y)(x - y)$ and find its value*

81. (**b**) We know, if $(a + b + c + d)$ is constant, then value of $a^2 + b^2 + c^2 + d^2$ is minimum at $a = b = c = d$.

Given, $a + b + c + d = 4m + 1$

Then, $a = b = c = d = m + 0.25$, but $m + 0.25$ is not an integer.

Thus, we should take their values in integers as close as possible.

Such values are

$m + 0, m + 0, m + 0, m + 1.$

Now, minimum value of

$a^2 + b^2 + c^2 + d^2$

$= (m + 0)^2 + (m + 0)^2 + (m + 0)^2 + (m + 1)^2$

$= 4m^2 + 2m + 1$

Hence, option (b) is correct.

82. Given, $(a + 3)^2 : b^2 = 9 : 1$

$\Rightarrow \quad (a + 3) : b = \pm 3$

$\Rightarrow \quad a + 3 = 3b$ and $a + 3 = -3b$

Also given, $(a - 1)^2 : (b - 1)^2 = 4 : 1$

$\Rightarrow \quad (a - 1) : (b - 1) = \pm 2$

$\Rightarrow a + 1 = 2b$ and $a - 3 = -2b$

Now, there may be four sets of two linear equations in two variables.

Set (i) $a + 3 = 3b$ and $a + 1 = 2b$, solution $\Rightarrow a = 3, b = 2$

Set (ii) $a + 3 = 3b$ and $a - 3 = -2b$, solution $\Rightarrow a = \frac{3}{5}, b = \frac{6}{5}$

Set (iii) $a + 3 = -3b$ and $a + 1 = 2b$, solution $\Rightarrow a = -5, b = -2$

Set (iv) $a + 3 = -3b$ and $a - 3 = -2b$, solution $\Rightarrow a = 15, b = -6$

Clearly, solutions of set(iv) satisfy the conditions that a and b are integers of opposite signs.

Hence, $a^2 : b^2 = 15^2 : (-6)^2 = 25 : 4$

83. The value of the given expression will be minimum if each term has minimum value and it is possible only when values of a, b, c, d are maximum.

Clearly, maximum possible values of (a, b, c, d) be $(8, 8, 7, 7)$, values are interchangeable.

Then, minimum value of the expression

$= (8 - 8)^2 + (8 - 7)^2 + (8 - 7)^2 = 2$

84. Given, $\sqrt{\frac{x}{y}} = \frac{24}{5} + \sqrt{\frac{y}{x}}$

$\Rightarrow \quad \sqrt{\frac{x}{y}} - \sqrt{\frac{y}{x}} = \frac{24}{5}$

$\Rightarrow \quad \frac{x}{y} + \frac{y}{x} - 2 = \frac{576}{25}$

$\Rightarrow \quad \frac{x^2 + y^2}{xy} = \frac{626}{5}$

$\Rightarrow \quad \frac{(x + y)^2 - 2xy}{xy} = \frac{626}{25}$

$\Rightarrow \quad \frac{(26)^2 - 2xy}{xy} = \frac{626}{25}$

$\Rightarrow \quad \frac{676}{xy} = \frac{676}{25}$

$\Rightarrow \quad xy = 25$

Alternatively

You may proceed through options.

(a) $xy = 5$, not possible as $(x + y)$ will always be less than 26.

(b) $xy = 15$, not possible as $(x + y)$ will always be less than 26.

(c) $xy = 25$, possible as $(x + y = 26)$ is satisfied at $x = 25$ and $y = 1$.

Since at $x = 25$ and $b = 1$,

$\sqrt{\frac{x}{y}} = \frac{24}{5} + \sqrt{\frac{y}{x}}$, option (c) is correct.

85. Given, $PQ = 1$, then $Q = \frac{1}{P}$

$= \frac{1}{7 + 4\sqrt{3}} = 7 - 4\sqrt{3}$

Now, $\frac{1}{P^2} + \frac{1}{Q^2} = \frac{P^2 + Q^2}{P^2 Q^2}$

$= \frac{(P + Q)^2 - 2PQ}{P^2 Q^2}$

$= \frac{(7 + 4\sqrt{3} + 7 - 4\sqrt{3})^2 - 2PQ}{P^2 Q^2}$

$= \frac{196 - 2}{1} = 194$

86. Expression

$\frac{x^2(y + z) + y^2(x + z) + z^2(x + y)}{xyz}$

$= \frac{x(y + z)}{yz} + \frac{y(x + z)}{xz} + \frac{z(x + y)}{xy}$

$= \frac{x}{z} + \frac{x}{y} + \frac{y}{z} + \frac{y}{x} + \frac{z}{y} + \frac{z}{x}$

$= \left(\frac{x}{z} + \frac{z}{x}\right) + \left(\frac{x}{y} + \frac{y}{x}\right) + \left(\frac{y}{z} + \frac{z}{y}\right)$

$= \left(\frac{x^2 + z^2}{zx}\right) + \left(\frac{x^2 + y^2}{xy}\right) + \left(\frac{y^2 + z^2}{yz}\right)$

We know, AM > GM, if all the given numbers are not equal.

$\Rightarrow \quad \frac{x^2 + z^2}{2} > \sqrt{x^2 z^2} \Rightarrow \frac{x^2 + z^2}{zx} > 2$

Similarly, $\quad \frac{x^2 + y^2}{xy} > 2$

and $\quad \frac{y^2 + z^2}{yz} > 2$

Hence, value of the expression would be greater than 6.

Alternatively

Take any value of x, y and z as per your convenience, such that the denominator must not be zero and then solve.

87. Given, $x = \frac{\sqrt{a + b} - \sqrt{a - b}}{\sqrt{a + b} + \sqrt{a - b}}$

$\Rightarrow \quad x = \frac{\sqrt{a + b} - \sqrt{a - b}}{\sqrt{a + b} + \sqrt{a - b}} \times \frac{\sqrt{a + b} - \sqrt{a - b}}{\sqrt{a + b} - \sqrt{a - b}}$

$= \frac{a + b + a - b + 2\sqrt{a^2 - b^2}}{a + b - a + b}$

$$= \frac{a + \sqrt{a^2 - b^2}}{b}$$

$\Rightarrow \quad bx - a = \sqrt{a^2 - b^2}$

$\Rightarrow \quad b^2 x^2 + a^2 - 2abx = a^2 - b^2$ (squaring both sides)

$\Rightarrow \quad b^2 x^2 - 2abx + b^2 = 0$

$\Rightarrow \quad bx^2 - 2ax + b = 0$

Alternatively

Let $a = 5$ and $b = 4$ (you are free to assume any value)

Then, $x = \dfrac{\sqrt{5+4} - \sqrt{5-4}}{\sqrt{5+4} + \sqrt{5-4}} = \dfrac{1}{2}$

Now, $bx^2 - 2ax + b = 4 \times \dfrac{1}{4} - 2 \times 5 \times \dfrac{1}{2} + 4 = 0$

88. Let $x = 1$, $y = 2$ and $z = -3$ (you are free to take any value of x, y, z satisfying $x + y + z = 0$. However, the value of denominator of any term at the assumed values must not be equal to 0, otherwise, the term will be undefined.)

Now, $\dfrac{3y^2 + x^2 + z^2}{2y^2 - xz}$

$$= \frac{3(2)^2 + 1^2 + (-3)^2}{2(2)^2 - 1(-3)} = 2$$

89. We have,

$(x + y + z)^2 = x^2 + y^2 + z^2 + 2(xy + yz + zx)$

$\Rightarrow \quad (22)^2 = x^2 + y^2 + z^2 + 2(35)$

$\Rightarrow \quad x^2 + y^2 + z^2 = 414$

Now, $(x - y)^2 + (y - z)^2 + (z - x)^2$

$= 2[x^2 + y^2 + z^2 - (xy + yz + zx)]$

$= 2(414 - 35) = 758$

90. Given, expression :

$4(x - 2)^2 + 4(y - 3)^2 - 2(x - 3)^2$

$= 4(y - 3)^2 + 4x^2 - 16x + 16 - 2x^2 + 12x - 18$

$= 4(y - 3)^2 + 2x^2 - 4x - 2$

$= 4(y - 3)^2 + 2(x - 1)^2 - 4$

Now, the least value of $(y - 3)^2$ be 0 at $y = 3$ and also the least value of $(x - 1)^2$ be 0 at $x = 1$.

Hence, the least value of the expression $= 4 \cdot 0 + 2 \cdot 0 - 4 = -4$

91. Given, $a - \dfrac{1}{a} = b$, $b - \dfrac{1}{b} = c$ and $c - \dfrac{1}{c} = a$

$\Rightarrow a + b + c = a + b + c - \left(\dfrac{1}{a} + \dfrac{1}{b} + \dfrac{1}{c}\right)$

$\Rightarrow \quad \left(\dfrac{1}{a} + \dfrac{1}{b} + \dfrac{1}{c}\right) = 0$

$\Rightarrow \quad \left(\dfrac{1}{a} + \dfrac{1}{b} + \dfrac{1}{c}\right)^2 = 0$

$\Rightarrow \quad \dfrac{1}{a^2} + \dfrac{1}{b^2} + \dfrac{1}{c^2} + 2\left(\dfrac{1}{ab} + \dfrac{1}{bc} + \dfrac{1}{ca}\right) = 0$

$\Rightarrow \quad (b^2 - a^2 + 2) + (c^2 - b^2 + 2) + (a^2 - c^2 + 2) = -2\left(\dfrac{1}{ab} + \dfrac{1}{bc} + \dfrac{1}{ca}\right)$

$\left(\because a - \dfrac{1}{a} = b \therefore a^2 + \dfrac{1}{a^2} - 2 = b^2\right.$

$\Rightarrow \dfrac{1}{a^2} = b^2 - a^2 + 2$.

Similar process for $\left.\dfrac{1}{b^2} \text{ and } \dfrac{1}{c^2}\right)$

$\Rightarrow \quad 6 = -2\left(\dfrac{1}{ab} + \dfrac{1}{bc} + \dfrac{1}{ca}\right)$

$\Rightarrow \quad \dfrac{1}{ab} + \dfrac{1}{bc} + \dfrac{1}{ca} = -3$

92. $x + \dfrac{1}{x} =$ linear coefficient with opposite sign $= -3$

Then, $x^2 + \dfrac{1}{x^2} = (-3)^2 - 2 = 7$

$\Rightarrow \quad \left(x^2 + \dfrac{1}{x^2}\right)^3 = 343$

$\Rightarrow$

$x^6 + \dfrac{1}{x^6} + 3 \cdot x^2 \cdot \dfrac{1}{x^2}\left(x^2 + \dfrac{1}{x^2}\right) = 343$

$\Rightarrow x^6 + \dfrac{1}{x^6} + 3 \cdot x^2 \cdot \dfrac{1}{x^2}(7) = 343$

$\Rightarrow \quad x^6 + \dfrac{1}{x^6} = 322$

93. Given, $\dfrac{a}{b+c} = \dfrac{b}{c+a} = \dfrac{c}{a+b}$

$\Rightarrow$ If $\dfrac{a}{b+c} + 1 = \dfrac{b}{c+a} + 1 = \dfrac{c}{a+b} + 1$

$\Rightarrow \dfrac{a+b+c}{b+c} = \dfrac{a+b+c}{c+a} = \dfrac{a+b+c}{a+b}$

$\Rightarrow \quad b + c = c + a = a + b$

or $\quad (a + b + c) = 0$

$\Rightarrow \quad a = b = c$ or $a = -(b + c)$

Then, value of each fraction $= \dfrac{a}{a+a}$ or $\dfrac{-(b+c)}{(b+c)}$ *i.e.* $\dfrac{1}{2}$ or -1

94. Given, $12x^2 - 21x + 1 = 0$

$\Rightarrow \quad 12x^2 + 1 = 21x$

$\Rightarrow \quad 4x\left(3x + \dfrac{1}{4x}\right) = 21x$

$\Rightarrow \quad \left(3x + \dfrac{1}{4x}\right) = \dfrac{21}{4}$

$\Rightarrow 9x^2 + (16x^2)^{-1} + \dfrac{3}{2} = \dfrac{225}{16}$ (squaring both sides)

$\Rightarrow \quad 9x^2 + (16x^2)^{-1} = \dfrac{417}{16}$

95. Given, $a^4 + \dfrac{1}{a^4} = 50$

Then, $a^2 + \dfrac{1}{a^2} = \sqrt{50 + 2} = 2\sqrt{13}$

and $\quad a + \dfrac{1}{a} = \sqrt{2\sqrt{13} + 2} = \sqrt{2(\sqrt{13} + 1)}$

Now, $a^3 + \dfrac{1}{a^3} = \left(\sqrt{2(\sqrt{13} + 1)}\right)^3 - 3 \cdot \left(\sqrt{2(\sqrt{13} + 1)}\right)$

$= 2(\sqrt{13} + 1) \times \sqrt{2(\sqrt{13} + 1)} - 3 \cdot \sqrt{2(\sqrt{13} + 1)}$

$= \sqrt{2(\sqrt{13} + 1)} \cdot (2\sqrt{13} - 1)$

Hence, option (a) is correct.

53

Part-II

Concept of Equations : Linear, Quadratic, Cubic and Others

Learning OBJECTIVES

After completing this chapter, aspirants will be able to :

- Understand the significance of this sub-section in diverse competitive exams;
- Comprehend concepts associated with the diverse types of equations;
- Be familiar with the general form of the equations;
- Understand the concept of solvability of simultaneous equations;
- Know the Sridharacharya's formula and cubic formula for obtaining roots of the equations;
- Compute the product of all roots of a polynomial equations and sum of product roots taken n together or sole;
- Know the techniques how to compare roots of two quadratic/cubic equations;
- Recognise the properties of a polynomial equation and its roots as well as nature of the roots;
- Realise themselves competent to find all roots of a cubic equation by a unique technqiue;
- Find the roots of a cubic equation if its one root is known;
- Establish the relation among quadratic coefficient, linear coefficient and constant of a quadratic equation when relation between its roots is known;
- Form a quadratic equation when its roots are known;
- Form a quadratic equation with the roots having certain relation with roots of another quadratic equations;
- Familiar with concept of common root(s) for two polynomial equations;
- Compute the minimum and maximum values of a quadratic expression and the value of the variable at which the quadratic equation is either maximum or minimum;

Inside

- ***Linear Equations :*** *Introduction, General form and Simultaneous Equations*
- ***Quadratic Equations :*** *(i) Introduction, their Properties, Sridharacharya's Formula (ii) Nature of Roots, their Sum and Product, (iii) Formation of Quadratic Equation in Diverse Cases, (iv) Maximum/Minimum Value of a Quadratic Equation, (v) Concept of Common Roots and their Number*
- ***Cubic and Other Equations :*** *(i) Critical Facts, How to Find Roots, (ii) Formation of Equation in Diverse Cases, (iii) Maximum Number of Positive and Negative, Real Roots of a Polynomial Equation (iv) Concept of Common Root(s)*
- ....
- .... *& Many More*
- *Work Out with Solutions*

- Compute maximum number of positive and negative real roots of a polynomial equation;
- Understand the philosophy of division of 'Illustrations' in two parts;
- Identify the problems which can easily be solved through options;
- Analyse the trend of the questions being asked in various competitive examinations;
- Solve the questions in moments;
- Learn many new concepts/fundamentals;
- Apply the pragmatic approaches in lieu of academic approaches; and
- Ponder on the alternative methods.

A. Linear Equations

Introduction

A lot of questions even pertaining to arithmetic chapters are based on the concept of linear equations directly or indirectly. An equation is a statement of equality which has one or more variables.

An equation having variable(s) with the highest power 1, is called linear equation. The equation is called linear because its graph is a straight line. When one variable is involved, it is called 'linear equation in one variable' and in case of two variables, 'linear equation in two variables'.

e.g.

$x + 9 = 0 \rightarrow$ an example of a linear equation in one variable

$2x + 3y - 13 = 0 \rightarrow$ an example of linear equation in two variables

$3x + 5y - 8z + 13 = 0 \rightarrow$ an example of linear equation in three variables

General form of Linear Equation

(i) General form of a linear equation in one variable (say, x) is $ax + b = 0$, where a and b are real numbers and $a \neq 0$.

(ii) General form of a linear equation in two variables (say, x and y) is $ax + by + c = 0$, where a, b and c are real numbers and $a \neq 0$, $b \neq 0$.

(iii) General form of a linear equation in three variables (say, x, y and z) is $ax + by + cz + d = 0$, where, a, b, c and d are real numbers and $a \neq 0$, $b \neq 0$, $c \neq 0$ and so on.

Critical Facts

(i) The value of variable(s) which satisfies(y) the equation is known as solution of the equation.

(ii) Moving a term from one side of the equation to its another side and changing its sign is called transposing the term. We also apply this transposition rule to solve the equation.

(iii) If we add the same number to both sides of an equation or subtract the same number from both sides or multiply both sides with the same number (non-zero) or divide both sides by the same non-zero number, there will be no effect on the equation as well as its solutions.

Take two examples to understand how to solve the equations.

Find the solutions of equations (a) $x + 9 = 0$ and (b) $2x + 3y - 13 = 0$

(a) $x + 9 = 0$

$\Rightarrow x = 0 - 9$ [$\because$ we moved 9 from left hand side to right hand side and also changed its sign]

Alternatively

We can subtract 9 from both sides to solve the equation

$$x + 9 - 9 = 0 - 9$$

$$\Rightarrow \quad x = -9$$

(b) $2x + 3y - 13 = 0$

$\Rightarrow 2x + 3y = 0 + 13$

[$\because$ we moved 13 from left hand side to right hand side and also changed its sign]

$\Rightarrow 2x = 13 - 3y$

[$\because$ we moved 3y from left hand side to right hand side and also changed its sign]

$$\Rightarrow \quad x = \frac{13 - 3y}{2}$$

If $y = 1$, then $x = 5$

If $y = 2$, then $x = 3.5$

If $y = 1.5$, then $x = 4.25$

Observation A linear equation in two variables has infinitely many solutions.

Simultaneous Equations

Simultaneous equations refer to a set of two or more equations, each having two or more variable, which are satisfied by the same values of the variables. It is also said that these equations form a 'system of equations'.

e.g. $\quad a_1 x + b_1 y + c_1 = 0$

and $\quad a_2 x + b_2 y + c_2 = 0$

form a system of two linear equations in variables x and y.

Keeping in view the trends of questions being asked in competitive exams, we will emphasis on a set of two linear equations in two variables.

For the above system of equations, the following facts are true with reference to solvability of these equations.

(i) If $\frac{a_1}{a_2} \neq \frac{b_1}{b_2}$, then the equations are said to be independent and the system of equations will have a unique solution.

e.g. $\quad 2x + 3y = 8$

and $4x + y = 6$

Clearly, $\frac{2}{4} \neq \frac{3}{1}$

It implies that the system of equations have unique solution. Now, learn how to solve the system of equations.

***Step* I** Multiply the given equation(s) by the suitable number(s) in order to make the coefficients of one of the variables equal.

***Step* II** If the coefficients, thus obtained, have same sign, subtract one equation from another. In case of opposite sign, add them. This is done to eliminate one term.

***Step* III** Try to solve the resulting linear equation to obtain the value of the variable inolved therein.

***Step* IV** Thereafter, put the value of the variable obtained in any one of the given equation and solve it to find the value of another variable.

$$2x + 3y = 8 \quad \ldots(i) \times 2$$
$$4x + y = 6 \quad \ldots(ii)$$

Eqs. (i) is multiplied by 2 in order to make coefficients of variable x equal.

Now, $4x + 6y = 16 \quad \ldots(i)$

$4x + y = 6 \quad \ldots(ii)$

Subtracting Eq. (ii) from Eq. (i) we get

$$5y = 10$$

$$\Rightarrow \quad y = \frac{10}{5} = 2$$

Now, put the value of y in any one of the equations.

Suppose we put value of y in Eqs. (i), then $2x + 3 \cdot 2 = 8$ $\Rightarrow x = 1$

Hence, the required solution is $x = 1$ and $y = 2$. It implies that the system of equations has unique solution.

Alternatively I

$$2x + 3y = 8 \quad \ldots(i)$$
$$4x + y = 6 \quad \ldots(ii) \times 3$$

Eq. (ii) is multiplied by 3 in order to make coefficients of variable y equal.

Now,

$$2x + 3y = 8 \quad \ldots(i)$$
$$12x + 3y = 18 \quad \ldots(ii)$$

Subtracting Eq. (ii) from Eq. (i) we get

$$-10x = -10$$

$$\Rightarrow \quad x = 1$$

Now, put the value of x in any one of the equations. Suppose we put value of x in Eq. (ii), then $4 \cdot 1 + y = 6$ $\Rightarrow y = 2$

Hence, the required solution is $x = 1$ and $y = 2$

Alternatively II

If the system of equations (in general form) is

$$a_1 x + b_1 y + c_1 = 0$$
$$a_2 x + b_2 y + c_2 = 0$$

Then, $x = \frac{b_1 c_2 - b_2 c_1}{a_1 b_2 - a_2 b_1}$ and $y = \frac{c_1 a_2 - c_2 a_1}{a_1 b_2 - a_2 b_1}$

As given, the system of equations is

$$2x + 3y + (-8) = 0$$

and $4x + 1y + (-6) = 0$

clearly, $a_1 = 2, a_2 = 4, b_1 = 3, b_2 = 1$ $c_1 = -8$ and $c_2 = -6$

Then, $x = \frac{b_1 c_2 - b_2 c_1}{a_1 b_2 - a_2 b_1}$

$$= \frac{3(-6) - 1(-8)}{2(1) - 4(3)} = \frac{-18 + 8}{2 - 12} = 1$$

and $y = \frac{c_1 a_2 - c_2 a_1}{a_1 b_2 - a_2 b_1} = \frac{-8(4) - (-6)2}{2(1) - 4(3)}$

$$= \frac{-32 + 12}{2 - 12} = 2$$

Note *Instead of finding the value of y with the help of this method, you should put the value of x in any one of the given equations to obtain the value of y. Obviously, it will save our time in examination hall.*

(ii) If $\frac{a_1}{a_2} = \frac{b_1}{b_2} = \frac{c_1}{c_2}$, then equations are said to be dependent and the system of equations has infinite solutions.

e.g. $4x - 3y - 5 = 0$

and $8x - 6y - 10 = 0$

Clearly, $\frac{4}{8} = \frac{-3}{-6} = \frac{-5}{-10}$

It implies that the system of equations has infinitely many solutions.

Remarks *If we multiply Eq. (i) by 2, both equations will be same and it is the fact that a linear equation in two variables has infinitely many solutions.*

(iii) If $\frac{a_1}{a_2} = \frac{b_1}{b_2} \neq \frac{c_1}{c_2}$, then equations are said to be inconsistent and the system of equations has no solutions.

e.g. $2x + y - 5 = 0$

and $6x + 3y - 10 = 0$

Cearly, $\frac{2}{6} = \frac{1}{3} \neq \frac{-5}{-10}$

It implies that the system of equations has no solution.

B. Quadratic Equations

Introduction

An equation of the form $ax^2 + bx + c = 0$, where a, b, c are real numbers and $a \neq 0$, is known as a quadratic equation in x.

Here, $x \to$ variable

$a \to$ quadratic coefficient

$b \to$ linear coefficient

$c \to$ constant

Note *In a quadratic equation, the linear coefficient or constant or both may be zero.*

The zeros of the quadratic polynomial $ax^2 + bx + c$ are known as roots or solutions of the quadratic equation $ax^2 + bx + c = 0$. If the real numbers α and β are two zeros of the quadratic polynomial $ax^2 + bx + c$, then α and β are said to be roots or solutions of the quadratic equation $ax^2 + bx + c = 0$.

How to find roots of a quadratic equation

Take two examples for building the concepts :

(a) Find the roots of $x^2 + 13x + 40 = 0$.

Or

Solve $x^2 + 13x + 40 = 0$

Let $x^2 + 13x + 40$ be a product of two linear expressions $(x + p)$ and $(x + q)$.

Then, $x^2 + 13x + 40 = (x + p)(x + q)$

$\Rightarrow x^2 + 13x + 40 = x^2 + (p + q)x + pq$

Clearly, $(p + q) = 13$ and $pq = 40$

Hence, $p = 8$ or 5 and $q = 5$ or 8

$\therefore \quad (x + 5)(x + 8) = 0$

$(x + 5) = 0$ and $(x + 8) = 0 \Rightarrow x = -5$ or $x = -8$

Thus, –5 and –8 are the roots of the quadratic equation $x^2 + 13x + 40 = 0$.

(b) Solve $2x^2 - 9x + 9 = 0$

Given equation : $2x^2 - 9x + 9 = 0$

$\Rightarrow x^2 - \frac{9}{2}x + \frac{9}{2} = 0$ [dividing by 2 both sides]

Let $x^2 - \frac{9}{2}x + \frac{9}{2} = (x + p)(x + q)$

$\Rightarrow \quad x^2 + \left(-\frac{9}{2}\right)x + \frac{9}{2} = x^2 + (p + q)x + pq$

Clearly, $p + q = -\frac{9}{2}$ and $pq = \frac{9}{2}$

These two conditions are satisfied if $p = (-3)$ or $\left(\frac{-3}{2}\right)$ and $q = \left(-\frac{3}{2}\right)$ or (-3).

$\therefore \quad [x + (-3)]\left[x + \left(-\frac{3}{2}\right)\right] = 0$

$\Rightarrow \quad (x - 3)\left(x - \frac{3}{2}\right) = 0$

$\Rightarrow \quad x = 3$ or $x = \frac{3}{2}$

Hence, 3 and $\frac{3}{2}$ are the roots of the quadratic equation $2x^2 - 9x + 9 = 0$.

Alternatively

(for $x^2 + 13x + 40 = 0$)

***Step* I** We multiply the quadratic coefficient and the constant with their signs. Here, $(+1) \times (+40) = +40$.

***Step* II** If the product obtained is positive, factorise the product in two terms such a way that their sum is equal to the linear coefficient (without its sign). In case the product is negative, then factorise it (without considering sign) in two terms such a way that their difference is equal to the linear coefficient (without its sign). Here, the product is positive. Then, $40 = 8 \times 5$ ($\because 8 + 5 =$ linear coefficient, 13)

***Step* III** Split the middle term considering the above factorisation minding signs.

Here, $13x = 8x + 5x$

***Step* IV** We get the desired factors by grouping the terms.

Clearly, $x^2 + 8x + 5x + 40 = 0$

$\Rightarrow \quad x(x + 8) + 5(x + 8) = 0$

$\Rightarrow \quad (x + 5)(x + 8) = 0$

$\Rightarrow \quad (x + 5) = 0$ and $(x + 8) = 0$

$\Rightarrow \quad x = -5$ or $x = -8$

Alternatively II

(for $2x^2 - 9x + 9 = 0$)

***Step* I** We multiply the quadratic coefficient and the constant with their signs. Here, $(+2) \times (+9) = +18$.

***Step* II** If the product obtained is positive, factorise the product in two terms such a way that their sum is equal to the absolute value of linear coefficient. In case the product is negative, then factorise it(without considering sign) in two terms such a way that their difference is equal to the absolute value of linear coefficient. We know that, an absolute value of ± 9 is 9.

Here, the product is positive. Then, $18 = 3 \times 6$ [$\because 3 + 6 =$ linear coefficient, 9]

***Step* III** Split the linear coefficient considering the above factorisation minding the signs.

Here, $-9 = -3 - 6$

***Step* IV** Divide both numbers i.e. (–3 and –6) by the quadratic coefficient. Hence, $\frac{-3}{2}$ and $\frac{-6}{2}$ i.e. $\frac{-3}{2}$ and –3.

***Step* V** Required roots will be the above results with opposite signs. Hence, roots are $\frac{3}{2}$ and 3.

Observations The aspirants are advised to adhere to the method envisaged under **Alternatively II** so as to solve the quadratic equation in moments without pen and paper.

Take other examples

(c) Solve $-x^2 + 2x + 24 = 0$

***Step* I** Product $= (-1) \times (+24) = -24$

***Step* II** Since product is negative, 24 is factorised into two terms such that their difference is 2 *viz.* $24 = 6 \times 4$

***Step* III** The linear coefficient (i.e. + 2) may be written as $+6 - 4$

***Step* IV** $\frac{6}{-1}$ and $\frac{-4}{-1}$ i.e. -6 and 4

***Step* V** The required roots = 6 and -4.

(d) Solve $5x^2 - \frac{17}{2}x + \frac{3}{2} = 0$

***Step* I** Product $= (+5) \times \left(+\frac{3}{2}\right) = +\frac{15}{2}$

***Step* II** $\frac{15}{2} = \frac{15}{2} \times 1$

***Step* III** $\frac{-17}{2} = \frac{-15}{2} - 1$

***Step* IV** $\frac{-15}{2 \times 5}$ and $\frac{-1}{5}$ i.e. $\frac{-3}{2}$ and $\frac{-1}{5}$

***Step* V** The required roots $= \frac{3}{2}$ and $\frac{1}{5}$

Alternatively

We may multiply both sides of the given equation by 2. Then, the equation be $10x^2 - 17x + 3 = 0$

Now, we can solve in easy way on the above pattern.

(e) Solve $x^2 + (\sqrt{5} + 1)x + \sqrt{5} = 0$

Here, the product of the quadratic coefficient and the constant is $+\sqrt{5}$. Now, we factorise this product in two terms such that their sum is equal to $(\sqrt{5} + 1)$,

viz. $\sqrt{5} = \sqrt{5} \times 1$

Then, $+(\sqrt{5} + 1) = +\sqrt{5} + 1$

$\therefore$ Required roots $= -\sqrt{5}, -1$

(f) Solve $x^2 - 6\sqrt{3}\,x + 24 = 0$

Here, the product of the quadratic coefficient and the constant is +24. Now, we factorise this product in two terms such that their sum is equal to $6\sqrt{3}$, *viz.* $24 = 2\sqrt{3} \times 4\sqrt{3}$

Then, $-6\sqrt{3} = -2\sqrt{3} - 4\sqrt{3}$

$\therefore$ Required roots $= 2\sqrt{3}$ and $4\sqrt{3}$

The factorisation of 24 under such condition seems to be somewhat difficult, hence, alternative approach is warranted.

Alternatively

Square of $\sqrt{3} = 3$

Now, divide the constant by 3. Here, the resultant is 8.

The equation can be written as $x^2 - 6x + 8 = 0$ [removing $\sqrt{3}$ from the linear coefficient]

Clearly, roots of the equation $x^2 - 6x + 8 = 0$ are 2 and 4.

Then, the required roots $= 2\sqrt{3}$ and $4\sqrt{3}$

Observation This approach is applicable when the linear coefficient is in a form of pure surd or mixed surd (mixed surd refers to the product of a rational number and a surd). In order to obtain a rational linear coefficient, we divide the linear coefficient by the given surd (say, $\sqrt{3}$). Also, we divide the constant by square of the surd. Thereafter, we solve the equation. The desired roots are $\sqrt{3}$ times the roots obtained.

It is to note that it is not always easy to factorise the product of the quadratic coefficient and the constant. Sometimes, it seems to be very difficult.

e.g. $x^2 - 6x + 4 = 0$

***Step* I** Product $= (+1) \times (+4) = +4$

***Step* II** 4 is to be factorised into two terms such that their sum is equal to linear coefficient 6. It is not easy to factorise 4 under such condition.

Hence, we will discuss about the quadratic formula with help of which roots of any quadratic equation can be computed.

Quadratic formula/Sridharacharya's formula

General form of a quadratic equation is $ax^2 + bx + c = 0$

$$\Rightarrow \quad x^2 + \frac{b}{a}x + \frac{c}{a} = 0 \quad \text{(on dividing both sides by } a\text{)}$$

$$\Rightarrow \quad x^2 + \frac{b}{a}x + \left(\frac{b}{2a}\right)^2 - \left(\frac{b}{2a}\right)^2 + \frac{c}{a} = 0$$

$$\Rightarrow \quad x^2 + 2\left(\frac{b}{2a}\right)x + \left(\frac{b}{2a}\right)^2 = \left(\frac{b}{2a}\right)^2 - \frac{c}{a}$$

$$\Rightarrow \quad \left(x + \frac{b}{2a}\right)^2 = \frac{b^2 - 4ac}{4a^2}$$

$$\Rightarrow \quad x + \frac{b}{2a} = \pm\sqrt{\frac{b^2 - 4ac}{4a^2}}$$

$$\Rightarrow \quad x = -\frac{b}{2a} \pm \frac{\sqrt{b^2 - 4ac}}{2a}$$

$$\Rightarrow \quad x = \frac{-b \pm \sqrt{b^2 - 4ac}}{2a}$$

$$\Rightarrow \quad x = \frac{-b + \sqrt{b^2 - 4ac}}{2a} \text{ or } x = \frac{-b - \sqrt{b^2 - 4ac}}{2a}$$

Hence, $\frac{-b + \sqrt{b^2 - 4ac}}{2a}$ and $\frac{-b - \sqrt{b^2 - 4ac}}{2a}$ are the roots of the quadratic equation $ax^2 + bx + c = 0$.

Note *$(b^2 - 4ac)$ is known as the discriminant (denoted by D) of the quadratic equation $ax^2 + bx + c = 0$. It is very clear from the above that if $D \geq 0$, then the equation will have real roots, otherwise it has no roots (technically, no real roots).*

Now, find the roots of $x^2 - 6x + 4 = 0$

Comparing the given equation $x^2 - 6x + 4 = 0$ with $ax^2 + bx + c = 0$, we get $a = +1, b = -6$ and $c = 4$.

Now, $D = b^2 - 4ac = (-6)^2 - 4 \cdot 1 \cdot 4 = 36 - 16 = 20$, which is greater than 0, it implies that the equation has real roots.

$$\therefore \text{Required roots} = \frac{-b + \sqrt{b^2 - 4ac}}{2a} \text{ and } \frac{-b - \sqrt{b^2 - 4ac}}{2a}$$

$$= \frac{-(-6) + \sqrt{20}}{2 \cdot 1} \text{ and } \frac{-(-6) - \sqrt{20}}{2 \cdot 1}$$

$$= \frac{(6) + 2\sqrt{5}}{2 \cdot 1} \text{ and } \frac{(6) - 2\sqrt{5}}{2 \cdot 1}$$

$$= 3 + \sqrt{5} \text{ and } 3 - \sqrt{5}$$

Sum and Product of roots of a quadratic equation

$$\text{Sum of roots} = \frac{-b + \sqrt{b^2 - 4ac}}{2a} + \frac{-b - \sqrt{b^2 - 4ac}}{2a}$$

$$= \frac{-b + \sqrt{b^2 - 4ac} - b - \sqrt{b^2 - 4ac}}{2a}$$

$$= \frac{-2b}{2a} = \frac{-b}{a}$$

$$\text{Products of roots} = \frac{-b + \sqrt{b^2 - 4ac}}{2a} \times \frac{-b - \sqrt{b^2 - 4ac}}{2a}$$

$$= \frac{b^2 - b^2 + 4ac}{4a^2} = \frac{c}{a}$$

Hence,

(i) Sum of roots $= \frac{-b}{a} = \left\{\frac{-(\text{linear coefficient})}{\text{quadratic coefficient}}\right\}$. It may be construed that if both the roots are equal in magnitude but opposite in sign, then $b = 0$.

(ii) Product of roots $= \frac{c}{a} = \left(\frac{\text{constant term}}{\text{quadratic coefficient}}\right)$. It may be interpreted that if both the roots are reciprocal to each other, then $a = c$.

Nature of roots

(i) If $D > 0$, then the equation has two distinct real roots.

(a) If D is perfect square, roots are rational and unequal.

(b) If D is not a perfect square, roots are irrational and a conjugate pair.

(ii) If $D < 0$, then the equation has no real roots.

(iii) If $D = 0$, then the equation has equal rational roots.

Properties of quadratic equation

(i) If one root of a quadratic equation with rational coefficients is $p + \sqrt{q}$, then another is its conjugate i.e. $p - \sqrt{q}$ and *vice-versa*.

Here, p is rational and $\sqrt{q}$ is a surd.

(ii) If α is a root of the quadratic equation $ax^2 + bx + c = 0$, then $(x - \alpha)$ is a factor of the quadratic expression.

(iii) If a quadratic equation $ax^2 + bx + c = 0$, has more than two roots, it is an identity i.e. $a = b = c = 0$.

(iv) In a quadratic equation $ax^2 + bx + c = 0$ if $a + b + c = 0$, then 1 is one of the roots of the equation.

(v) Since, a polynomial equation of n degree has n roots, a quadratic equation has two roots.

(vi) The graph of any quadratic equation will be a parabola.

Relation among quadratic coefficient, linear coefficient and constant of a quadratic equation when relation between its roots is known

Consider one case for building the concept :

"When one root of quadratic equation $ax^2 + bx + c = 0$ is m times the other"

Let one root be α and the other be $m\alpha$.

We have, sum of roots $= \frac{-b}{a}$

$$\Rightarrow \quad \alpha + m\alpha = \frac{-b}{a} \Rightarrow \alpha = \frac{-b}{a(1+m)}$$

Further, product of the roots $= \frac{c}{a} \Rightarrow \alpha \cdot m\alpha = \frac{c}{a}$

$$\Rightarrow \quad \left\{\frac{-b}{a(1+m)}\right\}^2 m = \frac{c}{a}$$

$$\Rightarrow \quad b^2 m = ac(1+m)^2$$

$$\Rightarrow \quad b^2 = ac\left(m + \frac{1}{m} + 2\right)$$

$$\Rightarrow \quad m + \frac{1}{m} = \frac{b^2 - 2ac}{ac}$$

On the same pattern, we can derive the relation under different circumstances. The relations which can be derived, are as below:

Cases	Circumstances	Relation among a, b, c of a quadratic equation $ax^2 + bx + c = 0$
(i)	When one root is m times the other	$m + \frac{1}{m} = \frac{b^2 - 2ac}{ac}$
(ii)	When ratio of roots is $p:q$	We may say that one root is $\frac{p}{q}$ times the other. Hence, we may put $m = \frac{p}{q}$ in the above relation. Otherwise, remember the result directly as $b^2pq = ac(p+q)^2$
(iii)	When one root is the square of the other	$a + c = \frac{b^3 - 3abc}{-ac}$
(iv)	When one root is m more than the other	$m^2 = \frac{b^2 - 4ac}{a^2}$
(v)	When sum of squares of both roots is 1	$b^2 = a^2 + 2ac$
(vi)	When one root is reciprocal to the other	$a = c$
(vii)	Both roots are of the same magnitude but opposite in signs	$b = 0$, hence, no first degree term.
(viii)	Both roots are equal	$b^2 = 4ac$

Formation of quadratic equations when roots are known

Suppose we have to find the quadratic equation whose roots are α and β.

Let the quadratic equation be $ax^2 + bx + c = 0$

$$\Rightarrow \quad x^2 + \frac{b}{a}x + \frac{c}{a} = 0 \quad \text{(On dividing by } a \text{ both sides)}$$

$$\Rightarrow \quad x^2 - \left(-\frac{b}{a}\right)x + \frac{c}{a} = 0$$

$$\Rightarrow \quad x^2 - (\text{sum of roots})\,x + \text{product of roots} = 0$$

Or, $\quad x^2 - (\alpha + \beta)x + \alpha\beta = 0$

Alternatively

$$\Rightarrow \quad x^2 - (\alpha + \beta)x + \alpha\beta = 0 \quad \text{(from the above)}$$

$$\Rightarrow \quad x^2 - \alpha x - \beta x + \alpha\beta = 0$$

$$\Rightarrow \quad x(x - \alpha) - \beta(x - \alpha) = 0$$

$$\Rightarrow \quad (x - \alpha)(x - \beta) = 0$$

e.g. Find the quadratic equation in x whose roots are 2 and -5.

Required equation is $x^2 -$ (sum of roots) x + product of roots $= 0$

$$\Rightarrow \quad x^2 - [2 + (-5)]x + 2(-5) = 0$$

$$\Rightarrow \quad x^2 + 3x - 10 = 0$$

Alternatively

$$[x - 2][x - (-5)] = 0 \Rightarrow (x - 2)(x + 5) = 0$$

$$\Rightarrow \quad x^2 + 3x - 10 = 0$$

Formation of a quadratic equation with the roots having certain relation with roots of another quadratic equation

Take few examples for concept-building

Write a quadratic equation whose roots are r more than the roots of a quadratic equation $ax^2 + bx + c = 0$.

Let α and β be roots of the given equation $ax^2 + bx + c = 0$

Then, $\quad \alpha + \beta = \frac{-b}{a}$ and $\alpha\beta = \frac{c}{a}$

Sum of the roots of the equation to be formed

$$= (\alpha + r) + (\beta + r) = \frac{-b}{a} + 2r$$

and product of roots of the equation to be formed

$$= (\alpha + r)\cdot(\beta + r) = \alpha\beta + r(\alpha + \beta) + r^2 = \frac{c}{a} + \frac{-br}{a} + r^2$$

Now, the new equation be

$$x^2 - (\text{sum of roots})\,x + \text{product of roots} = 0$$

$$\Rightarrow \quad x^2 - \left(\frac{-b}{a} + 2r\right)x + \left(\frac{c}{a} + \frac{-br}{a} + r^2\right) = 0$$

$$\Rightarrow \quad x^2 - 2xr + r^2 + \left(\frac{b}{a}\right)x + \left(\frac{c}{a}\right) + \frac{-br}{a} = 0$$

$$\Rightarrow \quad (x - r)^2 + \left(\frac{bx + c - br}{a}\right) = 0$$

$$\Rightarrow \quad a(x - r)^2 + b(x - r) + c = 0$$

Hence, new quadratic equation is

$$a(x - r)^2 + b(x - r) + c = 0$$

Similarly, we may form new quadratic equations with the roots having relation with roots of another equation in different way, depicted in the following chart.

Cases	How roots of the new quadratic equation are related with those of equation $ax^2 + bx + c = 0$.	New quadratic equation
(i)	greater by r (i.e. roots of newone is r greater than those of the given Eq.)	$a(x - r)^2 + b(x - r) + c = 0$
(ii)	less by r	$a(x + r)^2 + b(x + r) + c = 0$
(iii)	r times	$a\left(\frac{x}{r}\right)^2 + b\left(\frac{x}{r}\right) + c = 0$
(iv)	$1/r$ times	$a(xr)^2 + b(xr) + c = 0$
(v)	squares	$a(\sqrt{x})^2 + b\sqrt{x} + c = 0$ i.e. $ax + b\sqrt{x} + c = 0$
(vi)	cubes	$a(\sqrt[3]{x})^2 + b\sqrt[3]{x} + c = 0$
(vii)	reciprocals	$cx^2 + bx + a = 0$
(viii)	Same but opposite in sign	$ax^2 - bx + c = 0$

Maximum and Minimum value of a Quadratic Expression

Consider a quadratic expression $ax^2 + bx + c$.

Let the value of $ax^2 + bx + c$ is m.

$\Rightarrow \quad m = ax^2 + bx + c$

$\Rightarrow \quad ax^2 + bx + c - m = 0$

If x is real, then the discriminant of the equation $ax^2 + bx + c - m = 0$ must be ≥ 0.

$\Rightarrow \quad b^2 - 4 \cdot a \cdot (c - m) \geq 0$

$\Rightarrow \quad b^2 - 4ac + 4\,am \geq 0$

$\Rightarrow \quad 4\,am \geq 4ac - b^2$

We know, there is no effect on sign of inequality when both sides of an inequality are multiplied (or divided) by the same positive number. In case both sides are multiplied (or divided) by the same negative number, the sign of inequality is reversed.

Hence, $m \geq \dfrac{4ac - b^2}{4a}$ [on dividing by 4a, where a > 0] and

$m \leq \dfrac{4ac - b^2}{4a}$ [on dividing by 4a, where a < 0]

It may be construed from the above that

(i) For x be real, the minimum value of $ax^2 + bx + c$ is $\dfrac{4ac - b^2}{4a}$, if a > 0. However, we are not able to find the maximum value of the expression in case $a > 0$. It may be infinite.

(ii) For x be real, the maximum value of $ax^2 + bx + c$ is $\dfrac{4ac - b^2}{4a}$, if $a < 0$. However, we are not able to find the minimum value of the expression if $a < 0$.

Further, the minimum or maximum value of $ax^2 + bx + c = \dfrac{4ac - b^2}{4a}$

$\Rightarrow \dfrac{4ac - b^2}{4a} = ax^2 + bx + c$

$\Rightarrow 4ac - b^2 = 4a^2\,x^2 + 4abx + 4ac$

$\Rightarrow 4a^2\,x^2 + 4abx + b^2 = 0$

$\Rightarrow \quad (2ax + b)^2 = 0 \Rightarrow x = \dfrac{-b}{2a}$

It Implies that

(iii) At $x = \dfrac{-b}{2a}$, the value of $ax^2 + bx + c$ is minimum, which equals $\dfrac{4ac - b^2}{4a}$, if $a > 0$

(iv) At $x = \dfrac{-b}{2a}$, the value of $ax^2 + bx + c$ is maximum, which equals $\dfrac{4ac - b^2}{4a}$, if $a < 0$

Now, take some examples :

(a) For x be real what are the maximum and minimum values of $-x^2 + 4x + 9$?

Since the quadratic coefficient is less than 0, we cannot find the minimum value of the expression.

Now, the maximum value of the expression

$$= \frac{4ac - b^2}{4a}$$

$$= \frac{4 \cdot (-1) \cdot 9 - 4^2}{4(-1)} = 13$$

Alternatively

Given expression $-x^2 + 2 \cdot x \cdot 2 + 9$

$= -(x^2 - 2 \cdot x \cdot 2 + 2^2) + 13$

$= -(x - 2)^2 + 13$

$= 13 - (x - 2)^2$

Obviously, $13 - (x - 2)^2$ will have maximum value if the value of $(x - 2)^2$ is minimum and the minimum value of $(x - 2)^2$ is 0. Hence, the maximum value of the expression is 13.

Further, the value of $13 - (x - 2)^2$ will be minimum if the value of $(x - 2)^2$ is maximum. Since, the maximum value of $(x - 2)^2$ is not known, we cannot find the minimum value of $13 - (x - 2)^2$.

(b) For x be real, find the minimum value of $4x^2 + 4x + 9$ **[SSC (FCI-mains)-12]**

Since the quadratic coefficient is greater than 0, we can find the minimum value of the given expression.

The minimum value of the expression

$$= \frac{4ac - b^2}{4a}$$

$$= \frac{4 \cdot 4 \cdot 9 - 4^2}{4 \cdot 4} = 8$$

Alternatively

Given expression $4x^2 + 4x + 9$

$= (2x)^2 + 2 \cdot 2x \cdot 1 + 1^2 + 8$

$= (2x + 1)^2 + 8$

Obviously, $(2x + 1)^2 + 8$ will have minimum value if the value of $(2x + 1)^2$ is minimum and the minimum value of $(2x + 1)^2$ is 0. Hence, the minimum value of the expression is 8.

(c) For what value of x, a quadratic expression $-2x^2 + 4x - 6$ will have the maximum value ?

Obviously, at $x = \dfrac{-b}{2a}$, the expression will have maximum value.

$\therefore \quad x = \dfrac{-(4)}{2(-2)} = 1$

Concept of Common Root(s) between Quadratic Equations

Case I

When one root is common for two quadratic equations.

Let α be common root for two quadratic equations (i) $ax^2 + bx + c = 0$ and (ii) $dx^2 + ex + f = 0$

Since α is common, both the equations will be satisfied by this root.

Hence, $a\alpha^2 + b\alpha + c = 0$ ...(i)

and $d\alpha^2 + e\alpha + f = 0$ ...(ii)

We multiply Eq. (i) by d and Eq. (ii) by a.

Then, $da\alpha^2 + bd\alpha + cd = 0$ ...(i)

and $da\alpha^2 + ea\alpha + fa = 0$ ...(ii)

Subtracting Eq. (ii) from Eq. (i) we get,

$$\alpha(bd - ae) = -(cd - af)$$

$$\Rightarrow \quad \alpha = \frac{(af - cd)}{(bd - ae)}$$

It means we can find out the value of the common root by this result.

Further, the equations are as below when we multiply (i) by e and (ii) by b.

$ea\alpha^2 + eb\alpha + ec = 0$ ...(i)

and $bd\alpha^2 + be\alpha + bf = 0$ ...(ii)

Subtracting (ii) from (i) we get, $\alpha^2(ae - bd) + (ce - bf) = 0$

$$\Rightarrow \quad \alpha^2 = \frac{(ce - bf)}{(bd - ae)}$$

$$\Rightarrow \quad \left\{\frac{af - cd}{bd - ae}\right\}^2 = \frac{(ce - bf)}{(bd - ae)}$$

$$\Rightarrow \quad (af - cd)^2 = (ce - bf) \times (bd - ae)$$

$$\Rightarrow \quad (af - cd)^2 = (ae - bd) \times (bf - ce)$$

Further, $(af - cd)^2 = (ae - bd) \times (bf - ce)$

$$\Rightarrow \quad \frac{af - cd}{bd - ae} = \frac{ce - bf}{af - cd} \Rightarrow \alpha = \frac{ce - bf}{af - cd}$$

It is construed from the above that equations $ax^2 + bx + c = 0$ and $dx^2 + ex + f = 0$ will have a common root if $(af - cd)^2 = (ae - bd) \times (bf - ce)$. Also, if two equations $ax^2 + bx + c = 0$ and $dx^2 + ex + f = 0$ have one root common, then its value is either $\frac{(af - cd)}{(bd - ae)}$ or $\frac{ce - bf}{af - cd}$. No doubt, the value of $\frac{(af - cd)}{(bd - ae)}$ will be equal to $\frac{ce - bf}{af - cd}$.

In other words, both the given equations are satisfied at $x = \frac{(af - cd)}{(bd - ae)}$ or $\frac{ce - bf}{af - cd}$.

Case II

When both the roots are common for two quadratic equations

If $\frac{a}{d} = \frac{b}{e} = \frac{c}{f}$, then equations $ax^2 + bx + c = 0$ and $dx^2 + ex + f = 0$ will have both the roots common.

Another approach for conditionality for quadratic equations to have common roots(s) is detailed under 'Concept of common roots for polynomial equations, envisaged under Cubic and other equations'.

Number of common roots between two quadratic equations

Let the equations be $ax^2 + bx + c = 0$ and $dx^2 + ex + f = 0$. If

(i) $(af - cd)^2 = (ae - bd) \times (bf - ce)$, then number of common roots $= 1$

(ii) $\frac{a}{d} = \frac{b}{e} = \frac{c}{f}$, then number of common roots(s) = 2

(iii) None of $(af - cd)^2 = (ae - bd) \times (bf - ce)$ and $\frac{a}{d} = \frac{b}{e} = \frac{c}{f}$ is satisfied, then number of common root(s) = 0

Alternatively

You may solve the equations to check as to whether they have common root(s) or not, but it is not advisable as their solutions may take considerable time.

C. Cubic and other Equations

Critical Facts

(i) In a cubic equation $ax^3 + bx^2 + cx + d = 0$, if all of a, b, c and d are real numbers, then the equation has atleast one real root. It holds good for all odd degree polynomial equations.

(ii) Out of three roots of a cubic equation, two or even all three roots may be repeated.

(iii) In a cubic equation $ax^3 + bx^2 + cx + d = 0$, if $(a + c) = (b + d)$, then one root is –1. In case (a + c) = $-(b + d)$ then one root of the equation is 1.

(iv) If the coefficients of a polynomial are real, then the non-real roots are in pairs in form of $a \pm ib$.

Computation of Product of all roots and sum of product of roots taken n together or sole

Let an equation be $a_n x^n + a_{n-1} x^{n-1} + a_{n-2}x^{n-2}$ ....+ $a_2x^2 + a_1 x + a_0 = 0$, where a_n, a_{n-1}, ... a_2, a_1, a_0 are real numbers and n is non-negative integer.

Then, number of roots $= n$ [$\because$ degree of this polynomial equation is n]

$$\text{Product of roots} = (-1)^n \times \frac{\text{constant term}}{\text{coefficient of 1st term}}$$

$$= (-1)^n \times \frac{a_0}{a_n}$$

Sum of Product of roots taken n together at a time

$$= \frac{-[\text{coefficient of } (n+1)\text{th term}]}{a_n}, \text{ where } n \text{ is odd}$$

and $\frac{[\text{coefficient of } (n+1)\text{th term}]}{a_n}$, where n is even

e.g. Sum of roots $= \frac{-(a_{n-1})}{a_n}$

[$\because n = 1$ and coefficient of $(1+1)$th term is a_{n-1}]

and sum of pairwise product of roots $= \frac{a_{n-2}}{a_n}$ [$\because n = 2$ and coefficient of $(2+1)$th term is a_{n-2}]

In case of cubic equation

Let the roots of a cubic equation $ax^3 + bx^2 + cx + d = 0$ be α, β and γ. Then,

(i) $\alpha\beta\gamma = (-1)^3 \times \frac{d}{a} = \frac{-d}{a}$

(ii) $\alpha + \beta + \gamma = \frac{-b}{a}$

(iii) $\alpha\beta + \beta\gamma + \gamma\alpha = \frac{c}{a}$

In case of biquadratic (also known as quartic) equation

If the roots of a biquadratic equation $ax^4 + bx^3 + cx^2 + dx + e = 0$ are α, β, γ and δ, then,

(i) $\alpha\beta\gamma\delta = (-1)^4 \times \frac{e}{a} = \frac{e}{a}$

(ii) $\alpha + \beta + \gamma + \delta = \frac{-b}{a}$

(iii) $\alpha\beta + \alpha\gamma + \alpha\delta + \beta\gamma + \beta\delta + \gamma\delta = \frac{c}{a}$

(iv) $\alpha\beta\gamma + \beta\gamma\delta + \gamma\delta\alpha + \delta\alpha\beta = \frac{-d}{a}$

Note *If α is a root of a polynomial equation $a_n x^n + a_{n-1} x^{n-1} + \ldots\ldots + a_2 x^2 + a_1 x + a_0 = 0$, then $(x - \alpha)$ is a factor of the polynomial expression.*

Formation of equations when roots are known

(i) Let α, β and γ be roots of a cubic equation, then the equation be $x^3 - (\alpha + \beta + \gamma)x^2 + (\alpha\beta + \beta\gamma + \gamma\alpha)x - \alpha\beta\gamma = 0$ [mind the sign]

Or

$$(x - \alpha)(x - \beta)(x - \gamma) = 0$$

(ii) If α, β, γ and δ be roots of a biquadratic equation, then the equation be : $x^4 - x^3(\alpha + \beta + \gamma + \delta) + x^2(\alpha\beta + \alpha\gamma + \alpha\delta + \beta\gamma + \beta\delta + \gamma\delta) - x(\alpha\beta\gamma + \beta\gamma\delta + \gamma\delta\alpha + \delta\alpha\beta) + \alpha\beta\gamma\delta = 0$

Or

$$(x - \alpha)(x - \beta)(x - \gamma)(x - \delta) = 0$$

Maximum number of positive and negative real roots of a polynomial equation

(a) **For positive real roots** We count the number of sign changes in the given equation which is in form of $a_n x^n + a_{n-1} x^{n-1} + \ldots\ldots + a_2 x^2 + a_1 x + a_0 = 0$ (i.e. in descending powers of the variable).

The number of sign changes is the maximum number of positive real roots in the equation.

e.g. What is the maximum number of positive real roots in a polynomial equation

$$ax^5 - bx^4 + cx^3 - dx^2 + ex + f = 0$$

Here, the signs are $+ - + - + +$

Clearly, four signs change {i.e. 1st sign (from the left) + changes to –, 2nd sign (–) changes to +, 3rd sign (+) changes to – and 4th sign (–) changes to +.

Hence, maximum number of positive real roots = 4

Note *4 is the maximum no. of positive real roots, it may be 4, 2 or 0 (we count down in pairs).*

(b) **For negative real roots** Here, we also count the number of sign changes but in the new equation which is obtained with the negative variable.

e.g. Compute the maximum number of negative real roots in a polynomial equation

$ax^5 - bx^4 + cx^3 - dx^2 + ex + f = 0$

We 1st write a new equation with negative variable as below

$$a(-x)^5 - b(-x)^4 + c(-x)^3 - d(-x)^2 + e(-x) + f = 0$$

$$\Rightarrow \quad -ax^5 - bx^4 - cx^3 - dx^2 - ex + f = 0$$

Here, signs are $- - - - - +$

Clearly, one sign changes (i.e. 5th sign from the left changes to +).

Hence, maximum number of negative real roots = 1

Take another example for more clarity.

Computer the minimum number of negative real roots in a polynomial equation

$$-ax^6 + bx^5 + cx^4 + dx^3 + ex^2 + fx + g = 0$$

We write a new equation with the negative variable.

The new equation

$$-ax^6 - bx^5 + cx^4 - dx^3 + ex^2 - fx + g = 0$$

Clearly, number of sign changes = 5

Hence, maximum number of negative real roots = 5

However, there may be 5 or 3 or 1 (counting down in pairs) negative root(s).

It implies that atleast one negative real root is in the given equation.

Consider an equation which has only even power of the variable and coefficients and constant are of same sign. e.g. $ax^6 + bx^4 + cx^2 + d = 0$

Number of sign changes = 0

Hence, maximum number of positive real roots = 0

Further, the equation with $(-x)$ is the same as before. Then, maximum number of negative real roots = 0. Also, at $x = 0$, the equation is not satisfied.

As conclusion, we may say that in case of an equation with only even power of the variable wherein coefficients and constant are of same sign, there is no real root.

How to find roots of a cubic equation when one root is known

Take an example Find all roots of a cubic equation $x^3 + 2x^2 - 5x - 6 = 0$ if one root is 2.

We know, if α is a root of a polynomial equation ,

$a_n x^n + a_{n-1} x^{n-1} + \dots\dots\dots\dots + a_2 x^2 + a_1 x + a_0 = 0$, then $(x - \alpha)$ is a factor of the polynomial expression.

Hence, $(x - 2)$ is a factor of $x^2 + 2x^2 - 5x - 6$.

If we divide the expression by $(x - 2)$,we get $x^2 + 4x + 3$ as quotient.

Hence, $(x - 2)(x^2 + 4x + 3) = 0$

$\Rightarrow (x - 2)(x^2 + 3x + x + 3) = 0$

$\Rightarrow (x - 2)[x(x + 3) + 1(x + 3)] = 0$

$\Rightarrow (x - 2)(x + 1)(x + 3) = 0$

$\Rightarrow x = 2, -1 \text{ and } -3$

Alternatively (Through factor-method)

Clearly, $(x - 2)$ is a factor of $x^3 + 2x^2 - 5x - 6$.

Given equation $x^3 + 2x^2 - 5x - 6 = 0$

$\Rightarrow x^2(x - 2) + 4x^2 - 5x - 6 = 0$

$\Rightarrow x^2(x - 2) + 4x(x - 2) + 3x - 6 = 0$

$\Rightarrow x^2(x - 2) + 4x(x - 2) + 3(x - 2) = 0$

$\Rightarrow (x - 2)(x^2 + 4x + 3) = 0$

$\Rightarrow (x - 2)(x^2 + 3x + x + 3) = 0$

$\Rightarrow (x - 2)[x(x + 3) + 1(x + 3)] = 0$

$\Rightarrow (x - 2)(x + 1)(x + 3) = 0$

$\Rightarrow (x - 2) = 0, (x + 1) = 0 \text{ and } (x + 3) = 0$

$\Rightarrow x = 2, -1 \text{ and } -3.$

Another approach

Let a cubic equation be $ax^3 + bx^2 + cx + d = 0$ and one root is α. Then, find the other two.

The other two roots will be as same as roots of a quadratic equation

$$x^2 + (a\alpha + b)x + (a\alpha^2 + b\alpha + c) = 0$$

Now take the same example: Find all roots of a cubic equation $x^3 + 2x^2 - 5x - 6 = 0$ if one root is 2.

Required roots = roots of an quadratic equation

$$x^2 + (1 \cdot 2 + 2)x + (1 \cdot 2^2 + 2 \cdot 2 - 5) = 0$$

i.e. $$x^2 + 4x + 3 = 0$$

Now, solve $x^2 + 4x + 3 = 0$

Clearly, $+4 = +3 + 1$

[refer to, How to find roots of a quadratic equation]

Then, roots = – 3 and – 1

Hence, other two roots of the given cubic equation = – 3 and –1.

The above may be depicted as below.

Write all the coefficients including constant term with their signs in a row

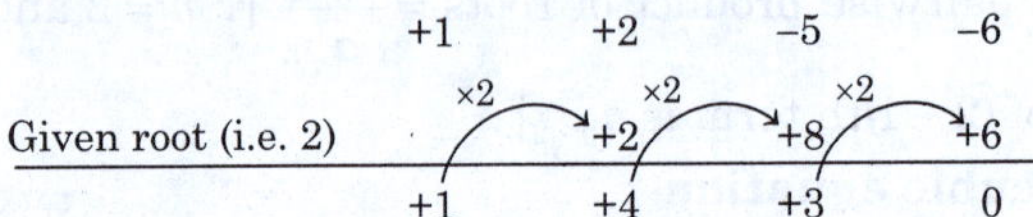

We write the leading coefficient i.e. 1st coefficient in the row vertically below the line. Then, 1st coefficient is multipled by the given root and the product is added to the 2nd coefficient. Thereafter, the sum so obtained is multiplied by the given root and the product obtained is added to the 3rd coefficient. Further, the sum so obtained is multiplied by the given root and the product obtained is added to the constant term. The sum so obtained is always zero. If the sum is not zero, we are committing a mistake in calculation for the above processing. In other words, obtaining 0 as sum is a confirmation of the fact that 2 is a root of the given cubic equation. We may apply this technique of confirmation in case of roots of any cubic equation.

Now, write a quadratic equation with the above results written below the line as (1 st result) x^2 + (2nd result) x + (3rd result) = 0

Hence, 1st result will be quadratic coefficient, 2nd be linear coefficient and 3rd result will be the constant term of this quadratic equation.

Clearly, the quadratic equation be $x^2 + 4x + 3 = 0$. The solutions of this quadratic equation will be other two roots of the given cubic equation.

Take another example

Find all roots of a cubic equation $2x^3 + x^2 - 13x + 6 = 0$ if one root is – 3.

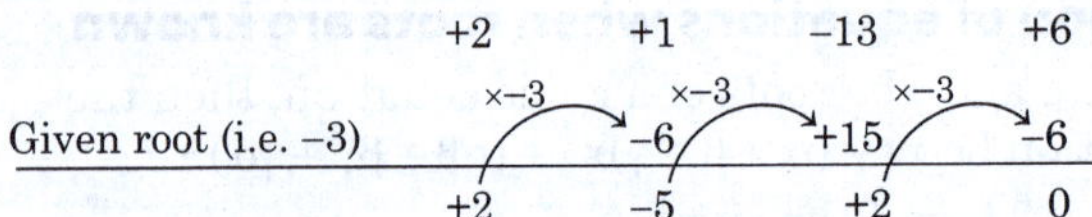

No need to write quadratic equation so as to save time. We may find directly the roots observing the structure of the quadratic equation.

Clearly, –5 = – 4 – 1

Then, roots $= \frac{4}{2}$ and $\frac{1}{2}$ i.e. 2 and $\frac{1}{2}$.

Take one more example, solve $\frac{x^3 - 6x^2 + 11x - 6}{(x - 1)} = 0$

If we divide $(x^3-6x^2+11x-6)$ by $(x-1)$, we get x^2-5x+6.

Hence, $x^2-5x+6=0$

Then, its roots = 2, 3

But the division is lengthy and time taking. Thus, we should adhere to another approach to save the time.

Another approach

In such type of questions, the denominator of the fraction of the LHS of the equation is a factor of its numerator. However, whether the denominator is a factor of numerator or not, may be checked by the factor-theorem.

Here, $x-1=0 \Rightarrow x=1$

At $x=1$, remainder $=1^3-6\cdot1^2+11\cdot1-6=0$

Since the remainder is zero, $(x-1)$ is a factor of $x^3-6x^2+11x-6$.

It is advisable not to check as to whether the denominator is a factor of the numerator or not as we have to save the time in examination hall.

Given equation be

$$\frac{x^3-6x^2+11x-6}{(x-1)}=0$$

$$\Rightarrow \quad x^3-6x^2+11x-6=0$$

Since $(x-1)$ is a factor of $x^3-6x^2+11x-6$, one root of $x^3-6x^2+11x-6=0$ is 1 $[\because x-1=0 \therefore x=1]$

Now, we may find the other two roots of $x^3-6x^2+11x-6=0$ on the above pattern as one is known.

	+2	−6	+11	−6
Given root (i.e. 1)		×1 → +1	×1 → −5	×1 → +6
	+1	−5	+6	0

No need to write quadratic equation. We may find directly the roots observing the structure of the quadratic equation.

Clearly, $-5=-2-3$

Then, roots = 2, 3

Hence, other two roots of the cubic equation $x^3-6x^2+11x-6=0$ are 2, 3. It implies that solutions of $\frac{x^3-6x^2+11x-6}{(x-1)}=0$ are 2, 3.

How to find all three roots of a cubic equation

One root of a cubic equation $ax^3+bx^2+cx+d=0$ is equal to

$$\frac{-b}{3a}+\sqrt[3]{\left(\frac{-b^3}{27a^3}+\frac{bc}{6a^2}-\frac{d}{2a}\right)+\sqrt{\left(\frac{-b^3}{27a^3}+\frac{bc}{6a^2}-\frac{d}{2a}\right)^2+\left(\frac{c}{3a}-\frac{b^2}{9a^2}\right)^3}}$$

$$+\sqrt[3]{\left(\frac{-b^3}{27a^3}+\frac{bc}{6a^2}-\frac{d}{2a}\right)-\sqrt{\left(\frac{-b^3}{27a^3}+\frac{bc}{6a^2}-\frac{d}{2a}\right)^2+\left(\frac{c}{3a}-\frac{b^2}{9a^2}\right)^3}}$$

Now, take one example: What are roots of an equation $2x^3+3x^2-11x-6=0$?

Comparing the given equation with $ax^3+bx^2+cx+d=0$, we get $a=2, b=3, c=-11$ and $d=-6$

Now, $\frac{-b^3}{27a^3}+\frac{bc}{6a^2}-\frac{d}{2a}=\frac{-1}{8}+\frac{-33}{24}-\frac{-6}{4}=0$

and $\frac{c}{3a}-\frac{b^2}{9a^2}=\frac{-11}{6}-\frac{9}{36}=\frac{-25}{12}$

Then, value of one root

$$=\frac{-3}{3\cdot2}+\sqrt[3]{0+\sqrt{(0)^2+\left(\frac{-25}{12}\right)^3}}+\sqrt[3]{0-\sqrt{(0)^2+\left(\frac{-25}{12}\right)^3}}$$

$$=\frac{-1}{2}+\sqrt[3]{\left(\frac{-25}{12}\right)^{\frac{3}{2}}}+\sqrt[3]{-1\times\left(\frac{-25}{12}\right)^{\frac{3}{2}}}$$

$$=\frac{-1}{2}+\left(\frac{-25}{12}\right)^{\frac{1}{2}}+(-1)^{\frac{1}{3}}\times\left(\frac{-25}{12}\right)^{\frac{1}{2}}$$

$$=\frac{-1}{2}+\left(\frac{-25}{12}\right)^{\frac{1}{2}}-\left(\frac{-25}{12}\right)^{\frac{1}{2}}=\frac{-1}{2}$$

We are well conversant with the method meant for finding all the roots of a cubic equation when one root is known. Write all the coefficients including constant term of the given equation with the signs in a row.

	+2	+3	−11	−6
Given root (i.e. $-\frac{1}{2}$)		$\times -\frac{1}{2}$ → −1	$\times -\frac{1}{2}$ → −1	$\times -\frac{1}{2}$ → +6
	+2	+2	−12	0

No need to write quadratic equation so as to save time. We may find directly the roots observing the structure of the quadratic equation.

Clearly, linear coefficient, $+2=+6-4$

Then, roots $=\frac{-6}{2}$ and $\frac{4}{2}$ i.e. −3 and 2.

Hence, all the three roots of the given cubic equation be −3, 2 and $-\frac{1}{2}$.

It is clear from the above that finding roots through the above cubic formula is very lengthy. Also, another approach in which x is replaced by $\left(y-\frac{b}{3a}\right)$ so as to eliminate the 2nd degree term, is very complicated. Hence, we ought to emphasis on the alternative approach.

Alternative approach

The given equation is $2x^3 + 3x^2 - 11x - 6 = 0$ and we have to find its roots.

We write the equation like

$2x^3 + 3x^2 - (11\times$ leading coefficient) $- 6$ (leading coefficient)$^2 = 0$

Hence, $2x^3 + 3x^2 - 22x - 24 = 0$ [$\because$ leading coefficient = 2]

Recall the concept of 'formation of equation'. When α, β and γ be roots of a cubic equations, then the equation be $x^3 - (\alpha + \beta + \gamma)x^2 + (\alpha\beta + \beta\gamma + \gamma\alpha)x - \alpha\beta\gamma = 0$

Now, factors of constant $24 = \pm 1, \pm 2, \pm 3, \pm 4, \pm 6, \pm 8, \pm 12, \pm 24$. We have to select any three factors of 24 whose product is −24 and sum is +3. We are free to take the repeated factors also, if required. The possible sets of factors are (2, −3, 4) and (1, 6, −4)

[***Hint*** Since the product is negative, either all three factors are negative or one of them is negative. If we select all three factors negative, then their sum will be negative, but the given sum is positive. Hence, we select one factor negative and the remaining two positive].

It is also fact that the cubic equation has only three roots, so only one set of factors is to be selected either (2, −3, 4) or (1, 6,−4). The coefficient associated with 3rd term is actually sum of pairwise product of roots. Here, it is −22. The 2nd set of factors i.e. (1, 6, −4) yielding − 22 as the sum of pairwise product of factors [say, $1\cdot 6 + 6\cdot(-4) + (-4)\cdot 1 = -22$]

Now, divide each factor of the 2nd set by the leading coefficient. The resultants with the opposite signs will be the roots of the given cubic equation .

Here, roots are $\frac{-1}{2}, \frac{-6}{2}$ and $\frac{4}{2}$ i.e. $\frac{-1}{2}$, −3 and 2.

Note

(i) The given equation may be written as $x^3 + \frac{3}{2}x^2 - \frac{11}{2}x - 3 = 0$. We may find its roots on the above pattern. However, in order to avoid possible complications arising from mathematical operations with fractions, we have taken 3 instead of $\frac{3}{2}$ as coefficient of 2nd term.

Coefficient of 2nd term refers to sum of roots. Accordingly, the coefficient associated with 3rd term as well as constant term have been modified.

(ii) In most of the cases, only one set of factors is possible. Thus, no need to observe the coefficient associated with the 3rd term However, in case of any doubt in selection of factors, there should be no hesitation to test as to whether the sum of pairwise product of factors selected, is equal to the 3rd term-coefficient or not.

(iii) The logic behind alteration in sign of the quotient obtained on dividing the factors of a set by the leading coefficient is that the 2nd term and the constant term both are negative when we form an equation from the given roots, whereas these are, in the standard form, positive [i.e. $x^3 - (\alpha + \beta + \gamma) x^2 + (\alpha\beta + \beta\gamma + \gamma\alpha)x - \alpha\beta\gamma = 0$ and $ax^3 + bx^2 + cx + d = 0$]

(iv) If the leading coefficient is 1, then there is no need to modify the coefficient associated with 3rd term as well as constant term because multiplication by 1 or its square does not affect the quantity. Also, no need to divide the set of factors selected, by the leading coefficient 1 as there is no effect if it is divided by 1.

(v) No need to mention that if one root of a cubic equation is known, the selection of other two factors of the constant term will be very easy. The given root with the opposite sign will be the one factor. We may find the other two roots either by this approach or by the approach as detailed under 'How to find roots of a cubic equation when one root is known'.

Take another example

What are the roots of $x^3 - 2x^2 - 5x + 6 = 0$?

Since the leading coefficient is 1, there is no need to modify coefficient associated with 3rd term as well as constant term.

Now, factors of constant term $6 = \pm 1, \pm 2, \pm 3, \pm 6$

We have to select any three factors of 6 such that their product is +6 and sum is −2.

The possible set of factors is (−1, 2, −3).

[***Hint*** Since the product is positive, either all three factors are positive or one of them is positive. If we select all three factors positive, then their sum will be positive, but the given sum is negative. Hence, we selected one factor positive and the remaining two negative.]

Then, the required roots are 1, −2 and 3.

Take one more example

What are the roots of $x^3 - 5x^2 + 8x - 4 = 0$?

Factors of $4 = \pm 1, \pm 2, \pm 4$

The possible set of factors is (−1, − 2, − 2). Sometimes, roots are repeated.

Then, required roots = 1, 2 and 2.

Alternatively

Here, $(a + c) = -(b + d)$. It implies that one root of the equation is 1.

Now, apply the method meant for computing the roots of a cubic equation when one root is known.

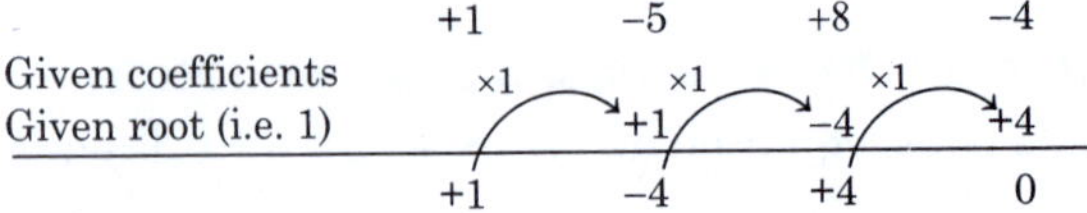

No need to write quadratic equation so as to save time. We may find directly the roots observing the structure of the quadratic equation.

Clearly, − 4 = − 2 − 2

Then, roots = + 2 , + 2

Hence, all three roots be 1, 2 and 2.

Concept of common roots for polynomial equations

(i) The HCF of two distinct polynomials is either their difference or its a factor or 1. Two polynomial equations have common root(s) if HCF of their corresponding polynomials is other than 1 and the values of the common roots are either same as roots of corresponding equations of difference between given polynomials or its subset. It is reiterated that the common roots must satisfy both the given equations.

(ii) Two polynomial equations have no common root if HCF of their corresponding polynomials is 1. It implies that the roots of the corresponding equation of difference between given polynomials or its subset shall not satisfy both the equations.

We may conclude from the above that if the roots of the corresponding equation of difference between given polynomials or its subset don't satisfy both the equations, the given equations have no common root and *vice-versa*. On the contrary, if roots of the corresponding equation of difference between given polynomials or its subset satisfy both the equations, the given equations have common root(s) and *vice-versa* and its/their values are same as value of the root(s) at which the equations are satisfied.

It is important to note that we should, if possible, equate the coefficients of the same degree term of the given equations so that the difference can be obtained in least terms.

Take one example for better understanding.

Total number of common roots between

$x^3 + 3x^2 + 4x + 5 = 0$

and $x^3 + 2x^2 + 7x + 3 = 0$ is ? **[CAT-03]**

Given equations are $x^3 + 3x^2 + 4x + 5 = 0$ and $x^3 + 2x^2 + 7x + 3 = 0$.

Difference between given polynomials $x^3 + 3x^2 + 4x + 5$ and $x^3 + 2x^2 + 7x + 3$ is $(x^2 - 3x + 2)$.

If they have common roots, the roots of $x^2 - 3x + 2 = 0$ or its subset must satisfy the both given polynomial equations. On the contrary, if they have no common root, the roots of $x^2 - 3x + 2 = 0$ or its subset must not satisfy the both given polynomial equations.

Now, roots of $x^2 - 3x + 2 = 0$ are 1 and 2 [we are familiar with technique for finding the roots of a quadratic equation]

At $x = 1$, $x^3 + 3x^2 + 4x + 5 = 13$ and $x^3 + 2x^2 + 7x + 3 = 13$

and at $x = 2$, $x^3 + 3x^2 + 4x + 5 = 33$

and $x^3 + 2x^2 + 7x + 3 = 33$

We observe that the given equations are not satisfied at any root of the corresponding equation of the difference of the given polynomials. Hence, the given equations have no common root.

Alternatively

We may find the roots of any one of the given equations and then we will check as to which root(s) obtained, is/are satisfying the another equation also. The root(s) satisfying both the equations will be the common root(s).

Unfortunately, we are unable to find roots of any one of the given equations with the help of the technique explained earlier. We may try this alternative approach in another case.

Take two more examples

- What is the total number of common roots between $2x^4 - 9x^3 + 6x^2 + 11x - 6 = 0$ and $x^3 - 3x^2 - 4x + 12 = 0$?

Since we are comfortable with finding the roots of a polynomial equation only upto 3rd degree, the difference of given polynomials must not be in degree higher than 3. Hence, the 4th degree term should be eliminated while subtracting.

Given, equations $2x^4 - 9x^3 + 6x^2 + 11x - 6 = 0$

and $2x^4 - 6x^3 - 8x^2 + 24x = 0$

[we multiplied 2nd equation by $2x$].

Now, difference between polynomials

$$= -3x^3 + 14x^2 - 13x - 6$$

Then, corresponding equation be

$$-3x^3 + 14x^2 - 13x - 6 = 0$$

$$\Rightarrow \quad 3x^3 - 14x^2 + 13x + 6 = 0$$

Solving this equation, we get $x = \frac{-1}{3}, 2, 3$

We know, if the given equations have common root(s), they are satisfied either at all three values of x or at its subset. Now, check as to which value(s) of x is/are satisfying the given equations.

At $x = 2, 3$ both the equations are satisfied.

Hence, common roots = 2,3

$\therefore$ Number of common roots = 2

Alternatively

We are capable to find the roots of the cubic equation $x^3 - 3x^2 - 4x + 12 = 0$.

On solving, we get $x = -2, 2, 3$

Now, check as to which value(s) of x is/are satisfying the another equation also.

At $x = 2, 3$ another equations $2x^4 - 9x^3 + 6x^2 + 11x - 6 = 0$ is also satisfied.

Since both the equations are satisfied at $x = 2, 3$, the common roots are 2 and 3.

$\therefore$ Number of common roots = 2

- What is the number of common roots between $2x^4 - 5x^3 + 5x - 2 = 0$ and $2x^4 - 7x^3 + 4x^2 + 7x - 6 = 0$?

Difference between respective polynomials

$$= 2x^3 - 4x^2 - 2x + 4$$

Then, corresponding equation be $2x^3 - 4x^2 - 2x + 4 = 0$

$\Rightarrow \quad x^3 - 2x^2 - x + 2 = 0$

On solving this equation, we get $x = -1, 1, 2$

At $x = -1, 1, 2$ both the given equations are satisfied.

Hence, values of common roots $x = -1, 1, 2$.

$\therefore$ Number of common roots = 3

[Since we are not able to find the roots of any one of the given equations with help of the technique elaborated before, the alternative approach does not work]

Note *No doubt, this concept will also cover the conditionality for quadratic equations to have common root(s).*

Illustrations

Problems based on Linear Equations

Problem 1 If $x + y - 7 = 0$ and $3x + y - 13 = 0$, then what is $4x^2 + y^2 + 4xy$ equals to

Sol. Given, $1x + 1y - 7 = 0$ ...(i)

and $3x + 1y - 13 = 0$ ...(ii)

Then, $x = \dfrac{1(-13) - 1(-7)}{1(1) - 3(1)} = \dfrac{-6}{-2} = 3$

Putting the value of x in Eq. (i), we get $y = 4$

Now, $4x^2 + y^2 + 4xy = (2x + y)^2$

$= (2 \cdot 3 + 4)^2 = 100$

Problem 2 The system of equations $3x + y - 4 = 0$ and $6x + 2y - 8 = 0$ has :

(a) a unique solution $x = 1, y = 1$

(b) a unique solution $x = 1, y = 2$

(c) no solution

(d) infinite solutions

Sol. Here, $\dfrac{3}{6} = \dfrac{1}{2} = \dfrac{-4}{-8}$

Since the given equations are dependent, the system of equations has infinite solutions.

Problem 3 For which of the following values of k will $3x + (k + 3)y = 1$ and $kx + 6y = 4$ have a unique solution? *[CAT-20, CDS-21]*

(a) 3 (b) – 6

(c) 6 (d) any value except 3 and – 6

Sol. We know, a system of equations $a_1x + b_1y + c_1 = 0$ and $a_2x + b_2y + c_2 = 0$ has a unique solution if $\dfrac{a_1}{a_2} \neq \dfrac{b_1}{b_2}$.

Hence, $\dfrac{3}{k} \neq \dfrac{k+3}{6} \Rightarrow k(k + 3) \neq 18$

If $k(k + 3) = 18$, then $k = 3$ and -6. It implies that at $k = 3, -6$ the system of equations has no unique solution. Hence, at any value of k other than 3 and – 6, the system has a unique solution.

Problem 4 What is the value of k for which the system of equations $x + 2y - 3 = 0$ and $5x + ky + 7 = 0$ has no solution?

Sol. We know, a system of equations $a_1x + b_1y + c_1 = 0$ and $a_2x + b_2y + c_2 = 0$ has no solution if $\dfrac{a_1}{a_2} = \dfrac{b_1}{b_2} \neq \dfrac{c_1}{c_2}$

Hence, $\dfrac{1}{5} = \dfrac{2}{k} \neq \dfrac{-3}{7}$

$\Rightarrow \quad k = 10$

Problem 5 The number of solutions of the equation $2x + y = 40$, where x and y are positive integers and $x \le y$, is *[CAT-06]*

Sol. Given, $2x + y = 40$

$\Rightarrow \quad y = 2(20 - x)$

Since x and y are positive integers, we may get the value of y in positive integer at any positive integer value of x between 0 and 20. Cleary, y will be less than x, if value of x is between 13 and 20.

Hence, at $x = 1, 2, 3, 4$............ $13, x \le y$.

It implies that number of solutions of the given equation = 13

Alternatively

Given, $2x + y = 40 \Rightarrow y = 2(20 - x)$

$\because \quad x \le y$

$\therefore \quad x \le 2(20 - x)$

$\Rightarrow \quad 3x \le 40 \Rightarrow x \le 13.33$

Since x is a positive integer, value of x is 1, 2, 3, 4 ....,13.

It implies that number of solutions of the given equation = 13

Problem 6 If $\dfrac{2x - 3y + 1}{2} = \dfrac{x + 4y + 8}{3} = \dfrac{4x - 7y + 2}{5}$, then what is $(x + y)$ equal to :

Sol. Given, $\dfrac{2x - 3y + 1}{2} = \dfrac{x + 4y + 8}{3}$

$\Rightarrow \quad 6x - 9y + 3 = 2x + 8y + 16$

$\Rightarrow \quad 4x - 17y - 13 = 0$ ...(i)

Also, $\dfrac{2x - 3y + 1}{2} = \dfrac{4x - 7y + 2}{5}$

[you may also take 2nd and 3rd term, no problem]

$\Rightarrow \quad 10x - 15y + 5 = 8x - 14y + 4$

$\Rightarrow \quad 2x - 1y + 1 = 0$ ...(ii)

Then, $x = \dfrac{-17(1) - (-1)(-13)}{4(-1) - 2(-17)} = \dfrac{-30}{30} = -1$

Putting the value of x in Eq. (i), we get $4(-1) - 17y - 13 = 0$

$\Rightarrow \quad -17y = 17$

$\Rightarrow \quad y = -1$

Now, $x + y = -1 - 1 = -2$

Problem 7 If $x + y = 5$, $y + z = 10$ and $z + x = 15$, then which one of the following is correct? *[CDS-14]*

(a) $z > x > y$ (b) $z > y > x$

(c) $x > y > z$ (d) $x > z > y$

Sol. Adding all the Eqs, we get $2(x+y+z)=30 \Rightarrow x+y+z=15$

Then, $x = 15-(y+z) = 15-10 = 5$

$y = 15-(z+x) = 15-15 = 0$

and $z = 15-(x+y) = 15-5 = 10$

Clearly, $z > x > y$

Problem 8 The total number of integer pairs (x, y) satisfying the equation $x+y=xy$ is/are : **[CAT-04]**

Sol. Given, $x+y=xy$

$\Rightarrow xy-x-y=0$

$\Rightarrow x(y-1)-y=0$

$\Rightarrow x(y-1)-1(y-1)=0+1$ [on adding 1 both sides]

$\Rightarrow (x-1)(y-1)=1$

Since x and y are integers, values of $(x-1)$ and $(y-1)$ must be integers.

Clearly, $(x-1)=1$ and $(y-1)=1$

$\Rightarrow x=2$ and $y=2$

Also, $(x-1)=-1$ and $(y-1)=-1$

$\Rightarrow x=0$ and $y=0$

Hence, number of integer pairs = 2

Problem 9 If $\frac{3}{x+y}+\frac{2}{x-y}=2$ and $\frac{9}{x+y}-\frac{4}{x-y}=1$, then what is the value of $\frac{x}{y}$?

Sol. Let $\frac{3}{x+y}=a$ and $\frac{2}{x-y}=b$

As given, $1a+1b-2=0$ ...(i)

and $3a-2b-1=0$ ...(ii)

Then, $a=\frac{1(-1)-(-2)(-2)}{1(-2)-3(1)}=\frac{-5}{-5}=1$

Putting the value of a in Eq. (i), we get $b=1$

$\because \frac{3}{x+y}=a$

$\therefore x+y=3$ ...(iii)

Also $\frac{2}{x-y}=b$

$\therefore x-y=2$ ...(iv)

On adding Eqs. (iii) and (iv), we get $2x=5$

$\Rightarrow x=2.5$

Putting the value of x in Eq. (iii) we get $y=0.5$

Now, $\frac{x}{y}=\frac{2.5}{0.5}=5$

Problem 10 If x and y are integers, then the equation $5x+19y=64$ has : **[CAT-03]**

(a) no solution for $x<300$ and $y<0$

(b) no solution for $x>250$ and $y>-100$

(c) a solution for $250<x<300$

(d) a solution for $-59<y<-56$

Sol. Given, $5x+19y=64 \Rightarrow 5x=64-19y$

$\Rightarrow 5x=60-15y+4-4y$

$\Rightarrow 5x=15(4-y)+4(1-y)$

Since, x and y are integers, $15(4-y)+4(1-y)$ must exactly be divisible by 5. Obviously, $15(4-y)$ is divisible by 5 and $4(1-y)$ will be divisible by 5 only when $(1-y)$ is divisible by 5. It implies that $1-y=5k$, where k is an integer constant.

$\because 1-y=5k \quad \therefore y=1-5k$

Hence, possible values of $y=(1-5\cdot1)$, $(1-5\cdot2)$, $(1-5\cdot3)$, ... i.e. $-4, -9, -14, -19, -24$, ... $-54, -59, -64, -69$.

Clearly, option (d) is eliminated.

The logic behind taking negative values of y is that negative value of y has been indicated in the options. Further, if the value of y is negative x will be positive and the positive value of x has been given in the options.

Since value of x in the option is near to 250. Then, y should be near to 60. [hint $(64-5\cdot250)\div19$]

If we take $y=-59$, x will be $[64+59\cdot19)\div5]$ i.e. 237

If we take $y=-64$, x will be 256 and at $y=-69$, x is 275 and so on.

Conclusion: The given equation has a solution for $250<x<300$.

Problems based on Quadratic Equations

[Word-based problems related to concept of quadratic equation have been discussed in respective chapters]

I. Determination of Number of Roots and Nature of Roots

Problem 11 The number of real solutions of the equation $|x|^2-3|x|+2=0$ are

Sol. Given, $|x|^2-3|x|+2=0$

Linear coefficient, $-3=-1-2$

Then, $|x|=1, 2$

$\therefore x=\pm1, \pm2$

Hence, number of real solutions – 4

Note *If $x^2-3|x|+2=0$, then values of x will be same as the above because $x^2-3|x|+2=0$ can be written as $|x|^2-3|x|+2=0$.*

Problem 12 Both the roots of the given equation $(x-a)(x-b)+(x-b)(x-c)+(x-c)(x-a)=0$ are always

(a) positive (b) negative (c) real (d) imaginary

Sol. Given equation $(x-a)(x-b)+(x-b)(x-c)+(x-c)(x-a)=0$

$\Rightarrow 3x^2-2(a+b+c)x+ab+bc+ca=0$

Now, discriminant $D=[2(a+b+c)]^2-4\cdot3\cdot(ab+bc+ca)$

$=4[a^2+b^2+c^2-ab-bc-ca]$

$=2\cdot2[a^2+b^2+c^2-ab-bc-ca]$

$=2[(a-b)^2+(b-c)^2+(c-a)^2]$

$[\because (a-b)^2+(b-c)^2+(c-a)^2 = 2(a^2+b^2+c^2-ab-bc-ca)]$

Obviously, $2[(a-b)^2+(b-c)^2+(c-a)^2]\geq0$ as square of any number is either 0 or greater than 0.

Since $D\geq0$, both the roots will be real.

Computation of Value of Algebric Expression Having Roots of a Quadratic Equation as Variables

Problem 13 If m and n $(m > n)$ are the roots of the equation $7(x + 2a)^2 + 3a^2 = 5a(7x + 23a)$, where, $a > 0$, then what is $3m - n$ equal to? *[CDS-14]*

(a) $12a$ (b) $14a$
(c) $15a$ (d) $18a$

Sol. Given, $7(x + 2a)^2 + 3a^2 = 5a(7x + 23a)$

$\Rightarrow 7(x^2 + 4ax + 4a^2) + 3a^2 = 35ax + 115a^2$

$\Rightarrow \quad 7x^2 - 7ax - 84a^2 = 0$

$\Rightarrow \quad x^2 - ax - 12a^2 = 0$

Linear coefficient, $-a = -4a + 3a$

[$\because$ product of quadratic coefficient and the constant is $-12a^2$]

The roots be $\dfrac{+4a}{\text{quadratic coefficient}}$ and $\dfrac{-3a}{\text{quadratic coefficient}}$

$= \dfrac{+4a}{1}$ and $\dfrac{-3a}{1} = 4a$ and $-3a$

[How to find roots has been discussed in length step by step in the beginning]

Clearly, $m = 4a$ and $n = -3a$ [$\because m > n$]

Now, $3m - n = 3(4a) - (-3a) = 15a$

Problem 14 If α and β are the roots of the equation $x^2 - x - 1 = 0$, then what is $\dfrac{\alpha^2 + \beta^2}{(\alpha^2 - \beta^2)(\alpha - \beta)}$ equal to

Sol. $\alpha + \beta = \dfrac{-(-1)}{1} = 1$ and $\alpha\beta = \dfrac{(-1)}{1} = -1$

Then, $(\alpha - \beta)^2 = (\alpha + \beta)^2 - 4\alpha\beta = (\alpha + \beta)^2 - 4\alpha\beta$

$= (1)^2 - 4(-1) = 5$

Now, $\dfrac{\alpha^2 + \beta^2}{(\alpha^2 - \beta^2)(\alpha - \beta)} = \dfrac{(\alpha + \beta)^2 - 2\alpha\beta}{(\alpha + \beta)(\alpha - \beta)(\alpha - \beta)} = \dfrac{(\alpha + \beta)^2 - 2\alpha\beta}{(\alpha + \beta)(\alpha - \beta)^2}$

$= \dfrac{(1)^2 - 2(-1)}{(1)\cdot 5} = \dfrac{3}{5}$

III. Relation Among Quadratic Coefficient, Linear Coefficient and Constant of a Quadratic Equation when Relation Between its Roots is Known

Problem 15 If one of the roots of the equation $px^2 + qx + r = 0$ is three times the other, then which one of the following relations is correct?

(a) $3q^2 = 16pr$ (b) $q^2 = 24pr$
(c) $p = q + r$ (d) $p + q + r = 1$

Sol. Let one root is α, then another be 3α.

Now, sum of roots $= \alpha + 3\alpha = \dfrac{-\text{(linear coefficient)}}{\text{quadratic coefficient}} = \dfrac{-q}{p}$

$\Rightarrow \quad \alpha = \dfrac{-q}{4p}$

Also, product of roots $= \alpha \cdot 3\alpha = \dfrac{\text{constant}}{\text{quadratic coefficient}} = \dfrac{r}{p}$

$\Rightarrow \quad 3\left(\dfrac{-q}{4p}\right)^2 = \dfrac{r}{p}$

$\Rightarrow \quad 3q^2 = 16pr$

Alternatively

We know, if one root of equation $ax^2 + bx + c = 0$ is m times the other then $m + \dfrac{1}{m} = \dfrac{b^2 - 2ac}{ac}$

$\Rightarrow \quad 3 + \dfrac{1}{3} = \dfrac{q^2 - 2pr}{pr}$ [$\because$ given Eq. is $px^2 + qx + r = 0$]

$\Rightarrow \quad 3q^2 = 16pr$

Problem 16 If roots of the equation $(a^2 - bc)x^2 + 2(b^2 - ac)x + (c^2 - ab) = 0$ are equal, where $b \neq 0$, then which one of the following is correct? *[CDS-14]*

(a) $a + b + c = abc$ (b) $a^2 + b^2 + c^2 = 0$
(c) $a^3 + b^3 + c^3 = 0$ (d) $a^3 + b^3 + c^3 = 3abc$

Sol. Since the roots are equal, the value of discriminant is 0.

$\Rightarrow [2(b^2 - ac)]^2 - 4\cdot(a^2 - bc)\cdot(c^2 - ab) = 0$

$\Rightarrow 4b^4 + 4a^2c^2 - 8b^2ac - 4a^2c^2 + 4a^3b + 4c^3b - 4b^2ac = 0$

$\Rightarrow 4b^4 - 12b^2ac + 4a^3b + 4c^3b = 0$

$\Rightarrow \quad 4b(b^3 - 3abc + a^3 + c^3) = 0$

$\Rightarrow \quad (b^3 - 3abc + a^3 + c^3) = 0$

$\Rightarrow \quad a^3 + b^3 + c^3 = 3abc$

Problem 17 If the roots of the equation $Ax^2 + Bx + C = 0$ are -1 and 1, then which one of the following is correct?

(a) A and C are both zero
(b) A and B are both positive
(c) A and C are both negative
(d) A and C are of opposite signs

Sol. Sum of the roots $= -1 + 1 = 0$

Product of the roots $= -1 \times 1 = -1$

Then, the actual equation is

$x^2 - (0)x + (-1) = 0$

$\Rightarrow x^2 - 1 = 0$ or $-x^2 + 1 = 0$

Hence, A and C are of opposite signs.

Alternatively

Given equation $Ax^2 + Bx + C = 0$

Sum of roots $= \dfrac{-B}{A} = -1 + 1$

$\Rightarrow \quad \dfrac{-B}{A} = 0 \Rightarrow B = 0$

Product of roots $= \dfrac{C}{A} = -1 \times 1$

$\Rightarrow \quad C = -A$ or $A = -C$

Hence, A and C are of opposite signs.

Problem 18 If one of the roots of the equation $x^2 - bx + c = 0$ is the square of the other, then which of the following option is correct? *[CDS-13]*

(a) $b^3 = 3bc + c^2 + c$ (b) $c^3 = 3bc + b^2 + b$
(c) $3bc = c^3 + b^2 + b$ (d) $3bc = c^3 + b^3 + b^2$

Sol. Let the roots be α and α^2.

Now, sum of roots $= \frac{-(-b)}{1}$

$\Rightarrow \alpha + \alpha^2 = b$

Product of roots $= \frac{c}{1} \Rightarrow \alpha \cdot \alpha^2 = c$

$\Rightarrow \alpha = c^{\frac{1}{3}}$

From the above, $\alpha + \alpha^2 = b$

$\Rightarrow c^{\frac{1}{3}} + c^{\frac{2}{3}} = b$

$\Rightarrow c + c^2 + 3\left(c^{\frac{1}{3}} \cdot c^{\frac{2}{3}}\right)\left(c^{\frac{1}{3}} + c^{\frac{2}{3}}\right) = b^3$ [on cubing both sides]

$\Rightarrow c + c^2 + 3cb = b^3$

$\Rightarrow b^3 = 3bc + c^2 + c$

Alternatively

We know, If ane root of equation $ax^2 + bx + c = 0$ is the square of the other, then $a + c = \frac{b^3 - 3abc}{ac}$

$\Rightarrow 1 + c = \frac{(-b)^3 - 3 \cdot 1(-b)c}{-1 \cdot c}$

[$\because$ given equation is $x^2 - bx + c = 0$]

$\Rightarrow -c - c^2 = -b^3 + 3bc$

$\Rightarrow b^3 = 3bc + c^2 + c$

Problem 19 If one root of the equation $\frac{x^2}{a} + \frac{x}{b} + \frac{1}{c} = 0$ is reciprocal to the other, then which one of the following is correct?

(a) $a = b$ (b) $b = c$
(c) $ac = 1$ (d) $a = c$

Sol. Let the roots be α and $\frac{1}{\alpha}$.

Now, product of roots $= \frac{\frac{1}{c}}{\frac{1}{a}}$

$\Rightarrow \alpha \cdot \frac{1}{\alpha} = \frac{a}{c} \Rightarrow a = c$

Alternatively

We know, if one root of equation $ax^2 + bx + c = 0$ is reciprocal of the other, then $a = c$

$\Rightarrow \frac{1}{a} = \frac{1}{c}$ [$\because$ given equation is $\frac{x^2}{a} + \frac{x}{b} + \frac{1}{c} = 0$]

$\Rightarrow a = c$

Problem 20 If $\sin\theta$ and $\cos\theta$ are the roots of the equation $ax^2 - bx + c = 0$, then which one of the following is correct?

(a) $a^2 + b^2 + 2ac = 0$ (b) $a^2 - b^2 + 2ac = 0$
(c) $a^2 + c^2 + 2ab = 0$ (d) $a^2 - b^2 - 2ac = 0$

Sol. Product of the roots $= \sin\theta \cdot \cos\theta = \frac{c}{a}$

Sum of the roots $= \sin\theta + \cos\theta = \frac{b}{a}$

$\Rightarrow \sin^2\theta + \cos^2\theta + 2 \cdot \sin\theta\cos\theta = \left(\frac{b}{a}\right)^2$ [squaring both sides]

$\Rightarrow 1 + 2 \cdot \frac{c}{a} = \frac{b^2}{a^2}$

$\Rightarrow \frac{2c}{a} = \frac{b^2 - a^2}{a^2}$

$\Rightarrow a^2 - b^2 + 2ac = 0$

Alternatively

Here, sum of squares of both roots is 1 as $\sin^2\theta + \cos^2\theta = 1$.

We know, if sum of squares of both roots of equation $ax^2 + bx + c = 0$ is 1, then $b^2 = a^2 + 2ac$

$\Rightarrow (-b)^2 = a^2 + 2ac$ [$\because$ given equation is $ax^2 - bx + c = 0$]

$\Rightarrow a^2 - b^2 + 2ac = 0$

IV. When One Root of a Quadratic Equation is known and Another is to be Found Out

Problem 21 If one root of the equation $ax^2 + x - 3 = 0$ is -1, then what is the other root?

Sol. Obviously, one value of x is -1.

Then, $a(-1)^2 + (-1)$ 3 0

$\Rightarrow a = 4$

Hence, $4x^2 + x - 3 = 0$

Product of roots of equation $4x^2 + x - 3 = 0$ is $-\frac{3}{4}$

$\Rightarrow -1 \times$ 2nd root $= -\frac{3}{4}$

$\Rightarrow$ 2nd root $= \frac{3}{4}$

V. Relation Among Quadratic Coefficient, Linear Coefficient and Constant of Two Quadratic Equations Whose Roots are Related Somehow

Problem 22 If m and n are the roots of $x^2 + ax + b = 0$ and m^2, n^2 are the roots of the equation $x^2 - cx + d = 0$, then which of the following is/are correct ?

I. $2b - a^2 = c$ II. $b^2 = d$

Select the correct answer using the codes given below :

(a) only I (b) Only II
(c) Both I and II (d) Neither I nor II

Sol. $m+n=-a$ and $mn=b$ [$\because$ m, n are roots of $x^2+ax+b=0$]

$\Rightarrow$ $(m+n)^2=a^2$

$\Rightarrow$ $(m^2+n^2)+2mn=a^2$

$\Rightarrow$ $c+2b=a^2$

$\Rightarrow$ $2b-a^2=-c$

Also, $m^2+n^2=c$ and $m^2n^2=d$

$\Rightarrow$ $d=(mn)^2 \Rightarrow d=b^2$

Hence, option (b) is correct.

Problem 23 If the ratio of the roots of $x^2+bx+c=0$ and $x^2+qx+r=0$ be same, then :

(a) $r^2c=b^2q$ (b) $r^2b=c^2q$

(c) $rb^2=cq^2$ (d) $rc^2=bq^2$

Sol. Let α, β are roots of $x^2+bx+c=0$ and m, n are roots of $x^2+qx+r=0$.

Then, $\alpha+\beta=-b$, $\alpha\beta=c$ and $m+n=-q$, $mn=r$

As given, $\frac{\alpha}{\beta}=\frac{m}{n}$

$\Rightarrow$ $\frac{\alpha+\beta}{\alpha-\beta}=\frac{m+n}{m-n}$

$\Rightarrow$ $\frac{(\alpha+\beta)^2}{(\alpha-\beta)^2}=\frac{(m+n)^2}{(m-n)^2}$

$\Rightarrow$ $\frac{(\alpha+\beta)^2}{(\alpha+\beta)^2-4\alpha\beta}=\frac{(m+n)^2}{(m+n)^2-4mn}$

$\Rightarrow$ $\frac{(-b)^2}{(-b)^2-4c}=\frac{(-q)^2}{(-q)^2-4r}$

$\Rightarrow$ $b^2q^2-4rb^2=b^2q^2-4cq^2$

$\Rightarrow$ $rb^2=cq^2$

VI. Number of Quadratic Equations or Quadratic Coefficient or Linear Coefficient or Constant of a Quadratic Equation When Nature of Root is Known

Problem 24 If both a and b belong to the set (1, 2, 3, 4), then the number of equation of the form $ax^2+bx+1=0$ having real roots, is **[CAT-96, 03]**

Sol. Since roots are real, $D\ge 0$, where D stands for Discriminant.

$\Rightarrow$ $b^2-4ac\ge 0$

$\Rightarrow$ $b^2-4a\ge 0$ [$\because c=1$]

$\Rightarrow$ $b^2\ge 4a$

The possible values of b^2 are $1^2, 2^2, 3^2$ and 4^2

For $b^2=1$, number of equations $=0$

[$\because b^2 \not\ge 4a$ at any value of a belonging to the set.]

For $b^2=4$, number of equations $=1$ [$\because b^2\ge 4a$ at $a=1$]

For $b^2=9$, number of equations $=2$ [$\because b^2\ge 4a$ at $a=1,2$]

For $b^2=16$, number of equations $=4$ [$\because b^2\ge 4a$ at $a=1,2,3,4$]

Hence, total number of equations $=0+1+2+4=7$

Problem 25 If the equation $x^2+2(1+k)x+k^2=0$ has equal roots, then what is the value of k? **[SSC (CGL-T-II)-12, CDS-21]**

Sol. Since roots are equal, the value of $D=0$, where D stands for Discriminant.

$\Rightarrow$ $b^2-4ac=0$

$\Rightarrow$ $(2+2k)^2-4\cdot 1\cdot k^2=0$

$\Rightarrow$ $4+8k=0$

$\Rightarrow$ $k=\frac{-1}{2}$

Problem 26 What is the condition that the equation $ax^2+bx+c=0$, where $a\ne 0$, has both the roots positive? **[CDS-11]**

(a) a, b and c are of the same sign

(b) a and b are of the same sign

(c) b and c have the same sign opposite to that of a

(d) a and c have the same sign opposite to that of b

Sol. If both roots are positive, their product (i.e. $\frac{c}{a}$) and sum (i.e. $\frac{-b}{a}$) both will be positive.

The value of $\frac{c}{a}$ will be positive only if a and c have same sign either positive or negative.

Also, $\frac{-b}{a}$ will be positive only if a is of negative sign i.e. sign of a is opposite to b.

From the above, it is clear that the signs of a and c should be same and sign of a should be opposite to that of b. It implies that a and c should have the same sign opposite to that of b.

Hence, option (d) is correct.

Problem 27 What is the least integral value of k for which the equation $x^2-2(k-1)x+(2k+1)=0$ has both the roots positive?

(a) 1 (b) $\frac{-1}{2}$

(c) 4 (d) 0

Sol. It is obvious that if both roots are positive, their product and sum both will be positive.

When product of roots >0

$\Rightarrow$ $\frac{2k+1}{1}>0$

$\Rightarrow$ $2k>-1$

$\Rightarrow$ $k>\frac{-1}{2}$

When sum of roots >0

$\Rightarrow$ $\frac{-[-2(k-1)]}{1}>0$

$\Rightarrow$ $2k-2>0$

$\Rightarrow$ $k>1$

Conclusion $k>1$

Hence, the option (c) is correct as others are not greater than 1.

Problem 28 For what value of k, will the roots of the equation $kx^2 - 5x + 6 = 0$ be in the ratio of 2 : 3 ?

Sol. Let α and β be roots of the given equation.

Then, $\alpha + \beta = \frac{5}{k}$ and $\alpha\beta = \frac{6}{k}$

Given, $\frac{\alpha}{\beta} = \frac{2}{3} \Rightarrow \alpha = \frac{2}{3}\beta$

Now, $\frac{2}{3}\beta + \beta = \frac{5}{k}$ $\quad [\because \alpha + \beta = \frac{5}{k}]$

$\Rightarrow \beta = \frac{3}{k} \Rightarrow \beta^2 = \frac{9}{k^2}$

Further, $\alpha\beta = \frac{6}{k} \Rightarrow \frac{2}{3}\beta \times \beta = \frac{6}{k} \Rightarrow \beta^2 = \frac{9}{k}$

From above, $\frac{9}{k^2} = \frac{9}{k} \Rightarrow k = 1$

Alternatively

You may proceed through options. Let options be 0, 1......

(a) $0 \Rightarrow 0 \cdot x^2 - 5x + 6 = 0 \Rightarrow$ only one root as it is a linear equation.

(b) $1 \Rightarrow 1 \cdot x^2 - 5x + 6 = 0 \Rightarrow x = 2$ and 3, Hence 2 : 3.

Thus, option(b) is correct.

Problem 29 Let u and v be the roots of the equation $x^2 - 2x + p = 0$ and let y and z be the roots of the equation $x^2 - 18x + q = 0$. If $u < v < y < z$ are in arithmetic progression, then p, q respectively equal to

(a) 8, 17 (b) 3, 7

(c) – 3, 11 (d) None of these

Sol. Let $u = a - 2d$, $v = a - d$, $y = a$ and $z = a + d$.

Sum of roots of the equation $x^2 - 2x + p = 0$ is 2.

$\Rightarrow (a - 2d) + (a - d) = 2$

$\Rightarrow 2a - 3d = 2 \quad \ldots(i)$

Also, sum of roots of the equation $x^2 - 18x + q = 0$ is 18.

$\Rightarrow a + (a + d) = 18$

$\Rightarrow 2a + d = 18 \quad \ldots(ii)$

Solving Eqs. (i) and (ii), we get $a = 7$ and $d = 4$

Now, product of roots of 1st equation

$= p = u \times v = (a - 2d)(a - d) = (7 - 2 \cdot 4)(7 - 4) = -3$

and product of roots of 2nd equation

$= q = yz = a(a + d) = 7(7 + 4) = 77$

Hence option (d) is correct.

Alternatively

You may proceed through the options.

(a) 8, 17 $\Rightarrow$ equation be $x^2 - 2x + 8 = 0$ and $x^2 - 18x + 17 = 0$.
Roots of 1st equation are imaginary and that of 2nd equation are 1, 17. Hence, they are not in consonance with the given condition.

(b) 3,7 $\Rightarrow$ Roots of $x^2 - 2x + 3 = 0$ and $x^2 - 18x + 7 = 0$ are not in consonance with the given condition.

(c) –3, 11 $\Rightarrow$ Roots of $x^2 - 2x - 3 = 0$ and $x^2 - 18x + 11 = 0$ are not in consonance with the given condition

Hence, option (d) is correct.

VII. Find the Quadratic Equation or its Roots, when its Roots with Changed Coefficients and Changed Constant are Known

Problem 30 In solving a problem, one student makes a mistake in the coefficient of the 1st degree term and obtains –9 and –1 for the roots. Another student makes a mistake in the constant term of the equation and obtains 8 and 2 for the roots. The correct equation was :

Sol. The equation having roots –9 and –1 is

x^2 – (sum of roots) x + product of roots = 0

$\Rightarrow x^2 - (-9 - 1)x + (-9)(-1) = 0$

$\Rightarrow x^2 + 10x + 9 = 0$

Further the equation having roots 8 and 2 is

$\Rightarrow x^2 - (8 + 2)x + 8 \cdot 2 = 0$

$\Rightarrow x^2 - 10x + 16 = 0$

Perusal of the question reveals that 1st student writes the product of the roots correct while another writes the sum of roots correct. It implies that the correct equation is $x^2 - 10x + 9 = 0$

Alternatively

1st student writes the coefficient of 2nd degree term and constant term correct whereas the another student writes coefficient of 2nd degree term as well as that of 1st degree term correct. It implies that 1st student writes the product of the roots correct while the another writes the sum of roots correct. Now, sum of roots of correct equation is (8 + 2) and product of roots of correct equation be (–9) (–1).

Hence, correct equation is

x^2 – (sum of roots) x + product of roots= 0

$\Rightarrow x^2 - (8 + 2)x + (-9)(-1) = 0$

$\Rightarrow x^2 - 10x + 9 = 0$

Problem 31 Two students A and B solve an equation of the form $x^2 + px + q = 0$. A starts with a wrong value of p and obtains the roots as 2 and 6. B starts with a wrong value of q and gets the roots as 2 and -9. What are the correct roots of the equation?

Sol. 1st student writes the coefficient of 2nd degree term as well as constant term correct whereas the another student writes coefficient of 2nd degree term as well as that of 1st degree term correct. It implies that 1st student writes the product of the roots correct while the another writes the sum of roots correct. Now, sum of roots of correct equation = 2 + (–9) = – 7

and product of roots of correct equation = 2 × 6 = 12

Hence, correct equation is

x^2 – (sum of roots) x + product of roots = 0

$\Rightarrow x^2 - (-7)x + 12 = 0$

$\Rightarrow x^2 + 7x + 12 = 0$

Since 7 = 3 + 4, $x = -3$ and –4

Thus, correct roots are –3 and –4.

VIII. Computation of Quadratic Coefficient or Linear Coefficient or Constant of a Quadratic Equation, when One of Its Roots is Known

Problem 32 If one of the roots of quadratic equation $7x^2 - 50x + k = 0$ is 7, then what is the value of k?

Sol. Product of roots $= \frac{k}{7}$

Then, value of 2nd root $= \frac{k}{7} \div 7 = \frac{k}{49}$

Now, sum of roots $= \frac{50}{7}$

$\Rightarrow \quad 7 + \frac{k}{49} = \frac{50}{7} \Rightarrow \frac{k}{49} = \frac{1}{7}$

$\Rightarrow \quad k = 7$

Alternatively

Clearly, one value of x is 7.

Then, $7 \cdot (7)^2 - 50 \cdot 7 + k = 0$

$\Rightarrow \quad k = 7$

Problem 33 If one of the roots of the equation $ax^2 + x - 3 = 0$ is -1.5, then what is the value of a? *[CDS-10]*

Sol. Clearly, one value of x is -1.5.

Then $a(-1.5)^2 + (-1.5) - 3 = 0$

$\Rightarrow \quad 2.25a = 4.5$

$\Rightarrow \quad a = 2$

Alternatively

You may proceed through options. Let options be 4, 2 ....

(a) $4 \Rightarrow 4(-1.5)^2 + (-1.5) - 3 \neq 0$

(b) $2 \Rightarrow 2(-1.5)^2 + (-1.5) - 3 = 0$

Hence, option (b) is correct.

IX. Problems Based on Concept of Common Root(s)

Problem 34 If the equations $2x^2 - 7x + 3 = 0$ and $4x^2 + ax - 3 = 0$ have a common root, then what is the value of a?

Sol. Given, $\quad 2x^2 - 7x + 3 = 0$

Clearly, $\quad -7 = -6 - 1$

Then, roots = 3 and $\frac{1}{2}$

If 3 is the common root, then $4 \cdot 3^2 + a \cdot 3 - 3 = 0 \Rightarrow a = -11$

If $\frac{1}{2}$ is the common root then $4 \cdot \left(\frac{1}{2}\right)^2 + a \cdot \left(\frac{1}{2}\right) - 3 = 0 \Rightarrow a = 4$

Hence, value of a is either -11 or 4.

Alternatively I

We know, if $(af - cd)^2 = (ae - bd) \times (bf - ce)$, then both the equations $ax^2 + bx + c = 0$ and $dx^2 + ex + f = 0$ will have a common root.

Given equations be $2x^2 - 7x + 3 = 0$ and $4x^2 + ax - 3 = 0$

Then, $[2(-3) - 3 \cdot 4]^2 = [2a - (-7)4] \times [(-7)(-3) - 3a]$

$\Rightarrow \quad 324 = (2a + 28)(21 - 3a)$

$\Rightarrow \quad 54 = (a + 14)(7 - a)$

$\Rightarrow \quad a^2 + 7a - 44 = 0$

Clearly, $7 = +11 - 4$

Then, $\quad a = -11$ or 4

Alternatively II

We know, the value of the common root for equations $ax^2 + bx + c = 0$ and $dx^2 + ex + f = 0$ is either $\frac{(af - cd)}{(bd - ae)}$ or $\frac{ce - bf}{af - cd}$.

Given equations be

$2x^2 - 7x + 3 = 0$

and $\quad 4x^2 + ax - 3 = 0$

Then, value of common root $= \frac{(-2) \cdot 3 - 3 \cdot 4}{(-7) \cdot 4 - 2 \cdot a} = \frac{9}{14 + a}$

Obviously, the common root will satisfy both the equations.

Roots of $2x^2 - 7x + 3 = 0$ are 3 and $\frac{1}{2}$ i.e. $x = 3, \frac{1}{2}$

In case $\frac{1}{2}$ is common, then, $\frac{1}{2} = \frac{9}{14 + a}$

$\Rightarrow \quad a = 4$

and in case 3 is common, then, $3 = \frac{9}{14 + a} \Rightarrow a = -11$

Hence, $a = -11$ or 4

Note *If we take value of the common root as $\frac{ce - bf}{af - cd}$ the answer will be same.*

Alternatively III

The 'concept of common roots for polynomial equations' detailed under cubic and other equations can be applied. Given equations : $2x^2 - 7x + 3 = 0$ and $4x^2 + ax - 3 = 0$

$\Rightarrow 4x^2 - 14x + 6 = 0$ and $4x^2 + ax - 3 = 0$

Difference between the corresponding polynomials $= -14x - ax + 9$

Now, corresponding equation of $-14x - ax + 9$ is $-14x - ax + 9 = 0$

Solving this equation, we get $x = \frac{9}{14 + a}$

We know, if roots of the corresponding equation of difference between given polynomials or its subset satisfy both the equations, the given equations have common root(s) and *vice-versa* and its/their values are same as value of the root(s) at which the equation are satisfied.

Roots of $2x^2 - 7x + 3 = 0$ are 3 and $\frac{1}{2}$ i.e. $x = 3, \frac{1}{2}$

In case $\frac{1}{2}$ is common, then, $\frac{1}{2} = \frac{9}{14 + a}$

$\Rightarrow \quad a = 4$

and in case 3 is common, then, $3 = \frac{9}{14 + a} \Rightarrow a = -11$

Hence, $a = -11$ or 4

Problem 35 If one of the roots of the equation $x^2 + ax + b = 0$ and $x^2 + bx + a = 0$ is coincident, then the numerical value of $(a + b)$ is :

Sol. Given equations : $x^2 + ax + b = 0$ and $x^2 + bx + a = 0$

Difference between the corresponding polynomials $= x(a - b) + (b - a)$

Now, corresponding equation of x(a – b) +(b – a) is x (a – b) + (b – a) = 0

$\Rightarrow \quad x(a-b) = (a-b) \Rightarrow x = 1$

We know, if roots of the corresponding equation of difference between given polynomials or its subset satisfy both the equations, the given equations have common root(s) and *vice-versa* and its/their values are same as value of the root(s) at which the equations are satisfied.

If we consider 1st equation, then $x^2 + ax + b = 0$

$\Rightarrow \quad (1)^2 + a(1) + b = 0$

$\Rightarrow \quad a + b = -1$

Note *Please try with other methods as explained under this heading.*

Problem 36 If a root of the equations $x^2 + px + q = 0$ and $x^2 + \alpha x + \beta = 0$ is common, then its value will be (where $p \neq \alpha$ and $q \neq \beta$)

(a) $\frac{q-\beta}{\alpha-p}$ (b) $\frac{p\beta-\alpha q}{q-\beta}$

(c) $\frac{q-\beta}{\alpha-p}$ or $\frac{p\beta-\alpha q}{q-\beta}$ (d) None of these

Sol. We know, if two equations $ax^2 + bx + c = 0$ and $dx^2 + ex + f = 0$ have one root common, then its value is either $\frac{(af-cd)}{(bd-ae)}$ or $\frac{ce-bf}{af-cd}$

Given equations be : $x^2 + px + q = 0$ and $x^2 + \alpha x + \beta = 0$

Then, value of the common root $= \frac{(\beta-q)}{(p-\alpha)}$ or $\frac{(q\alpha-p\beta)}{(\beta-q)}$

$\Rightarrow \quad \frac{q-\beta}{\alpha-p}$ or $\frac{p\beta-q\alpha}{q-\beta}$

Note *Please try with other method as explained above.*

Problem 37 If both the roots of $k(6x^2+3) + rx + 2x^2 - 1 = 0$ and $6k(2x^2+1) + px + 4x^2 - 2 = 0$ are common, then $2r - p$ is equal to :

(a) 0 (b) –1

(c) 2 (d) 1

Sol. Given equations are $k(6x^2+3) + rx + 2x^2 - 1 = 0$

and $\quad 6k(2x^2+1) + px + 4x^2 - 2 = 0$

$\Rightarrow \quad k6x^2 + 3k + rx + 2x^2 - 1 = 0$

and $\quad k12x^2 + 6k + px + 4x^2 - 2 = 0$

$\Rightarrow \quad x^2(6k+2) + rx + 3k - 1 = 0$

and $\quad x^2(12k+4) + px + 6k - 2 = 0$

We know that equations $ax^2 + bx + c = 0$ and $dx^2 + ex + f = 0$ will have both the roots common, if $\frac{a}{d} = \frac{b}{e} = \frac{c}{f}$.

Hence, $\frac{6k+2}{12k+4} = \frac{r}{p} = \frac{3k-1}{6k-2} \Rightarrow \frac{r}{p} = \frac{1}{2}$

$\Rightarrow \quad p = 2r$

$\Rightarrow \quad 2r - p = 0$

Problem 38 Two different quadratic equations have a common root. Let the three unique roots of the two equations be A, B and C – all are positive integers. If $(A + B + C) = 41$ and the product of the roots of one of the equation is 35, which of the following options is definitely correct? *[XAT-18]*

(a) The common root is 29

(b) The smallest among the roots is 1

(c) One of the roots is 5

(d) Product of the roots of the other equation is 5

(e) All of the above are possible, but none are definitely correct

Sol. Given, product of two roots of one equation = 35.

It implies that they may either be (1, 35) or (5, 7). If they are (1, 35), the 3rd root will be (41– 1 – 35) i.e. 5. Further in case they are (5 7), the 3rd root will be (41 – 5 – 7) i.e 29.

Hence, three unique roots are either (1, 35, 5) or (5, 7, 29).

Thus, it can definitely be said that one of the roots is 5.

X. Formation of a Quadratic Equation with the Roots Having Certain Relation with Roots of Another Quadratic Equation

Problem 39 If m and n are the roots of the equation $ax^2 + bx + c = 0$, then the equation whose roots are $\frac{m^2+1}{m}$ and $\frac{n^2+1}{n}$ is : *[CDS-14]*

(a) $acx^2 + (ab+bc)x + b^2 + (a-c)^2 = 0$

(b) $acx^2 + (ab-bc)x + b^2 + (a-c)^2 = 0$

(c) $acx^2 + (ab-bc)x + b^2 - (a-c)^2 = 0$

(d) $acx^2 + (ab+bc)x + b^2 - (a-c)^2 = 0$

Sol. Since m and n are the roots of the equation $ax^2 + bx + c = 0$, $m + n = \frac{-b}{a}$ and $mn = \frac{c}{a}$

Now, sum of the roots of the new equation $= \frac{m^2+1}{m} + \frac{n^2+1}{n}$

$$= m + n + \frac{1}{m} + \frac{1}{n} = (m+n) + \frac{m+n}{mn}$$

$$= \frac{-b}{a} + \frac{\frac{-b}{a}}{\frac{c}{a}} = \frac{-b}{a} + \frac{-b}{c} = \frac{-b(a+c)}{ac}$$

and product of roots $= \frac{m^2+1}{m} \times \frac{n^2+1}{n}$

$$= (m + \frac{1}{m})(n + \frac{1}{n})$$

$$= mn + \frac{m}{n} + \frac{n}{m} + \frac{1}{mn} = \frac{c}{a} + \frac{m^2+n^2}{mn} + \frac{a}{c}$$

$$= \frac{a^2+c^2}{ac} + \frac{(m+n)^2 - 2mn}{mn} = \frac{a^2+c^2}{ac} + \frac{\left(-\frac{b}{a}\right)^2}{\frac{c}{a}} - 2$$

$$= \frac{a^2+c^2}{ac} + \frac{b^2}{ac} - 2 = \frac{a^2+c^2-2ac+b^2}{ac}$$

$$= \frac{(a-c)^2 + b^2}{ac}$$

Now, the new equation be x^2 – (sum of roots) x + product of roots = 0

$$\Rightarrow x^2 - \left\{\frac{-b(a+c)}{ac}\right\}x + \frac{(a-c)^2 + b^2}{ac} = 0$$

$$\Rightarrow \quad cax^2 + (ab + bc)x + b^2 + (a-c)^2 = 0$$

Problem 40 If α and β are the roots of the equation $x^2 + px + q = 0$, then $-\alpha^{-1}, -\beta^{-1}$ are the roots of which one of the following equations ?

(a) $qx^2 - px + 1 = 0$ (b) $q^2 + px + 1 = 0$
(c) $x^2 + px - q = 0$ (d) $x^2 - px + q = 0$

Sol. Clearly, $\alpha + \beta = \frac{-p}{1} = -p$ and $\alpha\beta = q$

Now, $(-\alpha^{-1}) + (-\beta^{-1}) = -\left\{\frac{1}{\alpha} + \frac{1}{\beta}\right\} = -\left\{\frac{\alpha+\beta}{\alpha\beta}\right\} = -\left\{\frac{-p}{q}\right\} = \frac{p}{q}$

and $(-\alpha^{-1}) \times (-\beta^{-1}) = \frac{1}{\alpha\beta} = \frac{1}{q}$

Then, the required equation is

x^2 – (sum of roots) x + product of roots= 0

$$\Rightarrow \quad x^2 - \frac{p}{q}x + \frac{1}{q} = 0$$

$$\Rightarrow \quad qx^2 - px + 1 = 0$$

Alternatively

[Observe the chart given under 'Formation of a quadratic equation with the roots having certain relation with roots of another']

The new equation, if its roots are same as roots of equation $x^2 + px + q = 0$ but opposite in sign, is $x^2 - px + q = 0$.

Further, the equation whose roots are reciprocal of the roots of $x^2 - px + q = 0$ is $qx^2 - px + 1 = 0$

Hence, the required equation is $qx^2 - px + 1 = 0$

Problem 41 If α and β are the roots of the equation $x^2 + x - 1 = 0$, then the equation having roots α^5 and β^5 is **[SSC (CGL T-II)-17]**

Sol. Obviously, $\alpha + \beta = -1$ and $\alpha\beta = -1$

Now, $\alpha^2 + \beta^2 = (\alpha+\beta)^2 - 2\alpha\beta = (-1)^2 - 2\cdot(-1) = 3$ and

$\alpha^3 + \beta^3 = (\alpha+\beta)(\alpha^2 + \beta^2 - \alpha\beta) = (-1)(3+1) = -4$

Further, $(\alpha^2 + \beta^2)(\alpha^3 + \beta^3) = \alpha^5 + \beta^5 + \alpha^2\beta^3 + \alpha^3\beta^2$

$\Rightarrow 3 \times (-4) = \alpha^5 + \beta^5 + \alpha^2\beta^2(\alpha + \beta)$

$\Rightarrow \alpha^5 + \beta^5 = -12 - (-1)^2(-1) = -11$

Also, $\quad \alpha^5\beta^5 = -(-1)^5 = -1$

Now, the required equation be

x^2 – (sum of roots) x + product of roots = 0

$$\Rightarrow \quad x^2 - (-11)x + (-1) = 0$$

$$\Rightarrow \quad x^2 + 11x - 1 = 0$$

Observation We have consumed considerable time to compute the value of $\alpha^5 + \beta^5$. Instead, we can memorise the value of these types of expressions given below :

If α and β are roots of quadratic equation $x^2 - bx - c = 0$, then	
$\alpha + \beta = b$	$\alpha - \beta = \Delta$ (let)
$\alpha^2 + \beta^2 = b^2 + 2c$	$\alpha^2 - \beta^2 = b\Delta$
$\alpha^3 + \beta^3 = b^3 + 3bc$	$\alpha^3 - \beta^3 = (b^2 + c)\Delta$
$\alpha^4 + \beta^4 = b^4 + 4b^2c + 2c^2$	$\alpha^4 - \beta^4 = (b^3 + 2bc)\Delta$
$\alpha^5 + \beta^5 = b^5 + 5b^3c + 5bc^2$	$\alpha^5 - \beta^5 = (b^4 + 3b^2c + c^2)\Delta$
$\alpha^6 + \beta^6 = b^6 + 6b^4c + 9b^2c^2 + 2c^3$	$\alpha^6 - \beta^6 = (b^5 + 4b^3c + 3bc^2)\Delta$

Now, we may solve on base of the above table.

Given equation be $x^2 + x - 1 = 0$

$$\therefore \quad \alpha\beta = -1$$

Now, write the given equation in form of $x^2 - bx - c = 0$

i.e. $\quad x^2 - (-1)x - 1 = 0$

Hence, $b = -1$ and $c = 1$

Now, $\quad \alpha^5 + \beta^5 = b^5 + 5b^3c + 5bc^2$

$$= (-1)^5 + 5(-1)^3\cdot 1 + 5(-1)\cdot 1^2 = -11$$

Further, $\alpha^5\beta^5 = (-1)^5 = -1$

Then, required equation be

x^2 – (sum of roots) x + product of roots = 0

$$\Rightarrow \quad x^2 - (-11)x + (-1) = 0$$

$$\Rightarrow \quad x^2 + 11x - 1 = 0$$

XI. Problems Based on Concept of the Maximum and Minimum Value of a Quadratic Equation

Problem 42 The value of $x^2 - 4x + 11$ can never be less than

(a) 7 (b) 8
(c) 11 (d) 22

Sol. Minimum value of the expression

$$= \frac{4ac - b^2}{4a} = \frac{4\cdot 1\cdot 11 - (-4)^2}{4\cdot 1} = 7$$

Problem 43 A quadratic function $f(x)$ attains a maximum of 3 at $x = 1$. The value of the function at $x = 0$ is 1. What is the value of $f(x)$ at $x = 10$? **[CAT-07]**

Sol. Let the quadratic function be $ax^2 + bx + c$.

Given, value of the function at $x = 1$ is 3

$$\Rightarrow \quad a\cdot 1^2 + b\cdot 1 + c = 3$$

$$\Rightarrow \quad a + b + c = 3 \quad \ldots(i)$$

Further, value of the function at $x = 0$, will be $a\cdot 0^2 + b\cdot 0 + c$

$$\Rightarrow \quad 1 = c \quad \ldots(ii)$$

We know that a quadratic expression $ax^2 + bx + c$ has maximum value at $x = \frac{-b}{2a}$.

It is also fact that the coefficient of x^2 should be less than 0.

It implies that $\frac{-b}{2a} = 1$

$$\Rightarrow \quad -b = 2a \quad \ldots(iii)$$

From Eqs. (i) and (ii), we get $a + b = 2$

$$\Rightarrow \quad a = 2 - b$$

Then, Eq. (iii) becomes : $-b = 2(2 - b)$

$\Rightarrow \quad b = 4$

Then, $\quad a = 2 - 4 = -2$

It is true that the quadratic coefficient is less than 0.

Now the expression be

$$-2x^2 + 4x + 1$$

$\therefore$ Its value at $x = 10$ will be $-2 \cdot (10)^2 + 4 \cdot 10 + 1$; i.e. -159

XII. Comparison/Inequality Between Roots of the Given Quadratic Equations

In the following problems (44-50) two equations numbered I and II are given. You have to solve both the equations and mark the appropriate option.

Give answer

(a) If $x > y$ (b) If $x < y$

(c) If $x \geq y$ (d) If $x \leq y$

(e) If x = y or the relationship cannot be established.

Problem 44 I. $x^2 - 11x + 30 = 0$

II. $2y^2 - 9y + 10 = 0$

Problem 45 I. $15x^2 + 8x + 1 = 0$

II. $3y^2 + 14y + 8 = 0$

Problem 46 I. $4x^2 - 17x + 18 = 0$

II. $2y^2 - 21y + 40 = 0$

Problem 47 I. $6x^2 - 25x + 14 = 0$

II. $9y^2 - 9y + 2 = 0$

Problem 48 I. $8x^2 + 25x + 3 = 0$

II. $2y^2 + 17y + 30 = 0$

[SBI (PO-Pre)-16, 18, 19, IBPS (PO-Pre.)-18]

Problem 49 I. $x^2 + 10x + 24 = 0$

II. $y^2 - \sqrt{625} = 0$

Problem 50 I. $15x^2 - 11x + 2 = 0$

II. $10y^2 - 9y + 2 = 0$ *[IBPS (PO-Pre.)-16]*

44.

I. $x^2 - 11x + 30 = 0$	II. $2y^2 - 9y + 10 = 0$
$-11 = -5 - 6$	$-9 = -4 - 5$
$x = 5, 6$	$y = \frac{4}{2}, \frac{5}{2} = 2, 2.5$

Conclusion $x > y$ as each value of x is greater than both values of y.

Another Approach		
x	Relation	y
5	>	2
5	>	2.5
6	>	2
6	>	2.5
	Conclusion – sign : >	
Hence, $x > y$		

45.

I. $15x^2 + 8x + 1 = 0$	II. $3y^2 + 14y + 8 = 0$
$+8 = +5 + 3$	$+14 = +12 + 2$
$x = \frac{-5}{15}, \frac{-3}{15} = -0.33, -0.20$	$y = \frac{-12}{3}, \frac{-2}{3} = -4, -0.67$

Conclusion $x > y$ as each value of x is greater than both values of y.

Another Approach		
x	Relation	y
– 0.33	>	– 4
– 0.33	>	– 0.67
– 0.20	>	– 4
– 0.20	>	– 0.67
	Conclusion – sign : >	
Hence, $x > y$		

46.

I. $4x^2 - 17x + 18 = 0$	II. $2y^2 - 21y + 40 = 0$
$-17 = -9 - 8$	$-21 = -16 - 5$
$x = \frac{9}{4}, \frac{8}{4} = 2.25, 2$	$y = \frac{16}{2}, \frac{5}{2} = 8, 2.25$

Conclusion : $x < y$ as both values of x are less than each value of y.

Another Approach		
x	Relation	y
2.25	<	8
2.25	<	2.5
2	<	8
2	<	2.5
	Conclusion - sign : <	
Hence, $x < y$		

47.

I. $6x^2 - 25x + 14 = 0$	II. $9y^2 - 9y + 2 = 0$
$-25 = -21 - 4$	$-9 = -6 - 3$
$x = \frac{21}{6}, \frac{4}{6} = 3.5, 0.67$	$y = \frac{6}{9}, \frac{3}{9} = 0.67, 0.33$

Conclusion $x \geq y$ as one value of x is greater than both values of y and also another value of x is equal to one value of y and greater than another value of y.

Another Approach		
x	Relation	y
3.5	>	0.67
3.5	>	0.33
0.67	=	0.67
0.67	>	0.33
	Conclusion - sign : ≥	
Hence, $x \geq y$		

48.

I. $8x^2 + 25x + 3 = 0$	II. $2y^2 + 17y + 30 = 0$
$+25 = +24 + 1$	$+17 = +12 + 5$
$x = \frac{-24}{8}, \frac{-1}{8} = -3, -0.12$	$y = \frac{-12}{2}, \frac{-5}{2} = -6, -2.5$

Conclusion Relationship between x and y cannot be established as $-3 > -6$ and $-3 < -2.5$ i.e. $x > y$ and $x < y$ contradict each other (one value of x is greater than one value of y as well as the same value of x is less than another value of y)

Another Approach

x	Relation	y
– 3	>	– 6
– 3	<	– 2.5
– 0.12	>	– 6
– 0.12	>	– 2.5
	No conclusion as opposite inequality sings appear	
Hence, no relation can be establihsed between x and y.		

49.

I. $x^2 + 10x + 24 = 0$	II. $y^2 - \sqrt{625} = 0$
$+10 = +6 + 4$	$y^2 = \sqrt{625} = \pm 25$
$x = -6, -4$	$y = \pm 5$ (imaginary values have been ingnored)
Conclusion Relationship between x and y cannot be established as $-6 < +5, -5$ and $-4 > -5$ i.e. $x < y$ and $x > y$ contradicts each other. [one value of x is less than both values of y as well as the another value of x is greater than one value of y]	

x	Relation	y
– 6	<	+5
– 6	<	–5
– 4	<	+5
– 4	>	–5
	No conclusion as opposite inequality signs appear.	
Hence, no relation can be established between x and y.		

50.

I. $15x^2 - 11x + 2 = 0$	II. $10y^2 - 9y + 2 = 0$
$-11 = -5 - 6$	$-9 = -4 - 5$
$x = \frac{5}{15}, \frac{6}{15} = 0.33, 0.40$	$y = \frac{4}{10}, \frac{5}{10} = 0.40, 0.50$
Conclusion $x \le y$ as one value of x is less than both values of y and also another value of x is equal to one value of y and less than another value of y.	

Another Approach

X	Relation	Y
0.33	<	0.40
0.33	<	0.50
0.40	=	0.40
0.40	<	0.50
	Conclusion-sign : ≤	
Hence, $x \le y$		

Problems Based on Cubic and other Equations

Problem 51 If the roots of a cubic equation in variable x are 1, 2, –4, then what is the equation?

(a) $x^3 + x^2 - 10x + 8 = 0$ (b) $x^3 + 2x^2 - 4x + 8 = 0$
(c) $2x^3 + 4x^2 - x + 8 = 0$ (d) $2x^3 - 4x^2 - x + 8 = 0$

Sol. We know if α, β and γ be roots of a cubic equations then the equation be $x^3 - (\alpha + \beta + \gamma)x^2 + (\alpha\beta + \beta\gamma + \gamma\alpha)x - \alpha\beta\gamma = 0$

$\Rightarrow x^3 - (1 + 2 - 4)x^2 + (1\cdot 2 - 2\cdot 4 - 4\cdot 1)x + 1\cdot 2\cdot 4 = 0$

$\Rightarrow x^3 + x^2 - 10x + 8 = 0$

Note *It is not advisable to proceed through options.*

Problem 52 What is the maximum number of negative real roots in a polynomial equation $ax^4 + bx^3 - cx^2 + dx + e = 0$?

Sol. We 1st write a new equation with negative variable as below

$a(-x)^4 + b(-x)^3 - c(-x)^2 + d(-x) + e = 0$

$\Rightarrow ax^4 - bx^3 - cx^2 - dx + e = 0$

Here, signs are +, –, –, –, +

Cearly, two signs change.

Hence, maximum number of negative real roots = 2

Problem 53 What is the total number of common roots between $3x^4 + 5x^3 - 4x^2 - 2x + 3 = 0$ and $3x^4 + 4x^3 - 2x^2 - x + 1 = 0$

Sol. Deference between respective polynomials

$= x^3 - 2x^2 - x + 2$

Then, corresponding equation be $x^3 - 2x^2 - x + 2 = 0$

Solving this equation, we get $x = -1, 1, 2$

Both the given equations are not satisfied at any of the above values of x.

Hence, there is no common root.

Problem 54 What is the number of common roots between $x^5 + 2x^4 - 6x^3 - 8x^2 + 5x + 6 = 0$ and $x^5 + 2x^4 - 9x^3 - 14x^2 + 20x + 24 = 0$

Sol. Difference between respective polynomials

$= 3x^3 + 6x^2 - 15x - 18$

Then, corresponding equation be $3x^3 + 6x^2 - 15x - 18 = 0$

$\Rightarrow x^3 + 2x^2 - 5x - 6 = 0$

Solving this equation, we get $x = -1, -3, 2$

At $x = -1, -3, 2$ both the given equations are satisfied.

Hence, values of common root = –1, – 3, 2.

∴ Number of common roots = 3

Problem 55 If $x^4 - 8x^3 + ax^2 - bx + 16 = 0$ has positive real roots, then find $a - b$.

Sol. Let roots be α, β, γ and δ, then $\alpha + \beta + \gamma + \delta = \frac{-(-8)}{1} = 8$

and $\alpha\beta\gamma\delta = \frac{16}{1} = 16$

Clearly $\alpha = \beta = \gamma = \delta = 2$

$$\left\{\because AM = GM \Rightarrow \frac{\alpha + \beta + \gamma + \delta}{4} = \sqrt[4]{\alpha\beta\gamma\delta}\right\}$$

Further, $\alpha\beta + \alpha\gamma + \alpha\,\delta + \beta\gamma + \beta\delta + \gamma\delta = \frac{a}{1}$

$\Rightarrow 2\cdot 2 + 2\cdot 2 + 2\cdot 2 + 2\cdot 2 + 2\cdot 2 + 2\cdot 2 = a$

$\Rightarrow \quad a = 24$

Also, $\alpha\beta\gamma + \beta\gamma\delta + \gamma\delta\alpha + \delta\alpha\beta = \frac{-(-b)}{1}$

$\Rightarrow \quad 2\cdot 2\cdot 2 + 2\cdot 2\cdot 2 + 2\cdot 2\cdot 2 + 2\cdot 2\cdot 2 = b$

$\Rightarrow \quad b = 32$

Now, $\quad a - b = 24 - 32 = -8$

Problem 56 If the equation $x^3 - ax^2 + bx - a = 0$ has three real roots, then the following is true ?

(a) $a = 11$ (b) $a \neq 1$

(c) $b = 1$ (d) $b \neq 1$ ***[CAT 2000]***

Sol. Maximum number of positive real roots in the equation = 3 as three signs change (observe:+, −, + , −).

Maximum number of negative real roots = 0 as no sign changes when new equation is written with negative variable. (Observe −, −,−,−,). It may be construed that all the three roots are real and positive.

Let roots of the given equation be α,β and γ.

Then, $\alpha\beta\gamma = a$, $\alpha + \beta + \gamma = a$

and $\quad \alpha\beta + \beta\gamma + \gamma\alpha = b$

Now, $\quad \frac{\alpha + \beta + \gamma}{\alpha\beta\gamma} = \frac{a}{a} \Rightarrow \frac{1}{\beta\gamma} + \frac{1}{\alpha\gamma} + \frac{1}{\alpha\beta} = 1$

$\Rightarrow \quad \alpha\beta, \beta\gamma, \gamma\alpha > 1$

Hence, $\quad \alpha\beta + \beta\gamma + \gamma\alpha > 1 + 1 + 1$

$\Rightarrow \quad b > 3$

Hence, option (d) is correct.

Problem 57 For the following two equations, mark the appropriate option. Give answer

(a) If $x > y$ (b) If $x < y$

(c) If $x \geq y$ (d) If $x \leq y$

(e) If $x = y$ or the relationship cannot be established.

I. $\frac{x^3 - x^2 - 4x + 4}{(x - 1)} = 0$ II. $\frac{2y^3 - 9y^2 + 7y + 6}{(y - 2)} = 0$

Sol.

I. $\frac{x^3 - x^2 - 4x + 4}{(x - 1)} = 0$	II. $\frac{2y^3 - 9y^2 + 7y + 6}{(y-2)} = 0$
One root of $x^3 - x^2 - 4x + 4 = 0$ is 1 as $(x - 1)$ is a factor of $x^3 - x^2 - 4x + 4$	One root of $2y^3 - 9y^2 + 7y + 6 = 0$ is 2 as $(y - 2)$ is a factor of $2y^3 - 9y^2 + 7y + 6$
Recall 'how to find roots of a cubic equation when its one root is known'. $+1 \ -1 \ -4 \ +4$ KR (i.e. 1) $\ +1 \ 0 \ -4$ $+1 \ 0 \ -4 \ 0$ (KR refers to known root)	Recall 'how to find roots of a cubic equation when its one root is known'. $+2 \ -9 \ +7 \ +6$ KR (i.e. 2) $\ +4 \ -10 \ -6$ $+2 \ -5 \ -3 \ 0$ (KR refers to known root)
Accordingly, quadratic equation is $+1x^2 - 4 = 0$ ($\because$ coefficient of 2nd degree term is 0) Then, roots = + 2, −2	Accordingly, quadratic equation is $+ 2x^2 - 5x - 3 = 0$. Clearly, $-5 = -6 + 1$ Then, roots $= \frac{6}{2}, \frac{-1}{2}$ i.e. $3, -\frac{1}{2}$
Hence, solutions of the given equation are +2, −2 i.e. $x = +2, -2$	Hence, solutions of the given equation are $3, -\frac{1}{2}$ i.e. $y = 3, -\frac{1}{2}$

Now, for conclusion

x	Relation	y
+ 2	<	3
+ 2	>	$-\frac{1}{2}$
− 2	<	3
− 2	<	$-\frac{1}{2}$
	No conclusion as opposite inequality signs appear	
Hence, no relation can be established between x and y.		

WORKOUT

1. The values of a and b for which equation $x^4 - 4x^3 + ax^2 + bx + 1 = 0$ has four real roots :
(a) −6, −4 (b) −6, 5
(c) − 6, 4 (d) 6, −4

2. If x is real, the smallest value of the expression $3x^2 - 4x + 7$ is : *[IIFT-13]*
(a) $\frac{2}{3}$ (b) $\frac{3}{4}$
(c) $\frac{7}{9}$ (d) None of these

3. The system of equations $x + 2y = 3$ and $3x + 6y = 9$ has :
(a) a unique solution
(b) infinitely many solution
(c) no solution
(d) finite number of solutions

4. If p and q are the roots of the equation $x^2 - bx + c = 0$, then what is the equation if the roots are $(pq + p + q)$ and $(pq - p - q)$?
(a) $x^2 - 2cx + (c^2 - b^2) = 0$ *[CAT-16]*
(b) $x^2 - 2bx + (b^2 + c^2) = 0$
(c) $cx^2 - 2(b + c)\ x + c^2 = 0$
(d) $x^2 + 2bx - (c^2 - b^2) = 0$

5. Which of the following is a quadratic equation ? *[MAT-01, 99]*
(a) $x^{\frac{1}{2}} + 2x + 3 = 0$
(b) $(x - 1)(x + 4) = x^2 + 1$
(c) $x^4 - 3x + 5 = 0$
(d) $(2x + 1)(3x - 4) = 2x^2 + 3$

6. If $\frac{2}{x} + \frac{3}{y} = \frac{9}{xy}$ and $\frac{4}{x} + \frac{9}{y} = \frac{21}{xy}$ then what is $x + y$ equal to : *[CDS-11]*
(a) 2 (b) 3 (c) 4 (d) 8

7. The quadratic equation whose one root is $2 - \sqrt{3}$ will be
(a) $x^2 - 4x - 1 = 0$ (b) $x^2 - 4x + 1 = 0$
(c) $x^2 + 4x - 1 = 0$ (d) $x^2 + 4x + 1 = 0$

8. If x is a real number, y is greatest integer less than or equal to x, then $3|x| + 2 - y = 0$. Will the above equation have any real root ?
(a) yes
(b) no
(c) will have real roots for $x < 0$
(d) will have real roots for $x > 0$

9. If sum of the roots of the equation $x^2 + px + q = 0$ is three times their difference, then which one of the following is true ?
(a) $9p^2 = 2q$ (b) $2p^2 = 9q$
(c) $9p^2 = 4q$ (d) $9q^2 = 2p$

10. If $\frac{x}{2} + \frac{y}{3} = 4$ and $\frac{2}{x} + \frac{3}{y} = 1$, then what is $x + y$ equal to : *[CDS-13]*
(a) 11 (b) 10
(c) 9 (d) 8

11. For the given pair (x, y) of positive integers, such that $4x - 17y = 1$ and $x \leq 1000$, how many integer values of y satisfy the given conditions ? *[CAT-99]*
(a) 55 (b) 56
(c) 59 (d) 58

12. The number of roots of the quadratic equation $8\sec^2\theta - 6\sec\theta + 1 = 0$ is :
(a) ∞ (b) 2
(c) 0 (d) 1

13. If one root of $ax^2 + bx + c = 0$ where a, b and c are integers be $2 + \sqrt{5}$, then the other root is :
(a) ∞ (b) $2 - \sqrt{5}$
(c) 0
(d) cannot be determined

14. What is one of the roots of the equation $\sqrt{\frac{2x}{3-x}} - \sqrt{\frac{3-x}{2x}} = \frac{3}{2}$?
(a) 1 (b) 2
(c) 3 (d) 4

15. p is a prime and m is a positive integer. How many solutions exist for the equation $p^6 - p = (m^2 + m + 6)(p - 1)$? *[CAT-12]*
(b) 0 (b) 1
(c) 2 (d) infinite

16. The quadratic in t, such that AM of its root is A and GM is G, is :
(a) $t^2 - 2At + G^2 = 0$
(b) $t^2 - 2At - G^2 = 0$
(c) $t^2 + 2At + G^2 = 0$
(d) None of these

17. If $x^2 - kx - 21 = 0$ and $x^2 - 3kx + 35 = 0$ have one common root, then what is the value of k?
(a) +4 (b) −4
(c) ± 4 (d) None of these

18. If the roots, x_1 and x_2 of the quadratic equation $x^2 - 2x + c = 0$ also satisfy the equation $7x_2 - 4x_1 = 47$, then which of the following is true? *[CAT-97]*
(a) $c = -15$
(b) $x_1 = -5$, $x_2 = 3$
(c) $x_1 = 4\cdot 5$, $x_2 = -2.5$
(d) None of the above

19. If one of the roots of equation $x^2 + ax + 3 = 0$ is 3 and one of the roots of the equation $x^2 + ax + b = 0$ is three times the other root, then the value of b is equal to :
(a) 1 (b) 2
(c) 3 (d) 4

20. The number of real solutions of the equation $x^2 + 5|x| + 4 = 0$, are
(a) 0 (b) 2
(c) 3 (d) 4

21. If $ax^2 + bx + c = 0$ and $2a$, b and $2c$ are in arithmetic progression, which of the following are the roots of the equation? *[CAT-12]*
(a) a, c (b) $-a, -c$
(c) $-\frac{a}{2}, -\frac{c}{2}$ (d) $-\frac{c}{a}, -1$

22. If $\frac{a}{b} - \frac{b}{a} = \frac{x}{y}$ and $\frac{a}{b} + \frac{b}{a} = x - y$, then what is the value of x ?
(a) $\frac{a+b}{a}$ (b) $\frac{a+b}{b}$
(c) $\frac{a-b}{a}$ (d) None of these

23. The equation $2x^2 + 2(p + 1)x + p = 0$, where p is real, always has roots that are : *[CAT-10]*
(a) equal
(b) equal in magnitude but opposite in sign
(c) irrational
(d) real

24. The minimum value of $(x-2)(x-9)$ is : **[SSC (CGL-T-I)-13]**

(a) $-\frac{11}{4}$ (b) $\frac{49}{4}$ (c) 0 (d) $\frac{-49}{4}$

25. If the product of the roots of $x^2 - 3kx + 2k^2 - 1 = 0$ is 7 for a fixed k, then what is the nature of roots ?

(a) integral and positive
(b) integral and negative
(c) irrational
(d) rational but not integral

26. The number of real roots of the equation $\frac{A^2}{x} + \frac{B^2}{x-1} = 1$, where A and B are real numbers not equal to zero simultaneously, is : **[CAT-02]**

(a) None (b) 1
(c) 2 (d) 1 or 2

27. If $ax^2 + bx + c = 0$ and $bx^2 + cx + a = 0$ have a common root, $a \neq 0$, then $\frac{a^3 + b^3 + c^3}{abc} = ?$

(a) 1 (b) 2
(c) 3 (d) None of these

28. If x is real, then the minimum value of $x^2 - x + 1$ is : **[SSC(CGL-PT)-07]**

(a) $\frac{3}{4}$ (b) 0
(c) 1 (d) $\frac{1}{4}$

29. Given the quadratic equation $x^2 - (A-3)x - (A-7) = 0$, for what value of A will the sum of the squares of the roots be zero ? **[CAT-96]**

(a) –2 (b) 3
(c) 6 (d) None of these

30. The roots of $|x-2|^2 + |x-2| - 6 = 0$ are :

(a) 0,4 (b) –1, 3 (c) 4, 2 (d) 5,1

31. Number of solution of two equations $4x - y = 2$ and $2y - 8x + 4 = 0$ is : **[SSC (10+2, DEO/LDC)-13]**

(a) 0 (b) 1 (c) 2 (d) ∞

32. If the roots of the equation $(a^2 + b^2)x^2 + 2(b^2 + c^2)x + (b^2 + c^2) = 0$ are real, which of the following must hold true ? **[CAT-12]**

(a) $c^2 \geq a^2$
(b) $c^4 \geq a^2(b^2 + c^2)$
(c) $b^2 \geq a^2$
(d) $a^4 \leq b^2(a^2 + c^2)$

33. Under what condition do the equations $kx - y = 2$ and $6x - 2y = 3$, have a unique solution?

(a) $k = 3$
(b) $k \neq 3$
(c) $k = 0$
(d) $k \neq 0$

34. If one root of $x^2 - x - k = 0$ is square of the other, then value of k is ?

(a) $2 \pm \sqrt{3}$
(b) $2 \pm \sqrt{5}$
(c) $3 \pm \sqrt{2}$
(d) $5 \pm \sqrt{2}$

35. One root of $x^2 + kx - 8 = 0$ is square of the other, then the value of k is? **[CAT-95]**

(a) 2 (b) 8 (c) –8 (d) –2

36. If α and β are the roots of the equation $x^2 - 3x + 2 = 0$, then which equation has the roots $(\alpha + 1)$ and $(\beta + 1)$?

(a) $x^2 + 5x + 6 = 0$
(b) $x^2 - 5x - 6 = 0$
(c) $x^2 + 5x - 6 = 0$
(d) $x^2 - 5x + 6 = 0$

37. Let α and α^2 be the roots $x^2 + x + 1 = 0$, then the equation whose roots are α^{31}, α^{62} is

(a) $x^2 - x + 1 = 0$
(b) $x^2 + x - 1 = 0$
(c) $x^2 + x + 1 = 0$
(d) $x^{60} + x^{30} + 1 = 0$

38. The value of y which will satisfy the equations $2x^2 + 6x + 5y + 1 = 0$ and $2x + y + 3 = 0$ may be found by solving which one of the following equations ? **[CDS-08]**

(a) $y^2 + 14y - 7 = 0$
(b) $y^2 + 8y + 1 = 0$
(c) $y^2 + 10y - 7 = 0$
(d) $y^2 - 8y + 7 = 0$

39. How many roots the equation $x - \frac{2}{x-1} = 1 - \frac{2}{x-1}$ have?

(a) 1 (b) 2
(c) ∞ (d) 0

40. Let p and q be the roots of the quadratic equation $x^2 - (\alpha - 2)x - \alpha - 1 - 0$. What is minimum possible value of $p^2 + q^2$? **[CAT 03]**

(a) 0 (b) 4
(c) 5 (d) 3

41. When the roots of the quadratic equation $ax^2 + bx + c = 0$ are negative of reciprocals of each other, then which one of the following is correct **[CDS-10]**

(a) $b = 0$ (b) $c = 0$
(c) $a = c$ (d) $a = -c$

42. If roots of the quadratic equation $y^2 + My + N = 0$ are equal to N and M, then find the possible number of pairs of (M, N). **[CAT-09]**

(a) 0 (b) 1
(c) 2 (d) 3

43. If the roots of $x^2 - bx + c = 0$ are two consecutive integers, then $b^2 - 4c$ is equal to :

(a) 1 (b) 2
(c) 3 (d) 4

44. If $|x^2 + y^2| = 0.1$ and $|x - y| = 0.2$, then the value of $|x| + |y|$ is : **[CAT-99]**

(a) 0.6 (b) 0.2
(c) 0.36 (d) 0.4

45. Which one of the following is the quadratic equation whose roots are reciprocal to the roots of the quadratic equation $2x^2 - 3x - 4 = 0$?

(a) $3x^2 - 2x - 4 = 0$
(b) $4x^2 + 3x - 2 = 0$
(c) $3x^2 - 4x - 2 = 0$
(d) $4x^2 - 2x - 3 = 0$

46. If $x^2 + ax + 10 = 0$ and $x^2 + bx - 10 = 0$ have a common root then $a^2 - b^2$ is equal to ?

(a) 10 (b) 20
(c) 30 (d) 40

47. If α and β are the roots of the quadratic equation $x^2 - 10x + 15 = 0$, then find the quadratic equation whose roots are $\left(\alpha + \frac{\alpha}{\beta}\right)$ and $\left(\beta + \frac{\beta}{\alpha}\right)$. **[CAT-12]**

(a) $15x^2 + 71x + 210 = 0$
(b) $5x^2 - 22x + 56 = 0$
(c) $3x^2 - 44x + 78 = 0$
(d) cannot be determined

48. If α, β are the roots of $(x-a)(x-b) = c$, $c \neq 0$, then the roots of $(x-\alpha)(x-\beta) + c = 0$, shall be :

(a) a, c
(b) b, c
(c) a, b
(d) $a + c, b + c$

49. Which one of the following conditions must p, q and r satisfy so that the following system of linear simultaneous equations has at least one solution, such that $p+q+r \neq 0$?
$x+2y-3z=p, 2x+6y-11z=q$ and $x-2y+7z=r$ **[CAT-03]**
(a) $5p+2q+r=0$
(b) $5p-2q-r=0$
(c) $5p+2q-r=0$
(d) $5p-2q+r=0$

50. The roots of the equation $x^{\frac{2}{3}}+x^{\frac{1}{3}}-2=0$ are :
(a) 1, – 8 (b) 2, – 8
(c) 1, – 4 (d) 1, –2

51. Find the value of x for which the expression $2-3x-4x^2$ has the greatest value : **[SSC(CGL-T-I-13]**
(a) $-\frac{41}{16}$ (b) $\frac{3}{8}$
(c) $\frac{-3}{8}$ (d) $\frac{41}{16}$

52. If the roots of the equation $x^3-ax^2+bx-c=0$ are three consecutive integers, then what is the smallest possible value of b? **[CAT-08]**
(a) $-\frac{1}{\sqrt{3}}$ (b) –1
(c) 0 (d) 1
(e) $\frac{1}{\sqrt{3}}$

53. Product of real roots of the equation $t^2x^2+|x|+9=0$:
(a) is always positive
(b) is always negative
(c) does not exist
(d) None of the above

54. The number of non-negative real roots of $2^x-x-1=0$ equals : **[CAT-03]**
(a) 0 (b) 1
(c) 2 (d) 3

55. If one root of $x^2+px+12=0$ is 4, while the equation $x^2+px+q=0$ has equal roots, then the value of q is :
(a) $\frac{49}{4}$ (b) $\frac{4}{49}$
(c) 4 (d) $\frac{1}{4}$

56. The equation formed by decreasing each root of $ax^2+bx+c=0$ by 1 is $2x^2+8x+2=0$, then :
(a) $a=-b$ (b) $b=-c$
(c) $c=-a$ (d) $b=a+c$

57. If the equation $2x^2-7x+12=0$ has two roots α and β, then the value of $\frac{\alpha}{\beta}+\frac{\beta}{\alpha}$ is :
[SSC (CGL-T-I)-13, SSC (CGL-T-II)-19]
(a) $\frac{7}{2}$ (b) $\frac{1}{24}$
(c) $\frac{7}{24}$ (d) $\frac{97}{24}$

58. If one root of the equation $2x^2+3x+c=0$ is 0.5, then what is the value of c? **[CDS-11]**
(a) –1 (b) –2
(c) –3 (d) –4

59. If 3 is a root of $x^2+kx-24=0$, it is also a root of :
(a) $x^2+5x+k=0$
(b) $x^2-5x+k=0$
(c) $x^2-kx+6=0$
(d) $x^2+kx+24=0$

60. What is the total number of common roots between $2x^3-x^2-2x+1=0$ and $x^2+2x-3=0$?
(a) 0
(b) 1
(c) 2
(d) cannot be determined

61. The minimum possible value of the sum of the squares of the roots of the equation $x^2+(a+3)x-(a+5)=0$ is : **[CAT-17]**
(a) 1 (b) 2
(c) 3 (d) 4

62. If α and β are the roots of equation $x^2-2x+4=0$, then which equation will have roots $\frac{\alpha^3}{\beta^2}$ and $\frac{\beta^3}{\alpha^2}$?
(a) $x^2-4x+8=0$ **[SSC (CGL-T-II)-17]**
(b) $x^2-32x+4=0$
(c) $x^2-2x+4=0$
(d) $x^2-16x+4=0$

63. What is the positive value of m for which the roots of the equation $12x^2+mx+5=0$ are in the ratio 3 : 2? **[CDS-17]**
(a) $5\sqrt{10}$ (b) $\frac{5\sqrt{10}}{12}$
(c) $\frac{5}{12}$ (d) $\frac{12}{5}$

Directions *(Problem 64-67) Two equations number I and II are given You have to solve both the equations and mark the appropriate option*

Give answer
(a) if $x>y$ (b) if $x<y$
(c) if $x \geq y$ (d) if $x \leq y$
(e) $x=y$ or the relationship cannot be established

64. I. $x^2-6\sqrt{3}\,x+24=0$
II. $y^2-7\sqrt{3}\,y+30=0$

65. I. $x^2-7\sqrt{2}\,x+24=0$
II. $y^2+3\sqrt{3}\,y-12=0$

66. I. $\frac{x^3-x^2-4x+4}{(x+2)}=0$
II. $\frac{2y^3-9y^2+7y+6}{(2y+1)}=0$

67. I. $\frac{3x^3+\frac{7}{2}x^2-\frac{9}{2}x+1}{(3x-1)}=0$
II. $\frac{y^3-7y+6}{(y-1)}=0$

68. If $f_1(x)=x^2+11x+n$ and $f_2(x)=x$, then the largest positive integer n for which the equation $f_1(x)=f_2(x)$ has two distinct real roots, is : **[CAT-17]**
(a) 22 (b) – 2
(c) 9 (d) None of these

69. If α and β are the roots of the quadratic equation $2x^2+6x+k=0$, where $k<0$, then what is the maximum value of $\frac{\alpha}{\beta}+\frac{\beta}{\alpha}$?
(a) 2 (b) – 2
(c) 9 (d) –9

70. If one root of $(a^2-5a+3)x^2+(3a-1)x+2=0$ is twice the other, then what is the value of a?
(a) $\frac{2}{3}$ (b) $-\frac{2}{3}$
(c) $\frac{1}{3}$ (d) $-\frac{1}{3}$

71. If α and β are the roots of equation $x^2-x+1=0$, then which equation will have roots α^3 and β^3?
(a) $x^2+2x+1=0$ **[SSC(CGL-T-II)-17]**
(b) $x^2-2x-1=0$
(c) $x^2+3x-1=0$
(d) $x^2-3x+1=0$

72. The number of real solutions of the equation $|x^2 + 4x + 3| + 2x + 5 = 0$ is :

(a) 1 (b) 2
(c) 3 (d) 4

73. If p and q are the roots of the equation $x^2 - 15x + r = 0$ and $p - q = 1$, then what is the value of r ? ***[CDS-17]***

(a) 55 (b) 56
(c) 60 (d) 64

74. The quadratic equation $x^2 + bx + c = 0$ has two roots $4a$ and $3a$, where a is an integer. Which of the following is a possible value of $b^2 + c$? ***[CAT-19]***

(a) 3721 (b) 549
(c) 427 (d) 361

75. Consider the equation : $|x - 5|^2 + 5|x - 5| - 24 = 0$. The sum of all the real roots of this equation is : ***[IIFT-19]***

(a) 8 (b) 2
(c) 10 (d) 3

76. The number of solutions to the equation $|x|\ (6x^2 + 1) = 5x^2$: ***[CAT-19]***

(a) 2 (b) 3 (c) 4 (d) 5

Answer Key

1. (d)	2. (d)	3. (c)	4. (a)	5. (d)	6. (c)	7. (b)	8. (b)	9. (b)	10. (b)
11. (c)	12. (c)	13. (b)	14. (b)	15. (b)	16. (a)	17. (c)	18. (a)	19. (c)	20. (a)
21. (d)	22. (d)	23. (d)	24. (d)	25. (c)	26. (d)	27. (c)	28. (a)	29. (d)	30. (a)
31. (d)	32. (a)	33. (b)	34. (b)	35. (d)	36. (d)	37. (c)	38. (c)	39. (d)	40. (c)
41. (c)	42. (c)	43. (a)	44. (d)	45. (b)	46. (d)	47. (c)	48. (c)	49. (b)	50. (a)
51. (c)	52. (b)	53. (c)	54. (c)	55. (a)	56. (b)	57. (b)	58. (b)	59. (c)	60. (b)
61. (c)	62. (c)	63. (a)	64. (e)	65. (a)	66. (d)	67. (e)	68. (c)	69. (b)	70. (a)
71. (a)	72. (b)	73. (b)	74. (b)	75. (c)	76. (d)				

Hints and SOLUTIONS

1. Let roots be α ,β, γ and δ, then

$\alpha + \beta + \gamma + \delta = \dfrac{-(-4)}{1} = 4$

and $\alpha\beta\gamma\delta = \dfrac{1}{1} = 1$

Clearly, $\alpha = \beta = \gamma = \delta = 1$

$\left[\because \text{AM} = \text{GM} \Rightarrow \dfrac{\alpha + \beta + \gamma + \delta}{4} = \sqrt[4]{\alpha\beta\gamma\delta}\right]$

Further,

$\alpha\beta + \alpha\gamma + \alpha\delta + \beta\gamma + \beta\delta + \gamma\delta = \dfrac{a}{1}$

$\Rightarrow 1\cdot 1 + 1\cdot 1 + 1\cdot 1 + 1\cdot 1 + 1\cdot 1 + 1\cdot 1 = a$

$\Rightarrow \quad a = 6$

Also, $\alpha\beta\gamma + \beta\gamma\delta + \gamma\delta\alpha + \delta\alpha\beta = \dfrac{-b}{1}$

$\Rightarrow 1\cdot 1\cdot 1 + 1\cdot 1\cdot 1 + 1\cdot 1\cdot 1 + 1\cdot 1\cdot 1 = -b$

$\Rightarrow \quad b = -4$

Hence, $a = 6, b = -4$

2. Minimum value of the expression

$= \dfrac{4ac - b^2}{4a} = \dfrac{4\cdot 3\cdot 7 - (-4)^2}{4\cdot 3} = 5.66$

Hence, option (d) is correct.

3. Given, equations are

$1x + 2y - 3 = 0$

and $3x + 6y - 9 = 0$

Hence, $\dfrac{1}{3} = \dfrac{2}{6} = \dfrac{-3}{-9}$

Since the given equations are dependent, the system of equations has infinite solutions.

Hence, option (c) is correct.

4. Sum of roots of the given equation $= \dfrac{b}{1}$

$\Rightarrow \quad p + q = b$

and product of the roots $= \dfrac{c}{1}$

$\Rightarrow \quad pq = c$

Now, sum of the roots of the desired equation $= (pq + p + q) + (pq - p - q)$
$= 2pq = 2c$

and product of roots

$= (pq + p + q) \times (pq - p - q)$
$= (pq)^2 - (p + q)^2$
$= c^2 - b^2$

Now, the desired equation is $x^2 -$ (sum of roots) $x +$ (product of roots) $= 0$

$\Rightarrow x^2 - 2cx + (c^2 - b^2) = 0$

5. Proceed through options

(a) $x^{\frac{1}{2}} + 2x + 3 = 0 \Rightarrow$ Degree of the polynomial $\neq 2$

(b) $(x - 1)(x + 4) = x^2 + 3x - 4 = x^2 + 1$
$\Rightarrow 3x - 5 = 0$, a linear equation

(c) $x^4 - 3x + 5 = 0 \Rightarrow$ Degree of the polynomial $\neq 2$

(d) $(2x + 1)(3x - 4) = 2x^2 + 3$
$\Rightarrow 4x^2 - 5x - 7 = 0$, *a* quadratic equation.

Hence, option (d) is correct

6. Given, $\dfrac{2}{x} + \dfrac{3}{y} = \dfrac{9}{xy} \Rightarrow \dfrac{2y + 3x}{xy} = \dfrac{9}{xy}$

$\Rightarrow \quad 3x + 2y - 9 = 0 \quad$ ...(i)

Also given, $\dfrac{4}{x} + \dfrac{9}{y} = \dfrac{21}{xy}$

$\Rightarrow \quad 9x + 4y - 21 = 0 \quad$ ...(ii)

Then, $x = \dfrac{2(-21) - 4(-9)}{3(4) - 9(2)} = \dfrac{-6}{-6} = 1$

Putting the value of x in equation (i) we get, $3(1) + 2y - 9 = 0$

$\Rightarrow \quad 2y = 6 \therefore y = 3$

Now, $\quad x + y = 1 + 3 = 4$

7. We know, if one root of a quadratic equation with rational coefficients is $p+\sqrt{q}$, then another is its conjugate i.e. $p-\sqrt{q}$ and *vice-versa* where p is rational and $\sqrt{q}$ is a surd.

Then, 2nd root $= 2+\sqrt{3}$

Now, sum of roots
$$= (2-\sqrt{3}) + (2+\sqrt{3}) = 4$$
and product of roots
$$= (2-\sqrt{3})(2+\sqrt{3}) = 1$$
$\therefore$ Required equation : $x^2 -$ (sum of roots) $x +$ (product of roots) $= 0$
$$\Rightarrow \quad x^2 - 4x + 1 = 0$$

8. Given, $3\,|x| + 2 - y = 0$
$$\Rightarrow \quad 3\,|x| + 2 = y$$
Since y is the greatest integer less than or equal to x, the value of y shall not be greater than $|x|$ in any circumstances. It implies that the value of $3\,|x| + 2$ will always be greater than y.

Hence, $3|x| + 2 \neq y$ for real roots

$\therefore$ Option (b) is correct

9. Let α and β be the roots. Then, $\alpha+\beta = -p$ and $\alpha\beta = q$

As given, $\alpha+\beta = 3(\alpha-\beta)$
$$\Rightarrow \quad -p = 3(\alpha-\beta)$$
$$\Rightarrow \quad \alpha-\beta = \frac{-p}{3}$$
Now, $(\alpha-\beta)^2 = (\alpha+\beta)^2 - 4\alpha\beta$
$$\Rightarrow \quad \left(-\frac{p}{3}\right)^2 = (-p)^2 - 4q$$
$$\Rightarrow \quad 2p^2 = 9q$$

10. Let $\frac{x}{2} = a$ and $\frac{y}{3} = b$

As given, $a + b = 4$ and $\frac{1}{a} + \frac{1}{b} = 1$
$$\Rightarrow \quad \frac{a+b}{ab} = 1 \Rightarrow a + b = ab$$
Further, $(a-b)^2 = (a+b)^2 - 4ab$
$$\Rightarrow (a-b)^2 = (4)^2 - 4(a+b)$$
$$= (4)^2 - 4\cdot 4 = 0$$
$$\Rightarrow \quad a - b = 0 \Rightarrow a = b$$
$$\therefore \quad a = 2 \text{ and } b = 2$$
Now, $\frac{x}{2} = 2 \Rightarrow x = 4$

and $\frac{y}{3} = 2 \Rightarrow y = 6$

Then, $x + y = 4 + 6 = 10$

11. Given equation $4x - 17y = 1$
$$\Rightarrow \quad 4x = 1 + 17y$$
$$\Rightarrow \quad 4000 \geq 1 + 17y$$
$[\because x \leq 1000, \therefore 4x \leq 4000]$
$$\Rightarrow \quad 3999 \geq 17y \Rightarrow y \leq 235.23$$
Further, $4x = 1 + 17y$
$$\Rightarrow \quad 4x = 16y + (y+1)$$
Since x and y are positive integers, at $y = 3, 7, 11, \ldots\ldots$ x will be positive integer.

Number of integer values of y = number of terms in the A.P. series 3, 7, 11, ..., but the value of last term should not be greater than 235.

Let the last term be 235, Then,

number of terms $= \frac{235-3}{4} + 1 = 59$

Since number of term is in integer, the required answer is 59.

12. Linear coefficient $-6 = -4 - 2$

Then, $\sec\theta = \frac{4}{8}, \frac{2}{8}$ i.e. $\frac{1}{2}, \frac{1}{4}$

But it is the fact that $\sec\theta \geq 1$ or $\sec\theta \leq -1$

Hence, option (c) is correct.

13. We know, if one root of a quadratic equation with rational coefficients is $p+\sqrt{q}$, then another is its conjugate i.e. $p-\sqrt{q}$ and *vice-versa* where, p is rational and $\sqrt{q}$ is a surd

Hence, the required answer is $2-\sqrt{5}$.

14. Let $\sqrt{\frac{2x}{3-x}} = a$. Then, $a - \frac{1}{a} = \frac{3}{2}$
$$\Rightarrow \quad 2a^2 - 3a - 2 = 0$$
Also, 3 may be written as $-4+1$.

Hence, roots be 2 and $\frac{-1}{2}$

Then, $\sqrt{\frac{2x}{3-x}} = 2$ and $\sqrt{\frac{2x}{3-x}} = \frac{-1}{2}$

If $\sqrt{\frac{2x}{3-x}} = 2$
$$\Rightarrow \quad \frac{2x}{3-x} = 4 \text{ [on squaring both sides]}$$
$$\Rightarrow \quad 2x = 12 - 4x \Rightarrow x = 2$$
If $\sqrt{\frac{2x}{3-x}} = \frac{-1}{2}$
$$\Rightarrow \quad \frac{2x}{3-x} = \frac{1}{4} \Rightarrow x = \frac{1}{3}$$
Hence, option (b) is correct

[we may stop the process after we gets $x = 2$. The reason is that we have to find only one root and 2 is in the options.]

15. Given equation
$$p^6 - p = (m^2 + m + 6)(p-1)$$
$$\Rightarrow p(p^5 - 1) = (m^2 + m + 6)(p-1)$$
$$\Rightarrow p(p-1)(p^4 + p^3 + p^2 + p + 1)$$
$$= (m^2 + m + 6)(p-1)$$
$$\Rightarrow (p-1)(p^5 + p^4 + p^3 + p^2 + p)$$
$$= (m^2 + m + 6)(p-1)$$
$$\Rightarrow p^5 + p^4 + p^3 + p^2 + p$$
$$= m(m+1) + 6$$
The value of $m(m+1)$ is even as odd $\times$ even = even or even $\times$ odd = even.

Thus, value of $m(m+1) + 6$ is even.

Since LHS = RHS, the value of LHS is also even.

If p is an odd prime number, then the value of $p^5 + p^4 + p^3 + p^2 + p$ will be odd. Hence, p is not an odd prime number.

It is the fact that only 2 is an even prime number, Hence, $p = 2$

Now, $p^6 - p = (m^2 + m + 6)(p-1)$
$$\Rightarrow \; 2^6 - 2 = (m^2 + m + 6)(2-1)$$
$$\Rightarrow \; m^2 + m - 56 = 0$$
$$\Rightarrow \; m = -8 \text{ and } 7$$
Since m is positive integer, $m = 7$

Hence, only one solution exists at $p = 2$ and $m = 7$.

16. Let α and β be the roots.

Then, $A = \frac{\alpha+\beta}{2}$ and $G = \sqrt{\alpha\beta}$
$$\Rightarrow \alpha+\beta = 2A \text{ and } G^2 = \alpha\beta$$
Now, the required equation :
$$t^2 - (\alpha+\beta)t + \alpha\beta = 0$$
$$\Rightarrow \quad t^2 - 2At + G^2 = 0$$

17. We know, if $= (af-cd)^2 = (ae-bd)\times(bf-ce)$ then both the equations $ax^2 + bx + c = 0$ and $dx^2 + ex + f = 0$ will have a common root.

Given equations be $x^2 - kx - 21 = 0$ and $x^2 - 3kx + 35 = 0$

Then, $[1(35) + 21(1)]^2 = [1(-3k) - (-k)1] \times [(-k)35 - (-21)(-3k)]$
$$\Rightarrow \; 3136 = (-2k)(-98k)$$
$$\Rightarrow \quad 16 = k^2 \Rightarrow k = \pm 4$$

18. Given equation $x^2 - 2x + c = 0$

Then $\quad x_1 + x_2 = 2 \quad$ ...(i)

and $-4x_1 + 7x_2 = 47$ (as given) $\quad$ ...(ii)

Solving both Eqs. we get, $x_1 = -3$ and $x_2 = 5$

Now, product of roots $= -3 \times 5$
$$\Rightarrow \quad c = -15$$

19. The equation $x^2 + ax + 3 = 0$ is satisfied at $x = 3$, then
$$3^2 + 3a + 3 = 0$$
$$\Rightarrow a = -4$$

Let the roots of equation $x^2 + ax + b = 0$ be α and 3α.

Then, $\alpha + 3\alpha = -a$

$\Rightarrow \quad 4\alpha = 4$

$\Rightarrow \quad \alpha = 1$

Now, product of roots

$= \alpha \cdot 3\alpha = 1 \cdot 3 = 3$

$\Rightarrow \quad \frac{b}{1} = 3$

Hence, option (c) is correct

20. Given equation $x^2 + 5|x| + 4 = 0$ may be written as $|x|^2 + 5|x| + 4 = 0$

Clearly, the linear coefficient,

$5 = +1 + 4$

Then, $|x| = -1, -4$, which is not possible.

Hence, option (a) is correct.

21. Let $2a = x - d$, $b = x$ and $2c = x + d$

Then, $2a + 2c = 2x = 2b$

$\Rightarrow \quad b = a + c$

Now, roots

$$= \frac{-b \pm \sqrt{b^2 - 4ac}}{2a}$$

$$= \frac{-(a+c) \pm \sqrt{(a+c)^2 - 4ac}}{2a}$$

$$= \frac{-(a+c) \pm (a-c)}{2a}$$

$$= \frac{-(a+c) + (a-c)}{2a} \text{ and } = \frac{-(a+c) - (a-c)}{2a}$$

$$= -\frac{c}{a} \text{ and } -1$$

22. Given, $\frac{a}{b} - \frac{b}{a} = \frac{x}{y}$

$\Rightarrow \quad \frac{a^2 - b^2}{ab} = \frac{x}{y}$

Let $x = (a^2 - b^2)k$ and $y = abk$

Also, given $\frac{a}{b} + \frac{b}{a} = x - y$

$\Rightarrow \quad \frac{a^2 + b^2}{ab} = k(a^2 - b^2 - ab)$

$\Rightarrow k = \frac{a^2 + b^2}{ab(a^2 - b^2 - ab)}$

Then, $x = (a^2 - b^2)k =$

$$(a^2 - b^2)\frac{a^2 + b^2}{ab(a^2 - b^2 - ab)}$$

$$= \frac{(a-b)(a+b)(a^2+b^2)}{ab(a^2 - b^2 - ab)}$$

Hence, option (d) is correct.

23. Discriminant, $D = b^2 - 4ac$

$\Rightarrow D = [2(p+1)]^2 - 4 \cdot 2 \cdot p = 4p^2 + 4$

Since value of p^2 cannot be negative, the value of $4p^2 + 4 > 0$. We know, if $D > 0$, the roots are real.

Hence option (d) is correct.

24. Given quadratic expression

$= x^2 - 11x + 18$

Then, its minimum value

$$= \frac{4ac - b^2}{4a}$$

$$= \frac{4 \cdot 1 \cdot 18 - (-11)^2}{4 \cdot 1} = \frac{-49}{4}$$

25. Product of roots $= 2k^2 - 1 = 7$ (given)

$\Rightarrow \quad k = \pm 2$

Putting $k = +2$ in the given equation we get $x^2 - 6x + 7 = 0$ and putting $k = -2$ we get

$x^2 + 6x + 7 = 0$

Now, for equation $x^2 - 6x + 7 = 0$,

Discriminant $= (-6)^2 - 4 \cdot 1 \cdot 7 = 8$

and for equation $x^2 + 6x + 7 = 0$,

Discriminant $= 8$

Clearly, D is great than 0, but not a perfect suqare. It implies that roots are irrational.

26. If $A = 0$, then $\frac{0}{x} + \frac{B^2}{x-1} = 1$

$\Rightarrow x = B^2 + 1$

If $B = 0$, then $\frac{A^2}{x} + \frac{0}{x-1} = 1$

$\Rightarrow \ x = A^2$

If $A, B \neq 0$, then $\frac{A^2}{x} + \frac{B^2}{x-1} = 1$

$\Rightarrow \quad A^2x - A^2 + B^2x = x^2 - x$

$\Rightarrow x^2 - (A^2 + B^2 + 1)x + A^2 = 0$

Now, $D = [-(A^2 + B^2 + 1)]^2 - 4 \cdot 1 \cdot A^2$

$= A^4 + B^4 + 1 + 2A^2B^2 + 2B^2 - 2A^2$

$\Rightarrow A^4 + 1 - 2A^2 + B^4 + 2A^2B^2 + 2B^2$

$\Rightarrow (A^2 - 1)^2 + B^4 + 2A^2B^2 + 2B^2$

Obivously,

$(A^2 - 1)^2 + B^4 + 2A^2B^2 + 2B^2 > 0$ as $(A^2 - 1)^2$ cannot be negative.

Hence, the equation

$x^2 - (A^2 + B^2 + 1)x + A^2 = 0$ will have two real roots.

$\therefore$ Required answer is option (d).

27. We know, equations $ax^2 + bx + c = 0$ and $dx^2 + ex + f = 0$

will have a common root if

$(af - cd)^2 = (ae - bd) \times (bf - ce)$.

Then,

$(a \cdot a - cb)^2 = (ac - b \cdot b) \times (b \cdot a - c \cdot c)$

$\Rightarrow a^4 + b^2c^2 - 2a^2bc$

$= a^2bc - ac^3 - ab^3 + b^2c^2$

$\Rightarrow a^4 + ac^3 + ab^3 = 3a^2bc$

$\Rightarrow \quad a^3 + b^3 + c^3 = 3abc$

Then, $\frac{a^3 + b^3 + c^3}{abc} = \frac{3abc}{abc} = 3$

28. Since quadratic coefficient is positive, we can compute the minimum value of the given quadratic equation.

Its minimum value $= \frac{4ac - b^2}{4a}$

$$= \frac{4 \cdot 1 \cdot 1 - (-1)^2}{4 \cdot 1} = \frac{3}{4}$$

29. If α and β be the roots of the quadratic equation

$x^2 - (A-3)x - (A-7) = 0$

Then, $\alpha + \beta = (A - 3)$

and $\quad \alpha \cdot \beta = -(A - 7)$

As given, $\alpha^2 + \beta^2 = 0$

$\Rightarrow (\alpha + \beta)^2 - 2\alpha\beta = 0$

$\Rightarrow (A-3)^2 - 2[-(A-7)] = 0$

$\Rightarrow A^2 + 9 - 6A + 2A - 14 = 0$

$\Rightarrow \quad A^2 - 4A - 5 = 0$

Solving this equation we get

$A = -1, 5$

Hence, option (d) is correct.

30. Given equation :

$|x-2|^2 + |x-2| - 6 = 0$

Clearly, the linear coefficient

$1 = +3 - 2$

$\Rightarrow \ a = -3, 2 \quad$ [If $|x-2| = a$ (let)]

Then, $|x-2| = -3$, which is not possible

Also, $|x-2| = 2$

If $x - 2 > 0$, then $x - 2 = 2 \Rightarrow x = 4$

If $x - 2 < 0$, then $2 - x = 2 \Rightarrow x = 0$

Hence, option (a) is correct.

Alternatively

The best way is to check as to which option satisfies the given equation.

31. Given system of equations

$4x - y - 2 = 0$

and $\quad -8x + 2y + 4 = 0$

Here, $\quad \frac{4}{-8} = \frac{-1}{2} = \frac{-2}{4}$

We know, if $\frac{a_1}{a_2} = \frac{b_1}{b_2} = \frac{c_1}{c_2}$, then equations are said to be dependent and the system of equations has infinite solutions.

Hence, option (d) is correct.

32. If roots of a quadratic equation $ax^2 + bx + c = 0$ are real, then

$b^2 - 4ac \ge 0.$

$\Rightarrow [2(b^2 + c^2)]^2 - 4\cdot(a^2 + b^2)\cdot(b^2 + c^2) \ge 0$

$\Rightarrow (b^2 + c^2)(4b^2 + 4c^2 - 4a^2 - 4b^2) \ge 0$

$\Rightarrow 4(b^2 + c^2)(c^2 - a^2) \ge 0$

$\Rightarrow (c^2 - a^2) \ge 0$

$\Rightarrow c^2 \ge a^2$

33. We know, a system of equations $a_1 x + b_1 y + c_1 = 0$ and $a_2 x + b_2 y + c_2 = 0$ has a unique solution if, $\frac{a_1}{a_2} \ne \frac{b_1}{b_2}$

Hence, $\frac{kx}{6x} \ne \frac{-1}{-2}$

$\Rightarrow k \ne 3$

34. We know, if one root of equation $ax^2 + bx + c = 0$ is the square of the other, then $a + c = \frac{b^3 - 3abc}{-ac}$

$\Rightarrow 1 - k = \frac{(-1)^3 - 3\cdot1(-1)(-k)}{-1\cdot(-k)}$

[$\because$ Given equation is $x^2 - x - k = 0$]

$\Rightarrow k - k^2 = -1 - 3k$

$\Rightarrow k^2 - 4k - 1 = 0$

Solving this equation, we get

$k = 2 \pm \sqrt{5}$

35. Let the roots be α and α^2.

Now, sum of roots $= -k$

$\Rightarrow \alpha + \alpha^2 = -k$

Product of roots $= -8$

$\Rightarrow \alpha\cdot\alpha^2 = -8$

$\Rightarrow \alpha = -2$

From the above, $\alpha + \alpha^2 = -k$

$\Rightarrow -2 + (-2)^2 = -k$

$\Rightarrow 2 = -k$

$\Rightarrow k = -2$

36. We know, the equation whose roots are r greater than roots of an equation $ax^2 + bx + c = 0$ is $a(x - r)^2 + b(x - r) + c = 0.$

Hence, the required equation is

$(x - 1)^2 - 3(x - 1) + 2 = 0$

$\Rightarrow x^2 - 2x + 1 - 3x + 3 + 2 = 0$

$\Rightarrow x^2 - 5x + 6 = 0$

37. Given equation : $x^2 + x + 1 = 0$

Then, $\alpha + \alpha^2 = -1$ and $\alpha\cdot\alpha^2 = 1$

i.e. $\alpha^3 = 1$

Now, $\alpha^{31} + \alpha^{62} = \alpha^{30}\cdot\alpha(1 + \alpha\cdot\alpha^{30})$

$= (\alpha^3)^{10}\cdot\alpha[1 + \alpha\cdot(\alpha^3)^{10}]$

$= (1)^{10}\cdot\alpha[1 + \alpha\cdot(1)^{10}]$

$= \alpha(1 + \alpha) = \alpha + \alpha^2 = -1$

Also,

$\alpha^{31}\cdot\alpha^{62} = \alpha^{93} = (\alpha^3)^{31} = (1)^{31} = 1$

Since the sum and the product of roots of the required equation are same as those of the given equation, the required equation is

$x^2 + x + 1 = 0$

38. Given, $2x + y + 3 = 0$

$\Rightarrow x = \frac{-y-3}{2}$

[Since each of the equations given in the options has variable y, we took value of x in term of y]

Then, $2x^2 + 6x + 5y + 1 = 0$ will become

$2\left(\frac{-y-3}{2}\right)^2 + 6\left(\frac{-y-3}{2}\right) + 5y + 1 = 0$

$\Rightarrow \frac{y^2 + 9 + 6y}{2} - \frac{6y + 18}{2} + 5y + 1 = 0$

$\Rightarrow y^2 + 10y - 7 = 0$

39. Given, $x - \frac{2}{x-1} = 1 - \frac{2}{x-1}$

$\Rightarrow x = 1$

But at $x = 1$, the denominator of $\frac{2}{x-1}$ becomes zero and then the value of $\frac{2}{x-1}$ will be undefined. Hence, the given equation has no roots.

40. Given, $x^2 - (\alpha - 2)x - (\alpha + 1) = 0$

Sum of roots $= p + q = \frac{\alpha - 2}{1} = \alpha - 2$

and product of roots $= pq = -(\alpha + 1)$

Now, $p^2 + q^2 = (p + q)^2 - 2pq$

$= (\alpha - 2)^2 + 2(\alpha + 1)$

$= \alpha^2 - 4\alpha + 4 + 2\alpha + 2$

$= \alpha^2 - 2\alpha + 1 + 5 = (\alpha - 1)^2 + 5$

Since the minimum value of $(\alpha - 1)^2$ be 0, the minimum value of $(a - 1)^2 + 5$ is 5.

Hence, option (c) is correct.

41. Clearly, roots are $-\alpha$ and $-\frac{1}{\alpha}$. Then,

their product $= \frac{c}{a}$

$\Rightarrow -\alpha \times -\frac{1}{\alpha} = \frac{c}{a}$

$\Rightarrow a = c$

42. Sum of roots $= M + N = -M$

$\Rightarrow 2M = -N$

$\Rightarrow M = -N/2$

and product of roots $= MN = N$

$\Rightarrow M = 1$

$\because M = \frac{-N}{2}$

$\therefore 1 = \frac{-N}{2}$

$\Rightarrow N = -2$

Clearly, values of (M, N) are (1, –2).

Further, at $M = N = 0$, $M + N = -M$ and $MN = N$ both are satisfied.

Hence, number of pairs of $(M, N) = 2$ i.e. $(1, -2)$ and $(0, 0)$

43. Let roots be α and $(\alpha + 1)$.

Then, $\alpha + (\alpha + 1) = b$

and $\alpha(\alpha + 1) = c$

Now,

$b^2 - 4c = (2\alpha + 1)^2 - 4\alpha(\alpha + 1) = 1$

44. Given, $|x - y| = 0.2$

$\Rightarrow (x - y)^2 = 0.04$

Further, $|x^2 + y^2| = 0.1$

$\Rightarrow x^2 + y^2 = 0.1$

[$\because (x^2 + y^2)$ is always greater than 0]

$\Rightarrow (x - y)^2 + 2xy = 0.1$

$\Rightarrow 0.04 + 2xy = 0.1$

$\Rightarrow 2xy = 0.06$ ...(i)

Now, $(x + y)^2 = (x - y)^2 + 4xy$

$= 0.04 + 2(0.06) = 0.16$

$\Rightarrow x + y = \pm 0.4$ ...(ii)

Solving Eqs. (i) and (ii), we get

$x = \pm 0.3$ & $y = \pm 0.1$ or $x = \pm 0.1$ and $y = \pm 0.3$

Then, $|x| + |y| = |\pm 0.3| + |\pm 0.1| = 0.4$

or $|x| + |y| = |\pm 0.1| + |\pm 0.3| = 0.4$

Hance, option (d) is correct.

45. Observe the chart given under 'Formation of a quadratic equation with the roots having certain relation with roots of another'.

The equation whose roots are reciprocal to the roots of equation $ax^2 + bx + c = 0$ is $cx^2 + bx + a = 0$

Accordingly, the required equation is

$-4x^2 - 3x + 2 = 0$

$\Rightarrow 4x^2 + 3x - 2 = 0$

46. Given equations : $x^2 + ax + 10 = 0$ and $x^2 + bx - 10 = 0.$

Difference between the corresponding polynomials

$= x(a-b)+20$

Now, $x(a-b)+20=0$

$\Rightarrow \quad x = \dfrac{-20}{(a-b)}$

We know, if roots of the corresponding equation of difference between given polynomials or its subset satisfy both the equations, the given equations have common root(s) and *vice-versa* and its/their values are same as value of the root(s) at which the equations are satisfied.

If we consider 1st equation, then $x^2+ax+10=0$

$\Rightarrow \left(\dfrac{-20}{a-b}\right)^2 + a\left(\dfrac{-20}{a-b}\right)+10=0$

$\Rightarrow 40-2a(a-b)+(a-b)^2=0$

$\Rightarrow 40-2a^2+2ab+a^2-2ab+b^2=0$

$\Rightarrow b^2-a^2=-40$

$\Rightarrow a^2-b^2=40$

47. Clearly, $\alpha+\beta=10$ and $\alpha\beta=15$

Now, sum of the roots of the desired equation

$= \left(\alpha+\dfrac{\alpha}{\beta}\right)+\left(\beta+\dfrac{\beta}{\alpha}\right)$

$= \alpha+\beta+\left(\dfrac{\alpha^2+\beta^2}{\alpha\beta}\right)$

$= 10+\left(\dfrac{(\alpha+\beta)^2-2\alpha\beta}{\alpha\beta}\right)$

$= 10+\left(\dfrac{(10)^2-2\cdot15}{15}\right)=\dfrac{44}{3}$

and product of the roots of the desired equation

$= \left(\alpha+\dfrac{\alpha}{\beta}\right)\times\left(\beta+\dfrac{\beta}{\alpha}\right)=\alpha\beta+\beta+\alpha+1$

$= 15+10+1=26$

Now, the desired equation be x^2- (sum of roots) $x+$ product of roots $=0$

$\Rightarrow \quad x^2-\dfrac{44x}{3}+26=0$

$\Rightarrow \quad 3x^2-44x+78=0$

48. Given equation : $(x-a)(x-b)=c$

$\Rightarrow x^2-(a+b)x+ab-c=0$

Then, $\alpha+\beta=(a+b)$

and $\alpha\beta=ab-c$ (i.e. $\alpha\beta+c=ab$)

As given, $(x-\alpha)(x-\beta)+c=0$

$\Rightarrow \quad x^2-(\alpha+\beta)x+\alpha\beta+c=0$

Then, sum of roots $=\alpha+\beta$ and product of root $=\alpha\beta+c$

$\Rightarrow$ sum of roots $=a+b$ and product of root $=ab$

Hence, roots are a and b.

49. Proceed through options.

(a) $\Rightarrow 5(x+2y-3z)+2(2x+6y-11z)+(x-2y+7z)\neq 0$

(b) $\Rightarrow 5(x+2y-3z)-2(2x+6y-11z)-(x-2y+7z)=0$

Hence, option (b) is correct.

50. Let $x^{\frac{1}{3}}=a$

Then, $a^2+a-2=0$

Linear coefficient, $1=+2-1$

Then, $a=-2,+1$

Now, $x^{\frac{1}{3}}=-2, 1\Rightarrow x=-8, 1$

Alternatively

You may check as to which option satisfies the given equation.

51. Value of $x=\dfrac{-b}{2a}=\dfrac{-(-3)}{2(-4)}=\dfrac{-3}{8}$

52. Let the roots be $(\alpha-1)$, α and $(\alpha+1)$.

Then,

$(\alpha-1)\alpha+\alpha(\alpha+1)+(\alpha+1)(\alpha-1)$

$=\dfrac{b}{1}$

$\Rightarrow \alpha^2-\alpha+\alpha^2+\alpha+\alpha^2-1=b$

$\Rightarrow 3\alpha^2-1=b$

Now, the smallest value of $3\alpha^2=0$

as $\alpha^2\geq 0$

Then, the smallest value of

$3\alpha^2-1=0-1=-1$

Alternatively

The smallest value of

$3\alpha^2-1=\dfrac{4\cdot3(-1)-0^2}{4\cdot3}=-1$

$\left[\because \dfrac{4ac-b^2}{4a}\right.$ is the minimum value of a quadratic expression ax^2+bx+c, when $\left.a>0\right]$

53. Obviously, value of $t^2x^2\geq 0$ and value of $|x|\geq 0$

Then, $t^2x^2+|x|+9\geq 9$ but the given equation is $t^2x^2+|x|+9=0$

Hence, the given equation does not have real roots.

54. Given, $2^x-x-1=0$

$\Rightarrow 2^x=x+1$, it is possible only at $x=0, 1$.

Hence, number of non-negative real roots of $2^x-x-1=0$ is 2.

55. One root of $x^2+px+12=0$ is 4.

It means $4^2+p\cdot4+12=0$

$\Rightarrow \quad p=-7$

Since equation $x^2+px+q=0$ has equal roots, its discriminant is 0.

$\Rightarrow \quad p^2-4\cdot1\cdot q=0$

$\Rightarrow \quad (-7)^2-4\cdot1\cdot q=0$

$\Rightarrow \quad q=\dfrac{49}{4}$

56. Let roots of $ax^2+bx+c=0$ be α and β.

Then, $\alpha+\beta=\dfrac{-b}{a}$ and $\alpha\beta=\dfrac{c}{a}$.

As given, roots of $2x^2+8x+2=0$ be $(\alpha-1)$ and $(\beta-1)$

Now, sum of roots

$=(\alpha-1)+(\beta-1)=-4$

$\Rightarrow \quad \alpha+\beta=-2$

$\Rightarrow \quad \dfrac{-b}{a}=-2$

$\Rightarrow \quad b=2a$

and product of roots

$=(\alpha-1)(\beta-1)=1$

$\Rightarrow \alpha\beta-(\alpha+\beta)+1=1$

$\Rightarrow \quad \dfrac{c}{a}-(-2)+1=1$

$\Rightarrow \quad c=-2a$

$\Rightarrow \quad c=-b$

$\Rightarrow \quad b=-c$

57. Sum of roots $=\alpha+\beta=\dfrac{7}{2}$

and product of roots $=\alpha\beta=\dfrac{12}{2}=6$

Now,

$\dfrac{\alpha}{\beta}+\dfrac{\beta}{\alpha}=\dfrac{\alpha^2+\beta^2}{\alpha\beta}=\dfrac{(\alpha+\beta)^2-2\alpha\beta}{\alpha\beta}$

$=\dfrac{\left(\dfrac{7}{2}\right)^2-2\cdot6}{6}=\dfrac{1}{24}$

58. Clearly, one value of x is 0.5.

Then, $2\cdot(0.5)^2+3(0.5)+c=0$

$\Rightarrow \quad c=-2$

59. Obviously, $x^2+kx-24=0$ is satisfied at $x=3$.

Then, $3^2+3k-24=0$

$\Rightarrow \quad k=5$

Now, proceed through options and check as to which option is satisfied at $x=3$ and $k=5$.

$\therefore$ Option (c) is correct.

60. Solutions of $x^2+2x-3=0$ are $1, -3$

Another equation is satisfied only at $x=1$.

Hence. the value of common root is 1.

$\therefore$ Number of common roots = 1

61. Let the roots be α and β.

Then $\alpha + \beta = -(a+3)$ and $\alpha\beta = -(a+5)$

Now, $\alpha^2 + \beta^2 = (\alpha+\beta)^2 - 2\alpha\beta$

$= [-(a+3)]^2 - 2[-(a+5)] = a^2 + 8a + 19$

We know, for x be real, the minimum vale of $ax^2 + bx + c$ is $\frac{4ac - b^2}{4a}$, if $a > 0$.

Hence, the minimum value of $a^2 + 8a + 19$

$$= \frac{4 \cdot 1 \cdot 19 - 8^2}{4 \cdot 1} = 3$$

62. Obviously, $\alpha\beta = 4$

Given equation, $x^2 - 2x - (-4) = 0$

Now, $\frac{\alpha^3}{\beta^2} + \frac{\beta^3}{\alpha^2} = \frac{\alpha^5 + \beta^5}{\alpha^2\beta^2}$

$$= \frac{(2)^5 + 5(2)^3 \cdot (-4) + 5(2) \cdot (-4)^2}{4^2} = 2$$

and $\frac{\alpha^3}{\beta^2} \times \frac{\beta^3}{\alpha^2} = \alpha\beta = 4$

Then, the required equation be

$x^2 -$ (sum of roots) $x +$ product of roots = 0

$\Rightarrow \quad x^2 - 2x + 4 = 0$

63. Let α and β be roots of the given equation.

Then, $\alpha + \beta = \frac{-m}{12}$ and $\alpha\beta = \frac{5}{12}$

Given, $\frac{\alpha}{\beta} = \frac{3}{2} \Rightarrow \alpha = \frac{3}{2}\beta$

Now, $\frac{3}{2}\beta + \beta = \frac{-m}{12} \quad \left[\because \alpha + \beta = \frac{-m}{12}\right]$

$\Rightarrow \quad \beta = \frac{-m}{30} \Rightarrow \beta^2 = \frac{m^2}{900}$

Further, $\alpha\beta = \frac{5}{12}$

$\Rightarrow \quad \frac{3}{2}\beta \times \beta = \frac{5}{12} \Rightarrow \beta^2 = \frac{5}{18}$

From above,

$$\frac{m^2}{900} = \frac{5}{18} \Rightarrow m = \sqrt{250} = 5\sqrt{10}$$

64.

I. $x^2 - 6\sqrt{3}x + 24 = 0$	II. $y^2 - 7\sqrt{3}\,y + 30 = 0$
$x^2 - 6x + 8 = 0$ [on dividing the linear coefficient and the constant by $\sqrt{3}$ and its square respectively]	$y^2 - 7y + 10 = 0$ [on dividing the linear coefficient and the constant by $\sqrt{3}$ and its square respectively]
$-6 = -4 - 2$	$-7 = -5 - 2$
Roots of $x^2 - 6x + 8 = 0$ = 4, 2	Roots of $y^2 - 7y + 10 = 0$ = 5, 2
Required roots = $4\sqrt{3}, 2\sqrt{3}$ (on multiplying the roots of $x^2 - 6x + 8 = 0$ by $\sqrt{3}$)	$y = 5\sqrt{3}, 2\sqrt{3}$ (on multiplying the roots of $y^2 - 7y + 10 = 0$ by $\sqrt{3}$)

Now, for conclusion

x	Relation	y
$4\sqrt{3}$	<	$5\sqrt{3}$
$4\sqrt{3}$	>	$2\sqrt{3}$
$2\sqrt{3}$	<	$5\sqrt{3}$
$2\sqrt{3}$	=	$2\sqrt{3}$
	No conclusion as opposite inequality – signs appear.	

Hence, no relation can be established between x and y.

65.

I. $x^2 - 7\sqrt{2}\,x + 24 = 0$	II. $y^2 + 3\sqrt{3}\,y - 12 = 0$
$x^2 - 7x + 12 = 0$ [on dividing the linear coefficient and the constant by $\sqrt{2}$ and its square respectively]	$y^2 + 3y - 4 = 0$ [on dividing the linear coefficient and the constant by $\sqrt{3}$ and its square respectively]
$-7 = -4 - 3$	$+3 = +4 - 1$
Roots of $x^2 - 7x + 12 = 0$ = 4, 3	Roots of $y^2 + 3y - 4 = 0$ = -4, 1
Required roots = $4\sqrt{2}, 3\sqrt{2}$	Required roots = $-4\sqrt{3}, \sqrt{3}$

Now, for conclusion

x	Relation	y
$4\sqrt{2}$	>	$-4\sqrt{3}$
$4\sqrt{2}$	>	$\sqrt{3}$
$3\sqrt{2}$	>	$-4\sqrt{3}$
$3\sqrt{2}$	>	$\sqrt{3}$
	Conclusion – sign : >	

Hence, $x > y$

66.

I. $\frac{x^3 - x^2 - 4x + 4}{(x+2)} = 0$	II. $\frac{2y^3 - 9y^2 + 7y + 6}{(2y+1)} = 0$
One root of $x^3 - x^2 - 4x + 4 = 0$ is -2 as $(x+2)$ is a factor of $x^3 - x^2 - 4x + 4$	One root of $2y^3 - 9y^2 + 7y + 6 = 0$ is $\frac{-1}{2}$ as $(2y+1)$ is a factor of $2y^3 - 9y^2 + 7y + 6$
+1 −1 −4 +4 KR (i.e. −2) −2 +6 −4 +1 −3 +2 0 [KR refers to known root]	+2 −9 +7 +6 KR (i.e. $-\frac{1}{2}$) −1 +5 −6 +2 −10 +12 0 [KR refers to known root]
$-3 = -1 - 2$	$-10 = -4 - 6$
$x = 1, 2$	$y = \frac{4}{2}, \frac{6}{2}$ i.e. 2, 3

Now, for conclusion

x	Relation	y
1	<	2
1	<	3
2	=	2
2	<	3
	Conclusion –sign : $\le$	

Hence, $x \le y$

67.

I. $\dfrac{3x^3 + \frac{7}{2}x^2 - \frac{9}{2}x + 1}{(3x-1)} = 0$	II. $\dfrac{y^3 - 7y + 6}{(y-1)} = 0$
$\dfrac{6x^3 + 7x^2 - 9x + 2}{(3x-1)} = 0$ [For sake of convenience, we multiply both sides by 2. If you don't intend to do so, you may process accordingly]	No need to alter the given equation.
One root of $6x^3 + 7x^2 - 9x + 2 = 0$ is $\frac{1}{3}$ as $(3x - 1)$ is a factor of $6x^3 + 7x^2 - 9x + 2$	One root of $y^3 - 7y + 6 = 0$ is 1 as $(y-1)$ is a factor of $y^3 - 7y + 6$
+6 +7 –9 +2 KR (i.e. $\frac{1}{3}$) +2 +3 –2 +6 +9 –6 0 [KR refers to known root]	+1 0 –7 +6 KR (i.e.1) +1 +1 –6 +1 +1 –6 0 [Here, 2nd degree term is 0, so respective coefficient has been taken as 0]
$+9 = +12 - 3$	$+1 = +3 - 2$
$x = \frac{-12}{6}, \frac{3}{6}$ i.e. $-2, \frac{1}{2}$	$y = -3, 2$

Now, for conclusion

x	Relation	y
–2	>	–3
–2	<	2
$\frac{1}{2}$	>	–3
$\frac{1}{2}$	<	2
	No conclusion as opposite inequality – signs appear	

Hence, no relation can be established between x and y.

68. Given, $f_1(x) = f_2(x)$

$\Rightarrow \quad x^2 + 11x + n = x$

$\Rightarrow \quad x^2 + 10x + n = 0$

For two distinct real roots, the discriminant D must be greater than 0.

Hence, $10^2 - 4\cdot 1\cdot n > 0$

$\Rightarrow \quad n < 25$

Hence, the largest integer value of $n = 24$.

69. Sum of roots $= \alpha + \beta = \frac{-6}{2} = -3$

and product of roots $= \alpha\beta = \frac{k}{2}$

Now, $\frac{\alpha}{\beta} + \frac{\beta}{\alpha} = \frac{\alpha^2 + \beta^2}{\alpha\beta}$

$= \frac{(\alpha+\beta)^2 - 2\alpha\beta}{\alpha\beta} = \frac{(\alpha+\beta)^2}{\alpha\beta} - 2$

$= \frac{(-3)^2}{\frac{k}{2}} - 2 = \frac{18}{k} - 2$

Since k is negative, the maximum value of $\frac{18}{k}$ will be zero at $k = -\infty$.

Hence, required maximum value $= 0 - 2 = -2$

70. We know, if one root of equation $ax^2 + bx + c = 0$ is m times the other, then $m + \frac{1}{m} = \frac{b^2 - 2ac}{ac}$

$\Rightarrow \quad 2 + \frac{1}{2} = \frac{(3a-1)^2 - 2(a^2 - 5a + 3)2}{(a^2 - 5a + 3)2}$

$\Rightarrow \quad \frac{5}{2} = \frac{5a^2 + 14a - 11}{(a^2 - 5a + 3)2}$

$\Rightarrow \quad 5a^2 - 25a + 15 = 5a^2 + 14a - 11$

$\Rightarrow \quad -39a = -26 \Rightarrow a = \frac{2}{3}$

71. We know, the new equation whose roots are cubes of the roots to $ax^2 + bx + c = 0$

be $a(\sqrt[3]{x})^2 + b\sqrt[3]{x} + c = 0.$

Hence, the new equation be $1(\sqrt[3]{x})^2 - 1\sqrt[3]{x} + 1 = 0$

$\Rightarrow \quad x^{\frac{2}{3}} - x^{\frac{1}{3}} = -1$

$\Rightarrow \quad (x^{\frac{2}{3}} - x^{\frac{1}{3}})^3 = -1$ [cubing both sides]

$\Rightarrow x^2 - x - 3x^{\frac{2}{3}}x^{\frac{1}{3}}(x^{\frac{2}{3}} - x^{\frac{1}{3}}) = -1$

$\Rightarrow \quad x^2 - x - 3x(-1) = -1$

$\Rightarrow \quad x^2 + 2x + 1 = 0$

72. There may be two cases :

(i) $x^2 + 4x + 3 > 0$

(ii) $x^2 + 4x + 3 < 0$

Case (i) When $x^2 + 4x + 3 > 0$, then

$|x^2 + 4x + 3| + 2x + 5 = 0$ means

$x^2 + 4x + 3 + 2x + 5 = 0$

$\Rightarrow \quad x^2 + 6x + 8 = 0$

Linear coefficient, $+6 = +2 + 4$

Then, $x = -2, -4$, but at $x = -2$ the condition $x^2 + 4x + 3 > 0$ is not satisfied. Hence, only one solution in case (i).

Case (ii) When $x^2 + 4x + 3 < 0$, then $|x^2 + 4x + 3| + 2x + 5 = 0$ means $-(x^2 + 4x + 3) + 2x + 5 = 0$

$\Rightarrow \quad -x^2 - 2x + 2 = 0$

$\Rightarrow \quad x = -1 + \sqrt{3}, -1 - \sqrt{3}$

[on applying quadratic formula]

But at $x = -1 + \sqrt{3}$ the condition $x^2 + 4x + 3 < 0$ is not satisfied.

Hence, only one solution in case (ii).

Conclusion: Number of real solutions = 2

73. Sum of roots i.e.

$p + q = \frac{-(-15)}{1} = 15$ & $p - q = 1$ (as given)

Then, $p = 8$

Obviously the given equation will be satisfied at $x = 8$.

Then, $8^2 - 15 \cdot 8 + r = 0$

$\Rightarrow \quad r = 56$

74. Given equation : $x^2 + bx + c = 0$

Then, sum of roots $= \frac{-b}{1} = -b$ and

product of roots $= \frac{c}{1} = c$

As given, $-b = 4a + 3a$

$\Rightarrow \quad b = -7a$ and $c = 4a \times 3a$

$\Rightarrow \quad c = 12a^2$

Then, $b^2 + c = 49a^2 + 12a^2 = 61a^2$

Now, proceed through the given options.

(a) $61a^2 = 3721 \Rightarrow a^2 = 61$, hence $a \neq$ an integer

(b) $61a^2 = 549 \Rightarrow a^2 = 9 \Rightarrow a = \pm 3$, an integer.

Thus, option (b) is correct.

75. Given equation : $|x - 5|^2 + 5|x - 5| - 24 = 0$

Clearly, the linear coefficient $5 = +8 - 3$

$\Rightarrow a = -8, 3$ [if $|x - 5| = a$ (let)]

Then, $|x - 5| = -8$, which is not possible.

Also, $|x - 5| = 3$

If $x - 5 > 0$, then $x - 5 = 3 \Rightarrow x = 8$

If $x - 5 < 0$, then $5 - x = 3 \Rightarrow x = 2$

Now, sum of the roots $= 8 + 2 = 10$

76. There may be three cases viz. $x = 0$, $x > 0$ and $x < 0$.

Case (i) $x = 0$

At $x = 0$, the given equation is satisfied.

Case (ii) $x > 0$

Then, $x(6x^2 + 1) = 5x^2$

$\Rightarrow \quad 6x^2 - 5x + 1 = 0$

$\Rightarrow x = \frac{1}{2}, \frac{1}{3}$ (both are satisfying the given equation.)

Case (iii) $x < 0$

Then, $-x(6x^2 + 1) = 5x^2$

$\Rightarrow \quad 6x^2 + 5x + 1 = 0$

$\Rightarrow \quad x = \frac{-1}{2}, \frac{-1}{3}$

Hence, there are 5 solutions to the given equation.

54

Part-III

Concept of Inequality

Learning OBJECTIVES

After completing this sub-section, aspirants will be able to :

- Understand the significance of this sub-section in diverse competitive exams;
- Comprehend the diverse types of inequalities with illustrations;
- Familiar with the notation of range of values of solution set of inequalities;
- Find the techniques how to solve linear and quadratic inequalities;
- Find the techniques how to solve a system of linear inequalities in one variable and two variables;
- Find the techniques how to solve linear and quadratic inequalities having modulus;
- Determine the value of a quadratic equation;
- Familiar with some critical inequalities;
- Realise themselves competent to solve all types of problems based on inequalities;
- Understand the philosophy of division of 'Illustrations' in two parts;
- Identify the problems which can easily be solved through options;
- Realise themselves competent to solve the questions in unique way;
- Analyse the trend of the questions being asked in various competitive examinations;
- Solve the questions in moments;
- Learn many new concepts/fundamentals;
- Apply the pragmatic approaches in lieu of academic approaches; and
- Ponder on the alternative methods.

Inside

Introduction

An inequality is a relation between two different values, shown by >, <, ≥ or ≤. Values may be in form of real numbers or algebric expressions.

e.g. $7 > 6$, $3 < 5$ (example of numerical inequality)

$x \le 3, y \ge 2, z > 9, x < 1$ [examples of literal inequality]

An inequality having > or < is known as 'strict inequality' whereas an inequality with ≥ or ≤ is known as 'slack inequality'.

Example of

(i) linear inequality in one variable : $ax + b < 0$, where $a \ne 0$

(ii) linear inequality in two variables : $ax + by \ge c$, where $a \ne 0, b \ne 0$

(iii) Quadratic inequality in one variable : $ax^2 + bx + c \le 0$, where $a \ne 0$

Critical Properties of Inequalities

(i) There is no effect on the sign of inequality when the same number is added to (or subtracted from) the both sides of the inequality.

If $a > b$, then $(a + c) > (b + c)$ and $(a - c) > (b - c)$

If $a < b$, then $(a + c) < (b + c)$ and $(a - c) < (b - c)$

(ii) There is no effect on sign of inequality when both sides of an inequality are multiplied (or divided) by the same positive number. In case both sides are multiplied (or divided) by the same negative number, the sign of inequality is reversed.

If $a > b$, then $ac > bc$ and $\frac{a}{c} > \frac{b}{c}$

and if $a < b$, then $ac < bc$ and $\frac{a}{c} < \frac{b}{c}$

Also, if $a > b$, then $a(-c) < b(-c)$ and $\frac{a}{-c} < \frac{b}{-c}$

and if $a<b$, then $a(-c) > b(-c)$ and $\frac{a}{-c} > \frac{b}{-c}$

(iii) Any term of one side of an inequality can be transposed to another side with opposite sign. In doing so, the sign of inequality remains the same.

If $a + b > c$, then $a > c - b$

and $a + b < c$, then $a < c - b$

Also, if $a - b > c$, then $a > c + b$

and if $a - b < c$, then $a < c + b$

(iv) Two inequalities with the same sign may be added side-wise retaining the sign of inequality unchanged.

If $a > b$ and $c > d$, then $(a + c) > (b + d)$

and if $a < b$ and $c < d$, then $(a + c) < (b + d)$

In case of side-wise subtraction of one inequality from another inequality with opposite sign, the sign of the resultant inequality will be the same as the sign of the inequality from which another is subtracted.

If $a > b$ and $c < d$, then $(a - c) > (b - d)$

and if $a < b$ and $c > d$, then $(a - c) < (b - d)$

(v) If $a > b$, $a > 0$, $b > 0$, then $a^n > b^n$ for any natural number n and

If $a < b, a > 0, b > 0$, then $a^n < b^n$ for any natural number n.

Also, if $a > b$, $a > 0$, $b > 0$, then $a^{1/n} > b^{1/n}$ for any natural number n and

If $a < b, a > 0, b > 0$, then $a^{1/n} < b^{1/n}$ for any natural number n.

(vi) If $a > b$, then $-a < -b$ and if $a < b$, then $-a > -b$

It may be construed that if signs of all terms of an inequality is reversed, the sign of inequality is reversed.

(vii) If $a > b, a \ne 0, b \ne 0$, then $\frac{1}{a} < \frac{1}{b}$

and if $a < b$, then $\frac{1}{a} > \frac{1}{b}$

It may be construed that if we take reciprocals of both sides of an inequality, then the sign of inequality is reversed.

(viii) If $a > b$ and $b > c$, than $a > c$

and if $a < b$ and $b < c$, then $a < c$

Also, if $a = b$ and $b > c$, then $a > c$

and if $a = b$ and $b < c$, then $a < c$

Remarks

The above all the properties also hold good if > and < are replaced by ≥ and ≤ respectively. In addition to the above, the followings also hold :

(ix) If $a \ge b$ and $b > c$, then $a > c$

and if $a \le b$ and $b < c$, then $a < c$

(x) If one of a and b is positive and the other is negative, then :

If $a > b$, then $\frac{1}{a} > \frac{1}{b}$ and if $a < b$, then $\frac{1}{a} < \frac{1}{b}$

(xi) If $ab > 0$, then either both are positive or both are negative.

(xii) If $ab < 0$, then either $a > 0$ and $b < 0$ or $a < 0$ and $b > 0$.

(xiii) Two inequalities pointing in the same direction may be multiplied side-wise. Observe case-wise.

Case I When $a, b > 0$ and $c, d > 0$, all are real :

- If $a > b$ and $c > d$, then $ac > bd$
- If $a < b$ and $c < d$, then $ac < bd$
- If $a < b$ and $c \le d$, then $ac < bd$
- If $a \ge b$ and $c > d$, then $ac > bd$
- If $a < b$ and $c > d$, then no relation can be established between ac and bd.

Case II When $a, b > 0$ and $c > 0, d < 0$, all are real :

- If $a > b$ and $c > d$, then $ac > bd$
- If $a \geq b$ and $c > d$, then $ac > bd$

Case III When $a, b < 0$ and $c > 0, d < 0$, all are real :

- If $a >$ b and $c > d$, then $ac < bd$
- If $a \geq b$ and $c >$ d, then $ac < bd$

However, in the following cases, we cannot draw a conclusion about relation between ac and bd.

(a) $a > 0 > b$ and $c > 0 > d$

(b) $a > b > 0$ and $0 > c > d$

I. Linear Inequalities

(A) Solutions of linear inequalities in one variable

Solutions of an inequality in one variable refers to the values of the variable which make an inequality a true statement. Now, we will solve some linear inequalities applying the properties of inequalities.

Problem 1 Solve $3x + 8 < x + 14$

Sol. $3x - x < 14 - 8$ [property (iii)]

$\Rightarrow$ $2x < 6 \Rightarrow x < 3$ [property (ii)]

Alternatively

Let the options be $x > 4, x \leq 3, x \geq 5$ and $x < 3$.

Option (a) $\Rightarrow$ we take $x = 5$ as $x > 4$, but it does not satisfy the inequality.

Option (b) $\Rightarrow$ we take $x = 3$ as $x \leq 3$, but it does not satisfy the inequality.

Option (c) $\Rightarrow$ we take $x = 6$ as $x \geq 5$, but it does not satisfy the inequality.

Option (d) $\Rightarrow$ we take $x = 2$ as $x < 3$, and it satisfies the inequality.

Hence, option (d) is correct.

Problem 2 Solve $3x + 8 < 5x + 18$

Sol. $3x - 5x < 18 - 8$ [property (iii)]

$\Rightarrow$ $-2x < 10$

$\Rightarrow$ $-x < 5$ [property (ii)]

$\Rightarrow$ $x > -5$ [property (ii)]

Problem 3 Solve $2(x - 3) < 3(x + 2) - x - 10$

Sol. $2x - 6 < 3x + 6 - x - 10$

$\Rightarrow$ $2x - 6 < 2x - 4$

$\Rightarrow$ $0 \times x < 2$ [property (iii)]

Clearly, for all values of x, the inequality is true.

Problem 4 Solve $\dfrac{x}{5} - 2 \geq \dfrac{5 - 3x}{2}$

Sol. $\dfrac{x - 10}{5} \geq \dfrac{5 - 3x}{2}$

$\Rightarrow$ $2x - 20 \geq 25 - 15x$

[property (ii), both sides have been multiplied by 10]

$\Rightarrow$ $2x + 15x \geq 25 + 20$ [property (iii)]

$\Rightarrow$ $17x \geq 45$

$\Rightarrow$ $x \geq 45/17$

[property (ii), both sides have been divided by 17]

Observation All the four questions mentioned above can be solved through the given options.

(B) Notation of range of values of solution set of inequalites

The solution set of an inequality is the set of all solutions. We may write the solutions in the form of interval notation also. Interval notation expresses the inequality in form of interval having a pair of real numbers. 1st real number of the pair is starting point whereas 2nd number is endpoint.

Let the solutions of an inequality is $-1 < x < 3$. It implies that x is greater than -1 but less than 3. Clearly, -1 and 3 are not included in the solution set. We may write it in the interval notation as $(-1, 3)$. Observe the followings.

Solutions/ inequality	Interval notation	Remarks
$-1 < x < 3$	$(-1, 3)$	Starting and ending points both excluded
$-1 < x \leq 3$	$(-1, 3]$	Only ending point included
$-1 \leq x < 3$	$[-1, 3)$	Only starting point included
$-1 \leq x \leq 3$	$[-1, 3]$	Starting point and ending points both included
$x < 3$	$(-\infty, 3)$	
$x \leq 3$	$(-\infty, 3]$	
$x < -1$ or $x > 3$	$(-\infty, -1)$ U $(3, \infty)$	U stands for Union
$x < -1$ or $x \geq 3$	$(-\infty, -1)$ U $[3, \infty)$	
$-1 < x < 3$ or $4 \leq x < 7$	$(-1, 3)$ U $[4, 7)$	
$x < 0$ or $x > 3$	$(-\infty, 0)$ U$(3, \infty)$	

(C) Solutions of Linear Inequalities in Two Variables

Solutions of an inequality in two variables refer to the values of the variables (ordered pair) which make an inequality a true statement. Consider some examples

Problem 5 Whether $(-2, 1)$ is a solution of the linear inequality $x \geq 2y - 1$

Sol. Since $(-2, 1)$ is an ordered pair , $x = -2$ and $y = 1$

Now, substituting the values of variables in the given inequality we get $-2 \geq 2 \cdot 1 - 1$ i.e. $-2 \geq 1$, which is not true.

Hence, $(-2, 1)$ is not a solution of the linear inequality $x \geq 2y - 1$.

Problem 6 If $2x + 3y \leq 6, x \geq 0, y \geq 0$, then one of the solutions is : ***[CDS-13]***

(a) $x = -2$ and $y = 3$ (b) $x = -1$ and $y - -1$

(c) $x = 1$ and $y = 2$ (d) $x = 1$ and $y = 1$

Sol. Given, $2x + 3y \leq 6$

$\Rightarrow$ $2x \leq 6 - 3y \Rightarrow x \leq \dfrac{6 - 3y}{2}$

Now, proceed through options

Since, $x \geq 0$ and $y \geq 0$, options (a) and (b) are not possible.

Option (c) $\Rightarrow 1 \leq \frac{6-3\cdot 2}{2} \Rightarrow -1 \leq 0$ (not true)

Option (d) $\Rightarrow 1 \leq \frac{6-3\cdot 1}{2} \Rightarrow 1 \leq 1.5$ (true)

Hence, option (d) is correct.

(D) Finding Relation Between Any Two Elements if Relation Among Different Elements is Given in Form of the Inequality Statement(s)

Problem 7 If $A > B > C = D \geq E > F$, then what is the relation between A and F?

(a) $A \leq F$ (b) $A \geq F$ (c) $A = F$ (d) $A > F$

Sol. It is true that we cannot establish the relation between any two elements unless we have a common element which is related with both elements somehow (directly or indirectly). Here, A is related to B, B is related to C, C is related to D, D is related to E and E is related to F. It implies that there is also a relation between A and F. We can find the relation between A and C as B is related with both A and C. Thereafter, we can find relation between A and D as C would be common for both A and D. Then, relation can be established between A and E as D would be common for both A and E. Finally, we will be able to find relation between A and F when E will be common for both A and F. Now, proceed step by step :

Step (a) Given, $A>B>C$. It implies that $A>B$ and $B > C$. Then, $A > C$.

[Recall the property, if $a > b$ and $b > c$, then $a > c$]. This property can be written in brief as > and > ⇒ > for sake of convenience only where 1st > denotes relation between a and b, 2nd > denotes the relation between b and c and 3rd > does for relation between a and c. Please mind the order of elements in the statements.] All the properties to be applied under this section may be summerised as below :

Sign indicating how one element is related to the common element [say, a (inequality sign) b]	Sign indicating how the common element is related to the 2nd element [say b (inequality sign) c]	Conclusion sign (how 1st element is related to 2nd element). Hence, a (inequality sign) c.
$>$	$>$	$>$
e.g. If $a > b$ and $b > c$, then $a > c$, But if $a > b$ and $c > b$, then $a > c$ is not true as the elements of the inequality statement $c > b$ are not in order. It should be $b > c$ if we are inclined to conclude $a > c$.		
$>$	$=$	$>$
$=$	$>$	$>$
$>$	$\geq$	$>$
$\geq$	$>$	$>$
$=$	$\geq$	$\geq$
$\geq$	$=$	$\geq$
$\geq$	$\geq$	$\geq$
$=$	$=$	$=$
$<$	$<$	$<$
$<$	$=$	$<$
$=$	$<$	$<$
$<$	$\leq$	$<$
$\leq$	$<$	$<$
$=$	$\leq$	$\leq$
$\leq$	$=$	$\leq$
$\leq$	$\leq$	$\leq$
$>$	$<$	No conclusion hence we cannot establish relation between a and c.
$>$	$\leq$	No conclusion
$\geq$	$<$	No conclusion
$\geq$	$\leq$	No conclusion
$<$	$>$	No conclusion
$<$	$\geq$	No conclusion
$\leq$	$>$	No conclusion
$\leq$	$\geq$	No conclusion

How to remember the above facts

(a) Any sign + same sign ⇒ same sign

(b) Any sign + = ⇒ no change in sign, hence, the role of = is nothing.

(c) Any sign of slack inequality + corresponding sign of strict inequality ⇒ sign of strict inequality [**e.g.** ≥ + > ⇒ >]

Or Any sign of strict inequality + corresponding sign of slack inequality = sign of strict inequality

(d) Any sign + its reverse ⇒ no conclusion. The reverse means opposite sense of direction. [**e.g.** reverse of < is > as well as ≥]

Step (b) Now, $A > C$ and $C = D$, then $A > D$

Step (c) Since, $A > D$ and $D \geq E$, $A > E$

Step (d) $A > E$ and $E > F$, then $A > F$.

Hence, option (d) is correct.

Alternatively

Since we have to determine the relation between A and F, we will consider all the signs appeared between them in the inequality statement.

$$A > > = \geq > F$$
$$A\,(> >) = \geq > F$$
$$A\,(> =) \geq > F \qquad [\because > \text{ and } > \Rightarrow >]$$
$$A\,(> \geq) > F \qquad [\because > \text{ and } = \Rightarrow >]$$
$$A\,(> >)\,F \qquad [\because > \text{ and } \geq \Rightarrow >]$$
$$A > F \qquad [\because > \text{ and } > \Rightarrow >]$$

Observation The above process can be done in moments without pen and paper.

In the above question, what is the relation between B and E?

Given, $B > C = D \geq E$ then,

$$B > = \geq E$$
$$B\,(> =) \geq E$$
$$B\,(> \geq)\,E \qquad [\therefore > \text{ and } = \Rightarrow >]$$
$$B > E \qquad [\therefore > \text{ and } \geq \Rightarrow >]$$

Note *This approach is applicable for a single inequality statement or its part (as explained above). Also, it is applicable for a group of inequality statements wherein all the statements are arranged in such a way that the right side of one inequality is the same as left side of the succeeding inequality. Such arrangement of statements is hereinafter referred to as chain arrangement of statements.*

Problem 8 If $R \le M, N > M$ and $N = Q$, then what is the relation between R and Q?

Sol. Clearly, the given statements are not in chain arrangement. Hence, we should first do the same so as to make the above approach applicable.

Chain arrangement of statements is : $R \le M, M < N, N = Q$

Now, $R \le < = Q$

$R (\le <) = Q$

$R (< =) Q$ $[\because \le \text{and} < \Rightarrow <]$

$R < Q$ $[\because < \text{and} = \Rightarrow <]$

Problem 9 What is the relation between K and R, if $T < P \le U, L > U \le K$ and $P \ge R$?

(a) $K \le R$ (b) $K \ge R$ (c) $K = R$ (d) $K > R$

Sol. Here, three inequality statements have been given (i) $T < P \le U$ (ii) $L > U \le K$ (iii) $P \ge R$ and we have to find relation between two elements K and R, which lie in statements (ii) and (iii) respectively. It is obvious that we can't find the relation between them unless we have a common element which are related with both the elements K and R somehow. It is also fact that R is related with only P. It implies that only P can be the common element for K and R. Thus, we ought to establish the relation between K and P, thereafter we will be able to establish relation between K and R. Further, the relation between K and P has not been defined in one statement, thus, we should consider statements (i) and (ii) wherein they lie. In statements (i) and (ii), U is common element, which is also common for both P and K.

From statement (i); $P \le U$ and from statement (ii); $U \le K$. Since these are in chain arrangement, $P (\le \le) K$. Hence, $P \le K$.

Clearly, new statement $P \le K$ and statement (iii); $P \ge R$ are not in chain arrangement. Therefore, we should bring them in chain arrangement.

Chain arrangement of these statements is : $K \ge P, P \ge R$.

Then, $K \ge \ge R$

$K (\ge \ge) R$

$K \ge R$

Alternatively

Clearly, R is related with P, P is related with U and U is related with K.

Hence, $P \ge R$ (Statement III), $P \le U$ (statement I) and $U \le K$ (statement II)

Now, their chain arrangement : $R \le P, P \le U, U \le K$

Then, $R \le \le \le K$

$R (\le \le) \le K$

$R (\le \le) K$

$R \le K$

$K \ge R$

Directions (Problem 10 to 12) *In these questions a relationship between different elements is shown in statements. The statements are followed by two conclusions.*

Give answer :

(a) If only Conclusion I is true

(b) If only Conclusion II is true

(c) If either Conclusion I or II is true

(d) If neither Conclusion I nor II is true

(e) If both Conclusions I and II are true.

Problem 10 Statements : $H = I \le R; M \ge R < S$

Conclusions : I. $M = I$ II. $M > I$ **[Corporation Bank (PO)-11]**

Sol. Check for conclusions I and II.

Clearly, M and I are lying in statements (ii) and (i) respectively. In statements (i) and (ii), R is common element, which is also common for both M and I.

From statement (i), $I \le R$ and from statement (ii), $M \ge R$

Now, their chain arrangement : $I \le R, R \le M$

Then, $I \le \le M$

$I (\le \le) M$

$I \le M$

Hence, either $M = I$ or $M > I$

$\therefore$ Option (c) is correct.

Problem 11 Statements : $Q = Z, C \ge G, G \ge Q, Q \ge R$

Conclusions I. $G \ge Z$ II. $C \ge R$

Sol. **Check for conclusion I**

Clearly, G is in both the statements (ii) and (iii) and Z is in statement (i). However, we should take statements (iii) and (i) into consideration as common for the both G and Z i.e. Q is lying in statements (iii) and (i).

Chain arrangement of statements (iii) and (i) : $G \ge Q, Q = Z$

Then, $G \ge = Z$

$G (\ge =) Z$

$G \ge Z$

Alternatively

As given in statement (iii), $G \ge Q$

$\therefore$ $G \ge Z$ $[\because Q = Z]$

Check for conclusion II

Clearly, C is related to G, G is related to Q and Q is related to R. i.e. statements (ii), (iii) and (iv) respectively.

Now, chain arrangement of statements (ii), (iii) and (iv) : $C \ge G$, $G \ge Q, Q \ge R$.

Then, $C \ge \ge \ge R$

$C (\ge \ge) \ge R$

$C (\ge \ge) R$

$C \ge R$

Thus, option (e) is correct.

Problem 12 Statements : $A > B > C, D > E > F, D > C$.

Conclusions I. $E > C$ II. $F > B$

Sol. **Check for conclusion I**

Since D is common for both E and C, we should take statements(ii) and (iii) into consideration.

From statement (ii), $D > E$ and from statement (iii), $D > C$

Now, their chain arrangement : $E < D, D > C$

Then, $E (< >) C$.

We cannot find the relation between E and C as $< > \Rightarrow$ no conclusion.

Check for conclusion II

Clearly, B is related to C, C is related to D and D is related to F.

Hence, $B > C$ (statement I), $D > C$ (statement III) and $D > E > F$ (statement II)

Now, their chain arrangement : $B > C, C < D, D > E > F$

Then, $B > < > > F$

$$B\ (><) > > F$$

We cannot find the relation between B and F as $>\ < \Rightarrow$ no conclusion.

Alternatively

Since there in no relation between C and E (explained for conclusion I), we cannot find the relation between B and F because (C......E) is a part of (B ....F), if statements are in chain arrangement.

Hence, option (d) is correct.

Problem 13 Which of the following symbols should replace the question mark in the given expression in order to make the expressions '$K \le H$' and '$M > J$' definitely true?

$H \ge I = J\ ?\ K \le L < M$ *[Allahabad Bank (PO)-11]*

(a) $>$ (b) $\ge$
(c) $\le$ (d) either $<$ or $\le$
(e) $=$

Sol. Given expression : $H \ge I = J\ ?\ K \le L < M$

For 1st expression '$K \le H$'

$$H \ge =\ ?\ K$$
$$H\ (\ge =)\ ?\ K$$
$$H\ (\ge ?)\ K$$

If we replace ? by $\ge$ or $=$, we get $H \ge K$ i.e. 1st expression.

For 2nd expression '$M > J$'

$$J\ ? \le < M$$
$$J\ ?\ (\le <)\ M$$
$$J\ ? < M$$

If we replace ? by any one of $\le$, $<$ and $=$, we get $J < M$ i.e. 2nd expression.

From the above,

Replacement of ? is either $\ge$ or $=$ to make 1st expression true and replacement of ? is either $\le$ or $<$ or $=$ to make 2nd expression true.

It implies that if we replace ? by $=$ (i.e. common replacement), both the expressions will be true.

Hence, option (e) is correct.

(E) Solutions of a System of Linear Inequalities in One Variable

A system of linear inequalities in one variable consists of at least two linear inequalities in the same variable and the value of the variable for which each inequality in the system becomes a true statement, is known as the solution of a system of inequalities in one variable.

Take a few examples.

Problem 14 Solve the following system of inequalities :

$$3x + 1 > 4x - 2$$
$$5x + 1 > 3x - 2$$

Sol. Let's solve the 1st inequality

Given, $3x + 1 > 4x - 2$

$\Rightarrow$ $3x - 4x > -2 - 1$

$\Rightarrow$ $-x > -3$

$\Rightarrow$ $x < 3$

Now, solve the 2nd inequality

$5x + 1 > 3x - 2$

$\Rightarrow$ $5x - 3x > -2 - 1$

$\Rightarrow$ $2x > -3$

$\Rightarrow$ $x > -1.5$

Hence, solution of the system of inequalities is $-1.5 < x < 3$.

Problem 15 Solve the following system of inequalities :

$$3x - 1 \le 4x$$
$$x - 1 > 1$$

Sol. 1st inequality, $3x - 1 \le 4x$

$\Rightarrow$ $3x - 4x \le 1$

$\Rightarrow$ $-x \le 1$

$\Rightarrow$ $x \ge -1$

2nd inequality, $x - 1 > 1$

$\Rightarrow$ $x > 1 + 1 \Rightarrow x > 2$

Hence, solution of the system of inequalities is $x > 2$ because for this value of x, both the inequalities are true. In other words, $x > 2$ also covers $x \ge -1$. No need to mention that for $x \ge -1$, both the inequalities are not true.

> ***Observation*** If $x > a$ and $x > b$, $b > a$, then the conclusion is $x > b$.
> Also, if $x < a$ and $x < b$, $b > a$ then the conclusion is $x < a$.

Problem 16 Solve the following system of inequalities :

$$3x - 1 \le 4x$$
$$x + 1 < -1$$

Sol. 1st inequality, $3x - 1 \le 4x$

$\Rightarrow$ $3x - 4x \le 1$

$\Rightarrow$ $-x \le 1$

$\Rightarrow$ $x \ge -1$

2ns inequality, $x + 1 < -1$

$\Rightarrow$ $x < -1 - 1$

$\Rightarrow$ $x < -2$

Here, values of x are inconsistent. Hence, no solution.

Problem 17 Three inequalities in variable x, form a system of inequalities. If solutions of inequalities individually are $x \ge -2$, $x < 1$ and $x > -1$, then what is the solution of the system of inequalities in x?

Sol. We may conclude as $x > -1$ from the given solutions of inequalities $x \ge -2$ and $x > -1$.

Then, the required solution of the system : $-1 < x < 1$

Problem 18 Solve $9 \ge 2x - 1 > 3$

Sol. Clearly, there are two inequalities i.e $9 \ge 2x - 1$ and $2x - 1 > 3$. It may be said that these two form a system of inequalities.

1st inequality, $9 \ge 2x - 1$

$\Rightarrow$ $9 + 1 \ge 2x$

$\Rightarrow$ $x \le 5$

2nd inequality, $2x - 1 > 3$

$\Rightarrow$ $2x > 3 + 1 \Rightarrow x > 2$

Hence, solution of the system of inequalities is $2 < x \le 5$.

Alternatively

We may solve both the inequalities simultaneously.

We have, $9 \geq 2x - 1 > 3$

$\Rightarrow \quad 10 \geq 2x > 4$ [$\because$ on adding 1 to each term]

$\Rightarrow \quad 5 \geq x > 2$ [$\because$ on dividing by 2]

$\Rightarrow \quad 2 < x \leq 5$

(F) Solution of a System of Linear Inequalities in Two Variables

A system of linear inequalities in two variables consists of at least two linear inequalities in the same variables and the values of the variables (ordered pair) for which each inequality in the system becomes a true statement, is known as the solution of a system of inequalities in two variables.

e.g.

Problem 19 Is $(2, -1)$ is a solution of the given system of linear inequalities?

$y > x + 1$

$y \geq 2x - 5$

Sol. Since $(2, -1)$ is a ordered pair, $x = 2$ and $y = -1$.

Let's check the 1st inequality, $-1 > 2 + 1 \Rightarrow -1 > 3$ (False)

For 2nd inequality, $-1 \geq 2 \cdot 2 - 5 \Rightarrow -1 \geq -1$ (True)

Since $(2, -1)$ is not satisfying each of linear inequalities, it is not a solution of the given system of linear inequalities.

Problem 20 If $2x - 3y < 7$ and $x + 6y < 11$, then which one of the following is correct?

(a) $x + y < 5$ (b) $x + y < 6$

(c) $x + y \leq 5$ (d) $x + y \leq 6$ **[CDS-08]**

Sol. Given, $2x - 3y < 7$

and $\underline{x + 6y < 11}$

On adding, $3x + 3y < 18 \Rightarrow x + y < 6$

Hence, option (b) is correct.

Note *In the above question, finding the values of x and y individually through algebric method is much complicated. Thus, the solution of the system of linear inequalities is determined graphically.*

(G) Solution of linear inequalities having a modulus

Consider $|x| < a$

We are aware that the value of $|x|$ cannot be negative, hence, either be zero or positive. However, the value of x can either be negative or zero or positive. It is also fact that the value of a cannot be negative or zero as any number less than zero or a negative number is always negative which contradicts that the value of $|x|$ is non-negative.

Let a be equal to 6.

Then, values of x may be $\pm 5, \pm 4, \pm 3$ etc. Hence, absolute values of x (i.e. 5, 4, 3……) must be less than 6.

Clearly, $+5, +4, +3$ are less than 6 and $-5, -4, -3$ are greater than -6.

Hence, $x < 6$ or $x > -6$

If combined, $-6 < x < 6 \Rightarrow -a < x < a$

We may conclude that $|x| < a$, where $a > 0$ implies $-a < x < a$.

Now, consider, $|x| > a$

Here, values of a may be positive or zero.

(a) Let a be equal to 6, a positive number.

Then, values of x may be $\pm 7, \pm 8, \pm 9$ etc. Hence, absolute values of x (i.e. 7, 8, 9……) must be greater than 6.

Clearly, $+7, +8, +9$ are greater than 6 and $-7, -8, -9$ are less than -6.

Hence, $x > 6$ or $x < -6 \Rightarrow x > a$ or $x < -a$

We may conclude that $|x| >$ a, where $a > 0$ implies either $x > a$ or $x < -a$.

(b) If $a = 0$

Then, values of x may be $\pm 1, \pm 2, \pm 3$ etc. Hence, absolute values of x (i.e. 1, 2, 3……) must be greater than 0.

Clearly, $+1, +2, +3$ are greater than 0 and $-1, -2, -3$ are less than 0.

Hence, $x > 0$ or $x < 0$

We may conclude that $|x| > a$, where $a = 0$ implies x can be any real number except $x = 0$.

From above,

$|x| < a$, where $a > 0$ implies $-a < x < a$.

Similarly, $|x| \leq a$, where $a > 0$ implies $-a \leq x \leq a$.

Also,

$|x| > a$, where $a > 0$ implies either $x > a$ or $x < -a$.

and

$|x| \geq a$, where $a > 0$ implies either $x \geq a$ or $x \leq -a$.

It is also to note that :

(i) $|x| = 0$ implies $x = 0$.

(ii) $|x| > 0$ implies x can be any real number except $x = 0$

(iii) $|x| \geq 0$ implies x can be any real number.

(iv) $|x| < 0$ is not possible. Also, $|x| \leq 0$ is not possible, only one part $|x| = 0$ is possible.

Take a few examples.

Problem 21 Solve $|3x - 5| < 7$.

Sol. We know, $|x| < a$, where $a > 0$ implies $-a < x < a$.

Hence, $\quad -7 < 3x - 5 < 7$

$\Rightarrow \quad -2 < 3x < 12$ [$\because$ on adding 5 to each term]

$\Rightarrow \quad \frac{-2}{3} < x < 4$ [$\because$ on dividing by 3]

Problem 22 Solve $|3x - 2| \geq 4$.

Sol. We know, $|x| \geq a$, where $a > 0$ implies either $x \geq a$ or $x \leq -a$.

Hence, $3x - 2 \geq 4$ or $3x - 2 \leq -4 \Rightarrow 3x \geq 6$ or $3x \leq -2$

$\Rightarrow \quad x \geq 2$ or $x \leq \frac{-2}{3}$

Problem 23 Solve $||x-2|-3|>1$.

Sol. We know $|x|>a$, where $a>0$ implies either $x>a$ or $x<-a$.

Hence, $|x-2|-3>1$ or $|x-2|-3<-1$

$\Rightarrow \quad |x-2|>4$ or $|x-2|<2$

$\Rightarrow \quad x-2>4$ or $x-2<-4$ or $-2<x-2<2$

$\Rightarrow \quad x>6$ or $x<-2$ or $0<x<4$

Problem 24 Solve $\dfrac{2|x|}{|x+1|}\geq 0$.

Sol. Clearly, $2|x|\geq 0$ and $|x+1|>0$ or $2|x|\leq 0$ or $|x+1|<0$, as denominator cannot be equal to 0, hence, no sign of $\geq$ or $\leq$ with it.

Solution of $2|x|\geq 0$ is any real value of x

and $|x \div 1|>0$ implies x can be any real number except $x=-1$ i.e. $x<-1$ or $x>-1$.

Further, there is no solution of $|x|\leq 0$ (solution of only one part $|x|=0$ is possible at $x=0$) and also no solution of $|x+1|<0$.

Then, we may conclude from the above that $x<-1$ or $x>-1$

Now, solve the above inequality with changed sign i.e.

$\dfrac{2|x|}{|x+1|}>0$

Clearly, $2|x|>0$ and $|x+1|>0$ or $2|x|<0$ or $|x+1|<0$

Solution of $2|x|>0$ is any real value of x except $x=0$.

and $|x+1|>0$ implies x can be any real number except $x=-1$ i.e. $x<-1$ or $x>-1$.

Further, there are no solutions of

$|x|<0$ and $|x+1|<0$.

Now, we may conclude from the above that x can be any real number except $x=-1, 0$.

Hence, the required answer is $x<-1$ or $-1<x<0$ or $x>0$.

Problem 25 Solve $|2x-1|<|x+2|$.

Sol. We may solve it by the above method as below,

$-|x+2|<(2x-1)<|x+2|$

($\because |x|<a$, where $a>0$ implies $-a<x<a$)

Alternatively

Since both sides have absolute values, we can remove the modulus sign by just squaring both sides.

$\Rightarrow \quad 4x^2-4x+1<x^2+4x+4$

$\Rightarrow \quad 3x^2-8x-3<0$

Discriminant $D=b^2-4ac=(-8)^2-4\cdot 3(-3)=100$

Hence, $D>0$.

Also, quadratic coefficient >0. It implies that the value of the given quadratic expression is negative only if $\alpha<x<\beta$, where α and β are roots of corresponding equation and $\alpha<\beta$.

Solving the equation,

we get $\alpha=-1/3$ & 3

Hence, for $-1/3<x<3$, $3x^2-8x-3<0$

$\therefore$ The required answer is $-1/3<x<3$

[The above concept is given under next topic Quadratic inequalities.]

II. Quadratic Inequalities

(A) Value of a quadratic expression

We know that the value of a quadratic expression ax^2+bx+c is always zero at $x=\alpha,\beta$ where α and β are roots of its corresponding equation.

However, in case of other values of x, the quadratic expression is either greater than or less than zero, which is determined by the following rules :

S. No.	Cases	Value of quadratic expression	Remarks (How to memorise)
1	Discriminant $D<0$	$ax^2+bx+c>0$ for all real values of x, if $a>0$ (Value of ax^2+bx+c cannot be negative at any value of x, if $a>0$)	Sign of value of the expression is same as that of quadratic coefficient
		$ax^2+bx+c<0$ for all real values of x, if $a<0$ (Value of ax^2+bx+c cannot be positive at any value of x, if $a<0$)	
		$ax^2+bx+c\geq 0$ for all real values of x, if $a>0$ (Value of ax^2+bx+c cannot be negative at any value of x, if $a>0$)	Sign (> or <) of value of the expression is same as that of quadratic coefficient for all values of x except $x=\dfrac{-b}{2a}$. No need to mention that at $x=\dfrac{-b}{2a}$ the expression = 0.
2.	$D=0$	$ax^2+bx+c\leq 0$ for all real values of x, if $a<0$ (Value of ax^2+bx+c cannot be positive at any value of x, if $a<0$)	
3.		(a) $ax^2+bx+c>0$ for all real values of $x(x<\alpha$ or $x>\beta)$ if $a>0$ [α and β are roots of corresponding equation and $\alpha<\beta$] (b) $ax^2+bx+c<0$ for all real values of $x(\alpha<x<\beta)$ if $a>0$	Sign of value of the expression is same as that of quadratic coefficient for all real values of x except $(\alpha<x<\beta)$. In case of $(\alpha<x<\beta)$, the sign is reversed.
	$D>0$	(a) $ax^2+bx+c<0$ for all real values of $x(x<\alpha$ or $x>\beta)$ if $a<0$ (b) $ax^2+bx+c>0$ for all real values of $x(\alpha<x<\beta)$ if $a<0$	

(B) Solution of quadratic inequalities

Quadratic inequalities are in the following forms :

(i) $ax^2 + bx + c > 0$
(ii) $ax^2 + bx + c \geq 0$
(iii) $ax^2 + bx + c < 0$
(iv) $ax^2 + bx + c \leq 0$

We may take some questions for concept building.

Problem 26 What values of x satisfy an inequality $x^2 - x - 6 > 0$? *[Andhra Bank (SO)-02, CDS-19]*

Sol. Given inequality : $x^2 - x - 6 > 0$

We factorise the quadratic expression as below :

$\Rightarrow \quad x^2 - 3x + 2x - 6 > 0$

$\Rightarrow \quad x(x-3) + 2(x-3) > 0$

$\Rightarrow \quad (x-3)(x+2) > 0$

It is very clear that factor $(x-3)$ and $(x+2)$ both either be positive or be negative so as to make the product $(x-3)(x+2)$ positive i.e. greater than 0.

***Case (i) :* Both are positive**

Then, $(x-3) > 0$ and $(x+2) > 0$

$\Rightarrow \quad x > 3$ and $x > -2$

We may conclude that $x > 3$.

***Case (ii) :* Both are negative**

Then, $(x-3) < 0$ and $(x+2) < 0$

$\Rightarrow \quad x < 3$ and $x < -2$

We may conclude that $x < -2$.

Hence, required answer is $x > 3$ or $x < -2$.

Alternatively I

Given inequality, $x^2 - x - 6 > 0$

We solve the corresponding quadratic equation $x^2 - x - 6 = 0$ as below

Step I Product $= (+1) \times (-6) = -6$

Step II Since product is negative, 6 is factorised into two terms such that their difference is equal to absolute value of linear coefficient. *viz.* $6 = 3 \times 2$

Step III The linear coefficient (i.e.-1) may be written as $-3 + 2$

Step IV $\frac{-3}{1}$ and $\frac{+2}{1}$ i.e. -3 and 2

Step V The required roots $= 3$ and-2

[Please refer to 'Quadratic Equation' under which the above process has been detailed.]

Now, proceed step by step :

Step (i) We draw a number line indicating the values of x in ascending order.

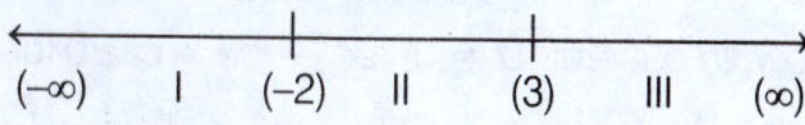

It appears three intervals

Interval I $\rightarrow$ [containing values less than -2 like $-3, -4, -5, \ldots$]

Interval II $\rightarrow$ [containing values between -2 and 3 like $-1, 0, 1, \ldots$]

Interval III $\rightarrow$ [containing values greater than 3 like $4, 5, 6, \ldots$]

Step (ii) Take one value from each interval preferably integers *viz.*

From interval I $\rightarrow -3$

From interval II $\rightarrow 0$

From interval III $\rightarrow 4$

Step (iii) Put the above values in the quadratic expression and test as to which values satisfy the inequality.

For $x = -3$, value of expression $= (-3)^2 - (-3) - 6 = 6$, which is greater than 0. Hence, $x = -3$ satisfies the inequality.

For $x = 0$, value of expression $= (0)^2 - (0) - 6 = -6$, which is less than 0. Hence, $x = 0$ does not satisfy the inequality.

For $x = 4$, value of expression $= (4)^2 - (4) - 6 = 6$, which is greater than 0. Hence, $x = 4$ satisfies the inequality.

Since intervals I and III satisfy the given inequality, the required answer is $x < -2$ or $x > 3$.

Alternatively II

Discriminant $D = b^2 - 4ac = (-1)^2 - 4 \cdot 1(-6) = 25$

Hence, $D > 0$.

Also, quadratic coefficient > 0. It implies that the value of the given quadratic expression is positive only if $x < \alpha$ or $x > \beta$, where α and β are roots of corresponding equation and $\alpha < \beta$

Solving the equation, we get $\alpha = -2, \beta = 3$.

Hence, the required answer is $x < -2$ or $x > 3$.

Alternatively III

Proceed through options. Let options be (a) $-2 < x < 3$ (b)$x < -2$ or $x > 3$ (c) $-3 < x < 2$ (d) $x > -2$ or $x > 3$ (e) None of these

(a) $-2 < x < 3 \Rightarrow$ we take $x = 0$ and at $x = 0$, $(0)^2 - (0) - 6 \not> 0$

(b) $x < -2$ or $x > 3 \Rightarrow$ we take $x = -3$ and 4 and at both values of x, the value of expression > 0, hence, these values satisfy the inequality.

Hence, option (b) is correct.

Problem 27 What values of x satisfy an inequality $x^2 - x + 7 \geq 13$?

Sol. Given inequality, $x^2 - x + 7 \geq 13$

$\Rightarrow \quad x^2 - x + 7 - 13 \geq 0 \Rightarrow x^2 - x - 6 \geq 0$

The corresponding quadratic equation of the inequality $x^2 - x - 6 \geq 0$ is $x^2 - x - 6 = 0$.

Solving this equation, we get $x = -2$ and 3.

The number line will be as below :

(−∞) I (−2) II (3) III (∞)

As seen in the previous question, intervals I and III satisfy the inequality $x^2 - x - 6 > 0$ (not $x^2 - x - 6 \geq 0$).

We are also aware that $x = -2$ and $x = 3$ makes the quadratic expression zero. It implies that values contained in interval I and III including -2 and 3 will satisfy the inequality $x^2 - x - 6 \geq 0$.

Thus, required answer is $x \geq 3$ or $x \leq -2$.

> ***Observation*** In case of inequality with sign $\geq$ or $\leq$, we solve its corresponding strict inequality and thereafter, we replace $>$ by $\geq$ and $<$ by $\leq$.

Alternatively II

Discriminant $D = b^2 - 4ac = (-1)^2 - 4 \cdot 1(-6) = 25$

Hence, $D > 0$.

Also, quadratic coefficient > 0. It implies that the value of the given quadratic expression is positive only if $x<\alpha$ or $x>\beta$, where α and β are roots of corresponding equation and $\alpha<\beta$.

Solving the equation, we get $\alpha=-2, \beta=3$

Hence, for $x<-2$ or $x>3$, $x^2-x-6>0$.

Obviously, in order to satisfy the inequality $x^2-x-6\geq 0$, we should also include values of α and β at which the value of the expression is 0.

Thus, the required answer is $x\leq -2$ or $x\geq 3$.

Problem 28 What values of x satisfy an inequality $2x^2-7x+6\leq 0$?

Sol. The corresponding strict inequality is $2x^2-7x+6<0$ and the corresponding quadratic equation of the inequality $2x^2-7x+6<0$ is $2x^2-7x+6=0$.

Solving this equation, we get $x=2$ and 1.5.

The number line will be as under :

(−∞) I (1.5) II (2) III (∞)

If we take $x=1$ (i.e. a value from interval I), $2x^2-7x+6\not<0$

If we take $x=1.6$(i.e. a value from interval II), $2x^2-7x+6<0$

If we take $x=3$ (i.e. a value from interval III), $2x^2-7x+6\not<0$

Hence, interval II i.e. $1.5<x<2$ is satisfying the inequality $2x^2-7x+6<0$.

Then, the required answer is $1.5\leq x\leq 2$.

Alternatively II

Discriminant $D=b^2-4ac=(-7)^2-4\cdot 2\cdot 6=1$

Hence, $D>0$.

Also, quadratic coefficient > 0. It implies that the value of the given quadratic expression is negative only if $\alpha<x<\beta$, where α and β are roots of corresponding equation and $\alpha<\beta$.

Solving the equation, we get $\alpha=1.5, \beta=2$.

Hence, for $1.5<x<2$, $2x^2-7x+6<0$.

Obviously, in order to satisfy the inequality $2x^2-7x+6\leq 0$, we should also include values of α and β at which the value of the expression is 0.

Thus, the required answer is $1.5\leq x\leq 2$.

Problem 29 What values of x satisfy an inequality $x^2+10x+25<0$?

Sol. The corresponding quadratic equation of the inequality is $x^2+10x+25=0$

Solving this equation, we get $x=-5$

The number line will be as under :

(−∞) I (-5) II (∞)

If we take $x=-6$ (i.e. a value from interval I), $x^2+10x+25\not<0$

If we take $x=0$ (i.e. a value from interval II), $x^2+10x+25\not<0$

Since both intervals are not satisfying the inequality, there is no solution of this inequality.

Alternatively

Discriminant $D=b^2-4ac=10^2-4\cdot 1\cdot 25=0$

and quadratic coefficient > 0.

Further, we know that when $D=0$, the value of ax^2+bx+c cannot be negative at any value of x, if the quadratic coefficient > 0. The given inequality contradicts the same. Hence, no solution of this inequality.

Problem 30 What values of x satisfy an inequality $x^2+10x+40>0$?

Sol. Discriminant $D=b^2-4ac=10^2-4\cdot 1\cdot 40=-60$

Hence, $D<0$.

The corresponding quadratic equation of the inequality is $x^2+10x+40=0$. Since discriminant $D<0$, the equation has no real roots. Then, it is not possible to indicate the values of x on the number line and process accordingly.

However, we know that when $D<0$ and the quadratic coefficient > 0, the value of the quadratic expression is greater than 0 for all real values of the variable. Hence, all real values of x satisfy the given inequality.

Problem 31 What values of x satisfy an inequality $-2x^2+5x-10\geq 0$?

Sol. Discriminant $D=b^2-4ac=5^2-4\cdot(-2)(-10)$

$=25-80=-55$

Hence, $D<0$ and quadratic coefficient < 0.

We are aware that when $D<0$ and the quadratic coefficient < 0, the value of the quadratic expression is less than 0 for all real values of the variable.

Here, the value of the given quadratic expression ≥ 0, hence, no solution of the given inequality.

Remarks

(i) *All the 1st four questions can be solved by the conventional method too. 1st question has been solved by this method. Try for the remaining.*

(ii) *All the questions can be solved by the given options. 1st question has been solved by this method.*

Problem 32 If the roots of the equation $x^2-2ax+a^2+a-3=0$ are real and less than 3, then which one of the following is correct ?

(a) $2<a<3$ (b) $a<2$

(c) $3<a<4$ (d) $a>4$ *[CDS-12]*

Sol. Since roots are real, $D\geq 0$.

$\Rightarrow \quad (-2a)^2-4\cdot 1(a^2+a-3)\geq 0$

$\Rightarrow \quad -4a+12\geq 0$

$\Rightarrow \quad -a+3\geq 0$

$\Rightarrow \quad a\geq -3 \Rightarrow a\leq 3 \quad$ ...(i)

We know, in case of $D=0$, $ax^2+bx+c\geq 0$ for all real values of x, if $a>0$.

Also, in case of $D>0$, $ax^2+bx+c>0$ for all real values of x ($x<\alpha$ or $x>\beta$) if $a>0$ [α and β are roots of corresponding equation and $\alpha<\beta$.]

Here, β is less than 3. Then, at $x\geq 3$, $ax^2+bx+c>0$.

As conclusion, $ax^2+bx+c>0$ at $x=3$.

$\therefore \quad 3^2-2a\cdot 3+a^2+a-3>0$

$\Rightarrow \quad a^2-5a+6>0$

Discriminant $D = b^2 - 4ac = (-5)^2 - 4 \cdot 1 \cdot 6 = 1$

Hence, $D > 0$.

Also, quadratic coefficient > 0. It implies that the value of the given quadratic expression is positive only if $a < \alpha$ or $a > \beta$, where α and β are roots of corresponding equation and $\alpha < \beta$.

Solving the equation, we get $\alpha = 2, \beta = 3$.

$\therefore$ Required answer is $a < 2$ or $a > 3$ ...(ii)

From Eqs. (i) and (ii), we get $a < 2$

Hence, option (b) is correct.

Alternatively

Proceed through options.

(a) $2 < a < 3 \Rightarrow$ let $a = 2.5$, then equation be

$$x^2 - 2 \times 2.5 \times x + (2.5)^2 + 2.5 - 3 = 0$$

$$\Rightarrow x^2 - 5x + 5.75 = 0$$

$$\Rightarrow x = \frac{5 \pm \sqrt{(-5)^2 - 4 \times 1 \times (5.75)}}{2.1} = \frac{5 \pm \sqrt{2}}{2.1} = 3.21 \text{ and } 1.79$$

Here, both the roots are not less than 3.

(b) $a < 2 \Rightarrow$ let $a = 1$, then equation be

$$x^2 - 2 \cdot 1x + 1^2 + 1 - 3 = 0$$

$$\Rightarrow x^2 - 2x - 1 = 0$$

$$\Rightarrow x = \frac{2 \pm \sqrt{(-2)^2 - 4 \times 1 \times (-1)}}{2.1} = \frac{2 \pm \sqrt{8}}{2.1} = 1 + \sqrt{2} \text{ and } 1 - \sqrt{2}$$

which are less than 3.

Hence, option (b) is correct.

Problem 33 Solve $x^2 - 4 > 0$.

Sol. Given, $x^2 - 4 > 0 \Rightarrow x^2 - 0x - 4 > 0$

Here, $D > 0$.

Also, quadratic coefficient > 0. It implies that the value of the given quadratic expression is positive only if $x < \alpha$ or $x > \beta$, where α and β are roots of corresponding equation and $\alpha < \beta$.

Solving the equation, we get $\alpha = -2, \beta = 2$

Hence, for $x < -2$ or $x > 2$, $x^2 - 4 > 0$.

(C) Solution of quadratic inequalities having modulus

Take a few examples for concept building :

Problem 34 Solve $|x^2 - 5x + 3| < 3$.

Sol. We know, $|x| < a$, where $a > 0$ implies $-a < x < a$.

$$\Rightarrow -3 < (x^2 - 5x + 3) < 3$$

$$\Rightarrow 0 < (x^2 - 5x + 6) < 6 \quad \text{[on adding 3 to each term]}$$

$$\Rightarrow x^2 - 5x + 6 > 0 \text{ and } x^2 - 5x < 0$$

For $\underline{x^2 - 5x + 6 > 0}$

Solution is $x < 2$ or $x > 3$.

For $\underline{x^2 - 5x < 0}$

$\Rightarrow x(x-5) < 0 \Rightarrow x > 0$ and $(x-5) < 0$ or $x < 0$ and $(x-5) > 0$

$\Rightarrow x > 0$ and $x < 5$ or $x < 0$ and $x > 5$.

$\Rightarrow 0 < x < 5$ as $x < 0$ and $x > 5$ are inconsistent.

We may conclude from the above solutions that $0 < x < 2$ or $3 < x < 5$. Technically, the required solution is the intersection of the both solutions. We may observe the same as below :

Solution of $x^2 - 5x + 6 > 0$

(0) (2) (3) (5)
(−∞) (∞)

Solution of $x^2 - 5x < 0$

(0) (2) (3) (5)
(−∞) (∞)

Clearly, common portions are (0 to 2) and (3 to 5).

Alternatively

The best way is to proceed through options. The option satisfying the given inequality is the required solution.

Problem 35 Solve $|x^2 - 4x| \geq 3$.

Sol. We know, $|x| \geq a$, where $a > 0$ implies either $x \geq a$ or $x \leq -a$

Hence, $x^2 - 4x \geq 3$ or $x^2 - 4x \leq -3$

$\Rightarrow x^2 - 4x - 3 \geq 0$ or $x^2 - 4x + 3 \leq 0$

Solution of $x^2 - 4x - 3 \geq 0$ is $x \leq -0.64$ or $x \geq 4.64$

and solution of $x^2 - 4x + 3 \leq 0$ is $1 \leq x \leq 3$

Hence, required answer is $x \leq -0.64$ or $x \geq 4.64$ or $1 \leq x \leq 3$.

Problem 36 The number of real values of x for which the equality $|3x^2 + 12x + 6| = 5x + 16$ holds good is :

(a) 2 (b) 3 (c) 4 (d) 1

Sol. There are 3 cases possible

Case (i) When $3x^2 + 12x + 6 > 0$

Case (ii) When $3x^2 + 12x + 6 < 0$

Case (iii) When $3x^2 + 12x + 6 = 0$

Case (i) $3x^2 + 12x + 6 > 0$

Since $D > 0$ and quadratie coefficient > 0, the roots will be $x < \alpha$ or $x > \beta$ [α and β are roots of corresponding equation and $\alpha < \beta$.]

Hence, $x < -2 - \sqrt{2}$ or $x > -2 + \sqrt{2}$.

Case (ii) $3x^2 + 12x + 6 < 0$

Since $D > 0$ and quadratic coefficient > 0, the roots will be $\alpha < x < \beta$.

Hence, $-2 - \sqrt{2} < x < -2 + \sqrt{2}$.

Case (iii) $3x^2 + 12x + 6 = 0$

Clearly, $x = -2 - \sqrt{2}, -2 + \sqrt{2}$

Consider case (i) and (iii). If $3x^2 + 12x + 6 > 0$, then $3x^2 + 12x + 6 = 5x + 16$. Also, if $3x^2 + 12x + 6 = 0$, then $3x^2 + 12x + 6 = 5x + 16$

$$\Rightarrow 3x^2 + 7x - 10 = 0 \Rightarrow x = \frac{-10}{3}, 1$$

But, none of $x < -2 - \sqrt{2}$ or $x > -2 + \sqrt{2}$ is satisfied at $x = \frac{-10}{3}$, however, $x > -2 + \sqrt{2}$ is satisfied at $x = 1$.

Further, consider case (ii). If $3x^2 + 12x + 6 < 0$, then $3x^2 + 12x + 6 = -(5x + 16)$

$$\Rightarrow 3x^2 + 17x + 22 = 0 \Rightarrow x = \frac{-11}{3}, -2$$

But $-2 - \sqrt{2} < x < -2 + \sqrt{2}$ is not satisfied at $\frac{-11}{3}$, however, it is satisfied at $x = -2$.

It may be construed from the above that the values of $x = 1, -2$.
Hence, option (a) is correct.

Alternatively

Given : $|3x^2 + 12x + 6| = 5x + 16$

It can be inferred that in case $3x^2 + 12x + 6 \geq 0$,

$$3x^2 + 12x + 6 = 5x + 16$$

$$\Rightarrow \quad 3x^2 + 7x - 10 = 0$$

$$\Rightarrow \quad x = \frac{-10}{3}, 1$$

But, at $x = \frac{-10}{3}$, the value of $5x + 16$ is negative, which is not possible. Hence, $x = 1$.

Further, if $3x^2 + 12x + 6 < 0$, $3x^2 + 12x + 6 = -(5x + 16)$

$$\Rightarrow \quad 3x^2 + 17x + 22 = 0 \Rightarrow x = \frac{-11}{3}, -2$$

But at $x = \frac{-11}{3}$, the value of $5x + 16$ is negative, which is not possible. Hence, $x = -2$. Conclusion $x = 1, -2$

$\therefore$ Option (a) is correct.

Problem 37 The product of the distinct roots of $|x^2 - x - 6| = x + 2$ is : ***[CAT-19]***

Sol. Given, $|x^2 - x - 6| = x + 2$

It can be inferred that in case $x^2 - x - 6 \geq 0$, $x^2 - x - 6 = x + 2$

$$\Rightarrow \quad x^2 - 2x - 8 = 0$$

$\Rightarrow x = -2, +4$ (both are satisfying the equation.)

Futher, if $x^2 - x - 6 < 0$, $x^2 - x - 6 = -(x + 2)$

$$\Rightarrow \quad x^2 = 4$$

$\Rightarrow \quad x = -2, +2$ (both are satisfying the equation.)

Hence, distinct roots $= -2, +4, +2$

Then, their product $= -16$

Some Critical Inequalities

(i) $(1 + x)^n \geq 1 + nx$, where $n = 1, 2, 3, \ldots.$ and $x > -1$
e.g. $(1 + x)^2 \geq 1 + 2x$

(ii) For real positive numbers, Root Mean Square $\geq$ Arithmetic Mean $\geq$ Geometric Mean $\geq$ their Harmonic mean $\left[\text{i.e. } \sqrt{\frac{x_1^2 + x_2^2 + x_3^2 + \ldots + x_n^2}{n}} \geq \frac{x_1 + x_2 + x_3 + \ldots x_n}{n} \geq \sqrt[n]{x_1\, x_2\, x_3 \ldots x_n} \geq \frac{n}{\frac{1}{x_1} + \frac{1}{x_2} + \frac{1}{x_3} + \ldots + \frac{1}{x_n}}\right]$. It is important to note that RMS = AM = GM = HM, when $x_1 = x_2 = x_3 = \ldots = x_n$.

e.g. (a) $\frac{a + b}{2} \geq \sqrt{ab} \geq \frac{2ab}{a + b}$, where $a, b > 0$

(b) $\frac{a + b + c}{3} \geq \sqrt[3]{abc} \geq \frac{3abc}{ab + bc + ca}$, where $a, b, c > 0$,

etc.

(iii) On the above pattern, $\frac{a^2 + b^2}{2} \geq \sqrt{a^2b^2}$. Hence, $a^2 + b^2 \geq 2ab$.
$\left[\text{or } \frac{a}{b} + \frac{b}{a} \geq 2\right]$, where a, b are real positive numbers.

(iv) $\frac{a}{b + c} + \frac{b}{c + a} + \frac{c}{a + b} \geq \frac{3}{2}$, where a, b, c are positive real numbers.

(v) $a^2 + b^2 + c^2 \geq ab + bc + ca$, where, a, b, c are real numbers.

(vi) $a^3 + b^3 \geq ab(a + b)$, where $a, b > 0$.

(vii) $|a + b| \leq |a| + |b|$, where a, b are real numbers.

(viii) $|a - b| \geq |a| - |b|$, where a, b are real numbers.

Note *The above inequalities (ii) to (viii) become equality if the given numbers are equal.*

(ix) The product of n equal positive numbers > product of n (not all equal) positive numbers [provided sum of n equal numbers equals sum of n unequal numbers]. This inequality is just an extension of an inequality, 'Arithmetic mean of positive real numbers $\geq$ their geometric mean'.
e.g. $10 \times 10 \times 10 > 9 \times 10 \times 11$
[here, $10 + 10 + 10 = 9 + 10 + 11$]
Also, $(10 \times 10 \times 10)$ is the greatest among all products of any three positive numbers provided their sum ≤ 30.
[That is why the area of a square is greater than that of a rectangle with the same perimeter, an equilateral triangle has maximum area in comparison to other types of triangles with the same perimeter, etc.]
It also holds, product of n closest positive numbers > product of n closer numbers, if sum of closest numbers $\geq$ sum of closer numbers.
e.g. (a) $10 \times 10 \times 10$ (extreme closest i.e. equal) > $9 \times 10 \times 11 > 9 \times 9 \times 12 > 8 \times 11 \times 11 > 8 \times 10 \times 12, \ldots$
(here, sum of all factors in each product is 30)
(b) $(10 \times 10) > (9 \times 11) > (8 \times 12) > (7 \times 13) > (6 \times 14) \ldots$
(here, sum of all factors in each product is 20)
Some questions based on this inequality have been given under 'Number System'.

WORKOUT

1. Which of the following symbols should be placed in the blank spaces respectively (in the same order from left to right) in order to complete the given expression in such a manner that '$N < K$' definitely holds true?

$K - L - M - N$

[Allahabad Bank (PO)-11]

(a) $\geq, =, >$
(b) $\geq, =, <$
(c) $>, \geq, <$
(d) $\leq, <, =$
(e) None of these

Directions (Q.Nos. 2-6) *In these questions a relationship between different elements is shown in statements. The statements are followed by two conclusions.*
Give answers :

(a) If only Conclusion I is true
(b) If only Conclusion II is true
(c) If either Conclusion I or II is true
(d) If neither Conclusion I nor II is true
(e) If both Conclusions I and II are true.

[IBPS(SO-Marketing)-16]

2. Statements $K \geq S = U \leq R$; $L < U \leq J$

Conclusions I. $R > J$ II. $L = K$

3. Statements $D > W \geq C \leq X$; $C \geq L$; $W < K$

Conclusions I. $X > K$ II. $L \leq W$

4. Statements $R \leq A < M \geq T \geq Y$; $M \leq S$

Conclusions I. $S > R$ II. $Y \leq S$

5. Statements $D > W \geq C \leq X$; $C \geq L$; $W < K$

Conclusions I. $D > W$ II. $L > D$

6. Statements $S > B = K \geq L$; $B = P \leq U$

Conclusions I. $L < U$ II. $U = L$

7. Solve the inequality $\dfrac{2x+3}{x-1} > 1$

(a) $(0, \infty)$
(b) $(-1, 4)$
(c) $(-\infty, -4)$ U $(1, \infty)$
(d) None of the above

8. The smallest value of $x^2 - 3x + 3$ in the interval $\left(-3, \dfrac{3}{2}\right)$ is :

(a) 3/4 (b) 5
(c) –15 (d) –20

9. If three positive real numbers x, y, z satisfy $y - x = z - y$ and $xyz = 4$, then what is the minimum possible value of y? ***[CAT-03]***

(a) $2^{1/3}$ (b) $2^{2/3}$
(c) $2^{1/4}$ (d) $2^{3/4}$

10. If a, b and c are real numbers such that $a - b + c < 1$, $a + b + c > -1$ and $9a + 3b + c < -4$, then what is the sign of 'a'? ***[MAT-05]***

(a) + only (b) – only
(c) + or – (d) Cannot be said

11. The sum of possible values of x in the equation $|x + 7| + |x - 8| = 16$ is : ***[XAT-14]***

(a) 0 (b) 1
(c) 2 (d) 3
(e) None of these

12. Solve $7 \leq 3x + 1 < 13$

(a) $0 \leq x < 4$ (b) $2 \leq x < 4$
(c) $2 < x < 4$ (d) None of these

13. If x is real, then maximum and minimum values of expression $\dfrac{x^2 + 14x + 9}{x^2 + 2x + 3}$ will be :

(a) 4, – 5 (b) 5, – 4
(c) – 4, 5 (d) – 4, – 5

14. Solve $x^2 - |3x + 5| \geq 0$

(a) $x \leq \dfrac{3 - \sqrt{29}}{2}$ (b) $x \geq \dfrac{3 + \sqrt{29}}{2}$
(c) (a) or (b) (d) None of these

15. If a, b, c and d are four positive real numbers such that $abcd = 1$, what is the minimum value of $(1 + a)(1 + b)(1 + c)(1 + d)$? ***[CAT-01]***

(a) 4 (b) 1
(c) 16 (d) 18

16. What is the maximum value of xy if $x + y = 8$ and x and y are real numbers ***[SSC(CGLT-I)-12]***

(a) 12 (b) 15
(c) 16 (d) 18

17. For what values of x is the function $\sqrt{x^2 + 6x - 40}$ defined in the real domain? ***[CAT-16]***

(a) $10 < x < 4$
(b) $-4 < x < 10$
(c) x does not lie between the closed interval –10 and 4
(d) x does not lie between the open interval –4 and 10

18. Solve $2x + 7 \geq 5x - 11$.

(a) $x < 4$ (b) $x < 6$
(c) $x < -6$ (d) $x \leq 6$

19. If α and $\beta (\alpha < \beta)$ are the roots of the equation $x^2 + bx + c = 0$, where $c < 0 < b$, then

(a) $0 < \alpha < \beta$ (b) $\alpha < 0 < \beta < |\alpha|$
(c) $\alpha < \beta < 0$ (d) $\alpha < 0 < |\alpha| < \beta$

20. Which of the following values of x don't satisfy the inequality $x^2 - 3x + 2 > 0$ at all? ***[CAT-96]***

(a) $1 \leq x \leq 2$ (b) $-1 \geq x \geq -2$
(c) $0 \leq x \leq 2$ (d) $0 \geq x \geq -2$

21. x and y are real numbers satisfying the conditions $2 < x < 3$ and $-8 < y < -7$. Which of the following expressions will have the least value? ***[CAT-01]***

(a) x^2y (b) xy^2
(c) $5xy$ (d) None of these

22. Solve $\dfrac{|-x^2 + 6x + 16|}{|x^2 + 4x + 4|} > 0$

(a) all real values
(b) all real values except 0 and 8
(c) all real values except –2, 2, 8
(d) Cannot be determined

23. If x is real and satisfies $x + 2 > \sqrt{x + 4}$, then

(a) $x < -2$ (b) $x > 0$
(c) $-3 < x < 0$ (d) $-3 < x < 4$

24. $x > 2$, $y > -1$, then which of the following holds good ? ***[CAT-2000]***

(a) $xy > -2$ (b) $xy < -2$
(c) $x > \dfrac{-2}{y}$ (d) None of these

25. Solve $|2x - 1| \leq 2$

(a) (0,15] (b) (–0.5, 1.5)
(c) [–0.5, 1.5] (d) No solution

26. What values of x satisfy $x^{2/3}+x^{1/3}-2<0$? *[CAT-06]*
(a) $-8\le x\le 1$
(b) $-1\le x\le 8$
(c) $1<x<8$
(d) $1\le x\le 8$

27. The set of all real numbers of x for which $x^2-|x+2|+x>0$?
(a) $(-\infty,-2)\cup(2,\infty)$
(b) $(-\infty,-\sqrt{2})\cup(\sqrt{2},\infty)$
(c) $(-\infty,-1)\cup(1,\infty)$
(d) $(\sqrt{2},\infty)$

28. Solve $1+\dfrac{2}{x-2}\ge\dfrac{1}{x}$
(a) $x\ge 2$ (b) $x<0$ or $x>2$
(c) $x<-1$ or $x>2$ (d) None of these

29. If x, y and z are real numbers such that $x+y+z=5$ and $xy+yz+zx=3$. What is the largest value that x can have? *[CAT-02]*
(a) 5/3 (b) $\sqrt{19}$
(c) 13/3 (d) None of these

30. Solve $|2x-5|>0$
(a) all real values except 0
(b) all real values
(c) all real value except 2.5
(d) Cannot be determined

31. If n is such that $36\le n\le 72$, then $x=\dfrac{n^2+2\sqrt{n}(n+4)+16}{n+4\sqrt{n}+n}$ satisfies *[CAT-03]*
(a) $20<x<54$ (b) $23<x<58$
(c) $25<x<64$ (d) $28<x<60$

32. If $b>a$, then the equation $(x-a)(x-b)=1$ has :
(a) both roots in $[a,b]$
(b) both roots in $(-\infty,a)$
(c) both roots in (b,∞)
(d) one root in $(-\infty,a)$ and the other in (b,∞)

33. Solve $\dfrac{3}{|x-1|}>2$
(a) $x\le 1$ (b) $-\dfrac{1}{2}<x<1$
(c) $1<x<\dfrac{5}{2}$ (d) (b) or (c)

34. m is the smallest positive integer such that for any integer $n\le m$, the quantity $n^3-7n^2+11n-5$ is positive. What is the value of m? *[CAT-01]*
(a) 4 (b) 5
(c) 8 (d) None of these

Directions (Q. Nos. 35-39) *In these questions, relationship between different element is shown in the statements. The statements are followed by two conclusions. Study the conclusions based on the given statements and select the appropriate answer.*
Give answer :
(a) If either conclusion I or II is true
(b) If both conclusions I and II are true
(c) If only conclusion I is true
(d) If only conclusion II is true
(e) If neither conclusion I nor II is true

Statements for (Q. Nos. 35-36)
$L<A=M<P;\ A\le C<T;\ M\ge O>R$

35. Conclusions I. $O<P$ II. $C>L$

36. Conclusions I. $T<M$ II. $A>R$

37. Statement $D\ge U=S\ge T$; $O=P\le S$
Conclusions I. $D>P$ II. $P=D$

Statements for (Q. Nos. 38 and 39)
$T>I\ge L>E;\ I\le N<B;\ N\ge S>D$

38. Conclusions I. $N>E$ II. $T<B$

39. Conclusions I. $S\le S$ II. $L>D$ *[IBPS (PO)-16]*

40. If x is real, the expression $\dfrac{x+2}{2x^2+3x+6}$ takes all value in the interval
(a) $\left(\dfrac{1}{13},\dfrac{1}{3}\right)$ (b) $\left(-\dfrac{1}{3},\dfrac{1}{13}\right)$
(c) $\left(-\dfrac{1}{13},\dfrac{1}{3}\right)$ (d) None of these

41. The number of solutions (x,y,z) to the equation $x-y-z=25$, where x, y and z are positve intergers such that $x\le 40$, $y\le 12$ and $z\le 12$ is *[CAT-17]*
(a) 101 (b) 99
(c) 87 (d) 105

42. For how many integers n, will the inequality $(n-5)(n-10)-3(n-2)\le 0$ be satisfied?
(a) 10 (b) 11
(c) 12 (d) 13

43. How many different pairs (a,b) of positive integers are there such that $a\le b$ and $\dfrac{1}{a}+\dfrac{1}{b}=\dfrac{1}{9}$? *[CAT-17]*
(a) 1 (b) 2
(c) 3 (d) 4

44. For $x>0$ what is the minimum value of $x+\dfrac{x+2}{2x}$? *[CDS-19]*
(a) 1
(b) 2
(c) 2.5
(d) Cannot be determined

45. If $2\le|x-1|\times|y+3|\le 5$ and x and y are negative integers, find the number of possible combinations of x and y. *[XAT-18]*
(a) 4 (b) 5
(c) 6 (d) 8
(e) 10

46. The number of the real roots of the equation $2\cos x(x+1)=2^x+2^{-x}$ is : *[CAT-19]*
(a) 2 (b) 1
(c) infinite (d) 0

Answers Key

1. (a)	2. (d)	3. (b)	4. (e)	5. (a)	6. (c)	7. (c)	8. (a)	9. (b)	10. (b)
11. (b)	12. (b)	13. (a)	14. (c)	15. (c)	16. (c)	17. (d)	18. (d)	19. (b)	20. (a)
21. (c)	22. (c)	23. (b)	24. (d)	25. (c)	26. (a)	27. (b)	28. (b)	29. (c)	30. (c)
31. (d)	32. (d)	33. (d)	34. (d)	35. (b)	36. (d)	37. (a)	38. (c)	39. (e)	40. (c)
41. (b)	42. (b)	43. (c)	44. (c)	45. (e)	46. (b)				

Hints and SOLUTIONS

1. Each of options (b) and (c) will yield no conclusion as a sign and its reverse are present in both options. Hence, no need to consider.

Now, consider option (a)

$$K \geq = > N$$
$$K (\geq =) > N$$
$$K (\geq >) N$$
$$K > N \text{ i.e. } N < K$$

Hence, option (a) is correct.

2. Check for conclusion I

Clearly, element U is common for both R and J.

From statement (i) $U \leq R$ and from statement (ii) $U \leq J$

Now, their chain arrangement $R \geq U$, $U \leq J$

We cannot establish relation between R and J as ($\geq \leq$) yield no conclusion.

Check for conclusion II

Clearly, element U is common for both L and K.

From statement (i), $K \geq S = U$ and from statement (ii), $L < U$

Now, their chain arrangement

$$K \geq S = U;\ U > L$$

Then,
$$K \geq = > L$$
$$K (\geq =) > L$$
$$K (\geq >) L$$
$$K > L$$

Hence, option (d) is correct.

3. Check for conclusion I

Clearly, element W is common for both X and K.

From statement (i), $W \geq C \leq X$ and from statement (iii), $W < K$

Now, their chain arrangement $K > W$, $W \geq C \leq X$

We cannot establish relation between K and X as ($\geq \leq$) yield no conclusion.

Check for Conclusion II

Clearly, element C is common for both L and W.

From statement (i), $W \geq C$ and from statement (ii), $C \geq L$

These are already in chain arrangement.

Then, $W (\geq \geq) L$

$$W \geq L$$

Hence, option (b) is correct.

4. Check for conclusion I

Clearly, element M is common for both S and R.

From statement (i), $R \leq A < M$ and from statement (ii), $M \leq S$

Since these are in chain arrangement, $R \leq < \leq S$

$$R (\leq <) \leq S$$
$$R (< \leq) S$$
$$R < S$$
$$S > R$$

Check for Conclusion II

Clearly, element M is common for both Y and S.

From statement (i), $M \geq T \geq Y$ and from statement (ii), $M \leq S$

Now, their chain arrangement $S \geq M$, $M \geq T \geq Y$

Then,
$$S \geq \geq \geq Y$$
$$S (\geq \geq) \geq Y$$
$$S (\geq \geq) Y$$
$$S \geq Y$$
$$Y \leq S$$

Hence, option (e) is correct.

5. Check for Conclusion I

From statement (i), $D > W$

Check for Conclusion II

Clearly, the elements are in statement $D > W \geq C \leq X$; and $C \geq L$

Then, $D > \geq \leq \geq L$

We cannot establish relation between D and L as ($\geq \leq$) yield no conclusion.

Hence, option (a) is correct.

6. Check for conclusion I and II :

Clearly, element B is common for both L and U.

From statement (i), $B = K \geq L$ and from statement (ii), $B = P \leq U$.

Now, the chain arrangement, $U \geq P = B$, $B = K \geq L$

Then,
$$U \geq - - \geq L$$
$$U (\geq =) = \geq L$$
$$U (\geq =) \geq L$$
$$U (\geq \geq) L$$
$$U \geq L$$

It implies that U is either greater than or equal to L.

Hence, option (c) is correct.

7. $\dfrac{2x+3}{x-1} - 1 > 0$

$$\Rightarrow \quad \frac{2x+3-x+1}{x-1} > 0$$
$$\Rightarrow \quad \frac{x+4}{x-1} > 0$$
$$\Rightarrow \quad (x+4)(x-1) > 0$$

Clearly, either $(x+4)$ and $(x-1)$ are positive or both be negative.

In case both are positive, then $(x+4) > 0$ and $(x-1) > 0$

$$\Rightarrow \quad x > -4 \text{ and } x > 1$$

Conclusion : $x > 1$

Further, in case both are negative, then $(x+4) < 0$ and $(x-1) < 0$

$$\Rightarrow \quad x < -4 \text{ and } x < 1$$

Conclusion : $x < -4$

Hence, the solution of the inequality :

either $x > 1$ or $x < -4$.

Alternatively

Proceed through options.

(a) We take $x = 1$, which does not satisfy the inequality

(b) We take $x = 0$, which does not satisfy the inequality

(c) We take $x = -10$ and 10 from both intervals respectively and they satisfy the given inequality.

Hence, option (c) is correct.

8. The smallest value

$$= \frac{4ac - b^2}{4a} = \frac{4\cdot 1\cdot 3 - (-3)^2}{4\cdot 1} = \frac{3}{4},$$

which lies in the interval $\left(-3, \dfrac{3}{2}\right)$.

9. Given, $y - x = z - y \Rightarrow y = \dfrac{x+z}{2}$

Also, $xyz = 4 \Rightarrow x \times \dfrac{x+z}{2} \times z = 4$

$$\Rightarrow \quad xz(x+z) = 8$$

Let $x + z = a$

Then, $x(a - x)\, a = 8$

$\Rightarrow -ax^2 + xa^2 = 8 \Rightarrow ax^2 - a^2x + 8 = 0$

Since x is real, $D \geq O \Rightarrow a^4 - 32a \geq 0$

$\Rightarrow a^3 \geq 32 \Rightarrow a \geq (32)^{\frac{1}{3}}$

Minimum possible value of
$a = (32)^{1/3} = 2^{5/3}$

Then, minimum value of
$$y = \frac{x+z}{2} = \frac{2^{5/3}}{2} = 2^{2/3}$$

10. Given, $a - b + c < 1$ ...(i)
and $-a - b - c < 1$ ...(ii)
Subtracting (ii) from (i) we get,
$2a + 2c < 0$ ...(iii)
Also given, $9a + 3b + c < -4$ ....(iv)
and $-3a - 3b - 3c < 3$ ...(ii)
Adding (iv) and (ii), we get
$6a - 2c < -1$ ...(v)
and $2a + 2c < 0$ ...(iii)
If we add Eqs. (v) and (iii), we get
$8a < -1$
$\Rightarrow \quad a < \frac{-1}{8}$

Hence, option (b) is correct.

11. Given, $|x + 7| + |x - 8| = 16$
$\Rightarrow \quad (x+7)^2 + (x-8)^2 + 2(x+7)(x-8) = 256$
[squaring on both sides]

Note *We squared both sides in order to avoid writing the value of the terms under modulus with + and - as square of positive or negative term is always positive. However, value of $2(x+7)(x-8)$ may be either positive or negative, if $(x+7)$ and $(x-8)$ are with same sign, product is positive otherwise negative.*

$\Rightarrow \quad 4x^2 - 4x - 255 = 0$
$\Rightarrow \quad x = -7.5, 8.5$

Then, their sum $= -7.5 + 8.5 = 1$

When, $2(x+7)(x-8)$ is negative, then, $(x+7)^2 + (x-8)^2 - 2(x+7)(x-8) = 256$

Solving we get, $225 = 256$, which is not possible.

Hence,option (b) is correct.

12. We have $7 \le 3x + 1 < 13$
$\Rightarrow \quad 6 \le 3x < 12$
[on subtracting 1 from each term]
$\Rightarrow \quad 2 \le x < 4$ [on dividing by 3]

Alternatively

Proceed through options.

(a) At $x = 0$, the inequality is not satisfied.

(b) At $x = 2, 3$, the inequality is satisfied.

Hence, option (b) is correct.

13. Given expression, $\frac{x^2 + 14x + 9}{x^2 + 2x + 3}$

Let $a = \frac{x^2 + 14x + 9}{x^2 + 2x + 3}$

Then, $ax^2 + 2ax + 3a = x^2 + 14x + 9$
$\Rightarrow x^2(a-1) + x(2a - 14) + 3a - 9 = 0$
Since x is real, Discriminant ≥ 0
$\Rightarrow (2a - 14)^2 - 4(a-1)(3a-9) \ge 0$
$\Rightarrow 4a^2 + 196 - 56a - 12a^2 + 48a - 36 \ge 0$
$\Rightarrow \quad -8a^2 - 8a + 160 \ge 0$
$\Rightarrow \quad a^2 + a - 20 \le 0$

Now, D for corresponding equation $a^2 + a - 20 = 0$ is 81, hence, $D > 0$.

Also, quadratic coefficient > 0. It implies that the value of the quadratic expression $a^2 + a - 20$ is negative only if $\alpha < x < \beta$, where α and β are roots of corresponding equation and $\alpha < \beta$.

Solving the equation, we get
$\alpha = -5, \beta = 4$

Hence, for $-5 < a < 4$, $a^2 + a - 20 < 0$. Obviously, in order to satisfy the inequality $a^2 + a - 20 \le 0$, we should also include values of α and β at which the value of the expression is 0.

Thus, $-5 \le a \le 4$

It implies that the minimum value of a is -5 while the maximum value is 4.

$\therefore$ Option (a) is correct.

14. Clearly, $|3x + 5| \le x^2$

The strict version of this inequality, $|3x + 5| < x^2$
$\Rightarrow -x^2 < 3x + 5 < x^2$
$\Rightarrow x^2 + 3x + 5 > 0$ or $x^2 - 3x - 5 > 0$

Clearly, $x^2 + 3x + 5 > 0$ is satisfied with all real values of x as $D < 0$ and $a > 0$.

Also, at imaginary value of $x, x^2 + 3x + 5 = 0$.

Further, solution of $x^2 - 3x - 5 > 0$ is
$$x < \frac{3 - \sqrt{29}}{2} \text{ or } x > \frac{3 + \sqrt{29}}{2}$$

We may conclude from the above that the solution of $|3x + 5| < x^2$ is
$$x < \frac{3 - \sqrt{29}}{2} \text{ or } x > \frac{3 + \sqrt{29}}{2}$$

Then, required solution is
$$x \le \frac{3 - \sqrt{29}}{2} \text{ or } x \ge \frac{3 + \sqrt{29}}{2}$$

15. We are aware that arithmetic mean of positive real numbers $\ge$ their geometric mean.

Hence, $\frac{1+a}{2} \ge \sqrt{1a} \Rightarrow (1+a) \ge 2\sqrt{a}$

Similarly, $(1+b) \ge 2\sqrt{b}$, $(1+c) \ge 2\sqrt{c}$ and $(1+d) \ge 2\sqrt{d}$

Obviously, $(1+a)(1+b)(1+c)(1+d) \ge 2\sqrt{a} \cdot 2\sqrt{b} \cdot 2\sqrt{c} \cdot 2\sqrt{d}$
$\Rightarrow (1+a)(1+b)(1+c)(1+d) \ge 16\sqrt{abcd}$
$\Rightarrow (1+a)(1+b)(1+c)(1+d) \ge 16\sqrt{1}$

Thus, minimum value of
$(1+a)(1+b)(1+c)(1+d) = 16$

16. We know, xy takes maximum value when each of x and y takes equal value.

Hence, maximum value of
$xy = 4 \times 4 = 16$

17. The given function is defined in real domain only when $x^2 - 6x - 40 \ge 0$.

Discriminant, $D = b^2 - 4ac$
$= (-6)^2 - 4 \cdot 1(-40) = 196$

Hence, $D > 0$

Also, quadratic coefficient > 0.

It implies that the value of the given quadratic expression is positive only if $x < \alpha$ or $x > \beta$, where α and β are roots of corresponding equation and $\alpha < \beta$.

Solving the equation, we get
$\alpha = -4, \beta = 10$

Hence, for $x < -4$ or $x > 10, x^2 - 6x - 40 > 0$.

Obviously, in order to satisfy the inequality $x^2 - 6x - 40 \ge 0$, we should also include values of α and β at which the value of the expression is 0.

Thus, the required answer is $x \le -4$ or $x \ge 10$ i.e. option (d).

18. $2x - 5x \ge -11 - 7$
$\Rightarrow \quad -3x \ge -18$
$\Rightarrow \quad -x \ge -6$
$\Rightarrow \quad x \le 6$

Alternatively

Proceed through options.

Since this is slack inequality, solution will also be a slack inequality.

Hence, option (d) is correct.

19. Product of roots $= \alpha\beta = c$
$\Rightarrow \alpha\beta =$ negative value as c is negative. It implies that one root is negative and another is positive.

Also, sum of roots $= \alpha + \beta = -b$
$\Rightarrow \alpha + \beta =$ negative value as b is positive. It means, the numerical value of the negative root is greater than that of the positive root.

As given, $\alpha < \beta$, it implies that α is negative root whereas β is positive root.

It may be construed from the above that $\alpha < 0 < \beta < |\alpha|$

Hence, option (b) is correct.

20. Given inequality : $x^2 - 3x + 2 > 0$

Since $D > 0$ and quadratic coefficient > 0,

$ax^2 + bx + c$ will be positive for all real values of $x(x < \alpha \text{ or } x > \beta)$ [α and β are roots of corresponding equation and $\alpha < \beta$]

Solving the equation $x^2 - 3x + 2 = 0$, we get $\alpha = 1$ and $\beta = 2$

Hence, the given inequality stands satisfied at $x < 1$ or $x > 2$. It implies that it is not satisfied at $1 \le x \le 2$.

Hence, option (a) is correct.

21. Given, $2 < x < 3 \Rightarrow 4 < x^2 < 9$

(on squaring both sides)

Also, $\quad 10 < 5x < 15$

(on multiplying by 5)

Now, proceed through options.

The value of xy^2 is positives as x and y^2 both are positive.

Clearly, both x^2y and $5xy$ are negative.

Let us compare them.

$$5x > x^2$$

[$\because x^2$ is less than 9 whereas $5x$ is greater than 10]

$$\Rightarrow \quad 5xy < x^2y$$

[on multiplying by a negative number, i.e. by y]

Hence, option (c) is correct.

22. Clearly, $|-x^2 + 6x + 16| > 0$

and $\quad |x^2 - 4x + 4| > 0$

Each of both has neither negative nor zero values.

Now, solve $|-x^2 + 6x + 16| > 0$

$-x^2 + 6x + 16$ is satisfied with all real values of x except values of x satisfying $-x^2 + 6x + 16 = 0$.

Hence, $x \ne 8, -2$

Now, solve $|x^2 - 4x + 4| > 0$

$x^2 - 4x + 4$ is satisfied with all real values of x except values of x satisfying $x^2 - 4x + 4 = 0$.

$$\Rightarrow \quad x \ne 2$$

Hence, we may conclude that the required solution is

$x < -2$ or $-2 < x < 2$ or $2 < x < 8$

or $\quad x > 8$

Alternatively

Proceed through options.

(a) At $x = 8$, the inequality is not satisfied.

(b) At $x = 2$, the inequality is not satisfied.

(c) All real values except $-2, 2, 8$ satisfy the inequality.

Hence, option (c) is correct.

23. Given, $x + 2 > \sqrt{x + 4}$

$\Rightarrow (x + 2)^2 > x + 4$ [on squaring both sides]

$\Rightarrow \quad x^2 + 3x > 0$

$\Rightarrow \quad x(x + 3) > 0$

$\Rightarrow x > 0$ and $(x + 3) > 0$ or $x < 0$ and $(x + 3) < 0$

$\Rightarrow x > 0$ and $x > -3$ or $x < 0$ and $x < -3$

$\Rightarrow x > 0$ or $x < -3$

Hence, option (b) is correct.

Alternatively

Given, $x + 2 > \sqrt{x + 4}$

$\Rightarrow \quad (x + 2)^2 > x + 4$

(on squaring both sides)

$\Rightarrow \quad x^2 + 3x + 0 > 0$

Since $D > 0$ and quadratic coefficient > 0, the roots will be $x < \alpha$ or $x > \beta$

[α and β are roots of corresponding equation and $\alpha < \beta$]

Hence, $x < -3$ or $x > 0$

Hence, option (b) is correct.

24. It is the fact that we cannot draw a conclusion about relation between ac and bd if $a > b > 0$ and $0 > c > d$.

Here, it is possible that value of y may be negative like $-0.5, -0.3$, etc.

Hence, option (d) is correct.

25. We know, $|x| \le a$, where $a > 0$ implies $-a \le x \le a$

Hence, $\quad -2 \le 2x - 1 \le 2$

$\Rightarrow \quad -1 \le 2x \le 3$ (on adding 1)

$\Rightarrow \quad \frac{-1}{2} \le x \le \frac{3}{2}$

(on dividing by 2)

Alternatively

Proceed through options.

26. Let $x^{1/3} = a$

Then, $a^2 + a - 2 \le 0$

The strict version of this inequality is $a^2 + a - 2 < 0$

Here, $D > 0$ and quadratic coefficient > 0.

It implies that the value of the given quadratic expression is negative only if $\alpha < a < \beta$, where α and β are roots of corresponding equation and $\alpha < \beta$.

Solving the equation, we get

$$\alpha = -2, \beta = 1$$

Hence, for $-2 < a < 1$, $a^2 + a - 2 < 0$

Obviously, in order to satisfy the inequality $a^2 + a - 2 \le 0$, we should also include values of α and β at which the value of the expression is 0.

Thus, $a^2 + a - 2 \le 0$ is satisfied at $-2 \le a \le 1$.

$\Rightarrow \quad -2 \le x^{1/3} \le 1$

$\Rightarrow \quad -8 \le x \le 1$

27. Given, $x^2 - |x + 2| + x > 0$

$\Rightarrow \quad |x + 2| < x^2 + x$

It is obvious that $x^2 + x > 0$ as its value cannot be 0 or negative.

$\Rightarrow \quad -(x^2 + x) < x + 2 < (x^2 + x)$

$\Rightarrow \quad x^2 + 2x + 2 > 0$ or $x^2 - 2 > 0$

Clearly, $x^2 + 2x + 2 > 0$ is satisfied with all real values of x as $D < 0$ and $a > 0$.

Now, solution of $x^2 - 2 > 0$ is $x < -\sqrt{2}$ or $x > \sqrt{2}$

Conclusion from solutions of $x^2 + 2x + 2 > 0$ and $x^2 - 2 > 0$ is $x < -\sqrt{2}$ or $x > \sqrt{2}$.

Hence, option (b) is correct.

Alternatively

You may proceed through options.

28. $1 + \frac{2}{x - 2} - \frac{1}{x} \ge 0$

$$\Rightarrow \quad \frac{x^2 - 2x + 2x - x + 2}{x(x - 2)} \ge 0$$

$$\Rightarrow \quad \frac{x^2 - x + 2}{x(x - 2)} \ge 0$$

Obviously, $x^2 - x + 2 > 0$ and $x(x - 2) > 0$ or $x^2 - x + 2 = 0$ and $x(x - 2) > 0$

$x^2 - x + 2 > 0$ is satisfied with all real values of x as $D < 0$ and $a > 0$ and $x^2 - x + 2 = 0$ is satisfied with imaginary values of x as $D < 0$.

Further, $x(x - 2) > 0$

Clearly, $x > 0$ and $(x - 2) > 0$ or $x < 0$ and $(x - 2) < 0$

$\Rightarrow x > 0$ and $x > 2$ or $x < 0$ and $x < 2$

$\Rightarrow \quad x > 2$ or $x < 0$

We may conclude from the above that $x < 0$ or $x > 2$ is the required solution.

Alternatively

Proceed through options.

(a) $x \geq 2 \Rightarrow$ the denominator of $\frac{2}{x-2}$ becomes zero at $x = 2$, hence not correct.

(b) Values of x satisfy the inequaity.

Hence, option (b) is correct.

29. We have, $(x+y+z)^2$
$= x^2 + y^2 + z^2 + 2xy + 2yz + 2zx$

$\Rightarrow \quad 5^2 = x^2 + y^2 + z^2 + 2(3)$

$\Rightarrow \quad x^2 + y^2 + z^2 = 19$

Value of each of y and z should be 0 if we intend to get the maximum value of x. However, values of both cannot be zero simultaneously as $xy + yz + zx \neq 0$.

Hence, $x^2 < 19$

$\therefore$ Option (b) is eliminated.

If $x = 5/3, \quad x^2 < 19$

If $x = 13/3, \quad x^2 < 19$

Since 13/3 is greater than 5/3, option (c) is correct.

30. We know, $|x| > 0$ implies x can be any real number except $x = 0$

Clearly, the solution is any real number except the value of x which satisfies $2x - 5 = 0$.

If $2x - 5 = 0$, then $x = 5/2$

Hence, the solution is any real number except $5/2$.

31. At $n = 36$,

$$x = \frac{(36)^2 + 2\sqrt{36}(36+4) + 16}{36 + 4\sqrt{36} + 4} = 28$$

and at $n = 72$,

$$x = \frac{(72)^2 + 2\sqrt{72}(72+4) + 16}{72 + 4\sqrt{72} + 4} = 59.02$$

Hence, option (d) is most appropriate answer.

32. Given equation $(x-a)(x-b) = 1$

$\Rightarrow x^2 - (a+b)x + ab - 1 = 0$

Then, Discriminant $= [-(a+b)]^2$
$-4 \times 1 \times (ab - 1) = (b-a)^2 + 4$

Since, value of the discriminant is greater than 0, the roots will be real.

Let the roots be α and β.

Then, $\alpha = \dfrac{(a+b) + \sqrt{(b-a)^2 + 4}}{2 \cdot 1}$

and $\beta = \dfrac{(a+b) - \sqrt{(b-a)^2 + 4}}{2 \cdot 1}$

$\Rightarrow \alpha > \dfrac{(a+b) + \sqrt{(b-a)^2}}{2}$

and $\beta < \dfrac{(a+b) - \sqrt{(b-a)^2}}{2}$

$\Rightarrow \quad \alpha > \dfrac{(a+b) + (b-a)}{2}$

and $\beta < \dfrac{(a+b) - (b-a)}{2}$

$\Rightarrow \quad \alpha > b$ and $\beta < a$

Hence, option (d) is correct.

33. Given, $\dfrac{3}{|x-1|} - 2 > 0$

$\Rightarrow \quad \dfrac{3 - 2|x-1|}{|x-1|} > 0$

Clearly, $3 - 2|x-1| > 0$

and $\quad |x-1| > 0$

$\Rightarrow 2|x-1| < 3$ and $|x-1| > 0$

$\Rightarrow \quad |x-1| < \dfrac{3}{2}$ and $|x-1| > 0$

$\Rightarrow -\dfrac{3}{2} < x - 1 < \dfrac{3}{2}$ and $|x-1| > 0$

$\Rightarrow \quad -\dfrac{1}{2} < x < \dfrac{5}{2}$ and $|x-1| > 0$

It is obvious that $(x-1) > 0$ is satisfied with all real values of x except $x = 1$.

Hence, the required solution is

$$-\frac{1}{2} < x < 1 \text{ or } 1 < x < \frac{5}{2}.$$

Alternatively

Proceed through options.

(a) $x \leq 1 \Rightarrow$ the denominator becomes zero at $x = 1$, hence not correct.

(b) and (c) both are true.

Hence, option (d) is correct.

34. Given $n^3 - 7n^2 + 11n - 5 > 0$

Since sum of alternative coefficients (including constant) are equal in magnitude but with opposite sign, one root of $n^3 - 7n^2 + 11n - 5 = 0$ is 1. Hence, $(n-1)$ is a factor of $n^3 - 7n^2 + 11n - 5$.

[you may refer to 'Cubic equation' for conceptual clarity]

Now, $(n-1)(n^2 - 6n + 5) > 0$

$\Rightarrow \quad (n-1)(n-1)(n-5) > 0$

$\Rightarrow \quad (n-1)^2(n-5) > 0$

Clearly, $(n-1)^2$ is positive.

Then, $(n-5) > 0 \Rightarrow n > 5$

Since $\quad n \leq m, 5 < n \leq m$.

$\therefore 5 < m$

Hence, least integer value of $m = 6$

$\therefore$ Option (d) is correct.

35. Check for Conclusion I

Since M is common for both O and P, we should take statements (i) and (iii) into consideration.

From statement (i), $M < P$ and from statement (iii), $M \geq O$

Now, their chain arrangement :
$P > M, M \geq O$

Then, $P (> \geq) O \Rightarrow P > O$

Check for Conclusion II

Since A is common for both C and L, we should take statements (i) and (ii) into consideration.

From statement (i), $L < A$ and from statement (ii), $A \leq C$

Now, their chain arrangement :
$L < A, A \leq C$

Then, $\quad L (< \leq) C \Rightarrow L < C$

Hence, both the conclusions are correct.

$\therefore$ Option (b) is correct.

36. Check for Conclusion I

Since A is common for both T and M, we should take statement (i) and (ii) into consideration.

From statement (i), $A = M$ and from statement (ii), $A \leq C < T$ i.e. $A < T$

Now, their chain arrangement :

$M = A, A < T$

Then, $\quad M (= <) T$

$\Rightarrow \quad M < T$

Check for Conclusion II

Since M is common for both A and R, we should take statement (i) and (iii) into consideration.

From statement (i), $A = M$ and from statement (iii), $M \geq O > R$ i.e. $M > R$

Now, their chain arrangement:
$A = M, M > R$

Then, $A (= >R) \Rightarrow A > R$

Hence, only conclusion II is correct.

$\therefore$ Option (d) is correct.

37. Check for Conclusions I and II

Since S is common for both D and P, we should take both the statement (i) and (ii) into consideration.

From statement (i), $D \geq U = S$ i.e. $D \geq S$ and from statement (ii), $P \leq S$

Now, their chain arrangement :
$D \geq S, S \geq P$.

Then, $D (\geq \geq) P \Rightarrow D \geq P$

Hence, either $D > P$ or $D = P$

$\therefore$ Option (a) is correct.

38. Check for Conclusion I

Since I is common for both N and E, we should take statements (i) and (ii) into consideration.

From statement (i), $I \geq L > E$ i.e. $I > E$ and from statement (ii), $I \leq N$

Now, their chain arrangement :

$E < I, I \leq N$

Then, $E (< \leq) N \Rightarrow E < N$

Check for Conclusion II

Since I is common for both T and B, we should take statements (i) and (ii) into consideration.

From statement (i), $T > I$ and from statement (ii), $I \leq N < B$ i.e. $I < B$

Now, their chin arrangement :

$T > I, I < B$

Then, $T (> <) B$

$\Rightarrow$ No conclusion can be drawn.

Hence, only conclusion I is correct.

$\therefore$ Option (c) is correct.

39. Check for conclusion I

Only $S = S$ is possible, no other sign will define it.

Check for Conclusion II

No element is common for L and D in direct way. However, we may establish the relation between them taking all three statements into consideration.

Clearly, I is common for L and N. From statement (i), $I \geq L$ and from statement (ii), $I \leq N$

Now, their chain arrangement $L \leq I$, $I \leq N$

Then, $L (\leq \leq) N \Rightarrow L \leq N$

Further, from statement (iii) $N \geq S > D$ i.e. $N \geq D$

Now, $L \leq N, N \geq D$

Then, $L (\leq \geq) D$

Hence, no relation can be established between L and D.

$\therefore$ Option (e) is correct.

Note *All the above five questions can be processed mentally in moments without pen and paper. Please try yourself.*

40. Let $a = \dfrac{x+2}{2x^2 + 3x + 6}$

$\Rightarrow 2ax^2 + 3ax + 6a = x + 2$

$\Rightarrow 2ax^2 + (3a - 1)x + 6a - 2 = 0$

Since x is real, Discriminant $D \geq 0$

$\Rightarrow \quad (3a-1)^2 - 4 \cdot 2a(6a-2) \geq 0$

$\Rightarrow \quad 9a^2 + 1 - 6a - 48a^2 + 16a \geq 0$

$\Rightarrow \quad -39a^2 + 10a + 1 \geq 0$

$\Rightarrow \quad 39a^2 - 10a - 1 \leq 0$

Now, D for corresponding equation $39a^2 - 10a - 1 = 0$ is 256, hence, $D > 0$.

Also, quadratic coefficient > 0. It implies that the value of the quadratic expression $39a^2 - 10a - 1$ is negative only if $\alpha < x < \beta$, where α and β are roots of corresponding equation and $\alpha < \beta$.

Solving the equation, we get

$$\alpha = -\frac{1}{13}, \beta = \frac{1}{3}$$

Hence, for $-\dfrac{1}{13} < a < \dfrac{1}{3}$,

$39a^2 - 10a - 1 < 0$.

Obviously, in order to satisfy the inequality $39a^2 - 10a - 1 \leq 0$, we should also include values of α and β at which the value of the expression is 0.

Thus, $-\dfrac{1}{13} \leq a \leq \dfrac{1}{3}$

$\therefore$ Option (c) is correct.

41. Given, $x - y - z = 25$

$\Rightarrow \quad x = 25 + y + z$

The values of variables may be as under

x	y	z	Number of solutions
27	1	1	1
28	1, 2	2, 1	2
29	1, 2, 3	3, 2, 1	3
....	....	......	......
38	1, 2, 3 ...11, 12	12, 11, 10, ...2, 1	12
39	2, 3, 4 ......11, 12	12, 11, 10, ...3, 2	11
40	3, 4 ...... 11, 12	12, 11, 10, ... 4, 3	10

Hence, total number of solutions

$= (1 + 2 + 3 + ...12) + 11 + 10$

$= \dfrac{12(12+1)}{2} + 21 = 99$

42. Given, $(n-5)(n-10) - 3(n-2) \leq 0$

$\Rightarrow \quad n^2 - 15n + 50 - 3n + 6 \leq 0$

$\Rightarrow \quad n^2 - 18n + 56 \leq 0$

The corresponding strict inequality is $n^2 - 18n + 56 < 0$

Now, discriminat D

$= b^2 - 4ac = (-18)^2 - 4 \cdot 1 \cdot 56 = 100$

Hence, $D > 0$

Also, quadratic coefficient > 0. It implies that the value of the given quadratic expression is negative only if $\alpha < n < \beta$, where α and β are roots of corresponding equation and $\alpha < \beta$.

Solving the equation we get

$$\alpha = 4, \beta = 14$$

Hence, for $4 < n < 14, n^2 - 18n + 56 < 0$

Obviously, in order to satisfy the inequality $n^2 - 18n + 56 \leq 0$, we should also include values of α and β at which the value of the expression is 0.

Thus, at $4 \leq x \leq 14$, the given inequality is satisfied.

$\therefore$ Number of values of $n = 11$

43. Given, $\dfrac{1}{a} + \dfrac{1}{b} = \dfrac{1}{9}$

$\Rightarrow \quad ab = 9(a+b)$

$\Rightarrow \quad ab - 9a - 9b = 0$

$\Rightarrow \quad a(b-9) - 9(b-9) = 81$

$\Rightarrow \quad (a-9)(b-9) = 81$

Since a and b are positive integers, the values of $(a-9)$ and $(b-9)$ may be as under:

(i) If $a - 9 = 3$, then $b - 9 = 81$

Hence, $a = 10$ and $b = 90$

(ii) If $a - 9 = 3$, then $b - 9 = 27$. Hence, $a = 12$ and $b = 36$

(iii) If $a - 9 = 9$, then $b - 9 = 9$. Hence, $a = 18$ and $b = 18$

(iv) If $a - 9 = 27$, then $b - 9 = 3$, Hence, $a = 36$ and $b = 12$

(v) If $a - 9 = 81$, then $b - 9 = 1$. Hence, $a = 90$ and $b = 10$

There are 3 pairs satisfying $a \leq b$. There pairs are (10, 90), (12, 36) and (18, 18).

44. Given, expression

$$= x + \frac{x+2}{2x} = x + \frac{1}{x} + \frac{1}{2}$$

We know, Arithmetic Mean $\geq$ Geometric Mean

$$\Rightarrow \quad \frac{x + \frac{1}{x}}{2} \geq \sqrt{x \times \frac{1}{x}}$$

$$\Rightarrow \quad x + \frac{1}{x} \geq 2$$

Thus, minimum value of

$$x + \frac{1}{x} + \frac{1}{2} = 2 + \frac{1}{2} = 2.5$$

45. Clearly, the value of $|x-1| \times |y+3|$ is either 2 or 3 or 4 or 5.

Case I If $|x-1| = 1$, then $|y+3| = 2$ or 3 or 4 or 5

(OR *vice versa*)

Solutions for $|x-1| = 1$

If $x-1>0, x-1=1 \Rightarrow x=2 \neq$ a negative integer

If $x-1<0, x-1=-1 \Rightarrow x=0 \neq$ a negative integer

Solutions for $|y+3|=2$

If $y+3>0, y+3=2 \Rightarrow y=-1=$ a negative integer

If $y+3<0, y+3=-2 \Rightarrow y=-5=a$ negative integer

Similarly, for $|y+3|=3, y=0,-6$

For $|y+3|=4, y=1,-7$

For $|y+3|=5, y=2,-8$

Case II If $|x-1|=2$, then $|y+3|=1$ or 2 (OR *vice versa*)

For $|x-1|=2, x=3,-1$

For $|y+3|=1, y=-2,-4$

For $|y+3|=2, y=-1,-5$

Case III If $|x-1|=3$, then $|y+3|=1$ (OR *vice versa*)

For $|x-1|=3, x=4,-2$

For $|y+3|=1, y=-2,-4$

Case IV If $|x-1|=4$, then $|y+3|=1$ (OR *vice versa*)

For $|x-1|=4, x=5,-3$

For $|y+3|=1, y=-2,-4$

Case V If $|x-1|=5$, then $|y+3|=1$ (OR *vice versa*)

For $|x-1|=5, x=6,-4$

For $|y+3|=1, y=-2,-4$

Possible combinations of x, y (case-wise)

$= [\mathbf{0} + \mathbf{4}$ {i.e. $(-1,-2),(-1,-4),(-1,-1),(-1,-5)$} $+\mathbf{2}$ {i.e. $(-2,-2)(-2,-4)$} $+\mathbf{2}$ {i.e. $(-3,-2),(-3,-4)$} $+ \mathbf{2}$ {i.e. $(-4,-2),(-4,-4)$}]

$= 10$

46. Since Arithmetic Mean $\geq$ Geometric Mean, $\dfrac{2^x + 2^{-x}}{2} \geq \sqrt{2^x \times 2^{-x}}$

$\Rightarrow \quad 2^x + 2^{-x} \geq 2$

It implies that $2\cos x(x+1) \geq 2$. Further, since the value of $\cos\theta$ cannot exceed 1, $2\cos x(x+1) = 2$.

Thus, the equation is valid only if $2^x + 2^{-x} = 2$ i.e. $x = 0$.

It may be said that the equation has only one real solution.

Unit 10

Geometry

- Point, Line, Plane and Angles
- Polygons
- Triangles
- Quadrilaterals
- Circles

Geometry

Introduction

The word 'Geometry' is derived from the greek words 'geo' and 'metron'. Geo means earth and Metron means measuring. Thus, the word Geometry means 'Earth Measurement'. Geometry Section is basically concerned with points, lines, shapes and space.

Euclid, a greek mathematician is known as father of Geometry. The geometry of plane figure, which is our subject matter, is known as Euclid's Geometry. A mathematician who works in the field of geometry is known as geometer.

Almost all the competitive examinations like CAT, MAT, XAT, SSC, Banks, Railway, LIC etc. cover this topic and in my opinion, it is an easy chapter. Hence, the highest marks can be secured without much efforts.

This part of mathematics is a game of imagination, thus, you must be capable to visualise the geometrical figures. You can do well in this part only when you have grip on axiom, theorem and other related concepts. Axioms are the basic facts which are granted without proof, viz, one fourth of equals are equal, whereas theorem is basically the result that is proved through logical reasoning based on axioms.

For the sake of convenience this unit is divided into following parts :

(A) Point, Line, Plane and Angles

(B) Polygons

(C) Triangles

(D) Quadrilaterals

(E) Circles

55

Point, Line, Plane and Angle

Learning OBJECTIVES

After completing this chapter, aspirants will be able to :

- Understand the significance of this sub-section in diverse competitive exams;
- Understand the definition of Point, Line, Line segment, Ray and Plane;
- Understand the diverse terminology related to point, line and plane;
- Be familiar with their properties and relation between them;
- Be familiar with meaning of an angle and various system of its measurement;
- Recognise various types of Angles with their properties;
- Understand the relation between pairs of Angles;
- Realise themselves competent to solve the questions in unique way;
- Analyse the trend of the questions being asked in various competitive examinations;
- Solve the questions in moments;
- Learn many new concepts/fundamentals;
- Apply the pragmatic approaches in lieu of academic approaches; and
- Ponder on the alternative methods.

Inside

A. Point

The figure which has no length, breadth or height.

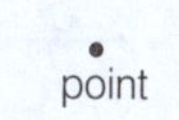

Critical facts

(i) Through a given point, infinite lines may pass.

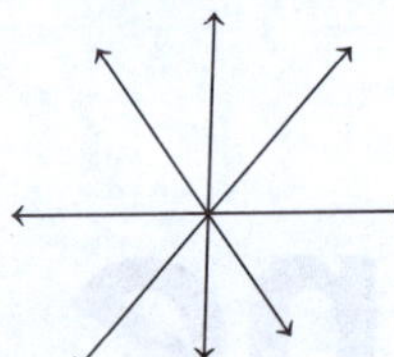

(ii) Given two distinct points, there is a unique line that passes through them.

(iii) When more than three points lie in the same plane, they are called as coplanar otherwise non-coplanar.

(iv) Three or more than three points are said to be collinear if there is a line which contains all of them.

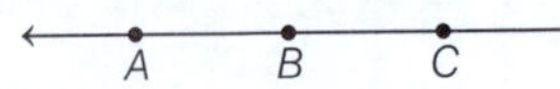

B. Line

A line is a set of points which has only length and can be extended in both directions. Line means straight line unless otherwise stated.

Critical facts

(i) A terminated line can be produced indefinitely.

(ii) Two distinct lines cannot have more than one point in common.

(iii) Two lines having a common point are called intersecting lines and this common point is known as point of intersection.

(iv) Three or more than three lines are said to be concurrent if there is a point which lies on all of them.

(v) When more than one line lie in the same plane, then these lines are called coplanar otherwise non-coplanar.

(vi) Two lines in a plane are said to be parallel, if they don't intersect when extended.

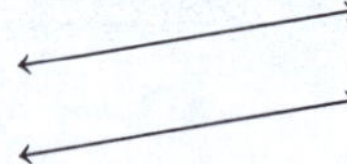

(vii) Two lines which lie in a single plane and intersect each other at 90° are called perpendicular lines.

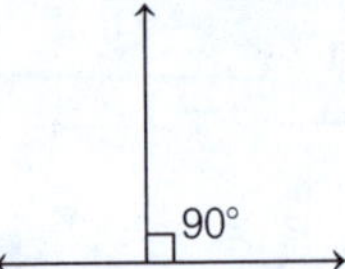

(viii) Two lines which are neither parallel nor intersecting are called skew lines.

(ix) Two lines which are both parallel to the same line, are parallel to each other.

(x) In a plane, when two different lines are perpendicular to a third line, they will obviously be parallel to each other.

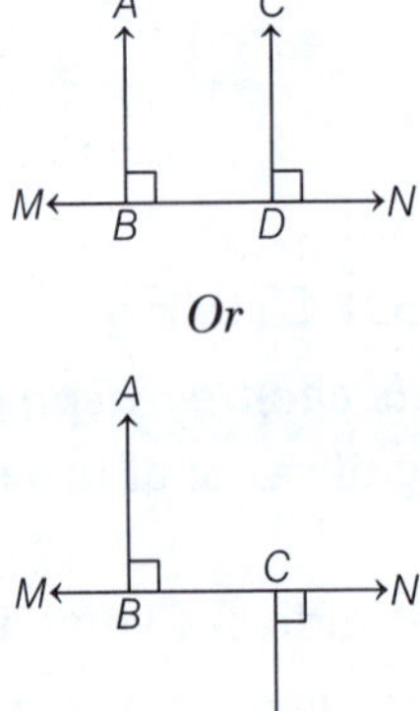

(xi) A line which intersects two or more given lines at distinct points is called transversal of the given lines.

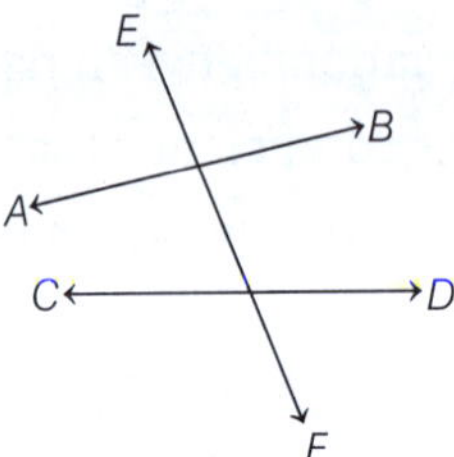

(xii) In case two or more parallel lines are intercepted by other transversal lines, the ratio of corresponding intercepts thus formed, are equal.

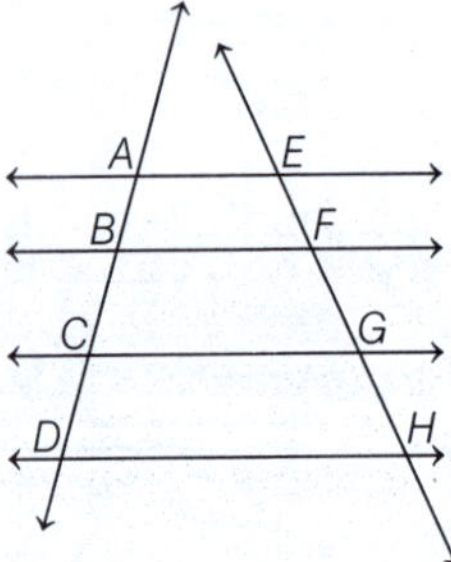

C. Line Segment

A line segment has two end points. We generally use word 'Line' for line-segment in Geometry.

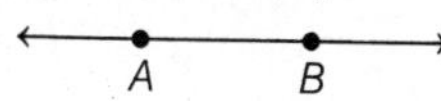

Critical facts

(i) Two line segments are said to be congruent if they have equal length.

(ii) Of all the segments that can be drawn to a given line from a point not lying on it, the perpendicular segment is the shortest. Here, PM is the shortest line.

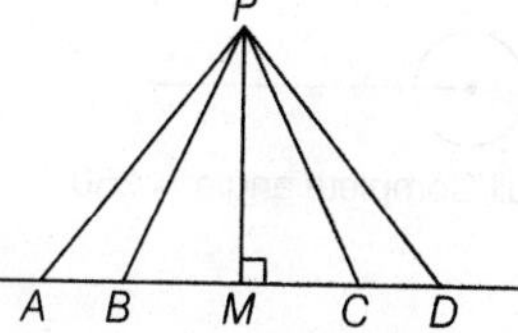

D. Ray

A Ray is a line with single endpoint that extends indefinitely in one direction.

E. Plane

A plane is a set of lines, it means it has infinite length and breadth but no thickness i.e. a two dimensional figure.

From the above, it may also be concluded that a solid has three dimensions, a surface (or plane) has two, a line has one and a point has none.

F. Angles

An angle is formed when two rays originate from the same end point. The rays making an angle are called arms of the angle and the end point is called the vertex of the angle.

The angle formed by two rays OA and OB is denoted as $\angle AOB$ or $\angle BOA$. We may write $\angle O$ instead of $\angle AOB$ or $\angle BOA$ provided there is only one angle with vertex O. Angles are said to be congruent if they have same size.

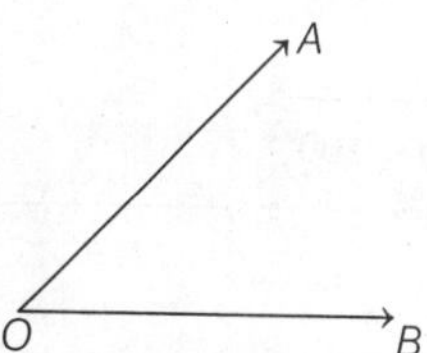

Measurement of Angles

(a) **Sexagesimal System (Base-60)** A right angle is divided into 90 equal parts known as degrees.

Each degree (°) is divided into 60 equal parts called minutes (′) and each minute is further divided into 60 equal parts called seconds (″).

Hence, 1 right angle = $90°$, $1° = 60'$ and $1' = 60''$

(b) **Centesimal System (Base-100)** A right angle is divided into 100 equal parts known as grades (g).

Hence, 1 right angle $= 100^g$ [i.e. 100 grade/gradian]

Also, $1^g = 100'$ [i.e. 100 centesimal minutes]

and $1' = 100''$ [i.e. 100 centesimal seconds]

(c) **Radian System** The angle subtended at centre of a circle by an arc (equal of radius in length) is defined as 1 radian (i.e. 1^c, c is meant for circular measure).

Hence, arc(i.e. equal to r) subtends 1 radian

Then, $2\pi r$ (i.e. circumference) will subtend 2π radian. Obviously, 2π radian = $360°$

$\therefore \pi$ radian = $180°$

Types of Angles

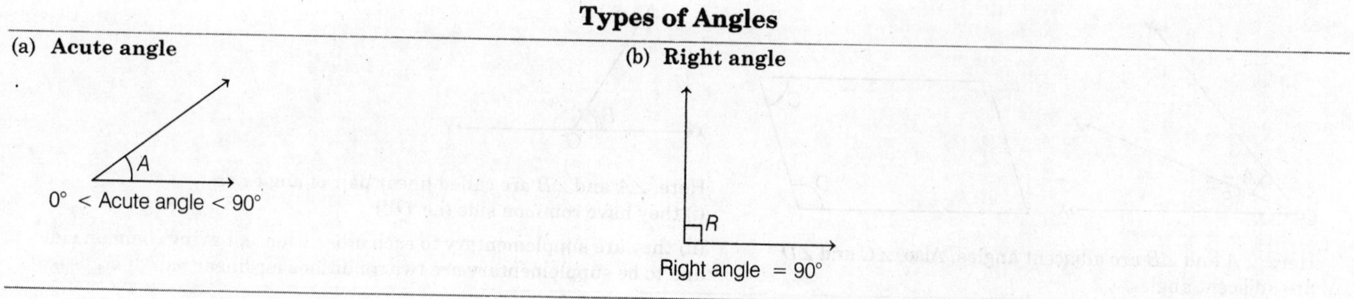

(c) **Obtuse angle**

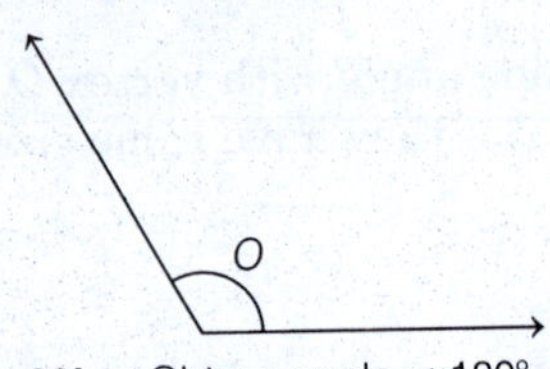

$90° <$ Obtuse angle $< 180°$

(d) **Straight angle**

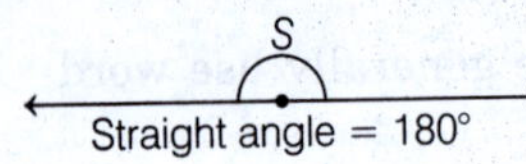

Straight angle = 180°

(e) **Reflex angle**

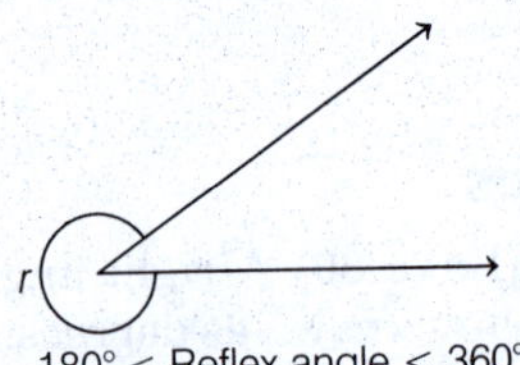

$180° <$ Reflex angle $< 360°$

(f) **Full/Complete angle**

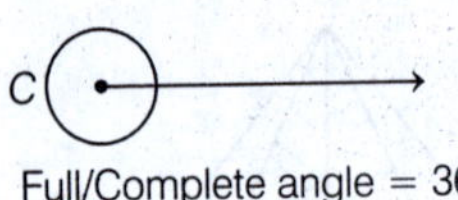

Full/Complete angle = 360°

(g) **Oblique angles** Angle that is not a right angle or a multiple of a right angle is called oblique angles. Ex. 120°, 135°, etc.

(h) **Congruent angles** Angles that have the same measure are said to be equal or congruent.

(i) **Complementary angles**

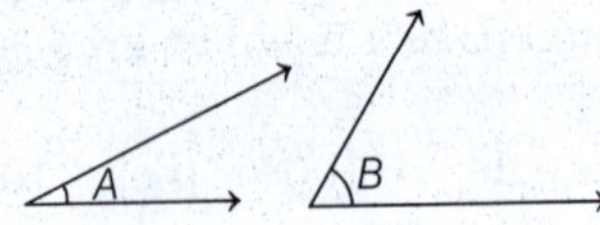

If $\angle A + \angle B = 90°$, then both are said to be complementary to each other.

(j) **Supplementary angles**

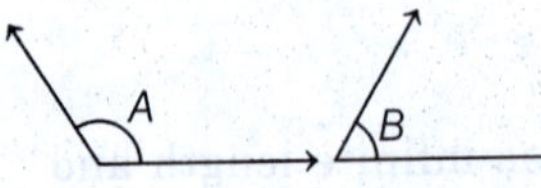

If $\angle A + \angle B = 180°$, then both are said to be supplementary to each other.

(k) **Explementary or Conjugate angles :**

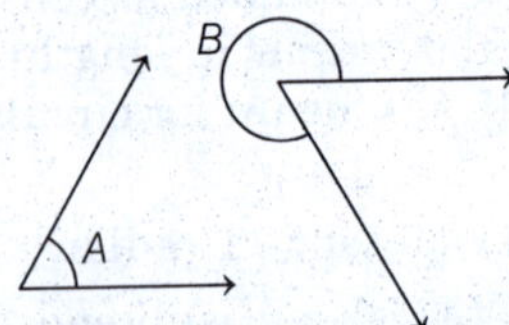

If $\angle A + \angle B = 360°$, then both are said to be Explementary or Conjugate to each other.

(l) **Reference angle** A reference angle is an acute version of any angle determined by either subtracting 180° or its multiple from that angle or subtracting that angle from 180°, as the case may be, so as to obtain the result between 0° and 90° (say acute).

If that angle is already an acute, its reference angle will be same as its value. e.g.

(i) Reference angle of $70° = 70°$

(ii) Reference angle of $160° = 180° - 160° = 20°$

(iii) Reference angle of $570° = 570° - (3 \times 180°) = 30°$

(m) **Adjacent angle** Two angles are adjacent if and only if they have common arm between them.

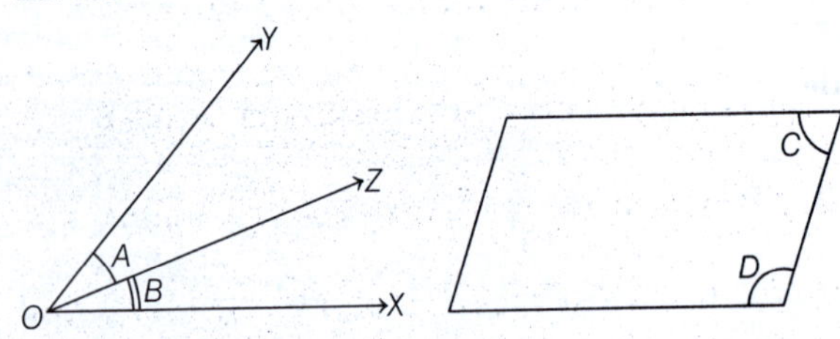

Here, $\angle A$ and $\angle B$ are adjacent angles. Also, $\angle C$ and $\angle D$ are adjacent angles.

(n) **Linear pair of angles**

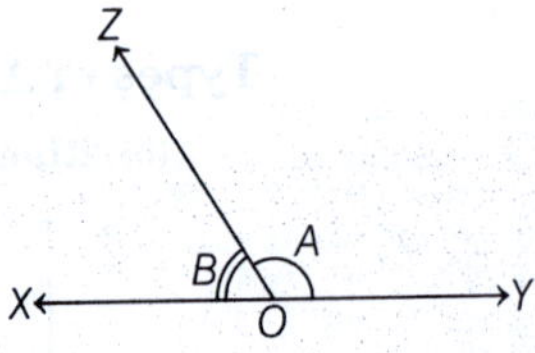

Here, $\angle A$ and $\angle B$ are called linear pair of angles as

(i) they have common side (i.e. OZ)

(ii) they are supplementary to each other. Hence, having common side and to be supplementary are two conditions for linear pair of angles.

(o) **Centre angle**

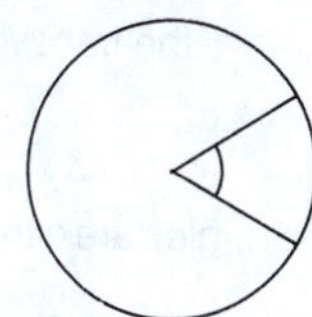

A centre angle is an angle whose vertex is the centre of a circle and whose sides pass through a pair of points on the circle. Centre angle is also known as Arc's angular distance.

(p) **Inscribed angle**

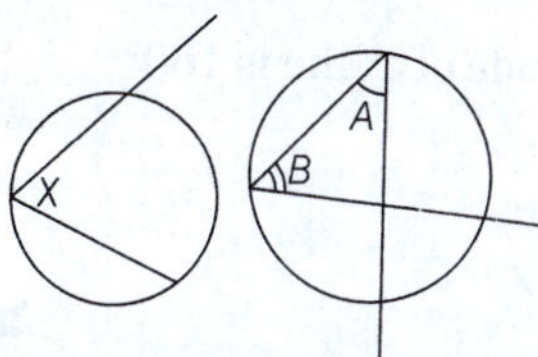

Inscribed angle is formed by two chords sharing an endpoint. [Relation between Inscribed and Centre angle will be discussed in detail in chapter 'Circle']

(q) **Dihedral or Torsion angle**

The angle between two planes is called dihedral or torsion angle

(r) **Vertically opposite angles**

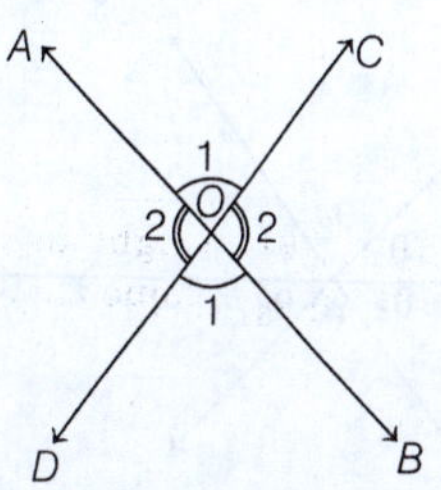

Vertically opposite angles formed when two lines (say AB and CD) intersect each other. Here, $\angle AOC$ and $\angle BOD$ and $\angle AOD$ and $\angle BOC$ are said to be vertically opposite angles. Also, they are equal in size. Hence, $\angle AOC = \angle BOD$ and $\angle AOD = \angle BOC$

Formation of pairs of angles when a transversal intersects two lines

Here, $\angle 1, \angle 2, \angle 7$ and $\angle 8$ are exterior angles whereas $\angle 3, \angle 4, \angle 5$ and $\angle 6$ are interior angles.

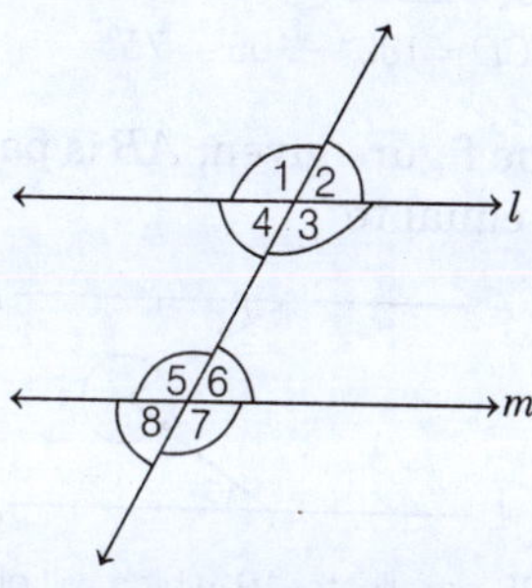

(a) **Pairs of Corresponding angles**

(i) $\angle 1$ and $\angle 5$ (ii) $\angle 2$ and $\angle 6$
(iii) $\angle 4$ and $\angle 8$ (iv) $\angle 3$ and $\angle 7$

(b) **Pairs of Alternate Interior angles**

(i) $\angle 4$ and $\angle 6$ (ii) $\angle 3$ and $\angle 5$

(c) **Pairs of Alternate Exterior angles**

(i) $\angle 1$ and $\angle 7$
(ii) $\angle 2$ and $\angle 8$

(d) **Pairs of Interior angles on the same side of the transversal**

(i) $\angle 4$ and $\angle 5$ (ii) $\angle 3$ and $\angle 6$

[These pairs are also known as Consecutive Interior angles or allied angles or co-interior angles].

(e) **Pairs of Vertically opposite angles**

(i) $\angle 1$ and $\angle 3$ (ii) $\angle 2$ and $\angle 4$
(iii) $\angle 5$ and $\angle 7$ (iv) $\angle 6$ and $\angle 8$

When a transversal intersects two parallel lines (assuming l and m be parallel in above figure), then

I. **Corresponding angles are equal,** *viz.*
$\angle 1 = \angle 5, \angle 2 = \angle 6, \angle 4 = \angle 8$ and $\angle 3 = \angle 7$

II. **Alternate Interior angles are equal,** *viz.*
$\angle 4 = \angle 6$ and $\angle 3 = \angle 5$

III. **Alternate Exterior angles are equal,** *viz.*
$\angle 1 = \angle 7$ and $\angle 2 = \angle 8$

IV. **Each pair of interior angles on the same side of the transversal is supplementary,** *viz.* $\angle 4 + \angle 5 = 180°$ and $\angle 3 + \angle 6 = 180°$

V. **Vertically opposite angles are equal,** *viz.*
$\angle 1 = \angle 3, \angle 2 = \angle 4, \angle 5 = \angle 7$ and $\angle 6 = \angle 8$

Note *If anyone of clauses from (I) to (IV) is surfaced when a transversal intersects any two lines, then it would be obvious that both the given lines are parallel.*

Critical Facts

(i) The sum of all the angles on one side of a line is 180°.

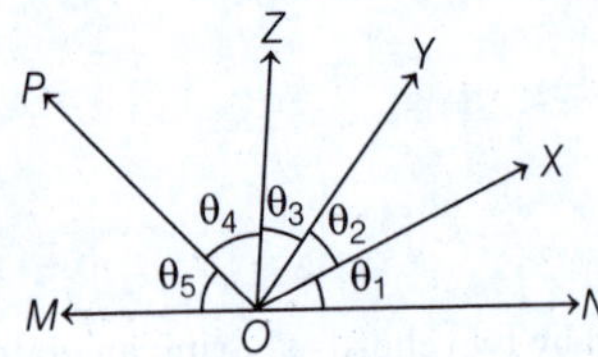

Here, $\theta_1 + \theta_2 + \theta_3 + \theta_4 + \theta_5 = 180°$

(ii) The sum of all the angles around a point is 360°.

Here, $\theta_1 + \theta_2 + \theta_3 + \theta_4 + \theta_5 + \theta_6 = 360°$

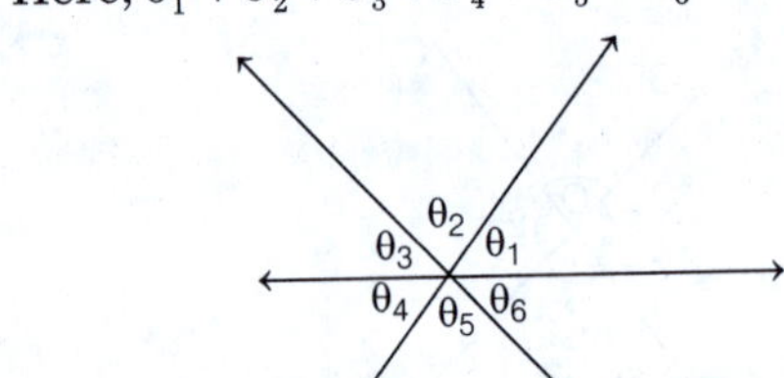

(iii) If the arms of one angle are respectively parallel to the arms of another angle, the two angles are either equal or supplementary.

(iv) If the both arms of one angle are perpendicular to the arms of another angle, the two angles are either equal or supplementary.

(v) If a transversal intersects two lines in such a way that bisectors of a pair of corresponding angles are parallel, then the two lines are parallel.

Problems Based on Lines and Angles

Problem 1 If the arms of one angle are respectively parallel to the arms of another angle, then what is the relation between the two angles? **[CDS-13]**

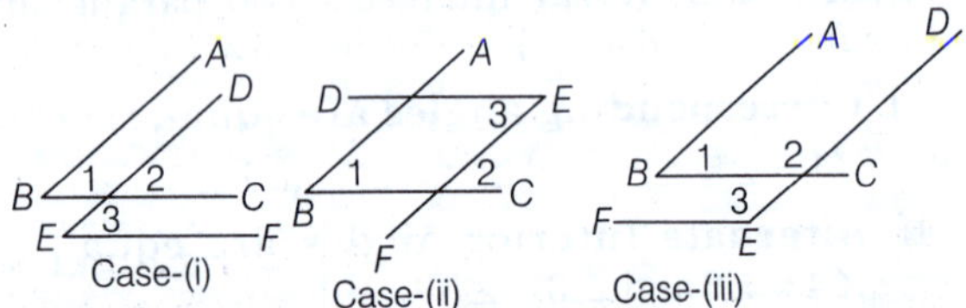

Sol. Case (i) : Clearly, arms of $\angle 1$ are respectively parallel to arms of $\angle 2$.

Then, $\angle 1 = \angle 2$ [∵ they are corresponding angles as AB II DE and BC is transversal]

Further, $\angle 2 = \angle 3$ [∵ they are corresponding angles as $BC \parallel EF$ and DE is transversal]

Hence, $\angle 1 = \angle 3$.

It means both the angles are equal.

Case (ii) : $\angle 1 = \angle 2$ [Corresponding angles as $AB \parallel EF$]

Also, $\angle 2 = \angle 3$ [Alternate Interior angles]

Thus, $\angle 1 = \angle 3$

Case (iii) : $\angle 1 + \angle 2 = 180°$ [∵ Interior angles on the same side of the transversal are supplementary]

Also, $\angle 2 = \angle 3$ [Corresponding angles]

Then, $\angle 1 + \angle 3 = 180°$

From the above, both the angles are either equal or supplement.

Problem 2 If given that $AB \parallel DE$, $\angle ABC = 115°$, $\angle CDE = 140°$, then what is the value of $\angle BCD$?

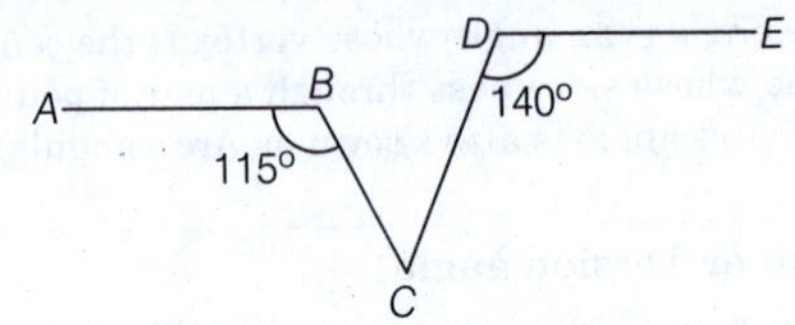

Sol. We draw a line MN through point C parallel to AB and DE.

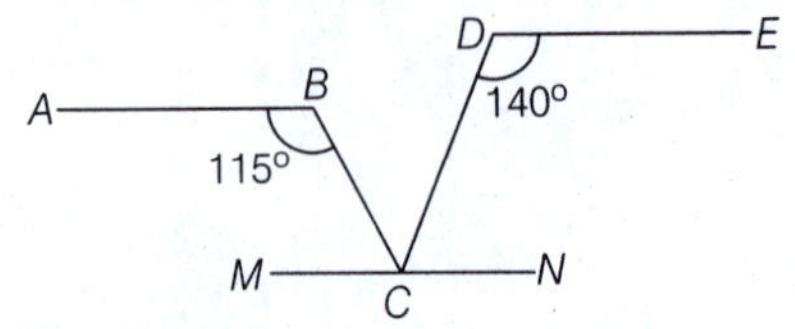

Then, $\angle ABC + \angle MCB = 180°$

[∵ Interior angles on the same side of the transversal are supplementary]

$\Rightarrow \quad \angle MCB = 180° - 115° = 65°$

Also, $\angle EDC + \angle NCD = 180°$

[∵ Interior angles on the same side of the transversal are supplementary]

$\Rightarrow \quad \angle NCD = 180° - 140° = 40°$

Clearly, $\angle MCB + \angle BCD + \angle NCD = 180°$

$\Rightarrow \quad 65° + \angle BCD + 40° = 180°$

$\Rightarrow \quad \angle BCD = 180° - 105° = 75°$

Problem 3 In the figure given, AB is parallel to LM. What is the $\angle a$ equal to **[CDS-10]**

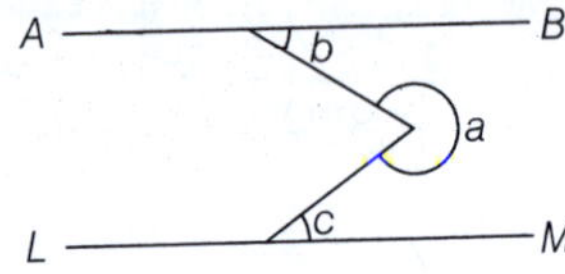

Sol. We draw a line RS parallel to AB, which will obviously be parallel to LM.

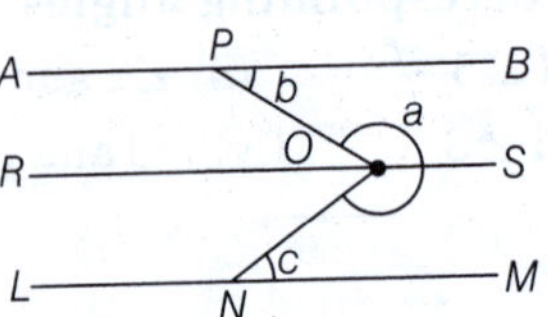

Clearly, $\angle BPO = \angle ROP$ [Alternate interior angles]

$\Rightarrow \quad \angle ROP = b$

Further, $\angle MNO = \angle RON$ [Alternate interior angles]

$\angle RON = c$

Then, $a = 360° - (\angle ROP + \angle RON)$

$= 2\pi - b - c$

Problem 4 In the figure given, $AC \parallel BD$ and $AE \parallel BF$. What is $\angle x$? **[CDS-07]**

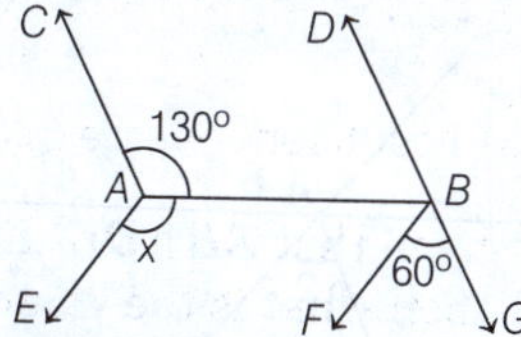

Sol. Clearly, $\angle CAB + \angle DBA = 180°$

[∵ Interior angles on the same side of the transversal are supplementary]

$\Rightarrow \quad \angle DBA = 180° - 130° = 50°$

Also, $x + \angle ABF = 180°$ [∵ Interior angles on the same side of the transversal are supplementary]

$\Rightarrow \quad x = 180° - \angle ABF$

$\Rightarrow \quad x = 180° - (180° - 60° - \angle DBA)$

$\Rightarrow \quad x = 180° - (180° - 60° - 50°) = 110°$

Problem 5 A wheel makes 12 revolutions per minute. The angle in radian described by a spoke of the wheel in 1 second is : **[CDS-12]**

Sol. Distance covered in 1 minute = 12 × circumference of the wheel

∴ Distance covered in one second

$= \frac{12 \times \text{circumference of the wheel}}{60}$

$= \frac{\text{circumference of the wheel}}{5}$

We are aware that $2\pi r$ (i.e. circumference) subtends 2π radian.

Then, $= \frac{\text{circumference of the wheel}}{5}$ will subtend $\frac{2\pi}{5}$ radian.

Problem 6 In the given figure, PQ is parallel to RS. What is $\angle NMS$ equal to?

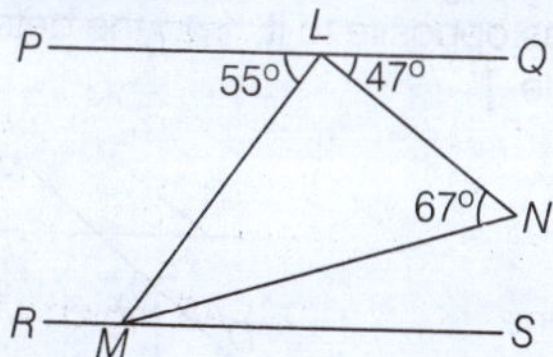

Sol. We draw a line AB parallel to PQ, which passes through point N. AB will obviously be parallel to RS.

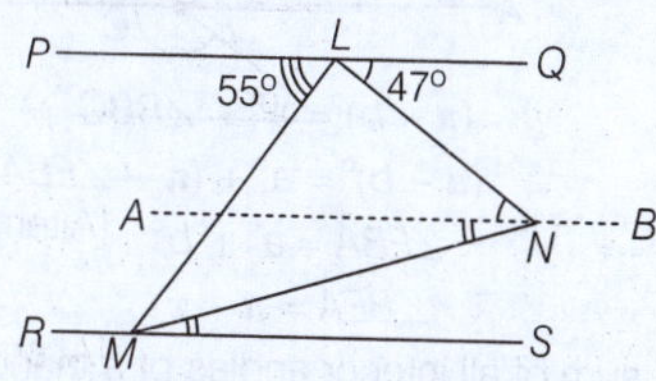

∴ $\quad \angle QLN = \angle LNA$ [Alternate interior angles]

$\Rightarrow \quad \angle LNA = 47°$

Also, $\quad \angle ANM = \angle NMS$ [Alternate interior angles]

$\Rightarrow \quad \angle LNM - \angle LNA = \angle NMS$

$\Rightarrow \quad 67° - 47° = \angle NMS$

∴ $\quad \angle NMS = 20°$

Remarks: For sake of convenience, you may indicate the angles in terms of numbers such as $\angle 1, \angle 2, \angle 3, \angle 4$ .... in lieu of $\angle QLN, \angle LNA, \angle NMS$,.....

Problem 7 In the given figure, LOM is a straight line. What is the value of x?

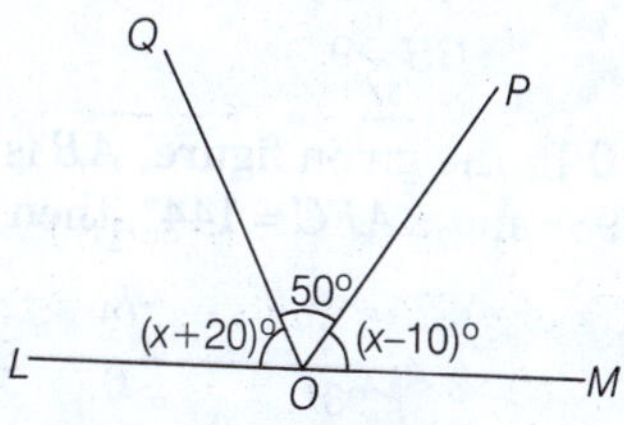

Sol. Clearly, $(x + 20)° + 50° + (x - 10)° = 180°$

[∵ sum of all the angles on one side of a line is 180°])

$\Rightarrow \quad 2x = 120°$

∴ $\quad x = 60°$

Problem 8 The line segments AB and CD intersect at O. OF is the internal bisector of obtuse $\angle BOC$ and OE is the internal bisector of acute $\angle AOC$. If $\angle BOC = 130°$, what is the measure of $\angle FOE$?

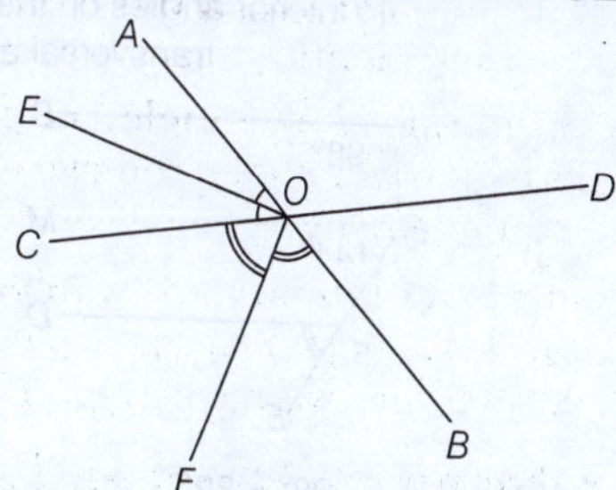

Sol. Clearly, $\quad \angle BOC + \angle AOC = 180°$

$\Rightarrow \quad \angle AOC = 180° - 130° = 50°$

Now, $\quad \angle FOE = \angle FOC + \angle EOC$

$\Rightarrow \quad \angle FOE = \frac{1}{2} \times \angle BOC + \frac{1}{2} \times \angle AOC$

$\Rightarrow \quad \angle FOE = \frac{1}{2} \times 130° + \frac{1}{2} \times 50° = 90°$

Problem 9 Two transversals S and T cut a set of distinct parallel lines. S cuts the parallel lines in points A, B, C, D and T cuts the parallel line in points E, F, G and H respectively. If $AB = 4$, $CD = 3$ and $EF = 12$, then what is the length of GH?

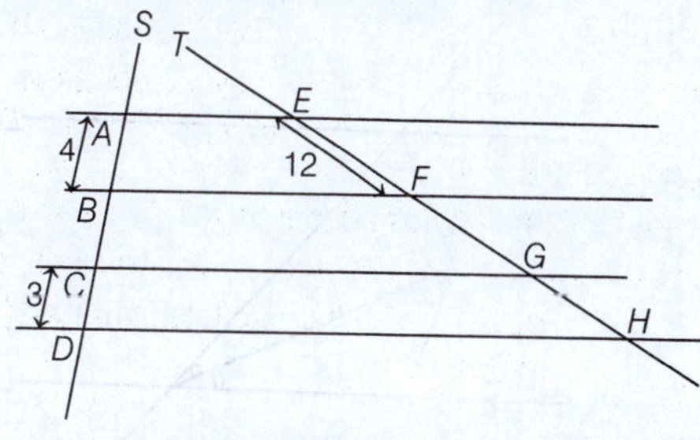

Sol. We are aware that in case two or more parallel lines are intercepted by other transversal lines, the ratio of corresponding intercepts thus formed, are equal.

Hence, $\frac{AB}{CD} = \frac{EF}{GH}$

$\Rightarrow \quad \frac{4}{3} = \frac{12}{GH}$

$\Rightarrow \quad GH = 9$

Problem 10 In the given figure, AB is parallel to CD. If $\angle BAF = 98°$ and $\angle AFC = 144°$, then what is $\angle ECD$ equal to? **[CDS-11]**

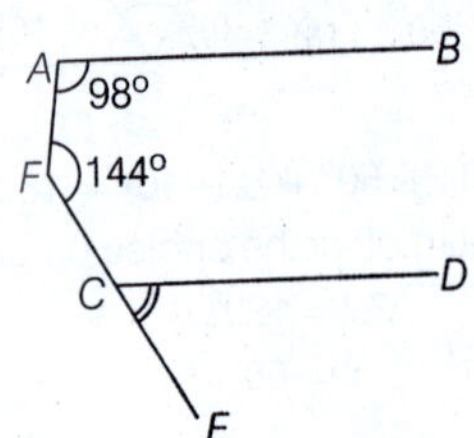

Sol. We draw a line FM parallel to AB, which will obviously be parallel to CD.

$\therefore \quad \angle BAF + \angle AFM = 180°$

[∵ Interior angles on the same side of the transversal are supplementary]

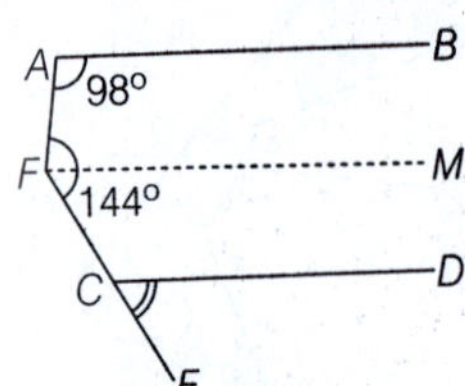

$\Rightarrow \quad \angle AFM = 180° - 98° = 82°$

Then, $\quad \angle CFM = 144° - 82° = 62°$

Now, $\quad \angle CFM = \angle ECD$ (Corresponding angles)

$\Rightarrow \quad \angle ECD = 62°$

Problem 11 The ratio of two complementary angles is 1 : 5, what is the difference between the two angles? **[CDS-12]**

Sol. We know, if $\angle A + \angle B = 90°$, then both are said to be complementary to each other.

Hence, $(1 + 5)$ units $= 90°$

Then, $(5 - 1)$ units $= 60°$

Problem 12 In the given figure, $L_1 \parallel L_2$, then $x = ?$

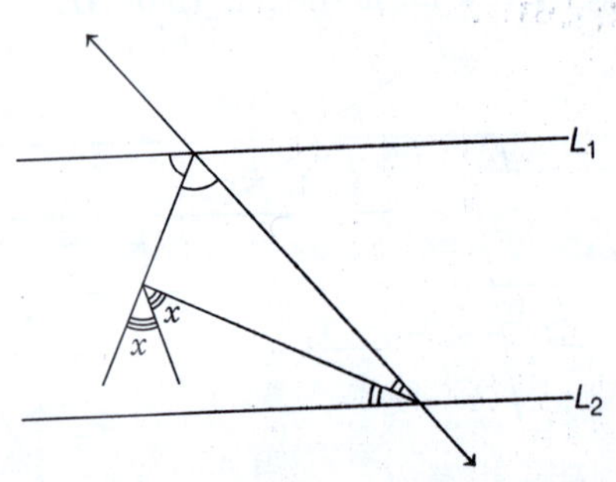

Sol. We have, $\angle 1 + \angle 2 + \angle 3 + \angle 4 = 180°$

[∵ each pair of interior angles on the same side of the transversal is supplementary.]

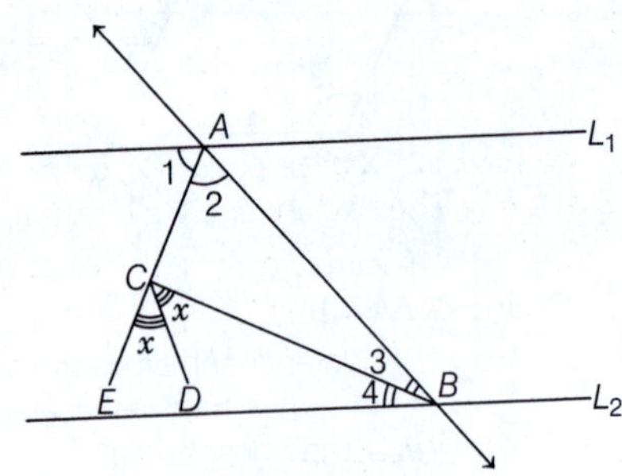

$\Rightarrow \quad \angle 2 + \angle 2 + \angle 3 + \angle 3 = 180°$

$\Rightarrow \quad \angle 2 + \angle 3 = 90°$

Further, $\quad x + x = \angle 2 + \angle 3$

[∵ the exterior angle of a triangle is equal to the sum of interior angles opposite to it]

$\Rightarrow \quad 2x = 90°$

$\Rightarrow \quad x = 45°$

Problem 13 The angles $x°$, $a°$, $c°$ and $(\pi - b)°$ are indicated in the figure given. What is value of x in terms of a, b and c? **[CDS-12]**

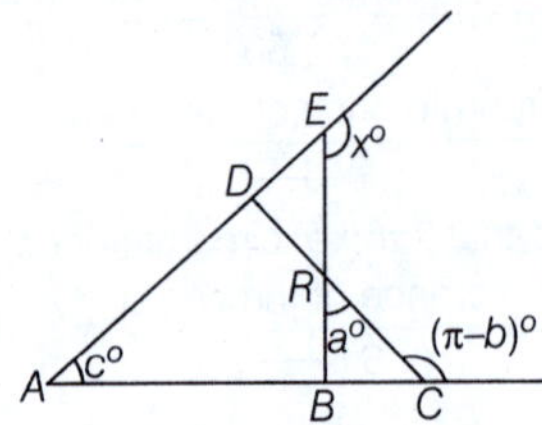

Sol. Clearly, $\angle RCN = \angle BRC + \angle RBC$

[∵ the exterior angle of a triangle is equal to the sum of two interior angles opposite to it, ... the details will be given in topic 'Triangle'.]

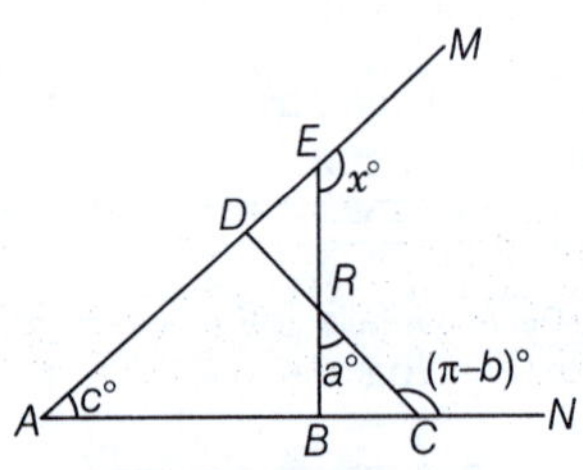

$\Rightarrow \quad (\pi - b)° = a° + \angle RBC$

$\Rightarrow \quad (\pi - b)° = a° + (\pi - \angle EBA)$

$\Rightarrow \quad \angle EBA = a° + b°$

Also, $\quad \angle BEA = \pi - x°$

We know, sum of all interior angles of a triangle is 180°.

Then, $\quad c° + \angle BEA + \angle EBA = 180°$

$\Rightarrow \quad c° + \pi - x° + a° + b° = \pi$

$\Rightarrow \quad x = a + b + c$

WORKOUT

1. In the given figure, AB is parallel to CD. If $\angle DCE = x$ and $\angle ABE = y$, then what is $\angle CEB$ equal to ? **[CDS-12]**

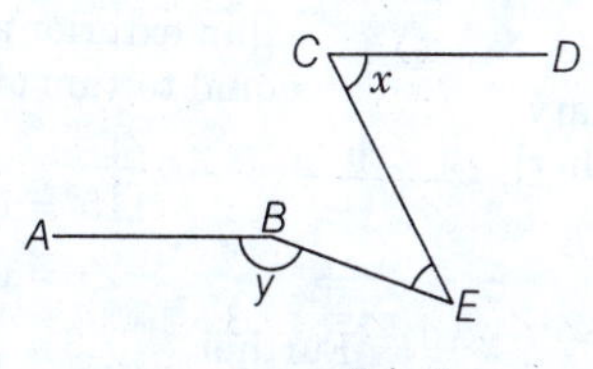

(a) $y - x$ (b) $\frac{x + y}{2}$

(c) $x + y - \frac{\pi}{2}$ (d) $x + y - \pi$

2. What is the angle which equals one-fifth of its supplement ? **[CDS-07]**

(a) 15° (b) 30° (c) 36° (d) 75°

3. What is the reference angle of 760° ?

(a) 40° (b) 90° (c) 140° (d) 100°

4. In the given figure, AB is parallel to CD. What is the value of $\angle XOY$? **[CDS-10]**

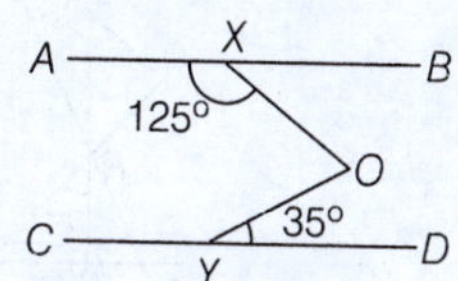

(a) 80° (b) 90° (c) 95° (d) 100°

5. The complement of an angle is equal to supplement of another angle. What is the value of the greater angle if both the angles are supplementary.

(a) 45° (b) 90° (c) 135° (d) 200°

6. What is the vlaue of x in the figure given? **[CDS-07]**

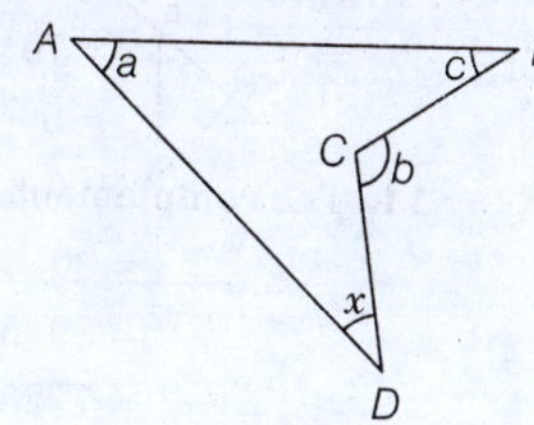

(a) $b - a - c$ (b) $b - a + c$

(c) $b + a - c$ (d) $\pi - (a + b + c)$

7. The angle which is 50° greater than its conjugate is ?

(a) 80° (b) 45°

(c) 120° (d) 205°

8. In the given figure, AB is parallel to CD, $\angle ABC = 65°$, $\angle CDE = 15°$ and $AB = AE$.
What is the value of $\angle AEF$? **[CDS-12]**

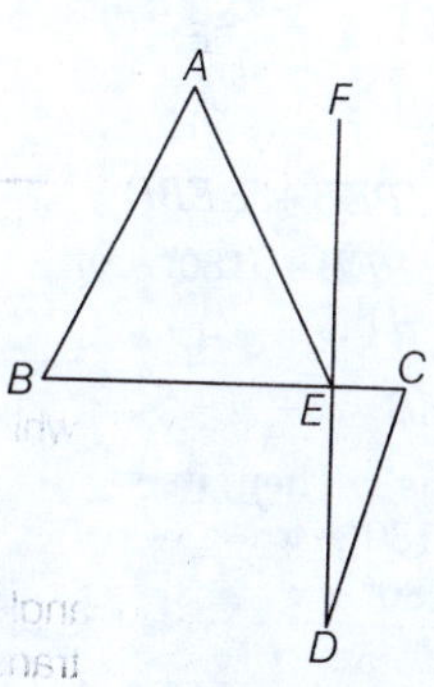

(a) 30° (b) 40°

(c) 35° (d) 45°

9. Two lines perpendicular to the same line are .... to each other.

(a) Perpendicular (b) Parallel

(c) Equal (d) Cannot say

10. In the given figure, $EC \mid\mid AB$, $\angle ECD = 70°$, $\angle BDO = 20°$ and $AB = AE$.
What is the value of $\angle OBD$? **[CDS-12]**

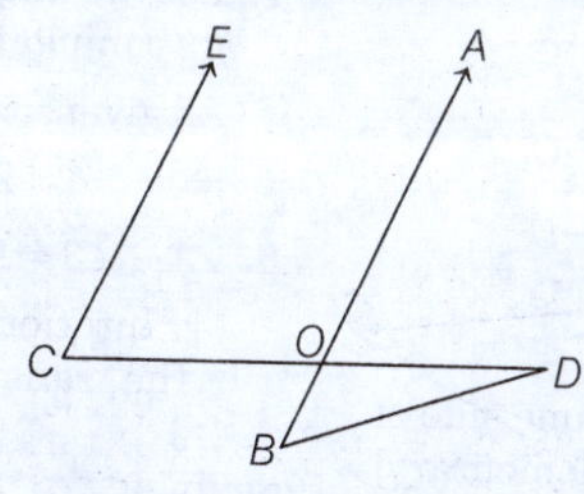

(a) 30° (b) 40°

(c) 20° (d) 50°

11. The complement of 82°36′ 44″ is :

(a) 8° 23′16″ (b) 7° 23′ 16″

(c) 7° 43′ 16″ (d) 7° 23′ 44″

Answer Key

1. (d)	2. (b)	3. (a)	4. (b)	5. (c)	6. (a)	7. (d)	8. (c)	9. (b)	10. (d)
11. (b)									

Hints and SOLUTIONS

1. We produce line AB to AR.

Then, $\angle EPR = \angle ECD$ [Corresponding angles]

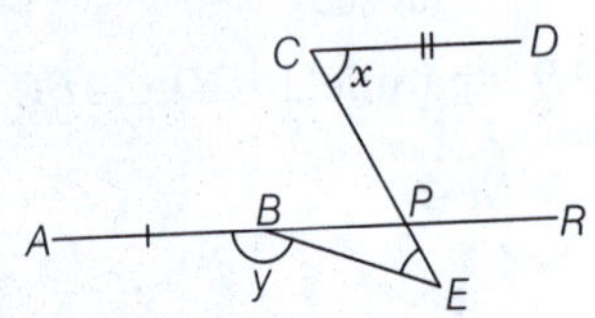

$\Rightarrow \quad \angle EPR = x$

Also, $\angle EPR = \angle PEB + \angle EBP$

$\Rightarrow \quad x = \angle PEB + (180^\circ - y)$

$\Rightarrow \quad \angle PEB = x + y - \pi$

$\Rightarrow \quad \angle CEB = x + y - \pi$

2. Let the angle be x. Then, its supplement $= 180^\circ - x$

As given, $x = \dfrac{180^\circ - x}{5}$

$\Rightarrow \quad 5x = 180^\circ - x$

$\Rightarrow \quad 6x = 180^\circ$

$\Rightarrow \quad x = 30^\circ$

Note *You may proceed through options.*

3. Reference angle of

$760^\circ = 760^\circ - (4 \times 180^\circ) = 40^\circ$

4. We draw a line PQ parallel to AB, which will obviously be parallel to CD.

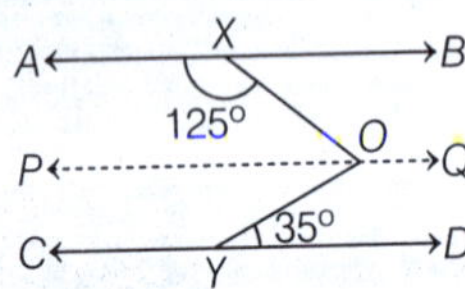

$\therefore \quad \angle AXO + \angle XOP = 180^\circ$

[$\because$ Interior angles on the same side of the transversal are supplementary]

$\Rightarrow \quad \angle XOP = 180^\circ - 125^\circ = 55^\circ$

Further, $\angle POY = \angle OYD$ [Alternate interior angles]

$\Rightarrow \quad \angle POY = 35^\circ$

Clearly, $\angle XOY = \angle XOP + \angle POY$

$= 55^\circ + 35^\circ = 90^\circ$

5. Let the angles be x and y.

According to the question,

$90^\circ - x = 180^\circ - y$

$\Rightarrow \quad 90^\circ - x = (x + y) - y$

[$\because x$ and y are supplementary to each other]

$\Rightarrow \quad 90^\circ = 2x$

$\Rightarrow \quad x = 45^\circ$

Then, $\quad y = 180^\circ - 45^\circ = 135^\circ$

Hence, the value of greater angle $= 135^\circ$

6. We join points A and C and also produce line AC to AM. Since an exterior angle of a triangle is equal to sum of two interior angles opposite to it, $\angle 4 = \angle 2 + c$ and $\angle 3 = \angle 1 + x$

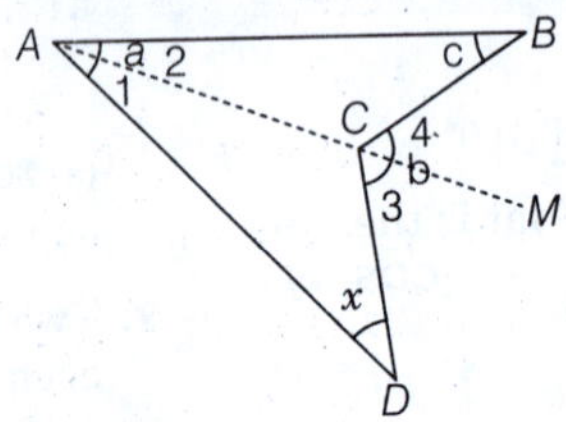

On addition,

$\angle 4 + \angle 3 = \angle 2 + c + \angle 1 + x$

$\Rightarrow \quad b = a + c + x$

$\Rightarrow \quad x = b - a - c$

7. Let the angle be x. Then, its conjugate $= 360^\circ - x$.

As given, $\quad x = (360^\circ - x) + 50^\circ$

$\Rightarrow \quad 2x = 410^\circ \Rightarrow x = 205^\circ$

8. $\angle 1 + \angle 3 = 180^\circ$

[$\because$ Interior angles on the same side of the transversal are supplementary]

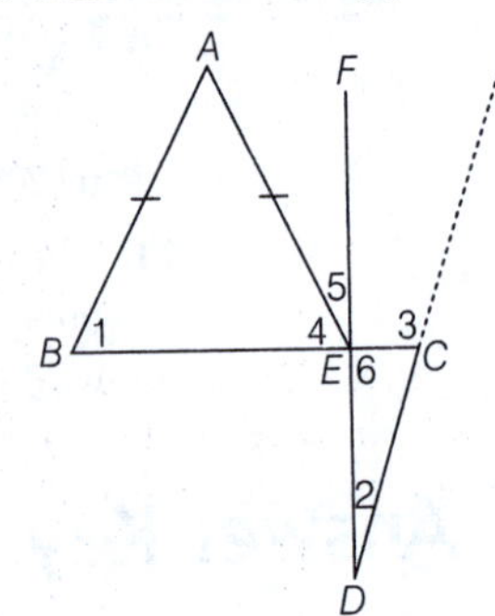

$\Rightarrow \quad \angle 3 = 180^\circ - 65^\circ$

$= 115^\circ$

Also, $\angle 3 = \angle 2 + \angle 6$

[an exterior angle of a triangle is equal to sum of two interior angles opposite to it]

$\Rightarrow \quad 115^\circ = 15^\circ + \angle 6$

$\Rightarrow \quad \angle 6 = 100^\circ$

Further, $\quad \angle 6 = \angle 4 + \angle 5$ [Vertically opposite angles]

$\Rightarrow \quad 100^\circ = \angle 1 + \angle 5$

[$\angle 4 = \angle 1$, as $AB = AE$, details in topic 'Triangle']

$\Rightarrow \quad 100^\circ = 65^\circ + \angle 5$

$\Rightarrow \quad \angle 5 = 35^\circ$

9. Two lines perpendicular to the same line are parallel to each other.

10. Clearly, $\angle AOD = \angle ECO$ [Corresponding angles]

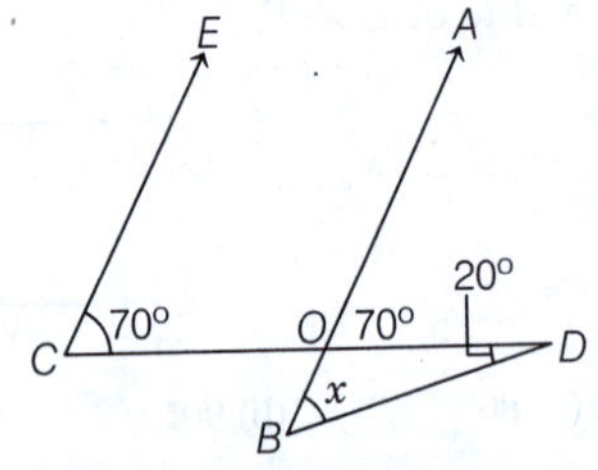

$\Rightarrow \quad \angle AOD = 70^\circ$

Also, $\angle AOD = x + 20^\circ$

[an exterior angle of a triangle is equal to sum of two interior angles opposite to it]

$\Rightarrow \quad x = 70^\circ - 20^\circ$

$= 50^\circ$

11. The complement of $82^\circ 36' 44''$ is

$= 90^\circ - 82^\circ 36' 44''$

$= 89^\circ 59' 60'' - 82^\circ 36' 44''$

$= 7^\circ 23' 16''$

56

Polygons

Learning OBJECTIVES

After completing this chapter, aspirants will be able to :

- Understand the significance of this sub-section in diverse competitive exams;
- Comprehend the meaning of polygon and its different types with their properties;
- Compute number of certain types of polygons in diverse cases;
- Apply the relevant formulae according to different situations;
- Realise themselves competent to solve the questions in unique way;
- Analyse the trend of the questions being asked in various competitive examinations;
- Recognise the types of question which can be easily solved through the given options;
- Solve the questions in moments;
- Learn many new concepts/fundamentals;
- Apply the pragmatic approaches in lieu of academic approaches; and
- Ponder on the alternative methods.

Inside

- *Introduction*
- *Various Types of Polygons with Their Properties*
- *Computation of Number of Certain Types of Polygons in Diverse Cases*
- *Critical Formulae*
- *Problems Based on Polygons*
- ....
- .... *& Many More*
- *Work Out with Solutions*

Introduction

Polygons are two-dimensional plane figures formed by a closed series of straight lines.

e.g. Triangles (Δ), Quadrilaterals (□), Hexagons, etc.

Types of polygons

(a) **Convex and Concave**

When all the diagonals of a polygon lie inside it, the polygon is Convex otherwise it is a concave polygon. It implies that all interior angle of Convex polygon are less than 180° and in Concave, atleast one interior angle is more than 180°.

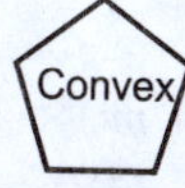

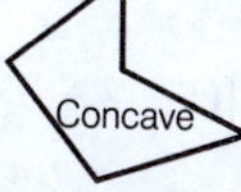

(b) **Regular and Irregular**

Polygon with all the sides equal and all the angles equal are known as Regular polygon, otherwise it is Irregular.

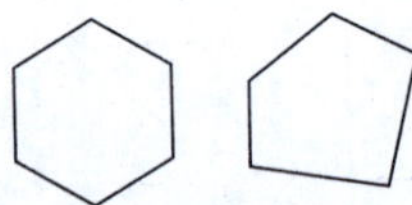

(c) **Equiangular and Equilateral**

Equiangular polygon is a polygon whose vertex angles are equal whereas a polygon with all sides equal is Equilateral polygon. It is important to note that a polygon with 5 or more sides can be equilateral without being convex.

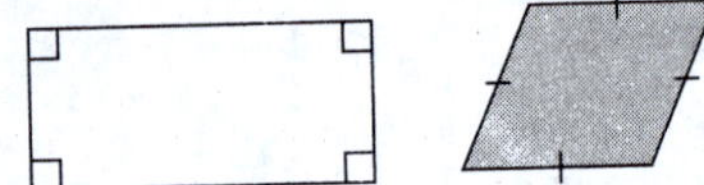

(d) **Tangential/Circumcyclic, Incyclic and Bicyclic**

A convex polygon which has an incircle touching all its sides is called Tangential polygon (also known as Circumcyclic polygon). The reason is that all sides of the polygon are tangent to the inscribed circle. On the other hand, a polygon having a circumcircle is called incyclic polygon. Either of these polygons may be called as Cyclic polygons.

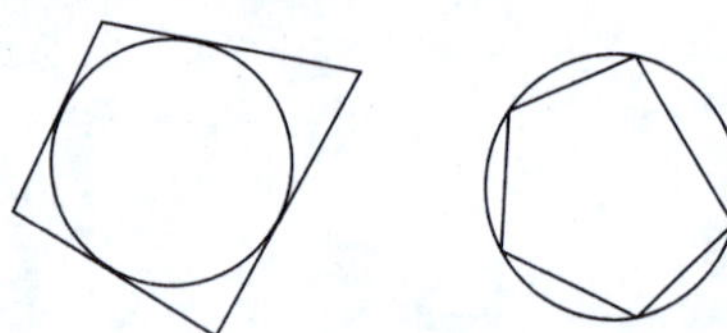

A polygon with incircle and circumcircle is known as Bicyclic polygon. All regular polygons, all triangles are Bicyclic.

Critical facts about Cyclic polygons

(i) All the angles of a cyclic polygon with odd number of sides are equal if the polygon is regular.

(ii) All the angles of a cyclic polygon with even number of sides are equal if the alternate sides are equal.

(e) **Rectilinear/Orthogonal**

A polygon all of whose edges meet at 90° is known as Rectilinear or Orthogonal polygon. It implies that interior angle at each vertex is either 90° or 270°.

In Rectilinear polygon, number of vertical and horizontal sides are equal. It implies that it has an even number of sides. Also, it is the fact that the number of 270° interior angles is 4 less than that of 90° interior angles in a simple orthogonal polygon. It implies that any orthogonal polygon has atleast four 90° interior angles.

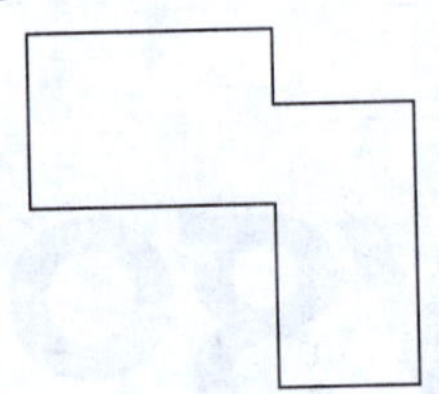

Computation of number of certain types of polygons in diverse cases

Case I In a square ['n' columns and 'n' rows]

(a) Number of squares $= 1^2 + 2^2 + \ldots\ldots + n^2$

$$= \frac{n(n+1)(2n+1)}{6}$$

(b) Number of rectangles

$= (1^3 + 2^3 + \ldots\ldots. + n^3) -$ total number of squares

$= \{1^3 + 2^3 + \ldots\ldots + n^3\} - \{1^2 + 2^2 + \ldots\ldots. + n^2\}$

$$= \left\{\frac{n(n+1)}{2}\right\}^2 - \frac{n(n+1)(2n+1)}{6}$$

e.g. Determine the number of squares and rectangles in the following square-grid.

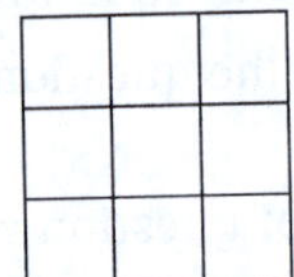

Clearly, number of column

= number of row = $n = 3$

Then, number of squares

$$= \frac{n(n+1)(2n+1)}{6} = \frac{3(3+1)(2\cdot 3+1)}{6} = 14$$

and number of rectangles

$$= \left\{\frac{n(n+1)}{2}\right\}^2 - \frac{n(n+1)(2n+1)}{6}$$

$$= \left\{\frac{3(3+1)}{2}\right\}^2 - 14 = 36 - 14 = 22$$

Case II In a rectangle ['m' column and 'n' rows]

(a) Number of squares

$= mn + (m-1)(n-1) + (m-2)(n-2) + \ldots 0$

(b) Number of rectangles

$= (1 + 2 + 3 + \ldots + m) \times (1 + 2 + 3 + \ldots + n)$

$-$ total number of squares

$= [(1 + 2 + 3 + \ldots. + m) \times (1 + 2 + 3 + \ldots. + n)$

$- mn + (m-1)(n-1) + (m-2)(n-2) + \ldots 0]$

Hence, the given regular polygon is a square, not a rectangle because in a regular polygon, all sides are equal and all angles are equal.

Problem 6 **Each interior angle of a regular polygon is three times its exterior angle, then the number of sides of the regular polygon is :** *[SSC (T-1)-12]*

Sol. Interior angle (I) + Exterior angle (E) = 180°

$\Rightarrow \quad 3E + E = 180° \Rightarrow E = 45°$

Then, number of sides $= \dfrac{360°}{45°} = 8$

Alternatively

We know, each interior angle $= \dfrac{(n-2)\times 180°}{n}$

and each exterior angle $= \dfrac{360°}{n}$

According to the question, $\dfrac{(n-2)\times 180°}{n} = 3 \times \dfrac{360°}{n}$

$\Rightarrow \quad 180° \times n = 1440° \Rightarrow n = 8$

Problem 7 **Measure of each interior angle of a regular polygon can never be :**

Sol. Given, options are 150°, 105°, 108° and 144°.

Proceed through options.

(a) 150° $\Rightarrow$ each exterior angle = 180° − 150° = 30°, then number of sides of the polygon = 360° ÷ 30° = 12

(b) 105° $\Rightarrow$ each exterior angle = 180° − 105° = 75°, then number of sides of the polygon = 360° ÷ 75° = 4.8

But, number of sides cannot be in fraction.

Hence, option (b) is correct.

Problem 8 **A polygon has 170 diagonals. What is the number of sides in that polygon ?**

Sol. Number of diagonals in n-sided polygons $= \dfrac{n(n-3)}{2}$

$\Rightarrow \quad 170 = \dfrac{n(n-3)}{2}$

$\Rightarrow \quad n^2 - 3n - 340 = 0$

$\Rightarrow \quad n^2 - 20n + 17n - 340 = 0$

$\Rightarrow \quad n(n-20) + 17(n-20) = 0$

$\Rightarrow \quad (n-20)(n+17) = 0$

$\Rightarrow \quad n = 20, n = -17$ (not possible)

Alternatively

The best way is to proceed through options.

Problem 9 **The greatest one of the interior angles of a convex polygon is 110°. What is the number of sides in the polygon provided all the interior angles are consecutive integers in term of degree.**

Sol. Given that the greatest interior angle is 110°, it means the smallest exterior angle is 70° as they are supplementary.

Obviously, interior angles are 110°, 109°, 108°, ...

Then, corresponding exterior angles will be 70°, 71°, 72°, ...

But the sum of all the exterior angles cannot exceed to 360°.

Hence, 70° + 71° + 72° + 73° + 74° = 360°

∴ Number of sides = 5

Alternatively

Let number of sides be n. Then,

$70° + (70° + 1°) + (70° + 2°) + \ldots\ldots [70° + (n-1°)] = 360°$

$\Rightarrow \quad 70n$ + sum of first $(n-1)$ natural numbers = 360°

$\Rightarrow \quad 70n + \dfrac{(n-1)(n-1+1)}{2} = 360°$

$\Rightarrow \quad 140n + n^2 - n = 720°$

$\Rightarrow \quad n^2 + 139n - 720 = 0$

$\Rightarrow \quad n^2 + 144n - 5n - 720 = 0$

$\Rightarrow \quad n(n+144) - 5(n+144) = 0$

$\Rightarrow \quad (n-5)(n+144) = 0$

$\Rightarrow \quad n = 5$ and $n = -144$ (not possible)

Hence, number of sides = 5

Problem 10 **Let A and B be two regular polygons having a and b sides, respectively. If $b = 2a$ and each interior angle of B is $\dfrac{3}{2}$ times each interior angle of A, then each interior angle, in degrees, of a regular polygon with $a + b$ sides is :** *[CAT-19]*

Sol. Each interior angle of polygon $A = \dfrac{(a-2)\times 180°}{a}$

and each interior angle of polygon $B = \dfrac{(b-2)\times 180°}{b}$

As given, $\dfrac{(b-2)\times 180°}{b} = \dfrac{3}{2} \times \dfrac{(a-2)\times 180°}{a}$

$\Rightarrow \quad \dfrac{(2a-2)}{2a} = \dfrac{3}{2} \times \dfrac{(a-2)}{a}$

$\Rightarrow \quad 2a - 2 = 3a - 6$

$\Rightarrow \quad a = 4$

Then, $b = 8$

Thus, number of sides in new polygon = 4 + 8 = 12

Then, its interior angle $= \dfrac{(12-2)\times 180°}{12} = 150°$

Problem 11 **All the interior angles of a convex regular polygon are integers in terms of degree. How many such polygons are possible ?**

(a) 24 (b) 20
(c) 22 (d) 16

Sol. Each interior angle $= 180° - \dfrac{360°}{n}$

Each interior angle can be integer only when the value of $\dfrac{360°}{n}$ be integer. No doubt, n, being number of sides of the polygon, must be integer.

Number of integers which can divide 360 exactly

$= (3+1)(2+1)(1+1) = 24$

$[\because 360 = 2^3 \times 3^2 \times 5^1$ – Recall the concept of factor]

It is also fact that 24 includes integers 1 and 2 also. But, no polygon can be formed either with 1 side or with 2 sides.

Hence, (24 − 2) convex regular polygons can be formed satisfying all the interior angles be integers.

WORKOUT

1. There are 10 interior angles of 270° in an orthogonal polygon. What is the number of interior angle with measure of 90° ?
(a) 14 (b) 12
(c) 18 (d) 6

2. Find the number of rectangles in the following figure.

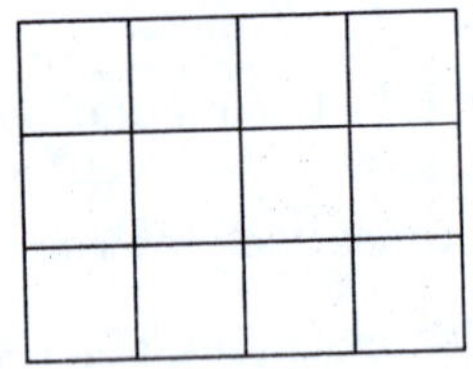

(a) 14 (b) 40
(c) 18 (d) 16

3. The ratio between interior angles of two regular polygons is 9 : 10 and that between number of their sides is 2 : 3. There sides are respectively *[SSC (CGL-T-II)-11]*
(a) 6, 9 (b) 8, 12
(c) 12, 18 (d) 4, 6

4. The interior angle of a regular polygon exceeds its exterior angle by 108°. The number of sides in the polygon is : *[SSC (CGL-T-II)-16]*
(a) 16 (b) 12
(c) 14 (d) 10

5. How many sides a regular polygon has with its interior angle eight times its exterior angle ? *[MAT-01]*
(a) 16 (b) 24 (c) 18 (d) 20

6. The sum of the interior angles of a polygon is 1620°. The number of sides of the polygon are : *[MAT-01, SSC (CGL-T-II)-15]*
(a) 9 (b) 11
(c) 15 (d) 12

7. Find the area of a regular hexagon with side 8 cm.
(a) $96\sqrt{3}$ cm^2 (b) $48\sqrt{3}$ cm^2
(c) $64\sqrt{2}$ cm^2 (d) $96\sqrt{8}$ cm^2

8. The area of a regular octagon is $50(1+\sqrt{2})$ cm^2. Find the length of its side.
(a) 4 cm (b) 6 cm (c) 8 cm (d) 5 cm

9. If a regular polygon has each of its angles equal to $\frac{3}{5}$ times of two right angles, then number of sides is : *[SSC (CGL-T-II)-12]*
(a) 3 (b) 5 (c) 6 (d) 8

10. Each interior angle of a regular octagon in radians is *[SSC (CGL-T-II)-13]*
(a) $\frac{\pi}{4}$ (b) $\frac{3\pi}{4}$ (c) $\frac{2\pi}{3}$ (d) $\frac{\pi}{3}$

11. ΔABC and ΔXYZ are equilateral triangles of 54 cm sides. All smaller triangles like ΔANM, ΔOCP, ΔQPX etc. are also equilateral triangles. Find the area of the shape $MNOPQRM$. *[XAT-16]*

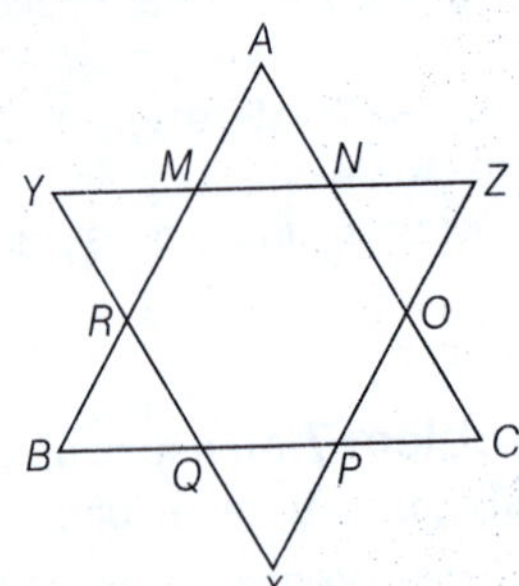

(a) $243\sqrt{3}$ cm^2
(b) $486\sqrt{3}$ cm^2
(c) $729\sqrt{3}$ cm^2
(d) None of these

12. If the area of a regular hexagon is equal to the area of an equilateral triangle of side 12 cm, then the length, (in cm), of each side of the hexagon is *[CAT-21]*
(a) $2\sqrt{6}$
(b) $4\sqrt{6}$
(c) $\sqrt{6}$
(d) $6\sqrt{6}$

Answer Key

1. (a)	2. (b)	3. (b)	4. (d)	5. (c)	6. (b)	7. (a)	8. (d)	9. (b)	10. (b)
11. (b)	12. (a)								

Hints and SOLUTIONS

1. We know, interior angle at each vertex of an orthogonal polygon is either 90° or 270°.

Also, sum of all interior angles

$= (n-2)\times 180°$

$\Rightarrow \quad (10\times 270°) + (x \times 90°)$

$= \quad (10 + x - 2)\times 180°$

[let number of 90° be x]

$\Rightarrow \quad 30 + x = 16 + 2x \Rightarrow x = 14$

Alternatively

We know that the number of 270° interior angles is 4 less than that of 90° interior angles in a simple orthogonal polygon.

Hence, required number of 90° interior angles = 10 + 4 = 14

2. Here, number of column = $m = 4$ and number of row = $n = 3$

Then, number of rectangles

$= (1+2+3+\dots+m)\times(1+2+3+\dots+n) - [mn + (m-1)(n-1) + (m-2)(n-2) + \dots 0)]$

$= (1+2+3+4)\times(1+2+3) - (4\cdot 3 + 3\cdot 2 + 2\cdot 1 + 1\cdot 0\}$

$= 60 - 20 = 40$

3. Let sides be $2n$ and $3n$.

According to the question,

$$\frac{\dfrac{(2n-2)\times 180°}{2n}}{\dfrac{(3n-2)\times 180°}{3n}} = \frac{9}{10}$$

$\left[\because \text{each interior angle} = \dfrac{(n-2)\times 180°}{n}\right]$

$\Rightarrow \quad 9n - 6 = 10n - 10$

$\Rightarrow \quad n = 4$

Hence, number of sides are (2×4) and (3×4) i.e. 8 and 12.

Alternatively

Proceed through options,

(a) 6, 9

$\Rightarrow \dfrac{(6-2)\times 180°}{6} : \dfrac{(9-2)\times 180°}{9} \neq 9:10$

(b) 8, 12

$\Rightarrow \dfrac{(8-2)\times 180°}{8} : \dfrac{(12-2)\times 180°}{12}$

$= 9:10$

Hence, option (b) is correct.

4. Interior angle (I) + Exterior angle (E) = 180°

$\Rightarrow (E + 108°) + E = 180°$

$\Rightarrow \quad E = 36°$

Then, number of sides $= \dfrac{360°}{36°} = 10$

5. Interior angle (I) + Exterior angle (E) = 180°

$\Rightarrow \quad 8E + E = 180° \Rightarrow E = 20°$

Then, number of sides $= \dfrac{360°}{20°} = 18$

6. Sum of interior angles $= (n-2)\times 180°$

$\Rightarrow \quad 1620° = (n-2)\times 180°$

$\Rightarrow \quad n\ 2 = 9$

$\Rightarrow \quad n = 11$

7. Area $= \dfrac{na^2}{4}\times \cot\left(\dfrac{180°}{n}\right)$

$= \dfrac{6\times 8^2}{4}\times \cot\left(\dfrac{180°}{6}\right)$

$= \dfrac{6\times 8^2}{4}\times \cot 30°$

$= 96\sqrt{3}\ \text{cm}^2$

Alternatively

Area = 6 × area of an equilateral triangle

$= 6\times \dfrac{\sqrt{3}\times 8^2}{4} = 96\sqrt{3}\ \text{cm}^2$

8. Area of regular octagon $= 2a^2(1+\sqrt{2})$

$\Rightarrow \quad 50(1+\sqrt{2}) = 2a^2(1+\sqrt{2})$

$\Rightarrow \quad a = 5\ \text{cm}$

9. Each interior angle of the polygon

$= \dfrac{(n-2)\times 180°}{n}$

$\Rightarrow \quad \dfrac{3}{5}\times(2\times 90°) = \dfrac{(n-2)\times 180°}{n}$

$\Rightarrow \quad 3n = 5n - 10 \Rightarrow n = 5$

10. Each interior angle of a regular octagon

$= \dfrac{(n-2)\times 180°}{n} = \dfrac{(8-2)\times \pi}{8}$

$[\because \pi \text{ radian} = 180°]$

$= \dfrac{3\pi}{4}$ radian

11. Obviously, *MNOPQRM* is a regular hexagon with side $\dfrac{54}{3}$ (i.e. 18) cm.

Then, its area = 6 × area of an equilateral triangle

$= 6\times \dfrac{\sqrt{3}\times 18^2}{4}$

$= 486\sqrt{3}\ \text{cm}^2$

12. Let the side of hexagon be a.

Then, its area $= 6\times \dfrac{\sqrt{3}}{4}\times a^2$

As given, $6\times \dfrac{\sqrt{3}}{4}\times a^2 = \dfrac{\sqrt{3}}{4}\times (12)^2$

$\Rightarrow \quad a^2 = 24 \Rightarrow a = 2\sqrt{6}$

57

Triangle

Learning OBJECTIVES

After completing this chapter, aspirants will be able to :

- Understand the significance of this sub-section in diverse competitive exams;
- Comprehend the definition of triangle and its different types with their properties;
- Be familiar with critical concepts/theorems;
- Understand diverse terms related to triangles and relation among them;
- Understand the conditions for congruency/similarity of triangles and their applications;
- Apply the relevant concepts according to different situations;
- Realise themselves competent to solve the questions in unique way;
- Analyse the trend of the questions being asked in various competitive examinations;
- Solve the questions in moments;
- Learn many new concepts/fundamentals;
- Apply the pragmatic approaches in lieu of academic approaches; and
- Ponder on the alternative methods.

Inside

Introduction

A triangle is a polygon having three sides. There are three interior angles, $\angle A$, $\angle B$ and $\angle C$ and three sides AB (i.e. c), AC (i.e. b) and BC (i.e. a).

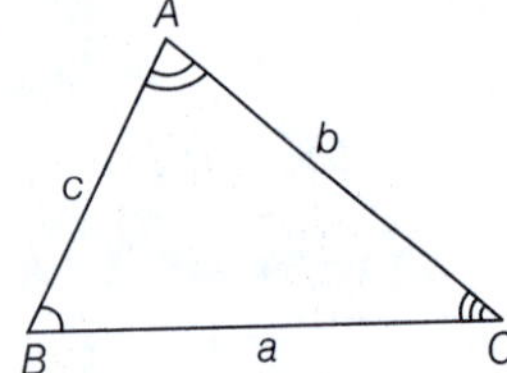

Types of Triangles

- **Based on interior angles**

  (i) **Acute angle triangle** Triangle with all three angles acute $(0° < \theta < 90°)$.

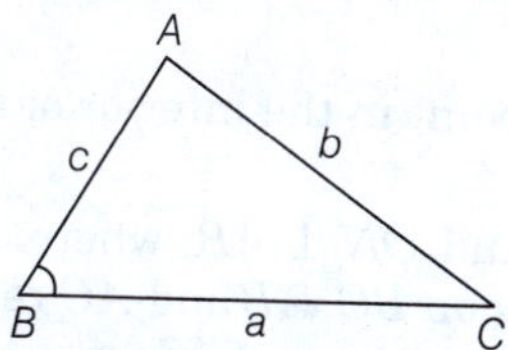

Here, $b^2 < a^2 + c^2$, if $\angle B$ is the greatest among all three.

(ii) **Right angled triangle** Triangle with one of the angles equal to 90°.

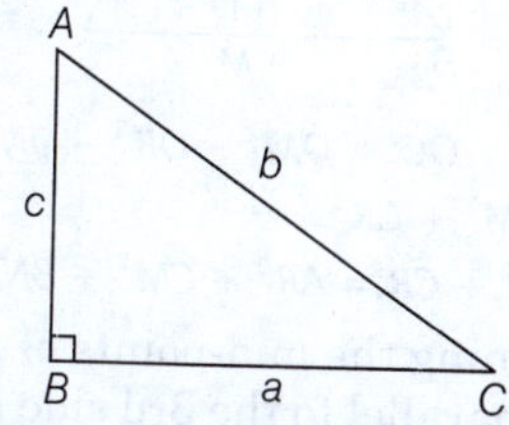

Here, $b^2 = a^2 + c^2$, if $\angle B$ is the greatest among all three.

(iii) **Obtuse angle triangle** Triangle with one of angles obtuse ($90° < \theta < 180°$).

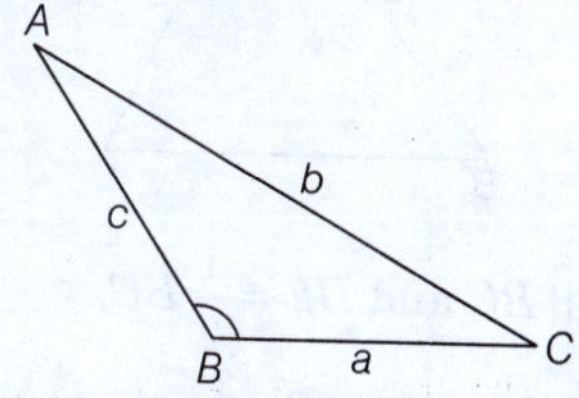

Here, $b^2 > a^2 + c^2$, if $\angle B$ is the greatest among all three.

- **Based on length of sides**

(i) **Equilateral triangle** Triangle with all sides equal.

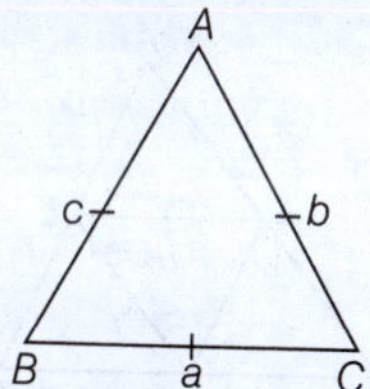

Here, $AB = BC = AC$ i.e. $c = a = b$

(ii) **Isosceles triangle** Triangle with two of its sides equal.

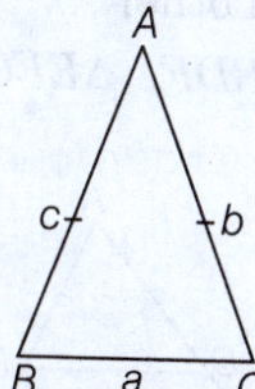

Here, $AB = AC$ i.e. $c = b$

(iii) **Scalene triangle** Triangle with none of the sides equal to any other side.

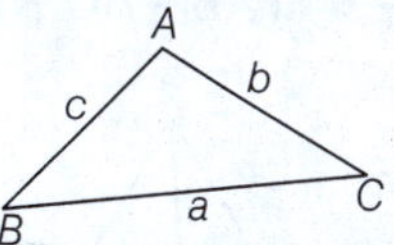

Here, $AB \neq BC \neq AC$ i.e. $c \neq a \neq b$

Fundamental properties of triangles

(i) Sum of the lengths of any two sides of a triangle is always greater than length of the 3rd side.

(ii) Difference between lengths of any two sides of a triangle is always less than length of the 3rd side.

(iii) Greater angle will have greater side opposite to it and smaller angle will have smaller side opposite to it.

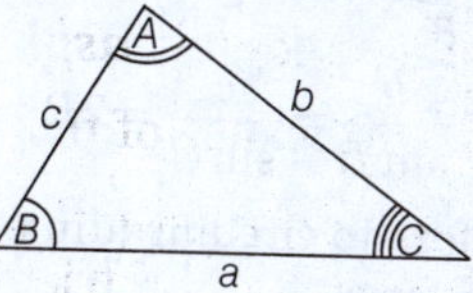

Here, $\angle B > \angle C$

$\therefore \quad b > c$

Also an angle opposite to the longest side is the greatest, which is greater than 60°.

(iv) There must be two acute angles in a triangles.

(v) The sum of all three interior angles is 180°.

Here, $\angle BAC + \angle ABC + \angle ACB = 180°$

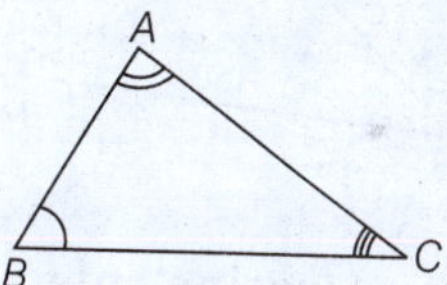

In brief, $\angle A + \angle B + \angle C = 180°$

(vi) The exterior angle of a triangle is equal to the sum of two interior angles opposite to it.

Here, $\angle 4 = \angle 1 + \angle 2$, $\angle 5 = \angle 2 + \angle 3$ and $\angle 6 = \angle 1 + \angle 3$

It may be construed that an exterior angle is always greater than each of the opposite interior angles.

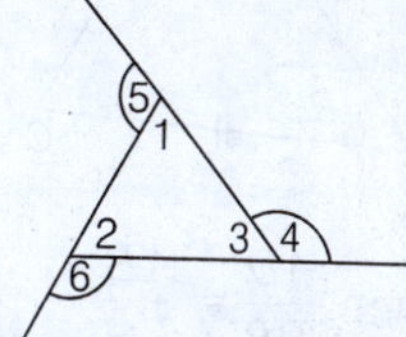

(vii) Sum of all exterior angles of a triangle is 360° and also, the sum of an interior angle and exterior angle adjacent to it is 180°. We have already discussed the same in the topic 'Polygon'.

(viii) If point P is lying in a triangle, then

$$AP + BP + CP > \frac{1}{2}(AB + BC + AC)$$

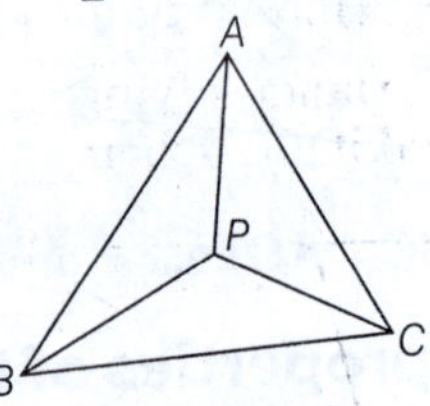

(ix) Here, $\dfrac{a}{\sin A} = \dfrac{b}{\sin B} = \dfrac{c}{\sin C}$

This is know as **sine-rule.**

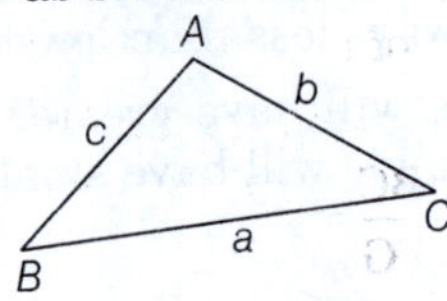

Also $\dfrac{a}{\sin A} = \dfrac{b}{\sin B} = \dfrac{c}{\sin C} = 2R$

where 'R' refers to circumradius. Circumradius will be discussed later.

(x) Here,

$b^2 = a^2 + c^2 - 2ac\cos B$

$a^2 = b^2 + c^2 - 2bc\cos A$

$c^2 = a^2 + b^2 - 2ab\cos C$

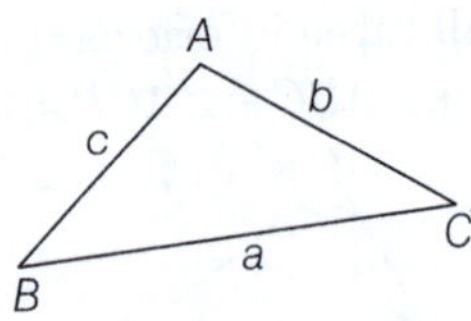

This is known as **'cosine rule.'**

(xi) If a line is drawn parallel to one side of a triangle to intersect the other two sides in distinct points, then the other two sides are divided in the same ratio. The converse is also true.

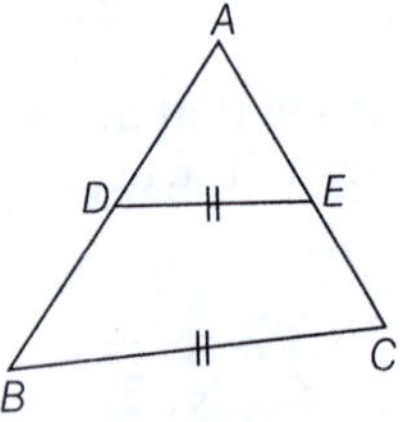

If $DE || BC$, then $\dfrac{AD}{DB} = \dfrac{AE}{EC}$

or $\dfrac{AD}{AB} = \dfrac{AE}{AC}$ or, $\dfrac{DB}{AB} = \dfrac{EC}{AC}$

It is also true that $\dfrac{AD}{DB} = \dfrac{AE}{BC} = \dfrac{DE}{BC}$ (by property of similarity.)

Note *In case ABC be an equilateral triangle, AD = AE and DB = EC.*

(xii) If O is the point in the interior of a ΔABC such that $OM \perp BC$,

$OR \perp AC$ and $ON \perp AB$, where M, N and R are points lying on BC, AB and AC respectively, then,

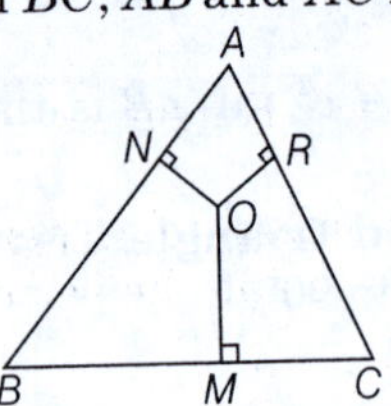

(a) $OA^2 + OB^2 + OC^2 - OM^2 - OR^2 - ON^2 = AN^2 + BM^2 + CR^2$

(b) $AN^2 + BM^2 + CR^2 = AR^2 + CM^2 + BN^2$

(xiii) The line joining the mid-points of any two sides of a triangle is parallel to the 3rd side and equals to half of the 3rd side.

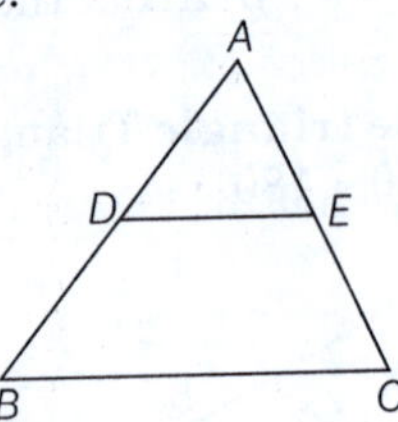

Here, $DE || BC$ and $DE = \dfrac{1}{2}BC$

It also implies that the perimeter of the triangle formed by joining the mid-points of the sides of a triangle is half the perimeter of the triangle.

Here, perimeter of $\Delta DEF = \dfrac{1}{2} \times$ perimeter of ΔABC

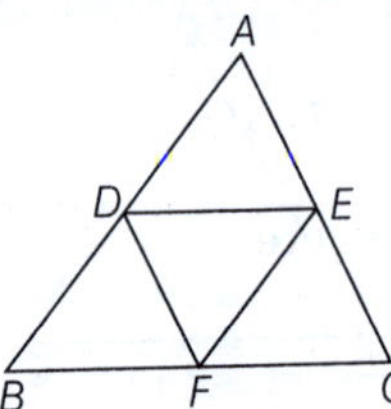

(xiv) If mid-points of all sides of a triangle are joined with each other, the triangles formed, 4 in number, are congruent to each other.

Here, ΔADE, ΔBDF, ΔEFC and ΔDEF all are congruent.

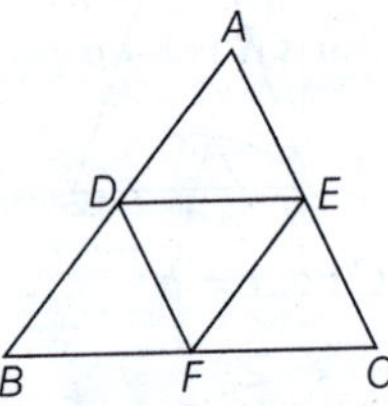

Further, each of the four small triangles is similar to the original triangle. Hence, if ΔABC is an equilateral triangle, all the four triangles will also be equilateral. In case of isosceles, all four will be isosceles and in case of scalene, all will be scalene.

(xv) In any triangle ΔABC, if D is a piont on BC, then $AB + BC + CA > 2AD$.

Area of triangles

(i) Area $= \frac{1}{2} \times$ Base $\times$ Height

Here, height refers to a perpendicular distance between the base and the vertex opposite to it.

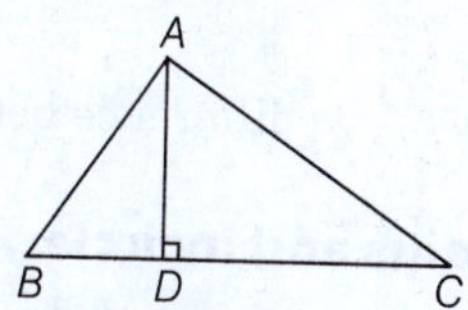

Area $= \frac{1}{2} \times BC \times AD$

(ii) Area $= \sqrt{s(s-a)(s-b)(s-c)}$, where $s = \frac{a+b+c}{2}$

(iii) Area $= \frac{1}{2} \times$ product of any two sides $\times$ sine of the included angle.

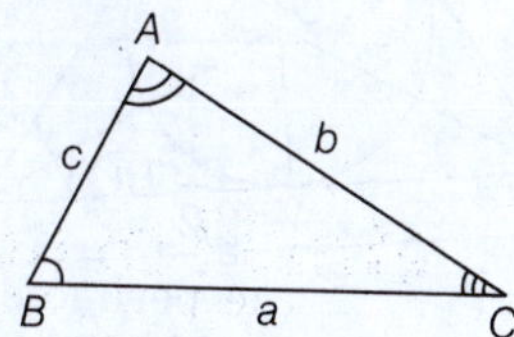

Here, area $= \frac{1}{2} \times ac \times \sin B$

$= \frac{1}{2} \times ab \times \sin C$

$= \frac{1}{2} \times bc \times \sin A$

(iv) Area $= rs$, where 'r' refers to inradius and $s = \frac{a+b+c}{2}$

[concept of inradius will be discussed later]

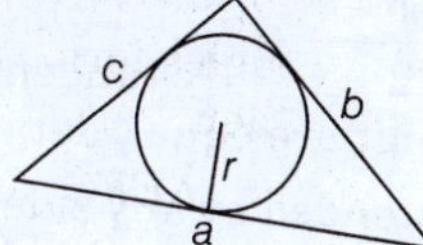

(v) Area $= \frac{abc}{4R}$, where R refers to circumradius

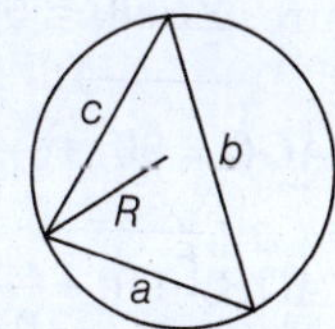

[concept of circumradius will be discussed later]

It is important to note that each of the above five approaches to compute area of a triangle is suitable for all types of triangles.

It also holds goods that

(a) The areas of triangles lying between same parallel lines are equal if all of them have same base.

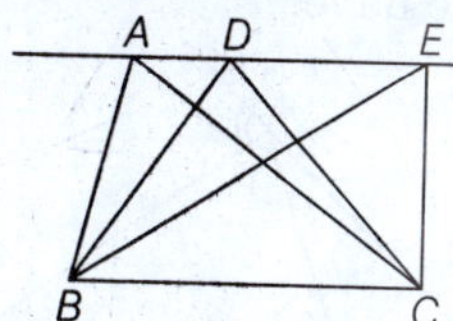

Here, area of $\Delta BAC =$ area of $\Delta BDC =$ area of ΔBEC

(b) Ratio of areas of triangles = Ratio of their heights [if their bases are equal or same]

Here, $\frac{\text{Area of } \Delta ABC}{\text{Area of } \Delta EFG} = \frac{AD}{GH}$ $[\because BC = EF]$

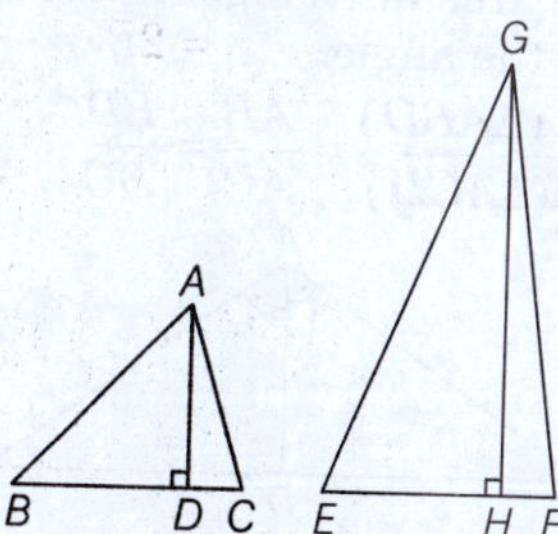

and Ratio of areas of triangles = Ratio of their bases [if their heights are equal or same]

Here, $\frac{\text{Area of } \Delta ABC}{\text{Area of } \Delta EFG} = \frac{BC}{EG}$ $[\because AD = FH]$

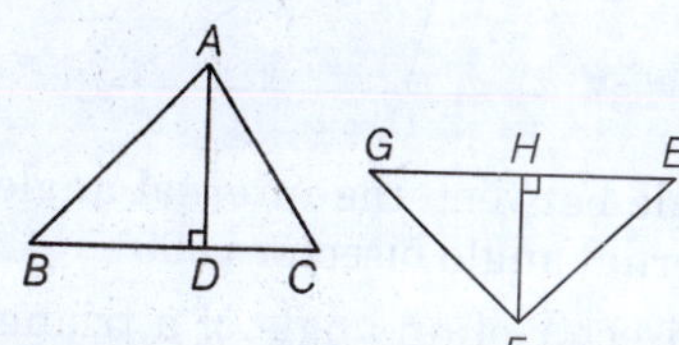

Critical terms related to triangle

A. Angle Bisector

A line which bisects the angle from where it originates, is called angle bisector.

Here, $\angle 1 = \angle 2 = \frac{\angle A}{2}$ and AD is the angle bisector.

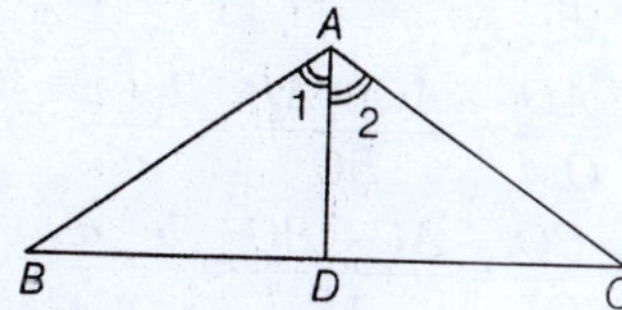

Interior Angle Bisector Theorem

The internal bisector of an angle of a triangle divides the opposite side internally in the same ratio as the sides containing the angle have.

In the above figure, $\frac{BD}{DC} = \frac{AB}{AC}$

Exterior Angle Bisector Theorem

The external bisector of an angle of a triangle divides the opposite side externally in the same ratio as the sides containing the angle have.

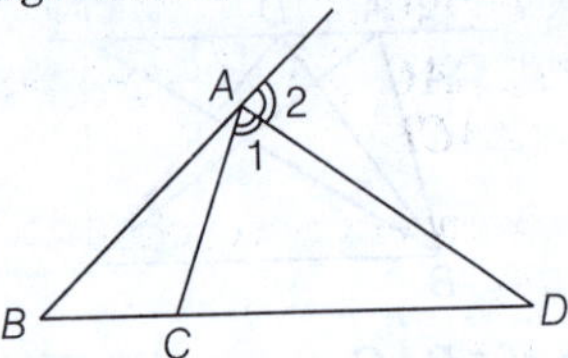

Here, AD is an exterior angle bisector. Then, $\frac{BD}{CD} = \frac{AB}{AC}$

It is also imperative to note that :

(i) The internal bisector of an angle of a triangle divides the area of triangle in the ratio of the sides containing the angle.

Here, $\frac{\text{Area }(\Delta ABD)}{\text{Area}(\Delta ACD)} = \frac{AB}{AC} = \frac{BD}{DC}$

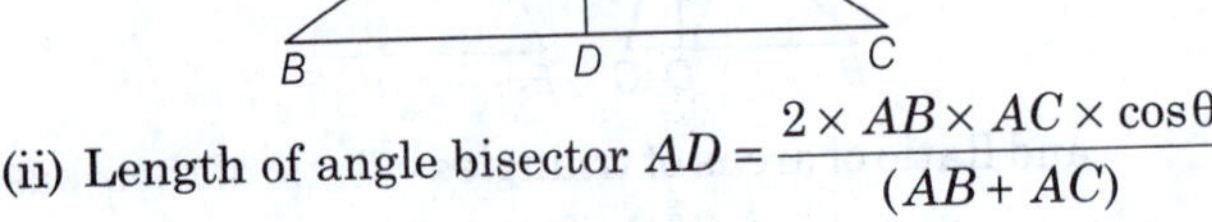

(ii) Length of angle bisector $AD = \frac{2 \times AB \times AC \times \cos\theta}{(AB + AC)}$

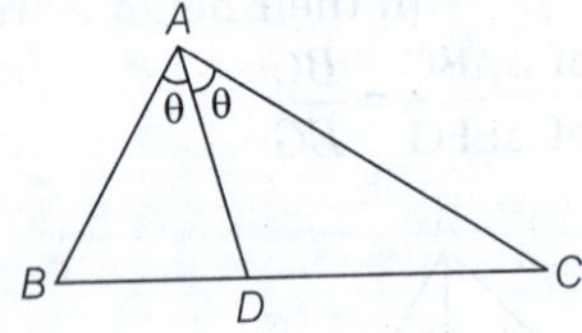

(iii) The angle between the internal angle bisector and the external angle bisector is 90°.

(iv) If the bisector of an angle of a triangle bisects the opposite side, the triangle is isosceles.

(v) If AM, BK and CL are internal angle bisectors as shown in the figure,

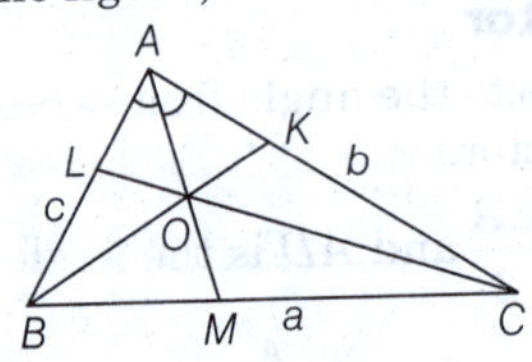

Then, $\frac{AO}{OM} = \frac{AC + AB}{BC} = \frac{b + c}{a}$

Similarly, $\frac{CO}{OL} = \frac{AC + BC}{AB} = \frac{b + a}{c}$

and $\frac{BO}{OK} = \frac{AB + BC}{AC} = \frac{c + a}{b}$

(vi) The angle between the angle bisector and the altitude originating from the same vertex of a triangle is equal to the half the difference between angles of other two vertices.

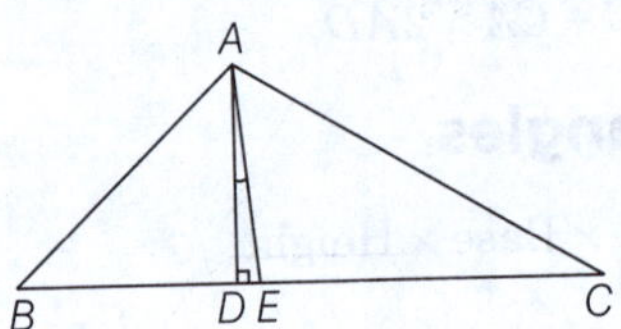

Here, AE is the angle bisector and AD is an altitude of the ΔABC.

Then, $\angle DAE = \frac{1}{2} \times$ difference between $\angle B$ and $\angle C$.

B. Incentre, inradius and incircle

The point of intersection of the angle bisectors of a triangle is called 'incentre' of the triangle and it is equidistant from all the sides of the triangle.

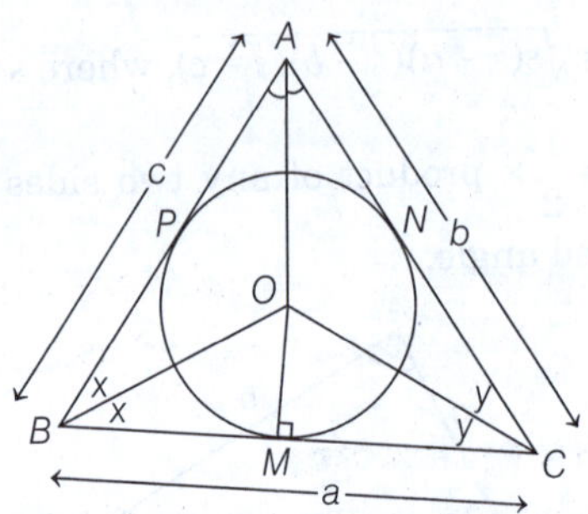

The perpendicular drawn from incentre to any side of triangle is called inradius and the circle drawn with inradius is known as incircle. No doubt, the circle will touch all the sides of the triangle.

Here, BO, AO and CO are angle bisectors. O is incentre and OM is inradius. Inradius is generally denoted by 'r'.

Points to be noted :

(i) In the above figure, $\left\{\begin{array}{lcl} BP = BM & = & s - b \\ AP = AN & = & s - a \\ CM = CN & = & s - c \end{array}\right\}$,

where $s = \frac{a + b + c}{2}$ [$\because$ tangents from an external point are equal, detailed in topic 'circle']

(ii) The angle subtended by any side at incentre is equal to the sum of right angle and half the angle opposite to that side.

In the above figure, $\angle BOC = 90° + \frac{\angle A}{2}$

$$\angle AOC = 90° + \frac{\angle B}{2}$$

$$\angle AOB = 90° + \frac{\angle C}{2}$$

(iii) In the given figure, BM and CM are external angle bisectors. Then,

$\angle BMC = 90^\circ - \dfrac{\angle A}{2}$. The similar result for the other two such angles too.

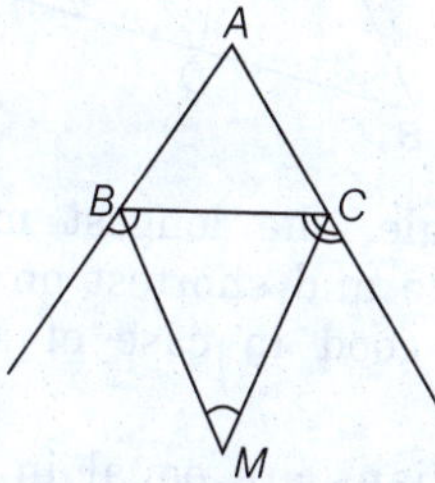

(iv) If sides of the triangle are divided by tangency points of the incircle into length of k and m, m and n and n and k, then the value of inradius $(r) = \sqrt{\dfrac{kmn}{k+m+n}}$

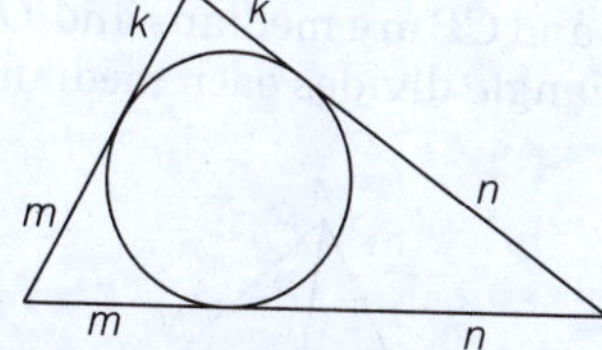

(v) If the altitudes of a triangles are k, m and n then, $\dfrac{1}{r} = \dfrac{1}{k} + \dfrac{1}{m} + \dfrac{1}{n}$ where, r = inradius.

C. Perpendicular bisector

A line which bisects a line/side of a triangle perpendicularly is said to be perpendicular bisector of that line/side. It is also fact that all points lying on the perpendicular bisector are equidistant from both ends of that line/side.

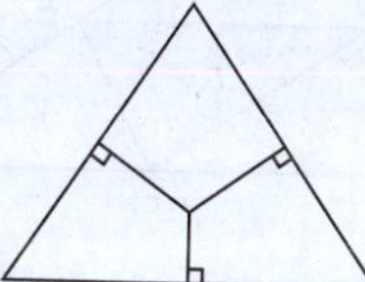

D. Circumcentre, circumradius and circumcircle

The point at which all three perpendicular bisectors of sides of a triangle meet, is known as circumcentre of the triangle.

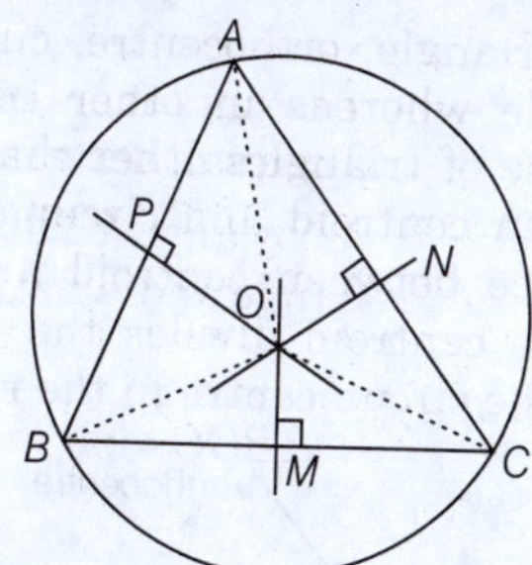

The line joining the circumcentre and any vertex of the triangle is called circumradius (R) and the circle drawn with the circumradius is called circumcircle of the triangle.

Here OM, ON and OP are the perpendicular bisectors. 'O' is circumcentre and OC, OB and OA are circumradii.

Points to Remember

(i) The angle subtended by any side of the triangle at circumcentre is twice the vertical angle opposite to that side.

In the above figure,

$\angle BOC = 2\angle BAC$, $\angle AOC = 2\angle ABC$ and $\angle AOB = 2\angle ACB$

Remarks *It may be a property of a circle, which will be discussed in topic 'Circle'.*

(ii) $\dfrac{a}{\sin A} = \dfrac{b}{\sin B} = \dfrac{c}{\sin C} = 2R$, also given under 'Fundamental properties of triangles'.

The circumradius (R) of a triangle cannot be smaller than twice its inradius (r). Hence, $R \geq 2r$

(iii) The distance between circumcentre and incentre of a triangle $= \sqrt{R(R-2r)}$, where R = circumradius and r = inradius

E. Height/Altitude and Orthocentre

A perpendicular drawn from any vertex to the opposite side of the triangle, is known as altitude of the triangle.

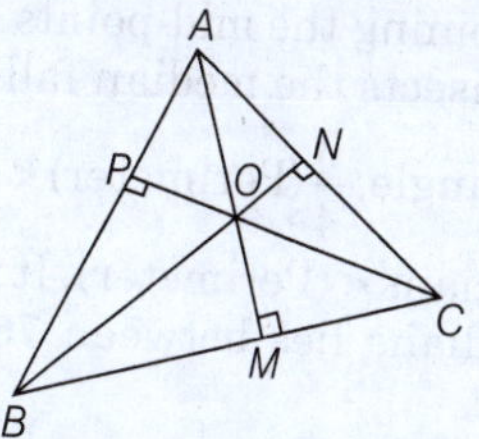

The meeting point of all three altitudes of a triangle is called 'orthocentre'.

Here, AM, BN and CP are altitudes and 'O' is orthocentre.

It is also true that the angle subtended by any side of a triangle on its orthocentre and the vertical angle opposite to that side are supplementary.

Hence, $\angle BOC + \angle BAC = 180^\circ$

$\angle AOC + \angle ABC = 180^\circ$

$\angle AOB + \angle ACB = 180^\circ$

Median

A line joining the mid-point of a side of a triangle to the vertex opposite to the side, is called median.

Here, AM, BN and CP are three medians of ΔABC.

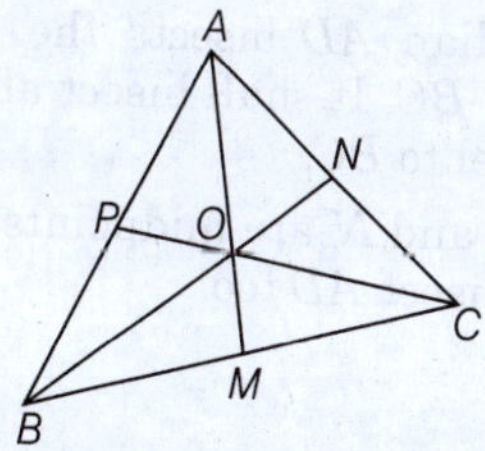

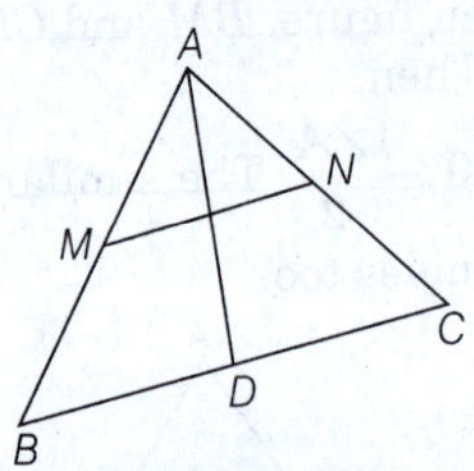

Critical facts

(i) A median divides the triangle in two parts having equal areas. It is also true that all the three medians divide a triangle into six smaller triangles of equal areas.

(ii) In any triangle, the sum of the squares of any two sides is equal to the sum of twice the square of median which bisects 3rd side and twice the square of half the 3rd side. This theorem is known as Apollonius theorem.

Hence, in the above figure,

$$(AB)^2 + (AC)^2 = 2(AM)^2 + 2\left(\frac{BC}{2}\right)^2$$

$$\Rightarrow \quad (AM)^2 = \frac{2(AB)^2 + 2(AC)^2 - (BC)^2}{4}$$

Similarly, $(BN)^2 = \dfrac{2(BC)^2 + 2(AB)^2 - (AC)^2}{4}$

and $\quad (CP)^2 = \dfrac{2(AC)^2 + 2(BC)^2 - (AB)^2}{4}$

It may also be construed from the above that we can find the length of each median if all the sides of a triangle are known.

(iii) The line joining the mid-points of any two sides of a triangle bisects the median falling on the 3rd side.

(iv) In any triangle, $\frac{3}{4}$(Perimeter) < sum of lengths of all three medians < (Perimeter). It implies that the sum of all medians lies between 75% and 100% of the perimeter.

(v) Sum of any two medians of a triangle is greater than the 3rd median.

(vi) $4[(AM)^2 + (BN)^2 + (CP)^2] = 3(a^2 + b^2 + c^2)$

(vii) $3[(AO)^2 + (BO)^2 + (CO)^2] = a^2 + b^2 + c^2$

(viii) Sum of any two sides of a triangle is greater than twice the median drawn to the 3rd side.

Here, $AB + BC > 2AD$

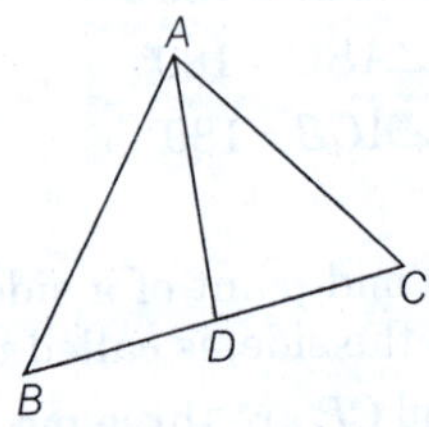

(ix) Here, median AD bisects the line MN, which is parallel to BC. It shall bisect all such lines, if they are parallel to BC.

Also, if M and N are midpoints of respective sides, MN will bisect AD too.

(x) In a triangle, the longest median falls on the shortest side and shortest on the longest side. It also holds good in case of altitudes and angle bisectors.

(xi) If two medians are equal in a triangle, it is an isosceles triangle.

F. Centroid

The point of intersection of all three medians is known as centroid.

Here, AM, BN and CP are medians and 'O' is centroid. The centroid of a triangle divides each median in the ratio 2 : 1 (from vertex).

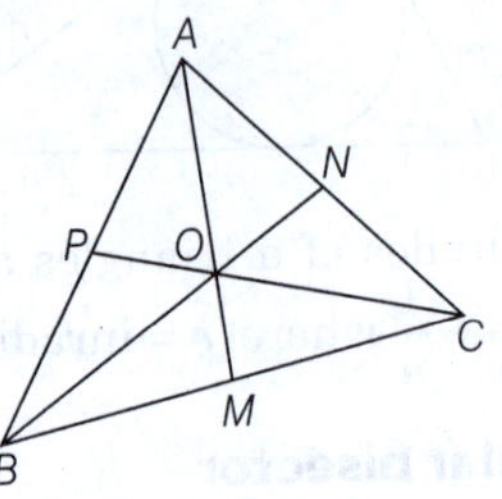

Hence, $AO : OM = BO : ON = CO : OP = 2 : 1$

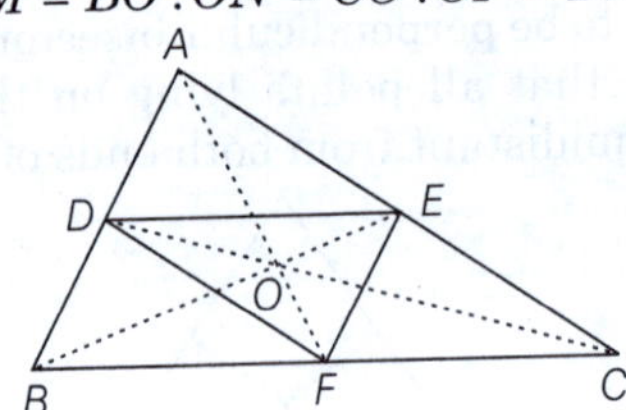

Also, a triangle and its medial triangle have the same centroid.

G. Relation among orthocentre, circumcentre and centroid

In equilateral triangle, orthocentre, circum- centre, and centroid coincide whereas in other triangles, they are collinear. In case of triangles other than equilateral, the distance between centroid and circumcentre is equal to half the distance between centroid and orthocentre. It implies that the centroid divides the distance from the orthocentre to the circumcentre in the ratio 2 : 1.

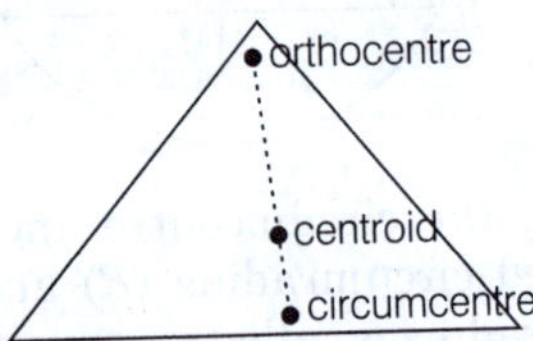

Triangle-wise properties

• Acute-angled triangle

In an acute angle triangle ABC, if AD is the altitude on BC from vertex A and $\angle ABC$ is the greatest angle among all three, then

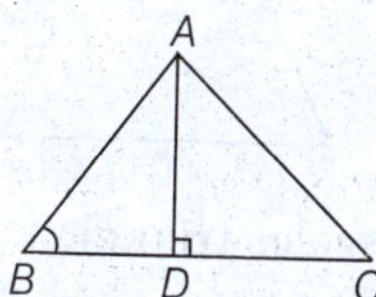

$$AC^2 = AB^2 + BC^2 - 2 \cdot BD \cdot BC$$

Hence, $AC^2 < AB^2 + BC^2$

Further, in an acute-angled triangle, each angle is less than the sum of the other two.

• Obtuse-angled triangle

In an obtuse angle triangle ABC, if AD is the altitude on BC produced from vertex A and $\angle ABC$ is the greatest angle among all three, then

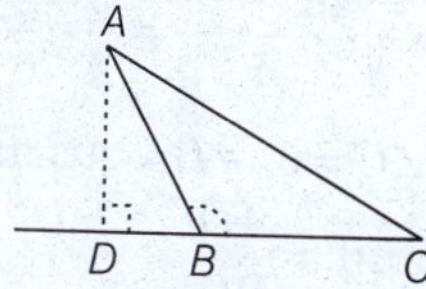

$$AC^2 = AB^2 + BC^2 + 2 \cdot BD \cdot BC$$

Hence, $AC^2 > AB^2 + BC^2$

It is also important to note that

(a) The circumcentre of an obtuse angled triangle lies outside the triangle.

(b) The greatest angle of an obtuse angled traingle is greater than the sum of the other two.

• Equilateral triangle

(i) Each angle is equal to $60°$.

(ii) All equilateral triangles are similar.

(iii) For the given perimeter, an equilateral triangle has maximum area in comparison to other types of triangles.

(iv) If areas of the triangles are same, then perimeter of the equilateral triangle will be minimum.

(v) An equilateral triangle inscribed in a circle will have maximum area in comparison to others.

(vi) Sum of perpendiculars drawn from a point P lying inside the equilateral triangle to all three sides is equal to the altitude of the triangle.

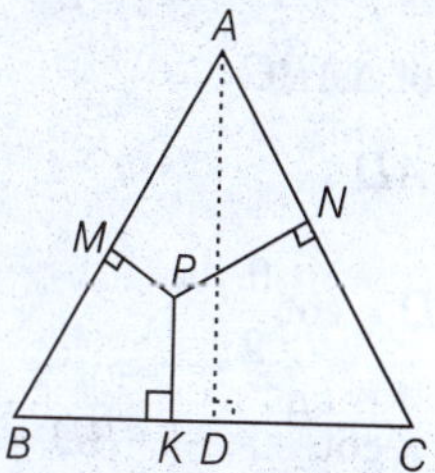

Here, $AD = PM + PK + PN$

(vii) If an equilateral triangle is inscribed in a circle, the line joining a vertex to any point lying on its oposite arc, is equal to sum of lines joining other two vertices to that point.

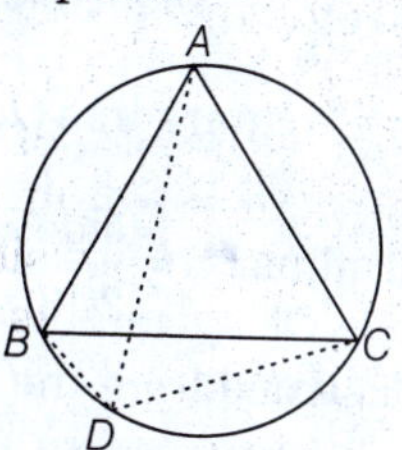

Here $AD = BD + CD$

(viii) In an equilateral triangle, altitude, median, angle-bisector and perpendicular bisector are all the same single line. Thus, orthocentre, centroid, incentre and circumcentre are all the same point.

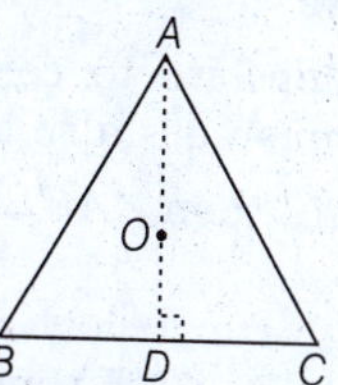

Here, $AD \Rightarrow$ altitude, median, angle-bisector and perpendicular bisector

and $O \Rightarrow$ orthocentre, centroid, incentre and circumcentre

(ix) In the above figure, $\angle B = 60°$

In ΔABD, $\sin B = \dfrac{AD}{AB}$

$$\Rightarrow \quad \sin 60° = \frac{AD}{AB} \Rightarrow \frac{\sqrt{3}}{2} = \frac{AD}{AB}$$

$$\Rightarrow \quad AD = \frac{\sqrt{3}}{2} \times AB$$

Hence, height of an equilateral triangle $= \dfrac{\sqrt{3}}{2} \times$ side

Further, $\sqrt{3}$ side $= 2 \times$ height

Then, $3 \times (\text{side})^2 = 4 \times (\text{height})^2$

(x) Area $= \dfrac{1}{2} \times \text{base} \times \text{height} = \dfrac{1}{2} \times \text{side} \times \left(\dfrac{\sqrt{3}}{2} \times \text{side}\right)$

Hence, area of equilateral triangle $= \dfrac{\sqrt{3}}{4} \times (\text{side})^2$

(xi) Since orthocentre, centroid, incentre and circumcentre of a equilateral triangle are all the same point and centroid divides the median in 2 : 1, length of $AO = \dfrac{2}{3} \times \text{median} \Rightarrow AO = \dfrac{2}{3} \times \text{altitude}$

[$\because$ median and altitude are the same line]

$$\Rightarrow \quad AO = \frac{2}{3} \times \frac{\sqrt{3}}{2} \times \text{side} = \frac{\text{side}}{\sqrt{3}}$$

Further, $OD = \frac{1}{2} \times AO$ $[\because AO : OD = 2 : 1]$

$\Rightarrow \quad OD = \frac{\text{side}}{2\sqrt{3}}$

Hence, AO = Circumradius $(R) = \frac{\text{side}}{\sqrt{3}}$

and OD = inradius $(r) = \frac{\text{side}}{2\sqrt{3}}$

(xii) It may be construed from the above that, $R = 2r$

(xiii) Three times the square of a side of an equilateral triangle is equal to four times the square of its altitude, as evident from point (ix).

(xiv) All the three medians together divide an equilateral triangle into six congruent smaller triangles.

- **Isosceles triangle**

(i) The angles opposite to congruent sides are also equal. The converse is also true.

Here, $AB = AC$, then $\angle ACB = \angle ABC$

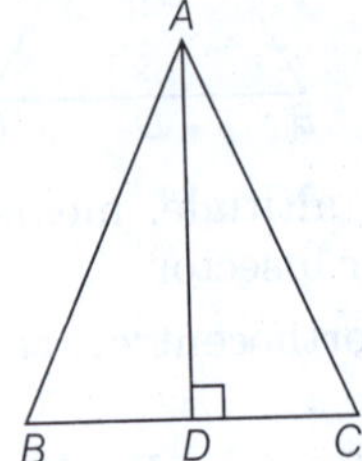

(ii) An isosceles triangle having one angle 90° is known as a right isosceles triangle.

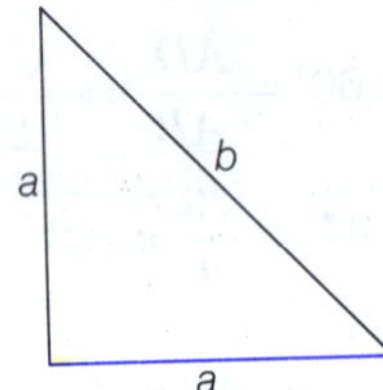

(iii) In an isosceles triangle (including right isosceles triangle) the altitude having the side other than congruent sides as its base has the mid-point of that side as its foot. Hence, this altitude is also a median and a perpendicular bisector as well. Also , this altitude will act as an angle bisector of the vertex.

In the above figure, AD is an altitude, median, angle bisector and perpendicular bisector. It implies that orthocentre, centroid, incentre and circumcentre of an isosceles triangle lies all on AD, hence they are collinear.

Also, AD is the line of symmetry of the triangle.

(iv) In the above figure, ΔABD is congruent to ΔACD.

(v) Area related properties.

Here, a = length of each of equal sides and b = length of the 3rd side

Area of any triangle $= \sqrt{s(s-a)(s-b)(s-c)}$,

where, $s = \frac{a+b+c}{2}$

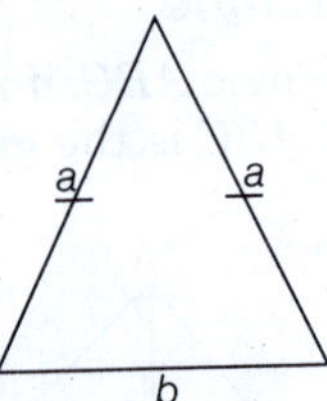

In case of isosceles triangle,

$$s = \frac{a+a+b}{2} = a + \frac{b}{2}$$

then, area of an isosceles triangle

$$= \sqrt{\left(a+\frac{b}{2}\right)\left(a+\frac{b}{2}-a\right)\left(a+\frac{b}{2}-a\right)\left(a+\frac{b}{2}-b\right)}$$

$$= \sqrt{\left(a+\frac{b}{2}\right)\left(\frac{b}{2}\right)\left(\frac{b}{2}\right)\left(a-\frac{b}{a}\right)}$$

$$= \left(\frac{b}{2}\right) \times \sqrt{\frac{4a^2-b^2}{4}} = \frac{b}{4}\sqrt{4a^2-b^2}$$

(vi) Here, $\angle B = \angle C = \theta$, $AB = AC$ and $BC = b$

In $\Delta\ ABD$, $\tan\theta = \frac{AD}{BD}$

$\Rightarrow \quad AD = BD \times \tan\theta$

Then, area of $\Delta ABC = \frac{1}{2} \times BC \times AD$

$= \frac{1}{2} \times b \times BD \times \tan\theta$

$= \frac{1}{2} \times b \times \frac{b}{2} \times \tan\theta$

$[\because BD = \frac{1}{2} BC]$

$= \frac{b^2}{4} \times \tan\theta$

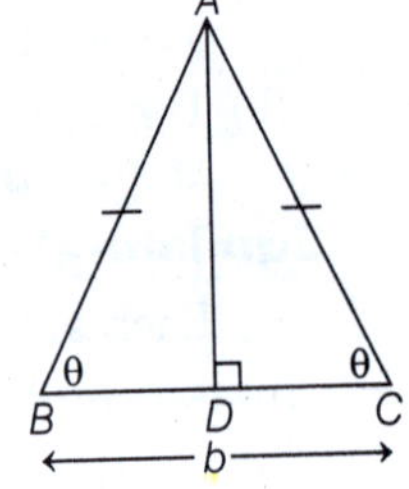

Hence, if each of the congruent angles is θ and the side other than the congruent sides is b, then area of the isosceles triangle $= \frac{b^2}{4} \times \tan\theta$

(vii) Here, $\angle A = \theta$ and $AB = AC$ and $BC = b$

In ΔABD, $\tan\frac{\theta}{2} = \frac{BD}{AD}$

$\Rightarrow \quad AD = BD \times \cot\frac{\theta}{2}$

Then, area of ΔABC

$= \frac{1}{2} \times BC \times AD$

$= \frac{1}{2} \times b \times BD \times \cot\frac{\theta}{2}$

$= \frac{1}{2} \times b \times \frac{b}{2} \times \cot\frac{\theta}{2}$ $\quad [\because BD = \frac{1}{2} BC]$

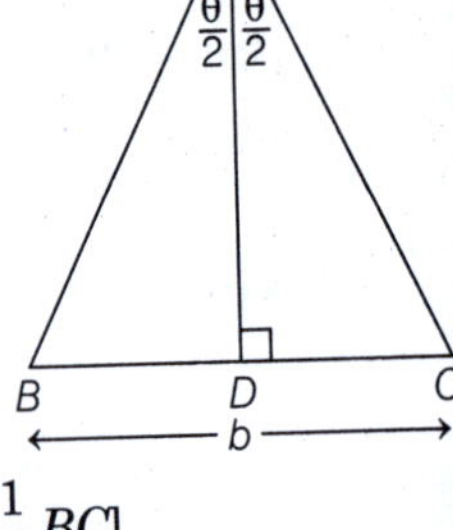

$$= \frac{b^2}{4}\cot\frac{\theta}{2}$$

Hence, if the angle other than the congruent angles is θ and the side other than the congruent sides is b, then area of the isosceles triangle.

$$= \frac{b^2}{4} \times \cot\frac{\theta}{2}$$

(viii) Here, $\angle A = \theta$ and $AB = AC = a$

In ΔABD, $\cos\frac{\theta}{2} = \frac{AD}{AB} \Rightarrow AD = a \times \cos\frac{\theta}{2}$

and $\quad \sin\frac{\theta}{2} = \frac{BD}{AB}$

$\Rightarrow \quad BD = a \times \sin\frac{\theta}{2}$

Then, area of $\Delta ABC = \frac{1}{2} \times BC \times AD$

$= \frac{1}{2} \times 2 \cdot BD \times a \times \cos\frac{\theta}{2}$

$= \frac{1}{2} \times 2 \times a \times \sin\frac{\theta}{2} \times a \times \cos\frac{\theta}{2}$

$= \frac{a^2}{2} \times 2\sin\frac{\theta}{2} \cdot \cos\frac{\theta}{2}$

$= \frac{a^2}{2} \times \sin\theta$

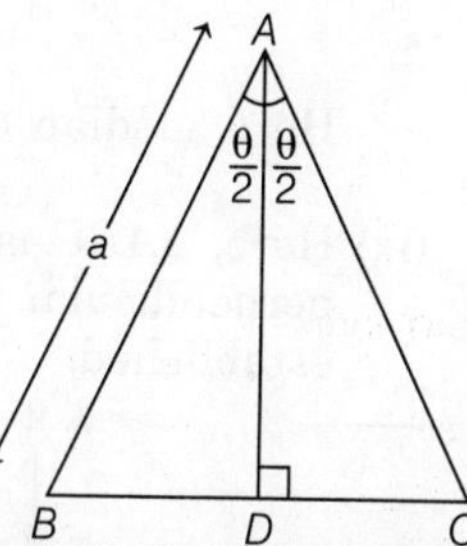

Hence, if the angle other than the congruent angles is θ and each of the congruent sides is a, then area of the isosceles triangle $= \frac{a^2}{2} \times \sin\theta$

(ix) Any n-sided regular polygon can be split into n isosceles triangle and measure of the angle other than the congruent angles will be $\left(\frac{360°}{n}\right)$.

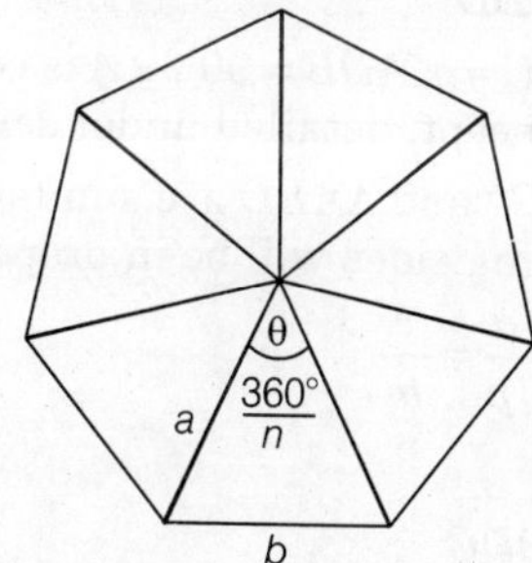

Then, area of n-sided polygon $= n \times$ area of one isosceles triangle $= n \times \frac{a^2}{2} \times \sin\left(\frac{360°}{n}\right)$

[$\because$ area of isosceles triangle $= \frac{a^2}{2} \times \sin\theta$]

$= \frac{1}{2} \times n \times \sin\left(\frac{360°}{n}\right) a^2$

Here, a = each of the congruent sides = length from centre to a corner of polygon.

Also,

Area of n-sides polygon $= n \times$ area of one isosceles triangle $= n \times \frac{b^2}{4}\cot\left(\frac{360°}{2n}\right)$

[$\because$ area of isosceles triangle $= \frac{b^2}{4}\cot\frac{\theta}{2}$]

$= \frac{nb^2}{4} \times \cot\left(\frac{180°}{n}\right)$

Here, b = side other than congruent sides = each side of the polygon

The above mentioned both formulae meant for the polygon have also been given under topic 'Polygon'.

(x) If two isosceles triangles have common base, the line joining their vertices bisects the base at right angle.

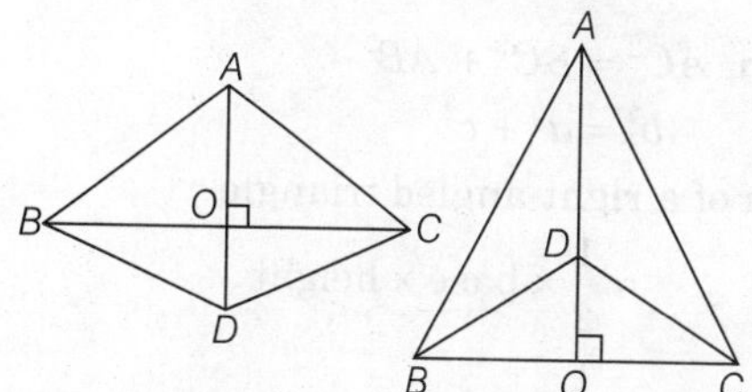

(xi) If the external bisector of the vertical angle of a triangle is parallel to its base, then the triangle is an isosceles triangle.

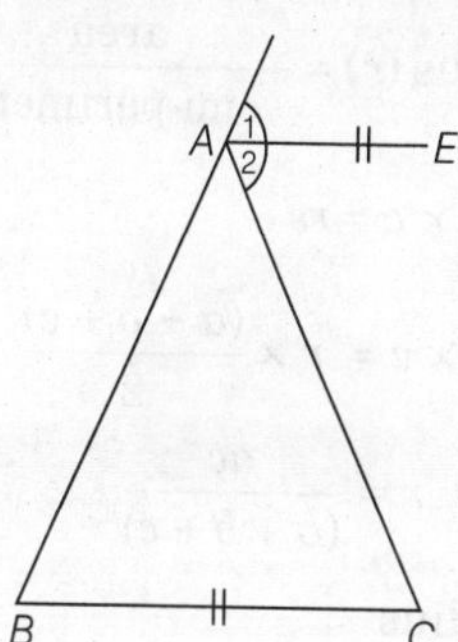

Here, $\angle 1 = \angle 2$ and $AE \parallel BC$

(xii) The line joining the mid-ponts of the congruent sides and the line joining the mid-point of the 3rd side and its opposite vertex (i.e. median or altitude or angle bisector or perpendicular bisector), bisect each other at right angle.

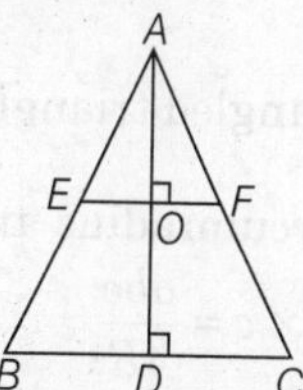

If D, E and F are mid-points of sides of an isosceles $\Delta\, ABC$ with $AB = AC$, then $\angle AOF = 90°$ and $AO = OD$ and $EO = OF$.

- **Right-angled Triangle**

(i) The side opposite to the right angle is called hypotenuse.

(ii) As per Pythagoras theorem,

$$(\text{Hypotenuse})^2 = (\text{Base})^2 + (\text{Height})^2$$

Here, AC is the hypotenuse and BC and AB are the perpendicular sides. BC may be termed as base or height and AB may be termed as height or base.

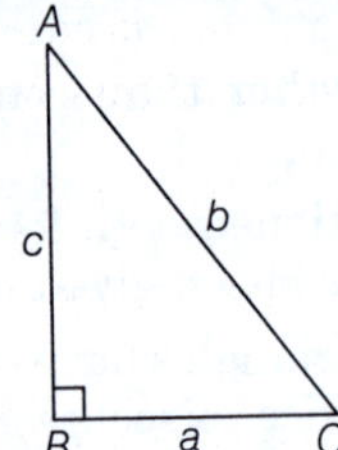

Then, $AC^2 = BC^2 + AB^2$

$\Rightarrow \quad b^2 = a^2 + c^2$

(iii) Area of a right-angled triangle

$$= \frac{1}{2} \times \text{base} \times \text{height}$$

$$= \frac{1}{2} ac$$

(iv) Area of a right angled triangle $= rs$

[where r = inradius and s = semi-perimeter]

Then, inradius $(r) = \dfrac{\text{area}}{\text{semi-perimeter}}$

Also, $\dfrac{1}{2} a \times c = rs$

$\Rightarrow \quad \dfrac{1}{2} \times a \times c = r \times \dfrac{(a+b+c)}{2}$

$\Rightarrow \quad r = \dfrac{ac}{(a+b+c)}$

Hence, inradius

$$(r) = \frac{\text{Product of perpendicular sides}}{\text{Perimeter}}$$

(v) Also, inradius (r)

$$= \frac{\text{Sum of perpendicular sides} - \text{hypotenuse}}{2}$$

$$= \frac{a+c-b}{2}$$

(vi) Area of a right-angled triangle $= \dfrac{abc}{4R}$

[where R = circumradius and a, b and c are sides]

$\Rightarrow \quad \dfrac{1}{2} \times a \times c = \dfrac{abc}{4R}$

$\Rightarrow \quad R = \dfrac{b}{2}$

Hence, circumradius $(R) = \dfrac{\text{hypotenuse}}{2}$

(vii) The orthocentre of a right angled triangle lies at the right anglular vertex.

(viii) The median drawn to the hypotenuse is equal to half the hypotenuse. It also holds good that if a median drawn to a side of a triangle is half the side, the triangle is a right-angled triangle.

Further, the above median divides the right angled triangle into two isosceles triangles.

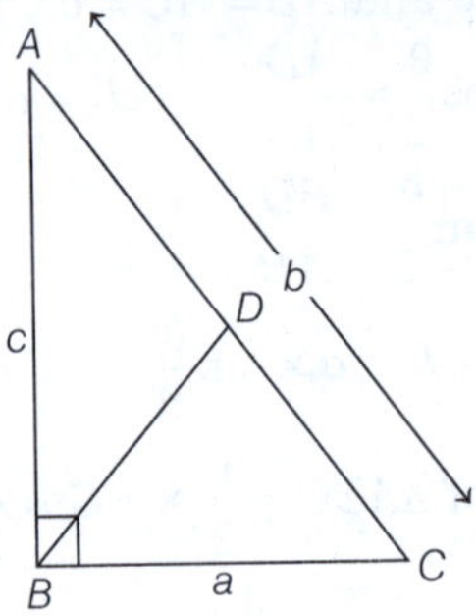

Here, median $BD = \dfrac{AC}{2}$

(ix) Here, ΔABC is a right-angled triangle and BD is perpendicular to AC. The following relations can be established.

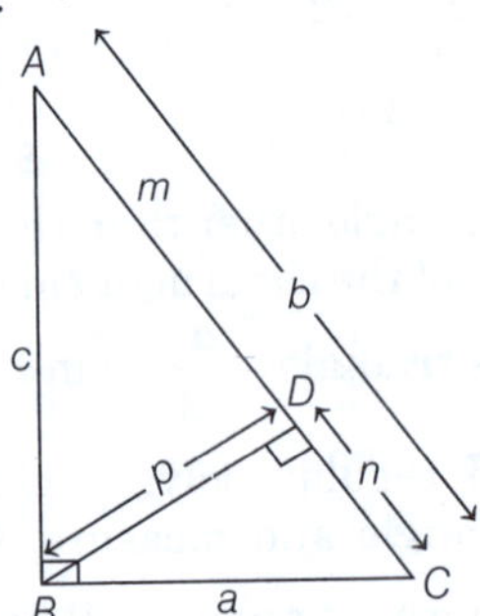

(a) $\Delta ABC \sim \Delta ABD$

[$\because \angle ABC = \angle ADB = 90°$, $\angle A$ is common $\Rightarrow A-A$ similarity concept, detailed under separate heading.]

Since ΔABC and ΔABD are similar, their corresponding sides will be in proportion.

Hence, $\dfrac{b}{c} = \dfrac{a}{p} = \dfrac{c}{m}$

$\Rightarrow c^2 = bm$

(b) $\Delta ABC \sim \Delta BDC$

[$\because \angle ABC = \angle BDC = 90°$, $\angle C$ is common $\Rightarrow A-A$ similarity concept, detailed under separate heading.]

Then, $\dfrac{b}{a} = \dfrac{c}{p} = \dfrac{a}{n} \Rightarrow a^2 = bn$

and $bp = ac$

(c) From the above, $\dfrac{a}{p} = \dfrac{c}{m}$ and $\dfrac{c}{p} = \dfrac{a}{n}$

On multiplication, $\frac{ac}{p^2} = \frac{ca}{mn} \Rightarrow p^2 = mn$

(d) From the above, $\frac{a}{p} = \frac{c}{m}$

$\Rightarrow \quad \frac{1}{p} = \frac{c}{am} \Rightarrow \frac{1}{p^2} = \frac{c}{a^2\, m^2}$

[squaring both sides]

$\Rightarrow \quad \frac{1}{p^2} = \frac{c^2}{a^2\left(\frac{c^2}{b}\right)^2} \quad [\because c^2 = bm \therefore m = \frac{c^2}{b}]$

$= \frac{c^2 \cdot b^2}{a^2 \cdot c^4} = \frac{b^2}{a^2 \cdot c^2}$

$= \frac{a^2 + c^2}{a^2 \cdot c^2} \quad \{\because b^2 = a^2 + c^2\}$

$= \frac{1}{c^2} + \frac{1}{a^2} \Rightarrow \frac{1}{p^2} = \frac{1}{c^2} + \frac{1}{a^2}$

(x) Here, ΔABC is a right-angled triangle and M and N are mid-points of sides AB and BC, respectively.

Then, $4(AN^2 + MC^2) = 5 \cdot AC^2 = 20 . MN^2$

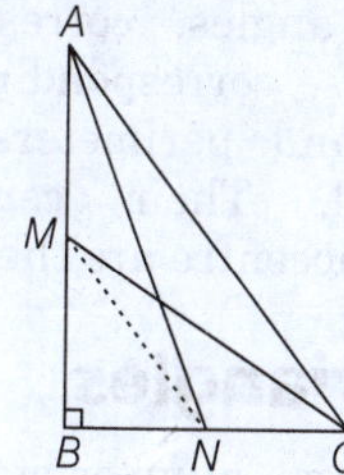

(xi) A triangle with angles 30°, 60° and 90° is termed as '30-60-90 triangle'.

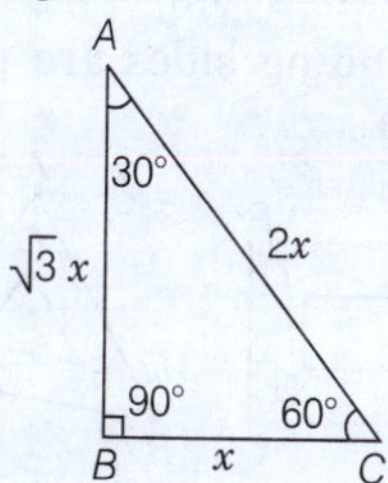

Here, $\tan 60° = \frac{AB}{BC}$

$\Rightarrow \quad AB = \sqrt{3}\, BC$ and $\tan 30° = \frac{BC}{AB}$

$\Rightarrow \quad AB = \sqrt{3}\, BC$

Then, $AC = \sqrt{AB^2 + BC^2} = \sqrt{(\sqrt{3}BC)^2 + BC^2} = 2BC$

It may be construed from the above that in a 30-60-90 triangle, the hypotenuse is twice the length of shorter perpendicular side and the longer perpendicular side is $\sqrt{3}$ times the shorter perpendicular side. This result helps us solve the questions quickly.

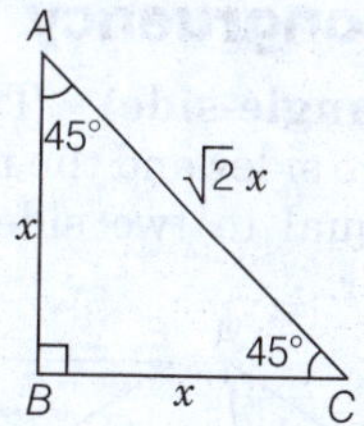

Also, in a '45-45-90 triangle', both the perpendicular sides are equal and the hypotenuse is equal to $\sqrt{2}$ times either of perpendicular sides. Hence, this triangle is an isosceles right-angled triangle.

In a 'right isosceles triangle' the altitutde to the hypotenuse is equal to half the hypotenuse.

Pythagorean triplets

A set of three numbers which satisfies the Pythagoras theorem is known as Pythagorean triplet.

$3^2 + 4^2 = 5^2$

[Here, (3, 4, 5) is termed as Pythagorean triplet.]

Examples of such triplets

(5, 12, 13), (7, 24, 25), (8,. 15, 17), (9, 40, 41), (11, 60, 61), (12, 35, 37), (16, 63, 65), (20, 21, 29), (28, 45, 53), (33, 56, 65) , etc.

No doubt, these triplets are very useful as many questions are based directly on them. It needs to be mentioned that (3, 4, 5) is only Pythagorean triplets which has consecutive integers.

When a Pythagorean triplet is either multiplied or divided by a real positive number, other triplets come out.

e.g.

5	12	13 ⇒ a triplet
↓×1.2	↓×1.2	↓×1.2
6	14.4	15.6 ⇒ another triplet

Or

5	12	13 ⇒ a triplet
↓÷1.2	↓÷1.2	↓÷1.2
2.5	6	6.5 ⇒ another triplet and so on.

Congruency of Triangles

Figures are said to be congruent if their shapes and sizes are both the same. In case of triangle, two triangles are said to be congruent if their corresponding sides are of the same length and also their corresponding angles are of the same measure. Hence, they have the same shape and size.

As we know, a triangle has three angles and three sides. In order to determine as to whether two triangles are congruent or not, we need not compare to all six components of one triangle with those of another. The same may be determined only by a set of three components, which may be regarded as criteria for the congruency. The notation for congruency is ≅.

Conditions for congruency

(a) **S-A-S (side-angle-side)** Two triangles are congruent, if two sides and the included angle of one triangle are equal to two sides and the included angle of another.

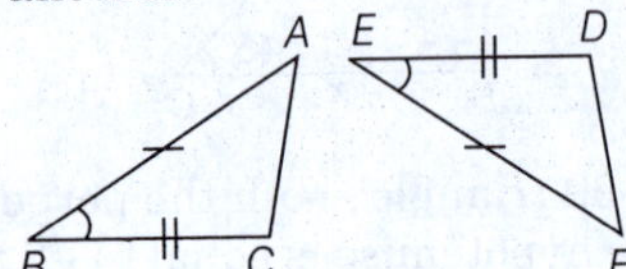

Here, $AB = EF$, $BC = DE$ and $\angle B = \angle E$.

Thus, ΔABC and ΔDEF are congruent. This may be written as $\Delta ABC \cong \Delta DEF$. It implies that both triangles have the same shape and equal size.

(b) **A-S-A (angle-side-angle)**

Two triangles are congruent, if two angles and the included side of one triangle are equal to two angles and the included side of another.

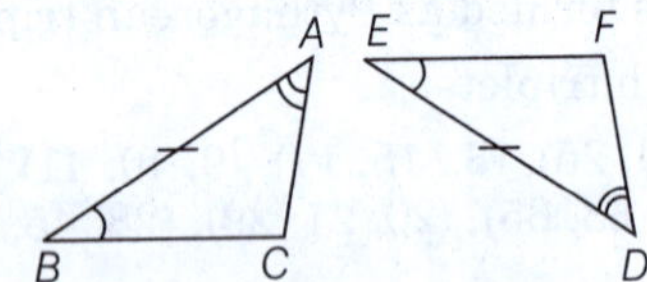

Here, $\angle A = \angle D$, $\angle B = \angle E$ and $AB = DE$. Hence, $\Delta ABC \cong \Delta DEF$.

(c) **A-A-S (angle-angle-side)** Two triangles are congruent, if two angles and a side other than included side of one triangle are equal to two angles and a corresponding side of another.

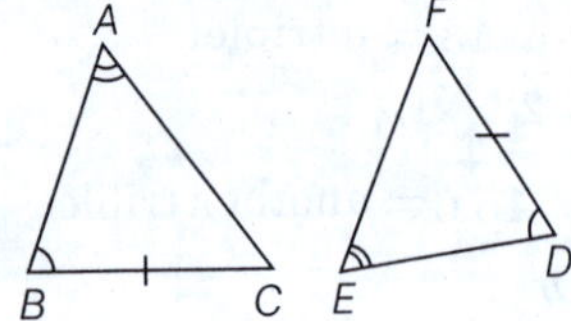

Here, $\angle A = \angle E$, $\angle B = \angle D$ and $BC = DF$. Hence, $\Delta ABC \cong \Delta DEF$.

(d) **S-S-S (side-side-side)** Two triangles are congruent, if three sides of one triangle are equal to three sides of another.

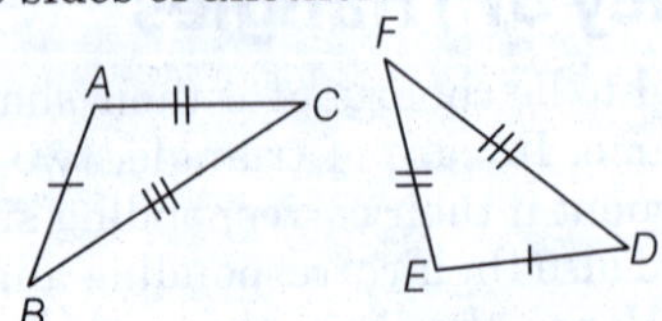

Here, $AB = DE, BC = DF, AC = EF$

Hence, $\Delta ABC \cong \Delta EDF$.

(e) **S-S-A (side-side-angle)/also known as A-S-S**

Two triangles are congruent, if two sides and a non-included angle of one triangle are equal to two sides and a corresponding non-included angle of another provided the side opposite to the angle is greater than or equal to the adjacent side.

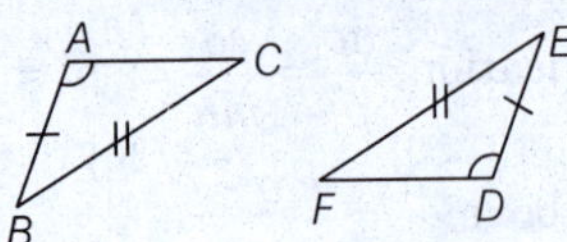

Here, $AB = DE$, $BC = EF$, $\angle A = \angle D$ and BC and EF are greater than or equal to their adjacent sides. Hence, $\Delta ABC \cong \Delta DEF$.

(f) **R-H-S (right angle-hypotenuse-side)** Two right-angled triangles are congruent, if hypotenuse and one side of one triangle are equal to hypotenuse and one side of another.

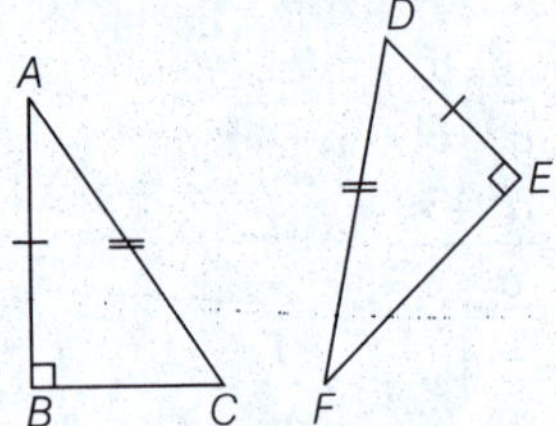

Here, hypotenuse $(AC) =$ hypotenuse (DF) and $AB = DE$. Hence, right-angled $\Delta ABC \cong$ right-angled ΔDEF.

Needless to say that in case of congruency, corresponding sides, corresponding angles, corresponding altitudes, corresponding medians, corresponding angle bisectors, circumradii, inradii and perimeters of two or more triangles are equal. Their centroids, incentres, circumcentres and orthocentre are the same point.

Similarity of Triangles

Two polygons of the same number of sides are said to be similar, if

(i) Their corresponding angles are equal; and

(ii) Their corresponding sides are proportional (i.e. in the same ratio).

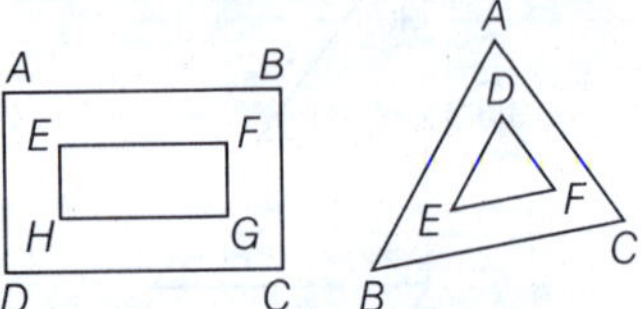

$\square ABCD$ and $\square EFGH$ are similar as

$$\frac{AB}{EF} = \frac{BC}{FG} = \frac{CD}{HG} = \frac{AD}{EH}$$

and $\angle A = \angle E, \angle B = \angle F, \angle C = \angle G$ and $\angle D = \angle H$

Also, ΔABC and ΔDEF are similar as

$\dfrac{AB}{DE} = \dfrac{AC}{DF} = \dfrac{BC}{EF}$ and $\angle A = \angle D, \angle B = \angle E$ and $\angle C = \angle F$.

However, we need not compare all the components of one triangle with those of another as to whether all angles are equal and sides be proportional or not. We test the same only by a defintie set of components, given below (say, conditions for similarity).

The notation for similarity is ~.

Conditions for Similarity

(a) **A-A (angle-angle)** Two triangles are similar if two angles of one triangle are respectively equal to two angles of another. No doubt, the 3rd angle becomes equal automatically.

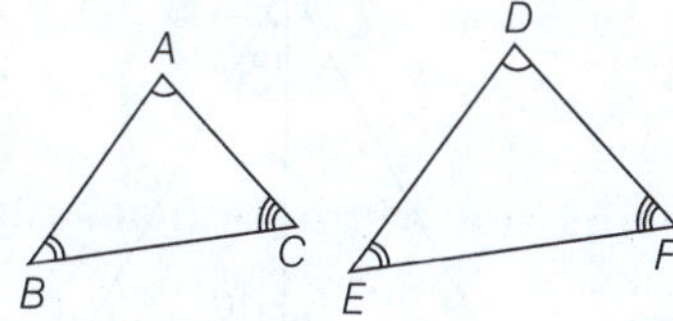

Here, $\angle A = \angle D, \angle B = \angle E$. Hence, $\Delta ABC \sim \Delta DEF$.

(b) **S-S-S (side-side-side)**

Two triangles are similar if the corresponding sides of two triangles are proportional.

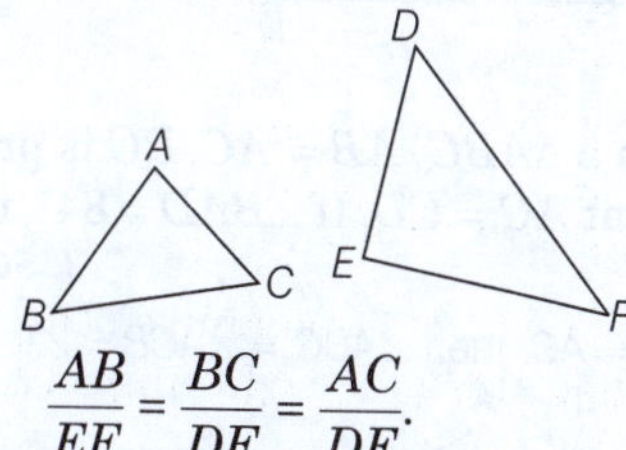

Here, $\dfrac{AB}{EF} = \dfrac{BC}{DF} = \dfrac{AC}{DE}$.

Hence, $\Delta ABC \sim \Delta DEF$.

(c) **S-A-S (side-angle-side)** Two triangles are similar if one angle of a triangle is equal to one angle of another and the sides containing these angles are proportional.

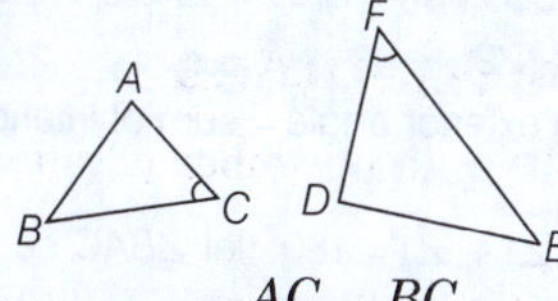

Here, $\angle C = \angle F$ and $\dfrac{AC}{DF} = \dfrac{BC}{EF}$.

Hence, $\Delta ABC \sim \Delta DEF$

(d) **R-H-S (right angle-hypotenuse-side)** Two right-angled triangles are similar if the hypotenuses and either pair of the two sides ae proportional.

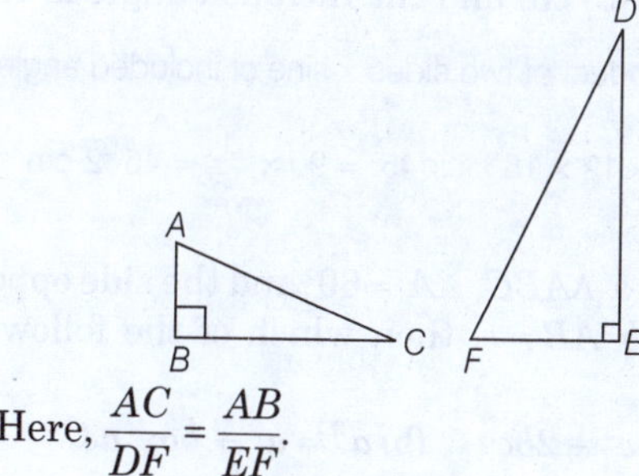

Here, $\dfrac{AC}{DF} = \dfrac{AB}{EF}$.

Hence, right-angled $\Delta ABC \sim$ right-angled ΔDEF.

In another way, two right-angled triangles are similar if ratio of the hypotenuse and one side of one triangle is equal to ratio of thc hypotenuse and one side of another.

Properties of Similar Triangles

(i) If two triangles are similar, then the followings hold good :

(a) Their corresponding sides are proportional.

(b) Their corresponding angles are equal.

Its converse is also true. It means, if the corresponding sides (all three pairs) are proportional or the corresponding angles (atleast two pairs) are equal, the two triangles are similar.

(ii) In case of similarity of two triangles, ratio of corresponding sides = ratio of corresponding altitudes = ratio of corresponding medians = ratio of corresponding angle bisectors = ratio of circumradii = ratio of inradii = ratio of perimeters.

(iii) Ratio of areas of two similar triangles = ratio of squares of corresponding sides/altitudes/ medians/ angle bisectors/circumradii/ inradii/ perimeters.

(iv) Two similar triangles are congruent if their areas are equal.

It is important to note that

(i) All the congruent figures/triangles are similar but the converse is not true. The reason is that shapes and sizes are same in the case of congruency whereas only shapes are same in case of similarity.

(ii) Each of the four triangles formed by the line segments joining the mid-points of sides of a triangle is similar to the original triangle. Thus, all these four must be similar. In fact all four are congruent. Here, ΔBDF, ΔCFE, ΔADE and ΔDEF are similar to ΔABC and also all smaller four triangles are congruent.

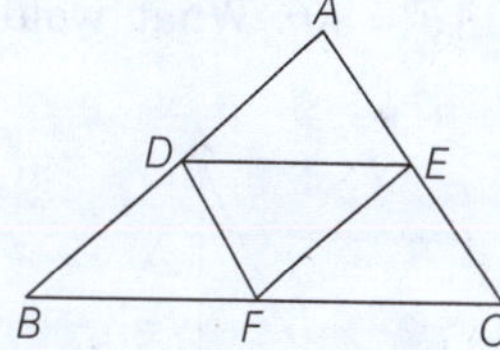

Problems Based on Types, Fundamental Properties and area of Triangles

Probem 1 Consider obtuse-angled triangles with sides 8 cm, 15 cm and x cm. If x is an integer, then how many such triangles exist? ***[CAT-08]***

Sol. We know, in an obutse-angled triangle,

$b^2 > a^2 + c^2$, if $\angle B$ is the greatest among all three.

There may be two possiblities.

(i) Side having length of x cm is the longest side i.e. x is greater than 15.

(ii) Side having length of 15 cm is the longest side i.e. x is smaller than 15.

For possibility (i), $x^2 > 8^2 + 15^2$

Here, the value of x must be greater than 17 as $17^2 = 8^2 + 15^2$. Hence, value of x = 18, 19, 20, ………

But, one side of a triangle is always less than sum of other two sides. Then, $x = 18, 19, 20, 21, 22$

For possibility (ii), $15^2 > 8^2 + x^2$

Here, the value of x must be less than 13 as $15^2 < 8^2 + 13^2$

Hence, value of $x = 12, 11, 10, \ldots\ldots$

But, one side of a triangle is always less than sum of other two sides. Then, $x = 12, 11, 10, 9, 8$.

Hence, there are 10 possible values of x. It implies that 10 such triangles exist.

Problem 2 **The side of a triangle are 5 cm, 7 cm and 3 cm name the triangle.**

Sol. Since, $7^2 > 5^2 + 3^2$, the triangle is an obtuse triangle.

Problem 3 **The side BC of ΔABC is produced to D. If $\angle ACD = 180°$ and $\angle B = \dfrac{\angle A}{2}$, then $\angle A = ?$** ***[SSC(CGL-T-1)-12]***

Sol. An exterior angle = sum of the opposite interior angles

$$\Rightarrow 108° = \angle A + \angle B \Rightarrow 108° = \angle A + \frac{\angle A}{2} \Rightarrow 108° = \frac{3\angle A}{2}$$

$$\Rightarrow \angle A = 72°$$

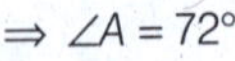

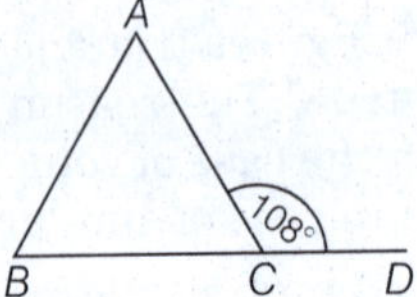

Remark

Certain questions based on the concept of exterior angles have also been given under previous chapter.

Problem 4 **In the given diagram, $CD = BF = 10$ units and $\angle CED = \angle BAF = 30°$. What would be the area of ΔAED?** ***[XAT-15]***

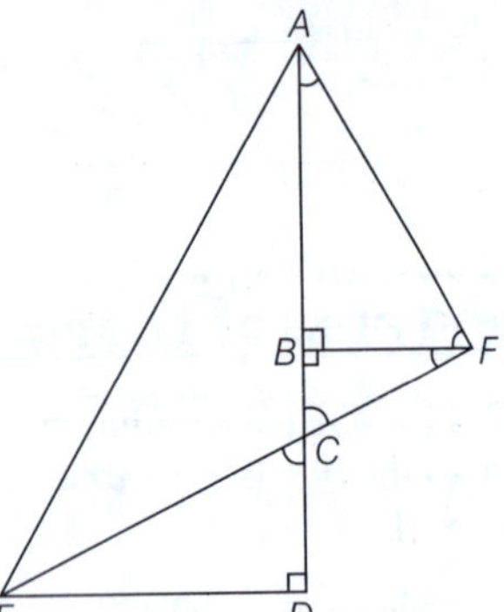

Sol. In ΔCED, $\tan 30° = \dfrac{CD}{ED} \Rightarrow \dfrac{1}{\sqrt{3}} = \dfrac{10}{ED} \Rightarrow ED = 10\sqrt{3}$

In ΔABF, $\tan 30° = \dfrac{FB}{AB} \Rightarrow \dfrac{1}{\sqrt{3}} = \dfrac{10}{AB} \Rightarrow AB = 10\sqrt{3}$

In ΔCBF, $\tan 30° = \dfrac{BC}{BF} \Rightarrow \dfrac{1}{\sqrt{3}} = \dfrac{BC}{10} \Rightarrow BC = \dfrac{10}{\sqrt{3}}$

Hence, $AD = AB + BC + CD = 10\sqrt{3} + \dfrac{10}{\sqrt{3}} + 10$

$$= 10\left(\sqrt{3} + \frac{1}{\sqrt{3}} + 1\right) = \frac{40 + 10\sqrt{3}}{\sqrt{3}}$$

Now, area of $\Delta AED = \dfrac{1}{2} \times ED \times AD$

$$\Rightarrow \frac{1}{2} \times 10\sqrt{3} \times \frac{40 + 10\sqrt{3}}{\sqrt{3}} = 50(\sqrt{3} + 4)$$

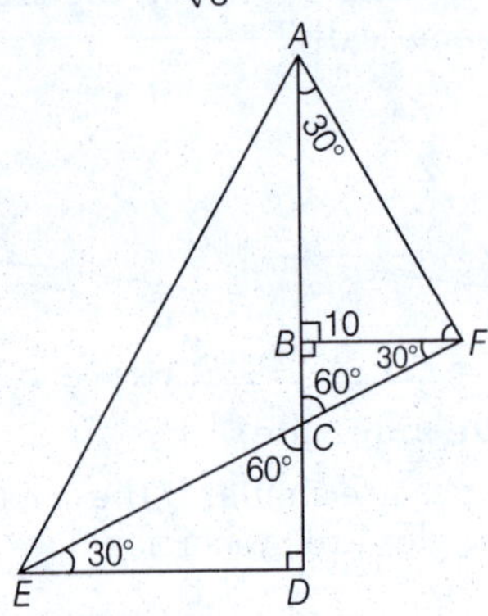

Problem 5 **In a ΔABC, $AB = AC$, BC is produced to D in such a way that $AC = CD$. If $\angle BAD = 84°$, then $\angle BAC = ?$** ***[SSC (FCI-Mains)-12]***

Sol. In ΔABC, $AB = AC$, then $\angle ABC = \angle ACB = \angle 1$

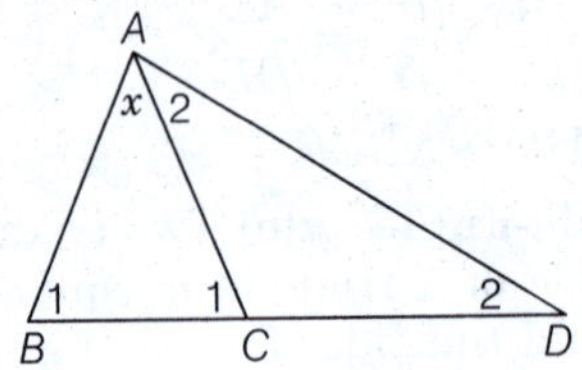

In ΔACD, $AC = CD$, then $\angle CAD = \angle ADC = \angle 2$

Now, $\angle 1 = \angle 2 + \angle 2$

[$\because$ an exterior angle = sum of interior opposite angles]

$\Rightarrow \angle 1 = 2\angle 2$

In $\Delta ABC = x + \angle 1 + \angle 1 = 180°$ (let $\angle BAC$ be x)

$\Rightarrow \quad x + 2\angle 2 + 2\angle 2 = 180°$

$\Rightarrow \quad x + 4\angle 2 = 180°$

$\Rightarrow \quad x + 4(84° - x) = 180°$ [$\because \angle BAD = x + \angle 2 = 84°$]

$\Rightarrow \quad 3x = 156° \Rightarrow x = 52°$

Problem 6 **Find the area of a triangle whose two sides are 12 cm and 15 cm and the included angle is 45°.**

Sol. Area $= \dfrac{1}{2} \times$ product of two sides $\times$ sine of included angle

$$= \frac{1}{2} \times 12 \times 15 \times \sin 45° = 90 \times \frac{1}{\sqrt{2}} = 45\sqrt{2} \text{ cm}^2$$

Problem 7 **In a ΔABC, $\angle A = 60°$ and the side opposite to it is a, $AC = b$, $AB = c$, then which of the followings is true?**

(a) $a^2 = b^2 + c^2 + 2bc$ (b) $a^2 = a^2 + bc - ac$

(c) $a^2 = b^2 + c^2 - bc$ (d) $a^2 = b^2 + c^2$

Sol. $a^2 = b^2 + c^2 - 2bc\cos A$ [Cosine rule]

$\Rightarrow a^2 = b^2 + c^2 - 2bc\cos 60°$

$\Rightarrow a^2 = b^2 + c^2 - 2bc.\dfrac{1}{2}$

$\Rightarrow a^2 = b^2 + c^2 - bc$

Problem 8 If A, B and C be the angles of a triangle, then of the following the incorrect relation is :

(a) $\sin\left(\frac{A+B}{2}\right)=\cos\frac{C}{2}$ (b) $\cos\left(\frac{A+B}{2}\right)=\sin\frac{C}{2}$

(c) $\tan\left(\frac{A+B}{2}\right)=\sec\frac{C}{2}$ (d) $\cot\left(\frac{A+B}{2}\right)=\tan\frac{C}{2}$

Sol. Clearly, $A+B+C=180°$

$\Rightarrow \frac{A+B+C}{2}=90° \Rightarrow \frac{A+B}{2}=90°-\frac{C}{2}$

Now, $\sin\left(\frac{A+B}{2}\right)=\sin\left(90°-\frac{C}{2}\right)$

$\Rightarrow \sin\left(\frac{A+B}{2}\right)=\cos\frac{C}{2}$

Further, $\cos\left(\frac{A+B}{2}\right)=\cos\left(90°-\frac{C}{2}\right)$

$\Rightarrow \cos\left(\frac{A+B}{2}\right)=\sin\frac{C}{2}$

Also, $\tan\left(\frac{A+B}{2}\right)=\tan\left(90°-\frac{C}{2}\right)$

$\Rightarrow \tan\left(\frac{A+B}{2}\right)=\cot\frac{C}{2}$

and $\cot\left(\frac{A+B}{2}\right)=\cot\left(90°-\frac{C}{2}\right)$

$\Rightarrow \cot\left(\frac{A+B}{2}\right)=\tan\frac{C}{2}$

Hence, option (c) is incorrect.

Problem 9 In a ΔABC, points D, E and F are lying on sides AB, BC and CA respectively in such a way that $\angle BDE=\angle DBE$ and $\angle CFE=\angle ECF$. What is the value of $\angle DEF$ if $\angle BAC=60°$?

Sol. Clearly, $\angle DEB=180°-\angle BDE-\angle DBE$

$=180°-2\angle DBE$ $[\because \angle BDE=\angle DBE]$

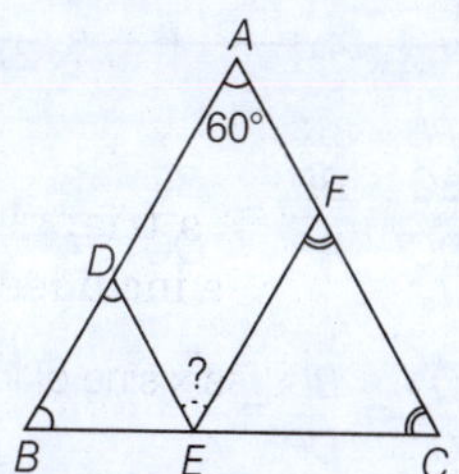

Similarly, $\angle FEC=180°-2\angle FCE$

Then, $\angle DEF=180°-\angle DEB-\angle FEC$

$=180°-(180°-2\angle DBE)-(180°-2\angle FCE)$

$=2\angle DBE-180°+2\angle FCE$

$=2(\angle DBE+\angle FCE)-180°$

$=2(180°-\angle A)-180°$

$=2(180°-60°)-180°=60°$

Alternatively

$\angle DEC=\angle BDE+\angle DBE$

[∵ an exterior angle = sum of interior opposite angles]

$\Rightarrow \angle DEC=2\angle DBE$ $[\because \angle BDE=\angle DBE]$

Similarly, $\angle FEB=2\angle FCE$

On addition,

$\angle DEC+\angle FEB=2(\angle DBE+\angle FCE)$

$\Rightarrow 180°+\angle DEF=2(180°-60°)$

$\Rightarrow \angle DEF=60°$

Problem 10 In the figure (not drawn to scale) given if $AD=CD=BC$ and $\angle BCE=96°$, how much is $\angle DBC$? **[CAT-03]**

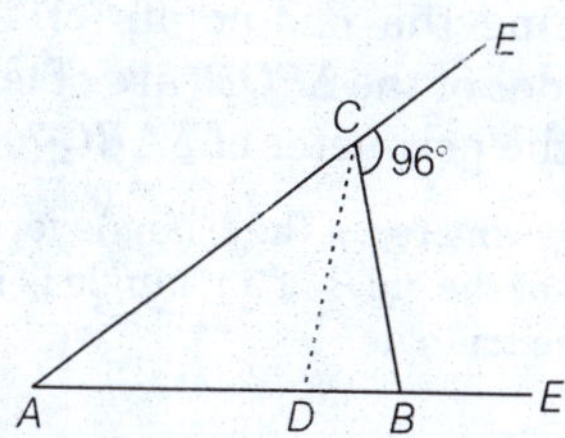

Sol. Here, $\angle 1=\angle 4+\angle 5$

[∵ exterior angle = sum of opposite interior angles]

$\Rightarrow \angle 1=\angle 4+\angle 4$ $[\because AD=CD \therefore \angle 4=\angle 5]$

$\Rightarrow \angle 1=2\angle 4$

$\Rightarrow \angle 2=2\angle 4$ $[\because CD=BC \therefore \angle 2=\angle 1]$

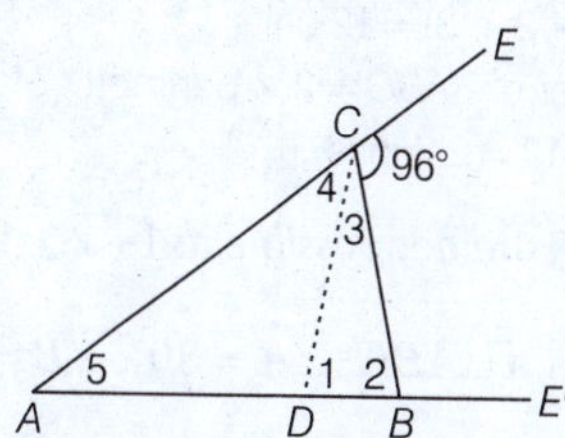

Clearly, $\angle 3+\angle 4=180°-96°=84°$ ...(i)

and $\angle 1+\angle 2+\angle 3=180°$

$\Rightarrow 2\angle 4+2\angle 4+\angle 3=180°$

$\Rightarrow 4\angle 4+\angle 3=180°$ ...(ii)

From Eqs. (i) and (ii), we get $\angle 4=32°$

Then, $\angle 2=2\angle 4=2\times 32°=64°$

Hence, $\angle DBC=64°$

Problem 11 In ΔABC, point D lies on side BC such that $BD=\frac{1}{4}DC$ and point M lies on line AD in such a way that $AM=\frac{MD}{3}$. Find the area of ΔABM, if area of $\Delta ABC=600\text{ cm}^2$.

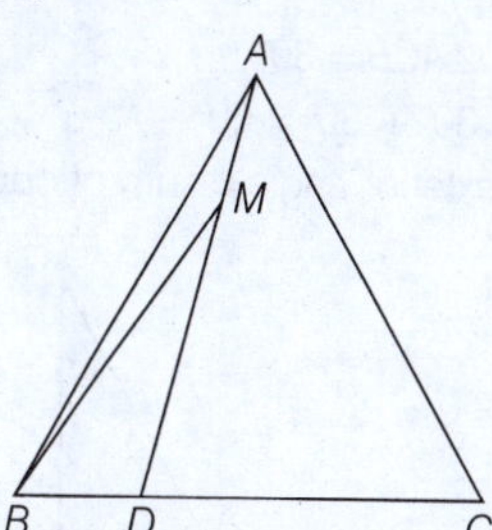

Sol. Clearly, ΔABD and ΔADC have same height but the base of ΔADC is four times that of ΔABD. Hence, ratio of areas of ΔABD and $\Delta ADC=1:4$

$\therefore$ Area of $\Delta ABD=600\times\frac{1}{5}=120$

Further, ratio of areas of ΔABM and $\Delta BDM = 1:3$ as they have same height and ratio of their bases is 1 : 3.

Then, area of $\Delta ABM = 120 \times \frac{1}{4} = 30\,\text{cm}^2$.

Problem 12 A triangle DEF is formed by joining the mid-points of the sides of ΔABC. Similarly, a ΔPQR is formed by joining the mid-points of the sides of the ΔDEF. If the sides of the ΔPQR are of lengths 1, 2 and 3 units, what is the perimeter of ΔABC? **[CDS-13]**

Sol. We know, the perimeter of the triangle formed by joining the mid-points of the sides of a triangle is half the perimeter of the triangle

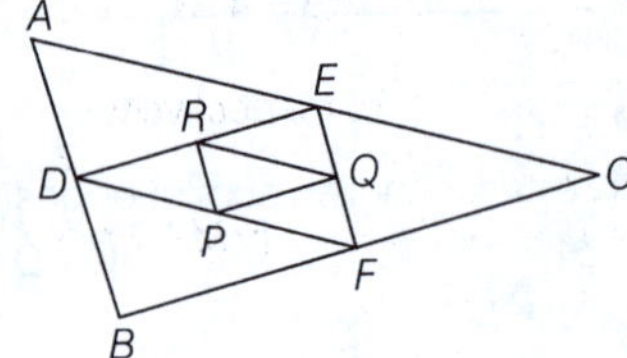

Hence, perimeter of $\Delta DEF = 2 \times$ perimeter of ΔPQR

$= 2(1 + 2 + 3) = 12$ units

Further, perimter of $\Delta ABC = 2 \times$ perimeter of ΔDEF

$= 2 \times 12 = 24$ units

Note *Side of ΔPQR are not possible as $1 + 2 \ngtr 3$*

Problem 13 In a ΔABC, $\angle A = 30°$, $\angle B = 60°$, then $a:b:c = ?$

Sol. Clearly, $\angle C = 90°$

We know, $\frac{a}{\sin A} = \frac{b}{\sin B} = \frac{c}{\sin C}$ (sine-rule)

$\Rightarrow \quad \frac{a}{\sin A} = \frac{b}{\sin B} = \frac{c}{\sin C} = k$ (let)

$\Rightarrow \quad a:b:c = k\sin A : k\sin B : k \cdot \sin C$

$\Rightarrow \quad \sin 30° : \sin 60° : \sin 90°$

$= \frac{1}{2} : \frac{\sqrt{3}}{2} : 1 = 1 : \sqrt{3} : 2$

Problem 14 ABC is an isosceles triangle with $AB = AC$. The side BA is produced to D such that $AB = AD$. If $\angle ABC = 30°$, then $\angle BCD$ is equal to **[SSC(10+2/LDC-DEO)-13, XAT-16]**

Sol. Clearly, $\angle ABC = \angle ACB = 30°$

Then, $\angle DAC = 30° + 30° = 60°$

[exterior angle = sum of opposite interior angles]

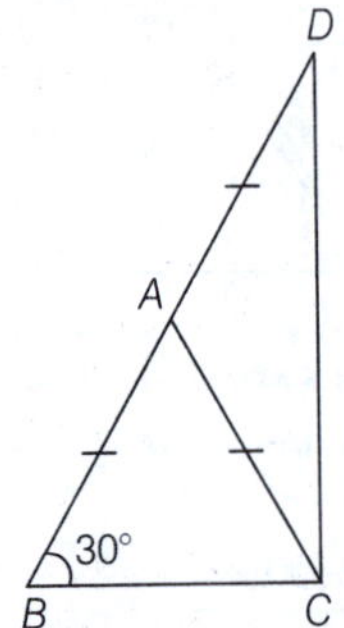

Further, $\angle ADC = \angle ACD \{\because AD = AC\}$

$\therefore \quad \angle ACD = \frac{180° - \angle DAC}{2} = \frac{180° - 60°}{2} = 60°$

Then, $\angle BCD = \angle ACB + \angle ACD$

$= 30° + 60° = 90°$

Problem 15 In a ΔABC, $\angle B = 45°$ and $b:c = 1:\sqrt{2}$, then $\angle A = ?$

Sol. We know, $\frac{b}{\sin B} = \frac{c}{\sin C}$ (sine-rule)

$\Rightarrow \quad \frac{\sin C}{\sin 45°} = \frac{c}{b} = \frac{\sqrt{2}}{1} \Rightarrow \frac{\sin C}{\frac{1}{\sqrt{2}}} = \frac{\sqrt{2}}{1}$

$\Rightarrow \quad \sin C = 1 \Rightarrow \sin C = \sin 90° \Rightarrow C = 90°$

Then, $\angle A = 180° - 45° - 90° = 45°$

Problem 16 In the figure (not drawn to scale) given P is a point on AB such that $AP:PB = 4:3$. PQ is parallel to AC and QD is parallel to CP. In ΔARC, $\angle ARC = 90°$ and in ΔPQS, $\angle PSQ = 90°$.

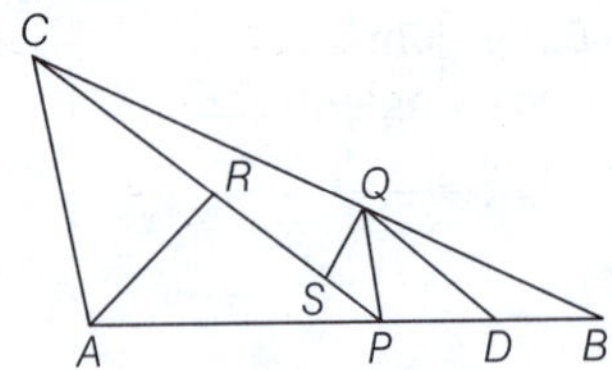

The length of $QS = 6$ cm. What is ratio $AP:PD$? **[CAT-03]**

Sol.

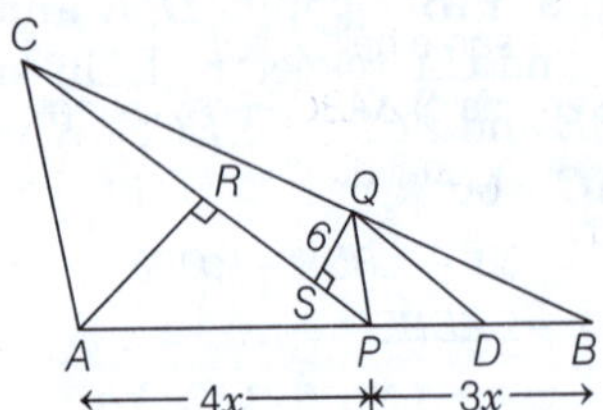

Let $AP = 4x$ and $PB = 3x$

Since, $PQ \parallel AC$, $\frac{BQ}{QC} = \frac{BP}{AP} \Rightarrow \frac{BQ}{QC} = \frac{3}{4}$

Also, $QD \parallel CP$

Then, $\frac{BD}{DP} = \frac{BQ}{QC} \Rightarrow \frac{BD}{DP} = \frac{3}{4}$

Let $BD = 3y$ and $PD = 4y$

Now, $PD + BD = 3x$

$\Rightarrow \quad 4y + 3y = 3x$

$\Rightarrow \quad 4y = \frac{12x}{7} \Rightarrow PD = \frac{12x}{7}$

Then, $AP : PD = 4x : \frac{12x}{7} = 7:3$

Problem 17 In a ΔABC, $a = 3$, $b = 5$ and $c = 7$, then value of the greatest angle = ?

Sol. Since the greatest side have the greatest angle in the triangle, $\angle C$ will be the greatest. Further, since we have to find $\angle C$, we should apply one of the cosine rules which has $\cos C$ in its term, i.e. $c^2 = a^2 + b^2 - 2ab\cos C$

$\Rightarrow \quad 7^2 = 3^2 + 5^2 - 2\cdot 3\cdot 5\cos C$

$\Rightarrow \quad \cos C = -\dfrac{1}{2}$

$\Rightarrow \quad \cos C = \cos(120°) \Rightarrow \angle C = 120°$

Problem 18 If a, b and c are the sides of a triangle and $a^2 + b^2 + c^2 = bc + ca + ab$, then the triangle is : **[CAT-2000]**

(a) equilateral (b) isosceles
(c) right-angled (d) obtuse-angled

Sol. Given, $a^2 + b^2 + c^2 = bc + ca + ab$

$\Rightarrow \quad 2a^2 + 2b^2 + 2c^2 = 2bc + 2ca + 2ab$

$\Rightarrow \quad 2a^2 + 2b^2 + 2c^2 - 2bc - 2ac - 2ab = 0$

$\Rightarrow \quad (a-b)^2 + (b-c)^2 + (c-a)^2 = 0$

It is possible only when each term is equal to 0 because value of any term cannot be negative.

Hence, $(a-b)^2 = (b-c)^2 = (c-a)^2 = 0 \Rightarrow a = b = c$

Thus, the given triangle is equilateral.

Problem 19 Side AB of rectangle $ABCD$ is divided into four equal parts by points x, y, z. The ratio of the area (Δxyc) and area (rectangle $ABCD$) is? **[SSC(CGL-T-1)-12]**

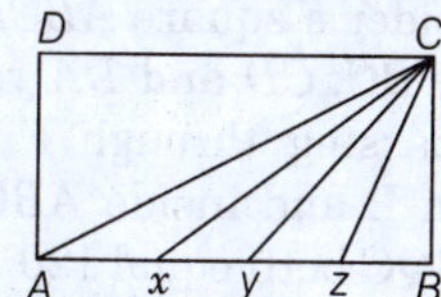

Sol. Ratio of areas of Δxyc and $\Delta ABC = 1:4$

[$\because$ they have same height but ratioof their bases is 1 : 4]

Further, ratio of areas of ΔABC and rectangle $ABCD = 1:2$

[$\because$ diagonal divides a rectangle in two equal parts]

Then, required ratio = 1 : 8

Problem 20 In the given figure, ΔABC is a right-angled triangle and $BMPN$ is a square. What is the area of ΔABC if $AP = 100$ cm and $PC = 75$ cm.

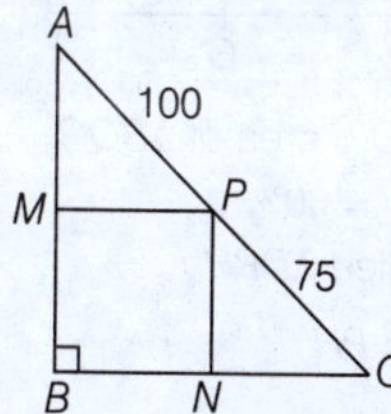

Sol. Since $BMPN$ is a square, $MP \parallel BN$. Thus, $MP \parallel BC$.

We know, if a line is drawn parallel to one side of a triangle to intersect the other two sides in distinct points, then the other two sides are divided in the same ratio. Hence, $\dfrac{AM}{MB} = \dfrac{AP}{PC}$

$\Rightarrow \dfrac{AM}{MB} = \dfrac{100}{75} \Rightarrow AM : MB = 4:3$

Let AM and MB be $4x$ and $3x$ respectively.

In ΔAMP, $AP^2 = AM^2 + MP^2$

$\Rightarrow 100^2 = (4x)^2 + (3x)^2$ [$\because MP = MB$, sides of a square]

$\Rightarrow 400 = x^2 \Rightarrow x = 20$

Then, $AB = 4x + 3x = 7x = 7 \times 20 = 140$

In ΔABC, $AC^2 = AB^2 + BC^2$

$\Rightarrow \quad 175^2 = 140^2 + BC^2$

$\Rightarrow \quad BC^2 = 11025 \Rightarrow BC = 105$

Now, area of $\Delta ABC = \dfrac{1}{2} \times BC \times AB = \dfrac{1}{2} \times 105 \times 140$

$= 7350 \text{ cm}^2$

Alternatively

Since $BMPN$ is a square, $MP \parallel BN$. Thus, $MP \parallel BC$

We know, if a line is drawn parallel to one side of a triangle to intersect the other two sides in distinct points, then the other two sides are divided in the same ratio. Hence, $\dfrac{AM}{MB} = \dfrac{AP}{PC}$

$\Rightarrow \quad \dfrac{AM}{MB} = \dfrac{100}{75} \Rightarrow AM : MB = 4:3$

Let AM and MB be $4x$ and $3x$ respectively.

In ΔAMP, $AP^2 = AM^2 + MP^2$

$\Rightarrow \quad 100^2 = (4x)^2 + (3x)^2$ [$\because MP = MB$, sides of a square]

$\Rightarrow \quad 400 = x^2 \Rightarrow x = 20$

Hence $MB = 3 \times 20 = 60$ and

$AB = 7x = 7 \times 20 = 140$

Since $PN \parallel AB$, $\dfrac{PC}{AP} = \dfrac{CN}{BN} \Rightarrow \dfrac{75}{100} = \dfrac{CN}{60}$

[$\because BN = MB$, sides of a square]

$\Rightarrow \quad CN = 45$

Then, $BC = BN + CN$

$= 60 + 45 = 105$

$\therefore$ Area of $\Delta ABC = \dfrac{1}{2} \times BC \times AB = \dfrac{1}{2} \times 105 \times 140 = 7350 \text{cm}^2$

Problem 21 An equilateral triangle BPC is drawn inside a square $ABCD$. What is the value of the angle APD in degrees ? **[CAT-06]**

Sol.

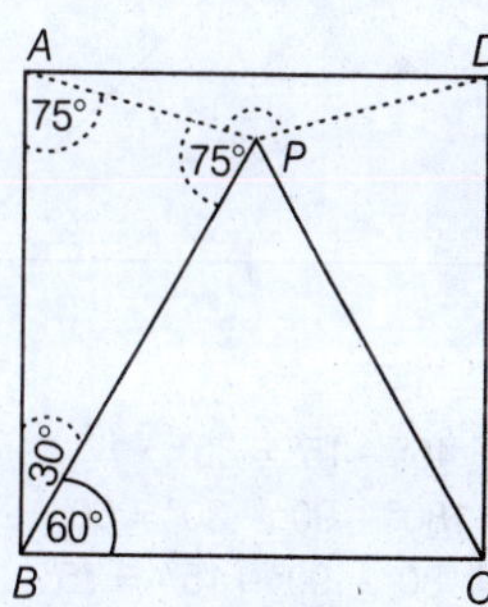

Here, $\angle PBC = \angle PCB = \angle BPC = 60°$ [$\because BPC$ is an equilateral Δ]

Then , $\angle ABP = 90° - 60° = 30°$

In ΔABP, $AB = BP$ [$\because BC = BP$ and $BC = AB$]

$\therefore \angle APB = \angle PAB$ {$\because$ angles opposite to equal sides are equal}

Hence, $\angle APB = \angle PAB = (180° - 30°)/2 = 75°$

Similarly, $\angle CPD = 75°$

Then, $\angle APD = 360° - \angle APB - \angle BPC - \angle CPD$

$= 360° - 75° - 60° - 75° = 150°$

Problem 22 In ΔABC, points M, N and R divide AB, AC and BC respectively in ratio $3:4$, $1:3$ and $2:3$. BM, BR and CN are smaller parts of their respective sides. Find the area of ΔABC, if area of $\Delta BMR = 72 \text{ cm}^2$.

Sol. Area of ΔBMR : Area of $\Delta MRC = 2 : 3$

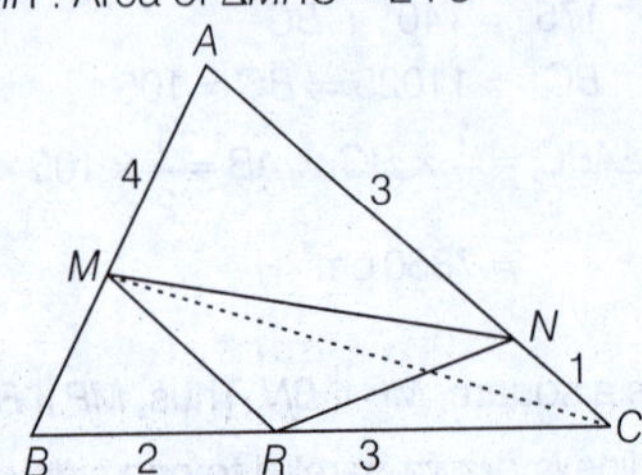

[$\because$ ratio of their bases = 2 : 3 and their heights are same]

$\therefore$ Area of $\Delta MRC = 72 \times 1.5 = 108$

Then, Area of ΔBMC = area of ΔBMR + area of ΔMRC

$= 72 + 108 = 180$

Further, area of ΔBMC : area of $\Delta AMC = 3 : 4$

[$\because$ ratio of their bases = 3 : 4 and their heights are same]

Then, area of $\Delta AMC = 180 \times \frac{4}{3} = 240$

Hence, area of ΔABC = area of ΔBMC + area of ΔAMC

$= 180 + 240 = 420 \text{ cm}^2$

Problem 23 An isosceles triangle ABC is right-angled at B. D is a point inside the triangle ABC. P and Q are the feet of the perpendiculars drawn from D on the side AB and AC respectively of ΔABC. If $AP = a$ cm, $AQ = b$ cm and $\angle BAD = 15°$, $\sin 75° = ?$ **[SSC(CGL-T-II)-13]**

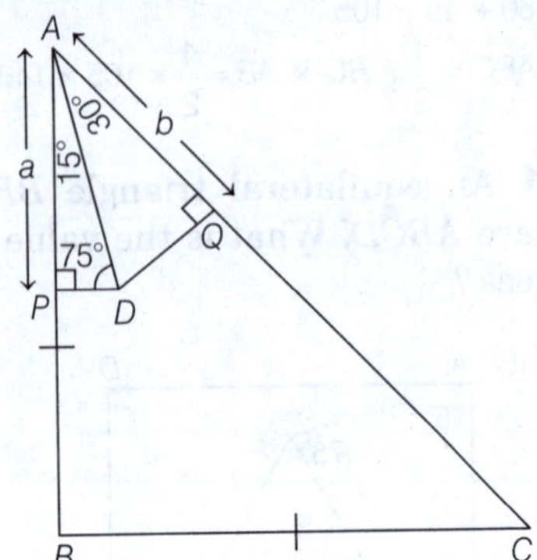

Sol. Clearly, $\angle A = \angle C = 45°$

Then, $\angle QAD = 45° - 15° = 30°$

$\therefore$ $\angle ADQ = 180° - 90° - 30° = 60°$

and $\angle ADP = 180° - 90° - 15° = 75°$

In ΔADQ, $\sin 60° = \frac{b}{AD}$

$\Rightarrow$ $AD = \frac{b}{\sin 60°}$

In ΔAPD, $\sin 75° = \frac{a}{AD}$

$= a \div \left(\frac{b}{\sin 60°}\right) = \frac{\sqrt{3}a}{2b}$

Problem 24 In ΔDEF, shown in the figure, points A, B and C are taken on DE, DF and EF respectively such that $EC = AC$ and $CF = BC$. $\angle D = 40°$, then what is $\angle ACB$ in degrees? **[CAT-01]**

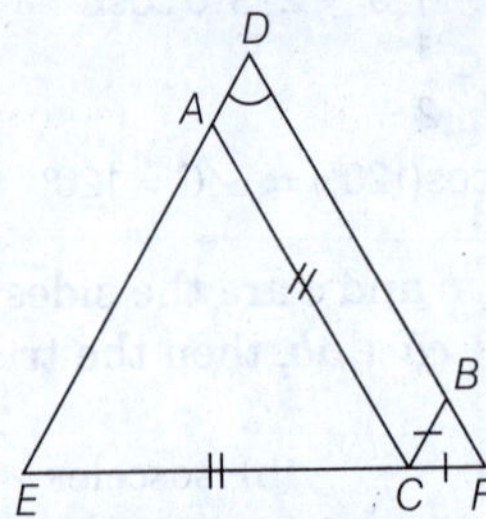

Sol. $\because$ $EC = AC$ $\therefore \angle EAC = \angle AEC$

Further, $CF = BC$ $\therefore \angle CBF = \angle BFC$

Clearly, $\angle D + \angle E + \angle F = 180°$

$\Rightarrow$ $40° + \angle AEC + \angle CBF = 180°$

$\Rightarrow$ $\angle EAC + \angle CBF = 140°$

$\Rightarrow$ $(180° - \angle DAC) + [180° - \angle CBD] = 140°$

$\Rightarrow$ $\angle DAC + \angle CBD = 220°$

Now, $\angle CAD + \angle ADB + \angle CBD + \angle BCA = 360°$

[$\because$ sum of all four angles in a quadrilateral be 360°]

$\Rightarrow \angle CAD + 40° + \angle CBD + \angle ACB = 360°$

$\Rightarrow$ $220° + 40° + \angle ACB = 360° \Rightarrow \angle ACB = 100°$

Problem 25 Consider a square $ABCD$ with mid-points E, F, G, H of AB, BC, CD and DA respectively. Let L denote the line passing through F and H. Consider points P and Q on L and inside $ABCD$, such that the angles APD and BQC both equal 120°. What is the ratio of the area of $ABQCDP$ to the remaining area insdie $ABCD$? **[CAT-08]**

Sol.

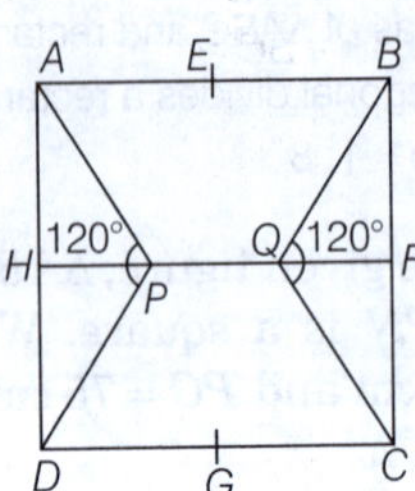

Clearly, area of ΔAPD = area of ΔBQC

Also, $\angle APH = \angle DPH = 60°$

Area of the right-angled ΔDPH

$= \frac{1}{2} \times HD \times PH$

$= \frac{1}{2} \times HD \times \frac{HD}{\sqrt{3}}$ [$\because \tan 60° = \frac{HD}{PH}$, $PH = HD/\sqrt{3}$]

Then, area (ΔAPD) = 2 × area (ΔDPH)

$= 2 \times \frac{1}{2} \times HD \times \frac{HD}{\sqrt{3}} = \frac{HD^2}{\sqrt{3}}$

Now, area of $ABQCDP$ = area of $ABCD$ – 2 (area of ΔAPD)

$= AD^2 - 2 \cdot \frac{HD^2}{\sqrt{3}} = (2HD)^2 - 2 \cdot \frac{HD^2}{\sqrt{3}}$

$\therefore$ Required ratio $= (2HD)^2 - 2 \cdot \frac{HD^2}{\sqrt{3}} : 2 \cdot \frac{HD^2}{\sqrt{3}}$

$= (2\sqrt{3} - 1) : 1$

Problem 26 In the given figure, $AB \perp BC, BD \perp AC$ and CE bisects $\angle C$, $\angle A = 30$. Then, what is $\angle CED$? **[CAT-95]**

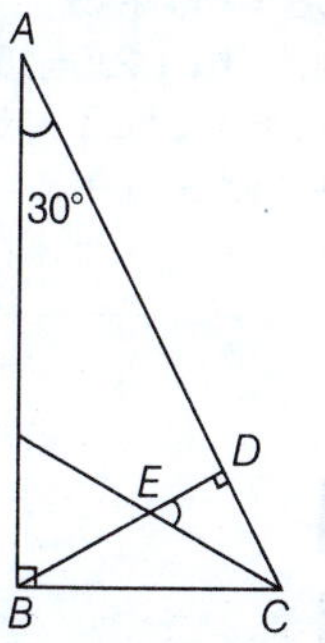

Sol. Obviously, $\angle BCA = 180° - \angle B - \angle A$

$= 180° - 90° - 30° = 60°$

Then, $\angle DCE = \dfrac{60°}{2} = 30°$ [$\because CE$ bisects $\angle C$]

In ΔDEC, $\angle DEC = 180° - \angle EDC - \angle DCE$

$= 180° - 90° - 30° = 60°$

Problem 27 In a right angled triangle, the product of two sides is equal to half of the square of the 3rd side i.e. hypotenuse. One of the acute angles must be : **[SSC(CGL-T-II)-13]**

Sol. Given, $AB \times BC = \dfrac{1}{2} \times AC^2$

$\Rightarrow AC^2 = 2 \times AB \times BC$

$\Rightarrow AB^2 + BC^2 = 2 \times AB \times BC$

$\Rightarrow AB^2 + BC^2 - 2 \times AB \times BC = 0$

$\Rightarrow (AB - BC)^2 = 0$

$\Rightarrow AB - BC = 0$

$\Rightarrow AB = BC$

Then, $\angle C = \angle A$ [$\because$ angles opposite to congruent sides are equal.]

Hence, $\angle C = \angle A = 45°$

Problem 28 In the given figure, $EADF$ is a rectangle and ABC is a triangle whose vertices lie on the sides of $EADF$. $AE = 22$, $BE = 6$, $CF = 16$ and $BF = 2$. Find the length of the line joining the mid points of the sides AB and BC. **[CAT-97]**

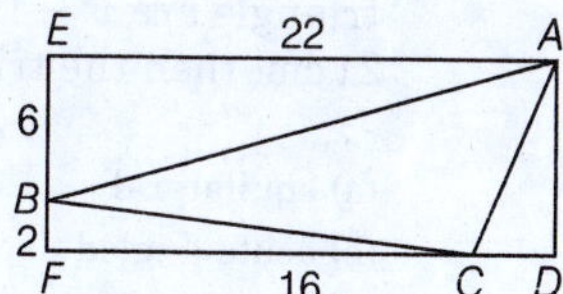

Sol. Clearly, $AD = EF = 6 + 2 = 8$ and $CD = FD - 16$

$= EA - 16 = 6$

ΔACD is right angled at D [$\because$ each angle of a rectangle is 90°]

Then, $AC = \sqrt{AD^2 + CD^2} = \sqrt{8^2 + 6^2} = 10$

We know, the line joining the mid-points of any two sides of a triangle is equal to half of the 3rd side.

Hence, required answer $= \dfrac{1}{2} \times AC = \dfrac{1}{2} \times 10 = 5$

Problem 29 The given figure shows two right angled triangles ΔOAB and ΔOQP with right angles at vertex A and P, respectively, having the common vertex O. The lengths of some of the sides are indicated in the figure. (Note that the figure is not drawn to scale.) AB and OP are parallel. What is $\angle QOB$? **[XAT-19]**

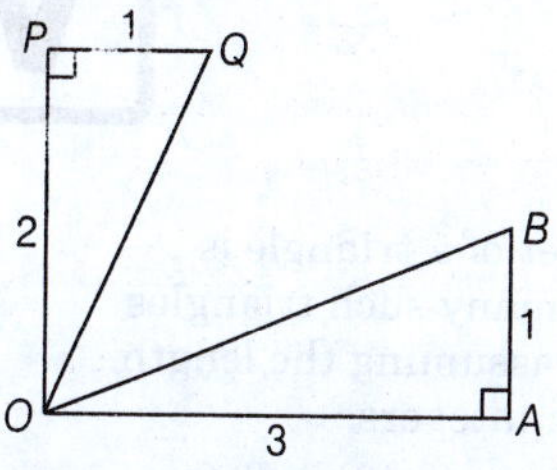

Sol. $OB = \sqrt{1^2 + 3^2} = \sqrt{10}$ and $OQ = \sqrt{1^2 = 2^2} = \sqrt{5}$

and $BQ = \sqrt{1^2 + 2^2} = \sqrt{5}$

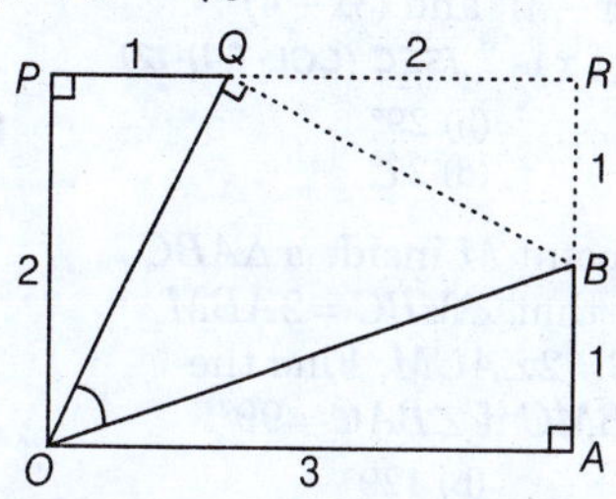

Since $OB^2 = OQ^2 + BQ^2$, ΔOQB is a right-angled triangle with right angle at vertex Q.

Then, $\angle QOB + \angle QBO = 180° - 90° = 90°$

$\Rightarrow \angle QOB + \angle QOB = 90°$

[$\because OQ = BQ$, the angles opposite to them are equal]

$\Rightarrow \angle QOB = 45°$

Problem 30 In the given figure, $AB = BC = CD = DE = EF = FG = GA$. Then, $\angle DAE$ is approximately : **[CAT-2000]**

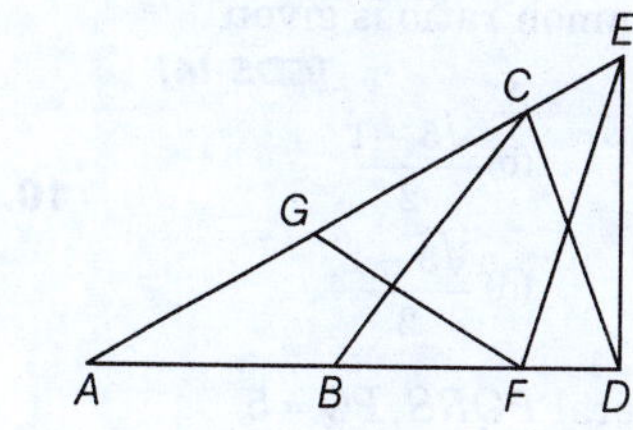

Sol.

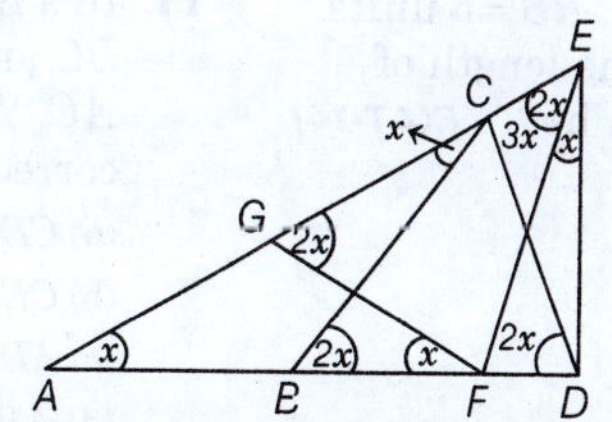

$\because AB = BC \therefore \angle BAC = \angle ACB = x$ (let)
Exterior $\angle CBD = x + x = 2x$
$\because BC = CD \therefore \angle CDB = 2x$
Clearly, $\angle BCD = 180° - 2x - 2x = 180° - 4x$
Then, $\angle DCE = 180° - x - (180° - 4x) = 3x$
$\therefore \angle DEC = 3x$ $[\because CD = DE]$
Further, $\angle GFA = x$
then exterior $\angle CGF = x + x = 2x$
$\therefore \angle GEF = 2x$ $[\because FG = EF]$
Then, $\angle FED = 3x - 2x = x$
Now, in ΔCED, $\angle CDE = 180° - 6x$
Then, $\angle EDF = 180° - 6x + 2x = 180° - 4x$
Clearly, $\angle EFD + \angle EDF + \angle FED = 180°$
$\Rightarrow 180° - 4x + 180° - 4x + x = 180°$
$\Rightarrow 180° - 7x = 0 \Rightarrow x = 25°$ (approx.)

WORKOUT

1. The perimeter of a triangle is 15 cm. How many such triangles are possible assuming the length of sides be in integers.
(a) 7 (b) 8 (c) 10 (d) 9

2. The angles of a triangle are $(x+5)°$, $(2x-3)°$ and $(3x+4)°$. The value of x is ? **[SSC (CGL-T-1)-12]**
(a) 30° (b) 29°
(c) 28° (d) 27°

3. There is a point M inside a ΔABC such a way that $\angle MBC = 2ABM$ and $\angle MCB = 2\angle ACM$. Find the value of $\angle BMC$ if $\angle BAC = 99°$.
(a) 126° (b) 129°
(c) 128° (d) 68°

4. The three sides of a triangle are 15, 25 and x units. Which one of the following is correct? **[CDS-14, 10]**
(a) $10 < x < 40$ (b) $10 \le x \le 40$
(c) $10 \le x < 40$ (d) $10 < x \le 40$

5. The sides of a triangle are in geometric progression with common ratio $r < 1$. If the triangle is a right angled triangle, the square of common ratio is given by : **[CDS-14]**
(a) $\frac{\sqrt{5}+1}{2}$ (b) $\frac{\sqrt{5}-1}{2}$
(c) $\frac{\sqrt{3}+1}{2}$ (d) $\frac{\sqrt{3}-1}{2}$

6. In quadrilateral $PQRS$, $PQ = 5$ units, $QR = 17$ units, $RS = 5$ units and $PS = 9$ units. The length of the diagonal QS can be **[XAT-14]**
(a) > 10 and < 12
(b) > 12 and < 14
(c) > 14 and < 16
(d) > 16 and < 18
(e) Cannot be determined

7. If the angles of a triangle are 90°, 60° and 30°, then what is the ratio of the sides opposite to these angles ?
(a) $\sqrt{3}:\sqrt{2}:1$
(b) $1:\sqrt{2}:2$
(c) $2:\sqrt{3}:1$
(d) $3:2:1$

8. A point D is taken from the side BC of a right-angled triangle ABC, where AB is hypotenuse. Then, **[SSC(CGL-T-II)-13]**
(a) $AB^2 + CD^2 = BC^2 + AD^2$
(b) $CD^2 + BD^2 = 2AD^2$
(c) $AB^2 + AC^2 = 2AD^2$
(d) $AB^2 = AD^2 + BD^2$

9. Consider the following statements :
A triangle can be constructed if its
I. two sides and the included angle are given.
II. three angles are given.
III. two angles and the included side are given.
Which of the statements given above are correct ?
(a) I and II (b) I and III
(c) II and III (d) All of these

10. In a ΔABC, $AB = AC$ and BA is produced to D such that $AC = AD$. Then $\angle BCD$ is equal to :
(a) 45° (b) 90° (c) 30° (d) 60°

11. In a ΔABC, AD is perpendicular to BC and BE is perpendicular to AC. Which of the following is correct ? **[CDS-14]**
(a) $CE \times CB = CA \times CD$
(b) $CE \times CA = CD \times CB$
(c) $AD \times BD = AE \times BE$
(d) $AB \times AC = AD \times BE$

12. If the sides of a triangle are extended in both the sides, then the sum of exterior angles so formed in both sides is : **[SSC (10 + 2/LDC-DEO)-14]**
(a) 360° (b) 720°
(c) 180° (d) 540°

13. In ΔABC, if $\angle B = 2\angle C = 2\angle A$, then what is the ratio of AC to AB?
(a) $\sqrt{2}:1$ (b) $\sqrt{3}:1$
(c) $1:1$ (d) $1:\sqrt{2}$

14. For a triangle, base is $6\sqrt{3}$ cm and two base angles are 30° and 60°. Then height of the triangle is
(a) $3\sqrt{3}$ cm (b) 4.5 cm
(c) $4\sqrt{3}$ cm (d) $2\sqrt{3}$ cm

15. The mid-points of sides of a triangle ABC are joined and the area of ΔPQR, thus formed, is 23 cm^2. What is the area of ΔABC ?
(a) 80 cm^2 (b) 69 cm^2
(c) 92 cm^2 (d) 115 cm^2

16. If the measure of the sides of a triangle are $x^2 - 1$, $x^2 + 1$ and $2x$ cm, then the triangle would be : **[SSC(CGL-T-I)-14]**
(a) equilateral
(b) acute-angled
(c) isosceles
(d) right-angled

17. In a square $PQRS$, A and B are two points on PS and SR such that $PA = 2\,AS$ and $RB = 2\,BS$. If $PQ = 6$, the area of the triangle ABQ is : **[XAT-13]**
(a) 6 (b) 8
(c) 10 (d) 12
(e) 14

18. In ΔABC, $\angle BAC = 90°$ and $AB = \frac{1}{2}$ BC, then measure of $\angle ACB$ is :

(a) 45° (b) 90°
(c) 30° (d) 60°

19. $ABCD$ is a square, F is mid-point of AB and E is a point on BC such that BE is 1/3rd of BC. If area of $\Delta FBE = 108\text{ m}^2$, then the length of AC is : ***[MAT-03]***

(a) 63 m
(b) $36\sqrt{2}$ m
(c) $63\sqrt{2}$ m
(d) $72\sqrt{2}$ m

20. By decreasing 15° of each angle of a triangle, the ratio of their angles are 2 : 3 : 5. The radian measure of the greatest angle is

(a) $\frac{11\pi}{24}$ radian (b) $\frac{13\pi}{24}$ radian
(c) $\frac{5\pi}{12}$ radian (d) $\frac{11\pi}{23}$ radian

21. If ΔFGH is isosceles and $FG < 3$ cm, $GH = 8$ cm, then of the following, the true relation is : ***[SSC(LDC/DEO)-13]***

(a) $GH = FH$
(b) $GF = GH$
(c) $FH > GH$
(d) $GH < GF$

22. Cosider the following inequalities in respect of any triangle ABC :

I. $AC - AB < BC$
II. $BC - AC < AB$
III. $AB - BC < AC$

Which of the above are correct? ***[CDS-19]***

(a) I and II (b) II and III
(c) I and III (d) All of these

23. In a ΔABC, $AB^2 + AC^2 = BC^2$ and $BC = \sqrt{2}AB$, then $\angle ABC$ is :

(a) 30° (b) 45°
(c) 60° (d) 90°

24. Let ABC be an isosceles triangle. Suppose that the sides AB and AC are equal and let the length of AB be x cm. Let b denotes the $\angle ABC$ and $\sin b = \frac{3}{5}$. If the area of the triangle ABC is M sq cm, then which of the following is true about M? ***[XAT-19]***

(a) $M < \frac{x^2}{4}$ (b) $\frac{x^2}{4} \le M < \frac{x^2}{2}$
(c) $\frac{3x^2}{4} \le M < x^2$ (d) $\frac{x^2}{2} \le M < \frac{3x^2}{4}$

25. From an interior point of an equilateral triangle, perpendiculars are drawn on all three sides. The sum of the lengths of the perpendiculars is s. Then, the area of the triangle is : ***[CAT-20]***

(a) $\frac{\sqrt{3}}{2}s^2$ (b) $\frac{s^2}{\sqrt{3}}$
(c) $\frac{2s^2}{\sqrt{3}}$ (d) $\frac{s^2}{2\sqrt{3}}$

Answer Key

1. (a)	2. (b)	3. (a)	4. (a)	5. (b)	6. (b)	7. (c)	8. (a)	9. (d)	10. (b)
11. (c)	12. (b)	13. (a)	14. (b)	15. (c)	16. (d)	17. (c)	18. (c)	19. (b)	20. (a)
21. (a)	22. (d)	23. (b)	24. (b)	25. (b)					

Hints and SOLUTIONS

1. We know that the sum of the length of any two sides of a triangle is always greater than length of the 3rd side. Also, the difference between lengths of any two sides of a triangle is always less than length of the 3rd side. Then, possible triangles are (1, 7, 7), (2, 6, 7), (3, 6, 6), (3, 5, 7), (4, 5, 6), (4, 4, 7), (5, 5, 5). Hence, number of triangles = 7

2. Sum of all interior angles of a triangle = 180°

$\Rightarrow (x+5)° + (2x-3)° + (3x+4)° = 180°$

$\Rightarrow 6x = 174° \Rightarrow x = 29°$

3. Given, $\angle A = 99°$, then

$\angle ABC + \angle ACB = 180° - 99° = 81°$

$\Rightarrow \quad (x + 2x) + (y + 2y) = 81°$

$\Rightarrow \quad 2x + 2y = 54°$

Then, $\angle BMC = 180° - 54° = 126°$

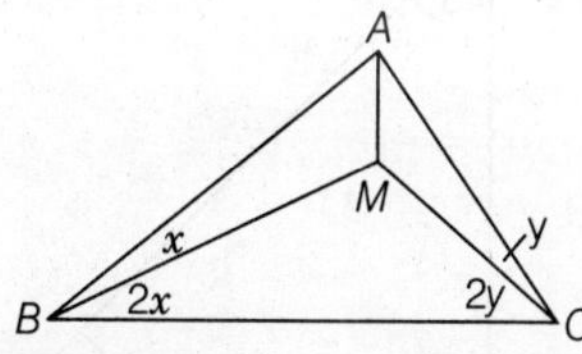

4. Since the sum of two sides is greater than the 3rd side, $x < 40$. Also, the difference between two sides must be less than the 3rd side, i.e. $10 < x$

Hence, $10 < x < 40$.

5. Let the sides be $\frac{a}{r}$, a and ar.

Since $r < 1$, $\frac{a}{r} > a > ar$. It implies that $\frac{a}{r}$ being the largest side, is the hypotenuse of the right-angled triangle.

Then, $\left(\frac{a}{r}\right)^2 = (a)^2 + (ar)^2$

$\Rightarrow \left(\frac{1}{r}\right)^2 = (1)^2 + (r)^2 \Rightarrow r^4 + r^2 - 1 = 0$

$\Rightarrow p^2 + p - 1 = 0$ [assuming $r^2 = p$]

$\Rightarrow p = \frac{-1 \pm \sqrt{1 - 4 \cdot 1 \cdot (-1)}}{2 \cdot 1}$

$= \frac{\sqrt{5}-1}{2}$ and $\frac{-\sqrt{5}-1}{2}$

Thus, option (b) is correct.

6. We know, one side of a triangle is always less than sum of other two sides. Then, in ΔPQS, QS must be less than $(PQ + PS)$ i.e. $QS < (5 + 9)$ and in ΔQRS, QR must be less than $(QS + SR)$ i.e. $17 < (QS + 5)$

$\Rightarrow QS > 12$

Hence, $14 > QS > 12$

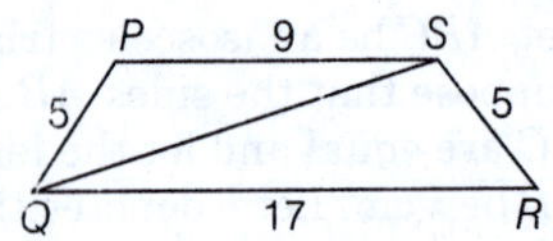

7. We know, $\frac{a}{\sin A} = \frac{b}{\sin B} = \frac{c}{\sin C}$ [Recall sine-rule]

$\Rightarrow \frac{a}{\sin 90^\circ} = \frac{b}{\sin 60^\circ} = \frac{c}{\sin 30^\circ}$

$\Rightarrow \frac{a}{1} = \frac{b}{\frac{\sqrt{3}}{2}} = \frac{c}{\frac{1}{2}} = k$ (let)

$\Rightarrow a = k, b = \frac{\sqrt{3}}{2}k$ and $c = \frac{1}{2}k$

Hence, $a : b : c = 2 : \sqrt{3} : 1$

8. In $\Delta ACB, AB^2 = AC^2 + BC^2$

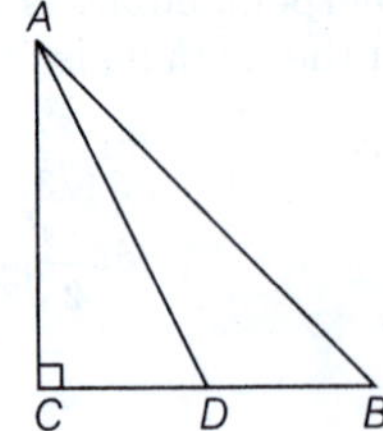

$\Rightarrow \quad AC^2 = AB^2 - BC^2$

and in $\Delta ACD, AD^2 = AC^2 + CD^2$

$\Rightarrow \; AD^2 = AB^2 - BC^2 + CD^2$

$\Rightarrow \; AB^2 + CD^2 = BC^2 + AD^2$

9. (*d*) $\because$ Clearly, option (d) is correct.

10. (*b*) $\because AB = AC \quad \therefore \angle ABC = \angle ACB$

Further, $AC = AD$, then

$\angle ADC = \angle ACD$

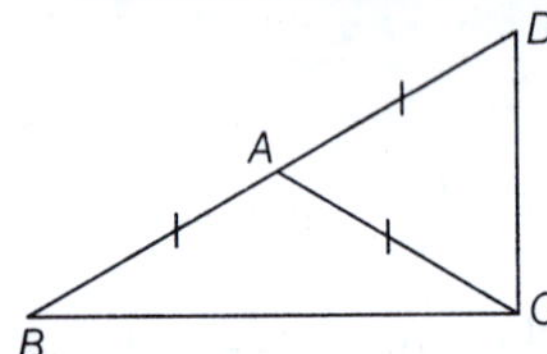

Now, $\angle CAD = \angle ABC + \angle ACB$

[$\because$ exterior angle = sum of opposite interior angles]

$\Rightarrow 180^\circ - \angle ADC - \angle ACD = \angle ACB + \angle ACB$

$\Rightarrow 180^\circ - \angle ACD - \angle ACD = \angle ACB + \angle ACB$

$\Rightarrow 180^\circ = 2(\angle ACD + \angle ACB)$

$\Rightarrow 180^\circ = 2\angle BCD \Rightarrow \angle BCD = 90^\circ$

11. Area of $\Delta AEB = \frac{1}{2} \times AE \times BE$

and area of $\Delta ABD = \frac{1}{2} \times BD \times AD$

Clearly, area of ΔAEB = area of ΔABD

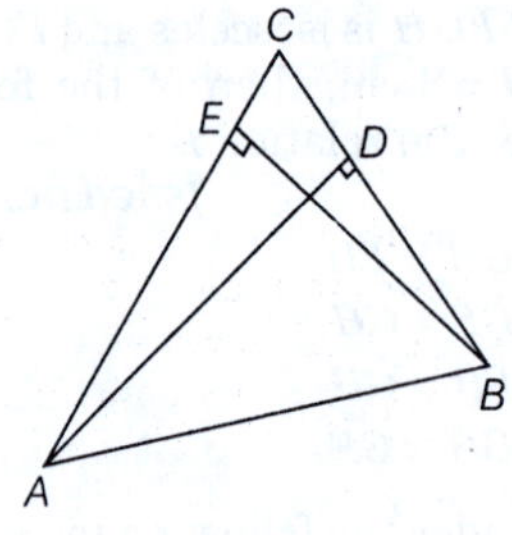

$\Rightarrow \frac{1}{2} \times AE \times BE = \frac{1}{2} \times BD \times AD$

$\Rightarrow \quad AE \times BE = BD \times AD$

12. Sum of of exterior angles in a polygon (here, Δ) = 360°

Hence, $\quad \angle 1 + \angle 2 + \angle 3 = 360^\circ$

Similarly, $\angle 4 + \angle 5 + \angle 6 = 360^\circ$

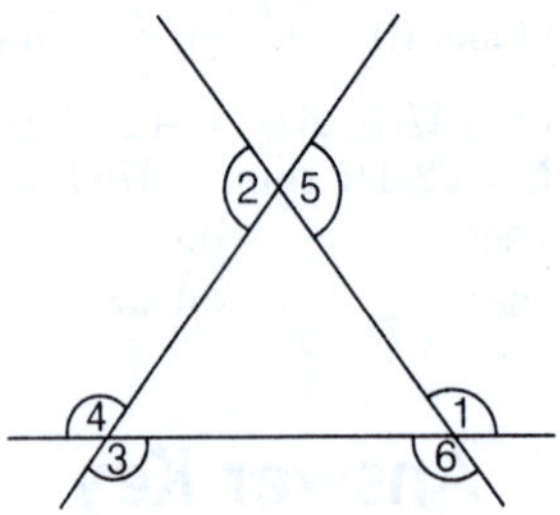

Then, required answer

$= 360^\circ + 360^\circ = 720^\circ$

13. Clearly, $\angle A + \angle B + \angle C = 180^\circ$

$\Rightarrow \angle A + 2\angle A + \angle A = 180^\circ$

$\Rightarrow \angle A = 45^\circ$

Then, $\angle C = 45^\circ$ and $\angle B = 90^\circ$

Hence, the given triangle is a right-angle triangle.

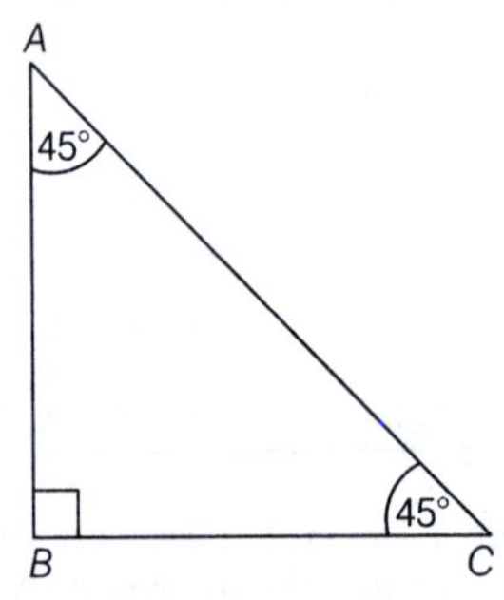

$\therefore \; \sin C = \frac{AB}{AC}$

$\Rightarrow \sin 45^\circ = \frac{AB}{AC} \Rightarrow \frac{1}{\sqrt{2}} = \frac{AB}{AC}$

$\Rightarrow AC : AB = \sqrt{2} : 1$

14. Clearly, $\angle A = 90^\circ$, Hence, the given triangle is right-angled.

In ΔABC, $\sin 30^\circ = \frac{AC}{BC}$

$\Rightarrow \frac{1}{2} = \frac{AC}{6\sqrt{3}} \Rightarrow AC = 3\sqrt{3}$

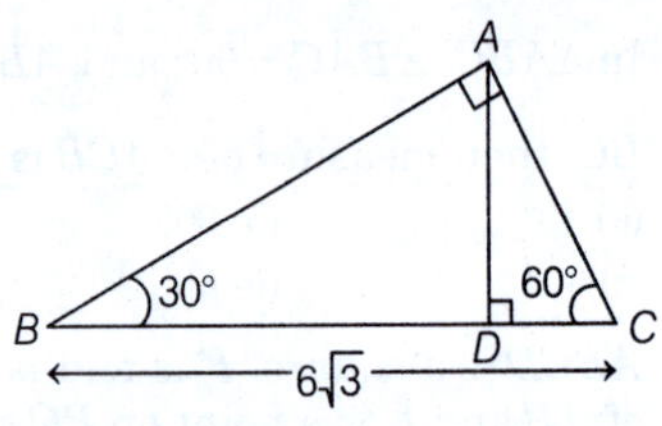

In ΔACD, $\sin 60^\circ = \frac{AD}{AC}$

$\Rightarrow \frac{\sqrt{3}}{2} = \frac{AD}{3\sqrt{3}} \Rightarrow AD = 4.5$ cm

15. Since all the four triangles formed by the line segments joining the mid-points of sides of a triangle are congruent, the area of $\Delta ABC = 4 \times 23 = 92 \text{ cm}^2$

16. Take any value of x (say, 2). Then, sides are 3, 5 and 4, hence, a Phythagorean triplet.

$\therefore$ The triangle is a right-angled triangle.

17. Here, $PQ = QR = RS = SP = 6$ and $AS = \frac{6}{3} = 2 = SB$

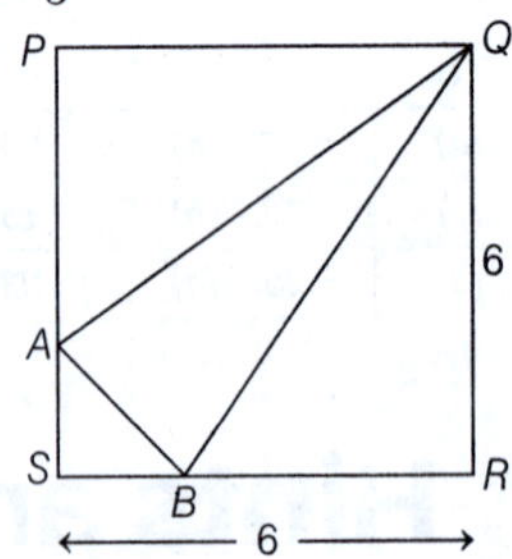

Area of ΔABQ = area of square $PQRS$ – area of ΔASB – area of ΔBRQ – area of ΔAPQ

$= 6 \times 6 - \frac{1}{2} \times 2 \times 2 - \frac{1}{2} \times (6-2) \times 6 - \frac{1}{2} \times (6-2) \times 6$

$= 36 - 2 - 12 - 12 = 10$

18. $\sin C = \frac{AB}{BC} = \frac{AB}{2AB}$ [$\because BC = 2AB$]

$\Rightarrow \quad \sin C = \frac{1}{2}$

$\Rightarrow \quad \sin C = \sin 30^\circ$

$\Rightarrow \quad C = 30^\circ$

Hence, $\angle ACB = 30^\circ$

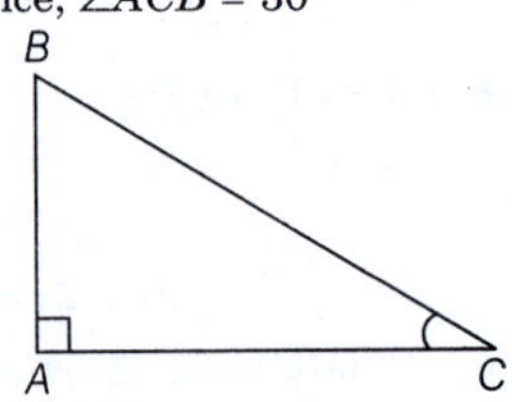

19. Area of $\Delta FBE = \frac{1}{2} \times BE \times FB$

$= \frac{1}{2} \times \frac{1}{3} BC \times \frac{1}{2} BC$

$[\because AB = BC]$

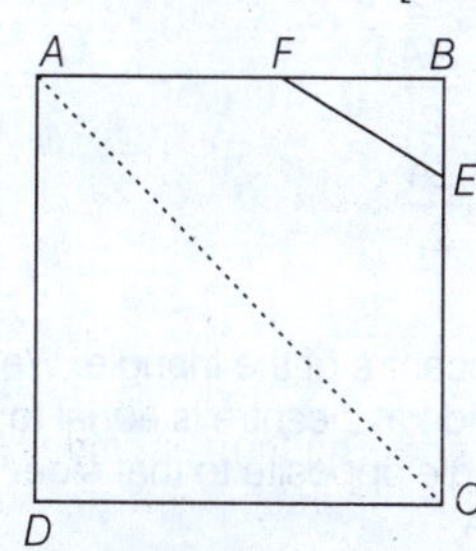

$\Rightarrow \quad 108 = \frac{1}{12} BC^2$

$\Rightarrow \quad BC = 12 \times 3 = 36$

Now, diagonal of the square

$= \sqrt{2} \times \text{side}$

$= 36\sqrt{2}$ m

20. Clearly, $(15° \times 3)$ has been reduced from 180°.

Then, $(2 + 3 + 5)$ units $= 180° - 45°$

$\Rightarrow \quad 5 \text{ units} = 67.5°$

Hence, the greatest angle

$= 67.5° + 15°$

$= 82.5°$

$\because \quad 180° = \pi$ radian

$\therefore \quad 82.5° = \frac{11\pi}{24}$ radian

21. Since ΔFGH is isosceles, any two sides are equal. The two sides are known. Then, the 3rd side i.e. FH is equal to either FG or GH.

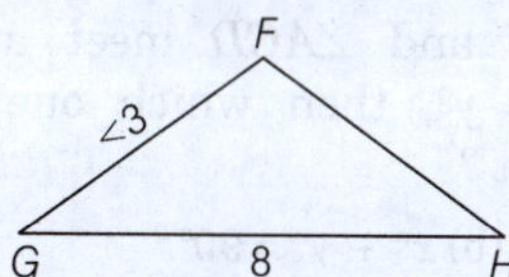

In case $FH = FG$, then sum of FH and FG will be lesser than the 3rd side, which contradicts the principle that the sum of any two sides of a triangle is greater than the 3rd side.

Hence, $FH = GH$.

22. Observe one of the fundamental properties of triangle, given in the beginning of this chapter, that the difference between lengths of any two sides of a triangle is always less than length of the 3rd side. Hence, option (d) is correct.

23. Since $AB^2 + AC^2 = BC^2$, $\angle A = 90°$

Furhter, $AB^2 + AC^2 = 2AB^2$

$[\because BC = \sqrt{2}AB]$

$\Rightarrow AB = AC \Rightarrow \angle B = \angle C$

[$\because$ angles opposite to equal sides are equal]

Clearly, $\angle B + \angle C$

$= 180° - 90° = 90°$

$\Rightarrow \angle B + \angle B = 90°$

$\Rightarrow \quad \angle B = 45°$

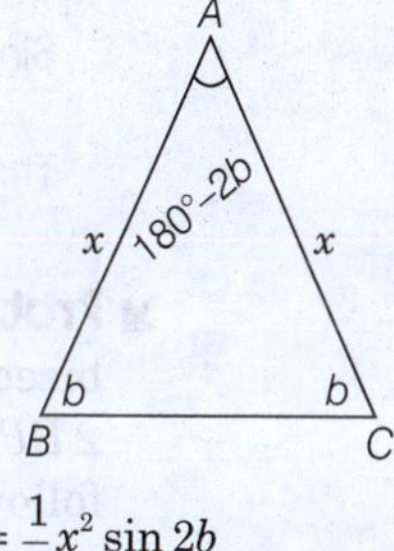

24. Area of ΔABC

$= \frac{1}{2} \times x \times x \times \sin(180° - 2b)$

$\Rightarrow M = \frac{1}{2} x^2 \sin 2b$

$\Rightarrow M = \frac{1}{2} x^2 \times 2 \sin b \times \cos b$

$\Rightarrow M = \frac{1}{2} x^2 \times 2 \times \frac{3}{5} \times \frac{4}{5}$

$\Rightarrow M = \frac{12x^2}{25} \Rightarrow M = 0.48x^2$

Hence, option (b) is correct.

Alternatively

Clearly, $\sin b = \frac{3}{5}$

$\Rightarrow \quad \frac{AD}{x} = \frac{3}{5} \Rightarrow AD = \frac{3x}{5}$

Then, $\quad BD = \sqrt{x^2 - \left(\frac{3x}{5}\right)^2}$

$= \frac{4x}{5}$

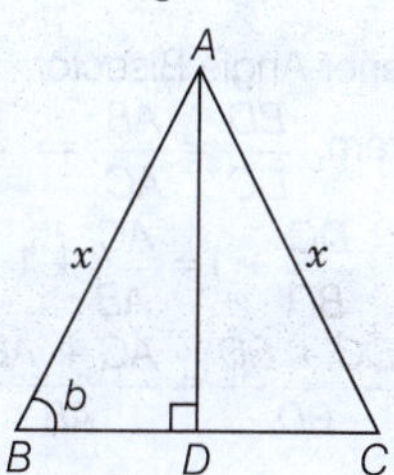

Since altitude of an isosceles triangle is also its median,

$BD = DC$

Now, area of

$\Delta ABC = \frac{1}{2} \times BC \times AD$

$\Rightarrow \quad M = \frac{1}{2} \times \left(\frac{4x}{5} + \frac{4x}{5}\right) \times \frac{3x}{5}$

$\Rightarrow \quad M = 0.48x^2$

Hence, option (b) is correct.

25. We know, sum of perpendiculars drawn from a point p lying inside the equilateral Δ to all three sides is equal to the altitude of the Δ (Discussed in beginning of this chapter).

Now, area of $\Delta = \frac{\sqrt{3}}{4} (\text{side})^2$

$= \left(\frac{\sqrt{3}}{2} \text{ side}\right)^2 \times \frac{1}{\sqrt{3}} = \frac{s^2}{\sqrt{3}}$

$\left[\because \text{altitude} = \frac{\sqrt{3}}{2} \times \text{side}\right]$

Problems based on the concept of Angle Bisector

Problem 31 If AD is the internal angle bisector of ΔABC with $AB = 3$ cm and $AC = 1$ cm, then what is $BD : BC$ equal to? **[CDS-14]**

Sol.

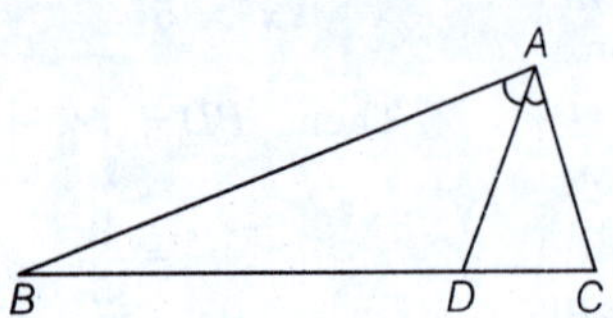

By Interior Angle Bisector Therorem, $\frac{BD}{DC} = \frac{AB}{AC} \Rightarrow \frac{DC}{BD} = \frac{AC}{AB}$

$\Rightarrow \quad \frac{DC}{BD} + 1 = \frac{AC}{AB} + 1$

$\Rightarrow \quad \frac{DC + BD}{BD} = \frac{AC + AB}{AB} \Rightarrow \frac{BC}{BD} = \frac{1+3}{3}$

$\Rightarrow \quad BD : BC = 3 : 4$

Problem 32 In a ΔDEF, EM and FN are bisectors of $\angle E$ and $\angle F$ respectively. If $DF = 23$ cm, then what is the value of DE provided $NM \parallel EF$.

Sol.

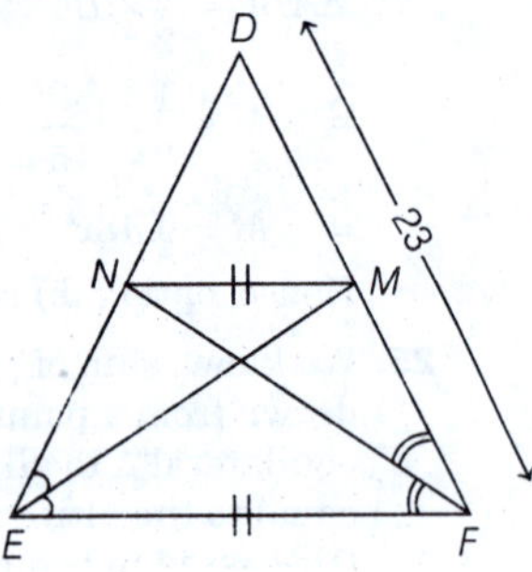

Since EM is the angle bisector of $\angle E$, $\frac{EF}{DE} = \frac{FM}{DM}$

Also, FN is the angle bisector of $\angle F$, $\frac{EF}{DF} = \frac{NE}{DN}$

We know, a line drawn parallel to one side divides the other two sides in the same ratio. Hence, $\frac{NE}{DN} = \frac{FM}{DM}$

Then, $\frac{EF}{DE} = \frac{EF}{DF} \Rightarrow DE = DF \Rightarrow DE = 23$ cm

Problem 33 In a ΔABC, the base angles at B and C are bisected by BO and CO respectively. Then $\angle BOC$ is : **[SSC(CGL-T-1)-12]**

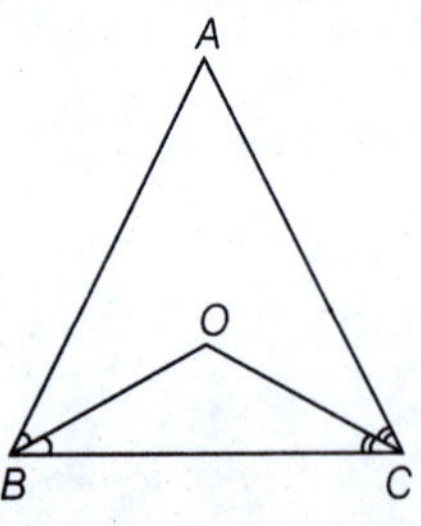

Sol. Clearly, $\angle B + \angle C = 180° - \angle A$

$\Rightarrow \quad \frac{\angle B}{2} + \frac{\angle C}{2} = \frac{1}{2}(180° - \angle A)$

$\Rightarrow \angle OBC + \angle OCB = 90° - \frac{\angle A}{2}$

$\Rightarrow \quad 180° - \angle BOC = 90° - \frac{\angle A}{2}$

$\Rightarrow \quad \angle BOC = 90° + \frac{\angle A}{2}$

Alternatively

Obviously, point O is the incentre of the triangle. We know, an angle subtended by any side at incentre is equal to the sum of right angle and half the angle opposite to that side.

Hence, $\angle BOC = 90° + \frac{\angle A}{2}$

Problem 34 ABC is a triangle such that $\frac{AB}{AC} = \frac{BD}{DC}$. $\angle B$ and $\angle C$ are 60° and 70° respectively. Find the value of $\angle DAB$ is point D lies on side BC.

Sol. Clearly, $\angle A = 180° - 60° - 70° = 50°$

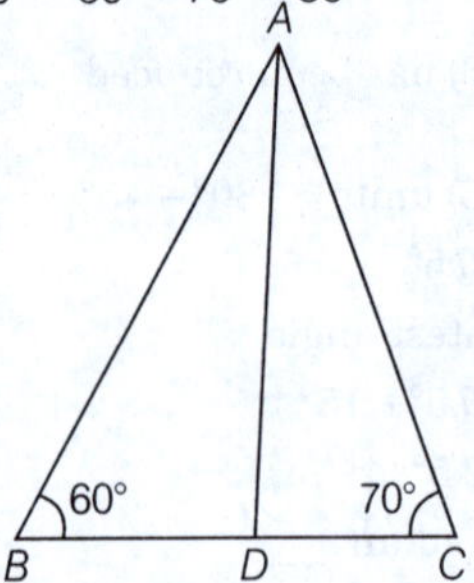

Since $\frac{AB}{AC} = \frac{BD}{DC}$, AD is the bisector of $\angle A$.

Then, $\angle DAB = \frac{\angle A}{2} = \frac{50°}{2} = 25°$

Problem 35 The side BC of a ΔABC is produced to D, bisectors of the $\angle ABC$ and $\angle ACD$ meet at P. If $\angle BPC = x°$ and $\angle BAC = y°$, then which one of the following option is correct ? **[CDS-13, 10]**

(a) $x° = y°$ (b) $x° + y° = 90°$
(c) $x° + y° = 180°$ (d) $2x° = y°$

Sol.

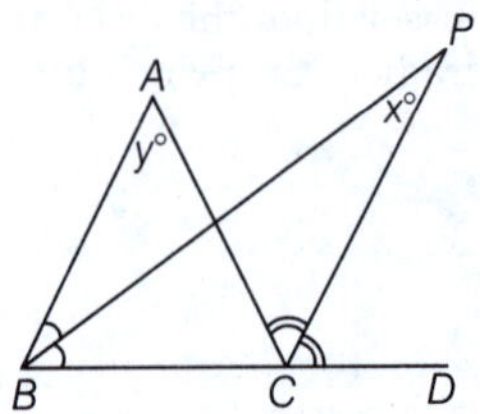

Here, $\angle ACP + \angle PCD = y° + \angle ABC$

[∵ exterior angle = sum of opposite interior angles]

$\Rightarrow \angle ACP + \angle PCD = y° + \angle ABP + \angle PBC$

$\Rightarrow \angle ACP + x° + \angle PBC = y° + \angle ABP + \angle PBC$

[∵ $\angle PCD = x° + \angle PBC$ (exterior angle = sum of opposite interior angles)]

$\Rightarrow \quad \angle ACP + x° = y° + \angle ABP$

$\Rightarrow \quad \angle PCD + x^\circ = y^\circ + \angle ABP$

[$\because \angle ACP = \angle PCD$ (CP bisects $\angle ACD$)]

$\Rightarrow x^\circ + \angle PBC + x^\circ = y^\circ + \angle ABP$

[$\because \angle PCD = x^\circ + \angle PBC$]

$\Rightarrow 2x^\circ = y^\circ$ [$\because \angle PBC = \angle ABP$ (BP bisects $\angle ABC$)]

Problem 36 The bisectors of $\angle A$, $\angle B$ and $\angle C$ of ΔABC intersect the circle circumscribed about the ΔABC at X, Y and Z respectively. If $\angle A = 50^\circ$, $\angle CZY = 30^\circ$, then $\angle BYZ = ?$

Sol.

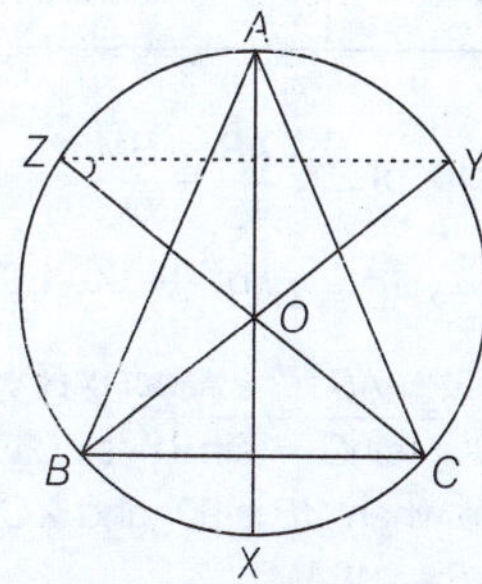

Since an angle subtended by any side at incentre is equal to the sum of right angle and half the angle opposite to that side,

$\angle BOC = 90^\circ + \frac{\angle A}{2}$ ($\because O$ is incentre)

Hence, $\angle BOC = 90^\circ + \frac{50^\circ}{2} = 115^\circ$

Further, $\angle BOC = \angle ZOY$ [Vertically opposite angles]

$\Rightarrow \quad \angle ZOY = 115^\circ$

$\therefore$ In ΔOYZ, $\angle ZOY + \angle OZY + \angle OYZ = 180^\circ$

$\Rightarrow \quad 115^\circ + 30^\circ + \angle OYZ = 180^\circ$

$\Rightarrow \quad \angle OYZ = 35^\circ$

Hence, $\quad \angle BYZ = 35^\circ$

Problem 37 In a triangle ABC, the internal bisector of the $\angle A$ meets BC at D. If $AB = 4$, $AC = 3$ and $\angle A = 60^\circ$, then the length of AD is : ***[CAT-02]***

Sol.

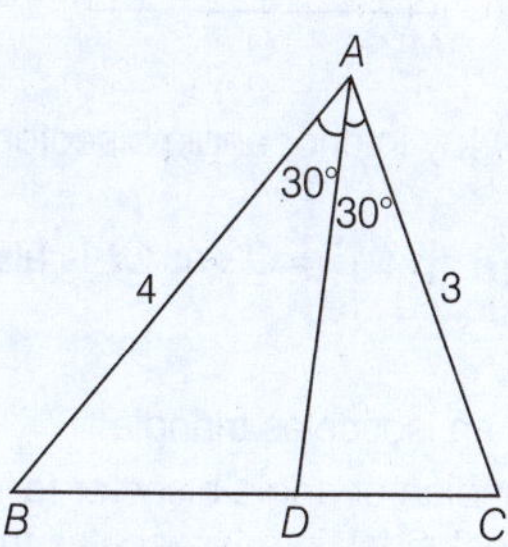

Here, $a^2 = b^2 + c^2 - 2bc\cos A$ [Recall cosine-rule]

$\Rightarrow \quad BC^2 = 3^2 + 4^2 - 2 \cdot 3 \cdot 4\cos 60^\circ = 13$

$\Rightarrow \quad BC = \sqrt{13}$

By Interior Angle Bisector Theorem,

$$\frac{AB}{AC} = \frac{BD}{DC}$$

$$\Rightarrow \quad \frac{BD}{DC} = \frac{4}{3}$$

$$\Rightarrow \frac{BD}{BD + DC} = \frac{4}{4+3} \Rightarrow \frac{BD}{BC} = \frac{4}{7}$$

$$\Rightarrow \quad BD = \sqrt{13} \times \frac{4}{7}$$

Now, in ΔABD, $BD^2 = AB^2 + AD^2 - 2 \cdot AB \cdot AD \cdot \cos 30^\circ$ [cosine-rule]

$$\Rightarrow \left(\sqrt{13} \times \frac{4}{7}\right)^2 = 4^2 + AD^2 - 2 \cdot 4 \cdot AD \cdot \cos 30^\circ$$

$$\Rightarrow \quad \frac{13 \cdot 16}{49} = 16 + AD^2 - 2 \cdot 4 \cdot AD \times \frac{\sqrt{3}}{2}$$

$\Rightarrow 49 \cdot 16 + 49x^2 - 4 \cdot 49\sqrt{3}x = 13 \cdot 16$ [let AD be x]

$\Rightarrow 49x^2 - 49 \cdot 4 \cdot \sqrt{3}x + 16 \cdot 36 = 0$

$$\Rightarrow x = \frac{196\sqrt{3} \pm \sqrt{196 \cdot 196 \cdot 3 - 4 \cdot 49 \cdot 16 \cdot 36}}{2 \cdot 49}$$

$$= \frac{196\sqrt{3} \pm \sqrt{196\,(588 - 576)}}{2 \cdot 49}$$

$$= \frac{196\sqrt{3} \pm \sqrt{196 \cdot 4 \cdot 3}}{2 \cdot 49} = \frac{196\sqrt{3} \pm 28\sqrt{3}}{2 \cdot 49}$$

$$= \frac{224\sqrt{3}}{2 \cdot 49} \text{ and } \frac{168\sqrt{3}}{2 \cdot 49}$$

$$= \frac{16\sqrt{3}}{7} \text{ and } \frac{12\sqrt{3}}{7}$$

Hence, option (b) is correct.

Alternatively

Clearly, Area of ΔABC = Area of ΔABD + Area of ΔACD

$$\Rightarrow \frac{1}{2} \times AB \times AC \times \sin 60^\circ$$

$$= \frac{1}{2} \times AB \times AD \times \sin 30^\circ + \frac{1}{2} \times AC \times AD \times \sin 30^\circ$$

$$\Rightarrow \frac{1}{2} \times 4 \times 3 \times \frac{\sqrt{3}}{2}$$

$$= \frac{1}{2} \times 4 \times AD \times \frac{1}{2} + \frac{1}{2} \times 3 \times AD \times \frac{1}{2}$$

$$\Rightarrow \quad 6\sqrt{3} = 2AD + \frac{3AD}{2}$$

$\Rightarrow \quad 7AD = 12\sqrt{3}$

$$\Rightarrow \quad AD = \frac{12\sqrt{3}}{7}$$

Problem 38 In ΔMNL, line NP bisects the $\angle MNL$. If $NP : NL = 2 : 3$ and $\angle MNL = 120^\circ$, then $NP : NL : MN$ is ***[IIFT-19]***

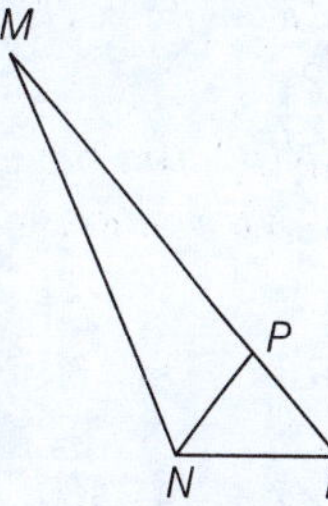

Sol. Angle bisector $NP = \frac{2 \times NL \times NM \times \cos 60^\circ}{NL + NM}$

$$\Rightarrow \quad 2NP = \frac{2 \times NL \times NM}{NL + NM}$$

$\Rightarrow 2 \times 2x = \dfrac{2 \times 3x \times NM}{3x + NM}$

[let NP and NL be $2x$ and $3x$ respectiely]

$\Rightarrow 4(3x + NM) = 6NM$

$\Rightarrow \quad NM = 6x$

$\therefore$ Required ratio $= 2x : 3x : 6x = 2 : 3 : 6$

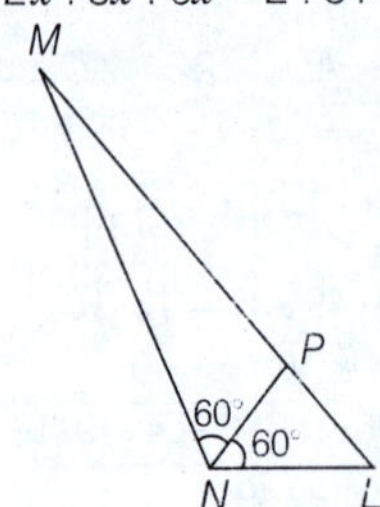

Problem 39 In ΔABC, $\angle C$ is an obtuse angle. The bisectors of the exterior angles at A and B meet BC and AC produced at D and E respectively. If $AB = AD = BE$, then $\angle ACB = ?$ **[SSC (CGL-T-I)-14]**

Sol. In ΔACB, $\angle MAC = \angle 4 + \angle 7$

[$\because$ an exterior angle = sum of interior opposite angles]

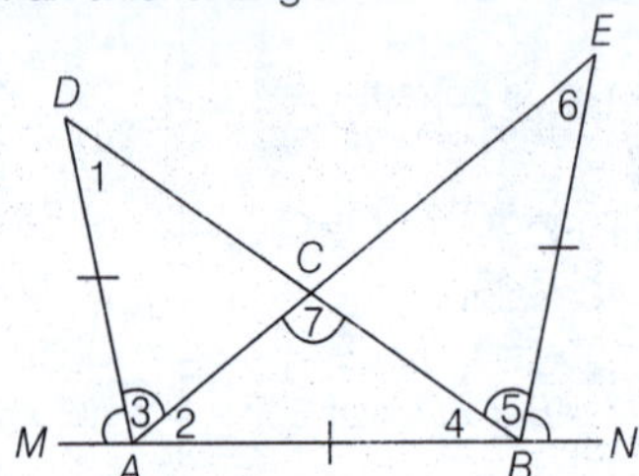

$\Rightarrow \quad 2\angle 3 = \angle 4 + \angle 7$

$\Rightarrow \quad \angle 3 = \dfrac{\angle 4}{2} + \dfrac{\angle 7}{2}$

$\Rightarrow \angle 7 - \angle 1 = \dfrac{\angle 4}{2} + \dfrac{\angle 7}{2}$ [$\because \angle 7 = \angle 1 + \angle 3$]

$\Rightarrow \quad \dfrac{\angle 7}{2} = \dfrac{3\angle 4}{2}$ [$\because AD = AB \;\therefore \angle 4 = \angle 1$]

$\Rightarrow \quad \angle 4 = \dfrac{\angle 7}{3}$

Also, in ΔACB, $\angle CBN = \angle 2 + \angle 7$

[$\because$ an exterior angle = sum of interior opposite angles]

$\Rightarrow \quad 2\angle 5 = \angle 2 + \angle 7$

$\Rightarrow \quad \angle 5 = \dfrac{\angle 2}{2} + \dfrac{\angle 7}{2}$

$\Rightarrow \angle 7 - \angle 6 = \dfrac{\angle 2}{2} + \dfrac{\angle 7}{2}$ [$\because \angle 7 = \angle 5 + \angle 6$]

$\Rightarrow \quad \dfrac{\angle 7}{2} = \dfrac{3\angle 2}{2}$ [$\because BE = AB \;\therefore \angle 2 = \angle 6$]

$\Rightarrow \quad \angle 2 = \dfrac{\angle 7}{3}$

Now, in ΔACB, $\angle 2 + \angle 7 + \angle 4 = 180°$

$\Rightarrow \quad \dfrac{\angle 7}{3} + \angle 7 + \dfrac{\angle 7}{3} = 180°$

$\Rightarrow \quad \angle 7 = 108°$

Problem 40 In ΔABC, $\angle BAC = 60°$ and a point M lies on BC such that $BM : MC = 1 : 2$. If AM bisects $\angle BAC$, then $\angle AMC = ?$

Sol.

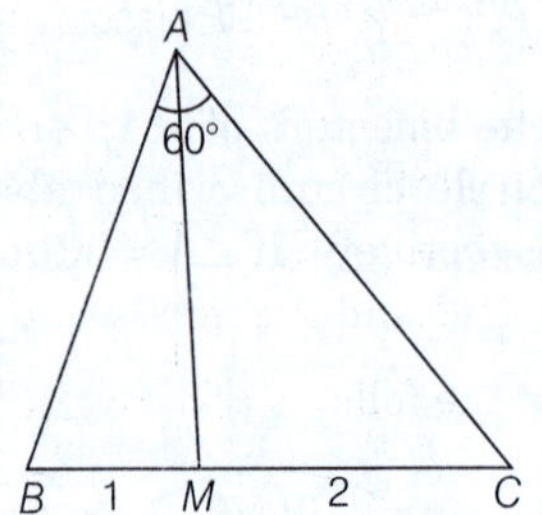

Since AM is bisector of $\angle A$, $\dfrac{AB}{AC} = \dfrac{BM}{MC}$

$\Rightarrow \quad \dfrac{AB}{AC} = \dfrac{1}{2} \Rightarrow AC = 2AB$

By sine-rule, $\dfrac{AC}{\sin B} = \dfrac{AB}{\sin C} \Rightarrow \dfrac{2AB}{\sin B} = \dfrac{AB}{\sin C} \Rightarrow \dfrac{\sin B}{\sin C} = \dfrac{2}{1}$

It may be possible when $\angle B$ is 90° and $\angle C = 30°$

Then, $\angle AMC = \angle B + \angle BAM$

[$\because$ exterior angle = sum of opposite interior angles]

$= 90° + \dfrac{60°}{2} = 120°$

Problem 41 In ΔLMN, LO is the median. Also, LO is the bisector of $\angle MLN$. If $LO = 3$ cm and $LM = 5$ cm, then find the area of ΔLMN. **[CAT-09]**

Sol.

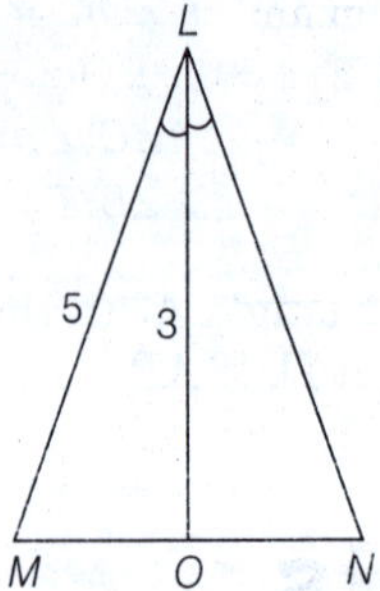

Here, $\dfrac{LM}{LN} = \dfrac{MO}{ON}$ [by Interior angle bisector theorem]

$\Rightarrow \quad \dfrac{LM}{LN} = \dfrac{MO}{MO}$ [$\because MO = ON$ as OL is also a median on MN]

$\Rightarrow \quad LM = LN$

Hence, ΔLMN is an isosceles triangle.

We know, the median or angle bisector falling on the side other than the congruent sides in an isosceles triangle, also acts as altitutde of the triangle. Hence, LO is altitutde too.

Then, in right-angled ΔLOM,

$$MO = \sqrt{5^2 - 3^2} = 4$$

Also, $MN = MO + ON = MO + MO = 4 + 4 = 8$

Now, area of $\quad \Delta LMN = \dfrac{1}{2} \times MN \times LO$

$= \dfrac{1}{2} \times 8 \times 3 = 12$ sq cm

WORKOUT

1. The side QR of ΔPQR is produced to S. The bisectors of $\angle PQR$ and $\angle PRS$ meet at T. If $\angle QPR = 80°$, then $\angle QTR = ?$ **[SSC(FCI-Mains)-12]**

(a) 60° (b) 40° (c) 30° (d) 75°

2. In the given figure, L is any point on the bisector of the acute angle ABC and the line ML is parallel to BC.

Which one of the following is correct ? **[CDS-10]**

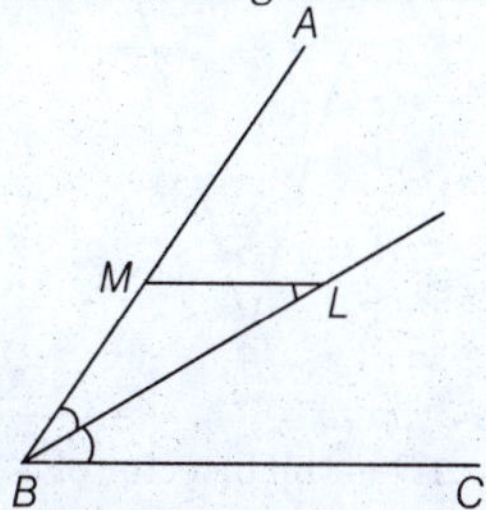

(a) The ΔBML is equilateral

(b) The ΔBML is isosceles but right angled

(c) The ΔBML is isosceles but not right-angled

(d) The ΔBML is not isosceles

3. The internal bisector of $\angle C$ of ΔABC intersects AB at point D and let $AB \neq AC$. E is such a point on CD that $AE = ED$. If $\angle ABC = 50°$. then $\angle CAE = ?$

(a) 60° (b) 45° (c) 30° (d) 50°

4. In a ΔABC, AD is an angle bisector. $AB = 6$ cm, $BC = 7$ cm and $AC = 8$ cm and O is incentre. Then, $AO : OD = ?$

(a) 1 : 2 (b) 2 : 1 (c) 3 : 5 (d) 5 : 3

5. The external bisector of $\angle ABC$ of ΔABC interesects the straight line through A and parallel to BC at the point D. If $\angle ABC = 50°$, then measure of $\angle ADB$ is : **[SSC (FCI-Mains)-12]**

(a) 60° (b) 65° (c) 30° (d) 75°

6. In ΔABC, $AB = AC$, O is a point on BC such that $BO = CO$ and OD is perpendicular to AB and OE is perpendicular to AC. If $\angle BOD = 30°$, then measure of $\angle AOE$ is : **[SSC(CGL-T-II)-14]**

(a) 60° (b) 45° (c) 30° (d) 75°

7. The external bisector of $\angle B$ and $\angle C$ of ΔABC {where AB and AC extended to E and F respectively} meet at point P. If $\angle BAC = 100°$, then the measure of $\angle BPC$ is : **[FCI (Asstt. Gr-III)-12, SSC(CGL-T-I)-12, SSC(CGL-T-I)-19]**

(a) 50° (b) 80° (c) 40° (d) 100°

8. In ΔABC, $\dfrac{AB}{AC} = \dfrac{BD}{DC}$, $\angle B = 70°$ and $\angle C = 50°$, then $\angle BAD = ?$ **[SSC (CPO)-14]**

(a) 60° (b) 30° (c) 20° (d) 50°

9. In ΔABC, $\angle B = 60°$, $\angle C = 40°$, AD is the bisector of $\angle A$ and AE is drawn perpendicular on BC from A. Then , the measure of $\angle EAD$ is **[SSC(10 + 2)/LDC-DEO-14]**

(a) 40° (b) 30° (c) 10° (d) 80°

10. In a triangle ABC with $\angle ABC = 65°$ and $\angle ACB = 50°$, AD is the angle bisector of $\angle BAC$ and $AE \perp BC$. Then, $\angle DAE = ?$

(a) 7.5° (b) 30° (c) 25° (d) 5°

Answer Key

1. (b)	2. (c)	3. (d)	4. (b)	5. (b)	6. (a)	7. (c)	8. (b)	9. (c)	10. (a)

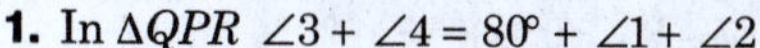

Hints and SOLUTIONS

1. In ΔQPR $\angle 3 + \angle 4 = 80° + \angle 1 + \angle 2$

[$\because$ an exterior angle = sum of interior opposite angles]

$\Rightarrow \dfrac{\angle 3 + \angle 4}{2} = \dfrac{80° + \angle 1 + \angle 2}{2}$

$\Rightarrow \quad \angle 4 = 40° + \angle 2 \quad [\because \angle 3 = \angle 4 \text{ and } \angle 1 = \angle 2]$

In ΔQTR, $\angle 4 = \angle 2 + \angle QTR$

[$\because$ an exterior angle = sum of interior opposite angles]

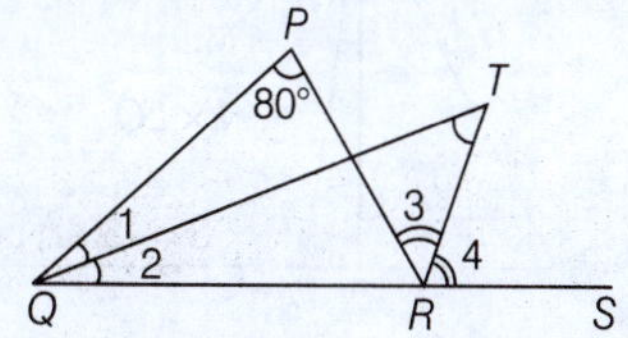

$\Rightarrow \quad 40° + \angle 2 = \angle 2 + \angle QTR$

$\Rightarrow \quad \angle QTR = 40°$

2. Clearly, $\angle MBL = \angle LBC$

Also, $\angle LBC = \angle MLB$ [Alternate interior angles]

$\therefore \quad \angle MBL = \angle MLB$. Then, $ML = MB$

Thus, ΔBML is an isosceles triangle.

3. In ΔADE, $AE = AD$. Then, $\angle ADE = \angle AED$

Also, $\quad \angle 1 = \angle 2 + x$

[$\because$ an exterior angle = sum of interior opposite angles]

In $\quad \Delta ABC$, $\angle A + \angle B + \angle C = 180°$

$\Rightarrow \quad \angle DAE + x + 50° + 2\angle 2 = 180°$

$\Rightarrow \; 180° - 2\angle 1 + x + 50° + 2\angle 2 = 180°$

$\Rightarrow\ -2(\angle 2 + x) + x + 50° + 2\angle 2 = 0° \Rightarrow\ x = 50°$

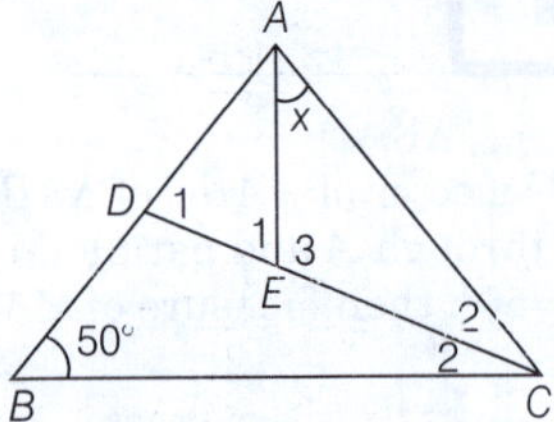

4. Clearly, $\frac{AO}{OD} = \frac{6+8}{7} \Rightarrow AO : OD = 2:1$

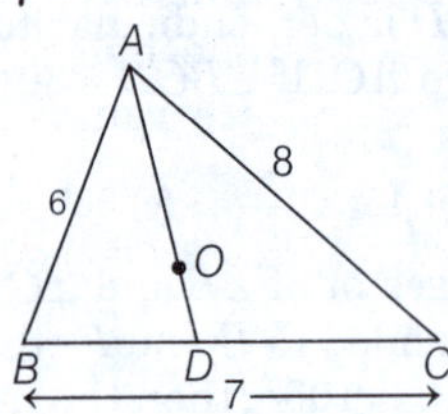

5. Clearly, $\angle ABY = 180° - 50° = 130°$

Then, $\angle YBD = \frac{130°}{2} = 65°$

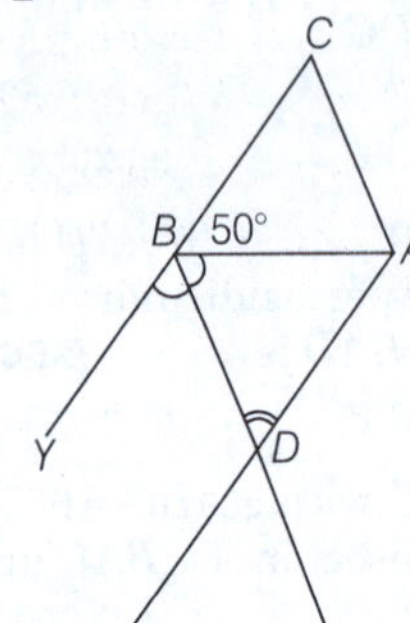

$\because CY \parallel AD$ and BD is transversal, then $\angle YBD = \angle ADB$ (Alternate interior angles)

Hence, $\angle ADB = 65°$

6. In ΔBOD, $\angle DBO = 180° - 90° - 30° = 60°$

$\Rightarrow \quad \angle ABC = 60° = \angle ACB$

$[\because AB = AC \therefore \angle ABC = \angle ACB]$

Then, $\angle BAC = 180° - 60° - 60° = 60°$

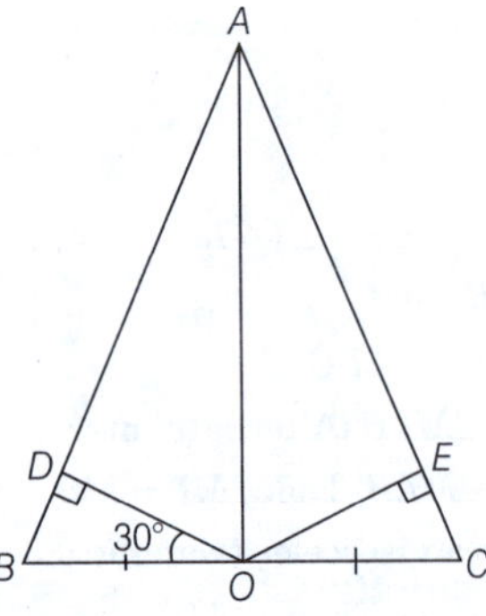

Since, each angle be 60°, ABC is an equilateral triangle, we know, in an equilateral triangle, medians also act as angle bisectors.

Hence, $\angle EAO = \frac{1}{2} \angle BAC = 30°$

Now, in ΔAOE, $\angle AOE = 180° - 90° - 30° = 60°$

7. Clearly, $\angle BPC = 90° - \frac{\angle A}{2}$

$\Rightarrow \quad \angle BPC = 90° - \frac{100°}{2} = 40°$

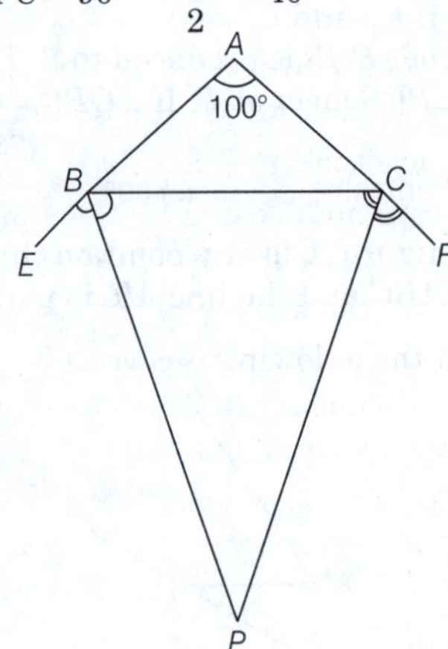

8. Since $\frac{AB}{AC} = \frac{BD}{DC}$, AD is the bisector of $\angle BAC$.

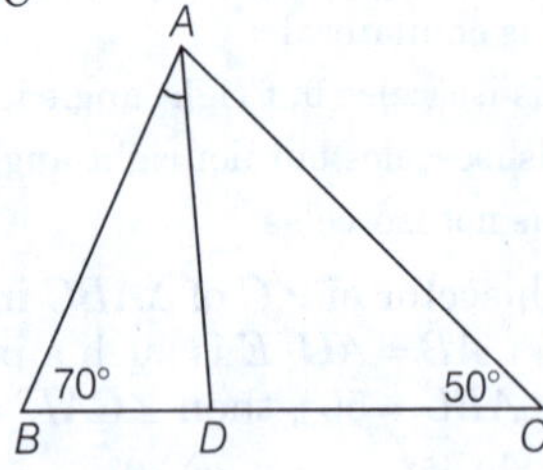

Then, $\angle BAD = \frac{1}{2} \angle BAC$

$= \frac{1}{2}(180° - 70° - 50°) = 30°$

9. In ΔBAC, $\angle BAC = 180° - 60° - 40° = 80°$

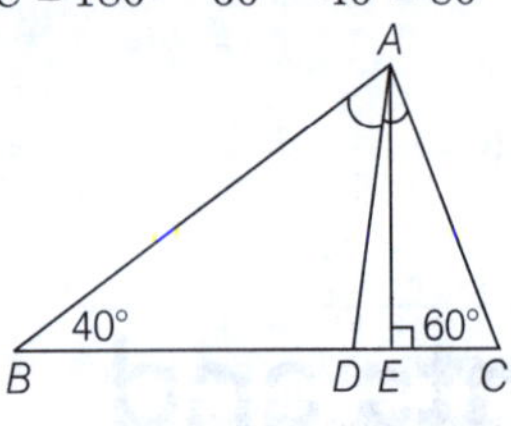

Then, $\angle BAD = \frac{80°}{2} = 40°$

In ΔAEB, $\angle EAB = 180° - 60° - 90° = 30°$

$\therefore \angle EAD = \angle BAD - \angle EAB = 40° - 30° = 10°$

10. We know, the angle between the angle bisector and the altitude originating from the same vertex of a triangle is equal to the half the difference between angles of other two vertices. Hence, required answer $= \frac{1}{2}(65° - 50°) = 7.5°$

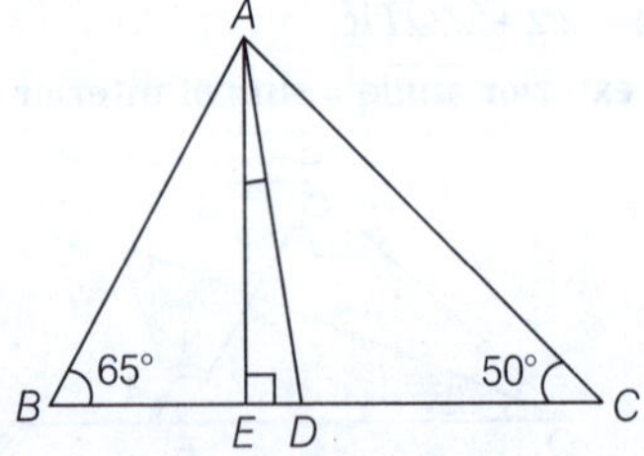

Problems based on Incircle/ Inradius/Incentre

Problem 42 If C_1 and C_2 and r_1 and r_2 are respectively the centroids and radii of incircles of two congruent triangles, then which one of the following is correct ?

(a) C_1 and C_2 are the same point and $r_1 = r_2$

(b) C_1 and C_2 are not necessarily the same point and $r_1 = r_2$

(c) C_1 and C_2 are the same point and r_1 is not necessarily equal to r_2

(d) C_1 and C_2 are not necessarily the same point and r_1 is not necessarily equal to r_2

Sol. Since both the triangles are congruent, their corresponding medians will be equal, hence, C_1 and C_2 are the same point. Further, their inradius will also be same. Thus, option (a) is correct.

Problem 43 I is the incentre of ΔABC, If $\angle ABC = 65°$ and $\angle ACB = 55°$, then the value of $\angle BIC$ is **[SSC(CGL-T-II)-12]**

Sol. Clearly, $\angle BAC = 180° - 65° - 55° = 60°$.

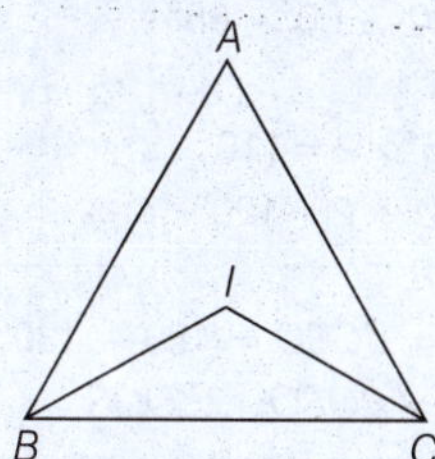

Since the angle subtended by any side at incentre is equal to the sum of right angle and half the angle opposite to that side, $\angle BIC = 90° + \frac{\angle BAC}{2}$. Hence, $\angle BIC = 90° + \frac{60°}{2} = 120°$

Alternatively

I is the incentre, it implies that IB and IC are angle bisectors of $\angle ABC$ and $\angle ACB$ respectively.

Then, $\angle IBC = 65°/2 = 32.5°$ and

$\angle ICB = 55°/2 = 27.5°$

$\therefore \quad \angle BIC = 180° - 32.5° - 27.5° = 120°$

Probem 44 In the given figure, ACB is a right-angled triangle. CD is the altitude. Circles are inscribed within the triangle ACD, BCD. P and Q are the centres of the circles. The distance PQ is : **[CAT-02]**

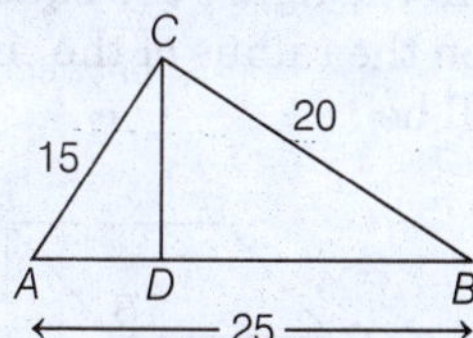

Sol. Area of $\Delta ABC = \frac{1}{2} \times 15 \times 20 = \frac{1}{2} \times 25 \times CD$

$\Rightarrow \quad CD = 12$

Clearly, $AD = 9$ because (9, 12, 15) is a Pythagorean triplet.

[In another way, $AD = \sqrt{AC^2 - CD^2} = \sqrt{15^2 - 12^2} = 9$]

Then, $BD = 25 - 9 = 16$

Inradius of ΔACD

$$= \frac{\text{Sum of perpendicular sides} - \text{hypotenuse}}{2}$$

$$= \frac{9 + 12 - 15}{2} = 3$$

and inradius of $\Delta BCD = \frac{16 + 12 - 20}{2} = 4$

Then, distance between radii of both the incircles = 3 + 4 = 7

Problem 45 ABC is a triangle with base $BC = 3$ cm and height $AD = 4$ cm. A circle has been inscribed in the triangle, then find out its radius if perimeter of the triangle is 15 cm.

Sol. Area of the triangle = rs [Where, r = inradius and s = semi perimeter]

$\Rightarrow \quad \frac{1}{2} \times 3 \times 4 = r \times \left(\frac{15}{2}\right)$

$\Rightarrow \quad r = 0.8$ cm

Problem 46 The tangency points of the incircle of a triangle divide the sides into lengths of 3 cm and 4 cm, 4 cm and 5 cm and 5 cm and 3 cm. What is inradius ?

Sol. We know, if sides of the triangle are divided by tangency points of the incircle into length of k and m, m and n and n and k, then the value of inradius (r)

$$= \sqrt{\frac{kmn}{k + m + n}}$$

Hence, $r = \sqrt{\frac{3 \cdot 4 \cdot 5}{3 + 4 + 5}} = \sqrt{5}$ cm

Problem 47 In a triangle two longest sides are 13 cm and 12 cm. The angles of the triangle are in arithmetic progression. The radius of the circle inscribed in this triangle is **[IIFT-19]**

Sol. Since angles are in AP, $(a - d) + (a) + (a + d) = 180°$

(here, d is common difference and a is intitial term)

$\Rightarrow \quad a = 60°$

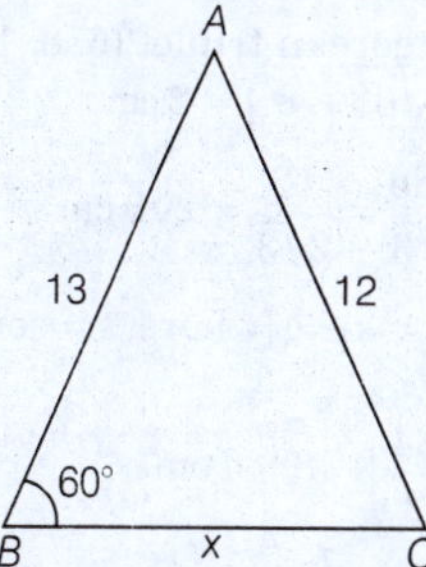

Hence, angles are $(60° - d)$, $(60°)$ and $(60° + d)$

We know that greater angle has greater side opposite to it and smaller angle has smaller side opposite to it.

Since 60° is the 2nd largest angle, it will be opposite to 2nd largest side (i.e. 12). It implies that $\angle ABC = 60°$

Now, $AC^2 = AB^2 + BC^2 - 2AB \cdot BC \cdot \cos 60°$

$\Rightarrow \quad 12^2 = 13^2 + x^2 \left(-2 \times 13 \times x \times \frac{1}{2}\right)$

$\Rightarrow x^2 - 13x + 25 = 0$

$\Rightarrow \quad x = 2.35$ and 10.65

Further, area of $\Delta ABC = rs$

[r is inradius and s is semi-perimeter of triangle]

$\Rightarrow \frac{1}{2} \times 13 \times 2.35 \times \sin 60° = r \times \frac{12 + 13 + 2.35}{2}$

$\Rightarrow \quad \frac{1}{2} \times 13 \times 2.35 \times \frac{\sqrt{3}}{2} = r \times \frac{12 + 13 + 2.35}{2}$

$\Rightarrow \quad r = 0.96$

Substituting $x = 10.65$ in the above equation we get, $r = 3.36$

Problem 48 The altitutdes of a triangle are 3 cm, 4 cm and 5 cm. Find the radius of the circle inscribed in the triangle.

Sol. $\frac{1}{r} = \frac{1}{3} + \frac{1}{4} + \frac{1}{5} \Rightarrow \frac{1}{r} = \frac{47}{60} \Rightarrow r = 1\frac{13}{47}$ cm

Workout

1. ABC is a right-angled triangle with $AB = 6$ cm and $BC = 8$ cm. A circle with centre O has been inscribed iside ΔABC. The radius of the circle is ***[SSC(CPO, DP-14]***
   (a) 1 cm (b) 2 cm (c) 3 cm (d) 4 cm
2. Find the radius of a circle inscribed in an equilateral triangle having side of 12 cm.
   (a) $2\sqrt{2}$ cm (b) $2\sqrt{5}$ cm (c) $2\sqrt{3}$ cm (d) $2\sqrt{7}$ cm
3. A circle is inscribed in a ΔABC such that it touches side AB at P, AC at N and BC at M. Find the value of CM if perimeter and AB are 24 cm and 8 cm respectively.
   (a) 1 cm (b) 2 cm (c) 3 cm (d) 4 cm

Answer Key

1. (b)	2. (c)	3. (d)		

SOLUTION

1. In the right angled traingle, inradius
   $= \frac{\text{sum of perpendicular sides} - \text{hypotenuse}}{2} = \frac{6 + 8 - 10}{2}$
   [Recal a Pythagorean triplet (6, 8, 10), otherwise hypotenuse $= \sqrt{6^2 + 8^2}$] = 2 cm
2. Inradius $= \frac{\text{side}}{2\sqrt{3}} = \frac{12}{2\sqrt{3}} = 2\sqrt{3}$ cm
3. We know, $CM = s - c$ [where, s = semi-perimeter and $c = AB$]
   $= \left(\frac{24}{2}\right) - 8 = 4$ cm

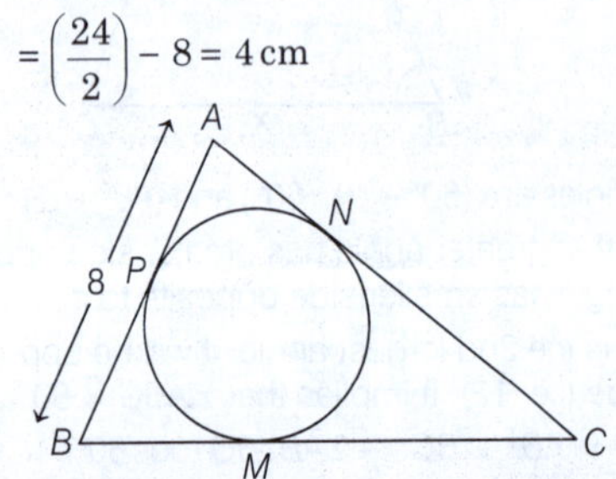

Problems Based on Circumcircle/ Circumradius/Circumcentre

Problem 49 O and C are respectively the orthocentre and circumcentre of an acute-angled ΔPQR. The points P and O are joined and produced to meet the side QR at S. If $\angle PQS = 60°$ and $\angle QCR = 130°$, then $\angle RPS$ = ? ***[SSC(DEO/LDC)-11]***

Sol. Since the angle subtended by any side of the triangle at circumcentre is twice the vertical angle opposite to that side,

$\angle QPR = \frac{130°}{2} = 65°$

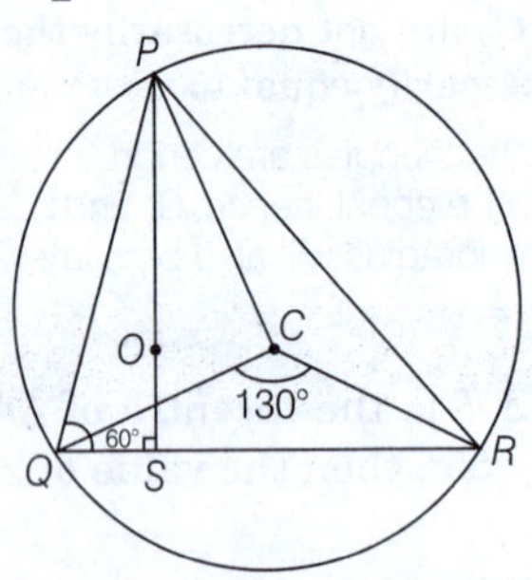

Then, $\angle PRQ = 180° - 60° - 65°$
$= 55°$

$\therefore \quad \angle PCQ = 55° \times 2 = 110°$

Also, $\angle PCR = 60° \times 2 = 120°$

In ΔQCR, $QC = CR$

$\therefore \quad \angle CQR = \angle CRQ = (180° - 130°)/2 = 25°$

In ΔQOS, $\angle COS = \angle OSQ + \angle OQS$

[Exterior angle = sum of opposite interior angles]

$\Rightarrow \quad \angle COS = 90° + 25° = 115°$

Further, in ΔPCO, $\angle COS = \angle CPO + \angle PCO$

$\Rightarrow \quad 115° = \angle CPO + 110°$

$\Rightarrow \quad \angle CPO = 5°$

Now, in ΔPCR, $PC = CR$ (radii). Then,

$\angle CPR = \angle CRP = (180° - \angle PCR)/2$
$= (180° - 120°)/2 = 30°$

Clearly, $\angle RPS = \angle CPR + \angle CPO = 30° + 5° = 35°$

Alternatively

$\angle QPS = 180° - 60° - 90° = 30°$

and $\angle QPR = 130°/2 = 65°$

$\Rightarrow \angle RPS = 65° - 30° = 35°$

Problem 50 If the lenth of diagonals DF, AG and CE of the cube shown in the figure are equal to the three sides of a triangle, then the radius of the circle circumscribing that triangle will be : ***[CAT-04]***

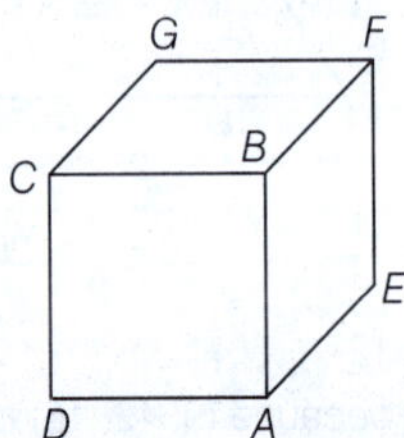

(a) equal to the side of the cube
(b) $\sqrt{3}$ times the side of the cube
(c) $\frac{1}{\sqrt{3}}$ times the side of the cube
(d) Impossible to find from the given information

Sol. (a) Diagonal of a cube = $\sqrt{3} \times$ a, where a is the side of the cube
Then, each side of the triangle = $\sqrt{3} \times a$
Now, circumradius of an equilateral triangle

$$= \frac{\text{side}}{\sqrt{3}} = \frac{\sqrt{3} \times a}{\sqrt{3}} = a$$

Hence, option (a) is correct.

Problem 51 The radius of the circumcircle of a right-angled triangle is 15 cm and the radius of its inscribed circle is 6 cm. Find the sum of perpendicular sides of the triangle.

Sol. In a right-angled triangle, circumradius = hypotenuse/2
$\Rightarrow$ hypotenuse = 2 × 15 = 30 cm

Further, inradius = $\frac{\text{sum of perpendicular sides} - \text{hypotenuse}}{2}$

$\Rightarrow \quad 6 = \frac{\text{Sum of perpendicular sides} - 30}{2}$

$\Rightarrow$ Sum of perpendicular sides = 42

Alternatively
Hypotenuse = 2 × 15 = 30 cm
Recall a Pythagorean triaplet 3, 4, 5. Then, another, triplet may be 3 × 6, 4 × 6, 5 × 6 considering the hypotenuse be 30.
Hence, sum of perpendicular sides = 18 + 24 = 42

Problem 52 In an isosceles ΔABC with $AB = BC = 8$ cm and $\angle B = 2\angle C$, what is the ratio of circumradius to inradius ?

Sol.

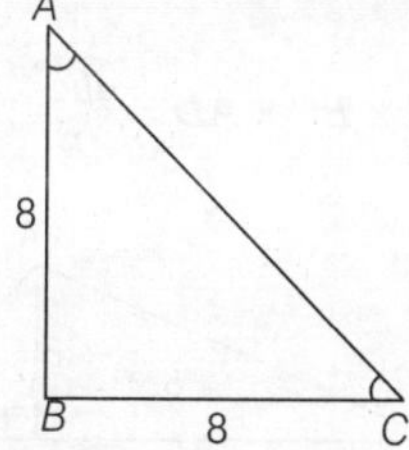

Since $\quad AB = BC, \angle A = \angle C$
Further, $\angle A + \angle B + \angle C = 180°$

$\Rightarrow \quad \frac{\angle B}{2} + \angle B + \frac{\angle B}{2} = 180° \Rightarrow \angle B = 90°$

Hence, ABC is a right-angled triangle,
Then, hypotenuse $(AC) = \sqrt{8^2 + 8^2} = 8\sqrt{2}$
Now, required ratio = $(8\sqrt{2})/2 : (8 + 8 - 8\sqrt{2})/2 = 1 : (\sqrt{2} - 1)$

Probem 53 The circumradius of an equilateral ΔABC be 8 cm, then the height of the triangle is :
[SSC(CGL-T-I)-13]

Sol. We know, circumradius $(R) = \frac{\text{side}}{\sqrt{3}} \Rightarrow 8 = \frac{\text{side}}{\sqrt{3}}$

$\Rightarrow \quad$ side = $8\sqrt{3}$

Then, height of the triangle = $\frac{\sqrt{3} \times \text{side}}{2} = \frac{\sqrt{3} \times 8\sqrt{3}}{2} = 12$ cm

Alternatively
In an equilateral triangle, centroid divides the median (or altitutde) in ratio 2 : 1 and also acts as circumcentre.
Hence, $\quad$ 2 units = 8 cm
$\therefore \quad$ (2 + 1) units = 12 cm

Problem 54 In ΔABC, $\angle B$ is 60° and the side opposite to $\angle B$ is 9 cm. Find the circumradius.

Sol. We have, $\frac{b}{\sin B} = 2R$

$\Rightarrow \quad \frac{9}{\sin 60°} = 2R \Rightarrow 9 \times \frac{2}{\sqrt{3}} = 2R$

$\Rightarrow \quad 6\sqrt{3} = 2R$

$\Rightarrow \quad R = 3\sqrt{3}$ cm

Problem 55 Find out the difference between the circumcentre and incentre of a triangle having sides of 10 cm, 24 cm and 26 cm.

Sol. Clearly the triangle is a right-angled (recall a Pythagorean triplet : (5, 12, 13. We can get the given figure by multiplying by 2)
Since 26 is the longest side, it is the hypotenuse.
Then, inradius (r)

$$= \frac{\text{sum of perpendicular sides} - \text{hypotenuse}}{2}$$

$$r = \frac{10 + 24 - 26}{2} \Rightarrow r = 4 \text{ cm}$$

R = hypotenuse/2 = 26/2 = 13 cm
Then, distance between R and r
$= \sqrt{R(R - 2r)}$
$= \sqrt{13(13 - 2 \cdot 4)} = \sqrt{65} = 8.06$ cm

Problem 56 If O is the circumcentre of ΔABC lying inside the triangle, then $\angle OBC + \angle BAC$ is equal to
[SSC(CGL-T-II)-15]

Sol.

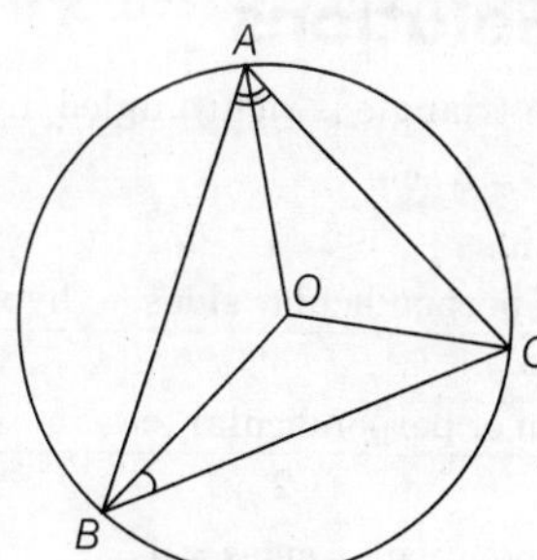

Clearly, $\angle OBC = \angle OCB$
[$\because OB = OC$ (radii) and angles opposite to equal sides are equal.]
Further, $\angle OBC + \angle OCB + \angle BOC = 180°$
$\Rightarrow \quad \angle OBC + \angle OBC + 2\angle BAC = 180°$
[$\because \angle BAC = 2\angle BOC$ as angle subtended by any side of the triangle at circumcentre is twice the vertical angle opposite to that side]
$\Rightarrow 2(\angle OBC + \angle BAC) = 180°$
$\Rightarrow \quad \angle OBC + BAC = 90°$

Workout

1. What is the area of a right-angled triangle whose inradius is 5 cm and circumradius = 18.5 cm.

(a) 92 cm^2 (b) 220 cm^2 (c) 184 cm^2 (d) 210 cm^2

2. 'O' is circumcentre of ΔABC. If $\angle BAC = 85°$, $\angle BCA = 75°$, then $\angle AOC$ is equal to *[SSC(FCI-Mains)-12, SSC(CGL-T-1)-12]*

(a) 60° (b) 70° (c) 20° (d) 50°

3. Find out the length of the circumradius of a triangle having sides of length 12 cm, 16 cm and 20 cm. *[SSC(CGL-T-1)-12]*

(a) 10 cm (b) 2 cm (c) 3 cm (d) 4 cm

4. The circumradius of an equilateral triangle be 10 cm, the measure of its inradius is?

(a) 5 cm (b) 2 cm (c) 3 cm (d) 4 cm

5. The altitude (AD) of ΔABC with $AB = 30$ cm, $AC = 12$ cm, is 8 cm. What is the radius of the circle circumscribed about the triangle ?

(a) 12 cm (b) 10.5 cm (c) 22.5 cm (d) 40 cm

Directions (Q. Nos. 6 and 7) *Consider the following for the next two items.*

"An equilateral ΔABC is inscribed in a circle of radius $20\sqrt{3}$ ".

6. What is the length of the side of the triangle?

(a) 30 cm (b) 40 cm (c) 50 cm (d) 60 cm

7. The centroid of the triangle is at a distance d from the vertex A. What is d equal to ? *[CDS-19]*

(a) 15 cm (b) 20 cm (c) $30\sqrt{3}$ cm (d) $20\sqrt{3}$ cm

Answer Key

1. (d) 2. (b) 3. (a) 4. (a) 5. (c) 6. (d) 7. (d)

Hints and Solutions

1. Since the given triangle is right-angled, hypotenuse = $2R$

$= 2 \times 18.5 = 37$ cm

Further, inradius

$$= \frac{\text{sum of perpendicular sides} - \text{hypotenuse}}{2}$$

$$\Rightarrow \quad 5 = \frac{\text{sum of perpendicular sides} - 37}{2}$$

$\Rightarrow$ sum of perpendicular sides = 47

Then, perimeter = sum of perpendicular sides + hypotenuse

$= 47 + 37 = 84$

$\therefore$ Semi-perimeter $(s) = \frac{84}{2} = 42$ cm

Now, area $= rs = 5 \times 42 = 210\text{ cm}^2$

2. Since an angle subtended by any side of the triangle at circumcentre is twice the verticle angle opposite to that side,

$\angle AOC = 2\angle ABC.$

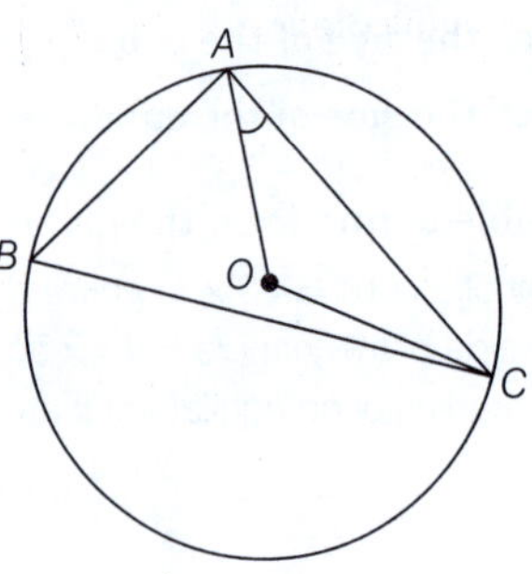

$\Rightarrow \quad \angle AOC = 2(180° - 85° - 75°) = 40°$

In ΔAOC, $\angle AOC + \angle OAC + \angle OCA = 180°$

$\Rightarrow \quad 40° + \angle OAC + \angle OAC = 180°$

[$\because$ radius OA = radius OC $\therefore$ $\angle OCA = \angle OAC$]

$\Rightarrow \quad 2\angle OAC = 140°$

$\Rightarrow \quad \angle OAC = 70°$

3. The triangle is a right-angled triangle.

[Recall a Pythagorean triplet (3, 4, 5) $\Rightarrow$ (3·4, 4·4, 5·4)]

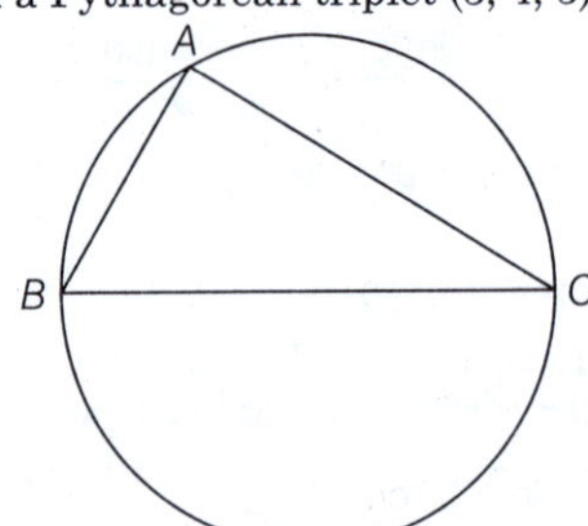

Then, circumradius $= \frac{\text{hypotenuse}}{2} = \frac{20}{2} = 10$ cm

4. Inradius of an equilateral triangle = circumradius/2

$= \frac{10}{2} = 5$ cm

5. Area of $\Delta ABC = \frac{1}{2} \times BC \times AD = \frac{abc}{4R}$

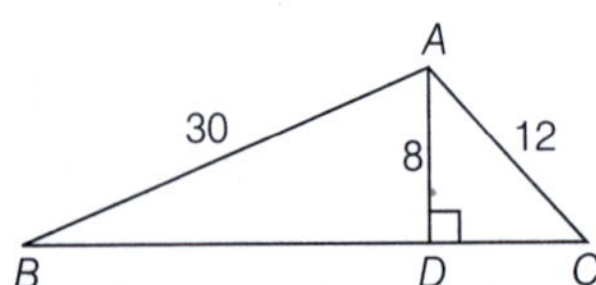

$\Rightarrow \frac{1}{2} \times BC \times AD = \frac{AB \times BC \times AC}{4R}$

$\Rightarrow \quad \frac{1}{2} \times BC \times 8 = \frac{30 \times BC \times 12}{4R}$

$\Rightarrow \quad 16R = 30 \times 12$

$\Rightarrow \quad R = 22.5$ cm

6. We know, circumradius $(R) = \frac{\text{side}}{\sqrt{3}}$

$\Rightarrow \quad 20\sqrt{3} = \frac{\text{side}}{\sqrt{3}}$

$\Rightarrow \quad$ side = 60 cm

7. (d) In equilateral triangle, circumcentre and centroid both are the same point. Hence, d will be the circumradius i.e. $20\sqrt{3}$ cm.

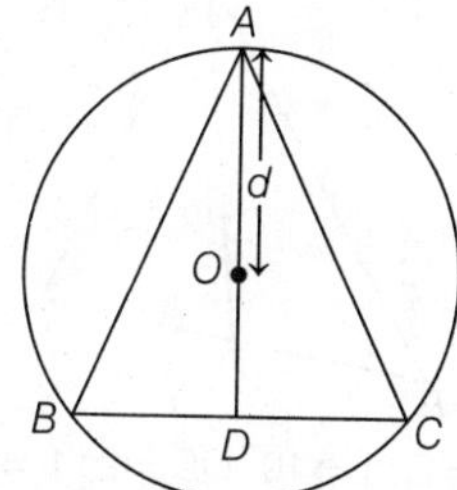

Problem Based on Altitude/Orthocentre

Problem 57 The sides of a triangle are 50 cm, 78 cm and 112 cm. The smallest altitude is

Sol. Semi-perimeter $(s) = \frac{50 + 78 + 112}{2} = 120$

Then, area of the triangle

$$= \sqrt{120(120-50)(120-78)(120-112)}$$

$$\sqrt{120 \cdot 70 \cdot 42 \cdot 8} = \sqrt{8 \cdot 15 \cdot 5 \cdot 14 \cdot 3 \cdot 14 \cdot 8}$$

$$= 8 \times 14 \times 15 = 1680$$

It is the fact that the smallest altitutde lie on the biggest side.

Then, area of triangle

$$= \frac{1}{2} \times \text{base (biggest side)} \times \text{height (smallest)}$$

$$\Rightarrow \quad 1680 = \frac{1}{2} \times 112 \times \text{smallest altitude}$$

$\Rightarrow$ Smallest altitude = 30 cm

Problem 58 PQR is an equilateral triangle, O is the point of intersection of altitutdes PL, QM and RN. If $OP = 8$ cm, then what is the perimeter of the ΔPQR? **[CDS-13]**

Sol.

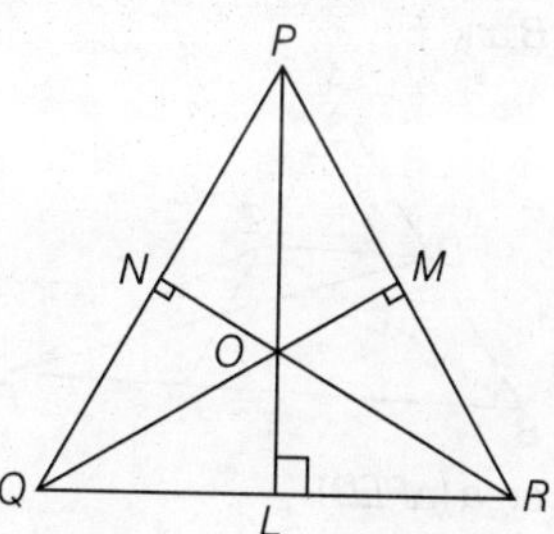

Since the triangle is equilateral, othrocentre 'O' will also be centroid and we know that centroid divides the median (here, altitude) in 2 : 1.

Clearly, $OP : OL = 2 : 1$

$$\Rightarrow \quad 8 : OL = 2 : 1 \Rightarrow OL = 4 \text{ cm}$$

Then, $\quad PL = 8 + 4 = 12$ cm

We know, altitude of an equilateral triangle $= \frac{\sqrt{3}}{2} \times$ side

$$\Rightarrow \quad 12 = \frac{\sqrt{3}}{2} \times \text{side}$$

$$\Rightarrow \quad \text{side} = 8\sqrt{3}$$

Then, perimeter of the triangle $= 3 \times 8\sqrt{3} = 24\sqrt{3}$ cm

Problem 59 In an obtuse angled ΔABC, $\angle A$ is the obtuse angle and O is the orthocentre. If $\angle BOC = 54°$, then $\angle BAC$ is : **[SSC(CGL-T-1)-12]**

Sol. We know, an angle subtended by any side of a triangle on its orthocentre and the vertical angle opposite to that side are supplementary.

Hence, $\angle BAC = 180° - 54° = 126°$

Workout

1. ΔABC is an isosceles triangle and $\overline{AB} = \overline{AC} = 2a$ unit, $\overline{BC} = a$ unit. Draw $\overline{AD} \perp \overline{BC}$ and find the length of $\overline{AD}$. **[SSC(CGL-T-II)-13]**

(a) $\sqrt{15}\,a$ unit (b) $\sqrt{17}\,a$ unit
(c) $\frac{\sqrt{15}}{2}\,a$ unit (d) $\frac{\sqrt{17}}{2}\,a$ unit

2. The ratio of altitutdes of a triangle is 5 : 2 : 3, then what is the ratio of their sides on which they are falling respectively?

(a) 6 : 15 : 10 (b) 6 : 9 : 10 (c) 3 : 4 : 10 (d) 3 : 4 : 5

Answer Key

1. (c) 2. (a)

Solutions

1. (c) Here, the altitude AD is also a median as ΔABC is an isosceles triangle.

Then, $BD = \frac{a}{2}$

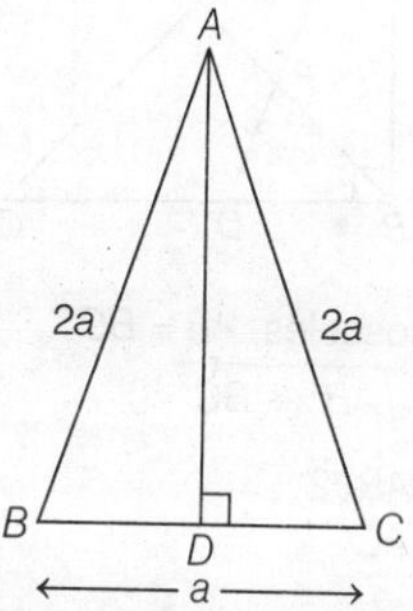

In right-angled ΔABD, $AD^2 = AB^2 - BD^2$

$$\Rightarrow \quad AD^2 = (2a)^2 - \left(\frac{a}{2}\right)^2 = \frac{15a^2}{4}$$

$$\Rightarrow \quad AD = \frac{\sqrt{15}}{2}\,a \text{ unit}$$

2. (a) Let the altitudes be $5k$, $2k$ and $3k$.

Area of the triangle

$$= \frac{1}{2} \times \text{base (corresponding)} \times 5k = \frac{1}{2} \times \text{base (corresponding)} \times 2k = \frac{1}{2} \times \text{base (corresponding)} \times 3k$$

[corresponding bases refer to sides on which altitudes are falling]

$= \text{base (corresponding)} \times 5k = \text{base (corresponding)} \times 2k$
$= \text{base (corresponding)} \times 3k$

$\Rightarrow$ Required ratio = 6 : 15 : 10

Problems Based on Median/Centroid

Problem 60 If G is the centroid and AD, BE, CF are three medians of ΔABC with area 72 cm^2, then the area of ΔBDG is

Sol.

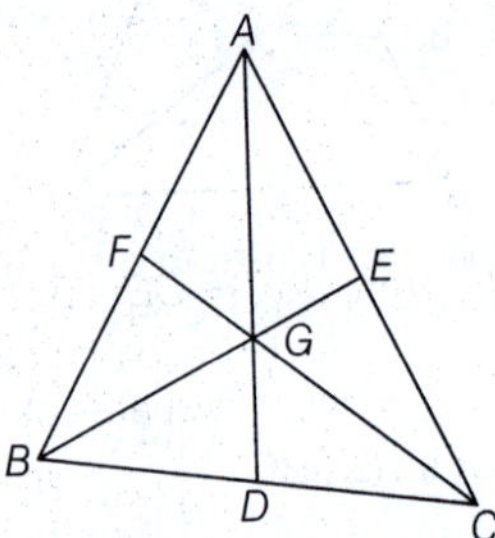

We know, all the three medians divide a triangle into six smaller triangles of equal areas.

Then, area of $\Delta BDG = \frac{72}{6} = 12\text{ cm}^2$

Problem 61 The median (BE) to the hypotenuse of an issosceles triangle ABC, right-angled at B, intersects the bisector (AD) of $\angle CAB$ at point P. If $BE = (10 + 5\sqrt{2})$ cm, then $BP = ?$

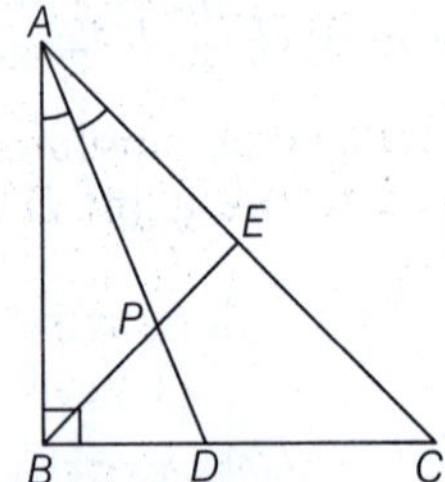

Sol. Since triangle is isosceles, $AB = BC$

Then, $AC = \sqrt{AB^2 + BC^2}$

$\Rightarrow \sqrt{AB^2 + AB^2} = AB\sqrt{2}$

Also, $AE = \frac{AC}{2}$ [$\because BE$ is a median]

$\Rightarrow AE = \frac{AB\sqrt{2}}{2} = \frac{AB}{\sqrt{2}}$

By interior angle bisector theorem,

$\frac{AB}{AE} = \frac{BP}{PE}$ [$\because AD$ bisects $\angle CAB$]

$\Rightarrow \frac{AB}{\left(\frac{AB}{\sqrt{2}}\right)} = \frac{BP}{PE} \Rightarrow \frac{\sqrt{2}}{1} = \frac{BP}{PE}$

$\Rightarrow BP : PE = \sqrt{2} : 1$

Then, $BP = (10 + 5\sqrt{2}) \times \frac{\sqrt{2}}{(1 + \sqrt{2})}$

[$\because BP + PE = BE = (10 + 5\sqrt{2})$]

$= 10$ cm

Problem 62 AD is the median of a ΔABC and O is the centroid such that $AO = 10$ cm. The length of OD is **[SSC(CGL-T-1)-12]**

Sol. We know the centroid of a triangle divides each median in the ratio 2 : 1 (from vertex).

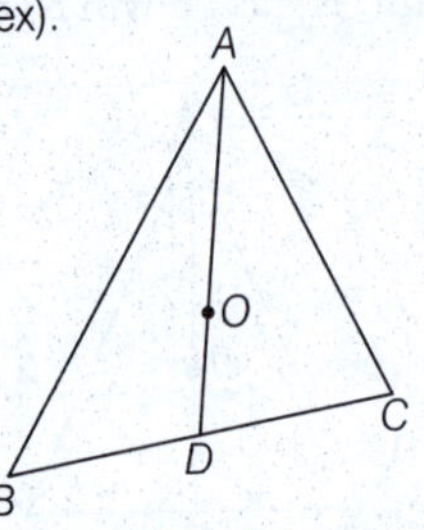

Hence, $AO : OD = 2 : 1 \Rightarrow 10 : OD = 2 : 1 \Rightarrow OD = 5$ cm

Problem 63 In the given figure, $\angle PQR = 90°$ and QL, is a median, $PQ = 5$ cm and $QR = 12$ cm. Then, QL is equal to

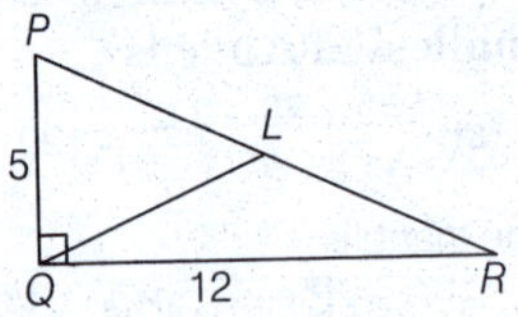

[CDS-13]

Sol. In a right-angled triangle, the median drawn to the hypotenuse is equal to half the hypotenuse.

Clearly, $PR = \sqrt{5^2 + 12^2} = 13$

[You may recall the examples of Pythagorean triplets (5, 12, 13) to avoid the calculation]

Then, $QL = \frac{13}{2} = 6.5$ cm

Problem 64 D and E are the mid-points of AB and AC of ΔABC; BC is produced to any point P; DE, DP and EP are joined. Then,

(a) $\Delta PED = \Delta BEC$ (b) $\Delta BDE = \Delta BEC$
(c) $\Delta ADE = \Delta BEC$ (d) $\Delta PED = \frac{1}{4}\Delta ABC$

Sol. We join points B and E.

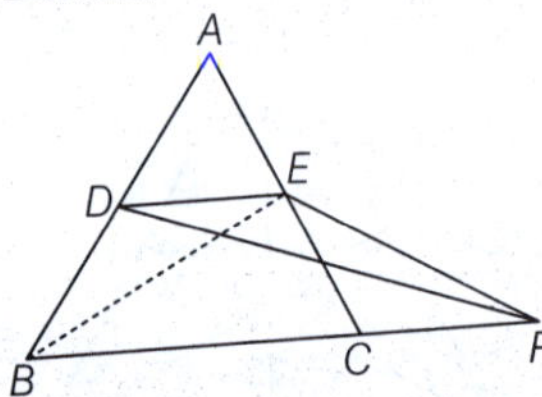

area (ΔDEB) = area (ΔPED)

[having same base and lying between same parallel lines]

Also, DE is a median of ΔAEB,

then area (ΔADE) = area (ΔDEB)

Since BE is a median of ΔABC, area (ΔBEC) = area (ΔABE)

[$\because$ a median divides th triangle in two parts having equal areas]

$\Rightarrow$ area (ΔBEC) = area (ΔDEB) + area (ΔADE)

$\Rightarrow$ area (ΔBEC) = area (ΔDEB) + area (ΔDEB)

$\Rightarrow$ area (ΔBEC) = 2 × area (ΔDEB)

$\Rightarrow \frac{1}{2}$ area (ΔABC) = 2 × area (ΔPED)

$\Rightarrow \Delta PED = \frac{1}{4}\Delta ABC$

Problem 65 If G is the centroid of ΔABC and $AG = BC$, then $\angle BGC = ?$

Sol.

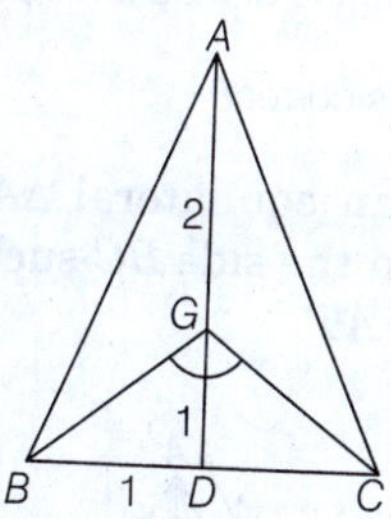

Clearly, $\frac{AG}{GD} = \frac{2}{1}$

As given, $AG = BC \Rightarrow \frac{AG}{2} = \frac{BC}{2}$

$\Rightarrow \quad GD = BD$ or CD

In ΔBGD, $GD = BD$, then $\angle GBD = \angle BGD$

In ΔCGD, $GD = CD$, then $\angle GCD = \angle CGD$

On addition,

$\angle GBD + \angle GCD = \angle BGD + \angle CGD$

$\Rightarrow \quad 180° - \angle BGC = \angle BGC$

$\Rightarrow \quad \angle BGC = 90°$

Problem 66 Let BE and CF be the two medians of a ΔABC and G be their intersection. Also, let EF cut AG at O. Then, $AO : OG$ is? ***[SSC(CGL-T-1)-12, 14, SSC(MTS)-17]***

Sol. We know, the line joining the mid-points of any two sides of a triangle bisects the median falling on the third side.

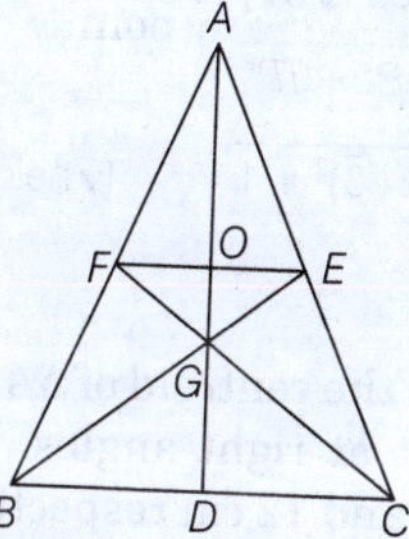

Hence, $AO = OD$

$\Rightarrow \quad AO = OG + GD$

$\Rightarrow \quad AO = OG + \frac{AG}{2}$ $\quad [\because AG : GD = 2 : 1]$

$\Rightarrow \quad AO = OG + \frac{AO + OG}{2}$

$\Rightarrow \quad 2AO = AO + 3OG$

$\Rightarrow \quad AO = 3OG$

$\Rightarrow AO : OG = 3 : 1$

Problem 67 ABC is a triangle and the medians CD and BE intersect each other at O. Then, $\Delta ODE : \Delta ABC = ?$

Sol. Clearly, $\Delta ODE \sim \Delta OBC$ (by AA-criterion)

$[\because DE \parallel BC$, then $\angle EDO = \angle OCB$, $\angle DEO = \angle OBC$ (alternate interior angles)]

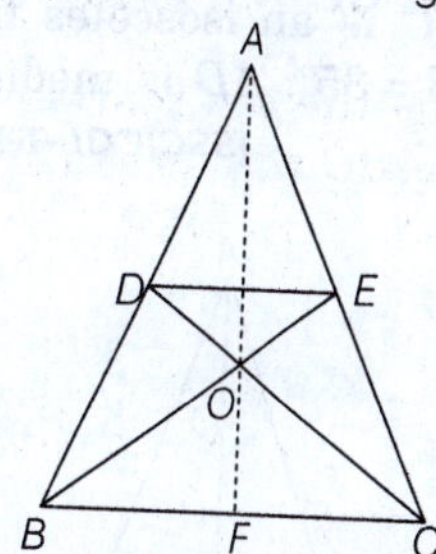

Then, area (ΔODE); area (OBC) $= DE^2 : BC^2$

[$\because$ Ratio of areas of similar triangles = ratio of squares of corresponding sides]

$\Rightarrow$ area (ΔODE) : area (ΔOBC)

$= \left(\frac{BC}{2}\right)^2 : BC^2 = 1 : 4$

[$\because$ line joining the mid-points of any two sides is half the third side]

$\Rightarrow$ area (ΔODE) : {area (ΔBOF) + area (ΔCOF)} = 1 : 4

$\Rightarrow$ [area (ΔBOF) + area (ΔCOF)] = 4 × area (ΔODE)

$\Rightarrow$ [$\frac{1}{6}$ of area (ΔABC) + $\frac{1}{6}$ of area (ΔABC)]

= 4 × area (ΔODE)

$\Rightarrow \frac{1}{3}$ of area (ΔABC) = 4 × area (ΔODE)

$\Rightarrow$ area (ΔODE) : area (ΔABC) = 1 : 12

Problem 68 Let ABC be an equilateral triangle. If the side BC is produced to the point D so that $BC = 2CD$, then AD^2 is equal to

(a) $3\,CD^2$
(b) $4\,CD^2$
(c) $5\,CD^2$
(d) $7\,CD^2$

Sol. We draw a median AE in ΔABC and join points A and D. Then, $BE = EC$

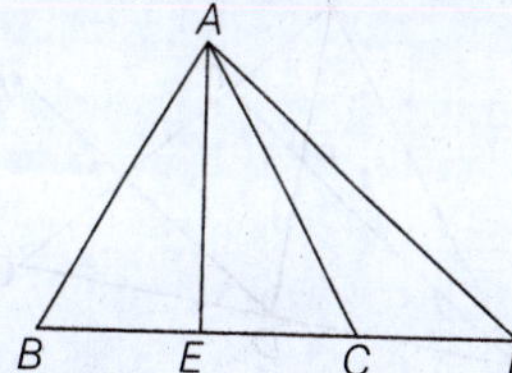

Since the median also acts as altitude of an equilateral triangle, ΔAEC and ΔAED are both right-angled triangles.

In ΔAEC, $AC^2 = AE^2 + EC^2$

$\Rightarrow \quad AC^2 = AE^2 + CD^2$

$[\because EC = \frac{1}{2}BC$ and $CD = \frac{1}{2}BC \therefore EC = CD]$

$\Rightarrow \quad \frac{4}{3}AE^2 = AE^2 + CD^2$ $\quad [\because \frac{\sqrt{3}}{2} \times AC = AE \therefore AC^2 = \frac{4}{3}AE^2]$

$\Rightarrow \quad AE^2 = 3CD^2$

In ΔAED, $AD^2 = AE^2 + ED^2$

$\Rightarrow \quad AD^2 = 3CD^2 + (2CD)^2$

$\Rightarrow \quad AD^2 = 7CD^2$

Problem 69 ABC is an isosceles triangles such that $AB = AC$ and $\angle B = 35°$. AD is median to the base BC. Then $\angle BAD$ is : **[SSC(CGL-T-I)-13, SSC(CGL-T-II)-14]**

Sol.

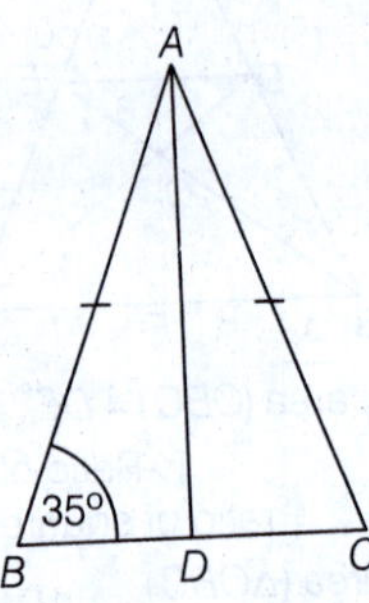

Here, $\angle C = 35°$ $\{\because AB = AC \therefore \angle B = \angle C\}$

Then, $\angle BAC = 180° - 35° - 35° = 110°$

We know, the median to the side other than the congruent sides is also an angle bisector.

Hence, $\angle BAD = \angle CAD$

$\because$ $\angle BAD = \frac{110°}{2} = 55°$

Alternatively

In an isosceles triangle, the median to the side other than the congruent sides is also an altitude. Hence, $\angle ADB = 90°$.

Then, $\angle BAD = 180° - 35° - 90° = 55°$

Problem 70 In ΔABC, AD is the median through A and E is the mid-point of AD and BE produced meets AC at. F. Then, AF is equal to **[CDS-14, SSC (10 + 2/DEO-LDC)-15]**

(a) $\frac{AC}{5}$ (b) $\frac{AC}{4}$ (c) $\frac{AC}{3}$ (d) $\frac{AC}{2}$

Sol. Draw a line $DM \parallel BF$ intersecting AC at point N.

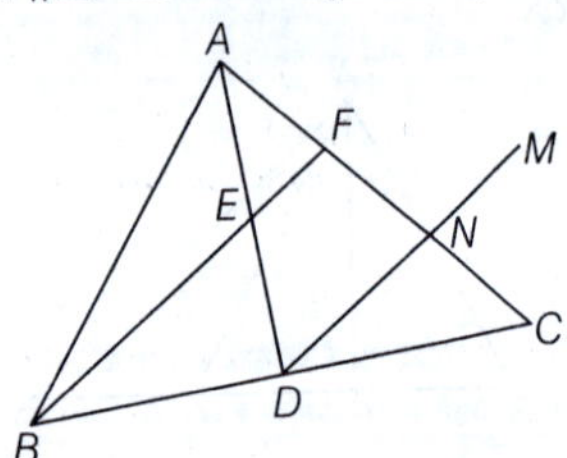

In ΔADN, $EF \parallel DN$ and E is the mid-point of AD, then F will be the mid-point of AN.

Further, in ΔBFC, $DN \parallel BF$ and D is the mid-point of BC, then N will be the mid-point of FC.

Now, $AC = AF + FN + NC$

$\Rightarrow$ $AC = AF + AF + AF$ $[\because FN = AF \text{ and } FN = NC]$

$\Rightarrow$ $AF = \frac{AC}{3}$

Problem 71 In ΔABC, AD, BE and CF are three medians. Then the ratio $(AD + BE + CF) : (AB + AC + BC)$ is :

(a) equal to $\frac{3}{4}$ (b) less than $\frac{3}{4}$

(c) greater than $\frac{3}{4}$ (d) equal to $\frac{1}{2}$

Sol. We know, in a triangle, $\frac{3}{4}$(perimeter) < sum of lengths of medians

$\Rightarrow$ $\frac{3}{4}$ < (sum of lengths of medians : perimeter)

Hence, option (c) is correct.

Problem 72 In an equilateral ΔABC with side 4 cm, point T is lying on the side BC such that $BT : TC = 1:3$. Find the value of AT.

Sol.

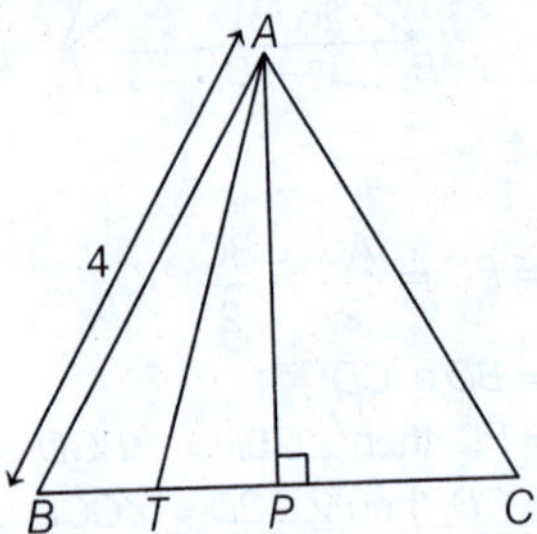

Draw $AP \perp BC$.

Since the altitutde of an equilateral triangle also acts as median,

$BP = \frac{BC}{2} = \frac{4}{2} = 2$ cm

Further, $BT = 4 \times \frac{1}{(1+3)}$ $[\because BT : TC = 1 : 3 \text{ and } BC = 4]$

$\Rightarrow$ $BT = 1$

Then, $TP = BP - BT = 2 - 1 = 1$

Now, in right-angled ΔATP,

$AT = \sqrt{AP^2 + TP^2}$

$\Rightarrow$ $AT = \sqrt{(2\sqrt{3})^2 + 1^2}$ $\left[\because \text{height } (AP) = \frac{\sqrt{3}}{2} \times 4 = 2\sqrt{3}\right]$

$\Rightarrow$ $AT = \sqrt{13}$

Problem 73 G is the centroid of ΔABC. The medians AD and BE intersect at right angles. If the lengths of AD and BE are 9 cm and 12 cm respectively, then the length of AB (in cm) is :

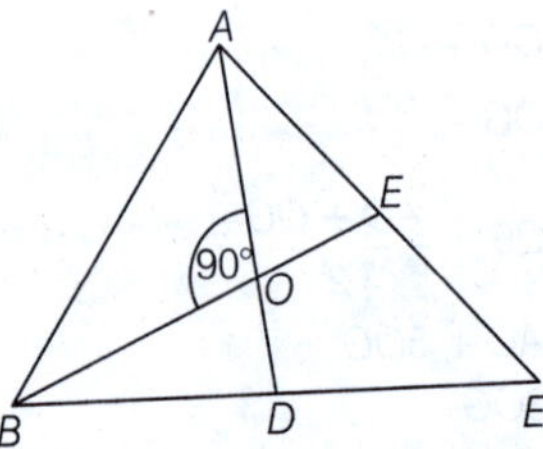

Sol. Clearly, $BG : GE = 2 : 1$

Then, $BG = 12 \times \frac{2}{3} = 8$ cm

Also, $AG = 9 \times \frac{2}{3} = 6$ cm

In ΔABG, $AB^2 = AG^2 + BG^2$

$= 6^2 + 8^2 = 100$

$\Rightarrow$ $AB = 10$ cm

Problem 74 In ΔABC, BE and CF are two medians perpendicular to each other and if $AB = 19$ cm and $AC = 22$ cm, then the length of BC is **[SSC(CPO, DP)-15]**

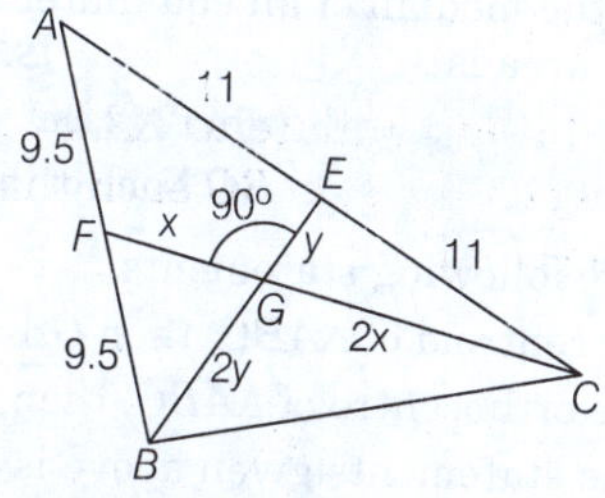

Sol. Here, $CG : GF = BG : GE = 2 : 1$ [$\because G$ is centroid]

Let $CG = 2x$,

$GF = x$, $BG = 2y$ and $GE = y$

In ΔBGF, $9.5^2 = x^2 + (2y)^2$

In ΔCEG, $11^2 = y^2 + (2x)^2$

On addition, $9.5^2 + 11^2 = 5(x^2 + y^2)$

$\Rightarrow 5(x^2 + y^2) = 211.25$

$\Rightarrow (x^2 + y^2) = 42.25$

In ΔBCG, $BC^2 = (2x)^2 + (2y)^2$

$= 4(x^2 + y^2) = 4 \times 42.25$

$\Rightarrow BC = \sqrt{4 \times 42.25} = 2 \times 6.5 = 13$ cm

Problem 75 In an isosceles triangle ABC with $AB = AC$, M, N and R are mid-points of AB, AC and BC respectively. Find the value of AC if $BC = 12\sqrt{3}$ cm and its area is $48\sqrt{3}$ cm^2.

Sol. We join points M and N and points A and R.

Since ABC is an isosceles triangle, median AR will also act as altitude of the triangle.

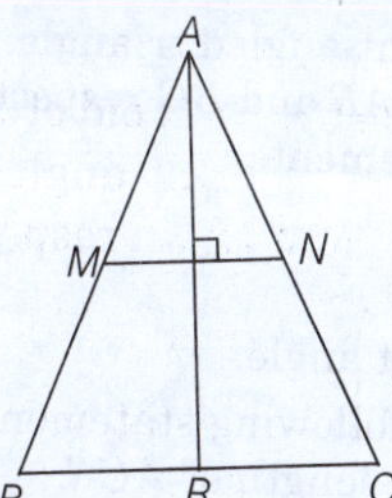

Now,

Area of $\Delta ABC = \frac{1}{2} \times BC \times AR$

$\Rightarrow 48\sqrt{3} = \frac{1}{2} \times 12\sqrt{3} \times AR$

$\Rightarrow AR = 8$ cm

We know, in an isosceles triangle, the line joining the mid-points of the congruent sides and the line joining the mid-point of the third side and its opposite vertex (i.e. median or altitude or angle bisector or perpendicular bisector), bisect each other at right angle.

Hence, $AT = \frac{1}{2} \times AR$ and $TN = \frac{1}{2} MN$

and $\angle ATN = 90°$

Then, $AT = \frac{1}{2} \times 8 = 4$ cm

Further, $MN = \frac{1}{2} BC$ [$\because$ the line joining the mid-points of any two sides of a triangle is equal to half of the third side]

$\Rightarrow MN = \frac{1}{2} \times 12\sqrt{3} = 6\sqrt{3}$

Then, $TN = \frac{1}{2} \times 6\sqrt{3} = 3\sqrt{3}$

Now, in right-angled ΔATN,

$AN = \sqrt{AT^2 + TN^2} = \sqrt{4^2 + (3\sqrt{3})^2} = \sqrt{43}$

$\therefore AC = 2 \times AN = 2 \times \sqrt{43} = 13.11$ cm

Problem 76 BL and CM are medians of ΔABC right-angled at A and $BC = 5$ cm. If $BL = \frac{3\sqrt{5}}{2}$ cm, then the length of CM is

Sol. We know, $(BL)^2 = \frac{2(AB)^2 + 2(BC)^2 - (AC)^2}{4}$

$\Rightarrow \left(\frac{3\sqrt{5}}{2}\right)^2 = \frac{2(AB)^2 + 2(5)^2 - \{(BC)^2 - (AB)^2\}}{4}$

$\Rightarrow 45 = 2(AB)^2 + 2(5)^2 - (5)^2 + (AB)^2$

$\Rightarrow 45 = 25 + 3(AB)^2$

$\Rightarrow 3(AB)^2 = 20$

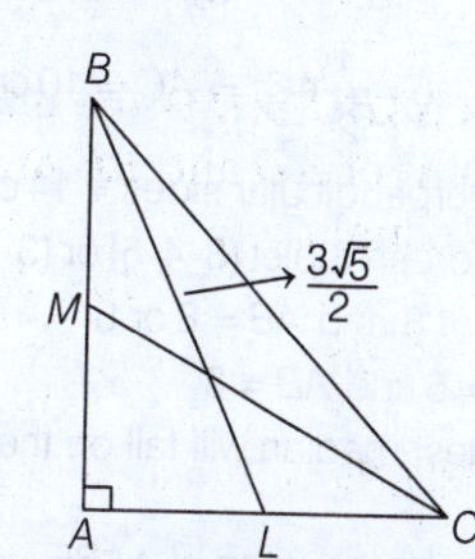

Further, $(CM)^2 = \frac{2(AC)^2 + 2(BC)^2 - (AB)^2}{4}$

$= \frac{2[(BC)^2 - (AB)^2] + 2(BC)^2 - (AB)^2}{4}$

$= \frac{2[(5)^2 - (AB)^2] + 2(5)^2 - (AB)^2}{4}$

$= \frac{100 - 3(AB)^2}{4} = \frac{100 - 20}{4} = 20$

$\Rightarrow CM = \sqrt{20} = 2\sqrt{5}$

Problem 77 In ΔABC, medians AD and BE are perpendicular to each other and have lengths 12 cm and 9 cm respectively. Then, the area of ΔABC, in sq cm, is **[CAT-19]**

Sol.

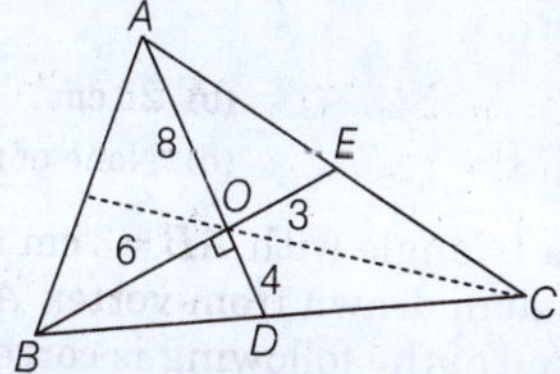

Since centroid divides the median in ratio 2 : 1, $BO = 6$ and $OE = 3$. Also, $AO = 8$ and $OD = 4$

Now, area of right-angled ΔBOD

$$= \frac{1}{2} \times 4 \times 6 = 12$$

We know that all the three medians divide a triangle into six smaller triangles of equal areas.

Thus, area of $\Delta ABC = 6 \times 12 = 72 \text{ cm}^2$

Problem 78 **The shortest median in a right-angled triangle is of 5 cm and the sum of perpendicular sides is 14 cm. Find the area of the triangle whose hypotenuse is the greatest median of the former triangle.**

Sol. Since the shortest median falls on the longest side (here, hypotenuse), the shortest median is BD.

Also, $BD = \frac{1}{2} AC$ [$\because$ median to the hypotenuse is half the length of hypotenuse]

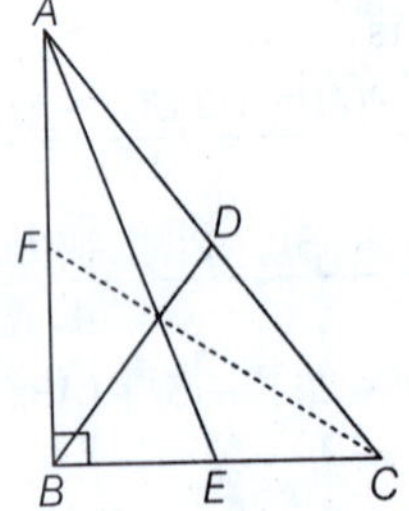

$\Rightarrow \quad 5 = \frac{1}{2} AC \Rightarrow AC = 10 \text{ cm}$

Given, sum of perpendicular sides = 14 cm

Recall a Pythagorean triplet (3, 4, 5) or (3·2, 4·2, 5·2).

Hence, $BC = 6$ or 8 and $AB = 8$ or 6

Case (i) If $BC = 6$ and $AB = 8$

Clearly, the largest median will fall on the shortest side i.e. on BC.

Hence, AE is the hypotenuse of ΔABE.

$\therefore$ Area of $\Delta ABE = \frac{1}{2} \times BE \times AB$

$$= \frac{1}{2} \times \left(\frac{6}{2}\right) \times 8 = 12 \text{ cm}^2$$

Case (ii) If $AB = 6$ and $BC = 8$

Area of $\Delta BCF = \frac{1}{2} \times BF \times BC$

$$= \frac{1}{2} \times \left(\frac{6}{2}\right) \times 8 = 12 \text{ cm}^2$$

Hence, required answer is 12 cm^2.

Workout

1. If G be the centroid of ΔABC and the area of ΔGBD is 6 cm^2 where D is the mid-point of side BC, then the area of ΔABC is : **[SSC(CGL-T-1)-12]**

(a) 45 cm^2 (b) 25 cm^2
(c) 36 cm^2 (d) None of these

2. Let ABC be a triangle with $AB = 3$ cm and $AC = 5$ cm. If AD is a median drawn from vertex A to the side BC, then which one of the following is correct ?

(a) AD is always greater than 4 cm but less than 5 cm
(b) AD is always greater than 5 cm
(c) AD is always less than 4 cm
(d) None of the above

3. If length of the median of an equailateral triangle be 6 cm, then its area is : **[SSC(FCI-mains)-12]**

(a) $12\sqrt{3} \text{ cm}^2$ (b) $12\sqrt{2} \text{ cm}^2$
(c) 36 cm^2 (d) None of these

4. Consider the following statements :

I. If G is the centroid of ΔABC, then $GA = GB = GC$
II. If H is the orthocentre of ΔABC, then $HA = HB = HC$

Which of the statements given above is/are correct? **[CDS-13]**

(a) Only I (b) Only II
(c) Both I and II (d) Neither I nor II

5. In ΔABC, $\angle A = 90°$, BP and CQ are two medians. Then, the value of $\frac{BP^2 + CQ^2}{BC^2}$ is **[SSC (10+2) DEO-LDC-14]**

(a) $\frac{4}{5}$ (b) $\frac{5}{4}$ (c) $\frac{3}{4}$ (d) $\frac{3}{5}$

6. If the medians of two equilateral triangles are in the ratio 3 : 2, then what is the ratio of their sides ?

(a) 1 : 1 (b) 2 : 3 (c) 3 : 2 (d) $\sqrt{3} : \sqrt{2}$

7. If the length of the three sides of a triangles are 6 cm, 8 cm and 10 cm, then the length of the median to its greatest side is :

(a) 8 cm (b) 6 cm (c) 5 cm (d) 4.8 cm

8. In the ΔABC, $AB = 2$ cm, $BC = 3$ cm and $AC = 4$ cm, D is the middle point of AC. If a square is constructed on the side BD. What is the area of the square ?

(a) 4.5 cm^2 (b) 2.5 cm^2
(c) 6.35 cm^2 (d) None of these

9. ΔABC is a right-angled triangle. D and E are mid-points of AB and BC respectively. Read the following statements :

I. $AE = 19$
II. $CD = 22$
III. $\angle B$ is a right angle.

Which of the following statement would be sufficient to determine the length of AC ? **[XAT-14]**

(a) Statements I and II (b) Statements I and III
(c) Statements II nd III (d) Statements III alone
(e) All three

10. In ΔABC, $\angle A = 90°$, M is the midpoint of BC such that $AD \perp BC$. If $AB = 7$ cm and $AC = 24$ cm, then $AD : AM$ is equal to **[SSC(CGL-T-1)-19]**

(a) 168 : 275 (b) 24 : 25
(c) 32 : 43 (d) 336 : 625

Answer Key

1. (c) 2. (c) 3. (a) 4. (d) 5. (b) 6. (c)
7. (c) 8. (b) 9. (e) 10. (d)

Hints and Solutions

1. Area of $\Delta ABC = 6 \times$ area of $\Delta GBD = 6 \times 6 = 36\,\text{cm}^2$

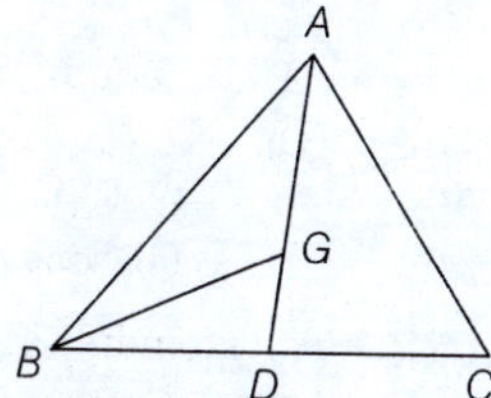

2. Here, $(3 + 5) > 2\text{AD}$ {∵ sum of any two sides of a triangle is greater than twice the median drawn to the 3rd side.}

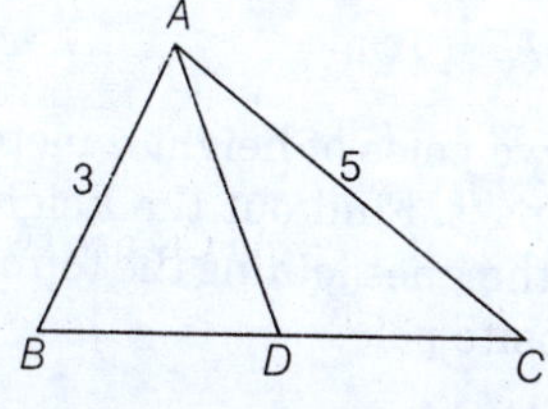

$\Rightarrow \quad 4 > AD$

Hence, AD is always less than 4 cm.

3. We know the median of an equilateral triangle is its height.

Hence, $\dfrac{\sqrt{3}}{2} \times \text{side} = 6$

$\Rightarrow \quad \text{side} = 4\sqrt{3}$

Then, area of triangle $= \dfrac{\sqrt{3}}{4} \times (4\sqrt{3})^2 = 12\sqrt{3}\,\text{cm}^2$

4. Both the statements are correct if the triangle is the equailateral triangle. But, here it is not given that the triangle is an equilateral triangle.

Hence, option (d) is correct.

5. In ΔABP, $BP^2 = AB^2 + AP^2$

and in ΔACQ, $CQ^2 = AQ^2 + AC^2$

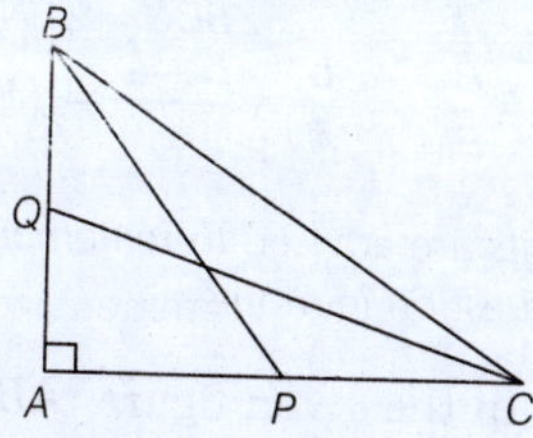

On addition, $BP^2 + CQ^2$

$$= AB^2 + AP^2 + AQ^2 + AC^2$$

$$\Rightarrow \frac{BP^2 + CQ^2}{BC^2} = \frac{AB^2 + AP^2 + AQ^2 + AC^2}{AB^2 + AC^2}$$

$$\Rightarrow \frac{BP^2 + CQ^2}{BC^2} = 1 + \frac{AP^2 + AQ^2}{AB^2 + AC^2}$$

$$\Rightarrow \frac{BP^2 + CQ^2}{BC^2} = 1 + \frac{AP^2 + AQ^2}{(2AQ)^2 + (2AP)^2} = 1 + \frac{1}{4} = \frac{5}{4}$$

6. Since both the triangles are equilateral, they are similar. Further, we know that in case of similarity, ratio of corresponding sides = ratio of medians. Then, ratio of their sides = 3 : 2

7. 6, 8 and 10 is a Pythagorean triplet, hence, the given triangle is a right-angled triangle. We know, the median drawn to the hypotenuse is equal to half the hypotenuse and the fact theat hypotenuse is the greatest side of a right-angled triangle. Then, required answer $= \dfrac{10}{2} = 5\,\text{cm}$

8. Side of the square = median AD

Recall the Apollonius theorem,

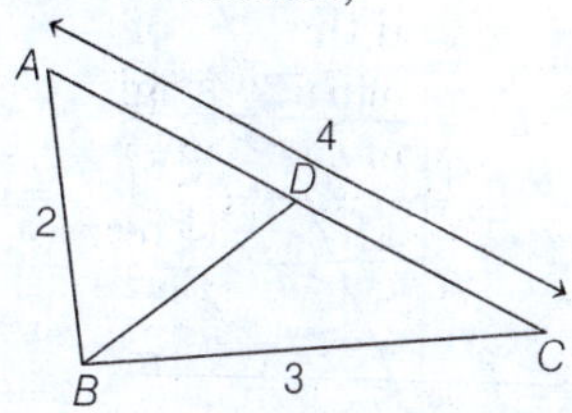

$$(AD)^2 = \frac{2(BC)^2 + 2(AB)^2 - (AC)^2}{4}$$

$$\Rightarrow (AD)^2 = \frac{2(3)^2 + 2(2)^2 - (4)^2}{4}$$

$$\Rightarrow \text{Area of the square} = \frac{2(3)^2 + 2(2)^2 - (4)^2}{4} = 2.5\,\text{cm}^2$$

9. Option (a) : We cannot determine the length of AC if two medians AE and CD are known.

(b) and (c) : We cannot determine the length of AC if one median either AE or CD is known and which angle, is the right angle, is known.

(d) It is not sufficient to determine the length of AC.

(e) We can find the length of AC if all three statements are given.

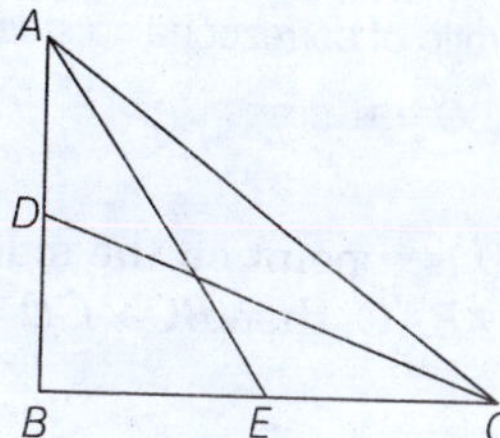

Hence, option (e) is correct.

[Try to determine the length of AC using the facts given in all three statements. Asnwer is 26]

10. Clearly, $BC = 25$ {∵ (7, 24, 25) is a Pythagorean triplet}

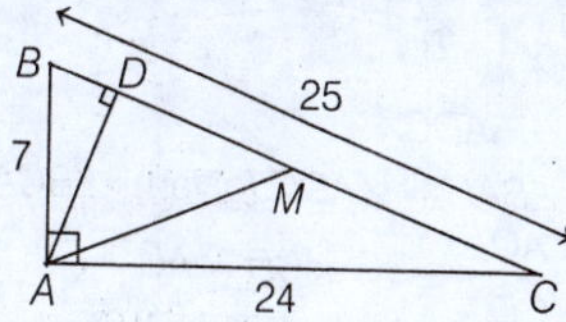

Now, area of $\Delta ABC = \dfrac{1}{2} \times 7 \times 24$

$\Rightarrow \dfrac{1}{2} \times 25 \times AD = 84 \Rightarrow AD = \dfrac{168}{25}$

Further, since the median drawn to the hypotenuse is equal to half the hypotenuse. Hence, $AM = \dfrac{25}{2}$

Now $AD : AM = \dfrac{168}{25} : \dfrac{25}{2} = 336 : 625$

Problems based on Similarity of Triangles

Problem 79 In the given figure, QA and PB are perpendicular to AB. If $AO = 10$ cm, $BO = 6$ cm and $PB = 9$ cm, then find AQ.

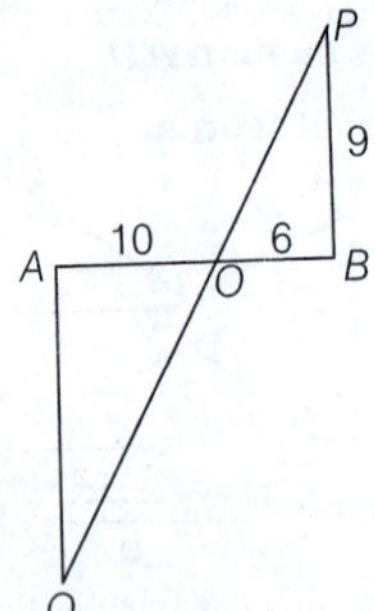

Sol. In ΔAQO and ΔPBO, $\angle QAO = \angle PBO = 90°$ and $\angle AOQ = \angle POB$ (vertically opposite angles)

$\therefore \Delta AQO \sim \Delta PBO$ (by AA-criterion)

Then, $\dfrac{AO}{OB} = \dfrac{AQ}{PB}$

$\Rightarrow \dfrac{10}{6} = \dfrac{AQ}{9} \Rightarrow AQ = 15$ cm

Problem 80 The perimeters of two similar triangles ABC and PQR are respectively 36 cm and 24 cm. If $PQ = 10$ cm, find AB. *[SSC(CGL-T-1)12]*

Sol. Clearly, $\dfrac{AB}{PQ} = \dfrac{36}{24}$

[∵ ratio of corresponding sides = ratio of perimeters]

$\Rightarrow \dfrac{AB}{10} = \dfrac{36}{24} \Rightarrow BC = 15$ cm

Problem 81 D is a point on the side BC of ΔABC such that $\angle ADC = \angle BAC$, then $BC \times CD = ?$

Sol.

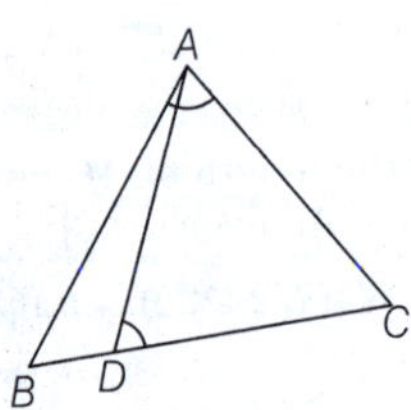

Here, $\Delta ABC \sim \Delta ADC$

{∵ $\angle BAC = \angle ADC$ (given), $\angle BCA = \angle ACD$ (common)}

Then, $\dfrac{BC}{AC} = \dfrac{AC}{CD} \Rightarrow BC \times CD = AC^2$

Problem 82 AD is perpendicular to the internal bisector of $\angle ABC$ of ΔABC. DE is drawn through D and parallel to BC to meet AC at E. If the length of AC is 12 cm, then the length of AE (in cm) is : *[SSC (CGL-T-II)-15]*

Sol. AD extended meets BC at point G.

In ΔABD and ΔBDG, $\angle ADB = \angle BDG = 90°$

and $\angle ABD = \angle GBD$ [∵ BD is bisector of $\angle ABC$]

Hence, $\Delta ABD \sim \Delta BDG$

$\therefore \dfrac{AD}{DG} = \dfrac{BD}{BD} \Rightarrow AD = DG$

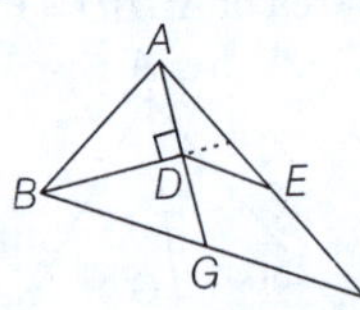

Clearly, $\Delta ADE \sim \Delta AGC$ [∵ $\angle A = \angle A$(common) and $\angle ADE = \angle AGC$ (corresponding angles)]

Then, $\dfrac{AD}{AG} = \dfrac{AE}{AC} \Rightarrow \dfrac{AD}{AD + DG} = \dfrac{AE}{12} \Rightarrow \dfrac{AD}{AD + AD} = \dfrac{AE}{12}$

$\Rightarrow AE = 6$ cm

Problem 83 Two poles of height a meters and b meters are p meters apart. Find out the height of the point of intersection of the lines joining the top of each pole to the foot of the opposite pole.

Sol. Here, $\Delta ABM \sim \Delta MCD$

[∵ $\angle ABM = \angle MCD$ (alternate interior angles), $\angle AMB = \angle CMD$ (vertically opposite angles)]

$\therefore \dfrac{AB}{CD} = \dfrac{BM}{MC} \Rightarrow \dfrac{a}{b} = \dfrac{BM}{MC} \Rightarrow \dfrac{b}{a} = \dfrac{MC}{BM}$

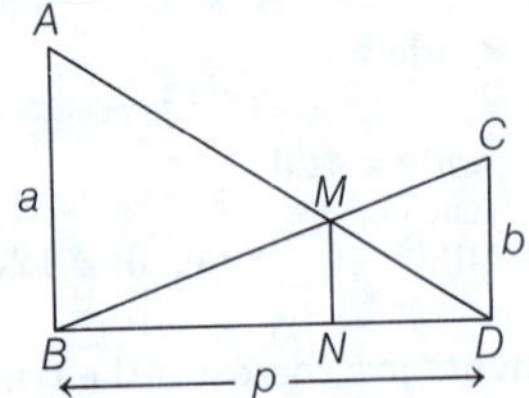

$\Rightarrow \dfrac{b}{a} + 1 = \dfrac{MC}{BM} + 1 \Rightarrow \dfrac{b + a}{a} = \dfrac{MC + BM}{BM}$

$\Rightarrow \dfrac{b + a}{a} = \dfrac{BC}{BM}$

Further, $\Delta BCD \sim \Delta BMN$ [∵ $\angle B = \angle B$ (common), $\angle BDC = \angle BNM$ (each equal to 90°)]

Then, $\dfrac{CD}{MN} = \dfrac{BC}{BM} \Rightarrow \dfrac{b}{MN} = \dfrac{b + a}{a} \Rightarrow MN = \dfrac{ab}{a + b}$

Remarks *Aspirants are advised to remember this result so as to solve the question in moments.*

Problem 84 In the given figure, ABC is an equilateral triangle of side length of 30 cm. XY is parallel to BC, XP is parallel to AC and YQ is parallel to AB. If $(XY + XP + YQ)$ is 40 cm, then what is PQ equal to ? *[CDS-13]*

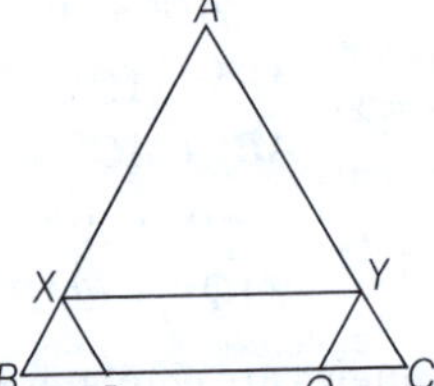

Sol. Since $XP \| AC$ and $YQ \| AB$, $\angle XPB = \angle QCY$ and $\angle XBP = \angle YQC = 60°$ (corresponding angles)

$\therefore$ ΔXBP and ΔYCQ are equilateral triangles.

Clearly, $\Delta AXY \sim ABC$. Then, $\frac{XY}{BC} = \frac{AX}{AB}$

$\Rightarrow$ $XY = Ax$ [$\because AB = BC$]

As given, $XY + XP + YQ = 40$

$\Rightarrow$ $AX + XB + YQ = 40$ [$\because XP = XB$]

$\Rightarrow$ $30 + YQ = 40$

$\Rightarrow$ $YQ = 10 \Rightarrow QC = 10$ [$\because \Delta YQC$ is equilateral]

Further, $\frac{AY}{YC} = \frac{AX}{XB}$ [$\because XY \parallel BC$]

$\Rightarrow$ $\frac{AY}{YQ} = \frac{AX}{XB}$

$\Rightarrow$ $\frac{30-10}{10} = \frac{AX}{XB} \Rightarrow AX : XB = 2 : 1$

$\therefore$ $XB = 10$ [$\because AX + XB = 30$]

Then, $BP = 10$

Hence, $PQ = BC - BP - QC$

$= 30 - 10 - 10 = 10$ cm

Problem 85 ABC is a triangle in which $AB = AC$ and D is a point on AC such that $BC^2 = AC \times CD$. If $BD = 7$ cm, then BC = ?

Sol. Given, $BC^2 = AC \times CD \Rightarrow \frac{BC}{CD} = \frac{AC}{BC}$

In ΔABC and ΔBCD, $\angle C = \angle C$ (common) and $\frac{BC}{CD} = \frac{AC}{BC}$

Hence, $\Delta ABC \sim \Delta BCD$ [by S-A-S criterion]

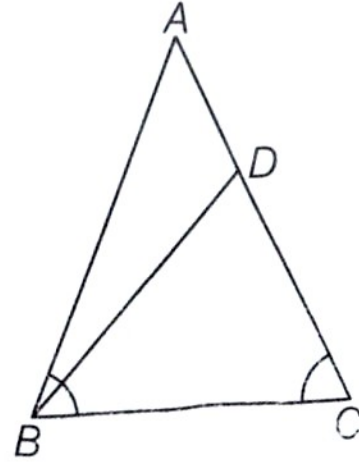

$\Rightarrow$ $\frac{AB}{BD} = \frac{BC}{CD} \Rightarrow \frac{AC}{BD} = \frac{BC}{CD}$ [$\because AB = AC$ (given)]

$\Rightarrow$ $\frac{AC}{BD} = \frac{AC}{BC}$ $\left[\because \frac{BC}{CD} = \frac{AC}{BC}\right]$

$\Rightarrow$ $BC = BD = 7$ cm

Problem 86 A rectangle inscribed in a triangle has its base coinciding with the base b of the triangle. If the altitude of the triangle is h and the altitude x of the rectangle is half the base of the rectangle, then : **[CAT-16]**

(a) $x = \frac{1}{2}h$ (b) $x = \frac{bh}{h+b}$

(c) $x = \frac{bh}{2h+b}$ (d) $x = \sqrt{\frac{bh}{2}}$

Sol. Given, altitutde of the rectangle $= DG = x$. Then, $DE = 2x$

Here, $\Delta BDG \sim \Delta BAH$

[$\because \angle B = \angle B$ (common) and $\angle BGD = \angle BHA$ (corresponding angles as $DG \parallel AH$ or equal to 90°]

Then, $\frac{BG}{BH} = \frac{DG}{AH} \Rightarrow \frac{BG}{BH} = \frac{x}{h}$

$\Rightarrow$ $BG = \frac{x}{h} \times BH$

Also, $\Delta EFC \sim \Delta AHC$

Then, $\frac{CF}{CH} = \frac{EF}{AH} \Rightarrow \frac{CF}{CH} = \frac{x}{h}$

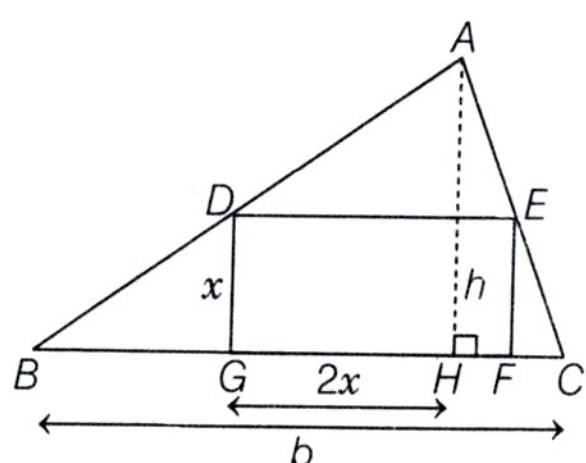

$\Rightarrow$ $CF = \frac{x}{h} \times CH$

Now, $BC = BG + FG + CF$

$\Rightarrow$ $b = \frac{x}{h} \times BH + 2x + \frac{x}{h} \times CH$

$\Rightarrow$ $b = \frac{x}{h}(BH + CH) + 2x$

$\Rightarrow$ $b = \frac{x}{h} \times BC + 2x \Rightarrow b = \frac{x}{h}b + 2x$

$\Rightarrow$ $b = x\left(\frac{b}{h} + 2\right) \Rightarrow x = \frac{bh}{2h+b}$

Problem 87 D and E are two points on the sides AC and BC respectively of ΔABC such that $DE = 18$ cm, $CE = 5$ cm and $\angle DEC = 90°$. If $\tan \angle ABC = 3.6$, then $AC : CD$ = ? **[SSC(CGL-T-II)-13]**

(a) $BC : 2CE$ (b) $2CE : BC$

(c) $2BC : CE$ (d) $CE : 2BC$

Sol.

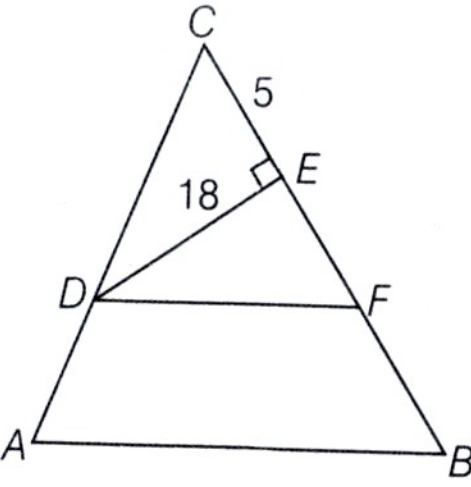

Draw DF parallel to AB.

Then, $\angle F = \angle B$ [corresponding angles]

$\therefore$ $\tan\angle B = \tan\angle F = 3.6$

$\Rightarrow$ $\frac{DE}{EF} = 3.6 \Rightarrow \frac{18}{EF} = 3.6 \Rightarrow EF = 5$

Hence, $CE = EF$

Also, $\Delta CDF \sim \Delta ABC$ [$\because \angle C = \angle C$ (common), $\angle F = \angle B$ (corresponding angles)]

Then, $\frac{AC}{CD} = \frac{BC}{CF} \Rightarrow \frac{AC}{CD} = \frac{BC}{CE + EF}$

$\Rightarrow$ $\frac{AC}{CD} = \frac{BC}{CE + CE} \Rightarrow \frac{AC}{CD} = \frac{BC}{2CE}$

$\Rightarrow$ $AC : CD = BC : 2CE$

Problem 88 In a ΔABC, point D and E are lying on sides AB and AC respectively such that $DE \parallel BC$ and $AD : DB = 5 : 4$, then find out the ratio of the areas of ΔDEF and ΔCFB, if the point F is the intersection of line DC and BE.

Sol. Here, $\Delta ADE \sim \Delta ABC$

[$\because \angle A = \angle A$(common) and $\angle AED = \angle ACB$] (corresponding angles) as $DE \parallel BC$)]

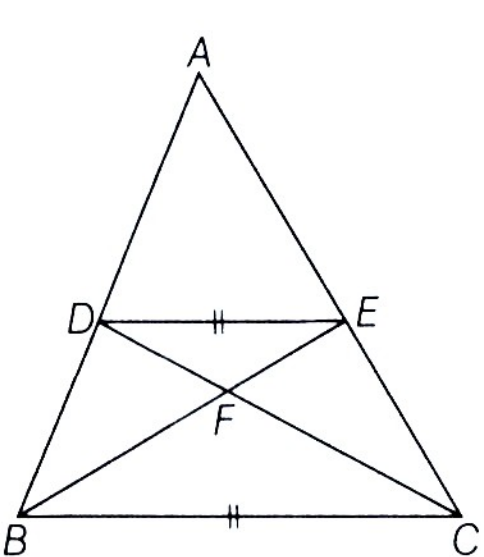

Then, $\frac{AD}{AB} = \frac{DE}{BC} \Rightarrow \frac{5}{(5+4)} = \frac{DE}{BC}$

$\Rightarrow \frac{DE}{BC} = \frac{5}{9}$

Further, $\Delta DEF \sim \Delta CFB$

[$\because \angle DEF = \angle FBC$ (alternate interior angles) and $\angle EDF = \angle FCB$ (alternate interior angles) as $DE \parallel BC$)]

Then, area of ΔDEF : area of $\Delta CFB = DE^2 : BC^2$

[$\because$ ratio of areas of similar triangles = ratio of squares of corresponding sides]

$= DE^2 : (\frac{9}{5} DE)^2 = 25 : 81$

Problem 89 Consider a rectangle $ABCD$ of area 90 units. The point P and Q trisect AB, and R bisects CD. The diagonal AC intersects the line segments PR and QR at M and N respectively. What is the area of the quadrilateral $PQMN$? ***[XAT-14]***

(a) > 9.5 and ≤ 10 (b) > 10 and ≤ 10.5
(c) > 10.5 and ≤ 11 (d) > 11 and ≤ 11.5
(e) > 11.5

Sol. Here, $\Delta NCR \sim \Delta ANQ$

[$\because \angle RNC = \angle ANQ$ and $\angle RCN = \angle QAN$ (alternate interior angles)]

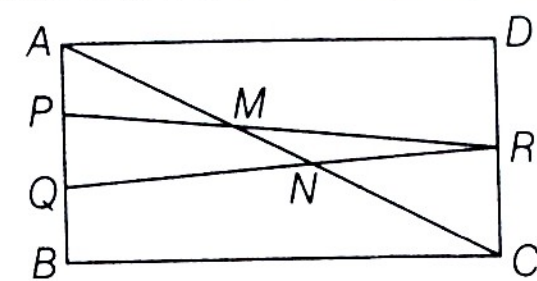

Then, $\frac{CR}{AQ} = \frac{RN}{NQ} \Rightarrow \frac{\frac{1}{2}AB}{\frac{2}{3}AB} = \frac{RN}{NQ}$

$\Rightarrow \frac{3}{4} = \frac{RN}{NQ} \Rightarrow \frac{RN}{RQ} = \frac{3}{3+4} \Rightarrow RN = \frac{3}{7}RQ$

Also, $\Delta CMR \sim \Delta AMP$

[$\because \angle CMR = \angle AMP$ and $\angle RCM = \angle PAM$ (alternate interior angles)]

Then, $\frac{CR}{AP} = \frac{RM}{MP} \Rightarrow \frac{\frac{1}{2}AB}{\frac{1}{3}AB} = \frac{RM}{MP}$

$\Rightarrow \frac{3}{2} = \frac{RM}{MP} \Rightarrow \frac{RM}{RP} = \frac{3}{2+3}$

$\Rightarrow RM = \frac{3}{5}RP$

Now, area of ΔMRN : area of ΔPRQ

$= \frac{1}{2} \times RM \times RN \times \sin \angle MRN$

$: \frac{1}{2} \times RP \times RQ \times \sin \angle PRQ$

$= \frac{3}{5}RP \times \frac{3}{7}RQ : RP \times RQ = 9 : 35$

$\therefore$ area of $PMNQ$: area of $\Delta PRQ = (35 - 9) : 35 = 26 : 35$

Further, area of $\Delta PRQ = \frac{1}{2} \times PQ \times BC$

$= \frac{1}{2} \times \frac{1}{3} AB \times BC = \frac{1}{6} \times 90 = 15$

It implies that 35 units = 15

Then, 26 units = 11.14 sq unit

Hence, option (d) is correct.

Problem 90 In a ΔABC, X and Y are two points on AB and BC respectively such that $XY \parallel AC$ and XY divides triangular region ABC into two parts equal in area. Find the value of $AX : AB$.

Sol.

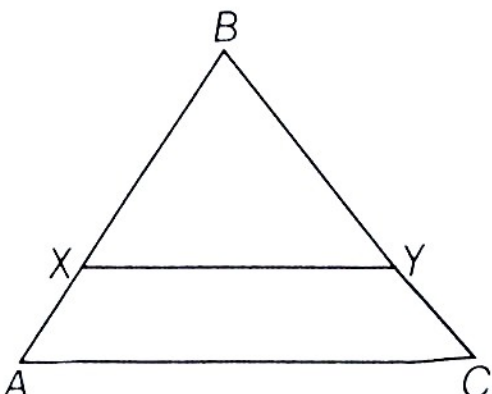

Area of ΔBAC = area of ΔBXY + area of $\square AXYC$

$\Rightarrow$ Area of ΔBAC = area of ΔBXY + area of ΔBXY

[$\because$ area of ΔBXY = area of $\square AXYC$, given]

$\Rightarrow$ area of ΔBXY : area of $\Delta BAC = 1 : 2$

Also, $\Delta BXY \sim \Delta BAC$

[$\because \angle B = \angle B$ (common) and $\angle BYX = \angle BCA$ (corresponding angles) as $XY \parallel AC$)]

Then, area of ΔBXY : area of $\Delta BAC = BX^2 : BA^2$

[$\because$ ratio of areas of similar triangles = ratio of squares of corresponding sides]

$\Rightarrow 1 : 2 = BX^2 : BA^2$

$\Rightarrow BX : BA = 1 : \sqrt{2}$

$\therefore AX = BA - BX = \sqrt{2} - 1$

Then, $AX : AB = (\sqrt{2} - 1) : \sqrt{2}$

Problem 91 In the figure given, $\angle MON = \angle MPO = \angle NQO = 90^\circ$ and OQ is the bisector of $\angle MON$ and $QN = 10, OR = \frac{40}{7}$. Find OP. **[CAT-12]**

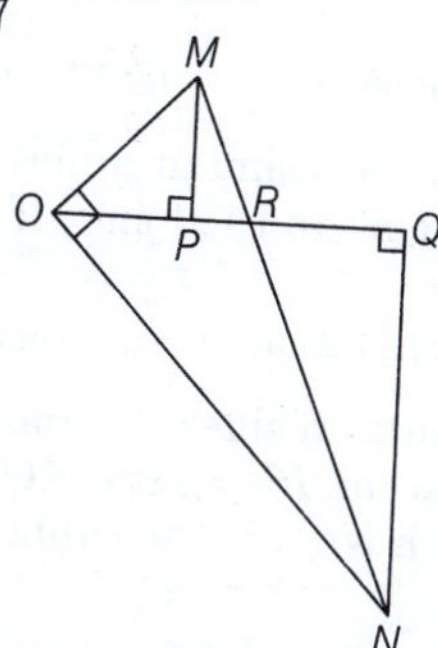

Sol.

$\because OQ$ is the bisector of $\angle MON, \angle MOP = \frac{90^\circ}{2} = 45^\circ$

Then, $\angle OMP = 180^\circ - 90^\circ - 45^\circ = 45^\circ$

Then, $MP = OP$

[$\because$ sides opposite to equal angles are equal]

Further, $QN = OQ$ [$\because$ in ΔOQN, $\angle QON = \angle QNO = 45^\circ$]

Here, $\Delta MPR \sim RQN$ {$\because \angle MPR = \angle NQR = 90^\circ, \angle PMR = \angle QNR$ (alternate interior angles)}

Then, $\frac{MP}{QN} = \frac{PR}{QR} \Rightarrow \frac{OP}{10} = \frac{PR}{OQ - OR}$

$\Rightarrow \frac{OP}{10} = \frac{OR - OP}{QN - OR} \Rightarrow \frac{OP}{10} = \frac{\frac{40}{7} - OP}{10 - \frac{40}{7}}$

$\Rightarrow \frac{30}{7}OP = \frac{400}{7} - 10 \cdot OP \Rightarrow \frac{100}{7}OP = \frac{400}{7}$

$\Rightarrow OP = 4$

Problem 92 Let ABC be an acute-angled triangle and CD be the altitude through C. If $AB = 8$ and $CD = 6$, then the distance between the mid-points of AD and BC is : **[MAT-02]**

Sol. Let the mid-points of AD and BC be M and N respectively. We draw a perpendicular NP to AB.

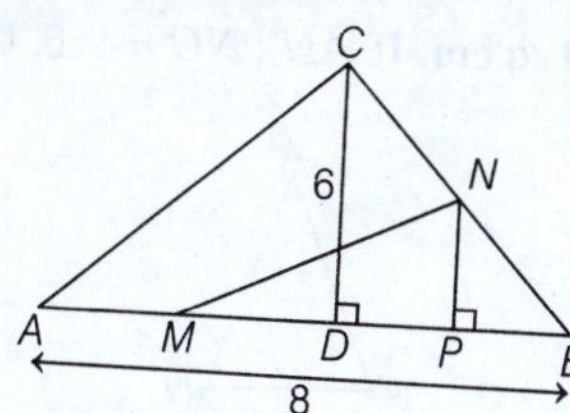

Here, $\Delta BDC \sim \Delta BPN$

[$\because \angle BDC = \angle BPN = 90^\circ$ and $\angle B = \angle B$ (common)]

Then, $\frac{BD}{BP} = \frac{BC}{BN}$

$\Rightarrow \frac{BD}{BP} - 1 = \frac{BC}{BN} - 1$

$\Rightarrow \frac{PD}{BP} = \frac{NC}{BN} = \frac{1}{1}$ {$\because N$ is the mid-point of BC}

$\Rightarrow PD = BP$

Now, $AM + MD + DP + PB = 8$ (given)

$\Rightarrow MD + MD + DP + DP = 8$ [$\because AM = MD$ and $DP = PB$]

$\Rightarrow MD + DP = 4$

$\Rightarrow MP = 4$

Further, $NP = \frac{1}{2}CD$

{$\because$ line joining the mid-points of any two sides is half the third side}

$\Rightarrow NP = \frac{1}{2} \times 6 = 3$

In ΔMNP, $MN^2 = MP^2 + NP^2$

$\Rightarrow MN^2 = 4^2 + 3^2 \Rightarrow MN = 5$

Problem 93 In a ΔABC, M and N are two points on AB and AC respectively such that $AM : MB = 2:3$ and $AN : NC = 2:1$. Find the area of ΔAMN if area of ΔBNC is 24 cm^2.

Sol.

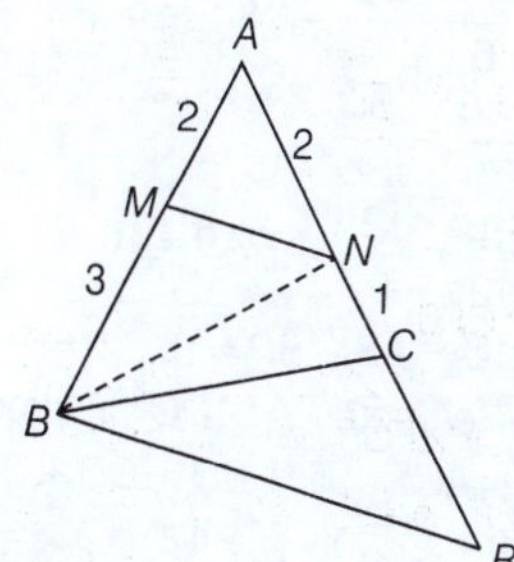

Produce AC to R such that $\frac{AM}{MB} = \frac{AN}{NR}$.

Hence, $MN \parallel BR$

Then, $\Delta AMN \sim \Delta ABR$ [$\because \angle A = \angle A$ (common), $\angle AMN = \angle ABR$ (corresponding angles)]

$\therefore \frac{AM}{AB} = \frac{MN}{BR} \Rightarrow \frac{2}{2+3} = \frac{MN}{BR}$

Since $\Delta AMN \sim \Delta ABR$, ratio of their heights = ratio of corresponding sides

$\therefore$ Ratio of their heights = 2 : (2 + 3) = 2 : 5

Then, height of ΔAMN : height of ΔBNR = 2 : (5 − 2) = 2 : 3

and base of ΔAMN : base of ΔBNR = 2 : 3

Hence, area of ΔAMN : area of ΔBNR

$= (2 \times 2) : (3 \times 3) = 4 : 9$

Now, $\frac{AM}{MB} = \frac{AN}{NR}$

$\Rightarrow \frac{2}{3} = \frac{2}{1 + CR}$

$\Rightarrow CR = 2$

It means $CR = 2NC$.

Since the base of ΔBCR is twice that of ΔBNC and their heights are same, the area of ΔBCR will be twice that of ΔBNC.

$\therefore$ Area of $\Delta BCR = 2 \times 24 = 48$

and area of $\Delta BNR = \Delta BCR + \Delta BNC = 48 + 24 = 72$

Also, area of ΔAMN : area of $\Delta BNR = 4 : 9$

$\Rightarrow$ area of $\Delta AMN : 72 = 4 : 9$

$\Rightarrow$ Area of $\Delta AMN = 32 \text{ cm}^2$

Problem 94 Consider the ΔABC shown in the adjacent figure where $BC = 12$ cm, $DB = 9$ cm, $CD = 6$ cm and $\angle BCD = \angle BAC$. What is the ratio of the perimeter of the triangle ADC to that of the triangle BDC? **[CAT-05]**

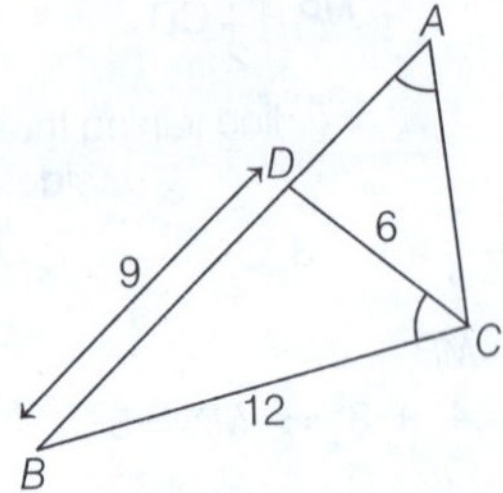

Sol. Here, $\Delta ABC \sim \Delta BCD$

$[\because \angle B = \angle B \text{ (common)}, \angle BCD = \angle BAC \text{ (given)}]$

Then, $\frac{AC}{CD} = \frac{BC}{BD}$

$\Rightarrow \frac{AC}{6} = \frac{12}{9} \Rightarrow AC = 8$

Also, $\frac{AB}{BC} = \frac{BC}{BD}$

$\Rightarrow \frac{AB}{12} = \frac{12}{9} \Rightarrow AB = 16$

Clearly, $AD = AB - BD = 16 - 9 = 7$

Required ratio = Perimeter of ΔADC : Perimeter of ΔBDC

$= (AD + AC + CD) : (BD + CD + BC)$

$= (7 + 8 + 6) : (9 + 6 + 12)$

$= 21 : 27 = 7 : 9$

Workout

1. At 8.00 AM, a tree 15 m long casts a shadow 24 m longh on the grond and a building casts 32 m long shadow. What is the height of the building?

(a) 18 m (b) 16 m (c) 15 m (d) 20 m

2. Two triangles BAC and BDC, right angled at A and D respectively, are drawn on the same base BC and on the same side of BC. If AC and DB intresect at P, then $AP \times PC = DP \times ?$

(a) BP (b) AC (c) PD (d) PC

3. In the figure given, P is a point on AB and PQ is parallel to AC. What is the number of pairs of distinct similar triangles in the figure? **[CDS-09]**

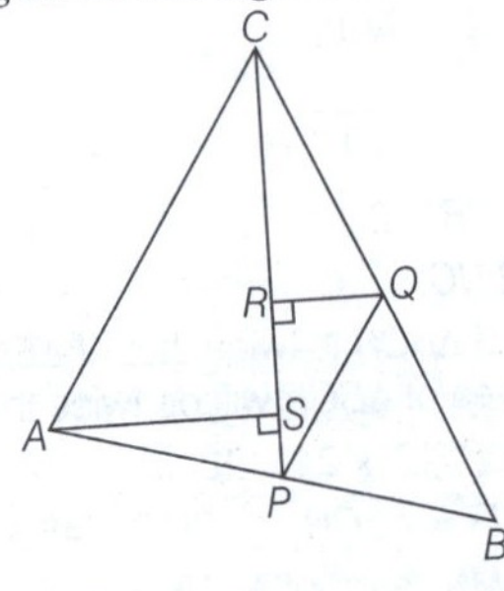

(a) 1 (b) 2
(c) 3 (d) 4

4. Find out the area of an equilateral triangle described on one side of a square as base if the area of the equailateral triangle described on its diagnonal as base is 140 cm^2.

(a) 70 cm^2 (b) 35 cm^2 (c) 170 cm^2 (d) 90 cm^2

5. ΔABC and ΔDEF are similar and their areas be respectively 64 cm^2 and 121 cm^2. If $EF = 15.4$ cm, BC is? **[SSC(CGL-T-1)-12]**

(a) 18 cm (b) 11.2 cm (c) 15 cm (d) 20 cm

6. P and Q are points on sides AB and AC respectively of ΔABC. If $AP = 3$ cm, $PB = 6$ cm, $AQ = 5$ cm, and $QC = 10$ cm, then what is the value of BC in term of PQ?

(a) PQ (b) $3PQ$ (c) $2PQ$ (d) $\frac{PQ}{2}$

7. AB, EF and CD are parallel lines. If $EG = 5$ cm, $GC = 10$ cm, $AB = 15$ cm and $DC = 18$ cm, then what is the value of AC? **[CDS-08]**

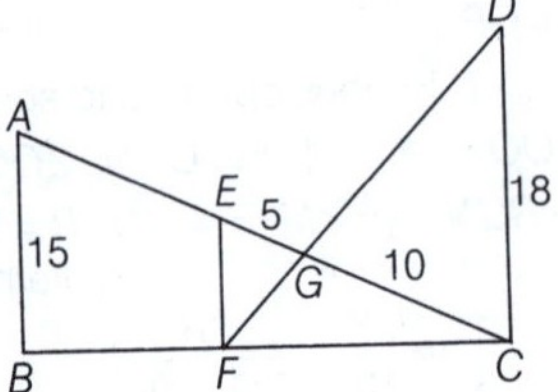

(a) 20 cm (b) 24 cm
(c) 25 cm (d) 28 cm

8. In the given trapzium $ABCD$, $AB \| CD$ and $CD = 2AB$. If the area of $\Delta COD = 84$ cm^2, what is the area of ΔAOB?

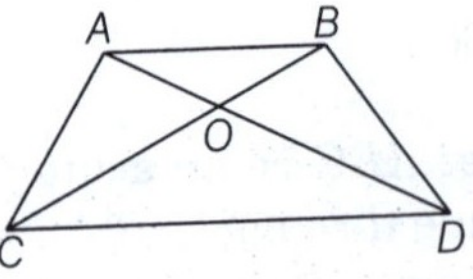

(a) 21 cm^2 (b) 42 cm^2
(c) 63 cm^2 (d) None of these

9. In ΔABC, $MN \| BC$, the area of quadrilateral $MBCN = 130$ sq cm. If $AN : NC = 4 : 5$, then the area of ΔMAN is : **[SSC(CGL-T-I)-19]**

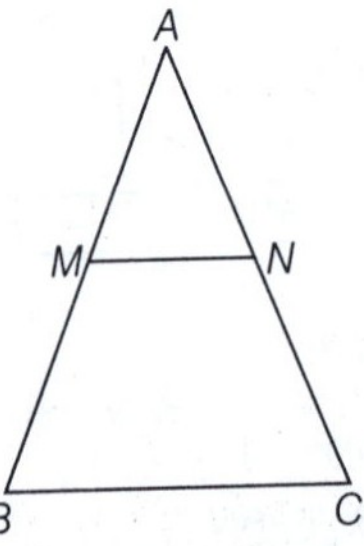

(a) 45 cm^2 (b) 65 cm^2 (c) 32 cm^2 (d) 40 cm^2

Answer Key

1. (d) 2. (a) 3. (b) 4. (a) 5. (b) 6. (b)
7. (c) 8. (a) 9. (c)

Hints and Solutions

1. Let AB be the tree and BC be its shadow and DE is the building and EF is its shadow.

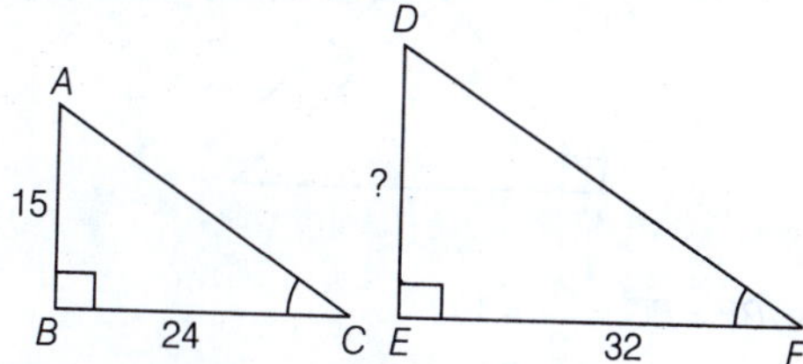

In ΔABC and ΔDEF, $\angle B = \angle E = 90^\circ$ and $\angle C = \angle F$ (Angular elevation of the sun)

$\therefore \Delta ABC \sim \Delta DEF$ (by AA-criterion)

Then, $\frac{AB}{DE} = \frac{BC}{EF} \Rightarrow \frac{15}{DE} = \frac{24}{32}$

$\Rightarrow \quad BC = 20\,\text{cm}$

2. Here, $\Delta APB \sim \Delta DPC$ {$\because \angle A = \angle D$ (90°, given), $\angle APB = \angle DPC$ (vertically opposite angles)}

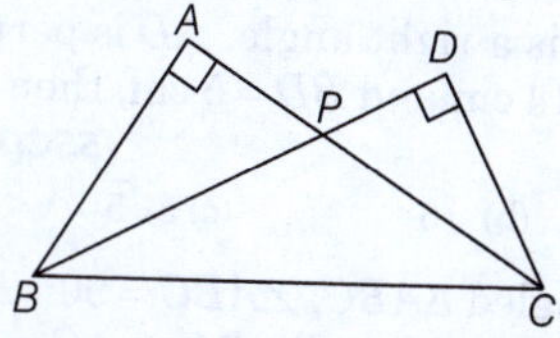

Then, $\frac{AP}{DP} = \frac{BP}{PC} \Rightarrow AP \times PC = DP \times BP$

3. In ΔBQP and ΔABC, $\angle B = \angle B$ (common) and $\angle BQP = \angle BCA$ (corresponding angles as $PQ \parallel AC$)

$\therefore \quad \Delta BQP \sim \Delta ABC$

Further, in ΔPQR and ΔASC, $\angle R = \angle S$ (each equal to 90°) and $\angle QPR = \angle ACS$ (Alternate interior angles as $PQ \parallel AC$)

$\therefore \ \Delta PQR \sim \Delta ASC$

Thus, two pairs of distinct similar triangles appear in the figure.

4. Since both the triangles are equilateral, they are equiangular.

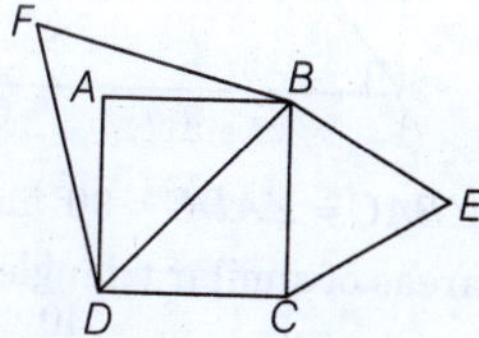

Hence, $\Delta BCE \sim \Delta BDF$

Then, area of ΔBCE : area of $\Delta BDF = BC^2 : BD^2$

[$\because$ Ratio of areas of similar triangles = ratio of squares of corresponding sides]

$\Rightarrow$ area of $\Delta BCE : 140 = BC^2 : (\sqrt{2}\,BC)^2$

[$\because$ diagonal of a square $= \sqrt{2} \times$ side]

$\Rightarrow$ area of $\Delta BCE : 140 = 1 : 2$

$\Rightarrow \quad$ area of $\Delta BCE = 70\,\text{cm}^2$

5. Since ratio of areas of similar triangles is equal to the ratio of squares of corresponding sides,

$\frac{64}{121} = \frac{BC^2}{EF^2} \Rightarrow 8EF = 11BC$

$\Rightarrow \quad BC = \frac{8 \times 15.4}{11} = 11.2\,\text{cm}$

6. Here, $\frac{AP}{PB} = \frac{AQ}{QC}$ $\left\{\because \frac{3}{6} = \frac{5}{10}\right\}$

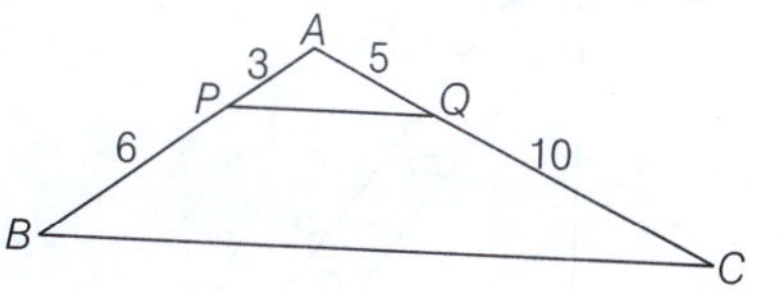

$\therefore \ PQ \parallel BC$

Further, $\Delta APQ \sim \Delta ABC$

[$\because \angle A = \angle A$ (common), $\angle APQ = \angle ABC$ (corresponding angles)]

$\therefore \quad \frac{PQ}{BC} = \frac{AQ}{AC}$

$\Rightarrow \quad \frac{PQ}{BC} = \frac{5}{(5+10)} \Rightarrow BC = 3PQ$

7. Here, $\Delta EFG \sim \Delta CDG$ {$\angle EGF = \angle CGD$ (common) and $\angle EFG = \angle CDG$ (Alternate interior angles)}

Then, $\frac{EG}{GC} = \frac{EF}{CD} \Rightarrow \frac{5}{10} = \frac{EF}{18}$

$\Rightarrow \quad EF = 9\,\text{cm}$

Further, $\Delta CEF \sim \Delta ABC$ {$\angle C = \angle C$ (common) and $\angle CEF = \angle CAB$ (corresponding angles as $EF \parallel AB$)}

Then, $\frac{EC}{AC} = \frac{EF}{AB}$

$\Rightarrow \quad \frac{15}{AC} = \frac{9}{15} \Rightarrow AC = 25\,\text{cm}$

8. Here, $\Delta AOB \sim \Delta COD$ {$\because \angle AOB = \angle COD$ (vertically opposite angles) and $\angle BAO = \angle ODC$ (alternate interior angles as $AB \parallel CD$)}

Then, area of ΔAOB : area of ΔCOD $= AB^2 : CD^2$

{$\because$ ratio of areas of similar triangles = ratio of squares of corresponding sides}

$\Rightarrow$ area of $\Delta AOB : 84 = AB^2 : CD^2$

$\Rightarrow$ area of $\Delta AOB : 84 = AB^2 : (2AB)^2$

$\Rightarrow$ area of $\Delta AOB : 84 = 1 : 4$

$\Rightarrow \quad$ area of $\Delta AOB = 21\,\text{cm}^2$

9. Clearly, $\Delta MAN \sim \Delta ABC$ ($\angle AMN = \angle ABC$ and $\angle ANM = \angle ACB \rightarrow$ corresponding angles)

Then, $\frac{\text{area of } \Delta MAN}{\text{area of } \Delta ABC} = \frac{4^2}{(4+5)^2}$

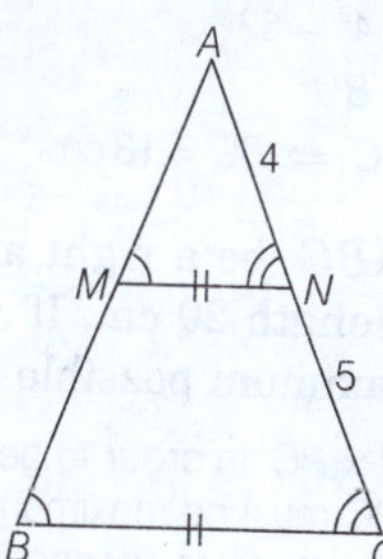

$\Rightarrow$ area of $\Delta\, MAN$: area of $\Delta ABC = 16 : 81$

As given, $\quad (81 - 16)$ units $= 130$

$\therefore \quad$ 16 units $= 32\,\text{cm}^2$

Problems based on the Results Related to Figure, given below

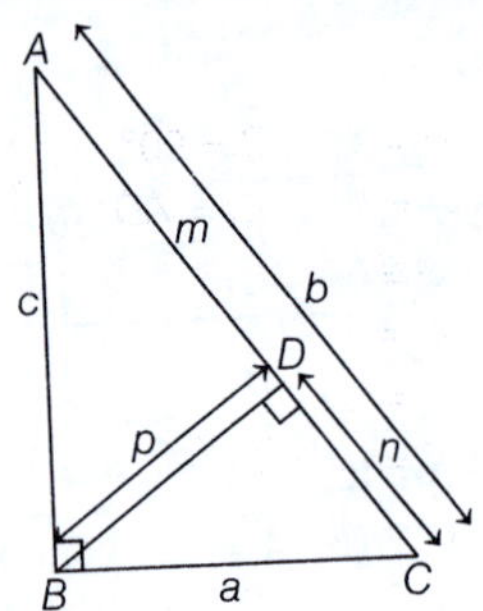

[Details of the reuslts have been given under heading 'triangle - wise properties', sub heading 'Right-angled triangle'. Results are

$c^2 = bm,\ a^2 = bn,\ bp = ac,\ p^2 = mn$ and $\frac{1}{p^2} = \frac{1}{c^2} + \frac{1}{a^2}$]

problem 95 In a ΔABC, $\angle BCA = 90^\circ$ and CD is perpendicular to AB. If $AD = 4$ cm and $BD = 9$ cm, then the value of DC will be **[CDS-13, 21, SSC(CGL-T-II)-15, 19]**

Sol. In a right-angled triangle, $CD^2 = AD \times DB$

$\Rightarrow CD^2 = 4 \times 9 \Rightarrow CD = 6$ cm

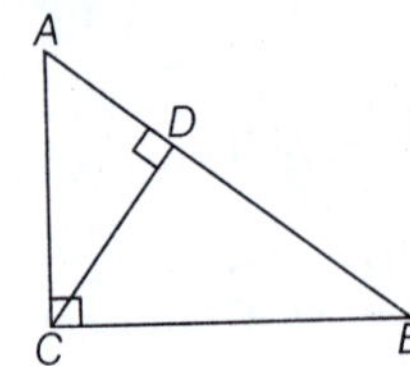

Problem 96 In a ΔABC, $\angle BAC = 90^\circ$ and AD is perpendicular to BC. If $AD = 6$ cm and $BD = 4$ cm, then the length of BC is :

Sol. In right-angled ΔABD, $AB^2 = AD^2 + BD^2$

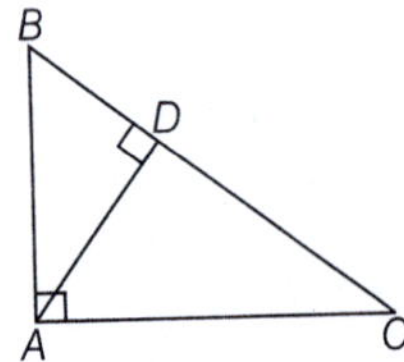

$\Rightarrow \quad AB^2 = 6^2 + 4^2 = 52$

Further, $AB^2 = BD \times BC$

$\Rightarrow \quad 52 = 4 \times BC \Rightarrow BC = 13$ cm

Problem 97 Let ABC be a right-angled triangle with hypotenuse BC of length 20 cm. If AP is perpendicular on BC, then the maximum possible lenght of AP, is :

Sol. We know, $AP^2 = BP \times PC$. In order to get AP maximum, the product of BP and PC must be maximum. Obviously, the product will be maximum when BP and PC are equal under given their sum.

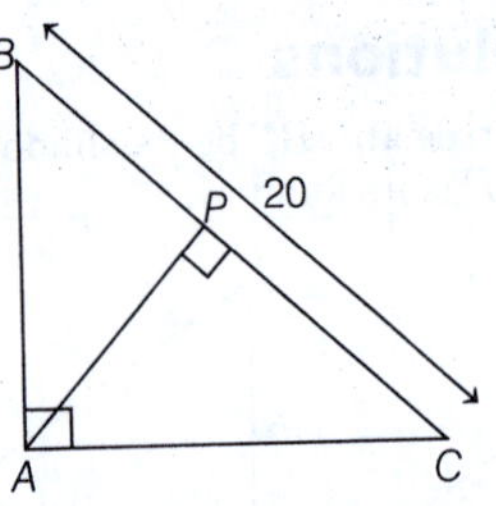

Hence, $BP = PC = \frac{20}{2} = 10$

Then, maximum value of $AP = \sqrt{10 \times 10} = 10$

Workout

1. Suppose ΔABC be a right-angled triangle where $\angle A = 90^\circ$ and $AD \perp BC$. If $\Delta ABC = 40\ \text{cm}^2$, $\Delta ACD = 10\ \text{cm}^2$ and $AC = 9$ cm, then the lenght of BC is : **[SSC(CGL-T-II)-12]**

(a) 12 cm (b) 18 cm (c) 4 cm (d) 6 cm

2. $\angle A$ of ΔABC is a right angle. AD is perpendicular on BC. If $BC = 14$ cm and $BD = 5$ cm, then measure of AD is : **[SSC(10+2/DEO-LDC)-15]**

(a) $2\sqrt{5}$ (b) $\sqrt{5}$ (c) $3\sqrt{5}$ (d) $3.5\sqrt{5}$

3. In a right-angled ΔABC, $\angle ABC = 90^\circ$, BN is perpendicular to AC. $AB = 6$ cm, $AC = 10$ cm. Then, $AN : NC$ is :

(a) $9 : 16$ (b) $4 : \sqrt{5}$ (c) $3\sqrt{5} : 1$ (d) $16 : 25$

Answer Key

1. (b) 2. (c) 3. (a)

Hints and Solutions

1. (*b*) Here, $\Delta ABC \sim \Delta ACD$

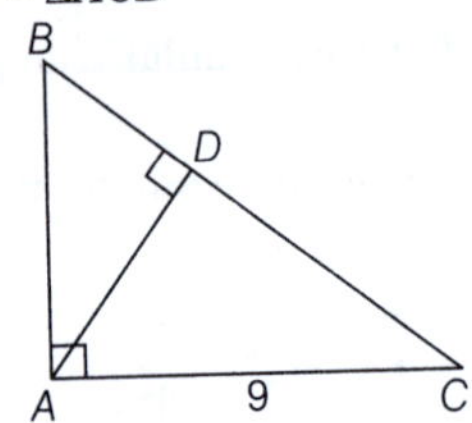

[$\because \angle BAC = \angle ADC = 90^\circ$ and $\angle B = \angle B$ (common)]

Since ratio of areas of similar triangles is equal to the ratio of square of corresponding sides, $\frac{40}{10} = \frac{BC^2}{AC^2} \Rightarrow \frac{40}{10} = \frac{BC^2}{9^2}$

$\Rightarrow \quad BC = 18$ cm

2. (c) Clearly, $AD^2 = BD \times DC$

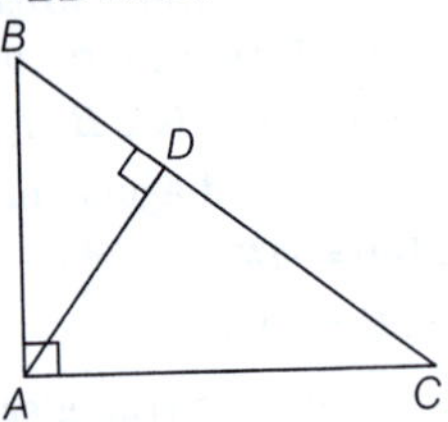

$\Rightarrow \quad 5 \times (14 - 5) = 45 \Rightarrow AD = \sqrt{45} = 3\sqrt{5}$ cm

3. (*a*) Clearly, $BC = 8\,\text{cm}$

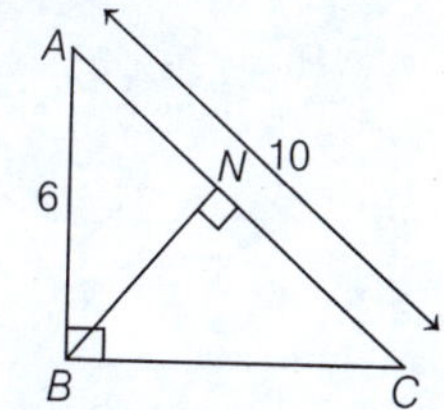

Now, $AB^2 = AN \times AC$

$\Rightarrow \quad 6 \times 6 = AN \times 10$

Further, $BC^2 = NC \times AC$

$\Rightarrow \quad 8 \times 8 = NC \times 10$

Then, $(AN \times 10) : (NC \times 10) = 6 \times 6 : 8 \times 8 = 9 : 16$

Problems Based on the concept of Congruency of Triangles

Problem 98 D is a point on the side BC of a ΔABC such that $AD \perp BC$, E is a point on AD for which $AE : ED = 5 : 1$. If $\angle BAD = 30^\circ$ and $\tan \angle ACB = 6 \tan \angle DBE$, then $\angle ACB = ?$ **[SSC(CGL-T-II)-13]**

Sol. Given, $\tan\angle ACB = 6\tan\angle DBE$

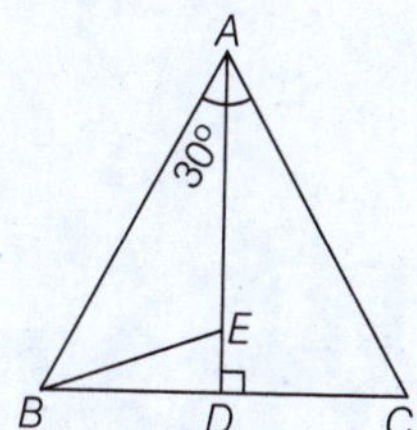

$$\Rightarrow \quad \frac{AD}{DC} = 6 \times \frac{ED}{BD} \Rightarrow \frac{AD}{ED} = 6 \times \frac{DC}{BD}$$

$$\Rightarrow \quad \frac{(5+1)}{1} = 6 \times \frac{DC}{BD} \Rightarrow DC = BD$$

In ΔABD and ΔACD,

$\angle ADB = \angle ADC = 90^\circ$,

$BD = DC$, $AD = AD$ (common)

Hence, $\Delta ABD \cong \Delta ACD$ [by S-A-S criterion]

Then, $\angle ABD = \angle ACD$

$\Rightarrow 180^\circ - 90^\circ - 30^\circ = \angle ACD$

$\Rightarrow \quad \angle ACB = 60^\circ$

Workout

1. In ΔABC and ΔDEF, $AB = DE$ and $BC = EF$. Then one can infer that $\Delta ABC \cong \Delta DEF$, when **[SSC (CGL-T-I)-13]**

(a) $\angle BAC = \angle EDF$

(b) $\angle ACB = \angle EDF$

(c) $\angle ACB = \angle DFE$

(d) $\angle ABC = \angle DEF$

1. (*d*) We know, two triangles are congruent, if two sides and the included angle of one triangle are equal to two sides and the included angle of another (S-A-S criterion).

Hence, option (d) is correct.

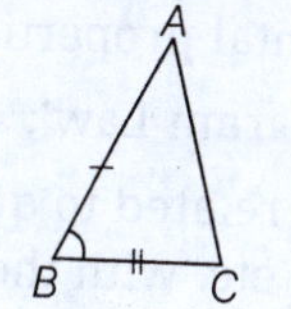

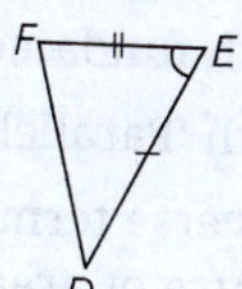

58

Quadrilaterals

Learning OBJECTIVES

After completing this chapter, aspirants will be able to :

- Understand the significance of this sub-section in diverse competitive exams;
- Comprehend the definition of quadrilaterals and its different types with their properties;
- Recognise how the concept of triangles works in case of quadrilaterals;
- Be familiar with fundamental properties of quadrilaterals;
- Be familiar with 'Parallelogram Law';
- Understand diverse terms related to quadrilaterals like median, bimedians, centre of area, etc. with the techniques to compute their values/locations;
- Comprehend the properties of cyclic and tangential quadrilaterals;
- Apply the relevant formulae/results according to different situations;
- Realise themselves competent to solve the questions in unique way;
- Analyse the trend of the questions being asked in various competitive examinations;
- Solve the questions in moments;
- Learn many new concepts/fundamentals;
- Apply the pragmatic approaches in lieu of academic approaches; and
- Ponder on the alternative methods.

Inside

Introduction

A quadrilateral is a closed plane figure bounded by four straight lines.

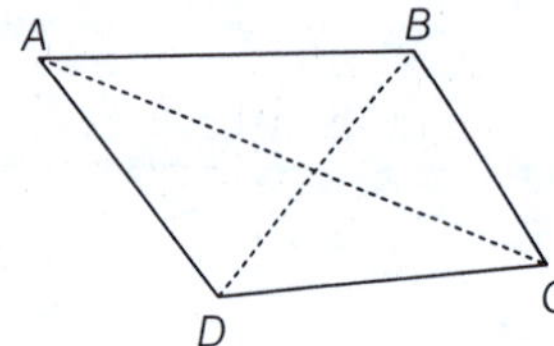

It has four sides (AB, BC, CD, DA), four angles ($\angle A$, $\angle B$, $\angle C$, $\angle D$), four vertices (A, B, C, D) and two diagonals (AC, BD).

General Properties of Quadrilaterals

(i) The sum of any three sides of a quadrilateral is greater than the fourth side.

(ii) Sum of all the interior angles of a quadrilateral is 360°. Also, sum of all exterior angles is 360°, already discussed under sub-section 'Polygon'.

(iii) The perimeter of a quadrilateral is greater than sum of their diagonals.

(iv) The sum of diagonals of a quadrilateral is greater than sum of two opposite sides.

(v) The line segments joining the mid-points of opposite sides of a convex quadrilateral are known as 'Bimedians'. Hence, two bimedians exist in a quadrilateral. Also, they bisect each other.

Here, EF and GH are bimedians and $EO = OF$ and $GO = OH$.

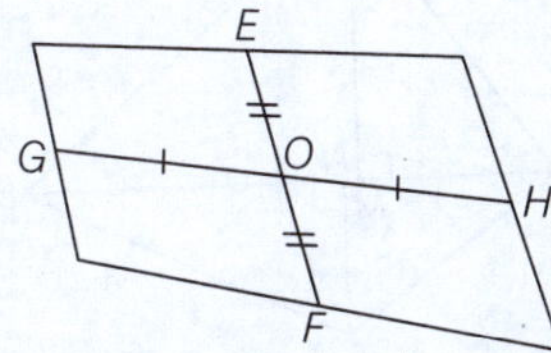

Point 'O' is the centroid of the vertices of the quadrilateral.

(vi) A cyclic or inscribed quadrilateral is a quadrilateral whose vertices all lie on a single circle. The circle is called the circumcircle or said to be circumscribed about a quadrilateral. The vertices are said to be concyclic.

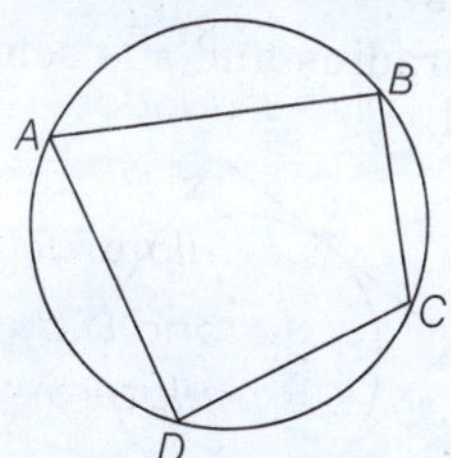

Here, $ABCD$ is a cyclic quadrilateral.

(a) In a cyclic quadrilateral, the sum of opposite angles is 180°. Hence, $\angle A + \angle C = 180°$ and $\angle B + \angle D = 180°$.

Also, if the sum of opposite angles is 180°, the quadrilateral is cyclic.

(b) The quadrilateral formed by angle bisectors of a cyclic quadrilateral is also cyclic.

Quadrilaterals $ABCD$ and $PQRS$ both are cyclic.

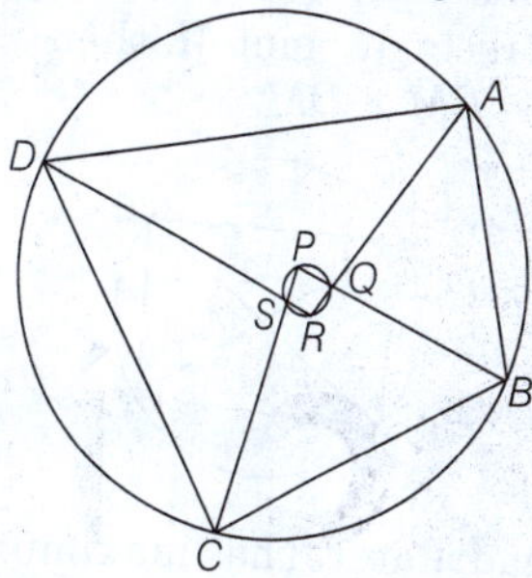

(c) The bisectors of the angles formed by producing opposite sides (if they are not parallel) of a cyclic quadrilateral, intersect each other at 90°.

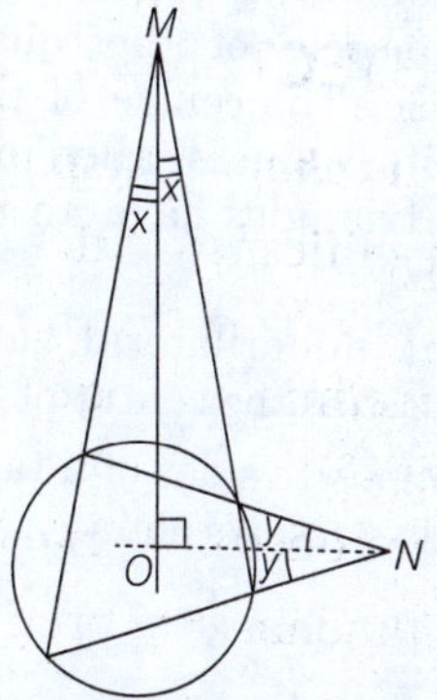

Here, $\angle MON = 90°$

(d) If a side of a cyclic quadrilateral is produced, then the exterior angle is equal to the interior opposite angle.

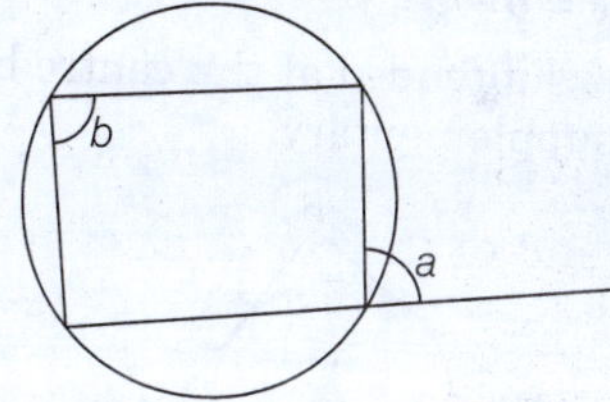

Here, $\angle a = \angle b$

(e) In a cyclic quadrilateral, the product of the diagonals is equal to the sum of the products of the opposite sides.

Here, $AC \times BD = ac + bd$

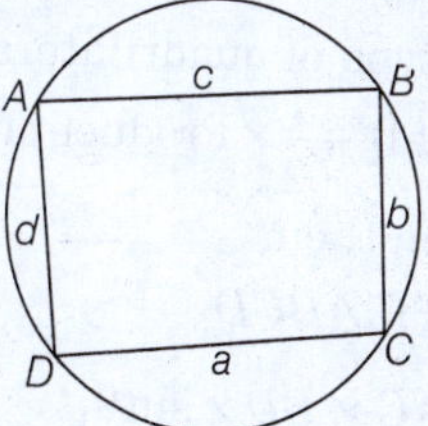

(f) If the diagonals BD and AC of a cyclic quadrilateral $ABCD$ intersect at point M, then $AM \times CM = BM \times DM$.

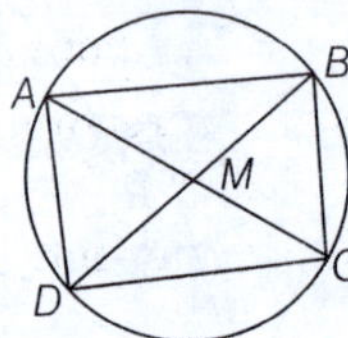

(g) A cyclic quadrilateral has maximum area among all types of quadrilaterals if they have same sequence of length of sides.

(vii) The quadrilateral circumscribed about a circle is termed as 'tangential quadrilateral'. The circle is called the incircle of the quadrilateral or its inscribed circle. The centre of this circle is called incentre and its radius is called inradius. It is also to note that all triangles have an incircle but not all quadrilaterals.

(a) In tangential quadrilateral, all the four angle bisectors concur at the centre of the incircle.

(b) The sum of opposite sides of a tangential quadrilateral is always equal.

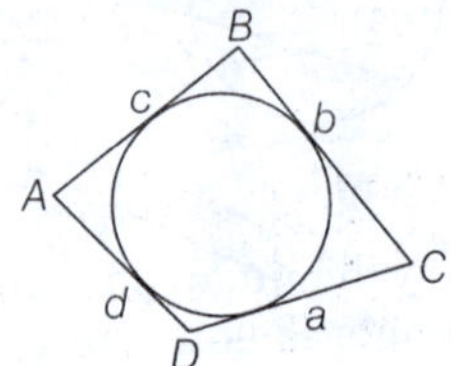

Here, $a + c = b + d$

(c) The angles subtended at the centre by the opposite sides are supplementary.

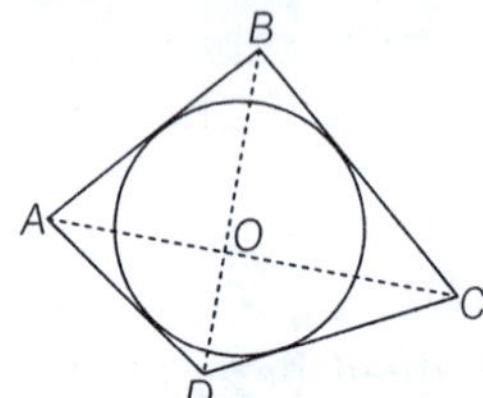

Here, $\angle AOB + \angle COD = \angle AOD + \angle BOC = 180°$

Area and diagonal of quadrilaterals

(i) Area of any type of quadrilaterals (including cyclic and tangential) $= \frac{1}{2} \times$ Product of diagonals $\times$ Sine of angle between them

Here, area of $\square ABCD$

$$= \frac{1}{2} \times AC \times BD \times \sin\theta_1$$

$$= \frac{1}{2} \times AC \times BD \times \sin\theta_2$$

$[\because \sin\theta_1 = \sin\theta_2 \text{ as } \sin\theta_1 = \sin(180° - \theta_1)]$

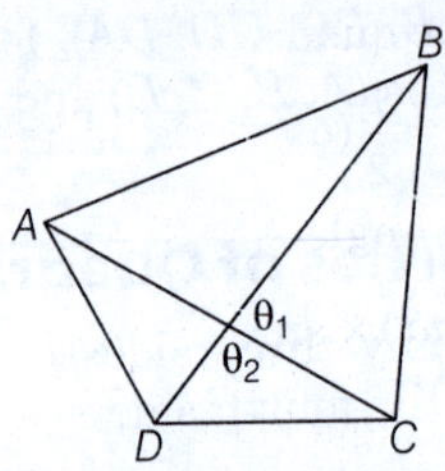

(ii) Area of any type of quadrilaterals (including cyclic and tangential)

$= $ area of ΔABD + area of ΔBCD

$= \frac{1}{2} \times BD \times AM + 1/2 \times BD \times CN$

$= \frac{1}{2} \times BD\,(AM + CN)$

$= \frac{1}{2} \times$ one of the diagonals $\times$ sum of perpendiculars drawn to it from opposite vertices

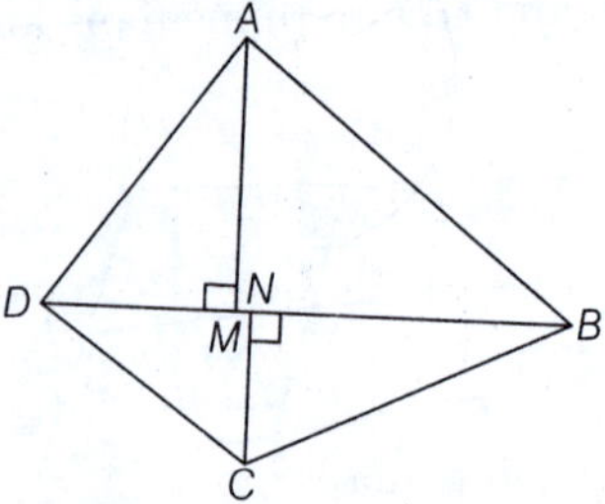

(iii) Area of any type of quadrilaterals (including cyclic and tangential) $= \sqrt{(s-a)(s-b)(s-c)(s-d)}$

where, $s = \dfrac{a+b+c+d}{2}$

(iv) (a) Area of a tangential quadrilateral $= \sqrt{abcd}$

(b) Area of a tangential quadrilateral $= rs$, where r is inradius and s is semi-perimeter of the quadrilateral.

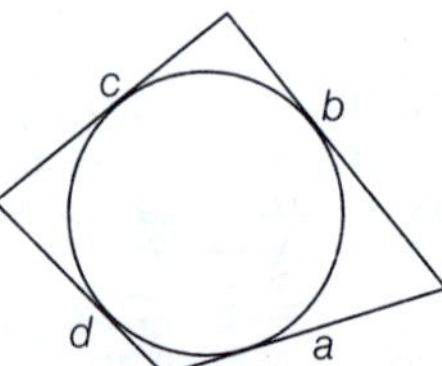

(v) (a) Area of a cyclic quadrilateral

$$= \frac{\sqrt{(ac+bd)(bc+ad)(ab+cd)}}{4R}$$

where, R is circumradius.

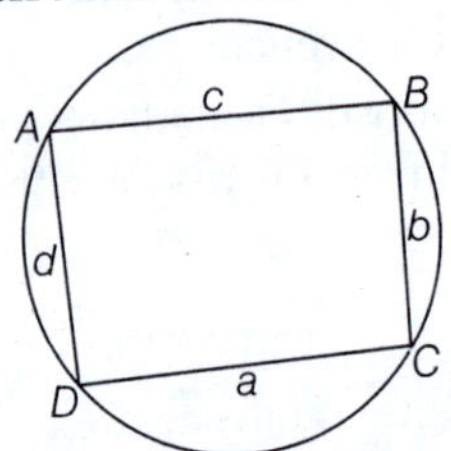

(b) Area of a cyclic quadrilateral

$$= \frac{1}{2}(ab + cd) \times \sin\theta,$$

where θ is the angle between a and b.

or $\frac{1}{2}(bc + ad) \times \sin\theta,$

where θ is the angle between b and c.

(c) Also, circumradius of cyclic quadrilateral $ABCD$ = circumradius of ΔABC = circumradius of ΔBCD = circumradius of ΔACD = circumradius of ΔABD

(d) Area of a cyclic quadrilateral $\leq 2R^2$, where R is circumradius. In case of squares, area = $2R^2$.

(vi) (a) For the adjoining cyclic quadrilateral,

$$\text{Diagonal } AC = \sqrt{\frac{(ab + cd)}{(ad + bc)} \times (ac + bd)}$$

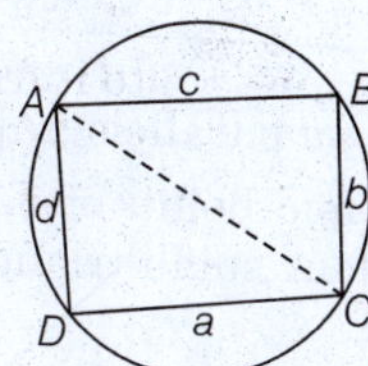

$$\text{and diagonal } BD = \sqrt{\frac{(ad + bc)}{(ab + cd)} \times (ac + bd)}$$

(b) For the adjoining quadrilateral,

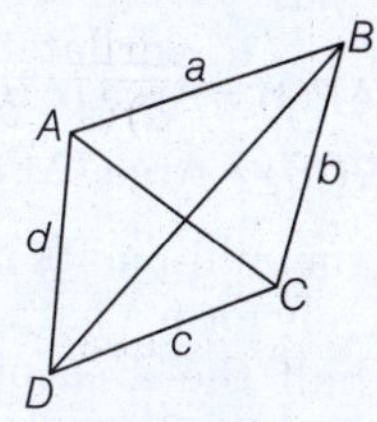

$$\text{Diagonal } BD = \sqrt{a^2 + d^2 - 2ad \cos A}$$
$$= \sqrt{b^2 + c^2 - 2bc \cos C}$$

$$\text{and diagonal } AC = \sqrt{c^2 + d^2 - 2cd \cos D}$$
$$= \sqrt{a^2 + b^2 - 2ab \cos B}$$

[Recall cosine-rule, discussed in sub-section 'Triangle']

We can compute the length of diagonals of any types of quadrilaterals(including cyclic and tangential) with help of the above formulae.

Types of Quadrilaterals with their Properties

Parallelogram

A parallelogram is a quadrilateral with opposite sides parallel. Parallelogram also includes rhomboid, rectangle, square, rhombus. Rhomboid refers to a quadrilateral whose opposite sides are parallel and adjacent sides are unequal and whose angles are not right angles.

Properties

(i) Opposite sides are equal.

(ii) Opposite angles are equal.

(iii) Diagonals may or may not be equal.

(iv) A parallelogram is a rectangle or a square if its diagonals are equal.

(v) Diagonals may or may not bisect angles at the vertices.

(vi) A parallelogram is a rhombus or a square if its diagonals bisect angles at the vertices.

(vii) Diagonals may or may not bisect at 90°.

(viii) A parallelogram is a rhombus or a square if its diagonals bisect at 90°.

(ix) If a diagonal of a parallelogram bisects one of the angles of the parallelogram, it also bisects the angle opposite to it and then the two diagonals are perpendicular to each other.

(x) Each diagonal divides a parallelogram into two congruent triangles.

(xi) The sum of the squares of the sides is equal to the sum of the squares of the diagonals. It is called 'Parallelogram Law'.

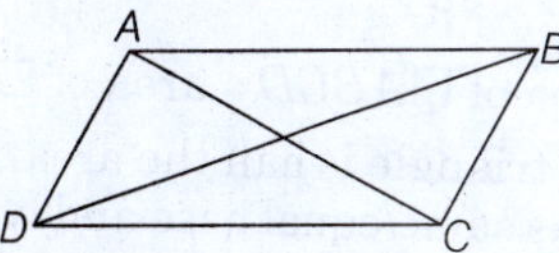

Here, $(AC)^2 + (BD)^2$

$= (AB)^2 + (BC)^2 + (CD)^2 + (DA)^2$

$= 2(AB)^2 + 2(BC)^2$ $[\because AB = CD \text{ and } BC = DA]$

$= 2[(AB)^2 + (BC)^2]$

Hence, the sum of squares of diagonals equals twice the sum of square of any two adjacent sides.

(xii) Bisectors of angles of a parallelogram form a rectangle.

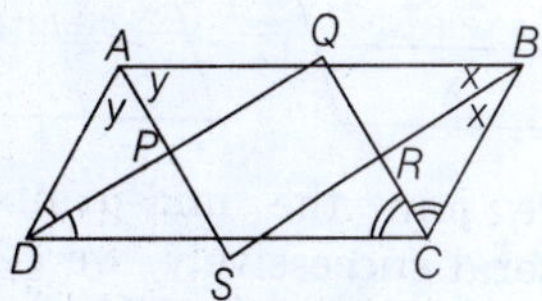

Here, $PQRS$ is a rectangle.

(xiii) Sum of any two adjacent angles is 180° as opposite sides are parallel. It implies that bisectors of two consecutive angles meet at 90°.

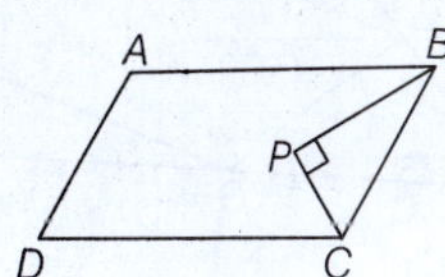

Here, $\angle BPC = 90°$, if BP and CP are bisectors of $\angle B$ and $\angle C$ respectively.

(xiv)

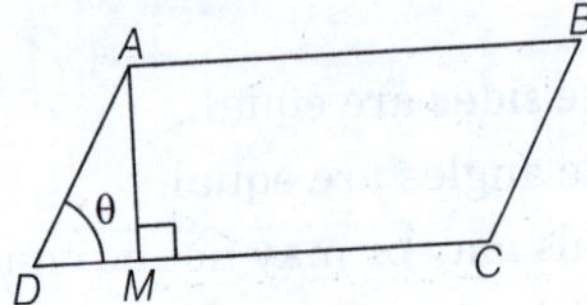

Area of parallelogram $ABCD$ = Base × Height

$= DC \times AM$

$= DC \times AD \times \sin\theta$ $\left[\because \sin\theta = \dfrac{AM}{AD}\right]$

= Product of any two adjacent sides × Sine of the included angle

Perimeter $= AB + BC + CD + DA$

$= 2AB + 2BC$ $[\because AB = CD \text{ and } BC = DA]$

$= 2(AB + BC)$

= 2 × Sum of any two adjacent sides

(xv) Parallelograms between same parallel lines and having same base are equal in area.

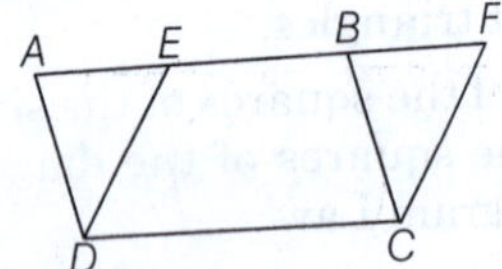

Here, area of □$ABCD$ = area of □$DEFC$

(xvi) Area of a triangle is half the area of a parallelogram if both has same/equal base and lying between same parallel lines.

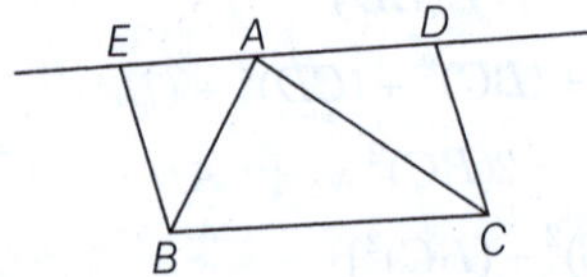

Here, area of $\Delta ABC = 1/2 \times$ area of □$EDCB$

(xvii) The line joining the mid-points of opposite sides of a parallelogram bisects the area of the parallelogram.

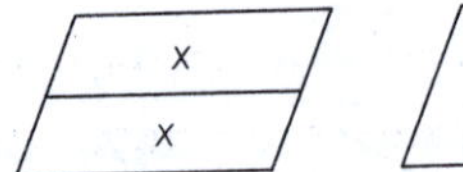

(xviii) When we join the mid-points of sides of a quadrilateral successively, we get a parallelogram (known as varignon parallelogram). Here, $EFGH$ is a varignon parallelogram. Then,

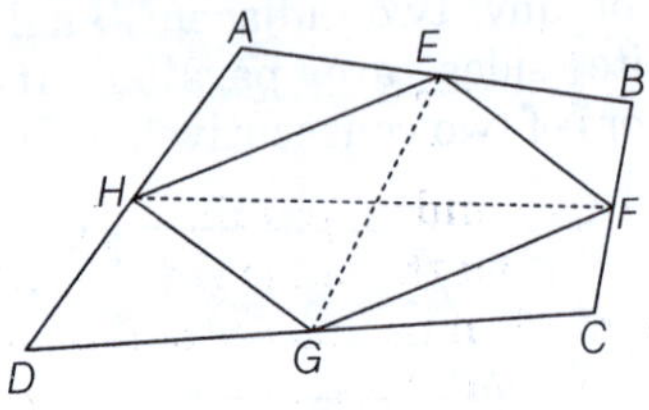

(a) $EH = GF = \dfrac{1}{2} BD$ and

$EF = HG = \dfrac{1}{2} BC$. It implies that perimeter of varignon parallelogram is equal to sum of the diagonals of original quadrilateral.

(b) The area of varignon parallelogram is equal to half the area to original quadrilateral.

(c) The diagonals of the varignon parallelogram are bimedians of the original quadrilateral.

(d) Lines joining the mid-points of sides of the varignon parallelogram form a parallelogram.

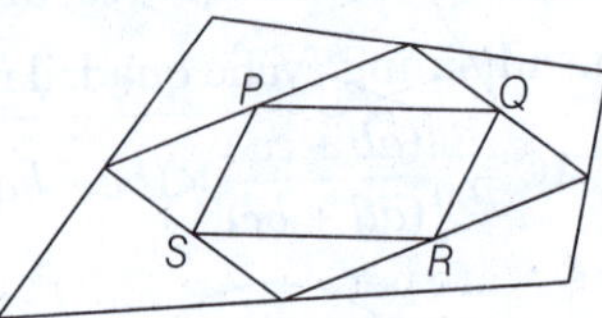

Here, $PQRS$ is a parallelogram.

(xix) When we join mid-points of sides of a parallelogram successively, four small triangles appear which are equal in area.

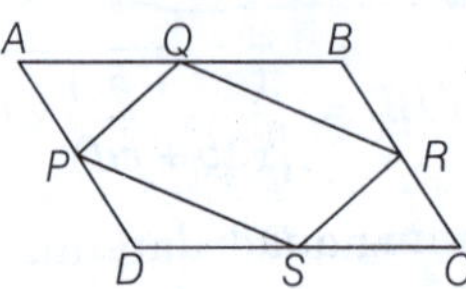

Here, area (ΔAPQ) = Area (ΔBQR)

= Area (ΔRCS) = Area (ΔPDS)

(xx) Lines joining the mid-points of sides of a parallelogram successively form a parallelogram. The area of the latter is half the area of the former and the perimeter of the latter equals sum of the diagonals of the former. Here, $PQRS$ is also a parallelogram.

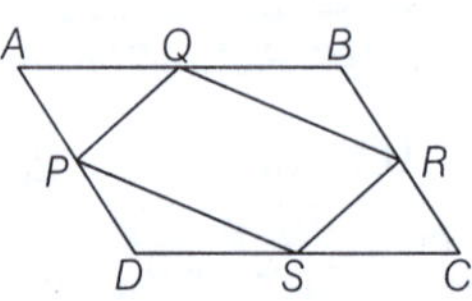

(xxi) The parallelogram inscribed in a circle is a rectangle. In other words, any cyclic parallelogram having unequal adjacent sides is necessarily a rectangle.

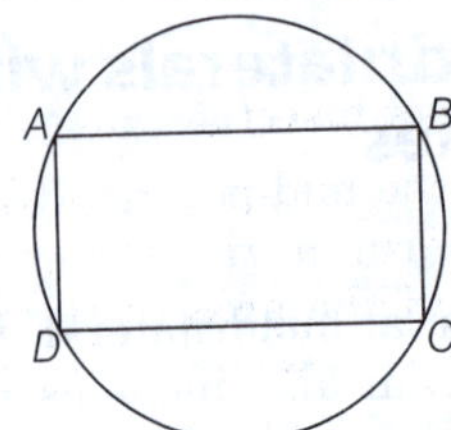

(xxii) The parallelogram circumscribed about a circle is a rhombus.

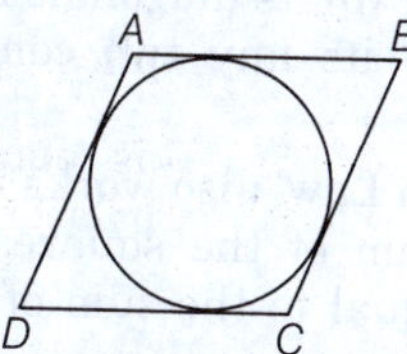

(xxiii) If M and N are mid-points of sides AD and BC respectively of a parallelogram $ABCD$, then $DP = PQ = QB$.

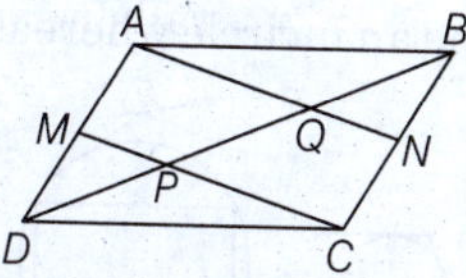

(xxiv) If AM and CN are perpendiculars to BD in the given parallelogram $ABCD$, then $AM = CN$.

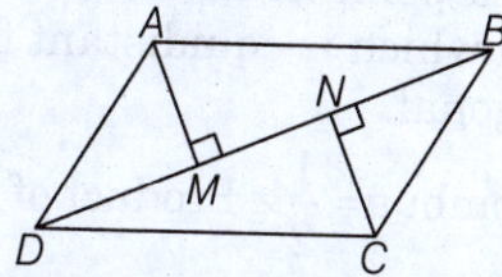

(xxv) Here, $ABCD$ is a parallelogram and E is the mid-point of BC. If DE and AB meet at point F, then $AF = 2AB$ and $DF = 2DE$.

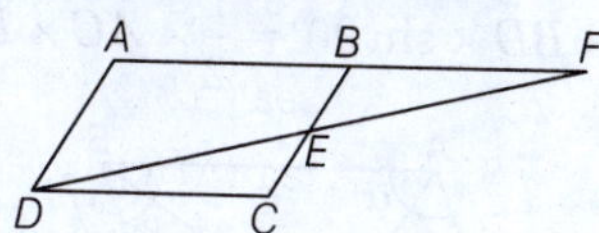

Rectangle

A rectangle is a parallelogram with each angle 90°. Hence, all rectangles are parallelogram but the converse is not true.

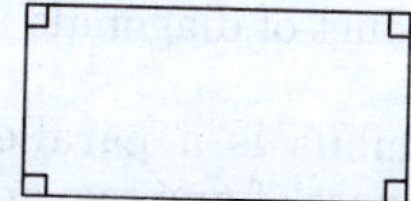

Properties

(i) Opposite sides are parallel and equal.

(ii) Each angle equals 90°.

(iii) Diagonals are equal and bisect each other but not at 90°.

(iv) Diagonals do not bisect angles at the vertices.

(v) Lines joining the mid-points of sides of a rectangle successively form a rhombus. The area of this rhombus is half the area of the rectangle and the perimeter of the rhombus equals twice the diagonal of the rectangle.

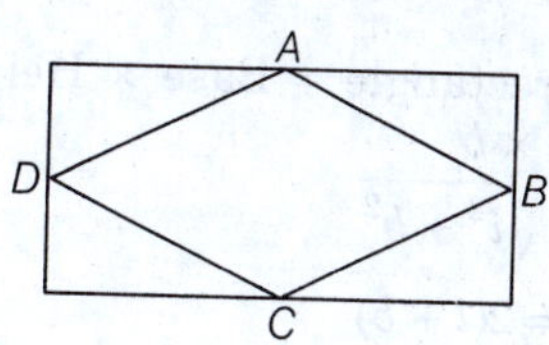

Here, $ABCD$ is a rhombus.

(vi) Bisectors of the angles of a rectangle form an another rectangle.

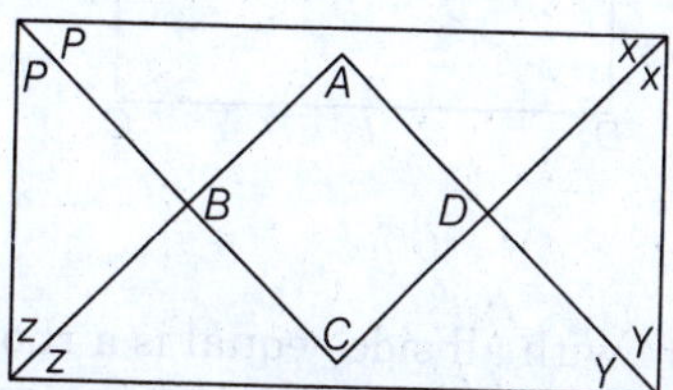

Here, $ABCD$ is a rectangle.

(vii) When a circle touches lengths of a rectangle internally, the breadth of that rectangle is diameter of the circle.

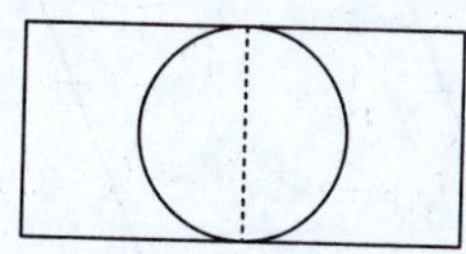

(viii) When a rectangle is inscribed in a circle, the diagonal of the rectangle is the diameter of the circle.

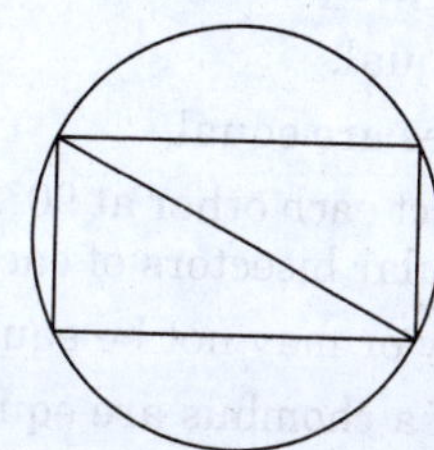

(ix) For the given perimeter, the area of a rectangle is less than that of a square.

(x) Bisectors of two consecutive angles meet at 90°.

(xi) 'O' is any point inside a rectangle $ABCD$. Then, $(OB)^2 + (OD)^2 = (AO)^2 + (OC)^2$

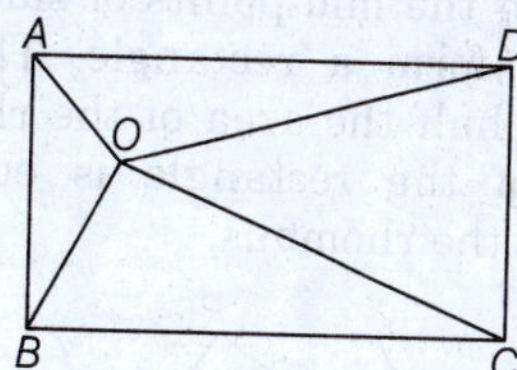

(xii) If a rectangle and a parallelogram stand on the same base and on the same side of the base with the same height, then perimeter of the parallelogram is greater than that of rectangle.

(xiii) Area of a rectangle = Base × Height = Length × Breadth = $l \times b$

Diagonal = $\sqrt{l^2 + b^2}$

Perimeter = $2(l + b)$

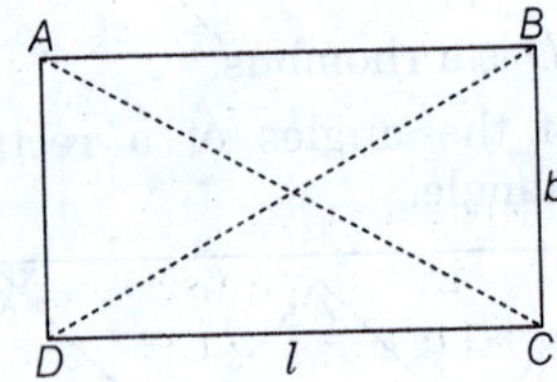

Rhombus

A parallelogram with all sides equal is a rhombus. Hence, all rhombuses are parallelogram but the converse is not true. Also, all squares are rhombus, but the converse is not true.

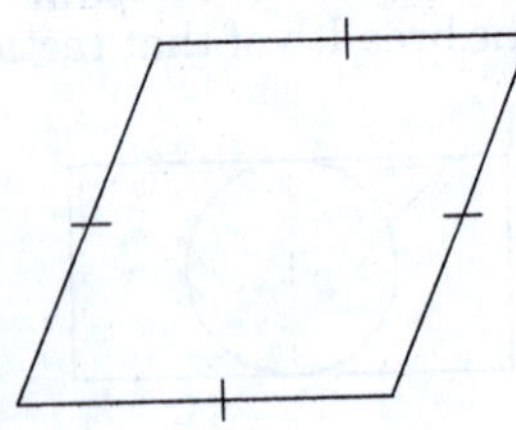

Properties

(i) Opposite sides are parallel.

(ii) All sides are equal.

(iii) Opposite angles are equal.

(iv) Diagonals bisect each other at 90°. Hence, diagonals are perpendicular bisectors of each other.

(v) Diagonals may or may not be equal.

(vi) If diagonals of a rhombus are equal, it is a square.

(vii) Diagonals bisect the vertex angles.

(viii) Each diagonal divides the rhombus into two congruent triangles.

(ix) Diagonals form four congruent triangles.

(x) Bisectors of two consecutive angles meet at 90°.

(xi) Lines joining the mid-points of sides of a rhombus successively form a rectangle. The area of this rectangle is half the area of the rhombus and the perimeter of the rectangle is equal to sum of diagonals of the rhombus.

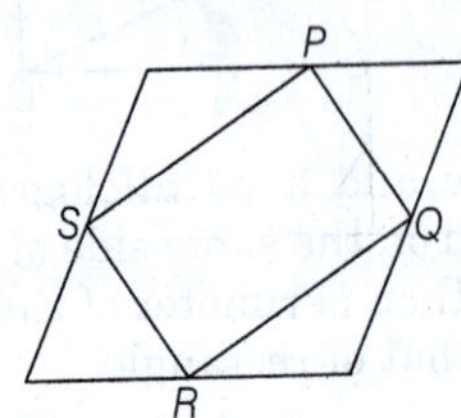

(xii) A line joining mid-points of opposite sides, bisects the area of a rhombus.

(xiii) A parallelogram is a rhombus. If (a) its diagonals bisect at 90° or (b) its diagonals bisect angles at the vertices or (c) its any two consecutive sides are equal.

(xiv) 'Parallelogram Law' also works in case of rhombus. Hence, the sum of the squares of the sides of a rhombus is equal to the sum of the squares of the diagonals .

(xv) Every rhombus is a kite and any quadrilateral which is both a kite and a parallelogram, is a rhombus.

(xvi) A rhombus has an incircle whereas a rectangle has a circumcircle.

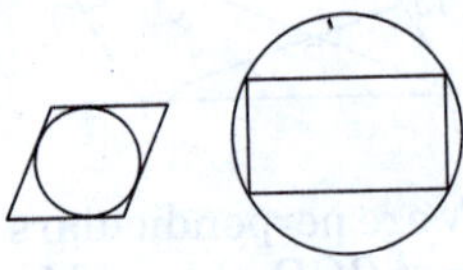

(xvii) The locus of a point in rhombus is a fixed point on one diagonal which is equidistant from both ends of another diagonal.

(xviii) Area of a rhombus = $\frac{1}{2}$ × Product of diagonals × Sine of included angle

$= \frac{1}{2} \times AC \times BD \times \sin\theta_1 (\text{or } \sin\theta_2)$ $[\because \theta_1 = \theta_2 = 90°]$

$= \frac{1}{2} \times AC \times BD \times \sin 90° = \frac{1}{2} \times AC \times BD \times 1$

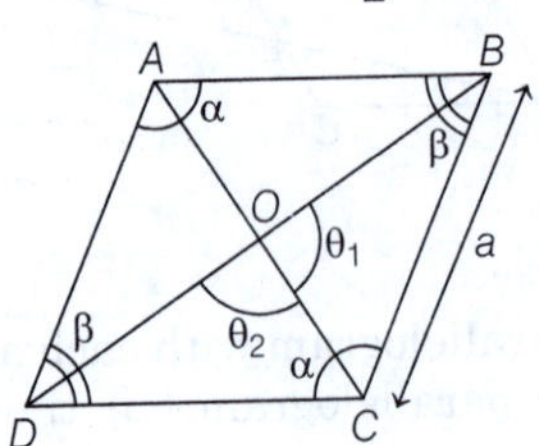

Area = $\frac{1}{2}$ × Product of diagonals

Since the rhombus is a parallelogram, area of a rhombus = product of any two adjacent sides × sine of included angle

$\therefore$ Area = $a^2 \times \sin\alpha = a^2 \times \sin\beta$

Also, Area = $2ar$, where r is radius.

(xix) From above, $r = \frac{\text{Area}}{2a}$

$\Rightarrow r = \frac{\frac{1}{2} \times \text{product of diagonals}}{2a} \Rightarrow r = \frac{d_1 \times d_2}{4a}$

[Let d_1 and d_2 be diagonals]

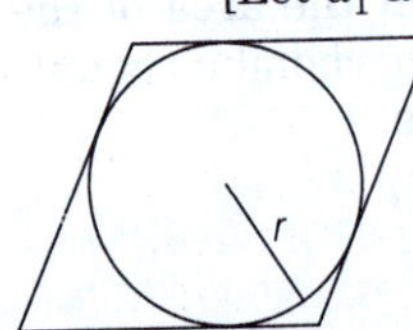

$$\Rightarrow \quad r = \frac{d_1 \times d_2}{4\sqrt{(d_1/2)^2 + (d_2/2)^2}}$$

[$\because$ In right-angled ΔBOC,
$a^2 = (d_1/2)^2 + (d_2/2)^2$]

$$\Rightarrow \quad r = \frac{d_1 \times d_2}{2\sqrt{(d_1)^2 + (d_2)^2}}$$

(xx) $BD^2 = BC^2 + CD^2 - 2 \cdot BC \cdot CD \cdot \cos C$

[Recall cosine-rule]

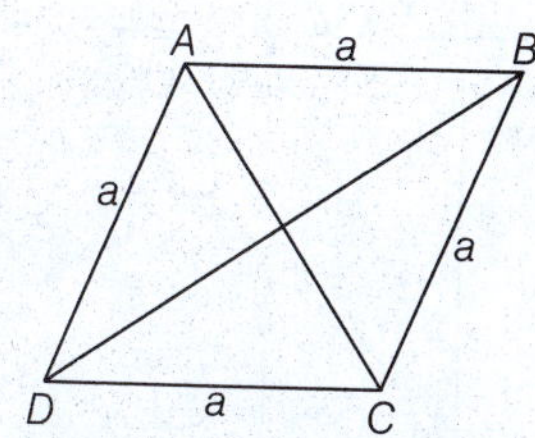

$$\Rightarrow \quad BD = \sqrt{a^2 + a^2 - 2a^2 \cos C}$$
$$= \sqrt{2} \cdot a\sqrt{1 - \cos C}$$

Hence, diagonal $BD = \sqrt{2} \cdot a\sqrt{1 - \cos C}$

or $\quad \sqrt{2} \cdot a\sqrt{1 - \cos A}$

and diagonal $AC = \sqrt{2} \cdot a\sqrt{1 - \cos D}$

or $\quad \sqrt{2} \cdot a\sqrt{1 - \cos B}$

(xxi) Perimeter = 4 × side = $4a$

Square

A square is a rectangle with two adjacent sides equal.

Or

A rhombus with each angle 90°.

Or

A parallelogram with a right angle and two adjacent sides equal.

Or

A quadrilateral with all sides equal and each angle 90°.

Properties

(i) All sides are equal.

(ii) Opposite sides are parallel.

(iii) Each angle is 90°.

(iv) Diagonals are equal and bisect each other at 90°.

(v) Diagonals bisects angles at the vertices.

(vi) Each diagonal divides the square into two congruent triangles.

(vii) Diagonals form four congruent triangles.

(viii) Bisectors of two consecutive angles meet at 90°.

(ix) Lines joining the mid-points of sides of a square successively form a square. The area of the latter is half the area of the former. Also, perimeter of the latter equals twice the diagonal of the former.

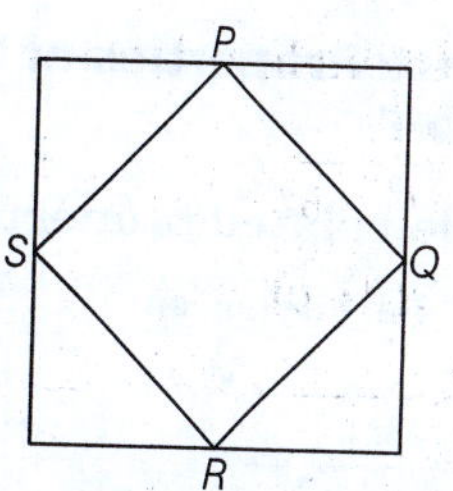

Here, $PQRS$ is a square.

(x) Diameter of a circle circumscribing a square is equal to the diagonal of the square.

(xi) Diameter of a circle inscribed in a square is equal to the side of the square.

(xii) If a rhombus and a square stand on the same base and on the same side of the base with the same height, then perimeter of the rhombus is greater than that of square.

(xiii) Area = Base × Height = $a \times a = a^2$

Area = (side)^2

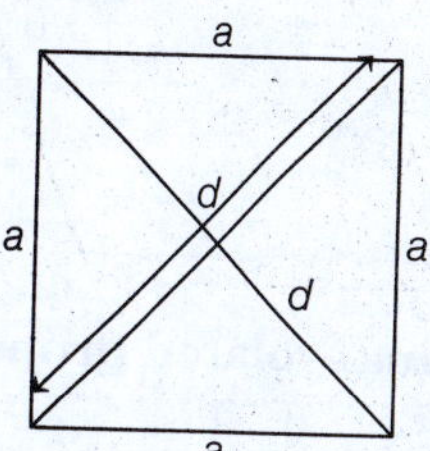

Also, area $= \frac{1}{2}$ × product of diagonals × sine of included angle

= 1/2 × first diagonal × second diagonal × sin 90°

$$\Rightarrow \quad \text{Area} = \frac{1}{2} \times (\text{diagonal})^2$$

[$\because$ both diagonals are equal]

Diagonal $= \sqrt{a^2 + a^2} = \sqrt{2}\,a$

Diagonal $= \sqrt{2} \times$ side

Perimeter = 4 × side

Inradius $(r) = \frac{a}{2} \Rightarrow r = \frac{\text{side}}{2}$

Circumradius $(R) = \frac{\text{diagonal}}{2}$

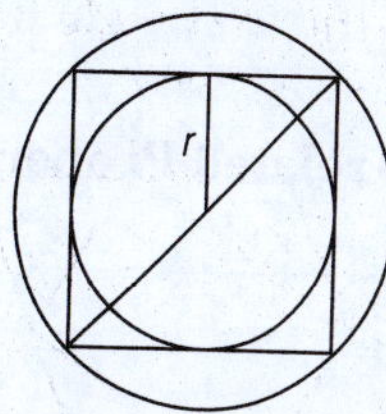

$$= \frac{\sqrt{2}a}{2} = \frac{a}{\sqrt{2}} \Rightarrow R = \frac{\text{side}}{\sqrt{2}}$$

(xiv) A square has a larger area than any other quadrilaterals with the same perimeter.

Comparative chart: Properties of Quadrilaterals 'for Quick Revision'

Side-related properties

	Parallelogram	Rectangle
Opposite sides are parallel	✓	✓
Opposite sides are equal	✓	✓

Angle-related properties

Opposite angles are equal	✓	✓	✓	✓
All four angles are equal, i.e. each 90°	×	✓	×	✓
Sum of any two adjacent angles is 180°	✓	✓	✓	✓
Bisectors of any two adjacent angles meet at 90°	✓	✓	✓	✓
Bisectors of angles form a rectangle	✓	✓	No rectangle is formed as bisectors are diagonals	

Diagonal-related properties

Diagonals are equal	×	✓	×	✓
Diagonals bisect each other	✓	✓	✓	✓
Diagonals bisect each other at 90°	×	×	✓	✓
Diagonals bisect vertex angles	×	×	✓	✓
Diagonals form four triangles of equal area	✓	✓	✓	✓
Diagonals form four congruent triangles	×	×	✓	✓
Each diagonal divides into two congruent triangles	✓	✓	✓	✓
Length of diagonal $=\sqrt{a^2+b^2-2ab\cos B}$, where, B is included angle between two adjacent sides a and b.	✓	✓(on simplification, $\sqrt{a^2+b^2}$)	✓	✓ (on simplification, $\sqrt{2}\cdot a$)

Area-related Properties

Area = Base × Height	✓	✓	✓	✓
Area = 1/2 × product of diagonals × sine of angle between them	✓	✓	✓	✓
Area = 1/2 × one diagonal × sum of perpendiculars on it from the both opposite vertices	✓	✓	✓	✓
Area = Product of any two adjacent sides × sine of included angle	✓	✓	✓	✓
They have equal areas if lying between same parallel lines and having same base	✓	✓	✓	✓
Area of a triangle is half the area of a figure (i.e. parallelogram, rectangle, rhombus, square) if both have same base and lying between same parallel lines	✓	✓	✓	✓
The line joining the mid-points of opposite sides bisects the area	✓	✓	✓	✓
Areas of four triangles, formed when we join mid-points of sides successively, are equal	✓	✓	✓	✓
Area of the figure formed, when we join mid-points of sides successively, is half the area of original figure. Also, its perimeter is equal to sum of the diagonals of the original figure.	✓	✓	✓	✓

Properties related to inscribed or circumscribed figures

Diameter of a circle circumscribing the figures	$\sqrt{a^2+b^2}$ (a parallelogram inscribed in a circle is a rectangle)	$\sqrt{a^2+b^2}$	Figure not possible	$\sqrt{2}\cdot a$
Diameter of a circle inscribed in the figures	Figure not possible	Figure not possible	Area/side	Length of side

Trapezium

A trapezium is a quadrilateral with only one pair of sides parallel.

Since $AB \parallel CD$, $ABCD$ is a trapezium.

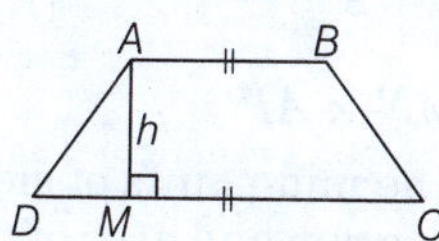

Properties

(i) Any line parallel to parallel sides of a trapezium divides the non-parallel sides proportionally.

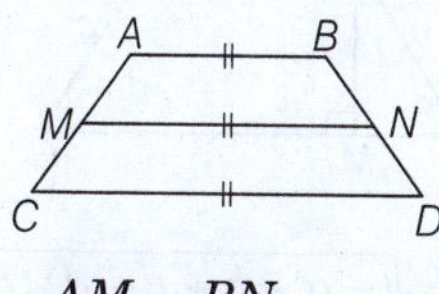

Here, $\dfrac{AM}{MC} = \dfrac{BN}{ND}$

(ii) If two parallel lines are cut by two distinct transversals, then the quadrilateral formed by the four lines is always a trapezium.

(iii) The line joining the mid-points of the non-parallel sides is called median or midline or midsegment and its length is half the sum of parallel sides. Also, median is equidistant from the parallel sides. In other words, median is parallel to parallel sides.

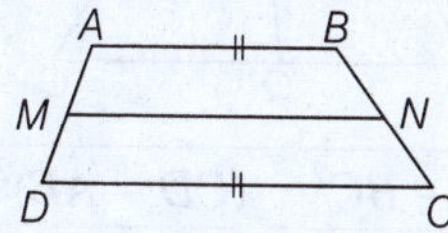

Here, median $(MN) = 1/2\,(AB + DC)$

(iv) The length of any line (say PQ) between and parallel to parallel sides is given by

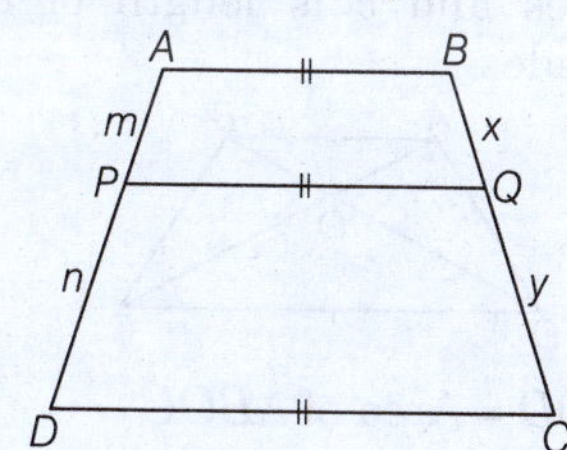

$$PQ = \frac{(m \times DC) + (n \times AB)}{AD}$$

$$= \frac{(x \times DC) + (y \times AB)}{BC}$$

where, $AP = m$, $PD = n$,

$BQ = x$ and $QC = y$

(v) If a line EF passes through intersection of diagonals and parallel to parallel sides, then

$$\frac{1}{EF} = \frac{1}{2}\left(\frac{1}{AB} + \frac{1}{DC}\right)$$

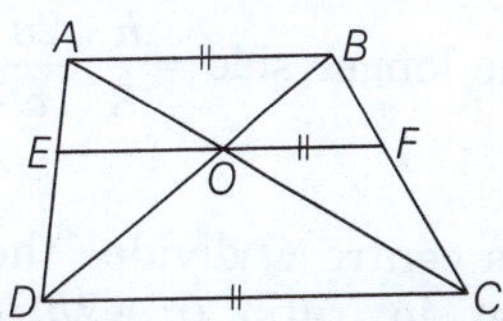

(vi) The sum of squares of diagonals of a trapezium is equal to the sum of squares of non-parallel sides plus twice the product of parallel sides.

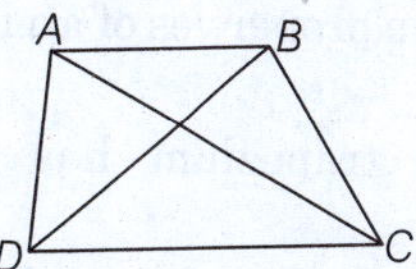

Here, $AC^2 + BD^2 = BC^2 + AD^2 + 2CD \cdot AB$

(vii) The line joining the mid-points of the diagonals of a trapezium is parallel to each of parallel sides and is half the difference between parallel sides.

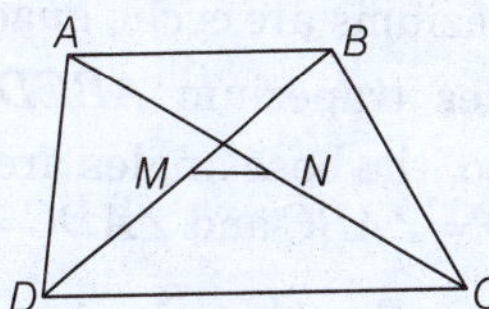

Here, $MN \parallel AB$ and DC

$\therefore \quad MN = \dfrac{1}{2}(DC - AB)$

(viii) The point of intersection of diagonals lies on the line joining mid-points of the parallel sides of a trapezium.

(ix) Lines joining the mid-points of sides of a trapezium form a parallelogram.

(x) Diagonals intersect each other in ratio equal to the ratio of parallel sides.

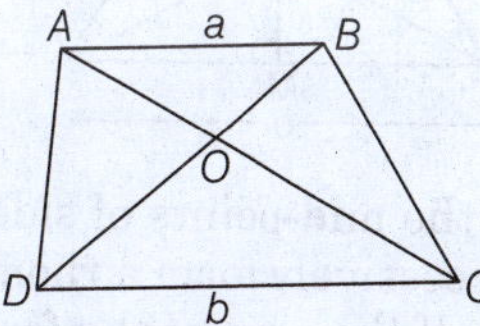

Hence, $\dfrac{AO}{OC} = \dfrac{BO}{OD} = \dfrac{a}{b}$

(xi) The centre of area lies on the line joining mid-points of the parallel sides of the trapezium.

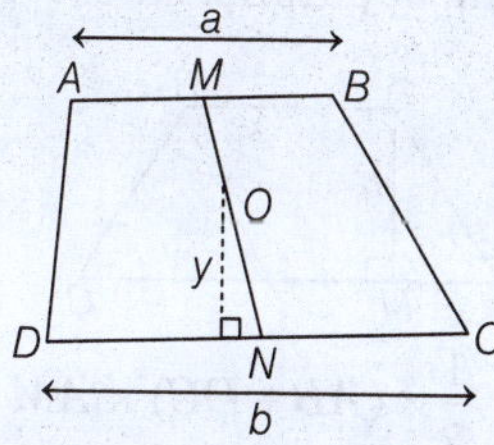

Here, $b > a$

Then, perpendicular distance [y] of centre of the area from the longer side $= \frac{h}{3}\left(\frac{2a+b}{a+b}\right)$, where h is height.

Further, this centre O divides the line MN (from shorter side) in ratio $(a+2b):(2a+b)$. Hence, $MO:ON = (a+2b):(2a+b)$

(xii) If the non-parallel sides of trapezium are equal in length, then this trapezium is called 'isosceles trapezium'. The properties of an isosceles trapezium are as below

(a) The isosceles trapezium has only one line of symmetry.

(b) Isosceles trapeziums have diagonals equal in length.

(c) If a trapezium is inscribed in a circle, then it is an isosceles trapezium.

(d) Opposite angles are supplementary. It implies that isosceles trapeziums are cyclic quadrilaterals.

(e) In an isosceles trapezium $ABCD$, $AO = BO$ and $DO = CO$. Also, the base angles are equal pairwise, hence, $\angle BAD = \angle ABC$ and $\angle ADC = \angle BCD$.

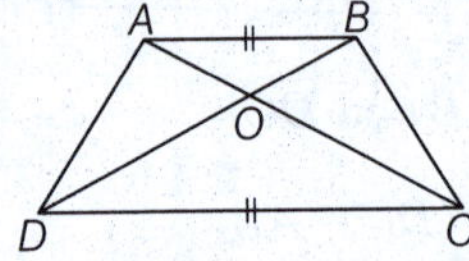

(f) The perpendicular distance of point of intersection of diagonals from the longer parallel side i.e. $OM = \frac{ah}{a+b}$, where h is the height of the isosceles trapezium.

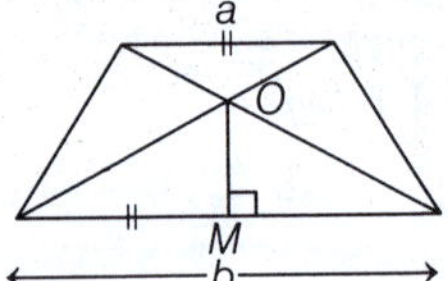

(g) Lines joining the mid-points of sides of an isosceles trapezium successively form a rhombus. The area of the latter is half the area of the former.

Also, perimeter of the latter equals twice the diagonal of the former.

(xiii) (a) Area of any type of trapezium

$$= \frac{1}{2} \times \text{Sum of parallel sides} \times \text{Height}$$

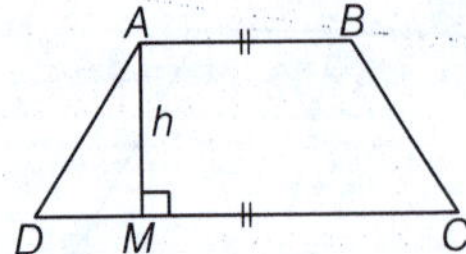

Hence, area $= \frac{1}{2} \times (AB + DC) \times AM$

(b) Area of any type of trapezium = Median × Height

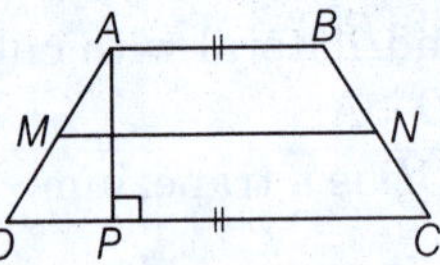

Here, area $= MN \times AP$

(c) Lines joining the mid-points of sides of a trapezium form a parallelogram and area of this parallelogram is half the area of trapezium. Also, perimeter of the parallelogram equals the sum of diagonals of the trapezium.

(xiv)

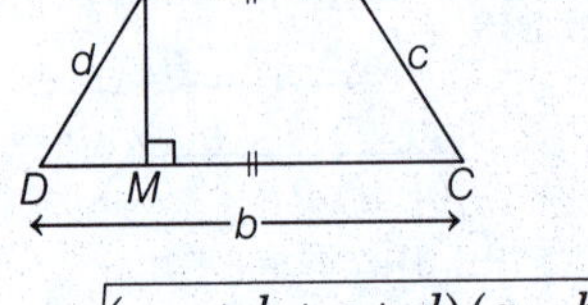

Height, $AM = \frac{\sqrt{(-a+b+c+d)(a-b+c+d)(a-b+c-d)(a-b-c+d)}}{2|b-a|}$

where, $a \neq b$ and c and d are unparallel sides.

(xv) Here, AB and DC are parallel sides.

Then, length of the line (say, PQ) joining their mid-points

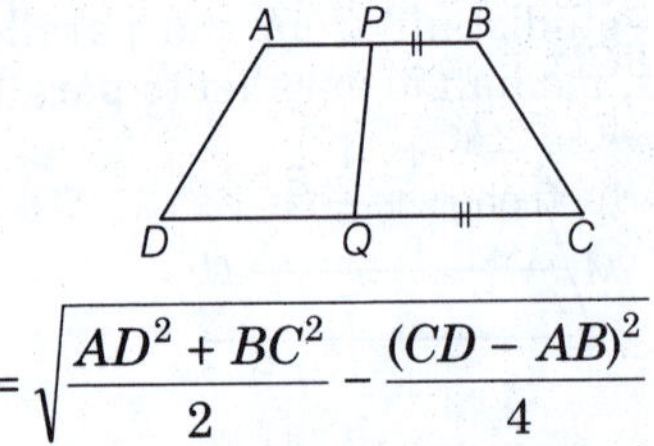

$$= \sqrt{\frac{AD^2 + BC^2}{2} - \frac{(CD - AB)^2}{4}}$$

(xvi) Length of each of diagonals of an isosceles trapezium $= \sqrt{ab + c^2}$, where a and b are length of parallel sides and c is length of each of equal unparallel sides.

(xvii)

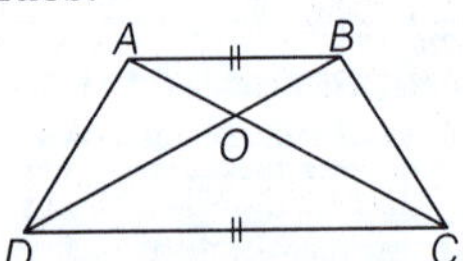

(a) Area of ΔAOD = Area of ΔBOC

(b) Area of ΔABD × Area of ΔBCD = Area of ΔABC × Area of ΔACD

(c) The ratio of the areas of each pair of adjacent triangles is equal to the ratio of parallel sides.

It means, $\frac{\text{Area of } \Delta AOD}{\text{Area of } \Delta COD} = \frac{AB}{DC}$

$$= \frac{\text{Area of } \Delta AOB}{\text{Area of } \Delta AOD}$$

$$= \frac{\text{Area of } \Delta AOB}{\text{Area of } \Delta BOC} = \frac{\text{Area of } \Delta BOC}{\text{Area of } \Delta COD}$$

(xviii) Here, P is any point on the line joining mid-points of parallel sides AB and DC, then

Area of ΔAPD = Area of ΔBPC.

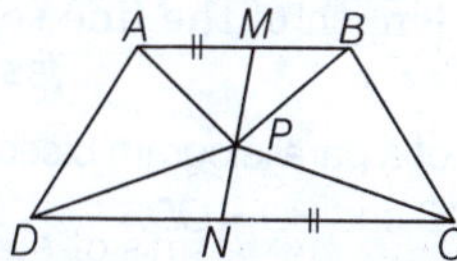

Illustrations

Problems Based on General Properties of Quadrilaterals

Problem 1 Points E, F, G and H are mid-points of sides AB, BC, CD and DA of a quadrilateral $ABCD$, respectively. Find the area of quadrilateral $ABCD$ if area of $\square EFGH$ is $25\sqrt{3}$ cm^2.

Sol. In such cases, area($\square ABCD$) = 2 × area($\square EFGH$)

$$= 2 \times 25\sqrt{3} = 50\sqrt{3} \text{ cm}^2$$

Problem 2 $ABCD$ is a quadrilateral such that $BC = BA$ and $CD > AD$. Which one of the following is correct ? ***[CDS-13]***

(a) $\angle BAD = \angle BCD$ (b) $\angle BAD < \angle BCD$

(c) $\angle BAD > \angle BCD$ (d) $2\angle BAD = \angle BCD$

Sol. (c) In ΔABC, $\angle BAC = \angle ACB$

[$\because BC = AB$ and angles opposite to equal sides are equal]

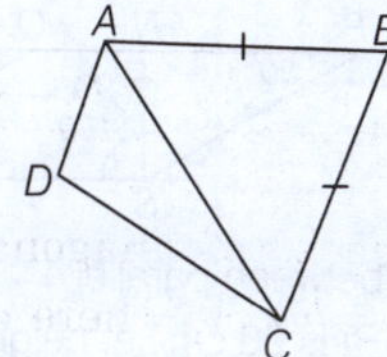

In ΔADC, $\angle DAC > \angle DCA$ [$\because$ greater side has greater angle]

On addition, $\angle BAC + \angle DAC > \angle ACB + \angle DCA$

$\Rightarrow \angle BAD > \angle BCD$

Problem 3 In a quadrilateral $ABCD$, $\angle B = 90^\circ$ and $AD^2 = AB^2 + BC^2 + CD^2$, then find the value of $\angle ACD$. ***[MAT-03]***

Sol. In ΔABC, $AC^2 = AB^2 + BC^2$

As given, $AD^2 = AC^2 + CD^2$

It is possible only when $\angle C = 90^\circ$, hence, $\angle ACD = 90^\circ$.

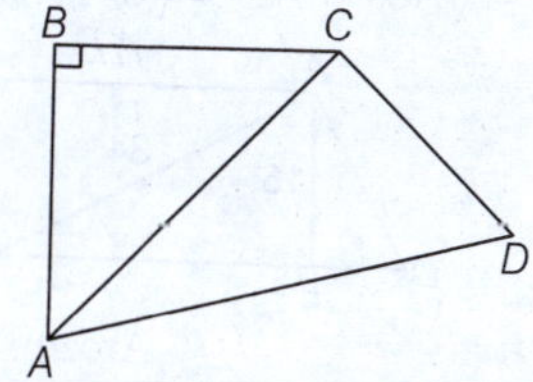

Problem 4 In the given figure, $EFCD$ is a cyclic quadrilateral and AO and BO are bisectors of $\angle EAF$ and $\angle DBE$, respectively. What is the length of AB if $AO = 8$ cm and $BO = 15$ cm?

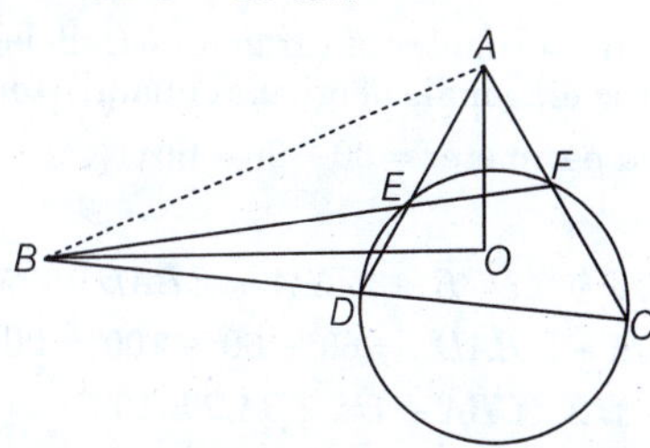

Sol. We know, the bisectors of the angles formed by producing opposite sides (if they are not parallel) of a cyclic quadrilateral, intersect each other at 90°.

Hence, $\angle AOB = 90^\circ$

Then, $AB = \sqrt{8^2 + 15^2} = 17$ cm

Work-out

1. What is the length of the diagonal of a quadrilateral opposite to the angle containing two sides 10 cm and 8 cm long and the angle between them is 60°?

(a) $2\sqrt{21}$ cm (b) $2\sqrt{7}$ cm

(c) $3\sqrt{2}$ cm (d) $2\sqrt{2}$ cm

2. What is the area of a quadrilateral whose diagonals are of 15 cm and 20 cm and they intersect at 60°?

(a) $25\sqrt{3}$ cm^2 (b) $45\sqrt{3}$ cm^2

(c) $75\sqrt{3}$ cm^2 (d) None of these

3. The mid-points of sides of a quadrilateral are joined successively. What is the perimeter of quadrilateral, thus formed if the diagonals of the parent quadrilateral are 30 cm and 36 cm long ?

(a) 44 cm (b) 76 cm

(c) 66 cm (d) 42 cm

4. $ABCD$ is a quadrilateral in which measures of $\angle D$ and $\angle C$ are 60° and 100°, respectively. If the internal bisectors of $\angle A$ and $\angle B$ meet at P, then measure of $\angle APB$ is ***[CLAT-18]***

(a) 80° (b) 90° (c) 100° (d) 110°

Answer Key

1. (a) 2. (c) 3. (c) 4. (a)

Hints and Solutions

1. Diagonal

$$AC = \sqrt{10^2 + 8^2 - 2\cdot 10\cdot 8\cdot \cos 60^\circ} = \sqrt{84} = 2\sqrt{21} \text{ cm}$$

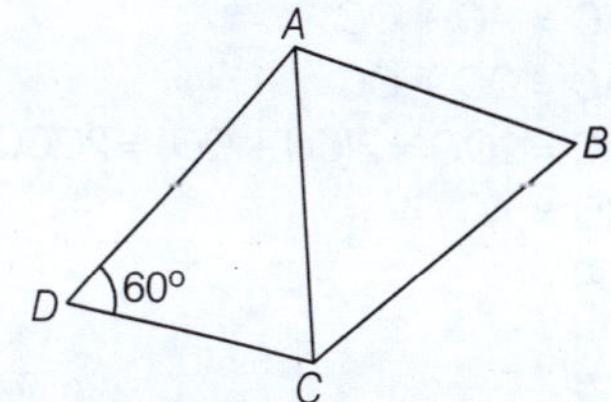

2. Area of the quadrilateral = 1/2 × Product of diagonals × Sine of angle between them

$$= 1/2 \times 15 \times 20 \times \sin 60^\circ = 150 \times \frac{\sqrt{3}}{2} = 75\sqrt{3}\ \text{cm}^2$$

3. We know, the perimeter of varignon parallelogram is equal to sum of the diagonals of original quadrilateral.

$\therefore$ Required perimeter = 30 + 36 = 66 cm

4. We have,

$$\angle ADC + \angle DCB + \angle CBA + \angle BAD = 360^\circ$$
$$\Rightarrow \angle CBA + \angle BAD = 360^\circ - 60^\circ - 100^\circ = 200^\circ$$
$$\Rightarrow 1/2\,\angle CBA + 1/2\,\angle BAD = 100^\circ$$
$$\Rightarrow \angle ABP + \angle BAP = 100^\circ$$

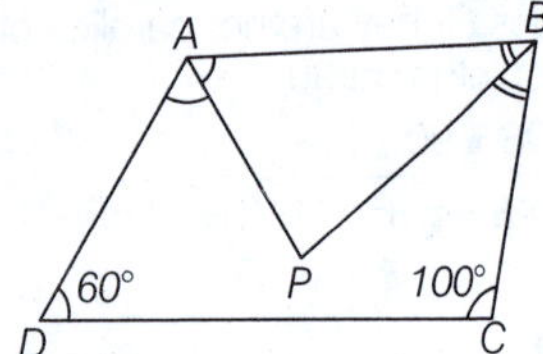

Now, for ΔABP,

$$\angle APB + \angle ABP + \angle BAP = 180^\circ$$
$$\Rightarrow \angle APB = 180^\circ - 100^\circ = 80^\circ$$

Problems based on Parallelograms

Problem 5 $ABCD$ is a parallelogram. P and R are the mid-points of DC and BC respectively. The line PR intersects the diagonal AC at Q. The distance CQ will be: **[CDS-14]**

(a) $AC/4$ (b) $BD/3$
(c) $BD/4$ (d) $AC/3$

Sol. (a) We join B and D. Let AC and BD intersect at point O.

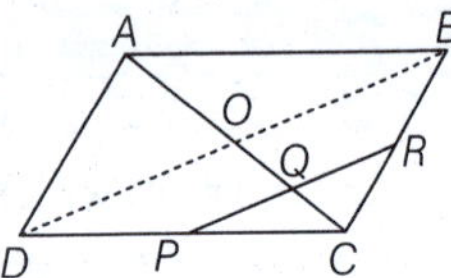

We know, the line joining the mid-points of two sides of a triangle is parallel to the 3rd side. Then, $PR \| BD$. Also, $PQ \| DO$

Now, in ΔCOD, $PQ \| DO$.

Then, $\frac{CQ}{QO} = \frac{CP}{PD}$

$\Rightarrow \frac{CQ}{QO} = \frac{1}{1}$ [$\because P$ is mid-point of DC]

$\Rightarrow CQ = QO$

Further, the diagonals of a parallelogram bisects each other.

Then, $AO = OC$

Clearly, $AC = AO + OC$

$\Rightarrow AC = OC + OC$

$\Rightarrow AC = 2OC = 2(CQ + QO) = 2(2CQ) = 4CQ$

$\Rightarrow CQ = AC/4$

Problem 6 The length of the diagonal BD of the parallelogram $ABCD$ is 18 cm. If P and Q are the centroid of the ΔABC and ΔADC respectively, then find the value of the length of the line segment PQ **[SSC (10 + 2-LDC/DEO)-11]**

Sol. Since diagonals of a parallelogram bisects each other,

$AO = OC$ and $BO = OD$.

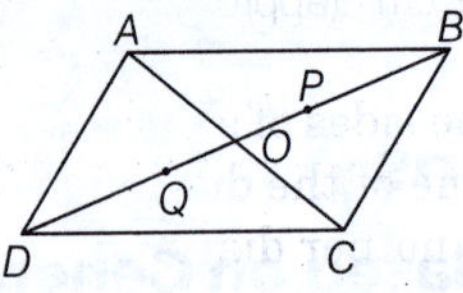

Then, $OB = OD = 9$ cm

Clearly, centroids P and Q will lie on OB and OD respectively as OB and OD are medians.

Now, $OP = 1/3 \times OB$

[$\because$ centroid divides the median in ratio 2 : 1 (from vertex)]

$\Rightarrow OP = 1/3 \times 9 = 3$ cm

Similarly, $OQ = 3$ cm

Then, $PQ = OP + OQ = 3 + 3 = 6$ cm

Problem 7 An obtuse angle made by a side of parallelogram $PQRS$ with other pair of parallel sides is 150°. If the perpendicular distance between these parallel sides (PQ and SR) is 20 cm, what is the length of the side RQ?

Sol. Since the sum of any two adjacent angles is 180° in a parallelogram,

$$\angle SPN = 180^\circ - 150^\circ = 30^\circ$$

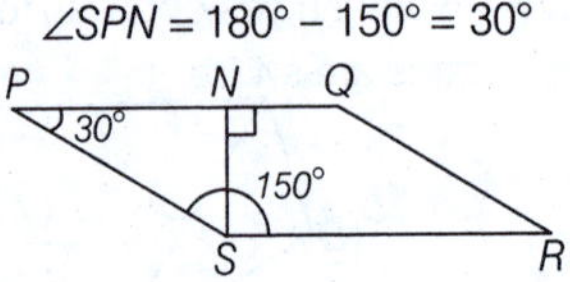

Given, distance between parallel sides (PQ and SR) is 20 cm i.e. height (NS) = 20 cm

Now, in ΔPNS, $\sin 30^\circ = \frac{NS}{PS}$

$\Rightarrow \frac{1}{2} = \frac{20}{PS} \Rightarrow PS = 40$

Since opposite sides are equal in a parallelogram,

$RQ = PS = 40$ cm

Problem 8 The altitude to DC and one diagonal AC of a parallelogram $ABCD$ are 16 cm and 34 cm respectively. What is the area of the parallelogram if its $\angle ADC = 60^\circ$?

Sol. As given, $AE = 16$ cm

Then, $CE = \sqrt{AC^2 - AE^2} = \sqrt{34^2 - 16^2} = 30$

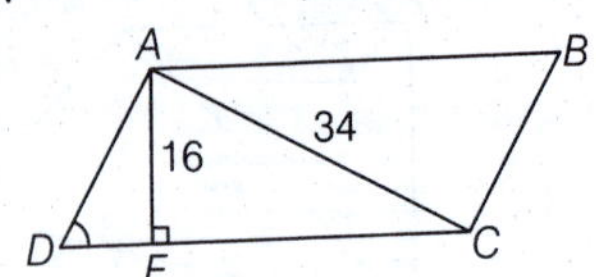

In ΔADE, $\tan 60° = AE/DE$

$\Rightarrow \quad \sqrt{3} = 16/DE$

$\Rightarrow \quad DE = 16/\sqrt{3}$

Clearly, $DC = DE + EC$

$= 16/\sqrt{3} + 30$

Now, area of parallelogram $ABCD = DC \times AE$

$= (16/\sqrt{3} + 30) \times 16$

$= 628\ \text{cm}^2$ (approx.)

Problem 9 **The sides of a parallelogram are 12 cm and 8 cm long and one of the diagonals is 10 cm long. What is the length of another diagonal?**

Sol. As per Parallelogram Law, the sum of the squares of the sides of a parallelogram is equal to the sum of the squares of the diagonals .

Hence, $10^2 + (\text{another diagonal})^2 = 2(12^2 + 8^2)$

[∵ opposite sides are equal]

$\Rightarrow \quad (\text{another diagonal})^2 = 316$

$\Rightarrow \quad \text{another diagonal} = 17.78\ \text{cm}$

Problem 10 **Two similar parallelograms have corresponding sides in the ratio $1 : k$. What is the ratio of their areas?**

Sol. Since their corresponding sides are in ratio $1 : k$, ratio of their bases and ratio of heights both will be in ratio $1 : k$.

Then, ratio of their areas $= (1 \times 1) : (k \times k)$

[∵ area of a parallelogram = base × height]

$= 1 : k^2$

Alternatively

Since parallelograms are similar, the corresponding angles will be same.

Then, ratio of their areas

$= (1 \times 1 \times \sin\theta) : (k \times k \times \sin\theta)$

[∵ area of a parallelogram = product of any two adjacent sides × sine of included angle]

$= 1 : k^2$

Work-out

1. The side AB of a parallelogram $ABCD$ is produced to E in such a way that $BE = AB$. DE intersects BC at Q. The point Q divides BC in the ratio :

(a) 1 : 2 (b) 1 : 1
(c) 2 : 3 (d) 2 : 1

2. In the figure given, $ABCD$ is a parallelogram. P is a point in BC such that $PB : PC = 1 : 2$. DP produced meets AB produced at Q. If the area of the ΔBPQ is 20 sq units, what is the area of ΔDCP? ***[CDS-10]***

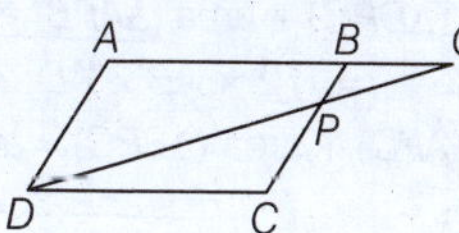

(a) 20 sq units (b) 30 sq units
(c) 40 sq units (d) None of these

3. M and N are points on diagonal BD of a parallelogram $ABCD$ such that $BM = ND$. Then, what is the shape of quadrilateral $AMCN$?

(a) Parallelogram (b) Rhombus
(c) Cannot determine (d) None of the above

4. A parallelogram $ABCD$ has sides $AB = 24$ cm and $AD = 16$ cm. The distance between the sides AB and DC is 10 cm. Find the distance between the sides AD and BC.

(a) 24 cm (b) 15 cm (c) 36 cm (d) 42 cm

5. A triangle and a parallelogram are constructed on the same base such that their areas are equal. If the altitude of the parallelogram is 100 m, then the altitude of the triangle is : ***[MAT-2003]***

(a) 100 m (b) 200 m (c) $100\sqrt{2}$ m (d) $10\sqrt{2}$ m

6. In a parallelogram $PQRS$, $\angle P$ is four times $\angle Q$, then the measure of $\angle R$ is: ***[SSC (CGL-T-I)-15]***

(a) 72° (b) 36° (c) 144° (d) 130°

7. The sides AB and CD of a trapezium $ABCD$ are paralle, with AB being the smaller side. P is the mid-point of CD and $ABPD$ is a parallelogram. If the difference between the areas of the parallelogram $ABPD$ and the triangle BPC is 10 sq cm, then the area, in sq cm, of the trapezium $ABCD$ is ***[CAT-21]***

(a) 30 (b) 40
(c) 25 (d) 20

Answer Key

1. (b) 2. (d) 3. (a) 4. (b) 5. (b) 6. (c) 7. (a)

Hints and Solutions

1. $\Delta CDQ \sim \Delta BQE$ [∵ $\angle CDQ = \angle BEQ$ and $\angle CQD = \angle BQE$]

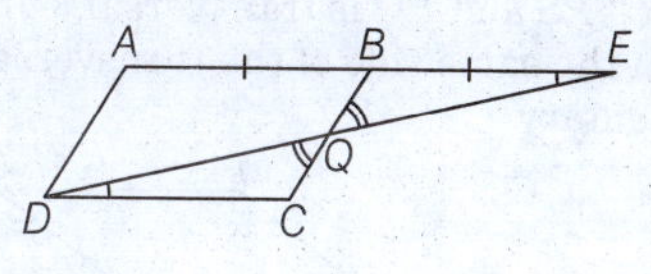

Then, $\dfrac{CQ}{BQ} = \dfrac{DC}{BE}$

$\Rightarrow \quad \dfrac{CQ}{BQ} = \dfrac{DC}{AB}$ [∵ $BE = AB$]

$\Rightarrow \quad \dfrac{CQ}{BQ} = \dfrac{AB}{AB}$

[∵ opposite sides are equal in a parallelogram]

$\Rightarrow \quad CQ : BQ = 1 : 1$

2. $\Delta BPQ \sim \Delta DCP$

[∵ $\angle BQP = \angle PDC$ (alternate interior angles as $AQ \parallel DC$) and $\angle BPQ = \angle DPC$ (vertically opposite angles)]

We know, ratio of areas of two similar triangles is equal to the ratio of square of their corresponding sides.

Then, $\dfrac{\text{area}(\Delta BPQ)}{\text{area}(\Delta DCP)} = \dfrac{PB^2}{PC^2} \Rightarrow \dfrac{20}{\text{Area}(\Delta DCP)} = \dfrac{1^2}{2^2}$

$\Rightarrow \quad \text{Area}(\Delta DCP) = 80\ \text{sq units}$

3. Since diagonals of a parallelogram bisect each other, $BO = OD$

$\Rightarrow \quad BO - BM = OD - ND \quad [\because BM = ND]$

$\Rightarrow \quad MO = ON$

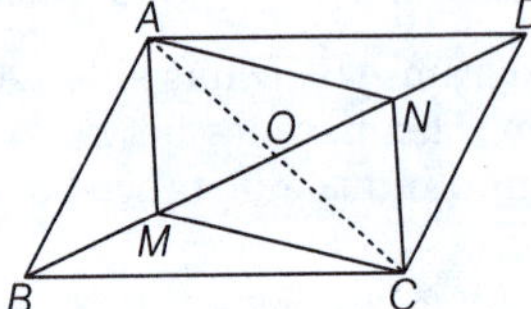

In $\square AMCN$, $MO = ON$ and $AO = OC$

$\because$ Diagonals of $\square AMCN$ are bisecting each other, it is a parallelogram.

4. Area = base × height

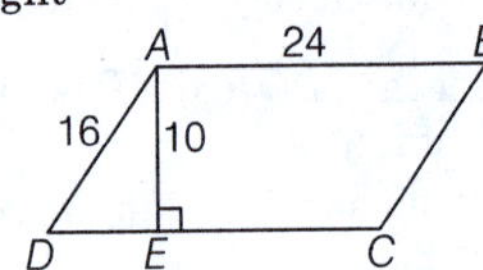

$\Rightarrow AB \times 10 = AD \times$ perpendicular distance between AD and BC

$\Rightarrow 24 \times 10 = 16 \times$ perpendicular distance between AD and BC

$\Rightarrow$ Perpendicular distance between AD and BC

$= \frac{24 \times 10}{16} = 15\text{ cm}$

5. Given, area of the triangle = area of the parallelogram

$\Rightarrow 1/2 \times$ base × altitude of triangle = base × altitude of parallelogram

$\Rightarrow 1/2 \times$ base × altitude of triangle = base × 100

$\Rightarrow 1/2 \times$ altitude of triangle = 100 [$\because$ both have same base]

Hence, altitude of the triangle = 200 m

6. Since $QR \parallel PS$ and PQ is transversal, sum of interior angles on the same side of the transversal is supplementary.

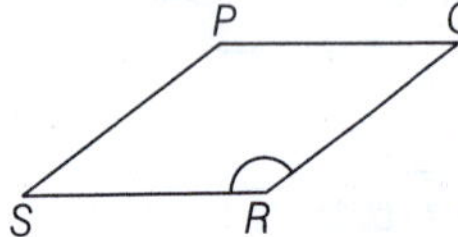

Hence, $\angle P + \angle Q = 180° \Rightarrow \angle P + \angle P/4 = 180° \Rightarrow \angle P = 144°$

$\because$ Opposite angles are equal in a parallelogram, $\angle P = \angle R$.

Then, $\angle R = 144°$

7.

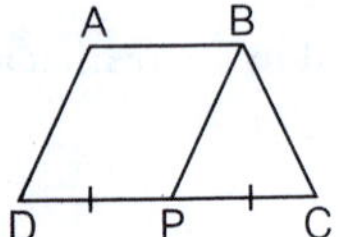

We know that the area of a triangle is half the area of a parallelogram if both has same/equal base and lying between same parallel lines. Here, $DP = PC$ and $AB \parallel DC$.

Hence, area of parallelogram $ABPD = 2 \times$ Area of ΔBPC.

As given, the difference between area of the parallelogram $ABPD$ and that of the ΔBPC is 10 cm^2. It means the area of

$\Delta BPC = 10\text{ cm}^2$

Now, area of the trapezium $ABCD = 3 \times 10 = 30\text{ cm}^2$

Problems Based on Rectangles

Problem 11 If $PQRS$ be a rectangle such that $PQ = \sqrt{3}\, QR$. Then, what is $\angle PRS$ equal to? **[CDS-14]**

Sol. In right-angled ΔPQR, $\tan \angle QPR = \frac{QR}{PQ}$

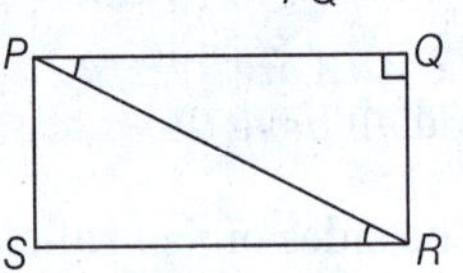

$\Rightarrow \quad \tan \angle QPR = \frac{QR}{\sqrt{3}\,QR}$

$\Rightarrow \quad \tan \angle QPR = \tan 30°$

$\Rightarrow \quad \angle QPR = 30°$

$\therefore \quad \angle PRS = 30°$

[$\because \angle PRS = \angle QPR$ (alternate interior angles) as $PQ \parallel SR$]

Problem 12 Q is a point in the interior of a rectangle $ABCD$. If $QA = 3$ cm, $QB = 4$ cm and $QC = 5$ cm, then find the length of QD (in cm).

Sol. We know that, if Q is any point inside a rectangle ABCD, then, $(QB)^2 + (QD)^2 = (AQ)^2 + (QC)^2$

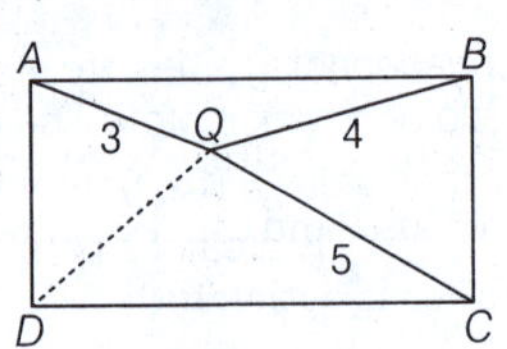

$\Rightarrow \quad (4)^2 + (QD)^2 = (3)^2 + (5)^2$

$\Rightarrow \quad (QD)^2 = 18 \Rightarrow QD = 3\sqrt{2}$ cm

Problem 13 A point P is inside a rectangle $ABCD$ and it is joined to the all vertices. What is the area of ΔAPB provided area $(\Delta APD + \Delta BPC) = 100\text{ cm}^2$ and area $(\Delta DPC) = 68\text{ cm}^2$?

Sol. Area $(\Delta APD) = 1/2 \times AD \times PM$

Area $(\Delta BPC) = 1/2 \times BC \times PN$

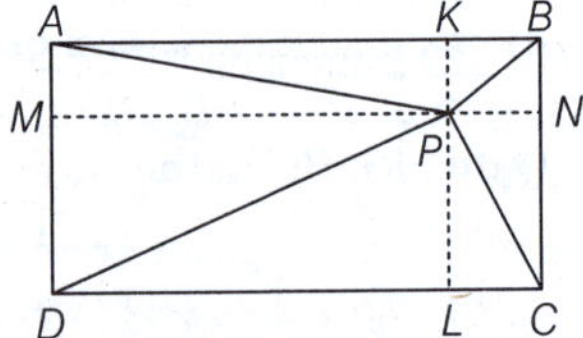

$\therefore$ Area(ΔAPD) + Area (BPC)

$= 1/2 \times AD\,(PM + PN) \quad [\because AD = BC]$

$= 1/2 \times AD \times AB$

Similarly, area (ΔDPC) + area (ΔAPB)

$= 1/2 \times AB \times AD$

Hence, area(ΔAPD) + area (ΔBPC) = area (ΔDPC) + area (ΔAPB)

$\Rightarrow \quad 100 = 68 + \text{area}(\Delta APB)$

$\Rightarrow \quad \text{area}(\Delta APB) = 32\text{ cm}^2$

Work-out

1. $ABCD$ is a rectangle, where the ratio of the length of AB and BC is 3 : 2. If P is the mid-point of AB, then the value of $\sin \angle CPB$ is: **[SSC (CGL-T-II)-13]**
 (a) 3/5 (b) 2/5 (c) 3/4 (d) 4/5

2. If a parallelogram with area P, a rectangle with area R and a triangle with area T are all constructed on the same base and all have the same altitude, then a false statement is **[MAT-99]**
 (a) $P = 2T$ (b) $T = R/2$ (c) $P = R$ (d) None of these

Answer Key

1. (d) 2. (d)

Hints and Solutions

1. Let AB and BC be $3x$ and $2x$ respectively.

Now, $PC = \sqrt{(BP)^2 + (BC)^2}$

$= \sqrt{(3x/2)^2 + (2x)^2} = \dfrac{5x}{2}$

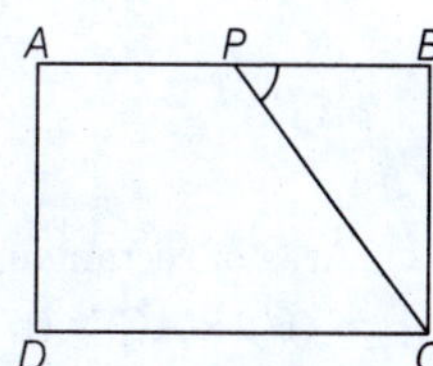

Then, $\sin \angle CPB = \dfrac{BC}{PC} = \dfrac{2x}{5x/2} = \dfrac{4}{5}$

2. Clearly, P = base × altitude of parallelogram

R = base × altitude of rectangle

T = 1/2 × base × altitude of triangle

Since bases are same and altitudes are same, $P = R = 2T$.
Hence, the answer is option(d).

Problems based on Rhombuses

Problem 14 $ABCD$ is a rhombus. A straight line through C cuts AD produced at P and AB produced at Q. If $DP = 1/2\,AB$, then the ratio of the length of BQ and AB is

Sol.

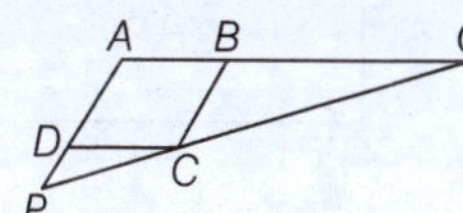

Here, $\Delta APQ \sim \Delta BCQ$

[∵ $\angle Q = \angle Q$ (common), $\angle QAP = \angle QBC$ (corresponding angles as $BC \parallel AP$)]

Then, $\dfrac{QA}{QB} = \dfrac{AP}{BC} \Rightarrow \dfrac{AB + BQ}{QB} = \dfrac{AD + DP}{BC}$

$\Rightarrow \dfrac{AB + BQ}{QB} = \dfrac{AB + \frac{1}{2}AB}{AB}$

[∵ $AD = AB = BC$ (sides of rhombus)]

$\Rightarrow \dfrac{AB}{QB} = \dfrac{3}{2} - 1 \Rightarrow BQ : AB = 2 : 1$

Problem 15 A rhombus $OABC$ is drawn inside a circle whose centre is at O in such a way that the vertices A, B and C of the rhombus are on the circle. If the area of the rhombus is $32\sqrt{3}$ m^2, then find the radius of the circle. **[MAT-03]**

Sol. In right-angled ΔOMC, $OC^2 = OM^2 + MC^2$

[∵ diagonals bisect each other at 90°]

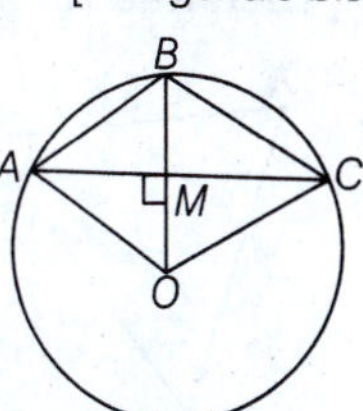

$\Rightarrow OB^2 = \left(\dfrac{OB}{2}\right)^2 + \left(\dfrac{AC}{2}\right)^2$

[∵ $OC = OB$ = radius and $OM = OB/2$ and $MC = AC/2$ as diagonals bisect each other]

$\Rightarrow 3 \cdot OB^2 = AC^2 \Rightarrow \sqrt{3}OB = AC$

Now, area of the rhombus = 1/2 × OB × AC

$\Rightarrow 32\sqrt{3} = 1/2 \times OB \times \sqrt{3}OB$

$\Rightarrow OB = 8$

Hence, radius of the circle is 8 cm.

Problem 16 $ABCD$ is a rhombus. AB is produced to F and BA is produced to E such that $AB = AE = BF$. Then, **[SSC (CGL-T-I)-15]**
 (a) $ED > CF$ (b) $ED \perp CF$
 (c) $ED^2 + CF^2 = EF^2$ (d) $ED \parallel CF$

Sol. (b) In ΔBED, $\dfrac{BA}{AE} = \dfrac{BO}{OD}$ [∵ diagonals bisect each other at 90°]

Hence, $AO \parallel ED$. Also, $OC \parallel DG$

Further, in ΔACF, $\dfrac{AB}{BF} = \dfrac{AO}{OC}$

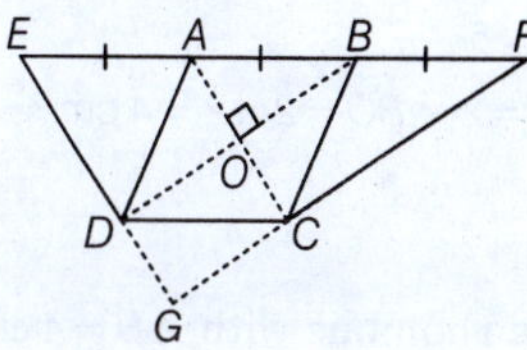

Hence, $OB \parallel CF$. Also, $DO \parallel CG$

Since, $OC \parallel DG$ and $DO \parallel CG$, $DOCG$ is a parallelogram.

Then, $\angle DOC = \angle DGC \Rightarrow \angle DGC = 90°$

Thus, $DG \perp CG$, i.e. $ED \perp CF$

Problem 17 What is the radius of a circle inscribed in a rhombus having diagonals of 8 cm and 6 cm?

Sol.

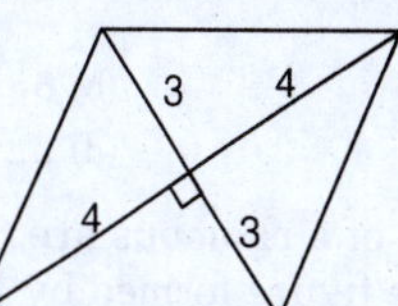

Area = $2ar$, where a is the side and r is in radius.

$\Rightarrow$ 1/2 × product of diagonals $= 2 \times \sqrt{3^2 + 4^2} \times r$

$\Rightarrow 1/2 \times 8 \times 6 = 10r \Rightarrow r = 2.4$ cm

Alternatively

Radius $= \dfrac{d_1 \times d_2}{2\sqrt{(d_1)^2 + (d_2)^2}} = \dfrac{8 \times 6}{2\sqrt{8^2 + 6^2}} = 2.4$ cm

Problem 18 $ABCD$ is a rhombus whose side $AB = 4$ cm and $\angle ABC = 120°$, then find the length of diagonal BD.

Sol.

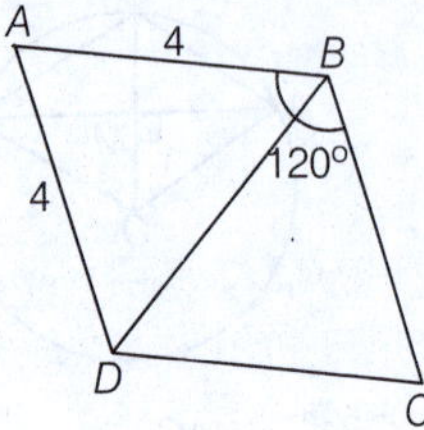

Here, $\angle ABC + \angle BCD = 180°$

[$\because$ adjacent angles are supplementary]

$\therefore$ $\angle BCD = 180° - 120° = 60°$

Now, diagonal $BD = \sqrt{2} \cdot a\sqrt{1-\cos C}$

$= \sqrt{2} \cdot 4\sqrt{1-\cos 60°} = 4$

Alternatively

Since diagonals of a rhombus bisect vertex angles,

$\angle OBC = 120°/2 = 60°$

ΔBOC is a right-angled triangle as diagonals of a rhombus bisect each other at 90°.

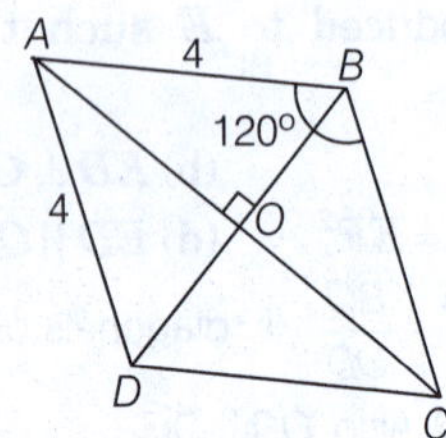

Now, $\cos 60° = \frac{BO}{BC} \Rightarrow 1/2 = \frac{BO}{4}$

$\Rightarrow$ $BO = 2$ cm

$\therefore$ $BD = 2 \times BO = 2 \times 2 = 4$ cm

Work-out

1. If $ABCD$ is a rhombus with $AB = 4$ cm, then $AC^2 + BD^2 = ?$
   (a) 24 cm
   (b) 15 cm
   (c) 36 cm
   (d) 64 cm
2. A circle is inscribed in a rhombus with diagonals 12 cm and 16cm. The ratio of the area of circle to the area of rhombus is: ***[CAT-20]***
   (a) $5\pi/18$ (b) $6\pi/25$
   (c) $3\pi/25$ (d) $2\pi/15$
3. The diagonals of a rhombus are 8 cm and 6 cm, then the area of the figure formed by joining the mid-points of its sides successively would be
   (a) 24 cm^2
   (b) 14 cm^2
   (c) 12 cm^2
   (d) None of the above

Answer Key

1. (d) 2. (b) 3. (c)

Hints and Solutions

1. As per Parallelogram law, the sum of the squares of the sides of a rhombus is equal to the sum of the squares of the diagonals .
   Hence, $AC^2 + BD^2 = 4 \times (\text{side}^2) = 4 \times (4)^2 = 64$ cm
2. We know, $r = \frac{d_1 \times d_2}{2\sqrt{(d_1)^2 + (d_2)^2}}$
   $\Rightarrow \quad \pi r^2 = \frac{\pi \times (12 \times 16)^2}{4[(12)^2 + (16)^2]} = \frac{36 \times 16}{25}\pi$
   Then, required ratio $= \frac{36 \times 16}{25}\pi : \frac{1}{2} \times 12 \times 16 = 6\pi/25$
3. Clearly, the figure formed is a rectangle.

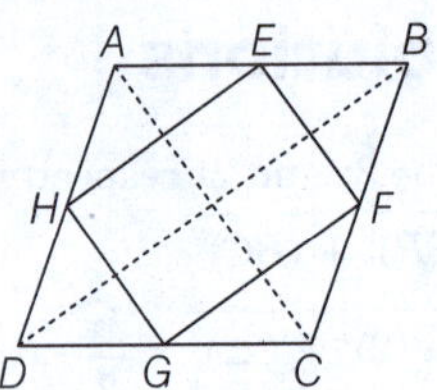

Area of rectangle $EFGH = EF \times EH$

$= (1/2 \times AC) \times (1/2 \times BD) = (1/2 \times 6) \times (1/2 \times 8) = 12$ cm^2

Problems based on Squares

Problem 19 Let $WXYZ$ be a square. If P, Q and R be the mid-points of WX, XY and ZW respectively and K, L be the mid-points of PQ and PR, respectively. Then, what is the value of area ΔPKL/area of square $WXYZ$?

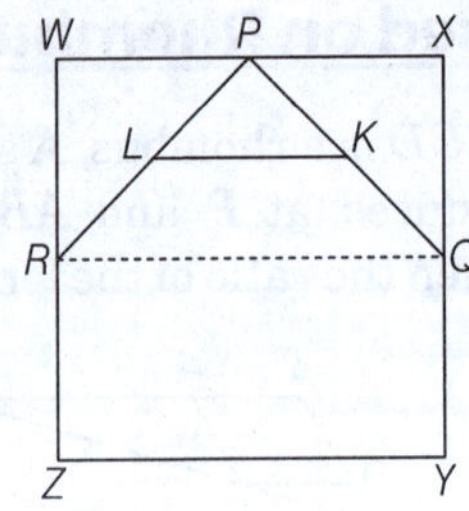

Sol. We know that the line joining the mid-points of opposite sides bisects the area. Then, area $(WXQR) = 1/2 \times$ area $(WXYZ)$

Further, area of a triangle is half the area of a square, if both have same base and lying between same parallel lines.

Then, area $(\Delta PRQ) = 1/2 \times$ area $(WXQR) = 1/2 \times 1/2 \times$ area$(WXYZ)$

Also, $\Delta PLK \sim \Delta PRQ$

[$\because \angle P = \angle P$ (common) and $\angle PLK = \angle PRQ$ (corresponding angles as $LK \parallel RQ$)]

Then, area (ΔPLK) : area $(\Delta PRQ) = PL^2 : PR^2$

$= PL^2 : (2\,PL)^2 = 1 : 4$

From above, area $(\Delta PRQ) = 1/2 \times 1/2 \times$ area $(WXYZ)$

$\Rightarrow$ $4 \times$ area $(\Delta PLK) = 1/4 \times$ area $(WXYZ)$

$\Rightarrow$ area (ΔPLK)/area$(WXYZ) = 1/16$

Remarks *You should write directly as area (ΔPLK) : area (ΔPRQ) = 1: 4 to save time in examination hall. The same has been discussed in sub-section 'Triangle'*

Problem 20 There are two squares S_1 and S_2 with areas 8 and 9 units, respectively. S_1 is inscribed within S_2, with one corner of S_1 on each side of S_2. The corners of the smaller square divides the sides of the bigger square into two segments. One of length a and the other of length b, where $b > a$. A possible value of b/a is ***[XAT-14]***

(a) ≥ 5 and < 8 (b) ≥ 8 and < 11

(c) ≥ 11 and < 14 (d) ≥ 14 and < 17

(e) > 17

Sol. (d) Side of the bigger square $S_2 = \sqrt{9} = 3$

i.e. $a + b = 3 \Rightarrow a = 3 - b$

Further, each side of the smaller square $S_1 = \sqrt{a^2 + b^2}$

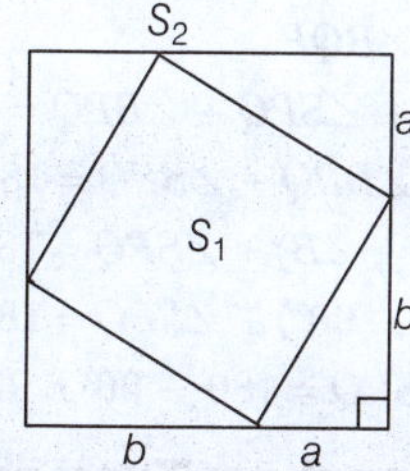

$\Rightarrow$ area of the smaller square $S_1 = a^2 + b^2$

$\Rightarrow \quad 8 = (3 - b^2) + b^2$

$\Rightarrow \quad 8 = 9 + 2b^2 - 6b$

$\Rightarrow \quad 2b^2 - 6b + 1 = 0$

$\Rightarrow \quad b = \dfrac{-(-6) \pm \sqrt{(-6)^2 - 4 \cdot 2 \cdot 1}}{2 \cdot 2}$

$= \dfrac{6 \pm \sqrt{28}}{2 \cdot 2}$

$\Rightarrow b = \dfrac{3 + \sqrt{7}}{2}$ and $= \dfrac{3 - \sqrt{7}}{2} = 2.82$ and 0.18

Then, $a = (3 - 2.82)$ and $(3 - 0.18) = 0.18$ and 2.82

Since, $b > a, b = 2.82$

and $a = 0.18$

Now, $\dfrac{b}{a} = \dfrac{2.82}{0.18} = 15.67$

Hence, option (d) is correct.

Problem 21 Points E, F, G, H lie on the sides AB, BC, CD and DA respectively, of a square $ABCD$. If $EFGH$ is also a square whose area is 62.5% of that of $ABCD$ and CG is longer than EB, then the ratio of length EB to that of CG is: ***[CAT-18]***

Sol.

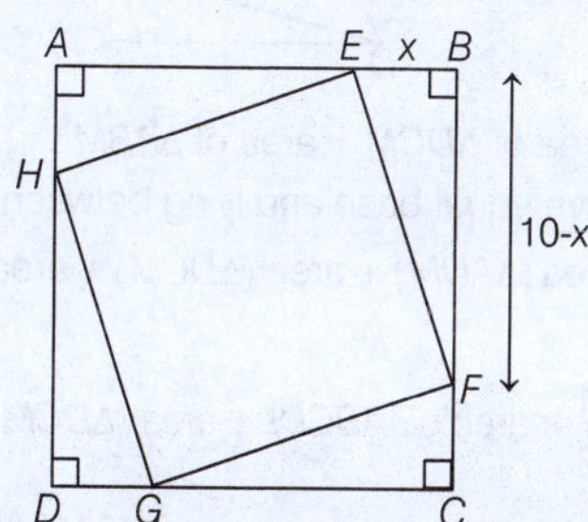

Let area of the bigger square be 100 and EB be x.

Then, side $BC = \sqrt{100} = 10$ and $BF = 10 - x$

$[\because EB = CF = DG = AH]$

Then, $\quad EF^2 = EB^2 + BF^2$

$\Rightarrow$ area of a $\square EFGH = x^2 + (10 - x)^2 = 62.5\%$ of 100

$\Rightarrow \quad 2x^2 - 20x + 37.5 = 0$

$\Rightarrow \quad x = 2.5$ and 7.5

Then, $\quad CG = BF = 7.5$ and 2.5

Since CG is longer than EB

$EB : CG = 2.5 : 7.5 = 1 : 3$

Problem 22 In the figure given, $ABCD$ is a square in which $AO = AX$. What is $\angle XOB$? ***[CDS-09]***

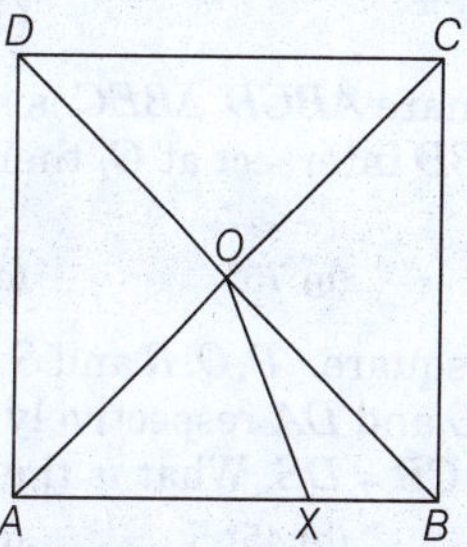

Sol. Since diagonals of square bisect vertex angles,

$\angle OBA = \dfrac{90°}{2} = \angle OAB$

Further, exterior $\angle OXA = \angle XOB + \angle OBX = \angle XOB + 45°$

In ΔAOX, $\angle AOX + \angle OXA + \angle XAO = 180°$

$\Rightarrow \quad \angle OXA + \angle OXA + 45° = 180°$

$[\because \angle AOX = \angle OXA$ as $AX = AO]$

$\Rightarrow \quad 2(\angle XOB + 45°) = 135°$

$\Rightarrow \quad \angle XOB = 22.5°$

Problem 23 $ABCD$ is a square. The diagonals AC and BD meet at O. Let K, L be the points on AB such that $AO = AK$ and $BO = BL$. If $\theta = \angle LOK$, then what is the value of $\tan \theta$?

Sol.

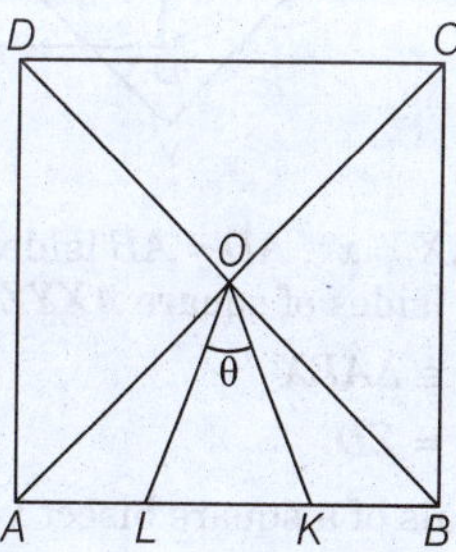

Since diagonals of a square bisect vertex angles,

$\angle OBA = \dfrac{90°}{2} = \angle OAB$

Further, exterior $\angle OKA = \angle KOB + \angle OBK = \angle KOB + 45°$

In ΔAOK, $\angle AOK + \angle OKA + \angle KAO = 180°$

$[\because \angle AOK = \angle OKA$ as $AK = AO]$

$\Rightarrow \quad 2(\angle KOB + 45°) = 135°$

$\Rightarrow \quad \angle KOB = 22.5°$

Similarly, $\quad \angle AOL = 22.5°$

Since diagonals of a square bisect each other at 90°, $\angle AOB = 90°$.

Clearly, $\angle KOB + \theta + \angle AOL = 90°$

$\Rightarrow \quad 22.5° + \theta + 22.5° = 90°$

$\Rightarrow \quad \theta = 45°$

$\Rightarrow \quad \tan\theta = \tan 45°$

$\Rightarrow \quad \tan\theta = 1$

Work-out

1. Let X be any point within a square $ABCD$. On AX, a square $AXYZ$ is described such that D is within it. Which one of the following is correct ? **[CDS-12]**
   (a) $AX = DZ$ (b) $\angle ADZ = \angle BAX$
   (c) $AD = DZ$ (d) $BX = DZ$
2. Inside a square $ABCD$, ΔBEC is an equilateral triangle. If CE and BD intersect at O, then $\angle BOC$ is equal to **[SSC (CGL-T-I)-13]**
   (a) 60° (b) 75° (c) 90° (d) 120°
3. $ABCD$ is a square. P, Q, R and S are points on the sides AB, BC, CD and DA respectively such that $AP = BQ = CR = DS$. What is the value of $\angle SPQ$?
   (a) 60° (b) 45° (c) 30° (d) 90°

Answer Key

1. (d) **2.** (b) **3.** (d)

Hints and Solutions

1. Let $\angle BAX$ be $x°$.
   Then, $\angle DAX = 90° - x°$
   $\therefore \quad \angle ZAD = 90° - (90° - x°) = x°$
   In ΔZAD and ΔABX,

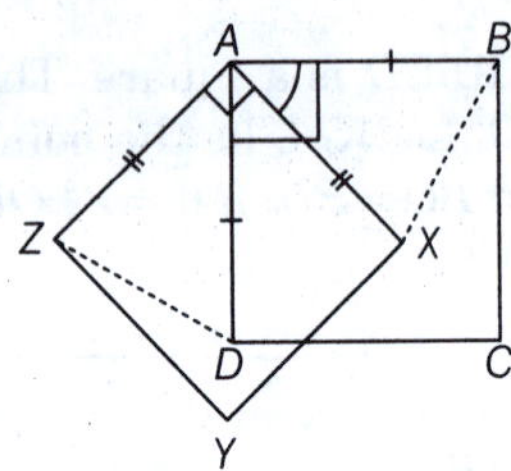

$\angle ZAD = \angle BAX = x°$, $AD = AB$ [sides of square $ABCD$] and $AZ = AX$ [sides of square $AXYZ$]

Hence, $\Delta ZAD \cong \Delta ABX$ [by criterion SAS]

$\therefore \quad BX = ZD$

2. Since, diagonals of a square bisect vertex angles,
   $\angle OBC = \frac{90°}{2} = 45°$

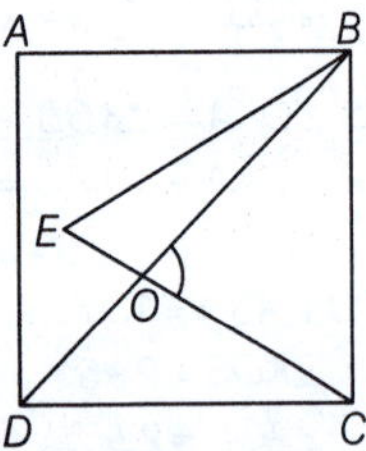

Further, $\angle ECB = 60°$
[$\because$ each angle of an equilateral triangle is 60°]

Then, $\angle BOC = 180° - \angle OBC - \angle OCB$
$= 180° - 45° - 60° = 75°$

3. Here, $\Delta APS \cong \Delta PBQ$

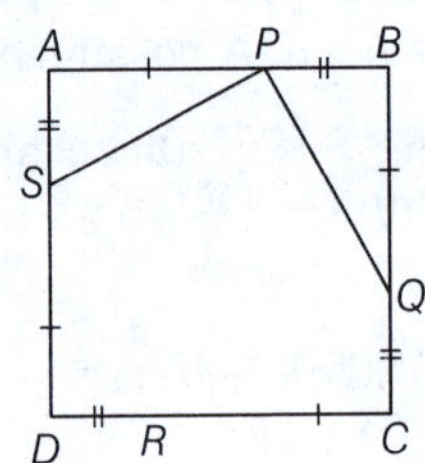

[$\because \angle A = \angle B = 90°$, $AP = BQ$ and $AS = PB$ (SAS criterion)]

Then, $\angle APS = \angle BQP$

Clearly, $\angle APS + \angle SPQ + \angle BPQ = 180°$

$\Rightarrow \quad \angle BQP + \angle BPQ + \angle SPQ = 180°$

$\Rightarrow \quad (180° - \angle B) + \angle SPQ = 180°$

$\Rightarrow \quad (180° - 90°) + \angle SPQ = 180°$

$\Rightarrow \quad \angle SPQ = 180° - 90° = 90°$

Problems based on Trapeziums

[Certain questions have also been discussed under 'Problems based on Similarity of triangles' under sub-section 'Triangle'.]

Problem 24 $ABCD$ is a trapezium, in which AB is parallel to CD. Let M be the mid-point of BC, then consider the following statements :

I. 'Area of ΔADM + Area of ΔDCM' is equal to 3/4th of the area of trapezium $ABCD$, if $AB = CD$.

II. 'Area of ΔABM + Area of ΔDCM' is always greater than half of the area of trapezium $ABCD$, if $AB = CD$.

Which of the above statement(s) is/are correct?

(a) Only I (b) Only II
(c) Both I and II (d) Neither I nor II

Sol. I. If $AB = CD$, the trapezium will become a parallelogram.

We know, area of a triangle is half the area of a parallelogram, if both has same base and lying between same parallel lines.

Hence, area of $\Delta ADM = \frac{1}{2} \times$ area of $\square ABCD$

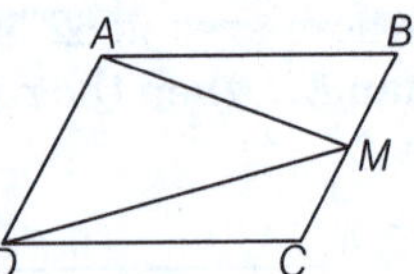

Further, area of ΔDCM = area of ΔABM

[$\because$ both have equal base and lying between same parallel lines]

Clearly, area (ΔADM) + area (ΔDCM) + area ΔABM = area ($\square ABCD$)

$\Rightarrow \quad \frac{1}{2} \times$ area ($\square ABCD$) + area (ΔDCM)

+ area (ΔABM) = area ($\square ABCD$)

$\Rightarrow \quad \text{area}(\Delta DCM) + \text{area}(\Delta DCM) = \frac{1}{2} \times \text{area}(\square ABCD)$

$\Rightarrow \quad \text{area}(\Delta DCM) = \frac{1}{4} \times \text{area}(\square ABCD)$

Then, $\text{area}(\Delta ADM) + \text{area}(\Delta DCM) = \frac{1}{2} \times \text{area}(\square ABCD)$

$$+ \frac{1}{4} \times \text{area}(\square ABCD) = \frac{3}{4} \times \text{area}(\square ABCD)$$

Hence, statement I is correct.

II. From the above, area (ΔDCM) + area (ΔABM)

$$= \frac{1}{2} \times \text{area}(\square ABCD)$$

Hence, statement II is incorrect.

Thus, option (a) is correct.

Problem 25 In an isosceles trapezium $AB \parallel DC$ and each of the equal legs is 5 cm. Find the diagonal, if $AB = 6$ cm and $DC = 10$ cm.

Sol. Legs refer to unparallel sides.

Each diagonal $= \sqrt{ab + c^2}$, where a and b are parallel sides and c is length of each of equal unparallel sides.

$$= \sqrt{(6 \times 10) + 5^2} = 9.2 \text{cm}$$

Problem 26 The parallel sides of a trapezoid $ABCD$ are in the ratio of 4 : 5. $ABCD$ is divided into an isosceles ΔABP and a parallelogram $PBCD$ (as shown). $ABCD$ has a perimeter equal to 1120 m and $PBCD$ has a perimeter equal to 1000 m. Find $\sin \angle ABC$, given that $2\angle DAB = \angle BCD$. ***[XAT-15]***

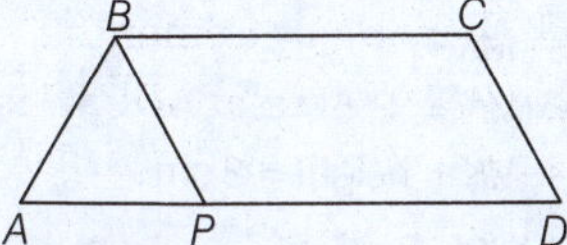

Sol. Let $AP = BP$ and $\angle BAP = \theta$.

Since ΔABP is isosceles,

$$\angle ABP = \angle BAP = \theta$$

Then, $\angle BCD = 2\theta$

Now, external $\angle BPD = \theta + \theta = 2\theta$

$\because BPDC$ is a parallelogram

$\therefore \quad \angle CBP + \angle BPD = 180°$

[$\because$ sum of any two adjacent angles is 180°]

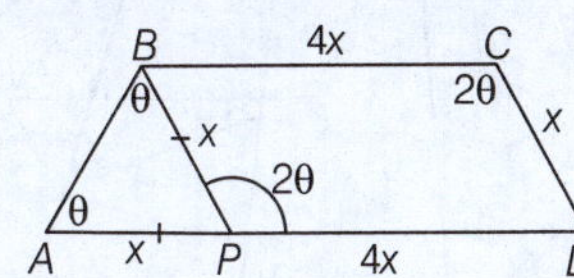

Then, $\angle CBP = 180° - 2\theta$

Clearly,

$\angle ABC = \angle ABP + \angle CBP = \theta + (180° - 2\theta) = 180° - \theta$

Let parallel sides BC and AD be $4x$ and $5x$, respectively.

Then, $AP = (AD - PD) = (5x - 4x)$

[$\because BC = PD$ as $PBCD$ is a parallelogram]

$\Rightarrow \quad AP = x$

Then, $BP = x$

As given, $AB + BC + CD + DP + PA = 1120$

and also, $BC + CD + DP + BP$ (or AP) $= 1000$

Then, $AB = 1120 - 1000 = 120$

Further, $2(x + 4x) = 1000$

$\Rightarrow \quad x = 100$

$\Rightarrow \quad AP = BP = 100$

Now, in ΔABP,

$$AP^2 = AB^2 + BP^2 - 2 \cdot AB \cdot BP \cdot \cos \angle ABP$$

[recall cosine rule]

$\Rightarrow \quad 100^2 = 120^2 + 100^2 - 2 \cdot 120 \cdot 100 \cdot \cos\theta$

$\Rightarrow \quad \cos\theta = \frac{3}{5}$

$\Rightarrow \quad \sin\theta = \sqrt{1 - \left(\frac{3}{5}\right)^2} = \frac{4}{5}$

$\Rightarrow \quad \sin(180° - \theta) = \frac{4}{5}$ $\quad [\sin\theta = \sin(180° - \theta)]$

$\Rightarrow \quad \sin \angle ABC = \frac{4}{5}$

Problem 27 In the given figure, $ABCD$ is a trapezium with $AB = 3$ cm and $DC = 5$ cm. $AB \parallel DC$ and O is the centre of the area. If the height of the trapezium is 4 cm, then OP = ?

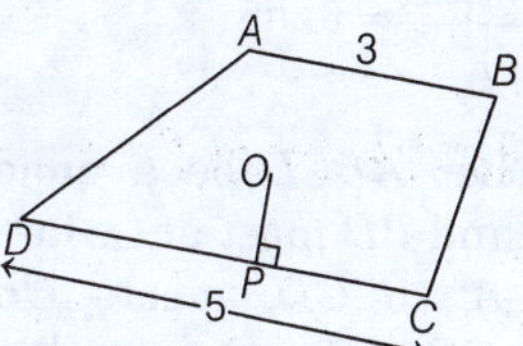

Sol. Clearly, $OP = \frac{\text{height}}{3}\left(\frac{2AB + DC}{AB + DC}\right) = \frac{4}{3}\left(\frac{2 \cdot 3 + 5}{3 + 5}\right) = 1.83 \text{ cm}$

Problem 28 $ABCD$ is a trapezium with parallel sides $AB = 2$ cm and $DC = 3$ cm. E and F are the mid-points of the non-parallel sides. The ratio of area of $ABFE$ to area of $EFCD$ is: ***(CDS-13)***

Sol.

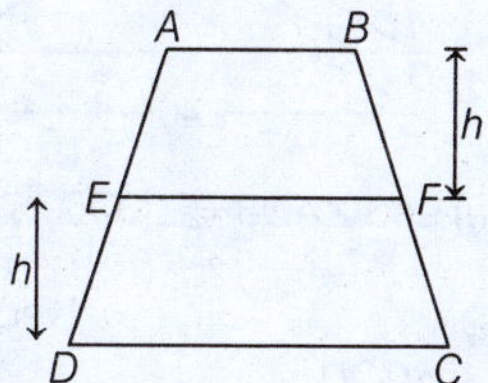

Since length of median is the half the sum of parallel sides,

$EF = \frac{1}{2} \times (2 + 3) = 2.5 \text{ cm}$

Obviously, the height of trapezium $ABFE$ and the height of trapezium $EFCD$ will be equal.

Then, ratio of area of $ABFE$ to area of $EFCD = \frac{1}{2} \times \text{height}$

$(AB + EF) : \frac{1}{2} \times \text{height} \times (EF + DC) = (2 + 2.5) : (3 + 2.5) = 9 : 11$

Problem 29 In a trapezium $ABCD$ with $AB = 4$ cm and $DC = 8$ cm, AB is parallel to CD. A line, parallel to DC, passing through intersection of the diagonals meets the unparallel sides at points E and F. What is the length of line EF ?

Sol. We know, $\frac{1}{EF} = \frac{1}{2}\left(\frac{1}{AB} + \frac{1}{DC}\right)$

$\Rightarrow \quad \frac{1}{EF} = \frac{1}{2}\left(\frac{1}{4} + \frac{1}{8}\right)$

$\Rightarrow \quad \frac{1}{EF} = \frac{3}{16}$

$\Rightarrow \quad EF = 5.33$ cm

Problem 30 In the given figure, what is the value of EF, if $AB \parallel EF \parallel DC$ and $AB = 5$ cm, $DC = 8$ cm, $BF = 2$ cm and $FC = 4$ cm.

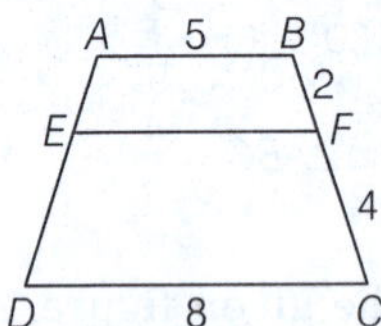

Sol. $EF = \frac{(BF \times DC) + (FC \times AB)}{BC}$

$= \frac{(2 \times 8) + (4 \times 5)}{(2 + 4)} = 6$ cm

Problem 31 Let $ABCD$ be a quadrilateral. Let the diagonals AC and BD meet at O. Let the perpendicular drawn from A to CD, meets CD at E. Further, $AO : OC = BO : OD$, $AB = 30$ cm, $CD = 40$ cm and the area of the quadrilateral $ABCD$ is 1050 cm^2. Then, what is the value of BE ?

Sol.

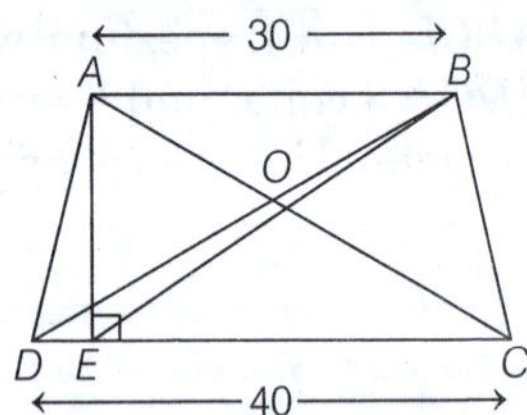

$\frac{AO}{OC} = \frac{BO}{OD}$ (given) and $\angle AOB = \angle COD$

[vertically opposite angles]

Then, $\quad \Delta AOB \sim \Delta COD$ [by criterion SAS]

$\therefore$ Their corresponding angles will be equal, i.e.

$\angle ABO = \angle ODC$

Since $\angle ABO$ and $\angle ODC$ are alternate interior angles, $AB \parallel CD$.

Hence, $ABCD$ is a trapezium.

Now, area of the trapezium $ABCD$

$= \frac{1}{2} \times AE \times (30 + 40)$

$\Rightarrow \quad 1050 = \frac{1}{2} \times AE \times 70$

$\Rightarrow \quad AE = 30$ cm

In right-angled ΔAEB, hypotenuse $(BE) = \sqrt{AB^2 + AE^2}$

$\Rightarrow \quad BE = \sqrt{30^2 + 30^2} = 30\sqrt{2}$ cm

Problem 32 Trapezium $ABCD$ has side $AB = 8$ cm and $CD = 10$ cm parallel to each other. If the area of trapezium is 27 cm^2, then how far is the point of intersection of the diagonals from CD ? ***[CAT-16]***

Sol. We draw a line PK perpendicular to parallel lines through point M.

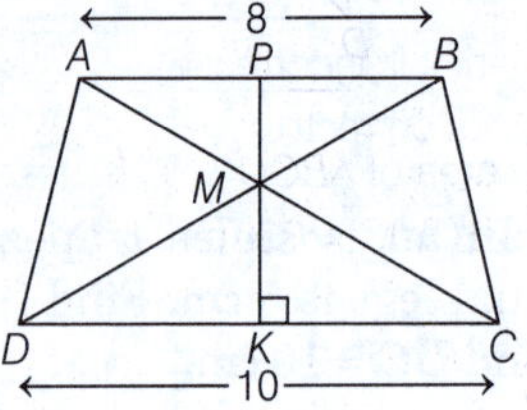

Given, area = 27 cm^2

$\Rightarrow \quad \frac{1}{2} \times$ height $\times$ (sum of parallel sides) = 27

$\Rightarrow \quad \frac{1}{2} \times$ height $\times (8 + 10) = 27$

$\Rightarrow \quad$ height = 3 cm

Here, $\Delta ABM \sim \Delta DMC$

[$\because \angle AMB = \angle DMC$ (vertically opposite angles) and $\angle BAM = \angle MCD$ (alternate interior angles)]

Then, $\quad \frac{AB}{DC} = \frac{PM}{MK}$

$\Rightarrow \quad \frac{8}{10} = \frac{PM}{MK}$

$\Rightarrow \quad PM : MK = 4 : 5$

Also, $\quad PM + MK$ = height = 3 cm

$\therefore \quad MK = 3 \times \frac{5}{(4+5)} = \frac{5}{3}$ cm

Problem 33 In the given figure, $ABCD$ is a trapezium with $AB = 3$ cm and $DC = 5$ cm. O is the centre of the area and M and N are mid-points of parallel sides AB and CD respectively, then what is length of ON if $MN = 6$ cm ?

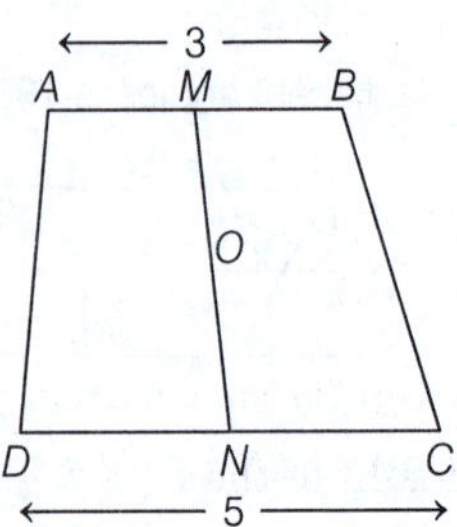

Sol. Clearly, $MO : ON = (3 + 2 \cdot 5) : (2 \cdot 3 + 5) = 13 : 11$

$\therefore \quad ON = 6 \times \frac{11}{(13 + 11)} = 2.75$ cm

Problem 34 $ABCD$ is a trapezium such that $AB : DC = 3:5$ and $AB || CD$. If diagonals intersect at point O and area of $\Delta AOB = 210$ cm^2, then area of $\Delta BOC = ?$

Sol. Clearly, area ΔAOB : area $\Delta BOC = AB : DC$

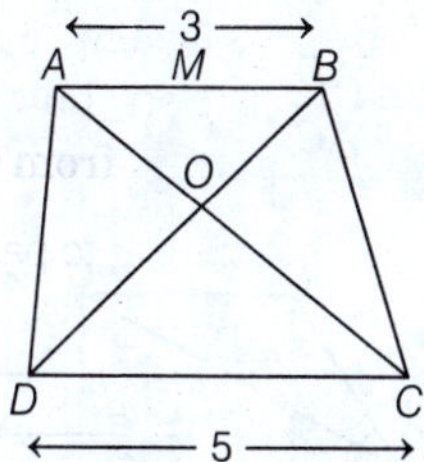

$\Rightarrow$ 210 : area of $\Delta BOC = 3:5$

Hence, area of $\Delta BOC = 350$ cm^2

Work Out

1. In a trapezium $ABCD$, $AB || DC$ and $AB = 5$ cm and $DC = 8$ cm. Find the length of the line joining mid-points of unparallel sides.

(a) 13 cm (b) 9 cm (c) 6.5 cm (d) 8 cm

2. A quadrilateral $ABCD$ is inscribed in a circle. If AB is parallel to CD and $AC = BD$, then the quadrilateral must be a ***[CDS 13]***

(a) parallelogram (b) rhombus
(c) trapezium (d) None of the above

3. In trapezium $ABCE$, $AB || EC$. The mid-point of longer side EC is D, which is joined to vertices A and B so as to form a parallelogram $ABCD$. What is the area of $ABCE$ if area of ΔABD is 57 cm^2.

(a) 114 cm^2 (b) 171 cm^2
(c) 85.5 cm^2 (d) None of these

4. $ABCD$ is a trapezium such that $AB = CD$ and $AD || BC$. $AD = 5$ cm, $BC = 9$ cm. If area of $ABCD$ is 35 sq cm, then CD is ***[SSC (CGL-T-I)-12]***

(a) $\sqrt{28}$ cm (b) 15 cm (c) $\sqrt{29}$ cm (d) 64 cm

5. In the given figure, if M and N are mid-points of diagonals of the trapezium $ABCD$ and $AB = 5$ cm and $DC = 12$ cm, then find the value of MN. ***[SSC (CGL-T-II)-20]***

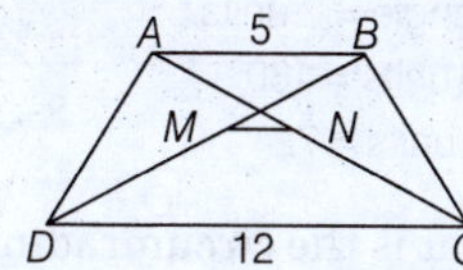

(a) 2.4 cm (b) 1.5 cm (c) 3.6 cm (d) 3.5 cm

6. What is the height of the given trapezium($AB || DC$)?

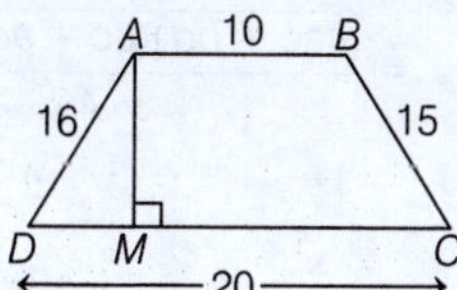

(a) 12.4 cm (b) 15.2 cm (c) 16.66 cm (d) 14.6 cm

7. In the figure given, $ABCD$ is a trapezium, EF is parallel to AD and BC. $\angle y$ is equal to: ***[CDS-13]***

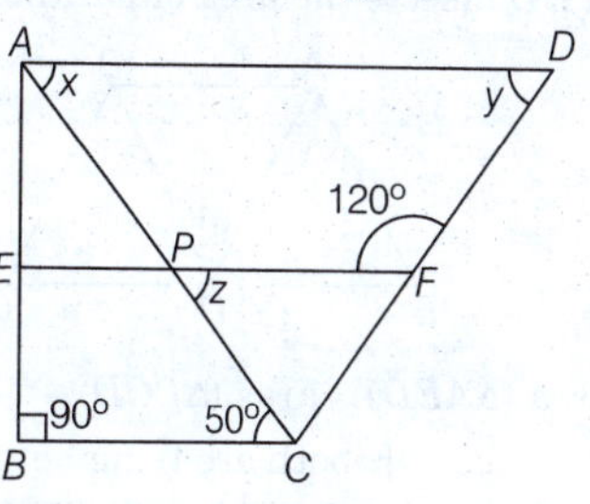

(a) 30° (b) 45° (c) 60° (d) 65°

8. In the trapezium $ABCD$, the sides AB and CD are parallel. The value of $\frac{\sin \angle BAC}{\sin \angle BAD}$ is : ***[XAT-19]***

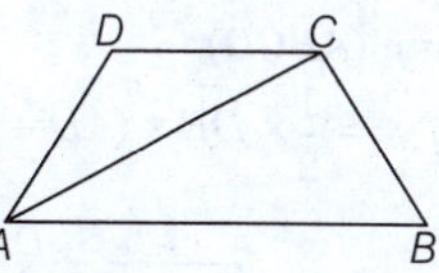

(a) $\frac{AB}{CD}$ (b) $\frac{AC}{CD}$
(c) $\frac{AD}{AC}$ (d) $\frac{BC}{AD}$
(e) $\frac{AC}{AD}$

9. In a trapezium $ABCD$, AB is paralled to DC, BC is perpendicular to DC and $\angle BAD = 45°$. If $DC = 5$ cm, $BC = 4$ cm, the area of the trapezium in sq. cm is : ***[CAT-20]***

(a) 48 (b) 28 (c) 40 (d) 42

Answer Key

1. (c)	**2.** (c)	**3.** (b)	**4.** (c)	**5.** (d)	**6.** (d)
7. (c)	**8.** (c)	**9.** (b)			

Hints and Solutions

1. Let E and F be mid-points of unparallel sides AD and BC, respectively.

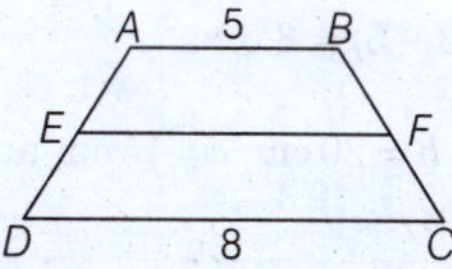

Then, $EF = \frac{1}{2}(AB + DC)$

$\Rightarrow$ $EF = \frac{1}{2}(5 + 8) = 6.5$ cm

2. Parallelogram is not possible as its diagonals are not necessarily equal. In case of rhombus, diagonals are not equal. In trapezium (isosceles trapezium), diagonals are equal. Further, isosceles trapeziums are cyclic quadrilaterals. It is also given that one pair of sides is parallel.

Hence, $ABCD$ must be a isosceles trapezium.

$\therefore$ Option (c) is correct.

3. Here, area (ΔBDC) = area (ΔABD)

[$\because$ diagonal BD bisects the area of parallelogram $ABCD$]

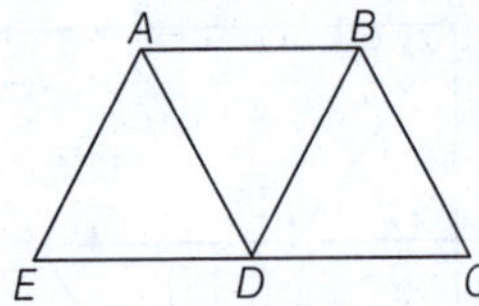

Further, area (ΔAED) = area (ΔBCD)

[$\because$ both are lying between same parallel lines and having equal bases $(ED = DC)$]

Hence, area (ΔBDC) = area (ΔABD) = area $(\Delta AED) = 57$

Then, area of trapezium $ABCE$ = area (ΔBDC) + area (ΔABD) + area of (ΔAED)

$= 3 \times 57 = 171 \text{cm}^2$

4. Area of trapezium $(ABCD)$

$= \frac{1}{2} \times DE \times (AD + BC)$

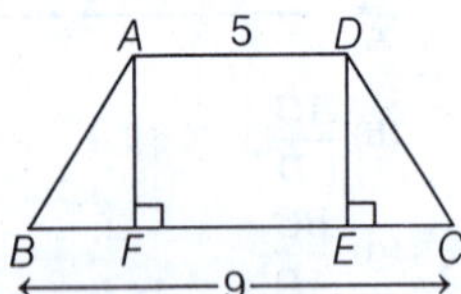

$\Rightarrow \quad 35 = \frac{1}{2} \times DE \times (5 + 9)$

$\Rightarrow \quad DE = 5$

Here, $\Delta ABF \cong \Delta DEC$

[$\because AB = DC$ (given), $AF = DE$ (altitudes) by RHS criterion]

Then, $\quad BF = EC$

Clearly, $\quad AD = FE$

$\therefore \quad BF = EC = \frac{1}{2}(9 - 5) = 2$

Now, in ΔDEC, $DC = \sqrt{DE^2 + EC^2}$

$= \sqrt{5^2 + 2^2} = \sqrt{29}$ cm

5. $MN = \frac{1}{2}(DC - AB)$

$= \frac{1}{2}(12 - 5) = 3.5$ cm

6. Here, $a = 10$ cm, $b = 20$ cm, $c = 15$ cm and $d = 16$ cm

Then, height (AM)

$$= \frac{\sqrt{(-a + b + c + d)(a - b + c + d)(a - b + c - d)(a - b - c + d)}}{2|b - a|}$$

where, c and d are legs of the trapezium

$$= \frac{\sqrt{(-10 + 20 + 15 + 16)(10 - 20 + 15 + 16)(10 - 20 + 15 - 16)(10 - 20 - 15 + 16)}}{2|20 - 10|}$$

$$= \frac{\sqrt{41 \times 21 \times (-11)(-9)}}{2|20 - 10|} = \frac{291.95}{20} = 14.6 \text{ cm}$$

7. In a trapezium $ADFP$, $AD \parallel PF$,

then $\angle y + 120° = 180°$

[$\because$ sum of interior angles on the same side of the transversal is supplementary]

$\Rightarrow \quad \angle y = 60°$

8. $\frac{\sin \angle BAC}{\sin \angle BAD} = \frac{\frac{h}{AC}}{\frac{h}{AD}} = \frac{AD}{AC}$

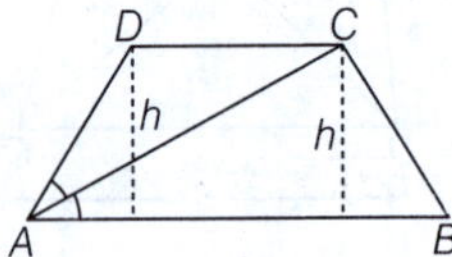

9. Draw $DE \perp AB$.

In ΔADE, $\tan 45° = \frac{ED}{AE}$

$\Rightarrow \quad AE = ED = 4$

Now, required area

= Area of rectangle □ $EBCD$ + Area (ΔAED)

$= (5 \times 4) + \frac{1}{2} \times 4 \times 4 = 28$ sq. cm

Problems based on cyclic quadrilaterals

Problem 35 The three successive angles of a cyclic quadrilateral are in the ratio 1 : 3 : 4, find the measure of the fourth angle.

Sol. In a cyclic quadrilateral, the sum of opposite angles is 180°.

Then, (1 + 4) units = 180°

$\therefore$ 3 units = 108°

Then, measure of the fourth angle

= 180° − 108° = 72°

Alternatively

(1 + 4) units = 180° = (3 + ?) units

$\therefore$ Fourth angle = 2 units

$\because$ (1 + 4) units = 180°

$\therefore$ 2 units = 72°

Problem 36 What is the circumradius of a quadrilateral with consecutive sides 10 cm, 15 cm, 20 cm and 25 cm ?

Sol. Let sides $a = 10$ cm, $b = 15$ cm, $c = 20$ cm and $d = 25$ cm

Area of a cyclic quadrilateral

$$= \frac{\sqrt{(ac + bd)(bc + ad)(ab + cd)}}{4R}$$

where, R is circumradius

$$\Rightarrow \quad \sqrt{(s-a)(s-b)(s-c)(s-d)} \quad \left(\text{where, } s = \frac{a+b+c+d}{2}\right)$$

$$= \frac{\sqrt{(ac+bd)(bc+ad)(ab+cd)}}{4R}$$

$$\Rightarrow \sqrt{(35-10)(35-15)(35-20)(35-25)}$$

$$= \frac{\sqrt{(200+375)(300+250)(150+500)}}{4R}$$

$$\Rightarrow \sqrt{25 \times 20 \times 15 \times 10}$$

$$= \frac{\sqrt{(200+375)(300+250)(150+500)}}{4R}$$

$$\Rightarrow \quad 50\sqrt{30} \times 4R = 14337.45$$

$$\Rightarrow \quad R = 13.1 \text{ cm}$$

Alternatively

$$\text{Diagonal } AC = \sqrt{\frac{(ad+bc)}{(ab+cd)} \times (ac+bd)}$$

$$= \sqrt{\frac{(10 \cdot 25 + 15 \cdot 20)}{(10 \cdot 15 + 20 \cdot 25)} \times (10 \cdot 20 + 15 \cdot 25)} = 22 \text{ (approx.)}$$

Now, find the circumradius of ΔABC

[Hint : area = product of sides/$4R$].

The circumradius of the ΔABC will be the circumradius of the quadrilateral .

Problem 37 A cyclic quadrilateral $ABCD$ is such that $AB = BC$, $AD = DC$, $AC \perp BD$, $\angle CAD = \theta$. Then, Find the value of $\angle ABC$.

Sol. In ΔADC, $\angle DAC = \angle DCA$ $(\because AD = DC)$

Then, $\angle ADC = 180° - \angle DAC - \angle DCA$

$= 180° - \theta - \theta = 180° - 2\theta$

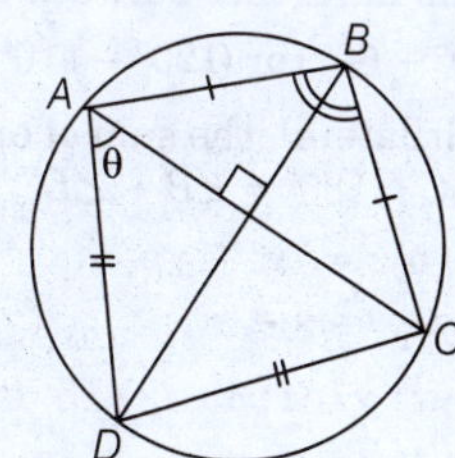

Further, $\angle ADC + \angle ABC = 180°$

[$\because$ sum of opposite angles in a cyclic quadrilateral is 180°]

$\Rightarrow 180° - 2\theta + \angle ABC = 180°$

$\Rightarrow \quad \angle ABC = 2\theta$

Problem 38 One diagonal of a cyclic quadrilateral with consecutive sides 10 cm, 6 cm, 8 cm and 9 cm, is 13.4 cm, find the another diagonal.

Sol. Since the product of the diagonals of a cyclic quadrilateral is equal to the sum of the products of the opposite sides,

Product of diagonals = $(10 \times 8) + (6 \times 9)$

$\Rightarrow 13.4 \times$ another diagonal = 134

Hence, another diagonal = 10 cm

Problem 39 $ABCD$ is a cyclic quadrilateral. The side AB is extended to E in such a way that $BE = BC$. If $\angle ADC = 70°$, $\angle BAD = 95°$, then $\angle DCE$ is equal to

Sol. We know, if a side of a cyclic quadrilateral is produced, then the exterior angle is equal to the interior opposite angle.

Hence, $\angle EBC = 70°$

Now, in ΔBEC, $EB = BC$

$\therefore \quad \angle BCE = \angle BEC$

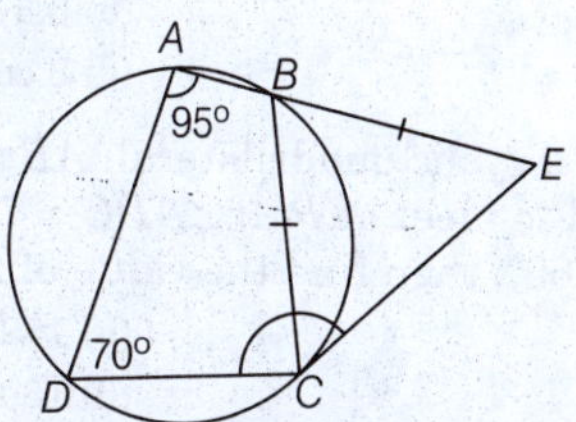

Clearly, $\angle EBC + \angle BCE + \angle BEC = 180°$

$\Rightarrow \quad 70° + \angle BCE + \angle BCE = 180° \Rightarrow \angle BCE = 55°$

Further, $\angle BCD = 180° - 95° = 85°$

[$\because$ in a cyclic quadrilateral, the sum of opposite angles is 180°]

Obviously, $\angle DCE = \angle BCD + \angle BCE = 85° + 55° = 140°$

Problem 40 $PQRS$ is a cyclic quadrilateral in which $PQ = x$, $QR = 16.8$ cm, $RS = 14$ cm, $PS = 25.2$ cm and PR bisects QS. What is the value of x? **[SSC (CGL-T-I)-19]**

Sol. $\angle PQS = \angle PRS$ and $\angle QPR = \angle RSQ$

[$\because$ angles in the same segment of a circle are equal]

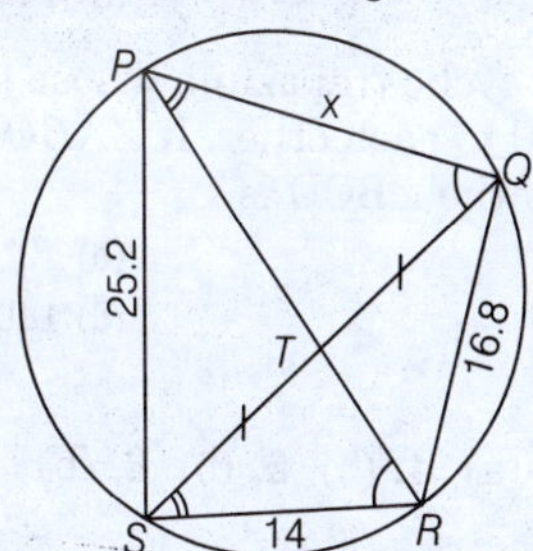

Hence, $\Delta PTQ \sim \Delta STR$

$$\Rightarrow \quad \frac{x}{14} = \frac{PT}{TS}$$

On the same analogy, $\Delta PTS \sim \Delta QTR$

Then, $\dfrac{25.2}{16.8} = \dfrac{PT}{QT} = \dfrac{PT}{TS}$ $[\because QT = TS]$

$$\Rightarrow \quad \frac{3}{2} = \frac{x}{14} \Rightarrow \quad x = 21$$

Work-out

1. *ABCD* is a cyclic parallelogram. The angle *B* is equal to

[SSC (FCI-Gr-III)-12]

(a) 30° (b) 45°
(c) 60° (d) 90°

2. In the given figure, if $AM = 6$ cm, $BM = 8$ cm and $DM = 5$ cm, then find the value of *CM*.

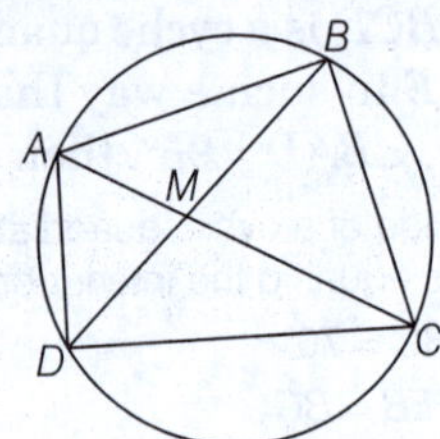

(a) 13 cm (b) 6.67 cm
(c) 6.5 cm (d) 8 cm

3. *ABCD* is a cyclic quadrilateral, *AB* and *DC* are produced to meet at *P*. If $\angle ADC = 70°$ and $\angle DAB = 60°$, then find the value of $\angle PBC + \angle PCB$.

(a) 130° (b) 150°
(c) 155° (d) 180°

4. In a cyclic quadrilateral *ABCD*, $\angle A$ is double its opposite angle and the difference between the other two angles is 1/3rd of $\angle A$. The minimum differences between any two angles of this quadrilateral is

[MAT-10, 16, 19]

(a) 30° (b) 10°
(c) 20° (d) 40°

5. If *ABCD* be a cyclic quadrilateral in which $\angle A = 4x°$, $\angle B = 7x°$, $\angle C = 5y°$ and $\angle D = y°$, then $x : y$ is :

[SSC (CGL-T-II)-14]

(a) 3 : 4 (b) 4 : 3
(c) 5 : 4 (d) 4 : 5

6. *ABCD* is a cyclic trapezium whose sides *AD* and *BC* are parallel to each other. If $\angle ABC = 72°$, then the measure of the $\angle BCD$ is :

(a) 144° (b) 72°
(c) 108° (d) 140°

Answer key

1. (d) 2. (b) 3. (a) 4. (b) 5. (b) 6. (b)

Hints and Solutions

1. We know, the parallelogram inscribed in a circle is a rectangle. Then, $\angle B$ will be equal to 90°.

Alternatively

In a cyclic quadrilateral, the sum of opposite angles is 180°.

$\therefore \quad \angle B + \angle D = 180°$

$\Rightarrow \quad \angle B + \angle B = 180°$

[$\because$ In a parallelogram, opposite angles are equal]

$\Rightarrow \quad \angle B = 90°$

2. We know, if the diagonals *BD* and *AC* of a cyclic quadrilateral *ABCD* intersect at point *M*, then $AM \times CM = BM \times DM$.

Then, $6 \times CM = 8 \times 5$

$\Rightarrow \quad CM = 6.67$ cm

3. In ΔPAD, $\angle P = 180° - 60° - 70° = 50°$

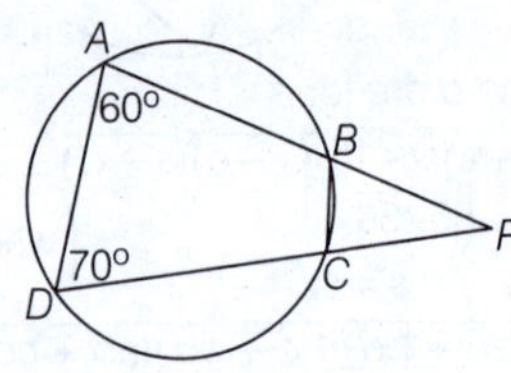

Now, in ΔPBC,

$\angle PBC + \angle PCB = 180° - \angle BPC$

$= 180° - 50° = 130°$

4. As given, $\angle A = 2\angle C$

Also, $\angle A + \angle C = 180°$

[$\because$ sum of opposite angles in a cyclic quadrilateral is 180°.]

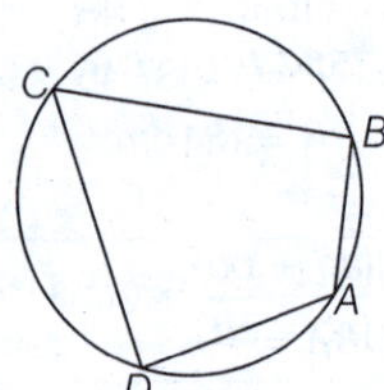

$\Rightarrow \quad \angle A + \frac{\angle A}{2} = 180°$

$\Rightarrow \quad \angle A = 120°$

Then, $\quad \angle B = 60°$

Further, $\quad \angle D - \angle B = \frac{\angle A}{3}$

$\Rightarrow \quad \angle D - \angle B = \frac{120°}{3} = 40° \quad$ ...(i)

and $\quad \angle D + \angle B = 180° \quad$ ...(ii)

From Eqs. (i) and (ii), we get

$\angle D = 110°$ and $\angle B = 70°$

Hence, minimum difference between any two angles

$= (70° - 60°)$ or $(120° - 110°)$ i.e. 10°.

5. In a cyclic quadrilateral, the sum of opposite angles is 180°.
Hence, $\angle A + \angle C = 180° = \angle B + \angle D$

$\Rightarrow \quad 4x° + 5y° = 7x° + y°$

$\Rightarrow \quad 4y° = 3x°$

$\Rightarrow \quad x : y = 4 : 3$

6. In a cyclic quadrilateral, the sum of opposite angles is 180°.

$\therefore \quad \angle B + \angle D = 180°$

$\Rightarrow \quad \angle D = 180° - 72° = 108°$

Further, $AD \parallel BC$, then $\angle D + \angle C = 180°$

[$\because$ sum of interior angles on the same side of the transversal is supplementary]

$\Rightarrow \quad 108° + \angle C = 180° \Rightarrow \angle BCD = 72°$

Problems based on Tangential Quadrilaterals

Problem 41 The sides of a quadrilateral circumscribing about a circle are 10,15, 20 and 25 cm. What is the radius of the quadrilateral?

Sol. We know, area of a tangential quadrilateral = $\sqrt{abcd}$,
where a, b, c and d are four sides.

$\Rightarrow \quad \text{area} = \sqrt{10\times15\times20\times25}$

$\Rightarrow \quad rs = 50\sqrt{30}$,

where r = radius and s = semi- perimeter

$\Rightarrow \quad r\times\dfrac{(10+15+20+25)}{2} = 50\sqrt{30}$

$\Rightarrow \quad r = \dfrac{50\sqrt{30}}{35}$

$= 7.82$ cm

Work-Out

1. A quadrilateral $ABCD$ circumscribes a circle and $AB = 6$ cm, $CD = 5$ cm and $AD = 7$ cm. The length of side BC is **[SSC (10+2/DEO-CDC)-13, SSC (CGL-T-I)-19]**

(a) 4 cm (b) 5 cm
(c) 3 cm (d) 6 cm

2. In the given figure, O is centre of the circle. If $\angle OAB = 60°$, $\angle OBA = 70°$, then $\angle COD$ = ?

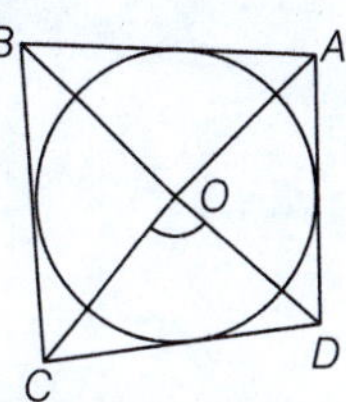

(a) 150° (b) 130° (c) 155° (d) 180°

Hints and Solutions

1. We know, the sum of opposite sides of a tangential quadrilateral is always equal.

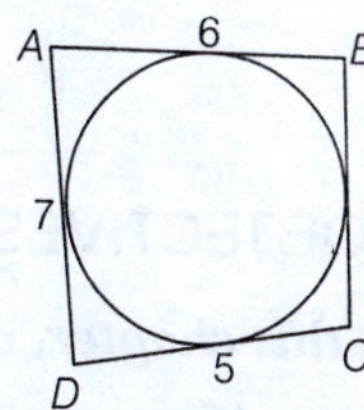

Then, $AB + DC = AD + BC$

$\Rightarrow \quad 6+5 = 7 + BC \Rightarrow BC = 4$ cm

2. In ΔAOB, $\angle AOB = 180°-60°-70° = 50°$

We know, the angles subtended by opposite sides of a tangential quadrilateral are supplementary.

Hence, $\angle AOB + \angle COD = 180°$

$\Rightarrow \quad \angle COD = 180°-50° = 130°$

59

Circle

Learning OBJECTIVES

After completing this chapter, aspirants will be able to :

- Understand the significance of this sub -section in diverse competitive exams;
- Comprehend the definition of circle and its components;
- Know the condition for the congruency of circles;
- Be familiar with the diverse critical terms related to the circle along with associated properties;
- Differentiate between major and minor arc, major and minor segment and major and minor sector of a circle;
- Recognise how the measure of central angle affects the length of an arc as well as area of the sector of the circle;
- Differentiate between the secant and the tangent to a circle;
- Be familiar with fundamental properties associated with the secant and the tangent of a circle;
- Apply the 'Alternate Segment Theorem' in different cases;
- Differentiate between Direct Common Tangent and Transverse Common Tangent;
- Be familiar with the techniques meant for the measure of the common tangent;
- Understand the philosophy of division of 'Illustrations' in two parts;
- Apply the relevant formulae/results according to different situations;
- Realise themselves competent to solve the questions in unique way;
- Analyse the trend of the questions being asked in various competitive examinations;
- Solve the questions in moments;
- Learn many new concepts/fundamentals;
- Apply the pragmatic approaches in lieu of academic approaches; and
- Ponder on the alternative methods.

Inside

- *Introduction*
- *Critical Terms Related to Circle Along with Associated Properties [Arc, Sector, Segment, Chord, Common Chord, Secant, Tangent, Common Tangent]*
- *Illustrations (A) : Based on Properties of Circle (Excluding those Related to Secant and Tangent)*
- *Illustrations (B) : Based on the Concepts Associated with Secant and Tangent*
- ....
- .... *& Many More*
- *Work Out with Solutions*

Introduction

A circle is a set of points on a plane which are at a constant distance (radius) from a fixed point (centre). Only one circle can pass through three or more non-collinear points.

However, an infinite number of circles can be drawn through a given pair of points. The diameter of the circle is the longest distance across a circle and it equals twice the radius.

Here, $O \Rightarrow$ centre, $r \Rightarrow$ radius,
$d \Rightarrow$ diameter of the circle

Area of a circle = πr^2

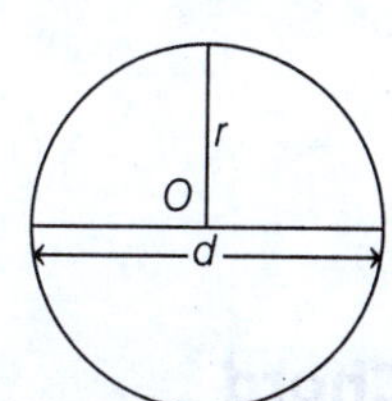

Circumference of the circle
(i.e. distance around a circle) = $2\pi r$

It is also to note that two circles are said to be congruent if they have equal radii. Also, circles having the same centre are known as 'concentric circles'.

Critical terms related to the circle

Arc of a circle

Arc is a part of the circumference of a circle. Any two points on the circumference form the major arc and the minor arc.

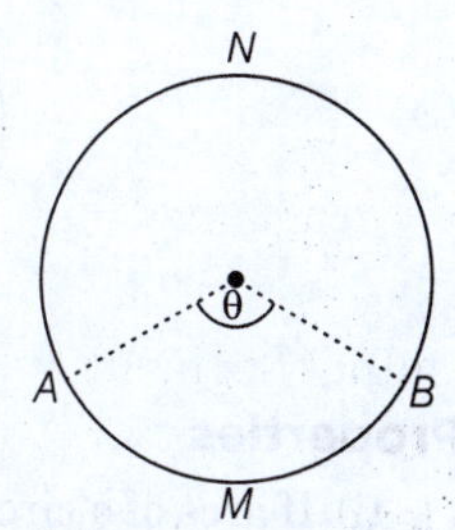

Here, AMB is minor arc whereas ANB is major arc.

Length of arc $AMB = 2\pi r \times \frac{\theta}{360°}$

[$\because$ 360° corresponds circumference $2\pi r$

$\therefore \theta$ will correspond $2\pi r \times \frac{\theta}{360°}$]

and Length of arc $ANB = 2\pi r \times \frac{(360° - \theta)}{360°}$

Properties

(a) Any angle subtended by the major arc at any point of the circle in the alternate segment is obtuse and the angle subtended by the minor arc in the alternate segment is acute.

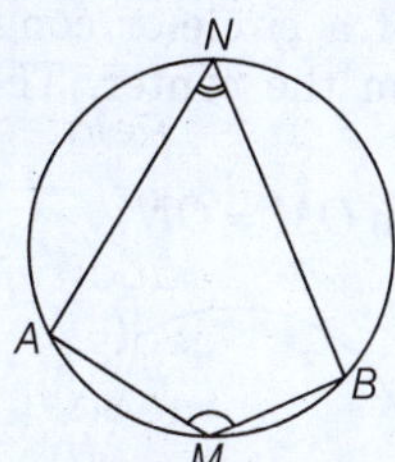

Here, $\angle AMB$ is obtuse and $\angle ANB$ is acute.

Also, $AMBN$ is a cyclic quadrilateral.

Then, $\angle M + \angle N = \angle A + \angle B = 180°$

(b) The angle subtended by an arc at the centre is twice the angle subtended by it at any point of the circle in the alternate segment of the circle. The angle at centre is actually the degree measure of the arc.

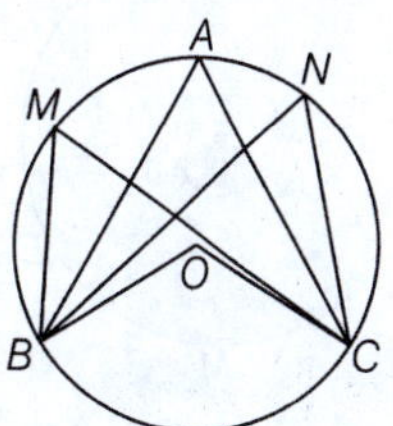

Here, $\angle BOC = 2\angle BAC = 2\angle BMC = 2\angle BNC$

Clearly, $\angle BAC = \angle BMC = \angle BNC$

It implies that angles in the same segment of a circle are equal.

(c) When an arc of a circle subtends a right angle at any point of a circle in the alternate or same segment, the arc is a semi-circle.

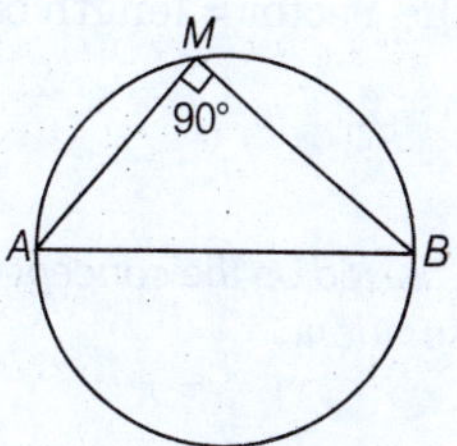

Here, arc AMB is a semi circle if $\angle AMB = 90°$.

Conversely, an angle in a semicircle is a right angle. If MN is a diameter, then $\angle MPN = 90°$.

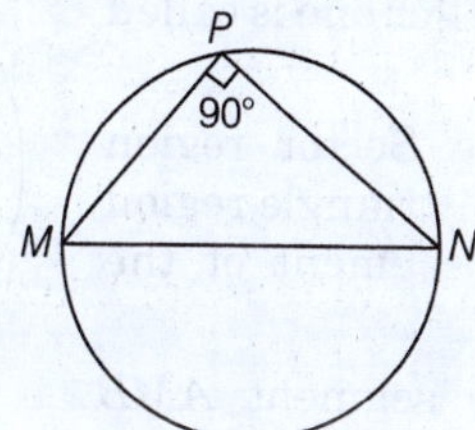

(d) Equal arcs of a circle (or of congruent circles) subtend equal angles at the centre.

If arc AB = arc CD, then $\angle AOB = \angle COD$

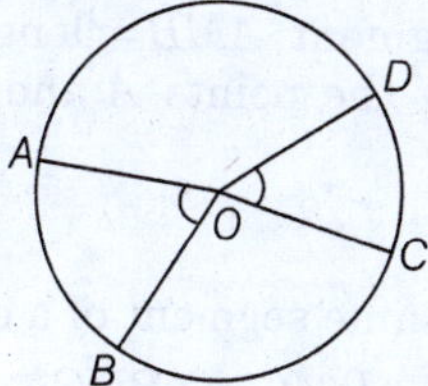

Sector

Sector of a circle is a part of the area of the circle between two radii.

Here, shaded region $OAMB$ is the minor sector and non-shaded region $OANB$ is the major sector.

[Recall that an arc is a part of circumference whereas a sector is a part of area]

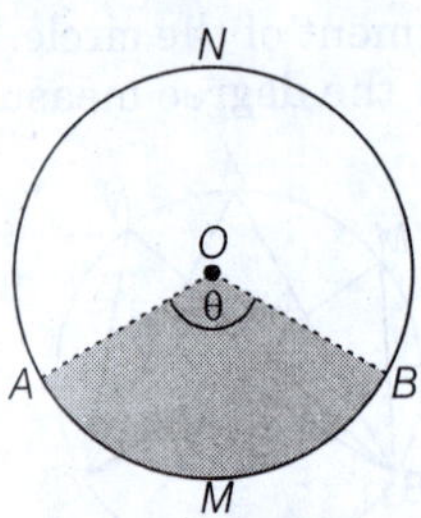

(i) Area of the Sector of a circle $= \dfrac{\pi r^2}{360°}\theta$

[where θ is the angle of the sector]

(ii) Area of Sector of a circle $= \dfrac{1}{2} \times$ length of arc $\times$ radius

(iii) Length of the arc of the sector $= \dfrac{2\pi r}{360°}\theta$

(iv) Perimeter of the Sector = length of the arc $+ 2r$

$$= \frac{2\pi r}{360°}\theta + 2r$$

Remarks *The problems based on the concept of sector have been covered in Mensuration.*

Segment

A line joining any two points of a circle divides the circle into two regions.

The bigger one is called major segment and the smaller one is called minor segment.

In other words, the Sector region formed by two radii – triangle region formed by them = segment of the circle.

Hence, Area of the segment AMB (shaded portion)= area of Sector $OAMB$ – area of ΔOAB

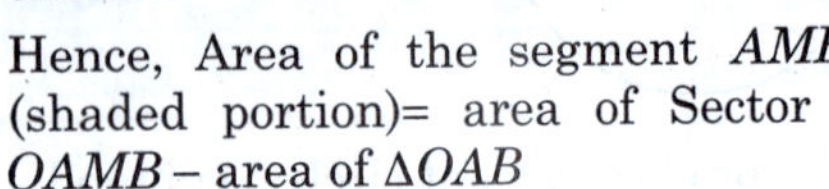

$$= \frac{\pi r^2}{360°}\theta - \frac{1}{2}r^2 \sin\theta$$

and Perimeter of segment AMB= length of arc AMB + length of line joining the points A and B (i.e. length of chord AB)

Properties

(i) Angles in the same segment of a circle are equal.

Here, $\angle BAC = \angle BMC = \angle BNC$

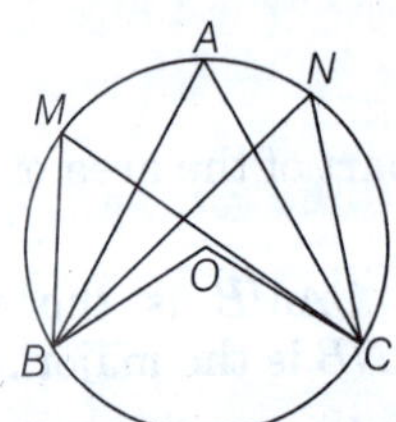

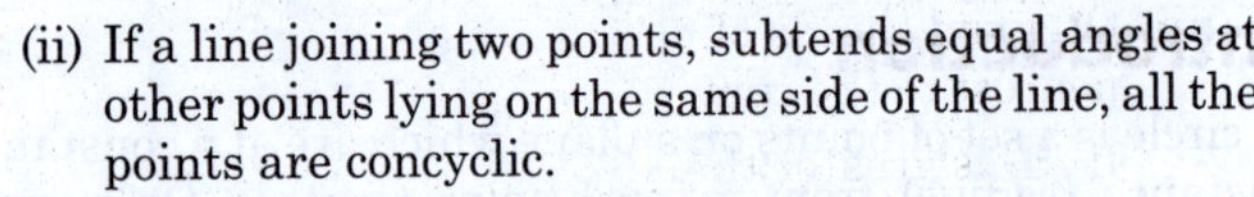

(ii) If a line joining two points, subtends equal angles at other points lying on the same side of the line, all the points are concyclic.

If $\theta_1 = \theta_2$, then points A, B, C and D are concyclic i.e. they all lie on the same circle.

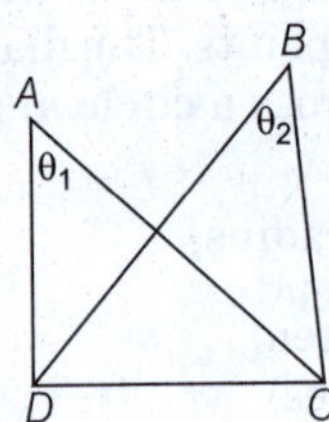

Chord

A line segment whose end points lie on the circle is known as chord of the circle. The diameter is the largest chord of a circle. Length of the chord $AB = 2r \sin\left(\dfrac{\theta}{2}\right)$

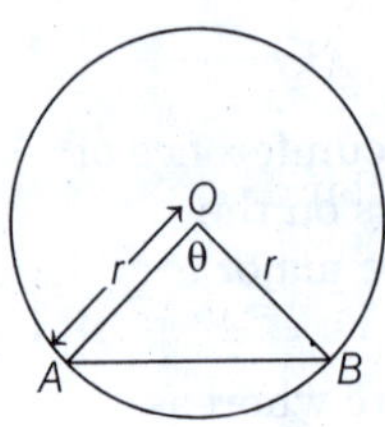

Properties

(i) If arcs of a circle (or congruent circles) are equal, the corresponding chords are equal. The converse is also true.

If arc AMB = arc CND, then chord AB = chord CD

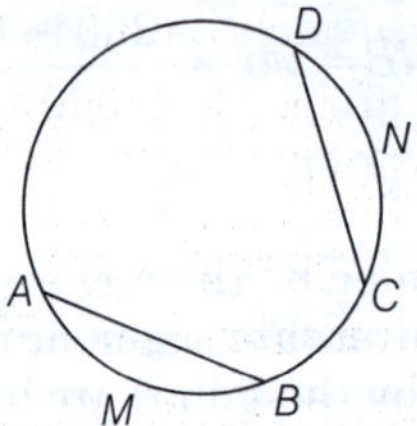

(ii) Equal chords of a circle(or congruent circles) are equidistant from the center. The converse is also true.

If $AB = CD$, then $OM = ON$

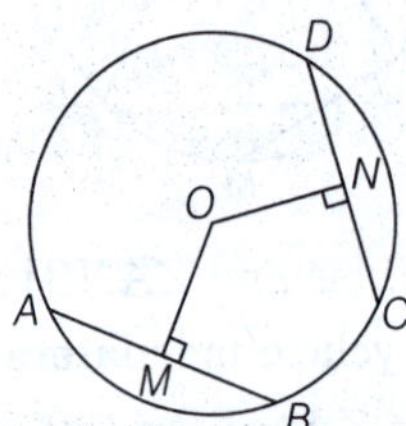

(iii) Of given chords of a circle, the one which is greatest is nearest to the centre. The converse is also true.

If AB is greater than CD, AB is nearer to the centre.

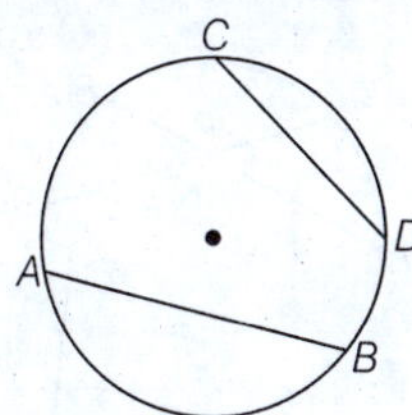

(iv) The angles subtended by equal chords of a circle (or congruent circles) at the center, are equal. The converse is also true.

If $AB = CD$, then $\angle AOB = \angle COD$

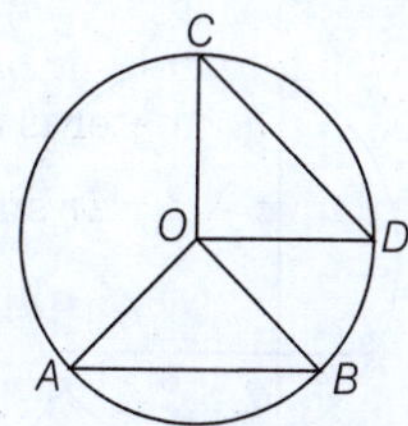

(v) The perpendicular drawn from the centre of a circle to a chord, bisects the chord. The converse is also true.

If $OM \perp AB$, then $AM = MB$

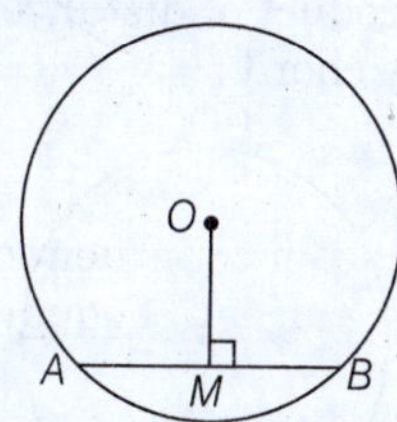

(vi) The perpendicular bisectors of chords of a circle intersect at the centre of the circle.

If OM and ON are perpendicular bisectors of chords AB and CD respectively, they will intersect at center 'O'.

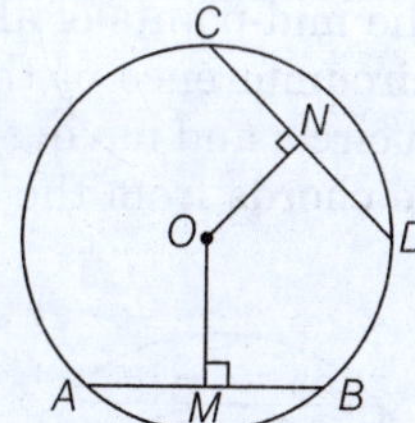

(vii) If two circles intersect each other in two points, then the line joining their centres is perpendicular bisector of their common chord.

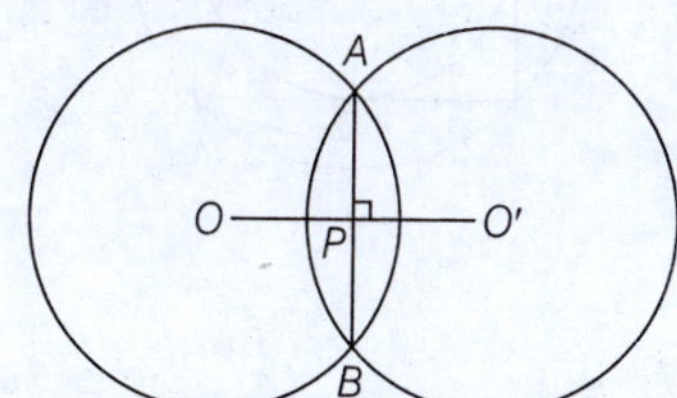

Here, AB is the common chord.

Then, $OO' \perp AB$ and $AP = PB$

Note *If two circles with same radius intersect each other in two points, the line joining their centres and the common chord bisect each other at right-angle.*

(a) Area of $\Delta AOO' = \frac{1}{2} \times OO' \times AP$

$$\Rightarrow \quad AP = \frac{2 \times \text{area } (\Delta\, AOO')}{OO'}$$

We have, $AB = 2 \times AP$

$$= 2 \times \frac{2 \times \text{area } (\Delta\, AOO')}{OO'}$$

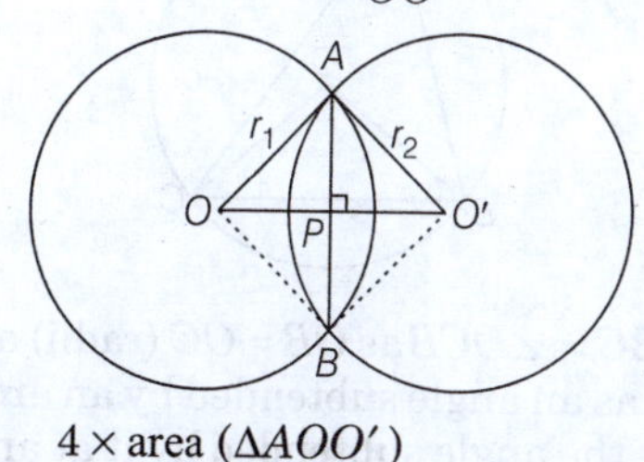

$$= \frac{4 \times \text{area } (\Delta AOO')}{OO'}$$

i.e. length of the common chord

$$= \frac{4 \times \text{area of the } \Delta \text{ having sides } r_1, r_2 \text{ and } d}{d},$$

Where, r_1 and r_2 are radii of circles and d is the distance between their centres.

(b) In case two circles with same radius r pass through each other's centre, then $\Delta AOO'$ will be equilateral.

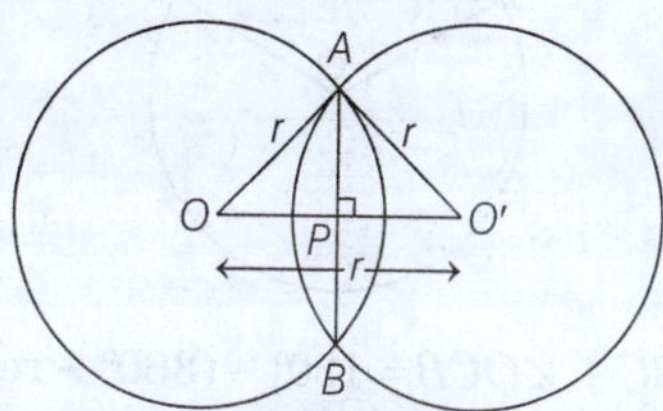

Then, length of the common chord

$$= \frac{4 \times \text{area } (\Delta\, AOO')}{OO'}$$

$$= \frac{4 \times \frac{\sqrt{3}}{4} \times r^2}{r} = \sqrt{3} \cdot r$$

Hence, if two circles with same radius pass through each other's centre, the common chord is equal to $\sqrt{3}$ times the radius.

(viii) If two chords of a circle are equal, then the bisector of angle between them passes through the centre of the circle. The converse is also true.

If $AB = BC$, then BE being angle bisector of $\angle ABC$, passes through centre 'O'.

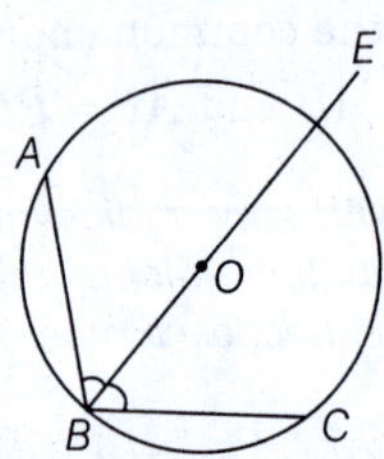

(ix) A circle is symmetrical about any diameter,

(x) In ΔBOC, $\angle OBC + \angle OCB = 180° - \angle BOC$

$\Rightarrow \quad 2\angle OBC = 180° - 2\angle BAC$

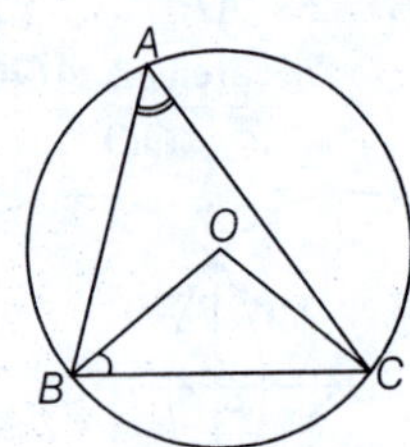

[$\because \angle OBC = \angle OCB$ as $OB = OC$ (radii) and $\angle BOC = 2\angle BAC$, as an angle subtended by an arc at the centre is twice the angle subtended by it at any point of the circle in the alternate segment]

$\Rightarrow \quad \angle BAC + \angle OBC = 90°$

Hence, $\angle A + \angle OBC = 90°$,

when point A is on major arc.

(xi) In ΔBOC,

$\angle OBC + \angle OCB = 180° - \angle BOC$

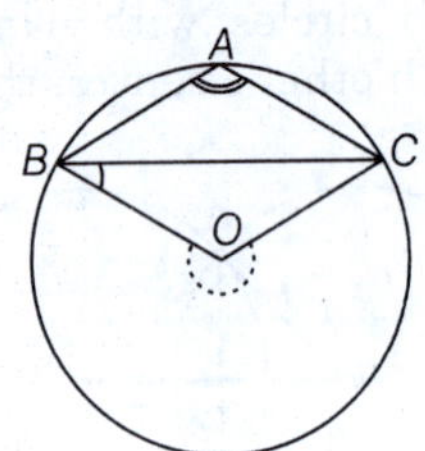

$\Rightarrow \angle OBC + \angle OCB = 180° - (360° - \text{reflex } \angle BOC)$

$\Rightarrow 2\angle OBC = 180° - (360° - 2\angle BAC)$

[$\because \angle OBC = \angle OCB$ as $OB = OC$ (radii) and reflex $\angle BOC = 2\angle BAC$]

$\Rightarrow \quad 2\angle BAC - 2\angle OBC = 180°$

Hence, $\angle A - \angle OBC = 90°$,

when point A is on minor arc.

(xii) If two chords of a circle intersect inside or outsid of the circle when produced, then $PA \times PB = PC \times PD$.

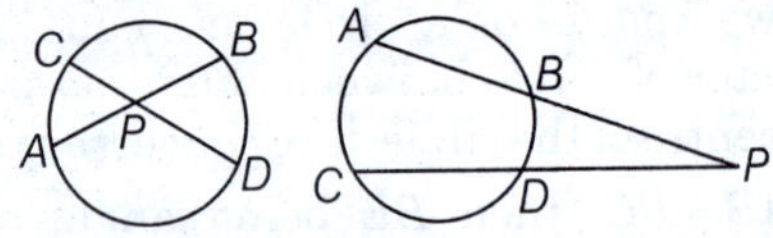

(xiii) Here, $PA \times PB = PC \times PD = r^2 - k^2$, where r is radius and k is distance from centre to the intersection point.

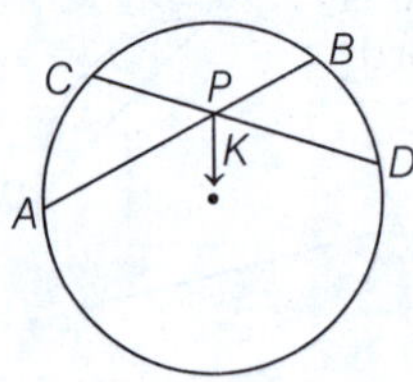

(xiv) If two chords intersect at 90° and the intersection point divides one into lengths a and b and another into lengths c and d, then $a^2 + b^2 + c^2 + d^2 =$ (diameter)^2

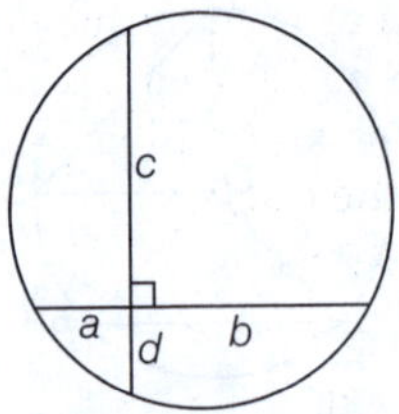

(xv) The product of the diameter and the perpendicular distance from a point on a circle to the given chord is equal to the product of distances from that point to the ends of the chord.

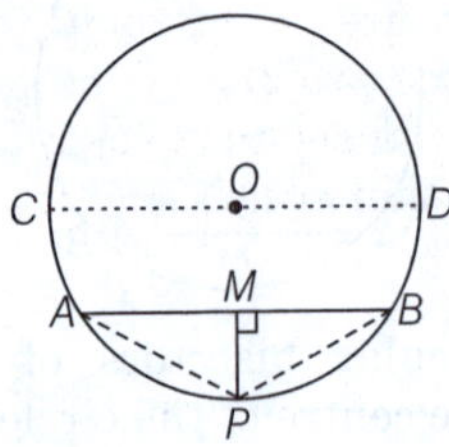

Here, $PM \times CD = PA \times PB$

(xvi) The locus of the mid-points of all equal chords in a circle is the circumference of the circle concentric with the given circle and having radius equal to the distance of the chords from the centre of the given circle.

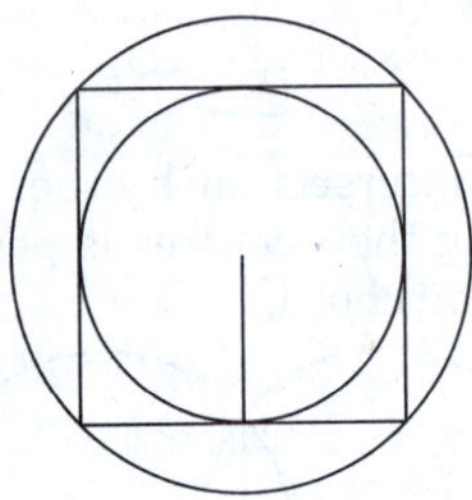

Secant

The line intersecting a circle in two distinct points is called a secant of the circle. The secant of a circle is basically an extension of the chord.

Here, line AB is a secant of the circle.

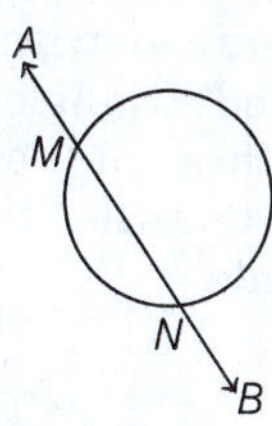

Properties

(i) If two secants of a circle meet outside the circle at point P, then $\angle P$ is equal to half the difference between angles subtended by the enclosed arcs at the centre of the circle.

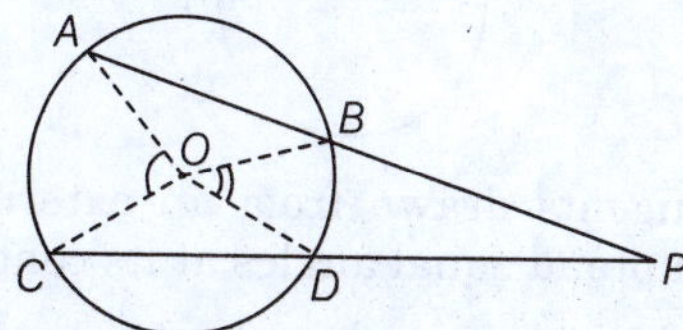

Here, $\angle APC = \frac{1}{2} \times |\angle AOC - \angle BOD|$

(ii) If two secants are drawn to a circle from an exterior point, then the product of the length of one secant and its external segment is equal to the product of length of another secant and its external segment.

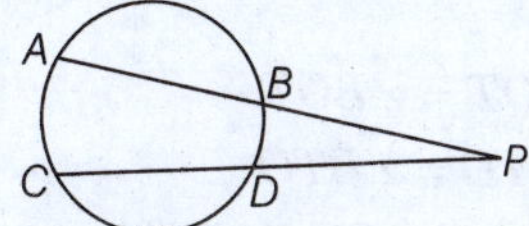

If two secants PBA and PDC are drawn to a circle from an exterior point , P then $PA \times PB = PC \times PD$.

Hence, $(AB + PB) \cdot PB = (CD + PD) \cdot PD$

Tangent

The line intersecting a circle in exactly one point is called tangent to the circle at that point.

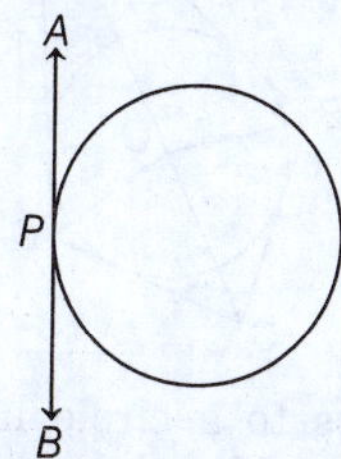

Here, line AB is tangent to the circle at point P. P is called the 'point of contact' of the tangent with the circle.

It is important to note that no tangent can be drawn to a circle through a point lying inside the circle. If a point lies on the circle, only one tangent to the circle can be drawn through that point. In case a point is outside the circle, two tangents can be drawn.

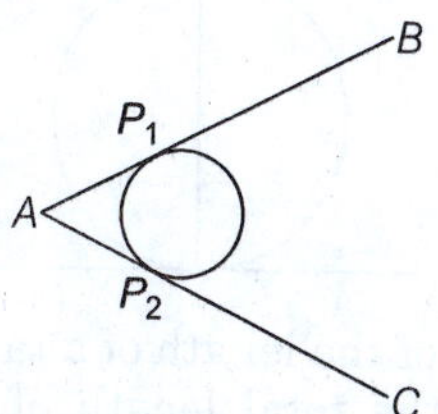

Here, AB and AC are two tangents drawn from a point P lying outside the circle. Their points of contact are P_1 and P_2 respectively.

Properties

(i) The tangent at any point of a circle is perpendicular to the radius through the point of contact. Conversely, if a line drawn through the end of a radius is perpendicular to the radius, it is a tangent to the circle. Here, tangent AB is perpendicular to the radius r through point of contact P.

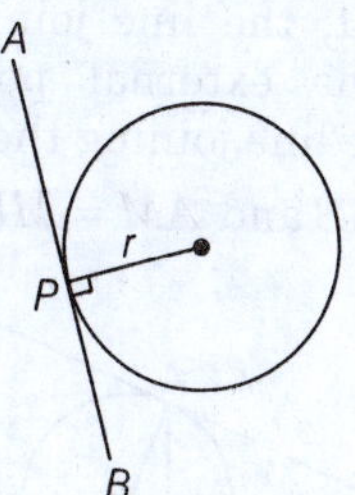

(ii) The tangents drawn from an external point to a circle are equal in length.

Here, $PA = PB$

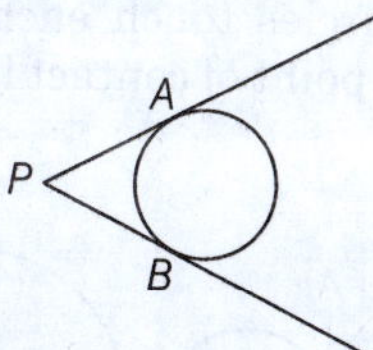

(iii) The centre of the circle lies on the bisector of the angle between the two tangents drawn to it from an external point.

Here, the center 0 lies on the bisector of $\angle APB$.

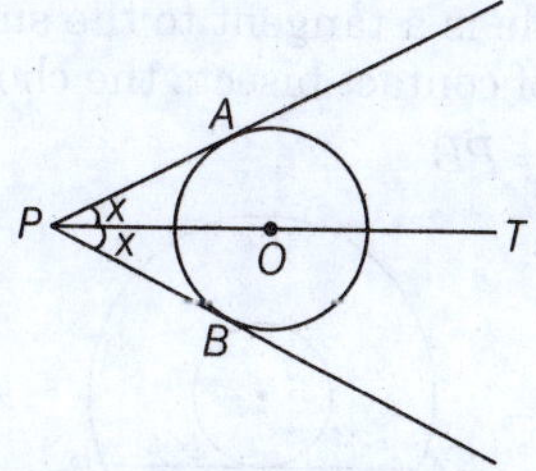

(iv) The tangents with end-points of the diameter of a circle as points of contact are parallel.

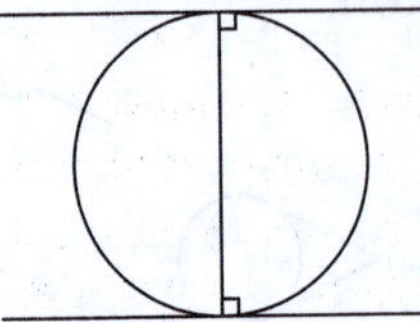

(v) The square of the length of a tangent is equal to the product of the total length of the secant and the external secant segment, if both the secant and the tangent are drawn from the same exterior point.

If PAB is a secant to the circle and PT is a tangent, then $PA \times PB = PT^2$

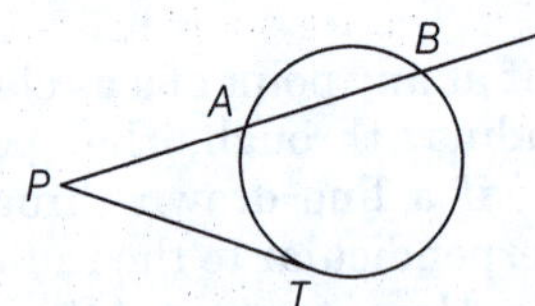

(vi) If two tangents to a circle are drawn from an external point, the line joining the centre of the circle and the external point is perpendicular bisector of the line joining the points of contact.

Here, $OP \perp AB$ and $AM = MB$

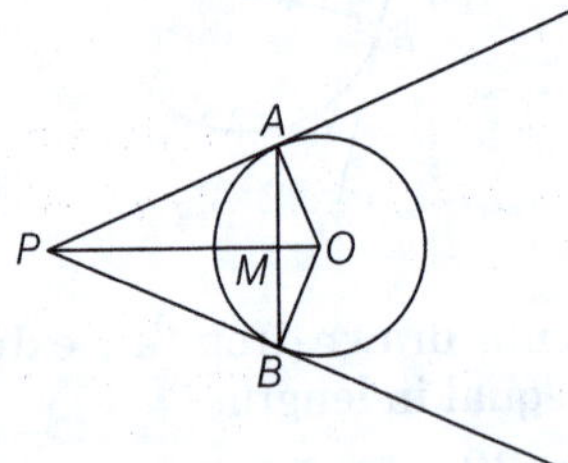

(vii) In case two circles touch each other externally or internally the point of contact lies on the line joining their centres.

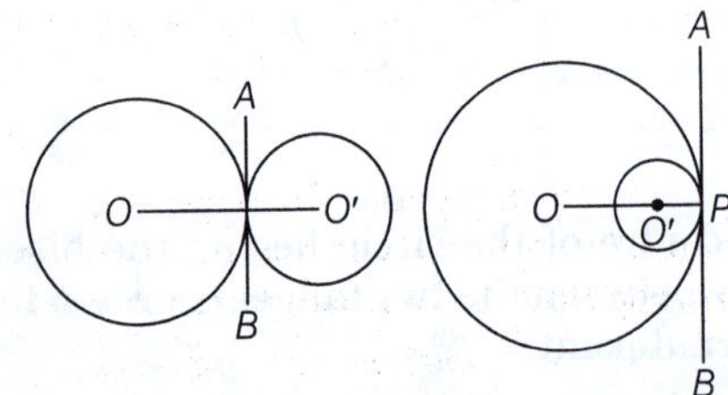

(viii) In case of two concentric circles, if a chord of the bigger circle is a tangent to the smaller circle, then the point of contact bisects the chord.

Here, $AP = PB$

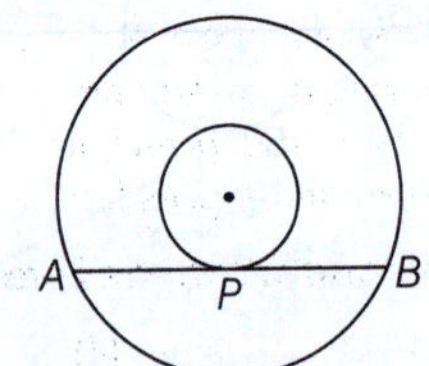

Also, length of $AB = 2\sqrt{a^2 - b^2}$,

where a and b are radii of the bigger and the smaller circles respectively. It also holds that all chords of the bigger circle which are tangents to the smaller circle are equal in length.

(ix) When a circle is drawn with one perpendicular side of a right-angled triangle as its diameter, the tangent drawn through the point of intersection of the hypotenuse and the circle bisects the another perpendicular side.

Here, $PN = NQ$

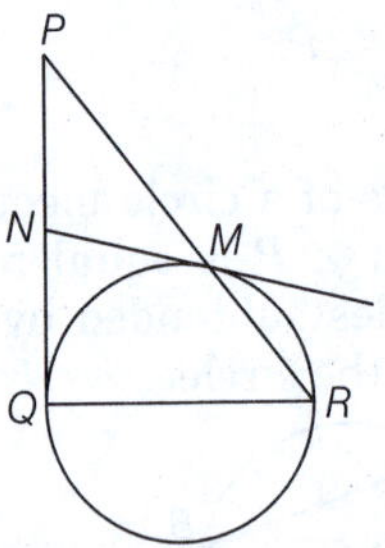

(x) The tangents drawn from an external point to a circle subtend equal angles at its centre.

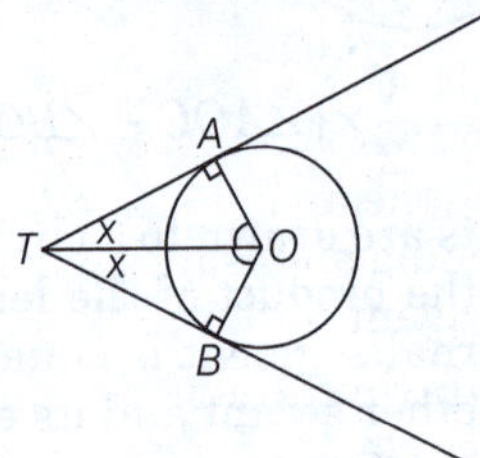

Here, $\angle AOT = \angle BOT$

Also, $\angle ATO = \angle BTO$

[$\because OT$ is the bisector of $\angle ATB$]

(xi) If all four sides of a quadrilateral are tangents to a circle, then the angles subtended at its centre by the opposite sides are supplementary.

Here$\angle AOB + \angle COD = \angle BOC + AOD = 180°$

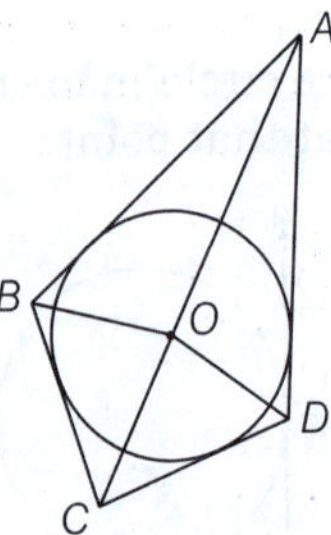

(xii) If two tangents to a circle having centre 0, with points of contact M and N, intersect at an exterior point P, then all the four points i.e. M, O, N and P are concyclic. Since M, O, N and P are concyclic,

$\angle OMP + \angle ONP = \angle MON + \angle MPN = 180°$

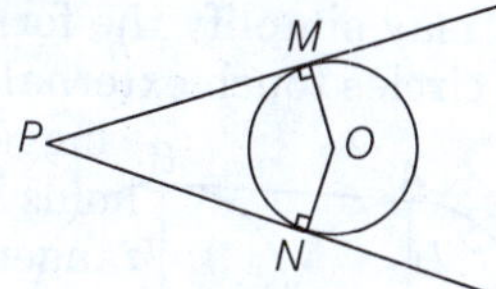

(xiii) If a chord is drawn through the point of contact of a tangent to a circle, the angles made by this chord with the given tangent are equal respectively to the angles subtended by the chord in the corresponding alternate segments. In brief, the angle between a tangent and a chord at the point of contact is equal to the angle in the alternate segment.

Here, $\angle BPQ = \angle PNQ$ and $\angle APQ = \angle PMQ$

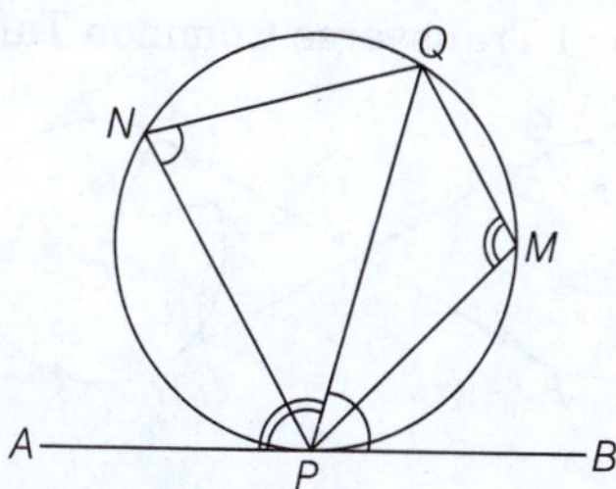

This is known as'Alternate Segment Theorem' or 'Tangent-Chord Theorem'.

Conversely, if the angles made by a chord of a circle with a line drawn through its end point are equal respectively to the angles subtended by the chord in the corresponding alternate segments, then the line drawn is a tangent to the circle.

(xiv) If a chord is drawn through the point of contact of a tangent to a circle, the acute (or right) angle made by this chord with the given tangent is half the angle subtended by the chord at the centre of the circle.

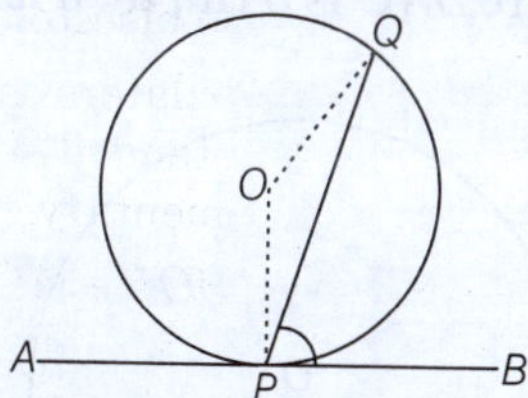

Here, acute $\angle QPB = \frac{1}{2}\angle POQ$ but obtuse

$\angle APQ \neq \frac{1}{2}\angle POQ$

Concept of Common Tangent

(i) A line that is tangent to two circles is called 'common tangent' to the circles. Now, observe number of common tangents in different cases.

Case	Diagram of position of two circles	No. of common tangents	Desriptions/Facts
(i)		0	One circle lies completely inside the other without having common point. No common tangent is possible in this case
(ii)		1	Two circles touch internally. There is only one common point.
(iii)		2	Two circles interect each other in two distinct points. i.e. they have two common points.
(iv)		3	Two circles touch each other externally. In this case, there is only one common point
(v)		4	One lies outside the other without touching or intersecting. There is no common point.

Note *If two circles have more than two common points, then both are the same circle.*

In cases (iii), (iv) and (v), common tangents AB and CD are equal and they are known as **'Direct Common Tangents'**. In case (v), common tangents EF and GH are also equal and known as **'Transverse Common Tangents'**. It is clear from the diagram associated with case(v) that transverse common tangents are possible only when distance between centres of two circles is greater than sum of their radii.

(ii) Direct Common Tangents (DCT) meet on the line joining centres of the circles.

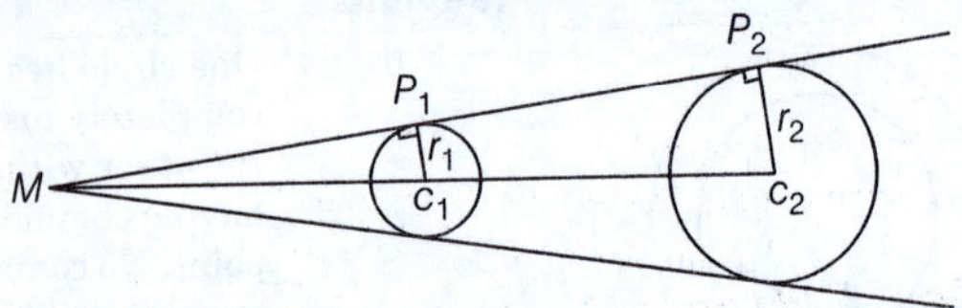

Here, DCTs meet at point M. It also holds that

$$\frac{MC_1}{MC_2} = \frac{MP_1}{MP_2} = \frac{r_1}{r_2} \quad [\because \Delta MC_1P_1 \sim \Delta MC_2P_2]$$

It may also be said that the point of intersection of DCTs divides each tangent as well as the line joining the centres of the circles externally in ratio $r_1 : r_2$. In other words, the point of intersection of the Direct common tangent(s) and the line joining the centres of the circles divides the line as well as the tangent(s) externally in ratio $r_1 : r_2$.

(iii) Transverse Common Tangents (TCT) also meet on the line joining centres of the circles.

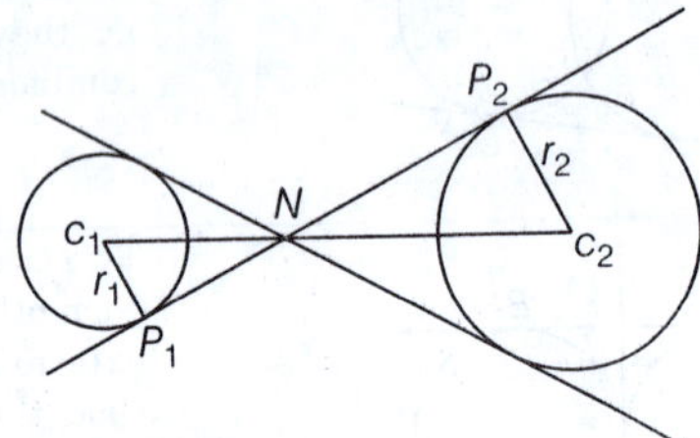

Here, TCTs meet at point N. It also holds that

$$\frac{NC_1}{NC_2} = \frac{NP_1}{NP_2} = \frac{r_1}{r_2} \quad [\Delta NC_1P_1 \sim \Delta NC_2P_2]$$

It may also be said that the point of intersection of TCTs divides each tangent as well as the line joining the centres of the circles internally in ratio $r_1 : r_2$. In other words, the point of intersection of the transverse common tangent(s) and the line joining the centres of the circles divides the line as well as the tangent(s) internally in ratio $r_1 : r_2$.

(iv) Length of Common Tangent

(a) Length of Direct Common tangent

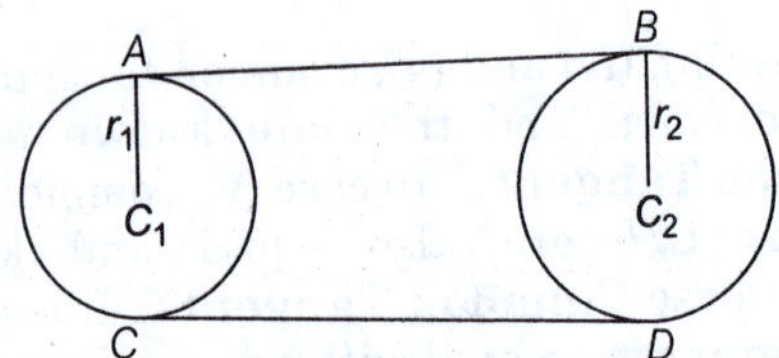

$$AB = CD = \sqrt{\begin{array}{l}\text{(distance between centres)}^2 - \\ \text{(difference between radii)}^2\end{array}}$$

$$\Rightarrow \quad AB = CD = \sqrt{(c_1c_2)^2 - (r_1 - r_2)^2}$$

Note *This formula is applicable for all three cases, viz. cases (iii), (iv) and (v).*

However, we may simplify the formula for case(iv) in which two circles touch externally.

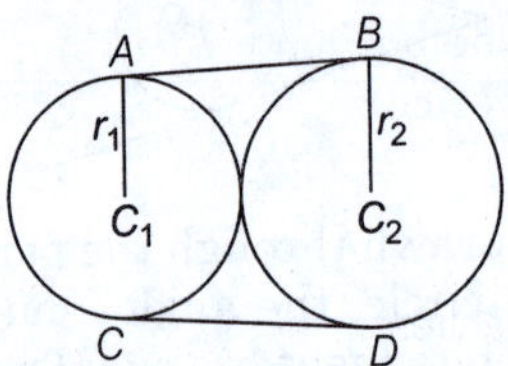

$$AB = CD = \sqrt{(c_1c_2)^2 - (r_1 - r_2)^2}$$
$$= \sqrt{(r_1 + r_2)^2 - (r_1 - r_2)^2}$$
$$= \sqrt{4.r_1.r_2}$$

(b) Length of Transverse Common Tangent

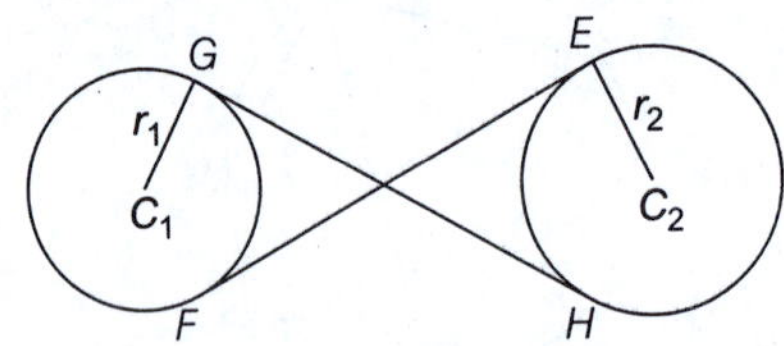

EF = GH

$$= \sqrt{\text{(distance between centres)}^2 - \text{(sum of radii)}^2}$$
$$= \sqrt{(c_1c_2)^2 - (r_1 + r_2)^2}$$

Illustrations: Part (A)

Based on properties of a circle (excluding those related to secant and tangent)

Problems Based on 'Length of the Arc'

Problem 1 In the given figure, OAC is a triangle right angled at O, the centre of the circle and $\angle AOB = 75°$. If the length of arc BNC is 3 cm, then length of the arc AMB is equal to

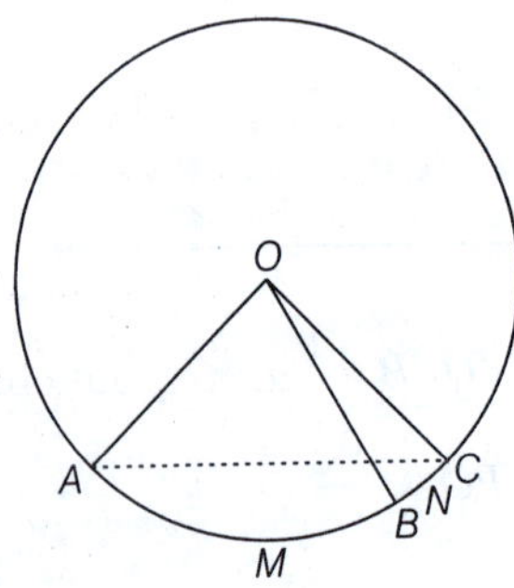

Sol. Obviously, $\angle BOC = \angle AOC - \angle AOB = 90° - 75° = 15°$

$\because$ 15° is corresponding to 3 cm

$\therefore$ 75° will correspond to 15 cm

Hence, length of the arc AMB = 15 cm.

[Alternatively, you may say that $2\pi r \times \frac{15°}{360°} = 3$

$\therefore \quad 2\pi r \times \frac{75°}{360°} = 15$]

Problem 2 In the given figure, two circular curves create 60° and 90° angles with their respective centres. If the length of the bottom curve Y is 10π, then find the length of the other curve. **[XAT-16]**

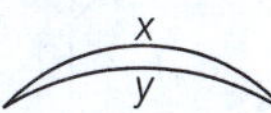

Sol. Let O and A be centres of the bigger and smaller circles and their radii are R and r respectively.

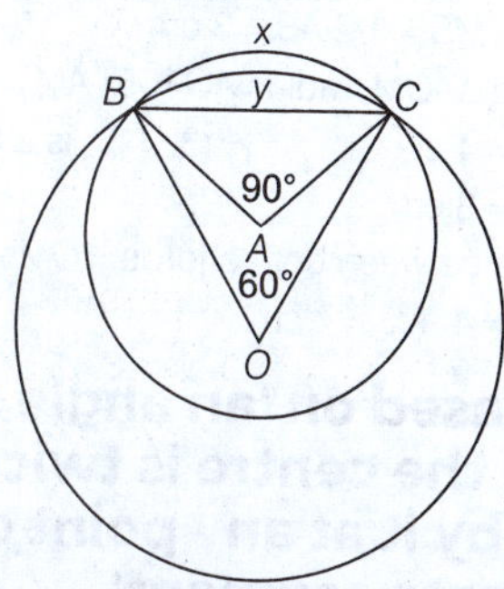

As given, $\angle BOC = 60°$ and $\angle BAC = 90°$

Length of the arc BYC of the bigger

circle $= 2\pi R \times \frac{60°}{360°}$

$\Rightarrow \quad 10\pi = 2\pi R \times \frac{60°}{360°}$

$\Rightarrow \quad R = 30 \Rightarrow OB = OC = 30$ cm

Further, length of the chord BC of the bigger circle

$= 2R \times \sin(60°/2)$

$= 2 \times 30 \times \sin 30° = 30$ cm

Now, in right-angled ΔABC, $AB = AC$ [radii of the smaller Δ]

Then, $\quad BC^2 = AB^2 + AC^2$

$\Rightarrow \quad 30^2 = AB^2 + AB^2 \Rightarrow AB = 15\sqrt{2}$

i.e. radius of the smaller circle $= 15\sqrt{2}$ cm

Then, length of the arc BXY of the smaller circle

$= 2\pi r \times \frac{90°}{360°} = 2\pi \times 15\sqrt{2} \times \frac{1}{4} = \frac{15\pi}{\sqrt{2}}$ cm

Problems based on 'Equal Chords of a Circle are Equidistant from the Center'

Problem 3 Two chords PQ and MN of a circle with centre O intersect each other at point R. What is the length of chord PQ if $MN = 17$ cm and RO is the bisector of $\angle NRQ$.

Sol. We draw perpendiculars OL and OK to chords MN and PQ respectively.

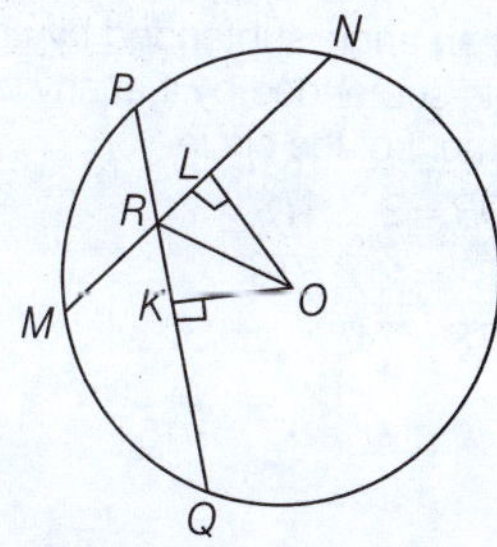

$\Delta RKO \cong \angle RLO$ [by criterion A-A-S as $\angle RKO = \angle RLO = 90°$, $RO = RO$ (common) and $\angle KRO = \angle LRO$ (given)]

$\therefore \quad OK = OL$

Since both the chords are equidistant from the centre, they will be equal in length. Hence, $PQ = MN = 17$ cm.

Problem 4 PQ and MN are two equal chords of a circle with centre O. OA and OB are perpendiculars to PQ and MN respectively. If $\angle AOB = 80°$, then what is $\angle OAB$ equal to ?

Sol.

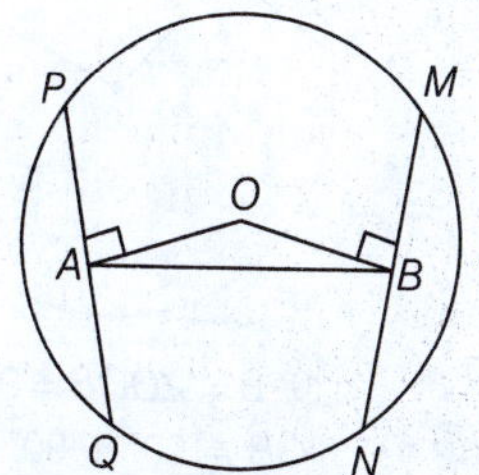

Since both the given chords are equal, they will be equidistant from the centre.

Hence, $\quad OA = OB$.

$\Rightarrow \quad \angle OBA = \angle OAB$

[∵ angles opposite to equal sides are equal]

Now, in ΔOAB,

$\angle OBA + \angle OAB + \angle AOB = 180°$

$\Rightarrow \quad \angle OAB + \angle OAB + 80° = 180°$

$\Rightarrow \quad \angle OAB = 50°$

Problems based on 'a Perpendicular drawn from the Centre to a Chord Bisects it'

Problem 5 O is the circumcentre of ΔABC with circumradius 13 cm. Let $BC = 24$ cm and OD is perpendicular to BC. Then, the length of OD is ?

[SSC (CGL T-I) 12], [SSC (CGL T-II) 19]

Sol. Since, the perpendicular drawn from the centre of a circle to a chord bisects the chord, $BD = DC$

Then, $\quad BD = DC = 1/2\ BC = 12$ cm

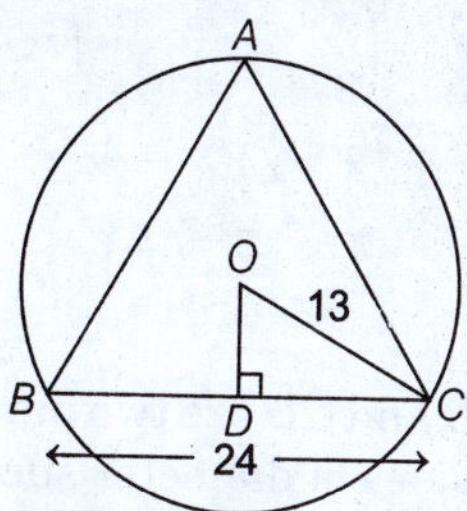

Now, in right-angled ΔODC,

$OD^2 = OC^2 - DC^2$

$\Rightarrow \quad OD = \sqrt{13^2 - 12^2} = 5$

[Note: You may write $OD = 5$ directly without computing as above because (5, 12, 13) is a Pythagorean triplet]

Problem 6 AB and BC are two chords of a circle with centre O. If P and Q are the mid-points of AB and BC respectively, then the quadrilateral $OQBP$ must be

(a) rectangle (b) square
(c) rhombus (d) cyclic

Sol. Here, OP and OQ bisect AB and BC respectively. It implies that $OP \perp AB$ and $OQ \perp BC$.

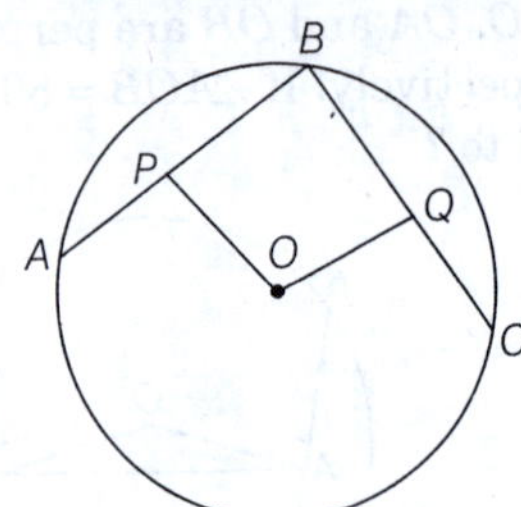

Hence, $\angle OPB = \angle OQB = 90°$

Then, $\angle OPB + \angle OQB = 90° + 90° = 180°$

Since sum of opposite angles of $\square OQBP$ is 180°, the quadrilateral is cyclic.

Problem 7 If the chord of an arc of a circle is of length x, the height of the arc is y and the radius of the circle is z. Then, which one of the following is correct ?

(a) $y(2z - y) = x^2$ (b) $y(2z - y) = 4x^2$
(c) $2y(2z - y) = x^2$ (d) $4y(2z - y) = x^2$ **[CDS-14]**

Sol. Given, radius $OB = z$, chord $AB = x$ and height of the arc $MC = y$. Since, the perpendicular drawn from the centre of a circle to a chord bisects the chord, $AM = MB = x/2$.

In right-angled ΔOMB, $OM^2 + MB^2 = OB^2$

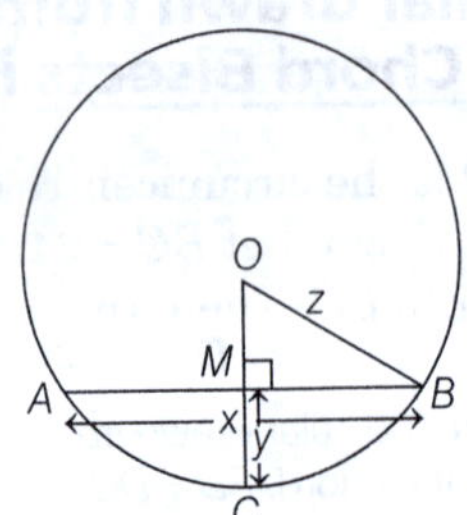

$\Rightarrow \quad (z - y)^2 + \left(\frac{x}{2}\right)^2 = z^2 \quad [\because OC = OB = z \therefore OM = z - y]$

$\Rightarrow \quad 4z^2 + 4y^2 - 8zy + x^2 = 4z^2$

$\Rightarrow \quad -4y^2 + 8zy = x^2$

$\Rightarrow \quad 4y(2z - y) = x^2$

Problem 8 AB and CD are two parallel chords drawn on two opposite sides of diameter such that $AB = 6$ cm and $CD = 8$ cm. If the radius of the circle is 5 cm, the distance between the chords is

Sol. We know that the perpendicular drawn from the centre of a circle to a chord bisects the chord. Hence, $AM = MB$.

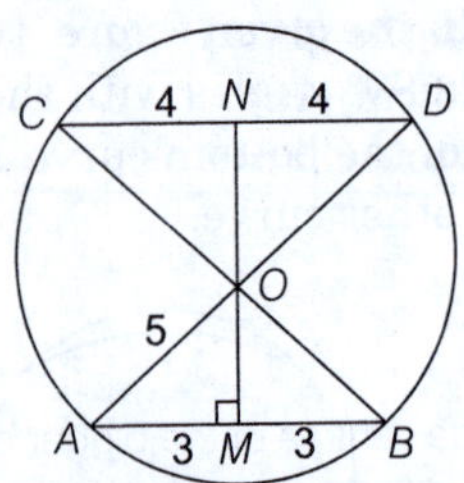

Then, $AM = MB = 1/2\, AB = 3$ cm

In right-angled ΔAOM, radius $AO = 5$, $AM = 3$

Then, $OM = 4$ [$\because$ (3, 4 ,5) is a Pythagorean triplet]

Similarly, $ON = 3$ cm

Now, distance between both chords $= OM + ON$

$= 4 + 3 = 7$ cm

Problems based on 'an angle subtended by an arc at the centre is twice the angle subtended by it at any point of the circle in the alternate segment'

Note *Problems under this heading may contain the properties discussed under preceding headings. The same is true for all the succeeding headings even in Part(B) of the illustrations.*

Problem 9 P and Q are the middle points of two chords (not diameters) AB and AC respectively of a circle with centre at a point O. The line OP and OQ are produced to meet the circle respectively at points R and S. T is any point on the major arc between the points R and S of the circle. If $\angle BAC = 32°$, $\angle RTS = ?$

Sol. Here, OP and OQ bisect AB and AC respectively.

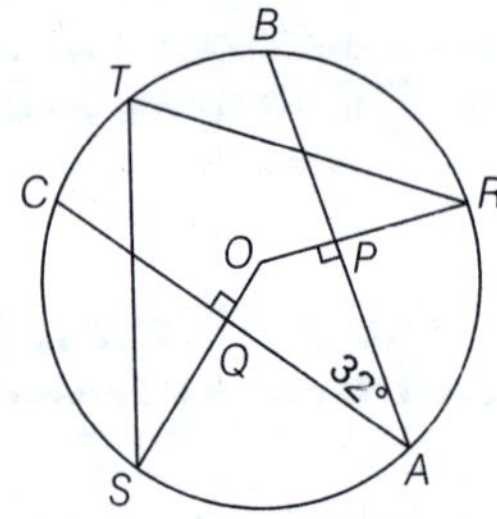

It implies that $OP \perp AB$ and $OQ \perp AC$.

Hence, $\angle OPA = \angle OQA = 90°$

In $\square OPAQ$, $\angle POQ = 360° - 90° - 90° - 32°$

$= 148°$

Hence, $\angle SOR = 148°$

We know that an angle subtended by an arc at the centre is twice the angle subtended by it at any point of the circle in the alternate segment of the circle.

Then, $\angle SOR = 2\angle RTS$

$\Rightarrow \quad 148° = 2\angle RTS$

$\Rightarrow \quad \angle RTS = 74°$

Problem 10 In the given figure, O is the centre of the circle, if $OA = 3$ cm, $AC = 3$ cm and OM is perpendicular to AC, then what is $\angle ABC$ equal to

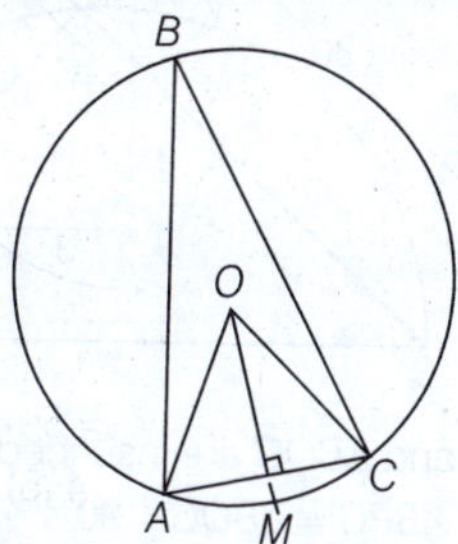

Sol. Since the perpendicular drawn from the centre of a circle to a chord bisects the chord, $AM = MC = AC/2 = 3/2 = 1.5$

In right-angled ΔAOM, $\sin \angle AOM = \frac{AM}{AO} = \frac{1.5}{3} = 1/2$

$\Rightarrow \quad \sin \angle AOM = \sin 30^\circ \Rightarrow \angle AOM = 30^\circ$

Similarly, $\angle COM = 30^\circ$

Then, $\angle AOC = 30^\circ + 30^\circ = 60^\circ$

Further, an angle subtended by an arc at the centre is twice the angle subtended by it at any point of the circle in the alternate segment of the circle.

Then, $\angle AOC = 2\angle ABC \Rightarrow 60^\circ = 2\angle ABC$

$\therefore \quad ABC = 30^\circ$

Alternatively I

$AO = OC = 3$ (radii)

In ΔOAC, $AC^2 = OA^2 + OC^2 - 2.\ OA\ .\ OC.\ \cos \angle AOC$

[cosine-rule]

$\Rightarrow \quad 3^2 = 3^2 + 3^2 - 2 \cdot 3 \cdot 3 \cos \angle AOC$

$\Rightarrow \quad \cos \angle AOC = \frac{1}{2} \Rightarrow \cos \angle AOC = \cos 60^\circ$

$\Rightarrow \quad \angle AOC = 60^\circ$

Then, $\angle ABC = 30^\circ$

Alternatively II

Clearly, $AO = OC = AC = 3$ cm

Hence, ΔOAC is equilateral. Then, its each angle will be of 60°.

$\therefore \quad \angle AOC = 60^\circ$

Then, $\angle ABC = 30^\circ$

Problem 11 Two circles with centre P and Q intersect at B and C. A, D are points on the circle such that A, C, D are collinear. If $\angle APB = 130^\circ$ and $\angle BQD = x^\circ$, then the value of x is:

Sol.

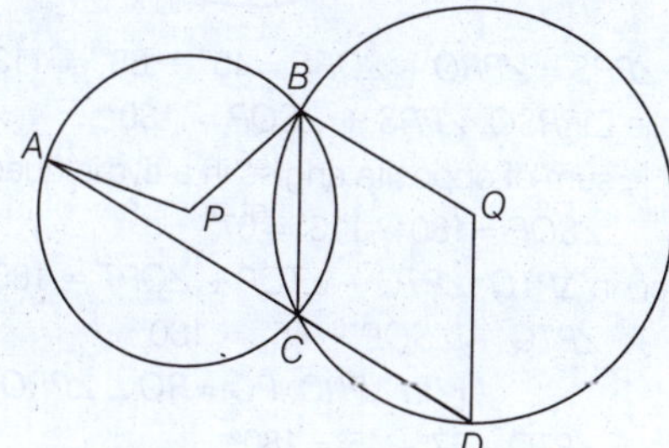

We know, an angle subtended by an arc at the centre is twice the angle subtended by it at any point of the circle in the alternate segment of the circle.

$\therefore \quad \angle ACB = 1/2 \times 130^\circ = 65^\circ$

Clearly, $\angle BCD = 180^\circ - \angle ACB = 180^\circ - 65^\circ = 115^\circ$

Then, reflex $\angle BQD = 2 \times 115^\circ = 230^\circ$

$\therefore$ Obtuse $\angle BQD = 360^\circ - 230^\circ = 130^\circ$

Problem 12 Chords AC and BD of a circle with centre O intersect at right angle at E. If $\angle OAB = 25^\circ$, then the value of $\angle EBC$ is **[SSC(Delhi Police SI)- 14]**

Sol. Given, $\angle OAB = 25^\circ$and $\angle BEC = 90^\circ$

In ΔABO, $OA = OB$ (radii)

$\therefore \quad \angle ABO = \angle OAB = 25^\circ$

Then, $\angle AOB = 180^\circ - 25^\circ - 25^\circ = 130^\circ$

$\therefore \quad \angle ACB = 1/2\ \angle AOB$

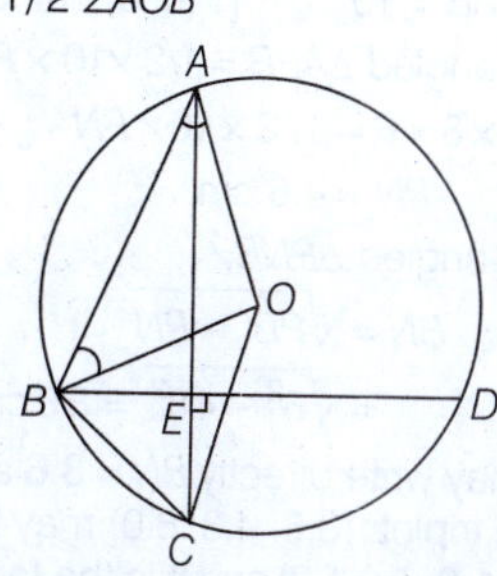

[$\because$ an angle subtended by an arc at the centre is twice the angle subtended by it at any point of the circle in the alternate segment]

$\Rightarrow \quad \angle ACB = 1/2 \times 130^\circ = 65^\circ$

Now, in ΔEBC, $\angle EBC = 180^\circ - \angle BEC - \angle ECB$

$\Rightarrow \angle EBC = 180^\circ - 90^\circ - \angle ACB = 180^\circ - 90^\circ - 65^\circ = 25^\circ$

Problems based on 'Angles in the same Segment are Equal'

Problem 13 ABC is an equilateral triangle inscribed in a circle. D is any point on the arc BC. What is $\angle ADB$ equal to: **[CDS-13]**

Sol.

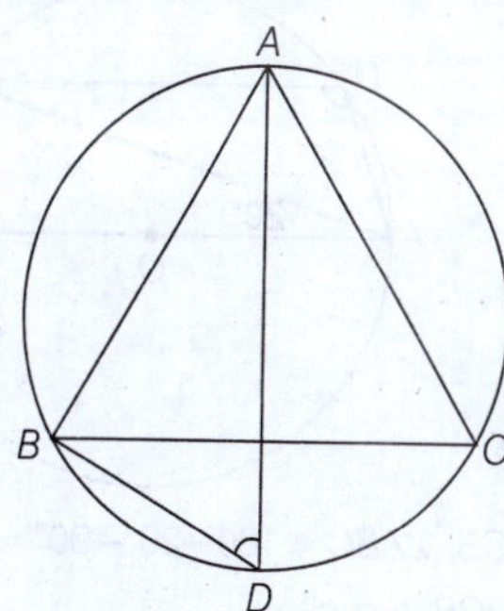

$\because \Delta ABC$ is equilateral. $\therefore \angle ACB = 60^\circ$

Since the angles in the same segment of a circle are equal, $\angle ACB = \angle ADB = 60^\circ$

Problems based on 'an Angle in a Semicircle is a Right Angle'

Problem 14 AB is a diameter of the circumcircle of ΔAPB. N is the foot of the perpendicular drawn from the point P on AB. If $AP = 8$ cm and $BP = 6$ cm, then the length of BN is :

Sol.

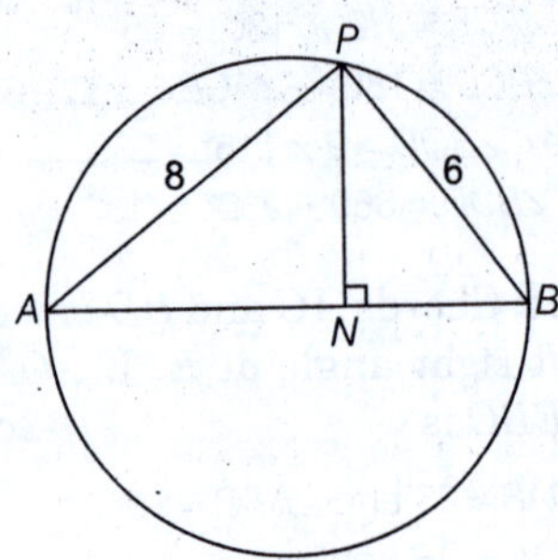

Since an angle in a semicircle is a right angle,

$\angle APB = 90^\circ$

Then, $AB = 10$ [$\because$ (6, 8, 10) is a Pythagorean triplet]

Area of right-angled $\Delta APB = 1/2 \times 10 \times PN$

$\Rightarrow \quad 1/2 \times 8 \times 6 = 1/2 \times 10 \times PN$

$\Rightarrow \quad PN = 4.8$ cm

Now, in right-angled ΔPNB,

$BN = \sqrt{PB^2 - PN^2}$

$= \sqrt{6^2 - 4.8^2} = 3.6$ cm

[Note: you may write directly $BN = 3.6$ as (3.6, 4.8, 6.0) is a Pythagorean triplet. (3.6, 4.8, 6.0) may be written as $(3 \times 1.2 \times 4 \times 1.2, 5 \times 1.2)$ and it is the fact that (3, 4,5) is a Pythagorean triplet]

Alternatively

Since $PN \perp AB$ in a right-angled ΔAPB, $PB^2 = AB \times BN$

$\Rightarrow \quad 6^2 = 10 \times BN$

$\Rightarrow \quad BN = 3.6$ cm

Problem 15 AB is a diameter of a circle with centre at O. DC is a chord of it such that $DC \parallel AB$. If $\angle BAC = 20^\circ$, then $\angle ADC = ?$ ***[SSC(CGL-T-I)- 12]***

Sol. $\angle ACB = 90^\circ$ [$\because$ an angle in a semicircle is a right angle]

and $\angle BAC = 20^\circ$ [given]

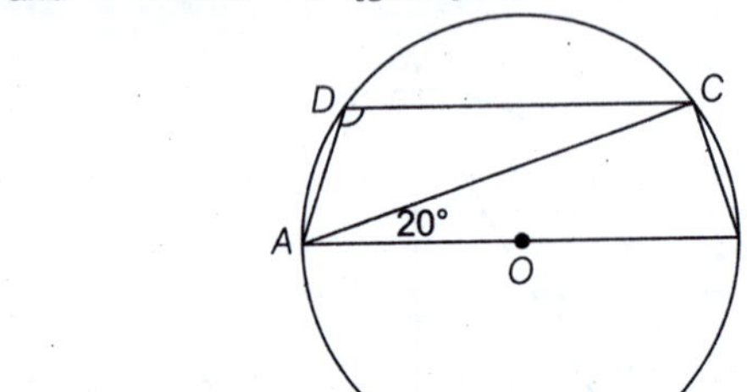

Then, in ΔACB, $\angle ABC = 180^\circ - 20^\circ - 90^\circ = 70^\circ$

Since $\square ADCB$ is cyclic,

$\angle ADC + \angle ABC = 180^\circ$

$\Rightarrow \quad \angle ADC + 70^\circ = 180^\circ$

$\Rightarrow \quad \angle ADC = 110^\circ$

Problem 16 On a triangle ABC, a circle BC is drawn, intersecting AB and AC at points P and Q respectively. If the lengths of AB, AC and CP are 30 cm, 25 cm and 20 cm respectively, then the length of BQ, in cm, is: ***[CAT 18]***

Sol.

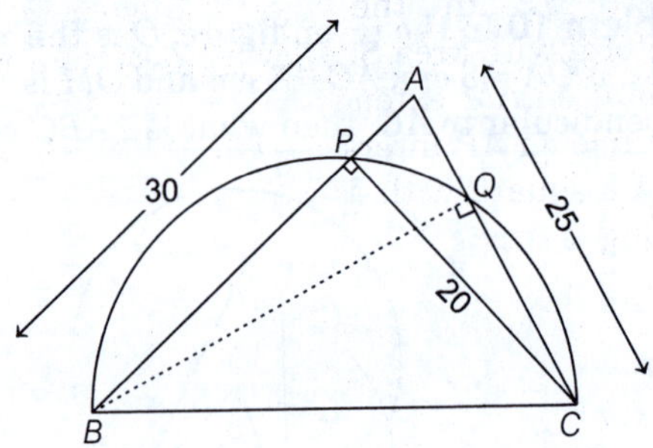

Since, ΔBPC and ΔBQC are inscribed in a semi circle,

$\angle BPC = \angle BQC = 90^\circ$

Now, area of $\Delta ABC = 1/2 \times AC \times BQ = 1/2 \times AB \times PC$

$\Rightarrow \quad 1/2 \times 25 \times BQ = 1/2 \times 30 \times 20 \Rightarrow BQ = 24$ cm

Problem 17 In the given figure, PQ is a diameter of the circle whose centre is at O. If $\angle ROS = 44^\circ$ and OR is a bisector of $\angle PRQ$, then what is the value of $\angle RTS$?

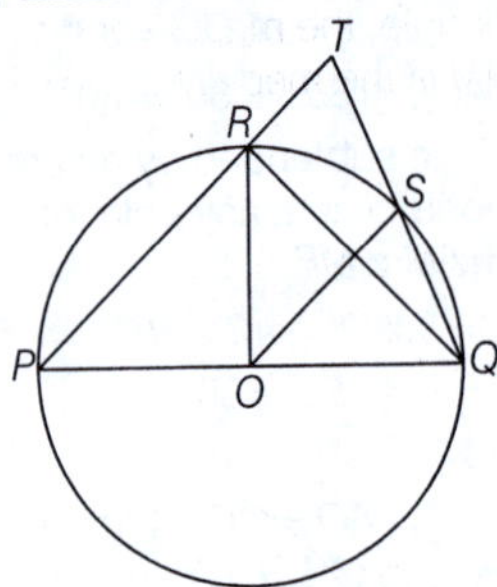

Sol. Since an angle in a semicircle is a right angle, $\angle PRQ = 90^\circ$

Then, $\angle PRO = \angle QRO = 45^\circ$ [$\because OR$ is a bisector of $\angle PRQ$]

In ΔROS, $RO = OS$ (radii)

Then, $\angle ORS = \angle OSR = \dfrac{180^\circ - 44^\circ}{2} = 68^\circ$

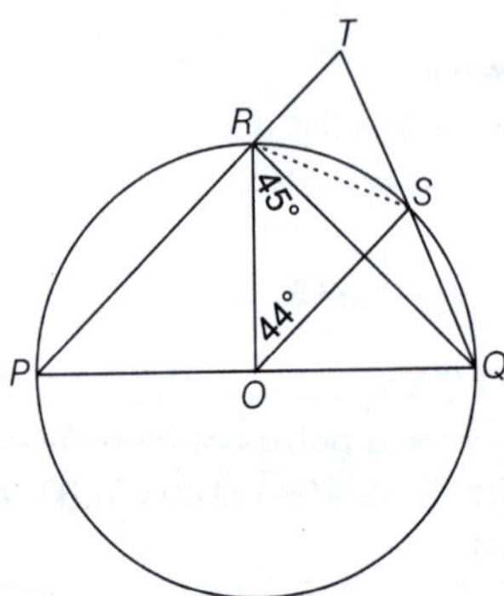

Now, $\angle PRS = \angle PRO + \angle ORS = 45^\circ + 68^\circ = 113^\circ$

In cyclic $\square PRSQ$, $\angle PRS + \angle SQP = 180^\circ$

[$\because$ sum of opposite angles in a cyclic quadrilateral is 180°]

$\Rightarrow \quad \angle SQP = 180^\circ - 113^\circ = 67^\circ$

Further, in ΔPTQ, $\angle PTQ + \angle TQP + \angle QPT = 180^\circ$

$\Rightarrow \quad \angle PTQ + \angle SQP + 45^\circ = 180^\circ$

[$\because$ in ΔPRO, $PO = RO \therefore \angle PRO = \angle RPO = 45^\circ$]

$\Rightarrow \quad \angle PTQ + 67^\circ + 45^\circ = 180^\circ$

$\Rightarrow \quad \angle PTQ = 68^\circ \Rightarrow \angle RTS = 68^\circ$

Problem 18 In the figure alongside, ΔABC is equilateral with area S. M is the mid-point of BC and P is a point on AM extended such that $MP = BM$. If the semi-circle on AP intersects CB extended at Q and the area of a square with MQ as a side is T, which of the following is true ?

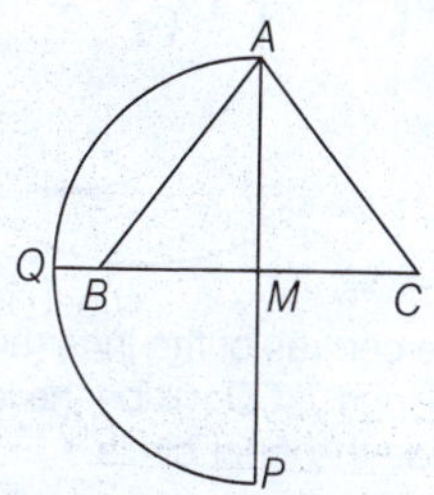

(a) $T = \sqrt{2}\,S$ (b) $T = S$
(c) $T = \sqrt{3}S$ (d) $T = 2S$

Sol. Since an angle in a semicircle is a right angle, $\angle PQA = 90°$.
In equilateral triangle, the median and the altitude are the same line. It means AM is the median as well as the altitude of the ΔABC.
In right-angled ΔAQP,
$$QM^2 = AM \times MP$$
[Recall the properties of a right-angled triangle detailed in sub-section 'Triangle']

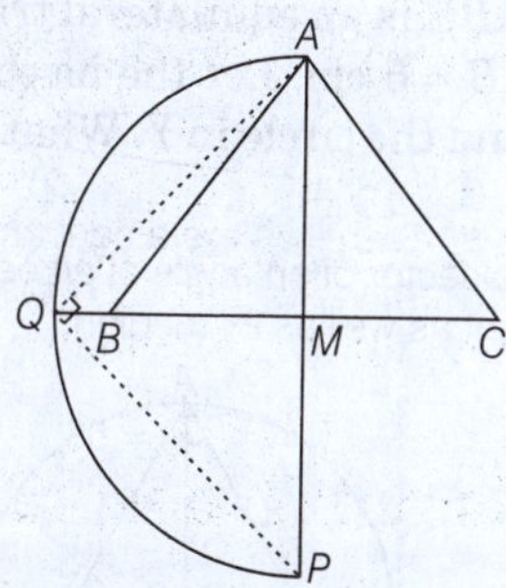

$QM^2 = AM \times BM$ [$\because$ MP = BM as given]
$\Rightarrow$ Area of the square $= AM \times 1/2\ BC$
$\Rightarrow$ $T = 1/2 \times BC \times AM$
$\Rightarrow$ T = area of ΔABC
$\Rightarrow$ $T = S$

Problems based on 'the Concept Associated with the Common Chord'

Problem 19 The length of the common chord of two circles of radii 15 cm and 20 cm whose centres are 25 cm apart, is:

Sol.

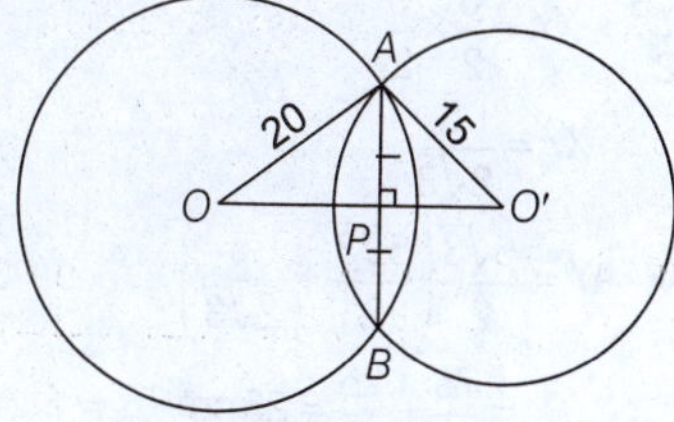

Given, $AO = 20$, $AO' = 15$ and $OO' = 25$

Since (15, 20, 25) is a Pythagorean triplet, AOO' is a triangle, right-angled at A.
We know, if two circles intersect each other in two points, then the line joining their centres is perpendicular bisector of their common chord. Hence, $AP \perp OO'$ and $AP = PB$.
Now, area of $\Delta AOO' = 1/2 \times AP \times OO'$
$\Rightarrow$ $1/2 \times 15 \times 20 = 1/2 \times AP \times 25 \Rightarrow AP = 12$
Then, $AB = 2AP = 2 \times 12 = 24$ cm
[Note: You may directly write as altitude AP
$$= \frac{\text{Product of perpendiculr sides}}{\text{hypotenuse}}$$
$$= \frac{15 \times 20}{\text{hypotenuse } 25} = 12 \text{ cm}]$$

Alternatively
Clearly, $r_1 = 20$, $r_2 = 15$ and distance between their centres i.e. $d = 25$
$\therefore$ Length of the common chord
$$= \frac{4 \times \text{area of the } \Delta \text{ having sides } r_1, r_2 \text{ and d}}{d}$$
$$= \frac{4 \times \text{area of the } \Delta \text{ having sides 15, 20 and 25}}{25}$$
$$= \frac{4 \times \frac{1}{2} \times 15 \times 20}{25}$$
[$\because$ the triangle is right-angled as (15, 20, 25) is a Pythagorean triplet]
$= 24$ cm

Problem 20 The length of the common chord of two intersecting circles is 24 cm. If the diameter of the circles are 30 cm and 26 cm, then the distance between the centres (in cm) is ***[SSC (CGL-T-II)-12, 20]***

Sol. Clearly, radii of the circles are 15 cm and 13 cm . Let radius OA = 15 cm and radius $O'A$ = 13 cm.

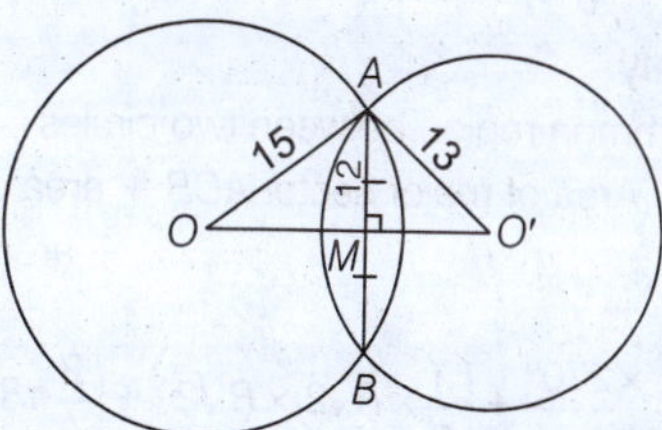

We know, if two circles intersect each other in two points, then the line joining their centres is perpendicular bisector of their common chord.
$\therefore$ $AM = MB = 12$ cm and $\angle AMO = \angle AMO' = 90°$
In right-angled ΔAMO, $OM = 9$
[$\because$ (9, 12, 15) is a Pythagorean triplet]
Also, in right-angled $\Delta AMO'$, $MO' = 5$
[$\because$ (5, 12, 13) is a Pythagorean triplet]
Then, $OO' = 9 + 5 = 14$ cm

Problem 21 Two circles with radius $2R$ and $\sqrt{2}\,R$ intersect each other at points A and B. The centres of both the circles are on the same side of AB. O is the centre of the bigger circle and $\angle AOB$ is 60°. Find the area of the common region between two circles. ***[XAT-18]***

Sol.

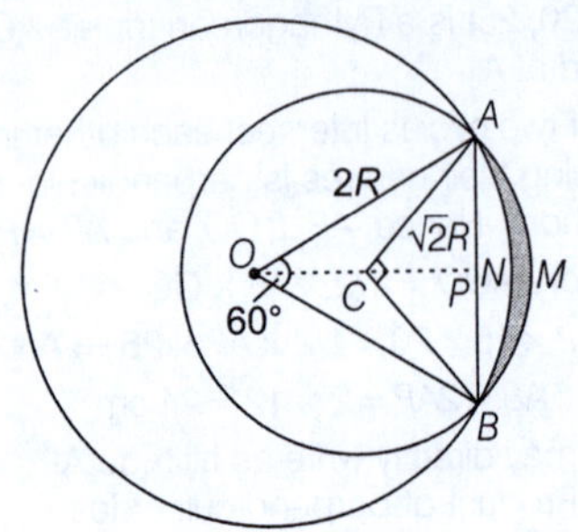

Given, $\angle AOB = 60°$ and side OA = side OB [$\because OA = OB$ = radii]

Hence, ΔAOB is an equilateral triangle.

Then, $AB = 2R$

$\therefore$ In ΔACB, $AB^2 = AC^2 + BC^2$

It implies that $\angle ACB = 90°$

Now, area of segment ABM = area of sector $ACBM$ – area of ΔACB

$$= \frac{\pi(R\sqrt{2})^2}{4} - \frac{1}{2} \times R\sqrt{2} \times R\sqrt{2} = \frac{\pi R^2}{2} - R^2$$

and area of segment ABN

$$= \frac{\pi(2R)^2}{6} - \frac{1}{2} \times 2R \times 2R \times \sin 60°$$

$$= \frac{2}{3}\pi R^2 - \sqrt{3}R^2$$

$\therefore$ Area of shaded portion = segment ABM – segment ABN

$$= \frac{\pi R^2}{2} - R^2 - \frac{2}{3}\pi R^2 + \sqrt{3}R^2$$

Clearly, area of common region between two circles = area of smaller circle - area of shaded portion

$$= \pi(R\sqrt{2})^2 - \frac{\pi R^2}{2} + R^2 + \frac{2}{3}\pi R^2 - \sqrt{3}R^2$$

$$= \frac{13\pi R^2}{6} + R^2 - \sqrt{3}R^2 = \left(\frac{13\pi}{6} + 1 - \sqrt{3}\right)R^2$$

Alternatively

Area of common region between two circles

= Area of major sector ACB + area of ΔACB + area of segment ABN

$$= \frac{\pi(R\sqrt{2})^2 \times 270°}{360°} + \left(\frac{1}{2} \times R\sqrt{2} \times R\sqrt{2}\right) + \left(\frac{2}{3}\pi R^2 - \sqrt{3}R^2\right)$$

$$= \left(\frac{13\pi}{6} + 1 - \sqrt{3}\right)R^2.$$

Problems based on 'when Two Chords of a Circle Intersect Inside the Circle, the Products of their Segments are Equal'

Cases of intersection of chords outside the circle, have been discussed under illustrations based on 'secant and tangent'

Problem 22 Two circles touch each other internally. Their radii are 4 cm and 6 cm. What is the length of the longest chord of the outer circle which is outside the inner circle? ***[CDS-11, SSC (CGL-T-I)-13]***

Sol.

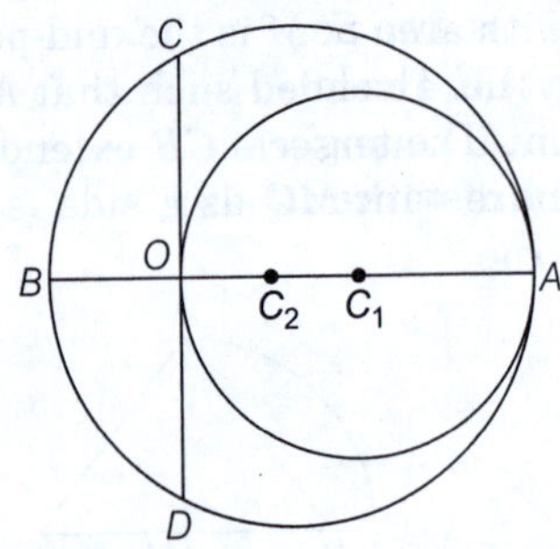

Let C_1 and C_2 be centres of the inner and the outer circles respectively. Obviously, CD will be the longest chord of the outer circle outside the inner circle.

We know, when two chords of a circle intersect each other inside the circle, the products of their segments are equal.

Here, chords AB and CD of the outer circle intersect at O.

Then, $BO \times OA = CO \times OD$

$\Rightarrow \quad (AB - OA) \times OA = CO \times CO$

[$\because OD = CO$ as the perpendicular (say C_2O) drawn from centre to a chord, bisects the chord]

$\Rightarrow \quad (12 - 8) \times 8 = CO \times CO$

$\Rightarrow \quad CO = \sqrt{32} = 4\sqrt{2}$

Then, $CD = 2 \times 4\sqrt{2} = 8\sqrt{2}$ cm

Problem 23 ABC is an equilateral triangle inscribed in a circle with $AB = 5$ cm. Let the bisector of the angle A meet BC in X and the circle in Y. What is the value of $AX \cdot AY$?

Sol. We know, the bisector of an angle of an equilateral triangle also acts as its height as well as its median.

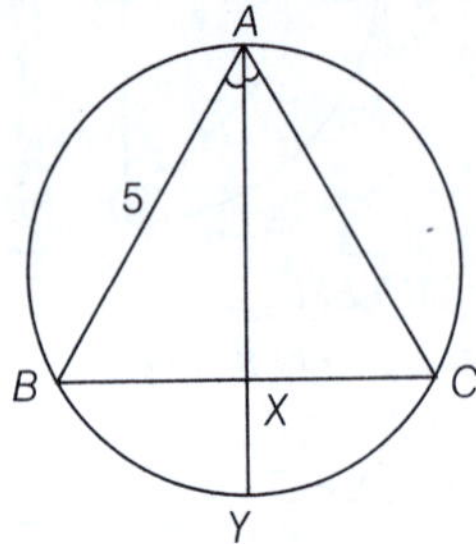

Hence, $BX = XC = \frac{5}{2}$

and height $AX = \frac{\sqrt{3}}{2} \times 5 = \frac{5\sqrt{3}}{2}$

We know, when two chords of a circle intersect each other inside the circle, the products of their segments are equal.

$\therefore \quad AX \times XY = BX \times XC$

$\Rightarrow \quad \frac{5\sqrt{3}}{2} \times XY = \frac{5}{2} \times \frac{5}{2}$

$\Rightarrow \quad XY = \frac{5}{2\sqrt{3}}$

Now, $\quad AX \cdot AY = \frac{5\sqrt{3}}{2}\left(\frac{5\sqrt{3}}{2} + \frac{5}{2\sqrt{3}}\right)$

$$= \frac{75}{4} + \frac{25}{4} = 25 \text{ cm}^2$$

Problem 24 The radius of a circle with centre O is $\sqrt{50}$ cm. A and C are two points on the circle and B is a point inside the circle. The length of AB is 6 cm and the length of BC is 2 cm. The angle ABC is a right angle. Find the square of the distance OB. **[XAT-13]**

Sol.

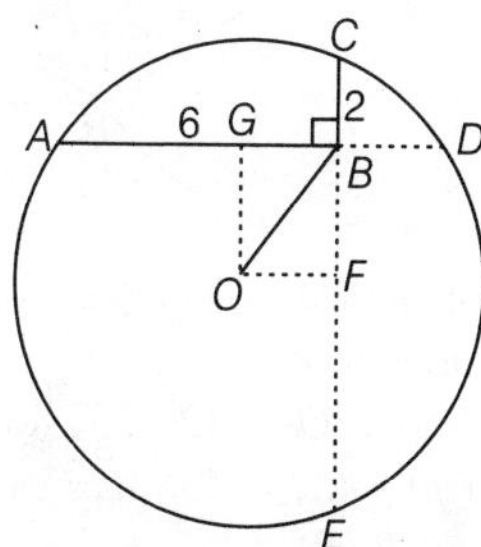

We know, when two chords of a circle intersect each other inside the circle, the products of their segments are equal.

$\therefore \quad AB \times BD = CB \times BE$

$\Rightarrow \quad 6 \times BD = 2 \times BE$

$\Rightarrow \quad BE = 3BD$

Further, we know, if two chords intersect at 90° and the intersection point divides one into lengths a and b and another into lengths c and d, then $a^2 + b^2 + c^2 + d^2 = (\text{diameter})^2$

$\Rightarrow \quad 6^2 + 2^2 + BD^2 + (3BD)^2 = (2\sqrt{50})^2$

$\Rightarrow \quad 6^2 + 2^2 + BD^2 + (3BD)^2 = 200$

$\Rightarrow \quad BD = 4$

Then, $\quad BE = 3 \times 4 = 12$

We drop perpendiculars OG and OF from the centre to AD and CE respectively. Then, points G and F will be mid-points of AD and CE respectively as the perpendicular drawn from the centre to a chord bisects the chord.

Then, $AG = AD/2 = (AB + BD)/2 = (6 + 4)/2 = 5$

$\therefore \quad BG = AB - AG = 6 - 5 = 1$

Further, $\quad CF = CE/2 = (CB + BE)/2 = (2 + 12)/2 = 7$

$\therefore \quad BF = CF - CB = 7 - 2 = 5$

Then, $\quad OG = BF = 5$

[$\because$ OGBF is a rectangle as each angle is 90°]

Now, in right-angled ΔOGB,

$OB^2 = BG^2 + OG^2 = 1^2 + 5^2 = 26$

Problem 25 Chords AB and CD of a circle intersect at E and are perpendicular to each other. Segments AE, EB and ED are of lengths 2 cm, 6 cm and 3 cm respectively. Then, the length of the diameter of the circle (in cm) is

Sol.

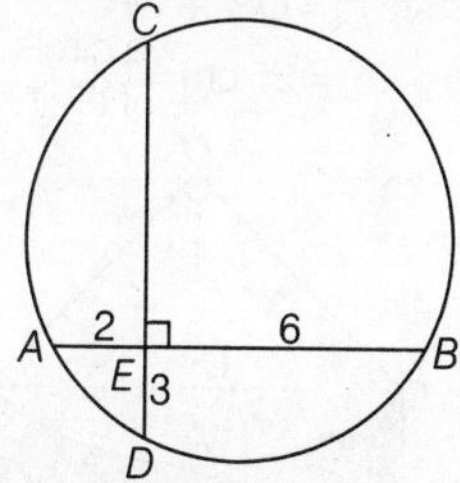

We know, when two chords of a circle intersect each other inside the circle, the products of their segments are equal.

$\therefore \quad AE \times EB = CE \times ED$

$\Rightarrow \quad 2 \times 6 = CE \times 3$

$\Rightarrow \quad CE = 4$

Further, we know, if two chords intersect at 90° and the intersection point divides one into lengths a and b another into length c and d, then

$a^2 + b^2 + c^2 + d^2 = (\text{diameter})^2$

$\Rightarrow \quad 2^2 + 3^2 + 6^2 + 4^2 = (\text{diameter})^2$

$\Rightarrow \quad \text{diameter} = \sqrt{65}$ cm

Problem 26 In a circle of radius 11 cm, CD is a diameter and AB is a chord of length 20.5 cm. If AB and CD intersect at point E inside the circle and CE has length 7 cm, then the difference of the lengths of BE and AE, in cm, is **[CAT 19]**

Sol.

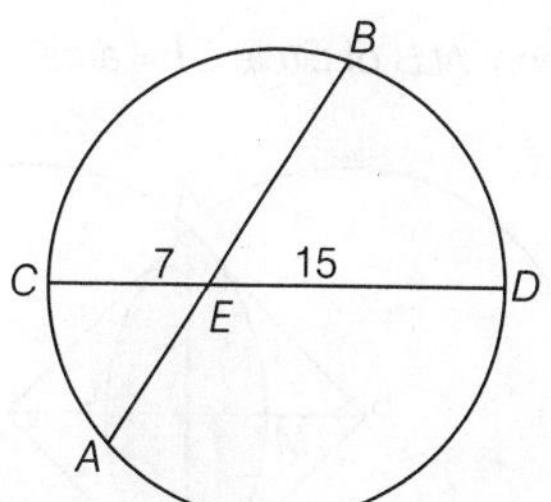

Diameter $CD = 2 \times 11 = 22$ cm

We know, $CE \times DE = AE \times BE$

$\Rightarrow \quad 7 \times 15 = AE \times BE = 105$

Now, $(AE - BE)^2 = (AE + BE)^2 - 4 \cdot AE \times BE$

$\Rightarrow \quad (AE - BE)^2 = (20.5)^2 - 4 \cdot 105 = 0.25$

$\Rightarrow \quad AE - BE = 0.5$

Problems based on Sector/ Segment of a Circle

A lot of questions have been discussed in Mensuration, so only four questions are being given here

Problem 27 A chord AB of length $3\sqrt{2}$ unit subtends a right angle at the centre O of a circle. Area of the minor segment AB is ?

Sol.

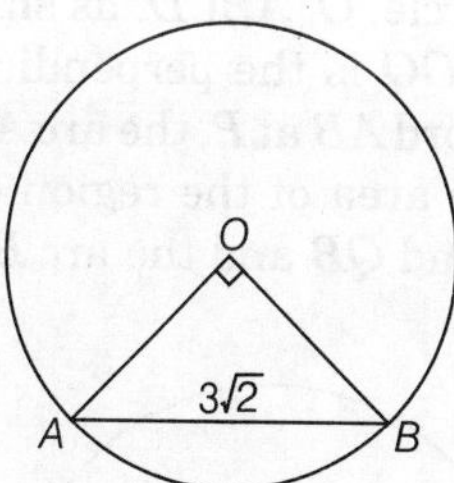

Let the radius of the circle be r.

Then, length of the chord $AB = 2r \sin(\theta/2)$

$\Rightarrow \quad 3\sqrt{2} = 2r \sin(90°/2)$

$\Rightarrow \quad 3\sqrt{2} = 2r \times \dfrac{1}{\sqrt{2}} \Rightarrow r = 3$

Then area of sector $AOB = \dfrac{\pi r^2}{360°}\theta$

$= \dfrac{\pi \times 3^2}{360°} \times 90° = \dfrac{9\pi}{4}$

Further, area of $\Delta AOB = \frac{1}{2} \times OA \times OB \times \sin\theta$

$$= \frac{1}{2} \times r \times r \times \sin 90^\circ = \frac{3^2}{2} = 4.5$$

$\therefore$ area of the minor segment

$$AB = \frac{9\pi}{4} - 4.5$$

$$= 2.57 \text{ sq units}$$

Problem 28 Two identical circles intersect so that their centres and the points at which they intersect form a square of 1 cm. The area in sq cm of the portion that is common to the two circles is ***[CAT 05]***

Sol. Area of segment ALB of circle $C1$ = area of sector PAB – area of ΔPAB

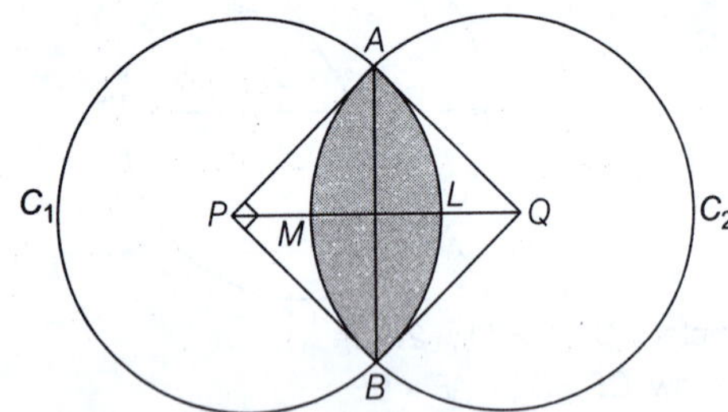

$$= \frac{\pi r^2}{360^\circ} \times 90^\circ - (1/2 \times 1 \times 1)$$

$$= \frac{\pi \times 1^2}{4} - \frac{1}{2} = \frac{\pi}{4} - \frac{1}{2}$$

Obviously, the area of segment AMB of circle C_2 will be same as that of segment ALB of circle C_1.

Hence, total common area $= 2\left(\frac{\pi}{4} - \frac{1}{2}\right)$

$$= \left(\frac{\pi}{2} - 1\right) \text{ sq cm}$$

Problem 29 In the given figure, the circle has a chord AB of length 12 cm, which makes an angle of 60° at the centre of the circle, O. $ABCD$, as shown in the diagram, is a rectangle. OQ is the perpendicular bisector of AB, intersecting chord AB at P, the arc AB at M and CD at Q. $OM = MQ$. The area of the region enclosed by the line segments AQ and QB and the arc BMA, is closest to (in sq cm) ***[XAT-20]***

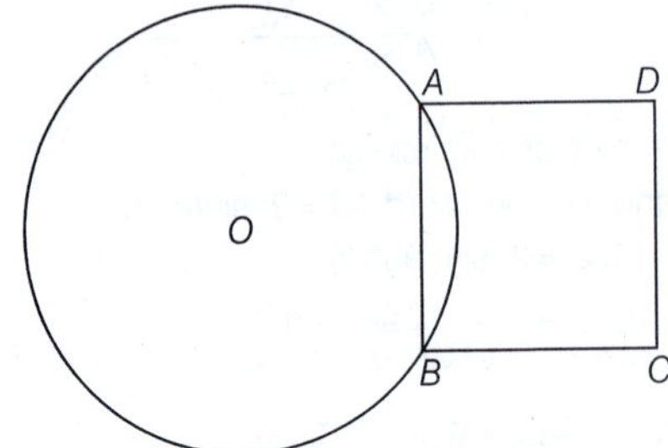

Sol. Given, $\angle AOB = 60^\circ$ Further, sides OA and OB of ΔAOB are equal as they are radii. It implies that $\angle OBA = \angle OAB = 60^\circ$.

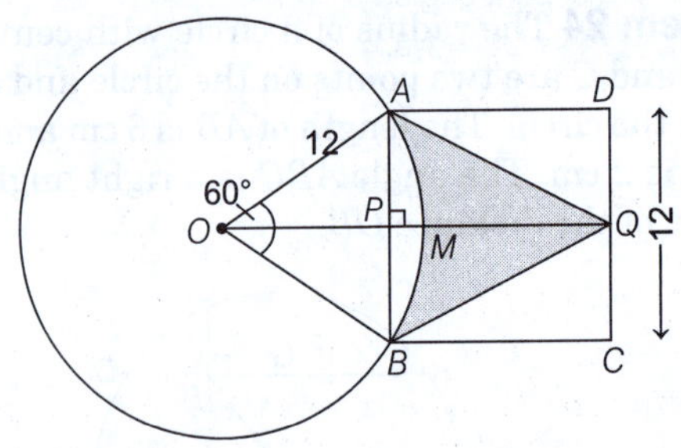

Thus, ΔAOB is an equilateral triangle.

Then, $OA = OB = AB = 12$

$\because$ Radius OM = Radius $OA = 12 = MQ$ (given)

Now, $OQ = OM + MQ = 12 + 12 = 24$

Clearly, $PQ = OQ - OP$

$$= 24 - \frac{\sqrt{3}}{2} \times 12 = 13.6$$

[$\because OP$ is height of equilateral ΔAOB]

Now, area of $\Delta ABQ = 1/2 \times AB \times PQ = 1/2 \times 12 \times 13.6 = 81.6$

and area of segment ABM = area of sector AOB – area of ΔAOB

$$= \frac{\pi \times 12^2}{6} - \frac{\sqrt{3} \times 12^2}{4} = 13.04$$

$\therefore$ Aarea of shaded region $= 81.6 – 13.04 \approx 69 \text{ cm}^2$

Problem 30 In the given figure, $PA = QB$ and PRQ is the arc of the circle, centre of which is O such that $\angle POQ = 90^\circ$. If $AB = 25\sqrt{2}$ cm and the perpendicular distance of AB from centre O is 30 cm, find the area of the shaded region. ***IIFT-18***

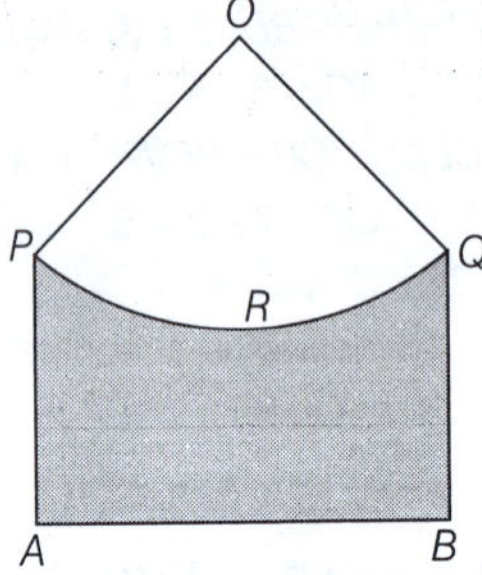

Sol. Here, $PQ = AB = 25\sqrt{2}$ [given]

Now, in right-angled ΔPOQ,

$$PQ^2 = OP^2 + OQ^2$$

$\Rightarrow$ $(25\sqrt{2})^2 = OP^2 + OP^2$ [$\because OP = OQ$ = radii]

$\Rightarrow$ $OP = 25 \text{ cm} = OQ$

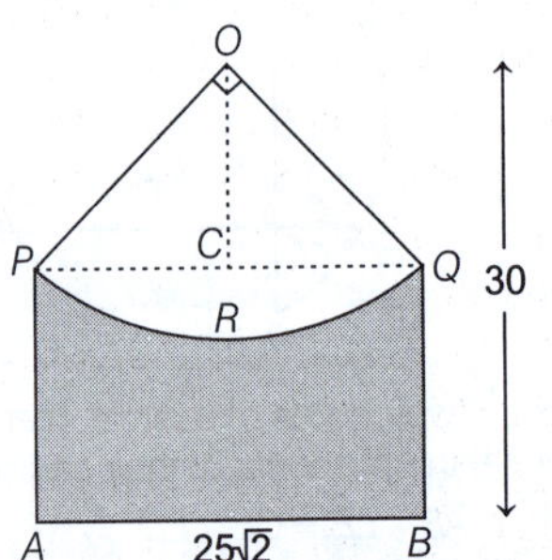

Now, area of right-angled $\Delta POQ = 1/2 \times OP \times OQ$

$\Rightarrow 1/2 \times PQ \times OC = 1/2 \times 25 \times 25$

$\Rightarrow 1/2 \times 25\sqrt{2} \times OC = 1/2 \times 25 \times 25$

$\Rightarrow OC = 25/\sqrt{2}$

Further, area of segment PRQ

$= \text{area of sector } OPRQ - \text{area of } \Delta OPQ$

$= \dfrac{\pi \times (25)^2}{4} - \dfrac{1 \times (25)^2}{2}$

Clearly,

Area of shaded region = area of $PABQ$ − area of segment PRQ

$= (AB \times BQ) - \dfrac{\pi \times (25)^2}{4} + \dfrac{1 \times (25)^2}{2}$

$= (25\sqrt{2} \times (30 - 25\sqrt{2}) - \dfrac{\pi \times (25)^2}{4} + \dfrac{1 \times (25)^2}{2}$

$= 750\sqrt{2} - 625\left(\dfrac{1}{2} + \dfrac{\pi}{4}\right)$

Miscellaneous (Including Other Remaining Properties)

Problem 31 One chord of a circle is known to be 10.1 cm. The radius of this circle must be

(a) 5 cm

(b) greater than 5 cm

(c) greater than or equal to 5 cm

(d) less than 5 cm

Sol. We know that the diameter of a circle is its longest chord. If the given chord is the diameter, the radius will be 10.1/2. i.e. 5.05 cm. Hence, option(b) is correct.

Problem 32 In the given figure, a circle with centre D is touching all three semicircles with centres A, E and F. Find its radius if A is the mid-point of MN and $MN = 12$ cm.

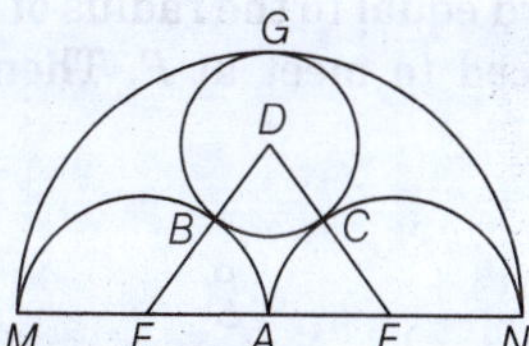

Sol. Given, A is the mid-point of MN

$\therefore \quad AM = AN = 6$ cm

and $\quad BE = FC = 3$ cm (radii)

Now, $\quad ED = (3 + r), DF = (3 + r)$ [let the desired radius be r]

$\because \quad ED = DF$

It implies that DEF is an isosceles triangle and DA is one of its medians as A is the mid-point of EF.

We know that the median of an isosceles triangle falling on the side other than congruent sides also acts as its altitude

$\therefore \quad \angle DAF = 90°$

In right-angled ΔDAF, $DF^2 = DA^2 + AF^2$

$\Rightarrow \quad (3 + r)^2 = (AG - DG)^2 + 3^2$

$\Rightarrow \quad (3 + r)^2 = (6 - r)^2 + 3^2$

[$\because AG = AN$ = (radii of the biggest semi-circle]

$\Rightarrow \quad 9 + 6r = 36 - 12r + 9 \Rightarrow r = 2$ cm

Remarks

It is clear from the above figure that the semicircles having centres E and F are congruent and the radius of the biggest semicircle is equal to the diameter of each of these congruent semicircles. In such type of cases, it may be construed from the above result that the radius of the circle touching all three semicircles is either equal to one-sixth of the diameter of the biggest semicircle or equal to two-third of the radius of each of congruent semicircles.)

Now, take one more example

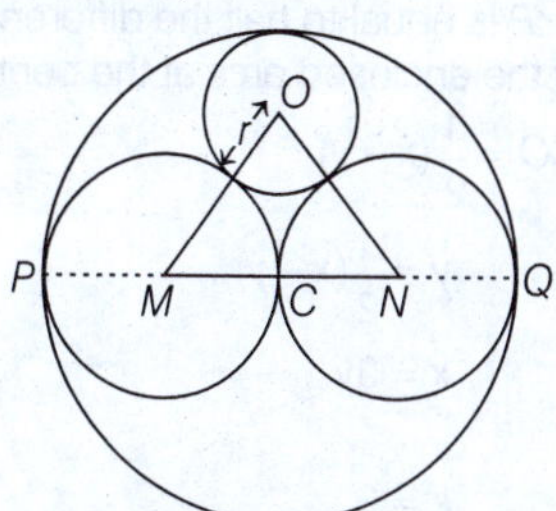

Here, radius of the smallest circle i.e. r

$= \dfrac{1}{6} \times PQ$

$= \dfrac{2}{3} \times CM$

$= \dfrac{2}{3} \times CN$

Problem 33 In the given figure, AB is the chord of a circle with centre O. AB is extended to C such that $BC = OB$. The straight line CO is produced to meet the circle at D. If $\angle ACD = y$ degrees and $\angle AOD = x$ degrees such that $x = ky$, then the value of k is **[CAT-03]**

(a) 3 (b) 2

(c) 1 (d) None of these

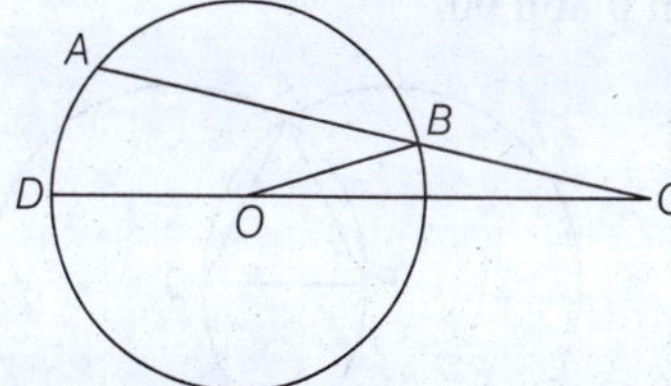

Sol. In ΔOBC, $OB = BC$ [given]

$\therefore \quad \angle BOC = BCO = y$

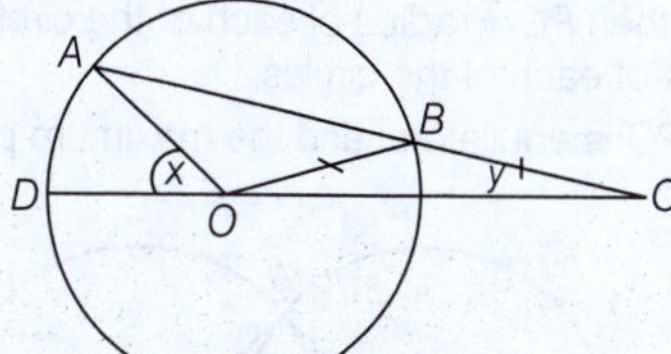

Exterior $\angle ABO$ of ΔOBC = sum of opposite interior angles

$= y + y = 2y$

In ΔAOB, $OA = OB$(radii)

$\therefore \quad \angle OAB = \angle ABO = 2y$

Clearly, $x + \angle AOB + \angle BOC = 180°$

$\Rightarrow \quad x + (180° - \angle OAB - \angle ABO) + y = 180°$

$\Rightarrow \quad x + (180° - 2y - 2y) + y = 180°$

$\Rightarrow \quad x = 3y$

But $x = ky$ [as given]

$\therefore \quad k = 3$

Alternatively

In ΔOBC, $OB = BC$ [given]

$\therefore \quad \angle BOC = \angle BCO = y$

We know, if two secants of a circle meet outside the circle at point P, then $\angle P$ is equal to half the difference between angles subtended by the enclosed arcs at the centre of the circle.

$\therefore \quad \angle BCO = \frac{1}{2}(x - y)$ [assumed $x > y$]

$\Rightarrow \quad y = \frac{1}{2}(x - y)$

$\Rightarrow \quad x = 3y$

But $x = ky$ [as given]

$\therefore \quad k = 3$

If we assume $x < y$, then $y = \frac{1}{2}(y - x)$

$\Rightarrow \quad y = -x$ (not possible)

Hence, $k = 3$ is correct answer.

Problem 34 Two circles with centres P and Q cut each other at two distinct points A and B. The circles have the same radii and neither P nor Q falls within the intersection of the circles. What is the smallest range that includes all possible values of the angle AQP in degrees ?

(a) Between 0 and 30
(b) Between 0 and 60
(c) Between 0 and 75
(d) Between 0 and 45
(e) Between 0 and 90 **[CAT 07]**

Sol.

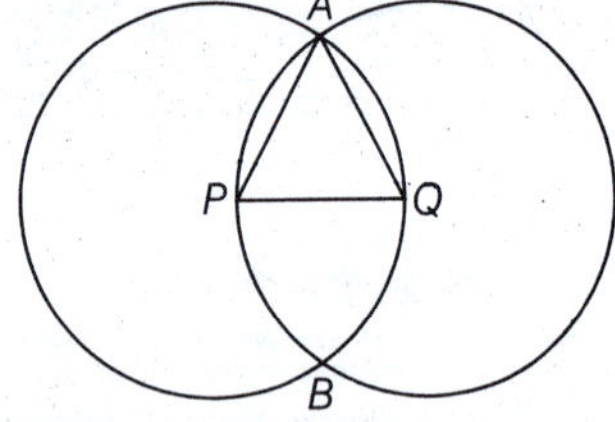

Suppose the centres P and Q lie on the arcs AB (as shown in the figure), then PQ = radius of each of the circles. Also, $AP = AQ$ = radius of each of the circles.

Hence, ΔAPQ is equilateral and the maximum possible value of $\angle APQ = 60°$

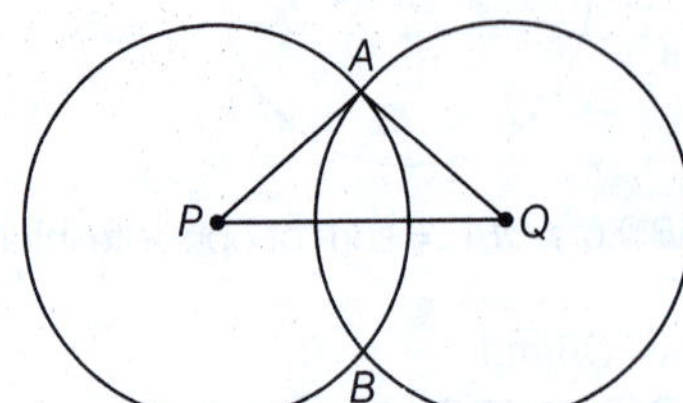

Now, consider the 2nd figure wherein neither P nor Q falls within the intersection of the circles. Here, length of PQ increases in comparison to the former as shown in the 1st, figure. Therefore, the value of $\angle PAQ$ will increase and the values of $\angle APQ$ and $\angle AQP$ will decrease in comparison to the former. The logic behind it is that an angle opposite to the greater side is the greater and that opposite to the smaller side is the smaller. Thus, the possible values of $\angle APQ$ is between 0° and 60°.

Problem 35 In the given figure, if $\angle BAC = 70°$, then $\angle OBC = ?$

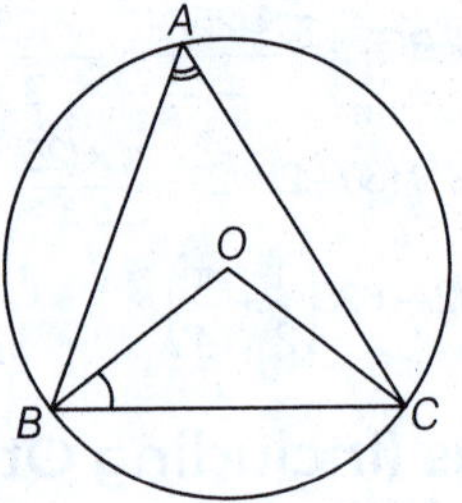

Sol. In ΔBOC,

$\angle OBC + \angle OCB = 180° - \angle BOC$

$\Rightarrow \quad 2\angle OBC = 180° - 2\angle BAC$

[$\because \angle OBC = \angle OCB$ as $OB = OC$ (radii)

and $\quad \angle BOC = 2\angle BAC$

as an angle subtended by an arc at the centre is twice the angle subtended by it at any point of the circle in the alternate segment]

$\Rightarrow \quad 2\angle OBC = 180° - (2 \times 70°)$

$\Rightarrow \quad \angle OBC = 20°$

[Alternatively, we know, $\angle A + \angle OBC = 90°$ when point A is on major arc

$\therefore \quad 70° + \angle OBC = 90° \Rightarrow \angle OBC = 20°$

Problem 36 AB is a diameter of a circle with centre O. CD is a chord equal to the radius of the circle. AC and BD are produced to meet at P. Then, the measure of $\angle APB = ?$

Sol.

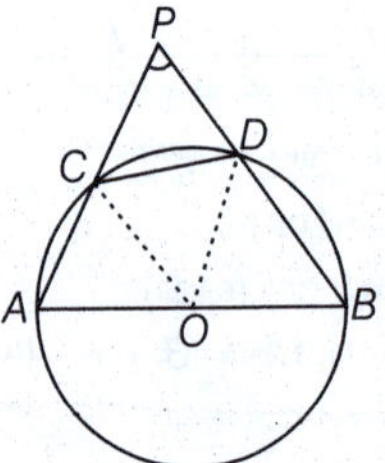

Sol. In ΔOCD, $OC = OD$ (radii) and also, CD is equal to radius (as given)

Hence, ΔOCD is equilateral.

Then, $\quad \angle OCD = \angle ODC = \angle COD = 60°$

In $\quad \Delta AOC$, $OA = OC$ [radii]

$\therefore \quad \angle OCA = \angle OAC$

Also, $\angle AOC + \angle OCA + \angle OAC = 180°$

$\Rightarrow \quad \angle AOC = 180° - 2\angle OAC$

Similarly, in ΔBOD, $\angle BOD = 180° - 2\angle OBD$

Clearly, $\angle AOC + \angle BOD + \angle COD = 180°$

$\Rightarrow 180° - 2\angle OAC + 180° - 2\angle OBD + 60° = 180°$

$\Rightarrow \angle OAC + \angle OBD = 120°$

Now, in ΔAPB, $\angle A + \angle B + \angle P = 180°$

$\Rightarrow \angle OAC + \angle OBD + \angle APB = 180°$

$\Rightarrow 120° + \angle APB = 180°$

$\Rightarrow \angle APB = 60°$

Problem 37 *AC* and *BC* are two equal chords of a circle. *BA* is produced to any point *P* and *CP*, when joined cuts the circle at *T*, then

(a) $CT : TP = AB : CA$ (b) $CT : TP = CA : AB$
(c) $CT : CB = CA : CP$ (d) $CT : CB = CP : CA$

Sol.

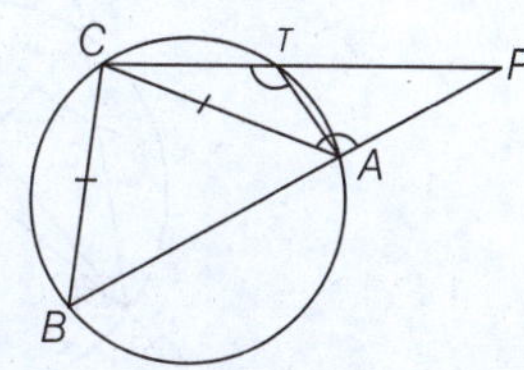

In cyclic □CTAB,

$\angle CBA + \angle CTA = 180°$

$\Rightarrow \angle CAB + \angle CTA = 180°$ [$\because BC = AC \therefore \angle CBA = \angle CAB$]

$\Rightarrow \angle CTA = 180° - \angle CAB$

Also, $\angle CAP = 180° - \angle CAB$

Hence, $\angle CTA = \angle CAP$

Now, $\Delta CAP \sim \Delta CTA$

[$\because \angle ACP = \angle ACT$ (common) and $\angle CAP = \angle CTA$}

Then, $\frac{CP}{AC} = \frac{AC}{CT} \Rightarrow \frac{CP}{AC} = \frac{BC}{CT}$ [$\because AC = BC$]

$\Rightarrow CT : CB = CA : CP$

Problem 38 *O* is the centre of the circle passing through points *A*, *B* and *C* such that $\angle BAO = 30°$, $\angle BCO = 40°$ and $\angle AOC = x°$, what is the value of *x* ?

Sol. In ΔAOB, $AO = BO$ (radii)

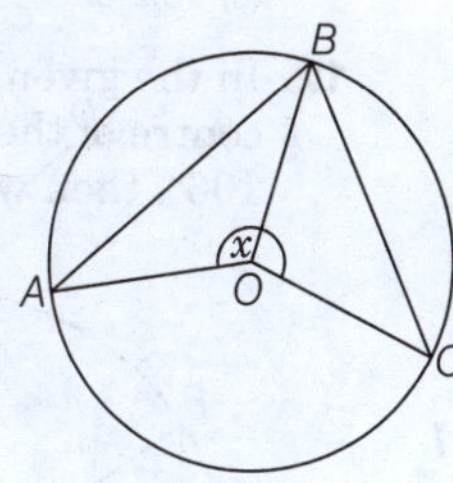

$\therefore \angle ABO = \angle BAO = 30°$

Then, $\angle AOB = 180° - 30° - 30° = 120°$

In ΔBOC, $BO = CO$ [radii]

$\therefore \angle BCO = \angle OBC = 40°$

Then, $\angle BOC = 180° - 40° - 40° = 100°$

Clearly, $\angle AOC = \angle AOB + \angle BOC$

$= 120° + 100° = 220°$

Or, minor $\angle AOC = 360° - 220° = 140°$

Problem 39 *AB* is the chord of a circle with centre *O* and *DOC* is a line segment originating from a point *D* on the circle and intersecting *AB* produced at *C* such that *BC* = *OD*. If $\angle BCD = 20°$, then $\angle AOD$ = ?

Sol.

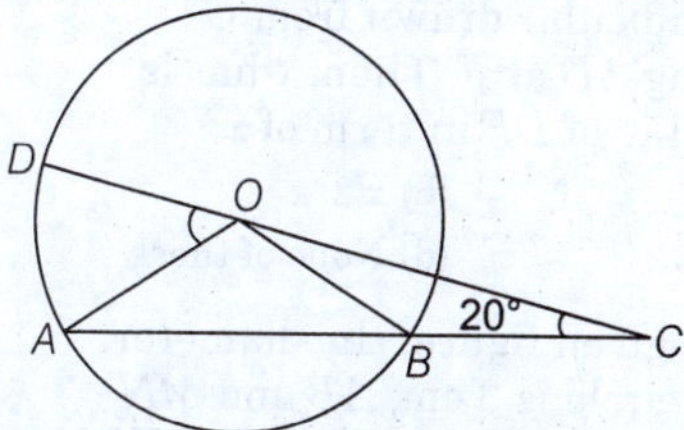

In ΔOBC, $OB = BC$

[$\because BC = OD$ (as given) and $OD = OB$ (radii)]

$\therefore \angle BOC = \angle OCB = 20°$

Exterior $\angle OBA = \angle BOC + \angle OCB$

$= 20° + 20° = 40°$

In ΔAOB, $\angle AOB = 180° - \angle OBA - \angle OAB$

$= 180° - \angle OBA - \angle OBA$

[$\because \angle OBA = \angle OAB$ as $OA = OB$ (radii)]

$= 180° - 40° - 40° = 100°$

Since, *DOC* is a straight line, $\angle AOD + \angle AOB + \angle BOC = 180°$

$\Rightarrow \angle AOD + 100° + 20° = 180°$

$\Rightarrow \angle AOD = 60°$

Problem 40 *A*, *B* and *C* are three points on the circumference of a circle and if $AB = AC = 5\sqrt{2}$ cm and $\angle BAC = 90°$, find the radius. **[SSC(CGL-T-I) -14]**

Sol. We know, when an arc of a circle subtends a right angle at any point of a circle in the alternate or same segment, the arc is a semi-circle. Hence, *BC* is the diameter.

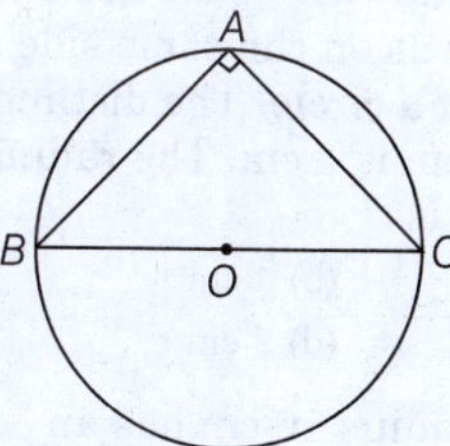

Clearly, $BC^2 = (5\sqrt{2})^2 + (5\sqrt{2})^2$

$\Rightarrow BC = 10$

$\therefore$ Radius $= \frac{10}{2} = 5$ cm

WORKOUT

1. Let ABC be a triangle in which $\angle ACB = 60°$ and $AC = x$. Let the circle with centre at C and radius x meets BC at D. Let CF be the perpendicular drawn from C meeting AD at F. Then, what is the value of DF in term of x ?

(a) x (b) $x/2$
(c) $2x$ (d) None of these

2. In the given figure, the diameter of the circle is 3 cm. AB and MN are two diameters such that MN is perpendicular to AB. In addition, CG is perpendicular to AB such that $AE : EB = 1 : 2$ and DF is perpendicular to MN such that $NL : LM = 1 : 2$. The length of DH in cm is **[CAT-05]**

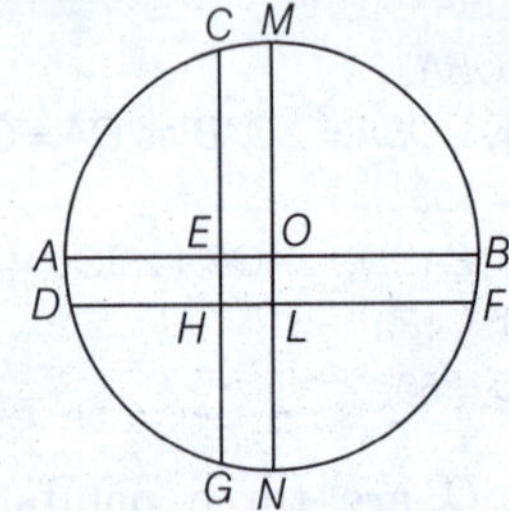

(a) $2\sqrt{2}-1$ (b) $\frac{(2\sqrt{2}-1)}{2}$
(c) $\frac{(3\sqrt{2}-1)}{2}$ (d) $\frac{(2\sqrt{2}-1)}{3}$

3. $AB = 8$ cm and $CD = 6$ cm are two parallel chords on the same side of the centre of a circle. The distance between them is 1 cm. The radius of the circle is

(a) 5 cm (b) 4 cm
(c) 3 cm (d) 2 cm

4. A circle of radius 10 cm has an equilateral triangle inscribed in it. The length of the perpendicular drawn from the centre to any side of the triangle is **[CDS-14]**

(a) $2.5\sqrt{3}$
(b) $5\sqrt{3}$
(c) $10\sqrt{3}$
(d) None of the above

5. In the given figure, the diameter of the larger circle is 20 cm and the smaller circle touches internally the larger circle at P and passes through O, the centre of the larger circle. Chord SP cuts the smaller circle at R and OR is equal to 8 cm. What is the length of chord SP ? **[CAT-13]**

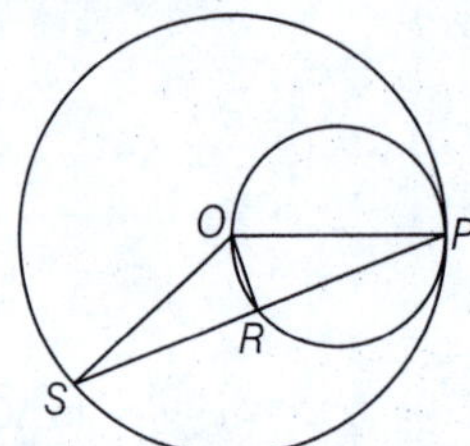

(a) 9 cm (b) 6 cm
(c) 14 cm d) 12 cm

6. Two equal circles pass through each other's centre. If the radius of each circle is 5 cm, what is length of the common chord ? **[SSC (CGL-T-1)-12]**

(a) $5\sqrt{3}$
(b) $10\sqrt{3}$
(c) $15\sqrt{3}$
(d) $2\sqrt{3}$

7. AC is the diameter of the circumcircle of the cyclic quadrilateral $ABCD$. If $\angle BDC = 42°$, then what is $\angle ACB$ equal to ?

(a) 42° (b) 45°
(c) 48° (d) 58°

8. P and Q are centres of two circles with radii 9 cm and 2 cm respectively, where $PQ = 17$ cm. R is the centre of another circle of radius x cm, which touches each of the above two circles externally. If $\angle PRQ = 90°$, then the value of x is **[SSC(CGL-T-1I)-12, SSC(10+2-DEO/LDC)- 13]**

(a) 4 cm
(b) 6 cm
(c) 7 cm
(d) 8 cm

9. In the given figure, O is the centre of the circle. AC and BD intersect at P. If $\angle AOB = 100°$ and $\angle DAP = 30°$, then what is $\angle APB$? **[CDS 10]**

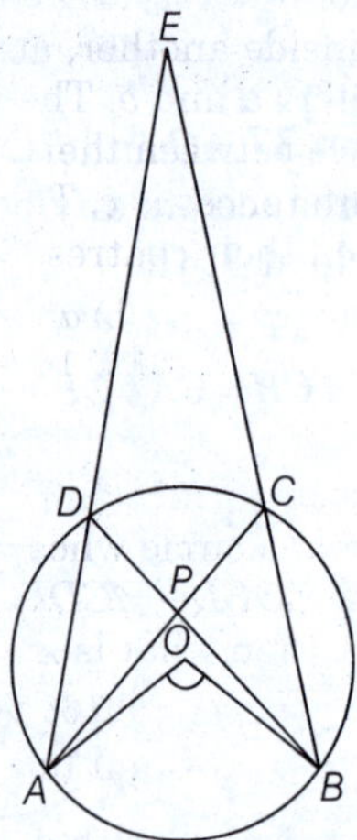

(a) 77° (b) 80°
(c) 85° (d) 90°

10. A semi-circle is drawn with AB as its diameter. From C, a point on AB, a line perpendicular to AB is drawn meeting the circumference of the semi-circle at D. Given that $AC = 2$ cm and $CD = 6$ cm, the radius of the semi-circle will be

(a) 15 cm (b) 14 cm
(c) 10 cm (d) 12 cm

11. In a circle of radius 21 cm, an arc subtends an angle of 72° at the centre. The length of the arc is

(a) 21.6 cm (b) 26.4 cm
(c) 13.2 cm (d) 19.8 cm

12. In the given figure, if O is the centre of the circle and $\angle AOD = 106°$, then what is $\angle BCD$ equal to **[CDS-09]**

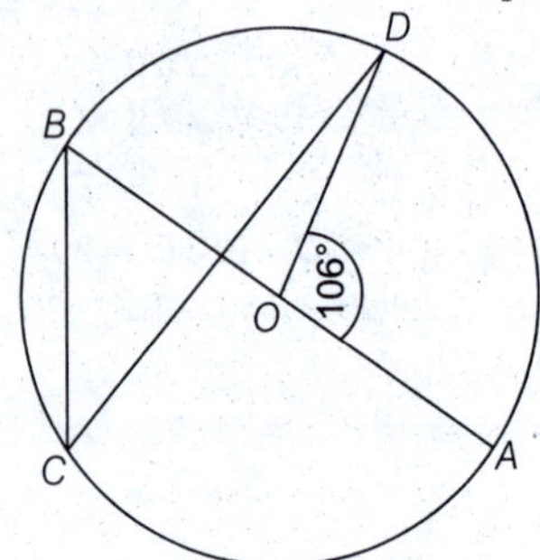

(a) 53° (b) 43°
(c) 40° (d) 37°

13. The angle subtended by a chord at its centre is 60°, then the ratio between chord and radius is ***[SSC(CGL-T-I 14)]***

(a) 1 : 2 (b) 1 : 1
(c) $\sqrt{2}$: 1 (d) 2 : 1

14. Two non-intersecting circles, one lying inside another, are of diameters a and b. The minimum distance between their circumferences is c. The distance between their centres is: ***[MAT-98]***

(a) $a - b - c$ (b) $a + b - c$
(c) $\frac{1}{2}(a - b - c)$ (d) $\frac{1}{2}(a - b) - c$

15. A, B, C and D are four distinct points on a circle whose centre is at O. If $\angle OBD - \angle CDB = \angle CBD - \angle ODB$, then what is $\angle$A equal to

(a) 45° (b) 60°
(c) 120° (d) 135°

16. On a semicircle with diameter AD, chord BC is parallel to the diameter. Further, each of the chords AB and CD has length 2 while AD has length 8. What is the length of BC? ***[CAT-04]***

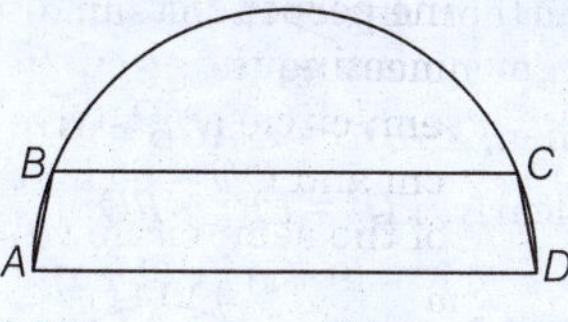

(a) 7.5 (b) 7
(c) 7.75 (d) NT

17. Two circles intersect each other at P and Q. PA and PB are two diameters. Then $\angle AQB$ = ? ***[SSC (CGL-T-I)-12]***

(a) 45° (b) 160°
(c) 180° (d) 135°

18. In the figure given, C and D are points on the semi-circle described on AB as diameter. If $\angle ABD = 75°$ and $\angle DAC = 35°$, then what is the $\angle BDC$?

(a) 130° (b) 110° (c) 90° (d) 100°

19. AB is a diameter of a circle with centre O. DC is a chord such that DC || AB. If $\angle BAC = 20°$, then measure of $\angle COD$ = ? ***[SSC(FCI-Mains)-12]***

(a) 45° (b) 60° (c) 120° (d) 100°

20. In the given figure, the $\angle AOC$ is 100°, where O is the centre of the circle. What is the $\angle ABC$?

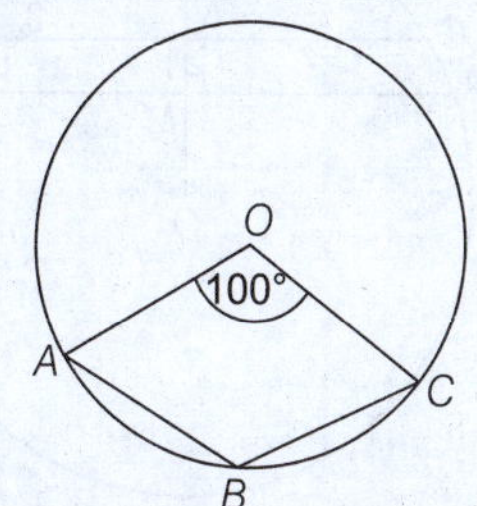

(a) 130° (b) 120° (c) 80° (d) 100°

21. In the given figure, chord ED is parallel to the diameter AC of the circle. If $\angle CBE = 65°$, then what is the value of $\angle DEC$? ***[CAT-04]***

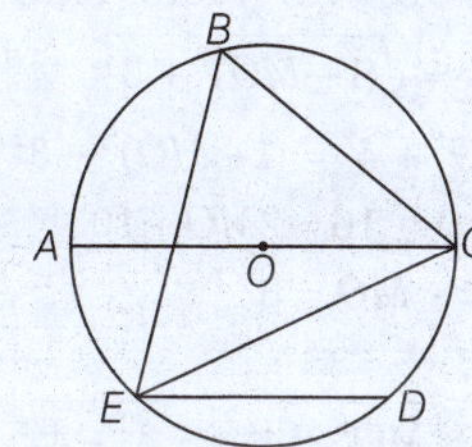

(a) 35° (b) 55° (c) 45° (d) 25°

22. Each of the circles of equal radii with centres A and B pass through the centre of one another. They cut at C and D, then , $\angle DBC$ is equal to:

(a) 60° (b) 100° (c) 120° (d) 140°

23. Two circles with centres A and B touch each other internally, as shown in the figure given below. Their radii are 5 and 3 units respectively. Perpendicular bisector of AB meets the bigger circle in P and Q. What is the length of PQ? ***[CDS-07 SSC(Delhi Police-SI-14]***

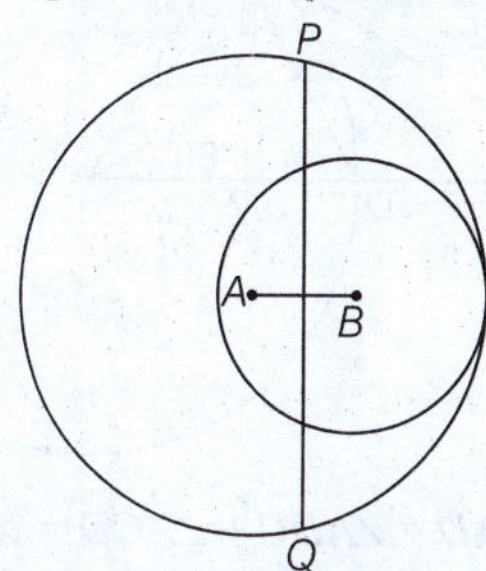

(a) $2\sqrt{6}$ (b) $\sqrt{34}$
(c) $4\sqrt{6}$ (d) $6\sqrt{2}$

24. A, B and C are three points on a circle such that the angles subtended by the chord AB and AC at the centre O are 110° and 130° respectively. Then, the value of $\angle BAC$ is : ***[SSC(CGL-T-I-19]***

(a) 45° (b) 60°
(c) 55° (d) 65°

25. A chord of length 5 cm subtends an angle of 60° at the centre of a circle. The length, in cm, of a chord that subtends an angle of 120° at the centre of the same circle is : ***[CAT-18]***

(a) 2π (b) $5\sqrt{3}$
(c) $6\sqrt{2}$ (d) 8

26. AB is a diameter of a circle of radius 5 cm. Let P and Q be two points on the circle so that the length of PB is 6 cm, and the length of AP is twice that of AQ. Then, the length, in cm, of QB is nearest to: ***[CAT-19]***

(a) 9.1 (b) 7.8
(c) 8.5 (d) 9.3

Answer Key

1. (b)	2. (b)	3. (a)	4. (d)	5. (d)	6. (a)	7. (c)	8. (b)	9. (b)	10. (c)
11. (b)	12. (d)	13. (b)	14. (d)	15. (b)	16. (b)	17. (c)	18. (a)	19. (d)	20. (a)
21. (d)	22. (c)	23. (c)	24. (b)	25. (b)	26. (a)				

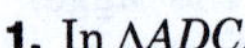

Hints and SOLUTIONS

1. In ΔADC,

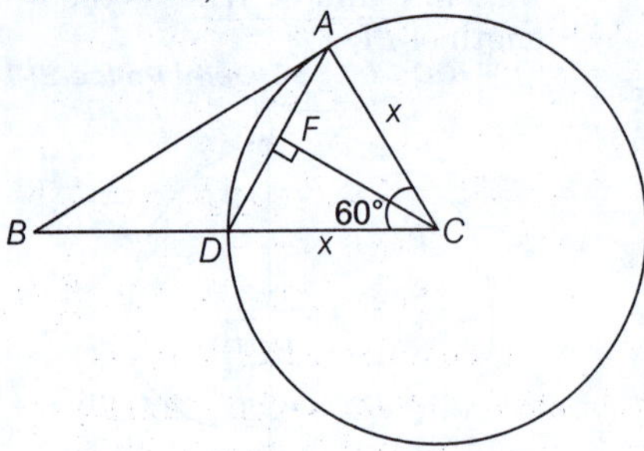

$\angle CAD + \angle ADC + \angle ACD = 180°$

$\Rightarrow \angle CAD + \angle ADC + 60° = 180°$

$\Rightarrow 2\angle CAD + 60° = 180°$

[$\because AC = DC$ (radii)

$\therefore \angle CAD = \angle ADC$]

$\Rightarrow \quad \angle CAD = 60°$

Hence, $\angle CAD = \angle ADC = \angle ACD = 60°$

It means ΔADC is equilateral.

Then, $AD = AC = CD = x$

Since a perpendicular drawn from the centre of a circle to a chord, bisects the chord, $DF = AF = x/2$.

2. Given, $AB = MN =$ diameter $= 3$ cm

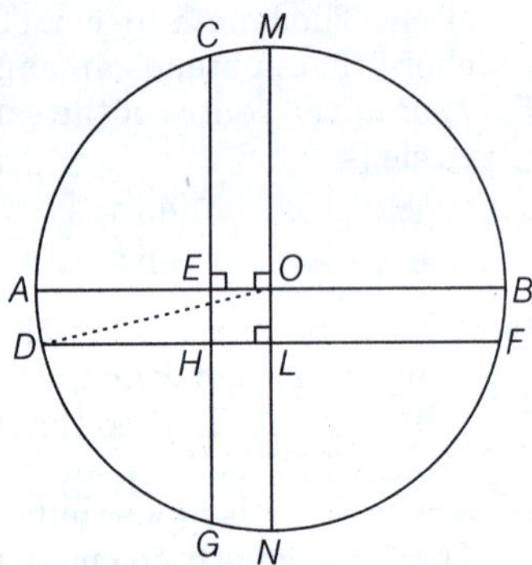

Then, $AE = 1$ cm and $EB = 2$ cm

[$\because AE : EB = 1 : 2$ as given]

and $NL = 1$ cm and $LM = 2$ cm

[$\because NL : LM = 1 : 2$ as given]

In right-angled ΔDOL,

$DL = \sqrt{OD^2 - OL^2} = \sqrt{(\text{radius})^2 - OL^2}$

$= \sqrt{(1.5)^2 - (0.5)^2} = \sqrt{2}$

[$\because$ radius $NO = 1.5$ and $NL = 1$

$\therefore OL = 0.5$]

Clearly, radius $AO = 1.5$ and $AE = 1$.

Then, $\quad EO = 0.5 = HL$

Hence, $DH = DL - HL = \sqrt{2} - 0.5$

$= \dfrac{(2\sqrt{2} - 1)}{2}$ cm

3. We draw a perpendicular on chord AB from the centre. Obviously, this perpendicular will bisect AB as well as CD as AB is parallel to CD.

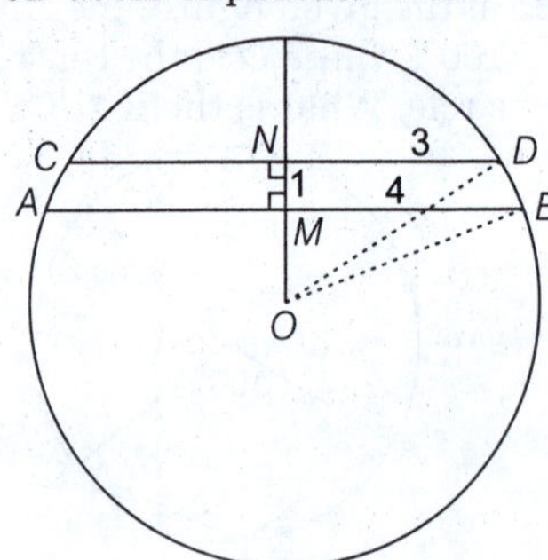

In right-angled ΔMOB,

$OB = \sqrt{MO^2 + MB^2} = \sqrt{MO^2 + 4^2}$

Also, in right-angled ΔNOD,

$OD = \sqrt{NO^2 + ND^2} = \sqrt{(1 + MO)^2 + 3^2}$

Since $OB = OD$ (radii), $\sqrt{MO^2 + 4^2}$

$= \sqrt{(1 + MO)^2 + 3^2}$

$\Rightarrow MO^2 + 4^2 = (1 + MO)^2 + 3^2$

$\Rightarrow \quad 16 = 2MO + 10$

$\Rightarrow \quad MO = 3$

Then,

$OB = \sqrt{MO^2 + 4^2} = \sqrt{3^2 + 4^2}$

$= 5$ cm

4. In equilateral triangle, circumradius $= 2 \times$ inradius

$\Rightarrow \quad 10 = 2 \times$ inradius

$\Rightarrow$ inradius $= 5$ cm

In an equilateral triangle, the radius is a part of its height.

Hence, required length = 5 cm

5. $\because$ Diameter of the larger circle = 20 cm

$\therefore$ Radius $OP = 10$ cm

Since the perpendicular drawn from the centre of a circle to a chord bisects the chord, $SR = RP$

Now, in right-angled ΔORP, $RP = 6$ as (6, 8, 10) is a Pythagorean triplet.

Then, $SP = SR + RP = 6 + 6 = 12$ cm

6. We know, if two circles with same radius pass through each other's centre, the common chord is equal to $\sqrt{3}$ times the radius. Then, length of the common chord $= 5\sqrt{3}$ cm.

7. Since an angle in a semicircle is a right angle, $\angle ADC = 90°$.

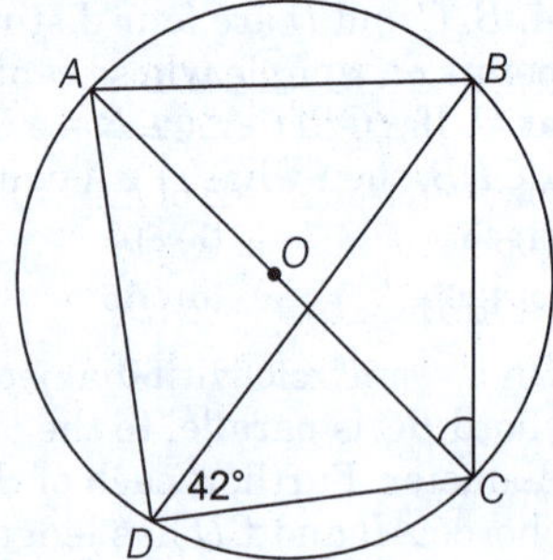

Then, $\angle ADB = 90° - 42° = 48°$

Further, angles in the same segments are equal.

Then, $\angle ADB = \angle ACB = 48°$

8. Clearly, $PQ^2 = PR^2 + RQ^2$

$\Rightarrow \quad 17^2 = (9 + x)^2 + (2 + x)^2$

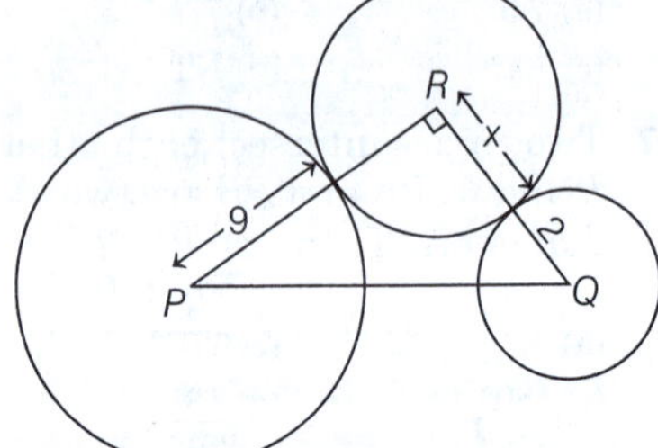

Since, (8, 15, 17) is a Pythagorean triplet,$(9 + x)$ and $(2 + x)$ will be equal to 8 and 15 or 15 and 8 respectively. Obviously, $(9 + x)$ will not be equal to 8 as x cannot be negative, hence, it equals 15.

$\Rightarrow \quad 9 + x = 15 \Rightarrow x = 6$

Further, $(2 + x) = 8 \Rightarrow x = 6$

Thus, the value of $x = 6$ cm

9. We know, an angle subtended by an arc at the centre is twice the angle subtended by it at any point of the circle in the alternate segment of the circle.

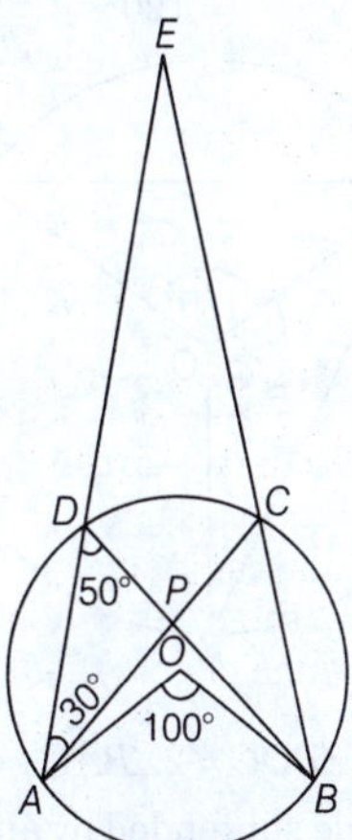

Hence, $\angle AOB = 2\angle ADB$

$\Rightarrow \quad ADB = 100°/2 = 50°$

Now, exterior $\angle APB$ of ΔADP = Sum of opposite interior angles

$\Rightarrow \angle APB = \angle DAP + \angle ADP$

$= 30° + 50° = 80°$

10. Since an angle in a semicircle is a right angle, $\angle ADB = 90°$

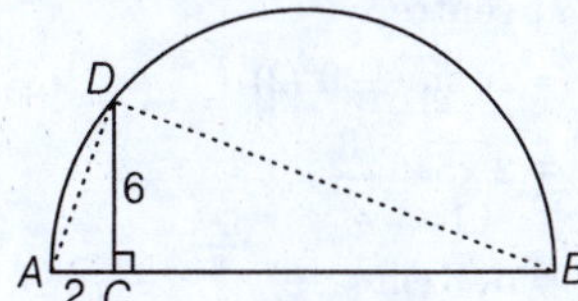

In right-angled ΔADB, $CD \perp AB$, then $CD^2 = AC \times BC$

[Recall the properties of a right-angled triangle detailed in sub-section 'Triangle']

$\Rightarrow \quad 6^2 = 2 \times BC \Rightarrow BC = 18$

Then, $AB = 2 + 18 = 20$ cm

$\therefore$ Radius of the semi-circle $= 20/2 = 10$ cm

11. Length of arc $= 2\pi r \times \dfrac{\theta}{360°}$

$= 2\pi \cdot 21 \times \dfrac{72°}{360°} = 26.4\,\text{cm}$

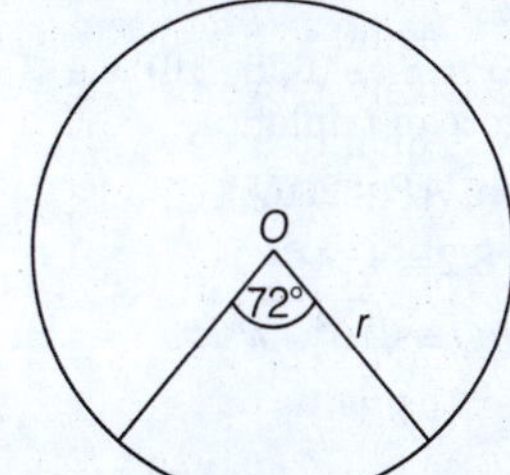

12. Clearly, $\angle BOD = 180° - 106° = 74°$

We know, the angle subtended by an arc at the centre is twice the angle subtended by it at any point of the circle in the alternate segment.

$\therefore \angle BOD = 2\,\angle BCD$ [$\because$ both angles have been made by arc AB]

$\Rightarrow 74° = 2\,\angle BCD \Rightarrow BCD = 37°$

13. In ΔAOB,

$\angle AOB + \angle OAB + \angle OBA = 180°$

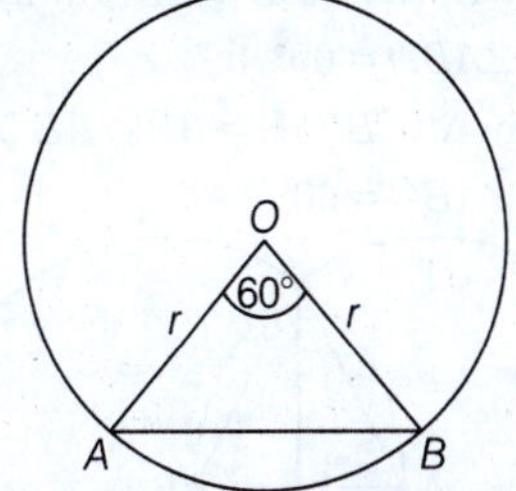

$\Rightarrow 60° + \angle OAB + \angle OAB = 180°$

[$\because \angle OAB = \angle OBA$ as $OA = OB$ (radii)]

$\Rightarrow \angle OAB = 60°$

$\Rightarrow \angle AOB = \angle OAB = \angle OBA = 60°$

Hence, ΔAOB is equilateral, hence, its each side is equal.

$\therefore \quad AB = OA = OB \Rightarrow AB = r$

Then, required ratio = 1 : 1

Alternatively

Length of chord $AB = 2r \times \sin(\theta/2)$

$= 2r \times \sin(60°/2) = r$

Hence, required ratio = 1 : 1

14. Let centres of the bigger and the smaller circles be O and B respectively. As given,

$OD = a/2$, $BC = b/2$ and $CD = c$

Clearly, $OD = DB + BO$

$\Rightarrow \quad \dfrac{a}{2} = (DC + CB) + BO$

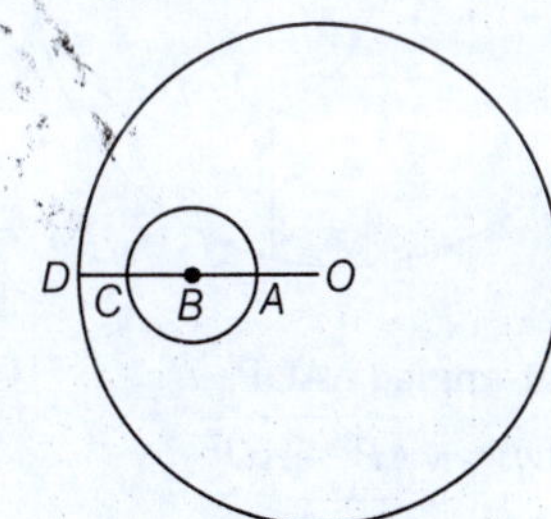

$\Rightarrow \quad \dfrac{a}{2} = c + \dfrac{b}{2} + BO$

$\Rightarrow \quad BO = \dfrac{1}{2}(a - b) - c$

Hence, distance between their centres i.e. $OB = \dfrac{1}{2}(a - b) - c$

15. Given, $\angle OBD - \angle CDB$

$= \angle CBD - \angle ODB$

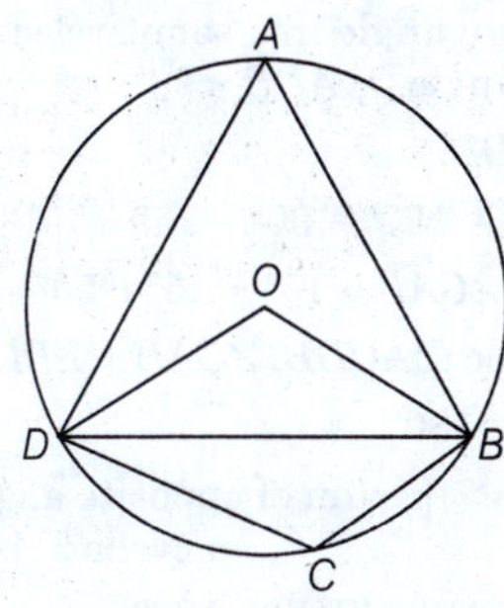

$\Rightarrow \angle OBD + \angle ODB = \angle CBD + \angle CDB$

$\Rightarrow 180° - \angle BOD = 180° - \angle BCD$

$\Rightarrow \quad \angle BOD = \angle BCD$

$\Rightarrow \quad 2\,\angle BAD = \angle BCD$

Since $\square ABCD$ is cyclic, $\angle BAD + \angle BCD = 180°$

[$\because$ sum of opposite angles is 180°]

$\Rightarrow \angle BAD + 2\,\angle BAD = 180°$

$\Rightarrow \angle BAD = 60°$

16. We draw the perpendiculars BP and CQ to AD. Since an angle in a semicircle is a right angle, $\angle ABD = 90° = \angle ACD$

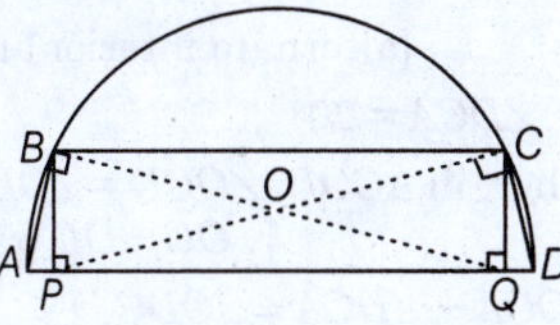

Now, in right-angled ΔABD,

$AB^2 = AP \times AD$

[Recall the properties of a right-angled Δ detailed in sub-section 'Triangle']

$\Rightarrow \quad 2^2 = AP \times 8$

$\Rightarrow \quad AP = 1/2$

Also, in right-angled ΔACD,

$CD^2 = QD \times AD$

$\Rightarrow \quad 2^2 = QD \times 8 \Rightarrow QD = \dfrac{1}{2}$

$\therefore \quad PQ = 8 - 1/2 - 1/2 = 7$

Hence, $BC = 7$ cm

17. We join P and Q.

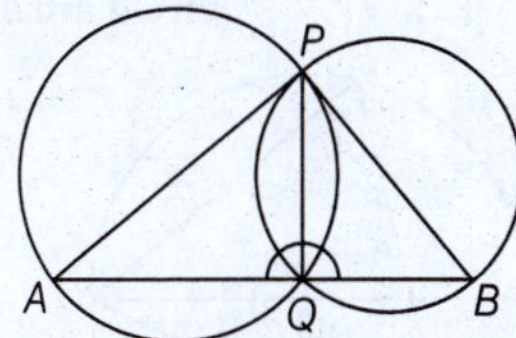

Since PQB and PQA are two semicircles, $\angle PQB = \angle PQA = 90°$

Then, $\angle AQB = 90° + 90° = 180°$

18. Since an angle in a semicircle is a right angle, $\angle ADB = 90°$

In ΔADB,

$\angle DAB = 180° - 90° - 75° = 15°$

Then, $\angle CAB = 15° + 35° = 50°$

In cyclic $\square ACDB$, $\angle CAB + \angle BDC = 180°$

[$\because$ sum of opposite angles in a cyclic $\square$ is 180°]

$\Rightarrow \quad 50° + \angle BDC = 180°$

$\Rightarrow \quad \angle BDC = 130°$

19. In ΔAOC, $AO = OC$ (radii)

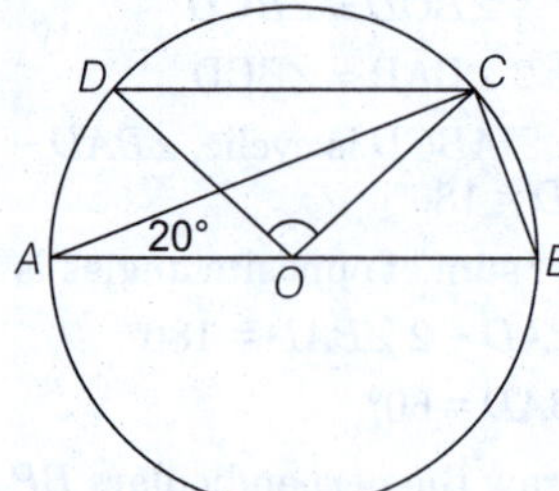

Then, $\angle OAC = \angle OCA = 20°$

$\because DC \mid\mid AB \therefore \angle DCA = \angle OAC$

[alternate interior angles]

$\Rightarrow \quad \angle DCA = 20°$

Further, in ΔCOD, $\angle OCD = \angle ODC$

[$\because OC = OD$ (radii)]

$\Rightarrow \angle OCA + \angle DCA = \angle ODC$

$\Rightarrow 20° + 20° = \angle ODC$

Hence, $\angle OCD = \angle ODC = 40°$

Then, $\angle COD = 180° - 40° - 40° = 100°$

20. Reflex $\angle AOC = 360° - 100° = 260°$

We know, the angle subtended by an arc at the centre is twice the angle subtended by it at any point of the circle in the alternate segment.

Hence, reflex $\angle AOC = 2\angle ABC$

$\Rightarrow \quad 260° = 2\angle ABC \Rightarrow \angle ABC = 130°$

21. (*d*) Since an angle in a semicircle is a right angle, $\angle AEC = 90°$

Also, $\angle EAC = \angle EBC = 65°$

[$\because$ angles in the same segment of a circle are equal]

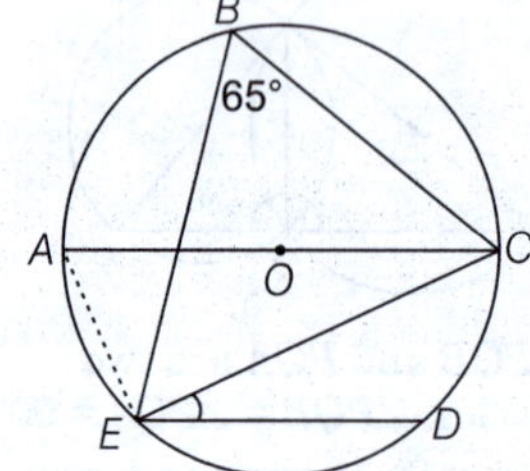

Now, in ΔAEC,

$\angle ACE = 180° - \angle EAC - \angle AEC$

$= 180° - 65° - 90° = 25°$

Given, $AC \mid\mid ED \therefore \angle ACE = \angle DEC$

(Alternate interior angles)

$\Rightarrow DEC = 25°$

22. In ΔADB, $AD = BD = AB$ (radii)

$\therefore \quad \angle ABD = 60°$

Also, in ΔACB, $AC = BC = AB$ [radii]

$\therefore \quad \angle ABC = 60°$

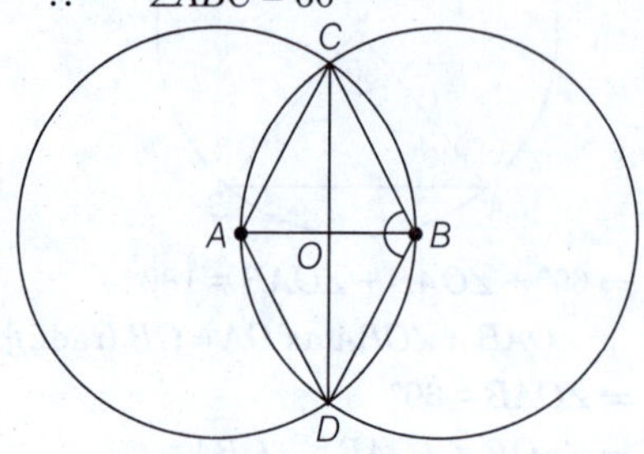

Then, $\angle DBC = \angle ABD + \angle ABC$

$= 60° + 60°$

$= 120°$

23. Clearly, $AB = 5 - 3 = 2$ units

Given, PQ is a perpendicular bisector of AB.

$\therefore \quad AO = OB = 1$ unit

Further, $AP = 5$ units

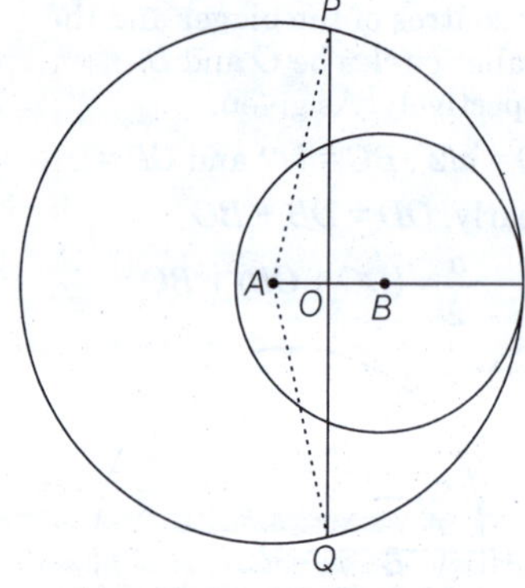

In right-angled ΔAOP,

$OP = \sqrt{AP^2 - AO^2}$

$OP = \sqrt{5^2 - 1^2}$

$= 2\sqrt{6}$

Since the perpendicular drawn from the centre to a chord, bisects the chord, $OP = OQ$.

Then, $PQ = 2OP$

$= 2 \times 2\sqrt{6}$

$= 4\sqrt{6}$ cm

24. Clearly, $\angle BOC = 360° - 110° - 130°$

$= 120°$

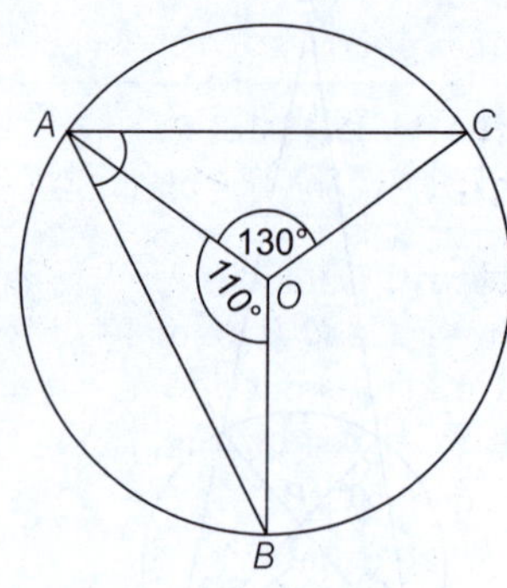

Further, $\angle BOC = 2\angle BAC$

[$\because$ the angle subtended by an arc at the centre is twice the angle subtended by it at any point of the circle in the alternate segment of the circle]

$\Rightarrow \quad \angle BAC = 60°$

25. Length of the chord $= 2r \sin(\theta/2)$

$\Rightarrow \quad 5 = 2r \times \sin(60°/2)$

$\Rightarrow \quad r = 5$

Now, length of the chord subtending 120° at centre

$= 2r \sin(120°/2)$

$= 2 \times 5 \times \frac{\sqrt{3}}{2}$

$= 5\sqrt{3}$ cm

26. Since, an angle in a semicircle is a right angle, ΔAQB and ΔAPB are right-angled triangles.

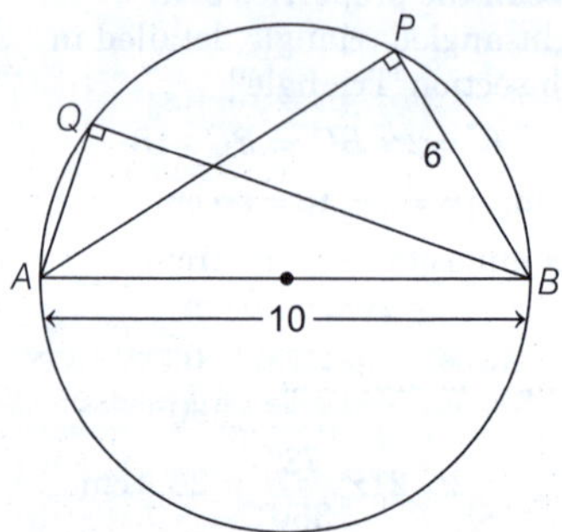

Now, AP = 8 as (6, 8 , 10) is a Pythagorean triplet.

As given, $AP = 2AQ$

$\therefore AQ = 8/2 = 4$

Then, $BQ = \sqrt{10^2 - 4^2}$

≈ 9.1

Illustrations Part B

Based on properties of a circle related to secant and tangent

Problems based on the concepts associated with the secant

Problem 1 Consider a circle with centre at O and radius r. Points A and B lie on its circumference and a point M lies outside of it such that M, A and O lie on the same straight line. Then, the ratio of MA to MB is

(a) equal to 1 (b) equal to r

(c) greater than 1 (d) less than 1

Sol. Clearly, two secants NM and BM of a circle meet outside the circle at point M. Then, $NM \times AM = BM \times KM$

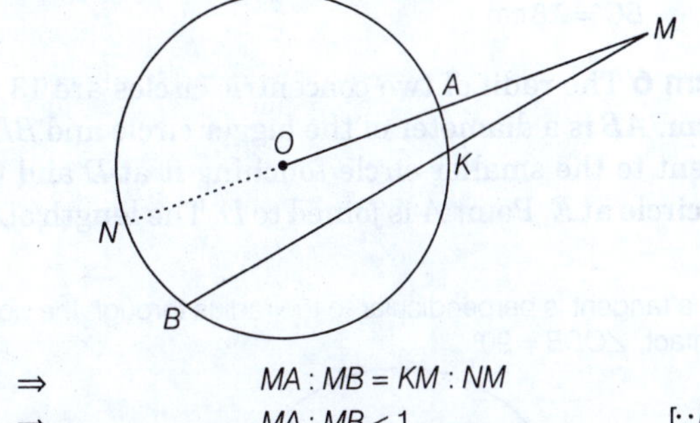

$\Rightarrow \quad MA : MB = KM : NM$

$\Rightarrow \quad MA : MB < 1 \quad [\because KM < NM]$

Problems based on the Concepts Associated with the Tangent

(Excluding Alternate Segment Theorem and Common Tangent)

Problem 2 A circular ring with centre O is kept in the vertical position by two weightless thin strings TP and TQ attached to the ring at P and Q. The line OT meets the the ring at E whereas a tangential string at E meets TP and TQ at A and B respectively. If the radius of ring is 5 cm and $OT = 13$ cm, then what is the length of AB? **[CDS-12, 08]**

Sol. Clearly, TP and TQ are tangents to the circle at points of contact P and Q respectively. We know, the tangent at and point of a circle is perpendicular to the radius through the point of contact.
Hence, $\angle OPT = 90°$ and $\angle OEA = 90°$

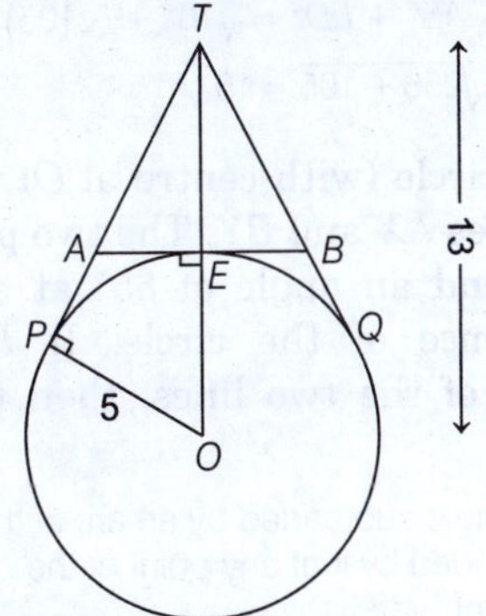

Now, in right-angle ΔTPO, $TP = 12$

$[\because (5, 12, 13)$ is a Pythagorean triplet$]$

In right-angle ΔTAE,

$TE = TO - OE = 13 - 5 = 8$

$TA = TP - AP = 12 - AP$

Then, $TA^2 = AE^2 + TE^2$

$\Rightarrow \quad (12 - AP)^2 = AE^2 + 8^2$

$\Rightarrow \quad (12 - AE)^2 = AE^2 + 8^2$

[$\because AP = AE$ as the tangents drawn from an external point to a circle are equal in length]

$\Rightarrow \quad 144 + AE^2 - 24AE = AE^2 + 64$

$\Rightarrow \quad AE = \frac{10}{3}$

On the above pattern, if we join O and Q and process accordingly, we get $EB = \frac{10}{3}$

$\therefore \quad AB = AE + EB = \frac{10}{3} + \frac{10}{3} = \frac{20}{3}$ cm

Alternatively

From the above, $AE = \frac{10}{3}$

Further, the centre of the circle lies on the bisector of the angle between the two tangents drawn from an external point.

$\therefore \quad \angle ATE = \angle BTE$

and $\angle TEA = \angle TEB = 90°$ and $TE = TE$ (common)

Thus, $\Delta ATE \cong \Delta BTE$ [by criterion A-S-A]

Then, $AE = EB = 10/3$

$\therefore \quad AB = AE + EB = \frac{10}{3} + \frac{10}{3} = \frac{20}{3}$ cm

Problem 3 Chords AB and CD of a circle intersect externally at P. If $AB = 6$ cm, $CD = 3$ cm and $PD = 5$ cm, then the length of PB is **[SSC (Delhi police-SI)-12, SSC (CGL-T-I)-19]**

Sol. We know, if two secants are drawn to a circle from an exterior point, then the product of the length of one secant and its external segment is equal to the product of length of another secant and its external segment.

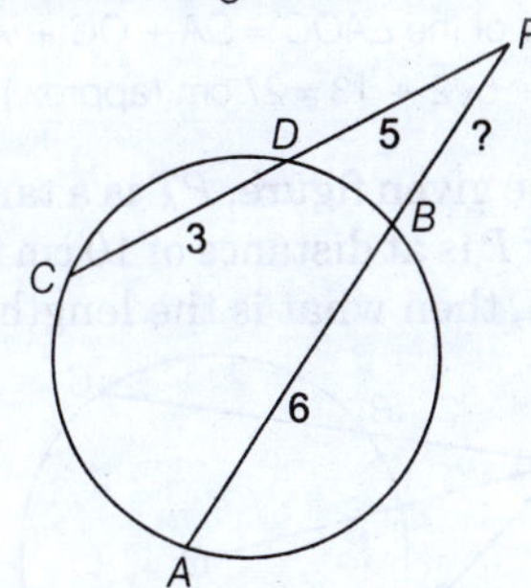

$\therefore \quad PC \times PD = PA \times PB$

$\Rightarrow \quad (3 + 5)\cdot 5 = (6 + x)\cdot x \quad$ [let PB be x]

$\Rightarrow \quad x^2 + 6x - 40 = 0 \Rightarrow x^2 + 10x - 4x - 40 = 0$

$\Rightarrow \quad x(x + 10) - 4(x + 10) = 0 \Rightarrow (x - 4)(x + 10) = 0$

$\Rightarrow \quad x = 4$ and -10

$\Rightarrow \quad x = 4$ as the value of x cannot be negative.

Hence, $PB = 4$ cm

Problem 4 In the given diagram, CT is tangent at C, making an angle of 45° with CD. O is the centre of circle, $CD = 10$ cm. What is the perimeter of the shaded region (ΔAOC) approximately? ***[CAT-13]***

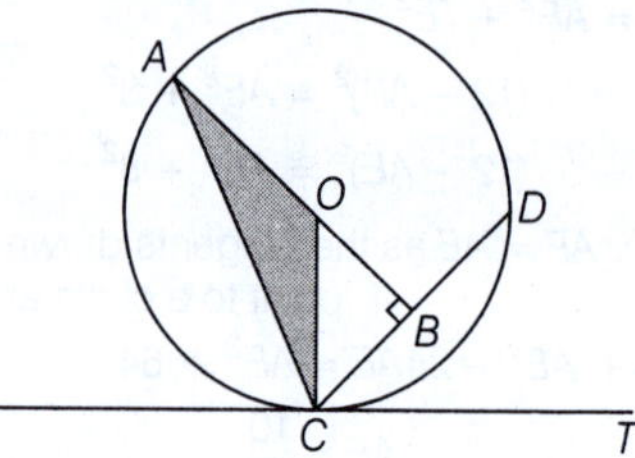

Sol. Since the perpendicular drawn from the centre of a circle to a chord, bisects the chord, $CB = BD = \frac{10}{2} = 5$ cm.

Futher, since a tangent is perpendicular to the radius through the point of contact, $\angle OCT = 90°$

Then, $\angle OCB = \angle OCT - \angle DCT = 90° - 45°$ [$\because \angle DCT = 45°$ as given]

$\Rightarrow \angle OCB = 45°$

In right-angled ΔOBC,

$\angle COB = 180° - \angle OBC - \angle OCB$
$= 180° - 90° - 45° = 45°$

In right-angle ΔOBC, $\angle COB = \angle OCB$

$\therefore CB = OB = 5$

Then, $OC = \sqrt{5^2 + 5^2} = 5\sqrt{2}$ cm

Further, $OC = OA$ [radii]

$\therefore OA = 5\sqrt{2}$

In right-angle ΔABC,

$AC = \sqrt{AB^2 + BC^2}$
$= \sqrt{(AO + OB)^2 + BC^2}$
$= \sqrt{(5\sqrt{2} + 5)^2 + 5^2}$
$= \sqrt{100 + 50\sqrt{2}}$
$= 13$ (approx.)

Now, premimeter of the $\Delta AOC = OA + OC + AC$
$= 5\sqrt{2} + 5\sqrt{2} + 13 = 27$ cm (approx.)

Problem 5 In the given figure, PT is a tangent to a circle of radius 6 cm. If P is at distance of 10 cm from the centre O and $PB = 5$ cm, then what is the length of chord BC?

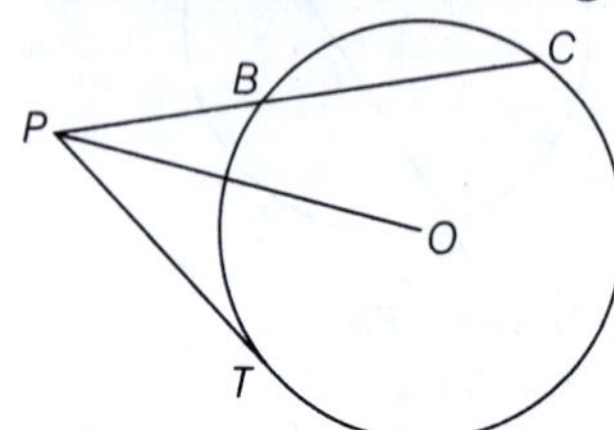

Sol. We know, the tangent at any point of a circle is perpendicular to the radius through the point of centact.

Hence, $\angle PTO = 90°$

Then, $PT = 8$ as (6, 8, 10) is Pythagorean triplet.

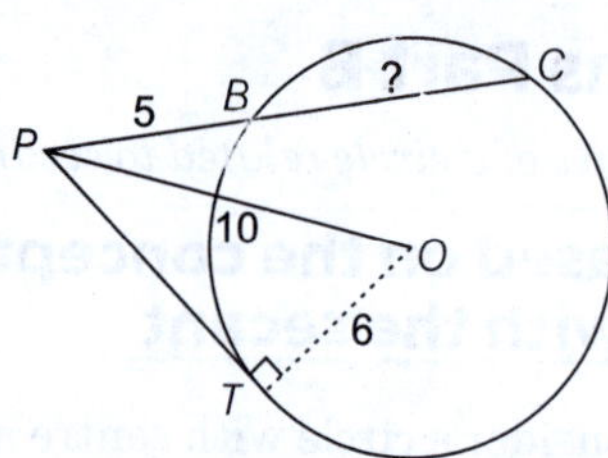

Further, the square of the length of a tangent is equal to the product of the total length of the secant and the external secant segment, if both the secant and the tangent are drawn from the same exterior point.

Hence, $PT^2 = PB \times PC$

$\Rightarrow 8^2 = 5 \times (PB + BC)$

$\Rightarrow 64 = 5(5 + BC)$

$\Rightarrow BC = 7.8$ cm

Problem 6 The radii of two concentric circles are 13 cm and 8 cm. AB is a diameter of the bigger circle and BD is a tangent to the smaller circle touching it at D and the bigger circle at E. Point A is joined to D. The length of AD is:

Sol. Since a tangent is perpendicular to the radius through the point of contact, $\angle ODB = 90°$

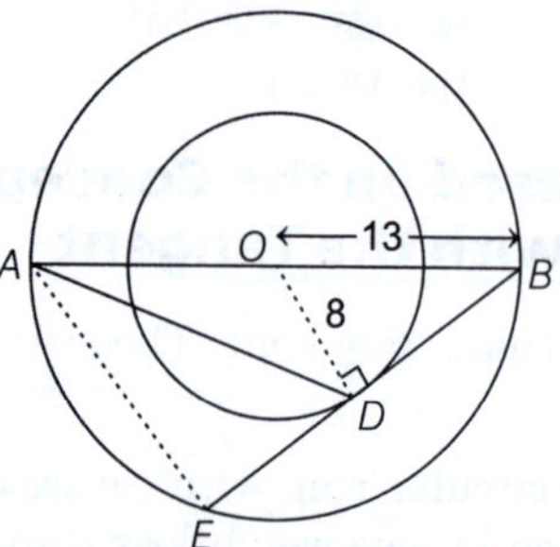

In right-angled ΔODB,

$DB = \sqrt{13^2 - 8^2} = \sqrt{105}$

Clearly, O is the mid-point of AB and D is the mid-point of EB as the perpendicular OD drawn from the centre divides the chord BE.

Since the line joining the mid-points of any two sides is equal to half the third side, $AE = 2OD$. Then, $AE = 2 \times 8 = 16$ cm.

$\angle AEB = 90°$ as an angle in a semicircle is a right angle.

Now, in right-angled ΔAED,

$AD = \sqrt{AE^2 + ED^2} = \sqrt{16^2 + (\sqrt{105})^2}$ $\{\because DB = ED\}$

$\Rightarrow AD = \sqrt{256 + 105} = 19$ cm

Problem 7 A circle (with centre at O) is touching two intersecting lines AX and BY. The two points of contact A and B subtend an angle of 65° at any point C on the circumference of the circle. If P is the point of intersection of the two lines, then the measure of $\angle APO$ is:

Sol. We know, the angle subtended by an arc at the centre is twice the angle subtended by it at any point of the circle in the alternate segment.

Hence, $\angle BOA = 2\angle BCA = 2 \times 65° = 130°$

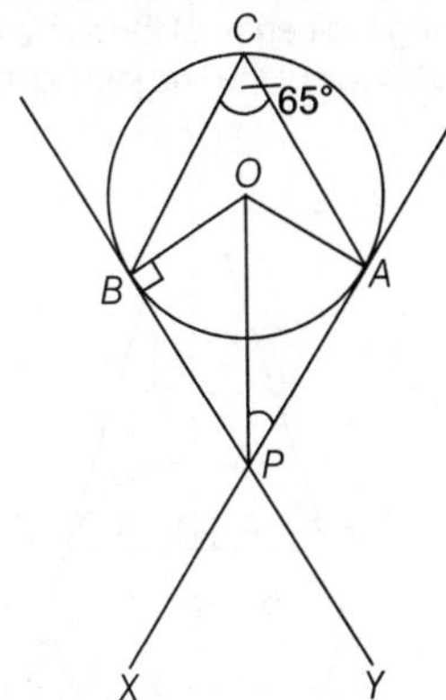

Since a tangent is perpendicular to the radius through the point of contact, $\angle OBP = 90° = \angle OAP$

Now, in $\square BOAP$,

$$\angle BPA = 360° - 130° - 90° - 90° = 50°$$

Also, it is the fact that the centre of a circle lies on the bisector of the angle between the two tangents drawn to it from an external point. It implies that OP is the angle bisector of $\angle BPA$.

$\therefore \quad \angle APO = \frac{1}{2}\angle BPA = \frac{1}{2} \times 50° = 25°$

Alternatively

We know, the angle subtended by an arc at the centre is twice the angle subtended by it at any point of the circle in the alternate segment.

$\therefore \quad \angle BOA = 2\angle BCA = 2 \times 65° = 130°$

Since a tangent is perpendicular to the radius through the point of contact, $\angle OBP = 90° = \angle OAP$

Now, in ΔBOP and ΔAOP,

$OB = OA$ (radii), $OP = OP$ (common)

and $BP = AP$

[$\because$ tangents drawn from an external point are equal]

Hence, $\Delta BOP \cong \Delta AOP$ [by criterion S-S-S]

Then, $\angle BOP = \angle AOP = \frac{130°}{2} = 65°$

In ΔOAP, $\angle APO = 180° - \angle OAP - \angle AOP$

$\Rightarrow \quad \angle APO = 180° - 90° - 65° = 25°$

Problem 8 **The centre of a circle inside a triangle is at a distance of 625 cm from each of the vertices of the triangle. If the diameter of the circle is 350 cm and the circle is touching only two sides of the triangle, then find the area of the triangle.** ***[XAT-15]***

Sol. As given, $AO = BO = CO = 625$ cm

and radius of the circle (i.e. OD and OE)

$$= \frac{350}{2} = 175$$

Also, $\angle ADO = \angle AEO = 90°$ as the tangent is perpendicular to the radius through point of contact.

In right-angled ΔADO,

$$AD = \sqrt{AO^2 - OD^2}$$
$$= \sqrt{625^2 - 175^2} = 600$$

Right-angled $\Delta ADO \cong$ right-angled ΔBDO [by criterion R-H-S as hypotenuse AO = hypotenuse BO and $OD = OD$ (common)]

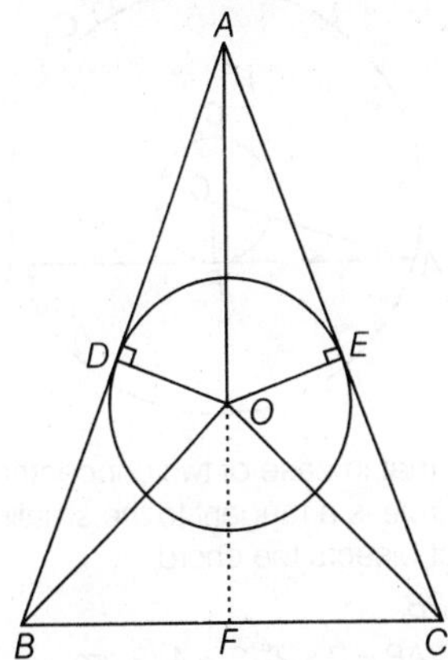

Then, $AD = BD$

$\Rightarrow \quad AB = AD + BD = 600 + 600 = 1200$

Also, $\Delta AOE \cong \Delta COE$

[by criterion R-H-S as hypotenuse AO = hypotenuse CO and $OE = OE$ (common)].

Then, $AE = EC = \sqrt{625^2 - 175^2} = 600$

Then, $AC = 600 + 600 = 1200$ cm

Now, we produce AO to point F which lies on BC. Since, the centre of a circle lies on the bisector of the angle between the two tangents drawn to it from an external point, AF is the angle bisector of $\angle BAC$. Also, it is fact that the bisector of the angle between two congruent sides of an isosceles triangle also acts as its median as well as altitude.

Hence, $\angle AFB = 90°$ and $BF = FC$

Further, $\Delta ADO \sim \Delta ABF$

[$\because \angle ADO = \angle AFB = 90°$, $\angle DAO = \angle BAF$ (common)]

$$\therefore \quad \frac{DO}{BF} = \frac{AO}{AB} = \frac{AD}{AF}$$

$$\Rightarrow \quad \frac{175}{BF} = \frac{625}{1200} = \frac{600}{AF}$$

$$\Rightarrow \quad BF = \frac{175 \times 1200}{625}$$

$$\text{and} \quad AF = \frac{600 \times 1200}{625}$$

$\Rightarrow \quad BF = 336$ and $AF = 1152$

Now, area of $\Delta ABC = \frac{1}{2} \times BC \times AF$

$$= \frac{1}{2} \times 2BF \times AF$$
$$= 336 \times 1152$$
$$= 387072 \text{ cm}^2$$

Problem 9 **A chord AB of a circle $C1$ of radius $(\sqrt{3}+1)$ touches a circle $C2$ which is concentric to $C1$. If radius of $C2$ is $(\sqrt{3}-1)$ cm, the length of AB is**

Sol. Since a tangent is perpendicular to the radius through the point of contact, $\angle OPA = 90°$

Clearly, $AP = \sqrt{AO^2 - OP^2}$

$$= \sqrt{(\sqrt{3}+1)^2 - (\sqrt{3}-1)^2}$$

$= \sqrt{4\sqrt{3}} = 2\sqrt[4]{3}$ cm

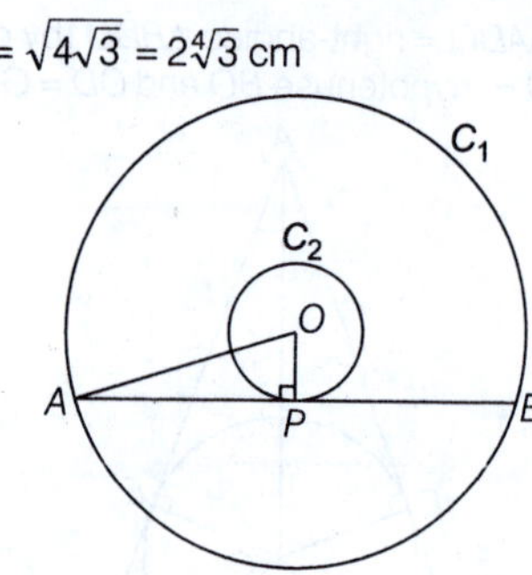

Also, we know that in case of two concentric circules, if a chord of the bigger circle is a tangent to the smaller circle, then the point of contact bisects the chord.

Hence, $AP = PB$

$\therefore \quad AB = 2AP = 2 \times 2\sqrt[4]{3} = 4\sqrt[4]{3}$ cm

Problem 10 A circle is touching the side BC of ΔABC at P and touching AB and AC produced at Q and R respectively. Then, length of the AQ is equal to:

(a) $\frac{1}{2} \times$ perimeter of ΔABC

(b) perimeter of ΔABC

(c) $\frac{1}{4} \times$ perimeter of ΔABC

(d) None of the above

Sol. Since tangent drawn from an exernal point to a circle are equal in length, $AQ = AR$, $BQ = BP$ and $CR = CP$.

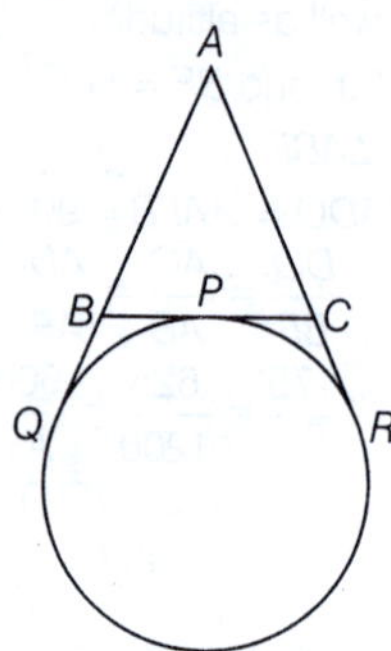

Now, perimeter of $\Delta ABC = AB + BC + CA$

$= AB + BP + CP + CA$

$= AQ - BQ + BP + CP + AR - CR$

$= AQ - BQ + BQ + CP + AR - CP$

$[\because BQ = BP \text{ and } CR = CP]$

$= AQ + AR$

$= AQ + AQ \quad [\because AR = AQ]$

$\Rightarrow$ perimeter of $\Delta ABC = 2AQ$

$\Rightarrow \quad AQ = \frac{1}{2} \times$ perimeter of ΔABC

Problem 11 PQ is a chord of length 8 cm, of a circle with centre O and of radius 5 cm. The tangents at P and Q intersect at point T. The length of TP is

[SSC (10 + 2 –DEO/∠DC)-13]

Sol. We know, if two tangents of a circle are drawn from an external point, the line joining the centre of the circle and the external point is perpendicular bisector of the line joining the points of contact.

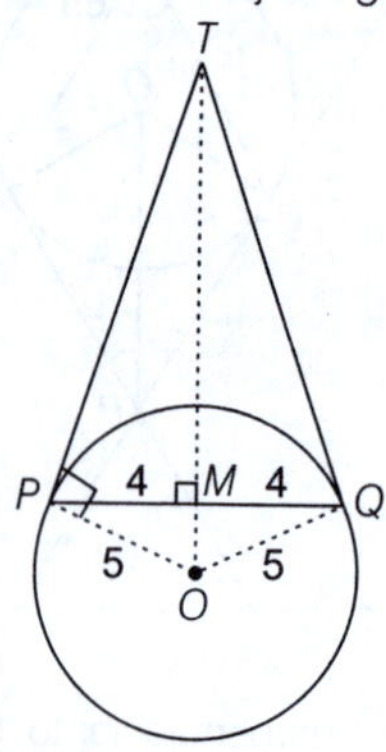

$\therefore \quad PM = MQ = \frac{8}{2} = 4$ cm

In right-angled ΔPMO, $OM = 3$ [$\because$ (3, 4, 5) is a Pythagorean triplet]

Since a tangent is perpendicular to the radius through the point of contact, $\angle TPO = 90°$

Then, in right-angled ΔTPO,

$PM^2 = TM \times OM$

[Recall the properties of a right-angle triangle detailed in sub-section 'Triangle']

$\Rightarrow \quad 4^2 = TM \times 3$

$\Rightarrow \quad TM = \frac{16}{3}$

Now, in right-angled ΔPTM,

$PT = \sqrt{PM^2 + TM^2}$

$= \sqrt{4^2 + \left(\frac{16}{3}\right)^2} = \frac{20}{3}$ cm

Problem 12 Two tangents are drawn from a point P on the circle with centre at O, touching the circle at point Q and T respectively. Another tangent AB touches the circle at point S. If angle $QPT = 55°$, find the angle AOB.

[IIFT-17]

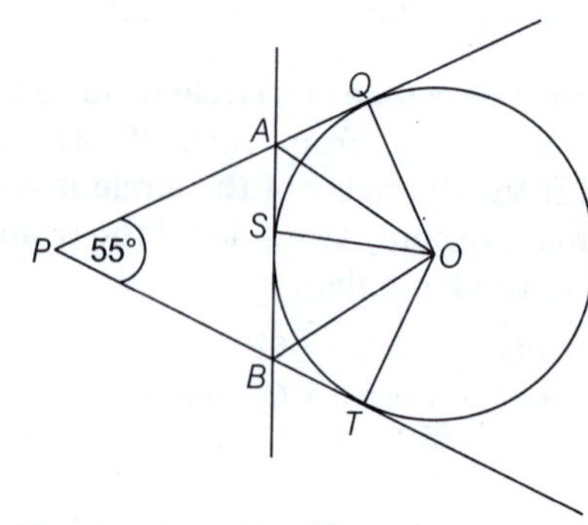

Sol. Since $PQOT$ is a quadrilateral,

$\angle QOT = 360° - \angle PQO - \angle PTO - \angle TPO$

$\Rightarrow \quad \angle QOT = 360° - 90° - 90° - 55° = 125°$

[$\because$ the tangent at any point of a circle is perpendicular to the radius through the point of contact]

Further, since AQ and AS are two tangents from the same point A, they will subtend equal angles at centre of the circle.

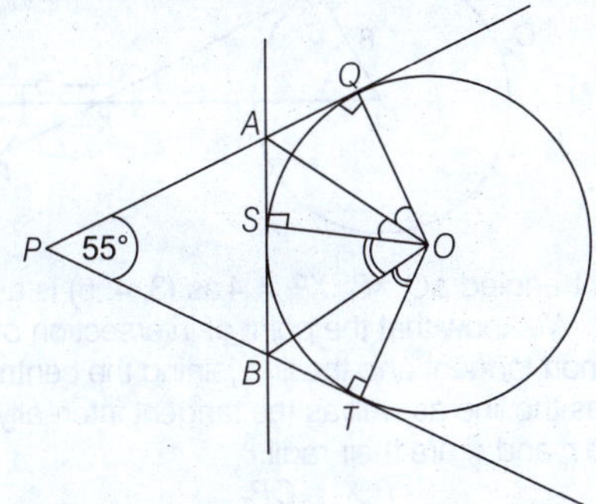

Hence, $\angle AOQ = \angle AOS$

Similarly, $\angle BOT = \angle BOS$

From the above, $\angle QOT = 125°$

$\Rightarrow \angle AOQ + \angle AOS + \angle BOT + \angle BOS = 125°$

$\Rightarrow \quad 2(\angle AOS + \angle BOS) = 125°$

$\Rightarrow \quad \angle AOB = \dfrac{125°}{2} = 62.5°$

Problems Based on Alternate Segment Theorem

Problem 13 PRT is a tangent to a circle with centre O, at the point R on it. Diameter SQ of the circle is produced to meet the tangent at P and QR is joined. If $\angle QRP = 28°$, then the measure of $\angle SPR$ is

[SSC (CGL-T-I)-19]

Sol. Since an angle in a semi-circle is a right angle, $\angle QRS = 90°$.

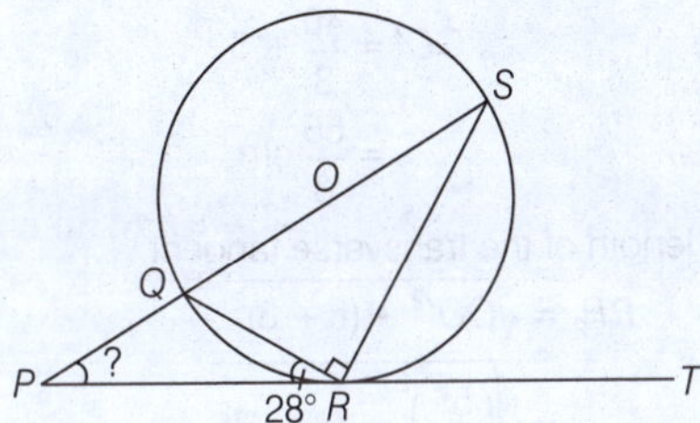

Further, since the angle between a tangent and a chord at the point of contact is equal to the angle in the alternate segment,

$\angle QRP = \angle QSR = 28°$

In ΔQRS, $\angle SQR = 180° - \angle QRS - \angle QSR$

$= 180° - 90° - 28° = 62°$

Now, exterior $\angle SQR = \angle QRP + ?$

[exterior angle of a triangle is equal to the sum of two interior angles opposite to it]

$\Rightarrow \quad ? = 62° - 28° = 34°$

Problem 14 $ABCD$ is concyclic quadrilateral. The tangents at A and C intersect each other at P. If $\angle ABC = 100°$, then what is $\angle APC$ equal to?

Sol. $\angle ADC + \angle ABC = 180°$

[∵ sum of opposite angles in a cyclic quadrilateral is 180°]

$\Rightarrow \quad \angle ADC = 180° - 100° = 80°$

Now, $\angle CAP = \angle ADC$ and $\angle ACP = \angle ADC$

[∵ the angle between a tangent and a chord at the point of contact is equal to the angle in the alternate segment.]

$\therefore \quad \angle CAP = \angle ACP = \angle ADC = 80°$

In ΔACP, $\angle APC = 180° - \angle CAP - \angle ACP$

$= 180° - 80° - 80° = 20°$

Problem 15 In the figure given, O is the centre of the circumcircle of the ΔXYZ. Tangents at X and Y intersect at T. If $\angle XTY = 80°$, then what is the value of $\angle ZXY$?

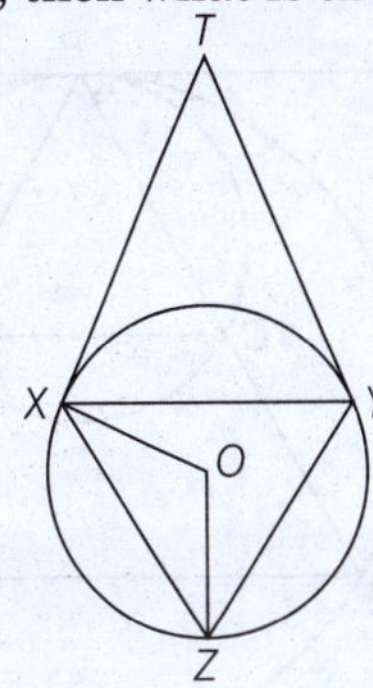

Sol. Since, the tangents drawn from an external point to a circle are equal in length, $XT = YT$.

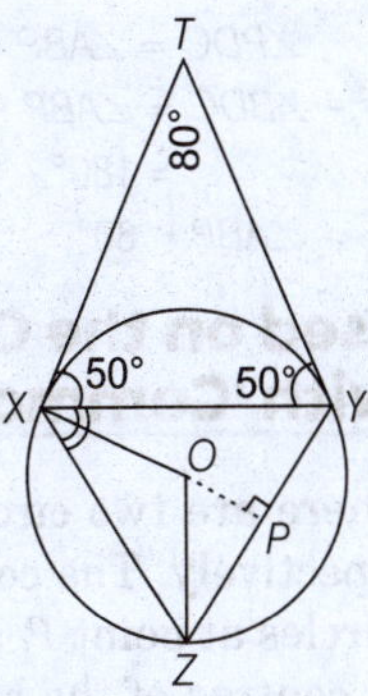

Then, $\angle TXY = \angle TYX$

Now in ΔIXY,

$\angle TXY + \angle TYX + \angle XTY = 180°$

$\Rightarrow \quad \angle TXY + \angle TYX + 80° = 180°$

$\Rightarrow \quad \angle TXY + \angle TXY = 100°$

$\Rightarrow \quad \angle TXY = 50°$

$\Rightarrow \quad \angle TXY = \angle TYX = 50°$

Further, since, a tangent is perpendicular to the radius through the point of contact, $\angle TXO = 90°$

$\Rightarrow \quad \angle TXY + \angle YXP = 90°$

$\Rightarrow \quad \angle YXP = 90° - 50° = 40°$

Obviously, $OP \perp ZY$ as perpendicular bisectors of sides of a triangle meet at circumcentre.

Then, in ΔXPY, $\angle XYP = 180° - \angle XPY - \angle YXP$

$= 180° - 90° - 40° = 50°$

Also, $\angle XZY = \angle TYX$

[$\because$ the angle between a tangent and a chord at the point of contact is equal to the angle in the alternate segment.]

$\Rightarrow \quad \angle XZY = 50°$

In ΔXYZ, $\angle XYZ + \angle YZX + \angle ZXY = 180°$

$\Rightarrow \quad \angle XYP + 50° + \angle ZXY = 180°$

$\Rightarrow \quad 50° + 50° + \angle ZXY = 180°$

$\Rightarrow \quad \angle ZXY = 60°$

Problem 16 **Two circles *C1* and *C2* touch each other internally at *P*. Two lines *PCA* and *PDB* meet the circles *C1* in *C, D* and *C2* in *A, B* respectively. If $\angle BDC = 120°$, then the value of $\angle ABP$ is equal to**

Sol. Since the angle between a tangent and a chord at the point of contact is equal to the angle in the alternate segment,

$\angle CPM = \angle PDC$ and $\angle APM = \angle ABP$

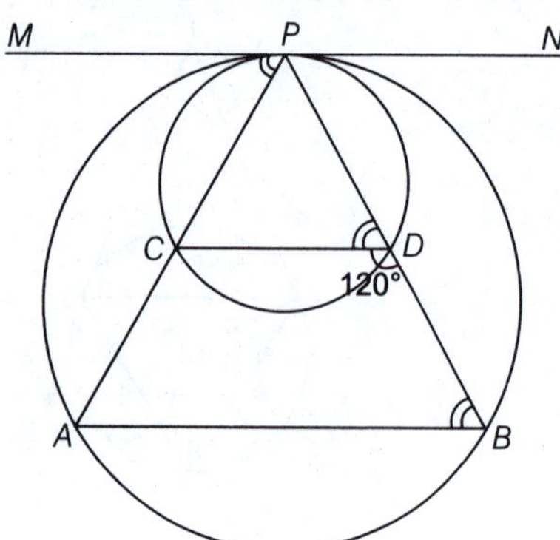

$\because \quad \angle CPM = \angle APM$

$\therefore \quad \angle PDC = \angle ABP$

$\Rightarrow \quad 180° - \angle BDC = \angle ABP$

$= 180° - 120° = \angle ABP$

$\therefore \quad \angle ABP = 60°$

Problems Based on the Concepts Associated with 'Common Tangent'

Problem 17 **There are two circles C_1 and C_2 of radii 3 and 8 units respectively. The common internal tangent *T* touches the circles at point P_1 and P_2 respectively. The line joining the centres of the circles intersects *T* at *X*. The distance of *X* from the centre of the smaller circle is 5 units. What is the length of the line segment $P_1 P_2$?**

(a) ≤ 13 (b) > 13 and ≤ 14 **[XAT-14]**

(c) > 14 and ≤ 15 (d) > 15 and ≤ 16

(e) > 16

Sol.

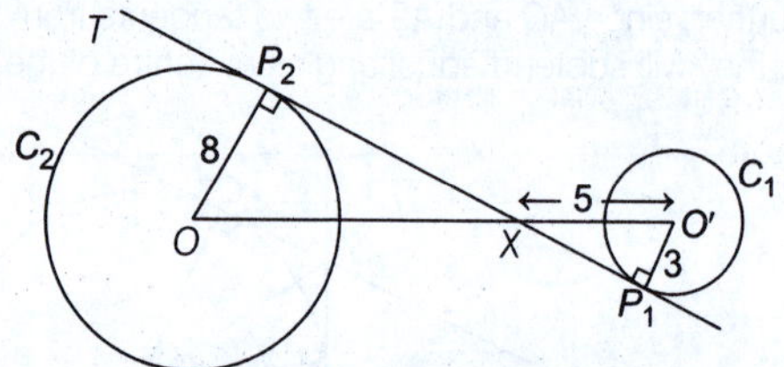

In right-angled $\Delta O'XP_1$, $XP_1 = 4$ as (3, 4, 5) is a Pythagorean triplet. We know that the point of intersection of the transverse common tangent and the line joining the centres of the circles divides the line as well as the tangent internally in ratio $r_1 : r_2$ where r_1 and r_2 are their radii.

$\therefore \quad \frac{P_2X}{P_1X} = \frac{OP_2}{O'P_1}$

$\Rightarrow \quad \frac{P_2X}{P_1X} = \frac{8}{3}$

$\Rightarrow \quad \frac{P_2X}{4} = \frac{8}{3}$

$\Rightarrow \quad P_2X = 10.66$

$\therefore \quad P_1P_2 = XP_1 + P_2X$

$= 4 + 10.66$

$= 14.66$ cm

Alternatively

We know that the point of intersection of the transverse common tangent and the line joining the centres of the circles divides the line internally in ratio $r_1 : r_2$, where r_1 and r_2 are their radii.

$\therefore \quad OX : XO' = 8 : 3$

$\Rightarrow \quad OX : 5 = 8 : 3$

$\Rightarrow \quad OX = \frac{40}{3}$ cm

Then, $\quad OO' = \frac{40}{3} + 5$

$= \frac{55}{3}$ cm

Now, length of the transverse tangent

$P_1P_2 = \sqrt{OO'^2 - (8+3)^2}$

$= \sqrt{\left(\frac{55}{3}\right)^2 - (8+3)^2}$

$= 14.66$ cm

Problem 18 **The centres of two circles of radii 7 cm and 2 cm are 13 cm apart. Then the length of direct common tangent to the circles between the points of contact is**

Sol. Length of the direct common tangent = $AB = CD$

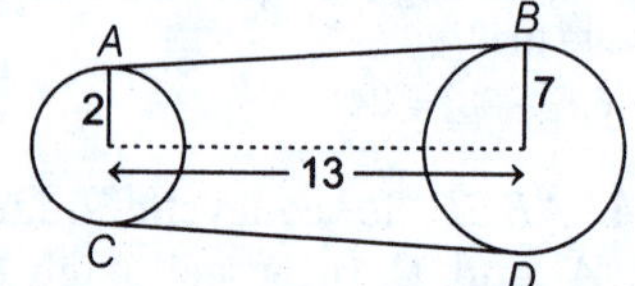

$= \sqrt{(\text{distance between centres})^2 - (\text{difference between radii})^2}$

$= \sqrt{13^2 - (7-2)^2}$

$= 12$ cm

Problem 19 Two circles with radii 4 cm and 9 cm touch each other externally. Find the radius of a circle touching the both circles externally and their direct common tangent.

Sol.

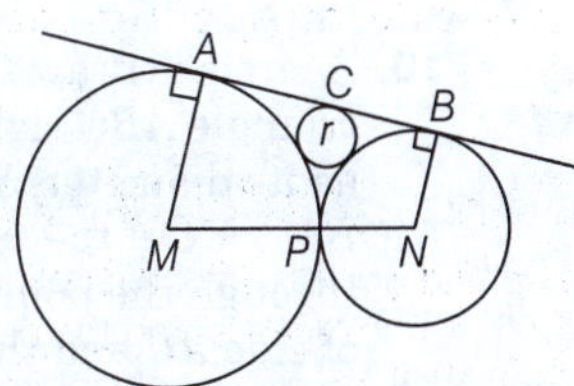

Let the desired radius be r. Length of common tangent

$$AB = \sqrt{4 \cdot r_1 \cdot r_2}$$

where r_1 and r_2 are radii of the circles.

$\Rightarrow \quad AB = \sqrt{4 \cdot 9 \cdot 4} = 12$ cm

Further, length of common tangent AC

$= \sqrt{4 \cdot 9 \cdot r} = 6\sqrt{r}$

and length of common tangent BC

$= \sqrt{4 \cdot 4 \cdot r} = 4\sqrt{r}$

Clearly, $AB = AC + BC$

$\Rightarrow \quad 12 = 6\sqrt{r} + 4\sqrt{r}$

$\Rightarrow \quad \sqrt{r} = 1.2$

$\Rightarrow \quad r = 1.44$ cm

Problem 20 AC is transverse common tangent to two circles with centres P and Q and radii 6 cm and 3 cm at the point A and C respectively. If AC cuts PQ at the point B and $AB = 8$ cm, then the length of PQ is : ***[SSC (CGL-T-T)-15]***

Sol.

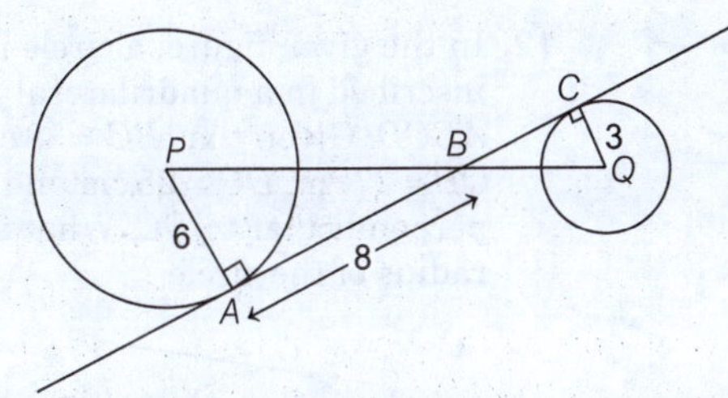

In right-angled ΔPAB, $PB = 10$ cm

[$\because$ (6, 8, 10) is a Pythogorean triplet]

We know that the point of intersection of the transverse common tangent and the line joining the centres of the circles divides the line internally in ratio $r_1 : r_2$, where r_1 and r_2 are their radii.

$\therefore \quad PB : BQ = 6 : 3$

$\Rightarrow \quad 10 : BQ = 6 : 3$

$\Rightarrow \quad BQ = 5$

Then, $\quad PQ = 10 + 5 = 15$ cm

Directions (Problems 21 to 23) *Answer the questions on the basis of the information given below.*

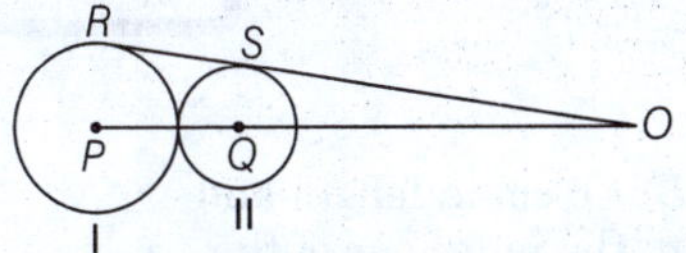

In the given figure, I and II are circles with centres P and Q respectively. The two circles touch each other and have a common tangent that touches them at point R and S respectively. This common tangent meets the line joining P and Q at O. The diameters of I and II are in the ratio 4 : 3. It is also known that the length of PO is 28 cm.

Problem 21 What is the ratio of the length of PQ to that of QO ?

Problem 22 What is the radius of the circle II ?

Problem 23 The length of SO is ***[CAT-04]***

Sol. Here, $r_1 : r_2 = 4 : 3$ as ratio of diameters of circle-I and II is 4 : 3

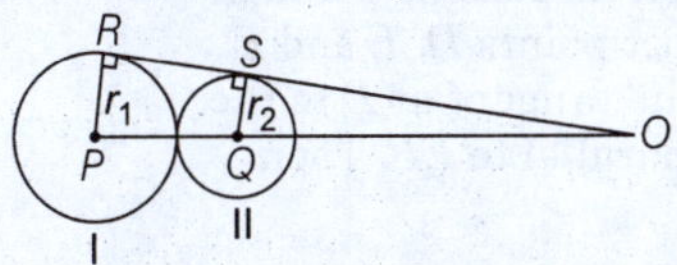

21. We know that the point of intersection of the Direct common tangent and the line joining the centres of the circles divides the line externally in ratio $r_1 : r_2$, where r_1 and r_2 are their radii.

$\therefore \quad PO : OQ = r_1 : r_2$

$\Rightarrow \quad PO : OQ = 4 : 3$

$\Rightarrow (PO - OQ) : QO = (4 - 3) : 3$

$\Rightarrow \quad PQ : QO = 1 : 3$

22. From above, $PQ : QO = 1 : 3$

As given, 4 units $\Rightarrow$ 28 cm [$\because PO = PQ + QO$]

$\therefore$ 1 unit $\Rightarrow$ 7 cm

i.e. $PQ = 7$ cm

$\Rightarrow PQ = r_1 + r_2 = 7$ cm

$\therefore \quad r_2 = 3$ cm [$\because r_1 : r_2 = 4 : 3$]

23. From above, $PO : OQ = 4 : 3$

$\Rightarrow \quad 28 : OQ = 4 : 3$

$\Rightarrow \quad OQ = 21$

Since, the tangent at any point of a circle is perpendicular to the radius through the point of contact, $\angle QSO = 90°$

In right-angled ΔQSO,

$$SO = \sqrt{OQ^2 - (r_2)^2} = \sqrt{21^2 - 3^2}$$

$$= \sqrt{24 \times 18} = 12\sqrt{3} \text{ cm}$$

WORKOUT

1. The line AB is 6 cm in length and is tangent to the inner one of the two concentric circles at point C. It is known that the radii of the two circles are integers. The radius of the outer circle is ...... cm. ***[CAT-15]***

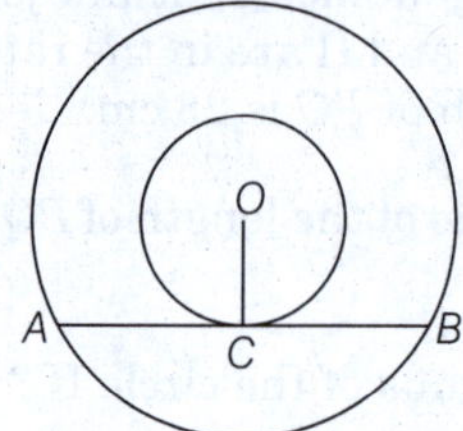

(a) 8 (b) 5 (c) 6 (d) 7

2. Let C be a circle passing through three distinct points D, E and F such that the tangent at D to the circle C is parallel to EF. Then, ΔDEF is

(a) isosceles
(b) equilateral
(c) scalene
(d) cannot be determined

3. The diameter of two circles are 18 cm and 8 cm. The distance between their centres is 13 cm. What is the number of common tangents ?

(a) 1 (b) 2
(c) 3 (d) None of these

4. In a ΔABC, $AB = AC$. A circle passing through B touches AC at D and intersects AB at P. If D is the mid-point of AC, then which one of the following is correct ? ***[CDS-07, SSC (CGL-T-1) 19]***

(a) $AB = 2AP$ (b) $AB = 3AP$
(c) $AB = 4AP$ (d) $AB = 5AP$

5. Two circles intersect at A and B. P is a point on produced BA. PT and PQ are tangents to the circles. The relation of PT and PQ is: ***[SSC (CGL-T-I)-13, SSC (CGL-T-II)-13]***

(a) $PT = 2PQ$ (b) $PT < PQ$
(c) $PT > 3PQ$ (d) $PT = PQ$

6. In the figure given, O is the centre of the circle and AC be the diameter. The line FEG is tangent to the circle at E. If $\angle GEC = 52°$, find the value of $\angle e + \angle c$ ***[CAT-11]***

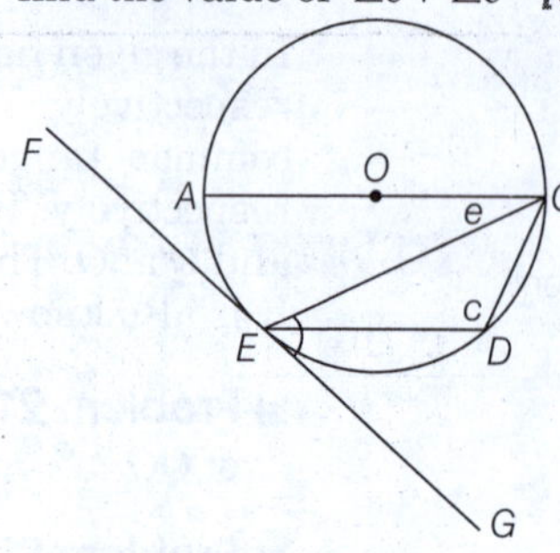

(a) 154° (b) 156°
(c) 166° (d) 180°

7. In the given figure (not drawn to scale) A, B and C are three points on a circle with centre O. The chord BA is extended to a point T such that CT becomes a tangent to the circle at point C. If $\angle ATC = 30°$ and $\angle ACT = 50°$, then the angle $\angle BOA$ is ***[CAT-03]***

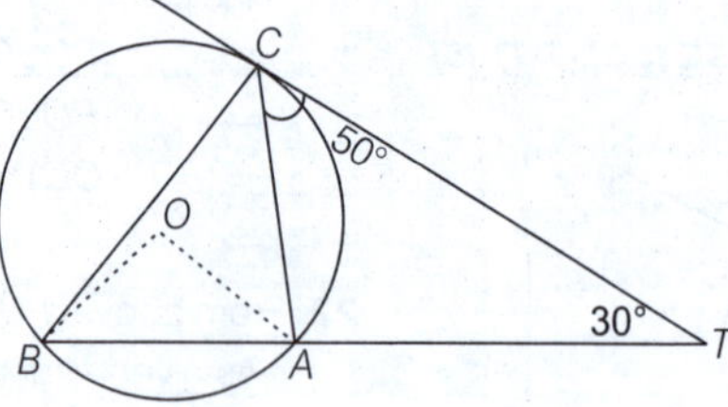

(a) 100° (b) 150°
(c) 80°
(d) not possible to determine

8. The distance between the centres of two circles having radii 4.5 cm and 3.5 cm respectively, is 10 cm. What is the length of the transverse common tangent of these circles ? ***[CDS-11]***

(a) 8 cm (b) 7 cm
(c) 6 cm
(d) None of these

9. If the length of a chord of a circle, which makes an angle 45° with the tangent drawn at one end point of the chord, is 6 cm, then the radius of the circle is:

(a) 8 cm (b) 7 cm
(c) 6 cm (d) None of these

10. A circle is drawn with side AB of a triangle ABC right angled at B, as its diameter. What is ratio of $BM : MC$ if the tangent drawn through the point of intersection of side AC and the circle, meets at point M on side BC.

(a) 1 : 1
(b) 1 : 2
(c) 1 : 4
(d) 2 : 3

11. In the given figure, l and m are two parallel tangents at A and B. The tangent at C makes an intercept DE between l and m . What is $\angle DFE$ equal to

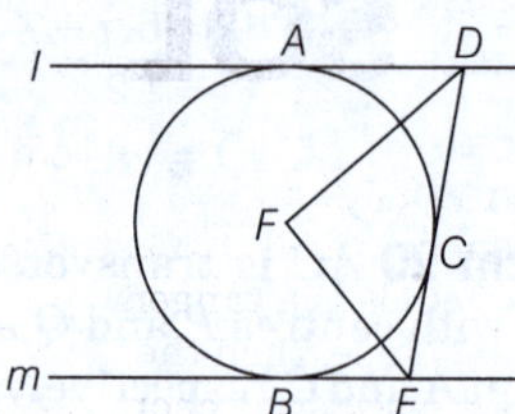

(a) 90° (b) 75°
(c) 45° (d) 120°

12. In the given figure, a circle is inscribed in a quadrilateral $ABCD$. Given that $BC = 38$ cm, $QB = 27$ cm, $DC = 25$ cm and AD is perpendicular to DC. What is the radius of the circle ?

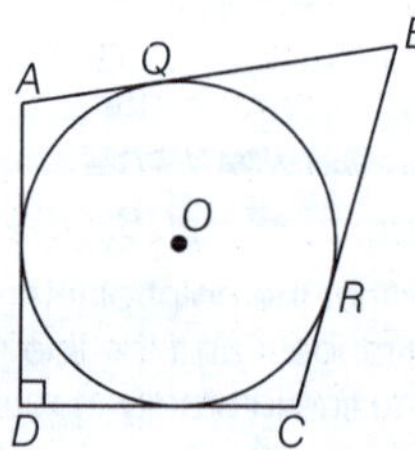

(a) 11 cm (b) 14 cm
(c) 15 cm (d) 16 cm

13. PQ is a common chord of two circles. APB is a secant line joining A and B on the two circles. Two tangents AC and BC are drawn. If $\angle ACB = 45°$, then what is $\angle AQB$ equal to: ***[CDS-09]***

(a) 75° (b) 90°
(c) 120° (d) 135°

14. Two circles touch externally at P. QR is a common tangent of the circles touching the circles at Q and R. Then, measure of $\angle QPR$ is: **[SSC (10 + 2-DEO/LDC)-14]**

(a) 30° (b) 60°
(c) 90° (d) 45°

15. In the given figure, AP and BP are tangents to a circle with centre O. If $\angle APB = 62°$, then the measure of $\angle AQB$ is **[SSC (CGL-T-1)-19]**

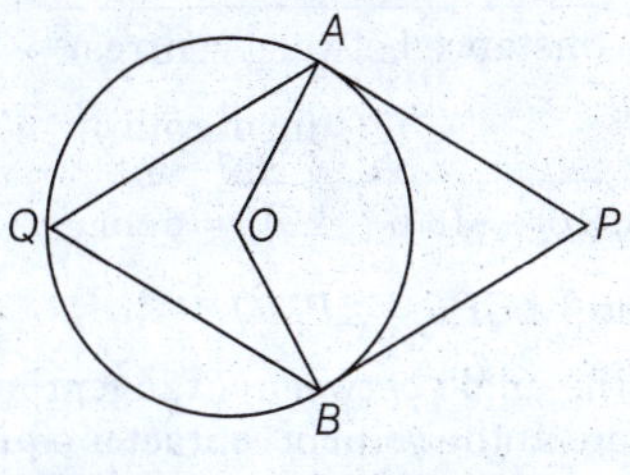

(a) 28° (b) 118°
(c) 31° (d) 59°

16. Two circles, each of radius 4 cm, touch externally. Each of these two circles is touched externally by a third circle. If these three circles have a common tangent, then the radius of the third circle, in cm, is **[CAT-19]**

(a) $\sqrt{2}$ (b) $\frac{\pi}{3}$
(c) $\frac{1}{\sqrt{2}}$ (d) 1

Answer Key

1. (b)	2. (a)	3. (c)	4. (c)	5. (d)	6. (c)	7. (a)	8. (c)	9. (d)	10. (a)
11. (a)	12. (b)	13. (d)	14. (c)	15. (d)	16. (d)				

Hints and SOLUTIONS

1. We know that in case of two concentric circles, if a chord of the bigger circle is a tangent to the smaller circle, then the point of contact bisects the chord.

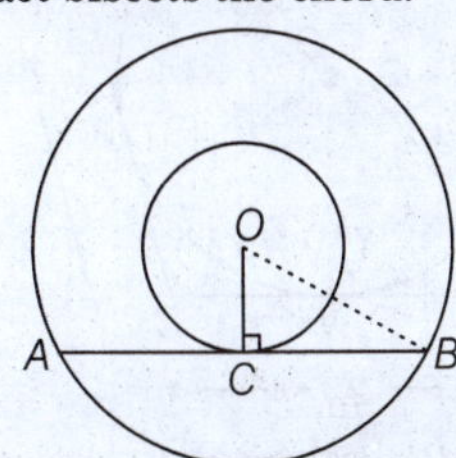

Hence, $AC = CB = 3\text{ cm}$

Since, a tangent is perpendicular to the radius through the point of contact, $\angle OCB = 90°$.

In right-angled ΔOCB, $CB = 3\text{ cm}$, then OB must be 5 as (3, 4, 5) is only Pythagorean triplet which has 3 as one of its components and all three are integers.

2. Here, $\angle FED = \angle EDM$ [alternate interior angles as $EF \parallel MN$]

We know, the angle between a tangent and a chord at the point of contact is equal to the angle in the alternate segment.

$\therefore \quad \angle EDM = \angle EFD$

Then, $\angle FED = \angle EFD$

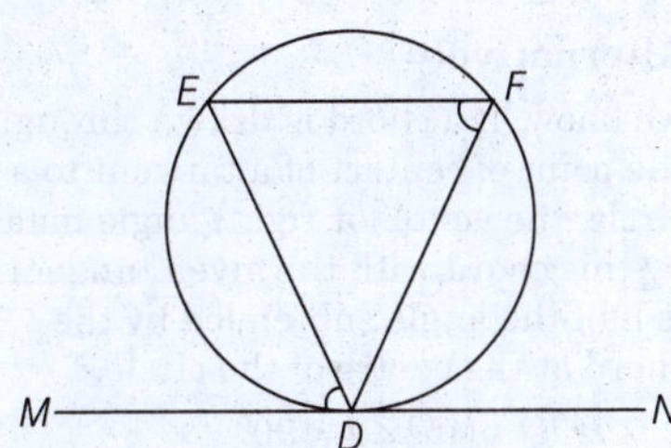

Hence, ΔDEF is isosceles.

3. Clearly, their radii are 9 cm and 4 cm.

Since the distance between their centres is (9 + 4) cm, they are touching externally. In this case, three common tangents are possible.

4. Here, AB is a secant and AD is a tangent.

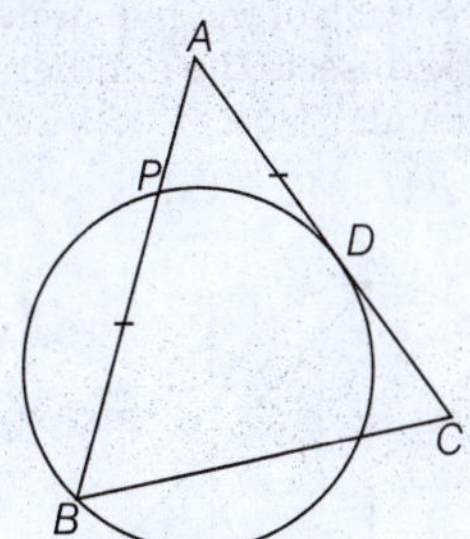

We know, the square of the length of a tangent is equal to the product of the total length of the secant and the external secant segment, if both the secant and the tangent are drawn from the same exterior point.

Then, $AD^2 = AB \times AP$

$\Rightarrow \left(\frac{AC}{2}\right)^2 = AC \times AP$

[$\because AB = AC$ and D is the mid-point of AC, as given]

$\Rightarrow \quad AC = 4AP$

$\Rightarrow \quad AB = 4AP$

5. We know, the square of the length of a tangent is equal to the product of the total length of the secant and the external secant segment, if both the secant and the tangent are drawn from the same exterior point.

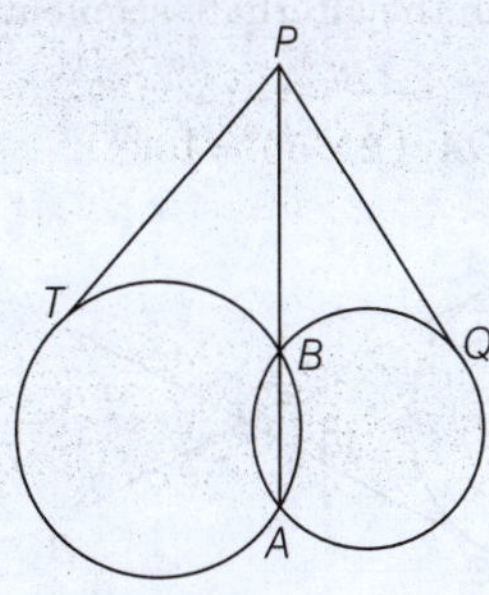

$\therefore \quad PT^2 = PA \times PB$

and $PQ^2 = PA \times PB$

Hence, $PT = PQ$

6. We know, the angle between a tangent and a chord at the point of contact is equal to the angle in the alternate segment.

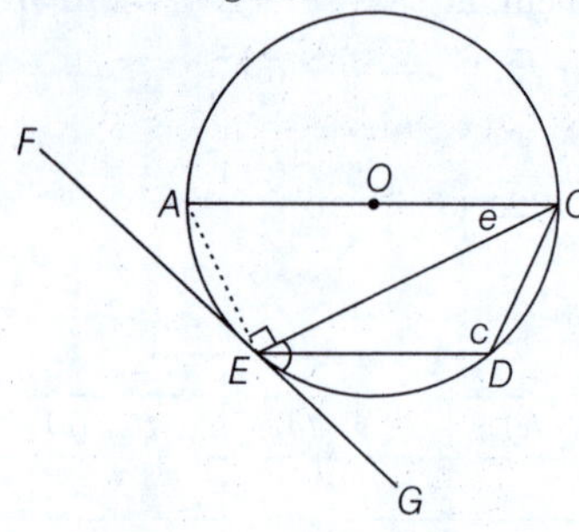

$\therefore \quad \angle GEC = \angle CAE = 52°$

Also, $\angle AEC = 90°$ as an angle in a semicircle is a right angle.

In ΔAEC. $\angle e = 180° - \angle AEC - \angle CAE$

$= 180° - 90° - 52° = 38°$

Further, $\angle CAE + \angle c = 180°$

[$\because$ $AEDC$ is a cyclic quadrilateral and sum of opposite angles of a cyclic quadrilateral is 180°]

$\Rightarrow \quad 52° + \angle c = 180°$

$\Rightarrow \quad \angle c = 128°$

Hence, $\angle e + \angle c = 38° + 128° = 166°$

7. In ΔACT,

$\angle CAT = 180° - 50° - 30° = 100°$

Exterior $\angle CAT$ of ΔACB = Sum of opposite angles

$\Rightarrow \quad 100° = \angle ABC + \angle ACB$

$\Rightarrow \quad 100° = 50° + \angle ACB$

[$\because$ the angle between a tangent and a chord at the point of contact is equal to the angle in the alternate segment i.e. $\angle ACT = \angle ABC$] $\Rightarrow \angle ACB = 50°$

Since an angle subtended by an arc at the centre is twice the angle subtended by it at any point of the circle in the alternate segment,

$\angle BOA = 2\angle BCA$

$\Rightarrow \angle BOA = 2 \times 50° = 100°$

8.

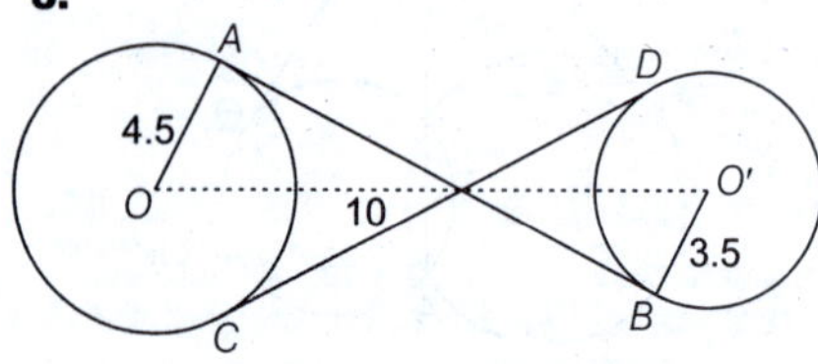

Length of the transverse common tangent = $AB = CD$

$= \sqrt{(\text{distance between centres})^2 - (\text{sum of radii})^2}$

$= \sqrt{(10)^2 - (3.5 + 4.5)^2} = 6\text{ cm}$

9. Here, $\angle QPB = \angle PMQ$

[$\because$ the angle between a tangent and a chord at the point of contact is equal to the angle in the alternate segment]

$\Rightarrow \quad \angle PMQ = 45°$

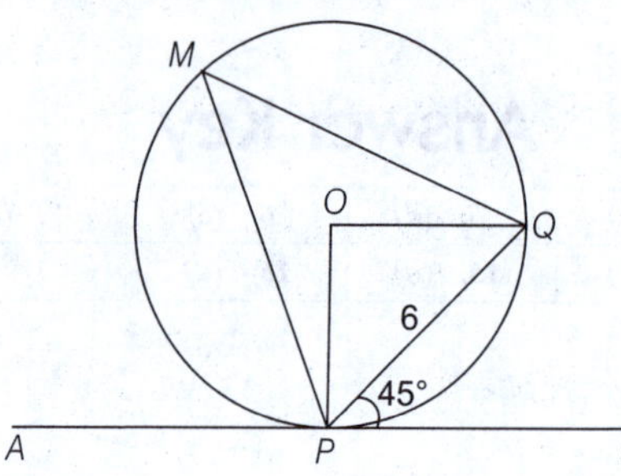

Since an angle subtended by an arc at the centre is twice the angle subtended by it at any point of the circle in the alternate segment,

$\angle POQ = 2\angle PMQ$

$\Rightarrow \quad \angle POQ = 2 \times 45° = 90°$

Alternatively

we know, if a chord is drawn through the point of contact of a tangent to a circle, the acute (or right) angle made by this chord with the given tangent is half the angle subtended by the chord at the centre of the circle.

$\therefore \quad \angle POQ = 45° \times 2 = 90°$

Now, length of chord PQ

$= 2r \sin\left(\frac{90°}{2}\right)$

$\Rightarrow \quad 6 = 2 \times r \times \frac{1}{\sqrt{2}} \Rightarrow r = 3\sqrt{2}\text{ cm}$

10. We know that when a circle is drawn with one perpendicular side of a right-angled triangle as its diameter, the tangent drawn through the point of intersection of the hypotenuse and the circle, bisects the another perpendicular side.

Hence, $BM : MC = 1 : 1$

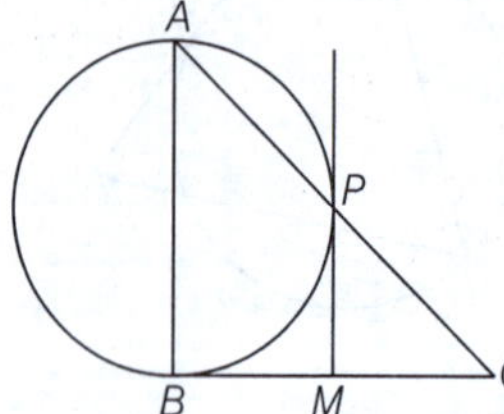

11. Since the centre of a circle lies on the bisector of the angle between the two tangents drawn to it from an external point, DF is the angle bisector of $\angle ADC$.

i.e. $\angle CDF = \frac{1}{2} \times \angle ADC$.

Similarly, $\angle FEC = \frac{1}{2} \times \angle BEC$

$\because \; l \parallel m \therefore \angle ADC + \angle BEC = 180°$

[$\because$ sum of interior angles on the same side of the transversal is 180°]

$\Rightarrow \frac{1}{2}(\angle ADC + \angle BEC) = \frac{1}{2} \times 180°$

$\Rightarrow \frac{1}{2}\angle ADC + \frac{1}{2}\angle BEC = 90°$

$\Rightarrow \quad \angle FDE + \angle FED = 90°$

Now, in ΔDFE,

$\angle DFE = 180° - (\angle FDE + \angle FED)$

$= 180° - 90° = 90°$

12. We join points of contact M and N to the centre O. We know that the tangents drawn from an external point to a circle are equal in length.

Then, $BQ = BR$ and $CR = CN$ and $DN = DM$

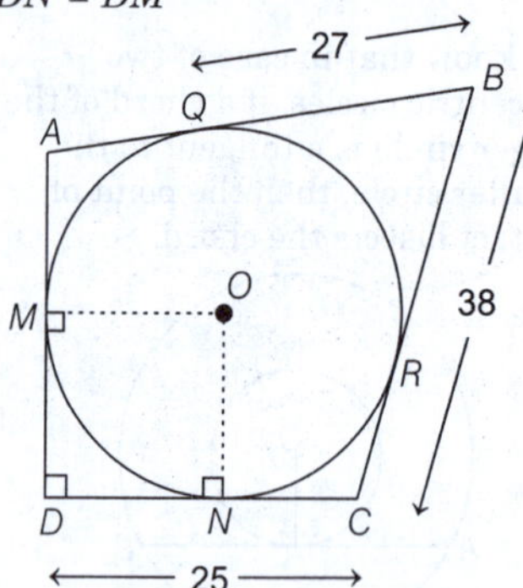

$\therefore \quad BR = 27\text{ cm}$

Then, $CN = CR = 38 - 27 = 11\text{cm}$

$\therefore \; DM = DN = 25 - 11 = 14\text{ cm}$

In $\square$ $MOND$, $\angle D$, $\angle M$ and $\angle N$ are 90° each

[$\because$ a tangent at any point of a circle is perpendicular to the radius through the point of contact]

Also, adjacent side OM and ON and DM and DN are equal. It implies that $MOND$ is a square.

Then, $DN = OM = 14\text{ cm}$

Hence, radius of the circle $OM = 14\text{ cm}$

13. Here, $\angle BAC = \angle PQA$ and $\angle ABC = \angle PQB$

[$\because$ the angle between a tangent and a chord at the point of contact is equal to the angle in the alternate segment.]

Then,

$\angle PQA + \angle PQB = \angle BAC + \angle ABC$

$\Rightarrow \angle AQB = 180° - \angle ACB$

$\therefore\ \angle AQB = 180° - 45° = 135°$

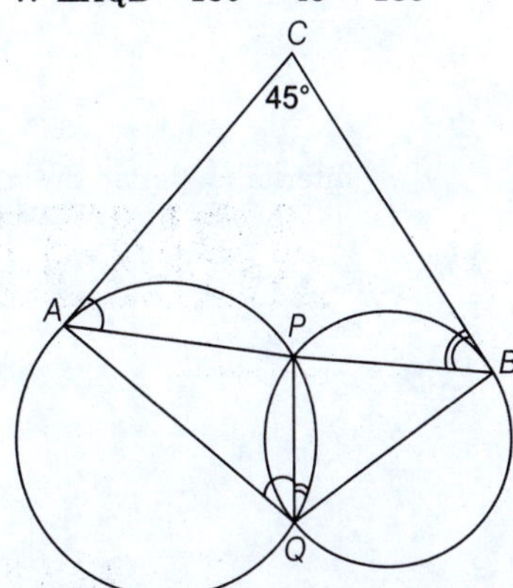

14. (*c*) We know, the tangents drawn to a circle from an external point are equal.

Hence, $MQ = MP$ and $MP = MR$

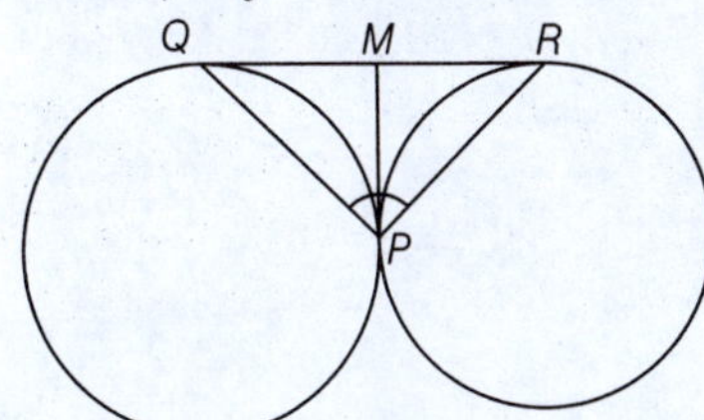

In ΔMPQ, $\angle MQP = \angle MPQ$

$[\because MQ = MP]$

Then,

$\angle QMP = 180° - \angle MQP - \angle MPQ$

$= 180° - 2\angle MPQ$

Now, exterior $\angle QMP$ of ΔMPR = sum of opposites angles

$\Rightarrow 180° - 2\angle MPQ = \angle MRP + \angle MPR$

$\Rightarrow 180° - 2\angle MPQ = \angle MPR + \angle MPR$

$[\because \angle MRP = \angle MPR$ as $MP = MR]$

$\Rightarrow 2(\angle MPQ + \angle MPR) = 180°$

$\Rightarrow \angle QPR = 90°$

15. Since $APBO$ is a quadrilateral,

$\angle AOB = 360° - \angle OAP - \angle OBP - \angle APB$

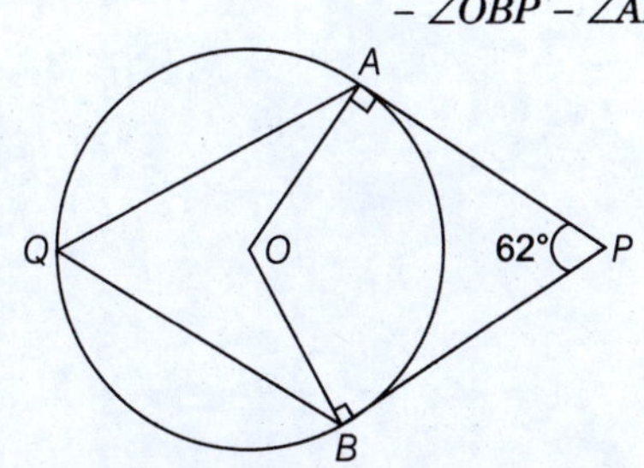

$\Rightarrow \angle AOB = 360° - 90° - 90° - 62°$

$= 118°$

[$\because$ the tangent at any point of a circle is perpendicular to the radius through the point of contact]

Further, $\angle AOB = 2\angle AQB$

[the angle subtended by an arc at centre is twice the angle subtended by it at the circle in the alternate segment of the circle.]

$\Rightarrow\ 118° = 2\angle AQB$

$\Rightarrow\ \angle AQB = 59°$

16. (*d*) Let the desired radius be r.

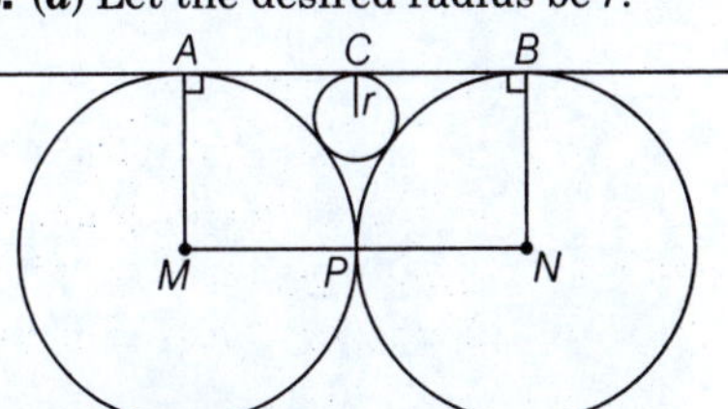

Length of common tangent

$AB = \sqrt{4 \cdot r_1 \cdot r_2}$

where r_1 and r_2 are radii of the circles.

$\Rightarrow\ AB = \sqrt{4 \cdot 4 \cdot 4} = 8\,\text{cm}$

Further, length of common tangent

$AC = \sqrt{4 \cdot 4 \cdot r} = 4\sqrt{r}$

and length of common tangent

$BC = \sqrt{4 \cdot 4 \cdot r} = 4\sqrt{r}$

Clearly, $AB = AC + BC$

$\Rightarrow\ 8 = 4\sqrt{r} + 4\sqrt{r}$

$\Rightarrow\ \sqrt{r} = 1$

$\Rightarrow\ r = 1\,\text{cm}$

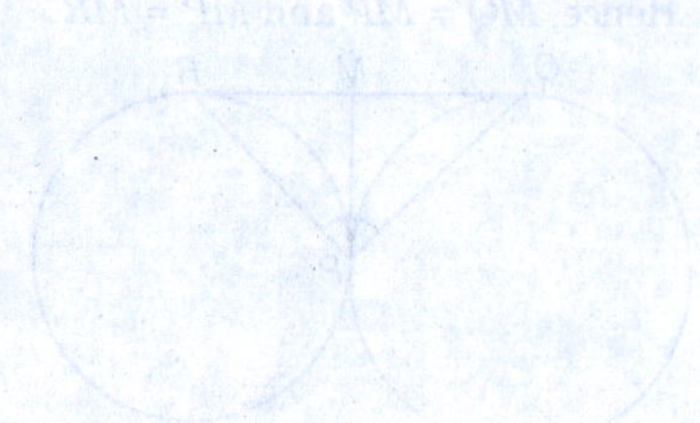

Unit 11

Fundamentals of Coordinate Geometry

60

Coordinate Geometry

Learning OBJECTIVES

Upon completion of this chapter, aspirants will be able to

- Understand the significance of this section in diverse competitive exams;
- Be familiar with the terminology of coordinate geometry;
- Be familiar with distance formula and section formula with their applications;
- Find the techniques how to find coordinates of geometric points of a triangle like incentre, orthocentre, etc;
- Be familiar with the concept of slope and forms of equations of a straight line under different conditions;
- Compute angle with two lines and also to compute area of a triangles or other shape whose sides are in equation forms;
- Learn how to compute the distance between parallel lines, distance between a point and a line;
- Establish relation between two lines observing their slopes;
- Realize themselves competent to solve all types of problems based on coordinate geometry;
- Understand the philosophy of division of 'Illustrations' in two parts;
- Identify the problems which can easily be solved through options;
- Realize themselves competent to solve the questions in unique way;
- Analyze the trend of the questions being asked in various competitive examinations;
- Solve the questions in moments;
- Learn many new concepts/fundamentals;
- Apply the pragmatic approaches in lieu of academic approaches; and
- Ponder on the alternative methods;

Inside

- *Introduction*
- *Distance and Section Formulae and Area of a Polygon with given Vertices*
- *Coordinates of Centroid, Incentre, Circumcentre, Orthocentre of a Δ*
- *Concept of Slope of a Straight line with its all aspects*
- *Angle between two lines*
- *Equation of lines under diverse conditions and relation among them*
- *Area of Δ formed by lines/axis.*
- *Perpendicular distance between lines/line and point*
- *Problems based on above all aspects*
- ....
- .... *& Many More*
- *Work Out with Solutions*

Introduction

Coordinate Geometry is a critical branch of mathematics, which provides a link between Geometry and Algebra through graphs of lines and curves. Almost all the competitive examinations like CAT, XAT, IIFT, MAT, SSC, Banks, Railway, LIC etc. cover this topic and in my opinion, it is very easy chapter at the level of understanding. Hence, the highest marks can be secured without much efforts.

Coordinate Geometry is basically a system of geometry in which a point is described by ordered pair of numbers. Let a point is lying in plane (say •) as below. The given point is described with reference to distances of the point from the two perpendicular lines as (*b*, *a*).

Here, *a* is the perpendicular distance of the given point from the horizontal line and *b* is its distance from the vertical line.

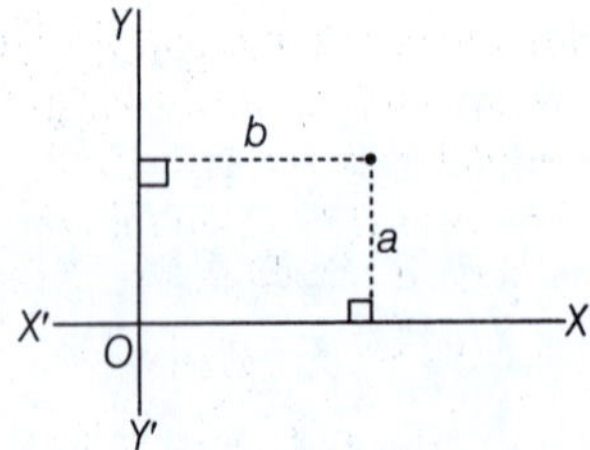

The plane where the object/point is lying is called the Cartesian or Coordinate Plane. → will be referred as 'is known as' in the followings.

- Perpendicular lines → Coordinate axes
- Horizontal line → *X*-axis
- Vertical line → *Y*-axis
- Point of intersection of both *X*-axis and *Y*-axis → origin (i.e. letter *O*)
- Distance of the point from *Y*-axis (i.e. *b*) → *x*-coordinate or abscissa
- Distance of the point from *X*-axis (i.e. *a*) → *y*-coordinate or ordinate
- (*b*, *a*) → Coordinates of the point

If *x* is abscissa of the point and *y* is its ordinate, then (*x*, *y*) → coordinates of the point. It implies that coordinates of the point are always expressed in form of (abscissa, ordinate). The coordinates of a point lying on origin will be (0, 0).

Obviously, the coordinate axes shall divide the plane into four parts called quadrants.

On the *X*-axis, values to the right of origin are positive and those to the left are negative whereas on the *Y*-axis, values above the origin and below the origin are positive and negative respectively. That is why,

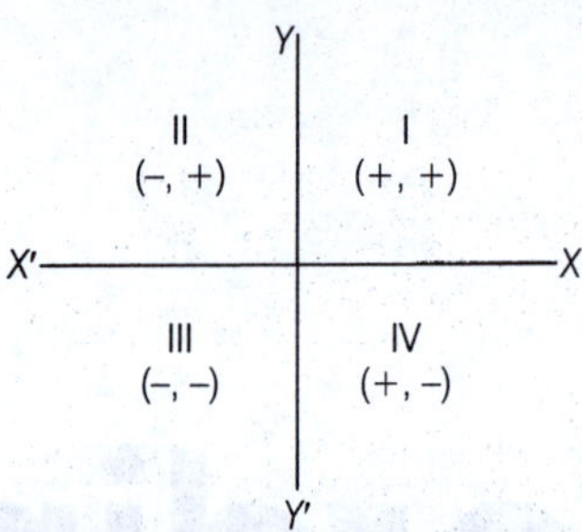

Coordinates of a point in 1st quadrant are in form of : (+,+)
Coordinates of a point in 2nd quadrant are in form of: (–,+)
Coordinates of a point in 3rd quadrant are in form of: (–,–)
Coordinates of a point in 4th quadrant are in form of: (+,–)

A. Distance between two points whose coordinates are known (Distance Formula)

Case I Let two points be $A(x_1, y_1)$ and $B(x_2, y_2)$.

Then, distance between points *A* and *B*

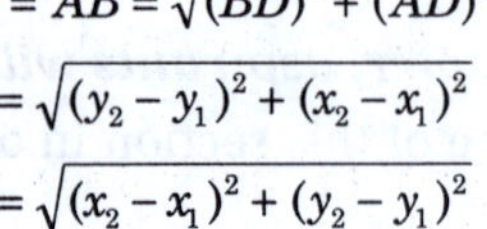

$$= AB = \sqrt{(BD)^2 + (AD)^2}$$
$$= \sqrt{(y_2 - y_1)^2 + (x_2 - x_1)^2}$$
$$= \sqrt{(x_2 - x_1)^2 + (y_2 - y_1)^2}$$

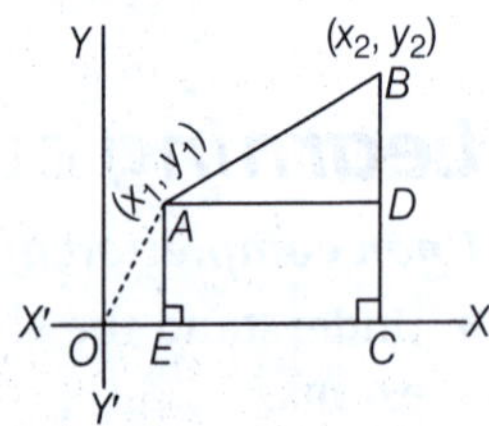

Case II Distance between point $A(x_1, y_1)$ and origin.

We know, the coordinates of the origin = (0,0)

Then, distance, $OA = \sqrt{(x_1 - 0)^2 + (y_1 - 0)^2}$

$$= \sqrt{(x_1)^2 + (y_1)^2}$$

Remarks *Three points P, Q and R are said to be collinear if PQ + QR = PR.*

B. Coordinates of a point dividing a line segment in a certain ratio (Section Formula)

Case I When a line segment joining points $A(x_1, y_1)$ and $B(x_2, y_2)$ is divided by a point p(x, y) internally in ratio m : n.

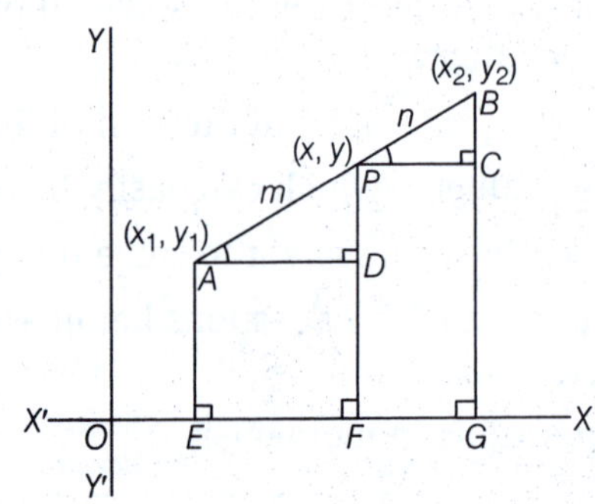

Clearly, $\Delta APD \sim \Delta PBC$

Then, $\frac{PA}{PB} = \frac{PD}{BC} = \frac{AD}{PC}$

$\Rightarrow \frac{m}{n} = \frac{y - y_1}{y_2 - y} = \frac{x - x_1}{x_2 - x} \Rightarrow \frac{m}{n} = \frac{y - y_1}{y_2 - y}$ and $\frac{m}{n} = \frac{x - x_1}{x_2 - x}$

$\Rightarrow y = \frac{my_2 + ny_1}{m + n}$ and $x = \frac{mx_2 + nx_1}{m + n}$

Thus, coordinates of point P

$$= (x, y) = \left(\frac{mx_2 + nx_1}{m + n}, \frac{my_2 + ny_1}{m + n}\right)$$

Case II When a line segment joining points $A(x_1, y_1)$ and $B(x_2, y_2)$ is divided by a point $P(x, y)$ externally in ratio m:n.

On the above pattern, the coordinates of point P dividing the line segment externally $= \left(\frac{mx_2 - nx_1}{m - n}, \frac{my_2 - ny_1}{m - n}\right)$

Case III When a line segment joining points $A(x_1, y_1)$ and $B(x_2, y_2)$ is divided by a point $P(x, y)$ internally in ratio 1:1

It implies that P is mid-point of the line segment.

Then, coordinates of point P

$$= \left(\frac{1 \cdot x_2 + 1 \cdot x_1}{1 + 1}, \frac{1 \cdot y_2 + 1 \cdot y_1}{1 + 1}\right) = \left(\frac{x_1 + x_2}{2}, \frac{y_1 + y_2}{2}\right)$$

C. How to compute area of a triangle or any polygon of which vertices are known

Let $A(x_1, y_1), B(x_2, y_2)$ and $C(x_3, y_3)$ be vertices of a ΔABC.

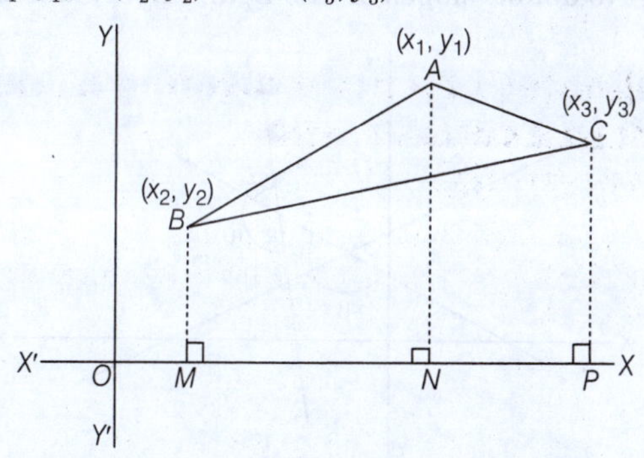

Area of ΔABC = Area of trapezium $ABMN$ + Area of trapezium $ACPN$ – Area of trapezium $BMPC$

$$= \frac{1}{2} \times MN(BM + NA) + \frac{1}{2} \times PN(PC + NA) - \frac{1}{2} \times MP(BM + PC)$$

$$= \frac{1}{2}(x_1 - x_2)(y_2 + y_1) + \frac{1}{2}(x_3 - x_1)(y_3 + y_1) - \frac{1}{2}(x_3 - x_2)(y_2 + y_3)$$

$$= \frac{1}{2}[x_1(y_2 - y_3) + x_2(y_3 - y_1) + x_3(y_1 - y_2)]$$

Since the area of a triangle cannot be negative, area of the triangle

$$= \frac{1}{2}|[x_1(y_2 - y_3) + x_2(y_3 - y_1) + x_3(y_1 - y_2)]|$$

Alternatively,

x_1, y_1
$\times \} \ (x_1y_2 - x_2y_1)$
x_2, y_2
$\times \} \ (x_2y_3 - x_3y_2)$
x_3, y_3
$\times \} \ (x_3y_1 - x_1y_3)$
x_1, y_1

$$\text{Area} = \frac{1}{2}|[(x_1y_2 - x_2y_1) + (x_2y_3 - x_3y_2) + (x_3y_1 - x_1y_3)]|$$

Remarks

(i) We may start with any vertex and end with the same vertex.

(ii) This alternative approach is applicable also for other polygons like quadrilaterals, hexagons, etc.

(iii) In case the area of a triangle is 0, all the three points are collinear. Hence, in case of collinearity of three points,

$$\frac{1}{2}|[x_1(y_2 - y_3) + x_2(y_3 - y_1) + x_3(y_1 - y_2)]| = 0$$

$$\Rightarrow x_1(y_2 - y_3) + x_2(y_3 - y_1) + x_3(y_1 - y_2) = 0$$

This is the condition for collinearity of three points.

D. Coordinates of centroid, incentre, orthocentre and circumcentre of a triangle of which vertices are known

Let $A(x_1, y_1), B(x_2, y_2)$ and $C(x_3, y_3)$ be vertices of a ΔABC. Then,

(i) Coordinates of centroid $= \left(\frac{x_1 + x_2 + x_3}{3}, \frac{y_1 + y_2 + y_3}{3}\right)$

(ii) Coordinates of incentre

$$= \left(\frac{ax_1 + bx_2 + cx_3}{a + b + c}, \frac{ay_1 + by_2 + cy_3}{a + b + c}\right)$$

where, $a = BC, b = AC$ and $c = AB$.

(iii) **For computation of coordinates of circumcentre** We find distances between circumcentre and all three given vertices taking assumed value of coordinates of the circumcentre. These distances are, indeed, circumradii. Since circumradii are equal, we solve the equations accordingly. The solution yields coordinates of the circumcentre.

(iv) **For computation of coordinates of orthocentre** We first find the equations of two sides, thereafter the equations of perpendicular lines passing through their opposite vertices. The coordinates of the point of intersection of perpendicular lines are the coordinates of the orthocentre. No need to mention that solutions of the equations of the perpendicular lines are the coordinates of the point of intersection.

Note *A triangle having all its vertices in integral coordinates, cannot be equilateral.*

E. Concept of slope of a straight line

If θ is inclination of a straight line, then $\tan\theta$ is called its slope or gradient. Here, inclination of the line means the angle which a line makes with the positive direction of X-axis and measured anti-clockwise. The slope of the straight line is denoted by letter 'm'. Hence, $m = \tan\theta$.

It is important to note:

(i) The slope of a line is positive when it makes an acute angle with the positive direction of X-axis and measured anti-clockwise.

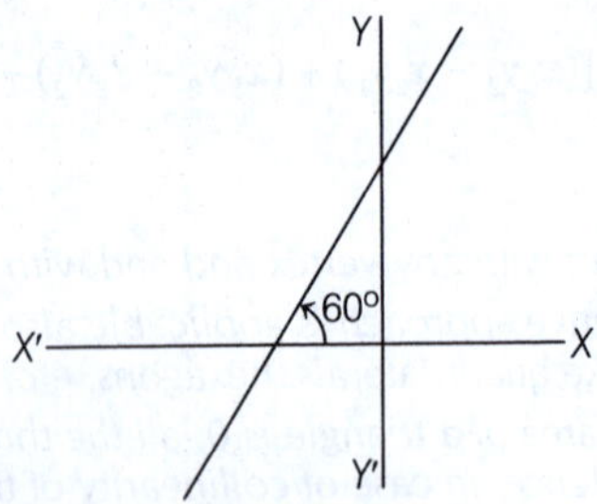

Clearly, $m = \tan\theta = \tan 60° = \sqrt{3}$

(ii) The slope of a line is negative when it makes an obtuse angle with the positive direction of X-axis and measured anti-clockwise.

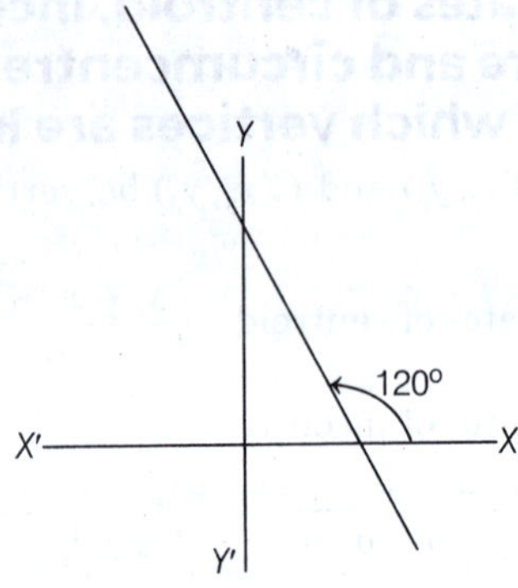

Here, $m = \tan\theta = \tan 120°$
$= \tan(180° - 60°) = -\tan 60°$
$= -\sqrt{3}$

(iii) If the line is vertical, then its slope is undefined as $\tan 90° = \infty$.

(iv) If the line is parallel to X-axis, then its slope is 0 as $\tan 0° = 0$.

(v) If slopes of two lines are equal, these lines are parallel.

(vi) Two lines are perpendicular to each other, if the product of their slopes is -1.

(vii) Three points A, B and C are collinear if the slope of AB is equal to the slope of BC.

Slope of a line when coordinates of any two points on the line are known

Let $P(x_1, y_1)$ and $Q(x_2, y_2)$ be two points on a line. Then,

slope of the line $= \dfrac{\text{rise}}{\text{run}} = \dfrac{y_2 - y_1}{x_2 - x_1}$

F. Angles between two lines

If α and β are the adjacent angles between two lines having slopes m_1 and m_2 and α is an acute angle, then:

$$\tan\alpha = \left|\frac{m_2 - m_1}{1 + m_1 \cdot m_2}\right|, \text{ where } 1 + m_1 \cdot m_2 \neq 0.$$

Obviously, $\beta = (180° - \alpha)$

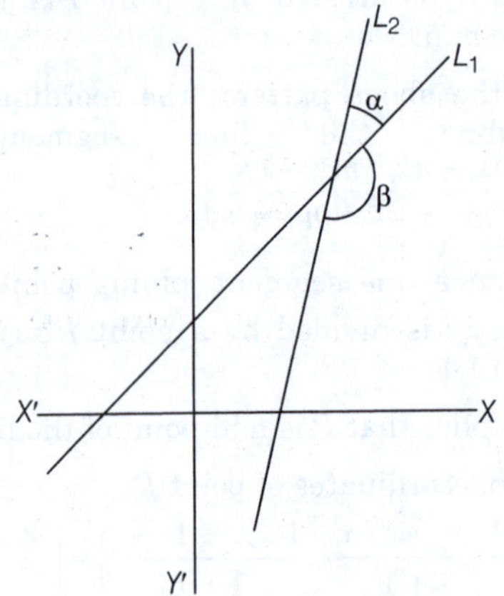

It is important to note that two lines with the different slopes may make equal angles with the another line.

Here, each of lines L_2 and L_3 makes angles α and β with the line L_1. No doubt, slopes of line L_2 and L_3 are different.

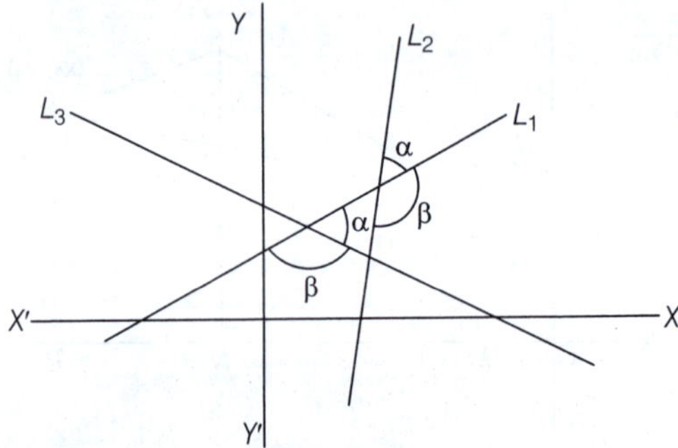

G. Forms of equation of a straight line under different conditions

(i) **General form of equation of a line** We are aware of the fact that the general form of a linear equation in two variables (say x and y) is $ax + by + c = 0$, where a, b and c are real numbers and $a \neq 0, b \neq 0$. Similarly, the general form of equation of a line is $ax + by + c = 0$, where a and b are not zero simultaneously.

(ii) **Equation of a line parallel to X-axis** Equation of a line parallel to X-axis and at a distance of b from the X-axis will be either $y = b$ or $y = -b$ as y-coordinate of every point lying on the line is either b or $-b$. If the line is above the X-axis, $y = b$ and if it is below the X-axis, $y = -b$.

Obviously, the equation of X-axis will be $y = 0$.

(iii) **Equation of a line parallel to Y-axis** Equation of a line parallel to Y-axis and at a distance of a from the Y-axis will be either $x = a$ or $x = -a$ as x-coordinate of every point lying on the line is either a or $-a$. If the line is in right side of the Y-axis, $x = a$ and if it is in left side, $x = -a$.

Obviously, the equation of Y-axis will be $x = 0$.

(iv) **'Point-slope form' of equation of a line** The equation of a line which has the slope m and passes through a point (x_1, y_1) is $(y - y_1) = m(x - x_1)$. This form of equation is termed as 'point-slope form'.

When a line with slope m passes through $(0, c)$, then equation of the line be $(y - c) = m(x - 0)$

$\Rightarrow \quad y = mx + c$

Hence, when a line with slope m makes y-intercept c, then equation of the line is $y = mx + c$. This form of equation is known as 'slope-intercept form'.

Note *When a line cuts Y-axis at a distance k from the origin, k is called y-intercept of the line. Also, when a line cuts X-axis at a distance L from the origin, L is known as x-intercept of the line.*

Similarly, the equation of a line which has the slope m and makes x-intercept d is $(y - 0) = m(x - d)$

$\Rightarrow \quad y = m(x - d)$

In view of the above, the equation of a line which has slope m and passes through the origin shall be

$$y = mx + 0 \text{ or } y = m(x - 0) \qquad [\because c = d = 0]$$

Hence, $y = mx$

(v) **'Two-point form' of equation of a line** The equation of a line passing through two points (x_1, y_1) and (x_2, y_2) is $(y - y_1) = \dfrac{y_2 - y_1}{x_2 - x_1}(x - x_1)$.

Let a line makes x-intercept a and y-intercept b on the axes. It is obvious that the line will obviously pass through $(0, b)$ and $(a, 0)$. Then, the equation of the line shall be $(y - b) = \dfrac{0 - b}{a - 0}(x - 0)$

$\Rightarrow \quad ay + bx = ab$

$\Rightarrow \quad \dfrac{x}{a} + \dfrac{y}{b} = 1$

This form of equation is known as 'intercept-form'. Notice that the slope of the line making x-intercept a and y-intercept b is $\dfrac{-b}{a}$.

The above forms of equation may be summarised as below:

Cases	Known	Known	Equation of the Line
I	Parallel to X-axis	a point through which the line passes	y = ordinate of the given point
II	Parallel to Y-axis	a point through which the line passes	x = abscissa of the given point
III	Slope (i.e.m)	a point (x_1, y_1) through which the line passes	$(y - y_1) = m(x - x_1)$
IV	Slope (i.e.m)	y- intercept (say c)	$y = mx + c$
V	Slope (i.e. m)	x-intercept (say d)	$y = m(x - d)$
VI	1st point (x_1, y_1) through which the line passes	2nd point (x_2, y_2) through which the line passes	$(y - y_1) = \dfrac{y_2 - y_1}{x_2 - x_1}(x - x_1)$ [Recall, $\dfrac{y_2 - y_1}{x_2 - x_1}$ is the slope of the line]
VII	x-intercept	y- intercept	$\dfrac{x}{x - \text{intercept}} + \dfrac{y}{y - \text{intercept}} = 1$ or x(y-intercept) + y(x-intercept) = (x-intercept) × (y-intercept)

H. Comparison of general form of equation of a line with other forms

(i) General form of the equation is $ax + by + c = 0$

$\Rightarrow \quad by = -ax - c$

$\Rightarrow \quad y = \dfrac{-a}{b}x + \dfrac{-c}{b}$

Also, the slope-intercept form of equation is $y = mx + c$, where c is y-intercept of the line.

Comparing both forms of the equation, we get

slope, $m = \dfrac{-a}{b}$ and y-intercept $= \dfrac{-c}{b}$

(ii) General form of the equation is $ax + by + c = 0$.

$\Rightarrow \quad \dfrac{ax}{-c} + \dfrac{by}{-c} - 1 = 0$

$\Rightarrow \quad \dfrac{x}{\left(\dfrac{-c}{a}\right)} + \dfrac{y}{\left(\dfrac{-c}{b}\right)} = 1$

Also, intercept-form of the equation is $\dfrac{x}{a} + \dfrac{y}{b} = 1$, where a and b are x-intercept and y-intercept of the line respectively.

Comparing both forms of the equation, we get x-intercept $= \left(\dfrac{-c}{a}\right)$ and y-intercept $= \left(\dfrac{-c}{b}\right)$.

Conclusion If $ax + by + c = 0$ is an equation of a line, then its slope $= \left(\frac{-a}{b}\right)$, x-intercept $= \left(\frac{-c}{a}\right)$ and y-intercept $= \left(\frac{-c}{b}\right)$.

These are very useful tools for solving the questions.

Alternative approach for deriving the above results

Given equation, $ax + by + c = 0$

$$\Rightarrow \quad by = -ax - c$$

$$\Rightarrow \quad y = \frac{-a}{b}x + \frac{-c}{b}$$

If the line passes through (L, 0), then $0 = \frac{-a}{b}.L + \frac{-c}{b}$

$$\Rightarrow \quad \frac{a}{b}.L = \frac{-c}{b} \Rightarrow L = \left(\frac{-c}{a}\right)$$

Hence, x-intercept of the line $= \left(\frac{-c}{a}\right)$

Further, if the line passes through (0, k),

then $$k = \frac{-a}{b}.0 + \frac{-c}{b} = \frac{-c}{b}$$

Hence, y-intercept of the line $= \left(\frac{-c}{b}\right)$

It is obvious that the line is passing through (L, 0) and (0, k), then slope of the line $= \frac{k-0}{0-L} = \frac{k}{-L} = \frac{\left(\frac{-c}{b}\right)}{-\left(\frac{-c}{a}\right)} = \left(\frac{-a}{b}\right)$

I. Area of a triangle formed by the line(s) and the coordinate axes(axis) or by the lines only

(i) Area of a triangle formed by the line $ax + by + c = 0$ and the coordinate axes:

The x-intercept and y-intercept of the line are $\frac{-c}{a}$ and $\frac{-c}{b}$, respectively.

Now, area of ΔAOB

$$= \frac{1}{2} \times OB \times OA = \frac{1}{2} \times \frac{-c}{a} \times \frac{-c}{b} = \frac{c^2}{2ab}.$$

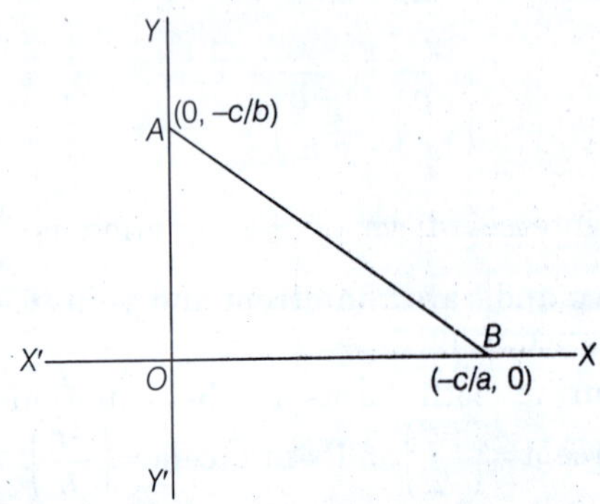

Since the area of a triangle cannot be negative, the area of the triangle formed by a line $ax + by + c = 0$ and the coordinate axes shall be $\frac{c^2}{|2ab|}$.

(ii) Area of triangle formed by two equations $y = m_1x + c_1$, $y = m_2x + c_2$ and Y-axis:

y-intercept of the line $y = m_1x + c_1$ is c_1 and y-intercept of the line $y = m_2x + c_2$ is c_2.

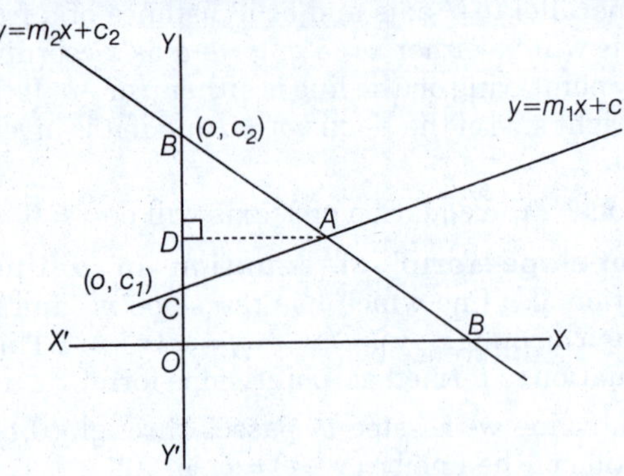

Clearly, two vertices of the triangle be $(0, c_1)$ and $(0, c_2)$. The third vertex will be the point of intersection of both the lines and its coordinates will be the solutions of the equations.

On solving both equations, we get

$$x = \frac{c_2 - c_1}{m_1 - m_2} \text{ and } y = \frac{m_1c_2 - m_2c_1}{m_1 - m_2}$$

Now, area of the ΔABC =

$$\frac{1}{2}\left|0\left(c_2 - \frac{m_1c_2 - m_2c_1}{m_1 - m_2}\right) + 0\left(\frac{m_1c_2 - m_2c_1}{m_1 - m_2} - c_1\right) + \frac{c_2 - c_1}{m_1 - m_2}(c_1 - c_2)\right|$$

$$= \frac{1}{2}\left|0 + 0 - \frac{c_2 - c_1}{m_1 - m_2}(c_2 - c_1)\right| = \frac{(c_2 - c_1)^2}{2|m_1 - m_2|}$$

Hence, the area of triangle formed by two equations $y = m_1x + c_1$, $y = m_2x + c_2$ and Y-axis $= \frac{(c_2 - c_1)^2}{2|m_1 - m_2|}$.

Or

Area of triangle formed by two equations $a_1x + b_1y + c_1 = 0$ and $a_2x + b_2y + c_2 = 0$ and Y-axis

$$= \frac{(\text{difference between } y\text{ - intercepts})^2}{2\,|\text{difference between slopes}|}$$

{If $ax + by + c = 0$ is an equation of a line, then its slope $= \left(\frac{-a}{b}\right)$ and y-intercept $= \left(\frac{-c}{b}\right)$}

Or Area of $\Delta ABC = \frac{1}{2} \times BC \times AD$

$$= \frac{1}{2} \times |\text{difference between } y\text{-intercepts}| \times |x\text{-coordinate of point of intersection of lines}|$$

(iii) Area of triangle formed by two equations $a_1x + b_1y + c_1 = 0$ and $a_2x + b_2y + c_2 = 0$ and X-axis

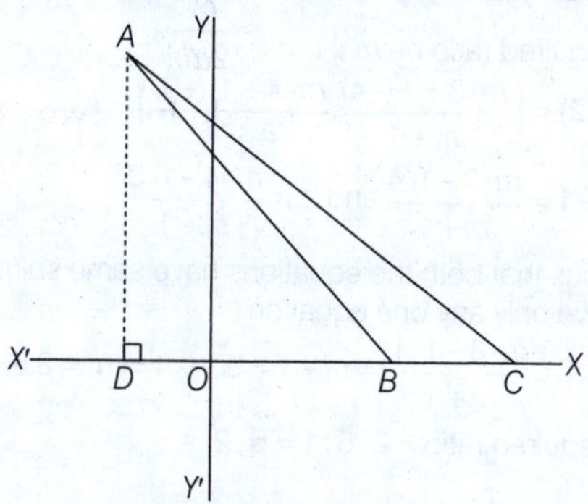

Area of $\Delta ABC = \frac{1}{2} \times BC \times AD$

$= \frac{1}{2} \times$ | difference between x-intercepts | $\times$

| y-coordinate of point of intersection of lines |

(iv) Area of triangle formed by three lines, whose equations are known

Basically, the point of intersection of any two lines is a vertex of the triangle and indeed, the coordinates of the point of intersection are the solutions of equations of two lines which intersect each other. Therefore, we first obtain all three vertices by solving two equations at a time and then apply the formula meant for computing the area of the triangle when all vertices are known.

J. Relation between two lines whose equations are known

Two lines $a_1x + b_1y + c_1 = 0$ and $a_2x + b_2y + c_2 = 0$ are:

(i) Coincident if $\frac{a_1}{a_2} = \frac{b_1}{b_2} = \frac{c_1}{c_2}$

(ii) Intersecting if $\frac{a_1}{a_2} \neq \frac{b_1}{b_2}$

[Recall the concept of a system of linear equations in 'algebra']

(iii) Parallel if $\frac{a_1}{a_2} = \frac{b_1}{b_2}$ $\quad [\because m_1 = m_2 \Rightarrow \frac{-a_1}{b_1} = \frac{-a_2}{b_2}]$

(iv) Perpendicular if $a_1a_2 + b_1b_2 = 0$

$[\because m_1m_2 = -1 \Rightarrow \frac{-a_1}{b_1} \times \frac{-a_2}{b_2} = -1]$

Note *Three lines are concurrent if solutions of equations of any two lines also satisfy the equation of the third line.*

K. Perpendicular distance of a line from a given point

The perpendicular distance of a line $ax + by + c = 0$ from a point $(x_1, y_1) = \frac{|ax_1 + by_1 + c|}{\sqrt{a^2 + b^2}}$

In case the point is the origin, the perpendicular distance $= \frac{|c|}{\sqrt{a^2 + b^2}}$

L. Distance between two parallel lines

Let two lines be $y = mx + c_1$ and $y = mx + c_2$.

Since the slopes of the parallel lines are equal, both equations have the slopes as m.

Then, perpendicular distance between them $= \frac{|c_1 - c_2|}{\sqrt{1 + m^2}}$

In case two lines are in form of $ax + by + c_1 = 0$ and $ax + by + c_2 = 0$, the perpendicular distance between them $= \frac{|c_1 - c_2|}{\sqrt{a^2 + b^2}}$

Illustrations

Problems Based on the Concept of Quadrants

Problem 1 If $ab > 0$ and the point (a, b) lies in the third quadrant, the quadrant in which the point $(-b, -a)$ lies is: ***SSC (CGL-T-I) 12***

Sol.

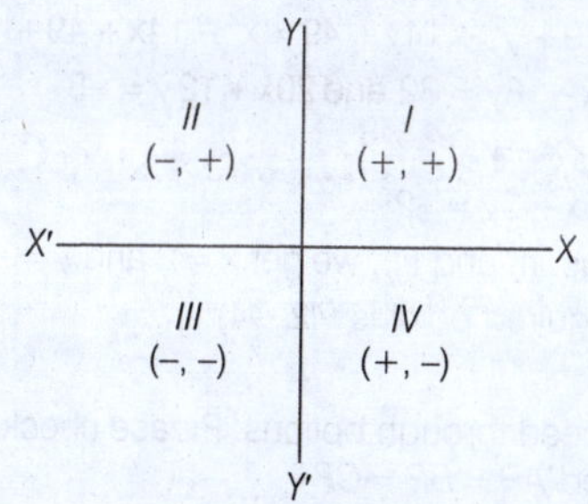

$ab > 0$, it means either both are positives or both are negatives.
Since (a, b) lies in third quadrant, both are negatives.
Then, $(-b, -a)$ will change to (b, a), which will obviously be in 1st quadrant.

Problems Based on Distance Formula

Problem 2 The distance between the points (0,0) and the intersecting point of the graphs of $x = 3$ and $y = 4$ is : **[SSC (CGL-T-I)-12]**

Sol. Distance between point (0, 0) and the intersecting point = hypotenuse $AO = \sqrt{(3)^2 + (4)^2} = 5$ units

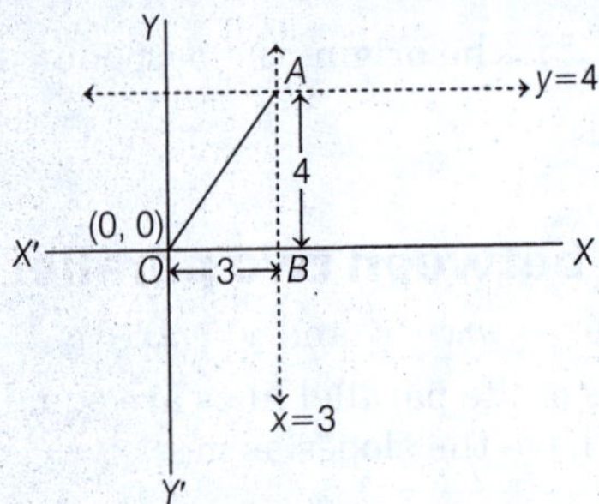

Alternatively

Clearly, coordinates of the intersecting point are (3, 4).

Then, distance between point (0, 0) and point (3, 4)

$$= \sqrt{(3-0)^2 + (4-0)^2} = 5 \text{ units}$$

Problem 3 Find the coordinates of the point equidistant from three given points $A(5,1)$, $B(-3,7)$ and $C(7,-1)$.

Sol. Let the required point be $P(x, y)$.

As per question, $AP = BP = CP$

$\Rightarrow \quad (AP)^2 = (BP)^2 = (CP)^2$

$\Rightarrow \quad (AP)^2 = (BP)^2$ and $(BP)^2 = (CP)^2$

$\Rightarrow \quad (x-5)^2 + (y-1)^2 = (x+3)^2 + (y+7)^2$ and $(x+3)^2 + (y+7)^2 = (x-7)^2 + (y+1)^2$

$\Rightarrow \quad x^2 - 10x + 25 + y^2 - 2y + 1 = x^2 + 6x + 9 + y^2 + 14y + 49$

and $x^2 + 6x + 9 + y^2 + 14y + 49 = x^2 - 14x + 49 + y^2 + 2y + 1$

$\Rightarrow \quad -16x - 16y = 32$ and $20x + 12y = -8$

$\Rightarrow \quad x + y = -2 \quad$ ...(i)

and $\quad 5x + 3y = -2 \quad$ ...(ii)

On solving Eqs. (i) and (ii), we get $x = 2$ and $y = -4$

Hence, the required point is $P(2, -4)$

Alternatively

You may proceed through options. Please check as to which option satisfied $AP = BP = CP$

Problems Based on Section Formula

Problem 4 Find the coordinates of the point which divides the line segment joining the points $P(3,-1)$ and $Q(5,4)$ in the ratio 1:2.

Sol. Coordinates of the point dividing the line segment

$$= \left(\frac{1\cdot 5 + 2\cdot 3}{1+2}, \frac{1\cdot 4 + 2\cdot(-1)}{1+2}\right)$$

$$= \left(\frac{11}{3}, \frac{2}{3}\right)$$

Problem 5 In what ratio does a point (1, 2) divide the line segment joining points $P(-4,-3)$ and $Q(3,4)$?

Sol. Let the required ratio be $m : 1$.

Then, $(1, 2) = \left(\frac{m\cdot 3 + 1\cdot(-4)}{m+1}, \frac{m\cdot 4 + 1\cdot(-3)}{m+1}\right)$

$\Rightarrow \quad 1 = \frac{m\cdot 3 - 1\cdot 4}{m+1}$ and $2 = \frac{m\cdot 4 - 1\cdot 3}{m+1}$

It is obvious that both the equations have same solution. Thus, solve only any one equation.

$$1 = \frac{m\cdot 3 - 1\cdot 4}{m+1} \Rightarrow m + 1 = 3m - 4 \Rightarrow m = 2.5$$

Hence, required ratio $= 2\cdot 5 : 1 = 5 : 2$

Note *It is not advisable to proceed through options.*

Problem 6 The line segment joining the points $(3,-4)$ and $(1,2)$ is trisected at the points $P(p,-2)$ and $Q\left(\frac{5}{3}, q\right)$, find the value of p and q.

K .... P .... Q L
(3, –4) $\quad$ (p, –2) $\quad$ $\left(\frac{5}{3}, q\right)$ $\quad$ (1, 2)

Sol. Clearly, the point $P(p,-2)$ divides KL in ratio 1 : 2

Then, $\quad (p, -2) = \left(\frac{1\cdot 1 + 2\cdot 3}{1+2}, \ldots\ldots\right)$

[no need to write the y-coordinate as we have to find value of only p]

$\therefore \quad p = \frac{1\cdot 1 + 2\cdot 3}{1+2} \Rightarrow p = \frac{7}{3}$

Hence, the coordinates of point P are $\left(\frac{7}{3}, -2\right)$

Also, $Q\left(\frac{5}{3}, q\right)$ is the mid-point of PL.

$$\Rightarrow \quad \left(\frac{5}{3}, q\right) = \left(\frac{\frac{7}{3}+1}{2}, \frac{-2+2}{2}\right)$$

[no need to write the x-coordinate as we have to find value of only q. However, it has been written here for only conceptual clarity.]

Problem 7 In what ratio will the X-axis divide a line segment joining $P(5,5)$ and $Q(1,-4)$?

Sol. Let a point $R(x, 0)$ lying on X-axis divides the line segment in ratio $m : 1$. Since the point R lies on X-axis, its ordinate will be 0.

Then, $\quad (x, 0) = \left(\frac{m\cdot 1 + 1\cdot 5}{m+1}, \frac{m(-4) + 1\cdot 5}{m+1}\right)$

$\Rightarrow \quad 0 = \frac{m\cdot(-4) + 1\cdot 5}{m+1}$

$\Rightarrow \quad -4m + 5 = 0 \Rightarrow m = \frac{5}{4}$

$\therefore \quad m : 1 = \frac{5}{4} : 1 = 5 : 4$

Problem 8 In the above question, what are the coordinates of the point of intersection of line segment PQ and the X-axis ?

Sol. From the above,

$$x\text{-coordinate of the point } R = \frac{m\cdot 1+1\cdot 5}{m+1} = \frac{\frac{5}{4}\cdot 1+1\cdot 5}{\frac{5}{4}+1} = \frac{25}{9}$$

Hence, coordinates of the point $R = \left(\frac{25}{9}, 0\right)$

Problem 9 A line $x + 2y - 6 = 0$ divides the line segment joining $P(5,5)$ and $Q(1,-4)$, what is the ratio of division ?

Sol. Let the line $x+2y-6=0$ divides the line segment at point $T(x,y)$ in ratio $m:1$

Then, $(x,y) = \left(\frac{m\cdot 1+1\cdot 5}{m+1}, \frac{m\cdot(-4)+1\cdot 5}{m+1}\right)$

$\Rightarrow \quad x = \frac{m+5}{m+1}$ and $y = \frac{5-4m}{m+1}$

Since the point T lies on the line $x+2y-6=0$, coordinates of the point T will also satisfy the equation.

$\Rightarrow \quad \frac{m+5}{m+1} + \frac{2(5-4m)}{m+1} - 6 = 0$

$\Rightarrow \quad -7m+15 = 6(m+1)$

$\Rightarrow \quad m = \frac{9}{13}$

Then, the required ratio $= \frac{9}{13}:1 = 9:13$

Problems based on Area, Median, Centroid, Incentre, Circumcentre, Orthocentre etc. of a triangle

Problem 10 Find the coordinates of the circumcentre of ΔABC with vertices at $A(5,1)$, $B(-3,-7)$ and $C(7,-1)$.

Sol. Let the circumcentre be $P(x,y)$.

Since the circumcentre of a triangle is equidistant from all three vertices, $AP = BP = CP$

$\Rightarrow \quad (AP)^2 = (BP)^2 = (CP)^2$

$\Rightarrow \quad (AP)^2 = (BP)^2$ and $(BP)^2 = (CP)^2$

$\Rightarrow \quad (x-5)^2 + (y-1)^2 = (x+3)^2 + (y+7)^2$

and $\quad (x+3)^2 + (y+7)^2 = (x-7)^2 + (y+1)^2$

$\Rightarrow x^2 - 10x + 25 + y^2 - 2y + 1 = x^2 + 6x + 9 + y^2 + 14y + 49$

and $x^2 + 6x + 9 + y^2 + 14y + 49 = x^2 - 14x + 49 + y^2 + 2y + 1$

$\Rightarrow \quad -16x - 16y = 32$

and $\quad 20x + 12y = -8$

$\Rightarrow \quad x + y = -2 \quad$ ...(i)

and $\quad 5x + 3y = -2 \quad$ ...(ii)

On solving Eqs. (i) and (ii), we get $x = 2$ and $y = -4$

Hence, the circumcentre is $P(2,-4)$

Alternatively

You may proceed through options. Please check as to which option satisfied $AP = BP = CP$.

Problem 11 Find the length of the median falling on the side AB of a triangle with vertices $A(4,1)$, $B(-2,5)$ and $C(3,-2)$.

Sol. Obviously, the distance between the mid-point of side AB and the vertex C of the triangle will be the length of the required median. Coordinates of the midpoint of side $AB = \left(\frac{4-2}{2}, \frac{1+5}{2}\right)$

$= (1,3)$

Now, the distance between the point (1, 3) and vertex c (3,–2)

$= \sqrt{(3-1)^2 + (-2-3)^2} = \sqrt{29}$

Problem 12 Two vertices of a ΔABC are given by $A(6,4)$ and $B(-2,2)$ and its centroid is $G(3,4)$. Find the coordinates of the third vertex C of ΔABC.

Sol. Let the third vertex be $C(x,y)$.

Then, coordinates of the centroid $= \left(\frac{6-2+x}{3}, \frac{4+2+y}{3}\right)$

$\Rightarrow \quad (3,4) = \left(\frac{6-2+x}{3}, \frac{4+2+y}{3}\right)$

$\Rightarrow \quad 3 = \frac{6-2+x}{3}$ and $4 = \frac{4+2+y}{3}$

$\Rightarrow \quad x = 5$ and $y = 6$

Hence, the third vertex is $C(5,6)$.

Problem 13 Find the coordinates of the incentre of a triangle whose vertices are $A(0,0)$, $B(0,6)$ and $C(8,0)$.

Sol. $BC = a = \sqrt{(8-0)^2 + (0-6)^2} = 10$

$AC = b = \sqrt{(8-0)^2 + (0-0)^2} = 8$

$AB = c = \sqrt{(0-0)^2 + (6-0)^2} = 6$

We know, coordinates of incentre of a triangle with vertices $A(x_1, y_1)$, $B(x_2, y_2)$ and $C(x_3, y_3)$ are $\left(\frac{ax_1 + bx_2 + cx_3}{a+b+c}, \frac{ay_1 + by_2 + cy_3}{a+b+c}\right)$, where, $a = BC$, $b = AC$ and $c = AB$.

Then, coordinates of incentre

$$= \left(\frac{10\cdot 0 + 8\cdot 0 + 6\cdot 8}{10+8+6}, \frac{10\cdot 0 + 8\cdot 6 + 6\cdot 0}{10+8+6}\right) = (2,2)$$

Problem 14 What is the area of a triangle whose vertices are $A(4,2)$, $B(0,5)$ and $C(-2,3)$

Sol. Area $= \frac{1}{2}|\{x_1(y_2 - y_3) + x_2(y_3 - y_1) + x_3(y_1 - y_2)\}|$

$= \frac{1}{2}|\{4(5-3) + 0(3-2) - 2(2-5)\}|$

$= \frac{1}{2}|\{8 + 0 + 6\}| = 7$ sq units.

Alternatively I

$$\begin{matrix} 4 & 2 \\ 0 & 5 \end{matrix} \;\; \rangle\rangle (20-0) = 20$$

$$\begin{matrix} 0 & 5 \\ -2 & 3 \end{matrix} \;\; \rangle\rangle (0+10) = 10$$

$$\begin{matrix} -2 & 3 \\ 4 & 2 \end{matrix} \;\; \rangle\rangle (-4-12) = -16$$

Then, Area $= \frac{1}{2}|20 + 10 - 16| = 7$ sq units

Alternatively II

Side $AB = \sqrt{(5-2)^2 + (0-4)^2} = 5$

Equation of the line passing through points $A(4,2)$ and $B(0,5)$ is

$$y - 2 = \frac{5-2}{0-4} \cdot (x-4)$$

$$\Rightarrow \quad -4y + 8 = 3x - 12 \Rightarrow 3x + 4y - 20 = 0$$

We know, the perpendicular distance of a line $ax + by + c = 0$ from a point $(x_1, y_1) = \frac{|ax_1 + by_1 + c|}{\sqrt{a^2 + b^2}}$

Then, the perpendicular distance between side AB and vertex

$$(-2,3) = \frac{|3\cdot(-2) + 4\cdot 3 - 20|}{\sqrt{3^2 + 4^2}} = \frac{14}{5}$$

Now, area of $\Delta ABC = \frac{1}{2} \times AB \times \text{height} = \frac{1}{2} \times 5 \times \frac{14}{5} = 7$ sq units

[this approach is not advisable for the examination hall, however, it is given here for conceptual clarity]

Problem 15 **The area of a triangle having vertices (2, 1), (−3, 4) is 10 sq units. What is the third vertex if it lies on the line $7x + 5y + 1 = 0$.**

Sol. Let the third vertex be (a, b).

Then,

2 1
⟩⟩$(8 + 3) = 11$
−3 4
⟩⟩$(-3b - 4a)$
a b
⟩⟩$(a - 2b)$
2 1

$\therefore$ Area of $\Delta = \frac{1}{2}|11 - 3b - 4a + a - 2b| = 10$

$$\Rightarrow \quad 10 = \frac{1}{2}|11 - 5b - 3a|$$

$$\Rightarrow \quad 20 = |11 - 5b - 3a|$$

If $11 - 5b - 3a > 0$, then, $20 = 11 - 5b - 3a$

$$\Rightarrow \quad 3a = -5b - 9 \Rightarrow a = \frac{-5b}{3} - 3$$

If $11 - 5b - 3a < 0$, then, $20 = -11 + 5b + 3a$

$$\Rightarrow \quad 3a = -5b + 31 \Rightarrow a = \frac{-5b}{3} + \frac{31}{3}$$

Obviously, the point (a, b) will satisfy the given equation $7x + 5y + 1 = 0$ as it lies on the equation.

$$\therefore \quad 7a + 5b + 1 = 0 \Rightarrow 7a = -1 - 5b$$

At $a = \frac{-5b}{3} - 3$, $7\left(\frac{-5b}{3} - 3\right) = -1 - 5b$

$$\Rightarrow \quad b = -3, \text{ then, } a = 2$$

At $a = \frac{-5b}{3} + \frac{31}{3}$, $7\left(\frac{-5b}{3} + \frac{31}{3}\right) = -1 - 5b$

$$\Rightarrow \quad b = 11, \text{ then, } a = -8$$

Hence, the third vertex is either $(2, -3)$ or $(-8, 11)$. In both cases, area of triangle will be 10 sq. units.

Alternatively

The best way is to proceed through options. Please check as to with which option, the area of the triangle is 10.

Problem 16 **Find the value of p for which the points $A(-1,3)$, $B(2, p)$ and $C(5, -1)$ are collinear.**

Sol. We know, three points $A(x_1, y_1)$, $B(x_2, y_2)$ and $C(x_3, y_3)$ are collinear if $x_1(y_2 - y_3) + x_2(y_3 - y_1) + x_3(y_1 - y_2) = 0$

$$\therefore \quad -1(p+1) + 2(-1-3) + 5(3-p) = 0$$

$$\Rightarrow \quad -p - 1 - 8 + 15 - 5p = 0 \Rightarrow p = 1$$

Alternatively

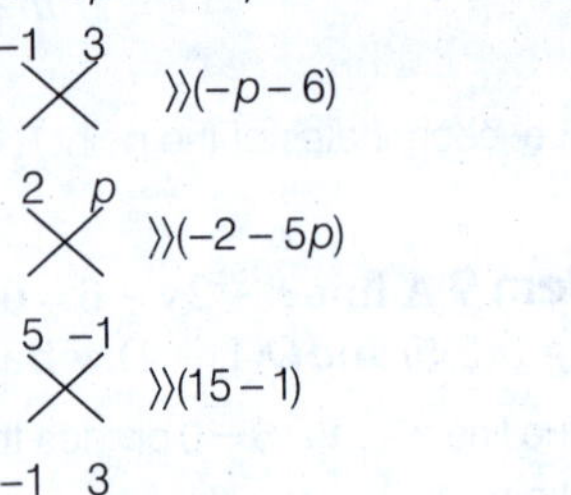

For collinearity of these three points, the value of $(-p-6) + (-2-5p) + (15-1)$ must be 0.

$\Rightarrow -p - 6 - 2 - 5p + 14 = 0 \Rightarrow p = 1$

Problem 17 **If the coordinates of orthocentre and the centroid of a Δ ABC are $(-5, 7)$ and $(5, 5)$, then the circumcentre of the Δ ABC is :** ***[IIFT-19]***

Sol. We know, the centroid divides the distance from the orthocentre to the circumcentre in ratio 2:1.

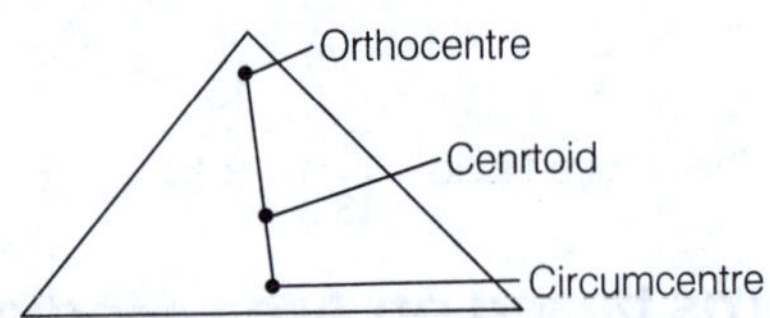

Let the coordinates of circumcentre be (x, y).

Then, $(5,5) = \left(\frac{2\cdot x + 1\cdot(-5)}{2+1}, \frac{2\cdot y + 1\cdot 7}{2+1}\right)$

[applying Section formula]

$$\Rightarrow \quad 5 = \frac{2x-5}{3} \text{ and } 5 = \frac{2y+7}{3} \Rightarrow x = 10 \text{ and } y = 4$$

Hence, coordinates of the circumcentre = (10, 4)

Problem 18 **Consider a triangle drawn on the X-Y plane with its three vertices at (41, 0), (0, 41) and (0, 0) each vertex being represented by its (X, Y) coordinates. The number of points with integer coordinates inside the triangle (excluding all the points on the boundary) is:** ***[CAT-05]***

Sol. The graph will be as below

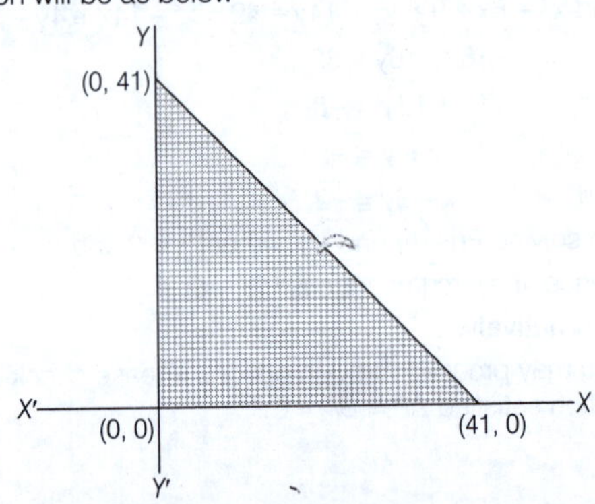

Integer coordinates of the points inside the triangle will be like (1, 39), (1, 38), (2, 38), (1, 37), (2, 37), (3, 37), (1, 36) (2,36), (3, 36), (4, 36) ......

Then total number of points $= 1+2+3+4+ ..... +39$

$$= \frac{39 \times 40}{2}$$

$$= 780$$

Problem 19 What are the coordinates of the orthocentre of a triangle whose vertices are $A(-2,1), B(3,2)$ and $C(1,-3)$.

Sol. Equation of the line passing through points $A(-2,1)$ and $B(3,2)$ is

$$(y-1) = \frac{2-1}{3+2}\cdot(x+2)$$

$$\Rightarrow \quad 5y-5 = x+2$$

$$\Rightarrow \quad x-5y+7=0$$

and equation of the line passing through points $B(3,2)$ and $C(1,-3)$ is

$$(y-2) = \frac{-3-2}{1-3}\cdot(x-3)$$

$$\Rightarrow \quad -2y+4 = -5x+15$$

$$\Rightarrow \quad 5x-2y-11=0$$

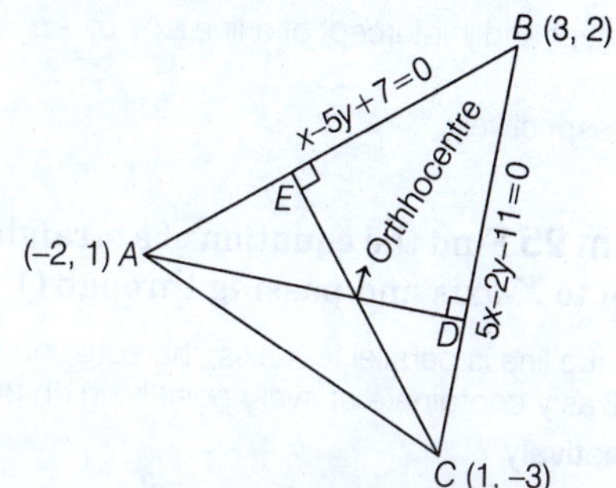

The slope of the line AD perpendicular to the line $5x-2y-11=0$

$$= \frac{-1}{\frac{5}{2}} = \frac{-2}{5}$$

[∵ product of slopes of two lines in case of perpendicularity = −1 and slope of the line $5x-2y-11=0$ is $\frac{-5}{-2}$ i.e. $\frac{5}{2}$]

Then, equation of line AD which passes through $(-2,1)$ is

$$(y-1) = \frac{-2}{5}(x+2)$$

$$\Rightarrow \quad 5y-5 = -2x-4$$

$$\Rightarrow \quad 2x+5y-1=0$$

Similarly, the equation of the line CE which is perpendicular to line $x-5y+7=0$ and passes through $(1,-3)$ is

$$5x+y-2=0$$

On solving both the equations

$$2x+5y-1-0$$

and $\quad 5x+y-2=0,$

we get $\quad x = \frac{9}{23}, y = \frac{1}{23}.$

Hence, coordinates of the orthocentre $= \left(\frac{9}{23}, \frac{1}{23}\right)$

Problems Based on Quadrilateral

Problem 20 The vertices of a polygon are $A(1,2)$, $B(5,4), C(3,8)$ and $D(-1,6)$. Then, $ABCD$ is:

(a) Rectangle (b) Rhombus

(c) Square (d) Cannot be determined

Sol. $AB = \sqrt{(5-1)^2+(4-2)^2} = 2\sqrt{5}$

Similarly, $BC = CD = DA = 2\sqrt{5}$

Now, diagonal $AC = \sqrt{(3-1)^2+(8-2)^2} = 2\sqrt{10}$

Similarly, $BD = 2\sqrt{10}$

Since the sides of quadrilateral $ABCD$ are equal and also, its diagonals are equal, $ABCD$ is a square.

Note *In order to solve such types of questions, we ought to know the properties of various types of quadrilaterals, already discussed in the section 'Geometry'. However, it is given for quick revision:*

(a) A quadrilateral is a parallelogram if its opposite sides are equal

(b) A quadrilateral is a rectangle if its opposite sides are equal and also its diagonals are equal.

(c) A quadrilateral is a square if its all sides are equal and also its diagonals are equal.

(d) A quadrilateral is a rhombus if its all sides are equal but its diagonals are not equal

Problem 21 If $(-2,-1), (a,0), (4,b)$ and $(1,2)$ are the vertices of a parallelogram, find the values of a and b.

Sol. Let $A(-2,-1), B(a,0), C(4,b)$ and $D(1,2)$ are vertices of the parallelogram.

We know, diagonals of a parallelogram bisect each other.

Then, coordinates of the midpoint of diagonal AC

$$= \left(\frac{-2+4}{2}, \frac{-1+b}{2}\right)$$

and coordinates of the midpoint of diagonal BD $= \left(\frac{a+1}{2}, \frac{0+2}{2}\right)$

Obviously, the midpoints of both the diagonals are the same.

$$\therefore \quad \frac{-2+4}{2} = \frac{a+1}{2} \text{ and } \frac{-1+b}{2} = \frac{0+2}{2}$$

$$\Rightarrow \quad a = 1 \text{ and } b = 3$$

Problem 22 Find the area of a quadrilateral whose vertices are $A(4,2), B(0,5), C(-2,3)$ and $D(2,1)$.

Sol.

4 2
⨯ ⟩⟩(20 − 0) = 20
0 5
⨯ ⟩⟩(0 + 10) = 10
−2 3
⨯ ⟩⟩(−2 − 6) = −8
2 1
⨯ ⟩⟩(4 − 4) = 0
4 2

Now, area $= \frac{1}{2}|20+10-8| = 11$ sq units.

Alternatively

You may split the quadrilateral in two triangles and add their areas. But, it is not advisable as it is a time consuming process.

Determination of equation of a line under different conditions

Problem 23 Find the equation of a line passing through the points (2, 4) and (–8, – 1).

Sol. We know, $y = mx + c$, where m is the slope of the line and c is the y-intercept.

$\Rightarrow \quad 4 = m\cdot 2 + c$ [if we consider the point (2, 4)]

$\Rightarrow \quad 4 = \frac{1}{2}\cdot 2 + c$ $\left[\because m = \frac{-1-4}{-8-2} = \frac{1}{2}\right]$

$\Rightarrow \quad c = 3$

Then, the required equation is $y = mx + c$

$\Rightarrow \quad y = \frac{1}{2}\cdot x + 3 \Rightarrow x - 2y + 6 = 0$

Alternatively I

If we consider the second point (–8, – 1), then $-1 = \frac{1}{2}\cdot(-8) + c$

$\Rightarrow \quad -1 = -4 + c \Rightarrow c = 3$

Then, the required equation is $y = \frac{1}{2}\cdot x + 3$

$\Rightarrow \quad x - 2y + 6 = 0$

Alternatively II

Let an another point (x, y) lies on the line.

Then, the slope m_1 of the line joining the points (x, y) and (2, 4) $= \frac{4-y}{2-x}$

Further, the slope m_2 of the line joining the points (2, 4) and $(-8,-1) = \frac{-1-4}{-8-2} = \frac{1}{2}$

Since slopes of the same line are equal, $\frac{4-y}{2-x} = \frac{1}{2}$. In other words, all the three points are collinear, then $m_1 = m_2$

$\Rightarrow \quad 8 - 2y = 2 - x \Rightarrow x - 2y + 6 = 0$

Alternatively III

The required equation

$$(y - y_1) = \frac{y_2 - y_1}{x_2 - x_1}(x - x_1)$$

$$\Rightarrow \quad (y-4) = \left(\frac{-1-4}{-8-2}\right)(x-2)$$

$$\Rightarrow \quad 2y - 8 = x - 2$$

$$\Rightarrow \quad x - 2y + 6 = 0$$

Alternatively IV

Proceed through options. Please check as to which option is satisfied by both the points.

Problem 24 Find the equation of a line which makes intercepts 2, – 3 on X and Y-axes respectively.

Sol. Clearly, the line is passing through points (2, 0) and (0, – 3). Now, get the equation on the above pattern. The answer is $3x - 2y - 6 = 0$.

In addition to the above first four approaches, the other approaches are as below:

Alternatively IV

We know, $y = mx + c$, where m is the slope of the line and c is the y-intercept

$\Rightarrow \quad y = mx - 3$ [$\because$ y-intercept is –3]

$\Rightarrow \quad 0 = m\cdot 2 - 3$

[considering the coordinates of the points (2, 0)]

$\Rightarrow \quad m = \frac{3}{2}$

Then, the required equation is : $y = \frac{3}{2}x - 3$

$\Rightarrow \quad 3x - 2y - 6 = 0.$

Alternatively V

The required equation:

x(y-intercept) + y(x-intercept) = (x-intercept) × (y-intercept)

$\Rightarrow \quad -3x + 2y = 2\cdot(-3)$

$\Rightarrow \quad 3x - 2y - 6 = 0$

Alternatively VI

Proceed through options. Please check as to which option yields 2 and –3 as x-intercept and y-intercept respectively. The x-intercept and y-intercept of a line $ax + by + c = 0$ are $\left(\frac{-c}{a}\right)$ and $\left(\frac{-c}{b}\right)$ respectively.

Problem 25 Find the equation of a straight line which is parallel to X-axis and passing through (1, – 3)

Sol. Since the line is parallel to X-axis, the equation of the line will be $y = -3$ as y-coordinate of every point lying on the line is –3 .

Alternatively

Slope of the line parallel to X-axis is 0.

Then, equation of the line: $y = 0\cdot x + c$

$\Rightarrow \quad -3 = 0\cdot 1 + c \Rightarrow c = -3$

Then, the required equation: $y = 0\cdot x - 3 \Rightarrow y = -3$

Problem 26 What is the equation of a line for which y-intercept is 2 and $\tan\theta = \frac{1}{2}$, where θ is the inclination of the line.

Sol. Given that the y-intercept is 2. It implies that the line passes through a point (0, 2).

Then, equation of the line: $y = mx + c$

$\Rightarrow \quad 2 = m\cdot 0 + c \Rightarrow c = 2$

Hence, the required equation is $y = \frac{1}{2}x + 2$ as $m = \tan\theta = \frac{1}{2}$

$\Rightarrow \quad x - 2y + 4 = 0$

Alternatively I

$y = mx + c$ is the 'slope-intercept form' of the equation of the line where, c is the y-intercept and m is the slope of the line.

$\Rightarrow \quad y = \frac{1}{2}x + 2$

$\Rightarrow \quad x - 2y + 4 = 0$

Alternatively II

Let the line passes through a point (x, y) also.

Hence, the line is passing through (x, y) and (0, 2).

Now, $m = \frac{2-y}{0-x}$

As given, $\frac{2-y}{0-x} = \frac{1}{2}$

$\Rightarrow \quad 4-2y = -x$

$\Rightarrow \quad x-2y+4 = 0$

Alternatively III

Proceed through options. Please check as to which option yields $\frac{1}{2}$ as slope of the line and is also satisfied with the coordinates of the point (0, 2). We know that the slope of a line $ax+by+c=0$ is $\left(\frac{-a}{b}\right)$.

Problem 27 **A line passing through a point $(-2, 4)$ makes intercepts on the axes equal in magnitude but opposite in sign. What is the equation of the line ?**

Sol. Let x-intercept be a, then y-intercept $= -a$

$\therefore$ Equation of the line $= \frac{x}{a} + \frac{y}{-a} = 1$

$\Rightarrow \quad x-y = a$

Since the line also passes through a point $(-2, 4)$, the coordinates of this point shall satisfy the equation of the line.

Then, $-2-4 = a \Rightarrow a = -6$

Hence, the equation of line is $x-y = -6$

$\Rightarrow \quad x-y+6 = 0$

Try yourself with other approaches.

Note *The best way is to proceed through options. Please check as to which option yields x-intercept and y-intercept of a line* $ax+by+c=0$ *are* $\left(\frac{-c}{a}\right)$ *and* $\left(\frac{-c}{b}\right)$ *respectively.*

Problems Based on Triangle whose Side(s) is/are in Equation Forms

Problem 28 **The area of the triangle formed by the graph of $ax + by = c$ (where a, b are two positive real numbers) and the coordinate axes, is:**

[SSC (FCI-Mains)-12]

Sol. Let the line $ax+by=c$ meets X-axis and Y-axis at points $B(x_1, y_1)$ and $A(x_2, y_2)$ respectively.

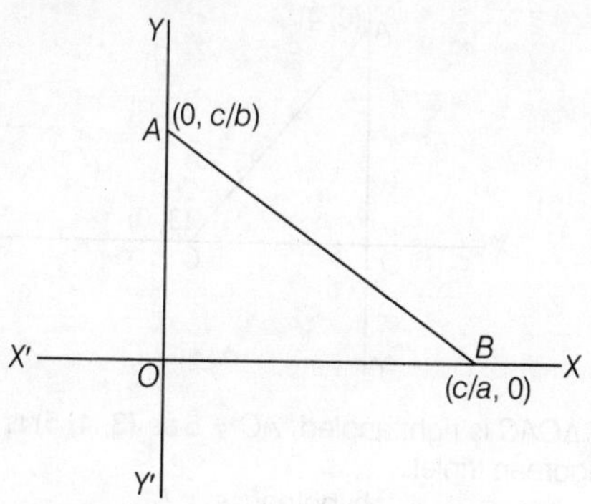

It is obvious that coordinates of point B will satisfy the given equation, however, the ordinate of the point B will be 0.

It means, $ax_1 + b \cdot 0 = c \Rightarrow x_1 = c/a$

Also, coordinates of point A will satisfy the given equation, however, the abscissa of the point A will be 0.

It means, $a \cdot 0 + b \cdot y_2 = c \Rightarrow y_2 = c/b$

Hence, OB $= c/a$ and OA $= c/b$

Then, area of the triangle $= \frac{1}{2} \times OB \times OA = \frac{1}{2} \times \frac{c}{a} \times \frac{c}{b} = \frac{c^2}{2ab}$

Alternatively

The equation of the line be $ax+by-c=0$

Then, the area of triangle $= \frac{(-c)^2}{|2ab|} = \frac{c^2}{2ab}$

[$\because$ a and b are positive real numbers as given]

Problem 29 **Area of the trapezium formed by X-axis, Y-axis and the lines $3x + 4y = 12$ and $6x + 8y = 60$, is:**

[SSC (CGL-T-I)-12]

Sol. The x-intercept and the y-intercept of the line $3x+4y-12=0$ are respectively $\frac{12}{3}$ and $\frac{12}{4}$ i.e. 4 and 3.

and x-intercept and y-intercept of the line $6x+8y-60=0$ are respectively $\frac{60}{6}$ and $\frac{60}{8}$ i.e. 10 and 7.5.

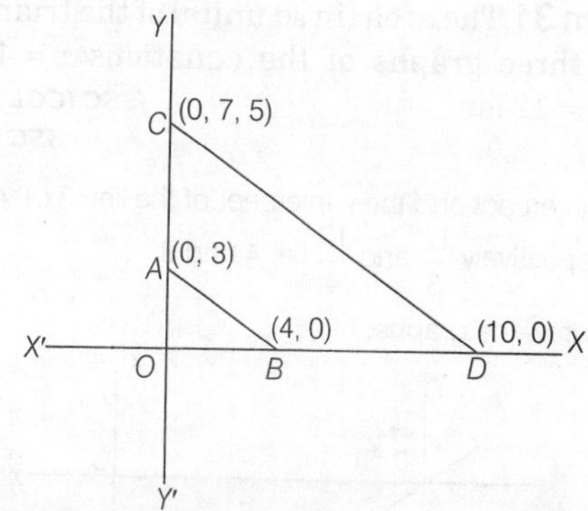

Now, area of $\Delta AOB = \frac{1}{2} \times 4 \times 3$

$= 6$ sq units

and area of $\Delta COD = \frac{1}{2} \times 10 \times 7.5$

$= 37.5$ sq units

$\therefore$ Area of trapezium $ABDC = 37.5 - 6 = 31.5$ sq units.

Alternatively

The x-intercept of the line $3x+4y-12=0$ is $\frac{12}{3}$ i.e. 4 and

x-intercept of the line $6x+8y-60=0$ is $\frac{60}{6}$ i.e. 10

It means, the parallel sides of the trapezium are in the same quadrant as depicted in the figure.

Now, area of triangle formed by line $3x+4y-12=0$ and coordinate axes $= \frac{c^2}{|2ab|} = \frac{(-12)^2}{|2 \cdot 3 \cdot 4|} = 6$ sq units.

and area of triangle formed by line $6x+8y-60=0$ and coordinate axes $= \frac{c^2}{|2ab|} = \frac{(-60)^2}{|2 \cdot 6 \cdot 8|} = 37.5$ sq units.

Then, area of the trapezium $= 37 \cdot 5 - 6 = 31 \cdot 5$ sq units

Problem 30 Find the area of a triangle formed by the lines $5x - 2y - 11 = 0$, $4x + 3y + 5 = 0$ and $x - 5y + 7 = 0$

Sol. Basically, the point of intersection of any two lines is a vertex of the triangle and indeed, the coordinates of the point of intersection are the solutions of equations of two lines which intersect each other.

On solving Eqs. (i) and (ii), we get, $x = 1, y = -3$

On solving Eqs. (ii) and (iii), we get, $x = -2, y = 1$

On solving Eqs. (iii) and (i), we get, $x = 3, y = 2$

Hence, vertices of the triangle are $(1, -3), (-2, 1)$ and $(3, 2)$.

For computation of area of the triangle:

1 -3 ╳))(1 − 6) = −5

−2 1 ╳))(−4 − 3) = −7

3 2 ╳))(−9 − 2) = −11

1 −3

Hence, area of the triangle $= \frac{1}{2}|-5-7-11| = \frac{23}{2}$ sq units

Problem 31 The area (in sq units) of the triangle formed by the three graphs of the equations $x = 4$, $y = 3$ and $3x + 4y = 12$ is: *[SSC (CGL T-I)-12, 14, 15, SSC (CGL T-II)-15]*

Sol. The x-intercept and the y-intercept of the line $3x + 4y - 12 = 0$ are respectively $\frac{12}{3}$ and $\frac{12}{4}$ i.e. 4 and 3.

Now, draw the graphs.

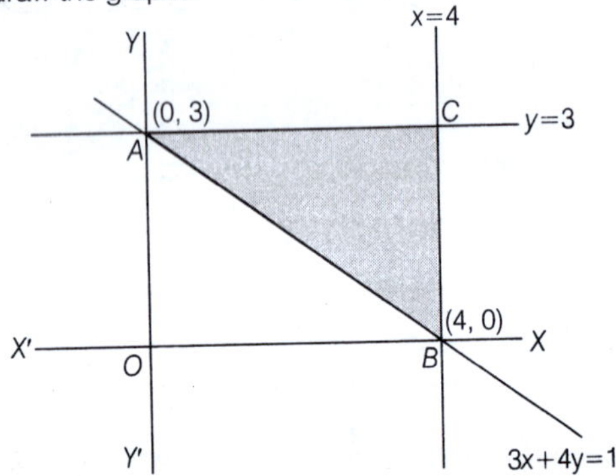

Clearly, area of $\Delta ABC = \frac{1}{2} \times 4 \times 3 = 6$ sq units

Problem 32 The area of the triangle formed by the lines $5x + 7y = 35$, $4x + 3y = 12$ and X-axis is: *[SSC (CGL-T-I)-12, SSC (T-II)-14, 20]*

Sol. Clearly, ordinate of the point of intersection of lines $5x + 7y = 35$ and $4x + 3y = 12$ shall be the altitude of ΔABC.

On solving both equation, we get

$$x = \frac{-21}{13} \text{ and } y = \frac{80}{13}$$

Hence, the altitude $AM = \frac{80}{13}$

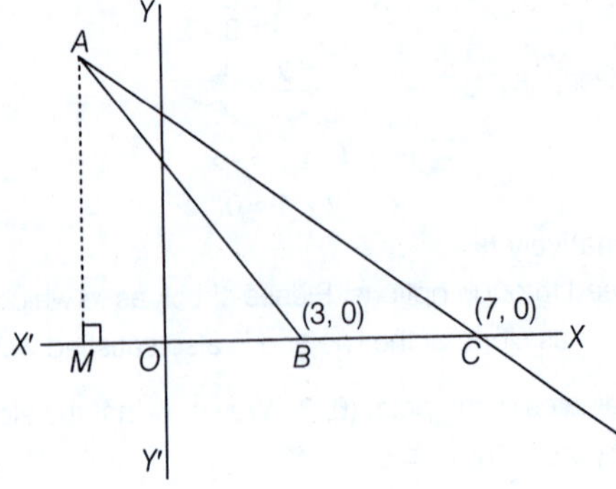

Now, x-intercept of the line $5x + 7y = 35$ is $\frac{-(-35)}{5} = 7$ and x-intercept of the line $4x + 3y = 12$ is $\frac{-(-12)}{4} = 3$

Hence, $BC = 7 - 3 = 4$ units.

Then, area of $\Delta ABC = \frac{1}{2} \times \text{base} \times \text{altitude}$

$$= \frac{1}{2} \times 4 \times \frac{80}{13} = 12.30 \text{ sq units.}$$

Alternatively

Area of the triangle $= \frac{1}{2} \times |$ difference between x-intercepts $| \times |$ y-coordinate of point of intersection of lines $|$

$$= \frac{1}{2} \times \left|\frac{-(-35)}{5} - \frac{-(-12)}{4}\right| \times \left|\frac{80}{13}\right| = 12.30 \text{ sq units.}$$

[recall the concept of a system of linear equations $a_1x + b_1y + c_1 = 0$ and $a_2x + b_2y + c_2 = 0$.

Value of $y = \frac{c_1a_2 - c_2a_1}{a_1b_2 - a_2b_1}$. We may find the value of y-coordinate through this method quickly]

Problem 33 Find the radius of the circumcircle of the triangle formed by X-axis, Y-axis and $4x + 3y = 12$. *[SSC (CGL-T-I)-12, CAT-19]*

Sol. The x-intercept and the y-intercept of the line $4x + 3y - 12 = 0$ are respectively $\frac{12}{4}$ and $\frac{12}{3}$ i.e. 3 and 4.

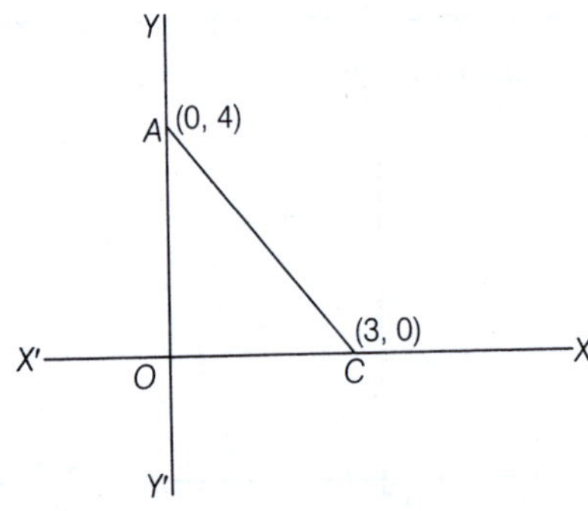

Since ΔOAC is right angled, $AC = 5$ as (3, 4, 5) is a Pythagorean triplet.

Then, circumradius $= \frac{\text{hypotenuse}}{2} = \frac{5}{2} = 2.5$ units.

Problem 34 What is the area of a triangle formed by lines $3x+4y-12=0$, $x-y-5=0$ and $x=0$.

Sol. $x=0$ means Y-axis.

We know, the area of triangle formed by two equations $a_1x+b_1y+c_1=0$ and $a_2x+b_2y+c_2=0$ and Y-axis

$$=\frac{(\text{difference between } y-\text{intercepts})^2}{2|\text{difference between slopes}|}$$

Slope and y-intercept of the line $3x+4y-12=0$ are $-3/4$ and 3 respectively

and Slope and y-intercept of the line $x-y-5=0$ are 1 and -5 respectively

Now, area of the triangle $=\dfrac{\{3-(-5)\}^2}{2\left|-\frac{3}{4}-1\right|}$

$$=\frac{32\times 4}{7}=18.29 \text{ sq units.}$$

Alternatively I

We may apply the formula after the equations are converted in slope-intercept form.

Alternatively II

Area of the triangle $=\frac{1}{2}\times$ | difference between y-intercepts | × | x-coordinate of point of intersection of lines |

$$=\frac{1}{2}\times\left|\frac{-(-12)}{4}-\frac{-(-5)}{-1}\right|\times\left|\frac{32}{7}\right|$$

$$=\frac{32\times 4}{7}=18.29 \text{ sq units.}$$

[recall the concept of a system of linear equations $a_1x+b_1y+c_1=0$ and $a_2x+b_2y+c_2=0$ Value of $x=\dfrac{b_1c_2-b_2c_1}{a_1b_2-a_2b_1}$.

We may find the value of x-coordinate through this method quickly]

Problems Based on Angle between Lines

Problem 35 If a straight line L makes an angle $\theta(\theta>90°)$ with the positive direction of X-axis, then the acute angle made by a straight line L_1, perpendicular to L, with the Y-axis is ***[SSC (CGL-T-I)-12]***

Sol. In quadrilateral AOCD,

$$\angle BAO+\angle AOC+\angle OCB+\angle CBA=360°$$

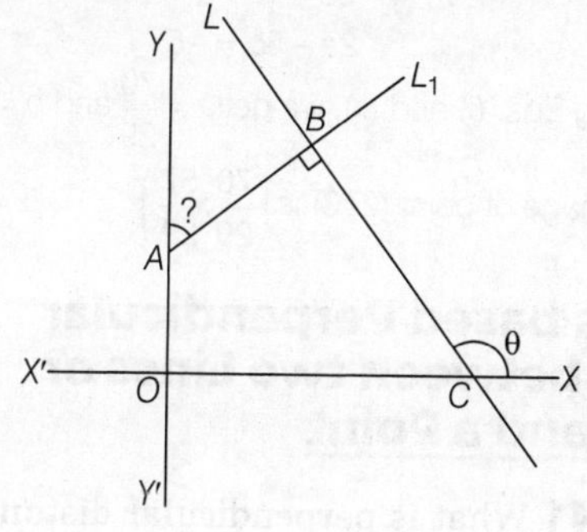

$\Rightarrow\quad \angle BAO+90°+(180°-\theta)+90°=360°$

$\Rightarrow\quad \angle BAO=\theta, \theta>90°$ as given.

$\therefore$ Acute $\angle BAY=(180°-\theta)$

Problem 36 If the angle between two straight lines is 45° and slope of one line is $\frac{1}{3}$, then what is the slope of another line?

Sol. We know, if α and β are the adjacent angles between two lines having slopes m_1 and m_2 and α is an acute angle, then

$$\tan\alpha=\left|\frac{m_2-m_1}{1+m_1\cdot m_2}\right|, \text{ where } 1+m_1\cdot m_2\neq 0$$

$$\Rightarrow\quad \tan 45°=\left|\frac{m_2-\frac{1}{3}}{1+\frac{1}{3}\cdot m_2}\right|$$

$$\Rightarrow\quad 1=\left|\frac{m_2-\frac{1}{3}}{1+\frac{1}{3}\cdot m_2}\right|$$

$$\Rightarrow\quad 1=\left(\frac{m_2-\frac{1}{3}}{1+\frac{1}{3}\cdot m_2}\right) \text{ or } 1=-\left(\frac{m_2-\frac{1}{3}}{1+\frac{1}{3}\cdot m_2}\right)$$

$$\Rightarrow\quad m_2=2 \text{ or } m_2=\frac{-1}{2}$$

Slope of another line is either 2 or $\frac{-1}{2}$. It implies that two lines with slopes 2 and $\frac{-1}{2}$ may make an angle of 45° with the given line. For conceptual clarity, observe the figure:

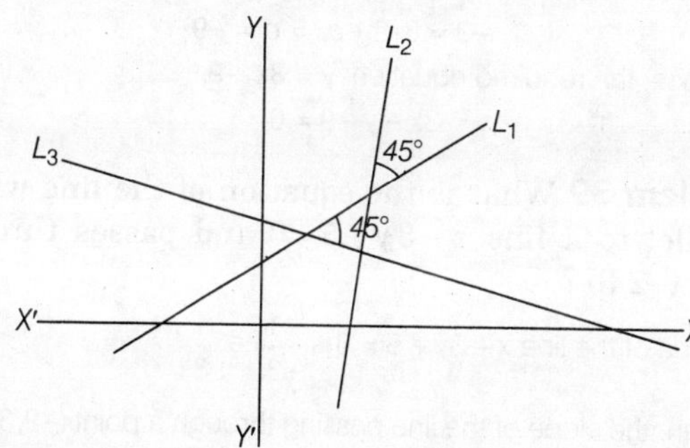

Here, each of lines L_2 and L_3 makes an angle 45° with the line L_1. The slopes of line L_2 and L_3 are different.

Alternatively

You may proceed through options. Please put the value of option in place of m_2 in the expression $\left|\dfrac{m_2-\frac{1}{3}}{1+\frac{1}{3}\cdot m_2}\right|$, where $m_1=\frac{1}{3}$.

The option with which the value of expression is 1, will be the correct answer.

Problems Based on angle between a Line and an Axis

Problem 37 The angle between the graph of the linear equation $239x - 239y + 5 = 0$ and X-axis is: *[SSC (CPO)-15]*

Sol. We know, if $ax + by + c = 0$ is an equation of a line, then its slope $= \left(\frac{-a}{b}\right)$

Here, slope of the line $= \frac{-239}{-239} = 1$

$\Rightarrow \quad \tan\theta = \tan 45°$

$\Rightarrow \quad \theta = 45°$

Problems Based on Relation between Two Lines like Relation of Perpendicularity, Parallelism, etc.

Problem 38 What is the equation of the line which passes through $(2, -3)$ and is perpendicular to the line joining the points $(-1, 5)$ and $(2, 4)$.

Sol. Slope of the line joining the points $(-1,5)$ and $(2,4) = \frac{4-5}{2+1} = \frac{-1}{3}$

Then, slope of the another line = 3

[$\because$ product of slopes of two lines in case of perpendicularity = −1]

Now, equation of the line which has slope-1 and passes through a point $(2, -3)$ be

$$(y+3) = 3(x-2) \Rightarrow 3x - y - 9 = 0$$

Alternatively

Equation of the line passing through $(2, -3)$ is $y = mx + c$

$\Rightarrow \quad -3 = 3 \cdot 2 + c \Rightarrow c = -9$

Hence, the required equation: $y = 3x - 9$

$\Rightarrow \quad 3x - y - 9 = 0$

Problem 39 What is the equation of the line which is parallel to a line $x - 3y + 5 = 0$ and passes through a point $(-2, 3)$?

Sol. Slope of the line $x - 3y + 5 = 0$ is $\frac{-1}{-3}$ i.e. $\frac{1}{3}$

Then, the slope of the line passing through a point $(-2,3) = \frac{1}{3}$ as parallel lines have equal slopes.

Now, its equations is $(y-3) = \frac{1}{3}(x+2)$

$\Rightarrow \quad 3y - 9 = x + 2$

$\Rightarrow \quad x - 3y + 11 = 0$

Alternatively

The equation of a line parallel to the line $ax + by + c = 0$ is $ax + by + c_1 = 0$ where c_1 is arbitrary constant.

It implies that the line $x - 3y + c_1 = 0$ passes through the point $(-2, 3)$

$\Rightarrow \quad -2 - 3 \times 3 + c_1 = 0$

$\Rightarrow \quad c_1 = 11$

Hence, the required equation be $x - 3y + 11 = 0$

Try yourself with other approaches.

Note *The best way is to proceed through options. Please check as to which option has the slope equal to the slope of the given equation $x - 3y + 5 = 0$ and is also satisfied with the coordinates of the given point.*

Problem 40 Let the line $2x - 5y + 8 = 0$ acts as the plane mirror, then what is the image of the point $(2, 3)$ in this line?

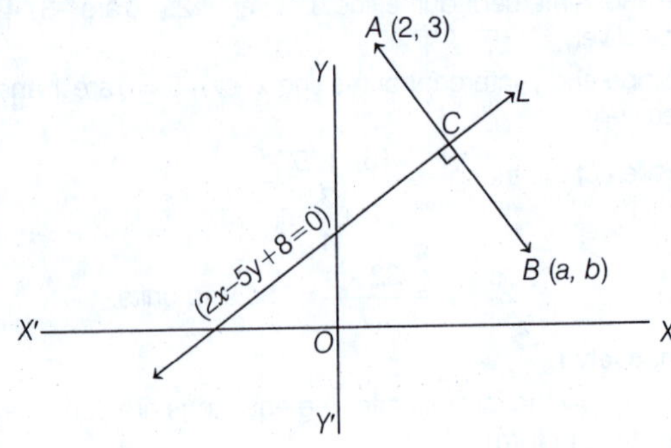

Sol. Let the given point be $A(2, 3)$ and its image be $B(a, b)$.

It is obvious that line L will bisect the line AB perpendicularly at point C.

We know, product of slopes of two lines in case of perpendicularity is −1.

$\Rightarrow$ Slope of line $AB \times$ slope of line $L = -1$

$\Rightarrow$ slope of line $AB \times \frac{-2}{-5} = -1$

$\Rightarrow \quad$ slope of line $AB = \frac{-5}{2}$

$\Rightarrow \quad \frac{b-3}{a-2} = \frac{-5}{2} \Rightarrow 2b - 6 = -5a + 10$

$\Rightarrow \quad 5a + 2b = 16 \quad$ ...(i)

Now, coordinates of point $C = \left(\frac{a+2}{2}, \frac{b+3}{2}\right)$ [by mid-point formula]

Obviously, point $C\left(\frac{a+2}{2}, \frac{b+3}{2}\right)$ will also satisfy the equation of line L as it lies on the line L too.

Then, $\quad 2\left(\frac{a+2}{2}\right) - 5\left(\frac{b+3}{2}\right) + 8 = 0$

$\Rightarrow \quad 2a + 4 - 5b - 15 = -16$

$\Rightarrow \quad 2a - 5b = -5 \quad$ ...(ii)

On solving Eqs. (i) and (ii), we get $a = \frac{70}{29}$ and $b = \frac{57}{29}$

Hence, image of point (2, 3) is $\left(\frac{70}{29}, \frac{57}{29}\right)$.

Problems based Perpendicular Distance between two Lines or one Line and a Point

Problem 41 What is perpendicular distance between a line $3x + 4y - 12 = 0$ and a point $(3, 5)$.

Sol. The x-intercept and the y-intercept of the line $3x + 4y - 12 = 0$ are respectively $\frac{12}{3}$ and $\frac{12}{4}$ i.e. 4 and 3.

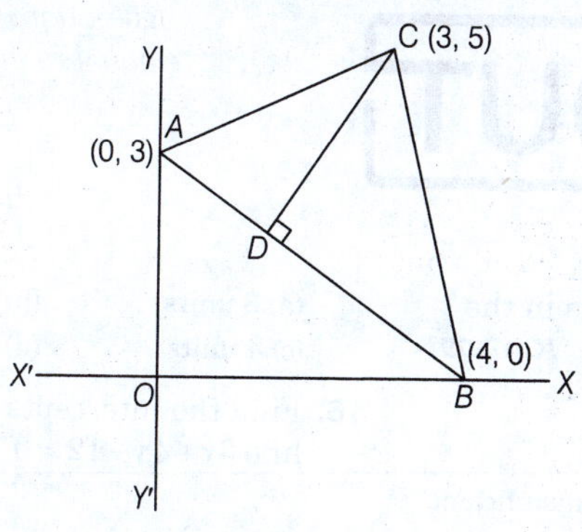

For computation of area of ΔABC,

$$\begin{matrix} 0 & 3 & \\ 4 & 0 & \rangle\rangle\ (0-12) = -12 \\ 3 & 5 & \rangle\rangle\ (20-0) = 20 \\ 0 & 3 & \rangle\rangle\ (9-0) = 9 \end{matrix}$$

Hence, area of $\Delta ABC = \frac{1}{2}|-12+20+9|$

$= \frac{17}{2}$ sq units

Now, area of $\Delta ABC = \frac{1}{2} \times AB \times CD$

[here, CD is the height, in other words, the perpendicular distance between point (3, 5) and the given line]

$\Rightarrow \frac{17}{2} = \frac{1}{2} \times 5 \times CD$ [$\because$ (3, 4, 5) is a Pythagorean triplet]

$\Rightarrow CD = \frac{17}{5} = 3.4$ units

Alternatively

We know, the perpendicular distance of a line $ax + by + c = 0$ from a point $(x_1, y_1) = \frac{|ax_1 + by_1 + c|}{\sqrt{a^2 + b^2}}$

Hence, the required distance $= \frac{|3 \times 3 + 4 \times 5 - 12|}{\sqrt{3^2 + 4^2}} = \frac{17}{5} = 3.4$ units

➘ Problem 42 What is the distance between two parallel lines $2x + y - 5 = 0$ and $2x + y - 8 = 0$

Sol. The x-intercept of the line $2x + y - 5 = 0$ is $\frac{5}{2}$ i.e. 2.5

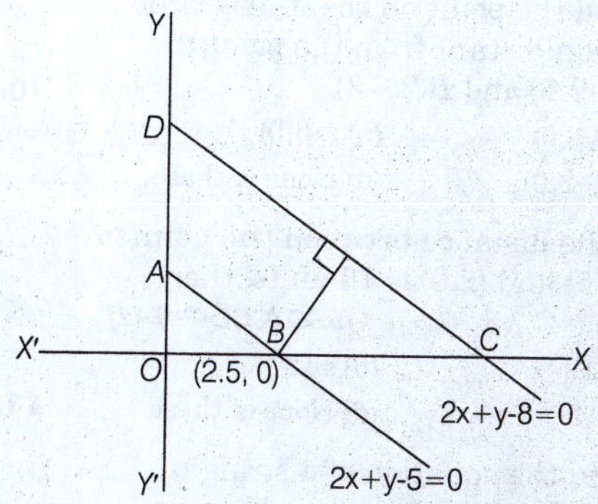

Now, the distance between a point (2.5, 0) and the line $2x + y - 8 = 0$ is $\frac{|2 \cdot 2.5 + 1 \cdot 0 - 8|}{\sqrt{2^2 + 1^2}} = \frac{3}{\sqrt{5}}$ units.

Alternatively

In case two lines are in form of $ax + by + c_1 = 0$ and $ax + by + c_2 = 0$, the perpendicular distance between them $= \frac{|c_1 - c_2|}{\sqrt{a^2 + b^2}}$

Hence, the required distance $= \frac{|-5-(-8)|}{\sqrt{2^2 + 1^2}} = \frac{3}{\sqrt{5}}$ units

WORKOUT

1. The graphs of $x = a$ and $y = b$ intersect at **[SSC (CGL-T-I)-14]**
(a) (a, b) (b) (b, a)
(c) $(-a, b)$ (d) $(a, -b)$

2. Find the point on the X-axis which is equidistant from the points $A(-2,5)$ and $B(2,-3)$.
(a) (3, 0) (b) (–3,0)
(c) (–2,0) (d) None of these

3. If the distance between two points (0, 5) and $(x,0)$ is 13 units, then $x = ?$ **[SSC (CGL-T-I)-12]**
(a) 12 (b) ±12
(c) 13 (d) None of these

4. Find the equation of a straight line that has slope 4 and passes through a point $(2,-3)$.
(a) $4x - y - 11 = 0$ (b) $4x - 3y - 11 = 0$
(c) $x - 4y - 15 = 0$ (d) None of these

5. The straight line $2x + 3y = 12$ passes through **[SSC (CGL-T-I)-13]**
(a) 1st, 2nd and 3rd quadrant
(b) 1st, 2nd and 4th quadrant
(c) 2nd, 3rd and 4th quadrant
(d) 1st, 3rd and 4th quadrant

6. In what ratio is the line segment joining the points $(-2,-3)$ and (3, 7) divided by the Y-axis?
(a) 3:2 (b) 2:3
(c) 3:4 (d) None of these

7. A polynomial $ax^3 + bx^2 + cx + d$ intersects X-axis at 1 and –1 and Y-axis at 2. The value of b is : **[XAT-14]**
(a) –2 (b) 0
(c) 1 (d) 2
(e) cannot be determined

8. An equation of the form $ax + by + c = 0$ where $a \neq 0$, $b \neq 0, c = 0$ represents a straight line which passes through: **[SSC (CGL-T-II)-12]**
(a) (0, 0) (b) (3, 2)
(c) (2, 4) (d) None of these

9. Coordinates of the points X, Y and Z are $X \equiv (6,4), Y \equiv (-3,5)$ and $Z \equiv (2,-4)$. Find the coordinates of a point which divides the medians from all the three vertices in the ratio 2 : 1. **[CAT-09]**
(a) $\left(\frac{5}{3}, \frac{5}{3}\right)$ (b) $\left(\frac{5}{3}, \frac{5}{2}\right)$
(c) $\left(\frac{7}{4}, \frac{3}{2}\right)$ (d) Data insufficient

10. If $A(5,-1), B(-3,-2)$ and $C(-1,8)$ are the vertices of ΔABC, find the length of the median through A and the coordinates of the centriod.
(a) $\sqrt{65}$ (b) 9
(c) $\sqrt{63}$ (d) None of these

11. The graph of the equation $4x - 5y = 20$ intersects the X-axis at the point **[SSC(Delhi Police-SI)-12]**
(a) (2, 0) (b) (5, 0)
(c) (4, 5) (d) (0, 5)

12. Circle C has a radius of 3 units. The line segment PQ is the only diameter of the circle which is parallel to the X-axis. P and Q are points on curves given by the equations $y = a^x$ and $y = 2a^x$ respectively, where a < 1. The value of a is: **[XAT-14]**
(a) $\frac{1}{\sqrt[6]{2}}$ (b) $\frac{1}{\sqrt[6]{3}}$
(c) $\frac{1}{\sqrt[3]{6}}$ (d) $\frac{1}{\sqrt{6}}$
(e) None of these

13. The length of intercept of the graph of the equation $9x - 12y = 108$ between two axes is **[SSC (CGL-T-II)-12, SSC (CGL-T-I)-15]**
(a) 15 units (b) 9 units
(c) 12 units (d) 18 units

14. The line joining the points $A(2,1)$ and $B(5,-8)$ is trisected at the points P and Q. If the point P lies on the line $2x - y + k = 0$, find the value of k.
(a) 10 (b) 12
(c) –8 (d) 6

15. A triangle is formed by the X-axis and the lines $2x + y = 4$ and $x - y + 1 = 0$ as three sides. Taking the side along X-axis as its base, the corresponding altitude of the triangle is: **[SSC (CGL-T-I)-12]**
(a) 3 units (b) 1 unit
(c) 4 units (d) 2 units

16. Find the intercepts made by the line $3x + 4y - 12 = 0$ on the axes. **[SNAP-11]**
(a) 2 and 3
(b) 4 and 3
(c) 3 and 5
(d) None of the above

17. In X-Y plane, the area of the region bounded by the graph of $|x + y| + |x - y| = 4$ is: **[CAT-05]**
(a) 8 (b) 12
(c) 16 (d) 20

18. Equation of the straight line parallel to X-axis and also 3 units below X-axis is: **[SSC (CGL-T-I)-13]**
(a) $x = -3$ (b) $y = 3$
(c) $y = -3$ (d) $x = 3$

19. Find the third vertex of a ΔABC if two of its vertices are $B(-3,1)$ and $C(0,-2)$ and its centroid is at the origin.
(a) (0, 0) (b) (3, 1)
(c) (2, 4) (d) None of these

20. The total area (in sq units) of the triangles formed by the graph of $4x + 5y = 40$, X-axis, Y-axis and $x = 5$ and $y = 4$ is: **[SSC (CGL-T-I)-14]**
(a) 10 (b) 20
(c) 30 (d) 40

21. Find the perpendicular distance between origin and the line $x - 3y + 21 = 0$.
(a) $\frac{21}{\sqrt{10}}$ (b) $\frac{21}{\sqrt{8}}$
(c) $\frac{21}{\sqrt{11}}$ (d) None of these

22. Consider the following two curves in the x-y plane $y = x^3 + x^2 + 5$ and $y = x^2 + x + 5$. Which of the following statements is true for $-2 \leq x \leq 2$? **[CAT-03]**
(a) The two curves intersect once
(b) The two curves intersect twice
(c) The two curves intersect thrice
(d) The two curves don't intersect

23. In what ratio does the line $x - y - 2 = 0$ divide the line segment joining $(3, -1)$ and $(8, 9)$?

(a) 3 : 2 (b) 2 : 3
(c) 3 : 4 (d) None of these

24. The area bounded by the lines $x = 0, y = 0, x + y = 1, 2x + 3y = 6$ (in sq units) is: ***[SSC (CGL-T-II)-12]***

(a) 2 (b) 3
(c) $2\frac{1}{3}$ (d) $2\frac{1}{2}$

25. What is the equation of a line for which x-intercept is -3 and $\tan\theta = \frac{1}{2}$, where θ is the inclination of the line.

(a) $3x - 2y + 3 = 0$
(b) $x - 4y + 3 = 0$
(c) $5x - 2y + 3 = 0$
(d) $x - 2y + 3 = 0$

26. Area of the triangle formed by the graph of the line $2x - 3y + 6 = 0$ along with the coordinate axes is:

(a) 1.5 sq units (b) 3 sq units
(c) 6 sq units (d) 0.5 sq units

27. Find the angle between lines $3x - y + 9 = 0$ and $x - 2y + 4 = 0$?

(a) either 45° or 135°
(b) 50°
(c) 60° (d) None of these

28. If a linear equation is of the form $x = k$ where k is a constant, then graph of the equation will be: ***[SSC (10+2)-LDC/DEO-12]***

(a) a line parallel to X-axis
(b) a line cutting both the axes
(c) a line making positive acute angle with X-axis
(d) a line parallel to Y-axis

29. A line makes intercepts on the axes such that the sum of their distances from the origin is 12 and it passes through a point (4, 2). What is the equation of the line?

(a) $x + 2y - 8 = 0$
(b) $x + y - 8 = 0$
(c) $2x + 3y - 12 = 0$
(d) $2x - 3y - 12 = 0$

30. The linear equation such that each point on its graph has an ordinate four times its abscissa is ***[SSC (CPO)-13]***

(a) $y + 4x = 0$ (b) $y = 4x$
(c) $x = 4y$ (d) $x + 4y = 0$

31. A triangle ABC has area 32 sq units and its side BC of length 8 units lies on the line $x = 4$. Then, the shortest possible distance between A and the point (0, 0) is: ***[CAT-18]***

(a) 4 units
(b) $4\sqrt{2}$ units
(c) $2\sqrt{2}$ units
(d) 8 units

32. The coordinates of a triangle ABC are $A(1, 5), B(-2, 3)$ and $C(0, -4)$, find the equation of the median AD. ***[IIFT-17]***

(a) $7x - 3y - 8 = 0$
(b) $5x - 4y + 15 = 0$
(c) $11x - 4y + 9 = 0$
(d) $x + 3y - 16 = 0$

33. Let P be the point of intersection of the lines $3x + 4y = 2a$ and $7x + 2y = 2018$ and Q be the point of intersection of the lines $3x + 4y = 2018$ and $5x + 3y = 1$. If the line through P and Q has slope 2, the value of a is ***[XAT-19]***

(a) 1 (b) $\frac{1}{2}$
(c) 4035 (d) 3026

Answers Key

1. (a)	2. (c)	3. (b)	4. (a)	5. (b)	6. (b)	7. (a)	8. (a)	9. (a)	10. (a)
11. (b)	12. (a)	13. (a)	14. (c)	15. (d)	16. (b)	17. (c)	18. (a)	19. (b)	20. (b)
21. (a)	22. (c)	23. (b)	24. (d)	25. (d)	26. (b)	27. (a)	28. (d)	29. (a)	30. (b)
31. (a)	32. (c)	33. (d)							

1. Obviously, values of a and b will act as abscissa and ordinate of the point of intersection.

Hence, option (a) is correct.

2. Let the point be $P(x, 0)$. Since the point P lies on X-axis, its ordinate will be 0.

Then, $AP = BP$

$\Rightarrow (AP)^2 = (BP)2$

$\Rightarrow (x+2)^2 + (0-5)^2$

$= (x-2)^2 + (0+3)^2$

$\Rightarrow x^2 + 4x + 29 = x^2 - 4x + 13$

$\Rightarrow 8x = -16$

$\Rightarrow x = -2$

Hence, the point is $P(-2, 0)$.

3. Clearly, $13 = \sqrt{(x-0)^2 + (0-5)^2}$

$\Rightarrow 169 = x^2 + 25$

$\Rightarrow x^2 = 144$

$\Rightarrow x = \sqrt{144} = \pm 12$

Alternatively

You may proceed through options.

4. Equation of line, $y = mx + c$

$\Rightarrow -3 = 4 \times 2 + c \Rightarrow c = -11$

Hence, the required equation is

$y = 4x - 11 \Rightarrow 4x - y - 11 = 0$

Alternatively

Check as to which option yields 4 as slope of the line and is also satisfied with the coordinates of the point (2 – 3). We know that the slope of a line $ax + by + c = 0$ is $\left(\frac{-a}{b}\right)$. It is the best way to solve the question in moments.

5. The x-intercept and the y-intercept of the line $2x + 3y - 12 = 0$ are respectively $\frac{12}{2}$ and $\frac{12}{3}$ i.e. 6 and 4.

Then, the graph will be as below:

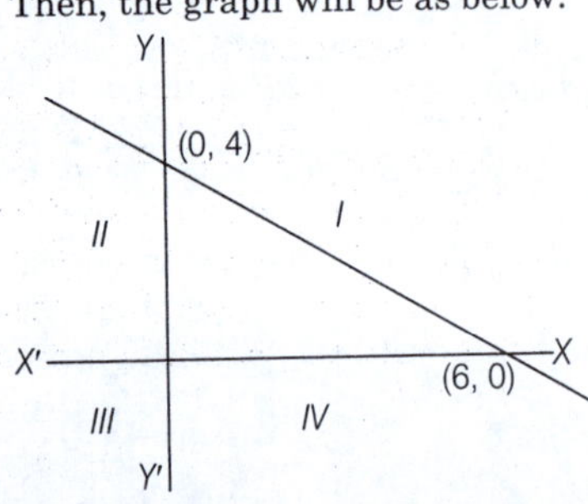

Clearly, the line passes through 1st, 2nd and 4th quadrant.

6. Let a point R lying on Y-axis divides the line segment in ratio m : 1

Then,

$$(0, y) = \left(\frac{m \cdot 3 + 1(-2)}{m+1}, \frac{m \cdot 7 + 1 \cdot (-3)}{m+1}\right)$$

$$\Rightarrow 0 = \frac{m \cdot 3 + 1 \cdot (-2)}{m+1}$$

$$\Rightarrow 3m - 2 = 0$$

$$\Rightarrow m = \frac{2}{3}$$

$$\therefore m : 1 = \frac{2}{3} : 1$$

$$= 2 : 3$$

7. Let $y = ax^3 + bx^2 + cx + d$

Clearly, coordinates of points of intersection are (1, 0), (–1, 0) and (0, 2)

At $x = 1$ and $y = 0$, the equation is

$$a + b + c + d = 0 \quad \text{...(i)}$$

At $x = -1$ and $y = 0$, the equation is

$$-a + b - c + d = 0 \quad \text{...(ii)}$$

At $x = 0$ and $y = 2$, the equation is

$$d = 2 \quad \text{...(iii)}$$

Solving Eqs. (i) and (ii), we get

$$b + d = 0$$

$$\Rightarrow b + 2 = 0$$

$$\Rightarrow b = -2$$

8. Given, $ax + by + c = 0$

$\Rightarrow ax + by + 0 = 0$

$\Rightarrow ax + by = 0$

It is possible at (0, 0).

Hence, option (a) is correct.

9. Coordinates of the centroid

$$= \left(\frac{6 - 3 + 2}{3}, \frac{4 + 5 - 4}{3}\right) = \left(\frac{5}{3}, \frac{5}{3}\right)$$

10. Coordinates of the centroid

$$= \left(\frac{5 - 3 - 1}{3}, \frac{-1 - 2 + 8}{3}\right) = \left(\frac{1}{3}, \frac{5}{3}\right)$$

Distance between centroid and vertex $A = \sqrt{\left(\frac{1}{3} - 5\right)^2 + \left(\frac{5}{3} + 1\right)^2}$

$$= \frac{2\sqrt{65}}{3}$$

We know, the centroid divides the median in ratio 2:1 (from vertex).

Hence, 2 units $\Rightarrow = \frac{2\sqrt{65}}{3}$ units

$\because$ 3 units $\Rightarrow \sqrt{65}$ units

Hence, the length of the required median

$= \sqrt{65}$ units.

11. The x-intercept of the line

$$4x - 5y - 20 = \frac{-c}{a} = \frac{20}{4} = 5$$

Hence, coordinates of point of intersection = (5, 0)

12. It is clear that one curve is passing through P and another is passing through Q and the distance between P and Q is 6 units.

Let the coordinates of point P be (R, T). Then, coordinates of point Q be $(R + 6, T)$ as the diameter is parallel to X-axis and point Q is 6 units far away from the point Q.

The curve passing through point P will be $T = a^R$

and the curve passing through point Q will be $T = 2a^{R+6}$

Hence, $a^R = 2a^{R+6}$

$\Rightarrow a^R = 2 \cdot a^R a^6$

$\Rightarrow a^6 = \frac{1}{2}$

$\Rightarrow a = \frac{1}{\sqrt[6]{2}}$ units

13. The x-intercept and the y-intercept of the line $9x - 12y - 108 = 0$ are respectively $\frac{108}{9}$ and $\frac{108}{12}$ i.e. 12 and 9.

Then, required length = 15 as (9, 12, 15) is a Pythagorean triplet.

14. Clearly, the point P divides the line segment in ratio 1 : 2.

Then, coordinates of point

$$P = \left(\frac{1 \cdot 5 + 2 \cdot 2}{1 + 2}, \frac{1(-8) + 2 \cdot 1}{1 + 2}\right)$$

$$= (3, -2)$$

Since the point P lies on the line $2x - y + k = 0$, coordinates of the point P will also satisfy the equation.

$$\Rightarrow 2 \times 3 - (-2) + k = 0$$

$$\Rightarrow k = -8$$

15.

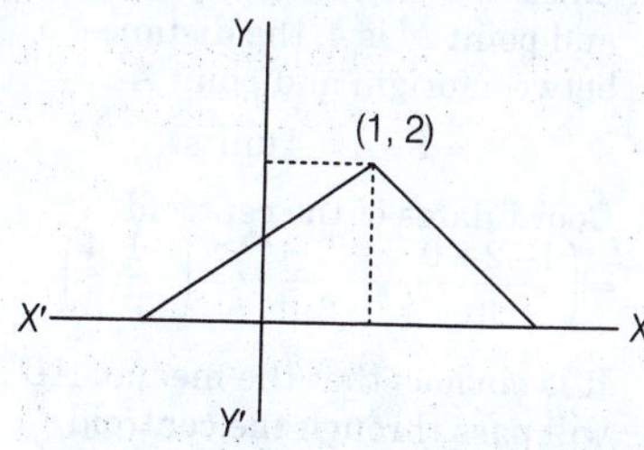

Clearly, ordinate of the point of intersection of lines $2x + y = 4$ and $x - y + 1 = 0$ shall be the required altitude.

On solving both equation, we get $x = 1$ and $y = 2$.

Hence, the required altitude = 2 units.

16. We know, if $ax + by + c = 0$ is an equation of a line, then x-intercept $= \left(\frac{-c}{a}\right)$ and y-intercept $= \left(\frac{-c}{b}\right)$

Then, x-intercept and y-intercept of the line $3x + 4y - 12 = 0$ are respectively $\frac{12}{3}$ and $\frac{12}{4}$ i.e. 4 and 3.

17. Given that $|x + y| + |x - y| = 4$

$\Rightarrow (|x + y|)^2 = (4 - |x - y|)^2$

$\Rightarrow x^2 + y^2 + 2xy$

$= 16 + x^2 + y^2 - 2xy - 8|x - y|$

$\Rightarrow \quad xy = 4 - 2|x - y|$

$\Rightarrow \quad xy = 4 - 2x + 2y,$

if $\quad x - y > 0$ ...(i)

and $\quad xy = 4 + 2x - 2y,$

if $\quad x - y < 0$ ...(ii)

Putting the value of $x = 0$ in Eqs. ...(i) and (ii), we get $y = -2$ and $y = 2$ respectively.

Putting the value of $y = 0$ in Eqs. (i) and (ii), we get $x = 2$ and $x = -2$ respectively.

Hence, points are (0, 2), (0, – 2), (2, 0) and (–2, 0)

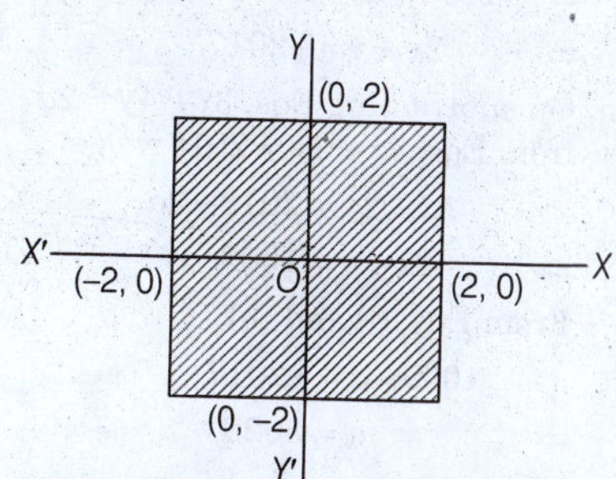

∴ Area of the region bounded by these points = $4 \times 4 = 16$ sq units.

18. The required equation be $x = -3$.

19. Let the third vertex be $A(x, y)$.

Then, coordinates of the centroid

$$= \left(\frac{-3 + 0 + x}{3}, \frac{1 - 2 + y}{3}\right)$$

$$\Rightarrow (0, 0) = \left(\frac{-3 + 0 + x}{3}, \frac{1 - 2 + y}{3}\right)$$

$$\Rightarrow \quad 0 = \frac{x - 3}{3} \text{ and } 0 = \frac{y - 1}{3}$$

$$\Rightarrow \quad x = 3 \text{ and } y = 1$$

Hence, the third vertex is $A(3, 1)$.

20.

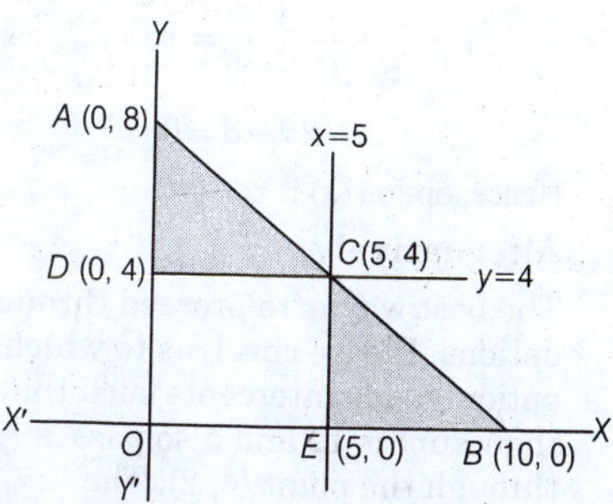

The x-intercept and the y-intercept of the line $4x + 5y - 40 = 0$ are respectively $\frac{40}{4}$ and $\frac{40}{5}$ i.e. 10 and 8.

Area of shaded region = area of ΔAOB – area of rectangle $OECD$

$$= \left(\frac{1}{2} \times 10 \times 8\right) - (5 \times 4)$$

$$= 20 \text{ sq units.}$$

21. We know, the perpendicular distance of a line $ax + by + c = 0$ from a point $(x_1, y_1) = \frac{|ax_1 + by_1 + c|}{\sqrt{a^2 + b^2}}$

Here, the point is the origin i.e. (0, 0).

Then, the required distance

$$= \frac{|1 \cdot 0 - 3 \cdot 0 + 21|}{\sqrt{1^2 + (-3)^2}} = \frac{21}{\sqrt{10}} \text{ units.}$$

22. Clearly, $x^3 + x^2 + 5 = x^2 + x + 5$

$$\Rightarrow \quad x^3 - x = 0$$

$$\Rightarrow \quad x(x^2 - 1) = 0$$

$$\Rightarrow \quad x(x - 1)(x + 1) = 0$$

$$\Rightarrow \quad x = 0, 1, -1$$

Since each of these three solutions lies in the given range $-2 \le x \le 2$, both curves intersect thrice.

23. Let the line $x - y - 2 = 0$ divides the line segment at point $T(x, y)$ in ration $m : 1$

Then, $(x, y) = \left(\frac{m \cdot 8 + 1 \cdot 3}{m + 1}, \frac{m \cdot 9 - 1 \cdot 1}{m + 1}\right)$

$$\Rightarrow \quad x = \frac{8m + 3}{m + 1}$$

and $\quad y = \frac{9m - 1}{m + 1}$

Since the point T lies on the line $x - y - 2 = 0$, coordinates of the point T will also satisfy the equation.

$$\Rightarrow \quad \frac{8m + 3}{m + 1} - \frac{9m - 1}{m + 1} - 2 = 0$$

$$\Rightarrow \quad -m + 4 = 2(m + 1)$$

$$\Rightarrow \quad m = \frac{2}{3}$$

∴ Required ratio $= \frac{2}{3} : 1 = 2 : 3$

24. The x-intercept and the y-intercept of the line $x + y - 1 = 0$ are respectively $\frac{1}{1}$ and $\frac{1}{1}$ i.e. 1 and 1.

and x-intercept and y-intercept of the line $2x + 3y - 6 = 0$ are respectively $\frac{6}{2}$ and $\frac{6}{3}$ i.e. 3 and 2.

further, $x = 0$ means the Y-axis and $y = 0$ means the X-axis.

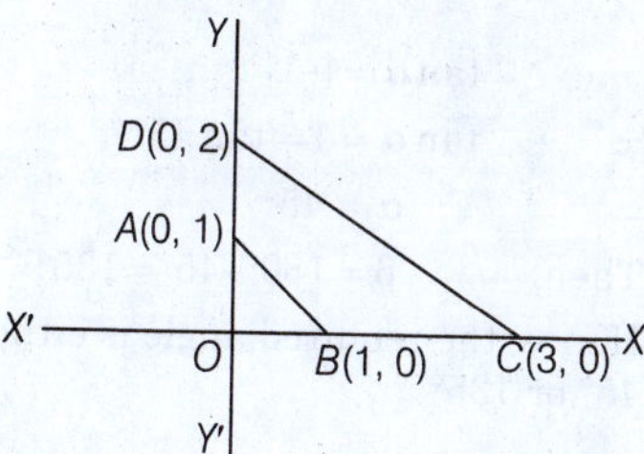

The required area of quadrilateral $ABCD$ = area of ΔOCD – area of ΔOAB

$$= \frac{1}{2} \times 3 \times 2 - \frac{1}{2} \times 1 \times 1$$

$$= 2.5 \text{ sq units}$$

Alternatively

$$\text{Required area} = \frac{(-6)^2}{|2 \cdot 2 \cdot 3|} - \frac{(-1)^2}{|2 \cdot 1 \cdot 1|}$$

$$= 2.5 \text{ sq units.}$$

25. The required equation,

$$y = m\{x - (-3)\}$$

$$\Rightarrow \quad y = \frac{1}{2}(x + 3)$$

$$\Rightarrow \quad x - 2y + 3 = 0$$

Alternatively

Check as to which option yields $\frac{1}{2}$ as slope of the line and is also satisfied with the coordinates of the point (–3, 0). We know that the slope of a line $ax + by + c = 0$ is $\left(\frac{-a}{b}\right)$. It is the best way to solve the question in moments.

26. Given equation, $2x - 3y + 6 = 0$

Comparing with $ax + by + c = 0$, we get $a = 2$, $b = -3$ and $c = 6$

Now, area of triangle

$$= \frac{c^2}{|2ab|} = \frac{6^2}{|2 \cdot 2(-3)|} = 3 \text{ sq units.}$$

27. Slope of the line $3x - y + 9 = 0$ (say m_1) is $\frac{-3}{-1}$ i.e. 3 and slope of the line $x - 2y + 4 = 0$ (say m_2) is $\frac{-1}{-2}$ i.e. $\frac{1}{2}$.

We know, if α and β are the adjacent angles between two lines having slopes m_1 and m_2 and α is an acute angle, then

$$\tan\alpha = \left|\frac{m_2 - m_1}{1 + m_1 . m_2}\right|,$$

where $1 + m_1 . m_2 \neq 0$ and $\beta = (180° - \alpha)$

$$\Rightarrow \quad \tan\alpha = \left|\frac{\frac{1}{2} - 3}{1 + 3 \times \frac{1}{2}}\right|$$

$$\Rightarrow \quad \tan\alpha = |-1|$$

$$\Rightarrow \quad \tan\alpha = 1 = \tan 45°$$

$$\Rightarrow \quad \alpha = 45°$$

Then, $\beta = 180° - 45° = 135°$

Hence, the required angle is either 45° or 135°

28.

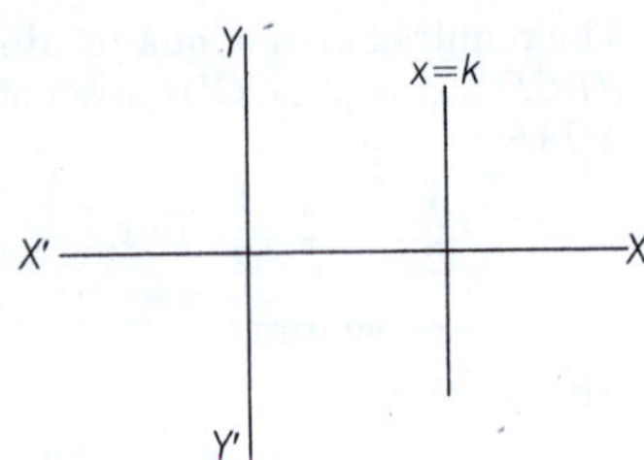

The abscissa of each point lying on the line will be k, but its ordinate will take any value.

Hence, the line is parallel to Y-axis.

29. Let x-intercept be a, then y-intercept $= (12 - a)$

$\therefore$ Equation of the line

$$= \frac{x}{a} + \frac{y}{(12 - a)} = 1$$

Since the line also passes through a point (4, 2), the coordinates of this point shall satisfy the equation of the line.

Then, $\frac{4}{a} + \frac{2}{(12 - a)} = 1$

$$\Rightarrow \quad 48 - 4a + 2a = 12a - a^2$$

$$\Rightarrow \quad a^2 - 14a + 48 = 0$$

$$\Rightarrow \quad a = 6, 8$$

$\therefore$ At $a = 6$, the equation of the line be

$$\frac{x}{6} + \frac{y}{(12 - 6)} = 1$$

$$\Rightarrow \quad x + y - 6 = 0$$

and at $a = 8$, the equation of the line be

$$\frac{x}{8} + \frac{y}{(12 - 8)} = 1$$

$$\Rightarrow \quad x + 2y - 8 = 0$$

Hence, option (a) is correct.

Alternatively

The best way is to proceed through options. Please check as to which option yields intercepts such that their sum is 12 and also passes through the point (4, 2). The x-intercept and y-intercept of a line $ax + by + c = 0$ are $\left(\frac{-c}{a}\right)$ and $\left(\frac{-c}{b}\right)$ respectively.

30. Clearly, $y = 4x$.

31. Obviously, the altitude of the ΔABC through vertex A will be parallel to X-axis. Since, we have to find the shortest distance between the origin and the vertex A, the vertex A must lie on the left side of the line $x = 4$ and also the altitude through vertex A must lie on the X-axis. Then, the triangle ABC may be as shown in figure or like ΔAMB or ΔAMC, altitudes of which lie on X-axis.

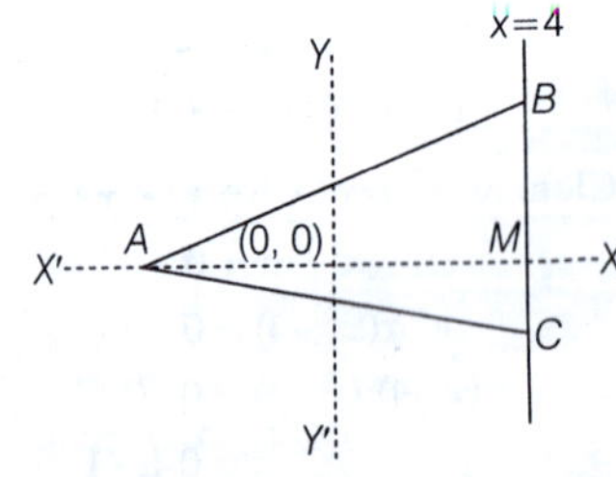

Now, Area of $\Delta ABC = \frac{1}{2} \times BC \times AM$

$$\Rightarrow \quad 32 = \frac{1}{2} \times 8 \times AM$$

$$\Rightarrow \quad AM = 8$$

Since, the distance between origin and point M is 4, the distance between origin and point A

$$= 8 - 4 = 4 \text{ units.}$$

32. Coordinates of the centroid

$$= \left(\frac{1 - 2 + 0}{3}, \frac{5 + 3 - 4}{3}\right) = \left(\frac{-1}{3}, \frac{4}{3}\right)$$

It is obvious that the median AD will pass through the centroid.

Hence, we have to find the equation of a line passing through vertex $A(1, 5)$ and centroid $\left(\frac{-1}{3}, \frac{4}{3}\right)$.

We know, the equation of a line passing through two points (x_1, y_1) and (x_2, y_2) is

$$(y - y_1) = \frac{y_2 - y_1}{x_2 - x_1}(x - x_1)$$

Then, $(y - 5) = \dfrac{\frac{4}{3} - 5}{\frac{-1}{3} - 1}(x - 1)$

$$\Rightarrow \quad (y - 5) = \frac{\frac{-11}{3}}{\frac{-4}{3}}(x - 1)$$

$$\Rightarrow \quad 4y - 20 = 11x - 11$$

$$\Rightarrow 11x - 4y + 9 = 0$$

Thus, equation of median AD

$$= 11x - 4y + 9 = 0$$

33. Solving Eqs. $3x + 4y = 2018$ and $5x + 3y = 1$ we get, $x = -550$ and $y = 917$, hence, coordinates of point $Q = (-550, 917)$.

Let coordinates of point P be (x, y).

Then, slope of the line passing through points P and Q

$$= \frac{917 - y}{-550 - x}$$

$$\Rightarrow \quad 2 = \frac{917 - y}{-550 - x}$$

$$\Rightarrow \quad -1100 - 2x = 917 - y$$

$$\Rightarrow \quad 2x - y = -2017 \quad \text{...(i)}$$

On subtracting Eqs. $3x + 4y = 2a$ from Eqs. $7x + 2y = 2018$ we get,

$$4x - 2y = 2018 - 2a$$

$$\Rightarrow \quad 2x - y = 1009 - a \quad \text{...(ii)}$$

From Eqs. (i) and (ii),

$$1009 - a = -2017$$

$$\Rightarrow \quad a = 3026$$